3. *Sum of two cubes:*

$$A^3 + B^3 = (A + B)(A^2 - AB + B^2)$$

4. *Difference of two cubes:*

$$A^3 - B^3 = (A - B)(A^2 + AB + B^2)$$

Variation

English Statement	*Equation*
y varies directly as x.	$y = kx$
y varies directly as x^n.	$y = kx^n$
y varies inversely as x.	$y = \frac{k}{x}$
y varies inversely as x^n.	$y = \frac{k}{x^n}$
y varies jointly as x and z.	$y = kxz$

Exponents

Definitions of Rational Exponents

1. $a^{\frac{1}{n}} = \sqrt[n]{a}$
2. $a^{\frac{m}{n}} = (\sqrt[n]{a})^m$ or $\sqrt[n]{a^m}$
3. $a^{-\frac{m}{n}} = \frac{1}{a^{\frac{m}{n}}}$

Properties of Rational Exponents

If m and n are rational exponents, and a and b are real numbers for which the following expressions are defined, then

1. $b^m \cdot b^n = b^{m+n}$
2. $\frac{b^m}{b^n} = b^{m-n}$
3. $(b^m)^n = b^{mn}$
4. $(ab)^n = a^n b^n$
5. $\left(\frac{a}{b}\right)^n = \frac{a^n}{b^n}$

Radicals

1. If n is even, then $\sqrt[n]{a^n} = |a|$.
2. If n is odd, then $\sqrt[n]{a^n} = a$.
3. The product rule: $\sqrt[n]{a} \cdot \sqrt[n]{b} = \sqrt[n]{ab}$
4. The quotient rule: $\frac{\sqrt[n]{a}}{\sqrt[n]{b}} = \sqrt[n]{\frac{a}{b}}$

Complex Numbers

1. The imaginary unit i is defined as

$$i = \sqrt{-1}, \quad \text{where} \quad i^2 = -1.$$

The set of numbers in the form $a + bi$ is called the set of complex numbers. If $b = 0$, the complex number is a real number. If $b \neq 0$, the complex number is an imaginary number.

The multiplication of conjugates results in a real number.

3. To simplify powers of i, rewrite the expression in terms of i^2. Then replace i^2 by -1 and simplify.

Quadratic Equations and Functions

1. The solutions of a quadratic equation in standard form

$$ax^2 + bx + c = 0, \quad a \neq 0,$$

are given by the quadratic formula

$$x = \frac{-b \pm \sqrt{b^2 - 4ac}}{2a}.$$

2. The discriminant, $b^2 - 4ac$, of the quadratic equation $ax^2 + bx + c = 0$ determines the number and type of solutions.

Discriminant	Solutions
Positive perfect square with a, b, and c rational numbers	2 rational solutions
Positive and not a perfect square	2 irrational solutions
Zero, with a, b, and c rational numbers	1 rational solution
Negative	2 imaginary solutions

3. The graph of the quadratic function

$$f(x) = a(x - h)^2 + k, \quad a \neq 0,$$

is called a parabola. The vertex, or turning point, is (h, k). The graph opens upward if a is positive and downward if a negative. The axis of symmetry is a vertical line passing through the vertex. The graph can be obtained using the vertex, x-intercepts, if any, [set $f(x)$ equal to zero], and the y-intercept (set $x = 0$).

4. A parabola whose equation is in the form

$$f(x) = ax^2 + bx + c, \quad a \neq 0,$$

has its vertex at

$$\left(-\frac{b}{2a}, f\left(-\frac{b}{2a}\right)\right).$$

If $a > 0$, then f has a minimum that occurs at $x = -\frac{b}{2a}$. If $a < 0$, then f has a maximum that occurs at $x = -\frac{b}{2a}$.

INTERMEDIATE ALGEBRA & TRIGONOMETRY

Taken from

Intermediate Algebra for College Students, Third Edition
By Robert Blitzer

Trigonometry: Enhanced with Graphing Utilities, Third Edition
By Michael Sullivan and Michael Sullivan, III

CUSTOM EDITION FOR CORNING COMMUNITY COLLEGE

PEARSON Custom Publishing

PEARSON Prentice Hall

Cover Art: *Matisse Dream*, by George Herman.

Taken from:

Intermediate Algebra for College Students, Third Edition
By Robert Blitzer

A Pearson Education Company
Upper Saddle River, New Jersey 07458

Trigonometry: A Unit Circle Approach, Seventh Edition
By Michael Sullivan

Published by Prentice-Hall, Inc.

This special edition published in cooperation with Pearson Custom Publishing.

Printed in the United States of America

10 9 8 7 6 5 4 3 2 1

ISBN 0-536-84045-8

2004360402

KK

Please visit our web site at *www.pearsoncustom.com*

PEARSON CUSTOM PUBLISHING
75 Arlington Street, Suite 300, Boston, MA 02116
A Pearson Education Company

Contents

Taken from *Intermediate Algebra for College Students,* Third Edition
By Robert Blitzer

CHAPTER 3 Systems of Linear Equations 153

CHAPTER 4 Inequalities and Problem Solving 229

CHAPTER 5 Polynomials, Polynomial Functions, and Factoring 275

CHAPTER 6 Rational Expressions, Functions, and Equations 369

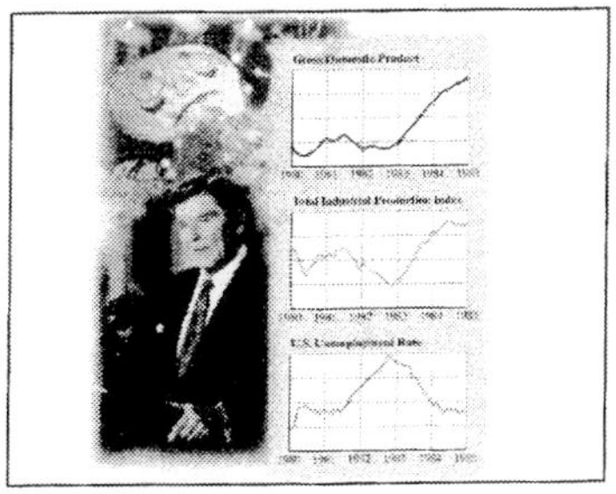

CHAPTER 7 Radicals, Radical Functions, and Rational Exponents 463

CHAPTER 8 Quadratic Equations and Functions 541

Taken from *Trigonometry: A Unit Circle Approach*, Seventh Edition
By Michael Sullivan

Taken from *Intermediate Algebra for College Students*, Third Edition
By Robert Blitzer

Taken from *Trigonometry: A Unit Circle Approach*, Seventh Edition
By Michael Sullivan

Preface

Intermediate Algebra for College Students, Third Edition, provides comprehensive, in-depth coverage of the topics required in a one-term course in intermediate algebra. The book is written for college students who have had a course in introductory algebra. The primary goals of the Third Edition are to help students acquire a solid foundation in intermediate algebra and to show how algebra can model and solve authentic real-world problems.

Writing the Third Edition

A source of frustration for me and my colleagues is that very few students read their textbook. When I ask students why they do not take full advantage of the text, their responses generally fall into two categories:

- "I cannot follow the explanations."
- "The applications are not interesting."

I thought about both of these objections in writing every page of the Third Edition.

"I can't follow the explanations." For many of my students, textbook explanations are too compressed. The chapters in the Third Edition have been extensively rewritten to make them more accessible. I have paid close attention to ensuring that the amount of detail and depth of coverage is appropriate for an intermediate college algebra course. Every section has been rewritten to contain a better range of simple, intermediate, and challenging examples. Voice balloons allow for more specific annotations in examples, further clarifying procedures and concepts. A more open format gives the book a less crowded look than the Second Edition.

"The applications are not interesting." One of the things I enjoy most about teaching in a large urban community college is the diversity of who my students are and what interests them. Real-world data that celebrate this variety are used to bring relevance to examples, discussions, and applications. Most data from the previous edition have been replaced to include data that extend as far up to the present as possible. Updated real-world data were selected on the basis of being interesting and intriguing to students. By connecting algebra to the whole spectrum of their interests, it is my intent to show students that their world is profoundly mathematical and, indeed, pi is in the sky.

New to the Third Edition

The Third Edition is a significant revision of the Second Edition, with increased emphasis on the relevance of algebra in everyday aspects of students' lives. In addition to the book's new open look, the expanded explanations, and the updated real-world data, you will find the following new features in the Third Edition.

- **Chapter-Opening and Section-Opening Scenarios.** Every chapter *and every section* opens with a compelling image that supports a scenario presenting a unique application of algebra in students' lives outside the classroom. Each scenario is revisited later in the chapter or section.
- **Check Point Examples.** Each worked example is followed by a similar matched problem for the student to work while reading the material. This actively involves the student in the learning process and gives students the opportunity to work with a concept as soon as they have learned it. Answers to all Check Points are given in the answer section.
- **Updated Real-World Data.** Real-world data are used to bring relevance to examples, discussions, and applications. Real-world data from the previous edition have been replaced to include data that extend as far up to the present as possible.
- **Reorganized Exercise Sets.** An extensive collection of exercises is included in an exercise set at the end of each section. The Third Edition organizes exercises by level within six category types: Practice Exercises, Application Exercises, Writing in Mathematics, Technology Exercises, Critical Thinking Exercises, and Review Exercises. This format makes it easy to create well-rounded homework assignments. Many new exercises have been added, with attention paid to making sure that the practice and application exercises are appropriate for the level and graded in difficulty.
- **Rewritten Exercise Sets.** In order to update applications and take them to a new level, most application problems from the previous edition have been replaced with new exercises. At the same time, applications were carefully chosen to avoid readers becoming overwhelmed by an excessive number of options. Expanded writing exercises offer students the opportunity to explain every concept covered in each section, providing reinforcement of the objectives and encouraging acquisition of a mathematical vocabulary. Additional writing exercises ask students to discuss, interpret, and give opinions about data. Each review exercises now contains the section number and example number of a similar worked-out example, giving students a pattern for problem solving.
- **Expanded Supplements Package.** The Third Edition is supported by a wealth of supplements designed for added effectiveness and efficiency. These items are described on pages x through xi.
- **Chapter Review Grids.** Each chapter contains a review chart that summarizes the definitions and concepts in every section of the chapter. Examples that illustrate these key concepts are also included in the chart. Like the summary grid, review exercises are now organized by each section of the chapter.
- **Separate Chapter on Inequalities.** In the Second Edition, inequalities were discussed in Chapter 2 (Functions, Linear Functions, and Inequalities)

and in Chapter 3 (Systems of Linear Equations and Inequalities). The Third Edition contains a separate chapter on inequalities that follows Chapters 2 and 3. This allows for shorter, more cohesive, chapters on linear functions and systems of linear equations, in addition to a more integrated problem-solving approach to the discussion of inequalities.

Preserved and Expanded from the Second Edition. The features described below that helped make the Second Edition so popular continue in the Third Edition.

- **Graphing and Functions.** Graphing is introduced in Chapter 1 and functions are introduced in Chapter 2, with an integrated graphing functional approach emphasized throughout the book. Graphs and functions that model data appear in nearly every section and exercise set. Examples and exercises use graphs of functions to explore relationships between data and to provide ways of visualizing a problem's solution.
- **Thorough, Yet Optional Technology.** Although the use of graphing utilities is optional, they are utilized in Using Technology boxes to enable students to visualize algebraic concepts. The use of graphing utilities is also reinforced in the technology exercises appearing in the exercise sets for those who want this option. With the book's early introduction to graphing and functions, students can look at the calculator screens in the Using Technology boxes and gain an increased understanding of an example's solution even if they are not using a graphing utility in the course.
- **Section Objectives.** Learning objectives open every section. The objectives are stated in the margin at their point of use.
- **Detailed Illustrative Examples.** Each illustrative example is titled, making clear the purpose of the example. Examples are clearly written and provide students with detailed step-by-step solutions. No steps are omitted and each step is explained.
- **Enrichment Essays.** Enrichment essays provide historical, interdisciplinary, and otherwise interesting connections throughout the text.
- **Study Tips.** Study Tip boxes offer suggestions for problem solving, point out common student errors, and provide informal tips and suggestions. These invaluable hints appear in abundance throughout the book.
- **Discovery.** Discover for Yourself boxes, found throughout the text, encourage students to further explore algebraic concepts. These explorations are optional and their omission does not interfere with the continuity of the topic under consideration.
- **Chapter Projects.** At the end of each chapter are collaborative activities that give students the opportunity to work cooperatively as they think and talk about mathematics. Many of these exercises should result in interesting group discussions.
- **End-of-Chapter Materials.** The new review grids provide a focused summary and illustrative examples for each section in the chapter. A omprehensive collection of review exercises for each of the chapter's sections follows the review grid. This is followed by a chapter test. Beginning with Chapter 2, each chapter concludes with a comprehensive collection of cumulative review exercises.

Supplements for the Instructor

Printed Resources

Annotated Instructor's Edition (0–13–061452–1)

- Answers to exercises on the same text page or in Graphing Answer Section.
- Graphing Answer Section contains answers to exercises requiring graphical solutions.

Instructor's Solutions Manual (0–13–034291–2)

- Step-by-step solutions for every even-numbered section exercise.
- Step-by-step solutions for every (even and odd) Check Point exercise, Chapter Review exercise, Chapter Test and Cumulative Review exercise.

Instructor's Resource Manual (0–13–034280–7)

- Notes to the Instructor
- Eight Chapter Tests per chapter (5 free response, 3 multiple choice)
- Eight Final Exams (4 free response, 4 multiple choice)
- Twenty additional exercises per section for added test exercises or worksheets.
- Answers to all items

Media Resources

TestGen-EQ with QuizMaster-EQ (CD ROM for IBM and Macintosh 0–13–061936–1)

- Algorithmically driven, text specific testing program.
- Networkable for administering tests and capturing grades on-line.
- Edit or add your own questions to create a nearly unlimited number of tests and worksheets.
- Use the new "Function Plotter" to create graphs.
- Tests can be easily exported to HTML so they can be posted to the Web.

Computerized Tutorial Software Course Management System

MathPro Explorer 4.0

- Network version for IBM and Macintosh
- Enables instructors to create either customized or algorithmically generated practice quizzes from any section of a chapter.
- Includes an e-mail function for networked users, enabling instructors to send a message to a specific student or to an entire group.
- Network based reports and summaries for a class or student and for cumulative or selected scores are available.

MathPro 5

- Anytime. Anywhere.
- Online tutorial with enhanced class and student management features.
- Integration of TestGen-EQ allows for testing to operate within the tutorial environment.
- Course management tracking of both tutorial and testing activity.

Online Options for Distance Learning

WebCT/Blackboard/CourseCompass

- Prentice Hall offers three different on-line interactivity and delivery options for a variety of distance learning needs. Instructors may access or adopt these in conjunction with this text.

Supplements for the Student

Printed Resources

Student Solutions Manual (0–13–034289–0)

- Step-by-step solutions for every odd-numbered section exercise.
- Step-by-step solutions for every (even and odd) Check Point exercise, Chapter Review exercise, Chapter Test and Cumulative Review exercise.

How to Study Mathematics

- Have your instructor contact the local Prentice Hall sales representative.

Math on the Internet: A Student's Guide

- Have your instructor contact the local Prentice Hall sales representative.

Media Resources

Computerized Tutorial Software

MathPro Explorer 4.0

- Keyed to each section of the text for text-specific tutorial exercises and instruction.
- Warm-up exercises and graded Practice Problems.
- Video clips show a problem being explained and worked out on the board.
- Algorithmically generated exercises. On-line help, glossary and summary of scores.

MathPro 5–Anytime. Anywhere.

- Enhanced, Internet-based version of Prentice Hall's popular tutorial software.

Lecture Videos

- Keyed to each section of the text.

Digitized Lecture Videos on CD.

- Have your instructor contact the local Prentice Hall sales representative.

Prentice Hall Tutor Center

- Provides one-on-one tutorial assistance by phone, e-mail, or fax.

Companion Website

- Offers Warm-ups, Real World Activities and Chapter Quizzes.
- E-mail results to your instructor.
- Destination links provide additional opportunities to explore other related sites.

Acknowledgments

I wish to express my thanks to all the users of the first and second editions of this text whose comments contributed to this new edition. In addition, I would like to express my appreciation to all the reviewers for their helpful criticisms and suggestions, frequently transmitted with wit, humor and intelligence. In particular, I would like to thank:

Thomas B. Clark	*Trident Technical College*
Bettyann Daley	*University of Delaware*
Marion K. Glasby	*Anne Arundel Community College*
Kristi Laird	*Jackson Community College*
Terri Moser	*Austin Community College*
Jeff Parent	*Oakland Community College*
Margaret Williamson	*Milwaukee Area Technical College*

Reviewers of Previous Editions:

Howard Anderson	*Skagit Valley College*
John Anderson	*Illinois Valley Community College*
Michael H. Andreoli	*Miami Dade Community College—North Campus*
Warren J. Burch	*Brevard Community College*
Alice Burstein	*Middlesex Community College*
Sandra Pryor Clarkson	*Hunter College*
Robert A. Davies	*Cuyahoga Community College*
Ben Divers, Jr.	*Ferrum College*
Irene Doo	*Austin Community College*
Charles C. Edgar	*Onondaga Community College*
Susan Forman	*Bronx Community College*
Donna Gerken	*Miami-Dade Community College*
Jay Graening	*University of Arkansas*
Robert B. Hafer	*Brevard Community College*
Mary Lou Hammond	*Spokane Community College*
Donald Herrick	*Northern Illinois University*
Beth Hooper	*Golden West College*
Tracy Hoy	*College of Lake County*
Judy Kasabian	*Lansing Community College*
Gary Knippenberg	*Lansing Community College*
Mary Kochler	*Cuyahoga Community College*
Hank Martel	*Broward Community College*
Diana Martelly	*Miami-Dade Community College*
John Robert Martin	*Tarrant County Junior College*
Irwin Metviner	*State University of New York at Old Westbury*
Kamilia Nemri	*Spokane Community College*
Allen R. Newhart	*Parkersburg Community College*
Scott W. Satake	*Eastern Washington University*
Kathy Shepard	*Monroe County Community College*

Gayle Smith	*Lane Community College*
Linda Smoke	*Central Michigan University*
Dick Spangler	*Tacoma Community College*
Janette Summers	*University of Arkansas*
Robert Thornton	*Loyola University*
Lucy C. Thrower	*Francis Marion College*
Andrew Walker	*North Seattle Community College*

Additional acknowledgments are extended to Jacquelyn White for creating the dynamic video tape series covering every section of the book; Penny Arnold for solving all the book's exercises; Lea Rosenberry for the Herculean task of preparing the solutions manuals, as well as serving as accuracy checker; the team at Laurel Technical Services for preparing the answer section and also serving as accuracy checker; Henry Small, research assistant, for obtaining much of the book's data and information for the new section openers; Melinda Alexander, photo researcher, for obtaining the book's photographs; Richard Blair for granting us permission to use one of his outstanding photographs from *Point Reyes Visions* for the book's cover; Brian Morris and the team of graphic artists and mathematicians at Scientific Illustrators, whose superb illustrations and graphs provide visual support to the verbal portions of the text; Prepare Inc., the book's compositor, for inputting hundreds of pages with hardly an error; and Bayani Mendoza de Leon, whose talents as production editor contributed to the book's wonderful look.

I would like to thank my editor at Prentice Hall, Karin Wagner, who guided and coordinated the many details of this project, and my development editor, Don Gecewicz. My thanks also to Eilish Main, Executive Marketing Manager, for her innovative marketing efforts, as well as to the entire Prentice Hall team for their confidence in and enthusiasm for my book.

Robert Blitzer

To the Student

I've written this book so that you can learn about the power of algebra and how it relates directly to your life outside the classroom. All concepts are carefully explained, important definitions and procedures are set off in boxes, and worked-out examples that present solutions in a step-by-step manner appear in every section. Each example is followed by a similar matched problem, called a Check Point, for you to try so that you can actively participate in the learning process as you read the book. (Answers to all Check Points appear in the back of the book.) Study Tips offer hints and suggestions and often point out common errors to avoid. A great deal of attention has been given to applying algebra to your life to make your learning experience both interesting and relevant.

As you begin your studies, I would like to offer some specific suggestions for using this book and for being successful in this course:

1. Attend all lectures. No book is intended to be a substitute for valuable insights and interactions that occur in the classroom. In addition to arriving for lecture on time and being prepared, you will find it useful to read the section before it is covered in lecture. This will give you a clear idea of the new material that will be discussed.

2. Read the book. Read each section with pen (or pencil) in hand. Move through the illustrative examples with great care. These worked-out examples provide a model for doing exercises in the exercise sets. As you proceed through the reading, do not give up if you do not understand every single word. Things will become clearer as you read on and see how various procedures are applied to specific worked-out examples.

3. Work problems every day and check your answers. The way to learn mathematics is by doing mathematics, which means working the Check Points and assigned exercises in the exercise sets. The more exercises you work, the better you will understand the material.

4. Prepare for chapter exams. After completing a chapter, study the summary chart, work the exercises in the Chapter Review, and work the exercises in the Chapter Test. Answers to all these exercises are given in the back of the book.

5. Use the supplements available with this book. A solutions manual containing worked-out solutions to the book's odd-numbered exercises, all review exercises, and all Check Points, a dynamic web page, and video tapes created for every section of the book are among the supplements created to help you tap into the power of mathematics. Ask your instructor or bookstore what supplements are available and where you can find them.

I wrote this book in Point Reyes National Seashore, 40 miles north of San Francisco. The park consists of 75,000 acres with miles of pristine surf-washed beaches, forested ridges, and bays bordered by white cliffs. It was my hope to convey the beauty and excitement of mathematics using nature's unspoiled beauty as a source of inspiration and creativity. Enjoy the pages that follow as you empower yourself with the algebra needed to succeed in college, your career, and in your life.

Regards,

Bob

Robert Blitzer

About the Author

Bob Blitzer is a native of Manhattan and received a Bachelor of Arts degree with dual majors in mathematics and psychology (minor: English literature) from the City College of New York. His unusual combination of academic interests led him toward a Master of Arts in mathematics from the University of Miami and a doctorate in behavioral sciences from Nova University. Bob is most energized by teaching mathematics and has taught a variety of mathematics courses at Miami-Dade Community College for nearly 30 years. He has received numerous teaching awards, including Innovator of the Year from the League for Innovations in the Community College, and was among the first group of recipients at Miami-Dade Community College for an endowed chair based on excellence in the classroom. In addition to *Intermediate Algebra for College Students*, Bob has written *Introductory Algebra for College Students, Introductory and Intermediate Algebra for College Students, Algebra for College Students*, *Thinking Mathematically*, *College Algebra*, *Algebra and Trigonometry*, and *Precalculus*, all published by Prentice Hall.

In the dramatic arts, ours is the era of the movies. As individuals and as a nation, we've grown up with them. Our images of love, war, family, country—even of things that terrify us—owe much to what we've seen on screen.

Take a rapid sequence of still photographs of a moving scene and project them onto a screen at thirty shots a second or faster. Our eyes see the result as continuous motion. The small difference between one frame and the next cannot be detected by the human visual system. In this chapter, you will see how the notion of a function and the steepness of a line can analyze motion and change by "freezing the frame" of a continuously changing process, instant by instant. Understanding the concept of a function will give you a new perspective on many situations, including using mathematics in a way that is similar to making a movie.

New Chapter Openers

Every chapter opens with a new photo and real-world "theme" application that is revisited later in the chapter.

page 90

New Section Openers

Each section opens with a new photo and application drawn from daily life to provide a real-life application of section topics.

▸ SECTION 2.1 *Introduction to Functions*

Objectives

1. Find the domain and range of a relation.
2. Determine whether a relation is a function.
3. Evaluate a function.
4. Graph functions by plotting points.
5. Use the vertical line test to identify functions.
6. Obtain information about a function from its graph.

SSM PH Tutor Center CD-ROM Video

Enjoy talking on the phone? In 2000, nearly 105 million Americans were chatting up a storm on their mobile phones, an increase of over 200% from 1995, when only 34 million Americans were using mobile phones. The graph in Figure 2.1 shows the increase in cellular telephone subscriptions in the United States from 1995 through 2000. With video mobile phones by 2020, there seems to be no limit to the ways in which we keep in touch.

1 Find the domain and range of a relation.

page 92

Emphasis on Problem Solving

The text has been reorganized to emphasize problem solving.
Section 2.5 is now "An Introduction to Problem Solving," and Chapter 3, "Problem Solving," fully explores problem-solving strategies.

1 Solve algebraic word problems using linear equations.

A Strategy for Solving Word Problems Using Equations

Problem solving is an important part of algebra. The problems in this book are presented in English. We must *translate* from the ordinary language of English into the language of algebraic equations. To translate, however, we must understand the English prose and be familiar with the forms of algebraic language. Here are some general steps we will follow in solving word problems:

Strategy for Solving Word Problems

Step 1 Read the problem carefully. Attempt to state the problem in your own words and state what the problem is looking for. Let x (or any variable) represent one of the quantities in the problem.

Step 2 If necessary, write expressions for any other unknown quantities in the problem in terms of x.

Step 3 Write an equation in x that describes the verbal conditions of the problem.

Step 4 Solve the equation and answer the problem's question.

Step 5 Check the solution *in the original wording* of the problem, not in the equation obtained from the words.

page 50

Voice Balloons

New *Voice Balloons* call out key problem-solving tips and observations to clarify the problem-solving process.

For example,

$$(T - S)(2000) = T(2000) - S(2000) = \$10{,}500 - \$3400 = \$7100.$$

In 2000, the difference between total tax and state and local tax was \$7100. This is the per capita federal tax.

Figure 2.10 on page 107 illustrates that information involving differences of functions often appears in graphs seen in newspapers and magazines. Like numbers and algebraic expressions, two functions can be added, subtracted, multiplied, or divided as long as there are numbers common to the domains of both functions. The common domain for functions T and S in Figure 2.10 is

$$\{1900, 1901, 1902, 1903, \ldots, 2000\}.$$

page 108

Revised and Relevant Applications

Over 90% of the *applications* and many of the *examples* have been rewritten to incorporate current, real-world data drawn from familiar sources, such as the Statistical Abstract published by the U.S. Census Bureau.

The graph shows annual cigarette consumption per U.S. adult as a function of time from 1925 through 2000. In Exercises 59–64, find the average rate of change in cigarette consumption per year for each time period.

Cigarette Consumption per U.S. Adult

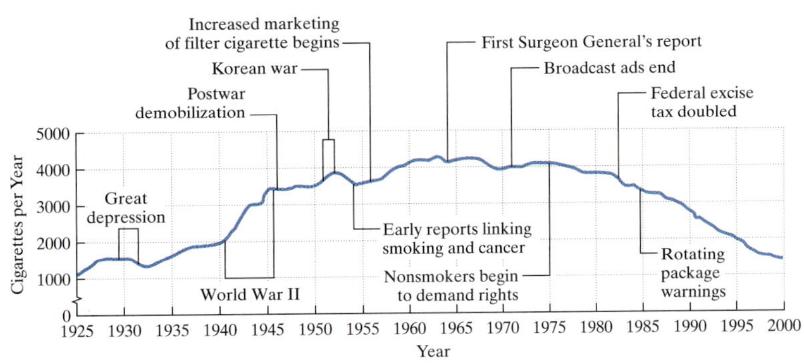

Source: U.S. Department of Health and Human Services

page 145

Critical Thinking Exercises

A wider array of exercises help students check concept mastery.

Critical Thinking Exercises

94. Which one of the following is true?

a. A linear function with nonnegative slope has a graph that rises from left to right.

b. Every line in the rectangular coordinate system has an equation that can be expressed in slope-intercept form.

page 130

Writing in Mathematics

65. Describe how to write the equation of a line if its slope and a point along the line are known.

66. Describe how to write the equation of a line if two points along the line are known.

page 141

Writing in Mathematics Exercises

The new editions provide even more exercises that ask students to explain concepts in their own words—providing reinforcement and encouraging acquisition of a mathematical vocabulary.

Discover for Yourself

If you are given two points on a line, you can use either point for (x_1, y_1) when you write its point-slope equation. Rework Example 2 using $(-2, 6)$ for (x_1, y_1). Once you solve for y, you should obtain the same slope-intercept equation as the one shown in the last line of the solution to Example 2.

page 133

Discover for Yourself Boxes

Discover for Yourself boxes encourage students to actively participate in the learning process as they read the book.

Study Tip

When computing slope, it makes no difference which point you call (x_1, y_1) and which point you call (x_2, y_2). If we let $(x_1, y_1) = (-2, 4)$ and $(x_2, y_2) = (-3, -1)$, the slope is still 5:

$$m = \frac{y_2 - y_1}{x_2 - x_1} = \frac{-1 - 4}{-3 - (-2)} = \frac{-5}{-1} = 5.$$

However, you should not subtract in one order in the numerator $(y_2 - y_1)$ and then in a different order in the denominator $(x_1 - x_2)$. The slope is *not*

$$\frac{-1 - 4}{-2 - (-3)} = \frac{-5}{1} = -5. \quad \text{Incorrect!}$$

page 119

Study Tips

Study Tips make mathematical content more accessible to the student.

Check Points

Each example is followed by an exercise that provides for a more interactive text and gives students the opportunity to work with a concept as soon as they have learned it.

EXAMPLE 6 Slope as Average Rate of Change

When a person receives a drug injected into a muscle, the concentration of the drug in the body, measured in milligrams per 100 milliliters, is a function of the time elapsed since the injection, measured in hours. Figure 2.24 shows the graph of such a function, where x represents hours after the injection and $f(x)$ is the drug's concentration at time x. Find the average rate of change in the drug's concentration between 3 and 7 hours.

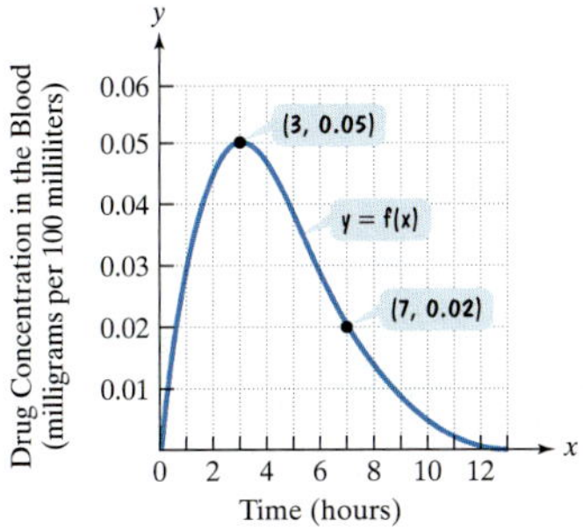

Figure 2.24 Concentration of a drug as a function of time

Solution At 3 hours, the drug's concentration is 0.05 and at 7 hours, the concentration is 0.02. The average rate of change in its concentration between 3 and 7 hours is the slope of the line connecting the points (3, 0.05) and (7, 0.02).

$$m = \frac{\text{Change in } y}{\text{Change in } x} = \frac{0.02 - 0.05}{7 - 3} = \frac{-0.03}{4} = -0.0075$$

The average rate of change is −0.0075. This means that the drug's concentration is decreasing at an average rate of 0.0075 milligrams per 100 milliliters per hour. ■

✔ **CHECK POINT 6** Use Figure 2.24 to find the average change in the drug's concentration between 1 hour and 3 hours.

page 126

CHAPTER 2 GROUP PROJECTS

Group members should consult an almanac, newspaper, magazine, or the Internet and return to the group with as much data as possible for one of the following situations:

1. Life expectancy by birth year, gender, ethnicity, and other categories of interest
2. Median annual income by level of education, broken down by gender, ethnicity, and other categories of interest
3. Global warming
4. Population growth
5. Drug use, including alcohol and cigarettes, by America's high school seniors or by adults in the U.S.
6. Any other area of interest in which data lie approximately on or near a line

Working by hand or using a graphing utility, group members should construct scatter plots for the data that were assembled. If working by hand, draw a line that approximately fits the data in each scatter plot and then write its equation as a function in slope-intercept form. If using a graphing utility, obtain the equation of each regression line. Then use each linear function's equation to make predictions about what might occur in the future. Are there circumstances that might affect the accuracy of the prediction? List some of these circumstances.

7. Consult an almanac, newspaper, magazine, or the Internet to find data displayed in the style of Figure 2.10 on page 107 or in Exercises 53–56 in Exercise Set 2.2 on page 114. Using the two graphs that group members find most interesting, introduce two or more functions that are related to the graphs. Then write and solve a problem involving function addition and function subtraction for each selected graph.

Chapter Projects and Group Activities

Extended applications conclude each chapter. Some activities feature related Websites for student research and exploration.

page 142

Review Exercises

78. If $f(x) = 3x^2 - 8x + 5$, find $f(-2)$. (Section 2.1, Example 3)

79. If $f(x) = x^2 - 3x + 4$ and $g(x) = 2x - 5$, find $(fg)(-1)$. (Section 2.2, Example 4)

80. The sum of the angles of a triangle is 180°. Find the three angles of a triangle if one angle is 20° greater than the smallest angle and the third angle is twice the smallest angle. (Section 1.5, Example 1)

page 142

Review Exercises

Review Exercises in section exercise sets are cross-referenced back to an example in the text—giving students a pattern for problem solving.

CHAPTER SUMMARY, REVIEW, AND TEST

SUMMARY

DEFINITIONS AND CONCEPTS	EXAMPLES
Section 2.1 Introduction to Functions	
A relation is any set of ordered pairs. The set of first components of the ordered pairs is the domain and the set of second components is the range. A function is a relation in which each member of the domain corresponds to exactly one member of the range. No two ordered pairs of a function can have the same first components and different second components.	The domain of the relation $\{(1, 2), (3, 4), (3, 7)\}$ is $\{1, 3\}$. The range is $\{2, 4, 7\}$. The relation is not a function: 3, in the domain, corresponds to both 4 and 7 in the range.
If a function is defined as an equation, the notation $f(x)$, read "f of x" or "f at x," describes the value of the function at the number, or input, x.	If $f(x) = 7x - 5$, then $f(a + 2) = 7(a + 2) - 5 = 7a + 14 - 5 = 7a + 9$.
The graph of a function is the graph of its ordered pairs. ***The Vertical Line Test for Functions*** If any vertical line intersects a graph in more than one point, the graph does not define y as a function of x. At the left or right of a function's graph, you will often find closed dots, open dots, or arrows. A closed dot shows the graph ends and the point belongs to the graph. An open dot shows the graph ends and the point does not belong to the graph. An arrow indicates the graph extends indefinitely.	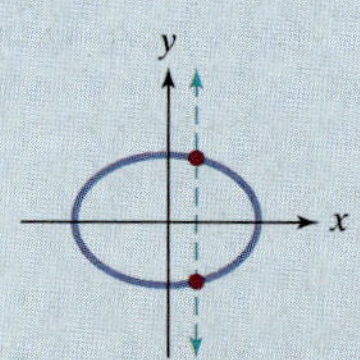 Not the graph of a function 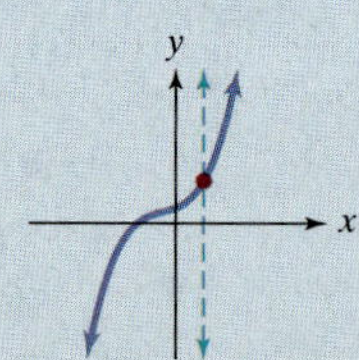 The graph of a function 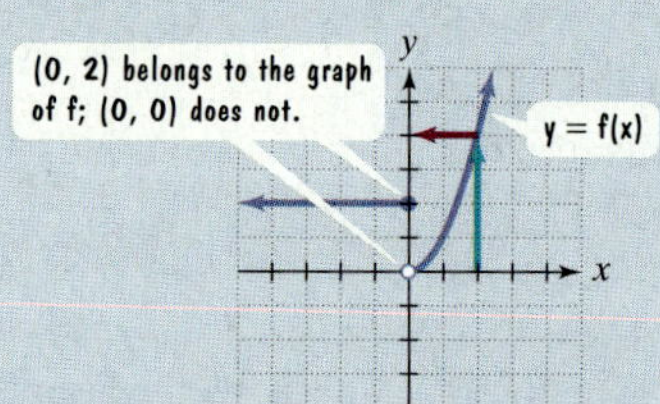 To find $f(2)$, locate 2 on the x-axis. The graph shows $f(2) = 4$.

page 143

Applications Index

D

E

F

Intermediate Algebra for College Students

For fish lovers, the Pacific salmon is in a class by itself. However, the drastic decline in their population has raised alarm and become one of the most important conservation issues in the Pacific Northwest.

The salmon fishing industry is one of the major industries of the states along the Pacific coast. It provides 60,000 jobs and more than $1 billion in income per year for the region. The formula

$$P = -0.22t + 9.6$$

in which P is the Pacific salmon population, in millions, t years after 1960 tells of a severely threatened population. Formulas can be used to describe what happened in the past and to make predictions about what might occur in the future. Knowing how to use and create formulas will help you recognize patterns, logic, and order in a world that can appear chaotic to the untrained eye. In many ways, algebra will provide you with a new way of looking at your world.

Chapter 1

Algebra, Mathematical Models, and Problem Solving

▸ SECTION 1.1 Algebraic Expressions and Real Numbers

Objectives

1. Translate English phrases to algebraic expressions.
2. Evaluate algebraic expressions.
3. Use mathematical models.
4. Recognize the sets that make up the real numbers.
5. Use set-builder notation.
6. Use the symbols $\in$ and $\notin$.
7. Use inequality symbols.

Insatiable killer. That's the reputation the gray wolf acquired in the United States in the nineteenth and early twentieth centuries. Although the label was undeserved, an estimated 2 million wolves were shot, trapped, or poisoned. By 1960, the population was reduced to 800. In this section, you will learn how the special language of algebra describes your world, including the increasing wolf population in the continental United States following the Endangered Species Act of 1973.

Algebraic Expressions

Algebra uses letters, such as x and y, to represent numbers. If a letter is used to represent various numbers, it is called a **variable**. For example, imagine that you are basking in the sun on the beach. We can let x represent the number of minutes that you can stay in the sun without burning with no sunscreen. With a number 6 sunscreen, exposure time without burning is six times as long, or 6 times x. This can be written $6 \cdot x$, but it is usually expressed as $6x$. Placing a number and a letter next to one another indicates multiplication.

Notice that $6x$ combines the number 6 and the variable x using the operation of multiplication. A combination of variables and numbers using the operations of addition, subtraction, multiplication, or division, as well as powers or roots, is called an **algebraic expression**. Here are some examples of algebraic expressions:

$$x + 6, \quad x - 6, \quad 6x, \quad \frac{x}{6}, \quad 3x + 5, \quad \sqrt{x} + 7.$$

Is every letter in algebra a variable? No. Some letters stand for a particular number. Such a letter is called a **constant**. For example, let d = the number of days in a week. The letter d represents just one number, namely 7, and is a constant.

1 Translate English phrases to algebraic expressions.

Translating English Phrases to Algebraic Expressions

Problem solving in algebra involves translating English phrases to algebraic expressions. Here is a list of words and phrases for four basic operations.

Addition	Subtraction	Multiplication	Division
sum	difference	product	quotient
plus	minus	times	divide
increased by	decreased by	of (used with fractions)	per
more than	less than	twice	ratio

EXAMPLE 1 Translating English Phrases to Algebraic Expressions

Write each English phrase as an algebraic expression. Let x represent the number:

a. Nine less than six times a number.

b. The quotient of five and a number, increased by twice the number.

Solution

a. Nine less than | six times a number

$$6x - 9$$

b. The quotient of five and a number, | increased by | twice the number

$$\frac{5}{x} + 2x$$

■

Study Tip

After working each Check Point, check your answer in the answer section before continuing your reading.

✔ **CHECK POINT 1** Write each English phrase as an algebraic expression. Let x represent the number:

a. Five more than 8 times a number.

b. The quotient of a number and 7, decreased by twice the number.

2 Evaluate algebraic expressions.

Evaluating Algebraic Expressions

Evaluating an algebraic expression means to find the value of the expression for a given value of the variable.

EXAMPLE 2 Evaluating an Algebraic Expression

The algebraic expression $81 - 0.6x$ approximates the percentage of American adults who smoked cigarettes x years after 1900. Evaluate the expression for $x = 100$. Describe what the answer means in practical terms.

Solution We begin by substituting 100 for x. Because $x = 100$, we will be finding the percentage of Americans who smoked cigarettes 100 years after 1900, in the year 2000.

$$81 - 0.6x$$

Replace x by 100.

$$= 81 - 0.6(100) = 81 - 60 = 21$$

Thus, in 2000, approximately 21% of American adults smoked cigarettes. ■

✔ **CHECK POINT 2** Evaluate $81 - 0.6x$ for $x = 50$. Describe what the answer means in practical terms.

Many algebraic expressions involve *exponents*. For example, the algebraic expression

$$0.72x^2 + 9.4x + 783$$

approximates the gray wolf population in the United States x years after 1960. The expression x^2 means $x \cdot x$, and is read "x to the second power" or "x squared." The exponent, 2, indicates that the base, x, appears as a factor two times.

Exponential Notation

If n is a counting number (1, 2, 3, and so on),

Exponent or Power

$$b^n = \underbrace{b \cdot b \cdot b \cdot \cdots \cdot b}_{b \text{ appears as a factor } n \text{ times.}}$$

Base

b^n is read "the nth power of b" or "b to the nth power." Thus, the nth power of b is defined as the product of n factors of b.

Furthermore, $b^1 = b$.

For example,

$$8^2 = 8 \cdot 8 = 64, \quad 5^3 = 5 \cdot 5 \cdot 5 = 125, \quad \text{and} \quad 2^4 = 2 \cdot 2 \cdot 2 \cdot 2 = 16.$$

When evaluating algebraic expressions, here are the rules for determining the order in which operations should be performed:

Order of Operations

1. First, perform all operations within grouping symbols.
2. Next, evaluate all exponential expressions.
3. Next, do all multiplications and divisions in the order in which they occur, working from left to right.
4. Finally, do all additions and subtractions in the order in which they occur, working from left to right.

Using Technology

You can use a calculator to evaluate exponential expressions. For example, to evaluate 5^3, press the following keys:

Scientific Calculator

5 [y^x] 3 [=]

Graphing Calculator

5 [∧] 3 [ENTER]

Although calculators have special keys to evaluate powers of ten and squaring bases, you can always use one of the sequences shown here.

EXAMPLE 3 Evaluating an Algebraic Expression

Evaluate $7 + 5(x - 4)^3$ for $x = 6$.

Solution

$7 + 5(x - 4)^3 = 7 + 5(6 - 4)^3$	Replace x with 6.
$= 7 + 5(2)^3$	First work inside parentheses: $6 - 4 = 2$.
$= 7 + 5(8)$	Evaluate the exponential expression: $2^3 = 2 \cdot 2 \cdot 2 = 8$.
$= 7 + 40$	Multiply: $5(8) = 40$.
$= 47$	Add.

✔ **CHECK POINT 3** Evaluate $8 + 6(x - 3)^2$ for $x = 13$.

3 Use mathematical models.

Formulas and Mathematical Models

An **equation** is formed when an equals sign is placed between two algebraic expressions. One aim of algebra is to provide a compact, symbolic description of your world. These descriptions involve the use of **formulas**, statements of equality expressing a relationship among two or more variables. Many formulas give an approximate description of actual data and are called **mathematical models**. We say that such formulas **model**, or approximately describe, the data.

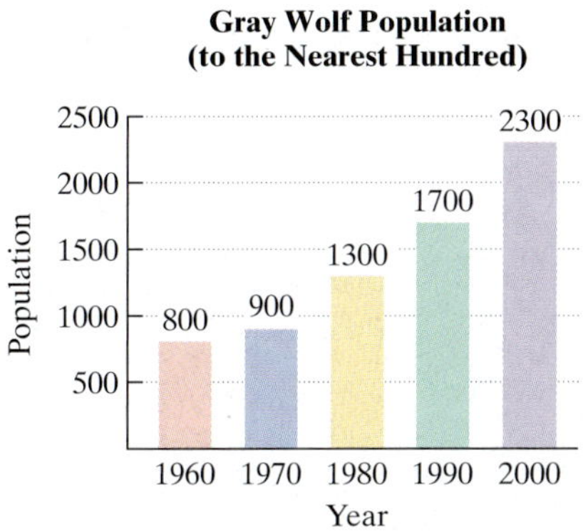

Figure 1.1 *Source*: U.S. Department of the Interior

EXAMPLE 4 Modeling the Gray Wolf Population

The formula

$$P = 0.72x^2 + 9.4x + 783$$

models the gray wolf population in the United States, P, x years after 1960. Use the formula to find the population in 1990. How well does the formula model the actual data shown in the bar graph in Figure 1.1?

Solution Because 1990 is 30 years after 1960, we substitute 30 for x in the given formula. Then we use the order of operations to find P, the gray wolf population in 1990.

$P = 0.72x^2 + 9.4x + 783$ — This is the given mathematical model.

$P = 0.72(30)^2 + 9.4(30) + 783$ — Replace each occurrence of x with 30.

$P = 0.72(900) + 9.4(30) + 783$ — Evaluate the exponential expression: $30^2 = 30 \cdot 30 = 900$.

$P = 648 + 282 + 783$ — Multiply from left to right: $0.72(900) = 648$ and $9.4(30) = 282$.

$P = 1713$ — Add.

The formula indicates that in 1990, the gray wolf population in the United States was 1713. The number given in Figure 1.1 is 1700, so the formula models the data quite well. ■

✔ **CHECK POINT 4** Use the formula in Example 3 to find the gray wolf population in 2000. How well does the formula model the data in Figure 1.1?

"As far as mathematical models refer to reality, they are not certain, and as far as they are certain, they do not refer to reality."
Albert Einstein
(Frances Broomfield's original paintings are available through Portal Gallery Ltd., London, England.)

Do mathematical models always provide an estimate that is a good approximation of actual data? No. Sometimes a mathematical model gives an estimate that is not a good approximation or is extended too far into the future, resulting in a prediction that does not make sense. In these cases, we say that **model breakdown** has occurred.

4 Recognize the sets that make up the real numbers.

The Set of Real Numbers

Before we describe the set of real numbers, let's be sure you are familiar with some basic ideas about sets. A **set** is a collection of objects whose contents can be clearly determined. The objects in a set are called the **elements** of the set. For example, the set of numbers used for counting can be represented by

$$\{1, 2, 3, 4, 5, \ldots\}.$$

The braces, { }, indicate that we are representing a set. This form of representation, called the **roster method**, uses commas to separate the elements of the set. The three dots after the 5 indicate that there is no final element and that the listing goes on forever.

Three common sets of numbers are the *natural numbers*, the *whole numbers*, and the *integers*.

Natural Numbers, Whole Numbers, and Integers

The Set of Natural Numbers

$$\{1, 2, 3, 4, 5, \ldots\}$$

These are the numbers that we use for counting.

The Set of Whole Numbers

$$\{0, 1, 2, 3, 4, 5, \ldots\}$$

The whole numbers add 0 to the set of natural numbers.

The Set of Integers

$$\{\ldots, -5, -4, -3, -2, -1, 0, 1, 2, 3, 4, 5, \ldots\}$$

The integers add the negatives of the natural numbers to the set of whole numbers.

5 Use set-builder notation.

A set can also be written in **set-builder notation**. In this notation, the elements of the set are described, but not listed. Here is an example:

$$\{x \mid x \text{ is a natural number less than } 6\}.$$

The set of all x | such that | x is a natural number less than 6.

The same set written using the roster method is

$$\{1, 2, 3, 4, 5\}.$$

6 Use the symbols $\in$ and $\notin$.

The symbol $\in$ is used to indicate that an element is in a particular set. The symbol $\in$ is read "is an element of." Here is an example:

$$7 \in \{1, 2, 3, 4, 5, \ldots\}.$$

7 | is an element of | the set of natural numbers.

The symbol $\notin$ is used to indicate that an element is not in a particular set. The symbol $\notin$ is read "is not an element of." Here is an example:

$$\frac{1}{2} \notin \{1, 2, 3, 4, 5, \ldots\}.$$

$\frac{1}{2}$ | is not an element of | the set of natural numbers.

EXAMPLE 5 Using the Symbols $\in$ and $\notin$

Determine whether each statement is true or false:

a. $100 \in \{x \mid x \text{ is an integer}\}$ **b.** $20 \notin \{5, 10, 15\}$.

Solution

a. Because 100 is an integer, the statement

$$100 \in \{x \mid x \text{ is an integer}\}$$

is true. The number 100 is an element of the set of integers.

b. Because 20 is not an element of $\{5, 10, 15\}$, the statement $20 \notin \{5, 10, 15\}$ is true. ■

✔ **CHECK POINT 5** Determine whether each statement is true or false:

a. $13 \in \{x \mid x \text{ is an integer}\}$

b. $6 \notin \{7, 8, 9, 10\}$.

Another common set is the set of *rational numbers*. These numbers can be expressed as a quotient of two integers, with the denominator not 0.

Rational Numbers

Rational numbers can be expressed as an integer divided by a nonzero integer.

The Set of Rational Numbers

$$\left\{\frac{a}{b} \,\middle|\, a \text{ and } b \text{ are integers and } b \neq 0\right\}$$

Two examples of rational numbers are:

$$\frac{1}{4} \quad \text{and} \quad 5 = \frac{5}{1}.$$

$a = 1$, $b = 4$ (for $\frac{1}{4}$); $a = 5$, $b = 1$ (for $\frac{5}{1}$)

Can you see that integers are also rational numbers because they can be written in terms of division by 1?

Rational numbers can be expressed in fraction or decimal notation. To express the fraction $\frac{a}{b}$ as a decimal, divide the denominator, b, into the numerator, a. In decimal notation, rational numbers either terminate (stop) or have a digit, or block of digits, that repeat. For example,

$$\frac{3}{8} = 3 \div 8 = 0.375 \quad \text{and} \quad \frac{7}{11} = 7 \div 11 = 0.6363\ldots = 0.\overline{63}.$$

The decimal stops; it is a terminating decimal.

This is a repeating decimal. The bar is written over the repeating part.

Some numbers cannot be expressed as terminating or repeating decimals. An example of such a number is $\sqrt{2}$, the square root of 2. The number $\sqrt{2}$ is a number that can be squared to give 2. No terminating or repeating decimal can be squared to get 2. However, some approximations have squares that come close to 2. We use the symbol $\approx$, which means "is approximately equal to."

- $\sqrt{2} \approx 1.4$ because $(1.4)^2 = (1.4)(1.4) = 1.96$.
- $\sqrt{2} \approx 1.41$ because $(1.41)^2 = (1.41)(1.41) = 1.9881$.
- $\sqrt{2} \approx 1.4142$ because $(1.4142)^2 = 1.99996164$.

$\sqrt{2}$ is an example of an *irrational number*.

ENRICHMENT ESSAY

Elephants and Irrational Numbers

The height of a fully grown elephant, measured from its foot to its shoulder, is approximately $2\pi \cdot$ the diameter of its foot.

Source: *The Joy of* π

Irrational Numbers

The set of **irrational numbers** is the set of numbers whose decimal representations neither terminate nor repeat. Irrational numbers cannot be expressed as quotients of integers.

Examples of irrational numbers include:

$$\sqrt{3} \approx 1.73205 \quad \text{and} \quad \pi(\text{pi}) \approx 3.141593.$$

Not all square roots are irrational. For example, $\sqrt{25} = 5$ because $5^2 = 5 \cdot 5 = 25$. Thus, $\sqrt{25}$ is a natural number, a whole number, an integer, and a rational number $\left(\sqrt{25} = \frac{5}{1}\right)$.

The set of *real numbers* is formed by combining the sets of rational numbers and irrational numbers. Thus, every real number is either rational or irrational, as shown in Figure 1.2.

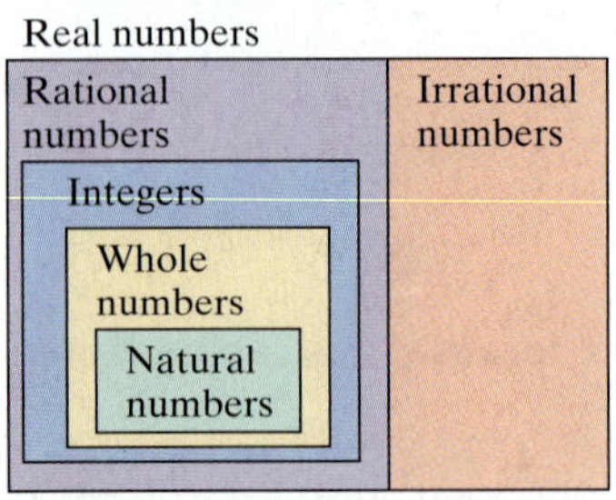

Figure 1.2 Every real number is rational or irrational.

Real Numbers

The set of **real numbers** is the set of numbers that are rational or irrational:

$$\{x \mid x \text{ is rational or } x \text{ is irrational}\}.$$

The Real Number Line

The **real number line** is a graph used to represent the set of real numbers. An arbitrary point, called the **origin**, is labeled 0. Select a point to the right of 0 and label it 1. The distance from 0 to 1 is called the **unit distance**; units to the right of the origin are **positive** and units to the left of the origin are **negative**.

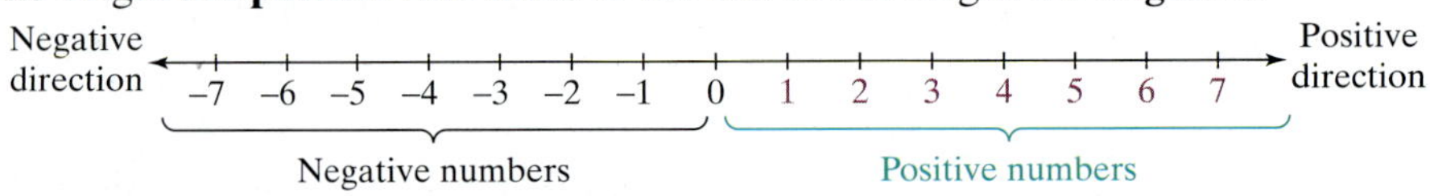

Real numbers are **graphed** on a number line by placing a dot at the correct location for each number. The integers are easiest to locate. In the following figure, we've graphed the integers −3, 0, and 4.

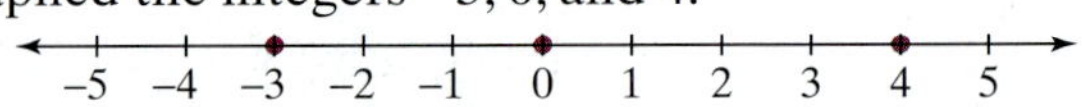

Every real number corresponds to a point on the number line and every point on the number line corresponds to a real number.

7 Use inequality symbols.

Ordering the Real Numbers

On the real number line, the real numbers increase from left to right. The lesser of two real numbers is the one farther to the left on a number line. The greater of two real numbers is the one farther to the right on a number line.

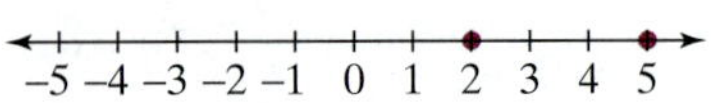

Figure 1.3

Look at the number line in Figure 1.3. The integers 2 and 5 are graphed. Observe that 2 is to the left of 5 on the number line. This means that 2 is less than 5:

$2 < 5$: 2 is less than 5 because 2 is to the *left* of 5 on the number line.

In Figure 1.3, we can also observe that 5 is to the right of 2 on the number line. This means that 5 is greater than 2:

$5 > 2$: 5 is greater than 2 because 5 is to the right of 2 on the number line.

Study Tip

The symbols < and > always point to the lesser of the two real numbers when the inequality is true.

$2 < 5$ The symbol points to 2, the lesser number.

$5 > 2$ The symbol points to 2, the lesser number.

The symbols < and > are called **inequality symbols**. They may be combined with an equal sign, as shown in the following table.

Symbols	Meaning	Example	Explanation
$a \le b$	a is less than or equal to b.	$3 \le 7$	Because $3 < 7$
		$7 \le 7$	Because $7 = 7$
$b \ge a$	b is greater than or equal to a.	$7 \ge 3$	Because $7 > 3$
		$-5 \ge -5$	Because $-5 = -5$

EXAMPLE 6 Using Inequality Symbols

Write out the meaning of each inequality. Then determine whether the inequality is true or false:

a. $-5 < -1$ **b.** $6 > -2$ **c.** $-6 \le 3$
d. $10 \ge 10$ **e.** $-9 \ge 6$.

Solution

Inequality	Meaning
a. $-5 < -1$	"-5 is less than -1." Because -5 is to the left of -1 on a number line, the inequality is true.
b. $6 > -2$	"6 is greater than -2." Because 6 is to the right of -2 on a number line, the inequality is true.
c. $-6 \le 3$	"-6 is less than or equal to 3." Because $-6 < 3$ is true, the inequality is true.
d. $10 \ge 10$	"10 is greater than or equal to 10." Because $10 = 10$ is true, the inequality is true.
e. $-9 \ge 6$	"-9 is greater than or equal to 6." Because neither $-9 > 6$ nor $-9 = 6$ is true, the inequality is false.

■

✔ **CHECK POINT 6** Write out the meaning of each inequality. Then determine whether the inequality is true or false:

a. $-8 < -2$
b. $7 > -3$
c. $-1 \le -4$
d. $5 \ge 5$
e. $2 \ge -14$.

Study Tip

Here are three similar English phrases that have very different translations:
- 7 minus a number: $7 - x$.
- 7 less than a number: $x - 7$.
- 7 is less than a number: $7 < x$.

Think carefully about what is expressed in English before you translate into the language of algebra.

EXERCISE SET 1.1

Practice Exercises

In Exercises 1–14, write each English phrase as an algebraic expression. Let x represent the number.

1. Five more than a number

2. A number increased by six

3. Four less than a number

4. Nine less than a number

5. Four times a number

6. Twice a number

7. Ten more than twice a number

8. Four more than five times a number

9. The difference of six and half of a number

10. The difference of three and half of a number

11. Two less than the quotient of four and a number

12. Three less than the quotient of five and a number

13. The quotient of three and the difference of five and a number

14. The quotient of six and the difference of ten and a number

In Exercises 15–26, evaluate each algebraic expression for the given value or values of the variable(s).

15. $7 + 5x$, for $x = 10$

16. $8 + 6x$, for $x = 5$

17. $6x - y$, for $x = 3$ and $y = 8$

18. $8x - y$, for $x = 3$ and $y = 4$

19. $x^2 + 3x$, for $x = 8$

20. $x^2 + 5x$, for $x = 6$

21. $x^2 - 6x + 3$, for $x = 7$

22. $x^2 - 7x + 4$, for $x = 8$

23. $4 + 5(x - 7)^3$, for $x = 9$

24. $6 + 5(x - 6)^3$, for $x = 8$

25. $x^2 - 3(x - y)$, for $x = 8$ and $y = 2$

26. $x^2 - 4(x - y)$, for $x = 8$ and $y = 3$

In Exercises 27–34, use the roster method to list the elements in each set.

27. $\{x \mid x$ is a natural number less than $5\}$

28. $\{x \mid x$ is a natural number less than $4\}$

29. $\{x \mid x$ is an integer between -8 and $-3\}$

30. $\{x \mid x$ is an integer between -7 and $-2\}$

31. $\{x \mid x$ is a natural number greater than $7\}$

32. $\{x \mid x$ is a natural number greater than $9\}$

33. $\{x \mid x$ is an odd whole number less than $11\}$

34. $\{x \mid x$ is an odd whole number less than $9\}$

In Exercises 35–48, use the meaning of the symbols $\in$ and $\notin$ to determine whether each statement is true or false.

35. $7 \in \{x \mid x$ is an integer$\}$

36. $9 \in \{x \mid x$ is a integer$\}$

37. $7 \in \{x \mid x$ is a rational number$\}$

38. $9 \in \{x \mid x$ is a rational number$\}$

39. $7 \in \{x \mid x$ is an irrational number$\}$

40. $9 \in \{x \mid x$ is an irrational number$\}$

41. $3 \notin \{x \mid x$ is an irrational number$\}$

42. $5 \notin \{x \mid x$ is an irrational number$\}$

43. $\frac{1}{2} \notin \{x \mid x$ is a rational number$\}$

44. $\frac{1}{4} \notin \{x \mid x$ is a rational number$\}$

45. $\sqrt{2} \notin \{x \mid x$ is a rational number$\}$

46. $\pi \notin \{x \mid x$ is a rational number$\}$

47. $\sqrt{2} \notin \{x \mid x$ is a real number$\}$

48. $\pi \notin \{x \mid x$ is a real number$\}$

In Exercises 49–64, write out the meaning of each inequality. Then determine whether the inequality is true or false.

49. $-6 < -2$

50. $-7 < -3$

51. $5 > -7$

52. $3 > -8$

53. $0 < -4$

54. $0 < -5$

55. $-4 \leq 1$

56. $-5 \leq 1$

57. $-2 \leq -6$

58. $-3 \leq -7$

59. $-2 \leq -2$

60. $-3 \leq -3$

61. $-2 \geq -2$

62. $-3 \geq -3$

63. $2 \leq -\frac{1}{2}$

64. $4 \leq -\frac{1}{2}$

Application Exercises

The graph shows life expectancy in the United States by year of birth. The data for U.S. women can be modeled by the formula

$$E = 0.22t + 71$$

where E represents life expectancy for women born t years after 1950. Use this formula to solve Exercises 65–66.

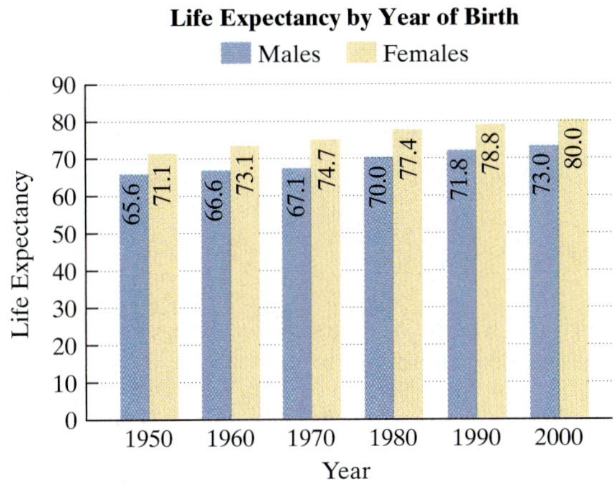

Source: U.S. Bureau of the Census

65. According to the formula, what is the life expectancy for women born in 2000? How well does the formula model the data shown in the bar graph?

66. According to the formula, what is the life expectancy for women born in 1990? How well does the formula model the data shown in the bar graph?

As the use of the Internet increases, so has the number of computer infections from viruses. The data in the bar graph can be modeled by the formula

$$N = 0.2x^2 - 1.2x + 2$$

where N represents the number of infections per month for every 1000 computers x years after 1990. Use this formula to solve Exercises 67–68.

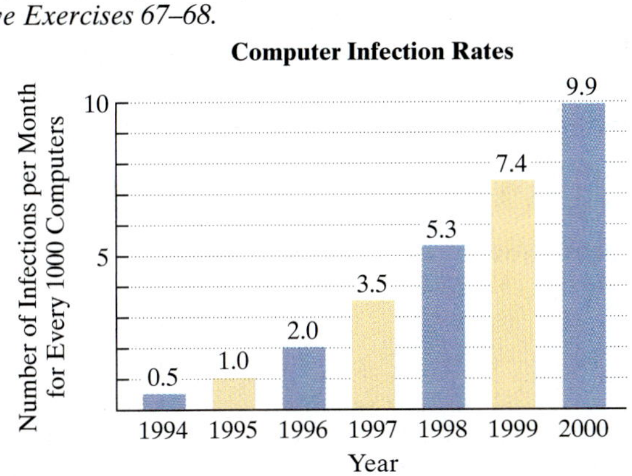

Source: Jupiter Communications

67. According to the formula, how many infections per month for every 1000 computers were there in 2000? How well does the formula model the data shown in the bar graph?

68. According to the formula, how many infections per month for every 1000 computers were there in 1998? How well does the formula model the data shown in the bar graph?

The formula

$$C = \frac{5}{9}(F - 32)$$

expresses the relationship between Fahrenheit temperature, F, and Celsius temperature, C. In Exercises 69–70, use the formula to convert the given Fahrenheit temperature to its equivalent temperature on the Celsius scale.

69. 50°F

70. 86°F

A football is kicked straight up from a height of 4 feet with an initial speed of 60 feet per second. The formula

$$h = 4 + 60t - 16t^2$$

describes the ball's height above the ground, h, in feet, t seconds after it was kicked. Use this formula to solve Exercises 71–72.

71. What is the ball's height 2 seconds after it is kicked?

72. What is the ball's height 3 seconds after it is kicked?

Writing in Mathematics

Writing about mathematics will help you to learn mathematics. For all writing exercises in this book, use complete sentences to respond to the question. Some writing exercises can be answered in a sentence; others require a paragraph or two. You can decide how much you need to write as long as your writing clearly and directly answers the question in the exercise. Standard references such as a dictionary and a thesaurus should be helpful.

73. What is a variable?

74. What is an algebraic expression? Give an example with your explanation.

75. If n is a natural number, what does b^n mean? Give an example with your explanation.

76. What does it mean when we say that a formula models data?

77. What is model breakdown?

78. What is a set?

79. Describe the roster method for representing a set.

80. What are the natural numbers?

81. What are the whole numbers?

82. What are the integers?

83. Describe the rational numbers.

84. Describe the difference between a rational number and an irrational number.

85. What are the real numbers?

86. What is set-builder notation?

87. Describe the meaning of the symbols $\in$ and $\notin$. Provide an example showing the correct use of each symbol.

88. What is the real number line?

89. If you are given two real numbers, explain how to determine which one is the lesser.

90. The mathematical model in Exercises 67–68 predicts a sharp increase in computer infections from viruses. What might occur in the future to cause this model to undergo model breakdown?

Critical Thinking Exercises

91. Which one of the following statements is true?

a. Every rational number is an integer.
b. Some whole numbers are not integers.
c. Some rational numbers are not positive.
d. Irrational numbers cannot be negative.

92. A bird lover visited a pet shop where there were twice 4 and 20 parrots. The bird lover purchased $\frac{1}{7}$ of the birds. English, of course, can be ambiguous and "twice 4 and 20" can mean $2(4 + 20)$ or $2 \cdot 4 + 20$. Explain how the conditions of the situation determine if "twice 4 and 20" means $2(4 + 20)$ or $2 \cdot 4 + 20$.

In Exercises 93–94, insert parentheses to make each statement true.

93. $2 \cdot 3 + 3 \cdot 5 = 45$

94. $8 + 2 \cdot 4 - 3 = 10$

95. Between which two consecutive integers is $-\sqrt{26}$? Do not use a calculator.

▸ SECTION 1.2 Operations with Real Numbers and Simplifying Algebraic Expressions

Objectives

1. Find a number's absolute value.
2. Add real numbers.
3. Find opposites.
4. Subtract real numbers.
5. Multiply real numbers.
6. Evaluate exponential expressions.
7. Divide real numbers.
8. Use the order of operations.
9. Use commutative, associative, and distributive properties.
10. Simplify algebraic expressions.

SSM PH Tutor Center CD-ROM Video

We Americans are often surprised at the generous social services provided by many European governments. However, European consumers often snap at contributing a tankful of taxes at the gas pump to help pay for these services. Is their anger justified? How does the price per gallon of premium gasoline in European countries compare to what we pay in the United States? In this section, we review operations with real numbers so that you can make some comparisons.

Absolute Value

Absolute value is used to describe how to operate with positive and negative numbers.

Absolute Value

The **absolute value** of a real number a, denoted by $|a|$, is the distance from 0 to a on the number line. This distance is always taken to be nonnegative.

1 Find a number's absolute value.

EXAMPLE 1 Finding Absolute Value

Find the absolute value:

a. $|-4|$ **b.** $|3.5|$ **c.** $|0|$.

Solution

a. $|-4| = 4$ The absolute value of -4 is 4 because -4 is 4 units from 0.

b. $|3.5| = 3.5$ 3.5 is 3.5 units from 0.

c. $|0| = 0$ 0 is 0 units from itself.

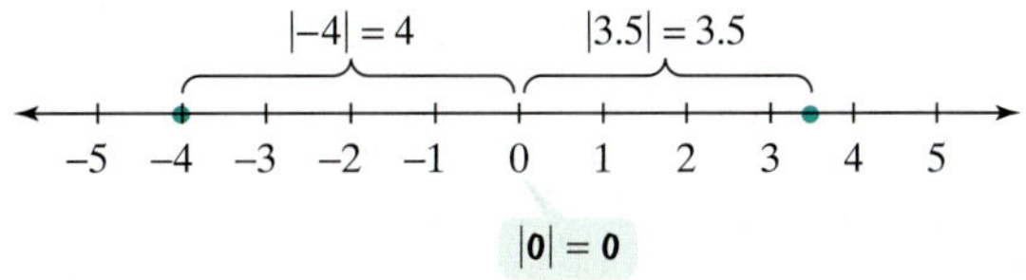

Can you see that the absolute value of a real number is either positive or zero? Zero is the only real number whose absolute value is 0:

$$|0| = 0.$$

The absolute value of a real number is never negative.

✔ **CHECK POINT 1** Find the absolute value:

a. $|-6|$ **b.** $|4.5|$ **c.** $|0|$.

2 Add real numbers.

Adding Real Numbers

Table 1.1 reviews how to add real numbers.

Table 1.1 Adding Real Numbers

Rule	Examples
To add two real numbers with the same sign, add their absolute values. Use the common sign as the sign of the sum.	$(-7) + (-4) = -(\|-7\| + \|-4\|)$ $= -(7 + 4)$ $= -11$
To add two real numbers with different signs, subtract the smaller absolute value from the greater absolute value. Use the sign of the number with the greater absolute value as the sign of the sum.	$7 + (-15) = -(\|-15\| - \|7\|)$ $= -(15 - 7)$ $= -8$

EXAMPLE 2 Adding Real Numbers

Add: **a.** $-12 + (-14)$ **b.** $-0.3 + 0.7$ **c.** $-\frac{3}{4} + \frac{1}{2}$.

Solution

a. $-12 + (-14) = -26$ Add absolute values: $12 + 14 = 26$.

Use the common sign.

b. $-0.3 + 0.7 = 0.4$ Subtract absolute values: $0.7 - 0.3 = 0.4$.

Use the sign of the number with the greater absolute value. The sign of the sum is assumed to be positive.

c. $-\frac{3}{4} + \frac{1}{2} = -\frac{1}{4}$ Subtract absolute values: $\frac{3}{4} - \frac{1}{2} = \frac{3}{4} - \frac{2}{4} = \frac{1}{4}$.

Use the sign of the number with the greater absolute value.

✔ **CHECK POINT 2** Add:

a. $-10 + (-18)$ **b.** $-0.2 + 0.9$ **c.** $-\frac{3}{5} + \frac{1}{2}$.

If one number in a sum is zero, the sum is the other number. For example,

$$-3 + 0 = -3 \quad \text{and} \quad 0 + 2 = 2.$$

In general,

$$a + 0 = a \quad \text{and} \quad 0 + a = a.$$

We call 0 the **identity of addition** or the **additive identity**. Thus, the additive identity can be deleted from a sum.

3 Find opposites.

Numbers with different signs but the same absolute value are called **opposites** or **additive inverses**. For example, 3 and −3 are additive inverses. When additive inverses are added, their sum is 0. For example,

$$3 + (-3) = 0 \quad \text{and} \quad -3 + 3 = 0.$$

Inverse Property of Addition

The sum of a real number and its additive inverse gives 0, the additive identity.

$$a + (-a) = 0 \quad \text{and} \quad (-a) + a = 0$$

The symbol "−" is used to name the opposite, or additive inverse, of a. When a is a negative number, $-a$, its opposite, is positive. For example, if a is −4, its opposite is 4. Thus,

$$-(-4) = 4.$$

The opposite of −4 is 4.

In general, if a is any real number,

$$-(-a) = a.$$

EXAMPLE 3 Finding Opposites

Find $-x$ if: **a.** $x = -6$ **b.** $x = \frac{1}{2}$.

Solution

a. If $x = -6$, then $-x = -(-6) = 6$. The opposite of −6 is 6.

b. If $x = \frac{1}{2}$, then $-x = -\frac{1}{2}$. The opposite of $\frac{1}{2}$ is $-\frac{1}{2}$.

✔ **CHECK POINT 3** Find $-x$ if: **a.** $x = -8$ **b.** $x = \frac{1}{3}$.

We can define the absolute value of the real number a using opposites without referring to a number line. The algebraic definition of the absolute value of a is given as follows:

Definition of Absolute Value

$$|a| = \begin{cases} a & \text{if } a \geq 0 \\ -a & \text{if } a < 0 \end{cases}$$

If a is nonnegative (that is $a \geq 0$), the absolute value of a is the number itself. For example,

$$|5| = 5 \qquad |\pi| = \pi \qquad \left|\frac{1}{3}\right| = \frac{1}{3} \qquad |0| = 0.$$

Zero is the only number whose absolute value is 0.

If a is a negative number (that is, $a < 0$), the absolute value of a is the opposite of a. This makes the absolute value positive. For example,

$$|-3| = -(-3) = 3 \qquad |-\pi| = -(-\pi) = \pi \qquad \left|-\frac{1}{3}\right| = -\left(-\frac{1}{3}\right) = \frac{1}{3}.$$

This middle step is usually omitted.

4 Subtract real numbers.

Subtracting Real Numbers

Subtraction of real numbers is defined in terms of addition.

Definition of Subtraction

If a and b are real numbers,

$$a - b = a + (-b).$$

To subtract a real number, add its opposite or additive inverse.

Thus, to subtract real numbers:

1. Change the subtraction to addition.
2. Change the sign of the number being subtracted.
3. Add, using one of the rules for adding numbers with the same sign or different signs.

EXAMPLE 4 Subtracting Real Numbers

Subtract:

a. $6 - 13$ **b.** $5.1 - (-4.2)$ **c.** $-10 - (-2)$.

Solution

a. $6 - 13 = 6 + (-13) = -7$

Change the subtraction to addition. Replace 13 with its opposite.

b. $5.1 - (-4.2) = 5.1 + 4.2 = 9.3$

Change the subtraction to addition. Replace −4.2 with its opposite.

c. $-10 - (-2) = -10 + 2 = -8$

Change the subtraction to addition. Replace -2 with its opposite.

✔ **CHECK POINT 4** Subtract:

a. $7 - 10$ **b.** $4.3 - (-6.2)$ **c.** $-9 - (-3)$.

5 Multiply real numbers.

Multiplying Real Numbers

You can think of multiplication as repeated addition or subtraction that starts at 0. For example,

$$3(-4) = 0 + (-4) + (-4) + (-4) = -12$$

The numbers have different signs and the product is negative.

and

$$(-3)(-4) = 0 - (-4) - (-4) - (-4) = 0 + 4 + 4 + 4 = 12.$$

The numbers have the same sign and the product is positive.

Table 1.2 reviews how to multiply real numbers.

Table 1.2 Multiplying Real Numbers

Rule	Examples
The product of two real numbers with different signs is found by multiplying their absolute values. The product is negative.	$7(-5) = -35$
The product of two real numbers with the same sign is found by multiplying their absolute values. The product is positive.	$(-6)(-11) = 66$
The product of 0 and any real number is 0: $a \cdot 0 = 0$ and $0 \cdot a = 0$.	$-17(0) = 0$
If no number is 0, a product with an odd number of negative factors is found by multiplying absolute values. The product is negative.	$-2(-3)(-5) = -30$ Three (odd) negative factors
If no number is 0, a product with an even number of negative factors is found by multiplying absolute values. The product is positive.	$-2(3)(-5) = 30$ Two (even) negative factors

6 Evaluate exponential expressions.

Because exponents indicate repeated multiplication, rules for multiplying real numbers can be used to evaluate exponential expressions.

EXAMPLE 5 Evaluating Exponential Expressions

Evaluate:

a. $(-6)^2$ **b.** -6^2 **c.** $(-5)^3$ **d.** $(-2)^4$.

Solution

a. $(-6)^2 = (-6)(-6) = 36$

Base is −6. Same signs give positive product.

b. $-6^2 = -(6 \cdot 6) = -36$

Base is 6. The negative is not inside parentheses and is not taken to the second power.

c. $(-5)^3 = (-5)(-5)(-5) = -125$

An odd number of negative factors gives a negative product.

d. $(-2)^4 = (-2)(-2)(-2)(-2) = 16$

An even number of negative factors gives a positive product.

CHECK POINT 5 Evaluate:

a. $(-5)^2$ **b.** -5^2 **c.** $(-4)^3$ **d.** $(-3)^4$.

7 Divide real numbers.

Dividing Real Numbers

If a and b are real numbers and b is not 0, then the quotient of a and b is defined by:

$$a \div b = a \cdot \frac{1}{b} \quad \text{or} \quad \frac{a}{b} = a \cdot \frac{1}{b}.$$

Thus, to find the quotient of a and b, we can divide by b or multiply by $\frac{1}{b}$. The nonzero real numbers b and $\frac{1}{b}$ are called **reciprocals**, or **multiplicative inverses**, of one another. When reciprocals are multiplied, their product is 1:

$$b \cdot \frac{1}{b} = 1.$$

Because division is defined in terms of multiplication, the sign rules for dividing numbers are the same as the sign rules for multiplying them.

Dividing Real Numbers

The quotient of two numbers with different signs is negative. The quotient of two numbers with the same sign is positive. The quotient is found by dividing absolute values.

EXAMPLE 6 Dividing Real Numbers

Divide:

a. $\frac{20}{-5}$ **b.** $-\frac{3}{4} \div \left(-\frac{5}{9}\right)$.

Solution

a. $\frac{20}{-5} = -4$ Divide absolute values: $\frac{20}{5} = 4$.

Different signs: negative quotient

b. $-\frac{3}{4} \div \left(-\frac{5}{9}\right) = \frac{27}{20}$ Divide absolute values: $\frac{3}{4} \div \frac{5}{9} = \frac{3}{4} \cdot \frac{9}{5} = \frac{27}{20}$.

Same sign: positive quotient

■

✔ **CHECK POINT 6** Divide:

a. $\frac{32}{-4}$ **b.** $-\frac{2}{3} \div \left(-\frac{5}{4}\right)$.

We must be careful with division when 0 is involved. Any nonzero number divided into 0 is 0. For example,

$$\frac{0}{-5} = 0.$$

Can you see why $\frac{0}{-5}$ must be 0? The definition of division tells us that

$$\frac{0}{-5} = 0 \cdot \left(-\frac{1}{5}\right)$$

and the product of 0 and any real number is 0. By contrast, what happens if we divide −5 by 0. The answer must be a number that, when multiplied by 0, gives −5. However, any number multiplied by 0 is 0. Thus, we cannot divide −5, or any other real number, by 0.

Division by Zero

Division by zero is not allowed; it is undefined. A real number can never have a denominator of 0.

8 Use the order of operations.

Order of Operations

The rules for order of operations can be applied to positive and negative real numbers. Recall that if no grouping symbols are present, we:

- Evaluate exponential expressions.
- Multiply and divide from left to right.
- Add and subtract from left to right.

EXAMPLE 7 Using the Order of Operations

Simplify: $4 - 7^2 + 8 \div 2(-3)^2$.

Solution

$4 - 7^2 + 8 \div 2(-3)^2$

$= 4 - 49 + 8 \div 2(9)$ Evaluate exponential expressions: $7^2 = 7 \cdot 7 = 49$ and $(-3)^2 = (-3)(-3) = 9$.

$= 4 - 49 + 4(9)$ Divide: $8 \div 2 = 4$.

$= 4 - 49 + 36$ Multiply: $4(9) = 36$.

$= -45 + 36$ Subtract: $4 - 49 = 4 + (-49) = -45$.

$= -9$ Add. ■

✔ CHECK POINT 7 Simplify: $3 - 5^2 + 12 \div 2(-4)^2$.

If an expression contains grouping symbols, we perform operations in these symbols first. Common grouping symbols are parentheses, brackets, and braces. Other grouping symbols include fraction bars, absolute value symbols, and radical symbols, such as square root signs $(\sqrt{\;})$.

EXAMPLE 8 Using the Order of Operations

Simplify: $\dfrac{13 - 3(-2)^4}{3 - (6 - 10)}$.

Solution Simplify the numerator and the denominator separately. Then divide.

$\dfrac{13 - 3(-2)^4}{3 - (6 - 10)}$

$= \dfrac{13 - 3(16)}{3 - (-4)}$ Evaluate the exponential expression in the numerator: $(-2)^4 = 16$. Subtract inside parentheses in the denominator: $6 - 10 = 6 + (-10) = -4$.

$= \dfrac{13 - 48}{7}$ Multiply in the numerator: $3(16) = 48$. Subtract in the denominator: $3 - (-4) = 3 + 4 = 7$.

$= \dfrac{-35}{7}$ Subtract in the numerator: $13 - 48 = 13 + (-48) = -35$.

$= -5$ Divide. ■

✔ CHECK POINT 8 Simplify: $\dfrac{4 + 3(-2)^3}{2 - (6 - 9)}$.

9 Use commutative, associative, and distributive properties.

The Commutative, Associative, and Distributive Properties

Basic algebraic properties enable us to write *equivalent algebraic expressions*. Two algebraic expressions that have the same value for all replacements are called **equivalent algebraic expressions**. In Section 1.4, you will use such expressions to solve equations.

In arithmetic, when two numbers are added or multiplied, the order in which the numbers are written does not affect the answer. These facts are called **commutative properties**.

The Commutative Properties

Let a, and b represent real numbers, variables, or algebraic expressions.

Addition: $a + b = b + a$

Multiplication: $ab = ba$

Changing order when adding or multiplying does not affect a sum or product.

EXAMPLE 9 Using the Commutative Properties

Write an algebraic expression equivalent to $3x + 7$ using each of the commutative properties.

Solution

Commutative of Addition

$$3x + 7 = 7 + 3x$$

Change the order of the addition.

Commutative of Multiplication

$$3x + 7 = x \cdot 3 + 7$$

Change the order of the multiplication. ■

✔ **CHECK POINT 9** Write an algebraic expression equivalent to $4x + 9$ using each of the commutative properties.

The **associative properties** enable us to form equivalent expressions by regrouping.

The Associative Properties

Let a, b, and c represent real numbers, variables, or algebraic expressions.

Addition: $(a + b) + c = a + (b + c)$

Multiplication: $(ab)c = a(bc)$

Changing grouping when adding or multiplying does not affect a sum or product.

EXAMPLE 10 Using the Associative Properties

Use an associative property to write an equivalent expression and simplify:

a. $7 + (3 + x)$ **b.** $-6(5x)$.

Solution

a. $7 + (3 + x) = (7 + 3) + x = 10 + x$

b. $-6(5x) = (-6 \cdot 5)x = -30x$ ■

✔ **CHECK POINT 10** Use an associative property to write an equivalent expression and simplify:

a. $6 + (12 + x)$ **b.** $-7(4x)$.

ENRICHMENT ESSAY

The Associative Property and the English Language

In the English language, phrases and sentences can take on different meanings depending on the way the words are associated with commas. Here are two examples.

- *What's the latest dope?*
  What's the latest, dope?
- *Population of New York City broken down by age and sex*
  Population of New York City, broken down by age and sex

The **distributive property** allows us to rewrite the product of a number and a sum as the sum of two products.

The Distributive Property

Let a, b, and c represent real numbers, variables, or algebraic expressions.

$$a(b + c) = ab + ac$$

Multiplication distributes over addition.

EXAMPLE 11 Using the Distributive Property

Use the distributive property to write an equivalent expression:

$$-2(3x + 5).$$

Solution

$$-2(3x + 5) = -2 \cdot 3x + (-2) \cdot 5 = -6x + (-10) = -6x - 10$$ ■

✔ **CHECK POINT 11** Use the distributive property to write an equivalent expression: $-4(7x + 2)$.

Study Tip

When using a distributive property to remove parentheses, be sure to multiply *each term* inside the parentheses by the factor outside.

INCORRECT

$$5(3y + 7) = 5 \cdot 3y + 7$$
$$= 15y + 7$$

Table 1.3 shows a number of other forms of the distributive property.

Table 1.3 Other Forms of the Distributive Property

Property	Meaning	Example
$a(b - c) = ab - ac$	Multiplication distributes over subtraction.	$5(4x - 3) = 5 \cdot 4x - 5 \cdot 3$ $= 20x - 15$
$a(b + c + d) = ab + ac + ad$	Multiplication distributes over three or more terms in parentheses.	$4(x + 10 + 3y)$ $= 4x + 4 \cdot 10 + 4 \cdot 3y$ $= 4x + 40 + 12y$
$(b + c)a = ba + ca$	Multiplication on the right distributes over addition (or subtraction).	$(x + 7)9 = x \cdot 9 + 7 \cdot 9$ $= 9x + 63$

10 Simplify algebraic expressions.

Combining Like Terms and Simplifying Algebraic Expressions

The **terms** of an algebraic expression are those parts that are separated by addition. For example, the algebraic expression $7x + 3$ contains two terms, namely $7x$ and 3. Notice that a term is a number, a variable, or a number multiplied by one or more variables.

The numerical part of a term is called its **numerical coefficient**. In the term $7x$, the 7 is the numerical coefficient. If a term containing one or more variables is written without a numerical coefficient, the numerical coefficient is understood to be 1. Thus, x means $1x$ and ab means $1ab$.

The parts of each term that are multiplied are called the **factors** of the term. The factors of the term $7x$ are 7 and x.

Like terms are terms that have exactly the same variable factors. For example, $3x$ and $7x$ are like terms. The distributive property in the form

$$ab + ac = (b + c)a$$

enables us to add or subtract like terms. For example,

$$3x + 7x = (3 + 7)x = 10x$$

$$7y^2 - y^2 = 7y^2 - 1y^2 = (7 - 1)y^2 = 6y^2.$$

This process is called **combining like terms**.

An algebraic expression is **simplified** when grouping symbols have been removed and like terms have been combined.

Study Tip

To combine like terms mentally, add or subtract the numerical coefficients of the terms. Use this result as the numerical coefficient of the terms' variable factor(s).

EXAMPLE 12 Simplifying an Algebraic Expression

Simplify: $7x + 12x^2 + 3x + x^2$.

Solution

$$7x + 12x^2 + 3x + x^2$$

$$= (7x + 3x) + (12x^2 + x^2)$$ Rearrange terms and group like terms using commutative and associative properties. This step is often done mentally.

$x^2 = 1x^2$

$$= (7 + 3)x + (12 + 1)x^2$$ Apply the distributive property.

$$= 10x + 13x^2$$ Simplify.

Using the commutative property of addition, we can write this simplified expression as $13x^2 + 10x$. ■

✔ **CHECK POINT 12** Simplify: $3x + 14x^2 + 11x + x^2$.

EXAMPLE 13 Simplifying an Algebraic Expression

Simplify: $4(7x - 3) - 10x$.

Solution

$$4(7x - 3) - 10x$$

$$= 4 \cdot 7x - 4 \cdot 3 - 10x$$ Use the distributive property to remove the parentheses.

$$= 28x - 12 - 10x$$ Multiply.

$$= (28x - 10x) - 12$$ Group like terms.

$$= (28 - 10)x - 12$$ Apply the distributive property.

$$= 18x - 12$$ Simplify. ■

✔ **CHECK POINT 13** Simplify: $8(2x - 5) - 4x$.

It is not uncommon to see algebraic expressions with parentheses preceded by a negative sign or subtraction. An expression of the form $-(b + c)$ can be simplified as follows:

$$-(b + c) = -1(b + c) = (-1)b + (-1)c = -b + (-c) = -b - c.$$

Do you see a fast way to obtain the simplified expression on the right? **If a negative sign or a subtraction symbol appears outside parentheses, drop the parentheses and change the sign of every term within the parentheses**. For example,

$$-(3x^2 - 7x - 4) = -3x^2 + 7x + 4.$$

EXAMPLE 14 Simplifying an Algebraic Expression

Simplify: $8x + 2[5 - (x - 3)]$.

Solution

$8x + 2[5 - (x - 3)]$

$= 8x + 2[5 - x + 3]$ Drop parentheses and change the sign of each term in parentheses: $-(x - 3) = -x + 3$.

$= 8x + 2[8 - x]$ Simplify inside brackets: $5 + 3 = 8$.

$= 8x + 16 - 2x$ Apply the distributive property: $2[8 - x] = 2 \cdot 8 - 2x = 16 - 2x$.

$= (8x - 2x) + 16$ Group like terms.

$= (8 - 2)x + 16$ Apply the distributive property.

$= 6x + 16$ Simplify. ■

CHECK POINT 14 Simplify: $6 + 4[7 - (x - 2)]$.

EXERCISE SET 1.2

Practice Exercises

In Exercises 1–10, find each absolute value.

1. $|-7|$ **2.** $|-10|$

3. $|4|$ **4.** $|13|$

5. $|-7.6|$ **6.** $|-8.3|$

7. $\left|\frac{\pi}{2}\right|$ **8.** $\left|\frac{\pi}{3}\right|$

9. $|-\sqrt{2}|$ **10.** $|-\sqrt{3}|$

In Exercises 11–24, add as indicated.

11. $-3 + (-8)$ **12.** $-5 + (-10)$

13. $-14 + 10$ **14.** $-15 + 6$

15. $-6.8 + 2.3$ **16.** $-7.9 + 2.4$

17. $\frac{11}{15} + \left(-\frac{3}{5}\right)$ **18.** $\frac{7}{10} + \left(-\frac{4}{5}\right)$

19. $-3.7 + (-4.5)$ **20.** $-6.2 + (-5.9)$

21. $0 + (-12.4)$ **22.** $0 + (-15.3)$

23. $12.4 + (-12.4)$ **24.** $15.3 + (-15.3)$

In Exercises 25–30, find $-x$ for the given value of x.

25. $x = 11$ **26.** $x = 13$

27. $x = -5$ **28.** $x = -9$

29. $x = 0$ **30.** $x = -\sqrt{2}$

In Exercises 31–42, subtract as indicated.

31. $3 - 15$ **32.** $4 - 20$

33. $8 - (-10)$ **34.** $7 - (-13)$

35. $-20 - (-5)$ **36.** $-30 - (-10)$

37. $\frac{1}{4} - \frac{1}{2}$ **38.** $\frac{1}{10} - \frac{2}{5}$

39. $-2.3 - (-7.8)$ **40.** $-4.3 - (-8.7)$

41. $0 - (-\sqrt{2})$ **42.** $0 - (-\sqrt{3})$

In Exercises 43–54, multiply as indicated.

43. $9(-10)$ **44.** $8(-10)$

45. $(-3)(-11)$ **46.** $(-7)(-11)$

47. $\frac{15}{13}(-1)$ **48.** $\frac{11}{13}(-1)$

49. $-\sqrt{2} \cdot 0$ **50.** $-\sqrt{3} \cdot 0$

51. $(-4)(-2)(-1)$ **52.** $(-5)(-3)(-2)$

53. $2(-3)(-1)(-2)(-4)$ **54.** $3(-2)(-1)(-5)(-3)$

In Exercises 55–64, evaluate each exponential expression.

55. $(-10)^2$

56. $(-8)^2$

57. -10^2

58. -8^2

59. $(-2)^3$

60. $(-3)^3$

61. $(-1)^4$

62. $(-4)^4$

63. $(-1)^{33}$

64. $(-1)^{35}$

In Exercises 65–76, divide as indicated or state that the division is undefined.

65. $\frac{12}{-4}$

66. $\frac{30}{-5}$

67. $\frac{-90}{-2}$

68. $\frac{-55}{-5}$

69. $\frac{0}{-4.6}$

70. $\frac{0}{-5.3}$

71. $-\frac{4.6}{0}$

72. $-\frac{5.3}{0}$

73. $-\frac{1}{2} \div \left(-\frac{7}{9}\right)$

74. $-\frac{1}{2} \div \left(-\frac{3}{5}\right)$

75. $6 \div \left(-\frac{2}{5}\right)$

76. $8 \div \left(-\frac{2}{9}\right)$

In Exercises 77–90, use the order of operations to simplify each expression.

77. $4(-5) - 6(-3)$

78. $8(-3) - 5(-6)$

79. $3(-2)^2 - 4(-3)^2$

80. $5(-3)^2 - 2(-2)^2$

81. $8^2 - 16 \div 2^2 \cdot 4 - 3$

82. $10^2 - 100 \div 5^2 \cdot 2 - 3$

83. $\frac{4^2 + 3^3}{5^2 - (-18)}$

84. $\frac{2^2 + (-3)^3}{5^2 - (-21)}$

85. $8 - 3[-2(2 - 5) - 4(8 - 6)]$

86. $8 - 3[-2(5 - 7) - 5(4 - 2)]$

87. $\frac{2(-2) - 4(-3)}{5 - 8}$

88. $\frac{6(-4) - 5(-3)}{9 - 10}$

89. $15 - \sqrt{3 - (-1)} + 12 \div 2 \cdot 3$

90. $17 - |5 - (-2)| + 12 \div 2 \cdot 3$

In Exercises 91–94, write an algebraic expression equivalent to the given expression using each of the commutative properties.

91. $4x + 10$

92. $5x + 30$

93. $7x - 5$

94. $3x - 7$

In Exercises 95–100, use an associative property to write an algebraic expression equivalent to each expression and simplify.

95. $4 + (6 + x)$

96. $12 + (3 + x)$

97. $-7(3x)$

98. $-10(5x)$

99. $-\frac{1}{3}(-3y)$

100. $-\frac{1}{4}(-4y)$

In Exercises 101–106, use the distributive property to write an equivalent expression.

101. $3(2x + 5)$

102. $5(4x + 7)$

103. $-7(2x + 3)$

104. $-9(3x + 2)$

105. $-(3x - 6)$

106. $-(6x - 3)$

In Exercises 107–120, simplify each algebraic expression.

107. $7x + 5x$

108. $8x + 10x$

109. $6x^2 - x^2$

110. $9x^2 - x^2$

111. $6x + 10x^2 + 4x + 2x^2$

112. $9x + 5x^2 + 3x + 4x^2$

113. $8(3x - 5) - 6x$

114. $7(4x - 5) - 8x$

115. $5(3y - 2) - (7y + 2)$

116. $4(5y - 3) - (6y + 3)$

117. $7 - 4[3 - (4y - 5)]$

118. $6 - 5[8 - (2y - 4)]$

119. $18x^2 + 4 - [6(x^2 - 2) + 5]$

120. $14x^2 + 5 - [7(x^2 - 2) + 4]$

Application Exercises

The graph compares the average price per gallon of premium gasoline in seven countries. The cost in Britain is \$4.29 per gallon. Use this information to solve Exercises 121–124 by filling in the missing dollar amount.

Average Price Per Gallon of Premium Gasoline

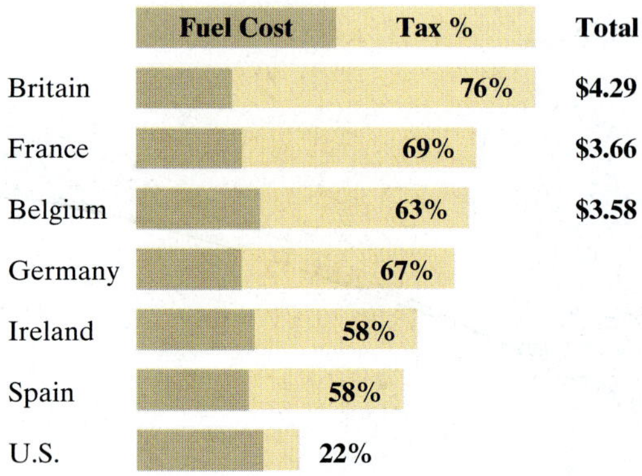

Source: Energy Information Administration; International Energy Agency

	Country	Price per Gallon of Premium Garoline	Difference from Britain
121.	France	\$3.66	____
122.	Belgium	\$3.58	____
123.	United States	____	\$2.55
124.	Spain	____	\$1.43

125. The bar graph shows that in 1995, Social Security had an annual cash surplus of \$233 billion. By 2020, this amount is expected to be a negative number—a deficit of \$244 billion. What is the difference between the 1995 surplus and the projected 2020 deficit?

Social Security Annual Operating Budget

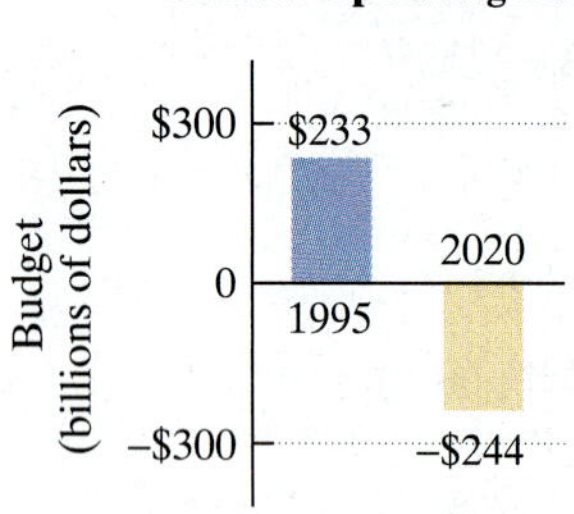

Source: U.S. Office of Management and Budget

126. The peak of Mount Everest is 8848 meters above sea level. The Marianas Trench, on the floor of the Pacific Ocean, is 10,915 meters below sea level. What is the difference in elevation between the peak of Mount Everest and the Marianas Trench?

127. The optimum heart rate is the rate that a person should achieve during exercise for the exercise to be most beneficial. The algebraic expression

$$0.6(220 - a)$$

describes a person's optimum heart rate, in beats per minute, where a represents the age of the person, in years.

a. Use the distributive property to rewrite the algebraic expression without parentheses.

b. Use each form of the algebraic expression to determine the optimum heart rate for a 20-year-old runner.

Writing in Mathematics

128. What is the meaning of $|a|$ in terms of a number line?

129. Explain how to add two numbers with the same sign. Give an example with your explanation.

130. Explain how to add two numbers with different signs. Give an example with your explanation.

131. What are opposites, or additive inverses? What happens when finding the sum of a number and its opposite?

132. Explain how to subtract real numbers.

133. Explain how to multiply two numbers with different signs. Give an example with your explanation.

134. Explain how to multiply two numbers with the same sign. Give an example with your explanation.

135. Explain how to determine the sign of a product that involves more than two numbers.

136. Explain how to divide real numbers.

137. Why is $\frac{0}{4} = 0$, although $\frac{4}{0}$ is undefined?

138. What are equivalent algebraic expressions?

139. State a commutative property and give an example of how it is used to write equivalent algebraic expressions.

140. State an associative property and give an example of how it is used to write equivalent algebraic expressions.

141. State a distributive property and give an example of how it is used to write equivalent algebraic expressions.

142. What are the terms of an algebraic expression? How can you tell if terms are like terms?

143. What does it mean to simplify an algebraic expression?

144. If a negative sign appears outside parentheses, explain how to simplify the expression. Give an example.

145. Use the graph in Exercises 121–124 to describe how fuel costs in the United States, excluding taxes, compare to fuel costs in the other six countries.

Critical Thinking Exercises

146. Which one of the following is true?

a. $16 \div 4 \cdot 2 = 16 \div 8 = 2$

b. $6 - 2(4 + 3) = 4(4 + 3) = 4(7) = 28$

c. $3 + 7x = 10x$

d. None of the above is true.

In Exercises 147–148, insert parentheses to make each statement true.

147. $8 - 2 \cdot 3 - 4 = 14$

148. $2 \cdot 5 - \frac{1}{2} \cdot 10 \cdot 9 = 45$

149. Simplify: $\dfrac{9[4 - (1 + 6)] - (3 - 9)^2}{5 + \dfrac{12}{5 - \dfrac{6}{2 + 1}}}$.

Review Exercises

From here on, each exercise set will contain three review exercises. It is important to review previously covered topics to improve your understanding of the topics and to help maintain your mastery of the material. If you are not certain how to solve a review exercise, turn to the section and the illustrative example given in parentheses at the end of each exercise.

150. Write the following English phrase as an algebraic expression: "The quotient of ten and a number, decreased by four times the number." Let x represent the number. (Section 1.1, Example 1)

151. Evaluate $10 + 2(x - 5)^4$ for $x = 7$. (Section 1.1, Example 3)

152. Determine whether the following statement is true or false: $\frac{1}{2} \notin \{x \mid x \text{ is an irrational number}\}$. (Section 1.1, Example 5)

▸ SECTION 1.3 Graphing Equations

Objectives

1. Plot points in the rectangular coordinate system.
2. Graph equations in the rectangular coordinate system.
3. Use the rectangular coordinate system to visualize relationships between variables.
4. Interpret information about a graphing utility's viewing rectangle.

SSM PH Tutor Center CD-ROM Video

Singers Ricky Martin and Jennifer Lopez, television journalist Soledad O'Brien, boxer Oscar De La Hoya, Internet entrepreneur Carlos Cardona, actor John Leguizamo, and political organizer Luigi Crespo are members of a generation of young Hispanics that is changing the way America looks, feels, thinks, eats, dances, and votes. There are 31 million Hispanics in the United States, pumping \$300 billion a year into the economy. By 2050, the Hispanic population is expected to reach 96 million—an increase of more than 200 percent. Hispanics are also younger than the rest of the nation: A third are under 18. The graph in Figure 1.4 on page 30 shows the average age of whites, African Americans, and Americans of Hispanic origin.

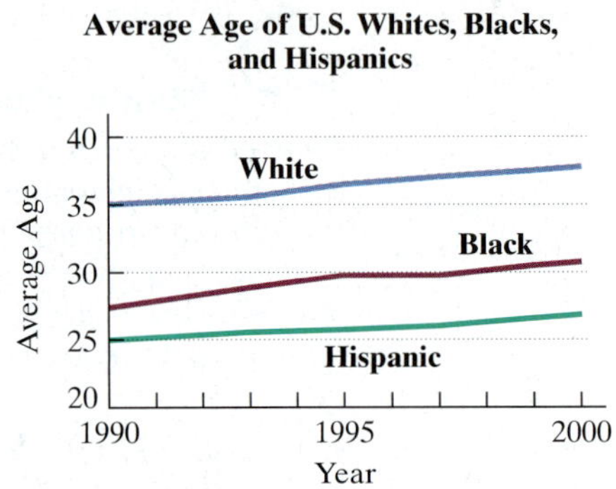

Source: U.S. Census Bureau

Figure 1.4

Magazines and newspapers often display information using **line graphs** like those in Figure 1.4. Line graphs are often used to illustrate trends over time. Some measure of time, such as months or years, frequently appears on the horizontal axis. Amounts are generally listed on the vertical axis.

A line graph displays information based on an idea developed by the French philosopher and mathematician René Descartes (1596–1650). Descartes developed a new branch of mathematics that brought together algebra and geometry in a unified way—a way that visualized numbers as points on a graph, equations as geometric figures, and geometric figures as equations. We begin this section by looking at Descartes's deceptively simple idea, called the **rectangular coordinate system** or (in his honor) the **Cartesian coordinate system**.

1 Plot points in the rectangular coordinate system.

Points and Ordered Pairs

Descartes used two number lines that intersect at right angles at their zero points, as shown in Figure 1.5. The horizontal number line is the ***x*-axis**. The vertical number line is the ***y*-axis**. The point of intersection of these axes is their zero points, called the **origin**. Positive numbers are shown to the right and above the origin. Negative numbers are shown to the left and below the origin. The axes divide the plane into four quarters, called **quadrants**. The points located on the axes are not in any quadrant.

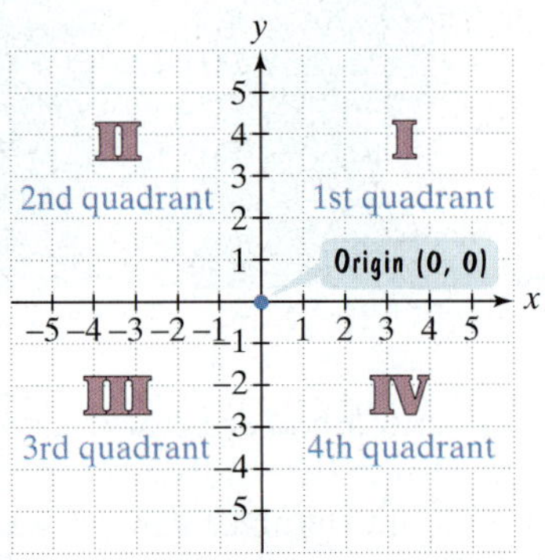

Figure 1.5 The rectangular coordinate system

Each point in the rectangular coordinate system corresponds to an **ordered pair** of real numbers, (x, y). Examples of such pairs are $(4, 2)$ and $(-5, -3)$. The first number in each pair, called the ***x*-coordinate**, denotes the distance and direction from the origin along the x-axis. The second number, called the ***y*-coordinate**, denotes vertical distance and direction along a line parallel to the y-axis or along the y-axis itself.

Figure 1.6 shows how we **plot**, or locate, the points corresponding to the ordered pairs (4, 2) and (−5, −3). We plot (4, 2) by going 4 units from 0 to the right along the x-axis. Then we go 2 units up parallel to the y-axis. We plot (−5, −3) by going 5 units from 0 to the left along the x-axis and 3 units down parallel to the y-axis. The phrase "the point corresponding to the ordered pair (−5, −3)" is often abbreviated as "the point (−5, −3)."

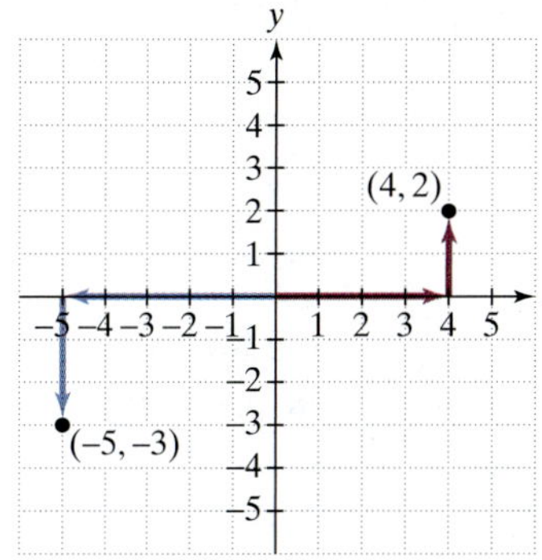

Figure 1.6 Plotting (4, 2) and (−5, −3)

EXAMPLE 1 Plotting Points in the Rectangular Coordinate System

Plot the points: $A(-3, 5)$, $B(2, -4)$, $C(5, 0)$, $D(-5, -3)$, $E(0, 4)$, and $F(0, 0)$.

Solution See Figure 1.7. We plot the points in the following way:

$A(-3, 5)$: 3 units left, 5 units up

$B(2, -4)$: 2 units right, 4 units down

$C(5, 0)$: 5 units right, 0 units up or down

$D(-5, -3)$: 5 units left, 3 units down

$E(0, 4)$: 0 units right or left, 4 units up

$F(0, 0)$: 0 units right or left, 0 units up or down

The origin is represented by (0, 0).

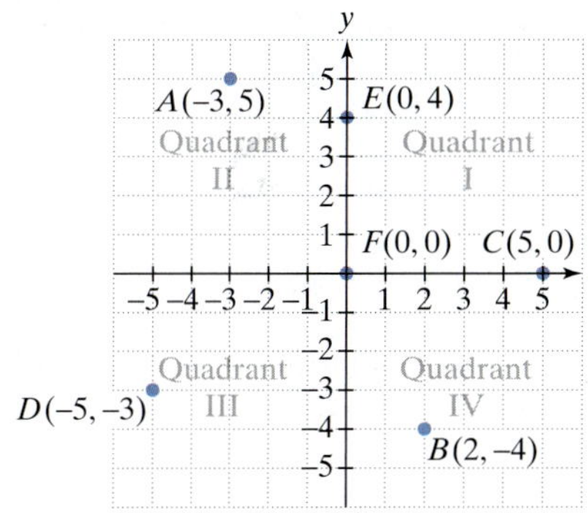

Figure 1.7 Plotting points

CHECK POINT 1 Plot the points:

$A(-2, 4)$, $B(4, -2)$, $C(-3, 0)$, and $D(0, -3)$.

2 Graph equations in the rectangular coordinate system.

Graphs of Equations

A relationship between two quantities can be expressed as an **equation in two variables**, such as

$$y = x^2 - 4.$$

A **solution** to this equation is an ordered pair of real numbers with the following property: When the x-coordinate is substituted for x and the y-coordinate is substituted for y in the equation, we obtain a true statement. For example, if we let $x = 3$, then $y = 3^2 - 4 = 9 - 4 = 5$. The ordered pair (3, 5) is a solution to the equation $y = x^2 - 4$. We also say that (3, 5) **satisfies** the equation.

We can generate as many ordered-pair solutions as desired to $y = x^2 - 4$ by substituting numbers for x and then finding the values for y. The **graph of the equation** is the set of all points whose coordinates satisfy the equation.

One method for graphing an equation such as $y = x^2 - 4$ is the **point-plotting method**. First, we find several ordered pairs that are solutions to the equation. Next, we plot these ordered pairs as points in the rectangular coordinate system. Finally, we connect the points with a smooth curve or line. This often gives us a picture of all ordered pairs that satisfy the equation.

EXAMPLE 2 Graphing an Equation Using the Point-Plotting Method

Graph $y = x^2 - 4$. Select integers for x, starting with -3 and ending with 3.

Solution For each value of x we find the corresponding value for y.

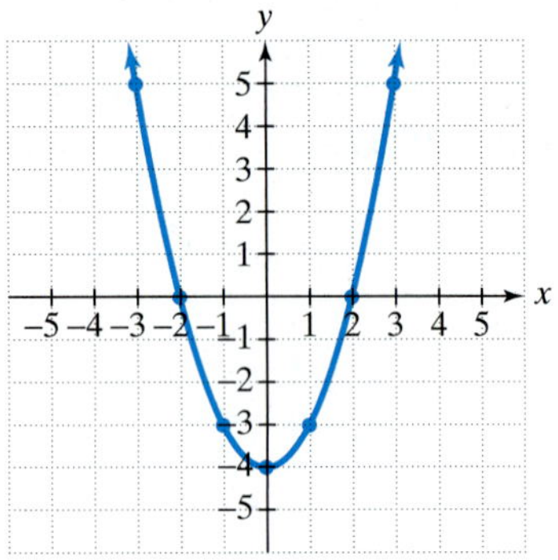

Figure 1.8 The graph of $y = x^2 - 4$

x	$y = x^2 - 4$	Ordered Pair (x, y)
-3	$y = (-3)^2 - 4 = 9 - 4 = 5$	$(-3, 5)$
-2	$y = (-2)^2 - 4 = 4 - 4 = 0$	$(-2, 0)$
-1	$y = (-1)^2 - 4 = 1 - 4 = -3$	$(-1, -3)$
0	$y = 0^2 - 4 = 0 - 4 = -4$	$(0, -4)$
1	$y = 1^2 - 4 = 1 - 4 = -3$	$(1, -3)$
2	$y = 2^2 - 4 = 4 - 4 = 0$	$(2, 0)$
3	$y = 3^2 - 4 = 9 - 4 = 5$	$(3, 5)$

Now we plot the seven points and join them with a smooth curve, as shown in Figure 1.8. The graph of $y = x^2 - 4$ is a curve where the part of the graph to the right of the y-axis is a reflection of the part to the left of it and vice versa. The arrows on the left and the right of the curve indicate that it extends indefinitely in both directions. ■

✔ **CHECK POINT 2** Graph $y = x^2 - 1$. Select integers for x, starting with -3 and ending with 3.

EXAMPLE 3 Graphing an Equation Using the Point-Plotting Method

Graph $y = |x|$. Select integers for x, starting with -3 and ending with 3.

Solution For each value of x we find the corresponding value for y.

x	$y = \|x\|$	Ordered Pair (x, y)
-3	$y = \|-3\| = 3$	$(-3, 3)$
-2	$y = \|-2\| = 2$	$(-2, 2)$
-1	$y = \|-1\| = 1$	$(-1, 1)$
0	$y = \|0\| = 0$	$(0, 0)$
1	$y = \|1\| = 1$	$(1, 1)$
2	$y = \|2\| = 2$	$(2, 2)$
3	$y = \|3\| = 3$	$(3, 3)$

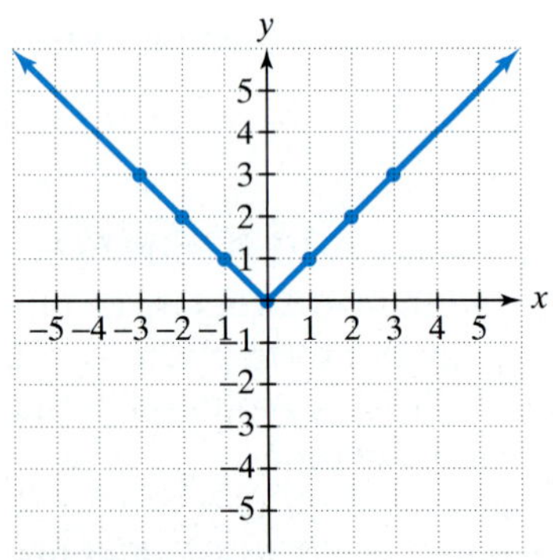

Figure 1.9 The graph of $y = |x|$

We plot the points and connect them, resulting in the graph shown in Figure 1.9. The graph is V-shaped and centered at the origin. For every point (x, y) on the graph, the point $(-x, y)$ is also on the graph. This shows that the absolute value of a positive number is the same as the absolute value of its opposite. ■

✔ **CHECK POINT 3** Graph $y = |x + 1|$. Select integers for x, starting with -4 and ending with 2.

3 Use the rectangular coordinate system to visualize relationships between variables.

Applications

The rectangular coordinate system allows us to visualize relationships between two variables by connecting any equation in two variables with a graph. Graphs in the rectangular coordinate system can also be used to tell a story.

EXAMPLE 4 Telling a Story with a Graph

Too late for that flu shot now! It's only 8 A.M. and you're feeling lousy. Fascinated by the way that algebra models the world (your author is projecting a bit here), you construct a graph showing your body temperature from 8 A.M. through 3 P.M. You decide to let x represent the number of hours after 8 A.M. and y your body temperature at time x. The graph is shown in Figure 1.10. The symbol ⌇ on the y-axis shows that there is a break in values between 0 and 98. Thus, the first tick mark on the y-axis represents a temperature of 98°F.

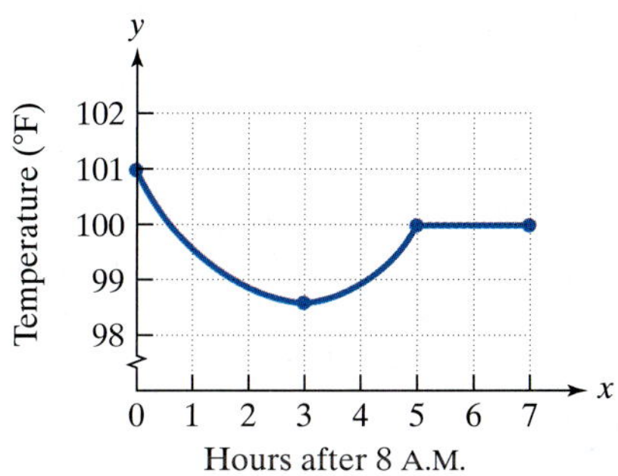

Figure 1.10 Body temperature from 8 A.M. through 3 P.M.

a. What is your temperature at 8 A.M.?

b. During which period of time is your temperature decreasing?

c. Estimate your minimum temperature during the time period shown. How many hours after 8 A.M. does this occur? At what time does this occur?

d. During which period of time is your temperature increasing?

e. Part of the graph is shown as a horizontal line segment. What does this mean about your temperature and when does this occur?

Solution

a. Because x is the number of hours after 8 A.M., your temperature at 8 A.M. corresponds to $x = 0$. Locate 0 on the horizontal axis and look at the point on the graph above 0. The figure on the right shows that your temperature at 8 A.M. is 101°F.

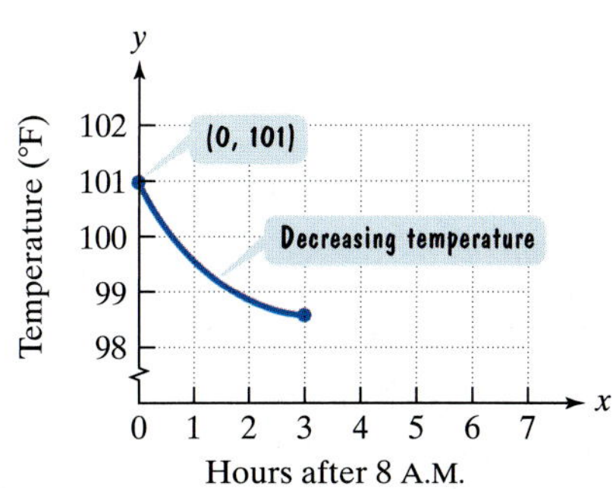

b. Your temperature is decreasing when the graph falls from left to right. This occurs between $x = 0$ and $x = 3$, shown in the preceding figure. Because x represents the number of hours after 8 A.M., your temperature is decreasing between 8 A.M. and 11 A.M.

c. Your minimum temperature can be found by locating the lowest point on the graph. This point lies above 3 on the horizontal axis, shown in the figure on the right. The y-coordinate of this point falls more than midway between 98 and 99, at approximately 98.6. The lowest point on the graph, (3, 98.6), shows that your minimum temperature, 98.6°F, occurs 3 hours after 8 A.M., at 11 A.M.

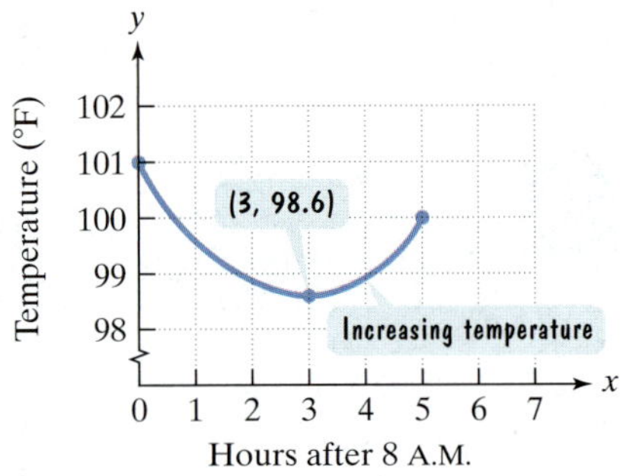

d. Your temperature is increasing when the graph rises from left to right. This occurs between $x = 3$ and $x = 5$, shown in the preceding figure. Because x represents the number of hours after 8 A.M., your temperature is increasing between 11 A.M. and 1 P.M.

e. The horizontal line segment shown in the figure on the right indicates that your temperature is neither increasing nor decreasing. Your temperature remains the same, 100°F, between $x = 5$ and $x = 7$. Thus, your temperature is at a constant 100°F between 1 P.M. and 3 P.M.

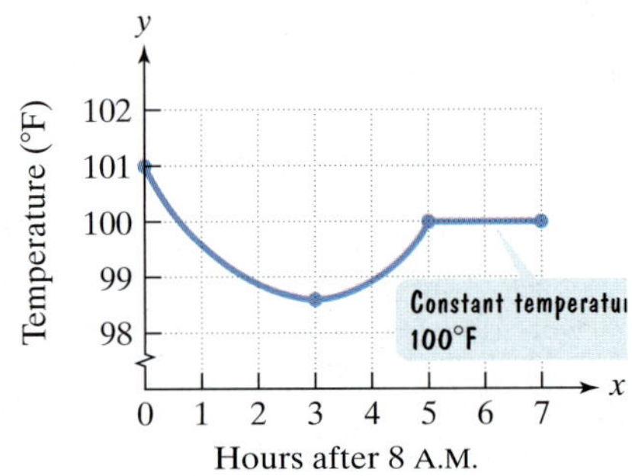

✔ **CHECK POINT 4** When a person receives a drug injected into a muscle, the concentration of the drug in the body, measured in milligrams per 100 milliliters, depends on the time elapsed after the injection, measured in hours. The figure shows the graph of drug concentration over time, where x represents hours after the injection and y represents the drug concentration at time x.

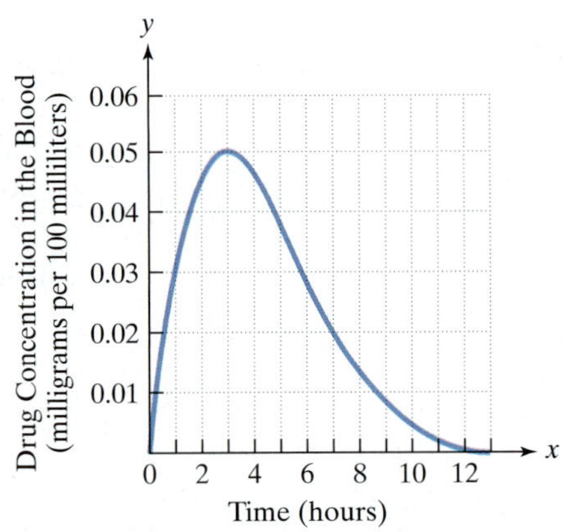

a. During which period of time is the drug concentration increasing?

b. During which period of time is the drug concentration decreasing?

c. What is the drug's maximum concentration and when does this occur?

d. What happens by the end of 13 hours?

4 Interpret information about a graphing utility's viewing rectangle.

Graphing Equations Using a Graphing Utility

Graphing calculators or graphing software packages for computers are referred to as **graphing utilities** or graphers. A graphing utility is a powerful tool that quickly generates the graph of an equation in two variables. Figure 1.11 shows two such graphs for the equations in Examples 2 and 3.

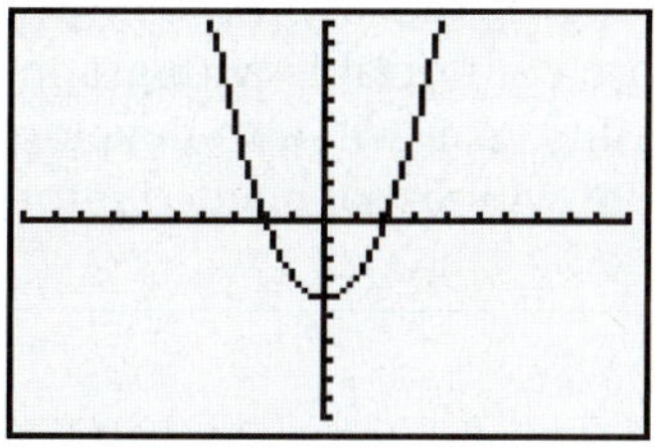

Figure 1.11(a) The graph of $y = x^2 - 4$

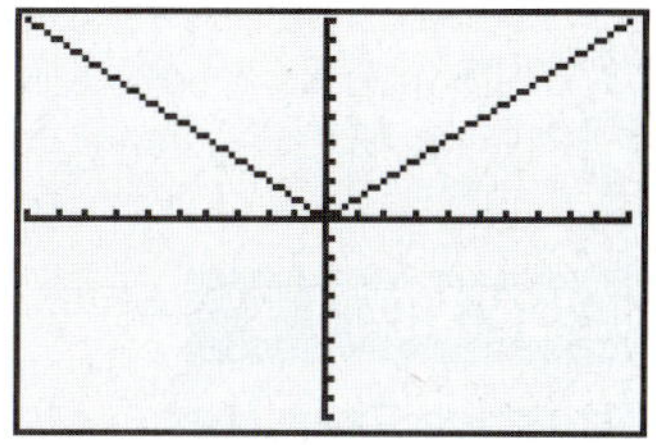

Figure 1.11(b) The graph of $y = |x|$

Study Tip

Even if you are not using a graphing utility in the course, read this part of the section. Knowing about viewing rectangles will enable you to understand the graphs that we display in the technology boxes throughout the book.

What differences do you notice between these graphs and the graphs that we drew by hand? They do seem a bit "jittery." Arrows do not appear on the left and right ends of the graphs. Furthermore, numbers are not given along the axes. For both graphs in Figure 1.11, the x-axis extends from -10 to 10 and the y-axis also extends from -10 to 10. The distance represented by each consecutive tick mark is one unit. We say that the **viewing rectangle** is $[-10, 10, 1]$ by $[-10, 10, 1]$.

$[-10,$ $10,$ $1]$ by $[-10,$ $10,$ $1]$

- -10: The minimum x-value along the x-axis is -10.
- 10: The maximum x-value along the x-axis is 10.
- 1: Distance between consecutive tick marks on the x-axis is one unit.
- -10: The minimum y-value along the y-axis is -10.
- 10: The maximum y-value along the y-axis is 10.
- 1: Distance between consecutive tick marks on the y-axis is one unit.

To graph an equation in x and y using a graphing utility, enter the equation and specify the size of the viewing rectangle. The size of the viewing rectangle sets minimum and maximum values for both the x- and y-axes. Enter these values, as well as the values between consecutive tick marks, on the respective axes. The $[-10, 10, 1]$ by $[-10, 10, 1]$ viewing rectangle used in Figure 1.11 is called the **standard viewing rectangle**.

EXAMPLE 5 Understanding the Viewing Rectangle

What is the meaning of a $[-2, 3, 0.5]$ by $[-10, 20, 5]$ viewing rectangle?

Solution We begin with $[-2, 3, 0.5]$, which describes the x-axis. The minimum x-value is -2, and the maximum x-value is 3. The distance between consecutive tick marks is 0.5.

Next, consider $[-10, 20, 5]$, which describes the y-axis. The minimum y-value is -10, and the maximum y-value is 20. The distance between consecutive tick marks is 5.

Figure 1.12 illustrates a $[-2, 3, 0.5]$ by $[-10, 20, 5]$ viewing rectangle. To make things clearer, we've placed numbers by each tick mark. These numbers do not appear on the axes when you use a graphing utility to graph an equation. ■

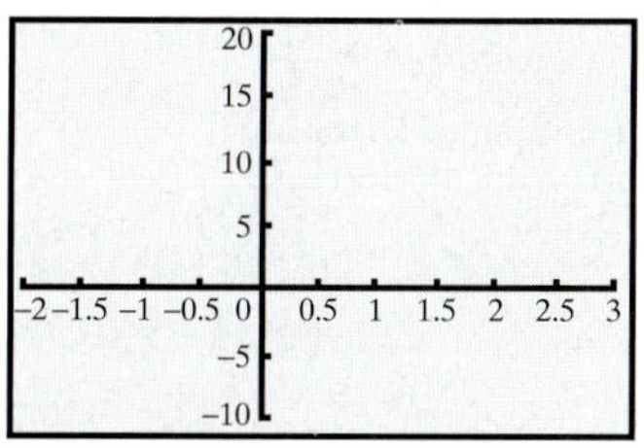

Figure 1.12 A $[-2, 3, 0.5]$ by $[-10, 20, 5]$ viewing rectangle

CHECK POINT 5 What is the meaning of a $[-100, 100, 50]$ by $[-100, 100, 10]$ viewing rectangle? Create a figure like the one in Figure 1.12 that illustrates this viewing rectangle.

On most graphing utilities, the display screen is two-thirds as high as it is wide. By using a square setting, you can make the x and y tick marks be equally spaced. (This does not occur in the standard viewing rectangle.) Graphing utilities can also *zoom in* and *zoom out*. When you zoom in, you see a smaller portion of the graph, but you do so in greater detail. When you zoom out, you see a larger portion of the graph. Thus, zooming out may help you to develop a better understanding of the overall character of the graph. With practice, you will become more comfortable with graphing equations in two variables using your graphing utility. You will also develop a better sense of the size of the viewing rectangle that will reveal needed information about a particular graph.

EXERCISE SET 1.3

Practice Exercises

In Exercises 1–10, plot the given point in a rectangular coordinate system.

1. (1, 4) **2.** (2, 5)
3. (−2, 3) **4.** (−1, 4)
5. (−3, −5) **6.** (−4, −2)
7. (4, −1) **8.** (3, −2)
9. (−4, 0) **10.** (0, −3)

Graph each equation in Exercises 11–26. Let x = −3, −2, −1, 0, 1, 2, and 3.

11. $y = x^2 - 2$ **12.** $y = x^2 + 2$
13. $y = x - 2$ **14.** $y = x + 2$
15. $y = 2x + 1$ **16.** $y = 2x - 4$
17. $y = -\frac{1}{2}x$ **18.** $y = -\frac{1}{2}x + 2$
19. $y = |x| + 1$ **20.** $y = |x| - 1$
21. $y = 2|x|$ **22.** $y = -2|x|$
23. $y = 4 - x^2$ **24.** $y = 9 - x^2$
25. $y = x^3$ **26.** $y = x^3 - 1$

In Exercises 27–30, match the viewing rectangle with the correct figure. Then label the tick marks in the figure to illustrate this viewing rectangle.

27. [−5, 5, 1] by [−5, 5, 1]
28. [−10, 10, 2] by [−4, 4, 2]
29. [−20, 80, 10] by [−30, 70, 10]
30. [−40, 40, 20] by [−1000, 1000, 100]

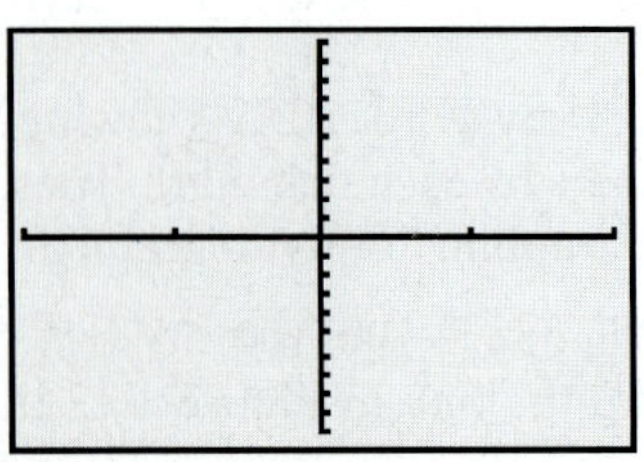
(a)

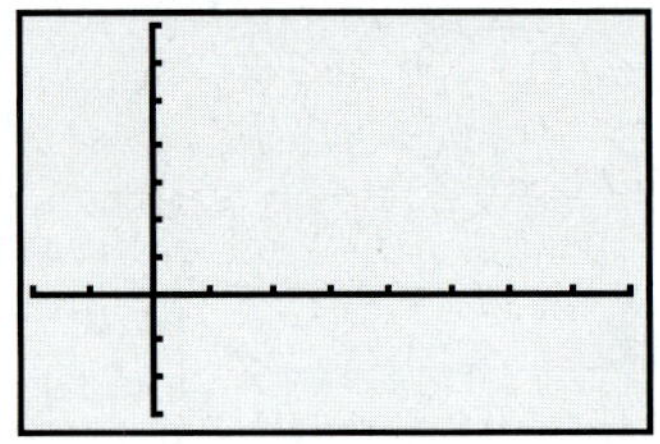
(b)

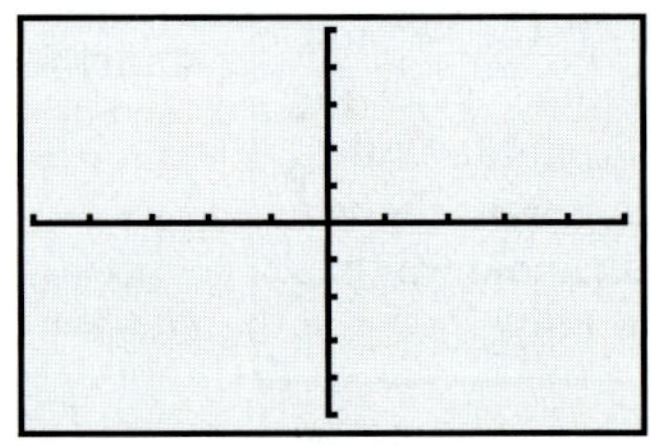
(c)

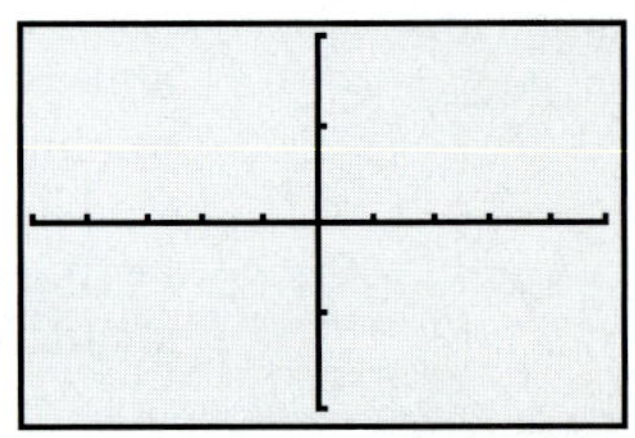
(d)

Application Exercises

The graph shows the percent distribution of divorces in the United States by number of years of marriage. Use the graph to solve Exercises 31–34.

Percent Distribution of Divorces by Number of Years of Marriage

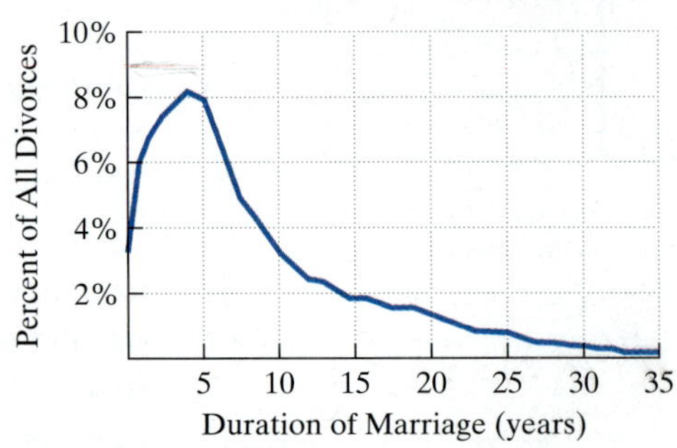

Source: Divorce Center

31. During which years of marriage is the chance of divorce increasing?
32. During which years of marriage is the chance of divorce decreasing?
33. During which year of marriage is the chance of divorce the highest? Estimate, to the nearest tenth of a percent, the percentage of divorces that occur during this year.

33. During which year of marriage is the chance of divorce the lowest? Estimate, to the nearest tenth of a percent, the percentage of divorces that occur during this year.

One effect of the presidential election of 2000 was to convince Americans that their votes count. Although turnout is crucial in any democracy, since the 1960s, the percentage of Americans who actually vote has been decreasing. Use the graph showing the percentage of Americans who vote in presidential elections to solve Exercises 35–38. Each tic mark on the horizontal axis represents 4 years.

Percentage of Americans Voting in Presidential Elections

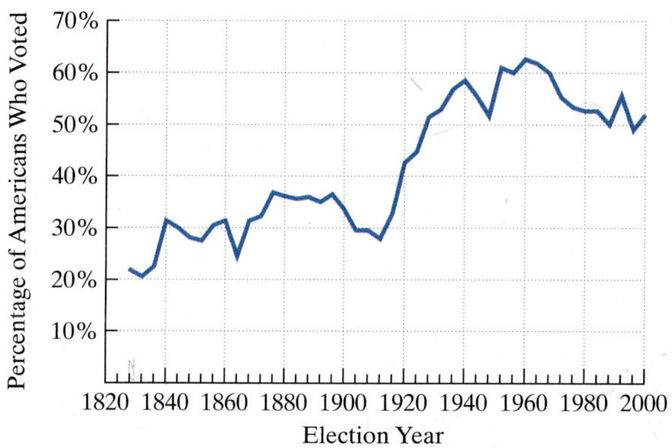

Source: National Journal

35. During which year was the percentage of Americans voting the lowest? What percentage of Americans voted in that year?
36. During which year was the percentage of Americans voting the highest? What percentage, to the nearest percent, voted in that year?
37. Approximately what percentage of Americans voted in the 2000 presidential election? Round to the nearest percent.
38. Approximately what percentage of Americans voted in the first presidential election? Round to the nearest percent.

In Exercises 39–42, match the story with the correct figure. The figures appear at the top of the next column and are labeled (a), (b), (c), and (d)..

39. As the blizzard got worse, the snow fell harder and harder.
40. The snow fell more and more softly.
41. It snowed hard, but then it stopped. After a short time, the snow started falling softly.
42. It snowed softly, and then it stopped. After a short time, the snow started falling hard.

(a)

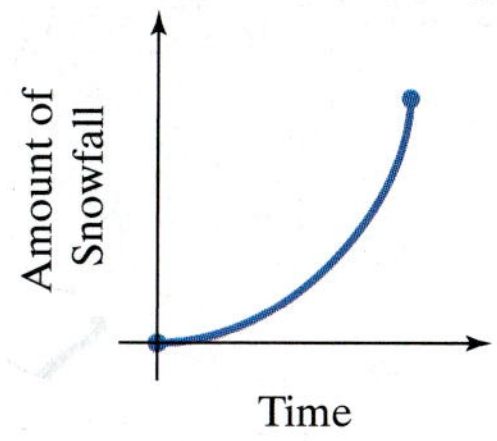

(b)

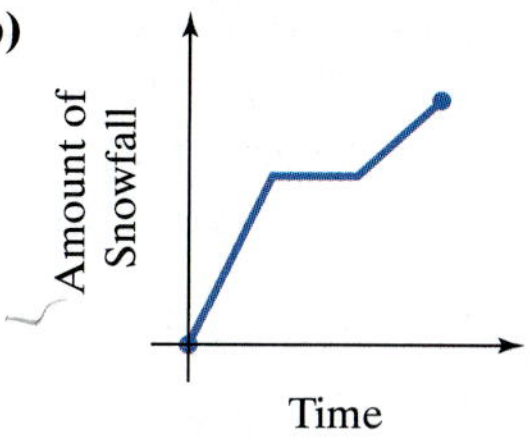

(c)

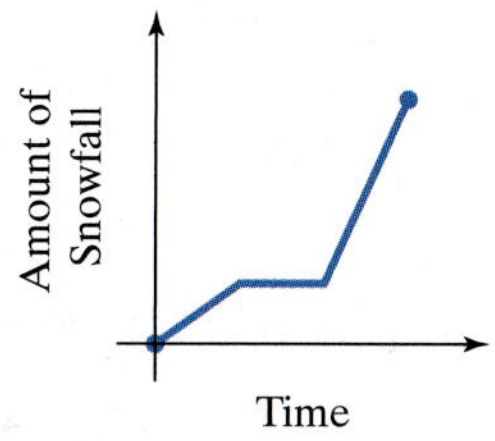

(d)

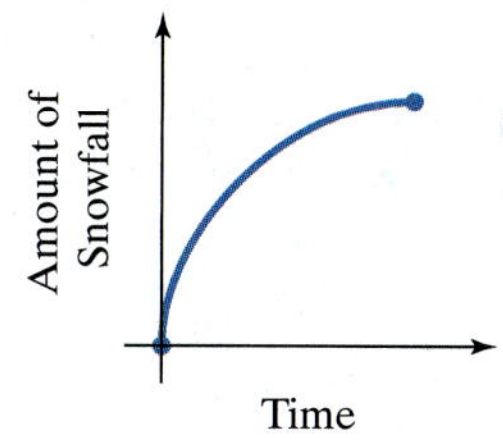

In Exercises 43–44, select the graph that best illustrates each story.

43. An airplane flew from Miami to San Francisco.

a.

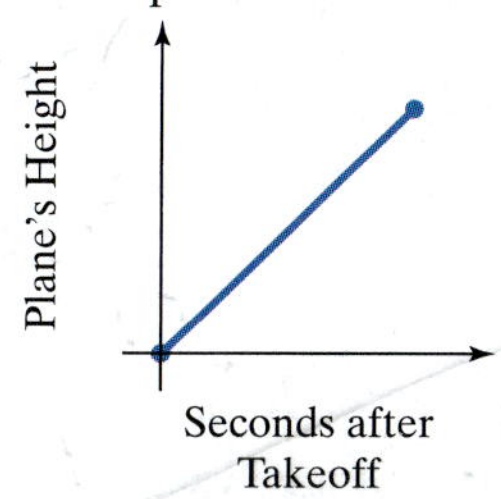

b.

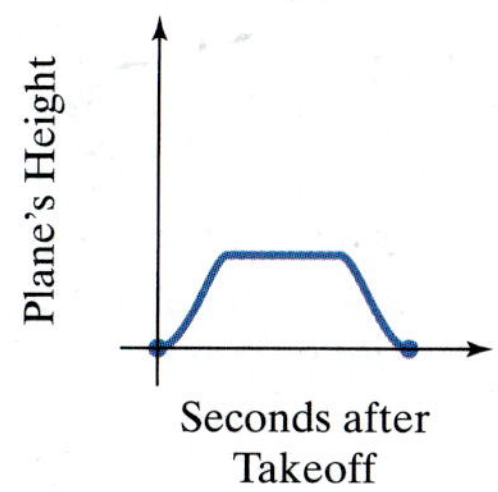

c.

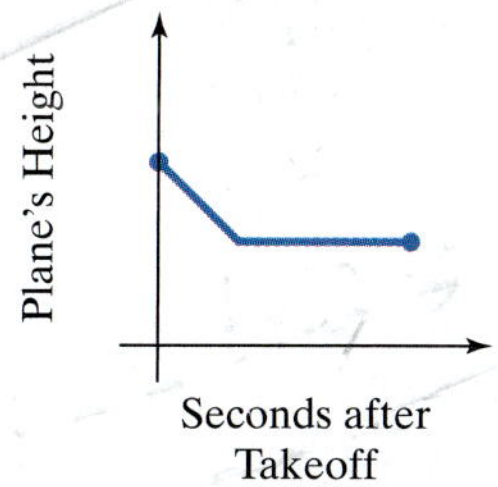

d.

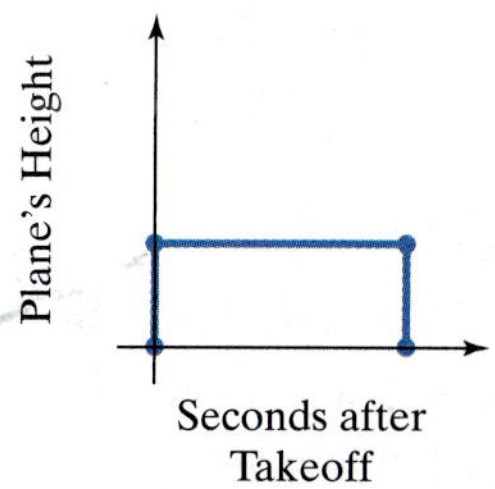

44. At noon, you begin to breathe in.

a.

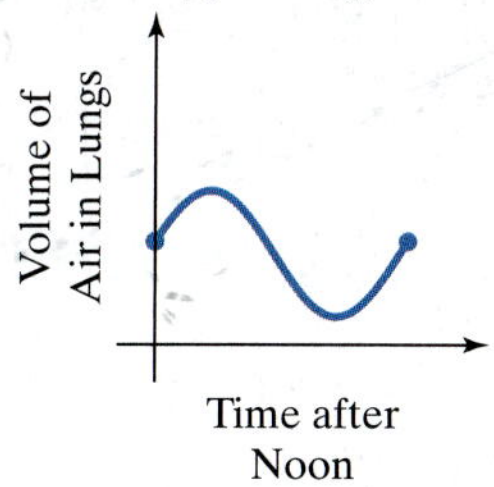

b.

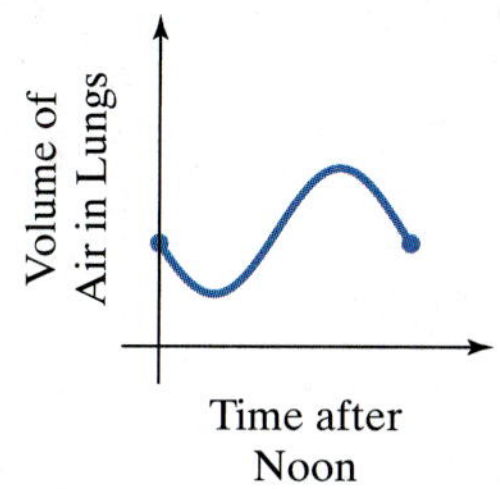

c.

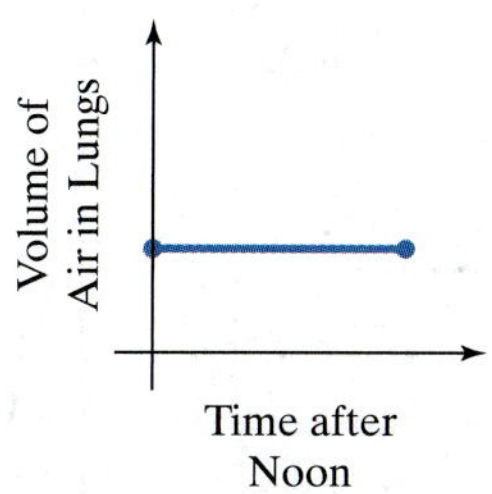

d.

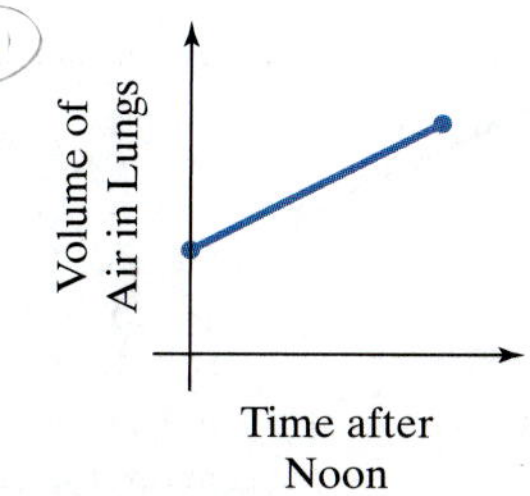

A portion of the rectangular coordinate system can be used to show bands of healthy weights for different heights. The figure shows such a weight chart for men between the ages of 20 and 65. Use the information shown to solve Exercises 45–48. Round all answers to the nearest pound.

Weight Chart for Men Ages 20 Through 65

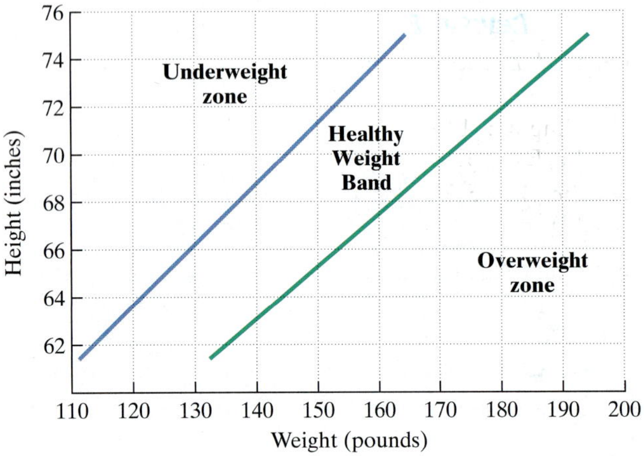

Source: United States Department of Agriculture

45. What is the maximum healthy weight for a man who is 6 feet (72 inches) tall?

46. What is the minimum healthy weight for a man who is 6 feet (72 inches) tall?

47. A man who is 70 inches tall needs to gain ten pounds. Approximately how much does this man weigh?

48. A man who is 70 inches tall needs to lose ten pounds. Approximately how much does this man weigh?

Writing in Mathematics

49. What is the rectangular coordinate system?

50. Explain how to plot a point in the rectangular coordinate system. Give an example with your explanation.

51. Explain why $(5, -2)$ and $(-2, 5)$ do not represent the same point.

52. Explain how to graph an equation in the rectangular coordinate system.

53. What does a $[-20, 2, 1]$ by $[-4, 5, 0.5]$ viewing rectangle mean?

In Exercises 54–55, write a story, or description, to match each graph.

54.

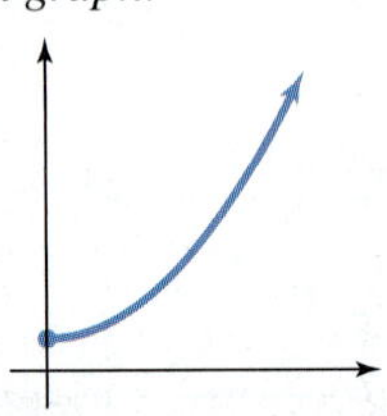

55.

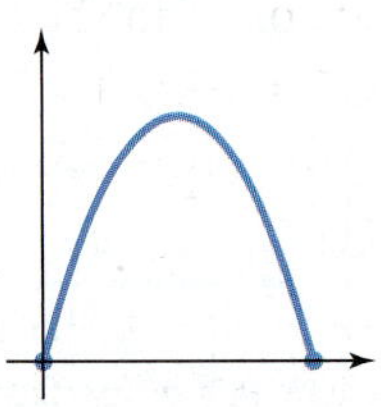

56. The graph for Exercises 31–34 shows that the chance of divorce is greatest during the fourth year of marriage. What explanation can you offer for this trend?

Technology Exercises

57. Use a graphing utility to verify each of your hand-drawn graphs in Exercises 11–26. Experiment with the range setting to make the graph displayed by the graphing utility resemble your hand-drawn graph as much as possible.

58. The stated intent of the 1994 "don't ask, don't tell" policy was to reduce the number of discharges of gay men and lesbians from the military. The formula

$$y = 45.48x^2 - 334.35x + 1237.9$$

models the number of gay service members, y, discharged from the military for homosexuality x years after 1990. Graph the equation in a $[0, 10, 1]$ by $[0, 2200, 100]$ viewing rectangle. Then describe something about the relationship between x and y that is revealed by looking at the graph that is not obvious from the equation. What does the graph reveal about the success or lack of success of "don't ask, don't tell"?

A graph of an equation is a complete graph if it shows all of the important features of the graph. Use a graphing utility to graph the equations in Exercises 59–62 in each of the given viewing rectangles. Then choose which viewing rectangle gives a complete graph.

59. $y = x^2 + 10$
- **a.** $[-5, 5, 1]$ by $[-5, 5, 1]$
- **b.** $[-10, 10, 1]$ by $[-10, 10, 1]$
- **c.** $[-10, 10, 1]$ by $[-50, 50, 1]$

60. $y = 0.1x^4 - x^3 + 2x^2$
- **a.** $[-5, 5, 1]$ by $[-8, 2, 1]$
- **b.** $[-10, 10, 1]$ by $[-10, 10, 1]$
- **c.** $[-8, 16, 1]$ by $[-16, 8, 1]$

61. $y = \sqrt{x} + 18$
- **a.** $[-10, 10, 1]$ by $[-10, 10, 1]$
- **b.** $[-50, 50, 10]$ by $[-10, 10, 1]$
- **c.** $[-10, 10, 1]$ by $[-50, 50, 10]$

62. $y = x^3 - 30x + 20$
- **a.** $[-10, 10, 1]$ by $[-10, 10, 1]$
- **b.** $[-10, 10, 1]$ by $[-50, 50, 10]$
- **c.** $[-10, 10, 1]$ by $[-50, 100, 10]$

Critical Thinking Exercises

63. Which one of the following is true?
- **a.** If the product of a point's coordinates is positive, the point must be in quadrant I.
- **b.** If a point is on the x-axis, it is neither up nor down, so $x = 0$.
- **c.** A graphing utility displays numbers along the x- and y-axes.
- **d.** None of the above is true.

The graph shows the height, y, in feet, of a ball dropped from the top of the Empire State Building at various times during its fall to the ground. Use the graph to solve Exercises 64–65.

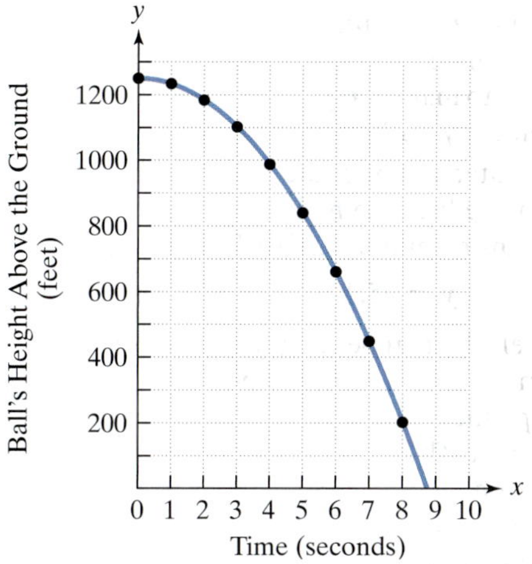

64. What is a reasonable estimate of the height of the Empire State Building?

65. What is a reasonable estimate, to the nearest tenth of a second, of how long it takes for the ball to hit the ground?

Review Exercises

66. Find the absolute value: $|-14.3|$. (Section 1.2, Example 1.)

67. Simplify: $[12 - (13 - 17)] - [9 - (6 - 10)]$. (Section 1.2, Examples 7 and 8)

68. Simplify: $6x - 5(4x + 3) - 10$. (Section 1.2, Example 13)

▸ SECTION 1.4 Solving Linear Equations

Objectives

1. Solve linear equations.
2. Recognize identities, conditional equations, and inconsistent equations.
3. Solve applied problems using mathematical models.

SSM PH Tutor Center CD-ROM Video

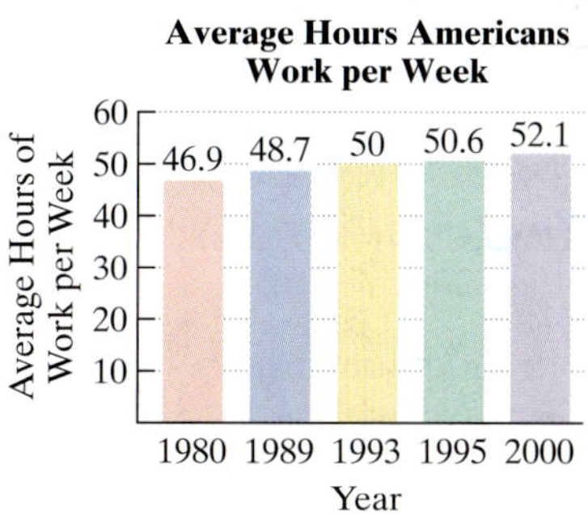

Figure 1.13 *Source: U.S.A. Today*

You would think that technological advances in the workplace have made it possible for us to work less and have more recreational time. Wrong. The bar graph in Figure 1.13 shows that the average number of hours that Americans work per week is gradually increasing. The data can be modeled by the formula

$$W = 0.3x + 46.6$$

where W represents the average number of hours that Americans work per week x years after 1980. So when will our workoholic nation average 55 (gasp) hours of work per week? Substitute 55 for W in the formula:

$$55 = 0.3x + 46.6.$$

We can reverse the two sides of this equation, expressing it as

$$0.3x + 46.6 = 55.$$

The form of this equation is $ax + b = c$, with $a = 0.3$, $b = 46.6$, and $c = 55$. Any equation of this form is called a **linear equation in one variable**. The exponent on the variable in such an equation is 1. In this section, we will study how to solve linear equations.

1 Solve linear equations.

Solving Linear Equations in One Variable

We begin by stating the definition of a linear equation in one variable.

Definition of a Linear Equation

A **linear equation in one variable** x is an equation that can be written in the form

$$ax + b = 0$$

where a and b are real numbers, and $a \neq 0$ (a is not equal to 0).

An example of a linear equation in one variable is

$$4x + 12 = 0.$$

Solving an equation in x involves determining all values of x that result in a true statement when substituted into the equation. Such values are **solutions** or **roots** of the equation. For example, substitute -3 for x in $4x + 12 = 0$. We obtain

$$4(-3) + 12 = 0, \quad \text{or} \quad -12 + 12 = 0.$$

This simplifies to the true statement $0 = 0$. Thus, -3 is a solution of the equation $4x + 12 = 0$. We also say that -3 **satisfies** the equation $4x + 12 = 0$, because when we substitute -3 for x, a true statement results. The set of all such solutions is called the equation's **solution set**. For example, the solution set of the equation $4x + 12 = 0$ is $\{-3\}$.

Equations that have the same solution set are called **equivalent equations**. For example, the equations

$$4x + 12 = 0 \quad \text{and} \quad 4x = -12 \quad \text{and} \quad x = -3$$

are equivalent equations because the solution set for each is $\{-3\}$. To solve a linear equation in x, we transform the equation into an equivalent equation one or more times. Our final equivalent equation should be of the form

$$x = \text{a number}.$$

The solution set of this equation is the set consisting of the number.

To generate equivalent equations, we will use the following properties:

The Addition and Multiplication Properties of Equality

The Addition Property of Equality

The same real number or algebraic expression may be added to both sides of an equation without changing the equation's solution set.

$$a = b \text{ and } a + c = b + c \text{ are equivalent equations.}$$

The Multiplication Property of Equality

The same nonzero real number or algebraic expression may multiply both sides of an equation without changing the equation's solution set.

$$a = b \text{ and } ac = bc \text{ are equivalent equations as long as } c \neq 0.$$

Because subtraction is defined in terms of addition, the addition property also lets us subtract the same number from both sides of an equation without changing the equation's solution set. Similarly, because division is defined in terms of multiplication, the multiplication property of equality can be used to divide both sides of an equation by a nonzero number to obtain an equivalent equation.

Table 1.4 illustrates how these properties are used to isolate x to obtain an equation of the form x = a number.

Table 1.4 Using Properties of Equality to Solve Linear Equations

Equation	How to Isolate x	Solving the Equation	The Equation's Solution Set
$x - 3 = 8$	Add 3 to both sides.	$x - 3 + 3 = 8 + 3$ $x = 11$	$\{11\}$
$x + 7 = -15$	Subtract 7 from both sides.	$x + 7 - 7 = -15 - 7$ $x = -22$	$\{-22\}$
$6x = 30$	Divide both sides by 6 (or multiply both sides by $\frac{1}{6}$).	$\frac{6x}{6} = \frac{30}{6}$ $x = 5$	$\{5\}$
$\frac{x}{5} = 9$	Multiply both sides by 5.	$5 \cdot \frac{x}{5} = 9 \cdot 5$ $x = 45$	$\{45\}$

EXAMPLE 1 Solving a Linear Equation

Solve and check: $2x + 3 = 17$.

Solution Our goal is to obtain an equivalent equation with x isolated on one side and a number on the other side.

$$2x + 3 = 17 \quad \text{This is the given equation.}$$

$$2x + 3 - 3 = 17 - 3 \quad \text{Subtract 3 from both sides.}$$

$$2x = 14 \quad \text{Simplify.}$$

$$\frac{2x}{2} = \frac{14}{2} \quad \text{Divide both sides by 2.}$$

$$x = 7 \quad \text{Simplify.}$$

Now we check the proposed solution, 7, by replacing x with 7 in the original equation.

$$2x + 3 = 17 \quad \text{This is the original equation.}$$

$$2 \cdot 7 + 3 \stackrel{?}{=} 17 \quad \text{Substitute 7 for x. The question mark indicates we do not yet know if the two sides are equal.}$$

$$14 + 3 \stackrel{?}{=} 17 \quad \text{Multiply: } 2 \cdot 7 = 14.$$

This statement is true. $$17 = 17 \quad \text{Add: } 14 + 3 = 17.$$

Because the check results in a true statement, we conclude that the solution of the given equation is 7, or the solution set is $\{7\}$. ■

✔ **CHECK POINT 1** Solve and check: $4x + 5 = 29$.

Here is a step-by-step procedure for solving a linear equation in one variable. Not all of these steps are necessary to solve every equation.

Solving a Linear Equation

1. Simplify the algebraic expression on each side.
2. Collect all the variable terms on one side and all the numbers, or constant terms, on the other side.
3. Isolate the variable and solve.
4. Check the proposed solution in the original equation.

EXAMPLE 2 Solving a Linear Equation

Solve and check: $2x - 7 + x = 3x + 1 + 2x$.

Solution

Step 1 Simplify the algebraic expression on each side.

$$2x - 7 + x = 3x + 1 + 2x$$ This is the given equation.

$$3x - 7 = 5x + 1$$ Combine like terms: $2x + x = 3x$ and $3x + 2x = 5x$.

Discover for Yourself

Solve the equation in Example 2 by collecting terms with the variable on the right and numerical terms on the left. What do you observe?

Step 2 Collect variable terms on one side and constant terms on the other side. We will collect variable terms on the left by subtracting $5x$ from both sides. We will collect the numbers on the right by adding 7 to both sides.

$$3x - 5x - 7 = 5x - 5x + 1$$ Subtract $5x$ from both sides.

$$-2x - 7 = 1$$ Simplify.

$$-2x - 7 + 7 = 1 + 7$$ Add 7 to both sides.

$$-2x = 8$$ Simplify.

Step 3 Isolate the variable and solve. We isolate x by dividing both sides by -2.

$$\frac{-2x}{-2} = \frac{8}{-2}$$ Divide both sides by -2.

$$x = -4$$ Simplify.

Step 4 Check the proposed solution in the original equation. Substitute -4 for x in the original equation.

$$2x - 7 + x = 3x + 1 + 2x$$ This is the original equation.

$$2(-4) - 7 + (-4) \stackrel{?}{=} 3(-4) + 1 + 2(-4)$$ Substitute -4 for x.

$$-8 - 7 + (-4) \stackrel{?}{=} -12 + 1 + (-8)$$ Multiply: $2(-4) = -8$, $3(-4) = -12$, and $2(-4) = -8$.

$$-15 + (-4) \stackrel{?}{=} -11 + (-8)$$ Add or subtract from left to right: $-8 - 7 = -15$ and $-12 + 1 = -11$.

$$-19 = -19$$ Add.

The true statement $-19 = -19$ verifies that -4 is the solution, or the solution set is $\{-4\}$. ■

✔ **CHECK POINT 2** Solve and check: $2x - 12 + x = 6x - 4 + 5x$.

EXAMPLE 3 Solving a Linear Equation

Solve and check: $4(2x + 1) - 29 = 3(2x - 5)$.

Solution

Step 1 Simplify the algebraic expression on each side.

$$4(2x + 1) - 29 = 3(2x - 5) \quad \text{This is the given equation.}$$

$$8x + 4 - 29 = 6x - 15 \quad \text{Use the distributive property.}$$

$$8x - 25 = 6x - 15 \quad \text{Simplify.}$$

Step 2 Collect variable terms on one side and constant terms on the other side. We will collect the variable terms on the left by subtracting $6x$ from both sides. We will collect the numbers on the right by adding 25 to both sides.

$$8x - 6x - 25 = 6x - 6x - 15 \quad \text{Subtract } 6x \text{ from both sides.}$$

$$2x - 25 = -15 \quad \text{Simplify.}$$

$$2x - 25 + 25 = -15 + 25 \quad \text{Add 25 to both sides.}$$

$$2x = 10 \quad \text{Simplify.}$$

Step 3 Isolate the variable and solve. We isolate x by dividing both sides by 2.

$$\frac{2x}{2} = \frac{10}{2} \quad \text{Divide both sides by 2.}$$

$$x = 5 \quad \text{Simplify.}$$

Step 4 Check the proposed solution in the original equation. Substitute 5 for x in the original equation.

$$4(2x + 1) - 29 = 3(2x - 5) \quad \text{This is the original equation.}$$

$$4(2 \cdot 5 + 1) - 29 \stackrel{?}{=} 3(2 \cdot 5 - 5) \quad \text{Substitute 5 for x.}$$

$$4(11) - 29 \stackrel{?}{=} 3(5) \quad \text{Simplify inside parentheses: } 2 \cdot 5 + 1 = 10 + 1 = 11 \text{ and } 2 \cdot 5 - 5 = 10 - 5 = 5.$$

$$44 - 29 \stackrel{?}{=} 15 \quad \text{Multiply: } 4(11) = 44 \text{ and } 3(5) = 15.$$

$$15 = 15 \quad \text{Subtract.}$$

The true statement $15 = 15$ verifies that 5 is the solution, or the solution set is $\{5\}$. ■

✔ **CHECK POINT 3** Solve and check: $2(x - 3) - 17 = 13 - 3(x + 2)$.

Using Technology

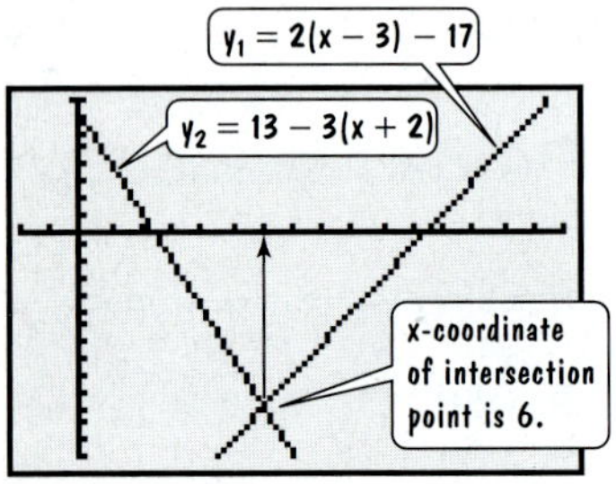

$[-2, 16, 1]$ by $[-14, 8, 1]$

You can use a graphing utility to check the solution to a linear equation in one variable. **Graph the left side and graph the right side. The solution is the x-coordinate of the point where the graphs intersect.** For example, to verify that 6 is the solution of

$$2(x-3)-17=13-3(x+2),$$

graph these two equations in the same viewing rectangle:

$$y_1=2(x-3)-17$$

$$\text{and}\quad y_2=13-3(x+2).$$

Choose a large enough range setting so that you can see where the graphs intersect. The viewing rectangle on the left shows that the x-coordinate of the intersection point is 6, verifying that $\{6\}$ is the solution set for the given equation. Is this the solution set you found in Check Point 2?

Linear Equations with Fractions

Equations are easier to solve when they do not contain fractions. How do we solve equations involving fractions? We begin by multiplying both sides of the equation by the least common denominator of all fractions in the equation. The least common denominator is the smallest number that all denominators will divide into. Multiplying every term on both sides of the equation by the least common denominator will eliminate the fractions in the equation. Example 4 shows how we "clear an equation of fractions."

EXAMPLE 4 Solving a Linear Equation Involving Fractions

Solve: $\dfrac{2x+5}{5}+\dfrac{x-7}{2}=\dfrac{3x+1}{2}$.

Solution The denominators are 5, 2, and 2. The smallest number that is divisible by 5, 2, and 2 is 10. We begin by multiplying both sides of the equation by the least common denominator.

$$\frac{2x+5}{5}+\frac{x-7}{2}=\frac{3x+1}{2}$$ This is the given equation.

$$10\left(\frac{2x+5}{5}+\frac{x-7}{2}\right)=10\left(\frac{3x+1}{2}\right)$$ Multiply both sides by 10.

$$\frac{10}{1}\cdot\frac{2x+5}{5}+\frac{10}{1}\cdot\frac{x-7}{2}=\frac{10}{1}\cdot\frac{3x+1}{2}$$ Use the distributive property and multiply each term by 10.

$$\frac{\overset{2}{\cancel{10}}}{1}\cdot\frac{2x+5}{\underset{1}{\cancel{5}}}+\frac{\overset{5}{\cancel{10}}}{1}\cdot\frac{x-7}{\underset{1}{\cancel{2}}}=\frac{\overset{5}{\cancel{10}}}{1}\cdot\frac{3x+1}{\underset{1}{\cancel{2}}}$$ Divide out common factors in each multiplication.

$$2(2x+5)+5(x-7)=5(3x+1)$$ The fractions are now cleared.

At this point, we have an equation similar to those we previously solved. Use the distributive property to begin simplifying each side.

$$4x + 10 + 5x - 35 = 15x + 5$$ Use the distributive property.

$$9x - 25 = 15x + 5$$ Combine like terms on the left side: $4x + 5x = 9x$ and $10 - 35 = -25$.

For variety, let's collect variable terms on the right and the numbers on the left.

$$9x - 9x - 25 = 15x - 9x + 5$$ Subtract 9x from both sides.

$$-25 = 6x + 5$$ Simplify.

$$-25 - 5 = 6x + 5 - 5$$ Subtract 5 from both sides.

$$-30 = 6x$$ Simplify.

Isolate x on the right side by dividing both sides by 6.

$$\frac{-30}{6} = \frac{6x}{6}$$ Divide both sides by 6.

$$-5 = x$$ Simplify.

Check the proposed solution in the original equation. Substitute -5 for x in the original equation. You should obtain $-7 = -7$. This true statement verifies that -5 is the solution, or the solution set is $\{-5\}$. ■

CHECK POINT 4 Solve: $\frac{x + 5}{7} + \frac{x - 3}{4} = \frac{5}{14}$.

2 Recognize identities, conditional equations, and inconsistent equations.

Types of Equations

We tend to place things in categories, allowing us to order and structure the world. For example, you can categorize yourself by your age group, your ethnicity, your academic major, or your gender. Equations can be placed into categories that depend on their solution sets.

An equation that is true for all real numbers for which both sides are defined is called an **identity**. An example of an identity is

$$x + 3 = x + 2 + 1.$$

Every number plus 3 is equal to that number plus 2 plus 1. Therefore, the solution set to this equation is the set of all real numbers. This set is written either as

$$\{x \mid x \text{ is a real number}\} \quad \text{or} \quad \mathbb{R}.$$

An equation that is not an identity, but that is true for at least one real number, is called a **conditional equation**. The equation $2x + 3 = 17$ is an example of a conditional equation. The equation is not an identity and is true only if x is 7.

An **inconsistent equation**, also called a **contradiction**, is an equation that is not true for even one real number. An example of an inconsistent equation is

$$x = x + 7.$$

There is no number that is equal to itself plus 7. This equation has no solution. Its solution set is written either as

$$\{\ \} \quad \text{or} \quad \varnothing.$$

These symbols stand for the empty set, a set with no elements.

It you attempt to solve an identity or an inconsistent equation, you will eliminate the variable. A true statement such as $6 = 6$ or a false statement such as $2 = 3$ will be the result. If a true statement results, the equation is an identity that is true for all real numbers. If a false statement results, the equation is an inconsistent equation with no solution.

EXAMPLE 5 Categorizing an Equation

Solve and determine whether the equation

$$2(x + 1) = 2x + 3$$

is an identity, a conditional equation, or an inconsistent equation.

Solution Begin by applying the distributive property on the left side. We obtain

$$2x + 2 = 2x + 3.$$

Does something look strange? Can doubling a number and increasing the product by 2 give the same result as doubling the same number and increasing the product by 3? No. Let's continue solving the equation by subtracting $2x$ from both sides.

$$2x - 2x + 2 = 2x - 2x + 3$$
$$2 = 3$$

The original equation is equivalent to the statement $2 = 3$, which is false for every value of x. The equation is inconsistent and has no solution. You can express this by writing "no solution" or using one of the symbols for the empty set, { } or ∅. ■

Using Technology

The graphs of $y_1 = 2(x + 1)$ and $y_2 = 2x + 3$ are parallel lines with no intersection point. This shows that the equation

$$2(x + 1) = 2x + 3$$

has no solution and is inconsistent.

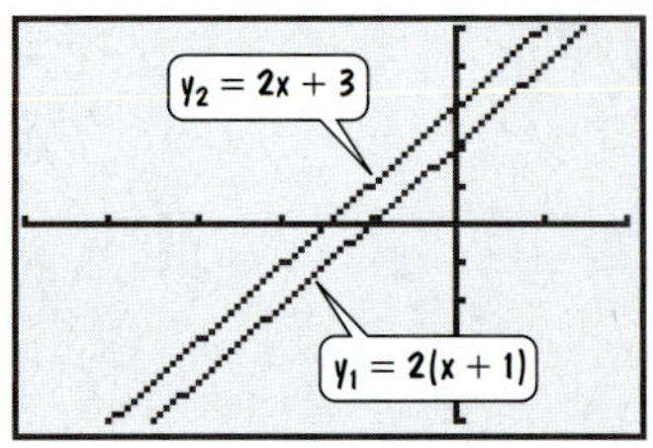

[−5, 2, 1] by [−5, 5, 1]

✔ **CHECK POINT 5** Solve and determine whether the equation

$$4x - 7 = 4(x - 1) + 3$$

is an identity, a conditional equation, or an inconsistent equation.

EXAMPLE 6 Categorizing an Equation

Solve and determine whether the equation

$$4x + 6 = 6(x + 1) - 2x$$

is an identity, a conditional equation, or an inconsistent equation.

ENRICHMENT ESSAY

The Empty Set in the Arts

In Example 5 we saw that

$$2(x + 1) = 2x + 3$$

has ∅, the empty set, as its solution set. The empty set denotes the notion of nothingness, such as the set of people in an empty room.

Nothingness appears not only in solution sets of equations but also in the arts. John Cage's *4′33″* is a musical work written for the piano that requires 4 minutes and 33 seconds of total silence as a musician sits frozen at a piano stool. In the opera *Porgy and Bess*, Ira Gershwin wrote, "I've got plenty of nothing." And the painting shown below, a 60″ by 60″ all-black canvas by the American abstract painter Ad Reinhardt, is a visual depiction of nothingness on permanent display at New York's Museum of Modern Art.

Ad Reinhardt "Abstract Painting" 1960–1961. Oil on canvans, 60 × 60 in. (152.4 × 152.4 cm). The Museum of Modern Art, New York. Purchase (by exchange). Photograph ©2000 The Museum of Modern Art, New York.

Solution

$$4x + 6 = 6(x + 1) - 2x \quad \text{This is the given equation.}$$

$$4x + 6 = 6x + 6 - 2x \quad \text{Apply the distributive property on the right side.}$$

$$4x + 6 = 4x + 6 \quad \text{Combine like terms on the right side: } 6x - 2x = 4x.$$

Can you see that the equation $4x + 6 = 4x + 6$ is true for every value of x? Let's continue solving the equation by subtracting $4x$ from both sides.

$$4x - 4x + 6 = 4x - 4x + 6$$

$$6 = 6$$

The original equation is equivalent to the statement $6 = 6$, which is true for every value of x. The equation is an identity, and all real numbers are solutions. You can express this by writing "all real numbers" or using one of the following notations:

$$\{x \mid x \text{ is a real number}\} \quad \text{or} \quad \mathbb{R}.$$

✔ **CHECK POINT 6** Solve and determine whether the equation

$$7x + 9 = 9(x + 1) - 2x$$

is an identity, a conditional equation, or an inconsistent equation.

3 Solve applied problems using mathematical models.

Applications

Our next example shows how the procedure for solving linear equations can be used to find the value of a variable in a mathematical model.

EXAMPLE 7 Working More and More

The formula

$$W = 0.3x + 46.6$$

models the average number of hours that Americans work per week, W, x years after 1980. When will we average 55 hours of work per week?

Solution We are interested in when we will work 55 hours, so substitute 55 for W in the formula and solve for x, the number of years after 1980.

$$W = 0.3x + 46.6 \quad \text{This is the given formula.}$$

$$55 = 0.3x + 46.6 \quad \text{Replace } W \text{ with 55.}$$

$$55 - 46.6 = 0.3x + 46.6 - 46.6 \quad \text{Subtract 46.6 from both sides.}$$

$$8.4 = 0.3x \quad \text{Simplify.}$$

$$\frac{8.4}{0.3} = \frac{0.3x}{0.3} \quad \text{Divide both sides by 0.3.}$$

$$28 = x \quad \text{Simplify.}$$

The formula indicates that 28 years after 1980, or in 2008, we will average 55 hours of work per week.

✔ **CHECK POINT 7** Use the formula in Example 7 to find when we will average 55.9 hours of work per week.

EXERCISE SET 1.4

Practice Exercises

In Exercises 1–22, solve and check each linear equation.

1. $5x + 3 = 18$
2. $3x + 8 = 50$
3. $6x - 3 = 63$
4. $5x - 8 = 72$
5. $14 - 5x = -41$
6. $25 - 6x = -83$
7. $11x - (6x - 5) = 40$
8. $5x - (2x - 8) = 35$
9. $2x - 7 = 6 + x$
10. $3x + 5 = 2x + 13$
11. $7x + 4 = x + 16$
12. $8x + 1 = x + 43$
13. $8y - 3 = 11y + 9$
14. $5y - 2 = 9y + 2$
15. $3(x - 2) + 7 = 2(x + 5)$
16. $2(x - 1) + 3 = x - 3(x + 1)$
17. $3(x - 4) - 4(x - 3) = x + 3 - (x - 2)$
18. $2 - (7x + 5) = 13 - 3x$
19. $16 = 3(x - 1) - (x - 7)$
20. $5x - (2x + 2) = x + (3x - 5)$
21. $7(x + 1) = 4[x - (3 - x)]$
22. $2[3x - (4x - 6)] = 5(x - 6)$

Exercises 23–36 contain equations with fractions. Solve each equation by multiplying both sides by the least common denominator, thereby clearing fractions.

23. $\frac{x}{3} = \frac{x}{2} - 2$
24. $\frac{x}{5} = \frac{x}{6} + 1$
25. $20 - \frac{x}{3} = \frac{x}{2}$
26. $\frac{x}{5} - \frac{1}{2} = \frac{x}{6}$
27. $\frac{3x}{5} = \frac{2x}{3} + 1$
28. $\frac{x}{2} = \frac{3x}{4} + 5$
29. $\frac{3x}{5} - x = \frac{x}{10} - \frac{5}{2}$
30. $2x - \frac{2x}{7} = \frac{x}{2} + \frac{17}{2}$
31. $\frac{x + 3}{6} = \frac{2}{3} + \frac{x - 5}{4}$
32. $\frac{x + 1}{4} = \frac{1}{6} + \frac{2 - x}{3}$
33. $\frac{x}{4} = 2 + \frac{x - 3}{3}$
34. $5 + \frac{x - 2}{3} = \frac{x + 3}{8}$
35. $\frac{x + 1}{3} = 5 - \frac{x + 2}{7}$
36. $\frac{3x}{5} - \frac{x - 3}{2} = \frac{x + 2}{3}$

In Exercises 37–48, solve each equation. Then state whether the equation is an identity, a conditional equation, or an inconsistent equation.

37. $5x + 9 = 9(x + 1) - 4x$
38. $4x + 7 = 7(x + 1) - 3x$
39. $3(y + 2) = 7 + 3y$
40. $4(y + 5) = 21 + 4y$
41. $10x + 3 = 8x + 3$
42. $5x + 7 = 2x + 7$
43. $4x + 5x = 8x$
44. $8x + 2x = 9x$
45. $-4x - 3(2 - 2x) = 7 + 2x$
46. $3x - 3(2 - x) = 6(x - 1)$
47. $y + 3(4y + 2) = 6(y + 1) + 5y$
48. $9y - 3(6 - 5y) = y - 2(3y + 9)$

Application Exercises

Medical researchers have found that the desirable heart rate, R, in beats per minute, for beneficial exercise is approximated by the formulas

$$R = 143 - 0.65A \quad \text{for women}$$

$$R = 165 - 0.75A \quad \text{for men}$$

where A is the person's age. Use these formulas to solve Exercises 49–50.

49. If the desirable heart rate for a woman is 117 beats per minute, how old is she? How is the solution shown on the line graph?

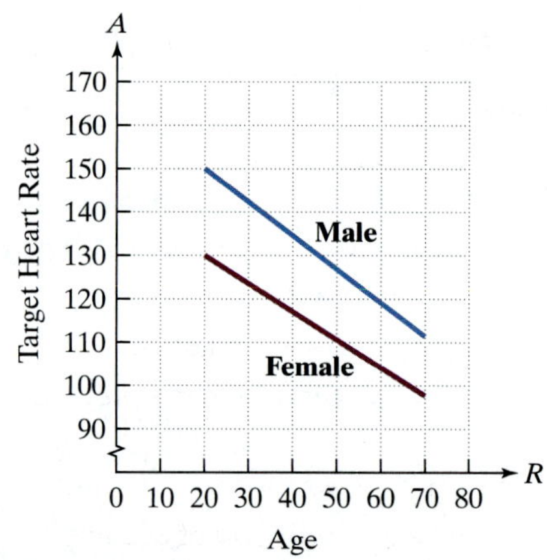

50. If the desirable heart rate for a man is 147 beats per minute, how old is he? How is the solution shown on the line graph?

How long can you expect to live? Life expectancies for men, M, and women, W, born t years after 1900 can be modeled by the formulas

$$M = 0.19t + 54.91$$

$$W = 0.11t + 68.41.$$

Use these formulas to solve Exercises 51–52.

51. Determine the year of birth for which U.S. men can expect to live 73.91 years.
52. Determine the year of birth for which U.S. women can expect to live 80.51 years.

Growth in human populations and economic activity threatens the continued existence of salmon in the Pacific Northwest. The bar graph shows the Pacific salmon population for various years. The data can be modeled by the formula

$$P = -0.22t + 9.6$$

in which P is the salmon population, in millions, t years after 1960. Use the formula to solve Exercises 53–54. Round to the nearest year.

Pacific Salmon Population

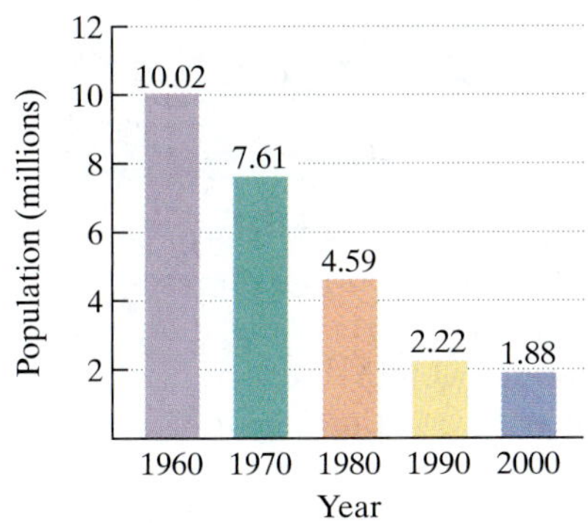

Source: U.S. Department of the Interior

53. When will the salmon population be reduced to 0.5 million?

54. When will there be no Pacific salmon?

Writing in Mathematics

55. What is a linear equation in one variable? Give an example of this type of equation.

56. What does it mean to solve an equation?

57. How do you determine whether a number is a solution of an equation?

58. What are equivalent equations? Give an example.

59. What is the addition property of equality?

60. What is the multiplication property of equality?

61. Explain how to clear an equation of fractions.

62. What is an identity? Give an example.

63. What is a conditional equation? Give an example.

64. What is an inconsistent equation? Give an example.

65. Describe a person or a group that would be interested in the formulas for life expectancy given in Exercises 51–52. What is the reason for this interest?

66. Describe one negative consequence of the dwindling Pacific salmon population modeled in Exercises 53–54.

67. What explanation can you offer for the fact that the average number of hours Americans work per week is gradually increasing?

Technology Exercises

For Exercises 68–71, use your graphing utility to graph each side of the equations in the same viewing rectangle. Based on the resulting graph, label each equation as conditional, inconsistent, or an identity. If the equation is conditional, use the x-coordinate of the intersection point to find the solution. Verify this value by direct substitution into the equation.

68. $2(x - 6) + 3x = x + 6$

69. $9x + 3 - 3x = 2(3x + 1)$

70. $2\left(x + \frac{1}{2}\right) = 5x + 1 - 3x$

71. $\frac{2x - 1}{3} - \frac{x - 5}{6} = \frac{x - 3}{4}$

Critical Thinking Exercises

72. Which one of the following is true?

a. The equation $-7x = x$ has no solution.

b. The equations $\frac{x}{x - 4} = \frac{4}{x - 4}$ and $x = 4$ are equivalent.

c. The equations $3y - 1 = 11$ and $3y - 7 = 5$ are equivalent.

d. If a and b are any real numbers, then $ax + b = 0$ always has only one number in its solution set.

73. Solve for x: $ax + b = c$.

74. Write three equations that are equivalent to $x = 5$.

75. If x represents a number, write an English sentence about the number that results in an inconsistent equation.

76. Find b such that $\frac{7x + 4}{b} + 13 = x$ will have a solution set given by $\{-6\}$.

Review Exercises

In Exercises 77–78, perform the indicated operations.

77. $-\frac{1}{5} - \left(-\frac{1}{2}\right)$ (Section 1.2, Example 4)

78. $4(-3)(-1)(-5)$ (Section 1.2, Examples in Table 1.2)

79. Graph $y = 4 - x^2$. Let $x = -3, -2, -1, 0, 1, 2$, and 3. (Section 1.3, Example 2)

▸ SECTION 1.5 Problem Solving and Using Formulas

Objectives

1. Solve algebraic word problems using linear equations.
2. Solve for a variable in a formula.

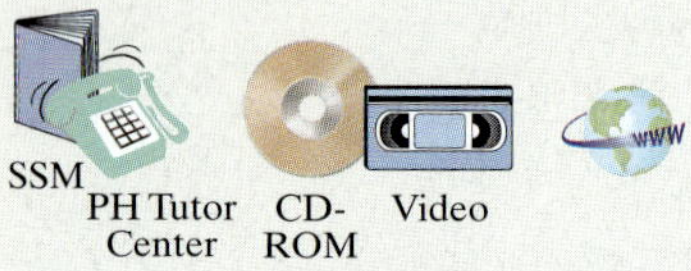

The human race is undeniably becoming a faster race. Since the beginning of the past century, track-and-field records have fallen in everything from sprints to miles to marathons. The performance arc is clearly rising, but no one knows how much higher it can climb. At some point, even the best-trained body simply has to up and quit. The question is, just where is that point, and is it possible for athletes, trainers, and genetic engineers to push it higher? In this section, you will learn a problem-solving strategy that uses linear equations to determine if anyone will ever run a 3-minute mile.

1 Solve algebraic word problems using linear equations.

A Strategy for Solving Word Problems Using Equations

Problem solving is an important part of algebra. The problems in this book are presented in English. We must *translate* from the ordinary language of English into the language of algebraic equations. To translate, however, we must understand the English prose and be familiar with the forms of algebraic language. Here are some general steps we will follow in solving word problems:

Strategy for Solving Word Problems

Step 1 Read the problem carefully. Attempt to state the problem in your own words and state what the problem is looking for. Let x (or any variable) represent one of the quantities in the problem.

Step 2 If necessary, write expressions for any other unknown quantities in the problem in terms of x.

Step 3 Write an equation in x that describes the verbal conditions of the problem.

Step 4 Solve the equation and answer the problem's question.

Step 5 Check the solution *in the original wording* of the problem, not in the equation obtained from the words.

EXAMPLE 1 Steps Forward: Granting Women the Right to Vote

From 1970 through 1999, 23 countries granted women the right to vote. The number of countries in the 1980s exceeded the number in the 1990s by 3. The number in the 1970s exceeded the number in the 1990s by 11. Determine the number of countries granting women the right to vote for each of these three decades.

Number of Countries Granting Women the Right to Vote

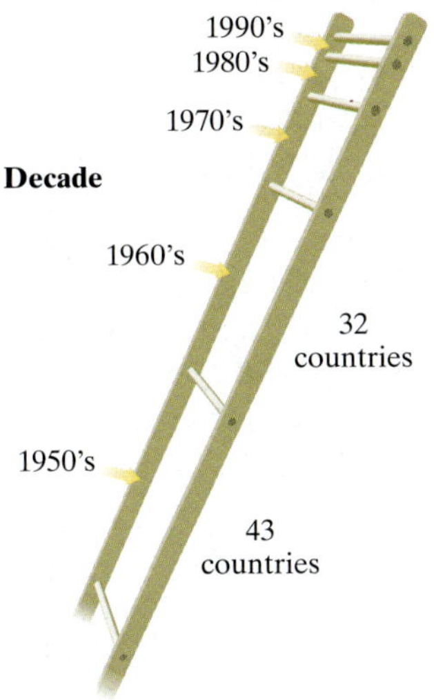

Source: The United Nations

Solution

Step 1 Let x represent one of the quantities. We know something about the number of countries in the 1980s and the 1970s; the numbers exceeded that of the 1990s by 3 and 11, respectively. We will let

$x =$ the number of countries granting women the right to vote in the 1990s.

Step 2 Represent other quantities in terms of x. Because the number of countries in the 1980s exceeded the number in the 1990s by 3, let

$x + 3 =$ the number of countries granting women the right to vote in the 1980s.

Because the number of countries in the 1970s exceeded the number in the 1990s by 11, let

$x + 11 =$ the number of countries granting women the right to vote in the 1970s.

Step 3 Write an equation in x that describes the conditions. From 1970 through 1999, 23 countries granted women the right to vote.

The number in the 1970s	plus	the number in the 1980s	plus	the number in the 1990s	equals	23.
$x + 11$	$+$	$x + 3$	$+$	x	$=$	23

Step 4 Solve the equation and answer the question.

$$x + 11 + x + 3 + x = 23$$ This is the equation for the problem's conditions.

$$3x + 14 = 23$$ Combine like terms.

$$3x = 9$$ Subtract 14 from both sides.

$$x = 3$$ Divide both sides by 3.

Thus,

the number of countries in the 1990s $= x = 3$.

the number of countries in the 1980s $= x + 3 = 3 + 3 = 6$.

the number of countries in the 1970s $= x + 11 = 3 + 11 = 14$.

There were 3 countries granting women the right to vote in the 1990s, 6 in the 1980s, and 14 in the 1970s.

Step 5 Check the proposed solution in the original wording of the problem. The problem states that from 1970 through 1999, 23 countries granted women the right to vote. By adding 3, 6, and 14, the numbers that we found for the 1990s, the 1980s, and the 1970s, respectively, we obtain

$$3 + 6 + 14, \quad \text{or} \quad 23,$$

as specified by the problem's conditions. ■

✔ CHECK POINT 1 We love to go to the movies, and a good soundtrack means we can take them home with us, too. The top film soundtracks are *The Bodyguard*, *Saturday Night Fever*, and *Purple Rain*. *Saturday Night Fever* sold 2 million copies more than *Purple Rain*, and sales of *The Bodyguard* exceeded sales of *Purple Rain* by 3 million. Combined, the three soundtracks have sold 44 million copies. Determine the sales, in millions, for each of these three soundtracks. (*Source*: Recording Industry Association of America)

EXAMPLE 2 Will Anyone Ever Run a Three-Minute Mile?

Mile records			
1886	4:12.3	1958	3:54.5
1923	4:10.4	1966	3:51.3
1933	4:07.6	1979	3:48.9
1945	4:01.3	1985	3:46.3
1954	3:59.4	1999	3:43.1

Source: U.S.A Track and Field

One yardstick for measuring how steadily—if slowly—athletic performance has improved is the mile run. In 1923, the record for the mile was a comparatively sleepy 4 minutes, 10.4 seconds. In 1954, Roger Bannister of Britain cracked the 4-minute mark, coming in at 3 minutes, 59.4 seconds. In the half-century since, about 0.3 second per year has been shaved off Bannister's record. If this trend continues, by which year will someone run a 3-minute mile?

Solution In solving this problem, we will express time for the mile run in seconds. Our interest is in a time of 3 minutes, or 180 seconds.

Step 1 Let x represent one of the quantities. The critical information in the problem is:

- In 1954, the record was 3 minutes, 59.4 seconds, or 239.4 seconds.
- The record has decreased by 0.3 second per year since then.

We are interested in when the record will be 180 seconds. Let

x = the number of years after 1954 when someone will run a 3-minute mile.

Step 2 Represent other quantities in terms of x. There are no other unknown quantities to find, so we can skip this step.

Step 3 Write an equation in x that describes the conditions.

The 1954 record time	decreased by	0.3 second per year for x years	equals	the 3-minute, or 180-second, mile.
239.4	$-$	$0.3x$	$=$	180

Step 4 Solve the equation and answer the question.

$$239.4 - 0.3x = 180$$ This is the equation for the problem's conditions.

$$239.4 - 239.4 - 0.3x = 180 - 239.4$$ Subtract 239.4 from both sides.

$$-0.3x = -59.4$$ Simplify.

$$\frac{-0.3x}{-0.3} = \frac{-59.4}{-0.3}$$ Divide both sides by -0.3.

$$x = 198$$ Simplify.

Using current trends, by 198 years (gasp!) after 1954, or in 2152, someone will run a 3-minute mile.

Step 5 Check the proposed solution in the original wording of the problem. The problem states that the record time should be 180 seconds. Do we obtain 180 seconds if we decrease the 1954 record time, 239.4 seconds, by 0.3 second per year for 198 years, our proposed solution?

$$239.4 - 0.3(198) = 239.4 - 59.4 = 180$$

This verifies that, using current trends, the 3-minute mile will be run 198 years after 1954. ■

ENRICHMENT ESSAY

A Poky Species

For a species that prides itself on its athletic prowess, human beings are a pretty poky group. Lions can sprint at up to 50 miles per hour; cheetahs move even faster, flooring it to a sizzling 70 miles per hour. But most humans—with our willowy spines and awkward, upright gait—would have trouble cracking 25 miles per hour with a tail wind, a flat track, and a good pair of running shoes.

✔ **CHECK POINT 2** In 2000, the population of the United States was approximately 275 million, increasing at a rate of 2.35 million people per year. If this trend continues, when will the population be 298.5 million?

EXAMPLE 3 Selecting a Long-Distance Carrier

You are choosing between two long-distance telephone plans. Plan A has a monthly fee of $20 with a charge of $0.05 per minute for all long-distance calls. Plan B has a monthly fee of $5 with a charge of $0.10 per minute for all long-distance calls. For how many minutes of long-distance calls will the costs for the two plans be the same?

Solution

Step 1 Let x represent one of the quantities. Let

x = the number of minutes of long-distance calls for the two plans to cost the same.

Step 2 Represent other unknown quantities in terms of x. There are no other unknown quantities, so we can skip this step.

Step 3 Write an equation in x that describes the conditions. The monthly cost for plan A is the monthly fee, $20, plus the per minute charge, $0.05, times the number of minutes of long-distance calls, x. The monthly cost for plan B is the monthly fee, $5, plus the per-minute charge, $0.10, times the number of minutes of long-distance calls, x.

The monthly cost for plan A | must equal | the monthly cost for plan B.

$$20 + 0.05x \quad = \quad 5 + 0.10x$$

Step 4 Solve the equation and answer the question.

$$20 + 0.05x = 5 + 0.10x$$ This is the equation for the problem's conditions.

$$20 = 5 + 0.05x$$ Subtract 0.05x from both sides.

$$15 = 0.05x$$ Subtract 5 from both sides.

$$\frac{15}{0.05} = \frac{0.05x}{0.05}$$ Divide both sides by 0.05.

$$300 = x$$ Simplify.

Because x represents the number of minutes of long-distance calls for the two plans to cost the same, the costs will be the same with 300 minutes of long-distance calls.

Step 5 Check the proposed solution in the original wording of the problem. The problem states that the costs for the two plans should be the same. Let's see if they are with 300 minutes of long distance calls:

$$\text{Cost for plan A} = \$20 + \$0.05(300) = \$20 + \$15 = \$35$$

Monthly fee | Per-minute charge

$$\text{Cost for plan B} = \$5 + \$0.10(300) = \$5 + \$30 = \$35.$$

With 300 minutes, or 5 hours, of long-distance chatting, both plans cost $35 for the month. Thus, the proposed solution, 300 minutes, satisfies the problem's conditions. ■

CHECK POINT 3 You are choosing between two long-distance telephone plans. Plan A has a monthly fee of \$15 with a charge of \$0.08 per minute for all long-distance calls. Plan B has a monthly fee of \$3 with a charge of \$0.12 per minute for all long-distance calls. For how many minutes of long-distance calls will the costs for the two plans be the same?

EXAMPLE 4 A Price Reduction on a Digital Camera

Your local computer store is having a terrific sale on digital cameras. After a 40% price reduction, you purchase a digital camera for \$276. What was the camera's price before the reduction?

Solution

Step 1 Let x represent one of the quantities. We will let

x = the original price of the digital camera prior to the reduction.

Step 2 Represent other quantities in terms of x. There are no other unknown quantities to find, so we can skip this step.

Step 3 Write an equation in x that describes the conditions. The camera's original price minus the 40% reduction is the reduced price, \$276.

Study Tip

Observe that the original price, x, reduced by 40% is $x - 0.4x$ and *not* $x - 0.4$.

Original price	minus	the reduction (40% of the original price)	is	the reduced price, \$276.
x	$-$	$0.4x$	$=$	276

Step 4 Solve the equation and answer the question.

$$x - 0.4x = 276$$ This is the equation for the problem's conditions.

$$0.6x = 276$$ Combine like terms: $x - 0.4x = 1x - 0.4x = 0.6x$.

$$\frac{0.6x}{0.6} = \frac{276}{0.6}$$ Divide both sides by 0.6.

$$x = 460$$ Simplify: $0.6\overline{)276.0}$ = 460.

The digital camera's price before the reduction was \$460.

Step 5 Check the proposed solution in the original wording of the problem. The price before the reduction, \$460, minus the 40% reduction should equal the reduced price given in the original wording, \$276:

$$460 - 40\% \text{ of } 460 = 460 - 0.4(460) = 460 - 184 = 276.$$

This verifies that the digital camera's price before the reduction was \$460. ■

✔ CHECK POINT 4 After a 30% price reduction, you purchase a new computer for \$980. What was the computer's price before the reduction?

Solving geometry problems usually requires a knowledge of basic geometric ideas and formulas. Formulas for area, perimeter, and volume are given in Table 1.5.

Table 1.5 Common Formulas for Area, Perimeter, and Volume

Square	Rectangle	Circle	Triangle	Trapezoid
$A = s^2$	$A = lw$	$A = \pi r^2$	$A = \frac{1}{2}bh$	$A = \frac{1}{2}h(a + b)$
$P = 4s$	$P = 2l + 2w$	$C = 2\pi r$		

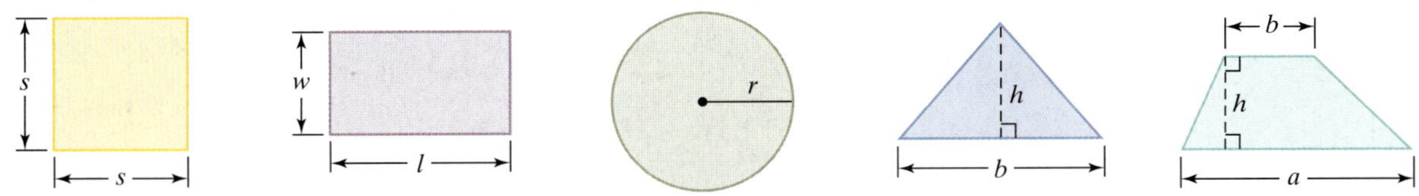

Cube	Rectangular Solid	Circular Cylinder	Sphere	Cone
$V = s^3$	$V = lwh$	$V = \pi r^2 h$	$V = \frac{4}{3}\pi r^3$	$V = \frac{1}{3}\pi r^2 h$

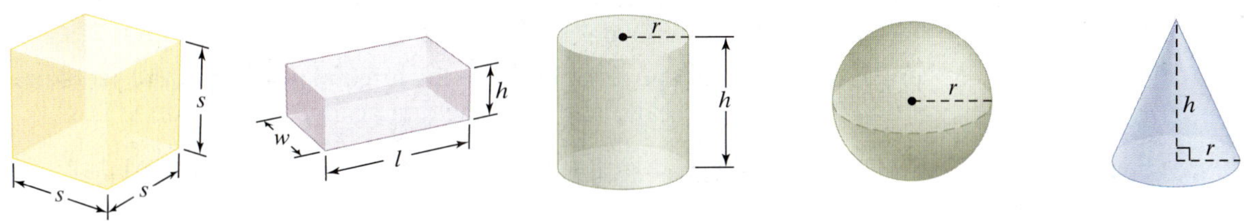

We will be using the formula for the perimeter of a rectangle, $P = 2l + 2w$, in our next example. A helpful verbal model for this formula is 2 times length plus 2 times width is a rectangle's perimeter.

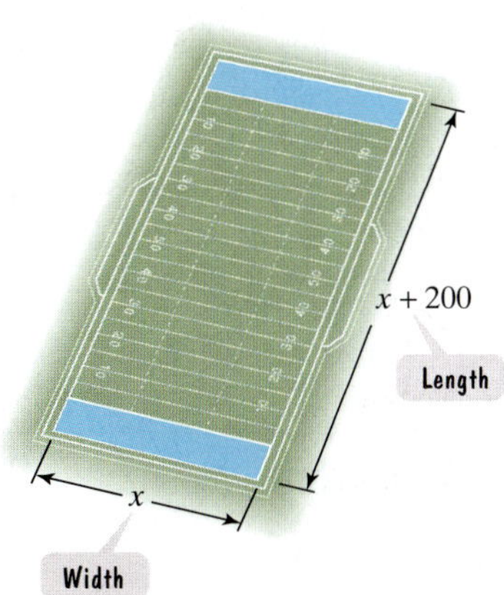

Figure 1.14 An American football field

EXAMPLE 5 Finding the Dimensions of an American Football Field

The length of an American football field is 200 feet more than the width. If the perimeter of the field is 1040 feet, what are its dimensions?

Solution

Step 1 Let x represent one of the quantities. We know something about the length; the length is 200 feet more than the width. We will let

$$x = \text{the width.}$$

Step 2 Represent other quantities in terms of x. Because the length is 200 feet more than the width, let

$$x + 200 = \text{the length.}$$

Figure 1.14 illustrates an American football field and its dimensions.

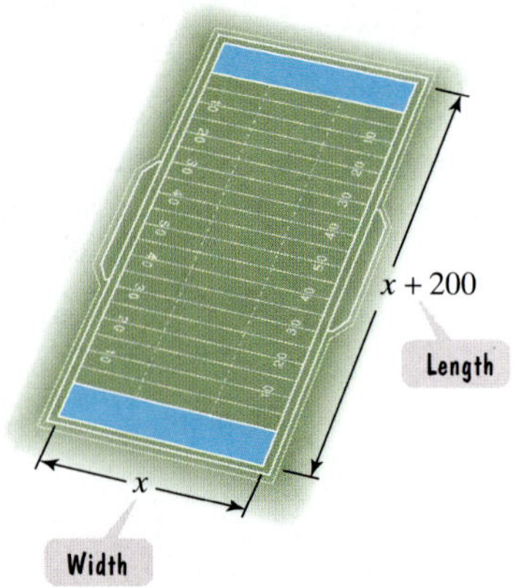

Figure 1.14, repeated

Step 3 Write an equation in x that describes the conditions. Because the perimeter of the field is 1040 feet,

Twice the length	plus	twice the width	is	the perimeter.
$2(x + 200)$	$+$	$2 \cdot x$	$=$	$1040.$

Step 4 Solve the equation and answer the question.

$$2(x + 200) + 2x = 1040$$ This is the equation for the problem's conditions.

$$2x + 400 + 2x = 1040$$ Apply the distributive property.

$$4x + 400 = 1040$$ Combine like terms: $2x + 2x = 4x$.

$$4x = 640$$ Subtract 400 from both sides.

$$x = 160$$ Divide both sides by 4.

Thus,

$$\text{width} = x = 160.$$
$$\text{length} = x + 200 = 160 + 200 = 360.$$

The dimensions of an American football field are 160 feet by 360 feet. (The 360-foot length is usually described as 120 yards.)

Step 5 Check the proposed solution in the original wording of the problem. The perimeter of the football field using the dimensions that we found is:

$$2(160 \text{ feet}) + 2(360 \text{ feet}) = 320 \text{ feet} + 720 \text{ feet} = 1040 \text{ feet}.$$

Because the problem's wording tells us that the perimeter is 1040 feet, our dimensions are correct. ■

✔ **CHECK POINT 5** The length of a rectangular basketball court is 44 feet more than the width. If the perimeter of the basketball court is 288 feet, what are its dimensions?

2 Solve for a variable in a formula.

Solving for a Variable in a Formula

When working with formulas, such as the geometric formulas shown in Table 1.5 on page 55, it is often necessary to solve for a specified variable. This is done by isolating the specified variable on one side of the equation. Begin by isolating all terms with the specified variable on one side of the equation and all terms without the specified variable on the other side. The next example shows how to do this.

EXAMPLE 6 Solving for a Variable in a Formula

Solve the formula $2l + 2w = P$ for w.

Solution First, isolate $2w$ on the left by subtracting $2l$ from both sides. Then solve for w by dividing both sides by 2.

We need to isolate w.

$$2l + 2w = P$$ This is the given formula.

$$2l - 2l + 2w = P - 2l$$ Isolate $2w$ by subtracting $2l$ from both sides.

$$2w = P - 2l$$ Simplify.

$$\frac{2w}{2} = \frac{P - 2l}{2}$$ Solve for w by dividing both sides by 2.

$$w = \frac{P - 2l}{2}$$ Simplify.

■

✔ **CHECK POINT 6** Solve the formula $2l + 2w = P$ for l.

Circular Cylinder

$V = \pi r^2 h$

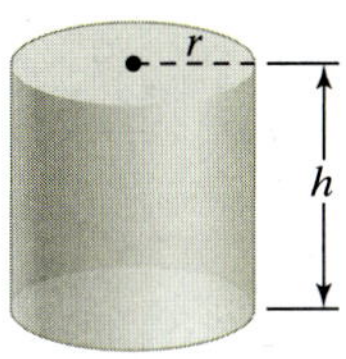

EXAMPLE 7 Solving for a Variable in a Formula

Table 1.5 shows that the volume of a circular cylinder is given by the formula

$$V = \pi r^2 h$$

where r is the radius of the circle at either end and h is the height. Solve this formula for h.

Solution Our goal is to get h by itself on one side of the formula. There is only one term with h, $\pi r^2 h$, and it is already isolated on the right side. We isolate h on the right by dividing both sides by πr^2.

We need to isolate h.

$$V = \pi r^2 h$$ This is the given formula.

$$\frac{V}{\pi r^2} = \frac{\pi r^2 h}{\pi r^2}$$ Isolate h by dividing both sides by πr^2.

$$\frac{V}{\pi r^2} = h$$ Simplify: $\frac{\pi r^2 h}{\pi r^2} = \frac{\pi r^2}{\pi r^2} \cdot h = 1h = h$.

Equivalently,

$$h = \frac{V}{\pi r^2}.$$

■

✔ **CHECK POINT 7** The volume of a rectangular solid is the product of its length, width, and height:

$$V = lwh.$$

Solve this formula for h.

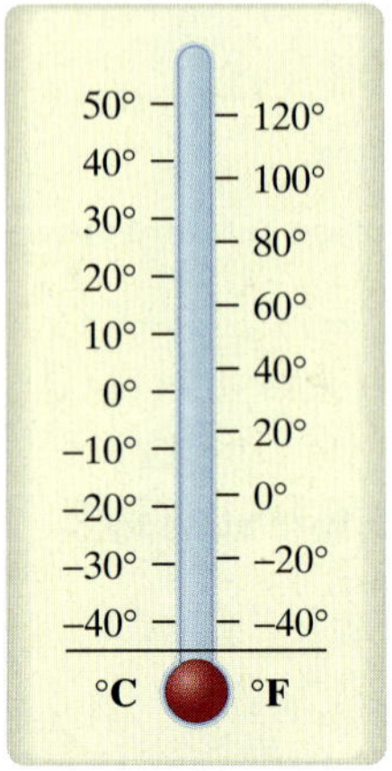

The Celsius scale is on the left and the Fahrenheit scale is on the right.

You'll be leaving the cold of winter for a Hawaii vacation. CNN International reports a temperature in Hawaii of 30°C. Should you cancel the trip? You can convert from Celsius temperature, C, to Fahrenheit temperature, F, using the formula

$$F = \frac{9}{5}C + 32.$$

A temperature of 30°C corresponds to a Fahrenheit temperature of

$$F = \frac{9}{5} \cdot 30 + 32 = \frac{9}{\cancel{5}_1} \cdot \frac{\cancel{30}^6}{1} + 32 = 54 + 32 = 86,$$

or a balmy 86°F. (Don't cancel the trip.)

Visitors to the United States are familiar with the Celsius temperature scale. For them, a useful formula is one that can be used to convert from Fahrenheit to Celsius. In Example 8, you will see how to obtain such a formula.

EXAMPLE 8 Solving for a Variable in a Formula

Solve the formula

$$F = \frac{9}{5}C + 32$$

for C.

Solution We begin by multiplying both sides of the formula by 5 to clear the fraction. Then we isolate the variable C.

$$F = \frac{9}{5}C + 32 \qquad \text{This is the given formula.}$$

$$5F = 5\left(\frac{9}{5}C + 32\right) \qquad \text{Multiply both sides by 5.}$$

$$5F = 5 \cdot \frac{9}{5}C + 5 \cdot 32 \qquad \text{Apply the distributive property.}$$

$$5F = 9C + 160 \qquad \text{Simplify.}$$

We need to isolate C.

$$5F - 160 = 9C + 160 - 160 \qquad \text{Subtract 160 from both sides.}$$

$$5F - 160 = 9C \qquad \text{Simplify.}$$

$$\frac{5F - 160}{9} = \frac{9C}{9} \qquad \text{Divide both sides by 9.}$$

$$\frac{5F - 160}{9} = C \qquad \text{Simplify.}$$

Using the distributive property, we can express $5F - 160$ as $5(F - 32)$. Thus,

$$C = \frac{5F - 160}{9} = \frac{5(F - 32)}{9}.$$

This formula, used to convert from Fahrenheit to Celsius, usually is given as

$$C = \frac{5}{9}(F - 32).$$

■

CHECK POINT 8 The formula

$$\frac{W}{2} - 3H = 53$$

models the recommended weight, W, in pounds, for a male, where H represents his height, in inches, over 5 feet. Solve this formula for W.

EXAMPLE 9 Solving for a Variable That Occurs Twice in a Formula

The formula

$$A = P + Prt$$

describes the amount, A, that a principal of P dollars is worth after t years when invested at a simple annual interest rate, r. Solve this formula for P.

Solution Notice that all the terms with P already occur on the right side of the formula.

$$A = P + Prt$$

We need to isolate P.

We can use the distributive property in the form $ab + ac = a(b + c)$ to convert the two occurrences of P into one.

$$A = P + Prt \quad \text{This is the given formula.}$$

$$A = P(1 + rt) \quad \text{Use the distributive property to obtain a single occurrence of } P.$$

$$\frac{A}{1 + rt} = \frac{P(1 + rt)}{1 + rt} \quad \text{Divide both sides by } 1 + rt.$$

$$\frac{A}{1 + rt} = P \quad \text{Simplify: } \frac{P(1 + rt)}{1(1 + rt)} = \frac{P}{1} = P.$$

Equivalently,

$$P = \frac{A}{1 + rt}.$$

■

Study Tip

You cannot solve $A = P + Prt$ for P by subtracting Prt from both sides and writing

$$A - Prt = P.$$

When a formula is solved for a specified variable, that variable must be isolated on one side. The variable P occurs on both sides of

$$A - Prt = P.$$

CHECK POINT 9 Solve the formula $P = C + MC$ for C.

ENRICHMENT ESSAY

Einstein's Famous Formula: $E = mc^2$

One of the most famous formulas in the world is $E = mc^2$ formulated by Albert Einstein. Einstein showed that any form of energy has mass and that mass itself is a form of energy. In this formula, E represents energy, in ergs, m represents mass, in grams, and c represents the speed of light. Because light travels at 30 billion centimeters per second, the formula indicates that 1 gram of mass will produce 900 billion ergs of energy.

Einstein's formula implies that the mass of a golf ball could provide the daily energy needs of the metropolitan Boston area. Mass and energy are equivalent, and the transformation of even a tiny amount of mass releases an enormous amount of energy. If this energy is released suddenly, a destructive force is unleashed, as in an atom bomb. When the release is gradual and controlled, the energy can be used to generate power.

The theoretical results implied by Einstein's formula $E = mc^2$ have not been realized because scientists have not yet developed a way of converting a mass completely to energy.

EXERCISE SET 1.5

Practice Exercises

Use the five-step strategy for solving word problems to find the number or numbers described in Exercises 1–10.

1. When five times a number is decreased by 4, the result is 26. What is the number?

2. When two times a number is decreased by 3, the result is 11. What is the number?

3. When a number is decreased by 20% of itself, the result is 20. What is the number?

4. When a number is decreased by 30% of itself, the result is 28. What is the number?

5. When 60% of a number is added to the number, the result is 192. What is the number?

6. When 80% of a number is added to the number, the result is 252. What is the number?

7. 70% of what number is 224?

8. 70% of what number is 252?

9. One number exceeds another by 26. The sum of the numbers is 64. What are the numbers?

10. One number exceeds another by 24. The sum of the numbers is 58. What are the numbers?

Application Exercises

In Exercises 11–40, use the five-step strategy for solving word problems.

11. The graph shows the five costliest natural disasters in U.S. history. The cost of the Northridge, California, earthquake exceeded Hurricane Hugo by $5.5 billion, and the cost of Hurricane Andrew exceeded Hugo by $13 billion. The combined costs of the three natural disasters was $39.5 billion. Determine the cost of each.

Costliest Natural Disasters in U.S. History

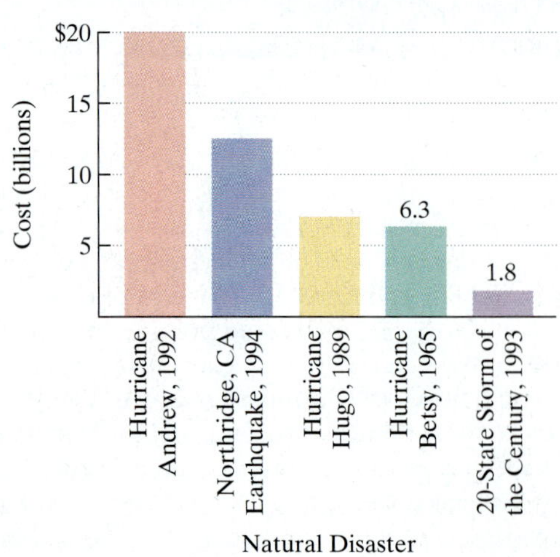

Source: Federal Emergency Management Agency

12. The graph shows the top five U.S. franchises, by number. There are 700 more 7-Elevens than Subways, and 1500 more McDonalds than Subways. With a total of 38,800 McDonalds, 7-Elevens, and Subways, find the number of each.

Top Five U.S. Franchises

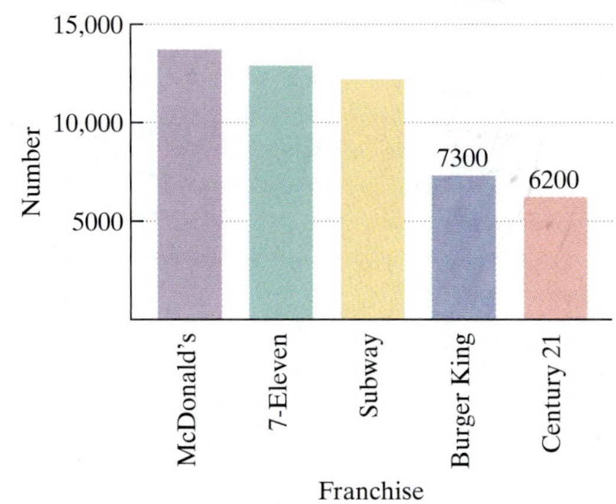

Source: Entrepreneur

Solve Exercises 13–14 using the fact that sum of the measures of the three angles of a triangle is 180°.

13. In a triangle, the measure of the first angle is twice the measure of the second angle. The measure of the third angle is 20° less than the second angle. What is the measure of each angle?

14. In a triangle, the measure of the first angle is three times the measure of the second angle. The measure of the third angle is 20° less than the second angle. What is the measure of each angle?

According to one mathematical model, the average life expectancy for American men born in 1900 was 55 years. Life expectancy has increased by about 0.2 year for each birth year after 1900. Use this information to solve Exercises 15–16.

15. If these trends continue, for which birth year will the average life expectancy be 85 years?

16. If these trends continue, for which birth year will the average life expectancy be 91 years?

According to the National Center for Health Statistics, in 1990, 28% of babies in the United States were born to parents who were not married. Throughout the 1990s, this increased by approximately 0.6% per year. Use this information to solve Exercises 17–18.

17. If this trend continues, in which year will 37% of babies be born out of wedlock?

18. If this trend continues, in which year will 40% of babies be born out of wedlock?

The graph shows the generally decreasing times in the Olympic 500-meter speed-skating event. In 1980, the winning time for women was 41.78 seconds. Since then, the winning time has decreased by about 0.19 second per year. Use this information to solve Exercises 19–20.

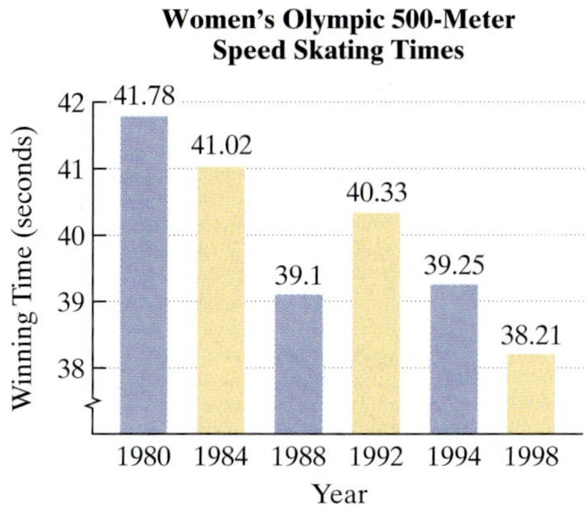

Source: *The Universal Almanac*

19. If this trend continues, in which year will the winning time be 37.22 seconds?

20. If this trend continues, in which year will the winning time be 34.94 seconds?

21. You are choosing between two plans at a discount warehouse. Plan A offers an annual membership fee of $100 and you pay 80% of the manufacturer's recommended list price. Plan B offers an annual membership fee of $40 and you pay 90% of the manufacturer's recommended list price. How many dollars of merchandise would you have to purchase in a year to pay the same amount under both plans? What will be the cost for each plan?

22. You are choosing between two plans at a discount warehouse. Plan A offers an annual membership fee of $300 and you pay 70% of the manufacturer's recommended list price. Plan B offers an annual membership fee of $40 and you pay 90% of the manufacturer's recommended list price. How many dollars of merchandise would you have to purchase in a year to pay the same amount under both plans? What will be the cost for each plan?

23. The bus fare in a city is $1.25. People who use the bus have the option of purchasing a monthly coupon book for $21.00. With the coupon book, the fare is reduced to $0.50. Determine the number of times in a month the bus must be used so that the total monthly cost without the coupon book is the same as the total monthly cost with the coupon book.

24. A coupon book for a bridge costs $21 per month. The toll for the bridge is normally $2.50, but it is reduced to $1 for people who have purchased the coupon book. Determine the number of times in a month the bridge must be crossed so that the total monthly cost without the coupon book is the same as the total monthly cost with the coupon book.

25. After a 20% reduction, you purchase a television for $336. What was the television's price before the reduction?

26. After a 30% reduction, you purchase a dictionary for $30.80. What was the dictionary's price before the reduction?

27. Including 8% sales tax, an inn charges $162 per night. Find the inn's nightly cost before the tax is added.

28. Including 5% sales tax, an inn charges $252 per night. Find the inn's nightly cost before the tax is added.

The graph shows median, or average, income by level of education. Exercises 29–30 use the information in the bar graph.

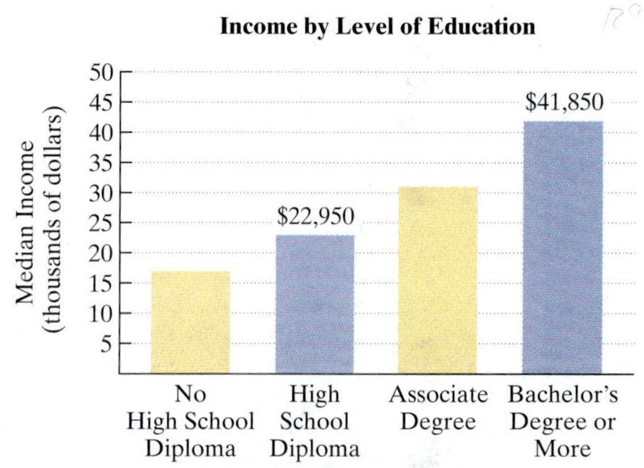

Source: U.S. Department of Commerce

29. The annual salary for people with a bachelor's degree or more is an increase of 35% over the annual salary for people with an associate degree. What is the annual salary with an associate degree?

30. The annual salary for high school graduates is an increase of 35% over the annual salary for people without a high school diploma. What is the average annual salary without a high school diploma?

Exercises 31–32 involve markup, the amount added to the dealer's cost of an item to arrive at the selling price of that item.

31. The selling price of a refrigerator is $584. If the markup is 25% of the dealer's cost, what is the dealer's cost of the refrigerator?

32. The selling price of a scientific calculator is $15. If the markup is 25% of the dealer's cost, what is the dealer's cost of the calculator?

33. A rectangular soccer field is twice as long as it is wide. If the perimeter of the soccer field is 300 yards, what are its dimensions?

34. A rectangular swimming pool is three times as long as it is wide. If the perimeter of the pool is 320 feet, what are its dimensions?

35. The length of the rectangular tennis court at Wimbledon is 6 feet longer than twice the width. If the court's perimeter is 228 feet, what are the court's dimensions?

36. The length of a rectangular pool is 6 meters less than twice the width. If the pool's perimeter is 126 meters, what are its dimensions?

37. The height of the bookcase in the figure is 3 feet longer than the length of a shelf. If 18 feet of lumber is available for the entire unit, find the length and height of the unit.

38. A bookcase is to be constructed as shown in the figure. The length is to be 3 times the height. If 60 feet of lumber is available for the entire unit, find the length and height of the bookcase.

39. An HMO pamphlet contains the following recommended weight for women: "Give yourself 100 pounds for the first 5 feet plus 5 pounds for every inch over 5 feet tall." Using this description, what height corresponds to an ideal weight of 135 pounds?

40. A job pays an annual salary of \$33,150, which includes a holiday bonus of \$750. If paychecks are issued twice a month, what is the gross amount for each paycheck?

In Exercises 41–66, solve each formula for the specified variable.

41. $A = lw$ for l

42. $A = lw$ for w

43. $A = \frac{1}{2}bh$ for b

44. $A = \frac{1}{2}bh$ for h

45. $I = Prt$ for P

46. $I = Prt$ for t

47. $T = D + pm$ for p

48. $P = C + MC$ for M

49. $A = \frac{1}{2}h(a + b)$ for a

50. $A = \frac{1}{2}h(a + b)$ for b

51. $V = \frac{1}{3}\pi r^2 h$ for h

52. $V = \frac{1}{3}\pi r^2 h$ for r^2

53. $y - y_1 = m(x - x_1)$ for m

54. $y_2 - y_1 = m(x_2 - x_1)$ for m

55. $V = \frac{d_1 - d_2}{t}$ for d_1

56. $z = \frac{x - u}{s}$ for x

57. $Ax + By = C$ for x

58. $Ax + By = C$ for y

59. $s = \frac{1}{2}at^2 + vt$ for v

60. $s = \frac{1}{2}at^2 + vt$ for a

61. $L = a + (n - 1)d$ for n

62. $L = a + (n - 1)d$ for d

63. $A = 2lw + 2lh + 2wh$ for l

64. $A = 2lw + 2lh + 2wh$ for h

65. $IR + Ir = E$ for I

66. $A = \frac{x_1 + x_2 + x_3}{n}$ for n

Writing in Mathematics

67. In your own words, describe a step-by-step approach for solving algebraic word problems.

68. Write an original word problem that can be solved using a linear equation. Then solve the problem.

69. Explain what it means to solve for a variable in a formula.

70. Did you have difficulties solving some of the problems that were assigned in this exercise set? Discuss what you did if this happened to you. Did your course of action enhance your ability to solve algebraic word problems?

71. The mile records in Example 2 on page 52 are a yardstick for measuring how athletes are getting better and better. Do you think that there is a limit to human performance? Explain your answer. If so, when might we reach it?

Technology Exercises

72. The formula $y = 55 + 0.2x$ models the average life expectancy, y, of American men born x years after 1900. Graph the formula in a $[0, 200, 20]$ by $[0, 100, 10]$ viewing rectangle. Then use the TRACE or ZOOM feature to verify your answer in Exercise 15 or 16.

73. The formula $y = 28 + 0.6x$ models the percentage, y, of U.S. babies born out of wedlock x years after 1990. Graph the formula in a $[0, 20, 1]$ by $[0, 50, 1]$ viewing rectangle. Then use the TRACE or ZOOM feature to verify your answer in Exercise 17 or 18.

Critical Thinking Exercises

74. Which one of the following is true?

a. If $I = prt$, then $t = I - pr$.

b. If $y = \frac{kx}{z}$, then $z = \frac{kx}{y}$.

c. It is not necessary to use the distributive property to solve $P = C + MC$ for C.

d. None of the above is true.

75. The price of a dress is reduced by 40%. When the dress still does not sell, it is reduced by 40% of the reduced price. If the price of the dress after both reductions is \$72, what was the original price?

76. Suppose that we agree to pay you 8¢ for every problem in this chapter that you solve correctly and fine you 5¢ for every problem done incorrectly. If at the end of 26 problems we do not owe each other any money, how many problems did you solve correctly?

77. It was wartime when the Ricardos found out Mrs. Ricardo was pregnant. Ricky Ricardo was drafted and made out a will, deciding that \$14,000 in a savings account was to be divided between his wife and his child-to-be. Rather strangely, and certainly with gender bias, Ricky stipulated that if the child were a boy, he would get twice the amount of the mother's portion. If it were a girl, the mother would get twice the amount the girl was to receive. We'll never know what Ricky was thinking of, for (as fate would have it) he did not return from war. Mrs. Ricardo gave birth to twins—a boy and a girl. How was the money divided?

78. A thief steals a number of rare plants from a nursery. On the way out, the thief meets three security guards, one after another. To each security guard, the thief is forced to give one-half the plants that he still has, plus 2 more. Finally, the thief leaves the nursery with 1 lone palm. How many plants were originally stolen?

79. Solve for C: $V = C - \frac{C - S}{L}N$.

Review Exercises

80. What does $-6 \le -6$ mean? Is the statement true or false? (Section 1.1, Example 6)

81. Simplify: $\frac{(2 + 4)^2 + (-1)^5}{12 \div 2 \cdot 3 - 3}$. (Section 1.2, Example 8)

82. Solve: $\frac{2x}{3} - \frac{8}{3} = x$. (Section 1.4, Example 4)

▸ SECTION 1.6 Properties of Integral Exponents

Objectives

1. Use the product rule.
2. Use the quotient rule.
3. Use the zero-exponent rule.
4. Use the negative-exponent rule.
5. Use the power rule.
6. Find the power of a product.
7. Find the power of a quotient.
8. Simplify exponential expressions.

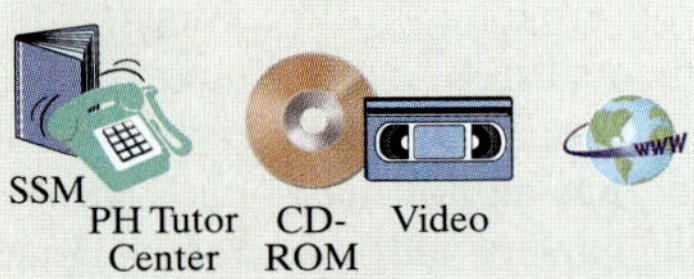

Horses were born with eight deformed legs, pigs with no eyes, and eggs contained several yolks. This was part of the grotesque aftermath of the 1986 explosion at the Chernobyl nuclear power plant in the former Soviet Union. Nearby cities were abandoned. Because of dangerous radiation levels for at least one hundred years, an exclusion zone was formed around the area. The spread of radioactive fallout from the world's worst nuclear accident is expected to result in at least 100,000 additional deaths from cancer in the Northern Hemisphere. In this section's exercise set, you will work with a formula that sadly explains why 335,000 people in cities surrounding Chernobyl were permanently displaced from their homes. The formula involves exponents, our topic for this section.

The Product and Quotient Rules

1 Use the product rule.

We have seen that exponents are used to indicate repeated multiplication. Now consider the multiplication of two exponential expressions, such as $b^4 \cdot b^3$. We are multiplying 4 factors of b and 3 factors of b. We have a total of 7 factors of b:

4 factors of b | 3 factors of b

$$b^4 \cdot b^3 = (b \cdot b \cdot b \cdot b) \cdot (b \cdot b \cdot b) = b^7.$$

Total: 7 factors of b

The product is exactly the same if we add the exponents:

$$b^4 \cdot b^3 = b^{4+3} = b^7.$$

This suggests the following rule:

The Product Rule

$$b^m \cdot b^n = b^{m+n}$$

When multiplying exponential expressions with the same base, add the exponents. Use this sum as the exponent of the common base.

EXAMPLE 1 Using the Product Rule

Multiply each expression using the product rule:

a. $b^8 \cdot b^{10}$ **b.** $(6x^4y^3)(5x^2y^7)$.

Solution

a. $b^8 \cdot b^{10} = b^{8+10} = b^{18}$

b. $(6x^4y^3)(5x^2y^7)$

$= 6 \cdot 5 \cdot x^4 \cdot x^2 \cdot y^3 \cdot y^7$ Use the associative and commutative properties. This step can be done mentally.

$= 30x^{4+2}y^{3+7}$

$= 30x^6y^{10}$ ■

CHECK POINT 1 Multiply each expression using the product rule:

a. $b^6 \cdot b^5$ **b.** $(4x^3y^4)(10x^2y^6)$.

2 Use the quotient rule.

Now, consider the division of two exponential expressions, such as the quotient of b^7 and b^3. We are dividing 7 factors of b by 3 factors of b.

$$\frac{b^7}{b^3} = \frac{b \cdot b \cdot b \cdot b \cdot b \cdot b \cdot b}{b \cdot b \cdot b} = \boxed{\frac{b \cdot b \cdot b}{b \cdot b \cdot b}} \cdot b \cdot b \cdot b \cdot b = 1 \cdot b \cdot b \cdot b \cdot b = b^4$$

This factor is equal to 1.

The quotient is exactly the same if we subtract the exponents:

$$\frac{b^7}{b^3} = b^{7-3} = b^4.$$

This suggests the following rule:

The Quotient Rule

$$\frac{b^m}{b^n} = b^{m-n}, \quad b \neq 0$$

When dividing exponential expressions with the same nonzero base, subtract the exponent in the denominator from the exponent in the numerator. Use this difference as the exponent of the common base.

EXAMPLE 2

Divide each expression using the quotient rule:

a. $\frac{b^{20}}{b^5}$ **b.** $\frac{30x^{12}y^9}{5x^3y^7}$.

Solution

a. $\frac{b^{20}}{b^5} = b^{20-5} = b^{15}$

b. $\frac{30x^{12}y^9}{5x^3y^7} = \frac{30}{5} \cdot \frac{x^{12}}{x^3} \cdot \frac{y^9}{y^7} = 6x^{12-3}y^{9-7} = 6x^9y^2$ ■

CHECK POINT 2 Divide each expression using the quotient rule:

a. $\frac{b^{15}}{b^3}$ **b.** $\frac{27x^{14}y^8}{3x^3y^5}$.

3 Use the zero-exponent rule.

Zero as an Exponent

A nonzero base can be raised to the 0 power. The quotient rule can be used to help determine what zero as an exponent should mean. Consider the quotient of b^4 and b^4, where b is not zero. We can determine this quotient in two ways.

$$\frac{b^4}{b^4} = 1 \qquad \frac{b^4}{b^4} = b^{4-4} = b^0$$

Any nonzero expression divided by itself is 1.

Use the quotient rule and subtract exponents.

This means that b^0 must equal 1.

The Zero-Exponent Rule

If b is any real number other than 0,

$$b^0 = 1.$$

EXAMPLE 3 Using the Zero-Exponent Rule

Use the zero-exponent rule to simplify each expression:

a. 8^0 **b.** $(-6)^0$ **c.** -6^0 **d.** $5x^0$ **e.** $(5x)^0$.

Solution

a. $8^0 = 1$ Any nonzero number raised to the 0 power is 1.

b. $(-6)^0 = 1$

c. $-6^0 = -(6^0) = -1$

Only 6 is raised to the 0 power.

d. $5x^0 = 5 \cdot 1 = 5$ Only x is raised to the 0 power.

e. $(5x)^0 = 1$ The entire expression, 5x, is raised to the 0 power. ■

CHECK POINT 3 Use the zero-exponent rule to simplify each expression:

a. 7^0 **b.** $(-5)^0$ **c.** -5^0 **d.** $10x^0$ **e.** $(10x)^0$.

4 Use the negative-exponent rule.

Negative Integers as Exponents

A nonzero base can be raised to a negative power. The quotient rule can be used to help determine what a negative integer as an exponent should mean. Consider the quotient of b^3 and b^5, where b is not zero. We can determine this quotient in two ways.

$$\frac{b^3}{b^5} = \frac{\cancel{b} \cdot \cancel{b} \cdot \cancel{b}}{\cancel{b} \cdot \cancel{b} \cdot \cancel{b} \cdot b \cdot b} = \frac{1}{b^2} \qquad \frac{b^3}{b^5} = b^{3-5} = b^{-2}$$

After cancelling, we have two factors of b in the denominator.

Use the quotient rule and subtract exponents.

Notice that $\frac{b^3}{b^5}$ equals both b^{-2} and $\frac{1}{b^2}$. This means that b^{-2} must equal $\frac{1}{b^2}$. This example is a special case of the **negative-exponent rule**.

The Negative-Exponent Rule

If b is any real number other than 0 and n is a natural number, then

$$b^{-n} = \frac{1}{b^n}.$$

EXAMPLE 4 Using the Negative-Exponent Rule

Use the negative-exponent rule to write each expression with a positive exponent. Simplify, if possible:

a. 7^{-2} **b.** $(-2)^{-4}$ **c.** $\frac{1}{6^{-2}}$ **d.** $7x^{-5}y^2$.

Solution

a. $7^{-2} = \frac{1}{7^2} = \frac{1}{49}$

b. $(-2)^{-4} = \frac{1}{(-2)^4} = \frac{1}{(-2)(-2)(-2)(-2)} = \frac{1}{16}$

Only the sign of the exponent, −4, changes. The base, −2, does not change sign.

c. $\frac{1}{6^{-2}} = \frac{1}{\frac{1}{6^2}} = 1 \cdot \frac{6^2}{1} = 6^2 = 36$

d. $7x^{-5}y^2 = 7 \cdot \frac{1}{x^5} \cdot y^2 = \frac{7y^2}{x^5}$ ■

CHECK POINT 4 Use the negative-exponent rule to write each expression with a positive exponent. Simplify, if possible:

a. 5^{-2} **b.** $(-3)^{-4}$ **c.** $\frac{1}{4^{-2}}$ **d.** $3x^{-6}y^4$.

In Example 4 and Check Point 4, did you notice that

$$\frac{1}{6^{-2}} = 6^2 \quad \text{and} \quad \frac{1}{4^{-2}} = 4^2?$$

In general, if a negative exponent appears in a denominator, an expression can be written with a positive exponent using

$$\frac{1}{b^{-n}} = b^n.$$

Negative Exponents in Numerators and Denominators

If b is any real number other than 0 and n is a natural number, then

$$b^{-n} = \frac{1}{b^n} \quad \text{and} \quad \frac{1}{b^{-n}} = b^n.$$

When a negative number appears as an exponent, switch the position of the base (from numerator to denominator or from denominator to numerator) and make the exponent positive. The sign of the base does not change.

EXAMPLE 5 Using Negative Exponents

Write each expression with positive exponents only. Then simplify, if possible:

a. $\frac{5^{-3}}{4^{-2}}$ **b.** $\frac{1}{6x^{-4}}$.

Solution

a. $\frac{5^{-3}}{4^{-2}} = \frac{4^2}{5^3} = \frac{4 \cdot 4}{5 \cdot 5 \cdot 5} = \frac{16}{125}$

Switch the position of each base to the other side of the fraction bar and change the sign of the exponent.

b. $\frac{1}{6x^{-4}} = \frac{x^4}{6}$ ■

✔ **CHECK POINT 5** Write each expression with positive exponents only. Then simplify, if possible:

a. $\frac{7^{-2}}{4^{-3}}$ **b.** $\frac{1}{5x^{-2}}$.

5 Use the power rule.

The Power Rule for Exponents (Powers to Powers)

The next property of exponents applies when an exponential expression is raised to a power. Here is an example:

$$(b^2)^4.$$

The exponential expression b^2 is raised to the fourth power.

There are 4 factors of b^2. Thus,

$$(b^2)^4 = b^2 \cdot b^2 \cdot b^2 \cdot b^2 = b^{2+2+2+2} = b^8$$

Add exponents when multiplying with the same base.

We can obtain the answer, b^8, by multiplying the exponents:

$$(b^2)^4 = b^{2 \cdot 4} = b^8.$$

This suggests the following rule:

The Power Rule (Powers to Powers)

$$(b^m)^n = b^{mn}$$

When an exponential expression is raised to a power, multiply the exponents. Place the product of the exponents on the base and remove the parentheses.

EXAMPLE 6 Using the Power Rule (Powers to Powers)

Simplify each expression using the power rule:

a. $(x^6)^4$ **b.** $(y^5)^{-3}$ **c.** $(b^{-4})^{-2}$.

Solution

a. $(x^6)^4 = x^{6 \cdot 4} = x^{24}$

b. $(y^5)^{-3} = y^{5(-3)} = y^{-15} = \frac{1}{y^{15}}$

c. $(b^{-4})^{-2} = b^{(-4)(-2)} = b^8$ ■

✔ **CHECK POINT 6** Simplify each expression using the power rule:

a. $(x^5)^3$ **b.** $(y^7)^{-2}$ **c.** $(b^{-3})^{-4}$.

6 Find the power of a product.

The Products-to-Powers Rule for Exponents

The next property of exponents applies when we are raising a product to a power. Here is an example:

$$(2x)^4.$$

The product 2x is raised to the fourth power.

There are four factors of $2x$. Thus,

$$(2x)^4 = 2x \cdot 2x \cdot 2x \cdot 2x = 2 \cdot 2 \cdot 2 \cdot 2 \cdot x \cdot x \cdot x \cdot x = 2^4x^4$$

We can obtain the answer, 2^4x^4, by raising each factor within the parentheses to the fourth power:

$$(2x)^4 = 2^4x^4.$$

This suggests the following rule:

Products to Powers

$$(ab)^n = a^nb^n$$

When a product is raised to a power, raise each factor to that power.

EXAMPLE 7 Using the Products-to-Powers Rule

Simplify each expression using the products-to-powers rule:

a. $(6x)^3$ **b.** $(-2y^2)^4$ **c.** $(-3x^{-1}y^3)^{-2}$.

Solution

a. $(6x)^3 = 6^3x^3$ Raise each factor to the third power.

$= 216x^3$ Simplify: $6^3 = 6 \cdot 6 \cdot 6 = 216$.

b. $(-2y^2)^4 = (-2)^4(y^2)^4$ Raise each factor to the fourth power.

$= (-2)^4y^{2\cdot4}$ With powers to powers, multiply exponents: $(b^m)^n = b^{mn}$.

$= 16y^8$ Simplify: $(-2)^4 = (-2)(-2)(-2)(-2) = 16$.

c. $(-3x^{-1}y^3)^{-2} = (-3)^{-2}(x^{-1})^{-2}(y^3)^{-2}$ Raise each factor to the −2 power.

$= (-3)^{-2}x^{(-1)(-2)}y^{3(-2)}$ Multiply powers.

$= (-3)^{-2}x^2y^{-6}$ Simplify.

$= \dfrac{1}{(-3)^2} \cdot x^2 \cdot \dfrac{1}{y^6}$ Apply $b^{-n} = \dfrac{1}{b^n}$ to the first and last factors.

$= \dfrac{x^2}{9y^6}$ Simplify: $(-3)^2 = (-3)(-3) = 9$. ■

✔ **CHECK POINT 7** Simplify each expression using the products-to-powers rule.

a. $(2x)^4$ **b.** $(-3y^2)^3$ **c.** $(-4x^5y^{-1})^{-2}$.

7 Find the power of a quotient.

The Quotients-to-Power Rule for Exponents

The following rule is used to raise a quotient to a power:

Quotients to Powers

If b is a nonzero real number, then

$$\left(\frac{a}{b}\right)^n = \frac{a^n}{b^n}.$$

When a quotient is raised to a power, raise the numerator to the power and divide by the denominator to the power.

EXAMPLE 8 Using the Quotients-to-Powers Rule

Simplify each expression using the quotients-to-powers rule.

a. $\left(\frac{x^2}{5}\right)^3$ **b.** $\left(\frac{2x^3}{y^{-4}}\right)^5$.

Solution

a. $\left(\frac{x^2}{5}\right)^3 = \frac{(x^2)^3}{5^3} = \frac{x^{2\cdot 3}}{5\cdot 5\cdot 5} = \frac{x^6}{125}$ Cube the numerator and the denominator.

b. $\left(\frac{2x^3}{y^{-4}}\right)^5 = \frac{(2x^3)^5}{(y^{-4})^5}$ Raise the numerator and the denominator to the fifth power.

$= \frac{2^5(x^3)^5}{(y^{-4})^5}$ Raise each factor in the numerator to the fifth power.

$= \frac{2^5 \cdot x^{3\cdot 5}}{y^{(-4)(5)}}$ Multiply exponents in both powers-to-powers expressions: $(b^m)^n = b^{mn}$.

$= \frac{32x^{15}}{y^{-20}}$ Simplify.

$= 32x^{15}y^{20}$ Move y^{-20} to the other side of the fraction bar and change -20 to 20: $\frac{1}{b^{-n}} = b^n$. ■

✔ **CHECK POINT 8** Simplify each expression using the quotients-to-powers rule:

a. $\left(\frac{x^5}{4}\right)^3$ **b.** $\left(\frac{2x^{-3}}{y^2}\right)^4$.

8 Simplify exponential expressions.

Simplifying Exponential Expressions

Properties of exponents are used to simplify exponential expressions. An exponential expression is **simplified** when:

- No parentheses appear.
- No powers are raised to powers.
- Each base occurs only once.
- No negative or zero exponents appear.

Simplifying Exponential Expressions

	Example
1. If necessary, remove parentheses by using $$(ab)^n = a^nb^n \quad \text{or} \quad \left(\frac{a}{b}\right)^n = \frac{a^n}{b^n}.$$	$(xy)^3 = x^3y^3$
2. If necessary, simplify powers to powers by using $$(b^m)^n = b^{mn}.$$	$(x^4)^3 = x^{4\cdot 3} = x^{12}$
3. If necessary, be sure that each base appears only once by using $$b^m \cdot b^n = b^{m+n} \quad \text{or} \quad \frac{b^m}{b^n} = b^{m-n}.$$	$x^4 \cdot x^3 = x^{4+3} = x^7$
4. If necessary, rewrite exponential expressions with zero powers as 1 $(b^0 = 1)$. Furthermore, write the answer with positive exponents by using $$b^{-n} = \frac{1}{b^n} \quad \text{or} \quad \frac{1}{b^{-n}} = b^n.$$	$\frac{x^5}{x^8} = x^{5-8} = x^{-3} = \frac{1}{x^3}$

The following example shows how to simplify exponential expressions. Throughout the example, assume that no variable in a denominator is equal to zero.

EXAMPLE 9 Simplifying Exponential Expressions

Simplify:

a. $(-2xy^{-14})(-3x^4y^5)^3$ **b.** $\left(\dfrac{25x^2y^4}{-5x^6y^{-8}}\right)^2$.

Solution

a. $(-2xy^{-14})(-3x^4y^5)^3$

$= (-2xy^{-14})(-3)^3(x^4)^3(y^5)^3$ Cube each factor in the second parentheses.

$= (-2xy^{-14})(-27)x^{12}y^{15}$ Multiply the exponents when raising a power to a power.

$= (-2)(-27)x^{1+12}y^{-14+15}$ Mentally rearrange factors and multiply like bases by adding the exponents.

$= 54x^{13}y$ Simplify.

b. $\left(\dfrac{25x^2y^4}{-5x^6y^{-8}}\right)^2$

$= (-5x^{2-6}y^{4-(-8)})^2$ Simplify inside the parentheses by subtracting the exponents when dividing.

$= (-5x^{-4}y^{12})^2$ Simplify.

$= (-5)^2(x^{-4})^2(y^{12})^2$ Square each factor in parentheses.

$= 25x^{-8}y^{24}$ Multiply the exponents when raising a power to a power.

$= \dfrac{25y^{24}}{x^8}$ Simplify x^{-8} using $b^{-n} = \dfrac{1}{b^n}$.

Study Tip

Try to avoid the following common errors that can occur when simplifying exponential expressions.

Correct	Incorrect	Description of Error
$b^3b^4 = b^7$	$b^3b^4 = b^{12}$	The exponents should be added, not multiplied.
$3^2 \cdot 3^4 = 3^6$	$3^2 \cdot 3^4 = 9^6$	The common base should be retained, not multiplied.
$\frac{5^{16}}{5^4} = 5^{12}$	$\frac{5^{16}}{5^4} = 5^4$	The exponents should be subtracted, not divided.
$(4a)^3 = 64a^3$	$(4a)^3 = 4a^3$	Both factors should be cubed.
$b^{-n} = \frac{1}{b^n}$	$b^{-n} = -\frac{1}{b^n}$	Only the exponent should change sign.
$(a + b)^{-1} = \frac{1}{a + b}$	$(a + b)^{-1} = \frac{1}{a} + \frac{1}{b}$	The exponent applies to the entire expression $a + b$.

CHECK POINT 9 Simplify:

a. $(-3x^{-6}y)(-2x^3y^4)^2$ **b.** $\left(\frac{10x^3y^5}{5x^6y^{-2}}\right)^2$.

ENRICHMENT ESSAY

Exponents and Large Numbers

- Although people do a great deal of talking, the total output since the beginning of gabble to the present day, including all baby talk, love songs, and congressional debates, only amounts to about 10^{16} (10 million billion) words.
- A googol is 10^{100}. The number of raindrops falling on Miami in a century is much less than a googol.
- A googol times a googol is $10^{100} \cdot 10^{100}$, or 10^{200}, which is much less than a googolplex! A googolplex is 10 to the googol power, or $10^{10^{100}}$, or 1 with a googol of zeros! If you tried to write a googolplex without using exponential notation, there would not be enough room to write it if you went to the farthest star, touring all the nebulae, and putting down zeros every inch of the way.
- The number of grains of sand on the beach at Coney Island, New York, is about 10^{20}.
- The number of snow crystals necessary to form the Ice Age was approximately 10^{30}.
- The number 10^{80} is greater than the total number of atoms in our entire universe.

EXERCISE SET 1.6

Practice Exercises

In Exercises 1–12, multiply each expression using the product rule.

1. $b^4 \cdot b^7$

2. $b^5 \cdot b^9$

3. $x \cdot x^3$

4. $x \cdot x^4$

5. $2^3 \cdot 2^2$

6. $2^4 \cdot 2^2$

7. $3x^4 \cdot 2x^2$

8. $5x^3 \cdot 3x^2$

9. $(-2y^{10})(-10y^2)$

10. $(-4y^8)(-8y^4)$

11. $(5x^3y^4)(20x^7y^8)$

12. $(4x^5y^6)(20x^7y^4)$

In Exercises 13–22, divide each expression using the quotient rule.

13. $\frac{b^{12}}{b^3}$

14. $\frac{b^{25}}{b^5}$

15. $\frac{15x^9}{3x^4}$

16. $\frac{18x^{11}}{3x^4}$

17. $\dfrac{x^9y^7}{x^4y^2}$

18. $\dfrac{x^9y^{12}}{x^2y^6}$

19. $\dfrac{50x^2y^7}{5xy^4}$

20. $\dfrac{36x^{12}y^4}{4xy^2}$

21. $\dfrac{-56a^{12}b^{10}c^8}{7ab^2c^4}$

22. $\dfrac{-66a^9b^7c^6}{6a^3bc^2}$

In Exercises 23–32, use the zero-exponent rule to simplify each expression.

23. 6^0

24. 9^0

25. $(-4)^0$

26. $(-2)^0$

27. -4^0

28. -2^0

29. $13y^0$

30. $17y^0$

31. $(13y)^0$

32. $(17y)^0$

In Exercises 33–50, write each expression with positive exponents only. Then simplify, if possible.

33. 3^{-2}

34. 4^{-2}

35. $(-5)^{-2}$

36. $(-7)^{-2}$

37. -5^{-2}

38. -7^{-2}

39. x^2y^{-3}

40. x^3y^{-4}

41. $8x^{-7}y^3$

42. $9x^{-8}y^4$

43. $\dfrac{1}{5^{-3}}$

44. $\dfrac{1}{2^{-5}}$

45. $\dfrac{1}{(-3)^{-4}}$

46. $\dfrac{1}{(-2)^{-4}}$

47. $\dfrac{x^{-2}}{y^{-5}}$

48. $\dfrac{x^{-3}}{y^{-7}}$

49. $\dfrac{a^{-4}b^7}{c^{-3}}$

50. $\dfrac{a^{-3}b^8}{c^{-2}}$

In Exercises 51–56, simplify each expression using the power rule.

51. $(x^6)^{10}$

52. $(x^3)^2$

53. $(b^4)^{-3}$

54. $(b^8)^{-3}$

55. $(7^{-4})^{-5}$

56. $(9^{-4})^{-5}$

In Exercises 57–70, simplify each expression using the products-to-powers rule.

57. $(4x)^3$

58. $(2x)^5$

59. $(-3x^7)^2$

60. $(-4x^9)^2$

61. $(2xy^2)^3$

62. $(3x^2y)^4$

63. $(-3x^2y^5)^2$

64. $(-3x^4y^6)^2$

65. $(-3x^{-2})^{-3}$

66. $(-2x^{-4})^{-3}$

67. $(5x^3y^{-4})^{-2}$

68. $(7x^2y^{-5})^{-2}$

69. $(-2x^{-5}y^4z^2)^{-4}$

70. $(-2x^{-4}y^5z^3)^{-4}$

In Exercises 71–82, simplify each expression using the quotients-to-powers rule.

71. $\left(\dfrac{2}{x}\right)^4$

72. $\left(\dfrac{y}{2}\right)^5$

73. $\left(\dfrac{x^3}{5}\right)^2$

74. $\left(\dfrac{x^4}{6}\right)^2$

75. $\left(-\dfrac{3x}{y}\right)^4$

76. $\left(-\dfrac{2x}{y}\right)^5$

77. $\left(\dfrac{x^4}{y^2}\right)^6$

78. $\left(\dfrac{x^5}{y^3}\right)^6$

79. $\left(\dfrac{x^3}{y^{-4}}\right)^3$

80. $\left(\dfrac{x^4}{y^{-2}}\right)^3$

81. $\left(\dfrac{a^{-2}}{b^3}\right)^{-4}$

82. $\left(\dfrac{a^{-3}}{b^5}\right)^{-4}$

In Exercises 83–106, simplify each exponential expression.

83. $\dfrac{x^3}{x^9}$

84. $\dfrac{x^6}{x^{10}}$

85. $\dfrac{20x^3}{-5x^4}$

86. $\dfrac{10x^5}{-2x^6}$

87. $\dfrac{16x^3}{8x^{10}}$

88. $\dfrac{15x^2}{3x^{11}}$

89. $\dfrac{20a^3b^8}{2ab^{13}}$

90. $\dfrac{72a^5b^{11}}{9ab^{17}}$

91. $x^3 \cdot x^{-12}$

92. $x^4 \cdot x^{-12}$

93. $(2a^5)(-3a^{-7})$

94. $(4a^2)(-2a^{-5})$

95. $\dfrac{6x^2}{2x^{-8}}$

96. $\dfrac{12x^5}{3x^{-10}}$

97. $\dfrac{x^{-7}}{x^3}$

98. $\dfrac{x^{-10}}{x^4}$

99. $\dfrac{30x^2y^5}{-6x^8y^{-3}}$

100. $\dfrac{24x^2y^{13}}{-2x^5y^{-2}}$

101. $\left(\dfrac{x^3}{x^{-5}}\right)^2$

102. $\left(\dfrac{x^4}{x^{-11}}\right)^3$

103. $\left(\dfrac{-15a^4b^2}{5a^{10}b^{-3}}\right)^3$

104. $\left(\dfrac{-30a^{14}b^8}{10a^{17}b^{-2}}\right)^3$

105. $\left(\dfrac{3a^{-5}b^2}{12a^3b^{-4}}\right)^0$

106. $\left(\dfrac{4a^{-5}b^3}{12a^3b^{-5}}\right)^0$

Application Exercises

The 1986 explosion at the Chernobyl nuclear power plant in the former Soviet Union sent about 1000 kilograms of radioactive cesium-137 into the atmosphere. The formula

$$y = 1000\left(\frac{1}{2}\right)^{x/30}$$

models the amount, y, in kilograms, of cesium-137 remaining in Chernobyl x years after 1986. If even 100 kilograms of cesium-137 remain in Chernobyl's atmosphere, the area is considered unsafe for human habitation. Use this information to solve Exercises 107–108.

107. Substitute 90 for x in the formula and determine if Chernobyl will be safe for human habitation 90 years after 1986, by 2076.

108. Substitute 60 for x in the formula and determine if Chernobyl will be safe for human habitation 60 years after 1986, by 2046.

The astronomical unit (AU) is often used to measure distances within the solar system. One AU is equal to the average distance between Earth and the sun, or 92,955,630 miles. The distance, d, of the nth planet from the sun is modeled by the formula

$$d = \frac{3(2^{n-2}) + 4}{10}$$

where d is measured in astronomical units. Use this formula to solve Exercises 109–112.

109. Substitute 1 for n and find the distance between Mercury and the sun.

110. Substitute 2 for n and find the distance between Venus and the sun.

111. Substitute 3 for n and evaluate the exponential expression. Describe what this means in practical terms.

112. Substitute 5 for n and evaluate the exponential expression. Describe what this means in practical terms.

Writing in Mathematics

113. Explain the product rule for exponents. Use $b^2 \cdot b^3$ in your explanation.

114. Explain the quotient rule for exponents. Use $\frac{b^8}{b^2}$ in your explanation.

115. Explain how to find any nonzero number to the 0 power.

116. Explain the negative-exponent rule and give an example.

117. Explain the power rule for exponents. Use $(b^2)^3$ in your explanation.

118. Explain how to simplify an expression that involves a product raised to a power. Give an example.

119. Explain how to simplify an expression that involves a quotient raised to a power. Give an example.

120. How do you know if an exponential expression is simplified?

Technology Exercises

121. Graph $y = 1000(0.5)^{x/30}$, the Chernobyl model described in Exercises 107–108, in a $[0, 110, 10]$ by $[0, 1000, 100]$ viewing rectangle. Enter the equation using

$y_1 = 1000$ [×] .5 [∧] [(] x [÷] 30 [)].

Then [TRACE] along the graph and determine when y is less than 100 kilograms of cesium-137. Approximately how many years after 1986 will Chernobyl be safe for human habitation?

122. The student-teacher ratio in U.S. elementary and secondary schools can be modeled by $y = 25.34(0.987)^x$, where x represents the number of years after 1959 and y represents the student-teacher ratio. Graph the equation in a $[1, 40, 1]$ by $[0, 26, 1]$ viewing rectangle. When did the student-teacher ratio become less than 21 students per teacher?

Critical Thinking Exercises

123. Which one of the following is true?

a. $2^2 \cdot 2^4 = 2^8$ **b.** $5^6 \cdot 5^2 = 25^8$
c. $2^3 \cdot 3^2 = 6^5$ **d.** None of these is true.

124. Which one of the following is true?

a. $\frac{1}{(-2)^3} = 2^{-3}$ **b.** $\frac{2^8}{2^{-3}} = 2^5$
c. $2^4 + 2^5 = 2^9$ **d.** None of these is true.

In Exercises 125–128, simplify the expression. Assume that all variables used as exponents represent integers and that all other variables represent nonzero real numbers.

125. $x^{n-1} \cdot x^{3n+4}$

126. $(x^{-4n} \cdot x^{n})^{-3}$

127. $\left(\dfrac{x^{3-n}}{x^{6-n}}\right)^{-2}$

128. $\left(\dfrac{x^{n}y^{3n+1}}{y^{n}}\right)^{3}$

Review Exercises

129. Graph $y = 2x - 1$ in a rectangular coordinate system. Let $x = -3, -2, -1, 0, 1, 2,$ and 3. (Section 1.3, Example 2)

130. Solve $Ax + By = C$ for y. (Section 1.5, Example 6)

131. The length of a rectangular playing field is 5 meters less than twice its width. If 230 meters of fencing enclose the field, what are its dimensions? (Section 1.5, Example 5)

▸ SECTION 1.7 Scientific Notation

Objectives

1. Convert from scientific to decimal notation.
2. Convert from decimal to scientific notation.
3. Perform computations with scientific notation.
4. Use scientific notation to solve problems.

Frozen yogurt appears to be losing its cool. The bar graph in Figure 1.15 shows sales are dropping as American waistwatchers shift to low-fat ice cream.

Retail Sales of Frozen Yogurt and Ice Cream, in Billions

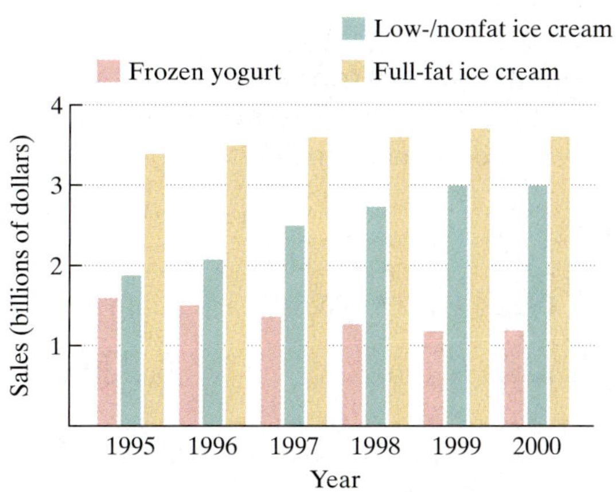

Source: FIND/SVP, Inc.

Figure 1.15

Table 1.6 Names of Large Numbers

10^{2}	hundred
10^{3}	thousand
10^{6}	million
10^{9}	billion
10^{12}	trillion
10^{15}	quadrillion
10^{18}	quintillion
10^{21}	sextillion
10^{24}	septillion
10^{27}	octillion
10^{30}	nonillion
10^{100}	googol

The bar on the right shows that in 2000, Americans spent approximately \$3.6 billion on full-fat ice cream. Because a billion is 10^{9} (see Table 1.6), this amount can be expressed as

$$3.6 \times 10^{9}.$$

The number 3.6×10^9 is written in a form called *scientific notation*. A number in **scientific notation** is expressed as a number greater than or equal to 1 and less than 10 multiplied by some power of 10. It is customary to use the multiplication symbol, ×, rather than a dot, in scientific notation. In this section, you will learn to use scientific notation to work with very large and very small numbers.

Conversions

Here are some additional examples of numbers in scientific notation:

$$6.4 \times 10^5 \quad \text{means} \quad 640{,}000.$$

$$2.17 \times 10^{-3} \quad \text{means} \quad 0.00217.$$

1 Convert from scientific to decimal notation.

Do you see that positive powers of 10 are used for large numbers and negative powers of 10 are used for small numbers? We can use the exponent on the 10 to mentally convert a number in scientific notation to decimal notation. If the exponent is *positive*, move the decimal point in the number to the *right* the same number of places as the exponent. If the exponent is *negative*, move the decimal point in the number to the *left* the same number of places as the exponent.

EXAMPLE 1 Converting from Scientific to Decimal Notation

Write each number in decimal notation:

a. 6.2×10^7 **b.** 2.019×10^{-8}.

Solution

a. We convert 6.2×10^7 to decimal notation by moving the decimal point in 6.2 seven places to the right. We need to add six zeros.

$$6.2 \times 10^7 = 62{,}000{,}000$$

b. We convert 2.019×10^{-8} to decimal notation by moving the decimal point in 2.019 eight places to the left. We need to add seven zeros to the right of the decimal point.

$$2.019 \times 10^{-8} = 0.00000002019$$

■

✔ **CHECK POINT 1** Write each number in decimal notation:

a. 2.6×10^9 **b.** 3.017×10^{-6}.

2 Convert from decimal to scientific notation.

To convert from decimal notation to scientific notation, we reverse the procedure of Example 1.

- Move the decimal point in the given number to obtain a number greater than or equal to 1 and less than 10.
- The number of places the decimal point moves gives the exponent on 10; the exponent is positive if the given number is greater than 10 and negative if the given number is between 0 and 1.

Using Technology

You can use your calculator's [EE] (enter exponent) or [EXP] key to convert from decimal to scientific notation. Here is how it's done for 0.00023:

Many Scientific Calculators

Keystrokes	Display
.00023 [EE] [=]	2.3 − 04

Many Graphing Calculators

Use the mode setting for scientific notation.

Keystrokes	Display
.00023 [ENTER]	2.3E − 4

EXAMPLE 2 Converting from Decimal Notation to Scientific Notation

Write each number in scientific notation:

a. 4,600,000 **b.** 0.00023.

Solution

Decimal point moves 6 places.

a. $4{,}600{,}000 = 4.6 \times 10^{?} \longrightarrow = 4.6 \times 10^{6}$

Decimal point moves 4 places.

b. $0.00023 = 2.3 \times 10^{-?} \longrightarrow = 2.3 \times 10^{-4}$ ■

✔ **CHECK POINT 2** Write each number in scientific notation.

a. 7,410,000,000 **b.** 0.000000092.

EXAMPLE 3 Expressing U.S. Personal Income in Scientific Notation

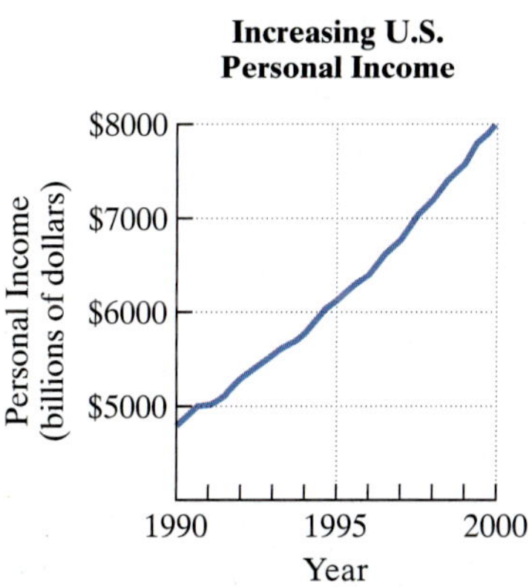

Figure 1.16 *Source:* Economagic

One result of the economic boom from 1990 through 2000 was a dramatic increase in U.S. personal income. The line graph in Figure 1.16 shows that in 2000, U.S. personal income was \$8000 billion. Express this figure in scientific notation.

Solution Because a billion in 10^9 (see Table 1.6 on page 75), U.S. personal income in 2000 can be expressed as

$$8000 \times 10^9.$$

This factor is not between 1 and 10, so the number is not in scientific notation.

The voice balloon indicates that we need to convert 8000 to scientific notation.

$$8000 \times 10^9 = (8 \times 10^3) \times 10^9 = 8 \times 10^{3+9} = 8 \times 10^{12}.$$

$8000 = 8 \times 10^3$

Because a trillion is 10^{12}, we see that U.S. personal income in 2000 was \$8 trillion. ■

✔ **CHECK POINT 3** Express 410×10^7 in scientific notation.

3 Perform computations with scientific notation.

Computations with Scientific Notation

We perform operations on numbers written in scientific notation by using properties of exponents:

$$10^m \cdot 10^n = 10^{m+n} \qquad \frac{10^m}{10^n} = 10^{m-n}.$$

Using Technology

$(6.1 \times 10^5)(4 \times 10^{-9})$
On a Calculator:

Many Scientific Calculators

6.1 [EE] 5 [×] 4 [EE] 9 [+/−] [=]

Display

2.44 − 03

Many Graphing Calculators

6.1 [EE] 5 [×] 4 [EE] [(−)] 9 [ENTER]

Display (in scientific notation mode)

2.44E − 3

EXAMPLE 4 Computations with Scientific Notation

Perform the indicated computations, writing the answers in scientific notation:

a. $(6.1 \times 10^5)(4 \times 10^{-9})$ **b.** $\dfrac{1.8 \times 10^4}{3 \times 10^{-2}}$.

Solution

a. $(6.1 \times 10^5)(4 \times 10^{-9})$

$= (6.1 \times 4) \times (10^5 \times 10^{-9})$ Regroup factors.

$= 24.4 \times 10^{5+(-9)}$ $10^m \cdot 10^n = 10^{m+n}$

$= 24.4 \times 10^{-4}$ Simplify.

$= 24.4 \times 10^1 \times 10^{-4}$ Convert 24.4 to scientific notation: $24.4 = 2.44 \times 10^1$

$= 2.44 \times 10^{-3}$ $10^1 \times 10^{-4} = 10^{1+(-4)} = 10^{-3}$

b. $\dfrac{1.8 \times 10^4}{3 \times 10^{-2}} = \dfrac{1.8}{3} \times \dfrac{10^4}{10^{-2}}$ Regroup factors.

$= 0.6 \times 10^{4-(-2)}$ $\dfrac{10^m}{10^n} = 10^{m-n}$

$= 0.6 \times 10^6$ Simplify: $4 - (-2) = 4 + 2 = 6$.

$= 6 \times 10^{-1} \times 10^6$ Convert 0.6 to scientific notation: $0.6 = 6 \times 10^{-1}$.

$= 6 \times 10^5$ $10^{-1} \times 10^6 = 10^{-1+6} = 10^5$ ■

✔ **CHECK POINT 4** Perform the indicated computations, writing answers in scientific notation:

a. $(7.1 \times 10^5)(5 \times 10^{-7})$ **b.** $\dfrac{1.2 \times 10^6}{3 \times 10^{-3}}$.

4 Use scientific notation to solve problems.

Applications: Putting Numbers in Perspective

We have seen that in 2000, Americans spent \$3.6 billion on full-fat ice cream. Example 5 shows how we can use scientific notation to comprehend the meaning of a number such as 3.6 billion.

EXAMPLE 5 Per Capita Spending On Full-Fat Ice Cream

In 2000, Americans spent 3.6×10^9 dollars on full-fat ice cream. At that time, the U.S. population was approximately 280 million, or 2.8×10^8. If ice cream spending is evenly divided, how much did each American spend?

Solution The amount that each of us spent is the total amount spent, 3.6×10^9 dollars, divided by the number of Americans, 2.8×10^8.

$$\frac{3.6 \times 10^9}{2.8 \times 10^8} = \left(\frac{3.6}{2.8}\right) \times \left(\frac{10^9}{10^8}\right) \approx 1.3 \times 10^{9-8} = 1.3 \times 10 = 130$$

We spent approximately \$130 per person on full-fat ice cream in 2000. ■

✔ **CHECK POINT 5** As of the end of 2000, the national debt was \$5.6 trillion, or 5.6×10^{12} dollars. If the national debt were evenly divided among every individual in the United States, how much would each citizen have to pay?

Many problems in algebra involve motion. Suppose that you ride your bike at an average speed of 12 miles per hour. What distance do you cover in 2 hours? Your distance is the product of your speed and the time that you travel:

$$\frac{12 \text{ miles}}{\cancel{\text{hour}}} \times 2 \cancel{\text{ hours}} = 24 \text{ miles}.$$

Your distance is 24 miles. Notice how the hour units cancel. The distance is expressed in miles.

In general, the distance covered by any moving body is the product of its average speed, or rate, and its time in motion.

A Formula for Motion

$$d = rt$$

Distance equals rate times time.

EXAMPLE 6 Using the Motion Formula

Light travels at a rate of approximately 1.86×10^5 miles per second. It takes light 5×10^2 seconds to travel from the sun to Earth. What is the distance between Earth and the sun?

Solution

$d = rt$ Use the motion formula.

$d = (1.86 \times 10^5) \times (5 \times 10^2)$ Substitute the given values.

$d = (1.86 \times 5) \times (10^5 \times 10^2)$ Rearrange factors.

$d = 9.3 \times 10^7$ Add the exponents when multiplying.

The distance between Earth and the sun is approximately 9.3×10^7 miles, or 93 million miles. ■

CHECK POINT 6 A futuristic spacecraft traveling at 1.55×10^3 miles per hour takes 20,000 hours (about 833 days) to travel from Venus to Mercury. What is the distance from Venus to Mercury?

EXERCISE SET 1.7

Practice Exercises

In Exercises 1–14, write each number in decimal notation without the use of exponents.

1. 3.8×10^2
2. 9.2×10^2
3. 6×10^{-4}
4. 7×10^{-5}
5. 7.16×10^6
6. 8.17×10^6
7. 1.4×10^0
8. 2.4×10^0
9. 7.9×10^{-1}
10. 6.8×10^{-1}
11. 4.15×10^{-3}
12. 3.14×10^{-3}
13. 6.00001×10^{10}
14. 7.00001×10^{10}

In Exercises 15–30, write each number in scientific notation.

15. 32,000
16. 64,000
17. 638,000,000,000,000,000
18. 579,000,000,000,000,000
19. 317
20. 326
21. 5716
22. 3829
23. 0.0027
24. 0.0083
25. 0.00000000504
26. 0.00000000405
27. 0.007
28. 0.005
29. 3.14159
30. 2.71828

In Exercises 31–50, perform the indicated computations. Write the answers in scientific notation. If necessary, round the decimal factor in your scientific notation answer to two decimal places.

31. $(3 \times 10^4)(2.1 \times 10^3)$
32. $(2 \times 10^4)(4.1 \times 10^3)$
33. $(1.6 \times 10^{15})(4 \times 10^{-11})$

34. $(1.4 \times 10^{15})(3 \times 10^{-11})$

35. $(6.1 \times 10^{-8})(2 \times 10^{-4})$

36. $(5.1 \times 10^{-8})(3 \times 10^{-4})$

37. $(4.3 \times 10^{8})(6.2 \times 10^{4})$

38. $(8.2 \times 10^{8})(4.6 \times 10^{4})$

39. $\dfrac{8.4 \times 10^8}{4 \times 10^5}$

40. $\dfrac{6.9 \times 10^8}{3 \times 10^5}$

41. $\dfrac{3.6 \times 10^4}{9 \times 10^{-2}}$

42. $\dfrac{1.2 \times 10^4}{2 \times 10^{-2}}$

43. $\dfrac{4.8 \times 10^{-2}}{2.4 \times 10^6}$

44. $\dfrac{7.5 \times 10^{-2}}{2.5 \times 10^6}$

45. $\dfrac{2.4 \times 10^{-2}}{4.8 \times 10^{-6}}$

46. $\dfrac{1.5 \times 10^{-2}}{3 \times 10^{-6}}$

47. $\dfrac{480{,}000{,}000{,}000}{0.00012}$

48. $\dfrac{282{,}000{,}000{,}000}{0.00141}$

49. $\dfrac{0.00072 \times 0.003}{0.00024}$

50. $\dfrac{66{,}000 \times 0.001}{0.003 \times 0.002}$

Application Exercises

The graph shows the projected number of people in the United States age 65 and over for the year 2000 and beyond. Use 10^6 for one million and the figures shown to solve Exercises 51–54. Express all answers in scientific notation.

U. S. Population Age 65 and Over

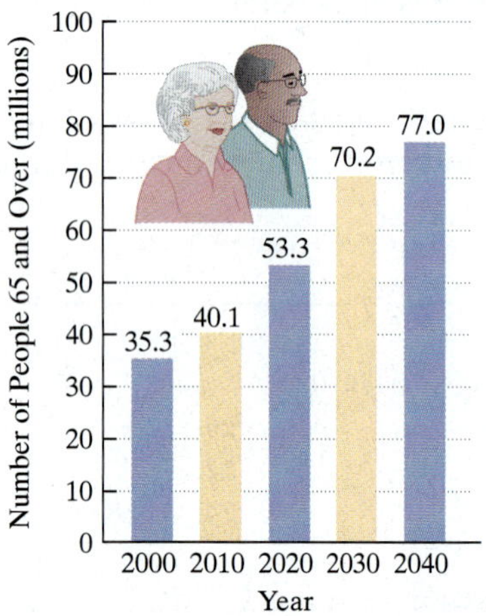

Source: U. S. Bureau of the Census

51. How many people 65 and over will there be in 2020?

52. How many people 65 and over will there be in 2030?

53. How many more people 65 and over will there be in 2040 than in 2000?

54. How many more people 65 and over will there be in 2030 than in 2000?

Use 10^{12} for one trillion and 2.8×10^8 for the U.S. population in 2000 to solve Exercises 55–57.

55. In 2000, the government collected approximately \$1.9 trillion in taxes. What was the per capita tax burden, or the amount that each U.S. citizen paid in taxes? Round to the nearest hundred dollars.

56. In 2000, U.S. personal income was \$8 trillion. What was the per capita income, or the income per U.S. citizen? Round to the nearest hundred dollars.

57. In the United States, we spend an average of \$4000 per person each year on health care—the highest in the world. What do we spend each year on health care nationwide? Express the answer in scientific notation.

58. Approximately 2×10^4 people run in the New York City Marathon each year. Each runner runs a distance of 26 miles. Write the total distance covered by all the runners (assuming that each person completes the marathon) in scientific notation.

59. The mass of one oxygen molecule is 5.3×10^{-23} gram. Find the mass of 20,000 molecules of oxygen. Express the answer in scientific notation.

60. The mass of one hydrogen atom is 1.67×10^{-24} gram. Find the mass of 80,000 hydrogen atoms. Express the answer in scientific notation.

Writing in Mathematics

61. How do you know if a number is written in scientific notation?

62. Explain how to convert from scientific to decimal notation and give an example.

63. Explain how to convert from decimal to scientific notation and give an example.

64. Describe one advantage of expressing a number in scientific notation over decimal notation.

65. Read Exercise 56. Does a fairly high per capita income indicate that the economic boom has reached all Americans? Explain your answer.

Technology Exercises

66. Use a calculator to check any three of your answers in Exercises 1–14.

67. Use a calculator to check any three of your answers in Exercises 15–30.

68. Use a calculator with an EE or EXP key to check any four of your computations in Exercises 31–50. Display the result of the computation in scientific notation.

Critical Thinking Exercises

69. Which one of the following is true?

a. $534.7 = 5.347 \times 10^3$

b. $\dfrac{8 \times 10^{30}}{4 \times 10^{-5}} = 2 \times 10^{25}$

c. $(7 \times 10^5) + (2 \times 10^{-3}) = 9 \times 10^2$
d. $(4 \times 10^3) + (3 \times 10^2) = 43 \times 10^2$

70. Our hearts beat about 70 times per minute. Express in scientific notation how many times the heart beats over a lifetime of 80 years. Round the decimal factor in your scientific notation answer to two decimal places.

71. Give an example of a number where there is no advantage in using scientific notation over decimal notation.

Review Exercises

72. Simplify: $9(10x - 4) - (5x - 10)$. (Section 1.2, Example 14)

73. Solve: $\frac{4x - 1}{10} = \frac{5x + 2}{4} - 4$. (Section 1.4, Example 4)

74. Simplify: $(8x^4y^{-3})^{-2}$. (Section 1.6, Example 7)

CHAPTER 1 GROUP PROJECTS

1. One of the best ways to learn how to *solve* a word problem in algebra is to *design* word problems of your own. Creating a word problem makes you very aware of precisely how much information is needed to solve the problem. You must also focus on the best way to present information to a reader and on how much information to give. As you write your problem, you gain skills that will help you solve problems created by others.

The group should design five different word problems that can be solved using an algebraic equation. All of the problems should be on different topics. For example, the group should not have more than one problem on a price reduction. The group should turn in both the problems and their algebraic solutions.

2. Each group member should create a graph that tells the story of a particular emotional experience. The experience can involve feelings of love, anger, sadness, or any other emotion you choose.

The horizontal axis should be labeled time and the vertical axis the emotion you are graphing. You will not be using your algebra skills to create your graph; however, you should try to make the graph as precise as possible. You may use negative numbers on the vertical axis, if appropriate. After each group member has created a graph, pool together all of the graphs and study them to see if there are any similarities in the graphs for a particular emotion or for all emotions.

3. Putting Numbers into Perspective. A large number, such as a trillion, is a staggering amount that is difficult to comprehend. If you counted to a trillion out loud, one number per second, it would take you 31,688 years. However, a large number can be put into perspective by comparing it with another number. For example, in Check Point 5 on page 78, you put the $5.6 trillion national debt into perspective by comparing it to the number of U.S. citizens. The total distance covered by all the runners in the New York City Marathon (Exercise 58 on page 80) can be put into perspective by comparing this distance with, say, the distance from New York City to San Francisco.

For this project, each group member should consult an almanac, a newspaper, or the World Wide Web to find a number greater than one million. Explain to other members of the group the context in which the large number is used. Express the number in scientific notation. Then put the number into perspective by comparing it with another number.

CHAPTER SUMMARY, REVIEW, AND TEST

SUMMARY

DEFINITIONS AND CONCEPTS	EXAMPLES
Section 1.1 Algebraic Expressions and Real Numbers	
Letters that represent numbers are called variables. An algebraic expression is a combination of variables, numbers, and operation symbols. English phrases can be translated into algebraic expressions: • Addition: sum, plus, increased by, more than • Subtraction: difference, minus, decreased by, less than • Multiplication: product, times, of, twice • Division: quotient, divide, per, ratio	Translate: Six less than the product of a number and five. 6 less than / the product of a number and 5 $5x - 6$ or $x \cdot 5 - 6$
Many algebraic expressions contain exponents. If b is a natural number, b^n, the nth power of b, is the product of n factors of b. Furthermore, $b^1 = b$. Evaluating an algebraic expression means to find the value of the expression for a given value of the variable.	Evaluate $6 + 5(x - 10)^3$ for $x = 12$. $= 6 + 5(12 - 10)^3$ $= 6 + 5 \cdot 2^3$ $= 6 + 5 \cdot 8$ $= 6 + 40 = 46$
An equation is a statement that two expressions are equal. Formulas are equations that express relationships among two or more variables. Formulas that approximately describe data are called mathematical models and are said to model the data.	The formula $h = -16t^2 + 200t + 4$ models the height, h, in feet, of fireworks t seconds after launch. What is the height after 2 seconds? $h = -16 \cdot 2^2 + 200 \cdot 2 + 4$ $= -16 \cdot 4 + 200 \cdot 2 + 4$ $= -64 + 400 + 4 = 340$ The height is 340 feet.
A set is a collection of objects, called elements, enclosed in braces. The roster method uses commas to separate the elements of the set. Set-builder notation describes the elements of a set, but does not list them. The symbol $\in$ means that an element is in a set; $\notin$ means that an element is not in a set. The set of real numbers is the set of all numbers that can be represented by points on the number line. Sets that make up the real numbers include: Natural numbers: $\{1, 2, 3, 4, \ldots\}$ Whole numbers: $\{0, 1, 2, 3, 4, \ldots\}$ Integers: $\{\ldots, -4, -3, -2, -1, 0, 1, 2, 3, 4, \ldots\}$ Rational numbers: $\{\frac{a}{b} \mid a \text{ and } b \text{ are integers and } b \neq 0\}$ Irrational numbers: $\{x \mid x \text{ is a real number and } x \text{ is not a rational number}\}$. In decimal form, rational numbers terminate or repeat. In decimal form, irrational numbers do neither.	• Use the roster method to list the elements of $\{x \mid x \text{ is a natural number less than } 6\}$. ***Solution*** $\{1, 2, 3, 4, 5\}$ • True or false: $\sqrt{2} \notin \{x \mid x \text{ is a rational number}\}$ ***Solution*** The statement is true: $\sqrt{2}$ is not a rational number. The decimal form of $\sqrt{2}$ neither terminates nor repeats. Thus, $\sqrt{2}$ is an irrational number.

DEFINITIONS AND CONCEPTS	EXAMPLES
Section 1.1 Algebraic Expressions and Real Numbers (continued)	
For any two real numbers, a and b, a is less than b if a is to the left of b on the number line. ***Inequality Symbols*** $<$: is less than $>$: is greater than $\le$: is less than or equal to $\ge$: is greater than or equal to	• $-1 < 5$, or -1 is less than 5, is true because -1 is to the left of 5 on a number line. • $-3 \ge 7$, -3 is greater than or equal to 7, is false. Neither $-3 > 7$ nor $-3 = 7$ is true.
Section 1.2 Operations with Real Numbers and Simplifying Algebraic Expressions	
Absolute Value $\lvert a\rvert = \begin{cases} a & \text{if } a \ge 0 \\ -a & \text{if } a < 0 \end{cases}$ The opposite, or additive inverse, of a is $-a$. When a is a negative number, $-a$ is positive.	• $\lvert 6.03\rvert = 6.03$ • $\lvert 0\rvert = 0$ • $\lvert -4.9\rvert = -(-4.9) = 4.9$
Adding Real Numbers To add two numbers with the same sign, add their absolute values and use their common sign. To add two numbers with different signs, subtract the smaller absolute value from the greater absolute value and use the sign of the number with the greater absolute value.	$-4.1 + (-6.2) = -10.3$ $-30 + 25 = -5$ $12 + (-8) = 4$
Subtracting Real Numbers $a - b = a + (-b)$	$-\frac{3}{4} - \left(-\frac{1}{2}\right) = -\frac{3}{4} + \frac{1}{2} = -\frac{3}{4} + \frac{2}{4} = -\frac{1}{4}$
Multiplying and Dividing Real Numbers The product or quotient of two numbers with the same sign is positive and with different signs is negative. If no number is 0, a product with an even number of negative factors is positive and with an odd number of negative factors is negative. Division by 0 is undefined.	• $2(-6)(-1)(-5) = -60$ Three (odd) negative factors give a negative product. • $(-2)^3 = (-2)(-2)(-2) = -8$ • $-\frac{1}{3}\left(-\frac{2}{5}\right) = \frac{2}{15}$ • $\frac{-14}{2} = -7$
Order of Operations **1.** Perform operations within grouping symbols, starting with the innermost grouping symbols. Grouping symbols include parentheses, brackets, fraction bars, absolute value symbols, and square root signs. **2.** Evaluate exponential expressions. **3.** Multiply and divide in order from left to right. **4.** Add and subtract in order from left to right.	Simplify: $\frac{6(8-10)^3 + (-2)}{(-5)^2(-2)}$. $= \frac{6(-2)^3 + (-2)}{(-5)^2(-2)} = \frac{6(-8) + (-2)}{25(-2)}$ $= \frac{-48 + (-2)}{-50} = \frac{-50}{-50} = 1$
Basic Algebraic Properties Commutative: $a + b = b + a$ and $ab = ba$ Associative: $(a + b) + c = a + (b + c)$ $(ab)c = a(bc)$ Distributive: $a(b + c) = ab + ac$ $a(b - c) = ab - ac$ $(b + c)a = ba + ca$	• Commutative $3x + 5 = 5 + 3x = 5 + x \cdot 3$ • Associative $-4(6x) = (-4 \cdot 6)x = -24x$ • Distributive $-4(9x + 3) = -4(9x) + (-4) \cdot 3$ $= -36x + (-12)$ $= -36x - 12$

DEFINITIONS AND CONCEPTS	EXAMPLES

Section 1.2 Operations with Real Numbers and Simplifying Algebraic Expressions (continued)

Simplifying Algebraic Expressions
Terms are separated by addition. Like terms have the same variable factors and are combined using the distributive property. An algebraic expression is simplified when grouping symbols have been removed and like terms have been combined.

Simplify: $7(3x - 4) - (10x - 5)$.

$$= 21x - 28 - 10x + 5$$
$$= 21x - 10x - 28 + 5$$
$$= 11x - 23$$

Section 1.3 Graphing Equations

The rectangular coordinate system consists of a horizontal number line, the x-axis, and a vertical number line, the y-axis, intersecting at their zero points, the origin. Each point in the system corresponds to an ordered pair of real numbers (x, y). The first number in the pair is the x-coordinate; the second number is the y-coordinate.

Plot: $(2, 3)$, $(-5, 4)$, $(-4, -3)$, and $(5, -2)$.

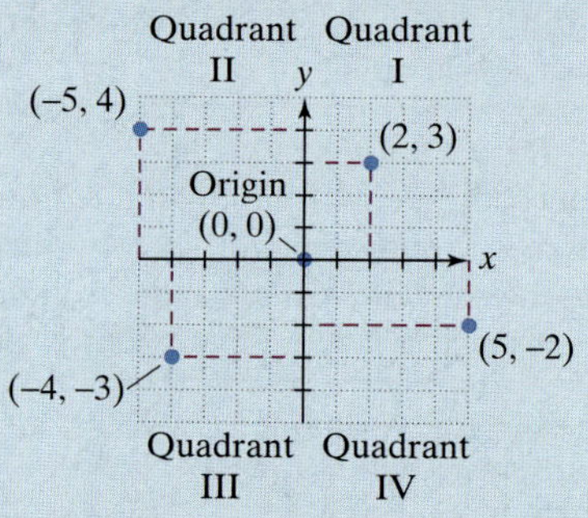

An ordered pair is a solution of an equation in two variables if replacing the variables by the corresponding coordinates results in a true statement. The ordered pair is said to satisfy the equation. The graph of the equation is the set of all points whose coordinates satisfy the equation. One method for graphing an equation is to plot ordered-pair solutions and connect them with a smooth curve or line.

Graph: $y = x^2 - 1$.

x	$y = x^2 - 1$
−2	$(-2)^2 - 1 = 3$
−1	$(-1)^2 - 1 = 0$
0	$0^2 - 1 = -1$
1	$1^2 - 1 = 0$
2	$2^2 - 1 = 3$

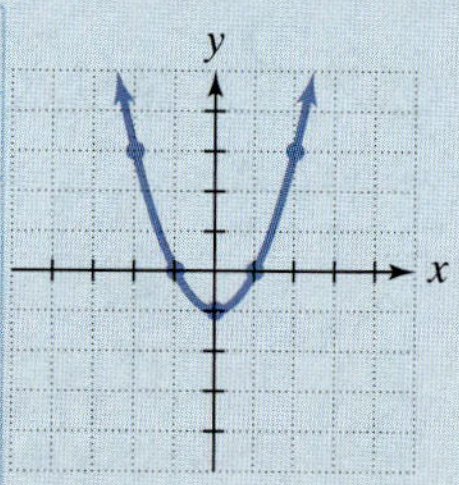

Section 1.4 Solving Linear Equations

A linear equation in one variable can be written in the form $ax + b = 0, a \neq 0$. A solution is a value of the variable that makes the equation a true statement. The set of all such solutions is the equation's solution set. Equivalent equations have the same solution set. To solve a linear equation:

1. Simplify each side.
2. Collect variable terms on one side and constant terms on the other side.
3. Isolate the variable and solve.
4. Check the proposed solution in the original equation.

Solve: $4(x - 5) = 2x - 14$.

$$4x - 20 = 2x - 14$$
$$4x - 2x - 20 = 2x - 2x - 14$$
$$2x - 20 = -14$$
$$2x - 20 + 20 = -14 + 20$$
$$2x = 6$$
$$\frac{2x}{2} = \frac{6}{2}$$
$$x = 3$$

Checking gives $-8 = -8$, so 3 is the solution, or $\{3\}$ is the solution set.

DEFINITIONS AND CONCEPTS | **EXAMPLES**

Section 1.4 Solving Linear Equations (continued)

Equations Containing Fractions

Multiply both sides (all terms) by the least common denominator. This clears the equation of fractions.

Solve: $\frac{x-2}{5} + \frac{x+2}{2} = \frac{x+4}{3}$.

$$30\left(\frac{x-2}{5} + \frac{x+2}{2}\right) = 30\left(\frac{x+4}{3}\right)$$

$$6(x-2) + 15(x+2) = 10(x+4)$$

$$6x - 12 + 15x + 30 = 10x + 40$$

$$21x + 18 = 10x + 40$$

$$11x = 22$$

$$x = 2$$

Checking gives $2 = 2$, so 2 is the solution, or $\{2\}$ is the solution set.

Types of Equations

An equation that is true for all real numbers, $\mathbb{R}$, is called an identity. When solving an identity, the variable is eliminated and a true statement, such as $3 = 3$, results. An equation that is not true for even one real number is called an inconsistent equation. A false statement, such as $3 = 7$, results when solving such an equation, whose solution set is $\varnothing$, the empty set. A conditional equation is not an identity, but is true for at least one real number.

Solve: $4x + 5 = 4(x + 2)$.

$$4x + 5 = 4x + 8$$

$$5 = 8, \quad \text{false}$$

The inconsistent equation has no solution: $\varnothing$.

Solve: $5x - 4 = 5(x + 1) - 9$

$$5x - 4 = 5x + 5 - 9$$

$$5x - 4 = 5x - 4$$

$$-4 = -4, \quad \text{true}$$

All real numbers satisfy the identity: $\mathbb{R}$.

Section 1.5 Problem Solving and Using Formulas

Strategy for Solving Algebraic Word Problems

1. Let x represent one of the quantities.
2. Represent other quantities in terms of x.
3. Write an equation that describes the conditions.
4. Solve the equation and answer the question.
5. Check the proposed solution in the original wording of the problem.

After a 60% reduction, a suit sold for \$32. What was the original price?

Let $x =$ the original price.

Original price	minus	60% reduction	=	reduced price
x	$-$	$0.6x$	$=$	32

$$0.4x = 32$$

$$\frac{0.4x}{0.4} = \frac{32}{0.4}$$

$$x = 80$$

The original price was \$80. Check this using the first sentence in the problem's conditions.

To solve for a variable in a formula, use the steps for solving a linear equation and isolate the specified variable on one side of the equation.

Solve for r:

$$E = I(R + r).$$

$$E = IR + Ir \quad \text{(We need to isolate } r\text{.)}$$

$$E - IR = Ir$$

$$\frac{E - IR}{I} = r$$

DEFINITIONS AND CONCEPTS	EXAMPLES
Section 1.6 Properties of Integral Exponents	
The Product Rule $b^m \cdot b^n = b^{m+n}$	$(-3x^{10})(5x^{20}) = -3 \cdot 5x^{10+20} = -15x^{30}$
The Quotient Rule $\frac{b^m}{b^n} = b^{m-n}, b \neq 0$	$\frac{5x^{20}}{10x^{10}} = \frac{5}{10} \cdot x^{20-10} = \frac{x^{10}}{2}$
Zero and Negative Exponents $b^0 = 1, b \neq 0$ $b^{-n} = \frac{1}{b^n}$ and $\frac{1}{b^{-n}} = b^n$	• $(3x)^0 = 1$ • $3x^0 = 3 \cdot 1 = 3$ • $\frac{2^{-3}}{4^{-2}} = \frac{4^2}{2^3} = \frac{16}{8} = 2$
Power Rule $(b^m)^n = b^{mn}$	$(x^5)^{-4} = x^{5(-4)} = x^{-20} = \frac{1}{x^{20}}$
Products to Powers $(ab)^n = a^nb^n$	$(5x^3y^{-4})^{-2} = 5^{-2} \cdot (x^3)^{-2} \cdot (y^{-4})^{-2}$ $= 5^{-2}x^{-6}y^8$ $= \frac{y^8}{5^2x^6} = \frac{y^8}{25x^6}$
Quotients to Powers $\left(\frac{a}{b}\right)^n = \frac{a^n}{b^n}$	$\left(\frac{2}{x^3}\right)^{-4} = \frac{2^{-4}}{(x^3)^{-4}} = \frac{2^{-4}}{x^{-12}}$ $= \frac{x^{12}}{2^4} = \frac{x^{12}}{16}$
An exponential expression is simplified when: • No parentheses appear. • No powers are raised to powers. • Each base occurs only once. • No negative or zero exponents appear.	Simplify: $\frac{-5x^{-3}y^2}{-20x^2y^{-6}}$. $= \frac{-5}{-20} \cdot x^{-3-2} \cdot y^{2-(-6)}$ $= \frac{1}{4}x^{-5}y^8 = \frac{y^8}{4x^5}$
Section 1.7 Scientific Notation	
A number in scientific notation is expressed as a number greater than or equal to 1 and less than 10 multiplied by some power of 10.	Write in decimal notation: 2.9×10^{-3}. $2.9 \times 10^{-3} = .0029 = 0.0029$ Write in scientific notation: 16,000. $16{,}000 = 1.6 \times 10^4$
Use $10^m \cdot 10^n = 10^{m+n}$ and $\frac{10^m}{10^n} = 10^{m-n}$ to perform computations with numbers in scientific notation.	$(5 \times 10^3)(4 \times 10^{-8})$ $= 5 \cdot 4 \times 10^{3+(-8)}$ $= 20 \times 10^{-5}$ $= 2 \times 10^1 \times 10^{-5} = 2 \times 10^{-4}$

Review Exercises

1.1 *In Exercises 1–3, write each English phrase as an algebraic expression. Let x represent the number.*

1. Ten less than twice a number
2. Four more than the product of six and a number
3. The quotient of nine and a number, increased by half of the number

In Exercises 4–6, evaluate each algebraic expression for the given value or values of the variable.

4. $x^2 - 7x + 4$, for $x = 10$
5. $6 + 2(x - 8)^3$, for $x = 11$
6. $x^4 - (x - y)$, for $x = 2$ and $y = 1$

In Exercises 7–8, use the roster method to list the elements in each set.

7. $\{x \mid x$ is a natural number less than $3\}$
8. $\{x \mid x$ is an integer greater than -4 and less than $2\}$

In Exercises 9–11, determine whether each statement is true or false.

9. $0 \in \{x \mid x$ is a natural number$\}$
10. $-2 \in \{x \mid x$ is a rational number$\}$
11. $\frac{1}{3} \notin \{x \mid x$ is an irrational number$\}$

In Exercises 12–14, write out the meaning of each inequality. Then determine whether the inequality is true or false.

12. $-5 < 2$
13. $-7 \geq -3$
14. $-7 \leq -7$
15. You are riding along an expressway traveling x miles per hour. The formula
$$S = 0.015x^2 + x + 10$$
models the recommended safe distance, S, in feet, between your car and other cars on the expressway. What is the recommended safe distance when your speed is 60 miles per hour?

1.2 *In Exercises 16–18, find each absolute value.*

16. $|-9.7|$
17. $|5.003|$
18. $|0|$

In Exercises 19–28, perform the indicated operation.

19. $-2.4 + (-5.2)$
20. $-6.8 + 2.4$
21. $-7 - (-20)$
22. $(-3)(-20)$
23. $4(-3)(-2)(-10)$
24. $(-2)^4$
25. -2^5
26. $-\frac{2}{3} \div \frac{8}{5}$
27. $\frac{-35}{-5}$
28. $\frac{54.6}{-6}$
29. Find $-x$ if $x = -7$.

In Exercises 30–36, use the order of operations to simplify each expression.

30. $-11 - [-17 + (-3)]$
31. $\left(-\frac{1}{2}\right)^3 \cdot 2^4$
32. $-3[4 - (6 - 8)]$
33. $8^2 - 36 \div 3^2 \cdot 4 - (-7)$
34. $\frac{(-2)^4 + (-3)^2}{2^2 - (-21)}$
35. $\frac{(7 - 9)^3 - (-4)^2}{2 + 2(8) \div 4}$
36. $4 - (3 - 8)^2 + 3 \div 6 \cdot 4^2$

In Exercises 37–41, simplify each algebraic expression.

37. $5(2x - 3) + 7x$
38. $5x + 7x^2 - 4x + 2x^2$
39. $3(4y - 5) - (7y + 2)$
40. $8 - 2[3 - (5x - 1)]$
41. $6(2x - 3) - 5(3x - 2)$

1.3 *In Exercises 42–44, plot the given point in a rectangular coordinate system.*

42. $(-1, 3)$
43. $(2, -5)$
44. $(0, -6)$

In Exercises 45–48, graph each equation. Let x = −3, −2, −1, 0, 1, 2, and 3.

45. $y = 2x - 2$
46. $y = x^2 - 3$
47. $y = x$
48. $y = |x| - 2$
49. What does a $[-20, 40, 10]$ by $[-5, 5, 1]$ viewing rectangle mean? Draw axes with tick marks and label the tick marks to illustrate this viewing rectangle.

The caseload of Alzheimer's disease in the United States is expected to explode as baby boomers head into their later years. The graph shows the percentage of Americans with the disease, by age. Use the graph to solve Exercises 50–52.

Alzheimer's Prevalence in the U.S., by Age

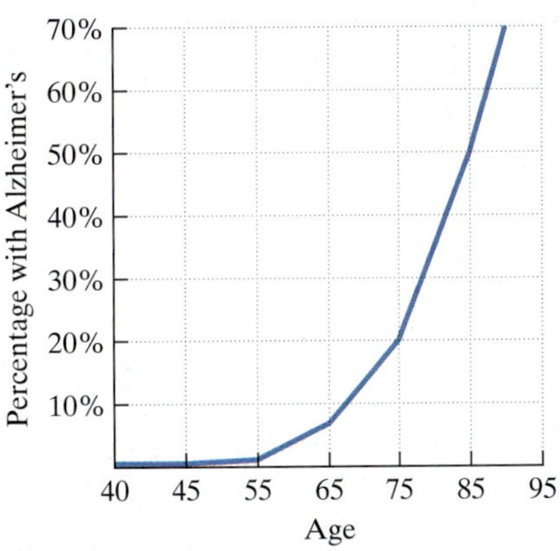

Source: Centers for Disease Control

50. What percentage of Americans who are 75 have Alzheimer's disease?
51. What age represents 50% prevalence of Alzheimer's disease?
52. Describe the trend shown by the graph.

53. Select the graph that best illustrates the following description: A train pulls into a station and lets off its passengers.

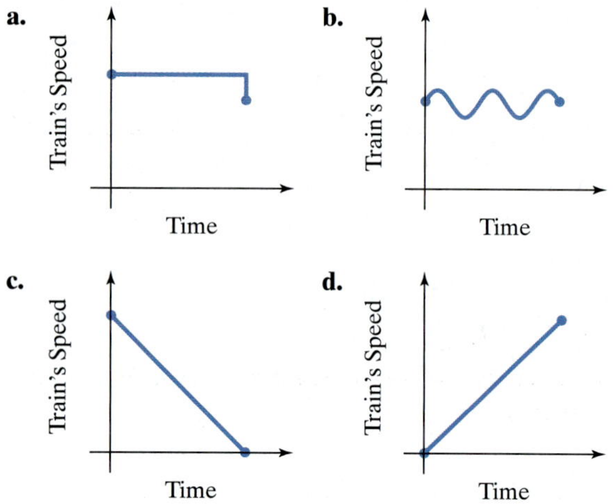

1.4 *In Exercises 54–59, solve and check each linear equation.*

54. $2x - 5 = 7$

55. $5x + 20 = 3x$

56. $7(x - 4) = x + 2$

57. $1 - 2(6 - x) = 3x + 2$

58. $2(x - 4) + 3(x + 5) = 2x - 2$

59. $2x - 4(5x + 1) = 3x + 17$

In Exercises 60–64, solve each equation by clearing fractions.

60. $\frac{2x}{3} = \frac{x}{6} + 1$

61. $\frac{x}{2} - \frac{1}{10} = \frac{x}{5} + \frac{1}{2}$

62. $\frac{2x}{3} = 6 - \frac{x}{4}$

63. $\frac{x}{4} = 2 + \frac{x - 3}{3}$

64. $\frac{3x + 1}{3} - \frac{13}{2} = \frac{1 - x}{4}$

In Exercises 65–67, solve each equation. Then state whether the equation is an identity, a conditional equation, or an inconsistent equation.

65. $7x + 5 = 5(x + 3) + 2x$

66. $7x + 13 = 4x - 10 + 3x + 23$

67. $7x + 13 = 3x - 10 + 2x + 23$

68. The formula $M = 420x + 720$ models the data for the amount of money lost to credit card fraud worldwide, M, expressed in millions of dollars, x years after 1989. In which year did losses amount to 4080 million dollars?

1.5 *In Exercises 69–74, use the five-step strategy for solving word problems.*

69. The cost of raising a child through the age of 17 varies by income group. The cost in middle-income families exceeds that of low-income families by \$63 thousand, and the cost in high-income families is \$3 thousand less than twice that of low-income families. Three children, one in a low-income family, one in a middle-income family, and one in a high-income family, will cost a total of \$756 thousand to raise through the age of 17. Find the cost of raising a child in each of the three income groups. (*Source*: *The World Almanac*; low annual income is less than \$36,800, middle is \$36,800–\$61,900, and high exceeds \$61,900.)

70. In 1965, women made up 1.2% of the U.S. military. With an increase at the rate of 0.4% per year, in which year will women make up 17.2% of the military?

71. You are choosing between two long-distance telephone plans. One plan has a monthly fee of \$15 with a charge of \$0.05 per minute. The other plan has a monthly fee of \$5 with a charge of \$0.07 per minute. For how many minutes of long-distance calls will the costs for the two plans be the same?

72. After a 20% price reduction, a cordless phone sold for \$48. What was the phone's price before the reduction?

73. A salesperson earns \$300 per week plus 5% commission of sales. How much must be sold to earn \$800 in a week?

74. The length of a rectangular field is 6 yards less than triple the width. If the perimeter of the field is 340 yards, what are its dimensions?

In Exercises 75–80, solve each formula for the specified variable.

75. $V = \frac{1}{3}Bh$ for h

76. $y - y_1 = m(x - x_1)$ for x

77. $E = I(R + r)$ for R

78. $C = \frac{5F - 160}{9}$ for F

79. $s = vt + gt^2$ for g

80. $T = gr + gvt$ for g

1.6 *In Exercises 81–93, simplify each exponential expression. Assume that no denominators are 0.*

81. $(-3x^7)(-5x^6)$

82. x^2y^{-5}

83. $\frac{3^{-2}x^4}{y^{-7}}$

84. $(x^3)^{-6}$

85. $(7x^3y)^2$

86. $\frac{16y^3}{-2y^{10}}$

87. $(-3x^4)(4x^{-11})$

88. $\frac{12x^7}{4x^{-3}}$

89. $\frac{-10a^5b^6}{20a^{-3}b^{11}}$

90. $(-3xy^4)(2x^2)^3$

91. $2^{-2} + \frac{1}{2}x^0$

92. $(5x^2y^{-4})^{-3}$

93. $(3x^4y^{-2})(-2x^5y^{-3})$

1.7 *In Exercises 94–95, write each number in decimal notation.*

94. 7.16×10^6 **95.** 1.07×10^{-4}

In Exercises 96–97, write each number in scientific notation.

96. 41,000,000,000,000

97. 0.00809

In Exercises 98–99, perform the indicated computations. Write the answers in scientific notation.

98. $(4.2 \times 10^{13})(3 \times 10^{-6})$

99. $\dfrac{5 \times 10^{-6}}{20 \times 10^{-8}}$

100. If the population of the United States is 2.8×10^8 and each person spends about \$150 per year going to the movies (or renting movies), express the total annual spending on movies in scientific notation.

Chapter 1 Test

1. Write the following English phrase as an algebraic expression:

   Five less than the product of a number and four.

   Let x represent the number.
2. Evaluate $8 + 2(x - 7)^4$, for $x = 10$.
3. Use the roster method to list the elements in the set:

   $\{x \mid x \text{ is a negative integer greater than } -5\}$.
4. Determine whether the following statement is true or false:

   $\frac{1}{4} \notin \{x \mid x \text{ is a natural number}\}$.
5. Write out the meaning of the inequality $-3 > -1$. Then determine whether the inequality is true or false.
6. The formula

   $$F = 24t^2 - 260t + 816$$

   models the number of police officers in the United States, F, convicted of felonies t years after 1990. How many convictions of police officers were there in 2000? (*Source*: FBI)
7. Find the absolute value: $|-17.9|$.

In Exercises 8–12, perform the indicated operation.

8. $-10.8 + 3.2$ **9.** $-\frac{1}{4} - \left(-\frac{1}{2}\right)$

10. $2(-3)(-1)(-10)$

11. $-\frac{1}{4}\left(-\frac{1}{2}\right)$ **12.** $\dfrac{-27.9}{-9}$

In Exercises 13–15, use the order of operations to simplify each expression.

13. $24 - 36 \div 4 \cdot 3$ **14.** $(5^2 - 2^4) + [9 \div (-3)]$

15. $\dfrac{(8 - 10)^3 - (-4)^2}{2 + 8(2) \div 4}$

In Exercises 16–17, simplify each algebraic expression.

16. $7x - 4(3x + 2) - 10$

17. $5(2y - 6) - (4y - 3)$

18. Plot $(-2, -4)$ in a rectangular coordinate system.

19. Graph $y = x^2 - 4$ in a rectangular coordinate system. Let $x = -3, -2, -1, 0, 1, 2,$ and 3.

In Exercises 20–22, solve each equation. If the solution set is $\varnothing$ or $\mathbb{R}$, classify the equation as an inconsistent equation or an identity.

20. $3(2x - 4) = 9 - 3(x + 1)$

21. $\dfrac{2x - 3}{4} = \dfrac{x - 4}{2} - \dfrac{x + 1}{4}$

22. $3(x - 4) + x = 2(6 + 2x)$

In Exercises 23–26, use the five-step strategy for solving word problems.

23. Find two numbers such that the second number is 3 more than twice the first number and the sum of the two numbers is 72.

24. You bought a new car for \$13,805. Its value is decreasing by \$1820 per year. After how many years will its value be \$4705?

25. After a 60% reduction, a jacket sold for \$20. What was the jacket's price before the reduction?

26. The length of a rectangular field exceeds the width by 260 yards. If the perimeter of the field is 1000 yards, what are its dimensions?

In Exercises 27–28, solve each formula for the specified variable.

27. $V = \frac{1}{3}lwh$ for h

28. $Ax + By = C$ for y

In Exercises 29–31, simplify each exponential expression.

29. $(-2x^5)(7x^{-10})$ **30.** $\dfrac{-10x^4y^3}{-40x^{-2}y^6}$

31. $(4x^{-5}y^2)^{-3}$

32. Write in decimal notation: 3.8×10^{-6}.

33. Write in scientific notation: 407,000,000,000.

34. Divide and write the answer in scientific notation:

$$\frac{4 \times 10^{-3}}{8 \times 10^{-7}}.$$

35. In 2000, the world population was approximately 6.08×10^9. By some projections, the world's population will double by 2040. Express the population at that time in scientific notation.

In the dramatic arts, ours is the era of the movies. As individuals and as a nation, we've grown up with them. Our images of love, war, family, country—even of things that terrify us—owe much to what we've seen on screen.

Take a rapid sequence of still photographs of a moving scene and project them onto a screen at thirty shots a second or faster. Our eyes see the result as continuous motion. The small difference between one frame and the next cannot be detected by the human visual system. In this chapter, you will see how the notion of a function and the steepness of a line can analyze motion and change by "freezing the frame" of a continuously changing process, instant by instant. Understanding the concept of a function will give you a new perspective on many situations, including using mathematics in a way that is similar to making a movie.

Chapter 2

Functions and Linear Functions

▸ SECTION 2.1 Introduction to Functions

Objectives

1. Find the domain and range of a relation.
2. Determine whether a relation is a function.
3. Evaluate a function.
4. Graph functions by plotting points.
5. Use the vertical line test to identify functions.
6. Obtain information about a function from its graph.

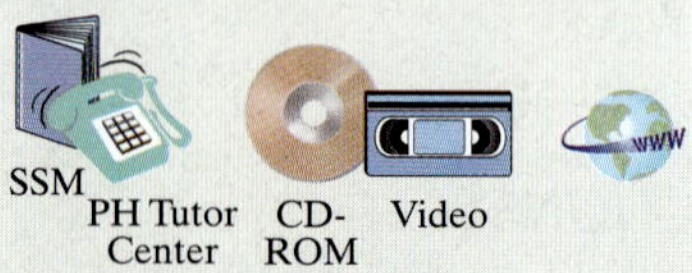

1 Find the domain and range of a relation.

Enjoy talking on the phone? In 2000, nearly 105 million Americans were chatting up a storm on their mobile phones, an increase of over 200% from 1995, when only 34 million Americans were using mobile phones. The graph in Figure 2.1 shows the increase in cellular telephone subscriptions in the United States from 1995 through 2000. With video mobile phones by 2020, there seems to be no limit to the ways in which we keep in touch.

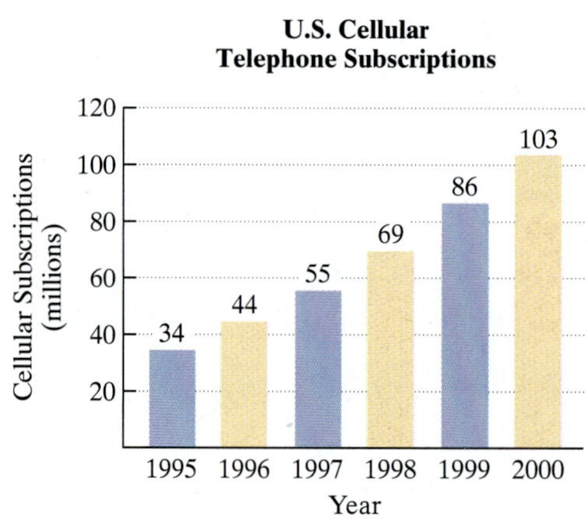

Figure 2.1 *Source:* Cellular Telephone Industry of America

If we let x represent a year and y the number of cellular telephone subscriptions, in millions, the bar graph in Figure 2.1 shows a correspondence between the two variables x and y. We can write this correspondence using a set of ordered pairs:

$$\{(1995, 34), (1996, 44), (1997, 55), (1998, 69), (1999, 86), (2000, 103)\}.$$

The mathematical term for a set of ordered pairs is a **relation**.

Definition of a Relation

A **relation** is any set of ordered pairs. The set of all first components of the ordered pairs is called the **domain** of the relation, and the set of all second components is called the **range** of the relation.

EXAMPLE 1 Analyzing U.S. Mobile-Phone Subscriptions as a Relation

Find the domain and range of the relation:

$$\{(1995, 34), (1996, 44), (1997, 55), (1998, 69), (1999, 86), (2000, 103)\}.$$

Solution The domain is the set of all first components. Thus, the domain is

$$\{1995, 1996, 1997, 1998, 1999, 2000\}.$$

The range is the set of all second components. Thus, the range is

$$\{34, 44, 55, 69, 86, 103\}.$$

■

✔ **CHECK POINT 1** Find the domain and the range of the relation:

$$\{(5, 12.8), (10, 16.2), (15, 18.9), (20, 20.7), (25, 21.8)\}.$$

As you worked Check Point 1, did you wonder if the numbers in each ordered pair represented anything? Think paid vacation days! The first number in each ordered pair is the number of years for full-time workers at medium to large U.S. companies. The second number is the average number of paid vacation days each year. Consider, for example, the ordered pair (15, 18.9).

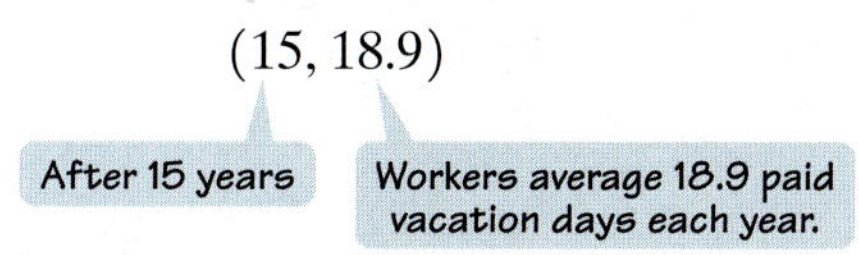

The relation in the vacation-days example can be pictured as follows:

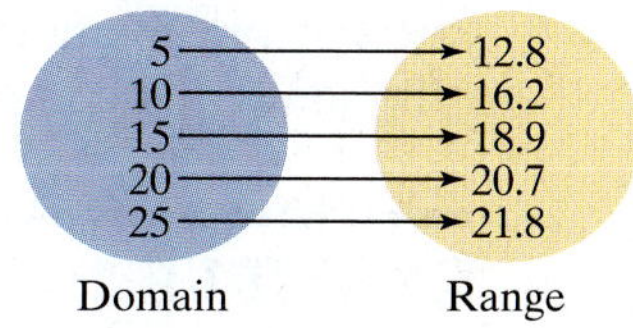

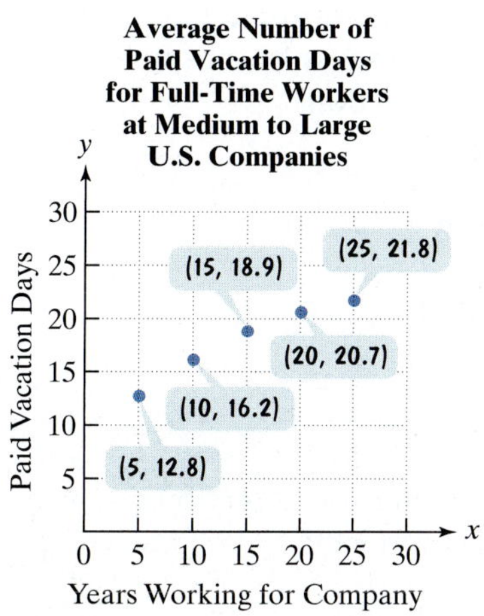

Figure 2.2 The graph of a relation showing a correspondence between years with a company and paid vacation days
Source: Bureau of Labor Statistics

The five points in Figure 2.2 are one way to visually represent the relation.

Functions

The SAT is the test that everyone loves to hate. The set of points in Figure 2.3 shows a relation indicating a correspondence between SAT scores and grade point averages for the first year in college for a group of randomly selected college students. The domain is the set of SAT scores for the students. The range is the set of their grade point averages. Is it possible for two students with the same SAT score to have different grade point averages? Look for two or more data points that are aligned vertically. We see that there are two students who have the same SAT score, 700, but their grade point averages are different. One student has a grade point average of approximately 2.4 and the other a grade point average of approximately 3.7. These students are represented by the following ordered pairs:

$$(700, 2.4) \qquad (700, 3.7).$$

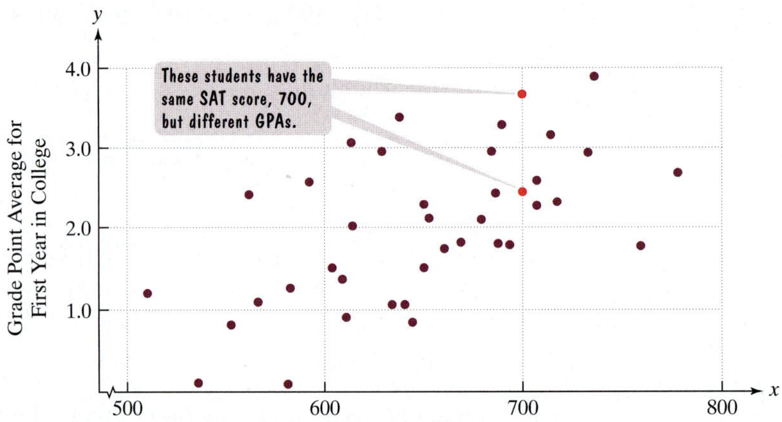

Figure 2.3

A relation in which each member of the domain corresponds to exactly one member of the range is a **function**. The relation in Figure 2.3 is not a function because at least one member of the domain corresponds to two members of the range.

$$(700, 2.4) \qquad (700, 3.7)$$

The member of the domain, 700, corresponds to two members of the range, 2.4 and 3.7. Because a function is a relation in which **no two ordered pairs have the same first component and different second components**, the ordered pairs (700, 2.4) and (700, 3.7) are not ordered pairs of a function.

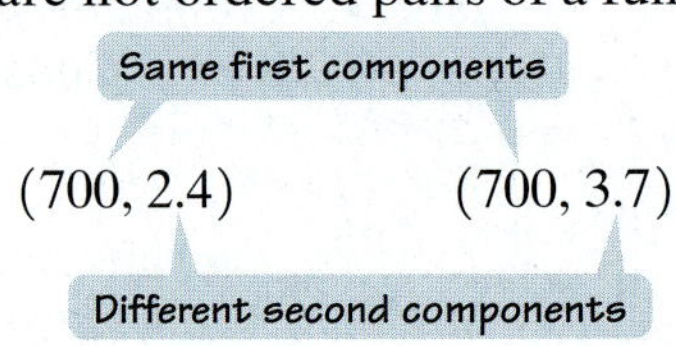

2 Determine whether a relation is a function.

Definition of a Function

A **function** is a relation in which each member of the domain corresponds to exactly one member of the range. No two ordered pairs of a function can have the same first component and different second components.

EXAMPLE 2 Determining Whether a Relation Is a Function

Determine whether each relation is a function:

a. $\{(1, 6), (2, 6), (3, 8), (4, 9)\}$ **b.** $\{(6, 1), (6, 2), (8, 3), (9, 4)\}$.

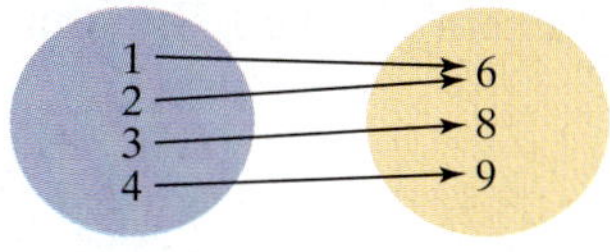

(a)

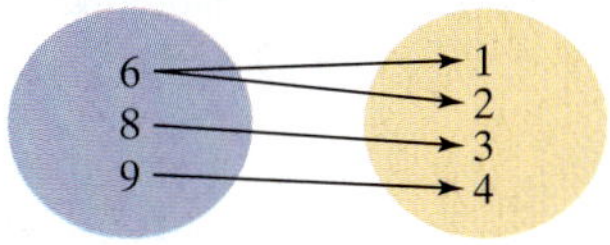

(b)

Figure 2.4

Solution We begin by making a figure for each relation that shows the domain and the range (Figure 2.4).

a. Figure 2.4(a) shows that every element in the domain corresponds to exactly one element in the range. The element 1 in the domain corresponds to the element 6 in the range. Furthermore, 2 corresponds to 6, 3 corresponds to 8, and 4 corresponds to 9. No two ordered pairs in the given relation have the same first component and different second components. Thus, the relation is a function.

b. Figure 2.4(b) shows that 6 corresponds to both 1 and 2. If any element in the domain corresponds to more than one element in the range, the relation is not a function. This relation is not a function; two ordered pairs have the same first component and different second components.

Same first components

$(6, 1)$ $(6, 2)$

Different second components

Look at Figure 2.4(a) again. The fact that 1 and 2 in the domain correspond to the same number, 6, in the range does not violate the definition of a function. A function can have two different first components with the same second component. By contrast, a relation is not a function when two different ordered pairs have the same first component and different second components. Thus, the relation in Example 2(b) is not a function.

CHECK POINT 2 Determine whether each relation is a function:

a. $\{(1, 2), (3, 4), (5, 6), (5, 8)\}$

b. $\{(1, 2), (3, 4), (6, 5), (8, 5)\}$.

Functions as Equations and Function Notation

Functions are usually given in terms of equations rather than as sets of ordered pairs. For example, here is an equation that models paid vacation days each year as a function of years working for a company:

$$y = -0.016x^2 + 0.93x + 8.5.$$

The variable x represents years working for a company. The variable y represents the average number of vacation days each year. The variable y is a function of the variable x. For each value of x, there is one and only one value of y. The variable x is called the **independent variable** because it can be assigned any value from the domain. Thus, x can be assigned any positive integer representing years working for a company. The variable y is called the **dependent variable** because its value depends on x. Paid vacation days depend on years working for a company. The value of the dependent variable, y, is calculated after selecting a value for the independent variable, x.

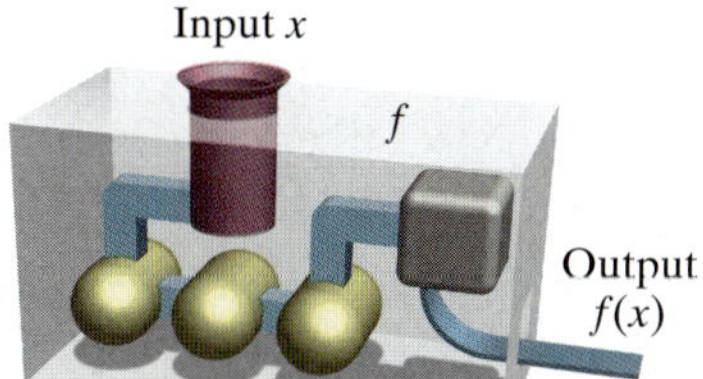

Figure 2.5 A "function machine" with inputs and outputs

If an equation in x and y gives one and only one value of y for each value of x, then the variable y is a function of the variable x. When an equation represents a function, the function is often named by a letter such as f, g, h, F, G, or H. Any letter can be used to name a function. Suppose that f names a function. Think of the domain as the set of the function's inputs and the range as the set of the function's outputs. As shown in Figure 2.5, the input is represented by x and the output by $f(x)$. The special notation $\boldsymbol{f(x)}$, read "f of x" or "f at x," represents the **value of the function at the number $\boldsymbol{x}$**.

3 Evaluate a function.

Let's make this clearer by considering a specific example. We know that the equation

$$y = -0.016x^2 + 0.93x + 8.5$$

defines y as a function of x. We'll name the function f. Now, we can apply our new function notation.

Input	Output	Equation	We read this equation as "f of x equals $-0.016x^2 + 0.93x + 8.5$."
x	$f(x)$	$f(x) = -0.016x^2 + 0.93x + 8.5$	

Suppose we are interested in finding $f(10)$, the function's output when the input is 10. To find the value of the function at 10, we substitute 10 for x. We are **evaluating the function** at 10.

$f(x) = -0.016x^2 + 0.93x + 8.5$ — This is the given function.

$f(10) = -0.016(10)^2 + 0.93(10) + 8.5$ — Replace each occurrence of x by 10.

$= -0.016(100) + 0.93(10) + 8.5$ — Evaluate the exponential expression: $10^2 = 100$.

$= -1.6 + 9.3 + 8.5$ — Perform the multiplications.

$= 16.2$ — Add from left to right.

Study Tip

The notation $f(x)$ does *not* mean "f times x." The notation describes the value of the function at x.

The statement $f(10) = 16.2$, read "f of 10 equals 16.2," tells us that the value of the function at 10 is 16.2. When the function's input is 10, its output is 16.2. (After 10 years, workers average 16.2 vacation days each year.) To find other function values, such as $f(15)$, $f(20)$, or $f(23)$, substitute the specified input values for x into the function's equation.

If a function is named f and x represents the independent variable, the notation $f(x)$ corresponds to the y-value for a given x. Thus,

$$f(x) = -0.016x^2 + 0.93x + 8.5 \quad \text{and} \quad y = -0.016x^2 + 0.93x + 8.5$$

define the same function. This function may be written as

$$y = f(x) = -0.016x^2 + 0.93x + 8.5.$$

EXAMPLE 3 Using Function Notation

Find the indicated function value:

a. $f(4)$ for $f(x) = 2x + 3$

b. $g(-2)$ for $g(x) = 2x^2 - 1$

c. $h(-5)$ for $h(r) = r^3 - 2r^2 + 5$

d. $F(a + h)$ for $F(x) = 5x + 7$.

Solution

a. $f(x) = 2x + 3$ — This is the given function.

$f(4) = 2 \cdot 4 + 3$ — To find f of 4, replace x by 4.

$= 8 + 3$

$= 11$ — Thus, f of 4 is 11.

b. $g(x) = 2x^2 - 1$ — This is the given function.

$g(-2) = 2(-2)^2 - 1$ — To find g of −2, replace x by −2.

$= 2(4) - 1$

$= 8 - 1$

$= 7$ — Thus, g of −2 is 7.

c. $h(r) = r^3 - 2r^2 + 5$ — The function's name is h and r represents the independent variable.

$h(-5) = (-5)^3 - 2(-5)^2 + 5$ — To find h of -5, replace each occurrence of r by -5.

$= -125 - 2(25) + 5$ — Evaluate exponential expressions.

$= -125 - 50 + 5$ — Multiply.

$= -170$ — Thus, h of -5 is -170.

d. $F(x) = 5x + 7$ — This is the given function.

$F(a + h) = 5(a + h) + 7$ — Replace x by $a + h$.

$= 5a + 5h + 7$ — Apply the distributive property. Thus, F of $a + h$ is $5a + 5h + 7$. ■

CHECK POINT 3 Find the indicated function value:

a. $f(6)$ for $f(x) = 4x + 5$

b. $g(-5)$ for $g(x) = 3x^2 - 10$

c. $h(-4)$ for $h(r) = r^2 - 7r + 2$

d. $F(a + h)$ for $F(x) = 6x + 9$.

Graphs of Functions and the Vertical Line Test

4 Graph functions by plotting points.

The **graph of a function** is the graph of its ordered pairs. For example, the graph of $f(x) = 2x$ is the set of points (x, y) in the rectangular coordinate system satisfying $y = 2x$. Similarly, the graph of $g(x) = 2x + 4$ is the set of points (x, y) in the rectangular coordinate system satisfying the equation $y = 2x + 4$. In the next example, we graph both of these functions in the same rectangular coordinate system.

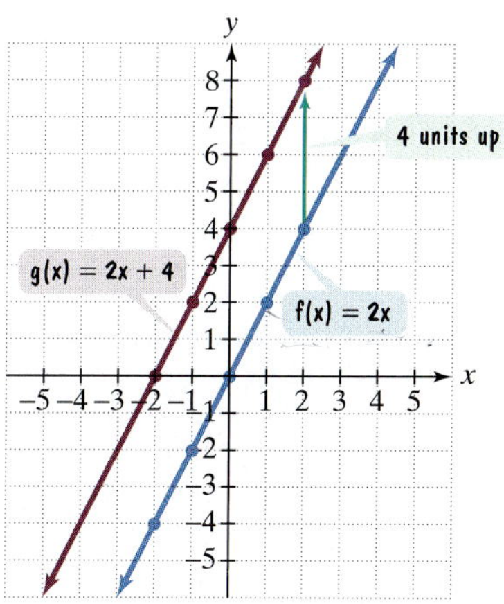

Figure 2.6

EXAMPLE 4 Graphing Functions

Graph the functions $f(x) = 2x$ and $g(x) = 2x + 4$ in the same rectangular coordinate system. Select integers for x, starting with -2 and ending with 2.

Solution We begin by setting up a partial table of coordinates for each function. Now, we plot the five points in each table and connect them, as shown in Figure 2.6. The graph of each function is a straight line. Do you see a relationship between the two graphs? The graph of g is the graph of f shifted vertically up by 4 units. ■

x	$f(x) = 2x$	(x, y) or $(x, f(x))$
-2	$f(-2) = 2(-2) = -4$	$(-2, -4)$
-1	$f(-1) = 2(-1) = -2$	$(-1, -2)$
0	$f(0) = 2 \cdot 0 = 0$	$(0, 0)$
1	$f(1) = 2 \cdot 1 = 2$	$(1, 2)$
2	$f(2) = 2 \cdot 2 = 4$	$(2, 4)$
Choose x.	Compute $f(x)$ by evaluating f at x.	Form the pair.

x	$g(x) = 2x + 4$	(x, y) or $(x, g(x))$
-2	$g(-2) = 2(-2) + 4 = 0$	$(-2, 0)$
-1	$g(-1) = 2(-1) + 4 = 2$	$(-1, 2)$
0	$g(0) = 2 \cdot 0 + 4 = 4$	$(0, 4)$
1	$g(1) = 2 \cdot 1 + 4 = 6$	$(1, 6)$
2	$g(2) = 2 \cdot 2 + 4 = 8$	$(2, 8)$
Choose x.	Compute $g(x)$ by evaluating g at x.	Form the pair.

The graphs in Example 4 are straight lines. All functions with equations of the form $f(x) = mx + b$ graph as straight lines. Such functions, called **linear functions**, will be discussed in detail in Sections 2.3–2.4.

Technology

Does your graphing utility have a TABLE feature? If so, you can use it to create tables of coordinates for a function. You will need to enter the equation of the function and specify the starting value for x, TBLSTART, and the increment between successive x-values, ΔTBL. For the tables of coordinates in Example 4, we start the table at $x = -2$ and increment by 1. Using the up- or down-arrow keys, you can scroll through the table and determine as many ordered pairs of the function as desired.

✔ **CHECK POINT 4** Graph the functions $f(x) = 2x$ and $g(x) = 2x - 3$ in the same rectangular coordinate system. Select integers for x, starting with -2 and ending with 2. How is the graph of g related to the graph of f?

5 Use the vertical line test to identify functions.

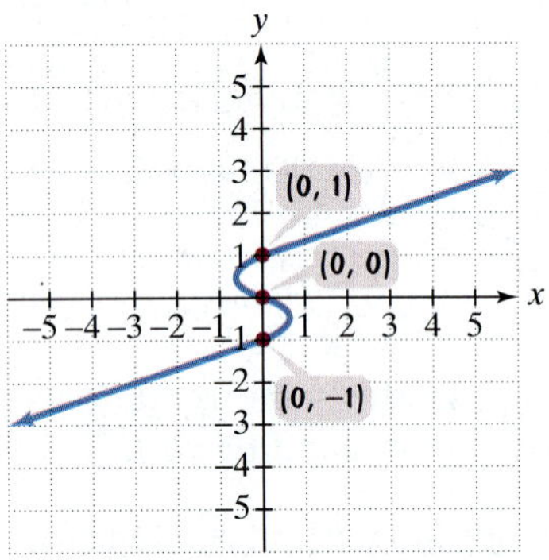

Figure 2.7 y is not a function of x because 0 is paired with three values of y, namely, 1, 0, and -1.

Not every graph in the rectangular coordinate system is the graph of a function. The definition of a function specifies that no value of x can be paired with two or more different values of y. Consequently, if a graph contains two or more different points with the same first coordinate, the graph cannot represent a function. This is illustrated in Figure 2.7. Observe that points sharing a common first coordinate are vertically above or below each other.

This observation is the basis of a useful test for determining whether a graph defines y as a function of x. The test is called the **vertical line test**.

The Vertical Line Test for Functions

If any vertical line intersects a graph in more than one point, the graph does not define y as a function of x.

EXAMPLE 5 Using the Vertical Line Test

Use the vertical line test to identify graphs in which y is a function of x.

a.

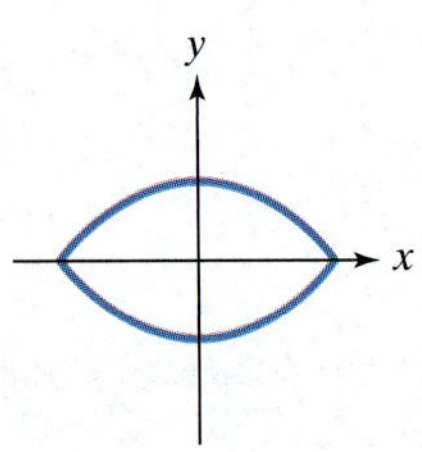

b.

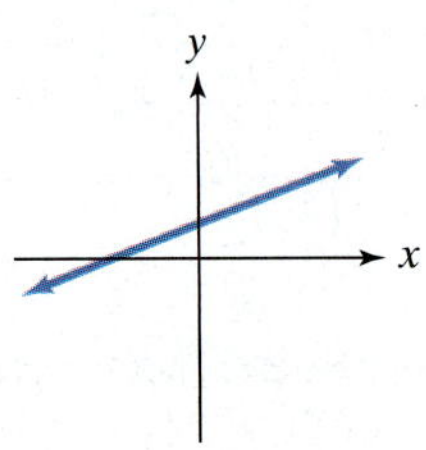

c.

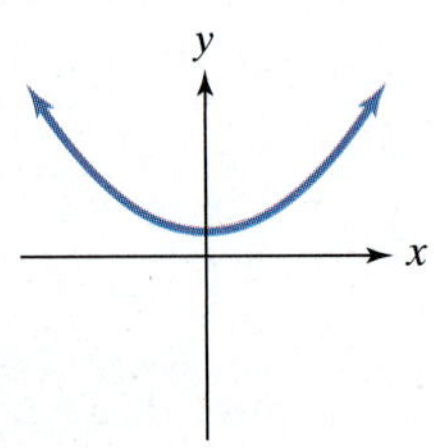

d.

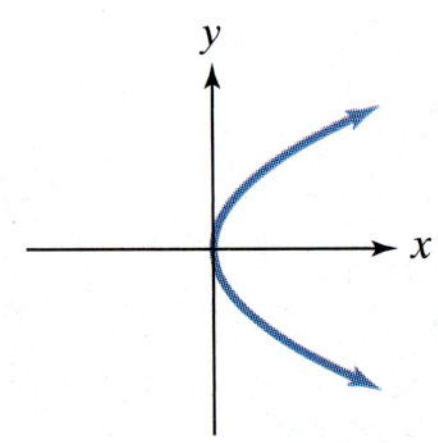

Solution y is a function of x for the graphs in (b) and (c).

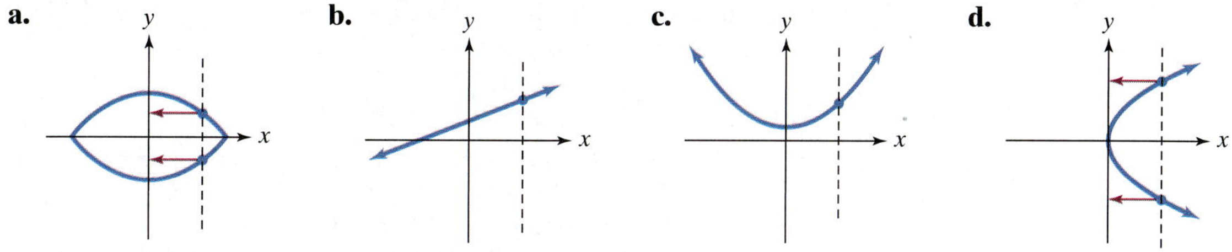

a. y **is not a function** of x. Two values of y correspond to an x-value.

b. y **is a function** of x.

c. y **is a function** of x.

d. y **is not a function** of x. Two values of y correspond to an x-value. ■

✔ CHECK POINT 5 Use the vertical line test to identify graphs in which y is a function of x.

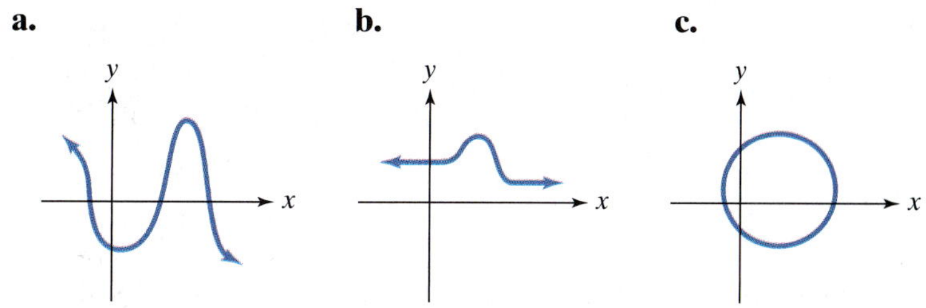

6 Obtain information about a function from its graph.

Obtaining Information from Graphs

You can obtain information about a function from its graph. At the right or left of a graph, you will often find closed dots, open dots, or arrows.

- A closed dot indicates that the graph does not extend beyond this point and the point belongs to the graph.
- An open dot indicates that the graph does not extend beyond this point and the point does not belong to the graph.
- An arrow indicates that the graph extends indefinitely in the direction in which the arrow points.

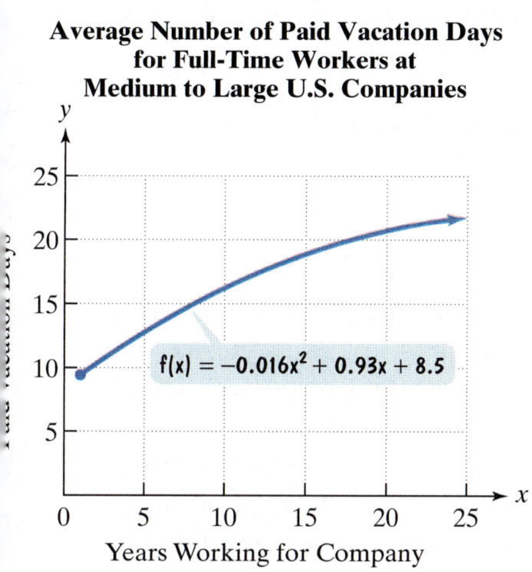

Figure 2.8 *Source:* Bureau of Labor Statistics

EXAMPLE 6 Analyzing the Graph of a Function

The function

$$f(x) = -0.016x^2 + 0.93x + 8.5$$

models the average number of paid vacation days each year, $f(x)$, for full-time workers at medium to large U.S. companies after x years. The graph of f is shown in Figure 2.8.

a. Explain why f represents the graph of a function.

b. Use the graph to find a reasonable estimate of $f(5)$.

c. For what value of x is $f(x) = 20$?

d. Describe the general trend shown by the graph.

Solution

a. No vertical line intersects the graph of f more than once. By the vertical line test, f represents the graph of a function.

b. To find $f(5)$, or f of 5, we locate 5 on the x-axis. The figure shows the point on the graph of f for which 5 is the first coordinate. From this point, we look to the y-axis to find the corresponding y-coordinate. A reasonable estimate of the y-coordinate is 13. Thus, $f(5) \approx 13$. After 5 years, a worker can expect approximately 13 paid vacation days.

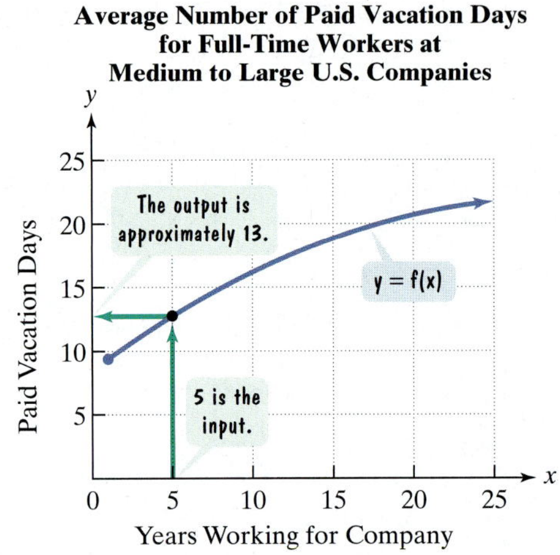

c. To find the value of x for which $f(x) = 20$, we locate 20 on the y-axis. The figure shows that there is one point on the graph of f for which 20 is the second coordinate. From this point, we look to the x-axis to find the corresponding x-coordinate. A reasonable estimate of the x-coordinate is 18. Thus, $f(x) = 20$ for $x \approx 18$. A worker with 20 paid vacation days has been with the company approximately 18 years.

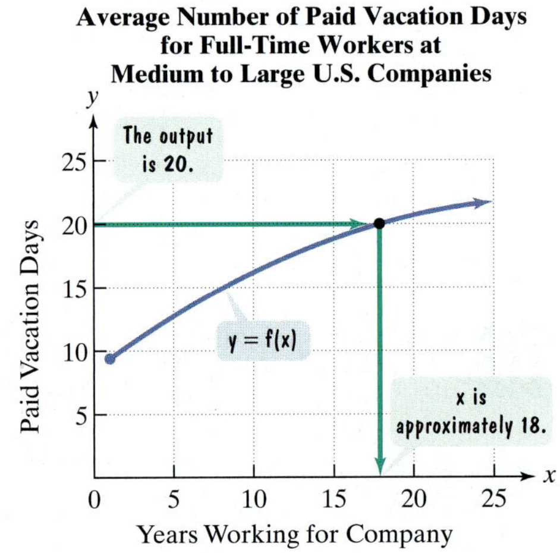

d. The graph of f is rising from left to right. This shows that paid vacation days increase as time with the company increases. However, the rate of increase is slowing down as the graph moves to the right. This means that the increase in paid vacation days takes place more slowly the longer an employee is with the company. ■

✔ CHECK POINT 6

a. Use the graph of f in Figure 2.8 on page 99 to find a reasonable estimate of $f(10)$.

b. For what value of x is $f(x) = 15$? Round to the nearest whole number.

Relationships between variables that you encounter in many everyday situations can be illustrated by functions and their graphs. For example, have you ever mailed a letter that seemed heavier than usual? Perhaps you worried that the letter would not have enough postage. Costs for mailing a letter weighing up to 5 ounces are given in Table 2.1 at the top of the next page. If your letter weighs an ounce or less, the cost is \$0.34. If your letter weighs 1.05 ounces, 1.50 ounces, 1.90 ounces, or 2.00 ounces, the cost "steps" to \$0.55. The cost does not take on any value between \$0.34 and \$0.55. If your letter weighs 2.05 ounces, 2.50 ounces, 2.90 ounces, or 3 ounces, the cost "steps" to \$0.76. Cost increases are \$0.21 per step.

Table 2.1 Cost of First-Class Mail (Effective January 7, 2001)

Weight Not Over	Cost
1 ounce	$0.34
2 ounces	0.55
3 ounces	0.76
4 ounces	0.97
5 ounces	1.18

Source: U.S. Postal Service

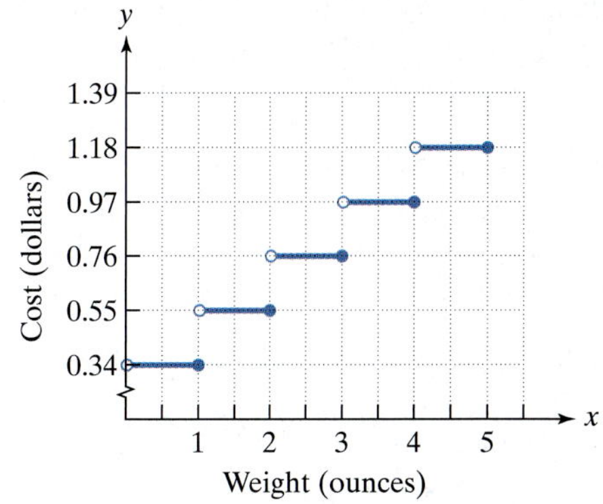

The graph of the function that models this situation is shown to the right of the table. Let

$$x = \text{the weight of the letter, in ounces, and}$$
$$y = f(x) = \text{the cost of mailing a letter weighing } x \text{ ounces.}$$

Notice how the graph of f consists of a series of steps that jump vertically 0.21 unit at each integer. The graph is constant between each pair of consecutive integers. The open dot on the left of each horizontal step shows that the left point is not included on the graph. By contrast, the closed dot on the right shows that the right point of each horizontal step is part of the graph.

EXERCISE SET 2.1

Practice Exercises

In Exercises 1–8, determine whether each relation is a function. Give the domain and range for each relation.

1. $\{(1, 2), (3, 4), (5, 5)\}$

2. $\{(4, 5), (6, 7), (8, 8)\}$

3. $\{(3, 4), (3, 5), (4, 4), (4, 5)\}$

4. $\{(5, 6), (5, 7), (6, 6), (6, 7)\}$

5. $\{(-3, -3), (-2, -2), (-1, -1), (0, 0)\}$

6. $\{(-7, -7), (-5, -5), (-3, -3), (0, 0)\}$

7. $\{(1, 4), (1, 5), (1, 6)\}$

8. $\{(4, 1), (5, 1), (6, 1)\}$

In Exercises 9–18, find the indicated function values.

9. $f(x) = x + 1$
a. $f(0)$ **b.** $f(5)$ **c.** $f(-8)$
d. $f(2a)$ **e.** $f(a + 2)$

10. $f(x) = x + 3$
a. $f(0)$ **b.** $f(5)$ **c.** $f(-8)$
d. $f(2a)$ **e.** $f(a + 2)$

11. $g(x) = 3x - 2$
a. $g(0)$ **b.** $g(-5)$ **c.** $g\left(\frac{2}{3}\right)$
d. $g(4b)$ **e.** $g(b + 4)$

12. $g(x) = 4x - 3$
a. $g(0)$ **b.** $g(-5)$ **c.** $g\left(\frac{3}{4}\right)$
d. $g(5b)$ **e.** $g(b + 5)$

13. $h(x) = 3x^2 + 5$
a. $h(0)$ **b.** $h(-1)$ **c.** $h(4)$
d. $h(-3)$ **e.** $h(4b)$

14. $h(x) = 2x^2 - 4$
a. $h(0)$ **b.** $h(-1)$ **c.** $h(5)$
d. $h(-3)$ **e.** $h(5b)$

15. $f(x) = 2x^2 + 3x - 1$
a. $f(0)$ **b.** $f(3)$ **c.** $f(-4)$
d. $f(b)$
e. $f(5a)$

16. $f(x) = 3x^2 + 4x - 2$
a. $f(0)$ **b.** $f(3)$ **c.** $f(-5)$
d. $f(b)$ **e.** $f(5a)$

17. $f(x) = \dfrac{2x - 3}{x - 4}$
a. $f(0)$ $\dfrac{3}{4}$ **b.** $f(3)$ **c.** $f(-4)$ $\dfrac{11}{8}$
d. $f(-5)$ $\dfrac{13}{9}$ **e.** $f(a + h)$ $\dfrac{2a + 2h - 3}{a + h - 4}$
f. Why must 4 be excluded from the domain of f?

18. $f(x) = \dfrac{3x - 1}{x - 5}$
a. $f(0)$ $\dfrac{1}{5}$ **b.** $f(3)$ **c.** $f(-3)$ $\dfrac{5}{4}$
d. $f(10)$ $\dfrac{29}{5}$ **e.** $f(a + h)$ $\dfrac{3a + 3h - 1}{a + h - 5}$
f. Why must 5 be excluded from the domain of f?

In Exercises 19–28, graph the given functions, f and g, in the same rectangular coordinate system. Select integers for x, starting with −2 and ending with 2. Once you have obtained your graphs, describe how the graph of g is related to the graph of f.

19. $f(x) = x,\ g(x) = x + 3$

20. $f(x) = x,\ g(x) = x - 4$

21. $f(x) = -2x,\ g(x) = -2x - 1$

22. $f(x) = -2x, g(x) = -2x + 3$

23. $f(x) = x^2,\ g(x) = x^2 + 1$

24. $f(x) = x^2,\ g(x) = x^2 - 2$

25. $f(x) = |x|,\ g(x) = |x| - 2$

26. $f(x) = |x|, g(x) = |x| + 1$

27. $f(x) = x^3,\ g(x) = x^3 + 2$

28. $f(x) = x^3,\ g(x) = x^3 - 1$

In Exercises 29–36, use the vertical line test to identify graphs in which y is a function of x.

29.
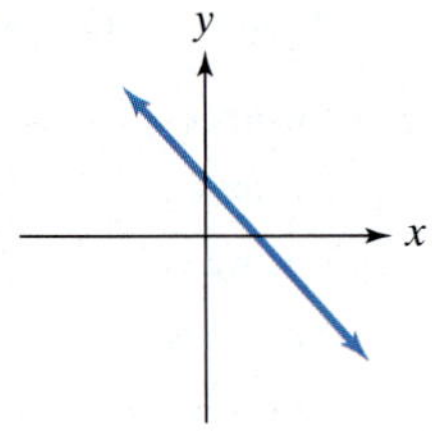

30.
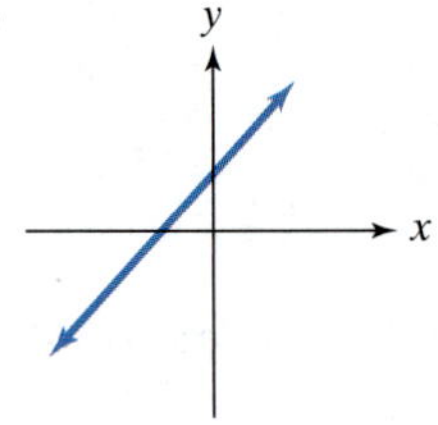

31.
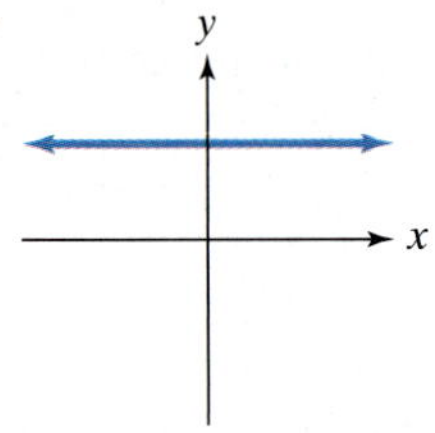

32.
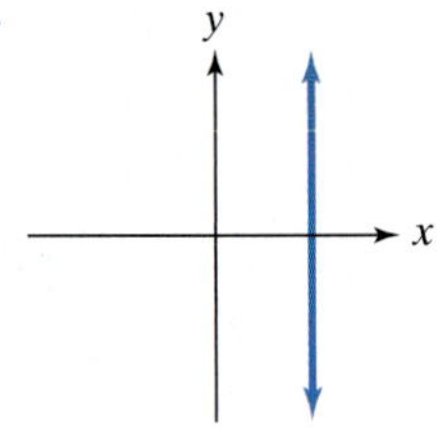

33.
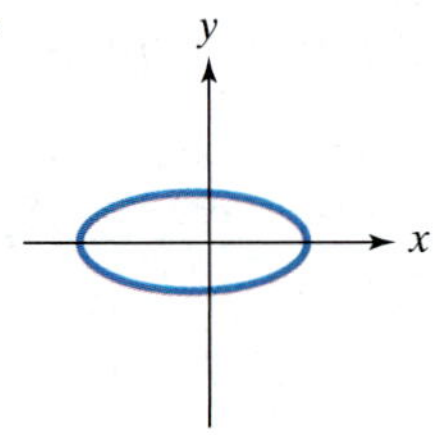

34.
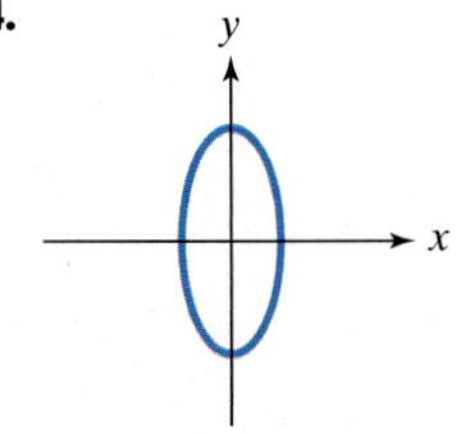

35.
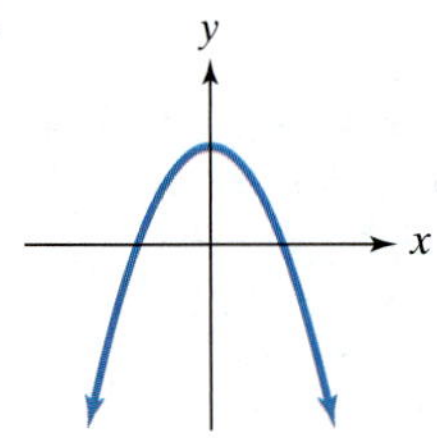

36.
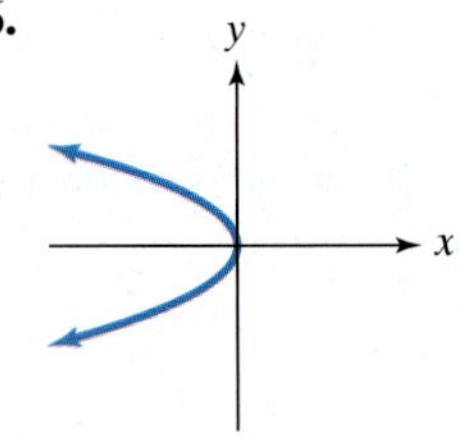

In Exercises 37–42, use the graph of f to find each indicated function value.

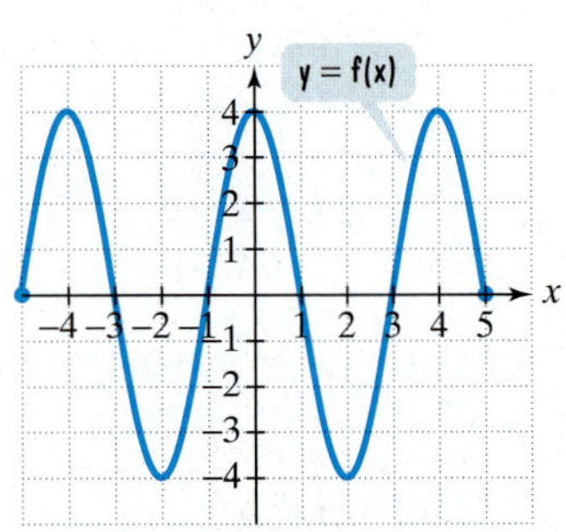

37. $f(-2)$

38. $f(2)$

39. $f(4)$

40. $f(-4)$

41. $f(-3)$

42. $f(-1)$

Use the graph of g to solve Exercises 43–48.

43. Find $g(-4)$.

44. Find $g(2)$.

45. Find $g(-10)$.

46. Find $g(10)$.

47. For what value of x is $g(x) = 1$?

48. For what value of x is $g(x) = -1$?

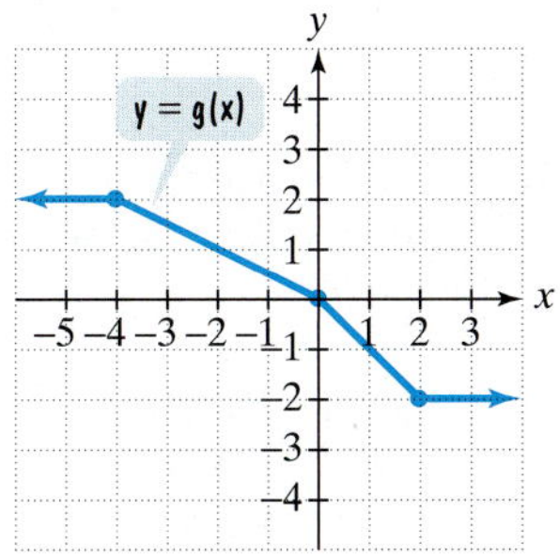

Application Exercises

49. The bar graph shows the percentage of people in the United States using the Internet by education level. Write five ordered pairs for

(education level, percentage using Internet)

as a relation. Find the domain and the range of the relation. Is this relation a function? Explain your answer.

Internet Use in the U.S. by Education Level

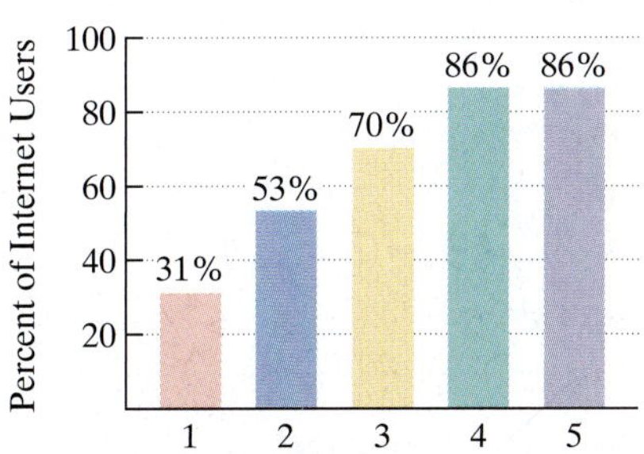

(1 = less than high school, 2 = high school graduate, 3 = some college, 4 = college graduate, 5 = advanced degree)

Source: U.C.L.A. Center for Communication Policy

50. The debate over the place of guns in our society rages on. The graph at the top of the next column shows gun deaths in eight selected countries. Each gun represents one death per 100,000 people. Write eight ordered pairs for

(country number, gun deaths per 100,000 people)

as a relation. Find the domain and the range of the relation. Is this relation a function? Explain your answer.

Gun Deaths per 100,000 People, by Country

Source: Centers for Disease Control and Prevention

The male minority? The graphs show enrollment in U.S. colleges, with projections from 2000 to 2009. The trend indicated by the graphs is among the hottest topics of debate among college-admissions officers. Some private liberal arts colleges have quietly begun special efforts to recruit men—including admissions preferences for them.

Enrollment in U.S. Colleges

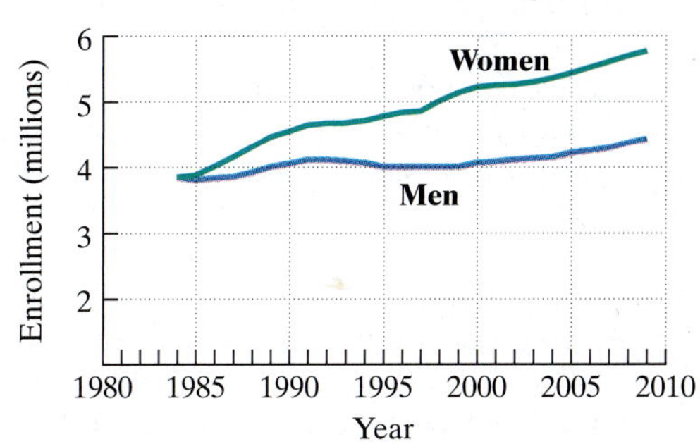

Source: Department of Education

The function

$$W(x) = 0.07x + 4.1$$

models the number of women, $W(x)$, in millions, enrolled in U.S. colleges x years after 1984. The function

$$M(x) = 0.01x + 3.9$$

models the number of men, $M(x)$, in millions, enrolled in U.S. colleges x years after 1984. Use these equations to solve Exercises 51–54.

51. Find and interpret $W(16)$. Identify this information as a point on the graph for women.

52. Find and interpret $M(16)$. Identify this information as a point on the graph for men.

53. Find and interpret $W(20) - M(20)$.

54. Find and interpret $W(25) - M(25)$.

The function

$$f(x) = 0.4x^2 - 36x + 1000$$

models the number of accidents, $f(x)$, per 50 million miles driven as a function of the driver's age, x, in years, where x includes drivers from ages 16 through 74. The graph of f is shown. Use the equation for f to solve Exercises 55–58.

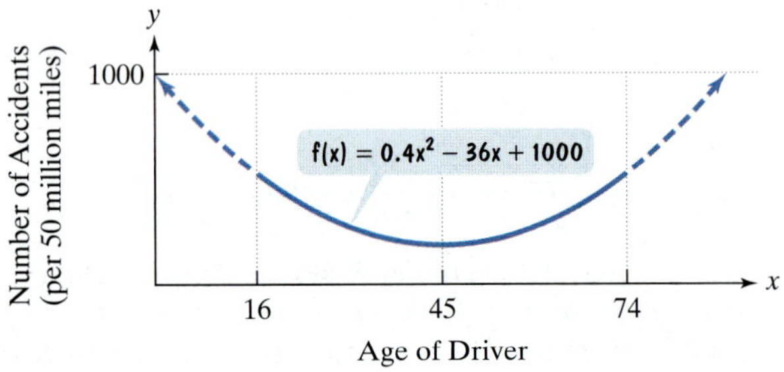

55. Find and interpret $f(20)$. Identify this information as a point on the graph of f.

56. Find and interpret $f(50)$. Identify this information as a point on the graph of f.

57. For what value of x does the graph reach its lowest point? Use the equation for f to find the minimum value of y. Describe the practical significance of this minimum value.

58. Use the graph to identify two different ages for which drivers have the same number of accidents. Use the equation for f to find the number of accidents for drivers at each of these ages.

The figure shows the percentage of the U.S. population, $f(x)$, made up of Jewish Americans as a function of time, x, where x is the number of years after 1900. Use the graph to solve Exercises 59–66.

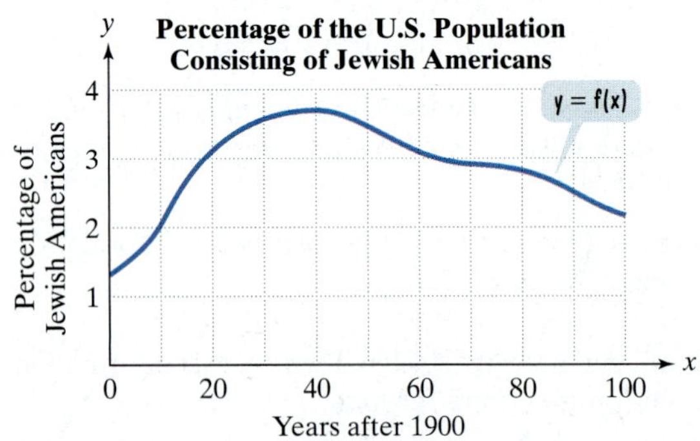

Source: American Jewish Yearbook

59. Use the graph to find a reasonable estimate of $f(60)$. What does this mean in terms of the variables in this situation?

60. Use the graph to find a reasonable estimate of $f(100)$. What does this mean in terms of the variables in this situation?

61. For what value or values of x is $f(x) = 3$? Round to the nearest year. What does this mean in terms of the variables in this situation?

62. For what value or values of x is $f(x) = 2.5$? Round to the nearest year. What does this mean in terms of the variables in this situation?

63. In which year did the percentage of Jewish Americans in the U.S. population reach a maximum? What is a reasonable estimate of the percentage for that year?

64. In which year was the percentage of Jewish Americans in the U.S. population at a minimum? What is a reasonable estimate of the percentage for that year?

65. Explain why f represents the graph of a function.

66. Describe the general trend shown by the graph.

The figure shows the cost of mailing a first-class letter, f(x), as a function of its weight, x, in ounces. Use the graph to solve Exercises 67–70.

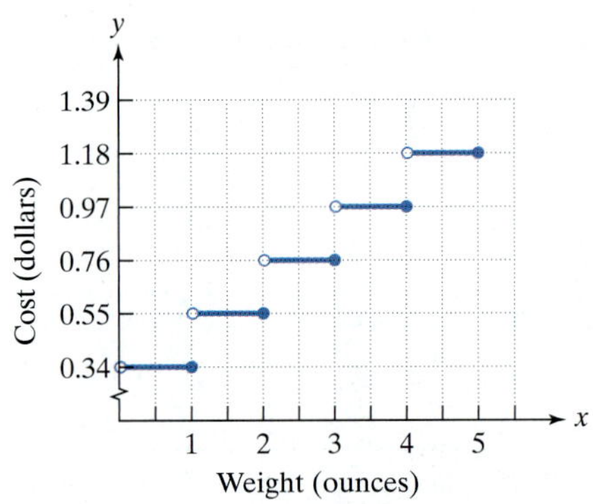

67. Find $f(3)$. What does this mean in terms of the variables in this situation?

68. Find $f(4)$. What does this mean in terms of the variables in this situation?

69. What is the cost of mailing a letter that weighs 1.5 ounces?

70. What is the cost of mailing a letter that weighs 1.8 ounces?

Writing in Mathematics

71. What is a relation? Describe what is meant by its domain and its range.

72. Explain how to determine whether a relation is a function. What is a function?

73. Does $f(x)$ mean f times x when referring to function f? If not, what does $f(x)$ mean? Provide an example with your explanation.

74. What is the graph of a function?

75. Explain how the vertical line test is used to determine whether a graph is a function.

76. For people filing a single return, federal income tax is a function of adjusted gross income because for each value of adjusted gross income there is a specific tax to be paid. By contrast, the price of a house is not a function of the lot size on which the house sits because houses on same-sized lots can sell for many different prices.

  a. Describe an everyday situation between variables that is a function.

  b. Describe an everyday situation between variables that is not a function.

77. Do you believe that the trend shown by the graphs for Exercises 51–54 should be reversed by providing admissions preferences for men? Explain your position on this issue.

Technology Exercises

78. Use a graphing utility to verify the pairs of graphs that you drew by hand in Exercises 19–28.

79. The function

$$f(x) = -0.00002x^3 + 0.008x^2 - 0.3x + 6.95$$

models the number of annual physician visits, $f(x)$, by a person of age x. Graph the function in a [0, 100, 5] by [0, 40, 2] viewing rectangle. What does the shape of the graph indicate about the relationship between one's age and the number of annual physician visits? Use the TRACE or minimum function capability to find the coordinates of the minimum point on the graph of the function. What does this mean?

Critical Thinking Exercises

80. Which one of the following is true?

  a. All relations are functions.

  b. No two ordered pairs of a function can have the same second component and different first components.

  c. The graph of every line is a function.

  d. A horizontal line can intersect the graph of a function in more than one point.

81. If $f(x) = 3x + 7$, find $\dfrac{f(a + h) - f(a)}{h}$.

82. Take another look at the cost of first-class mail and its graph (Table 2.1 and the figure to its right on page 101). Change the description of the heading in the left column of Table 2.1 so that the graph includes the point on the left of each horizontal step, but does not include the point on the right.

83. If $f(x + y) = f(x) + f(y)$ and $f(1) = 3$, find $f(2)$, $f(3)$, and $f(4)$. Is $f(x + y) = f(x) + f(y)$ for all functions?

Review Exercises

84. Simplify: $24 \div 4[2 - (5 - 2)]^2 - 6$. (Section 1.2, Example 7)

85. Simplify: $\left(\dfrac{3x^2y^{-2}}{y^3}\right)^{-2}$. (Section 1.6, Example 9)

86. Solve: $\dfrac{x}{3} = \dfrac{3x}{5} + 4$. (Section 1.4, Example 4)

▸ SECTION 2.2 The Algebra of Functions

Objectives

1. Find the domain of a function.
2. Find the sum of two functions.
3. Use the algebra of functions to combine functions and determine domains.

SSM PH Tutor Center CD-ROM Video

© Bettmann Corbis

America is a nation of immigrants. Since 1820, over 40 million people have immigrated to the United States from all over the world. They chose to come for various reasons, such as to live in freedom, to practice religion freely, to escape poverty or oppression, and to make better lives for themselves and their children. We open this section by looking at America's many faces from the perspective of functions. You will see that functions can be combined using procedures that will remind you of combining algebraic expressions.

1 Find the domain of a function.

The Domain of a Function

The early part of the twentieth century was the golden age of immigration in America. More than 13 million people migrated to the United States between 1900 and 1914. By 1910, foreign-born residents accounted for 15% of the total U.S. population. The graph in Figure 2.9 shows the percentage of Americans who were foreign born throughout the twentieth century.

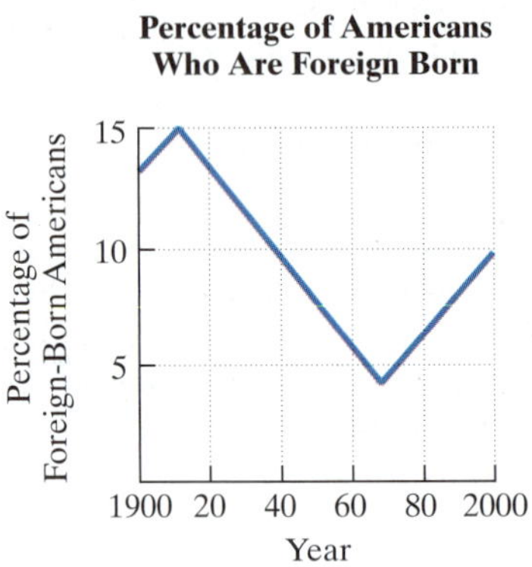

Source: U.S. Census Bureau

Figure 2.9

Although the graph in Figure 2.9 consists of three line segments, no vertical line can be drawn that intersects any one of these line segments more than once. This shows that the percentage of foreign-born Americans is a function of time. The domain of this function is

$$\{1900, 1901, 1902, 1903, \ldots, 2000\}.$$

Functions that model data often have their domains explicitly given on the horizontal axis of a coordinate system, as in Figure 2.9, or along with the function's equation. However, for most functions, only an equation is given, and the domain is not specified. In cases like this, the domain of function f is the largest set of real numbers for which the value of $f(x)$ is a real number. For example, consider the function

$$f(x) = \frac{1}{x - 3}.$$

Because division by 0 is undefined (and not a real number), the denominator $x - 3$ cannot be 0. Thus, x cannot equal 3. The domain of the function consists of all real numbers other than 3, represented by

$$\text{Domain of } f = \{x \mid x \text{ is a real number and } x \neq 3\}.$$

In Chapter 7, we will be studying square root functions such as

$$g(x) = \sqrt{x}.$$

The equation tells us to take the square root of x. Because only nonnegative numbers have square roots that are real numbers, the expression under the square root sign, x, must be nonnegative. Thus,

$$\text{Domain of } g = \{x \mid x \text{ is a nonnegative real number}\}.$$

Using Technology

You can graph a function and visually determine its domain. For example,

$$g(x) = \sqrt{x}, \text{ or } y = \sqrt{x}$$

appears only for x greater than or equal to 0: $x \geq 0$. This verifies that

$$\{x \mid x \text{ is a nonnegative real number}\}$$

is the domain.

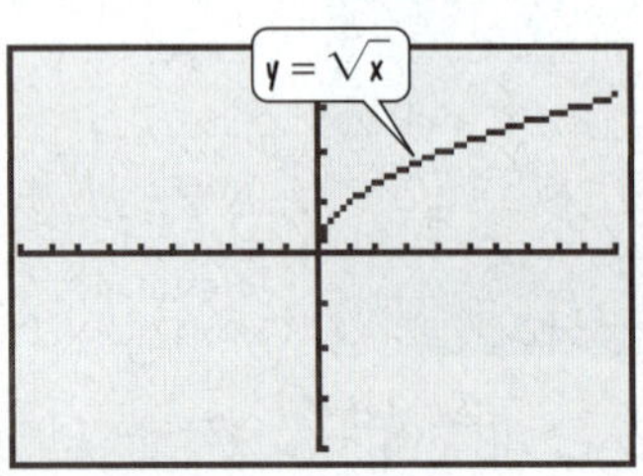

$[-10, 10, 1]$ by $[-4, 4, 1]$

Finding a Function's Domain

If a function f does not model data or verbal conditions, its domain is the largest set of real numbers for which the value of $f(x)$ is a real number. Exclude from a function's domain real numbers that cause division by zero and real numbers that result in a square root of a negative number.

EXAMPLE 1 Finding the Domain of a Function

Find the domain of each function:

a. $f(x) = 3x + 2$ **b.** $g(x) = \dfrac{3x + 2}{x + 1}$.

Solution

a. The function $f(x) = 3x + 2$ contains neither division nor a square root. For every real number, x, the algebraic expression $3x + 2$ is a real number. Thus, the domain of f is the set of all real numbers.

$$\text{Domain of } f = \{x \mid x \text{ is a real number}\}$$

b. The function $g(x) = \dfrac{3x + 2}{x + 1}$ contains division. Because division by 0 is undefined, we must exclude from the domain the value of x that causes $x + 1$ to be 0. Thus, x cannot equal -1.

$$\text{Domain of } g = \{x \mid x \text{ is a real number and } x \neq -1\}$$ ■

✔ **CHECK POINT 1** Find the domain of each function:

a. $f(x) = \frac{1}{2}x + 3$

b. $g(x) = \dfrac{7x + 4}{x + 5}$.

The Algebra of Functions

The United States was a low-tax country in the early part of the twentieth century. Figure 2.10 shows how the tax burden has grown since then.

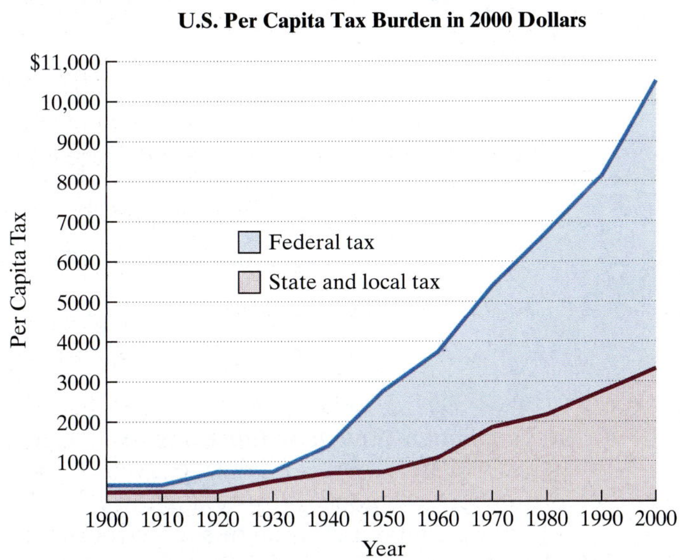

Figure 2.10 *Source:* Tax Foundation

Take a look at the information shown for the year 2000. The total per capita tax burden is approximately \$10,500. The per capita state and local tax is approximately \$3400. The per capita federal tax is the difference between these amounts:

$$\text{Per capita federal tax} = \$10{,}500 - \$3400 = \$7100.$$

We can think of this subtraction as the subtraction of function values. We do this by introducing the following functions:

$$\text{Let } T(x) = \text{total per capital tax in year } x.$$
$$\text{Let } S(x) = \text{per capita state and local tax in year } x.$$

Using Figure 2.10, we see that

$$T(2000) = \$10{,}500 \quad \text{and} \quad S(2000) = \$3400.$$

We can subtract these function values by introducing a new function, $T - S$, defined by the subtraction of $T(x)$ and $S(x)$. Thus,

$$(T - S)(x) = T(x) - S(x) = \text{total per capita tax in year } x \text{ minus state and local per capita tax in year } x.$$

For example,

$$(T - S)(2000) = T(2000) - S(2000) = \$10{,}500 - \$3400 = \$7100.$$

In 2000, the difference between total tax and state and local tax was \$7100. This is the per capita federal tax.

Figure 2.10 on page 107 illustrates that information involving differences of functions often appears in graphs seen in newspapers and magazines. Like numbers and algebraic expressions, two functions can be added, subtracted, multiplied, or divided as long as there are numbers common to the domains of both functions. The common domain for functions T and S in Figure 2.10 is

$$\{1900, 1901, 1902, 1903, \ldots, 2000\}.$$

2 Find the sum of two functions.

Because functions are usually given as equations, we perform operations by carrying out these operations with the algebraic expressions that appear on the right side of the equations. For example, we can combine the following two functions using addition:

$$f(x) = 2x + 1 \quad \text{and} \quad g(x) = x^2 - 4.$$

To do so, we add the terms to the right of the equal sign for $f(x)$ to the terms to the right of the equal sign for $g(x)$. Here is how it's done:

$$(f + g)(x) = f(x) + g(x)$$
$$= (2x + 1) + (x^2 - 4) \quad \text{Add terms for } f(x) \text{ and } g(x).$$
$$= 2x - 3 + x^2 \quad \text{Combine like terms.}$$
$$= x^2 + 2x - 3. \quad \text{Arrange terms in descending powers of } x.$$

The name of this new function is $f + g$. Thus, the sum $f + g$ is the function defined by $(f + g)(x) = x^2 + 2x - 3$. The domain of $f + g$ consists of the numbers x that are in the domain of f and in the domain of g. Because neither f nor g contains division or square roots, the domain of each function is the set of all real numbers. Thus, the domain of $f + g$ is also the set of all real numbers.

EXAMPLE 2 Finding the Sum of Two Functions

Let $f(x) = x^2 - 3$ and $g(x) = 4x + 5$. Find:

a. $(f + g)(x)$ **b.** $(f + g)(3)$.

Solution

a. $(f + g)(x) = f(x) + g(x) = (x^2 - 3) + (4x + 5) = x^2 + 4x + 2$

Thus,

$$(f + g)(x) = x^2 + 4x + 2.$$

b. We find $(f + g)(3)$ by substituting 3 for x in the equation for $f + g$.

$$(f + g)(x) = x^2 + 4x + 2 \quad \text{This is the equation for } f + g.$$

Substitute 3 for x.

$$(f + g)(3) = 3^2 + 4 \cdot 3 + 2 = 9 + 12 + 2 = 23$$ ■

✔ **CHECK POINT 2** Let $f(x) = 3x^2 + 4x - 1$ and $g(x) = 2x + 7$. Find:

a. $(f + g)(x)$

b. $(f + g)(4)$.

Here is a general definition for function addition:

The Sum of Functions

Let f and g be two functions. The **sum** $\boldsymbol{f + g}$ is the function defined by

$$(f + g)(x) = f(x) + g(x).$$

The domain of $f + g$ is the set of all real numbers that are common to the domain of f and the domain of g.

EXAMPLE 3 Adding Functions and Determining the Domain

Let $f(x) = \frac{4}{x}$ and $g(x) = \frac{3}{x + 2}$. Find:

a. $(f + g)(x)$

b. the domain of $f + g$.

Solution

a. $(f + g)(x) = f(x) + g(x) = \frac{4}{x} + \frac{3}{x + 2}$

(In Chapter 6, we will discuss how to combine these algebraic fractions. Perhaps you remember how to do so from your work in introductory algebra. For now, we will leave these fractions in the form shown.)

b. The domain of $f + g$ is the set of all real numbers that are common to the domain of f and the domain of g. Thus, we must find the domains of f and g. We will do so for f first.

Note that $f(x) = \frac{4}{x}$ is a function involving division. Because division by 0 is undefined, x cannot equal 0.

$$\text{Domain of } f = \{x \mid x \text{ is a real number and } x \neq 0\}$$

The function $g(x) = \frac{3}{x + 2}$ is also a function involving division. Because division by 0 is undefined, x cannot equal -2.

$$\text{Domain of } g = \{x \mid x \text{ is a real number and } x \neq -2\}$$

In order to find $f(x) + g(x)$, x must be in both domains listed. Thus,

$$\text{Domain of } f + g = \{x \mid x \text{ is a real number and } x \neq 0 \text{ and } x \neq -2\}.$$ ■

✔ **CHECK POINT 3** Let $f(x) = \frac{5}{x}$ and $g(x) = \frac{7}{x-8}$. Find:

a. $(f+g)(x)$

b. the domain of $f+g$.

3 Use the algebra of functions to combine functions and determine domains.

We can also combine functions using subtraction, multiplication, and division by performing operations with the algebraic expressions that appear on the right side of the equations. For example, the functions $f(x) = 2x$ and $g(x) = x - 1$ can be combined to form the difference, product, and quotient of f and g. Here's how it's done:

Difference: $f - g$

$$(f-g)(x) = f(x) - g(x)$$
$$= 2x - (x-1) = 2x - x + 1 = x + 1$$

Product: fg

$$(fg)(x) = f(x) \cdot g(x)$$
$$= 2x(x-1) = 2x^2 - 2x$$

Quotient: $\frac{f}{g}$

$$\left(\frac{f}{g}\right)(x) = \frac{f(x)}{g(x)} = \frac{2x}{x-1}, x \neq 1.$$

Just like the domain for $f + g$, the domain for each of these functions consists of all real numbers that are common to the domains of f and g. In the case of the quotient function $\frac{f(x)}{g(x)}$, we must remember not to divide by 0, so we add the further restriction that $g(x) \neq 0$.

The following definitions summarize our discussion.

The Algebra of Functions: Sum, Difference, Product, and Quotient of Functions

Let f and g be two functions. The **sum** $f + g$, the **difference** $f - g$, the **product** fg, and the **quotient** $\frac{f}{g}$ are functions whose domains are the set of all real numbers common to the domains of f and g, defined as follows:

1. Sum: $(f+g)(x) = f(x) + g(x)$
2. Difference: $(f-g)(x) = f(x) - g(x)$
3. Product: $(fg)(x) = f(x) \cdot g(x)$
4. Quotient: $\left(\frac{f}{g}\right)(x) = \frac{f(x)}{g(x)}$, provided $g(x) \neq 0$.

EXAMPLE 4 Using the Algebra of Functions

Let $f(x) = x^2 + x$ and $g(x) = x - 5$. Find:

a. $(f + g)(4)$

b. $(f - g)(x)$ and $(f - g)(-3)$

c. $\left(\frac{f}{g}\right)(x)$ and $\left(\frac{f}{g}\right)(7)$

d. $(fg)(-2)$.

Solution

a. We can find $(f + g)(4)$ using $f(4)$ and $g(4)$.

$$f(x) = x^2 + x \qquad g(x) = x - 5$$
$$f(4) = 4^2 + 4 = 20 \qquad g(4) = 4 - 5 = -1$$

Thus,

$$(f + g)(4) = f(4) + g(4) = 20 + (-1) = 19.$$

We can also find $(f + g)(4)$ by first finding $(f + g)(x)$ and then substituting 4 for x:

$$(f + g)(x) = f(x) + g(x)$$ This is the definition of the sum $f + g$.

$$= (x^2 + x) + (x - 5)$$ Substitute the given functions.

$$= x^2 + 2x - 5.$$ Simplify.

Using $(f + g)(x) = x^2 + 2x - 5$. we have

$$(f + g)(4) = 4^2 + 2 \cdot 4 - 5 = 16 + 8 - 5 = 19.$$

b. $(f - g)(x) = f(x) - g(x)$ This is the definition of the difference $f - g$.

$$= (x^2 + x) - (x - 5)$$ Substitute the given functions.

$$= x^2 + x - x + 5$$ Remove parentheses and change the sign of each term in the second set of parentheses.

$$= x^2 + 5$$ Simplify.

Using $(f - g)(x) = x^2 + 5$, we have

$$(f - g)(-3) = (-3)^2 + 5 = 9 + 5 = 14.$$

c. $\left(\frac{f}{g}\right)(x) = \frac{f(x)}{g(x)}$ This is the definition of the quotient $\frac{f}{g}$.

$$= \frac{x^2 + x}{x - 5}$$ Substitute the given functions.

Using $\left(\frac{f}{g}\right)(x) = \frac{x^2 + x}{x - 5}$, we have

$$\left(\frac{f}{g}\right)(7) = \frac{7^2 + 7}{7 - 5} = \frac{56}{2} = 28.$$

d. We can find $(fg)(-2)$ using the fact that

$$(fg)(-2) = f(-2) \cdot g(-2).$$

$$f(x) = x^2 + x \qquad g(x) = x - 5$$

$$f(-2) = (-2)^2 + (-2) = 4 - 2 = 2 \qquad g(-2) = -2 - 5 = -7$$

Thus,

$$(fg)(-2) = f(-2) \cdot g(-2) = 2(-7) = -14.$$

We could also have found $(fg)(-2)$ by multiplying $f(x) \cdot g(x)$ and then substituting -2 into the product. We will discuss how to multiply expressions such as $x^2 + x$ and $x - 5$ in Chapter 5. ■

CHECK POINT 4 Let $f(x) = x^2 - 2x$ and $g(x) = x + 3$. Find:

a. $(f + g)(5)$

b. $(f - g)(x)$ and $(f - g)(-1)$

c. $\left(\frac{f}{g}\right)(x)$ and $\left(\frac{f}{g}\right)(7)$

d. $(fg)(-4)$.

EXERCISE SET 2.2

Practice Exercises

In Exercises 1–10, find the domain of each function.

1. $f(x) = 3x + 5$
2. $f(x) = 4x + 7$
3. $g(x) = \dfrac{1}{x + 4}$
4. $g(x) = \dfrac{1}{x + 5}$
5. $f(x) = \dfrac{2x}{x - 3}$
6. $f(x) = \dfrac{4x}{x - 2}$
7. $g(x) = x + \dfrac{3}{5 - x}$
8. $g(x) = x + \dfrac{7}{6 - x}$
9. $f(x) = \dfrac{1}{x + 7} + \dfrac{3}{x - 9}$
10. $f(x) = \dfrac{1}{x + 8} + \dfrac{3}{x - 10}$

In Exercises 11–16, find **a.** $(f + g)(x)$ **b.** $(f + g)(5)$.

11. $f(x) = 3x + 1,\ g(x) = 2x - 6$
12. $f(x) = 4x + 2, g(x) = 2x - 9$
13. $f(x) = x - 5, g(x) = 3x^2$
14. $f(x) = x - 6, g(x) = 2x^2$
15. $f(x) = 2x^2 - x - 3, g(x) = x + 1$
16. $f(x) = 4x^2 - x - 3, g(x) = x + 1$

In Exercises 17–28, for each pair of functions, f and g, determine the domain of f + g.

17. $f(x) = 3x + 7,\ g(x) = 9x + 10$
18. $f(x) = 7x + 4,\ g(x) = 5x - 2$

19. $f(x) = 3x + 7,\ g(x) = \dfrac{2}{x - 5}$

20. $f(x) = 7x + 4,\ g(x) = \dfrac{2}{x - 6}$

21. $f(x) = \dfrac{1}{x},\ g(x) = \dfrac{2}{x - 5}$

22. $f(x) = \dfrac{1}{x},\ g(x) = \dfrac{2}{x - 6}$

23. $f(x) = \dfrac{8x}{x - 2},\ g(x) = \dfrac{6}{x + 3}$

24. $f(x) = \dfrac{9x}{x - 4},\ g(x) = \dfrac{7}{x + 8}$

25. $f(x) = \dfrac{8x}{x - 2},\ g(x) = \dfrac{6}{2 - x}$

26. $f(x) = \dfrac{9x}{x - 4},\ g(x) = \dfrac{7}{4 - x}$

27. $f(x) = x^2,\ g(x) = x^3$

28. $f(x) = x^2 + 1,\ g(x) = x^3 - 1$

In Exercises 29–48, let

$$f(x) = x^2 + 4x \quad \text{and} \quad g(x) = 2 - x.$$

Find each of the following.

29. $(f + g)(x)$ and $(f + g)(3)$

30. $(f + g)(x)$ and $(f + g)(4)$

31. $f(-2) + g(-2)$

32. $f(-3) + g(-3)$

33. $(f - g)(x)$ and $(f - g)(5)$

34. $(f - g)(x)$ and $(f - g)(6)$

35. $f(-2) - g(-2)$

36. $f(-3) - g(-3)$

37. $(fg)(-2)$

38. $(fg)(-3)$

39. $(fg)(5)$

40. $(fg)(6)$

41. $\left(\dfrac{f}{g}\right)(x)$ and $\left(\dfrac{f}{g}\right)(1)$

42. $\left(\dfrac{f}{g}\right)(x)$ and $\left(\dfrac{f}{g}\right)(3)$

43. $\left(\dfrac{f}{g}\right)(-1)$

44. $\left(\dfrac{f}{g}\right)(0)$

45. The domain of $f + g$

46. The domain of $f - g$

47. The domain of $\dfrac{f}{g}$

48. The domain of fg

Application Exercises

It seems that Phideau's medical bills are costing us an arm and a paw. The graph shows veterinary costs, in billions of dollars, for dogs and cats in five selected years. Let

$D(x)$ = veterinary costs, in billions of dollars, for dogs in year x

$C(x)$ = veterinary costs, in billions of dollars, for cats in year x.

Use the graph to solve Exercises 49–52.

Veterinary Costs in the U.S.

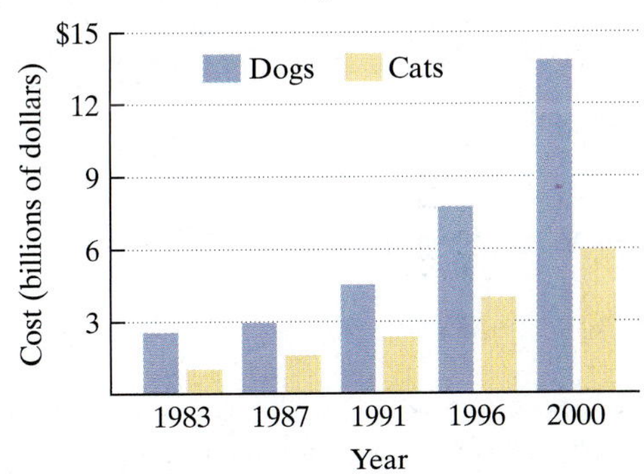

Source: American Veterinary Medical Association

49. Find an estimate of $(D + C)(2000)$. What does this mean in terms of the variables in this situation?

50. Find an estimate of $(D - C)(2000)$. What does this mean in terms of the variables in this situation?

51. Using the information shown in the graph, what is the domain of $D + C$?

52. Using the information shown in the graph, what is the domain of $D - C$?

Consider the following functions:

$f(x) =$ population of the world's more-developed regions in year x

$g(x) =$ the population of the world's less-developed regions in year x

$h(x) =$ total world population in year x.

Use these functions and the graphs shown to answer Exercises 53–56.

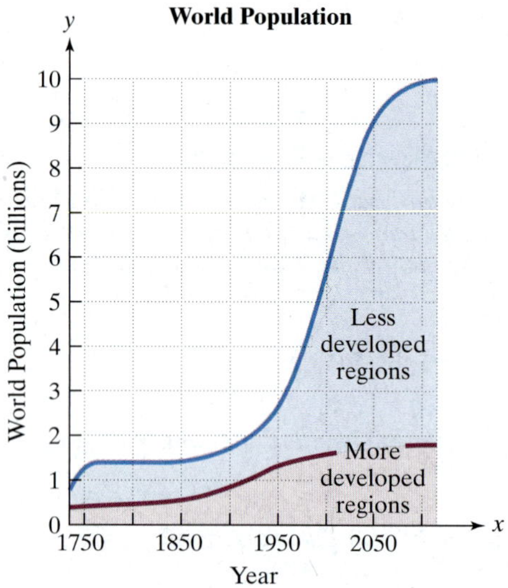

Source: Population Reference Bureau

53. What does the function $f + g$ represent?

54. What does the function $h - g$ represent?

55. Use the graph to estimate $(f + g)(2000)$.

56. Use the graph to estimate $(h - g)(2000)$.

57. A company that sells radios has yearly fixed costs of \$600,000. It costs the company \$45 to produce each radio. Each radio will sell for \$65. The company's costs and revenue are modeled by the following functions:

$C(x) = 600,000 + 45x$ — This function models the company's costs.

$R(x) = 65x$ — This function models the company's revenue.

Find and interpret $(R - C)(20{,}000)$, $(R - C)(30{,}000)$ and $(R - C)(40{,}000)$.

58. The function $f(t) = -0.14t^2 + 0.51t + 31.6$ models the U.S. population, $f(t)$, in millions, ages 65 and older t years after 1990. The function $g(t) = 0.54t^2 + 12.64t + 107.1$ models the total yearly cost of Medicare, $g(t)$, in billions of dollars, t years after 1990.

a. What does the function $\frac{g}{f}$ represent?

b. Find and interpret $\frac{g}{f}(10)$.

Writing in Mathematics

59. If a function is defined by an equation, explain how to find its domain.

60. If equations for functions f and g are given, explain how to find $f + g$.

61. If the equations of two functions are given, explain how to obtain the quotient function and its domain.

62. If equations for functions f and g are given, describe two ways to find $(f - g)(3)$.

63. Explain how to use the graphs in Figure 2.10 on page 107 to estimate the per capita federal tax for any one of the years shown on the horizontal axis.

Technology Exercises

64. Graph the function $\frac{g}{f}$ from Exercise 58 in a $[0, 15, 1]$ by $[0, 60, 1]$ viewing rectangle. What does the shape of the graph indicate about the per capita costs of Medicare for the U.S. population ages 65 and over with increasing time?

In Exercises 65–68, graph each of the three functions in the same $[-10, 10, 1]$ *by* $[-10, 10, 1]$ *viewing rectangle.*

65. $y_1 = 2x + 3$
$y_2 = 2 - 2x$
$y_3 = y_1 + y_2$

66. $y_1 = x - 4$
$y_2 = 2x$
$y_3 = y_1 - y_2$

67. $y_1 = x$
$y_2 = x - 4$
$y_3 = y_1 \cdot y_2$

68. $y_1 = x^2 - 2x$
$y_2 = x$
$y_3 = \frac{y_1}{y_2}$

69. In Exercise 68, use the TRACE feature to trace along y_3. What happens at $x = 0$? Explain why this occurs.

Critical Thinking Exercises

70. Which one of the following is false?
- **a.** If $(f + g)(a) = 0$, then $f(a)$ and $g(a)$ must be opposites, or additive inverses.
- **b.** If $(f - g)(a) = 0$, then $f(a)$ and $g(a)$ must be equal.
- **c.** If $\left(\dfrac{f}{g}\right)(a) = 0$, then $f(a)$ must be 0.
- **d.** If $(fg)(a) = 0$, then $f(a)$ must be 0.

71. Use the graphs given in Exercises 53–56 to create a graph that shows the population, in billions, of less-developed regions from 1950 through 2050.

Review Exercises

72. Solve for b: $R = 3(a + b)$. (Section 1.5, Example 6)

73. Solve: $3(6 - x) = 3 - 2(x - 4)$. (Section 1.4, Example 3)

74. If $f(x) = 6x - 4$, find $f(b + 2)$. (Section 2.1, Example 3)

▸ SECTION 2.3 Linear Functions and Slope

Objectives

1. Use intercepts to graph a linear function in standard form.
2. Compute a line's slope.
3. Find a line's slope and y-intercept from its equation.
4. Graph linear functions in slope-intercept form.
5. Graph horizontal or vertical lines.
6. Calculate average rate of change in applied situations.

Is there a relationship between literacy and child mortality? As the percentage of adult females who are literate increases, does the mortality of children under five decrease? Figure 2.11, based on data from the United Nations, indicates that this is, indeed, the case. Each point in the figure represents one country.

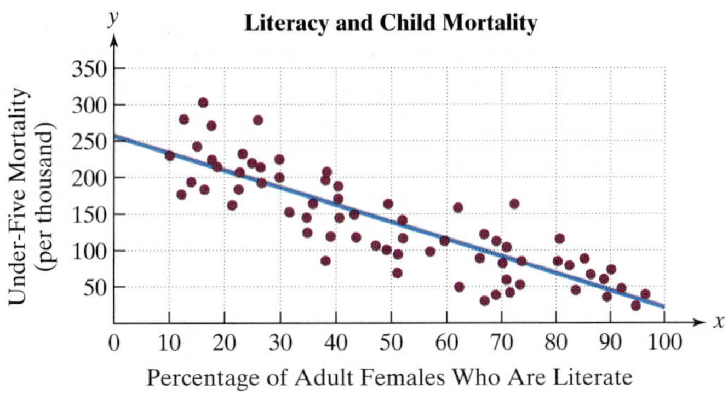

Figure 2.11 *Source*: United Nations

Data presented in a visual form as a set of points is called a **scatter plot**. Also shown in Figure 2.11 is a line that passes through or near the points. A line that best fits the data points in a scatter plot is called a **regression line**. By writing the equation of this line, we can obtain a model of the data and make predictions about child mortality based on the percentage of adult females in a country who are literate.

Data often fall on or near a line. In the remainder of this chapter, we will use equations to model such data and make predictions. We begin with a discussion of graphing linear functions using intercepts.

1 Use intercepts to graph a linear function in standard form.

Graphing Using Intercepts

The equation of the regression line in Figure 2.11 on page 115 is

$$y = -2.39x + 254.47.$$

The variable x represents the percentage of adult females in a country who are literate. The variable y represents child mortality, per thousand, for children under five in that country. Using function notation, we can rewrite the equation as

$$f(x) = -2.39x + 254.47.$$

A function such as this, whose graph is a straight line, is called a **linear function**. There is another way that we can write the function's equation

$$y = -2.39x + 254.47.$$

We will collect the x- and y-terms on the left side. This is done by adding $2.39x$ to both sides:

$$2.39x + y = 254.47.$$

The form of this equation is $Ax + By = C$.

$$2.39x \quad + \quad y = 254.47.$$

A, the coefficient of x, is 2.39.

B, the coefficient of y, is 1.

C, the constant on the right, is 254.47.

All equations of the form $Ax + By = C$ are straight lines when graphed, as long as A and B are not both zero. Such an equation is called the **standard form of the equation of a line**. To graph equations of this form, we will use two important points: the **intercepts**.

An ***x*-intercept** of a graph is the x-coordinate of a point where the graph intersects the x-axis. For example, look at the graph of $2x - 4y = 8$ in Figure 2.12. The graph crosses the x-axis at $(4, 0)$. Thus, the x-intercept is 4. **The *y*-coordinate corresponding to an *x*-intercept is always zero.**

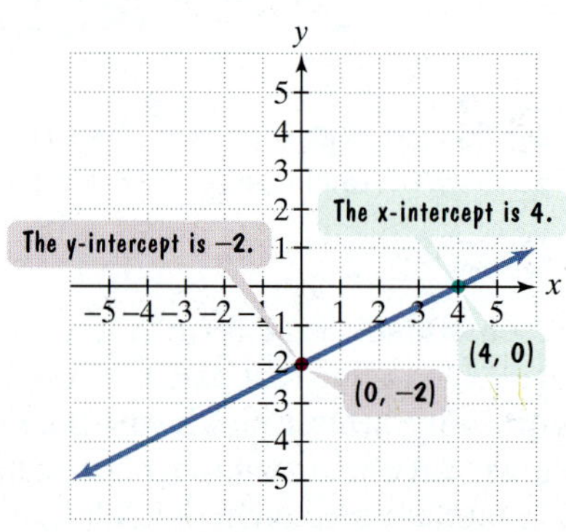

Figure 2.12 The graph of $2x - 4y = 8$

A ***y*-intercept** of a graph is the y-coordinate of a point where the graph intersects the y-axis. The graph of $2x - 4y = 8$ in Figure 2.12 shows that the graph crosses the y-axis at $(0, -2)$. Thus, the y-intercept is -2. **The *x*-coordinate corresponding to a *y*-intercept is always zero.**

When graphing using intercepts, it is a good idea to use a third point, a checkpoint, before drawing the line. A checkpoint can be obtained by selecting a value for either variable, other than 0, and finding the corresponding value for the other variable. The checkpoint should lie on the same line as the x- and y-intercepts. If it does not, recheck your work and find the error.

Using Intercepts to Graph $Ax + By = C$

1. Find the x-intercept. Let $y = 0$ and solve for x.
2. Find the y-intercept. Let $x = 0$ and solve for y.
3. Find a checkpoint, a third ordered-pair solution.
4. Graph the equation by drawing a line through the three points.

EXAMPLE 1 Using Intercepts to Graph a Linear Equation

Graph: $3x + 2y = 6$.

Solution

Step 1 Find the *x*-intercept. Let $y = 0$ and solve for x.

$$3x + 2 \cdot 0 = 6$$
$$3x = 6$$
$$x = 2$$

Step 2 Find the *y*-intercept. Let $x = 0$ and solve for y.

$$3 \cdot 0 + 2y = 6$$
$$2y = 6$$
$$y = 3$$

The x-intercept is 2, so the line passes through $(2, 0)$. The y-intercept is 3, so the line passes through $(0, 3)$.

Step 3 Find a checkpoint, a third ordered-pair solution.
For our checkpoint, we will let $x = 1$ and find the corresponding value for y.

$$3x + 2y = 6 \quad \text{This is the given equation.}$$
$$3(1) + 2y = 6 \quad \text{Substitute 1 for x.}$$
$$3 + 2y = 6 \quad \text{Simplify.}$$
$$2y = 3 \quad \text{Subtract 3 from both sides.}$$
$$y = \frac{3}{2} \quad \text{Divide both sides by 2.}$$

The checkpoint is the ordered pair $\left(1, \frac{3}{2}\right)$ or $(1, 1.5)$.

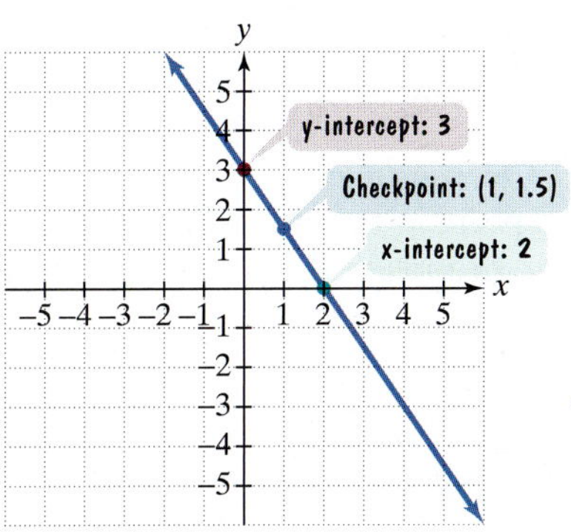

Figure 2.13 The graph of $3x + 2y = 6$

Step 4 Graph the equation by drawing a line through the three points. The three points in Figure 2.13 lie along the same line. Drawing a line through the three points results in the graph of $3x + 2y = 6$. ■

Using Technology

You can use a graphing utility to graph equations of the form $Ax + By = C$. Begin by solving the equation for y. For example, to graph $3x + 2y = 6$, solve the equation for y.

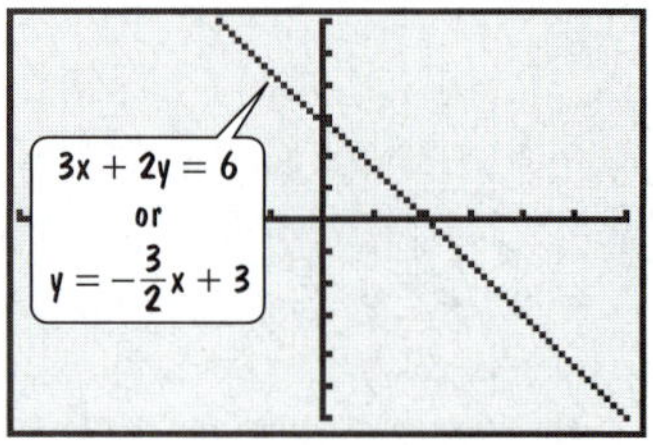

$$3x + 2y = 6$$ This is the equation to be graphed.

$$3x - 3x + 2y = -3x + 6$$ Subtract 3x from both sides.

$$2y = -3x + 6$$ Simplify.

$$\frac{2y}{2} = \frac{-3x + 6}{2}$$ Divide both sides by 2.

$$y = -\frac{3}{2}x + 3$$ Simplify.

This is the equation to enter in your graphing utility. The graph of $y = -\frac{3}{2}x + 3$ or, equivalently, $3x + 2y = 6$ is shown above in a $[-6, 6, 1]$ by $[-6, 6, 1]$ viewing rectangle.

CHECK POINT 1 Graph: $2x + 3y = 6$.

2 Compute a line's slope.

The Slope of a Line

Mathematicians have developed a useful measure of the steepness of a line, called the **slope** of the line. Slope compares the vertical change (the **rise**) to the horizontal change (the **run**) when moving from one fixed point to another along the line. To calculate the slope of a line, mathematicians use a ratio comparing the change in y (the rise) to the change in x (the run).

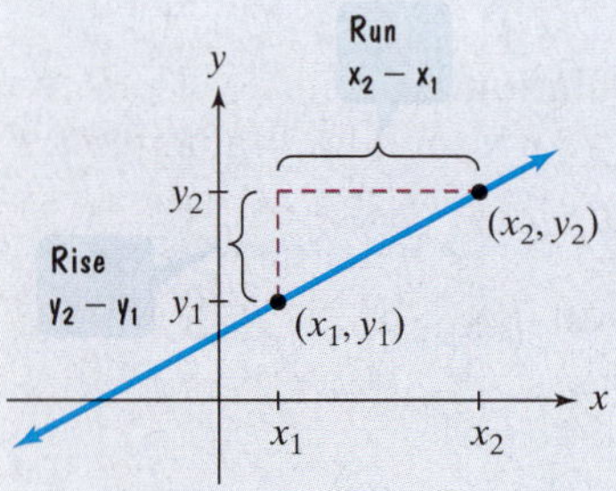

Definition of Slope

The **slope** of the line through the distinct points (x_1, y_1) and (x_2, y_2) is

$$\frac{\text{Change in } y}{\text{Change in } x} = \frac{\text{Rise}}{\text{Run}} = \frac{y_2 - y_1}{x_2 - x_1}$$

where $x_2 - x_1 \neq 0$.

It is common notation to let the letter m represent the slope of a line. The letter m is used because it is the first letter of the French verb *monter*, meaning to rise, or to ascend.

EXAMPLE 2 Using the Definition of Slope

Find the slope of the line passing through each pair of points:

a. $(-3, -1)$ and $(-2, 4)$ **b.** $(-3, 4)$ and $(2, -2)$.

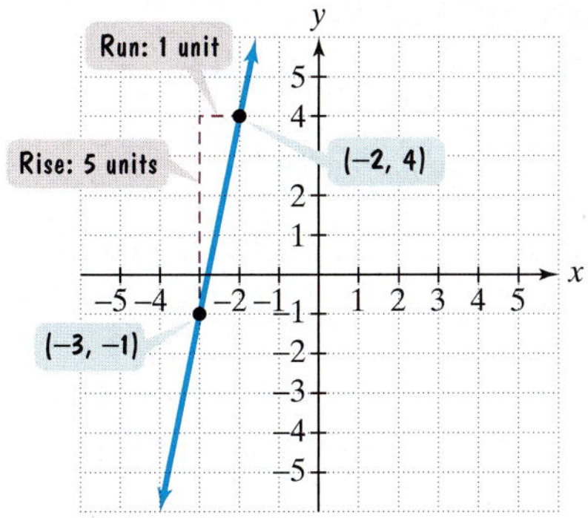

Figure 2.14 Visualizing a slope of 5

Solution

a. Let $(x_1, y_1) = (-3, -1)$ and $(x_2, y_2) = (-2, 4)$. We obtain a slope of

$$m = \frac{\text{Change in } y}{\text{Change in } x} = \frac{y_2 - y_1}{x_2 - x_1} = \frac{4 - (-1)}{-2 - (-3)} = \frac{5}{1} = 5.$$

The situation is illustrated in Figure 2.14. The slope of the line is 5, indicating that there is a vertical change, a rise, of 5 units for each horizontal change, a run, of 1 unit. The slope is positive, and the line rises from left to right.

Study Tip

When computing slope, it makes no difference which point you call (x_1, y_1) and which point you call (x_2, y_2). If we let $(x_1, y_1) = (-2, 4)$ and $(x_2, y_2) = (-3, -1)$, the slope is still 5:

$$m = \frac{y_2 - y_1}{x_2 - x_1} = \frac{-1 - 4}{-3 - (-2)} = \frac{-5}{-1} = 5.$$

However, you should not subtract in one order in the numerator $(y_2 - y_1)$ and then in a different order in the denominator $(x_1 - x_2)$. The slope is *not*

$$\frac{-1 - 4}{-2 - (-3)} = \frac{-5}{1} = -5. \quad \text{Incorrect!}$$

b. We can let $(x_1, y_1) = (-3, 4)$ and $(x_2, y_2) = (2, -2)$. The slope of the line shown in Figure 2.15 is computed as follows:

$$m = \frac{-2 - 4}{2 - (-3)} = \frac{-6}{5} = -\frac{6}{5}.$$

The slope of the line is $-\frac{6}{5}$. For every vertical change of −6 units (6 units down), there is a corresponding horizontal change of 5 units. The slope is negative and the line falls from left to right. ■

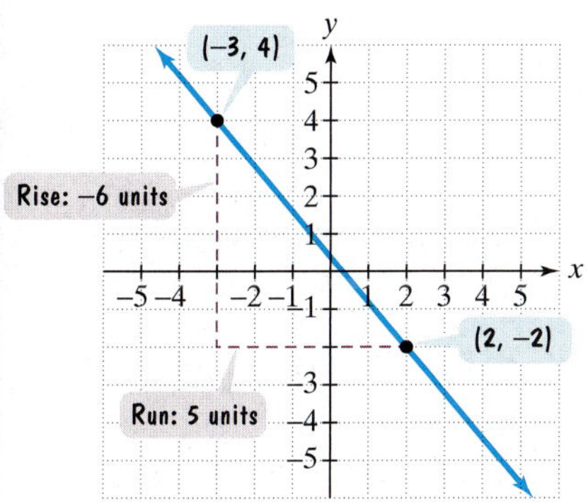

Figure 2.15 Visualizing a slope of $-\frac{6}{5}$

✔ **CHECK POINT 2** Find the slope of the line passing through each pair of points:

a. (−3, 4) and (−4, −2) **b.** (4, −2) and (−1, 5).

Example 2 illustrates that a line with a positive slope is rising from left to right and a line with a negative slope is falling from left to right. By contrast, a horizontal line neither rises nor falls and has a slope of zero. A vertical line has no horizontal change, so $x_2 - x_1 = 0$ in the formula for slope. Because we cannot divide by zero, the slope of a vertical line is undefined. This discussion is summarized in Table 2.2.

Table 2.2 Possibilities for a Line's Slope

Positive Slope	Negative Slope	Zero Slope	Undefined Slope
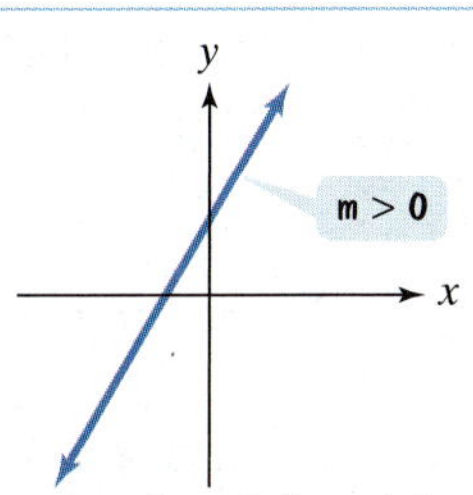	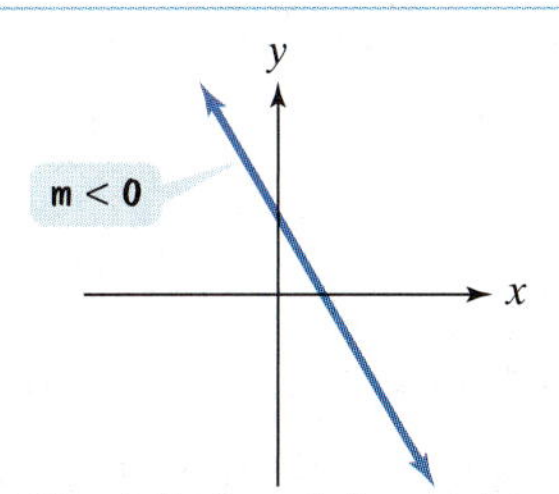	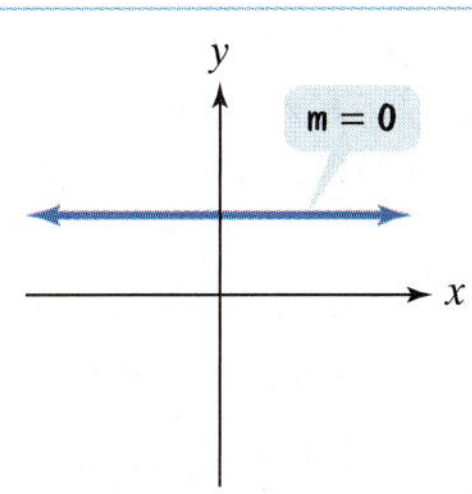	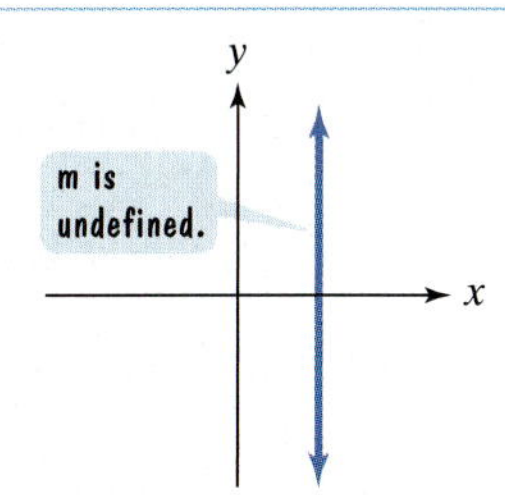
Line rises from left to right.	Line falls from left to right.	Line is horizontal.	Line is vertical.

The Slope-Intercept Form of the Equation of a Line

We opened this section with a linear function that modeled child mortality as a function of literacy. The function's equation can be expressed as

$$y = -2.39x + 254.47 \quad \text{or} \quad f(x) = -2.39x + 254.47.$$

What is the significance of -2.39, the x-coefficient, or of 254.47, the constant term? To answer this question, let's look at an equation in the same form with simpler numbers. In particular, consider the equation $y = 2x + 4$.

Figure 2.16 shows the graph of $y = 2x + 4$. Verify that the x-intercept is -2 by setting y equal to 0 and solving for x. Similarly, verify that the y-intercept is 4 by setting x equal to 0 and solving for y.

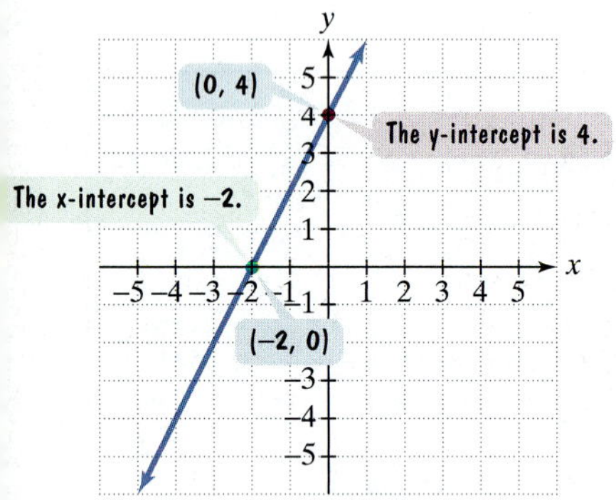

Figure 2.16 The graph of $y = 2x + 4$

Now that we have two points on the line, we can calculate the slope of the graph of $y = 2x + 4$.

$$\text{Slope} = \frac{\text{Change in } y}{\text{Change in } x}$$

$$= \frac{4 - 0}{0 - (-2)} = \frac{4}{2} = 2$$

We see that the slope of the line is 2, the same as the coefficient of x in the equation $y = 2x + 4$. The y-intercept is 4, the same as the constant in the equation $y = 2x + 4$.

$$y = 2x + 4$$

The slope is 2. The y-intercept is 4.

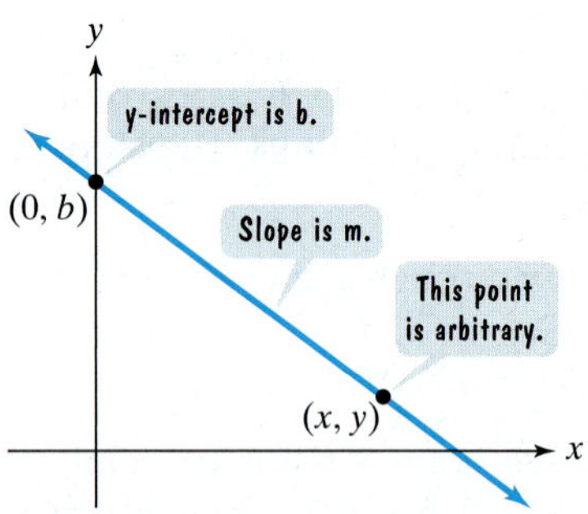

Figure 2.17 A line with slope m and y-intercept b

It is not merely a coincidence that the x-coefficient is the line's slope and the constant term is the y-intercept. Let's find the equation of any nonvertical line with slope m and y-intercept b. Because the y-intercept is b, the point $(0, b)$ lies on the line. Now, let (x, y) represent any other point on the line, shown in Figure 2.17. Keep in mind that the point (x, y) is arbitrary and is not in one fixed position. By contrast, the point $(0, b)$ is fixed.

Regardless of where the point (x, y) is located, the steepness of the line in Figure 2.17 remains the same. Thus, the ratio for slope stays a constant m. This means that for all points along the line,

$$m = \frac{\text{Change in } y}{\text{Change in } x} = \frac{y - b}{x - 0} = \frac{y - b}{x}.$$

We can clear the fraction by multiplying both sides by x, the least common denominator.

$m = \dfrac{y - b}{x}$	This is the slope of the line in Figure 2.17.
$mx = \dfrac{y - b}{x} \cdot x$	Multiply both sides by x.
$mx = y - b$	Simplify: $\dfrac{y - b}{\cancel{x}} \cdot \cancel{x} = y - b$.
$mx + b = y - b + b$	Add b to both sides and solve for y.
$mx + b = y$	Simplify.

Now, if we reverse the two sides, we obtain the slope-intercept form of the equation of a line.

Slope-Intercept Form of the Equation of a Line

The **slope-intercept equation** of a nonvertical line with slope m and y-intercept b is

$$y = mx + b.$$

3 Find a line's slope and y-intercept from its equation.

The slope-intercept equation, $y = mx + b$, can be expressed in function notation by replacing y by $f(x)$:

$$f(x) = mx + b.$$

We have seen that functions in this form are called **linear functions**. Thus, in the equation of a linear function, the x-coefficient is the line's slope and the constant term is the y-intercept. Here are two examples.

Study Tip

The variables in $y = mx + b$ vary in different ways. The values for slope, m, and y-intercept, b, vary from one line's equation to another. However, they remain constant in the equation of a single line. By contrast, the variables x and y represent the infinitely many points, (x, y), on a single line. Thus, these variables vary in both the equation of a single line, as well as from one equation to another.

$$y = 2x - 4$$

The slope is 2. The y-intercept is −4.

$$f(x) = \frac{1}{2}x + 2$$

The slope is $\frac{1}{2}$. The y-intercept is 2.

If a linear function's equation is in standard form, $Ax + By = C$, do you see how we can identify the line's slope and y-intercept? Solve the equation for y and convert to slope-intercept form. For example, consider the equation

$$3x + 2y = 6.$$

We graph this equation with a graphing utility by first solving for y. We also solve for y to find the slope and y-intercept. This is shown in the Using Technology box on page 118. We obtain the slope-intercept equation

$$y = -\frac{3}{2}x + 3.$$

The slope is $-\frac{3}{2}$. The y-intercept is 3.

If a linear function's equation is in slope-intercept form, we can use the y-intercept and the slope to obtain its graph.

4 Graph linear functions in slope-intercept form.

Graphing $y = mx + b$ Using the Slope and y-Intercept

1. Plot the y-intercept on the y-axis. This is the point $(0, b)$.
2. Obtain a second point using the slope, m. Write m as a fraction, and use rise over run, starting at the point containing the y-intercept, to plot this point.
3. Use a straightedge to draw a line through the two points. Draw arrowheads at the ends of the line to show that the line continues indefinitely in both directions.

EXAMPLE 3 Graphing by Using the Slope and *y*-Intercept

Graph the line whose equation is $y = 4x - 3$.

Solution We write $y = 4x - 3$ in the form $y = mx + b$.

$$y = 4x + (-3)$$

The slope is 4. The *y*-intercept is −3.

Now that we have identified the slope and the *y*-intercept, we use the three steps in the box to graph the equation.

Step 1 Plot the *y*-intercept on the *y*-axis. The *y*-intercept is −3. We plot the point $(0, -3)$, shown in Figure 2.18(a).

Step 2 Obtain a second point using the slope, *m*. Write *m* as a fraction, and use rise over run, starting at the point containing the *y*-intercept, to plot this point. We express the slope, 4, as a fraction.

$$m = \frac{4}{1} = \frac{\text{Rise}}{\text{Run}}.$$

We plot the second point on the line by starting at $(0, -3)$, the first point. Based on the slope, we move 4 units *up* (the rise) and 1 unit to the *right* (the run). This puts us at a second point on the line, $(1, 1)$, shown in Figure 2.18(b).

Step 3 Use a straightedge to draw a line through the two points. The graph of $y = 4x - 3$ is shown in Figure 2.18(c).

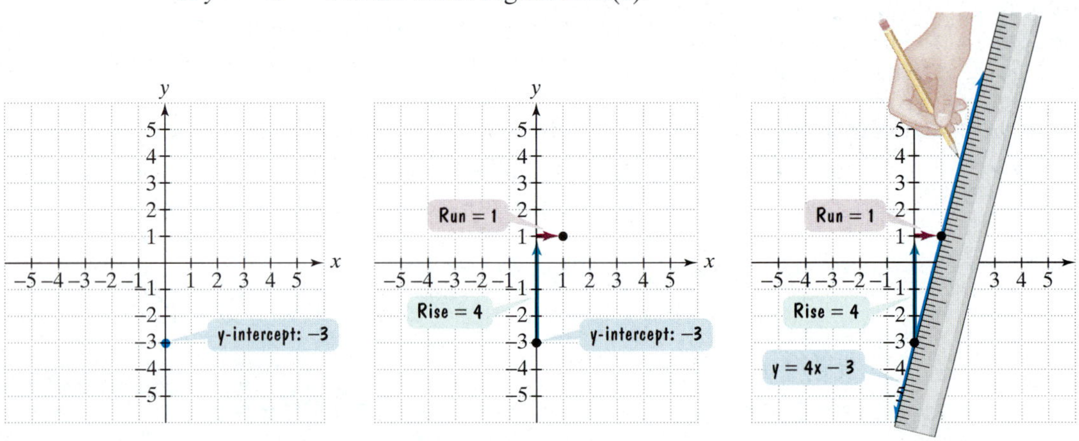

(a) The *y*-intercept is −3, so $(0, -3)$ is a point on the line.

(b) The slope is 4.

(c) The graph of $y = 4x - 3$

Figure 2.18 Graphing $y = 4x - 3$ using the *y*-intercept and slope

■

✔ **CHECK POINT 3** Graph the line whose equation is $y = 3x - 2$.

EXAMPLE 4 Graphing by Using the Slope and *y*-Intercept

Graph the linear function: $f(x) = \frac{2}{3}x + 2$.

Solution The equation of the line is in the form $f(x) = mx + b$. We can find the slope, *m*, by identifying the coefficient of *x*. We can find the *y*-intercept, *b*, by identifying the constant term.

$$f(x) = \frac{2}{3}x + 2$$

The slope is $\frac{2}{3}$.

The y-intercept is 2.

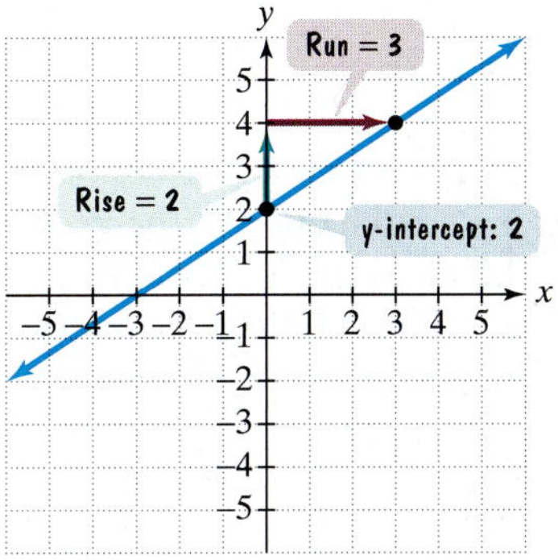

Figure 2.19 The graph of $f(x) = \frac{2}{3}x + 2$

Now that we have identified the slope and the y-intercept, we use the three-step procedure to graph the equation.

Step 1 Plot the y-intercept on the y-axis.
The y-intercept is 2. We plot $(0, 2)$, shown in Figure 2.19.

Step 2 Obtain a second point using the slope, m. Write m as a fraction, and use rise over run, starting at the point containing the y-intercept, to plot this point.
We express the slope, $\frac{2}{3}$, as a fraction.

$$m = \frac{2}{3} = \frac{\text{Rise}}{\text{Run}}$$

We plot the second point on the line by starting at $(0, 2)$, the first point. Based on the slope, we move 2 units *up* (the rise) and 3 units to the *right* (the run). This puts us at a second point on the line, $(3, 4)$, shown in Figure 2.19.

Step 3 Use a straightedge to draw a line through the two points. The graph of the linear function $f(x) = \frac{2}{3}x + 2$ is shown as a blue line in Figure 2.19. ■

✔ **CHECK POINT 4** Graph the linear function: $f(x) = \frac{3}{5}x + 1$.

5 Graph horizontal or vertical lines.

Equations of Horizontal and Vertical Lines

Some things change very little. For example, Figure 2.20 shows that the percentage of people in the United States satisfied with their lives remains relatively constant for all age groups. Shown in the figure is a horizontal line that passes near most tops of the six bars.

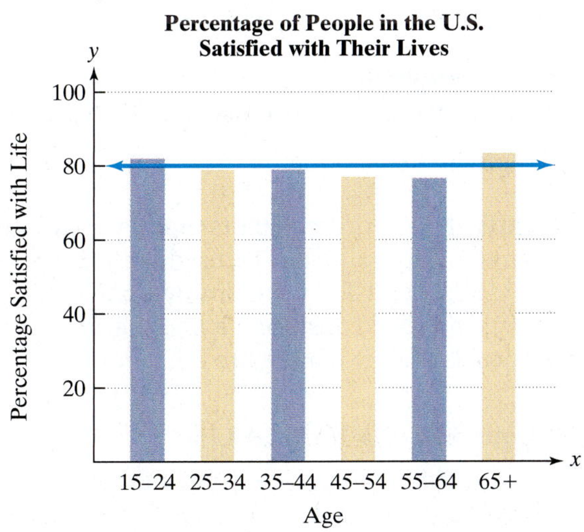

Source: *Culture Shift in Advanced Industrial Society*, Princeton University Press

Figure 2.20

We can use $y = mx + b$, the slope-intercept form of a line's equation, to write the equation of the horizontal line in Figure 2.20. We need the line's slope, m, and its y-intercept, b. Because the line is horizontal, $m = 0$. The line intersects the y-axis at 80, so $b = 80$. Thus, an equation that models the percentage of people at age x satisfied with their lives is

$$y = 0x + 80, \quad \text{or} \quad y = 80.$$

The percentage of people satisfied with their lives remains relatively constant in the United States for all age groups, at approximately 80 percent. Using function notation, we can write

$$f(x) = 80$$

For all age inputs, x, in the domain — the output is a constant 80%.

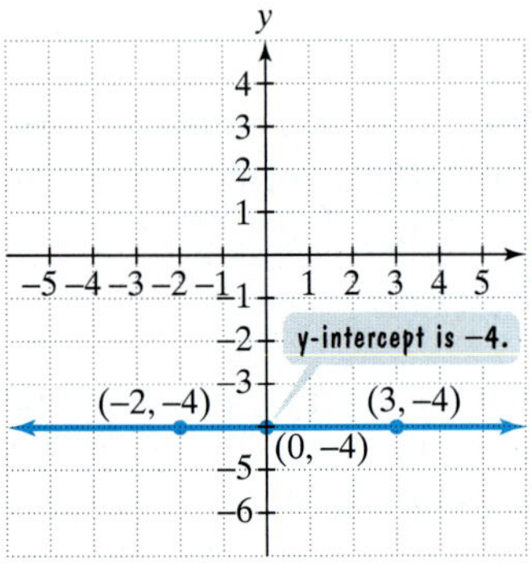

Figure 2.21 The graph of $y = -4$ or $f(x) = -4$

In general, if a line is horizontal, its slope is zero: $m = 0$. Thus, the equation $y = mx + b$ becomes $y = b$, where b is the y-intercept. For example, the graph of $y = -4$ is a horizontal line with a y-intercept of -4. The graph is shown in Figure 2.21. Three of the points along the line are shown and labeled. No matter what the x-coordinate is, the corresponding y-coordinate for every point on the line is -4.

Equation of a Horizontal Line

A horizontal line is given by an equation of the form

$$y = b$$

where b is the y-intercept.

Because any vertical line can intersect the graph of the horizontal line $y = b$ only once, a horizontal line is the graph of a function. Thus, we can express the equation $y = b$ as $f(x) = b$. This linear function is often called a **constant function**. The function modeling the percentage of people satisfied with their lives at age x, namely $f(x) = 80$, is an example of a constant function.

Next, let's see what we can discover about a vertical line by looking at an example.

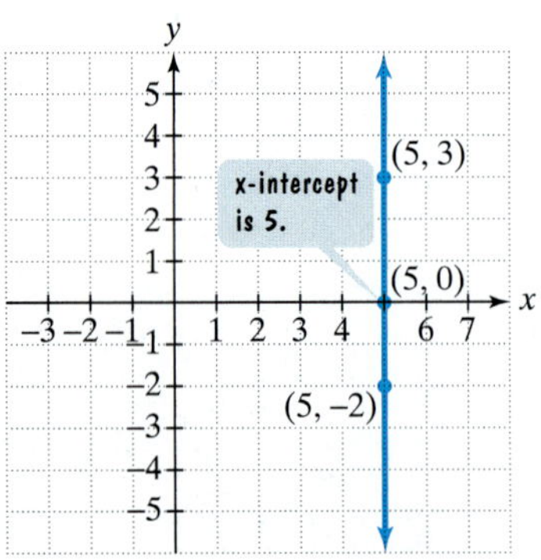

Figure 2.22 The graph of $x = 5$

EXAMPLE 5 Graphing a Vertical Line

Graph the linear equation: $x = 5$.

Solution All ordered pairs that are solutions of $x = 5$ have a value of x that is always 5. Any value can be used for y. Let us select three of the possible values for y: -2, 0, and 3. So, the ordered pairs that are solutions of $x = 5$ are $(5, -2)$, $(5, 0)$, and $(5, 3)$. Drawing a line that passes through the three points gives the vertical line shown in Figure 2.22. ■

Equation of a Vertical Line

A vertical line is given by an equation of the form

$$x = a$$

where a is the x-intercept.

Does a vertical line represent the graph of a linear function? No. Look at the graph of $x = 5$ in Figure 2.22. A vertical line drawn through 5 intersects the graph infinitely many times. This shows that associated with the input 5 are infinitely many outputs. **No vertical line is a linear function.**

✔ **CHECK POINT 5** Graph the linear equation: $x = -2$.

6 Calculate average rate of change in applied situations.

Applications: Slope as Average Rate of Change

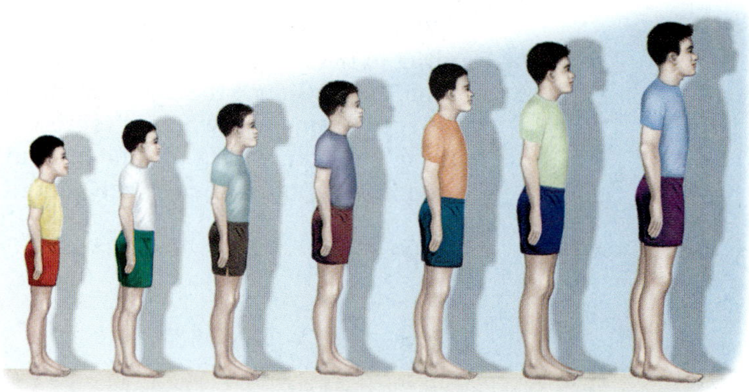

Can this possibly be my little cousin whom I haven't seen in five years? He's changed from a kid to an attractive young adult. I know that things change over time and that most changes occur at uneven rates, but this is a radical transformation.

Figure 2.23 shows the graph of your cousin's height, in inches, as a function of his age, in years. Two points on the graph are labeled: (13, 57) and (18, 76). At age 13, your cousin was 57 inches tall, and at age 18, he was 76 inches tall.

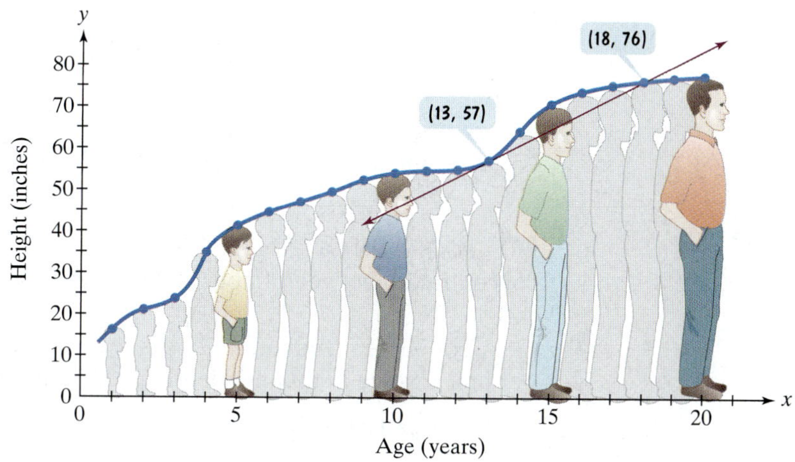

Figure 2.23 Height as a function of age

If the graph of a function is not a straight line, the **average rate of change** between any two points is the slope of the line containing the two points. The slope of the line in Figure 2.23 is

$$m = \frac{\text{Change in } y}{\text{Change in } x} = \frac{76 - 57}{18 - 13} = \frac{19}{5} = 3\frac{4}{5}.$$

Your cousin's average rate of change, or average growth rate, from 13 to 18 was $3\frac{4}{5}$ inches per year.

For any function, $y = f(x)$, the slope of the line between any two points is the average change in y per unit change in x.

EXAMPLE 6 Slope as Average Rate of Change

When a person receives a drug injected into a muscle, the concentration of the drug in the body, measured in milligrams per 100 milliliters, is a function of the time elapsed since the injection, measured in hours. Figure 2.24 shows the graph of such a function, where x represents hours after the injection and $f(x)$ is the drug's concentration at time x. Find the average rate of change in the drug's concentration between 3 and 7 hours.

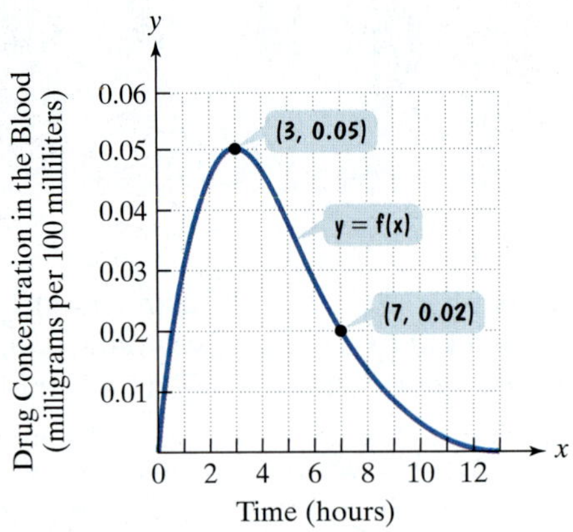

Figure 2.24 Concentration of a drug as a function of time

Solution At 3 hours, the drug's concentration is 0.05 and at 7 hours, the concentration is 0.02. The average rate of change in its concentration between 3 and 7 hours is the slope of the line connecting the points $(3, 0.05)$ and $(7, 0.02)$.

$$m = \frac{\text{Change in } y}{\text{Change in } x} = \frac{0.02 - 0.05}{7 - 3} = \frac{-0.03}{4} = -0.0075$$

The average rate of change is -0.0075. This means that the drug's concentration is decreasing at an average rate of 0.0075 milligrams per 100 milliliters per hour. ■

✔ **CHECK POINT 6** Use Figure 2.24 to find the average change in the drug's concentration between 1 hour and 3 hours.

For a linear function, slope is the rate of change of the dependent variable per unit change in the independent variable. This is illustrated in Table 2.3.

Table 2.3 Interpreting Slope

Linear Function	What the Function Models	Interpretation of Slope
$W = 0.3x + 46.9$ (Slope)	Average number of hours, W, Americans work per week x years after 1980	The rate of change of hours worked per week is 0.3 hour per year. Each year, hours worked per week increase by 0.3.
$y = -2.39x + 254.47$ (Slope)	Child mortality, y, per thousand, for children under five in a country where x percent of adult females are literate	The rate of change in child mortality is -2.39 per unit change in literacy. For each 1% increase in literacy, child mortality decreases by 2.39 per thousand.

ENRICHMENT ESSAY

How Calculus Studies Change

Calculus allows motion and change to be analyzed by "freezing the frame" of a continuous changing process, instant by instant. For example, Figure 2.25 shows a male's changing height over intervals of time. Over the period of time from P to D, his average rate of growth is his change in height—that is, his height at time D minus his height at time P—divided by the change in time from P to D. This is the slope of line PD.

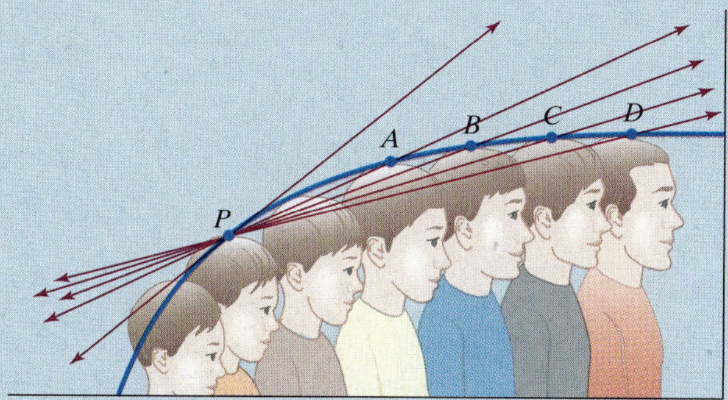

Figure 2.25 Analyzing continuous growth over intervals of time and at an instant in time

The lines PD, PC, PB, and PA shown in Figure 2.25 have slopes that show average growth rates for successively shorter periods of time. Calculus makes these time frames so small that they approach a single point—that is, a single instant in time. This point is shown as point P in Figure 2.25. The slope of the line that touches the graph at P gives the male's growth rate at one instant in time, P.

EXERCISE SET 2.3

Practice Exercises

In Exercises 1–14, use intercepts and a checkpoint to graph each linear function.

1. $x + y = 5$
2. $x + y = 3$
3. $3x + y = 6$
4. $x + 2y = 4$
5. $2x - 6y = 12$
6. $9x - 6y = 18$
7. $-x + 3y = 6$
8. $-x + 4y = 8$
9. $3x - y = 9$
10. $2x - y = 6$
11. $3x = 2y + 6$
12. $5x = 3y - 15$
13. $6x - 3y = 15$
14. $8x - 2y = 12$

In Exercises 15–24, find the slope of the line passing through each pair of points or state that the slope is undefined. Then indicate whether the line through the points rises, falls, is horizontal, or is vertical.

15. $(4, 7)$ and $(8, 10)$
16. $(2, 1)$ and $(3, 4)$
17. $(-2, 1)$ and $(2, 2)$
18. $(-1, 3)$ and $(2, 4)$
19. $(4, -2)$ and $(3, -2)$
20. $(4, -1)$ and $(3, -1)$
21. $(-2, 4)$ and $(-1, -1)$
22. $(6, -4)$ and $(4, -2)$
23. $(5, 3)$ and $(5, -2)$
24. $(3, -4)$ and $(3, 5)$

In Exercises 15–34, find the slope and the y-intercept. Then graph the equation.

25. $y = 2x + 1$
26. $y = 3x + 2$
27. $y = -2x + 1$
28. $y = -3x + 2$
29. $f(x) = \frac{3}{4}x - 2$
30. $f(x) = \frac{3}{4}x - 3$
31. $f(x) = -\frac{3}{5}x + 7$
32. $f(x) = -\frac{2}{5}x + 6$
33. $y = -\frac{1}{2}x$
34. $y = -\frac{1}{3}x$

In Exercises 35–42,

a. *Rewrite the given equation in slope-intercept form by solving for y.*
b. *Give the slope and y-intercept.*
c. *Use the slope and y-intercept to graph the linear function.*

35. $2x + y = 0$
36. $3x + y = 0$

37. $4y = 5x$ **38.** $3y = 4x$

39. $3x + y = 4$ **40.** $2x + y = 3$

41. $5x + 3y = 15$ **42.** $7x + 2y = 14$

In Exercises 43–56, graph each equation in the rectangular coordinate system.

43. $y = 3$
44. $y = 5$
45. $y = -2$
46. $y = -4$
47. $3y = 18$
48. $5y = -30$
49. $f(x) = 2$
50. $f(x) = 1$
51. $x = 5$
52. $x = 4$
53. $3x = -12$
54. $4x = -12$
55. $x = 0$
56. $y = 0$

Application Exercises

In 2000, world population was approximately 6 billion. The figure at the top of the next column shows that world population is expected to reach somewhere between 7.3 billion and 10.7 billion by 2050. Use this information to solve Exercises 57–58.

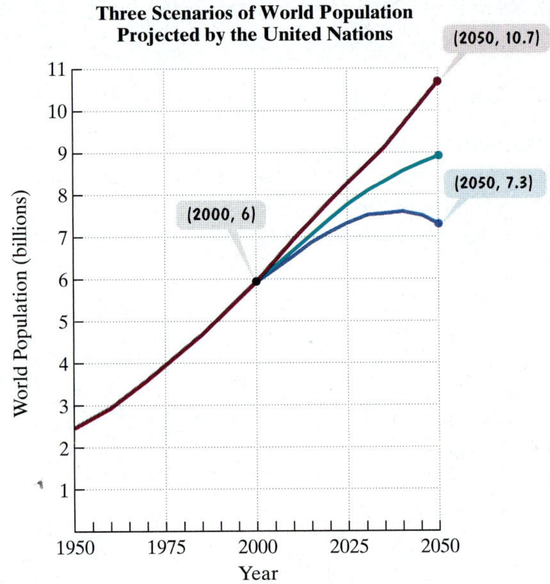

Source: United Nations

57. Use the points (2000, 6) and (2050, 10.7) in the rapid-growth scenario to find the projected average rate of change of population per year from 2000 projected through 2050.

58. Use the points (2000, 6) and (2050, 7.3) in the slow-growth scenario to find the projected average rate of change of population per year from 2000 through 2050.

The graph shows annual cigarette consumption per U.S. adult as a function of time from 1925 through 2000. In Exercises 59–64, find the average rate of change in cigarette consumption per year for each time period.

Cigarette Consumption per U.S. Adult

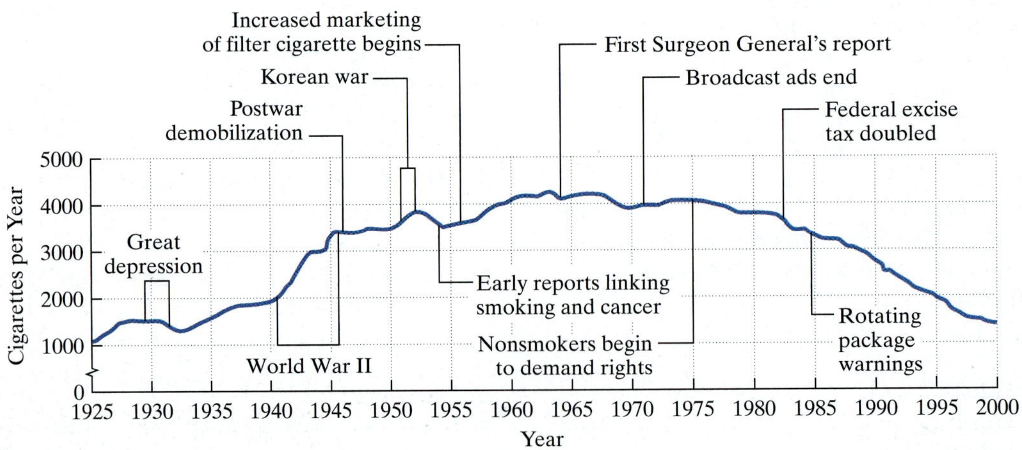

Source: U.S. Department of Health and Human Services

59. from 1975 through 2000; Use the points (1975, 4100) and (2000, 1400).

60. from 1980 through 2000; Use the points (1980, 3800) and (2000, 1400).

Solve Exercises 61–64 by estimating coordinates of appropriate points on the function's graph.

61. from 1940 through 1996

62. from 1935 through 2000

63. from 1940 through 1945

64. from 1955 through 1960

65. The scatter plot shows that from 1985 to 2000, the number of Americans participating in downhill skiing has remained relatively constant. Write an equation that models the number of participants in downhill skiing, y, in millions, for this period.

Number of U.S. Participants in Downhill Skiing

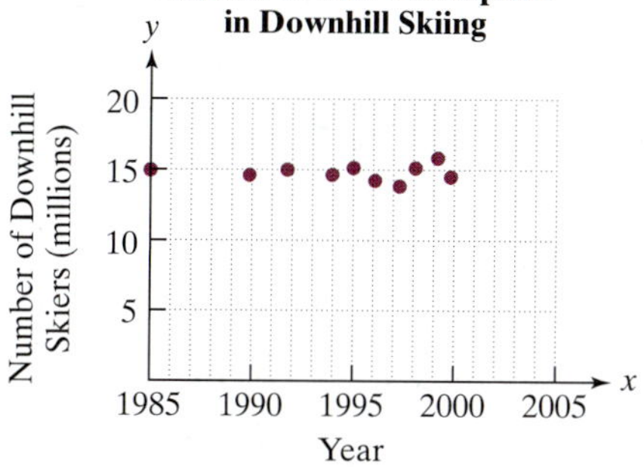

Source: National Ski Areas Association

The graphs of functions f and g should help you to get a reading on book sales in the United States. Use the information in the graphs to solve Exercises 66–67.

Dollar Sales and Unit Sales for Books Sold in the U.S.

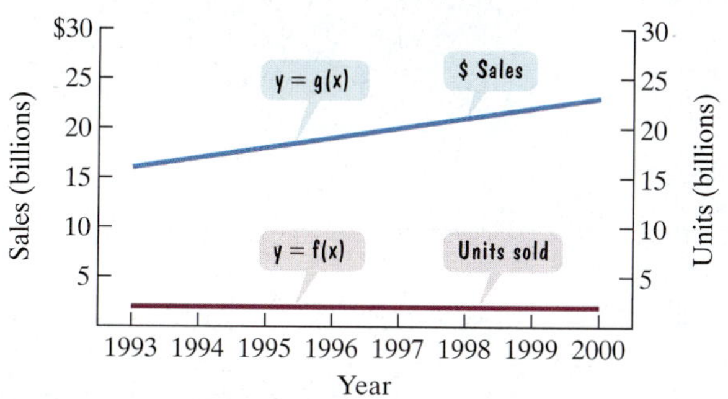

Source: Book Industry Study Group

66. The units, or number of books, sold from 1993 to 2000 remained constant. Write an equation that models the units sold, y, in billions, for this period.

67. The dollar sales from books steadily increased from 1993 through 2000. Approximate the rate of increase in sales per year, in billions of dollars, for this period.

In Exercises 68–71, a linear function that models data is described. Find the slope of each model. Then describe what this means in terms of the rate of change of the dependent variable per unit change in the independent variable.

68. The linear function $y = -0.4x + 38$ models the percentage of U.S. men, y, smoking cigarettes x years after 1980.

69. The linear function $y = 2x + 24$ models the average cost in dollars of a retail drug prescription in the United States, y, x years after 1991.

70. The linear function $f(x) = 0.01x + 57.7$ models the global average temperature of Earth, $f(x)$, in degrees Fahrenheit, x years after 1995.

71. The linear function $f(x) = -0.7x + 60$ models the percentage of U.S. adults, $f(x)$, who read a newspaper x years after 1995.

Writing in Mathematics

72. What is a scatter plot?

73. What is a regression line?

74. What is the standard form of the equation of a line?

75. What is an x-intercept of a graph?

76. What is a y-intercept of a graph?

77. If you are given the standard form of the equation of a line, explain how to find the x-intercept.

78. If you are given the standard form of the equation of a line, explain how to find the y-intercept.

79. What is the slope of a line?

80. Describe how to calculate the slope of a line passing through two points.

81. What does it mean if the slope of a line is zero?

82. What does it mean if the slope of a line is undefined?

83. Describe how to find the slope of a line whose equation is given.

84. Describe how to graph a line using the slope and y-intercept. Provide an original example with your description.

85. Describe the graph of $y = b$.

86. Describe the graph of $x = a$.

87. We saw that the percentage of people satisfied with their lives remains relatively constant for all age groups. Exercise 65 showed that the number of skiers in the United States has remained relatively constant over time. Give another example of a real-world phenomenon that has remained relatively constant. Try writing an equation that models this phenomenon.

88. A linear function in the form $f(x) = mx + b$ models the percentage of high school seniors in the United States, $f(x)$, who admit to using marijuana, where x is the number of years since 1992. Would you expect m to be positive, negative, or zero? Explain your answer.

Technology Exercises

89. Use a graphing utility to verify any three of your hand-drawn graphs in Exercises 1–14. Solve the equation for y before entering it.

In Exercises 90–93, use a graphing utility to graph each linear function. Then use the TRACE *feature to trace along the line and find the coordinates of two points. Use these points to compute the line's slope. Check your result by using the coefficient of x in the line's equation.*

90. $y = 2x + 4$

91. $y = -3x + 6$

92. $f(x) = -\frac{1}{2}x - 5$

93. $f(x) = \frac{3}{4}x - 2$

Critical Thinking Exercises

94. Which one of the following is true?

a. A linear function with nonnegative slope has a graph that rises from left to right.

b. Every line in the rectangular coordinate system has an equation that can be expressed in slope-intercept form.

c. The graph of the linear function $5x + 6y = 30$ is a line passing through the point $(6, 0)$ with slope $-\frac{5}{6}$.

d. The graph of $x = 7$ in the rectangular coordinate system is the single point $(7, 0)$.

In Exercises 95–96, find the coefficients that must be placed in each shaded area so that the function's graph will be a line satisfying the specified conditions.

95. $\square x + \square y = 12$; x-intercept $= -2$; y-intercept $= 4$

96. $\square x + \square y = 12$; y-intercept $= -6$; slope $= \frac{1}{2}$

97. Use the graphs of functions f and g in Exercises 66–67 to find the rate of increase in the average price of a book per year from 1993 through 2000.

Review Exercises

98. Simplify: $\left(\frac{4x^2}{y^{-3}}\right)^2$. (Section 1.6, Example 9)

99. Multiply and write the answer in scientific notation:

$$(8 \times 10^{-7})(4 \times 10^3).$$

(Section 1.7, Example 4)

100. Simplify: $5 - [3(x - 4) - 6x]$. (Section 1.2, Example 14)

▸ SECTION 2.4 The Point-Slope Form of the Equation of a Line

Objectives

1. Use the point-slope form to write equations of a line.
2. Model data with linear functions and make predictions.
3. Find slopes and equations of parallel and perpendicular lines.

SSM PH Tutor Center CD-ROM Video

If present trends continue, is it possible that our descendants could live to be 200 years of age? To answer this question, we need to develop a function that models life expectancy by birth year. In this section, you will learn to use another form of a line's equation to obtain functions that model data.

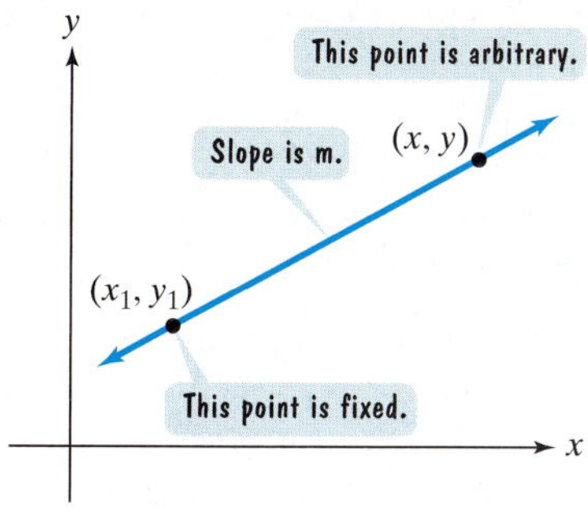

Figure 2.26 A line passing through (x_1, y_1) with slope m

Point-Slope Form

We can use the slope of a line to obtain another useful form of the line's equation. Consider a nonvertical line that has a slope of m and contains the point (x_1, y_1). Now, let (x, y) represent any other point on the line, shown in Figure 2.26. Keep in mind that the point (x, y) is arbitrary and is not in one fixed position. By contrast, the point (x_1, y_1) is fixed.

Regardless of where the point (x, y) is located, the steepness of the line in Figure 2.26 remains the same. Thus, the ratio for slope stays a constant m. This means that for all points along the line,

$$m = \frac{\text{Change in } y}{\text{Change in } x} = \frac{y - y_1}{x - x_1}.$$

We can clear the fraction by multiplying both sides by $x - x_1$, the least common denominator.

$$m = \frac{y - y_1}{x - x_1}$$ This is the slope of the line in Figure 2.26.

$$m(x - x_1) = \frac{y - y_1}{x - x_1} \cdot x - x_1$$ Multiply both sides by $x - x_1$.

$$m(x - x_1) = y - y_1$$ Simplify: $\frac{y - y_1}{\cancel{x - x_1}} \cdot \cancel{x - x_1} = y - y_1$.

Now, if we reverse the two sides, we obtain the **point-slope form** of the equation of a line.

Point-Slope Form of the Equation of a Line

The **point-slope equation** of a nonvertical line with slope m that passes through the point (x_1, y_1) is

$$y - y_1 = m(x - x_1).$$

For example, the point-slope equation of the line passing through $(1, 5)$ with a slope of 2 $(m = 2)$ is

$$y - 5 = 2(x - 1).$$

1 Use the point-slope form to write equations of a line.

Using the Point-Slope Form to Write a Line's Equation

If we know the slope of a line and a point through which the line passes, the point-slope form is the equation that we should use. Once we have obtained this equation, it is customary to solve for y and write the equation in slope-intercept form. Examples 1 and 2 illustrate these ideas.

EXAMPLE 1 Writing the Point-Slope Form and the Slope-Intercept Form

Write the point-slope form and the slope-intercept form of the equation of the line with slope 4 that passes through the point $(-1, 3)$.

Solution We begin with the point-slope equation of a line with $m = 4$, $x_1 = -1$, and $y_1 = 3$.

$$y - y_1 = m(x - x_1)$$ This is the point-slope form of the equation.

$$y - 3 = 4[x - (-1)]$$ Substitute the given values.

$$y - 3 = 4(x + 1)$$ We now have the point-slope form of the equation for the given line.

Now we solve this equation for y and write an equivalent equation in slope-intercept form ($y = mx + b$).

$$y - 3 = 4(x + 1)$$ This is the point-slope equation.

$$y - 3 = 4x + 4$$ Use the distributive property.

$$y = 4x + 7$$ Add 3 to both sides.

The slope-intercept form of the line's equation is $y = 4x + 7$. Using function notation, the equation is $f(x) = 4x + 7$. ■

✔ **CHECK POINT 1** Write the point-slope form and the slope-intercept form of the equation of the line with slope 6 that passes through the point $(2, -5)$.

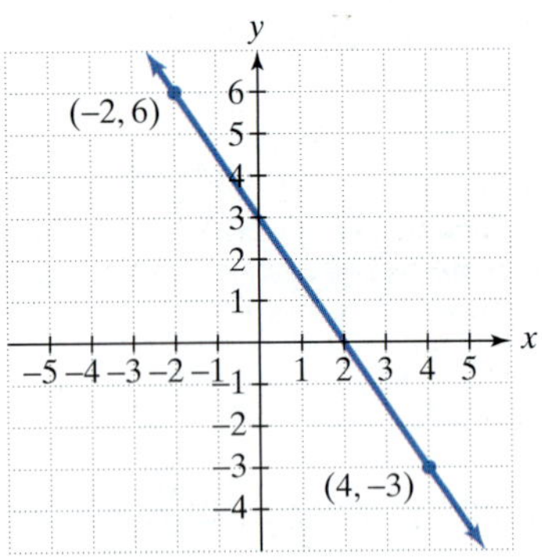

Figure 2.27 Line passing through $(4, -3)$ and $(-2, 6)$

EXAMPLE 2 Writing the Point-Slope Form and the Slope-Intercept Form

A line passes through the points $(4, -3)$ and $(-2, 6)$. (See Figure 2.27). Find an equation of the line:

a. in point-slope form

b. in slope-intercept form.

Solution

a. To use the point-slope form, we need to find the slope. The slope is the change in the y-coordinates divided by the corresponding change in the x-coordinates.

$$m = \frac{6 - (-3)}{-2 - 4} = \frac{9}{-6} = -\frac{3}{2}$$ This is the definition of slope using $(4, -3)$ and $(-2, 6)$.

We can take either point on the line to be (x_1, y_1). Let's use $(x_1, y_1) = (4, -3)$. Now, we are ready to write the point-slope equation.

$$y - y_1 = m(x - x_1)$$ This is the point-slope form of the equation.

$$y - (-3) = -\frac{3}{2}(x - 4)$$ Substitute: $(x_1, y_1) = (4, -3)$, and $m = -\frac{3}{2}$.

$$y + 3 = -\frac{3}{2}(x - 4)$$ Simplify.

This equation is the point-slope form of the equation of the line shown in Figure 2.27.

b. Now, we solve this equation for y and write an equivalent equation in slope-intercept form ($y = mx + b$).

$$y + 3 = -\frac{3}{2}(x - 4)$$ This is the point-slope equation.

$$y + 3 = -\frac{3}{2}x + 6$$ Use the distributive property.

$$y = -\frac{3}{2}x + 3$$ Subtract 3 from both sides.

This equation is the slope-intercept form of the equation of the line shown in Figure 2.27. Using function notation, the equation is $f(x) = -\frac{3}{2}x + 3$. ■

Discover for Yourself

If you are given two points on a line, you can use either point for (x_1, y_1) when you write its point-slope equation. Rework Example 2 using $(-2, 6)$ for (x_1, y_1). Once you solve for y, you should obtain the same slope-intercept equation as the one shown in the last line of the solution to Example 2.

CHECK POINT 2 A line passes through the points $(-2, -1)$ and $(-1, -6)$. Find an equation of the line

a. in point-slope form

b. in slope-intercept form

Here is a summary of the various forms for equations of lines:

Equations of Lines

1. Standard form:	$Ax + By = C$
2. Slope-intercept form:	$y = mx + b$ or $f(x) = mx + b$
3. Horizontal line:	$y = b$
4. Vertical line:	$x = a$
5. Point-slope form:	$y - y_1 = m(x - x_1)$

In Examples 1 and 2, we eventually write a line's equation in slope-intercept form, or in function notation. But where do we start our work?

Starting with $y = mx + b$	Starting with $y - y_1 = m(x - x_1)$
Begin with the slope-intercept form if you know: 1. The slope of the line and the y-intercept.	Begin with the point-slope form if you know: 1. The slope of the line and a point on the line or 2. Two points on the line.

2 Model data with linear functions and make predictions.

Applications

Linear functions are useful for modeling data that fall on or near a line. For example, Table 2.4 gives the life expectancy for an American born in the indicated year. The data are displayed in the scatter plot in Figure 2.28.

Table 2.4 Life Expectancy

Year of Birth	x (Years after 1980)	y (Life Expectancy)
1980	0	73.7
1985	5	74.7
1990	10	75.4
1993	13	75.5
1996	16	76.1
1997	17	76.5
1998	18	76.7

Source: U.S. Bureau of the Census

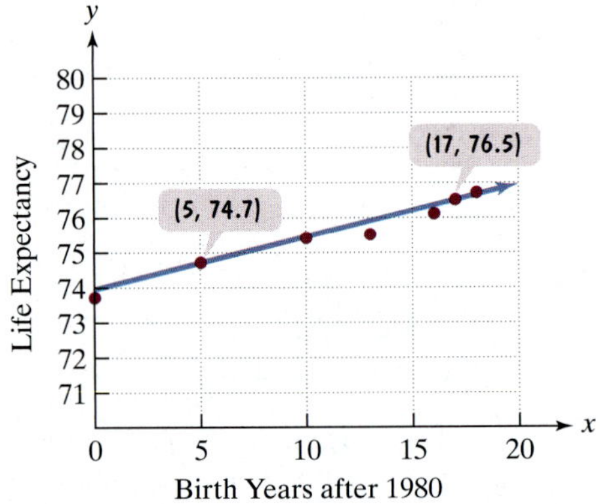

Figure 2.28

Also birth years shown in Figure 2.28 is a line that passes through or near the seven points. By writing the equation of this line, we can obtain a model for life expectancy and make predictions about how long Americans will live in the future.

Using Technology

You can use a graphing utility to obtain a model for a scatter plot in which the data points fall on or near a straight line. After entering the data in Table 2.4, a graphing utility displays a scatter plot of the data and the regression line, that is, the line that best fits the data.

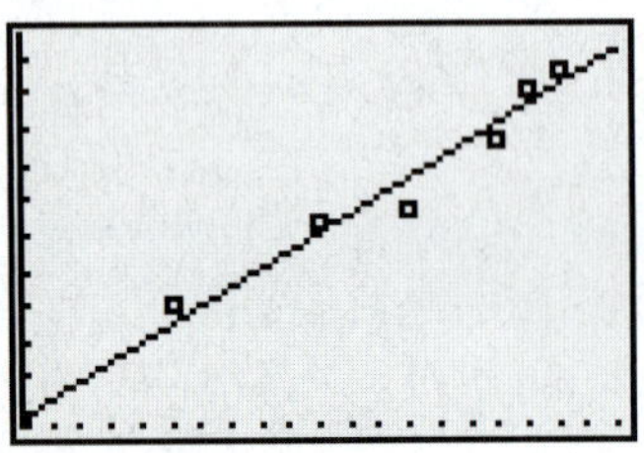

[0, 20, 1] by [73.7, 77.0, 0.3]

Also displayed is the regression line's equation.

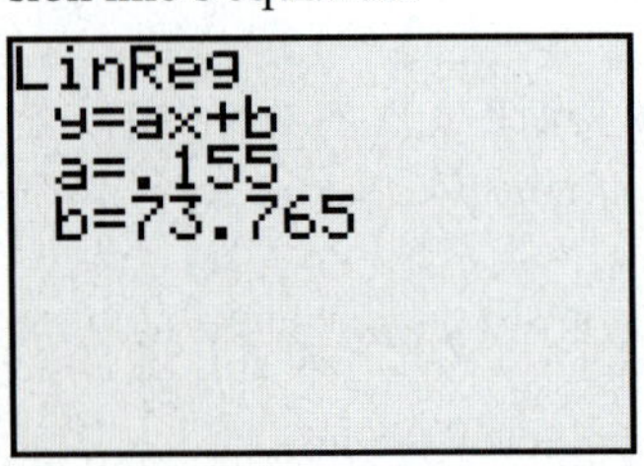

EXAMPLE 3 Modeling Life Expectancy

Write the slope-intercept equation of the line shown in Figure 2.28. Use the equation to predict the life expectancy of an American born in 2020.

Solution The line in Figure 2.28 passes through (5, 74.7) and (17, 76.5). We start by finding the slope.

$$m = \frac{\text{Change in } y}{\text{Change in } x} = \frac{76.5 - 74.7}{17 - 5} = \frac{1.8}{12} = 0.15$$

Now we write the line's slope-intercept equation.

$y - y_1 = m(x - x_1)$ — Begin with the point-slope form.

$y - 74.7 = 0.15(x - 5)$ — Either ordered pair can be (x_1, y_1). Let $(x_1, y_1) = (5, 74.7)$. From above, $m = 0.15$.

$y - 74.7 = 0.15x - 0.75$ — Apply the distributive property on the right side.

$y = 0.15x + 73.95$ — Add 74.7 to both sides and solve for y.

A linear function that models life expectancy for Americans born x years after 1980 is

$$f(x) = 0.15x + 73.95.$$

Now let's use this function to predict the life expectancy of an American born in 2020. Because 2020 is 40 years after 1980, substitute 40 for x and evaluate the function at 40.

$$f(40) = 0.15(40) + 73.95 = 79.95$$

Our model predicts that Americans born in 2020 will have a life expectancy of 79.95, or approximately 80 years. ■

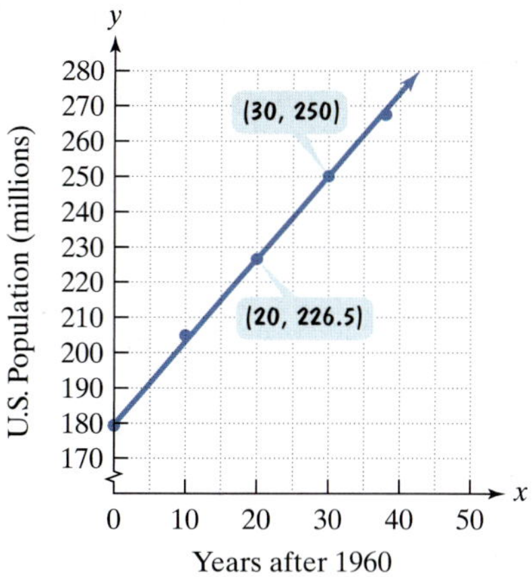

Figure 2.29

✔ **CHECK POINT 3** The five points in the scatter plot in Figure 2.29 show the population of the United States, in millions, for five selected years. Write the slope-intercept equation of the line shown in the figure. Then use the linear function to predict U.S. population in 2010.

Parallel and Perpendicular Lines

Two nonintersecting lines that lie in the same plane are **parallel**. If two lines do not intersect, the ratio of the vertical change to the horizontal change is the same for each line. Because two parallel lines have the same "steepness," they must have the same slope.

Slope and Parallel Lines

1. If two nonvertical lines are parallel, then they have the same slope.
2. If two distinct nonvertical lines have the same slope, then they are parallel.
3. Two distinct vertical lines, both with undefined slopes, are parallel.

3 Find slopes and equations of parallel and perpendicular lines.

EXAMPLE 4 Writing Equations of a Line Parallel to a Given Line

Write an equation of the line passing through $(-3, 2)$ and parallel to the line whose equation is $y = 2x + 1$. Express the equation in point-slope form and slope-intercept form.

Solution The situation is illustrated in Figure 2.30. We are looking for the equation of the line shown on the left. How do we obtain this equation? Notice that the line passes through the point $(-3, 2)$. Using the point-slope form of the line's equation, we have $x_1 = -3$ and $y_1 = 2$.

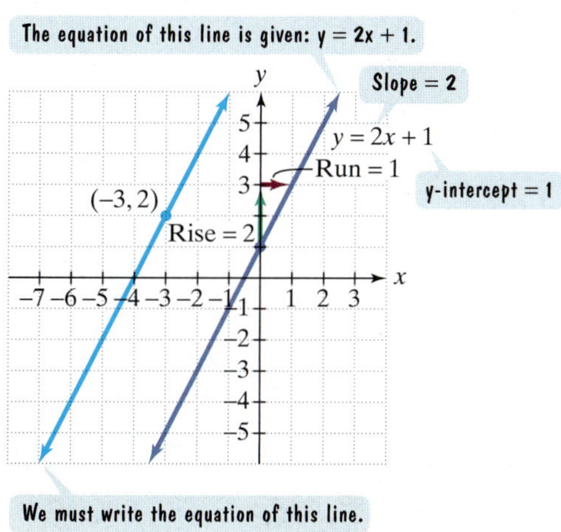

Figure 2.30

$$y - y_1 = m(x - x_1)$$

$y_1 = 2$ $\quad$ $x_1 = -3$

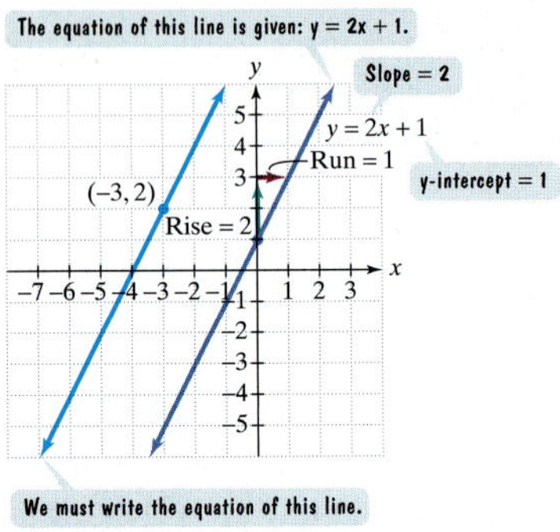

Figure 2.30, repeated

Now the only thing missing from the equation is m, the slope of the line on the left. Do we know anything about the slope of either line in Figure 2.30? The answer is yes; we know the slope of the line on the right, whose equation is given.

$$y = 2x + 1$$

The slope of the line on the right in Figure 2.30 is 2.

Parallel lines have the same slope. Because the slope of the line with the given equation is 2, $m = 2$ for the line whose equation we must write.

$$y - y_1 = m(x - x_1)$$

$y_1 = 2$ $m = 2$ $x_1 = -3$

The point-slope form of the line's equation is

$$y - 2 = 2[x - (-3)] \text{ or}$$
$$y - 2 = 2(x + 3).$$

Solving for y, we obtain the slope-intercept form of the equation.

$y - 2 = 2x + 6$ Apply the distributive property.

$y = 2x + 8$ Add 2 to both sides. This is the slope-intercept form, $y = mx + b$, of the equation. Using function notation, the equation is $f(x) = 2x + 8$. ■

✔ **CHECK POINT 4** Write an equation of the line passing through $(-2, 5)$ and parallel to the line whose equation is $y = 3x + 1$. Express the equation in point-slope form and slope-intercept form.

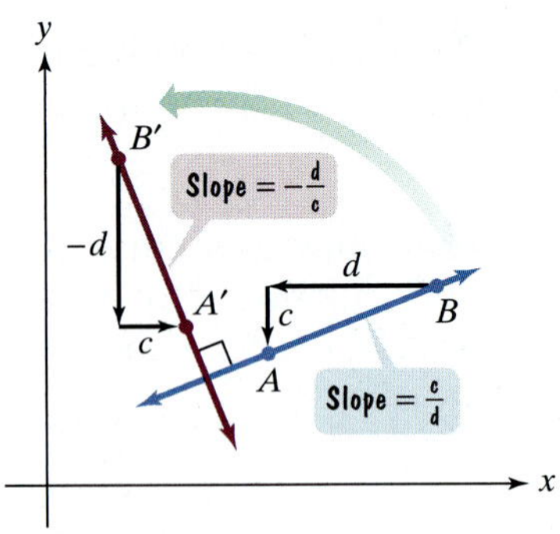

Figure 2.31 Slopes of perpendicular lines

Two lines that intersect at a right angle (90°) are said to be **perpendicular**, shown in Figure 2.31. The relationship between the slopes of perpendicular lines is not as obvious as the relationship between parallel lines. Figure 2.31 shows line AB, with a slope of $\frac{c}{d}$. Rotate line AB through 90° to the left to obtain line $A'B'$, perpendicular to line AB. The figure indicates that the rise and the run of the new line are reversed from the original line, but the rise is now negative. This means that the slope of the new line is $-\frac{d}{c}$. Notice that the product of the slopes of the two perpendicular lines is -1:

$$\left(\frac{c}{d}\right)\left(-\frac{d}{c}\right) = -1.$$

This relationship holds for all perpendicular lines and is summarized in the following box.

Slope and Perpendicular Lines

1. If two nonvertical lines are perpendicular, then the product of their slopes is -1.
2. If the product of the slopes of two lines is -1, then the lines are perpendicular.
3. A horizontal line having zero slope is perpendicular to a vertical line having undefined slope.

An equivalent way of stating this relationship is to say that one line is perpendicular to another line if its slope is the *negative reciprocal* of the slope of the other. For example, if a line has slope 5, any line having slope $-\frac{1}{5}$ is perpendicular to it. Similarly, if a line has slope $-\frac{3}{4}$, any line having slope $\frac{4}{3}$ is perpendicular to it.

EXAMPLE 5 Finding the Slope of a Line Perpendicular to a Given Line

Find the slope of any line that is perpendicular to the line whose equation is $x + 4y - 8 = 0$.

Solution We begin by writing the equation of the given line in slope-intercept form. Solve for y.

$x + 4y - 8 = 0$ This is the given equation.

$4y = -x + 8$ To isolate the y-term, subtract x and add 8 on both sides.

$y = -\frac{1}{4}x + 2$ Divide both sides by 4.

Slope is $-\frac{1}{4}$.

The given line has slope $-\frac{1}{4}$. Any line perpendicular to this line has a slope that is the negative reciprocal of $-\frac{1}{4}$. Thus, the slope of any perpendicular line is 4. ■

CHECK POINT 5 Find the slope of any line that is perpendicular to the line whose equation is $x + 3y - 12 = 0$.

EXERCISE SET 2.4

Practice Exercises

Write the point-slope form of the line satisfying each of the conditions in Exercises 1–28. Then use the point-slope form of the equation to write the slope-intercept form of the equation in function notation.

1. Slope $= 2$, passing through $(3, 5)$

2. Slope $= 4$, passing through $(1, 3)$

3. Slope $= 6$, passing through $(-2, 5)$

4. Slope $= 8$, passing through $(4, -1)$

5. Slope $= -3$, passing through $(-2, -3)$

6. Slope $= -5$, passing through $(-4, -2)$

7. Slope $= -4$, passing through $(-4, 0)$

8. Slope $= -2$, passing through $(0, -3)$

9. Slope $= -1$, passing through $\left(-\frac{1}{2}, -2\right)$

10. Slope $= -1$, passing through $\left(-4, -\frac{1}{4}\right)$

11. Slope $= \frac{1}{2}$, passing through the origin

12. Slope $= \frac{1}{3}$, passing through the origin

13. Slope $= -\frac{2}{3}$, passing through $(6, -2)$

14. Slope $= -\frac{3}{5}$, passing through $(10, -4)$

15. Passing through $(1, 2)$ and $(5, 10)$

16. Passing through $(3, 5)$ and $(8, 15)$

17. Passing through $(-3, 0)$ and $(0, 3)$

18. Passing through $(-2, 0)$ and $(0, 2)$

19. Passing through $(-3, -1)$ and $(2, 4)$

20. Passing through $(-2, -4)$ and $(1, -1)$

21. Passing through $(-3, -2)$ and $(3, 6)$

22. Passing through $(-3, 6)$ and $(3, -2)$

23. Passing through $(-3, -1)$ and $(4, -1)$

24. Passing through $(-2, -5)$ and $(6, -5)$

25. Passing through $(2, 4)$ with x-intercept $= -2$

26. Passing through $(1, -3)$ with x-intercept $= -1$

27. x-intercept $= -\frac{1}{2}$ and y-intercept $= 4$

28. x-intercept $= 4$ and y-intercept $= -2$

In Exercises 29–44, the equation of a line is given. Find the slope of a line that is (a) parallel to the line with the given equation; and (b) perpendicular to the line with the given equation.

29. $y = 5x$

30. $y = 3x$

31. $y = -7x$

32. $y = -9x$

33. $y = \frac{1}{2}x + 3$

34. $y = \frac{1}{4}x - 5$

35. $y = -\frac{2}{5}x - 1$

36. $y = -\frac{3}{7}x - 2$

37. $4x + y = 7$

38. $8x + y = 11$

39. $2x + 4y - 8 = 0$

40. $3x + 2y - 6 = 0$

41. $2x - 3y - 5 = 0$

42. $3x - 4y + 7 = 0$

43. $x = 6$

44. $y = 9$

In Exercises 45–48, write an equation for line L in point-slope form and slope-intercept form.

45.

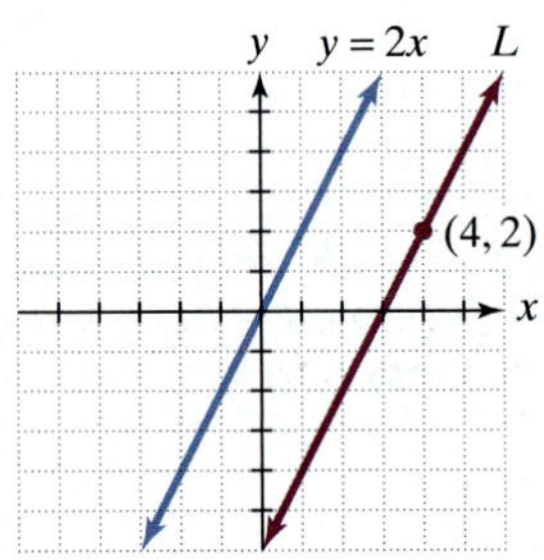

L is parallel to $y = 2x$.

46.

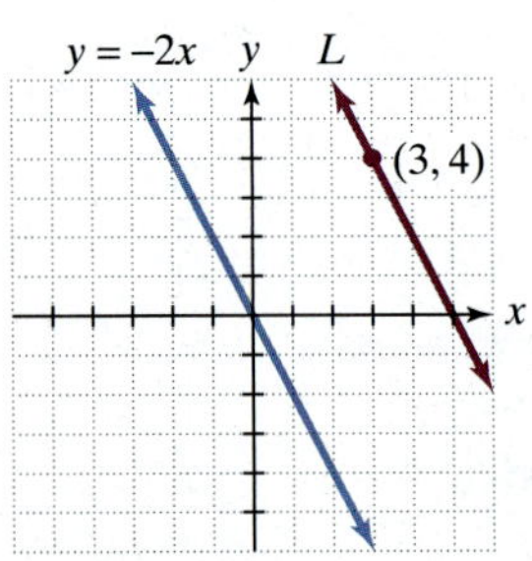

L is parallel to $y = -2x$.

47.

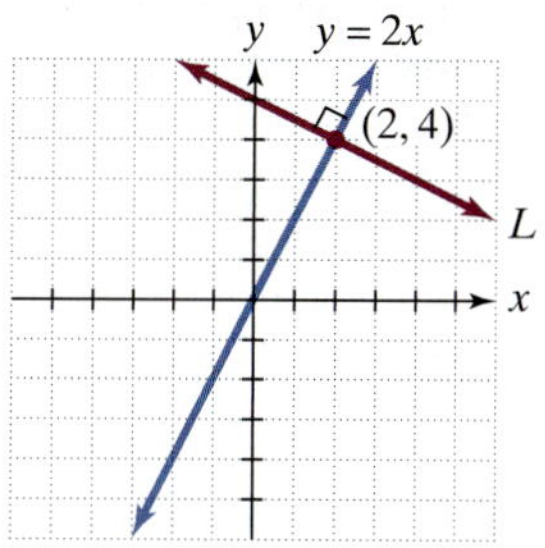

L is perpendicular to $y = 2x$.

48.

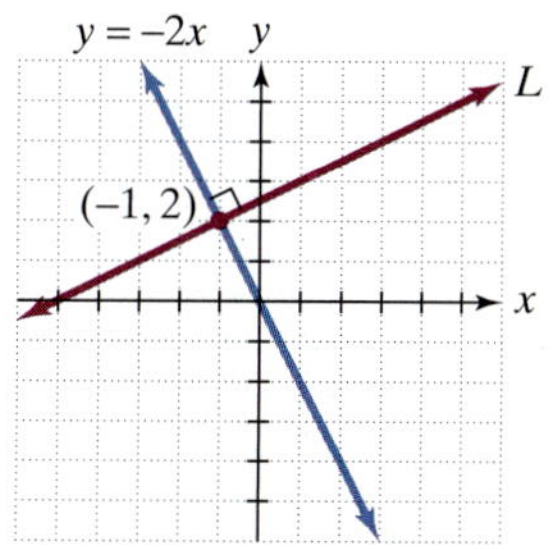

L is perpendicular to $y = -2x$.

In Exercises 49–56, use the given conditions to write an equation for each line in point-slope form and slope-intercept form.

49. Passing through $(-8, -10)$ and parallel to the line whose equation is $y = -4x + 3$

50. Passing through $(-2, -7)$ and parallel to the line whose equation is $y = -5x + 4$

51. Passing through $(2, -3)$ and perpendicular to the line whose equation is $y = \frac{1}{5}x + 6$

52. Passing through $(-4, 2)$ and perpendicular to the line whose equation is $y = \frac{1}{3}x + 7$

53. Passing through $(-2, 2)$ and parallel to the line whose equation is $2x - 3y - 7 = 0$

54. Passing through $(-1, 3)$ and parallel to the line whose equation is $3x - 2y - 5 = 0$

55. Passing through $(4, -7)$ and perpendicular to the line whose equation is $x - 2y - 3 = 0$

56. Passing through $(5, -9)$ and perpendicular to the line whose equation is $x + 7y - 12 = 0$

Application Exercises

57. There are approximately 170 million automobile drivers in the United States and more than 100 million cell phones—a sometimes dangerous combination. The scatter plot shows the number of cellular phone–related automobile accidents in North Carolina from 1994 through 2000. Also shown is a line that passes through or near the points.

North Carolina Cellular Phone-Related Automobile Accidents Per Year

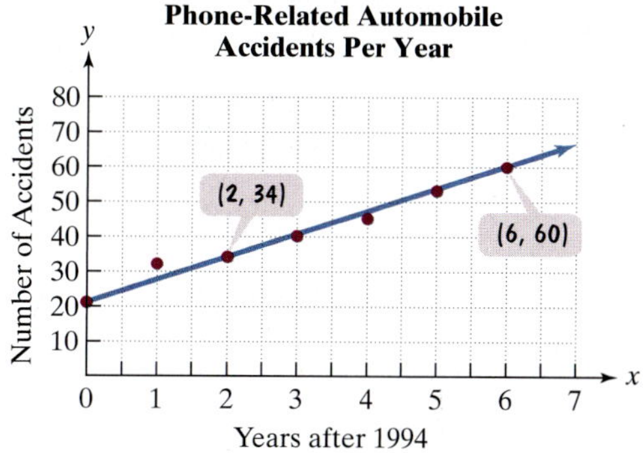

Source: U.S. Department of Transportation

a. Use the two points whose coordinates are shown by the voice balloons to find the point-slope equation of the line that models the number of cellular phone–related automobile accidents, y, in North Carolina x years after 1994.

b. Write the equation in part (a) in slope-intercept form. Use function notation.

c. Use the linear function to predict the number of cellular phone–related accidents in North Carolina in 2010.

58. Corporations in the United States are doing quite well, thank you. The scatter plot shows corporate profits, in billions of dollars, from 1990 through 2000. Also shown is a line that passes through or near the points.

Corporate Profits

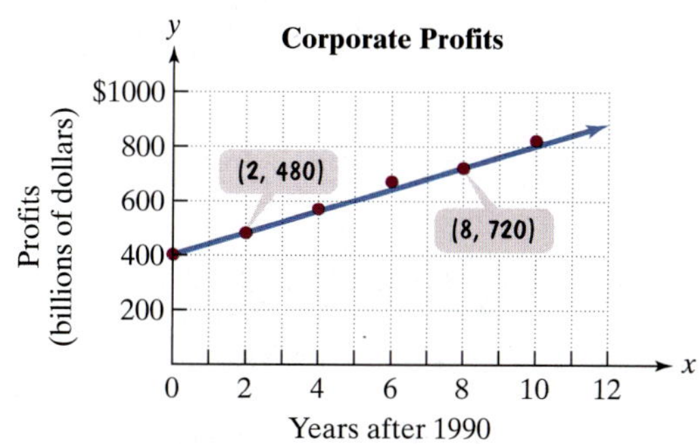

Source: U.S. Department of Labor

a. Use the two points whose coordinates are shown by the voice balloons on page 139 to find the point-slope equation of the line that models corporate profits, y, in billions of dollars, x years after 1990.

b. Write the equation in part (a) in slope-intercept form. Use function notation.

c. Use the linear function to predict corporate profits in 2010.

59. The graph shows the increasing U.S. prison population during the last two decades of the twentieth century. Let x represent the number of years after 1980 and let y represent the prison population, in millions. Estimate the coordinates of two points on the graph. Then use the two points to find a linear equation in point-slope form that models the data. Use your model to predict U.S. prison population in 2010.

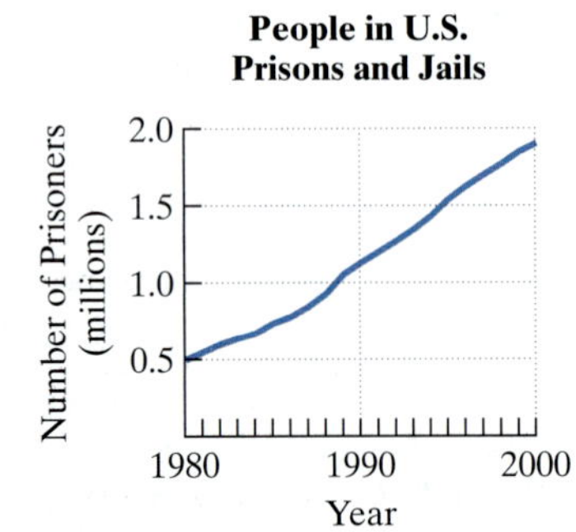

Source: Bureau of Justice Statistics

60. The graph shows the increase in the number of people in the United States using the Internet. Let x represent the number of years after 1997 and let y represent the number of Internet users, in millions. Estimate the coordinates of two points on the graph. Then use the two points to find a linear equation in point-slope form that models the data. Use your model to predict the number of Internet users in 2010.

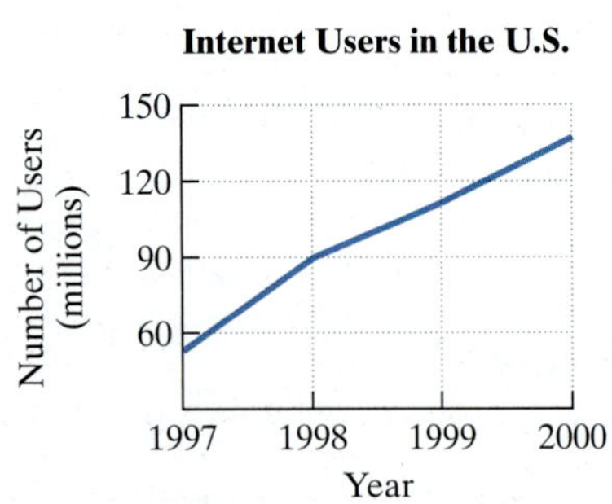

Source: Computer Industry Almanac

The data show the percentage of America's high school seniors who have ever used marijuana or cocaine. Use this information to solve Exercises 61–62.

	Class of 1995	Class of 1996	Class of 1997	Class of 1998	Class of 1999
Percentage of seniors who ever used marijuana	41.7	44.9	49.6	49.1	49.7
Percentage of seniors who ever used cocaine	6.0	7.1	8.7	9.3	9.8

Source: National Institute on Drug Abuse

61. a. Let x represent the number of years after 1995 and let y represent the percentage of seniors who used marijuana. Graph the five data points (0, 41.7), (1, 44.9), (2, 49.6), (3, 49.1), and (4, 49.7) and create a scatter plot for the percentage of seniors who have ever used marijuana.

b. Draw a line that fits the data so that the spread of the data points around the line is as small as possible.

c. Use the coordinates of two points along your line to write its point-slope and slope-intercept equations.

d. Use the equation in part (c) to predict the percentage of high school seniors who will try marijuana in the class of 2010.

62. a. Let x represent the number of years after 1995 and let y represent the percentage of seniors who used cocaine. Graph the five data points (0, 6.0), (1, 7.1), (2, 8.7), (3, 9.3), and (4, 9.8) and create a scatter plot for the percentage of seniors who have ever used cocaine.

b. Draw a line that fits the data so that the spread of the data points around the line is as small as possible.

c. Use the coordinates of two points along your line to write its point-slope and slope-intercept equations.

d. Use the equation in part (c) to predict the percentage of high school seniors who will try cocaine in the class of 2010.

63. The line graph shows the number of people in the United States projected to live alone through 2010 by gender.

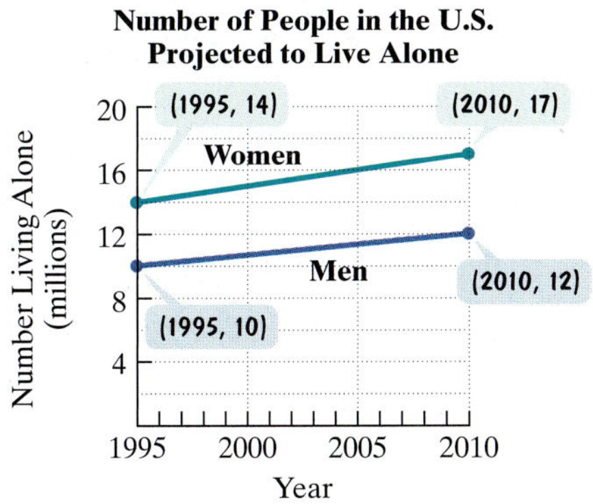

Source: Forrester Research

a. Find the slope of the line for U.S. women.
b. Find the slope of the line for U.S. men.
c. Are the lines parallel? What does this mean in terms of the yearly increase for women and the yearly increase for men?

64. Shown, again, is the scatter plot that indicates a relationship between the percentage of adult females in a country who are literate and the mortality of children under five. Also shown is a line that passes through or near the points. Find a function that models the data by finding the slope-intercept equation of the line. Use the function to make a prediction about child mortality based on the percentage of adult females in a country who are literate.

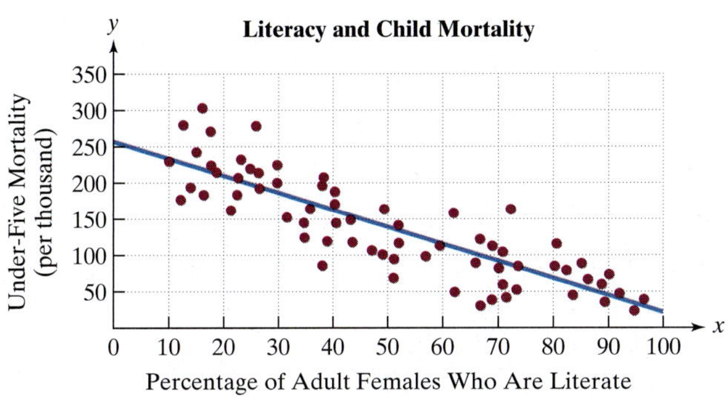

Source: United Nations

Writing in Mathematics

65. Describe how to write the equation of a line if its slope and a point along the line are known.

66. Describe how to write the equation of a line if two points along the line are known.

67. If two lines are parallel, describe the relationship between their slopes.

68. If two lines are perpendicular, describe the relationship between their slopes.

69. If you know a point on a line and you know the equation of a line parallel to this line, explain how to write the line's equation.

70. In Example 3 on page 134, we developed a model that predicted Americans born in 2020 will have a life expectancy of approximately 80 years. Describe something that might occur that would make this prediction inaccurate.

71. Take a second look at the scatter plot in Exercise 64. Although there is a relationship between literacy and child mortality, we cannot conclude that increased literacy causes child mortality to decrease. Offer two or more possible explanations for the data in the scatter plot.

Technology Exercises

72. The lines whose equations are $y = \frac{1}{3}x + 1$ and $y = -3x - 2$ are perpendicular because the product of their slopes, $\frac{1}{3}$ and -3, respectively, is -1.

a. Use a graphing utility to graph the equations in a $[-10, 10, 1]$ by $[-10, 10, 1]$ viewing rectangle. Do the lines appear to be perpendicular?
b. Now use the zoom square feature of your utility. Describe what happens to the graphs. Explain why this is so.

73. a. Use the statistical menu of your graphing utility to enter the five data points shown in Exercise 61.
b. Use the DRAW menu and the scatter plot capability to draw a scatter plot of the data points.
c. Select the linear regression option. Use your utility to obtain values for a and b for the equation of the regression line, $y = ax + b$. Compare this equation to the one that you obtained by hand in Exercise 61. You may also be given a **correlation coefficient**, r. Values of r close to 1 indicate that the points can be modeled by a linear function and the regression line has a positive slope. Values of r close to -1 indicate that the points can be modeled by a linear function and the regression line has a negative slope. Values of r close to 0 indicate no linear relationship between the variables. In this case, a linear model does not accurately describe the data.
d. Use the appropriate sequence (consult your manual) to graph the regression equation on top of the points in the scatter plot.

74. Repeat Exercise 73 using the five data points shown in Exercise 62.

Critical Thinking Exercises

In Exercises 75–76, use the given conditions to write an equation for each line in point-slope form and slope-intercept form.

75. Having an x-intercept of -3 and perpendicular to the line passing through $(0, 0)$ and $(6, -2)$

76. Perpendicular to $3x - 2y = 4$ with the same y-intercept

77. Prove that the equation of a line passing through $(a, 0)$ and $(0, b)$ $(a \neq 0, b \neq 0)$ can be written in the form $\frac{x}{a} + \frac{y}{b} = 1$. Why is this called the *intercept form* of a line?

Review Exercises

78. If $f(x) = 3x^2 - 8x + 5$, find $f(-2)$. (Section 2.1, Example 3)

79. If $f(x) = x^2 - 3x + 4$ and $g(x) = 2x - 5$, find $(fg)(-1)$. (Section 2.2, Example 4)

80. The sum of the angles of a triangle is 180°. Find the three angles of a triangle if one angle is 20° greater than the smallest angle and the third angle is twice the smallest angle. (Section 1.5, Example 1)

CHAPTER 2 GROUP PROJECTS

Group members should consult an almanac, newspaper, magazine, or the Internet and return to the group with as much data as possible for one of the following situations:

1. Life expectancy by birth year, gender, ethnicity, and other categories of interest
2. Median annual income by level of education, broken down by gender, ethnicity, and other categories of interest
3. Global warming
4. Population growth
5. Drug use, including alcohol and cigarettes, by America's high school seniors or by adults in the U.S.
6. Any other area of interest in which data lie approximately on or near a line

Working by hand or using a graphing utility, group members should construct scatter plots for the data that were assembled. If working by hand, draw a line that approximately fits the data in each scatter plot and then write its equation as a function in slope-intercept form. If using a graphing utility, obtain the equation of each regression line. Then use each linear function's equation to make predictions about what might occur in the future. Are there circumstances that might affect the accuracy of the prediction? List some of these circumstances.

7. Consult an almanac, newspaper, magazine, or the Internet to find data displayed in the style of Figure 2.10 on page 107 or in Exercises 53–56 in Exercise Set 2.2 on page 114. Using the two graphs that group members find most interesting, introduce two or more functions that are related to the graphs. Then write and solve a problem involving function addition and function subtraction for each selected graph.

CHAPTER SUMMARY, REVIEW, AND TEST

SUMMARY

DEFINITIONS AND CONCEPTS	EXAMPLES
Section 2.1 Introduction to Functions	
A relation is any set of ordered pairs. The set of first components of the ordered pairs is the domain and the set of second components is the range. A function is a relation in which each member of the domain corresponds to exactly one member of the range. No two ordered pairs of a function can have the same first components and different second components.	The domain of the relation $\{(1, 2), (3, 4), (3, 7)\}$ is $\{1, 3\}$. The range is $\{2, 4, 7\}$. The relation is not a function: 3, in the domain, corresponds to both 4 and 7 in the range.
If a function is defined as an equation, the notation $f(x)$, read "f of x" or "f at x," describes the value of the function at the number, or input, x.	If $f(x) = 7x - 5$, then $f(a + 2) = 7(a + 2) - 5 = 7a + 14 - 5 = 7a + 9.$
The graph of a function is the graph of its ordered pairs. ***The Vertical Line Test for Functions*** If any vertical line intersects a graph in more than one point, the graph does not define y as a function of x. At the left or right of a function's graph, you will often find closed dots, open dots, or arrows. A closed dot shows the graph ends and the point belongs to the graph. An open dot shows the graph ends and the point does not belong to the graph. An arrow indicates the graph extends indefinitely.	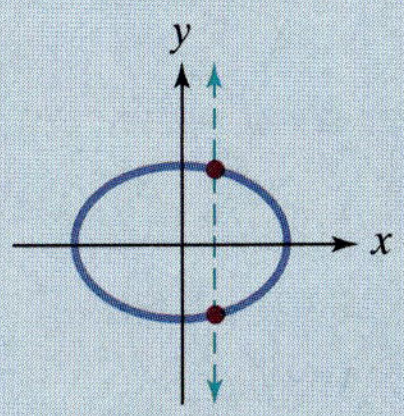 Not the graph of a function 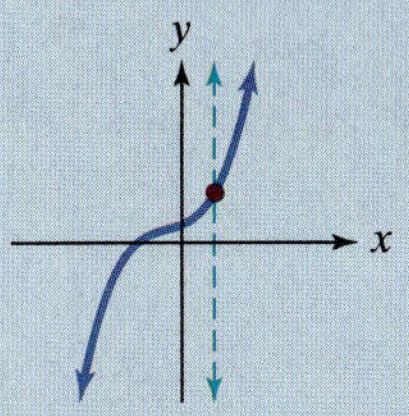 The graph of a function 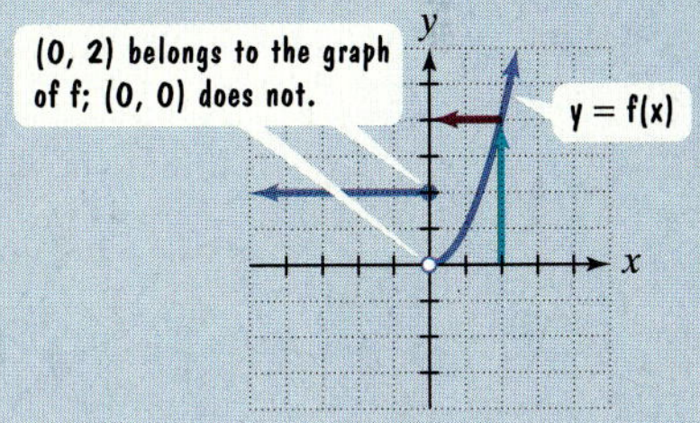 To find $f(2)$, locate 2 on the x-axis. The graph shows $f(2) = 4$.
Section 2.2 The Algebra of Functions	
A Function's Domain If a function f does not model data or verbal conditions, its domain is the largest set of real numbers for which the value of $f(x)$ is a real number. Exclude from a function's domain real numbers that cause division by zero and real numbers that result in a square root of a negative number.	$f(x) = 7x + 13$ Domain of $f = \{x \mid x \text{ is a real number}\}$ $g(x) = \frac{7x}{12 - x}$ Domain of $g = \{x \mid x \text{ is a real number and } x \neq 12\}$

DEFINITIONS AND CONCEPTS	EXAMPLES
Section 2.2 The Algebra of Functions (continued)	
The Algebra of Functions Let f and g be two functions. The sum $f + g$, the difference $f - g$, the product fg, and the quotient $\frac{f}{g}$ are functions whose domains are the set of all real numbers common to the domains of f and g, defined as follows: **1.** Sum: $(f + g)(x) = f(x) + g(x)$ **2.** Difference: $(f - g)(x) = f(x) - g(x)$ **3.** Product: $(fg)(x) = f(x) \cdot g(x)$ **4.** Quotient: $\left(\frac{f}{g}\right)(x) = \frac{f(x)}{g(x)}, g(x) \neq 0.$	Let $f(x) = x^2 + 2x$ and $g(x) = 4 - x$. • $(f + g)(x) = (x^2 + 2x) + (4 - x) = x^2 + x + 4$ $(f + g)(-2) = (-2)^2 + (-2) + 4 = 4 - 2 + 4 = 6$ • $(f - g)(x) = (x^2 + 2x) - (4 - x) = x^2 + 2x - 4 + x$ $= x^2 + 3x - 4$ $(f - g)(5) = 5^2 + 3 \cdot 5 - 4 = 25 + 15 - 4 = 36$ • $(fg)(1) = f(1) \cdot g(1) = (1^2 + 2 \cdot 1)(4 - 1)$ $= 3(3) = 9$ • $\left(\frac{f}{g}\right)(x) = \frac{x^2 + 2x}{4 - x}, x \neq 4$ $\left(\frac{f}{g}\right)(3) = \frac{3^2 + 2 \cdot 3}{4 - 3} = \frac{9 + 6}{1} = 15$
Section 2.3 Linear Functions and Slope	
Data presented in a visual form as a set of points is called a scatter plot. A line that best fits the data points is called a regression line. A function whose graph is a straight line is called a linear function. All linear functions can be written in the form $f(x) = mx + b$.	$f(x) = 3x + 10$ is a linear function. $g(x) = 3x^2 + 10$ is not a linear function.
If a graph intersects the x-axis at $(a, 0)$, then a is an x-intercept. If a graph intersects the y-axis at $(0, b)$, then b is a y-intercept. The standard form of the equation of a line, $$Ax + By = C,$$ can be graphed using intercepts and a checkpoint.	Graph using intercepts: $4x + 3y = 12$. x-intercept: (Set $y = 0$.) $4x = 12$, $x = 3$ y-intercept: (Set $x = 0$.) $3y = 12$, $y = 4$ checkpoint: Let $x = 2$. $4 \cdot 2 + 3y = 12$ $8 + 3y = 12$ $3y = 4$ $y = \frac{4}{3}$ 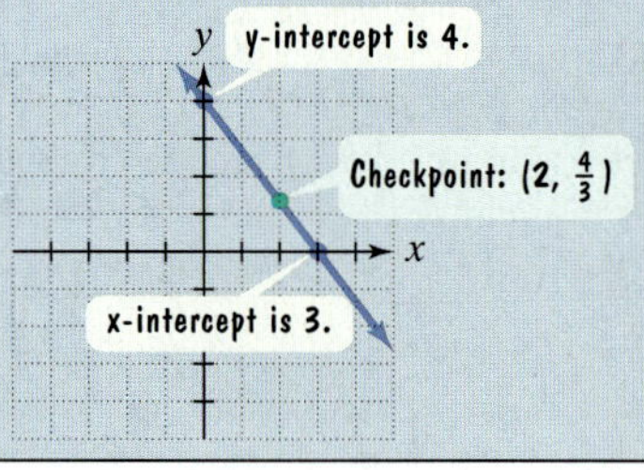

DEFINITIONS AND CONCEPTS | **EXAMPLES**

Section 2.3 Linear Functions and Slope (continued)

The slope, m, of the line through the points (x_1, y_1) and (x_2, y_2) is

$$m = \frac{y_2 - y_1}{x_2 - x_1}, \quad x_2 - x_1 \neq 0.$$

If the slope is positive, the line rises from left to right. If the slope is negative, the line falls from left to right. The slope of a horizontal line is 0. The slope of a vertical line is undefined.

For points $(-7, 2)$ and $(3, -4)$, the slope of the line through the points is:

$$m = \frac{\text{Change in } y}{\text{Change in } x} = \frac{-4 - 2}{3 - (-7)} = \frac{-6}{10} = -\frac{3}{5}.$$

Slope is negative, so the line falls.

For points $(2, -5)$ and $(2, 16)$, the slope of the line through the points is:

$$m = \frac{\text{Change in } y}{\text{Change in } x} = \frac{16 - (-5)}{2 - 2} = \frac{21}{0}.$$

undefined

Slope is undefined, so the line is vertical.

The slope-intercept equation of a nonvertical line with slope m and y-intercept b is

$$y = mx + b.$$

Using function notation, the equation is

$$f(x) = mx + b.$$

Graph: $f(x) = -\frac{3}{4}x + 1.$

Slope is $-\frac{3}{4}$.

y-intercept is 1.

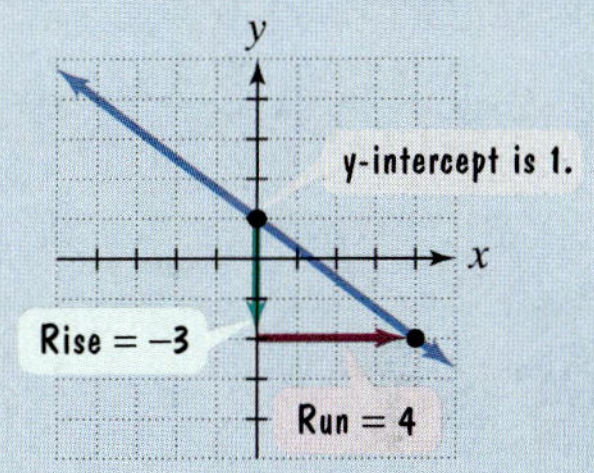

Horizontal and Vertical Lines

The graph of $y = b$, or $f(x) = b$, is a horizontal line. The y-intercept is b. The linear function $f(x) = b$ is called a constant function.

The graph of $x = a$ is a vertical line. The x-intercept is a. A vertical line is not a linear function.

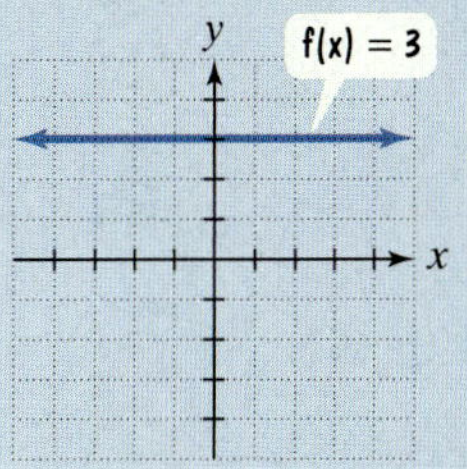

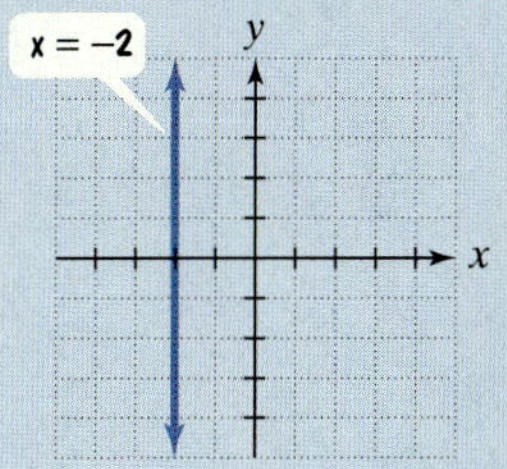

If the graph of a function is not a straight line, the average rate of change between any two points is the slope of the line containing the two points.

For a linear function, slope is the rate of change of the dependent variable per unit change of the independent variable.

The function

$$p(t) = -0.59t + 80.75$$

slope

models the percentage, $p(t)$, of Americans smoking cigarettes t years after 1900. Slope, -0.59, shows that the percentage of smokers is decreasing by 0.59% per year.

<table>
<tr><th>DEFINITIONS AND CONCEPTS</th><th>EXAMPLES</th></tr>
<tr><td colspan="2">Section 2.4 The Point-Slope Form of the Equation of a Line</td></tr>
<tr><td>

The point-slope equation of a nonvertical line with slope m that passes through the point (x_1, y_1) is

$$y - y_1 = m(x - x_1).$$

</td><td>

Slope = −3, passing through (−1, 5)

$m = -3$ $x_1 = -1$ $y_1 = 5$

The line's point-slope equation is

$$y - 5 = -3[x - (-1)].$$

Simplify:

$$y - 5 = -3(x + 1).$$

</td></tr>
<tr><td>

To write the point-slope form of the line passing through two points, begin by using the points to compute the slope, m. Use either given point as (x_1, y_1) and write the point-slope equation:

$$y - y_1 = m(x - x_1).$$

Solving this equation for y gives the slope-intercept form of the line's equation.

</td><td>

Write an equation in point-slope form and slope-intercept form of the line passing through (−1, −3) and (4, 2).

$$m = \frac{2 - (-3)}{4 - (-1)} = \frac{2 + 3}{4 + 1} = \frac{5}{5} = 1$$

Using (4, 2) as (x_1, y_1), the point-slope equation is

$$y - 2 = 1(x - 4).$$

Solve for y to obtain the slope-intercept form.

$$y - 2 = x - 4$$

$$y = x - 2$$

In function notation,

$$f(x) = x - 2.$$

</td></tr>
<tr><td>

Nonvertical parallel lines have the same slope. If the product of the slopes of two lines is −1, then the lines are perpendicular. One line is perpendicular to another line if its slope is the negative reciprocal of the slope of the other.

</td><td>

Write point-slope and slope-intercept equations of the line passing through (2, −1)

x_1 y_1

and perpendicular to $y = -\frac{1}{5}x + 6$.

slope

Slope, m, of perpendicular line is 5, the negative reciprocal of $-\frac{1}{5}$.

$$y - (-1) = 5(x - 2)$$ Point-slope equation

$$y + 1 = 5(x - 2)$$

$$y + 1 = 5x - 10$$ Slope-intercept equation

$$y = 5x - 11 \quad \text{or} \quad f(x) = 5x - 11$$

</td></tr>
</table>

Review Exercises

2.1 *In Exercises 1–3, determine whether each relation is a function. Give the domain and range for each relation.*

1. $\{(3, 10), (4, 10), (5, 10)\}$

2. $\{(1, 12), (2, 100), (3, \pi), (4, -6)\}$

3. $\{(13, 14), (15, 16), (13, 17)\}$

In Exercises 4–5, find the indicated function values.

4. $f(x) = 7x - 5$

a. $f(0)$ **b.** $f(3)$ **c.** $f(-10)$
d. $f(2a)$ **e.** $f(a + 2)$

5. $g(x) = 3x^2 - 5x + 2$

a. $g(0)$ **b.** $g(5)$ **c.** $g(-4)$
d. $g(b)$ **e.** $g(4a)$

In Exercises 6–7, graph the given functions, f and g, in the same rectangular coordinate system. Select integers for x, starting with −2 and ending with 2. Once you have obtained your graphs, describe how the graph of g is related to the graph of f.

6. $f(x) = x^2, \quad g(x) = x^2 - 1$

7. $f(x) = |x|, \quad g(x) = |x| + 2$

In Exercises 8–11, use the vertical line test to identify graphs in which y is a function of x.

8.

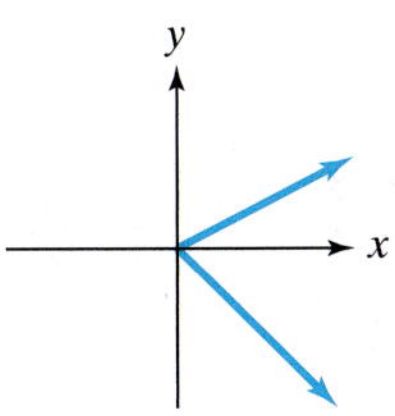

9.

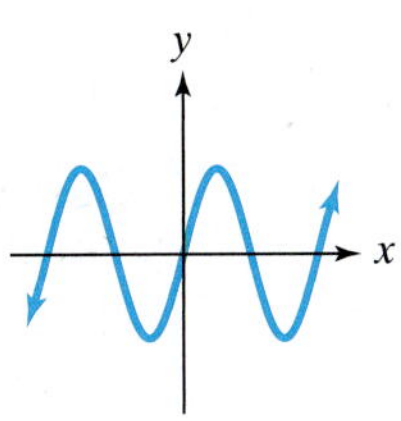

10.

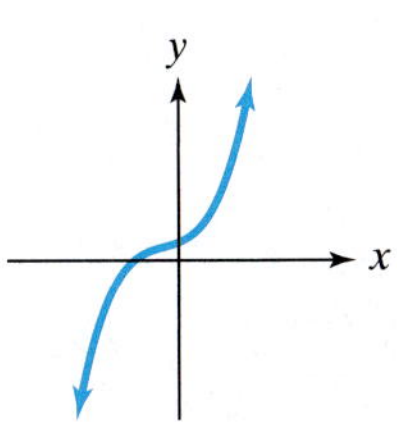

11.

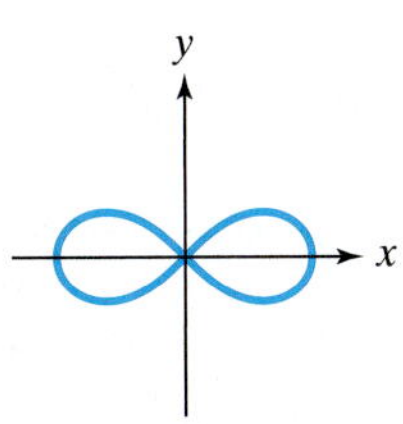

Use the graph of f to solve Exercises 12–14.

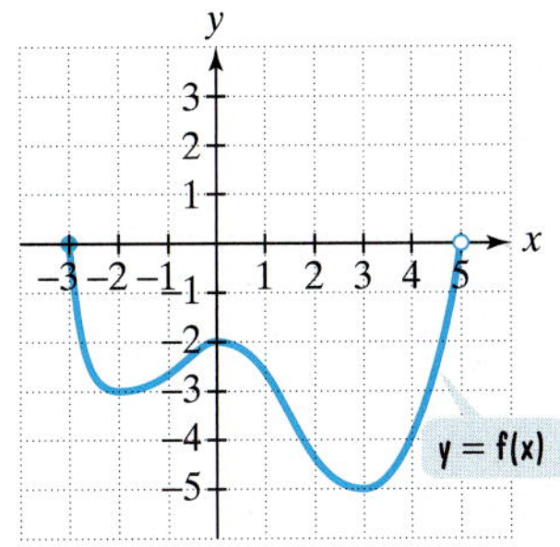

12. Find $f(-2)$.

13. Find $f(0)$.

14. For what value of x is $f(x) = -5$?

15. The graph shows the height, in meters, of a vulture in terms of its time, in seconds, in flight.

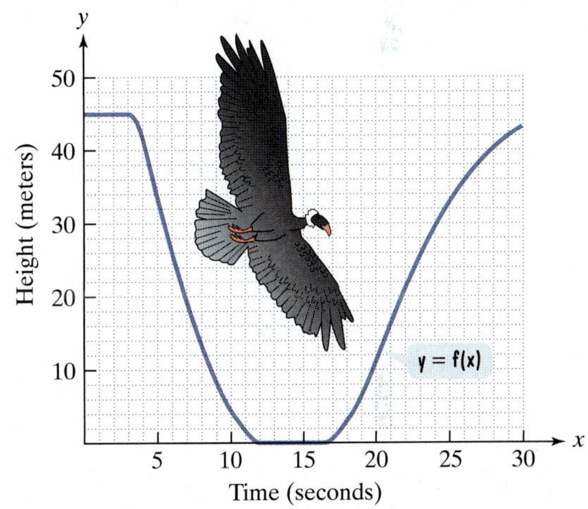

a. Use the graph to explain why the vulture's height is a function of its time in flight.

b. Find $f(15)$. Describe what this means in practical terms.

c. What is a reasonable estimate of the vulture's maximum height?

d. For what values of x is $f(x) = 20$? Describe what this means in practical terms.

e. Use the graph of the function to write a description of the vulture's flight.

2.2 *In Exercises 16–18, find the domain of each function.*

16. $f(x) = 7x - 3$

17. $g(x) = \dfrac{1}{x + 8}$

18. $f(x) = x + \dfrac{3x}{x - 5}$

In Exercises 19–20, find **a.** $(f + g)(x)$ *and* **b.** $(f + g)(3)$.

19. $f(x) = 4x - 5, \quad g(x) = 2x + 1$

20. $f(x) = 5x^2 - x + 4, \quad g(x) = x - 3$

In Exercises 21–22, for each pair of functions, f and g, determine the domain of f + g.

21. $f(x) = 3x + 4, \quad g(x) = \dfrac{5}{4 - x}$

22. $f(x) = \dfrac{7x}{x + 6}, \quad g(x) = \dfrac{4}{x + 1}$

In Exercises 23–30, let

$$f(x) = x^2 - 2x \quad \text{and} \quad g(x) = x - 5.$$

Find each of the following.

23. $(f + g)(x)$ and $(f + g)(-2)$

24. $f(3) + g(3)$

25. $(f - g)(x)$ and $(f - g)(1)$

26. $f(4) - g(4)$

27. $(fg)(-3)$

28. $\left(\dfrac{f}{g}\right)(x)$ and $\left(\dfrac{f}{g}\right)(4)$

29. The domain of $f - g$

30. The domain of $\dfrac{f}{g}$

2.3 *In Exercises 31–33, use intercepts and a checkpoint to graph each linear function.*

31. $x + 2y = 4$

32. $2x - 3y = 12$

33. $4x = 8 - 2y$

In Exercises 34–37, find the slope of the line passing through each pair of points or state that the slope is undefined. Then indicate whether the line through the points rises, falls, is horizontal, or is vertical.

34. $(5, 2)$ and $(2, -4)$

35. $(-2, 3)$ and $(7, -3)$

36. $(3, 2)$ and $(3, -1)$

37. $(-3, 4)$ and $(-1, 4)$

In Exercises 38–40, give the slope and y-intercept of each line whose equation is given. Then graph the linear function.

38. $y = 2x - 1$

39. $f(x) = -\dfrac{1}{2}x + 4$

40. $y = \dfrac{2}{3}x$

In Exercises 41–43, rewrite the equation in slope-intercept form. Give the slope and y-intercept.

41. $2x + y = 4$

42. $-3y = 5x$

43. $5x + 3y = 6$

In Exercises 44–48, graph each equation in the rectangular coordinate system.

44. $y = 2$

45. $7y = -21$

46. $f(x) = -4$

47. $x = 3$

48. $2x = -10$

49. The graph shows annual spending per uniformed member of the U.S. military in inflation-adjusted dollars. Find the average rate of change of spending per year from 1955 through 2000.

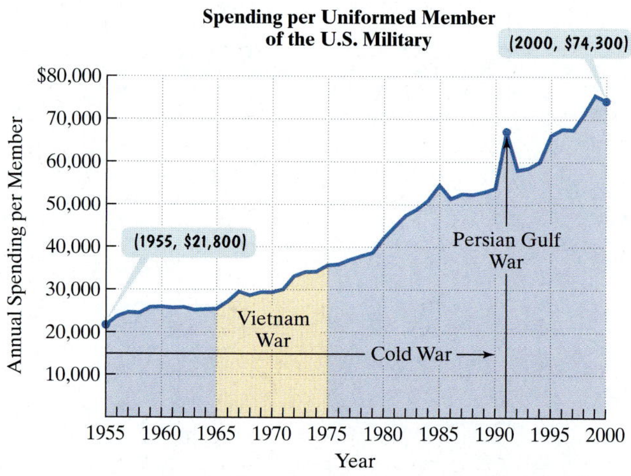

Source: Center for Strategic and Budgetary Assessments

50. The function $f(t) = -0.27t + 70.45$ models record time, $f(t)$, in seconds, for the women's 400-meter run t years after 1900. What is the slope of this model? Describe what this means in terms of rate of change.

2.4 *In Exercises 51–54, use the given conditions to write an equation for each line in point-slope form and slope-intercept form.*

51. Passing through $(-3, 2)$ with a slope of -6

52. Passing through $(1, 6)$ and $(-1, 2)$

53. Passing through $(4, -7)$ and parallel to the line whose equation is $3x + y - 9 = 0$

54. Passing through $(-2, 6)$ and perpendicular to the line whose equation is $y = \frac{1}{3}x + 4$.

55. We seem to be fed up with being lectured about our waistlines. The points in the graph show the average weight of Americans adults from 1990 through 2000. Also shown is a line that passes through or near the points.

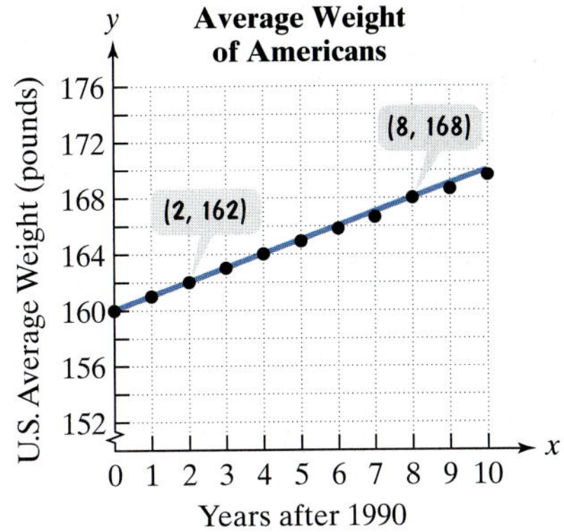

Source: Diabetes Care

a. Use the two points whose coordinates are shown by the voice balloons to find the point-slope equation of the line that models average weight of Americans, y, in pounds, x years after 1990.

b. Write the equation in part (a) in slope-intercept form. Use function notation.

c. Use the linear function to predict the average weight of Americans in 2005.

Chapter 2 Test

In Exercises 1–2, determine whether each relation is a function. Give the domain and range for each relation.

1. $\{(1, 2), (3, 4), (5, 6), (6, 6)\}$

2. $\{(2, 1), (4, 3), (6, 5), (6, 6)\}$

3. If $f(x) = 3x - 2$, find $f(a + 4)$.

4. If $f(x) = 4x^2 - 3x + 6$, find $f(-2)$.

5. Graph $f(x) = x^2 - 1$ and $g(x) = x^2 + 1$ in the same rectangular coordinate system. Select integers for x, starting with -2 and ending with 2. Once you have obtained your graphs, describe how the graph of g is related to the graph of f.

In Exercises 6–7, identify the graph or graphs in which y is a function of x.

6.

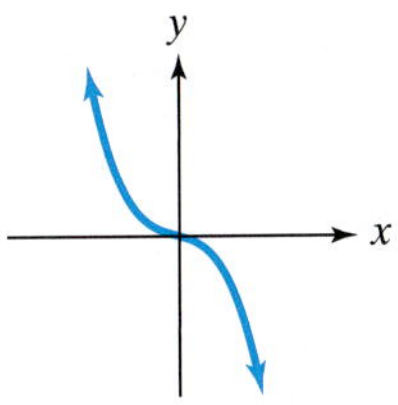

7.

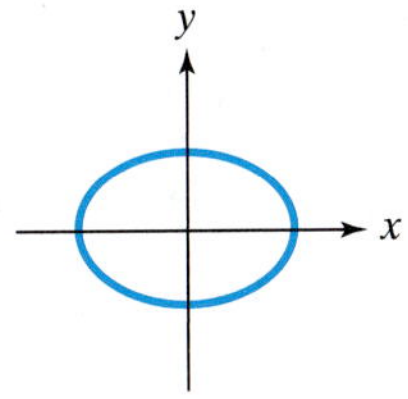

Use the graph of f to solve Exercises 8–9.

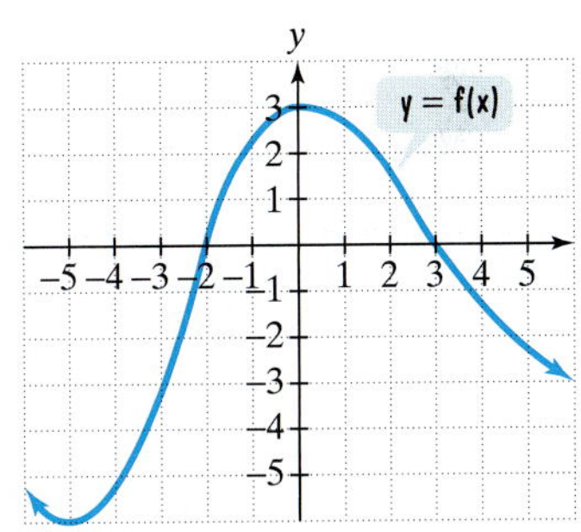

$f(-2) = ?\quad f(6) = ?$

8. Find $f(6)$.

9. List a value of x for which $f(x) = -6$.

10. Find the domain of $f(x) = \dfrac{6}{10 - x}$.

In Exercises 11–15, let

$$f(x) = x^2 + 4x \quad \textit{and} \quad g(x) = x + 2.$$

Find each of the following.

11. $(f + g)(x)$ and $(f + g)(3)$

12. $(f - g)(x)$ and $(f - g)(-1)$

13. $(fg)(-5)$

14. $\left(\frac{f}{g}\right)(x)$ and $\left(\frac{f}{g}\right)(2)$

15. The domain of $\frac{f}{g}$

In Exercises 16–18, graph each linear function.

16. $4x - 3y = 12$

17. $f(x) = -\frac{1}{3}x + 2$

18. $f(x) = 4$

In Exercises 19–20, find the slope of the line passing through each pair of points or state that the slope is undefined. Then indicate whether the line through the points rises, falls, is horizontal, or is vertical.

19. $(5, 2)$ and $(1, 4)$

20. $(4, 5)$ and $(4, -5)$

The function $V(t) = 3.6t + 140$ models the number of Super Bowl viewers, in millions, t years after 1995. Use the model to solve Exercises 21–22.

21. Find $V(10)$. Describe what this means in terms of the variables in the model.

22. What is the slope of this model? Describe what this means in terms of rate of change.

In Exercises 23–24, use the given conditions to write an equation for each line in point-slope form and slope-intercept form.

23. Passing through $(-1, -3)$ and $(4, 2)$

24. Passing through $(-2, 3)$ and perpendicular to the line whose equation is $y = -\frac{1}{2}x - 4$

25. Strong demand plus higher fuel and labor costs are driving up the price of flying. The graph shows the national averages for one-way fares. Also shown is a line that models the data.

National Averages for One-Way Airline Fares: Business Travel

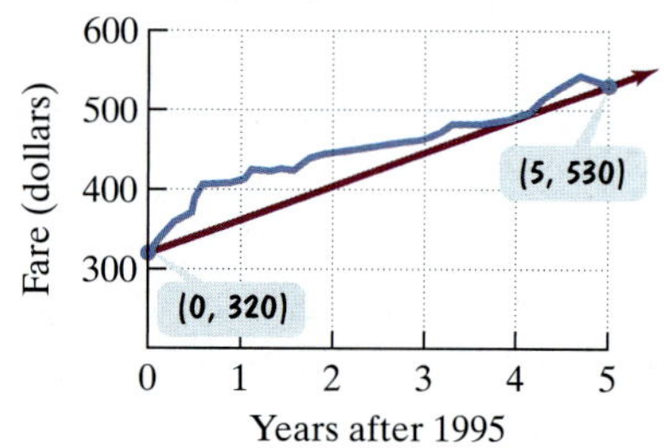

Source: American Express

a. Use the two points whose coordinates are shown by the voice balloons to find the point-slope equation of the line that models the average one-way fare, y, in dollars, x years after 1995.

b. Write the equation in part (a) in slope-intercept form. Use function notation.

c. Use the linear function to predict the national average for one-way fares in 2008.

Cumulative Review Exercises (Chapters 1–2)

1. Use the roster method to list the elements in the set:

$$\{x \mid x \text{ is a whole number less than } 4\}.$$

2. Determine whether the following statement is true or false:

$$\pi \notin \{x \mid x \text{ is an irrational number}\}.$$

In Exercises 3–4, use the order of operations to simplify each expression.

3. $\dfrac{8 - 3^2 \div 9}{|-5| - [5 - (18 \div 6)]^2}$

4. $4 - (2 - 9)^0 + 3^2 \div 1 + 3$

5. Simplify: $3 - [2(x - 2) - 5x]$.

In Exercises 6–8, solve each equation. If the solution set is $\varnothing$ or $\mathbb{R}$, classify the equation as an inconsistent equation or an identity.

6. $2 + 3x - 4 = 2(x - 3)$

7. $4x + 12 - 8x = -6(x - 2) + 2x$

8. $\dfrac{x - 2}{4} = \dfrac{2x + 6}{3}$

9. After a 20% reduction, a computer sold for $1800. What was the computer's price before the reduction?

10. Solve for t: $A = p + prt$.

In Exercises 11–12, simplify each exponential expression.

11. $(3x^4y^{-5})^{-2}$ **12.** $\left(\dfrac{3x^2y^{-4}}{x^{-3}y^2}\right)^2$

13. Multiply and write the answer in scientific notation:

$$(7 \times 10^{-8})(3 \times 10^2).$$

14. Is $\{(1,5),(2,5),(3,5),(4,5),(6,5)\}$ a function? Give the relation's domain and range.

15. Graph $f(x) = |x| - 1$ and $g(x) = |x| + 2$ in the same rectangular coordinate system. Select integers for x, starting with -2 and ending with 2. Once you have obtained your graphs, describe how the graph of g is related to the graph of f.

16. Find the domain of $f(x) = \dfrac{1}{15 - x}$.

17. If $f(x) = 3x^2 - 4x + 2$ and $g(x) = x^2 - 5x - 3$, find $(f - g)(x)$ and $(f - g)(-1)$.

In Exercises 18–19, graph each linear function.

18. $f(x) = -2x + 4$

19. $x - 2y = 6$

20. Write equations in point-slope form and slope-intercept form for the line passing through $(3, -5)$ and parallel to the line whose equation is $y = 4x + 7$.

Traffic jams getting you down? Powerful computers, able to solve systems of equations with hundreds of thousands of variables in a single bound, may promise a gridlock-free future. The computer in your car could be linked to a central computer that manages traffic flow by controlling traffic lights, rerouting you away from traffic congestion, issuing weather reports, and selecting the best route to your destination. New technologies could eventually drive your car at a steady 75 miles per hour along automated highways as you comfortably nap.

Real-world problems often involve solving thousands of equations, sometimes containing a million variables. The methods that you will learn in this chapter provide the foundation for solving the kinds of complex systems needed to create a future free of traffic jams.

Chapter 3

Systems of Linear Equations

▸ SECTION 3.1 Systems of Linear Equations in Two Variables

Objectives

1. Determine whether an ordered pair is a solution of a system of linear equations.
2. Solve systems of linear equations by graphing.
3. Solve systems of linear equations by substitution.
4. Solve systems of linear equations by addition.
5. Select the most efficient method for solving a system of linear equations.
6. Identify systems that do not have exactly one ordered-pair solution.

Women now attain higher levels of education than ever before, and a majority of women choose careers in the labor force rather than homemaking. One of the most important developments in the work force has been the dramatic increase in the number of women, at approximately $\frac{1}{2}\%$ per year. By contrast, the percentage of men is decreasing by $\frac{1}{4}\%$ per year. These changes are illustrated in Figure 3.1. If current trends continue, the graphs will intersect. The intersection point reveals when the percentage of women and men in the labor force will be equal.

Projections frequently involve graphs that intersect. In this section, you will learn three methods for finding intersection points. With these methods, you can use mathematical models to determine when variable quantities, such as the percentage of women and men in the work force, will be equal.

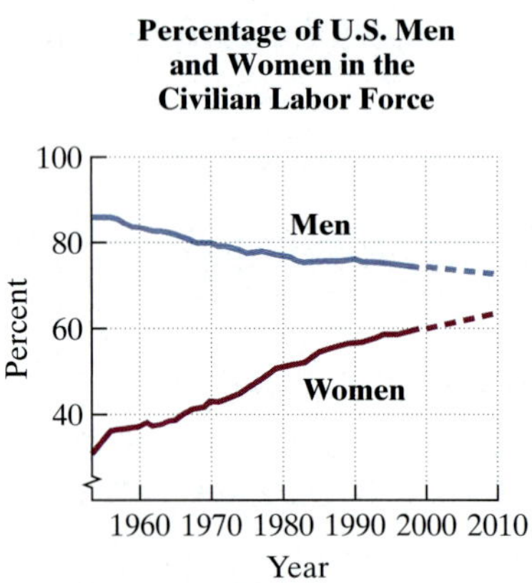

Figure 3.1 *Source:* U.S. Department of Labor

Systems of Linear Equations and Their Solutions

We have seen that all equations in the form $Ax + By = C$ are straight lines when graphed. Two such equations are called a **system of linear equations**. A **solution to a system of linear equations** is an ordered pair that satisfies all equations in the system. For example, (3, 4) satisfies the system

$$x + y = 7 \quad \text{(3 + 4 is, indeed, 7.)}$$
$$x - y = -1. \quad \text{(3 − 4 is, indeed, −1.)}$$

Thus, (3, 4) satisfies both equations and is a solution of the system. The solution can be described by saying that $x = 3$ and $y = 4$. The solution can also be described using set notation. The solution set to the system is $\{(3, 4)\}$—that is, the set consisting of the ordered pair (3, 4).

A system of linear equations can have exactly one solution, no solution, or infinitely many solutions. We begin with systems with exactly one solution.

1 Determine whether an ordered pair is a solution of a system of linear equations.

EXAMPLE 1 Determining Whether an Ordered Pair Is a Solution of a System

Determine whether (−5, −6) is a solution of the system

$$2x - y = -4$$
$$3x - 5y = 15.$$

Solution Because −5 is the x-coordinate and −6 is the y-coordinate of (−5, −6), we replace x by −5 and y by −6.

$$\begin{aligned} 2x - y &= -4 \\ 2(-5) - (-6) &\stackrel{?}{=} -4 \\ -10 + 6 &\stackrel{?}{=} -4 \\ -4 &= -4, \quad \text{true} \end{aligned} \qquad \begin{aligned} 3x - 5y &= 15 \\ 3(-5) - 5(-6) &\stackrel{?}{=} 15 \\ -15 + 30 &\stackrel{?}{=} 15 \\ 15 &= 15, \quad \text{true} \end{aligned}$$

The pair $(-5, -6)$ satisfies both equations: It makes each equation true. Thus, the pair is a solution of the system. ■

CHECK POINT 1 Determine whether $(-2, -3)$ is a solution of the system

$$2x - 4y = 8$$
$$2x - y = -1.$$

2 Solve systems of linear equations by graphing.

Solving Linear Systems by Graphing

The solution of a system of two linear equations in two variables can be found by graphing both of the equations in the same rectangular coordinate system. For a system with one solution, **the coordinates of the point of intersection give the system's solution.**

Study Tip

When solving linear systems by graphing, neatly drawn graphs are essential for determining points of intersection.

- Use rectangular coordinate graph paper.
- Use a ruler or straightedge.
- Use a pencil with a sharp point.

Solving Systems of Two Linear Equations in Two Variables, x and y, by Graphing

1. Graph the first equation.
2. Graph the second equation on the same set of axes.
3. If the lines representing the two graphs intersect at a point, determine the coordinates of this point of intersection. The ordered pair is the solution to the system.
4. Check the solution in both equations.

EXAMPLE 2 Solving a Linear System by Graphing

Solve by graphing:

$$y = -x - 1$$
$$4x - 3y = 24.$$

Solution

Step 1 Graph the first equation. We use the y-intercept and slope to graph $y = -x - 1$.

$$y = -x - 1$$

The slope is −1. The y-intercept is −1.

The graph of the linear function is shown in Figure 3.2.

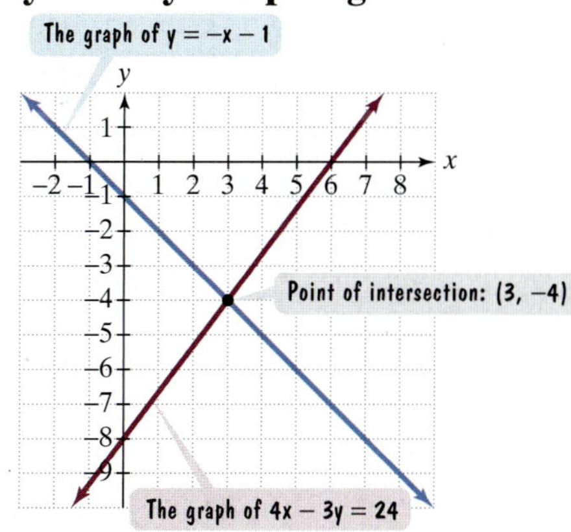

Figure 3.2

Step 2 Graph the second equation. We use intercepts to graph $4x - 3y = 24$.

x-intercept (Set $y = 0$.)	**y-intercept (Set $x = 0$.)**
$4x - 3 \cdot 0 = 24$	$4 \cdot 0 - 3y = 24$
$4x = 24$	$-3y = 24$
$x = 6$	$y = -8$

The x-intercept is 6, so the line passes through $(6, 0)$. The y-intercept is -8, so the line passes through $(0, -8)$. The graph of $4x - 3y = 24$ is shown in Figure 3.2.

Step 3 Determine the coordinates of the intersection point. This ordered pair is the system's solution. Using Figure 3.2, it appears that the lines intersect at $(3, -4)$. The "apparent" solution of the system is $(3, -4)$.

Using Technology

A graphing utility can be used to solve the system in Example 2. Graph each equation and use the intersection feature. The utility displays the solution $(3, -4)$ as $x = 3$, $y = -4$.

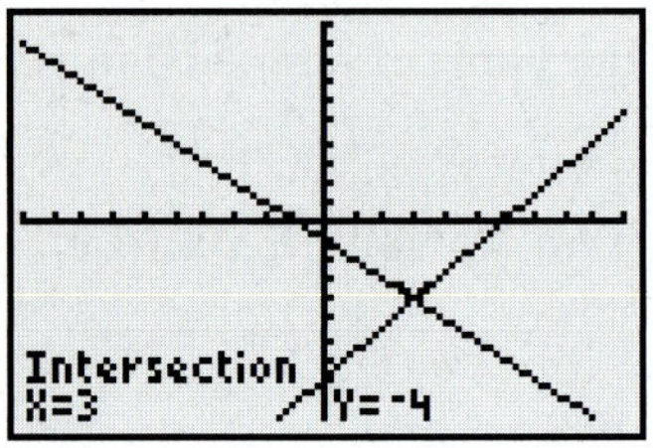

Step 4 **Check the solution in both equations.**

Check $(3, -4)$ in $y = -x - 1$:

$$-4 \stackrel{?}{=} -3 - 1$$

$$-4 = -4, \quad \text{true}$$

Check $(3, -4)$ in $4x - 3y = 24$:

$$4(3) - 3(-4) \stackrel{?}{=} 24$$

$$12 + 12 \stackrel{?}{=} 24$$

$$24 = 24, \quad \text{true}$$

Because both equations are satisfied, $(3, -4)$ is the solution and $\{(3, -4)\}$ is the solution set. ■

✔ **CHECK POINT 2** Solve by graphing:

$$y = -2x + 6$$

$$2x - y = -2.$$

3 Solve systems of linear equations by substitution.

Eliminating a Variable Using the Substitution Method

Finding the solution to a linear system by graphing equations may not be easy to do. For example, a solution of $\left(-\frac{2}{3}, \frac{157}{29}\right)$ would be difficult to "see" as an intersection point on a graph.

Let's consider a method that does not depend on finding a system's solution visually: the substitution method. This method involves converting the system to one equation in one variable by an appropriate substitution.

EXAMPLE 3 Solving a System by Substitution

Solve by the substitution method:

$$y = -2x + 4$$

$$7x - 2y = 3.$$

Solution

Step 1 **Solve either of the equations for one variable in terms of the other.** This step has already been done for us. The first equation, $y = -2x + 4$, has y solved in terms of x.

Step 2 **Substitute the expression from step 1 into the other equation.** We substitute the expression $-2x + 4$ for y in the other equation:

$$y = \boxed{-2x + 4} \qquad 7x - 2\boxed{y} = 3. \qquad \text{Substitute } -2x + 4 \text{ for } y.$$

This gives us an equation in one variable, namely

$$7x - 2(-2x + 4) = 3.$$

The variable y has been eliminated.

Step 3 Solve the resulting equation containing one variable.

$$7x - 2(-2x + 4) = 3$$ This is the equation containing one variable.

$$7x + 4x - 8 = 3$$ Apply the distributive property.

$$11x - 8 = 3$$ Combine like terms.

$$11x = 11$$ Add 8 to both sides.

$$x = 1$$ Divide both sides by 11.

Step 4 Back-substitute the obtained value into one of the original equations. We now know that the x-coordinate of the solution is 1. To find the y-coordinate, we back-substitute the x-value into either original equation. We will use

$$y = -2x + 4.$$

Substitute 1 for x.

$$y = -2 \cdot 1 + 4 = -2 + 4 = 2$$

With $x = 1$ and $y = 2$, the proposed solution is $(1, 2)$.

Step 5 Check the proposed solution in both of the system's given equations. Replace x with 1 and y with 2.

$y = -2x + 4$	$7x - 2y = 3$
$2 \stackrel{?}{=} -2 \cdot 1 + 4$	$7(1) - 2(2) \stackrel{?}{=} 3$
$2 \stackrel{?}{=} -2 + 4$	$7 - 4 \stackrel{?}{=} 3$
$2 = 2$, true	$3 = 3$, true

The pair $(1, 2)$ satisfies both equations. The solution is $(1, 2)$ and the system's solution set is $\{(1, 2)\}$. ■

CHECK POINT 3 Solve by the substitution method:

$$y = 3x - 7$$
$$5x - 2y = 8.$$

Before considering additional examples, let's summarize the steps used in the substitution method.

Study Tip

In step 1, if possible, solve for a variable whose coefficient is 1 or -1 to avoid working with fractions.

Solving Linear Systems by Substitution

1. Solve either of the equations for one variable in terms of the other. (If one of the equations is already in this form, you can skip this step.)
2. Substitute the expression found in step 1 into the other equation. This will result in an equation in one variable.
3. Solve the equation obtained in step 2.
4. Back-substitute the value found in step 3 into one of the original equations. Simplify and find the value of the remaining variable.
5. Check the proposed solution in both of the system's given equations.

EXAMPLE 4 Solving a System by Substitution

Solve by the substitution method:

$$5x + 2y = 1$$
$$x - 3y = 7.$$

Solution

Step 1 Solve either of the equations for one variable in terms of the other. We begin by isolating one of the variables in either of the equations. By solving for x in the second equation, which has a coefficient of 1, we can avoid fractions.

$$x - 3y = 7 \quad \text{This is the second equation in the given system.}$$
$$x = 3y + 7 \quad \text{Solve for } x \text{ by adding } 3y \text{ to both sides.}$$

Step 2 Substitute the expression from step 1 into the other equation. We substitute $3y + 7$ for x in the first equation.

$$x = \boxed{3y + 7} \qquad 5\boxed{x} + 2y = 1$$

This gives us an equation in one variable, namely

$$5(3y + 7) + 2y = 1.$$

The variable x has been eliminated.

Step 3 Solve the resulting equation containing one variable.

$$5(3y + 7) + 2y = 1 \quad \text{This is the equation containing one variable.}$$
$$15y + 35 + 2y = 1 \quad \text{Apply the distributive property.}$$
$$17y + 35 = 1 \quad \text{Combine like terms.}$$
$$17y = -34 \quad \text{Subtract 35 from both sides.}$$
$$y = -2 \quad \text{Divide both sides by 17.}$$

Step 4 Back-substitute the obtained value into one of the original equations. We back-substitute -2 for y into one of the original equations to find x. Let's use both equations to show that we obtain the same value for x in either case.

Using the first equation:	**Using the second equation:**
$5x + 2y = 1$	$x - 3y = 7$
$5x + 2(-2) = 1$	$x - 3(-2) = 7$
$5x - 4 = 1$	$x + 6 = 7$
$5x = 5$	$x = 1$
$x = 1$	

With $x = 1$ and $y = -2$, the proposed solution is $(1, -2)$.

Step 5 Check. Take a moment to show that $(1, -2)$ satisfies both given equations. The solution is $(1, -2)$ and the solution set is $\{(1, -2)\}$. ■

Study Tip

Get into the habit of checking ordered-pair solutions in *both* equations of the system.

✔ CHECK POINT 4 Solve by the substitution method:

$$5x - 4y = 9$$
$$x - 2y = -3.$$

4 Solve systems of linear equations by addition.

Eliminating a Variable Using the Addition Method

The substitution method is most useful if one of the given equations has an isolated variable. A third method for solving a linear system is the addition method. Like the substitution method, the addition method involves eliminating a variable and ultimately solving an equation containing only one variable. However, this time we eliminate a variable by adding the equations.

For example, consider the following equations:

$$3x - 4y = 11$$
$$-3x + 2y = -7.$$

When we add these two equations, the x-terms are eliminated. This occurs because the coefficients of the x-terms, 3 and -3, are opposites (additive inverses) of each other:

$$\begin{aligned} 3x - 4y &= 11 \\ \underline{-3x + 2y} &= \underline{-7} \\ \text{Add: } 0x - 2y &= 4 \\ -2y &= 4 \\ y &= -2. \quad \text{Solve for } y\text{, dividing both sides by } -2. \end{aligned}$$

Now we can back-substitute -2 for y into one of the original equations to find x. It does not matter which equation you use; you will obtain the same value for x in either case. If we use either equation, we can show that $x = 1$ and the solution $(1, -2)$ satisfies both equations in the system.

When we use the addition method, we want to obtain two equations whose sum is an equation containing only one variable. The key step is to obtain, for one of the variables, coefficients that differ only in sign. To do this, we may need to multiply one or both equations by some nonzero number so that the coefficients of one of the variables, x or y, become opposites. Then when the two equations are added, this variable is eliminated.

EXAMPLE 5 Solving a System by the Addition Method

Solve by the addition method:

$$3x + 4y = -10$$
$$5x - 2y = 18.$$

Solution We must rewrite one or both equations in equivalent forms so that the coefficients of the same variable (either x or y) are opposites of each other. Consider the terms in y in each equation, that is, $4y$ and $-2y$. To eliminate y, we can multiply each term of the second equation by 2 and then add equations.

$$\begin{aligned} 3x + 4y = -10 \xrightarrow{\text{No change}} \quad 3x + 4y &= -10 \\ 5x - 2y = \ \ 18 \xrightarrow{\text{Multiply by 2.}} \underline{10x - 4y} &= \underline{\ \ 36} \\ \text{Add: } 13x + 0y &= \ \ 26 \\ 13x &= \ \ 26 \\ x &= \ \ 2 \quad \text{Solve for } x\text{, dividing both sides by 13.} \end{aligned}$$

Thus, $x = 2$. We back-substitute this value into either one of the given equations. We'll use the first one.

$$3x + 4y = -10 \quad \text{This is the first equation in the given system.}$$

$$3(2) + 4y = -10 \quad \text{Substitute 2 for } x.$$

$$6 + 4y = -10 \quad \text{Multiply.}$$

$$4y = -16 \quad \text{Subtract 6 from both sides.}$$

$$y = -4 \quad \text{Divide both sides by 4.}$$

We see that $x = 2$ and $y = -4$. The ordered pair $(2, -4)$ can be shown to satisfy both equations in the system. Consequently, the solution is $(2, -4)$ and the solution set is $\{(2, -4)\}$. ■

Solving Linear Systems by Addition

1. If necessary, rewrite both equations in the form $Ax + By = C$.
2. If necessary, multiply either equation or both equations by appropriate nonzero numbers so that the sum of the x-coefficients or the sum of the y-coefficients is 0.
3. Add the equations in step 2. The sum is an equation in one variable.
4. Solve the equation in one variable.
5. Back-substitute the value obtained in step 4 into either of the given equations and solve for the other variable.
6. Check the solution in both of the original equations.

✔ **CHECK POINT 5** Solve by the addition method:

$$4x + 5y = 3$$
$$2x - 3y = 7.$$

EXAMPLE 6 Solving a System by the Addition Method

Solve by the addition method:

$$7x = 5 - 2y$$
$$3y = 16 - 2x.$$

Solution

Step 1 Rewrite both equations in the form $Ax + By = C$. We first arrange the system so that variable terms appear on the left and constants appear on the right. We obtain

$$7x + 2y = 5 \quad \text{Add } 2y \text{ to both sides of the first equation.}$$

$$2x + 3y = 16. \quad \text{Add } 2x \text{ to both sides of the second equation.}$$

Step 2 If necessary, multiply either equation or both equations by appropriate numbers so that the sum of the x-coefficients or the sum of the y-coefficients is 0. We can eliminate x or y. Let's eliminate y by multiplying the first equation by 3 and the second equation by -2.

$$\begin{array}{rcl} 7x + 2y = 5 & \xrightarrow{\text{Multiply by 3.}} & 21x + 6y = 15 \\ 2x + 3y = 16 & \xrightarrow{\text{Multiply by } -2.} & -4x - 6y = -32 \end{array}$$

Step 3 Add the equations.

$$\text{Add: } 17x + 0y = -17$$
$$17x = -17$$

Step 4 Solve the equation in one variable. We solve $17x = -17$ by dividing both sides by 17.

$$\frac{17x}{17} = \frac{-17}{17} \quad \text{Divide both sides by 17.}$$

$$x = -1 \quad \text{Simplify.}$$

Step 5 Back-substitute and find the value for the other variable. We can back-substitute -1 for x into either one of the given equations. We'll use the second one.

$$3y = 16 - 2x \quad \text{This is the second equation in the given system.}$$
$$3y = 16 - 2(-1) \quad \text{Substitute } -1 \text{ for } x.$$
$$3y = 16 + 2 \quad \text{Multiply.}$$
$$3y = 18 \quad \text{Add.}$$
$$y = 6 \quad \text{Divide both sides by 3.}$$

With $x = -1$ and $y = 6$, the proposed solution is $(-1, 6)$.

Step 6 Check. Take a moment to show that $(-1, 6)$ satisfies both given equations. The solution is $(-1, 6)$ and the solution set is $\{(-1, 6)\}$. ■

✔ **CHECK POINT 6** Solve by the addition method:

$$3x = 2 - 4y$$
$$5y = -1 - 2x.$$

Some linear systems have solutions that are not integers. If the value of one variable turns out to be a "messy" fraction, back-substitution might lead to cumbersome arithmetic. If this happens, you can return to the original system and use addition to find the value of the other variable.

EXAMPLE 7 Solving a System by the Addition Method

Solve by the addition method:

$$\frac{x}{2} - 5y = 32$$
$$\frac{3x}{2} - 7y = 45.$$

Solution

Step 1 Rewrite both equations in the form $Ax + By = C$. Although each equation is already in this form, the coefficients of x are not integers. There is less chance for error if the coefficients for x and y in $Ax + By = C$ are integers. Consequently, we begin by clearing fractions. Multiply both sides of each equation by 2.

$$\frac{x}{2} - 5y = 32 \xrightarrow{\text{Multiply by 2.}} x - 10y = 64$$

$$\frac{3x}{2} - 7y = 45 \xrightarrow{\text{Multiply by 2.}} 3x - 14y = 90$$

Step 2 If necessary, multiply either equation or both equations by appropriate numbers so that the sum of the x-coefficients or the sum of the y-coefficients is 0. We will eliminate x. Multiply the first equation with integral coefficients by −3 and leave the second equation unchanged.

$$\begin{aligned} x - 10y &= 64 \xrightarrow{\text{Multiply by } -3.} & -3x + 30y &= -192 \\ 3x - 14y &= 90 \xrightarrow{\text{No change}} & 3x - 14y &= 90 \end{aligned}$$

Step 3 Add the equations.

$$\begin{aligned} 0x + 16y &= -102 \\ 16y &= -102 \end{aligned}$$

Step 4 Solve the equation in one variable. We solve $16y = -102$ by dividing both sides by 16.

$$\frac{16y}{16} = \frac{-102}{16} \qquad \text{Divide both sides by 16.}$$

$$y = -\frac{102}{16} = -\frac{51}{8} \qquad \text{Simplify.}$$

Step 5 Back-substitute and find the value of the other variable. Back-substitution of $-\frac{51}{8}$ for y into either of the given equations results in cumbersome arithmetic. Instead, let's use the addition method on the system with integral coefficients to find the value of x. Thus, we eliminate y by multiplying the first equation by −7 and the second equation by 5.

$$\begin{aligned} x - 10y &= 64 \xrightarrow{\text{Multiply by } -7.} & -7x + 70y &= -448 \\ 3x - 14y &= 90 \xrightarrow{\text{Multiply by 5.}} & 15x - 70y &= 450 \\ & \text{Add:} & 8x &= 2 \\ & & x &= \frac{2}{8} = \frac{1}{4} \end{aligned}$$

With $x = \frac{1}{4}$ and $y = -\frac{51}{8}$, the proposed solution is $\left(\frac{1}{4}, -\frac{51}{8}\right)$.

Step 6 Check. For this system, a calculator is helpful in showing that $\left(\frac{1}{4}, -\frac{51}{8}\right)$ satisfies both of the original equations of the system. The solution is $\left(\frac{1}{4}, -\frac{51}{8}\right)$ and the solution set is $\left\{\left(\frac{1}{4}, -\frac{51}{8}\right)\right\}$. ■

✔ **CHECK POINT 7** Solve by the addition method:

$$\frac{3x}{2} - 2y = \frac{5}{2}$$

$$x - \frac{5y}{2} = -\frac{3}{2}.$$

5 Select the most efficient method for solving a system of linear equations.

Comparing the Three Solution Methods

The following chart compares the graphing, substitution, and addition methods for solving linear systems of equations in two variables. With increased practice, you will find it easier to select the best method for solving a particular linear system.

Method	Advantages	Disadvantages
Graphing	You can see the solutions.	If the solution does not involve integers or is too large to be seen on the graph, it's impossible to tell exactly what the solution is.
Substitution	Gives exact solutions. Easy to use if a variable is on one side by itself.	Solutions cannot be seen. Introduces extensive work with fractions when no variable has a coefficient of 1 or −1.
Addition	Gives exact solutions. Easy to use if no variable has a coefficient of 1 or −1.	Solutions cannot be seen.

6 Identify systems that do not have exactly one ordered-pair solution.

Linear Systems Having No Solution or Infinitely Many Solutions

We have seen that a system of linear equations in two variables represents a pair of lines. The lines either intersect at one point, are parallel, or are identical. Thus, there are three possibilities for the number of solutions to a system of two linear equations.

The Number of Solutions to a System of Two Linear Equations

The number of solutions to a system of two linear equations in two variables is given by one of the following. (See Figure 3.3.)

Number of Solutions	What This Means Graphically
Exactly one ordered-pair solution	The two lines intersect at one point.
No solution	The two lines are parallel.
Infinitely many solutions	The two lines are identical.

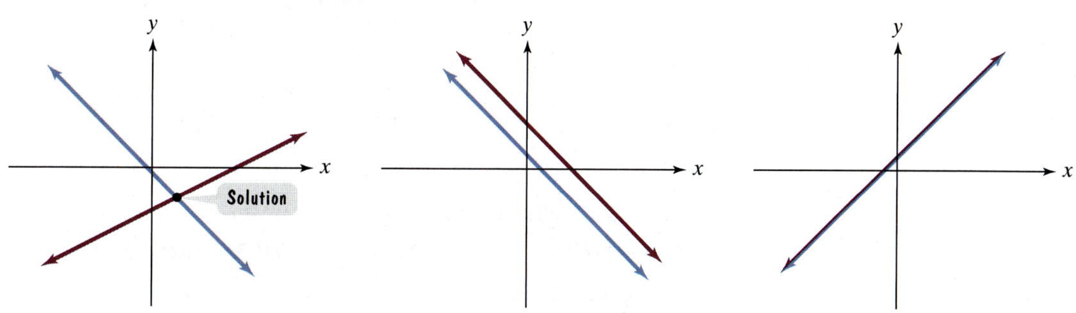

Exactly one solution — No solution (parallel lines) — Infinitely many solutions (lines coincide)

Figure 3.3 Possible graphs for a system of two linear equations in two variables

A linear system with no solution is called an **inconsistent system**. If you attempt to solve such a system by substitution or addition, you will eliminate both variables. A false statement such as $0 = 17$ will be the result.

EXAMPLE 8 A System with No Solution

Solve the system:

$$4x + 6y = 12.$$
$$6x + 9y = 12.$$

Solution Because no variable is isolated, we will use the addition method. To obtain coefficients of x that differ only in sign, we multiply the first equation by 3 and multiply the second equation by -2.

$$\begin{array}{rcl} 4x + 6y = 12 & \xrightarrow{\text{Multiply by 3.}} & 12x + 18y = 36 \\ 6x + 9y = 12 & \xrightarrow{\text{Multiply by } -2.} & -12x - 18y = -24 \\ & \text{Add:} & 0 = 12 \end{array}$$

There are no values of x and y for which $0 = 12$.

The false statement $0 = 12$ indicates that the system is inconsistent and has no solution. The solution set is the empty set, $\varnothing$. ■

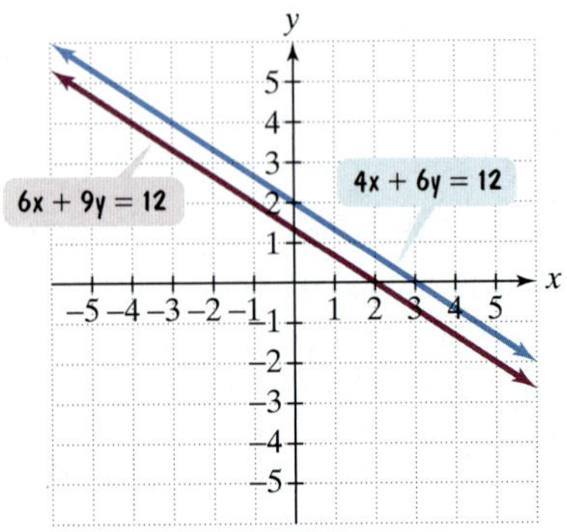

Figure 3.4 The graph of an inconsistent system

The lines corresponding to the two equations in Example 6 are shown in Figure 3.4. The lines are parallel and have no point of intersection.

Discover for Yourself

Show that the graphs of $4x + 6y = 12$ and $6x + 9y = 12$ must be parallel lines by solving each equation for y. What is the slope and y-intercept for each line? What does this mean? If a linear system is inconsistent, what must be true about the slopes and y-intercepts for the system's graphs?

✔ **CHECK POINT 8** Solve the system:

$$x + 2y = 4$$
$$3x + 6y = 13.$$

A linear system that has at least one solution is called a **consistent system**. Lines that intersect and lines that coincide both represent consistent systems. If the lines coincide, then the consistent system has infinitely many solutions, represented by every point on the coinciding lines.

The equations in a linear system with infinitely many solutions are called **dependent**. If you attempt to solve such a system by substitution or addition, you will eliminate both variables. However, a true statement such as $0 = 0$ will be the result.

EXAMPLE 9 A System with Infinitely Many Solutions

Solve the system:

$$y = 3 - 2x$$
$$4x + 2y = 6.$$

Solution Because the variable y is isolated in the first equation, we can use the substitution method. We substitute the expression for y in the second equation.

$$y = \boxed{3 - 2x} \qquad 4x + 2\boxed{y} = 6$$ Substitute $3 - 2x$ for y.

$$4x + 2y = 6$$ This is the second equation in the given system.

$$4x + 2(3 - 2x) = 6$$ Substitute $3 - 2x$ for y.

$$4x + 6 - 4x = 6$$ Apply the distributive property.

$$6 = 6$$ Simplify. This statement is true for all values of x and y.

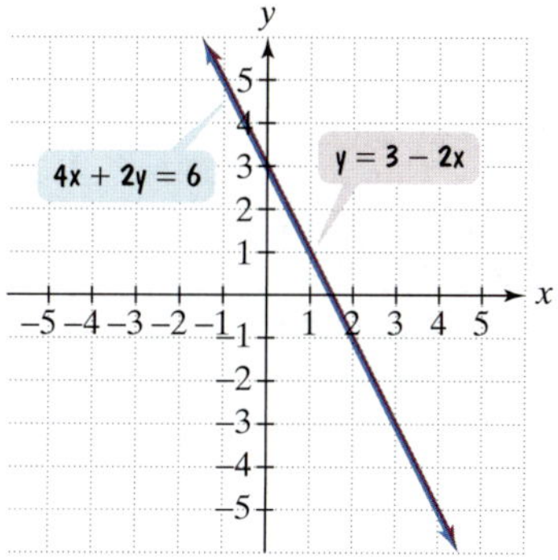

Figure 3.5 The graph of a system with infinitely many solutions

In our final step, both variables have been eliminated, and the resulting statement $6 = 6$ is true. This true statement indicates that the system has infinitely many solutions. The solution set consists of all points (x, y) lying on either of the coinciding lines, $y = 3 - 2x$ or $4x + 2y = 6$, as shown in Figure 3.5.

We express the solution set for the system in one of two equivalent ways:

$\{(x, y) \mid y = 3 - 2x\}$ The set of all ordered pairs (x, y) such that $y = 3 - 2x$

or $\{(x, y) \mid 4x + 2y = 6\}$. The set of all ordered pairs (x, y) such that $4x + 2y = 6$ ■

CHECK POINT 9 Solve the system:

$$y = 4x - 4$$
$$8x - 2y = 8.$$

EXERCISE SET 3.1

Practice Exercises

In Exercises 1–6, determine whether the given ordered pair is a solution of the system.

1. $(7, -5)$
$x - y = 12$
$x + y = 2$

2. $(-3, 1)$
$x - y = -4$
$2x + 10y = 4$

3. $(2, -1)$
$3x + 4y = 2$
$2x + 5y = 1$

4. $(4, 2)$
$2x - 5y = -2$
$3x + 4y = 18$

5. $(5, -3)$
$y = 2x - 13$
$4x + 9y = -7$

6. $(-3, -4)$
$y = 3x + 5$
$5x - 2y = -7$

In Exercises 7–24, solve each system by graphing. Identify systems with no solution and systems with infinitely many solutions, using set notation to express their solution sets.

7. $x + y = 4$
$x - y = 2$

8. $x + y = 6$
$x - y = -4$

9. $2x + y = 4$
$y = 4x + 1$

10. $x + 2y = 4$
$y = -2x - 1$

11. $3x - 2y = 6$
$x - 4y = -8$

12. $4x + y = 4$
$3x - y = 3$

13. $2x + 3y = 6$
$4x = -6y + 12$

14. $3x - 3y = 6$
$2x = 2y + 4$

15. $y = 2x - 2$
$y = -5x + 5$

16. $y = -x + 1$
$y = 3x + 5$

17. $3x - y = 4$
$6x - 2y = 4$

18. $2x - y = -4$
$4x - 2y = 6$

19. $2x + y = 4$
$4x + 3y = 10$

20. $4x - y = 9$
$x - 3y = 16$

21. $x - y = 2$
$y = 1$

22. $x + 2y = 1$
$x = 3$

23. $3x + y = 3$
$6x + 2y = 12$

24. $2x - 3y = 6$
$4x - 6y = 24$

In Exercises 25–42, solve each system by the substitution method. Identify inconsistent systems and systems with dependent equations, using set notation to express their solution sets.

25. $x + y = 6$
$y = 2x$

26. $x + y = 10$
$y = 4x$

27. $2x + 3y = 9$
$x = y + 2$

28. $3x - 4y = 18$
$y = 1 - 2x$

29. $y = -3x + 7$
$5x - 2y = 8$

30. $x = 3y + 8$
$2x - y = 6$

31. $4x + y = 5$
$2x - 3y = 13$

32. $x - 3y = 3$
$3x + 5y = -19$

33. $x - 2y = 4$
$2x - 4y = 5$

34. $x - 3y = 6$
$2x - 6y = 5$

35. $2x + 5y = -4$
$3x - y = 11$

36. $2x + 5y = 1$
$-x + 6y = 8$

37. $2(x - 1) - y = -3$
$y = 2x + 3$

38. $x + y - 1 = 2(y - x)$
$y = 3x - 1$

39. $\frac{x}{4} - \frac{y}{4} = -1$
$x + 4y = -9$

40. $\frac{x}{6} - \frac{y}{2} = \frac{1}{3}$
$x + 2y = -3$

41. $y = \frac{2}{5}x - 2$
$2x - 5y = 10$

42. $y = \frac{1}{3}x + 4$
$3y = x + 12$

In Exercises 43–58, solve each system by the addition method. Identify inconsistent systems and systems with dependent equations, using set notation to express their solution sets.

43. $x + y = 7$
$x - y = 3$

44. $2x + y = 3$
$x - y = 3$

45. $12x + 3y = 15$
$2x - 3y = 13$

46. $4x + 2y = 12$
$3x - 2y = 16$

47. $x + 3y = 2$
$4x + 5y = 1$

48. $x + 2y = -1$
$2x - y = 3$

49. $6x - y = -5$
$4x - 2y = 6$

50. $x - 2y = 5$
$5x - y = -2$

51. $3x - 5y = 11$
$2x - 6y = 2$

52. $4x - 3y = 12$
$3x - 4y = 2$

53. $2x - 5y = 13$
$5x + 3y = 17$

54. $4x + 5y = -9$
$6x - 3y = -3$

55. $2x + 6y = 8$
$3x + 9y = 12$

56. $x - 3y = -6$
$3x - 9y = 9$

57. $2x - 3y = 4$
$4x + 5y = 3$

58. $4x - 3y = 8$
$2x - 5y = -14$

In Exercises 59–82, solve each system by the method of your choice. Identify inconsistent systems and systems with dependent equations, using set notation to express solution sets.

59. $3x - 7y = 1$
$2x - 3y = -1$

60. $2x - 3y = 2$
$5x + 4y = 51$

61. $x = y + 4$
$3x + 7y = -18$

62. $y = 3x + 5$
$5x - 2y = -7$

63. $9x + \frac{4y}{3} = 5$
$4x - \frac{y}{3} = 5$

64. $\frac{x}{6} - \frac{y}{5} = -4$
$\frac{x}{4} - \frac{y}{6} = -2$

65. $\frac{1}{4}x - \frac{1}{9}y = \frac{2}{3}$
$\frac{1}{2}x - \frac{1}{3}y = 1$

66. $\frac{1}{16}x - \frac{3}{4}y = -1$
$\frac{3}{4}x + \frac{5}{2}y = 11$

67. $x = 3y - 1$
$2x - 6y = -2$

68. $x = 4y - 1$
$2x - 8y = -2$

69. $y = 2x + 1$
$y = 2x - 3$

70. $y = 2x + 4$
$y = 2x - 1$

71. $0.4x + 0.3y = 2.3$
$0.2x - 0.5y = 0.5$

72. $0.2x - y = -1.4$
$0.7x - 0.2y = -1.6$

73. $5x - 40 = 6y$
$2y = 8 - 3x$

74. $4x - 24 = 3y$
$9y = 3x - 1$

75. $3(x + y) = 6$
$3(x - y) = -36$

76. $4(x - y) = -12$
$4(x + y) = -20$

77. $3(x - 3) - 2y = 0$
$2(x - y) = -x - 3$

78. $5x + 2y = -5$
$4(x + y) = 6(2 - x)$

79. $x + 2y - 3 = 0$
$12 = 8y + 4x$

80. $2x - y - 5 = 0$
$10 = 4x - 2y$

81. $3x + 4y = 0$
$7x = 3y$

82. $5x + 8y = 20$
$4y = -5x$

Application Exercises

83. In the United States, deaths from car accidents, per 100,000 persons, are decreasing at a faster rate than deaths from gunfire, shown by the scatter plots and the regression lines. The formula $y = -0.4x + 28$ models deaths from car accidents, y, per 100,000 persons, x years after 1965. The formula $0.07x + y = 15$ models deaths from gunfire, y, per 100,000 persons, x years after 1965. Use these models to project when the number of deaths from gunfire will equal the number of deaths from car accidents. Round to the nearest year. How many annual deaths, per 100,000 persons, will there be from gunfire and from car accidents at that time? Describe how this is illustrated by the regression lines in the figure shown.

Annual Deaths in the U.S. from Car Accidents and Gunfire

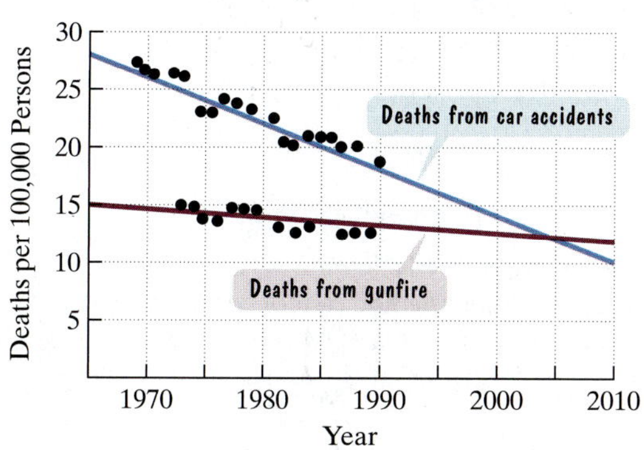

Source: Journal of the American Medical Association

84. The formula $y = 0.52x + 35.7$ models the percentage, y, of U.S. women in the work force x years after 1955. The formula $0.25x + y = 85.4$ models the percentage, y, of U.S. men in the work force x years after 1955. Use these models to determine when the percentage of women in the work force will be the same as the percentage of men. Round to the nearest year. What percentage of women and what percentage of men will be in the work force at that time?

85. The June 7, 1999 issue of *Newsweek* presented statistics showing progress African Americans have made in education, health, and finance. Infant mortality for blacks is decreasing at a faster rate than it is for whites, shown by the graphs. Infant mortality for blacks can be modeled by $M = -0.41x + 22$ and for whites by $M = -0.18x + 10$. In both models, x is the number of years after 1980 and M is infant mortality, measured in deaths per 1000 live births. Use these models to project when infant mortality for blacks and whites will be the same. What is infant mortality for both groups at that time?

Infant Mortality

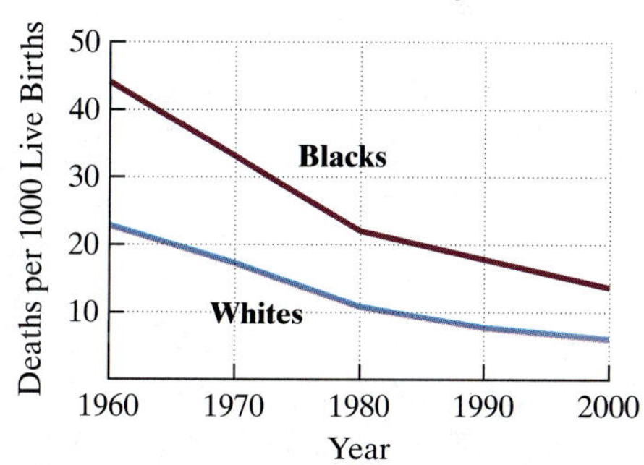

Source: National Center for Health Statistics

86. Life expectancies for women have been longer than life expectancies for men for many years. However, life expectancies for men are now increasing at a faster rate than those for women. Life expectancy for men can be modeled by $E = 0.19x + 54.91$ and for women by $E = 0.11x + 68.41$. In both models, x is the number of birth years after 1900 and E is life expectancy, in years. Use these models to project when life expectancies for women and men will be the same. Round to the nearest year. What are the life expectancies for women and men at that time? Do you feel confident about this prediction? Explain why or why not.

An important application of systems of equations arises in connection with supply and demand. As the price of a product increases, the demand for that product decreases. However, at higher prices suppliers are willing to produce greater quantities of the product. Exercises 87–88 involve supply and demand.

87. A chain of video stores specializes in cult films. The weekly demand and supply models for *The Rocky Horror*

Picture Show are given by

$$N = -13p + 760 \quad \text{Demand model}$$
$$N = 2p + 430 \quad \text{Supply model}$$

in which p is the price of the video and N is the number of copies of the video sold or supplied each week to the chain of stores.

a. How many copies of the video can be sold and supplied at \$18 per copy?

b. Find the price at which supply and demand are equal. At this price, how many copies of *Rocky Horror* can be supplied and sold each week?

88. At a price of p dollars per ticket, the number of tickets to a rock concert that can be sold is given by the demand model $N = -25p + 7500$. At a price of p dollars per ticket, the number of tickets that the concert's promoters are willing to make available is given by the supply model $N = 5p + 6000$.

a. How many tickets can be sold and supplied for \$40 per ticket?

b. Find the ticket price at which supply and demand are equal. At this price, how many tickets will be supplied and sold?

Writing in Mathematics

89. What is a system of linear equations? Provide an example with your description.

90. What is a solution to a system of linear equations?

91. Explain how to determine if an ordered pair is a solution of a system of linear equations.

92. Explain how to solve a system of linear equations by graphing.

93. Explain how to solve a system of equations using the substitution method. Use $y = 3 - 3x$ and $3x + 4y = 6$ to illustrate your explanation.

94. Explain how to solve a system of equations using the addition method. Use $5x + 8y = -1$ and $3x + y = 7$ to illustrate your explanation.

95. When is it easier to use the addition method rather than the substitution method to solve a system of equations?

96. When using the addition or substitution method, how can you tell if a system of linear equations has no solution? What is the relationship between the graphs of the two equations?

97. When using the addition or substitution method, how can you tell if a system of linear equations has infinitely many solutions? What is the relationship between the graphs of the two equations?

98. The formula $y = 0.94x + 5.64$ models annual U.S. consumption of chicken, y, in pounds per person, x years after 1950. The formula $0.74x + y = 146.76$ models annual U.S. consumption of red meat, y, in pounds per person, x years after 1950. What is the most efficient method for solving this system? What does the solution mean in terms of the variables in the formulas? (It is not necessary to solve the system.)

99. In Exercise 98, find the slope of each model. Describe what this means in terms of the rate of change of chicken consumption and the rate of change of red meat consumption. Why must the graphs have an intersection point? What happens to the right of the intersection point?

Technology Exercises

100. Verify your solutions to any five exercises from 7 through 24 by using a graphing utility to graph the two equations in the system in the same viewing rectangle. Then use the intersection feature to display the solution.

101. Some graphing utilities can give the solution to a linear system of equations. (Consult your manual for details.) This capability is usually accessed with the [SIMULT] (simultaneous equations) feature. First, you will enter 2, for two equations in two variables. With each equation in $Ax + By = C$ form, you will then enter the coefficients for x and y and the constant term, one equation at a time. After entering all six numbers, press [SOLVE]. The solution will be displayed on the screen. (The x-value may be displayed as $x_1 =$ and the y-value as $x_2 =$.) Use this capability to verify the solutions to any five of the exercises you solved in the practice exercises of this exercise set. Describe what happens when you use your graphing utility on a system with no solution or infinitely many solutions.

Critical Thinking Exercises

102. Which one of the following is true?

a. The addition method cannot be used to eliminate either variable in a system of two equations in two variables.

b. The solution set of the system

$$5x - y = 1$$
$$10x - 2y = 2$$

is $\{(2, 9)\}$.

c. A linear system of equations can have a solution set consisting of precisely two ordered pairs.

d. The solution set of the system

$$y = 4x - 3$$
$$y = 4x + 5$$

is the empty set.

103. Write a system of equations having $\{(-2, 7)\}$ as a solution set. (More than one system is possible.)

104. Solve the system for x and y in terms of a_1, b_1, c_1, a_2, b_2, and c_2:

$$a_1x + b_1y = c_1$$
$$a_2x + b_2y = c_2.$$

Review Exercises

105. Solve: $6x = 10 + 5(3x - 4)$. (Section 1.4, Example 3)

106. Simplify: $(4x^2y^4)^2(-2x^5y^0)^3$. (Section 1.6, Example 9)

107. If $f(x) = x^2 - 3x + 7$, find $f(-1)$. (Section 2.1, Example 3)

▸ SECTION 3.2 Problem Solving and Business Applications Using Systems of Equations

Objectives

1. Solve problems using systems of equations.
2. Use functions to model revenue, cost, and profit.
3. Perform a break-even analysis.

SSM | PH Tutor Center | CD-ROM | Video | www

Driving through your neighborhood, you see kids selling lemonade. Would it surprise you to know that this activity can be analyzed using functions and systems of equations? By doing so, you will view profit and loss in the business world in a new way. In this section, we use systems of equations to solve problems and model business ventures.

1 Solve problems using systems of equations.

A Strategy for Solving Word Problems Using Systems of Equations

When we solved problems in Chapter 1, we let x represent a quantity that was unknown. Problems in this section involve two unknown quantities. We will let x and y represent these quantities. We then translate from the verbal conditions of the problem to a *system* of linear equations.

EXAMPLE 1 Solving a Problem Involving Energy Efficiency of Building Materials

A heat-loss survey by an electric company indicated that a wall of a house containing 40 square feet of glass and 60 square feet of plaster lost 1920 Btu (British thermal units) of heat. A second wall containing 10 square feet of glass and 100 square feet of plaster lost 1160 Btu of heat. Determine the heat lost per square foot for the glass and for the plaster.

Solution

Step 1 Use variables to represent unknown quantities.
Let x = the heat lost per square foot for the glass.
Let y = the heat lost per square foot for the plaster.

Step 2 Write a system of equations describing the problem's conditions. The heat loss for each wall is the heat lost by the glass plus the heat lost by the plaster. One wall containing 40 square feet of glass and 60 square feet of plaster lost 1920 Btu of heat.

Heat lost by the glass + heat lost by the plaster = total heat lost.

$$\begin{pmatrix}\text{Number}\\ \text{of ft}^2\end{pmatrix}\cdot\begin{pmatrix}\text{heat lost}\\ \text{per ft}^2\end{pmatrix} + \begin{pmatrix}\text{number}\\ \text{of ft}^2\end{pmatrix}\cdot\begin{pmatrix}\text{heat lost}\\ \text{per ft}^2\end{pmatrix} = \text{total heat lost.}$$

$$40 \cdot x + 60 \cdot y = 1920$$

A second wall containing 10 square feet of glass and 100 square feet of plaster lost 1160 Btu of heat.

$$\begin{pmatrix}\text{Number}\\ \text{of ft}^2\end{pmatrix}\cdot\begin{pmatrix}\text{heat lost}\\ \text{per ft}^2\end{pmatrix} + \begin{pmatrix}\text{number}\\ \text{of ft}^2\end{pmatrix}\cdot\begin{pmatrix}\text{heat lost}\\ \text{per ft}^2\end{pmatrix} = \text{total heat lost.}$$

$$10 \cdot x + 100 \cdot y = 1160$$

Step 3 Solve the system and answer the problem's question. The system

$$40x + 60y = 1920$$
$$10x + 100y = 1160$$

can be solved by addition. We'll multiply the second equation by −4 and then add equations to eliminate x.

$$\begin{aligned} 40x + 60y &= 1920 \xrightarrow{\text{No change}} & 40x + 60y &= 1920\\ 10x + 100y &= 1160 \xrightarrow{\text{Multiply by } -4.} & -40x - 400y &= -4640\\ &\text{Add:} & -340y &= -2720\\ && y &= \frac{-2720}{-340} = 8 \end{aligned}$$

Now we can find the value of x by back-substituting 8 for y in either of the system's equations.

$$10x + 100y = 1160 \quad \text{We'll use the second equation.}$$
$$10x + 100(8) = 1160 \quad \text{Back-substitute 8 for } y.$$
$$10x + 800 = 1160 \quad \text{Multiply.}$$
$$10x = 360 \quad \text{Subtract 800 from both sides.}$$
$$x = 36 \quad \text{Divide both sides by 10.}$$

We see that $x = 36$ and $y = 8$. Because x represents heat lost per square foot for the glass and y for the plaster, the glass lost 36 Btu of heat per square foot and the plaster lost 8 Btu per square foot.

Step 4 Check the proposed solution in the original wording of the problem. The problem states that the wall with 40 square feet of glass and 60 square feet of plaster lost 1920 Btu.

$$40(36) + 60(8) = 1440 + 480 = 1920 \text{ Btu of heat}$$

Proposed solution is 36 Btu per ft^2 for glass and 8 Btu per ft^2 for plaster.

Our proposed solution checks with the first statement. The problem also states that the wall with 10 square feet of glass and 100 square feet of plaster lost 1160 Btu.

$$10(36) + 100(8) = 360 + 800 = 1160 \text{ Btu of heat}$$

Our proposed solution also checks with the second statement. ■

✔ CHECK POINT 1 One pan pizza and two beef burritos provide 1980 calories. Two pan pizzas and one beef burrito provide 2670 calories. Find the number of calories in each item.

Next, we will solve problems involving investments, mixtures, and motion with systems of equations. We will continue using our four-step problem-solving strategy. We will also use tables to help organize the information in the problems.

Dual Investments with Simple Interest

Simple interest involves interest calculated only on the amount of money that we invest, called the **principal**. The formula $I = Pr$ is used to find the simple interest, I, earned for one year when the principal, P, is invested at an annual interest rate, r. Dual investment problems involve different amounts of money in two or more investments, each paying a different rate.

EXAMPLE 2 Solving a Dual Investment Problem

Your grandmother needs your help. She has \$50,000 to invest. Part of this money is to be invested in noninsured bonds paying 15% annual interest. The rest of this money is to be invested in a government-insured certificate of deposit paying 7% annual interest. She told you that she requires \$6000 per year in extra income from both of these investments. How much money should be placed in each investment?

Solution

Step 1 Use variables to represent unknown quantities.
Let x = the amount invested in the 15% noninsured bonds.
Let y = the amount invested in the 7% certificate of deposit.

Step 2 Write a system of equations describing the problem's conditions. Because Grandma has \$50,000 to invest,

The amount invested at 15%	plus	the amount invested at 7%	equals	\$50,000.
x	$+$	y	$=$	50,000

Furthermore, Grandma requires \$6000 in total interest. We can use a table to organize the information in the problem and obtain a second equation.

	Principal (amount invested)	×	Interest rate	=	Interest earned
15% Investment	x		0.15		$0.15x$
7% Investment	y		0.07		$0.07y$

The interest for the two investments combined must be \$6000.

Interest from the 15% investment	plus	interest from the 7% investment	is	\$6000.
$0.15x$	$+$	$0.07y$	$=$	6000

Step 3 Solve the system and answer the problem's question. The system

$$x + y = 50{,}000$$
$$0.15x + 0.07y = 6000$$

can be solved by substitution or addition. Substitution works well because both variables in the first equation have coefficients of 1. Addition also works well; if we multiply the first equation by -0.15 or -0.07, adding equations will eliminate a variable. We will use addition.

$$\begin{array}{rcl} x + y = 50{,}000 & \xrightarrow{\text{Multiply by } -0.07.} & -0.07x - 0.07y = -3500 \\ 0.15x + 0.07y = 6000 & \xrightarrow{\text{No change}} & \underline{0.15x + 0.07y = 6000} \\ & \text{Add:} & 0.08x = 2500 \\ & & x = \dfrac{2500}{0.08} = 31{,}250 \end{array}$$

Because x represents the amount that should be invested at 15%, Grandma should place \$31,250 in 15% noninsured bonds. Now we can find y, the amount that she should place in the 7% certificate of deposit. We do so by back-substituting 31,250 for x in either of the system's equations.

$$x + y = 50{,}000 \quad \text{We'll use the first equation.}$$
$$31{,}250 + y = 50{,}000 \quad \text{Back-substitute 31,250 for x.}$$
$$y = 18{,}750 \quad \text{Subtract 31,250 from both sides.}$$

Because $x = 31{,}250$ and $y = 18{,}750$, Grandma should invest \$31,250 at 15% and \$18,750 at 7%.

Step 4 Check the proposed answers in the original wording of the problem. Has Grandma invested \$50,000?

$$\$31{,}250 + \$18{,}750 = \$50{,}000$$

Yes, all her money was placed in the dual investments. Can she count on \$6000 interest? The interest earned on \$31,250 at 15% is (\$31,250)(0.15), or \$4687.50. The interest earned on \$18,750 at 7% is (\$18,750)(0.07), or \$1312.50. The total interest is \$4687.50 + \$1312.50, or \$6000, exactly as it should be. You've made your grandmother happy. (Now if you would just visit her more often . . .) ■

✔ **CHECK POINT 2** You inherited \$16,000 with the stipulation that for the first year the money had to be invested in two accounts paying 6% and 8% annual interest. How much did you invest at each rate if the total interest earned for the year was \$1180?

Problems Involving Mixtures

Chemists and pharmacists often have to change the concentration of solutions and other mixtures. In these situations, the amount of a particular ingredient in the solution or mixture is expressed as a percentage of the total.

EXAMPLE 3 Solving a Mixture Problem

A chemist working on a flu vaccine needs to mix a 10% sodium-iodine solution with a 60% sodium-iodine solution to obtain 50 milliliters of a 30% sodium-iodine solution. How many milliliters of the 10% solution and of the 60% solution should be mixed?

Solution

Step 1 Use variables to represent unknown quantities.
Let x = the number of milliliters of the 10% solution to be used in the mixture.
Let y = the number of milliliters of the 60% solution to be used in the mixture.

Step 2 Write a system of equations describing the problem's conditions. The situation is illustrated in Figure 3.6.

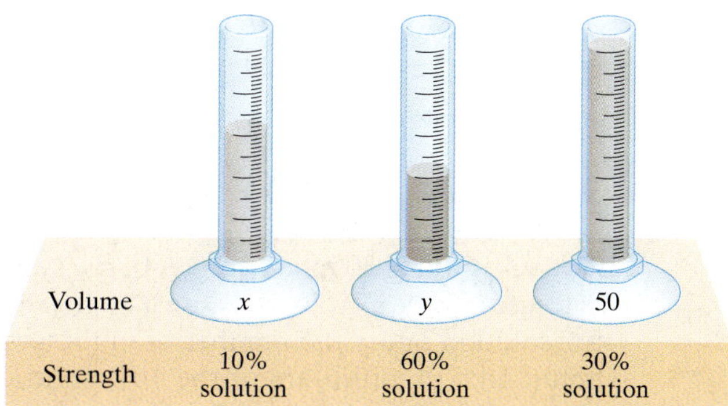

Figure 3.6

The chemist needs 50 milliliters of a 30% sodium-iodine solution. We form a table that shows the amount of sodium-iodine in each of the three solutions.

Solution	Number of milliliters	×	Percent of Sodium-Iodine	=	Amount of Sodium-Iodine
10% Solution	x		$10\% = 0.1$		$0.1x$
60% Solution	y		$60\% = 0.6$		$0.6y$
30% Mixture	50		$30\% = 0.3$		$0.3(50) = 15$

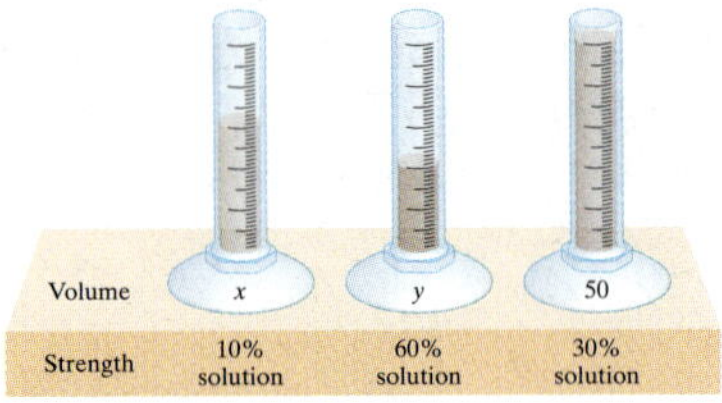

Figure 3.6, repeated

The chemist needs to obtain a 50-milliliter mixture.

The number of milliliters used of the 10% solution	plus	The number of milliliters used of the 60% solution	must equal	50 milliliters.
x	$+$	y	$=$	50

The 50-milliliter mixture must be 30% sodium-iodine. The amount of sodium-iodine must be 30% of 50, or $(0.3)(50) = 15$ milliliters.

Amount of sodium-iodine in the 10% solution	plus	amount of sodium-iodine in the 60% solution	equals	amount of sodium-iodine in the mixture.
$0.1x$	$+$	$0.6y$	$=$	15

Step 3 Solve the system and answer the problem's question. The system

$$x + y = 50$$
$$0.1x + 0.6y = 15$$

can be solved by substitution or addition. Let's use substitution. Solving the first equation for y, we obtain $y = 50 - x$.

$$y = \boxed{50 - x} \qquad 0.1x + 0.6\boxed{y} = 15$$

We substitute $50 - x$ for y in the second equation. This gives us an equation in one variable.

$$0.1x + 0.6(50 - x) = 15 \quad \text{This equation contains one variable, } x.$$
$$0.1x + 30 - 0.6x = 15 \quad \text{Apply the distributive property.}$$
$$-0.5x + 30 = 15 \quad \text{Combine like terms.}$$
$$-0.5x = -15 \quad \text{Subtract 30 from both sides.}$$
$$x = \frac{-15}{-0.5} = 30 \quad \text{Divide both sides by } -0.5.$$

Back-substituting 30 for x in either of the system's equations ($x + y = 50$ is easier to use) gives $y = 20$. Because x represents the number of milliliters of the 10% solution and y the number of milliliters of the 60% solution, the chemist should mix 30 milliliters of the 10% solution with 20 milliliters of the 60% solution.

Step 4 Check the proposed solution in the original wording of the problem. The problem states that the chemist needs 50 milliliters of a 30% sodium-iodine solution. The amount of sodium-iodine in this mixture is 0.3(50), or 15 milliliters. The amount of sodium-iodine in 30 milliliters of the 10% solution is 0.1(30), or 3 milliliters. The amount of sodium-iodine in 20 milliliters of the 60% solution is $0.6(20) = 12$ milliliters. The amount of sodium-iodine in the two solutions used in the mixture is 3 milliliters + 12 milliliters, or 15 milliliters, exactly as it should be. ■

✔ **CHECK POINT 3** A chemist needs to mix an 18% acid solution with a 45% acid solution to obtain 12 liters of a 36% acid solution. How many liters of each of the acid solutions must be used?

Problems Involving Motion

We have seen that the distance, d, covered by any moving body is the product of its average rate, r, and its time in motion, t:

$$d = rt. \qquad \text{Distance equals rate times time.}$$

Wind speed and water current have the effect of increasing or decreasing a traveler's rate.

EXAMPLE 4 Solving a Motion Problem

When a small airplane flies with the wind, it can travel 450 miles in 3 hours. When the same airplane flies in the opposite direction against the wind, it takes 5 hours to fly the same distance. Find the speed of the plane in still air and the speed of the wind.

Solution

Step 1 Use variables to represent unknown quantities.
Let x = the speed of the plane in still air.
Let y = the speed of the wind.

Step 2 Write a system of equations describing the problem's conditions. As it travels with the wind, the plane's speed is increased. The net rate is its speed in still air, x, plus the speed of the wind, y, given by the expression $x + y$. As it travels against the wind, the plane's speed is decreased. The net rate is its speed in still air, x, minus the speed of the wind, y, given by the expression $x - y$. Here is a chart that summarizes the problem's information and includes the increased and decreased rates.

	Rate	×	Time	=	Distance
Trip with the Wind	$x + y$		3		$3(x + y)$
Trip against the Wind	$x - y$		5		$5(x - y)$

The problem states that the distance in each direction is 450 miles. We use this information to write our system of equations.

The distance of the trip with the wind is 450 miles.

$$3(x + y) = 450$$

The distance of the trip against the wind is 450 miles.

$$5(x - y) = 450$$

Step 3 Solve the system and answer the problem's question. We can simplify the system by dividing both sides of the equations by 3 and 5, respectively.

$$\begin{aligned} 3(x + y) &= 450 \xrightarrow{\text{Divide by 3.}} x + y = 150 \\ 5(x - y) &= 450 \xrightarrow{\text{Divide by 5.}} x - y = 90 \end{aligned}$$

Solve the system on the right by the addition method.

$$\begin{aligned} x + y &= 150 \\ x - y &= 90 \\ \text{Add: } 2x \quad &= 240 \\ x &= 120 \quad \text{Divide both sides by 2.} \end{aligned}$$

Back-substituting 120 for x in either of the system's equations gives $y = 30$. Because $x = 120$ and $y = 30$, the speed of the plane in still air is 120 miles per hour and the speed of the wind is 30 miles per hour.

Step 4 Check the proposed solution in the original wording of the problem. The problem states that the distance in each direction is 450 miles. The speed of the plane with the wind is $120 + 30 = 150$ miles per hour. In 3 hours, it travels $150 \cdot 3$, or 450 miles, which checks with the stated condition. Furthermore, the speed of the plane against the wind is $120 - 30 = 90$ miles per hour. In 5 hours, it travels $90 \cdot 5 = 450$ miles, which is the stated distance. ■

✔ **CHECK POINT 4** With the current, a motorboat can travel 84 miles in 2 hours. Against the current, the same trip takes 3 hours. Find the speed of the boat in still water and the speed of the current.

2 Use functions to model revenue, cost, and profit.

Functions of Business: Break-Even Analysis

As a young entrepreneur, did you ever try selling lemonade in your front yard? Suppose that you charged 55 cents for each cup and you sold 45 cups. Your **revenue** is your income from selling these 45 units, or $\$0.55(45) = \$24.75$. Your *revenue function* from selling x cups is

$$R(x) = 0.55x.$$

This is the unit price: 55¢ for each cup. This is the number of units sold.

For any business, the **revenue function**, $R(x)$, is the money generated by selling x units of the product:

$$R(x) = px.$$

Price per unit x units sold

Back to selling lemonade and energizing the neighborhood with white sugar: Is your revenue for the afternoon also your profit? No. We need to consider the cost of the business. You estimate that the lemons, white sugar, and bottled water cost 5 cents per cup. Furthermore, mommy dearest is charging you a \$10 rental fee for use of your (her?) front yard. Your *cost function* for selling x cups of lemonade is

$$C(x) = 10 + 0.05x.$$

This is your \$10 fixed cost. This is your variable cost: 5¢ for each cup produced.

For any business, the **cost function**, $C(x)$, is the cost of producing x units of the product:

$$C(x) = (\text{fixed cost}) + cx.$$

Cost per unit x units produced

The term on the right, cx, represents **variable cost**, because it varies based on the number of units produced. Thus, the cost function is the sum of the fixed cost and the variable cost.

What does every entrepreneur, from a kid selling lemonade to Donald Trump, want to do? Generate profit, of course. The *profit* made is the money taken in, or the revenue, minus the money spent, or the cost. This relationship between revenue and cost allows us to define the *profit function*, $P(x)$.

The Profit Function

The profit, $P(x)$, generated after producing and selling x units of a product is given by the **profit function**

$$P(x) = R(x) - C(x)$$

where R and C are the revenue and cost functions, respectively.

Figure 3.7 shows the graphs of the revenue and cost functions for the lemonade business. Similar graphs and models apply no matter how small or large a business venture may be.

$$R(x) = 0.55x \qquad\qquad C(x) = 10 + 0.05x$$

Revenue is 55¢ times the number of cups sold.

Cost is $10 plus 5¢ times the number of cups produced.

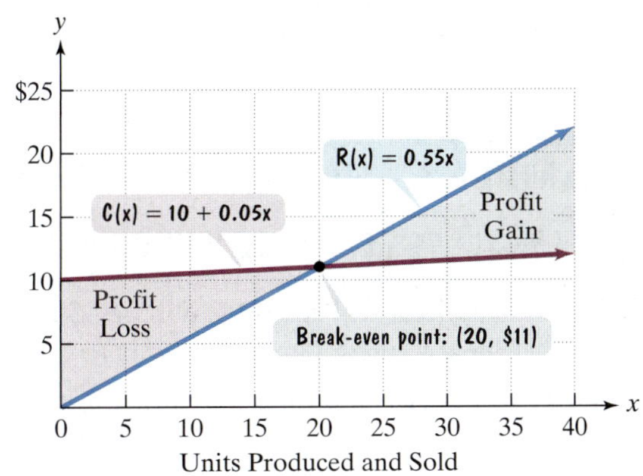

Figure 3.7

The lines intersect at the point (20, 11). This means that when 20 cups are produced and sold, both cost and revenue are $11. In business, this point of intersection is called the **break-even point**. At the break-even point, the money coming in is equal to the money going out. Can you see what happens for x-values less than 20? The cost graph is above the revenue graph. The cost is greater than the revenue and the business is losing money. Thus, if you sell fewer than 20 cups of lemonade, the result is a *profit loss*. By contrast, look at what happens for x-values greater than 20. The revenue graph is above the cost graph. The revenue is greater than the cost and the business is making money. Thus, if you sell more than 20 cups of lemonade, the result is a *profit gain*.

EXAMPLE 5 Writing a Profit Function

Use the revenue and cost functions shown in Figure 3.7,

$$R(x) = 0.55x \quad \text{and} \quad C(x) = 10 + 0.05x,$$

to write the profit function for producing and selling x cups of lemonade.

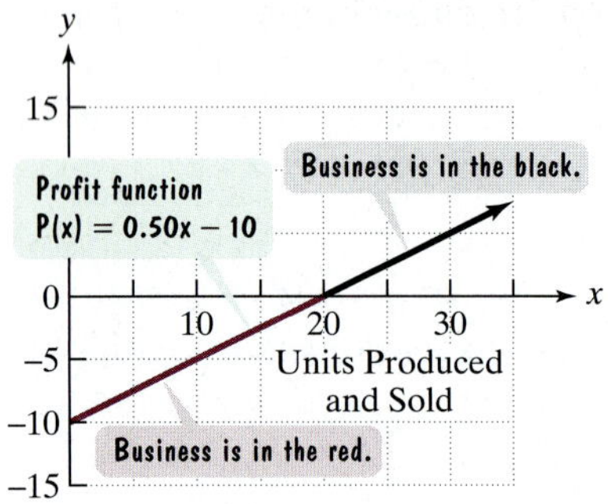

Figure 3.8 In the red and in the black

Solution The profit function is the difference between the revenue function and the cost function.

$$
\begin{aligned}
P(x) &= R(x) - C(x) && \text{This is the definition of the profit function.}\\
&= 0.55x - (10 + 0.05x) && \text{Substitute the given functions.}\\
&= 0.55x - 10 - 0.05x && \text{Distribute } -1 \text{ to each term in parentheses.}\\
&= 0.50x - 10 && \text{Simplify.}
\end{aligned}
$$

The profit function is $P(x) = 0.50x - 10$. ■

The graph of the profit function, which can be expressed as $y = 0.50x - 10$, is shown in Figure 3.8. The red portion of the graph lies below the x-axis and shows a profit loss when fewer than 20 units are sold. The lemonade business is "in the red." The black portion of the graph lies above the x-axis and shows a profit gain when more than 20 units are sold. The lemonade business is "in the black."

✔ **CHECK POINT 5** Use the revenue and cost functions

$$R(x) = 200x$$
$$C(x) = 160{,}000 + 75x$$

to write the profit function for producing and selling x units.

3 Perform a break-even analysis.

EXAMPLE 6 Finding a Break-Even Point

Technology is now promising to bring light, fast, and beautiful wheelchairs to millions of disabled people. A company is planning to manufacture these radically different wheelchairs. Fixed cost will be \$500,000 and it will cost \$400 to produce each wheelchair. Each wheelchair will be sold for \$600.

a. Write the cost function, C, of producing x wheelchairs.

b. Write the revenue function, R, from the sale of x wheelchairs.

c. Determine the break-even point. Describe what this means.

Solution

a. The cost function is the sum of the fixed cost and variable cost.

Fixed cost of \$500,000 | plus | Variable cost: \$400 for each chair produced

$$C(x) = 500{,}000 + 400x$$

b. The revenue function is the money generated from the sale of x wheelchairs.

Revenue per chair, \$600, times | the number of chairs sold

$$R(x) = 600x$$

c. The break-even point occurs where the graphs of C and R intersect. Thus, we find this point by solving the system

$$
\begin{aligned}
C(x) &= 500{,}000 + 400x \\
R(x) &= 600x
\end{aligned}
\quad \text{or} \quad
\begin{aligned}
y &= 500{,}000 + 400x \\
y &= 600x.
\end{aligned}
$$

Using Technology

The graphs of the wheelchair company's cost and revenue functions

$$C(x) = 500{,}000 + 400x$$

and $R(x) = 600x$
are shown in a

$$[0, 5000, 1000] \text{ by}$$

$$[0, 3{,}000{,}000, 1{,}000{,}000]$$

viewing rectangle. The intersection point confirms that the company breaks even by producing and selling 2500 wheelchairs.

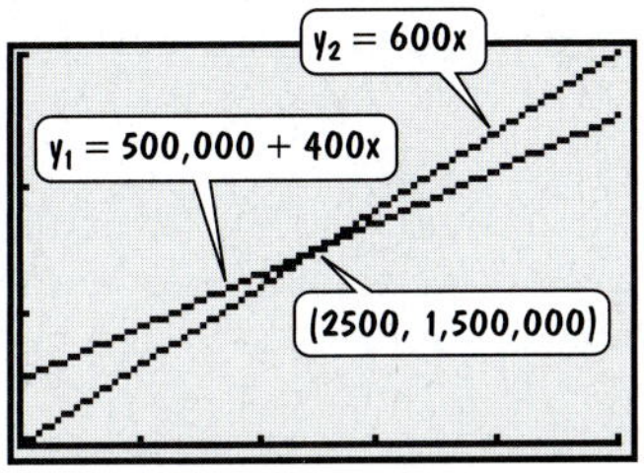

Using substitution, we can substitute $600x$ for y in the first equation.

$$600x = 500{,}000 + 400x \quad \text{Substitute } 600x \text{ for } y.$$

$$200x = 500{,}000 \quad \text{Subtract } 400x \text{ from both sides.}$$

$$x = 2500 \quad \text{Divide both sides by 200.}$$

Back-substituting 2500 for x in either of the system's equations (or functions), we obtain

$$R(2500) = 600(2500) = 1{,}500{,}000.$$

We used $R(x) = 600x$.

The break-even point is (2500, 1,500,000). This means that the company will break even if it produces and sells 2500 wheelchairs. At this level, the money coming in is equal to the money going out: $1,500,000. ■

✔ CHECK POINT 6 A company that manufactures running shoes has a fixed cost of $300,000. Additionally, it costs $30 to produce each pair of shoes. They are sold at $80 per pair.

a. Write the cost function, C, of producing x pairs of running shoes.

b. Write the revenue function, R, from the sale of x pairs of running shoes.

c. Determine the break-even point. Describe what this means.

EXERCISE SET 3.2

Practice Exercises

In Exercises 1–4, let x represent one number and let y represent the other number. Use the given conditions to write a system of equations. Solve the system and find the numbers.

1. The sum of two numbers is 7. If one number is subtracted from the other, the result is −1. Find the numbers.

2. The sum of two numbers is 2. If one number is subtracted from the other, the result is 8. Find the numbers.

3. Three times a first number decreased by a second number is 1. The first number increased by twice the second number is 12. Find the numbers.

4. The sum of three times a first number and twice a second number is 8. If the second number is subtracted from twice the first number, the result is 3. Find the numbers.

In Exercises 5–8, cost and revenue functions for producing and selling x units of a product are given. Cost and revenue are expressed in dollars.

a. *Write the profit function from producing and selling x units of the product.*

b. *Find the number of units that must be produced and sold to break even. At this level, what is the dollar amount coming in and going out?*

5. $C(x) = 25{,}500 + 15x$
$R(x) = 32x$

6. $C(x) = 15{,}000 + 12x$
$R(x) = 32x$

7. $C(x) = 105x + 70{,}000$
$R(x) = 245x$

8. $C(x) = 1.2x + 1500$
$R(x) = 1.7x$

Application Exercises

In Exercises 9–36, use the four-step strategy to solve each problem.

9. The graph shows the number of divorced Americans, in millions, by age group. Combined, there are 8.5 million divorced Americans in the 40–44 and 45–54 age groups. The difference between the number of divorces for the two groups is 2.3 million. Find the number, in millions, of divorced Americans in each of these two age groups.

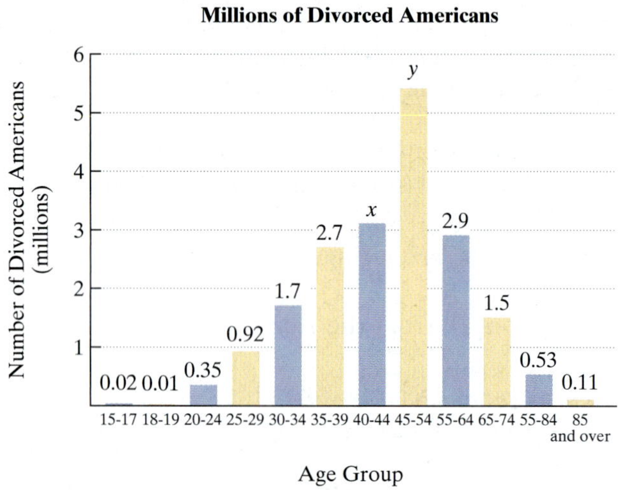

Source: National Center for Health Statistics

10. People in the United States produce more garbage than people in any other country. The graph shows the pounds of trash per person per day in the top five garbage-producing nations. Combined, one American and one Canadian produce 8.1 pounds of trash per day. The difference between the waste per person per day in the United States and Canada is 0.5 pound. Find the number of pounds of trash per person per day in the United States and Canada.

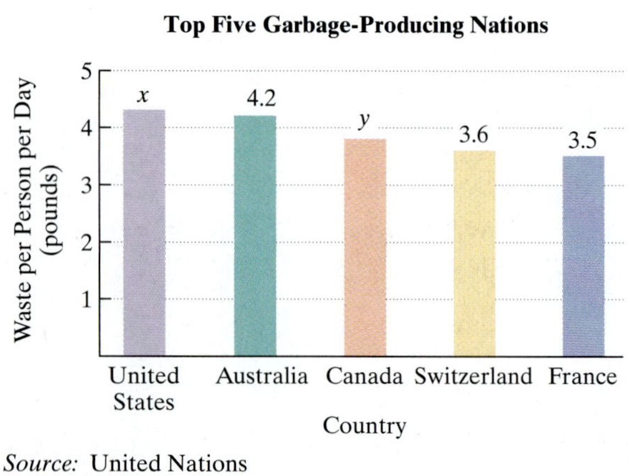

Source: United Nations

11. The verdict is in: After years of research, the nation's health experts agree that high cholesterol in the blood is a major contributor to heart disease. Cholesterol intake should be limited to 300 milligrams or less each day. Cake can vary in cholesterol content. Four slices of sponge cake and 2 slices of pound cake contain 784 milligrams of cholesterol. One slice of sponge cake and 3 slices of pound cake contain 366 milligrams of cholesterol. Find the cholesterol content in each item.

12. How do the Quarter Pounder and Whopper with cheese measure up in the calorie department? Actually, not too well. Two Quarter Pounders and three Whoppers with cheese provide 2607 calories. Even one of each provide enough calories to bring tears to Jenny Craig's eyes—9 calories in excess of what is allowed on a 1000 calorie-a-day diet. Find the calories in each item.

13. You invest \$20,000 in two accounts paying 7% and 9% annual interest, respectively. If the total interest earned for the year is \$1550, how much was invested at each rate?

14. You invest \$20,000 in two accounts paying 7% and 8% annual interest, respectively. If the total interest earned for the year is \$1520, how much was invested at each rate?

15. A bank loaned out \$250,000, part of it at the rate of 8% annual mortgage interest and the rest at the rate of 18% annual credit card interest. The interest received on both loans totaled \$23,000. How much was loaned at each rate?

16. A bank loaned out \$120,000, part of it at the rate of 8% annual mortgage interest and the rest at the rate of 18% annual credit card interest. The interest received on both loans totaled \$10,000. How much was loaned at each rate?

17. Things did not go quite as planned. You invested \$8000, part of it in a stock that paid 12% annual interest. However, the rest of the money suffered a 5% loss. If the total annual income from both investments was \$620, how much was invested at each rate?

18. Things did not go quite as planned. You invested \$12,000, part of it in a stock that paid 14% annual interest. However, the rest of the money suffered a 6% loss. If the total annual income from both investments was \$680, how much was invested at each rate?

19. A chef needs to mix a 45% fat content cheese with a 20% fat content cheese to obtain 30 grams of a cheese mixture that is 30% fat. How many grams of each kind of cheese must be used?

20. A candy company needs to mix a 30% fat content chocolate with a 12% fat content chocolate to obtain 50 pounds of a 20% fat content chocolate. How many pounds of each kind of chocolate must be used?

21. At the north campus of a small liberal arts college, 10% of the students are women. At the south campus, 50% of the students are women. The campuses are merged into one east campus. If 40% of the 1200 students at the east campus are women, how many students did the north and south campuses have before the merger?

22. At the north campus of a performing arts school, 10% of the students are music majors. At the south campus, 90% of the students are music majors. The campuses are merged into one east campus. If 42% of the 1000 students at the east campus are music majors, how many students did the north and south campuses have before the merger?

23. A grocer needs to mix tea worth \$6.00 per pound with tea worth \$8.00 per pound to obtain 144 pounds of a tea mixture worth \$7.50 per pound. How many pounds of each kind of tea must be used?

24. A grocer needs to mix cashews worth \$6 per pound with peanuts worth \$2 per pound to obtain 10 pounds of a mixture worth \$3 per pound. How many pounds of each kind of nut must be used?

25. When a small plane flies with the wind, it can travel 800 miles in 5 hours. When the plane flies in the opposite direction, against the wind, it takes 8 hours to fly the same distance. Find the speed of the plane in still air and the speed of the wind.

26. When a plane flies with the wind, it can travel 4200 miles in 6 hours. When the plane flies in the opposite direction, against the wind, it takes 7 hours to fly the same distance. Find the speed of the plane in still air and the speed of the wind.

27. A boat's crew rowed 16 kilometers downstream, with the current, in 2 hours. The return trip upstream, against the current, covered the same distance, but took 4 hours. Find the crew's rowing rate in still water and the rate of the current.

28. A motorboat traveled 36 miles downstream, with the current, in 1.5 hours. The return trip upstream, against the current, covered the same distance, but took 2 hours. Find the boat's rate in still water and the rate of the current.

29. With the current, you can canoe 24 miles in 4 hours. Against the same current, you can canoe only $\frac{3}{4}$ of this distance in 6 hours. Find your speed in still water and the speed of the current.

30. With the current, you can row 24 miles in 3 hours. Against the same current, you can row only $\frac{2}{3}$ of this distance in 4 hours. Find your rowing speed in still water and the speed of the current.

31. A rectangular lot whose perimeter is 360 feet is fenced along three sides. An expensive fencing along the lot's length costs \$20 per foot, and an inexpensive fencing along the two side widths costs only \$8 per foot. The total cost of the fencing along the three sides comes to \$3280. What are the lot's dimensions?

32. A rectangular lot whose perimeter is 320 feet is fenced along three sides. An expensive fencing along the lot's length costs \$16 per foot, and an inexpensive fencing along the two side widths costs only \$5 per foot. The total cost of the fencing along the three sides comes to \$2140. What are the lot's dimensions?

33. A hotel has 200 rooms. Those with kitchen facilities rent for \$100 per night and those without kitchen facilities rent for \$80 per night. On a night when the hotel was completely occupied, revenues were \$17,000. How many of each type of room does the hotel have?

34. In a new development, 50 one- and two-bedroom condominiums were sold. Each one-bedroom condominium sold for \$120 thousand and each two-bedroom condominium sold for \$150 thousand. If sales totaled \$7050 thousand, how many of each type of unit was sold?

35. The calorie-nutrient information for an apple and an avocado is given in the table. How many of each should be eaten to get exactly 1000 calories and 100 grams of carbohydrates?

	One Apple	One Avocado
Calories	100	350
Carbohydrates (grams)	24	14

36. Nutritional information for macaroni and broccoli is given in the table. How many servings of each would it take to get exactly 14 grams of protein and 48 grams of carbohydrates?

	Macaroni	Broccoli
Protein (grams/serving)	3	2
Carbohydrates (grams/serving)	16	4

The figure shows the graphs of the cost and revenue functions for a company that manufactures and sells small radios. Use the information in the figure to solve Exercises 37–42.

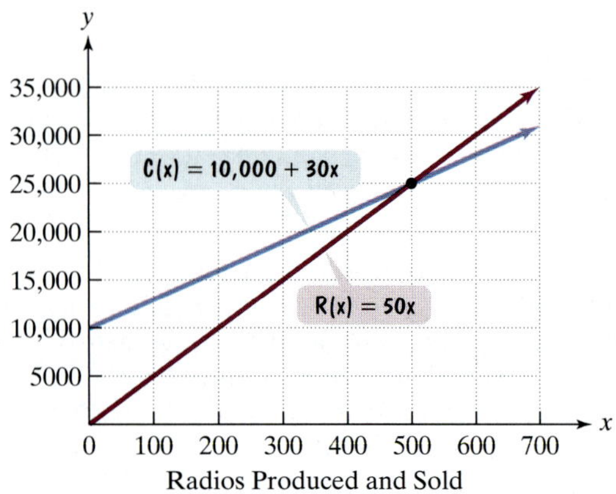

37. How many radios must be produced and sold for the company to break even?

38. More than how many radios must be produced and sold for the company to have a profit gain?

39. Use the formulas shown in the voice balloons to find $R(200) - C(200)$. Describe what this means for the company.

40. Use the formulas shown in the voice balloons to find $R(300) - C(300)$. Describe what this means for the company.

41. Use the formulas shown in the voice balloons to find the company's profit gain if 10,000 radios are produced and sold.

42. Use the formulas shown in the voice balloons to find the company's profit gain if 20,000 radios are produced and sold.

Exercises 43–46 describe a number of business ventures. For each exerise:

a. *Write the cost function, C.*
b. *Write the revenue function, R.*
c. *Determine the break-even point. Describe what this means.*

43. A company that manufactures small canoes has a fixed cost of \$18,000. It costs \$20 to produce each canoe. The selling price is \$80 per canoe. (In solving this exercise, let x represent the number of canoes produced and sold.)

44. A company that manufactures bicycles has a fixed cost of \$100,000. It costs \$100 to produce each bicycle. The selling price is \$300 per bike. (In solving this exercise, let x represent the number of bicycles produced and sold.)

45. You invest in a new play. The cost includes an overhead of \$30,000, plus production costs of \$2500 per performance. A sold-out performance brings in \$3125. (In solving this exercise, let x represent the number of sold-out performances.)

46. You invested \$30,000 and started a business writing greeting cards. Supplies cost 2¢ per card and you are selling each card for 50¢. (In solving this exercise, let x represent the number of cards produced and sold.)

Writing in Mathematics

47. Describe the conditions in a problem that enable it to be solved using a system of linear equations.

48. Write a word problem that can be solved by translating to a system of linear equations. Then solve the problem.

49. Describe a revenue function for a business venture.

50. Describe a cost function for a business venture. What are the two kinds of costs that are modeled by this function?

51. What is the profit function for a business venture and how is it determined?

52. Describe the break-even point for a business.

53. In Exercise 10, what explanation can you offer for the particular countries shown as the world's top five garbage producers?

54. The law of supply and demand states that, in a free market economy, a commodity tends to be sold at its equilibrium price. At this price, the amount that the seller will supply is the same amount that the consumer will buy. Explain how graphs can be used to determine the equilibrium price.

Technology Exercises

In Exercises 55–56, graph the revenue and cost functions in the same viewing rectangle. Then use the intersection feature to determine the break-even point.

55. $R(x) = 50x, \quad C(x) = 20x + 180$

56. $R(x) = 92.5x, \quad C(x) = 52x + 1782$

57. Use the procedure in Exercises 55–56 to verify your work for any one of the break-even points that you found in Exercises 43–46.

Critical Thinking Exercises

58. A marching band has 52 members, and there are 24 in the pom-pom squad. They wish to form several hexagons and squares like those diagrammed below. Can it be done with no people left over?

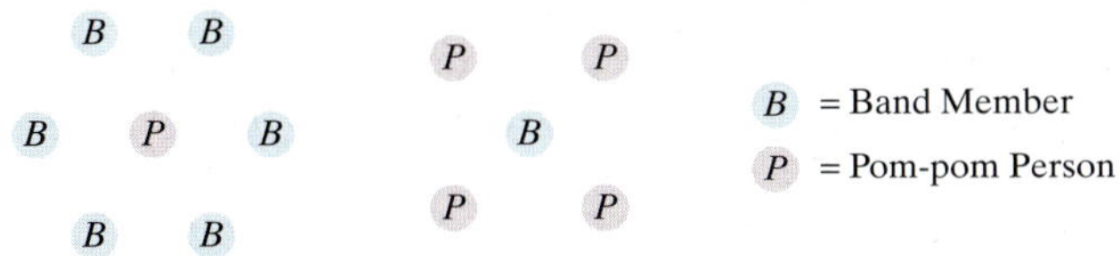

59. Find the measures of the angles marked $x°$ and $y°$ in the figure.

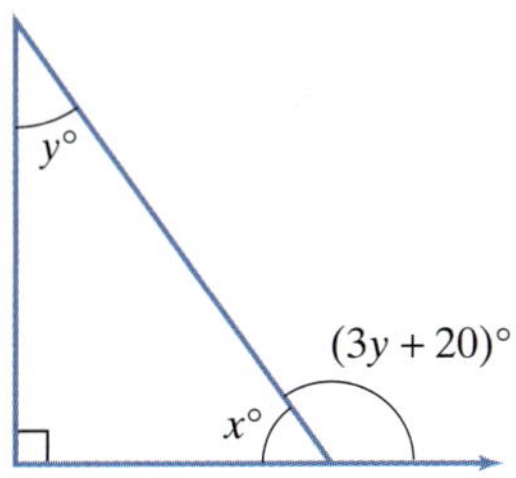

60. A boy has as many brothers as he has sisters. Each of his sisters has twice as many brothers as she has sisters. How many boys and girls are in this family?

61. A dealer paid a total of \$67 for mangos and avocados. The mangos were sold at a profit of 20% on the dealer's cost, but the avocados started to spoil, resulting in a selling price of a 2% loss on the dealer's cost. The dealer made a profit of \$8.56 on the total transaction. How much did the dealer pay for the mangos and for the the avocados?

Review Exercises

In Exercises 62–63, use the given conditions to write an equation for each line in point-slope form and slope-intercept form.

62. Passing through $(-2, 5)$ and $(-6, 13)$
(Section 2.4, Example 2)

63. Passing through $(-3, 0)$ and parallel to the line whose equation is $-x + y = 7$
(Section 2.4, Example 4)

64. Find the domain of $g(x) = \frac{x - 2}{3 - x}$.
(Section 2.2, Example 1)

▸ SECTION 3.3 Systems of Linear Equations in Three Variables

Objectives

1. Verify the solution of a system of linear equations in three variables.
2. Solve systems of linear equations in three variables.
3. Identify inconsistent and dependent systems.
4. Solve problems using systems in three variables.

SSM PH Tutor Center CD-ROM Video

All animals sleep, but the length of time they sleep varies widely: Cattle sleep for only a few minutes at a time. We humans seem to need more sleep than other animals, up to eight hours a day. Without enough sleep, we have difficulty concentrating, make mistakes in routine tasks, lose energy, and feel bad-tempered. There is a relationship between hours of sleep and death rate per year per 100,000 people. How many hours of sleep will put you in the group with the minimum death rate? In this section you will learn how to solve systems of linear equations with more than two variables in order to answer this question.

1 Verify the solution of a system of linear equations in three variables.

Systems of Linear Equations in Three Variables and Their Solutions

An equation such as $x + 2y - 3z = 9$ is called a **linear equation in three variables**. In general, any equation of the form

$$Ax + By + Cz = D$$

where A, B, C, and D are real numbers such that A, B, and C are not all 0, is a linear equation in the variables x, y, and z. The graph of this linear equation in three variables is a plane in three-dimensional space.

The process of solving a system of three linear equations in three variables is geometrically equivalent to finding the point of intersection (assuming that there is one) of three planes in space (see Figure 3.9). A **solution** to a system of linear equations in three variables is an ordered triple of real numbers that satisfies all equations of the system. The **solution set** of the system is the set of all its solutions.

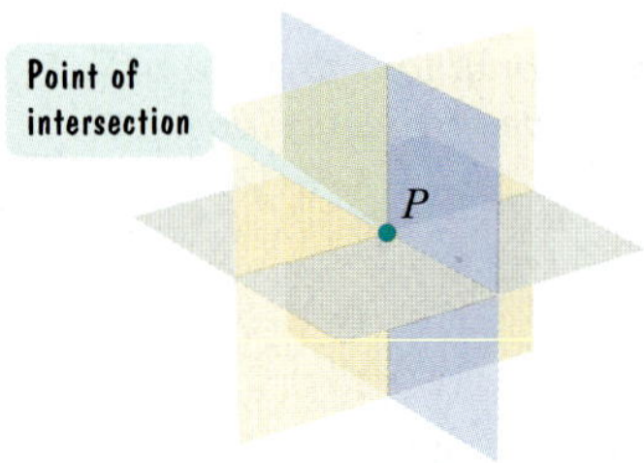

Figure 3.9

EXAMPLE 1 Determining Whether an Ordered Triple Satisfies a System

Show that the ordered triple $(-1, 2, -2)$ is a solution of the system:

$$x + 2y - 3z = 9$$
$$2x - y + 2z = -8$$
$$-x + 3y - 4z = 15.$$

Solution Because -1 is the x-coordinate, 2 is the y-coordinate, and -2 is the z-coordinate of $(-1, 2, -2)$, we replace x by -1, y by 2, and z by -2 in each of the three equations.

$$x + 2y - 3z = 9 \qquad 2x - y + 2z = -8 \qquad -x + 3y - 4z = 15$$

$$-1 + 2(2) - 3(-2) \stackrel{?}{=} 9 \qquad 2(-1) - 2 + 2(-2) \stackrel{?}{=} -8 \qquad -(-1) + 3(2) - 4(-2) \stackrel{?}{=} 15$$

$$-1 + 4 + 6 \stackrel{?}{=} 9 \qquad -2 - 2 - 4 \stackrel{?}{=} -8 \qquad 1 + 6 + 8 \stackrel{?}{=} 15$$

$$9 = 9, \text{ true} \qquad -8 = -8, \text{ true} \qquad 15 = 15, \text{ true}$$

The ordered triple $(-1, 2, -2)$ satisfies the three equations: It makes each equation true. Thus, the ordered triple is a solution of the system. ■

✔ **CHECK POINT 1** Show that the ordered triple $(-1, -4, 5)$ is a solution of the system:

$$x - 2y + 3z = 22$$
$$2x - 3y - z = 5$$
$$3x + y - 5z = -32.$$

2 Solve systems of linear equations in three variables.

Solving Systems of Linear Equations in Three Variables by Eliminating Variables

The method for solving a system of linear equations in three variables is similar to that used on systems of linear equations in two variables. We use addition to eliminate any variable, reducing the system to two equations in two variables.

Once we obtain a system of two equations in two variables, we use addition or substitution to eliminate a variable. The result is a single equation in one variable. We solve this equation to get the value of the remaining variable. Other variable values are found by back-substitution.

Solving Linear Systems in Three Variables by Eliminating Variables

1. Reduce the system to two equations in two variables. This is usually accomplished by taking two different pairs of equations and using the addition method to eliminate the same variable from each pair.
2. Solve the resulting system of two equations in two variables using addition or substitution. The result is an equation in one variable that gives the value of that variable.
3. Back-substitute the value of the variable found in step 2 into either of the equations in two variables to find the value of the second variable.
4. Use the values of the two variables from steps 2 and 3 to find the value of the third variable by back-substituting into one of the original equations.
5. Check the proposed solution in each of the original equations.

EXAMPLE 2 Solving a System in Three Variables

Solve the system:

$$\begin{aligned} 5x - 2y - 4z &= 3 && \text{Equation 1} \\ 3x + 3y + 2z &= -3 && \text{Equation 2} \\ -2x + 5y + 3z &= 3. && \text{Equation 3} \end{aligned}$$

Solution There are many ways to proceed. Because our initial goal is to reduce the system to two equations in two variables, **the central idea is to take two different pairs of equations and eliminate the same variable from each pair**.

Step 1 Reduce the system to two equations in two variables. We choose any two equations and use the addition method to eliminate a variable. Let's eliminate z from Equations 1 and 2. We do so by multiplying Equation 2 by 2. Then we add equations.

$$\begin{aligned} &\text{(Equation 1)}\quad 5x - 2y - 4z = 3 \xrightarrow{\text{No change}} && 5x - 2y - 4z = 3 \\ &\text{(Equation 2)}\quad 3x + 3y + 2z = -3 \xrightarrow{\text{Multiply by 2.}} && \underline{6x + 6y + 4z = -6} \\ &&\text{Add:}\quad & 11x + 4y \qquad = -3 \quad \text{(Equation 4)} \end{aligned}$$

Now we must eliminate the *same* variable from another pair of equations. We can eliminate z from Equations 2 and 3. First, we multiply Equation 2 by -3. Next, we multiply Equation 3 by 2. Finally, we add equations.

$$\begin{aligned} &\text{(Equation 2)}\quad 3x + 3y + 2z = -3 \xrightarrow{\text{Multiply by } -3.} && -9x - 9y - 6z = 9 \\ &\text{(Equation 3)}\quad -2x + 5y + 3z = 3 \xrightarrow{\text{Multiply by 2.}} && \underline{-4x + 10y + 6z = 6} \\ &&\text{Add:}\quad & -13x + y \qquad = 15 \quad \text{(Equation 5)} \end{aligned}$$

Equations 4 and 5 give us a system of two equations in two variables.

Step 2 Solve the resulting system of two equations in two variables. We will use the addition method to solve Equations 4 and 5 for x and y. To do so, we multiply Equation 5 on both sides by -4 and add this to Equation 4.

(Equation 4) $11x + 4y = -3$ $\xrightarrow{\text{No change}}$ $11x + 4y = -3$

(Equation 5) $-13x + y = 15$ $\xrightarrow{\text{Multiply by } -4.}$ $52x - 4y = -60$

Add: $63x = -63$

$x = -1$ Divide both sides by 63.

Step 3 Use back-substitution in one of the equations in two variables to find the value of the second variable. We back-substitute -1 for x in either Equation 4 or 5 to find the value of y.

$-13x + y = 15$ Equation 5

$-13(-1) + y = 15$ Substitute -1 for x.

$13 + y = 15$ Multiply.

$y = 2$ Subtract 13 from both sides.

Step 4 Back-substitute the values found for two variables into one of the original equations to find the value of the third variable. We can now use any one of the original equations and back-substitute the values of x and y to find the value for z. We will use Equation 2.

$3x + 3y + 2z = -3$ Equation 2

$3(-1) + 3(2) + 2z = -3$ Substitute -1 for x and 2 for y.

$3 + 2z = -3$ Multiply and then add: $3(-1) + 3(2) = -3 + 6 = 3$.

$2z = -6$ Subtract 3 from both sides.

$z = -3$ Divide both sides by 2.

With $x = -1$, $y = 2$, and $z = -3$, the proposed solution is the ordered triple $(-1, 2, -3)$.

Step 5 Check. Check the proposed solution, $(-1, 2, -3)$, by substituting the values for x, y, and z into each of the three original equations. These substitutions yield three true statements. Thus, the solution is $(-1, 2, -3)$ and the solution set is $\{(-1, 2, -3)\}$. ■

✔ **CHECK POINT 2** Solve the system:

$$\begin{aligned} x + 4y - z &= 20 \\ 3x + 2y + z &= 8 \\ 2x - 3y + 2z &= -16. \end{aligned}$$

In some examples, one of the variables is already eliminated from a given equation. In this case, the same variable should be eliminated from the other two equations, thereby making it possible to omit one of the elimination steps. We illustrate this idea in Example 3.

EXAMPLE 3 Solving a System of Equations with a Missing Term

Solve the system:

$x + z = 8$ Equation 1

$x + y + 2z = 17$ Equation 2

$x + 2y + z = 16.$ Equation 3

Solution

Step 1 Reduce the system to two equations in two variables. Because Equation 1 contains only x and z, we could eliminate y from Equations 2 and 3. This will give us two equations in x and z. To eliminate y from Equations 2 and 3, we multiply Equation 2 by -2 and add Equation 3.

$$\begin{array}{lrcl} \text{(Equation 2)} & x + y + 2z = 17 & \xrightarrow{\text{Multiply by } -2.} & -2x - 2y - 4z = -34 \\ \text{(Equation 3)} & x + 2y + z = 16 & \xrightarrow{\text{No change}} & x + 2y + z = 16 \\ & & \text{Add:} & -x \qquad - 3z = -18 \quad \text{(Equation 4)} \end{array}$$

Equation 4 and the given Equation 1 provide us with a system of two equations in two variables.

Step 2 Solve the resulting system of two equations in two variables. We will solve Equations 1 and 4 for x and z.

$$\begin{array}{rrll} & x + z = & 8 & \text{Equation 1} \\ & -x - 3z = & -18 & \text{Equation 4} \\ \text{Add:} & -2z = & -10 & \\ & z = & 5 & \text{Divide both sides by } -2. \end{array}$$

Step 3 Use back-substitution in one of the equations in two variables to find the value of the second variable. To find x,we back-substitute 5 for z in either Equation 1 or 4. We will use Equation 1.

$$\begin{array}{rl} x + z = 8 & \text{Equation 1} \\ x + 5 = 8 & \text{Substitute 5 for } z. \\ x = 3 & \text{Subtract 5 from both sides.} \end{array}$$

Step 4 Back-substitute the values found for two variables into one of the original equations to find the value of the third variable. To find y, we back-substitute 3 for x and 5 for z into Equation 2 or 3. We can't use Equation 1 because y is missing in this equation. We will use Equation 2.

$$\begin{array}{rl} x + y + 2z = 17 & \text{Equation 2} \\ 3 + y + 2(5) = 17 & \text{Substitute 3 for } x \text{ and 5 for } z. \\ y + 13 = 17 & \text{Multiply and add.} \\ y = 4 & \text{Subtract 13 from both sides.} \end{array}$$

We found that $z = 5$, $x = 3$, and $y = 4$. Thus, the proposed solution is the ordered triple $(3, 4, 5)$.

Step 5 Check. Substituting 3 for x, 4 for y, and 5 for z into each of the three original equations yields three true statements. Consequently, the solution is $(3, 4, 5)$ and the solution set is $\{(3, 4, 5)\}$. ■

✔ **CHECK POINT 3** Solve the system:

$$\begin{aligned} 2y - z &= 7 \\ x + 2y + z &= 17 \\ 2x - 3y + 2z &= -1. \end{aligned}$$

3 Identify inconsistent and dependent systems.

Inconsistent and Dependent Systems

A system of linear equations in three variables represents three planes. The three planes need not intersect at one point. The planes may have no common point of intersection and represent an **inconsistent system** with no solution. Figure 3.10 illustrates some of the geometric possibilities for inconsistent systems.

Three planes are parallel with no common intersection point.

Two planes are parallel with no common intersection point.

Planes intersect two at a time. There is no intersection point common to all three planes.

Figure 3.10 Three planes may have no common point of intersection.

If you attempt to solve an inconsistent system algebraically, at some point in the solution process you will eliminate all three variables. A false statement such as $0 = 17$ will be the result. For example, consider the system

$$\begin{aligned} 2x + 5y + z &= 12 & \text{Equation 1} \\ x - 2y + 4z &= 10 & \text{Equation 2} \\ -3x + 6y - 12z &= 20. & \text{Equation 3} \end{aligned}$$

Suppose we reduce the system to two equations in two variables by eliminating x. To eliminate x in Equations 2 and 3, we multiply Equation 2 by 3 and add Equation 3:

$$\begin{array}{rcl} x - 2y + 4z = -10 & \xrightarrow{\text{Multiply by 3.}} & 3x - 6y + 12z = -30 \\ -3x + 6y - 12z = 20 & \xrightarrow{\text{No change}} & \underline{-3x + 6y - 12z = 20} \\ & \text{Add:} & 0 = -10. \end{array}$$

There are no values of x, y, and z for which $0 = -10$. The false statement $0 = -10$ indicates that the system is inconsistent and has no solution. The solution set is the empty set, $\varnothing$.

We have seen that a linear system that has at least one solution is called a **consistent system**. Planes that intersect at one point and planes that intersect at infinitely many points both represent consistent systems. Figure 3.11 illustrates planes that intersect at infinitely many points. The equations in these linear systems with infinitely many solutions are called **dependent**.

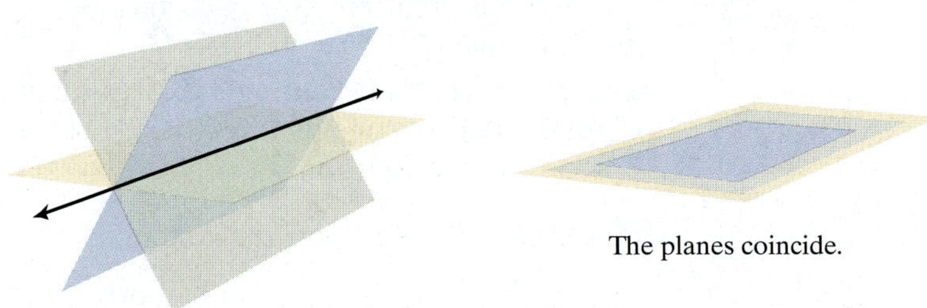

The planes intersect along a common line.

The planes coincide.

Figure 3.11 Three planes may intersect at infinitely many points.

If you attempt to solve a system with dependent equations algebraically, at some point in the solution process you will eliminate all three variables. A true statement such as $0 = 0$ will be the result. If this occurs as you are solving a linear system, simply state that the equations are dependent.

4 Solve problems using systems in three variables.

Applications

Systems of equations may allow us to find models for data without using a graphing utility. Three data points that do not lie on or near a line determine the graph of a function of the form

$$y = ax^2 + bx + c, a \neq 0.$$

Such a function is called a **quadratic function**. Its graph has a cuplike shape, making it ideal for modeling situations in which values of y are decreasing and then increasing. In Chapter 8, we'll have lots of interesting things to tell you about quadratic functions and their graphs.

The process of determining a function whose graph contains given points is called **curve fitting**. In our next example, we fit three data points to the curve whose equation is $y = ax^2 + bx + c$. Using a system of equations, we find values for a, b, and c.

EXAMPLE 4 Modeling Data Relating Sleep and Death Rate

In a study relating sleep and death rate, the following data were obtained. Use the function $y = ax^2 + bx + c$ to model the data.

x (Average Number of Hours of Sleep)	y (Death Rate per Year per 100,000 Males)
4	1682
7	626
9	967

Solution We need to find values for a, b, and c. We can do so by solving a system of three linear equations in a, b, and c. We obtain the three equations by using the values of x and y from the data as follows:

$\boldsymbol{y = ax^2 + bx + c}$ Use the quadratic function to model the data.

When $x = 4, y = 1682$: $1682 = a \cdot 4^2 + b \cdot 4 + c$ or $16a + 4b + c = 1682$

When $x = 7, y = 626$: $626 = a \cdot 7^2 + b \cdot 7 + c$ or $49a + 7b + c = 626$

When $x = 9, y = 967$: $967 = a \cdot 9^2 + b \cdot 9 + c$ or $81a + 9b + c = 967$.

The easiest way to solve this system is to eliminate c from two pairs of equations, obtaining two equations in a and b. Solving this system gives $a = 104.5$, $b = -1501.5$, and $c = 6016$. We now substitute the values for a, b, and c into $y = ax^2 + bx + c$. The function that models the given data is

$$y = 104.5x^2 - 1501.5x + 6016.$$ ■

We can use the model that we obtained in Example 4 to find the death rate of males who average, say, 6 hours of sleep. First, write the model in function notation:

$$f(x) = 104.5x^2 - 1501.5x + 6016.$$

Substitute 6 for x:

$$f(6) = 104.5(6)^2 - 1501.5(6) + 6016 = 769.$$

According to the model, the death rate for males who average 6 hours of sleep is 769 deaths per 100,000 males.

Using Technology

The graph of

$$y = 104.5x^2 - 1501.5x + 6016$$

is displayed in a $[3, 12, 1]$ by $[500, 2000, 100]$ viewing rectangle. The minimum function feature shows that the lowest point on the graph is approximately (7.2, 622.5). Men who average 7.2 hours of sleep are in the group with the lowest death rate, approximately 622.5 per 100,000.

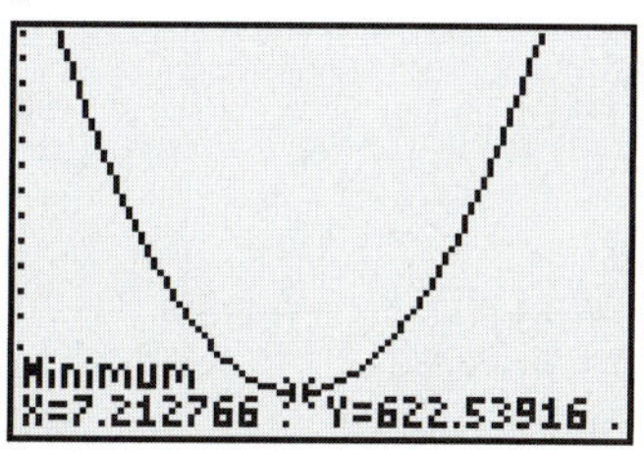

CHECK POINT 4 Find the quadratic function $y = ax^2 + bx + c$ whose graph passes through the points $(1, 4)$, $(2, 1)$, and $(3, 4)$.

Problems involving three unknowns can be solved using the same strategy for solving problems with two unknown quantities. You can let x, y, and z represent the unknown quantities. We then translate from the verbal conditions of the problem to a system of three equations in three variables. Problems of this type are included in the exercise set that follows.

EXERCISE SET 3.3

Practice Exercises

In Exercises 1–4, determine if the given ordered triple is a solution of the system.

1. $\begin{aligned} x + y + z &= 4 \\ x - 2y - z &= 1 \\ 2x - y - z &= -1 \end{aligned}$
$(2, -1, 3)$

2. $\begin{aligned} x + y + z &= 0 \\ x + 2y - 3z &= 5 \\ 3x + 4y + 2z &= -1 \end{aligned}$
$(5, -3, -2)$

3. $\begin{aligned} x - 2y &= 2 \\ 2x + 3y &= 11 \\ y - 4z &= -7 \end{aligned}$
$(4, 1, 2)$

4. $\begin{aligned} x - 2z &= -5 \\ y - 3z &= -3 \\ 2x - z &= -4 \end{aligned}$
$(-1, 3, 2)$

Solve each system in Exercises 5–22. If there is no solution or if there are infinitely many solutions and a system's equations are dependent, so state.

5. $\begin{aligned} x + y + 2z &= 11 \\ x + y + 3z &= 14 \\ x + 2y - z &= 5 \end{aligned}$

6. $\begin{aligned} 2x + y - 2z &= -1 \\ 3x - 3y - z &= 5 \\ x - 2y + 3z &= 6 \end{aligned}$

7. $\begin{aligned} 4x - y + 2z &= 11 \\ x + 2y - z &= -1 \\ 2x + 2y - 3z &= -1 \end{aligned}$

8. $\begin{aligned} x - y + 3z &= -8 \\ 3x + y - 2z &= -2 \\ 2x + 4y + z &= 0 \end{aligned}$

9. $\begin{aligned} 3x + 2y - 3z &= -2 \\ 2x - 5y + 2z &= -2 \\ 4x - 3y + 4z &= 10 \end{aligned}$

10. $\begin{aligned} 2x + 3y + 7z &= 13 \\ 3x + 2y - 5z &= -22 \\ 5x + 7y - 3z &= -28 \end{aligned}$

11. $\begin{aligned} 2x - 4y + 3z &= 17 \\ x + 2y - z &= 0 \\ 4x - y - z &= 6 \end{aligned}$

12. $\begin{aligned} x + z &= 3 \\ x + 2y - z &= 1 \\ 2x - y + z &= 3 \end{aligned}$

13. $\begin{aligned} 2x + y &= 2 \\ x + y - z &= 4 \\ 3x + 2y + z &= 0 \end{aligned}$

14. $\begin{aligned} x + 3y + 5z &= 20 \\ y - 4z &= -16 \\ 3x - 2y + 9z &= 36 \end{aligned}$

15. $\begin{aligned} x + y &= -4 \\ y - z &= 1 \\ 2x + y + 3z &= -21 \end{aligned}$

16. $\begin{aligned} x + y &= 4 \\ x + z &= 4 \\ y + z &= 4 \end{aligned}$

17. $\begin{aligned} 2x + y + 2z &= 1 \\ 3x - y + z &= 2 \\ x - 2y - z &= 0 \end{aligned}$

18. $\begin{aligned} 3x + 4y + 5z &= 8 \\ x - 2y + 3z &= -6 \\ 2x - 4y + 6z &= 8 \end{aligned}$

19. $\begin{aligned} 5x - 2y - 5z &= 1 \\ 10x - 4y - 10z &= 2 \\ 15x - 6y - 15z &= 3 \end{aligned}$

20. $\begin{aligned} x + 2y + z &= 4 \\ 3x - 4y + z &= 4 \\ 6x - 8y + 2z &= 8 \end{aligned}$

21. $\begin{aligned} 3(2x + y) + 5z &= -1 \\ 2(x - 3y + 4z) &= -9 \\ 4(1 + x) &= -3(z - 3y) \end{aligned}$

22. $\begin{aligned} 7z - 3 &= 2(x - 3y) \\ 5y + 3z - 7 &= 4x \\ 4 + 5z &= 3(2x - y) \end{aligned}$

In Exercises 23–26, find the quadratic function $y = ax^2 + bx + c$ whose graph passes through the given points.

23. $(-1, 6), (1, 4), (2, 9)$

24. $(-2, 7), (1, -2), (2, 3)$

25. $(-1, -4), (1, -2), (2, 5)$

26. $(1, 3), (3, -1), (4, 0)$

In Exercises 27–28, let x represent the first number, y the second number, and z the third number. Use the given conditions to write a system of equations. Solve the system and find the numbers.

27. The sum of three numbers is 16. The sum of twice the first number, 3 times the second number, and 4 times the third number is 46. The difference between 5 times the first number and the second number is 31. Find the three numbers.

28. Three numbers are unknown. Three times the first number plus the second number plus twice the third number is 5. If 3 times the second number is subtracted from the sum of the first number and 3 times the third number, the result is 2. If the third number is subtracted from 2 times the first number and 3 times the second number, the result is 1. Find the numbers.

Application Exercises

29. The bar graph shows that the percentage of the U.S. population that was foreign-born decreased between 1960 and 1970 and then increased between 1970 and 1980.

Percentage of U.S. Population That Is Foreign-Born, 1900-1999

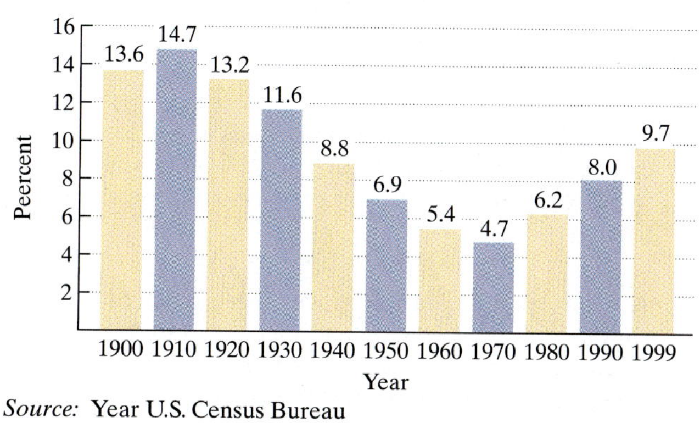

Source: Year U.S. Census Bureau

a. Write the data for 1960, 1970, and 1980 as ordered pairs (x, y), where x is the number of years after 1960 and y is the percentage of the U.S. population that was foreign-born in that year.

b. The three data points in part (a) can be modeled by the quadratic function $y = ax^2 + bx + c$. Substitute each ordered pair into this function, one ordered pair at a time, and write a system of linear equations in three variables that can be used to find values for a, b, and c. It is not necessary to solve the system.

30. The bar graph shows that the U.S. divorce rate increased between 1970 and 1985 and then decreased between 1985 and 1998.

U.S. Divorce Rates: Number of Divorces per 1000 People

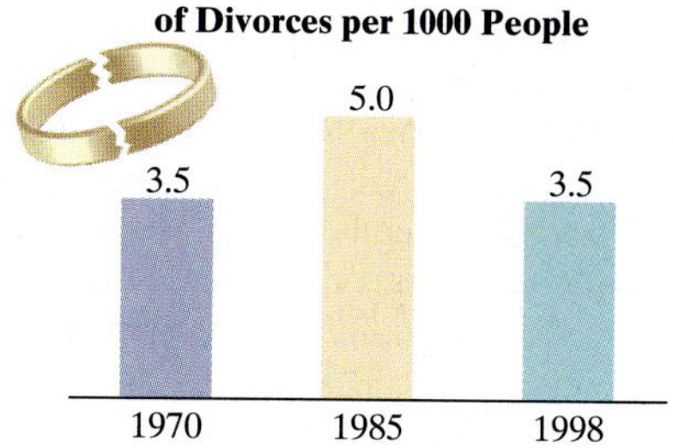

Source: U.S. Census Bureau

a. Write the data for 1970, 1985, and 1998 as ordered pairs (x, y), where x is the number of years after 1970 and y is that year's divorce rate.

b. The three data points in part (a) can be modeled by the quadratic function $y = ax^2 + bx + c$. Substitute each ordered pair into this function, one ordered pair at a time, and write a system of linear equations in three variables that can be used to find values for a, b, and c. It is not necessary to solve the system.

31. You throw a ball straight up from a rooftop. The ball misses the rooftop on its way down and eventually strikes the ground. A mathematical model can be used to describe the relationship for the ball's height above the ground, y, after x seconds. Consider the following data.

x, seconds after the ball is thrown	y, ball's height, in feet, above the ground
1	224
3	176
4	104

a. Find the quadratic function $y = ax^2 + bx + c$ whose graph passes through the given points.

b. Use the function in part (a) to find the value for y when $x = 5$. Describe what this means.

32. A mathematical model can be used to describe the relationship between the number of feet a car travels once the brakes are applied, y, and the number of seconds the car is in motion after the brakes are applied, x. A research firm collects the following data.

x, seconds in motion after brakes are applied	y, feet car travels once the brakes are applied
1	46
2	84
3	114

a. Find the quadratic function $y = ax^2 + bx + c$ whose graph passes through the given points.

b. Use the function in part (a) to find the value for y when $x = 6$. Describe what this means.

In Exercises 33–40, use the four-step strategy to solve each problem. Use x, y, and z to represent unknown quantities. Then translate from the verbal conditions of the problem to a system of three equations in three variables.

33. The bar graph shows the average starting salaries for the five top-paying fields for college graduates. If we add the average starting salaries for college graduates who are chemical, mechanical, and electrical engineers, the total is \$121,421. The difference between the starting salaries for chemical and mechanical engineers is \$2906. The difference between the starting salaries for mechanical engineers and electrical engineers is \$1041. Find the average starting salaries for chemical, mechanical, and electrical engineers.

Average Starting Salaries for the Five Top Paying Fields for College Graduates in 1999

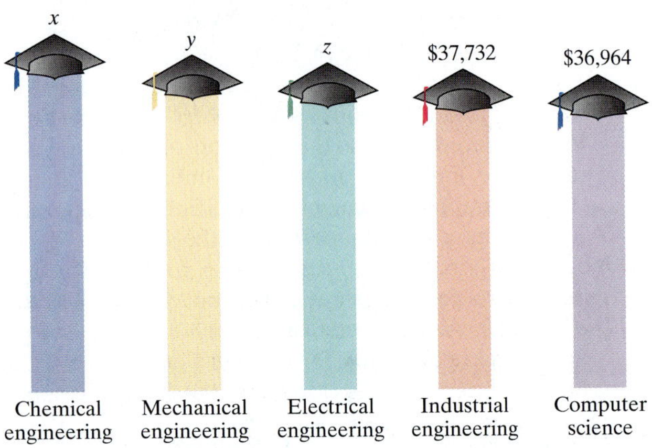

Source: Michigan State University

34. The circle graph indicates computers in use for the United States and the rest of the world. The percentage of the world's computers in Europe and Japan combined is 13% less than the percentage of the world's computers in the United States. If the percentage of the world's computers in Europe is doubled, it is only 3% more than the percentage of the world's computers in the United States. Find the percentage of the world's computers in the United States, Europe, and Japan.

Percentage of the World's Computers: U.S. and the World

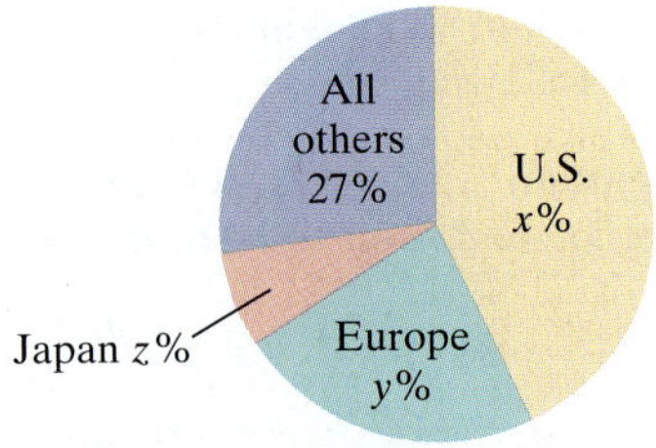

Source: Jupiter Communications

35. A person invested \$6700 for one year, part at 8%, part at 10%, and the remainder at 12%. The total annual income from these investments was \$716. The amount of money invested at 12% was \$300 more than the amount invested at 8% and 10% combined. Find the amount invested at each rate.

36. A person invested \$17,000 for one year, part at 10%, part at 12%, and the remainder at 15%. The total annual income from these investments was \$2110. The amount of money invested at 12% was \$1000 less than the amount invested at 10% and 15% combined. Find the amount invested at each rate.

37. At a college production of *Evita*, 400 tickets were sold. The ticket prices were \$8, \$10, and \$12, and the total income from ticket sales was \$3700. How many tickets of each type were sold if the combined number of \$8 and \$10 tickets sold was 7 times the number of \$12 tickets sold?

38. A certain brand of razor blades comes in packages of 6, 12, and 24 blades, costing \$2, \$3, and \$4 per package, respectively. A store sold 12 packages containing a total of 162 razor blades and took in \$35. How many packages of each type were sold?

39. Three foods have the following nutritional content per ounce.

	Calories	Protein (in grams)	Vitamin C (in milligrams)
Food *A*	40	5	30
Food *B*	200	2	10
Food *C*	400	4	300

If a meal consisting of the three foods allows exactly 660 calories, 25 grams of protein, and 425 milligrams of vitamin C, how many ounces of each kind of food should be used?

40. A furniture company produces three types of desks: a children's model, an office model, and a deluxe model. Each desk is manufactured in three stages: cutting, construction, and finishing. The time requirements for each model and manufacturing stage are given in the following table.

	Children's model	Office model	Deluxe model
Cutting	2 hr	3 hr	2 hr
Construction	2 hr	1 hr	3 hr
Finishing	1 hr	1 hr	2 hr

Each week the company has available a maximum of 100 hours for cutting, 100 hours for construction, and 65 hours for finishing. If all available time must be used, how many of each type of desk should be produced each week?

Writing in Mathematics

41. What is a system of linear equations in three variables?

42. How do you determine whether a given ordered triple is a solution of a system in three variables?

43. Describe in general terms how to solve a system in three variables.

44. Describe what happens when using algebraic methods to solve an inconsistent system.

45. Describe what happens when using algebraic methods to solve a system with dependent equations.

46. AIDS is taking a deadly toll on southern Africa. Describe how to use the techniques that you learned in this section to obtain a model for African life span using projections with AIDS. Let x represent the number of years after 1985 and let y represent African life span in that year.

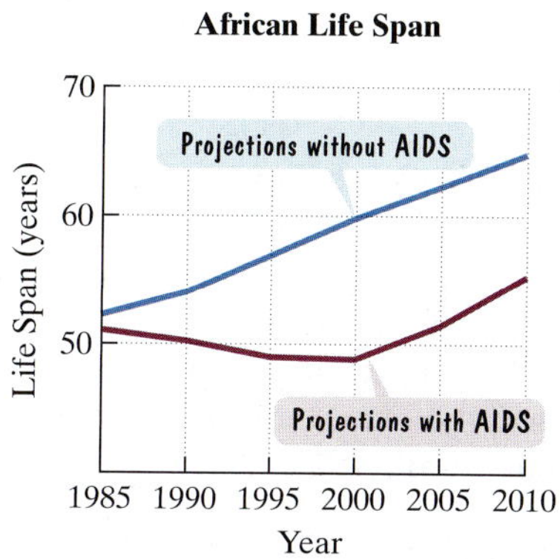

Source: United Nations

Technology Exercises

47. Does your graphing utility have a feature that allows you to solve linear systems by entering coefficients and constant terms? If so, use this feature to verify the solutions to any five exercises that you worked by hand from Exercises 5–16.

48. Verify your results in Exercises 23–26 by using a graphing utility to graph the quadratic function. Trace along the curve and convince yourself that the three points given in the exercise lie on the function's graph.

Critical Thinking Exercises

49. Which one of the following is true?

- **a.** The ordered triple (2, 15, 14) is the only solution to the equation $x + y - z = 3$.
- **b.** The equation $x - y - z = -6$ is satisfied by $(2, -3, 5)$.
- **c.** If two equations in a system are $x + y - z = 5$ and $2x + 2y - 2z = 7$, then the system must be inconsistent.
- **d.** An equation with four variables, such as $x + 2y - 3z + 5w = 2$, cannot be satisfied by real numbers.

50. Describe how the system

$$\begin{aligned} x + y - z - 2w &= -8 \\ x - 2y + 3z + w &= 18 \\ 2x + 2y + 2z - 2w &= 10 \\ 2x + y - z + w &= 3 \end{aligned}$$

could be solved. Is it likely that in the near future a graphing utility will be available to provide a geometric solution (using intersecting graphs) to this system? Explain.

51. A modernistic painting consists of triangles, rectangles, and pentagons, all drawn so as to not overlap or share sides. Within each rectangle are drawn 2 red roses, and each pentagon contains 5 carnations. How many triangles, rectangles, and pentagons appear in the painting if the painting contains a total of 40 geometric figures, 153 sides of geometric figures, and 72 flowers?

52. Find the values of the angles marked $x°$, $y°$, and $z°$ in the following triangle.

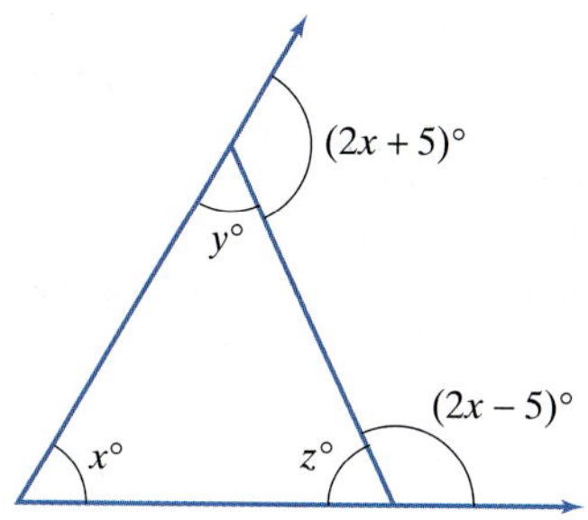

Review Exercises

In Exercises 53–55, graph each linear function.

53. $f(x) = -\frac{3}{4}x + 3$. (Section 2.3, Example 4)

54. $-2x + y = 6$ (Section 2.3, Example 1)

55. $f(x) = -5$ (Section 2.3, Figure 2.21)

▸ SECTION 3.4 Matrix Solutions to Linear Systems

Objectives

1. Write the augmented matrix for a linear system.
2. Perform matrix row operations.
3. Use matrices to solve systems.
4. Use matrices to identify inconsistent and dependent systems.

SSM PH Tutor Center CD-ROM Video

Yes, we overindulged, but it was delicious. Anyway, a few hours of moderate activity and we'll just burn off those extra calories. The following chart should help. We see that the number of calories burned per hour depends on our weight. Four hours of tennis and we'll be as good as new!

How Fast You Burn Off Calories

	Weight (pounds)					
	110	**132**	**154**	**176**	**187**	**209**
Activity	**Calories Burned per Hour**					
Housework	175	210	245	285	300	320
Cycling	190	215	245	270	280	295
Tennis	335	380	425	470	495	520
Watching TV	60	70	80	85	90	95

The 24 numbers inside the red brackets are arranged in four rows and six columns. This rectangular array of 24 numbers, arranged in rows and columns and placed in brackets, is an example of a **matrix** (plural: **matrices**). The numbers inside the brackets are called **elements** of the matrix. Matrices are used to display information and to solve systems of linear equations.

1 Write the augmented matrix for a linear system.

Solving Linear Systems by Using Matrices

A matrix gives us a shortened way of writing a system of equations. The first step in solving a system of linear equations using matrices is to write the augmented matrix. An **augmented matrix** has a vertical bar separating the columns of the matrix into two groups. The coefficients of each variable are placed to the left of the vertical line, and the constants are placed to the right. If any variable is missing, its coefficient is 0. Here are two examples.

System of Linear Equations	Augmented Matrix
$\begin{aligned} x + 3y &= 5 \\ 2x - y &= -4 \end{aligned}$	$\left[\begin{array}{cc\|c} 1 & 3 & 5 \\ 2 & -1 & -4 \end{array}\right]$
$\begin{aligned} x + 2y - 5z &= -19 \\ y + 3z &= 9 \\ z &= 4 \end{aligned}$	$\left[\begin{array}{ccc\|c} 1 & 2 & -5 & -19 \\ 0 & 1 & 3 & 9 \\ 0 & 0 & 1 & 4 \end{array}\right]$

Notice how the matrix at the bottom of page 194 contains 1s down the diagonal from upper left to lower right and 0s below the 1s. This arrangement makes it easy to find the solution of the system of equations, as Example 1 shows.

EXAMPLE 1 Solving a System Using a Matrix

Write the solution for a system of equations represented by the matrix

$$\left[\begin{array}{ccc|c} 1 & 2 & -5 & -19 \\ 0 & 1 & 3 & 9 \\ 0 & 0 & 1 & 4 \end{array}\right].$$

Solution The system represented by the given matrix is

$$\left[\begin{array}{ccc|c} 1 & 2 & -5 & -19 \\ 0 & 1 & 3 & 9 \\ 0 & 0 & 1 & 4 \end{array}\right] \rightarrow \begin{array}{l} 1x + 2y - 5z = -19 \\ 0x + 1y + 3z = 9 \\ 0x + 0y + 1z = 4 \end{array}.$$

This system can be simplified as follows.

$$x + 2y - 5z = -19 \quad \text{Equation 1}$$
$$y + 3z = 9 \quad \text{Equation 2}$$
$$z = 4 \quad \text{Equation 3}$$

The value of z is known. We can find y by back-substitution.

$$y + 3z = 9 \quad \text{Equation 2}$$
$$y + 3(4) = 9 \quad \text{Substitute 4 for z.}$$
$$y + 12 = 9 \quad \text{Multiply.}$$
$$y = -3 \quad \text{Subtract 12 from both sides.}$$

With values for y and z, we can now use back-substitution to find x.

$$x + 2y - 5z = -19 \quad \text{Equation 1}$$
$$x + 2(-3) - 5(4) = -19 \quad \text{Substitute −3 for y and 4 for z.}$$
$$x - 6 - 20 = -19 \quad \text{Multiply.}$$
$$x - 26 = -19 \quad \text{Add.}$$
$$x = 7 \quad \text{Add 26 to both sides.}$$

We see that $x = 7$, $y = -3$, and $z = 4$. The solution is $(7, -3, 4)$ and the solution set for the system is $\{(7, -3, 4)\}$. ■

✔ **CHECK POINT 1** Write the solution for a system of equations represented by the matrix

$$\left[\begin{array}{ccc|c} 1 & -1 & 1 & 8 \\ 0 & 1 & -12 & -15 \\ 0 & 0 & 1 & 1 \end{array}\right].$$

Our goal in solving a linear system using matrices is to produce a matrix with 1s down the main diagonal and 0s below the 1s. In general, the matrix will be one of the following forms.

This is the desired form for systems with two equations.

$$\left[\begin{array}{cc|c} 1 & a & b \\ 0 & 1 & c \end{array}\right] \qquad \left[\begin{array}{ccc|c} 1 & a & b & c \\ 0 & 1 & d & e \\ 0 & 0 & 1 & f \end{array}\right]$$

This is the desired form for systems with three equations.

The last row of these matrices gives us the value of one variable. The other variables can then be found by back-substitution.

2 Perform matrix row operations.

A matrix with 1s down the main diagonal and 0s below the 1s is said to be in **triangular form** or **row-echelon form**. How do we produce a matrix in this form? We use **row operations** on the augmented matrix. These row operations are just like what you did when solving a linear system by the addition method. The difference is that we no longer write the variables, usually represented by x, y, and z.

Matrix Row Operations

These row operations produce matrices that represent systems with the same solution.

1. Two rows of a matrix may be interchanged. This is the same as interchanging two equations in the linear system.
2. The elements in any row may be multiplied by a nonzero number. This is the same as multiplying both sides of an equation by a nonzero number.
3. The elements in any row may be multiplied by a nonzero number, and these products may be added to the corresponding elements in any other row. This is the same as multiplying both sides of an equation by a nonzero number and then adding equations to eliminate a variable.

Two matrices are **row equivalent** if one can be obtained from the other by a sequence of row operations.

Study Tip

When performing the row operation

$$kR_i + R_j$$

you use row i to find the products. However, **elements in row i do not change. It is the elements in row j that change:** Add k times the elements in row i to the corresponding elements in row j. Replace elements in row j by these sums.

Each matrix row operation in the preceding box can be expressed symbolically as follows:

1. Interchange the elements in the ith and jth rows: $R_i \leftrightarrow R_j$.
2. Multiply each element in the ith row by k: kR_i.
3. Add k times the elements in row i to the corresponding elements in row j: $kR_i + R_j$.

EXAMPLE 2 Performing Matrix Row Operations

Use the matrix

$$\left[\begin{array}{ccc|c} 3 & 18 & -12 & 21 \\ 1 & 2 & -3 & 5 \\ -2 & -3 & 4 & -6 \end{array}\right]$$

and perform each indicated row operation:

a. $R_1 \leftrightarrow R_2$ **b.** $\frac{1}{3}R_1$ **c.** $2R_2 + R_3$.

Study Tip

As you read the solution, keep looking back at the given matrix

$$\left[\begin{array}{ccc|c} 3 & 18 & -12 & 21 \\ 1 & 2 & -3 & 5 \\ -2 & -3 & 4 & -6 \end{array}\right].$$

Solution

a. The notation $R_1 \leftrightarrow R_2$ means to interchange the elements in row 1 and row 2. This results in the row-equivalent matrix

$$\left[\begin{array}{ccc|c} 1 & 2 & -3 & 5 \\ 3 & 18 & -12 & 21 \\ -2 & -3 & 4 & -6 \end{array}\right].$$

This was row 2; now it's row 1.

This was row 1; now it's row 2.

b. The notation $\frac{1}{3} R_1$ means to multiply each element in row 1 by $\frac{1}{3}$. This results in the row-equivalent matrix

$$\left[\begin{array}{ccc|c} \frac{1}{3}(3) & \frac{1}{3}(18) & \frac{1}{3}(-12) & \frac{1}{3}(21) \\ 1 & 2 & -3 & 5 \\ -2 & -3 & 4 & -6 \end{array}\right] = \left[\begin{array}{ccc|c} 1 & 6 & -4 & 7 \\ 1 & 2 & -3 & 5 \\ -2 & -3 & 4 & -6 \end{array}\right].$$

c. The notation $2R_2 + R_3$ means to add 2 times the elements in row 2 to the corresponding elements in row 3. Replace the elements in row 3 by these sums. First, we find 2 times the elements in row 2:

$$2(1) \text{ or } 2, \qquad 2(2) \text{ or } 4, \qquad 2(-3) \text{ or } -6, \qquad 2(5) \text{ or } 10.$$

Now we add these products to the corresponding elements in row 3. Although we use row 2 to find the products, row 2 does not change. It is the elements in row 3 that change, resulting in the row-equivalent matrix

$$\left[\begin{array}{ccc|c} 3 & 18 & -12 & 21 \\ 1 & 2 & -3 & 5 \\ -2+2=0 & -3+4=1 & 4+(-6)=-2 & -6+10=4 \end{array}\right] = \left[\begin{array}{ccc|c} 3 & 18 & -12 & 21 \\ 1 & 2 & -3 & 5 \\ 0 & 1 & -2 & 4 \end{array}\right].$$ ■

✔ CHECK POINT 2 Use the matrix

$$\left[\begin{array}{ccc|c} 4 & 12 & -20 & 8 \\ 1 & 6 & -3 & 7 \\ -3 & -2 & 1 & -9 \end{array}\right]$$

and perform each indicated row operation:

a. $R_1 \leftrightarrow R_2$

b. $\frac{1}{4} R_1$

c. $3R_2 + R_3$.

3 Use matrices to solve systems.

The process that we use to solve linear systems using matrix row operations is often called **Gaussian elimination**, after the German mathematician Carl Friedrich Gauss (1777–1835). Here are the steps used in solving systems with matrices.

Solving Linear Systems Using Matrices

1. Write the augmented matrix for the system.
2. Use matrix row operations to simplify the matrix to one with 1s down the diagonal from upper left to lower right, and 0s below the 1s.
3. Write the system of linear equations corresponding to the matrix in step 2, and use back-substitution to find the system's solution.

EXAMPLE 3 Using Matrices to Solve a Linear System

Use matrices to solve the system:

$$\begin{aligned} 4x - 3y &= -15 \\ x + 2y &= -1. \end{aligned}$$

Solution

Step 1 Write the augmented matrix for the system.

Linear System	Augmented Matrix
$\begin{aligned} 4x - 3y &= -15 \\ x + 2y &= -1 \end{aligned}$	$\left[\begin{array}{cc\|c} 4 & -3 & -15 \\ 1 & 2 & -1 \end{array}\right]$

Step 2 Use matrix row operations to simplify the matrix to one with 1s down the diagonal from upper left to lower right, and 0s below the 1s. Our goal is to obtain a matrix of the form

$$\left[\begin{array}{cc|c} 1 & a & b \\ 0 & 1 & c \end{array}\right].$$

Our first step in achieving this goal is to get 1 in the top position of the first column.

We want 1 in this position.

$$\left[\begin{array}{cc|c} 4 & -3 & -15 \\ 1 & 2 & -1 \end{array}\right]$$

To get 1 in this position, we interchange rows 1 and 2: $R_1 \leftrightarrow R_2$.

$$\left[\begin{array}{cc|c} 1 & 2 & -1 \\ 4 & -3 & -15 \end{array}\right]$$

This was row 2; now it's row 1.
This was row 1; now it's row 2.

Now we want 0 below the 1 in the first column.

We want 0 in this position.

$$\left[\begin{array}{cc|c} 1 & 2 & -1 \\ 4 & -3 & -15 \end{array}\right]$$

Let's get a 0 where there is now a 4. If we multiply the top row of numbers by −4 and add these products to the second row of numbers, we will get 0 in this position: $-4R_1 + R_2$. *We change only row 2.*

$$\left[\begin{array}{cc|c} 1 & 2 & -1 \\ -4(1) + 4 & -4(2) + (-3) & -4(-1) + (-15) \end{array}\right] = \left[\begin{array}{cc|c} 1 & 2 & -1 \\ 0 & -11 & -11 \end{array}\right]$$

row 1 element, row 2 element; row 1 element, row 2 element; row 1 element, row 2 element

We move on to the second column. We want 1 in the second row, second column.

We want 1 in this position.
$$\left[\begin{array}{cc|c} 1 & 2 & -1 \\ 0 & -11 & -11 \end{array}\right]$$

To get 1 in the desired position, we multiply -11 by its multiplicative inverse, or reciprocal, $-\frac{1}{11}$. Therefore, we multiply all the numbers in the second row by $-\frac{1}{11}$: $-\frac{1}{11}R_2$.

$$\left[\begin{array}{cc|c} 1 & 2 & -1 \\ -\frac{1}{11}(0) & -\frac{1}{11}(-11) & -\frac{1}{11}(-11) \end{array}\right] = \left[\begin{array}{cc|c} 1 & 2 & -1 \\ 0 & 1 & 1 \end{array}\right]$$

We now have the desired matrix with 1s down the diagonal and 0 below the 1.

Step 3 **Write the system of linear equations corresponding to the matrix in step 2, and use back-substitution to find the system's solution.** The system represented by the matrix in step 2 is

$$\left[\begin{array}{cc|c} 1 & 2 & -1 \\ 0 & 1 & 1 \end{array}\right] \rightarrow \begin{aligned} 1x + 2y &= -1 \\ 0x + 1y &= 1 \end{aligned} \quad \text{or} \quad \begin{aligned} x + 2y &= -1 \\ y &= 1 \end{aligned}.$$

We immediately see that the value for y is 1. To find x, we back-substitute 1 for y in the first equation.

$$x + 2y = -1 \quad \text{Equation 1}$$
$$x + 2 \cdot 1 = -1 \quad \text{Substitute 1 for } y.$$
$$x = -3 \quad \text{Subtract 2 from both sides.}$$

With $x = -3$ and $y = 1$, the proposed solution is $(-3, 1)$. Take a moment to show that $(-3, 1)$ satisfies both equations. The solution is $(-3, 1)$ and the solution set is $\{(-3, 1)\}$. ■

✔ **CHECK POINT 3** Use matrices to solve the system:

$$2x - y = -4$$
$$x + 3y = 5.$$

EXAMPLE 4 Using Matrices to Solve a Linear System

Use matrices to solve the system:

$$\begin{aligned} 3x + y + 2z &= 31 \\ x + y + 2z &= 19 \\ x + 3y + 2z &= 25. \end{aligned}$$

Solution

Step 1 **Write the augmented matrix for the system.**

Linear System

$$\begin{aligned} 3x + y + 2z &= 31 \\ x + y + 2z &= 19 \\ x + 3y + 2z &= 25 \end{aligned}$$

Augmented Matrix

$$\left[\begin{array}{ccc|c} 3 & 1 & 2 & 31 \\ 1 & 1 & 2 & 19 \\ 1 & 3 & 2 & 25 \end{array}\right]$$

Step 2 Use matrix row operations to simplify the matrix to one with 1s down the diagonal from upper left to lower right, and 0s below the 1s. Our goal is to obtain a matrix of the form

$$\left[\begin{array}{ccc|c} 1 & a & b & c \\ 0 & 1 & d & e \\ 0 & 0 & 1 & f \end{array}\right].$$

Our first step in achieving this goal is to get 1 in the top position of the first column.

We want 1 in this position.

$$\left[\begin{array}{ccc|c} 3 & 1 & 2 & 31 \\ 1 & 1 & 2 & 19 \\ 1 & 3 & 2 & 25 \end{array}\right]$$

To get 1 in this position, we interchange rows 1 and 2: $R_1 \leftrightarrow R_2$. (We could also interchange rows 1 and 3 to attain our goal.)

$$\left[\begin{array}{ccc|c} 1 & 1 & 2 & 19 \\ 3 & 1 & 2 & 31 \\ 1 & 3 & 2 & 25 \end{array}\right]$$

This was row 2; now it's row 1.

This was row 1; now it's row 2.

Now we want to get 0s below the 1 in the first column.

We want 0 in these positions.

$$\left[\begin{array}{ccc|c} 1 & 1 & 2 & 19 \\ 3 & 1 & 2 & 31 \\ 1 & 3 & 2 & 25 \end{array}\right]$$

To get a 0 where there is now a 3, multiply the top row of numbers by -3 and add these products to the second row of numbers: $-3R_1 + R_2$. To get a 0 where there is now a 1, multiply the top row of numbers by -1 and add these products to the third row of numbers: $-1R_1 + R_3$. Although we are using row 1 to find the products, the numbers in row 1 do not change.

$$\begin{array}{l} \\ -3R_1 + R_2 \\ -1R_1 + R_3 \end{array}\left[\begin{array}{ccc|c} 1 & 1 & 2 & 19 \\ -3(1) + 3 & -3(1) + 1 & -3(2) + 2 & -3(19) + 31 \\ -1(1) + 1 & -1(1) + 3 & -1(2) + 2 & -1(19) + 25 \end{array}\right]$$

$$= \left[\begin{array}{ccc|c} 1 & 1 & 2 & 19 \\ 0 & -2 & -4 & -26 \\ 0 & 2 & 0 & 6 \end{array}\right]$$

We move on to the second column. We want 1 in the second row, second column.

We want 1 in this position.

$$\left[\begin{array}{ccc|c} 1 & 1 & 2 & 19 \\ 0 & -2 & -4 & -26 \\ 0 & 2 & 0 & 6 \end{array}\right]$$

To get 1 in the desired position, we multiply -2 by its reciprocal, $-\frac{1}{2}$. Therefore, we multiply all the numbers in the second row by $-\frac{1}{2}$: $-\frac{1}{2}R_2$.

$$\left[\begin{array}{ccc|c} 1 & 1 & 2 & 19 \\ -\frac{1}{2}(0) & -\frac{1}{2}(-2) & -\frac{1}{2}(-4) & -\frac{1}{2}(-26) \\ 0 & 2 & 0 & 6 \end{array}\right] = \left[\begin{array}{ccc|c} 1 & 1 & 2 & 19 \\ 0 & 1 & 2 & 13 \\ 0 & 2 & 0 & 6 \end{array}\right].$$

We want 0 in this position.

Study Tip

You can carry out step 2 on the right using the following order. Start with the augmented matrix.

$$\left[\begin{array}{ccc|c} * & * & * & * \\ * & * & * & * \\ * & * & * & * \end{array}\right]$$

Get 1 in the upper left-hand corner.

$$\left[\begin{array}{ccc|c} 1 & * & * & * \\ * & * & * & * \\ * & * & * & * \end{array}\right]$$

Get 0s in the first column beneath the 1.

$$\left[\begin{array}{ccc|c} 1 & * & * & * \\ 0 & * & * & * \\ 0 & * & * & * \end{array}\right]$$

Get 1 in the second row/second column position.

$$\left[\begin{array}{ccc|c} 1 & * & * & * \\ 0 & 1 & * & * \\ 0 & * & * & * \end{array}\right]$$

Get 0 below the 1 in the second column.

$$\left[\begin{array}{ccc|c} 1 & * & * & * \\ 0 & 1 & * & * \\ 0 & 0 & * & * \end{array}\right]$$

Get 1 in the third row/third column position.

$$\left[\begin{array}{ccc|c} 1 & * & * & * \\ 0 & 1 & * & * \\ 0 & 0 & 1 & * \end{array}\right]$$

We are not yet done with the second column. The voice balloon shows that we want to get a 0 where there is now a 2. If we multiply the second row of numbers by -2 and add these products to the third row of numbers, we will get 0 in this position: $-2R_2 + R_3$. Although we are using the numbers in row 2 to find the products, the numbers in row 2 do not change.

$-2R_2 + R_3$

$$\left[\begin{array}{ccc|c} 1 & 1 & 2 & 19 \\ 0 & 1 & 2 & 13 \\ -2(0)+0 & -2(1)+2 & -2(2)+0 & -2(13)+6 \end{array}\right] = \left[\begin{array}{ccc|c} 1 & 1 & 2 & 19 \\ 0 & 1 & 2 & 13 \\ 0 & 0 & -4 & -20 \end{array}\right]$$

We move on to the third column. We want 1 in the third row, third column.

We want 1 in this position.

$$\left[\begin{array}{ccc|c} 1 & 1 & 2 & 19 \\ 0 & 1 & 2 & 13 \\ 0 & 0 & -4 & -20 \end{array}\right]$$

To get 1 in the desired position, we multiply -4 by its reciprocal, $-\frac{1}{4}$. Therefore, we multiply all the numbers in the third row by $-\frac{1}{4}$: $-\frac{1}{4}R_3$.

$$\left[\begin{array}{ccc|c} 1 & 1 & 2 & 19 \\ 0 & 1 & 2 & 13 \\ -\frac{1}{4}(0) & -\frac{1}{4}(0) & -\frac{1}{4}(-4) & -\frac{1}{4}(-20) \end{array}\right] = \left[\begin{array}{ccc|c} 1 & 1 & 2 & 19 \\ 0 & 1 & 2 & 13 \\ 0 & 0 & 1 & 5 \end{array}\right].$$

We now have the desired matrix with 1s down the diagonal and 0s below the 1s.

Step 3 Write the system of linear equations corresponding to the matrix in step 2, and use back-substitution to find the system's solution. The system represented by the matrix in step 2 is

$$\left[\begin{array}{ccc|c} 1 & 1 & 2 & 19 \\ 0 & 1 & 2 & 13 \\ 0 & 0 & 1 & 5 \end{array}\right] \rightarrow \begin{aligned} 1x + 1y + 2z &= 19 \\ 0x + 1y + 2z &= 13 \\ 0x + 0y + 1z &= 5 \end{aligned} \quad \text{or} \quad \begin{aligned} x + y + 2z &= 19 \\ y + 2z &= 13. \\ z &= 5 \end{aligned}$$

We immediately see that the value for z is 5. To find y, we back-substitute 5 for z in the second equation.

$y + 2z = 13$ Equation 2

$y + 2(5) = 13$ Substitute 5 for z.

$y = 3$ Solve for y.

Finally, back-substitute 3 for y and 5 for z in the first equation.

$x + y + 2z = 19$ Equation 1

$x + 3 + 2(5) = 19$ Substitute 3 for y and 5 for z.

$x + 13 = 19$ Multiply and add.

$x = 6$ Subtract 13 both sides.

The solution for the original system is $(6, 3, 5)$ and the solution set is $\{(6, 3, 5)\}$. Check to see that the solution satisfies all three equations in the given system. ■

✔ **CHECK POINT 4** Use matrices to solve the system

$$\begin{aligned} 2x + y + 2z &= 18 \\ x - y + 2z &= 9 \\ x + 2y - z &= 6. \end{aligned}$$

Modern supercomputers are capable of solving systems with more than 600,000 variables. The augmented matrices for such systems are huge, but the solution using matrices is exactly like what we did in Example 4. Work with the augmented matrix, one column at a time. First, get 1 in the desired position. Then get 0s below the 1.

4 Use matrices to identify inconsistent and dependent systems.

Inconsistent Systems and Systems with Dependent Equations

When solving a system using matrices, you might obtain a matrix with a row in which the numbers to the left of the vertical bar are all zeros but a nonzero number number appears on the right. In such a case, the system is inconsistent and has no solution. For example, a system of equations that yields the following matrix is an inconsistent system:

$$\left[\begin{array}{cc|c} 1 & -2 & 3 \\ 0 & 0 & -4 \end{array}\right].$$

The second row of the matrix represents the equation $0x + 0y = -4$, which is false for all values of x and y.

If you obtain a matrix in which a 0 appears across an entire row, the system contains dependent equations and has infinitely many solutions. This row of zeros represents $0x + 0y = 0$ or $0x + 0y + 0z = 0$. These equations are satisfied by infinitely many ordered pairs or triples.

EXERCISE SET 3.4

Practice Exercises

In Exercises 1–8, write the system of linear equations represented by the augmented matrix. Use x, y, and, if necessary, z, for the variables. Once the system is written, use back-substitution to find its solution.

1. $\left[\begin{array}{cc|c} 1 & -3 & 11 \\ 0 & 1 & -3 \end{array}\right]$

2. $\left[\begin{array}{cc|c} 1 & 3 & 5 \\ 0 & 1 & 2 \end{array}\right]$

3. $\left[\begin{array}{cc|c} 1 & -3 & 1 \\ 0 & 1 & -1 \end{array}\right]$

4. $\left[\begin{array}{cc|c} 1 & 2 & 13 \\ 0 & 1 & 4 \end{array}\right]$

5. $\left[\begin{array}{ccc|c} 1 & 0 & -4 & 5 \\ 0 & 1 & -12 & 13 \\ 0 & 0 & 1 & -\frac{1}{2} \end{array}\right]$

6. $\left[\begin{array}{ccc|c} 1 & 2 & 1 & 0 \\ 0 & 1 & 0 & -2 \\ 0 & 0 & 1 & 3 \end{array}\right]$

7. $\left[\begin{array}{ccc|c} 1 & \frac{1}{2} & 1 & \frac{11}{2} \\ 0 & 1 & \frac{3}{2} & 7 \\ 0 & 0 & 1 & 4 \end{array}\right]$

8. $\left[\begin{array}{ccc|c} 1 & 1 & 0 & 3 \\ 0 & 1 & \frac{3}{2} & -2 \\ 0 & 0 & 1 & 0 \end{array}\right]$

In Exercises 9–22, perform each matrix row operation and write the new matrix.

9. $\left[\begin{array}{cc|c} 2 & 2 & 5 \\ 1 & -\frac{3}{2} & 5 \end{array}\right] R_1 \leftrightarrow R_2$

10. $\left[\begin{array}{cc|c} -6 & 9 & 4 \\ 1 & -\frac{3}{2} & 4 \end{array}\right] R_1 \leftrightarrow R_2$

11. $\left[\begin{array}{cc|c} -6 & 8 & -12 \\ 3 & 5 & -2 \end{array}\right] -\frac{1}{6}R_1$

12. $\left[\begin{array}{cc|c} -2 & 3 & -10 \\ 4 & 2 & 5 \end{array}\right] -\frac{1}{2}R_1$

13. $\left[\begin{array}{cc|c} 1 & -3 & 5 \\ 2 & 6 & 4 \end{array}\right] -2R_1 + R_2$

14. $\left[\begin{array}{cc|c} 1 & -3 & 1 \\ 2 & 1 & -5 \end{array}\right] -2R_1 + R_2$

15. $\left[\begin{array}{cc|c} 1 & -\frac{3}{2} & \frac{7}{2} \\ 3 & 4 & 2 \end{array}\right] -3R_1 + R_2$

16. $\left[\begin{array}{cc|c} 1 & -\frac{2}{5} & \frac{3}{4} \\ 4 & 2 & -1 \end{array}\right] -4R_1 + R_2$

17. $\left[\begin{array}{ccc|c} 2 & -6 & 4 & 10 \\ 1 & 5 & -5 & 0 \\ 3 & 0 & 4 & 7 \end{array}\right] \frac{1}{2}R_1$

18. $\left[\begin{array}{ccc|c} 3 & -12 & 6 & 9 \\ 1 & -4 & 4 & 0 \\ 2 & 0 & 7 & 4 \end{array}\right] \frac{1}{3}R_1$

19. $\left[\begin{array}{ccc|c} 1 & -3 & 2 & 0 \\ 3 & 1 & -1 & 7 \\ 2 & -2 & 1 & 3 \end{array}\right] -3R_1 + R_2$

20. $\left[\begin{array}{ccc|c} 1 & -1 & 5 & -6 \\ 3 & 3 & -1 & 10 \\ 1 & 3 & 2 & 5 \end{array}\right] -3R_1 + R_2$

21. $\left[\begin{array}{ccc|c} 1 & 1 & -1 & 6 \\ 2 & -1 & 1 & -3 \\ 3 & -1 & -1 & 4 \end{array}\right] \begin{array}{c} -2R_1 + R_2 \\ \text{and} \\ -3R_1 + R_3 \end{array}$

22. $\left[\begin{array}{ccc|c} 1 & 2 & 1 & 2 \\ -2 & -1 & 2 & 5 \\ 1 & 3 & -2 & -8 \end{array}\right] \begin{array}{c} 2R_1 + R_2 \\ \text{and} \\ -1R_3 + R_3 \end{array}$

In Exercises 23–46, solve each system using matrices. If there is no solution or if there are infinitely many solutions and a system's equations are dependent, so state.

23. $x + y = 6$
$x - y = 2$

24. $x + 2y = 11$
$x - y = -1$

25. $2x + y = 3$
$x - 3y = 12$

26. $3x - 5y = 7$
$x - y = 1$

27. $5x + 7y = -25$
$11x + 6y = -8$

28. $3x - 5y = 22$
$4x - 2y = 20$

29. $4x - 2y = 5$
$-2x + y = 6$

30. $-3x + 4y = 12$
$6x - 8y = 16$

31. $x - 2y = 1$
$-2x + 4y = -2$

32. $3x - 6y = 1$
$2x - 4y = \frac{2}{3}$

33. $x + y - z = -2$
$2x - y + z = 5$
$-x + 2y + 2z = 1$

34. $x - 2y - z = 2$
$2x - y + z = 4$
$-x + y - 2z = -4$

35. $x + 3y = 0$
$x + y + z = 1$
$3x - y - z = 11$

36. $3y - z = -1$
$x + 5y - z = -4$
$-3x + 6y + 2z = 11$

37. $2x + 2y + 7z = -1$
$2x + y + 2z = 2$
$4x + 6y + z = 15$

38. $3x + 2y + 3z = 3$
$4x - 5y + 7z = 1$
$2x + 3y - 2z = 6$

39. $x + y + z = 6$
$x - z = -2$
$y + 3z = 11$

40. $x + y + z = 3$
$-y + 2z = 1$
$-x + z = 0$

41. $x - y + 3z = 4$
$2x - 2y + 6z = 7$
$3x - y + 5z = 14$

42. $3x - y + 2z = 4$
$-6x + 2y - 4z = 1$
$5x - 3y + 8z = 0$

43.
$$\begin{aligned} x - 2y + z &= 4 \\ 5x - 10y + 5z &= 20 \\ -2x + 4y - 2z &= -8 \end{aligned}$$

44.
$$\begin{aligned} x - 3y + z &= 2 \\ 4x - 12y + 4z &= 8 \\ -2x + 6y - 2z &= -4 \end{aligned}$$

45.
$$\begin{aligned} x + y &= 1 \\ y + 2z &= -2 \\ 2x - z &= 0 \end{aligned}$$

46.
$$\begin{aligned} x + 3y &= 3 \\ y + 2z &= -8 \\ x - z &= 7 \end{aligned}$$

Application Exercises

47. The table shows the number of inmates in federal and state prisons in the United States for three selected years.

x (Number of Years after 1980)	1	5	10
y (Number of Inmates, in thousands)	344	480	740

a. Use the quadratic function $y = ax^2 + bx + c$ to model the data. Solve the system of linear equations involving a, b, and c using matrices.

b. Predict the number of inmates in the year 2010.

c. List one factor that would change the accuracy of this model for the year 2010.

48. A football is kicked straight upward. The graph shows the ball's height, y, in feet, after x seconds.

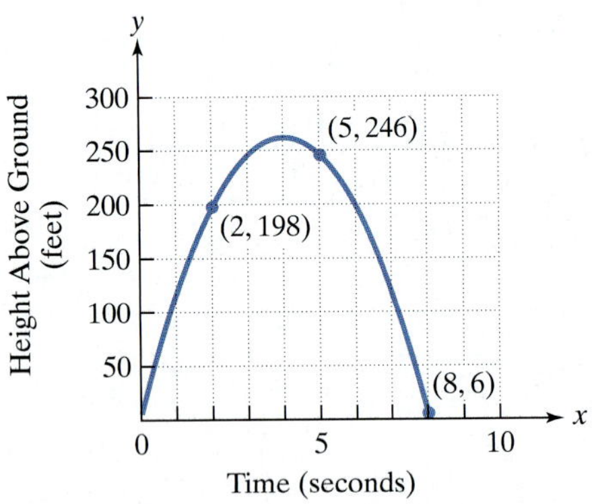

a. Find the quadratic function

$$y = ax^2 + bx + c$$

whose graph passes through the three points labeled on the graph. Solve the system of linear equations involving a, b, and c using matrices.

b. Use the function in part (a) to find the value for y when $x = 7$. Describe what this means.

Write a system of linear equations in three variables to solve Exercises 49–50. Then use matrices to solve the system.

49. The circle graph indicates the ages of the 40 million online users in the United States. The percentage of online users in the youngest (under 30) and oldest (50 and over) age groups combined exceeds the percentage in the 30–49 age group by 2%. If the percentage of users in the oldest age group is doubled, it is 3% less than the percentage of users in the youngest age group. Find the percentage of online users in each of the three age groups.

Age of U.S. Online Users

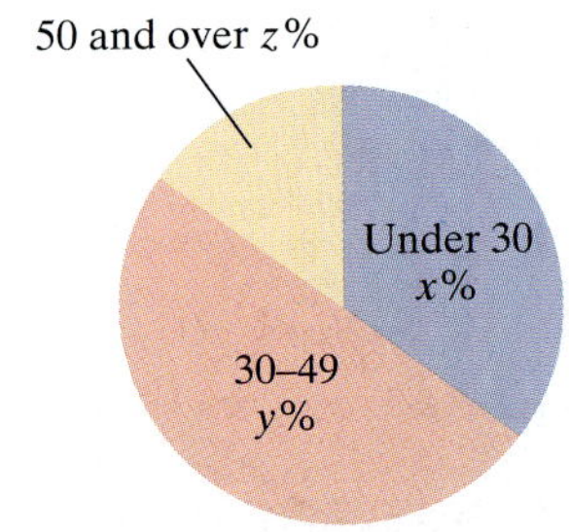

Source: U.S. Census Bureau

50. From *shlemiel* to *shlemazal*, evocative Yiddish words have entered our vocabulary. Yiddish, Thai, and Persian are spoken by 621 thousand people in the United States. The difference between the number of people who speak Yiddish and the number who speak Thai is 7 thousand. The difference between the number of people who speak Thai and the number who speak Persian is 4 thousand. Find the thousands of people in the United States who speak Yiddish, Thai, and Persian.

Writing in Mathematics

51. What is a matrix?

52. Describe what is meant by the augmented matrix of a system of linear equations.

53. In your own words, describe each of the three matrix row operations. Give an example of each of the operations.

54. Describe how to use matrices and row operations to solve a system of linear equations.

55. When solving a system using matrices, how do you know if the system has no solution?

56. When solving a system using matrices, how do you know if the system has infinitely many solutions?

Technology Exercises

57. Most graphing utilities can perform row operations on matrices. Consult the owner's manual for your graphing utility to learn proper keystroke sequences for performing these operations. Then duplicate the row operations of any three exercises that you solved from Exercises 11–20.

58. The final augmented matrix that we obtain when using Gaussian elimination is said to be in **row-echelon form**. For systems of linear equations with unique solutions, this form results when each entry in the main diagonal is 1 and all entries below the main diagonal are 0s. Some graphing utilities can transform a matrix to row-echelon form. Consult your owner's manual. If your utility has this capability, use it to verify of any five of the matrices you obtained in Exercises 23–46.

Critical Thinking Exercises

59. Which one of the following is true?

a. A matrix row operation such as $-\frac{4}{5}R_1 + R_2$ is not permitted because of the negative fraction.

b. The augmented matrix for the system

$$\begin{aligned} x - 3y &= 5 \\ y - 2z &= 7 \\ 2x + z &= 4 \end{aligned} \quad \text{is} \quad \left[\begin{array}{cc|c} 1 & -3 & 5 \\ 1 & -2 & 7 \\ 2 & 1 & 4 \end{array}\right].$$

c. In solving a linear system of three equations in three variables, we begin with the augmented matrix and use row operations to obtain a row-equivalent matrix with 0s down the diagonal from left to right and 1s below each 0.

d. None of the above is true.

60. Matrices can be used to solve systems involving more than three equations. Write the augmented matrix for the following system. Then use matrix row operations to obtain a matrix with 1s down the diagonal from upper left to lower right, and 0s below the 1s. Use this matrix and back-substitution to solve the system.

$$\begin{aligned} 2x + y + 3z - w &= 6 \\ x - y + 2z - 2w &= -1 \\ x - y - z + w &= -4 \\ -x + 2y - 2z - w &= -7 \end{aligned}$$

61. The vitamin content per ounce for three foods is given in the following table.

	Milligrams per Ounce		
	Thiamin	**Riboflavin**	**Niacin**
Food A	3	7	1
Food B	1	5	3
Food C	3	8	2

a. Use matrices to show that no combination of these foods can provide exactly 14 milligrams of thiamin, 32 milligrams of riboflavin, and 9 milligrams of niacin.

b. Use matrices to describe in practical terms what happens if the riboflavin requirement is increased by 5 mg and the other requirements stay the same.

Review Exercises

62. If $f(x) = -3x + 10$, find $f(2a - 1)$. (Section 2.1, Example 3)

63. If $f(x) = 3x$ and $g(x) = 2x - 3$, find $(fg)(x)$ and $(fg)(-1)$. (Section 2.2, Example 4)

64. Simplify: $\dfrac{-4x^8y^{-12}}{12x^{-3}y^{24}}$. (Section 1.6, Example 9)

▸ SECTION 3.5 Determinants and Cramer's Rule

Objectives

1. Evaluate a second-order determinant.
2. Solve a system of linear equations in two variables using Cramer's rule.
3. Evaluate a third-order determinant.
4. Solve a system of linear equations in three variables using Cramer's rule.
5. Use determinants to identify inconsistent and dependent systems.

SSM PH Tutor Center CD-ROM Video

A portion of Charles Babbage's unrealized Difference Engine

As cyberspace absorbs more and more of our work, play, shopping, and socializing, where will it all end? Which activities will still be offline in 2025?

Our technologically transformed lives can be traced back to the English inventor Charles Babbage (1792–1871). Babbage knew of a method for solving linear systems called *Cramer's rule*, in honor of the Swiss geometer Gabriel Cramer (1704–1752). Cramer's rule was simple, but involved numerous multiplications for large systems. Babbage designed a machine, called the "difference engine," that consisted of toothed wheels on shafts for performing these multiplications. Despite the fact that only one-seventh of the functions ever worked, Babbage's invention demonstrated how complex calculations could be handled mechanically. In 1944, scientists at IBM used the lessons of the difference engine to create the world's first computer.

Those who invented computers hoped to relegate the drudgery of repeated computation to a machine. In this section, we look at a method for solving linear systems that played a critical role in this process. The method uses arrays of numbers called *determinants*. As with matrix methods, solutions are obtained by writing down the coefficients and constants of a linear system and performing operations with them.

1 Evaluate a second-order determinant.

The Determinant of a 2 × 2 Matrix

A matrix of **order $m \times n$** has m rows and n columns. If $m = n$, a matrix has the same number of rows as columns and is called a **square matrix**. Associated with every square matrix is a real number, called its **determinant**. The determinant for a 2×2 square matrix is defined as follow:

Study Tip

To evaluate a 2×2 determinant, find the difference of the products of the two diagonals.

$$\begin{vmatrix} a_1 & b_1 \\ a_2 & b_2 \end{vmatrix} = a_1b_2 - a_2b_1$$

Definition of the Determinant of a 2 × 2 Matrix

The determinant of the matrix $\begin{bmatrix} a_1 & b_1 \\ a_2 & b_2 \end{bmatrix}$ is denoted by $\begin{vmatrix} a_1 & b_1 \\ a_2 & b_2 \end{vmatrix}$ and is defined by

$$= \begin{vmatrix} a_1 & b_1 \\ a_2 & b_2 \end{vmatrix} = a_1b_2 - a_2b_1.$$

We also say that the **value** of the **second-order determinant** $\begin{vmatrix} a_1 & b_1 \\ a_2 & b_2 \end{vmatrix}$ is $a_1b_2 - a_2b_1$.

Example 1 illustrates that the determinant of a matrix may be positive or negative. The determinant can also have 0 as its value.

EXAMPLE 1 Evaluating the Determinant of a 2 × 2 Matrix

Evaluate the determinant of:

a. $\begin{bmatrix} 5 & 6 \\ 7 & 3 \end{bmatrix}$ **b.** $\begin{bmatrix} 2 & 4 \\ -3 & -5 \end{bmatrix}$.

Solution We multiply and subtract as indicated.

a. $\begin{vmatrix} 5 & 6 \\ 7 & 3 \end{vmatrix} = 5 \cdot 3 - 7 \cdot 6 = 15 - 42 = -27$ The value of the second-order determinant is −27.

b. $\begin{vmatrix} 2 & 4 \\ -3 & -5 \end{vmatrix} = 2(-5) - (-3)(4) = -10 + 12 = 2$ The value of the second-order determinant is 2. ■

Discover for Yourself

Write and then evaluate three determinants, one whose value is positive, one whose value is negative, and one whose value is 0.

CHECK POINT 1 Evaluate the determinant of:

a. $\begin{bmatrix} 10 & 9 \\ 6 & 5 \end{bmatrix}$

b. $\begin{bmatrix} 4 & 3 \\ -5 & -8 \end{bmatrix}$.

2 Solve a system of linear equations in two variables using Cramer's rule.

Solving Linear Systems of Equations in Two Variables Using Determinants

Determinants can be used to solve a linear system in two variables. In general, such a system appears as

$$a_1x + b_1y = c_1$$
$$a_2x + b_2y = c_2.$$

Let's first solve this system for x using the addition method. We can solve for x by eliminating y from the equations. Multiply the first equation by b_2 and the second equation by $-b_1$. Then add the two equations.

$$\begin{aligned} a_1x + b_1y = c_1 &\xrightarrow{\text{Multiply by } b_2.} & a_1b_2x + b_1b_2y &= c_1b_2 \\ a_2x + b_2y = c_2 &\xrightarrow{\text{Multiply by } -b_1.} & -a_2b_1x - b_1b_2y &= -c_2b_1 \\ &\text{Add:} & (a_1b_2 - a_2b_1)x &= c_1b_2 - c_2b_1 \\ & & x &= \frac{c_1b_2 - c_2b_1}{a_1b_2 - a_2b_1} \end{aligned}$$

Because

$$\begin{vmatrix} c_1 & b_1 \\ c_2 & b_2 \end{vmatrix} = c_1b_2 - c_2b_1 \quad \text{and} \quad \begin{vmatrix} a_1 & b_1 \\ a_2 & b_2 \end{vmatrix} = a_1b_2 - a_2b_1$$

we can express our answer for x as the quotient of two determinants:

$$x = \frac{\begin{vmatrix} c_1 & b_1 \\ c_2 & b_2 \end{vmatrix}}{\begin{vmatrix} a_1 & b_1 \\ a_2 & b_2 \end{vmatrix}}.$$

In a similar way, we could use the addition method to solve our system for y, again expressing y as the quotient of two determinants. This method of using determinants to solve the linear system, called **Cramer's rule**, is summarized in the box.

Solving a Linear System in Two Variables Using Determinants

Cramer's Rule

If

$$a_1x + b_1y = c_1$$
$$a_2x + b_2y = c_2$$

then

$$x = \frac{\begin{vmatrix} c_1 & b_1 \\ c_2 & b_2 \end{vmatrix}}{\begin{vmatrix} a_1 & b_1 \\ a_2 & b_2 \end{vmatrix}} \quad \text{and} \quad y = \frac{\begin{vmatrix} a_1 & c_1 \\ a_2 & c_2 \end{vmatrix}}{\begin{vmatrix} a_1 & b_1 \\ a_2 & b_2 \end{vmatrix}}$$

where

$$\begin{vmatrix} a_1 & b_1 \\ a_2 & b_2 \end{vmatrix} \neq 0.$$

Here are some helpful tips when solving

$$a_1x + b_1y = c_1$$
$$a_2x + b_2y = c_2$$

using determinants:

1. Three different determinants are used to find x and y. The determinants in the denominators for x and y are identical. The determinants in the numerators for x and y differ. In abbreviated notation, we write

$$x = \frac{D_x}{D} \quad \text{and} \quad y = \frac{D_y}{D} \text{ where } D \neq 0.$$

2. The elements of D, the determinant in the denominator, are the coefficients of the variables in the system.

$$D = \begin{vmatrix} a_1 & b_1 \\ a_2 & b_2 \end{vmatrix}$$

3. D_x, the determinant in the numerator of x,is obtained by replacing the x-coefficients in D, a_1 and a_2, with the constants on the right side of the equations, c_1 and c_2.

$$D = \begin{vmatrix} a_1 & b_1 \\ a_2 & b_2 \end{vmatrix} \quad \text{and} \quad D_x = \begin{vmatrix} c_1 & b_1 \\ c_2 & b_2 \end{vmatrix}$$

Replace the column with a_1 and a_2 with the constants c_1 and c_2 to get D_x.

4. D_y, the determinant in the numerator for y, is obtained by replacing the y-coefficients in D, b_1 and b_2,with the constants on the right side of the equations, c_1 and c_2.

$$D = \begin{vmatrix} a_1 & b_1 \\ a_2 & b_2 \end{vmatrix} \quad \text{and} \quad D_y = \begin{vmatrix} a_1 & c_1 \\ a_2 & c_2 \end{vmatrix}$$

Replace the column with b_1 and b_2 with the constants c_1 and c_2 to get D_y.

Example 2 illustrates the use of Cramer's rule.

EXAMPLE 2 Using Cramer's Rule to Solve a Linear System

Use Cramer's rule to solve the system:

$$\begin{aligned} 5x - 4y &= 2 \\ 6x - 5y &= 1. \end{aligned}$$

Solution Because

$$x = \frac{D_x}{D} \quad \text{and} \quad y = \frac{D_y}{D},$$

we will set up and evaluate the three determinants D, D_x, and D_y.

1. D, the determinant in both denominators, consists of the x- and y-coefficients.

$$D = \begin{vmatrix} 5 & -4 \\ 6 & -5 \end{vmatrix} = (5)(-5) - (6)(-4) = -25 + 24 = -1$$

Because this determinant is not zero, we continue to use Cramer's rule to solve the system.

2. D_x, the determinant in the numerator for x, is obtained by replacing the x-coefficients in D, 5 and 6, by the constants on the right side of the equation, 2 and 1.

$$D_x = \begin{vmatrix} 2 & -4 \\ 1 & -5 \end{vmatrix} = (2)(-5) - (1)(-4) = -10 + 4 = -6$$

3. D_y, the determinant in the numerator for y, is obtained by replacing the y-coefficients in D, -4 and -5, by the constants on the right side of the equation, 2 and 1.

$$D_y = \begin{vmatrix} 5 & 2 \\ 6 & 1 \end{vmatrix} = (5)(1) - (6)(2) = 5 - 12 = -7$$

4. Thus,

$$x = \frac{D_x}{D} = \frac{-6}{-1} = 6 \quad \text{and} \quad y = \frac{D_y}{D} = \frac{-7}{-1} = 7.$$

As always, the ordered pair (6, 7) can be checked by substituting these values into the original equations. The solution is (6, 7) and the solution set is $\{(6, 7)\}$. ■

✔ **CHECK POINT 2** Use Cramer's rule to solve the system:

$$\begin{aligned} 5x + 4y &= 12 \\ 3x - 6y &= 24. \end{aligned}$$

3 Evaluate a third-order determinant.

The Determinant of a 3 × 3 Matrix

Associated with every square matrix is a real number called its determinant. The determinant for a 3×3 matrix is defined in terms of second-order determinants.

Definition of the Determinant of a 3 × 3 Matrix

A third-order determinant is defined by

Subtract. Add.

$$\begin{vmatrix} a_1 & b_1 & c_1 \\ a_2 & b_2 & c_2 \\ a_3 & b_3 & c_3 \end{vmatrix} = a_1 \begin{vmatrix} b_2 & c_2 \\ b_3 & c_3 \end{vmatrix} - a_2 \begin{vmatrix} b_1 & c_1 \\ b_3 & c_3 \end{vmatrix} + a_3 \begin{vmatrix} b_1 & c_1 \\ b_2 & c_2 \end{vmatrix}.$$

The a's on the right come from the first column.

Here are some tips that may be helpful when evaluating the determinant of a 3 × 3 matrix:

Evaluating the Determinant of a 3 × 3 Matrix

1. Each of the three terms in the definition contains two factors—a numerical factor and a second-order determinant.
2. The numerical factor in each term is an element from the first column of the third-order determinant.
3. The minus sign precedes the second term.
4. The second-order determinant that appears in each term is obtained by crossing out the row and the column containing the numerical factor.

$$a_1 \begin{vmatrix} b_2 & c_2 \\ b_3 & c_3 \end{vmatrix} - a_2 \begin{vmatrix} b_1 & c_1 \\ b_3 & c_3 \end{vmatrix} + a_3 \begin{vmatrix} b_1 & c_1 \\ b_2 & c_2 \end{vmatrix}$$

$$\downarrow \qquad\qquad \downarrow \qquad\qquad \downarrow$$

$$\begin{vmatrix} \not a_1 & \not b_1 & \not c_1 \\ \not a_2 & b_2 & c_2 \\ \not a_3 & b_3 & c_3 \end{vmatrix} \quad \begin{vmatrix} \not a_1 & b_1 & c_1 \\ \not a_2 & \not b_2 & \not c_2 \\ \not a_3 & b_3 & c_3 \end{vmatrix} \quad \begin{vmatrix} \not a_1 & b_1 & c_1 \\ \not a_2 & b_2 & c_2 \\ \not a_3 & \not b_3 & \not c_3 \end{vmatrix}$$

The **minor** of an element is the determinant that remains after deleting the row and column of that element. For this reason, we call this method **expansion by minors**.

EXAMPLE 3 Evaluating the Determinant of a 3 × 3 Matrix

Evaluate the determinant of:

$$\begin{bmatrix} 4 & 1 & 0 \\ -9 & 3 & 4 \\ -3 & 8 & 1 \end{bmatrix}.$$

Solution We know that each of the three terms in the determinant contains a numerical factor and a second-order determinant. The numerical factors are from the first column of the determinant of the given matrix. They are shown in red in the following matrix:

$$\begin{bmatrix} 4 & 1 & 0 \\ -9 & 3 & 4 \\ -3 & 8 & 1 \end{bmatrix}.$$

We find the minor for each numerical factor by deleting the row and column of that element:

$$\begin{bmatrix} 4 & 1 & 0 \\ -9 & 3 & 4 \\ -3 & 8 & 1 \end{bmatrix} \quad \begin{bmatrix} 4 & 1 & 0 \\ -9 & 3 & 4 \\ -3 & 8 & 1 \end{bmatrix} \quad \begin{bmatrix} 4 & 1 & 0 \\ -9 & 3 & 4 \\ -3 & 8 & 1 \end{bmatrix}.$$

The minor for 4 is $\begin{vmatrix} 3 & 4 \\ 8 & 1 \end{vmatrix}$.

The minor for −9 is $\begin{vmatrix} 1 & 0 \\ 8 & 1 \end{vmatrix}$.

The minor for −3 is $\begin{vmatrix} 1 & 0 \\ 3 & 4 \end{vmatrix}$.

Now we have three numerical factors, 4, −9, and −3, and three second-order determinants. We multiply each numerical factor by its second-order determinant to find the three terms of the third-order determinant:

$$4\begin{vmatrix} 3 & 4 \\ 8 & 1 \end{vmatrix}, \quad -9\begin{vmatrix} 1 & 0 \\ 8 & 1 \end{vmatrix}, \quad -3\begin{vmatrix} 1 & 0 \\ 3 & 4 \end{vmatrix}.$$

Based on the preceding definition, we subtract the second term from the first term and add the third term:

Don't forget to supply the minus sign.

$$\begin{vmatrix} 4 & 1 & 0 \\ -9 & 3 & 4 \\ -3 & 8 & 1 \end{vmatrix} = 4\begin{vmatrix} 3 & 4 \\ 8 & 1 \end{vmatrix} - (-9)\begin{vmatrix} 1 & 0 \\ 8 & 1 \end{vmatrix} - 3\begin{vmatrix} 1 & 0 \\ 3 & 4 \end{vmatrix}$$

$$= 4(3 \cdot 1 - 8 \cdot 4) + 9(1 \cdot 1 - 8 \cdot 0) - 3(1 \cdot 4 - 3 \cdot 0)$$

$$= 4(3 - 32) + 9(1 - 0) - 3(4 - 0)$$ Evaluate the three second-order determinants.

$$= 4(-29) + 9(1) - 3(4)$$ Subtract within parentheses.

$$= -116 + 9 - 12$$ Multiply.

$$= -119$$ Add and subtract as indicated. ■

Using Technology

Verify the result of Example 3 by using your graphing utility. First enter the given matrix:

$$A = \begin{bmatrix} 4 & 1 & 0 \\ -9 & 3 & 4 \\ -3 & 8 & 1 \end{bmatrix}.$$

Then enter

det [A] [ENTER].

The result should be −119.

✔ **CHECK POINT 3** Evaluate the determinant of:

$$\begin{bmatrix} 2 & 1 & 7 \\ -5 & 6 & 0 \\ -4 & 3 & 1 \end{bmatrix}.$$

4 Solve a system of linear equations in three variables using Cramer's rule.

Solving Linear Systems of Equations in Three Variables Using Determinants

Cramer's rule can be applied to solving systems of linear equations in three variables. The determinants in the numerator and denominator of all variables are third-order determinants.

Solving Three Equations in Three Variables Using Determinants

Cramer's Rule

If

$$a_1x + b_1y + c_1z = d_1$$
$$a_2x + b_2y + c_2z = d_2$$
$$a_3x + b_3y + c_3z = d_3$$

then

$$x = \frac{D_x}{D}, \; y = \frac{D_y}{D}, \text{ and } z = \frac{D_z}{D}.$$

These four third-order determinants are given by:

$$D = \begin{vmatrix} a_1 & b_1 & c_1 \\ a_2 & b_2 & c_2 \\ a_3 & b_3 & c_3 \end{vmatrix}$$ These are the coefficients of the variables x, y, and z, $D \neq 0$.

$$D_x = \begin{vmatrix} d_1 & b_1 & c_1 \\ d_2 & b_2 & c_2 \\ d_3 & b_3 & c_3 \end{vmatrix}$$ Replace x-coefficients in D with the constants at the right of the three equations.

$$D_y = \begin{vmatrix} a_1 & d_1 & c_1 \\ a_2 & d_2 & c_2 \\ a_3 & d_3 & c_3 \end{vmatrix}$$ Replace y-coefficients in D with the constants at the right of the three equations.

$$D_z = \begin{vmatrix} a_1 & b_1 & d_1 \\ a_2 & b_2 & d_2 \\ a_3 & b_3 & d_3 \end{vmatrix}.$$ Replace z-coefficients in D with the constants at the right of the three equations.

EXAMPLE 4 Using Cramer's Rule to Solve a Linear System in Three Variables

Use Cramer's rule to solve:

$$x + 2y - z = -4$$
$$x + 4y - 2z = -6$$
$$2x + 3y + z = 3.$$

Solution Because

$$x = \frac{D_x}{D}, \; y = \frac{D_y}{D}, \quad \text{and} \quad z = \frac{D_z}{D},$$

we need to set up and evaluate four determinants.

Step 1 Set up the determinants.

1. D, the determinant in all three denominators, consists of the x-, y-, and z-coefficients.

$$D = \begin{vmatrix} 1 & 2 & -1 \\ 1 & 4 & -2 \\ 2 & 3 & 1 \end{vmatrix}$$

ENRICHMENT ESSAY

Why Software Programs for Solving Linear Systems Do Not Use Cramer's Rule

Supercomputers can perform one trillion (10^{12}) multiplications per second. To solve a linear system with a "mere" 20 equations using Cramer's rule requires over 5.1×10^{19} multiplications. This would take a supercomputer more than 590 days. The cost for this venture? At $2200 per hour, typical of the costs for supercomputer time, the computations required by Cramer's rule would cost more than $30 million!

2. D_x, the determinant in the numerator for x, is obtained by replacing the x-coefficients in D, 1, 1, and 2, with the constants on the right side of the equation, −4, −6, and 3.

$$D_x = \begin{vmatrix} -4 & 2 & -1 \\ -6 & 4 & -2 \\ 3 & 3 & 1 \end{vmatrix}$$

3. D_y, the determinant in the numerator for y, is obtained by replacing the y-coefficients in D, 2, 4, and 3, with the constants on the right side of the equation, −4, −6, and 3.

$$D_y = \begin{vmatrix} 1 & -4 & -1 \\ 1 & -6 & -2 \\ 2 & 3 & 1 \end{vmatrix}$$

4. D_z, the determinant in the numerator for z, is obtained by replacing the z-coefficients in D, −1, −2, and 1, with the constants on the right side of the equation, −4, −6, and 3.

$$D_z = \begin{vmatrix} 1 & 2 & -4 \\ 1 & 4 & -6 \\ 2 & 3 & 3 \end{vmatrix}$$

Step 2 Evaluate the four determinants.

$$D = \begin{vmatrix} 1 & 2 & -1 \\ 1 & 4 & -2 \\ 2 & 3 & 1 \end{vmatrix} = 1\begin{vmatrix} 4 & -2 \\ 3 & 1 \end{vmatrix} - 1\begin{vmatrix} 2 & -1 \\ 3 & 1 \end{vmatrix} + 2\begin{vmatrix} 2 & -1 \\ 4 & -2 \end{vmatrix}$$

$$= 1(4 + 6) - 1(2 + 3) + 2(-4 + 4)$$

$$= 1(10) - 1(5) + 2(0) = 5$$

Using the same technique to evaluate each determinant, we obtain

$$D_x = -10, \quad D_y = 5, \quad \text{and} \quad D_z = 20.$$

Step 3 Substitute these four values and solve the system.

$$x = \frac{D_x}{D} = \frac{-10}{5} = -2$$

$$y = \frac{D_y}{D} = \frac{5}{5} = 1$$

$$z = \frac{D_z}{D} = \frac{20}{5} = 4$$

The ordered triple (−2, 1, 4) can be checked by substitution into the original three equations. The solution is (−2, 1, 4) and the solution set is $\{(-2, 1, 4)\}$. ■

✔ **CHECK POINT 4** Use Cramer's rule to solve the system:

$$\begin{aligned} 3x - 2y + z &= 16 \\ 2x + 3y - z &= -9. \\ x + 4y + 3z &= 2 \end{aligned}$$

5 Use determinants to identify inconsistent and dependent systems.

Cramer's Rule with Inconsistent and Dependent Systems

If D, the determinant in the denominator, is 0, the variables described by the quotient of determinants are not real numbers. However, when $D = 0$, this indicates that the system is inconsistent or contains dependent equations. This gives rise to the following two situations:

Determinants: Inconsistent and Dependent Systems

1. If $D = 0$ and at least one of the determinants in the numerator is not 0, then the system is inconsistent. The solution set is $\varnothing$.
2. If $D = 0$ and all the determinants in the numerators are 0, then the equations in the system are dependent. The system has infinitely many solutions.

Although we have focused on applying determinants to solve linear systems, they have other applications, some of which we consider in the exercise set that follows.

EXERCISE SET 3.5

Practice Exercises

Evaluate each determinant in Exercises 1–10.

1. $\begin{vmatrix} 5 & 7 \\ 2 & 3 \end{vmatrix}$

2. $\begin{vmatrix} 4 & 8 \\ 5 & 6 \end{vmatrix}$

3. $\begin{vmatrix} -4 & 1 \\ 5 & 6 \end{vmatrix}$

4. $\begin{vmatrix} 7 & 9 \\ -2 & -5 \end{vmatrix}$

5. $\begin{vmatrix} -7 & 14 \\ 2 & -4 \end{vmatrix}$

6. $\begin{vmatrix} 1 & -3 \\ -8 & 2 \end{vmatrix}$

7. $\begin{vmatrix} -5 & -1 \\ -2 & -7 \end{vmatrix}$

8. $\begin{vmatrix} \frac{1}{5} & \frac{1}{6} \\ -6 & 5 \end{vmatrix}$

9. $\begin{vmatrix} \frac{1}{2} & \frac{1}{2} \\ \frac{1}{8} & -\frac{3}{4} \end{vmatrix}$

10. $\begin{vmatrix} \frac{2}{3} & \frac{1}{3} \\ -\frac{1}{2} & \frac{3}{4} \end{vmatrix}$

For Exercises 11–26, use Cramer's rule to solve each system or to determine that the system is inconsistent or contains dependent equations.

11. $x + y = 7$
$x - y = 3$

12. $2x + y = 3$
$x - y = 3$

13. $12x + 3y = 15$
$2x - 3y = 13$

14. $x - 2y = 5$
$5x - y = -2$

15. $4x - 5y = 17$
$2x + 3y = 3$

16. $3x + 2y = 2$
$2x + 2y = 3$

17. $x - 3y = 4$
$3x - 4y = 12$

18. $2x - 9y = 5$
$3x - 3y = 11$

19. $3x - 4y = 4$
$2x + 2y = 12$

20. $3x = 7y + 1$
$2x = 3y - 1$

21. $2x = 3y + 2$
$5x = 51 - 4y$

22. $y = -4x + 2$
$2x = 3y + 8$

23. $3x = 2 - 3y$
$2y = 3 - 2x$

24. $x + 2y - 3 = 0$
$12 = 8y + 4x$

25. $4y = 16 - 3x$
$6x = 32 - 8y$

26. $2x = 7 + 3y$
$4x - 6y = 3$

Evaluate each determinant in Exercises 27–32.

27. $\begin{vmatrix} 3 & 0 & 0 \\ 2 & 1 & -5 \\ 2 & 5 & -1 \end{vmatrix}$

28. $\begin{vmatrix} 4 & 0 & 0 \\ 3 & -1 & 4 \\ 2 & -3 & 5 \end{vmatrix}$

29. $\begin{vmatrix} 3 & 1 & 0 \\ -3 & 4 & 0 \\ -1 & 3 & -5 \end{vmatrix}$

30. $\begin{vmatrix} 2 & -4 & 2 \\ -1 & 0 & 5 \\ 3 & 0 & 4 \end{vmatrix}$

31. $\begin{vmatrix} 1 & 1 & 1 \\ 2 & 2 & 2 \\ -3 & 4 & -5 \end{vmatrix}$

32. $\begin{vmatrix} 1 & 2 & 3 \\ 2 & 2 & -3 \\ 3 & 2 & 1 \end{vmatrix}$

In Exercises 33–40, use Cramer's rule to solve each system.

33. $\begin{aligned} x + y + z &= 0 \\ 2x - y + z &= -1 \\ -x + 3y - z &= -8 \end{aligned}$

34. $\begin{aligned} x - y + 2z &= 3 \\ 2x + 3y + z &= 9 \\ -x - y + 3z &= 11 \end{aligned}$

35. $\begin{aligned} 4x - 5y - 6z &= -1 \\ x - 2y - 5z &= -12 \\ 2x - y &= 7 \end{aligned}$

36. $\begin{aligned} x - 3y + z &= -2 \\ x + 2y &= 8 \\ 2x - y &= 1 \end{aligned}$

37. $\begin{aligned} x + y + z &= 4 \\ x - 2y + z &= 7 \\ x + 3y + 2z &= 4 \end{aligned}$

38. $\begin{aligned} 2x + 2y + 3z &= 10 \\ 4x - y + z &= -5 \\ 5x - 2y + 6z &= 1 \end{aligned}$

39. $\begin{aligned} x + 2z &= 4 \\ 2y - z &= 5 \\ 2x + 3y &= 13 \end{aligned}$

40. $\begin{aligned} 3x + 2z &= 4 \\ 5x - y &= -4 \\ 4y + 3z &= 22 \end{aligned}$

Application Exercises

Determinants are used to find the area of a triangle whose vertices are given by three points in a rectangular coordinate system. The area of a triangle with vertices (x_1, y_1), (x_2, y_2), and (x_3, y_3) is:

$$\text{Area} = \pm\frac{1}{2}\begin{vmatrix} x_1 & y_1 & 1 \\ x_2 & y_2 & 1 \\ x_3 & y_3 & 1 \end{vmatrix}.$$

where the symbol $(\pm)$ indicates that the appropriate sign should be chosen to yield a positive area. Use this information to work Exercises 41–42.

41. Use determinants to find the area of the triangle whose vertices are $(3, -5)$, $(2, 6)$, and $(-3, 5)$.

42. Use determinants to find the area of the triangle whose vertices are $(1, 1)$, $(-2, -3)$, and $(11, -3)$.

Determinants are used to show that three points lie on the same line (are collinear). If

$$\begin{vmatrix} x_1 & y_1 & 1 \\ x_2 & y_2 & 1 \\ x_3 & y_3 & 1 \end{vmatrix} = 0$$

then the points (x_1, y_1), (x_2, y_2), and (x_3, y_3) are collinear. If the determinant does not equal 0, then the points are not collinear. Use this information to work Exercises 43–44.

43. Are the points $(3, -1)$, $(0, -3)$, and $(12, 5)$ collinear?

44. Are the points $(-4, -6)$, $(1, 0)$, and $(11, 12)$ collinear?

Determinants are used to write an equation of a line passing through two points. An equation of the line passing through the distinct points (x_1, y_1) and (x_2, y_2) is given by

$$\begin{vmatrix} x & y & 1 \\ x_1 & y_1 & 1 \\ x_2 & y_2 & 1 \end{vmatrix} = 0.$$

Use this information to work Exercises 45–46.

45. Use the determinant to write an equation for the line passing through $(3, -5)$ and $(-2, 6)$. Then expand the determinant, expressing the line's equation in slope-intercept form.

46. Use the determinant to write an equation for the line passing through $(-1, 3)$ and $(2, 4)$. Then expand the determinant, expressing the line's equation in slope-intercept form.

Writing in Mathematics

47. Explain how to evaluate a second-order determinant.

48. Describe the determinants D, D_x, and D_y in terms of the coefficients and constants in a system of two equations in two variables.

49. Explain how to evaluate a third-order determinant.

50. When expanding a determinant by minors, when is it necessary to supply a minus sign?

51. Without going into too much detail, describe how to solve a linear system in three variables using Cramer's rule.

52. In applying Cramer's rule, what does it mean if $D = 0$?

53. The process of solving a linear system in three variables using Cramer's rule can involve tedious computation. Is there a way of speeding up this process, perhaps using Cramer's rule to find the value for only one of the variables? Describe how this process might work, presenting a specific example with your description. Remember that your goal is still to find the value for each variable in the system.

54. If you could use only one method to solve linear systems in three variables, which method would you select? Explain why this is so.

Technology Exercises

55. Use the feature of your graphing utility that evaluates the determinant of a square matrix to verify any five of the determinants that you evaluated by hand in Exercises 1–10, or 27–32.

56. What is the fastest method for solving a linear system with your graphing utility?

Critical Thinking Exercises

57. Which one of the following is true?

a. If $D = 0$, then every variable has a value of 0.

b. Using Cramer's rule, we use $\dfrac{D}{D_y}$ to get the value of y.

c. Because there are different determinants in the numerators of x and y, if a system is solved using Cramer's rule, x and y cannot have the same value.

d. Only one 2×2 determinant is needed to evaluate

$$\begin{vmatrix} 2 & 3 & -2 \\ 0 & 1 & 3 \\ 0 & 4 & -1 \end{vmatrix}.$$

58. Solve for x: $\begin{vmatrix} 2x + 1 & 5 \\ x - 1 & -3 \end{vmatrix} = 3x - 1.$

59. What happens to the value of a second-order determinant if the two columns are interchanged?

60. Consider the system

$$\begin{aligned} a_1x + b_1y &= c_1 \\ a_2x + b_2y &= c_2. \end{aligned}$$

Use Cramer's rule to prove that if the first equation of the system is replaced by the sum of the two equations, the resulting system has the same solution as the original system.

61. Show that the equation of a line through (x_1, y_1) and (x_2, y_2) is given by the determinant

$$\begin{vmatrix} x & y & 1 \\ x_1 & y_1 & 1 \\ x_2 & y_2 & 1 \end{vmatrix} = 0.$$

Review Exercises

62. Solve: $6x - 4 = 2 + 6(x - 1)$. (Section 1.4, Example 6)

63. Solve for y: $-2x + 3y = 7$. (Section 1.5, Example 6)

64. Solve: $\dfrac{4x + 1}{3} = \dfrac{x - 3}{6} + \dfrac{x + 5}{6}$. (Section 1.4, Example 4)

CHAPTER 3 GROUP PROJECTS

1. The group is going into business for the next year. Your first task is to determine the product that you plan to manufacture and sell. Choose something unique, but realistic, reflecting the abilities and interests of group members. How much will it cost to produce each unit of your product? What will be the selling price for each unit? What is a reasonable estimate of your fixed cost for the entire year? Be sure to include utilities, labor, materials, marketing, and anything else that is relevant to your business venture. Once you have determined the product and these three figures:
   a. Write the cost function, C, of producing x units of your product.
   b. Write the revenue function, R, from the sale of x units of your product.
   c. Determine the break-even point. Does this seem realistic in terms of your product and its target market?
   d. Graph the cost function and the revenue function in the same rectangular coordinate system. Label the figure just like Figure 3.7 on page 177. Indicate the regions that show the profit loss and profit gain for your business venture.
   e. Write the profit function, P, from producing and selling x units of your product. Graph the profit function, in the same style as Figure 3.8 on page 178. Indicate where the business is in the red and where it is in the black.
   f. MTV's *The Real World* has offered to pay the fixed cost for the business venture, videotaping all group interactions for its forthcoming *Real World: Profit and Loss*. Group members need to determine whether or not to accept the offer.
2. Group members should consult an almanac, newspaper, magazine, or the Internet and return to the group with data that have a variable that is first decreasing and then increasing over time, or vice versa. The group should select the two most interesting data sets. For each set selected:
   a. Identify three data points and use the function $y = ax^2 + bx + c$ to model the data. Let x represent the number of years after the first year in the data set.
   b. Use the quadratic function to make a prediction about what might occur in the future.
3. The group should write four original word problems that can be solved using a system of linear equations in two or three variables. Each problem should be on a different topic. The group should turn in the four problems and their algebraic solutions.
4. Turn on your computer and read your e-mail or write a paper. When you need to do research, use the Internet to browse through art museums and photography exhibits. When you need a break, load a flight simulator program and fly through a photorealistic computer world. As different as these experiences may be, thay all share one thing —you're looking at images based on matrices. Matrices have applications in numerous fields, including the new technology of digital photography in which pictures are represented by numbers rather than film. Members of the group should research applications of matrices that they find intriguing. The group should then present a seminar to the class about these applications.

CHAPTER SUMMARY, REVIEW, AND TEST

SUMMARY

DEFINITIONS AND CONCEPTS | **EXAMPLES**

Section 3.1 Systems of Linear Equations in Two Variables

A system of linear equations in two variables, x and y, consists of two equations that can be written in the form $Ax + By = C$. The solution set is the set of all ordered pairs that satisfy both equations. Using the graphing method, a solution of a linear system is a point common to the graphs of both equations in the system.

Solve by graphing:

$$2x - y = 6$$
$$x + y = 6.$$

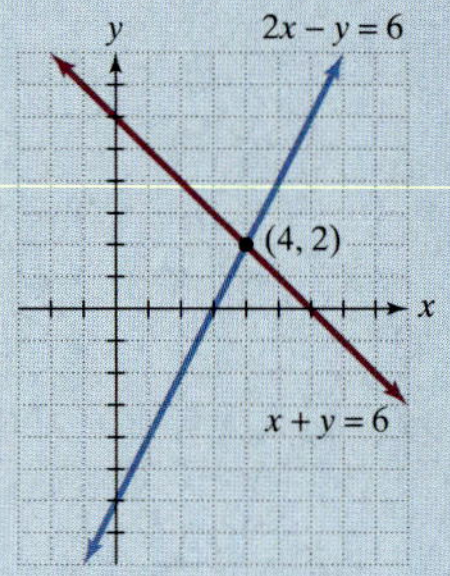

The intersection point gives the solution: (4, 2). The solution set is $\{(4, 2)\}$.

To solve a linear system by the substitution method:

1. Solve one equation for a variable.
2. Substitute the expression for that variable into the other equation.
3. Solve the equation in one variable.
4. Back-substitute the value of the variable into one of the original equations and find the value of the other variable.
5. Check the proposed solution in both equations.

Solve by the substitution method:

$$y = 2x + 3$$
$$7x - 5y = -18.$$

Substitute $2x + 3$ for y in the second equation.

$$7x - 5(2x + 3) = -18$$
$$7x - 10x - 15 = -18$$
$$-3x - 15 = -18$$
$$-3x = -3$$
$$x = 1$$

Find y. Substitute 1 for x in $y = 2x + 3$.

$$y = 2 \cdot 1 + 3 = 2 + 3 = 5$$

The ordered pair (1, 5) checks. The solution is (1, 5) and $\{(1, 5)\}$ is the solution set.

To solve a linear system by the addition method:

1. Write equations in $Ax + By = C$ form.
2. Multiply one or both equations by nonzero numbers so that coefficients of a variable are opposites.
3. Add equations.
4. Solve the resulting equation for a variable.
5. Back-substitute the value of the variable in either original equation and find the value of the remaining variable.
6. Check the proposed solution in both equations.

Solve by the addition method:

$$3x + y = -11$$
$$6x - 2y = -2.$$

Eliminate y. Multiply both sides of the first equation by 2.

$$6x + 2y = -22$$
$$6x - 2y = -2$$
$$\text{Add: } 12x = -24$$
$$x = -2$$

Find y. Back-substitute -2 for x. Use the first equation.

$$3(-2) + y = -11$$
$$-6 + y = -11$$
$$y = -5$$

The ordered pair $(-2, -5)$, checks. The solution is $(-2, -5)$ and $\{(-2, -5)\}$ is the solution set.

DEFINITIONS AND CONCEPTS	EXAMPLES

Section 3.1 Systems of Linear Equations in Two Variables (continued)

A linear system with at least one solution is a consistent system. A system that has no solution, with $\varnothing$ as its solution set, is an inconsistent system. A linear system with infinitely many solutions has dependent equations. Solving inconsistent systems by substitution or addition leads to a false statement, such as $0 = 3$. Solving systems with dependent equations leads to a true statement, such as $7 = 7$.

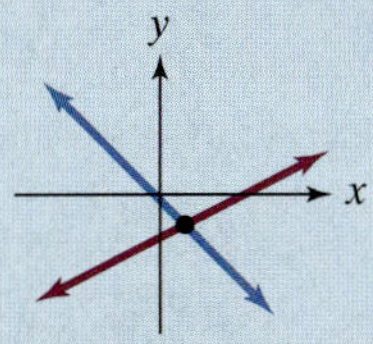

One solution: consistent

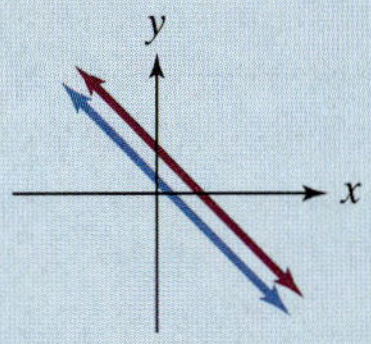

No solution: inconsistent

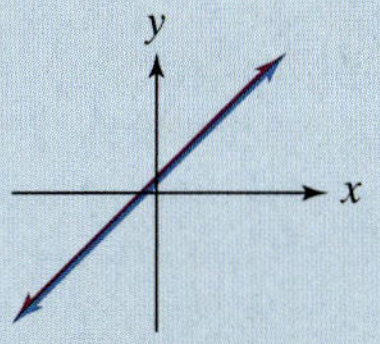

Infinitely many solutions: dependent and consistent

Section 3.2 Problem Solving and Business Applications Using Systems of Equations

A Problem-Solving Strategy

1. Use variables, usually x and y, to represent unknown quantities.
2. Write a system of equations describing the problem's conditions.
3. Solve the system and answer the problem's question.
4. Check proposed answers in the problem's wording.

You invested \$14,000 in two stocks paying 7% and 9% interest. Total year-end interest was \$1180. How much was invested at each rate?

Let x = amount invested at 7% and
y = amount invested at 9%.

amount invested at 7% — amount invested at 9%

$$x + y = 14{,}000$$

Interest from 7% investment — Interest from 9% investment

$$0.07x + 0.09y = 1180$$

Solving by substitution or addition, $x = 4000$ and $y = 10{,}000$. Thus, \$4000 was invested at 7% and \$10,000 at 9%.

Functions of Business

A company produces and sells x units of a product.

Revenue Function

$$R(x) = (\text{price per unit})x$$

Cost Function

$$C(x) = \text{fixed cost} + (\text{cost per unit produced})x$$

Profit Function

$$P(x) = R(x) - C(x)$$

The point of intersection of the graphs of R and C is the break-even point. The point reveals the number of units that a company must produce and sell so that the money coming in, the revenue, is equal to the money going out, the cost.

A company that manufactures lamps has a fixed cost of \$80,000 and it costs \$20 to produce each lamp. Lamps are sold for \$70.

a. Write the cost function.

$$C(x) = 80{,}000 + 20x$$

Fixed cost — Variable cost: \$20 per lamp.

b. Write the revenue function.

$$R(x) = 70x$$

Revenue per lamp, \$70, times number of lamps sold

c. Find the break-even point.

Solve
$$y = 80{,}000 + 20x$$
$$y = 70x$$
by substitution. Solving
$$70x = 80{,}000 + 20x$$
yields $x = 1600$. Back-substituting, $y = 112{,}000$. The break-even point is $(1600, 112{,}000)$: The company breaks even if it sells 1600 lamps. At this level, money coming in equals money going out: \$112,000.

DEFINITIONS AND CONCEPTS	EXAMPLES
Section 3.3 Systems of Linear Equations in Three Variables	
A system of linear equations in three variables, x, y, and z, consists of three equations of the form $Ax + By + Cz = D$. The solution set is the set of ordered triples that satisfy all three equations. The solution represents the point of intersection of three planes in space.	Is $(2, -1, 3)$ a solution of $3x + 5y - 2z = -5$ $2x + 3y - z = -2$ $2x + 4y + 6z = 18$? Replace x by 2, y by -1, and z by 3. Using Equation 1, we obtain: $3 \cdot 2 + 5(-1) - 2(3) \stackrel{?}{=} -5$ $6 - 5 - 6 \stackrel{?}{=} -5$ $-5 = -5$, true The ordered triple $(2, -1, 3)$ satisfies the first equation. In a similar manner, it satisfies the other two equations and is a solution.
To solve a linear system in three variables by eliminating variables: **1.** Reduce the system to two equations in two variables. **2.** Solve the resulting system of two equations in two variables. **3.** Use back-substitution in one of the equations in two variables to find the value of the second variable. **4.** Back-substitute the values for two variables into one of the original equations to find the value of the third variable. **5.** Check. If all variables are eliminated and a false statement results, the system is inconsistent and has no solution. If a true statement results, the system contains dependent equations and has infinitely many solutions.	Solve: $2x + 3y - 2z = 0$ [1] $x + 2y - z = 1$ [2] $3x - y + z = -15.$ [3] Add equations [2] and [3] to eliminate z. $4x + y = -14$ [4] Eliminate z again. Multiply equation [3] by 2 and add to equation [1]. $8x + y = -30$ [5] Multiply equation [4] by -1 and add to equation [5]. $-4x - y = 14$ $8x + y = -30$ Add: $4x = -16$ $x = -4$ Substitute -4 for x in [4]. $4(-4) + y = -14$ $y = 2$ Substitute -4 for x and 2 for y in [3]. $3(-4) - 2 + z = -15$ $-14 + z = -15$ $z = -1$ Checking verifies that $(-4, 2, -1)$ is the solution and $\{(-4, 2, -1)\}$ is the solution set.

DEFINITIONS AND CONCEPTS	EXAMPLES

Section 3.3 Systems of Linear Equations in Three Variables (continued)

Curve Fitting

Curve fitting is determining a function whose graph contains given points. Three points that do not lie on a line determine the graph of a quadratic function

$$y = ax^2 + bx + c.$$

Use the three given points to create a system of three equations. Solve the system to find a, b, and c.

Find the quadratic function $y = ax^2 + bx + c$ whose graph passes through the points $(-1, 2)$, $(1, 8)$, and $(2, 14)$.

Use $y = ax^2 + bx + c$.

When $x = -1, y = 2$: $2 = a(-1)^2 + b(-1) + c$

When $x = 1, y = 8$: $8 = a \cdot 1^2 + b \cdot 1 + c$

When $x = 2, y = 14$: $14 = a \cdot 2^2 + b \cdot 2 + c$

Solving,

$$\begin{aligned} a - b + c &= 2 \\ a + b + c &= 8 \\ 4a + 2b + c &= 14, \end{aligned}$$

$a = 1$, $b = 3$, and $c = 4$. The quadratic function, $y = ax^2 + bx + c$, is $y = x^2 + 3x + 4$.

Section 3.4 Matrix Solutions to Linear Systems

A matrix is a rectangular array of numbers. The augmented matrix of a linear system is obtained by writing the coefficients of each variable, a vertical bar, and the constants of the system.

$$\begin{aligned} x + 4y &= 9 \\ 3x + y &= 5 \end{aligned}$$

Augmented matrix is $\left[\begin{array}{cc|c} 1 & 4 & 9 \\ 3 & 1 & 5 \end{array}\right]$.

The following row operations produce matrices that represent systems with the same solution. Two matrices are row equivalent if one can be obtained from the other by a sequence of these row operations.

1. Interchange the elements in the ith and jth rows: $R_i \leftrightarrow R_j$.
2. Multiply each element in the ith row by k: kR_i.
3. Add k times the elements in row i to the corresponding elements in row j: $kR_i + R_j$.

Find $-4R_1 + R_2$:

$$\left[\begin{array}{ccc|c} 1 & 0 & -2 & 5 \\ 4 & -1 & 2 & 6 \\ 3 & -7 & 9 & 10 \end{array}\right].$$

Add -4 times the elements in row 1 to the corresponding elements in row 2.

$$\left[\begin{array}{ccc|c} 1 & 0 & -2 & 5 \\ -4(1) + 4 & -4(0) + (-1) & -4(-2) + 2 & -4(5) + 6 \\ 3 & -7 & 9 & 10 \end{array}\right]$$

$$= \left[\begin{array}{ccc|c} 1 & 0 & -2 & 5 \\ 0 & -1 & 10 & -14 \\ 3 & -7 & 9 & 10 \end{array}\right]$$

Solving Linear Systems Using Matrices

1. Write the augmented matrix for the system.
2. Use matrix row operations to simplify the matrix to one with 1s down the diagonal from upper left to lower right, and 0s below the 1s.
3. Write the system of linear equations corresponding to the matrix in step 2, and use back-substitution to find the system's solution.

If you obtain a matrix with a row containing 0s to the left of the vertical bar and a nonzero number on the right, the system is inconsistent. If 0s appear across an entire row, the system contains dependent equations.

Solve using matrices:

$$\begin{aligned} 3x + y &= 5 \\ x + 4y &= 9. \end{aligned}$$

$$\left[\begin{array}{cc|c} 3 & 1 & 5 \\ 1 & 4 & 9 \end{array}\right] \xrightarrow{R_1 \leftrightarrow R_2} \left[\begin{array}{cc|c} 1 & 4 & 9 \\ 3 & 1 & 5 \end{array}\right]$$

$$\xrightarrow{-3R_1 + R_2} \left[\begin{array}{cc|c} 1 & 4 & 9 \\ 0 & -11 & -22 \end{array}\right] \xrightarrow{-\frac{1}{11}R_2} \left[\begin{array}{cc|c} 1 & 4 & 9 \\ 0 & 1 & 2 \end{array}\right]$$

$$\rightarrow \begin{aligned} x + 4y &= 9 \\ y &= 2 \end{aligned}$$

When $y = 2$, $x + 4 \cdot 2 = 9$, so $x = 1$. The solution is $(1, 2)$ and the solution set is $\{(1, 2)\}$.

DEFINITIONS AND CONCEPTS	EXAMPLES
Section 3.5 Determinants and Cramer's Rule	
A square matrix has the same number of rows as columns. A determinant is a real number associated with a square matrix. The determinant is denoted by placing vertical bars about the array of numbers. The value of a second-order determinant is $\begin{vmatrix} a_1 & b_1 \\ a_2 & b_2 \end{vmatrix} = a_1b_2 - a_2b_1.$	Evaluate: $\begin{vmatrix} 2 & -1 \\ 3 & 4 \end{vmatrix} = 2(4) - 3(-1) = 8 + 3 = 11.$
Cramer's Rule for Two Linear Equations in Two Variables. If $a_1x + b_1y = c_1$, $a_2x + b_2y = c_2$ then $x = \frac{\begin{vmatrix} c_1 & b_1 \\ c_2 & b_2 \end{vmatrix}}{\begin{vmatrix} a_1 & b_1 \\ a_2 & b_2 \end{vmatrix}} = \frac{D_x}{D}$ and $y = \frac{\begin{vmatrix} a_1 & c_1 \\ a_2 & c_2 \end{vmatrix}}{\begin{vmatrix} a_1 & b_1 \\ a_2 & b_2 \end{vmatrix}} = \frac{D_y}{D}, \quad D \neq 0.$ If $D = 0$ and any numerator is not zero, the system is inconsistent and has no solution. If all determinants are 0, the system contains dependent equations and has infinitely many solutions.	Solve by Cramer's rule: $5x + 3y = 7$, $-x + 2y = 9.$ $D = \begin{vmatrix} 5 & 3 \\ -1 & 2 \end{vmatrix} = 5(2) - (-1)(3) = 10 + 3 = 13$ $D_x = \begin{vmatrix} 7 & 3 \\ 9 & 2 \end{vmatrix} = 7 \cdot 2 - 9 \cdot 3 = 14 - 27 = -13$ $D_y = \begin{vmatrix} 5 & 7 \\ -1 & 9 \end{vmatrix} = 5(9) - (-1)(7) = 45 + 7 = 52$ $x = \frac{D_x}{D} = \frac{-13}{13} = -1,\ y = \frac{D_y}{D} = \frac{52}{13} = 4$ The solution is $(-1, 4)$ and the solution set is $\{(-1, 4)\}$.
The value of a third-order determinant is $\begin{vmatrix} a_1 & b_1 & c_1 \\ a_2 & b_2 & c_2 \\ a_3 & b_3 & c_3 \end{vmatrix} = a_1\begin{vmatrix} b_2 & c_2 \\ b_3 & c_3 \end{vmatrix} - a_2\begin{vmatrix} b_1 & c_1 \\ b_3 & c_3 \end{vmatrix} + a_3\begin{vmatrix} b_1 & c_1 \\ b_2 & c_2 \end{vmatrix}.$ Each 2×2 determinant is called a minor.	Evaluate: $\begin{vmatrix} 1 & -2 & 1 \\ 3 & 1 & -2 \\ 5 & 5 & 3 \end{vmatrix}.$ $= 1\begin{vmatrix} 1 & -2 \\ 5 & 3 \end{vmatrix} - 3\begin{vmatrix} -2 & 1 \\ 5 & 3 \end{vmatrix} + 5\begin{vmatrix} -2 & 1 \\ 1 & -2 \end{vmatrix}$ $= 1(3 - (-10)) - 3(-6 - 5) + 5(4 - 1)$ $= 1(13) - 3(-11) + 5(3)$ $= 13 + 33 + 15 = 61$
Cramer's Rule for Three Linear Equations in Three Variables If $a_1x + b_1y + c_1z = d_1$, $a_2x + b_2y + c_2z = d_2$, $a_3x + b_3y + c_3z = d_3$	Solve by Cramer's rule: $x - 2y + z = 4$, $3x + y - 2z = 3$, $5x + 5y + 3z = -8.$

DEFINITIONS AND CONCEPTS	EXAMPLES
Section 3.5 Determinants and Cramer's Rule (continued)	
then $x = \frac{D_x}{D}, \quad y = \frac{D_y}{D}, \quad z = \frac{D_z}{D}.$ $D = \begin{vmatrix} a_1 & b_1 & c_1 \\ a_2 & b_2 & c_2 \\ a_3 & b_3 & c_3 \end{vmatrix} \neq 0, \quad D_x = \begin{vmatrix} d_1 & b_1 & c_1 \\ d_2 & b_2 & c_2 \\ d_3 & b_3 & c_3 \end{vmatrix}$ $D_y = \begin{vmatrix} a_1 & d_1 & c_1 \\ a_2 & d_2 & c_2 \\ a_3 & d_3 & c_3 \end{vmatrix}, \quad D_z = \begin{vmatrix} a_1 & b_1 & d_1 \\ a_2 & b_2 & d_2 \\ a_3 & b_3 & d_3 \end{vmatrix}$ If $D = 0$ and any numerator is not zero, the system is inconsistent. If all determinants are 0, the system contains dependent equations.	$D = \begin{vmatrix} 1 & -2 & 1 \\ 3 & 1 & -2 \\ 5 & 5 & 3 \end{vmatrix} = 61$ shown on page 222 $D_x = \begin{vmatrix} 4 & -2 & 1 \\ 3 & 1 & -2 \\ -8 & 5 & 3 \end{vmatrix} = 61$ $D_y = \begin{vmatrix} 1 & 4 & 1 \\ 3 & 3 & -2 \\ 5 & -8 & 3 \end{vmatrix} = -122$ $D_z = \begin{vmatrix} 1 & -2 & 4 \\ 3 & 1 & 3 \\ 5 & 5 & -8 \end{vmatrix} = -61$ $x = \frac{D_x}{D} = \frac{61}{61} = 1, \quad y = \frac{D_y}{D} = \frac{-122}{61} = -2,$ $z = \frac{D_z}{D} = \frac{-61}{61} = -1.$ The solution is $(1, -2, -1)$ and the solultion set is $\{(1, -2, -1)\}$.

Review Exercises

3.1 *In Exercises 1–2, determine whether the given ordered pair is a solution of the system.*

1. $(4, 2)$

$$2x - 5y = -2$$
$$3x + 4y = 4$$

2. $(-5, 3)$

$$-x + 2y = 11$$
$$y = -\frac{x}{3} + \frac{4}{3}$$

In Exercises 3–6, solve each system by graphing. Identify systems with no solution and systems with infinitely many solutions, using set notation to express their solution sets.

3. $x + y = 5$

$3x - y = 3$

4. $3x - 2y = 6$

$6x - 4y = 12$

5. $y = \frac{3}{5}x - 3$

$2x - y = -4$

6. $y = -x + 4$

$3x + 3y = -6$

In Exercises 7–13, solve each system by the substitution method or the addition method. Identify systems with no solution and systems with infinitely many solutions, using set notation to express their solution sets.

7. $2x - y = 2$

$x + 2y = 11$

8. $y = -2x + 3$

$3x + 2y = -17$

9. $3x + 2y = -8$

$2x + 5y = 2$

10. $5x - 2y = 14$

$3x + 4y = 11$

11. $y = 4 - x$

$3x + 3y = 12$

12. $\frac{x}{8} + \frac{3y}{4} = \frac{19}{8}$

$-\frac{x}{2} + \frac{3y}{4} = \frac{1}{2}$

13. $x - 2y + 3 = 0$

$2x - 4y + 7 = 0$

3.2 *In Exercises 14–18, use the four-step strategy to solve each problem.*

14. An appliance store is having a sale on small TVs and stereos. One day a salesperson sells 3 of the TVs and 4 stereos for \$2530. The next day the same employee sells 4 of the same TVs and 3 of the same stereos for \$2510. What are the prices of a TV and a stereo?

15. You invested \$10,000 in two stocks paying 8% and 10% annual interest, respectively. At the end of the year, the total interest from these investments was \$940. How much was invested at each rate?

16. A chemist needs to mix a solution that is 34% silver nitrate with one that is 4% silver nitrate to obtain 100 milliliters of a mixture that is 7% silver nitrate. How many milliliters of each of the solutions must be used?

17. When a plane flies with the wind, it can travel 2160 miles in 3 hours. When the plane flies in the opposite direction, against the wind, it takes 4 hours to fly the same distance. Find the speed of the plane in still air and the speed of the wind.

18. The perimeter of a rectangular table tennis top is 28 feet. The difference between 4 times the length and 3 times the width is 21 feet. Find the dimensions.

The cost and revenue functions for producing and selling x units of a new graphing calculator are

$$C(x) = 22{,}500 + 40x \quad \text{and} \quad R(x) = 85x.$$

Use these functions to solve Exercises 19–21.

19. Find the profit loss or the profit gain from selling 400 graphing calculators.

20. Write the profit function, P, from producing and selling x graphing calculators.

21. Determine the break-even point. Describe what this means.

22. A company is planning to manufacture computer desks. The fixed cost will be \$60,000 and it will cost \$200 to produce each desk. Each desk will be sold for \$450.

a. Write the cost function, C, of producing x desks.

b. Write the revenue function, R, from the sale of x desks.

c. Determine the break-even point. Describe what this means.

3.3

23. Is $(-3, -2, 5)$ a solution of the system

$$\begin{aligned} x + y + z &= 0 \\ 2x - 3y + z &= 5 \\ 4x + 2y + 4z &= 3? \end{aligned}$$

Solve each system in Exercises 24–26 by eliminating variables using the addition method. If there is no solution or if there are infinitely many solutions and a system's equations are dependent, so state.

24. $$\begin{aligned} 2x - y + z &= 1 \\ 3x - 3y + 4z &= 5 \\ 4x - 2y + 3z &= 4 \end{aligned}$$

25. $$\begin{aligned} x + 2y - z &= 5 \\ 2x - y + 3z &= 0 \\ 2y + z &= 1 \end{aligned}$$

26. $$\begin{aligned} 3x - 4y + 4z &= 7 \\ x - y - 2z &= 2 \\ 2x - 3y + 6z &= 5 \end{aligned}$$

27. Find the quadratic function $y = ax^2 + bx + c$ whose graph passes through the points $(1, 4)$, $(3, 20)$, and $(-2, 25)$.

28. The graph shows a low savings rate in the United States compared to that of many industrialized countries. Adding the rates for Japan, Germany, and France gives 45%. The savings rate in Japan exceeds that for Germany by 1% and is 12% less than twice that for France. Find the savings rates for Japan, Germany, and France.

Comparative Savings Rates

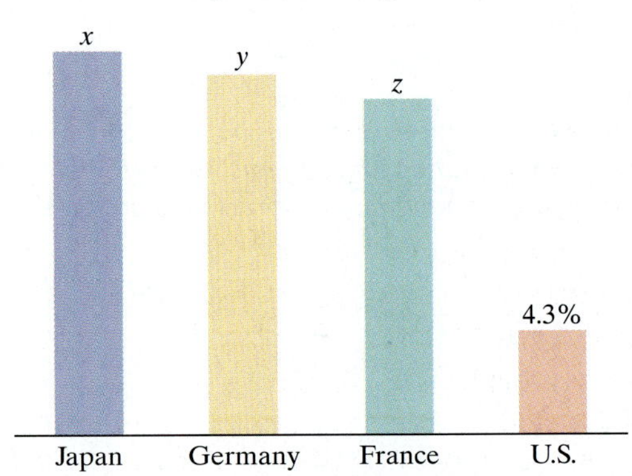

Source: Office of Management and Budget

3.4 *In Exercises 29–30, write the system of linear equations represented by the augmented matrix. Use x, y, and, if necessary, z, for the variables. Once the system is written, use back-substitution to find its solution.*

29. $\left[\begin{array}{cc|c} 2 & 3 & -10 \\ 0 & 1 & -6 \end{array}\right]$

30. $\left[\begin{array}{ccc|c} 1 & 1 & 3 & 12 \\ 0 & 1 & -2 & -4 \\ 0 & 0 & 1 & 3 \end{array}\right]$

In Exercises 31–34, perform each matrix row operation and write the new matrix.

31. $\left[\begin{array}{cc|c} 1 & -8 & 3 \\ 0 & 7 & -14 \end{array}\right] \frac{1}{7}R_2$

32. $\left[\begin{array}{cc|c} 1 & -3 & 1 \\ 2 & 1 & -5 \end{array}\right] -2R_1 + R_2$

33. $\left[\begin{array}{ccc|c} 2 & -2 & 1 & -1 \\ 1 & 2 & -1 & 2 \\ 6 & 4 & 3 & 5 \end{array}\right] \frac{1}{2}R_1$

34. $\left[\begin{array}{ccc|c} 1 & 2 & 2 & 2 \\ 0 & 1 & -1 & 2 \\ 0 & 5 & 4 & 1 \end{array}\right] -5R_2 + R_3$

In Exercises 35–38, solve each system using matrices. If there is no solution or if a system's equations are dependent, so state.

35. $x + 4y = 7$
$3x + 5y = 0$

36. $2x - 3y = 8$
$-6x + 9y = 4$

37. $x + 2y + 3z = -5$
$2x + y + z = 1$
$x + y - z = 8$

38. $x - 2y + z = 0$
$y - 3z = -1$
$2y + 5z = -2$

3.5 *In Exercises 39–42, evaluate each determinant.*

39. $\begin{vmatrix} 3 & 2 \\ -1 & 5 \end{vmatrix}$

40. $\begin{vmatrix} -2 & -3 \\ -4 & -8 \end{vmatrix}$

41. $\begin{vmatrix} 2 & 4 & -3 \\ 1 & -1 & 5 \\ -2 & 4 & 0 \end{vmatrix}$

42. $\begin{vmatrix} 4 & 7 & 0 \\ -5 & 6 & 0 \\ 3 & 2 & -4 \end{vmatrix}$

In Exercises 43–46, use Cramer's rule to solve each system. If there is no solution or if a system's equations are dependent, so state.

43. $x - 2y = 8$
$3x + 2y = -1$

44. $7x + 2y = 0$
$2x + y = -3$

45. $x + 2y + 2z = 5$
$2x + 4y + 7z = 19$
$-2x - 5y - 2z = 8$

46. $2x + y = -4$
$y - 2z = 0$
$3x - 2z = -11$

47. Use the quadratic function $y = ax^2 + bx + c$ to model the following data:

x (Age of a Driver)	*y* (Average Number of Automobile Accidents per Day in the United States)
20	400
40	150
60	400

Use Cramer's rule to determine values for *a*, *b*, and *c*. Then use the model to write a statement about the average number of automobile accidents in which 30-year-olds and 50-year-olds are involved daily.

Chapter 3 Test

1. Solve by graphing:

$$x + y = 6$$
$$4x - y = 4.$$

In Exercises 2–4, solve each system by the substitution method or the addition method. Identify systems with no solution and systems with infinitely many solutions, using set notation to express their solution sets.

2. $5x + 4y = 10$
$3x + 5y = -7$

3. $x = y + 4$
$3x + 7y = -18$

4. $4x = 2y + 6$
$y = 2x - 3$

In Exercises 5–8, use the four-step strategy to solve each problem.

5. Health experts agree that cholesterol intake should be limited to 300 milligrams or less each day. Three ounces of shrimp and 2 ounces of scallops contain 156 milligrams of cholesterol. Five ounces of shrimp and 3 ounces of scallops contain 45 milligrams of cholesterol less than the suggested maximum daily intake. Determine the cholesterol content in an ounce of each item.

6. You invested \$6000 in two stocks paying 9% and 6% annual interest, respectively. At the end of the year, the total interest from these investments was \$480. How much was invested at each rate?

7. A chemist needs to mix a 20% acid solution with a 50% acid solution to obtain 60 ounces of a 30% acid solution. How many ounces of each of the solutions must be used?

8. When a plane flies with the wind, it can travel 1600 kilometers in 2 hours. When the plane flies in the opposite direction, against the wind, it takes 3 hours to travel 1950 kilometers. Find the speed of the plane in still air and the speed of the wind.

9. The cost and revenue functions for producing and selling x units of a toaster oven are

$$C(x) = 40x + 350{,}000 \quad \text{and} \quad R(x) = 125x.$$

Write the profit function, P, from producing and selling x toaster ovens.

Use this information to solve Exercises 10–12: A company is planning to produce and sell a new line of computers. The fixed cost will be \$360,000, and it will cost \$850 to produce each computer. Each computer will be sold for \$1150.

10. Write the cost function, C, of producing x computers.

11. Write the revenue function, R, from the sale of x computers.

12. Determine the break-even point. Describe what this means.

13. Solve by eliminating variables using the addition method:

$$\begin{aligned} x + y + z &= 6 \\ 3x + 4y - 7z &= 1 \\ 2x - y + 3z &= 5. \end{aligned}$$

14. Perform the indicated matrix row operation and write the new matrix.

$$\left[\begin{array}{ccc|c} 1 & 0 & -4 & 5 \\ 6 & -1 & 2 & 10 \\ 2 & -1 & 4 & -3 \end{array}\right] \quad -6R_1 + R_2$$

In Exercises 15–16, solve each system using matrices.

15. $2x + y = 6$
$3x - 2y = 16$

16. $x - 4y + 4z = -1$
$2x - y + 5z = 6$
$-x + 3y - z = 5$

In Exercises 17–18, evaluate each determinant.

17. $\begin{vmatrix} -1 & -3 \\ 7 & 4 \end{vmatrix}$

18. $\begin{vmatrix} 3 & 4 & 0 \\ -1 & 0 & -3 \\ 4 & 2 & 5 \end{vmatrix}$

In Exercises 19–20, use Cramer's rule to solve each system.

19. $4x - 3y = 14$
$3x - y = 3$

20. $2x + 3y + z = 2$
$3x + 3y - z = 0$
$x - 2y - 3z = 1$

Cumulative Review Exercises (Chapters 1–3)

1. Simplify: $\dfrac{6(8 - 10)^3 + (-2)}{(-5)^2(-2)}$.

2. Simplify: $7x - [5 - 2(4x - 1)]$.

In Exercises 3–5, solve each equation.

3. $5 - 2(3 - x) = 2(2x + 5) + 1$

4. $\dfrac{3x}{5} + 4 = \dfrac{x}{3}$

5. $3x - 4 = 2(3x + 2) - 3x$

6. For a summer sales job, you are choosing between two pay arrangements: a weekly salary of \$200 plus 5% commission on sales, or a straight 15% commission. For how many dollars of sales will the earnings be the same regardless of the pay arrangement?

7. Simplify: $\dfrac{-5x^6y^{-10}}{20x^{-2}y^{20}}$.

8. If $f(x) = -4x + 5$, find $f(a + 2)$.

9. Find the domain of $f(x) = \dfrac{4}{x + 3}$.

10. If $f(x) = 2x^2 - 5x + 2$ and $g(x) = x^2 - 2x + 3$, find $(f - g)(x)$ and $(f - g)(3)$.

In Exercises 11–12, graph each linear function.

11. $f(x) = -\frac{2}{3}x + 2$ **12.** $2x - y = 6$

In Exercises 13–14, use the given conditions to write an equation for each line in point-slope form and slope-intercept form.

13. Passing through $(2, 4)$ and $(4, -2)$

14. Passing through $(-1, 0)$ and parallel to the line whose equation is $3x + y = 6$

In Exercises 15–16, solve each system by eliminating variables using the addition method.

15.
$$\begin{aligned} 3x + 12y &= 25 \\ 2x - 6y &= 12 \end{aligned}$$

16.
$$\begin{aligned} x + 3y - z &= 5 \\ -x + 2y + 3z &= 13 \\ 2x - 5y - z &= -8 \end{aligned}$$

17. If two pads and 19 pens are sold for \$5.40 and 7 of the same pads and 4 of the same pens sell for \$6.40, find the cost of one pad and one pen.

18. Evaluate: $\begin{vmatrix} 0 & 1 & -2 \\ -7 & 0 & -4 \\ 3 & 0 & 5 \end{vmatrix}$.

19. Solve using matrices:
$$\begin{aligned} 2x + 3y - z &= -1 \\ x + 2y + 3z &= 2 \\ 3x + 5y - 2z &= -3. \end{aligned}$$

20. Solve using Cramer's rule (determinants):
$$\begin{aligned} 3x + 4y &= -1 \\ -2x + y &= 8. \end{aligned}$$

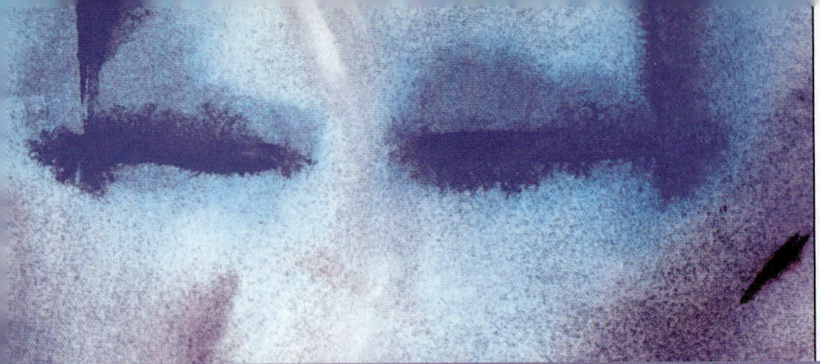

Had a good workout lately? If so, could you tell if you were overdoing it or not pushing yourself hard enough?

Experts suggest a method for determining the pulse rate you should maintain during aerobic activities, such as running, biking, or swimming. Subtract your age from 220; find 72% of that figure for the low end of the range and 87% for the high end. Your pulse rate, in beats per minute, should stay between these two figures if you want to get the most cardiovascular benefit from your training.

The phrase "should stay between" indicates an algebraic inequality, a symbolic statement containing an inequality symbol such as $<$ (is less than). In this chapter, you will learn methods for modeling your world with inequalities. You will even see how inequalities are used to describe some of the most magnificent places in our nation's landscape.

Chapter 4

Inequalities and Problem Solving

▸ SECTION 4.1 Solving Linear Inequalities

Objectives

1. Graph an inequality's solution set.
2. Use set-builder and interval notations.
3. Use properties of inequalities to solve inequalities.
4. Recognize inequalities with no solution or infinitely many solutions.
5. Solve problems using linear inequalities.

SSM PH Tutor Center CD-ROM Video

You can go online and obtain a list of telecommunication companies that provide residential long-distance phone service. The list contains the monthly fee, the monthly minimum, and the rate per minute for each service provider. You've chosen a plan that has a monthly fee of $15 with a charge of $0.08 per minute for all long-distance calls. Suppose you are limited by how much money you can spend for the month: You can spend at most $35. If we let x represent the number of minutes of long-distance calls in a month, we can write an inequality that describes the given conditions:

The monthly fee of $15	plus	the charge of $0.08 per minute for x minutes	must be less than or equal to	$35.
15	$+$	$0.08x$	$\leq$	$35.$

Using the commutative property of addition, we can express this inequality as

$$0.08x + 15 \leq 35.$$

The form of this inequality is $ax + by \leq c$, with $a = 0.08$, $b = 15$, and $c = 35$. Any inequality in this form is called **a linear inequality in one variable**. The greatest exponent on the variable in such an inequality is 1. The symbol between $ax + b$ and c can be $\leq$ (is less than or equal to), $<$ (is less than), $\geq$ (is greater than or equal to), or $>$ (is greater than).

In this section, we will study how to solve linear inequalities such as $0.08x + 15 \leq 35$. **Solving an inequality** is the process of finding the set of numbers that make the inequality a true statement. These numbers are called the **solutions** of the inequality, and we say that they **satisfy** the inequality. The set of all solutions is called the **solution set** of the inequality. We begin by discussing how to graph and how to represent these solution sets.

1 Graph an inequality's solution set.

Graphs of Inequalities; Interval Notation

There are infinitely many solutions to the inequality $x > -4$, namely all real numbers that are greater than -4. Although we cannot list all the solutions, we can make a drawing on a number line that represents these solutions. Such a drawing is called the **graph of the inequality**.

Graphs of solutions to linear inequalities are shown on a number line by shading all points representing numbers that are solutions. For example, the solutions of $x > 4$ are all real numbers that are greater than 4. They are graphed on a number line by shading all points to the right of 4. The open dot at 4 indicates that 4 is not a solution:

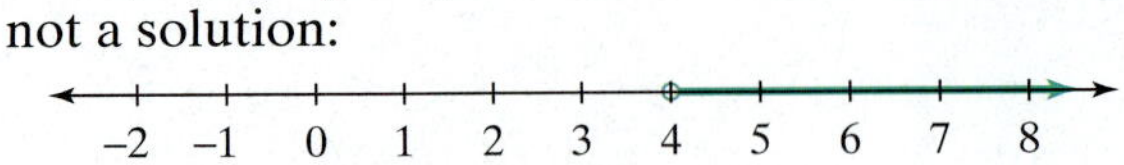

.

In this book, a convenient notation, called *interval notation*, will be used to represent solution sets of inequalities. To help understand this notation, a different graphing notation will be used. Parentheses, rather than open dots, indicate endpoints that are not solutions. Thus, we graph the solutions of $x > 4$ as

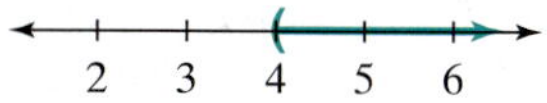 rather than as 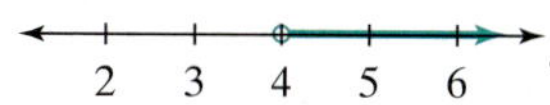.

Square brackets, rather than closed dots, indicate endpoints that are solutions. Thus, we graph the solutions of $x \geq 4$ as

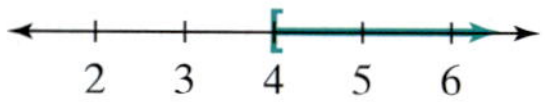 rather than as 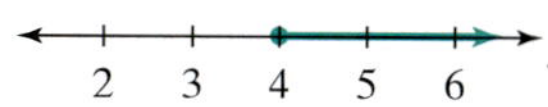.

EXAMPLE 1 Graphing Inequalities

Graph the solutions of:

a. $x < 3$ **b.** $x \geq -1$ **c.** $-1 < x \leq 3$.

Solution

a. The solutions of $x < 3$ are all real numbers that are less than 3. They are graphed on a number line by shading all points to the left of 3. The parenthesis at 3 indicates that 3 is not a solution, but numbers such as 2.9999 and 2.6 are. The arrow shows that the graph extends indefinitely to the left.

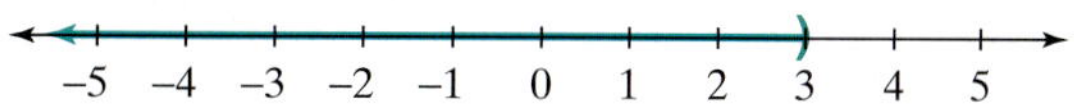

Study Tip

Because an inequality symbol points to the smaller number, $x < 3$ (x is less than 3) may be expressed as $3 > x$ (3 is greater than x).

b. The solutions of $x \geq -1$ are all real numbers that are greater than or equal to -1. We shade all points to the right of -1 and the point for -1 itself. The bracket at -1 shows that -1 is a solution for the given inequality. The arrow shows that the graph extends indefinitely to the right.

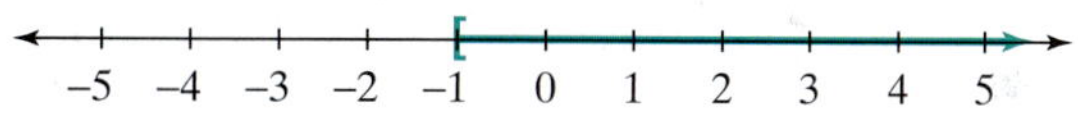

c. The inequality $-1 < x \leq 3$ is read "-1 is less than x *and* x is less than or equal to 3," or "x is greater than -1 *and* less than or equal to 3." The solutions of $-1 < x \leq 3$ are all real numbers between -1 and 3, not including -1 but including 3. The parenthesis at -1 indicates that -1 is not a solution. By contrast, the bracket at 3 shows that 3 is a solution. Shading indicates the other solutions.

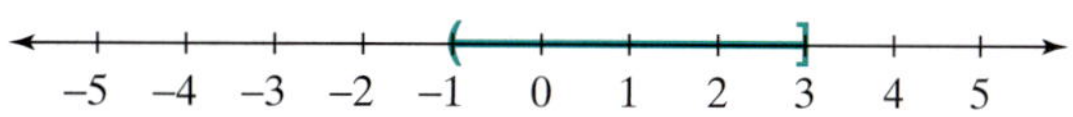

CHECK POINT 1 Graph the solutions of:

a. $x \leq 2$ **b.** $x > -4$ **c.** $2 \leq x < 6$

2 Use set-builder and interval notations.

Now that we know how to graph the solution set for an inequality such as $x > -4$, let's see how to represent the solution set. One method is with **set-builder notation**. Using this method, the solution set for $x > -4$ can be expressed as

$$\{x \mid x > -4\}.$$

The set of all x — such that

We read this as "the set of all real numbers x such that x is greater than -4."

As previously mentioned, another method used to represent solution sets for inequalities is **interval notation**. Using this notation, the solution set for $x > -4$ is expressed as $(-4, \infty)$. The parenthesis at -4 indicates that -4 is not included in the interval. The infinity symbol, ∞, does not represent a real number. It indicates that the interval extends indefinitely to the right.

Table 4.1 lists nine possible types of intervals used to describe sets of real numbers.

Table 4.1 Intervals on the Real Number Line

Let a and b be real numbers such that $a < b$.

Interval Notation	Set-Builder Notation	Graph
(a, b)	$\{x \mid a < x < b\}$	
$[a, b]$	$\{x \mid a \le x \le b\}$	
$[a, b)$	$\{x \mid a \le x < b\}$	
$(a, b]$	$\{x \mid a < x \le b\}$	
(a, ∞)	$\{x \mid x > a\}$	
$[a, \infty)$	$\{x \mid x \ge a\}$	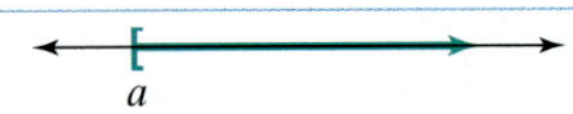
$(-\infty, b)$	$\{x \mid x < b\}$	
$(-\infty, b]$	$\{x \mid x \le b\}$	
$(-\infty, \infty)$	$\mathbb{R}$ (set of all real numbers)	

EXAMPLE 2 Intervals and Inequalities

Express the intervals in terms of inequalities and graph:

a. $(-1, 4]$ **b.** $[2.5, 4]$ **c.** $(-4, \infty)$.

Solution

a. $(-1, 4] = \{x \mid -1 < x \le 4\}$

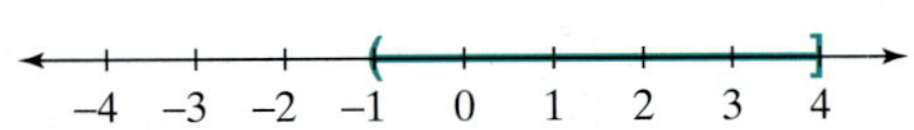

b. $[2.5, 4] = \{x \mid 2.5 \le x \le 4\}$

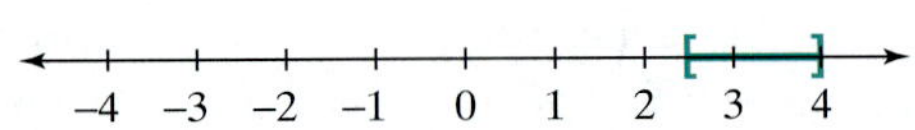

c. $(-4, \infty) = \{x \mid x > -4\}$

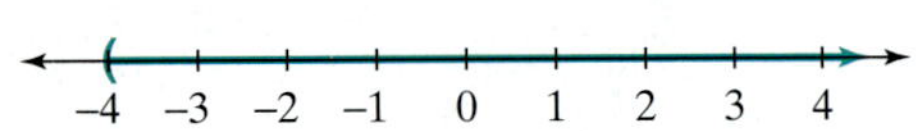

■

✔ **CHECK POINT 2** Express the intervals in terms of inequalities and graph:

a. $[-2, 5)$ **b.** $[1, 3.5]$ **c.** $(-\infty, -1)$.

3 Use properties of inequalities to solve inequalities.

Solving Linear Inequalities

Let's return to the question that opened this section: How many minutes of long-distance calls can you make in a month if you can spend at most \$35? We answer the question by solving

$$0.08x + 15 \leq 35$$

for x. The solution procedure is nearly identical to that for solving

$$0.08x + 15 = 35.$$

Our goal is to get x by itself on the left side. We do this by isolating 0.08: subtracting 15 from both sides:

$$0.08x + 15 \leq 35$$

$$0.08x + 15 - 15 \leq 35 - 15$$

$$0.08x \leq 20.$$

Finally, we isolate x from $0.08x$ by dividing both sides of the inequality by 0.08:

$$\frac{0.08x}{0.08} \leq \frac{20}{0.08}$$

$$x \leq 250.$$

With at most \$35 per month to spend, you can make no more than 250 minutes of long-distance calls each month.

We started with the inequality $0.08x + 15 \leq 35$ and obtained the inequality $x \leq 250$ in the final step. Both of these inequalities have the same solution set, namely $\{x \mid x \leq 250\}$. Inequalities such as these, with the same solution set, are said to be **equivalent**.

We isolated x from $0.08x$ by dividing both sides of $0.08x \leq 20$ by 0.08, a positive number. Let's see what happens if we divide both sides of an inequality by a negative number. Consider the inequality $10 < 14$. Divide 10 and 14 by -2:

$$\frac{10}{-2} = -5 \quad \text{and} \quad \frac{14}{-2} = -7.$$

Because -5 lies to the right of -7 on the number line, -5 is greater than -7:

$$-5 > -7.$$

Notice that the direction of the inequality symbol is reversed:

$$10 < 14$$

$$-5 > -7.$$

In general, **when we multiply or divide both sides of an inequality by a negative number, the direction of the inequality symbol is reversed**. When we reverse the direction of the inequality symbol, we say that we change the *sense* of the inequality.

We can isolate a variable in a linear inequality the same way we can isolate a variable in a linear equation. The following properties are used to create equivalent inequalities:

Properties of Inequalities

Property	The Property in Words	Example
The Addition Property of Inequality If $a < b$, then $a + c < b + c$. If $a < b$, then $a - c < b - c$.	If the same quantity is added to or subtracted from both sides of an inequality, the resulting inequality is equivalent to the original one.	$2x + 3 < 7$ Subtract 3: $2x + 3 - 3 < 7 - 3$. Simplify: $2x < 4$.
The Positive Multiplication Property of Inequality If $a < b$ and c is positive, then $ac < bc$. If $a < b$ and c is positive, then $\frac{a}{c} < \frac{b}{c}$.	If we multiply or divide both sides of an inequality by the same positive quantity, the resulting inequality is equivalent to the original one.	$2x < 4$ Divide by 2: $\frac{2x}{2} < \frac{4}{2}$. Simplify: $x < 2$.
The Negative Multiplication Property of Inequality If $a < b$ and c is negative, then $ac > bc$. If $a < b$ and c is negative, then $\frac{a}{c} > \frac{b}{c}$.	If we multiply or divide both sides of an inequality by the same negative quantity and reverse the direction of the inequality symbol, the result is an equivalent inequality.	$-4x < 20$ Divide by -4 and reverse the sense of the inequality: $\frac{-4x}{-4} > \frac{20}{-4}$. Simplify: $x > -5$.

If an inequality does not contain fractions, it can be solved using the following procedure. Notice, again, how similar this procedure is to the procedure for solving an equation.

Solving a Linear Inequality

1. Simplify the algebraic expression on each side.
2. Use the addition property of inequality to collect all the variable terms on one side and all the constant terms on the other side.
3. Use the multiplication property of inequality to isolate the variable and solve. Reverse the sense of the inequality when multiplying or dividing both sides by a negative number.
4. Express the solution set in set-builder or interval notation and graph the solution set on a number line.

EXAMPLE 3 Solving a Linear Inequality

Solve and graph the solution set on a number line:

$$3x - 5 > -17.$$

Solution

Step 1 Simplify each side. Because each side is already simplified, we can skip this step.

Step 2 Collect variable terms on one side and constant terms on the other side. The variable term, $3x$, is already on the left. We will collect constant terms on the right by adding 5 to both sides.

$$3x - 5 > -17 \quad \text{This is the given inequality.}$$
$$3x - 5 + 5 > -17 + 5 \quad \text{Add 5 to both sides.}$$
$$3x > -12 \quad \text{Simplify.}$$

Step 3 Isolate the variable and solve. We isolate the variable, x, by dividing both sides by 3. Because we are dividing by a positive number, we do not reverse the inequality symbol.

$$\frac{3x}{3} > \frac{-12}{3} \quad \text{Divide both sides by 3.}$$
$$x > -4 \quad \text{Simplify.}$$

Discover for Yourself

As a partial check, select one number from the solution set for the inequality in Example 3. Substitute that number into the original inequality. Perform the resulting computations. You should obtain a true statement.

Is it possible to perform a partial check using a number that is not in the solution set? What should happen in this case? Try doing this.

Step 4 Express the solution set in set-builder or interval notation and graph the set on a number line. The solution set consists of all real numbers that are greater than -4, expressed as $\{x \mid x > -4\}$ in set-builder notation. The interval notation for this solution set is $(-4, \infty)$. The graph of the solution set is shown as follows:

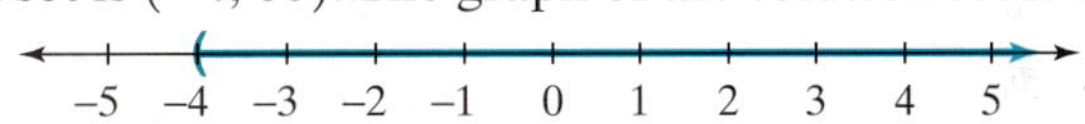

CHECK POINT 3 Solve and graph the solution set on a number line:

$$4x - 3 > -23.$$

EXAMPLE 4 Solving a Linear Inequality

Solve and graph the solution set on a number line:

$$7x + 15 \geq 13x + 51.$$

Solution

Step 1 Simplify each side. Because each side is already simplified, we can skip this step.

Step 2 Collect variable terms on one side and constant terms on the other side. We will collect variable terms on the left and constant terms on the right.

$$7x + 15 \geq 13x + 51 \quad \text{This is the given inequality.}$$
$$7x + 15 - 13x \geq 13x + 51 - 13x \quad \text{Subtract 13x from both sides.}$$
$$-6x + 15 \geq 51 \quad \text{Simplify.}$$
$$-6x + 15 - 15 \geq 51 - 15 \quad \text{Subtract 15 from both sides.}$$
$$-6x \geq 36 \quad \text{Simplify.}$$

Step 3 Isolate the variable and solve. We isolate the variable, x, by dividing both sides by -6. Because we are dividing by a negative number, we must reverse the inequality symbol.

$$\frac{-6x}{-6} \leq \frac{36}{-6} \quad \text{Divide both sides by } -6 \text{ and reverse the sense of the inequality.}$$
$$x \leq -6 \quad \text{Simplify.}$$

Study Tip

You can solve

$$7x + 15 \geq 13x + 51$$

by isolating x on the right side. Subtract $7x$ from both sides.

$$7x + 15 - 7x \geq 13x + 51 - 7x$$
$$15 \geq 6x + 51$$

Now subtract 51 from both sides.

$$15 - 51 \geq 6x + 51 - 51$$
$$-36 \geq 6x$$

Finally, divide both sides by 6.

$$\frac{-36}{6} \geq \frac{6x}{6}$$
$$-6 \geq x$$

This last inequality means the same thing as

$$x \leq -6.$$

Step 4 Express the solution set in set-builder or interval notation and graph the set on a number line. The solution set consists of all real numbers that are less than or equal to −6, expressed in set-builder notation as $\{x \mid x \leq -6\}$. The interval notation for this solution set is $(-\infty, -6]$. The graph of the solution set is shown as follows:

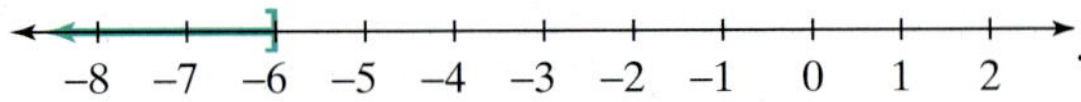

■

Using Technology

You can use a graphing utility to verify that $(-\infty, -6]$ is the solution set for

$$7x + 15 \geq 13x + 51.$$

For what values of x does the graph of y = 7x + 15 | lie above or on | the graph of y = 13x + 51?

The graphs are shown below in a $[-10, 2, 1]$ by $[-40, 5, 5]$ viewing rectangle. Notice that the graph of $y = 7x + 15$ lies above or on the graph of $y = 13x + 51$ when $x \leq -6$.

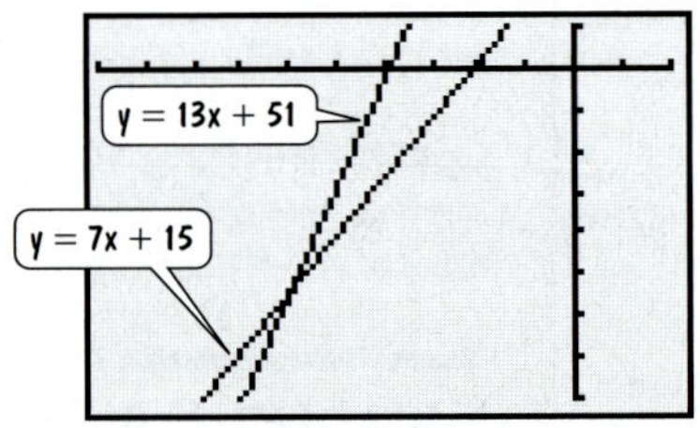

CHECK POINT 4 Solve and graph the solution set: $6 - 3x \geq 5x - 2$.

If an inequality contains fractions, begin by multiplying both sides by the least common denominator. This will clear the inequality of fractions.

EXAMPLE 5 Solving a Linear Inequality

Solve and graph the solution set on a number line:

$$\frac{x+3}{4} > \frac{x-2}{3} + \frac{1}{4}.$$

Solution The denominators are 4, 3, and 4. The least common denominator is 12. We begin by multiplying both sides of the inequality by 12.

$$\frac{x+3}{4} > \frac{x-2}{3} + \frac{1}{4}$$ This is the given inequality.

$$12\left(\frac{x+3}{4}\right) > 12\left(\frac{x-2}{3} + \frac{1}{4}\right)$$ Multiply both sides by 12. Multiplying by a positive number preserves the sense of the inequality.

$$\frac{12}{1} \cdot \frac{x+3}{4} > \frac{12}{1} \cdot \frac{x-2}{3} + \frac{12}{1} \cdot \frac{1}{4}$$ Use the distributive property and multiply each term by 12.

$$\frac{\overset{3}{\cancel{12}}}{1} \cdot \frac{x+3}{\underset{1}{\cancel{4}}} > \frac{\overset{4}{\cancel{12}}}{1} \cdot \frac{x-2}{\underset{1}{\cancel{3}}} + \frac{\overset{3}{\cancel{12}}}{1} \cdot \frac{1}{\underset{1}{\cancel{4}}}$$ Divide out common factors in each multiplication.

$$3(x+3) > 4(x-2) + 3$$ The fractions are now cleared.

Now that fractions are cleared, we follow the four steps that we used in the previous examples.

Step 1 Simplify each side.

$$3(x+3) > 4(x-2) + 3$$ This is the inequality with the fractions cleared.

$$3x + 9 > 4x - 8 + 3$$ Use the distributive property.

$$3x + 9 > 4x - 5$$ Simplify.

Step 2 Collect variable terms on one side and constant terms on the other side. We will collect variable terms on the left and constant terms on the right.

$$3x + 9 - 4x > 4x - 5 - 4x$$ Subtract 4x from both sides.

$$-x + 9 > -5$$ Simplify.

$$-x + 9 - 9 > -5 - 9$$ Subtract 9 from both sides.

$$-x > -14$$ Simplify.

Step 3 Isolate the variable and solve. To isolate x, we must eliminate the negative sign in front of the x. Because $-x$ means $-1x$, we can do this by multiplying (or dividing) both sides of the inequality by -1. We are multiplying by a negative number. Thus, we must reverse the inequality symbol.

$$(-1)(-x) < (-1)(-14)$$ Multiply both sides by -1 and reverse the sense of the inequality.

$$x < 14$$ Simplify.

Step 4 Express the solution set in set-builder or interval notation and graph the set on a number line. The solution set consists of all real numbers that are less than 14, expressed in set-builder notation as $\{x \mid x < 14\}$. The interval notation for this solution set is $(-\infty, 14)$. The graph of the solution set is shown as follows:

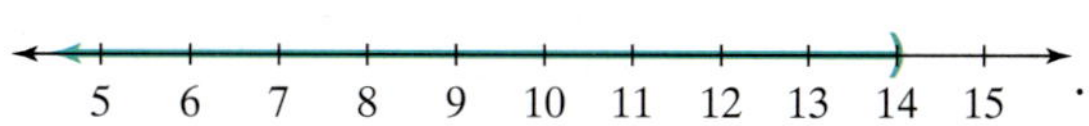

CHECK POINT 5 Solve and graph the solution set on a number line:

$$\frac{x-4}{2} > \frac{x-2}{3} + \frac{5}{6}.$$

4 Recognize inequalities with no solution or infinitely many solutions.

Inequalities with Unusual Solution Sets

We have seen that some equations have no solution. This is also true for some inequalities. An example of such an inequality is

$$x > x + 1.$$

There is no number that is greater than itself plus 1. This inequality has no solution and its solution set is $\varnothing$, the empty set.

By contrast, some inequalities are true for all real numbers. An example of such an inequality is

$$x < x + 1.$$

Every real number is less than itself plus 1. The solution set is $\{x \mid x \text{ is a real number}\}$ or $\mathbb{R}$. In interval notation, the solution set is $(-\infty, \infty)$.

If you attempt to solve an inequality that has no solution, you will eliminate the variable and obtain a false statement, such as $0 > 1$. If you attempt to solve an inequality that is true for all real numbers, you will eliminate the variable and obtain a true statement, such as $0 < 1$.

Using Technology

The graphs of

$y_1 = 2(x + 4)$ and $y_2 = 2x + 3$

are parallel lines. The graph of y_1 is always above the graph of y_2. Every value of x satisfies the inequality $y_1 > y_2$. Thus, the solution set of the inequality

$$2(x + 4) > 2x + 3$$

is $\mathbb{R}$.

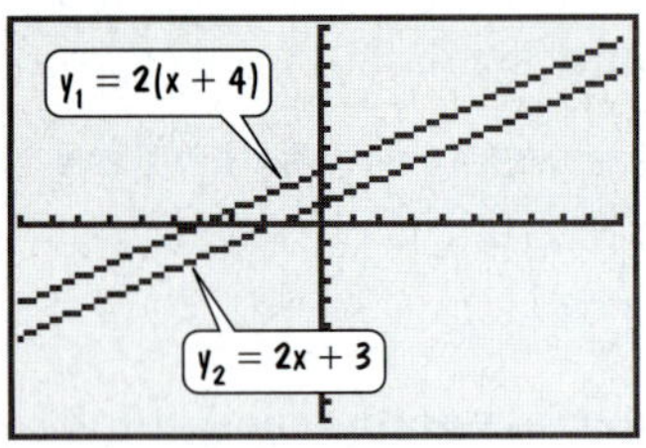

$[-10, 10, 1]$ by $[-30, 30, 3]$

EXAMPLE 6 Solving Linear Inequalities

Solve each inequality:

a. $2(x + 4) > 2x + 3$ **b.** $x + 7 \le x - 2$.

Solution

a.

$2(x + 4) > 2x + 3$	This is the given inequality.
$2x + 8 > 2x + 3$	Apply the distributive property.
$2x + 8 - 2x > 2x + 3 - 2x$	Subtract 2x from both sides.
$8 > 3$	Simplify. The statement $8 > 3$ is true.

The inequality $8 > 3$ is true for all values of x. Because this inequality is equivalent to the original inequality, the original inequality is true for all real numbers. The solution set is

$$\{x \mid x \text{ is a real number}\} \text{ or } \mathbb{R} \text{ or } (-\infty, \infty).$$

b.

$x + 7 \le x - 2$	This is the given inequality.
$x + 7 - x \le x - 2 - x$	Subtract x from both sides.
$7 \le -2$	Simplify. The statement $7 \le -2$ is false.

The inequality $7 \le -2$ is false for all values of x. Because this inequality is equivalent to the original inequality, the original inequality has no solution. The solution set is $\varnothing$. ■

✔ **CHECK POINT 6** Solve each inequality:

a. $3(x + 1) > 3x + 2$

b. $x + 1 \le x - 1$.

5 Solve problems using linear inequalities.

Applications

Commonly used English phrases such as "at least" and "at most" indicate inequalities. Table 4.2 lists sentences containing these phrases and their algebraic translations into inequalities.

Table 4.2 English Sentences and Inequalities

English Sentence	Inequality
x is at least 5.	$x \ge 5$
x is at most 5.	$x \le 5$
x is between 5 and 7.	$5 < x < 7$
x is no more than 5.	$x \le 5$
x is no less than 5.	$x \ge 5$

Our next example shows how to use an inequality to select the better deal between two pricing options. We use our strategy for solving word problems, translating from the verbal conditions of the problem to a linear inequality.

EXAMPLE 7 Selecting the Better Deal

Acme Car rental agency charges \$4 a day plus \$0.15 a mile, whereas Interstate rental agency charges \$20 a day and \$0.05 a mile. How many miles must be driven to make the daily cost of an Acme rental a better deal than an Interstate rental?

Solution

Step 1 Let x represent one of the quantities. We are looking for the number of miles that must be driven in a day to make Acme the better deal. Thus,

let x = the number of miles driven in a day.

Step 2 Represent other quantities in terms of x. We are not asked to find another quantity, so we can skip this step.

Step 3 Write an inequality in x that describes the conditions. Acme is a better deal than Interstate if the daily cost of Acme is less than the daily cost of Interstate.

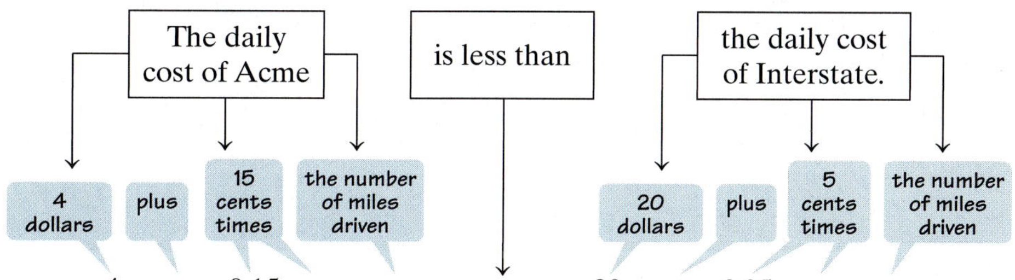

$$4 + 0.15 \cdot x < 20 + 0.05 \cdot x$$

Step 4 Solve the inequality and answer the question.

$4 + 0.15x < 20 + 0.05x$ — This is the inequality that describes the verbal conditions.

$4 + 0.15x - 0.05x < 20 + 0.05x - 0.05x$ — Subtract $0.05x$ from both sides.

$4 + 0.1x < 20$ — Simplify.

$4 + 0.1x - 4 < 20 - 4$ — Subtract 4 from both sides.

$0.1x < 16$ — Simplify.

$\dfrac{0.1x}{0.1} < \dfrac{16}{0.1}$ — Divide both sides by 0.1.

$x < 160$ — Simplify.

Thus, driving fewer than 160 miles per day makes Acme the better deal.

Step 5 Check the proposed solution in the original wording of the problem. One way to do this is to take a mileage less than 160 miles per day to see if Acme is the better deal. Suppose that 150 miles are driven in a day.

$$\text{Cost for Acme} = 4 + 0.15(150) = 26.50$$

$$\text{Cost for Interstate} = 20 + 0.05(150) = 27.50$$

Acme has a lower daily cost, making Acme the better deal. ■

Using Technology

The graphs of the daily cost models for the car rental agencies

$$y_1 = 4 + 0.15x$$

$$\text{and } y_2 = 20 + 0.05x$$

are shown in a [0, 300, 10] by [0, 40, 4] viewing rectangle. The graphs intersect at (160, 28). To the left of $x = 160$, the graph of Acme's daily cost lies below that of Interstate's daily cost. This shows that for fewer than 160 miles per day, Acme offers the better deal.

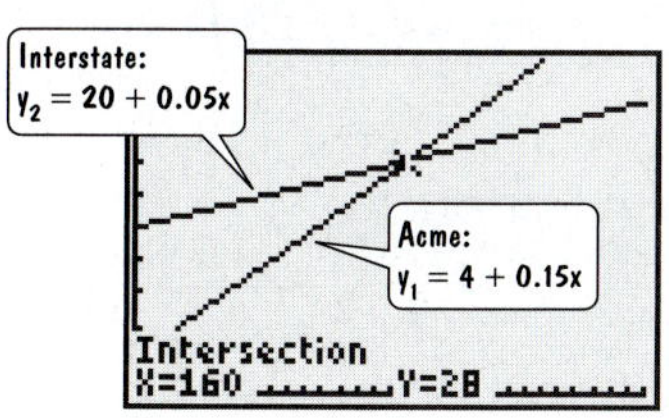

CHECK POINT 7 A car can be rented from Basic Rental for \$260 per week with no extra charge for mileage. Continental charges \$80 per week plus 25 cents for each mile driven to rent the same car. How many miles must be driven in a week to make the rental cost for Basic Rental a better deal than Continental's?

EXERCISE SET 4.1

Practice Exercises

In Exercises 1–12, graph the solutions of each inequality on a number line.

1. $x > 6$

2. $x > -2$

3. $x < -4$

4. $x < 0$

5. $x \geq -3$

6. $x \geq -5$

7. $x \leq 4$

8. $x \leq 7$

9. $-2 < x \leq 5$

10. $-3 \leq x < 7$

11. $-1 < x < 4$

12. $-7 \leq x \leq 0$

In Exercises 13–26, express each interval in terms of an inequality and graph the interval on a number line.

13. $(1, 6]$

14. $(-2, 4]$

15. $[-5, 2)$

16. $[-4, 3)$

17. $[-3, 1]$

18. $[-2, 5]$

19. $(2, \infty)$

20. $(3, \infty)$

21. $[-3, \infty)$

22. $[-5, \infty)$

23. $(-\infty, 3)$

24. $(-\infty, 2)$

25. $(-\infty, 5.5)$

26. $(-\infty, 3.5]$

In Exercises 27–52, solve each linear inequality. Other than ∅, graph the solution set on a number line.

27. $5x + 11 < 26$

28. $2x + 5 < 17$

29. $3x - 8 \geq 13$

30. $8x - 2 \geq 14$

31. $-9x \geq 36$

32. $-5x \leq 30$

33. $8x - 11 \leq 3x - 13$

34. $18x + 45 \leq 12x - 8$

35. $4(x + 1) + 2 \geq 3x + 6$

36. $8x + 3 > 3(2x + 1) + x + 5$

37. $2x - 11 < -3(x + 2)$

38. $-4(x + 2) > 3x + 20$

39. $1 - (x + 3) \geq 4 - 2x$

40. $5(3 - x) \leq 3x - 1$

41. $\frac{x}{4} - \frac{1}{2} \leq \frac{x}{2} + 1$

42. $\frac{3x}{10} + 1 \geq \frac{1}{5} - \frac{x}{10}$

43. $1 - \frac{x}{2} > 4$

44. $7 - \frac{4}{5}x < \frac{3}{5}$

45. $\frac{x - 4}{6} \geq \frac{x - 2}{9} + \frac{5}{18}$

46. $\frac{4x - 3}{6} + 2 \geq \frac{2x - 1}{12}$

47. $4(3x - 2) - 3x < 3(1 + 3x) - 7$

48. $3(x - 8) - 2(10 - x) < 5(x - 1)$

49. $8(x + 1) \leq 7(x + 5) + x$

50. $4(x - 1) \geq 3(x - 2) + x$

51. $3x < 3(x - 2)$

52. $5x < 5(x - 3)$

Application Exercises

The bar graph shows how we spend our leisure time. Let x represent the percentage of the population regularly participating in an activity. In Exercises 53–60, write the name or names of the activity described by the given inequality or interval.

Percentage of U.S. Population Participating in Each Activity on a Regular Basis

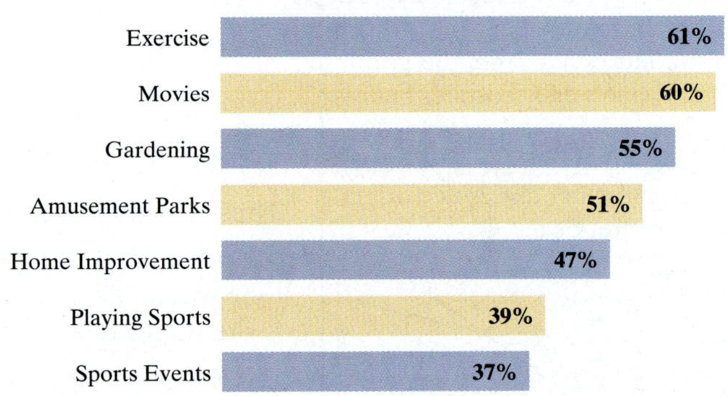

Source: U.S. Census Bureau

53. $x < 40\%$

54. $x < 50\%$

55. $[51\%, 61\%]$

56. $[47\%, 60\%]$

57. $(51\%, 61\%)$

58. $(47\%, 60\%)$

59. $(39\%, 55\%]$

60. $(37\%, 47\%]$

61. The percentage, P, of U.S. adults who read the daily newspaper can be modeled by the formula

$$P = -0.7x + 80$$

where x is the number of years after 1965. In which years will fewer than 52% of U.S. adults read the daily newspaper?

62. The formula

$$W = 0.3x + 22.1$$

models the average number of hours, W, that Americans work per week x years after 1900. In which years will Americans average more than 55.1 hours of work per week?

The Olympic 500-meter speed skating times have generally been decreasing over time. The formulas

$$W = -0.19t + 57 \quad \text{and} \quad M = -0.15t + 50$$

model the winning times, in seconds, for women, W, and men, M, t years after 1900. Use these models to solve Exercises 63–64.

63. Find values of t such that $W < M$. Describe what this means in terms of winning times.

64. Find values of t such that $W > M$. Describe what this means in terms of winning times.

In Exercises 65–72, use the strategy for solving word problems, translating from the verbal conditions of the problem to a linear inequality.

65. A truck can be rented from Basic Rental for \$50 a day plus \$0.20 per mile. Continental charges \$20 per day plus \$0.50 per mile to rent the same truck. How many miles must be driven in a day to make the rental cost for Basic Rental a better deal than Continental's?

66. You are choosing between two long-distance telephone plans. Plan A has a monthly fee of \$15 with a charge of \$0.08 per minute for all long-distance calls. Plan B has a monthly fee of \$3 with a charge of \$0.12 per minute for all long-distance calls. How many minutes of long-distance calls in a month make plan A the better deal?

67. A city commission has proposed two tax bills. The first bill requires that a homeowner pay \$1800 plus 3% of the assessed home value in taxes, The second bill requires taxes of \$200 plus 8% of the assessed home value. What price range of home assessment would make the first bill a better deal?

68. A local bank charges \$8 per month plus 5¢ per check. The credit union charges \$2 per month plus 8¢ per check. How many checks should be written each month to make the credit union a better deal?

69. A company manufactures and sells blank audiocassette tapes. The weekly fixed cost is \$10,000 and it costs \$0.40 to produce each tape. The selling price is \$2.00 per tape. How many tapes must be produced and sold each week for the company to have a profit gain?

70. A company manufactures and sells personalized stationery. The weekly fixed cost is \$3000 and it costs \$3.00 to produce each package of stationery. The selling price is \$5.50 per package. How many packages of stationery must be produced and sold each week for the company to have a profit gain?

71. An elevator at a construction site has a maximum capacity of 2800 pounds. If the elevator operator weighs 265 pounds and each cement bag weighs 65 pounds, how many bags of cement can be safely lifted on the elevator in one trip?

72. An elevator at a construction site has a maximum capacity of 3000 pounds. If the elevator operator weighs 245 pounds and each cement bag weighs 95 pounds, how many bags of cement can be safely lifted on the elevator in one trip?

Writing in Mathematics

73. When graphing the solutions of an inequality, what does a parenthesis signify? What does a bracket signify?

74. When solving an inequality, when is it necessary to change the sense of the inequality? Give an example.

75. Describe ways in which solving a linear inequality is similar to solving a linear equation.

76. Describe ways in which solving a linear inequality is different from solving a linear equation.

77. When solving a linear inequality, describe what happens if the solution set is $(-\infty, \infty)$.

78. When solving a linear inequality, describe what happens if the solution set is $\varnothing$.

79. The formula

$$V = 3.5x + 120$$

models Super Bowl viewers, V, in millions, x years after 1990. Use the formula to write a word problem that can be solved using a linear inequality. Then solve the problem.

Technology Exercises

In Exercises 80–81, solve each inequality using a graphing utility. Graph each side separately. Then determine the values of x for which the graph on the left side lies above the graph on the right side.

80. $-3(x - 6) > 2x - 2$

81. $-2(x + 4) > 6x + 16$

Use the same technique employed in Exercises 80–81 to solve each inequality in Exercises 82–83. In each case, what conclusion can you draw? What happens if you try solving the inequalities algebraically?

82. $12x - 10 > 2(x - 4) + 10x$

83. $2x + 3 > 3(2x - 4) - 4x$

84. A bank offers two checking account plans. Plan A has a base service charge of \$4.00 per month plus 10¢ per check. Plan B charges a base service charge of \$2.00 per month plus 15¢ per check.

a. Write models for the total monthly costs for each plan if x checks are written.

b. Use a graphing utility to graph the models in the same $[0, 50, 1]$ by $[0, 10, 1]$ viewing rectangle.

c. Use the graphs (and the TRACE or intersection feature) to determine for what number of checks per month plan A will be better than plan B.

d. Verify the result of part (c) algebraically by solving an inequality.

Critical Thinking Exercises

85. Which one of the following is true?

a. The inequality $3x > 6$ is equivalent to $2 > x$.

b. The smallest real number in the solution set of $2x > 6$ is 4.

c. If x is at least 7, then $x > 7$.

d. The inequality $-3x > 6$ is equivalent to $-2 > x$.

86. What's wrong with this argument? Suppose x and x represent two real numbers, where $x > y$.

$2 > 1$	This is a true statement.
$2(y - x) > 1(y - x)$	Multiply both sides by $y - x$.
$2y - 2x > y - x$	Use the distributive property.
$y - 2x > -x$	Subtract y from both sides.
$y > x$	Add $2x$ to both sides.

The final inequality, $y > x$, is impossible because we were initially given $x > y$.

87. The graphs of $y = 6$, $y = 3(-x - 5) - 9$, and $y = 0$ are shown in the figure. The graph was obtained using a graphing utility and a $[-12, 1, 1]$ by $[-2, 8, 1]$ viewing rectangle. Use the graphs to write the solution set for the inequality $0 < 3(-x - 5) - 9 < 6$.

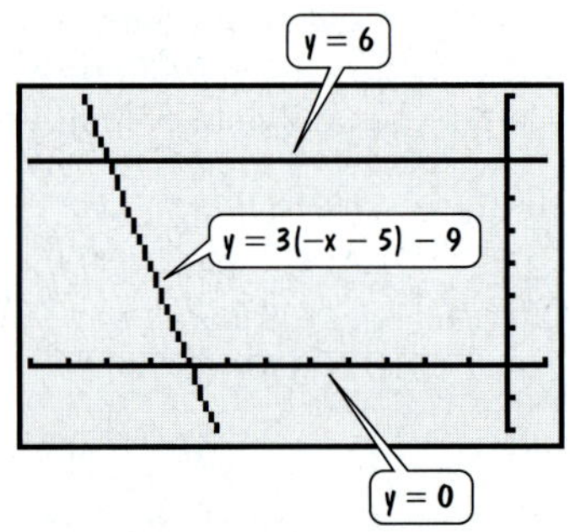

Review Exercises

88. If $f(x) = x^2 - 2x + 5$, find $f(-4)$. (Section 2.1, Example 3)

89. Solve the system:

$$\begin{aligned} 2x - y - z &= -3 \\ 3x - 2y - 2z &= -5 \\ -x + y + 2z &= 4. \end{aligned}$$

(Section 3.3, Example 2)

90. Simplify: $\left(\dfrac{2x^4y^{-2}}{4xy^3}\right)^3$ (Section 1.6, Example 9)

SECTION 4.2 Compound Inequalities

Objectives

1. Find the intersection of two sets.
2. Solve compound inequalities involving "and."
3. Find the union of two sets.
4. Solve compound inequalities involving "or."

SSM   PH Tutor Center   CD-ROM   Video

Boys and girls differ in their toy preferences. These differences are functions of our gender stereotypes—that is, our widely shared beliefs about males' and females' abilities, personality traits, and social behavior. Generally, boys have less leeway to play with "feminine" toys than girls do with "masculine" toys.

Which toys are requested by more than 40% of boys *and* more than 10% of girls? Which toys are requested by more than 40% of boys *or* more than 10% of girls? These questions are not the same. One involves inequalities joined by the word "and"; the other involves inequalities joined by the word "or." A **compound inequality** is formed by joining two inequalities with the word *and* or *or*.

Examples of Compound Inequalities

$$x - 3 < 5 \quad \text{and} \quad 2x + 4 < 14$$

$$3x - 5 \leq 13 \quad \text{or} \quad 5x + 2 > -3$$

In this section, you will learn to solve compound inequalities. With this skill, you will analyze childrens' toy preferences in the exercise set.

1 Find the intersection of two sets.

Compound Inequalities Involving "And"

You need to determine whether there is sufficient support on campus to have a blood drive. You take a survey to obtain information, asking students the following:

Would you be willing to donate blood?

Would you be willing to help serve a free breakfast to blood donors?

Part of your report involves the number of students willing to both donate blood *and* serve breakfast.

You must focus on the set of all students who are common to both the set of donors and the set of breakfast servers. This set is called the *intersection* of the two sets.

Definition of the Intersection of Sets

The **intersection** of sets A and B, written $A \cap B$, is the set of elements common to both set A and set B. This definition can be expressed in set-builder notation as follows:

$$A \cap B = \{x \mid x \in A \text{ AND } x \in B\}.$$

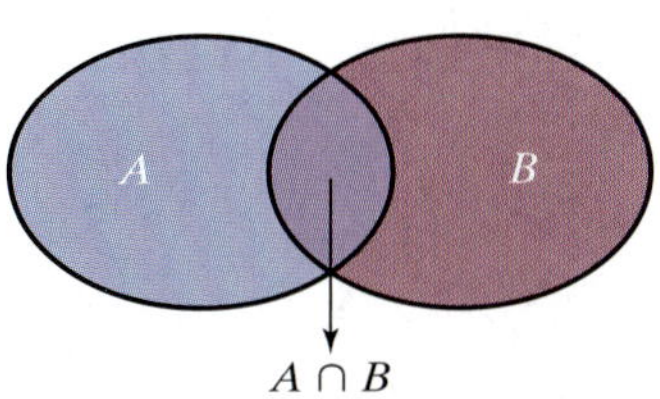

Figure 4.1 Picturing the intersection of two sets

Figure 4.1 shows a useful way of picturing the intersection of sets A and B. The figure indicates that $A \cap B$ contains those elements that belong to both A and B at the same time.

EXAMPLE 1 Finding the Intersection of Two Sets

Find the intersection: $\{7, 8, 9, 10, 11\} \cap \{6, 8, 10, 12\}$.

Solution The elements common to $\{7, 8, 9, 10, 11\}$ and $\{6, 8, 10, 12\}$ are 8 and 10. Thus,

$$\{7, 8, 9, 10, 11\} \cap \{6, 8, 10, 12\} = \{8, 10\}.$$

✔ CHECK POINT 1 Find the intersection: $\{3, 4, 5, 6, 7\} \cap \{3, 7, 8, 9\}$.

2 Solve compound inequalities involving "and."

A number is a **solution of a compound inequality** formed by the word "and" if it is a solution of both inequalities. For example, the solution set of the compound inequality

$$x \leq 6 \quad \text{and} \quad x \geq 2$$

is the set of values of x that satisfy $x \leq 6$ and $x \geq 2$. Thus, the solution set is the intersection of the solution sets of the two inequalities.

What are the numbers that satisfy both $x \leq 6$ and $x \geq 2$? These numbers are easier to see if we graph the solution set to each inequality on a number line. These graphs are shown in Figure 4.2. The intersection is shown in the third graph.

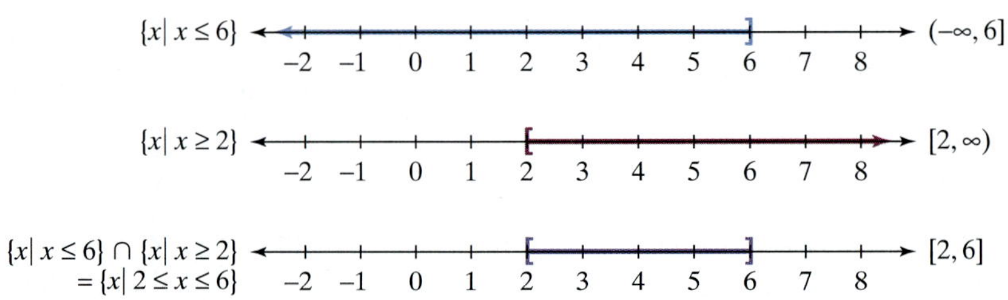

Figure 4.2 Numbers satisfying both $x \leq 6$ and $x \geq 2$

The numbers common to both sets are those that are less than or equal to 6 and greater than or equal to 2. This set is $\{x | 2 \leq x \leq 6\}$, or, in interval notation, $[2, 6]$.

Here is a procedure for finding the solution set of a compound inequality containing the word "and."

Solving Compound Inequalities Involving "And"

1. Solve each inequality separately.
2. Graph the solution set to each inequality on a number line and take the intersection of these solution sets.

EXAMPLE 2 Solving a Compound Inequality with "And"

Solve: $x - 3 < 5$ and $2x + 4 < 14$.

Solution

Step 1 Solve each inequality separately.

$$\begin{aligned} x - 3 &< 5 \quad &\text{and} \quad 2x + 4 &< 14 \\ x &< 8 & 2x &< 10 \\ & & x &< 5 \end{aligned}$$

Step 2 Take the intersection of the solution sets of the two inequalities. We graph the solution sets of $x < 8$ and $x < 5$. The intersection is shown in the third graph.

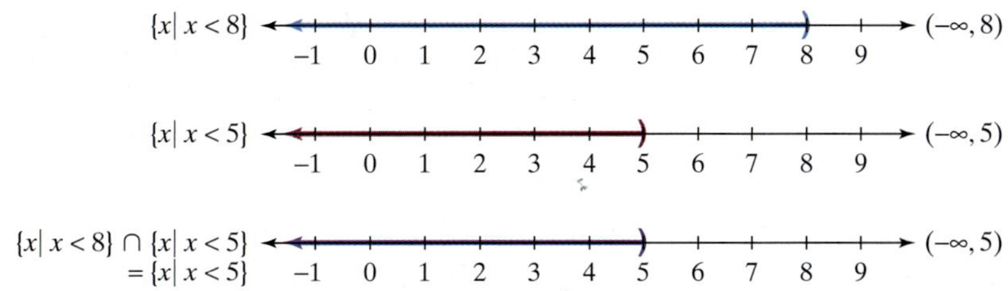

The numbers common to both sets are those that are less than 5. The solution set is $\{x \mid x < 5\}$, or, in interval notation, $(-\infty, 5)$. Take a moment to check that any number in $(-\infty, 5)$ satisfies both of the original inequalities. ■

✔ **CHECK POINT 2** Solve: $x + 2 < 5$ and $2x - 4 < -2$.

EXAMPLE 3 Solving a Compound Inequality with "And"

Solve: $2x - 7 > 3$ and $5x - 4 < 6$.

Solution

Step 1 Solve each inequality separately.

$$2x - 7 > 3 \quad \text{and} \quad 5x - 4 < 6$$
$$2x > 10 \qquad\qquad 5x < 10$$
$$x > 5 \qquad\qquad x < 2$$

Step 2 Take the intersection of the solution sets of the two inequalities. We graph the solution sets of $x > 5$ and $x < 2$. We use these graphs to find their intersection.

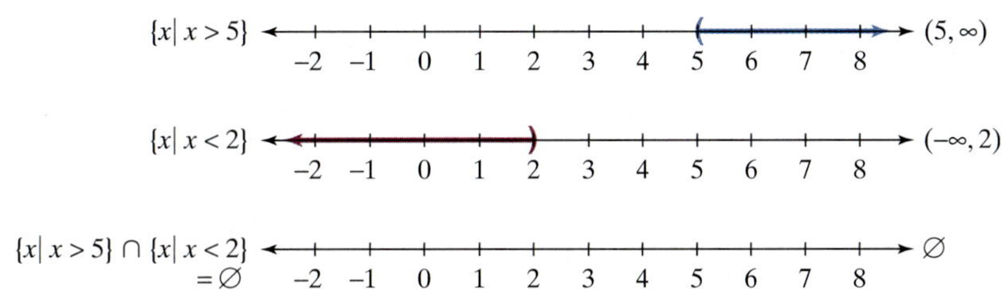

There is no number that is both greater than 5 and less than 2. Thus, the solution set is the empty set, $\varnothing$. ■

✔ **CHECK POINT 3** Solve: $4x - 5 > 7$ and $5x - 2 < 3$.

If $a < b$, the compound inequality

$$a < x \text{ and } x < b$$

can be written in the shorter form

$$a < x < b.$$

For example, the compound inequality

$$-3 < 2x + 1 \text{ and } 2x + 1 < 3$$

can be abbreviated

$$-3 < 2x + 1 < 3.$$

The word "and" does not appear when the inequality is written in the shorter form, although it is implied. The shorter form enables us to solve both inequalities at once. By performing the same operations on all three parts of the inequality, our goal is to **isolate *x* in the middle**.

EXAMPLE 4 Solving a Compound Inequality

Solve and graph the solution set:

$$-3 < 2x + 1 \le 3.$$

Solution We would like to isolate x in the middle. We can do this by first subtracting 1 from all three parts of the compound inequality. Then we isolate x from $2x$ by dividing all three parts of the inequality by 2.

$-3 < 2x + 1 \le 3$	This is the given inequality.
$-3 - 1 < 2x + 1 - 1 \le 3 - 1$	Subtract 1 from all three parts.
$-4 < 2x \le 2$	Simplify.
$\frac{-4}{2} < \frac{2x}{2} \le \frac{2}{2}$	Divide each part by 2.
$-2 < x \le 1$	Simplify.

The solution set consists of all real numbers greater than -2 and less than or equal to 1, represented by $\{x \mid -2 < x \le 1\}$ in set-builder notation and $(-2, 1]$ in interval notation. The graph is shown as follows:

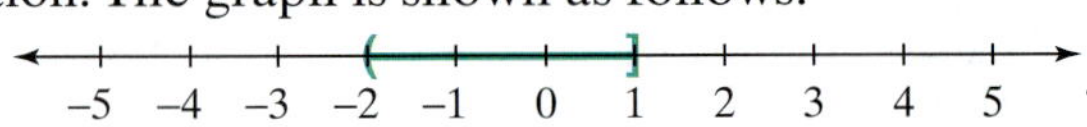

Using Technology

To check Example 4, graph each part of

$$-3 < 2x + 1 \le 3.$$

The figure shows that the graph of $y_2 = 2x + 1$ lies above the graph of $y_1 = -3$ and on or below the graph of $y_3 = 3$ when x is in the interval $(-2, 1]$.

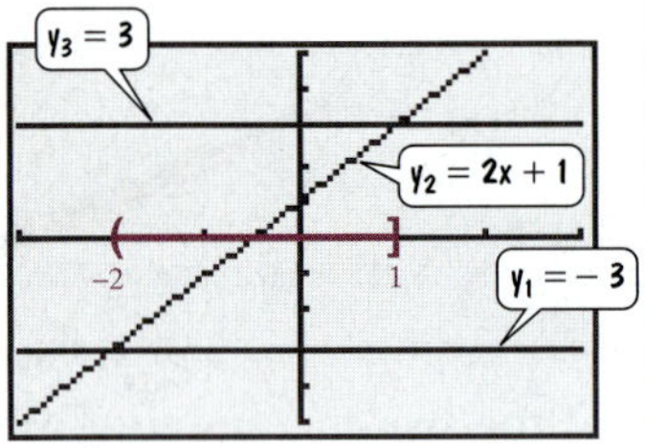

$[-3, 3, 1]$ by $[-5, 5, 1]$

✔ **CHECK POINT 4** Solve and graph the solution set: $1 \le 2x + 3 < 11$.

3 Find the union of two sets.

Compound Inequalities Involving "Or"

You continue to sort the results of the survey. Part of your report involves the number of students willing to donate blood *or* serve breakfast *or* do both. You must focus on the set containing donors or the set of breakfast servers or both of these sets. This set is called the *union* of two sets.

Definition of the Union of Sets

The **union** of sets A and B, written $A \cup B$, is the set of elements that are members of set A or of set B or of both sets. This definition can be expressed in set-builder notation as follows:

$$A \cup B = \{x \mid x \in A \; OR \; x \in B\}.$$

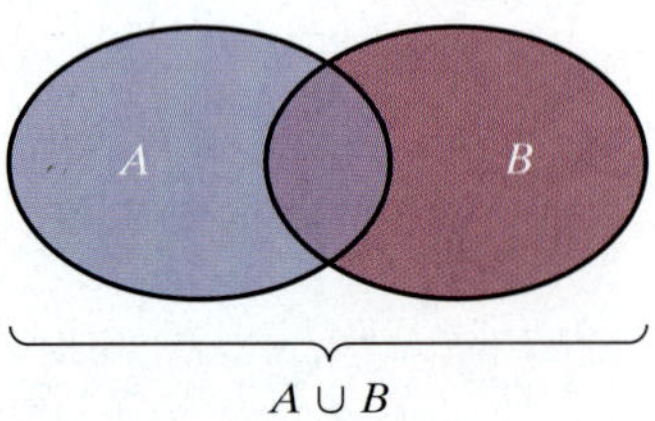

Figure 4.3 Picturing the union of two sets

Figure 4.3 shows a useful way of picturing the union of sets A and B. The figure indicates that $A \cup B$ is formed by joining the sets together.

We can find the union of set A and set B by listing the elements of set A. Then, we include any elements of set B that have not already been listed. Enclose all elements that are listed with braces. This shows that the union of two sets is also a set.

Study Tip

When finding the union of two sets, do not list twice any elements that appear in both sets.

EXAMPLE 5 Finding the Union of Two Sets "Or"

Find the union: $\{7, 8, 9, 10, 11\} \cup \{6, 8, 10, 12\}$.

Solution To find $\{7, 8, 9, 10, 11\} \cup \{6, 8, 10, 12\}$, start by listing all the elements from the first set, namely 7, 8, 9, 10, and 11. Now list all the elements from the second set that are not in the first set, namely 6 and 12. The union is the set consisting of all these elements. Thus,

$$\{7, 8, 9, 10, 11\} \cup \{6, 8, 10, 12\} = \{6, 7, 8, 9, 10, 11, 12\}.$$

CHECK POINT 5 Find the union: $\{3, 4, 5, 6, 7\} \cup \{3, 7, 8, 9\}$.

4 Solve compound inequalities involving "or."

A number is a **solution of a compound inequality** formed by the word "or" if it is a solution of either inequality. Thus, the solution set of a compound inequality formed by the word "or" is the union of the solution sets of the two inequalities.

Solving Compound Inequalities Involving "Or"

1. Solve each inequality separately.
2. Graph the solution set to each inequality on a number line and take the union of these solution sets.

EXAMPLE 6 Solving a Compound Inequality with "Or"

Solve: $2x - 3 < 7$ or $35 - 4x \leq 3$.

Solution

Step 1 Solve each inequality separately.

$$\begin{aligned} 2x - 3 &< 7 & \text{or} \quad 35 - 4x &\leq 3 \\ 2x &< 10 & -4x &\leq -32 \\ x &< 5 & x &\geq 8 \end{aligned}$$

Step 2 Take the union of the solution sets of the two inequalities. We graph the solution sets of $x < 5$ and $x \geq 8$. We use these graphs to find their union.

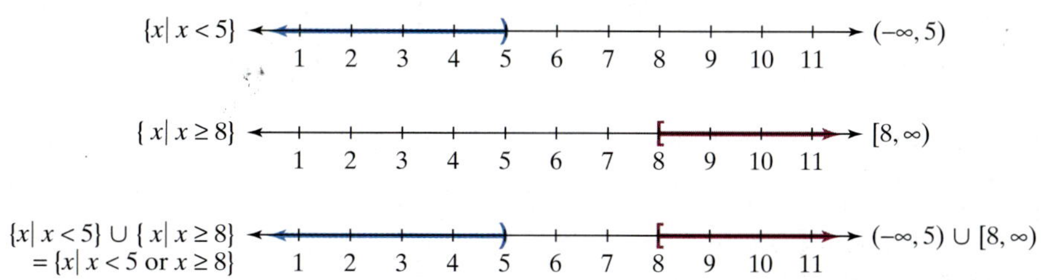

The solution set consists of all numbers that are less than 5 or greater than or equal to 8. The solution set is $\{x \mid x < 5 \text{ or } x > 8\}$, or, in interval notation, $(-\infty, 5) \cup [8, \infty)$. There is no shortcut way to express the solution of this union when interval notation is used.

CHECK POINT 6 Solve: $3x - 5 \leq -2$ or $10 - 2x < 4$.

EXAMPLE 7 Solving a Compound Inequality with "Or"

Solve: $3x - 5 \le 13$ or $5x + 2 > -3$.

Solution

Step 1 Solve each inequality separately.

$$\begin{aligned} 3x - 5 &\le 13 \quad \text{or} \quad & 5x + 2 &> -3 \\ 3x &\le 18 & 5x &> -5 \\ x &\le 6 & x &> -1 \end{aligned}$$

Step 2 Take the union of the solution sets of the two inequalities. We graph the solution sets of $x \le 6$ and $x > -1$. We use these graphs to find their union.

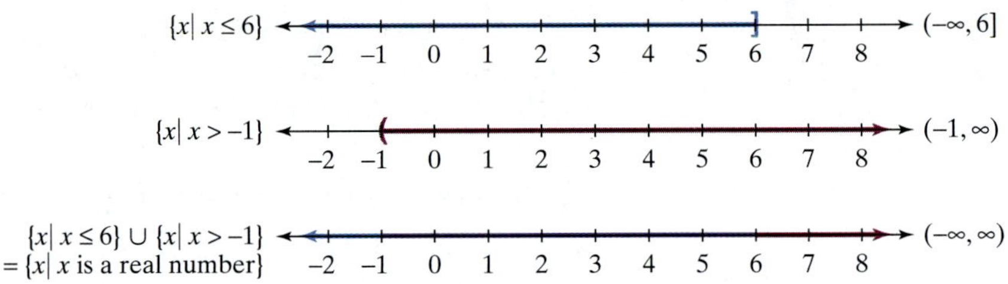

Because all real numbers are less than or equal to 6 or greater than −1, the union of the two sets fills the entire number line. Thus, the solution set is $\{x \mid x \text{ is a real number}\}$, or $\mathbb{R}$. The solution set in interval notation is $(-\infty, \infty)$. Any real number that you select will satisfy either or both of the original inequalities. ■

✔ **CHECK POINT 7** Solve: $2x + 5 \ge 3$ or $2x + 3 < 3$.

EXERCISE SET 4.2

Practice Exercises

In Exercises 1–4, find the intersection of the sets.

1. $\{1, 2, 3, 4\} \cap \{2, 4, 5\}$
2. $\{1, 3, 7\} \cap \{2, 3, 8\}$
3. $\{1, 3, 5, 7\} \cap \{2, 4, 6, 8, 10\}$
4. $\{0, 1, 3, 5\} \cap \{-5, -3, -1\}$

In Exercises 5–22, solve each compound inequality. Use graphs to show the solution set to each of the two given inequalities, as well as a third graph that shows the solution set of the compound inequality. Except for the empty set, express the solution set in both set-builder and interval notations.

5. $x > 3$ and $x > 6$
6. $x > 2$ and $x > 4$
7. $x \le 5$ and $x \le 1$
8. $x \le 6$ and $x \le 2$
9. $x < 2$ and $x \ge -1$
10. $x < 3$ and $x \ge -1$
11. $x > 2$ and $x < -1$
12. $x > 3$ and $x < -1$
13. $5x < -20$ and $3x > -18$
14. $3x \le 15$ and $2x > -6$
15. $x - 4 \le 2$ and $3x + 1 > -8$
16. $3x + 2 > -4$ and $2x - 1 < 5$
17. $2x > 5x - 15$ and $7x > 2x + 10$
18. $6 - 5x > 1 - 3x$ and $4x - 3 > x - 9$
19. $4(1 - x) < -6$ and $\dfrac{x - 7}{5} \le -2$
20. $5(x - 2) > 15$ and $\dfrac{x - 6}{4} \le -2$
21. $x - 1 \le 7x - 1$ and $4x - 7 < 3 - x$
22. $2x + 1 > 4x - 3$ and $x - 1 \ge 3x + 5$

In Exercises 23–30, solve each inequality and graph the solution set on a number line. Express the solution set in both set-builder and interval notations.

23. $6 < x + 3 < 8$

24. $7 < x + 5 < 11$

25. $-3 \le x - 2 < 1$

26. $-6 < x - 4 \le 1$

27. $-11 < 2x - 1 \le -5$

28. $3 \le 4x - 3 < 19$

29. $-3 \le \frac{2x}{3} - 5 < -1$

30. $-6 \le \frac{x}{2} - 4 < -3$

In Exercises 31–34, find the union of the sets.

31. $\{1, 2, 3, 4\} \cup \{2, 4, 5\}$

32. $\{1, 3, 7, 8\} \cup \{2, 3, 8\}$

33. $\{1, 3, 5, 7\} \cup \{2, 4, 6, 8, 10\}$

34. $\{0, 1, 3, 5\} \cup \{2, 4, 6\}$

In Exercises 35–50, solve each compound inequality. Use graphs to show the solution set to each of the two given inequalities, as well as a third graph that shows the solution set of the compound inequality. Express the solution set in both set-builder and interval notations.

35. $x > 3$ or $x > 6$

36. $x > 2$ or $x > 4$

37. $x \le 5$ or $x \le 1$

38. $x \le 6$ or $x \le 2$

39. $x < 2$ or $x \ge -1$

40. $x < 3$ or $x \ge -1$

41. $x \ge 2$ or $x < -1$

42. $x \ge 3$ or $x < -1$

43. $3x > 12$ or $2x < -6$

44. $3x < 3$ or $2x > 10$

45. $3x + 2 \le 5$ or $5x - 7 \ge 8$

46. $2x - 5 \le -11$ or $5x + 1 \ge 6$

47. $4x + 3 < -1$ or $2x - 3 \ge -11$

48. $2x + 1 < 15$ or $3x - 4 \ge -1$

49. $-2x + 5 > 7$ or $-3x + 10 > 2x$

50. $16 - 3x \ge -8$ or $13 - x > 4x + 3$

Application Exercises

As a result of cultural expectations about what is appropriate behavior for each gender, boys and girls differ substantially in their toy preferences. The graph shows the percentage of boys and girls asking for various types of toys in letters to Santa Claus. Use the information in the graph to solve Exercises 51–56.

Toys Requested by Children

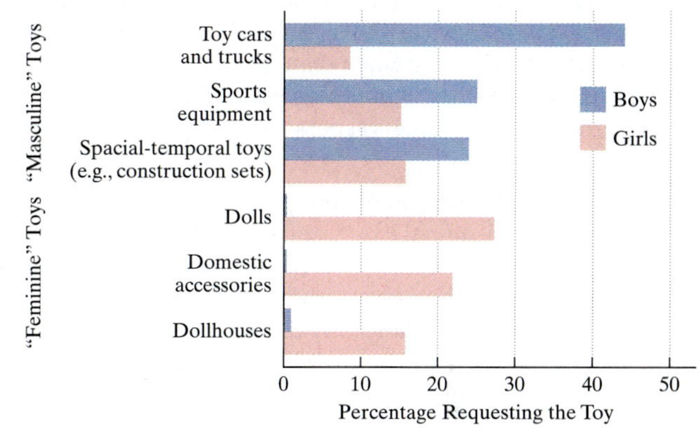

Source: Richardson, J. G., & Simpson, C. H. (1982). Children, gender and social structure: An analysis of the contents of letters to Santa Claus. *Child Development, 53*, 429–436.

51. Which toys were requested by more than 10% of the boys *and* less than 20% of the girls?

52. Which toys were requested by fewer than 5% of the boys *and* fewer than 20% of the girls?

53. Which toys were requested by more than 10% of the boys *or* less than 20% of the girls?

54. Which toys were requested by fewer than 5% of the boys *or* fewer than 20% of the girls?

55. Which toys were requested by more than 40% of the boys *and* more than 10% of the girls?

56. Which toys were requested by more than 40% of the boys *or* more than 10% of the girls?

57. The formula

$$T = 0.01x + 56.7$$

models the global mean temperature, T, in degrees Fahrenheit, of Earth x years after 1905. For which range of years was the global mean temperature at least 52°F and at most 57.2°F?

58. The formula for converting Fahrenheit temperature, F, to Celsius temperature, C, is

$$C = \frac{5}{9}(F - 32).$$

If Celsius temperature ranges from 15° to 35°, inclusive, what is the range for the Fahrenheit temperature? Use interval notation to express this range.

59. On the first of four exams, your grades are 70, 75, 87, and 92. There is still one more exam, and you are hoping to earn a B in the course. This will occur if the average of your five exam grades is greater than or equal to 80 and less than 90. What range of grades on the fifth exam will result in earning a B? Use interval notation to express this range.

60. On the first of four exams, your grades are 82, 75, 80, and 90. There is still a final exam, and it counts as two grades. You are hoping to earn a B in the course: This will occur if the average of your six exam grades is greater than or equal to 80 and less than 90. What range of grades on the final exam will result in earning a B? Use interval notation to express this range.

61. The toll to a bridge is \$3.00. A three-month pass costs \$7.50 and reduces the toll to \$0.50. A six-month pass costs \$30 and permits crossing the bridge for no additional fee. How many crossings per three-month period does it take for the three-month pass to be the best deal?

62. Parts for an automobile repair cost \$175. The mechanic charges \$34 per hour. If you receive an estimate for at least \$226 and at most \$294 for fixing the car, what is the time interval that the mechanic will be working on the job?

Writing in Mathematics

63. Describe what is meant by the intersection of two sets. Give an example.

64. Explain how to solve a compound inequality involving "and."

65. Why is $1 < 2x + 3 < 9$ a compound inequality? Where are the two inequalities and what is the word that joins them?

66. Explain how to solve $1 < 2x + 3 < 9$.

67. Describe what is meant by the union of two sets. Give an example.

68. Explain how to solve a compound inequality involving "or."

69. How many Christmas trees are sold each holiday season? The function $T(x) = 0.9x + 32$ models the number of trees sold, $T(x)$, in millions, x years after 1990. Use the function to write a problem that can be solved using a compound inequality. Then solve the problem.

Technology Exercises

In Exercises 70–73, solve each inequality using a graphing utility. Graph each of the three parts of the inequality separately in the same viewing rectangle. The solution set consists of all values of x for which the graph of the linear function in the middle lies between the graphs of the constant functions on the left and the right.

70. $1 < x + 3 < 9$

71. $-1 < \frac{x + 4}{2} < 3$

72. $1 \le 4x - 7 \le 3$

73. $2 \le 4 - x \le 7$

Critical Thinking Exercises

74. Which one of the following is true?

a. $(-\infty, -1] \cap [-4, \infty) = [-4, -1]$

b. $(-\infty, 3) \cup (-\infty, -2) = (-\infty, -2)$

c. The union of two sets can never give the same result as the intersection of those sets.

d. The solution set of $x < a$ and $x > a$ is the set of all real numbers excluding a.

In Exercises 75–76, solve each compound inequality. Express solution sets in interval notation.

75. $-7 \le 8 - 3x \le 20$ and $-7 < 6x - 1 < 41$

76. $2x - 3 < 3x + 1 < 6x + 2$

77. Graph the linear function $f(x) = -x + 4$. Then use your graph to solve $2 < 4 - x < 7$.

78. Graph the linear function $f(x) = x + 3$. Then use your graph to solve $x + 3 > 3$ or $x + 3 < 2$.

The graphs of $f(x) = \sqrt{4 - x}$ and $g(x) = \sqrt{x + 1}$ are shown in a $[-3, 10, 1]$ by $[-2, 5, 1]$ viewing rectangle.

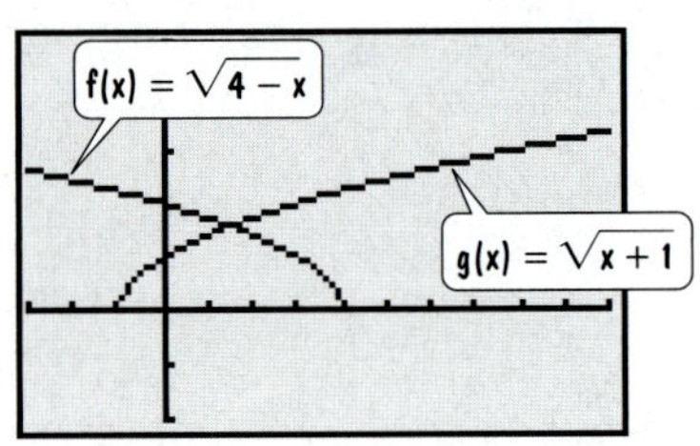

In Exercises 79–82, use the graphs and interval notation to express the domain of the given function.

79. The domain of f

80. The domain of g

81. The domain of $f + g$

82. The domain of $\frac{f}{g}$

83. At the end of the day, the change machine at a laundrette contained at least \$3.20 and at most \$5.45 in nickels. dimes, and quarters. There were 3 fewer dimes than twice the number of nickels and 2 more quarters than twice the number of nickels. What was the least possible number and the greatest possible number of nickels?

Review Exercises

84. If $f(x) = x^2 - 3x + 4$ and $g(x) = 2x - 5$, find $(g - f)(x)$ and $(g - f)(-1)$. (Section 2.2, Example 4)

85. Use function notation to write the equation of the line passing through $(4, 2)$ and perpendicular to the line whose equation is $4x - 2y = 8$. (Section 2.4, Examples 4 and 5)

86. Simplify: $4 - [2(x - 4) - 5]$. (Section 1.2, Example 14)

▸ SECTION 4.3 Equations and Inequalities Involving Absolute Value

Objectives

1. Solve equations involving absolute value.
2. Solve inequalities involving absolute value.
3. Recognize absolute value inequalities with no solution or infinitely many solutions.
4. Solve problems using absolute value inequalities.

SSM PH Tutor Center CD-ROM Video www

*M*A*S*H* took place in the early 1950s during the Korean War. By the final episode, the show had lasted four times as long as the Korean War.

At the end of the twentieth century, there were 94 million households in the United States with television sets. The television program viewed by the greatest percentage of such households in that century was the final episode of *M*A*S*H*. Over 50 million American households watched this program.

Numerical information, such as the number of households watching a television program, is often given with a margin of error. Inequalities involving absolute value are used to describe errors in polling, as well as errors of measurement in manufacturing, engineering, science, and other fields. In this section, you will learn to solve equations and inequalities containing absolute value. With these skills, you will analyze the percentage of households that watched the final episode of *M*A*S*H*.

1 Solve equations involving absolute value.

Equations Involving Absolute Value

We have seen that the absolute value of a, $|a|$, is the distance from 0 to a on the number line. Now consider **absolute value equations**, such as

$$|x| = 2.$$

This means that we must determine real numbers whose distance from the origin on the number line is 2. Figure 4.4 shows that there are two numbers such that $|x| = 2$, namely, 2 or -2. We write $x = 2$ or $x = -2$. This observation can be generalized as follows:

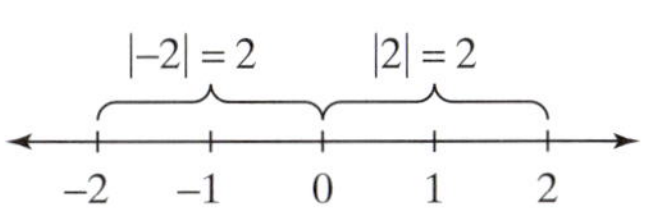

Figure 4.4 If $|x| = 2$, then $x = 2$ or $x = -2$.

Rewriting an Absolute Value Equation without Absolute Value Bars

If c is a positive real number and X represents any algebraic expression, then $|X| = c$ is equivalent to $X = c$ or $X = -c$.

Using Technology

You can use a graphing utility to verify the solution set of an absolute value equation. Consider, for example,

$$|2x - 3| = 11.$$

Graph $y_1 = |2x - 3|$ and $y_2 = 11$. The graphs are shown in a $[-10, 10, 1]$ by $[-1, 15, 1]$ viewing rectangle. The x-coordinates of the intersection points are -4 and 7, verifying that $\{-4, 7\}$ is the solution set.

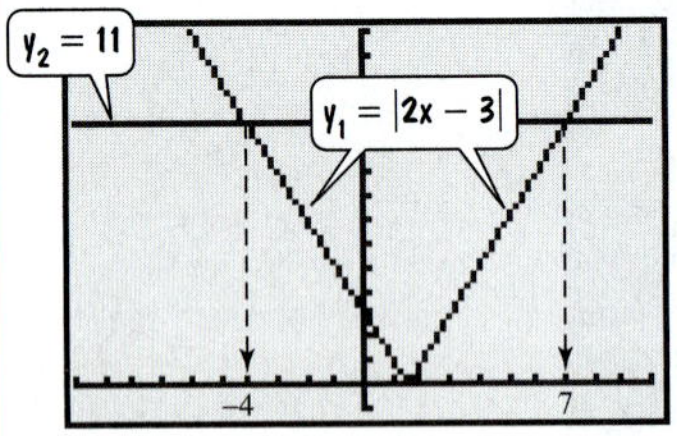

$[-10, 10, 1]$ by $[-1, 15, 1]$

EXAMPLE 1 Solving an Equation Involving Absolute Value

Solve: $|2x - 3| = 11$.

Solution

$|2x - 3| = 11$		This is the given equation.
$2x - 3 = 11$ or	$2x - 3 = -11$	Rewrite the equation without absolute value bars.
$2x = 14$	$2x = -8$	Add 3 to both sides of each equation.
$x = 7$	$x = -4$	Divide both sides of each equation by 2.

Check 7:	**Check −4:**	
$|2x - 3| = 11$	$|2x - 3| = 11$	This is the original equation.
$|2(7) - 3| \stackrel{?}{=} 11$	$|2(-4) - 3| \stackrel{?}{=} 11$	Substitute the proposed solutions.
$|14 - 3| \stackrel{?}{=} 11$	$|-8 - 3| \stackrel{?}{=} 11$	Perform operations inside the absolute value bars.
$|11| \stackrel{?}{=} 11$	$|-11| \stackrel{?}{=} 11$	
$11 = 11$, true	$11 = 11$, true	These true statements indicate that 7 and −4 are solutions.

The solutions are −4 and 7. We can also say that the solution set is $\{-4, 7\}$. ■

✔ **CHECK POINT 1** Solve: $|2x - 1| = 5$.

The absolute value of a number is never negative. Thus, if X is an algebraic expression and c is a negative number, then $|X| = c$ has no solution. For example, the equation $|3x - 6| = -2$ has no solution because $|3x - 6|$ cannot be negative. The solution set is $\varnothing$, the empty set.

The absolute value of 0 is 0. Thus, if X is an algebraic expression and $|X| = 0$, the solution is found by solving $X = 0$. For example, the solution of $|x - 2| = 0$ is obtained by solving $x - 2 = 0$. The solution is 2 and the solution set is $\{2\}$.

To solve some absolute value equations, it is necessary to first isolate the expression containing the absolute value symbols. For example, consider the equation

$$3|2x - 3| - 8 = 25.$$

We need to isolate $|2x - 3|$.

How can we isolate $|2x - 3|$? Add 8 to both sides of the equation and then divide both sides by 3.

$3|2x - 3| - 8 = 25$	This is the given equation.
$3|2x - 3| = 33$	Add 8 to both sides.
$|2x - 3| = 11$	Divide both sides by 3.

This results in the equation we solved in Example 1.

Some equations have two absolute value expressions, such as

$$|3x - 1| = |x + 5|.$$

These absolute value expressions are equal when the expressions inside the absolute value bars are equal to or opposites of each other.

Rewriting an Absolute Value Equation with Two Absolute Values without Absolute Value Bars

If $|X_1| = |X_2|$, then $X_1 = X_2$ or $X_1 = -X_2$.

EXAMPLE 2 Solving an Absolute Value Equation with Two Absolute Values

Solve: $|3x - 1| = |x + 5|$.

Solution We rewrite the equation without absolute value bars.

$|X_1| = |X_2|$ means $X_1 = X_2$ or $X_1 = -X_2$.

$|3x - 1| = |x + 5|$ means $3x - 1 = x + 5$ or $3x - 1 = -(x + 5)$.

We now solve the two equations that do not contain absolute value bars.

$$
\begin{aligned}
3x - 1 &= x + 5 &\quad\text{or}\quad 3x - 1 &= -(x + 5)\\
2x - 1 &= 5 & 3x - 1 &= -x - 5\\
2x &= 6 & 4x - 1 &= -5\\
x &= 3 & 4x &= -4\\
& & x &= -1
\end{aligned}
$$

Take a moment to complete the solution process by checking the two proposed solutions in the original equation. The solutions are −1 and 3, and the solution set is $\{-1, 3\}$. ■

✔ **CHECK POINT 2** Solve: $|2x - 7| = |x + 3|$.

2 Solve inequalities involving absolute value.

Inequalities Involving Absolute Value

Absolute value can also arise in inequalities. Consider, for example,

$$|x| < 2.$$

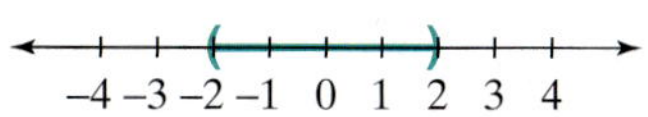

Figure 4.5 $|x| < 2$, so $-2 < x < 2$.

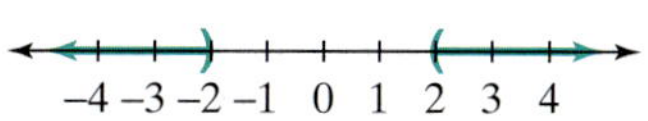

Figure 4.6 $|x| > 2$, so $x < -2$ or $x > 2$.

This means that the distance of x from 0 is *less than* 2, as shown in Figure 4.5. The interval shows values of x that lie less than 2 units from 0. Thus, x can lie between −2 and 2. That is, x is greater than −2 and less than 2: $-2 < x < 2$.

Some absolute value inequalities use the "greater than" symbol. For example, $|x| > 2$ means that the distance of x from 0 is *greater than* 2, as shown in Figure 4.6. Thus, x can be less than −2 *or* greater than 2: $x < -2$ or $x > 2$.

These observations suggest the following principles for solving inequalities with absolute value.

Solving an Absolute Value Inequality

If X is an algebraic expression and c is a positive number:

1. The solutions of $|X| < c$ are the numbers that satisfy $-c < X < c$.
2. The solutions of $|X| > c$ are the numbers that satisfy $X < -c$ or $X > c$.

These rules are valid if $<$ is replaced by $\le$ and $>$ is replaced by $\ge$.

EXAMPLE 3 Solving an Absolute Value Inequality with <

Solve and graph the solution set on a number line: $|x - 4| < 3$.

Solution We rewrite the inequality without absolute value bars.

$|X| < c$ means $-c < X < c$.

$$|x - 4| < 3 \quad \text{means} \quad -3 < x - 4 < 3.$$

We solve the compound inequality by adding 4 to all three parts.

$$-3 < x - 4 < 3$$
$$-3 + 4 < x - 4 + 4 < 3 + 4$$
$$1 < x < 7$$

The solution set is all real numbers greater than 1 and less than 7, denoted by $\{x|1 < x < 7\}$ or $(1, 7)$. The graph of the solution set is shown as follows:

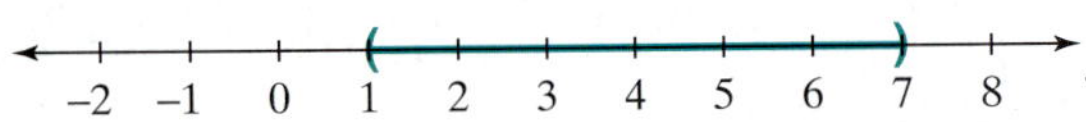

■

We can use the rectangular coordinate system to visualize the solution set of

$$|x - 4| < 3.$$

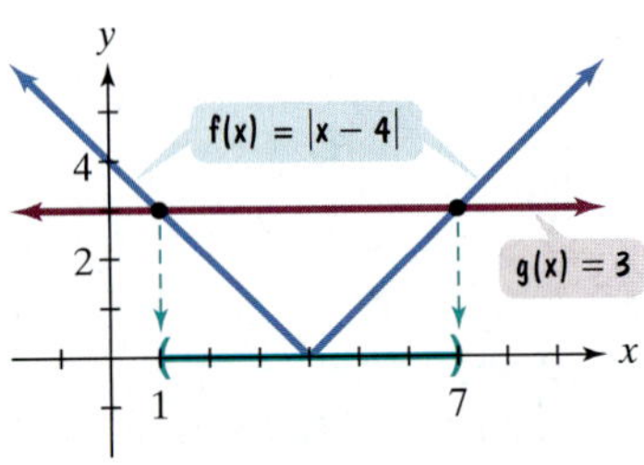

Figure 4.7 The solution set of $|x - 4| < 3$ is $(1, 7)$.

Figure 4.7 shows the graphs of $f(x) = |x - 4|$ and $g(x) = 3$. The solution set of $|x - 4| < 3$ consists of all values of x for which the blue graph of f lies below the red graph of g. These x-values make up the interval $(1, 7)$, which is the solution set.

✔ **CHECK POINT 3** Solve and graph the solution set on a number line: $|x - 2| < 5$.

EXAMPLE 4 Solving an Absolute Value Inequality with ≥

Solve and graph the solution set on a number line: $|2x + 3| \geq 5$.

Solution We rewrite the inequality without absolute value bars.

$|X| \geq c$ means $X \leq -c$ or $X \geq c$.

$$|2x + 3| \geq 5 \quad \text{means} \quad 2x + 3 \leq -5 \quad \text{or} \quad 2x + 3 \geq 5.$$

We solve this compound inequality by solving each of these inequalities separately. Then we take the union of their solution sets.

$2x + 3 \leq -5$	or	$2x + 3 \geq 5$	*These are the inequalities without absolute value bars.*
$2x \leq -8$	or	$2x \geq 2$	*Subtract 3 from both sides.*
$x \leq -4$	or	$x \geq 1$	*Divide both sides by 2.*

The solution set consists of all numbers that are less than or equal to -4 or greater than or equal to 1. The solution set is $\{x|x \leq -4 \text{ or } x \geq 1\}$, or, in interval notation, $(-\infty, -4] \cup [1, \infty)$. The graph of the solution set is shown as follows:

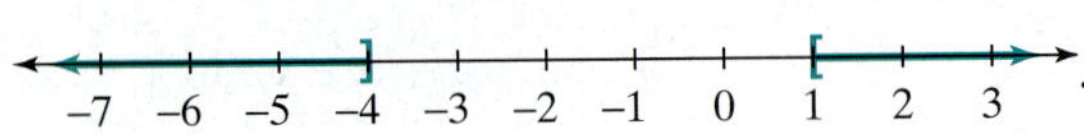

■

✔ **CHECK POINT 4** Solve and graph the solution set on a number line: $|2x - 5| \geq 3$.

3 Recognize absolute value inequalities with no solution or infinitely many solutions.

Absolute Value Inequalities with Unusual Solution Sets

If c is a positive number, the solutions of $|X| < c$ comprise a single interval. By contrast, the solutions of $|X| > c$ make up two intervals.

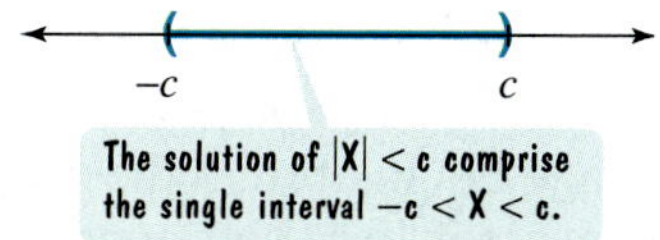

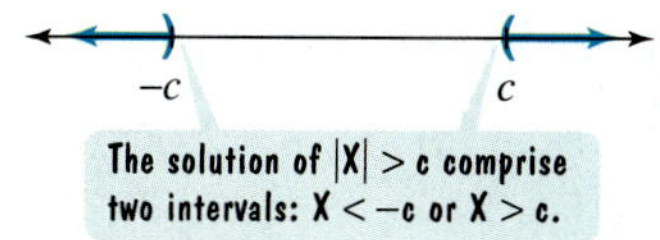

Now let's see what happens to these inequalities if c is a negative number. Consider, for example, $|x| < -2$. Because $|x|$ always has a value that is greater than or equal to 0, there is no number whose absolute value is less than -2. The inequality $|x| < -2$ has no solution. The solution set is $\varnothing$.

Now consider the inequality $|x| > -2$. Because $|x|$ is never negative, all numbers have an absolute value that is greater than -2. All real numbers satisfy the inequality $|x| > -2$. The solution set is $(-\infty, \infty)$.

Absolute Value Inequalities with Unusual Solution Sets

If X is an algebraic expression and c is a negative number:

1. The inequality $|X| < c$ has no solution.
2. The inequality $|X| > c$ is true for all real numbers for which X is defined.

4 Solve problems using absolute value inequalities.

Applications

When you were between the ages of 6 and 14, how would you have responded to this question:

What is bad about being a kid?

Table 4.3 What Is Bad About Being a Kid?

Kids Say	
Getting bossed around	17%
School, homework	15%
Can't do everything I want	11%
Chores	9%
Being grounded	9%

Source: Penn, Schoen, and Berland using 1172 interviews with children ages 6 to 14 from May 14 to June 1, 1999, Margin of error: ±2.9%

Note the margin of error.

In a random sample of 1172 children ages 6 to 14, 17% of the children responded, "Getting bossed around." The problem is that this is a single random sample. Do 17% of kids in the entire population of children ages 6 to 14 think that getting bossed around as a kid is a bad thing?

If you look at the results of a poll like the one in Table 4.3, you will observe that a **margin of error** is reported. The margin of error is ±2.9%. This means that the actual percentage of children who feel getting bossed around is a bad thing is at most 2.9% greater than or less than 17%. If x represents the percentage of children in the population who think that getting bossed around is a bad thing, then the poll's margin of error can be expressed as an absolute value inequality:

$$|x - 17| \leq 2.9.$$

EXAMPLE 5 Analyzing a Poll's Margin of Error

The inequality

$$|x - 9| \leq 2.9$$

describes the percentage of children in the population who think that being grounded is a bad thing about being a kid. (See Table 4.3.) Solve the inequality and interpret the solution.

Solution We rewrite the inequality without absolute value bars.

$|X| \le c$ means $-c \le X \le c.$

$$|x - 9| \le 2.9 \quad \text{means} \quad -2.9 \le x - 9 \le 2.9.$$

We solve the compound inequality by adding 9 to all three parts.

$$-2.9 \le x - 9 \le 2.9$$
$$-2.9 + 9 \le x - 9 + 9 \le 2.9 + 9$$
$$6.1 \le x \le 11.9$$

The percentage of children in the population who think that being grounded is a bad thing is somewhere between a low of 6.1% and a high of 11.9%. Notice that these percents are always within 2.9%, the poll's margin of error, of 9%. ■

✔ **CHECK POINT 5** Solve the inequality:

$$|x - 11| \le 2.9.$$

Interpret the solution in terms of the information in Table 4.3 on page 255.

EXERCISE SET 4.3

Practice Exercises

In Exercises 1–38, find the solution set for each equation.

1. $|x| = 8$
2. $|x| = 6$
3. $|x - 2| = 7$
4. $|x + 1| = 5$
5. $|2x - 1| = 7$
6. $|2x - 3| = 11$
7. $\left|\frac{4x - 2}{3}\right| = 2$
8. $\left|\frac{3x - 1}{5}\right| = 1$
9. $|x| = -8$
10. $|x| = -6$
11. $|x + 3| = 0$
12. $|x + 2| = 0$
13. $2|y + 6| = 10$
14. $3|y + 5| = 12$
15. $3|2x - 1| = 21$
16. $2|3x - 2| = 14$
17. $|6y - 2| + 4 = 32$
18. $|3y - 1| + 10 = 25$
19. $7|5x| + 2 = 16$
20. $7|3x| + 2 = 16$
21. $|x + 1| + 5 = 3$
22. $|x + 1| + 6 = 2$
23. $|4y + 1| + 10 = 4$
24. $|3y - 2| + 8 = 1$
25. $|2x - 1| + 3 = 3$
26. $|3x - 2| + 4 = 4$
27. $|5x - 8| = |3x + 2|$
28. $|4x - 9| = |2x + 1|$
29. $|2x - 4| = |x - 1|$
30. $|6x| = |3x - 9|$
31. $|2x - 5| = |2x + 5|$
32. $|3x - 5| = |3x + 5|$
33. $|x - 3| = |5 - x|$
34. $|x - 3| = |6 - x|$
35. $|2y - 6| = |10 - 2y|$
36. $|4y + 3| = |4y + 5|$
37. $\left|\frac{2x}{3} - 2\right| = \left|\frac{x}{3} + 3\right|$
38. $\left|\frac{x}{2} - 2\right| = \left|x - \frac{1}{2}\right|$

In Exercises 39–70, solve and graph the solution set on a number line.

39. $|x| < 3$
40. $|x| < 5$
41. $|x - 2| < 1$
42. $|x - 1| < 5$
43. $|x + 2| \le 1$
44. $|x + 1| \le 5$
45. $|2x - 6| < 8$
46. $|3x + 5| < 17$
47. $|x| > 3$
48. $|x| > 5$
49. $|x + 3| > 1$
50. $|x - 2| > 5$
51. $|x - 4| \ge 2$
52. $|x - 3| \ge 4$
53. $|3x - 8| > 7$
54. $|5x - 2| > 13$
55. $|2(x - 1) + 4| \le 8$
56. $|3(x - 1) + 2| \le 20$

57. $\left|\frac{2y+6}{3}\right| < 2$

58. $\left|\frac{3x-3}{4}\right| < 6$

59. $\left|\frac{2x+2}{4}\right| \geq 2$

60. $\left|\frac{3x-3}{9}\right| \geq 1$

61. $\left|3 - \frac{2x}{3}\right| > 5$

62. $\left|3 - \frac{3x}{4}\right| > 9$

63. $|x - 2| < -1$

64. $|x - 3| < -2$

65. $|x + 6| > -10$

66. $|x + 4| > -12$

67. $|x + 2| + 9 \leq 16$

68. $|x - 2| + 4 \leq 5$

69. $2|2x - 3| + 10 > 12$

70. $3|2x - 1| + 2 > 8$

Application Exercises

The three television programs viewed by the greatest percentage of U.S. households in the twentieth century are shown in the table. The data are from a random survey of 4000 TV households by Nielsen Media Research. In Exercises 71–72, let x represent the actual viewing percentage in the U.S. population.

TV Programs with the Greatest U.S. Audience Viewing Percentage of the Twentieth Century

Program	Viewing Percentage in Survey
1. "M*A*S*H" Feb. 28, 1983	60.2%
2. "Dallas" Nov. 21, 1980	53.3%
3. "Roots" Part 8 Jan. 30, 1977	51.1%

Source: Nielsen Media Research

71. Solve the inequality: $|x - 60.2| \leq 1.6$. Interpret the solution in terms of the information in the table. What is the margin of error?

72. Solve the inequality: $|x - 51.1| \leq 1.6$. Interpret the solution in terms of the information in the table. What is the margin of error?

73. The inequality $|T - 57| \leq 7$ describes the range of monthly average temperature, T, in degrees Fahrenheit, for San Francisco, California. Solve the inequality and interpret the solution.

74. The inequality $|T - 50| \leq 22$ describes the range of monthly average temperature, T, in degrees Fahrenheit, for Albany, New York. Solve the inequality and interpret the solution.

The specifications for machine parts are given with tolerance limits that describe a range of measurements for which the part is acceptable. In Exercises 75–76, x represents the length of a machine part, in centimeters. The tolerance limit is 0.01 centimeter.

75. Solve: $|x - 8.6| \leq 0.01$. If the length of the machine part is supposed to be 8.6 centimeters, interpret the solution.

76. Solve: $|x - 9.4| \leq 0.01$. If the length of the machine part is supposed to be 9.4 centimeters, interpret the solution.

77. It a coin is tossed 100 times, we would expect approximately 50 of the outcomes to be heads. It can be demonstrated that a coin is unfair if h, the number of outcomes that result in heads, satisfies $\left|\frac{h-50}{5}\right| \geq 1.645$. Describe the number of outcomes that determine an unfair coin that is tossed 100 times.

Writing in Mathematics

78. Explain how to solve an equation containing one absolute value expression.

79. Explain why the procedure that you described in Exercise 78 does not apply to the equation $|x - 5| = -3$. What is the solution set of this equation?

80. Describe how to solve an absolute value equation with two absolute values.

81. Describe how to solve an absolute value inequality involving the symbol $<$. Give an example.

82. Explain why the procedure that you described in Exercise 81 does not apply to the inequality $|x - 5| < -3$. What is the solution set of this inequality?

83. Describe how to solve an absolute value inequality involving the symbol $>$. Give an example.

84. Explain why the procedure that you described in Exercise 83 does not apply to the inequality $|x - 5| > -3$. What is the solution set of this inequality?

85. The final episode of *M*A*S*H* was viewed by more than 58% of U.S. television households. Is it likely that a popular television series in the twenty-first century will achieve a 58% market share? Explain your answer.

Technology Exercises

In Exercises 86–88, solve each equation using a graphing utility. Graph each side separately in the same viewing rectangle. The solutions are the x-coordinates of the intersection points.

86. $|x + 1| = 5$

87. $|3(x + 4)| = 12$

88. $|2x - 3| = |9 - 4x|$

In Exercises 89–91, solve each inequality using a graphing utility. Graph each side separately in the same viewing rectangle. The solution set consists of all values of x for which the the graph of the left side lies below the graph of the right side.

89. $|2x + 3| < 5$

90. $\left|\frac{2x - 1}{3}\right| < \frac{5}{3}$

91. $|x + 4| < -1$

In Exercises 92–94, solve each inequality using a graphing utility. Graph each side separately in the same viewing rectangle. The solution set consists of all values of x for which the graph of the left side lies above the graph of the right side.

92. $|2x - 1| > 7$

93. $|0.1x - 0.4| + 0.4 > 0.6$

94. $|x + 4| > -1$

95. Use a graphing utility to verify the solution sets for any five equations or inequalities that you solved by hand in Exercises 1–70.

Critical Thinking Exercises

96. Which one of the following is true?

a. All absolute value equations have two solutions.

b. The equation $|x| = -6$ is equivalent to $x = 6$ or $x = -6$.

c. We can rewrite the inequality $x > 5$ or $x < -5$ more compactly as $-5 < x < 5$.

d. Absolute value inequalities in the form $|ax + b| < c$ translate into *and* compound statements, which may be written as three-part inequalities.

In Exercises 97–98, use the graph of $f(x) = |4 - x|$ to solve each equation or inequality.

97. $|4 - x| = 1$

98. $|4 - x| < 5$

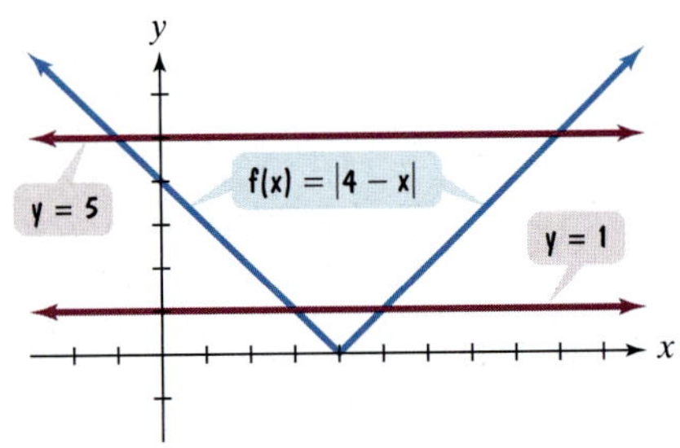

99. The percentage, p, of defective products manufactured by a company is given as $|p - 0.3\%| \le 0.2\%$. If 100,000 products are manufactured and the company offers a \$5 refund for each defective product, describe the company's cost for refunds.

100. Solve: $|2x + 5| = 3x + 4$.

Review Exercises

In Exercises 101–103, graph each linear function.

101. $3x - 5y = 15$ (Section 2.3, Example 1)

102. $f(x) = -\frac{2}{3}x$ (Section 2.3, Example 4)

103. $f(x) = -2$ (Section 2.3, Figure 2.21)

SECTION 4.4 Linear Inequalities in Two Variables

Objectives

1. Graph a linear inequality in two variables.
2. Graph a system of linear inequalities.
3. Solve applied problems involving systems of inequalities.

SSM PH Tutor Center CD-ROM Video

This book was written in Point Reyes National Seashore, 40 miles north of San Francisco. The park consists of 75,000 acres with miles of pristine surf-washed beaches, forested ridges, and bays bordered by white cliffs.

Like your author, many people are kept inspired and energized surrounded by nature's unspoiled beauty. In this section, you will see how systems of inequalities model whether a region's natural beauty manifests itself in forests, grasslands, or deserts.

Linear Inequalities in Two Variables and Their Solution

We have seen that equations in the form $Ax + By = C$ are straight lines when graphed. If we change the $=$ sign to $>, <, \geq$, or $\leq$, we obtain a **linear inequality in two variables**. Some examples of linear inequalities in two variables are $x + y > 2$, $3x - 5y \leq 15$, and $2x - y < 4$.

A **solution of an inequality in two variables**, x and y, is an ordered pair of real numbers with the following property: When the x-coordinate is substituted for x and the y-coordinate is substituted for y in the inequality, we obtain a true statement. For example, $(3, 2)$ is a solution of the inequality $x + y > 1$. When 3 is substituted for x and 2 is substituted for y, we obtain the true statement $3 + 2 > 1$, or $5 > 1$. Because there are infinitely many pairs of numbers that have a sum greater than 1, the inequality $x + y > 1$ has infinitely many solutions. Each ordered pair solution is said to **satisfy** the inequality. Thus, $(3, 2)$ satisfies the inequality $x + y > 1$.

1 Graph a linear inequality in two variables.

The Graph of a Linear Inequality in Two Variables

We know that the graph of an equation in two variables is the set of all points whose coordinates satisfy the equation. Similarly, the **graph of an inequality in two variables** is the set of all points whose coordinates satisfy the inequality.

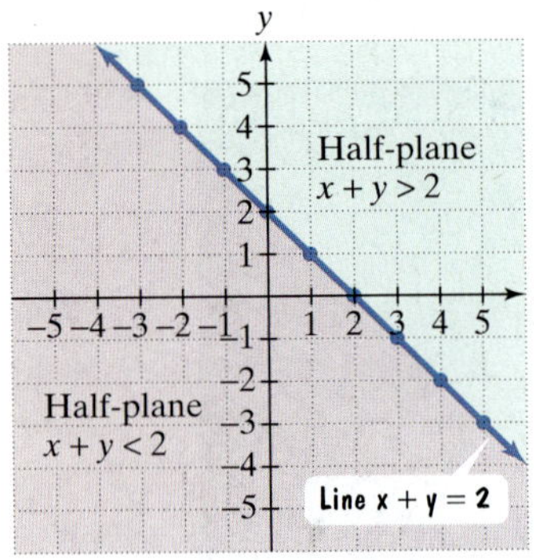

Figure 4.8

Let's use Figure 4.8 to get an idea of what the graph of a linear inequality in two variables looks like. Part of the figure shows the graph of the linear equation $x + y = 2$. The line divides the points in the rectangular coordinate system into three sets. First, there is the set of points along the line, satisfying $x + y = 2$. Next, there is the set of points in the green region above the line. Points in the green region satisfy the linear inequality $x + y > 2$. Finally, there is the set of points in the pink region below the line. Points in the pink region satisfy the linear inequality $x + y < 2$.

A **half-plane** is the set of all the points on one side of a line. In Figure 4.8, the green region is a half-plane. The pink region is also a half-plane. A half-plane is the graph of a linear inequality that involves $>$ or $<$. The graph of an inequality that involves $\geq$ or $\leq$ is a half-plane and a line. A solid line is used to show that the line is part of the graph. A dashed line is used to show that a line is not part of a graph.

Graphing a Linear Inequality in Two Variables

1. Replace the inequality symbol with an equal sign and graph the corresponding linear equation. Draw a solid line if the original inequality contains a $\leq$ or $\geq$ symbol. Draw a dashed line if the original inequality contains a $<$ or $>$ symbol.
2. Choose a test point in one of the half-planes that is not on the line. Substitute the coordinates of the test point into the inequality.
3. If a true statement results, shade the half-plane containing this test point. If a false statement results, shade the half-plane not containing this test point.

EXAMPLE 1 Graphing a Linear Inequality in Two Variables

Graph: $3x - 5y \geq 15$.

Solution

Step 1 Replace the inequality symbol by = and graph the linear equation. We need to graph $3x - 5y = 15$. We can use intercepts to graph this line.

We set $y = 0$ to find the x-intercept:	**We set $x = 0$ to find the y-intercept:**
$3x - 5y = 15$	$3x - 5y = 15$
$3x - 5 \cdot 0 = 15$	$3 \cdot 0 - 5y = 15$
$3x = 15$	$-5y = 15$
$x = 5.$	$y = -3.$

The x-intercept is 5, so the line passes through $(5, 0)$. The y-intercept is -3, so the line passes through $(0, -3)$. Using the intercepts, the line is shown in Figure 4.9 as a solid line. This is because the inequality $3x - 5y \geq 15$ contains a $\geq$ symbol, in which equality is included.

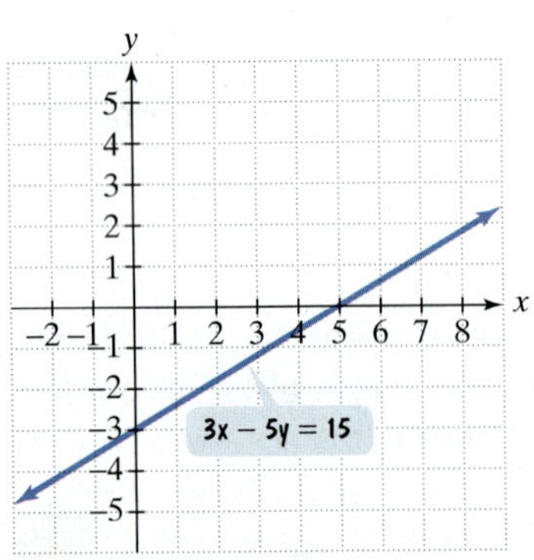

Figure 4.9 Preparing to graph $3x - 5y \geq 15$

Step 2 Choose a test point in one of the half-planes that is not on the line. Substitute its coordinates into the inequality. The line $3x - 5y = 15$ divides the plane into three parts—the line itself and two half-planes. The points in one half-plane satisfy $3x - 5y > 15$. The points in the other half-plane satisfy $3x - 5y < 15$. We need to find which half-plane belongs to the solution. To do so, we test a point from either half-plane. The origin, $(0, 0)$, is the easiest point to test.

$$3x - 5y \geq 15 \quad \text{This is the given inequality.}$$
$$3 \cdot 0 - 5 \cdot 0 \overset{?}{\geq} 15 \quad \text{Test (0, 0) by substituting 0 for } x \text{ and 0 for } y.$$
$$0 - 0 \overset{?}{\geq} 15 \quad \text{Multiply.}$$
$$0 \geq 15 \quad \text{This statement is false.}$$

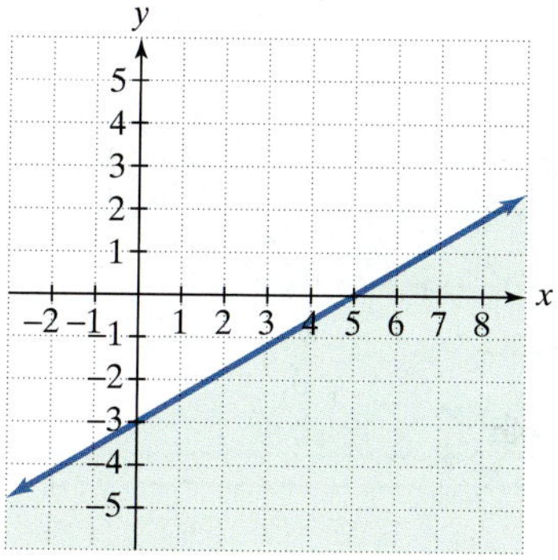

Figure 4.10 The graph of $3x - 5y \geq 15$

Step 3 If a false statement results, shade the half-plane not containing the test point. Because 0 is not greater than or equal to 15, the test point, (0, 0), is not part of the solution set. Thus, the half-plane below the solid line $3x - 5y = 15$ is part of the solution set. The solution set is the line and the half-plane that does not contain the point (0, 0), indicated by shading this half-plane. The graph is shown using green shading and a blue line in Figure 4.10. ■

✔ **CHECK POINT 1** Graph: $2x - 4y \geq 8$.

When graphing a linear inequality, test a point that lies in one of the half-planes and *not on the line dividing the half-planes*. The test point (0, 0) is convenient because it is easy to calculate when 0 is substituted for each variable. However, if (0, 0) lies on the dividing line and not in a half-plane, a different test point must be selected.

EXAMPLE 2 Graphing a Linear Inequality in Two Variables

Graph: $y > -\frac{2}{3}x$.

Solution

Step 1 Replace the inequality symbol by = and graph the linear equation. We need to graph $y = -\frac{2}{3}x$. We can use the slope and the y-intercept to graph this linear function.

$$y = -\frac{2}{3}x + 0$$

Slope $= \frac{-2}{3} = \frac{\text{rise}}{\text{run}}$ y-intercept $= 0$

The y-intercept is 0, so the line passes through (0, 0). Using the y-intercept and the slope, the line is shown in Figure 4.11 as a dashed line. This is because the inequality $y > -\frac{2}{3}x$ contains a $>$ symbol, in which equality is not included.

Step 2 Choose a test point in one of the half-planes that is not on the line. Substitute its coordinates into the inequality. We cannot use (0, 0) as a test point because it lies on the line and not in a half-plane. Let's use (1, 1), which lies in the half-plane above the line.

$$y > -\frac{2}{3}x \quad \text{This is the given inequality.}$$
$$1 \overset{?}{>} -\frac{2}{3} \cdot 1 \quad \text{Test (1, 1) by substituting 1 for } x \text{ and 1 for } y.$$
$$1 > -\frac{2}{3} \quad \text{This statement is true.}$$

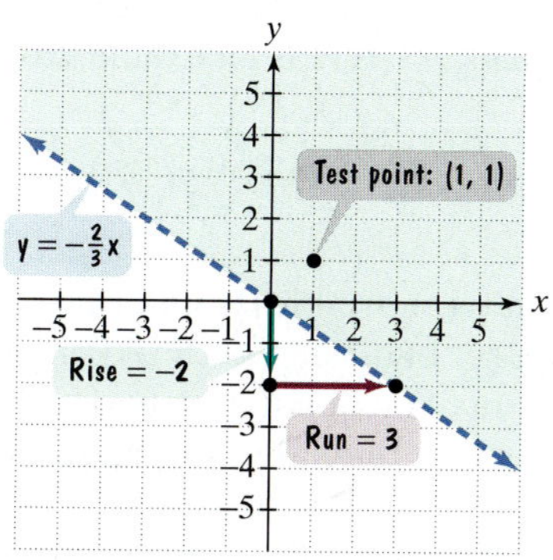

Figure 4.11 The graph of $y > -\frac{2}{3}x$

Step 3 If a true statement results, shade the half-plane containing the test point. Because 1 is greater than $-\frac{2}{3}$, the test point (1, 1) is part of the solution set. All the points on the same side of the line $y = -\frac{2}{3}x$ as the point (1, 1) are members of the solution set. The solution set is the half-plane that contains the point (1, 1), indicated by shading this half-plane. The graph is shown using green shading and a dashed blue line in Figure 4.11. ■

Using Technology

Most graphing utilities can graph inequalities in two variables with the SHADE feature. The procedure varies by model, so consult your manual. For most graphing utilities, you must first solve for y if it is not already isolated. The figure shows the graph of $y > -\frac{2}{3}x$. Most displays do not distinguish between dashed and solid boundary lines.

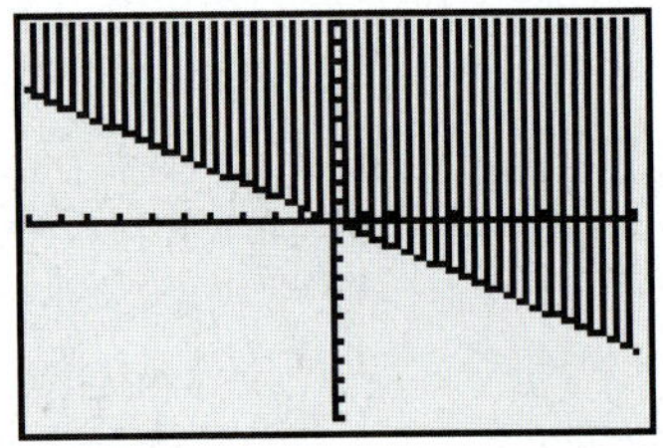

CHECK POINT 2 Graph: $y > -\frac{3}{4}x$.

You can graph inequalities in the form $y > mx + b$ or $y < mx + b$ without using test points. The inequality symbol indicates which half-plane to shade.

- If $y > mx + b$, shade above the line $y = mx + b$.
- If $y < mx + b$, shade below the line $y = mx + b$.

In Section 2.3, we learned that $y = b$, the constant function, graphs as a horizontal line, where b is the y-intercept. Similarly, the graph of $x = a$ is a vertical line, where a is the x-intercept. Half-planes can be separated by horizontal or vertical lines. For example, Figure 4.12 shows the graph of $y \leq 2$. Because $(0, 0)$ satisfies this inequality ($0 \leq 2$ is true), the graph consists of the half-plane below the line $y = 2$ and the line. Similarly, Figure 4.13 shows the graph of $x < 4$.

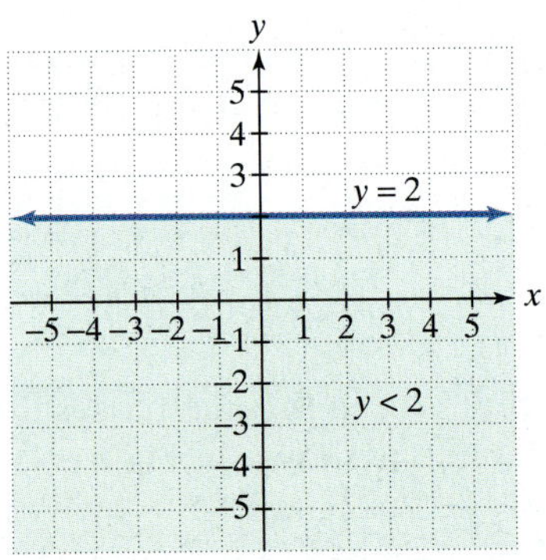

Figure 4.12 The graph of $y \leq 2$

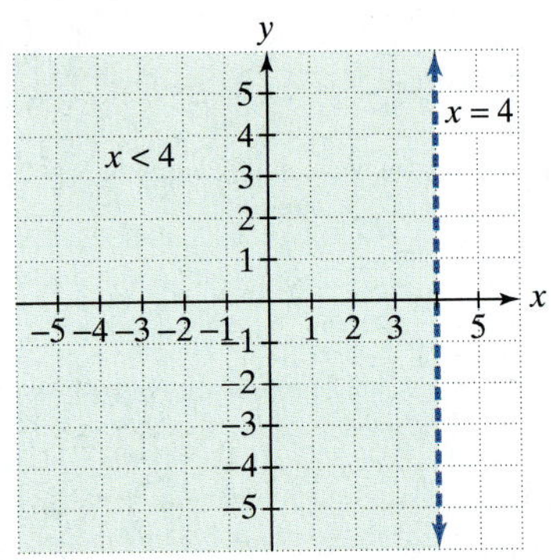

Figure 4.13 The graph of $x < 4$

2 Graph a system of linear inequalities.

Systems of Linear Inequalities

Just as two linear equations make up a system of linear equations, two or more linear inequalities make up a **system of linear inequalities**. Here is an example of a system of linear inequalities:

$$2x - y < 4$$
$$x + y \geq -1.$$

A **solution of a system of linear inequalities** in two variables is an ordered pair that satisfies each inequality in the system. The set of all such ordered pairs is the **solution set** of the system. Thus, to graph a system of inequalities in two variables, begin by graphing each individual inequality in the same rectangular coordinate system. Then find the region, if there is one, that is common to every graph in the system. This region of intersection gives a picture of the system's solution set.

EXAMPLE 3 Graphing a System of Linear Inequalities

Graph the solution set of the system:

$$2x - y < 4$$
$$x + y \geq -1.$$

Solution We begin by graphing $2x - y < 4$. Because the inequality contains a $<$ symbol, rather than $\leq$, we graph $2x - y = 4$ as a dashed line. (If $x = 0$, then $y = -4$, and if $y = 0$, then $x = 2$. The x-intercept is 2 and the y-intercept is -4.) Because $(0, 0)$ makes the inequality $2x - y < 4$ true, we shade the half-plane containing $(0, 0)$, shown in yellow in Figure 4.14.

Now we graph $x + y \geq -1$ in the same rectangular coordinate system. Because the inequality contains a $\geq$ symbol, in which equality is included, we graph $x + y = -1$ as a solid line. (If $x = 0$, then $y = -1$, and if $y = 0$, then $x = -1$. The x-intercept and y-intercept are both -1.) Because $(0, 0)$ makes the inequality true, we shade the half-plane containing $(0, 0)$. This is shown in Figure 4.15 using green vertical shading above the solid red line.

The solution set of the system is shown graphically by the intersection (the overlap) of the two half-planes. This is shown in Figure 4.15 as the region in which the yellow shading and the green vertical shading overlap. The solutions of the system are shown again in Figure 4.16. ■

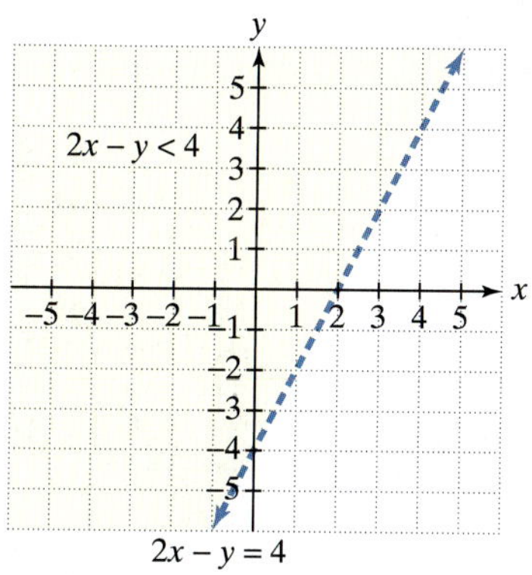

Figure 4.14 The graph of $2x - y < 4$

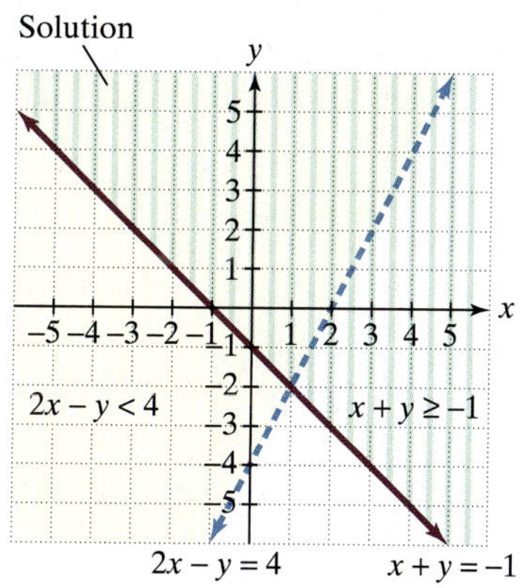

Figure 4.15 Adding the graph of $x + y \geq -1$

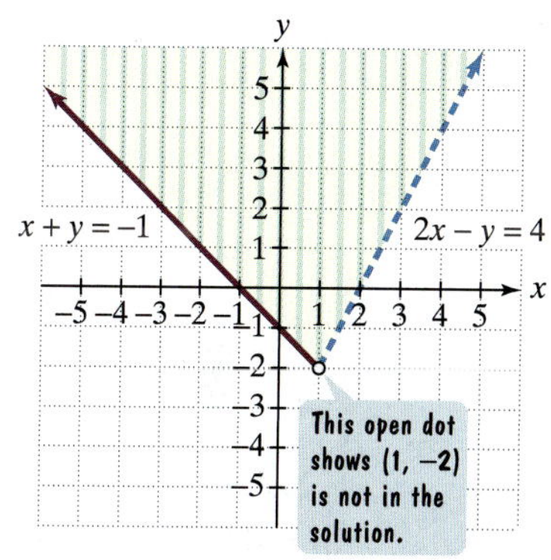

Figure 4.16 The graph of $2x - y < 4$ and $x + y \geq -1$

CHECK POINT 3 Graph the solution set of the system:

$$x + 2y > 4$$
$$2x - 3y \leq -6.$$

A system of inequalities has no solution if there are no points in the rectangular coordinate system that simultaneously satisfy each inequality in the system. For example, the system

$$2x + 3y \geq 6$$
$$2x + 3y \leq 0$$

whose separate graphs are shown in Figure 4.17 has no overlapping region. Thus, the system has no solution. The solution set is $\varnothing$, the empty set.

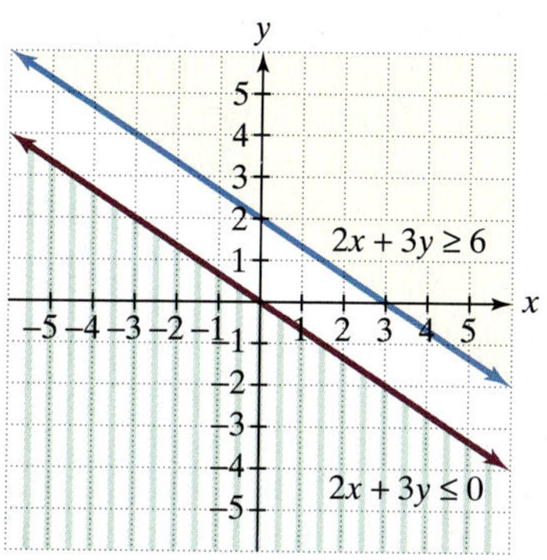

Figure 4.17 A system of inequalities with no solution.

EXAMPLE 4 Graphing a System of Inequalities

Graph the solution set:

$$\begin{aligned} x - y &< 2 \\ -2 \le x &< 4 \\ y &< 3. \end{aligned}$$

Solution We begin by graphing $x - y < 2$, the first given inequality. The line $x - y = 2$ has an x-intercept of 2 and a y-intercept of -2. The test point $(0, 0)$ makes the inequality $x - y < 2$ true, and its graph is shown in Figure 4.18.

Now, let's consider the second given inequality, $-2 \le x < 4$. Replacing the inequality symbols by $=$, we obtain $x = -2$ and $x = 4$, graphed as vertical lines. The line of $x = 4$ is not included. Using $(0, 0)$ as a test point and substituting the x-coordinate, 0, into $-2 \le x < 4$, we obtain the true statement $-2 \le 0 < 4$. We therefore shade the region between the vertical lines. We've added this region to Figure 4.18, intersecting the region between the vertical lines with the yellow region in Figure 4.18. The resulting region is shown in yellow and green vertical shading in Figure 4.19.

Finally, let's consider the third given inequality, $y < 3$. Replacing the inequality symbol by $=$, we obtain $y = 3$, which graphs as a horizontal line. Because $(0, 0)$ satisfies $y < 3$ $(0 < 3$ is true$)$, the graph consists of the half-plane below the line $y = 3$. We've added this half-plane to the region in Figure 4.19, intersecting the half-plane with this region. The resulting region is shown in yellow and green vertical shading in Figure 4.20. This region represents the graph of the solution set of the given system. ■

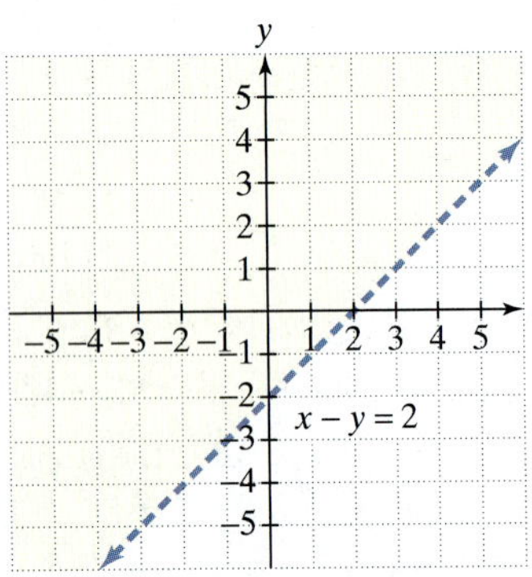

Figure 4.18 The graph of $x - y < 2$

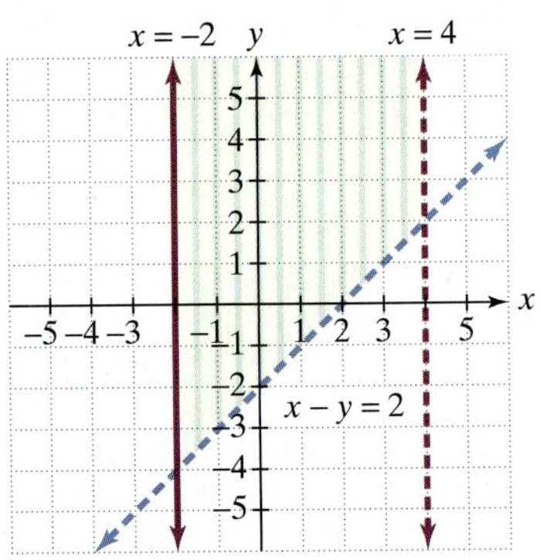

Figure 4.19 The graph of $x - y < 2$ and $-2 \le x < 4$

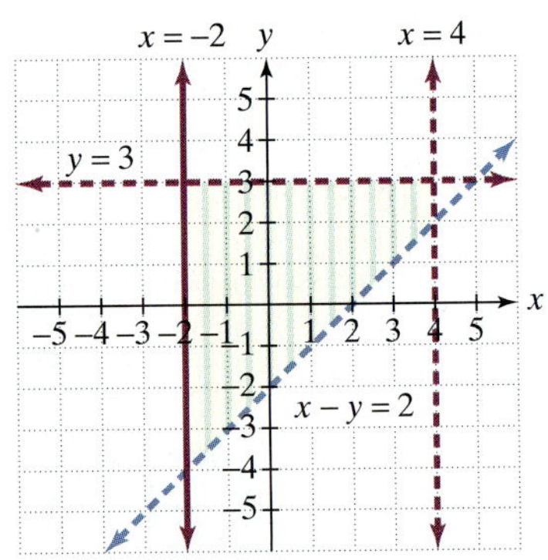

Figure 4.20 The graph of $x - y < 2$ and $-2 \le x < 4$ and $y < 3$

✔ **CHECK POINT 4** Graph the solution set:

$$\begin{aligned} x + y &< 2 \\ -2 \le x &< 1 \\ y &> -3. \end{aligned}$$

3 Solve applied problems involving systems of inequalities.

Applications

Temperature and precipitation affect whether or not trees and forests can grow. At certain levels of precipitation and temperature, only grasslands and deserts will exist. Figure 4.21 shows three kinds of regions—deserts, grasslands, and forests—that result from various ranges of temperature, T, and precipitation, P.

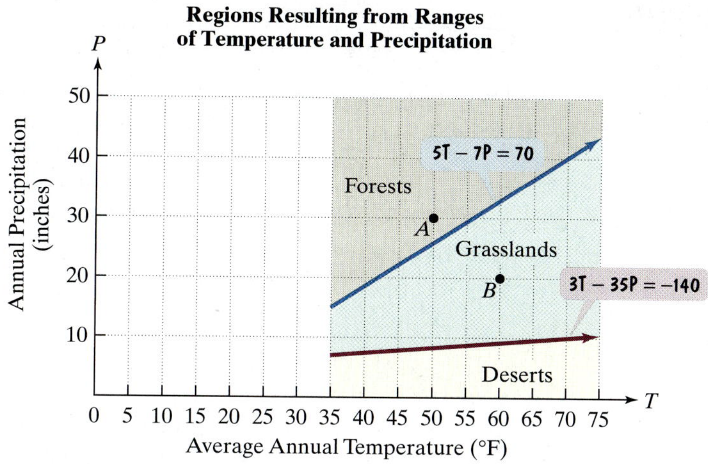

Figure 4.21 *Source*: A. Miller and J. Thompson, *Elements of Meteorology*

Systems of inequalities can be used to describe where forests, grasslands, and deserts occur. Because these regions occur when the average annual temperature, T, is 35°F or greater, each system contains the inequality $T \geq 35$.

Forests occur if	Grasslands occur if	Deserts occur if
$T \geq 35$	$T \geq 35$	$T \geq 35$
$5T - 7P < 70.$	$5T - 7P \geq 70$	$3T - 35P > -140.$
	$3T - 35P \leq -140.$	

EXAMPLE 5 Forests and Systems of Inequalities

Show that point A in Figure 4.21 is a solution of the system of inequalities that describes where forests occur.

Solution Point A has coordinates (50, 30). This means that if a region has an average annual temperature of 50°F and an average annual precipitation of 30 inches, a forest occurs. We can show that (50, 30) satisfies the system of inequalities for forests by substituting 50 for T and 30 for P in each inequality in the system.

$$T \geq 35 \qquad\qquad 5T - 7P < 70$$

$$50 \geq 35, \text{ true} \qquad\qquad 5 \cdot 50 - 7 \cdot 30 \stackrel{?}{<} 70$$

$$250 - 210 \stackrel{?}{<} 70$$

$$40 < 70, \text{ true}$$

The coordinates (50, 30) make each inequality true. Thus, (50, 30) satisfies the system for forests. ■

✔ **CHECK POINT 5** Show that point B in Figure 4.21 is a solution of the system of inequalities that describes where grasslands occur.

EXERCISE SET 4.4

Practice Exercises

In Exercises 1–22, graph each inequality.

1. $x + y \geq 3$

2. $x + y \geq 2$

3. $x - y < 5$

4. $x - y < 6$

5. $x + 2y > 4$

6. $2x + y > 6$

7. $3x - y \leq 6$

8. $x - 3y \leq 6$

9. $\frac{x}{2} + \frac{y}{3} < 1$

10. $\frac{x}{4} + \frac{y}{2} < 1$

11. $y > \frac{1}{3}x$

12. $y > \frac{1}{4}x$

13. $y \leq 3x + 2$

14. $y \leq 2x - 1$

15. $y < -\frac{1}{4}x$

16. $y < -\frac{1}{3}x$

17. $x \leq 2$

18. $x \leq -4$

19. $y > -4$

20. $y > -2$

21. $y \geq 0$

22. $x \leq 0$

In Exercises 23–46 graph the solution set of each system of inequalities or indicate that the system has no solution.

23. $3x + 6y \leq 6$
$2x + y \leq 8$

24. $x - y \geq 4$
$x + y \leq 6$

25. $2x - 5y \leq 10$
$3x - 2y > 6$

26. $2x - y \leq 4$
$3x + 2y > -6$

27. $y > 2x - 3$
$y < -x + 6$

28. $y < -2x + 4$
$y < x - 4$

29. $x + 2y \leq 4$
$y \geq x - 3$

30. $x + y \leq 4$
$y \geq 2x - 4$

31. $x \leq 2$
$y \geq -1$

32. $x \leq 3$
$y \leq -1$

33. $-2 \leq x < 5$

34. $-2 < y \leq 5$

35. $x - y \leq 1$
$x \geq 2$

36. $4x - 5y \geq -20$
$x \geq -3$

37. $x + y > 4$
$x + y < -1$

38. $x + y > 3$
$x + y < -2$

39. $x + y > 4$
$x + y > -1$

40. $x + y > 3$
$x + y > -2$

41. $x - y \leq 2$
$x \geq -2$
$y \leq 3$

42. $3x + y \leq 6$
$x \geq -2$
$y \leq 4$

43. $x \geq 0$
$y \geq 0$
$2x + 5y \leq 10$
$3x + 4y \leq 12$

44. $x \geq 0$
$y \geq 0$
$2x + y \leq 4$
$2x - 3y \leq 6$

45. $3x + y \leq 6$
$2x - y \leq -1$
$x \geq -2$
$y \leq 4$

46. $2x + y \leq 6$
$x + y \geq 2$
$1 \leq x \leq 2$
$y \leq 3$

Application Exercises

For people between ages 10 and 70, inclusive, the target zone for aerobic exercise is given by the following system of inequalities in which a represents one's age and p is one's pulse rate.

$$10 \leq a \leq 70$$
$$2a + 3p \geq 450$$
$$a + p \leq 190$$

The graph of this target zone is shown in the figure. Use this information to solve Exercises 47–48.

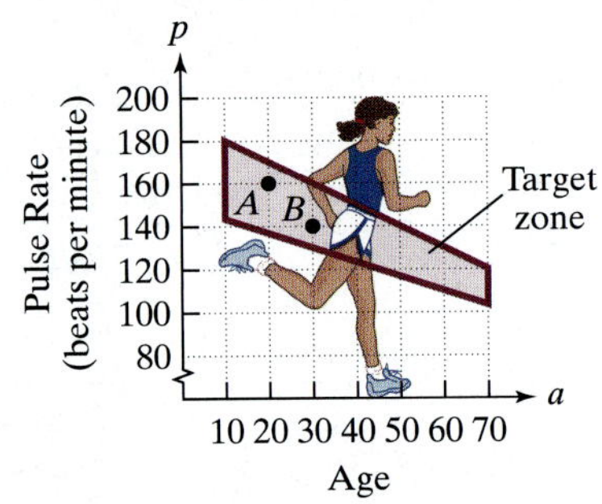

47. a. What are the coordinates of point A and what does this mean in terms of age and pulse rate?

b. Show that point A is a solution of the system of inequalities.

48. a. What are the coordinates of point B and what does this mean in terms of age and pulse rate?

b. Show that point B is a solution of the system of inequalities.

49. Use Figure 4.21 on page 265 to identify the coordinates of a point that lies on the boundary line between grasslands and forests. Then show that this point is a solution of the system of inequalities that describes where grasslands occur.

50. Use Figure 4.21 on page 265 to identify the coordinates of a point that lies on the boundary line between grasslands and deserts. Then show that this point is a solution of the system of inequalities that describes where grasslands occur.

The shaded region in the figure shows recommended weight and height combinations based on information from the Department of Agriculture. Points along the blue and red lines also belong to the healthy weight region. Use this region to solve Exercises 51–55.

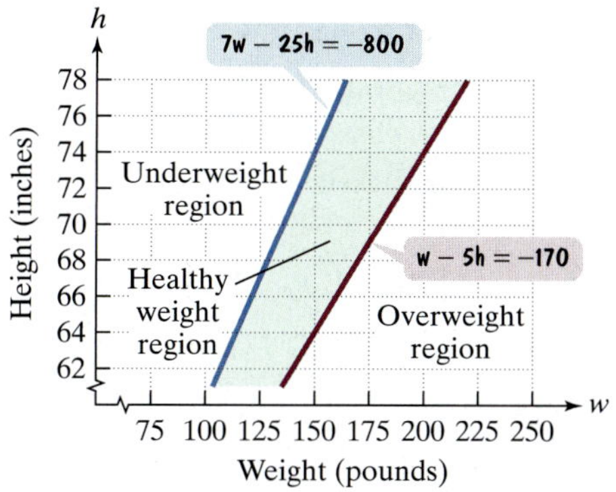

51. Estimate the maximum healthy weight for a person who is 6 feet tall.

52. Estimate the minimum healthy weight for a person who is 6 feet tall.

53. Write an inequality that describes the underweight region. Then identify a point in this region and show that it is a solution of the inequality.

54. Write an inequality that describes the overweight region. Then identify a point in this region and show that it is a solution of the inequality.

55. Write a system of inequalities that describes the healthy weight region. Then identify a point in this region and show that it is a solution of the system.

Writing in Mathematics

56. What is a linear inequality in two variables? Provide an example with your description.

57. How do you determine whether an ordered pair is a solution of an inequality in two variables, x and y?

58. What is a half-plane?

59. What does a solid line mean in the graph of an inequality?

60. What does a dashed line mean in the graph of an inequality?

61. Explain how to graph $x - 2y < 4$.

62. What is a system of linear inequalities?

63. What is a solution of a system of linear inequalities?

64. Explain how to graph the solution set of a system of inequalities.

65. What does it mean if a system of linear inequalities has no solution?

66. Look at the shaded region showing recommended weight and height combinations in the figure for Exercises 51–55. Describe why a system of inequalities, rather than a linear equation in w and h, is better suited to give the recommended combinations.

Technology Exercises

Graphing utilities can be used to shade regions in the rectangular coordinate system, thereby graphing an inequality in two variables. Read the section of the user's manual for your graphing utility that describes how to shade a region. Then use your graphing utility to graph the inequalities in Exercises 67–70.

67. $y \leq 4x + 4$

68. $y \geq \frac{2}{3}x - 2$

69. $2x + y \leq 6$

70. $3x - 2y \geq 6$

71. Does your graphing utility have any limitations in terms of graphing inequalities? If so, what are they?

72. Use a graphing utility with a SHADE feature to verify any five of the graphs that you drew by hand in Exercises 1–22.

73. Use a graphing utility with a SHADE feature to verify any five of the graphs that you drew by hand for the systems in Exercises 23–46.

Critical Thinking Exercises

74. Write a linear inequality in two variables whose graph is shown.

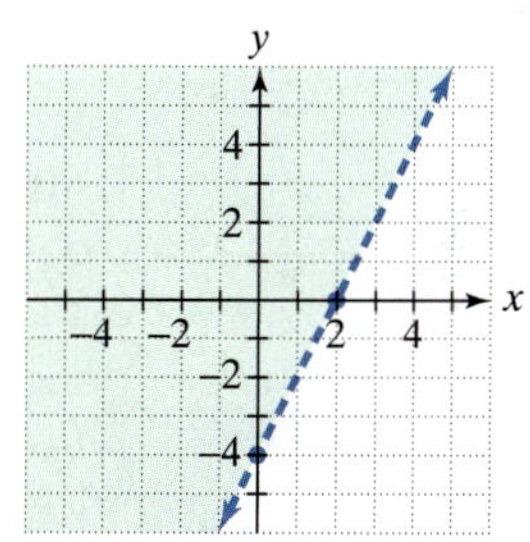

75. Write a system of inequalities whose solution set includes every point in the rectangular coordinate system.

76. A person with no more than \$15,000 to invest plans to place the money in two investments. One investment is high risk, high yield; the other is low risk, low yield. At least \$2000 is to be placed in the high-risk investment. Furthermore, the amount invested at low risk should be at least three times the amount invested at high risk. Find and graph a system of inequalities that describes all possibilities for placing the money in the high- and low-risk investments.

77. Sketch the graph of the solution set for the following system of inequalities:

$$y \geq nx + b \ (n < 0, b > 0)$$
$$y \leq mx + b \ (m > 0, b > 0).$$

Review Exercises

78. Solve using matrices:

$$3x - y = 8$$
$$x - 5y = -2.$$

(Section 3.4, Example 3)

79. Solve by graphing:

$$y = 3x - 2$$
$$y = -2x + 8.$$

(Section 3.1, Example 2)

80. Evaluate:

$$\begin{vmatrix} 8 & 2 & -1 \\ 3 & 0 & 5 \\ 6 & -3 & 4 \end{vmatrix}.$$

(Section 3.5, Example 3)

▸ SECTION 4.5 Linear Programming

Objectives

1. Write an objective function describing a quantity that must be maximized or minimized.
2. Use inequalities to describe limitations in a situation.
3. Use linear programming to solve problems.

SSM PH Tutor Center CD-ROM Video

West Berlin children at Tempelhof airport watch fleets of U.S. airplanes bringing in supplies to circumvent the Russian blockade. The airlift began June 28, 1948 and continued for 15 months.

The Berlin Airlift (1948–1949) was an operation by the United States and Great Britain. It was a response to military action by the former Soviet Union: Soviet troops closed all roads and rail lines between West Germany and Berlin, cutting off supply routes to the city. The Allies used a mathematical technique developed during World War II to maximize the amount of supplies transported. During the 15-month airlift, 278,228 flights provided basic necessities to blockaded Berlin, saving one of the world's great cities.

In this section, we will look at an important application of systems of linear inequalities. Such systems arise in **linear programming**, a method for solving problems in which a particular quantity that must be maximized or minimized is limited. Linear programming is one of the most widely used tools in management science. It helps businesses allocate resources to manufacture products in a way that will maximize profit. Linear programming accounts for more than 50% and perhaps as much as 90% of all computing time used for management decisions in business. The Allies used linear programming to save Berlin.

1 Write an objective function describing a quantity that must be maximized or minimized.

Objective Functions in Linear Programming

Many problems involve quantities that must be maximized or minimized. Businesses are interested in maximizing profit. An operation in which bottled water and medical kits are shipped to earthquake victims needs to maximize the number of victims helped by this shipment. An **objective function** is an algebraic expression in two or more variables describing a quantity that must be maximized or minimized.

EXAMPLE 1 Writing an Objective Function

Bottled water and medical supplies are to be shipped to victims of an earthquake by plane. Each container of bottled water will serve 10 people and each medical kit will aid 6 people. If x represents the number of bottles of water to be shipped and y represents the number of medical kits, write the objective function that describes the number of people that can be helped.

Solution Because each bottle of water serves 10 people and each medical kit aids 6 people, we have

The number of people helped	is	10 times the number of bottles of water	plus	6 times the number of medical kits.
	$=$	$10x$	$+$	$6y$.

Using z to represent the objective function, we have

$$z = 10x + 6y.$$

Unlike the functions that we have seen so far, the objective function is an equation in three variables. For a value of x and a value of y, there is one and only one value of z. Thus, z is a function of x and y. ■

✔ **CHECK POINT 1** A company manufactures bookshelves and desks for computers. Let x represent the number of bookshelves manufactured daily and y the number of desks manufactured daily. The company's profits are \$25 per bookshelf and \$55 per desk. Write the objective function that describes the company's total daily profit, z, from x bookshelves and y desks. (Check Points 1 through 4 are related to this situation, so keep track of your answers.)

2 Use inequalities to describe limitations in a situation.

Constraints in Linear Programming

Ideally, the number of earthquake victims helped in Example 1 should increase without restriction so that every victim receives water and medical kits. However, the planes that ship these supplies are subject to weight and volume restrictions. In linear programming problems, such restrictions are called **constraints**. Each constraint is expressed as a linear inequality. The list of constraints forms a system of linear inequalities.

EXAMPLE 2 Writing a Constraint

Each plane can carry no more than 80,000 pounds. The bottled water weighs 20 pounds per container and each medical kit weighs 10 pounds. If x represents the number of bottles of water to be shipped and y represents the number of medical kits, write an inequality that describes this constraint.

Solution Because each plane can carry no more than 80,000 pounds, we have

The total weight of the water bottles	plus	the total weight of the medical kits	must be less than or equal to	80,000 pounds.
$20x$	$+$	$10y$	$\leq$	80,000.
Each bottle weighs 20 pounds.		Each kit weighs 10 pounds.		

The plane's weight constraint is described by the inequality

$$20x + 10y \leq 80{,}000.$$ ■

✔ **CHECK POINT 2** To maintain high quality, the company in Check Point 1 should not manufacture more than 80 bookshelves and desks per day. Write an inequality that describes this constraint.

In addition to a weight constraint on its cargo, each plane has a limited amount of space in which to carry supplies. Example 3 demonstrates how to express this constraint.

EXAMPLE 3 Writing a Constraint

Planes can carry a total volume for supplies that does not exceed 6000 cubic feet. Each water bottle is 1 cubic foot and each medical kit also has a volume of 1 cubic foot. With x still representing the number of water bottles and y the number of medical kits, write an inequality that describes this second constraint.

Solution Because each plane can carry a volume of supplies that does not exceed 6000 cubic feet, we have

The total volume of the water bottles	plus	the total volume of the medical kits	must be less than or equal to	6000 cubic feet.
$1x$	$+$	$1y$	$\leq$	$6000.$
Each bottle is 1 cubic foot.		Each kit is 1 cubic foot.		

The plane's volume constraint is described by the inequality $x + y \leq 6000.$ ■

In summary, here's what we have described in this aid-to-earthquake-victims situation:

$$z = 10x + 6y$$

This is the objective function describing the number of people helped with x bottles of water and y medical kits.

$$20x + 10y \leq 80{,}000$$
$$x + y \leq 6000.$$

These are the constraints based on each plane's weight and volume limitations.

✔ **CHECK POINT 3** To meet customer demand, the company in Check Point 1 must manufacture between 30 and 80 bookshelves per day inclusive. Furthermore, the company must manufacture at least 10 and no more than 30 desks per day, inclusive. Write an inequality that describes each of these sentences. Then summarize what you have described about this company by writing the objective function for its profits, and the three constraints.

3 Use linear programming to solve problems.

Solving Problems with Linear Programming

The problem in the earthquake situation described previously is to maximize the number of victims who can be helped, subject to the planes' weight and volume constraints. The process of solving this problem is called linear programming, based on a theorem that was proven during World War II.

Solving a Linear Programming Problem

Let $z = ax + by$ be an objective function that depends on x and y. Furthermore, z is subject to a number of constraints on x and y. If a maximum or minimum value of z exists, it can be determined as follows:

1. Graph the system of inequalities representing the constraints.
2. Find the value of the objective function at each corner, or **vertex**, of the graphed region. The maximum and minimum of the objective function occur at one or more of the corner points.

EXAMPLE 4 Solving a Linear Programming Problem

Determine how many bottles of water and how many medical kits should be sent on each plane to maximize the number of earthquake victims who can be helped.

Solution We must maximize $z = 10x + 6y$ subject to the constraints:

$$20x + 10y \le 80{,}000$$
$$x + y \le 6000.$$

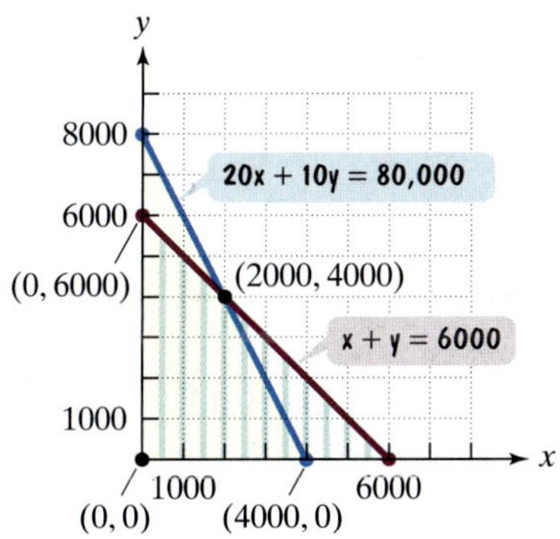

Figure 4.22 The region in quadrant I representing the constraints
$20x + 10y \le 80{,}000$
$x + y \le 6000$

Step 1 Graph the system of inequalities representing the constraints. Because x (the number of bottles of water per plane) and y (the number of medical kits per plane) must be nonnegative, we need to graph the system of inequalities in quadrant I and its boundary only. To graph the inequality $20x + 10y \le 80{,}000$, we graph the equation $20x + 10y = 80{,}000$ as a solid blue line (Figure 4.22). Setting $y = 0$, the x-intercept is 4000 and setting $x = 0$, the y-intercept is 8000. Using $(0, 0)$ as a test point, the inequality is satisfied, so we shade below the blue line, as shown in yellow in Figure 4.22. Now we graph $x + y \le 6000$ by first graphing $x + y = 6000$ as a solid line. Setting $y = 0$, the x-intercept is 6000. Setting $x = 0$, the y-intercept is 6000. Using $(0, 0)$ as a test point, the inequality is satisfied, so we shade below the red line, as shown using green vertical shading in Figure 4.22.

We use the addition method to find where the lines $20x + 10y = 80{,}000$ and $x + y = 6000$ intersect.

$$\begin{array}{rcrcl} 20x + 10y = 80{,}000 & \xrightarrow{\text{No change}} & 20x + 10y & = & 80{,}000 \\ x + y = 6000 & \xrightarrow{\text{Multiply by } -10.} & -10x - 10y & = & -60{,}000 \\ & \text{Add:} & 10x & = & 20{,}000 \\ & & x & = & 2000 \end{array}$$

Back-substituting 2000 for x in $x + y = 6000$, we find $y = 4000$, so the intersection point is $(2000, 4000)$.

The system of inequalities representing the constraints is shown by the region in which the yellow shading and the green vertical shading overlap in Figure 4.22. The graph of the system of inequalities is shown again in Figure 4.23. The red and blue line segments are included in the graph.

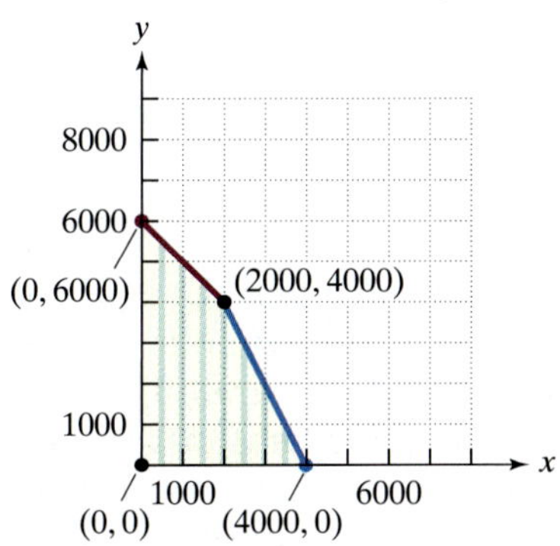

Figure 4.23

Step 2 Find the value of the objective function at each corner of the graphed region. The maximum and minimum of the objective function occur at one or more of the corner points. We must evaluate the objective function, $z = 10x + 6y$, at the four corners, or vertices, of the region in Figure 4.23.

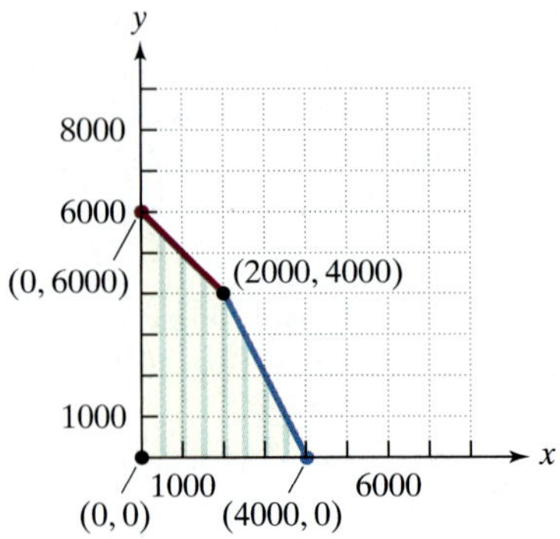

Figure 4.23, repeated

Corner (x, y)	**Objective Function** $z = 10x + 6y$
(0, 0)	$z = 10(0) + 6(0) = 0$
(4000, 0)	$z = 10(4000) + 6(0) = 40{,}000$
(2000, 4000)	$z = 10(2000) + 6(4000) = 44{,}000$ ← maximum
(0, 6000)	$z = 10(0) + 6(6000) = 36{,}000$

Thus, the maximum value of z is 44,000 and this occurs when $x = 2000$ and $y = 4000$. In practical terms, this means that the maximum number of earthquake victims who can be helped with each plane shipment is 44,000. This can be accomplished by sending 2000 water bottles and 4000 medical kits per plane. ■

✔ **CHECK POINT 4** For the company in Check Points 1–3, how many bookshelves and how many desks should be manufactured per day to obtain maximum profit? What is the maximum daily profit?

EXAMPLE 5 Solving a Linear Programming Problem

Find the maximum value of the objective function

$$z = 2x + y$$

subject to the constraints:

$$x \geq 0,\ y \geq 0$$
$$x + 2y \leq 5$$
$$x - y \leq 2.$$

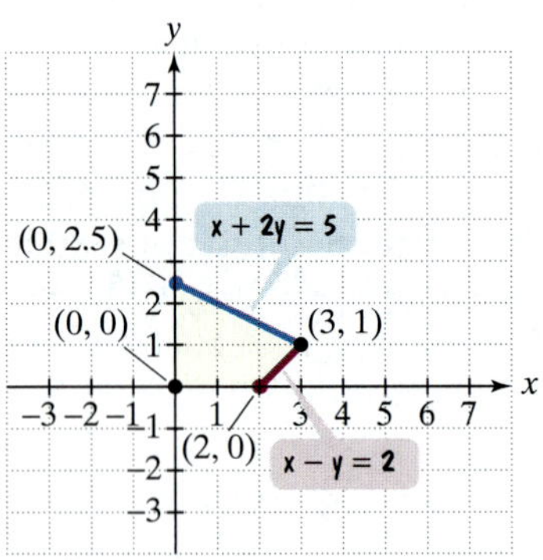

Figure 4.24 The graph of $x + 2y \leq 5$ and $x - y \leq 2$ in quadrant I

Solution We begin by graphing the region in quadrant I ($x \geq 0$, $y \geq 0$) formed by the constraints. The graph is shown in Figure 4.24.

Now we evaluate the objective function at the four vertices of this region.

Objective function: $z = 2x + y$

At (0, 0): $z = 2 \cdot 0 + 0 = 0$

At (2, 0): $z = 2 \cdot 2 + 0 = 4$

At (3, 1): $z = 2 \cdot 3 + 1 = 7$ — Maximum value of z

At (0, 2.5): $z = 2 \cdot 0 + 2.5 = 2.5$

Thus, the maximum value of z is 7, and this occurs when $x = 3$ and $y = 1$. ■

We can see why the objective function in Example 5 has a maximum value that occurs at a vertex by solving the equation for y.

$z = 2x + y$ — This is the objective function of Example 5.

$y = -2x + z$ — Solve for y. Recall that the slope-intercept form of a line is $y = mx + b$.

Slope $= -2$; y-intercept $= z$

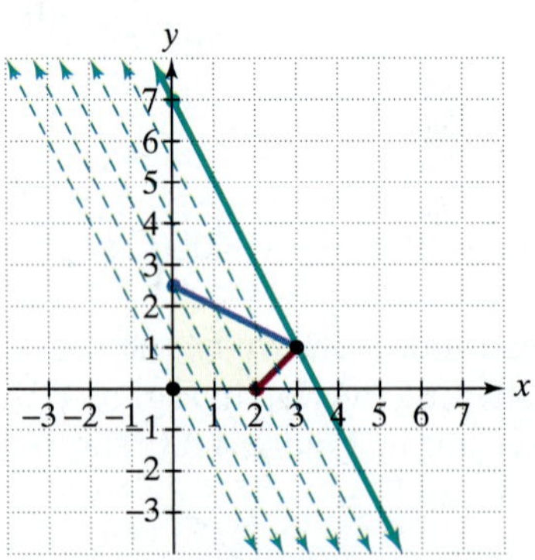

Figure 4.25 The line with slope −2 with the greatest y-intercept that intersects the shaded region passes through one of its vertices.

In this form, z represents the y-intercept of the objective function. The equation describes infinitely many parallel lines, each with a slope of −2. The process in linear programming involves finding the maximum z-value for all lines that intersect the region determined by the constraints. Of all the lines whose slope is −2, we're looking for the one with the greatest y-intercept that intersects the given region. As we see in Figure 4.25, such a line will pass through one (or possibly more) of the vertices of the region.

CHECK POINT 5 Find the maximum value of the objective function $z = 3x + 5y$ subject to the constraints $x \geq 0, y \geq 0, x + y \geq 1, x + y \leq 6$.

ENRICHMENT ESSAY

Faster and Faster

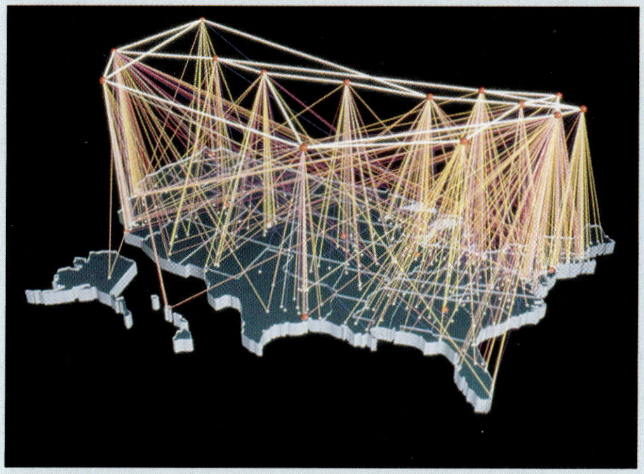

The network of computer linkages in the United States

The problems we solve nowadays have thousands of equations, sometimes a million variables. One of the things that still amazes me is to see a program run on the computer—and to see the answer come out. If we think of the number of combinations of different solutions that we're trying to choose the best of, it's akin to the stars in the heavens. Yet we solve them in a matter of moments. This, to me, is staggering. Not that we can solve them—but that we can solve them so rapidly and efficiently.

—George Dantzig
Inventor of a linear programming method

Problems in linear programming can involve objective functions with thousands of variables subject to thousands of constraints. Several nongeometric linear programming methods are available on software for solving such problems. And we continue to search for faster and faster linear programming methods. This area of applied mathematics has a direct impact on the efficiency and profitability of numerous industries, including telephone and computer communications, and the airlines.

EXERCISE SET 4.5

Practice Exercises

In Exercises 1–4, find the value of the objective function at each corner of the graphed region. What is the maximum value of the objective function? What is the minimum value of the objective function?

1. Objective Function $z = 5x + 6y$

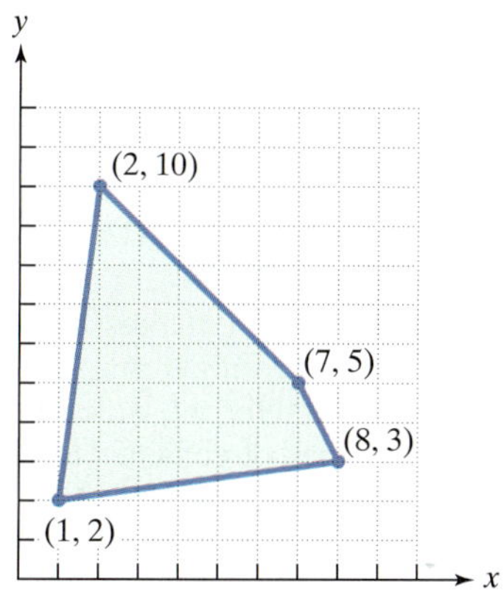

2. Objective Function $z = 3x + 2y$

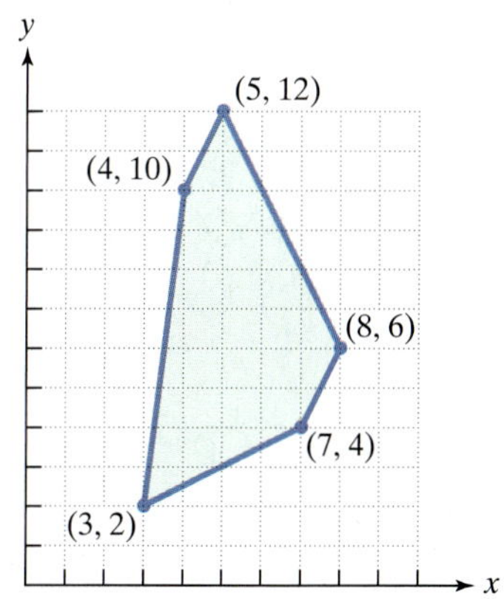

3. Objective Function $z = 40x + 50y$

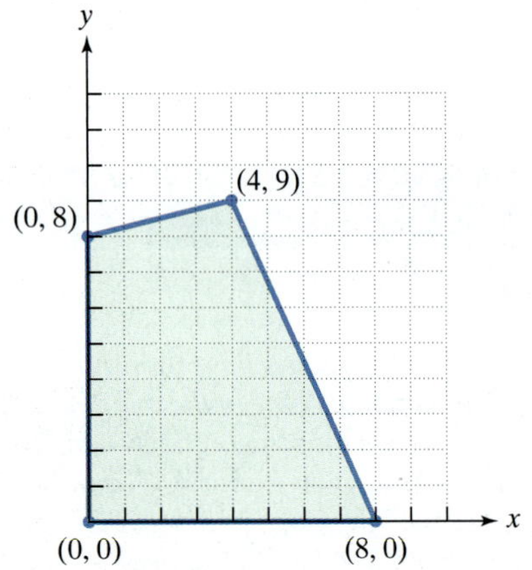

4. Objective Function $z = 30x + 45y$

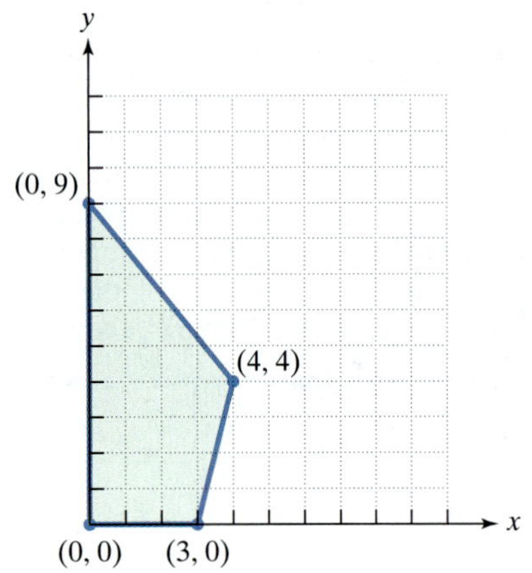

In Exercises 5–14, an objective function and a system of linear inequalities representing constraints are given.

a. *Graph the system of inequalities representing the constraints.*
b. *Find the value of the objective function at each corner of the graphed region.*
c. *Use the values in part (b) to determine the maximum value of the objective function and the values of x and y for which the maximum occurs.*

5. Objective Function $z = 3x + 2y$
Constraints $x \geq 0, y \geq 0$
$2x + y \leq 8$
$x + y \geq 4$

6. Objective Function $z = 2x + 3y$
Constraints $x \geq 0, y \geq 0$
$2x + y \leq 8$
$2x + 3y \leq 12$

7. Objective Function $z = 4x + y$
Constraints $x \geq 0, y \geq 0$
$2x + 3y \leq 12$
$x + y \geq 3$

8. Objective Function $z = x + 6y$
Constraints $x \geq 0, y \geq 0$
$2x + y \leq 10$
$x - 2y \geq -10$

9. Objective Function $z = 3x - 2y$
Constraints $1 \leq x \leq 5$
$y \geq 2$
$x - y \geq -3$

10. Objective Function $z = 5x - 2y$
Constraints $0 \leq x \leq 5$
$0 \leq y \leq 3$
$x + y \geq 2$

11. Objective Function $z = 4x + 2y$
Constraints $x \geq 0, y \geq 0$
$2x + 3y \leq 12$
$3x + 2y \leq 12$
$x + y \geq 2$

12. Objective Function $z = 2x + 4y$
Constraints $x \geq 0, y \geq 0$
$x + 3y \geq 6$
$x + y \geq 3$
$x + y \leq 9$

13. Objective Function $z = 10x + 12y$
Constraints $x \geq 0, y \geq 0$
$x + y \leq 7$
$2x + y \leq 10$
$2x + 3y \leq 18$

14. Objective Function $z = 5x + 6y$
Constraints $x \geq 0, y \geq 0$
$2x + y \geq 10$
$x + 2y \geq 10$
$x + y \leq 10$

Application Exercises

15. A television manufacturer makes console and wide-screen televisions. The profit per unit is \$125 for the console televisions and \$200 for the wide-screen televisions.

a. Let x = the number of consoles manufactured in a month and y = the number of wide-screens manufactured in a month. Write the objective function that describes the total monthly profit.

b. The manufacturer is bound by the following constraints:

1. Equipment in the factory allows for making at most 450 console televisions in one month.
2. Equipment in the factory allows for making at most 200 wide-screen televisions in one month.
3. The cost to the manufacturer per unit is $600 for the console televisions and $900 for the wide-screen televisions. Total monthly costs cannot exceed $360,000.

Write a system of three inequalities that describes these constraints.

c. Graph the system of inequalities in part (b). Use only the first quadrant and its boundary, because x and y must both be nonnegative.

d. Evaluate the objective function for total monthly profit at each of the five vertices of the graphed region. (The vertices should occur at $(0, 0)$, $(0, 200)$, $(300, 200)$, $(450, 100)$, and $(450, 0)$.)

e. Complete the missing portions of this statement: The television manufacturer will make the greatest profit by manufacturing ____ console televisions each month and ____ wide-screen televisions each month. The maximum monthly profit is $______.

16. a. A student earns $10 per hour for tutoring and $7 per hour as a teacher's aid. Let x = the number of hours each week spent tutoring, and y = the number of hours each week spent as a teacher's aid. Write the objective function that describes total weekly earnings.

b. The student is bound by the following constraints:

- To have enough time for studies, the student can work no more than 20 hours a week.
- The tutoring center requires that each tutor spend at least three hours a week tutoring.
- The tutoring center requires that each tutor spend no more than eight hours a week tutoring.

Write a system of three inequalities that describes these constraints.

c. Graph the system of inequalities in part (b). Use only the first quadrant and its boundary, because x and y are nonnegative.

d. Evaluate the objective function for total weekly earnings at each of the four vertices of the graphed region. (The vertices should occur at $(3, 0)$, $(8, 0)$, $(3, 17)$, and $(8, 12)$.)

e. Complete the missing portions of this statement: The student can earn the maximum amount per week by tutoring for __ hours per week and working as a teacher's aid for ___ hours per week. The maximum amount that the student can earn each week is $____.

Use the two steps for solving a linear programming problem, given in the box on page 271, to solve the problems in Exercises 17–23.

17. A manufacturer produces two models of mountain bicycles. The times (in hours) required for assembling and painting each model are given in the following table.

	Model *A*	Model *B*
Assembling	5	4
Painting	2	3

The maximum total weekly hours available in the assembly department and the paint department are 200 hours and 108 hours, respectively. The profits per unit are $25 for model *A* and $15 for model *B*. How many of each type should be produced to maximize profit?

18. A large institution is preparing lunch menus containing foods A and B. The specifications for the two foods are given in the following table.

Food	Units of Fat per Ounce	Units of Carbohydrates per Ounce	Units of Protein per Ounce
A	1	2	1
B	1	1	1

Each lunch must provide at least 6 units of fat per serving, no more than 7 units of protein, and at least 10 units of carbohydrates. The institution can purchase food A for $0.12 per ounce and food B for $0.08 per ounce. How many ounces of each food should a serving contain to meet the dietary requirements at the least cost?

19. Food and clothing are shipped to victims of a natural disaster. Each carton of food will feed 12 people, while each carton of clothing will help 5 people. Each 20-cubic-foot box of food weighs 50 pounds and each 10-cubic-foot box of clothing weighs 20 pounds. The commercial carriers transporting food and clothing are bound by the following constraints:

1. The total weight per carrier cannot exceed 19,000 pounds.
2. The total volume must be less than 8000 cubic feet.

How many cartons of food and clothing should be sent with each plane shipment to maximize the number of people who can be helped?

20. On June 24, 1948, the former Soviet Union blocked all land and water routes through East Germany to Berlin. A gigantic airlift was organized using American and British planes to bring food, clothing, and other supplies to the more than 2 million people in West Berlin. The cargo capacity was 30,000 cubic feet for an American plane and 20,000 cubic feet for a British plane. To break the Soviet blockade, the Western Allies had to maximize cargo capacity, but were subject to the following restrictions:

1. No more than 44 planes could be used.
2. The larger American planes required 16 personnel per flight, double that of the requirement for the British planes. The total number of personnel available could not exceed 512.
3. The cost of an American flight was \$9000 and the cost of a British flight was \$5000. Total weekly costs could not exceed \$300,000.

Find the number of American and British planes that were used to maximize cargo capacity.

21. A theater is presenting a program on drinking and driving for students and their parents. The proceeds will be donated to a local alcohol information center. Admission is \$2.00 for parents and \$1.00 for students. However, the situation has two constraints: The theater can hold no more than 150 people and every two parents must bring at least one student. How many parents and students should attend to raise the maximum amount of money?

22. You are about to take a test that contains computation problems worth 6 points each and word problems worth 10 points each. You can do a computation problem in 2 minutes and a word problem in 4 minutes. You have 40 minutes to take the test and may answer no more than 12 problems. Assuming you answer all the problems attempted correctly, how many of each type of problem must you do to maximize your score? What is the maximum score?

23. In 1978, a ruling by the Civil Aeronautics Board allowed Federal Express to purchase larger aircraft. Federal Express's options included 20 Boeing 727s that United Airlines was retiring and/or the French-built Dassault Fanjet Falcon 20. To aid in their decision, executives at Federal Express analyzed the following data:

	Boeing 727	**Falcon 20**
Direct Operating Cost	\$1400 per hour	\$500 per hour
Payload	42,000 pounds	6000 pounds

Federal Express was faced with the following constraints:

1. Hourly operating cost was limited to \$35,000.
2. Total payload had to be at least 672,000 pounds.
3. Only twenty 727s were available.

Given the constraints, how many of each kind of aircraft should Federal Express have purchased to maximize the number of aircraft?

Writing in Mathematics

24. What kinds of problems are solved using the linear programming method?

25. What is an objective function in a linear programming problem?

26. What is a constraint in a linear programming problem? How is a constraint represented?

27. In your own words, describe how to solve a linear programming problem.

28. Describe a situation in your life in which you would like to maximize something, but are limited by at least two constraints. Can linear programming be used in this situation? Explain your answer.

Critical Thinking Exercises

29. Suppose that you inherit \$10,000. The will states how you must invest the money. Some (or all) of the money must be invested in stocks and bonds. The requirements are that at least \$3000 be invested in bonds, with expected returns of \$0.08 per dollar, and at least \$2000 be invested in stocks, with expected returns of \$0.12 per dollar. Because the stocks are medium risk, the final stipulation requires that the investment in bonds should never be less than the investment in stocks. How should the money be invested so as to maximize your expected returns?

30. Consider the objective function $z = Ax + By$ ($A > 0$ and $B > 0$) subject to the following constraints: $2x + 3y \le 9$, $x - y \le 2$, $x \ge 0$, and $y \ge 0$. Prove that the objective function will have the same maximum value at the vertices $(3, 1)$ and $(0, 3)$ if $A = \frac{2}{3}B$.

Review Exercises

31. Simplify: $(2x^4y^3)(3xy^4)^3$. (Section 1.6, Example 9)

32. Solve for L: $3P = \dfrac{2L - W}{4}$. (Section 1.5, Example 8)

33. If $f(x) = x^3 + 2x^2 - 5x + 4$, find $f(-1)$. (Section 2.1, Example 3)

CHAPTER 4 GROUP PROJECTS

1. Each group member should research one situation that provides two different pricing options. These can involve areas such as public transportation options (with or without coupon books) or long-distance telephone plans or anything of interest. Be sure to bring in all the details for each option. At the group meeting, select the two pricing situations that are most interesting and relevant. Using each situation, write a word problem about selecting the better of the two options. The word problem should be one that can be solved using a linear inequality. The group should turn in the two problems and their solutions.

2. For this activity, group members will conduct interviews with a random sample of students on campus. Each student is to be asked, "What is the worst thing about being a student?" One response should be recorded for each student.

   a. Each member should interview enough students so that there are at least 50 randomly selected students in the sample.

   b. After all responses have been recorded, the group should organize the four most common answers. For each answer, compute the percentage of students in the sample who felt that this is the worst thing about being a student.

   c. If there are n students in your sample, the poll's margin of error is given by $\pm\frac{1}{\sqrt{n}}$. Compute the margin of error for your poll. Express the margin of error as a percent, correct to the nearest tenth of a percent.

   d. For each of the four most common answers, write a statement about the percentage of all students, x, on your campus who feel that this is the worst thing about being a student. Use the inequality

$$|x - \text{percentage in the sample}| \leq \text{margin of error.}$$

3. Group members should choose a particular field of interest. Research how linear programming is used to solve problems in that field. If possible, investigate the solution of a specific practical problem. Present a report on your findings, including the contributions of George Dantzig, Narendra Karmarkar, and L.G. Khachion to linear programming.

4. Members of the group should interview a business executive who is in charge of deciding the product mix for a business. How are production policy decisions made? Are other methods used in conjunction with linear programming? What are these methods? What sort of academic background, particularly in mathematics, does this executive have? Present a group report addressing these questions, emphasizing the role of linear programming for the business.

CHAPTER SUMMARY, REVIEW, AND TEST

SUMMARY

DEFINITIONS AND CONCEPTS	EXAMPLES
Section 4.1 Solving Linear Inequalities	
A linear inequality in one variable can be written in the form $ax + b < c$, $ax + b \le c$, $ax + b > c$, or $ax + b \ge c$. The set of all numbers that make the inequality a true statement is its solution set. Graphs of solution sets are shown on a number line by shading all points representing numbers that are solutions. Parentheses indicate endpoints that are not solutions. Square brackets indicate endpoints that are solutions.	• $(-2, 1] = \{x \mid -2 < x \le 1\}$ 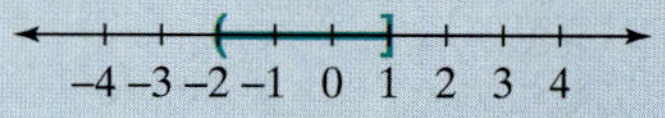 • $[-2, \infty) = \{x \mid x \ge -2\}$ 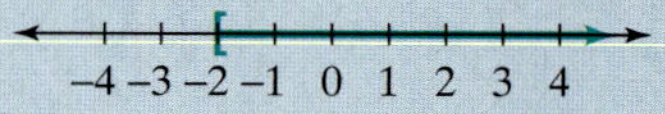
Solving a Linear Inequality **1.** Simplify each side. **2.** Collect variable terms on one side and constant terms on the other side. **3.** Isolate the variable and solve. If an inequality is multiplied or divided by a negative number, the inequality symbol must be reversed.	Solve: $2(x + 3) - 5x \le 15.$ $2x + 6 - 5x \le 15$ $-3x + 6 \le 15$ $-3x \le 9$ $\frac{-3x}{-3} \ge \frac{9}{-3}$ $x \ge -3$ Solution set: $\{x \mid x \ge -3\}$ or $[-3 \infty)$ 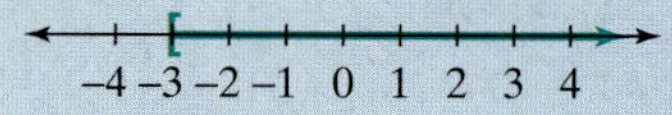
Section 4.2 Compound Inequalities	
Intersection $(\cap)$ ***and Union*** $(\cup)$ $A \cap B$ is the set of elements common to both set A and set B. $A \cup B$ is the set of elements that are members of set A or set B or of both sets.	$\{1, 3, 5, 7\} \cap \{5, 7, 9, 11\} = \{5, 7\}$ $\{1, 3, 5, 7\} \cup \{5, 7, 9, 11\} = \{1, 3, 5, 7, 9, 11\}$
A compound inequality is formed by joining two inequalities with the word *and* or *or*. When the connecting word is *and*, graph each inequality separately and take the intersection of their solution sets.	Solve: $x + 1 > 3$ and $x + 4 \le 8.$ $x > 2$ and $x \le 4$ 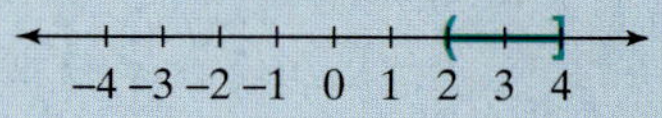 Solution set: $\{x \mid 2 < x \le 4\}$ or $(2, 4]$
The compound inequality $a < x < b$ means $a < x$ and $x < b$. Solve by isolating the variable in the middle.	Solve: $-1 < \frac{2x + 1}{3} \le 2.$ $-3 < 2x + 1 \le 6$ Multiply by 3. $-4 < 2x \le 5$ Subtract 1. $-2 < x \le \frac{5}{2}$ Divide by 2. Solution set: $\left\{x \mid -2 < x \le \frac{5}{2}\right\}$ or $\left(-2, \frac{5}{2}\right]$ 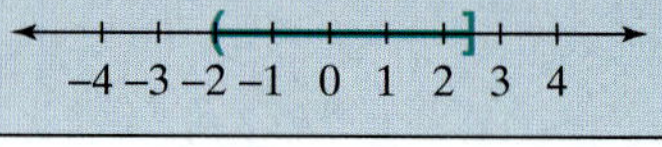

DEFINITIONS AND CONCEPTS | **EXAMPLES**

Section 4.2 Compound Inequalities (continued)

When the connecting word in a compound inequality is *or*, graph each inequality separately and take the union of their solution sets.

Solve: $x - 2 > -3$ or $2x \le -6$.

$$x > -1 \quad \text{or} \quad x \le -3$$

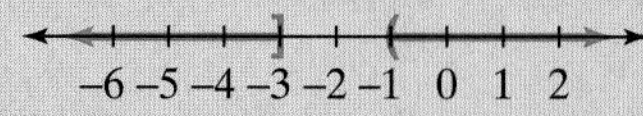

Solution set: $\{x \mid x \le -3 \text{ or } x > -1\}$ or $(-\infty, -3] \cup (-1, \infty)$

Section 4.3 Equations and Inequalities Involving Absolute Value

Absolute Value Equations

1. If $c > 0$, then $|X| = c$ means $X = c$ or $X = -c$.
2. If $c < 0$, then $|X| = c$ has no solution.
3. If $c = 0$, then $|X| = 0$ means $X = 0$.

Solve: $|2x - 7| = 3$.

$$2x - 7 = 3 \quad \text{or} \quad 2x - 7 = -3$$
$$2x = 10 \qquad\qquad 2x = 4$$
$$x = 5 \qquad\qquad x = 2$$

The solution set is $\{2, 5\}$.

Absolute Value Equations with Two Absolute Value Bars

If $|X_1| = |X_2|$, then $X_1 = X_2$ or $X_1 = -X_2$.

Solve: $|x - 6| = |2x + 1|$.

$$x - 6 = 2x + 1 \quad \text{or} \quad x - 6 = -(2x + 1)$$
$$-x - 6 = 1 \qquad\qquad x - 6 = -2x - 1$$
$$-x = 7 \qquad\qquad 3x - 6 = -1$$
$$x = -7 \qquad\qquad 3x = 5$$
$$x = \frac{5}{3}$$

The solutions are -7 and $\frac{5}{3}$, and the solution set is $\{-7, \frac{5}{3}\}$.

If c is a positive number, then to solve $|X| < c$, solve the compound inequality $-c < X < c$. If c is negative, $|X| < c$ has no solution.

Solve: $|x - 4| < 3$.

$$-3 < x - 4 < 3$$
$$1 < x < 7$$

The solution set is $\{x \mid 1 < x < 7\}$ or $(1, 7)$.

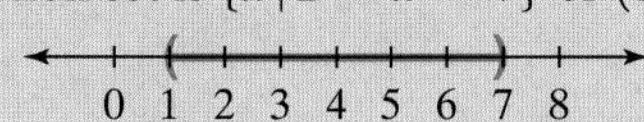

If c is a positive number, then to solve $|X| > c$, solve the compound inequality $X < -c$ *or* $X > c$. If c is negative, $|X| > c$ is true for all real numbers for which X is defined.

Solve: $\left|\frac{x}{3} - 1\right| \ge 2$.

$$\frac{x}{3} - 1 \le -2 \quad \text{or} \quad \frac{x}{3} - 1 \ge 2$$
$$x - 3 \le -6 \quad \text{or} \quad x - 3 \ge 6 \qquad \text{Multiply by 3.}$$
$$x \le -3 \quad \text{or} \quad x \ge 9 \qquad \text{Add 3.}$$

The solution set is $\{x \mid x \le -3 \text{ or } x \ge 9\}$ or $(-\infty, -3] \cup [9, \infty)$.

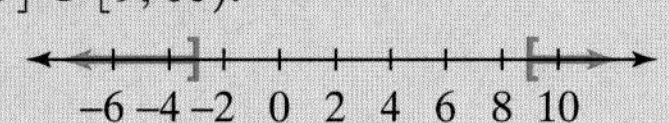

DEFINITIONS AND CONCEPTS	EXAMPLES

Section 4.4 Linear Inequalities in Two Variables

If the equal sign in $Ax + By = C$ is replaced with an inequality symbol, the result is a linear inequality in two variables. Its graph is the set of all points whose coordinates satisfy the inequality. To obtain the graph:

1. Replace the inequality symbol with an equal sign and graph the boundary line. Use a solid line for $\leq$ or $\geq$ and a dashed line for $<$ or $>$.
2. Choose a test point not on the line and substitute its coordinates into the inequality.
3. If a true statement results, shade the half-plane containing the test point. If a false statement results, shade the half-plane not containing the test point.

Graph: $x - 2y \leq 4$.

1. Graph $x - 2y = 4$. Use a solid line because the inequality symbol is $\leq$.
2. Test $(0, 0)$.

$$0 - 2 \cdot 0 \stackrel{?}{=} 4$$
$$0 \leq 4, \quad \textbf{true}$$

3. The inequality is true. Shade the half-plane containing $(0, 0)$.

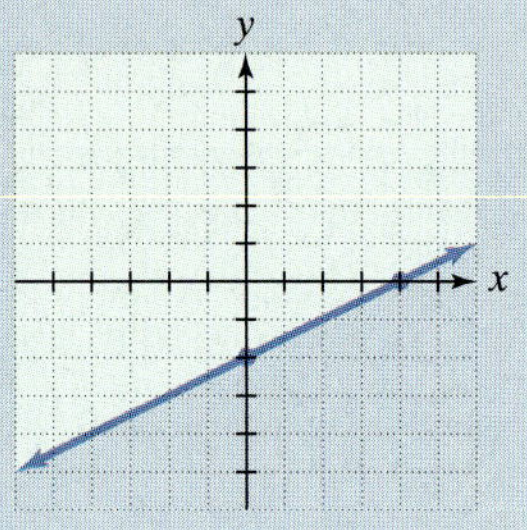

Two or more linear inequalities make up a system of linear inequalities. A solution is an ordered pair satisfying all inequalities in the system. To graph a system of inequalities, graph each inequality in the system. The overlapping region represents the solutions of the system.

Graph the solutions of the system:

$$y \leq -2x$$
$$x - y \geq 3.$$

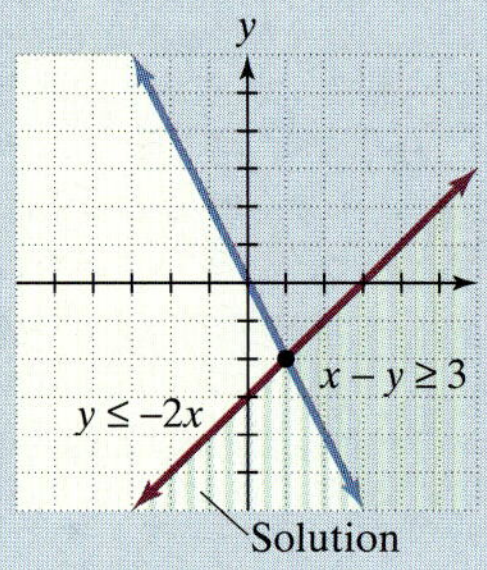

Section 4.5 Linear Programming

Linear programming is a method for solving problems in which a particular quantity that must be maximized or minimized is limited. An objective function is an algebraic expression in three variables describing a quantity that must be maximized or minimized. Constraints are restrictions, expressed as linear inequalities.

Solving a Linear Programming Problem

1. Graph the system of inequalities representing the constraints.
2. Find the value of the objective function at each corner, or vertex, of the graphed region. The maximum and minimum of the objective function occur at one or more vertices.

Find the maximum value of the objective function $z = 3x + 2y$ subject to the following constraints: $x \geq 0$, $y \geq 0$, $2x + 3y \leq 18$, $2x + y \leq 10$.

1. Graph the system of inequalities representing the constraints.

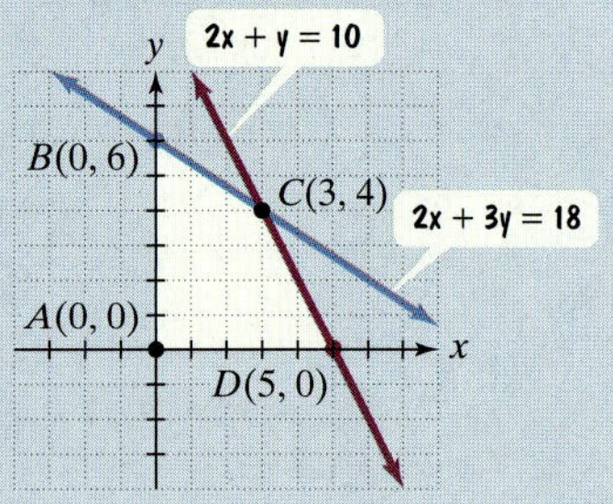

2. Evaluate the objective function at each vertex.

Vertex	$z = 3x + 2y$
$A(0, 0)$	$z = 3(0) + 2(0) = 0$
$B(0, 6)$	$z = 3(0) + 2(6) = 12$
$C(3, 4)$	$z = 3(3) + 2(4) = 17$
$D(5, 0)$	$z = 3(5) + 2(0) = 15$

The maximum value of the objective function is 17.

Review Exercises

4.1 *In Exercises 1–3, graph the solutions of each inequality on a number line.*

1. $x > 5$ **2.** $x \le 1$ **3.** $-3 \le x < 0$

In Exercises 4–6, express each interval in terms of an inequality and graph the interval on a number line.

4. $(-2, 3]$
5. $[-1.5, 2]$
6. $(-1, \infty)$

In Exercises 7–12, solve each linear inequality. Other than $\varnothing$, graph the solution set on a number line. Express the solution set in both set-builder and interval notations.

7. $-6x + 3 \le 15$
8. $6x - 9 \ge -4x - 3$
9. $\frac{x}{3} - \frac{3}{4} - 1 > \frac{x}{2}$
10. $6x + 5 > -2(x - 3) - 25$
11. $3(2x - 1) - 2(x - 4) \ge 7 + 2(3 + 4x)$
12. $2x + 7 \le 5x - 6 - 3x$
13. A person can choose between two charges on a checking account. The first method involves a fixed cost of $11 per month plus 6¢ for each check written. The second method involves a fixed cost of $4 per month plus 20¢ for each check written. How many checks should be written to make the first method a better deal?
14. A salesperson earns $500 per month plus a commission of 20% of sales. Describe the sales needed to receive a total income that exceeds $3200 per month.

4.2 *In Exercises 15–18, let $A = \{a, b, c\}$, $B = \{a, c, d, e\}$ and $C = \{a, d, f, g\}$. Find the indicated set.*

15. $A \cap B$ **16.** $A \cap C$
17. $A \cup B$ **18.** $A \cup C$

In Exercises 19–29, solve each compound inequality. Except for the empty set, express the solution set in both set-builder and interval notations. Graph the solution set on a number line.

19. $x \le 3$ and $x < 6$
20. $x \le 3$ or $x < 6$
21. $-2x < -12$ and $x - 3 < 5$
22. $5x + 3 \le 18$ and $2x - 7 \le -5$
23. $2x - 5 > -1$ and $3x < 3$
24. $2x - 5 > -1$ or $3x < 3$
25. $x + 1 \le -3$ or $-4x + 3 < -5$
26. $5x - 2 \le -22$ or $-3x - 2 > 4$
27. $5x + 4 \ge -11$ or $1 - 4x \ge 9$
28. $-3 < x + 2 \le 4$
29. $-1 \le 4x + 2 \le 6$
30. On the first of four exams, your grades are 72, 73, 94, and 80. There is still one more exam, and you are hoping to earn a B in the course. This will occur if the average of your five exam grades is greater than or equal to 80 and less than 90. What range of grades on the fifth exam will result in receiving a B? Use interval notation to express this range.

4.3 *In Exercises 31–34, find the solution set for each equation.*

31. $|2x + 1| = 7$ **32.** $|3x + 2| = -5$
33. $2|x - 3| - 7 = 10$
34. $|4x - 3| = |7x + 9|$

In Exercises 35–38, solve and graph the solution set on a number line. Except for the empty set, express the solution set in both set-builder and interval notations.

35. $|2x + 3| \le 15$
36. $\left|\frac{2x + 6}{3}\right| > 2$
37. $|2x + 5| - 7 < -6$
38. $|2x - 3| + 4 \ge -10$
39. Approximately 90% of the population sleeps h hours daily, where h is modeled by the inequality $|h - 6.5| \le 1$. Write a sentence describing the range for the number of hours that most people sleep. Do *not* use the phrase "absolute value" in your description.

4.4 *In Exercises 40–45, graph each inequality in a rectangular coordinate system.*

40. $3x - 4y > 12$ **41.** $x - 3y \le 6$
42. $y \le -\frac{1}{2}x + 2$ **43.** $y > \frac{3}{5}x$
44. $x \le 2$ **45.** $y > -3$

In Exercises 46–54, graph the solution set of each system of inequalities or indicate that the system has no solution.

46. $3x - y \le 6$, $x + y \ge 2$
47. $y < -x + 4$, $y > x - 4$
48. $-3 \le x < 5$
49. $-2 < y \le 6$
50. $x \ge 3$, $y \le 0$
51. $2x - y > -4$, $x \ge 0$
52. $x + y \le 6$, $y \ge 2x - 3$
53. $3x + 2y \ge 4$, $x - y \le 3$, $x \ge 0, y \ge 0$
54. $2x - y > 2$, $2x - y < -2$

4.5

55. Find the value of the objective function $z = 2x + 3y$ at each corner of the graphed region shown. What is the maximum value of the objective function? What is the minimum value of the objective function?

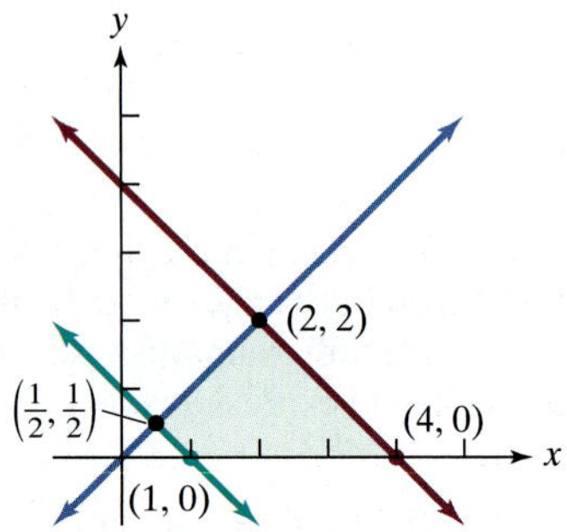

In Exercises 56–58, graph the region determined by the constraints. Then find the maximum value of the given objective function, subject to the constraints.

56. Objective Function $z = 2x + 3y$
Constraints $x \geq 0, y \geq 0$
$x + y \leq 8$
$3x + 2y \geq 6$

57. Objective Function $z = x + 4y$
Constraints $0 \leq x \leq 5, 0 \leq y \leq 7$
$x + y \geq 3$

58. Objective Function $z = 5x + 6y$
Constraints $x \geq 0, y \geq 0$
$y \leq x$
$2x + y \leq 12$
$2x + 3y \geq 6$

59. A paper manufacturing company converts wood pulp to writing paper and newsprint. The profit on a unit of writing paper is \$500 and the profit on a unit of newsprint is \$350.

a. Let x represent the number of units of writing paper produced daily. Let y represent the number of units of newsprint produced daily. Write the objective function that models total daily profit.

b. The manufacturer is bound by the following constraints:

1. Equipment in the factory allows for making at most 200 units of paper (writing paper and newsprint) in a day.

2. Regular customers require at least 10 units of writing paper and at least 80 units of newsprint daily.

Write a system of inequalities that models these constraints.

c. Graph the inequalities in part (b). Use only the first quadrant, because x and y must both be positive. (*Suggestion*: Let each unit along the x- and y-axes represent 20.)

d. Evaluate the objective profit function at each of the three vertices of the graphed region.

e. Complete the missing portions of this statement: The company will make the greatest profit by producing ___ units of writing paper and __ units of newsprint each day. The maximum daily profit is \$ _____.

60. A manufacturer of lightweight tents makes two models whose specifications are given in the following table.

	Cutting Time per Tent	Assembly Time per Tent
Model *A*	0.9 hour	0.8 hour
Model *B*	1.8 hours	1.2 hours

On a monthly basis, the manufacturer has no more than 864 hours of labor available in the cutting department and at most 672 hours in the assembly division. The profits come to \$25 per tent for model A and \$40 per tent for model B. How many of each should be manufactured monthly to maximize the profit?

Chapter 4 Test

In Exercises 1–2, express each interval as an inequality and graph the interval on a number line.

1. $[-3, 2)$

2. $(-\infty, -1]$

In Exercises 3–4, solve and graph the solution set on a number line. Express the solution set in both set-builder and interval notations.

3. $3(x + 4) \geq 5x - 12$

4. $\frac{x}{6} + \frac{1}{8} \leq \frac{x}{2} - \frac{3}{4}$

5. You are choosing between two telephone plans for local calls. Plan A charges \$25 per month for unlimited calls. Plan B has a monthly fee of \$13 with a charge of \$0.06 per local call. How many local telephone calls in a month make plan A the better deal?

6. Find the intersection: $\{2, 4, 6, 8, 10\} \cap \{4, 6, 12, 14\}$.

7. Find the union: $\{2, 4, 6, 8, 10\} \cup \{4, 6, 12, 14\}$.

In Exercises 8–12, solve each compound inequality. Except for the empty set, express the solution set in both set-builder and interval notations. Graph the solution set on a number line.

8. $2x + 4 < 2$ and $x - 3 > -5$

9. $x + 6 \geq 4$ and $2x + 3 \geq -2$

10. $2x - 3 < 5$ or $3x - 6 \leq 4$

11. $x + 3 \leq -1$ or $-4x + 3 < -5$

12. $-3 \le \frac{2x + 5}{3} < 6$

In Exercises 13–14, find the solution set for each equation.

13. $|5x + 3| = 7$

14. $|6x + 1| = |4x + 15|$

In Exercises 15–16, solve and graph the solution set on a number line. Express the solution set in both set-builder and interval notations.

15. $|2x - 1| < 7$

16. $|2x - 3| \ge 5$

17. The inequality $|T - 74| \le 8$ describes the range of monthly average temperature, T, in degrees Fahrenheit, for Miami, Florida. Solve the inequality and interpret the solution.

In Exercises 18–20, graph each inequality in a rectangular coordinate system.

18. $3x - 2y < 6$

19. $y \ge \frac{1}{2}x - 1$

20. $y \le -1$

In Exercises 21–23, graph the solution set of each system of inequalities.

21. $\begin{aligned} x + y &\ge 2 \\ x - y &\ge 4 \end{aligned}$

22. $\begin{aligned} 3x + y &\le 9 \\ 2x + 3y &\ge 6 \\ x \ge 0, y &\ge 0 \end{aligned}$

23. $-2 < x \le 4$

24. Find the maximum value of the objective function $z = 3x + 5y$ subject to the following constraints: $x \ge 0$, $y \ge 0$, $x + y \le 6$, $x \ge 2$.

25. A manufacturer makes two types of jet skis, regular and deluxe. The profit on a regular jet ski is \$200 and the profit on the deluxe model is \$250. To meet customer demand, the company must manufacture at least 50 regular jet skis per week and at least 75 deluxe models. To maintain high quality, the total number of both models of jet skis manufactured by the company should not exceed 150 per week. How many jet skis of each type should be manufactured per week to obtain maximum profit? What is the maximum weekly profit?

Cumulative Review Exercises (Chapters 1–4)

In Exercises 1–2, solve each equation.

1. $5(x + 1) + 2 = x - 3(2x + 1)$

2. $\frac{2(x + 6)}{3} = 1 + \frac{4x - 7}{3}$

3. Simplify: $\frac{-10x^2y^4}{15x^7y^{-3}}$.

4. If $f(x) = x^2 - 3x + 4$, find $f(-3)$ and $f(2a)$.

5. If $f(x) = 3x^2 - 4x + 1$ and $g(x) = x^2 - 5x - 1$, find $(f - g)(x)$ and $(f - g)(2)$.

6. Use function notation to write the equation of the line passing through $(2, 3)$ and perpendicular to the line whose equation is $y = 2x - 3$.

In Exercises 7–10, graph each equation or inequality in a rectangular coordinate system.

7. $f(x) = 2x + 1$

8. $y > 2x$

9. $2x - y \ge 6$

10. $f(x) = -1$

11. Solve the system:

$$\begin{aligned} 3x - y + z &= -15 \\ x + 2y - z &= 1. \\ 2x + 3y - 2z &= 0 \end{aligned}$$

12. Solve using matrices:

$$\begin{aligned} 2x - y &= -4 \\ x + 3y &= 5 \end{aligned}.$$

13. Evaluate: $\begin{vmatrix} 4 & 3 \\ -1 & -5 \end{vmatrix}$.

14. A motel with 60 rooms charges \$90 per night for rooms with kitchen facilities and \$80 per night for rooms without kitchen facilities. When all rooms are occupied, the nightly revenue is \$5260. How many rooms of each kind are there?

15. Which of the following are functions?

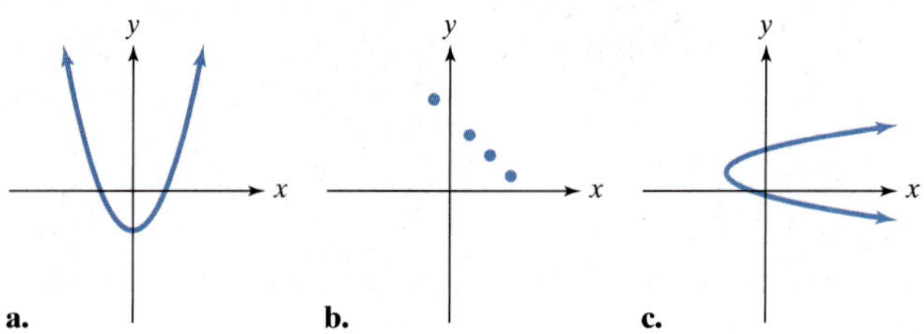

In Exercises 16–20, solve and graph the solution set on a number line. Express the solution set in both set-builder and interval notations.

16. $\frac{x}{4} - \frac{3}{4} - 1 \le \frac{x}{2}$

17. $2x + 5 \le 11$ and $-3x > 18$

18. $x - 4 \ge 1$ or $-3x + 1 \ge -5 - x$

19. $|2x + 3| \le 17$

20. $|3x - 8| > 7$

Coffee, our culture's most thoroughly approved drug, has become relevant and contemporary. Coffee bars, providing everything from personal computers so that customers can surf the Net to matchmaking services, have made coffee an important part of social gathering places. There is a wide variety of coffee offerings—from size, flavor, roast, preparation, and toppings—and plenty of gourmet shops to provide them.

Many of the 110 million daily coffee drinkers in the United States are surprised to learn that caffeine is a classically addictive drug. Many regular users cannot think clearly in the morning until they have had their first cup. Coffee is most notorious for causing insomnia, irritability, and nervousness. There is a function that models the percentage of coffee drinkers, $P(x)$, who are x years old who become irritable if they do not have coffee at their regular time:

$$P(x) = -0.05x^2 + 4.2x - 26.$$

The algebraic expression on the right side contains variables to powers that are whole numbers and is an example of a **polynomial**. Much of what we do in algebra involves polynomials and polynomial functions. In this chapter, we study polynomials, polynomial functions, and their applications.

Chapter 5

Polynomials, Polynomial Functions, and Factoring

▸ SECTION 5.1 Introduction to Polynomials and Polynomial Functions

Objectives

1. Use the vocabulary of polynomials.
2. Evaluate polynomial functions.
3. Determine end behavior.
4. Add polynomials.
5. Subtract polynomials.

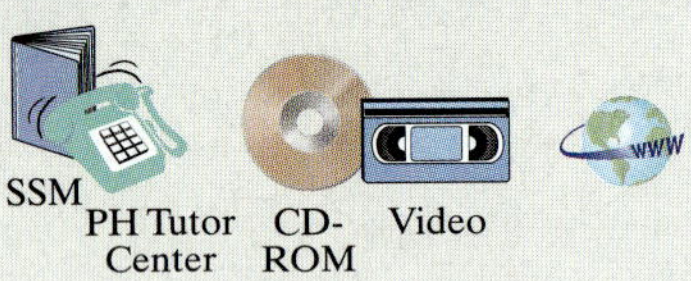

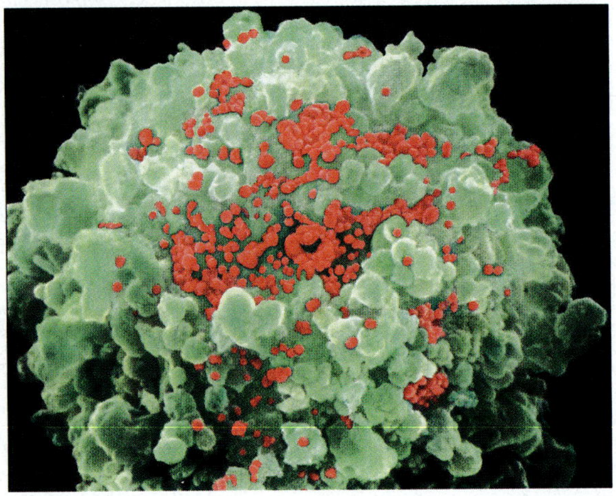

Magnified 6000 times, this color-scanned image shows a T-lymphocyte blood cell (green) infected with the HIV virus (red). Depletion of the number of T-cells causes destruction of the immune system.

In 1980, U.S. doctors diagnosed 41 cases of a rare form of cancer, Kaposi's sarcoma, that involved skin lesions, pneumonia, and severe immunological deficiencies. All cases involved gay men ranging in age from 26 to 51. By the end of 2000, approximately 775,000 Americans, straight and gay, male and female, old and young, were infected with the HIV virus.

Modeling AIDS-related data and making predictions about the epidemic's havoc is serious business. Changing circumstances and unforeseen events have resulted in models that are not particularly useful over long periods of time. For example, the function

$$f(x) = -143x^3 + 1810x^2 - 187x + 2331$$

models the number of AIDS cases diagnosed in the United States x years after 1983. The model was obtained using cases diagnosed from 1983 through 1991. Figure 5.1 shows the graph of f from 1983 through 1991 in a $[0, 8, 1]$ by $[0, 50{,}000, 5000]$ viewing rectangle. The function used to describe what was happening with new HIV infections over a limited period of time contains an algebraic expression that is an example of a *polynomial*. A **polynomial** is a single term or the sum of two or more terms containing variables in the numerator with whole number exponents. Functions containing polynomials are used in such diverse areas as science, business, medicine, psychology, and sociology. In this section, we present basic ideas about polynomials and polynomial functions. We will use our knowledge of combining like terms to find sums and differences of polynomials.

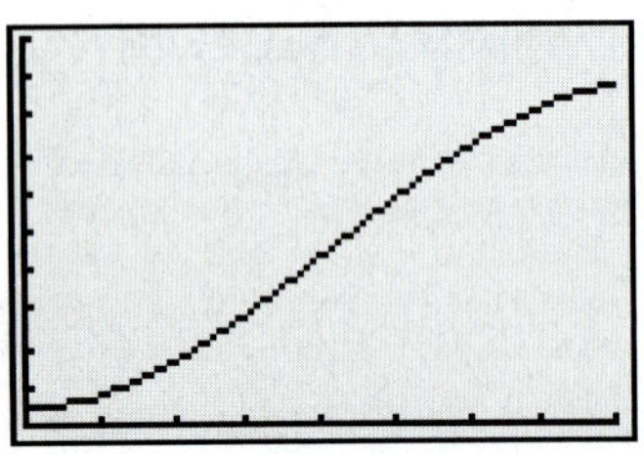

Figure 5.1 The graph of a function modeling the number of AIDS cases from 1983 through 1991

1 Use the vocabulary of polynomials.

How We Describe Polynomials

Consider the polynomial

$$-143x^3 + 1810x^2 - 187x + 2331.$$

We can express this polynomial as

$$-143x^3 + 1810x^2 + (-187x) + 2331.$$

This polynomial contains four terms. It is customary to write the terms in the order of descending powers of the variable. This is the **standard form** of a polynomial.

> **Study Tip**
>
> We can express 0 in many ways, including $0x$, $0x^2$, and $0x^3$. It is impossible to assign a single exponent on the variable. This is why 0 has no defined degree.

Some polynomials contain only one variable. Each term of such a polynomial in x is of the form ax^n. If $a \neq 0$, the **degree** of ax^n is n. For example, the degree of the term $-143x^3$ is 3.

The Degree of ax^n

If $a \neq 0$, the degree of ax^n is n. The degree of a nonzero constant is 0. The constant 0 has no defined degree.

Here is an example of a polynomial and the degree of each of its four terms:

$$-143x^3 + 1810x^2 - 187x + 2331.$$

degree 3 | degree 2 | degree 1 | degree of nonzero constant: 0

Notice that the exponent on x for the term $-187x$ is understood to be 1: $-187x^1$. For this reason, the degree of $-187x$ is 1. You can think of 2331 as $2331x^0$; thus, its degree is 0.

A polynomial with exactly one term is called a **monomial**. A **binomial** is a polynomial that has two terms, each with a different exponent. A **trinomial** is a polynomial with three terms, each with a different exponent. Polynomials with four or more terms have no special names.

Some polynomials contain two or more variables. Here is an example of a polynomial in two variables, x and y:

$$7x^2y^3 - 17x^4y^2 + xy - 6y^2 + 9.$$

A polynomial in two variables, x and y, contains the sum of one or more monomials of the form ax^ny^m. The constant a is the **coefficient**. The exponents, n and m, represent whole numbers. The **degree of the term** ax^ny^m is the sum of the exponents of the variables, $n + m$.

The **degree of a polynomial** is the highest of the degrees of its terms. If there is precisely one term of the highest degree, it is called the **leading term**. Its coefficient is called the **leading coefficient**.

EXAMPLE 1 Using the Vocabulary of Polynomials

Determine the coefficient of each term, the degree of each term, the degree of the polynomial, the leading term, and the leading coefficient of the polynomial

$$7x^2y^3 - 17x^4y^2 + xy - 6y^2 + 9.$$

Solution

Term	Coefficient	Degree (Sum of Exponents on the Variables)
$7x^2y^3$	7	$2 + 3 = 5$
$-17x^4y^2$	−17	$4 + 2 = 6$
xy	1	$1 + 1 = 2$
$-6y^2$	−6	$0 + 2 = 2$
9	9	$0 + 0 = 0$

Think of xy as $1x^1y^1$.

Think of $-6y^2$ as $-6x^0y^2$.

Think of 9 as $9x^0y^0$.

The degree of the polynomial is the highest degree of all its terms, which is 6. The leading term is the term of the highest degree, which is $-17x^4y^2$. Its coefficient, −17, is the leading coefficient. ■

✔ **CHECK POINT 1** Determine the coefficient of each term, the degree of each term, the degree of the polynomial, the leading term, and the leading coefficient of the polynomial

$$8x^4y^5 - 7x^3y^2 - x^2y - 5x + 11.$$

If a polynomial contains three or more variables, the degree of a term is the sum of the exponents of the variables. Here is an example of a polynomial in three variables, x, y, and z:

The coefficients are $\frac{1}{4}$, −2, 6, and 5.

$$\frac{1}{4}xy^2z^4 - 2xyz + 6x^2 + 5.$$

Degree: $1 + 2 + 4 = 7$ | Degree: $1 + 1 + 1 = 3$ | Degree: 2 | Degree: 0

The degree of this polynomial is the highest degree of all its terms, which is 7.

2 Evaluate polynomial functions.

Polynomial Functions

The expression $4x^3 - 5x^2 + 3$ is a polynomial. If we write

$$f(x) = 4x^3 - 5x^2 + 3$$

then we have a **polynomial function**. In a polynomial function, the expression that defines the function is a polynomial. How do we evaluate a polynomial function? Use substitution, just as we did to evaluate other functions in Chapter 2.

EXAMPLE 2 Evaluating a Polynomial Function

The bar graph in Figure 5.2 shows the cost of Medicare, in billions of dollars, projected through 2005. Using the regression feature of a graphing utility, these data can be modeled by the polynomial function

$$f(x) = 0.08x^3 - 0.06x^2 + 20.08x + 178.32$$

in which x represents the number of years after 1995 and $f(x)$ represents Medicare spending, in billions of dollars, for that year. Find $f(10)$ and describe what this means in practical terms.

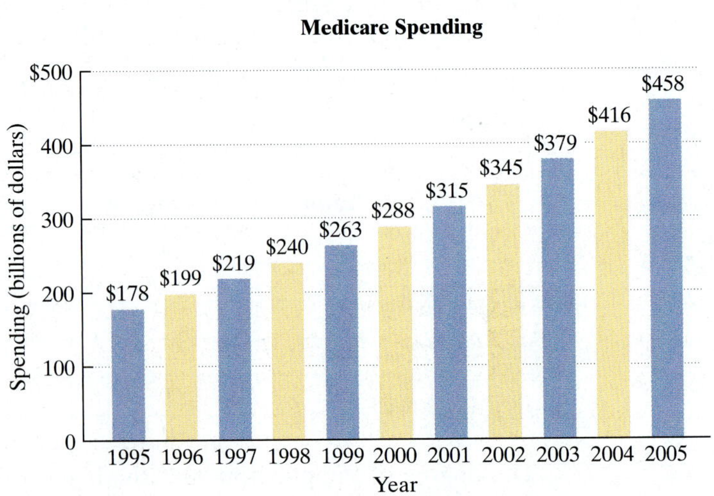

Figure 5.2 *Source*: Congressional Budget Office

Solution

$$f(x) = 0.08x^3 - 0.06x^2 + 20.08x + 178.32$$ This is the given function.

$$f(10) = 0.08(10)^3 - 0.06(10)^2 + 20.08(10) + 178.32$$ To find f of 10, replace x by 10.

$$= 0.08(1000) - 0.06(100) + 20.08(10) + 178.32$$ Evaluate exponential expressions.

$$= 80 - 6 + 200.8 + 178.32$$ Multiply.

$$= 453.12$$ Simplify.

Thus, $f(10) = 453.12$. This means that 10 years after 1995, in 2005, Medicare spending is projected to be \$453.12 billion. ■

✔ **CHECK POINT 2** For the polynomial function

$$f(x) = 4x^3 - 3x^2 - 5x + 6,$$

find $f(2)$.

Smooth, Continuous Graphs

Polynomial functions of degree 2 or higher have graphs that are *smooth* and *continuous*. By **smooth**, we mean that the graph contains only rounded curves with no sharp corners. By **continuous**, we mean that the graph has no breaks and can be drawn without lifting your pencil from the rectangular coordinate system. These ideas are illustrated in Figure 5.3.

Graphs of Polynomial Functions **Not Graphs of Polynomial Functions**

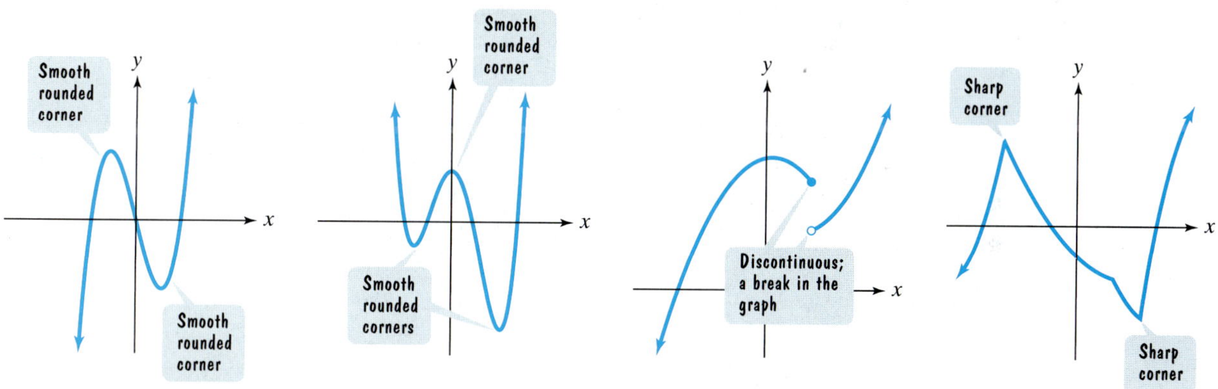

Figure 5.3 Recognizing graphs of polynomial functions

3 Determine end behavior.

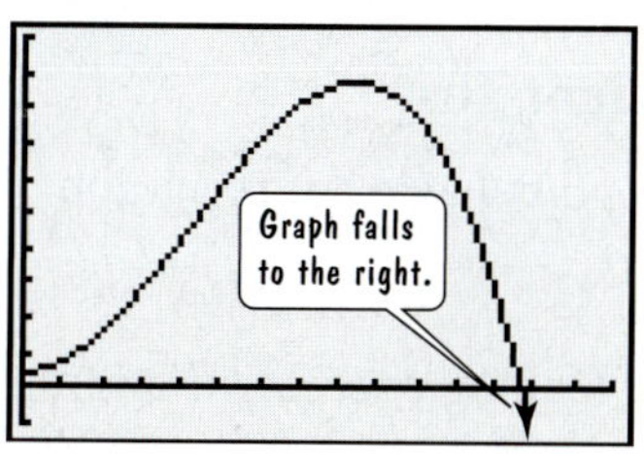

Figure 5.4 By extending the viewing rectangle, we see that y is eventually negative and the function no longer models the number of AIDS cases.

End Behavior of Polynomial Functions

Figure 5.4 shows the graph of the function

$$f(x) = -143x^3 + 1810x^2 - 187x + 2331,$$

which models U.S. AIDS cases from 1983 through 1991. Look what happens to the graph when we extend the year up through 1998 with a [0, 15, 1] by [−5000, 50,000, 5000] viewing rectangle. By year 13 (1996), the values of y are negative and the function no longer models AIDS cases. We've added an arrow to the graph at the far right to emphasize that it continues to decrease without bound. It is this far-right *end behavior* of the graph that makes it inappropriate for modeling AIDS cases into the future.

The behavior of a graph of a function to the far left or the far right is called its **end behavior**. Although the graph of a polynomial function may have intervals where it increases or decreases, the graph will eventually rise or fall without bound as it moves far to the left or far to the right.

How can you determine whether the graph of a polynomial function goes up or down at each end? The end behavior depends upon the leading term. In particular, the sign of the leading coefficient and the degree of the polynomial reveal the graph's end behavior. In terms of end behavior, only the leading term—that is, the term of the highest degree counts, summarized by the **Leading Coefficient Test**.

The Leading Coefficient Test

As x increases or decreases without bound, the graph of a polynomial function eventually rises or falls. In particular,

1. For odd-degree polynomials:

If the leading coefficient is positive, the graph falls to the left and rises to the right.

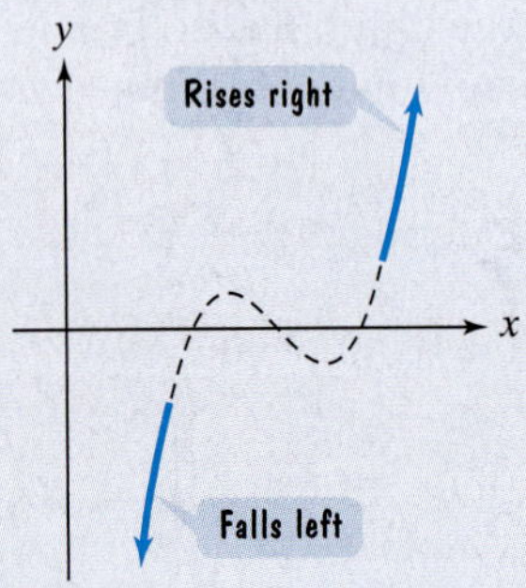

Figure 5.5(a)

If the leading coefficient is negative, the graph rises to the left and falls to the right.

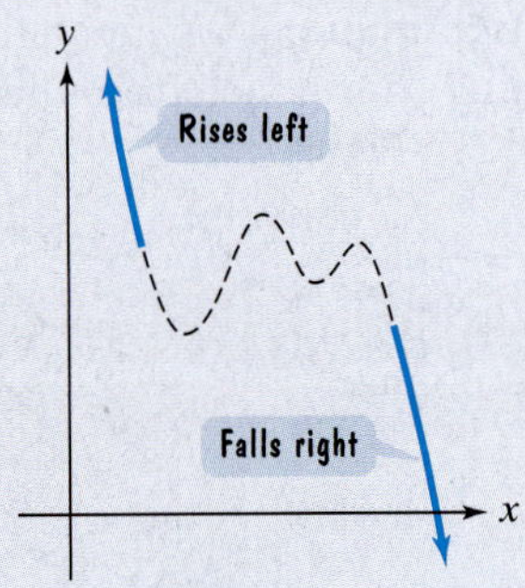

Figure 5.5(b)

2. For even-degree polynomials:

If the leading coefficient is positive, the graph rises to the left and to the right.

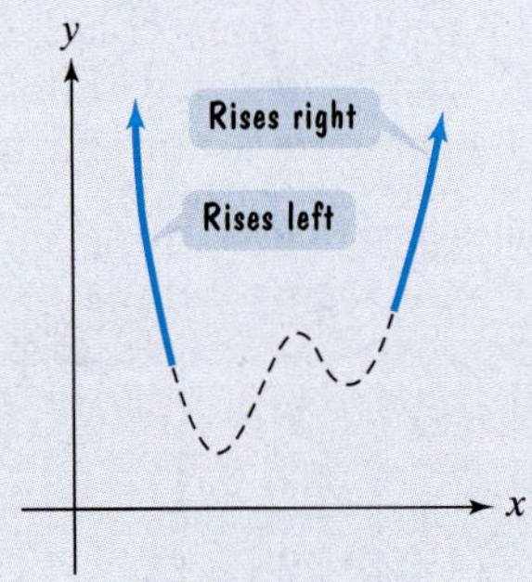

Figure 5.5(c)

If the leading coefficient is negative, the graph falls to the left and to the right.

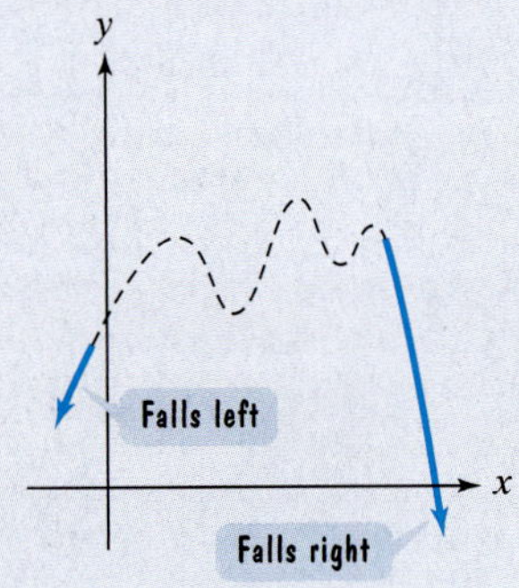

Figure 5.5(d)

Study Tip

Odd-degree polynomial functions have graphs with opposite behavior at each end. Even-degree polynomial functions have graphs with the same behavior at each end.

EXAMPLE 3 Using the Leading Coefficient Test

Use the Leading Coefficient Test to determine the end behavior of the graph of

$$f(x) = x^3 + 3x^2 - x - 3.$$

Solution We begin by identifying the sign of the leading coefficient and the degree of the polynomial.

$$f(x) = x^3 + 3x^2 - x - 3$$

The leading coefficient, 1, is positive. The degree of the polynomial, 3, is odd.

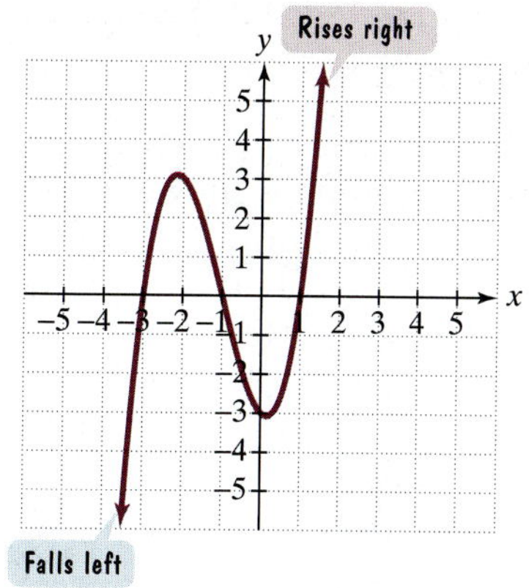

Figure 5.6 The graph of $f(x) = x^3 + 3x^2 - x - 3$

The degree of the function f is 3, which is odd. Odd-degree polynomial functions have graphs with opposite behavior at each end. The leading coefficient, 1, is positive. Thus, the graph falls to the left and rises to the right. This behavior is shown in Figure 5.5(a) in the box on page 290 that describes the Leading Coefficient Test. The graph of f is shown in Figure 5.6. ■

✔ **CHECK POINT 3** Use the Leading Coefficient Test to determine the end behavior of the graph of $f(x) = x^4 - 4x^2$.

EXAMPLE 4 Using the Leading Coefficient Test

Use end behavior to explain why

$$f(x) = -143x^3 + 1810x^2 - 187x + 2331$$

is only an appropriate model for AIDS cases for a limited time period.

Solution We begin by identifying the sign of the leading coefficient and the degree of the polynomial.

$$f(x) = -143x^3 + 1810x^2 - 187x + 2331$$

The leading coefficient, −143, is negative. The degree of the polynomial, 3, is odd.

The degree of f is 3, which is odd. Odd-degree polynomial functions have graphs with opposite behavior at each end. The leading coefficient, −143, is negative. Thus, the graph rises to the left and falls to the right. This behavior is shown in Figure 5.5(b) in the box on page 290 that describes the Leading Coefficient Test. The fact that the graph falls to the right indicates that at some point the number of AIDS cases will be negative, an impossibility. If a function has a graph that decreases without bound over time, it will not be capable of modeling nonnegative phenomena over long time periods. ■

✔ **CHECK POINT 4** The polynomial function

$$f(x) = -0.27x^3 + 9.2x^2 - 102.9x + 400$$

models the ratio of students to computers in U.S. public schools x years after 1980. Use end behavior to determine whether this function could be an appropriate model for computers in the classroom well into the twenty-first century. Explain your answer.

If you use a graphing utility to graph a polynomial function, it is important to select a viewing rectangle that accurately reveals the graph's end behavior. If the viewing rectangle is too small, it may not accurately show end behavior.

EXAMPLE 5 Using the Leading Coefficient Test

The graph of $f(x) = -x^4 + 8x^3 + 4x^2 + 2$ was obtained with a graphing utility using a $[-8, 8, 1]$ by $[-10, 10, 1]$ viewing rectangle. The graph is shown in Figure 5.7(a). Does the graph show the end behavior of the function?

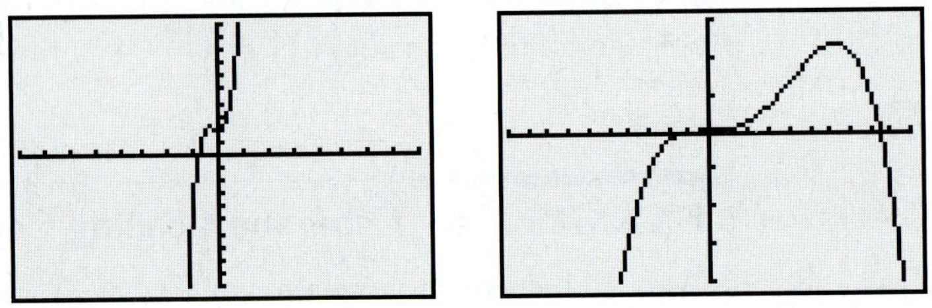

Figure 5.7 **(a)** $[-8, 8, 1]$ by $[-10, 10, 1]$ **(b)** $[-10, 10, 1]$ by $[-1000, 750, 250]$

Solution We begin by identifying the sign of the leading coefficient and the degree of the polynomial.

$$f(x) = -x^4 + 8x^3 + 4x^2 + 2$$

The leading coefficient, −1, is negative. The degree of the polynomial, 4, is even.

The degree of f is 4, which is even. Even-degree polynomial functions have graphs with the same behavior at each end. The leading coefficient, −1, is negative. Thus, the graph should fall to the left and fall to the right. (See Figure 5.5(d) in the box on page 290 on the Leading Coefficient Test.) The graph in Figure 5.7(a) is falling to the left, but it is not falling to the right. Therefore, the graph is not complete enough to show end behavior. A more complete graph of the function is shown in a larger viewing rectangle in Figure 5.7(b). ■

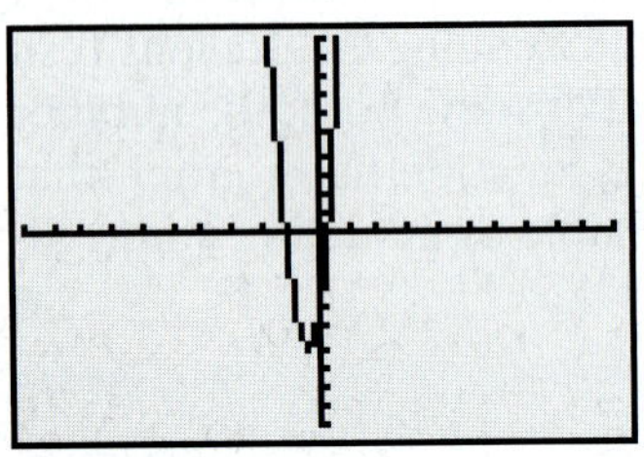

Figure 5.8

✔ **CHECK POINT 5** The graph of $f(x) = x^3 + 13x^2 + 10x - 4$ is shown in a standard viewing rectangle in Figure 5.8. Use the Leading Coefficient Test to determine whether the graph shows the end behavior of the function. Explain your answer.

4 Add polynomials.

Adding Polynomials

Polynomials are added by combining like terms. Here are two examples that illustrate the use of the distributive property in adding monomials and combining like terms:

$$-9x^3 + 13x^3 = (-9 + 13)x^3 = 4x^3$$

Add coefficients and keep the same variable factor(s).

$$-7x^3y^2 + 4x^3y^2 = (-7 + 4)x^3y^2 = -3x^3y^2.$$

EXAMPLE 6 Adding Polynomials

Add: $(-6x^3 + 5x^2 + 4) + (2x^3 + 7x^2 - 10)$.

Solution

$$(-6x^3 + 5x^2 + 4) + (2x^3 + 7x^2 - 10)$$

$$= -6x^3 + 5x^2 + 4 + 2x^3 + 7x^2 - 10$$ Remove the parentheses.

$$= \underbrace{-6x^3 + 2x^3}_{-4x^3} + \underbrace{5x^2 + 7x^2}_{+12x^2} + \underbrace{4 - 10}_{-6}$$ Rearrange the terms so that like terms are adjacent.

$$= -4x^3 + 12x^2 - 6$$ Combine like terms. ■

✔ CHECK POINT 6 Add: $(-7x^3 + 4x^2 + 3) + (4x^3 + 6x^2 - 13)$.

EXAMPLE 7 Adding Polynomials

Add: $(5x^3y - 4x^2y - 7y) + (2x^3y + 6x^2y - 4y - 5)$.

Solution

$$(5x^3y - 4x^2y - 7y) + (2x^3y + 6x^2y - 4y - 5)$$ The given problem involves adding polynomials in two variables.

$$= 5x^3y - 4x^2y - 7y + 2x^3y + 6x^2y - 4y - 5$$ Remove the parentheses.

$$= \underbrace{5x^3y + 2x^3y}_{7x^3y} \underbrace{- 4x^2y + 6x^2y}_{+2x^2y} \underbrace{- 7y - 4y}_{-11y} \underbrace{- 5}_{-5}$$ Rearrange the terms so that like terms are adjacent.

$$= 7x^3y + 2x^2y - 11y - 5$$ Combine like terms. ■

Polynomials can be added by arranging like terms in columns. Then combine like terms, column by column. Here's the solution to Example 7 using columns and a vertical format:

$$\begin{array}{r} 5x^3y - 4x^2y - \ \ 7y \\ 2x^3y + 6x^2y - \ \ 4y - 5 \\ \hline 7x^3y + 2x^2y - 11y - 5 \end{array}$$

✔ CHECK POINT 7 Add: $(7xy^3 - 5xy^2 - 3y) + (2xy^3 + 8xy^2 - 12y - 9)$.

5 Subtract polynomials.

Subtracting Polynomials

We subtract real numbers by adding the opposite, or additive inverse, of the number being subtracted. For example,

$$8 - 3 = 8 + (-3) = 5.$$

Similarly, we subtract one polynomial from another by adding the opposite of the polynomial being subtracted.

Subtracting Polynomials

To subtract two polynomials, change the sign of every term of the second polynomial. Add this result to the first polynomial.

Study Tip

You can also subtract polynomials using a vertical format. Here's the solution to Example 8 using a vertical format:

$$\begin{array}{r} 7x^3 - 8x^2 + 9x - 6 \\ -(2x^3 - 6x^2 - 3x + 9) \\ \hline \end{array}$$

$$\begin{array}{r} 7x^3 - 8x^2 + 9x - 6 \\ + -2x^3 + 6x^2 + 3x - 9. \\ \hline 5x^3 - 2x^2 + 12x - 15 \end{array}$$

EXAMPLE 8 Subtracting Polynomials

Subtract: $(7x^3 - 8x^2 + 9x - 6) - (2x^3 - 6x^2 - 3x + 9)$.

Solution

$(7x^3 - 8x^2 + 9x - 6) - (2x^3 - 6x^2 - 3x + 9)$

$= (7x^3 - 8x^2 + 9x - 6) + (-2x^3 + 6x^2 + 3x - 9)$ Change the sign of each term of the second polynomial and add the two polynomials.

$= \underbrace{7x^3 - 2x^3}\ \underbrace{- 8x^2 + 6x^2}\ \underbrace{+ 9x + 3x}\ \underbrace{- 6 - 9}$ Rearrange terms.

$= \quad 5x^3 \quad -2x^2 \quad +12x \quad -15$ Combine like terms. ■

✔ **CHECK POINT 8** Subtract: $(14x^3 - 5x^2 + x - 9) - (4x^3 - 3x^2 - 7x + 1)$.

Study Tip

Be careful of the order in Example 9. For example, subtracting 2 from 5 is equivalent to $5 - 2$. In general, subtracting B from A translates as $A - B$. The order of the resulting problem is not the same as the order in English.

EXAMPLE 9 Subtracting Polynomials

Subtract $-2x^5y^2 - 3x^3y + 7$ from $3x^5y^2 - 4x^3y - 3$.

Solution

$(3x^5y^2 - 4x^3y - 3) - (-2x^5y^2 - 3x^3y + 7)$

$= 3x^5y^2 - 4x^3y - 3 + 2x^5y^2 + 3x^3y - 7$ Remove the parentheses, changing the sign of every term after the subtraction sign.

$= \underbrace{3x^5y^2 + 2x^5y^2}\ \underbrace{- 4x^3y + 3x^3y}\ \underbrace{- 3 - 7}$ Rearrange terms.

$= \quad 5x^5y^2 \quad -x^3y \quad -10$ Combine like terms. ■

✔ **CHECK POINT 9** Subtract $-7x^2y^5 - 4xy^3 + 2$ from $6x^2y^5 - 2xy^3 - 8$.

EXERCISE SET 5.1

Practice Exercises

In Exercises 1–10, determine the coefficient of each term, the degree of each term, the degree of the polynomial, the leading term, and the leading coefficient of the polynomial.

1. $-x^4 + x^2$

2. $x^3 - 4x^2$

3. $5x^3 + 7x^2 - x + 9$

4. $11x^3 - 6x^2 + x + 3$

5. $3x^2 - 7x^4 - x + 6$

6. $2x^2 - 9x^4 - x + 5$

7. $x^3y^2 - 5x^2y^7 + 6y^2 - 3$

8. $12x^4y - 5x^3y^7 - x^2 + 4$

9. $x^5 + 3x^2y^4 + 7xy + 9x - 2$

10. $3x^6 + 4x^4y^4 - x^3y + 4x^2 - 5$

In Exercises 11–20, let

$$f(x) = x^2 - 5x + 6 \quad \text{and} \quad g(x) = 2x^3 - x^2 + 4x - 1.$$

Find the indicated function values.

11. $f(3)$ **12.** $f(4)$

13. $f(-1)$ **14.** $f(-2)$

15. $g(3)$ **16.** $g(2)$

17. $g(-2)$ **18.** $g(-3)$

19. $g(0)$ **20.** $f(0)$

In Exercises 21–24, identify which graphs are not those of polynomial functions.

21.

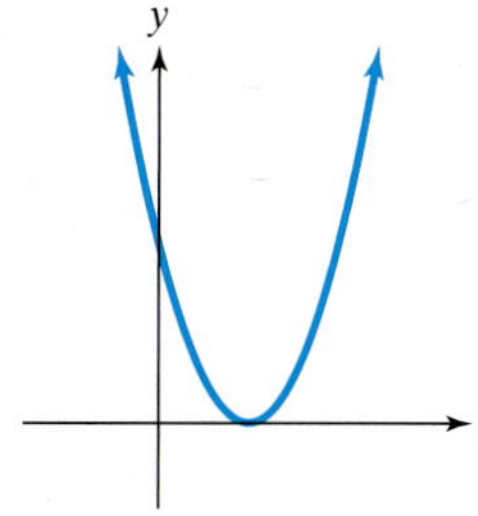

22.

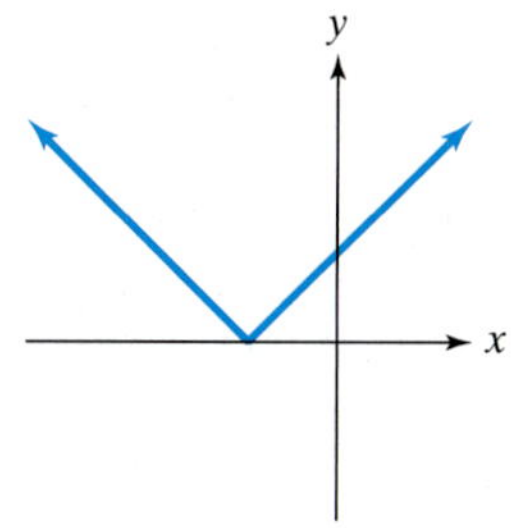

23.

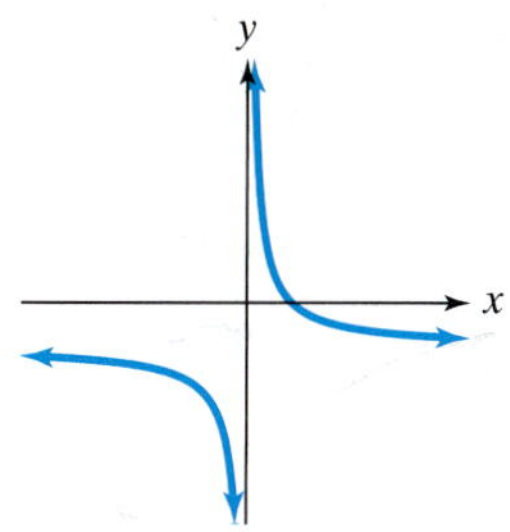

24.

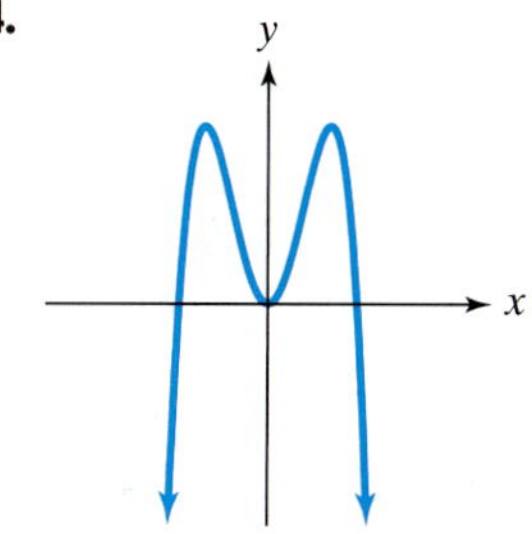

In Exercises 25–28, use the Leading Coefficient Test to determine the end behavior of the graph of the given polynomial function. Then use this end behavior to match the polynomial function with its graph. [The graphs are labeled (a) through (d).]

25. $f(x) = -x^4 + x^2$

26. $f(x) = x^3 - 4x^2$

27. $f(x) = x^2 - 6x + 9$

28. $f(x) = -x^3 - x^2 + 5x - 3$

(a)

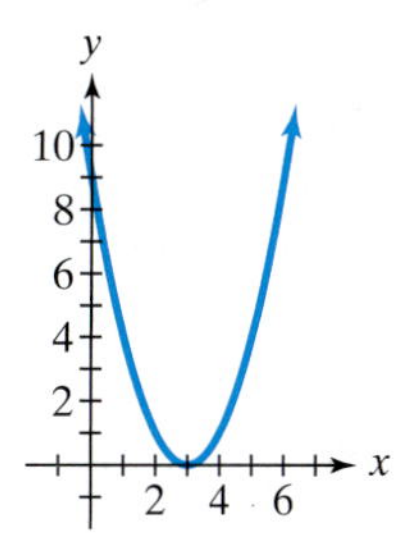

(b)

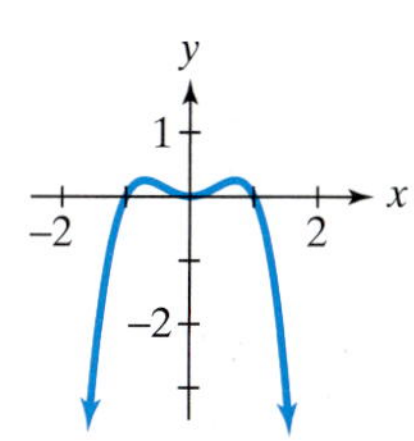

(c)

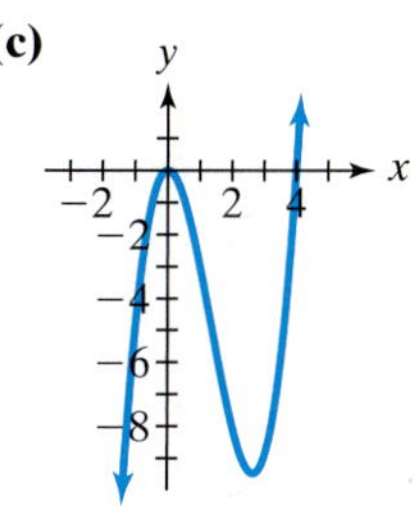

(d)

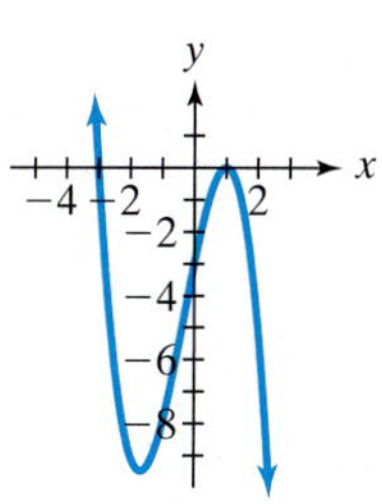

In Exercises 29–38, add the polynomials.

29. $(-6x^3 + 5x^2 - 8x + 9) + (17x^3 + 2x^2 - 4x - 13)$

30. $(-7x^3 + 6x^2 - 11x + 13) + (19x^3 - 11x^2 + 7x - 17)$

31. $\left(\frac{2}{5}x^4 + \frac{2}{3}x^3 + \frac{5}{8}x^2 + 7\right) + \left(-\frac{4}{5}x^4 + \frac{1}{3}x^3 - \frac{1}{4}x^2 - 7\right)$

32. $\left(\frac{1}{5}x^4 + \frac{1}{3}x^3 + \frac{3}{8}x^2 + 6\right) + \left(-\frac{3}{5}x^4 + \frac{2}{3}x^3 - \frac{1}{2}x^2 - 6\right)$

33. $(7x^2y - 5xy) + (2x^2y - xy)$

34. $(-4x^2y + xy) + (7x^2y + 8xy)$

35. $(5x^2y + 9xy + 12) + (-3x^2y + 6xy + 3)$

36. $(8x^2y + 12xy + 14) + (-2x^2y + 7xy + 4)$

37. $(9x^4y^2 - 6x^2y^2 + 3xy) + (-18x^4y^2 - 5x^2y - xy)$

38. $(10x^4y^2 - 3x^2y^2 + 2xy) + (-16x^4y^2 - 4x^2y - xy)$

In Exercises 39–46, subtract the polynomials.

39. $(17x^3 - 5x^2 + 4x - 3) - (5x^3 - 9x^2 - 8x + 11)$

40. $(18x^3 - 2x^2 - 7x + 8) - (9x^3 - 6x^2 - 5x + 7)$

41. $(13y^5 + 9y^4 - 5y^2 + 3y + 6) - (-9y^5 - 7y^3 + 8y^2 + 11)$

42. $(12y^5 + 7y^4 - 3y^2 + 6y + 7) - (-10y^5 - 8y^3 + 3y^2 + 14)$

43. $(x^3 + 7xy - 5y^2) - (6x^3 - xy + 4y^2)$

44. $(x^4 - 7xy - 5y^3) - (6x^4 - 3xy + 4y^3)$

45. $(3x^4y^2 + 5x^3y - 3y) - (2x^4y^2 - 3x^3y - 4y + 6x)$

46. $(5x^4y^2 + 6x^3y - 7y) - (3x^4y^2 - 5x^3y - 6y + 8x)$

47. Subtract $-5a^2b^4 - 8ab^2 - ab$ from $3a^2b^4 - 5ab^2 + 7ab$.

48. Subtract $-7a^2b^4 - 8ab^2 - ab$ from $13a^2b^4 - 17ab^2 + ab$.

49. Subtract $-4x^3 - x^2y + xy^2 + 3y^3$ from $x^3 + 2x^2y - y^3$.

50. Subtract $-6x^3 + x^2y - xy^2 + 2y^3$ from $x^3 + 2xy^2 - y^3$.

Application Exercises

The polynomial function

$$f(x) = 0.036x^2 - 2.8x + 58.14$$

models the number of deaths per year per thousand people, $f(x)$, for people who are x years old, $40 \le x \le 60$. Use this function to solve Exercises 51–52.

51. Find and interpret $f(50)$.

52. Find and interpret $f(40)$.

The graph of the polynomial function in Exercises 51–52 is shown in the figure. Use the graph to solve Exercises 53–56.

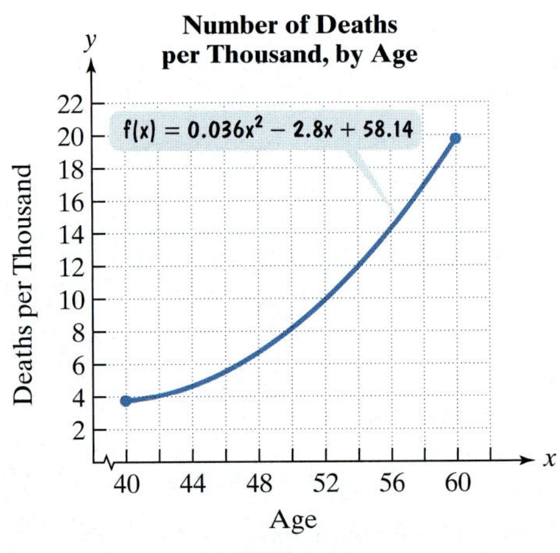

Source: U.S. Department of Health and Human Services

53. Identify your answer from Exercise 51 as a point on the graph.

54. Identify your answer from Exercise 52 as a point on the graph.

55. At which age, to the nearest year, do approximately 16 people per thousand die each year?

56. At which age, to the nearest year, do approximately 6 people per thousand die each year?

The polynomial function

$$f(x) = -0.001618x^4 + 0.077326x^3 - 1.2367x^2 + 11.460x + 2.914$$

models the age in human years, $f(x)$, of a dog that is x years old, where $x \geq 1$. The coefficients make it difficult to use this function when computing by hand. However, a graph of the function makes approximations possible. Use the graph to solve Exercises 57–60. Round all answers to the nearest year.

Dog's Age in Human Years

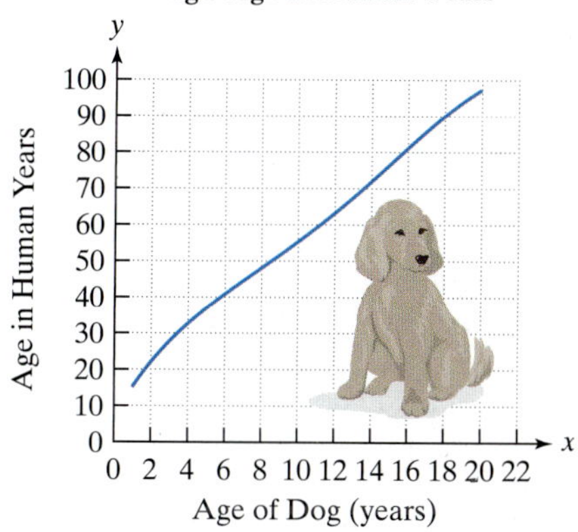

Source: U.C. Davis

57. If your dog is 10 years old, estimate the equivalent age in human years.

58. If your dog is 12 years old, estimate the equivalent age in human years.

59. If you are 20, what is the equivalent age for dogs?

60. If you are 90, what is the equivalent age for dogs?

61. The common cold is caused by a rhinovirus. After x days of invasion by the viral particles, the number of particles in our bodies, $f(x)$, in billions, can be modeled by the polynomial function

$$f(x) = -0.75x^4 + 3x^3 + 5.$$

Use the Leading Coefficient Test to determine the graph's end behavior to the right. What does this mean about the number of viral particles in our bodies over time?

62. The polynomial function

$$f(x) = -0.87x^3 + 0.35x^2 + 81.62x + 7684.94$$

models the number of thefts, $f(x)$, in thousands, in the United States x years after 1987. Will this function be useful in modeling the number of thefts over an extended period of time? Explain your answer.

63. A herd of 100 elk is introduced to a small island. The number of elk, $f(x)$, after x years is modeled by the polynomial function

$$f(x) = -x^4 + 21x^2 + 100.$$

Use the Leading Coefficient Test to determine the graph's end behavior to the right. What does this mean about what will eventually happen to the elk population?

Writing in Mathematics

64. What is a polynomial?

65. Explain how to determine the degree of each term of a polynomial.

66. Explain how to determine the degree of a polynomial.

67. Explain how to determine the leading coefficient of a polynomial.

68. What is a polynomial function?

69. What do we mean when we describe the graph of a polynomial function as smooth and continuous?

70. What is meant by the end behavior of a polynomial function?

71. Explain how to use the Leading Coefficient Test to determine the end behavior of a polynomial function.

72. Why is a polynomial function of degree 3 with a negative leading coefficient not appropriate for modeling nonnegative real-world phenomena over a long period of time?

73. Explain how to add polynomials.

74. Explain how to subtract polynomials.

75. In a favorable habitat and without natural predators, a population of reindeer is introduced to an island preserve. The reindeer population t years after their introduction is modeled by the polynomial function $f(t) = -0.125t^5 + 3.125t^4 + 4000$. Discuss the growth and decline of the reindeer population. Describe the factors that might contribute to this population model.

Technology Exercises

Write a polynomial function that imitates the end behavior of each graph in Exercises 76–79. The dashed portions of the graphs indicate that you should focus only on imitating the left and right end behavior of the graph and can be flexible about what occurs between the left and right ends. Then use your graphing utility to graph the polynomial function and verify that you imitated the end behavior shown in the given graph.

76.

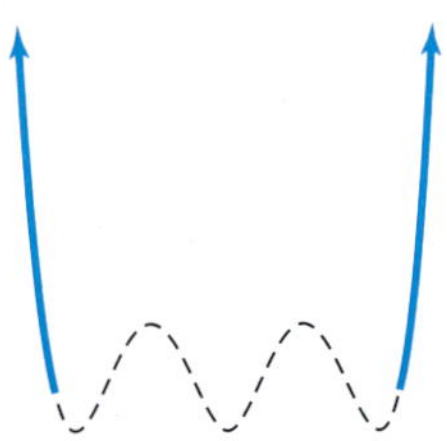

77.

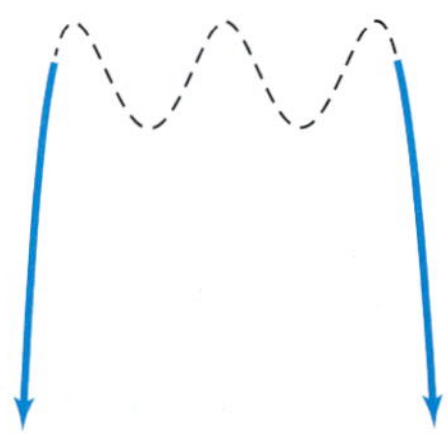

78.

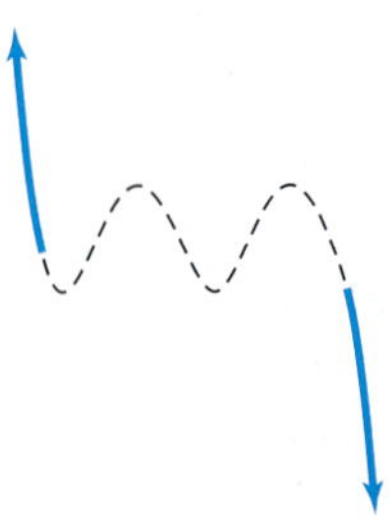

79.

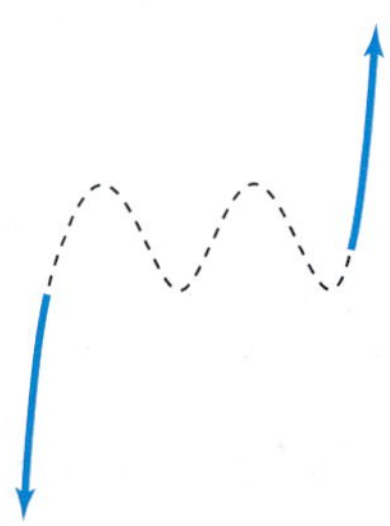

In Exercises 80–83, use a graphing utility with a viewing rectangle large enough to show end behavior to graph each polynomial function.

80. $f(x) = x^3 + 13x^2 + 10x - 4$

81. $f(x) = -2x^3 + 6x^2 + 3x - 1$

82. $f(x) = -x^4 + 8x^3 + 4x^2 + 2$

83. $f(x) = -x^5 + 5x^4 - 6x^3 + 2x + 20$

For Exercises 84–85, use a graphing utility to graph f and g in the same viewing rectangle. Then use the ZOOM OUT *feature to show that f and g have identical end behavior.*

84. $f(x) = x^3 - 6x + 1, \quad g(x) = x^3$

85. $f(x) = -x^4 + 2x^3 - 6x, \quad g(x) = -x^4$

Critical Thinking Exercises

86. Which one of the following is true?

a. $4x^3 + 7x^2 - 5x + \frac{2}{x}$ is a polynomial containing four terms.

b. If two polynomials of degree 2 are added, their sum must be a polynomial of degree 2.

c. $(x^2 - 7x) - (x^2 - 4x) = -11x$ for any values of x.

d. All terms of a polynomial are monomials.

In Exercises 87–88, perform the indicated operations. Assume that exponents represent whole numbers.

87. $(x^{2n} - 3x^n + 5) + (4x^{2n} - 3x^n - 4) - (2x^{2n} - 5x^n - 3)$

88. $(y^{3n} - 7y^{2n} + 3) - (-3y^{3n} - 2y^{2n} - 1) + (6y^{3n} - y^{2n} + 1)$

89. From what polynomial must $4x^2 + 2x - 3$ be subtracted to obtain $5x^2 - 5x + 8$?

Review Exercises

90. Solve: $9(x - 1) = 1 + 3(x - 2)$. (Section 1.4, Example 3)

91. Graph: $2x - 3y < -6$. (Section 4.4, Example 1)

92. Write the point-slope and slope-intercept equations of a line passing through the point $(-2, 5)$ and parallel to the line whose equation is $3x - y = 9$. (Section 2.4, Example 4)

▸ SECTION 5.2 Multiplication of Polynomials

Objectives

1. Multiply monomials.
2. Multiply a monomial and a polynomial.
3. Multiply polynomials when neither is a monomial.
4. Use FOIL in polynomial multiplication.
5. Square binomials.
6. Multiply the sum and difference of two terms.
7. Find the product of functions.
8. Use polynomial multiplication to evaluate functions.

SSM PH Tutor Center CD-ROM Video

The rising and setting of the sun suggest the obvious: Things change over time. Calculus is the study of rates of change, allowing the motion of the rising sun to be measured by "freezing the frame" at one instant in time. Calculus reveals a function's rate of change at any "frozen" instant. In this section, you will encounter an algebraic expression that plays an important role in this process. You will be able to work with this expression by learning how polynomials are multiplied.

Multiplying Monomials

To multiply monomials, begin by multiplying the coefficients. Then multiply the variables. Use the product rule for exponents to multiply the variables: Retain the variable and add the exponents.

1 Multiply monomials.

EXAMPLE 1 Multiplying Monomials

Multiply:

a. $(5x^3y^4)(-6x^7y^8)$ **b.** $(4x^3y^2z^5)(2x^5y^2z^4)$.

Solution

a. $(5x^3y^4)(-6x^7y^8) = 5(-6)x^3 \cdot x^7 \cdot y^4 \cdot y^8$ Rearrange factors. This step is usually done mentally.

$= -30x^{3+7}y^{4+8}$ Multiply coefficients and add exponents.

$= -30x^{10}y^{12}$ Simplify.

b. $(4x^3y^2z^5)(2x^5y^2z^4) = 4 \cdot 2 \cdot x^3 \cdot x^5 \cdot y^2 \cdot y^2 \cdot z^5 \cdot z^4$ Rearrange factors.

$= 8x^{3+5}y^{2+2}z^{5+4}$ Multiply coefficients and add exponents.

$= 8x^8y^4z^9$ Simplify. ■

✔ **CHECK POINT 1** Multiply:

a. $(6x^5y^7)(-3x^2y^4)$

b. $(10x^4y^3z^6)(3x^6y^3z^2)$.

2 Multiply a monomial and a polynomial.

Multiplying a Monomial and a Polynomial That Is Not a Monomial

We use the distributive property to multiply a monomial and a polynomial that is not a monomial. For example,

$$3x^2(2x^3 + 5x) = 3x^2 \cdot 2x^3 + 3x^2 \cdot 5x = 3 \cdot 2x^{2+3} + 3 \cdot 5x^{2+1} = 6x^5 + 15x^3.$$

Monomial | Binomial | Multiply coefficients and add exponents.

To multiply a monomial and a polynomial, multiply each term of the polynomial by the monomial. Once the monomial factor is distributed, we multiply the resulting monomials using the procedure shown in Example 1.

EXAMPLE 2 Multiplying a Monomial and a Trinomial

Multiply:

a. $4x^3(6x^5 - 2x^2 + 3)$ **b.** $5x^3y^4(2x^7y - 6x^4y^3 - 3)$.

Solution

a. $4x^3(6x^5 - 2x^2 + 3) = 4x^3 \cdot 6x^5 - 4x^3 \cdot 2x^2 + 4x^3 \cdot 3$ Use the distributive property.

$= 24x^8 - 8x^5 + 12x^3$ Multiply coefficients and add exponents.

b. $5x^3y^4(2x^7y - 6x^4y^3 - 3)$

$= 5x^3y^4 \cdot 2x^7y - 5x^3y^4 \cdot 6x^4y^3 - 5x^3y^4 \cdot 3$ Use the distributive property.

$= 10x^{10}y^5 - 30x^7y^7 - 15x^3y^4$ Multiply coefficients and add exponents. ■

✔ **CHECK POINT 2** Multiply:

a. $6x^4(2x^5 - 3x^2 + 4)$

b. $2x^4y^3(5xy^6 - 4x^3y^4 - 5)$.

3 Multiply polynomials when neither is a monomial.

Multiplying Polynomials When Neither Is a Monomial

How do we multiply two polynomials if neither is a monomial? For example, consider

$$(2x + 3)(x^2 + 4x + 5).$$

binomial | trinomial

One way to perform this multiplication is to distribute $2x$ throughout the trinomial

$$2x(x^2 + 4x + 5)$$

and 3 throughout the trinomial

$$3(x^2 + 4x + 5).$$

Then combine the like terms that result. In general, the product of two polynomials is the polynomial obtained by multiplying each term of one polynomial by each term of the other polynomial and then combining like terms.

EXAMPLE 3 Multiplying a Binomial and a Trinomial

Multiply: $(2x + 3)(x^2 + 4x + 5)$.

Solution

$(2x + 3)(x^2 + 4x + 5)$

$= 2x(x^2 + 4x + 5) + 3(x^2 + 4x + 5)$ Multiply the trinomial by each term of the binomial.

$= 2x \cdot x^2 + 2x \cdot 4x + 2x \cdot 5 + 3x^2 + 3 \cdot 4x + 3 \cdot 5$ Use the distributive property.

$= 2x^3 + 8x^2 + 10x + 3x^2 + 12x + 15$ Multiply monomials: Multiply coefficients and add exponents.

$= 2x^3 + 11x^2 + 22x + 15$ Combine like terms: $8x^2 + 3x^2 = 11x^2$ and $10x + 12x = 22x$. ■

✔ **CHECK POINT 3** Multiply: $(5x + 2)(x^2 - 4x + 3)$.

Another method for solving Example 3 is to use a vertical format similar to that used for multiplying whole numbers.

$$\begin{array}{rrrrl} & x^2 + & 4x + & 5 & \\ & & 2x + & 3 & \\ \hline & 3x^2 + & 12x + & 15 & \leftarrow 3(x^2 + 4x + 5) \\ 2x^3 + & 8x^2 + & 10x & & \leftarrow 2x(x^2 + 4x + 5) \\ \hline 2x^3 + & 11x^2 + & 22x + & 15 & \text{Combine like terms.} \end{array}$$

Write like terms in the same column.

EXAMPLE 4 Multiplying a Binomial and a Trinomial

Multiply: $(2x^2y + 3y)(5x^4y - 4x^2y + y)$.

Solution

$(2x^2y + 3y)(5x^4y - 4x^2y + y)$

$= 2x^2y(5x^4y - 4x^2y + y) + 3y(5x^4y - 4x^2y + y)$ Multiply the trinomial by each term of the binomial.

$= 2x^2y \cdot 5x^4y - 2x^2y \cdot 4x^2y + 2x^2y \cdot y + 3y \cdot 5x^4y - 3y \cdot 4x^2y + 3y \cdot y$
Use the distributive property

$= 10x^6y^2 - 8x^4y^2 + 2x^2y^2 + 15x^4y^2 - 12x^2y^2 + 3y^2$ Multiply coefficients and add exponents.

$= 10x^6y^2 + 7x^4y^2 - 10x^2y^2 + 3y^2$

Combine like terms: $-8x^4y^2 + 15x^4y^2 = 7x^4y^2$ and $2x^2y^2 - 12x^2y^2 = -10x^2y^2$. ■

✔ **CHECK POINT 4** Multiply: $(4xy^2 + 2y)(3xy^4 - 2xy^2 + y)$.

4 Use FOIL in polynomial multiplication.

The Product of Two Binomials: FOIL

Frequently we need to find the product of two binomials. We can use a method called FOIL, which is based on the distributive property, to do so. For example, we can find the product of the binomials $3x + 2$ and $4x + 5$ as follows:

$$(3x + 2)(4x + 5) = 3x(4x + 5) + 2(4x + 5)$$

First, distribute $3x$ over $4x + 5$. Then distribute 2.

$$= 3x(4x) + 3x(5) + 2(4x) + 2(5)$$

$$= 12x^2 + 15x + 8x + 10.$$

Two binomials can be quickly multiplied using the FOIL method, in which *F* represents the product of the **first** terms in each binomial, *O* represents the product of the **outside** terms, *I* represents the product of the two **inside** terms, and *L* represents the product of the **last**, or second, terms in each binomial.

first, last, inside, outside

F O I L

$$(3x + 2)(4x + 5) = 12x^2 + 15x + 8x + 10$$

$$= 12x^2 + 23x + 10$$

Combine like terms.

In general, here's how to use the FOIL method to find the product of $ax + b$ and $cx + d$:

Using the FOIL Method to Multiply Binomials

first, last, inside, outside

$$(ax + b)(cx + d) = ax \cdot cx + ax \cdot d + b \cdot cx + b \cdot d$$

Product of First terms; Product of Outside terms; Product of Inside terms; Product of Last terms

ENRICHMENT ESSAY

Visualizing Polynomial Multiplication

Rectangles and their areas make it possible to visualize certain polynomial multiplications. What does the FOIL product

$$(x + 3)(x + 2) = x^2 + 5x + 6$$

look like? Use a rectangle with dimensions $x + 2$ and $x + 3$.

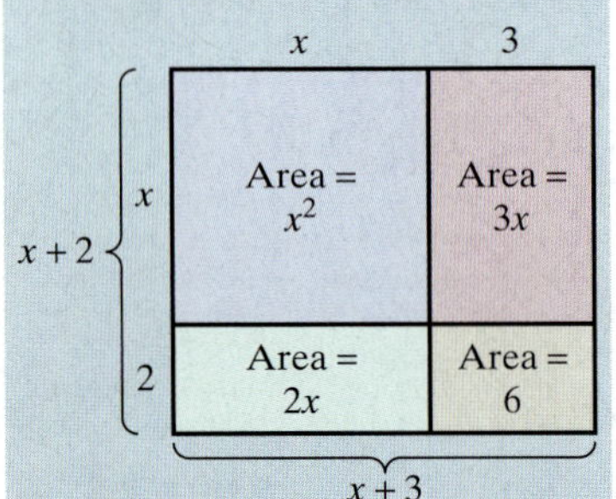

The area of the large rectangle is

$$(x + 3)(x + 2).$$

The sum of areas of the four smaller rectangles is

$$x^2 + 3x + 2x + 6$$
$$= x^2 + 5x + 6.$$

The area of the large rectangle equals the sum of the areas of the four smaller rectangles:

$$(x + 3)(x + 2) = x^2 + 5x + 6.$$

EXAMPLE 5 Using the FOIL Method

Multiply:

a. $(x + 3)(x + 2)$ **b.** $(3x + 5y)(x - 2y)$ **c.** $(5x^3 - 6)(4x^3 - x)$.

Solution

first, last, inside, outside

F O I L

a. $(x + 3)(x + 2) = x \cdot x + x \cdot 2 + 3 \cdot x + 3 \cdot 2$

$$= x^2 + 2x + 3x + 6$$

$$= x^2 + 5x + 6$$

Combine like terms.

first, last, inside, outside

b. $(3x + 5y)(x - 2y) = 3x \cdot x + 3x(-2y) + 5y \cdot x + 5y(-2y)$
F O I L

$= 3x^2 - 6xy + 5xy - 10y^2$

$= 3x^2 - xy - 10y^2$ Combine like terms.

first, last, inside, outside

c. $(5x^3 - 6)(4x^3 - x) = 5x^3 \cdot 4x^3 + 5x^3(-x) + (-6)(4x^3) + (-6)(-x)$
F O I L

$= 20x^6 - 5x^4 - 24x^3 + 6x$ There are no like terms to combine. ■

✔ CHECK POINT 5 Multiply:

a. $(x + 5)(x + 3)$ **b.** $(7x + 4y)(2x - y)$ **c.** $(4x^3 - 5)(x^3 - 3x)$.

5 Square binomials.

The Square of a Binomial

Let us find $(A + B)^2$, the square of a binomial sum. To do so, we begin with the FOIL method and look for a general rule.

$$(A + B)^2 = (A + B)(A + B) = A \cdot A + A \cdot B + A \cdot B + B \cdot B$$
F O I L

$$= A^2 + 2AB + B^2$$

This result implies the following rule, which is often called a **special-product formula:**

Study Tip

Caution! The square of a sum is *not* the sum of the squares.

$(A + B)^2 \neq A^2 + B^2$

The middle term 2AB is missing.

$(x + 3)^2 \neq x^2 + 9$

Incorrect

Show that $(x + 3)^2$ and $x^2 + 9$ are not equal by substituting 5 for x in each expression and simplifying.

The Square of a Binomial Sum

$(A + B)^2$	$=$	A^2	$+$	$2AB$	$+$	B^2
The square of a binomial sum	is	first term squared	plus	2 times the product of the terms	plus	last term squared.

EXAMPLE 6 Finding the Square of a Binomial Sum

Multiply:

a. $(x + 5)^2$ **b.** $(3x + 2y)^2$.

Solution Use the special-product formula shown.

$(A + B)^2 = A^2 + 2AB + B^2$

	(First Term)^2	+	2 · Product of the Terms	+	(Last Term)^2	= Product
a. $(x + 5)^2 =$	x^2	+	$2 \cdot x \cdot 5$	+	5^2	$= x^2 + 10x + 25$
b. $(3x + 2y)^2 =$	$(3x)^2$	+	$2 \cdot 3x \cdot 2y$	+	$(2y)^2$	$= 9x^2 + 12xy + 4y^2$

■

✔ CHECK POINT 6 Multiply:

a. $(x + 10)^2$ **b.** $(4x + 5y)^2$.

Using the FOIL method on $(A - B)^2$, the square of a binomial difference, we obtain the following rule:

The Square of a Binomial Difference

$$(A - B)^2 = A^2 - 2AB + B^2$$

The square of a binomial difference | is | first term squared | minus | 2 times the product of the terms | plus | last term squared.

EXAMPLE 7 Finding the Square of a Binomial Difference

Multiply:

a. $(x - 8)^2$ **b.** $\left(\frac{1}{2}x - 4y^3\right)^2$.

Solution Use the special-product formula shown.

$$(A - B)^2 = A^2 - 2AB + B^2$$

	(First Term)^2	−	2 · Product of the Terms	+	(Last Term)^2	= Product
a. $(x - 8)^2 =$	x^2	−	$2 \cdot x \cdot 8$	+	8^2	$= x^2 - 16x + 64$
b. $\left(\frac{1}{2}x - 4y^3\right)^2 =$	$\left(\frac{1}{2}x\right)^2$	−	$2 \cdot \frac{1}{2}x \cdot 4y^3$	+	$(4y^3)^2$	$= \frac{1}{4}x^2 - 4xy^3 + 16y^6$

■

✔ **CHECK POINT 7** Multiply:

a. $(x - 5)^2$ **b.** $(2x - 6y^4)^2$.

6 Multiply the sum and difference of two terms.

Multiplying the Sum and Difference of Two Terms

We can use the FOIL method to multiply $A + B$ and $A - B$ as follows:

$$(A + B)(A - B) = \underset{F}{A^2} - \underset{O}{AB} + \underset{I}{AB} - \underset{L}{B^2} = A^2 - B^2.$$

Notice that the outside and inside products have a sum of 0 and the terms cancel. The FOIL multiplication provides us with a quick rule for multiplying the sum and difference of two terms, which is another example of a special-product formula.

The Product of the Sum and Difference of Two Terms

$$(A + B)(A - B) = A^2 - B^2$$

The product of the sum and the difference of the same two terms | is | the square of the first term minus the square of the second term.

EXAMPLE 8 Finding the Product of the Sum and Difference of Two Terms

Multiply:

a. $(x + 8)(x - 8)$ **b.** $(9x + 5y)(9x - 5y)$

c. $(6a^2b - 3b)(6a^2b + 3b)$.

Solution Use the special-product formula shown.

$$(A + B)(A - B) = A^2 - B^2$$

First term squared − Second term squared = Product

a. $(x + 8)(x - 8) = x^2 - 8^2 = x^2 - 64$

b. $(9x + 5y)(9x - 5y) = (9x)^2 - (5y)^2 = 81x^2 - 25y^2$

c. $(6a^2b - 3b)(6a^2b + 3b) = (6a^2b)^2 - (3b)^2 = 36a^4b^2 - 9b^2$ ■

✔ **CHECK POINT 8** Multiply:

a. $(x + 3)(x - 3)$

b. $(5x + 7y)(5x - 7y)$

c. $(5ab^2 - 4a)(5ab^2 + 4a)$.

Special products can sometimes be used to find the products of certain trinomials, as illustrated in Example 9.

EXAMPLE 9 Using the Special Products

Multiply: **a.** $(7x + 5 + 4y)(7x + 5 - 4y)$ **b.** $(3x + y + 1)^2$.

Solution

a. By grouping the first two terms within each of the parentheses, we can find the product using the form for the sum and difference of two terms.

$$(A + B) \cdot (A - B) = A^2 - B^2$$

$$[(7x + 5) + 4y] \cdot [(7x + 5) - 4y] = (7x + 5)^2 - (4y)^2$$

$$= (7x)^2 + 2 \cdot 7x \cdot 5 + 5^2 - (4y)^2$$

$$= 49x^2 + 70x + 25 - 16y^2$$

b. We can group the terms so that the formula for the square of a binomial can be applied.

$$(A + B)^2 = A^2 + 2 \cdot A \cdot B + B^2$$

$$[(3x + y) + 1]^2 = [(3x + y)^2 + 2 \cdot (3x + y) \cdot 1 + 1^2]$$

$$= 9x^2 + 6xy + y^2 + 6x + 2y + 1$$ ■

✔ **CHECK POINT 9** Multiply:

a. $(3x + 2 + 5y)(3x + 2 - 5y)$

b. $(2x + y + 3)^2$.

ENRICHMENT ESSAY

Visualizing a Special-Product Formula

The formula

$$(A + B)^2 = A^2 + 2AB + B^2$$

can be interpreted geometrically.

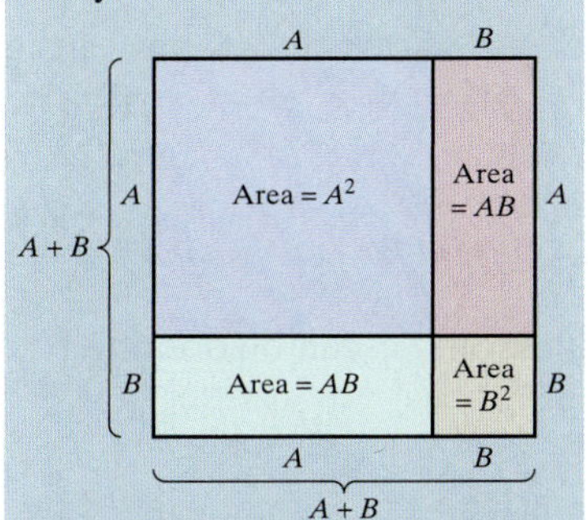

The area of the large rectangle is

$$(A + B)(A + B) = (A + B)^2.$$

The sum of the areas of the four smaller rectangles is

$$A^2 + AB + AB + B^2$$
$$= A^2 + 2AB + B^2.$$

The area of the large rectangle equals the sum of the areas of the four smaller rectangles:

$$(A + B)^2 = A^2 + 2AB + B^2.$$

7 Find the product of functions.

Multiplication of Polynomial Functions

In Chapter 2, we developed an algebra of functions, defining the product of functions f and g as follows:

$$(fg)(x) = f(x) \cdot g(x).$$

Now that we know how to multiply polynomials, we can find the product of functions.

EXAMPLE 10 Using the Algebra of Functions

Let $f(x) = x - 5$ and $g(x) = x - 2$. Find:

a. $(fg)(x)$ **b.** $(fg)(1)$.

Solution

a. $(fg)(x) = f(x) \cdot g(x)$ This is the definition of the product function, fg.

$= (x - 5)(x - 2)$ Substitute the given functions.

F O I L

$= x^2 - 2x - 5x + 10$ Multiply by the FOIL method.

$= x^2 - 7x + 10$ Combine like terms.

Thus,

$$(fg)(x) = x^2 - 7x + 10.$$

b. We use the product function to find $(fg)(1)$—that is, the function fg at 1. Replace x by 1.

$$(fg)(1) = 1^2 - 7 \cdot 1 + 10 = 4$$ ■

Curious about what all this looks like? Figure 5.9 shows the graphs of the three functions $f(x) = x - 5$, $g(x) = x - 2$, and $(fg)(x) = x^2 - 7x + 10$. The three points indicated by the voice balloons show that

$$f(1) = -4,\ g(1) = -1, \text{ and } (fg)(1) = 4.$$

The y-coordinate of every point on the product function, fg, can be obtained by multiplying the y-coordinates of functions f and g.

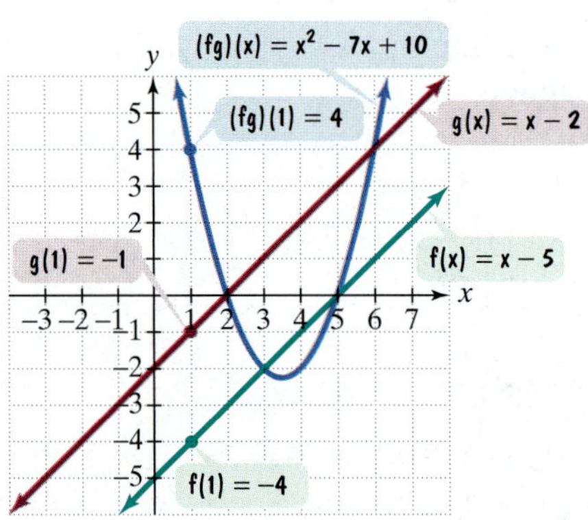

Figure 5.9

In Chapter 3, we introduced quadratic functions, functions of the form

$$f(x) = ax^2 + bx + c \text{ or } y = ax^2 + bx + c, a \neq 0.$$

Functions f and g in Figure 5.9 are linear functions of the form $f(x) = mx + b$. By contrast, the product function is a quadratic function of the form $f(x) = ax^2 + bx + c$. Notice the cuplike shape of the graph of the quadratic function.

Discover for Yourself

Describe how the graphs of linear functions f and g in Figure 5.9 exhibit the end behavior expected of a polynomial function. If the slope of a linear function is negative, how does this affect its end behavior?

Are linear functions and quadratic functions also polynomial functions? The answer is yes. They are defined in terms of first-degree and second-degree polynomials. Thus, a linear function is also a first-degree polynomial function. A quadratic function is also a second-degree polynomial function.

Can you see that the graph of the product function in Figure 5.9 exhibits the end behavior expected of a polynomial function? The degree is 2, which is even. Even-degree polynomial functions have graphs with the same behavior at each end. The leading coefficient, 1, is positive. Thus, the graph should rise to the left and rise to the right, and it does.

✔ **CHECK POINT 10** Let $f(x) = x - 3$ and $g(x) = x - 7$. Find:

a. $(fg)(x)$

b. $(fg)(2)$.

8 Use polynomial multiplication to evaluate functions.

If you are given a function, f, calculus can reveal how it is changing at any instant in time. The algebraic expression

$$\frac{f(a + h) - f(a)}{h}$$

plays an important role in this process. Our work with polynomial multiplication can be used to evaluate the numerator of this expression.

EXAMPLE 11 Using Polynomial Multiplication to Evaluate Functions

Given $f(x) = x^2 - 7x + 3$, find and simplify:

a. $f(a + 4)$ **b.** $f(a + h) - f(a)$.

Solution

a. We find $f(a + 4)$, read "f at a plus 4," by replacing x by $a + 4$ each time that x appears in the polynomial.

$$f(x) = x^2 - 7x + 3$$

Replace x by $a + 4$. Replace x by $a + 4$. Replace x by $a + 4$. Copy the 3. There is no x in this term.

$$f(a + 4) = (a + 4)^2 - 7(a + 4) + 3$$

$$= a^2 + 8a + 16 - 7a - 28 + 3 \quad \text{Multiply as indicated.}$$

$$= a^2 + a - 9 \quad \text{Combine like terms: } 8a - 7a = a \text{ and } 16 - 28 + 3 = -9.$$

b. To find $f(a + h) - f(a)$, we first replace each occurrence of x by $a + h$ and then replace each occurrence of x by a. Then we multiply and simplify.

$$= \overbrace{(a+h)^2 - 7(a+h) + 3}^{f(a+h)} \overset{-}{\downarrow} \overbrace{(a^2 - 7a + 3)}^{f(a)}$$

Use $f(x) = x^2 - 7x + 3$, replacing x by $a + h$ and a, respectively.

$$= a^2 + 2ah + h^2 - 7a - 7h + 3 - a^2 + 7a - 3$$

Multiply as indicated.

$$= 2ah + h^2 - 7h$$

Combine like terms. Observe that $a^2 - a^2$, $-7a + 7a$, and $3 - 3$ all have zero sums. ■

CHECK POINT 11 Given $f(x) = x^2 - 5x + 4$, find and simplify:

a. $f(a + 3)$

b. $f(a + h) - f(a)$.

EXERCISE SET 5.2

Practice Exercises

In Exercises 1–6, multiply the monomials.

1. $(3x^2)(5x^4)$

2. $(4x^2)(6x^4)$

3. $(3x^2y^4)(5xy^7)$

4. $(6x^4y^2)(3x^7y)$

5. $(-3xy^2z^5)(2xy^7z^4)$

6. $(11x^2yz^4)(-3xy^5z^6)$

In Exercises 7–18, multiply the monomial and the polynomial.

7. $4x^2(3x + 2)$

8. $5x^2(6x + 7)$

9. $2y(y^2 - 5y)$

10. $3y(y^2 - 4y)$

11. $5x^3(2x^5 - 4x^2 + 9)$

12. $6x^3(3x^5 - 5x^2 + 7)$

13. $4xy(7x + 3y)$

14. $5xy(8x + 3y)$

15. $3ab^2(6a^2b^3 + 5ab)$

16. $5ab^2(10a^2b^3 + 7ab)$

17. $-4x^2y(3x^4y^2 - 7xy^3 + 6)$

18. $-3x^2y(10x^2y^4 - 2xy^3 + 7)$

In Exercises 19–30, find each product using either a horizontal or a vertical format.

19. $(x - 3)(x^2 + 2x + 5)$

20. $(x + 4)(x^2 - 5x + 8)$

21. $(x - 1)(x^2 + x + 1)$

22. $(x - 2)(x^2 + 2x + 4)$

23. $(a - b)(a^2 + ab + b^2)$

24. $(a + b)(a^2 - ab + b^2)$

25. $(x^2 + 2x - 1)(x^2 + 3x - 4)$

26. $(x^2 - 2x + 3)(x^2 + x + 1)$

27. $(x - y)(x^2 - 3xy + y^2)$

28. $(x - y)(x^2 - 4xy + y^2)$

29. $(xy + 2)(x^2y^2 - 2xy + 4)$

30. $(xy + 3)(x^2y^2 - 2xy + 5)$

In Exercises 31–48, use the FOIL method to multiply the binomials.

31. $(x + 4)(x + 7)$

32. $(x + 5)(x + 8)$

33. $(y + 5)(y - 6)$

34. $(y + 5)(y - 8)$

35. $(5x + 3)(2x + 1)$

36. $(4x + 3)(5x + 1)$

37. $(3y - 4)(2y - 1)$

38. $(5y - 2)(3y - 1)$

39. $(3x - 2)(5x - 4)$

40. $(2x - 3)(4x - 5)$

41. $(x - 3y)(2x + 7y)$

42. $(3x - y)(2x + 5y)$

43. $(7xy + 1)(2xy - 3)$

44. $(3xy - 1)(5xy + 2)$

45. $(x - 4)(x^2 - 5)$

46. $(x - 5)(x^2 - 3)$

47. $(8x^3 + 3)(x^2 - 5)$

48. $(7x^3 + 5)(x^2 - 2)$

In Exercises 49–60, multiply using one of the rules for the square of a binomial.

49. $(x + 3)^2$
50. $(x + 4)^2$
51. $(y - 5)^2$
52. $(y - 6)^2$
53. $(2x + y)^2$
54. $(4x + y)^2$
55. $(5x - 3y)^2$
56. $(3x - 4y)^2$
57. $(2x^2 + 3y)^2$
58. $(4x^2 + 5y)^2$
59. $(4xy^2 - xy)^2$
60. $(5xy^2 - xy)^2$

In Exercises 61–72, multiply using the rule for the product of the sum and difference of two terms.

61. $(x + 4)(x - 4)$
62. $(x + 5)(x - 5)$
63. $(5x + 3)(5x - 3)$
64. $(3x + 2)(3x - 2)$
65. $(4x + 7y)(4x - 7y)$
66. $(8x + 7y)(8x - 7y)$
67. $(y^3 + 2)(y^3 - 2)$
68. $(y^3 + 3)(y^3 - 3)$
69. $(1 - y^5)(1 + y^5)$
70. $(2 - y^5)(2 + y^5)$
71. $(7xy^2 - 10y)(7xy^2 + 10y)$
72. $(3xy^2 - 4y)(3xy^2 + 4y)$

In Exercises 73–83, find each product.

73. $[(2x + 3) + 4y][(2x + 3) - 4y]$

74. $[(3x + 2) + 5y][(3x + 2) - 5y]$

75. $(x + y + 3)(x + y - 3)$
76. $(x + y + 4)(x + y - 4)$
77. $(5x + 7y - 2)(5x + 7y + 2)$
78. $(7x + 5y - 2)(7x + 5y + 2)$
79. $[5y + (2x + 3)][5y - (2x + 3)]$

80. $[8y + (3x + 2)][8y - (3x + 2)]$

81. $(x + y + 1)^2$
82. $(x + y + 2)^2$
83. $(x + 1)(x - 1)(x^2 + 1)$
84. $(x + 2)(x - 2)(x^2 + 4)$
85. Let $f(x) = x - 2$ and $g(x) = x + 6$. Find each of the following.
a. $(fg)(x)$
b. $(fg)(-1)$
c. $(fg)(0)$

86. Let $f(x) = x - 4$ and $g(x) = x + 10$. Find each of the following.
a. $(fg)(x)$
b. $(fg)(-1)$
c. $(fg)(0)$

87. Let $f(x) = x - 3$ and $g(x) = x^2 + 3x + 9$. Find each of the following.
a. $(fg)(x)$
b. $(fg)(-2)$
c. $(fg)(0)$

88. Let $f(x) = x + 3$ and $g(x) = x^2 - 3x + 9$. Find each of the following.
a. $(fg)(x)$
b. $(fg)(-2)$
c. $(fg)(0)$

In Exercises 89–92, find each of the following and simplify:

a. $f(a + 2)$ **b.** $f(a + h) - f(a)$.

89. $f(x) = x^2 - 3x + 7$

90. $f(x) = x^2 - 4x + 9$

91. $f(x) = 3x^2 + 2x - 1$

92. $f(x) = 4x^2 + 5x - 1$

Application Exercises

In Exercises 93–94, find the area of the large rectangle in two ways:

a. *Find the sum of the areas of the four smaller rectangles.*
b. *Multiply the length and the width of the large rectangle using the FOIL method. Compare this product with your answer to part (a).*

93.

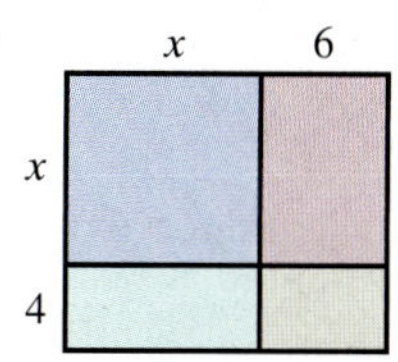

94.

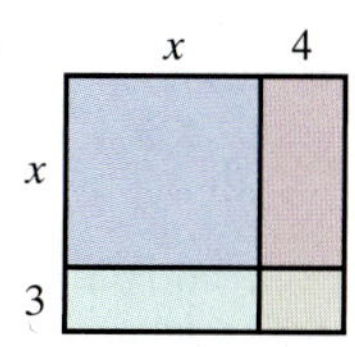

In Exercises 95–96, express each polynomial in standard form—that is, in descending powers of x.

a. *Write a polynomial that represents the area of the large rectangle.*

b. *Write a polynomial that represents the area of the small, unshaded rectangle.*

c. *Write a polynomial that represents the area of the shaded region.*

95.

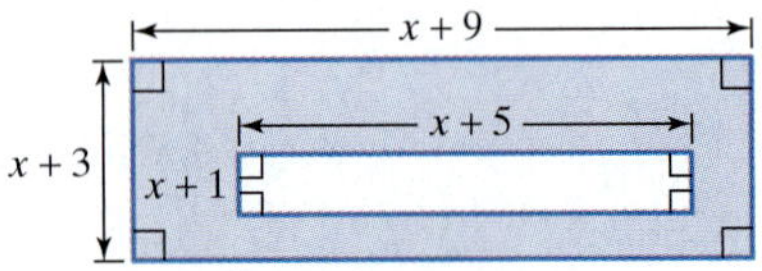

96.

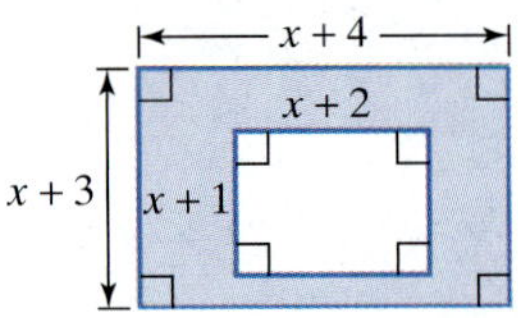

In Exercises 97–98, express each polynomial in standard form.

a. *Write a polynomial that represents the area of of the rectangular base of the open box.*

b. *Write a polynomial that represents the volume of the open box.*

97.

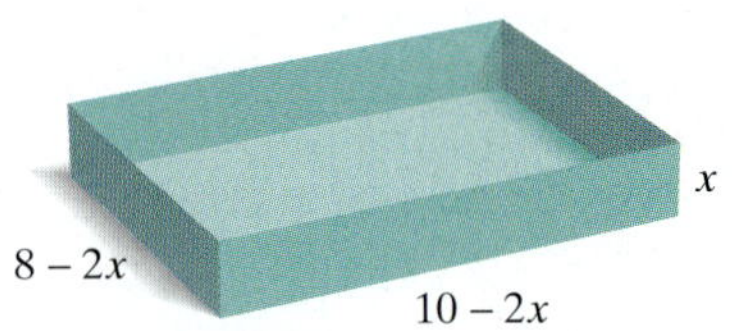

98.

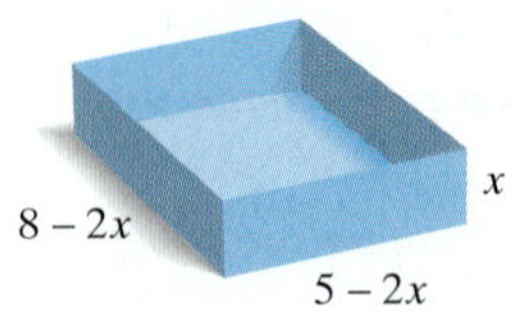

Writing in Mathematics

99. Explain how to multiply monomials. Give an example.

100. Explain how to multiply a monomial and a polynomial that is not a monomial. Give an example.

101. Explain how to multiply a binomial and a trinomial.

102. What is the FOIL method and when is it used? Give an example of the method.

103. Explain how to square a binomial sum. Give an example.

104. Explain how to square a binomial difference. Give an example.

105. Explain how to find the product of the sum and difference of two terms. Give an example with your explanation.

106. How can the graph of function fg be obtained from the graphs of functions f and g?

107. Explain how to find $f(a + h) - f(a)$ for a given function f.

Technology Exercises

In Exercises 108–111, use a graphing utility to graph the functions y_1 and y_2. Select a viewing rectangle that is large enough to show the end behavior of y_2. What can you conclude? Verify your conclusions using polynomial multiplication.

108. $y_1 = (x - 2)^2$
$y_2 = x^2 - 4x + 4$

109. $y_1 = (x - 4)(x^2 - 3x + 2)$
$y_2 = x^3 - 7x^2 + 14x - 8$

110. $y_1 = (x - 1)(x^2 + x + 1)$
$y_2 = x^3 - 1$

111. $y_1 = (x + 1.5)(x - 1.5)$
$y_2 = x^2 - 2.25$

112. Graph $f(x) = x + 4$, $g(x) = x - 2$, and the product function, fg, in a $[-6, 6, 1]$ by $[-10, 10, 1]$ viewing rectangle. Trace along the curves and show that $(fg)(1) = f(1) \cdot g(1)$.

Critical Thinking Exercises

113. Which one of the following is true?

a. If f is a polynomial function, then
$$f(a + h) - f(a) = f(a) + f(h) - f(a) = f(h).$$

b. $(x - 5)^2 = x^2 - 5x + 25$

c. $(x + 1)^2 = x^2 + 1$

d. Suppose a square garden has an area represented by $9x^2$ square feet. If one side is made 7 feet longer and the other side is made 2 feet shorter, then the trinomial that represents the area of the larger garden is $9x^2 + 15x - 14$ square feet.

114. Express the area of the plane figure shown as a polynomial in standard form.

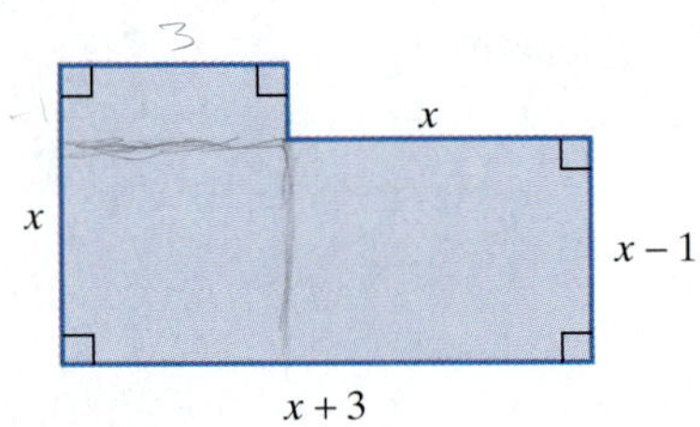

In Exercises 115–116, represent the volume of each figure as a polynomial in standard form.

115.

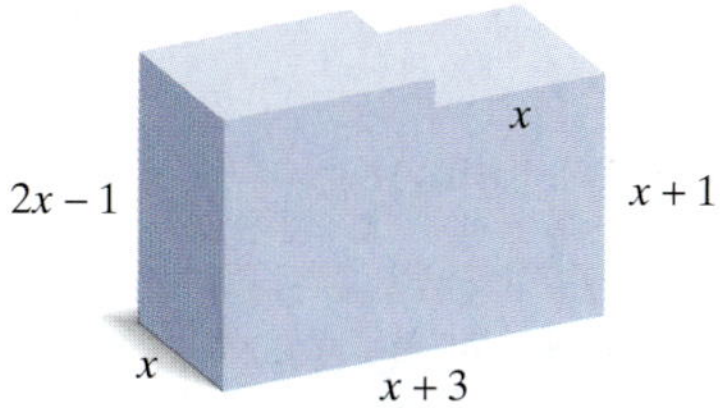

116.

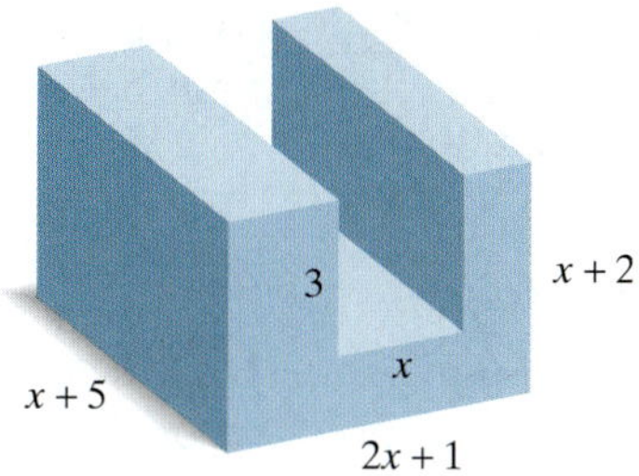

117. Simplify: $(y^n + 2)(y^n - 2) - (y^n - 3)^2$.

Review Exercises

118. Solve: $|3x + 4| \geq 10$. (Section 4.3, Example 1)

119. Solve: $2 - 6x \leq 20$. (Section 4.1, Example 3)

120. Write in scientific notation: 8,034,000,000. (Section 1.7, Example 2)

▸ SECTION 5.3 Greatest Common Factors and Factoring by Grouping

Objectives

1. Factor out the greatest common factor of a polynomial.
2. Factor out a common factor with a negative coefficient.
3. Factor by grouping.

SSM PH Tutor Center CD-ROM Video

The inability to understand numbers and their meanings is called *innumeracy* by mathematics professor John Allen Paulos. Paulos has written a book about mathematical illiteracy. Entitled *Innumeracy*, the book seeks to explain why so many people are numerically inept and to show how the problem can be corrected.

Jasper Johns *0 Through 9*, 1961. Oil on canvas, 137 × 105 cm. The Saatchi Collection, courtesy of the Leo Castelli Gallery. © JasperHohns/Licensed by VAGA/New York, NY.

Did you know that one of the most common ways that you are given numerical information is with percents? Unfortunately, many people are innumerate when it comes to this topic. For example, a computer whose price has been reduced by 40% and then another 40% is not selling at 20% of its original price.

To cure this bout of innumeracy (see Exercise 63 in Exercise Set 5.3), we turn to a process that reverses polynomial multiplication. For example, we can multiply polynomials and show that

$$5x(2x + 3) = 10x^2 + 15x.$$

We can also reverse this process and express the resulting polynomial as

$$10x^2 + 15x = 5x(2x + 3).$$

Factoring is the process of writing a polynomial as the product of two or more polynomials. The **factors** of $10x^2 + 15x$ are $5x$ and $2x + 3$.

In this chapter, we will be factoring over the set of integers, meaning that the coefficients in the factors are integers. Polynomials that cannot be factored using integer coefficients are called **prime polynomials**.

1 Factor out the greatest common factor of a polynomial.

Factoring Out the Greatest Common Factor

In any factoring problem, the first step is to look for the *greatest common factor*. The **greatest common factor**, abbreviated GCF, is an expression of the highest degree that divides each term of the polynomial. Can you see that $5x$ is the greatest common factor of $10x^2 + 15x$? 5 is the greatest integer that divides 10 and 15. Furthermore, x is the greatest expression that divides x^2 and x.

The variable part of the greatest common factor always contains the smallest power of a variable that appears in all terms of the polynomial. For example, consider the polynomial

$$10x^2 + 15x.$$

x^1, or x, is the variable raised to the smallest exponent.

We see that x is the variable part of the greatest common factor, $5x$.

When factoring a monomial from a polynomial, determine the greatest common factor of all terms in the polynomial. Sometimes there may not be a GCF other than 1. When a GCF other than 1 exists, we use the following procedure:

Factoring a Monomial from a Polynomial

1. Determine the greatest common factor of all terms in the polynomial.
2. Express each term as the product of the GCF and its other factor.
3. Use the distributive property to factor out the GCF.

EXAMPLE 1 Factoring Out the Greatest Common Factor

Factor: $10x^2 + 15x$.

Solution The GCF of the two terms of the polynomial is $5x$.

$$10x^2 + 15x$$

$$= 5x(2x) + 5x(3)$$ Express each term as the product of the GCF and its other factor.

$$= 5x(2x + 3)$$ Factor out the GCF.

We can check this factorization by multiplying $5x$ and $2x + 3$, obtaining the original polynomial as the answer. ■

✔ **CHECK POINT 1** Factor: $20x^2 + 30x$.

EXAMPLE 2 Factoring Out the Greatest Common Factor

Factor:

a. $9x^5 + 15x^3$ **b.** $16x^2y^3 - 24x^3y^4$ **c.** $12x^5y^4 - 4x^4y^3 + 2x^3y^2$.

Solution

a. First, determine the greatest common factor.

3 is the greatest integer that divides 9 and 15.

$$9x^5 + 15x^3$$

x^3 is the variable raised to the smallest exponent.

The GCF of the two terms of the polynomial is $3x^3$.

$$9x^5 + 15x^3$$

$$= 3x^3 \cdot 3x^2 + 3x^3 \cdot 5$$ Express each term as the product of the GCF and its other factor.

$$= 3x^3(3x^2 + 5)$$ Factor out the GCF.

b. Begin by determining the greatest common factor.

8 is the greatest integer that divides 16 and 24.

$$16x^2y^3 - 24x^3y^4$$

The variables raised to the smallest exponents are x^2 and y^3.

The GCF of the two terms of the polynomial is $8x^2y^3$.

$$16x^2y^3 - 24x^3y^4$$

$$= 8x^2y^3 \cdot 2 - 8x^2y^3 \cdot 3xy$$ Express each term as the product of the GCF and its other factor.

$$= 8x^2y^3(2 - 3xy)$$ Factor out the GCF.

c. First, determine the greatest common factor of the three terms.

2 is the greatest integer that divides 12, 4, and 2.

$$12x^5y^4 - 4x^4y^3 + 2x^3y^2.$$

The variables raised to the smallest exponents are x^3 and y^2.

The GCF is $2x^3y^2$.

$$12x^5y^4 - 4x^4y^3 + 2x^3y^2$$

$$= 2x^3y^2 \cdot 6x^2y^2 - 2x^3y^2 \cdot 2xy + 2x^3y^2 \cdot 1$$ Express each term as the product of the GCF and its other factor.

$$= 2x^3y^2(6x^2y^2 - 2xy + 1)$$ Factor out the GCF. ■

Because factoring reverses the process of multiplication, all factoring results can be checked by multiplying. Take a few minutes to check each of the three factorizations in Example 2. Use the distributive property to multiply the factors. This should give you the original polynomial.

✔ **CHECK POINT 2** Factor:

a. $9x^4 + 21x^2$

b. $15x^3y^2 - 25x^4y^3$

c. $16x^4y^5 - 8x^3y^4 + 4x^2y^3$.

2 Factor out a common factor with a negative coefficient.

When the leading coefficient of a polynomial is negative, it is often desirable to factor out a common factor with a negative coefficient. The common factor is the GCF preceded by a negative sign.

EXAMPLE 3 Using a Common Factor with a Negative Coefficient

Factor: $-3x^3 + 12x^2 - 15x$.

Solution The GCF is $3x$. Because the leading coefficient, -3, is negative, we factor out a common factor with a negative coefficient. We will factor out the negative of the GCF, or $-3x$.

$$-3x^3 + 12x^2 - 15x$$

$$= -3x(x^2) - 3x(-4x) - 3x(5)$$ Express each term as the product of the common factor and its other factor.

$$= -3x(x^2 - 4x + 5)$$ Factor out the GCF. ■

✔ **CHECK POINT 3** Factor out a common factor with a negative coefficient: $-2x^3 + 10x^2 - 6x$.

Factoring by Grouping

Up to now, we have factored a monomial from a polynomial. By contrast, in our next example, the greatest common factor of the polynomial is a binomial.

EXAMPLE 4 Factoring Out the Greatest Common Binomial Factor

Factor: **a.** $2(x - 7) + 9a(x - 7)$ **b.** $5y(a - b) - (a - b)$.

Solution Let's identify the common binomial factor in each part of the problem.

$2(x - 7) + 9a(x - 7)$ — The GCF, a binomial, is $x - 7$.

$5y(a - b) - (a - b)$ — The GCF, a binomial, is $a - b$.

We factor out each common binomial factor as follows.

a. $2(x - 7) + 9a(x - 7)$

$= (x - 7)2 + (x - 7)9a$ This step, usually omitted, shows each term as the product of the GCF and its other factor, in that order.

$= (x - 7)(2 + 9a)$ Factor out the GCF.

b. $5y(a - b) - (a - b)$

$= 5y(a - b) - 1(a - b)$ Write $-(a - b)$ as $-1(a - b)$ to aid in the factoring.

$= (a - b)(5y - 1)$ Factor out the GCF. ■

CHECK POINT 4 Factor:

a. $3(x - 4) + 7a(x - 4)$

b. $7x(a + b) - (a + b)$.

3 Factor by grouping.

Some polynomials have only a greatest common factor of 1. However, by a suitable grouping of the terms, it still may be possible to factor. This process, called **factoring by grouping**, is illustrated in Example 5.

EXAMPLE 5 Factoring by Grouping

Factor: $x^3 - 5x^2 + 3x - 15$.

Solution There is no factor other than 1 common to all terms. However, we can group terms that have a common factor:

$$\boxed{x^3 - 5x^2} + \boxed{3x - 15}$$

Common factor is x^2. Common factor is 3.

We now factor the given polynomial as follows.

$x^3 - 5x^2 + 3x - 15$

$= (x^3 - 5x^2) + (3x - 15)$ Group terms with common factors.

$= x^2(x - 5) + 3(x - 5)$ Factor out the greatest common factor from the grouped terms. The remaining two terms have $x - 5$ as a common binomial factor.

$= (x - 5)(x^2 + 3)$ Factor out the GCF.

Thus, $x^3 - 5x^2 + 3x - 15 = (x - 5)(x^2 + 3)$. Check the factorization by multiplying the right side of the equation using the FOIL method. If the factorization is correct, you will obtain the original polynomial. ■

Discover for Yourself

In Example 5, group the terms as follows:

$$x^2 + 3x - 5x^2 - 15.$$

Factor out the common factor from each group and complete the factoring process. Describe what happens. What can you conclude?

CHECK POINT 5 Factor: $x^3 - 4x^2 + 5x - 20$.

Factoring by Grouping

1. Group terms that have a common monomial factor. There will usually be two groups. Sometimes the terms must be rearranged.
2. Factor out the common monomial factor from each group.
3. Factor out the remaining common binomial factor (if one exists).

EXAMPLE 6 Factoring by Grouping

Factor: $3x^2 + 12x - 2xy - 8y$.

Solution There is no factor other than 1 common to all terms. However, we can group terms that have a common factor:

$$\boxed{3x^2 + 12x} + \boxed{-2xy - 8y}.$$

Common factor is 3x: $3x^2 + 12x = 3x(x + 4)$.

Use $-2y$, rather than $2y$, as the common factor: $-2xy - 8y = -2y(x + 4)$. In this way, the common binomial factor, $x + 4$, appears.

The voice balloons illustrate that it is sometimes necessary to use a factor with a negative coefficient to obtain a common binomial factor for the two groupings. We now factor the given polynomial as follows.

$$3x^2 + 12x - 2xy - 8y$$

$$= (3x^2 + 12x) + (-2xy - 8y) \quad \text{Group terms with common factors.}$$

$$= 3x(x + 4) - 2y(x + 4) \quad \text{Factor out the common factors from the grouped terms.}$$

$$= (x + 4)(3x - 2y) \quad \text{Factor out the GCF.}$$

Thus, $3x^2 + 12x - 2xy - 8y = (x + 4)(3x - 2y)$. Using the commutative property of multiplication, the factorization can also be expressed as $(3x - 2y)(x + 4)$. Verify the factorization by showing that, regardless of the order, FOIL multiplication gives the original polynomial. ■

CHECK POINT 6 Factor: $4x^2 + 20x - 3xy - 15y$.

Factoring by grouping sometimes requires that the terms be rearranged before the groupings are made. For example, consider the polynomial

$$3x^2 - 8y + 12x - 2xy.$$

The first two terms have no common factor other than 1. We must rearrange the terms and try a different grouping. Example 6 showed one such rearrangement and two groupings.

EXERCISE SET 5.3

Practice Exercises

In Exercises 1–18, factor the greatest common factor from each polynomial.

1. $10x^2 + 4x$
2. $12x^2 + 9x$
3. $y^2 - 4y$
4. $y^2 - 7y$
5. $x^3 + 5x^2$
6. $x^3 + 7x^2$
7. $12x^4 - 8x^2$
8. $20x^4 - 8x^2$
9. $32x^4 + 2x^3 + 8x^2$
10. $9x^4 + 18x^3 + 6x^2$
11. $4x^2y^3 + 6xy$
12. $6x^3y^2 + 9xy$
13. $30x^2y^3 - 10xy^2$
14. $27x^2y^3 - 18xy^2$
15. $12xy - 6xz + 4xw$
16. $14xy - 10xz + 8xw$
17. $15x^3y^6 - 9x^4y^4 + 12x^2y^5$
18. $15x^4y^6 - 3x^3y^5 + 12x^4y^4$

In Exercises 19–30, factor out the negative of the greatest common factor.

19. $-4x + 12$
20. $-5x + 20$
21. $-8x - 48$
22. $-7x - 63$
23. $-2x^2 + 6x - 14$
24. $-2x^2 + 8x - 12$
25. $-5y^2 + 40x$
26. $-9y^2 + 45x$
27. $-4x^3 + 32x^2 - 20x$
28. $-5x^3 + 50x^2 - 10x$
29. $-x^2 - 7x + 5$
30. $-x^2 - 8x + 8$

In Exercises 31–40, factor the greatest common binomial factor from each polynomial.

31. $4(x + 3) + a(x + 3)$
32. $5(x + 4) + a(x + 4)$
33. $x(y - 6) - 7(y - 6)$
34. $x(y - 9) - 5(y - 9)$
35. $3x(x + y) - (x + y)$
36. $7x(x + y) - (x + y)$
37. $4x^2(3x - 1) + 3x - 1$
38. $6x^2(5x - 1) + 5x - 1$
39. $(x + 2)(x + 3) + (x - 1)(x + 3)$
40. $(x + 4)(x + 5) + (x - 1)(x + 5)$

In Exercises 41–60, factor by grouping.

41. $x^2 + 3x + 5x + 15$
42. $x^2 + 2x + 4x + 8$
43. $x^2 + 7x - 4x - 28$
44. $x^2 + 3x - 5x - 15$
45. $x^3 - 3x^2 + 4x - 12$
46. $x^3 - 2x^2 + 5x - 10$
47. $xy - 6x + 2y - 12$
48. $xy - 5x + 9y - 45$
49. $xy + x - 7y - 7$
50. $xy + x - 5y - 5$
51. $10x^2 - 12xy + 35xy - 42y^2$
52. $3x^2 - 6xy + 5xy - 10y^2$
53. $4x^3 - x^2 - 12x + 3$
54. $3x^3 - 2x^2 - 6x + 4$
55. $x^2 - ax - bx + ab$
56. $x^2 + ax - bx - ab$
57. $x^3 - 12 - 3x^2 + 4x$
58. $2x^3 - 10 + 4x^2 - 5x$
59. $ay - by + bx - ax$
60. $cx - dx + dy - cy$

Application Exercises

61. A ball is thrown straight upward. The function

$$f(t) = -16t^2 + 40t$$

describes the ball's height above the ground, $f(t)$, in feet, t seconds after it was thrown.

a. Find and interpret $f(2)$.

b. Find and interpret $f(2.5)$.

c. Factor the polynomial $-16t^2 + 40t$ and write the function in factored form.

d. Use the factored form of the function to find $f(2)$ and $f(2.5)$. Do you get the same answers as you did in parts (a) and (b)? If so, does this prove that your factorization is correct? Explain.

62. An explosion causes debris to rise vertically. The function

$$f(t) = -16t^2 + 72t$$

describes the height of the debris above the ground, $f(t)$, in feet, t seconds after the explosion.

a. Find and interpret $f(2)$.

b. Find and interpret $f(4.5)$.

c. Factor the polynomial $-16t^2 + 72t$ and write the function in factored form.

d. Use the factored form of the function to find $f(2)$ and $f(4.5)$. Do you get the same answers as you did in parts (a) and (b)? If so, does this prove that your factorization is correct? Explain.

63. Your computer store is having an incredible sale. The price on one model is reduced by 40%. Then the sale price is reduced by another 40%. If x is the computer's original price, the sale price can be represented by

$$(x - 0.4x) - 0.4(x - 0.4x).$$

a. Factor out $(x - 0.4x)$ from each term. Then simplify the resulting expression.

b. Use the simplified expression from part (a) to answer these questions. With a 40% reduction followed by a 40% reduction, is the computer selling at 20% of its original price? If not, at what percentage of the original price is it selling?

64. Your local electronics store is having an end-of-the-year sale. The price on a large-screen television had been reduced by 30%. Now the sale price is reduced by another 30%. If x is the television's original price, the sale price can be represented by

$$(x - 0.3x) - 0.3(x - 0.3x).$$

a. Factor out $(x - 0.3x)$ from each term. Then simplify the resulting expression.

b. Use the simplified expression from part (a) to answer these questions. With a 30% reduction followed by a 30% reduction, is the television selling at 40% of its original price? If not, at what percentage of the original price is it selling?

Exercises 65–66 involve compound interest. **Compound interest** *is interest computed on your original savings as well as on any accumulated interest.*

65. After 2 years, the balance, A, in an account with principal P and interest rate r compounded annually is given by the formula

$$A = P + Pr + (P + Pr)r.$$

Use factoring by grouping to express the formula as $A = P(1 + r)^2$.

66. After 3 years, the balance, A, in an account with principal P and interest rate r compounded annually is given by the formula

$$A = P(1 + r)^2 + P(1 + r)^2 r.$$

Use factoring by grouping to express the formula as $A = P(1 + r)^3$.

67. The area of the skating rink with semicircular ends shown is $A = \pi r^2 + 2rl$. Express the area, A, in factored form.

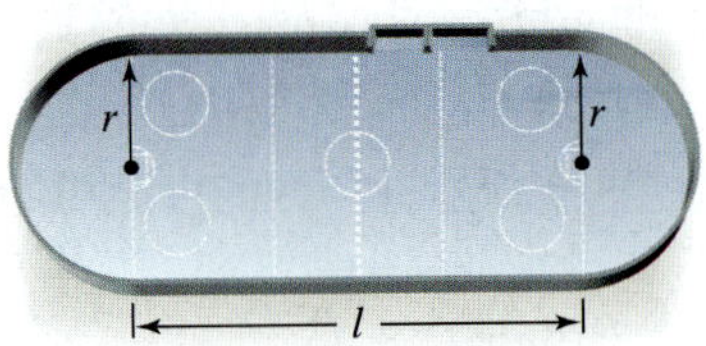

68. The amount of sheet metal needed to manufacture a cylindrical tin can, that is, its surface area, S, is $S = 2\pi r^2 + 2\pi rh$. Express the surface area, S, in factored form.

Writing in Mathematics

69. What is factoring?

70. If a polynomial has a greatest common factor other than 1, explain how to find its GCF.

71. Using an example, explain how to factor out the greatest common factor of a polynomial.

72. Suppose that a polynomial contains four terms and can be factored by grouping. Explain how to obtain the factorization.

73. Use two different groupings to factor

$$ac - ad + bd - bc$$

in two ways. Then explain why the two factorizations are the same.

74. Write a sentence that uses the word "factor" as a noun. Then write a sentence that uses the word "factor" as a verb.

Technology Exercises

In Exercises 75–79, use a graphing utility to graph the function on each side of the equation in the same viewing rectangle. Use end behavior to show a complete picture of the polynomial function on the left side. Do the graphs coincide? If so, this means that the polynomial on the left side has been factored correctly. If not, factor the polynomial correctly and then use your graphing utility to verify the factorization.

75. $x^2 - 4x = x(x - 4)$

76. $x^2 - 2x + 5x - 10 = (x - 2)(x - 5)$

77. $x^2 + 2x + x + 2 = x(x + 2) + 1$

78. $x^3 - 3x^2 + 4x - 12 = (x^2 + 4)(x - 3)$

79. $x^3 + x^2 - x - 1 = (x^2 + 1)(x - 1)$

Critical Thinking Exercises

80. Which one of the following is true?

a. If all terms of a polynomial contain the same variable raised to different powers, the exponent on the variable of the GCF is the highest power that appears in all the terms.

b. Because the GCF of $9x^3 + 6x^2 + 3x$ is $3x$, it is not necessary to write the 1 when $3x$ is factored from the last term.

c. Some polynomials with four terms, such as $x^3 + x^2 + 4x - 4$, cannot be factored by grouping.

d. None of the above is true.

In Exercises 81–83, factor each polynomial. Assume that all variable exponents represent whole numbers.

81. $x^{4n} + x^{2n} + x^{3n}$

82. $3x^{3m}y^{m} - 6x^{2m}y^{2m}$

83. $8y^{2n+4} + 16y^{2n+3} - 12y^{2n}$

In Exercises 84–85, write a polynomial that fits the given description. Do not use a polynomial that appears in this section or in the exercise set.

84. The polynomial has three terms and can be factored using a greatest common factor that has both a negative coefficient and a variable.

85. The polynomial has four terms and can be factored by grouping.

Review Exercises

86. Solve by Cramer's rule:

$$3x - 2y = 8$$
$$2x - 5y = 10.$$

(Section 3.5, Example 2)

87. Determine whether each relation is a function.

a. $\{(0, 5), (3, -5), (5, 5), (7, -5)\}$

b. $\{(1, 2), (3, 4), (5, 5), (5, 6)\}$ (Section 2.1, Example 2)

88. The length of a rectangle is 2 feet greater than twice its width. If the rectangle's perimeter is 22 feet, find the length and width. (Section 1.5, Example 5)

▸ SECTION 5.4 Factoring Trinomials

Objectives

1. Factor a trinomial whose leading coefficient is 1.
2. Factor using a substitution.
3. Factor a trinomial whose leading coefficient is not 1.
4. Factor trinomials by grouping.

SSM PH Tutor Center CD-ROM Video

We have a long history of throwing things. Prior to 400 B.C., the Greeks competed in games that included discus throwing. In the seventeenth century, English soldiers organized cannonball-throwing competitions. In 1827, a Yale University student, disappointed over failing an exam, took out his frustrations at the passing of a collection plate in chapel. Seizing the monetary tray, he flung it in the direction of a large open space on campus. Yale students see this act of frustration as the origin of the Frisbee.

How can we use mathematical models to analyze the where and when of the things that we enjoy throwing? The first step is to learn techniques for factoring trinomials. We will use these techniques in Section 5.7 to model and visualize motion.

1 Factor a trinomial whose leading coefficient is 1.

Factoring a Trinomial Whose Leading Coefficient Is 1

In Section 5.2, we used the FOIL method to multiply two binomials. The product was often a trinomial. The following are some examples.

Factored Form		F O I L		Trinomial Form
$(x + 3)(x + 4)$	=	$x^2 + 4x + 3x + 12$	=	$x^2 + 7x + 12$
$(x - 3)(x - 4)$	=	$x^2 - 4x - 3x + 12$	=	$x^2 - 7x + 12$
$(x + 3)(x - 5)$	=	$x^2 - 5x + 3x - 15$	=	$x^2 - 2x - 15$

Observe that each trinomial is of the form $x^2 + bx + c$, where the coefficient of the squared term is 1. Our goal in this part of the section is to start with the trinomial form and, assuming that it is factorable, return to the factored form.

Let's start by multiplying $(x + 3)(x + 4)$ using the FOIL method to obtain $x^2 + 7x + 12$.

$$(x + 3)(x + 4) = x^2 + 7x + 12$$

Now, we can make several important observations.

$$x^2 + 7x + 12 = (x + 3)(x + 4)$$

The product of the First terms is $x \cdot x = x^2$.

$$x^2 + 7x + 12 = (x + 3)(x + 4)$$

The product of the Last terms is $3 \cdot 4 = 12$.

$$x^2 + 7x + 12 = (x + 3)(x + 4)$$

I:$3x$

O:$4x$

The sum of the Outside and Inside products is $4x + 3x = 7x$.

These observations provide us with a procedure for factoring $x^2 + bx + c$.

A Strategy for Factoring $x^2 + bx + c$

1. Enter x as the first term of each factor.

$$(x \qquad)(x \qquad) = x^2 + bx + c$$

2. List pairs of factors of the constant c.
3. Try various combinations of these factors. Select the combination in which the sum of the *O*utside and *I*nside products is equal to bx.

$$(x + \square)(x + \square) = x^2 + bx + c$$

I

O

Sum of $O + I$

4. Check your work by multiplying the factors using the FOIL method. You should obtain the original trinomial.

If none of the possible combinations yield an *O*utside product and an *I*nside product whose sum is equal to bx, the trinomial cannot be factored using integers and is called **prime**.

EXAMPLE 1 Factoring a Trinomial Whose Leading Coefficient Is 1

Factor: $x^2 + 5x + 6$.

Solution

Step 1 Enter x as the first term of each factor.

$$x^2 + 5x + 6 = (x \qquad)(x \qquad)$$

To find the second term of each factor, we must find two integers whose product is 6 and whose sum is 5.

Step 2 List pairs of factors of the constant, 6.

Factors of 6	6, 1	3, 2	−6, −1	−3, −2

Step 3 Try various combinations of these factors. The correct factorization of $x^2 + 5x + 6$ is the one in which the sum of the *O*utside and *I*nside products is equal to $5x$. Here is a list of the possible factorizations.

Possible Factorizations of $x^2 + 5x + 6$	Sum of *O*utside and *I*nside Products (Should Equal $5x$)
$(x + 6)(x + 1)$	$x + 6x = 7x$
$(x + 3)(x + 2)$	$2x + 3x = 5x$
$(x - 6)(x - 1)$	$-x - 6x = -7x$
$(x - 3)(x - 2)$	$-2x - 3x = -5x$

This is the required middle term.

Using Technology

The polynomial functions

$$y_1 = x^2 + 5x + 6$$

and

$$y_2 = (x + 3)(x + 2)$$

have identical graphs. This verifies that

$$x^2 + 5x + 6 = (x + 3)(x + 2).$$

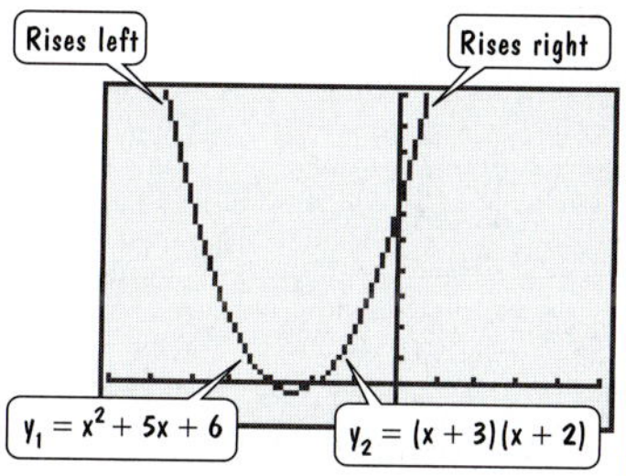

$[-7, 5, 1]$ by $[-1, 10, 1]$

Notice the cuplike shape of the graph of the quadratic function. The graph of the even-degree quadratic function exhibits the same behavior at each end. This graph rises to the left and to the right.

Thus,

$$x^2 + 5x + 6 = (x + 3)(x + 2).$$

Check this result by multiplying the right side using the FOIL method. You should obtain the original trinomial. Because of the commutative property, the factorization can be expressed as

$$x^2 + 5x + 6 = (x + 2)(x + 3).$$

■

In factoring a trinomial of the form $x^2 + bx + c$, you can speed things up by listing the factors of c, and then finding their sums. We are interested in a sum of b. For example, in factoring $x^2 + 5x + 6$, we are interested in the factors of 6 whose sum is 5.

Factors of 6	6, 1	3, 2	−6, −1	−3, −2
Sum of Factors	7	5	−7	−5

This is the desired sum.

Thus, $x^2 + 5x + 6 = (x + 3)(x + 2)$.

✔ **CHECK POINT 1** Factor: $x^2 + 6x + 8$.

EXAMPLE 2 Factoring a Trinomial Whose Leading Coefficient Is 1

Factor: $x^2 - 14x + 24$.

Solution

Step 1 Enter x as the first term of each factor.

$$x^2 - 14x + 24 = (x \quad\quad)(x \quad\quad)$$

To find the second term of each factor, we must find two integers whose product is 24 and whose sum is -14.

Step 2 List pairs of factors of the constant, 24. Because the desired sum, −14, is negative, we will list only the negative pairs of factors of 24.

Negative Factors of 24	−24, −1	−12, −2	−8, −3	−6, −4

Step 3 Try various combinations of these factors. We are interested in factors whose sum is −14.

Negative Factors of 24	−24, −1	−12, −2	−8, −3	−6, −4
Sum of Factors	−25	−14	−11	−10

This is the desired sum.

Thus, $x^2 - 14x + 24 = (x - 12)(x - 2)$. ■

Study Tip

To factor $x^2 + bx + c$ when c is positive, find two numbers with the same sign as the middle term.

$x^2 + 5x + 6 = (x + 3)(x + 2)$

Same signs

$x^2 - 14x + 24 = (x - 12)(x - 2)$

Same signs

✔ **CHECK POINT 2** Factor: $x^2 - 9x + 20$.

EXAMPLE 3 Factoring a Trinomial Whose Leading Coefficient Is 1

Factor: $y^2 + 7y - 60$

Solution

Step 1 Enter y as the first term of each factor.

$$y^2 + 7y - 60 = (y \qquad)(y \qquad)$$

To find the second term of each factor, we must find two integers whose product is −60 and whose sum is 7.

Steps 2 and 3 List pairs of factors of the constant, − 60, and try various combinations of these factors. Because the desired sum, 7, is positive, the positive factor of −60 must be farther from 0 than the negative factor is. Thus, we will only list pairs of factors of −60 in which the positive factor has the larger absolute value.

Some Factors of − 60	60, −1	30, −2	20, −3	15, −4	12, −5	10, −6
Sum of Factors	59	28	17	11	7	4

This is the desired sum.

Thus, $y^2 + 7y - 60 = (y + 12)(y - 5)$. ■

Study Tip

To factor $x^2 + bx + c$ when c is negative, find two numbers with opposite signs whose sum is the coefficient of the middle term.

$y^2 + 7y - 60 = (y + 12)(y - 5)$

Negative   Opposite signs

✔ **CHECK POINT 3** Factor: $y^2 + 19y - 66$.

EXAMPLE 4 Factoring a Trinomial in Two Variables

Factor: $x^2 - 4xy - 21y^2$.

Solution

Step 1 Enter x as the first term of each factor. Because the last term of the trinomial contains y^2, the second term of each factor must contain y.

$$x^2 - 4xy - 21y^2 = (x \quad ?y)(x \quad ?y)$$

The question marks indicate that we are looking for the coefficients of y in each factor. To find these coefficients, we must find two integers whose product is -21 and whose sum is -4.

Steps 2 and 3 List pairs of factors of the coefficient of the last term, -21, and try various combinations of these factors. We are interested in factors whose sum is -4.

Factors of -21	$1, -21$	$3, -7$	$-1, 21$	$-3, 7$
Sum of Factors	-20	-4	20	4

This is the desired sum.

Thus, $x^2 - 4xy - 21y^2 = (x + 3y)(x - 7y)$ or $(x - 7y)(x + 3y)$.

Step 4 Verify the factorization using the FOIL method.

$$(x + 3y)(x - 7y) = x^2 - 7xy + 3xy - 21y^2 = x^2 - 4xy - 21y^2$$

Because the product of the factors is the original polynomial, the factorization is correct. ■

✔ **CHECK POINT 4** Factor: $x^2 - 5xy + 6y^2$.

Can every trinomial be factored? The answer is no. For example, consider

$$x^2 + x - 5 = (x \quad\quad)(x \quad\quad).$$

To find the second term of each factor, we must find two integers whose product is -5 and whose sum is 1. Because no such integers exist, $x^2 + x - 5$ cannot be factored. This trinomial is prime.

To factor some polynomials, more than one technique must be used. **Always begin by trying to factor out the greatest common factor.** A polynomial is **factored completely** when it is written as the product of prime polynomials.

EXAMPLE 5 Factoring Completely

Factor: $8x^3 - 40x^2 - 48x$.

Solution The GCF of the three terms of the polynomial is $8x$. We begin by factoring out $8x$. Then we factor the remaining trinomial.

$$8x^3 - 40x^2 - 48x$$

$$= 8x(x^2 - 5x - 6)$$ Factor out the GCF.

$$= 8x(x \quad\quad)(x \quad\quad)$$ Begin factoring $x^2 - 5x - 6$. Find two integers whose product is -6 and whose sum is -5.

$$= 8x(x - 6)(x + 1)$$ The integers are -6 and 1.

Thus,

$$8x^3 - 40x^2 - 48x = 8x(x - 6)(x + 1).$$

Be sure to include the GCF in the factorization.

You can check this factorization by multiplying the binomials using the FOIL method. Then use the distributive property and multiply each term in this product by $8x$. Try doing this now. If the factorization is correct, you should obtain the original polynomial. ■

Some trinomials, such as $-x^2 + 5x + 6$, have a leading coefficient of -1. Because it is easier to factor a trinomial with a positive leading coefficient, begin by factoring out -1. For example,

$$-x^2 + 5x + 6 = -1(x^2 - 5x - 6) = -(x - 6)(x + 1).$$

✔ **CHECK POINT 5** Factor: $3x^3 - 15x^2 - 42x$.

2 Factor using a substitution.

In some trinomials, the highest power is greater than 2, and the exponent in one of the terms is half that of the other term. By letting t equal the variable to the smaller power, the trinomial can be written in a form that makes its possible factorization more obvious. Here are some examples.

Given Trinomial	Substitution	New Trinomial
$x^6 - 8x^3 + 15$ or $(x^3)^2 - 8x^3 + 15$	$t = x^3$	$t^2 - 8t + 15$
$x^4 - 8x^2 - 9$ or $(x^2)^2 - 8x^2 - 9$	$t = x^2$	$t^2 - 8t - 9$

In each case, we factor the given trinomial by working with the new trinomial on the right. If a factorization is found, we replace all occurrences of t by the substitution shown in the middle column.

EXAMPLE 6 Factoring by Substitution

Factor: $x^6 - 8x^3 + 15$.

Solution Notice that the exponent on x^3 is half that of the exponent on x^6. We will let t equal the variable to the power that is half of 6. Thus, let $t = x^3$.

$(x^3)^2 - 8x^3 + 15$ — This is the given polynomial, with x^6 written as $(x^3)^2$.

$= t^2 - 8t + 15$ — Let $t = x^3$. Rewrite the trinomial in terms of t.

$= (t - 5)(t - 3)$ — Factor.

$= (x^3 - 5)(x^3 - 3)$ — Now substitute x^3 for t.

Thus, the given trinomial can be factored as

$$x^6 - 8x^3 + 15 = (x^3 - 5)(x^3 - 3).$$

Check this result using FOIL multiplication on the right. ■

✔ **CHECK POINT 6** Factor: $x^6 - 7x^3 + 10$.

3 Factor a trinomial whose leading coefficient is not 1.

Factoring Trinomials Whose Leading Coefficient Is Not 1

How do we factor a trinomial such as $5x^2 - 14x + 8$? Notice that the leading coefficient is 5. We must find two binomials whose product is $5x^2 - 14x + 8$. The product of the *F*irst terms must be $5x^2$:

$$(5x \qquad)(x \qquad).$$

From this point on, the factoring strategy is exactly the same as the one we use to factor trinomials whose leading coefficient is 1.

A Strategy for Factoring $ax^2 + bx + c$

Assume, for the moment, that there is no greatest common factor.

1. Find two **F**irst terms whose product is ax^2:

$$(\square x + \qquad)(\square x + \qquad) = ax^2 + bx + c.$$

2. Find two **L**ast terms whose product is c:

$$(\square x + \square)(\square x + \square) = ax^2 + bx + c.$$

3. By trial and error, perform steps 1 and 2 until the sum of the **O**utside product and **I**nside product is bx:

$$(\square x + \square)(\square x + \square) = ax^2 + bx + c.$$

I

O

Sum of $O + I$

If no such combinations exist, the polynomial is called prime.

EXAMPLE 7 Factoring a Trinomial Whose Leading Coefficient Is Not 1

Factor: $5x^2 - 14x + 8$.

Solution

Step 1 Find two *F*irst terms whose product is $5x^2$.

$$5x^2 - 14x + 8 = (5x \qquad)(x \qquad)$$

Step 2 Find two *L*ast terms whose product is 8. The number 8 has pairs of factors that are either both positive or both negative. Because the middle term, $-14x$, is negative, both factors must be negative. The negative factorizations of 8 are $-1(-8)$ and $-2(-4)$.

Step 3 Try various combinations of these factors. The correct factorization of $5x^2 - 14x + 8$ is the one in which the sum of the *O*utside and *I*nside products is equal to $-14x$. Here is a list of the possible factorizations.

Possible Factorizations of $5x^2 - 14x + 8$	Sum of *O*utside and *I*nside Products (Should Equal $-14x$)
$(5x - 1)(x - 8)$	$-40x - x = -41x$
$(5x - 8)(x - 1)$	$-5x - 8x = -13x$
$(5x - 2)(x - 4)$	$-20x - 2x = -22x$
$(5x - 4)(x - 2)$	$-10x - 4x = -14x$

This is the required middle term.

Study Tip

With practice, you will find that it is not necessary to list every possible factorization of the trinomial. As you practice factoring, you will be able to narrow down the list of possible factorizations to just a few. When it comes to factoring, practice makes perfect. (Sorry about the cliché!)

Thus,

$$5x^2 - 14x + 8 = (5x - 4)(x - 2).$$

Show that this factorization is correct by multiplying the factors using the FOIL method. You should obtain the original trinomial. ■

✔ **CHECK POINT 7** Factor: $3x^2 - 20x + 28$.

EXAMPLE 8 Factoring a Trinomial Whose Leading Coefficient Is Not 1

Factor: $8x^6 - 10x^5 - 3x^4$.

Solution The GCF of the three terms of the polynomial is x^4. We begin by factoring out x^4.

$$8x^6 - 10x^5 - 3x^4 = x^4(8x^2 - 10x - 3)$$

Now we factor the remaining trinomial, $8x^2 - 10x - 3$.

Step 1 Find two *F*irst terms whose product is $8x^2$.

$$8x^2 - 10x - 3 \stackrel{?}{=} (8x \quad\quad)(x \quad\quad)$$
$$8x^2 - 10x - 3 \stackrel{?}{=} (4x \quad\quad)(2x \quad\quad)$$

Step 2 Find two *L*ast terms whose product is -3. The possible factorizations are $1(-3)$ and $-1(3)$.

Step 3 Try various combinations of these factors. The correct factorization of $8x^2 - 10x - 3$ is the one in which the sum of the *O*utside and *I*nside products is equal to $-10x$. Here is a list of the possible factorizations.

Possible Factorizations of $8x^2 - 10x - 3$	Sum of *O*utside and *I*nside Products (Should Equal $-10x$)
$(8x + 1)(x - 3)$	$-24x + x = -23x$
$(8x - 3)(x + 1)$	$8x - 3x = 5x$
$(8x - 1)(x + 3)$	$24x - x = 23x$
$(8x + 3)(x - 1)$	$-8x + 3x = -5x$
$(4x + 1)(2x - 3)$	$-12x + 2x = -10x$
$(4x - 3)(2x + 1)$	$4x - 6x = -2x$
$(4x - 1)(2x + 3)$	$12x - 2x = 10x$
$(4x + 3)(2x - 1)$	$-4x + 6x = 2x$

This is the required middle term.

Study Tip

Here are some suggestions for reducing the list of possible factorizations for $ax^2 + bx + c$.

1. If b is relatively small, avoid the larger factors of a.
2. If c is positive, the signs in both binomial factors must match the sign of b.
3. If the trinomial has no common factor, no binomial factor can have a common factor.
4. Reversing the signs in the binomial factors reverses the sign of bx, the middle term.

The factorization of $8x^2 - 10x - 3$ is $(4x + 1)(2x - 3)$. Now we include the GCF in the complete factorization of the given polynomial. Thus,

$$8x^6 - 10x^5 - 3x^4 = x^4(8x^2 - 10x - 3) = x^4(4x + 1)(2x - 3).$$ ■

This is the complete factorization with the GCF, x^4, included.

✔ **CHECK POINT 8** Factor: $6x^6 + 19x^5 - 7x^4$.

We have seen that not every trinomial can be factored. For example, consider

$$6x^2 + 14x + 7 = (6x + \square)(x + \square)$$

$$6x^2 + 14x + 7 = (3x + \square)(2x + \square)$$

The possible factors for the last term are 1 and 7. However, regardless of how these factors are placed in the boxes shown, the sum of the *O*utside and *I*nside products is not equal to $14x$. Thus, the trinomial $6x^2 + 14x + 7$ cannot be factored and is prime.

EXAMPLE 9 Factoring a Trinomial in Two Variables

Factor: $3x^2 - 13xy + 4y^2$.

Solution

Step 1 Find two *F*irst terms whose product is $3x^2$.

$$3x^2 - 13xy + 4y^2 = (3x \quad ?y)(x \quad ?y)$$

The question marks indicate that we are looking for the coefficients of y in each factor.

Steps 2 and 3 List pairs of factors of the coefficient of the last term, 4, and try various combinations of these factors. The correct factorization is the one in which the sum of the *O*utside and *I*nside products is equal to $-13xy$. Because of the negative coefficient, -13, we will consider only the negative pairs of factors of 4. The possible factorizations are $-1(-4)$ or $-2(-2)$.

Possible Factorizations of $3x^2 - 13xy + 4y^2$	Sum of *O*utside and *I*nside Products (Should Equal $-13xy$)
$(3x - y)(x - 4y)$	$-12xy - xy = -13xy$
$(3x - 4y)(x - y)$	$-3xy - 4xy = -7xy$
$(3x - 2y)(x - 2y)$	$-6xy - 2xy = -8xy$

This is the required middle term.

Thus,

$$3x^2 - 13xy + 4y^2 = (3x - y)(x - 4y).$$ ■

✔ **CHECK POINT 9** Factor: $2x^2 - 7xy + 3y^2$.

EXAMPLE 10 Factoring by Substitution

Factor: $6y^4 + 13y^2 + 6$.

Solution

$6(y^2)^2 + 13y^2 + 6$	This is the given polynomial, with y^4 written as $(y^2)^2$.
$= 6t^2 + 13t + 6$	Let $t = y^2$. Rewrite the trinomial in terms of t.
$= (3t + 2)(2t + 3)$	Factor the trinomial.
$= (3y^2 + 2)(2y^2 + 3)$	Now substitute y^2 for t.

Therefore, $6y^4 + 13y^2 + 6 = (3y^2 + 2)(2y^2 + 3)$. Check using FOIL multiplication. ■

✔ **CHECK POINT 10** Factor: $3y^4 + 10y^2 - 8$.

4 Factor trinomials by grouping.

Factoring Trinomials by Grouping

A second method for factoring $ax^2 + bx + c$, $a \neq 0$, is called the **grouping method**. The method involves both trial and error, as well as grouping. The trial and error in factoring $ax^2 + bx + c$ depends on finding two numbers, p and q, for which $p + q = b$. Then we factor $ax^2 + px + qx + c$ using grouping.

Let's see how this works by looking at a particular factorization.

$$8y^2 - 10y - 3 = (2y - 3)(4y + 1)$$

If we multiply using FOIL on the right, we obtain

$$(2y - 3)(4y + 1) = 8y^2 + 2y - 12y - 3.$$

In this case, the desired numbers, p and q, are $p = 2$ and $q = -12$. Compare these numbers to ac and b in the given polynomial.

$$8y^2 - 10y - 3$$

$a = 8$ $b = -10$ $c = -3$

$$ac = 8(-3) = -24$$

Can you see that p and q, 2 and -12, are factors of ac, or -24? Furthermore, p and q have a sum of b, namely -10. By expressing the middle term, $-10y$, in terms of p and q, we can factor by grouping as follows:

$8y^2 - 10y - 3$	
$= 8y^2 + (2y - 12y) - 3$	Rewrite $-10y$ as $2y - 12y$.
$= (8y^2 + 2y) + (-12y - 3)$	Group terms.
$= 2y(4y + 1) - 3(4y + 1)$	Factor from each group.
$= (4y + 1)(2y - 3)$	Factor out the common binomial factor.

Factoring $ax^2 + bx + c$ Using Grouping ($a \neq 1$)

1. Multiply the leading coefficient, a, and the constant, c.
2. Find the factors of ac whose sum is b.
3. Rewrite the middle term, bx, as a sum or difference using the factors from step 2.
4. Factor by grouping.

EXAMPLE 11 Factoring by Grouping

Factor by grouping: $12x^2 - 5x - 2$.

Solution The trinomial is of the form $ax^2 + bx + c$.

$$12x^2 - 5x - 2$$

$a = 12$ $b = -5$ $c = -2$

Step 1 Multiply the leading coefficient, a, and the constant, c. Using $a = 12$ and $c = -2$,

$$ac = 12(-2) = -24.$$

Step 2 Find the factors of ac whose sum is b. We want the factors of -24 whose sum is b, or -5. The factors of -24 whose sum is -5 are -8 and 3.

Step 3 Rewrite the middle term, $-5x$, as a sum or difference using the factors from step 2, -8 and 3.

$$12x^2 - 5x - 2 = 12x^2 - 8x + 3x - 2$$

Step 4 Factor by grouping.

$$= 4x(3x - 2) + 1(3x - 2)$$
$$= (3x - 2)(4x + 1)$$

Thus,

$$12x^2 - 5x - 2 = (3x - 2)(4x + 1).$$

Discover for Yourself

In step 2, we found that the desired numbers were -8 and 3. We wrote $-5x$ as $-8x + 3x$. What happens if we write $-5x$ as $3x - 8x$? Use factoring by grouping on

$$12x^2 - 5x - 2$$
$$= 12x^2 + 3x - 8x - 12.$$

Describe what happens.

✔ **CHECK POINT 11** Factor by grouping: $8x^2 - 22x + 5$.

EXERCISE SET 5.4

Practice Exercises

In Exercises 1–30, factor each trinomial, or state that the trinomial is prime. Check each factorization using FOIL multiplication.

1. $x^2 + 5x + 6$
2. $x^2 + 10x + 9$
3. $x^2 + 8x + 12$
4. $x^2 + 8x + 15$
5. $x^2 + 9x + 20$
6. $x^2 + 11x + 24$
7. $y^2 + 10y + 16$
8. $y^2 + 9y + 18$
9. $x^2 - 8x + 15$
10. $x^2 - 5x + 6$
11. $y^2 - 12y + 20$
12. $y^2 - 25y + 24$
13. $a^2 + 5a - 14$
14. $a^2 + a - 12$
15. $x^2 + x - 30$
16. $x^2 + 14x - 32$
17. $x^2 - 3x - 28$
18. $x^2 - 4x - 21$
19. $y^2 - 5y - 36$
20. $y^2 - 3y - 40$
21. $x^2 - x + 7$
22. $x^2 + 3x + 8$
23. $x^2 - 9xy + 14y^2$
24. $x^2 - 8xy + 15y^2$

25. $x^2 - xy - 30y^2$

26. $x^2 - 3xy - 18y^2$

27. $x^2 + xy + y^2$

28. $x^2 - xy + y^2$

29. $a^2 - 18ab + 80b^2$

30. $a^2 - 18ab + 45b^2$

In Exercises 31–38, factor completely.

31. $3x^2 + 3x - 18$

32. $4x^2 - 4x - 8$

33. $2x^3 - 14x^2 + 24x$

34. $2x^3 + 6x^2 + 4x$

35. $3y^3 - 15y^2 + 18y$

36. $4y^3 + 12y^2 - 72y$

37. $2x^4 - 26x^3 - 96x^2$

38. $3x^4 + 54x^3 + 135x^2$

In Exercises 39–44, factor by introducing an appropriate substitution.

39. $x^6 - x^3 - 6$

40. $x^6 + x^3 - 6$

41. $x^4 - 5x^2 - 6$

42. $x^4 - 4x^2 - 5$

43. $(x + 1)^2 + 6(x + 1) + 5$ (Let $t = x + 1$.)

44. $(x + 1)^2 + 8(x + 1) + 7$ (Let $t = x + 1$.)

In Exercises 45–68, use the method of your choice to factor each trinomial, or state that the trinomial is prime. Check each factorization using FOIL multiplication.

45. $3x^2 + 8x + 5$

46. $2x^2 + 9x + 7$

47. $5x^2 + 56x + 11$

48. $5x^2 - 16x + 3$

49. $3y^2 + 22y - 16$

50. $5y^2 + 33y - 14$

51. $4y^2 + 9y + 2$

52. $8y^2 + 10y + 3$

53. $10x^2 + 19x + 6$

54. $6x^2 + 19x + 15$

55. $8x^2 - 18x + 9$

56. $4x^2 - 27x + 18$

57. $6y^2 - 23y + 15$

58. $16y^2 - 6y - 27$

59. $6y^2 + 14y + 3$

60. $4y^2 + 22y - 5$

61. $3x^2 + 4xy + y^2$

62. $2x^2 + 3xy + y^2$

63. $6x^2 - 7xy - 5y^2$

64. $6x^2 - 5xy - 6y^2$

65. $15x^2 - 31xy + 10y^2$

66. $15x^2 + 11xy - 14y^2$

67. $3a^2 - ab - 14b^2$

68. $15a^2 - ab - 6b^2$

In Exercises 69–82, factor completely.

69. $15x^3 - 25x^2 + 10x$

70. $10x^3 + 24x^2 + 14x$

71. $24x^4 + 10x^3 - 4x^2$

72. $15x^4 - 39x^3 + 18x^2$

73. $15y^5 - 2y^4 - y^3$

74. $10y^5 - 17y^4 + 3y^3$

75. $24x^2 + 3xy - 27y^2$

76. $12x^2 + 10xy - 8y^2$

77. $6a^2b - 2ab - 60b$

78. $8a^2b + 34ab - 84b$

79. $12x^2y - 34xy^2 + 14y^3$

80. $12x^2y - 46xy^2 + 14y^3$

81. $13x^3y^3 + 39x^3y^2 - 52x^3y$

82. $4x^3y^5 + 24x^2y^5 - 64xy^5$

In Exercises 83–92, factor by introducing an appropriate substitution.

83. $2x^4 - x^2 - 3$

84. $5x^4 + 2x^2 - 3$

85. $2x^6 + 11x^3 + 15$

86. $2x^6 + 13x^3 + 15$

87. $2y^{10} + 5y^5 + 3$

88. $5y^{10} + 29y^5 - 42$

89. $5(x + 1)^2 + 12(x + 1) + 7$ (Let $t = x + 1$.)

90. $3(x + 1)^2 - 5(x + 1) + 2$ (Let $t = x + 1$.)

91. $2(x - 3)^2 - 5(x - 3) - 7$

92. $3(x - 2)^2 - 5(x - 2) - 2$

Application Exercises

93. A diver jumps directly upward from a board that is 32 feet high. The function

$$f(t) = -16t^2 + 16t + 32$$

describes the diver's height above the water, $f(t)$, in feet, after t seconds.

a. Find and interpret $f(1)$.

b. Find and interpret $f(2)$.

c. Factor the expression for $f(t)$ and write the function in completely factored form.

d. Use the factored form of the function to find $f(1)$ and $f(2)$.

94. The function $V(x) = 3x^3 - 2x^2 - 8x$ describes the volume, $V(x)$, in cubic inches, of the box shown whose height is x inches.

a. Find and interpret $V(4)$.

b. Factor the expression for $V(x)$ and write the function in completely factored form.

c. Use the factored form of the function to find $V(4)$ and $V(5)$.

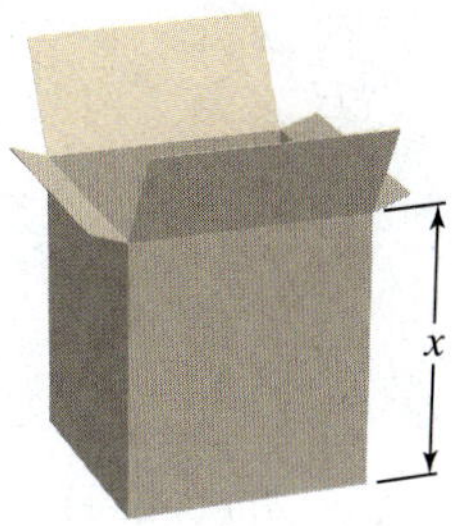

95. Find the area of the large rectangle shown below in two ways:

a. Find the sum of the areas of the six smaller rectangles and squares.

b. Express the area of the large rectangle as the product of its length and width.

c. Explain how the figure serves as a geometric model for the factorization of the sum that you wrote in part (a).

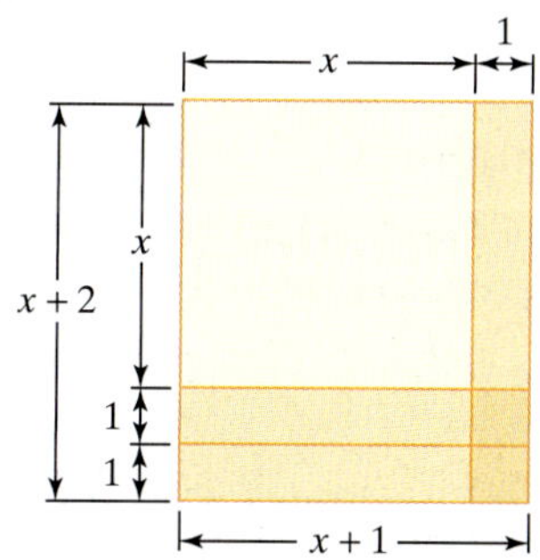

96. If x represents a positive integer, factor $x^3 + 3x^2 + 2x$ to show that the trinomial represents the product of three consecutive integers.

Writing in Mathematics

97. Explain how to factor $x^2 + 8x + 15$.

98. Give two helpful suggestions for factoring $x^2 - 5x + 6$.

99. In factoring $x^2 + bx + c$, describe how the last term in each factor is related to b and c.

100. Describe the first thing that you should try doing when factoring a polynomial.

101. What does it mean to factor completely?

102. Explain how to factor $x^6 - 7x^3 + 10$ by substitution.

103. Is it possible to factor $x^6 - 7x^3 + 10$ without using substitution? How might this be done?

104. Explain how to factor $2x^2 - x - 1$.

Technology Exercises

In Exercises 105–108, use a graphing utility to graph the function on each side of the equation in the same viewing rectangle. Use end behavior to show a complete picture of the polynomial function on the left side. Do the graphs coincide? If so, this means that the polynomial on the left side has been factored correctly. If not, factor the polynomial correctly and then use your graphing utility to verify the factorization.

105. $x^2 + 7x + 12 = (x + 4)(x + 3)$

106. $x^2 - 7x + 6 = (x - 2)(x - 3)$

107. $6x^3 + 5x^2 - 4x = x(3x + 4)(2x - 1)$

108. $x^4 - x^2 - 20 = (x^2 + 5)(x^2 - 4)$

Critical Thinking Exercises

109. Which one of the following is true?

a. Once a GCF is factored from $6y^6 - 19y^5 + 10y^4$, the remaining trinomial factor is prime.

b. One factor of $8y^2 - 51y + 18$ is $8y - 3$.

c. We can immediately tell that $6x^2 - 11xy - 10y^2$ is prime because 11 is a prime number and the polynomial contains two variables.

d. A factor of $12x^2 - 19xy + 5y$ is $4x - y$.

In Exercises 110–111, find all integers b so that the trinomial can be factored.

110. $4x^2 + bx - 1$

111. $3x^2 + bx + 5$

In Exercises 112–113, factor each polynomial. Assume that all variable exponents represent whole numbers.

112. $9x^{2n} + x^n - 8$

113. $4x^{2n} - 9x^n + 5$

Review Exercises

114. Solve: $-2x \le 6$ and $-2x + 3 < -7$. (Section 4.2, Example 2)

115. Solve the system:

$$\begin{aligned} 2x - y - 2z &= -1 \\ x - 2y - z &= 1 \\ x + y + z &= 4. \end{aligned}$$

(Section 3.3, Example 2)

116. Graph the solution set of the system:

$$\begin{aligned} x - y &\ge -4 \\ x + 2y &\ge 2. \end{aligned}$$

(Section 4.4, Example 3)

▸ SECTION 5.5 Factoring Special Forms

Objectives

1. Factor the difference of two squares.
2. Factor perfect square trinomials.
3. Use grouping to obtain the difference of two squares.
4. Factor the sum and difference of two cubes.

SSM PH Tutor Center CD-ROM Video

Bees use honeycombs to store honey and house larvae. They construct honey storage cells from wax. Each cell has the shape of a six-sided figure whose sides are all the same length and whose angles all have the same measure, called a regular hexagon. The cells fit together perfectly, preventing dirt or predators from entering. Squares or equilateral triangles would fit equally well, but regular hexagons provide the largest storage space for the amount of wax used.

In this section, we develop factoring techniques by reversing the formulas for special products discussed in Section 5.2. Like the construction of honeycombs, these factorizations can be visualized by perfectly fitting together "cells" of squares and rectangles to form larger rectangles.

1 Factor the difference of two squares.

Factoring the Difference of Two Squares

A method for factoring the difference of two squares is obtained by reversing the special product for the sum and difference of two terms.

The Difference of Two Squares

If A and B are real numbers, variables, or algebraic expressions, then

$$A^2 - B^2 = (A + B)(A - B).$$

In words: The difference of the squares of two terms factors as the product of a sum and a difference of those terms.

EXAMPLE 1 Factoring the Difference of Two Squares

Factor:

a. $9x^2 - 100$ **b.** $36y^6 - 49x^4$.

Solution We must express each term as the square of some monomial. Then we use the formula for factoring $A^2 - B^2$.

a. $9x^2 - 100 = (3x)^2 - 10^2 = (3x + 10)(3x - 10)$

$$A^2 \quad - \quad B^2 \quad = \quad (A \quad + \quad B) \quad (A \quad - \quad B)$$

b. $36y^6 - 49x^4 = (6y^3)^2 - (7x^2)^2 = (6y^3 + 7x^2)(6y^3 - 7x^2)$ ■

ENRICHMENT ESSAY

Visualizing the Factoring for the Difference of Two Squares

Shaded Area: $A^2 - B^2$

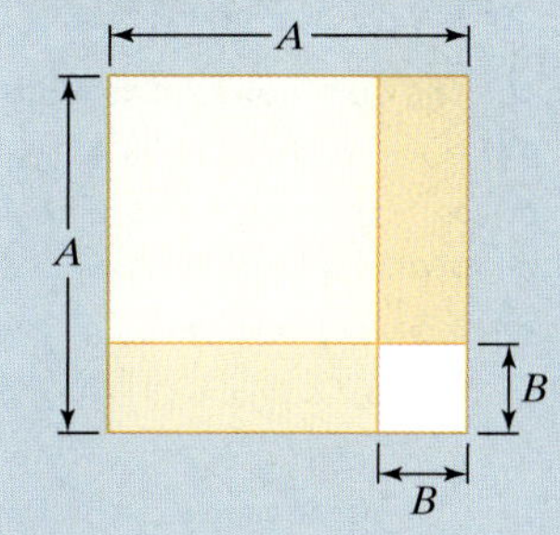

Shaded Area: $(A + B)(A - B)$

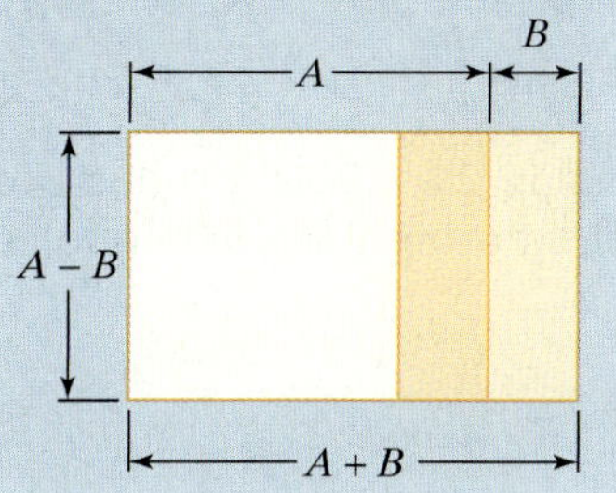

The shaded areas are the same. Thus,

$$A^2 - B^2 = (A + B)(A - B).$$

✔ CHECK POINT 1 Factor:

a. $16x^2 - 25$ **b.** $100y^6 - 9x^4$.

Can $x^2 - 5$ be factored using integers and the formula for factoring $A^2 - B^2$? No. The number 5 is not the square of an integer. Thus, $x^2 - 5$ is prime.

When factoring, always check first for common factors. If there are common factors, factor out the GCF and then factor the resulting polynomial.

EXAMPLE 2 Factoring Out the GCF and Then Factoring the Difference of Two Squares

Factor: $3y - 3x^6y^5$.

Solution

$$3y - 3x^6y^5 = 3y(1 - x^6y^4) = 3y[1^2 - (x^3y^2)^2] = 3y(1 + x^3y^2)(1 - x^3y^2)$$ ■

Factor out the GCF.

$A^2 - B^2 = (A + B)(A - B)$

✔ CHECK POINT 2 Factor: $6y - 6x^2y^7$.

We have seen that a polynomial is factored completely when it is written as the product of prime polynomials. To be sure that you have factored completely, check to see whether the factors can be factored.

EXAMPLE 3 A Repeated Factorization

Factor completely: $81x^4 - 16$.

Solution

$81x^4 - 16 = (9x^2)^2 - 4^2$ — Express as the difference of two squares.

$= (9x^2 + 4)(9x^2 - 4)$ — The factors are the sum and difference of the expressions being squared.

$= (9x^2 + 4)[(3x)^2 - 2^2]$ — The factor $9x^2 - 4$ is the difference of two squares and can be factored.

$= (9x^2 + 4)(3x + 2)(3x - 2)$ — The factors of $9x^2 - 4$ are the sum and difference of the expressions being squared. ■

Study Tip

Factoring $81x^4 - 16$ as

$$(9x^2 + 4)(9x^2 - 4)$$

is not a complete factorization. The second factor, $9x^2 - 4$, is itself a difference of two squares and can be factored.

Are you tempted to further factor $9x^2 + 4$, the sum of two squares, in Example 3? Resist the temptation! **The sum of two squares, $A^2 + B^2$, with no common factor other than 1 is a prime polynomial.**

✔ CHECK POINT 3 Factor completely: $16x^4 - 81$.

In our next example, we begin with factoring by grouping. We can then factor further using the difference of squares.

EXAMPLE 4 Factoring Completely

Factor completely: $x^3 + 5x^2 - 9x - 45$.

Solution

$$\begin{aligned} & x^3 + 5x^2 - 9x - 45 \\ &= (x^3 + 5x^2) + (-9x - 45) && \text{Group terms with common factors.} \\ &= x^2(x + 5) - 9(x + 5) && \text{Factor out the common factor from each group.} \\ &= (x + 5)(x^2 - 9) && \text{Factor out } x + 5\text{, the common binomial factor, from both terms.} \\ &= (x + 5)(x + 3)(x - 3) && \text{Factor } x^2 - 3^3\text{, the difference of two squares.} \end{aligned}$$

■

2 Factor perfect square trinomials.

✔ **CHECK POINT 4** Factor completely: $x^3 + 7x^2 - 4x - 28$.

ENRICHMENT ESSAY

Visualizing the Factoring for a Perfect Square Trinomial

Area: $(A + B)^2$

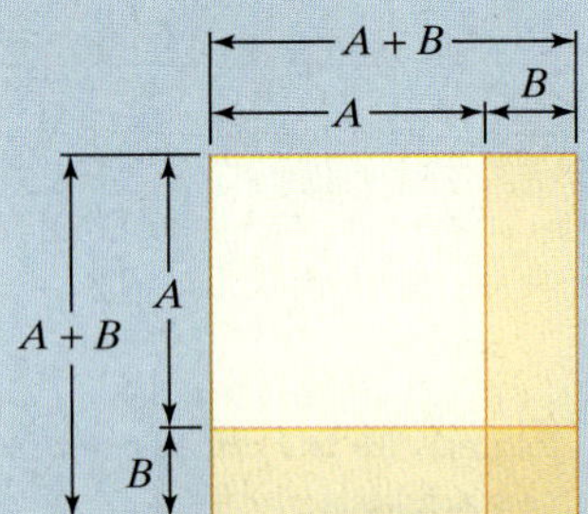

Sum of Areas:
$A^2 + 2AB + B^2$

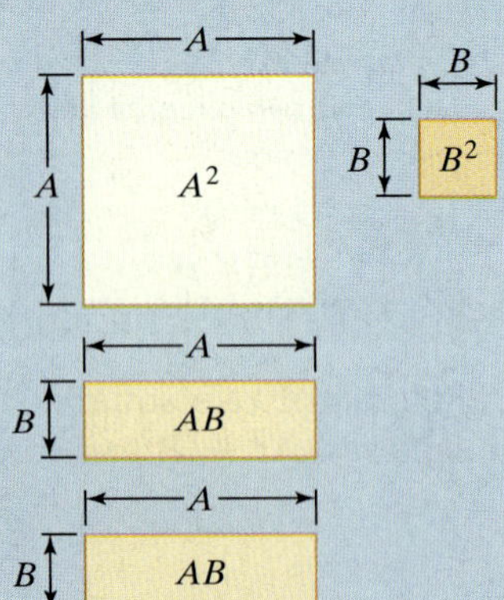

Thus,
$A^2 + 2AB + B^2 = (A + B)^2$.

Factoring Perfect Square Trinomials

Our next factoring technique is obtained by reversing the special products for squaring binomials. The trinomials that are factored using this technique are called **perfect square trinomials**.

Factoring Perfect Square Trinomials

Let A and B be real numbers, variables, or algebraic expressions.

1. $A^2 + 2AB + B^2 = (A + B)^2$
   Same sign
2. $A^2 - 2AB + B^2 = (A - B)^2$
   Same sign

The two items in the box show that perfect square trinomials come in two forms: one in which the middle term is positive and one in which the middle term is negative. Here's how to recognize a perfect square trinomial:

1. The first and last terms are squares of monomials or integers.
2. The middle term is twice the product of the expressions being squared in the first and last terms.

EXAMPLE 5 Factoring Perfect Square Trinomials

Factor:

a. $x^2 + 14x + 49$ **b.** $4x^2 + 12xy + 9y^2$ **c.** $9y^4 - 12y^2 + 4$.

Solution

a. $x^2 + 14x + 49 = x^2 + 2 \cdot x \cdot 7 + 7^2 = (x + 7)^2$ The middle term has a positive sign.

$A^2 + 2AB + B^2 = (A + B)^2$

b. We suspect that $4x^2 + 12xy + 9y^2$ is a perfect square trinomial because $4x^2 = (2x)^2$ and $9y^2 = (3y)^2$. The middle term can be expressed as twice the product of $2x$ and $3y$.

$$4x^2 + 12xy + 9y^2 = (2x)^2 + 2 \cdot 2x \cdot 3y + (3y)^2 = (2x + 3y)^2$$

$$A^2 + 2AB + B^2 = (A + B)^2$$

c. $9y^4 - 12y^2 + 4 = (3y^2)^2 - 2 \cdot 3y^2 \cdot 2 + 2^2 = (3y^2 - 2)^2$

$$A^2 - 2AB + B^2 = (A - B)^2$$

The middle term has a negative sign. ■

✔ **CHECK POINT 5** Factor:

a. $x^2 + 6x + 9$

b. $16x^2 + 40xy + 25y^2$

c. $4y^4 - 20y^2 + 25$.

3 Use grouping to obtain the difference of two squares.

Using Special Forms When Factoring by Grouping

If a polynomial contains four terms, try factoring by grouping. In the next example, we group the terms to obtain the difference of two squares. One of the squares is a perfect square trinomial.

EXAMPLE 6 Using Grouping to Obtain the Difference of Two Squares

Factor: $x^2 - 8x + 16 - y^2$.

Solution

$x^2 - 8x + 16 - y^2$

$= (x^2 - 8x + 16) - y^2$ — Group as a perfect square trinomial minus y^2 to obtain a difference of squares.

$= (x - 4)^2 - y^2$ — Factor the perfect square trinomial.

$= (x - 4 + y)(x - 4 - y)$ — Factor the difference of squares. The factors are the sum and difference of the expressions being squared. ■

✔ **CHECK POINT 6** Factor: $x^2 + 10x + 25 - y^2$.

EXAMPLE 7 Using Grouping to Obtain the Difference of Two Squares

Factor: $a^2 - b^2 + 10b - 25$.

Solution Grouping into two groups of two terms does not result in a common binomial factor. Let's look for a perfect square trinomial. Can you see that the perfect square trinomial is the expression being subtracted from a^2?

$$a^2 - b^2 + 10b - 25$$

$$= a^2 - (b^2 - 10b + 25)$$ Factor out −1 and group as a^2 minus a perfect square trinomial to obtain a difference of squares.

$$= a^2 - (b - 5)^2$$ Factor the perfect square trinomial.

$$= [a + (b - 5)][a - (b - 5)]$$ Factor the difference of squares. The factors are the sum and difference of the expressions being squared.

$$= (a + b - 5)(a - b + 5)$$ Simplify. ■

✔ **CHECK POINT 7** Factor: $a^2 - b^2 + 4b - 4$.

4 Factor the sum and difference of two cubes.

Factoring the Sum and Difference of Two Cubes

Here are two multiplications that lead to factoring formulas for the sum of two cubes and the difference of two cubes.

$$(A + B)(A^2 - AB + B^2) = A(A^2 - AB + B^2) + B(A^2 - AB + B^2)$$
$$= A^3 - A^2B + AB^2 + A^2B - AB^2 + B^3$$
$$= A^3 + B^3 \quad \text{Combine like terms.}$$

and

$$(A - B)(A^2 + AB + B^2) = A(A^2 + AB + B^2) - B(A^2 + AB + B^2)$$
$$= A^3 + A^2B + AB^2 - A^2B - AB^2 - B^3$$
$$= A^3 - B^3 \quad \text{Combine like terms.}$$

By reversing the two sides of these equations, we obtain formulas that allow us to factor a sum or difference of two cubes. These formulas should be memorized.

Factoring the Sum and Difference of Two Cubes

1. Factoring the Sum of Two Cubes

$$A^3 + B^3 = (A + B)(A^2 - AB + B^2)$$

Same sign   Opposite signs

2. Factoring the Difference of Two Cubes

$$A^3 - B^3 = (A - B)(A^2 + AB + B^2)$$

Same sign   Opposite signs

Study Tip

When factoring the sum or difference of cubes, it is helpful to recognize the following cubes:

$8 = 2^3$

$27 = 3^3$

$64 = 4^3$

$125 = 5^3$

$216 = 6^3$

$1000 = 10^3$.

EXAMPLE 8 Factoring the Sum of Two Cubes

Factor:

a. $x^3 + 125$ **b.** $x^6 + 64y^3$.

Solution We must express each term as the cube of some monomial. Then we use the formula for factoring $A^3 + B^3$.

a. $x^3 + 125 = x^3 + 5^3 = (x + 5)(x^2 - x \cdot 5 + 5^2) = (x + 5)(x^2 - 5x + 25)$

$A^3 + B^3 = (A + B)\ (A^2 - A\ B + B^2)$

b. $x^6 + 64y^3 = (x^2)^3 + (4y)^3 = (x^2 + 4y)[(x^2)^2 - x^2 \cdot 4y + (4y)^2]$

$A^3 + B^3 = (A + B)\ (A^2 - A\ B + B^2)$

$= (x^2 + 4y)(x^4 - 4x^2y + 16y^2)$ ■

✔ **CHECK POINT 8** Factor:

a. $x^3 + 27$
b. $x^6 + 1000y^3$.

EXAMPLE 9 Factoring the Difference of Two Cubes

Factor:

a. $x^3 - 216$ **b.** $8 - 125x^3y^3$.

Solution We must express each term as the cube of some monomial. Then we use the formula for factoring $A^3 - B^3$.

a. $x^3 - 216 = x^3 - 6^3 = (x - 6)(x^2 + x \cdot 6 + 6^2) = (x - 6)(x^2 + 6x + 36)$

$A^3 - B^3 = (A - B)\ (A^2 + A\ B + B^2)$

b. $8 - 125x^3y^3 = 2^3 - (5xy)^3 = (2 - 5xy)[2^2 + 2 \cdot 5xy + (5xy)^2]$

$A^3 - B^3 = (A - B)\ (A^2 + A\ B + B^2)$

$= (2 - 5xy)(4 + 10xy + 25x^2y^2)$ ■

✔ **CHECK POINT 9** Factor:

a. $x^3 - 8$
b. $1 - 27x^3y^3$.

EXERCISE SET 5.5

Practice Exercises

In Exercises 1–18, factor each difference of two squares.

1. $x^2 - 4$
2. $x^2 - 16$
3. $9x^2 - 25$
4. $4x^2 - 9$
5. $9 - 25y^2$
6. $16 - 49y^2$
7. $36x^2 - 49y^2$
8. $64x^2 - 25y^2$
9. $x^2y^2 - 1$
10. $x^2y^2 - 100$
11. $9x^4 - 25y^6$
12. $25x^4 - 9y^6$
13. $x^{14} - y^4$
14. $x^4 - y^{10}$
15. $(x - 3)^2 - y^2$
16. $(x - 6)^2 - y^2$
17. $a^2 - (b - 2)^2$
18. $a^2 - (b - 3)^2$

In Exercises 19–44, factor completely, or state that the polynomial is prime.

19. $2x^3 - 8x$

20. $2x^3 - 72x$

21. $50 - 2y^2$

22. $72 - 2y^2$

23. $8x^2 - 8y^2$

24. $6x^2 - 6y^2$

25. $2x^3y - 18xy$

26. $2x^3y - 32xy$

27. $a^3b^2 - 49ac^2$

28. $4a^3c^2 - 16ax^2y^2$

29. $5y - 5x^2y^7$

30. $2y - 2x^6y^3$

31. $8x^2 + 8y^2$

32. $6x^2 + 6y^2$

33. $x^2 + 25y^2$

34. $x^2 + 36y^2$

35. $x^4 - 16$

36. $x^4 - 1$

37. $81x^4 - 1$

38. $1 - 81x^4$

39. $2x^5 - 2xy^4$

40. $3x^5 - 3xy^4$

41. $x^3 + 3x^2 - 4x - 12$

42. $x^3 + 3x^2 - 9x - 27$

43. $x^3 - 7x^2 - x + 7$

44. $x^3 - 6x^2 - x + 6$

In Exercises 45–60, factor any perfect square trinomials, or state that the polynomial is prime.

45. $x^2 + 4x + 4$

46. $x^2 + 2x + 1$

47. $x^2 - 10x + 25$

48. $x^2 - 14x + 49$

49. $x^4 - 4x^2 + 4$

50. $x^4 - 6x^2 + 9$

51. $9y^2 + 6y + 1$

52. $4y^2 + 4y + 1$

53. $64y^2 - 16y + 1$

54. $25y^2 - 10y + 1$

55. $x^2 - 12xy + 36y^2$

56. $x^2 + 16xy + 64y^2$

57. $x^2 - 8xy + 64y^2$

58. $x^2 - 9xy + 81y^2$

59. $9x^2 + 48xy + 64y^2$

60. $16x^2 - 40xy + 25y^2$

In Exercises 61–70, factor by grouping to obtain the difference of two squares.

61. $x^2 - 6x + 9 - y^2$

62. $x^2 - 12x + 36 - y^2$

63. $x^2 + 20x + 100 - x^4$

64. $x^2 + 16x + 64 - x^4$

65. $9x^2 - 30x + 25 - 36y^2$

66. $25x^2 - 20x + 4 - 81y^2$

67. $x^4 - x^2 - 2x - 1$

68. $x^4 - x^2 - 6x - 9$

69. $z^2 - x^2 + 4xy - 4y^2$

70. $z^2 - x^2 + 10xy - 25y^2$

In Exercises 71–88, factor using the formula for the sum or difference of two cubes.

71. $x^3 + 64$

72. $x^3 + 1$

73. $x^3 - 27$

74. $x^3 - 1000$

75. $8y^3 + 1$

76. $27y^3 + 1$

77. $125x^3 - 8$

78. $27x^3 - 8$

79. $x^3y^3 + 27$

80. $x^3y^3 + 64$

81. $64x - x^4$

82. $216x - x^4$

83. $x^6 + 27y^3$

84. $x^6 + 8y^3$

85. $125x^6 - 64y^6$

86. $125x^6 - y^6$

87. $x^9 + 1$

88. $x^9 - 1$

Application Exercises

In Exercises 89–92, find the formula for the area of the shaded region and express it in factored form.

89.

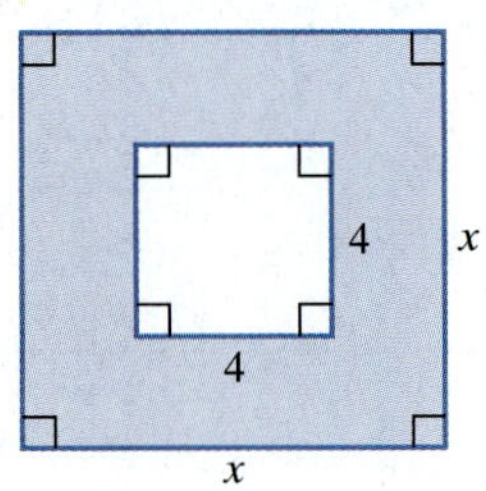

90.

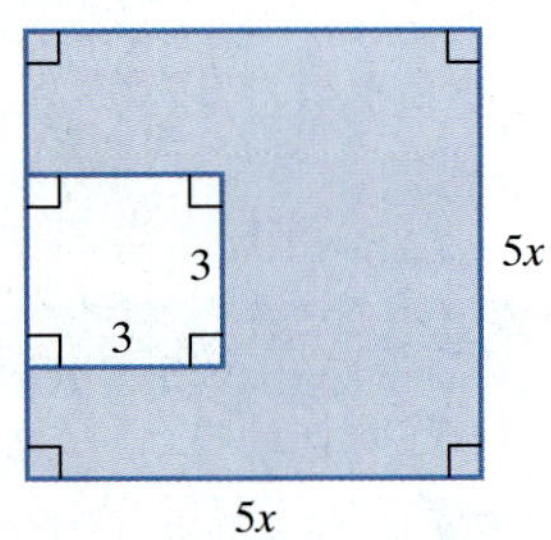

91.

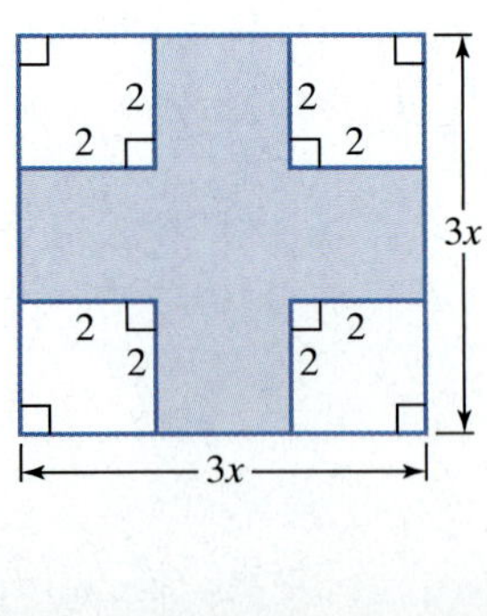

92.

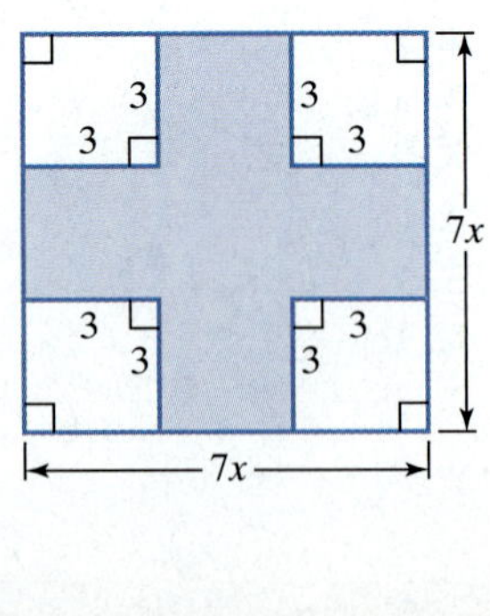

In Exercises 93–94, find the formula for the volume of the region outside the smaller rectangular solid and inside the larger rectangular solid. Then express the volume in factored form.

93.

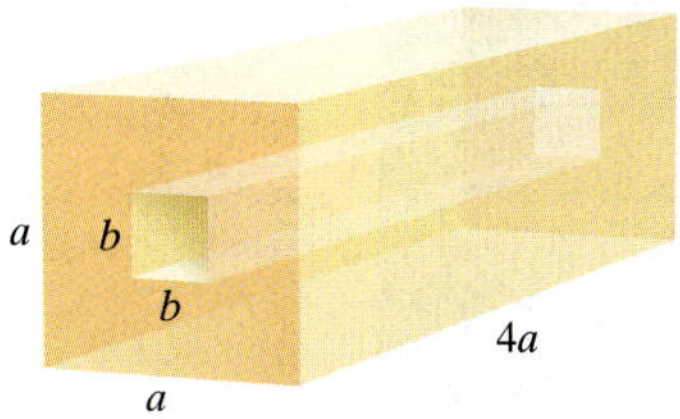

94.

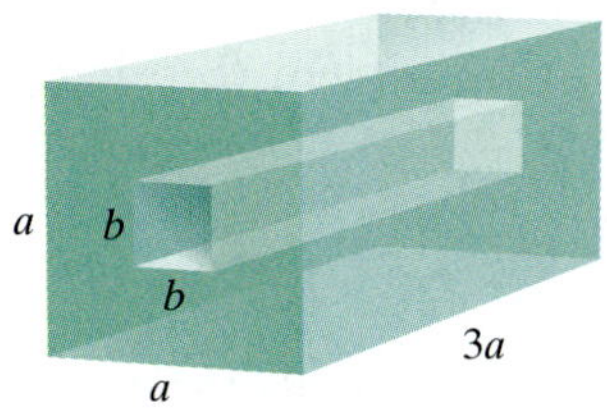

Writing in Mathematics

95. Explain how to factor the difference of two squares. Provide an example with your explanation.

96. What is a perfect square trinomial and how is it factored?

97. Explain how to factor $x^2 - y^2 + 8x - 16$. Should the expression be grouped into two groups of two terms? If not, why not, and what sort of grouping should be used?

98. Explain how to factor $x^3 + 1$.

Technology Exercises

In Exercises 99–106, use a graphing utility to graph the function on each side of the equation in the same viewing rectangle. Use end behavior to show a complete picture of the polynomial function on the left side. Do the graphs coincide? If so, this means that the polynomial on the left side has been factored correctly. If not, factor the polynomial correctly and then use your graphing utility to verify the factorization.

99. $9x^2 - 4 = (3x + 2)(3x - 2)$

100. $x^2 + 4x + 4 = (x + 4)^2$

101. $9x^2 + 12x + 4 = (3x + 2)^2$

102. $25 - (x^2 + 4x + 4) = (x + 7)(x - 3)$

103. $(2x + 3)^2 - 9 = 4x(x + 3)$

104. $(x - 3)^2 + 8(x - 3) + 16 = (x - 1)^2$

105. $x^3 - 1 = (x - 1)(x^2 - x + 1)$

106. $(x + 1)^3 + 1 = (x + 1)(x^2 + x + 1)$

Critical Thinking Exercises

107. Which one of the following is true?
- **a.** $9x^2 + 15x + 25 = (3x + 5)^2$
- **b.** $x^3 - 27 = (x - 3)(x^2 + 6x + 9)$
- **c.** $x^3 - 64 = (x - 4)^3$
- **d.** None of the above is true.

In Exercises 108–111, factor each polynomial completely. All variable exponents represent whole numbers.

108. $y^3 + x + x^3 + y$

109. $36x^{2n} - y^{2n}$

110. $x^{3n} + y^{12n}$

111. $4x^{2n} + 20x^n y^m + 25y^{2m}$

112. Factor $x^6 - y^6$ first as the difference of squares and then as the difference of cubes. From these two factorizations, determine a factorization for $x^4 + x^2y^2 + y^4$.

In Exercises 113–114, find all integers k so that the trinomial is a perfect square trinomial.

113. $kx^2 + 8xy + y^2$

114. $64x^2 - 16x + k$

Review Exercises

115. Solve: $2x + 2 \geq 12$ and $\frac{2x - 1}{3} \leq 7$.
(Section 4.2, Example 2)

116. Solve using matrices:

$$3x - 2y = -8$$
$$x + 6y = 4.$$

(Section 3.4, Example 3)

117. Simplify: $\frac{4^{-2}x^{-3}y^1}{x^2y^{-5}}$.
(Section 1.6, Example 9)

▸ SECTION 5.6 A General Factoring Strategy

Objectives

1. Recognize the appropriate method for factoring a polynomial.
2. Use a general strategy for factoring polynomials.

SSM PH Tutor Center CD-ROM Video

Successful problem solving involves understanding the problem, devising a plan for solving it, and then carrying out the plan. In this section, you will learn a step-by-step strategy that provides a plan and direction for solving factoring problems.

A Strategy for Factoring Polynomials

It is important to practice factoring a wide variety of polynomials so that you can quickly select the appropriate technique. The polynomial is factored completely when all its polynomial factors, except possibly for monomial factors, are prime. Because of the commutative property, the order of the factors does not matter.

1 Recognize the appropriate method for factoring a polynomial.

A Strategy for Factoring a Polynomial

1. If there is a common factor, factor out the GCF or factor out a common factor with a negative coefficient.
2. Determine the number of terms in the polynomial and try factoring as follows:
   a. If there are two terms, can the binomial be factored by one of the following special forms?

      Difference of two squares: $A^2 - B^2 = (A + B)(A - B)$

      Sum of two cubes: $A^3 + B^3 = (A + B)(A^2 - AB + B^2)$

      Difference of two cubes: $A^3 - B^3 = (A - B)(A^2 + AB + B^2)$

   b. If there are three terms, is the trinomial a perfect square trinomial? If so, factor by one of the following special forms:

      $$A^2 + 2AB + B^2 = (A + B)^2$$
      $$A^2 - 2AB + B^2 = (A - B)^2.$$

      If the trinomial is not a perfect square trinomial, try factoring by trial and error or grouping.

   c. If there are four or more terms, try factoring by grouping.
3. Check to see if any factors with more than one term in the factored polynomial can be factored further. If so, factor completely.

Remember to check the factored form by multiplying or by using a graphing utility.

2 Use a general strategy for factoring polynomials.

The following examples and those in the exercise set are similar to the previous factoring problems. However, these factorizations are not all of the same type. They are intentionally mixed to promote the development of a general factoring strategy.

Using Technology

The polynomial functions

$$y_1 = 2x^3 + 8x^2 + 8x$$

and

$$y_2 = 2x(x + 2)^2$$

have identical graphs. This verifies that

$$2x^3 + 8x^2 + 8x = 2x(x + 2)^2.$$

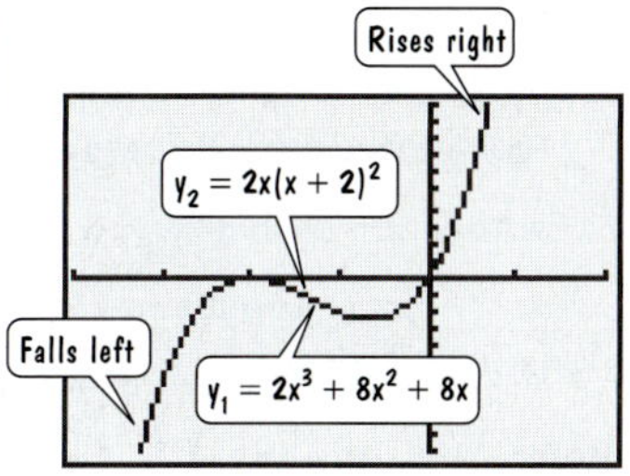

[−4, 2, 1] by [−10, 10, 1]

The degree of y_1 is 3, which is odd. Odd-degree polynomial functions have graphs with opposite behavior at each end. The leading coefficient, 2, is positive. The graph should fall to the left and rise to the right. The viewing rectangle used is complete enough to show this end behavior.

EXAMPLE 1 Factoring a Polynomial

Factor: $2x^3 + 8x^2 + 8x$.

Solution

Step 1 If there is a common factor, factor out the GCF. Because $2x$ is common to all terms, we factor it out.

$$2x^3 + 8x^2 + 8x = 2x(x^2 + 4x + 4) \qquad \text{Factor out the GCF.}$$

Step 2 Determine the number of terms and factor accordingly. The factor $x^2 + 4x + 4$ has three terms and is a perfect square trinomial. We factor using $A^2 + 2AB + B^2 = (A + B)^2$.

$$\begin{aligned} 2x^3 + 8x^2 + 8x &= 2x(x^2 + 4x + 4) \\ &= 2x(x^2 + 2 \cdot x \cdot 2 + 2^2) \\ &\qquad A^2 + 2AB + B^2 \\ &= 2x(x + 2)^2 \qquad A^2 + 2AB + B^2 = (A + B)^2 \end{aligned}$$

Step 3 Check to see if factors can be factored further. In this problem, they cannot. Thus,

$$2x^3 + 8x^2 + 8x = 2x(x + 2)^2.$$

■

✔ CHECK POINT 1 Factor: $3x^3 - 30x^2 + 75x$.

EXAMPLE 2 Factoring a Polynomial

Factor: $4x^2y - 16xy - 20y$.

Solution

Step 1 If there is a common factor, factor out the GCF. Because $4y$ is common to all terms, we factor it out.

$$4x^2y - 16xy - 20y = 4y(x^2 - 4x - 5) \qquad \text{Factor out the GCF.}$$

Step 2 Determine the number of terms and factor accordingly. The factor $x^2 - 4x - 5$ has three terms, but it is not a perfect square trinomial. We factor it using trial and error.

$$4x^2y - 16xy - 20y = 4y(x^2 - 4x - 5) = 4y(x + 1)(x - 5)$$

Step 3 Check to see if factors can be factored further. In this case, they cannot, so we have factored completely. ■

✔ CHECK POINT 2 Factor: $3x^2y - 12xy - 36y$.

EXAMPLE 3 Factoring a Polynomial

Factor: $9b^2x - 16y - 16x + 9b^2y$.

Solution

Step 1 If there is a common factor, factor out the GCF. Other than 1 or -1, there is no common factor.

Step 2 Determine the number of terms and factor accordingly. There are four terms. We try factoring by grouping. Notice that the first and last terms have a common factor of $9b^2$ and the two middle terms have a common factor of -16. Thus, we begin by rearranging the terms.

$$9b^2x - 16y - 16x + 9b^2y$$

$$= (9b^2x + 9b^2y) + (-16x - 16y)$$ Rearrange terms and group terms with common factors.

$$= 9b^2(x + y) - 16(x + y)$$ Factor from each group.

$$= (x + y)(9b^2 - 16)$$ Factor out the common binomial factor, $x + y$.

Step 3 Check to see if factors can be factored further. We note that $9b^2 - 16$ is the difference of two squares, $(3b)^2 - 4^2$, so we continue factoring.

$$9b^2x - 16y - 16x + 9b^2y$$

$$= (x + y)[(3b)^2 - 4^2]$$ Express $9b^2 - 16$ as the difference of squares.

$$= (x + y)(3b + 4)(3b - 4)$$ The factors of $9b^2 - 16$ are the sum and difference of the expression being squared. ■

✔ **CHECK POINT 3** Factor: $16a^2x - 25y - 25x + 16a^2y$.

EXAMPLE 4 Factoring a Polynomial

Factor: $x^2 - 25a^2 + 8x + 16$.

Solution

Step 1 If there is a common factor, factor out the GCF. Other than 1 or -1, there is no common factor.

Step 2 Determine the number of terms and factor accordingly. There are four terms. We try factoring by grouping. Grouping into two groups of two terms does not result in a common binomial factor. Let's try grouping as a difference of squares.

$$x^2 - 25a^2 + 8x + 16$$

$$= (x^2 + 8x + 16) - 25a^2$$ Rearrange terms and group as a perfect square trinomial minus $25a^2$ to obtain a difference of squares.

$$= (x + 4)^2 - (5a)^2$$ Factor the perfect square trinomial.

$$= (x + 4 + 5a)(x + 4 - 5a)$$ Factor the difference of squares. The factors are the sum and difference of the expressions being squared.

Step 3 Check to see if factors can be factored further. In this case, they cannot, so we have factored completely. ■

CHECK POINT 4 Factor: $x^2 - 36a^2 + 20x + 100$.

EXAMPLE 5 Factoring a Polynomial

Factor: $3x^{10} + 3x$.

Solution

Step 1 If there is a common factor, factor out the GCF. Because $3x$ is common to both terms, we factor it out.

$$3x^{10} + 3x = 3x(x^9 + 1) \quad \text{Factor out the GCF.}$$

Step 2 Determine the number of terms and factor accordingly. The factor $x^9 + 1$ has two terms. This binomial can be expressed as $(x^3)^3 + 1^3$, so it can be factored as the sum of two cubes.

$$3x^{10} + 3x = 3x(x^9 + 1)$$
$$= 3x[(x^3)^3 + 1^3] = 3x(x^3 + 1)[(x^3)^2 - x^3 \cdot 1 + 1^2]$$
$$A^3 + B^3 = (A + B)\ (A^2 - AB + B^2)$$
$$= 3x(x^3 + 1)(x^6 - x^3 + 1) \quad \text{Simplify.}$$

Step 3 Check to see if factors can be factored further. We note that $x^3 + 1$ is the sum of two cubes, $x^3 + 1^3$, so we continue factoring.

$$3x^{10} + 3x$$
$$= 3x(x^3 + 1)(x^6 - x^3 + 1) \quad \text{This is our factorization in the previous step.}$$
$$= 3x(x + 1)(x^2 - x + 1)(x^6 - x^3 + 1) \quad \text{Factor completely by factoring } x^3 + 1^3 \text{, the sum of cubes.}$$

CHECK POINT 5 Factor: $x^{10} + 512x$. *Hint*: $512 = 8^3$.

EXERCISE SET 5.6

Practice Exercises

In Exercises 1–60, factor completely, or state that the polynomial is prime.

1. $x^3 - 16x$
2. $x^3 - x$
3. $3x^2 + 18x + 27$
4. $8x^2 + 40x + 50$
5. $81x^3 - 3$
6. $24x^3 - 3$
7. $x^2y - 16y + 32 - 2x^2$
8. $12x^2y - 27y - 4x^2 + 9$
9. $4a^2b - 2ab - 30b$
10. $32y^2 - 48y + 18$
11. $ay^2 - 4a - 4y^2 + 16$
12. $ax^2 - 16a - 2x^2 + 32$
13. $11x^5 - 11xy^2$
14. $4x^9 - 400x$

15. $4x^5 - 64x$

16. $7x^5 - 7x$

17. $x^3 - 4x^2 - 9x + 36$

18. $x^3 - 5x^2 - 4x + 20$

19. $2x^5 + 54x^2$

20. $3x^5 + 24x^2$

21. $3x^4y - 48y^5$

22. $32x^4y - 2y^5$

23. $12x^3 + 36x^2y + 27xy^2$

24. $18x^3 + 48x^2y + 32xy^2$

25. $x^2 - 12x + 36 - 49y^2$

26. $x^2 - 10x + 25 - 36y^2$

27. $4x^2 + 25y^2$

28. $16x^2 + 49y^2$

29. $12x^3y - 12xy^3$

30. $9x^2y^2 - 36y^2$

31. $6bx^2 + 6by^2$

32. $6x^2 - 66$

33. $x^4 - xy^3 + x^3y - y^4$

34. $x^3 - xy^2 + x^2y - y^3$

35. $x^2 - 4a^2 + 12x + 36$

36. $x^2 - 49a^2 + 14x + 49$

37. $5x^3 + x^6 - 14$

38. $6x^3 + x^6 - 16$

39. $4x - 14 + 2x^3 - 7x^2$

40. $3x^3 + 8x + 9x^2 + 24$

41. $54x^3 - 16y^3$

42. $54x^3 - 250y^3$

43. $x^2 + 10x - y^2 + 25$

44. $x^2 + 6x - y^2 + 9$

45. $x^8 - y^8$

46. $x^8 - 1$

47. $x^3y - 16xy^3$

48. $x^3y - 100xy^3$

49. $x + 8x^4$

50. $x + 27x^4$

51. $16y^2 - 4y - 2$

52. $32y^2 + 4y - 6$

53. $14y^3 + 7y^2 - 10y$

54. $5y^3 - 45y^2 + 70y$

55. $27x^2 + 36xy + 12y^2$

56. $125x^2 + 50xy + 5y^2$

57. $12x^3 + 3xy^2$

58. $3x^4 + 27x^2$

59. $x^6y^6 - x^3y^3$

60. $x^3 - 2x^2 - x + 2$

Application Exercises

In Exercises 61–64:

a. *Write an expression for the area of the shaded region.*
b. *Write the expression in factored form.*

61.

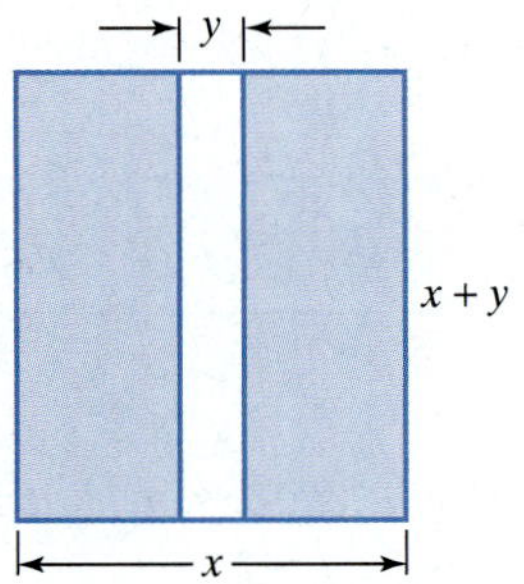

62.

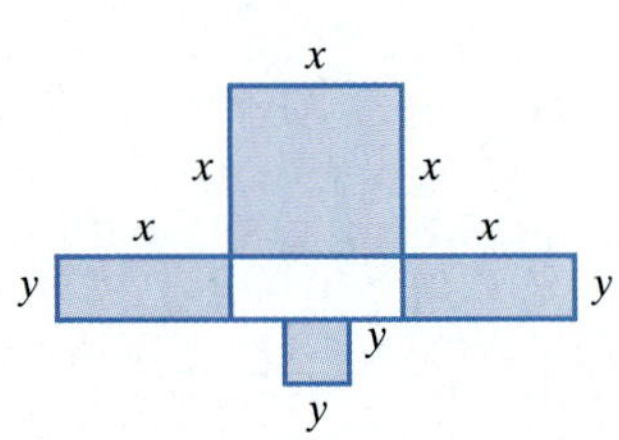

63.

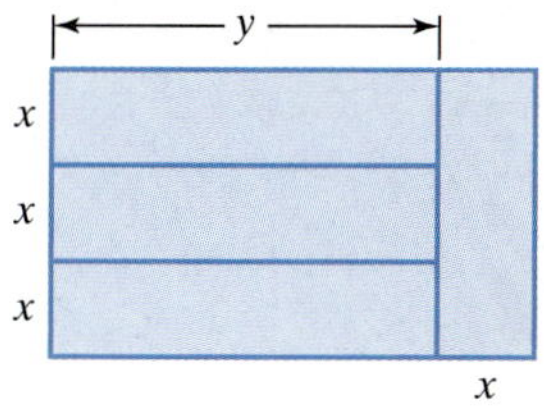

64.

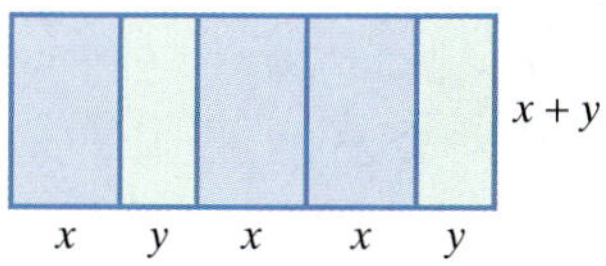

Writing in Mathematics

65. Describe a strategy that can be used to factor polynomials.

66. Describe some of the difficulties in factoring polynomials. What suggestions can you offer to overcome these difficulties?

Technology Exercises

In Exercises 67–70, use a graphing utility to graph the function on each side of the equation in the same viewing rectangle. Use end behavior to show a complete picture of the polynomial on the left side. Do the graphs coincide? If so, the factorization is correct. If not, factor correctly and then use your graphing utility to verify the factorization.

67. $4x^2 - 12x + 9 = (4x - 3)^2$

68. $2x^3 + 10x^2 - 2x - 10 = 2(x + 5)(x^2 + 1)$

69. $x^4 - 16 = (x^2 + 4)(x + 2)(x - 2)$

70. $x^3 + 1 = (x + 1)^3$

Critical Thinking Exercises

71. Which one of the following is true?

a. $x^4 - 16$ is factored completely as $(x^2 + 4)(x^2 - 4)$.

b. The trinomial $x^2 - 4x - 4$ is a prime polynomial.

c. $x^2 + 36 = (x + 6)^2$

d. $x^3 - 64 = (x + 4)(x^2 + 4x - 16)$

In Exercises 72–76, factor completely. All variable exponents represent whole numbers.

72. $(x^2 - 3)^2 - 7(x^2 - 3) + 6$

73. $(x^2 - 8)^2 - 4(x^2 - 8) - 32$

74. $5x^3y^2 - 5xy^4 - 5x^2y^2 + 5xy^3$

75. $x^{4n+1} - xy^{4n}$

76. $x^4 - y^4 - 2x^3y + 2xy^3$

77. In certain circumstances, the sum of two perfect squares can be factored by adding and subtracting the same perfect square. For example,

$$x^4 + 4 = x^4 + 4x^2 + 4 - 4x^2. \quad \text{Add and subtract } 4x^2.$$

Use this first step to factor $x^4 + 4$.

78. Express $x^3 + x + 2x^4 + 4x^2 + 2$ as the product of two polynomials of degree 2.

Review Exercises

79. Solve: $\frac{3x - 1}{5} + \frac{x + 2}{2} = -\frac{3}{10}$.
(Section 1.4, Example 4)

80. Simplify: $(4x^3y^{-1})^2(2x^{-3}y)^{-1}$.
(Section 1.6, Example 9)

81. Evaluate: $\begin{vmatrix} 0 & -3 & 2 \\ 1 & 5 & 3 \\ -2 & 1 & 4 \end{vmatrix}$.
(Section 3.5, Example 3)

SECTION 5.7 Polynomial Equations and Their Applications

Objectives

1. Solve quadratic equations by factoring.
2. Solve higher-degree polynomial equations by factoring.
3. Solve problems using polynomial equations.

SSM PH Tutor Center CD-ROM Video

If you are a coffee drinker, do you become a tad unnerved if you do not get your caffeine rush at a regular time? We opened the chapter with a polynomial function that models the percentage of coffee drinkers, $P(x)$, who are x years old who become irritable if they do not have coffee at their regular time:

$$P(x) = -0.05x^2 + 4.2x - 26.$$

How can we use this function to find the age at which 36% of coffee drinkers become irritable if they do not have coffee at their regular time? Substitute 36 for $P(x)$ and solve for x:

$$36 = -0.05x^2 + 4.2x - 26.$$

Do you see how this equation differs from a linear equation? The highest exponent on x is 2. Solving such an equation involves finding the numbers that will make the equation a true statement. In this section, we use factoring for solving equations in the form $ax^2 + bx + c = 0$. We also look at applications of these equations.

ENRICHMENT ESSAY

Trends in U.S. Coffee Consumption

- 54% of adults drink coffee every day, representing 110 million daily drinkers.
- Another 25% of adults, or 51 million people, drink coffee occasionally.
- Twenty million adults drink gourmet coffee beverages every day.
- Coffee drinkers consume, on average, 3.1 cups of coffee per day.

Source: National Coffee Association

The Standard Form of a Quadratic Equation

We begin by defining a quadratic equation.

Definition of a Quadratic Equation

A **quadratic equation** in x is an equation that can be written in the **standard form**

$$ax^2 + bx + c = 0$$

where a, b, and c are real numbers with $a \neq 0$. A quadratic equation in x is also called a **second-degree polynomial equation** in x.

An example of a quadratic equation in standard form is $x^2 - 7x + 10 = 0$. The coefficient of x^2 is $1 (a = 1)$, the coefficient of x is $-7 (b = -7)$, and the constant term is $10 (c = 10)$.

1 Solve quadratic equations by factoring.

Solving Quadratic Equations by Factoring

We can factor the left side of the quadratic equation $x^2 - 7x + 10 = 0$. We obtain $(x - 5)(x - 2) = 0$. If a quadratic equation has zero on one side and a factored expression on the other side, it can be solved using the **zero-product principle**.

The Zero-Product Principle

If the product of two algebraic expressions is zero, then at least one of the factors is equal to zero.

If $AB = 0$, then $A = 0$ or $B = 0$.

For example, consider the equation $(x - 5)(x - 2) = 0$. According to the zero-product principle, this product can be zero only if at least one of the factors is zero. We set each individual factor equal to zero and solve each resulting equation for x.

$$(x - 5)(x - 2) = 0$$

$$x - 5 = 0 \quad \text{or} \quad x - 2 = 0$$

$$x = 5 \qquad\qquad x = 2$$

The solutions of the original quadratic equation, $x^2 - 7x + 10 = 0$, are 5 and 2. The solution set is $\{5, 2\}$.

Solving a Quadratic Equation by Factoring

1. If necessary, rewrite the equation in the form $ax^2 + bx + c = 0$, moving all terms to one side, thereby obtaining zero on the other side.
2. Factor.
3. Apply the zero-product principle, setting each factor equal to zero.
4. Solve the equations in step 3.
5. Check the solutions in the original equation.

EXAMPLE 1 Solving a Quadratic Equation by Factoring

Solve: $2x^2 + 7x = 4$.

Solution

Step 1 Move all terms to one side and obtain zero on the other side. Subtract 4 from both sides and write the equation in standard form.

$$2x^2 + 7x - 4 = 4 - 4$$

$$2x^2 + 7x - 4 = 0$$

Study Tip

Do not confuse factoring a polynomial with solving a quadratic equation by factoring.

INCORRECT

Factor: $2x^2 + 7x - 4$.

$$(2x - 1)(x + 4)$$

$$2x - 1 = 0 \quad \text{or} \quad x + 4 = 0$$

$$x = \frac{1}{2} \qquad\qquad x = -4$$

Step 2 Factor.

$$2x^2 + 7x - 4 = 0$$

$$(2x - 1)(x + 4) = 0$$

Steps 3 and 4 Set each factor equal to zero and solve each resulting equation.

$$2x - 1 = 0 \quad \text{or} \quad x + 4 = 0$$

$$2x = 1 \quad \text{or} \quad x = -4$$

$$x = \frac{1}{2}$$

Step 5 Check the solutions in the original equation.

Check $\frac{1}{2}$:

$$2x^2 + 7x = 4$$
$$2\left(\frac{1}{2}\right)^2 + 7\left(\frac{1}{2}\right) \stackrel{?}{=} 4$$
$$\frac{1}{2} + \frac{7}{2} \stackrel{?}{=} 4$$
$$4 = 4, \text{ true}$$

Check -4:

$$2x^2 + 7x = 4$$
$$2(-4)^2 + 7(-4) \stackrel{?}{=} 4$$
$$32 + (-28) \stackrel{?}{=} 4$$
$$4 = 4, \text{ true}$$

The solutions are -4 and $\frac{1}{2}$, and the solution set is $\left\{\frac{1}{2}, -4\right\}$. ■

✔ **CHECK POINT 1** Solve: $2x^2 - 9x = 5$.

There is an important relationship between a quadratic equation in standard form, such as

$$2x^2 + 7x - 4 = 0$$

and a quadratic function, such as

$$y = 2x^2 + 7x - 4.$$

The solutions of $ax^2 + bx + c = 0$ correspond to the x-intercepts of the graph of the quadratic function $y = ax^2 + bx + c$. For example, you can visualize the solutions of $2x^2 + 7x - 4 = 0$ by looking at the x-intercepts of the graph of the quadratic function $y = 2x^2 + 7x - 4$. The cuplike graph is shown in Figure 5.10. The solutions of the equation $2x^2 + 7x - 4 = 0$, -4 and $\frac{1}{2}$, appear as the graph's x-intercepts.

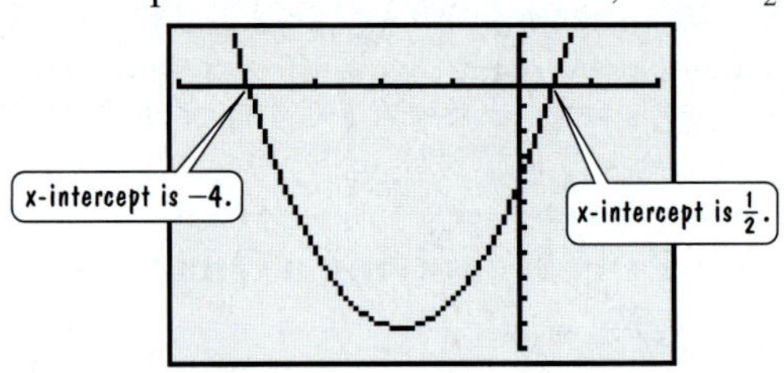

[−5, 2, 1] by [−12, 2, 1]

Figure 5.10

EXAMPLE 2 Solving Quadratic Equations by Factoring

Solve:

a. $5x^2 = 20x$ **b.** $x^2 + 4 = 8x - 12$ **c.** $(x - 7)(x + 5) = -20$.

Solution

a.

$5x^2 = 20x$	This is the given equation.
$5x^2 - 20x = 0$	Subtract $20x$ from both sides, and write the equation in standard form.
$5x(x - 4) = 0$	Factor.
$5x = 0$ or $x - 4 = 0$	Set each factor equal to 0.
$x = 0$ or $x = 4$	Solve the resulting equations.

Check by substituting 0 and 4 into the given equation. The graph of $y = 5x^2 - 20x$, obtained with a graphing utility, is shown in Figure 5.11. The x-intercepts are 0 and 4. This verifies that the solutions are 0 and 4, and the solution set is $\{0, 4\}$.

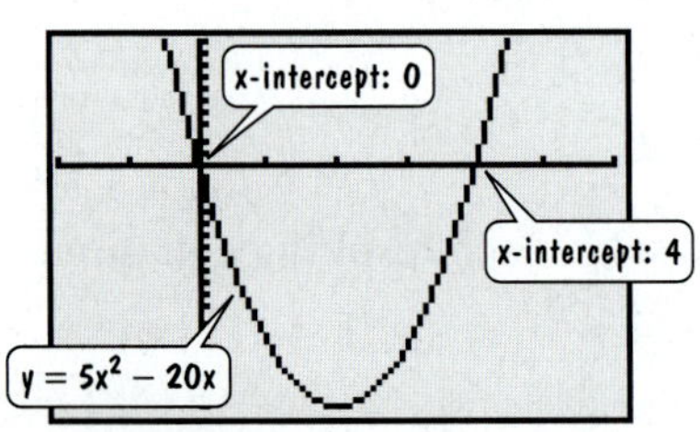

[−2, 6, 1] by [−20, 10, 1]

Figure 5.11 The solution set of $5x^2 - 20x = 0$ is $\{0, 4\}$.

b. $x^2 + 4 = 8x - 12$ — This is the given equation.

$x^2 - 8x + 16 = 0$ — Write the equation in standard form by subtracting 8x and adding 12 on both sides.

$(x - 4)(x - 4) = 0$ — Factor.

$x - 4 = 0$ or $x - 4 = 0$ — Set each factor equal to 0.

$x = 4$ or $x = 4$ — Solve the resulting equations.

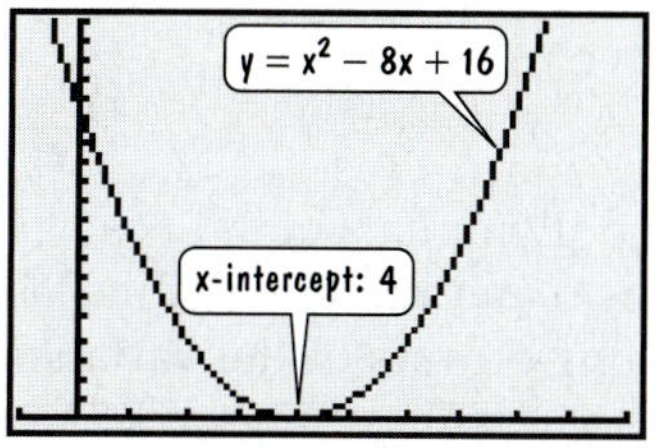

[−1, 10, 1] by [0, 20, 1]

Figure 5.12 The solution set of $x^2 - 8x + 16 = 0$ is $\{4\}$.

Notice that there is only one solution (or, if you prefer, a repeated solution.) The trinomial $x^2 - 8x + 16$ is a perfect square trinomial that could have been factored as $(x - 4)^2$. The graph of $y = x^2 - 8x + 16$, obtained with a graphing utility, is shown in Figure 5.12. The graph has only one x-intercept at 4. This verifies that the equation's solution is 4, and the solution set is $\{4\}$.

c. Be careful! Although the left side is factored in $(x - 7)(x + 5) = -20$, we cannot use the zero product principle. Why not? The right side of the equation is not 0. So we begin by multiplying the factors on the left side of the equation. Then we add 20 to both sides to obtain 0 on the right side.

$(x - 7)(x + 5) = -20$ — This is the given equation.

$x^2 - 2x - 35 = -20$ — Use the FOIL method to multiply on the left side.

$x^2 - 2x - 15 = 0$ — Add 20 to both sides.

$(x + 3)(x - 5) = 0$ — Factor.

$x + 3 = 0$ or $x - 5 = 0$ — Set each factor equal to 0.

$x = -3$ or $x = 5$ — Solve the resulting equations.

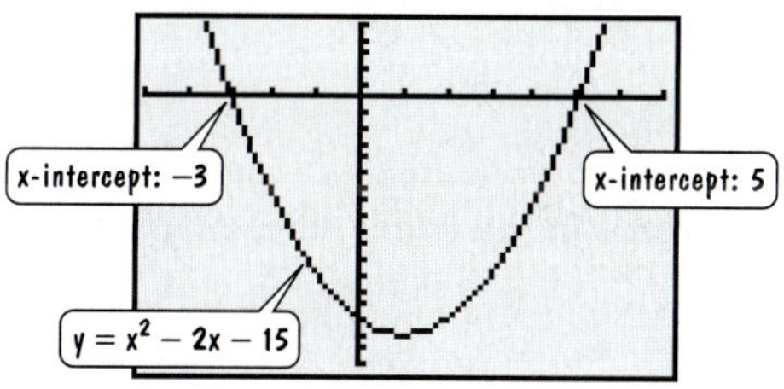

[−5, 7, 1] by [−18, 5, 1]

Figure 5.13 The solution set of $x^2 - 2x - 15 = 0$ is $\{-3, 5\}$.

Check by substituting −3 and 5 into the given equation. The graph of $y = x^2 - 2x - 15$, obtained with a graphing utility, is shown in Figure 5.13. The x-intercepts are −3 and 5. This verifies that the solutions are −3 and 5, and the solution set is $\{-3, 5\}$. ■

Study Tip

Avoid the following errors:

$$5x^2 = 20x$$
$$\frac{5x^2}{x} = \frac{20x}{x}$$
$$5x = 20$$
$$x = 4$$

Never divide both sides of an equation by x. Division by zero is undefined and x may be zero. Indeed, the solutions for this equation (Example 2a) are 0 and 4. Dividing both sides by x does not permit us to find both solutions.

$$(x - 7)(x + 5) = -20$$
$$x - 7 = -20 \quad \text{or} \quad x + 5 = -20$$
$$x = -13 \quad \text{or} \quad x = -25$$

The zero product principle cannot be used because the right side of the equation is not equal to 0.

CHECK POINT 2 Solve:

a. $3x^2 = 2x$

b. $x^2 + 7 = 10x - 18$

c. $(x - 2)(x + 3) = 6.$

2 Solve higher-degree polynomial equations by factoring.

Polynomial Equations

A **polynomial equation** is the result of setting two polynomials equal to each other. The equation is in **standard form** if one side is 0 and the polynomial on the other side is in standard form, that is, in descending powers of the variable. The **degree of a polynomial equation** is the same as the highest degree of any term in the equation. Here are examples of three polynomial equations:

$$3x + 5 = 14 \qquad 2x^2 + 7x = 4 \qquad x^3 + x^2 = 4x + 4.$$

This equation is of degree 1 because 1 is the highest degree.

This equation is of degree 2 because 2 is the highest degree.

This equation is of degree 3 because 3 is the highest degree.

Notice that a polynomial equation of degree 1 is a linear equation. A polynomial equation of degree 2 is a quadratic equation.

Some polynomial equations of degree 3 or higher can be solved by moving all terms to one side, thereby obtaining 0 on the other side. Once the equation is in standard form, factor and then set each factor equal to 0.

EXAMPLE 3 Solving a Polynomial Equation by Factoring

Solve by factoring: $x^3 + x^2 = 4x + 4.$

Solution

Step 1 Move all terms to one side and obtain zero on the other side. Subtract $4x$ and subtract 4 from both sides.

$$x^3 + x^2 - 4x - 4 = 4x + 4 - 4x - 4$$
$$x^3 + x^2 - 4x - 4 = 0$$

Step 2 Factor. Use factoring by grouping. Group terms that have a common factor.

$$\boxed{x^3 + x^2} + \boxed{-4x - 4} = 0$$

Common factor is x^2. Common factor is -4.

$$x^2(x + 1) - 4(x + 1) = 0$$ Factor x^2 from the first two terms and -4 from the last two terms.

$$(x + 1)(x^2 - 4) = 0$$ Factor out the common binomial, $x + 1$, from each term.

$$(x + 1)(x + 2)(x - 2) = 0$$ Factor completely by factoring $x^2 - 4$ as the difference of squares.

Steps 3 and 4 Set each factor equal to zero and solve the resulting equations.

$$x + 1 = 0 \quad \text{or} \quad x + 2 = 0 \quad \text{or} \quad x - 2 = 0$$
$$x = -1 \qquad\qquad x = -2 \qquad\qquad x = 2$$

Step 5 Check the solutions in the original equation. Check the three solutions, $-1, -2$, and 2, by substituting them into the original equation. Can you verify that the solutions are $-1, -2$, and 2, and the solution set is $\{-2, -1, 2\}$? ■

Using Technology

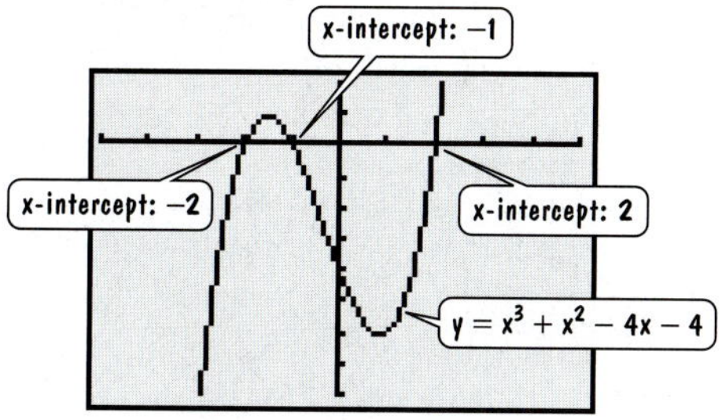

[−5, 5, 1] by [−8, 2, 1]

You can use a graphing utility to check the solutions to $x^3 + x^2 - 4x - 4 = 0$. Graph $y = x^3 + x^2 - 4x - 4$, as shown on the left. Is the graph complete? Because the degree, 3, is odd, and the leading coefficient, 1, is positive, it should fall to the left and rise to the right. The graph shows this end behavior and is therefore complete. The x-intercepts are −2, −1, and 2, corresponding to the equation's solutions.

✔ **CHECK POINT 3** Solve by factoring: $2x^3 + 3x^2 = 8x + 12$.

3 Solve problems using polynomial equations.

Applications of Polynomial Equations

Solving polynomial equations by factoring can be used to answer questions about variables contained in mathematical models.

EXAMPLE 4 Modeling Motion

You throw a ball straight up from a rooftop 160 feet high with an initial speed of 48 feet per second. The function

$$s(t) = -16t^2 + 48t + 160$$

describes the ball's height above the ground, $s(t)$, in feet, t seconds after you threw it. The ball misses the rooftop on its way down and eventually strikes the ground. How long will it take for the ball to hit the ground?

Solution The ball hits the ground when $s(t)$, its height above the ground, is 0 feet. Thus, we substitute 0 for $s(t)$ in the given function and solve for t.

$$s(t) = -16t^2 + 48t + 160$$ This is the function that models the ball's height.

$$0 = -16t^2 + 48t + 160$$ Substitute 0 for $s(t)$.

$$0 = -16(t^2 - 3t - 10)$$ Factor out −16.

$$0 = -16(t - 5)(t + 2)$$ Factor the trinomial.

Do not set the constant, −16, equal to zero: $-16 \neq 0$.

$$t - 5 = 0 \quad \text{or} \quad t + 2 = 0$$ Set each variable factor equal to 0.

$$t = 5 \quad \text{or} \quad t = -2$$ Solve for t.

Because we begin describing the ball's height at $t = 0$, we discard the solution $t = -2$. The ball hits the ground after 5 seconds. ■

Figure 5.14 shows the graph of the function $s(t) = -16t^2 + 48t + 160$. The horizontal axis is labeled t, for the ball's time in motion. The vertical axis is labeled $s(t)$, for the ball's height above the ground at time t. Because time and height are both positive, the function is graphed in quadrant I only.

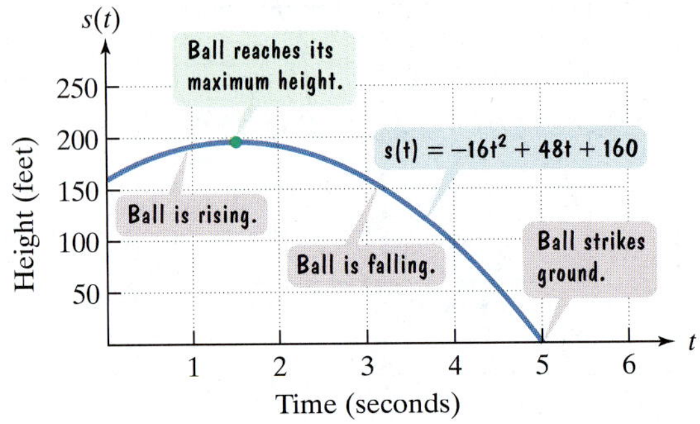

Figure 5.14

The graph visually shows what we discovered algebraically: The ball hits the ground after 5 seconds. The graph also reveals that the ball reaches its maximum height, nearly 200 feet, after 1.5 seconds. Then the ball begins to fall.

✔ **CHECK POINT 4** Use the function $s(t) = -16t^2 + 48t + 160$ to determine when the ball's height is 192 feet. Identify your solutions as points on the graph in Figure 5.14.

ENRICHMENT ESSAY

Visualizing Irritability by Age

The polynomial function

$$P(x) = -0.05x^2 + 4.2x - 26$$

models the percentage of coffee drinkers, $P(x)$, who are x years old who become irritable if they do not have coffee at their regular time. Figure 5.15 shows the graph of the function. The graph reveals that 62.2% of 42-year-old coffee drinkers become irritable. This is the maximum percentage for any age, x, in the function's domain.

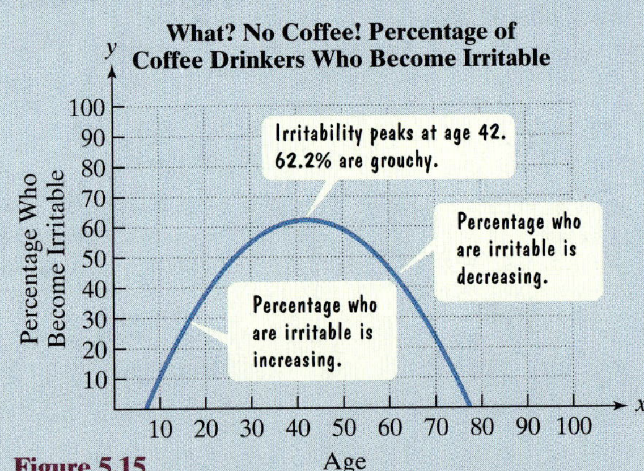

Figure 5.15

Source: LMK Associates

In our next example, we use our five-step strategy for solving word problems.

EXAMPLE 5 Solving a Problem Involving Landscape Design

A rectangular garden measures 80 feet by 60 feet. A large path of uniform width is to be added along both shorter sides and one longer side of the garden. The landscape designer doing the work wants to double the garden's area with the addition of this path. How wide should the path be?

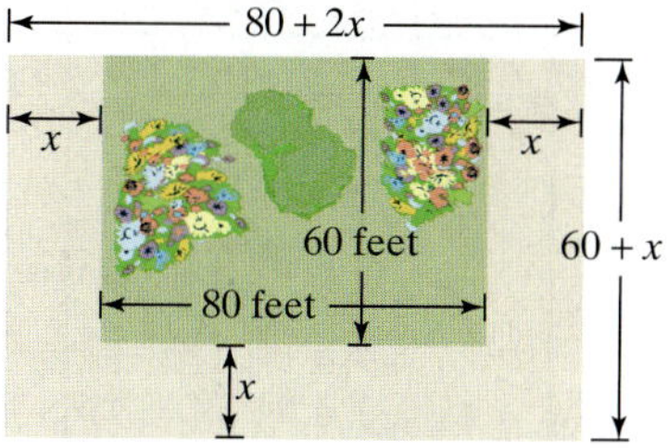

Figure 5.16 The garden's area is to be doubled by adding the path.

Solution

Step 1 Let x represent one of the quantities. We will let

$$x = \text{the width of the path.}$$

The situation is illustrated in Figure 5.16. The figure shows the original 80-by-60 foot rectangular garden and the path of width x added along both shorter sides and one longer side.

Step 2 Represent other quantities in terms of x. Because the path is added along both shorter sides and one longer side, Figure 5.16 shows that

$$80 + 2x = \text{the length of the new, expanded, rectangle}$$
$$60 + x = \text{the width of the new, expanded, rectangle.}$$

Step 3 Write an equation that describes the conditions. The area of the rectangle must be doubled by the addition of the path.

The area, or length times width, of the new, expanded rectangle	must be	twice that of	the area of the garden.
$(80 + 2x)(60 + x)$	$=$	$2 \cdot$	$80 \cdot 60$

Step 4 Solve the equation and answer the question.

$$(80 + 2x)(60 + x) = 2 \cdot 80 \cdot 60$$ This is the equation for the problem's conditions.

$$4800 + 200x + 2x^2 = 9600$$ Multiply, using FOIL on the left side.

$$2x^2 + 200x - 4800 = 0$$ Subtract 9600 from both sides and write the equation in standard form.

$$2(x^2 + 100x - 2400) = 0$$ Factor out 2, the GCF.

$$2(x - 20)(x + 120) = 0$$ Factor the trinomial.

$$x - 20 = 0 \quad \text{or} \quad x + 120 = 0$$ Set each variable factor equal to 0.

$$x = 20 \quad \text{or} \quad x = -120$$ Solve for x.

The path cannot have a negative width. Because -120 is geometrically impossible, we use $x = 20$. The width of the path should be 20 feet.

Step 5 Check the proposed solution in the original wording of the problem. Has the landscape architect doubled the garden's area with the 20-foot-wide path? The area of the garden is 80 feet times 60 feet, or 4800 square feet. Because $80 + 2x$ and $60 + x$ represent the length and width of the expanded rectangle,

$$80 + 2x = 80 + 2 \cdot 20 = 120 \text{ feet is the expanded rectangle's length.}$$
$$60 + x = 60 + 20 = 80 \text{ feet is the expanded rectangle's width.}$$

The area of the expanded rectangle is 120 feet times 80 feet, or 9600 square feet. This is double the area of the garden, 4800 square feet, as specified by the problem's conditions. ■

✔ **CHECK POINT 5** A rectangular garden measures 16 feet by 12 feet. A path of uniform width is to be added so as to surround the entire garden. The landscape artist doing the work wants the garden and path to cover an area of 320 square feet. How wide should the path be?

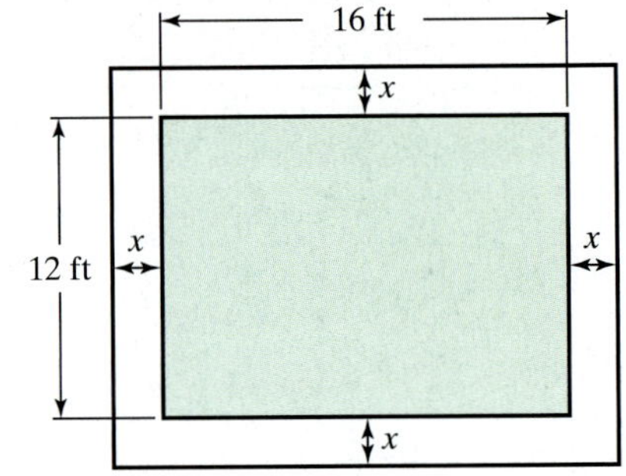

The solution to our next problem relies on knowing the **Pythagorean Theorem.** The theorem relates the lengths of the three sides of a **right triangle**, a triangle with one angle measuring 90°. The side opposite the 90° angle is called the **hypotenuse.** The other sides are called **legs.** The legs form the two sides of the right angle.

The Pythagorean Theorem

The sum of the squares of the lengths of the legs of a right triangle equals the square of the length of the hypotenuse.

If the legs have lengths a and b, and the hypotenuse has length c, then

$$a^2 + b^2 = c^2.$$

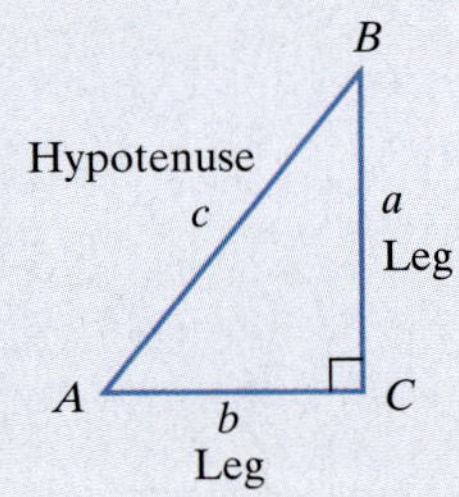

EXAMPLE 6 Using the Pythagorean Theorem to Obtain a Polynomial Equation

Figure 5.17 shows a tent with wires attached to help stabilize it. The length of the wire is 8 feet greater than the height from the ground where the wire is attached to the tent. Find the length of the wire.

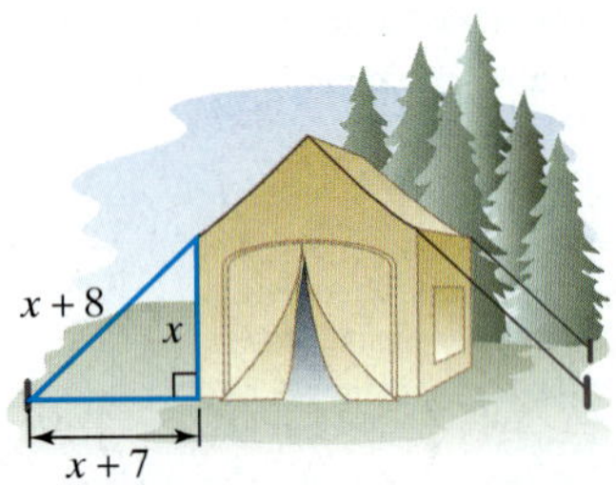

Figure 5.17

Solution Figure 5.17 shows a right triangle. The length of the legs are x and $x + 7$. The length of the hypotenuse, $x + 8$, represents the length of the wire. We use the Pythagorean Theorem to find this length.

$$\underbrace{x^2}_{\text{leg}^2} + \underbrace{(x+7)^2}_{\text{leg}^2} = \underbrace{(x+8)^2}_{\text{hypotenuse}^2}$$ This is the equation arising from the Pythagorean Theorem.

$x^2 + x^2 + 14x + 49 = x^2 + 16x + 64$ Square $x + 7$ and $x + 8$.

$2x^2 + 14x + 49 = x^2 + 16x + 64$ Combine like terms: $x^2 + x^2 = 2x^2$.

$x^2 - 2x - 15 = 0$ Subtract $x^2 + 16x + 64$ from both sides and write the quadratic equation in standard form.

$(x - 5)(x + 3) = 0$ Factor the trinomial.

$x - 5 = 0$ or $x + 3 = 0$ Set each factor equal to 0.

$x = 5$ or $x = -3$ Solve for x.

Because x represents the height from the ground where the wire is attached, x cannot be negative. Thus, we only use $x = 5$. Figure 5.17 on page 354 shows that the length of the wire is $x + 8$ feet. The length of the wire is $5 + 8$ feet, or 13 feet.

We can check to see that the lengths of the three sides of the right triangle, x, $x + 7$, and $x + 8$, satisfy the Pythagorean Theorem when $x = 5$. The lengths are 5 feet, 12 feet, and 13 feet.

$$\text{leg}^2 + \text{leg}^2 = \text{hypotenuse}^2$$
$$5^2 + 12^2 \stackrel{?}{=} 13^2$$
$$25 + 144 \stackrel{?}{=} 169$$
$$169 = 169, \text{ true}$$

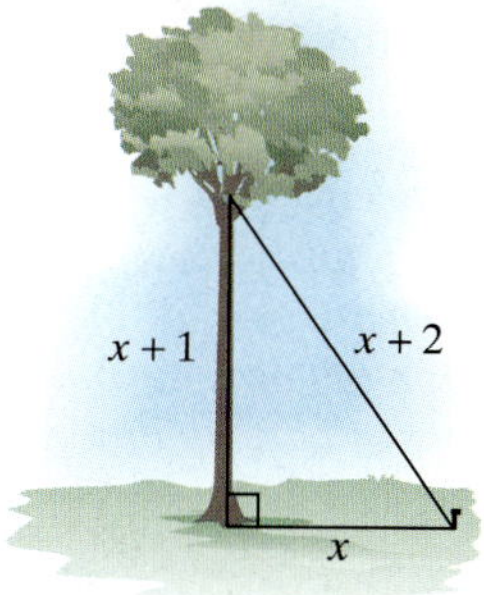

Figure 5.18

CHECK POINT 6 A guy wire is attached to a tree to help it grow straight. The situation is illustrated in Figure 5.18. The length of the wire is 2 feet greater than the distance from the base of the tree to the stake. Find the length of the wire.

EXERCISE SET 5.7

Practice Exercises

In Exercises 1–36, use factoring to solve each quadratic equation. Check by substitution or by using a graphing utility and identifying x-intercepts.

1. $x^2 + x - 12 = 0$

2. $x^2 - 2x - 15 = 0$

3. $x^2 + 6x = 7$

4. $x^2 - 4x = 45$

5. $3x^2 + 10x - 8 = 0$

6. $2x^2 - 5x - 3 = 0$

7. $5x^2 = 8x - 3$

8. $7x^2 = 30x - 8$

9. $3x^2 = 2 - 5x$

10. $5x^2 = 2 + 3x$

11. $x^2 = 8x$

12. $x^2 = 4x$

13. $3x^2 = 5x$

14. $2x^2 = 5x$

15. $x^2 + 4x + 4 = 0$

16. $x^2 + 6x + 9 = 0$

17. $x^2 = 14x - 49$

18. $x^2 = 12x - 36$

19. $9x^2 = 30x - 25$

20. $4x^2 = 12x - 9$

21. $x^2 - 25 = 0$

22. $x^2 - 49 = 0$

23. $9x^2 = 100$

24. $4x^2 = 25$

25. $x(x - 3) = 18$

26. $x(x - 4) = 21$

27. $(x - 3)(x + 8) = -30$

28. $(x - 1)(x + 4) = 14$

29. $x(x + 8) = 16(x - 1)$

30. $x(x + 9) = 4(2x + 5)$

31. $(x + 1)^2 - 5(x + 2) = 3x + 7$

32. $(x + 1)^2 = 2(x + 5)$

33. $x(8x + 1) = 3x^2 - 2x + 2$

34. $2x(x + 3) = -5x - 15$

35. $\frac{x^2}{18} + \frac{x}{2} + 1 = 0$

36. $\frac{x^2}{4} - \frac{5x}{2} + 6 = 0$

In Exercises 37–46, use factoring to solve each polynomial equation. Check by substitution or by using a graphing utility and identifying x-intercepts.

37. $x^3 + 4x^2 - 25x - 100 = 0$

38. $x^3 - 2x^2 - x + 2 = 0$

39. $x^3 - x^2 = 25x - 25$

40. $x^3 + 2x^2 = 16x + 32$

41. $3x^4 - 48x^2 = 0$

42. $5x^4 - 20x^2 = 0$

43. $2x^4 = 16x$

44. $3x^4 = 81x$

45. $2x^3 + 16x^2 + 30x = 0$

46. $3x^3 - 9x^2 - 30x = 0$

In Exercises 47–50, determine the x-intercepts of the graph of each quadratic function. Then match the function with its graph, labeled (a)–(d). Each graph is shown in a $[-10, 10, 1]$ *by* $[-10, 10, 1]$ *viewing rectangle.*

47. $y = x^2 - 6x + 8$

48. $y = x^2 - 2x - 8$

49. $y = x^2 + 6x + 8$

50. $y = x^2 + 2x - 8$

a.

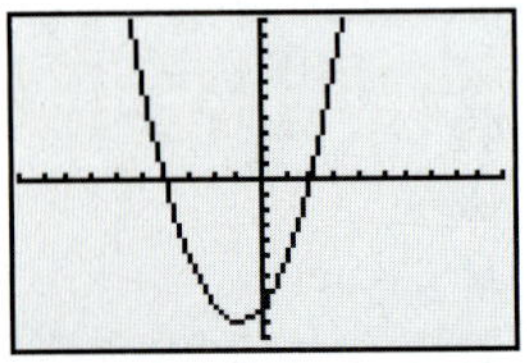

b.

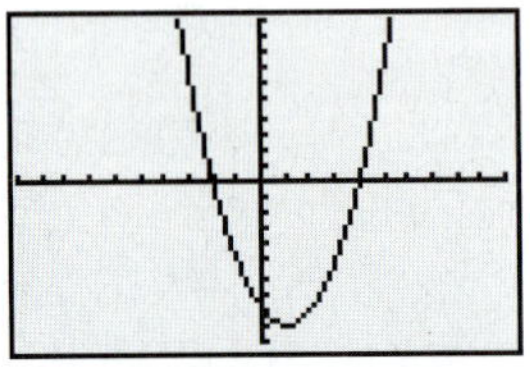

c.

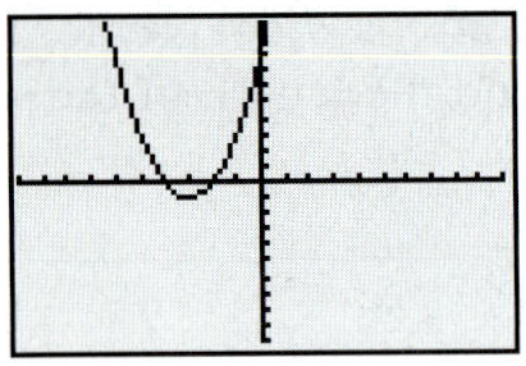

d.

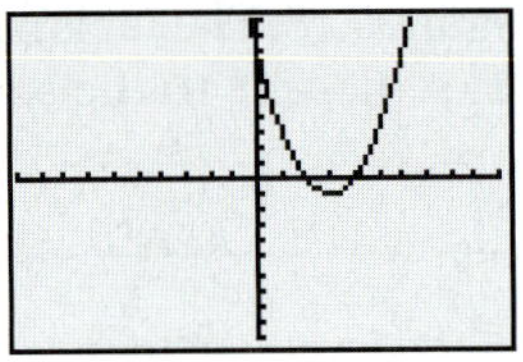

Application Exercises

A ball is thrown straight up from a rooftop 300 feet high. The function

$$s(t) = -16t^2 + 20t + 300$$

describes the ball's height above the ground, s(t), in feet, t seconds after it was thrown. The ball misses the rooftop on its way down and eventually strikes the ground. The graph of the function is shown, with tick marks omitted along the horizontal axis. Use the function to solve Exercises 51–53.

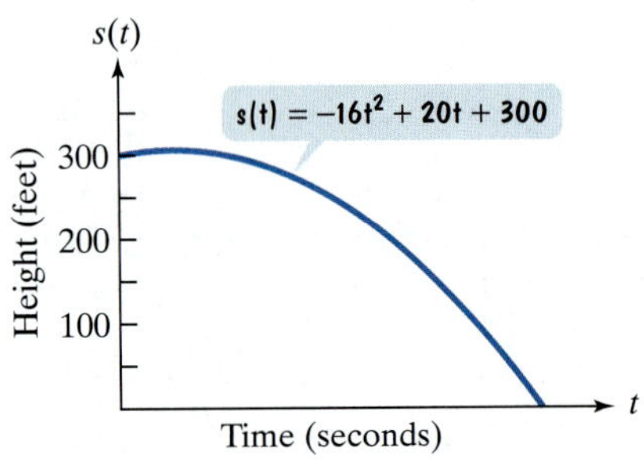

51. How long will it take for the ball to hit the ground? Use this information to provide tick marks with appropriate numbers along the horizontal axis in the figure shown.

52. When will the ball's height be 304 feet? Identify the solution as one or more points on the graph.

53. When will the ball's height be 276 feet? Identify the solution as one or more points on the graph.

54. A model rocket is launched from the top of a cliff 80 feet above the ocean. The function

$$s(t) = -16t^2 + 64t + 80$$

describes the rocket's height above the water, $s(t)$, in feet, t seconds after it was launched. The rocket misses the edge of the cliff on its way down and eventually lands in the ocean. How long will it take for the rocket to hit the water?

The Food Stamp Program is America's first line of defense against hunger for millions of families. Over half of all participants are children; one out of six is a low-income older adult. Exercises 55–58 involve the number of participants in the program from 1990 through 2000.

The function

$$f(x) = -\frac{1}{2}x^2 + 4x + 19$$

models the number of people, f(x), in millions, receiving food stamps x years after 1990. Use the function to solve Exercises 55–56.

55. In which year did 27 million people receive food stamps?

56. In which years did 19 million people receive food stamps?

The graph of the quadratic function in Exercises 55–56 is shown. Use the graph to solve Exercises 57–58.

Number of People Receiving Food Stamps

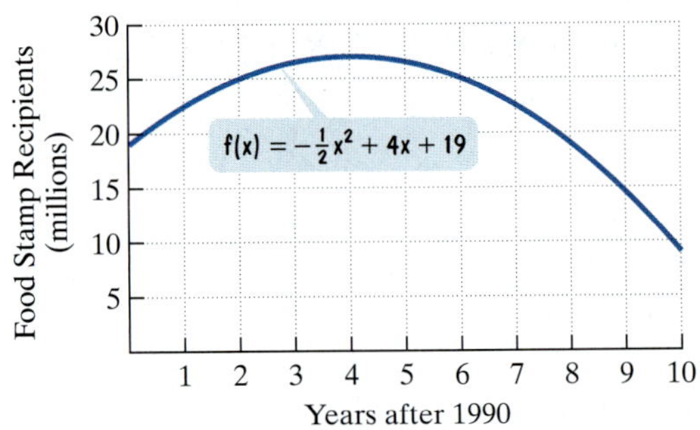

Source: New York Times

57. Identify your solution in Exercise 55 as a point on the graph. Describe what is significant about this point.

58. Identify your solution in Exercise 56 as one or more points on the graph. Then describe the trend shown by the graph.

59. The length of a rectangular sign is 3 feet longer than the width. If the sign's area is 54 square feet, find its length and width.

60. A rectangular parking lot has a length that is 3 yards greater than the width. The area of the parking lot is 180 square yards. Find the length and the width.

61. Each side of a square is lengthened by 3 inches. The area of this new, larger square is 64 square inches. Find the length of a side of the original square.

62. Each side of a square is lengthened by 2 inches. The area of this new, larger square is 36 square inches. Find the length of a side of the original square.

63. A pool measuring 10 meters by 20 meters is surrounded by a path of uniform width, as shown in the figure. If the area of the pool and the path combined is 600 square meters, what is the width of the path?

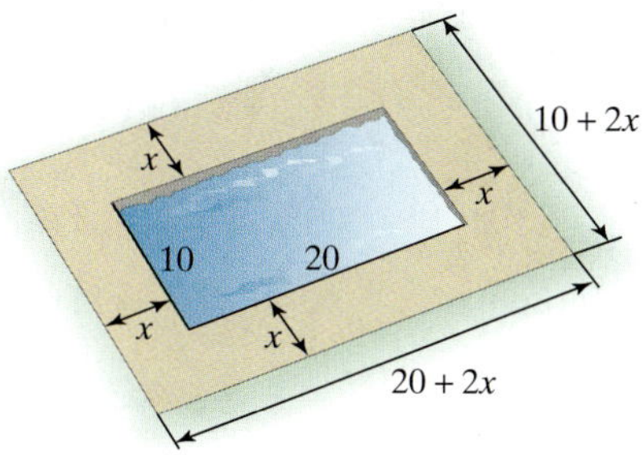

64. A vacant rectangular lot is being turned into a community vegetable garden measuring 15 meters by 12 meters. A path of uniform width is to surround the garden. If the area of the lot is 378 square meters, find the width of the path surrounding the garden.

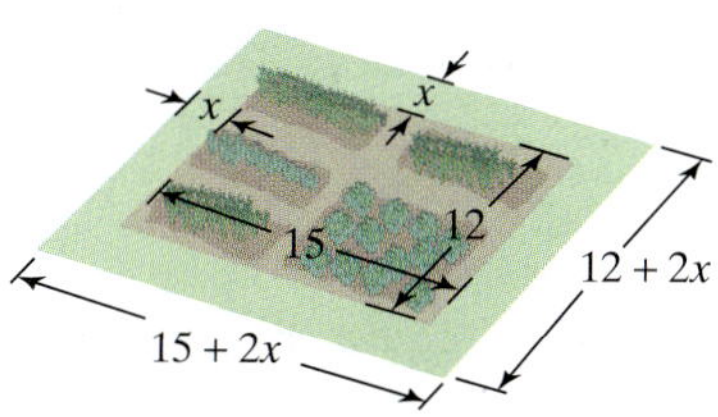

65. As part of a landscaping project, you put in a flower bed measuring 10 feet by 12 feet. You plan to surround the bed with a uniform border of low-growing plants.

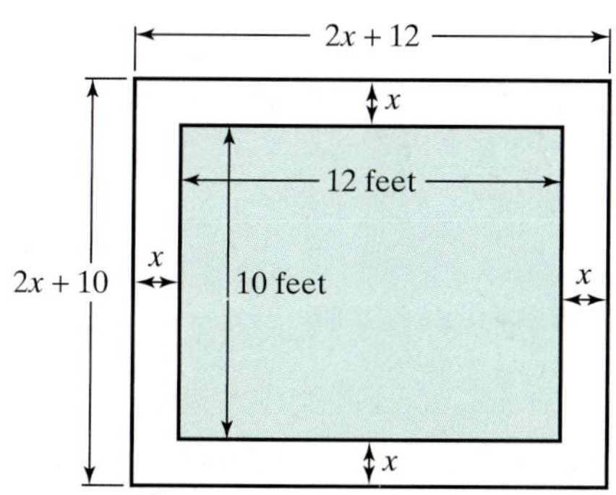

a. Write a polynomial that describes the area of the uniform border that surrounds your flower bed. *Hint:* The area of the border is the area of the large rectangle shown in the figure minus the area of the flower bed.

b. The low growing plants surrounding the flower bed require 1 square foot each when mature. If you have 168 of these plants, how wide a strip around the flower bed should you prepare for the border?

66. As part of a landscaping project, you put in a flower bed measuring 20 feet by 30 feet. To finish off the project, you are putting in a uniform border of pine bark around the outside of the rectangular garden. You have enough pine bark to cover 336 square feet. How wide should the border be?

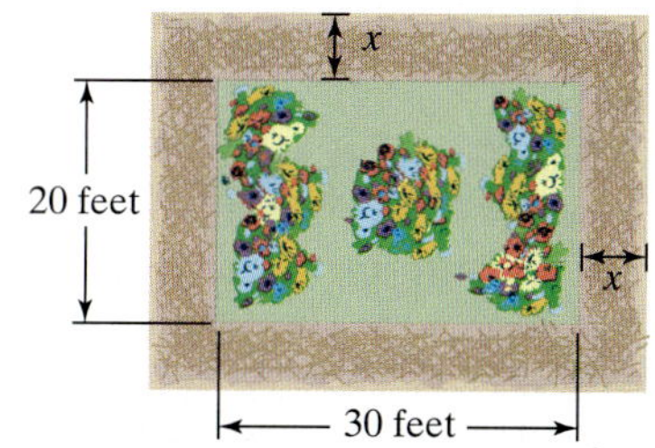

67. A machine produces open boxes using square sheets of metal. The figure illustrates that the machine cuts equal-sized squares measuring 2 inches on a side from the corners and then shapes the metal into an open box by turning up the sides. If each box must have a volume of 200 cubic inches, find the length and width of the open box.

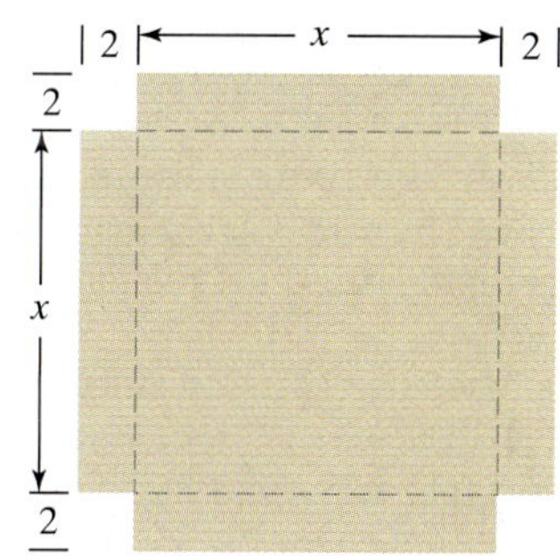

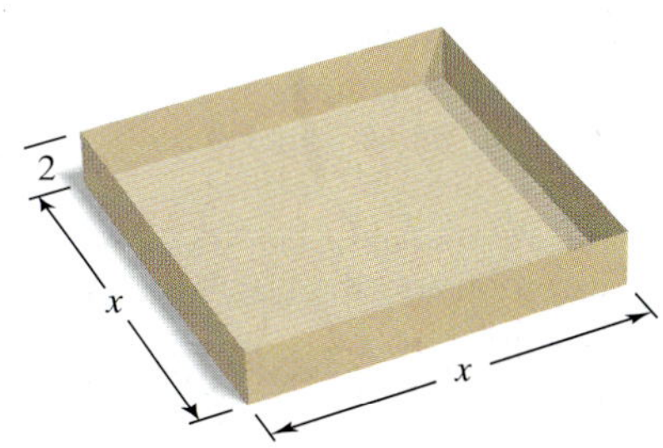

68. A machine produces open boxes using square sheets of metal. The machine cuts equal-sized squares measuring 3 inches on a side from the corners and then shapes the metal into an open box by turning up the sides. If each box must have a volume of 75 cubic inches, find the length and width of the open box.

69. The rectangular floor of a closet is divided into two right triangles by drawing a diagonal, as shown in the figure. One leg of the right triangle is 2 feet more than twice the other leg. The hypotenuse is 13 feet. Determine the closet's length and width.

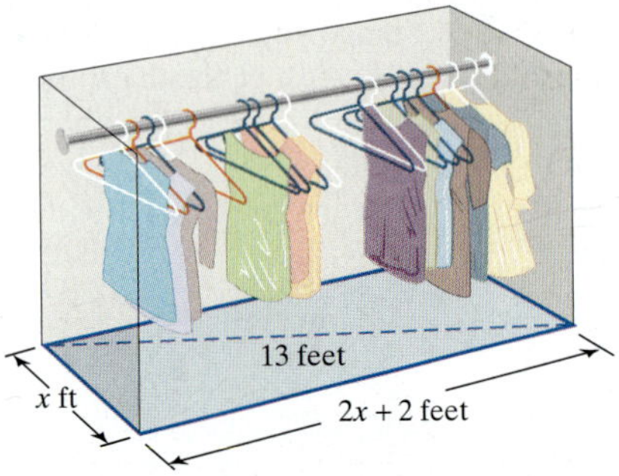

70. A piece of wire measuring 20 feet is attached to a telephone pole as a guy wire. The distance along the ground from the bottom of the pole to the end of the wire is 4 feet greater than the height where the wire is attached to the pole. How far up the pole does the guy wire reach?

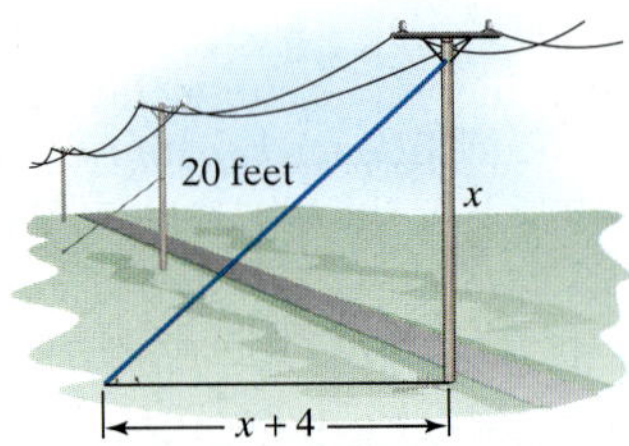

71. A tree is supported by a wire anchored in the ground 15 feet from its base. The wire is 4 feet longer than the height that it reaches on the tree. Find the length of the wire.

72. A tree is supported by a wire anchored in the ground 5 feet from its base. The wire is 1 foot longer than the height that it reaches on the tree. Find the length of the wire.

Writing in Mathematics

73. What is a quadratic equation?

74. What is the zero-product principle?

75. Explain how to solve $x^2 - x = 6$.

76. Describe the relationship between the solutions of a quadratic equation and the graph of the corresponding quadratic function.

77. What is a polynomial equation? When is it in standard form?

78. What is the degree of a polynomial equation? What are polynomial equations of degree 1 and degree 2, respectively, called?

79. Explain how to solve $x^3 + x^2 = x + 1$.

80. If something is thrown straight up, or possibly dropped, describe a situation in which it is important to know how long it will take the object to hit the ground or possibly the water.

81. A toy rocket is launched vertically upward. Using a quadratic equation, we find that the rocket will reach a height of 220 feet at 2.5 seconds and again at 5.5 seconds. How can this be?

82. Describe a situation in which a landscape designer might use polynomials and polynomial equations.

83. In your own words, state the Pythagorean Theorem.

Technology Exercises

In Exercises 84–87, use a graphing utility with a viewing rectangle large enough to show end behavior to graph each polynomial function. Then use the x-intercepts for the graph to solve the polynomial equation. Check by substitution.

84. Use the graph of $y = x^2 + 3x - 4$ to solve $x^2 + 3x - 4 = 0$.

85. Use the graph of $y = x^3 + 3x^2 - x - 3$ to solve $x^3 + 3x^2 - x - 3 = 0$.

86. Use the graph of $y = 2x^3 - 3x^2 - 11x + 6$ to solve $2x^3 - 3x^2 - 11x + 6 = 0$.

87. Use the graph of $y = -x^4 + 4x^3 - 4x^2$ to solve $-x^4 + 4x^3 - 4x^2 = 0$.

Critical Thinking Exercises

88. Which one of the following is true?

a. Quadratic equations solved by factoring always have two different solutions.

b. If $4x(x^2 + 49) = 0$, then

$4x = 0$ or $x^2 + 49 = 0$

$x = 0$ or $x = 7$ or $x = -7$.

c. If -4 is a solution to $7y^2 + (2k - 5)y - 20 = 0$, then k must equal 14.

d. Some quadratic equations have more than two solutions.

In Exercises 89–90, find the domain of the given function f.

89. $f(x) = \dfrac{3}{x^2 + 4x - 45}$

90. $f(x) = \dfrac{7}{x^2 - 3x - 28}$

91. Write a quadratic equation in standard form whose solutions are -3 and 7.

92. Solve: $|x^2 + 2x - 36| = 12$.

Review Exercises

93. Solve: $|3x - 2| = 8$. (Section 4.3, Example 1)

94. Simplify: $3(5 - 7)^2 + \sqrt{16} + 12 \div (-3)$. (Section 1.2, Example 7)

95. You invest \$3000 in two accounts paying 5% and 8% annual interest, respectively. If the total interest earned for the year is \$189, how much was invested at each rate? (Section 3.2, Example 2)

CHAPTER 5 GROUP PROJECTS

1. Divide the group in half. Without looking at any factoring problems in the book, each group should create five factoring problems. Make sure that some of your problems require at least two factoring strategies. Next, exchange problems with the other half of the group. Work to factor the five problems. After completing the factorizations, evaluate the factoring problems that you were given. Are they too easy? Too difficult? Can the polynomials really be factored? Share your responses with the half of the group that wrote the problems. Finally, grade each other's work in factoring the polynomials. Each factoring problem is worth 20 points. You may award partial credit. If you take off points, explain why points are deducted and how you decided to take off a particular number of points for the error(s) that you found.

2. The graph shows the number of applicants to U.S. medical schools from 1985 through 2000. The smooth, continuous graph suggests modeling the data with a polynomial function. Because the graph rises to the left and falls to the right, an odd-degree polynomial function with a negative leading coefficient is an appropriate choice.

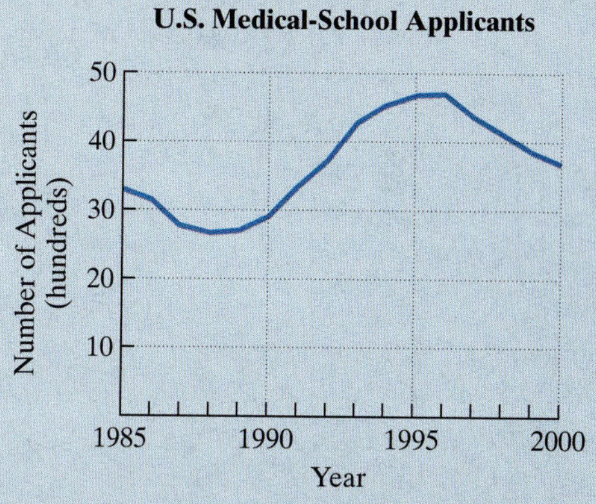

   a. Each group member should bring in a graph like the one shown with data that can be modeled using a polynomial function.

   b. For each graph, group members should describe the degree (even or odd) and the leading coefficient (positive or negative) of a polynomial function that would be a good choice for modeling the data.

   c. Group members should describe what is happening to the data over time. Offer explanations for these trends.

3. Divisible only by themselves and the number 1, **prime numbers** are the building blocks in the mathematics of natural numbers. All other natural numbers, known as **composite numbers**, can be written as the product of unique prime factors. Many topics from number theory involve factoring large composite numbers into primes. The following topics from number theory are appropriate for either individual or group research projects. A report should be given to the class on the researched topic. Useful references include liberal arts mathematics textbooks, books about numbers and number theory, books whose purpose is to excite the reader about mathematics, books about the history of mathematics, encyclopedias, and the World Wide Web.

   a. Euclide and Number Theory
   b. An Unsolved Problem from Number Theory
   c. Perfect Number
   d. Mersenne Primes
   e. Deficient and Abundant Numbers
   f. Formulas that Yield Primes
   g. The Sieve of Eratosthenes

CHAPTER SUMMARY, REVIEW, AND TEST

SUMMARY

DEFINITIONS AND CONCEPTS | **EXAMPLES**

Section 5.1 Introduction to Polynomials and Polynomial Functions

A polynomial is a single term or the sum of two or more terms containing variables in the numerator with whole number exponents. A monomial is a polynomial with exactly one term; a binomial has exactly two terms; a trinomial has exactly three terms. If $a \neq 0$, the degree of ax^n is n and the degree of ax^ny^m is $n + m$. The degree of a nonzero constant is 0. The constant 0 has no defined degree. The degree of a polynomial is the highest of the degrees of its terms. The leading term is the term of highest degree. Its coefficient is called the leading coefficient.

$$7x^3y - 4x^5y^4 - 2x^4y$$

Degree is $3 + 1 = 4$. Degree is $5 + 4 = 9$. Degree is $4 + 1 = 5$.

The degree of the polynomial is 9. The leading term is $-4x^5y^4$. The leading coefficient is -4. This polynomial is a trinomial.

In a polynomial function, the expression that defines the function is a polynomial. Polynomial functions have graphs that are smooth and continuous. The behavior of the graph of a polynomial function to the far left or the far right is called its end behavior.

The Leading Coefficient Test

1. Odd-degree polynomial functions have graphs with opposite behavior at each end. If the leading coefficient is positive, the graph falls to the left and rises to the right. If the leading coefficient is negative, the graph rises to the left and falls to the right.
2. Even-degree polynomial functions have graphs with the same behavior at each end. If the leading coefficient is positive, the graph rises to the left and rises to the right. If the leading coefficient is negative, the graph falls to the left and falls to the right.

Describe the end behavior of the graph of each polynomial function.

- $f(x) = -2x^3 + 3x^2 + 11x - 6$
  The degree, 3, is odd. The leading coefficient -2, is negative. Graph rises to the left and falls to the right.

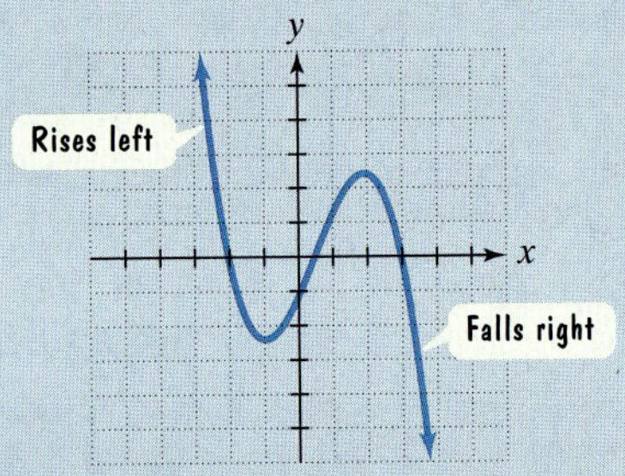

- $f(x) = x^2 - 4$
  The degree, 2, is even. The leading coefficient, 1, is positive. Graph rises to the left and rises to the right.

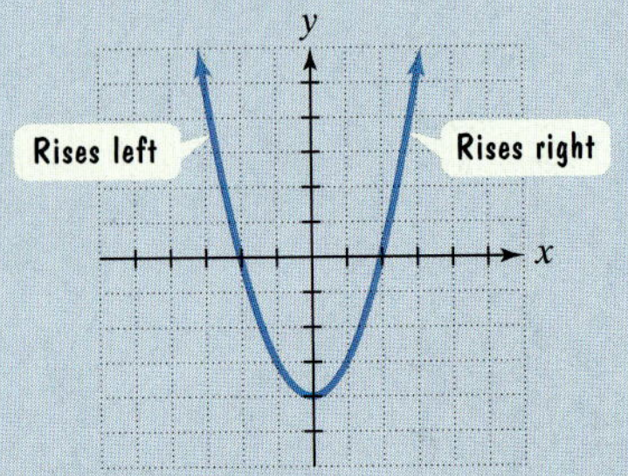

To add polynomials, add like terms.

$$(6x^3y + 5x^2y - 7y) + (-9x^3y + x^2y + 6y)$$
$$= (6x^3y - 9x^3y) + (5x^2y + x^2y) + (-7y + 6y)$$
$$= -3x^3y + 6x^2y - y$$

DEFINITIONS AND CONCEPTS	EXAMPLES
Section 5.1 Introduction to Polynomials and Polynomial Functions (continued)	
To subtract two polynomials, change the sign of every term of the second polynomial. Add this result to the first polynomial.	$(5y^3 - 9y^2 - 4) - (3y^3 - 12y^2 - 5)$ $= (5y^3 - 9y^2 - 4) + (-3y^3 + 12y^2 + 5)$ $= (5y^3 - 3y^3) + (-9y^2 + 12y^2) + (-4 + 5)$ $= 2y^3 + 3y^2 + 1$
Section 5.2 Multiplication of Polynomials	
To multiply monomials, multiply coefficients and add exponents.	$(-2x^2y^4)(-3x^3y)$ $= (-2)(-3)x^{2+3}y^{4+1} = 6x^5y^5$
To multiply a monomial and a polynomial, multiply each term of the polynomial by the monomial.	$7x^2y(4x^3y^5 - 2xy - 1)$ $= 7x^2y \cdot 4x^3y^5 - 7x^2y \cdot 2xy - 7x^2y \cdot 1$ $= 28x^5y^6 - 14x^3y^2 - 7x^2y$
To multiply polynomials if neither is a monomial, multiply each term of one by each term of the other.	$(x^3 + 2x)(5x^2 - 3x + 4)$ $= x^3(5x^2 - 3x + 4) + 2x(5x^2 - 3x + 4)$ $= 5x^5 - 3x^4 + 4x^3 + 10x^3 - 6x^2 + 8x$ $= 5x^5 - 3x^4 + 14x^3 - 6x^2 + 8x$
The FOIL method may be used when multiplying two binomials: First terms multiplied. Outside terms multiplied. Inside terms multiplied. Last terms multiplied.	F O I L $(5x - 3y)(2x + y) = 5x \cdot 2x + 5x \cdot y + (-3y) \cdot 2x + (-3y) \cdot y$ $= 10x^2 + 5xy - 6xy - 3y^2$ $= 10x^2 - xy - 3y^2$
The Square of a Binomial Sum $(A + B)^2 = A^2 + 2AB + B^2$	$(x^2 + 6)^2 = (x^2)^2 + 2 \cdot x^2 \cdot 6 + 6^2$ $= x^4 + 12x^2 + 36$
The Square of a Binomial Difference $(A - B)^2 = A^2 - 2AB + B^2$	$(4x - 5)^2 = (4x)^2 - 2 \cdot 4x \cdot 5 + 5^2$ $= 16x^2 - 40x + 25$
The Product of the Sum and Difference of Two Terms $(A + B)(A - B) = A^2 - B^2$	• $(3x + 7y)(3x - 7y) = (3x)^2 - (7y)^2$ $= 9x^2 - 49y^2$ • $[(x + 2) - 4y][(x + 2) + 4y]$ $= (x + 2)^2 - (4y)^2 = x^2 + 4x + 4 - 16y^2$

DEFINITIONS AND CONCEPTS	EXAMPLES
Section 5.3 Greatest Common Factors and Factoring by Grouping	
Factoring is the process of writing a polynomial as the product of two or more polynomials. Polynomials that cannot be factored using integer coefficients are called prime polynomials. The greatest common factor, GCF, is an expression that divides every term of the polynomial. The GCF is the product of the largest common numerical factor and the variable of lowest degree common to every term of the polynomial. To factor a monomial from a polynomial, express each term as the product of the GCF and its other factor. Then use the distributive property to factor out the GCF. When the leading coefficient of a polynomial is negative, it is often desirable to factor out a common factor with a negative coefficient.	Factor: $4x^4y^2 - 12x^2y^3 + 20xy^2$. (GCF is $4xy^2$.) $= 4xy^2 \cdot x^3 - 4xy^2 \cdot 3xy + 4xy^2 \cdot 5$ $= 4xy^2(x^3 - 3xy + 5)$ Factor: $-25x^3 + 10x^2 - 15x$. (Use $-5x$ as a common factor.) $= -5x(5x^2) - 5x(-2x) - 5x(3)$ $= -5x(5x^2 - 2x + 3)$
To factor by grouping, factor out the GCF from each group. Then factor out the remaining factor.	$xy + 5x - 3y - 15$ $= x(y + 5) - 3(y + 5)$ $= (y + 5)(x - 3)$
Section 5.4 Factoring Trinomials	
To factor a trinomial of the form $x^2 + bx + c$, find two numbers whose product is c and whose sum is b. The factorization is $(x + \text{one number})(x + \text{other number})$.	Factor: $x^2 + 9x + 20$. Find two numbers whose product is 20 and whose sum is 9. The numbers are 4 and 5. $x^2 + 9x + 20 = (x + 4)(x + 5)$
In some trinomials, the highest power is greater than 2, and the exponent in one of the terms is half that of the other term. Factor by introducing a substitution. Let t equal the variable to the smaller power.	Factor: $x^6 - 7x^3 + 12$. $= (x^3)^2 - 7x^3 + 12$ $= t^2 - 7t + 12$ Let $t = x^3$. $= (t - 4)(t - 3)$ $= (x^3 - 4)(x^3 - 3)$
To factor $ax^2 + bx + c$ by trial and error, try various combinations of factors of ax^2 and c until a middle term of bx is obtained for the sum of outside and inside products.	Factor: $3x^2 + 7x - 6$. Factors of $3x^2$: $3x$, x Factors of -6: 1 and -6, -1 and 6, 2 and -3, -2 and 3. $(3x - 2)(x + 3)$ Sum of outside and inside products should equal $7x$. $9x - 2x = 7x$ Thus, $3x^2 + 7x - 6 = (3x - 2)(x + 3)$.
To factor $ax^2 + bx + c$ by grouping, find the factors of ac whose sum is b. Write bx using these factors. Then factor by grouping.	Factor: $3x^2 + 7x - 6$. Find the factors of $3(-6)$, or -18, whose sum is 7. They are 9 and -2. $3x^2 + 7x - 6$ $= 3x^2 + 9x - 2x - 6$ $= 3x(x + 3) - 2(x + 3) = (x + 3)(3x - 2)$

DEFINITIONS AND CONCEPTS	EXAMPLES
Section 5.5 Factoring Special Forms	
The Difference of Two Squares $A^2 - B^2 = (A + B)(A - B)$	$9x^2 - 25y^2$ $= (3x)^2 - (5y)^2 = (3x + 5y)(3x - 5y)$
Perfect Square Trinomials $A^2 + 2AB + B^2 = (A + B)^2$ $A^2 - 2AB + B^2 = (A - B)^2$	$x^2 + 16x + 64 = x^2 + 2 \cdot x \cdot 8 + 8^2 = (x + 8)^2$ $25x^2 - 30x + 9 = (5x)^2 - 2 \cdot 5x \cdot 3 + 3^2 = (5x - 3)^2$
Sum and Difference of Cubes $A^3 + B^3 = (A + B)(A^2 - AB + B^2)$ $A^3 - B^3 = (A - B)(A^2 + AB + B^2)$	$8x^3 - 125 = (2x)^3 - 5^3$ $= (2x - 5)[(2x)^2 + 2x \cdot 5 + 5^2]$ $= (2x - 5)(4x^2 + 10x + 25)$
When using factoring by grouping, terms can sometimes be grouped to obtain the difference of two squares. One of the squares is a perfect square trinomial.	$\underbrace{x^2 + 18x + 81} - 25y^2$ $= (x + 9)^2 - (5y)^2$ $= (x + 9 + 5y)(x + 9 - 5y)$
Section 5.6 A General Factoring Strategy	
A Factoring Strategy **1.** Factor out the GCF or a common factor with a negative coefficient. **2. a.** If two terms, try $A^2 - B^2 = (A + B)(A - B)$ $A^3 + B^3 = (A + B)(A^2 - AB + B^2)$ $A^3 - B^3 = (A - B)(A^2 + AB + B^2)$. **b.** If three terms, try $A^2 + 2AB + B^2 = (A + B)^2$ $A^2 - 2AB + B^2 = (A - B)^2$. If not a perfect square trinomial, try trial and error or grouping. **c.** If four terms, try factoring by grouping. **3.** See if any factors can be factored further.	Factor: $2x^4 + 10x^3 - 8x^2 - 40x$. The GCF is $2x$. $2x^4 + 10x^3 - 8x^2 - 40x$ $= 2x(x^3 + 5x^2 - 4x - 20)$ Four terms: Try grouping. $= 2x[x^2(x + 5) - 4(x + 5)]$ $= 2x(x + 5)(x^2 - 4)$ This can be factored further. $= 2x(x + 5)(x + 2)(x - 2)$
Section 5.7 Polynomial Equations and Their Applications	
A quadratic equation in x can be written in the standard form $ax^2 + bx + c, a \neq 0.$ A polynomial equation is the result of setting two polynomials equal to each other. The equation is in standard form if one side is 0 and the polynomial on the other side is in standard form, that is, in descending powers of the variable. Its degree is the highest degree of any term in the equation. A polynomial equation of degree 1 is a linear equation and of degree 2 a quadratic equation. Some polynomial equations can be solved by writing the equation in standard form, factoring, and then using the Zero-Product Principle: If a product is 0, then at least one of the factors is equal to 0.	Solve: $4x^2 + 9x = 9$. $4x^2 + 9x - 9 = 0$ $(4x - 3)(x + 3) = 0$ $4x - 3 = 0$ or $x + 3 = 0$ $4x = 3$ $x = -3$ $x = \frac{3}{4}$ The solutions are -3 and $\frac{3}{4}$, the solution set is $\{-3, \frac{3}{4}\}$. (The solutions are the x-intercepts of the graph of $y = 4x^2 + 9x - 9$.)

Review Exercises

5.1 *In Exercises 1–2, determine the coefficient of each term, the degree of each term, the degree of the polynomial, the leading term, and the leading coefficient of the polynomial.*

1. $-5x^3 + 7x^2 - x + 2$

2. $8x^4y^2 - 7xy^6 - x^3y$

3. If $f(x) = x^3 - 4x^2 + 3x - 1$, find $f(-2)$.

4. The polynomial function $f(x) = 0.26x^2 + 2.09x + 16.4$ models the number of people in the United States, $f(x)$, in thousands, waiting for organ transplants x years after 1990. Find and interpret $f(10)$.

In Exercises 5–8, use the Leading Coefficient Test to determine the end behavior of the graph of the given polynomial function. Then use this end behavior to match the polynomial function with its graph. [The graphs are labeled (a) through (d).]

5. $f(x) = -x^3 + 12x^2 - x$

6. $f(x) = x^6 - 6x^4 + 9x^2$

7. $f(x) = x^5 - 5x^3 + 4x$

8. $f(x) = -x^4 + 1$

a.

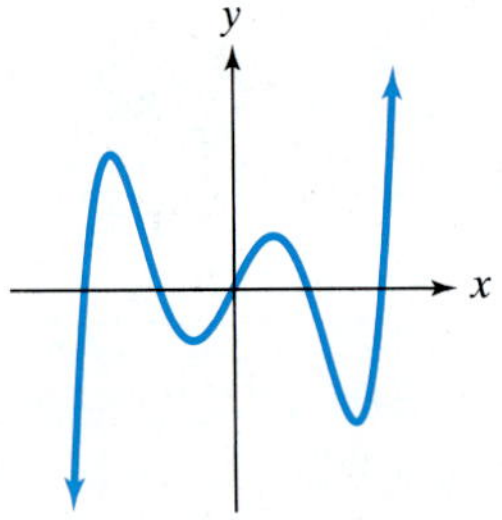

b.

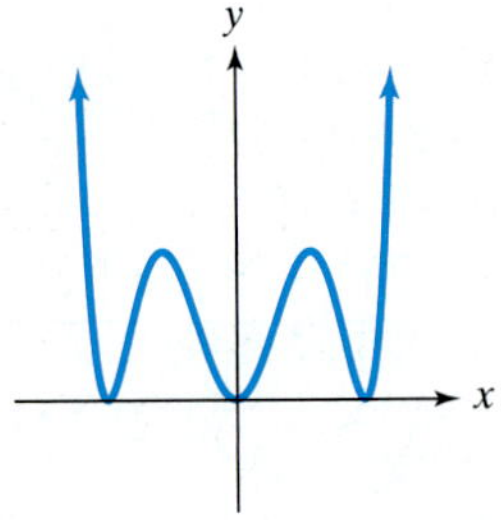

c.

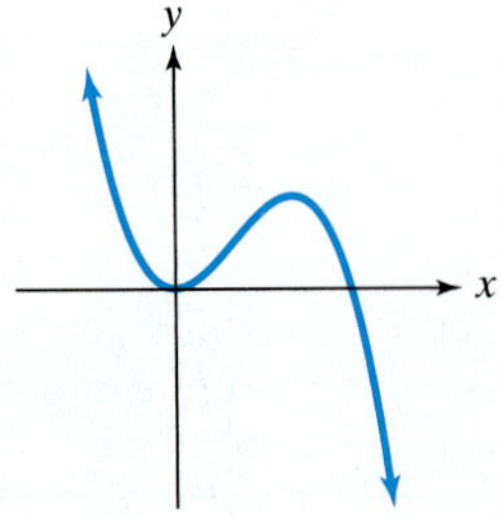

d.

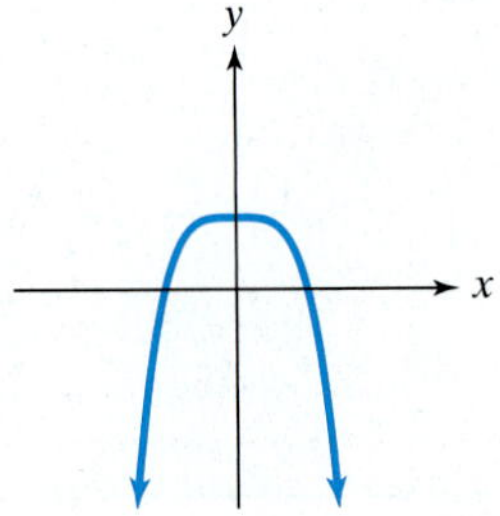

9. The function $f(x) = -0.0013x^3 + 0.78x^2 - 1.43x + 18.1$ models the percentage of U.S. families below the poverty level x years after 1960. Use end behavior to explain why the model is valid only for a limited period of time.

In Exercises 10–11, add the polynomials.

10. $(-8x^3 + 5x^2 - 7x + 4) + (9x^3 - 11x^2 + 6x - 13)$

11. $(7x^3y - 13x^2y - 6y) + (5x^3y + 11x^2y - 8y - 17)$

In Exercises 12–13, subtract the polynomials.

12. $(7x^3 - 6x^2 + 5x - 11) - (-8x^3 + 4x^2 - 6x - 3)$

13. $(4x^3y^2 - 7x^3y - 4) - (6x^3y^2 - 3x^3y + 4)$

14. Subtract $-2x^3 - x^2y + xy^2 + 7y^3$ from $x^3 + 4x^2y - y^3$.

5.2 *In Exercises 15–27, multiply the polynomials.*

15. $(4x^2yz^5)(-3x^4yz^2)$

16. $6x^3\left(\frac{1}{3}x^5 - 4x^2 - 2\right)$

17. $7xy^2(3x^4y^2 - 5xy - 1)$

18. $(2x + 5)(3x^2 + 7x - 4)$

19. $(x^2 + x - 1)(x^2 + 3x + 2)$

20. $(4x - 1)(3x - 5)$

21. $(3xy - 2)(5xy + 4)$

22. $(3x + 7y)^2$

23. $(x^2 - 5y)^2$

24. $(2x + 7y)(2x - 7y)$

25. $(3xy^2 - 4x)(3xy^2 + 4x)$

26. $[(x + 3) + 5y][(x + 3) - 5y]$

27. $(x + y + 4)^2$

28. Let $f(x) = x - 3$ and $g(x) = 2x + 5$. Find $(fg)(x)$ and $(fg)(-4)$.

29. Let $f(x) = x^2 - 7x + 2$. Find each of the following and simplify:

a. $f(a - 1)$

b. $f(a + h) - f(a)$.

5.3 *In Exercises 30–33, factor the greatest common factor from each polynomial.*

30. $16x^3 + 24x^2$

31. $2x - 36x^2$

32. $21x^2y^2 - 14xy^2 + 7xy$

33. $18x^3y^2 - 27x^2y$

In Exercises 34–35, factor out a common factor with a negative coefficient.

34. $-12x^2 + 8x - 48$

35. $-x^2 - 11x + 14$

In Exercises 36–38, factor by grouping.

36. $x^3 - x^2 - 2x + 2$

37. $xy - 3x - 5y + 15$

38. $5ax - 15ay + 2bx - 6by$

5.4 *In Exercises 39–47, factor each trinomial completely, or state that the trinomial is prime.*

39. $x^2 + 8x + 15$

40. $x^2 + 16x - 80$

41. $x^2 + 16xy - 17y^2$

42. $3x^3 - 36x^2 + 33x$

43. $3x^2 + 22x + 7$

44. $6x^2 - 13x + 6$

45. $5x^2 - 6xy - 8y^2$

46. $6x^3 + 5x^2 - 4x$

47. $2x^2 + 11x + 15$

In Exercises 48–51, factor by introducing an appropriate substitution.

48. $x^6 + x^3 - 30$

49. $x^4 - 10x^2 - 39$

50. $(x + 5)^2 + 10(x + 5) + 24$

51. $5x^6 + 17x^3 + 6$

5.5 *In Exercises 52–55, factor each difference of two squares.*

52. $4x^2 - 25$

53. $1 - 81x^2y^2$

54. $x^8 - y^6$

55. $(x - 1)^2 - y^2$

In Exercises 56–60, factor any perfect square trinomials, or state that the polynomial is prime.

56. $x^2 + 16x + 64$

57. $9x^2 - 6x + 1$

58. $25x^2 + 20xy + 4y^2$

59. $49x^2 + 7x + 1$

60. $25x^2 - 40xy + 16y^2$

In Exercises 61–62, factor by grouping to obtain the difference of two squares.

61. $x^2 + 18x + 81 - y^2$

62. $z^2 - 25x^2 + 10x - 1$

In Exercises 63–65, factor using the formula for the sum or difference of two cubes.

63. $64x^3 + 27$

64. $125x^3 - 8$

65. $x^3y^3 + 1$

5.6 *In Exercises 66–86, factor completely, or state that the polynomial is prime.*

66. $15x^2 + 3x$

67. $12x^4 - 3x^2$

68. $20x^4 - 24x^3 + 28x^2 - 12x$

69. $x^3 - 15x^2 + 26x$

70. $-2y^4 + 24y^3 - 54y^2$

71. $9x^2 - 30x + 25$

72. $5x^2 - 45$

73. $2x^3 - x^2 - 18x + 9$

74. $6x^2 - 23xy + 7y^2$

75. $2y^3 + 12y^2 + 18y$

76. $x^2 + 6x + 9 - 4a^2$

77. $8x^3 - 27$

78. $x^5 - x$

79. $x^4 - 6x^2 + 9$

80. $x^2 + xy + y^2$

81. $4a^3 + 32$

82. $x^4 - 81$

83. $ax + 3bx - ay - 3by$

84. $27x^3 - 125y^3$

85. $10x^3y + 22x^2y - 24xy$

86. $6x^6 + 13x^3 - 5$

In Exercises 87–88:

a. *Write an expression for the area of the shaded region.*

b. *Write the expression in factored form.*

87.

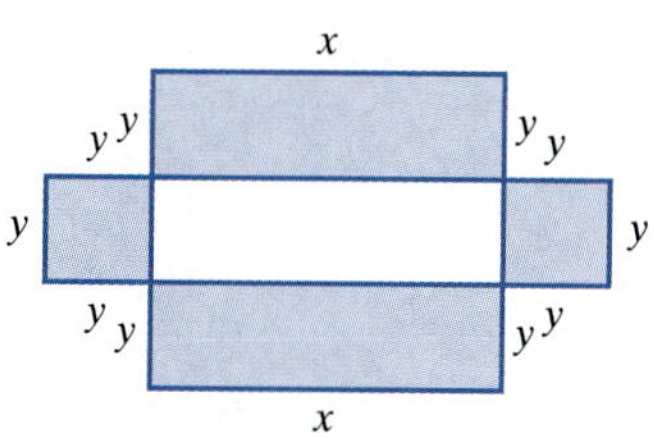

88.

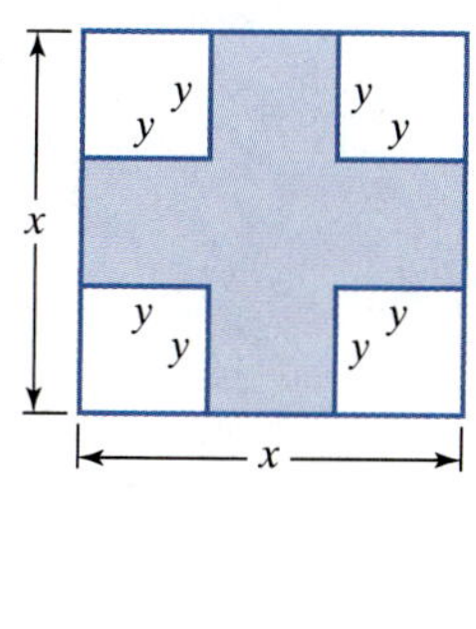

5.7 *In Exercises 89–93, use factoring to solve each polynomial equation.*

89. $x^2 + 6x + 5 = 0$

90. $3x^2 = 22x - 7$

91. $(x + 3)(x - 2) = 50$

92. $3x^2 = 12x$

93. $x^3 + 5x^2 = 9x + 45$

94. A model rocket is launched from the top of a cliff 144 feet above sea level. The function

$$s(t) = -16t^2 + 128t + 144$$

describes the rocket's height above the water, $s(t)$, in feet, t seconds after it was launched. The rocket misses the edge of the cliff on its way down and eventually lands in the ocean. How long will it take for the rocket to hit the water?

95. The function

$$N(x) = 2x^2 + 22x + 320$$

models the number of inmates, $N(x)$, in thousands, in U.S. state and federal prisons x years after 1980. In which year were there 740 thousand inmates in U.S. state and federal prisons?

96. The length of a rectangular sign is 3 feet longer than the width. If the sign has space for 54 square feet of advertising, find its length and its width.

97. A painting measuring 10 inches by 16 inches is surrounded by a frame of uniform width. If the combined area of the painting and frame is 280 square inches, determine the width of the frame.

98. A lot is in the shape of a right triangle. The longer leg of the triangle is 20 yards longer than twice the length of the shorter leg. The hypotenuse is 30 yards longer than twice the length of the shorter leg. What are the lengths of the three sides?

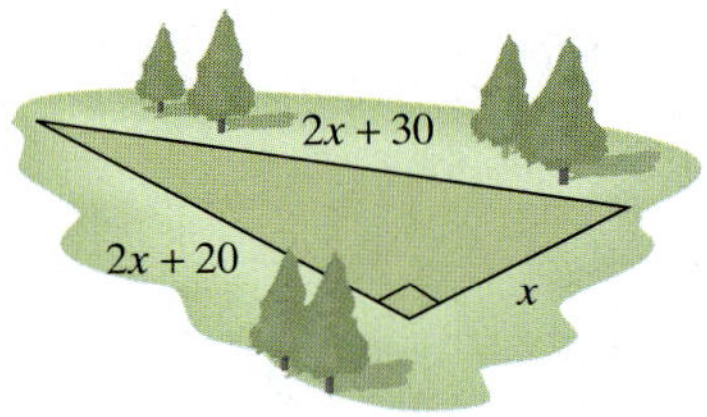

Chapter 5 Test

In Exercises 1–2, give the degree and the leading coefficient of the polynomial.

1. $7x - 5 + x^2 - 6x^3$

2. $4xy^3 + 7x^4y^5 - 3xy^4$

3. If $f(x) = 3x^3 + 5x^2 - x + 6$, find $f(0)$ and $f(-2)$.

In Exercises 4–5, use the Leading Coefficient Test to describe the end behavior of the graph of the polynomial function.

4. $f(x) = -16x^2 + 160x$

5. $f(x) = 4x^3 + 12x^2 - x - 3$

In Exercises 6–13, perform the indicated operations.

6. $(4x^3y - 19x^2y - 7y) + (3x^3y + x^2y + 6y - 9)$

7. $(6x^2 - 7x - 9) - (-5x^2 + 6x - 3)$

8. $(-7x^3y)(-5x^4y^2)$

9. $(x - y)(x^2 - 3xy - y^2)$

10. $(7x - 9y)(3x + y)$

11. $(2x - 5y)(2x + 5y)$

12. $(4y - 7)^2$

13. $[(x + 2) + 3y][(x + 2) - 3y]$

14. Let $f(x) = x + 2$ and $g(x) = 3x - 5$. Find $(fg)(x)$ and $(fg)(-5)$.

15. Let $f(x) = x^2 - 5x + 3$. Find $f(a + h) - f(a)$ and simplify.

In Exercises 16–33, factor completely, or state that the polynomial is prime.

16. $14x^3 - 15x^2$

17. $81y^2 - 25$

18. $x^3 + 3x^2 - 25x - 75$

19. $25x^2 - 30x + 9$

20. $x^2 + 10x + 25 - 9y^2$

21. $x^4 + 1$

22. $y^2 - 16y - 36$

23. $14x^2 + 41x + 15$

24. $5x^3 - 5$

25. $12x^2 - 3y^2$

26. $12x^2 - 34x + 10$

27. $3x^4 - 3$

28. $x^8 - y^8$

29. $12x^2y^4 + 8x^3y^2 - 36x^2y$

30. $x^6 - 12x^3 - 28$

31. $x^4 - 2x^2 - 24$

32. $12x^2y - 27xy + 6y$

33. $y^4 - 3y^3 + 2y^2 - 6y$

In Exercises 34–37, solve each polynomial equation.

34. $3x^2 = 5x + 2$

35. $(5x + 4)(x - 1) = 2$

36. $15x^2 - 5x = 0$

37. $x^3 - 4x^2 - x + 4 = 0$

38. A baseball is thrown straight up from a rooftop 448 feet high. The function

$$s(t) = -16t^2 + 48t + 448$$

describes the ball's height above the ground, $s(t)$, in feet, t seconds after it was thrown. How long will it take for the ball to hit the ground?

39. An architect is allowed 15 square yards of floor space to add a small bedroom to a house. Because of the room's design in relationship to the existing structure, the width of the rectangular floor must be 7 yards less than two times the length. Find the length and width of the rectangular floor that the architect is permitted.

40. Find the lengths of the three sides of the right triangle in the figure shown.

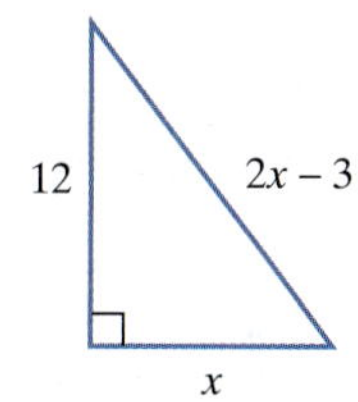

Cumulative Review Exercises (Chapters 1–5)

In Exercises 1–7, solve each equation, inequalities or system of equations.

1. $8(x + 2) - 3(2 - x) = 4(2x + 6) - 2$

2. $2x + 4y = -6$
$x = 2y - 5$

3. $2x - y + 3z = 0$
$2y + z = 1$
$x + 2y - z = 5$

4. $2x + 4 < 10$ and $3x - 1 > 5$

5. $|2x - 5| \geq 9$

6. $2x^2 = 7x - 5$

7. $2x^3 + 6x^2 = 20x$

8. Solve for x: $x = \dfrac{ax + b}{c}$.

9. Use function notation to write the equation of the line passing through $(-2, -3)$ and $(2, 5)$.

10. In a campuswide election for student government president, 2800 votes were cast for the two candidates. If the winner had 160 more votes than the loser, how many votes were cast for each candidate?

In Exercises 11–13, graph each equation or inequality in a rectangular coordinate system.

11. $f(x) = -\frac{1}{3}x + 1$

12. $4x - 5y < 20$

13. $y \leq -1$

14. Simplify: $-\dfrac{8x^3y^6}{16x^9y^{-4}}$.

15. Write in scientific notation: 0.0000706.

In Exercises 16–17, perform the indicated operations.

16. $(3x^2 - y)^2$

17. $(3x^2 - y)(3x^2 + y)$

In Exercises 18–20, factor completely.

18. $x^3 - 3x^2 - 9x + 27$

19. $x^6 - x^2$

20. $14x^3y^2 - 28x^4y^2$

During the 1980s, the label "supply-side economics" was applied to the theory that lower tax rates would improve private sector incentives, leading to higher employment, productivity and output in the economy. George Bush, father of president George W. Bush and an opponent of Ronald Reagan in the 1980 primaries, referred to an extreme version of this theory espoused by Reagan as "voodoo economics." In this version, a cut in tax rates was predicted to result in an increase in tax revenue, and thus not increase the government deficit.

Budget deficits exploded in the 1980s after tax rates were cut in 1981. In this chapter, you will see how a function consisting of the quotient of two polynomials models the most controversial version of supply-side economics. Functions involving fractional expressions will give you new insights into phenomena as diverse as the cost of environmental cleanup, the relationship between high production levels and competitively low prices, the percentage of deaths from a disease caused by smoking cigarettes, and even our ongoing processes of learning and forgetting.

Chapter 6

Rational Expressions, Functions, and Equations

▸ SECTION 6.1 Rational Expressions and Functions: Multiplying and Dividing

Objectives

1. Evaluate rational functions.
2. Find the domain of a rational function.
3. Interpret information given by the graph of a rational function.
4. Simplify rational expressions.
5. Multiply rational expressions.
6. Divide rational expressions.

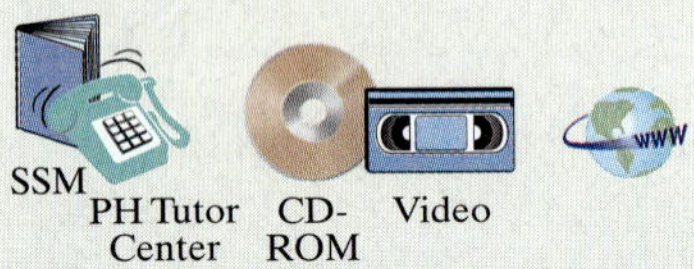

Environmental scientists and municipal planners often make decisions using **cost-benefit models.** These mathematical models estimate the cost of removing a pollutant from the atmosphere as a function of the percentage of pollutant removed. What kinds of functions describe the cost of reducing environmental pollution? In this section, we introduce this new category of functions, called *rational functions*.

Rational Expressions

A **rational expression** consists of a polynomial divided by a nonzero polynomial. Here are some examples of rational expressions:

$$\frac{120x}{100-x}, \quad \frac{2x+1}{2x^2-x-1}, \quad \frac{3x^2+12xy-15y^2}{6x^3-6xy^2}.$$

1 Evaluate rational functions.

Rational Functions

A **rational function** is defined by a formula that is a rational expression.

EXAMPLE 1 Using a Cost-Benefit Model

The rational function

$$f(x) = \frac{120x}{100-x}$$

models the cost, $f(x)$, in thousands of dollars, to remove $x\%$ of the pollutants that a city has discharged into a lake. Find and interpret:

a. $f(20)$ **b.** $f(80)$.

Solution We use substitution to evaluate a rational function, just as we did to evaluate other functions in Chapter 2.

a. $f(x) = \frac{120x}{100-x}$ This is the given rational function.

$f(20) = \frac{120(20)}{100-20}$ To find f of 20, replace x by 20.

$= \frac{2400}{80} = 30$ Perform the indicated operations.

Thus, $f(20) = 30$. This means that the cost to remove 20% of the lake's pollutants is \$30 thousand. Figure 6.1 illustrates the solution by the point (20, 30) on the graph of the rational function.

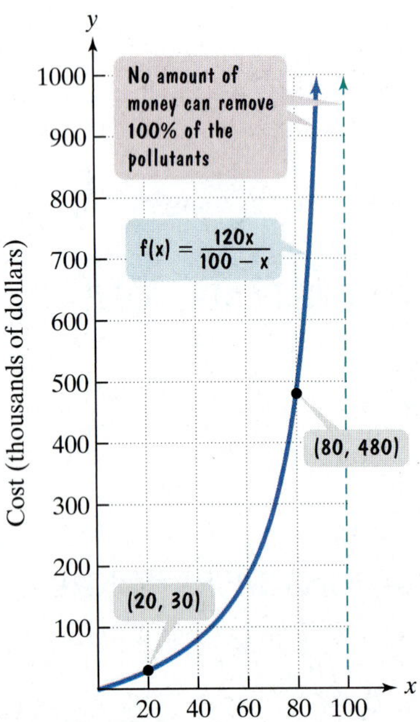

Figure 6.1 Percentage of pollutants removed

b. $f(x) = \dfrac{120x}{100 - x}$ This is the given rational function.

$$f(80) = \frac{120(80)}{100 - 80} = \frac{9600}{20} = 480$$

Thus, $f(80) = 480$. This means that the cost to remove 80% of the lake's pollutants is \$480 thousand. Figure 6.1 illustrates the solution by the point (80, 480) on the graph of the cost-benefit model. The graph illustrates that costs spiral upward as the percentage of pollutants removed increases. ■

✔ **CHECK POINT 1** Use the rational function in Example 1 to find and interpret: **a.** $f(40)$ **b.** $f(60)$. Identify each solution as a point on the graph of the rational function in Figure 6.1.

2 Find the domain of a rational function.

The Domain of a Rational Function

Does the cost-benefit model

$$f(x) = \frac{120x}{100 - x}$$

indicate that the city can clean up its lake completely? To do this, the city must remove 100% of the pollutants. The problem is that the rational function is undefined for $x = 100$.

$$f(x) = \frac{120x}{100 - x}$$

If $x = 100$, the value of the denominator is 0.

Notice how the graph of the rational function in Figure 6.1 approaches, but never touches, the dashed green vertical line drawn through $x = 100$. The graph continues to rise more and more steeply, visually showing the escalating costs. By never touching the dashed vertical line, the graph illustrates that no amount of money will be enough to remove all pollutants from the lake.

The **domain of a rational function** is the set of all real numbers except those for which the denominator is zero. We can find the domain by determining when the denominator is zero. For the cost-benefit model, the denominator is zero when $x = 100$. Furthermore, for this model, negative values of x and values of x greater than 100 are not meaningful. The domain of the function is $[0, 100)$ and excludes 100.

Inspection can sometimes be used to find a rational function's domain. Here are two examples.

$$f(x) = \frac{4}{x - 2} \qquad g(x) = \frac{x}{(x + 1)(x - 1)}$$

This denominator would equal zero if $x = 2$.

This factor would equal zero if $x = -1$.

This factor would equal zero if $x = 1$.

The domain of f can be expressed in set-builder or interval notation:

$$\text{Domain of } f = \{x \mid x \text{ is a real number and } x \neq 2\}$$

$$\text{Domain of } f = (-\infty, 2) \cup (2, \infty).$$

Likewise, the domain of g can be expressed in set-builder or interval notation:

$$\text{Domain of } g = \{x \mid x \text{ is a real number and } x \neq -1 \text{ and } x \neq 1\}$$

$$\text{Domain of } g = (-\infty, -1) \cup (-1, 1) \cup (1, \infty).$$

Ask your professor if a particular notation is preferred in the course.

EXAMPLE 2 Finding the Domain of a Rational Function

Find the domain of f if

$$f(x) = \frac{2x + 1}{2x^2 - x - 1}.$$

Solution The domain of f is the set of all real numbers except those for which the denominator is zero. We can identify such numbers by setting the denominator equal to zero and solving for x.

$$2x^2 - x - 1 = 0 \quad \text{Set the denominator equal to 0.}$$

$$(2x + 1)(x - 1) = 0 \quad \text{Factor.}$$

$$2x + 1 = 0 \quad \text{or} \quad x - 1 = 0 \quad \text{Set each factor equal to 0.}$$

$$2x = -1 \qquad x = 1 \quad \text{Solve the resulting equations.}$$

$$x = -\frac{1}{2}$$

Because $-\frac{1}{2}$ and 1 make the denominator zero, these are the values to exclude. Thus,

$$\text{Domain of } f = \left\{x \,\middle|\, x \text{ is a real number and } x \neq -\frac{1}{2} \text{ and } x \neq 1\right\}$$

or $$\text{Domain of } f = \left(-\infty, -\frac{1}{2}\right) \cup \left(-\frac{1}{2}, 1\right) \cup (1, \infty).$$

■

✔ CHECK POINT 2 Find the domain of f if

$$f(x) = \frac{x - 5}{2x^2 + 5x - 3}.$$

Does every rational function have values to exclude? The answer is no. For example, consider

$$f(x) = \frac{2}{x^2 + 1}.$$

No values of x cause this denominator to equal zero.

The domain of f consists of the set of all real numbers. Recall that this can be expressed in three different ways:

$$\text{Domain of } f = \{x \mid x \text{ is a real number}\} \text{ or } \mathbb{R} \text{ or } (-\infty, \infty).$$

3 Interpret information given by the graph of a rational function.

What Graphs of Rational Functions Look Like

In everyday speech, a continuous process is one that goes on without interruption and without abrupt changes. In mathematics, a continuous function has much the same meaning. The graph of a continuous function does not have interrupting breaks, such as holes, gaps, or jumps. Thus, the graph of a continuous function can be drawn without lifting a pencil off the paper.

Most rational functions are not continuous functions. For example, the graph of the rational function

$$f(x) = \frac{2x + 1}{2x^2 - x - 1}$$

Using Technology

When using a graphing utility to graph a rational function, you might not be pleased with the quality of the display. The graph of the rational function

$$y = \frac{2x + 1}{2x^2 - x - 1}$$

is shown using the DOT mode in a $[-6, 6, 1]$ by $[-6, 6, 1]$ viewing rectangle. In this mode, the utility displays unconnected points that have been calculated. Would you agree that the graph's behavior is better illustrated by the hand-drawn graph in Figure 6.2?

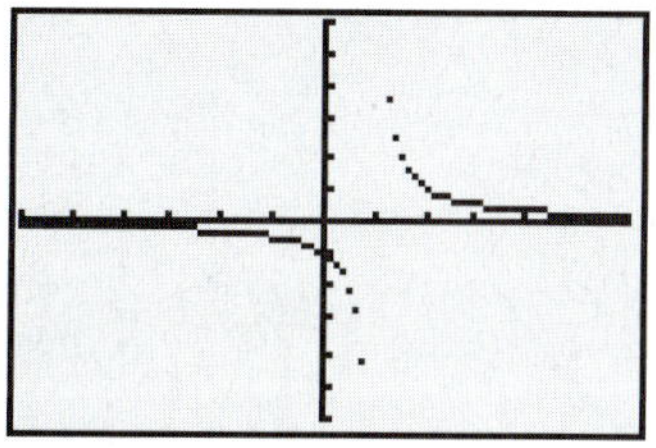

is shown in Figure 6.2. In Example 2, we excluded $-\frac{1}{2}$ and 1 from this function's domain. Unlike the graph of a polynomial function, this graph has two breaks in it—one at each of the excluded values. At $-\frac{1}{2}$, a hole in the graph appears. The graph is composed of two distinct branches. Each branch approaches, but never touches, the dashed vertical line drawn at 1, the other excluded value.

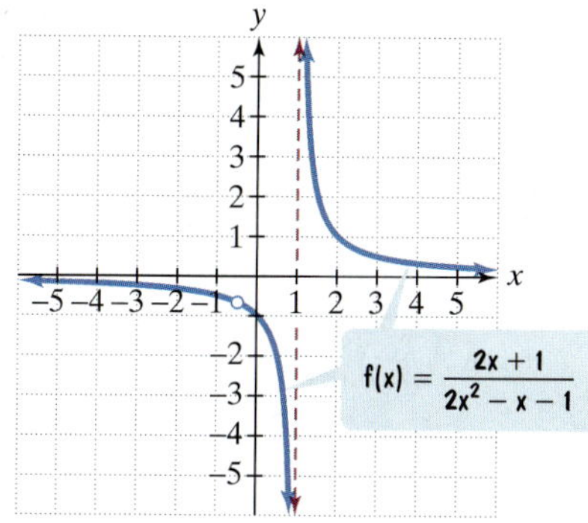

Figure 6.2

A vertical line that the graph of a function approaches, but does not touch, is said to be a **vertical asymptote** of the graph. In Figure 6.2, the line $x = 1$ is a vertical asymptote of the graph of f. A rational function may have no vertical asymptotes, one vertical asymptote, or several vertical asymptotes. The graph of a rational function never intersects a vertical asymptote.

Unlike the graph of a polynomial function, the graph of the rational function in Figure 6.2 does not go up or down at each end. At the far left and the far right, the graph is getting closer to, but not actually reaching, the x-axis. This shows that as x increases or decreases without bound, the function values are approaching 0. The line $y = 0$ (that is, the x-axis) is a **horizontal asymptote** of the graph. Many, but not all, rational functions have horizontal asymptotes.

For the remainder of this section, we will focus on the rational expressions that define rational functions. Operations with these expressions should remind you of those performed in arithmetic with fractions.

4 Simplify rational expressions.

Simplifying Rational Expressions

A rational expression is **simplified** if its numerator and denominator have no common factors other than 1 or -1. The following procedure can be used to simplify rational expressions:

Simplifying Rational Expressions

1. Factor the numerator and denominator completely.
2. Divide both the numerator and denominator by the common factors.

EXAMPLE 3 Simplifying a Rational Expression

Simplify: $\frac{x^2 + 4x + 3}{x + 1}$.

Solution

$$\frac{x^2 + 4x + 3}{x + 1} = \frac{(x + 1)(x + 3)}{1(x + 1)}$$ Factor the numerator and denominator.

$$= \frac{\cancel{(x + 1)}(x + 3)}{1\cancel{(x + 1)}}$$ Divide out the common factor, $x + 1$.

$$= x + 3$$

■

Simplifying a rational expression can change the numbers that make it undefined. For example, we just showed that

$$\frac{x^2 + 4x + 3}{x + 1} = x + 3.$$

This is undefined for $x = -1$.

This simplified form is defined for all real numbers.

Thus, to equate the two expressions, we must restrict the values for x in the simplified expression to exclude -1. We can write

$$\frac{x^2 + 4x + 3}{x + 1} = x + 3, x \neq -1.$$

Without this restriction, the expressions are not equal. The slight difference between them is illustrated using the graphs of

$$f(x) = \frac{x^2 + 4x + 3}{x + 1} \quad \text{and} \quad g(x) = x + 3$$

in Figure 6.3.

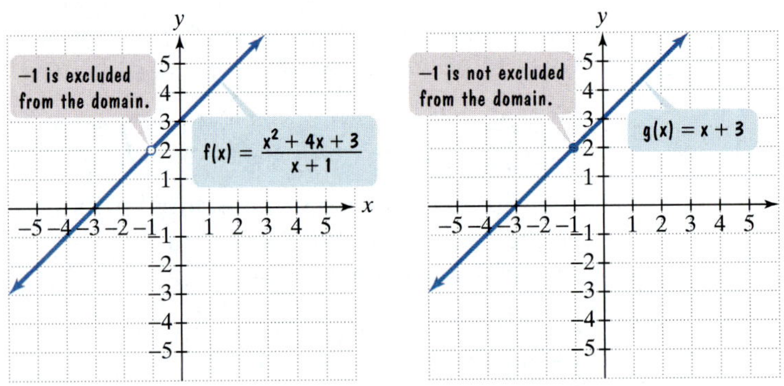

Figure 6.3 Visualizing the difference between $\frac{x^2 + 4x + 3}{x + 1}$ and $x + 3$

Hereafter, we will assume that a simplified rational expression is equal to the original rational expression for all real numbers except those for which either denominator is 0.

✔ **CHECK POINT 3** Simplify: $\frac{x^2 + 7x + 10}{x + 2}$.

EXAMPLE 4 Simplifying Rational Expressions

Simplify:

a. $\frac{x^2 - 7x - 18}{2x^2 + 3x - 2}$ **b.** $\frac{3x^2 + 12xy - 15y^2}{6x^3 - 6xy^2}$.

Solution

a. $\dfrac{x^2 - 7x - 18}{2x^2 + 3x - 2} = \dfrac{(x - 9)(x + 2)}{(2x - 1)(x + 2)}$ Factor the numerator and denominator.

$= \dfrac{(x - 9)\cancel{(x + 2)}}{(2x - 1)\cancel{(x + 2)}}$ Divide out the common factor, $x + 2$.

$= \dfrac{x - 9}{2x - 1}$

b. $\dfrac{3x^2 + 12xy - 15y^2}{6x^3 - 6xy^2} = \dfrac{3(x^2 + 4xy - 5y^2)}{6x(x^2 - y^2)}$ Factor out the GCF in the numerator and denominator.

$= \dfrac{3(x + 5y)(x - y)}{6x(x + y)(x - y)}$ Factor the numerator and denominator completely.

$= \dfrac{\overset{1}{\cancel{3}}(x + 5y)\cancel{(x - y)}}{\underset{2}{\cancel{6}}x(x + y)\cancel{(x - y)}}$ Divide out the common factor, $3(x - y)$.

$= \dfrac{x + 5y}{2x(x + y)}$ It is not necessary to carry out the multiplication in the denominator. ■

CHECK POINT 4 Simplify:

a. $\dfrac{x^2 - 2x - 15}{3x^2 + 8x - 3}$

b. $\dfrac{3x^2 + 9xy - 12y^2}{9x^3 - 9xy^2}$.

Study Tip

When simplifying rational expressions, you can only divide out, or cancel, factors common to the numerator and denominator. **It is incorrect to divide out common terms from the numerator and denominator.**

INCORRECT

$\dfrac{\cancel{x} + 4}{\cancel{x}} = 4 \qquad \dfrac{x^2 - \cancel{4}}{\cancel{4}} = x^2 - 1 \qquad \dfrac{\overset{x}{\cancel{x^2}} - \overset{3}{\cancel{9}}}{\underset{1}{\cancel{x}} - \underset{1}{\cancel{3}}} = x - 3$

The first two expressions, $\dfrac{x + 4}{x}$ and $\dfrac{x^2 - 4}{4}$, have no common factors in their numerators and denominators. Only when expressions are multiplied can they be factors. If you can't factor, then don't try to cancel. The third rational expression shown above, $\dfrac{x^2 - 9}{x - 3}$, can be simplified as follows:

CORRECT

$$\frac{x^2 - 9}{x - 3} = \frac{(x + 3)\cancel{(x - 3)}}{1\cancel{(x - 3)}} = x + 3.$$

Divide out the common factor, $x - 3$.

5 Multiply rational expressions.

Multiplying Rational Expressions

The product of two rational expressions is the product of their numerators divided by the product of their denominators. For example,

$$\frac{x^2}{y+3} \cdot \frac{x+5}{y-7} = \frac{x^2(x+5)}{(y+3)(x-7)}.$$

Multiply numerators.

Multiply denominators.

Here is a step-by-step procedure for multiplying rational expressions. Before multiplying, divide out any factors common to both a numerator and a denominator.

Multiplying Rational Expressions

1. Factor all numerators and denominators completely.
2. Divide numerators and denominators by common factors.
3. Multiply the remaining factors in the numerators and multiply the remaining factors in the denominators.

EXAMPLE 5 Multiplying Rational Expressions

Multiply: $\frac{x+3}{x-4} \cdot \frac{x^2-2x-8}{x^2-9}$.

Solution

$$\frac{x+3}{x-4} \cdot \frac{x^2-2x-8}{x^2-9}$$

$$= \frac{1(x+3)}{1(x-4)} \cdot \frac{(x-4)(x+2)}{(x+3)(x-3)}$$ Factor all numerators and denominators completely.

$$= \frac{1\cancel{(x+3)}}{1\cancel{(x-4)}} \cdot \frac{\cancel{(x-4)}(x+2)}{\cancel{(x+3)}(x-3)}$$ Divide numerators and denominators by common factors.

$$= \frac{x+2}{x-3}$$ Multiply the remaining factors in the numerators and denominators. ■

✔ **CHECK POINT 5** Multiply: $\frac{x+4}{x-7} \cdot \frac{x^2-4x-21}{x^2-16}$.

Some rational expressions contain factors in the numerator and denominator that are opposites, or additive inverses. Here is an example of such an expression:

$$\frac{(2x+5)(2x-5)}{3(5-2x)}.$$

The factors $2x - 5$ and $5 - 2x$ are opposites. They differ only in sign.

Although you can factor out -1 from the numerator or the denominator and then divide out the common factor, there is an even faster way to simplify this rational expression.

Simplifying Rational Expressions with Opposite Factors in the Numerator and Denominator

The quotient of two polynomials that have opposite signs and are additive inverses is -1.

For example,

$$\frac{(2x+5)(2x-5)}{3(5-2x)} = \frac{(2x+5)\overset{(-1)}{\cancel{(2x-5)}}}{3\cancel{(5-2x)}} = \frac{-(2x+5)}{3} \quad \text{or} \quad -\frac{2x+5}{3} \quad \text{or} \quad \frac{-2x-5}{3}.$$

EXAMPLE 6 Multiplying Rational Expressions

Multiply: $\dfrac{5x+5}{7x-7x^2} \cdot \dfrac{2x^2+x-3}{4x^2-9}$.

Solution

$$\frac{5x+5}{7x-7x^2} \cdot \frac{2x^2+x-3}{4x^2-9}$$

$$= \frac{5(x+1)}{7x(1-x)} \cdot \frac{(2x+3)(x-1)}{(2x+3)(2x-3)}$$ Factor all numerators and denominators completely.

$$= \frac{5(x+1)}{7x\cancel{(1-x)}} \cdot \frac{\cancel{(2x+3)}\overset{(-1)}{\cancel{(x-1)}}}{\cancel{(2x+3)}(2x-3)}$$ Divide numerators and denominators by common factors. Because $1-x$ and $x-1$ are opposites, their quotient is -1.

$$= \frac{-5(x+1)}{7x(2x-3)} \quad \text{or} \quad -\frac{5(x+1)}{7x(2x-3)}$$ Multiply the remaining factors in the numerators and denominators.

■

✔ **CHECK POINT 6** Multiply: $\dfrac{4x+8}{6x-3x^2} \cdot \dfrac{3x^2-4x-4}{9x^2-4}$.

6 Divide rational expressions.

Dividing Rational Expressions

The quotient of two rational expressions is the product of the first expression and the multiplicative inverse, or reciprocal, of the second expression. The reciprocal is found by interchanging the numerator and the denominator.

Dividing Rational Expressions

If P, Q, R, and S are polynomials, where $Q \neq 0$, $R \neq 0$, and $S \neq 0$, then

$$\frac{P}{Q} \div \frac{R}{S} = \frac{P}{Q} \cdot \frac{S}{R} = \frac{PS}{QR}.$$

Change division to multiplication.

Replace $\frac{R}{S}$ by its reciprocal by interchanging numerator and denominator.

Thus, **we find the quotient of two rational expressions by inverting the divisor and multiplying**. For example,

$$\frac{x}{7} \div \frac{6}{y} = \frac{x}{7} \cdot \frac{y}{6} = \frac{xy}{42}.$$

Change the division to multiplication.

Replace $\frac{6}{y}$ with its reciprocal by interchanging numerator and denominator.

Study Tip

When performing operations with rational expressions; if a rational expression is written without a denominator, it is helpful to write the expression with a denominator of 1. In Example 7(a), we wrote

$$4x^2 - 25 \text{ as } \frac{4x^2 - 25}{1}.$$

EXAMPLE 7 Dividing Rational Expressions

Divide:

a. $(4x^2 - 25) \div \frac{2x + 5}{14}$ **b.** $\frac{x^2 + 3x - 10}{2x} \div \frac{x^2 - 5x + 6}{x^2 - 3x}$.

Solution

a. $(4x^2 - 25) \div \frac{2x + 5}{14}$

$= \frac{4x^2 - 25}{1} \div \frac{2x + 5}{14}$ Write $4x^2 - 25$ with a denominator of 1.

$= \frac{4x^2 - 25}{1} \cdot \frac{14}{2x + 5}$ Invert the divisor and multiply.

$= \frac{(2x + 5)(2x - 5)}{1} \cdot \frac{14}{1(2x + 5)}$ Factor.

$= \frac{\cancel{(2x + 5)}(2x - 5)}{1} \cdot \frac{14}{1\cancel{(2x + 5)}}$ Divide the numerator and denominator by the common factor, $2x + 5$.

$= 14(2x - 5)$ Multiply the remaining factors in the numerators and denominators.

b. $\frac{x^2 + 3x - 10}{2x} \div \frac{x^2 - 5x + 6}{x^2 - 3x}$

$= \frac{x^2 + 3x - 10}{2x} \cdot \frac{x^2 - 3x}{x^2 - 5x + 6}$ Invert the divisor and multiply.

$= \frac{(x + 5)(x - 2)}{2x} \cdot \frac{x(x - 3)}{(x - 3)(x - 2)}$ Factor.

$= \frac{(x + 5)\cancel{(x - 2)}}{2\cancel{x}} \cdot \frac{\cancel{x}\cancel{(x - 3)}}{\cancel{(x - 3)}\cancel{(x - 2)}}$ Divide numerators and denominators by common factors.

$= \frac{x + 5}{2}$ Multiply the remaining factors in the numerators and denominators.

■

✔ **CHECK POINT 7** Divide:

a. $(9x^2 - 49) \div \frac{3x - 7}{9}$

b. $\frac{x^2 - x - 12}{5x} \div \frac{x^2 - 10x + 24}{x^2 - 6x}$.

EXERCISE SET 6.1

Practice Exercises

In Exercises 1–6, use the given rational function to find the indicated function values. If a function value does not exist, so state.

1. $f(x) = \dfrac{x^2 - 9}{x + 3}$; $f(-2), f(0), f(5)$
2. $f(x) = \dfrac{x^2 - 16}{x + 4}$; $f(-2), f(0), f(5)$
3. $f(x) = \dfrac{x^2 - 2x - 3}{4 - x}$; $f(-1), f(2), f(6)$
4. $f(x) = \dfrac{x^2 - 3x - 4}{3 - x}$; $f(-1), f(3), f(5)$
5. $g(t) = \dfrac{2t^3 - 5}{t^2 + 1}$; $g(-1), g(0), g(2)$
6. $g(t) = \dfrac{2t^3 - 1}{t^2 + 4}$; $g(-1), g(0), g(2)$

In Exercises 7–16, use inspection or set the denominator equal to zero to find the domain of the given rational function. Use the notation of your choice (set-builder or interval) or the notation requested by your professor.

7. $f(x) = \dfrac{x - 2}{x - 5}$
8. $f(x) = \dfrac{x - 3}{x - 6}$
9. $f(x) = \dfrac{x - 4}{(x - 1)(x + 3)}$
10. $f(x) = \dfrac{x - 5}{(x - 2)(x + 4)}$
11. $f(x) = \dfrac{2x}{(x + 5)^2}$
12. $f(x) = \dfrac{2x}{(x + 7)^2}$
13. $f(x) = \dfrac{3x}{x^2 - 8x + 15}$
14. $f(x) = \dfrac{3x}{x^2 - 13x + 36}$
15. $f(x) = \dfrac{(x - 1)^2}{3x^2 - 2x - 8}$
16. $f(x) = \dfrac{(x - 1)^2}{4x^2 - 13x + 3}$

The graph of a rational function, f, is shown in the figure. Use the graph to solve Exercises 17–26.

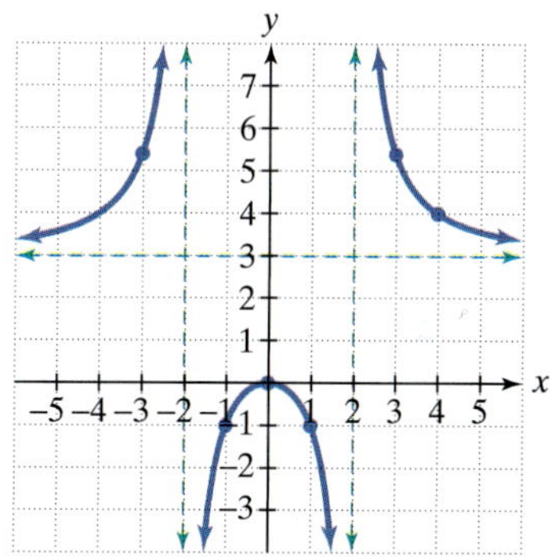

17. Find $f(4)$.
18. Find $f(1)$.
19. What is the domain of f?
20. What are the vertical asymptotes of the graph of f?
21. Describe the end behavior of the graph at the far left. What is the equation of the horizontal asymptote?
22. Describe the end behavior of the graph at the far right. What is the equation of the horizontal asymptote?
23. Explain how the graph shows that $f(-2)$ does not exist.
24. Explain how the graph shows that $f(2)$ does not exist.
25. How can you tell that this is not the graph of a polynomial function?
26. List two real numbers that are not function values of f.

In Exercises 27–50, simplify each rational expression. If the rational expression cannot be simplified, so state.

27. $\dfrac{x^2 - 4}{x - 2}$
28. $\dfrac{x^2 - 25}{x - 5}$
29. $\dfrac{x + 2}{x^2 - x - 6}$
30. $\dfrac{x + 1}{x^2 - 2x - 3}$

31. $\frac{4x + 20}{x^2 + 5x}$

32. $\frac{5x + 30}{x^2 + 6x}$

33. $\frac{4y - 20}{y^2 - 25}$

34. $\frac{6y - 42}{y^2 - 49}$

35. $\frac{3x - 5}{25 - 9x^2}$

36. $\frac{5x - 2}{4 - 25x^2}$

37. $\frac{y^2 - 49}{y^2 - 14y + 49}$

38. $\frac{y^2 - 9}{y^2 - 6y + 9}$

39. $\frac{x^2 + 7x - 18}{x^2 - 3x + 2}$

40. $\frac{x^2 - 4x - 5}{x^2 + 5x + 4}$

41. $\frac{3x + 7}{3x + 10}$

42. $\frac{2x + 3}{2x + 5}$

43. $\frac{x^2 - x - 12}{16 - x^2}$

44. $\frac{x^2 - 7x + 12}{9 - x^2}$

45. $\frac{x^2 + 3xy - 10y^2}{3x^2 - 7xy + 2y^2}$

46. $\frac{x^2 + 2xy - 3y^2}{2x^2 + 5xy - 3y^2}$

47. $\frac{x^3 - 8}{x^2 - 4}$

48. $\frac{x^3 - 1}{x^2 - 1}$

49. $\frac{x^3 + 4x^2 - 3x - 12}{x + 4}$

50. $\frac{x^3 - 2x^2 + x - 2}{x - 2}$

In Exercises 51–68, multiply as indicated.

51. $\frac{x - 3}{x + 7} \cdot \frac{3x + 21}{2x - 6}$

52. $\frac{x - 2}{x + 3} \cdot \frac{2x + 6}{5x - 10}$

53. $\frac{x^2 - 49}{x^2 - 4x - 21} \cdot \frac{x + 3}{x}$

54. $\frac{x^2 - 25}{x^2 - 3x - 10} \cdot \frac{x + 2}{x}$

55. $\frac{x^2 - 9}{x^2 - x - 6} \cdot \frac{x^2 + 5x + 6}{x^2 + x - 6}$

56. $\frac{x^2 - 1}{x^2 - 4} \cdot \frac{x^2 - 5x + 6}{x^2 - 2x - 3}$

57. $\frac{x^2 + 4x + 4}{x^2 + 8x + 16} \cdot \frac{(x + 4)^3}{(x + 2)^3}$

58. $\frac{x^2 - 2x + 1}{x^2 - 4x + 4} \cdot \frac{(x - 2)^3}{(x - 1)^3}$

59. $\frac{8y + 2}{y^2 - 9} \cdot \frac{3 - y}{4y^2 + y}$

60. $\frac{6y + 2}{y^2 - 1} \cdot \frac{1 - y}{3y^2 + y}$

61. $\frac{y^3 - 8}{y^2 - 4} \cdot \frac{y + 2}{2y}$

62. $\frac{y^2 + 6y + 9}{y^3 + 27} \cdot \frac{1}{y + 3}$

63. $(x - 3) \cdot \frac{x^2 + x + 1}{x^2 - 5x + 6}$

64. $(x + 1) \cdot \frac{x + 2}{x^2 + 7x + 6}$

65. $\frac{x^2 + xy}{x^2 - y^2} \cdot \frac{4x - 4y}{x}$

66. $\frac{x^2 - y^2}{x} \cdot \frac{x^2 + xy}{x + y}$

67. $\frac{x^2 + 2xy + y^2}{x^2 - 2xy + y^2} \cdot \frac{4x - 4y}{3x + 3y}$

68. $\frac{2x^2 - 3xy - 2y^2}{3x^2 - 4xy + y^2} \cdot \frac{3x^2 - 2xy - y^2}{x^2 + xy - 6y^2}$

In Exercises 69–86, divide as indicated.

69. $\frac{x + 5}{7} \div \frac{4x + 20}{9}$

70. $\frac{x + 1}{3} \div \frac{3x + 3}{7}$

71. $\frac{4}{y - 6} \div \frac{40}{7y - 42}$

72. $\frac{7}{y - 5} \div \frac{28}{3y - 15}$

73. $\frac{x^2 - 2x}{15} \div \frac{x - 2}{5}$

74. $\frac{x^2 - x}{15} \div \frac{x - 1}{5}$

75. $\frac{y^2 - 25}{2y - 2} \div \frac{y^2 + 10y + 25}{y^2 + 4y - 5}$

76. $\frac{y^2 + y}{y^2 - 4} \div \frac{y^2 - 1}{y^2 + 5y + 6}$

77. $(x^2 - 16) \div \frac{x^2 + 3x - 4}{x^2 + 4}$

78. $(x^2 + 4x - 5) \div \frac{x^2 - 25}{x + 7}$

79. $\frac{y^2 - 4y - 21}{y^2 - 10y + 25} \div \frac{y^2 + 2y - 3}{y^2 - 6y + 5}$

80. $\frac{y^2 + 4y - 21}{y^2 + 3y - 28} \div \frac{y^2 + 14y + 48}{y^2 + 4y - 32}$

81. $\frac{8x^3 - 1}{4x^2 + 2x + 1} \div \frac{x - 1}{(x - 1)^2}$

82. $\frac{x^2 - 9}{x^3 - 27} \div \frac{x^2 + 6x + 9}{x^2 + 3x + 9}$

83. $\frac{x^2 - 4y^2}{x^2 + 3xy + 2y^2} \div \frac{x^2 - 4xy + 4y^2}{x + y}$

84. $\frac{xy - y^2}{x^2 + 2x + 1} \div \frac{2x^2 + xy - 3y^2}{2x^2 + 5xy + 3y^2}$

85. $\frac{x^4 - y^8}{x^2 + y^4} \div \frac{x^2 - y^4}{3x^2}$

86. $\frac{(x - y)^3}{x^3 - y^3} \div \frac{x^2 - 2xy + y^2}{x^2 - y^2}$

Application Exercises

The rational function

$$f(x) = \frac{130x}{100 - x}$$

models the cost, $f(x)$, in millions of dollars, to inoculate $x\%$ of the population against a particular strain of flu. The graph of the rational function is shown. Use the function's equation to solve Exercises 87–90.

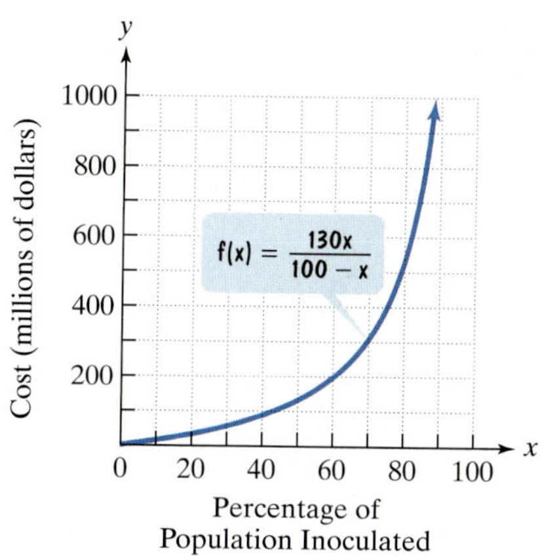

87. Find and interpret $f(60)$. Identify your solution as a point on the graph.

88. Find and interpret $f(80)$. Identify your solution as a point on the graph.

89. What value of x must be excluded from the rational function's domain? Does the cost model indicate that we can inoculate all of the population against the flu? Explain.

90. What happens to the cost as x approaches 100%? How is this shown by the graph? Explain what this means.

The rational function

$$P(x) = \frac{72{,}900}{100x^2 + 729}$$

models the percentage of people in the United States, $P(x)$, with x years of education who are unemployed. The graph of the rational function is shown. Use the function's equation to solve Exercises 91–94.

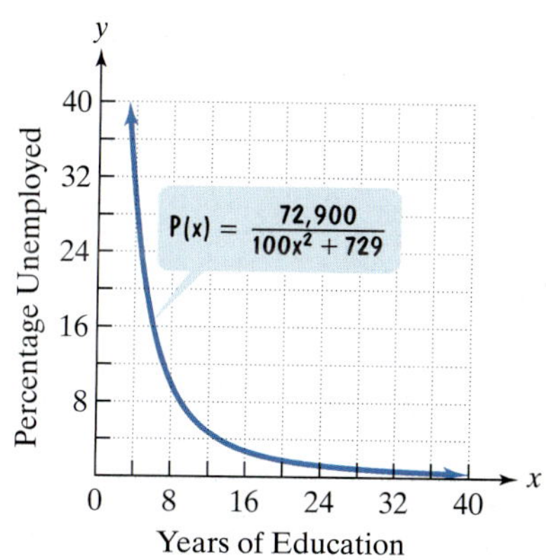

91. Find and interpret $P(10)$. Round to the nearest whole percent. Identify your solution as a point on the graph.

92. Find and interpret $P(20)$. Round to the nearest whole percent. Identify your solution as a point on the graph.

93. Describe the end behavior at the far right of the graph. Is there an education level that leads to guaranteed employment? If not, how is this indicated by the horizontal asymptote, the x-axis?

94. Describe the behavior of the graph as x approaches 0. What does this mean in terms of the percentage unemployed with decreasing levels of education?

Among all deaths from a particular disease, the percentage that are smoking related (21–39 cigarettes per day) is a function of the disease's ***incidence ratio****. The incidence ratio describes the number of times more likely smokers are than nonsmokers to die from the disease. The following table shows the incidence ratios for heart disease and lung cancer for two age groups.*

Incidence Ratios

	Heart Disease	Lung Cancer
Ages 55–64	1.9	10
Ages 65–74	1.7	9

Source: Alexander M. Walker, *Observations and Inference, page 20.*

For example, the incidence ratio of 9 in the table means that smokers between the ages of 65 and 74 are 9 times more likely than nonsmokers in the same age group to die from lung cancer. The rational function

$$P(x) = \frac{100(x - 1)}{x}$$

models the percentage of smoking-related deaths among all deaths from a disease, $P(x)$, in terms of the disease's incidence ratio, x. The graph of the rational function is shown. Use this function to solve Exercises 95–98.

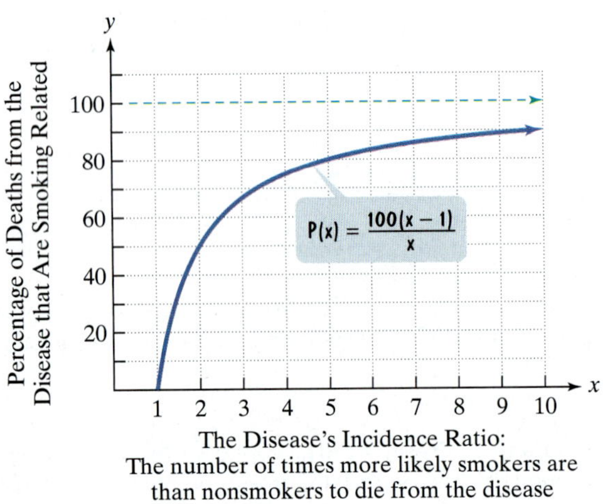

95. Find $P(10)$. Describe what this means in terms of the incidence ratio, 10, given in the table. Identify your solution as a point on the graph.

96. Find $P(9)$. Round to the nearest percent. Describe what this means in terms of the incidence ratio, 9, given in the table. Identify your solution as a point on the graph.

97. What is the horizontal asymptote of the graph? Describe what this means about the percentage of deaths caused by smoking with increasing incidence ratios.

98. According to the model and its graph, is there a disease for which all deaths are caused by smoking? Explain your answer.

Writing in Mathematics

99. What is a rational expression? Give an example with your explanation.

100. What is a rational function? Provide an example.

101. What is the domain of a rational function?

102. If you are given the equation of a rational function, explain how to determine its domain.

103. Describe two ways the graph of a rational function differs from the graph of a polynomial function.

104. What is a vertical asymptote?

105. What is a horizontal asymptote?

106. Explain how to simplify a rational expression.

107. Explain how to simplify a rational expression with opposite factors in the numerator and denominator.

108. Explain how to multiply rational expressions.

109. Explain how to divide rational expressions.

110. Although your friend has a family history of heart disease, he smokes, on average, 25 cigarettes per day. He sees the table showing incidence ratios for heart disease (see Exercises 95–98) and feels comfortable that they are less than 2, compared to 9 and 10 for lung cancer. He claims that all family deaths have been from heart disease, and decides not to give up smoking. Use the given function and its graph to describe some additional information not given in the table that might influence his decision.

Technology Exercises

In Exercises 111–114, determine if the multiplication or division has been performed correctly by graphing the function on each side of the equation in the same viewing rectangle. If the graphs do not coincide, correct the expression on the right side and then verify your correction using the graphing utility.

111. $\frac{x^2 + x}{3x} \cdot \frac{6x}{x + 1} = 2x$

112. $\frac{x^3 - 25x}{x^2 - 3x - 10} \cdot \frac{x + 2}{x} = x + 5$

113. $\frac{x^2 - 9}{x + 4} \div \frac{x - 3}{x + 4} = x - 3$

114. $(x - 5) \div \frac{2x^2 - 11x + 5}{4x^2 - 1} = 2x - 1$

115. **a.** Graph $f(x) = \frac{x^2 - x - 2}{x - 2}$ and $g(x) = x + 1$ in the same viewing rectangle. What do you observe?

b. Simplify the formula in the definition of function f. Do f and g represent exactly the same function? Explain.

c. Display the graphs of f and g separately in the utility's viewing rectangle. TRACE along each of the curves until you get to $x = 2$. What difference do you observe? What does this mean?

116. The rational function

$$f(x) = \frac{27{,}725(x - 14)}{x^2 + 9} - 5x$$

models the number of arrests, $f(x)$, per 100,000 drivers, for driving under the influence of alcohol, as a function of a driver's age, x.

a. Graph the function in a $[0, 70, 5]$ by $[0, 400, 20]$ viewing rectangle.

b. Describe the trend shown by the graph.

c. Use the ZOOM and TRACE features or the maximum function feature of your graphing utility to find the age that corresponds to the greatest number of arrests. How many arrests, per 100,000 drivers, are there for this age group?

Critical Thinking Exercises

117. Which one of the following is true?

a. $\frac{x^2 - 25}{x - 5} = x - 5$

b. $\frac{x^2 + 7}{7} = x^2 + 1$

c. The domain of $f(x) = \frac{x - 3}{x^2 + 1}$ is $(-\infty, 3) \cup (3, \infty)$.

d. None of the above is true.

118. Graph: $f(x) = \frac{x^2 - x - 2}{x - 2}$.

119. Let $f(x) = x^2 - 5x + 3$. Find

$$\frac{f(a + h) - f(a)}{h}$$

and simplify.

120. Simplify: $\frac{6x^{3n} + 6x^{2n}y^n}{x^{2n} - y^{2n}}$.

121. Divide: $\frac{y^{2n} - 1}{y^{2n} + 3y^n + 2} \div \frac{y^{2n} + y^n - 12}{y^{2n} - y^n - 6}$.

Review Exercises

122. Graph: $4x - 5y \geq 20$. (Section 4.4, Example 1)

123. Multiply: $(2x - 5)(x^2 - 3x - 6)$. (Section 5.2, Example 3)

124. Factor completely: $x^4 - 3x^3 + 2x^2 - 6x$. (Section 5.6, Example 3)

▸ SECTION 6.2 Adding and Subtracting Rational Expressions

Objectives

1. Add rational expressions with the same denominators.
2. Subtract rational expressions with the same denominators.
3. Find the least common denominators.
4. Add and subtract rational expressions with different denominators.
5. Add and subtract rational expressions with opposite denominators.

SSM PH Tutor Center CD-ROM Video

Mothers Against Drunk Driving (MADD) volunteer Janet Priewe of Wisconsin holds up picture posters of drunk driving victims during a 20th anniversary rally outside the U.S. Capitol, September 6, 2000 in Washington.

Did you know that people in California are at far greater risk of injury or death from drunk drivers than from earthquakes? According to the U.S. Bureau of Justice statistics, half the arrests for driving under the influence of alcohol involve drivers ages 25 through 34. The rational function

$$f(x) = \frac{27{,}725(x - 14)}{x^2 + 9} - 5x$$

models the number of arrests for driving under the influence, $f(x)$, per 100,000 drivers, as a function of a driver's age, x.

The formula for function f involves the subtraction of two expressions. It is possible to perform this subtraction and express the function's formula as a single rational expression. In this section, we will draw on your experience from arithmetic to add and subtract rational expressions.

1 Add rational expressions with the same denominators.

Addition and Subtraction When Denominators Are the Same

To add rational numbers having the same denominators, such as $\frac{2}{9}$ and $\frac{5}{9}$, we add the numerators and place the sum over the common denominator:

$$\frac{2}{9} + \frac{5}{9} = \frac{2+5}{9} = \frac{7}{9}.$$

We add rational expressions with the same denominators identically.

Adding Rational Expressions with Common Denominators

If $\frac{P}{R}$ and $\frac{Q}{R}$ are rational expressions, then

$$\frac{P}{R} + \frac{Q}{R} = \frac{P+Q}{R}.$$

To add rational expressions with the same denominators, add numerators and place the sum over the common denominator. If possible, simplify the final result.

EXAMPLE 1 Adding Rational Expressions When Denominators Are the Same

Add: $\dfrac{x^2 + 2x - 2}{x^2 + 3x - 10} + \dfrac{5x + 12}{x^2 + 3x - 10}.$

Solution

$$\frac{x^2+2x-2}{x^2+3x-10}+\frac{5x+12}{x^2+3x-10}$$

$$=\frac{x^2+2x-2+5x+12}{x^2+3x-10}$$ Add numerators. Place this sum over the common denominator.

$$=\frac{x^2+7x+10}{x^2+3x-10}$$ Combine like terms: $2x+5x=7x$ and $-2+12=10$.

$$=\frac{(x+2)\cancel{(x+5)}}{(x-2)\cancel{(x+5)}}$$ Factor and simplify by dividing out the common factor, $x+5$.

$$=\frac{x+2}{x-2}$$

■

✔ **CHECK POINT 1** Add: $\frac{x^2-5x-15}{x^2+5x+6}+\frac{2x+5}{x^2+5x+6}$.

The following box shows how to subtract rational expressions with the same denominators:

2 Subtract rational expressions with the same denominators.

Subtracting Rational Expressions with Common Denominators

If $\frac{P}{R}$ and $\frac{Q}{R}$ are rational expressions, then

$$\frac{P}{R}-\frac{Q}{R}=\frac{P-Q}{R}.$$

To subtract rational expressions with the same denominators, subtract numerators and place the difference over the common denominator. If possible, simplify the final result.

EXAMPLE 2 Subtracting Rational Expressions When Denominators Are the Same

Subtract: $\frac{3y^3-5x^3}{x^2-y^2}-\frac{4y^3-6x^3}{x^2-y^2}$.

Solution

$$\frac{3y^3-5x^3}{x^2-y^2}-\frac{4y^3-6x^3}{x^2-y^2}=\frac{3y^3-5x^3-(4y^3-6x^3)}{x^2-y^2}$$ Subtract numerators and include parentheses to indicate that both terms are subtracted. Place this difference over the common denominator.

$$=\frac{3y^3-5x^3-4y^3+6x^3}{x^2-y^2}$$ Remove parentheses and then change the sign of each term.

$$=\frac{x^3-y^3}{x^2-y^2}$$ Combine like terms: $-5x^3+6x^3=x^3$ and $3y^3-4y^3=-y^3$.

$$=\frac{\cancel{(x-y)}(x^2+xy+y^2)}{(x+y)\cancel{(x-y)}}$$ Factor and simplify by dividing out the common factor, $x-y$.

$$=\frac{x^2+xy+y^2}{x+y}$$

■

Study Tip

When a numerator is being subtracted, be sure to **subtract every term in that expression.**

The − sign applies to the entire numerator, $4y^3 - 6x^3$.

Insert parentheses to indicate this.

The sign of every term of $4y^3 - 6x^3$ changes.

$$\frac{3y^3 - 5x^3}{x^2 - y^2} - \frac{4y^3 - 6x^3}{x^2 - y^2} = \frac{3y^3 - 5x^3 - (4y^3 - 6x^3)}{x^2 - y^2} = \frac{3y^3 - 5x^3 - 4y^3 + 6x^3}{x^2 - y^2}$$

The entire numerator of the second rational expression must be subtracted. Avoid the common error of subtracting only the first term.

INCORRECT!

$$\frac{3y^3 - 5x^3}{x^2 - y^2} - \frac{4y^3 - 6x^3}{x^2 - y^2} = \frac{3y^3 - 5x^3 - 4y^3 - 6x^3}{x^2 - y^2}$$

$-6x^3$ must also be subtracted.

✔ **CHECK POINT 2** Subtract: $\dfrac{5x - y}{x^2 - y^2} - \dfrac{4x - 2y}{x^2 - y^2}$.

3 Find the least common denominator.

Finding the Least Common Denominator

We can gain insight into adding rational expressions with different denominators by looking closely at what we do when adding fractions with different denominators. For example, suppose that we want to add $\frac{1}{2}$ and $\frac{2}{3}$. We must first write the fractions with the same denominator. We look for the smallest number that contains both 2 and 3 as factors. This number, 6, is then used as the *least common denominator*, or LCD.

The **least common denominator** of a group of rational expressions is a polynomial of least degree whose factors include all the factors of the denominators in the group.

Finding the Least Common Denominator

1. Factor each denominator completely.
2. List the factors of the first denominator.
3. Add to the list in step 2 any factors of the second denominator that do not appear in the list.
4. Form the product of each different factor from the list in step 3. This product is the least common denominator.

EXAMPLE 3 Finding the Least Common Denominator

Find the LCD of

$$\frac{3}{10x^2} \text{ and } \frac{7}{15x}.$$

Solution

Step 1 Factor each denominator completely.

$$10x^2 = 5 \cdot 2x^2 \quad (\text{or } 5 \cdot 2 \cdot x \cdot x)$$
$$15x = 5 \cdot 3x$$

Step 2 List the factors of the first denominator.

$$5, 2, x^2 \quad (\text{or } 5, 2, x, x)$$

Step 3 Add any unlisted factors from the second denominator. Two factors from $5 \cdot 3x$ are already in our list. These factors include 5 and x. We add the unlisted factor, 3, to our list. We have

$$3, 5, 2, x^2.$$

Step 4 The least common denominator is the product of all factors in the final list. Thus,

$$3 \cdot 5 \cdot 2 \cdot x^2$$

or $30x^2$ is the least common denominator. ■

✔ **CHECK POINT 3** Find the LCD of

$$\frac{7}{6x^2} \text{ and } \frac{2}{9x}.$$

EXAMPLE 4 Finding the Least Common Denominator

Find the LCD of

$$\frac{9}{7x^2 + 28x} \text{ and } \frac{11}{x^2 + 8x + 16}.$$

Solution

Step 1 Factor each denominator completely.

$$7x^2 + 28x = 7x(x + 4)$$
$$x^2 + 8x + 16 = (x + 4)^2$$

Step 2 List the factors of the first denominator.

$$7, x, (x + 4)$$

Step 3 Add any unlisted factors from the second denominator. The second denominator is $(x + 4)^2$, or $(x + 4)(x + 4)$. One factor of $x + 4$ is already in our list, but the other factor is not. We add a second factor of $x + 4$ to the list. We have

$$7, x, (x + 4), (x + 4).$$

Step 4 The least common denominator is the product of all factors in the final list. Thus,

$$7x(x + 4)(x + 4), \text{ or } 7x(x + 4)^2,$$

is the least common denominator. ■

CHECK POINT 4 Find the LCD of

$$\frac{7}{5x^2 + 15x} \quad \text{and} \quad \frac{9}{x^2 + 6x + 9}.$$

4 Add and subtract rational expressions with different denominators.

Addition and Subtraction When Denominators Are Different

Finding the least common denominator for two (or more) rational expressions is the first step needed to add or subtract the expressions. For example, to add $\frac{1}{2}$ and $\frac{2}{3}$, we first determine that the LCD is 6. Then we write each fraction in terms of the LCD.

$$\frac{1}{2} + \frac{2}{3} = \frac{1}{2} \cdot \frac{3}{3} + \frac{2}{3} \cdot \frac{2}{2}$$

Multiply the numerator and denominator of each fraction by whatever extra factors are required to form 6, the LCD.

$\frac{3}{3} = 1$ and $\frac{2}{2} = 1$. Multiplying by 1 does not change a fraction's value.

$$= \frac{3}{6} + \frac{4}{6}$$

$$= \frac{3 + 4}{6}$$

Add numerators. Place this sum over the LCD.

$$= \frac{7}{6}$$

We follow the same steps in adding or subtracting rational expressions with different denominators.

Adding and Subtracting Rational Expressions That Have Different Denominators

1. Find the LCD of the rational expressions.
2. Rewrite each rational expression as an equivalent expression whose denominator is the LCD. To do so, multiply the numerator and the denominator of each rational expression by any factor(s) needed to convert the denominator into the LCD.
3. Add or subtract numerators, placing the resulting expression over the LCD.
4. If necessary, simplify the resulting rational expression.

EXAMPLE 5 Adding Rational Expressions with Different Denominators

Add: $\frac{3}{10x^2} + \frac{7}{15x}$.

Solution

Step 1 Find the least common denominator. In Example 3, we found that the LCD for these rational expressions is $30x^2$.

Step 2 Write equivalent expressions with the LCD as denominators. We must rewrite each rational expression with a denominator of $30x^2$.

$$\frac{3}{10x^2} \cdot \frac{3}{3} = \frac{9}{30x^2} \qquad \frac{7}{15x} \cdot \frac{2x}{2x} = \frac{14x}{30x^2}$$

Multiply the numerator and denominator by 3 to get $30x^2$, the LCD.

Multiply the numerator and denominator by $2x$ to get $30x^2$, the LCD.

Because $\frac{3}{3} = 1$ and $\frac{2x}{2x} = 1$, we are not changing the value of either rational expression, only its appearance. In summary, we have

$$\frac{3}{10x^2} + \frac{7}{15x} \qquad \text{The LCD is } 30x^2.$$

$$= \frac{3}{10x^2} \cdot \frac{3}{3} + \frac{7}{15x} \cdot \frac{2x}{2x} \qquad \text{Write equivalent expressions with the LCD.}$$

$$= \frac{9}{30x^2} + \frac{14x}{30x^2}.$$

Steps 3 and 4 Add numerators, putting this sum over the LCD. Simplify, if possible.

$$= \frac{9 + 14x}{30x^2} \quad \text{or} \quad \frac{14x + 9}{30x^2}$$

The numerator is prime and further simplification is not possible. ■

Study Tip

It is incorrect to add rational expressions by adding numerators and adding denominators. Avoid this common error.

INCORRECT!

$$\frac{3}{10x^2} + \frac{7}{15x}$$

$$= \frac{3 + 7}{10x^2 + 15x}$$

$$= \frac{10}{10x^2 + 15x}$$

✔ **CHECK POINT 5** Add: $\frac{7}{6x^2} + \frac{2}{9x}$.

EXAMPLE 6 Adding Rational Expressions with Different Denominators

Add: $\frac{x}{x - 3} + \frac{x - 1}{x + 3}$.

Solution

Step 1 Find the least common denominator. Start by factoring the denominators.

$$x - 3 = 1(x - 3)$$

$$x + 3 = 1(x + 3)$$

The factors of the first denominator are 1 and $x - 3$. The only factor from the second denominator that is unlisted is $x + 3$. Thus, the least common denominator is $1(x - 3)(x + 3)$, or $(x - 3)(x + 3)$.

Step 2 Write equivalent expressions with the LCD as denominators.

$$\frac{x}{x - 3} + \frac{x - 1}{x + 3}$$

$$= \frac{x(x + 3)}{(x - 3)(x + 3)} + \frac{(x - 1)(x - 3)}{(x - 3)(x + 3)}$$

Multiply each numerator and denominator by the extra factor required to form $(x - 3)(x + 3)$, the LCD.

Steps 3 and 4 Add numerators, putting this sum over the LCD. Simplify, if possible.

$$= \frac{x(x+3) + (x-1)(x-3)}{(x-3)(x+3)}$$

$$= \frac{x^2 + 3x + x^2 - 4x + 3}{(x-3)(x+3)}$$

Perform the multiplications using the distributive property and FOIL.

$$= \frac{2x^2 - x + 3}{(x-3)(x+3)}$$

Combine like terms: $x^2 + x^2 = 2x^2$ and $3x - 4x = -x$.

The numerator is prime and further simplification is not possible. ■

✔ **CHECK POINT 6** Add: $\frac{x}{x-4} + \frac{x-2}{x+4}$.

EXAMPLE 7 Subtracting Rational Expressions with Different Denominators

Subtract: $\frac{x-1}{x^2+x-6} - \frac{x-2}{x^2+4x+3}$.

Solution

Step 1 Find the least common denominator. Start by factoring the denominators.

$$x^2 + x - 6 = (x+3)(x-2)$$

$$x^2 + 4x + 3 = (x+3)(x+1)$$

The factors of the first denominator are $x + 3$ and $x - 2$. The only factor from the second denominator that is unlisted is $x + 1$. Thus, the least common denominator is $(x+3)(x-2)(x+1)$.

Step 2 Write equivalent expressions with the LCD as denominators.

$$\frac{x-1}{x^2+x-6} - \frac{x-2}{x^2+4x+3}$$

$$= \frac{x-1}{(x+3)(x-2)} - \frac{x-2}{(x+3)(x+1)}$$

Factor denominators. The LCD is $(x+3)(x-2)(x+1)$.

$$= \frac{(x-1)(x+1)}{(x+3)(x-2)(x+1)} - \frac{(x-2)(x-2)}{(x+3)(x-2)(x+1)}$$

Multiply each numerator and denominator by the extra factor required to form $(x+3)(x-2)(x+1)$, the LCD.

Steps 3 and 4 Subtract numerators, putting this difference over the LCD. Simplify if possible.

$$= \frac{(x-1)(x+1)-(x-2)(x-2)}{(x+3)(x-2)(x+1)}$$

$$= \frac{x^2-1-(x^2-4x+4)}{(x+3)(x-2)(x+1)}$$ Perform the multiplications in the numerator. Don't forget the parentheses.

$$= \frac{x^2-1-x^2+4x-4}{(x+3)(x-2)(x+1)}$$ Remove parentheses and change the sign of each term in parentheses.

$$= \frac{4x-5}{(x+3)(x-2)(x+1)}$$ Combine like terms.

The numerator is prime and further simplification is not possible. ■

✔ **CHECK POINT 7** Subtract: $\frac{2x-3}{x^2-5x+6} - \frac{x+4}{x^2-2x-3}$.

EXAMPLE 8 Adding and Subtracting Rational Expressions with Different Denominators

Perform the indicated operations:

$$\frac{3y+2}{y-5} + \frac{4}{3y+4} - \frac{7y^2+24y+28}{3y^2-11y-20}.$$

Solution

Step 1 Find the least common denominator. Start by factoring the denominators.

$$y-5 = 1(y-5)$$
$$3y+4 = 1(3y+4)$$
$$3y^2-11y-20 = (3y+4)(y-5)$$

The factors of the first denominator are 1 and $y - 5$. The only factor from the second denominator that is unlisted is $3y + 4$. Adding this factor to our list, we have 1, $y - 5$, and $3y + 4$. We have listed all factors from the third denominator. Thus, the least common denominator is $1(y - 5)(3y + 4)$, or $(y - 5)(3y + 4)$.

Step 2 Write equivalent expressions with the LCD as denominators.

$$\frac{3y+2}{y-5} + \frac{4}{3y+4} - \frac{7y^2+24y+28}{3y^2-11y-20}$$

$$= \frac{3y+2}{y-5} + \frac{4}{3y+4} - \frac{7y^2+24y+28}{(3y+4)(y-5)}$$

Factor denominators. The LCD is $(y - 5)(3y + 4)$.

$$= \frac{(3y+2)(3y+4)}{(y-5)(3y+4)} + \frac{4(y-5)}{(y-5)(3y+4)} - \frac{7y^2+24y+28}{(3y+4)(y-5)}$$

Multiply the first two numerators and denominators by the extra factor required to form the LCD.

Step 3 Add and subtract numerators, putting this result over the LCD. Simplify if possible.

$$= \frac{(3y + 2)(3y + 4) + 4(y - 5) - (7y^2 + 24y + 28)}{(y - 5)(3y + 4)}$$

$$= \frac{9y^2 + 18y + 8 + 4y - 20 - 7y^2 - 24y - 28}{(y - 5)(3y + 4)}$$

Perform multiplications. Remove parentheses and change the sign of each term.

$$= \frac{2y^2 - 2y - 40}{(y - 5)(3y + 4)}$$

Combine like terms: $9y^2 - 7y^2 = 2y^2$, $18y + 4y - 24y = -2y$, and $8 - 20 - 28 = -40$.

$$= \frac{2(y^2 - y - 20)}{(y - 5)(3y + 4)}$$

Factor out the GCF in the numerator.

$$= \frac{2(y + 4)\cancel{(y - 5)}}{\cancel{(y - 5)}(3y + 4)}$$

Factor completely and simplify.

$$= \frac{2(y + 4)}{3y + 4}$$

■

✔ **CHECK POINT 8** Perform the indicated operations:

$$\frac{y - 1}{y - 2} + \frac{y - 6}{y^2 - 4} - \frac{y + 1}{y + 2}.$$

5 Add and subtract rational expressions with opposite denominators.

In some situations, we need to add or subtract rational expressions with denominators that are opposites, or additive inverses. Multiply the numerator and the denominator of either of the rational expressions by -1. Then they will have the same denominators.

EXAMPLE 9 Adding Rational Expressions When Denominators Are Opposites

Add: $\frac{4x - 16y}{x - 5y} + \frac{x - 6y}{5y - x}$.

Solution

$$\frac{4x - 16y}{x - 5y} + \frac{x - 6y}{5y - x}$$

The denominators, $x - 5y$ and $5y - x$, are opposites, or additive inverses.

$$= \frac{4x - 16y}{x - 5y} + \frac{(-1)}{(-1)} \cdot \frac{x - 6y}{5y - x}$$

Multiply the numerator and denominator of the second rational expression by -1.

$$= \frac{4x - 16y}{x - 5y} + \frac{-x + 6y}{-5y + x}$$

Perform the multiplications by -1.

$$= \frac{4x - 16y}{x - 5y} + \frac{-x + 6y}{x - 5y}$$

Rewrite $-5y + x$ as $x - 5y$. Both rational expressions now have the same denominators.

$$= \frac{4x - 16y - x + 6y}{x - 5y}$$ Add numerators. Place this sum over the common denominator.

$$= \frac{3x - 10y}{x - 5y}$$ Combine like terms.

✔ **CHECK POINT 9** Add: $\frac{4x - 7y}{x - 3y} + \frac{x - 2y}{3y - x}$.

EXERCISE SET 6.2

Practice Exercises

In Exercises 1–16, perform the indicated operations. These exercises involve addition and subtraction when denominators are the same. Simplify the result, if possible.

1. $\frac{7}{9x} + \frac{5}{9x}$
2. $\frac{5}{6x} + \frac{4}{6x}$
3. $\frac{x}{x - 4} + \frac{9x + 7}{x - 4}$
4. $\frac{x}{x - 3} + \frac{11x + 5}{x - 3}$
5. $\frac{x^2 - 2x}{x^2 + 3x} + \frac{x^2 + x}{x^2 + 3x}$
6. $\frac{x^2 + 7x}{x^2 - 5x} + \frac{x^2 - 4x}{x^2 - 5x}$
7. $\frac{y^2}{y^2 - 9} + \frac{9 - 6y}{y^2 - 9}$
8. $\frac{y^2}{y^2 - 25} + \frac{25 - 10y}{y^2 - 25}$
9. $\frac{3x}{4x - 3} - \frac{2x - 1}{4x - 3}$
10. $\frac{3x}{7x - 4} - \frac{2x - 1}{7x - 4}$
11. $\frac{x^2 - 2}{x^2 + 6x - 7} - \frac{19 - 4x}{x^2 + 6x - 7}$
12. $\frac{x^2 + 6x + 2}{x^2 + x - 6} - \frac{2x - 1}{x^2 + x - 6}$
13. $\frac{20y^2 + 5y + 1}{6y^2 + y - 2} - \frac{8y^2 - 12y - 5}{6y^2 + y - 2}$
14. $\frac{y^2 + 3y - 6}{y^2 - 5y + 4} - \frac{4y - 4 - 2y^2}{y^2 - 5y + 4}$
15. $\frac{2x^3 - 3y^3}{x^2 - y^2} - \frac{x^3 - 2y^3}{x^2 - y^2}$
16. $\frac{4y^3 - 3x^3}{y^2 - x^2} - \frac{3y^3 - 2x^3}{y^2 - x^2}$

In Exercises 17–26, find the least common denominator of the rational expressions.

17. $\frac{11}{25x^2}$ and $\frac{14}{35x}$
18. $\frac{7}{15x^2}$ and $\frac{9}{24x}$
19. $\frac{2}{x - 5}$ and $\frac{3}{x^2 - 25}$
20. $\frac{2}{x + 3}$ and $\frac{5}{x^2 - 9}$
21. $\frac{7}{y^2 - 100}$ and $\frac{13}{y(y - 10)}$
22. $\frac{7}{y^2 - 4}$ and $\frac{15}{y(y + 2)}$
23. $\frac{8}{x^2 - 16}$ and $\frac{x}{x^2 - 8x + 16}$
24. $\frac{3}{x^2 - 25}$ and $\frac{x}{x^2 - 10x + 25}$
25. $\frac{7}{y^2 - 5y - 6}$ and $\frac{y}{y^2 - 4y - 5}$
26. $\frac{3}{y^2 - y - 20}$ and $\frac{y}{2y^2 + 7y - 4}$

In Exercises 27–62, perform the indicated operations. These exercises involve addition and subtraction when denominators are different. Simplify the result, if possible.

27. $\frac{3}{5x^2} + \frac{10}{x}$
28. $\frac{7}{2x^2} + \frac{4}{x}$
29. $\frac{3}{x - 2} + \frac{4}{x + 3}$
30. $\frac{2}{x - 1} + \frac{3}{x + 2}$
31. $\frac{x}{x^2 + 5x + 6} + \frac{4}{x^2 - x - 6}$
32. $\frac{x}{x^2 + 2x + 1} + \frac{4}{x^2 + 5x + 4}$

33. $\frac{y-7}{y+4}+\frac{y+4}{y-7}$

34. $\frac{y-5}{y+3}+\frac{y+3}{y-5}$

35. $\frac{8x}{x^2-16}-\frac{5}{x+4}$

36. $\frac{5x}{x^2-9}-\frac{4}{x+3}$

37. $\frac{3y+7}{y^2-5y+6}-\frac{3}{y-3}$

38. $\frac{2y+9}{y^2-7y+12}-\frac{2}{y-3}$

39. $\frac{x^2-6}{x^2+9x+18}-\frac{x-4}{x+6}$

40. $\frac{x^2-39}{x^2+3x-10}-\frac{x-7}{x-2}$

41. $\frac{4x+1}{x^2+7x+12}+\frac{2x+3}{x^2+5x+4}$

42. $\frac{3x-2}{x^2-x-6}+\frac{4x-3}{x^2-9}$

43. $\frac{x+4}{x^2-x-2}-\frac{2x+3}{x^2+2x-8}$

44. $\frac{2x+1}{x^2-7x+6}-\frac{x+3}{x^2-5x-6}$

45. $4+\frac{1}{x-3}$

46. $7+\frac{1}{x-5}$

47. $\frac{y-7}{y^2-16}+\frac{7-y}{16-y^2}$

48. $\frac{y-3}{y^2-25}+\frac{y-3}{25-y^2}$

49. $\frac{x+3}{3x+6}+\frac{x}{4-x^2}$

50. $\frac{x+7}{4x+12}+\frac{x}{9-x^2}$

51. $\frac{2x}{x-4}+\frac{64}{x^2-16}-\frac{2x}{x+4}$

52. $\frac{x}{x-3}+\frac{x+2}{x^2-2x-3}-\frac{4}{x+1}$

53. $\frac{7x}{x^2-y^2}-\frac{3}{y-x}$

54. $\frac{3x}{x^2-y^2}-\frac{2}{y-x}$

55. $\frac{3}{5x+6}-\frac{4}{x-2}+\frac{x^2-x}{5x^2-4x-12}$

56. $\frac{x-1}{x^2+2x+1}-\frac{3}{2x-2}+\frac{x}{x^2-1}$

57. $\frac{3x-y}{x^2-9xy+20y^2}+\frac{2y}{x^2-25y^2}$

58. $\frac{x+2y}{x^2+4xy+4y^2}-\frac{2x}{x^2-4y^2}$

59. $\frac{3x}{x^2-4}+\frac{5x}{x^2+x-2}-\frac{3}{x^2-4x+4}$

60. $\frac{1}{x}+\frac{4}{x^2-4}-\frac{2}{x^2-2x}$

61. $\frac{6a+5b}{6a^2+5ab-4b^2}-\frac{a+2b}{9a^2-16b^2}$

62. $\frac{5a-b}{a^2+ab-2b^2}-\frac{3a+2b}{a^2+5ab-6b^2}$

Application Exercises

You plan to drive from Miami, Florida, to Atlanta, Georgia. Your trip involves approximately 470 miles of travel in Florida and 250 miles in Georgia. The speed limit is 70 miles per hour in Florida and 65 miles per hour in Georgia. If you average x miles per hour over these speed limits, the total driving time, $T(x)$, in hours, is given by the function

$$T(x)=\frac{470}{x+70}+\frac{250}{x+65}.$$

The graph of T is shown in the figure. Use the function's equation to solve Exercises 63–68.

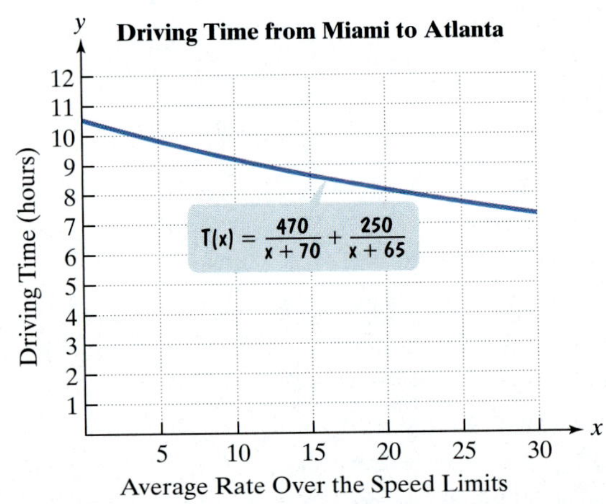

63. Find and interpret $T(0)$. Round to the nearest hour. Identify your solution as a point on the graph.

64. Find and interpret $T(5)$. Round to the nearest hour. Identify your solution as a point on the graph.

65. Find a simplified form of $T(x)$ by adding the rational expressions in the function's formula. Then use this form of the function to find $T(0)$.

66. Find a simplified form of $T(x)$ by adding the rational expressions in the function's formula. Then use this form of the function to find $T(5)$.

67. Use the graph to answer this question. If you want the driving time to be 9 hours, how much over the speed limits do you need to drive? Round to the nearest mile per hour. Does this seem like a realistic driving time for the trip? Explain your answer.

68. Use the graph to answer this question. If you want the driving time to be 8 hours, how much over the speed limits do you need to drive? Round to the nearest mile per hour. Does this seem like a realistic driving time for the trip? Explain your answer.

In Exercises 69–70, express the perimeter of each rectangle as a single rational expression.

69.

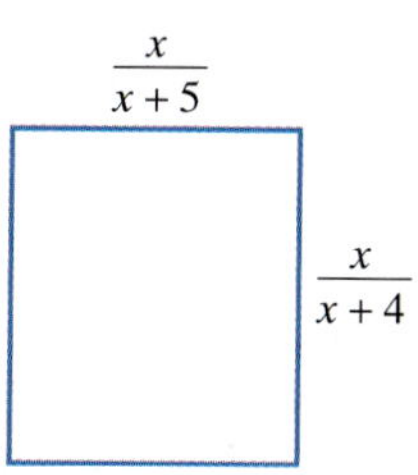

70.

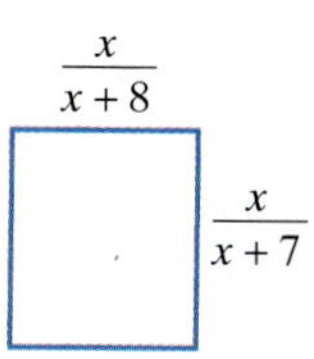

71. In the section opener, we saw that the rational function

$$f(x) = \frac{27{,}725(x - 14)}{x^2 + 9} - 5x$$

models the number of arrests, $f(x)$, per 100,000 drivers, for driving under the influence of alcohol as a function of a driver's age, x. Perform the subtraction and express the function as a single rational expression.

Writing in Mathematics

72. Explain how to add rational expressions when denominators are the same. Give an example with your explanation.

73. Explain how to subtract rational expressions when denominators are the same. Give an example with your explanation.

74. Explain how to find the least common denominator for denominators of $x^2 - 100$ and $x^2 - 20x + 100$.

75. Explain how to add rational expressions that have different denominators. Use $\frac{5}{x + 1} + \frac{3}{x + 4}$ in your explanation.

76. Explain how to add rational expressions when denominators are opposites. Use an example to support your explanation.

Explain the error in Exercises 77–78. Then rewrite the right side of the equation to correct the error that now exists.

77. $\frac{1}{a} + \frac{1}{b} = \frac{1}{a + b}$

78. $\frac{1}{x} + \frac{3}{7} = \frac{4}{x + 7}$

Technology Exercise

79. Graph the given rational function and the function expressed as a single rational expression in Exercise 71 in a $[0, 70, 10]$ by $[0, 400, 50]$ viewing rectangle. The graphs should be identical. What trend do the graphs show in terms of the relationship between a driver's age and the number of arrests per 100,000 drivers for driving under the influence of alcohol?

Critical Thinking Exercises

80. Which one of the following is true?

a. $\frac{a}{b} + \frac{a}{c} = \frac{a}{b + c}$

b. $6 + \frac{1}{x} = \frac{7}{x}$

c. $\frac{1}{x + 3} + \frac{x + 3}{2} = \frac{1}{\cancel{(x + 3)}} + \frac{\cancel{(x + 3)}}{2} = 1 + \frac{1}{2} = \frac{3}{2}$

d. None of the above is true.

In Exercises 81–83, perform the indicated operations.

81. $\left(\frac{2x + 3}{x + 1} \cdot \frac{x^2 + 4x - 5}{2x^2 + x - 3}\right) - \frac{2}{x + 2}$

82. $\left(\frac{5}{x - 5} - \frac{2}{x + 3}\right) \div (3x + 25)$

83. $(x - y)^{-1} + (x - y)^{-2}$

In Exercises 84–87, let

$$f(x) = \frac{2x}{x^2 - 9} \quad \text{and} \quad g(x) = \frac{6x^2}{x^2 + 3x - 18}.$$

84. Find $(f + g)(x)$.

85. Find $(f - g)(x)$.

86. Find $\left(\frac{f}{g}\right)(x)$.

87. Find $\left(\frac{g}{f}\right)(x)$.

Review Exercises

88. Simplify: $\left(\frac{3x^2y^{-2}}{y^3}\right)^{-2}$. (Section 1.6, Example 9)

89. Solve: $|3x - 1| \le 14$. (Section 4.3, Example 3)

90. Factor completely: $50x^3 - 18x$. (Section 5.6, Example 5)

▸ SECTION 6.3 Complex Rational Expressions

Objectives

1. Simplify complex rational expressions by multiplying by one.
2. Simplify complex rational expressions by dividing.

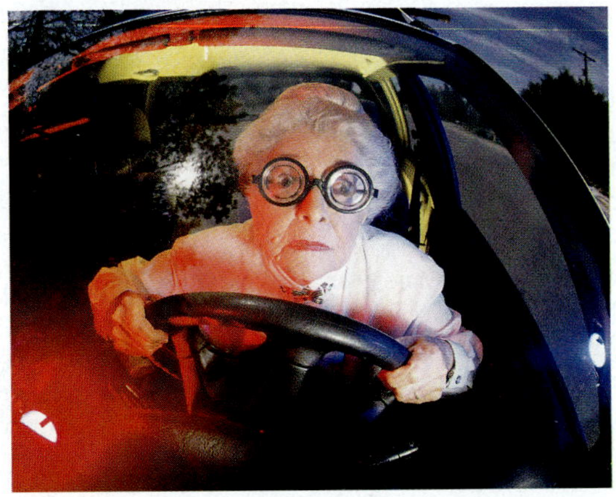

One area in finance of great interest to us ordinary folks when we buy a new car is the amount of each monthly payment. If P is the principal, or the amount borrowed, i is the monthly interest rate, and n is the number of monthly payments, then the amount, A, of each monthly payment is

$$A = \frac{Pi}{1 - \frac{1}{(1 + i)^n}}.$$

Do you notice anything unusual about the rational expression for the amount of each payment? It has a separate rational expression in its denominator.

$$\frac{Pi}{1 - \frac{1}{(1 + i)^n}}$$

Numerator: Pi — Main fraction bar — Denominator: $1 - \frac{1}{(1+i)^n}$

A separate rational expression occurs in the denominator.

Complex rational expressions, also called **complex fractions**, have numerators or denominators containing one or more rational expressions. Here is another example of such an expression:

$$\frac{\frac{1}{x} + \frac{y}{x^2}}{\frac{1}{y} + \frac{x}{y^2}}.$$

Numerator — Main fraction bar — Denominator

Separate rational expressions occur in the numerator and the denominator.

In this section, we study two methods for simplifying complex rational expressions.

Simplifying Complex Rational Expressions by Multiplying by One

One method for simplifying a complex rational expression is to find the least common denominator of all the rational expressions in its numerator and denominator. Then multiply each term in its numerator and denominator by this least common denominator. Because we are multiplying by a form of 1, we will obtain an equivalent expression that does not contain fractions in its numerator or denominator.

1 Simplify complex rational expressions by multiplying by one.

Simplifying a Complex Rational Expression by Multiplying by One in the Form $\frac{\text{LCD}}{\text{LCD}}$

1. Find the LCD of all rational expressions within the complex rational expression.
2. Multiply both the numerator and the denominator of the complex rational expression by this LCD.
3. Use the distributive property and multiply each term in the numerator and denominator by this LCD. Simplify. No fractional expressions should remain.
4. If possible, factor and simplify.

EXAMPLE 1 Simplifying a Complex Rational Expression

Simplify:

$$\frac{\frac{1}{x}+\frac{y}{x^2}}{\frac{1}{y}+\frac{x}{y^2}}.$$

Solution The denominators in the complex rational expression are x, x^2, y, and y^2. The LCD is x^2y^2. Multiply both the numerator and the denominator of the complex rational expression by x^2y^2.

$$\frac{\frac{1}{x}+\frac{y}{x^2}}{\frac{1}{y}+\frac{x}{y^2}} = \frac{x^2y^2}{x^2y^2}\cdot\frac{\left(\frac{1}{x}+\frac{y}{x^2}\right)}{\left(\frac{1}{y}+\frac{x}{y^2}\right)}$$

$$= \frac{x^2y^2\cdot\frac{1}{x}+x^2y^2\cdot\frac{y}{x^2}}{x^2y^2\cdot\frac{1}{y}+x^2y^2\cdot\frac{x}{y^2}}$$ Use the distributive property.

$$= \frac{xy^2+y^3}{x^2y+x^3}$$ Simplify.

$$= \frac{y^2(x+y)}{x^2(y+x)}$$ Factor and simplify.

$$= \frac{y^2}{x^2}$$

■

✔ **CHECK POINT 1** Simplify:

$$\frac{\frac{x}{y} - 1}{\frac{x^2}{y^2} - 1}.$$

Complex rational expressions are often written with negative exponents. For example,

$$\frac{x^{-1} + x^{-2}y}{y^{-1} + xy^{-2}} \quad \text{means} \quad \frac{\frac{1}{x} + \frac{y}{x^2}}{\frac{1}{y} + \frac{x}{y^2}}.$$

This is the expression that we simplified in Example 1. If an expression contains negative exponents, first rewrite it as an equivalent expression with positive exponents. Then simplify by multiplying the numerator and the denominator by the LCD.

EXAMPLE 2 Simplifying a Complex Rational Expression

Simplify:

$$\frac{\frac{1}{x+h} - \frac{1}{x}}{h}.$$

Solution The denominators in the complex rational expression are $x + h$ and x. The LCD is $x(x + h)$. Multiply both the numerator and the denominator of the complex rational expression by $x(x + h)$.

$$\frac{\frac{1}{x+h} - \frac{1}{x}}{h} = \frac{x(x+h)}{x(x+h)} \cdot \frac{\left(\frac{1}{x+h} - \frac{1}{x}\right)}{h}$$

$$= \frac{x(x+h) \cdot \frac{1}{x+h} - x(x+h) \cdot \frac{1}{x}}{x(x-h)h}$$

Use the distributive property in the numerator.

$$= \frac{x - (x+h)}{x(x+h)h}$$

Simplify: $x\cancel{(x+h)} \cdot \frac{1}{\cancel{(x+h)}} = x$ and $\cancel{x}(x+h) \cdot \frac{1}{\cancel{x}} = x + h$

$$= \frac{x - x - h}{x(x+h)h}$$

Remove parentheses and change the sign of each term.

$$= \frac{-h}{x(x+h)h}$$

Simplify.

$$= \frac{-\cancel{h}}{x(x+h)\cancel{h}}$$

Divide the numerator and the denominator by the common factor, h.

$$= -\frac{1}{x(x+h)}$$

■

✔ **CHECK POINT 2** Simplify:

$$\frac{\frac{1}{x+7}-\frac{1}{x}}{7}.$$

2 Simplify complex rational expressions by dividing.

Simplifying Complex Rational Expressions by Dividing

A second method for simplifying a complex rational expression is to combine its numerator into a single expression and combine its denominator into a single expression. Then perform the division by inverting the denominator and multiplying.

Simplifying a Complex Rational Expression by Dividing

1. If necessary, add or subtract to get a single rational expression in the numerator.
2. If necessary, add or subtract to get a single rational expression in the denominator.
3. Perform the division indicated by the main fraction bar: Invert the denominator of the complex rational expression and multiply.
4. If possible, simplify.

EXAMPLE 3 Simplifying a Complex Rational Expression

Simplify:

$$\frac{\frac{x+1}{x}+\frac{x+1}{x-1}}{\frac{x+2}{x}-\frac{2}{x-1}}.$$

Solution

Step 1 Add to get a single rational expression in the numerator.

$$\frac{x+1}{x}+\frac{x+1}{x-1}$$

The LCD is $x(x-1)$.

$$=\frac{(x+1)(x-1)}{x(x-1)}+\frac{x(x+1)}{x(x-1)}=\frac{(x+1)(x-1)+x(x+1)}{x(x-1)}=\frac{x^2-1+x^2+x}{x(x-1)}=\frac{2x^2+x-1}{x(x-1)}$$

Step 2 Subtract to get a single rational expression in the denominator.

$$\frac{x+2}{x}-\frac{2}{x-1}$$

The LCD is $x(x-1)$.

$$=\frac{(x+2)(x-1)}{x(x-1)}-\frac{2x}{x(x-1)}=\frac{(x+2)(x-1)-2x}{x(x-1)}=\frac{x^2+x-2-2x}{x(x-1)}=\frac{x^2-x-2}{x(x-1)}$$

Steps 3 and 4 **Perform the division indicated by the main fraction bar: Invert and multiply. If possible, simplify.**

$$\frac{\dfrac{x+1}{x}+\dfrac{x+1}{x-1}}{\dfrac{x+2}{x}-\dfrac{2}{x-1}}=\frac{\dfrac{2x^2+x-1}{x(x-1)}}{\dfrac{x^2-x-2}{x(x-1)}}$$

These are the single rational expressions from steps 1 and 2.

$$=\frac{2x^2+x-1}{x(x-1)}\cdot\frac{x(x-1)}{x^2-x-2}=\frac{(2x-1)\cancel{(x+1)}}{\cancel{x(x-1)}}\cdot\frac{\cancel{x(x-1)}}{(x-2)\cancel{(x+1)}}=\frac{2x-1}{x-2}$$

Invert and multiply.

■

✔ CHECK POINT 3 Simplify:

$$\frac{\dfrac{x+1}{x-1}-\dfrac{x-1}{x+1}}{\dfrac{x-1}{x+1}+\dfrac{x+1}{x-1}}.$$

Which of the two methods do you prefer? Let's try them both in Example 4.

EXAMPLE 4 Simplifying a Complex Rational Expression: Comparing Methods

Simplify:

$$\frac{1-x^{-2}}{1-4x^{-1}+3x^{-2}}.$$

Solution First rewrite the expression without negative exponents.

$$\frac{1-x^{-2}}{1-4x^{-1}+3x^{-2}}=\frac{1-\dfrac{1}{x^2}}{1-\dfrac{4}{x}+\dfrac{3}{x^2}}$$

Method 1 Multiplying by One

$$\frac{1-\dfrac{1}{x^2}}{1-\dfrac{4}{x}+\dfrac{3}{x^2}}=\frac{x^2}{x^2}\cdot\frac{\left(1-\dfrac{1}{x^2}\right)}{\left(1-\dfrac{4}{x}+\dfrac{3}{x^2}\right)}$$

Multiply the numerator and denominator by x^2, the LCD of all fractions.

$$=\frac{x^2\cdot 1-x^2\cdot\dfrac{1}{x^2}}{x^2\cdot 1-x^2\cdot\dfrac{4}{x}+x^2\cdot\dfrac{3}{x^2}}$$

Apply the distributive property.

$$=\frac{x^2-1}{x^2-4x+3}$$

Simplify.

$$=\frac{(x+1)\cancel{(x-1)}}{(x-3)\cancel{(x-1)}}$$

Factor and simplify.

$$=\frac{x+1}{x-3}$$

Method 2 Dividing

$$\frac{1-\dfrac{1}{x^2}}{1-\dfrac{4}{x}+\dfrac{3}{x^2}}=\frac{\dfrac{x^2}{x^2}-\dfrac{1}{x^2}}{\dfrac{x^2}{x^2}-\dfrac{4}{x}\cdot\dfrac{x}{x}+\dfrac{3}{x^2}}$$

Get a single rational expression in the numerator and denominator

$$=\frac{\dfrac{x^2-1}{x^2}}{\dfrac{x^2-4x+3}{x^2}}$$

$$=\frac{x^2-1}{x^2}\cdot\frac{x^2}{x^2-4x+3}$$

Invert and multiply.

$$=\frac{(x+1)\cancel{(x-1)}}{\cancel{x^2}}\cdot\frac{\cancel{x^2}}{\cancel{(x-1)}(x-3)}$$

Factor and simplify.

$$=\frac{x+1}{x-3}$$

■

✔ **CHECK POINT 4** Simplify by the method of your choice:

$$\frac{1 - 4x^{-2}}{1 - 7x^{-1} + 10x^{-2}}.$$

EXERCISE SET 6.3

Practice Exercises

In Exercises 1–38, simplify each complex rational expression by the method of your choice.

1. $\dfrac{4 + \frac{2}{x}}{1 - \frac{3}{x}}$

2. $\dfrac{5 - \frac{2}{x}}{3 + \frac{1}{x}}$

3. $\dfrac{\frac{3}{x} + \frac{x}{3}}{\frac{x}{3} - \frac{3}{x}}$

4. $\dfrac{\frac{x}{5} - \frac{5}{x}}{\frac{1}{5} + \frac{1}{x}}$

5. $\dfrac{\frac{1}{x} + \frac{1}{y}}{\frac{1}{x} - \frac{1}{y}}$

6. $\dfrac{\frac{x}{y} + \frac{1}{x}}{\frac{y}{x} + \frac{1}{x}}$

7. $\dfrac{8x^{-2} - 2x^{-1}}{10x^{-1} - 6x^{-2}}$

8. $\dfrac{12x^{-2} - 3x^{-1}}{15x^{-1} - 9x^{-2}}$

9. $\dfrac{\frac{1}{x - 2}}{1 - \frac{1}{x - 2}}$

10. $\dfrac{\frac{1}{x + 2}}{1 + \frac{1}{x + 2}}$

11. $\dfrac{\frac{1}{x + 5} - \frac{1}{x}}{5}$

12. $\dfrac{\frac{1}{x + 6} - \frac{1}{x}}{6}$

13. $\dfrac{\frac{4}{x + 4}}{\frac{1}{x + 4} - \frac{1}{x}}$

14. $\dfrac{\frac{7}{x + 7}}{\frac{1}{x + 7} - \frac{1}{x}}$

15. $\dfrac{\frac{1}{x - 1} + 1}{\frac{1}{x + 1} - 1}$

16. $\dfrac{\frac{1}{x + 1} - 1}{\frac{1}{x - 1} + 1}$

17. $\dfrac{x^{-1} + y^{-1}}{(x + y)^{-1}}$

18. $(x^{-1} + y^{-1})^{-1}$

19. $\dfrac{\frac{x + 2}{x - 2} - \frac{x - 2}{x + 2}}{\frac{x - 2}{x + 2} + \frac{x + 2}{x - 2}}$

20. $\dfrac{\frac{x + 1}{x - 1} - \frac{x - 1}{x + 1}}{\frac{x - 1}{x + 1} + \frac{x + 2}{x - 1}}$

21. $\dfrac{\frac{2}{x^3y} + \frac{5}{xy^4}}{\frac{5}{x^3y} - \frac{3}{xy}}$

22. $\dfrac{\frac{3}{xy^2} + \frac{2}{x^2y}}{\frac{1}{x^2y} + \frac{2}{xy^3}}$

23. $\dfrac{\frac{3}{x + 2} - \frac{3}{x - 2}}{\frac{5}{x^2 - 4}}$

24. $\dfrac{\frac{3}{x + 1} - \frac{3}{x - 1}}{\frac{5}{x^2 - 1}}$

25. $\dfrac{3a^{-1} + 3b^{-1}}{4a^{-2} - 9b^{-2}}$

26. $\dfrac{5a^{-1} - 2b^{-1}}{25a^{-2} - 4b^{-2}}$

27. $\dfrac{\frac{4x}{x^2 - 4} - \frac{5}{x - 2}}{\frac{2}{x - 2} + \frac{3}{x + 2}}$

28. $\dfrac{\frac{2}{x + 3} + \frac{5x}{x^2 - 9}}{\frac{4}{x + 3} + \frac{2}{x - 3}}$

29. $\dfrac{\frac{2y}{y^2 + 4y + 3}}{\frac{1}{y + 3} + \frac{2}{y + 1}}$

30. $\dfrac{\frac{5y}{y^2 - 5y + 6}}{\frac{3}{y - 3} + \frac{2}{y - 2}}$

31. $\dfrac{\frac{2}{a^2} - \frac{1}{ab} - \frac{1}{b^2}}{\frac{1}{a^2} - \frac{3}{ab} + \frac{2}{b^2}}$

32. $\dfrac{\dfrac{2}{b^2} - \dfrac{5}{ab} - \dfrac{3}{a^2}}{\dfrac{2}{b^2} + \dfrac{7}{ab} + \dfrac{3}{a^2}}$

33. $\dfrac{\dfrac{2x}{x^2 - 25} + \dfrac{1}{3x - 15}}{\dfrac{5}{x - 5} + \dfrac{3}{4x - 20}}$

34. $\dfrac{\dfrac{7x}{2x - 2} + \dfrac{x}{x^2 - 1}}{\dfrac{4}{x + 1} - \dfrac{1}{3x + 3}}$

35. $\dfrac{\dfrac{3}{x + 2y} - \dfrac{2y}{x^2 + 2xy}}{\dfrac{3y}{x^2 + 2xy} + \dfrac{5}{x}}$

36. $\dfrac{\dfrac{1}{x^3 - y^3}}{\dfrac{1}{x - y} - \dfrac{1}{x^2 + xy + y^2}}$

37. $\dfrac{x}{1 - \dfrac{1}{1 + \dfrac{1}{x}}}$

38. $\dfrac{\dfrac{1}{x + 1}}{x - \dfrac{1}{x + \dfrac{1}{x}}}$

Application Exercises

39. How much are your monthly payments on a loan? If P is the principal, or amount borrowed, i is the monthly interest rate, and n is the number of monthly payments, then the amount, A, of each monthly payments is

$$A = \frac{Pi}{1 - \dfrac{1}{(1 + i)^n}}.$$

a. Simplify the complex rational expression for the amount of each payment

b. You purchase a \$20,000 automobile at 1% monthly interest to be paid over 48 months. How much do you pay each month? Use the simplified rational expression from part (a) and a calculator. Round to the nearest dollar.

40. The average speed on a round-trip commute having a one-way distance d is given by the complex rational expression

$$\frac{2d}{\dfrac{d}{r_1} + \dfrac{d}{r_2}}$$

in which r_1 and r_2 are the speeds on the outgoing and return trips, respectively.

a. Simplify the complex rational expression.

b. Find your average speed if you drive to campus averaging 30 miles per hour and return home on the same route averaging 40 miles per hour. Use the simplified rational expression from part (a).

41. If three resistors with resistances R_1, R_2, and R_3 are connected in parallel, their combined resistance, R, is given by the formula

$$R = \frac{1}{\dfrac{1}{R_1} + \dfrac{1}{R_2} + \dfrac{1}{R_3}}.$$

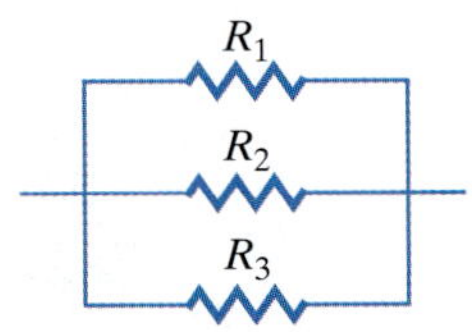

Simplify the complex rational expression on the right side of the formula. Then find R when R_1 is 4 ohms, R_2 is 8 ohms, and R_3 is 12 ohms.

42. A camera lens has a measurement called its focal length, f. When an object is in focus, its distance from the lens, p, and its image distance from the lens, q, satisfy the formula

$$f = \frac{1}{\dfrac{1}{p} + \dfrac{1}{q}}.$$

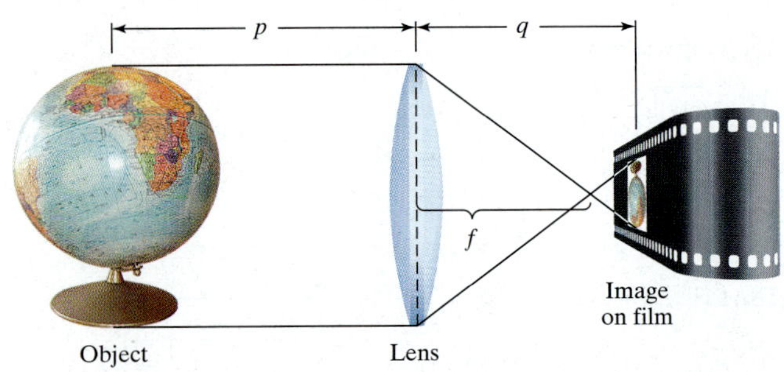

Simplify the complex rational expression on the right side of the formula.

Writing in Mathematics

43. What is a complex rational expression? Give an example with your explanation.

44. Describe two ways to simplify $\dfrac{\dfrac{2}{x}+\dfrac{2}{y}}{\dfrac{2}{x}-\dfrac{2}{y}}$.

45. Which method do you prefer for simplifying complex rational expressions? Why?

46. Of the four complex rational expressions in Exercises 39–42, which one do you find most useful? Explain how you might use this expression in a practical situation.

Technology Exercises

In Exercises 47–50, use a graphing utility to determine if the simplification is correct by graphing the function on each side of the equation in the same viewing rectangle. If the graphs do not coincide, correct the expression on the right side and then verify your correction using the graphing utility.

47. $\dfrac{x-\dfrac{1}{2x+1}}{1-\dfrac{x}{2x+1}} = 2x - 1$

48. $\dfrac{\dfrac{1}{x}+1}{\dfrac{1}{x}} = 2$

49. $\dfrac{\dfrac{1}{x}+\dfrac{1}{3}}{\dfrac{1}{3x}} = x + \dfrac{1}{3}$

50. $\dfrac{\dfrac{x}{3}}{\dfrac{2}{x+1}} = \dfrac{3(x+1)}{2}$

Critical Thinking Exercises

51. Simplify:

$$\frac{\dfrac{x+h}{x+h+1}-\dfrac{x}{x+1}}{h}.$$

52. Let $f(x) = \dfrac{1}{x+1}$. Find $f\left(\dfrac{1}{a+1}\right)$ and simplify.

53. Simplify:

$$x + \cfrac{1}{x + \cfrac{1}{x + \cfrac{1}{x}}}.$$

In Exercises 54–55, use the given rational function to find and simplify

$$\frac{f(a+h)-f(a)}{h}.$$

54. $f(x) = \dfrac{3}{x}$

55. $f(x) = \dfrac{1}{x^2}$

Review Exercises

56. Solve: $x^2 + 27 = 12x$. (Section 5.7, Example 2)

57. Multiply: $(4x^2 - y)^2$. (Section 5.2, Example 7)

58. Solve: $-4 < 3x - 7 < 8$. (Section 4.2, Example 4)

▸ SECTION 6.4 Division of Polynomials

Objectives

1. Divide a polynomial by a monomial.
2. Use long division to divide by a polynomial containing more than one term.

SSM PH Tutor Center CD-ROM Video

© Roger Ressmeyer/CORBIS

During the 1980s, the controversial economist Arthur Laffer promoted the idea that tax *increases* lead to a *reduction* in government revenue. Called supply-side economics, the theory uses rational functions as models. One such function

$$f(x) = \frac{80x - 8000}{x - 110}, \; 30 \le x \le 100$$

models the government tax revenue, $f(x)$, in tens of billions of dollars, as a function of the tax rate, x. The graph of the rational function is shown in Figure 6.4. The graph shows tax revenue decreasing quite dramatically as the tax rate increases. At a tax rate of (gasp) 100%, the government takes all our money and no one has an incentive to work. With no income earned, zero dollars in tax revenue is generated.

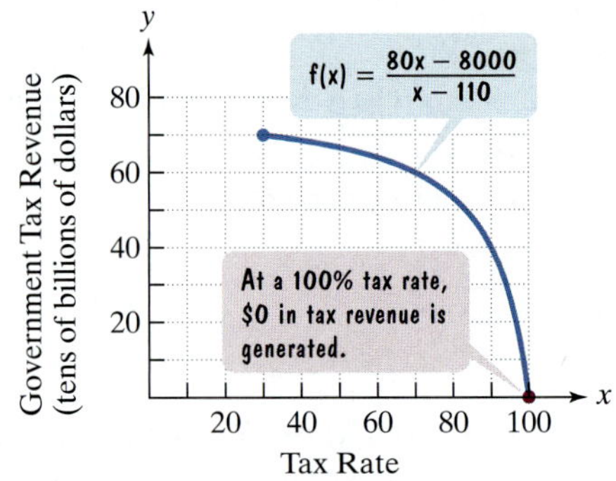

Figure 6.4

Like all rational functions, the Laffer model consists of the quotient of polynomials. Although the rational expression in the model cannot be simplified, it is possible to perform the division, thereby expressing the function in another form. In this section, you will learn to divide polynomials.

1 Divide a polynomial by a monomial.

Dividing a Polynomial by a Monomial

We have seen that to divide monomials, we divide the coefficients and subtract the exponents when bases are the same. For example,

$$\frac{25x^{12}}{5x^4} = \frac{25}{5}x^{12-4} = 5x^8 \quad \text{and} \quad \frac{60x^4y^3}{-30x^2y} = \frac{60}{-30}x^{4-2}y^{3-1} = -2x^2y^2.$$

How do we divide a polynomial that is not a monomial by a monomial? We divide each term of the polynomial by the monomial. For example,

$$\frac{10x^8 + 15x^6}{5x^3} = \frac{10x^8}{5x^3} + \frac{15x^6}{5x^3} = \frac{10}{5}x^{8-3} + \frac{15}{5}x^{6-3} = 2x^5 + 3x^3.$$

Polynomial dividend: $10x^8 + 15x^6$. Monomial divisor: $5x^3$. Divide the first term by $5x^3$. Divide the second term by $5x^3$.

Dividing a Polynomial That Is Not a Monomial by a Monomial

To divide a polynomial by a monomial, divide each term of the polynomial by the monomial.

EXAMPLE 1 Dividing a Polynomial by a Monomial

Divide: $(15x^3 - 5x^2 + x + 5) \div (5x)$.

Solution

$$\frac{15x^3 - 5x^2 + x + 5}{5x}$$ Rewrite the division in a vertical format.

$$= \frac{15x^3}{5x} - \frac{5x^2}{5x} + \frac{x}{5x} + \frac{5}{5x}$$ Divide each term of the polynomial by the monomial.

$$= 3x^2 - x + \frac{1}{5} + \frac{1}{x}$$ Simplify each quotient. ■

✔ **CHECK POINT 1** Divide: $(16x^3 - 32x^2 + 2x + 4) \div 4x$.

EXAMPLE 2 Dividing a Polynomial by a Monomial

Divide $8x^4y^5 - 10x^4y^3 + 12x^2y^3$ by $4x^3y^2$.

Solution

$$\frac{8x^4y^5 - 10x^4y^3 + 12x^2y^3}{4x^3y^2}$$ Express the division in a vertical format.

$$= \frac{8x^4y^5}{4x^3y^2} - \frac{10x^4y^3}{4x^3y^2} + \frac{12x^2y^3}{4x^3y^2}$$ Divide each term of the polynomial by the monomial.

$$= 2xy^3 - \frac{5}{2}xy + \frac{3y}{x}$$ Simplify each quotient. ■

✔ **CHECK POINT 2** Divide $15x^4y^5 - 5x^3y^4 + 10x^2y^2$ by $5x^2y^3$.

2 Use long division to divide by a polynomial containing more than one term.

Dividing by a Polynomial Containing More Than One Term

We now look at division by a polynomial containing more than one term, such as

$$x + 3\overline{)x^2 + 10x + 21}.$$

Divisor has two terms. Dividend has three terms.

When a divisor has more than one term, the four steps used to divide whole numbers—**divide, multiply, subtract, bring down the next term**—form the repetitive procedure for polynomial long division.

EXAMPLE 3 Dividing a Polynomial by a Binomial

Divide $x^2 + 10x + 21$ by $x + 3$.

Solution The following steps illustrate how polynomial division is very similar to numerical division.

$$x + 3\overline{)x^2 + 10x + 21}$$

Arrange the terms of the dividend $(x^2 + 10x + 21)$ and the divisor $(x + 3)$ in descending powers of x.

$$\begin{array}{r} x \\ x + 3\overline{)x^2 + 10x + 21} \end{array}$$

DIVIDE x^2 (the first term in the dividend) by x (the first term in the divisor): $\frac{x^2}{x} = x$. Align like terms.

$$\begin{array}{r} x \\ x + 3\overline{)x^2 + 10x + 21} \\ x^2 + 3x \end{array}$$

times, equals

MULTIPLY each term in the divisor $(x + 3)$ by x aligning terms of the product under like terms in the dividend.

$$\begin{array}{r} x \\ x + 3\overline{)x^2 + 10x + 21} \\ \underline{x^2 + 3x} \\ 7x \end{array}$$

SUBTRACT $x^2 + 3x$ from $x^2 + 10x$ by changing the sign of each term in the lower expression and adding.

$$\begin{array}{r} x \\ x + 3\overline{)x^2 + 10x + 21} \\ \underline{x^2 + 3x \downarrow} \\ 7x + 21 \end{array}$$

BRING DOWN 21 from the original dividend and add algebraically to form a new dividend.

$$\begin{array}{r} x + 7 \\ x+3 \overline{\smash{\big)} x^2 + 10x + 21} \\ \underline{x^2 + 3x } \downarrow \\ 7x + 21 \end{array}$$

Find the second term of the quotient. DIVIDE the first term of $7x + 21$ by x, the first term of the divisor: $\frac{7x}{x} = 7$.

times

$$\begin{array}{r} x + 7 \\ x+3 \overline{\smash{\big)} x^2 + 10x + 21} \\ \underline{x^2 + 3x } \\ 7x + 21 \\ \underline{7x + 21} \\ 0 \end{array}$$

equals

MULTIPLY the divisor $(x + 3)$ by 7, aligning under like terms in the new dividend. Then subtract to obtain the remainder of 0.

Remainder

The quotient is $x + 7$ and the remainder is 0. We will not list a remainder of 0 in the answer. Thus,

$$(x^2 + 10x + 21) \div (x + 3) = x + 7.$$

After performing polynomial long division, the answer can be checked. Find the product of the divisor and the quotient and add the remainder. If the result is the dividend, the answer to the division problem is correct. For example, let's check our work in Example 3.

$$(x^2 + 10x + 21) \div (x + 3) = x + 7$$

Dividend — Divisor — Quotient we wish to check

Multiply the divisor and the quotient and add the remainder, 0:

$$(x + 3)(x + 7) + 0 = x^2 + 7x + 3x + 21 + 0 = x^2 + 10x + 21.$$

Divisor — Quotient — Remainder — This is the dividend.

Because we obtained the dividend, the quotient is correct.

✔ **CHECK POINT 3** Divide $x^2 + 14x + 45$ by $x + 9$.

Before considering additional examples, let's summarize the general procedure for dividing by a polynomial that contains more than one term.

Long Division of Polynomials

1. **Arrange the terms** of both the dividend and the divisor in descending powers of any variable.
2. **Divide** the first term in the dividend by the first term in the divisor. The result is the first term of the quotient.
3. **Multiply** every term in the divisor by the first term in the quotient. Write the resulting product beneath the dividend with like terms lined up.
4. **Subtract** the product from the dividend.
5. **Bring down** the next term in the original dividend and write it next to the remainder to form a new dividend.
6. Use this new expression as the dividend and repeat this process until the remainder can no longer be divided. This will occur when the degree of the remainder (the highest exponent on a variable in the remainder) is less than the degree of the divisor.

In our next long division, we will obtain a nonzero remainder.

EXAMPLE 4 Long Division of Polynomials

Divide $4 - 5x - x^2 + 6x^3$ by $3x - 2$.

Solution We begin by writing the dividend in descending powers of x.

$$4 - 5x - x^2 + 6x^3 = 6x^3 - x^2 - 5x + 4$$

Multiply.

$$\begin{array}{r} 2x^2 \phantom{{}- x^2 - 5x + 4} \\ 3x - 2\overline{)6x^3 - x^2 - 5x + 4} \\ \ominus 6x^3 \oplus 4x^2 \phantom{{}- 5x + 4} \\ \hline 3x^2 - 5x \phantom{{}+ 4} \end{array}$$

Change signs of the polynomial being subtracted.

Divide: $\frac{6x^3}{3x} = 2x^2$.

Multiply: $2x^2(3x - 2) = 6x^3 - 4x^2$.

Subtract $6x^3 - 4x^2$ from $6x^3 - x^2$ and bring down $-5x$.

Now we divide $3x^2$ by $3x$ to obtain x, multiply x and the divisor, and subtract.

Multiply.

$$\begin{array}{r} 2x^2 + x \phantom{{}- 5x + 4} \\ 3x - 2\overline{)6x^3 - x^2 - 5x + 4} \\ 6x^3 - 4x^2 \phantom{{}- 5x + 4} \\ \hline 3x^2 - 5x \phantom{{}+ 4} \\ \ominus 3x^2 \oplus 2x \phantom{{}+ 4} \\ \hline -3x + 4 \end{array}$$

Change signs of the polynomial being subtracted.

Divide: $\frac{3x^2}{3x} = x$.

Multiply: $x(3x - 2) = 3x^2 - 2x$.

Subtract $3x^2 - 2x$ from $3x^2 - 5x$ and bring down 4.

Now we divide $-3x$ by $3x$ to obtain -1, multiply -1 and the divisor, and subtract.

Multiply.

$$\begin{array}{r} 2x^2 + x - 1 \\ 3x - 2\overline{)6x^3 - x^2 - 5x + 4} \\ 6x^3 - 4x^2 \phantom{{}- 5x + 4} \\ \hline 3x^2 - 5x \phantom{{}+ 4} \\ 3x^2 - 2x \phantom{{}+ 4} \\ \hline -3x + 4 \\ \oplus 3x \ominus 2 \\ \hline 2 \end{array}$$

Change signs of the polynomial being subtracted.

Remainder

Divide: $\frac{-3x}{3x} = -1$.

Multiply: $-1(3x - 2) = -3x + 2$.

Subtract $-3x + 2$ from $-3x + 4$, leaving a remainder of 2.

The quotient is $2x^2 + x - 1$ and the remainder is 2. When there is a nonzero remainder, as in this example, list the quotient, plus the remainder above the divisor. Thus,

$$\frac{6x^3 - x^2 - 5x + 4}{3x - 2} = 2x^2 + x - 1 + \frac{2}{3x - 2}.$$

Quotient

Remainder above divisor

Check this result by showing that the product of the divisor and the quotient,

$$(3x - 2)(2x^2 + x - 1),$$

plus the remainder, 2, is the dividend, $6x^3 - x^2 - 5x + 4$. ■

CHECK POINT 4 Divide $-9 + 7x - 4x^2 + 4x^3$ by $2x - 1$.

If a power of x is missing in either a dividend or a divisor, add that power of x with a coefficient of 0 and then divide. In this way, like terms will be aligned as you carry out the long division.

EXAMPLE 5 Long Division of Polynomials

Divide $6x^4 + 5x^3 + 3x - 5$ by $3x^2 - 2x$.

Solution We write the dividend, $6x^4 + 5x^3 + 3x - 5$, as $6x^4 + 5x^3 + 0x^2 + 3x - 5$ to keep all like terms aligned.

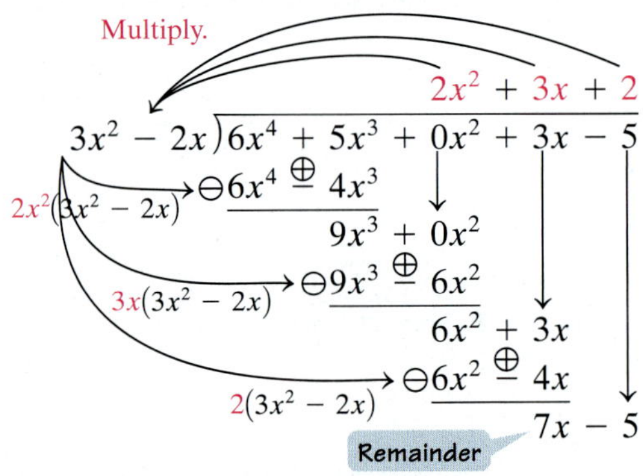

The division process is finished because the degree of $7x - 5$, which is 1, is less than the degree of the divisor $3x^2 - 2x$, which is 2. The answer is

$$\frac{6x^4 + 5x^3 + 3x - 5}{3x^2 - 2x} = 2x^2 + 3x + 2 + \frac{7x - 5}{3x^2 - 2x}.$$

CHECK POINT 5 Divide $2x^4 + 3x^3 - 7x - 10$ by $x^2 - 2x$.

EXERCISE SET 6.4

Practice Exercises

In Exercises 1–12, divide the polynomial by the monomial.

1. $\dfrac{25x^7 - 15x^5 + 10x^3}{5x^3}$

2. $\dfrac{49x^7 - 28x^5 + 14x^3}{7x^3}$

3. $\dfrac{18x^3 + 6x^2 - 9x - 6}{3x}$

4. $\dfrac{25x^3 + 50x^2 - 40x - 10}{5x}$

5. $(28x^3 - 7x^2 - 16x) \div (4x^2)$

6. $(70x^3 - 10x^2 - 14x) \div (7x^2)$

7. $(25x^8 - 50x^7 + 3x^6 - 40x^5) \div (-5x^5)$

8. $(18x^7 - 9x^6 + 20x^5 - 10x^4) \div (-2x^4)$

9. $(18a^3b^2 - 9a^2b - 27ab^2) \div (9ab)$

10. $(12a^2b^2 + 6a^2b - 15ab^2) \div (3ab)$

11. $(36x^4y^3 - 18x^3y^2 - 12x^2y) \div (6x^3y^3)$

12. $(40x^4y^3 - 20x^3y^2 - 50x^2y) \div (10x^3y^3)$

In Exercises 13–36, divide as indicated. Check at least five of your answers by showing that the product of the divisor and the quotient, plus the remainder, is the dividend.

13. $(x^2 + 8x + 15) \div (x + 5)$

14. $(x^2 + 3x - 10) \div (x - 2)$

15. $(x^3 - 2x^2 - 5x + 6) \div (x - 3)$

16. $(x^3 + 5x^2 + 7x + 2) \div (x + 2)$

17. $(x^2 - 7x + 12) \div (x - 5)$

18. $(2x^2 + x - 9) \div (x - 2)$

19. $(2x^2 + 13x + 5) \div (2x + 3)$

20. $(8x^2 + 6x - 25) \div (4x + 9)$

21. $(x^3 + 3x^2 + 5x + 4) \div (x + 1)$

22. $(x^3 + 6x^2 - 2x + 3) \div (x - 1)$

23. $(4y^3 + 12y^2 + 7y - 3) \div (2y + 3)$

24. $(6y^3 + 7y^2 + 12y - 5) \div (3y - 1)$

25. $(9x^3 - 3x^2 - 3x + 4) \div (3x + 2)$

26. $(2x^3 + 13x^2 + 9x - 6) \div (2x + 3)$

27. $(4x^3 - 6x - 11) \div (2x - 4)$

28. $(2x^3 + 6x - 4) \div (x + 4)$

29. $(4y^3 - 5y) \div (2y - 1)$

30. $(6y^3 - 5y) \div (2y - 1)$

31. $(4y^4 - 17y^2 + 14y - 3) \div (2y - 3)$

32. $(2y^4 - y^3 + 16y^2 - 4) \div (2y - 1)$

33. $(4x^4 + 3x^3 + 4x^2 + 9x - 6) \div (x^2 + 3)$

34. $(3x^5 - x^3 + 4x^2 - 12x - 8) \div (x^2 - 2)$

35. $(15x^4 + 3x^3 + 4x^2 + 4) \div (3x^2 - 1)$

36. $(18x^4 + 9x^3 + 3x^2) \div (3x^2 + 1)$

Application Exercises

In the section opener, we saw that

$$f(x) = \frac{80x - 8000}{x - 110}, \quad 30 \le x \le 100$$

models the government tax revenue, $f(x)$, in tens of billions of dollars, as a function of the tax rate percentage, x. Use this function to solve Exercises 37–40. Round to the nearest ten billion dollars.

37. Find and interpret $f(30)$. Identify the solution as a point on the graph of the function in Figure 6.4 on page 404.

38. Find and interpret $f(40)$. Identify the solution as a point on the graph of the function in Figure 6.4 on page 404.

39. Rewrite the function by using long division to perform

$$(80x - 8000) \div (x - 110).$$

Then use this new form of the function to find $f(30)$. Do you obtain the same answer as you did in Exercise 37? Which form of the function do you find easier to use?

40. Rewrite the function by using long division to perform

$$(80x - 8000) \div (x - 110).$$

Then use this new form of the function to find $f(40)$. Do you obtain the same answer as you did in Exercise 38? Which form of the function do you find easier to use?

Writing in Mathematics

41. Explain how to divide a polynomial that is not a monomial by a monomial. Give an example.

42. In your own words, explain how to divide by a polynomial containing more than one term. Use $\frac{x^2 + 4}{x + 2}$ in your explanation.

43. When performing polynomial long division, explain when to stop dividing.

44. After performing polynomial long division, explain how to check the answer.

45. When performing polynomial long division with missing terms, explain the advantage of writing the missing terms with zero coefficients.

46. The idea of supply-side economics is that an increase in the tax rate may actually reduce government revenue. What explanation can you offer for this theory?

Technology Exercises

In Exercises 47–50, use a graphing utility to determine if the division has been performed correctly. Graph the function on each side of the equation in the same viewing rectangle. If the graphs do not coincide, correct the expression on the right side by using polynomial long division. Then verify your correction using the graphing utility.

47. $(6x^2 + 16x + 8) \div (3x + 2) = 2x + 4$

48. $(4x^3 + 7x^2 + 8x + 20) \div (2x + 4) = 2x^2 - \frac{1}{2}x + 3$

49. $(3x^2 - 4x + 6) \div (x - 2) = 3x + 2 + \frac{10}{x - 2}$

50. $(3x^4 + 4x^3 - 32x^2 - 5x - 20) \div (x + 4) = 3x^3 - 8x^2 + 5$

Critical Thinking Exercises

51. Which one of the following is true?

a. All long-division problems can be done by the alternative method of factoring the dividend and canceling identical factors in the dividend and the divisor.

b. Polynomial long division always shows that the quotient of two polynomials is a polynomial.

c. The long division process should be continued until the degree of the remainder is the same as the degree of the divisor.

d. If a polynomial long-division problem results in a remainder that is zero, then the divisor is a factor of the dividend.

In Exercises 52–54, divide as indicated.

52. $(x^{3n} - 4x^{2n} - 2x^n - 12) \div (x^n - 5)$

53. $(x^{3n} + 1) \div (x^n + 1)$

54. When $2x^2 - 7x + 9$ is divided by a polynomial, the quotient is $2x - 3$ and the remainder is 3. Find the polynomial.

55. Find k so that the remainder is 0:
$(20x^3 + 23x^2 - 10x + k) \div (4x + 3)$.

Review Exercises

56. Solve: $|2x - 3| > 4$. (Section 4.3, Example 4)

57. Write 40,610,000 in scientific notation. (Section 1.7, Example 2)

58. Simplify: $2x - 4[x - 3(2x + 1)]$. (Section 1.2, Example 14)

▸ SECTION 6.5 Synthetic Division and the Remainder Theorem

Objectives

1. Divide polynomials using synthetic division.
2. Use the Remainder Theorem to evaluate a polynomial function.
3. Use the Remainder Theorem to show that a number is a solution of a polynomial equation.

SSM PH Tutor Center CD-ROM Video

A moth has moved into your closet. She appeared in your bedroom at night, but somehow her relatively stout body escaped your clutches. Within a few weeks swarms of moths in your tattered wardrobe suggest that Mama Moth was in the family way. There must be at least 200 critters nesting in every crevice of your clothing.

Two hundred plus moth-tykes from one female moth; is this possible? Indeed it is. The number of eggs, $f(x)$, in a female moth is a function of her abdominal width, x, in millimeters, modeled by

$$f(x) = 14x^3 - 17x^2 - 16x + 34$$

for $1.5 \le x \le 3.5$. Because there are 200 moths feasting on your favorite sweaters, Mama's abdominal width can be estimated by finding the solutions of the polynomial equation

$$14x^3 - 17x^2 - 16x + 34 = 200.$$

With mathematics present even in your quickly disappearing attire, we move from rags to a shortcut for long division, called *synthetic division*. In the exercise set, you will use this shortcut to find Mama Moth's abdominal width.

1 Divide polynomials using synthetic division.

Dividing Polynomials Using Synthetic Division

We can use **synthetic division** to divide polynomials if the divisor is of the form $x - c$. This method provides a quotient more quickly than long division. Let's compare the two methods showing $x^3 + 4x^2 - 5x + 5$ divided by $x - 3$.

Long Division

$$\begin{array}{r}
x^2 + 7x + 16 \\
x - 3\,\overline{)\,x^3 + 4x^2 - 5x + 5} \\
\ominus x^3 \oplus 3x^2 \qquad\qquad \\
\hline
7x^2 - 5x \qquad \\
\ominus 7x^2 \oplus 21x \qquad \\
\hline
16x + 5 \\
\ominus 16x \oplus 48 \\
\hline
53
\end{array}$$

Quotient: $x^2 + 7x + 16$

Divisor: $x - c$; $c = 3$

Dividend: $x^3 + 4x^2 - 5x + 5$

Remainder: 53

Synthetic Division

$$\begin{array}{r|rrrr}
3 & 1 & 4 & -5 & 5 \\
 & & 3 & 21 & 48 \\
\hline
 & 1 & 7 & 16 & 53
\end{array}$$

Notice the relationship between the polynomials in the long division process and the numbers that appear in synthetic division.

These are the coefficients of the dividend $x^3 + 4x^2 - 5x + 5$.

The divisor is $x - 3$. This is 3, or c in $x - c$.

$$\begin{array}{r|rrrr} 3 & 1 & 4 & -5 & 5 \\ & & 3 & 21 & 48 \\ \hline & 1 & 7 & 16 & 53 \end{array}$$

These are the coefficients of the quotient $x^2 + 7x + 16$.

This is the remainder.

Now let's look at the steps involved in synthetic division.

Synthetic Division

To divide a polynomial by $x - c$:

Example

1. Arrange polynomials in descending powers, with a 0 coefficient for any missing term.

$$x - 3\overline{)x^3 + 4x^2 - 5x + 5}$$

2. Write c for the divisor, $x - c$. To the right, write the coefficients of the dividend.

$$\begin{array}{r|rrrr} 3 & 1 & 4 & -5 & 5 \end{array}$$

3. Write the leading coefficient of the dividend on the bottom row.

$$\begin{array}{r|rrrr} 3 & 1 & 4 & -5 & 5 \\ & \downarrow & & & \\ \hline & 1 & & & \end{array}$$

Bring down.

4. Multiply c (in this case, 3) times the value just written on the bottom row. Write the product in the next column in the second row.

$$\begin{array}{r|rrrr} 3 & 1 & 4 & -5 & 5 \\ & & 3 & & \\ \hline & 1 & & & \end{array}$$

Multiply by 3: $3 \cdot 1 = 3$.

5. Add the values in this new column, writing the sum in the bottom row.

$$\begin{array}{r|rrrr} 3 & 1 & 4 & -5 & 5 \\ & & 3 & & \\ \hline & 1 & 7 & & \end{array}$$

Add.

6. Repeat this series of multiplications and additions until all columns are filled in.

$$\begin{array}{r|rrrr} 3 & 1 & 4 & -5 & 5 \\ & & 3 & 21 & \\ \hline & 1 & 7 & 16 & \end{array}$$

Add.

Multiply by 3: $3 \cdot 7 = 21$.

$$\begin{array}{r|rrrr} 3 & 1 & 4 & -5 & 5 \\ & & 3 & 21 & 48 \\ \hline & 1 & 7 & 16 & 53 \end{array}$$

Add.

Multiply by 3: $3 \cdot 16 = 48$.

7. Use the numbers in the last row to write the quotient and remainder in fractional form. **The degree of the first term of the quotient is one less than the degree of the first term of the dividend.** The final value in this row is the remainder.

Written from the last row of the synthetic division.

$$\begin{array}{r} 1x^2 + 7x + 16 + \dfrac{53}{x-3} \\ x-3\,\overline{)\,x^3 + 4x^2 - 5x + 5} \end{array}$$

EXAMPLE 1 Using Synthetic Division

Use synthetic division to divide $5x^3 + 6x + 8$ by $x + 2$.

Solution The divisor must be in the form $x - c$. Thus, we write $x + 2$ as $x - (-2)$. This means that $c = -2$. Writing a 0 coefficient for the missing x^2-term in the dividend, we can express the division as follows:

$$x - (-2)\,\overline{)\,5x^3 + 0x^2 + 6x + 8}.$$

Now we are ready to set up the problem so that we can use synthetic division.

Use the coefficients of the dividend $5x^3 + 0x^2 + 6x + 8$ in descending powers of x.

This is c in $x - (-2)$.

$$\underline{-2|}\quad 5 \quad 0 \quad 6 \quad 8$$

We begin the synthetic division process by bringing down 5. This is followed by a series of multiplications and additions.

1. Bring down 5.

$$\begin{array}{r|rrrr} -2 & 5 & 0 & 6 & 8 \\ \hline & 5 & & & \end{array}$$

2. Multiply: $-2(5) = -10$.

$$\begin{array}{r|rrrr} -2 & 5 & 0 & 6 & 8 \\ & & -10 & & \\ \hline & 5 & & & \end{array}$$

Multiply by −2.

3. Add: $0 + (-10) = -10$.

$$\begin{array}{r|rrrr} -2 & 5 & 0 & 6 & 8 \\ & & -10 & & \\ \hline & 5 & -10 & & \end{array}$$

Add.

4. Multiply: $-2(-10) = 20$.

$$\begin{array}{r|rrrr} -2 & 5 & 0 & 6 & 8 \\ & & -10 & 20 & \\ \hline & 5 & -10 & & \end{array}$$

Multiply by −2.

5. Add: $6 + 20 = 26$.

$$\begin{array}{r|rrrr} -2 & 5 & 0 & 6 & 8 \\ & & -10 & 20 & \\ \hline & 5 & -10 & 26 & \end{array}$$

Add.

6. Multiply: $-2(26) = -52$.

$$\begin{array}{r|rrrr} -2 & 5 & 0 & 6 & 8 \\ & & -10 & 20 & -52 \\ \hline & 5 & -10 & 26 & \end{array}$$

Multiply by −2.

7. Add: $8 + (-52) = -44$.

$$\begin{array}{r|rrrr} -2 & 5 & 0 & 6 & 8 \\ & & -10 & 20 & -52 \\ \hline & 5 & -10 & 26 & -44 \end{array}$$

Add.

The numbers in the last row represent the coefficients of the quotient and the remainder. The degree of the first term of the quotient is one less than that of the dividend. Because the degree of the dividend, $5x^3 + 6x + 8$, is 3, the degree of the quotient is 2. This means that the 5 in the last row represents $5x^2$.

$$\begin{array}{r|rrrr} -2 & 5 & 0 & 6 & 8 \\ & & -10 & 20 & -52 \\ \hline & 5 & -10 & 26 & -44 \end{array}$$

The quotient is $5x^2 - 10x + 26$.

The remainder is -44.

Thus,

$$x + 2 \overline{)5x^3 + 6x + 8} \quad \text{quotient: } 5x^2 - 10x + 26 - \frac{44}{x+2}$$

CHECK POINT 1 Use synthetic division to divide $x^3 - 7x - 6$ by $x + 2$.

2 Use the Remainder Theorem to evaluate a polynomial function.

The Remainder Theorem

We have seen that the answer to a long division problem can be checked: Find the product of the divisor and the quotient and add the remainder. The result should be the dividend. If the divisor is $x - c$, we can express this idea symbolically:

$$f(x) = (x - c)q(x) + r.$$

Dividend | Divisor | Quotient | The remainder, r, is a constant when dividing by $x - c$.

Now let's evaluate f at c.

$f(c) = (c - c)q(c) + r$ — Find $f(c)$, letting $x = c$. This will give an expression for r.

$f(c) = 0 \cdot q(c) + r$ — $c - c = 0$ and $0 \cdot q(c) = 0$.

$f(c) = r$ — On the right, $0 + r = r$.

What does this last equation mean? If a polynomial is divided by $x - c$, the remainder is the value of the polynomial at c. This result is called the **Remainder Theorem**.

> **The Remainder Theorem**
> If the polynomial $f(x)$ is divided by $x - c$, then the remainder is $f(c)$.

Example 2 shows how we can use the Remainder Theorem to evaluate a polynomial function at 2. Rather than substituting 2 for x, we divide the function by $x - 2$. The remainder is $f(2)$.

EXAMPLE 2 Using the Remainder Theorem to Evaluate a Polynomial Function

Given $f(x) = x^3 - 4x^2 + 5x + 3$, use the Remainder Theorem to find $f(2)$.

Solution By the Remainder Theorem, if $f(x)$ is divided by $x - 2$, then the remainder is $f(2)$. We'll use synthetic division to divide.

$$\begin{array}{r|rrrr} 2 & 1 & -4 & 5 & 3 \\ & & 2 & -4 & 2 \\ \hline & 1 & -2 & 1 & 5 \end{array}$$

Remainder

The remainder, 5, is the value of $f(2)$. Thus, $f(2) = 5$. We can verify that this is correct by evaluating $f(2)$ directly. Using $f(x) = x^3 - 4x^2 + 5x + 3$, we obtain

$$f(2) = 2^3 - 4 \cdot 2^2 + 5 \cdot 2 + 3 = 8 - 16 + 10 + 3 = 5.$$

■

✔ **CHECK POINT 2** Given $f(x) = 3x^3 + 4x^2 - 5x + 3$, use the Remainder Theorem to find $f(-4)$.

3 Use the Remainder Theorem to show that a number is a solution of a polynomial equation.

If the polynomial $f(x)$ is divided by $x - c$ and the remainder is zero, then $f(c) = 0$. This means that c is a solution of the polynomial equation $f(x) = 0$.

EXAMPLE 3 Using the Remainder Theorem

Show that 3 is a solution of the equation

$$2x^3 - 3x^2 - 11x + 6 = 0.$$

Then solve the polynomial equation.

Solution One way to show that 3 is a solution is to substitute 3 for x in the equation and obtain 0. An easier way is to use synthetic division and the Remainder Theorem.

Proposed solution

$$\begin{array}{r|rrrr} 3 & 2 & -3 & -11 & 6 \\ & & 6 & 9 & 6 \\ \hline & 2 & 3 & -2 & 0 \end{array}$$

Remainder

$$x - 3\overline{)2x^3 - 3x^2 - 11x + 6} \quad \text{quotient: } 2x^2 + 3x - 2$$

Equivalently,

$$2x^3 - 3x^2 - 11x + 6 = (x - 3)(2x^2 + 3x - 2).$$

Because the remainder is 0, the polynomial has a value of 0 when $x = 3$. Thus, 3 is a solution of the given equation.

The synthetic division also shows that $x - 3$ divides the polynomial with a zero remainder. Thus, $x - 3$ is a factor of the polynomial, as shown to the right of the synthetic division. The other factor is the quotient found in the last row of the synthetic division. Now we can solve the polynomial equation.

$2x^3 - 3x^2 - 11x + 6 = 0$ This is the given equation.

$(x - 3)(2x^2 + 3x - 2) = 0$ Factor using the result from the synthetic division.

$(x - 3)(2x - 1)(x + 2) = 0$ Factor the trinomial.

$x - 3 = 0$ or $2x - 1 = 0$ or $x + 2 = 0$ Set each factor equal to 0.

$x = 3$ $\quad x = \frac{1}{2}$ $\quad x = -2$ Solve for x.

The solutions are $-2, \frac{1}{2}$, and 3, and the solution set is $\{-2, \frac{1}{2}, 3\}$. ■

✔ **CHECK POINT 3** Use synthetic division to show that -1 is a solution of the equation

$$15x^3 + 14x^2 - 3x - 2 = 0.$$

Then solve the polynomial equation.

Using Technology

Because the solution set of

$$2x^3 - 3x^2 - 11x + 6 = 0$$

is $\{-2, \frac{1}{2}, 3\}$, this implies that the polynomial function

$$f(x) = 2x^3 - 3x^2 - 11x + 6$$

has x-intercepts at $-2, \frac{1}{2}$, and 3. This is verified by the graph of f.

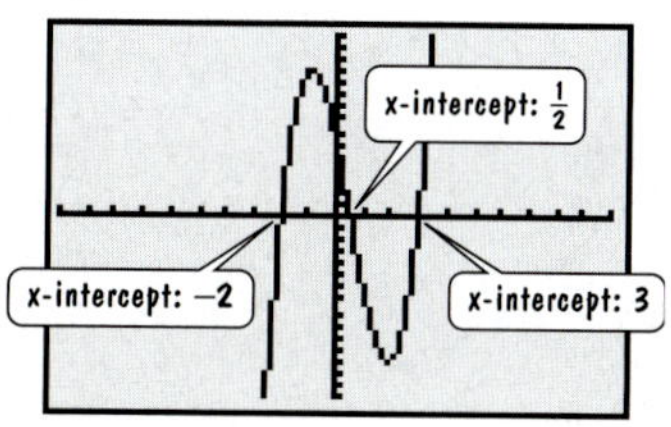

$[-10, 10, 1]$ by $[-15, 15, 1]$

EXERCISE SET 6.5

Practice Exercises

In Exercises 1–18, divide using synthetic division. In the first two exercises, begin the process as shown.

1. $(2x^2 + x - 10) \div (x - 2)$

   $2\rfloor \quad 2 \quad 1 \quad -10$

2. $(x^2 + x - 2) \div (x - 1)$

   $1\rfloor \quad 1 \quad 1 \quad -2$

3. $(3x^2 + 7x - 20) \div (x + 5)$

4. $(5x^2 - 12x - 8) \div (x + 3)$

5. $(4x^3 - 3x^2 + 3x - 1) \div (x - 1)$

6. $(5x^3 - 6x^2 + 3x + 11) \div (x - 2)$

7. $(6x^5 - 2x^3 + 4x^2 - 3x + 1) \div (x - 2)$

8. $(x^5 + 4x^4 - 3x^2 + 2x + 3) \div (x - 3)$

9. $(x^2 - 5x - 5x^3 + x^4) \div (5 + x)$

10. $(x^2 - 6x - 6x^3 + x^4) \div (6 + x)$

11. $(3x^3 + 2x^2 - 4x + 1) \div \left(x - \frac{1}{3}\right)$

12. $(2x^4 - x^3 + 2x^2 - 3x + 1) \div \left(x - \frac{1}{2}\right)$

13. $\dfrac{x^5 + x^3 - 2}{x - 1}$

14. $\dfrac{x^7 + x^5 - 10x^3 + 12}{x + 2}$

15. $\dfrac{x^4 - 256}{x - 4}$

16. $\dfrac{x^7 - 128}{x - 2}$

17. $\dfrac{2x^5 - 3x^4 + x^3 - x^2 + 2x - 1}{x + 2}$

18. $\dfrac{x^5 - 2x^4 - x^3 + 3x^2 - x + 1}{x - 2}$

In Exercises 19–26, use synthetic division and the Remainder Theorem to find the indicated function value.

19. $f(x) = 2x^3 - 11x^2 + 7x - 5;\quad f(4)$

20. $f(x) = x^3 - 7x^2 + 5x - 6;\quad f(3)$

21. $f(x) = 3x^3 - 7x^2 - 2x + 5;\quad f(-3)$

22. $f(x) = 4x^3 + 5x^2 - 6x - 4;\quad f(-2)$

23. $f(x) = x^4 + 5x^3 + 5x^2 - 5x - 6;\quad f(3)$

24. $f(x) = x^4 - 5x^3 + 5x^2 + 5x - 6;\quad f(2)$

25. $f(x) = 7x^4 - 3x^3 + 6x + 9;\quad f(-5)$

26. $f(x) = 3x^4 + 6x^3 - 2x + 4;\quad f(-4)$

In Exercises 27–32, use synthetic division to show that the number given to the right of each equation is a solution of the equation. Then solve the polynomial equation.

27. $x^3 - 4x^2 + x + 6 = 0;\quad -1$

28. $x^3 - 2x^2 - x + 2 = 0;\quad -1$

29. $2x^3 - 5x^2 + x + 2 = 0;\quad 2$

30. $2x^3 - 3x^2 - 11x + 6 = 0;\quad -2$

31. $6x^3 + 25x^2 - 24x + 5 = 0;\quad -5$

32. $3x^3 + 7x^2 - 22x - 8 = 0;\quad -4$

Application Exercises

33. a. Use synthetic division to show that 3 is a solution of the polynomial equation

$$14x^3 - 17x^2 - 16x - 177 = 0.$$

b. Use the solution from part (a) to solve this problem. The number of eggs, $f(x)$, in a female moth is a function of her abdominal width, x, in millimeters, modeled by

$$f(x) = 14x^3 - 17x^2 - 16x + 34.$$

What is the abdominal width when there are 211 eggs?

34. a. Use synthetic division to show that 2 is a solution of the polynomial equation

$$2h^3 + 14h^2 - 72 = 0.$$

b. Use the solution from part (a) to solve this problem. The width of a rectangular box is twice the height and the length is 7 inches more than the height. If the volume is 72 cubic inches, find the dimensions of the box.

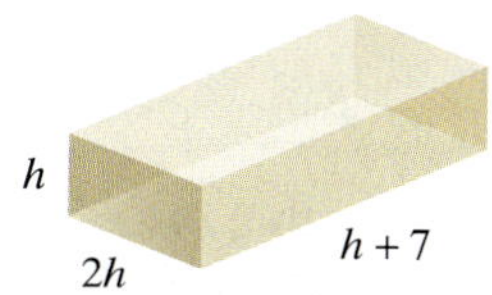

Writing in Mathematics

35. Explain how to perform synthetic division. Use the division problem

$$(2x^3 - 3x^2 - 11x + 7) \div (x - 3)$$

to support your explanation.

36. State the Remainder Theorem.

37. Explain how the Remainder Theorem can be used to find $f(-6)$ if $f(x) = x^4 + 7x^3 + 8x^2 + 11x + 5$. What advantage is there to using the Remainder Theorem in this situation rather than evaluating $f(-6)$ directly?

38. How can the Remainder Theorem and synthetic division be used to determine whether -4 is a solution of the equation

$$5x^3 + 22x^2 + x - 28 = 0?$$

Technology Exercise

39. For each equation that you solved in Exercises 27–32, use a graphing utility to graph the polynomial function on the left side of the equation. Use end behavior to obtain a complete graph. Then use the graph's x-intercepts to verify your solutions.

Critical Thinking Exercises

40. Synthetic division is a process for dividing a polynomial by $x - c$. The coefficient of x is 1. How might synthetic division be used if you are dividing by $2x - 4$?

41. Use synthetic division to show that 5 is a solution of

$$x^4 - 4x^3 - 9x^2 + 16x + 20 = 0.$$

Then solve the polynomial equation. *Hint:* Use factoring by grouping when working with the quotient factor.

Review Exercises

42. Solve: $4x + 3 - 13x - 7 < 2(3 - 4x)$. (Section 4.1, Example 5)

43. Solve: $2x(x + 3) + 6(x - 3) = -28$. (Section 5.7, Example 2)

44. Solve by Cramer's rule:

$$\begin{aligned} 7x - 6y &= 17 \\ 3x + y &= 18. \end{aligned}$$

(Section 3.5, Example 2)

▸ SECTION 6.6 Rational Equations

Objectives

1 Solve rational equations.

2 Solve problems involving rational functions that model applied situations.

The lake is in one of the city's favorite parks, and the time has come to clean it up. Voters in the city have committed \$80 thousand for the cleanup. We know that

$$f(x) = \frac{120x}{100 - x}$$

models the cost, $f(x)$, in thousands of dollars, to remove $x\%$ of the lake's pollutants. What percentage of pollutants can be removed for \$80 thousand?

To determine the percentage, we use the given cost-benefit model. Voters have committed \$80 thousand, so substitute 80 for $f(x)$:

$$80 = \frac{120x}{100 - x}.$$

This equation contains a rational expression.

Now we need to solve the equation and find the value for x. This variable represents the percentage of pollutants that can be removed for \$80 thousand.

A **rational equation**, also called a **fractional equation**, is an equation containing one or more rational expressions. The equation shown above is an example of a rational equation. Do you see that there is a variable in a denominator? This is a characteristic of many rational equations. In this section, you will learn a procedure for solving such equations.

1 Solve rational equations.

Solving Rational Equations

We have seen that the LCD is used to add and subtract rational expressions. By contrast, when solving rational equations, **the LCD is used as a multiplier that clears an equation of fractions.**

EXAMPLE 1 Solving a Rational Equation

Solve: $\frac{x + 4}{2x} + \frac{x + 20}{3x} = 3.$

Solution Notice that the variable x appears in two of the denominators. We must avoid any values of the variable that make a denominator zero.

$$\frac{x + 4}{2x} + \frac{x + 20}{3x} = 3$$

This denominator would equal zero if $x = 0$.

This denominator would equal zero if $x = 0$.

We see that x cannot equal zero.

The denominators are $2x$ and $3x$. The least common denominator is $6x$. We begin by multiplying both sides of the equation by $6x$. We will also write the restriction that x cannot equal zero to the right of the equation.

$$\frac{x+4}{2x} + \frac{x+20}{3x} = 3, \; x \neq 0 \qquad \text{This is the given equation.}$$

$$6x\left(\frac{x+4}{2x} + \frac{x+20}{3x}\right) = 6x \cdot 3 \qquad \text{Multiply both sides by } 6x\text{, the LCD.}$$

$$\frac{6x}{1} \cdot \frac{x+4}{2x} + \frac{6x}{1} \cdot \frac{x+20}{3x} = 18x \qquad \text{Use the distributive property.}$$

$$3(x+4) + 2(x+20) = 18x \qquad \text{Divide out common factors in the multiplications.}$$

Observe that the equation is now cleared of fractions.

$$3x + 12 + 2x + 40 = 18x \qquad \text{Use the distributive property.}$$

$$5x + 52 = 18x \qquad \text{Combine like terms.}$$

$$52 = 13x \qquad \text{Subtract } 5x \text{ from both sides.}$$

$$4 = x \qquad \text{Divide both sides by 4.}$$

The proposed solution, 4, is not part of the restriction $x \neq 0$. It should check in the original equation.

Check 4:

$$\frac{x+4}{2x} + \frac{x+20}{3x} = 3$$

$$\frac{4+4}{2 \cdot 4} + \frac{4+20}{3 \cdot 4} \stackrel{?}{=} 3$$

$$\frac{8}{8} + \frac{24}{12} \stackrel{?}{=} 3$$

$$1 + 2 \stackrel{?}{=} 3$$

$$3 = 3, \qquad \text{true}$$

This true statement verifies that the solution is 4 and the solution set is $\{4\}$. ■

✔ **CHECK POINT 1** Solve: $\dfrac{x+6}{2x} + \dfrac{x+24}{5x} = 2.$

The following steps may be used to solve a rational equation.

Solving Rational Equations

1. List restrictions on the variable. Avoid any values of the variable that make a denominator zero.
2. Clear the equation of fractions by multiplying both sides by the LCD of all rational expressions in the equation.
3. Solve the resulting equation.
4. Reject any proposed solution that is in the list of restrictions on the variable. Check other proposed solutions in the original equation.

Study Tip

Do not confuse simplifying rational expressions and solving rational equations. We simplify expressions by adding or subtracting:

$$\frac{x}{2}+\frac{x}{3} \quad \text{The LCD is 6.}$$

$$=\frac{x\cdot 3}{2\cdot 3}+\frac{x\cdot 2}{3\cdot 2}=\frac{3x}{6}+\frac{2x}{6}=\frac{5x}{6}.$$

By contrast, we solve equations.

Solve: $\frac{x}{2}+\frac{x}{3}=5.$

$$6\left(\frac{x}{2}+\frac{x}{3}\right)=6\cdot 5$$

$$3x+2x=30$$

$$5x=30$$

$$x=6$$

The solution is 6 and the solution set is $\{6\}$.

A rational expression does not contain an equal sign and cannot be cleared of fractions. A rational equation contains an equal sign and can be cleared of fractions.

EXAMPLE 2 Solving a Rational Equation

Solve: $\frac{x+1}{x+10}=\frac{x-2}{x+4}$.

Solution

Step 1 List restrictions on the variable.

$$\frac{x+1}{x+10}=\frac{x-2}{x+4}$$

This denominator would equal zero if $x = -10$. This denominator would equal zero if $x = -4$.

The restrictions are $x \neq -10$ and $x \neq -4$.

Step 2 Multiply both sides by the LCD. The denominators are $x + 10$ and $x + 4$. Thus, the LCD is $(x + 10)(x + 4)$.

$$\frac{x+1}{x+10}=\frac{x-2}{x+4},\quad x\neq -10,\ x\neq -4 \qquad \text{This is the given equation.}$$

$$(x+10)(x+4)\cdot\left(\frac{x+1}{x+10}\right)=(x+10)(x+4)\cdot\left(\frac{x-2}{x+4}\right) \qquad \text{Multiply both sides by the LCD.}$$

$$(x+4)(x+1)=(x+10)(x-2) \qquad \text{Simplify.}$$

Step 3 Solve the resulting equation.

$$(x+4)(x+1)=(x+10)(x-2) \qquad \text{This is the equation cleared of fractions.}$$

$$x^2+5x+4=x^2+8x-20 \qquad \text{Use FOIL multiplication on each side.}$$

$$5x+4=8x-20 \qquad \text{Subtract } x^2 \text{ from both sides.}$$

$$-3x+4=-20 \qquad \text{Subtract 8x from both sides.}$$

$$-3x=-24 \qquad \text{Subtract 4 from both sides.}$$

$$x=8 \qquad \text{Divide both sides by } -3.$$

Step 4 Check the proposed solution in the original equation. The proposed solution, 8, is not part of the restriction that $x \neq -10$ and $x \neq -4$. Substitute 8 for x in the given equation. You should obtain the true statement $\frac{1}{2}=\frac{1}{2}$. The solution is 8 and the solution set is $\{8\}$. ■

✔ **CHECK POINT 2** Solve: $\frac{x-3}{x+1}=\frac{x-2}{x+6}$.

EXAMPLE 3 Solving a Rational Equation

Solve: $\frac{x}{x-3}=\frac{3}{x-3}+9$.

Solution

Step 1 List restrictions on the variable.

$$\frac{x}{x-3}=\frac{3}{x-3}+9$$

These denominators are zero if $x = 3$.

The restriction is $x \neq 3$.

Step 2 Multiply both sides by the LCD. The LCD is $x - 3$.

$$\frac{x}{x-3} = \frac{3}{x-3} + 9, \quad x \neq 3$$ This is the given equation.

$$(x-3) \cdot \frac{x}{x-3} = (x-3)\left(\frac{3}{x-3} + 9\right)$$ Multiply both sides by the LCD.

$$\cancel{(x-3)} \cdot \frac{x}{\cancel{x-3}} = \cancel{(x-3)} \cdot \frac{3}{\cancel{x-3}} + 9(x-3)$$ Use the distributive property on the right side.

$$x = 3 + 9(x-3)$$ Simplify.

Step 3 Solve the resulting equation.

$$x = 3 + 9(x-3)$$ This is the equation cleared of fractions.

$$x = 3 + 9x - 27$$ Use the distributive property on the right side.

$$x = 9x - 24$$ Combine numerical terms.

$$-8x = -24$$ Subtract 9x from both sides.

$$x = 3$$ Divide both sides by −8.

Study Tip

Reject any proposed solution that causes any denominator in a rational equation to equal 0.

Step 4 Check proposed solutions. The proposed solution, 3, is *not* a solution because of the restriction that $x \neq 3$. Notice that 3 makes two of the denominators zero in the original equation. There is no solution to this equation. The solution set is $\varnothing$, the empty set. ■

✔ **CHECK POINT 3** Solve: $\frac{8x}{x+1} = 4 - \frac{8}{x+1}$.

Examples 4 and 5 involve rational equations that become quadratic equations after clearing fractions.

EXAMPLE 4 Solving a Rational Equation

Solve: $\frac{x}{3} + \frac{9}{x} = 4$.

Solution

Step 1 List restrictions on the variable.

$$\frac{x}{3} + \frac{9}{x} = 4$$

This denominator would equal zero if $x = 0$.

The restriction is $x \neq 0$.

Step 2 Multiply both sides by the LCD. The denominators are 3 and x. Thus, the LCD is $3x$.

$$\frac{x}{3} + \frac{9}{x} = 4, \quad x \neq 0 \qquad \text{This is the given equation.}$$

$$3x\left(\frac{x}{3} + \frac{9}{x}\right) = 3x \cdot 4 \qquad \text{Multiply both sides by the LCD.}$$

$$3x \cdot \frac{x}{3} + 3x \cdot \frac{9}{x} = 12x \qquad \text{Use the distributive property on the left side.}$$

$$x^2 + 27 = 12x \qquad \text{Simplify.}$$

Step 3 Solve the resulting equation. Can you see that we have a quadratic equation? Write the equation in standard form and solve for x.

$$x^2 + 27 = 12x \qquad \text{This is the equation cleared of fractions.}$$

$$x^2 - 12x + 27 = 0 \qquad \text{Subtract 12x from both sides.}$$

$$(x - 9)(x - 3) = 0 \qquad \text{Factor.}$$

$$x - 9 = 0 \quad \text{or} \quad x - 3 = 0 \qquad \text{Set each factor equal to 0.}$$

$$x = 9 \qquad\qquad x = 3 \qquad \text{Solve the resulting equations.}$$

Step 4 Check proposed solutions in the original equation. The proposed solutions, 9 and 3, are not part of the restriction that $x \neq 0$. Substitute 9 for x, and then 3 for x, in the given equation. In each case, you should obtain the true statement $4 = 4$. The solutions are 3 and 9, and the solution set is $\{3, 9\}$. ■

Using Technology

The graphs of

$$y_1 = \frac{x}{3} + \frac{9}{x}$$

and $\quad y_2 = 4$

have two intersection points. The x-coordinates of the points are 3 and 9. This verifies that

$$\frac{x}{3} + \frac{9}{x} = 4$$

has both 3 and 9 as solutions.

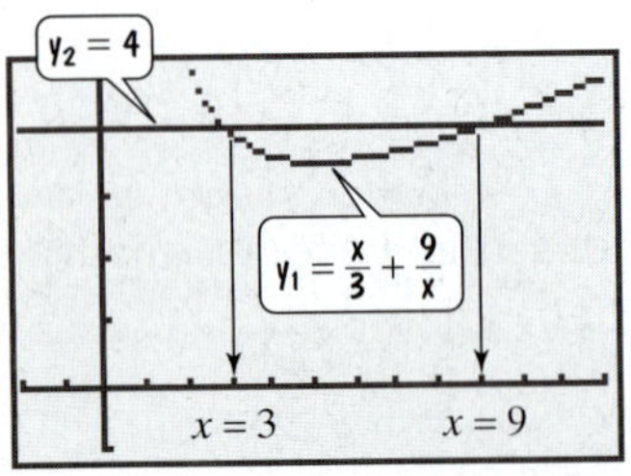

$[-2, 12, 1]$ by $[-1, 5, 1]$

✔ CHECK POINT 4 Solve: $\frac{x}{2} + \frac{12}{x} = 5.$

EXAMPLE 5 Solving a Rational Equation

Solve: $\frac{2x}{x-3} + \frac{6}{x+3} = -\frac{28}{x^2 - 9}$.

Solution

Step 1 List restrictions on the variable. By factoring denominators, it makes it easier to see values that make denominators zero.

$$\frac{2x}{x-3} + \frac{6}{x+3} = -\frac{28}{(x+3)(x-3)}$$

This denominator is zero if $x = 3$. This denominator is zero if $x = -3$. This denominator is zero if $x = -3$ or $x = 3$.

The restrictions are $x \neq -3$ and $x \neq 3$.

Step 2 Multiply both sides by the LCD. The LCD is $(x+3)(x-3)$.

$$\frac{2x}{x-3}+\frac{6}{x+3}=-\frac{28}{(x+3)(x-3)},\quad x\neq-3, x\neq 3$$

This is the given equation with a denominator factored.

$$(x+3)(x-3)\left(\frac{2x}{x-3}+\frac{6}{x+3}\right)=(x+3)(x-3)\left(-\frac{28}{(x+3)(x-3)}\right)$$

Multiply both sides by the LCD.

$$(x+3)\cancel{(x-3)}\cdot\frac{2x}{\cancel{x-3}}+\cancel{(x+3)}(x-3)\cdot\frac{6}{\cancel{x+3}}=\cancel{(x+3)}\cancel{(x-3)}\left(-\frac{28}{\cancel{(x+3)}\cancel{(x-3)}}\right)$$

Use the distributive property on the left side.

$$2x(x+3)+6(x-3)=-28$$

Simplify.

Step 3 Solve the resulting equation.

$$2x(x+3)+6(x-3)=-28$$ This is the equation cleared of fractions.

$$2x^2+6x+6x-18=-28$$ Use the distributive property twice on the left side.

$$2x^2+12x-18=-28$$ Combine like terms.

$$2x^2+12x+10=0$$ Add 28 to both sides and write the quadratic equation in standard form.

$$2(x^2+6x+5)=0$$ Factor out the GCF.

$$2(x+5)(x+1)=0$$ Factor the trinomial.

$$x+5=0 \quad\text{or}\quad x+1=0$$ Set each variable factor equal to 0.

$$x=-5 \qquad\qquad x=-1$$ Solve for x.

Step 4 Check the proposed solutions in the original equation. The proposed solutions, -5 and -1, are not part of the restriction that $x\neq-3$ and $x\neq 3$. Substitute -5 for x, and then -1 for x, in the given equation. The resulting true statements verify that -5 and -1 are the solutions, and $\{-5,-1\}$ is the solution set. ■

✔ **CHECK POINT 5** Solve: $\frac{3}{x-3}+\frac{5}{x-4}=\frac{x^2-20}{x^2-7x+12}$.

2 Solve problems involving rational functions that model applied situations.

Applications of Rational Equations

Solving rational equations can be used to answer questions about variables contained in rational functions.

EXAMPLE 6 Using a Cost-Benefit Model

The function

$$f(x)=\frac{120x}{100-x}$$

models the cost, $f(x)$, in thousands of dollars, to remove $x\%$ of a lake's pollutants. If voters commit \$80 thousand for this project, what percentage of pollutants can be removed?

Solution Substitute 80, the cost in thousands of dollars, for $f(x)$ and solve the resulting rational equation for x.

$$80 = \frac{120x}{100 - x}$$ The LCD is $100 - x$.

$$(100 - x)80 = (100 - x) \cdot \frac{120x}{100 - x}$$ Multiply both sides by the LCD.

$$80(100 - x) = 120x$$ Simplify.

$$8000 - 80x = 120x$$ Use the distributive property on the left side.

$$8000 = 200x$$ Add $80x$ to both sides.

$$40 = x$$ Divide both sides by 200.

If voters commit $80 thousand, 40% of the lake's pollutants can be removed. ■

CHECK POINT 6 Use the cost-benefit model in Example 6 to answer this question: If voters in the city commit $120 thousand for the project, what percentage of the lake's pollutants can be removed?

EXERCISE SET 6.6

Practice Exercises

In Exercises 1–32, solve each rational equation. If an equation has no solution, so state.

1. $\frac{1}{x} + 2 = \frac{3}{x}$
2. $\frac{1}{x} - 3 = \frac{4}{x}$
3. $\frac{5}{x} + \frac{1}{3} = \frac{6}{x}$
4. $\frac{4}{x} + \frac{1}{2} = \frac{5}{x}$
5. $\frac{x - 2}{2x} + 1 = \frac{x + 1}{x}$
6. $\frac{7x - 4}{5x} = \frac{9}{5} - \frac{4}{x}$
7. $\frac{3}{x + 1} = \frac{5}{x - 1}$
8. $\frac{6}{x + 3} = \frac{4}{x - 3}$
9. $\frac{x - 6}{x + 5} = \frac{x - 3}{x + 1}$
10. $\frac{x + 2}{x + 10} = \frac{x - 3}{x + 4}$
11. $\frac{x + 6}{x + 3} = \frac{3}{x + 3} + 2$
12. $\frac{3x + 1}{x - 4} = \frac{6x + 5}{2x - 7}$
13. $1 - \frac{4}{x + 7} = \frac{5}{x + 7}$
14. $5 - \frac{2}{x - 5} = \frac{3}{x - 5}$
15. $\frac{4x}{x + 2} + \frac{2}{x - 1} = 4$
16. $\frac{3x}{x + 1} + \frac{4}{x - 2} = 3$
17. $\frac{8}{x^2 - 9} + \frac{4}{x + 3} = \frac{2}{x - 3}$
18. $\frac{32}{x^2 - 25} = \frac{4}{x + 5} + \frac{2}{x - 5}$
19. $x + \frac{7}{x} = -8$
20. $x + \frac{6}{x} = -7$
21. $\frac{6}{x} - \frac{x}{3} = 1$
22. $\frac{x}{2} - \frac{12}{x} = 1$
23. $\frac{x + 6}{3x - 12} = \frac{5}{x - 4} + \frac{2}{3}$
24. $\frac{1}{5x + 5} = \frac{3}{x + 1} - \frac{7}{5}$
25. $\frac{1}{x - 1} + \frac{1}{x + 1} = \frac{2}{x^2 - 1}$
26. $\frac{1}{x - 2} + \frac{1}{x + 2} = \frac{4}{x^2 - 4}$
27. $\frac{5}{x + 4} + \frac{3}{x + 3} = \frac{12x + 19}{x^2 + 7x + 12}$
28. $\frac{2x - 1}{x^2 + 2x - 8} + \frac{2}{x + 4} = \frac{1}{x - 2}$

29. $\frac{4x}{x+3} - \frac{12}{x-3} = \frac{4x^2+36}{x^2-9}$

30. $\frac{2}{x+3} - \frac{5}{x+1} = \frac{3x+5}{x^2+4x+3}$

31. $\frac{4}{x^2+3x-10} + \frac{1}{x^2+9x+20} = \frac{2}{x^2+2x-8}$

32. $\frac{4}{x^2+3x-10} - \frac{1}{x^2+x-6} = \frac{3}{x^2-x-12}$

Application Exercises

The function

$$f(x) = \frac{250x}{100-x}$$

models the cost, $f(x)$, in millions of dollars, to remove $x\%$ of a river's pollutants. Use this function to solve Exercises 33–34.

33. If the government commits $375 million for this project, what percentage of pollutants can be removed?

34. If the government commits $750 million for this project, what percentage of pollutants can be removed?

In an experiment on memory, students in a language class are asked to memorize 40 vocabulary words in Latin, a language with which the students are not familiar. After studying the words for one day, students are tested each day after to see how many words they remember. The class average is then found. The function

$$f(x) = \frac{5x+30}{x}$$

models the average number of Latin words remembered by the students, $f(x)$, after x days. The graph of the rational function is shown. Use the function to solve Exercises 35–38.

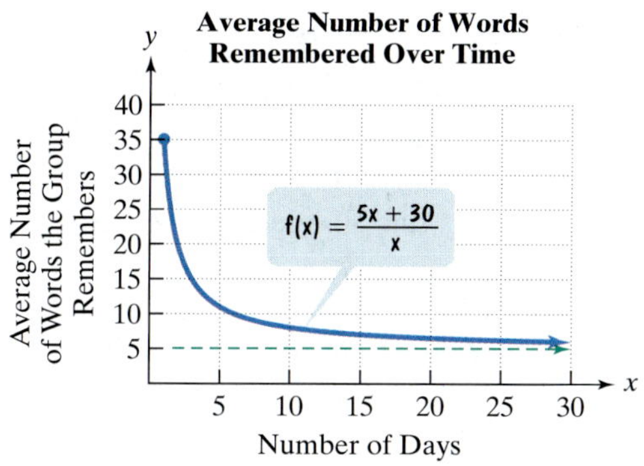

35. After how many days do the students remember 8 words? Identify your solution as a point on the graph.

36. After how many days do the students remember 7 words? Identify your solution as a point on the graph.

37. What is the horizontal asymptote of the graph? Describe what this means about the average number of Latin words remembered by the students over an extended period of time.

38. According to the graph, between which two days do students forget the most? Describe the trend shown by the memory function's graph.

Rational functions can be used to model learning. Many of these functions model the proportion of correct responses as a function of the number of trials of a particular task. One such model, called a learning curve, is

$$f(x) = \frac{0.9x-0.4}{0.9x+0.1}$$

where $f(x)$ is the proportion of correct responses after x trials. If $f(x) = 0$, there are no correct responses. If $f(x) = 1$, all responses are correct. The graph of the rational function is shown. Use the function to solve Exercises 39–42.

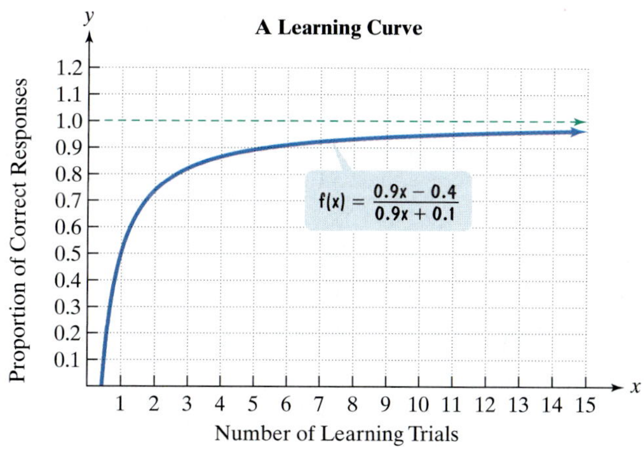

39. How many learning trials are necessary for 0.95 of the responses to be correct? Identify your solution as a point on the graph.

40. How many learning trials are necessary for 0.5 of the responses to be correct? Identify your solution as a point on the graph.

41. Describe the trend shown by the graph in terms of learning new tasks. What happens initially and what happens as time increases?

42. What is the horizontal asymptote of the graph? Once the performance level approaches peak efficiency, what effect does additional practice have on performance? Describe how this is shown by the graph.

A grocery store sells 4000 cases of canned soup per year. By averaging costs to purchase soup and pay storage costs, the owner has determined that if x cases are ordered at a time, the yearly inventory cost, $f(x)$, can be modeled by

$$f(x) = \frac{10{,}000}{x} + 3x.$$

Use this information to solve Exercises 43–44.

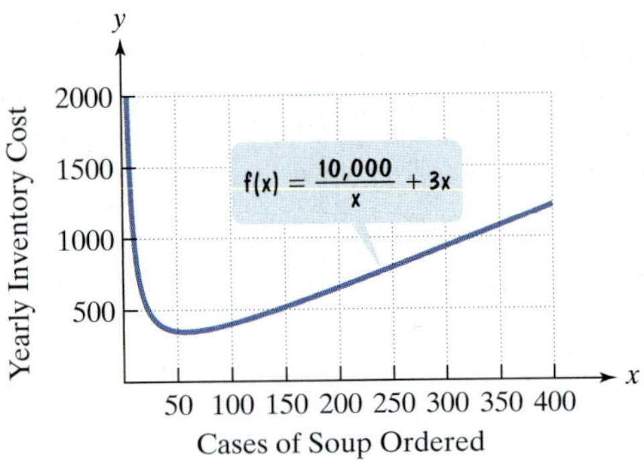

43. How many cases should be ordered at a time for yearly inventory costs to be \$350? Identify your solutions as points on the graph.

44. How many cases should be ordered at a time for yearly inventory costs to be \$790? Identify your solutions as points on the graph.

Writing in Mathematics

45. What is a rational equation?

46. Explain how to solve a rational equation.

47. Explain how to find restrictions on the variable in a rational equation.

48. Why should restrictions on the variable in a rational equation be listed before you begin solving the equation?

49. Describe similarities and differences between the procedures needed to solve the following problems:

Add: $\frac{2}{x} + \frac{3}{4}$ Solve: $\frac{2}{x} + \frac{3}{4} = 1.$

50. Rational functions model learning and forgetting. Use the graphs in Exercises 33–42 to describe one similarity and one difference between learning and forgetting over time.

51. Does the graph of the learning curve shown in Exercises 39–42 indicate that practice makes perfect? Explain. Does this have anything to do with what psychologists call "the curse of perfection"?

Technology Exercises

In Exercises 52–56, use a graphing utility to solve each rational equation. Graph each side of the equation in the given viewing rectangle. The solution is the first coordinate of the point(s) of intersection. Check by direct substitution.

52. $\frac{x}{2} + \frac{x}{4} = 6$
$[-5, 10, 1]$ by $[-5, 10, 1]$

53. $\frac{50}{x} = 2x$
$[-10, 10, 1]$ by $[-20, 20, 2]$

54. $x + \frac{6}{x} = -5$
$[-10, 10, 1]$ by $[-7, 10, 1]$

55. $\frac{2}{x} = x + 1$
$[-5, 5, 1]$ by $[-5, 5, 1]$

56. $\frac{3}{x} - \frac{x + 21}{3x} = \frac{5}{3}$
$[-5, 5, 1]$ by $[-5, 5, 1]$

Critical Thinking Exercises

57. Which one of the following is true?

a. Once a rational equation is cleared of fractions, all solutions of the resulting equation are also solutions of the rational equation.

b. We find

$$\frac{4}{x} - \frac{2}{x + 1}$$

by multiplying each term by the LCD, $x(x + 1)$, thereby clearing fractions.

c. All real numbers satisfy the equation $\frac{7}{x} - \frac{2}{x} = \frac{5}{x}$.

d. None of the above is true.

58. If $f(x) = \frac{5}{3x} + 1$ find all values of a such that $f(a) = \frac{7}{6}$.

59. Write an original rational equation that has no solution.

60. Solve

$$\left(\frac{4}{x - 1}\right)^2 + 2\left(\frac{4}{x - 1}\right) + 1 = 0$$

by introducing the substitution $t = \frac{4}{x - 1}$.

Review Exercises

61. Graph the solution set:

$$x + 2y \geq 2$$
$$x - y \geq -4.$$

(Section 4.4, Example 3)

62. Solve:

$$\frac{x-4}{2} - \frac{1}{5} = \frac{7x+1}{20}.$$

(Section 1.4, Example 4)

63. Solve for F:

$$C = \frac{5F - 160}{9}.$$

(Section 1.5, Example 8)

▸ SECTION 6.7 Formulas and Applications of Rational Equations

Objectives

1. Solve for a variable in a formula with a rational expression.
2. Solve business problems involving average cost.
3. Solve problems involving time in motion.
4. Solve problems involving work.

SSM PH Tutor Center CD-ROM Video

The edges between reality and virtual reality are becoming blurred. The people who built your computer talk of "bandwidth out the wazoo" that will give you the visual experience, in high-definition 3-D format, of being in the same room with a person who is actually in another city. It is already possible to "experience" an exciting activity such as skiing down a mountain.

As computers become more powerful, technology is now promising to make the virtual experience even more real and bring it to our home computer. The cost of manufacturing computers with massive bandwidth can be modeled by rational functions. In this section, you will see how high production levels of "tele-immersion" computers can eventually make them affordable for home use. Get ready to feel the wind in your hair, the frost on your eyebrows, and the gentle heat of the sun on your face, as well as the shudder through your ski boots as you race down mountains without ever leaving your home.

1 Solve for a variable in a formula with a rational expression.

Solving for a Variable in a Formula

Formulas and mathematical models frequently contain rational expressions. We solve for a specified variable using the procedure for solving rational equations. The goal is to get the specified variable alone on one side of the equation. To do so, collect all terms with this variable on one side and all other terms on the other side. It is sometimes necessary to factor out the variable you are solving for.

EXAMPLE 1 Solving for a Variable in a Formula

If you wear glasses, did you know that each lens has a measurement called its focal length, f? When an object is in focus, its distance from the lens, p, and the distance from the lens to your retina, q, satisfy the formula

$$\frac{1}{p} + \frac{1}{q} = \frac{1}{f}.$$

Figure 6.5

(See Figure 6.5.) Solve this formula for p.

Solution Our goal is to isolate the variable p. We begin by multiplying both sides by the least common denominator, pqf, to clear the equation of fractions.

$$\frac{1}{p} + \frac{1}{q} = \frac{1}{f}$$ This is the given formula.

$$pqf\left(\frac{1}{p} + \frac{1}{q}\right) = pqf\left(\frac{1}{f}\right)$$ Multiply both sides by pqf, the LCD.

$$\cancel{p}qf\left(\frac{1}{\cancel{p}}\right) + p\cancel{q}f\left(\frac{1}{\cancel{q}}\right) = pq\cancel{f}\left(\frac{1}{\cancel{f}}\right)$$ Use the distributive property on the left side.

$$qf + pf = pq$$ Simplify.

Observe that the formula is now cleared of fractions. Collect terms with p, the specified variable, on one side of the equation. To do so, subtract pf from both sides.

$$qf + pf = pq$$ This is the equation cleared of fractions.

$$qf = pq - pf$$ Subtract pf from both sides.

$$qf = p(q - f)$$ Factor out p, the specified variable.

$$\frac{qf}{q - f} = \frac{p(q - f)}{q - f}$$ Divide both sides by $q - f$ and solve for p.

$$\frac{qf}{q - f} = p$$ Simplify

✔ **CHECK POINT 1** Solve $\frac{1}{x} + \frac{1}{y} = \frac{1}{z}$ for x.

2 Solve business problems involving average cost.

Business Problems Involving Average Cost

We have seen that the cost function for a business is the sum of its fixed and variable costs:

$$C(x) = (\text{fixed cost}) + cx.$$

Cost per unit times the number of units produced, x

The **average cost** per unit for a company to produce x units is the sum of its fixed and variable costs divided by the number of units produced. The **average cost function** is a rational function that is denoted by $\overline{C}$. Thus,

$$\overline{C}(x) = \frac{(\text{fixed cost}) + cx}{x}.$$

Cost of producing x units: fixed plus variable costs

Number of units produced

EXAMPLE 2 Average Cost for a Business

The time: the not-too-distant future. A new company is hoping to replace traditional computers and two-dimensional monitors with its virtual reality system. The fixed monthly cost will be \$600,000 and it will cost \$500 to produce each system.

a. Write the cost function, C, of producing x virtual reality systems.

b. Write the average cost function, $\overline{C}$, of producing x virtual reality systems.

c. How many virtual reality systems must be produced each month for the company to have an average cost of \$2000 per system?

Solution

a. The cost function, C, is the sum of the fixed cost and and the variable cost.

Fixed cost is \$600,000. Variable cost: \$500 for each system produced

$$C(x) = 600{,}000 + 500x$$

b. The average cost function, $\overline{C}$, is the sum of fixed and variable costs divided by the number of virtual reality systems produced.

$$\overline{C}(x) = \frac{600{,}000 + 500x}{x}$$

c. We are interested in the company's production level that results in an average cost of \$2000 per virtual reality system. Substitute 2000, the average cost, for $\overline{C}(x)$ and solve the resulting rational equation for x.

$$2000 = \frac{600{,}000 + 500x}{x} \quad \text{Substitute 2000 for } \overline{C}(x).$$

$$2000x = 600{,}000 + 500x \quad \text{Multiply both sides by the LCD, } x.$$

$$1500x = 600{,}000 \quad \text{Subtract } 500x \text{ from both sides.}$$

$$x = 400 \quad \text{Divide both sides by 1500.}$$

The company must produce 400 virtual reality systems each month for an average cost of \$2000 per system. ■

Figure 6.6 shows the graph of the average cost function in Example 2. As the production level increases, the average cost of producing each virtual reality system decreases. The horizontal asymptote, $y = 500$, is also shown in the figure. This means that the more systems produced each month, the closer the average cost per system for the company comes to \$500. The least possible cost per virtual reality system is approaching \$500. Competitively low prices take place with high production levels, posing a major problem for small businesses.

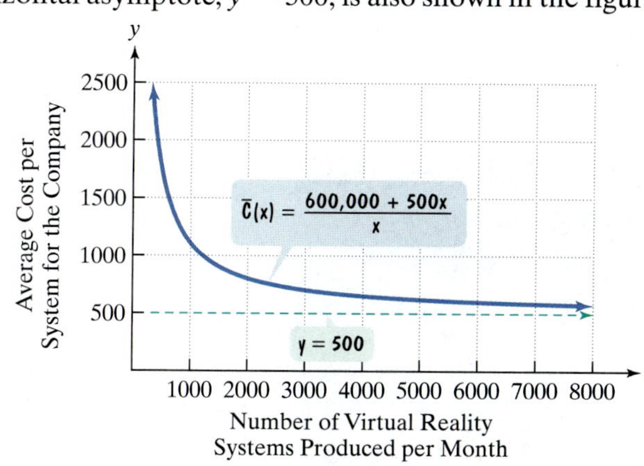

Figure 6.6

CHECK POINT 2 A company is planning to manufacture wheelchairs that are light, fast, and beautiful. Fixed monthly cost will be \$500,000, and it will cost \$400 to produce each radically innovative chair.

a. Write the cost function, C, of producing x wheelchairs.

b. Write the average cost function, $\overline{C}$, of producing x wheelchairs.

c. How many wheelchairs must be produced each month for the company to have an average cost of \$450 per chair?

3 Solve problems involving time in motion.

Problems Involving Motion

We have seen that the distance, d, covered by any moving body is the product of its average rate, r, and its time in motion, t: $d = rt$. Rational expressions appear in motion problems when the conditions of the problem involve the time traveled. We can obtain an expression for t, the time traveled, by dividing both sides of $d = rt$ by r.

$$d = rt \quad \text{Distance equals rate times time.}$$

$$\frac{d}{r} = \frac{rt}{r} \quad \text{Divide both sides by } r.$$

$$\frac{d}{r} = t \quad \text{Simplify.}$$

ENRICHMENT ESSAY

Reducing Time in Motion: Space Planes

Engineers are now working on a new kind of plane that combines flight and space technology. Space planes will fly high into the atmosphere where they will go into orbit, reaching speeds of more than 10,000 miles per hour. This new aircraft will be capable of flying you to the other end of the Earth in an hour or so.

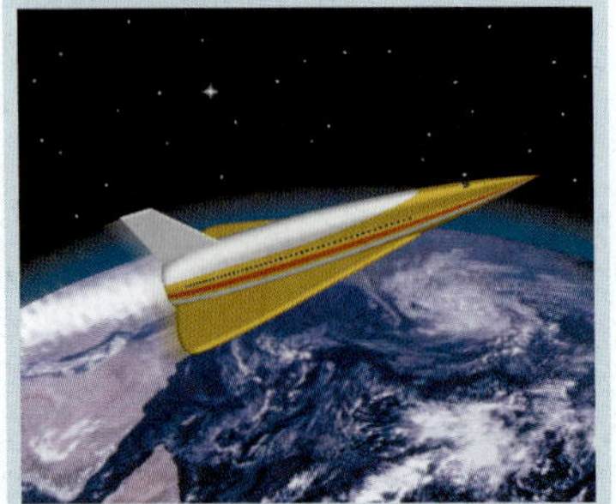

Time in Motion

$$t = \frac{d}{r}$$

$$\text{Time traveled} = \frac{\text{Distance traveled}}{\text{Rate of travel}}$$

EXAMPLE 3 A Motion Problem Involving Time

You commute to work a distance of 40 miles and return on the same route at the end of the day. Your average rate on the return trip is 30 miles per hour faster than your average rate on the outgoing trip. If the round trip takes 2 hours, what is your average rate on the outgoing trip to work?

Solution

Step 1 Let x represent one of the quantities. Let

$$x = \text{the rate on the outgoing trip.}$$

Step 2 Represent other quantities in terms of x. Because the average rate on the return trip is 30 miles per hour faster than the average rate on the outgoing trip, let

$$x + 30 = \text{the rate on the return trip.}$$

Step 3 Write an equation that describes the conditions. By reading the problem again, we discover that the crucial idea is that the time for the round trip is 2 hours. Thus, the time on the outgoing trip plus the time on the return trip is 2 hours.

	Distance	Rate	Time $= \dfrac{\text{Distance}}{\text{Rate}}$
Outgoing Trip	40	x	$\dfrac{40}{x}$
Return Trip	40	$x + 30$	$\dfrac{40}{x+30}$

The sum of these times is 2 hours.

We are now ready to write an equation that describes the problem's conditions.

Time on the outgoing trip plus time on the return trip equals 2 hours.

$$\frac{40}{x} + \frac{40}{x+30} = 2$$

Step 4 Solve the equation and answer the question.

$$\frac{40}{x} + \frac{40}{x+30} = 2 \quad \text{This is the equation for the problem's conditions.}$$

$$x(x+30)\left(\frac{40}{x} + \frac{40}{x+30}\right) = 2x(x+30) \quad \text{Multiply both sides by the LCD, } x(x+30).$$

$$\cancel{x}(x+30)\cdot\frac{40}{\cancel{x}} + x\cancel{(x+30)}\cdot\frac{40}{\cancel{x+30}} = 2x^2 + 60x \quad \text{Use the distributive property on each side}$$

$$40(x+30) + 40x = 2x^2 + 60x \quad \text{Simplify.}$$

$$40x + 1200 + 40x = 2x^2 + 60x \quad \text{Use the distributive property.}$$

$$80x + 1200 = 2x^2 + 60x \quad \text{Combine like terms: } 40x + 40x = 80x.$$

$$0 = 2x^2 - 20x - 1200 \quad \text{Subtract } 80x + 1200 \text{ from both sides.}$$

$$0 = 2(x^2 - 10x - 600) \quad \text{Factor out the GCF.}$$

$$0 = 2(x-30)(x+20) \quad \text{Factor completely.}$$

$$x - 30 = 0 \quad \text{or} \quad x + 20 = 0 \quad \text{Set each variable factor equal to 0.}$$

$$x = 30 \quad \text{or} \quad x = -20 \quad \text{Solve for } x.$$

Because x represents the rate on the outgoing trip, we reject the negative value, -20. The rate on the outgoing trip is 30 miles per hour.

Step 5 Check the proposed solution in the original wording of the problem. Does the round trip take 2 hours? Because the rate on the return trip is 30 miles

Using Technology

The graph of the rational function for the time for the round trip

$$f(x) = \frac{40}{x} + \frac{40}{x+30}$$

is shown in a $[0, 60, 30]$ by $[0, 10, 1]$ viewing rectangle.

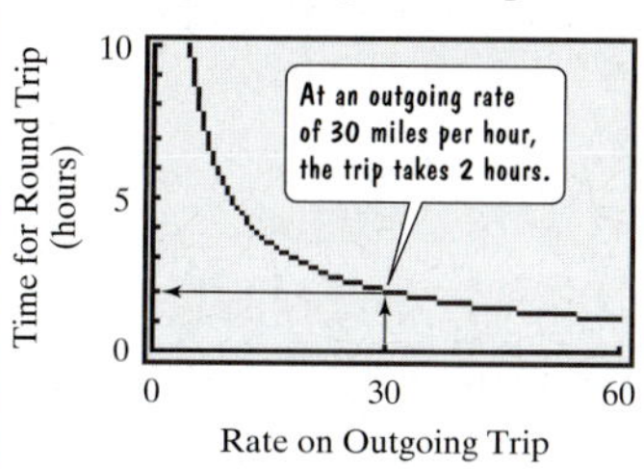

The graph is falling from left to right. This shows that the time for the round trip decreases as the rate increases.

per hour faster than the rate on the outgoing trip, the rate on the return trip is 30 + 30, or 60 miles per hour.

$$\text{Time on the outgoing trip} = \frac{\text{Distance}}{\text{Rate}} = \frac{40}{30} = \frac{4}{3} \text{ hours}$$

$$\text{Time on the return trip} = \frac{\text{Distance}}{\text{Rate}} = \frac{40}{60} = \frac{2}{3} \text{ hour}$$

The total time for the round trip is $\frac{4}{3} + \frac{2}{3} = \frac{6}{3}$, or 2 hours. This checks with the original conditions of the problem. ■

✔ **CHECK POINT 3** After riding at a steady speed for 40 miles, a bicyclist had a flat tire and walked 5 miles to a repair shop. The cycling rate was 4 times faster than the walking rate. If the time spent cycling and walking was 5 hours, at what rate was the cyclist riding?

4 Solve problems involving work.

Problems Involving Work

You are thinking of designing your own Web site You estimate that it will take 30 hours to do the job. In 1 hour, $\frac{1}{30}$ of the job is completed. In 2 hours, $\frac{2}{30}$, or $\frac{1}{15}$, of the job is completed. In 3 hours, the fractional part of the job done is $\frac{3}{30}$, or $\frac{1}{10}$. In x hours, the fractional part of the job that you can complete is $\frac{x}{30}$.

Your friend, who has experience developing Web sites, took 20 hours working on her own to design an impressive site. You wonder about the possibility of working together. How long would it take both of you to design your Web site?

Problems involving work usually have two (or more) people or machines working together to complete a job. The amount of time it takes each person to do the job working alone is frequently known. The question deals with how long it will take both people working together to complete the job.

In work problems, **the number 1 represents one whole job completed**. For example, the completion of your Web site is represented by 1. Equations in work problems are often based on the following condition:

Fractional part of the job done by the first person	+	fractional part of the job done by the second person	=	1 (one whole job completed).

EXAMPLE 4 Solving a Problem Involving Work

You can design a Web site in 30 hours. Your friend can design the same site in 20 hours. How long will it take to design the Web site if you both work together?

Solution

Step 1 Let x represent one of the quantities. Let

x = the time, in hours, for you and your friend working together to design the Web site.

Step 2 Represent other quantities in terms of x. Because there are no other unknown quantities, we can skip this step.

Step 3 Write an equation that describes the conditions. We construct a table to help find the fractional part of the task completed by you and your friend in x hours.

	Fractional part of job completed in 1 hour	Time working together	Fractional part of the job completed in x hours
You	$\frac{1}{30}$	x	$\frac{x}{30}$
Your friend	$\frac{1}{20}$	x	$\frac{x}{20}$

You can design the site in 30 hours.

Your friend can design the site in 20 hours.

Fractional part of the job done by you + fractional part of the job done by your friend = one whole job.

$$\frac{x}{30} + \frac{x}{20} = 1$$

Step 4 Solve the equation and answer the question.

$$\frac{x}{30} + \frac{x}{20} = 1$$
This is the equation for the problem's conditions.

$$60\left(\frac{x}{30} + \frac{x}{20}\right) = 60 \cdot 1$$
Multiply both sides by 60, the LCD.

$$60 \cdot \frac{x}{30} + 60 \cdot \frac{x}{20} = 60$$
Use the distributive property on the left side.

$$2x + 3x = 60$$
Simplify: $\frac{\overset{2}{\cancel{60}}}{1} \cdot \frac{x}{\underset{1}{\cancel{30}}} = 2x$ and $\frac{\overset{3}{\cancel{60}}}{1} \cdot \frac{x}{\underset{1}{\cancel{20}}} = 3x$.

$$5x = 60$$
Combine like terms.

$$x = 12$$
Divide both sides by 5.

If you both work together, you can design your Web site in 12 hours.

Step 5 Check the proposed solution in the original wording of the problem. Will you both complete the job in 12 hours? In 12 hours, you can complete $\frac{12}{30}$, or $\frac{2}{5}$ of the job. In 12 hours, your friend can complete $\frac{12}{20}$, or $\frac{3}{5}$ of the job. Notice that $\frac{2}{5} + \frac{3}{5} = 1$, which represents the completion of the entire job, or one whole job. ■

Study Tip

Let

a = the time it takes person A to do a job working alone, and

b = the time it takes person B to do the same job working alone.

If x represents the time it takes for A and B to complete the entire job working together, then the situation can be modeled by the rational equation

$$\frac{x}{a} + \frac{x}{b} = 1.$$

✔ **CHECK POINT 4** One person can paint the outside of a house in 8 hours. A second person can do it in 4 hours. How long will it take them to do the job if they work together?

EXAMPLE 5 Solving a Problem Involving Work

After designing your Web site, you and your friend decide to go into business setting up sites for others. With lots of practice, you can now work together and design a modest site in 4 hours. Your friend is still a faster worker. Working alone, you require 6 more hours than she to design a site for a client. How many hours does it take your friend to design a Web site if she works alone?

Solution

Step 1 Let x represent one of the quantities. Let

$x =$ the time, in hours, for your friend to design a Web site working alone.

Step 2 Represent other quantities in terms of x. Because you require 6 more hours than your friend to do the job, let

$x + 6 =$ the time, in hours, for you to design a Web site working alone.

Step 3 Write an equation that describes the conditions. Working together, you and your friend can complete the job in 4 hours. We construct a table to find the fractional part of the task completed by you and your friend in 4 hours.

	Fractional part of job completed in 1 hour	Time working together	Fractional part of the job completed in 4 hours
Your friend (Your friend can design the site in x hours.)	$\frac{1}{x}$	4	$\frac{4}{x}$
You (You can design the site in x + 6 hours.)	$\frac{1}{x+6}$	4	$\frac{4}{x+6}$

Because you can both complete the job in 4 hours,

Fractional part of the job done by your friend + fractional part of the job done by you = one whole job.

$$\frac{4}{x} + \frac{4}{x+6} = 1.$$

Step 4 Solve the equation and answer the question.

$$\frac{4}{x} + \frac{4}{x+6} = 1$$ This is the equation for the problem's conditions.

$$x(x+6)\left(\frac{4}{x} + \frac{4}{x+6}\right) = x(x+6) \cdot 1$$ Multiply both sides by $x(x+6)$, the LCD.

$$x(x+6) \cdot \frac{4}{x} + x(x+6) \cdot \frac{4}{(x+6)} = x^2 + 6x$$ Use the distributive property on each side.

$$4(x+6) + 4x = x^2 + 6x$$ Simplify.

$$4x + 24 + 4x = x^2 + 6x$$ Use the distributive property.

$$8x + 24 = x^2 + 6x$$ Combine like terms: $4x + 4x = 8x$.

$$0 = x^2 - 2x - 24$$ Subtract $8x + 24$ from both sides and write the quadratic equation in standard form.

$$0 = (x-6)(x+4)$$ Factor.

$$x - 6 = 0 \quad \text{or} \quad x + 4 = 0$$ Set each factor equal to 0.

$$x = 6 \qquad\qquad x = -4$$ Solve for x.

Because x represents the time for your friend to design a Web site working alone, we reject the negative value, -4. Your friend can design a Web site in 6 hours.

Step 5 Check the proposed solution in the original wording of the problem. Will you both complete the job in 4 hours? Working alone, your friend takes 6 hours. Because you require 6 more hours than your friend, you require 12 hours to complete the job on your own.

In 4 hours, your friend completes $\frac{4}{6} = \frac{2}{3}$ of the job.

In 4 hours, you complete $\frac{4}{12} = \frac{1}{3}$ of the job.

Notice that $\frac{2}{3} + \frac{1}{3} = 1$, which represents the completion of the entire job, or the design of one Web site. ■

✔ **CHECK POINT 5** An experienced carpenter can panel a room 3 times faster than an apprentice can. Working together, they can panel the room in 6 hours. How long would it take each person working alone to do the job?

EXERCISE SET 6.7

Practice Exercises

In Exercises 1–14, solve each formula for the specified variable.

1. $\frac{V_1}{V_2} = \frac{P_2}{P_1}$ for P_1 (Chemistry)

2. $\frac{V_1}{V_2} = \frac{P_2}{P_1}$ for V_2 (Chemistry)

3. $\frac{1}{p} + \frac{1}{q} = \frac{1}{f}$ for f (Optics)

4. $\frac{1}{p} + \frac{1}{q} = \frac{1}{f}$ for q (Optics)

5. $P = \frac{A}{1 + r}$ for r (Investment)

6. $S = \frac{a}{1 - r}$ for r (Mathematics)

7. $F = \frac{Gm_1m_2}{d^2}$ for m_1 (Physics)

8. $F = \frac{Gm_1m_2}{d^2}$ for m_2 (Physics)

9. $z = \frac{x - \bar{x}}{s}$ for x (Statistics)

10. $z = \frac{x - \bar{x}}{s}$ for s (Statistics)

11. $I = \frac{E}{R + r}$ for R (Electronics)

12. $I = \frac{E}{R + r}$ for r (Electronics)

13. $f = \frac{f_1f_2}{f_1 + f_2}$ for f_1 (Optics)

14. $f = \frac{f_1f_2}{f_1 + f_2}$ for f_2 (Optics)

Application Exercises

The figure shows the graph of the average cost function for the company described in Check Point 2 that manufactures wheelchairs. Use the graph to solve Exercises 15–18.

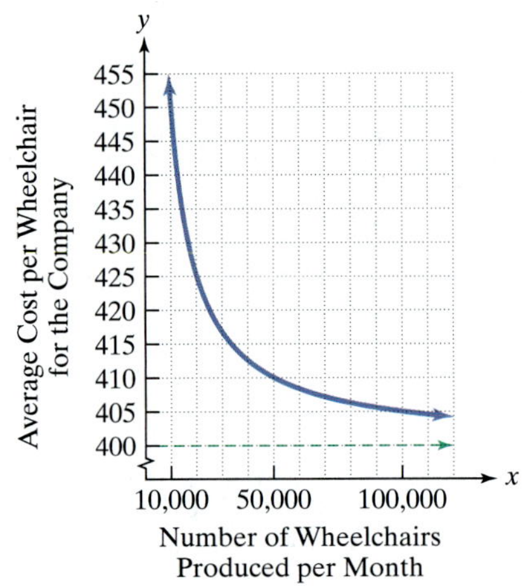

15. How many wheelchairs must be produced each month for the company to have an average cost of \$410 per chair?

16. How many wheelchairs must be produced each month for the company to have an average cost of \$425 per chair?

17. What is the horizontal asymptote of the graph? What is the meaning of the horizontal asymptote as production level increases?

18. Describe the end behavior of the graph at the far right. Is there a production level that results in an average cost of \$400 per chair? Explain your answer.

19. A company is planning to manufacture mountain bikes. Fixed monthly cost will be \$100,000 and it will cost \$100 to produce each bicycle.
    a. Write the cost function, C, of producing x mountain bikes.
    b. Write the average cost function, $\overline{C}$, of producing x mountain bikes.
    c. How many mountain bikes must be produced each month for the company to have an average cost of \$300 per bike?

20. A company is planning to manufacture small canoes. Fixed monthly cost will be $20,000 and it will cost $20 to produce each canoe.

a. Write the cost function, C, of producing x canoes.

b. Write the average cost function, $\overline{C}$, of producing x canoes.

c. How many canoes must be produced each month for the company to have an average cost of $40 per canoe?

It's vacation time. You drive 90 miles along a scenic highway and then take a 5-mile walk along a hiking trail. Your driving time is nine times that of your walking rate. The graph shows the total time you spend driving and hiking, $f(x)$, as a function of your walking rate, x. Use the graph of this rational function to solve Exercises 21–24.

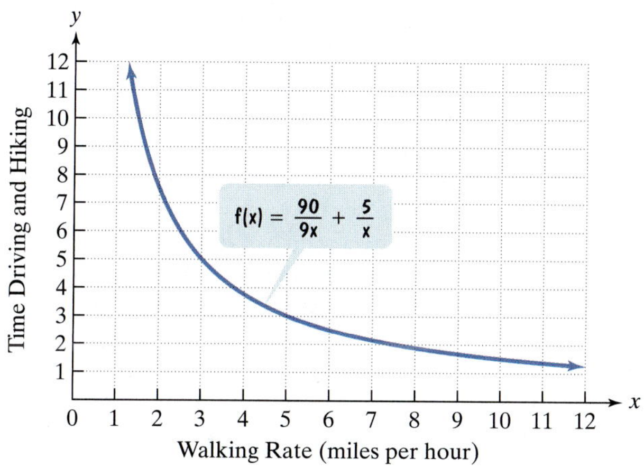

21. If the total time for driving and hiking is 3 hours, what is your walking rate?

22. If the total time for driving and hiking is 5 hours, what is your walking rate?

23. Describe the behavior of the graph as x approaches 0. What does this show about the time driving and hiking as your walking rate is close to zero miles per hour (a stupefied crawl)?

24. The graph is falling from left to right. What does this show?

Use a rational equation to solve Exercises 25–36. Each exercise is a problem involving motion.

25. You can travel 40 miles on motorcycle in the same time that it takes to travel 15 miles on bicycle. If your motorcycle's rate is 20 miles per hour faster than your bicycle's, find the average rate for each.

26. How bad is the heavy traffic? You can walk 10 miles in the same time that it takes to travel 15 miles by car. If the car's rate is 3 miles per hour faster than your walking rate, find the average rate for each.

27. You ride your bike to campus a distance of 5 miles and return home on the same route. Going to campus, you ride mostly downhill and average 9 miles per hour faster than on your return trip home. If the round trip takes one hour and ten minutes—that is $\frac{7}{6}$ hours—what is your average rate on the return trip?

28. An engine pulls a train 140 miles. Then a second engine, whose average rate is 5 miles per hour faster than the first engine, takes over and pulls the train 200 miles. The total time required for both engines is 9 hours. Find the average rate of each engine.

29. In still water, a boat averages 8 miles per hour. It takes the same amount of time to travel 15 miles downstream, with the current, as 9 miles upstream, against the current. What is the rate of the water's current?

	Distance	Rate	Time = $\frac{\text{Distance}}{\text{Rate}}$
With the current	15	$8 + x$	
Against the current	9	$8 - x$	

These times are equal.

30. In still water, a canoeist paddles at an average rate of 3 miles per hour. It takes the same amount of time to travel 10 miles downstream, with the current, as 2 miles upstream, against the current. What is the rate of the water's current?

	Distance	Rate	Time = $\frac{\text{Distance}}{\text{Rate}}$
With the current	10	$3 + x$	
Against the current	2	$3 - x$	

These times are equal.

31. The water's current is 2 miles per hour. A canoe can travel 6 miles downstream, with the current, in the same amount of time it travels 2 miles upstream, against the current. What is the canoe's average rate in still water?

32. The water's current is 2 miles per hour. A boat can travel 6 miles downstream, with the current, in the same amount of time it travels 4 miles upstream, against the current. What is the boat's average rate in still water?

33. A moving sidewalk at an airport moves at a rate of 1.8 feet per second. Walking on the moving sidewalk, you travel 100 feet forward in the same time it takes to travel 40 feet in the opposite direction. Find your walking speed on a nonmoving sidewalk.

34. A moving sidewalk at an airport moves at a rate of 1.8 feet per second. Walking on the moving sidewalk, you travel 105 feet forward in the same time it takes to travel 50 feet in the opposite direction. Find your walking speed on a nonmoving sidewalk. Round to the nearest tenth.

35. Two runners, one averaging 8 miles per hour and the other 6 miles per hour, start at the same place and run along the same trail. The slower runner arrives at the end of the trail a half hour after the faster runner. How far did each person run?

36. Two sailboats, one averaging 20 miles per hour and the other 18 miles per hour, start at the same place and follow the same course. The slower boat arrives at the end of the course $\frac{1}{6}$ of an hour after the faster sailboat. How far did each boat travel?

Use a rational equation to solve Exercises 37–48. Each exercise is a problem involving work.

37. You promised your parents that you would wash the family car. You have not started the job, and they are due home in 16 minutes. You can wash the car in 40 minutes, and your sister claims she can do it in 30 minutes. If you work together, how long will it take to do the job? Will this give you enough time before your parents return?

38. You must leave for campus in 10 minutes, or you will be late for class. Unfortunately, you are snowed in. You can shovel the driveway in 20 minutes, and your brother claims he can do it in 15 minutes. If you shovel together, how long will it take to clear the driveway? Will this give you enough time before you have to leave?

39. A pool can be filled by one pipe in 3 hours and by a second pipe in 6 hours. How long will it take using both pipes to fill the pool?

40. A pool can be filled by one pipe in 4 hours and by a second pipe in 6 hours. How long will it take using both pipes to fill the pool?

41. Working with your cousin, you can refinish a table in 3 hours. Working alone, your cousin can complete the job in 4 hours. How long would it take you to refinish the table working alone?

42. Working with your cousin, you can split a cord of firewood in 5 hours. Working alone, your cousin can complete the job in 7 hours. How long would it take you to split the firewood working alone?

43. An earthquake strikes and an isolated area is without food or water. Three crews arrive. One can dispense needed supplies in 20 hours, a second in 30 hours, and a third in 60 hours. How long will it take all three crews working together to dispense food and water?

44. A hurricane strikes, and a rural area is without food or water. Three crews arrive. One can dispense needed supplies in 10 hours, a second in 15 hours, and a third in 20 hours. How long will it take all three crews working together to dispense food and water?

45. Working together, Lou and Bud can paint a room in 6 hours. Working alone, it takes Lou 5 hours longer than Bud to do the job. How long would it take Bud to paint the room working alone?

46. Working together, Shlemiel and Shlemazal can clear a lot in 12 hours. Working alone, it takes Shlemiel 10 hours longer than Shlemazal to do the job. How long would it take Shlemazal to clear the lot working alone?

47. A faucet can fill a sink in 5 minutes. It takes twice that long for the drain to empty the sink. How long will it take to fill the sink if the drain is open and the faucet is on?

48. A pool can be filled by a pipe in 3 hours. It takes 3 times as long for another pipe to empty the pool. How long will it take to fill the pool if both pipes are open?

Writing in Mathematics

49. Without showing the details, explain how to solve the formula

$$\frac{1}{R} = \frac{1}{R_1} + \frac{1}{R_2}$$

for R_1. (The formula is used in electronics.)

50. Explain how to find the average cost function for a business.

51. How does the average cost function illustrate a problem for small businesses?

52. What is the relationship among time traveled, distance traveled, and rate of travel?

53. If you know how many hours it takes for you to do a job, explain how to find the fractional part of the job you can complete in x hours.

54. If you can do a job in 6 hours and your friend can do the same job in 3 hours, explain how to find how long it takes to complete the job working together. It is not necessary to solve the problem.

55. When two people work together to complete a job, describe one factor that can result in more or less time than the time given by the rational equations we have been using.

Technology Exercises

56. For Exercises 19–20, use a graphing utility to graph the average cost function described by the problem's conditions. Then TRACE along the curve and find the point that visually shows the solution in part (c).

57. For Exercises 45–46, use a graphing utility to graph the function representing the sum of the fractional parts of the job done by the two people. Then TRACE along the curve and find the point that visually shows the problem's solution.

58. A boat can travel 10 miles per hour in still water. The boat travels 24 miles upstream, against the current, and then 24 miles downstream, with the current.
   a. Let x = the rate of the current. Write a function in terms of x that models the total time for the boat to travel upstream and downstream.
   b. Use a graphing utility to graph the rational function in part (a).
   c. TRACE along the curve and determine the current's rate if the trip's time is 5 hours. Then verify this result algebraically.

Critical Thinking Exercises

59. Which one of the following is true?
   a. As production level increases, the average cost for a company to produce each unit of its product also increases.
   b. To solve $qf + pf = pq$ for p, subtract qf from both sides and then divide by f.
   c. If you can clean the house in 3 hours and your sloppy friend can completely mess it up in 6 hours, then $\frac{x}{3} - \frac{x}{6} = 1$ can be used to find how long it takes to clean the house if you both "work" together.
   d. None of the above is true.

60. Solve $\frac{1}{s} = f + \frac{1 - f}{p}$ for f.

61. A new schedule for a train requires it to travel 351 miles in $\frac{1}{4}$ hour less time than before. To accomplish this, the rate of the train must be increased by 2 miles per hour. What should the average rate of the train be so that it can keep on the new schedule?

62. It takes Mr. Todd 4 hours longer to prepare an order of pies than it takes Mrs. Lovett. They bake together for 2 hours when Mrs. Lovett leaves. Mr. Todd takes 7 additional hours to complete the work. Working alone, how long does it take Mrs. Lovett to prepare the pies?

Review Exercises

63. Factor: $x^2 + 4x + 4 - 9y^2$. (Section 5.6, Example 4)

64. Solve using matrices:
$$\begin{aligned} 2x + 5y &= -5 \\ x + 2y &= -1. \end{aligned}$$
(Section 3.4, Example 3)

65. Solve the system:
$$\begin{aligned} x + y + z &= 4 \\ 2x + 5y &= 1 \\ x - y - 2z &= 0. \end{aligned}$$
(Section 3.3, Example 3)

▸ SECTION 6.8 Modeling Using Variation

Objectives

1. Solve direct variation problems.
2. Solve inverse variation problems.
3. Solve combined variation problems.
4. Solve problems involving joint variation.

SSM PH Tutor Center CD-ROM Video

Have you ever wondered how telecommunication companies estimate the number of phone calls expected per day between two cities? The formula

$$N = \frac{400P_1P_2}{d^2}$$

shows that the daily number of phone calls, N, increases as the populations of the cities, P_1 and P_2, in thousands, increase and decreases as the distance, d, between the cities increases.

Certain formulas occur so frequently in applied situations that they are given special names. Variation formulas show how one quantity changes in relation to other quantities. Quantities can vary *directly, inversely*, or *jointly*. In this section, we look at situations that can be modeled by each of these kinds of variation. And think of this. The next time you get one of those "all-circuits-are-busy" messages, you will be able to use a variation formula to estimate how many other callers you're competing with for those precious 5-cent minutes.

1 Solve direct variation problems.

Direct Variation

Because light travels faster than sound, during a thunderstorm we see lightning before we hear thunder. The formula

$$d = 1080t$$

describes the distance, in feet, of the storm's center if it takes t seconds to hear thunder after seeing lightning. Thus,

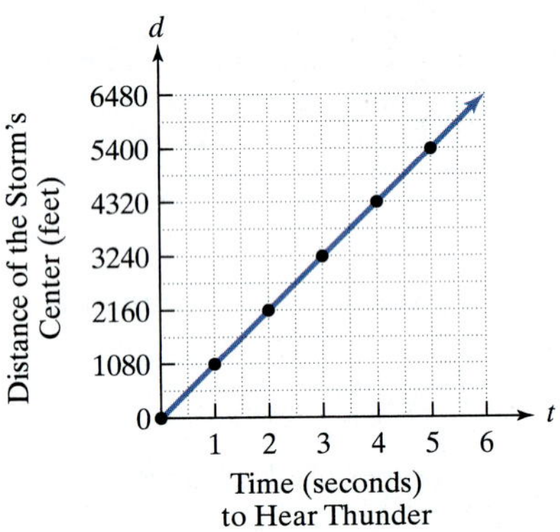

The graph of $d = 1080t$. Distance to a storm's center varies directly as the time it takes to hear thunder.

If $t = 1$, $d = 1080 \cdot 1 = 1080$. If it takes 1 second to hear thunder, the storm's center is 1080 feet away.

If $t = 2$, $d = 1080 \cdot 2 = 2160$. If it takes 2 seconds to hear thunder, the storm's center is 2160 feet away.

If $t = 3$, $d = 1080 \cdot 3 = 3240$. If it takes 3 seconds to hear thunder, the storm's center is 3240 feet away.

As the formula $d = 1080t$ illustrates, the distance to the storm's center is a constant multiple of how long it takes to hear the thunder. When the time is doubled, the storm's distance is doubled; when the time is tripled, the storm's distance is tripled; and so on. Because of this, the distance is said to **vary directly** as the time. The **equation of variation** is

$$d = 1080t.$$

Generalizing, we obtain the following statement:

Direct Variation

If a situation is described by an equation in the form

$$y = kx$$

where k is a constant, we say that **y varies directly as x.** The number k is called the **constant of variation.**

Problems involving direct variation can be solved using the following procedure. This procedure applies to direct variation problems as well as to the other kinds of variation problems that we will discuss.

Solving Variation Problems

1. Write an equation that describes the given English statement.
2. Substitute the given pair of values into the equation in step 1 and find the value of k.
3. Substitute the value of k into the equation in step 1.
4. Use the equation from step 3 to answer the problem's question.

EXAMPLE 1 Solving a Direct Variation Problem

The amount of garbage, G, varies directly as the population, P. Allegheny County, Pennsylvania, has a population of 1.3 million and creates 26 million pounds of garbage each week. Find the weekly garbage produced by New York City with a population of 7.3 million.

Solution

Step 1 Write an equation. We know that y varies directly as x is expressed as

$$y = kx.$$

By changing letters, we can write an equation that describes the following English statement: Garbage production, G, varies directly as the population, P.

$$G = kP$$

Step 2 Use the given values to find k. Allegheny County has a population of 1.3 million and creates 26 million pounds of garbage weekly. Substitute 26 for G and 1.3 for P in the direct variation equation. Then solve for k.

$$G = kP$$

$$26 = k \cdot 1.3$$

$$\frac{26}{1.3} = \frac{k \cdot 1.3}{1.3} \quad \text{Divide both sides by 1.3.}$$

$$20 = k \quad \text{Simplify.}$$

Step 3 Substitute the value of k into the equation.

$$G = kP \quad \text{Use the equation from step 1.}$$

$$G = 20P \quad \text{Replace } k\text{, the constant of variation, with 20.}$$

Step 4 Answer the problem's question. New York City has a population of 7.3 million. To find its weekly garbage production, substitute 7.3 for P in $G = 20P$ and solve for G.

$$G = 20P \quad \text{Use the equation from step 3.}$$

$$G = 20(7.3) \quad \text{Substitute 7.3 for } P.$$

$$G = 146$$

The weekly garbage produced by New York City weighs approximately 146 million pounds. ■

✔ **CHECK POINT 1** The pressure, P, of water on an object below the surface varies directly as its distance, D, below the surface. If a submarine experiences a pressure of 25 pounds per square inch 60 feet below the surface, how much pressure will it experience 330 feet below the surface?

The direct variation equation $y = kx$ is a linear function. If $k > 0$, then the slope of the line is positive. Consequently, as x increases, y also increases.

A direct variation situation can involve variables to higher powers. For example, y can vary directly as x^2 $(y = kx^2)$ or as x^3 $(y = kx^3)$.

Direct Variation with Powers

y* varies directly as the *n*th power of *x if there exists some nonzero constant k such that

$$y = kx^n.$$

Direct variation with whole number powers is modeled by polynomial functions. In our next example, the graph of the variation equation is the graph of a quadratic function.

EXAMPLE 2 Solving a Direct Variation Problem

The distance, s, that a body falls from rest varies directly as the square of the time, t, of the fall. If skydivers fall 64 feet in 2 seconds, how far will they fall in 4.5 seconds?

Solution

Step 1 Write an equation. We know that y varies directly as the square of x is expressed as

$$y = kx^2.$$

By changing letters, we can write an equation that describes the following English statement: Distance, s, varies directly as the square of time, t, of the fall.

$$s = kt^2$$

Step 2 Use the given values to find k. Skydivers fall 64 feet in 2 seconds. Substitute 64 for s and 2 for t in the direct variation equation. Then solve for k.

$$s = kt^2$$

$$64 = k \cdot 2^2$$

$$64 = 4k$$

$$\frac{64}{4} = \frac{4k}{4}$$

$$16 = k$$

Step 3 Substitute the value of k into the equation.

$$s = kt^2 \quad \text{Use the equation from step 1.}$$

$$s = 16t^2 \quad \text{Replace } k \text{, the constant of variation, with 16.}$$

Step 4 Answer the problem's question. How far will the skydivers fall in 4.5 seconds? Substitute 4.5 for t in $s = 16t^2$ and solve for s.

$$s = 16(4.5)^2 = 16(20.25) = 324$$

Thus, in 4.5 seconds, skydivers will fall 324 feet. ■

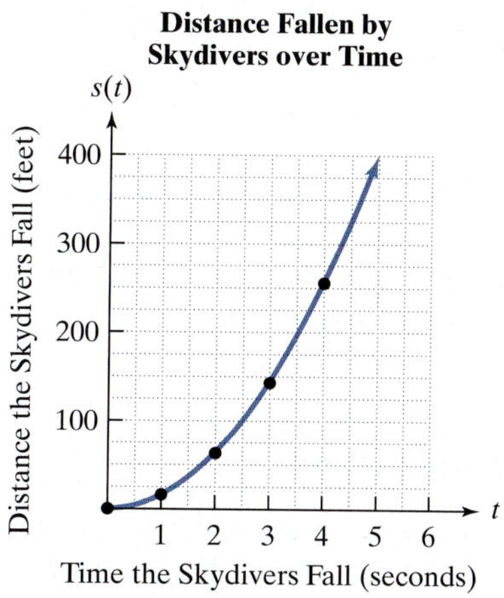

Figure 6.7 The graph of $s(t) = 16t^2$

We can express the variation equation from Example 2 in function notation, writing

$$s(t) = 16t^2.$$

The distance that a body falls from rest is a function of the time, t, of the fall. The graph of this quadratic function is shown in Figure 6.7. The graph increases rapidly from left to right, showing the effects of the acceleration of gravity.

✔ **CHECK POINT 2** The distance required to stop a car varies directly as the square of its speed. If 200 feet are required to stop a car traveling 60 miles per hour, how many feet are required to stop a car traveling 100 miles per hour?

2 Solve inverse variation problems.

Inverse Variation

The distance from Atlanta, Georgia, to Orlando, Florida, is 450 miles. The time that it takes to drive from Atlanta to Orlando depends on the rate at which one drives and is given by

$$\text{Time} = \frac{450}{\text{Rate}}.$$

For example, if you average 45 miles per hour, the time for the drive is

$$\text{Time} = \frac{450}{45} = 10,$$

or 10 hours. If you ignore speed limits and average 75 miles per hour, the time for the drive is

$$\text{Time} = \frac{450}{75} = 6,$$

or 6 hours. As your rate (or speed) increases, the time for the trip decreases and vice versa. This is illustrated by the graph in Figure 6.8.

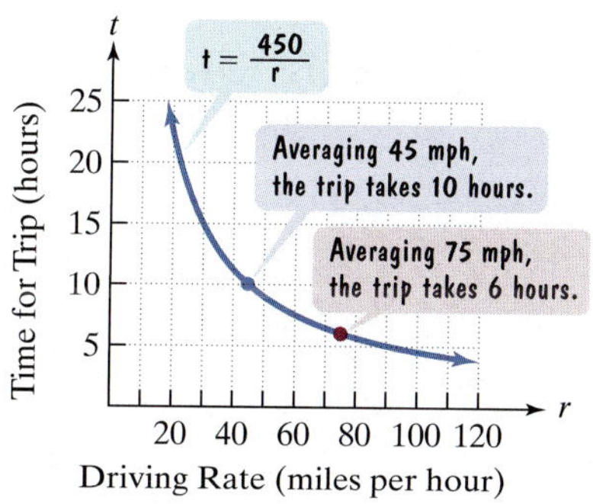

Figure 6.8

We can express the time for the Atlanta-Orlando trip using t for time and r for rate:

$$t = \frac{450}{r}.$$

This equation is an example of an **inverse variation** equation. Time, t, **varies inversely** as rate, r. When two quantities vary inversely, one quantity increases as the other decreases, and vice versa.

Generalizing, we obtain the following statement:

Inverse Variation

If a situation is described by an equation in the form

$$y = \frac{k}{x}$$

where k is a constant, we say that ***y* varies inversely as *x***. The number k is called the **constant of variation**.

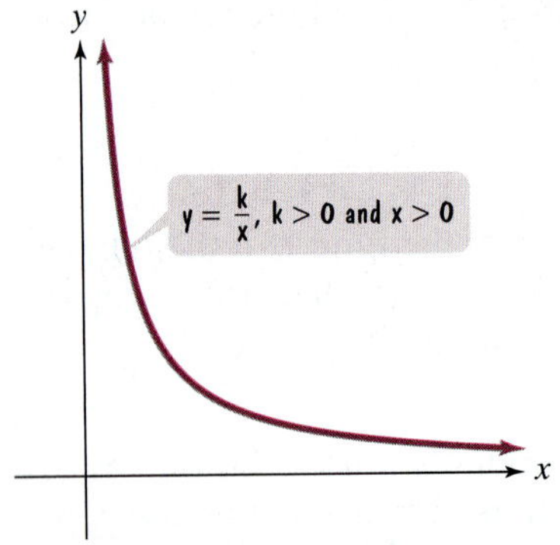

Figure 6.9 The graph of the inverse variation equation

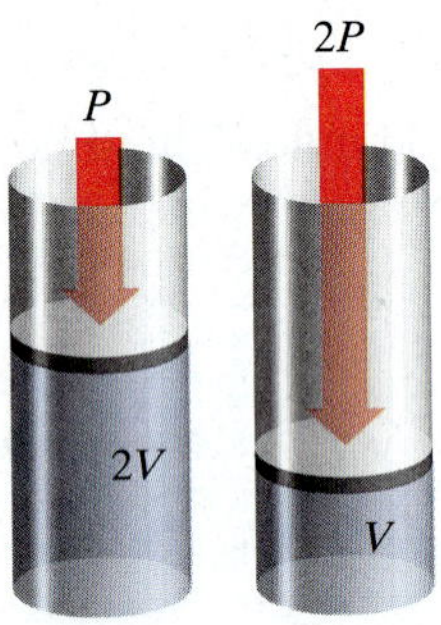

Doubling the pressure halves the volume.

Notice that the inverse variation equation

$$y = \frac{k}{x}, \quad \text{or} \quad f(x) = \frac{k}{x},$$

is a rational function. For $k > 0$ and $x > 0$, the graph of the function takes on the shape shown in Figure 6.9.

We use the same procedure to solve inverse variation problems as we did to solve direct variation problems. Example 3 illustrates this procedure.

EXAMPLE 3 Solving an Inverse Variation Problem

When you use a spray can and press the valve at the top, you decrease the pressure of the gas in the can. This decrease of pressure causes the volume of the gas in the can to increase. Because the gas needs more room than is provided in the can, it expands in spray form through the small hole near the valve. In general, if the temperature is constant, the pressure, P, of a gas in a container varies inversely as the volume, V, of the container. The pressure of a gas sample in a container whose volume is 8 cubic inches is 12 pounds per square inch. If the sample expands to a volume of 22 cubic inches, what is the new pressure of the gas?

Solution

Step 1 Write an equation. We know that y varies inversely as x is expressed as

$$y = \frac{k}{x}.$$

By changing letters, we can write an equation that describes the following English statement: The pressure, P, of a gas in a container varies inversely as the volume, V.

$$P = \frac{k}{V}.$$

Step 2 Use the given values to find k. The pressure of a gas sample in a container whose volume is 8 cubic inches is 12 pounds per square inch. Substitute 12 for P and 8 for V in the inverse variation equation. Then solve for k.

$$P = \frac{k}{V}$$

$$12 = \frac{k}{8}$$

$$12 \cdot 8 = \frac{k}{8} \cdot 8 \qquad \text{Multiply both sides by 8.}$$

$$96 = k \qquad \text{Simplify.}$$

Step 3 Substitute the value of k into the equation.

$$P = \frac{k}{V} \qquad \text{Use the equation from step 1.}$$

$$P = \frac{96}{V} \qquad \text{Replace } k\text{, the constant of variation, with 96.}$$

Step 4 **Answer the problem's question.** We need to find the pressure when the volume expands to 22 cubic inches. Substitute 22 for V and solve for P.

$$P = \frac{96}{V} = \frac{96}{22} = 4\frac{4}{11}$$

When the volume is 22 cubic inches, the pressure of the gas is $4\frac{4}{11}$ pounds per square inch. ■

CHECK POINT 3 The price, P, of oil varies inversely as the supply, S. An OPEC nation sells oil for \$19.50 per barrel when its daily production level is 4 million barrels. At what price will it sell oil if the daily production level is decreased to 3 million barrels?

3 Solve combined variation problems.

Combined Variation

In a **combined variation** situation, direct and inverse variation occur at the same time. For example, as the advertising budget, A, of a company increases, its monthly sales, S, also increase. Monthly sales vary directly as the advertising budget:

$$S = kA.$$

By contrast, as the price of the company's product, P, increases, its monthly sales, S, decrease. Monthly sales vary inversely as the price of the product:

$$S = \frac{k}{P}.$$

We can combine these two variation equations into one combined equation:

$$S = \frac{kA}{P}.$$

The following example illustrates the application of combined variation.

EXAMPLE 4 Solving a Combined Variation Problem

The owners of Rollerblades Now determine that the monthly sales, S, of its skates vary directly as its advertising budget, A, and inversely as the price of the skates, P. When \$60,000 is spent on advertising and the price of the skates is \$40, the monthly sales are 12,000 pairs of rollerblades.

a. Write an equation of variation that describes this situation.

b. Determine monthly sales if the amount of the advertising budget is increased to \$70,000.

Solution

a. Write an equation.

$$S = \frac{kA}{P}$$ Translate "sales vary directly as the advertising budget and inversely as the skates' price."

Use the given values to find k.

$$12{,}000 = \frac{k(60{,}000)}{40}$$ When \$60,000 is spent on advertising ($A = 60{,}000$) and the price is \$40 ($P = 40$), monthly sales are 12,000 units ($S = 12{,}000$).

$$12{,}000 = k \cdot 1500$$ Divide 60,000 by 40.

$$\frac{12{,}000}{1500} = \frac{k \cdot 1500}{1500}$$ Divide both sides of the equation by 1500.

$$8 = k$$ Simplify.

Therefore, the equation of variation that describes monthly sales is

$$S = \frac{8A}{P}.$$

b. The advertising budget is increased to \$70,000, so $A = 70{,}000$. The skates' price is still \$40, so $P = 40$.

$$S = \frac{8A}{P}$$ This is the equation from part (a).

$$S = \frac{8(70{,}000)}{40}$$ Substitute 70,000 for A and 40 for P.

$$S = 14{,}000$$

With a \$70,000 advertising budget and \$40 price, the company can expect to sell 14,000 pairs of rollerblades in a month (up from 12,000). ■

✔ **CHECK POINT 4** The number of minutes needed to solve an exercise set of variation problems varies directly as the number of problems and inversely as the number of people working to solve the problems. It takes 4 people 32 minutes to solve 16 problems. How many minutes will it take 8 people to solve 24 problems?

4 Solve problems involving joint variation.

Joint Variation

Joint variation is a variation in which a variable varies directly as the product of two or more other variables. Thus, the equation $y = kxz$ is read "y varies jointly as x and z."

Joint variation plays a critical role in Isaac Newton's formula for gravitation:

$$F = G\frac{m_1m_2}{d^2}.$$

The formula states that the force of gravitation, F, between two bodies varies jointly as the product of their masses, m_1 and m_2, and inversely as the square of the distance between them, d. (G is the gravitational constant.) The formula

indicates that gravitational force exists between any two objects in the universe, increasing as the distance between the bodies decreases. One practical result is that the pull of the moon on the oceans is greater on the side of Earth closer to the moon. This gravitational imbalance is what produces tides.

EXAMPLE 5 Modeling Centrifugal Force

The centrifugal force, C, of a body moving in a circle varies jointly with the radius of the circular path, r, and the body's mass, m, and inversely with the square of the time, t, it takes to move about one full circle. A 6-gram body moving in a circle with radius 100 centimeters at a rate of 1 revolution in 2 seconds has a centrifugal force of 6000 dynes. Find the centrifugal force of an 18-gram body moving in a circle with radius 100 centimeters at a rate of 1 revolution in 3 seconds.

Solution

$$C = \frac{krm}{t^2}$$ Translate "Centrifugal force, C, varies jointly with radius, r, and mass, m, and inversely with the square of time, t."

$$6000 = \frac{k(100)(6)}{2^2}$$ If $r = 100$, $m = 6$, and $t = 2$, then $C = 6000$.

$$6000 = 150k$$ Simplify.

$$40 = k$$ Divide both sides by 150 and solve for k.

$$C = \frac{40rm}{t^2}$$ Substitute 40 for k in the model for centrifugal force.

$$= \frac{40(100)(18)}{3^2}$$ Find C when $r = 100$, $m = 18$, and $t = 3$.

$$= 8000$$ Simplify.

The centrifugal force is 8000 dynes. ■

CHECK POINT 5 The volume of a cone, V, varies jointly as its height, h, and the square of its radius r. A cone with a radius measuring 6 feet and a height measuring 10 feet has a volume of 120π cubic feet. Find the volume of a cone having a radius of 12 feet and a height of 2 feet.

EXERCISE SET 6.8

Practice Exercises

Use the four-step procedure for solving variation problems given on page 441 to solve Exercises 1–8.

1. y varies directly as x. $y = 35$ when $x = 5$. Find y when $x = 12$.

2. y varies directly as x. $y = 55$ when $x = 5$. Find y when $x = 13$.

3. y varies inversely as x. $y = 10$ when $x = 5$. Find y when $x = 2$.

4. y varies inversely as x. $y = 5$ when $x = 3$. Find y when $x = 9$.

5. y varies directly as x and inversely as the square of z. $y = 20$ when $x = 50$ and $z = 5$. Find y when $x = 3$ and $z = 6$.

6. a varies directly as b and inversely as the square of c. $a = 7$ when $b = 9$ and $c = 6$. Find a when $b = 4$ and $c = 8$.

7. y varies jointly as x and z. $y = 25$ when $x = 2$ and $z = 5$. Find y when $x = 8$ and $z = 12$.

8. C varies jointly as A and T. $C = 175$ when $A = 2100$ and $T = 4$. Find C when $A = 2400$ and $T = 6$.

Application Exercises

9. A person's fingernail growth, G, in inches, varies directly as the number of weeks it has been growing, W.

a. Write an equation that expresses this relationship.

b. Fingernails grow at a rate of about 0.02 inch per week. Substitute 0.02 for k, the constant of variation, in the equation in part (a) and write the equation for fingernail growth.

c. Substitute 52 for W to determine your fingernail growth at the end of one year if for some bizarre reason you decided not to cut them and they did not break.

10. A person's wages, W, vary directly as the number of hours worked, h.

a. Write an equation that expresses this relationship.

b. For a 40-hour work week, Gloria earned \$1400. Substitute 1400 for W and 40 for h in the equation from part (a) and find k, the constant of variation.

c. Substitute the value of k into your equation in part (a) and write the equation that describes Gloria's wages in terms of the number of hours she works.

d. Use the equation from part (c) to find Gloria's wages for 25 hours of work.

Use the four-step procedure for solving variation problems given on page 441 to solve Exercises 11–29.

11. The cost, C, of an airplane ticket varies directly as the number of miles, M, in the trip. A 3000-mile trip costs \$400. What is the cost of a 450-mile trip?

12. An object's weight on the moon, M, varies directly as its weight on Earth, E. A person who weighs 55 kilograms on Earth weighs 8.8 kilograms on the moon. What is the moon weight of a person who weighs 90 kilograms on Earth?

13. The Mach number is a measurement of speed named after the man who suggested it, Ernst Mach (1838–1916). The speed of an aircraft varies directly as its Mach number. Shown here are two aircraft. Use the figures for the Concorde to determine the Blackbird's speed.

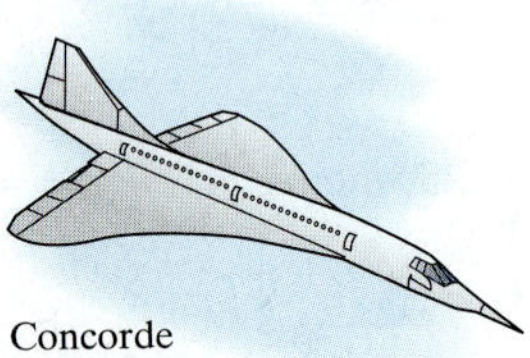

Concorde
Mach 2.03
Speed = 1502.2 miles per hour

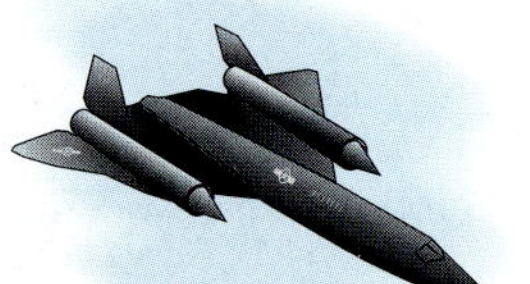

SR-71 Blackbird
Mach 3.3
Speed = ?

14. Do you still own records, or are you strictly a CD person? Record owners claim that the quality of sound on good vinyl surpasses that of a CD, although this is up for debate. This, however, is not debatable: The number of revolutions a record makes as it is being played varies directly as the time that it is on the turntable. A record that lasted 3 minutes made 135 revolutions. If a record takes 2.4 minutes to play, how many revolutions does it make?

15. If all men had identical body types, their weight would vary directly as the cube of their height. Shown is Robert Wadlow, who reached a record height of 8 feet 11 inches (107 inches) before his death at age 22. If a man who is 5 feet 10 inches tall (70 inches) with the same body type as Mr. Wadlow weighs 170 pounds, what was Robert Wadlow's weight shortly before his death?

16. The distance that an object falls varies directly as the square of the time it has been falling. An object falls 144 feet in 3 seconds. Find how far it will fall in 7 seconds.

17. The time that it takes you to get to campus varies inversely as your driving rate. Averaging 20 miles per hour in terrible traffic, it takes you 1.5 hours to get to campus. How long would the trip take averaging 60 miles per hour?

18. The weight that can be supported by a 2-inch by 4-inch piece of pine (called a 2-by-4) varies inversely as its length. A 10-foot 2-by-4 can support 500 pounds. What weight can be supported by a 5-foot 2-by-4?

19. The volume of a gas in a container at a constant temperature varies inversely as the pressure. If the volume is 32 cubic centimeters at a pressure of 8 pounds, find the pressure when the volume is 40 cubic centimeters.

20. The current in a circuit varies inversely as the resistance. The current is 20 amperes when the resistance is 5 ohms. Find the current for a resistance of 16 ohms.

21. A person's body-mass index is used to assess levels of fatness, with an index from 20 to 26 considered in the desirable range. The index varies directly as one's weight, in pounds, and inversely as one's height, in inches. A person who weighs 150 pounds and is 70 inches tall has an index of 21. What is the body-mass index of a person who weighs 240 pounds and is 74 inches tall? Because the index is rounded to the nearest whole number, do so and then determine if this person's level of fatness is in the desirable range.

22. The volume of a gas varies directly as its temperature and inversely as its pressure. At a temperature of 100 Kelvin and a pressure of 15 kilograms per square meter, the gas occupies a volume of 20 cubic meters. Find the volume at a temperature of 150 Kelvin and a pressure of 30 kilograms per square meter.

23. The intensity of illumination on a surface varies inversely as the square of the distance of the light source from the surface. The illumination from a source is 25 foot-candles at a distance of 4 feet. What is the illumination when the distance is 6 feet?

24. The gravitational force with which Earth attracts an object varies inversely as the square of the distance from the center of Earth. A gravitational force of 160 pounds acts on an object 400 miles from Earth's center. Find the force of attraction on an object 6000 miles from the center of Earth.

25. Kinetic energy varies jointly as the mass and the square of the velocity. A mass of 8 grams and velocity of 3 centimeters per second has a kinetic energy of 36 ergs. Find the kinetic energy for a mass of 4 grams and velocity of 6 centimeters per second.

26. The electrical resistance of a wire varies directly as its length and inversely as the square of its diameter. A wire of 720 feet with $\frac{1}{4}$-inch diameter has a resistance of $1\frac{1}{2}$ ohms. Find the resistance for 960 feet of the same kind of wire if its diameter is doubled.

27. The average number of phone calls between two cities in a day varies jointly as the product of their populations and inversely as the square of the distance between them. The population of Minneapolis is 2538 thousand, and the population of Cincinnati is 1818 thousand. The average number of telephone calls per day between the two cities, which are 608 miles apart, is 158,233. Find the average number of telephone calls per day between Orlando, Florida (population 1225 thousand) and Seattle, Washington (population 2970 thousand), two cities that are 3403 miles apart.

28. The force of attraction between two bodies varies jointly as the product of their masses and inversely as the square of the distance between them. Two 1-kilogram masses separated by 1 meter exert a force of attraction of 6.67×10^{-11} newton. What is the gravitational force exerted by Earth on a 1000-kilogram satellite orbiting at an altitude of 300 kilometers? (The mass of Earth is 5.98×10^{24} kilograms and its radius is 6400 kilometers.. Consequently, the distance between Earth and the satellite is $6400 + 300$, or 6700 kilometers.)

29. The force of wind blowing on a window positioned at a right angle to the direction of the wind varies jointly as the area of the window and the square of the wind's speed. It is known that a wind of 30 miles per hour blowing on a window measuring 4 feet by 5 feet exerts a force of 150 pounds. During a storm with winds of 60 miles per hour, should hurricane shutters be placed on a window that measures 3 feet by 4 feet and is capable of withstanding 300 pounds of force?

30. The table of values shows the values for the current, I, in an electric circuit and the resistance, R, of the circuit.

***I* (amperes)**	0.5	1.0	1.5	2.0	2.5	3.0	4.0	5.0
***R* (ohms)**	12	6.0	4.0	3.0	2.4	2.0	1.5	1.2

a. Graph the ordered pairs in the table of values, with values of I along the x-axis and values of R along the y-axis. Connect the eight points with a smooth curve.

b. Does current vary directly or inversely as resistance? Use your graph and explain how you arrived at your answer.

c. Write an equation of variation for I and R, using one of the ordered pairs in the table to find the constant of variation. Then use your variation equation to verify the other seven ordered pairs in the table.

Writing in Mathematics

31. What does it mean if two quantities vary directly?

32. In your own words, explain how to solve a variation problem.

33. What does it mean if two quantities vary inversely?

34. Explain what is meant by combined variation. Give an example with your explanation.

35. Explain what is meant by joint variation. Give an example with your explanation.

In Exercises 36–37, describe in words the variation shown by the given equation.

36. $z = \frac{k\sqrt{x}}{y^2}$

37. $z = kx^2\sqrt{y}$

38. We have seen that the daily number of phone calls between two cities varies jointly as their populations and inversely as the square of the distance between them. This model, used by telecommunication companies to estimate the line capacities needed among various cities, is called the *gravity model*. Compare the model to Newton's formula for gravitation on page 446 and describe why the name *gravity model* is appropriate.

Technology Exercise

39. Use a graphing utility to graph any three of the variation equations in Exercises 11–20. Then TRACE along each curve and identify the point that corresponds to the problem's solution.

Critical Thinking Exercises

40. In a hurricane, the wind pressure varies directly as the square of the wind velocity. If wind pressure is a measure of a hurricane's destructive capacity, what happens to this destructive power when the wind speed doubles?

41. The illumination from a light source varies inversely as the square of the distance from the light source. If you raise a lamp from 15 inches to 30 inches over your desk, what happens to the illumination?

42. The heat generated by a stove element varies directly as the square of the voltage and inversely as the resistance. If the voltage remains constant, what needs to be done to triple the amount of heat generated?

43. Galileo's telescope brought about revolutionary changes in astronomy. A comparable leap in our ability to observe the universe took place as a result of the Hubble Space Telescope. The space telescope can see stars and galaxies whose brightness is $\frac{1}{50}$ of the faintest objects now observable using ground-based telescopes. Use the fact that the brightness of a point source, such as a star, varies inversely as the square of its distance from an observer to show that the space telescope can see about seven times farther than a ground-based telescope.

Review Exercises

44. Evaluate:

$$\begin{vmatrix} -1 & 2 \\ 3 & -4 \end{vmatrix}.$$

(Section 3.5, Example 1)

45. Factor completely:

$$x^2y - 9y - 3x^2 + 27.$$

(Section 5.6, Example 3)

46. Find the degree of

$$7xy + x^2y^2 - 5x^3 - 7.$$

(Section 5.1, Example 1)

CHAPTER 6 GROUP PROJECTS

1. A cost-benefit analysis compares the estimated costs of a project with the benefits that will be achieved. Costs and benefits are given monetary values and compared using a benefit-cost ratio. As shown in the figure, a favorable ratio for a project means that the benefits outweigh the costs, and the project is cost-effective. As a group, select an environmental project that was implemented in your area of the country. Research the cost and benefit graphs that resulted in the implementation of this project. How were the benefits converted into monetary terms? Is there an equation for either the cost model or the benefit model? Group members may need to interview members of environmental groups and businesses that were part of this project. You may wish to consult an environmental science textbook to find out more about cost-benefit analyses. After doing your research, the group should write or present a report explaining why the cost-benefit analysis resulted in the project's implementation.

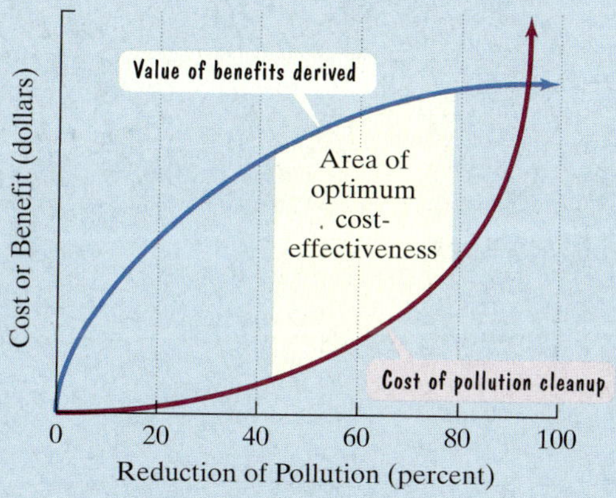

2. Group members are to write a helpful list of items for a pamphlet called "Errors That Occur when Working with Rational Expressions and How to Avoid Them." This pamphlet will be used by students who use the math lab because they are having problems simplifying, multiplying, dividing, adding, and subtracting rational expressions, as well as simplifying complex rational expressions and solving rational equations. What helpful guidelines for avoiding errors with rational expressions can you offer from the perspective of a student that you won't find in math books?
3. Begin by deciding on a product that interests the group because you are now in charge of advertising this product. Members were told that the demand for the product varies directly as the amount spent on advertising and inversely as the price of the product. However, as more money is spent on advertising, the price of your product rises. Under what conditions would members recommend an increased expense in advertising? Once you've determined what your product is, write formulas for the given conditions and experiment with hypothetical numbers. What other factors might you take into consideration in terms of your recommendation? How do these factors affect the demand for your product?

CHAPTER SUMMARY, REVIEW, AND TEST

SUMMARY

DEFINITIONS AND CONCEPTS	EXAMPLES

Section 6.1 Rational Expressions and Functions: Multiplying and Dividing

A rational expression consists of a polynomial divided by a nonzero polynomial. A rational function is defined by a formula that is a rational expression. The domain of a rational function is the set of all real numbers except those for which the denominator is zero. Graphs of rational functions often contain disconnected branches. The graphs often approach, but do not touch, vertical or horizontal lines, called vertical asymptotes and horizontal asymptotes, respectively.

Let $f(x) = \dfrac{x+2}{x^2+x-6}$. Find the domain of f.

$$x^2 + x - 6 = 0$$
$$(x+3)(x-2) = 0$$
$$x + 3 = 0 \quad \text{or} \quad x - 2 = 0$$
$$x = -3 \qquad\qquad x = 2$$

These values make the denominator zero. Exclude them from domain.
Domain of $f = (-\infty, -3) \cup (-3, 2) \cup (2, \infty)$.
The branches of the graph of f break at -3 and 2.

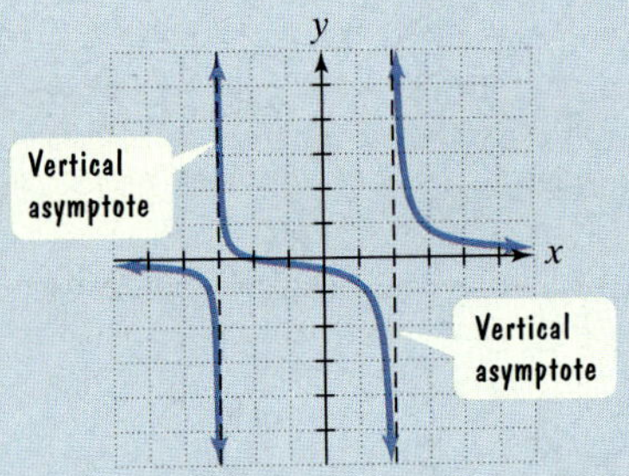

Simplifying Rational Expressions

1. Factor the numerator and denominator completely.
2. Divide both the numerator and denominator by the common factors.

Simplify: $\dfrac{3x+18}{x^2-36}$.

$$\frac{3x+18}{x^2-36} = \frac{3\cancel{(x+6)}}{\cancel{(x+6)}(x-6)} = \frac{3}{x-6}$$

Multiplying Rational Expressions

1. Factor completely.
2. Divide numerators and denominators by common factors.
3. Multiply remaining factors in the numerators and the denominators.

$$\frac{x^2+3x-10}{x^2-2x} \cdot \frac{x^2}{x^2-25}$$
$$= \frac{\cancel{(x+5)}\cancel{(x-2)}}{\cancel{x}\cancel{(x-2)}} \cdot \frac{\cancel{x} \cdot x}{\cancel{(x+5)}(x-5)}$$
$$= \frac{x}{x-5}$$

Dividing Rational Expressions
Invert the divisor and multiply.

$$\frac{3y+3}{(y+2)^2} \div \frac{y^2-1}{y+2}$$
$$= \frac{3y+3}{(y+2)^2} \cdot \frac{y+2}{y^2-1} = \frac{3\cancel{(y+1)}}{(y+2)\cancel{(y+2)}} \cdot \frac{\cancel{(y+2)}}{\cancel{(y+1)}(y-1)}$$
$$= \frac{3}{(y+2)(y-1)}$$

DEFINITIONS AND CONCEPTS	EXAMPLES

Section 6.2 Adding and Subtracting Rational Expressions

Adding or Subtracting Rational Expressions

If the denominators are the same, add or subtract the numerators. Place the result over the common denominator. If the denominators are different, write all rational expressions with the least common denominator. The LCD is a polynomial of least degree whose factors include all the factors of the denominators. Once all rational expressions are written in terms of the LCD, add or subtract as described above. Simplify the result, if possible.

$$\frac{x+1}{2x-2}-\frac{2x}{x^2+2x-3}$$

$$=\frac{x+1}{2(x-1)}-\frac{2x}{(x-1)(x+3)} \quad \text{LCD is } 2(x-1)(x+3).$$

$$=\frac{(x+1)(x+3)}{2(x-1)(x+3)}-\frac{2x\cdot 2}{2(x-1)(x+3)}$$

$$=\frac{x^2+4x+3-4x}{2(x-1)(x+3)}=\frac{x^2+3}{2(x-1)(x+3)}$$

Section 6.3 Complex Rational Expressions

Complex rational expressions have numerators or denominators containing one or more rational expressions. Complex rational expressions can be simplified by multiplying the numerator and denominator by the LCD. They can also be simplified by obtaining single expressions in the numerator and denominator and then dividing.

Simplify: $\dfrac{\frac{x+3}{x}}{x-\frac{9}{x}}$.

Multiplying by the LCD, x:

$$\frac{x}{x}\cdot\frac{\left(\frac{x+3}{x}\right)}{\left(x-\frac{9}{x}\right)}=\frac{x+3}{x^2-9}=\frac{\cancel{(x+3)}}{\cancel{(x+3)}(x-3)}=\frac{1}{x-3}.$$

Simplifying by dividing:

$$\frac{\frac{x+3}{x}}{x-\frac{9}{x}}=\frac{\frac{x+3}{x}}{\frac{x}{1}\cdot\frac{x}{x}-\frac{9}{x}}=\frac{\frac{x+3}{x}}{\frac{x^2-9}{x}}=\frac{x+3}{x}\cdot\frac{x}{x^2-9}$$

LCD is x.

$$=\frac{\cancel{x+3}}{\cancel{x}}\cdot\frac{\cancel{x}}{\cancel{(x+3)}(x-3)}=\frac{1}{x-3}.$$

Section 6.4 Division of Polynomials

To divide a polynomial by a monomial, divide each term of the polynomial by the monomial.

$$\frac{9x^3y^4-12x^3y^2-7x^2y^2}{3xy^2}$$

$$=\frac{9x^3y^4}{3xy^2}-\frac{12x^3y^2}{3xy^2}-\frac{7x^2y^2}{3xy^2}=3x^2y^2-4x^2-\frac{7x}{3}$$

To divide by a polynomial containing more than one term, use long division. If necessary, arrange the dividend in descending powers of the variable. If a power of a variable is missing, add that power with a coefficient of 0. Repeat the four steps of the long division process—divide, multiply, subtract, bring down the next term—until the degree of the remainder is less than the degree of the divisor.

Divide: $(2x^3-x^2-7)\div(x-2)$.

```
                2x² + 3x +  6
        x - 2 ) 2x³ -  x² + 0x -  7
                2x³ - 4x²
                      3x² + 0x
                      3x² - 6x
                            6x -  7
                            6x - 12
                                  5
```

The answer is $2x^2+3x+6+\dfrac{5}{x-2}$.

DEFINITIONS AND CONCEPTS	EXAMPLES

Section 6.5 Synthetic Division and the Remainder Theorem

A shortcut to long division, called synthetic division, can be used to divide a polynomial by a binomial of the form $x - c$.

Here is the division problem shown on page 453 using synthetic division.

$$(2x^3 - x^2 - 7) \div (x - 2)$$

Coefficients of the dividend, $2x^3 - x^2 + 0x - 7$

This is c in $x - c$. For $x - 2$, c is 2.

$$\begin{array}{r|rrrr} 2 & 2 & -1 & 0 & -7 \\ & \downarrow & 4 & 6 & 12 \\ \hline & 2 & 3 & 6 & 5 \end{array}$$

Coefficients of quotient

Remainder

The answer is $2x^2 + 3x + 6 + \dfrac{5}{x - 2}$.

The Remainder Theorem

If the polynomial $f(x)$ is divided by $x - c$, then the remainder is $f(c)$. This can be used to evaluate a polynomial function at c. Rather than substituting c for x, divide the function by $x - c$. The remainder is $f(c)$.

If $f(x) = x^4 - 3x^2 - 2x + 5$, use the Remainder Theorem to find $f(-2)$.

$$\begin{array}{r|rrrrr} -2 & 1 & 0 & -3 & -2 & 5 \\ & & -2 & 4 & -2 & 8 \\ \hline & 1 & -2 & 1 & -4 & 13 \end{array}$$

Remainder

Thus, $f(-2) = 13$.

Section 6.6 Rational Equations

A rational equation is an equation containing one or more rational expressions.

Solving Rational Equations

1. List restrictions on the variable.
2. Clear fractions by multiplying both sides by the LCD.
3. Solve the resulting equations.
4. Reject any proposed solution in the list of restrictions. Check other proposed solutions in the original equation.

Solve: $\dfrac{7x}{x^2 - 4} + \dfrac{5}{x - 2} = \dfrac{2x}{x^2 - 4}$.

$$\frac{7x}{(x + 2)(x - 2)} + \frac{5}{x - 2} = \frac{2x}{(x + 2)(x - 2)}$$

Denominators would equal 0 if $x = -2$ or $x = 2$. Restrictions: $x \neq -2$ and $x \neq 2$.

LCD is $(x + 2)(x - 2)$.

$$(x + 2)(x - 2)\left[\frac{7x}{(x + 2)(x - 2)} + \frac{5}{x - 2}\right] = (x + 2)(x - 2) \cdot \frac{2x}{(x + 2)(x - 2)}$$

$$7x + 5(x + 2) = 2x$$
$$7x + 5x + 10 = 2x$$
$$12x + 10 = 2x$$
$$10 = -10x$$
$$-1 = x$$

The proposed solution, -1, is not part of the restriction $x \neq -2$ and $x \neq 2$. It checks. The solution is -1 and the solution set is $\{-1\}$.

DEFINITIONS AND CONCEPTS	EXAMPLES

Section 6.7 Formulas and Applications of Rational Equations

To solve for a variable in a formula, get the specified variable alone on one side of the formula. When working with formulas containing rational expressions, it is sometimes necessary to factor out the variable you are solving for.

Solve $\frac{e}{E} = \frac{r}{r+R}$ for r.

$$E(r+R)\cdot\frac{e}{E} = E(r+R)\cdot\frac{r}{r+R}$$

LCD is $E(r+R)$.

$$e(r+R) = Er$$
$$er + eR = Er$$
$$eR = Er - er$$
$$eR = (E-e)r$$
$$\frac{eR}{E-e} = r$$

The average cost function, $\bar{C}$, for a business is a rational function representing the average cost for the company to produce each unit of its product. The function consists of the sum of fixed and variable costs divided by the number of units produced. As production level increases, the average cost to produce each unit of a product decreases.

A company has a fixed cost of \$80,000 monthly and it costs \$20 to produce each unit of its product. Its cost function, C, of producing x units is

$$C(x) = 80{,}000 + 20x.$$

Its average cost function, $\bar{C}$, is

$$\bar{C}(x) = \frac{80{,}000 + 20x}{x}.$$

Problems involving time in motion and work translate into rational equations. Motion problems involving time are solved using

$$\text{Time traveled} = \frac{\text{Distance traveled}}{\text{Rate of travel}}.$$

Work problems are solved using the following condition:

Fraction of job done by first + fraction of job done by second = 1.

It takes a cyclist who averages 16 miles per hour in still air the same time to travel 48 miles with the wind as 16 miles against the wind. What is the wind's rate?

x = wind's rate
$16 + x$ = cyclist's rate with wind
$16 - x$ = cyclist's rate against wind

	Distance	Rate	Time = $\frac{\text{Distance}}{\text{Rate}}$
With wind	48	$16 + x$	$\frac{48}{16+x}$
Against wind	16	$16 - x$	$\frac{16}{16-x}$

These times are equal.

$$(16+x)(16-x)\cdot\frac{48}{16+x} = \frac{16}{16-x}\cdot(16+x)(16-x)$$
$$48(16-x) = 16(16+x)$$

Solving this equation, $x = 8$. The wind's rate is 8 miles per hour.

DEFINITIONS AND CONCEPTS	EXAMPLES
Section 6.8 Modeling Using Variation	
English Statement / ***Equation*** y varies directly as x. $y = kx$ y varies directly as x^n. $y = kx^n$ y varies inversely as x. $y = \frac{k}{x}$ y varies inversely as x^n. $y = \frac{k}{x^n}$ y varies jointly as x and z. $y = kxz$	The time that it takes you to get to campus varies inversely as your driving rate. Averaging 20 miles per hour in terrible traffic, it takes you 1.5 hours to get to campus. How long would the trip take averaging 60 miles per hour? **1.** $t = \frac{k}{r}$ **2.** It takes 1.5 hours at 20 miles per hour: $1.5 = \frac{k}{20}$, $k = 1.5(20) = 30.$ **3.** $t = \frac{30}{r}$ **4.** How long at 60 miles per hour? $t = \frac{30}{60} = \frac{1}{2}$ hour

Review Exercises

6.1

1. If $f(x) = \dfrac{x^2 + 2x - 3}{x^2 - 4}$

find the following function values. If a function value does not exist, so state.

a. $f(4)$ **b.** $f(0)$

c. $f(-2)$ **d.** $f(-3)$

In Exercises 2–3, find the domain of the given rational function.

2. $f(x) = \dfrac{x - 6}{(x - 3)(x + 4)}$

3. $f(x) = \dfrac{x + 2}{x^2 + x - 2}$

In Exercises 4–8, simplify each rational expression. If the rational expression cannot be simplified, so state.

4. $\dfrac{5x^3 - 35x}{15x^2}$

5. $\dfrac{x^2 + 6x - 7}{x^2 - 49}$

6. $\dfrac{6x^2 + 7x + 2}{2x^2 - 9x - 5}$

7. $\dfrac{x^2 + 4}{x^2 - 4}$

8. $\dfrac{x^3 - 8}{x^2 - 4}$

In Exercises 9–14, multiply or divide as indicated.

9. $\dfrac{5x^2 - 5}{3x + 12} \cdot \dfrac{x + 4}{x - 1}$

10. $\dfrac{2x + 5}{4x^2 + 8x - 5} \cdot \dfrac{4x^2 - 4x + 1}{x + 1}$

11. $\dfrac{x^2 - 9x + 14}{x^3 + 2x^2} \cdot \dfrac{x^2 - 4}{x^2 - 4x + 4}$

12. $\dfrac{1}{x^2 + 8x + 15} \div \dfrac{3}{x + 5}$

13. $\dfrac{x^2 + 16x + 64}{2x^2 - 128} \div \dfrac{x^2 + 10x + 16}{x^2 - 6x - 16}$

14. $\dfrac{y^2 - 16}{y^3 - 64} \div \dfrac{y^2 - 3y - 18}{y^2 + 5y + 6}$

15. A number of deer are placed into a newly acquired habitat. The deer population over time is modeled by a rational function whose graph is shown in the figure. Use the graph to answer each of the following questions.
   a. How many deer were introduced into the habitat?
   b. What is the population after 10 years?
   c. What is the equation of the horizontal asymptote shown in the figure? What does this mean in terms of the deer population?

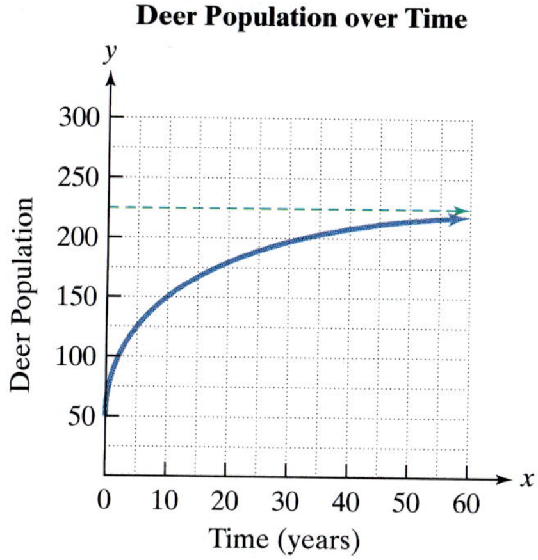

6.2 *In Exercises 16–18, add or subtract as indicated. Simplify the result, if possible.*

16. $\dfrac{4x+1}{3x-1} + \dfrac{8x-5}{3x-1}$

17. $\dfrac{2x-7}{x^2-9} - \dfrac{x-4}{x^2-9}$

18. $\dfrac{4x^2-11x+4}{x-3} - \dfrac{x^2-4x+10}{x-3}$

In Exercises 19–20, find the least common denominator of the rational expressions.

19. $\dfrac{7}{9x^3}$ and $\dfrac{5}{12x}$

20. $\dfrac{x+7}{x^2+2x-35}$ and $\dfrac{x}{x^2+9x+14}$

In Exercises 21–27, perform the indicated operations. Simplify the result, if possible.

21. $\dfrac{1}{x} + \dfrac{2}{x-5}$

22. $\dfrac{2}{x^2-5x+6} + \dfrac{3}{x^2-x-6}$

23. $\dfrac{x-3}{x^2-8x+15} + \dfrac{x+2}{x^2-x-6}$

24. $\dfrac{3x^2}{9x^2-16} - \dfrac{x}{3x+4}$

25. $\dfrac{y}{y^2+5y+6} - \dfrac{2}{y^2+3y+2}$

26. $\dfrac{x}{x+3} + \dfrac{x}{x-3} - \dfrac{9}{x^2-9}$

27. $\dfrac{3x^2}{x-y} + \dfrac{3y^2}{y-x}$

6.3 *In Exercises 28–32, simplify each complex rational expression.*

28. $\dfrac{\frac{3}{x}-3}{\frac{8}{x}-8}$

29. $\dfrac{\frac{5}{x}+1}{1-\frac{25}{x^2}}$

30. $\dfrac{3-\frac{1}{x+3}}{3+\frac{1}{x+3}}$

31. $\dfrac{\frac{4}{x+3}}{\frac{2}{x-2}-\frac{1}{x^2+x-6}}$

32. $\dfrac{x^{-2}+x^{-1}}{x^{-2}-x^{-1}}$

6.4 *In Exercises 33–34, divide the polynomial by the monomial.*

33. $(15x^3-30x^2+10x-2)\div(5x^2)$

34. $(36x^4y^3+12x^2y^3-60x^2y^2)\div(6xy^2)$

In Exercises 35–38, divide as indicated.

35. $(6x^2-5x+5)\div(2x+3)$

36. $(10x^3-26x^2+17x-13)\div(5x-3)$

37. $(x^6 + 3x^5 - 2x^4 + x^2 - 3x + 2) \div (x - 2)$

38. $(4x^4 + 6x^3 + 3x - 1) \div (2x^2 + 1)$

6.5 *In Exercises 39–41, divide using synthetic division.*

39. $(4x^3 - 3x^2 - 2x + 1) \div (x + 1)$

40. $(3x^4 - 2x^2 - 10x - 20) \div (x - 2)$

41. $(x^4 + 16) \div (x + 4)$

In Exercises 42–43, use synthetic division and the Remainder Theorem to find the indicated function value.

42. $f(x) = 2x^3 - 5x^2 + 4x - 1;\quad f(2)$

43. $f(x) = 3x^4 + 6x^3 - 2x + 4;\quad f(-4)$

In Exercises 44–45, use synthetic division to determine whether or not the number given to the right of each equation is a solution of the equation.

44. $2x^3 - x^2 - 8x + 4 = 0;\quad -2$

45. $x^4 - x^3 - 7x^2 + x + 6 = 0;\quad 4$

46. Use synthetic division to show that $\frac{1}{2}$ is a solution of

$$6x^3 + x^2 - 4x + 1 = 0.$$

Then solve the polynomial equation.

6.6 *In Exercises 46–52, solve each rational equation. If an equation has no solution, so state.*

47. $\frac{3}{x} + \frac{1}{3} = \frac{5}{x}$

48. $\frac{5}{3x + 4} = \frac{3}{2x - 8}$

49. $\frac{1}{x - 5} - \frac{3}{x + 5} = \frac{6}{x^2 - 25}$

50. $\frac{x + 5}{x + 1} - \frac{x}{x + 2} = \frac{4x + 1}{x^2 + 3x + 2}$

51. $\frac{2}{3} - \frac{5}{3x} = \frac{1}{x^2}$

52. $\frac{2}{x - 1} = \frac{1}{4} + \frac{7}{x + 2}$

53. $\frac{2x + 7}{x + 5} - \frac{x - 8}{x - 4} = \frac{x + 18}{x^2 + x - 20}$

54. The function

$$f(x) = \frac{4x}{100 - x}$$

models the cost, $f(x)$, in millions of dollars, to remove $x\%$ of pollution from a river due to pesticide runoff from area farms. What percentage of pollutants can be removed for $16 million?

6.7 *In Exercises 55–59, solve each formula for the specified variable.*

55. $P = \frac{R - C}{n}$ for C

56. $\frac{P_1V_1}{T_1} = \frac{P_2V_2}{T_2}$ for T_1

57. $T = \frac{A - P}{Pr}$ for P

58. $\frac{1}{R} = \frac{1}{R_1} + \frac{1}{R_2}$ for R

59. $I = \frac{nE}{R + nr}$ for n

60. A company is planning to manufacture affordable graphing calculators. Fixed monthly cost will be $50,000, and it will cost $25 to produce each calculator.

a. Write the cost function, C, of producing x graphing calculators.

b. Write the average cost function, $\bar{C}$, of producing x graphing calculators.

c. How many graphing calculators must be produced each month for the company to have an average cost of $35 per graphing calculator?

61. After riding at a steady rate for 60 miles, a bicyclist had a flat tire and walked 8 miles to a repair shop. The cycling rate was 3 times faster than the walking rate. If the time spent cycling and walking was 7 hours, at what rate was the cyclist riding?

62. The current of a river moves at 3 miles per hour. It takes a boat a total of 3 hours to travel 12 miles upstream against the current, and return the same distance traveling downstream, with the current. What is the boat's rate in still water?

63. Working alone, two people can clean their house in 3 hours and 6 hours, respectively. They have agreed to clean together so that they can finish in time to watch a TV program that begins in $1\frac{1}{2}$ hours. How long will it take them to clean the house working together? Can they finish before the program starts?

64. Working together, two crews can clear snow from the city's streets in 20 hours. Working alone, the faster crew requires 9 hours less time than the slower crew. How many hours would it take each crew to clear the streets working alone?

65. An inlet faucet can fill a small pond in 60 minutes. The pond can be emptied by an outlet pipe in 80 minutes. You begin filling the empty pond. By accident, the outlet pipe that empties the pond is left open. Under these conditions, how long will it take for the pond to fill?

6.8 *Solve the variation problems in Exercises 66–71.*

66. An electric bill varies directly as the amount of electricity used. The bill for 1400 kilowatts of electricity is $98. What is the bill for 2200 kilowatts of electricity?

67. The distance that a body falls from rest varies directly as the square of the time of the fall. If skydivers fall 144 feet in 3 seconds, how far will they fall in 10 seconds?

68. The time it takes to drive a certain distance varies inversely as the rate of travel. If it takes 4 hours at 50 miles per hour to drive the distance, how long will it take at 40 miles per hour?

69. The loudness of a stereo speaker, measured in decibels, varies inversely as the square of your distance from the speaker. When you are 8 feet from the speaker, the loudness is 28 decibels. What is the loudness when you are 4 feet from the speaker?

70. The time required to assemble computers varies directly as the number of computers assembled and inversely as the number of workers. If 30 computers can be assembled by 6 workers in 10 hours, how long would it take 5 workers to assemble 40 computers?

71. The volume of a pyramid varies jointly as its height and the area of its base. A pyramid with a height of 15 feet and a base with an area of 35 square feet has a volume of 175 cubic feet. Find the volume of a pyramid with a height of 20 feet and a base with an area of 120 square feet.

Chapter 6 Test

1. Find the domain of $f(x) = \dfrac{x^2 - 2x}{x^2 - 7x + 10}$. Then simplify the right side of the function's equation.

In Exercises 2–9, perform the indicated operations. Simplify where possible.

2. $\dfrac{x^2}{x^2 - 16} \cdot \dfrac{x^2 + 7x + 12}{x^2 + 3x}$

3. $\dfrac{x^3 + 27}{x^2 - 1} \div \dfrac{x^2 - 3x + 9}{x^2 - 2x + 1}$

4. $\dfrac{x^2 - 5x - 2}{6x^2 - 11x - 35} - \dfrac{x^2 - 7x + 5}{6x^2 - 11x - 35}$

5. $\dfrac{x}{x + 3} + \dfrac{5}{x - 3}$

6. $\dfrac{2}{x^2 - 4x + 3} + \dfrac{3x}{x^2 + x - 2}$

7. $\dfrac{5x}{x^2 - 4} - \dfrac{2}{x^2 + x - 2}$

8. $\dfrac{x - 4}{x - 5} - \dfrac{3}{x + 5} - \dfrac{10}{x^2 - 25}$

9. $\dfrac{1}{10 - x} + \dfrac{x - 1}{x - 10}$

In Exercises 10–11, simplify each rational expression.

10. $\dfrac{\dfrac{x}{4} - \dfrac{1}{x}}{1 + \dfrac{x+4}{x}}$

11. $\dfrac{\dfrac{1}{x} - \dfrac{3}{x+2}}{\dfrac{2}{x^2 + 2x}}$

In Exercises 12–14, divide as indicated.

12. $(12x^4y^3 + 16x^2y^3 - 10x^2y^2) \div (4x^2y)$

13. $(9x^3 - 3x^2 - 3x + 4) \div (3x + 2)$

14. $(3x^4 + 2x^3 - 8x + 6) \div (x^2 - 1)$

15. Divide using synthetic division:

$$(3x^4 + 11x^3 - 20x^2 + 7x + 35) \div (x + 5).$$

16. Given that

$$f(x) = x^4 - 2x^3 - 11x^2 + 5x + 34,$$

use synthetic division and the Remainder Theorem to find $f(-2)$.

17. Use synthetic division to decide whether -2 is a solution of $2x^3 - 3x^2 - 11x + 6 = 0$.

In Exercises 18–19, solve each rational equation.

18. $\dfrac{4}{x+1} = \dfrac{5}{2x+1}$

19. $\dfrac{x+1}{x^2 + 2x - 3} - \dfrac{1}{x+3} = \dfrac{1}{x-1}$

20. Park rangers introduce 50 elk into a wildlife preserve. The function

$$f(t) = \frac{250(3t + 5)}{t + 25}$$

models the elk population, $f(t)$, after t years. How many years will it take for the population to increase to 125 elk?

21. Solve $R = \dfrac{as}{a + s}$ for a.

22. A company is planning to manufacture pocket-sized televisions. Fixed monthly cost will be $300,000 and it will cost $10 to produce each television.

a. Write the cost function, C, of producing x televisions.

b. Write the average cost function, C, of producing x televisions.

c. How many pocket-sized televisions must be produced each month for the company to have an average cost of $25 per television?

23. It takes one pipe 3 hours to fill a pool and a second pipe 4 hours to drain the pool. The pool is empty and the first pipe begins to fill it. The second pipe is accidently left open, so the water is also draining out of the pool. Under these conditions, how long will it take to fill the pool?

24. A motorboat averages 20 miles per hour in still water. It takes the boat the same amount of time to travel 3 miles with the current as it does to travel 2 miles against the current. What is the current's rate?

25. The intensity of light received at a source varies inversely as the square of the distance from the source. A particular light has an intensity of 20 foot-candles at 15 feet. What is the light's intensity at 10 feet?

Cumulative Review Exercises (Chapters 1–6)

In Exercises 1–5, solve each equation, inequality, or system.

1. $2x + 5 \le 11$ and $-3x > 18$

2. $2x^2 = 7x + 4$

3.
$$\begin{aligned} 4x + 3y + 3z &= 4 \\ 3x \qquad + 2z &= 2 \\ 2x - 5y \qquad &= -4 \end{aligned}$$

4. $|3x - 4| \le 10$

5. $\dfrac{x}{x-8} + \dfrac{6}{x-2} = \dfrac{x^2}{x^2 - 10x + 16}$

6. Solve for s: $I = \dfrac{2R}{w + 2s}$.

7. Solve by graphing:

$$\begin{aligned} 2x - y &= 4 \\ x + y &= 5. \end{aligned}$$

8. Use function notation to write the equation of the line with slope -3 and passing through the point $(1, -5)$.

In Exercises 9–11, graph each equation, inequality, or system in a rectangular coordinate system.

9. $y = |x| + 2$

10. $y \geq 2x - 1$
$x \geq 1$

11. $2x - y < 4$

In Exercises 12–15, perform the indicated operations.

12. $[(x + 2) + 3y][(x + 2) - 3y]$

13. $\dfrac{2x^2 + x - 1}{2x^2 - 9x + 4} \div \dfrac{6x + 15}{3x^2 - 12x}$

14. $\dfrac{3x}{x^2 - 9x + 20} - \dfrac{5}{2x - 8}$

15. $(3x^2 + 10x + 10) \div (x + 2)$

In Exercises 16–17, factor completely.

16. $xy - 6x + 2y - 12$

17. $24x^3y + 16x^2y - 30xy$

18. A baseball is thrown straight up from a height of 64 feet. The function

$$s(t) = -16t^2 + 48t + 64$$

describes the ball's height above the ground, $s(t)$, in feet, t seconds after it was thrown. How long will it take for the ball to hit the ground?

19. The local cable television company offers two deals. Basic cable service with one movie channel costs \$35 per month. Basic service with two movie channels costs \$45 per month. Find the charge for the basic cable service and the charge for each movie channel.

20. A rectangular garden 10 feet wide and 12 feet long is surrounded by a rock border of uniform width. The area of the garden and rock border combined is 168 square feet. What is the width of the rock border?

You enjoy the single life. You have a terrific career, a support group of close friends, and your own home. Still, you feel the pressure to marry. Your parents belong to a generation that believes that you are not complete until you have found your "other half." They see your staying unattached as a "failure to marry." Perhaps you will someday enter into a permanent relationship, but at the moment you are satisfied with your life and career, and plan to stay single into the future.

An increase in the median age at which Americans first marry and the increased rate of divorce have contributed to the growth of our single population. Data indicate that the number of Americans living alone increased quite rapidly toward the end of the twentieth century. However, now this growth rate is beginning to slow down. In this chapter, you will see why radical functions with square roots are used to describe phenomena that are continuing to grow, but whose growth is leveling off. By learning about radicals and radical functions, you will have new algebraic tools for describing your world.

Chapter 7

Radicals, Radical Functions, and Rational Exponents

▸ SECTION 7.1 Radical Expressions and Functions

Objectives

1. Evaluate square roots.
2. Evaluate square root functions.
3. Find the domain of square root functions.
4. Use models that are square root functions.
5. Simplify expressions of the form $\sqrt{a^2}$.
6. Evaluate cube root functions.
7. Simplify expressions of the form $\sqrt[3]{a^3}$.
8. Find even and odd roots.
9. Simplify expressions of the form $\sqrt[n]{a^n}$.

SSM PH Tutor Center CD-ROM Video

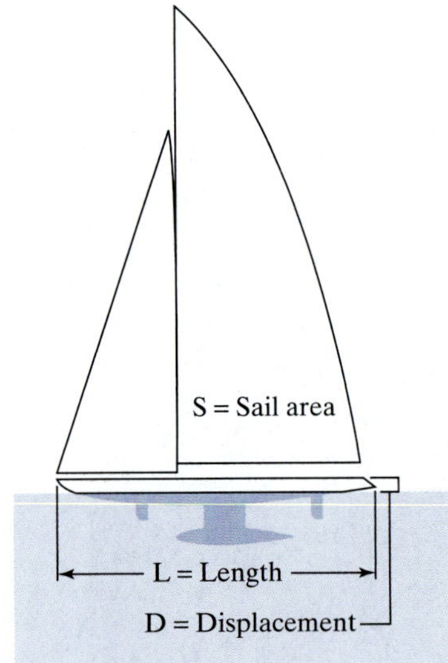

The America's Cup is the supreme event in ocean sailing. Competition is fierce and the costs are huge. Competitors look to mathematics to provide the critical innovation that can make the difference between winning and losing. The basic dimensions of competitors' yachts must satisfy an inequality containing square roots and cube roots:

$$L + 1.25\sqrt{S} - 9.8\sqrt[3]{D} \leq 16.296.$$

In the inequality, L is the yacht's length, in meters, S is its sail area, in square meters, and D is its displacement, in cubic meters.

In this section, we introduce a new category of expressions and functions that contain roots. You will see why square root functions are used to describe phenomena that are continuing to grow but whose growth is leveling off.

1 Evaluate square roots.

Square Roots

From our earlier work with exponents, we are aware that the square of both 5 and −5 is 25:

$$5^2 = 25 \quad \text{and} \quad (-5)^2 = 25.$$

The reverse operation of squaring a number is finding the *square root* of the number. For example,

- A square root of 25 is 5 because $5^2 = 25$.
- A square root of 25 is also −5 because $(-5)^2 = 25$.

In general, **if $b^2 = a$, then b is a square root of a.**

The symbol $\sqrt{\ }$ is used to denote the *positive* or *principal square root* of a number. For example,

- $\sqrt{25} = 5$ because $5^2 = 25$ and 5 is positive.
- $\sqrt{100} = 10$ because $10^2 = 100$ and 10 is positive.

The symbol $\sqrt{\ }$ that we use to denote the principal square root is called a **radical sign**. The number under the radical sign is called the **radicand**. Together we refer to the radical sign and its radicand as a **radical expression**.

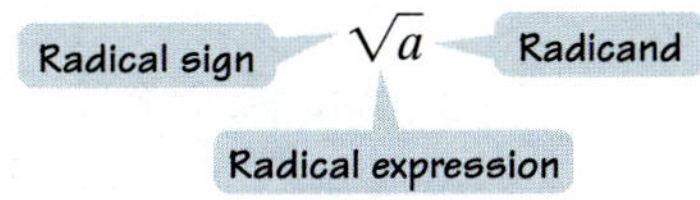

Definition of the Principal Square Root

If a is a nonnegative real number, the nonnegative number b such that $b^2 = a$, denoted by $b = \sqrt{a}$, is the **principal square root** of a.

The symbol $-\sqrt{\ }$ is used to denote the negative square root of a number. For example,

- $-\sqrt{25} = -5$ because $(-5)^2 = 25$ and -5 is negative.
- $-\sqrt{100} = -10$ because $(-10)^2 = 100$ and -10 is negative.

EXAMPLE 1 Evaluating Square Roots

Evaluate:

a. $\sqrt{81}$ **b.** $-\sqrt{9}$ **c.** $\sqrt{\frac{4}{49}}$

d. $\sqrt{0.0064}$ **e.** $\sqrt{36 + 64}$ **f.** $\sqrt{36} + \sqrt{64}$.

Solution

a. $\sqrt{81} = 9$ The principal square root of 81 is 9 because $9^2 = 81$.

b. $-\sqrt{9} = -3$ The negative square root of 9 is -3 because $(-3)^2 = 9$.

c. $\sqrt{\frac{4}{49}} = \frac{2}{7}$ The principal square root of $\frac{4}{49}$ is $\frac{2}{7}$ because $\left(\frac{2}{7}\right)^2 = \frac{4}{49}$.

d. $\sqrt{0.0064} = 0.08$ The principal square root of 0.0064 is 0.08 because $(0.08)^2 = (0.08)(0.08) = 0.0064$.

e. $\sqrt{36 + 64} = \sqrt{100}$ Simplify the radicand.

$= 10$ Take the principal square root of 100, which is 10.

f. $\sqrt{36} + \sqrt{64} = 6 + 8$ $\sqrt{36} = 6$ because $6^2 = 36$. $\sqrt{64} = 8$ because $8^2 = 64$.

$= 14$ ■

Study Tip

In Example 1, parts (e) and (f), observe that $\sqrt{36 + 64}$ is not equal to $\sqrt{36} + \sqrt{64}$. In general,

$$\sqrt{a + b} \neq \sqrt{a} + \sqrt{b}$$

and

$$\sqrt{a - b} \neq \sqrt{a} - \sqrt{b}.$$

✔ **CHECK POINT 1** Evaluate:

a. $\sqrt{64}$ **b.** $-\sqrt{49}$ **c.** $\sqrt{\frac{16}{25}}$

d. $\sqrt{0.0081}$ **e.** $\sqrt{9 + 16}$ **f.** $\sqrt{9} + \sqrt{16}$.

Let's see what happens to the radical expression $\sqrt{x}$ if x is a negative number. Is the square root of a negative number a real number? For example, consider $\sqrt{-25}$. Is there a real number whose square is -25? No. Thus, $\sqrt{-25}$ is not a real number. In general, **a square root of a negative number is not a real number**.

2 Evaluate square root functions.

Square Root Functions

Because each nonnegative real number, x, has precisely one principal square root, $\sqrt{x}$, there is a **square root function** defined by

$$f(x) = \sqrt{x}.$$

The domain of this function is $[0, \infty)$. We can graph $f(x) = \sqrt{x}$ by selecting nonnegative real numbers for x. It is easiest to choose perfect squares, numbers that have rational square roots. Table 7.1 shows five such choices for x and the calculations for the corresponding outputs. We plot these ordered pairs as points in the rectangular coordinate system and connect the points with a smooth curve. The graph of $f(x) = \sqrt{x}$ is shown in Figure 7.1.

Table 7.1

x	$f(x) = \sqrt{x}$	(x, y) or $(x, f(x))$
0	$f(0) = \sqrt{0} = 0$	(0, 0)
1	$f(1) = \sqrt{1} = 1$	(1, 1)
4	$f(4) = \sqrt{4} = 2$	(4, 2)
9	$f(9) = \sqrt{9} = 3$	(9, 3)
16	$f(16) = \sqrt{16} = 4$	(16, 4)

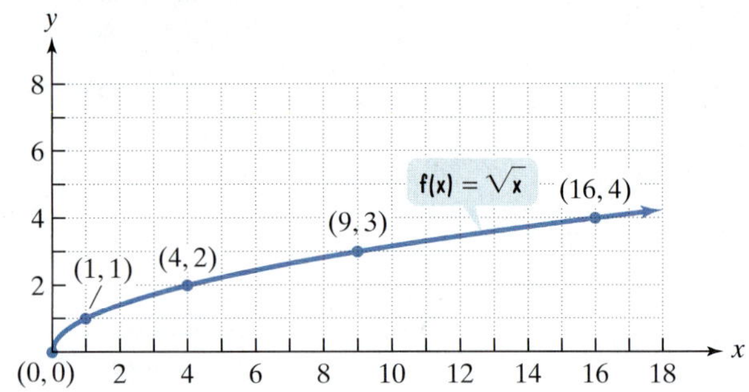

Figure 7.1 The graph of the square root function $f(x) = \sqrt{x}$

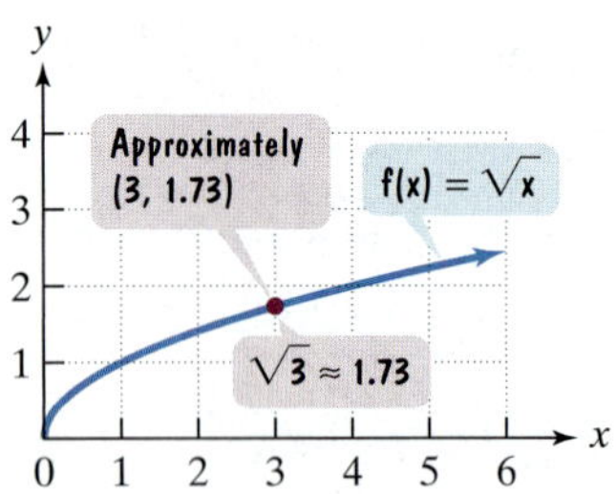

Figure 7.2 Visualizing $\sqrt{3}$ as a point on the graph of $f(x) = \sqrt{x}$

Is it possible to choose values of x in Table 7.1 that are not squares of integers, or perfect squares? Yes. For example, we can let $x = 3$. Thus, $f(3) = \sqrt{3}$. Because 3 is not a perfect square, $\sqrt{3}$ is an irrational number, one that cannot be expressed as a quotient of integers. We can use a calculator to find a decimal approximation of $\sqrt{3}$.

Most Scientific Calculators	**Most Graphing Calculators**
3 [√]	[√] 3 [ENTER]

Rounding the displayed number to two decimal places, $\sqrt{3} \approx 1.73$. This information is shown visually as a point, approximately (3, 1.73), on the graph of $f(x) = \sqrt{x}$ in Figure 7.2.

To evaluate a square root function, we use substitution, just as we do to evaluate other functions.

EXAMPLE 2 Evaluating Square Root Functions

For each function, find the indicated function value:

a. $f(x) = \sqrt{5x - 6}$; $f(2)$ **b.** $g(x) = -\sqrt{64 - 8x}$; $g(-3)$.

Solution

a. $f(2) = \sqrt{5 \cdot 2 - 6}$ Substitute 2 for x in $f(x) = \sqrt{5x - 6}$.

$= \sqrt{4} = 2$ Simplify the radicand and take the square root.

b. $g(-3) = -\sqrt{64 - 8(-3)}$ Substitute −3 for x in $g(x) = -\sqrt{64 - 8x}$.

$= -\sqrt{88} \approx -9.38$ Simplify the radicand: $64 - 8(-3) = 64 - (-24) = 64 + 24 = 88$. Then use a calculator to approximate $\sqrt{88}$. ■

✔ **CHECK POINT 2** For each function, find the indicated function value:

a. $f(x) = \sqrt{12x - 20}$; $f(3)$

b. $g(x) = -\sqrt{9 - 3x}$; $g(-5)$.

3 Find the domain of a square root function.

We have seen that the domain of a function f is the largest set of real numbers for which the value of $f(x)$ is a real number. Because only nonnegative numbers have real square roots, the domain of a square root function is the set of real numbers for which the radicand is nonnegative.

EXAMPLE 3 Finding the Domain of a Square Root Function

Find the domain of

$$f(x) = \sqrt{3x + 12}.$$

Solution The domain is the set of real numbers, x, for which the radicand, $3x + 12$, is nonnegative. We set the radicand greater than or equal to 0 and solve the resulting inequality.

$$3x + 12 \ge 0$$
$$3x \ge -12$$
$$x \ge -4$$

The domain of f is $\{x \mid x \ge -4\}$ or $[-4, \infty)$. ■

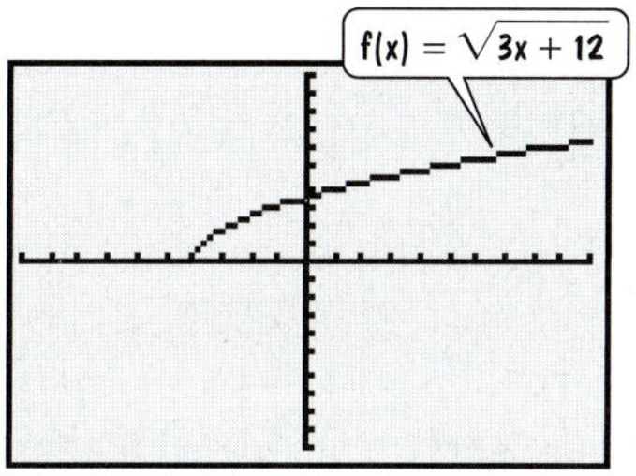

Figure 7.3

Figure 7.3 shows the graph of $f(x) = \sqrt{3x + 12}$ in a $[-10, 10, 1]$ by $[-10, 10, 1]$ viewing rectangle. The graph appears only for $x \ge -4$, verifying $[-4, \infty)$ as the domain. Can you see how the graph also illustrates this square root function's range? The graph only appears for nonnegative values of y. Thus, the range is $[0, \infty)$.

✔ **CHECK POINT 3** Find the domain of

$$f(x) = \sqrt{9x - 27}.$$

4 Use models that are square root functions.

The graph of the square root function $f(x) = \sqrt{x}$ is increasing from left to right. However, the rate of increase is slowing down as the graph moves to the right. This is why square root functions are often used to model growing phenomena with growth that is leveling off. The amount of money for new student loans, shown from 1993 through 2000 in Figure 7.4, displays this growth pattern.

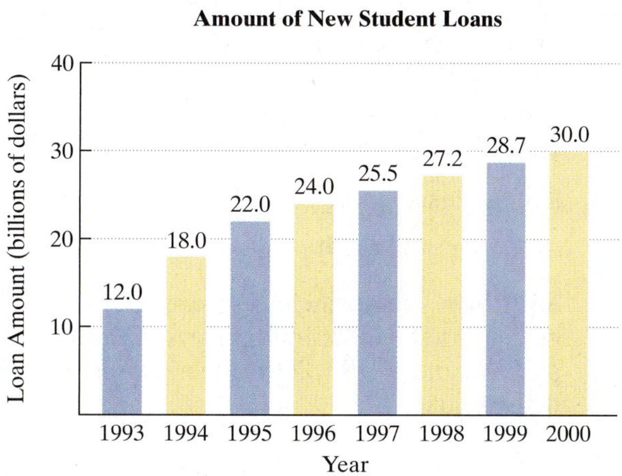

Figure 7.4 *Source:* U.S. Department of Education

Using Technology

The graph of

$$f(x) = 6.75\sqrt{x} + 12$$

is shown in a $[0, 7, 1]$ by $[0, 30, 3]$ viewing rectangle. Can you see how the graph shows that the amount for new student loans is growing, but leveling off?

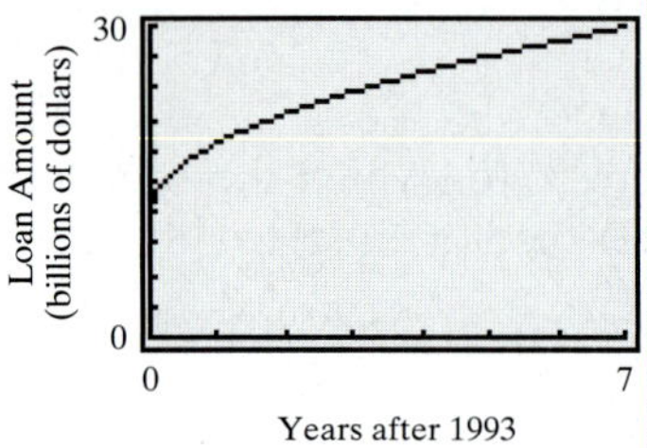

EXAMPLE 4 Modeling with a Square Root Function

The data in Figure 7.4 can be modeled by the function

$$f(x) = 6.75\sqrt{x} + 12$$

where $f(x)$ is the amount, in billions of dollars, of new student loans x years after 1993. According to the model, how much was loaned in 2000? Round to the nearest tenth of a billion. How well does the model describe the actual data?

Solution Because 2000 is 7 years after 1993, we substitute 7 for x and evaluate the function at 7.

$$f(x) = 6.75\sqrt{x} + 12 \quad \text{Use the given function.}$$

$$f(7) = 6.75\sqrt{7} + 12 \quad \text{Substitute 7 for } x.$$

$$\approx 29.9 \quad \text{Use a calculator.}$$

Rounding to the nearest tenth of a billion, the amount of money in new student loans in 2000 was \$29.9 billion. The amount given in the data in Figure 7.4 on the bottom of page 467 is \$30.0 billion. Thus, the model describes the actual data fairly well. ■

✔ **CHECK POINT 4** Use the square root function in Example 4 to find how much was loaned in 1997. Round to the nearest tenth of a billion. How well does the model describe the actual data?

5 Simplify expressions of the form $\sqrt{a^2}$.

Simplifying Expressions of the Form $\sqrt{a^2}$

You may think that $\sqrt{a^2} = a$. However, this is not necessarily true. Consider the following examples:

$$\sqrt{4^2} = \sqrt{16} = 4$$

$$\sqrt{(-4)^2} = \sqrt{16} = 4.$$

The result is not -4, but rather the absolute value of -4, or 4.

Here is a rule for simplifying expressions of the form $\sqrt{a^2}$:

Simplifying $\sqrt{a^2}$

For any real number a,

$$\sqrt{a^2} = |a|.$$

In words, the principal square root of a^2 is the absolute value of a.

Using Technology

The graphs of

$$f(x) = \sqrt{x^2} \text{ and } g(x) = |x|$$

are shown in a $[-10, 10, 1]$ by $[-2, 10, 1]$ viewing rectangle. The graphs are the same. Thus,

$$\sqrt{x^2} = |x|.$$

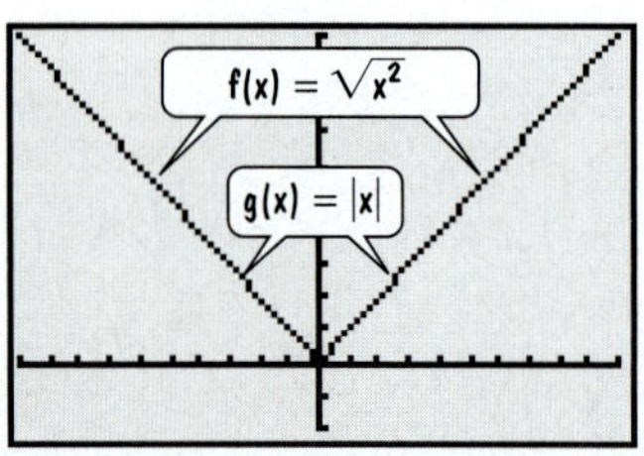

EXAMPLE 5 Simplifying Radical Expressions

Simplify each expression:

a. $\sqrt{(-6)^2}$ **b.** $\sqrt{(x + 5)^2}$ **c.** $\sqrt{25x^6}$ **d.** $\sqrt{x^2 - 4x + 4}$.

Solution The principal square root of an expression squared is the absolute value of that expression. In parts (a) and (b), we are given squared radicands. In parts (c) and (d), it will first be necessary to express the radicand as an expression that is squared.

a. $\sqrt{(-6)^2} = |-6| = 6$

b. $\sqrt{(x + 5)^2} = |x + 5|$

c. To simplify $\sqrt{25x^6}$, first write $25x^6$ as an expression that is squared: $25x^6 = (5x^3)^2$. Then simplify.

$$\sqrt{25x^6} = \sqrt{(5x^3)^2} = |5x^3|$$

d. To simplify $\sqrt{x^2 - 4x + 4}$, first write $x^2 - 4x + 4$ as an expression that is squared by factoring the perfect square trinomial: $x^2 - 4x + 4 = (x - 2)^2$. Then simplify.

$$\sqrt{x^2 - 4x + 4} = \sqrt{(x - 2)^2} = |x - 2|$$

■

✔ **CHECK POINT 5** Simplify each expression:

a. $\sqrt{(-7)^2}$

b. $\sqrt{(x + 8)^2}$

c. $\sqrt{49x^{10}}$

d. $\sqrt{x^2 - 6x + 9}$.

In some situations we are told that no radicands involve negative quantities to even powers. When the expression being squared is nonnegative, it is not necessary to use absolute value when simplifying $\sqrt{a^2}$. For example, assuming that no radicands contain negative quantities that are squared,

$$\sqrt{x^8} = \sqrt{(x^4)^2} = x^4$$

$$\sqrt{25x^2 + 10x + 1} = \sqrt{(5x + 1)^2} = 5x + 1.$$

6 Evaluate cube root functions.

Cube Roots and Cube Root Functions

Finding the square root of a number reverses the process of squaring a number. Similarly, finding the cube root of a number reverses the process of cubing a number. For example, $2^3 = 8$, and so the cube root of 8 is 2. The notation that we use is $\sqrt[3]{8} = 2$.

Definition of the Cube Root of a Number

The **cube root** of a real number a is written $\sqrt[3]{a}$.

$$\sqrt[3]{a} = b \quad \text{means that } b^3 = a.$$

Study Tip

Some cube roots occur so frequently that you might want to memorize them.

$$\sqrt[3]{1} = 1$$
$$\sqrt[3]{8} = 2$$
$$\sqrt[3]{27} = 3$$
$$\sqrt[3]{64} = 4$$
$$\sqrt[3]{125} = 5$$
$$\sqrt[3]{216} = 6$$
$$\sqrt[3]{1000} = 10$$

For example,

$$\sqrt[3]{64} = 4 \quad \text{because} \quad 4^3 = 64.$$

$$\sqrt[3]{-27} = -3 \quad \text{because} \quad (-3)^3 = -27.$$

By contrast to square roots, the cube root of a negative number is a real number. All real numbers have cube roots. The cube root of a positive number is positive. The cube root of a negative number is negative.

Because every real number, x, has precisely one cube root, $\sqrt[3]{x}$, there is a **cube root function** defined by

$$f(x) = \sqrt[3]{x}.$$

The domain of this function is the set of all real numbers. We can graph $f(x) = \sqrt[3]{x}$ by selecting perfect cubes, numbers that have rational cube roots, for x. Table 7.2 shows five such choices for x and the calculations for the corresponding outputs. We plot these ordered pairs as points in the rectangular coordinate system and connect the points with a smooth curve. The graph of $f(x) = \sqrt[3]{x}$ is shown in Figure 7.5.

Table 7.2

x	$f(x) = \sqrt[3]{x}$	(x, y) or $(x, f(x))$
-8	$f(-8) = \sqrt[3]{-8} = -2$	$(-8, -2)$
-1	$f(-1) = \sqrt[3]{-1} = -1$	$(-1, -1)$
0	$f(0) = \sqrt[3]{0} = 0$	$(0, 0)$
1	$f(1) = \sqrt[3]{1} = 1$	$(1, 1)$
8	$f(8) = \sqrt[3]{8} = 2$	$(8, 2)$

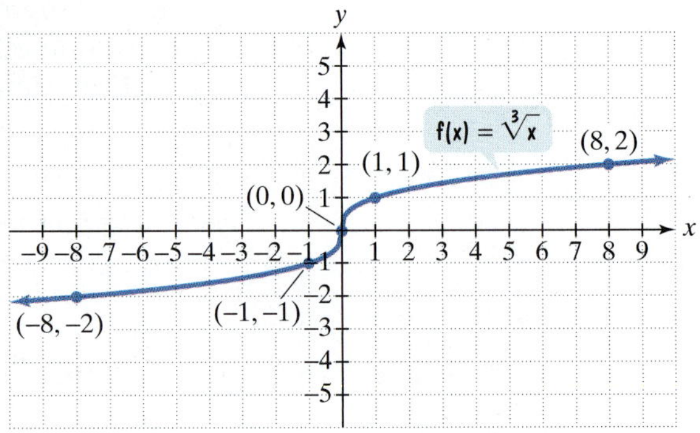

Figure 7.5 The graph of the cube root function $f(x) = \sqrt[3]{x}$

Notice that both the domain and range of $f(x) = \sqrt[3]{x}$ are sets of all real numbers.

EXAMPLE 6 Evaluating Cube Root Functions

For each function, find the indicated function value:

a. $f(x) = \sqrt[3]{x-2}$; $f(127)$ **b.** $g(x) = \sqrt[3]{8x-8}$; $g(-7)$.

Solution

a. $f(127) = \sqrt[3]{127-2}$ Substitute 127 for x in $f(x) = \sqrt[3]{x-2}$.

$= \sqrt[3]{125}$ Simplify the radicand.

$= 5$ $\sqrt[3]{125} = 5$ because $5^3 = 125$.

b. $g(-7) = \sqrt[3]{8(-7)-8}$ Substitute −7 for x in $g(x) = \sqrt[3]{8x-8}$.

$= \sqrt[3]{-64}$ Simplify the radicand: $8(-7)-8 = -56 - 8 = -64$.

$= -4$ $\sqrt[3]{-64} = -4$ because $(-4)^3 = -64$. ■

✔ **CHECK POINT 6** For each function, find the indicated function value:

a. $f(x) = \sqrt[3]{x-6}$; $f(33)$

b. $g(x) = \sqrt[3]{2x+2}$; $g(-5)$.

7 Simplify expressions of the form $\sqrt[3]{a^3}$.

Because the cube root of a positive number is positive and the cube root of a negative number is negative, absolute value is not needed to simplify expressions of the form $\sqrt[3]{a^3}$.

Simplifying $\sqrt[3]{a^3}$

For any real number a,

$$\sqrt[3]{a^3} = a.$$

In words, the cube root of any expression cubed is that expression.

EXAMPLE 7 Simplifying a Cube Root

Simplify: $\sqrt[3]{-64x^3}$.

Solution Begin by expressing the radicand as an expression that is cubed: $-64x^3 = (-4x)^3$. Then simplify.

$$\sqrt[3]{-64x^3} = \sqrt[3]{(-4x)^3} = -4x$$

We can check our answer by cubing $-4x$:

$$(-4x)^3 = (-4)^3x^3 = -64x^3.$$

By obtaining the radicand, we know that our simplification is correct. ■

✔ CHECK POINT 7 Simplify: $\sqrt[3]{-27x^3}$.

8 Find even and odd roots.

Even and Odd *n*th Roots

Up to this point, we have focused on square and cube roots. Other radical expressions have different roots. For example, the fifth root of a, written $\sqrt[5]{a}$, is the number b for which $b^5 = a$. Thus,

$$\sqrt[5]{32} = 2 \quad \text{because} \quad 2^5 = 2 \cdot 2 \cdot 2 \cdot 2 \cdot 2 = 32.$$

The radical expression $\sqrt[n]{a}$ represents the ***n*th root of *a***. The number n is called the **index**. An index of 2 represents a square root and is not written. An index of 3 represents a cube root.

If the index n in $\sqrt[n]{a}$ is an odd number, a root is said to be an **odd root**. A cube root is an odd root. Other odd roots have the same characteristics as cube roots. Every real number has exactly one real root when n is odd. The (odd) nth root of a, $\sqrt[n]{a}$, is the number b for which $b^n = a$. An odd root of a positive number is positive, and an odd root of a negative number is negative. For example,

$$\sqrt[5]{243} = 3 \quad \text{because} \quad 3^5 = 3 \cdot 3 \cdot 3 \cdot 3 \cdot 3 = 243$$

and $\sqrt[5]{-243} = -3$ because $(-3)^5 = (-3)(-3)(-3)(-3)(-3) = -243$.

If the index n in $\sqrt[n]{a}$ is an even number, a root is said to be an **even root**. A square root is an even root. Other even roots have the same characteristics as square roots. Every positive real number has two real roots when n is even. One root is positive and one is negative. The positive root, called the **principal *n*th root** and represented by $\sqrt[n]{a}$, is the nonnegative number b for which $b^n = a$. The symbol $-\sqrt[n]{a}$ is used to denote the negative nth root. **An even root of a negative number is not a real number.**

Study Tip

Some higher even and odd roots occur so frequently that you might want to memorize them.

Fourth Roots	Fifth Roots
$\sqrt[4]{1} = 1$	$\sqrt[5]{1} = 1$
$\sqrt[4]{16} = 2$	$\sqrt[5]{32} = 2$
$\sqrt[4]{81} = 3$	$\sqrt[5]{243} = 3$
$\sqrt[4]{256} = 4$	
$\sqrt[4]{625} = 5$	

EXAMPLE 8 Finding Even and Odd Roots

Find the indicated root, or state that the expression is not a real number:

a. $\sqrt[4]{81}$ **b.** $-\sqrt[4]{81}$ **c.** $\sqrt[4]{-81}$ **d.** $\sqrt[5]{-32}$.

Solution

a. $\sqrt[4]{81} = 3$ The principal fourth root of 81 is 3 because $3^4 = 3 \cdot 3 \cdot 3 \cdot 3 = 81$.

b. $-\sqrt[4]{81} = -3$ The negative fourth root of 81 is -3 because $(-3)^4 = (-3)(-3)(-3)(-3) = 81$.

c. $\sqrt[4]{-81}$ is not a real number because the index, 4, is even and the radicand, -81, is negative. No real number can be raised to the fourth power to give a negative result such as -81. Real numbers to even powers can only result in nonnegative numbers.

d. $\sqrt[5]{-32} = -2$ because $(-2)^5 = (-2)(-2)(-2)(-2)(-2) = -32$. An odd root of a negative real number is always negative. ■

✔ **CHECK POINT 8** Find the indicated root, or state that the expression is not a real number:

a. $\sqrt[4]{16}$ **b.** $-\sqrt[4]{16}$

c. $\sqrt[4]{-16}$ **d.** $\sqrt[5]{-1}$.

9 Simplify expressions of the form $\sqrt[n]{a^n}$.

Simplifying Expressions of the Form $\sqrt[n]{a^n}$

We have seen that

$$\sqrt{a^2} = |a| \quad \text{and} \quad \sqrt[3]{a^3} = a.$$

Expressions of the form $\sqrt[n]{a^n}$ can be simplified in the same manner. Unless a is known to be nonnegative, absolute value notation is needed when n is even. When the index is odd, absolute value bars are not necessary.

Simplifying $\sqrt[n]{a^n}$

For any real number a,

1. If n is even, $\sqrt[n]{a^n} = |a|$.
2. If n is odd, $\sqrt[n]{a^n} = a$.

EXAMPLE 9 Simplifying Radical Expressions

Simplify:

a. $\sqrt[4]{(x-3)^4}$ **b.** $\sqrt[5]{(2x+7)^5}$ **c.** $\sqrt[6]{(-5)^6}$.

Solution Each expression involves the nth root of a radicand raised to the nth power. Thus, each radical expression can be simplified. Absolute value bars are necessary in parts (a) and (c) because n is even.

a. $\sqrt[4]{(x-3)^4} = |x-3|$ — $\sqrt[n]{a^n} = |a|$ if n is even.
b. $\sqrt[5]{(2x+7)^5} = 2x+7$ — $\sqrt[n]{a^n} = a$ if n is odd.
c. $\sqrt[6]{(-5)^6} = |-5| = 5$ — $\sqrt[n]{a^n} = |a|$ if n is even. ■

✔ **CHECK POINT 9** Simplify:

a. $\sqrt[4]{(x+6)^4}$
b. $\sqrt[5]{(3x-2)^5}$
c. $\sqrt[6]{(-8)^6}$.

EXERCISE SET 7.1

Practice Exercises

In Exercises 1–16, evaluate each expression, or state that the expression is not a real number.

1. $\sqrt{36}$
2. $\sqrt{16}$
3. $-\sqrt{36}$
4. $-\sqrt{16}$
5. $\sqrt{-36}$
6. $\sqrt{-16}$
7. $\sqrt{\frac{1}{25}}$
8. $\sqrt{\frac{1}{49}}$
9. $-\sqrt{0.04}$
10. $-\sqrt{0.64}$
11. $\sqrt{25 - 16}$
12. $\sqrt{144 + 25}$
13. $\sqrt{25} - \sqrt{16}$
14. $\sqrt{144} + \sqrt{25}$
15. $\sqrt{16 - 25}$
16. $\sqrt{25 - 144}$

In Exercises 17–22, find the indicated function values for each function. If necessary, round to two decimal places. If the function value is not a real number and does not exist, so state.

17. $f(x) = \sqrt{x - 2};\quad f(18), f(3), f(2), f(-2)$
18. $f(x) = \sqrt{x - 3};\quad f(28), f(4), f(3), f(-1)$
19. $g(x) = -\sqrt{2x + 3};\quad g(11), g(1), g(-1), g(-2)$
20. $g(x) = -\sqrt{2x + 1};\quad g(4), g(1), g\left(-\frac{1}{2}\right), g(-1)$
21. $h(x) = \sqrt{(x - 1)^2};\quad h(5), h(3), h(0), h(-5)$
22. $h(x) = \sqrt{(x - 2)^2};\quad h(5), h(3), h(0), h(-5)$

In Exercises 23–28, find the domain of each square root function. Then use the domain to match the radical function with its graph. [The graphs are labeled (a) through (f) and are shown in $[-10, 10, 1]$ by $[-10, 10, 1]$ viewing rectangles.]

23. $f(x) = \sqrt{x - 3}$
24. $f(x) = \sqrt{x + 2}$
25. $f(x) = \sqrt{3x + 15}$
26. $f(x) = \sqrt{3x - 15}$
27. $f(x) = \sqrt{6 - 2x}$
28. $f(x) = \sqrt{8 - 2x}$

(a)

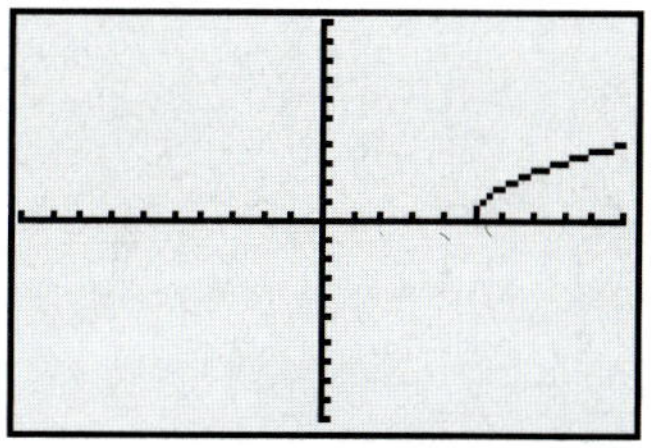

(b)

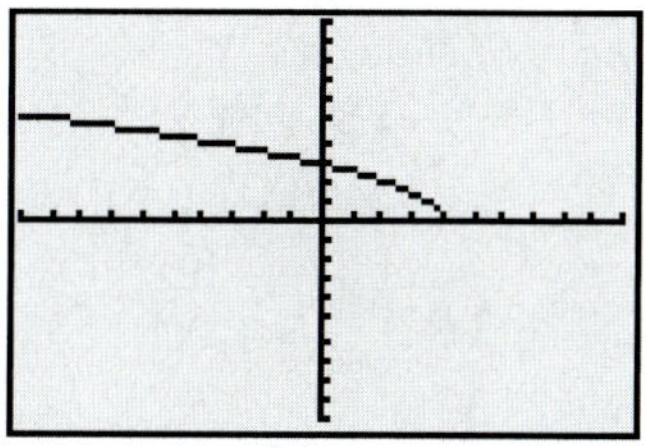

(c)

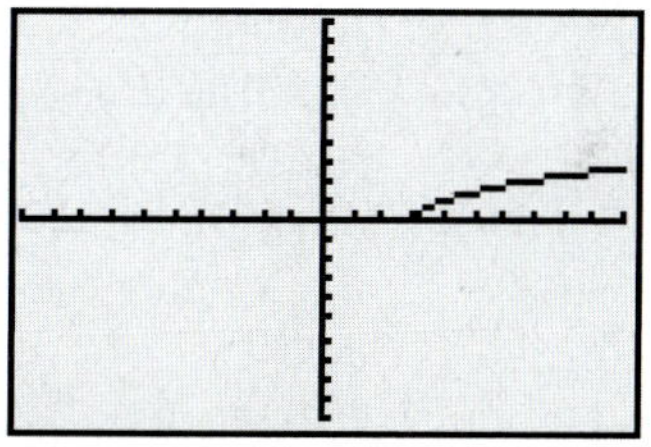

(d)

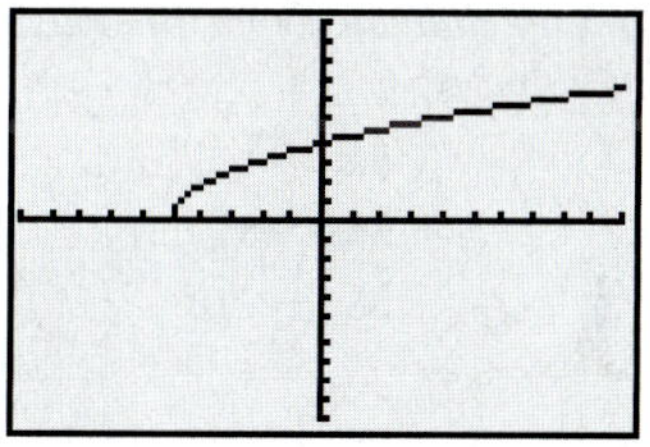

(e)

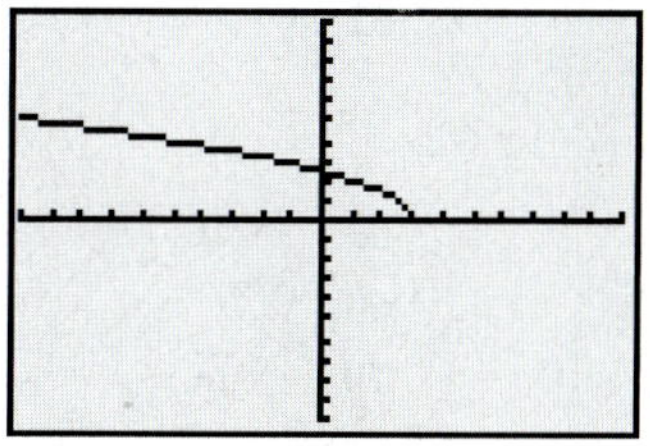

(f)

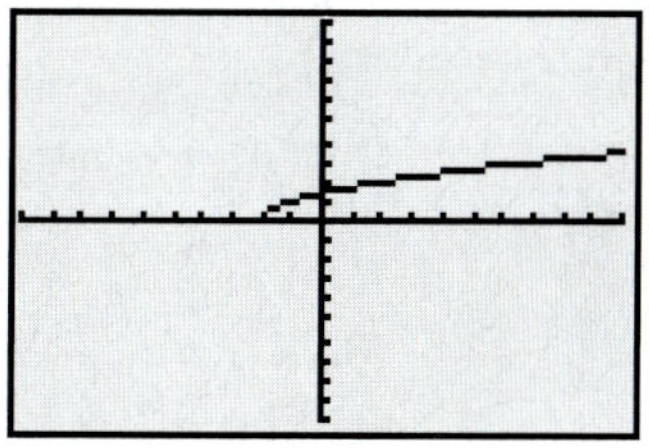

In Exercises 29–42, simplify each expression.

29. $\sqrt{5^2}$
30. $\sqrt{7^2}$

31. $\sqrt{(-4)^2}$

32. $\sqrt{(-10)^2}$

33. $\sqrt{(x-1)^2}$

34. $\sqrt{(x-2)^2}$

35. $\sqrt{36x^4}$

36. $\sqrt{81x^4}$

37. $-\sqrt{100x^6}$

38. $-\sqrt{49x^6}$

39. $\sqrt{x^2+12x+36}$

40. $\sqrt{x^2+14x+49}$

41. $-\sqrt{x^2-8x+16}$

42. $-\sqrt{x^2-10x+25}$

In Exercises 43–48, find each cube root.

43. $\sqrt[3]{27}$

44. $\sqrt[3]{64}$

45. $\sqrt[3]{-27}$

46. $\sqrt[3]{-64}$

47. $\sqrt[3]{\dfrac{1}{125}}$

48. $\sqrt[3]{\dfrac{1}{1000}}$

In Exercises 49–52, find the indicated function values for each function.

49. $f(x)=\sqrt[3]{x-1};\quad f(28), f(9), f(0), f(-63)$

50. $f(x)=\sqrt[3]{x-3};\quad f(30), f(11), f(2), f(-122)$

51. $g(x)=-\sqrt[3]{8x-8}; g(2), g(1), g(0)$

52. $g(x)=-\sqrt[3]{2x+1};\quad g(13), g(0), g(-63)$

In Exercises 53–70, find the indicated root, or state that the expression is not a real number.

53. $\sqrt[4]{1}$

54. $\sqrt[5]{1}$

55. $\sqrt[4]{16}$

56. $\sqrt[4]{81}$

57. $-\sqrt[4]{16}$

58. $-\sqrt[4]{81}$

59. $\sqrt[4]{-16}$

60. $\sqrt[4]{-81}$

61. $\sqrt[5]{-1}$

62. $\sqrt[7]{-1}$

63. $\sqrt[6]{-1}$

64. $\sqrt[8]{-1}$

65. $-\sqrt[4]{256}$

66. $-\sqrt[4]{10{,}000}$

67. $\sqrt[6]{64}$

68. $\sqrt[5]{32}$

69. $-\sqrt[5]{32}$

70. $-\sqrt[6]{64}$

In Exercises 71–84, simplify each expression. Include absolute value bars where necessary.

71. $\sqrt[3]{x^3}$

72. $\sqrt[5]{x^5}$

73. $\sqrt[4]{y^4}$

74. $\sqrt[6]{y^6}$

75. $\sqrt[3]{-8x^3}$

76. $\sqrt[3]{-125x^3}$

77. $\sqrt[3]{(-5)^3}$

78. $\sqrt[3]{(-6)^3}$

79. $\sqrt[4]{(-5)^4}$

80. $\sqrt[6]{(-6)^6}$

81. $\sqrt[4]{(x+3)^4}$

82. $\sqrt[4]{(x+5)^4}$

83. $\sqrt[5]{-32(x-1)^5}$

84. $\sqrt[5]{-32(x-2)^5}$

Application Exercises

The table shows the median, or average, heights for boys of various ages in the United States, from birth through 60 months, or five years. The data can be modeled by the radical function

$$f(x) = 2.9\sqrt{x} + 20.1$$

where $f(x)$ is the median height, in inches, of boys who are x months of age. Use the function to solve Exercises 85–86.

Boys' Median Heights

Age (months)	Height (inches)
0	20.5
6	27.0
12	30.8
18	32.9
24	35.0
36	37.5
48	40.8
60	43.4

Source: *The Portable Pediatrician for Parents*, by Laura Walther Nathanson, M.D., FAAP

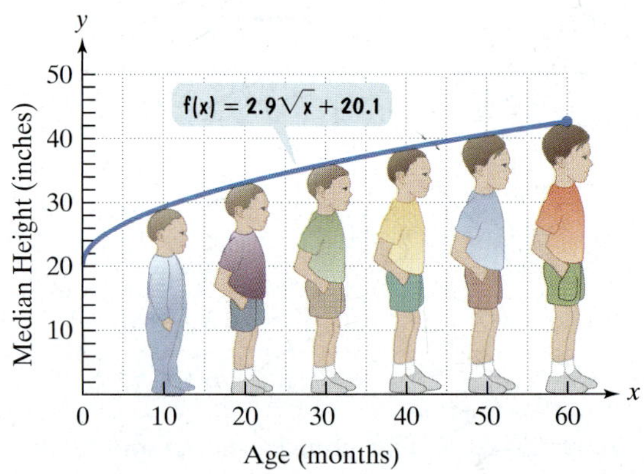

85. According to the model, what is the median height of boys who are 48 months, or four years, old? Use a calculator and round to the nearest tenth of an inch. How well does the model describe the actual median height shown in the table?

86. According to the model, what is the median height of boys who are 60 months, or five years, old? Use a calculator and round to the nearest tenth of an inch. How well does the model describe the actual median height shown in the table?

Police use the function $f(x) = \sqrt{20x}$ to estimate the speed of a car, $f(x)$, in miles per hour, based on the length, x, in feet, of its skid marks upon sudden braking on a dry asphalt road. Use the function to solve Exercises 87–88.

87. A motorist is involved in an accident. A police officer measures the car's skid marks to be 245 feet long. Estimate the speed at which the motorist was traveling before braking. If the posted speed limit is 50 miles per hour and the motorist tells the officer he was not speeding, should the officer believe him? Explain.

88. A motorist is involved in an accident. A police officer measures the car's skid marks to be 45 feet long. Estimate the speed at which the motorist was traveling before braking. If the posted speed limit is 35 miles per hour and the motorist tells the officer she was not speeding, should the officer believe her? Explain.

Writing in Mathematics

89. What are the square roots of 36? Explain why each of these numbers is a square root.

90. What does the symbol $\sqrt{}$ denote? Which of your answers in Exercise 89 is given by this symbol? Write the symbol needed to obtain the other answer.

91. Explain why $\sqrt{-1}$ is not a real number.

92. Explain how to find the domain of a square root function.

93. Explain how to simplify $\sqrt{a^2}$. Give an example with your explanation.

94. Explain why $\sqrt[3]{8}$ is 2. Then describe what is meant by the cube root of a real number.

95. Describe two differences between odd and even roots.

96. Explain how to simplify $\sqrt[n]{a^n}$ if n is even and if n is odd. Give examples with your explanations.

97. Explain the meaning of the words "radical," "radicand," and "index." Give an example with your explanation.

98. Describe the trend in a boy's growth from birth through five years, shown in the table for Exercises 85–86. Why is a square root function a useful model for the data?

99. The function $f(x) = 6.85\sqrt{x} + 19$ models the percentage, $f(x)$, of U.S. households online x years after 1997. Write a word problem using the function. Then solve the problem.

Technology Exercises

100. Use a graphing utility to graph $y_1 = \sqrt{x}$, $y_2 = \sqrt{x+4}$, and $y_3 = \sqrt{x-3}$ in the same $[-5, 10, 1]$ by $[0, 6, 1]$ viewing rectangle. Describe one similarity and one difference that you observe among the graphs. Use the word "shift" in your response.

101. Use a graphing utility to graph $y = \sqrt{x}$, $y = \sqrt{x} + 4$, and $y = \sqrt{x} - 3$ in the same $[-1, 10, 1]$ by $[-10, 10, 1]$ viewing rectangle. Describe one similarity and one difference that you observe among the graphs.

102. Use a graphing utility to graph $f(x) = \sqrt{x}$, $g(x) = -\sqrt{x}$, $h(x) = \sqrt{-x}$, and $k(x) = -\sqrt{-x}$ in the same $[-10, 10, 1]$ by $[-4, 4, 11]$ viewing rectangle. Use the graphs to describe the domain and range of functions f, g, h, and k.

103. Use a graphing utility to graph $y_1 = \sqrt{x^2}$ and $y_2 = -x$ in the same viewing rectangle.

a. For what values of x is $\sqrt{x^2} = -x$?

b. For what values of x is $\sqrt{x^2} \neq -x$?

Critical Thinking Exercises

104. Which one of the following is true?

a. The domain of $f(x) = \sqrt[3]{x-4}$ is $[4, 00)$.

b. If n is odd and b is negative, then $\sqrt[n]{b}$ is not a real number.

c. The expression $\sqrt[n]{4}$ represents increasingly larger numbers for $n = 2, 3, 4, 5, 6$, and so on.

d. None of the above is true.

105. Write a function whose domain is $(-\infty, 5]$.

106. Let $f(x) = \sqrt{x-3}$ and $g(x) = \sqrt{x+1}$. Find the domain of $f + g$ and $\frac{f}{g}$.

107. Simplify: $\sqrt{(2x+3)^{10}}$.

In Exercises 108–109, graph each function by hand. Then describe the relationship between the function that you graphed and the graph of $f(x) = \sqrt{x}$.

108. $g(x) = \sqrt{x} + 3$

109. $h(x) = \sqrt{x+3}$

Review Exercises

110. Simplify: $3x - 2[x - 3(x+5)]$. (Section 1.2, Example 14)

111. Simplify: $(-3x^{-4}y^3)^{-2}$. (Section 1.6, Example 7c)

112. Solve: $|3x - 4| > 11$. (Section 4.3, Example 4)

▸ SECTION 7.2 Rational Exponents

Objectives

1 Use the definition of $a^{\frac{1}{n}}$.

2 Use the definition of $a^{\frac{m}{n}}$.

3 Use the definition of $a^{-\frac{m}{n}}$.

4 Use models that are functions with rational exponents.

5 Simplify expressions with rational exponents.

6 Simplify radical expressions using rational exponents.

SSM PH Tutor Center CD-ROM Video

Marine iguanas of the Galapágos Islands.

The Galápagos Islands are a volcanic chain of islands lying 600 miles west of Ecuador. They are famed for over 5000 species of plants and animals, including a rare flightless cormorant, marine iguanas, and giant tortoises weighing more than 600 pounds. Early in 2001, the plants and wildlife that live in the Galápagos chain were at risk from a massive oil spill that flooded 150,000 gallons of toxic fuel into one of the world's most fragile ecosystems. The long-term danger of the accident is that fuel sinking to the ocean floor will destroy algae that is vital to the food chain. Any imbalance in the food chain could threaten the rare Galápagos plant and animal species that have evolved for thousands of years in isolation with little human intervention.

At risk on these ecologically vulnerable islands are unique flora and fauna that helped to inspire Charles Darwin's theory of evolution. Darwin made an enormous collection of the islands' plant species. The function

$$f(x) = 29x^{\frac{1}{3}}$$

models the number of plant species, $f(x)$, on the various islands of the Galápagos chain in terms of the area, x, in square miles, of a particular island. x to the *what* power? How can we interpret the information given by this function? In this section, we turn our attention to rational exponents such as $\frac{1}{3}$ and their relationship to roots of real numbers.

Defining Rational Exponents

We define rational exponents so that their properties are the same as the properties for integer exponents. For example, suppose that $x = 7^{\frac{1}{3}}$. We know that exponents are multiplied when an exponential expression is raised to a power. For this to be true,

$$x^3 = \left(7^{\frac{1}{3}}\right)^3 = 7^{\frac{1}{3}\cdot 3} = 7^1 = 7.$$

We see that $x^3 = 7$. This means that x is the number whose cube is 7. Thus, $x = \sqrt[3]{7}$. Remember that we began with $x = 7^{\frac{1}{3}}$. This means that

$$7^{\frac{1}{3}} = \sqrt[3]{7}.$$

We can generalize this idea with the following definition:

1 Use the definition of $a^{\frac{1}{n}}$.

The Definition of $a^{\frac{1}{n}}$

If $\sqrt[n]{a}$ represents a real number and $n \geq 2$ is an integer, then

$$a^{\frac{1}{n}} = \sqrt[n]{a}.$$

The denominator of the rational exponent is the radical's index.

If a is negative, n must be odd. If a is nonnegative, n can be any index.

EXAMPLE 1 Using the Definition of $a^{\frac{1}{n}}$

Use radical notation to rewrite each expression. Simplify, if possible:

a. $64^{\frac{1}{2}}$ **b.** $(-125)^{\frac{1}{3}}$ **c.** $(6x^2y)^{\frac{1}{5}}$.

Solution

a. $64^{\frac{1}{2}} = \sqrt{64} = 8$

The denominator is the index.

b. $(-125)^{\frac{1}{3}} = \sqrt[3]{-125} = -5$

c. $(6x^2y)^{\frac{1}{5}} = \sqrt[5]{6x^2y}$ ■

CHECK POINT 1 Use radical notation to rewrite each expression. Simplify, if possible:

a. $25^{\frac{1}{2}}$ **b.** $(-8)^{\frac{1}{3}}$ **c.** $(5xy^2)^{\frac{1}{4}}$.

In our next example, we begin with radical notation and rewrite the expression with rational exponents.

The radical's index becomes the exponent's denominator.

$$\sqrt[n]{a} = a^{\frac{1}{n}}$$

The radicand becomes the base.

EXAMPLE 2 Using the Definition of $a^{\frac{1}{n}}$

Rewrite with rational exponents:

a. $\sqrt[5]{13ab}$ **b.** $\sqrt[7]{\frac{xy^2}{17}}$.

Solution Parentheses are needed to show that the entire radicand becomes the base.

a. $\sqrt[5]{13ab} = (13ab)^{\frac{1}{5}}$

The index is the exponent's denominator.

b. $\sqrt[7]{\frac{xy^2}{17}} = \left(\frac{xy^2}{17}\right)^{\frac{1}{7}}$ ■

CHECK POINT 2 Rewrite with rational exponents:

a. $\sqrt[4]{5xy}$ **b.** $\sqrt[5]{\dfrac{a^3b}{2}}$.

2 Use the definition of $a^{\frac{m}{n}}$.

Can rational exponents have numerators other than 1? The answer is yes. If the numerator is some other integer, we still want to multiply exponents when raising a power to a power. For this reason,

$$a^{\frac{2}{3}} = (a^{\frac{1}{3}})^2 \quad \text{and} \quad a^{\frac{2}{3}} = (a^2)^{\frac{1}{3}}.$$

This means $(\sqrt[3]{a})^2$. This means $\sqrt[3]{a^2}$.

Thus,

$$a^{\frac{2}{3}} = (\sqrt[3]{a})^2 = \sqrt[3]{a^2}.$$

Do you see that the denominator, 3, of the rational exponent is the same as the index of the radical? The numerator, 2, of the rational exponent serves as an exponent in each of the two radical forms. We generalize these ideas with the following definition:

The Definition of $a^{\frac{m}{n}}$

If $\sqrt[n]{a}$ represents a real number, and $\dfrac{m}{n}$ is a positive rational number, $n \geq 2$, then

$$a^{\frac{m}{n}} = (\sqrt[n]{a})^m$$

and

$$a^{\frac{m}{n}} = \sqrt[n]{a^m}.$$

The first form of the definition shown in the box involves taking the root first. This form is often preferable because smaller numbers are involved. Notice that the rational exponent consists of two parts, indicated by the following voice balloons.

The numerator is the exponent.

$$a^{\frac{m}{n}} = (\sqrt[n]{a})^m$$

The denominator is the radical's index.

ENRICHMENT ESSAY

A Radical Curiosity

The expression

$$a^{\frac{m}{n}}$$

is not defined if $\sqrt[n]{a}$ is not a real number. This can lead to some strange results. For example,

$$(-125)^{\frac{2}{6}}$$

is not defined because $\sqrt[6]{-125}$ is not a real number. However,

$$(-125)^{\frac{1}{3}}$$

is defined because

$$\sqrt[3]{-125} = -5.$$

EXAMPLE 3 Using the Definition of $a^{\frac{m}{n}}$

Use radical notation to rewrite each expression and simplify:

a. $27^{\frac{2}{3}}$ **b.** $9^{\frac{3}{2}}$ **c.** $-32^{\frac{4}{5}}$.

Solution

a. $27^{\frac{2}{3}} = (\sqrt[3]{27})^2 = 3^2 = 9$

b. $9^{\frac{3}{2}} = (\sqrt{9})^3 = 3^3 = 27$

c. $-32^{\frac{4}{5}} = -(\sqrt[5]{32})^4 = -2^4 = -16$

The base is 32 and the negative sign is not affected by the exponent.

Technology

Here are the calculator keystroke sequences for $27^{\frac{2}{3}}$:

Scientific Calculator

27 [y^x] [(] 2 [÷] 3 [)] [=]

Graphing Calculator

27 [∧] [(] 2 [÷] 3 [)] [ENTER].

✔ **CHECK POINT 3** Use radical notation to rewrite each expression and simplify.

a. $27^{\frac{4}{3}}$ **b.** $4^{\frac{3}{2}}$ **c.** $-16^{\frac{3}{4}}$

In our next example, we begin with radical notation and rewrite the expression with rational exponents. When changing from radical form to exponential form, the index becomes the denominator of the rational exponent.

EXAMPLE 4 Using the Definition of $a^{\frac{m}{n}}$

Rewrite with rational exponents:

a. $\sqrt[3]{7^5}$ **b.** $(\sqrt[4]{13xy})^9$.

Solution

a. $\sqrt[3]{7^5} = 7^{\frac{5}{3}}$

The index is the exponent's denominator.

b. $(\sqrt[4]{13xy})^9 = (13xy)^{\frac{9}{4}}$ ■

✔ **CHECK POINT 4** Rewrite with rational exponents:

a. $\sqrt[3]{6^4}$ **b.** $(\sqrt[5]{2xy})^7$.

3 Use the definition of $a^{-\frac{m}{n}}$.

Can a rational exponent be negative? Yes. The following definition is similar to the way that negative integer exponents are defined.

The Definition of $a^{-\frac{m}{n}}$

If $a^{\frac{m}{n}}$ is a nonzero real number, then

$$a^{-\frac{m}{n}} = \frac{1}{a^{\frac{m}{n}}}.$$

EXAMPLE 5 Using the Definition of $a^{-\frac{m}{n}}$

Rewrite each expression with a positive exponent. Simplify, if possible:

a. $100^{-\frac{1}{2}}$ **b.** $27^{-\frac{1}{3}}$ **c.** $81^{-\frac{3}{4}}$ **d.** $(7xy)^{-\frac{4}{7}}$.

Solution

a. $100^{-\frac{1}{2}} = \frac{1}{100^{\frac{1}{2}}} = \frac{1}{\sqrt{100}} = \frac{1}{10}$

b. $27^{-\frac{1}{3}} = \frac{1}{27^{\frac{1}{3}}} = \frac{1}{\sqrt[3]{27}} = 3$

c. $81^{-\frac{3}{4}} = \frac{1}{81^{\frac{3}{4}}} = \frac{1}{(\sqrt[4]{81})^3} = \frac{1}{3^3} = \frac{1}{27}$

d. $(7xy)^{-\frac{4}{7}} = \frac{1}{(7xy)^{\frac{4}{7}}}$ ■

Using Technology

Here are the calculator keystroke sequences for $81^{-\frac{3}{4}}$:

Scientific calculator

81 [y^x] [(] 3 [+/−] [÷] 4 [)] [=]

Graphing Calculator

81 [∧] [(] [(−)] 3 [÷] 4 [)] [ENTER]

CHECK POINT 5 Rewrite each expression with a positive exponent. Simplify, if possible.

a. $25^{-\frac{1}{2}}$ **b.** $64^{-\frac{1}{3}}$

c. $32^{-\frac{4}{5}}$ **d.** $(3xy)^{-\frac{5}{9}}$.

4 Use models that are functions with rational exponents.

Applications

Now that you know the meaning of rational exponents, you can work with functions that contain these exponents.

EXAMPLE 6 Filing Tax Returns Early

The bar graph in Figure 7.6 shows the percentage of U.S. taxpayers who file their tax returns at least one month before the April 15 deadline. The function

$$P(x) = 152x^{-\frac{1}{5}}$$

models the percentage, $P(x)$, of taxpayers who are x years old who file early. What percentage of 32-year-olds are expected to file early?

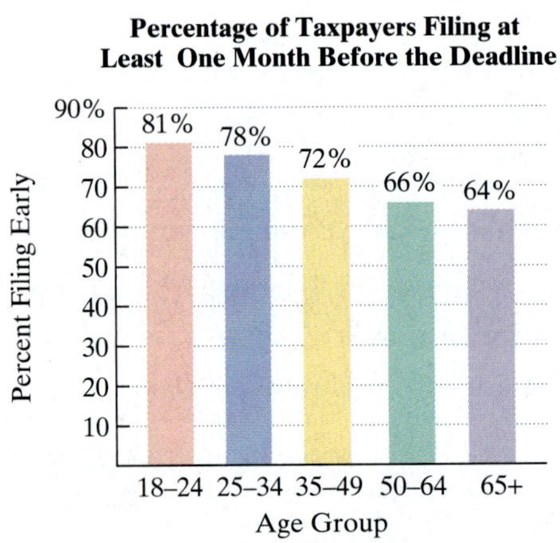

Source: Bruskin/Goldring Research

Figure 7.6

Solution Because we are interested in taxpayers of age 32, substitute 32 for x and evaluate the function at 32.

$$P(x) = 152x^{-\frac{1}{5}} \quad \text{This is the given function.}$$

$$P(32) = 152 \cdot 32^{-\frac{1}{5}} \quad \text{Substitute 32 for x.}$$

$$= \frac{152}{32^{\frac{1}{5}}} = \frac{152}{\sqrt[5]{32}} = \frac{152}{2} = 76$$

We see that $P(32) = 76$. This means that 76% of 32-year-olds are expected to file early. ■

CHECK POINT 6 The function $S(x) = 63.25x^{\frac{1}{4}}$ models the average sale price, $S(x)$, in thousands of dollars, of single-family homes in the U.S. Midwest x years after 1981. What was the average sale price in 1997?

Properties of Rational Exponents

The same properties apply to rational exponents as to integer exponents. The following is a summary of these properties:

Properties of Rational Exponents

If m and n are rational exponents, and a and b are real numbers for which the following expressions are defined, then

1. $b^m \cdot b^n = b^{m+n}$ — When multiplying exponential expressions with the same base, add the exponents.
2. $\dfrac{b^m}{b^n} = b^{m-n}$ — When dividing exponential expressions with the same base, subtract the exponents.
3. $(b^m)^n = b^{mn}$ — When an exponential expression is raised to a power, multiply the exponents.
4. $(ab)^n = a^n b^n$ — When a product is raised to a power, raise each factor to that power and multiply.
5. $\left(\dfrac{a}{b}\right)^n = \dfrac{a^n}{b^n}$ — When a quotient is raised to a power, raise the numerator and denominator to that power and divide.

5 Simplify expressions with rational exponents.

We can use these properties to simplify exponential expressions with rational exponents. As with integer exponents, an expression with rational exponents is **simplified** when:

- No parentheses appear.
- No powers are raised to powers.
- Each base occurs only once.
- No negative or zero exponents appear.

EXAMPLE 7 Simplifying Expressions with Rational Exponents

Simplify:

a. $6^{\frac{1}{7}} \cdot 6^{\frac{4}{7}}$ **b.** $\dfrac{32x^{\frac{1}{2}}}{16x^{\frac{3}{4}}}$ **c.** $\left(8.3^{\frac{3}{4}}\right)^{\frac{2}{3}}$ **d.** $\left(x^{-\frac{2}{5}}y^{\frac{1}{3}}\right)^{\frac{1}{2}}$

Solution

a. $6^{\frac{1}{7}} \cdot 6^{\frac{4}{7}} = 6^{\frac{1}{7}+\frac{4}{7}}$ — To multiply with the same base, add exponents.

$= 6^{\frac{5}{7}}$ — Simplify: $\frac{1}{7} + \frac{4}{7} = \frac{5}{7}$.

b. $\dfrac{32x^{\frac{1}{2}}}{16x^{\frac{3}{4}}} = \dfrac{32}{16}x^{\frac{1}{2}-\frac{3}{4}}$ — Divide coefficients. To divide with the same base, subtract exponents.

$= 2x^{\frac{2}{4}-\frac{3}{4}}$ — Write exponents in terms of the LCD, 4.

$= 2x^{-\frac{1}{4}}$ — Subtract: $\frac{2}{4} - \frac{3}{4} = -\frac{1}{4}$.

$= \dfrac{2}{x^{\frac{1}{4}}}$ — Rewrite with a positive exponent: $a^{-\frac{m}{n}} = \dfrac{1}{a^{\frac{m}{n}}}$.

c. $(8.3^{\frac{3}{4}})^{\frac{2}{3}} = 8.3^{\left(\frac{3}{4}\right)\left(\frac{2}{3}\right)}$ — To raise a power to a power, multiply exponents.

$= 8.3^{\frac{1}{2}}$ — Multiply: $\frac{3}{4} \cdot \frac{2}{3} = \frac{6}{12} = \frac{1}{2}$.

d. $(x^{-\frac{2}{5}}y^{\frac{1}{3}})^{\frac{1}{2}} = (x^{-\frac{2}{5}})^{\frac{1}{2}}(y^{\frac{1}{3}})^{\frac{1}{2}}$ — To raise a product to a power, raise each factor to the power.

$= x^{-\frac{1}{5}}y^{\frac{1}{6}}$ — Multiply: $-\frac{2}{5} \cdot \frac{1}{2} = -\frac{1}{5}$ and $\frac{1}{3} \cdot \frac{1}{2} = \frac{1}{6}$.

$= \frac{y^{\frac{1}{6}}}{x^{\frac{1}{5}}}$ — Rewrite with a positive exponent.

CHECK POINT 7 Simplify:

a. $7^{\frac{1}{2}} \cdot 7^{\frac{1}{3}}$

b. $\frac{50x^{\frac{1}{3}}}{10x^{\frac{4}{3}}}$

c. $(9.1^{\frac{2}{5}})^{\frac{3}{4}}$

d. $(x^{-\frac{3}{5}}y^{\frac{1}{4}})^{\frac{1}{3}}$.

6 Simplify radical expressions using rational exponents.

Using Rational Exponents to Simplify Radical Expressions

Some radical expressions can be simplified using rational exponents. We will use the following procedure:

Simplifying Radical Expressions Using Rational Exponents

1. Rewrite each radical expression as an exponential expression with a rational exponent.
2. Simplify using properties of rational exponents.
3. Rewrite in radical notation when rational exponents still appear.

EXAMPLE 8 Simplifying Radical Expressions Using Rational Exponents

Use rational exponents to simplify:

a. $\sqrt[10]{x^5}$ **b.** $\sqrt[3]{27a^{15}}$ **c.** $\sqrt[4]{x^6y^2}$ **d.** $\sqrt{x} \cdot \sqrt[3]{x}$ **e.** $\sqrt[3]{\sqrt{x}}$.

Solution

a. $\sqrt[10]{x^5} = x^{\frac{5}{10}}$ — Rewrite as an exponential expression.

$= x^{\frac{1}{2}}$ — Simplify the exponent.

$= \sqrt{x}$ — Rewrite in radical notation.

b. $\sqrt[3]{27a^{15}} = (27a^{15})^{\frac{1}{3}}$ — Rewrite as an exponential expression.

$= 27^{\frac{1}{3}}(a^{15})^{\frac{1}{3}}$ — Raise each factor in parentheses to the $\frac{1}{3}$ power.

$= \sqrt[3]{27} \cdot a^{15\left(\frac{1}{3}\right)}$ — To raise a power to a power, multiply exponents.

$= 3a^5$ — $\sqrt[3]{27} = 3$. Multiply exponents: $15 \cdot \frac{1}{3} = 5$.

c. $\sqrt[4]{x^6y^2} = (x^6y^2)^{\frac{1}{4}}$ Rewrite as an exponential expression.

$= (x^6)^{\frac{1}{4}}(y^2)^{\frac{1}{4}}$ Raise each factor in parentheses to the $\frac{1}{4}$ power.

$= x^{\frac{6}{4}}y^{\frac{2}{4}}$ To raise powers to powers, multiply.

$= x^{\frac{3}{2}}y^{\frac{1}{2}}$ Simplify.

$= (x^3y)^{\frac{1}{2}}$ $a^nb^n = (ab)^n$

$= \sqrt{x^3y}$ Rewrite in radical notation.

d. $\sqrt{x} \cdot \sqrt[3]{x} = x^{\frac{1}{2}} \cdot x^{\frac{1}{3}}$ Rewrite as exponential expressions.

$= x^{\frac{1}{2}+\frac{1}{3}}$ To multiply with the same base, add exponents.

$= x^{\frac{3}{6}+\frac{2}{6}}$ Write exponents in terms of the LCD, 6.

$= x^{\frac{5}{6}}$ Add: $\frac{3}{6} + \frac{2}{6} = \frac{5}{6}$.

$= \sqrt[6]{x^5}$ Rewrite in radical notation.

e. $\sqrt[3]{\sqrt{x}} = \sqrt[3]{x^{\frac{1}{2}}}$ Write the radicand as an exponential expression.

$= (x^{\frac{1}{2}})^{\frac{1}{3}}$ Write the entire expression in exponential form.

$= x^{\frac{1}{6}}$ To raise powers to powers, multiply: $\frac{1}{2} \cdot \frac{1}{3} = \frac{1}{6}$.

$= \sqrt[6]{x}$ Rewrite in radical notation. ■

CHECK POINT 8 Use rational exponents to simplify:

a. $\sqrt[6]{x^3}$ **b.** $\sqrt[3]{8a^{12}}$ **c.** $\sqrt[8]{x^4y^2}$ **d.** $\dfrac{\sqrt{x}}{\sqrt[3]{x}}$ **e.** $\sqrt{\sqrt[3]{x}}$.

ENRICHMENT ESSAY

Rational Exponents and Windchill

The way that we perceive the temperature on a cold day depends on both air temperature and wind speed. The **windchill temperature** is what the air temperature would have to be with no wind to achieve the same chilling effect on the skin. The formula that describes windchill temperature, W, in terms of the velocity of the wind, v, in miles per hour, and the actual air temperature, t, in degrees Fahrenheit, is

$$W = 91.4 - \frac{(10.5 + 6.7v^{\frac{1}{2}} - 0.45v)(457 - 5t)}{110}.$$

Use your calculator to describe how cold the air temperature feels (that is, the windchill temperature) when the temperature is 15° Fahrenheit and the velocity of the wind is 5 miles per hour. Contrast this with a temperature of 40° Fahrenheit and a wind blowing at 50 miles per hour.

EXERCISE SET 7.2

Practice Exercises

In Exercises 1–20, use radical notation to rewrite each expression. Simplify, if possible.

1. $49^{\frac{1}{2}}$ **2.** $100^{\frac{1}{2}}$

3. $(-27)^{\frac{1}{3}}$ **4.** $(-64)^{\frac{1}{3}}$

5. $-16^{\frac{1}{4}}$ **6.** $-81^{\frac{1}{4}}$

7. $(xy)^{\frac{1}{3}}$ **8.** $(xy)^{\frac{1}{4}}$

9. $(2xy^3)^{\frac{1}{5}}$ **10.** $(3xy^4)^{\frac{1}{5}}$

11. $81^{\frac{3}{2}}$ **12.** $25^{\frac{3}{2}}$

13. $125^{\frac{2}{3}}$ **14.** $1000^{\frac{2}{3}}$

15. $(-32)^{\frac{3}{5}}$ **16.** $(-27)^{\frac{2}{3}}$

17. $27^{\frac{2}{3}} + 16^{\frac{3}{4}}$ **18.** $4^{\frac{5}{2}} - 8^{\frac{2}{3}}$

19. $(xy)^{\frac{4}{7}}$ **20.** $(xy)^{\frac{4}{9}}$

In Exercises 21–38, rewrite each expression with rational exponents.

21. $\sqrt{7}$ **22.** $\sqrt{13}$

23. $\sqrt[3]{5}$ **24.** $\sqrt[3]{6}$

25. $\sqrt[5]{11x}$ **26.** $\sqrt[5]{13x}$

27. $\sqrt{x^3}$ **28.** $\sqrt{x^5}$

29. $\sqrt[5]{x^3}$ **30.** $\sqrt[7]{x^4}$

31. $\sqrt[5]{x^2y}$ **32.** $\sqrt[7]{xy^3}$

33. $(\sqrt{19xy})^3$ **34.** $(\sqrt{11xy})^3$

35. $(\sqrt[6]{7xy^2})^5$ **36.** $(\sqrt[6]{9x^2y})^5$

37. $2x\sqrt[3]{y^2}$ **38.** $4x\sqrt[5]{y^2}$

In Exercises 39–54, rewrite each expression with a positive rational exponent. Simplify, if possible.

39. $49^{-\frac{1}{2}}$ **40.** $9^{-\frac{1}{2}}$

41. $27^{-\frac{1}{3}}$ **42.** $125^{-\frac{1}{3}}$

43. $16^{-\frac{3}{4}}$ **44.** $81^{-\frac{5}{4}}$

45. $8^{-\frac{2}{3}}$ **46.** $32^{-\frac{4}{5}}$

47. $\left(\frac{8}{27}\right)^{-\frac{1}{3}}$ **48.** $\left(\frac{8}{125}\right)^{-\frac{1}{3}}$

49. $(-64)^{-\frac{2}{3}}$ **50.** $(-8)^{-\frac{2}{3}}$

51. $(2xy)^{-\frac{7}{10}}$ **52.** $(4xy)^{-\frac{4}{7}}$

53. $5xz^{-\frac{1}{3}}$ **54.** $7xz^{-\frac{1}{4}}$

In Exercises 55–78, use properties of rational exponents to simplify each expression. Assume that all variables represent positive numbers.

55. $3^{\frac{3}{4}} \cdot 3^{\frac{1}{4}}$ **56.** $5^{\frac{2}{3}} \cdot 5^{\frac{1}{3}}$

57. $\dfrac{16^{\frac{3}{4}}}{16^{\frac{1}{4}}}$ **58.** $\dfrac{100^{\frac{3}{4}}}{100^{\frac{1}{4}}}$

59. $x^{\frac{1}{2}} \cdot x^{\frac{1}{3}}$ **60.** $x^{\frac{1}{2}} \cdot x^{\frac{2}{3}}$

61. $\dfrac{x^{\frac{4}{5}}}{x^{\frac{1}{5}}}$ **62.** $\dfrac{x^{\frac{3}{7}}}{x^{\frac{1}{7}}}$

63. $\dfrac{x^{\frac{1}{3}}}{x^{\frac{3}{4}}}$ **64.** $\dfrac{x^{\frac{1}{4}}}{x^{\frac{3}{5}}}$

65. $(5^{\frac{2}{3}})^3$ **66.** $(3^{\frac{4}{5}})^5$

67. $(y^{-\frac{2}{3}})^{\frac{1}{4}}$ **68.** $(y^{-\frac{3}{4}})^{\frac{1}{6}}$

69. $(2x^{\frac{1}{5}})^5$ **70.** $(2x^{\frac{1}{4}})^4$

71. $(25x^4y^6)^{\frac{1}{2}}$ **72.** $(125x^9y^6)^{\frac{1}{3}}$

73. $(x^{\frac{1}{2}}y^{-\frac{3}{5}})^{\frac{1}{2}}$ **74.** $(x^{\frac{1}{4}}y^{-\frac{2}{5}})^{\frac{1}{3}}$

75. $\dfrac{3^{\frac{1}{2}} \cdot 3^{\frac{3}{4}}}{3^{\frac{1}{4}}}$ **76.** $\dfrac{5^{\frac{3}{4}} \cdot 5^{\frac{1}{2}}}{5^{\frac{1}{4}}}$

77. $\dfrac{(3y^{\frac{1}{4}})^3}{y^{\frac{1}{12}}}$ **78.** $\dfrac{(2y^{\frac{1}{5}})^4}{y^{\frac{3}{10}}}$

In Exercises 79–106, use rational exponents to simplify each expression. If rational exponents appear after simplifying, write the answer in radical notation. Assume that all variables represent positive numbers.

79. $\sqrt[8]{x^2}$ **80.** $\sqrt[10]{x^2}$

81. $\sqrt[3]{8a^6}$ **82.** $\sqrt[3]{27a^{12}}$

83. $\sqrt[5]{x^{10}y^{15}}$ **84.** $\sqrt[5]{x^{15}y^{20}}$

85. $(\sqrt[3]{xy})^{18}$

86. $(\sqrt[3]{xy})^{21}$

87. $\sqrt[10]{(3y)^2}$

88. $\sqrt[12]{(3y)^2}$

89. $(\sqrt[6]{2a})^4$

90. $(\sqrt[8]{2a})^6$

91. $\sqrt[9]{x^6y^3}$

92. $\sqrt[4]{x^2y^6}$

93. $\sqrt{2} \cdot \sqrt[3]{2}$

94. $\sqrt{3} \cdot \sqrt[3]{3}$

95. $\dfrac{\sqrt[4]{x}}{\sqrt[5]{x}}$

96. $\dfrac{\sqrt[3]{x}}{\sqrt[4]{x}}$

97. $\dfrac{\sqrt[3]{y^2}}{\sqrt[6]{y}}$

98. $\dfrac{\sqrt[5]{y^2}}{\sqrt[10]{y^3}}$

99. $\sqrt[4]{\sqrt{x}}$

100. $\sqrt[5]{\sqrt{x}}$

101. $\sqrt{\sqrt{x^2y}}$

102. $\sqrt{\sqrt{xy^2}}$

103. $\sqrt[4]{\sqrt[3]{2x}}$

104. $\sqrt[5]{\sqrt[3]{2x}}$

105. $(\sqrt[4]{x^3y^5})^{12}$

106. $(\sqrt[5]{x^4y^2})^{20}$

Application Exercises

The Galápagos Islands, lying 600 miles west of Ecuador, are famed for their extraordinary wildlife. The function

$$f(x) = 29x^{\frac{1}{3}}$$

models the number of plant species, $f(x)$, on the various islands of the Galápagos chain in terms of the area, x, in square miles, of a particular island. Use the function to solve Exercises 107–108.

107. How many species of plants are on a Galápagos island that has an area of 8 square miles?

108. How many species of plants are on a Galápagos island that has an area of 27 square miles?

The function

$$f(x) = 0.07x^{\frac{3}{2}}$$

models the duration of a storm, $f(x)$, in hours, in terms of the diameter of the storm, x, in miles. Use the function to solve Exercises 109–110. Round answers to the nearest tenth of an hour.

109. What is the duration of a storm with a diameter of 9 miles?

110. What is the duration of a storm with a diameter of 16 miles?

According to the American Management Association, the percentage of potential employees testing positive for illegal drugs is on the decline. The function

$$p(t) = \frac{73t^{\frac{1}{3}} - 28t^{\frac{2}{3}}}{t}$$

models the percentage, $p(t)$, of people applying for jobs who test positive t years after 1985. Use this model to solve Exercises 111–112.

111. What percentage of people applying for jobs tested positive for illegal drugs in 1993?

112. What percentage of people applying for jobs tested positive for illegal drugs in 2001? Use a calculator and round to nearest hundredth of a percent.

The function

$$f(x) = 70x^{\frac{3}{4}}$$

models the number of calories per day, $f(x)$, a person needs to maintain life in terms of that person's weight, x, in kilograms. (1 kilogram is approximately 2.2 pounds.) Use this model and a calculator to solve Exercises 113–114. Round answers to the nearest calorie.

113. How many calories per day does a person who weighs 80 kilograms (approximately 176 pounds) need to maintain life?

114. How many calories per day does a person who weighs 70 kilograms (approximately 154 pounds) need to maintain life?

Your job is to determine whether or not yachts are eligible for the America's Cup, the supreme event in ocean sailing. The basic dimensions of competitors' yachts must satisfy

$$L + 1.25\sqrt{S} - 9.8\sqrt[3]{D} \le 16.296,$$

where L is the yacht's length, in meters; S is its sail area, in square meters, and D is its displacement, in cubic meters. Use this information to solve Exercises 115–116.

115. a. Rewrite the inequality using rational exponents.

b. Use your calculator to determine if a yacht with length 20.85 meters, sail area 276.4 square meters, and displacement 18.55 cubic meters is eligible for the America's Cup.

116. a. Rewrite the inequality using rational exponents.

b. Use your calculator to determine if a yacht with length 22.85 meters, sail area 312.5 square meters, and displacement 22.34 cubic meters is eligible for the America's Cup.

Writing in Mathematics

117. What is the meaning of $a^{\frac{1}{n}}$? Give an example to support your explanation.

118. What is the meaning of $a^{\frac{m}{n}}$? Give an example.

119. What is the meaning of $a^{-\frac{m}{n}}$? Give an example.

120. Explain why $a^{\frac{1}{n}}$ is negative when n is odd and a is negative. What happens if n is even and a is negative? Why?

121. In simplifying $36^{\frac{3}{2}}$, is it better to use $a^{\frac{m}{n}} = \sqrt[n]{a^m}$ or $a^{\frac{m}{n}} = (\sqrt[n]{a})^m$? Explain.

122. How can you tell if an expression with rational exponents is simplified?

123. Explain how to simplify $\sqrt[3]{x} \cdot \sqrt{x}$.

124. Explain how to simplify $\sqrt[3]{\sqrt{x}}$.

125. In Exercises 111–112, you used a model for the percentage of people applying for jobs who test positive for illegal drugs. Drug testing is a controversial issue. Is testing of this kind an infringement of our right to privacy or an appropriate procedure for companies to screen potential new hires? Explain your position on this issue.

Technology Exercises

126. Use a scientific or graphing calculator to verify your results in Exercises 15–18.

127. Use a scientific or graphing calculator to verify your results in Exercises 45–50.

128. The graph of $y = x^{\frac{2}{3}}$ is shown in the figure in a $[-10, 10, 1]$ by $[-10, 10, 1]$ viewing rectangle.

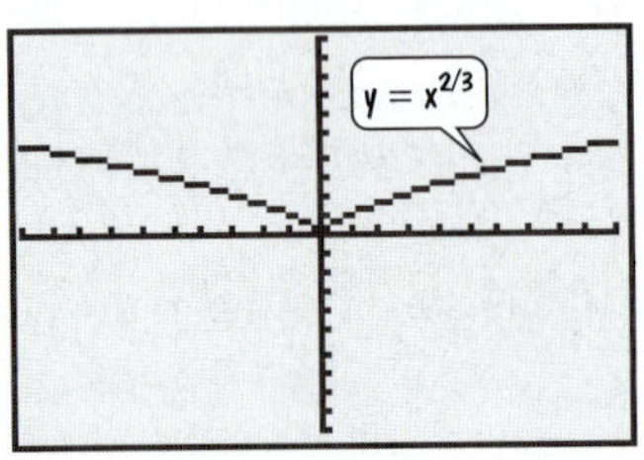

a. Use a graphing utility to obtain this graph by entering

$$Y_1 = X\boxed{\wedge}(2\boxed{\div}3).$$

Describe what happens.

b. Repeat part (a), but this time enter

$$Y_1 = (X\boxed{\wedge}(1\boxed{\div}3))\boxed{\wedge}2.$$

Describe what happens.

129. a. Graph $y = x^{\frac{2}{6}}$ by entering

$$Y_1 = (X\boxed{\wedge}(1\boxed{\div}6))\boxed{\wedge}2.$$

b. Graph $y = x^{\frac{1}{3}}$ by entering

$$Y_1 = X\boxed{\wedge}(1\boxed{\div}3).$$

c. Does your graphing utility produce the same graph in parts (a) and (b)? Can you explain what is happening?

Critical Thinking Exercises

130. Which one of the following is true?

a. If n is odd, then $(-b)^{\frac{1}{n}} = -b^{\frac{1}{n}}$.

b. $(a + b)^{\frac{1}{n}} = a^{\frac{1}{n}} + b^{\frac{1}{n}}$

c. $8^{-\frac{2}{3}} = -4$

d. None of the above is true.

131. A mathematics professor recently purchased a birthday cake for her son with the inscription

$$\text{Happy } (2^{\frac{5}{2}} \cdot 2^{\frac{3}{4}} \div 2^{\frac{1}{4}})\text{th Birthday.}$$

How old is the son?

132. The birthday boy in Exercise 131, excited by the inscription on the cake, tried to wolf down the whole thing. Professor Mom, concerned about the possible metamorphosis of her son into a blimp, exclaimed, "Hold on! It is your birthday, so why not take $\dfrac{8^{-\frac{4}{3}} + 2^{-2}}{16^{-\frac{3}{4}} + 2^{-1}}$ of the cake? I'll eat half of what's left over." How much of the cake did the professor eat?

133. Simplify: $\left[3 + (27^{\frac{2}{3}} + 32^{\frac{2}{5}})\right]^{\frac{3}{2}} - 9^{\frac{1}{2}}$.

134. Find the domain of $f(x) = (x - 3)^{\frac{1}{2}}(x + 4)^{-\frac{1}{2}}$.

Review Exercises

135. Write the equation of the linear function whose graph passes through $(5, 1)$ and $(4, 3)$. (Section 2.4, Example 2)

136. Graph, $y \le -\dfrac{3}{2}x + 3$.

(Section 4.4, Example 2)

137. Solve by Cramer's rule:

$$\begin{aligned} 5x - 3y &= 3 \\ 7x + y &= 25. \end{aligned}$$

(Section 3.5, Example 2)

▸ SECTION 7.3 Multiplying and Simplifying Radical Expressions

Objectives

1. Use the product rule to multiply radicals.
2. Use factoring and the product rule to simplify radicals.
3. Multiply radicals and then simplify.

SSM PH Tutor Center CD-ROM Video

George Tooker (b. 1920) "Mirror II" 1963, egg tempera on gesso panel, 1968.4. Gift of R.H. Donnelley Erdman (PA 1956). Addison Gallery of American Art, Phillips Academy, Andover, Massachusetts. All Rights Reserved.

A difficult challenge of middle adulthood is confronting the aging process. Middle-aged adults are forced to acknowledge their mortality as they witness the deaths of parents, colleagues, and friends. In addition, there are a number of physical transformations, including changes in vision that require glasses for reading, the onset of wrinkles and sagging skin, and a decrease in heart response. A change in heart response occurs fairly early; after 20, our hearts become less adept at accelerating in response to exercise. In this section, you will see how a radical function models changes in heart function throughout the aging process, as we turn to multiplying and simplifying radical expressions.

1 Use the product rule to multiply radicals.

The Product Rule for Radicals

A rule for multiplying radicals can be generalized by comparing $\sqrt{25} \cdot \sqrt{4}$ and $\sqrt{25 \cdot 4}$. Notice that

$$\sqrt{25} \cdot \sqrt{4} = 5 \cdot 2 = 10 \qquad \text{and} \qquad \sqrt{25 \cdot 4} = \sqrt{100} = 10.$$

Because we obtain 10 in both situations, the original radical expressions must be equal. That is,

$$\sqrt{25} \cdot \sqrt{4} = \sqrt{25 \cdot 4}.$$

This result is a special case of the **product rule for radicals** that can be generalized as follows:

The Product Rule for Radicals

If $\sqrt[n]{a}$ and $\sqrt[n]{b}$ are real numbers, then

$$\sqrt[n]{a} \cdot \sqrt[n]{b} = \sqrt[n]{ab}.$$

The product of two nth roots is the nth root of the product.

Study Tip

The product rule can be used only when the radicals have the same index. If indices differ, rational exponents can be used, as in $\sqrt{x} \cdot \sqrt[3]{x}$, which was Example 8(d) in the previous section.

EXAMPLE 1 Using the Product Rule for Radicals

Multiply:

a. $\sqrt{3} \cdot \sqrt{7}$ **b.** $\sqrt{x+7} \cdot \sqrt{x-7}$ **c.** $\sqrt[3]{7} \cdot \sqrt[3]{9}$ **d.** $\sqrt[8]{10x} \cdot \sqrt[8]{8x^4}$.

Solution In each problem, the indices are the same. Thus, we multiply by multiplying the radicands.

a. $\sqrt{3} \cdot \sqrt{7} = \sqrt{3 \cdot 7} = \sqrt{21}$

b. $\sqrt{x+7} \cdot \sqrt{x-7} = \sqrt{(x+7)(x-7)} = \sqrt{x^2 - 49}$

This is *not* equal to $\sqrt{x^2} - \sqrt{49}$.

c. $\sqrt[3]{7} \cdot \sqrt[3]{9} = \sqrt[3]{7 \cdot 9} = \sqrt[3]{63}$

d. $\sqrt[8]{10x} \cdot \sqrt[8]{8x^4} = \sqrt[8]{10x \cdot 8x^4} = \sqrt[8]{80x^5}$ ■

CHECK POINT 1 Multiply:

a. $\sqrt{5} \cdot \sqrt{11}$

b. $\sqrt{x+4} \cdot \sqrt{x-4}$

c. $\sqrt[3]{6} \cdot \sqrt[3]{10}$

d. $\sqrt[7]{2x} \cdot \sqrt[7]{6x^3}$.

2 Use factoring and the product rule to simplify radicals.

Using Factoring and the Product Rule to Simplify Radicals

A number that is the square of an integer is a **perfect square**. For example, 100 is a perfect square because $100 = 10^2$. A number is a **perfect cube** if it is the cube of an integer. Thus, 125 is a perfect cube because $125 = 5^3$. In general, a number is a **perfect *n*th power** if it is the nth power of an integer. Thus, p is a perfect nth power if there is an integer q such that $p = q^n$.

A radical of index n is **simplified** when its radicand has no factors other than 1 that are perfect nth powers. For example, $\sqrt{300}$ is not simplified because it can be expressed as $\sqrt{100 \cdot 3}$ and 100 is a perfect square. We can use the product rule in the form

$$\sqrt[n]{ab} = \sqrt[n]{a} \cdot \sqrt[n]{b}$$

to simplify $\sqrt[n]{ab}$ when $\sqrt[n]{a}$ or $\sqrt[n]{b}$ is a perfect nth power. Consider $\sqrt{300}$. To simplify, we factor 300 so that one of its factors is the largest perfect square possible.

$\sqrt{300} = \sqrt{100 \cdot 3}$ Factor 300. 100 is the largest perfect square factor.

$= \sqrt{100} \cdot \sqrt{3}$ Use the product rule: $\sqrt[n]{ab} = \sqrt[n]{a} \cdot \sqrt[n]{b}$.

$= 10\sqrt{3}$ Write $\sqrt{100}$ as 10. We read $10\sqrt{3}$ as "ten times the square root of three."

Using Technology

You can use a calculator to provide numerical support that $\sqrt{300} = 10\sqrt{3}$. For $\sqrt{300}$:

300 [√] ≈ 17.32

or

[√] 300 [ENTER] ≈ 17.32.

For $10\sqrt{3}$:

10 [×] 3 [√] ≈ 17.32

or

10 √ 3 [ENTER] ≈ 17.32.

Correct to two decimal places, $\sqrt{300} \approx 17.32$ and $10\sqrt{3} \approx 17.32$. This verifies that

$$\sqrt{300} = 10\sqrt{3}.$$

Use this technique to support the numerical results for the answers in this section.

Caution: A simplified radical does not mean a decimal approximation.

Simplifying Radical Expressions by Factoring

A radical expression whose index is n is **simplified** when its radicand has no factors that are perfect nth powers. To simplify:

1. Write the radicand as the product of two factors, one of which is the largest perfect nth power.
2. Use the product rule to take the nth root of each factor.
3. Find the nth root of the perfect nth power.

EXAMPLE 2 Simplifying Radicals by Factoring

Simplify by factoring:

a. $\sqrt{75}$ **b.** $\sqrt[3]{54}$ **c.** $\sqrt[5]{64}$ **d.** $\sqrt{500xy^2}$.

Solution

a. $\sqrt{75} = \sqrt{25 \cdot 3}$ — 25 is the largest perfect square that is a factor of 75.

$= \sqrt{25} \cdot \sqrt{3}$ — Take the square root of each factor: $\sqrt[n]{ab} = \sqrt[n]{a} \cdot \sqrt[n]{b}$.

$= 5\sqrt{3}$ — Write $\sqrt{25}$ as 5.

b. $\sqrt[3]{54} = \sqrt[3]{27 \cdot 2}$ — 27 is the largest perfect cube that is a factor of 54: $27 = 3^3$.

$= \sqrt[3]{27} \cdot \sqrt[3]{2}$ — Take the cube root of each factor: $\sqrt[n]{ab} = \sqrt[n]{a} \cdot \sqrt[n]{b}$.

$= 3\sqrt[3]{2}$ — Write $\sqrt[3]{27}$ as 3.

c. $\sqrt[5]{64} = \sqrt[5]{32 \cdot 2}$ — 32 is the largest perfect fifth power that is a factor of 64: $32 = 2^5$.

$= \sqrt[5]{32} \cdot \sqrt[5]{2}$ — Take the fifth root of each factor: $\sqrt[n]{ab} = \sqrt[n]{a} \cdot \sqrt[n]{b}$.

$= 2\sqrt[5]{2}$ — Write as 2.

d. $\sqrt{500xy^2} = \sqrt{100y^2 \cdot 5x}$ — $100y^2$ is the largest perfect square that is a factor of $500xy^2$: $100y^2 = (10y)^2$.

$= \sqrt{100y^2} \cdot \sqrt{5x}$ — Factor into two radicals.

$= 10|y|\sqrt{5x}$ — Take the square root of $100y^2$. ■

✔ **CHECK POINT 2** Simplify by factoring:

a. $\sqrt{80}$ **b.** $\sqrt[3]{40}$

c. $\sqrt[4]{32}$ **d.** $\sqrt{200x^2y}$.

EXAMPLE 3 Simplifying a Radical Function

If

$$f(x) = \sqrt{2x^2 + 4x + 2},$$

express the function, f, in simplified form.

Solution Begin by factoring the radicand. The GCF is 2. Simplification is possible if we obtain a factor that is a perfect square.

$f(x) = \sqrt{2x^2 + 4x + 2}$ — This is the given function.

$= \sqrt{2(x^2 + 2x + 1)}$ — Factor out the GCF.

$= \sqrt{2(x + 1)^2}$ — Factor the perfect square trinomial: $A^2 + 2AB + B^2 = (A + B)^2$.

$= \sqrt{2} \cdot \sqrt{(x + 1)^2}$ — Take the square root of each factor. The factor $(x + 1)^2$ is a perfect square.

$= \sqrt{2}|x + 1|$ — Take the square root of $(x + 1)^2$.

In simplified form,

$$f(x) = \sqrt{2}|x + 1|.$$

■

Using Technology

The graphs of

$$f(x) = \sqrt{2x^2 + 4x + 2}, \quad g(x) = \sqrt{2}|x + 1|, \quad \text{and} \quad h(x) = \sqrt{2}(x + 1)$$

are shown in three separate $[-5, 5, 1]$ by $[-5, 5, 1]$ viewing rectangles. The graphs in parts (a) and (b) are identical. This verifies that our simplification in Example 3 is correct. Now compare the graphs in parts (a) and (c). Can you see that they are not the same? This illustrates the importance of not leaving out absolute value bars.

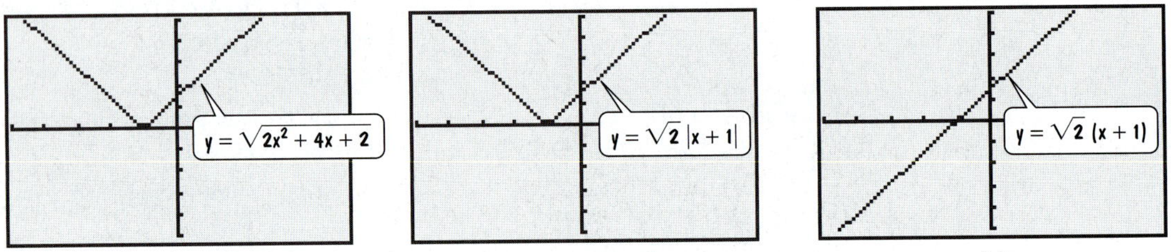

(a) (b) (c)

CHECK POINT 3 If $f(x) = \sqrt{3x^2 - 12x + 12}$, express the function, f, in simplified form.

For the remainder of this chapter, in situations that do not involve functions, we will **assume that no radicands involve negative quantities raised to even powers. Based upon this assumption, absolute value bars are not necessary when taking even routes.**

Simplifying When Variables to Even Powers in a Radicand Are Nonnegative Quantities

For any nonnegative real number a,

$$\sqrt[n]{a^n} = a.$$

In simplifying an nth root, how do we find variable factors in the radicand that are perfect nth powers? The **perfect nth powers have exponents that are divisible by n.** Simplification is possible by observation or by using rational exponents. Here are some examples:

- $\sqrt{x^6} = \sqrt{(x^3)^2} = x^3$ or $\sqrt{x^6} = (x^6)^{\frac{1}{2}} = x^3$

  6 is divisible by the index, 2. Thus, x^6 is a perfect square.

- $\sqrt[3]{y^{21}} = \sqrt[3]{(y^7)^3} = y^7$ or $\sqrt[3]{y^{21}} = (y^{21})^{\frac{1}{3}} = y^7$

  21 is divisible by the index, 3. Thus, y^{21} is a perfect cube.

- $\sqrt[6]{z^{24}} = \sqrt[6]{(z^4)^6} = z^4$ or $\sqrt[6]{z^{24}} = (z^{24})^{\frac{1}{6}} = z^4$.

  24 is divisible by the index, 6. Thus, z^{24} is a perfect 6th power.

EXAMPLE 4 Simplifying a Radical by Factoring

Simplify: $\sqrt{x^5y^{13}z^7}$.

Solution We write the radicand as the product of the largest perfect square factor and another factor. Because the index is 2, variables that have exponents that are divisible by 2 are part of the perfect square factor. We use the largest exponents that are divisible by 2.

$$\sqrt{x^5y^{13}z^7} = \sqrt{x^4 \cdot x \cdot y^{12} \cdot y \cdot z^6 \cdot z}$$ Use the largest even power of each variable.

$$= \sqrt{(x^4y^{12}z^6)(xyz)}$$ Group the perfect square factors.

$$= \sqrt{x^4y^{12}z^6} \cdot \sqrt{xyz}$$ Factor into two radicals.

$$= x^2y^6z^3\sqrt{xyz}$$ $\sqrt{x^4y^{12}z^6} = (x^4y^{12}z^6)^{\frac{1}{2}} = x^2y^6z^3$ ■

Discover for Yourself

Square the answer in Example 4 and show that it is correct. If it is a square root, you should obtain the given radicand, $x^5y^{13}z^7$.

CHECK POINT 4 Simplify: $\sqrt{x^9y^{11}z^3}$.

EXAMPLE 5 Simplifying a Radical by Factoring

Simplify: $\sqrt[3]{32x^8y^{16}}$.

Solution We write the radicand as the product of the largest perfect cube factor and another factor. Because the index is 3, variables that have exponents that are divisible by 3 are part of the perfect cube factor. We use the largest exponents that are divisible by 3.

$$\sqrt[3]{32x^8y^{16}} = \sqrt[3]{8 \cdot 4 \cdot x^6 \cdot x^2 \cdot y^{15} \cdot y}$$ Identify perfect cube factors.

$$= \sqrt[3]{(8x^6y^{15})(4x^2y)}$$ Group the perfect cube factors.

$$= \sqrt[3]{8x^6y^{15}} \cdot \sqrt[3]{4x^2y}$$ Factor into two radicals.

$$= 2x^2y^5\sqrt[3]{4x^2y}$$ $\sqrt[3]{8} = 2$ and $\sqrt[3]{x^6y^{15}} = (x^6y^{15})^{\frac{1}{3}} = x^2y^5$. ■

CHECK POINT 5 Simplify: $\sqrt[3]{40x^{10}y^{14}}$.

EXAMPLE 6 Simplifying a Radical by Factoring

Simplify: $\sqrt[5]{64x^3y^7z^{29}}$.

Solution We write the radicand as the product of the largest perfect 5th power and another factor. Because the index is 5, variables that have exponents that are divisible by 5 are part of the perfect fifth factor. We use the largest exponents that are divisible by 5.

$$\sqrt[5]{64x^3y^7z^{29}} = \sqrt[5]{32 \cdot 2 \cdot x^3 \cdot y^5 \cdot y^2 \cdot z^{25} \cdot z^4}$$ Identify perfect fifth factors.

$$= \sqrt[5]{(32y^5z^{25})(2x^3y^2z^4)}$$ Group the perfect fifth factors.

$$= \sqrt[5]{32y^5z^{25}} \cdot \sqrt[5]{2x^3y^2z^4}$$ Factor into two radicals.

$$= 2yz^5\sqrt[5]{2x^3y^2z^4}$$ $\sqrt[5]{32} = 2$ and $\sqrt[5]{y^5z^{25}} = (y^5z^{25})^{\frac{1}{5}} = yz^5$. ■

CHECK POINT 6 Simplify: $\sqrt[5]{32x^{12}y^2z^8}$.

3 Multiply radicals and then simplify.

Multiplying and Simplifying Radicals

We have seen how to use the product rule when multiplying radicals with the same index. Sometimes after multiplying, we can simplify the resulting radical.

EXAMPLE 7 Multiplying Radicals and then Simplifying

Multiply and simplify:

a. $\sqrt{15} \cdot \sqrt{3}$ **b.** $7\sqrt[3]{4} \cdot 5\sqrt[3]{6}$ **c.** $\sqrt[4]{8x^3y^2} \cdot \sqrt[4]{8x^5y^3}$.

Solution

a. $\sqrt{15} \cdot \sqrt{3} = \sqrt{15 \cdot 3}$ Use the product rule.

$= \sqrt{45} = \sqrt{9 \cdot 5}$ 9 is the largest perfect square factor of 45.

$= \sqrt{9} \cdot \sqrt{5} = 3\sqrt{5}$

b. $7\sqrt[3]{4} \cdot 5\sqrt[3]{6} = 35\sqrt[3]{4 \cdot 6}$ Use the product rule.

$= 35\sqrt[3]{24} = 35\sqrt[3]{8 \cdot 3}$ 8 is the largest perfect cube factor of 24.

$= 35\sqrt[3]{8} \cdot \sqrt[3]{3} = 35 \cdot 2 \cdot \sqrt[3]{3}$

$= 70\sqrt[3]{3}$

c. $\sqrt[4]{8x^3y^2} \cdot \sqrt[4]{8x^5y^3} = \sqrt[4]{8x^3y^2 \cdot 8x^5y^3}$ Use the product rule.

$= \sqrt[4]{64x^8y^5}$ Multiply.

$= \sqrt[4]{16 \cdot 4 \cdot x^8 \cdot y^4 \cdot y}$ Identify perfect fourth factors.

$= \sqrt[4]{(16x^8y^4)(4y)}$ Group the perfect fourth factors.

$= \sqrt[4]{16x^8y^4} \cdot \sqrt[4]{4y}$ Factor into two radicals.

$= 2x^2y\sqrt[4]{4y}$ $\sqrt[4]{16} = 2$ and

CHECK POINT 7 Multiply and simplify:

a. $\sqrt{6} \cdot \sqrt{2}$

b. $10\sqrt[3]{16} \cdot 5\sqrt[3]{2}$

c. $\sqrt[4]{4x^2y} \cdot \sqrt[4]{8x^6y^3}$.

EXERCISE SET 7.3

Practice Exercises

In Exercises 1–20, use the product rule to multiply.

1. $\sqrt{3} \cdot \sqrt{5}$
2. $\sqrt{7} \cdot \sqrt{5}$
3. $\sqrt[3]{2} \cdot \sqrt[3]{9}$
4. $\sqrt[3]{5} \cdot \sqrt[3]{4}$
5. $\sqrt[4]{11} \cdot \sqrt[4]{3}$
6. $\sqrt[5]{9} \cdot \sqrt[5]{3}$
7. $\sqrt{3x} \cdot \sqrt{11y}$
8. $\sqrt{5x} \cdot \sqrt{11y}$
9. $\sqrt[5]{6x^3} \cdot \sqrt[5]{4x}$
10. $\sqrt[4]{6x^2} \cdot \sqrt[4]{3x}$
11. $\sqrt{x+3} \cdot \sqrt{x-3}$
12. $\sqrt{x+6} \cdot \sqrt{x-6}$
13. $\sqrt[6]{x-4} \cdot \sqrt[6]{(x-4)^4}$
14. $\sqrt[6]{x-5} \cdot \sqrt[6]{(x-5)^4}$
15. $\sqrt{\frac{2x}{3}} \cdot \sqrt{\frac{3}{2}}$
16. $\sqrt{\frac{2x}{5}} \cdot \sqrt{\frac{5}{2}}$
17. $\sqrt[4]{\frac{x}{7}} \cdot \sqrt[4]{\frac{3}{y}}$
18. $\sqrt[4]{\frac{x}{3}} \cdot \sqrt[4]{\frac{7}{y}}$
19. $\sqrt[7]{7x^2y} \cdot \sqrt[7]{11x^3y^2}$
20. $\sqrt[9]{12x^2y^3} \cdot \sqrt[9]{3x^3y^4}$

In Exercises 21–32, simplify by factoring.

21. $\sqrt{50}$

22. $\sqrt{27}$

23. $\sqrt{45}$

24. $\sqrt{28}$

25. $\sqrt{75x}$

26. $\sqrt{40x}$

27. $\sqrt[3]{16}$

28. $\sqrt[3]{54}$

29. $\sqrt[3]{27x^3}$

30. $\sqrt[3]{250x^3}$

31. $\sqrt[3]{-16x^2y^3}$

32. $\sqrt[3]{-32x^2y^3}$

In Exercises 33–38, express the function, f, in simplified form. Assume that x can be any real number.

33. $f(x) = \sqrt{36(x + 2)^2}$

34. $f(x) = \sqrt{81(x - 2)^2}$

35. $f(x) = \sqrt[3]{32(x + 2)^3}$

36. $f(x) = \sqrt[3]{48(x - 2)^3}$

37. $f(x) = \sqrt{3x^2 - 6x + 3}$

38. $f(x) = \sqrt{5x^2 - 10x + 5}$

In Exercises 39–60, simplify by factoring. Assume that all variables in a radicand represent positive real numbers and no radicands involve negative quantities raised to even powers.

39. $\sqrt{x^7}$

40. $\sqrt{x^5}$

41. $\sqrt{x^8y^9}$

42. $\sqrt{x^6y^7}$

43. $\sqrt{48x^3}$

44. $\sqrt{40x^3}$

45. $\sqrt[3]{y^8}$

46. $\sqrt[3]{y^{11}}$

47. $\sqrt[3]{x^{14}y^3z}$

48. $\sqrt[3]{x^3y^{17}z^2}$

49. $\sqrt[3]{81x^8y^6}$

50. $\sqrt[3]{32x^9y^{17}}$

51. $\sqrt[3]{(x + y)^5}$

52. $\sqrt[3]{(x + y)^4}$

53. $\sqrt[5]{y^{17}}$

54. $\sqrt[5]{y^{18}}$

55. $\sqrt[5]{64x^6y^{17}}$

56. $\sqrt[5]{64x^7y^{16}}$

57. $\sqrt[4]{80x^{10}}$

58. $\sqrt[4]{96x^{11}}$

59. $\sqrt[4]{(x - 3)^{10}}$

60. $\sqrt[4]{(x - 2)^{14}}$

In Exercises 61–82, multiply and simplify. Assume that all variables in a radicand represent positive real numbers and no radicands involve negative quantities raised to even powers.

61. $\sqrt{12} \cdot \sqrt{2}$

62. $\sqrt{3} \cdot \sqrt{6}$

63. $\sqrt{5x} \cdot \sqrt{10y}$

64. $\sqrt{8x} \cdot \sqrt{10y}$

65. $\sqrt{12x} \cdot \sqrt{3x}$

66. $\sqrt{20x} \cdot \sqrt{5x}$

67. $\sqrt{50xy} \cdot \sqrt{4xy^2}$

68. $\sqrt{5xy} \cdot \sqrt{10xy^2}$

69. $2\sqrt{5} \cdot 3\sqrt{40}$

70. $3\sqrt{15} \cdot 5\sqrt{6}$

71. $\sqrt[3]{12} \cdot \sqrt[3]{4}$

72. $\sqrt[4]{4} \cdot \sqrt[4]{8}$

73. $\sqrt{5x^3} \cdot \sqrt{8x^2}$

74. $\sqrt{2x^7} \cdot \sqrt{12x^4}$

75. $\sqrt[3]{25x^4y^2} \cdot \sqrt[3]{5xy^{12}}$

76. $\sqrt[3]{6x^7y} \cdot \sqrt[3]{9x^4y^{12}}$

77. $\sqrt[4]{8x^2y^3z^6} \cdot \sqrt[4]{2x^4yz}$

78. $\sqrt[4]{4x^2y^3z^3} \cdot \sqrt[4]{8x^4yz^6}$

79. $\sqrt[5]{8x^4y^6} \cdot \sqrt[5]{8xy^7}$

80. $\sqrt[5]{8x^4y^3} \cdot \sqrt[5]{8xy^9}$

81. $\sqrt[3]{x - y} \cdot \sqrt[3]{(x - y)^7}$

82. $\sqrt[3]{x - 6} \cdot \sqrt[3]{(x - 6)^7}$

Application Exercises

Racing cyclists use the function $r(x) = 4\sqrt{x}$ to determine the maximum rate, $r(x)$, in miles per hour, to turn a corner of radius x, in feet, without tipping over. Use this function to solve Exercises 83–84.

83. What is the maximum rate that a cyclist should travel around a corner of radius 8 feet without tipping over? Write the answer in simplified radical form. Then use the simplified radical form and a calculator to express the answer to the nearest tenth.

84. What is the maximum rate that a cyclist should travel around a corner of radius 12 feet without tipping over? Write the answer in simplified radical form. Then use the simplified radical form and a calculator to express the answer to the nearest tenth.

The function

$$T(x) = \sqrt{\frac{x}{16}}$$

models the time, $T(x)$, in seconds, that it takes an object to fall x feet. Use this function to solve Exercises 85–86.

85. How long will it take a ball dropped from the top of a 320-foot building to hit the ground? Write the answer in simplified radical form. Then use the simplified radical form and a calculator to express the answer to the nearest tenth of a second.

86. How long will it take a ball dropped from the top of a 640-foot building to hit the ground? Write the answer in simplified radical form. Then use the simplified radical form and a calculator to express the answer to the nearest tenth of a second.

Your **cardiac index** *is your heart's output, in liters of blood per minute, divided by your body's surface area, in square meters. The cardiac index, $C(x)$, can be modeled by*

$$C(x) = \frac{7.644}{\sqrt[4]{x}}, \quad 10 \le x \le 80,$$

where x is an individual's age, in years. The graph of the function is shown. Use the function to solve Exercises 87–88.

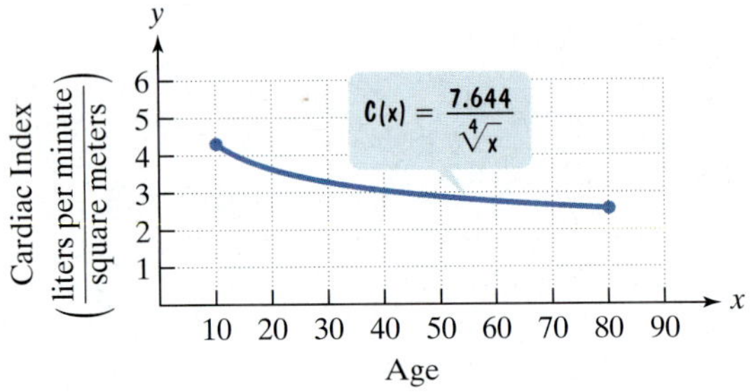

87. a. Find the cardiac index of a 32-year-old. Express the denominator in simplified radical form and reduce the fraction.

b. Use the form of the answer in part (a) and a calculator to express the cardiac index to the nearest hundredth. Identify your solution as a point on the graph.

88. a. Find the cardiac index of an 80-year-old. Express the denominator in simplified radical form and reduce the fraction.

b. Use the form of the answer in part (a) and a calculator to express the cardiac index to the nearest hundredth. Identify your solution as a point on the graph.

Writing in Mathematics

89. What is the product rule for radicals? Give an example to show how it is used.

90. Explain why $\sqrt{50}$ is not simplified. What do we mean when we say a radical expression is simplified?

91. In simplifying an *n*th root, explain how to find variable factors in the radicand that are perfect *n*th powers.

92. Without showing all the details, explain how to simplify $\sqrt[3]{16x^{14}}$.

93. As you get older, what would you expect to happen to your heart's output? Explain how this is shown in the graph for Exercises 87–88. Is this trend taking place progressively more rapidly or more slowly over the entire interval? What does this mean about this aspect of aging?

Technology Exercises

94. Use a calculator to provide numerical support for your simplifications in Exercises 21–24 and 27–28. In each case, find a decimal approximation for the given expression. Then find a decimal approximation for your simplified expression. The approximations should be the same.

In Exercises 95–98, determine if each simplification is correct by graphing the function on each side of the equation with your graphing utility. Use the given viewing rectangle. The graphs should be the same. If they are not, correct the right side of the equation and then use your graphing utility to verify the simplification.

95. $\sqrt{x^4} = x^2$
$[0, 5, 1]$ by $[0, 20, 1]$

96. $\sqrt{8x^2} = 4x\sqrt{2}$
$[-5, 5, 1]$ by $[0, 20, 1]$

97. $\sqrt{3x^2 - 6x + 3} = (x - 1)\sqrt{3}$
$[-5, 5, 1]$ by $[-5, 5, 1]$

98. $\sqrt[3]{2x} \cdot \sqrt[3]{4x^2} = 4x$
$[-10, 10, 1]$ by $[-10, 10, 1]$

Critical Thinking Exercises

99. Which one of the following is true?

a. $2\sqrt{5} \cdot 6\sqrt{5} = 12\sqrt{5}$

b. $\sqrt[3]{4} \cdot \sqrt[3]{4} = 4$

c. $\sqrt{12} = 2\sqrt{6}$

d. $\sqrt[5]{3^{25}} = 243$

100. If a number is tripled, what happens to its square root?

101. What must be done to a number so that its cube root is tripled?

102. If $f(x) = \sqrt[3]{2x}$ and $(fg)(x) = 2x$, find $g(x)$.

103. Graph $f(x) = \sqrt{(x - 1)^2}$ by hand.

Review Exercises

104. Solve: $2x - 1 \le 21$ and $2x + 2 \ge 12$.
(Section 4.2, Example 2)

105. Solve:

$$5x + 2y = 2$$
$$4x + 3y = -4.$$

(Section 3.1, Example 6)

106. Factor: $64x^3 - 27$.
(Section 5.5, Example 9)

▸ SECTION 7.4 Adding, Subtracting, and Dividing Radical Expressions

Objectives

1. Add and subtract radical expressions.
2. Use the quotient rule to simplify radical expressions.
3. Use the quotient rule to divide radical expressions.

SSM PH Tutor Center CD-ROM Video

The future is now: You have the opportunity to explore the cosmos in a starship traveling near the speed of light. The experience will enable you to understand the mysteries of the universe in deeply personal ways, transporting you to unimagined levels of knowing and being. The down side: According to Einstein's theory of relativity, close to the speed of light, your aging rate relative to friends on Earth is nearly zero. You will return from your two-year journey to a futuristic world in which friends and loved ones are long dead. Do you explore space or stay here on Earth? In this section, in addition to learning to perform various operations with radicals, you will see how they model your return to a world in which the people you knew have long departed.

1 Add and subtract radical expressions.

Adding and Subtracting Radical Expressions

Two or more radical expressions that have the same indices and radicands are called **like radicals**. Like radicals are combined in exactly the same way that we combine like terms. For example,

$$7\sqrt{11} + 6\sqrt{11} = (7 + 6)\sqrt{11} = 13\sqrt{11}.$$

7 square roots of 11 plus 6 square roots of 11 result in 13 square roots of 11.

EXAMPLE 1 Adding and Subtracting Like Radicals

Simplify (add or subtract) by combining like radical terms:

a. $7\sqrt{2} + 5\sqrt{2}$ **b.** $\sqrt[3]{5} - 4x\sqrt[3]{5} + 8\sqrt[3]{5}$

c. $8\sqrt[6]{5x} - 5\sqrt[6]{5x} + 4\sqrt[3]{5x}$.

Solution

a. $7\sqrt{2} + 5\sqrt{2}$

$= (7 + 5)\sqrt{2}$ Apply the distributive property.

$= 12\sqrt{2}$ Simplify.

b. $\sqrt[3]{5} - 4x\sqrt[3]{5} + 8\sqrt[3]{5}$

$= (1 - 4x + 8)\sqrt[3]{5}$ Apply the distributive property.

$= (9 - 4x)\sqrt[3]{5}$ Simplify.

c. $8\sqrt[6]{5x} - 5\sqrt[6]{5x} + 4\sqrt[3]{5x}$

$= (8 - 5)\sqrt[6]{5x} + 4\sqrt[3]{5x}$ Apply the distributive property to the two like radicals.

$= 3\sqrt[6]{5x} + 4\sqrt[3]{5x}$ The indices, 6 and 3, differ. These are not like terms and cannot be combined. ■

CHECK POINT 1 Simplify by combining like radical terms:

a. $8\sqrt{13} + 2\sqrt{13}$

b. $9\sqrt[3]{7} - 6x\sqrt[3]{7} + 12\sqrt[3]{7}$

c. $7\sqrt[4]{3x} - 2\sqrt[4]{3x} + 2\sqrt[3]{3x}$.

In some cases, radical expressions can be combined once they have been simplified. For example, to add $\sqrt{2}$ and $\sqrt{8}$, we can write $\sqrt{8}$ as $\sqrt{4 \cdot 2}$ because 4 is a perfect square factor of 8.

$$\sqrt{2} + \sqrt{8} = \sqrt{2} + \sqrt{4 \cdot 2} = 1\sqrt{2} + 2\sqrt{2} = (1 + 2)\sqrt{2} = 3\sqrt{2}$$

EXAMPLE 2 Combining Radicals That First Require Simplification

Simplify by combining like radical terms, if possible:

a. $7\sqrt{18} + 5\sqrt{8}$ **b.** $4\sqrt{27x} - 8\sqrt{12x}$ **c.** $7\sqrt{3} - 2\sqrt{5}$.

Solution

a. $7\sqrt{18} + 5\sqrt{8}$

$= 7\sqrt{9 \cdot 2} + 5\sqrt{4 \cdot 2}$	Factor the radicands using the largest perfect square factors.
$= 7\sqrt{9} \cdot \sqrt{2} + 5\sqrt{4} \cdot \sqrt{2}$	Take the square root of each factor.
$= 7 \cdot 3 \cdot \sqrt{2} + 5 \cdot 2 \cdot \sqrt{2}$	$\sqrt{9} = 3$ and $\sqrt{4} = 2$.
$= 21\sqrt{2} + 10\sqrt{2}$	Multiply.
$= (21 + 10)\sqrt{2}$	Apply the distributive property.
$= 31\sqrt{2}$	Simplify.

b. $4\sqrt{27x} - 8\sqrt{12x}$

$= 4\sqrt{9 \cdot 3x} - 8\sqrt{4 \cdot 3x}$	Factor the radicands using the largest perfect square factors.
$= 4\sqrt{9} \cdot \sqrt{3x} - 8\sqrt{4} \cdot \sqrt{3x}$	Take the square root of each factor.
$= 4 \cdot 3 \cdot \sqrt{3x} - 8 \cdot 2 \cdot \sqrt{3x}$	$\sqrt{9} = 3$ and $\sqrt{4} = 2$.
$= 12\sqrt{3x} - 16\sqrt{3x}$	Multiply.
$= (12 - 16)\sqrt{3x}$	Apply the distributive property.
$= -4\sqrt{3x}$	Simplify.

c. $7\sqrt{3} - 2\sqrt{5}$ cannot be simplified. The radical expressions have different radicands, namely 3 and 5, and are not like terms. ■

CHECK POINT 2 Simplify by combining like radical terms, if possible:

a. $3\sqrt{20} + 5\sqrt{45}$

b. $3\sqrt{12x} - 6\sqrt{27x}$

c. $8\sqrt{5} - 6\sqrt{2}$.

EXAMPLE 3 Adding and Subtracting with Higher Indices

Simplify by combining like radical terms, if possible:

a. $2\sqrt[3]{16} - 4\sqrt[3]{54}$ **b.** $5\sqrt[3]{xy^2} + \sqrt[3]{8x^4y^5}$.

Solution

a. $2\sqrt[3]{16} - 4\sqrt[3]{54}$

$= 2\sqrt[3]{8 \cdot 2} - 4\sqrt[3]{27 \cdot 2}$ Factor the radicands using the largest perfect cube factors.

$= 2\sqrt[3]{8} \cdot \sqrt[3]{2} - 4\sqrt[3]{27} \cdot \sqrt[3]{2}$ Take the cube root of each factor.

$= 2 \cdot 2 \cdot \sqrt[3]{2} - 4 \cdot 3 \cdot \sqrt[3]{2}$ $\sqrt[3]{8} = 2$ and $\sqrt[3]{27} = 3$.

$= 4\sqrt[3]{2} - 12\sqrt[3]{2}$ Multiply.

$= (4 - 12)\sqrt[3]{2}$ Apply the distributive property.

$= -8\sqrt[3]{2}$ Simplify.

b. $5\sqrt[3]{xy^2} + \sqrt[3]{8x^4y^5}$

$= 5\sqrt[3]{xy^2} + \sqrt[3]{(8x^3y^3)xy^2}$ Factor the second radicand using the largest perfect cube factor.

$= 5\sqrt[3]{xy^2} + \sqrt[3]{8x^3y^3} \cdot \sqrt[3]{xy^2}$ Take the cube root of each factor.

$= 5\sqrt[3]{xy^2} + 2xy\sqrt[3]{xy^2}$ $\sqrt[3]{8} = 2$ and $\sqrt[3]{x^3y^3} = (x^3y^3)^{\frac{1}{3}} = xy$.

$= (5 + 2xy)\sqrt[3]{xy^2}$ Apply the distributive property. ■

✔ **CHECK POINT 3** Simplify by combining like radical terms, if possible.

a. $3\sqrt[3]{24} - 5\sqrt[3]{81}$

b. $5\sqrt[3]{x^2y} + \sqrt[3]{27x^5y^4}$.

Dividing Radical Expressions

We have seen that the root of a product is the product of the roots. The root of a quotient can also be expressed as the quotient of roots. Here is an example:

$$\sqrt{\frac{64}{4}} = \sqrt{16} = 4 \quad \text{and} \quad \frac{\sqrt{64}}{\sqrt{4}} = \frac{8}{2} = 4.$$

This expression is the square root of a quotient.

This expression is the quotient of two square roots.

The two procedures produce the same result, 4. This is a special case of the **quotient rule for radicals.**

2 Use the quotient rule to simplify radical expressions.

The Quotient Rule for Radicals

If $\sqrt[n]{a}$ and $\sqrt[n]{b}$ are real numbers and $b \neq 0$, then

$$\sqrt[n]{\frac{a}{b}} = \frac{\sqrt[n]{a}}{\sqrt[n]{b}}.$$

The nth root of a quotient is the quotient of the nth roots.

We know that a radical is simplified when its radicand has no factors other than 1 that are perfect nth powers. The quotient rule can be used to simplify some radicals. Keep in mind that all variables in radicands represent positive real numbers.

EXAMPLE 4 Using the Quotient Rule to Simplify Radicals

Simplify using the quotient rule:

a. $\sqrt[3]{\frac{16}{27}}$ **b.** $\sqrt{\frac{x^2}{25y^6}}$ **c.** $\sqrt[4]{\frac{7y^5}{x^{12}}}$.

Solution We simplify each expression by taking the roots of the numerator and denominator. Then we use factoring to simplify the resulting radicals, if possible.

a. $\sqrt[3]{\frac{16}{27}} = \frac{\sqrt[3]{16}}{\sqrt[3]{27}} = \frac{\sqrt[3]{8 \cdot 2}}{3} = \frac{\sqrt[3]{8} \cdot \sqrt[3]{2}}{3} = \frac{2\sqrt[3]{2}}{3}$

b. $\sqrt{\frac{x^2}{25y^6}} = \frac{\sqrt{x^2}}{\sqrt{25y^6}} = \frac{x}{5(y^6)^{\frac{1}{2}}} = \frac{x}{5y^3}$

Try to do this step mentally.

c. $\sqrt[4]{\frac{7y^5}{x^{12}}} = \frac{\sqrt[4]{7y^5}}{\sqrt[4]{x^{12}}} = \frac{\sqrt[4]{y^4 \cdot 7y}}{\sqrt[4]{x^{12}}} = \frac{y\sqrt[4]{7y}}{x^3}$ ■

✔ **CHECK POINT 4** Simplify using the quotient rule:

a. $\sqrt[3]{\frac{24}{125}}$ **b.** $\sqrt{\frac{9x^3}{y^{10}}}$ **c.** $\sqrt[3]{\frac{8y^7}{x^{12}}}$.

3 Use the quotient rule to divide radical expressions.

By reversing the two sides of the quotient rule, we obtain a procedure for dividing radical expressions.

Dividing Radical Expressions

If $\sqrt[n]{a}$ and $\sqrt[n]{b}$ are real numbers and $b \neq 0$, then

$$\frac{\sqrt[n]{a}}{\sqrt[n]{b}} = \sqrt[n]{\frac{a}{b}}.$$

To divide two radical expressions with the same index, divide the radicands and retain the common index.

EXAMPLE 5 Dividing Radical Expressions

Divide and, if possible, simplify:

a. $\frac{\sqrt{48x^3}}{\sqrt{6x}}$ **b.** $\frac{\sqrt{45xy}}{2\sqrt{5}}$ **c.** $\frac{\sqrt[3]{16x^5y^2}}{\sqrt[3]{2xy^{-1}}}$.

Solution In each part of this problem, the indices in the numerator and the denominator are the same. Perform each division by dividing the radicands and retaining the common index.

a. $\frac{\sqrt{48x^3}}{\sqrt{6x}} = \sqrt{\frac{48x^3}{6x}} = \sqrt{8x^2} = \sqrt{4x^2 \cdot 2} = \sqrt{4x^2} \cdot \sqrt{2} = 2x\sqrt{2}$

b. $\frac{\sqrt{45xy}}{2\sqrt{5}} = \frac{1}{2} \cdot \sqrt{\frac{45xy}{5}} = \frac{1}{2} \cdot \sqrt{9xy} = \frac{1}{2} \cdot 3\sqrt{xy}$ or $\frac{3\sqrt{xy}}{2}$

c. $\frac{\sqrt[3]{16x^5y^2}}{\sqrt[3]{2xy^{-1}}} = \sqrt[3]{\frac{16x^5y^2}{2xy^{-1}}}$ Divide the radicands and retain the common index.

$= \sqrt[3]{8x^{5-1}y^{2-(-1)}}$ Divide factors in the radicand. Subtract exponents on common bases.

$= \sqrt[3]{8x^4y^3}$ Simplify.

$= \sqrt[3]{(8x^3y^3)x}$ Factor using the largest perfect cube factor.

$= \sqrt[3]{8x^3y^3} \cdot \sqrt[3]{x}$ Factor into two radicals.

$= 2xy\sqrt[3]{x}$ Simplify. ■

CHECK POINT 5 Divide and, if possible, simplify:

a. $\frac{\sqrt{40x^5}}{\sqrt{2x}}$ **b.** $\frac{\sqrt{50xy}}{2\sqrt{2}}$

c. $\frac{\sqrt[3]{48x^7y}}{\sqrt[3]{6xy^{-2}}}$.

EXERCISE SET 7.4

Practice Exercises

In this exercise set, assume that all variables represent positive real numbers.

In Exercises 1–10, add or subtract as indicated.

1. $8\sqrt{5} + 3\sqrt{5}$

2. $7\sqrt{3} + 2\sqrt{3}$

3. $9\sqrt[3]{6} - 2\sqrt[3]{6}$

4. $9\sqrt[3]{7} - 4\sqrt[3]{7}$

5. $4\sqrt[5]{2} + 3\sqrt[5]{2} - 5\sqrt[5]{2}$

6. $6\sqrt[5]{3} + 2\sqrt[5]{3} - 3\sqrt[5]{3}$

7. $3\sqrt{13} - 2\sqrt{5} - 2\sqrt{13} + 4\sqrt{5}$

8. $8\sqrt{17} - 5\sqrt{19} - 6\sqrt{17} + 4\sqrt{19}$

9. $3\sqrt{5} - \sqrt[3]{x} + 4\sqrt{5} + 3\sqrt[3]{x}$

10. $6\sqrt{7} - \sqrt[3]{x} + 2\sqrt{7} + 5\sqrt[3]{x}$

In Exercises 11–28, add or subtract as indicated. You will need to simplify terms to identify the like radicals.

11. $\sqrt{3} + \sqrt{27}$

12. $\sqrt{5} + \sqrt{20}$

13. $7\sqrt{12} + \sqrt{75}$

14. $5\sqrt{12} + \sqrt{75}$

15. $3\sqrt{32x} - 2\sqrt{18x}$

16. $5\sqrt{45x} - 2\sqrt{20x}$

17. $5\sqrt[3]{16} + \sqrt[3]{54}$

18. $3\sqrt[3]{24} + \sqrt[3]{81}$

19. $3\sqrt{45x^3} + \sqrt{5x}$

20. $8\sqrt{45x^3} + \sqrt{5x}$

21. $\sqrt[3]{54xy^3} + y\sqrt[3]{128x}$

22. $\sqrt[3]{24xy^3} + y\sqrt[3]{81x}$

23. $\sqrt[3]{54x^4} - \sqrt[3]{16x}$

24. $\sqrt[3]{81x^4} - \sqrt[3]{24x}$

25. $\sqrt{9x - 18} + \sqrt{x - 2}$

26. $\sqrt{4x - 12} + \sqrt{x - 3}$

27. $2\sqrt[3]{x^4y^2} + 3x\sqrt[3]{xy^2}$

28. $4\sqrt[3]{x^4y^2} + 5x\sqrt[3]{xy^2}$

In Exercises 29–44, simplify using the quotient rule.

29. $\sqrt{\frac{11}{4}}$

30. $\sqrt{\frac{19}{25}}$

31. $\sqrt[3]{\frac{19}{27}}$

32. $\sqrt[3]{\frac{11}{64}}$

33. $\sqrt{\frac{x^2}{36y^8}}$

34. $\sqrt{\frac{x^2}{144y^{12}}}$

35. $\sqrt{\frac{8x^3}{25y^6}}$

36. $\sqrt{\frac{50x^3}{81y^8}}$

37. $\sqrt[3]{\frac{x^4}{8y^3}}$

38. $\sqrt[3]{\frac{x^5}{125y^3}}$

39. $\sqrt[3]{\frac{50x^8}{27y^{12}}}$

40. $\sqrt[3]{\frac{81x^8}{8y^{15}}}$

41. $\sqrt[4]{\frac{9y^6}{x^8}}$

42. $\sqrt[4]{\frac{13y^7}{x^{12}}}$

43. $\sqrt[5]{\frac{64x^{13}}{y^{20}}}$

44. $\sqrt[5]{\frac{64x^{14}}{y^{15}}}$

In Exercises 45–62, divide and, if possible, simplify.

45. $\frac{\sqrt{40}}{\sqrt{5}}$

46. $\frac{\sqrt{200}}{\sqrt{10}}$

47. $\frac{\sqrt[3]{48}}{\sqrt[3]{6}}$

48. $\frac{\sqrt[3]{54}}{\sqrt[3]{2}}$

49. $\frac{\sqrt{54x^3}}{\sqrt{6x}}$

50. $\frac{\sqrt{72x^3}}{\sqrt{2x}}$

51. $\frac{\sqrt{x^5y^3}}{\sqrt{xy}}$

52. $\frac{\sqrt{x^7y^6}}{\sqrt{x^3y^2}}$

53. $\frac{\sqrt{200x^3}}{\sqrt{10x^{-1}}}$

54. $\frac{\sqrt{500x^3}}{\sqrt{10x^{-1}}}$

55. $\frac{\sqrt{72xy}}{2\sqrt{2}}$

56. $\frac{\sqrt{50xy}}{2\sqrt{2}}$

57. $\frac{\sqrt[3]{24x^3y^5}}{\sqrt[3]{3y^2}}$

58. $\frac{\sqrt[3]{250x^5y^3}}{\sqrt[3]{2x^3}}$

59. $\frac{\sqrt[4]{32x^{10}y^8}}{\sqrt[4]{2x^2y^{-2}}}$

60. $\frac{\sqrt[5]{96x^{12}y^{11}}}{\sqrt[5]{3x^2y^{-2}}}$

61. $\frac{\sqrt[3]{x^2 + 5x + 6}}{\sqrt[3]{x + 2}}$

62. $\frac{\sqrt[3]{x^2 + 7x + 12}}{\sqrt[3]{x + 3}}$

Application Exercises

What does travel in space have to do with radicals? Imagine that in the future we will be able to travel in starships at velocities approaching the speed of light (approximately 186,000 miles per second). According to Einstein's theory of relativity, time would pass more quickly on Earth than it would in the moving starship. The radical expression

$$R_f \frac{\sqrt{c^2 - v^2}}{\sqrt{c^2}}$$

gives the aging rate of an astronaut relative to the aging rate of a friend, R_f, on Earth. In the expression, v is the astronaut's velocity and c is the speed of light. Use the expression to solve Exercises 63–64. Imagine that you are the astronaut on the starship.

63. a. Use the quotient rule and simplify the expression that shows your aging rate relative to a friend on Earth. Working in a step-by-step manner, express your aging rate as

$$R_f\sqrt{1 - \left(\frac{v}{c}\right)^2}.$$

b. You are moving at velocities approaching the speed of light. Substitute c, the speed of light, for v in the simplified expression from part (a). Simplify completely. Close to the speed of light, what is your aging rate relative to a friend of Earth? What does this mean?

64. a. Use the quotient rule and simplify the expression that shows your aging rate relative to a friend on Earth. Working in a step-by-step manner, express your aging rate as

$$R_f\sqrt{1 - \left(\frac{v}{c}\right)^2}.$$

b. You are moving at 90% of the speed of light. Substitute $0.9c$ for v, your velocity, in the simplified expression from part (a). What is your aging rate, correct to two decimal places, relative to a friend on Earth? If you are gone for 44 weeks, approximately how many weeks have passed for your friend?

In Exercises 65–66, find the perimeter and area of each rectangle. Express answers in simplified radical form.

65.

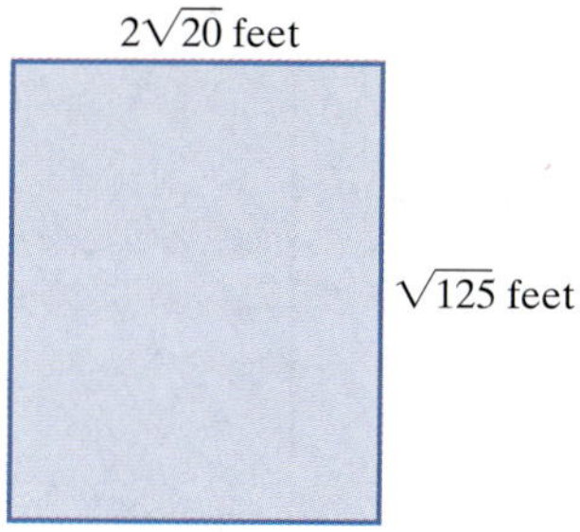

66.

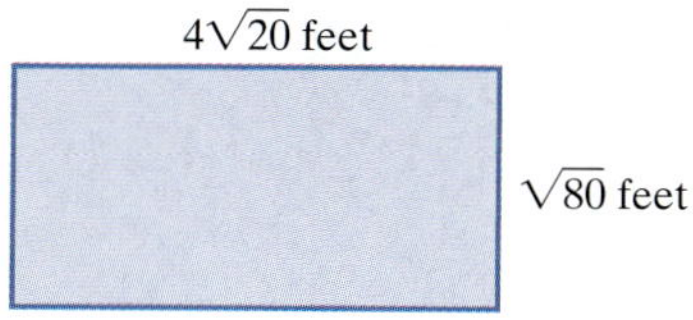

Writing in Mathematics

67. What are like radicals? Give an example with your explanation.

68. Explain how to add like radicals. Give an example with your explanation.

69. If only like radicals can be combined, why is it possible to add $\sqrt{2}$ and $\sqrt{8}$?

70. Explain how to simplify a radical expression using the quotient rule. Provide an example.

71. Explain how to divide radical expressions with the same index.

72. Answer the question posed in the section opener on page 495. What will you do: explore space or stay here on Earth? What are the reasons for your choice?

Technology Exercises

73. Use a calculator to provide numerical support to any four exercises that you worked from Exercises 1–62 that do not contain variables. Begin by finding a decimal approximation for the given expression. Then find a decimal approximation for your answer. The two decimal approximations should be the same.

In Exercises 74–76, determine if each operation is performed correctly by graphing the function on each side of the equation with your graphing utility. Use the given viewing rectangle. The graphs should be the same. If they are not, correct the right side of the equation and then use your graphing utility to verify the correction.

74. $\sqrt{4x} + \sqrt{9x} = 5\sqrt{x}$
$[0, 5, 1]$ by $[0, 10, 1]$

75. $\sqrt{16x} - \sqrt{9x} = \sqrt{7x}$
$[0, 5, 1]$ by $[0, 5, 1]$

76. $x\sqrt{8} + x\sqrt{2} = x\sqrt{10}$
$[-5, 5, 1]$ by $[-15, 15, 1]$

Critical Thinking Exercises

77. Which one of the following is true?

a. $\sqrt{5} + \sqrt{5} = \sqrt{10}$

b. $4\sqrt{3} + 5\sqrt{3} = 9\sqrt{6}$

c. If any two radical expressions are completely simplified, they can then be combined through addition or subtraction.

d. None of the above is true.

78. If an irrational number is decreased by $2\sqrt{18} - \sqrt{50}$, the result is $\sqrt{2}$. What is the number?

79. Simplify: $\dfrac{\sqrt{20}}{3} + \dfrac{\sqrt{45}}{4} - \sqrt{80}$.

80. Simplify: $\dfrac{6\sqrt{49xy}\sqrt{ab^2}}{7\sqrt{36x^{-3}y^{-5}}\sqrt{a^{-9}b^{-1}}}$.

Review Exercises

81. Solve: $2(3x - 1) - 4 = 2x - (6 - x)$.
(Section 1.4, Example 3)

82. Factor: $x^2 - 8xy + 12y^2$.
(Section 5.4, Example 4)

83. Add: $\dfrac{2}{x^2 + 5x + 6} + \dfrac{3x}{x^2 + 6x + 9}$.
(Section 6.2, Example 7)

▸ SECTION 7.5 Multiplying with More Than One Term and Rationalizing Denominators

Objectives

1. Multiply radicals with more than one term.
2. Use polynomial special products to multiply radicals.
3. Rationalize denominators containing one term.
4. Rationalize denominators containing two terms.
5. Rationalize numerators.

SSM PH Tutor Center CD-ROM Video

PEANUTS reprinted by permission of United Feature Syndicate, Inc.

The late Charles Schultz, creator of the "Peanuts" comic strip, transfixed 350 million readers worldwide with the joys and angst of his hapless Charlie Brown and Snoopy, a romantic self-deluded beagle. In 18,250 comic strips that spanned nearly 50 years, mathematics was often featured. Is the discussion of radicals shown above the real thing, or is it just an algebraic scam? By the time you complete this section on multiplying and dividing radicals, you will be able to decide.

1 Multiply radicals with more than one term.

Multiplying Radicals with More Than One Term

Radical expressions with more than one term are multiplied in much the same way that polynomials with more than one term are multiplied. Example 1 uses the distributive property and the FOIL method to perform multiplications.

EXAMPLE 1 Multiplying Radicals

Multiply:

a. $\sqrt{7}(x + \sqrt{2})$ **b.** $\sqrt[3]{x}(\sqrt[3]{6} - \sqrt[3]{x^2})$ **c.** $(5\sqrt{2} + 2\sqrt{3})(4\sqrt{2} - 3\sqrt{3})$.

Solution

a. $\sqrt{7}(x + \sqrt{2})$

$= \sqrt{7} \cdot x + \sqrt{7} \cdot \sqrt{2}$ Use the distributive property.

$= x\sqrt{7} + \sqrt{14}$ Multiply the radicals.

b. $\sqrt[3]{x}(\sqrt[3]{6} - \sqrt[3]{x^2})$

$= \sqrt[3]{x} \cdot \sqrt[3]{6} - \sqrt[3]{x} \cdot \sqrt[3]{x^2}$ Use the distributive property.

$= \sqrt[3]{6x} - \sqrt[3]{x^3}$ Multiply the radicals: $\sqrt[n]{a} \cdot \sqrt[n]{b} = \sqrt[n]{ab}$.

$= \sqrt[3]{6x} - x$ Simplify: $\sqrt[3]{x^3} = x$.

first last

c. $(5\sqrt{2} + 2\sqrt{3})(4\sqrt{2} - 3\sqrt{3})$ Use the FOIL method.

inside outside

F O I L

$= (5\sqrt{2})(4\sqrt{2}) + (5\sqrt{2})(-3\sqrt{3}) + (2\sqrt{3})(4\sqrt{2}) + (2\sqrt{3})(-3\sqrt{3})$

$= 20 \cdot 2 - 15\sqrt{6} + 8\sqrt{6} - 6 \cdot 3$ Multiply. Note that $\sqrt{2} \cdot \sqrt{2} = \sqrt{4} = 2$ and $\sqrt{3} \cdot \sqrt{3} = \sqrt{9} = 3$.

$= 40 - 15\sqrt{6} + 8\sqrt{6} - 18$ Complete the multiplications.

$= (40 - 18) + (-15\sqrt{6} + 8\sqrt{6})$ Group like terms. Try to do this step mentally.

$= 22 - 7\sqrt{6}$ Combine numerical terms and like radicals. ■

CHECK POINT 1 Multiply:

a. $\sqrt{6}(x + \sqrt{10})$

b. $\sqrt[3]{y}(\sqrt[3]{y^2} - \sqrt[3]{7})$

c. $(6\sqrt{5} + 3\sqrt{2})(2\sqrt{5} - 4\sqrt{2})$.

2 Use polynomial special products to multiply radicals.

Some radicals can be multiplied using the special products for multiplying polynomials.

Study Tip

In case you forgot them, special products for multiplying polynomials are summarized as follows:

Square of a Binomial:

$(A + B)^2 = A^2 + 2AB + B^2$

$(A - B)^2 = A^2 - 2AB + B^2$

Product of the Sum and Difference of Two Terms:

$(A + B)(A - B) = A^2 - B^2$.

EXAMPLE 2 Using Special Products to Multiply Radicals

Multiply:

a. $(\sqrt{3} + \sqrt{7})^2$ **b.** $(\sqrt{7} + \sqrt{3})(\sqrt{7} - \sqrt{3})$ **c.** $(\sqrt{a} - \sqrt{b})(\sqrt{a} + \sqrt{b})$.

Solution Use the special-product formulas shown.

a. $(A + B)^2 = A^2 + 2 \cdot A \cdot B + B^2$

$(\sqrt{3} + \sqrt{7})^2 = (\sqrt{3})^2 + 2 \cdot \sqrt{3} \cdot \sqrt{7} + (\sqrt{7})^2$ Use the special product for $(A + B)^2$.

$= 3 + 2\sqrt{21} + 7$ Multiply the radicals.

$= 10 + 2\sqrt{21}$ Simplify.

b. $(A + B) \cdot (A - B) = A^2 - B^2$

$(\sqrt{7} + \sqrt{3})(\sqrt{7} - \sqrt{3}) = (\sqrt{7})^2 - (\sqrt{3})^2$ Use the special product for $(A + B)(A - B)$.

$= 7 - 3$ Simplify: $(\sqrt{a})^2 = a$.

$= 4$

$(A - B) \cdot (A + B) = A^2 - B^2$

c. $(\sqrt{a} - \sqrt{b})(\sqrt{a} + \sqrt{b}) = (\sqrt{a})^2 - (\sqrt{b})^2 = a - b$ ■

Radical expressions that involve the sum and difference of the same two terms are called **conjugates**. For example,

$$\sqrt{7} + \sqrt{3} \qquad \text{and} \qquad \sqrt{7} - \sqrt{3}$$

are conjugates of each other. Parts (b) and (c) of Example 2 illustrate that the product of two radical expressions need not be a radical expression:

$$(\sqrt{7} + \sqrt{3})(\sqrt{7} - \sqrt{3}) = 4$$
$$(\sqrt{a} - \sqrt{b})(\sqrt{a} + \sqrt{b}) = a - b.$$

The product of conjugates does not contain a radical.

Later in this section, we will use conjugates to simplify quotients.

✔ **CHECK POINT 2** Multiply:

a. $(\sqrt{5} + \sqrt{6})^2$

b. $(\sqrt{6} + \sqrt{5})(\sqrt{6} - \sqrt{5})$

c. $(\sqrt{a} - \sqrt{7})(\sqrt{a} + \sqrt{7})$.

3 Rationalize denominators containing one term.

Rationalizing Denominators Containing One Term

You can use a calculator to compare the approximate values for $\frac{1}{\sqrt{3}}$ and $\frac{\sqrt{3}}{3}$. The two approximations are the same. This is not a coincidence:

$$\frac{1}{\sqrt{3}} = \frac{1}{\sqrt{3}} \cdot \boxed{\frac{\sqrt{3}}{\sqrt{3}}} = \frac{\sqrt{3}}{\sqrt{9}} = \frac{\sqrt{3}}{3}.$$

Any number divided by itself is 1. Multiplication by 1 does not change the value of $\frac{1}{\sqrt{3}}$.

This process involves rewriting a radical expression as an equivalent expression in which the denominator no longer contains a radical. The process is called **rationalizing the denominator**. When the denominator contains a single radical with an nth root, **multiply the numerator and denominator by a radical of index n that produces a perfect nth power in the denominator's radicand.**

EXAMPLE 3 Rationalizing Denominators

Rationalize each denominator:

a. $\frac{\sqrt{5}}{\sqrt{6}}$

b. $\sqrt[3]{\frac{7}{16}}$.

Solution

a. If we multiply the numerator and denominator of $\frac{\sqrt{5}}{\sqrt{6}}$ by $\sqrt{6}$, the denominator becomes $\sqrt{6} \cdot \sqrt{6} = \sqrt{36} = 6$. The denominator's radicand, 36, is a perfect square. The denominator no longer contains a radical. Therefore, we multiply by 1, choosing $\frac{\sqrt{6}}{\sqrt{6}}$ for 1.

$$\frac{\sqrt{5}}{\sqrt{6}} = \frac{\sqrt{5}}{\sqrt{6}} \cdot \frac{\sqrt{6}}{\sqrt{6}}$$ Multiply the numerator and denominator by $\sqrt{6}$ to remove the radical in the denominator.

$$= \frac{\sqrt{30}}{\sqrt{36}}$$ Multiply numerators and denominators. The denominator's radicand, 36, is a perfect square.

$$= \frac{\sqrt{30}}{6}$$ Simplify: $\sqrt{36} = 6$.

b. Using the quotient rule, we can express $\sqrt[3]{\frac{7}{16}}$ as $\frac{\sqrt[3]{7}}{\sqrt[3]{16}}$. We have cube roots, so we want the denominator's radicand to be a perfect cube. Right now, the denominator's radicand is 16 or 4^2. We know that $\sqrt[3]{4^3} = 4$. If we multiply the numerator and denominator of $\frac{\sqrt[3]{7}}{\sqrt[3]{16}}$ by $\sqrt[3]{4}$, the denominator becomes

$$\sqrt[3]{16} \cdot \sqrt[3]{4} = \sqrt[3]{4^2} \cdot \sqrt[3]{4} = \sqrt[3]{4^3} = 4.$$

The denominator's radicand, 4^3, is a perfect cube. The denominator no longer contains a radical. Therefore, we multiply by 1, choosing $\frac{\sqrt[3]{4}}{\sqrt[3]{4}}$ for 1.

$$\sqrt[3]{\frac{7}{16}} = \frac{\sqrt[3]{7}}{\sqrt[3]{16}}$$ Use the quotient rule and rewrite as the quotient of radicals.

$$= \frac{\sqrt[3]{7}}{\sqrt[3]{4^2}}$$ Write the denominators radicand as an exponential expression.

$$= \frac{\sqrt[3]{7}}{\sqrt[3]{4^2}} \cdot \frac{\sqrt[3]{4}}{\sqrt[3]{4}}$$ Multiply the numerator and denominator by $\sqrt[3]{4}$ to remove the radical in the denominator.

$$= \frac{\sqrt[3]{28}}{\sqrt[3]{4^3}}$$ Multiply numerators and denominators. The denominator's radicand, 4^3, is a perfect cube.

$$= \frac{\sqrt[3]{28}}{4}$$ Simplify: $\sqrt[3]{4^3} = 4$. ■

CHECK POINT 3 Rationalize each denominator:

a. $\frac{\sqrt{3}}{\sqrt{7}}$ **b.** $\sqrt[3]{\frac{2}{9}}$.

Example 3 showed that it is helpful to express the denominator's radicand using exponents. In this way, we can find the extra factor needed to produce a perfect *n*th power. For example, suppose that $\sqrt[5]{8}$ appears in the denominator. We want a perfect fifth power. By expressing $\sqrt[5]{8}$ as $\sqrt[5]{2^3}$, we would multiply the numerator and denominator by $\sqrt[5]{2^2}$ because

$$\sqrt[5]{2^3} \cdot \sqrt[5]{2^2} = \sqrt[5]{2^5} = 2.$$

EXAMPLE 4 Rationalizing Denominators

Rationalize each denominator:

a. $\sqrt{\dfrac{3x}{5y}}$ **b.** $\dfrac{\sqrt[3]{x}}{\sqrt[3]{36y}}$ **c.** $\dfrac{10y}{\sqrt[5]{4x^3y}}$.

Solution By examining each denominator, you can determine how to multiply by 1. Let's first look at the denominators. For the square root, we must produce exponents of 2 in the radicand. For the cube root, we need exponents of 3, and for the fifth root, we want exponents of 5.

- $\sqrt{5y}$

Multiply by $\sqrt{5y}$: $\sqrt{5y}\cdot\sqrt{5y}=\sqrt{25y^2}=5y$.

- $\sqrt[3]{36y}$ or $\sqrt[3]{6^2y}$

Multiply by $\sqrt[3]{6y^2}$: $\sqrt[3]{6^2y}\cdot\sqrt[3]{6y^2}=\sqrt[3]{6^3y^3}=6y$.

- $\sqrt[5]{4x^3y}$ or $\sqrt[5]{2^2x^3y}$

Multiply by $\sqrt[5]{2^3x^2y^4}$: $\sqrt[5]{2^2x^3y}\cdot\sqrt[5]{2^3x^2y^4}=\sqrt[5]{2^5x^5y^5}=2xy$.

a. $$\sqrt{\frac{3x}{5y}}=\frac{\sqrt{3x}}{\sqrt{5y}}=\frac{\sqrt{3x}}{\sqrt{5y}}\cdot\frac{\sqrt{5y}}{\sqrt{5y}}=\frac{\sqrt{15xy}}{\sqrt{25y^2}}=\frac{\sqrt{15xy}}{5y}$$

Multiply by 1. $25y^2$ is a perfect square.

b. $$\frac{\sqrt[3]{x}}{\sqrt[3]{36y}}=\frac{\sqrt[3]{x}}{\sqrt[3]{6^2y}}=\frac{\sqrt[3]{x}}{\sqrt[3]{6^2y}}\cdot\frac{\sqrt[3]{6y^2}}{\sqrt[3]{6y^2}}=\frac{\sqrt[3]{6xy^2}}{\sqrt[3]{6^3y^3}}=\frac{\sqrt[3]{6xy^2}}{6y}$$

Multiply by 1. 6^3y^3 is a perfect cube.

c. $$\frac{10y}{\sqrt[5]{4x^3y}}=\frac{10y}{\sqrt[5]{2^2x^3y}}=\frac{10y}{\sqrt[5]{2^2x^3y}}\cdot\frac{\sqrt[5]{2^3x^2y^4}}{\sqrt[5]{2^3x^2y^4}}$$

Multiply by 1.

$$=\frac{10y\sqrt[5]{2^3x^2y^4}}{\sqrt[5]{2^5x^5y^5}}=\frac{10y\sqrt[5]{8x^2y^4}}{2xy}=\frac{5\sqrt[5]{8x^2y^4}}{x}$$

$2^5x^5y^5$ is a perfect 5th power. Simplify: Divide numerator and denominator by $2y$.

CHECK POINT 4 Rationalize each denominator:

a. $\sqrt{\dfrac{2x}{7y}}$

b. $\dfrac{\sqrt[3]{x}}{\sqrt[3]{9y}}$

c. $\dfrac{6x}{\sqrt[5]{8x^2y^4}}$.

4 Rationalize denominators containing two terms.

Rationalizing Denominators Containing Two Terms

How can we rationalize a denominator if the denominator contains two terms with one or more square roots? Multiply the numerator and the denominator by the conjugate of the denominator. Here are three examples of such expressions:

- $\dfrac{8}{3\sqrt{2}+4}$ The conjugate of the denominator is $3\sqrt{2}-4$.
- $\dfrac{2+\sqrt{5}}{\sqrt{6}-\sqrt{3}}$ The conjugate of the denominator is $\sqrt{6}+\sqrt{3}$.
- $\dfrac{h}{\sqrt{x+h}-\sqrt{x}}$ The conjugate of the denominator is $\sqrt{x+h}+\sqrt{x}$.

The product of the denominator and its conjugate is found using the formula

$$(A+B)(A-B)=A^2-B^2.$$

The product will not contain a radical.

ENRICHMENT ESSAY

Golden Rectangles

The early Greeks believed that the most pleasing of all rectangles were **golden rectangles**, whose ratio of width to height is

$$\frac{w}{h}=\frac{2}{\sqrt{5}-1}.$$

The Parthenon at Athens fits into a golden rectangle once the triangular pediment is reconstructed. By working Exercise 108 in the exercise set, you will rationalize the denominator of this rational expression and obtain a better understanding about the ratio of adjacent sides in all golden rectangles.

EXAMPLE 5 Rationalizing a Denominator Containing Two Terms

Rationalize the denominator: $\dfrac{8}{3\sqrt{2}+4}$.

Solution The conjugate of the denominator is $3\sqrt{2}-4$. If we multiply the numerator and denominator by $3\sqrt{2}-4$, the denominator will not contain a radical. Therefore, we multiply by 1, choosing $\dfrac{3\sqrt{2}-4}{3\sqrt{2}-4}$ for 1.

$$\frac{8}{3\sqrt{2}+4}=\frac{8}{3\sqrt{2}+4}\cdot\frac{3\sqrt{2}-4}{3\sqrt{2}-4}$$ Multiply by 1.

$$=\frac{8(3\sqrt{2}-4)}{(3\sqrt{2})^2-4^2}$$ $(A+B)(A-B)=A^2-B^2$

$$=\frac{8(3\sqrt{2}-4)}{18-16}$$ $(3\sqrt{2})^2=9\cdot 2=18$

Leave the numerator in factored form. This helps simplify, if possible.

$$=\frac{8(3\sqrt{2}-4)}{2}$$ This expression can still be simplified.

$$=\frac{\overset{4}{\cancel{8}}(3\sqrt{2}-4)}{\underset{1}{\cancel{2}}}$$ Divide the numerator and denominator by 2.

$$=4(3\sqrt{2}-4)\quad\text{or}\quad 12\sqrt{2}-16$$ ■

✔ **CHECK POINT 5** Rationalize the denominator: $\dfrac{18}{2\sqrt{3}+3}$.

EXAMPLE 6 Rationalizing a Denominator Containing Two Terms

Rationalize the denominator: $\dfrac{2 + \sqrt{5}}{\sqrt{6} - \sqrt{3}}$.

Solution Multiplication of both the numerator and denominator by $\sqrt{6} + \sqrt{3}$ will rationalize the denominator.

$$\frac{2 + \sqrt{5}}{\sqrt{6} - \sqrt{3}} = \frac{2 + \sqrt{5}}{\sqrt{6} - \sqrt{3}} \cdot \frac{\sqrt{6} + \sqrt{3}}{\sqrt{6} + \sqrt{3}}$$

Multiply by 1.

F O I L

$$= \frac{2\sqrt{6} + 2\sqrt{3} + \sqrt{5} \cdot \sqrt{6} + \sqrt{5} \cdot \sqrt{3}}{(\sqrt{6})^2 - (\sqrt{3})^2}$$

Use FOIL on the numerator and $(A - B)(A + B) = A^2 - B^2$ in the denominator.

$$= \frac{2\sqrt{6} + 2\sqrt{3} + \sqrt{30} + \sqrt{15}}{6 - 3}$$

$$= \frac{2\sqrt{6} + 2\sqrt{3} + \sqrt{30} + \sqrt{15}}{3}$$

Further simplification is not possible. ■

✔ **CHECK POINT 6** Rationalize the denominator: $\dfrac{3 + \sqrt{7}}{\sqrt{5} - \sqrt{2}}$.

5 Rationalize numerators.

Rationalizing Numerators

We have seen that square root functions are often used to model growing phenomena with growth that is leveling off. Figure 7.7 shows a male's height as a function of his age. The pattern of his growth suggests modeling with a square root function.

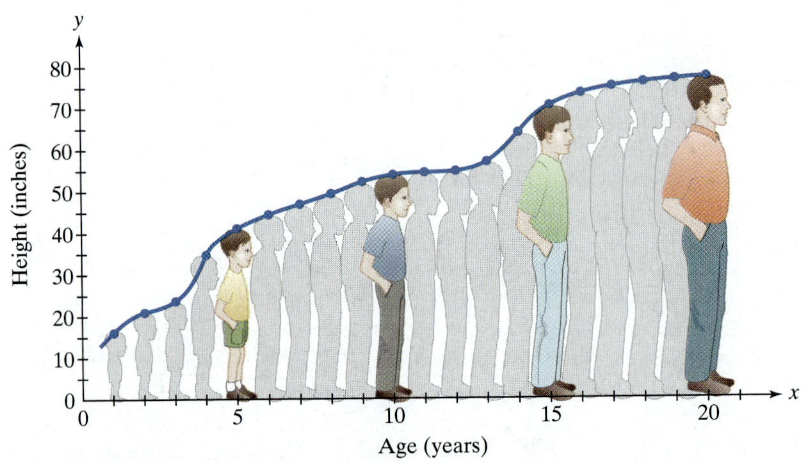

Figure 7.7

If we use $f(x) = \sqrt{x}$ to model height, $f(x)$, at age x, the expression

$$\frac{f(a+h) - f(a)}{h} = \frac{\sqrt{a+h} - \sqrt{a}}{h}$$

describes the man's average growth rate from age a to age $a + h$. Can you see that this expression is not defined if $h = 0$? However, to explore the man's average growth rates for successively shorter periods of time, we need to know what happens to the expression as h takes on values that get closer and closer to 0.

What happens to growth near the instant in time that the man is age a? The question is answered by first **rationalizing the numerator.** The procedure is similar to rationalizing the denominator. **To rationalize a numerator, multiply by 1 to eliminate the radical in the *numerator*.**

EXAMPLE 7 Rationalizing a Numerator

Rationalize the numerator:

$$\frac{\sqrt{a+h} - \sqrt{a}}{h}.$$

Solution The conjugate of the numerator is $\sqrt{a+h} + \sqrt{a}$. If we multiply the numerator and denominator by $\sqrt{a+h} + \sqrt{a}$, the numerator will not contain radicals. Therefore, we multiply by 1, choosing $\frac{\sqrt{a+h} + \sqrt{a}}{\sqrt{a+h} + \sqrt{a}}$ for 1.

$$\frac{\sqrt{a+h} - \sqrt{a}}{h} = \frac{\sqrt{a+h} - \sqrt{a}}{h} \cdot \frac{\sqrt{a+h} + \sqrt{a}}{\sqrt{a+h} + \sqrt{a}}$$ Multiply by 1.

$$= \frac{(\sqrt{a+h})^2 - (\sqrt{a})^2}{h(\sqrt{a+h} + \sqrt{a})}$$ $(A - B)(A + B) = A^2 - B^2$ Leave the denominator in factored form.

$$= \frac{a + h - a}{h(\sqrt{a+h} + \sqrt{a})}$$ $(\sqrt{a+h})^2 = a + h$ and $(\sqrt{a})^2 = a$.

$$= \frac{h}{h(\sqrt{a+h} + \sqrt{a})}$$ Simplify the numerator.

$$= \frac{1}{\sqrt{a+h} + \sqrt{a}}$$ Simplify by dividing the numerator and the denominator by h.

■

What happens to the rationalized expression in Example 7 as h is close to 0? The expression is close to

$$\frac{1}{\sqrt{a+0} + \sqrt{a}} = \frac{1}{\sqrt{a} + \sqrt{a}} = \frac{1}{2\sqrt{a}}.$$

The man's growth rate at one instant in time, age a, can be determined with this expression. This expression reveals how the square root function $f(x) = \sqrt{x}$ is changing, instant by instant.

✔ **CHECK POINT 7** Rationalize the numerator: $\frac{\sqrt{x+3} - \sqrt{x}}{3}$.

ENRICHMENT ESSAY

Roller Coasters and Instantaneous Change

Roller coaster rides give you the opportunity to spend a few hair-raising minutes plunging hundreds of feet, accelerating to 80 miles per hour in seven seconds, and enduring vertical loops that turn you upside-down. By finding a function that models your distance above the ground, techniques such as rationalizing numerators help determine the instant when your velocity is the greatest. As you experience the glorious agony of the roller coaster, this is your moment of peak terror.

EXERCISE SET 7.5

Practice Exercises

In this exercise set, assume that all variables represent positive real numbers.

In Exercises 1–38, multiply as indicated. If possible, simplify any radical expressions that appear in the product.

1. $\sqrt{2}(x + \sqrt{7})$
2. $\sqrt{5}(x + \sqrt{3})$
3. $\sqrt{6}(7 - \sqrt{6})$
4. $\sqrt{3}(5 - \sqrt{3})$
5. $\sqrt{3}(4\sqrt{6} - 2\sqrt{3})$
6. $\sqrt{6}(4\sqrt{6} - 3\sqrt{2})$
7. $\sqrt[3]{2}(\sqrt[3]{6} + 4\sqrt[3]{5})$
8. $\sqrt[3]{3}(\sqrt[3]{6} + 7\sqrt[3]{4})$
9. $\sqrt[3]{x}(\sqrt[3]{16x^2} - \sqrt[3]{x})$
10. $\sqrt[3]{x}(\sqrt[3]{24x^2} - \sqrt[3]{x})$
11. $(5 + \sqrt{2})(6 + \sqrt{2})$
12. $(7 + \sqrt{2})(8 + \sqrt{2})$
13. $(6 + \sqrt{5})(9 - 4\sqrt{5})$
14. $(4 + \sqrt{5})(10 - 3\sqrt{5})$
15. $(6 - 3\sqrt{7})(2 - 5\sqrt{7})$
16. $(7 - 2\sqrt{7})(5 - 3\sqrt{7})$
17. $(\sqrt{2} + \sqrt{7})(\sqrt{3} + \sqrt{5})$
18. $(\sqrt{3} + \sqrt{2})(\sqrt{10} + \sqrt{11})$
19. $(\sqrt{2} - \sqrt{7})(\sqrt{3} - \sqrt{5})$
20. $(\sqrt{3} - \sqrt{2})(\sqrt{10} - \sqrt{11})$
21. $(3\sqrt{2} - 4\sqrt{3})(2\sqrt{2} + 5\sqrt{3})$
22. $(3\sqrt{5} - 2\sqrt{3})(4\sqrt{5} + 5\sqrt{3})$
23. $(\sqrt{3} + \sqrt{5})^2$
24. $(\sqrt{2} + \sqrt{7})^2$
25. $(\sqrt{3x} - \sqrt{y})^2$
26. $(\sqrt{2x} - \sqrt{y})^2$
27. $(\sqrt{5} + 7)(\sqrt{5} - 7)$
28. $(\sqrt{6} + 2)(\sqrt{6} - 2)$
29. $(2 - 5\sqrt{3})(2 + 5\sqrt{3})$
30. $(3 - 5\sqrt{2})(3 + 5\sqrt{2})$
31. $(3\sqrt{2} + 2\sqrt{3})(3\sqrt{2} - 2\sqrt{3})$
32. $(4\sqrt{3} + 3\sqrt{2})(4\sqrt{3} - 3\sqrt{2})$
33. $(3 - \sqrt{x})(2 - \sqrt{x})$
34. $(4 - \sqrt{x})(3 - \sqrt{x})$
35. $(\sqrt[3]{x} - 4)(\sqrt[3]{x} + 5)$
36. $(\sqrt[3]{x} - 3)(\sqrt[3]{x} + 7)$
37. $(x + \sqrt[3]{y^2})(2x - \sqrt[3]{y^2})$
38. $(x - \sqrt[5]{y^3})(2x + \sqrt[5]{y^3})$

In Exercises 39–64, rationalize each denominator.

39. $\frac{\sqrt{2}}{\sqrt{5}}$
40. $\frac{\sqrt{7}}{\sqrt{3}}$
41. $\sqrt{\frac{11}{x}}$
42. $\sqrt{\frac{6}{x}}$
43. $\frac{9}{\sqrt{3y}}$
44. $\frac{12}{\sqrt{3y}}$
45. $\frac{1}{\sqrt[3]{2}}$
46. $\frac{1}{\sqrt[3]{3}}$
47. $\frac{6}{\sqrt[3]{4}}$
48. $\frac{10}{\sqrt[3]{5}}$
49. $\sqrt[3]{\frac{2}{3}}$
50. $\sqrt[3]{\frac{3}{4}}$
51. $\frac{4}{\sqrt[3]{x}}$
52. $\frac{7}{\sqrt[3]{x}}$
53. $\sqrt[3]{\frac{2}{y^2}}$
54. $\sqrt[3]{\frac{5}{y^2}}$
55. $\frac{7}{\sqrt[3]{2x^2}}$
56. $\frac{10}{\sqrt[3]{4x^2}}$
57. $\sqrt[3]{\frac{2}{xy^2}}$
58. $\sqrt[3]{\frac{3}{xy^2}}$
59. $\frac{3}{\sqrt[4]{x}}$
60. $\frac{5}{\sqrt[4]{x}}$
61. $\frac{6}{\sqrt[5]{8x^3}}$
62. $\frac{10}{\sqrt[5]{16x^2}}$
63. $\frac{2x^2y}{\sqrt[5]{4x^2y^4}}$
64. $\frac{3xy^2}{\sqrt[5]{8xy^3}}$

In Exercises 65–82, rationalize each denominator. Simplify, if possible.

65. $\frac{8}{\sqrt{5} + 2}$
66. $\frac{15}{\sqrt{6} + 1}$

67. $\dfrac{13}{\sqrt{11}-3}$

68. $\dfrac{17}{\sqrt{10}-2}$

69. $\dfrac{6}{\sqrt{5}+\sqrt{3}}$

70. $\dfrac{12}{\sqrt{7}+\sqrt{3}}$

71. $\dfrac{\sqrt{a}}{\sqrt{a}-\sqrt{b}}$

72. $\dfrac{\sqrt{b}}{\sqrt{a}-\sqrt{b}}$

73. $\dfrac{25}{5\sqrt{2}-3\sqrt{5}}$

74. $\dfrac{35}{5\sqrt{2}-3\sqrt{5}}$

75. $\dfrac{\sqrt{5}+\sqrt{3}}{\sqrt{5}-\sqrt{3}}$

76. $\dfrac{\sqrt{11}-\sqrt{5}}{\sqrt{11}+\sqrt{5}}$

77. $\dfrac{\sqrt{x}+1}{\sqrt{x}+3}$

78. $\dfrac{\sqrt{x}-2}{\sqrt{x}-5}$

79. $\dfrac{5\sqrt{3}-3\sqrt{2}}{3\sqrt{2}-2\sqrt{3}}$

80. $\dfrac{2\sqrt{6}+\sqrt{5}}{3\sqrt{6}-\sqrt{5}}$

81. $\dfrac{2\sqrt{x}+\sqrt{y}}{\sqrt{y}-2\sqrt{x}}$

82. $\dfrac{3\sqrt{x}+\sqrt{y}}{\sqrt{y}-3\sqrt{x}}$

In Exercises 83–94, rationalize each numerator. Simplify, if possible.

83. $\sqrt{\dfrac{3}{2}}$

84. $\sqrt{\dfrac{5}{3}}$

85. $\dfrac{\sqrt[3]{4x}}{\sqrt[3]{y}}$

86. $\dfrac{\sqrt[3]{2x}}{\sqrt[3]{y}}$

87. $\dfrac{\sqrt{x}+3}{\sqrt{x}}$

88. $\dfrac{\sqrt{x}+4}{\sqrt{x}}$

89. $\dfrac{\sqrt{a}+\sqrt{b}}{\sqrt{a}-\sqrt{b}}$

90. $\dfrac{\sqrt{a}-\sqrt{b}}{\sqrt{a}+\sqrt{b}}$

91. $\dfrac{\sqrt{x+5}-\sqrt{x}}{5}$

92. $\dfrac{\sqrt{x+7}-\sqrt{x}}{7}$

93. $\dfrac{\sqrt{x}+\sqrt{y}}{x^2-y^2}$

94. $\dfrac{\sqrt{x}-\sqrt{y}}{x^2-y^2}$

Application Exercises

The bar graph shows the percentage of U.S. households online from 1997 through 2001. The function

$$P(t) = 6.85\sqrt{t} + 19$$

models the percentage, $P(t)$, of U.S. households online t years after 1997. Use the function to solve Exercises 95–96.

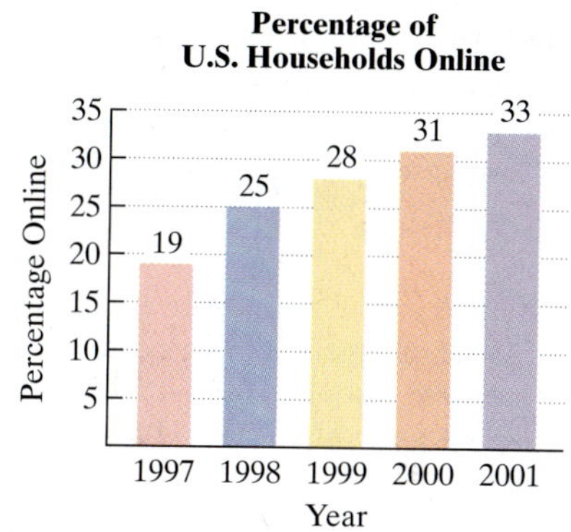

Source: Forrester Research

95. Find $P(4)$ and describe what this means. Round to the nearest percent. How well does your answer model the actual data shown in the graph?

96. Find $P(3)$ and describe what this means. Use a calculator and round to the nearest percent. How well does your answer model the actual data shown in the graph?

In Exercises 97–98, use the data shown in the bar graph for the percentage of U.S. households online.

97. The average rate of change in the percentage of households online from 1997 through 2001 is

$$\frac{\text{percentage online in 2001} - \text{percentage online in 1997}}{2001 - 1997}.$$

Change in percent

Change in time

Find the average yearly increase in the percentage of online households during this period.

98. The average rate of change in the percentage of households online from 1998 through 2001 is

$$\frac{\text{percentage online in 2001} - \text{percentage online in 1998}}{2001 - 1998}.$$

Change in percent

Change in time

Find the average yearly increase in the percentage of online households during this period.

The algebraic expression

$$6.85\left(\frac{\sqrt{t+h}-\sqrt{t}}{h}\right)$$

models the average rate of change in the percentage of households online from t years after 1997 to t + h years after 1997. Use the expression to solve Exercises 99–102.

99. Let $t = 0$ and $h = 4$. Evaluate the expression to find the average rate of change in the percentage of households online from 1997 through 2001—that is, between 0 years and 4 years after 1997. How well does your answer model the actual yearly increase in the percentage that you found in Exercise 97?

100. Let $t = 1$ and $h = 3$. Evaluate the expression to find the average rate of change in the percentage of households online from 1998 through 2001—that is, between 1 year and $1 + 3$, or 4, years after 1997. How well does your answer model the actual yearly increase in the percentage that you found in Exercise 98?

101. a. Rewrite the average rate of change as an equivalent expression by rationalizing the numerator of the rational expression in parentheses.

b. Substitute 0 for h in your expression from part (a) and simplify. The resulting expression gives the rate of change in the percentage of households online precisely t years after 1997.

c. Use the expression from part (b) to find the rate of change in the percentage of households online in 2001. Round to the nearest tenth of a percent.

102. Solve parts (a) and (b) of Exercise 101. Then find the rate of change in the percentage of households online in 1998. Round to the nearest tenth of a percent.

103. Do you expect to pay more taxes than were withheld? Would you be surprised to know that the percentages of taxpayers who receive a refund and the percentage of taxpayers who pay more taxes varies according to age? The function

$$p(x) = \frac{x(13+\sqrt{x})}{5\sqrt{x}}$$

models the percentage, $p(x)$, of taxpayers who are x years old who must pay more taxes.

a. What percentage of 25-year-olds must pay more taxes?

b. Rewrite the function by rationalizing the denominator.

104. Did you know that the U.S. infant mortality rate has been declining steadily? The function

$$f(x) = \frac{28.46}{\sqrt[3]{x^2}}$$

models the infant mortality rate, $f(x)$, in deaths per 1000 live births, x years after 1959.

a. Find and interpret $f(8)$.

b. Rewrite the function by rationalizing the denominator.

In Exercises 105–106, write expressions for the perimeter and area of each figure. Then simplify these expressions. Assume that all measures are given in inches.

105. $\sqrt{8}+1$

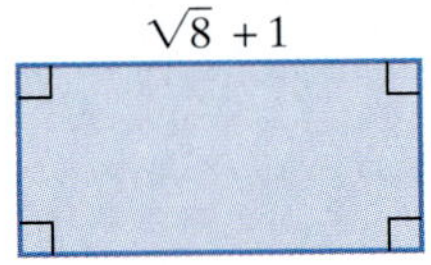

$\sqrt{8}-1$

106. $2\sqrt{3}+\sqrt{2}$

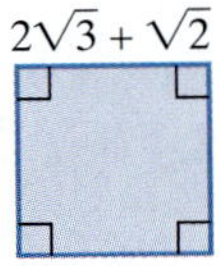

$2\sqrt{3}+\sqrt{2}$

107. In the "Peanuts" cartoon shown in the section opener on page 502, Woodstock appears to be working steps mentally. Fill in the missing steps that show how to go from $\frac{7\sqrt{2\cdot 2\cdot 3}}{6}$ to $\frac{7}{3}\sqrt{3}$.

108. Rationalize the denominator of the golden ratio

$$\frac{2}{\sqrt{5}-1},$$

discussed in the enrichment essay on page 507. Then use a calculator and find the ratio of width to height, correct to the nearest hundredth, in golden rectangles.

Writing in Mathematics

109. Explain how to perform this multiplication: $\sqrt{2}(\sqrt{7}+\sqrt{10})$.

110. Explain how to perform this multiplication: $(2+\sqrt{3})(4+\sqrt{3})$.

111. Explain how to perform this multiplication: $(2 + \sqrt{3})^2$.

112. What are conjugates? Give an example with your explanation.

113. Describe how to multiply conjugates.

114. Describe what it means to rationalize a denominator. Use both $\frac{1}{\sqrt{5}}$ and $\frac{1}{5 + \sqrt{5}}$ in your explanation.

115. When a radical expression has its denominator rationalized, we change the denominator so that it no longer contains a radical. Doesn't this change the value of the radical expression? Explain.

116. Suppose that you do not have a calculator and wish to obtain a decimal approximation for $\frac{5}{\sqrt{7}}$. From a table of square roots, you know that $\sqrt{7} \approx 2.6458$. Describe the advantage of rationalizing the denominator to obtain a decimal approximation for $\frac{5}{\sqrt{7}}$.

117. Square the real number $\frac{2}{\sqrt{3}}$. Observe that the radical is eliminated from the denominator. Explain whether this process is equivalent to rationalizing the denominator.

118. Describe the trend shown in the bar graph in Exercises 95–96 for the percentage of U.S. households online from 1997 through 2001. What explanation can you offer for this trend?

Technology Exercises

In Exercises 119–122, determine if each operation is performed correctly by graphing the function on each side of the equation with your graphing utility. Use the given viewing rectangle. The graphs should be the same. If they are not, correct the right side of the equation and then use your graphing utility to verify the correction.

119. $(\sqrt{x} - 1)(\sqrt{x} - 1) = x + 1$
$[0, 5, 1]$ by $[-1, 2, 1]$

120. $(\sqrt{x} + 2)(\sqrt{x} - 2) = x^2 - 4$ for $x = 0$
$[0, 10, 1]$ by $[-10, 10, 1]$

121. $(\sqrt{x} + 1)^2 = x + 1$
$[0, 8, 1]$ by $[0, 15, 1]$

122. $\frac{3}{\sqrt{x + 3} - \sqrt{x}} = \sqrt{x + 3} + \sqrt{x}$
$[0, 8, 1]$ by $[0, 6, 1]$

Critical Thinking Exercises

123. Which one of the following is true?

a. $\frac{\sqrt{3} + 7}{\sqrt{3} - 2} = -\frac{7}{2}$

b. $\frac{4}{\sqrt{x + y}} = \frac{4\sqrt{x - y}}{x - y}$

c. $\frac{4\sqrt{x}}{\sqrt{x} - y} = \frac{4x + 4y\sqrt{x}}{x - y^2}$

d. $(\sqrt{x} - 7)^2 = x - 49$

124. Simplify: $\sqrt{2} + \sqrt{\frac{1}{2}}$.

125. Simplify: $(\sqrt{2 + \sqrt{3}} + \sqrt{2 - \sqrt{3}})^2$.

126. Rationalize the denominator: $\frac{1}{\sqrt{2} + \sqrt{3} + \sqrt{4}}$.

127. Show that $\frac{3 + \sqrt{3}}{3}$ is a solution of $3x^2 + 2 = 6x$.

Review Exercises

128. Add: $\frac{2}{x - 2} + \frac{3}{x^2 - 4}$.
(Section 6.2, Example 6)

129. Solve: $\frac{2}{x - 2} + \frac{3}{x^2 - 4} = 0$.
(Section 6.6, Example 5)

130. If $f(x) = x^4 - 3x^2 - 2x + 5$, use synthetic division and the Remainder Theorem to find $f(-2)$. (Section 6.5, Example 2)

▸ SECTION 7.6 Radical Equations

Objectives

1 Solve radical equations.

2 Use models that are radical functions to solve problems.

SSM PH Tutor Center CD-ROM Video

A best guess at the look of our nation in the next decades indicates that the number of men and women living alone will increase each year. By 2010, approximately 28 million men and women are projected to be living alone. The function

$$f(x) = 2.6\sqrt{x} + 11$$

models the number of Americans living alone, $f(x)$, in millions, x years after 1970. How can we predict the year when, say, 29 million of us will live alone? Substitute 29 for $f(x)$ and solve for x:

$$29 = 2.6\sqrt{x} + 11.$$

The resulting equation contains a variable in the radicand and is called a *radical equation*. A **radical equation** is an equation in which the variable occurs in a square root, cube root, or any higher root. Some examples of radical equations are

$$\sqrt{2x+3} = 5, \quad \sqrt{3x+1} - \sqrt{x+4} = 1, \quad \text{and} \quad \sqrt[3]{3x-1} + 4 = 0.$$

Variables occur in radicands.

In this section, you will learn how to solve radical equations. Solving such equations will enable you to solve new kinds of problems using radical functions.

1 Solve radical equations.

Solving Radical Equations

Solving equations involving radicals involves raising both sides of the equation to a power equal to the radical's index. Unfortunately, if the index is even, all the solutions of the equation raised to the even power may not be solutions of the original equation. Consider, for example, the equation

$$x = 4.$$

If we square both sides, we obtain

$$x^2 = 16.$$

This new equation has two solutions, −4 and 4. By contrast, only 4 is a solution of the original equation, $x = 4$. For this reason, **when raising both sides of an equation to an even power, always check proposed solutions in the original equation.**

Here is a general method for solving radical equations with nth roots:

Solving Radical Equations Containing *n*th Roots

1. If necessary, arrange terms so that one radical is isolated on one side of the equation.
2. Raise both sides of the equation to the nth power to eliminate the nth root.
3. Solve the resulting equation. If this equation still contains radicals, repeat steps 1 and 2.
4. Check all proposed solutions in the original equation.

EXAMPLE 1 Solving a Radical Equation

Solve: $\sqrt{2x+3} = 5$.

Solution

Step 1 Isolate a radical on one side. The radical, $\sqrt{2x+3}$, is already isolated on the left side of the equation, so we can skip this step.

Step 2 Raise both sides to the *n*th power. Because n, the index, is 2, we square both sides.

$$\sqrt{2x+3} = 5 \quad \text{This is the given equation.}$$
$$\left(\sqrt{2x+3}\right)^2 = 5^2 \quad \text{Square both sides to eliminate the radical.}$$
$$2x + 3 = 25 \quad \text{Simplify.}$$

Step 3 Solve the resulting equation.

$$2x + 3 = 25 \quad \text{The resulting equation is a linear equation.}$$
$$2x = 22 \quad \text{Subtract 3 from both sides.}$$
$$x = 11 \quad \text{Divide both sides by 2.}$$

Step 4 Check the proposed, solution in the original equation. Because both sides were raised to an even power, this check is essential.

CHECK 11:

$$\sqrt{2x+3} = 5$$
$$\sqrt{2 \cdot 11 + 3} \stackrel{?}{=} 5$$
$$\sqrt{25} \stackrel{?}{=} 5$$
$$5 = 5, \quad \text{true}$$

The solution is 11 and the solution set is $\{11\}$. ■

✔ **CHECK POINT 1** Solve: $\sqrt{3x+4} = 8$.

EXAMPLE 2 Solving a Radical Equation

Solve: $\sqrt{x-3} + 6 = 5$.

Solution

Step 1 Isolate a radical on one side. The radical, $\sqrt{x-3}$, can be isolated by subtracting 6 from both sides. We obtain

$$\sqrt{x-3} = -1.$$

Does anything look strange about this equation? Can a principal square root be negative? It appears that this equation has no solution. However, we'll continue the solution procedure to see what happens.

Step 2 Raise both sides to the *n*th power. Because n, the index, is 2, we square both sides.

$$(\sqrt{x-3})^2 = (-1)^2$$

$$x - 3 = 1 \qquad \text{Simplify.}$$

Step 3 Solve the resulting equation.

$$x - 3 = 1 \qquad \text{The resulting equation is a linear equation.}$$

$$x = 4 \qquad \text{Add 3 to both sides.}$$

Step 4 Check the proposed solution in the original equation.

CHECK 4:

$$\sqrt{x-3} + 6 = 5$$

$$\sqrt{4-3} + 6 \stackrel{?}{=} 5$$

$$\sqrt{1} + 6 \stackrel{?}{=} 5$$

$$1 + 6 \stackrel{?}{=} 5$$

$$7 = 5, \quad \text{false}$$

This false statement indicates that 4 is not a solution. Thus, the equation has no solution. The solution set is $\varnothing$, the empty set. ■

Example 2 illustrates that extra solutions may be introduced when you raise both sides of a radical equation to an even power. Such solutions, which are not solutions of the given equation, are called **extraneous solutions**. Thus, 4 is an extraneous solution of $\sqrt{x-3} + 6 = 5$.

✔ CHECK POINT 2 Solve: $\sqrt{x-1} + 7 = 2$.

EXAMPLE 3 Solving a Radical Equation

Solve: $x + \sqrt{26 - 11x} = 4$.

Solution

Step 1 Isolate a radical on one side. We isolate the radical by subtracting x from both sides. We obtain

$$\sqrt{26 - 11x} = 4 - x.$$

Step 2 Square both sides.

$$(\sqrt{26-11x})^2 = (4-x)^2$$

$$26 - 11x = 16 - 8x + x^2 \qquad \text{Simplify. Use the formula } (A-B)^2 = A^2 - 2AB + B^2 \text{ to square } 4 - x.$$

Step 3 Solve the resulting equation. Because of the x^2-term, the resulting equation is a quadratic equation. We need to write this quadratic equation in standard form. We can obtain zero on the left side by subtracting 26 and adding $11x$ on both sides.

Study Tip

Be sure to square *both sides* of an equation. Do *not* square each term.

CORRECT:

$$(\sqrt{26-11})^2 = (4-x)^2$$

~~INCORRECT:~~

~~$(\sqrt{26-11})^2 = 4^2 - x^2$~~

$$26 - 26 - 11x + 11x = 16 - 26 - 8x + 11x + x^2$$

$$0 = x^2 + 3x - 10 \quad \text{Simplify.}$$

$$0 = (x + 5)(x - 2) \quad \text{Factor.}$$

$$x + 5 = 0 \quad \text{or} \quad x - 2 = 0 \quad \text{Set each factor equal to zero.}$$

$$x = -5 \qquad\qquad x = 2 \quad \text{Solve for } x.$$

Step 4 Check the proposed solutions in the original equation.

CHECK −5:	**CHECK 2:**
$x + \sqrt{26 - 11x} = 4$	$x + \sqrt{26 - 11x} = 4$
$-5 + \sqrt{26 - 11(-5)} \stackrel{?}{=} 4$	$2 + \sqrt{26 - 11 \cdot 2} \stackrel{?}{=} 4$
$-5 + \sqrt{81} \stackrel{?}{=} 4$	$2 + \sqrt{4} \stackrel{?}{=} 4$
$-5 + 9 \stackrel{?}{=} 4$	$2 + 2 \stackrel{?}{=} 4$
$4 = 4$, true	$4 = 4$, true

The solutions are -5 and 2, and the solution set is $\{-5, 2\}$. ■

Using Technology

The graph of

$$y = x + \sqrt{26 - 11x} - 4$$

is shown in a $[-10, 3, 1]$ by $[-4, 3, 1]$ viewing rectangle. The x-intercepts are -5 and 2. This verifies $\{-5, 2\}$ as the solution set of $x + \sqrt{26 - 11x} = 4$.

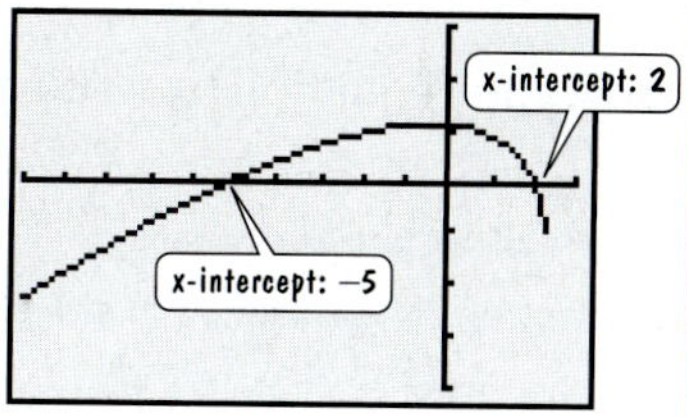

✔ **CHECK POINT 3** Solve: $\sqrt{6x + 7} - x = 2$.

The solution of radical equations with two or more square root expressions involves isolating a radical, squaring both sides, and then repeating this process. Let's consider an equation containing two square root expressions.

EXAMPLE 4 Solving an Equation Involving Two Radicals

Solve: $\sqrt{3x + 1} - \sqrt{x + 4} = 1$.

Solution

Step 1 Isolate a radical on one side. We can isolate the radical $\sqrt{3x + 1}$ by adding $\sqrt{x + 4}$ to both sides. We obtain

$$\sqrt{3x + 1} = \sqrt{x + 4} + 1.$$

Step 2 Square both sides.

$$\left(\sqrt{3x + 1}\right)^2 = \left(\sqrt{x + 4} + 1\right)^2$$

Squaring the expression on the right side of the equation can be a bit tricky. We need to use the formula

$$(A + B)^2 = A^2 + 2AB + B^2.$$

Focusing on just the right side, here is how the squaring is done.

$$(A + B)^2 = A^2 + 2 \cdot A \cdot B + B^2$$

$$\left(\sqrt{x + 4} + 1\right)^2 = \left(\sqrt{x + 4}\right)^2 + 2 \cdot \sqrt{x + 4} \cdot 1 + 1^2$$

This simplifies to $x + 4 + 2\sqrt{x + 4} + 1$. Thus, our equation can be written as follows:

$$3x + 1 = x + 4 + 2\sqrt{x + 4} + 1.$$

$$3x + 1 = x + 5 + 2\sqrt{x + 4} \quad \text{Combine numerical terms on the right.}$$

Can you see that the resulting equation still contains a radical, namely $\sqrt{x + 4}$? Thus, we need to repeat the first two steps.

Repeat Step 1 Isolate a radical on one side. We isolate $2\sqrt{x+4}$, the radical term, by subtracting $x + 5$ from both sides. We obtain

$$3x + 1 = x + 5 + 2\sqrt{x+4}$$ This is the equation on the bottom of page 517.

$$2x - 4 = 2\sqrt{x+4}.$$

Although we can simplify the equation by dividing both sides by 2, this sort of simplification is not always possible. Thus, we will work with the equation in this form.

Repeat Step 2 Square both sides.

$$(2x-4)^2 = (2\sqrt{x+4})^2$$

$$4x^2 - 16x + 16 = 4(x+4)$$

Subtract $x + 5$ from both sides. Square both the 2 and the $\sqrt{x+4}$ on the right side.

Step 3 Solve the resulting equation. We solve this quadratic equation by writing it in standard form.

$$4x^2 - 16x + 16 = 4x + 16$$ Use the distributive property.

$$4x^2 - 20x = 0$$ Subtract $4x + 16$ from both sides.

$$4x(x-5) = 0$$ Factor.

$$4x = 0 \quad \text{or} \quad x - 5 = 0$$ Set each factor equal to zero.

$$x = 0 \qquad\qquad x = 5$$ Solve for x.

Step 4 Check the proposed solutions in the original equation.

CHECK 0:

$$\sqrt{3x+1} - \sqrt{x+4} = 1$$
$$\sqrt{3\cdot 0+1} - \sqrt{0+4} \stackrel{?}{=} 1$$
$$\sqrt{1} - \sqrt{4} \stackrel{?}{=} 1$$
$$1 - 2 \stackrel{?}{=} 1$$
$$-1 = 1, \quad \text{false}$$

CHECK 5:

$$\sqrt{3x+1} - \sqrt{x+4} = 1$$
$$\sqrt{3\cdot 5+1} - \sqrt{5+4} \stackrel{?}{=} 1$$
$$\sqrt{16} - \sqrt{9} \stackrel{?}{=} 1$$
$$4 - 3 \stackrel{?}{=} 1$$
$$1 = 1, \quad \text{true}$$

The check indicates that 0 is not a solution. It is an extraneous solution brought about by squaring each side of the equation. The only solution is 5, and the solution set is $\{5\}$. ■

Using Technology

The graph of

$$y = \sqrt{3x+1} - \sqrt{x+4} - 1$$

has only one x-intercept at 5. This verifies that the solution set of

$$\sqrt{3x+1} - \sqrt{x+4} = 1$$

is $\{5\}$.

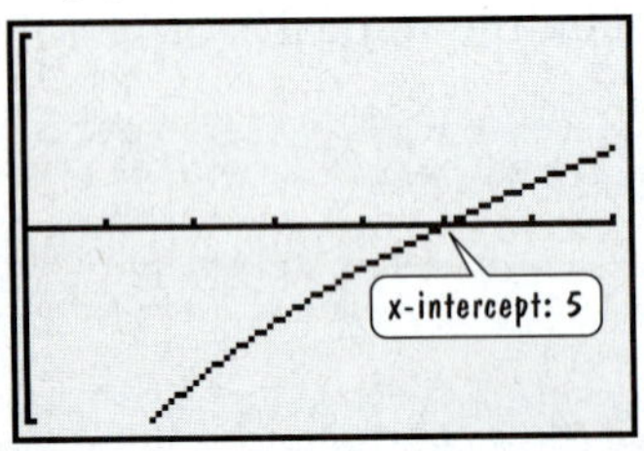

✔ **CHECK POINT 4** Solve: $\sqrt{x+5} - \sqrt{x-3} = 2$.

EXAMPLE 5 Solving a Radical Equation

Solve: $(3x-1)^{\frac{1}{3}} + 4 = 0$.

Solution Although we can rewrite the equation in radical form

$$\sqrt[3]{3x-1} + 4 = 0,$$

it is not necessary to do so. Because the equation involves a cube root, we isolate the radical term—that is, the term with the rational exponent—and cube both sides.

$$(3x-1)^{\frac{1}{3}} + 4 = 0$$ This is the given equation.

$$(3x-1)^{\frac{1}{3}} = -4$$ Subtract 4 from both sides.

$$\left[(3x-1)^{\frac{1}{3}}\right]^3 = (-4)^3$$ Cube both sides.

$$3x - 1 = -64$$ Multiply exponents on the left side and simplify.

$$3x = -63 \quad \text{Add 1 to both sides.}$$
$$x = -21 \quad \text{Divide both sides by 3.}$$

Because both sides were raised to an odd power, it is not essential to check the proposed solution, −21. However, checking is always a good idea. Do so now and verify that −21 is the solution and the solution set is $\{-21\}$. ■

Example 5 illustrates that a radical equation with rational exponents can be solved by:

1. isolating the expression with the rational exponent, and
2. raising both sides of the equation to a power that is the reciprocal of the rational exponent.

Keep in mind that it is essential to check when both sides have been raised to even powers. Thus, equations with rational exponents such as $\frac{1}{2}$ and $\frac{1}{4}$ must be checked.

CHECK POINT 5 Solve: $(2x - 3)^{\frac{1}{3}} + 3 = 0$.

2 Use models that are radical functions to solve problems.

Applications of Radical Equations

Radical equations can be solved to answer questions about variables contained in radical functions.

EXAMPLE 6 Living Alone

The function

$$f(x) = 2.6\sqrt{x} + 11$$

models the number of Americans living alone, $f(x)$, in millions, x years after 1970. The U.S. population in 2000 was approximately 280 million. According to the model, when will the number of people living alone be $\frac{1}{8}$ of the population in 2000?

Solution We are interested in $\frac{1}{8}$ of the total population, 280 million:

$$\frac{1}{8} \cdot 280 \text{ million} = 35 \text{ million.}$$

When will 35 million Americans live alone? Substitute 35 for $f(x)$ in the given function. Then solve for x, the number of years after 1970.

$$f(x) = 2.6\sqrt{x} + 11 \quad \text{This is the given function.}$$
$$35 = 2.6\sqrt{x} + 11 \quad \text{Substitute 35 for } f(x).$$
$$24 = 2.6\sqrt{x} \quad \text{Subtract 11 from both sides.}$$
$$\frac{24}{2.6} = \sqrt{x} \quad \text{Divide both sides by 2.6.}$$
$$\left(\frac{24}{2.6}\right)^2 = (\sqrt{x})^2 \quad \text{Square both sides.}$$
$$85 \approx x \quad \text{Use a calculator.}$$

The model indicates that 35 million Americans will live alone approximately 85 years after 1970. Because 1970 + 85 = 2055, this is predicted to occur in 2055. ■

CHECK POINT 6 Use the information in Example 6 to find in which year the number of Americans living alone will be $\frac{1}{10}$ of the population in 2000. Use a calculator and round to the nearest year.

ENRICHMENT ESSAY

The Percentage of Young People Who Remain Single

Although popular stereotypes suggest that being single is more difficult for women than for men, most studies find that single women are healthier and more satisfied with their lives than single men are. Evidence suggests that women get along better without men than men get along without women. (*Source*: L. Cargan, *Singles: Myths and Realities*)

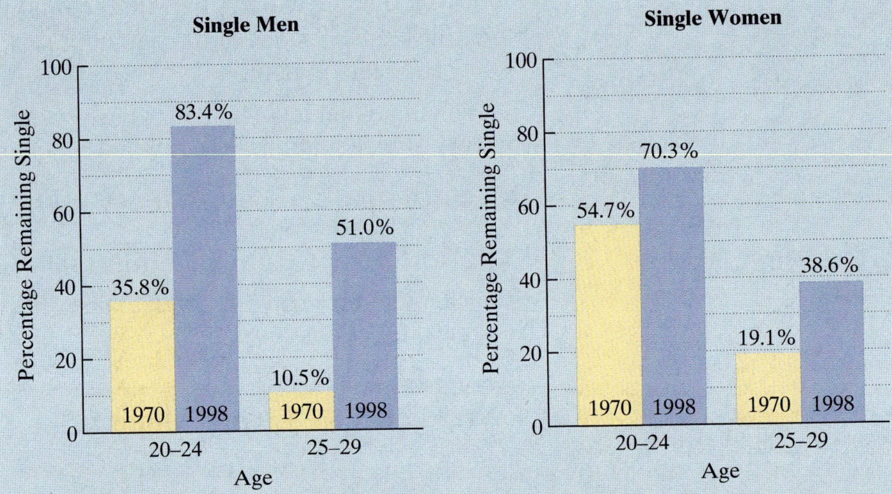

Source: U.S. Bureau of the Census

EXERCISE SET 7.6

Practice Exercises

In Exercises 1–36, solve each radical equation.

1. $\sqrt{3x-2}=4$
2. $\sqrt{5x-1}=8$
3. $\sqrt{5x-4}-9=0$
4. $\sqrt{3x-2}-5=0$
5. $\sqrt{3x+7}+10=4$
6. $\sqrt{2x+5}+11=6$
7. $x=\sqrt{7x+8}$
8. $x=\sqrt{6x+7}$
9. $\sqrt{5x+1}=x+1$
10. $\sqrt{2x+1}=x-7$
11. $x=\sqrt{2x-2}+1$
12. $x=\sqrt{3x+7}-3$
13. $x-2\sqrt{x-3}=3$
14. $3x-\sqrt{3x+7}=-5$
15. $\sqrt{2x-5}=\sqrt{x+4}$
16. $\sqrt{6x+2}=\sqrt{5x+3}$
17. $\sqrt[3]{2x+11}=3$
18. $\sqrt[3]{6x-3}=3$
19. $\sqrt[3]{2x-6}-4=0$
20. $\sqrt[3]{4x-3}-5=0$
21. $\sqrt{x-7}=7-\sqrt{x}$
22. $\sqrt{x-8}=\sqrt{x}-2$
23. $\sqrt{x+2}+\sqrt{x-1}=3$
24. $\sqrt{x-4}+\sqrt{x+4}=4$
25. $(2x+3)^{\frac{1}{3}}+4=6$
26. $(3x-6)^{\frac{1}{3}}+5=8$

27. $(3x + 1)^{\frac{1}{4}} + 7 = 9$

28. $(2x + 3)^{\frac{1}{4}} + 7 = 10$

29. $(x + 2)^{\frac{1}{2}} + 8 = 4$

30. $(x - 3)^{\frac{1}{2}} + 8 = 6$

31. $\sqrt{2x - 3} - \sqrt{x - 2} = 1$

32. $\sqrt{x + 2} + \sqrt{3x + 7} = 1$

33. $3x^{\frac{1}{3}} = (x^2 + 17x)^{\frac{1}{3}}$

34. $2(x - 1)^{\frac{1}{3}} = (x^2 + 2x)^{\frac{1}{3}}$

35. $(x + 8)^{\frac{1}{4}} = (2x)^{\frac{1}{4}}$

36. $(x - 2)^{\frac{1}{4}} = (3x - 8)^{\frac{1}{4}}$

Application Exercises

First the good news: The graph shows that U.S. seniors' scores in standard testing in science have improved since 1982. Now the bad news: The highest possible score is 500, and in 1970, the average test score was 304.

U.S. Seniors' Test Scores in Science

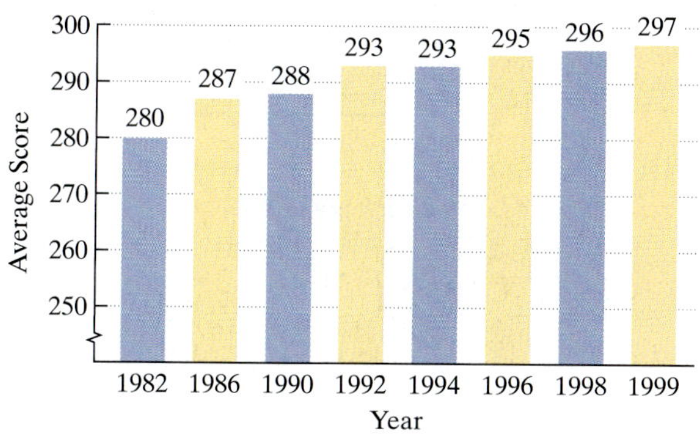

Source: National Assessment of Educational Progress

The function

$$f(x) = 4\sqrt{x} + 280$$

models the average science test score, f(x), x years after 1982. Use the model to solve Exercises 37–38.

37. When will the average science score return to the 1970 average of 304?

38. When will the average science test score be 300?

The function

$$f(x) = 6.75\sqrt{x} + 12$$

models the amount, f(x), in billions of dollars, of new student loans x years after 1993. Use the model to solve Exercises 39–40.

39. When is the amount loaned expected to reach $32.25 billion?

40. When is the amount loaned expected to reach $39 billion?

Out of a group of 50,000 births, the number of people, f(x), surviving to age x is modeled by the function

$$f(x) = 5000\sqrt{100 - x}.$$

The graph of the function is shown. Use the function to solve Exercises 41–42.

Number of Survivors, by Age, from a Group of 50,000 Births

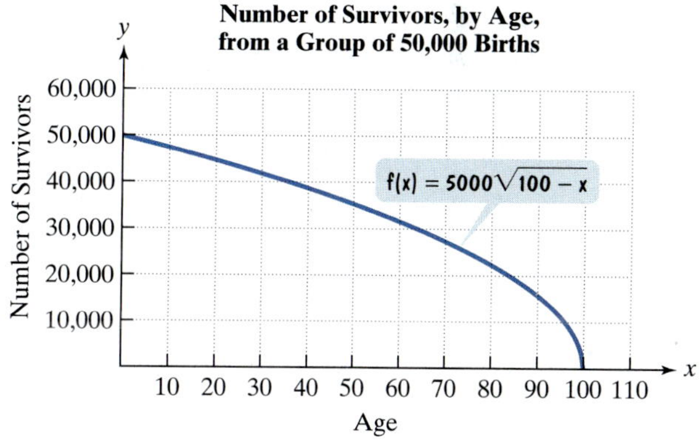

41. To what age will 40,000 people in the group survive? Identify the solution as a point on the graph of the function.

42. To what age will 35,000 people in the group survive? Identify the solution as a point on the graph of the function.

The function

$$f(x) = 29x^{\frac{1}{3}}$$

models the number of plant species, f(x), on the various islands of the Galápagos chain in terms of the area, x, in square miles, of a particular island. Use the function to solve Exercises 43–44.

43. What is the area of a Galápagos island that has 87 species of plants?

44. What is the area of a Galápagos island that has 58 species of plants?

For each planet in our solar system, its year is the time it takes the planet to revolve once around the sun. The function

$$f(x) = 0.2x^{\frac{3}{2}}$$

models the number of Earth days in a planet's year, $f(x)$, where x is the average distance of the planet from the sun, in millions of kilometers. Use the function to solve Exercises 45–46.

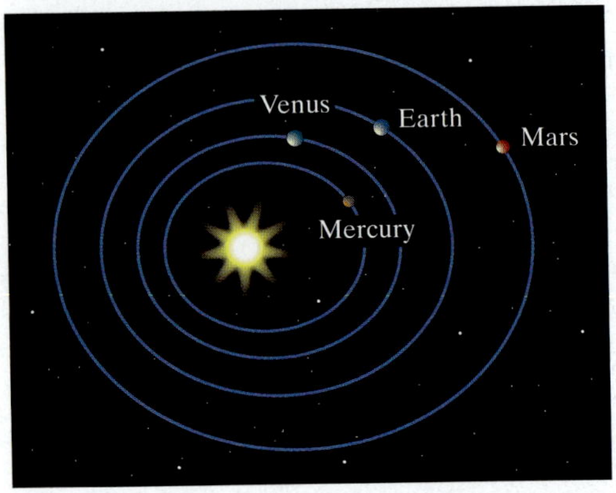

45. We, of course, have 365 Earth days in our year. What is the average distance of Earth from the sun? Use a calculator and round to the nearest million kilometers.

46. There are approximately 88 Earth days in the year of the planet Mercury. What is the average distance of Mercury from the sun? Use a calculator and round to the nearest million kilometers.

Writing in Mathematics

47. What is a radical equation?

48. In solving $\sqrt{2x-1} + 2 = x$, why is it a good idea to isolate the radical term? What if we don't do this and simply square each side? Describe what happens.

49. What is an extraneous solution to a radical equation?

50. Explain why $\sqrt{x} = -1$ has no solution.

51. Explain how to solve a radical equation with rational exponents.

52. The radical function in Example 6 on page 519 shows a gradual year-by-year increase in the number of Americans living alone. What explanations can you offer for this trend? Describe an event that might occur in the future that could cause this trend to change.

53. Describe the trend shown by the graph of f in Exercises 41–42. When is the rate of decrease most rapid? What does this mean about survival rate by age?

Technology Exercises

In Exercises 54–58, use a graphing utility to solve each radical equation. Graph each side of the equation in the given viewing rectangle. The equation's solution is given by the x-coordinate of the point(s) of intersection. Check by substitution.

54. $\sqrt{2x+2} = \sqrt{3x-5}$
$[-1, 10, 1]$ by $[-1, 5, 1]$

55. $\sqrt{x} + 3 = 5$
$[-1, 6, 1]$ by $[-1, 6, 1]$

56. $\sqrt{x^2+3} = x + 1$
$[-1, 6, 1]$ by $[-1, 6, 1]$

57. $4\sqrt{x} = x + 3$
$[-1, 10, 1]$ by $[-1, 14, 1]$

58. $\sqrt{x} + 4 = 2$
$[-2, 18, 1]$ by $[0, 10, 1]$

Critical Thinking Exercises

59. Which one of the following is true?

a. The first step in solving $\sqrt{x+6} = x + 2$ is to square both sides, obtaining $x + 6 = x^2 + 4$.

b. The equations $\sqrt{x+4} = -5$ and $x + 4 = 25$ have the same solution.

c. If $T = 2\pi\sqrt{\frac{L}{32}}$, then $L = \frac{8T^2}{\pi^2}$.

d. The equation $\sqrt{x^2 + 9x + 3} = -x$ has no solution because a principal square root is always nonnegative.

60. Find the length of the three sides of the right triangle shown in the figure.

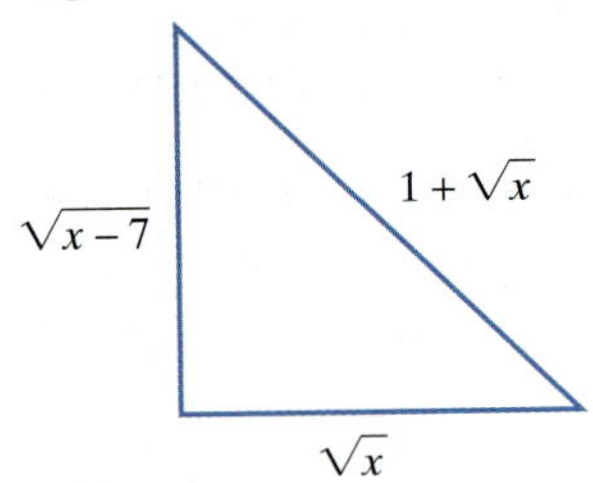

In Exercises 61–63, solve each equation.

61. $\sqrt[3]{x\sqrt{x}} = 9$

62. $\sqrt{\sqrt{x} + \sqrt{x+9}} = 3$

63. $(x-4)^{\frac{2}{3}} = 25$

Review Exercises

64. Divide using synthetic division:
$(4x^4 - 3x^3 + 2x^2 - x - 1) \div (x + 3)$.
(Section 6.5, Example 1)

65. Divide:

$$\frac{3x^2 - 12}{x^2 + 2x - 8} \div \frac{6x + 18}{x + 4}.$$

(Section 6.1, Example 7)

66. Factor: $y^2 - 6y + 9 - 25x^2$.
(Section 5.5, Example 6)

▸ SECTION 7.7 Complex Numbers

Objectives

1. Express square roots of negative numbers in terms of i.
2. Add and subtract complex numbers.
3. Multiply complex numbers.
4. Divide complex numbers.
5. Simplify powers of i.

SSM PH Tutor Center CD-ROM Video

Who is this kid warning us about our eyeballs turning black if we attempt to find the square root of −9? Don't believe what you hear on the street. Although square roots of negative numbers are not real numbers, they do play a significant role in algebra. In this section, we move beyond real numbers and discuss square roots with negative radicands.

1 Express square roots of negative numbers in terms of i.

The Imaginary Unit *i*

In Chapter 8, we will study equations whose solutions involve the square roots of negative numbers. Because the square of a real number is never negative, there is no real number x such that $x^2 = -1$. To provide a setting in which such equations have solutions, mathematicians invented an expanded system of numbers, the complex numbers. The *imaginary number i*, defined to be a solution to the equation $x^2 = -1$, is the basis of this new set.

The Imaginary Unit *i*

The **imaginary unit *i*** is defined as

$$i = \sqrt{-1}, \quad \text{where} \quad i^2 = -1.$$

Using the imaginary unit i, we can express the square root of any negative number as a real multiple of i. For example,

$$\sqrt{-25} = \sqrt{25(-1)} = \sqrt{25}\sqrt{-1} = 5i.$$

We can check that $\sqrt{-25} = 5i$ by squaring $5i$ and obtaining -25.

$$(5i)^2 = 5^2i^2 = 25(-1) = -25$$

The Square Root of a Negative Number

If b is a positive real number, then

$$\sqrt{-b} = \sqrt{b(-1)} = \sqrt{b}\sqrt{-1} = \sqrt{b}i \quad \text{or} \quad i\sqrt{b}.$$

EXAMPLE 1 Expressing Square Roots of Negative Numbers as Multiples of *i*

Write as a multiple of i: **a.** $\sqrt{-9}$ **b.** $\sqrt{-7}$ **c.** $\sqrt{-8}$.

Solution

a. $\sqrt{-9} = \sqrt{9(-1)} = \sqrt{9}\sqrt{-1} = 3i$

b. $\sqrt{-7} = \sqrt{7(-1)} = \sqrt{7}\sqrt{-1} = \sqrt{7}i$

Be sure not to write i under the radical.

c. $\sqrt{-8} = \sqrt{8(-1)} = \sqrt{8}\sqrt{-1} = \sqrt{4 \cdot 2}\sqrt{-1} = 2\sqrt{2}i$ ■

✔ **CHECK POINT 1** Write as a multiple of i:

a. $\sqrt{-16}$ **b.** $\sqrt{-5}$ **c.** $\sqrt{-50}$.

A new system of numbers, called **complex numbers**, is based on adding multiples of i, such as $5i$, to the real numbers.

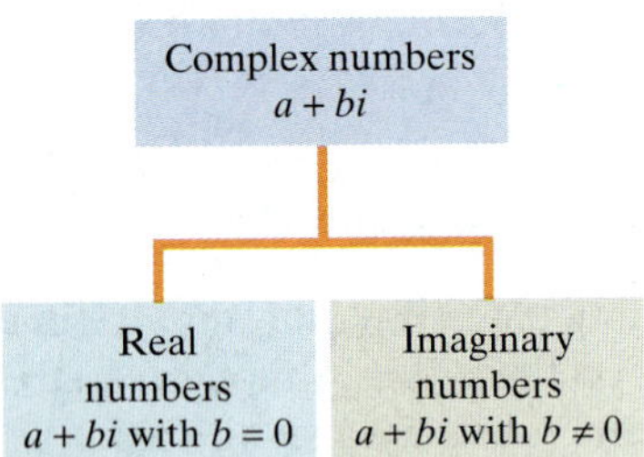

Figure 7.8 The complex number system

Complex Numbers and Imaginary Numbers

The set of all numbers in the form

$$a + bi$$

with real numbers a and b, and i, the imaginary unit, is called the set of **complex numbers**. The real number a is called the **real part**, and the real number b is called the **imaginary part**, of the complex number $a + bi$. If $b \neq 0$, then the complex number is called an **imaginary number** (Figure 7.8).

Here are some examples of complex numbers. Each number can be written in the form $a + bi$.

$$-4 + 6i \qquad 2i = 0 + 2i \qquad 3 = 3 + 0i$$

$-4 + 6i$	$2i = 0 + 2i$	$3 = 3 + 0i$
a, the real part, is −4. b, the imaginary part, is 6.	a, the real part, is 0. b, the imaginary part, is 2.	a, the real part, is 3. b, the imaginary part, is 0.

Can you see that b, the imaginary part, is not zero in the first two complex numbers? Because $b \neq 0$, these complex numbers are imaginary numbers. By contrast, the imaginary part of the complex number on the right is zero. This complex number is not an imaginary number. The number 3, or $3 + 0i$, is a real number.

2 Add and subtract complex numbers.

Adding and Subtracting Complex Numbers

The form of a complex number $a + bi$ is like the binomial $a + bx$. Consequently we can add, subtract, and multiply complex numbers using the same methods we used for binomials, remembering that $i^2 = -1$.

Adding and Subtracting Complex Numbers

1. $(a + bi) + (c + di) = (a + c) + (b + d)i$

   In words, this says that you add complex numbers by adding their real parts, adding their imaginary parts, and expressing the sum as a complex number.

2. $(a + bi) - (c + di) = (a - c) + (b - d)i$

   In words, this says that you subtract complex numbers by subtracting their real parts, subtracting their imaginary parts, and expressing the difference as a complex number.

EXAMPLE 2 Adding and Subtracting Complex Numbers

Perform the indicated operations, writing the result in the form $a + bi$:

a. $(5 - 11i) + (7 + 4i)$ **b.** $(-5 + 7i) - (-11 - 6i)$.

Study Tip

The following examples, using the same integers as in Example 2, show how operations with complex numbers are just like operations with polynomials.

a. $(5 - 11x) + (7 + 4x)$
$= 12 - 7x$

b. $(-5 + 7x) - (-11 - 6x)$
$= -5 + 7x + 11 + 6x$
$= 6 + 13x$

Solution

a. $(5 - 11i) + (7 + 4i)$

$= 5 - 11i + 7 + 4i$ Remove the parentheses.

$= 5 + 7 - 11i + 4i$ Group real and imaginary terms.

$= (5 + 7) + (-11 + 4)i$ Use the distributive property.

$= 12 - 7i$ Add real parts and add imaginary parts.

b. $(-5 + 7i) - (-11 - 6i)$

$= -5 + 7i + 11 + 6i$ Remove the parentheses.

$= -5 + 11 + 7i + 6i$ Group real and imaginary terms.

$= (-5 + 11) + (7 + 6)i$

$= 6 + 13i$ ■

✔ **CHECK POINT 2** Add or subtract as indicated:

a. $(5 - 2i) + (3 + 3i)$

b. $(2 + 6i) - (12 - 4i)$.

3 Multiply complex numbers.

Multiplying Complex Numbers

Multiplication of complex numbers is performed the same way as multiplication of polynomials, using the distributive property and the FOIL method. After completing the multiplication, we replace i^2 with -1. This idea is illustrated in the next example.

EXAMPLE 3 Multiplying Complex Numbers

Find the products: **a.** $4i(3 - 5i)$ **b.** $(7 - 3i)(-2 - 5i)$.

Solution

a. $4i(3 - 5i)$

$= 4i \cdot 3 - 4i \cdot 5i$ Distribute $4i$ throughout the parentheses.

$= 12i - 20i^2$ Multiply.

$= 12i - 20(-1)$ Replace i^2 with -1.

$= 20 + 12i$ Simplify to $12i + 20$ and write in $a + bi$ form.

b. $(7 - 3i)(-2 - 5i)$

F O I L

$$= -14 - 35i + 6i + 15i^2$$ Use the FOIL method.

$$= -14 - 35i + 6i + 15(-1)$$ $i^2 = -1$

$$= -14 - 15 - 35i + 6i$$ Group real and imaginary terms.

$$= -29 - 29i$$ Combine real and imaginary terms. ■

CHECK POINT 3 Find the products:

a. $7i(2 - 9i)$ **b.** $(5 + 4i)(6 - 7i)$.

Consider the multiplication problem

$$5i \cdot 2i = 10i^2 = 10(-1) = -10.$$

This problem can also be given in terms of square roots of negative numbers:

$$\sqrt{-25} \cdot \sqrt{-4}.$$

Because the product rule for radicals only applies to real numbers, multiplying radicands is incorrect. **When performing operations with square roots of negative numbers, begin by expressing all square roots in terms of *i*.** Then perform the indicated operation.

CORRECT:

$$\sqrt{-25} \cdot \sqrt{-4} = \sqrt{25}\sqrt{-1} \cdot \sqrt{4}\sqrt{-1}$$
$$= 5i \cdot 2i$$
$$= 10i^2 = 10(-1) = -10$$

INCORRECT:

$$\sqrt{-25} \cdot \sqrt{-4} = \sqrt{(-25)(-4)}$$
$$= \sqrt{100}$$
$$= 10$$

EXAMPLE 4 Multiplying Square Roots of Negative Numbers

Multiply: $\sqrt{-3} \cdot \sqrt{-5}$.

Solution

$$\sqrt{-3} \cdot \sqrt{-5} = \sqrt{3}\sqrt{-1} \cdot \sqrt{5}\sqrt{-1}$$

$$= \sqrt{3}i \cdot \sqrt{5}i$$ Express square roots in terms of i.

$$= \sqrt{15}i^2$$ $\sqrt{3} \cdot \sqrt{5} = \sqrt{15}$ and $i \cdot i = i^2$.

$$= \sqrt{15}(-1)$$ $i^2 = -1$

$$= -\sqrt{15}$$ ■

CHECK POINT 4 Multiply: $\sqrt{-5} \cdot \sqrt{-7}$.

4 Divide complex numbers.

Conjugates and Division

It is possible to multiply imaginary numbers and obtain a real number. Here is an example:

F O I L

$$(4 + 7i)(4 - 7i) = 16 - 28i + 28i - 49i^2$$

$$= 16 - 49i^2 = 16 - 49(-1) = 65.$$

Replace i^2 with -1.

ENRICHMENT ESSAY

Complex Numbers on a Postage Stamp

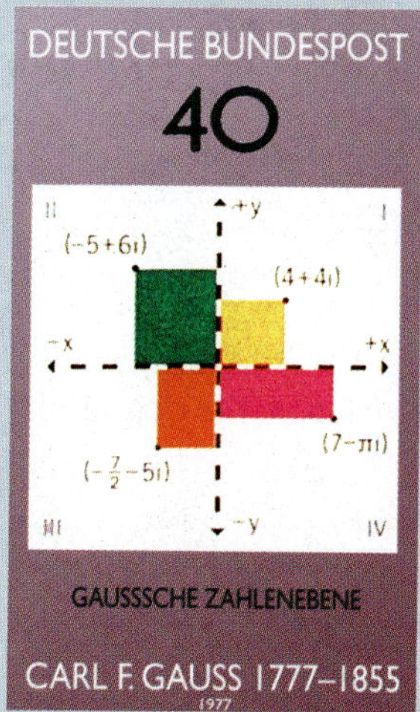

This stamp honors the work done by the German mathematician Carl Friedrich Gauss (1777–1855) with complex numbers. Gauss represented complex numbers as points in the plane. (Stamp from the private collection of Professor C.M. Lang, photography by Gary J. Shulfer, University of Wisconsin, Stevens Point. "Germany: #5"; Scott Standard Postage Stamp Catalogue, Scott Pub. Co., Sidney, Ohio)

You can also perform the multiplication using the formula

$$(A + B)(A - B) = A^2 - B^2.$$

A real number is obtained even faster:

$$(4 + 7i)(4 - 7i) = 4^2 - (7i)^2 = 16 - 49i^2 = 16 - 49(-1) = 65.$$

The **conjugate** of the complex number $a + bi$ is $a - bi$. The **conjugate** of the complex number $a - bi$ is $a + bi$. The multiplication problem that we just performed involved conjugates. The multiplication of conjugates always results in a real number:

$$(a + bi)(a - bi) = a^2 - (bi)^2 = a^2 - b^2i^2 = a^2 - b^2(-1) = a^2 + b^2$$

The product eliminates i.

Conjugates are used to divide complex numbers. By multiplying the numerator and the denominator of the division by the conjugate of the denominator, you will obtain a real number in the denominator. Here are two examples of such divisions:

- $\dfrac{7 + 4i}{2 - 5i}$ — The conjugate of the denominator is $2 + 5i$.
- $\dfrac{5i - 4}{3i}$ or $\dfrac{5i - 4}{0 + 3i}$. — The conjugate of the denominator is $0 - 3i$, or $-3i$.

The procedure for dividing complex numbers, illustrated in Examples 5 and 6, should remind you of rationalizing denominators.

EXAMPLE 5 Using Conjugates to Divide Complex Numbers

Divide and simplify to the form $a + bi$:

$$\frac{7 + 4i}{2 - 5i}.$$

Solution The conjugate of the denominator is $2 + 5i$. Multiplication of both the numerator and denominator by $2 + 5i$ will eliminate i from the denominator.

$$\frac{7 + 4i}{2 - 5i} = \frac{7 + 4i}{2 - 5i} \cdot \frac{2 + 5i}{2 + 5i}$$ Multiply by 1. (F O I L)

$$= \frac{14 + 35i + 8i + 20i^2}{2^2 - (5i)^2}$$ Use FOIL in the numerator and $(A - B)(A + B) = A^2 - B^2$ in the denominator.

$$= \frac{14 + 43i + 20i^2}{4 - 25i^2}$$ Simplify.

$$= \frac{14 + 43i + 20(-1)}{4 - 25(-1)}$$ $i^2 = -1$

$$= \frac{14 + 43i - 20}{4 + 25}$$ Perform the multiplications involving -1.

$$= \frac{-6 + 43i}{29}$$ Combine real terms in the numerator and denominator.

✔ **CHECK POINT 5** Divide and simplify to the form $a + bi$:

$$\frac{6 + 2i}{4 - 3i}.$$

EXAMPLE 6 Using Conjugates to Divide Complex Numbers

Divide and simplify to the form $a + bi$:

$$\frac{5i - 4}{3i}.$$

Solution The conjugate of the denominator, $0 + 3i$, is $0 - 3i$. Multiplication of both the numerator and denominator by $-3i$ will eliminate i from the denominator.

$$\frac{5i - 4}{3i} = \frac{5i - 4}{3i} \cdot \frac{-3i}{-3i} \quad \text{Multiply by 1.}$$

$$= \frac{-15i^2 + 12i}{-9i^2} \quad \text{Multiply. Use the distributive property in the numerator.}$$

$$= \frac{-15(-1) + 12i}{-9(-1)} \quad i^2 = -1$$

$$= \frac{15 + 12i}{9} \quad \text{Perform the multiplications involving } -1.$$

$$= \frac{15}{9} + \frac{12}{9}i \quad \text{Express the division in the form } a + bi.$$

$$= \frac{5}{3} + \frac{4}{3}i \quad \text{Simplify real and imaginary parts.}$$

■

✔ **CHECK POINT 6** Divide and simplify to the form $a + bi$:

$$\frac{3 - 2i}{4i}.$$

5 Simplify powers of i.

Powers of i

Cycles govern many aspects of life—heartbeats, sleep patterns, seasons, and tides all follow regular, predictable cycles. Surprisingly, so do powers of i. To see how this occurs, use the fact that $i^2 = -1$ and express each power of i in terms of i^2:

$$i$$
$$i^2 = -1$$
$$i^3 = i^2 \cdot i = (-1)i = -i$$
$$i^4 = (i^2)^2 = (-1)^2 = 1$$
$$i^5 = i^4 \cdot i = (i^2)^2 \cdot i = (-1)^2 \cdot i = i$$
$$i^6 = (i^2)^3 = (-1)^3 = -1.$$

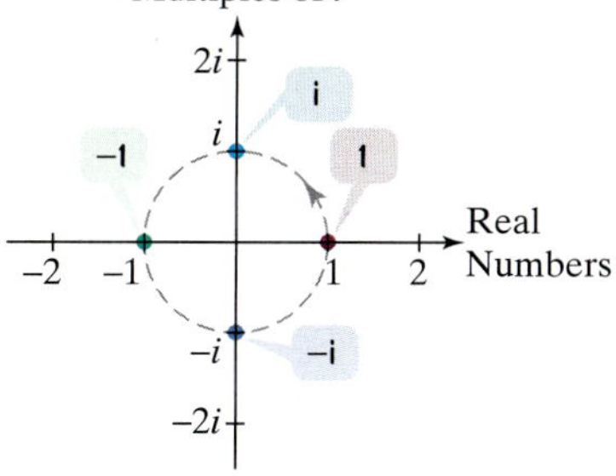

Figure 7.9

$$i^4 = (i^2)^2 = (-1)^2 = 1$$
$$i^5 = i^4 \cdot i = (i^2)^2 \cdot i = (-1)^2 \cdot i = i$$
$$i^6 = (i^2)^3 = (-1)^3 = -1.$$

Can you see that the powers of i are cycling through the values $i, -1, -i$, and 1? The cycle is illustrated in Figure 7.9. In the figure, tick marks on the horizontal axis represent real numbers and tick marks along the vertical axis represent multiples of i. Using this representation, powers of i are equally spaced at 90° intervals on a circle having radius 1.

Here is a procedure for simplifying powers of i:

Simplifying Powers of *i*

1. Express the given power of i in terms of i^2.
2. Replace i^2 by -1 and simplify. Use the fact that -1 to an even power is 1, and -1 to an odd power is -1.

EXAMPLE 7 Simplifying Powers of *i*

Simplify: **a.** i^{12} **b.** i^{39} **c.** i^{39}.

Solution

a. $i^{12} = (i^2)^6 = (-1)^6 = 1$

b. $i^{39} = i^{38}i = (i^2)^{19}i = (-1)^{19}i = (-1)i = -i$

ENRICHMENT ESSAY

The Patterns of Chaos

R.F. Voss "29-Fold M-set Seahorse" computer-generated image. © 1990 R.F. Voss/IBM Research.

One of the new frontiers of mathematics suggests that there is an underlying order in things that appear to be random, such as the hiss and crackle of background noises as you tune a radio. Irregularities in the heartbeat, some of them severe enough to cause a heart attack, or irregularities in our sleeping patterns, such as insomnia, are example of chaotic behavior. Chaos in the mathematical sense does not mean a complete lack of form or arrangement. In mathematics, chaos is used to describe something that appears to be random but is not actually random. The patterns of chaos appear in images like the one on the left, called the Mandelbrot set. Magnified portions of this image yield repetitions of the original structure, as well as new and unexpected patterns. The Mandelbrot set transforms the hidden structure of chaotic events into a source of wonder and inspiration.

The Mandelbrot set is made possible by opening up graphing to include complex numbers in the form $a + bi$. Although the details are beyond the scope of this text, the coordinate system that is used is shown in Figure 7.9. Plot certain complex numbers in this system, add color to the magnified boundary of the graph, and the patterns of chaos begin to appear.

EXERCISE SET 7.7

Practice Exercises

In Exercises 1–16, express each number in terms of i and simplify, if possible.

1. $\sqrt{-49}$
2. $\sqrt{-36}$
3. $\sqrt{-17}$
4. $\sqrt{-11}$
5. $\sqrt{-75}$
6. $\sqrt{-20}$
7. $\sqrt{-28}$
8. $\sqrt{-45}$
9. $-\sqrt{-150}$
10. $-\sqrt{-700}$
11. $7 + \sqrt{-16}$
12. $9 + \sqrt{-4}$
13. $5 + \sqrt{-5}$
14. $10 + \sqrt{-3}$
15. $6 - \sqrt{-18}$
16. $6 - \sqrt{-98}$

In Exercises 17–32, add or subtract as indicated. Write the result in the form a + bi.

17. $(3 + 2i) + (5 + i)$
18. $(6 + 5i) + (4 + 3i)$
19. $(7 + 2i) + (1 - 4i)$
20. $(-2 + 6i) + (4 - i)$
21. $(10 + 7i) - (5 + 4i)$
22. $(11 + 8i) - (2 + 5i)$
23. $(9 - 4i) - (10 + 3i)$
24. $(8 - 5i) - (6 + 2i)$
25. $(3 + 2i) - (5 - 7i)$
26. $(-7 + 5i) - (9 - 11i)$
27. $(-5 + 4i) - (-13 - 11i)$
28. $(-9 + 2i) - (-17 - 6i)$
29. $8i - (14 - 9i)$
30. $15i - (12 - 11i)$
31. $(2 + \sqrt{3}i) + (7 + 4\sqrt{3}i)$
32. $(4 + \sqrt{5}i) + (8 + 6\sqrt{5}i)$

In Exercises 33–62, find each product. Write the result in the form a + bi.

33. $2i(5 + 3i)$
34. $5i(4 + 7i)$
35. $3i(7i - 5)$
36. $8i(4i - 3)$
37. $-7i(2 - 5i)$
38. $-6i(3 - 5i)$
39. $(3 + i)(4 + 5i)$
40. $(4 + i)(5 + 6i)$
41. $(7 - 5i)(2 - 3i)$
42. $(8 - 4i)(3 - 2i)$
43. $(6 - 3i)(-2 + 5i)$
44. $(7 - 2i)(-3 + 6i)$
45. $(3 + 5i)(3 - 5i)$
46. $(2 + 7i)(2 - 7i)$
47. $(-5 + 3i)(-5 - 3i)$
48. $(-4 + 2i)(-4 - 2i)$
49. $(3 - \sqrt{2}i)(3 + \sqrt{2}i)$
50. $(5 - \sqrt{3}i)(5 + \sqrt{3}i)$
51. $(2 + 3i)^2$
52. $(3 + 2i)^2$
53. $(5 - 2i)^2$
54. $(5 - 3i)^2$
55. $\sqrt{-7} \cdot \sqrt{-2}$
56. $\sqrt{-7} \cdot \sqrt{-3}$
57. $\sqrt{-9} \cdot \sqrt{-4}$
58. $\sqrt{-16} \cdot \sqrt{-4}$
59. $\sqrt{-7} \cdot \sqrt{-25}$
60. $\sqrt{-3} \cdot \sqrt{-36}$
61. $\sqrt{-8} \cdot \sqrt{-3}$
62. $\sqrt{-9} \cdot \sqrt{-5}$

In Exercises 63–84, divide and simplify to the form a + bi.

63. $\dfrac{2}{3 + i}$
64. $\dfrac{3}{4 + i}$
65. $\dfrac{2i}{1 + i}$
66. $\dfrac{5i}{2 + i}$
67. $\dfrac{7}{4 - 3i}$
68. $\dfrac{9}{1 - 2i}$

69. $\frac{6i}{3 - 2i}$

70. $\frac{5i}{2 - 3i}$

71. $\frac{1 + i}{1 - i}$

72. $\frac{1 - i}{1 + i}$

73. $\frac{2 - 3i}{3 + i}$

74. $\frac{2 + 3i}{3 - i}$

75. $\frac{5 - 2i}{3 + 2i}$

76. $\frac{6 - 3i}{4 + 2i}$

77. $\frac{4 + 5i}{3 - 7i}$

78. $\frac{5 - i}{3 - 2i}$

79. $\frac{7}{3i}$

80. $\frac{5}{7i}$

81. $\frac{8 - 5i}{2i}$

82. $\frac{3 + 4i}{5i}$

83. $\frac{4 + 7i}{-3i}$

84. $\frac{5 + i}{-4i}$

In Exercises 85–100, simplify each expression.

85. i^{10}

86. i^{14}

87. i^{11}

88. i^{15}

89. i^{22}

90. i^{46}

91. i^{200}

92. i^{400}

93. i^{17}

94. i^{21}

95. $(-i)^4$

96. $(-i)^6$

97. $(-i)^9$

98. $(-i)^{13}$

99. $i^{24} + i^2$

100. $i^{28} + i^{30}$

Application Exercises

Complex numbers are used in electronics to describe the current in an electric circuit. Ohm's law relates the current in a circuit, I, in amperes, the voltage of the circuit, E, in volts, and the resistance of the circuit, R, in ohms, by the formula $E = IR$. Use this formula to solve Exercises 101–102.

101. Find E, the voltage of a circuit, if $I = (4 - 5i)$ amperes and $R = (3 + 7i)$ ohms.

102. Find E, the voltage of a circuit, if $I = (2 - 3i)$ amperes and $R = (3 + 5i)$ ohms.

103. The mathematician Girolamo Cardano is credited with the first use (in 1545) of negative square roots in solving the now-famous problem, "Find two numbers whose sum is 10 and whose product is 40." Show that the complex numbers $5 + \sqrt{15}i$ and $5 - \sqrt{15}i$ satisfy the conditions of the problem. (Cardano did not use the symbolism $\sqrt{15}i$ or even $\sqrt{-15}$. He wrote R.m 15 for $\sqrt{-15}$, meaning "radix minus 15." He regarded the numbers 5 + R.m 15 and 5 − R.m 15 as "fictitious" or "ghost numbers," and considered the problem "manifestly impossible." But in a mathematically adventurous spirit, he exclaimed, "Nevertheless, we will operate.")

Writing in Mathematics

104. What is the imaginary unit i?

105. Explain how to write $\sqrt{-64}$ as a multiple of i.

106. What is a complex number? Explain when a complex number is a real number and when it is an imaginary number. Provide examples with your explanation.

107. Explain how to add complex numbers. Give an example.

108. Explain how to subtract complex numbers. Give an example.

109. Explain how to find the product of $2i$ and $5 + 3i$.

110. Explain how to find the product of $2i + 3$ and $5 + 3i$.

111. Explain how to find the product of $2i + 3$ and $2i - 3$.

112. Explain how to find the product of $\sqrt{-1}$ and $\sqrt{-4}$. Describe a common error in the multiplication that needs to be avoided.

113. What is the conjugate of $2 + 3i$? What happens when you multiply this complex number by its conjugate?

114. Explain how to divide complex numbers. Provide an example with your explanation.

115. Explain each of the three jokes in the cartoon on page 523.

116. A stand-up comedian uses algebra in some jokes, including one about a telephone recording that announces "You have just reached an imaginary number. Please multiply by i and dial again." Explain the joke.

Explain the error in Exercises 117–118.

117. $\sqrt{-9} + \sqrt{-16} = \sqrt{-25} = \sqrt{25}i = 5i$

118. $(\sqrt{-9})^2 = \sqrt{-9} \cdot \sqrt{-9} = \sqrt{81} = 9$

Critical Thinking Exercises

119. Which one of the following is true?

a. Some irrational numbers are not complex numbers.

b. $(3 + 7i)(3 - 7i)$ is an imaginary number.

c. $\frac{7 + 3i}{5 + 3i} = \frac{7}{5}$

d. In the complex number system, $x^2 + y^2$ (the sum of two squares) can be factored as $(x + yi)(x - yi)$.

In Exercises 120–122, perform the indicated operations and write the result in the form $a + bi$.

120. $(8 + 9i)(2 - i) - (1 - i)(1 + i)$

121. $\frac{4}{(2 + i)(3 - i)}$

122. $\frac{1 + i}{1 + 2i} + \frac{1 - i}{1 - 2i}$

123. Evaluate $x^2 - 2x + 2$ for $x = 1 + i$.

Review Exercises

124. Simplify:

$$\frac{\frac{x}{y^2} + \frac{1}{y}}{\frac{y}{x^2} + \frac{1}{x}}.$$

(Section 6.3, Example 1)

125. Solve for x: $\frac{1}{x} + \frac{1}{y} = \frac{1}{z}$.
(Section 6.7, Example 1)

126. Solve:

$$2x - \frac{x - 3}{8} = \frac{1}{2} + \frac{x + 5}{2}.$$

(Section 1.4, Example 4)

CHAPTER 7 GROUP PROJECTS

1. Group members should consult an almanac, newspaper, magazine, or the Internet and return to the group with as much data as possible that show phenomena that are continuing to grow over time, but whose growth is leveling off. Select the five data sets that you find most intriguing. Let x represent the number of years after the first year in each data set. Model the data by hand using

$$f(x) = a\sqrt{x} + b.$$

Use the first and last data points to find values for a and b. The first data point corresponds to $x = 0$. Its second coordinate gives the value of b. To find a, substitute the second data point into $f(x) = a\sqrt{x} + b$, with the value that you obtained for b. Now solve the equation and find a. Substitute a and b into $f(x) = a\sqrt{x} + b$ to obtain a square root function that models each data set. Then use the function to make predictions about what might occur in the future. Are there circumstances that might affect the accuracy of the predictions? List some of these circumstances.

2. Group members should prepare and present a seminar on chaos. Include one or more of the following topics in your presentation: fractal images, the role of complex numbers in generating fractal images, algorithms, iterations, iteration number, and fractals in nature. Be sure to include visual images that will intrigue your audience.

3. The math lab has asked the group to prepare a pamphlet for students having difficulties with the 30-item Chapter 7 test on pages 538–539. The pamphlet's title should be "How To Ace the Test on Radicals, Radical Functions, and Rational Exponents." Offer guidelines for avoiding errors, as well as hints and suggestions for succeeding with each of the test items.

CHAPTER SUMMARY, REVIEW, AND TEST

SUMMARY

DEFINITIONS AND CONCEPTS	EXAMPLES
Section 7.1 Radical Expressions and Functions	
If $b^2 = a$, then b is a square root of a. The principal square root of a, designated $\sqrt{a}$, is the nonnegative number satisfying $b^2 = a$. The negative square root of a is written $-\sqrt{a}$. A square root of a negative number is not *a* real number. A radical function in x is a function defined by an expression containing a root of x. The domain of a square root function is the set of real numbers for which the radicand is nonnegative.	Let $f(x) = \sqrt{6 - 2x}$. $f(-15) = \sqrt{6 - 2(-15)} = \sqrt{6 + 30} = \sqrt{36} = 6$ $f(5) = \sqrt{6 - 2 \cdot 5} = \sqrt{6 - 10} = \sqrt{-4}$, not a real number Domain of f: Set the radicand greater than or equal to zero. $6 - 2x \geq 0$ $-2x \geq -6$ $x \leq 3$ Domain of $f = \{x \mid x \leq 3\}$ or $(-\infty, 3]$.
The cube root of a real number a is written $\sqrt[3]{a}$. $\sqrt[3]{a} = b$ means that $b^3 = a$. The nth root of a real number a is written $\sqrt[n]{a}$. The number n is the index. Every real number has one root when n is odd. The odd nth root of a, $\sqrt[n]{a}$, is the number b for which $b^n = a$. Every positive real number has two real roots when n is even. An even root of a negative number is not a real number. If n is even, then $\sqrt[n]{a^n} = \lvert a \rvert$. If n is odd, then $\sqrt[n]{a^n} = a$.	• $\sqrt[3]{-8} = -2$ because $(-2)^3 = -8$. • $\sqrt[4]{-16}$ is not a real number. • $\sqrt{x^2 - 14x + 49} = \sqrt{(x - 7)^2} = \lvert x - 7 \rvert$ • $\sqrt[3]{125(x + 6)^3} = 5(x + 6)$
Section 7.2 Rational Exponents	
• $a^{\frac{1}{n}} = \sqrt[n]{a}$ • $a^{\frac{m}{n}} = (\sqrt[n]{a})^m$ or $\sqrt[n]{a^m}$ • $a^{-\frac{m}{n}} = \dfrac{1}{a^{\frac{m}{n}}}$	• $16^{\frac{1}{2}} = \sqrt{16} = 4$ • $27^{\frac{1}{3}} = \sqrt[3]{27} = 3$ • $8^{\frac{5}{3}} = (\sqrt[3]{8})^5 = 2^5 = 32$ • $81^{-\frac{3}{4}} = \dfrac{1}{81^{\frac{3}{4}}} = \dfrac{1}{(\sqrt[4]{81})^3} = \dfrac{1}{3^3} = \dfrac{1}{27}$ • $(\sqrt[3]{3xy})^4 = (3xy)^{\frac{4}{3}}$
Properties of integer exponents are true for rational exponents. An expression with rational exponents is simplified when no parentheses appear, no powers are raised to powers, each base occurs once, and no negative or zero exponents appear.	Simplify: $(8x^{\frac{1}{3}}y^{-\frac{1}{2}})^{\frac{1}{3}}$. $= 8^{\frac{1}{3}}(x^{\frac{1}{3}})^{\frac{1}{3}}(y^{-\frac{1}{2}})^{\frac{1}{3}}$ $= 2x^{\frac{1}{9}}y^{-\frac{1}{6}} = \dfrac{2x^{\frac{1}{9}}}{y^{\frac{1}{6}}}$

DEFINITIONS AND CONCEPTS	EXAMPLES
Section 7.2 Rational Exponents (continued)	
Some radical expressions can be simplifed using rational exponents. Rewrite the expression using rational exponents, simplify, and rewrite in radical notation if rational exponents still appear.	• $\sqrt[9]{x^3} = x^{\frac{3}{9}} = x^{\frac{1}{3}} = \sqrt[3]{x}$ • $\sqrt[5]{x^2} \cdot \sqrt[4]{x} = x^{\frac{2}{5}} \cdot x^{\frac{1}{4}} = x^{\frac{2}{5}+\frac{1}{4}}$ $= x^{\frac{8}{20}+\frac{5}{20}} = x^{\frac{13}{20}} = \sqrt[20]{x^{13}}$
Section 7.3 Multiplying and Simplifying Radical Expressions	
The product rule for radicals can be used to multiply radicals with the same indices: $\sqrt[n]{a} \cdot \sqrt[n]{b} = \sqrt[n]{ab}.$	$\sqrt[3]{7x} \cdot \sqrt[3]{10y^2} = \sqrt[3]{7x \cdot 10y^2} = \sqrt[3]{70xy^2}$
The product rule for radicals can be used to simplify radicals: $\sqrt[n]{ab} = \sqrt[n]{a} \cdot \sqrt[n]{b}.$ A radical expression with index n is simplified when its radicand has no factors that are perfect nth powers. To simplify, write the radicand as the product of two factors, one of which is the largest perfect nth power. Then use the product rule to take the nth root of each factor. If all variables in a radicand are positive, then $\sqrt[n]{a^n} = a.$ Some radicals can be simplified after the multiplication is performed.	• Simplify: $\sqrt[3]{54x^7y^{11}}$. $= \sqrt[3]{27 \cdot 2 \cdot x^6 \cdot x \cdot y^9 \cdot y^2}$ $= \sqrt[3]{(27x^6y^9)(2xy^2)}$ $= \sqrt[3]{27x^6y^9} \cdot \sqrt[3]{2xy^2} = 3x^2y^3\sqrt[3]{2xy^2}$ • Assuming positive variables, multiply and simplify: $\sqrt[4]{4x^2y} \cdot \sqrt[4]{4xy^3}$. $= \sqrt[4]{4x^2y \cdot 4xy^3} = \sqrt[4]{16x^3y^4}$ $= \sqrt[4]{16y^4} \cdot \sqrt[4]{x^3} = 2y\sqrt[4]{x^3}$
Section 7.4 Adding, Subtracting, and Dividing Radical Expressions	
Like radicals have the same indices and radicands. Like radicals can be added or subtracted using the distributive property. In some cases, radicals can be combined once they have been simplified.	$4\sqrt{18} - 6\sqrt{50}$ $= 4\sqrt{9 \cdot 2} - 6\sqrt{25 \cdot 2} = 4 \cdot 3\sqrt{2} - 6 \cdot 5\sqrt{2}$ $= 12\sqrt{2} - 30\sqrt{2} = -18\sqrt{2}$
The quotient rule for radicals can be used to simplify radicals: $\sqrt[n]{\frac{a}{b}} = \frac{\sqrt[n]{a}}{\sqrt[n]{b}}.$	$\sqrt[3]{-\frac{8}{x^{12}}} = \frac{\sqrt[3]{-8}}{\sqrt[3]{x^{12}}} = -\frac{2}{x^4}$ $\sqrt[3]{x^{12}} = (x^{12})^{\frac{1}{3}} = x^4$
The quotient rule for radicals can be used to divide radicals with the same indices: $\frac{\sqrt[n]{a}}{\sqrt[n]{b}} = \sqrt[n]{\frac{a}{b}}.$ Some radicals can be simplified after the division is performed.	Assuming a positive variable, divide and simplify: $\frac{\sqrt[4]{64x^5}}{\sqrt[4]{2x^{-2}}} = \sqrt[4]{32x^{5-(-2)}} = \sqrt[4]{32x^7}$ $= \sqrt[4]{16 \cdot 2 \cdot x^4 \cdot x^3} = \sqrt[4]{16x^4} \cdot \sqrt[4]{2x^3}$ $= 2x\sqrt[4]{2x^3}.$

DEFINITIONS AND CONCEPTS	EXAMPLES

Section 7.5 Multiplying with More Than One Term and Rationalizing Denominators

Radical expressions with more than one term are multiplied in much the same way that polynomials with more than one term are multiplied.

- $\sqrt{5}(2\sqrt{6} - \sqrt{3}) = 2\sqrt{30} - \sqrt{15}$
- $(4\sqrt{3} - 2\sqrt{2})(\sqrt{3} + \sqrt{2})$

F O I L

$$= 4\sqrt{3} \cdot \sqrt{3} + 4\sqrt{3} \cdot \sqrt{2} - 2\sqrt{2} \cdot \sqrt{3} - 2\sqrt{2} \cdot \sqrt{2}$$
$$= 4 \cdot 3 + 4\sqrt{6} - 2\sqrt{6} - 2 \cdot 2$$
$$= 12 + 4\sqrt{6} - 2\sqrt{6} - 4 = 8 + 2\sqrt{6}$$

Radical expressions that involve the sum and difference of the same two terms are called conjugates. Use

$$(A + B)(A - B) = A^2 - B^2$$

to multiply conjugates.

$$(8 + 2\sqrt{5})(8 - 2\sqrt{5})$$
$$= 8^2 - (2\sqrt{5})^2 = 64 - 4 \cdot 5$$
$$= 64 - 20 = 44$$

The process of rewriting a radical expression as an equivalent expression without a radical in the denominator is called rationalizing the denominator. When the denominator contains a single radical with an nth root, multiply the numerator and denominator by a radical of index n that produces a perfect nth power in the denominator's radicand.

Rationalize the denominator: $\dfrac{7}{\sqrt[3]{2x}}$.

$$= \frac{7}{\sqrt[3]{2x}} \cdot \frac{\sqrt[3]{4x^2}}{\sqrt[3]{4x^2}} = \frac{7\sqrt[3]{4x^2}}{\sqrt[3]{8x^3}} = \frac{7\sqrt[3]{4x^2}}{2x}$$

If the denominator contains two terms, rationalize the denominator by multiplying the numerator and denominator by the conjugate of the denominator.

$$\frac{9}{7 - \sqrt{5}} = \frac{9}{7 - \sqrt{5}} \cdot \frac{7 + \sqrt{5}}{7 + \sqrt{5}}$$
$$= \frac{9(7 + \sqrt{5})}{7^2 - (\sqrt{5})^2}$$
$$= \frac{9(7 + \sqrt{5})}{49 - 5} = \frac{9(7 + \sqrt{5})}{44}$$

Section 7.6 Radical Equations

A radical equation is an equation in which the variable occurs in a radicand.

Solving Radical Equations Containing nth Roots

1. Isolate the radical.
2. Raise both sides to the nth power.
3. Solve the resulting equation.
4. Check proposed solutions in the original equation. Solutions of an equation to an even power that is radical-free, but not the original equation, are called extraneous solutions.

Solve: $\sqrt{2x + 1} - x = -7$.

$\sqrt{2x + 1} = x - 7$ — Isolate the radical.

$(\sqrt{2x + 1})^2 = (x - 7)^2$ — Square both sides.

$2x + 1 = x^2 - 14x + 49$

$0 = x^2 - 16x + 48$ — Subtract $2x + 1$ from both sides.

$0 = (x - 12)(x - 4)$ — Factor.

$x - 12 = 0$ or $x - 4 = 0$

$x = 12$ $\quad x = 4$

Check both proposed solutions. 12 checks, but 4 is extraneous. The solution is 12 and the solution set is $\{12\}$.

DEFINITIONS AND CONCEPTS	EXAMPLES
Section 7.7 Complex Numbers	
The imaginary unit i is defined as $$i = \sqrt{-1}, \text{ where } i^2 = -1.$$ The set of numbers in the form $a + bi$ is called the set of complex numbers; a is the real part and b is the imaginary part. If $b = 0$, the complex number is a real number. If $b \neq 0$, the complex number is an imaginary number.	• $\sqrt{-36} = \sqrt{36(-1)} = \sqrt{36}\sqrt{-1} = 6i$ • $\sqrt{-50} = \sqrt{50(-1)} = \sqrt{25 \cdot 2}\sqrt{-1} = 5\sqrt{2}i$
To add or subtract complex numbers, add or subtract their real parts and add or subtract their imaginary parts.	$(2 - 4i) - (7 - 10i)$ $= 2 - 4i - 7 + 10i$ $= (2 - 7) + (-4 + 10)i = -5 + 6i$
To multiply complex numbers, multiply as if they are polynomials. After completing the multiplication, replace i^2 with -1. When performing operations with square roots of negative numbers, begin by expressing all square roots in terms of i. Then multiply.	F O I L • $(2 - 3i)(4 + 5i) = 8 + 10i - 12i - 15i^2$ $= 8 + 10i - 12i - 15(-1)$ $= 23 - 2i$ • $\sqrt{-36} \cdot \sqrt{-100} = \sqrt{36(-1)} \cdot \sqrt{100(-1)}$ $= 6i \cdot 10i = 60i^2 = 60(-1) = -60$
The complex numbers $a + bi$ and $a - bi$ are conjugates. Conjugates can be multiplied using the formula $$(A + B)(A - B) = A^2 - B^2.$$ The multiplication of conjugates results in a real number.	$(3 + 5i)(3 - 5i)$ $= 3^2 - (5i)^2$ $= 9 - 25i^2$ $= 9 - 25(-1) = 34$
To divide complex numbers, multiply the numerator and the denominator by the conjugate of the denominator.	$\frac{5 + 2i}{4 - i} = \frac{5 + 2i}{4 - i} \cdot \frac{4 + i}{4 + i} = \frac{20 + 5i + 8i + 2i^2}{16 - i^2}$ $= \frac{20 + 13i + 2(-1)}{16 - (-1)}$ $= \frac{20 + 13i - 2}{16 + 1}$ $= \frac{18 + 13i}{17} = \frac{18}{17} + \frac{13}{17}i$
To simplify powers of i, rewrite the expression in terms of i^2. Then replace i^2 by -1 and simplify.	Simplify: i^{27}. $i^{27} = i^{26} \cdot i = (i^2)^{13} i$ $= (-1)^{13} i = (-1)i = -i$

Review Exercises

7.1 *In Exercises 1–5, find the indicated root, or state that the expression is not a real number.*

1. $\sqrt{81}$

2. $-\sqrt{\frac{1}{100}}$

3. $\sqrt[3]{-27}$

4. $\sqrt[4]{-16}$

5. $\sqrt[5]{-32}$

In Exercises 6–7, find the indicated function values for each function. If necessary, round to two decimal places. If the function value is not a real number and does not exist, so state.

6. $f(x) = \sqrt{2x - 5};\ \ f(15), f(4), f\left(\frac{5}{2}\right), f(1)$

7. $g(x) = \sqrt[3]{4x - 8};\ \ g(4), g(0), g(-14)$

In Exercises 8–9, find the domain of each square root function.

8. $f(x) = \sqrt{x - 2}$

9. $g(x) = \sqrt{100 - 4x}$

In Exercises 10–15, simplify each expression. Assume that each variable can represent any real number, so include absolute value bars where necessary.

10. $\sqrt{25x^2}$

11. $\sqrt{(x + 14)^2}$

12. $\sqrt{x^2 - 8x + 16}$

13. $\sqrt[3]{64x^3}$

14. $\sqrt[4]{16x^4}$

15. $\sqrt[5]{-32(x + 7)^5}$

7.2 *In Exercises 16–18, use radical notation to rewrite each expression. Simplify, if possible.*

16. $(5xy)^{\frac{1}{3}}$

17. $16^{\frac{3}{2}}$

18. $32^{\frac{4}{5}}$

In Exercises 19–20, rewrite each expression with rational exponents.

19. $\sqrt{7x}$

20. $(\sqrt[3]{19xy})^5$

In Exercises 21–22, rewrite each expression with a positive rational exponent. Simplify, if possible.

21. $8^{-\frac{2}{3}}$

22. $3x(ab)^{-\frac{4}{5}}$

In Exercises 23–26, use properties of rational exponents to simplify each expression.

23. $x^{\frac{1}{3}} \cdot x^{\frac{1}{4}}$

24. $\dfrac{5^{\frac{1}{2}}}{5^{\frac{1}{3}}}$

25. $(8x^6y^3)^{\frac{1}{3}}$

26. $(x^{-\frac{2}{3}}y^{\frac{1}{4}})^{\frac{1}{2}}$

In Exercises 27–31, use rational exponents to simplify each expression. If rational exponents appear after simplifying, write the answer in radical notation.

27. $\sqrt[3]{x^9y^{12}}$

28. $\sqrt[9]{x^3y^9}$

29. $\sqrt{x} \cdot \sqrt[3]{x}$

30. $\dfrac{\sqrt[3]{x^2}}{\sqrt[4]{x^2}}$

31. $\sqrt[5]{\sqrt[3]{x}}$

32. The function $f(x) = 350x^{\frac{2}{3}}$ models the expenditures, $f(x)$, in millions of dollars, for the U.S. National Park Service x years after 1985. According to this model, what will expenditures be in 2012?

7.3 *In Exercises 33–35, use the product rule to multiply.*

33. $\sqrt{3x} \cdot \sqrt{7y}$

34. $\sqrt[5]{7x^2} \cdot \sqrt[5]{11x}$

35. $\sqrt[6]{x - 5} \cdot \sqrt[6]{(x - 5)^4}$

36. If $f(x) = \sqrt{7x^2 - 14x + 7}$, express the function, f, in simplified form. Assume that x can be any real number.

In Exercises 37–39, simplify by factoring. Assume that all variables in a radicand represent positive real numbers.

37. $\sqrt{20x^3}$

38. $\sqrt[3]{54x^8y^6}$

39. $\sqrt[4]{32x^3y^{11}}$

In Exercises 40–43, multiply and simplify, if possible. Assume that all variables in a radicand represent positive real numbers.

40. $\sqrt{6x^3} \cdot \sqrt{4x^2}$

41. $\sqrt[3]{4x^2y} \cdot \sqrt[3]{4xy^4}$

42. $\sqrt[5]{2x^4y^3} \cdot \sqrt[5]{8xy^6}$

43. $\sqrt{x + 1} \cdot \sqrt{x - 1}$

7.4 *Assume that all variables represent positive real numbers.*

In Exercises 44–47, add or subtract as indicated.

44. $6\sqrt[3]{3} + 2\sqrt[3]{3}$

45. $5\sqrt{18} - 3\sqrt{8}$

46. $\sqrt[3]{27x^4} + \sqrt[3]{xy^6}$

47. $2\sqrt[3]{6} - 5\sqrt[3]{48}$

In Exercises 48–50, simplify using the quotient rule.

48. $\sqrt[3]{\dfrac{16}{125}}$

49. $\sqrt{\dfrac{x^3}{100y^4}}$

50. $\sqrt[4]{\dfrac{3y^5}{16x^{20}}}$

In Exercises 51–54, divide and, if possible, simplify.

51. $\dfrac{\sqrt{48}}{\sqrt{2}}$

52. $\dfrac{\sqrt[3]{32}}{\sqrt[3]{2}}$

53. $\dfrac{\sqrt[4]{64x^7}}{\sqrt[4]{2x^2}}$

54. $\dfrac{\sqrt{200x^3y^2}}{\sqrt{2x^{-2}y}}$

7.5 *Assume that all variables represent positive real numbers.*

In Exercises 55–62, multiply as indicated. If possible, simplify any radical expressions that appear in the product.

55. $\sqrt{3}(2\sqrt{6} + 4\sqrt{15})$

56. $\sqrt[3]{5}(\sqrt[3]{50} - \sqrt[3]{2})$

57. $(\sqrt{7} - 3\sqrt{5})(\sqrt{7} + 6\sqrt{5})$

58. $(\sqrt{x} - \sqrt{11})(\sqrt{y} - \sqrt{11})$

59. $(\sqrt{5} + \sqrt{8})^2$

60. $(2\sqrt{3} - \sqrt{10})^2$

61. $(\sqrt{7} + \sqrt{13})(\sqrt{7} - \sqrt{13})$

62. $(7 - 3\sqrt{5})(7 + 3\sqrt{5})$

In Exercises 63–75, rationalize each denominator. Simplify, if possible.

63. $\frac{4}{\sqrt{6}}$ **64.** $\sqrt{\frac{2}{7}}$

65. $\frac{12}{\sqrt[3]{9}}$ **66.** $\sqrt{\frac{2x}{5y}}$

67. $\frac{14}{\sqrt[3]{2x^2}}$ **68.** $\sqrt[4]{\frac{7}{3x}}$

69. $\frac{5}{\sqrt[5]{32x^4y}}$

70. $\frac{6}{\sqrt{3}-1}$ **71.** $\frac{\sqrt{7}}{\sqrt{5}+\sqrt{3}}$

72. $\frac{10}{2\sqrt{5}-3\sqrt{2}}$

73. $\frac{\sqrt{x}+5}{\sqrt{x}-3}$

74. $\frac{\sqrt{7}+\sqrt{3}}{\sqrt{7}-\sqrt{3}}$ **75.** $\frac{2\sqrt{3}+\sqrt{6}}{2\sqrt{6}+\sqrt{3}}$

In Exercises 76–79, rationalize each numerator. Simplify, if possible.

76. $\sqrt{\frac{2}{7}}$ **77.** $\frac{\sqrt[3]{3x}}{\sqrt[3]{y}}$

78. $\frac{\sqrt{7}}{\sqrt{5}+\sqrt{3}}$ **79.** $\frac{\sqrt{7}+\sqrt{3}}{\sqrt{7}-\sqrt{3}}$

7.6 *In Exercises 80–84, solve each radical equation.*

80. $\sqrt{2x+4} = 6$

81. $\sqrt{x-5} + 9 = 4$

82. $\sqrt{2x-3} + x = 3$

83. $\sqrt{x-4} + \sqrt{x+1} = 5$

84. $(x^2+6x)^{\frac{1}{3}} + 2 = 0$

85. The time, $f(x)$, in seconds, for a free-falling object to fall x feet is modeled by the function

$$f(x) = \sqrt{\frac{x}{16}}.$$

If a worker accidently drops a hammer from a building and it hits the ground after 4 seconds, from what height was the hammer dropped?

86. Out of a group of 50,000 births, the number of people, $f(x)$, surviving to age x is modeled by the function

$$f(x) = 5000\sqrt{100-x}.$$

To what age will 20,000 people in the group survive?

7.7 *In Exercises 87–89, express each number in terms of i and simplify, if possible.*

87. $\sqrt{-81}$ **88.** $\sqrt{-63}$

89. $-\sqrt{-8}$

In Exercises 90–99, perform the indicated operation. Write the result in the form a + bi.

90. $(7+12i)+(5-10i)$

91. $(8-3i)-(17-7i)$

92. $4i(3i-2)$

93. $(7-5i)(2+3i)$

94. $(3-4i)^2$

95. $(7+8i)(7-8i)$

96. $\sqrt{-8}\cdot\sqrt{-3}$

97. $\frac{6}{5+i}$

98. $\frac{3+4i}{4-2i}$

99. $\frac{5+i}{3i}$

In Exercises 100–101, simplify each expression.

100. i^{16} **101.** i^{23}

Chapter 7 Test

1. Let $f(x) = \sqrt{8-2x}$.

a. Find $f(-14)$.

b. Find the domain of f.

2. Evaluate: $27^{-\frac{4}{3}}$.

3. Simplify: $(25x^{-\frac{1}{2}}y^{\frac{1}{4}})^{\frac{1}{2}}$.

In Exercises 4–5, use rational exponents to simplify each expression. If rational exponents appear after simplifying, write the answer in radical notation.

4. $\sqrt[8]{x^4}$ **5.** $\sqrt[4]{x}\cdot\sqrt[5]{x}$

In Exercises 6–9, simplify each expression. Assume that each variable can represent any real number.

6. $\sqrt{75x^2}$ **7.** $\sqrt{x^2-10x+25}$

8. $\sqrt[3]{16x^4y^8}$ **9.** $\sqrt[5]{-\frac{32}{x^{10}}}$

In Exercises 10–17, perform the indicated operation and, if possible, simplify. Assume that all variables represent positive real numbers.

10. $\sqrt[3]{5x^2}\cdot\sqrt[3]{10y}$

11. $\sqrt[4]{8x^3y}\cdot\sqrt[4]{4xy^2}$

12. $3\sqrt{18} - 4\sqrt{32}$

13. $\sqrt[3]{8x^4} + \sqrt[3]{xy^6}$

14. $\dfrac{\sqrt[3]{16x^8}}{\sqrt[3]{2x^4}}$

15. $\sqrt{3}(4\sqrt{6} - \sqrt{5})$

16. $(5\sqrt{6} - 2\sqrt{2})(\sqrt{6} + \sqrt{2})$

17. $(7 - \sqrt{3})^2$

In Exercises 18–20, rationalize each denominator. Simplify, if possible. Assume all variables represent positive real numbers.

18. $\sqrt{\dfrac{5}{x}}$

19. $\dfrac{5}{\sqrt[3]{5x^2}}$

20. $\dfrac{\sqrt{2} - \sqrt{3}}{\sqrt{2} + \sqrt{3}}$

In Exercises 21–23, solve each radical equation.

21. $3 + \sqrt{2x - 3} = x$

22. $\sqrt{x + 9} - \sqrt{x - 7} = 2$

23. $(11x + 6)^{\frac{1}{3}} + 3 = 0$

24. The function

$$f(x) = 2.9\sqrt{x} + 20.1$$

models the average height, in inches, of boys who are x months of age, $0 \le x \le 60$. Find the age at which the average height of boys is 40.4 inches.

25. Express in terms of i and simplify: $\sqrt{-75}$.

In Exercises 26–29, perform the indicated operation. Write the result in the form $a + bi$.

26. $(5 - 3i) - (6 - 9i)$

27. $(3 - 4i)(2 + 5i)$

28. $\sqrt{-9} \cdot \sqrt{-4}$

29. $\dfrac{3 + i}{1 - 2i}$

30. Simplify: i^{35}.

Cumulative Review Exercises (Chapters 1–7)

In Exercises 1–5, solve each equation, inequality, or system.

1.
$$\begin{aligned} 2x - y + z &= -5 \\ x - 2y - 3z &= 6 \\ x + y - 2z &= 1 \end{aligned}$$

2. $3x^2 - 11x = 4$

3. $2(x + 4) < 5x + 3(x + 2)$

4. $\dfrac{1}{x + 2} + \dfrac{15}{x^2 - 4} = \dfrac{5}{x - 2}$

5. $\sqrt{x + 2} - \sqrt{x + 1} = 1$

6. Graph the solution set of the system:

$$\begin{aligned} x + 2y &< 2 \\ 2y - x &> 4. \end{aligned}$$

In Exercises 7–15, perform the indicated operations.

7. $\dfrac{8x^2}{3x^2 - 12} \div \dfrac{40}{x - 2}$

8. $\dfrac{x + \dfrac{1}{y}}{y + \dfrac{1}{x}}$

9. $(2x - 3)(4x^2 - 5x - 2)$

10. $\dfrac{7x}{x^2 - 2x - 15} - \dfrac{2}{x - 5}$

11. $7(8 - 10)^3 - 7 + 3 \div (-3)$

12. $\sqrt{80x} - 5\sqrt{20x} + 2\sqrt{45x}$

13. $\dfrac{\sqrt{3} - 2}{2\sqrt{3} + 5}$

14. $(2x^3 - 3x^2 + 3x - 4) \div (x - 2)$

15. $(2\sqrt{3} + 5\sqrt{2})(\sqrt{3} - 4\sqrt{2})$

In Exercises 16–17, factor completely.

16. $24x^2 + 10x - 4$

17. $16x^4 - 1$

18. The amount of light provided by a light bulb varies inversely as the square of the distance from the bulb. The illumination provided is 120 lumens at a distance of 10 feet. How many lumens are provided at a distance of 15 feet?

19. You invested \$6000 in two accounts paying 7% and 9% annual interest, respectively. At the end of the year, the total interest from these investments was \$510. How much was invested at each rate?

20. Although there are 2332 students enrolled in the college, this is 12% fewer students than there were last year. How many students were enrolled last year?

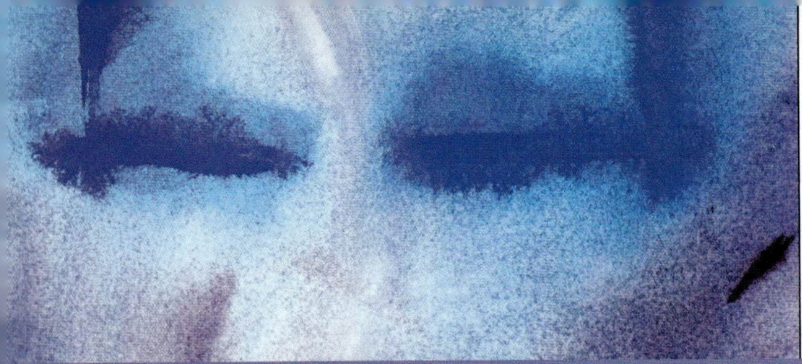

Thanks to the efficiency of U.S. farms, most Americans enjoy an abundance of inexpensive foods. Nearly one in five U.S. kids ages 9 through 13 say they've already been on a weight-loss diet. By contrast, worldwide, every three seconds a child dies of hunger. If all the world's undernourished people were gathered into a new country, it would be the third most populous nation, just behind China and India.

Source: The State of Food Insecurity in the World, 2000; 2000 BBC News

Because we tend to take our food abundance for granted, it might surprise you to know that the number of U.S. farms has declined since the 1920s, as individually owned family farms have been swallowed up by huge agribusinesses owned by corporations. Currently, there are about 1.8 acres of cropland to grow food for each American. If the current trends in population growth and loss of farmland continue, there will be only 0.6 acre per American in the year 2050. This is the rate that currently exists worldwide.

Like the number of farms and the amount of cropland in the United States, many phenomena follow trends that involve growth and decline, or vice versa. In this chapter, you will learn to solve equations and graph functions that provide new ways of looking at and understanding these phenomena.

Chapter 8

Quadratic Equations and Functions

8.1 THE SQUARE ROOT PROPERTY AND COMPLETING THE SQUARE

8.2 THE QUADRATIC FORMULA

8.3 QUADRATIC FUNCTIONS AND THEIR GRAPHS

8.4 EQUATIONS QUADRATIC IN FORM

8.5 QUADRATIC AND RATIONAL INEQUALITIES

▸ SECTION 8.1 The Square Root Property and Completing the Square

Objectives

1. Solve quadratic equations using the square root property.
2. Complete the square of a binomial.
3. Solve quadratic equations by completing the square.
4. Solve problems using the square root property.

Sandy Skoglund "Radioactive Cats" cibachrome, 30 × 40 inches.

Using Einstein's famous formula $E = mc^2$, it can be shown that the mass of a golf ball could provide the daily energy needs of a large metropolitan area. (See the enrichment essay on page 59.) However, physicists have not yet developed a way of converting a mass, m, completely to energy, E. The best that has been done in nuclear fission reactors is to convert about 0.1% of a mass to energy. With only 0.1% efficiency, it would take about 30 kilograms, or approximately 66 pounds, of mass to power a large metropolitan area for a day.

Numerous problems stand in the way of nuclear power expansion, not the least of which is what to do with nuclear waste. So far, scientists have not found a safe way to dispose of the dangerous radioactive wastes produced by nuclear reactors. By 2000, more than 25,000 tons of waste had accumulated around the 104 nuclear power plants in the United States.

The formula $E = mc^2$ shows that mass and energy are equivalent, and the transformation of even a tiny amount of mass releases an enormous amount of energy. Is there a way to work with the formula $E = mc^2$ and solve for c, thereby expressing the speed of light, c, in terms of energy, E, and mass, m? In this section, by learning how this is done, you will develop two new ways for solving quadratic equations. These new methods are called the *square root property* and *completing the square*.

Study Tip

Here is a summary of what we already know about quadratic equations and quadratic functions.

1. A **quadratic equation** in x can be written in the standard form
$$ax^2 + bx + c = 0, \quad a \neq 0.$$
2. Some quadratic equations can be solved by factoring.
$$\begin{aligned} \text{Solve:} \quad 2x^2 + 7x - 4 &= 0. \\ (2x - 1)(x + 4) &= 0 \\ 2x - 1 = 0 \quad \text{or} \quad x + 4 &= 0 \\ 2x = 1 \qquad\qquad x &= -4 \\ x = \tfrac{1}{2} \qquad\qquad & \end{aligned}$$
The solutions are -4 and $\frac{1}{2}$, and the solution set is $\{-4, \frac{1}{2}\}$.
3. A polynomial function of the form
$$f(x) = ax^2 + bx + c, \quad a \neq 0$$
is a **quadratic function**. Graphs of quadratic functions have cuplike shapes with the same behavior at each end.

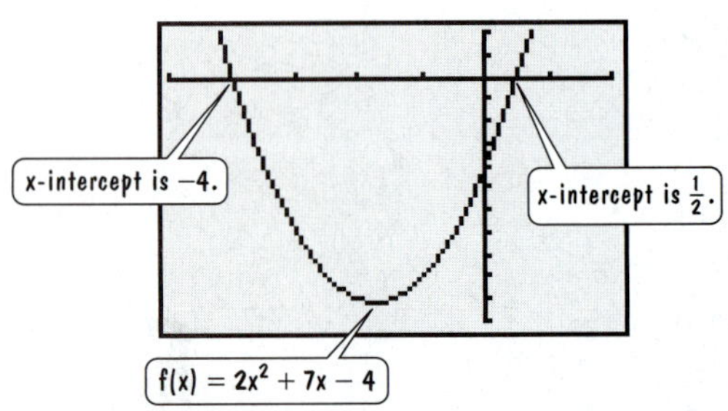

Figure 8.1

4. The real solutions of $ax^2 + bx + c = 0$ correspond to the x-intercepts for the graph of the quadratic function $f(x) = ax^2 + bx + c$. For example, the solutions of the equation $2x^2 + 7x - 4 = 0$ are -4 and $\frac{1}{2}$. Figure 8.1 shows that the solutions appear as x-intercepts on the graph of the quadratic function.

Now that we've summarized what we know, let's look at where we go. How do we solve a quadratic equation, $ax^2 + bx + c = 0$, if the trinomial $ax^2 + bx + c$ cannot be factored? Methods other than factoring are needed. In this section, we look at other ways of solving quadratic equations.

1 Solve quadratic equations using the square root property.

The Square Root Property

Let's begin with a relatively simple quadratic equation:

$$x^2 = 9.$$

The value of x must be a number whose square is 9. There are two numbers whose square is 9:

$$x = \sqrt{9} = 3 \quad \text{or} \quad x = -\sqrt{9} = -3.$$

Thus, the solutions of $x^2 = 9$ are 3 and -3. This is an example of the **square root property**.

Discover for Yourself

Solve $x^2 = 9$, or

$$x^2 - 9 = 0,$$

by factoring. What is the advantage of using the square root property?

The Square Root Property

If u is an algebraic expression and d is a nonzero real number, then $u^2 = d$ has exactly two solutions:

$$\text{If } u^2 = d, \quad \text{then } u = \sqrt{d} \text{ or } u = -\sqrt{d}.$$

Equivalently,

$$\text{If } u^2 = d, \quad \text{then } u = \pm\sqrt{d}.$$

Notice that $u = \pm\sqrt{d}$ is a shorthand notation to indicate that $u = \sqrt{d}$ or $u = -\sqrt{d}$. Although we usually read $u = \pm\sqrt{d}$ as "u equals plus or minus the square root of d," we actually mean that u is the positive square root of d or the negative square root of d.

EXAMPLE 1 Solving a Quadratic Equation by the Square Root Property

Solve: $4x^2 = 20$.

Solution To apply the square root property, we need a squared expression by itself on one side of the equation.

$$4x^2 = 20$$

We want x^2 by itself.

We can get x^2 by itself if we divide both sides by 4.

$4x^2 = 20$	This is the original equation.
$\dfrac{4x^2}{4} = \dfrac{20}{4}$	Divide both sides by 4.
$x^2 = 5$	Simplify.
$x = \sqrt{5}$ or $x = -\sqrt{5}$	Apply the square root property.

Now let's check these proposed solutions in the original equation.

Check $\sqrt{5}$:	**Check $-\sqrt{5}$:**
$4x^2 = 20$	$4x^2 = 20$
$4(\sqrt{5})^2 \stackrel{?}{=} 20$	$4(-\sqrt{5})^2 \stackrel{?}{=} 20$
$4 \cdot 5 \stackrel{?}{=} 20$	$4 \cdot 5 \stackrel{?}{=} 20$
$20 = 20$, true	$20 = 20$, true

The solutions are $-\sqrt{5}$ and $\sqrt{5}$. The solution set is $\{-\sqrt{5}, \sqrt{5}\}$ or $\{\pm\sqrt{5}\}$. ■

CHECK POINT 1 Solve: $5x^2 = 15$.

In this section, we will express irrational solutions in simplified radical form, rationalizing denominators when possible.

EXAMPLE 2 Solving a Quadratic Equation by the Square Root Property

Solve: $2x^2 - 7 = 0$.

Solution To solve by the square root property, we place the squared expression by itself on one side of the equation.

$$2x^2 - 7 = 0$$

We want x^2 by itself.

$$2x^2 - 7 = 0 \quad \text{This is the original equation.}$$

$$2x^2 = 7 \quad \text{Add 7 to both sides.}$$

$$x^2 = \frac{7}{2} \quad \text{Divide both sides by 2.}$$

$$x = \sqrt{\frac{7}{2}} \quad \text{or} \quad x = -\sqrt{\frac{7}{2}} \quad \text{Apply the square root property.}$$

Because the proposed solutions are opposites, we can rationalize both denominators at once:

$$\pm\sqrt{\frac{7}{2}} = \pm\frac{\sqrt{7}}{\sqrt{2}} \cdot \frac{\sqrt{2}}{\sqrt{2}} = \pm\frac{\sqrt{14}}{2}.$$

Substitute these values in the original equation and verify that the solutions are $-\frac{\sqrt{14}}{2}$ and $\frac{\sqrt{14}}{2}$. The solution set is $\left\{-\frac{\sqrt{14}}{2}, \frac{\sqrt{14}}{2}\right\}$ or $\left\{\pm\frac{\sqrt{14}}{2}\right\}$. ■

CHECK POINT 2 Solve: $2x^2 - 5 = 0$.

Some quadratic equations have solutions that are imaginary numbers.

EXAMPLE 3 Solving a Quadratic Equation by the Square Root Property

Solve: $9x^2 + 25 = 0$.

Solution

$$9x^2 + 25 = 0 \quad \text{This is the original equation.}$$

We need to isolate x^2.

$$9x^2 = -25 \quad \text{Subtract 25 from both sides.}$$

$$x^2 = -\frac{25}{9} \quad \text{Divide both sides by 9.}$$

$$x = \sqrt{-\frac{25}{9}} \quad \text{or} \quad x = -\sqrt{-\frac{25}{9}} \quad \text{Apply the square root property.}$$

$$x = \sqrt{\frac{25}{9}}\sqrt{-1} \qquad x = -\sqrt{\frac{25}{9}}\sqrt{-1}$$

$$x = \frac{5}{3}i \qquad x = -\frac{5}{3}i \quad \sqrt{-1} = i$$

Because the equation has an x^2-term and no x-term, we can check both proposed solutions, $\pm\frac{5}{3}i$, at once.

Using Technology

The graph of

$$f(x) = 9x^2 + 25$$

has no x-intercepts. This shows that

$$9x^2 + 25 = 0$$

has no real solutions. Example 3 algebraically establishes that the solutions are imaginary numbers.

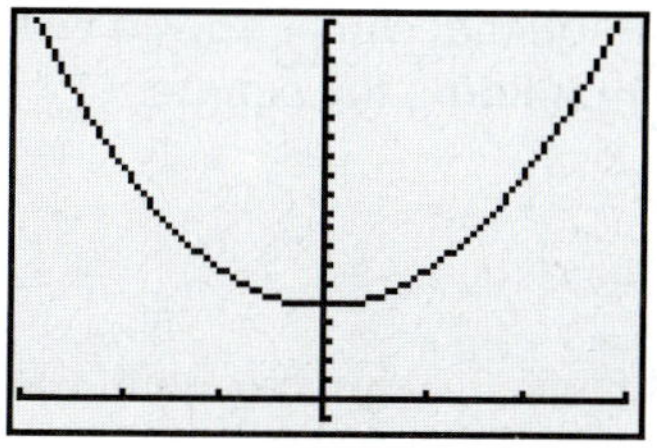

$[-3, 3, 1]$ by $[-5, 100, 5]$

Check $\frac{5}{3}i$ and $-\frac{5}{3}i$:

$$9x^2 + 25 = 0$$

$$9\left(\pm\frac{5}{3}i\right)^2 + 25 \stackrel{?}{=} 0$$

$$9\left(\frac{25}{9}i^2\right) + 25 \stackrel{?}{=} 0$$

$$25i^2 + 25 \stackrel{?}{=} 0$$

$i^2 = -1$

$$25(-1) + 25 \stackrel{?}{=} 0$$

$$0 = 0, \text{ true}$$

The solutions are $-\frac{5}{3}i$ and $\frac{5}{3}i$. The solution set is $\left\{-\frac{5}{3}i, \frac{5}{3}i\right\}$ or $\left\{\pm\frac{5}{3}i\right\}$. ■

✔ CHECK POINT 3 Solve: $4x^2 + 9 = 0$.

Can we solve an equation such as $(x - 1)^2 = 5$ using the square root property? Yes. The equation is in the form $u^2 = d$, where u^2, the squared expression, is by itself on the left side.

$$(x - 1)^2 \quad = \quad 5$$

This is u^2 in $u^2 = d$ with $u = x - 1$.

This is d in $u^2 = d$ with $d = 5$.

Discover for Yourself

Try solving

$$(x - 1)^2 = 5$$

by writing the equation in standard form and factoring. What problem do you encounter?

EXAMPLE 4 Solving a Quadratic Equation by the Square Root Property

Solve by the square root property: $(x - 1)^2 = 5$.

Solution

$$(x - 1)^2 = 5 \qquad \text{This is the original equation.}$$

$$x - 1 = \sqrt{5} \quad \text{or} \quad x - 1 = -\sqrt{5} \qquad \text{Apply the square root property.}$$

$$x = 1 + \sqrt{5} \qquad x = 1 - \sqrt{5} \qquad \text{Add 1 to both sides in each equation.}$$

Check $1 + \sqrt{5}$:

$$(x - 1)^2 = 5$$
$$(1 + \sqrt{5} - 1)^2 \stackrel{?}{=} 5$$
$$(\sqrt{5})^2 \stackrel{?}{=} 5$$
$$5 = 5, \text{ true}$$

Check $1 - \sqrt{5}$:

$$(x - 1)^2 = 5$$
$$(1 - \sqrt{5} - 1)^2 \stackrel{?}{=} 5$$
$$(-\sqrt{5})^2 \stackrel{?}{=} 5$$
$$5 = 5, \text{ true}$$

The solutions are $1 \pm \sqrt{5}$, and the solution set is $\{1 + \sqrt{5}, 1 - \sqrt{5}\}$ or $\{1 \pm \sqrt{5}\}$. ■

✔ **CHECK POINT 4** Solve: $(x - 2)^2 = 7$.

2 Complete the square of a binomial.

Completing the Square

We return to the question that opened this section: How do we solve a quadratic equation, $ax^2 + bx + c = 0$, if the trinomial $ax^2 + bx + c$ cannot be factored? We can convert the equation into an equivalent equation that can be solved using the square root property. This is accomplished by **completing the square**.

Completing the Square

If $x^2 + bx$ is a binomial, then by adding $\left(\frac{b}{2}\right)^2$, which is the square of half the coefficient of x, a perfect square trinomial will result. That is,

$$x^2 + bx + \left(\frac{b}{2}\right)^2 = \left(x + \frac{b}{2}\right)^2.$$

EXAMPLE 5 Completing the Square

Complete the square for each binomial. Then factor the resulting perfect square trinomial:

a. $x^2 + 8x$ **b.** $x^2 - 7x$ **c.** $x^2 + \frac{3}{5}x$.

Solution To complete the square, we must add a term to each binomial. The term that should be added is the square of half the coefficient of x.

$$x^2 + 8x \qquad x^2 - 7x \qquad x^2 + \frac{3}{5}x$$

Add $\left(\frac{8}{2}\right)^2 = 4^2$. Add 16 to complete the square.

Add $\left(\frac{-7}{2}\right)^2$, or $\frac{49}{4}$, to complete the square.

Add $\left(\frac{1}{2} \cdot \frac{3}{5}\right)^2 = \left(\frac{3}{10}\right)^2$. Add $\frac{9}{100}$ to complete the square.

a. The coefficient of the x-term in $x^2 + 8x$ is 8. Half of 8 is 4, and $4^2 = 16$. Add 16.

$$x^2 + 8x + 16 = (x + 4)^2$$

b. The coefficient of the x-term in $x^2 - 7x$ is -7. Half of -7 is $-\frac{7}{2}$, and $\left(-\frac{7}{2}\right)^2 = \frac{49}{4}$. Add $\frac{49}{4}$.

$$x^2 - 7x + \frac{49}{4} = \left(x - \frac{7}{2}\right)^2$$

c. The coefficient of the x-term in $x^2 + \frac{3}{5}x$ is $\frac{3}{5}$. Half of $\frac{3}{5}$ is $\frac{1}{2} \cdot \frac{3}{5}$, or $\frac{3}{10}$, and $\left(\frac{3}{10}\right)^2 = \frac{9}{100}$. Add $\frac{9}{100}$.

$$x^2 + \frac{3}{5}x + \frac{9}{100} = \left(x + \frac{3}{10}\right)^2$$ ■

ENRICHMENT ESSAY

Visualizing Completing the Square

This figure, with area $x^2 + 8x$, is not a complete square. The bottom-right corner is missing.

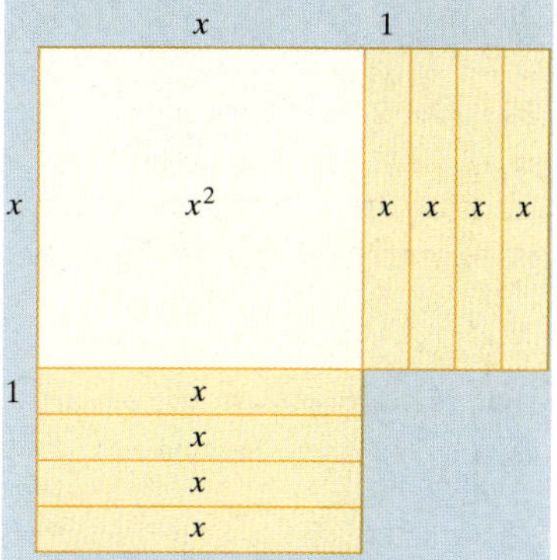

Area: $x^2 + 8x$

Add 16 square units to the missing portion and you, literally, complete the square.

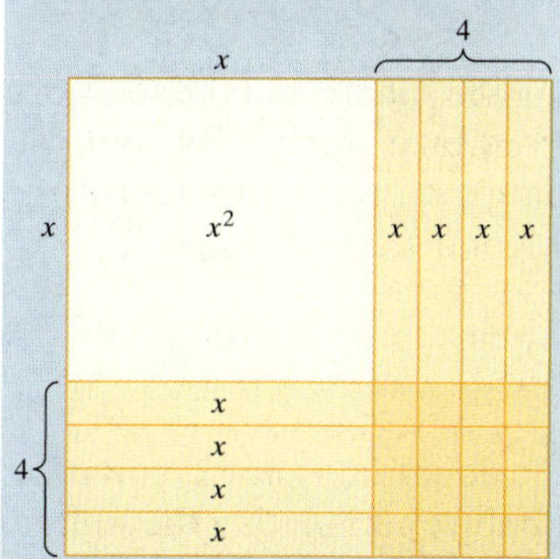

Area: $x^2 + 8x + 16 = (x + 4)^2$

Study Tip

You may not be accustomed to factoring perfect square trinomials in which fractions are involved. The constant in the factorization is always half the coefficient of x.

$$x^2 - 7x + \frac{49}{4} = \left(x - \frac{7}{2}\right)^2 \qquad x^2 + \frac{3}{5}x + \frac{9}{100} = \left(x + \frac{3}{10}\right)^2$$

Half the coefficient of x, -7, is $-\frac{7}{2}$.

Half the coefficient of x, $\frac{3}{5}$, is $\frac{3}{10}$.

✔ **CHECK POINT 5** Complete the square for each binomial. Then factor the resulting perfect square trinomial:

a. $x^2 + 10x$

b. $x^2 - 3x$

c. $x^2 + \frac{3}{4}x$.

3 Solve quadratic equations by completing the square.

Solving Quadratic Equations by Completing the Square

We can solve any quadratic equation by completing the square. If the coefficient of the x^2-term is one, we add the square of half the coefficient of x to both sides of the equation. **When you add a constant term to one side of the equation to complete the square, be certain to add the same constant to the other side of the equation.** These ideas are illustrated in Example 6.

EXAMPLE 6 Solving a Quadratic Equation by Completing the Square

Solve by completing the square: $x^2 - 6x + 2 = 0$.

Solution We begin by subtracting 2 from both sides. This is done to isolate the binomial $x^2 - 6x$ so that we can complete the square.

$$x^2 - 6x + 2 = 0 \qquad \text{This is the original equation.}$$

$$x^2 - 6x = -2 \qquad \text{Subtract 2 from both sides.}$$

Next, we complete the square. Find half the coefficient of the x-term and square it. The coefficient of the x-term is -6. Half of -6 is -3 and $(-3)^2 = 9$. Thus, we add 9 to both sides of the equation.

$$x^2 - 6x + 9 = -2 + 9 \qquad \text{Add 9 to both sides to complete the square.}$$

$$(x - 3)^2 = 7 \qquad \text{Factor and simplify.}$$

$$x - 3 = \sqrt{7} \quad \text{or} \quad x - 3 = -\sqrt{7} \qquad \text{Apply the square root property.}$$

$$x = 3 + \sqrt{7} \qquad x = 3 - \sqrt{7} \qquad \text{Add 3 to both sides in each equation.}$$

The solutions are $3 \pm \sqrt{7}$, and the solution set is $\{3 + \sqrt{7}, 3 - \sqrt{7}\}$ or $\{3 \pm \sqrt{7}\}$. ■

If you solve a quadratic equation by completing the square and the solutions are rational numbers, the equation can also be solved by factoring. By contrast, quadratic equations with irrational solutions cannot be solved by factoring. However, all quadratic equations can be solved by completing the square.

✔ **CHECK POINT 6** Solve by completing the square: $x^2 - 10x + 18 = 0$.

If the coefficient of the x^2-term in a quadratic equation is not 1, you must divide each side of the equation by this coefficient before completing the square. For example, to solve $2x^2 + 5x - 4 = 0$ by completing the square, first divide every term by 2:

$$\frac{2x^2}{2} + \frac{5x}{2} - \frac{4}{2} = \frac{0}{2}$$

$$x^2 + \frac{5}{2}x - 2 = 0.$$

Now that the coefficient of the x^2-term is 1, we can solve by completing the square.

EXAMPLE 7 Solving a Quadratic Equation by Completing the Square

Solve by completing the square: $2x^2 + 5x - 4 = 0$.

Using Technology

Obtain a decimal approximation for each solution of $2x^2 + 5x - 4 = 0$, the equation in Example 7.

$$\frac{-5 + \sqrt{57}}{4} \approx 0.6$$

$$\frac{-5 - \sqrt{57}}{4} \approx -3.1$$

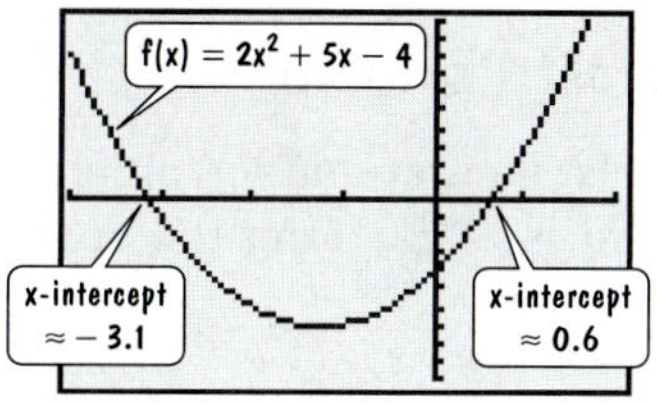

$[-4, 2, 1]$ by $[-10, 10, 1]$

The x-intercepts of $f(x) = 2x^2 + 5x - 4$ verify the solutions.

Solution

$$2x^2 + 5x - 4 = 0$$
This is the original equation.

$$x^2 + \frac{5}{2}x - 2 = 0$$
Divide both sides by 2.

$$x^2 + \frac{5}{2}x = 2$$
Add 2 to both sides to isolate the binomial.

$$x^2 + \frac{5}{2}x + \frac{25}{16} = 2 + \frac{25}{16}$$
Complete the square: Half of $\frac{5}{2}$ is $\frac{5}{4}$ and $\left(\frac{5}{4}\right)^2 = \frac{25}{16}$.

$$\left(x + \frac{5}{4}\right)^2 = \frac{57}{16}$$
Factor and simplify. On the right: $2 + \frac{25}{16} = \frac{32}{16} + \frac{25}{16} = \frac{57}{36}$.

$$x + \frac{5}{4} = \sqrt{\frac{57}{16}} \quad \text{or} \quad x + \frac{5}{4} = -\sqrt{\frac{57}{16}}$$
Apply the square root property.

$$x + \frac{5}{4} = \frac{\sqrt{57}}{4} \quad \text{or} \quad x + \frac{5}{4} = -\frac{\sqrt{57}}{4}$$
$\sqrt{\frac{57}{16}} = \frac{\sqrt{57}}{\sqrt{16}} = \frac{\sqrt{57}}{4}$

$$x = -\frac{5}{4} + \frac{\sqrt{57}}{4} \quad \text{or} \quad x = -\frac{5}{4} - \frac{\sqrt{57}}{4}$$
Solve the equations, subtracting $\frac{5}{4}$ from both sides.

$$x = \frac{-5 + \sqrt{57}}{4} \quad \text{or} \quad x = \frac{-5 - \sqrt{57}}{4}$$
Express solutions with a common denominator.

The solutions are $\frac{-5 \pm \sqrt{57}}{4}$, and the solution set is $\left\{\frac{-5 \pm \sqrt{57}}{4}\right\}$. ■

CHECK POINT 7 Solve by completing the square: $2x^2 - 10x - 1 = 0$.

4 Solve problems using the square root property.

Applications

We all want a wonderful life with fulfilling work, good health, and loving relationships. And let's be honest: Financial security wouldn't hurt! Achieving this goal depends on understanding how money in a savings account grows in remarkable ways as a result of *compound interest*. **Compound interest** is interest computed on your original investment as well as on any accumulated interest. For example, suppose you deposit \$2000, the principal, in a savings account at a rate of 6%. The interest at the end of the first year is

$$I = Pr = (\$2000)(0.06) = \$120.$$

Interest earned is the principal times the interest rate.

Assuming that you do not withdraw the \$120 interest, the amount in the account during year two is \$2000 + \$120, or \$2120. The interest paid at the end of the second year is computed using \$2120:

$$I = Pr = (\$2120)(0.06) = \$127.20.$$

The bank now adds \$127.20 to the \$2120 in your account. Thus, during year 3 the account contains \$2120 + \$127.20, or \$2247.20. The interest paid at the end of the third year is computed using \$2247.20:

$$I = Pr = (\$2247.20)(0.06) \approx \$134.83.$$

The amount in the savings account at the end of year three is \$2247.20 + \$134.83, or \$2382.03.

A faster way to determine the amount, A, in an account subject to compound interest is to use the following formula:

A Formula for Compound Interest

Suppose that an amount of money, P, is invested at interest rate r, compounded annually. In t years, the amount, A, or balance, in the account is given by the formula

$$A = P(1 + r)^t.$$

Some compound interest problems can be solved using quadratic equations.

EXAMPLE 8 Solving a Compound Interest Problem

You invested \$1000 in a savings account whose interest is compounded annually. After 2 years, the amount, or balance, in the account is \$1210. Find the annual interest rate.

Solution We are given that

P (the amount invested) = \$1000

t (the time of the investment) = 2 years

A (the amount, or balance, in the account) = \$1210.

We are asked to find the annual interest rate, r. We do this by substituting the three given values in the compound interest formula and solving for r.

$$A = P(1 + r)^t \quad \text{Use the compound interest formula.}$$

$$1210 = 1000(1 + r)^2 \quad \text{Substitute the given values.}$$

$$\frac{1210}{1000} = (1 + r)^2 \quad \text{Divide both sides by 1000.}$$

$$\frac{121}{100} = (1 + r)^2 \quad \text{Simplify the fraction.}$$

$$1 + r = \sqrt{\frac{121}{100}} \quad \text{or} \quad 1 + r = -\sqrt{\frac{121}{100}} \quad \text{Apply the square root property.}$$

$$1 + r = \frac{11}{10} \quad \text{or} \quad 1 + r = -\frac{11}{10} \quad \sqrt{\frac{121}{100}} = \frac{\sqrt{121}}{\sqrt{100}} = \frac{11}{10}$$

$$r = \frac{11}{10} - 1 \quad \text{or} \quad r = -\frac{11}{10} - 1 \quad \text{Subtract 1 from both sides.}$$

$$r = \frac{1}{10} \quad \text{or} \quad r = -\frac{21}{10} \quad \frac{11}{10} - 1 = \frac{11}{10} - \frac{10}{10} = \frac{1}{10} \text{ and } -\frac{11}{10} - 1 = -\frac{11}{10} - \frac{10}{10} = -\frac{21}{10}.$$

Because the interest rate cannot be negative, we reject $-\frac{21}{10}$. Thus, the annual interest rate is $\frac{1}{10} = 0.10 = 10\%$.

We can check this answer using the formula $A = P(1 + r)^t$. If \$1000 is invested for 2 years at 10% interest, compounded annually, the balance in the account is

$$A = \$1000(1 + 0.10)^2 = \$1000(1.10)^2 = \$1210.$$

Because this is precisely the amount given by the problem's conditions, the annual interest rate is, indeed, 10% compounded annually. ■

CHECK POINT 8 You invested \$3000 in an account whose interest is compounded annually. After 2 years, the amount, or balance, in the account is \$4320. Find the annual interest rate.

In Chapter 5, we solved problems using the Pythagorean Theorem. Recall that in a right triangle, the side opposite the 90° angle is the hypotenuse and the other sides are legs. The Pythagorean Theorem states that the sum of the squares of the lengths of the legs equals the square of the length of the hypotenuse. Some problems that involve the Pythagorean Theorem can be solved with the square root property.

EXAMPLE 9 Using the Pythagorean Theorem and the Square Root Property

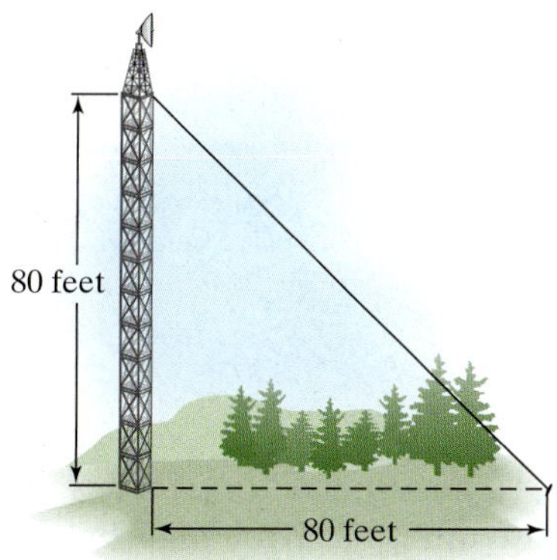

Figure 8.2

A supporting wire is to be attached to the top of an 80-foot antenna, as shown in Figure 8.2. Because of surrounding trees, the wire must be anchored 80 feet from the base of the antenna. What length of wire is required?

Solution Let w = the wire's length. We can find w in the right triangle using the Pythagorean Theorem.

(Leg)^2	plus	(Leg)^2	equals	(Hypotenuse)^2.
80^2	$+$	80^2	$=$	w^2.

We solve this equation using the square root property.

$$6400 + 6400 = w^2 \quad \text{Square 80.}$$

$$12{,}800 = w^2 \quad \text{Add.}$$

$$w = \sqrt{12{,}800} \quad \text{or} \quad w = -\sqrt{12{,}800} \quad \text{Apply the square root property.}$$

$$w = \sqrt{6400}\sqrt{2} \quad \text{or} \quad w = -\sqrt{6400}\sqrt{2} \quad \text{6400 is the largest perfect square factor of 12,800.}$$

$$w = 80\sqrt{2} \quad \text{or} \quad w = -80\sqrt{2}$$

Because w represents the wire's length, we reject the negative value. Thus, a wire of length $80\sqrt{2}$ feet is needed to support the antenna. Using a calculator, this is approximately 113.1 feet of wire. ■

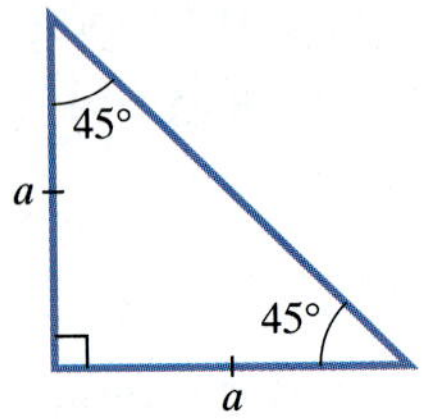

Figure 8.3 An isosceles right triangle has legs that are the same length and acute angles each measuring 45°.

Take a second look at the right triangle in Figure 8.2 on page 551. Can you see that both legs are the same length? If the lengths of both legs of a right triangle are the same, the triangle is called an **isosceles right triangle**. Figure 8.3 shows such a triangle. In Example 9, the wire's length was the length of a leg, 80 feet, times $\sqrt{2}$. We can use Figure 8.3 to show that the length of the hypotenuse in any isosceles right triangle is the length of a leg times $\sqrt{2}$.

$$c^2 = a^2 + a^2$$ Apply the Pythagorean Theorem to the isosceles right triangle in Figure 8.3.

$$c^2 = 2a^2$$ Combine like terms.

$$c = \sqrt{2a^2} \quad \text{or} \quad c = -\sqrt{2a^2}$$ Apply the square root property.

Reject because c cannot be negative.

$$c = \sqrt{2}a \quad \text{or} \quad a\sqrt{2}$$ Simplify: $\sqrt{2a^2} = \sqrt{2}\sqrt{a^2} = \sqrt{2}a$.

Lengths within Isosceles Right Triangles

The length of the hypotenuse in an isosceles right triangle is the length of a leg times $\sqrt{2}$.

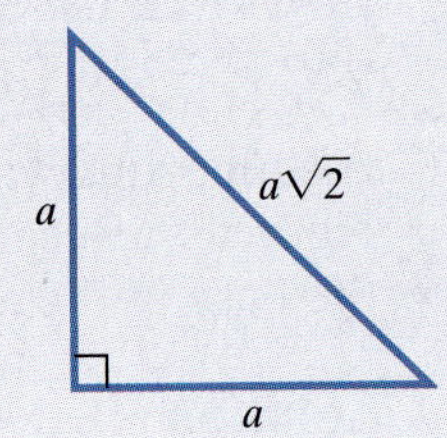

✔ **CHECK POINT 9** A 50-foot supporting wire is to be attached to an antenna. The wire is anchored 20 feet from the base of the antenna. How high up the antenna is the wire attached? Express the answer in simplified radical form. Then find a decimal approximation to the nearest tenth of a foot.

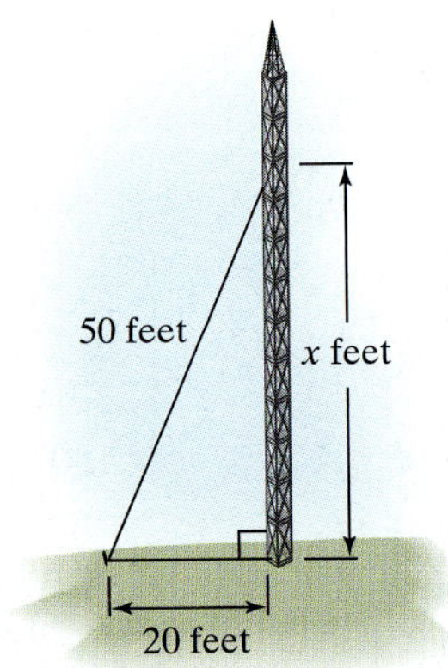

EXERCISE SET 8.1

Practice Exercises

In Exercises 1–22, solve each equation by the square root property. If possible, simplify radicals or rationalize denominators. Express imaginary solutions in the form a + bi.

1. $3x^2 = 75$

2. $5x^2 = 20$

3. $7x^2 = 42$

4. $8x^2 = 40$

5. $16x^2 = 25$

6. $4x^2 = 49$

7. $3x^2 - 2 = 0$

8. $3x^2 - 5 = 0$

9. $25x^2 + 16 = 0$

10. $4x^2 + 49 = 0$

11. $(x + 7)^2 = 9$

12. $(x + 3)^2 = 64$

13. $(x - 3)^2 = 5$

14. $(x - 4)^2 = 3$

15. $(x + 2)^2 = 8$

16. $(x + 2)^2 = 12$

17. $(x - 5)^2 = -9$

18. $(x - 5)^2 = -4$

19. $\left(x + \frac{3}{4}\right)^2 = \frac{11}{16}$

20. $\left(x + \frac{2}{5}\right)^2 = \frac{7}{25}$

21. $x^2 - 6x + 9 = 36$

22. $x^2 - 6x + 9 = 49$

In Exercises 23–34, complete the square for each binomial. Then factor the resulting perfect square trinomial.

23. $x^2 + 2x$

24. $x^2 + 4x$

25. $x^2 - 14x$

26. $x^2 - 10x$

27. $x^2 + 7x$

28. $x^2 + 9x$

29. $x^2 - \frac{1}{2}x$

30. $x^2 - \frac{1}{3}x$

31. $x^2 + \frac{4}{3}x$

32. $x^2 + \frac{4}{5}x$

33. $x^2 - \frac{9}{4}x$

34. $x^2 - \frac{9}{5}x$

In Exercises 35–56, solve each quadratic equation by completing the square.

35. $x^2 + 6x = -8$

36. $x^2 + 6x = 7$

37. $x^2 + 6x = -2$

38. $x^2 + 2x = 5$

39. $x^2 + 4x + 1 = 0$

40. $x^2 + 6x - 5 = 0$

41. $x^2 + 2x + 2 = 0$

42. $x^2 - 4x + 8 = 0$

43. $x^2 + 3x - 1 = 0$

44. $x^2 - 3x - 5 = 0$

45. $x^2 = 7x - 3$

46. $x^2 = 5x - 3$

47. $x^2 + x - 1 = 0$

48. $x^2 + 3x - 1 = 0$

49. $2x^2 - 3x + 1 = 0$

50. $2x^2 - x - 1 = 0$

51. $2x^2 + 10x + 11 = 0$

52. $2x^2 + 8x + 5 = 0$

53. $3x^2 - 2x - 4 = 0$

54. $4x^2 - 2x - 3 = 0$

55. $8x^2 - 4x + 1 = 0$

56. $9x^2 - 6x + 5 = 0$

Application Exercises

In Exercises 57–60, use the compound interest formula

$$A = P(1 + r)^t$$

to find the annual interest rate, r.

57. In 2 years, an investment of \$2000 grows to \$2880.

58. In 2 years, an investment of \$2000 grows to \$2420.

59. In 2 years, an investment of \$3125 grows to \$3360.

60. In 2 years, an investment of \$1280 grows to \$1445.

Cable-TV modems make it possible to quickly access the Internet and download files. The graph shows the number, in millions, of Internet users in the United States with this new technology. The data can be modeled by the function

$$f(x) = 0.4x^2 + 0.5,$$

where $f(x)$ represents millions of people in the United States using cable-TV modems x years after 1996. Use this function and the square root property to solve Exercises 61–62.

Number of People in the United States Using Cable-TV Modems

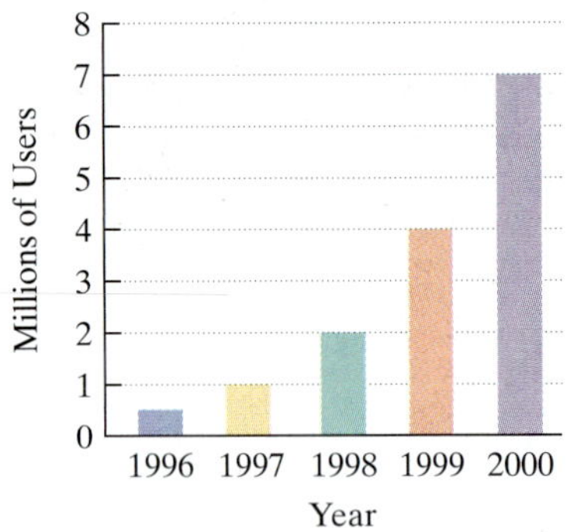

Source: *The New York Times*

61. In which year will 20 million Americans use cable-TV modems?

62. In which year did 4 million Americans use cable-TV modems? How well does the function model the actual number of users for that year shown in the graph?

The function $s(t) = 16t^2$ models the distance, $s(t)$, in feet, that an object falls in t seconds. Use this function and the square root property to solve Exercises 63–64. Express answers in simplified radical form. Then use your calculator and find a decimal approximation to the nearest tenth of a second.

63. A sky diver jumps from an airplane and falls for 4800 feet before opening a parachute. For how many seconds was the diver in a free fall?

64. A sky diver jumps from an airplane and falls for 3200 feet before opening a parachute. For how many seconds was the diver in a free fall?

Use the Pythagorean Theorem and the square root property to solve Exercises 65–70. Express answers in simplified radical form. Then find a decimal approximation to the nearest tenth.

65. A rectangular park is 6 miles long and 3 miles wide. How long is a pedestrian route that runs diagonally across the park?

66. A rectangular park is 4 miles long and 2 miles wide. How long is a pedestrian route that runs diagonally across the park?

67. The base of a 20-foot ladder is 15 feet from the house. How far up the house does the ladder reach?

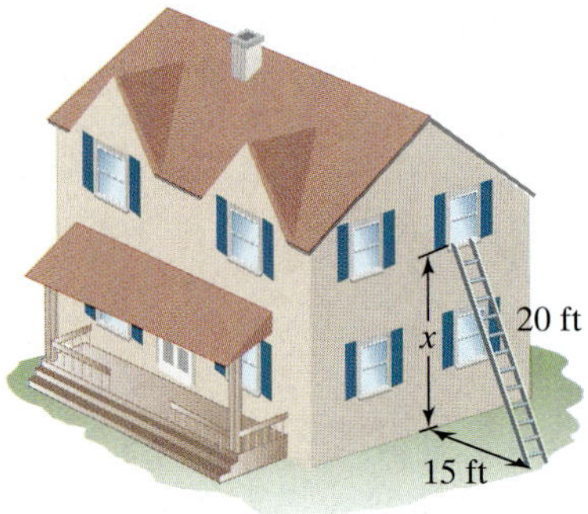

68. A baseball diamond is actually a square with 90-foot sides. What is the distance from home plate to second base?

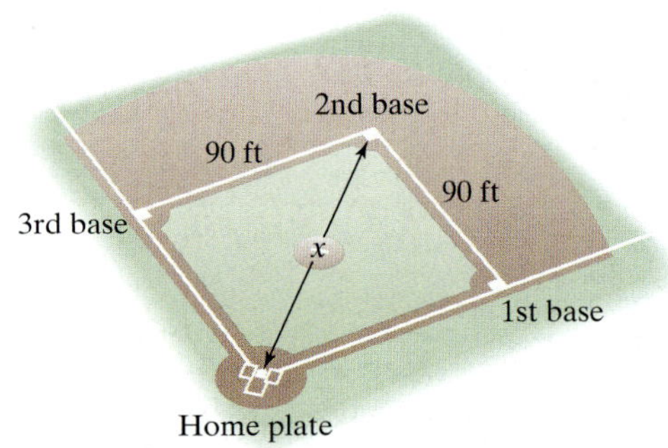

69. A supporting wire is to be attached to the top of a 50-foot antenna. If the wire must be anchored 50 feet from the base of the antenna, what length of wire is required?

70. A supporting wire is to be attached to the top of a 70-foot antenna. If the wire must be anchored 70 feet from the base of the antenna, what length of wire is required?

71. In the section opener, we looked at Einstein's famous formula

$$E = mc^2.$$

The formula implies that a mere pound of any kind of matter contains enough energy to send a large vessel on at least 100 ocean voyages. Energy, E, is equal to mass, m, multiplied by the square of the speed of light, c. Solve for c to express the speed of light in terms of energy and mass.

72. The surface area, A, of a sphere with radius r is given by the formula

$$A = 4\pi r^2.$$

Solve for r to express the radius in terms of the area.

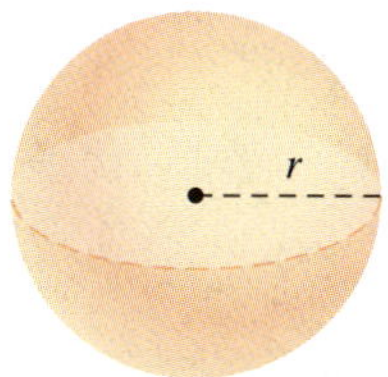

73. A square flower bed is to be enlarged by adding 2 meters on each side. If the larger square has an area of 144 square meters, what is the length of a side of the original square?

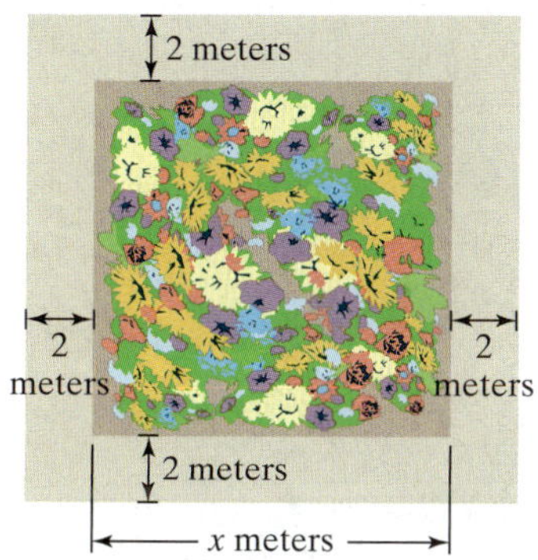

74. A square flower bed is to be enlarged by adding 3 feet on each side. If the larger square has an area of 169 square feet, what is the length of a side of the original square?

Writing in Mathematics

75. What is the square root property?

76. Explain how to solve $(x - 1)^2 = 16$ using the square root property.

77. Explain how to complete the square for a binomial. Use $x^2 + 6x$ to illustrate your explanation.

78. Explain how to solve $x^2 + 6x + 8 = 0$ by completing the square.

79. What is compound interest?

80. In your own words, describe the compound interest formula

$$A = P(1 + r)^t.$$

81. What is an isosceles right triangle? What is the relationship between the length of its hypotenuse and the length of a leg?

82. A basketball player's hang time is the time spent in the air when shooting a basket. The function $f(x) = 4x^2$ models the player's vertical leap, $f(x)$, in feet, in terms of the hang time, x, measured in seconds. Use this function to write a problem that can be solved by the square root property. Then solve the problem. Be realistic: It is unlikely that hang time can exceed 1 second.

83. In the section opener, we mentioned the possibility of the mass of a golf ball providing the daily energy needs of a large metropolitan area. Discuss some of the problems involved with gradual and controlled release of energy through nuclear power. What problems stand in the way of nuclear power expansion? Consult the Internet to help answer the question.

Technology Exercises

84. Use a graphing utility to solve $4 - (x + 1)^2 = 0$. Graph $y = 4 - (x + 1)^2$ in a $[-5, 5, 1]$ by $[-5, 5, 1]$ viewing rectangle. The equation's solutions are the graph's x-intercepts. Check by substitution in the given equation.

85. Use a graphing utility to solve $(x - 1)^2 - 9 = 0$. Graph $y = (x - 1)^2 - 9$ in a $[-5, 5, 1]$ by $[-9, 3, 1]$ viewing rectangle. The equation's solutions are the graph's x-intercepts. Check by substitution in the given equation.

86. Use a graphing utility and x-intercepts to verify any of the real solutions that you obtained for five of the quadratic equations in Exercises 35–56.

Critical Thinking Exercises

87. Which one of the following is true?

a. The equation $(x - 5)^2 = 12$ is equivalent to $x - 5 = 2\sqrt{3}$.

b. In completing the square for $2x^2 - 6x = 5$, we should add 9 to both sides.

c. Although not every quadratic equation can be solved by completing the square, they can all be solved by factoring.

d. The graph of $y = (x - 2)^2 + 3$ cannot have x-intercepts.

88. Solve for y: $\frac{x^2}{a^2} + \frac{y^2}{b^2} = 1$.

89. Solve by completing the square:

$$x^2 + x + c = 0.$$

90. Solve by completing the square:

$$x^2 + bx + c = 0.$$

91. Solve: $x^4 - 8x^2 + 15 = 0$.

Review Exercises

92. Simplify: $4x - 2 - 3[4 - 2(3 - x)]$. (Section 1.2, Example 14)

93. Factor: $1 - 8x^3$. (Section 5.5, Example 9)

94. Divide: $(x^4 - 5x^3 + 2x^2 - 6) \div (x - 3)$. (Section 6.5, Example 1)

▸ SECTION 8.2 The Quadratic Formula

Objectives

1. Solve quadratic equations using the quadratic formula.
2. Use the discriminant to determine number and type of solutions.
3. Determine the most efficient method to use when solving a quadratic equation.
4. Write quadratic equations from solutions.
5. Use the quadratic formula to solve problems.

SSM PH Tutor Center CD-ROM Video

"Serpico", 1973, starring Al Pacino, was a movie about police corruption

In 2000, a police scandal shocked Los Angeles. A police officer who had been convicted of stealing cocaine held as evidence described how members of his unit behaved in ways that resembled the gangs they were targeting, assaulting and framing innocent people.

Is police corruption on the rise? How many police officers are convicted of felonies each year, and is there a trend to the data? In this section, we use a quadratic function to model this upsetting phenomenon. We begin by deriving a formula that will enable you to solve quadratic equations more quickly than the method of completing the square. In the exercise set, you will use this formula to work with the function that models data for police officers convicted as felons.

1 Solve quadratic equations using the quadratic formula.

Solving Quadratic Equations Using the Quadratic Formula

We can use the method of completing the square to derive a formula that can be used to solve all quadratic equations. The derivation given here also shows a particular quadratic equation, $3x^2 - 2x - 4 = 0$, to specifically illustrate each of the steps.

Deriving the Quadratic Formula

Standard Form of a Quadratic Equation	Comment	A Specific Example
$ax^2 + bx + c = 0, a > 0$	This is the given equation.	$3x^2 - 2x - 4 = 0$
$x^2 + \frac{b}{a}x + \frac{c}{a} = 0$	Divide both sides by the coefficient of x^2.	$x^2 - \frac{2}{3}x - \frac{4}{3} = 0$
$x^2 + \frac{b}{a}x = -\frac{c}{a}$	Isolate the binomial by adding $-\frac{c}{a}$ on both sides.	$x^2 - \frac{2}{3}x = \frac{4}{3}$
$x^2 + \frac{b}{a}x + \left(\frac{b}{2a}\right)^2 = -\frac{c}{a} + \left(\frac{b}{2a}\right)^2$ (half)^2 $x^2 + \frac{b}{a}x + \frac{b^2}{4a^2} = -\frac{c}{a} + \frac{b^2}{4a^2}$	Complete the square. Add the square of half the coefficient of x to both sides.	$x^2 - \frac{2}{3}x + \left(-\frac{1}{3}\right)^2 = \frac{4}{3} + \left(-\frac{1}{3}\right)^2$ (half)^2 $x^2 - \frac{2}{3}x + \frac{1}{9} = \frac{4}{3} + \frac{1}{9}$
$\left(x + \frac{b}{2a}\right)^2 = -\frac{c}{a} \cdot \frac{4a}{4a} + \frac{b^2}{4a^2}$	Factor on the left side and obtain a common denominator on the right side.	$\left(x - \frac{1}{3}\right)^2 = \frac{4}{3} \cdot \frac{3}{3} + \frac{1}{9}$
$\left(x + \frac{b}{2a}\right)^2 = \frac{-4ac + b^2}{4a^2}$ $\left(x + \frac{b}{2a}\right)^2 = \frac{b^2 - 4ac}{4a^2}$	Add fractions on the right.	$\left(x - \frac{1}{3}\right)^2 = \frac{12 + 1}{9}$ $\left(x - \frac{1}{3}\right)^2 = \frac{13}{9}$
$x + \frac{b}{2a} = \pm\sqrt{\frac{b^2 - 4ac}{4a^2}}$	Apply the square root property.	$x - \frac{1}{3} = \pm\sqrt{\frac{13}{9}}$
$x + \frac{b}{2a} = \pm\frac{\sqrt{b^2 - 4ac}}{2a}$	Take the square root of the quotient, simplifying the denominator.	$x - \frac{1}{3} = \pm\frac{\sqrt{13}}{3}$
$x = \frac{-b}{2a} \pm \frac{\sqrt{b^2 - 4ac}}{2a}$	Solve for x by subtracting $\frac{b}{2a}$ from both sides.	$x = \frac{1}{3} \pm \frac{\sqrt{13}}{3}$
$x = \frac{-b \pm \sqrt{b^2 - 4ac}}{2a}$	Combine fractions on the right side.	$x = \frac{1 \pm \sqrt{13}}{3}$

The formula shown at the bottom of the left column is called the **quadratic formula**. A similar proof shows that the same formula can be used to solve quadratic equations if a, the coefficient of the x^2-term, is negative.

The Quadratic Formula

The solutions of a quadratic equation in standard form $ax^2 + bx + c = 0$, with $a \neq 0$, are given by the **quadratic formula**

$$x = \frac{-b \pm \sqrt{b^2 - 4ac}}{2a}.$$

x equals negative b, plus or minus the square root of $b^2 - 4ac$, all divided by 2a.

To use the quadratic formula, write the quadratic equation in standard form if necessary. Then determine the numerical values for a (the coefficient of the squared term), b (the coefficient of the x-term), and c (the constant term). Substitute the values of a, b, and c in the quadratic formula and evaluate the expression. The $\pm$ sign indicates that there are two solutions of the equation.

EXAMPLE 1 Solving a Quadratic Equation Using the Quadratic Formula

Solve using the quadratic formula: $8x^2 + 2x - 1 = 0$.

Solution The given equation is in standard form. Begin by identifying the values for a, b, and c.

$$8x^2 + 2x - 1 = 0$$

$a = 8$ $b = 2$ $c = -1$

Substituting these values into the quadratic formula and simplifying gives the equation's solutions.

$$x = \frac{-b \pm \sqrt{b^2 - 4ac}}{2a}$$ Use the quadratic formula.

$$x = \frac{-2 \pm \sqrt{2^2 - 4(8)(-1)}}{2(8)}$$ Substitute the values for a, b, and c: $a = 8$, $b = 2$, and $c = -1$.

$$= \frac{-2 \pm \sqrt{4 - (-32)}}{16}$$ $2^2 - 4(8)(-1) = 4 - (-32)$

$$= \frac{-2 \pm \sqrt{36}}{16}$$ $4 - (-32) = 4 + 32 = 36$

$$= \frac{-2 \pm 6}{16}$$ $\sqrt{36} = 6$

Now we will evaluate this expression in two different ways to obtain the two solutions. At left, we will *add* 6 to −2. At right, we will *subtract* 6 from −2.

$$x = \frac{-2 + 6}{16} = \frac{4}{16} = \frac{1}{4} \quad \text{or} \quad x = \frac{-2 - 6}{16} = \frac{-8}{16} = -\frac{1}{2}$$

The solutions are $-\frac{1}{2}$ and $\frac{1}{4}$, and the solution set is $\left\{-\frac{1}{2}, \frac{1}{4}\right\}$. ■

Using Technology

The graph of the quadratic function

$$y = 8x^2 + 2x - 1$$

has x-intercepts at $-\frac{1}{2}$ and $\frac{1}{4}$. This verifies that $\left\{-\frac{1}{2}, \frac{1}{4}\right\}$ is the solution set of the quadratic equation

$$8x^2 + 2x - 1 = 0.$$

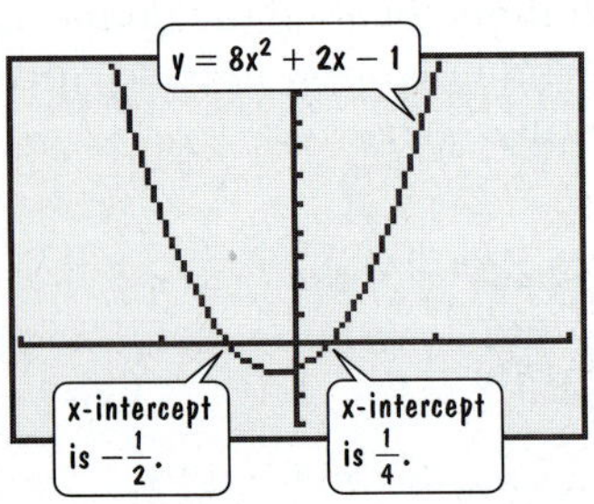

[−2, 2, 1] by [−3, 10, 1]

In Example 1, the solutions of $8x^2 + 2x - 1 = 0$ are rational numbers. This means that the equation can also be solved by factoring. The reason that the solutions are rational numbers is that $b^2 - 4ac$, the radicand in the quadratic formula, is 36, which is a perfect square. If a, b, and c are rational numbers, all quadratic equations for which $b^2 - 4ac$ is a perfect square have rational solutions.

✔ **CHECK POINT 1** Solve using the quadratic formula: $2x^2 + 9x - 5 = 0$.

ENRICHMENT ESSAY

To Die at Twenty

Can the equations

$7x^5 + 12x^3 - 9x + 4 = 0$

and

$8x^6 - 7x^5 + 4x^3 - 19 = 0$

be solved using a formula similar to the quadratic formula? The first equation has five solutions and the second has six solutions, but they cannot be found using a formula. How do we know? In 1832, a 20-year-old Frenchman, Evariste Galois, wrote down a proof showing that there is no general formula to solve equations when the exponent on the variable is 5 or greater. Galois was jailed as a political activist several times while still a teenager. The day after his brilliant proof he fought a duel over a woman. The duel was a political setup. As he lay dying, Galois told his brother, Alfred, of the manuscript that contained his proof: "Mathematical manuscripts are in my room. On the table. Take care of my work. Make it known. Important. Don't cry, Alfred. I need all my courage—to die at twenty." (Our source is Leopold Infeld's biography of Galois, *Whom the Gods Love*. Some historians, however, dispute the story of Galois's ironic death the very day after his algebraic proof. Mathematical truths seem more reliable than historical ones!)

EXAMPLE 2 Solving a Quadratic Equation Using the Quadratic Formula

Solve using the quadratic formula:

$$2x^2 = 4x + 1.$$

Solution The quadratic equation must be in standard form to identify the values for a, b, and c. To move all terms to one side and obtain zero on the right, we subtract $4x + 1$ from both sides. Then we can identify the values for a, b, and c.

$$2x^2 = 4x + 1$$ This is the given equation.

$$2x^2 - 4x - 1 = 0$$ Subtract $4x + 1$ from both sides.

$a = 2$ $\quad b = -4$ $\quad c = -1$

Substituting these values into the quadratic formula and simplifying gives the equation's solutions.

$$x = \frac{-b \pm \sqrt{b^2 - 4ac}}{2a}$$ Use the quadratic formula.

$$x = \frac{-(-4) \pm \sqrt{(-4)^2 - 4(2)(-1)}}{2(2)}$$ Substitute the values for a, b, and c: $a = 2$, $b = -4$, and $c = -1$.

$$= \frac{4 \pm \sqrt{16 - (-8)}}{4}$$ $(-4)^2 - 4(2)(-1) = 16 - (-8)$

$$= \frac{4 \pm \sqrt{24}}{4}$$ $16 - (-8) = 16 + 8 = 24$

$$= \frac{4 \pm 2\sqrt{6}}{4}$$ $\sqrt{24} = \sqrt{4 \cdot 6} = \sqrt{4} \cdot \sqrt{6} = 2\sqrt{6}$

$$= \frac{2(2 \pm \sqrt{6})}{4}$$ Factor out 2 from the numerator.

$$= \frac{2 \pm \sqrt{6}}{2}$$ Divide the numerator and denominator by 2.

The solutions are $\frac{2 \pm \sqrt{6}}{2}$, and the solution set is $\left\{\frac{2 + \sqrt{6}}{2}, \frac{2 - \sqrt{6}}{2}\right\}$ or $\left\{\frac{2 \pm \sqrt{6}}{2}\right\}$.

In Example 2, the solutions of $2x^2 = 4x + 1$ are irrational numbers. This means that the equation cannot be solved by factoring. The reason that the solutions are irrational numbers is that $b^2 - 4ac$, the radicand in the quadratic formula, is 24, which is not a perfect square. Notice, too, that the solutions, $\frac{2+\sqrt{6}}{2}$ and $\frac{2-\sqrt{6}}{2}$, are conjugates.

Study Tip

Many students use the quadratic formula correctly until the last step, where they make an error in simplifying the solutions. Be sure to factor the numerator before dividing the numerator and denominator by the greatest common factor.

$$\frac{4 \pm 2\sqrt{6}}{4} = \frac{2(2 \pm \sqrt{6})}{4} = \frac{\overset{1}{\cancel{2}}(2 \pm \sqrt{6})}{\underset{2}{\cancel{4}}} = \frac{2 \pm \sqrt{6}}{2}$$

You cannot divide just one term in the numerator and the denominator by their greatest common factor.

INCORRECT

$$\frac{\overset{1}{\cancel{4}} \pm 2\sqrt{6}}{\underset{1}{\cancel{4}}} = 1 \pm 2\sqrt{6} \qquad \frac{4 \pm \overset{1}{\cancel{2}}\sqrt{6}}{\underset{2}{\cancel{4}}} = \frac{4 \pm \sqrt{6}}{2}$$

Can all irrational solutions of quadratic equations be simplified? No. The following solutions cannot be simplified:

$$\frac{5 \pm 2\sqrt{7}}{2} \quad \text{and} \quad \frac{4 \pm 3\sqrt{7}}{2}.$$

Other than 1, terms in each numerator have no common factor.

✔ **CHECK POINT 2** Solve using the quadratic formula: $2x^2 = 6x - 1$.

EXAMPLE 3 Solving a Quadratic Equation Using the Quadratic Formula

Solve using the quadratic formula:

$$3x^2 + 2 = -4x.$$

Solution Begin by writing the quadratic equation in standard form.

$$3x^2 + 2 = -4x \qquad \text{This is the given equation.}$$

$$3x^2 + 4x + 2 = 0 \qquad \text{Add } 4x \text{ to both sides.}$$

$a = 3$ $b = 4$ $c = 2$

Substituting these values into the quadratic formula and simplifying gives the equation's solutions.

Using Technology

The graph of the quadratic function

$$y = 3x^2 + 4x + 2$$

has no x-intercepts. This verifies that the equation in Example 3

$$3x^2 + 2 = -4x, \quad \text{or}$$
$$3x^2 + 4x + 2 = 0$$

has imaginary solutions.

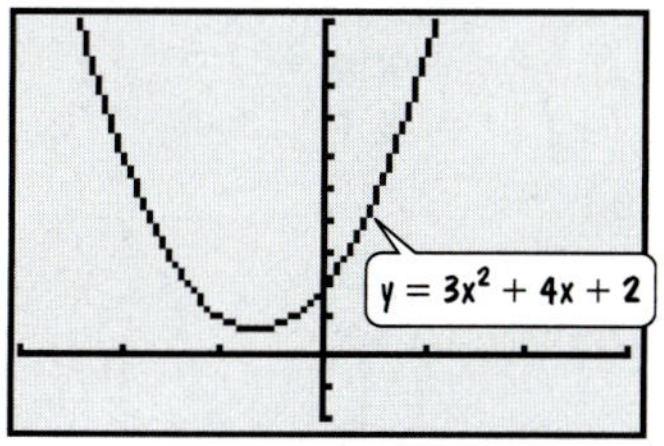

[−3, 3, 1] by [−2, 10, 1]

$$x = \frac{-b \pm \sqrt{b^2 - 4ac}}{2a}$$ Use the quadratic formula.

$$x = \frac{-4 \pm \sqrt{4^2 - 4 \cdot 3 \cdot 2}}{2 \cdot 3}$$ Substitute the values for a, b, and c: $a = 3$, $b = 4$, and $c = 2$.

$$= \frac{-4 \pm \sqrt{16 - 24}}{6}$$ Multiply under the radical.

$$= \frac{-4 \pm \sqrt{-8}}{6}$$ Subtract under the radical.

$$= \frac{-4 \pm 2\sqrt{2}i}{6}$$ $\sqrt{-8} = \sqrt{8(-1)} = \sqrt{8}\sqrt{-1} = \sqrt{8}i = \sqrt{4 \cdot 2}i = 2\sqrt{2}i$

$$= \frac{2(-2 \pm \sqrt{2}i)}{6}$$ Factor out 2 from the numerator.

$$= \frac{-2 \pm \sqrt{2}i}{3}$$ Divide the numerator and denominator by 2.

$$-\frac{2}{3} \pm \frac{\sqrt{2}i}{3}$$ Write in the form $A + b_i$.

The solutions are $-\frac{2}{3} \pm \frac{\sqrt{2}}{3}i$, and the solution set is $\left\{-\frac{2}{3} + \frac{\sqrt{2}}{3}i, -\frac{2}{3} - \frac{\sqrt{2}}{3}i\right\}$ or $\left\{-\frac{2}{3} \pm \frac{\sqrt{2}}{3}i\right\}$. ■

In Example 3, the solutions of $3x^2 + 2 = -4x$ are imaginary numbers. This means that the equation cannot be solved using factoring. The reason that the solutions are imaginary numbers is that $b^2 - 4ac$, the radicand in the quadratic formula, is −8, which is negative. Notice, too, that the solutions are complex conjugates.

✔ **CHECK POINT 3** Solve using the quadratic formula: $3x^2 + 5 = -6x$.

Some rational equations can be solved using the quadratic formula. For example, consider the equation

$$3 + \frac{4}{x} = -\frac{2}{x^2}.$$

The denominators are x and x^2. The least common denominator is x^2. We clear fractions by multiplying both sides of the equation by x^2. Notice that x cannot equal zero.

$$x^2\left(3 + \frac{4}{x}\right) = x^2\left(-\frac{2}{x^2}\right), \quad x \neq 0$$

$$3x^2 + \frac{4}{x} \cdot x^2 = x^2\left(-\frac{2}{x^2}\right)$$ Use the distributive property.

$$3x^2 + 4x = -2$$ Simplify.

By adding 2 to both sides, we obtain the standard form of the quadratic equation:

$$3x^2 + 4x + 2 = 0.$$

This is the equation that we solved in Example 3. The two imaginary solutions are not part of the restriction that $x \neq 0$.

2 Use the discriminant to determine the number and type of solutions.

The Discriminant

The quantity $b^2 - 4ac$, which appears under the radical sign in the quadratic formula, is called the **discriminant**. In Example 1, the discriminant was 36, a positive number that is a perfect square. The equation had two solutions that were rational numbers. In Example 2, the discriminant was 24, a positive number that is not a perfect square. The equation had two solutions that were irrational numbers. Finally, in Example 3, the discriminant was −8, a negative number. The equation had solutions that were imaginary numbers involving i. In this case, the graph of the corresponding quadratic function had no x-intercepts.

These observations are generalized in Table 8.1.

Study Tip

Checking irrational and imaginary solutions can be time-consuming. The solutions given by the quadratic formula are always correct, unless you have made a careless error. Checking for computational errors or errors in simplification is sufficient.

Table 8.1 The Discriminant and the Kinds of Solutions to $ax^2 + bx + c = 0$

Discriminant $b^2 - 4ac$	Kinds of Solutions to $ax^2 + bx + c = 0$	Graph of $y = ax^2 + bx + c$
$b^2 - 4ac > 0$	Two unequal real solutions; if a, b, and c are rational numbers and the discriminant is a perfect square, the solutions are rational. If the discriminant is not a perfect square, the solutions are irrational conjugates.	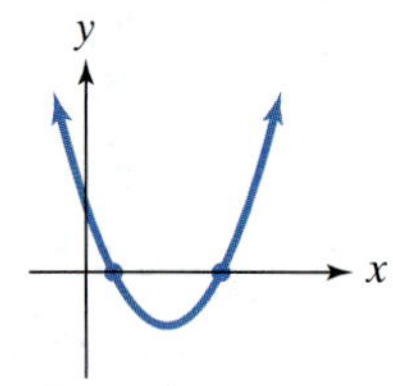 Two x-intercepts
$b^2 - 4ac = 0$	One real solution (a repeated solution) that is a real number; If a, b, and c are rational numbers, the repeated solution is also a rational number.	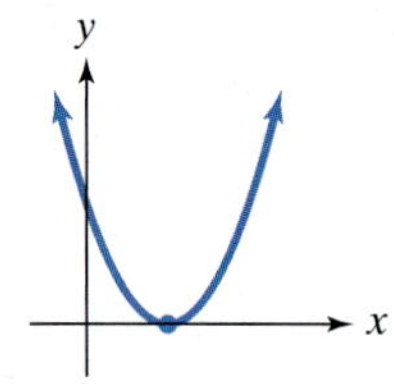 One x-intercept
$b^2 - 4ac < 0$	No real solution; two imaginary solutions; The solutions are complex conjugates.	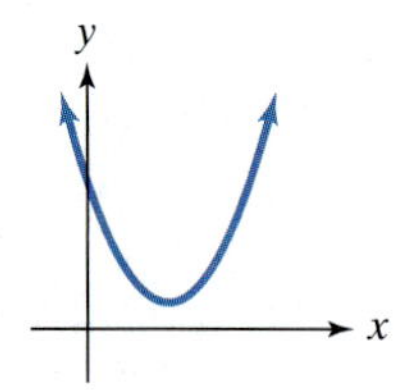 No x-intercepts

EXAMPLE 4 Using the Discriminant

For each equation, compute the discriminant. Then determine the number and type of solutions:

a. $3x^2 + 4x - 5 = 0$ **b.** $9x^2 - 6x + 1 = 0$ **c.** $3x^2 - 8x + 7 = 0$.

Solution Begin by identifying the values for a, b, and c in each equation. Then compute $b^2 - 4ac$, the discriminant.

a. $3x^2 + 4x - 5 = 0$

$a = 3$ $b = 4$ $c = -5$

Substitute and compute the discriminant:

$$b^2 - 4ac = 4^2 - 4 \cdot 3(-5) = 16 - (-60) = 16 + 60 = 76.$$

The discriminant, 76, is a positive number that is not a perfect square. Thus, there are two irrational solutions. (These solutions are conjugates of each other.)

b. $9x^2 - 6x + 1 = 0$

$a = 9$ $b = -6$ $c = 1$

Substitute and compute the discriminant:

$$b^2 - 4ac = (-6)^2 - 4 \cdot 9 \cdot 1 = 36 - 36 = 0.$$

The discriminant, 0, shows that there is only one solution. This solution is a rational number.

c. $3x^2 - 8x + 7 = 0$

$a = 3$ $b = -8$ $c = 7$

$$b^2 - 4ac = (-8)^2 - 4 \cdot 3 \cdot 7 = 64 - 84 = -20$$

The negative discriminant, −20, shows that there are two imaginary solutions. (These solutions are complex conjugates of each other.) ■

✔ **CHECK POINT 4** For each equation, compute the discriminant. Then determine the number and type of solutions:

a. $x^2 + 6x + 9 = 0$

b. $2x^2 - 7x - 4 = 0$

c. $3x^2 - 2x + 4 = 0.$

3 Determine the most efficient method to use when solving a quadratic equation.

Determining Which Method to Use

All quadratic equations can be solved by the quadratic formula. However, if an equation is in the form $u^2 = d$, such as $x^2 = 5$ or $(2x + 3)^2 = 8$, it is faster to use the square root property, taking the square root of both sides. If the equation is not in the form $u^2 = d$, write the quadratic equation in standard form $(ax^2 + bx + c = 0)$. Try to solve the equation by the factoring method. If $ax^2 + bx + c$ cannot be factored, then solve the quadratic equation by the quadratic formula.

Because we used the method of completing the square to derive the quadratic formula, we no longer need it for solving quadratic equations. However, we will use completing the square in Chapter 10 to help graph certain kinds of equations.

Table 8.2 summarizes our observations about which technique to use when solving a quadratic equation.

Table 8.2 Determining the Most Efficient Technique to Use When Solving a Quadratic Equation

Description and Form of the Quadratic Equation	Most Efficient Solution Method	Example
$ax^2 + bx + c = 0$ and $ax^2 + bx + c$ can be factored easily.	Factor and use the zero-product principle.	$3x^2 + 5x - 2 = 0$ $(3x - 1)(x + 2) = 0$ $3x - 1 = 0$ or $x + 2 = 0$ $x = \frac{1}{3}$ $\quad x = -2$
$ax^2 + c = 0$ The quadratic equation has no x-term. ($b = 0$)	Solve for x^2 and apply the square root property.	$4x^2 - 7 = 0$ $4x^2 = 7$ $x^2 = \frac{7}{4}$ $x = \pm\frac{\sqrt{7}}{2}$
$u^2 = d$; u is a first-degree polynomial.	Use the square root property.	$(x + 4)^2 = 5$ $x + 4 = \pm\sqrt{5}$ $x = -4 \pm \sqrt{5}$
$ax^2 + bx + c = 0$ and $ax^2 + bx + c$ cannot be factored or the factoring is too difficult.	Use the quadratic formula: $x = \frac{-b \pm \sqrt{b^2 - 4ac}}{2a}$.	$x^2 - 2x - 6 = 0$ $a = 1$ $\quad b = -2$ $\quad c = -6$ $x = \frac{-(-2) \pm \sqrt{(-2)^2 - 4(1)(-6)}}{2(1)}$ $= \frac{2 \pm \sqrt{4 - 4(1)(-6)}}{2(1)}$ $= \frac{2 \pm \sqrt{28}}{2} = \frac{2 \pm \sqrt{4}\sqrt{7}}{2}$ $= \frac{2 \pm 2\sqrt{7}}{2} = \frac{2(1 \pm \sqrt{7})}{2}$ $= 1 \pm \sqrt{7}$

4 Write quadratic equations from solutions.

Writing Quadratic Equations from Solutions

Using the zero-product principle, the equation $(x - 3)(x + 5) = 0$ has two solutions, 3 and -5. By applying the zero-product principle in reverse, we can find a quadratic equation that contains two given solutions.

The Zero-Product Principle in Reverse

If $A = 0$ or $B = 0$, then $AB = 0$.

EXAMPLE 5 Writing Equations from Solutions

Write a quadratic equation with the given solution set:

a. $\left\{-\frac{5}{3}, \frac{1}{2}\right\}$ **b.** $\{-5i, 5i\}$.

Jasper Johns, "Zero". © Jasper Johns/Licensed by VAGA, New York, NY.

The special properties of zero make it possible to write a quadratic equation from its solutions.

Solution

a. Because the solution set is $\left\{-\frac{5}{3}, \frac{1}{2}\right\}$, then

$$x = -\frac{5}{3} \quad \text{or} \quad x = \frac{1}{2}.$$

$$x + \frac{5}{3} = 0 \quad \text{or} \quad x - \frac{1}{2} = 0$$

Obtain zero on one side of each equation.

$$3x + 5 = 0 \quad \text{or} \quad 2x - 1 = 0$$

Clear fractions, multiplying by 3 and 2, respectively.

$$(3x + 5)(2x - 1) = 0$$

Use the zero-product principle in reverse: If $A = 0$ or $B = 0$, then $AB = 0$.

$$6x^2 - 3x + 10x - 5 = 0$$

Use the FOIL method to multiply.

$$6x^2 + 7x - 5 = 0.$$

Combine like terms.

Thus, the equation is $6x^2 + 7x - 5 = 0$. Many other quadratic equations have $\left\{-\frac{5}{3}, \frac{1}{2}\right\}$ for their solution sets. These equations can be obtained by multiplying both sides of $6x^2 + 7x - 5 = 0$ by any nonzero real number.

b. Because the solution set is $\{-5i, 5i\}$, then

$$x = -5i \quad \text{or} \quad x = 5i$$

$$x + 5i = 0 \quad \text{or} \quad x - 5i = 0$$

Obtain zero on one side of each equation.

$$(x + 5i)(x - 5i) = 0$$

Use the zero-product principle in reverse: If $A = 0$ or $B = 0$, then $AB = 0$.

$$x^2 - (5i)^2 = 0$$

Multiply conjugates using $(A + B)(A - B) = A^2 - B^2$.

$$x^2 - 25i^2 = 0$$

$(5i)^2 = 5^2i^2 = 25i^2$

$$x^2 - 25(-1) = 0$$

$i^2 = -1$

$$x^2 + 25 = 0.$$

This is the required equation. ■

✔ **CHECK POINT 5** Write a quadratic equation with the given solution set:

a. $\left\{-\frac{3}{5}, \frac{1}{4}\right\}$

b. $\{-7i, 7i\}$.

5 Use the quadratic formula to solve problems.

Applications

Quadratic equations can be solved to answer questions about variables contained in quadratic functions.

EXAMPLE 6 AIDS Deaths in the United States

The bar graph in Figure 8.4 shows the cumulative number of deaths from AIDS in the United States for each year from 1990 through 1999. The cumulative number of U.S. AIDS deaths, $f(x)$, in thousands, x years after 1990 can be modeled by the quadratic function

$$f(x) = -1.65x^2 + 51.8x + 111.44.$$

According to the model, in which year will the total number of U.S. AIDS deaths reach 500 thousand?

Cumulative Number of Deaths from AIDS in the United States

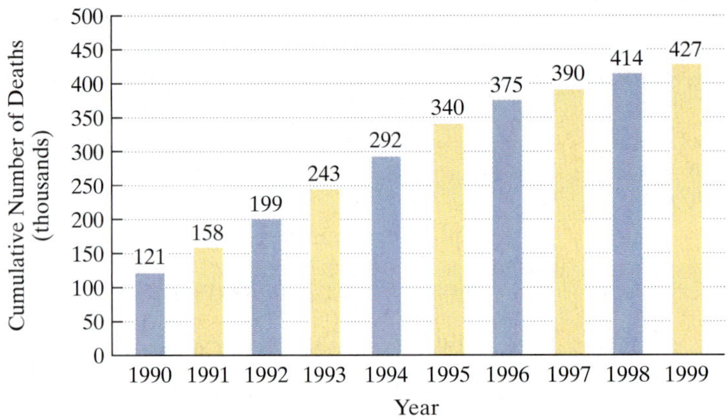

Figure 8.4 *Source*: Centers for Disease Control

Solution Because we are interested in when the number of deaths will reach 500 thousand, we substitute 500 for $f(x)$ in the given function. Then we solve for x, the number of years after 1990.

$$f(x) = -1.65x^2 + 51.8x + 111.44$$ This is the given function.

$$500 = -1.65x^2 + 51.8x + 111.44$$ Substitute 500 for $f(x)$.

$$0 = -1.65x^2 + 51.8x - 388.56$$ Subtract 500 from both sides and write the quadratic equation in standard form.

$a = -1.65$ $b = 51.8$ $c = -388.56$

Because the trinomial on the right side of the equation is not easily factored, if it can be factored at all, we solve using the quadratic formula.

$$x = \frac{-b \pm \sqrt{b^2 - 4ac}}{2a}$$ Use the quadratic formula.

$$= \frac{-51.8 \pm \sqrt{(51.8)^2 - 4(-1.65)(-388.56)}}{2(-1.65)}$$ Substitute the values for a, b, and c: $a = -1.65$, $b = 51.8$, $c = -388.56$.

$$= \frac{-51.8 \pm \sqrt{118.744}}{-3.3}$$ Use a calculator to simplify the radicand.

Using Technology

On most calculators, here is how to approximate

$$\frac{-51.8 + \sqrt{118.744}}{-3.3}.$$

Most Scientific Calculators:

[(] 51.8 [+/−] [+] 118.744

[√] [)] [÷] 3.3 [+/−] [=]

Most Graphing Calculators:

[(] [(−)] 51.8 [+] [√] 118.744

[)] [÷] [(−)] 3.3 [ENTER]

Similar keystrokes can be used to approximate the other irrational solution in Example 6.

$$x = \frac{-51.8 + \sqrt{118.744}}{-3.3} \quad \text{or} \quad x = \frac{-51.8 - \sqrt{118.744}}{-3.3}$$

$$x \approx 12 \qquad\qquad x \approx 19 \qquad \text{Use a calculator.}$$

According to the model, approximately 12 years after 1990, in 2002, cumulative U.S. AIDS deaths will reach 500 thousand. We must be careful not to extend the model too far into the future. The approximate solution, 19, makes no sense. The only way that cumulative deaths can reach 500 thousand in 2002 and again in 1990 + 19, or in 2009, is that no AIDS deaths occur between 2002 and 2009. This seems unlikely, indicating that model breakdown has occurred. ■

CHECK POINT 6 According to the model in Example 6, in which year did the cumulative number of deaths from AIDS in the United States reach 330 thousand? What was the actual number for that year? How well does the function describe the situation for that year?

EXERCISE SET 8.2

Practice Exercises

In Exercises 1–16, solve each equation using the quadratic formula. Simplify solutions, if possible.

1. $x^2 + 8x + 12 = 0$
2. $x^2 + 8x + 15 = 0$
3. $2x^2 - 7x = -5$
4. $5x^2 + 8x = -3$
5. $x^2 + 3x - 20 = 0$
6. $x^2 + 5x - 10 = 0$
7. $3x^2 - 7x = 3$
8. $4x^2 + 3x = 2$
9. $6x^2 = 2x + 1$
10. $2x^2 = -4x + 5$
11. $4x^2 - 3x = -6$
12. $9x^2 + x = -2$
13. $x^2 - 4x + 8 = 0$
14. $x^2 + 6x + 13 = 0$
15. $3x^2 = 8x - 7$
16. $3x^2 = 4x - 6$

In Exercises 17–28, compute the discriminant. Then determine the number and type of solutions for the given equation.

17. $x^2 + 8x + 3 = 0$
18. $x^2 + 7x + 4 = 0$
19. $x^2 + 6x + 8 = 0$
20. $x^2 + 2x - 3 = 0$
21. $2x^2 + x + 3 = 0$
22. $2x^2 - 4x + 3 = 0$
23. $2x^2 + 6x = 0$
24. $3x^2 - 5x = 0$

25. $5x^2 + 3 = 0$

26. $5x^2 + 4 = 0$

27. $9x^2 = 12x - 4$

28. $4x^2 = 20x - 25$

In Exercises 29–44, solve each equation by the method of your choice. Simplify solutions, if possible.

29. $3x^2 - 4x = 4$

30. $2x^2 - x = 1$

31. $x^2 - 2x = 1$

32. $2x^2 + 3x = 1$

33. $(2x - 5)(x + 1) = 2$

34. $(2x + 3)(x + 4) = 1$

35. $(3x - 4)^2 = 16$

36. $(2x + 7)^2 = 25$

37. $\frac{x^2}{2} + 2x + \frac{2}{3} = 0$

38. $\frac{x^2}{3} - x - \frac{1}{6} = 0$

39. $(3x - 2)^2 = 10$

40. $(4x - 1)^2 = 15$

41. $\frac{1}{x} + \frac{1}{x + 2} = \frac{1}{3}$

42. $\frac{1}{x} + \frac{1}{x + 3} = \frac{1}{4}$

43. $(2x - 6)(x + 2) = 5(x - 1) - 12$

44. $7x(x - 2) = 3 - 2(x + 4)$

In Exercises 45–58, write a quadratic equation in standard form with the given solution set.

45. $\{-3, 5\}$

46. $\{-2, 6\}$

47. $\left\{-\frac{2}{3}, \frac{1}{4}\right\}$

48. $\left\{-\frac{5}{6}, \frac{1}{3}\right\}$

49. $\{-6i, 6i\}$

50. $\{-8i, 8i\}$

51. $\{-\sqrt{2}, \sqrt{2}\}$

52. $\{-\sqrt{3}, \sqrt{3}\}$

53. $\{-2\sqrt{5}, 2\sqrt{5}\}$

54. $\{-3\sqrt{5}, 3\sqrt{5}\}$

55. $\{1 + i, 1 - i\}$

56. $\{2 + i, 2 - i\}$

57. $\{1 + \sqrt{2}, 1 - \sqrt{2}\}$

58. $\{1 + \sqrt{3}, 1 - \sqrt{3}\}$

Application Exercises

The bar graph shows the number of convictions of police officers throughout the United States for seven years. The number of police officers convicted of felonies, $f(x)$, x years after 1990 can be modeled by the quadratic function

$$f(x) = 23.4x^2 - 259.1x + 815.8.$$

Use the function to solve Exercises 59–60.

Convictions of Police Officers

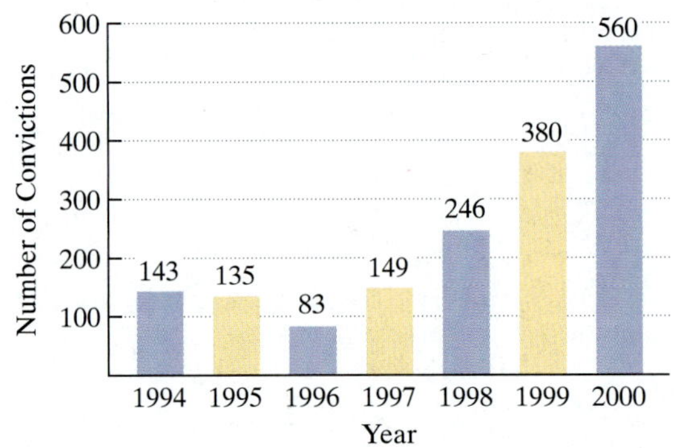

Source: F.B.I.

59. In which year will 1000 police officers be convicted of felonies?

60. In which year after 1994 were 250 police officers convicted of felonies? How well does the function model the actual number of convictions for that year shown in the bar graph?

A driver's age has something to do with his or her chance of getting into a fatal car crash. The bar graph shows the number of fatal vehicle crashes per 100 million miles driven for drivers of various age groups. For example, 25-year-old drivers are involved in 4.1 fatal crashes per 100 million miles driven. Thus, when a group of 25-year-old Americans have driven a total of 100 million miles, approximately 4 have been in accidents in which someone died.

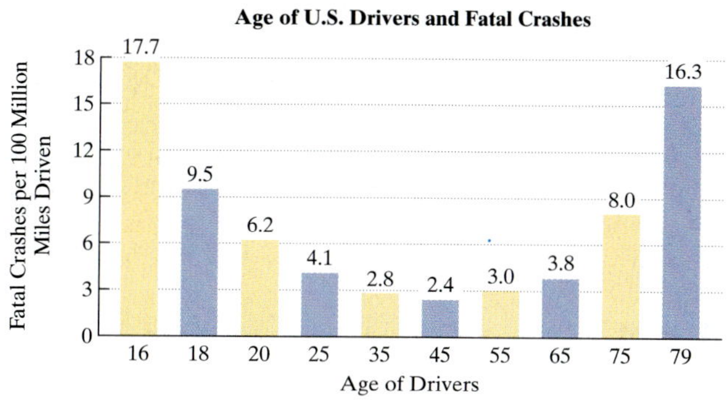

Source: Insurance Institute for Highway Safety

The number of fatal vehicle crashes per 100 million miles, $f(x)$, for drivers of age x can be modeled by the quadratic function

$$f(x) = 0.013x^2 - 1.19x + 28.24.$$

Use the function to solve Exercises 61–62.

61. What age groups are expected to be involved in 10 fatal crashes per 100 million miles driven? How well does the function model the trend in the actual data shown in the bar graph?

62. What age groups are expected to be involved in 3 fatal crashes per 100 million miles driven? How well does the function model the trend the actual data shown in the bar graph?

63. A football is kicked straight up from a height of 4 feet with an initial speed of 60 feet per second. The function

$$f(t) = -16t^2 + 60t + 4$$

models the ball's height above the ground, $f(t)$, in feet, t seconds after it was kicked. How long will it take for the football to hit the ground? Use a calculator and round to the nearest tenth of a second.

64. Standing on a platform 50 feet high, a person accidentally fires a gun straight into the air. The function

$$f(t) = -16t^2 + 100t + 50$$

models the bullet's height above the ground, $f(t)$, in feet, t seconds after the gun was fired. How long will it take for the bullet to hit the ground? Use a calculator and round to the nearest tenth of a second.

65. The length of a rectangle is 3 meters longer than the width. If the area is 36 square meters, find the rectangle's dimensions. Round to the nearest tenth of a meter.

66. The length of a rectangle is 2 meters longer than the width. If the area is 10 square meters, find the rectangle's dimensions. Round to the nearest tenth of a meter.

67. The hypotenuse of a right triangle is 4 feet long. One leg is 1 foot longer than the other. Find the lengths of the legs. Round to the nearest tenth of a foot.

68. The hypotenuse of a right triangle is 6 feet long. One leg is 1 foot shorter than the other. Find the lengths of the legs. Round to the nearest tenth of a foot.

69. A rain gutter is made from sheets of aluminum that are 20 inches wide. As shown in the figure, the edges are turned up to form right angles. Determine the depth of the gutter that will allow a cross-sectional area of 13 square inches. Show that there are two different solutions to the problem. Round to the nearest tenth of an inch.

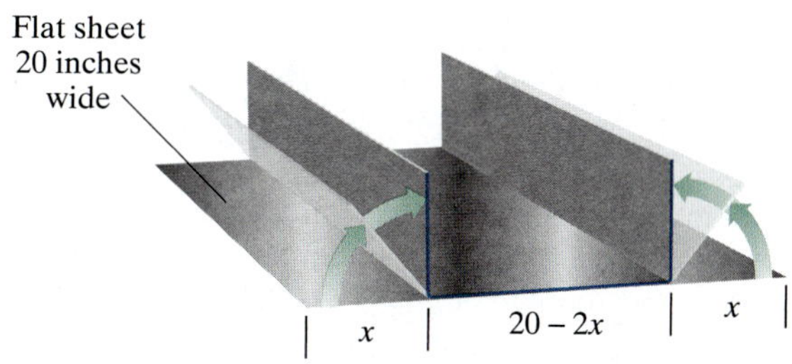

70. A piece of wire is 8 inches long. The wire is cut into two pieces and then each piece is bent into a square. Find the length of each piece if the sum of the areas of these squares is to be 2 square inches.

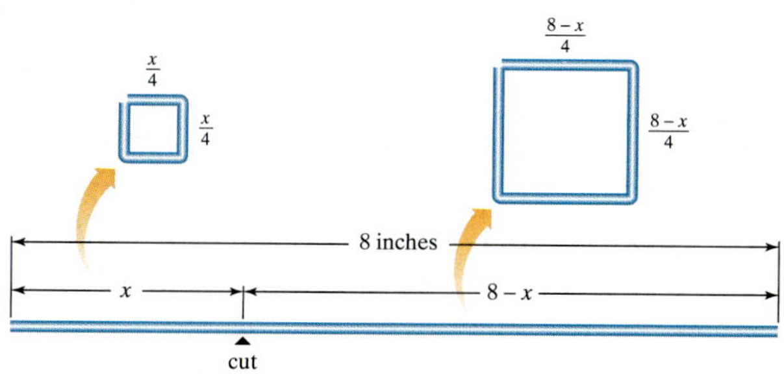

Writing in Mathematics

71. What is the quadratic formula and why is it useful?

72. Without going into specific details for every step, describe how the quadratic formula is derived.

73. Explain how to solve $x^2 + 6x + 8 = 0$ using the quadratic formula.

74. If a quadratic equation has imaginary solutions, how is this shown on the graph of the corresponding quadratic function?

75. What is the discriminant, and what information does it provide about a quadratic equation?

76. If you are given a quadratic equation, how do you determine which method to use to solve it?

77. Explain how to write a quadratic equation from its solution set. Give an example with your explanation.

78. Describe the trend shown by the data for the convictions of police officers in the graph for Exercises 59–60. Do you believe that this trend is likely to continue or might something occur that would make it impossible to extend the model into the future? Explain your answer.

Technology Exercises

79. Use a graphing utility to graph the quadratic function related to any five of the quadratic equations in Exercises 17–28. How does each graph illustrate what you determined algebraically using the discriminant?

80. Use a graphing utility to obtain a graph of the quadratic function in Exercise 63 or Exercise 64. [TRACE] along the curve and verify what you determined algebraically about when the football or bullet hits the ground.

81. Reread Exercise 69. The cross-sectional area of the gutter is given by the quadratic function

$$f(x) = x(20 - 2x).$$

Graph the function in a $[0, 10, 1]$ by $[0, 60, 5]$ viewing rectangle. Then [TRACE] along the curve or use the maximum function feature to determine the depth of the gutter that will maximize its cross-sectional area and allow the greatest amount of water to flow. What is the maximum area? Does the situation described in Exercise 69 take full advantage of the sheets of aluminum?

Critical Thinking Exercises

82. Which one of the following is true?

a. The quadratic formula is developed by applying factoring and the zero-product principle to the quadratic equation $ax^2 + bx + c = 0$.

b. In using the quadratic formula to solve the quadratic equation $5x^2 = 2x - 7$, we have $a = 5$, $b = 2$, and $c = -7$.

c. The quadratic formula cannot be used to solve the equation $x^2 - 9 = 0$.

d. Any quadratic equation that can be solved by completing the square can be solved by the quadratic formula.

83. Solve: $x^2 - 3\sqrt{2}x = -2$.

84. Solve: $10^{-4}x^2 + 2 \cdot 10^{-3}x + 10^{-2} = 0$.

85. A rectangular swimming pool is 12 meters long and 8 meters wide. A tile border of uniform width is to be built around the pool using 120 square meters of tile. The tile is from a discontinued stock (so no additional materials are available), and all 120 square meters are to be used. How wide should the border be? Round to the nearest tenth of a meter. If zoning laws require at least a 2-meter-wide border around the pool, can this be done with the available tile?

86. An owner of a large diving boat that can carry as many as 70 people charges \$10 per passenger to groups of between 15 and 20 people. If more than 20 divers charter the boat, the fee per passenger is decreased by 10 cents times the number of people exceeding 20. How many people must there be in a diving group to generate an income of \$323.90?

Review Exercises

87. Solve: $|5x + 2| = |4 - 3x|$. (Section 4.3, Example 2)

88. Solve: $\sqrt{2x - 5} - \sqrt{x - 3} = 1$. (Section 7.6, Example 4)

89. Rationalize the denominator: $\dfrac{5}{\sqrt{3} + x}$. (Section 7.5, Example 5)

SECTION 8.3 Quadratic Functions and Their Graphs

Objectives

1. Recognize characteristics of parabolas.
2. Graph parabolas in the form $f(x) = a(x - h)^2 + k$.
3. Graph parabolas in the form $f(x) = ax^2 + bx + c$.
4. Solve problems involving minimizing or maximizing quadratic functions.

The wage gap is used to desribe women's earnings relative to men's. The wage gap is expressed as a percent and is calculated by dividing the median annual earnings for women by the median annual earnings for men. In 1999, the wage gap was approximately 77%. Thus, women earned approximately 77% as much as men did. Median weekly earnings of female full-time workers were \$473, compared to \$618 for their male counterparts.

The quadratic function

$$f(x) = 0.022x^2 - 0.4x + 60.07$$

models the wage gap, $f(x)$, x years after 1960. By graphing the function, we can get a good idea of the trend for the wage gap, including the year in which women's earnings relative to men's were the lowest. In this section, we study quadratic functions and their graphs.

1 Recognize characteristics of parabolas.

Graphs of Quadratic Functions

The graph of the quadratic function

$$f(x) = ax^2 + bx + c, \quad a \neq 0,$$

is called a **parabola**. Parabolas are shaped like cups, as shown in Figure 8.5. If the coefficient of x^2 (the value of a in $ax^2 + bx + c$) is positive, the parabola opens upward. If the coefficient of x^2 is negative, the parabola opens downward. The **vertex** (or turning point) of the parabola is the lowest point on the graph when it opens upward and the highest point on the graph when it opens downward.

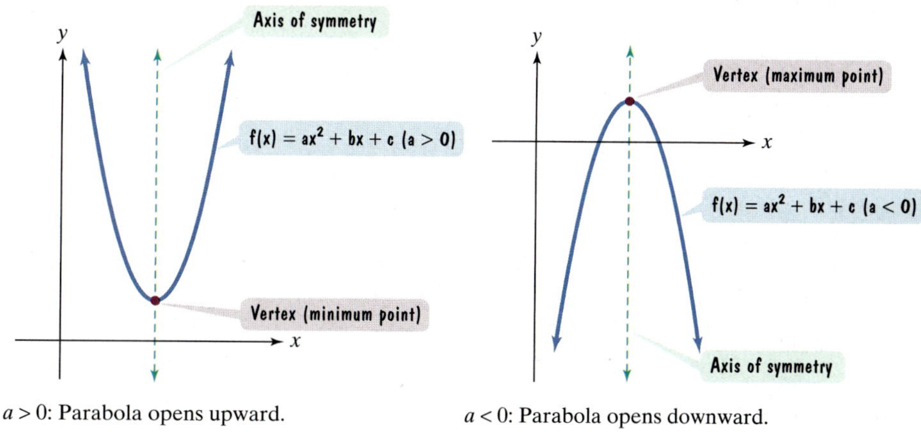

Figure 8.5 Characteristics of graphs of quadratic functions

Look at the unusual image of the word "mirror" shown below. The artist, Scott Kim, has created the image so that the two halves of the whole are mirror images of each other. A parabola shares this kind of symmetry, in which a line through the vertex divides the figure in half. Parabolas are symmetric with respect to this line, called the **axis of symmetry**. If a parabola is folded along its axis of symmetry, the two halves match exactly.

2 Graph parabolas in the form $f(x) = a(x - h)^2 + k$.

Graphing Quadratic Functions in the Form $f(x) = a(x - h)^2 + k$

One way to obtain the graph of a quadratic function is to use point plotting. Let's begin by graphing the functions $f(x) = x^2$, $g(x) = 2x^2$, and $h(x) = \frac{1}{2}x^2$ in the same rectangular coordinate system. Select integers for x, starting with −3 and ending with 3. The partial table of coordinates for each function is shown below. The three parabolas are shown in Figure 8.6.

x	$f(x) = x^2$	(x, y) or $(x, f(x))$
−3	$f(-3) = (-3)^2 = 9$	$(-3, 9)$
−2	$f(-2) = (-2)^2 = 4$	$(-2, 4)$
−1	$f(-1) = (-1)^2 = 1$	$(-1, 1)$
0	$f(0) = 0^2 = 0$	$(0, 0)$
1	$f(1) = 1^2 = 1$	$(1, 1)$
2	$f(2) = 2^2 = 4$	$(2, 4)$
3	$f(3) = 3^2 = 9$	$(3, 9)$

x	$g(x) = 2x^2$	(x, y) or $(x, g(x))$
−3	$g(-3) = 2(-3)^2 = 18$	$(-3, 18)$
−2	$g(-2) = 2(-2)^2 = 8$	$(-2, 8)$
−1	$g(-1) = 2(-1)^2 = 2$	$(-1, 2)$
0	$g(0) = 2 \cdot 0^2 = 0$	$(0, 0)$
1	$g(1) = 2 \cdot 1^2 = 2$	$(1, 2)$
2	$g(2) = 2 \cdot 2^2 = 8$	$(2, 8)$
3	$g(3) = 2 \cdot 3^2 = 18$	$(3, 18)$

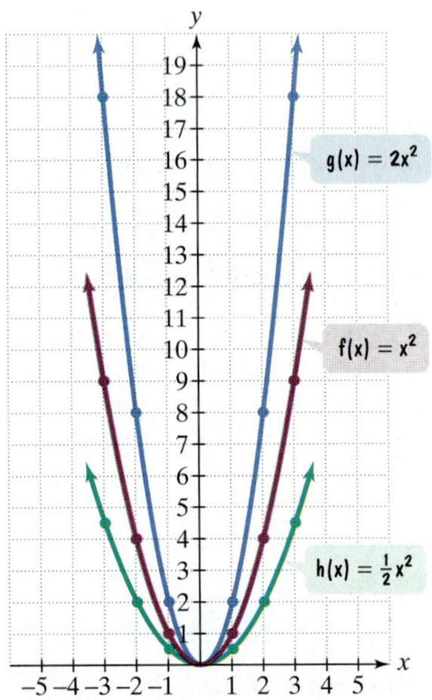

Figure 8.6

x	$h(x) = \frac{1}{2}x^2$	(x, y) or $(x, h(x))$
−3	$h(-3) = \frac{1}{2}(-3)^2 = \frac{9}{2}$	$\left(-3, \frac{9}{2}\right)$
−2	$h(-2) = \frac{1}{2}(-2)^2 = 2$	$(-2, 2)$
−1	$h(-1) = \frac{1}{2}(-1)^2 = \frac{1}{2}$	$\left(-1, \frac{1}{2}\right)$
0	$h(0) = \frac{1}{2} \cdot 0^2 = 0$	$(0, 0)$
1	$h(1) = \frac{1}{2} \cdot 1^2 = \frac{1}{2}$	$\left(1, \frac{1}{2}\right)$
2	$h(2) = \frac{1}{2} \cdot 2^2 = 2$	$(2, 2)$
3	$h(3) = \frac{1}{2} \cdot 3^2 = \frac{9}{2}$	$\left(3, \frac{9}{2}\right)$

Can you see that the graphs of f, g, and h all have the same vertex, $(0, 0)$? They also have the same axis of symmetry, the y-axis, or $x = 0$. This is true for all graphs of the form $f(x) = ax^2$. However, the blue graph of $g(x) = 2x^2$ is a narrower parabola than the red graph of $f(x) = x^2$. By contrast, the green graph of $h(x) = \frac{1}{2}x^2$ is a flatter parabola than the red graph of $f(x) = x^2$.

Is there a more efficient method than point plotting to obtain the graph of a quadratic function? The answer is yes. The method is based on comparing graphs of the form $g(x) = a(x - h)^2 + k$ to those of the form $f(x) = ax^2$.

In Figure 8.7(a), the graph of $f(x) = ax^2$ for $a > 0$ is shown in black. The parabola's vertex is $(0, 0)$ and it opens upward. In Figure 8.7(b), the graph of $f(x) = ax^2$ for $a < 0$ is shown in black. The parabola's vertex is $(0, 0)$ and it opens downward.

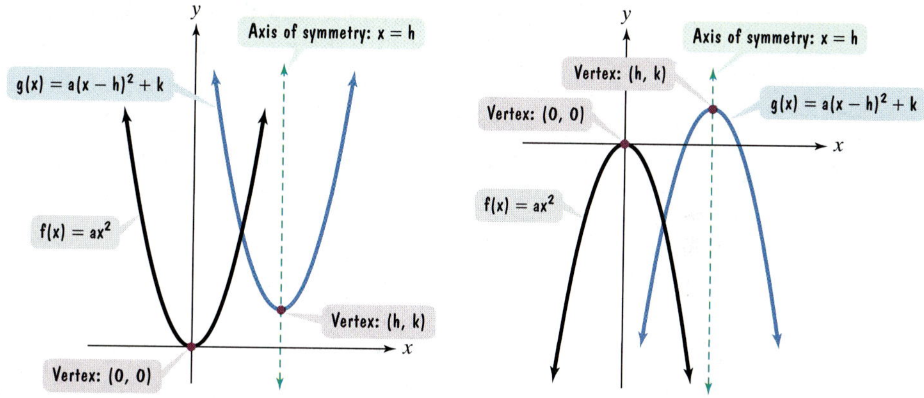

(a) $a > 0$: Parabola opens upward. (b) $a < 0$: Parabola opens downward.

Figure 8.7 Moving, or shifting, the graph of $f(x) = ax^2$

Figure 8.7 also shows the graph of $g(x) = a(x - h)^2 + k$ in blue. Compare these graphs to those of $f(x) = ax^2$. Observe that h causes a horizontal move, or shift, and k causes a vertical shift of the graph of $f(x) = ax^2$. Consequently, the vertex $(0, 0)$ on the graph of $f(x) = ax^2$ moves to the point (h, k) on the graph of $g(x) = a(x - h)^2 + k$. The axis of symmetry is the vertical line whose equation is $x = h$.

The form of the expression for g is convenient because it immediately identifies the vertex of the parabola as (h, k). The sign of a in $g(x) = a(x - h)^2 + k$ determines whether the parabola opens upward or downward. Furthermore, if $|a|$ is small, the parabola opens more flatly than if $|a|$ is large.

Graphing Quadratic Functions with Equations in the Form $f(x) = a(x - h)^2 + k$

To graph $f(x) = a(x - h)^2 + k$:

1. **Determine whether the parabola opens upward or downward.** If $a > 0$, it opens upward. If $a < 0$, it opens downward.
2. **Determine the vertex of the parabola.** The vertex is (h, k).
3. **Find any x-intercepts by replacing $f(x)$ with 0.** Solve the resulting quadratic equation for x.
4. **Find the y-intercept by replacing x with 0.**
5. **Plot the intercepts and vertex and additional points as necessary.** Connect these points with a smooth curve that is shaped like a cup.

EXAMPLE 1 Graphing a Quadratic Function in the Form $f(x) = a(x - h)^2 + k$

Graph the quadratic function $f(x) = -2(x - 3)^2 + 8$.

Solution We can graph this function by following the steps in the preceding box. We begin by identifying values for a, h, and k.

$$f(x) = a(x - h)^2 + k$$

$a = -2$ $h = 3$ $k = 8$

$$f(x) = -2(x - 3)^2 + 8$$

Step 1 Determine how the parabola opens. Note that a, the coefficient of x^2, is -2. Thus, $a < 0$; this negative value tells us that the parabola opens downward.

Step 2 Find the vertex. The vertex of the parabola is at (h, k). Because $h = 3$ and $k = 8$, the parabola has its vertex at $(3, 8)$.

Step 3 Find the x-intercepts. Replace $f(x)$ with 0 in $f(x) = -2(x - 3)^2 + 8$.

$0 = -2(x - 3)^2 + 8$		Find x-intercepts, setting f(x) equal to 0.
$2(x - 3)^2 = 8$		Solve for x. Add $2(x - 3)^2$ to both sides of the equation.
$(x - 3)^2 = 4$		Divide both sides by 2.
$x - 3 = \sqrt{4}$ or	$x - 3 = -\sqrt{4}$	Apply the square root property.
$x - 3 = 2$	$x - 3 = -2$	$\sqrt{4} = 2$
$x = 5$	$x = 1$	Add 3 to both sides in each equation.

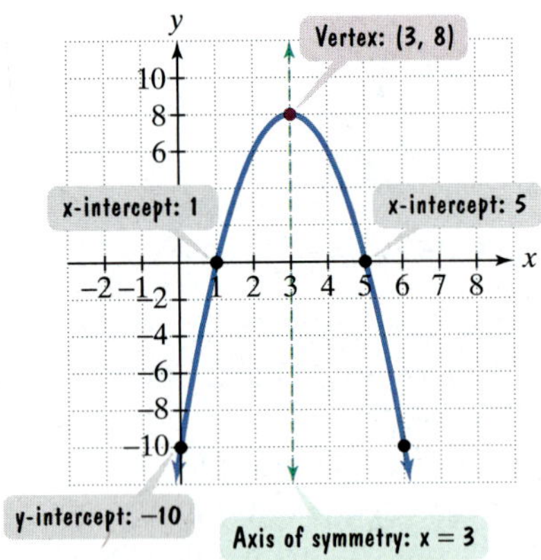

Figure 8.8 The graph of $f(x) = -2(x - 3)^2 + 8$

The x-intercepts are 5 and 1. The parabola passes through $(5, 0)$ and $(1, 0)$.

Step 4 Find the y-intercept. Replace x with 0 in $f(x) = -2(x - 3)^2 + 8$.

$$f(0) = -2(0 - 3)^2 + 8 = -2(-3)^2 + 8 = -2(9) + 8 = -10$$

The y-intercept is -10. The parabola passes through $(0, -10)$.

Step 5 Graph the parabola. With a vertex at $(3, 8)$, x-intercepts at 5 and 1, and a y-intercept at -10, the graph of f is shown in Figure 8.8. The axis of symmetry is the vertical line whose equation is $x = 3$. ■

✔ **CHECK POINT 1** Graph the quadratic function $f(x) = -(x - 1)^2 + 4$.

EXAMPLE 2 Graphing a Quadratic Function in the Form $f(x) = a(x - h)^2 + k$

Graph the quadratic function $f(x) = (x + 3)^2 + 1$.

Solution We begin by finding values for a, h, and k.

$$f(x) = a(x - h)^2 + k \quad \text{Form of given equation}$$

$$f(x) = (x + 3)^2 + 1 \quad \text{Given equation}$$

$$\text{or} \quad f(x) = 1(x - (-3))^2 + 1$$

$a = 1$ $h = -3$ $k = 1$

Step 1 Determine how the parabola opens. Note that a, the coefficient of x^2, is 1. Thus, $a > 0$; this positive value tells us that the parabola opens upward.

Step 2 Find the vertex. The vertex of the parabola is at (h, k). Because $h = -3$ and $k = 1$, the parabola has its vertex at $(-3, 1)$.

Step 3 Find the x-intercepts. Replace $f(x)$ with 0 in $f(x) = (x + 3)^2 + 1$. Because the vertex is at $(-3, 1)$, which lies above the x-axis, and the parabola opens upward, it appears that this parabola has no x-intercepts. We can verify this observation algebraically.

$$0 = (x + 3)^2 + 1$$ Find possible x-intercepts, setting $f(x)$ equal to 0.

$$-1 = (x + 3)^2$$ Solve for x. Subtract 1 from both sides.

$$x + 3 = \sqrt{-1} \quad \text{or} \quad x + 3 = -\sqrt{-1}$$ Apply the square root property.

$$x + 3 = i \qquad\qquad x + 3 = -i$$ $\sqrt{-1} = i$

$$x = -3 + i \qquad\qquad x = -3 - i$$ The solutions are $-3 \pm i$.

Because this equation has no real solutions, the parabola has no x-intercepts.

Step 4 Find the y-intercept. Replace x with 0 in $f(x) = (x + 3)^2 + 1$.

$$f(0) = (0 + 3)^2 + 1 = 3^2 + 1 = 9 + 1 = 10$$

The y-intercept is 10. The parabola passes through $(0, 10)$.

Step 5 Graph the parabola. With a vertex at $(-3, 1)$, no x-intercepts, and a y-intercept at 10, the graph of f is shown in Figure 8.9. The axis of symmetry is the vertical line whose equation is $x = -3$. ■

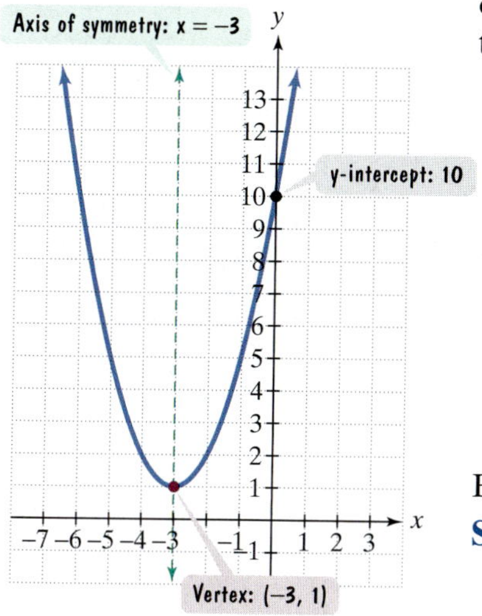

Figure 8.9 The graph of $f(x) = (x + 3)^2 + 1$

✔ CHECK POINT 2 Graph the quadratic function $f(x) = (x - 2)^2 + 1$.

3 Graph parabolas in the form $f(x) = ax^2 + bx + c$.

Graphing Quadratic Functions in the Form $f(x) = ax^2 + bx + c$

Quadratic functions are frequently expressed in the form $f(x) = ax^2 + bx + c$. How can we identify the vertex of a parabola whose equation is in this form? Completing the square provides the answer to this question.

$$f(x) = ax^2 + bx + c$$

$$= a\left(x^2 + \frac{b}{a}x\right) + c$$ Factor out a from $ax^2 + bx$.

$$= a\left(x^2 + \frac{b}{a}x + \frac{b^2}{4a^2}\right) + c - a\left(\frac{b^2}{4a^2}\right)$$

Complete the square by adding the square of half the coefficient of x.

By completing the square, we added $a \cdot \frac{b^2}{4a^2}$. To avoid changing the function's equation, we must subtract this term.

$$= a\left(x + \frac{b}{2a}\right)^2 + c - \frac{b^2}{4a}$$ Write the trinomial as the square of a binomial and simplify the constant term.

On the next page, we compare the form of this equation with a quadratic function in the form $f(x) = a(x - h)^2 + k$.

The form we know how to graph $f(x) = a(x - h)^2 + k$

$h = -\frac{b}{2a}$ $k = c - \frac{b^2}{4a}$

Equation under discussion $f(x) = a\left(x - \left(-\frac{b}{2a}\right)\right)^2 + c - \frac{b^2}{4a}$

The important part of this observation is that h, the x-coordinate of the vertex, is $-\frac{b}{2a}$. The y-coordinate can be found by evaluating the function at $-\frac{b}{2a}$.

The Vertex of a Parabola Whose Equation Is $f(x) = ax^2 + bx + c$

Consider the parabola defined by the quadratic function $f(x) = ax^2 + bx + c$. The parabola's vertex is at $\left(-\frac{b}{2a}, f\left(-\frac{b}{2a}\right)\right)$.

We can apply our five-step procedure and graph parabolas in the form $f(x) = ax^2 + bx + c$. The only step that is different is how we determine the vertex.

EXAMPLE 3 Graphing a Quadratic Function in the Form $f(x) = ax^2 + bx + c$

Graph the quadratic function $f(x) = -x^2 + 4x - 1$.

Solution

Step 1 Determine how the parabola opens. Note that a, the coefficient of x^2, is -1. Thus, $a < 0$; this negative value tells us that the parabola opens downward.

Step 2 Find the vertex. We know that the x-coordinate of the vertex is $x = -\frac{b}{2a}$. We identify a, b, and c in $f(x) = ax^2 + bx + c$.

$$f(x) = -x^2 + 4x - 1$$

$a = -1$ $b = 4$ $c = -1$

Substitute the values of a and b into the equation for the x-coordinate:

$$x = -\frac{b}{2a} = -\frac{4}{2(-1)} = \frac{-4}{-2} = 2.$$

The x-coordinate of the vertex is 2. We substitute 2 for x in the equation of the function to find the y-coordinate:

$$f(2) = -2^2 + 4 \cdot 2 - 1 = -4 + 8 - 1 = 3.$$

The vertex is at $(2, 3)$.

Step 3 Find the x-intercepts. Replace $f(x)$ with 0 in $f(x) = -x^2 + 4x - 1$. We obtain $0 = -x^2 + 4x - 1$ or $-x^2 + 4x - 1 = 0$. This equation cannot be solved by factoring. We will use the quadratic formula to solve it.

$$a = -1, \qquad b = 4, \qquad c = -1$$

$$x = \frac{-b \pm \sqrt{b^2 - 4ac}}{2a} = \frac{-4 \pm \sqrt{4^2 - 4(-1)(-1)}}{2(-1)} = \frac{-4 \pm \sqrt{16 - 4}}{-2}$$

$$x = \frac{-4 + \sqrt{12}}{-2} \approx 0.3 \quad \text{or} \quad x = \frac{-4 - \sqrt{12}}{-2} \approx 3.7$$

The x-intercepts are approximately 0.3 and 3.7. The parabola passes through $(0.3, 0)$ and $(3.7, 0)$.

Step 4 Find the y-intercept. Replace x with 0 in $f(x) = -x^2 + 4x - 1$.

$$f(0) = -0^2 + 4 \cdot 0 - 1 = -1$$

The y-intercept is -1. The parabola passes through $(0, -1)$.

Step 5 Graph the parabola. With a vertex at $(2, 3)$, x-intercepts at 0.3 and 3.7, and a y-intercept at -1, the graph of f is shown in Figure 8.10. The axis of symmetry is the vertical line whose equation is $x = 2$. ■

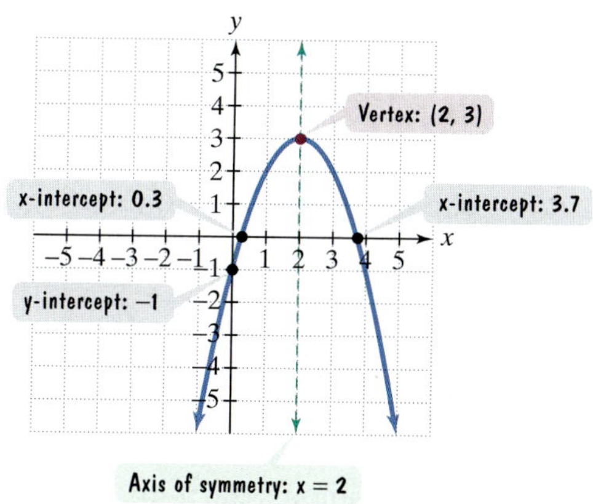

Figure 8.10 The graph of $f(x) = -x^2 + 4x - 1$

✔ **CHECK POINT 3** Graph the quadratic function $f(x) = x^2 - 2x - 3$.

4 Solve problems involving minimizing or maximizing quadratic functions.

Applications of Quadratic Functions

When were women's earnings as a percentage of men's at the lowest? What is the age of a driver having the least number of car accidents? How many units of a product should a business manufacture to maximize its profits? The answers to these questions involve finding the maximum or minimum value of quadratic functions.

Consider the quadratic function $f(x) = ax^2 + bx + c$. If $a > 0$, the parabola opens upward and the vertex is its lowest point. If $a < 0$, the parabola opens downward and the vertex is its highest point. The x-coordinate of the vertex is $-\frac{b}{2a}$. Thus, we can find the minimum or maximum value of f by evaluating the quadratic function at $x = -\frac{b}{2a}$.

Minimum and Maximum: Quadratic Functions

Consider $f(x) = ax^2 + bx + c$.

1. If $a > 0$, then f has a minimum that occurs at $x = -\frac{b}{2a}$.

2. If $a < 0$, then f has a maximum that occurs at $x = -\frac{b}{2a}$.

ENRICHMENT ESSAY

Variations in the Wage Gap

Gender differences in earnings vary among ethnic groups.

Women's Earnings as a Percentage of Men's in 1999

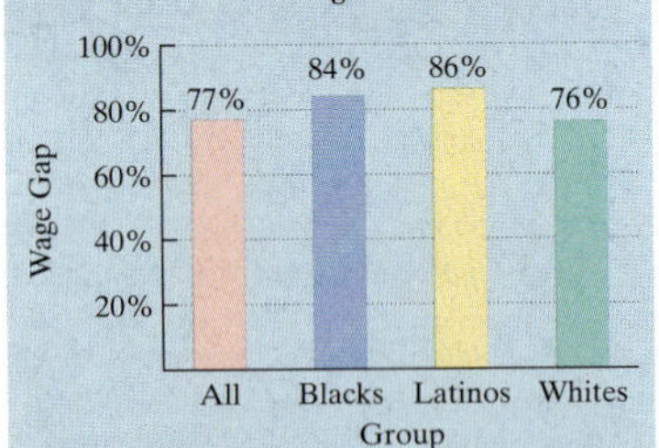

Source: U.S. Department of Labor

EXAMPLE 4 An Application: The Wage Gap

The function

$$f(x) = 0.022x^2 - 0.4x + 60.07$$

models women's earnings as a percentage of men's x years after 1960. In which year was this percentage at a minimum? What was the percentage for that year?

Solution The quadratic function is in the form $f(x) = ax^2 + bx + c$ with $a = 0.022$ and $b = -0.4$. With $a > 0$, the function has a minimum when $x = -\frac{b}{2a}$.

$$x = -\frac{b}{2a} = -\frac{(-0.4)}{2(0.022)} \approx 9$$

This means that women's earnings as a percentage of men's were at their lowest approximately 9 years after 1960, or in 1969. The percentage for that year was

$$f(9) = 0.022(9)^2 - 0.4(9) + 60.07 \approx 58.$$

In 1969, women earned approximately 58% as much as men. ■

✔ **CHECK POINT 4** The function $f(x) = 0.4x^2 - 36x + 1000$ models the number of accidents, $f(x)$, per 50 million miles driven, as a function of a driver's age, x, in years, where $16 \leq x \leq 74$. What is the age of a driver having the least number of car accidents? What is the minimum number of car accidents per 50 million miles driven?

Quadratic functions can also be formed from verbal conditions. Once we have obtained a quadratic function, we can then use the x-coordinate of the vertex to determine its maximum or minimum value. Here is a step-by-step strategy for solving these kinds of problems:

Strategy for Solving Problems Involving Maximizing or Minimizing Quadratic Functions

1. Read the problem carefully and decide which quantity is to be maximized or minimized.
2. Use the conditions of the problem to express the quantity as a function in one variable.
3. Rewrite the function in the form $f(x) = ax^2 + bx + c$.
4. If $a > 0$, f has a minimum at $x = -\frac{b}{2a}$. If $a < 0$, f has a maximum at $x = -\frac{b}{2a}$.
5. Answer the question posed in the problem.

EXAMPLE 5 Maximizing Area

You have 100 yards of fencing to enclose a rectangular region. Find the dimensions of the rectangle that maximize the enclosed area. What is the maximum area?

Solution

Step 1 Decide what must be maximized or minimized. We must maximize area. What we do not know are the rectangle's dimensions, x and y.

Step 2 Express this quantity as a function in one variable. Because we must maximize area, we have $A = xy$. We need to transform this into a function in which A is represented by one variable. Because you have 100 yards of fencing, the perimeter of the rectangle is 100 yards. This means that

$$2x + 2y = 100.$$

We can solve this equation for y in terms of x, substitute the result into $A = xy$, and obtain A as a function in one variable. We begin by solving for y.

$$2y = 100 - 2x \quad \text{Subtract } 2x \text{ from both sides.}$$

$$y = \frac{100 - 2x}{2} \quad \text{Divide both sides by 2.}$$

$$y = 50 - x \quad \text{Divide each term in the numerator by 2.}$$

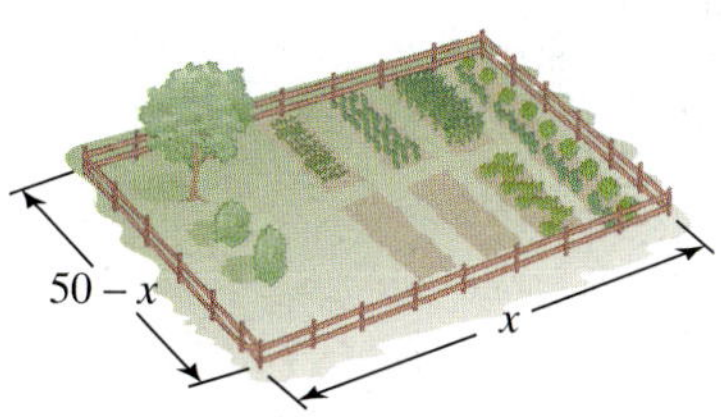

Figure 8.11 What value of x will maximize the rectangle's area?

Now we substitute $50 - x$ for y in $A = xy$.

$$A = xy = x(50 - x)$$

The rectangle and its dimensions are illustrated in Figure 8.11. Because A is now a function of x, we can write

$$A(x) = x(50 - x).$$

Step 3 Write the function in the form $f(x) = ax^2 + bx + c$. Applying the distributive property and rearranging terms, we obtain

$$A(x) = -x^2 + 50x.$$

Step 4 If $a < 0$, the function has a maximum at $x = -\frac{b}{2a}$. Using our formula for $A(x)$, we see that $a = -1$ and $b = 50$.

$$A(x) = -x^2 + 50x$$

$a = -1$ $b = 50$

The function has a maximum at $x = -\frac{b}{2a}$.

$$x = -\frac{b}{2a} = -\frac{50}{2(-1)} = 25$$

Step 5 Answer the question posed by the problem. We found that $x = 25$. Figure 8.11 shows that the rectangle's other dimension is $50 - x = 50 - 25 = 25$. The dimensions that maximize area are 25 feet by 25 feet. The rectangle that gives the maximum area is actually a square with an area of 25 feet · 25 feet, or 625 square feet. ■

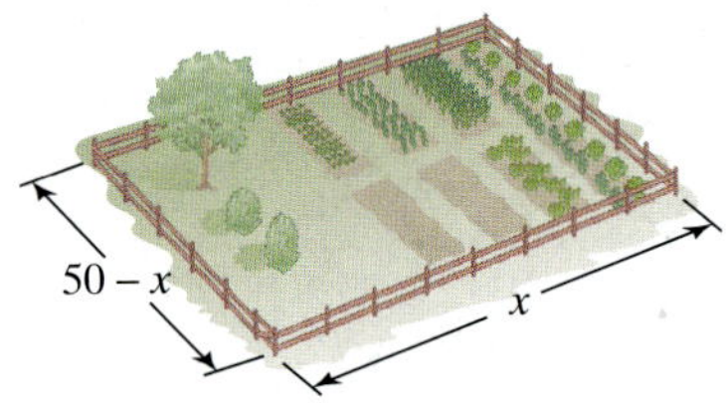

Figure 8.11, repeated

Using Technology

The graph of the area function

$$A(x) = x(50 - x)$$

was obtained with a graphing utility using a [0, 50, 2] by [0, 700, 25] viewing rectangle. The maximum function feature verifies that a maximum area of 625 square feet occurs when one of the dimensions is 25 feet.

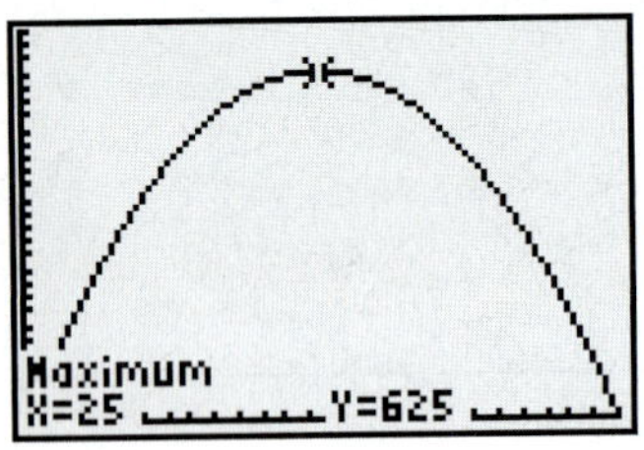

✔ CHECK POINT 5 You have 120 feet of fencing to enclose a rectangular region. Find the dimensions of the rectangle that maximize the enclosed area. What is the maximum area?

Using Technology

We've come a long way from the small nation of "embattled farmers" who launched the American Revolution. In the early days of our Republic, 95% of the population was involved in farming. The graph in Figure 8.12 shows the number of farms in the United States from 1850 through 2010 (projected). Because the graph is shaped like a cup, with an increasing number of farms from 1850 to 1910 and a decreasing number of farms from 1910 to 2010, a quadratic function is an appropriate model for the data. You can use the statistical menu of a graphing utility to enter the data in Figure 8.12. We entered the data using (number of decades after 1850, millions of U.S. farms). The data are shown to the right of Figure 8.12.

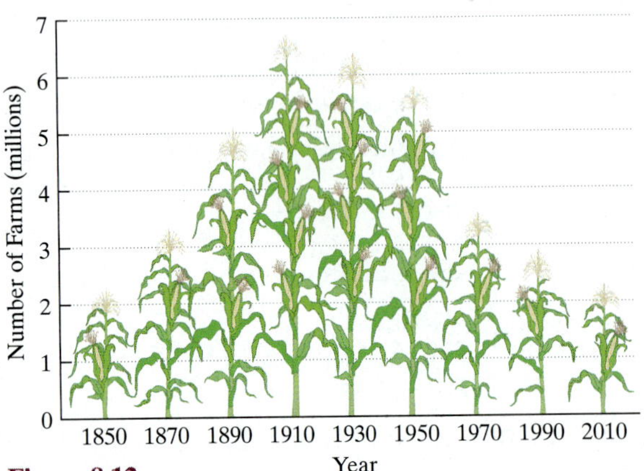

Figure 8.12

Source: U. S. Bureau of the Census

Data:
(0, 2.3), (2, 3.3), (4, 5.1), (6, 6.7), (8, 6.4), (10, 5.8), (12, 3.6), (14, 2.9), (16, 2.3)

```
QuadReg
 y=ax²+bx+c
 a=-.0643668831
 b=.9873701299
 c=2.203636364
```

Upon entering the QUADratic REGression program, we obtain the results shown in the screen. Thus, the quadratic function of best fit is

$$f(x) = -0.064x^2 + 0.99x + 2.2$$

where x represents the number of decades after 1850 and $f(x)$ represents the number of U.S. farms, in millions.

EXERCISE SET 8.3

Practice Exercises

In Exercises 1–4, the graph of a quadratic function is given. Write the function's equation, selecting from the following options:

$$f(x) = (x + 1)^2 - 1,\ g(x) = (x + 1)^2 + 1,$$
$$h(x) = (x - 1)^2 + 1,\ j(x) = (x - 1)^2 - 1.$$

1.

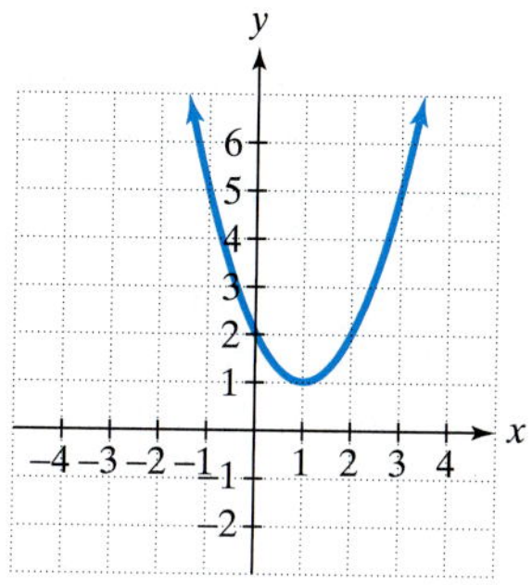

2.

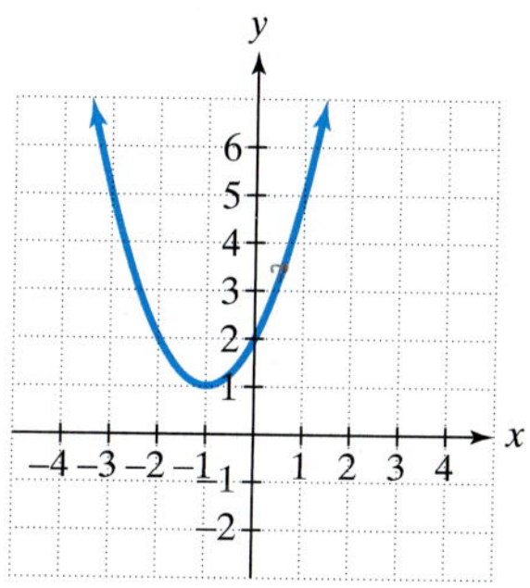

3.

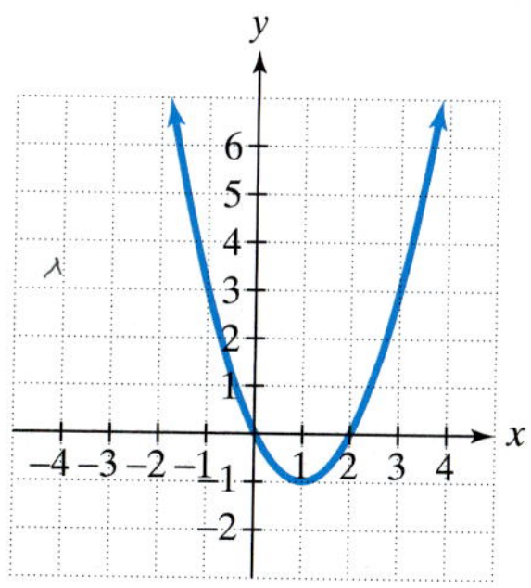

4.

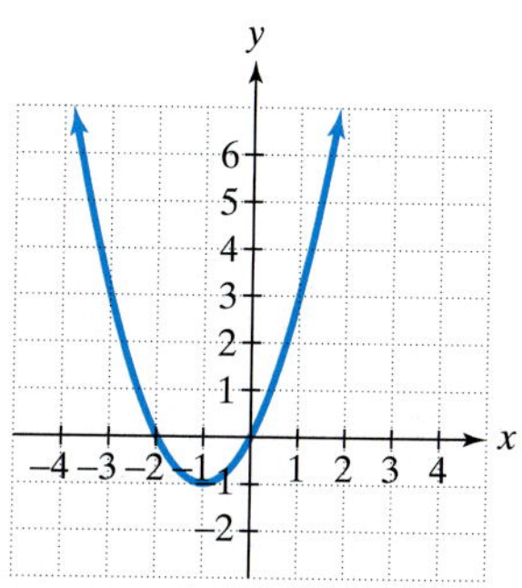

In Exercises 5–8, the graph of a quadratic function is given. Write the function's equation, selecting from the following options:

$$f(x) = x^2 + 2x + 1,\ g(x) = x^2 - 2x + 1,$$
$$h(x) = x^2 - 1,\ j(x) = -x^2 - 1.$$

5.

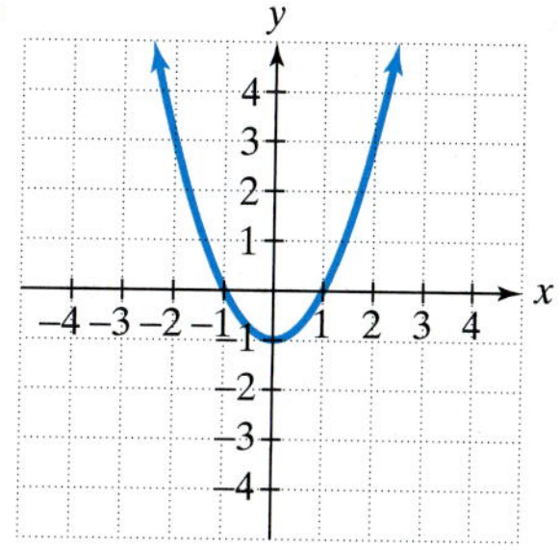

6.

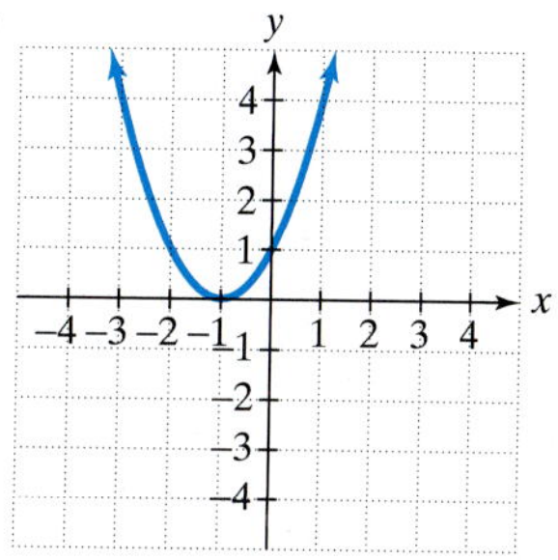

7.

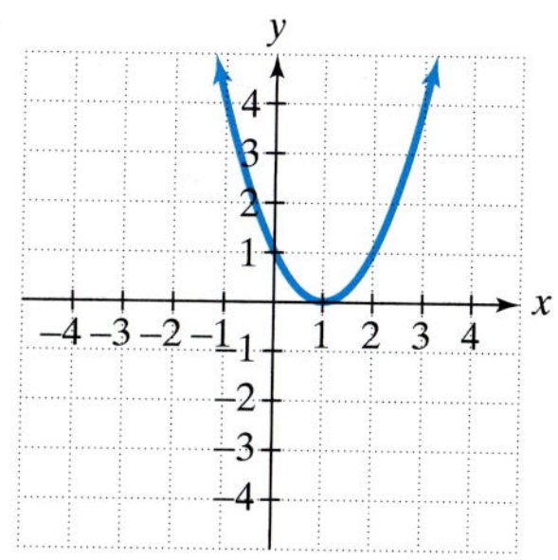

8.

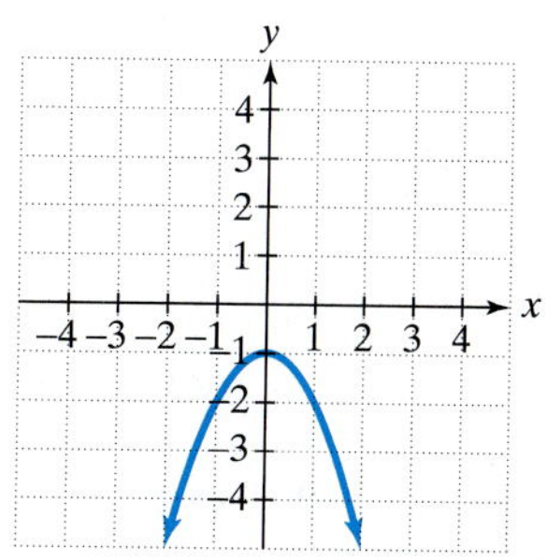

In Exercises 9–16, find the coordinates of the vertex for the parabola defined by the given quadratic function.

9. $f(x) = 2(x - 3)^2 + 1$
10. $f(x) = -3(x - 2)^2 + 12$
11. $f(x) = -2(x + 1)^2 + 5$
12. $f(x) = -2(x + 4)^2 - 8$
13. $f(x) = 2x^2 - 8x + 3$
14. $f(x) = 3x^2 - 12x + 1$
15. $f(x) = -x^2 - 2x + 8$
16. $f(x) = -2x^2 + 8x - 1$

In Exercises 17–34, use the vertex and intercepts to sketch the graph of each quadratic function. Give the equation for the parabola's axis of symmetry.

17. $f(x) = (x - 4)^2 - 1$
18. $f(x) = (x - 1)^2 - 2$
19. $f(x) = (x - 1)^2 + 2$
20. $f(x) = (x - 3)^2 + 2$
21. $y - 1 = (x - 3)^2$
22. $y - 3 = (x - 1)^2$
23. $f(x) = 2(x + 2)^2 - 1$
24. $f(x) = \frac{5}{4} - \left(x - \frac{1}{2}\right)^2$
25. $f(x) = 4 - (x - 1)^2$
26. $f(x) = 1 - (x - 3)^2$
27. $f(x) = x^2 + 2x - 3$
28. $f(x) = x^2 - 2x - 15$
29. $f(x) = x^2 + 3x - 10$
30. $f(x) = 2x^2 - 7x - 4$
31. $f(x) = 2x - x^2 + 3$
32. $f(x) = 5 - 4x - x^2$
33. $f(x) = 2x - x^2 - 2$
34. $f(x) = 6 - 4x + x^2$

In Exercises 35–40, determine, without graphing, whether the given quadratic function has a minimum value or a maximum value. Then find the coordinates of the minimum or the maximum point.

35. $f(x) = 3x^2 - 12x - 1$
36. $f(x) = 2x^2 - 8x - 3$
37. $f(x) = -4x^2 + 8x - 3$
38. $f(x) = -2x^2 - 12x + 3$
39. $f(x) = 5x^2 - 5x$
40. $f(x) = 6x^2 - 6x$

Application Exercises

41. The function

$$f(x) = -3.1x^2 + 51.4x + 4024.5$$

models the average annual per capita consumption of cigarettes, $f(x)$, by Americans 18 and older x years after 1960. According to this model, in which year did cigarette consumption per capita reach a maximum? What was the consumption for that year?

42. The function

$$f(x) = -0.0065x^2 + 0.23x + 8.47$$

models the American marriage rate, $f(x)$ (the number of marriages per 1000 people), x years after 1960. According to this model, in which year was the marriage rate the highest? What was the marriage rate for that year?

43. The function

$$f(x) = 104.5x^2 - 1501.5x + 6016$$

models the death rate per year per 100,000 males, $f(x)$, for U.S. men who average x hours of sleep each night. How many hours of sleep, to the nearest tenth of an hour, corresponds to the minimum death rate? What is this minimum death rate, to the nearest whole number?

44. There is a relationship between the amount of one's annual income, x, in thousands of dollars, and the percentage of this income, $P(x)$, that one contributes to charities. This relationship is modeled by the quadratic function

$$P(x) = 0.0014x^2 - 0.1529x + 5.855,$$

where $5 \le x \le 100$. What annual income corresponds to the minimum percentage given to charity? What is the minimum percentage?

45. A person standing close to the edge on the top of a 160-foot building throws a baseball vertically upward. The quadratic function

$$s(t) = -16t^2 + 64t + 160$$

models the ball's height above the ground, $s(t)$, in feet, t seconds after it was thrown.

a. After how many seconds does the ball reach its maximum height? What is the maximum height?

b. How many seconds does it take until the ball finally hits the ground? Round to the nearest tenth of a second.

c. Find $s(0)$ and describe what this means.

d. Use your results from parts (a) through (c) to graph the quadratic function. Begin the graph with $t = 0$ and end with the value of t for which the ball hits the ground.

46. A person standing close to the edge on the top of a 200-foot building throws a baseball vertically upward. The quadratic function

$$s(t) = -16t^2 + 64t + 200$$

models the ball's height above the ground, $s(t)$, in feet, t seconds after it was thrown.

a. After how many seconds does the ball reach its maximum height? What is the maximum height?

b. How many seconds does it take until the ball finally hits the ground? Round to the nearest tenth of a second.

c. Find $s(0)$ and describe what this means.

d. Use your results from parts (a) through (c) to graph the quadratic function. Begin the graph with $t = 0$ and end with the value of t for which the ball hits the ground.

47. Prices, as measured by the U.S. Consumer Price Index, have risen since World War II. What cost \$100 in 1967, the reference year, cost \$60 in 1920, \$116.30 in 1970, and \$511.50 by 2000. Suppose that a quadratic function is used to model the data shown in the graph with ordered pairs representing (number of years after 1920, Consumer Price Index).

Consumer Price Index, 1920–2000

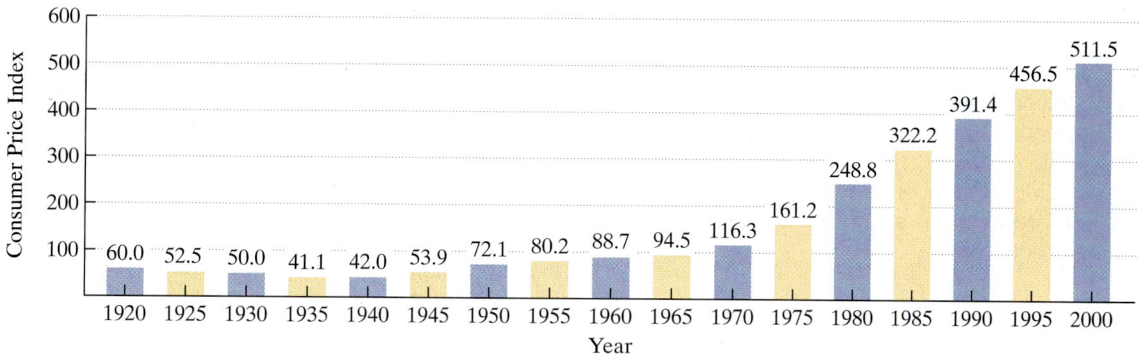

Source: U.S. Department of Labor

Determine, without obtaining an actual quadratic function that models the data, the coordinates of the vertex for the function's graph. Describe what this means in practical terms.

48. Suppose that a quadratic function is used to model the data shown in the graph with ordered pairs representing (number of years after 1971, millions of students enrolled in U.S. schools).

Millions of Students Enrolled in U.S. Schools

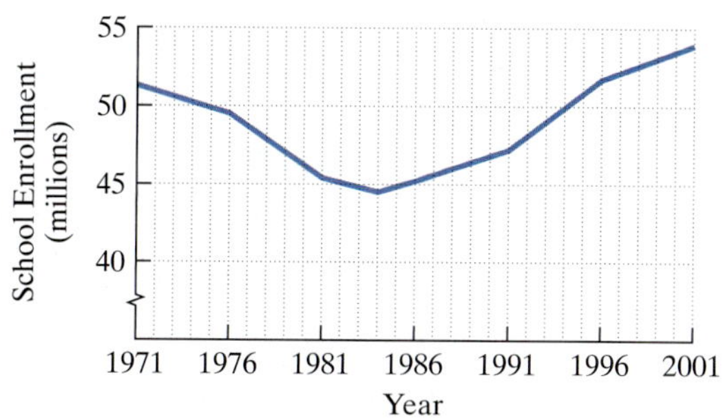

Source: National Education Association

Determine, without obtaining an actual quadratic function that models the data, the approximate coordinates of the vertex for the function's graph.

49. You have 120 feet of fencing to enclose a rectangular plot that borders on a river. If you do not fence the side along the river, find the length and width of the plot that will maximize the area. What is the largest area that can be enclosed?

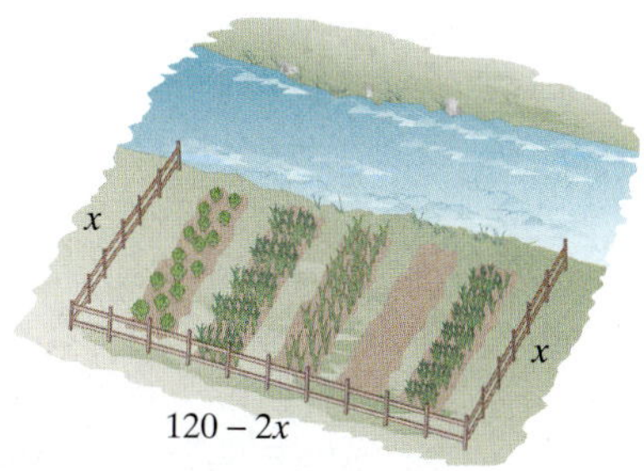

50. You have 600 feet of fencing to enclose a rectangular plot that borders on a river. If you do not fence the side along the river, find the length and width of the plot that will maximize the area. What is the largest area that can be enclosed?

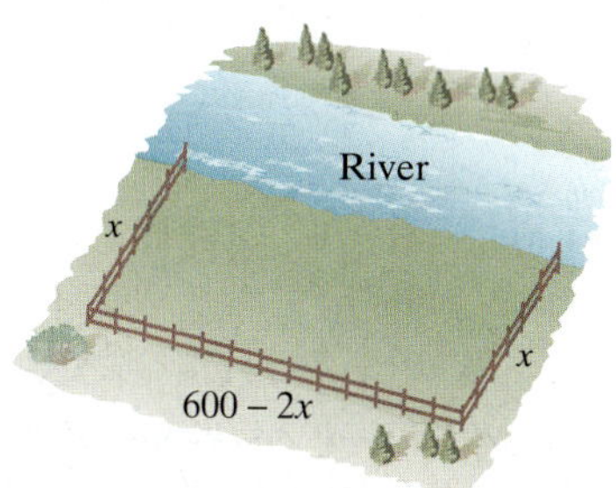

51. Among all pairs of numbers whose sum is 16, find a pair whose product is as large as possible. What is the maximum product?

52. Among all pairs of numbers whose sum is 20, find a pair whose product is as large as possible. What is the maximum product?

53. Among all pairs of numbers whose difference is 10, find a pair whose product is as small as possible. What is the minimum product?

54. Among all pairs of numbers whose difference is 8, find a pair whose product is as small as possible. What is the minimum product?

55. You have 50 yards of fencing to enclose a rectangular region. Find the dimensions of the rectangle that maximize the enclosed area. What is the maximum area?

56. You have 80 yards of fencing to enclose a rectangular region. Find the dimensions of the rectangle that maximize the enclosed area. What is the maximum area?

57. A rain gutter is made from sheets of aluminum that are 20 inches wide by turning up the edges to form right angles. Determine the depth of the gutter that will maximize its cross-sectional area and allow the greatest amount of water to flow. What is the maximum cross-sectional area?

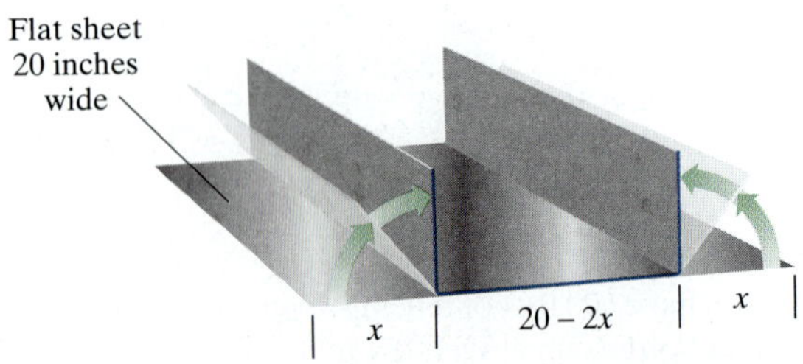

58. A rain gutter is made from sheets of aluminum that are 12 inches wide by turning up the edges to form right angles. Determine the depth of the gutter that will maximize its cross-sectional area and allow the greatest amount of water to flow. What is the maximum cross-sectional area?

In Chapter 3, we saw that the profit, $P(x)$, generated after producing and selling x units of a product is given by the function

$$P(x) = R(x) - C(x)$$

where R and C are the revenue and cost functions, respectively. Use these functions to solve Exercises 59–60.

59. Hunky Beef, a local sandwich store, has a fixed weekly cost of \$525.00, and variable costs for making a roast beef sandwich are \$0.55.

a. Let x represent the number of roast beef sandwiches made and sold each week. Write the weekly cost function, C, for Hunky Beef.

b. The function $R(x) = -0.001x^2 + 3x$ describes the money that Hunky Beef takes in each week from the sale of x roast beef sandwiches. Use this revenue function and the cost function from part (a) to write the store's weekly profit function, P.

c. Use the store's profit function to determine the number of roast beef sandwiches it should make and sell each week to maximize profit. What is the maximum weekly profit?

60. Virtual Fido is a company that makes electronic virtual pets. The fixed weekly cost is \$3000, and variable costs for each pet are \$20.

a. Let x represent the number of virtual pets made and sold each week. Write the weekly cost function, C, for Virtual Fido.

b. The function $R(x) = -x^2 + 1000x$ describes the money that Virtual Fido takes in each week from the sale of x virtual pets. Use this revenue function and the cost function from part (a) to write the weekly profit function, P.

c. Use the profit function to determine the number of virtual pets that should be made and sold each week to maximize profit. What is the maximum weekly profit?

Writing in Mathematics

61. What is a parabola? Describe its shape.

62. Explain how to decide whether a parabola opens upward or downward.

63. Describe how to find a parabola's vertex if its equation is in the form $f(x) = a(x - h)^2 + k$. Give an example.

64. Describe how to find a parabola's vertex if its equation is in the form $f(x) = ax^2 + bx + c$. Use $f(x) = x^2 - 6x + 8$ as an example.

65. A parabola that opens upward has its vertex at $(1, 2)$. Describe as much as you can about the parabola based on this information. Include in your discussion the number of x-intercepts (if any) for the parabola.

66. Use the Consumer Price Index shown in Exercise 47 to answer this question: Does a worker who earned \$40,000 in 1990 and \$50,000 in 2000 have more buying power in 2000 than in 1990? Explain your answer.

67. The quadratic function

$$f(x) = -0.018x^2 + 1.93x - 25.34$$

describes the miles per gallon of a Ford Taurus driven at x miles per hour. Suppose that you own a Ford Taurus. Describe how you can use this function to save money.

Technology Exercises

68. Use a graphing utility to verify any five of your hand-drawn graphs in Exercises 17–34.

69. a. Use a graphing utility to graph $y = 2x^2 - 82x + 720$ in a standard viewing rectangle. What do you observe?

b. Find the coordinates of the vertex for the given quadratic function.

c. The answer to part (b) is $(20.5, -120.5)$. Because the leading coefficient, 2, of the given function is positive, the vertex is a minimum point on the graph. Use this fact to help find a viewing rectangle that will give a relatively complete picture of the parabola. With an axis of symmetry at $x = 20.5$, the setting for x should extend past this, so try Xmin $= 0$ and Xmax $= 30$. The setting for y should include (and probably go below) the y-coordinate of the graph's minimum point, so try Ymin $= -130$. Experiment with Ymax until your utility shows the parabola's major features.

d. In general, explain how knowing the coordinates of a parabola's vertex can help determine a reasonable viewing rectangle on a graphing utility for obtaining a complete picture of the parabola.

In Exercises 70–73, find the vertex for each parabola. Then determine a reasonable viewing rectangle on your graphing utility and use it to graph the parabola.

70. $y = -0.25x^2 + 40x$

71. $y = -4x^2 + 20x + 160$

72. $y = 5x^2 + 40x + 600$

73. $y = 0.01x^2 + 0.6x + 100$

74. The quadratic function $f(x) = 0.013x^2 - 0.96x + 25.4$ describes the average yearly consumption of whole milk per person, $f(x)$, in the United States x years after 1970. The linear function $g(x) = 0.41x + 6.03$ describes the average yearly consumption of low-fat milk per person, $g(x)$, in the United States x years after 1970.

a. Use a graphing utility to graph each function in the same viewing rectangle for the years 1970 through 2000.

b. Use the graphs to describe the trend in consumption for both types of milk. What possible explanations are there for these consumption patterns?

75. The function $y = 0.011x^2 - 0.097x + 4.1$ models the number of people in the United States, y, in millions, holding more than one job x years after 1970. Use a graphing utility to graph the function in a $[0, 20, 1]$ by $[3, 6, 1]$ viewing rectangle. TRACE along the curve or use your utility's minimum value feature to approximate the coordinates of the parabola's vertex. Describe what this represents in practical terms.

76. The following data show fuel efficiency, in miles per gallon, for all U.S. automobiles in the indicated year.

x (Years after 1940)	y (Average Number of Miles per Gallon for U.S. Automobiles)
1940: 0	14.8
1950: 10	13.9
1960: 20	13.4
1970: 30	13.5
1980: 40	15.9
1990: 50	20.2
1998: 58	21.8

Source: U.S. Department of Transportation

a. Use a graphing utility to draw a scatter plot of the data. Explain why a quadratic function is appropriate for modeling these data.

b. Use the quadratic regression feature to find the quadratic function that best fits the data.

c. Use the equation in part (b) to determine the worst year for automobile fuel efficiency. What was the average number of miles per gallon for that year?

d. Use a graphing utility to draw a scatter plot of the data and graph the quadratic function of best fit on the scatter plot.

Critical Thinking Exercises

77. Which one of the following is true?

a. No quadratic functions have a range of $(-\infty, \infty)$.

b. The vertex of the parabola described by $f(x) = 2(x - 5)^2 - 1$ is at $(5, 1)$.

c. The graph of $f(x) = -2(x + 4)^2 - 8$ has one y-intercept and two x-intercepts.

d. The maximum value of y for the quadratic function $f(x) = -x^2 + x + 1$ is 1.

In Exercises 78–79, find the axis of symmetry for each parabola whose equation is given. Use the axis of symmetry to find a second point on the parabola whose y-coordinate is the same as the given point.

78. $f(x) = 3(x + 2)^2 - 5;\quad (-1, -2)$

79. $f(x) = (x - 3)^2 + 2;\quad (6, 11)$

80. A rancher has 1000 feet of fencing to construct six corrals, as shown in the figure. Find the dimensions that maximize the enclosed area. What is the maximum area?

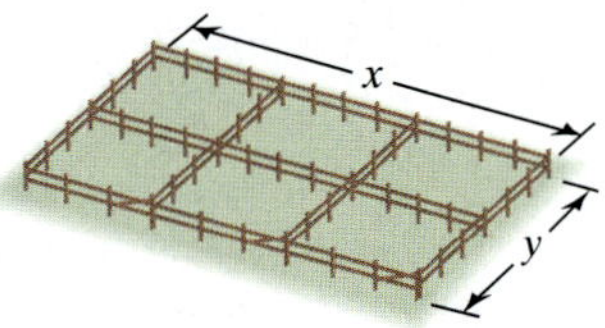

81. The annual yield per lemon tree is fairly constant at 320 pounds when the number of trees per acre is 50 or fewer. For each additional tree over 50, the annual yield per tree for all trees on the acre decreases by 4 pounds due to overcrowding. Find the number of trees that should be planted on an acre to produce the maximum yield. How many pounds is the maximum yield?

Review Exercises

82. Solve: $\frac{2}{x+5} + \frac{1}{x-5} = \frac{16}{x^2 - 25}$. (Section 6.6, Example 5)

83. Simplify: $\dfrac{1 + \frac{2}{x}}{1 - \frac{4}{x^2}}$. (Section 6.3, Example 1)

84. Solve using determinants (Cramer's Rule):

$$2x + 3y = 6 \qquad x - 4y = 14.$$

(Section 3.5, Example 2)

▸ SECTION 8.4 Equations Quadratic in Form

Objective

1. Solve equations that are quadratic in form.

SSM PH Tutor Center CD-ROM Video

"My husband asked me if we have any cheese puffs. Like he can't go and lift the couch cushion up himself."—Roseanne

How important is it for you to have a clean house? The percentage of people who find this to be quite important varies by age. In the exercise set, you will work with a function that models this phenomenon. Your work will be based on equations that are not quadratic, but that can be written as quadratic equations using an appropriate substitution. Here are some examples:

Given Equation	Substitution	New Equation
$x^4 - 10x^2 + 9 = 0$ or $(x^2)^2 - 10x^2 + 9 = 0$	$t = x^2$	$t^2 - 10t + 9 = 0$
$5x^{\frac{2}{3}} + 11x^{\frac{1}{3}} + 2 = 0$ or $5(x^{\frac{1}{3}})^2 + 11x^{\frac{1}{3}} + 2 = 0$	$t = x^{\frac{1}{3}}$	$5t^2 + 11t + 2 = 0$

An equation that is **quadratic in form** is one that can be expressed as a quadratic equation using an appropriate substitution. Both of the preceding given equations are quadratic in form.

1 Solve equations that are quadratic in form.

Equations that are quadratic in form contain an expression to a power, the same expression to that power squared, and a constant term. By letting t equal the expression to the power, a quadratic equation in t will result. Now it's easy. Solve this quadratic equation for t. Finally, use your substitution to find the values for the variable in the given equation. Example 1 shows how this is done.

EXAMPLE 1 Solving an Equation Quadratic in Form

Solve: $x^4 - 10x^2 + 9 = 0$.

Solution Notice that the equation contains an expression to a power, x^2, the same expression to that power squared, x^4 or $(x^2)^2$, and a constant term, 9. We let t equal the expression to the power. Thus,

$$\text{let } t = x^2.$$

Now we write the given equation as a quadratic equation in t and solve for t.

$$x^4 - 10x^2 + 9 = 0 \quad \text{This is the given equation.}$$

$$(x^2)^2 - 10x^2 + 9 = 0 \quad \text{The given equation contains } x^2 \text{ and } x^2 \text{ squared.}$$

$$t^2 - 10t + 9 = 0 \quad \text{Replace } x^2 \text{ with } t.$$

$$(t - 9)(t - 1) = 0 \quad \text{Factor.}$$

$$t - 9 = 0 \quad \text{or} \quad t - 1 = 0 \quad \text{Set each factor equal to zero.}$$

$$t = 9 \quad \text{or} \quad t = 1 \quad \text{Solve for } t.$$

We're not done! Why not? We were asked to solve for x and we have values for t. We use the original substitution, $t = x^2$, to solve for x. Replace t with x^2 in each equation shown.

$$x^2 = 9 \quad \text{or} \quad x^2 = 1$$

$$x = \pm\sqrt{9} \qquad x = \pm\sqrt{1} \quad \text{Apply the square root property.}$$

$$x = \pm 3 \qquad x = \pm 1$$

Substitute these values into the given equation and verify that the solutions are $-3, 3, -1$, and 1. The solution set is $\{-3, 3, -1, 1\}$. You may prefer to express the solution set in numerical order as $\{-3, -1, 1, 3\}$. ■

Using Technology

The graph of

$$y = x^4 - 10x^2 + 9$$

has x-intercepts at $-3, -1, 1$, and 3. This verifies that the solution set of

$$x^4 - 10x^2 + 9 = 0$$

is $\{-3, -1, 1, 3\}$.

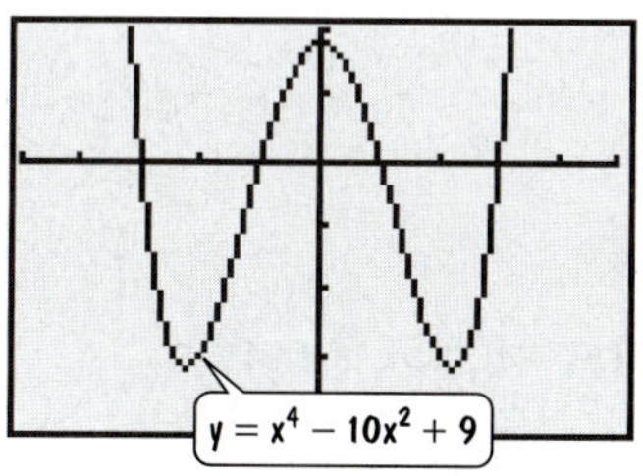

$[-5, 5, 1]$ by $[-20, 10, 5]$

✔ **CHECK POINT 1** Solve: $x^4 - 17x^2 + 16 = 0$.

If checking proposed solutions is not overly cumbersome, you should do so either algebraically or with a graphing utility. The Using Technology box shows a check of the four solutions in Example 1. Are there situations when solving equations quadratic in form where a check is essential? Yes. **If at any point in the solution process both sides of an equation are raised to an even power, a check is required**. Extraneous solutions that are not solutions of the given equations may have been introduced.

EXAMPLE 2 Solving an Equation Quadratic in Form

Solve: $2x - \sqrt{x} - 10 = 0$.

Solution In order to identify exponents on the terms, let's rewrite $\sqrt{x}$ as $x^{\frac{1}{2}}$. The equation can be expressed as

$$2x^1 - x^{\frac{1}{2}} - 10 = 0.$$

Notice that equation contains an expression to a power, $x^{\frac{1}{2}}$, the same expression to that power squared, x^1 or $\left(x^{\frac{1}{2}}\right)^2$, and a constant term, -10. We let t equal the expression to the power. Thus,

$$\text{let } t = x^{\frac{1}{2}}.$$

Now we write the given equation as a quadratic equation in t and solve for t.

$$2x^1 - x^{\frac{1}{2}} - 10 = 0 \quad \text{This is the given equation in exponential form.}$$

$$2\left(x^{\frac{1}{2}}\right)^2 - x^{\frac{1}{2}} - 10 = 0 \quad \text{The equation contains } x^{\frac{1}{2}} \text{ and } x^{\frac{1}{2}} \text{ squared.}$$

$$2t^2 - t - 10 = 0 \quad \text{Let } t = x^{\frac{1}{2}}.$$

$$(2t - 5)(t + 2) = 0 \quad \text{Factor.}$$

$$2t - 5 = 0 \quad \text{or} \quad t + 2 = 0 \quad \text{Set each factor equal to zero.}$$

$$t = \frac{5}{2} \quad \text{or} \quad t = -2 \quad \text{Solve for } t.$$

Use the original substitution, $t = x^{\frac{1}{2}}$, to solve for x. Replace t with $x^{\frac{1}{2}}$ in each of the preceding equations.

$$x^{\frac{1}{2}} = \frac{5}{2} \quad \text{or} \quad x^{\frac{1}{2}} = -2 \quad \text{Replace } t \text{ with } x^{\frac{1}{2}}.$$

$$\left(x^{\frac{1}{2}}\right)^2 = \left(\frac{5}{2}\right)^2 \qquad \left(x^{\frac{1}{2}}\right)^2 = (-2)^2 \quad \text{Solve for } x \text{ by squaring both sides of each equation.}$$

Both sides are raised to even powers. We must check.

$$x = \frac{25}{4} \qquad x = 4 \quad \text{Square } \frac{5}{2} \text{ and } -2.$$

It is essential to check both proposed solutions in the original equation.

Check $\frac{25}{4}$:

$$2x - \sqrt{x} - 10 = 0$$

$$2 \cdot \frac{25}{4} - \sqrt{\frac{25}{4}} - 10 \stackrel{?}{=} 0$$

$$\frac{25}{2} - \frac{5}{2} - 10 \stackrel{?}{=} 0$$

$$\frac{20}{2} - 10 \stackrel{?}{=} 0$$

$$0 = 0, \text{ true}$$

Check 4:

$$2x - \sqrt{x} - 10 = 0$$

$$2 \cdot 4 - \sqrt{4} - 10 \stackrel{?}{=} 0$$

$$8 - 2 - 10 \stackrel{?}{=} 0$$

$$6 - 10 \stackrel{?}{=} 0$$

$$-4 = 0, \text{ false}$$

The check indicates that 4 is not a solution. It is an extraneous solution brought about by squaring each side of the equation. The only solution is $\frac{25}{4}$, and the solution set is $\left\{\frac{25}{4}\right\}$.

CHECK POINT 2 Solve: $x - 2\sqrt{x} - 8 = 0$.

The equations in Examples 1 and 2 can be solved by methods other than using substitutions.

$$x^4 - 10x^2 + 9 = 0$$

This equation can be solved directly by factoring: $(x^2 - 9)(x^2 - 1) = 0$.

$$2x - \sqrt{x} - 10 = 0$$

This equation can be solved by isolating the radical term: $2x - 10 = \sqrt{x}$. Then square both sides.

In the examples that follow, solving the equations by methods other than first introducing a substitution becomes increasingly difficult.

EXAMPLE 3 Solving an Equation Quadratic in Form

Solve: $(x^2 - 5)^2 + 3(x^2 - 5) - 10 = 0$.

Solution This equation contains $x^2 - 5$ and $x^2 - 5$ squared. We let

$$t = x^2 - 5.$$

$(x^2 - 5)^2 + 3(x^2 - 5) - 10 = 0$	This is the given equation.
$t^2 + 3t - 10 = 0$	Let $t = x^2 - 5$.
$(t + 5)(t - 2) = 0$	Factor.
$t + 5 = 0$ or $t - 2 = 0$	Set each factor equal to zero.
$t = -5$ or $t = 2$	Solve for t.

Use the original substitution, $t = x^2 - 5$, to solve for x. Replace t by $x^2 - 5$ in each of the preceding equations.

$x^2 - 5 = -5$ or $x^2 - 5 = 2$	Replace t by $x^2 - 5$.
$x^2 = 0$ $\quad$ $x^2 = 7$	Solve for x by isolating x^2.
$x = 0$ $\quad$ $x = \pm\sqrt{7}$	Apply the square root property.

Although we did not raise both sides of an equation to an even power, checking the three proposed solutions in the original equation is a good idea. Do this now and verify that the solutions are $-\sqrt{7}$, 0, and $\sqrt{7}$, and the solution set is $\{-\sqrt{7}, 0, \sqrt{7}\}$.

CHECK POINT 3 Solve: $(x^2 - 4)^2 + (x^2 - 4) - 6 = 0$.

EXAMPLE 4 Solving an Equation Quadratic in Form

Solve: $10x^{-2} + 7x^{-1} + 1 = 0$.

Solution Notice that the equation contains on expression to a power, x^{-1}, the same expression to that power squared, x^{-2} or $(x^{-1})^2$, and a constant term, 1. We let t equal the expression to the power. Thus,

$$\text{let } t = x^{-1}.$$

Now we write the given equation as a quadratic equation in t and solve for t.

$$10x^{-2} + 7x^{-1} + 1 = 0 \quad \text{This is the given equation.}$$

$$10(x^{-1})^2 + 7x^{-1} + 1 = 0 \quad \text{The equation contains } x^{-1} \text{ and } x^{-1} \text{ squared.}$$

$$10t^2 + 7t + 1 = 0 \quad \text{Let } t = x^{-1}.$$

$$(5t + 1)(2t + 1) = 0 \quad \text{Factor.}$$

$$5t + 1 = 0 \quad \text{or} \quad 2t + 1 = 0 \quad \text{Set each factor equal to zero.}$$

$$5t = -1 \quad \text{or} \quad 2t = -1$$

$$t = -\frac{1}{5} \quad \text{or} \quad t = -\frac{1}{2} \quad \text{Solve for } t.$$

Use the original substitution, $t = x^{-1}$, to solve for x. Replace t with x^{-1} in each of the preceding equations.

$$x^{-1} = -\frac{1}{5} \quad \text{or} \quad x^{-1} = -\frac{1}{2} \quad \text{Replace } t \text{ with } x^{-1}.$$

$$(x^{-1})^{-1} = \left(-\frac{1}{5}\right)^{-1} \qquad (x^{-1})^{-1} = \left(-\frac{1}{2}\right)^{-1} \quad \text{Solve for } x \text{ by raising both sides of each equation to the } -1 \text{ power.}$$

$$x = -5 \qquad x = -2$$

$$\left(-\frac{1}{5}\right)^{-1} = \frac{1}{-\frac{1}{5}} = -5 \qquad \left(-\frac{1}{2}\right)^{-1} = \frac{1}{-\frac{1}{2}} = -2$$

We did not raise both sides of an equation to an even power. A check will show that both -5 and -2 are solutions of the original equation. The solution set is $\{-5, -2\}$. ■

✔ **CHECK POINT 4** Solve: $2x^{-2} + x^{-1} - 1 = 0$.

EXAMPLE 5 Solving an Equation Quadratic in Form

Solve: $5x^{\frac{2}{3}} + 11x^{\frac{1}{3}} + 2 = 0$.

Solution Notice that the equations contains on expression to a power, $x^{\frac{1}{3}}$, the same expression to that power squared, $x^{\frac{2}{3}}$ or $\left(x^{\frac{1}{3}}\right)^2$, and a constant term, 1. We let t equal the expression to the power. Thus,

$$\text{let } t = x^{\frac{1}{3}}.$$

Now we write the given equation as a quadratic equation in t and solve for t.

$$5x^{\frac{2}{3}} + 11x^{\frac{1}{3}} + 2 = 0 \quad \text{This is the given equation.}$$

$$5\left(x^{\frac{1}{3}}\right)^2 + 11\left(x^{\frac{1}{3}}\right) + 2 = 0 \quad \text{The given equation contains } x^{\frac{1}{3}} \text{ and } x^{\frac{1}{3}} \text{ squared.}$$

$$5t^2 + 11t + 2 = 0 \quad \text{Let } t = x^{\frac{1}{3}}.$$

$$(5t + 1)(t + 2) = 0 \quad \text{Factor.}$$

$$5t + 1 = 0 \quad \text{or} \quad t + 2 = 0 \quad \text{Set each factor equal to 0.}$$

$$t = -\tfrac{1}{5} \quad \text{or} \quad t = -2 \quad \text{Solve for } t.$$

Use the original substitution, $t = x^{\frac{1}{3}}$, to solve for x. Replace t with $x^{\frac{1}{3}}$ in each of the preceding equations.

$$x^{\frac{1}{3}} = -\frac{1}{5} \qquad x^{\frac{1}{3}} = -2 \qquad \text{Replace } t \text{ with } x^{\frac{1}{3}}.$$

$$\left(x^{\frac{1}{3}}\right)^3 = \left(-\frac{1}{5}\right)^3 \qquad \left(x^{\frac{1}{3}}\right)^3 = (-2)^3 \qquad \text{Solve for } x \text{ by cubing both sides of each equation.}$$

$$x = -\frac{1}{125} \qquad x = -8$$

We did not raise both sides of an equation to an even power. A check will show that both -8 and $-\frac{1}{125}$ are solutions of the original equation. The solution set is $\left\{-8, -\frac{1}{125}\right\}$. ■

✔ **CHECK POINT 5** Solve: $3x^{\frac{2}{3}} - 11x^{\frac{1}{3}} - 4 = 0$.

EXERCISE SET 8.4

Practice Exercises

In Exercises 1–32, solve each equation by making an appropriate substitution. If at any point in the solution process both sides of an equation are raised to an even power, a check is required.

1. $x^4 - 5x^2 + 4 = 0$

2. $x^4 - 13x^2 + 36 = 0$

3. $x^4 - 11x^2 + 18 = 0$

4. $x^4 - 9x^2 + 20 = 0$

5. $x^4 + 2x^2 = 8$

6. $x^4 + 4x^2 = 5$

7. $x + \sqrt{x} - 2 = 0$

8. $x + \sqrt{x} - 6 = 0$

9. $x - 4x^{\frac{1}{2}} - 21 = 0$

10. $x - 6x^{\frac{1}{2}} + 8 = 0$

11. $x - 13\sqrt{x} + 40 = 0$

12. $2x - 7\sqrt{x} - 30 = 0$

13. $(x - 5)^2 - 4(x - 5) - 21 = 0$

14. $(x + 3)^2 + 7(x + 3) - 18 = 0$

15. $(x^2 - 1)^2 - (x^2 - 1) = 2$

16. $(x^2 - 2)^2 - (x^2 - 2) = 6$

17. $(x^2 + 3x)^2 - 8(x^2 + 3x) - 20 = 0$

18. $(x^2 - 2x)^2 - 11(x^2 - 2x) + 24 = 0$

19. $x^{-2} - x^{-1} - 20 = 0$

20. $x^{-2} - x^{-1} - 6 = 0$

21. $2x^{-2} - 7x^{-1} + 3 = 0$

22. $20x^{-2} + 9x^{-1} + 1 = 0$

23. $x^{-2} - 4x^{-1} = 3$

24. $x^{-2} - 6x^{-1} = -4$

25. $x^{\frac{2}{3}} - x^{\frac{1}{3}} - 6 = 0$

26. $x^{\frac{2}{3}} + 2x^{\frac{1}{3}} - 3 = 0$

27. $x^{\frac{2}{5}} + x^{\frac{1}{5}} - 6 = 0$

28. $x^{\frac{2}{5}} + x^{\frac{1}{5}} - 2 = 0$

29. $2x^{\frac{1}{2}} - x^{\frac{1}{4}} = 1$

30. $2x^{\frac{1}{2}} - 5x^{\frac{1}{4}} = 3$

31. $\left(x - \frac{8}{x}\right)^2 + 5\left(x - \frac{8}{x}\right) - 14 = 0$

32. $\left(x - \frac{10}{x}\right)^2 + 6\left(x - \frac{10}{x}\right) - 27 = 0$

In Exercises 33–38, find the x-intercepts of the given function, f. Then use the x-intercepts to match each function with its graph. (The graphs are labeled (a) through (f).)

33. $f(x) = x^4 - 5x^2 + 4$

34. $f(x) = x^4 - 13x^2 + 36$

35. $f(x) = x^{\frac{1}{3}} + 2x^{\frac{1}{6}} - 3$

36. $f(x) = x^{-2} - x^{-1} - 6$

37. $f(x) = (x + 2)^2 - 9(x + 2) + 20$

38. $f(x) = 2(x + 2)^2 + 5(x + 2) - 3$

a. 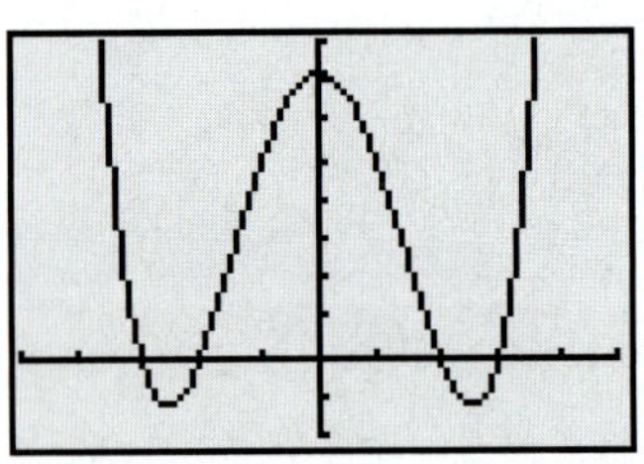
[−5, 5, 1] by [−10, 40, 5]

b. 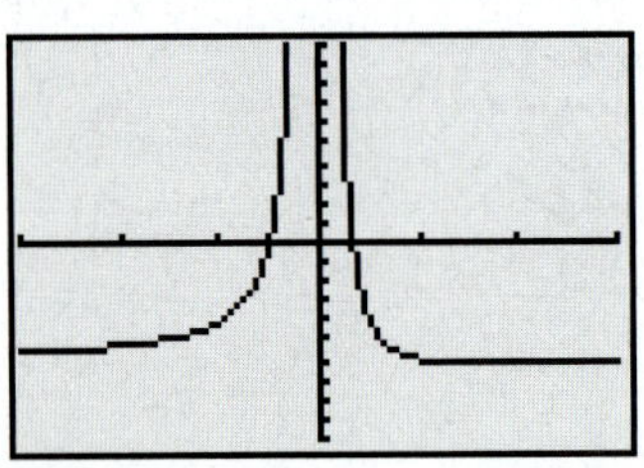
[−3, 3, 1] by [−10, 10, 1]

c. 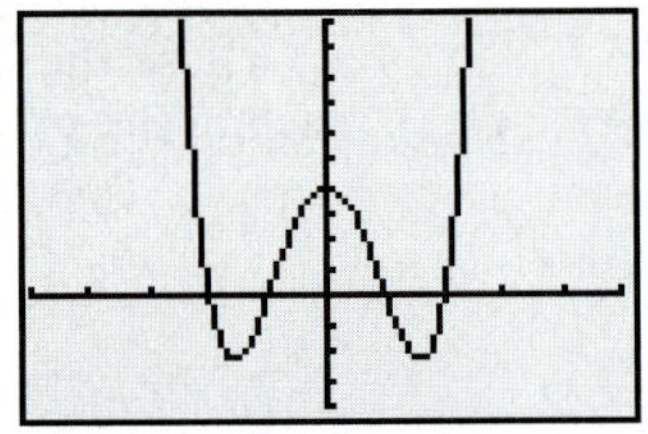
[−5, 5, 1] by [−4, 10, 1]

d. 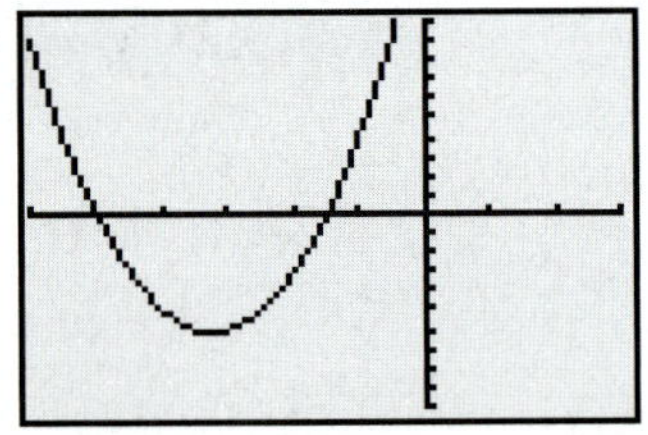
[−6, 3, 1] by [−10, 10, 1]

e. 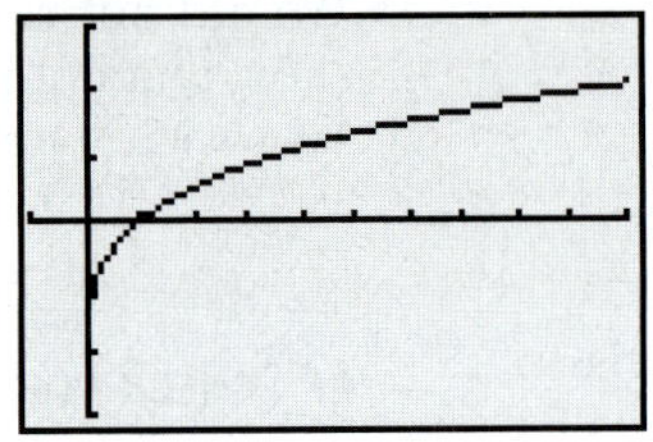
[−1, 10, 1] by [−3, 3, 1]

f. 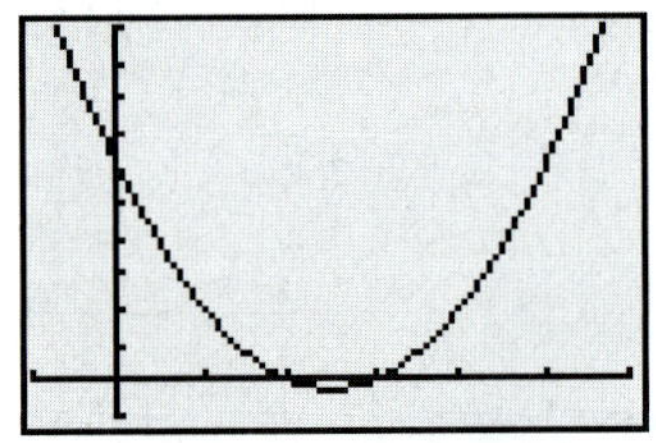
[−1, 6, 1] by [−1, 10, 1]

Application Exercises

How important is it for you to have a clean house? The bar graph indicates that the percentage of people who find this to be quite important varies by age. The percentage, P(x), who find having a clean house very important can be modeled by the function

$$P(x) = 0.04(x + 40)^2 - 3(x + 40) + 104$$

where x is the number of years a person's age is above or below 40. Thus, x is positive for people over 40 and negative for people under 40. Use the function to solve Exercises 39–40.

The Importance of Having a Clean House, by Age

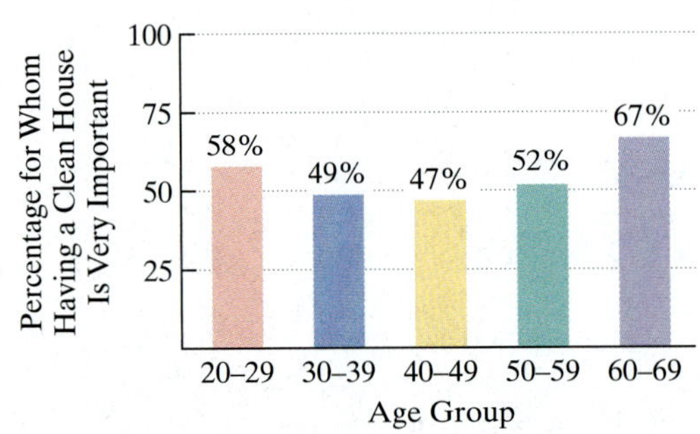

Source: Soap and Detergent Association

39. According to the model, at which ages do 60% of us feel that having a clean house is very important? Substitute 60 for $P(x)$ and solve the quadratic-in-form equation. How well does the function model the data shown in the bar graph?

40. According to the model, at which ages do 50% of us feel that having a clean house is very important? Substitute 50 for $P(x)$ and solve the quadratic-in-form equation. How well does the function model the data shown in the bar graph?

Writing in Mathematics

41. Explain how to recognize an equation that is quadratic in form. Provide two original examples with your explanation.

42. Describe two methods for solving this equation:

$$x - 5\sqrt{x} + 4 = 0.$$

43. In the twenty-first century, collecting and presenting data are big business. Because statisticians both record and influence our behavior, you should always ask yourself if the person or group presenting the data has any special case to make for or against the displayed information. Using this criterion, describe your impressions of the data in the bar graph for Exercises 39–40.

Technology Exercises

44. Use a graphing utility to verify the solutions of any five equations in Exercises 1–32 that you solved algebraically. The real solutions should appear as x-intercepts on the graph of the function related to the given equation.

Use a graphing utility to solve the equations in Exercises 45–52. Check by direct substitution.

45. $x^6 - 7x^3 - 8 = 0$

46. $3(x - 2)^{-2} - 4(x - 2)^{-1} + 1 = 0$

47. $x^4 - 10x^2 + 9 = 0$

48. $2x + 6\sqrt{x} = 8$

49. $2(x + 1)^2 = 5(x + 1) + 3$

50. $(x^2 - 3x)^2 + 2(x^2 - 3x) - 24 = 0$

51. $x^{\frac{1}{2}} + 4x^{\frac{1}{4}} = 5$

52. $x^{\frac{2}{3}} - 3x^{\frac{1}{3}} + 2 = 0$

Critical Thinking Exercises

53. Which one of the following is true?
   a. If an equation is quadratic in form, there is only one method that can be used to obtain its solution.
   b. An equation that is quadratic in form must have a variable factor in one term that is the square of the variable factor in another term.
   c. Because x^6 is the square of x^3, the equation $x^6 - 5x^3 + 6x = 0$ is quadratic in form.
   d. To solve $x - 9\sqrt{x} + 14 = 0$, we let $\sqrt{t} = x$.

In Exercises 54–56, use a substitution to solve each equation.

54. $x^4 - 5x^2 - 2 = 0$

55. $5x^6 + x^3 = 18$

56. $\sqrt{\dfrac{x + 4}{x - 1}} + \sqrt{\dfrac{x - 1}{x + 4}} = \dfrac{5}{2}$ $\left(\text{Let } t = \sqrt{\dfrac{x + 4}{x - 1}}.\right)$

Review Exercises

57. Simplify:

$$\frac{2x^2}{10x^3 - 2x^2}.$$

(Section 6.1, Example 4)

58. Divide: $\dfrac{2 + i}{1 - i}$. (Section 7.7, Example 5)

59. Solve using matrices:

$$\begin{aligned} 2x + y &= 6 \\ x - 2y &= 8. \end{aligned}$$

(Section 3.4, Example 3)

▸ SECTION 8.5 Quadratic and Rational Inequalities

Objectives

1. Solve quadratic inequalities.
2. Solve rational inequalities.
3. Solve problems using nonlinear inequalities.

SSM PH Tutor Center CD-ROM Video

People are going to live longer in the twenty-first century. This will put added pressure on the Social Security and Medicare systems. The bar graph in Figure 8.13 shows the cost of Medicare, in billions of dollars, projected through 2005.

Medicare Spending

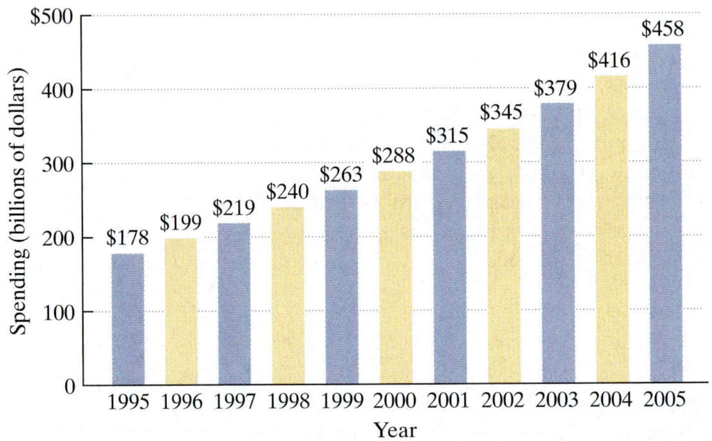

Figure 8.13 *Source*: Congressional Budget Office

Medicare spending, $f(x)$, in billions of dollars, x years after 1995 can be modeled by the quadratic function

$$f(x) = 1.2x^2 + 15.2x + 181.4.$$

To determine in which years Medicare spending will exceed \$500 billion, we must solve the inequality

$$1.2x^2 + 15.2x + 181.4 > 500.$$

Medicare spending | exceeds | \$500 billion.

We begin by subtracting 500 from both sides. This will give us zero on the right:

$$1.2x^2 + 15.2x + 181.4 - 500 > 500 - 500$$

$$1.2x^2 + 15.2x - 318.6 > 0.$$

The form of this inequality is $ax^2 + bx + c > 0$. Such an inequality is called a *quadratic inequality*.

Definition of a Quadratic Inequality

A **quadratic inequality** is any inequality that can be put in one of the forms

$$ax^2 + bx + c < 0 \qquad ax^2 + bx + c > 0$$

$$ax^2 + bx + c \le 0 \qquad ax^2 + bx + c \ge 0$$

where a, b, and c are real numbers and $a \neq 0$.

In this section we establish the basic techniques for solving quadratic inequalities. We will use these techniques to solve inequalities containing quotients, called **rational inequalities**.

1 Solve quadratic inequalities

Solving Quadratic Inequalities

Graphs can help us to estimate the solutions of quadratic inequalities. The graph of $y = x^2 - 7x + 10$ is shown in Figure 8.14. The x-intercepts, 2 and 5, are **boundary points** between where the graph lies above the x-axis, shown in blue, and where the graph lies below the x-axis, shown in red. These boundary points play a critical role in solving quadratic inequalities.

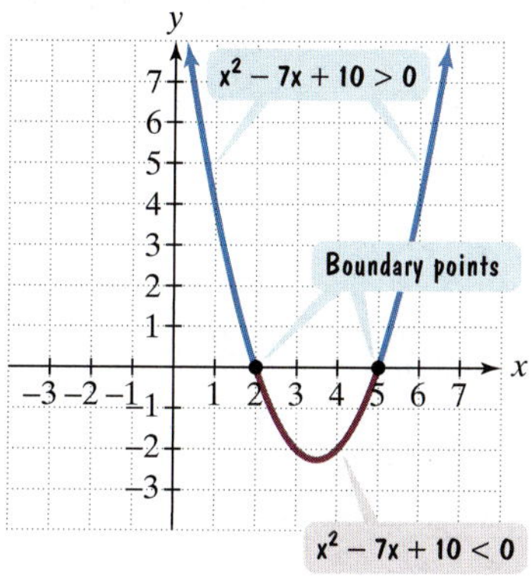

Figure 8.14 The graph of

$$y = x^2 - 7x + 10.$$

The blue parts of the graph lie above the x-axis:

$$x^2 - 7x + 10 > 0.$$

By contrast, the red part lies below the x-axis:

$$x^2 - 7x + 10 < 0.$$

Procedure for Solving Quadratic Inequalities

1. Express the inequality in the standard form
$$ax^2 + bx + c > 0 \quad \text{or} \quad ax^2 + bx + c < 0.$$
2. Solve the equation $ax^2 + bx + c = 0$. The real solutions are the **boundary points**.
3. Locate these boundary points on a number line, thereby dividing the number line into **test intervals**.
4. Choose one representative number within each test interval. If substituting that value into the original inequality produces a true statement, then all real numbers in the test interval belong to the solution set. If substituting that value into the original inequality produces a false statement, then no real number in the test interval belongs to the solution set.
5. Write the solution set, selecting the interval(s) that produced a true statement. The graph of the solution set on a number line usually appears as

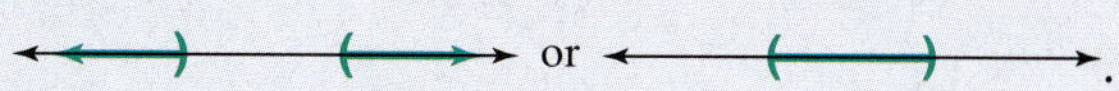

This procedure is valid if $<$ is replaced by $\le$ and $>$ is replaced by $\ge$.

EXAMPLE 1 Solving a Quadratic Inequality

Solve and graph the solution set on a real number line: $x^2 - 7x + 10 < 0$.

Solution

Step 1 Write the inequality in standard form. The inequality is given in this form, so this step has been done for us.

Step 2 Solve the related quadratic equation. This equation is obtained by replacing the inequality sign in $x^2 - 7x + 10 < 0$ by an equal sign. Thus, we will solve $x^2 - 7x + 10 = 0$.

$$x^2 - 7x + 10 = 0 \quad \text{This is the related quadratic equation.}$$

$$(x - 2)(x - 5) = 0 \quad \text{Factor.}$$

$$x - 2 = 0 \quad \text{or} \quad x - 5 = 0 \quad \text{Set each factor equal to 0.}$$

$$x = 2 \quad \text{or} \quad x = 5 \quad \text{Solve for } x.$$

The boundary points are 2 and 5.

Step 3 Locate the boundary points on a number line. The number line with the boundary points is shown as follows:

The boundary points divide the number line into three test intervals, namely $(-\infty, 2)$, $(2, 5)$, and $(5, \infty)$.

Step 4 Take one representative number within each test interval and substitute that number into the original inequality.

Test Interval	Representative Number	Substitute into $x^2 - 7x + 10 < 0$	Conclusion
$(-\infty, 2)$	0	$0^2 - 7 \cdot 0 + 10 \overset{?}{<} 0$ $10 < 0$, false	$(-\infty, 2)$ does not belong to the solution set.
$(2, 5)$	3	$3^2 - 7 \cdot 3 + 10 \overset{?}{<} 0$ $9 - 21 + 10 \overset{?}{<} 0$ $-2 < 0$, true	$(2, 5)$ belongs to the solution set.
$(5, \infty)$	6	$6^2 - 7 \cdot 6 + 10 \overset{?}{<} 0$ $36 - 42 + 10 \overset{?}{<} 0$ $4 < 0$, false	$(5, \infty)$ does not belong to the solution set.

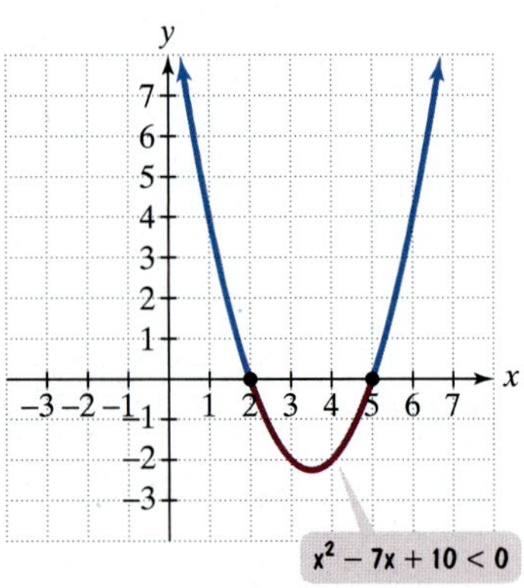

Figure 8.15 The graph lies below the x-axis between the boundary points 2 and 5, in the interval $(2, 5)$.

Step 5 The solution set is the interval that produced a true statement. Our analysis shows that the solution set is the interval $(2, 5)$, or $\{x \mid 2 < x < 5\}$. The graph in Figure 8.15 confirms that $x^2 - 7x + 10 < 0$ (lies below the x-axis) in this interval. The graph of the solution set on a number line is shown as follows:

■

✔ **CHECK POINT 1** Solve and graph the solution set:

$$x^2 + 2x - 3 < 0.$$

EXAMPLE 2 Solving a Quadratic Inequality

Solve and graph the solution set on a real number line: $2x^2 + x \geq 15$.

Solution

Step 1 Write the inequality in standard form. We can write $2x^2 + x \geq 15$ in standard form by subtracting 15 from both sides. This will give us zero on the right.

$$2x^2 + x - 15 \geq 15 - 15$$

$$2x^2 + x - 15 \geq 0$$

Step 2 Solve the related quadratic equation. This equation is obtained by replacing the inequality sign in $2x^2 + x - 15 \geq 0$ by an equal sign. Thus, we will solve $2x^2 + x - 15 = 0$.

$$2x^2 + x - 15 = 0 \quad \text{This is the related quadratic equation.}$$

$$(2x - 5)(x + 3) = 0 \quad \text{Factor.}$$

$$2x - 5 = 0 \quad \text{or} \quad x + 3 = 0 \quad \text{Set each factor equal to 0.}$$

$$x = \tfrac{5}{2} \quad \text{or} \quad x = -3 \quad \text{Solve for } x.$$

The boundary points are -3 and $\frac{5}{2}$.

Step 3 Locate the boundary points on a number line. The number line with the boundary points is shown as follows:

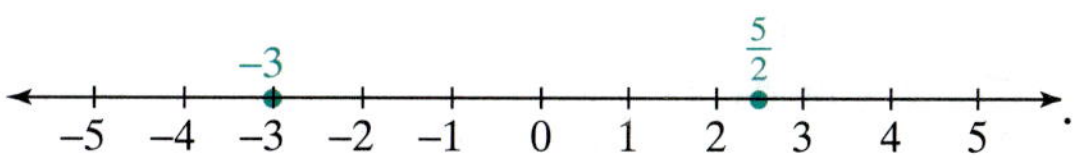

The boundary points divide the number line into three test intervals. Including the boundary points (because of the given greater than or *equal to* sign), the intervals are $(-\infty, -3]$, $[-3, \frac{5}{2}]$, and $[\frac{5}{2}, \infty)$.

Step 4 Take one representative number within each test interval and substitute that number into the original inequality.

Test Interval	Representative Number	Substitute into $2x^2 + x \geq 15$	Conclusion
$(-\infty, -3]$	-4	$2(-4)^2 + (-4) \overset{?}{\geq} 15$ $28 \geq 15$, true	$(-\infty, -3]$ belongs to the solution set.
$\left[-3, \frac{5}{2}\right]$	0	$2 \cdot 0^2 + 0 \overset{?}{\geq} 15$ $0 \geq 15$, false	$\left[-3, \frac{5}{2}\right]$ does not belong to the solution set.
$\left[\frac{5}{2}, \infty\right)$	3	$2 \cdot 3^2 + 3 \overset{?}{\geq} 15$ $21 \geq 15$, true	$\left[\frac{5}{2}, \infty\right)$ belongs to the solution set.

Using Technology

The solution set for

$$2x^2 + x \geq 15$$

or, equivalently,

$$2x^2 + x - 15 \geq 0$$

can be verified with a graphing utility. The graph of $y = 2x^2 + x - 15$ was obtained using a $[-10, 10, 1]$ by $[-16, 6, 1]$ viewing rectangle. The graph lies above or on the x-axis, representing $\geq$, for all x in $(-\infty, -3] \cup [\frac{5}{2}, \infty)$.

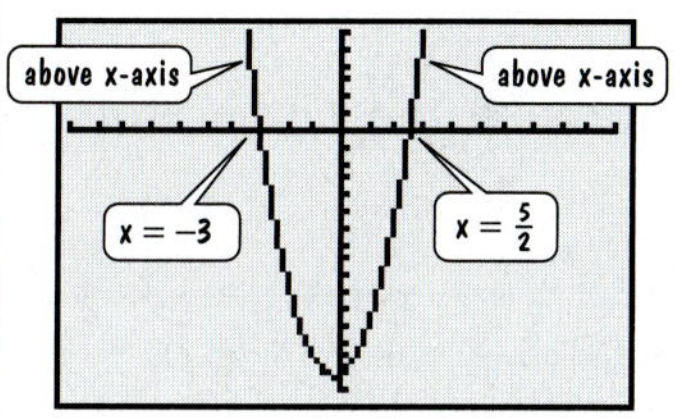

Step 5 The solution set is the union of intervals that produced a true statement. Our analysis on page 597 shows that the solution set is

$$(-\infty, -3] \cup \left[\frac{5}{2}, \infty\right)$$

$$\text{or } \left\{x \mid x \leq -3 \text{ or } x \geq \frac{5}{2}\right\}.$$

The graph of the solution set on a number line is shown as follows:

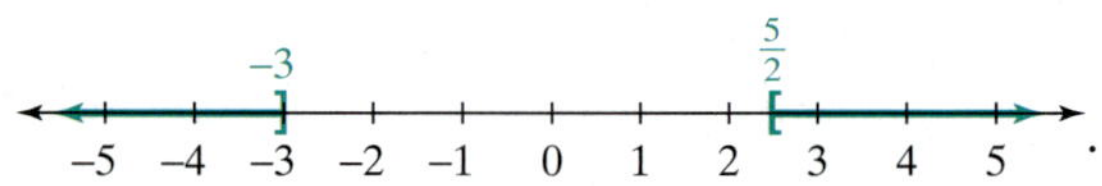

■

✔ **CHECK POINT 2** Solve and graph the solution set: $x^2 - x \geq 20$.

2 Solve rational inequalities.

Solving Rational Inequalities

Inequalities that involve quotients can be solved in the same manner as quadratic inequalities. For example, the inequalities

$$(x+3)(x-7) > 0 \quad \text{and} \quad \frac{x+3}{x-7} > 0$$

are similar in that both are positive under the same conditions. To be positive, each of these inequalities must have two positive linear expressions

$$x + 3 > 0 \quad \text{and} \quad x - 7 > 0$$

or two negative linear expressions

$$x + 3 < 0 \quad \text{and} \quad x - 7 < 0.$$

Consequently, we solve $\frac{x+3}{x-7} > 0$ using boundary points to divide the number line into test intervals. Then we select one representative number in each interval to determine whether that interval belongs to the solution set. Example 3 illustrates how this is done.

Study Tip

Many students want to solve

$$\frac{x+3}{x-7} > 0$$

by first multiplying both sides by $x - 7$ to clear fractions. This is incorrect. The problem is that $x - 7$ contains a variable and can be positive or negative, depending on the value of x. Thus, we do not know whether or not to reverse the sense of the inequality.

EXAMPLE 3 Using Test Numbers to Solve a Rational Inequality

Solve and graph the solution set: $\frac{x+3}{x-7} > 0.$

Solution We begin by finding values of x that make the numerator and denominator 0.

$$x + 3 = 0 \qquad x - 7 = 0$$ *Set the numerator and denominator equal to 0.*

$$x = -3 \qquad x = 7$$ *Solve.*

The boundary points are -3 and 7. We locate these numbers on a number line as follows:

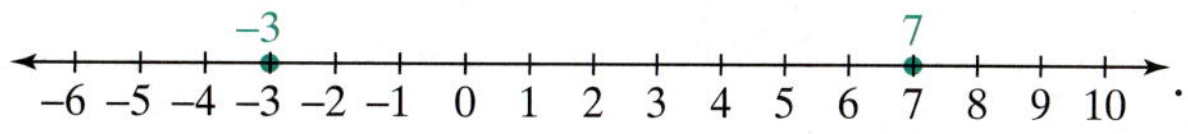

These boundary points divide the number line into three test intervals, namely $(-\infty, -3)$, $(-3, 7)$, and $(7, \infty)$. Now, we take one representative number from each test interval and substitute that number into the original inequality.

Test Interval	**Representative Number**	**Substitute into** $\frac{x+3}{x-7} > 0$	**Conclusion**
$(-\infty, -3)$	-4	$\frac{-4+3}{-4-7} \overset{?}{>} 0$ $\frac{-1}{-11} \overset{?}{>} 0$ $\frac{1}{11} > 0$, true	$(-\infty, -3)$ belongs to the solution set.
$(-3, 7)$	0	$\frac{0+3}{0-7} \overset{?}{>} 0$ $-\frac{3}{7} > 0$, false	$(-3, 7)$ does not belong to the solution set.
$(7, \infty)$	8	$\frac{8+3}{8-7} \overset{?}{>} 0$ $11 > 0$, true	$(7, \infty)$ belongs to the solution set.

Our analysis shows that the solution set is

$$(-\infty, -3) \cup (7, \infty)$$

$$\text{or } \{x \mid x < -3 \text{ or } x > 7\}.$$

The graph of the solution set on a number line is shown as follows:

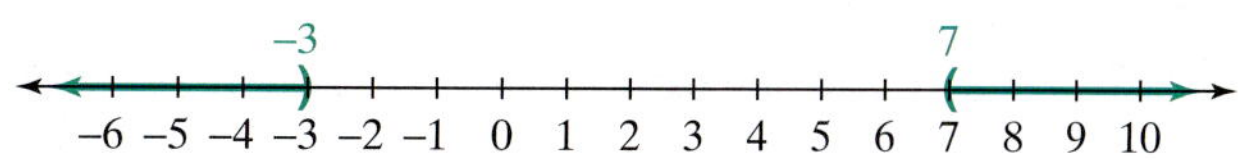

■

CHECK POINT 3 Solve and graph the solution set: $\frac{x-5}{x+2} > 0$.

The first step in solving a rational inequality is to bring all terms to one side, obtaining zero on the other side. Then express the nonzero side as a single quotient. At this point, we follow the same procedure as in Example 3, finding values of the variable that make the numerator and denominator 0.

Study Tip

Do not begin solving

$$\frac{x+1}{x+3} \le 2$$

by multiplying both sides by $x + 3$. We do not know if $x + 3$ is positive or negative. Thus, we do not know whether or not to reverse the sense of the inequality.

EXAMPLE 4 Solving a Rational Inequality

Solve and graph the solution set: $\dfrac{x+1}{x+3} \le 2.$

Solution

Step 1 Express the inequality so that one side is zero and the other side is a single quotient. We subtract 2 from both sides to obtain zero on the right.

$$\frac{x+1}{x+3} \le 2$$ This is the given inequality.

$$\frac{x+1}{x+3} - 2 \le 0$$ Subtract 2 from both sides, obtaining 0 on the right.

$$\frac{x+1}{x+3} - \frac{2(x+3)}{x+3} \le 0$$ The least common denominator is $x + 3$. Express 2 in terms of this denominator.

$$\frac{x+1-2(x+3)}{x+3} \le 0$$ Subtract rational expressions.

$$\frac{x+1-2x-6}{x+3} \le 0$$ Apply the distributive property.

$$\frac{-x-5}{x+3} \le 0$$ Simplify.

Step 2 Find boundary points by setting the numerator and the denominator equal to zero.

$$-x - 5 = 0 \qquad x + 3 = 0$$ Set the numerator and denominator equal to 0. These are the values that make the previous quotient zero or undefined.

$$x = -5 \qquad x = -3$$ Solve for x.

The boundary points are -5 and -3. Because equality is included in the given less-than-or-equal-to symbol, we include the value of x that causes the quotient $\dfrac{-x-5}{x+3}$ to be zero. Thus, -5 is included in the solution set. By contrast, we do not include -3 in the solution set because -3 makes the denominator zero.

Step 3 Locate boundary points on a number line. The number line, with the boundary points, is shown as follows:

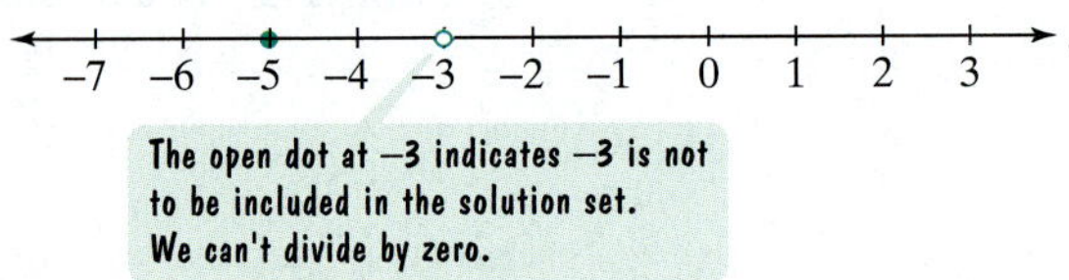

The boundary points divide the number line into three test intervals, namely $(-\infty, -5]$, $[-5, -3)$, and $(-3, \infty)$.

Step 4 Take one representative number within each test interval and substitute that number into the original inequality.

Test Interval	Representative Number	Substitute into $\frac{x+1}{x+3} \le 2$	Conclusion
$(-\infty, -5]$	-6	$\frac{-6+1}{-6+3} \overset{?}{\le} 2$ $\frac{5}{3} \le 2$, true	$(-\infty, -5]$ belongs to the solution set.
$[-5, -3)$	-4	$\frac{-4+1}{-4+3} \overset{?}{\le} 2$ $3 \le 2$, false	$[-5, -3)$ does not belong to the solution set.
$(-3, \infty)$	0	$\frac{0+1}{0+3} \overset{?}{\le} 2$ $\frac{1}{3} \le 2$, true	$(-3, \infty)$ belongs to the solution set.

Discover for Yourself

Because $(x+3)^2$ is positive, it is possible so solve

$$\frac{x+1}{x+3} \le 2$$

by first multiplying both sides by $(x+3)^2$ (where $x \ne -3$). This will not reverse the sense of the inequality and will clear the fraction. Try using this solution method and compare it to the solution on pages 600–601.

Step 5 The solution set consists of the intervals that produced a true statement. Our analysis shows that the solution set is

$$(-\infty, -5] \cup (-3, \infty)$$

$$\text{or } \{x \mid x \le -5 \text{ or } x > -3\}.$$

The graph of the solution set on a number line is shown as follows:

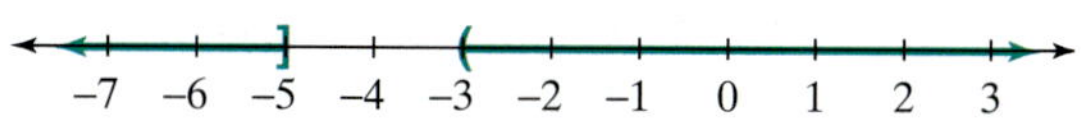

■

CHECK POINT 4 Solve and graph the solution set: $\frac{2x}{x+1} \le 1$.

3 Solve problems using nonlinear inequalities.

Applications

Quadratic inequalities can be solved to answer questions about variables contained in quadratic functions.

EXAMPLE 5 Modeling the Position of a Free-Falling Object

The Leaning Tower of Pisa is 176 feet high. Figure 8.16 on page 602 shows that a ball is thrown vertically upward from the top of the tower. The function

$$s(t) = -16t^2 + 96t + 176$$

models the ball's height above the ground, $s(t)$, in feet, t seconds after it was thrown. During which time period will the ball's height exceed that of the tower?

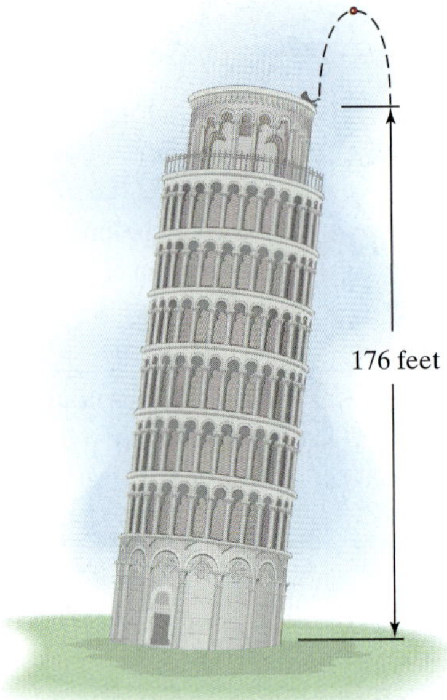

Figure 8.16 Throwing a ball from the top of the Leaning Tower of Pisa

Solution Using the problem's question and the given model for the ball's height, $s(t) = -16t^2 + 96t + 176$, we obtain a quadratic inequality.

When will the ball's height | exceed that | of the tower?

$$-16t^2 + 96t + 176 > 176$$

$$-16t^2 + 96t + 176 > 176 \quad \text{This is the inequality implied by the problem's question. We must find } t.$$

$$-16t^2 + 96t > 0 \quad \text{Subtract 176 from both sides.}$$

$$-16t^2 + 96t = 0 \quad \text{Solve the related quadratic equation.}$$

$$-16t(t - 6) = 0 \quad \text{Factor.}$$

$$-16t = 0 \quad \text{or} \quad t - 6 = 0 \quad \text{Set each factor equal to 0.}$$

$$t = 0 \qquad t = 6 \quad \text{Solve for } t. \text{ The boundary points are 0 and 6.}$$

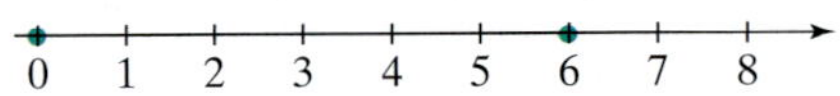

Locate these values on a number line, with $t \geq 0$.

The intervals are $(0, 6)$ and $(6, \infty)$, although the time interval should not extend to infinity but rather to the value of t when the ball hits the ground. (By setting $-16t^2 + 96t + 176$ equal to zero, we find $t \approx 7.47$; the ball hits the ground after approximately 7.47 seconds.)

We use $(0, 6)$ and $(6, 7.47)$ for our test intervals.

Test Interval	Representative Number	Substitute into $-16t^2 + 96t + 176 > 176$	Conclusion
$(0, 6)$	1	$-16 \cdot 1^2 + 96 \cdot 1 + 176 \stackrel{?}{>} 176$ $256 > 176$, true	$(0, 6)$ belongs to the solution set.
$(6, 7.47)$	7	$-16 \cdot 7^2 + 96 \cdot 7 + 176 \stackrel{?}{>} 176$ $64 > 176$, false	$(6, 7.47)$ does not belong to the solution set.

The ball's height exceeds that of the tower between 0 and 6 seconds, excluding $t = 0$ and $t = 6$. ■

CHECK POINT 5 An object is propelled straight up from ground level with an initial velocity of 80 feet per second. Its height at time t is modeled by

$$s(t) = -16t^2 + 80t,$$

where the height, $s(t)$, is measured in feet and the time, t, is measured in seconds. In which time interval will the object be more than 64 feet above the ground?

EXERCISE SET 8.5

Practice Exercises

Solve each quadratic inequality in Exercises 1–26, and graph the solution set on a real number line.

1. $(x - 4)(x + 2) > 0$
2. $(x + 3)(x - 5) > 0$
3. $(x - 7)(x + 3) \le 0$
4. $(x + 1)(x - 7) \le 0$
5. $x^2 - 5x + 4 > 0$
6. $x^2 - 4x + 3 < 0$
7. $x^2 + 5x + 4 > 0$
8. $x^2 + x - 6 > 0$
9. $x^2 - 6x + 8 \le 0$
10. $x^2 - 2x - 3 \ge 0$
11. $3x^2 + 10x - 8 \le 0$
12. $9x^2 + 3x - 2 \ge 0$
13. $2x^2 + x < 15$
14. $6x^2 + x > 1$
15. $4x^2 + 7x < -3$
16. $3x^2 + 16x < -5$
17. $x^2 - 4x \ge 0$
18. $x^2 + 2x < 0$
19. $2x^2 + 3x > 0$
20. $3x^2 - 5x \le 0$
21. $-x^2 + x \ge 0$
22. $-x^2 + 2x \ge 0$
23. $x^2 \le 4x - 2$
24. $x^2 \le 2x + 2$
25. $x^2 - 6x + 9 < 0$
26. $4x^2 - 4x + 1 \ge 0$

Solve each rational inequality in Exercises 27–42, and graph the solution set on a real number line.

27. $\dfrac{x - 4}{x + 3} > 0$
28. $\dfrac{x + 5}{x - 2} > 0$
29. $\dfrac{x + 3}{x + 4} < 0$
30. $\dfrac{x + 5}{x + 2} < 0$
31. $\dfrac{-x + 2}{x - 4} \ge 0$
32. $\dfrac{-x - 3}{x + 2} \le 0$
33. $\dfrac{4 - 2x}{3x + 4} \le 0$
34. $\dfrac{3x + 5}{6 - 2x} \ge 0$
35. $\dfrac{x}{x - 3} > 0$
36. $\dfrac{x + 4}{x} > 0$
37. $\dfrac{x + 1}{x + 3} < 2$
38. $\dfrac{x}{x - 1} > 2$
39. $\dfrac{x + 4}{2x - 1} \le 3$
40. $\dfrac{1}{x - 3} < 1$
41. $\dfrac{x - 2}{x + 2} \le 2$
42. $\dfrac{x}{x + 2} \ge 2$

Application Exercises

43. You throw a ball straight up from a rooftop 160 feet high with an initial speed of 48 feet per second. The function

$$s(t) = -16t^2 + 48t + 160$$

models the ball's height above the ground, $s(t)$, in feet, t seconds after it was thrown. During which time period will the ball's height exceed that of the rooftop?

44. A model rocket is launched from the top of a cliff 80 feet above sea level. The function

$$s(t) = -16t^2 + 64t + 80$$

models the rocket's height above the water, $s(t)$, in feet, t seconds after it was launched. During which time period will the rocket's height exceed that of the cliff?

The bar graph in Figure 8.13 on page 594 shows the cost of Medicare, in billions of dollars, projected through 2005. Using the regression feature of a graphing utility, these data can be modeled by

a linear function, $f(x) = 27x + 163$;

a quadratic function, $g(x) = 1.2x^2 + 15.2x + 181.4$.

In each function, x represents the number of years after 1995. Use these functions to solve Exercises 45–48.

45. The graph indicates that Medicare spending will reach \$379 billion in 2003. Find the amount predicted by each of the functions, f and g, for that year. How well do the functions model the value in the graph?

46. The graph indicates that Medicare spending will reach \$458 billion in 2005. Find the amount predicted by each of the functions, f and g, for that year. How well do the functions model the value in the graph? Which function serves as a better model for that year?

47. For which years does the quadratic model indicate that Medicare spending will exceed \$536.6 billion?

48. For which years does the quadratic model indicate that Medicare spending will exceed \$629.4 billion?

The United States has more people in prison, as well as more people in prison per capita, than any other western industrialized nation. The bar graph shows the number of inmates in U.S. state and federal prisons in seven selected years from 1980 through 1999. The data can be modeled by the function

$$f(x) = 1.3x^2 + 32x + 303,$$

in which $f(x)$ represents the number of inmates, in thousands, x years after 1980. Use this function to solve Exercises 49–52.

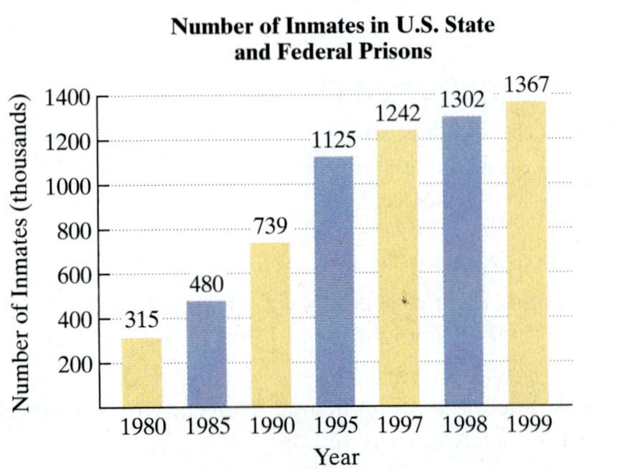

Source: U.S. Justice Department

49. Find the number of inmates in 1998 using the given function. How well does the function model the actual number shown for that year?

50. Find the number of inmates in 1999 using the given function. How well does the function model the actual number shown for that year?

51. For which years does the model indicate that U.S. inmate population will exceed 2433?

52. For which years does the model indicate that U.S. inmate population was no more than 1463 thousand?

A company manufactures wheelchairs. The average cost function, $\bar{C}$, of producing x wheelchairs per month is given by

$$\bar{C}(x) = \frac{500{,}000 + 400x}{x}.$$

The graph of the rational function is shown. Use the function to solve Exercises 53–54.

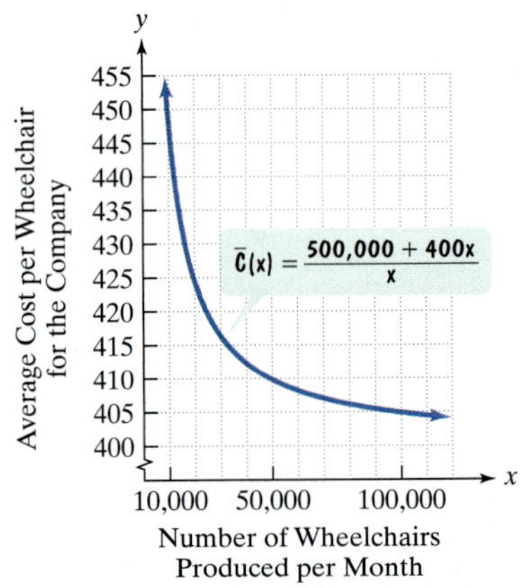

53. Describe the company's production level so that the average cost of producing each wheelchair does not exceed \$425. Use a rational inequality to solve the problem. Then explain how your solution is shown on the graph.

54. Describe the company's production level so that the average cost of producing each wheelchair does not exceed \$410. Use a rational inequality to solve the problem. Then explain how your solution is shown on the graph.

Writing in Mathematics

55. What is a quadratic inequality?

56. What is a rational inequality?

57. Describe similarities and differences between the solutions of

$$(x-2)(x+5) \geq 0 \quad \text{and} \quad \frac{x-2}{x+5} \geq 0.$$

58. What explanations can you offer for the trend shown in the bar graph for Exercises 49–52? Why do you think that the United States has more people in prison than any other industrialized nation?

Technology Exercises

Solve each inequality in Exercises 59–64 using a graphing utility.

59. $x^2 + 3x - 10 > 0$

60. $2x^2 + 5x - 3 \le 0$

61. $\dfrac{x - 4}{x - 1} \le 0$

62. $\dfrac{x + 2}{x - 3} \le 2$

63. $\dfrac{1}{x + 1} \le \dfrac{2}{x + 4}$

64. $x^3 + 2x^2 - 5x - 6 > 0$

Critical Thinking Exercises

65. Which one of the following is true?

a. The solution set to $x^2 > 25$ is $(5, \infty)$.

b. The inequality $\dfrac{x - 2}{x + 3} < 2$ can be solved by multiplying both sides by $x + 3$, resulting in the equivalent inequality $x - 2 < 2(x + 3)$.

c. $(x + 3)(x - 1) \ge 0$ and $\dfrac{x + 3}{x - 1} \ge 0$ have the same solution set.

d. None of the above statements is true.

66. Write a quadratic inequality whose solution set is $[-3, 5]$.

67. Write a rational inequality whose solution set is $(-\infty, -4) \cup [3, \infty)$.

In Exercises 68–71, use inspection to describe each inequality's solution set. Do not solve any of the inequalities.

68. $(x - 2)^2 > 0$

69. $(x - 2)^2 \le 0$

70. $(x - 2)^2 < -1$

71. $\dfrac{1}{(x - 2)^2} > 0$

72. The graphing calculator screen shows the graph of $y = 4x^2 - 8x + 7$.

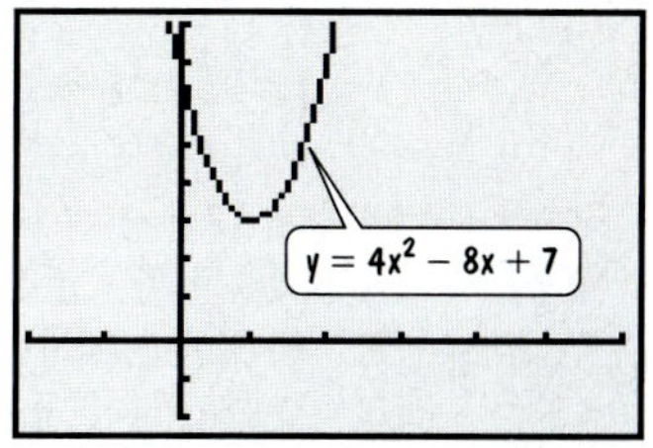

a. Use the graph to describe the solution set for $4x^2 - 8x + 7 > 0$.

b. Use the graph to describe the solution set for $4x^2 - 8x + 7 < 0$.

c. Use an algebraic approach to verify each of your descriptions in parts (a) and (b).

73. The graphing calculator screen shows the graph of $y = \sqrt{27 - 3x^2}$. Write and solve a quadratic inequality that explains why the graph only appears for $-3 \le x \le 3$.

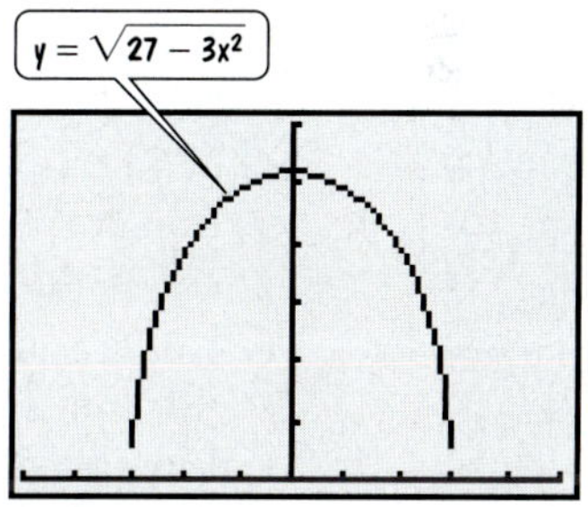

Review Exercises

74. Solve: $\left|\dfrac{x - 5}{3}\right| < 8$. (Section 4.3, Example 3)

75. Divide:

$$\frac{2x + 6}{x^2 + 8x + 16} \div \frac{x^2 - 9}{x^2 + 3x - 4}.$$

(Section 6.1, Example 7)

76. Factor completely: $x^4 - 16y^4$. (Section 5.5, Example 3)

CHAPTER 8 GROUP PROJECTS

1. Each group member should consult an almanac, newspaper, magazine, or the Internet to find data that can be modeled by a quadratic function. Group members should select the two sets of data that are most interesting and relevant. For each data set selected:

a. Use the quadratic regression feature of a graphing utility to find the quadratic function that best fits the data.

(*continues on next page*)

b. Use the equation of the quadratic function to make a prediction from the data. What circumstances might affect the accuracy of your prediction?

c. Use the equation of the quadratic function to write and solve a problem involving maximizing or minimizing the function.

2. Throughout the chapter, we have considered functions that model the position of free-falling objects. Any object that is falling, or vertically projected into the air, has its height above the ground, $s(t)$, in feet, after t seconds in motion, modeled by the quadratic function

$$s(t) = -16t^2 + v_o t + s_o,$$

where v_o is the original velocity (initial velocity) of the object, in feet per second, and s_o is the original height (initial height) of the object, in feet, above the ground. In this exercise, group members will be working with this position function. The exercise is appropriate for groups of three to five people.

a. Drop ball from a height of 3 feet, 6 feet, and 12 feet. Record the number of seconds it takes for the ball to hit the ground.

b. For each of the three initial positions, use the position function to determine the time required for the ball to hit the ground.

c. What factors might result in differences between the times that you recorded and the times indicated by the function?

d. What appears to be happening to the time required for a free-falling object to hit the ground as its initial height is doubled? Verify this observation algebraically and with a graphing utility.

e. Repeat part (a) using a sheet of paper rather than a ball. What differences do you observe? What factor seems to be ignored in the position function?

f. What is meant by the acceleration of gravity and how does this number appear in the position function for a free-falling object?

3. Members of your group act as the marketing research division for a large corporation. Your division estimates that at a price of x dollars per unit, the weekly cost, C, and revenue, R, in thousands of dollars, will be given by the functions

$$C(x) = 14 - x$$

$$\text{and} \quad R(x) = 8x - x^2.$$

Begin by finding a price range for x that describes when a profit will result. (A profit will result if revenue is greater than cost.) Once you find the price range, decide on the kind of product manufactured and sold by the corporation that will be consistent with this price range. Once the product is determined, present a written report to the corporation, heading the report as "Profit Analysis." Include the following in your report:

a. The price range for x, describing when a profit will result

b. A price that will result in the maximum profit

c. Graphs that the corporation might find helpful

d. Factors that might change the cost and revenue functions in the future (This will depend on the product chosen.)

e. A discussion of the possible limitations of the given functions

f. Anything else that you think might be helpful.

CHAPTER SUMMARY, REVIEW, AND TEST

SUMMARY

DEFINITIONS AND CONCEPTS | **EXAMPLES**

Section 8.1 The Square Root Property and Completing the Square

The Square Root Property

If u is an algebraic expression and d is a real number, then

$$\text{If } u^2 = d, \quad \text{then} \quad u = \sqrt{d} \quad \text{or} \quad u = -\sqrt{d}.$$

Equivalently,

$$\text{If } u^2 = d, \quad \text{then} \quad u = \pm\sqrt{d}.$$

Solve: $(x - 6)^2 = 50.$

$$x - 6 = \pm\sqrt{50}$$
$$x - 6 = \pm\sqrt{25 \cdot 2}$$
$$x - 6 = \pm 5\sqrt{2}$$
$$x = 6 \pm 5\sqrt{2}$$

The solutions are $6 \pm 5\sqrt{2}$ and the solution set is $\{6 \pm 5\sqrt{2}\}$.

Completing the Square

If $x^2 + bx$ is a binomial, then by adding $\left(\frac{b}{2}\right)^2$, the square of half the coefficient of x, a perfect square trinomial will result. That is,

$$x^2 + bx + \left(\frac{b}{2}\right)^2 = \left(x + \frac{b}{2}\right)^2.$$

Complete the square: $x^2 + \frac{2}{7}x.$

Half of $\frac{2}{7}$ is $\frac{1}{2} \cdot \frac{2}{7} = \frac{1}{7}$ and $\left(\frac{1}{7}\right)^2 = \frac{1}{49}$.

$$x^2 + \frac{2}{7}x + \frac{1}{49} = \left(x + \frac{1}{7}\right)^2$$

Solving Quadratic Equations by Completing the Square

1. If the coefficient of x^2 is not 1, divide both sides by this coefficient.
2. Isolate variable terms on one side.
3. Complete the square by adding the square of half the coefficient of x to both sides.
4. Factor the perfect square trinomial.
5. Solve by applying the square root property.

Solve by completing the square:

$$2x^2 + 12x - 4 = 0.$$
$$\frac{2x^2}{2} + \frac{12x}{2} - \frac{4}{2} = \frac{0}{2} \qquad \text{Divide by 2.}$$
$$x^2 + 6x - 2 = 0 \qquad \text{Simplify.}$$
$$x^2 + 6x = 2 \qquad \text{Add 2.}$$

The coefficient of x is 6. Half of 6 is 3 and $3^2 = 9$. Add 9 to both sides.

$$x^2 + 6x + 9 = 2 + 9$$
$$(x + 3)^2 = 11$$
$$x + 3 = \pm\sqrt{11}$$
$$x = -3 \pm \sqrt{11}$$

Section 8.2 The Quadratic Formula

The solutions of a quadratic equation in standard form

$$ax^2 + bx + c = 0, \quad a \neq 0,$$

are given by the quadratic formula

$$x = \frac{-b \pm \sqrt{b^2 - 4ac}}{2a}.$$

Solve using the quadratic formula:

$$2x^2 + 4x = 5.$$

First write the equation in standard form by subtracting 5 from both sides.

$$2x^2 + 4x - 5 = 0$$

$a = 2$ $b = 4$ $c = -5$

$$x = \frac{-4 \pm \sqrt{4^2 - 4 \cdot 2(-5)}}{2 \cdot 2} = \frac{-4 \pm \sqrt{16 - (-40)}}{4}$$
$$= \frac{-4 \pm \sqrt{56}}{4} = \frac{-4 \pm \sqrt{4 \cdot 14}}{4} = \frac{-4 \pm 2\sqrt{14}}{4}$$
$$= \frac{\cancel{2}(-2 \pm \sqrt{14})}{\cancel{2} \cdot 2} = \frac{-2 \pm \sqrt{14}}{2}$$

DEFINITIONS AND CONCEPTS	EXAMPLES

Section 8.2 The Quadratic Formula (continued)

The Discriminant

The discriminant, $b^2 - 4ac$, of the quadratic equation $ax + bx + c = 0$ determines the number and type of solutions.

Discriminant	Solutions
Positive perfect square,with a, b, and c rational numbers	2 rational solutions
Positive and not a perfect square	2 irrational solutions
Zero, with a, b, and c rational numbers	1 rational solution
Negative	2 imaginary solutions

- $2x^2 - 7x - 4 = 0$

$a = 2$ $b = -7$ $c = -4$

$$b^2 - 4ac = (-7)^2 - 4(2)(-4)$$
$$= 49 - (-32) = 49 + 32 = 81$$

Positive perfect square

The equations has 2 rational solutions.

Writing Quadratic Equations from Solutions

The zero-product principle in reverse makes it possible to write a quadratic equation from solutions:

$$\text{If } A = 0 \quad \text{or} \quad B = 0, \quad \text{then} \quad AB = 0.$$

Write a quadratic equation with the solution set $\{-2\sqrt{3}, 2\sqrt{3}\}$.

$$x = -2\sqrt{3} \quad \text{or} \quad x = 2\sqrt{3}$$
$$x + 2\sqrt{3} = 0 \quad \text{or} \quad x - 2\sqrt{3} = 0$$
$$(x + 2\sqrt{3})(x - 2\sqrt{3}) = 0$$
$$x^2 - (2\sqrt{3})^2 = 0$$
$$x^2 - 12 = 0$$

Section 8.3 Quadratic Functions and Their Graphs

The graph of the quadratic function

$$f(x) = a(x - h)^2 + k, \quad a \neq 0,$$

is called a parabola. The vertex, or turning point, is (h, k). The graph opens upward if a is positive and downward if a is negative. The axis of symmetry is a vertical line passing through the vertex. The graph can be obtained using the vertex, x-intercepts, if any, (set $f(x)$ equal to zero), and the y-intercept (set $x = 0$).

Graph: $f(x) = -(x + 3)^2 + 1$.

$$f(x) = -1(x - (-3))^2 + 1$$

$a = -1$ $h = -3$ $k = 1$

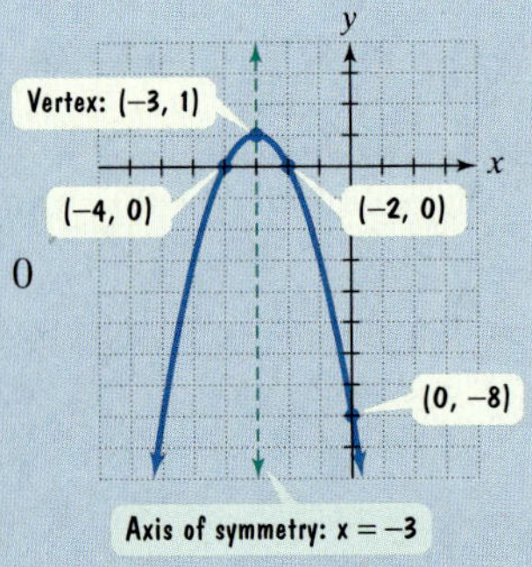

- Vertex: $(h, k) = (-3, 1)$.
- Opens downward because $a < 0$
- x-intercepts: Set $f(x) = 0$.

$$0 = -(x + 3)^2 + 1$$
$$(x + 3)^2 = 1$$
$$x + 3 = \pm\sqrt{1}$$
$$x + 3 = 1 \quad \text{or} \quad x + 3 = -1$$
$$x = -2 \qquad\qquad x = -4$$

- y-intercept: Set $x = 0$.

$$f(0) = -(0 + 3)^2 + 1 = -9 + 1 = -8$$

DEFINITIONS AND CONCEPTS	EXAMPLES

Section 8.3 Quadratic Functions and Their Graphs (continued)

A parabola whose equation is in the form

$$f(x) = ax^2 + bx + c, \quad a \neq 0,$$

has its vertex at

$$\left(-\frac{b}{2a}, f\left(-\frac{b}{2a}\right)\right).$$

The parabola is graphed as described in the left column at the bottom of page 608. The only difference is how we determine the vertex. If $a > 0$, then f has a minimum that occurs at $x = -\frac{b}{2a}$. If $a < 0$, then f has a maximum that occurs at $x = -\frac{b}{2a}$.

Graph: $f(x) = x^2 - 6x + 5$.

$a = 1$ $b = -6$ $c = 5$

- Vertex: $x = -\frac{b}{2a} = -\frac{-6}{2 \cdot 1} = 3$

  $f(3) = 3^2 - 6 \cdot 3 + 5 = -4$

  Vertex is at $(3, -4)$.
- Opens upward because $a > 0$.
- x-intercepts: Set $f(x) = 0$.

$$x^2 - 6x + 5 = 0$$
$$(x - 1)(x - 5) = 0$$
$$x = 1 \quad \text{or} \quad x = 5$$

- y-intercept: Set $x = 0$.

$$f(0) = 0^2 - 6 \cdot 0 + 5 = 5$$

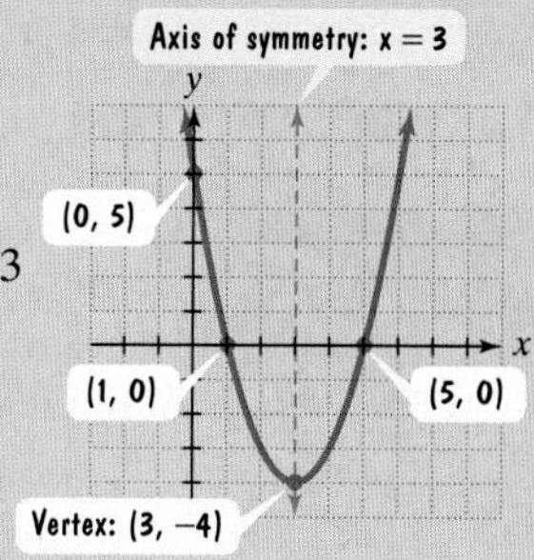

Section 8.4 Equations Quadratic in Form

An equation that is quadratic in form is one that can be expressed as a quadratic equation using an appropriate substitution. These equations contain an expression to a power, the same expression to that power squared, and a constant term. Let $t =$ the expression to the power. If at any point in the solution process both sides of an equation are raised to an even power, a check is required.

Solve: $x^{\frac{2}{3}} - 3x^{\frac{1}{3}} + 2 = 0$.

$$(x^{\frac{1}{3}})^2 - 3x^{\frac{1}{3}} + 2 = 0$$

Let $t = x^{\frac{1}{3}}$.

$$t^2 - 3t + 2 = 0$$
$$(t - 1)(t - 2) = 0$$
$$t - 1 = 0 \quad \text{or} \quad t - 2 = 0$$
$$t = 1 \qquad\qquad t = 2$$
$$x^{\frac{1}{3}} = 1 \qquad\qquad x^{\frac{1}{3}} = 2$$
$$(x^{\frac{1}{3}})^3 = 1^3 \qquad\qquad (x^{\frac{1}{3}})^3 = 2^3$$
$$x = 1 \qquad\qquad x = 8$$

The solutions are 1 and 8, and the solution set is $\{1, 8\}$.

Section 8.5 Quadratic and Rational Inequalities

Solving Quadratic Inequalities

1. Write the inequality in standard form:

   $ax^2 + bx + c > 0$ or $ax^2 + bx + c < 0$.
2. Solve the related quadratic equation.
3. Use solutions to locate boundary points on a number line, dividing the line into test intervals.
4. Substitute a test point from each interval into the inequality. If a true statement results, the interval belongs to the solution set.
5. The solution set is the union of all such intervals.

This procedure is valid if $<$ is replaced by $\leq$ and $>$ is replaced by $\geq$. In these cases, include the boundary points in the solution set.

Solve: $2x^2 + x - 6 > 0$.

$$2x^2 + x - 6 = 0$$
$$(2x - 3)(x + 2) = 0$$
$$2x - 3 = 0 \quad \text{or} \quad x + 2 = 0$$
$$x = \frac{3}{2} \qquad\qquad x = -2$$

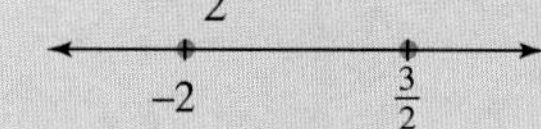

$x = -3$ makes the inequality true,
$x = 0$ makes it false, and
$x = 2$ makes it true.
The solution set is $\{x | x < -2 \text{ or } x > \frac{3}{2}\}$ or $(-\infty, -2) \cup (\frac{3}{2}, \infty)$.

DEFINITIONS AND CONCEPTS | **EXAMPLES**

Section 8.5 Quadratic and Rational Inequalities (continued)

Solving Rational Inequalities

1. Express the inequality so that one side is zero and the other side is a single quotient.
2. Find boundary points by setting the numerator and denominator equal to zero.
3. Locate boundary points on a number line, dividing the line into test intervals.
4. Substitute a test point from each interval into the inequality. If a true statement results, the interval belongs to the solution set.
5. The solution set is the union of all such intervals. Exclude any values that make the denominator zero.

Solve: $\frac{x}{x+4} \geq 2.$

$$\frac{x}{x+4} - \frac{2(x+4)}{x+4} \leq 0$$

$$\frac{-x-8}{x+4} \leq 0$$

$$-x - 8 = 0 \qquad x + 4 = 0$$

$$-8 = x \qquad x = -4$$

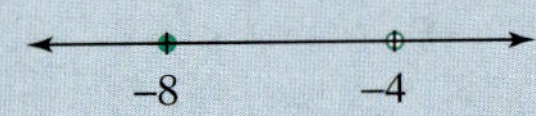

$x = -9$ makes the original inequality false,

$x = -7$ makes it true, and

$x = -3$ makes it false.

The solution set is $\{x|-8 \leq x < -4\}$ or $[-8, -4)$.

Review Exercises

8.1 *In Exercises 1–5, solve each equation by the square root property. If possible, simplify radicals or rationalize denominators. Express imaginary solutions in the form $a + bi$.*

1. $2x^2 - 3 = 125$

2. $3x^2 - 150 = 0$

3. $3x^2 - 2 = 0$

4. $(x - 4)^2 = 18$

5. $(x + 7)^2 = -36$

In Exercises 6–7, complete the square for each binomial. Then factor the resulting perfect square trinomial.

6. $x^2 + 20x$

7. $x^2 - 3x$

In Exercises 8–10, solve each quadratic equation by completing the square.

8. $x^2 - 12x + 27 = 0$

9. $x^2 - 7x - 1 = 0$

10. $2x^2 + 3x - 4 = 0$

11. In 2 years, an investment of \$2500 grows to \$2916. Use the compound interest formula

$$A = P(1 + r)^t$$

to find the annual interest rate, r.

12. The function $W(t) = 3t^2$ models the weight of a human fetus, $W(t)$, in grams, after t weeks, where $0 \leq t \leq 39$. After how many weeks does the fetus weigh 1200 grams?

13. A building casts a shadow that is double the length of the building's height. If the distance from the end of the shadow to the top of the building is 300 meters, how high is the building? Express the answer in simplified radical form. Then find a decimal approximation to the nearest tenth of a meter.

8.2 *In Exercises 14–16, solve each equation using the quadratic formula. Simplify solutions, if possible.*

14. $x^2 = 2x + 4$

15. $x^2 - 2x + 19 = 0$

16. $2x^2 = 3 - 4x$

In Exercises 17–19, compute the discriminant. Then determine the number and type of solutions for the given equation.

17. $x^2 - 4x + 13 = 0$

18. $9x^2 = 2 - 3x$

19. $2x^2 + 4x = 3$

In Exercises 20–26, solve each equation by the method of your choice. Simplify solutions, if possible.

20. $2x^2 - 11x + 5 = 0$

21. $(3x + 5)(x - 3) = 5$

22. $3x^2 - 7x + 1 = 0$

23. $x^2 - 9 = 0$

24. $(x - 3)^2 - 8 = 0$

25. $3x^2 - x + 2 = 0$

26. $\frac{5}{x+1} + \frac{x-1}{4} = 2$

In Exercises 27–29, write a quadratic equation in standard form with the given solution set.

27. $\left\{-\frac{1}{3}, \frac{3}{5}\right\}$

28. $\{-9i, 9i\}$

29. $\{-4\sqrt{3}, 4\sqrt{3}\}$

30. As the use of the Internet increases, so has the number of computer infections from viruses. The function

$$N(x) = 0.2x^2 - 1.2x + 2$$

models the number of infections per month for every 1000 PCs, $N(x)$, x years after 1990. In which year will the infection rate be 13 PCs per month for every 1000 PCs?

31. A baseball is hit by a batter. The function

$$s(t) = -16t^2 + 140t + 3$$

models the ball's height above the ground, $s(t)$, in feet, t seconds after it was hit. How long will it take for the ball to strike the ground? Round to the nearest tenth of a second.

8.3 *In Exercises 32–35, use the vertex and intercepts to sketch the graph of each quadratic function. Give the equation for the parabola's axis of symmetry.*

32. $f(x) = -(x + 1)^2 + 4$

33. $f(x) = (x + 4)^2 - 2$

34. $f(x) = -x^2 + 2x + 3$

35. $f(x) = 2x^2 - 4x - 6$

36. The function

$$f(x) = -0.02x^2 + x + 1$$

models the yearly growth of a young redwood tree, $f(x)$, in inches, with x inches of rainfall per year. How many inches of rainfall per year results in maximum tree growth? What is the maximum yearly growth?

37. A model rocket is launched upward from a platform 30 feet above the ground. The quadratic function

$$s(t) = -16t^2 + 400t + 40$$

models the rocket's height above the ground, $s(t)$, in feet, t seconds after it was launched. After how many seconds does the rocket reach its maximum height? What is the maximum height?

38. Suppose that a quadratic function is used to model the data shown in the graph using

(number of years after 1960, divorce rate per 1000 population).

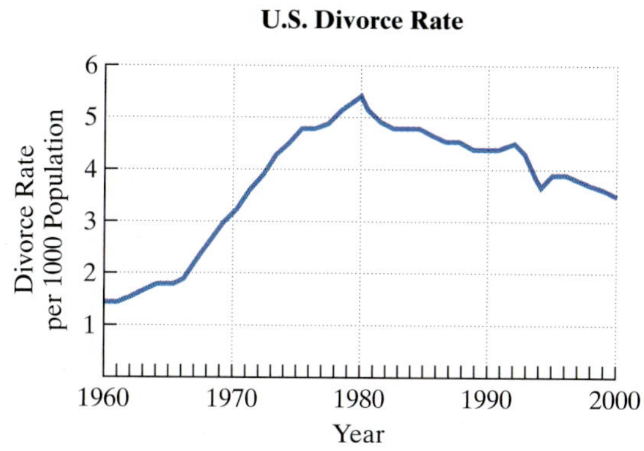

Source: U.S. Department of Health and Human Services

Determine, without obtaining an actual quadratic function that models the data, the approximate coordinates of the vertex for the function's graph. Describe what this means in practical terms.

39. A field bordering a straight stream is to be enclosed. The side bordering the stream is not to be fenced. If 1000 yards of fencing material is to be used, what are the dimensions of the largest rectangular field that can be fenced? What is the maximum area?

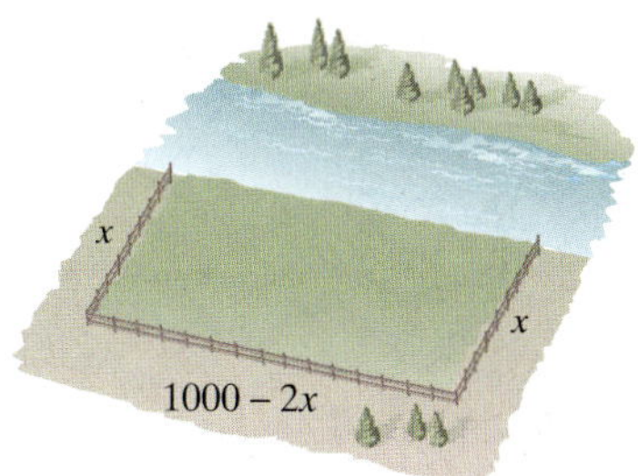

40. Among all pairs of numbers whose difference is 14, find a pair whose product is as small as possible. What is the minimum product?

8.4 *In Exercises 41–46, solve each equation by making an appropriate substitution. When necessary, check proposed solutions.*

41. $x^4 - 6x^2 + 8 = 0$

42. $x + 7\sqrt{x} - 8 = 0$

43. $(x^2 + 2x)^2 - 14(x^2 + 2x) = 15$

44. $x^{-2} + x^{-1} - 56 = 0$

45. $x^{\frac{2}{3}} - x^{\frac{1}{3}} - 12 = 0$

46. $x^{\frac{1}{2}} + 3x^{\frac{1}{4}} - 10 = 0$

8.5 *In Exercises 47–50, solve each inequality and graph the solution set on a real number line.*

47. $2x^2 + 5x - 3 < 0$

48. $2x^2 + 9x + 4 \geq 0$

49. $\frac{x-6}{x+2} > 0$

50. $\frac{x+3}{x-4} \leq 5$

51. A model rocket is launched from ground level. The function

$$s(t) = -16t^2 + 48t$$

models the rocket's height above the ground, $s(t)$, in feet, t seconds after it was launched. During which time period will the rocket's height exceed 32 feet?

52. The function

$$H(x) = \frac{15}{8}x^2 - 30x + 200$$

models heart rate, $H(x)$, in beats per minute, x minutes after a strenuous workout.

a. What is the heart rate immediately following the workout?

b. According to the model, during which intervals of time after the strenuous workout does the heart rate exceed 110 beats per minute? For which of these intervals has model breakdown occurred? Which interval provides a more realistic answer? How did you determine this?

Chapter 8 Test

Express solutions to all equations in simplified form. Rationalize denominators, if possible.

In Exercises 1–2, solve each equation by the square root property.

1. $2x^2 - 5 = 0$

2. $(x - 3)^2 = 20$

In Exercises 3–4, complete the square for each binomial. Then factor the resulting perfect square trinomial.

3. $x^2 - 16x$

4. $x^2 + \frac{2}{5}x$

5. Solve by completing the square: $x^2 - 6x + 7 = 0$.

6. Use the measurements determined by the surveyor to find the width of the pond. Express the answer in simplified radical form.

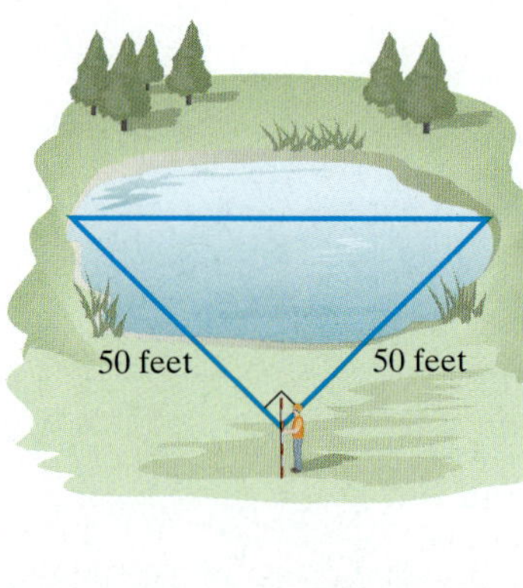

In Exercises 7–8, compute the discriminant. Then determine the number and type of solutions for the given equation.

7. $3x^2 + 4x - 2 = 0$

8. $x^2 = 4x - 8$

In Exercises 9–12, solve each equation by the method of your choice.

9. $2x^2 + 9x = 5$

10. $x^2 + 8x + 5 = 0$

11. $(x + 2)^2 + 25 = 0$

12. $2x^2 - 6x + 5 = 0$

In Exercises 13–14, write a quadratic equation in standard form with the given solution set.

13. $\{-3, 7\}$

14. $\{-10i, 10i\}$

15. The function

$$f(x) = -0.5x^2 + 4x + 19$$

models the number of people in the United States, $f(x)$, in millions, receiving food stamps x years after 1990. In which year(s) were 20 million people receiving food stamps? Use a calculator and round to the nearest year(s).

In Exercises 16–17, use the vertex and intercepts to sketch the graph of each quadratic function. Give the equation for the parabola's axis of symmetry.

16. $f(x) = (x + 1)^2 + 4$

17. $f(x) = x^2 - 2x - 3$

A baseball player hits a pop fly into the air. The function

$$s(t) = -16t^2 + 64t + 5$$

models the ball's height above the ground, $s(t)$, in feet, t seconds after it was hit. Use the function to solve Exercises 18–19.

18. When does the baseball reach its maximum height? What is that height?

19. After how many seconds does the baseball hit the ground? Round to the nearest tenth of a second.

20. The function $f(x) = -x^2 + 46x - 360$ models the daily profit, $f(x)$, in hundreds of dollars, for a company that manufactures x computers daily. How many computers should be manufactured each day to maximize profit? What is the maximum daily profit?

In Exercises 21–23, solve each equation by making an appropriate substitution. When necessary, check proposed solutions.

21. $(2x - 5)^2 + 4(2x - 5) + 3 = 0$

22. $x^4 - 13x^2 + 36 = 0$

23. $x^{\frac{2}{3}} - 9x^{\frac{1}{3}} + 8 = 0$

In Exercises 24–25, solve each inequality and graph the solution set on a real number line.

24. $x^2 - x - 12 < 0$

25. $\dfrac{2x + 1}{x - 3} \leq 3$

Cumulative Review Exercises (Chapters 1–8)

In Exercises 1–7, solve each equation, inequality, or system.

1. $8 - (4x - 5) = x - 7$

2. $\begin{aligned} 5x + 4y &= 22 \\ 3x - 8y &= -18 \end{aligned}$

3. $\begin{aligned} -3x + 2y + 4z &= 6 \\ 7x - y + 3z &= 23 \\ 2x + 3y + z &= 7 \end{aligned}$

4. $|x - 1| > 3$

5. $\sqrt{x + 4} - \sqrt{x - 4} = 2$

6. $x - 4 \geq 0$ and $-3x \leq -6$

7. $2x^2 = 3x - 2$

In Exercises 8–12, graph each function, equation, or inequality in a rectangular coordinate system.

8. $3x = 15 + 5y$

9. $2x - 3y > 6$

10. $f(x) = -\frac{1}{2}x + 1$

11. $f(x) = x^2 + 6x + 8$

12. $f(x) = (x - 3)^2 - 4$

13. Evaluate:

$$\begin{vmatrix} 3 & 1 & 0 \\ 0 & 5 & -6 \\ -2 & -1 & 0 \end{vmatrix}.$$

14. Solve for c: $A = \dfrac{cd}{c + d}$.

15. Write the equation of the linear function whose graph contains the point $(-2, 4)$ and is perpendicular to the line whose equation is $2x + y = 10$.

In Exercises 16–20, perform the indicated operations and simplify, if possible.

16. $\dfrac{-5x^3y^7}{15x^4y^{-2}}$

17. $(4x^2 - 5y)^2$

18. $(5x^3 - 24x^2 + 9) \div (5x + 1)$

19. $\dfrac{\sqrt[3]{32xy^{10}}}{\sqrt[3]{2xy^2}}$

20. $\dfrac{x + 2}{x^2 - 6x + 8} + \dfrac{3x - 8}{x^2 - 5x + 6}$

In Exercises 21–22, factor completely.

21. $x^4 - 4x^3 + 8x - 32$

22. $2x^2 + 12xy + 18y^2$

23. The length of a rectangular carpet is 4 feet greater than twice its width. If the area is 48 square feet, find the carpet's length and width.

24. Working alone, you can mow the lawn in 2 hours, and your sister can do it in 3 hours. How long will it take you to do the job if you work together?

25. Your motorboat can travel 15 miles per hour in still water. Traveling with the river's current, the boat can cover 20 miles in the same time it takes to go 10 miles against the current. Find the rate of the current.

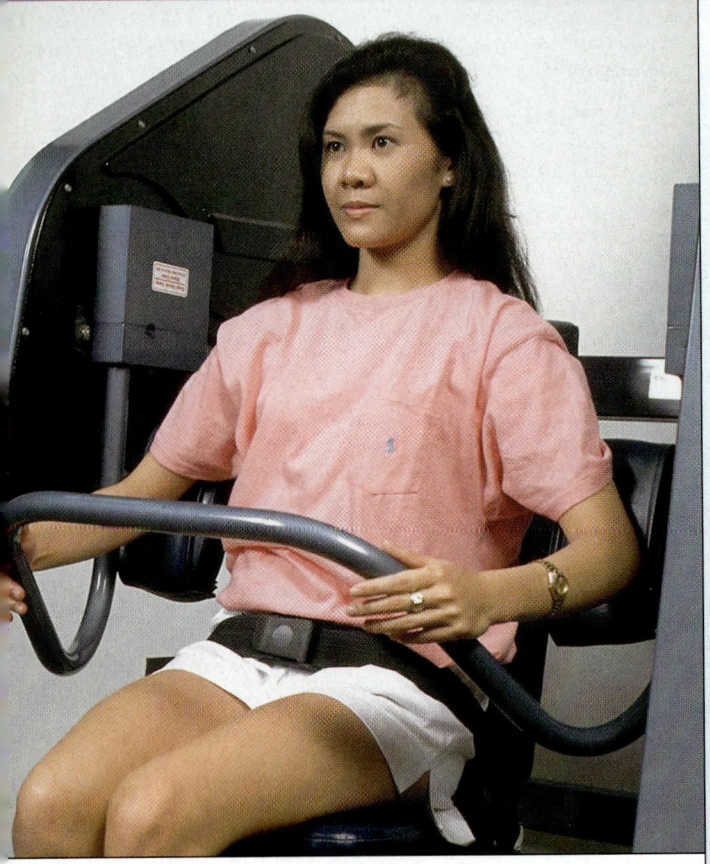

You've recently taken up weightlifting, recording the maximum number of pounds you can lift at the end of each week. At first your weight limit increases rapidly, but now you notice that this growth is beginning to level off. You wonder about a function that would serve as a mathematical model to predict the number of pounds you can lift as you continue the sport.

What went wrong on the space shuttle *Challenger*? Will population growth lead to a future without comfort or individual choice? Can I put aside a small amount of money and have millions for early retirement? Why did I feel I was walking too slowly on my visit to New York City? Why are people in California at far greater risk from drunk drivers than from earthquakes? What is the difference between earthquakes measuring 6 and 7 on the Richter scale? And what can I hope to accomplish in weightlifting?

The functions that you will be learning about in this chapter will provide you with the mathematics for answering these questions. You will see how these remarkable functions enable us to predict the future and rediscover the past.

Chapter 9

Exponential and Logarithmic Functions

▸ SECTION 9.1 Exponential Functions

Objectives

1. Evaluate exponential functions.
2. Graph exponential functions.
3. Evaluate functions with base e.
4. Use compound interest formulas.

SSM PH Tutor Center CD-ROM Video

The space shuttle *Challenger* exploded approximately 73 seconds into flight on January 28, 1986. The tragedy involved damage to O-rings, which were used to seal the connections between different sections of the shuttle engines. The number of O-rings damaged increases dramatically as Fahrenheit temperature falls.

The function

$$f(x) = 13.49(0.967)^x - 1$$

models the number of O-rings expected to fail when the temperature is x°F. Can you see how this function is different from polynomial functions? The variable x is in the exponent. Functions whose equations contain a variable in the exponent are called **exponential functions**. Many real-life situations, including population growth, growth of epidemics, radioactive decay, and other changes that involve rapid increase or decrease can be described using exponential functions.

> **Definition of the Exponential Function**
>
> The **exponential function f with base b** is defined by
>
> $$f(x) = b^x \quad \text{or} \quad y = b^x$$
>
> where b is a positive constant other than $1 (b > 0 \text{ and } b \neq 1)$ and x is any real number.

Here are some examples of exponential functions.

$$f(x) = 2^x \qquad g(x) = 10^x \qquad h(x) = 3^{x+1}$$

Base is 2. — Base is 10. — Base is 3.

Each of these functions has a constant base and a variable exponent. By contrast the following functions are not exponential.

$$F(x) = x^2 \qquad G(x) = 1^x \qquad H(x) = x^x$$

Variable is the base and not exponent. — The base of an exponential function must be a positive constant other than 1. — Variable is both the base and the exponent.

Why is $G(x) = 1^x$ not classified as an exponential function? The number 1 raised to any power is 1. Thus, the function G can be written as $G(x) = 1$, which is a constant function.

1 Evaluate exponential functions.

You will need a calculator to evaluate exponential expressions. Most scientific calculators have an $\boxed{x^y}$ key. Graphing calculators have a $\boxed{\wedge}$ key. To evaluate expressions of the form b^x, enter the base b, press $\boxed{x^y}$ or $\boxed{\wedge}$, enter the exponent x, and finally press $\boxed{=}$ or $\boxed{\text{ENTER}}$.

EXAMPLE 1 Evaluating an Exponential Function

The exponential function $f(x) = 13.49(0.967)^x - 1$ describes the number of O-rings expected to fail, $f(x)$, when the temperature is x°F. On the morning the *Challenger* was launched, the temperature was 31°F, colder than any previous experience. Find the number of O-rings expected to fail at this temperature.

Solution Because the temperature was 31°F, substitute 31 for x and evaluate the function at 31.

$$f(x) = 13.49(0.967)^x - 1 \quad \text{This is the given function.}$$

$$f(31) = 13.49(0.967)^{31} - 1 \quad \text{Substitute 31 for } x.$$

Use a scientific or graphing calculator to evaluate $(0.967)^{31}$. Press the following keys on your calculator to do this:

Scientific calculator: .967 $\boxed{x^y}$ 31 $\boxed{=}$

Graphing calculator: .967 $\boxed{\wedge}$ 31 $\boxed{\text{ENTER}}$.

The display should be approximately .353362693426. Multiplying this number by 13.49 and subtracting 1, we obtain

$$f(31) = 13.49(0.967)^{31} - 1 \approx 4.$$

Thus, four O-rings are expected to fail at a temperature of 31°F. ■

✔ **CHECK POINT 1** Use the function in Example 1 to find the number of O-rings expected to fail at a temperature of 60°F. Round to the nearest whole number.

2 Graph exponential functions.

Graphing Exponential Functions

We are familiar with expressions involving b^x where x is a rational number. For example,

$$b^{1.7} = b^{\frac{17}{10}} = \sqrt[10]{b^{17}} \quad \text{and} \quad b^{1.73} = b^{\frac{173}{100}} = \sqrt[100]{b^{173}}.$$

However, note that the definition of $f(x) = b^x$ includes all real numbers for the domain x. You may wonder what b^x means when x is an irrational number, such as $b^{\sqrt{3}}$ or b^{π}. Using the nonrepeating and nonterminating approximation 1.73205 for $\sqrt{3}$, we can think of $b^{\sqrt{3}}$ as the value that has the successively closer approximations

$$b^{1.7}, b^{1.73}, b^{1.732}, b^{1.73205}, \ldots.$$

In this way, we can graph the exponential function with no holes, or points of discontinuity, at the irrational domain values.

EXAMPLE 2 Graphing an Exponential Function

Graph: $f(x) = 2^x$.

Solution We begin by setting up a table of coordinates.

x	$f(x) = 2^x$
-3	$f(-3) = 2^{-3} = \frac{1}{8}$
-2	$f(-2) = 2^{-2} = \frac{1}{4}$
-1	$f(-1) = 2^{-1} = \frac{1}{2}$
0	$f(0) = 2^0 = 1$
1	$f(1) = 2^1 = 2$
2	$f(2) = 2^2 = 4$
3	$f(3) = 2^3 = 8$

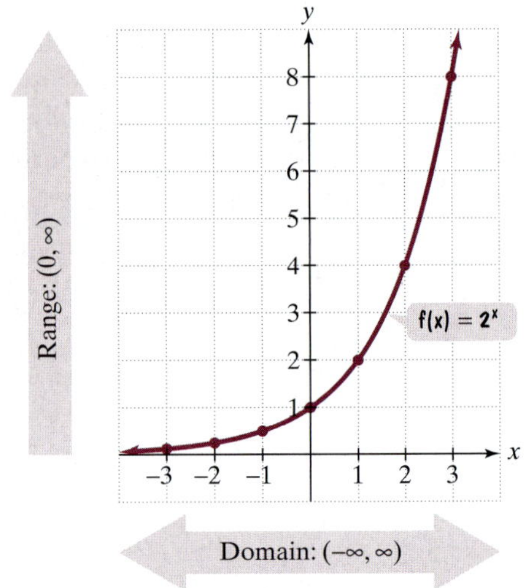

Figure 9.1 The graph of $f(x) = 2^x$

We plot these points, connecting them with a continuous curve. Figure 9.1 shows the graph of $f(x) = 2^x$. Observe that the graph approaches but never touches the negative portion of the x-axis. Thus, the x-axis is a horizontal asymptote. The range is the set of all positive real numbers. Although we used integers for x in our table of coordinates, you can use a calculator to find additional points. For example, $f(0.3) = 2^{0.3} \approx 1.231$, $f(0.95) = 2^{0.95} \approx 1.932$. The points $(0.3, 1.231)$ and $(0.95, 1.932)$ fit the graph. ■

✔ **CHECK POINT 2** Graph: $f(x) = 3^x$.

EXAMPLE 3 Graphing an Exponential Function

Graph: $f(x) = \left(\frac{1}{2}\right)^x$.

Solution We begin by setting up a table of coordinates. We compute the function values by noting that

$$f(x) = \left(\frac{1}{2}\right)^x = (2^{-1})^x = 2^{-x}.$$

We plot these points, connecting them with a continuous curve. Figure 9.2 on page 619 shows the graph of $f(x) = \left(\frac{1}{2}\right)^x$. This time the graph approaches but never touches the *positive* portion of the x-axis. Once again, the x-axis is a horizontal asymptote. The range consists of the set of all positive real numbers.

x	$f(x) = \left(\frac{1}{2}\right)^x$ or 2^{-x}
-3	$f(-3) = 2^{-(-3)} = 2^3 = 8$
-2	$f(-2) = 2^{-(-2)} = 2^2 = 4$
-1	$f(-1) = 2^{-(-1)} = 2^1 = 2$
0	$f(0) = 2^{-0} = 1$
1	$f(1) = 2^{-1} = \frac{1}{2^1} = \frac{1}{2}$
2	$f(2) = 2^{-2} = \frac{1}{2^2} = \frac{1}{4}$
3	$f(3) = 2^{-3} = \frac{1}{2^3} = \frac{1}{8}$

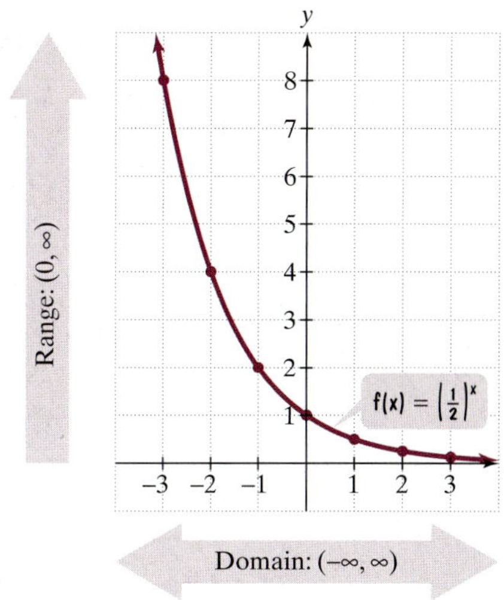

Figure 9.2 The graph of $f(x) = \left(\frac{1}{2}\right)^x$

Do you notice a relationship between the graphs of $f(x) = 2^x$ and $f(x) = \left(\frac{1}{2}\right)^x$ in Figures 9.1 and 9.2? The graph of $f(x) = \left(\frac{1}{2}\right)^x$ is a mirror image, or reflection, of the graph of $f(x) = 2^x$ about the y-axis.

CHECK POINT 3 Graph: $f(x) = \left(\frac{1}{3}\right)^x$. Note that $f(x) = \left(\frac{1}{3}\right)^x = (3^{-1})^x = 3^{-x}$.

Four exponential functions have been graphed in Figure 9.3. Compare the black and green graphs, where $b > 1$, to those in blue and red, where $b < 1$. When $b > 1$, the value of y increases as the value of x increases. When $b < 1$, the value of y decreases as the value of x increases. Notice that all four graphs pass through $(0, 1)$.

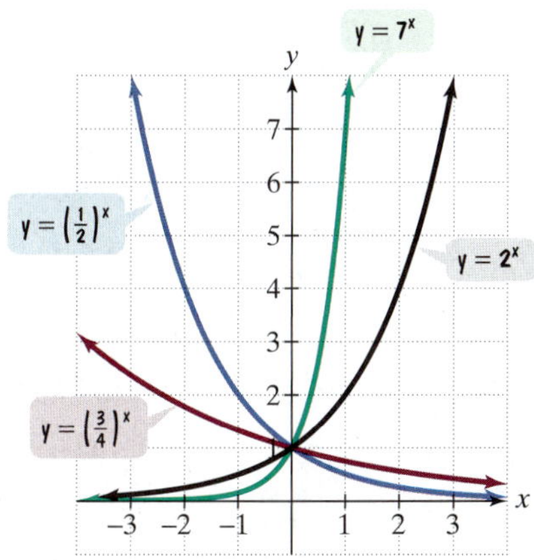

Figure 9.3 Graphs of four exponential functions

These graphs illustrate the following general characteristics of exponential functions, listed in the box on page 620.

Characteristics of Exponential Functions of the Form $f(x) = b^x$

1. The domain of $f(x) = b^x$ consists of all real numbers. The range of $f(x) = b^x$ consists of all positive real numbers.
2. The graphs of all exponential functions of the form $f(x) = b^x$ pass through the point $(0, 1)$ because $f(0) = b^0 = 1$ $(b \neq 0)$. The y-intercept is 1.
3. If $b > 1$, $f(x) = b^x$ has a graph that goes up to the right and is an increasing function. The greater the value of b, the steeper the increase.
4. If $0 < b < 1$, $f(x) = b^x$ has a graph that goes down to the right and is a decreasing function. The smaller the value of b, the steeper the decrease.
5. The graph of $f(x) = b^x$ approaches, but does not cross, the x-axis. The x-axis, or $y = 0$, is a horizontal asymptote.

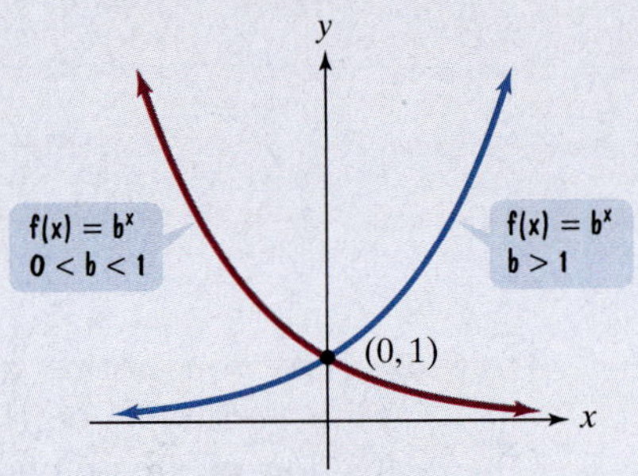

EXAMPLE 4 Graphing Exponential Functions

Graph $f(x) = 3^x$ and $g(x) = 3^{x+1}$ in the same rectangular coordinate system. How is the graph of g related to the graph of f?

Solution We begin by setting up a table showing some of the coordinates for f and g, selecting integers from -2 to 2 for x. Notice that $x + 1$ is the exponent for $g(x) = 3^{x+1}$.

x	$f(x) = 3^x$	$g(x) = 3^{x+1}$
-2	$f(-2) = 3^{-2} = \frac{1}{9}$	$g(-2) = 3^{-2+1} = 3^{-1} = \frac{1}{3}$
-1	$f(-1) = 3^{-1} = \frac{1}{3}$	$g(-1) = 3^{-1+1} = 3^0 = 1$
0	$f(0) = 3^0 = 1$	$g(0) = 3^{0+1} = 3^1 = 3$
1	$f(1) = 3^1 = 3$	$g(1) = 3^{1+1} = 3^2 = 9$
2	$f(2) = 3^2 = 9$	$g(2) = 3^{2+1} = 3^3 = 27$

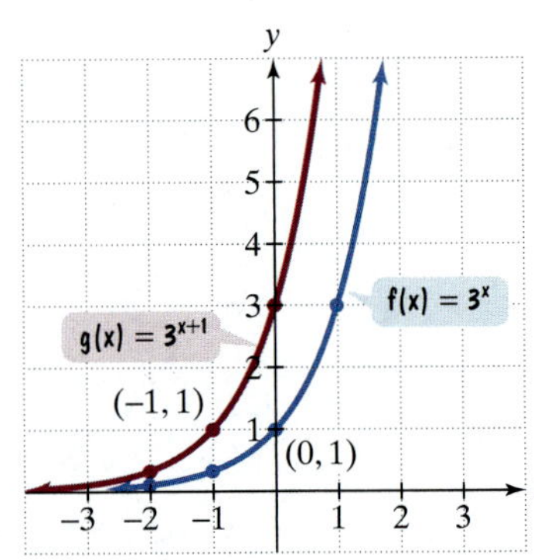

Figure 9.4

We plot the points for each function and connect them with a smooth curve. Because of the scale on the y-axis, the points on each function corresponding to $x = 2$ are not shown. Figure 9.4 shows the graphs of $f(x) = 3^x$ and $g(x) = 3^{x+1}$. The graph of g is the graph of f shifted one unit to the left. ■

CHECK POINT 4 Graph $f(x) = 3^x$ and $g(x) = 3^{x-1}$ in the same rectangular coordinate system. Select integers from -2 to 2 for x. How is the graph of g related to the graph of f?

EXAMPLE 5 Graphing Exponential Functions

Graph $f(x) = 2^x$ and $g(x) = 2^x - 3$ in the same rectangular coordinate system. How is the graph of g related to the graph of f?

Solution We begin by setting up a table showing some of the coordinates for f and g, selecting integers from -2 to 2 for x.

x	$f(x) = 2^x$	$g(x) = 2^x - 3$
-2	$f(-2) = 2^{-2} = \frac{1}{4}$	$g(-2) = 2^{-2} - 3 = \frac{1}{4} - 3 = -2\frac{3}{4}$
-1	$f(-1) = 2^{-1} = \frac{1}{2}$	$g(-1) = 2^{-1} - 3 = \frac{1}{2} - 3 = -2\frac{1}{2}$
0	$f(0) = 2^0 = 1$	$g(0) = 2^0 - 3 = 1 - 3 = -2$
1	$f(1) = 2^1 = 2$	$g(1) = 2^1 - 3 = 2 - 3 = -1$
2	$f(2) = 2^2 = 4$	$g(2) = 2^2 - 3 = 4 - 3 = 1$

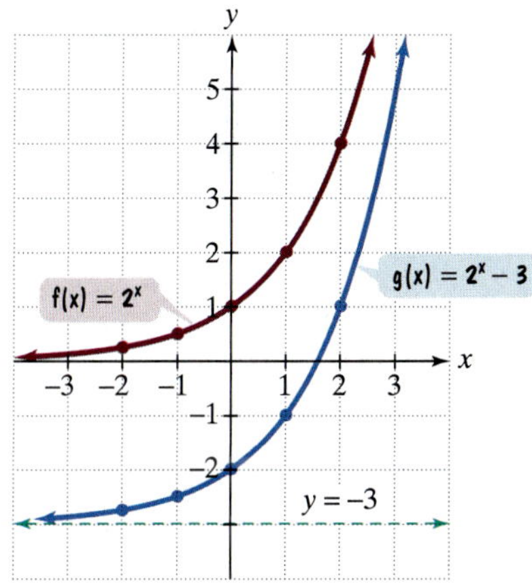

Figure 9.5

We plot the points for each function and connect them with a smooth curve. Figure 9.5 shows the graphs of $f(x) = 2^x$ and $g(x) = 2^x - 3$. The graph of g is the graph of f shifted down three units. As a result, $y = -3$ is the horizontal asymptote for g. ■

CHECK POINT 5 Graph $f(x) = 2^x$ and $g(x) = 2^x + 3$ in the same rectangular coordinate system. Select integers from -2 to 2 for x. How is the graph of g related to the graph of f?

3 Evaluate functions with base e.

The Natural Base e

An irrational number, symbolized by the letter e, appears as the base in many applied exponential functions. This irrational number is approximately equal to 2.72. More accurately,

$$e \approx 2.71828\ldots.$$

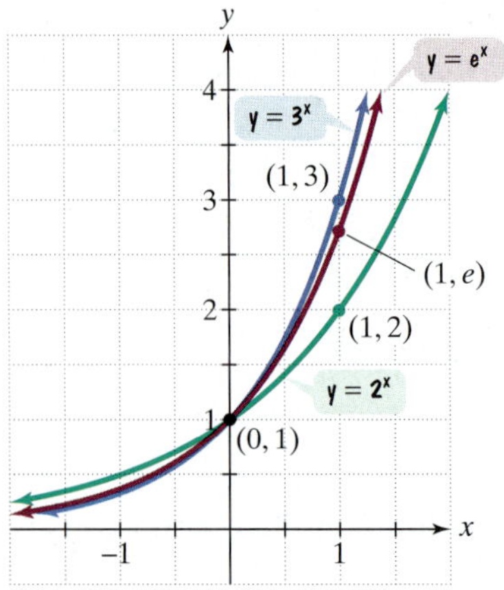

Figure 9.6 Graphs of three exponential functions

The number e is called the **natural base**. The function $f(x) = e^x$ is called the **natural exponential function**.

Use a scientific or graphing calculator with an $\boxed{e^x}$ key to evaluate e to various powers. For example, to find e^2, press the following keys on most calculators:

Scientific calculator: 2 $\boxed{e^x}$

Graphing calculator: $\boxed{e^x}$ 2 $\boxed{\text{ENTER}}$.

The display is approximately 7.389.

$$e^2 \approx 7.389$$

The number e lies between 2 and 3. Because $2^2 = 4$ and $3^2 = 9$, it makes sense that e^2, approximately 7.389, lies between 4 and 9.

Because $2 < e < 3$, the graph of $y = e^x$ is between the graphs of $y = 2^x$ and $y = 3^x$, shown in Figure 9.6.

EXAMPLE 6 World Population

In a report entitled *Resources and Man*, the U.S. National Academy of Sciences concluded that a world population of 10 billion "is close to (if not above) the maximum that an intensely managed world might hope to support with some degree of comfort and individual choice." At the time the report was issued in 1969, the world population was approximately 3.6 billion, with a growth rate of 2% per year. The function

$$f(x) = 3.6e^{0.02x}$$

describes world population, $f(x)$, in billions, x years after 1969. Use the function to find world population in the year 2020. Is there cause for alarm?

World Population in Billions

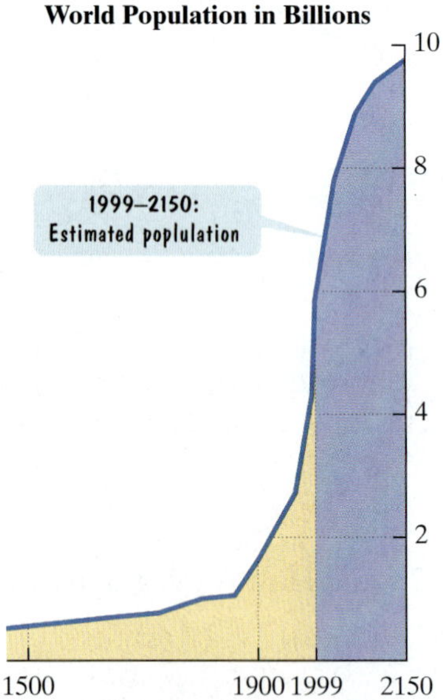

Source: U.N. Population Division

Growth rate in world population has slowed down since 1969. Current models project that world population will not be close to 10 billion until 2150.

Solution Because 2020 is 51 years after 1969, we substitute 51 for x:

$$f(51) = 3.6e^{0.02(51)}.$$

Although this computation can be done on your calculator in one step, we will break it down into smaller steps so that you can clearly see how we use the $\boxed{e^x}$ key. First find 0.02(51):

$$0.02(51) = 1.02.$$

Now find $e^{1.02}$:

Scientific calculator: 1.02 $\boxed{e^x}$

Graphing calculator: $\boxed{e^x}$ 1.02 $\boxed{\text{ENTER}}$.

The display is approximately 2.7731948. Multiplying this number by 3.6, we obtain

$$f(51) = 3.6e^{0.02(51)} = 3.6e^{1.02} \approx 9.98.$$

This indicates that world population in the year 2020 will be approximately 9.98 billion. Because this number is quite close to 10 billion, the given function suggests that there may be cause for alarm. ■

World population in 2000 was approximately 6 billion, but the growth rate was no longer 2%. It had slowed down to 1.3%. Using this current growth rate, exponential functions now predict a world population of 7.6 billion in the year 2020. Experts think the population may stabilize at 10 billion after 2200 if the deceleration in growth rate continues.

✔ **CHECK POINT 6** The function $f(x) = 6e^{0.013x}$ describes world population, $f(x)$, in billions, x years after 2000 subject to a growth rate of 1.3% annually. Use the function to find world population in 2050.

4 Use compound interest formulas.

Compound Interest

In Chapter 8, we saw that the amount of money, A, that a principal, P, will be worth after t years at interest rate r, compounded annually, is given by the formula

$$A = P(1 + r)^t.$$

Most savings institutions have plans in which interest is paid more than once a year. If compound interest is paid twice a year, the compounding period is six months. We say that the interest is **compounded semiannually**. When compound interest is paid four times a year, the compounding period is three months and the interest is said to be **compounded quarterly**. Some plans allow for monthly compounding or daily compounding.

In general, when compound interest is paid n times a year, we say that there are ***n* compounding periods per year**. The formula $A = P(1 + r)^t$ can be adjusted to take into account the number of compounding periods in a year. If there are n compounding periods per year, the formula becomes

$$A = P\left(1 + \frac{r}{n}\right)^{nt}.$$

Some banks use **continuous compounding**, where the number of compounding periods increases infinitely (compounding interest every trillionth of a second, every quadrillionth of a second, etc.). As n, the number of compounding periods in a year, increases without bound, the expression $\left(1 + \frac{1}{n}\right)^n$ approaches e. As a result, the formula for continuous compounding is $A = Pe^{rt}$. Although continuous compounding sounds terrific, it yields only a fraction of a percent more interest over a year than daily compounding.

n	$\left(1 + \frac{1}{n}\right)^n$
1	2
2	2.25
5	2.48832
10	2.59374246
100	2.704813829
1000	2.716923932
10,000	2.718145927
100,000	2.718268237
1,000,000	2.718280469
1,000,000,000	2.718281827

As n takes on increasingly large values, the expression $\left(1 + \frac{1}{n}\right)^n$ approaches e.

Formulas for Compound Interest

After t years, the balance, A, in an account with principal P and annual interest rate r (in decimal form) is given by the following formulas:

1. For n compoundings per year: $A = P\left(1 + \frac{r}{n}\right)^{nt}$
2. For continuous compounding: $A = Pe^{rt}$.

Using Technology

The graphs illustrate that as x increases, $\left(1 + \frac{1}{x}\right)^x$ approaches e.

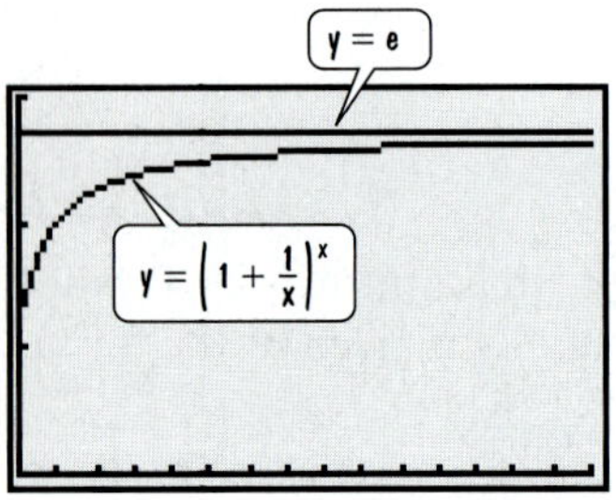

EXAMPLE 7 Choosing between Investments

You want to invest \$8000 for 6 years, and you have a choice between two accounts. The first pays 7% per year, compounded monthly. The second pays 6.85% per year, compounded continuously. Which is the better investment?

Solution The better investment is the one with the greater balance in the account after 6 years. Let's begin with the account with monthly compounding. We use the compound interest model with $P = 8000$, $r = 7\% = 0.07$, $n = 12$ (monthly compounding means 12 compoundings per year), and $t = 6$.

$$A = P\left(1 + \frac{r}{n}\right)^{nt} = 8000\left(1 + \frac{0.07}{12}\right)^{12 \cdot 6} \approx 12{,}160.84$$

The balance in this account after 6 years is \$12,160.84. For the second investment option, we use the model for continuous compounding with $P = 8000$, $r = 6.85\% = 0.0685$, and $t = 6$.

$$A = Pe^{rt} = 8000e^{0.0685(6)} \approx 12{,}066.60$$

The balance in this account after 6 years is \$12,066.60, slightly less than the previous amount. Thus, the better investment is the 7% monthly compounding option. ■

CHECK POINT 7 A sum of \$10,000 is invested at an annual rate of 8%. Find the balance in the account after 5 years subject to **a.** quarterly compounding and **b.** continuous compounding.

EXERCISE SET 9.1

Practice Exercises

In Exercises 1–10, approximate each number using a calculator. Round your answer to three decimal places.

1. $2^{3.4}$ **2.** $3^{2.4}$

3. $3^{\sqrt{5}}$ **4.** $5^{\sqrt{3}}$

5. $4^{-1.5}$ **6.** $6^{-1.2}$

7. $e^{2.3}$ **8.** $e^{3.4}$

9. $e^{-0.95}$ **10.** $e^{-0.75}$

In Exercises 11–16, set up a table of coordinates for each function. Select integers from −2 to 2 for x. Then use the table of coordinates to match the function with its graph. [The graphs are labeled (a) through (f).]

11. $f(x) = 3^x$ **12.** $f(x) = 3^{x-1}$

13. $f(x) = 3^x - 1$ **14.** $f(x) = -3^x$

15. $f(x) = 3^{-x}$ **16.** $f(x) = -3^{-x}$

(a)

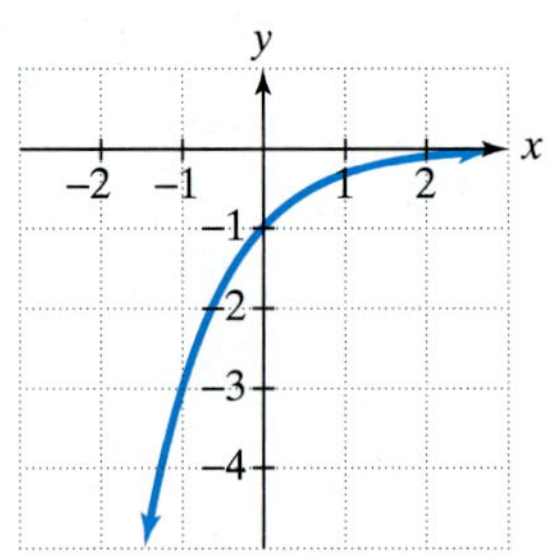

(b)

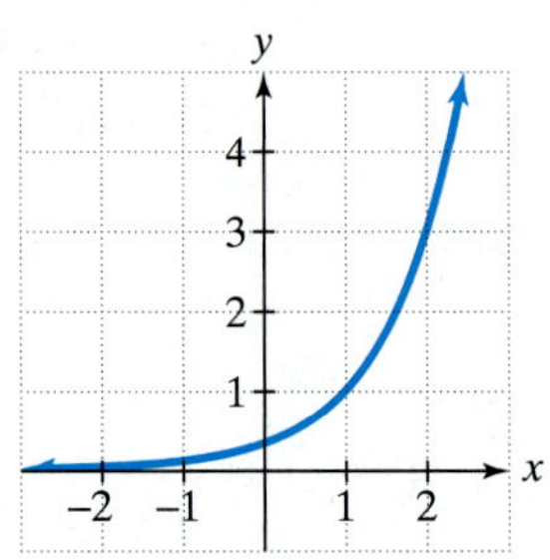

(c)

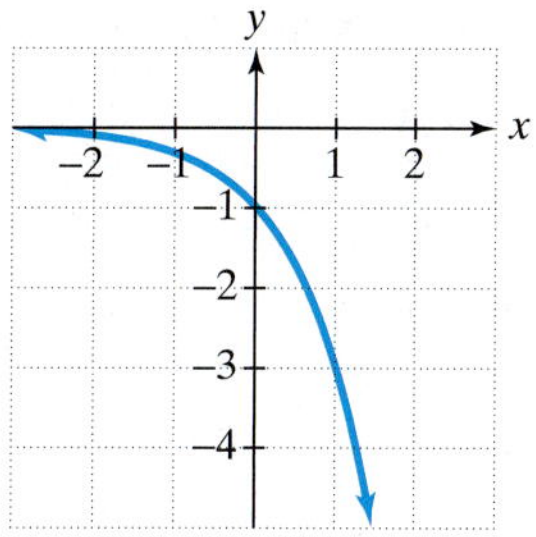

(d)

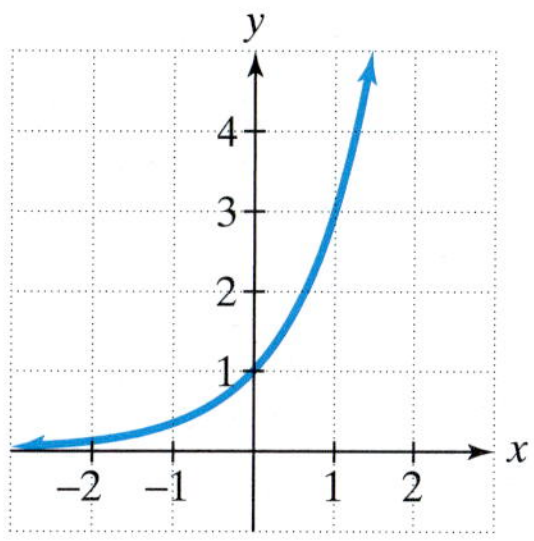

(e)

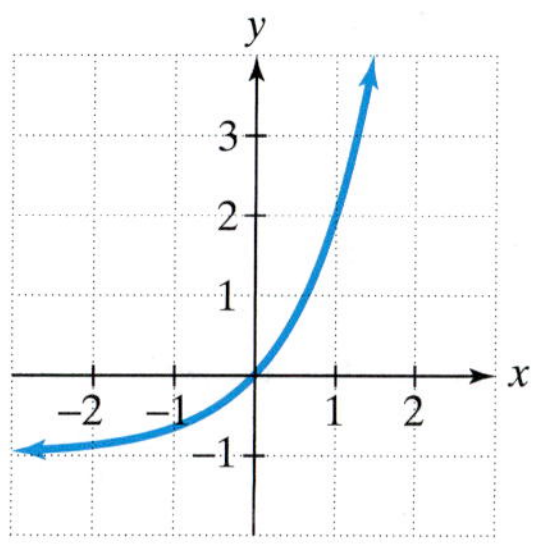

(f)

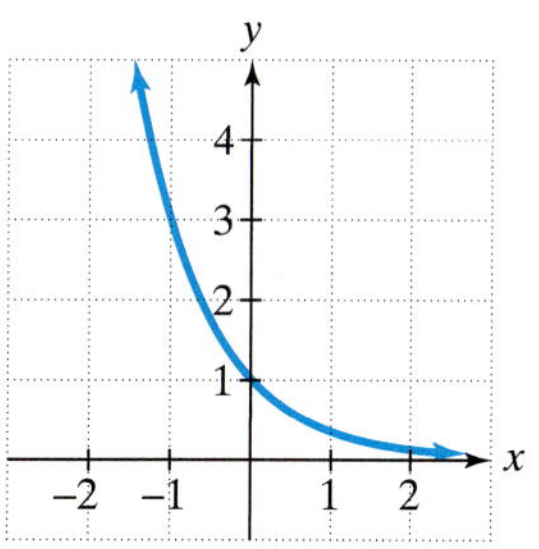

In Exercises 17–24, graph each function by making a table of coordinates. If applicable, use a graphing utility to confirm your hand-drawn graph.

17. $f(x) = 4^x$

18. $f(x) = 5^x$

19. $g(x) = \left(\frac{3}{2}\right)^x$

20. $g(x) = \left(\frac{4}{3}\right)^x$

21. $h(x) = \left(\frac{1}{2}\right)^x$

22. $h(x) = \left(\frac{1}{3}\right)^x$

23. $f(x) = (0.6)^x$

24. $f(x) = (0.8)^x$

In Exercises 25–38, graph functions f and g in the same rectangular coordinate system. Select integers from −2 to 2 for x. Then describe how the graph of g is related to the graph of f. If applicable, use a graphing utility to confirm your hand-drawn graphs.

25. $f(x) = 2^x$ and $g(x) = 2^{x+1}$

26. $f(x) = 2^x$ and $g(x) = 2^{x+2}$

27. $f(x) = 2^x$ and $g(x) = 2^{x-2}$

28. $f(x) = 2^x$ and $g(x) = 2^{x-1}$

29. $f(x) = 2^x$ and $g(x) = 2^x + 1$

30. $f(x) = 2^x$ and $g(x) = 2^x + 2$

31. $f(x) = 2^x$ and $g(x) = 2^x - 2$

32. $f(x) = 2^x$ and $g(x) = 2^x - 1$

33. $f(x) = 3^x$ and $g(x) = -3^x$

34. $f(x) = 3^x$ and $g(x) = 3^{-x}$

35. $f(x) = 2^x$ and $g(x) = 2^{x+1} - 1$

36. $f(x) = 2^x$ and $g(x) = 2^{x+1} - 2$

37. $f(x) = 3^x$ and $g(x) = \frac{1}{3} \cdot 3^x$

38. $f(x) = 3^x$ and $g(x) = 3 \cdot 3^x$

Use the compound interest formulas, $A = P\left(1 + \frac{r}{n}\right)^{nt}$ *and* $A = Pe^{rt}$*, to solve Exercises 39–42. Round answers to the nearest cent.*

39. Find the accumulated value of an investment of $10,000 for 5 years at an interest rate of 5.5% if the money is **a**. compounded semiannually; **b**. compounded monthly; **c**. compounded continuously.

40. Find the accumulated value of an investment of $5000 for 10 years at an interest rate of 6.5% if the money is **a**. compounded semiannually; **b**. compounded monthly; **c**. compounded continuously.

41. Suppose that you have $12,000 to invest. Which investment yields the greatest return over 3 years: 7% compounded monthly or 6.85% compounded continuously?

42. Suppose that you have $6000 to invest. Which investment yields the greatest return over 4 years: 8.25% compounded quarterly or 8.3% compounded semiannually?

Application Exercises

Use a calculator with an $\boxed{x^y}$ *key or a* $\boxed{\wedge}$ *key to solve Exercises 43–48.*

43. The exponential function $f(x) = 67.38(1.026)^x$ describes the population of Mexico, $f(x)$, in millions, x years after 1980.

a. Substitute 0 for x and, without using a calculator, find Mexico's population in 1980.

b. Substitute 27 for x and use your calculator to find Mexico's population in the year 2007 as predicted by this function.

c. Find Mexico's population in the year 2034 as predicted by this function.

d. Find Mexico's population in the year 2061 as predicted by this function.

e. What appears to be happening to Mexico's population every 27 years?

44. The 1986 explosion at the Chernobyl nuclear power plant in the former Soviet Union sent about 1000 kilograms of radioactive cesium-137 into the atmosphere. The function $f(x) = 1000(0.5)^{\frac{x}{30}}$ describes the amount, $f(x)$, in kilograms, of cesium-137 remaining in Chernobyl x years after 1986. If even 100 kilograms of cesium-137 remain in Chernobyl's atmosphere, the area is considered unsafe for human habitation. Find $f(80)$ and determine if Chernobyl will be safe for human habitation by 2066.

The formula $S = C(1 + r)^t$ *models inflation, where* $C =$ *the value today,* $r =$ *the annual inflation rate, and* $S =$ *the inflated value* t *years from now. Use this formula to solve Exercises 45–46.*

45. If the inflation rate is 6%, how much will a house now worth $65,000 be worth in 10 years?

46. If the inflation rate is 3%, how much will a house now worth $110,000 be worth in 5 years?

47. A decimal approximation for $\sqrt{3}$ is 1.7320508. Use a calculator to find $2^{1.7}$, $2^{1.73}$, $2^{1.732}$, $2^{1.73205}$, and $2^{1.7320508}$. Now find $2^{\sqrt{3}}$. What do you observe?

48. A decimal approximation for π is 3.141593. Use a calculator to find 2^3, $2^{3.1}$, $2^{3.14}$, $2^{3.141}$, $2^{3.1415}$, $2^{3.14159}$, and $2^{3.141593}$. Now find 2^{π}. What do you observe?

Use a calculator with an $\boxed{e^x}$ *key to solve Exercises 49–55. The function* $f(x) = 24{,}000e^{0.21x}$ *describes the number of AIDS cases in the United States among intravenous drug users* x *years after 1989. Use this function to solve Exercises 49–50.*

49. Evaluate $f(11)$ and describe what this means in practical terms.

50. Evaluate $f(31)$ and describe what this means in practical terms.

51. In college, we study large volumes of information—information that, unfortunately, we do not often retain for very long. The function

$$f(x) = 80e^{-0.5x} + 20$$

describes the percentage of information, $f(x)$, that a particular person remembers x weeks after learning the information.

a. Substitute 0 for x and, without using a calculator, find the percentage of information remembered at the moment it is first learned.

b. Substitute 1 for x and find the percentage of information that is remembered after 1 week.

c. Find the percentage of information that is remembered after 4 weeks.

d. Find the percentage of information that is remembered after one year (52 weeks).

52. In 1626, Peter Minuit convinced the Wappinger Indians to sell him Manhattan Island for $24. If the Native Americans had put the $24 into a bank account paying 5% interest, how much would the investment have been worth in the year 2000 if interest were compounded

a. monthly?

b. continuously?

The function

$$f(x) = \frac{90}{1 + 270e^{-0.122x}}$$

models the percentage, f(x) of people x years old with some coronary heart disease. Use this function to solve Exercises 53–54. Round answers to the nearest tenth of a percent.

53. Evaluate $f(30)$ and describe what this means in practical terms.

54. Evaluate $f(70)$ and describe what this means in practical terms.

55. The function

$$N(t) = \frac{30{,}000}{1 + 20e^{-1.5t}}$$

describes the number of people, $N(t)$, who become ill with influenza t weeks after its initial outbreak in a town with 30,000 inhabitants. The horizontal asymptote in the graph indicates that there is a limit to the epidemic's growth.

a. How many people became ill with the flu when the epidemic began? (When the epidemic began, $t = 0$.)

b. How many people were ill by the end of the third week?

c. Why can't the spread of an epidemic simply grow indefinitely? What does the horizontal asymptote shown in the graph indicate about the limiting size of the population that becomes ill?

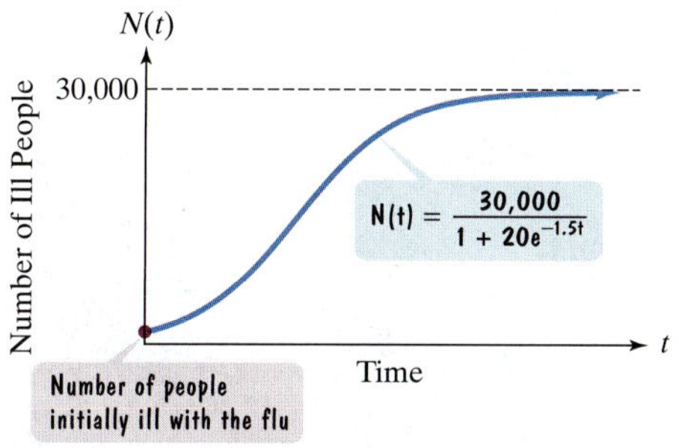

Writing in Mathematics

56. What is an exponential function?

57. What is the natural exponential function?

58. Use a calculator to evaluate $\left(1 + \frac{1}{x}\right)^x$ for $x = 10, 100,$ 1000, 10,000, 100,000, and 1,000,000. Describe what happens to the expression as x increases.

59. Write an example similar to Example 7 on page 624 in which continuous compounding at a slightly lower yearly interest rate is a better investment than compounding n times per year.

60. Describe how you could use the graph of $f(x) = 2^x$ to obtain a decimal approximation for $\sqrt{2}$.

61. In 2000, world population was 6 billion with an annual growth rate of 1.3%. Discuss two factors that would cause this growth rate to slow down over the next ten years.

Technology Exercises

62. Graph $y = 13.49(0.967)^x - 1$, the function for the number of O-rings expected to fail at x°F, in a $[0, 90, 10]$ by $[0, 20, 5]$ viewing rectangle. If NASA engineers had used this function and its graph, is it likely they would have allowed the *Challenger* to be launched when the temperature was 31°F? Explain.

63. The student–teacher ratio in U.S. elementary and secondary schools can be modeled by $y = 25.34(0.987)^x$, where x represents the number of years after 1959 and y represents the student–teacher ratio. Graph the function in a $[1, 40, 1]$ by $[0, 26, 1]$ viewing rectangle. When did the student–teacher ratio become less than 21 students per teacher?

64. You have \$10,000 to invest. One bank pays 5% interest compounded quarterly and the other pays 4.5% interest compounded monthly.

a. Use the formula for compound interest to write a function for the balance in each account at any time t, in years.

b. Use a graphing utility to graph both functions in an appropriate viewing rectangle. Based on the graphs, which bank offers the better return on your money?

65. **a.** Graph $y = e^x$ and $y = 1 + x + \frac{x^2}{2}$ in the same viewing rectangle.

b. Graph $y = e^x$ and $y = 1 + x + \frac{x^2}{2} + \frac{x^3}{6}$ in the same viewing rectangle.

c. Graph $y = e^x$ and $y = 1 + x + \frac{x^2}{2} + \frac{x^3}{6} + \frac{x^4}{24}$ in the same viewing rectangle.

d. Describe what you observe in parts (a)–(c). Try generalizing this observation.

Critical Thinking Exercises

66. Which one of the following is true?

a. As the number of compounding periods increases on a fixed investment, the amount of money in the account over a fixed interval of time will increase without bound.

b. The functions $f(x) = 3^{-x}$ and $g(x) = -3^x$ have the same graph.

c. $e = 2.718$.

d. The functions $f(x) = \left(\frac{1}{3}\right)^x$ and $g(x) = 3^{-x}$ have the same graph.

67. The graphs labeled (a)–(d) in the figure represent $y = 3^x$, $y = 5^x$, $y = \left(\frac{1}{3}\right)^x$, and $y = \left(\frac{1}{5}\right)^x$, but not necessarily in that order. Which is which? Describe the process that enables you to make this decision.

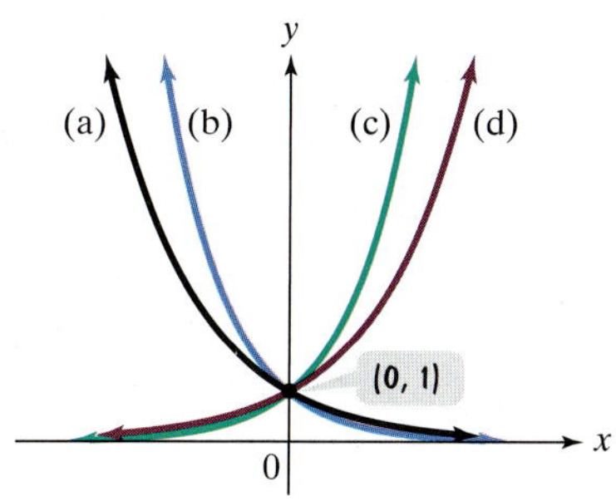

68. The hyperbolic cosine and hyperbolic sine functions are defined by

$$\cosh x = \frac{e^x + e^{-x}}{2} \quad \text{and} \quad \sinh x = \frac{e^x - e^{-x}}{2}.$$

Prove that $(\cosh x)^2 - (\sinh x)^2 = 1$.

Review Exercises

69. Solve for b: $D = \dfrac{ab}{a + b}$.
(Section 6.7, Example 1)

70. Evaluate: $\begin{vmatrix} 3 & -2 \\ 7 & -5 \end{vmatrix}$.
(Section 3.5, Example 1)

71. Solve: $x(x - 3) = 10$.
(Section 8.2, Example 1)

▸ SECTION 9.2 Composite and Inverse Functions

Objectives

1. Form composite functions.
2. Verify inverse functions.
3. Find the inverse of a function.
4. Use the horizontal line test to determine if a function has an inverse function.
5. Use the graph of a one-to-one function to graph its inverse function.

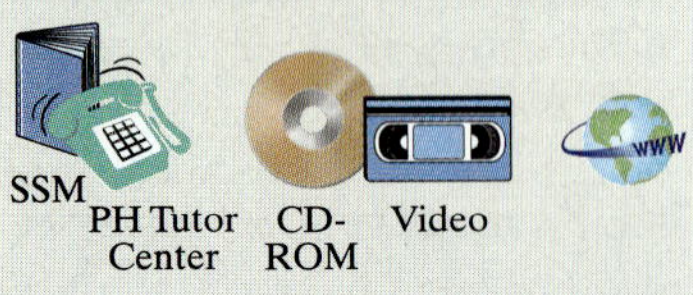

You are an over-40 student returning to college. Through consistent hard work, you've earned a near-perfect grade point average, feeling more empowered by knowledge with each academic success. Unfortunately, when it comes to technology, you are at a disadvantage compared to younger students. There were no computers in the classroom during your high school years, and you glaze over hearing students chatting about browsers, megabytes, RAM, and ROM. It's time to turn things around and become computer literate.

Luckily, your local computer store is having a sale. The models that are on sale cost either \$300 less than the regular price or 85% of the regular price. If x represents the computer's regular price, both discounts can be described with the following functions.

$$f(x) = x - 300 \qquad\qquad g(x) = 0.85x$$

The computer is on sale for \$300 less than its regular price.

The computer is on sale for 85% of its regular price.

At the store, you bargain with the salesperson. Eventually, she makes an offer you can't refuse: The sale price is 85% of the regular price followed by a \$300 reduction:

$$0.85x - 300.$$

85% of the regular price — followed by a \$300 reduction

In terms of functions f and g, this offer can be obtained by taking the output of $g(x) = 0.85x$, namely $0.85x$, and using it as the input of f:

$$f(x) = x - 300$$

Replace x with 0.85 x, the output of $g(x) = 0.85x$.

$$f(0.85x) = 0.85x - 300.$$

Because $0.85x$ is $g(x)$, we can write this last equation as

$$f(g(x)) = 0.85x - 300.$$

We read this equation as "f of g of x is equal to $0.85x - 300$." We call $f(g(x))$ the **composition of the function *f* with *g***, or a **composite function.** This composite function is written $f \circ g$. Thus,

$$(f \circ g)(x) = f(g(x)) = 0.85x - 300.$$

Like all functions, we can evaluate $f \circ g$ for a specified value of x in the function's domain. For example, here's how to find the value of this function at 1400:

$$(f \circ g)(x) = 0.85x - 300$$

This composite function describes the offer you cannot refuse.

Replace x with 1400.

$$(f \circ g)(1400) = 0.85(1400) - 300 = 1190 - 300 = 890.$$

This means that a computer that regularly sells for \$1400 is on sale for \$890 subject to both discounts.

In this section, we will focus on the composition of two functions. We will also study functions whose compositions have a special relationship.

1 Form composite functions.

Before you run out to buy a computer, let's generalize our discussion of the computer's double discount and define the composition of any two functions.

The Composition of Functions

The **composition of the function *f* with *g*** is denoted by $f \circ g$ and is defined by the equation

$$(f \circ g)(x) = f(g(x)).$$

The domain of the **composite function $f \circ g$** is the set of all x such that

1. x is in the domain of g and
2. $g(x)$ is in the domain of f.

The composition of f with g, $f \circ g$, is pictured as a machine with inputs and outputs in Figure 9.7. The diagram indicates that the output of g, or $g(x)$, becomes the input for "machine" f. If $g(x)$ is not in the domain of f, it cannot be input into machine f, and so $g(x)$ must be discarded.

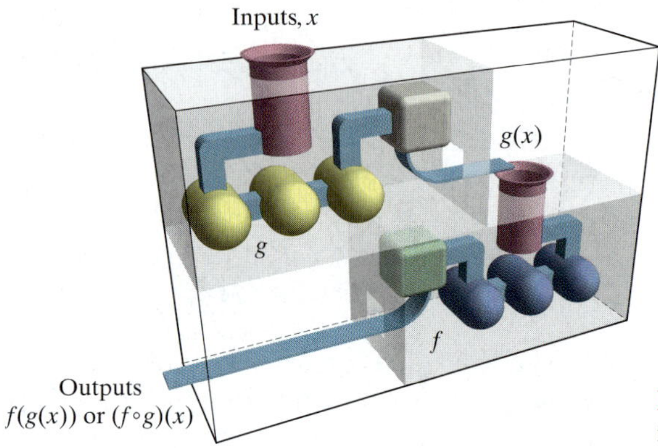

Figure 9.7 Inputting one function into a second function

EXAMPLE 1 Forming Composite Functions

Given $f(x) = 3x - 4$ and $g(x) = x^2 + 6$, find:

a. $(f \circ g)(x)$ **b.** $(g \circ f)(x)$.

Solution

a. We begin with $(f \circ g)(x)$, the composition of f with g. Because $(f \circ g)(x)$ means $f(g(x))$, we must replace each occurrence of x in the equation for f with $g(x)$.

$$f(x) = 3x - 4 \quad \text{This is the given equation for } f.$$

Replace x with $g(x)$.

$$(f \circ g)(x) = f(g(x)) = 3g(x) - 4$$

$$= 3(x^2 + 6) - 4 \quad \text{Because } g(x) = x^2 + 6, \text{ replace } g(x) \text{ with } x^2 + 6.$$

$$= 3x^2 + 18 - 4 \quad \text{Use the distributive property.}$$

$$= 3x^2 + 14 \quad \text{Simplify.}$$

Thus, $(f \circ g)(x) = 3x^2 + 14$.

b. Next, we find $(g \circ f)(x)$, the composition of g with f. Because $(g \circ f)(x)$ means $g(f(x))$, we must replace each occurrence of x in the equation for g with $f(x)$.

$$g(x) = x^2 + 6 \quad \text{This is the given equation for } g.$$

Replace x with $f(x)$.

$$(g \circ f)(x) = g(f(x)) = (f(x))^2 + 6$$

$$= (3x - 4)^2 + 6 \quad \text{Because } f(x) = 3x - 4, \text{ replace } f(x) \text{ with } 3x - 4.$$

$$= 9x^2 - 24x + 16 + 6 \quad \text{Use } (A - B)^2 = A^2 - 2AB + B^2 \text{ to square } 3x - 4.$$

$$= 9x^2 - 24x + 22 \quad \text{Simplify.}$$

Thus, $(g \circ f)(x) = 9x^2 - 24x + 22$. Notice that $f \circ g$ is not the same function as $g \circ f$. ■

CHECK POINT 1 Given $f(x) = 5x + 6$ and $g(x) = x^2 - 1$, find:

a. $(f \circ g)(x)$ **b.** $(g \circ f)(x)$.

Inverse Functions

Here are two functions that describe situations related to the price x of a computer:

$$f(x) = x - 300 \qquad g(x) = x + 300.$$

Function f subtracts \$300 from the computer's price and function g adds \$300 to the computer's price. Let's see what $f(g(x))$ does. Put $g(x)$ into f:

$f(x) = x - 300$ This is the given equation for f.

Replace x with $g(x)$.

$f(g(x)) = g(x) - 300$

$= x + 300 - 300$ Because $g(x) = x + 300$, replace $g(x)$ with $x + 300$.

$= x.$ This is the computer's original price.

Using $f(g(x))$, the computer's price, x, went through two changes: the first, an increase; the second, a decrease:

$$x + 300 - 300.$$

The final price of the computer, x, is identical to its starting price, x.

In general, if the changes made to x by function g are undone by the changes made by function f, then

$$f(g(x)) = x.$$

Assume, also, that this "undoing" takes place in the other direction:

$$g(f(x)) = x.$$

Under these conditions, we say that each function is the **inverse function** of the other. The fact that g is the inverse of f is expressed by renaming g as f^{-1}, read "f-inverse." For example, the inverse functions

$$f(x) = x - 300 \qquad g(x) = x + 300$$

are usually named as follows:

$$f(x) = x - 300 \qquad f^{-1}(x) = x + 300.$$

With these ideas in mind, we present the formal definition of the inverse of a function.

Study Tip

The notation f^{-1} represents the inverse function of f. The -1 is *not* an exponent. The notation f^{-1} does *not* mean $\frac{1}{f}$:

$$f^{-1} \neq \frac{1}{f}.$$

Definition of the Inverse of a Function

Let f and g be two functions such that

$$f(g(x)) = x \quad \text{for every } x \text{ in the domain of } g$$

and

$$g(f(x)) = x \quad \text{for every } x \text{ in the domain of } f.$$

The function g is the **inverse** of the function f, and is denoted by f^{-1} (read "f-inverse"). Thus, $f(f^{-1}(x)) = x$ and $f^{-1}(f(x)) = x$. The domain of f is equal to the range of f^{-1}, and vice versa.

2 Verify inverse functions.

EXAMPLE 2 Verifying Inverse Functions

Show that each function is an inverse of the other:

$$f(x) = 5x \quad \text{and} \quad g(x) = \frac{x}{5}.$$

Solution To show that f and g are inverses of each other, we must show that $f(g(x)) = x$ and $g(f(x)) = x$. We begin with $f(g(x))$.

$$f(x) = 5x$$ This is the given equation for f.

Replace x with $g(x)$.

$$f(g(x)) = 5g(x) = 5\left(\frac{x}{5}\right) = x$$

Next, we find $g(f(x))$.

$$g(x) = \frac{x}{5}$$ This is the given equation for g.

Replace x with $f(x)$.

$$g(f(x)) = \frac{f(x)}{5} = \frac{5x}{5} = x$$

Because g is the inverse of f (and vice versa), we can use inverse notation and write

$$f(x) = 5x \quad \text{and} \quad f^{-1}(x) = \frac{x}{5}.$$

Notice how f^{-1} undoes the change produced by f: f changes x by multiplying by 5 and f^{-1} undoes this by dividing by 5. ■

✔ **CHECK POINT 2** Show that each function is an inverse of the other:

$$f(x) = 7x \quad \text{and} \quad g(x) = \frac{x}{7}.$$

EXAMPLE 3 Verifying Inverse Functions

Show that each function is an inverse of the other:

$$f(x) = 3x + 2 \quad \text{and} \quad g(x) = \frac{x-2}{3}.$$

Solution To show that f and g are inverses of each other, we must show that $f(g(x)) = x$ and $g(f(x)) = x$. We begin with $f(g(x))$.

$$f(x) = 3x + 2$$ This is the equation for f.

Replace x with $g(x)$.

$$f(g(x)) = 3g(x) + 2 = 3\left(\frac{x-2}{3}\right) + 2 = x - 2 + 2 = x$$

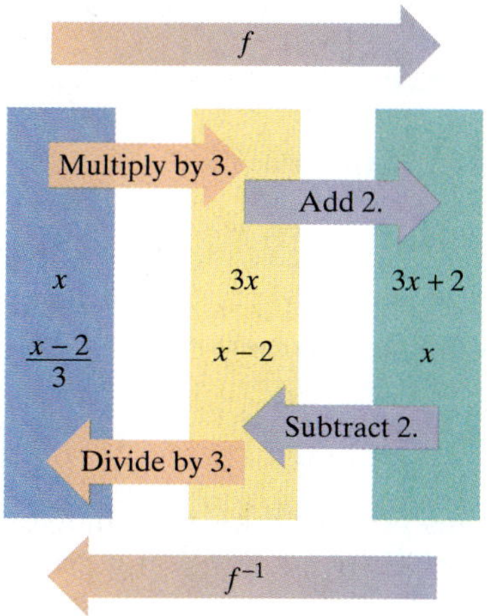

Figure 9.8 f^{-1} undoes the changes produced by f.

Next, we find $g(f(x))$.

$$g(x) = \frac{x-2}{3} \quad \text{This is the equation for } g.$$

Replace x with $f(x)$.

$$g(f(x)) = \frac{f(x)-2}{3} = \frac{(3x+2)-2}{3} = \frac{3x}{3} = x$$

Because g is the inverse of f (and vice versa), we can use inverse notation and write

$$f(x) = 3x + 2 \quad \text{and} \quad f^{-1}(x) = \frac{x-2}{3}.$$

Notice how f^{-1} undoes the changes produced by f: f changes x by *multiplying* by 3 and *adding* 2, and f^{-1} undoes this by *subtracting* 2 and *dividing* by 3. This "undoing" process is illustrated in Figure 9.8. ■

✔ **CHECK POINT 3** Show that each function is an inverse of the other:

$$f(x) = 4x - 7 \quad \text{and} \quad g(x) = \frac{x+7}{4}.$$

3 Find the inverse of a function.

Finding the Inverse of a Function

The definition of the inverse of a function tells us that the domain of f is equal to the range of f^{-1}, and vice versa. This means that if the function f is the set of ordered pairs (x, y), then the inverse of f is the set of ordered pairs (y, x). If a function is defined by an equation, we can obtain the equation for f^{-1}, the inverse of f, by interchanging the role of x and y in the equation for the function f.

Study Tip

The procedure for finding a function's inverse uses a *switch-and-solve* strategy. Switch x and y, then solve for y.

Finding the Inverse of a Function

The equation for the inverse of a function f can be found as follows:

1. Replace $f(x)$ with y in the equation for $f(x)$.
2. Interchange x and y.
3. Solve for y. If this equation does not define y as a function of x, the function f does not have an inverse function and this procedure ends. If this equation does define y as a function of x, the function f has an inverse function.
4. If f has an inverse function, replace y in step 3 with $f^{-1}(x)$. We can verify our result by showing that $f(f^{-1}(x)) = x$ and $f^{-1}(f(x)) = x$.

EXAMPLE 4 Finding the Inverse of a Function

Find the inverse of $f(x) = 7x - 5$.

Solution

Step 1 Replace $f(x)$ with y:

$$y = 7x - 5.$$

Step 2 Interchange x and y:

$$x = 7y - 5 \quad \text{This is the inverse function.}$$

Discover for Yourself

In Example 4, we found that if $f(x) = 7x - 5$, then

$$f^{-1}(x) = \frac{x+5}{7}.$$

Verify this result by showing that

$$f(f^{-1}(x)) = x$$

and

$$f^{-1}(f(x)) = x.$$

Step 3 Solve for *y*:

$$x + 5 = 7y \quad \text{Add 5 to both sides.}$$

$$\frac{x+5}{7} = y \quad \text{Divide both sides by 7.}$$

Step 4 Replace *y* with $f^{-1}(x)$:

$$f^{-1}(x) = \frac{x+5}{7} \quad \text{The equation is written with } f^{-1} \text{ on the left.}$$

Thus, the inverse of $f(x) = 7x - 5$ is $f^{-1}(x) = \dfrac{x+5}{7}$.

The inverse function, f^{-1}, undoes the changes produced by f. f changes x by multiplying by 7 and subtracting 5. f^{-1} undoes this by adding 5 and dividing by 7. ■

✔ **CHECK POINT 4** Find the inverse of $f(x) = 2x + 7$.

EXAMPLE 5 Finding the Equation of the Inverse

Find the inverse of $f(x) = x^3 + 1$.

Solution

Step 1 Replace $f(x)$ with *y*: $y = x^3 + 1$.

Step 2 Interchange *x* and *y*: $x = y^3 + 1$.

Step 3 Solve for *y*.

$$x - 1 = y^3$$

$$\sqrt[3]{x-1} = \sqrt[3]{y^3}$$

$$\sqrt[3]{x-1} = y.$$

Step 4 Replace *y* with $f^{-1}(x)$: $f^{-1}(x) = \sqrt[3]{x-1}$.

Thus, the inverse of $f(x) = x^3 + 1$ is $f^{-1}(x) = \sqrt[3]{x-1}$. ■

✔ **CHECK POINT 5** Find the inverse of $f(x) = 4x^3 - 1$.

4 Use the horizontal line test to determine if a function has an inverse function.

The Horizontal Line Test and One-to-One Functions

Let's see what happens if we try to find the inverse of the quadratic function $f(x) = x^2$.

Step 1 Replace $f(x)$ with *y*: $y = x^2$.

Step 2 Interchange *x* and *y*: $x = y^2$.

Step 3 Solve for *y*: We apply the square root property to solve $y^2 = x$ for y. We obtain

$$y = \pm\sqrt{x}.$$

The $\pm$ in this last equation shows that for certain values of x (all positive real numbers), there are two values of y. Because this equation does not represent y as a function of x, the quadratic function $f(x) = x^2$ does not have an inverse function.

Can we look at the graph of a function and tell if it represents a function with an inverse? Yes. The graph of the quadratic function $f(x) = x^2$ is shown in Figure 9.9. Four units above the x-axis, a horizontal line is drawn. This line intersects the graph at two of its points, $(-2, 4)$ and $(2, 4)$. Because inverse functions have ordered pairs with the coordinates reversed, let's see what happens if we reverse these coordinates. We obtain $(4, -2)$ and $(4, 2)$. A function provides exactly one output for each input. However, the input 4 is associated with two outputs, -2 and 2. The points $(4, -2)$ and $(4, 2)$ do not define a function.

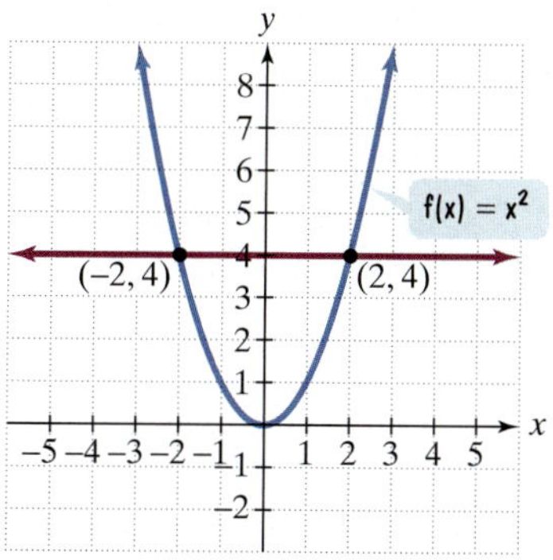

Figure 9.9 The horizontal line intersects the graph twice.

If any horizontal line, such as the one in Figure 9.9, intersects a graph at two or more points, the set of these points will not define a function when their coordinates are reversed. This suggests the **horizontal line test** for inverse functions.

The Horizontal Line Test For Inverse Functions

A function f has an inverse that is a function, f^{-1}, if there is no horizontal line that intersects the graph of the function f at more than one point.

Discover for Yourself

How might you restrict the domain of $f(x) = x^2$, graphed in Figure 9.9, so that the remaining portion of the graph passes the horizontal line test?

EXAMPLE 6 Applying the Horizontal Line Test

Which of the following graphs represent functions that have inverse functions?

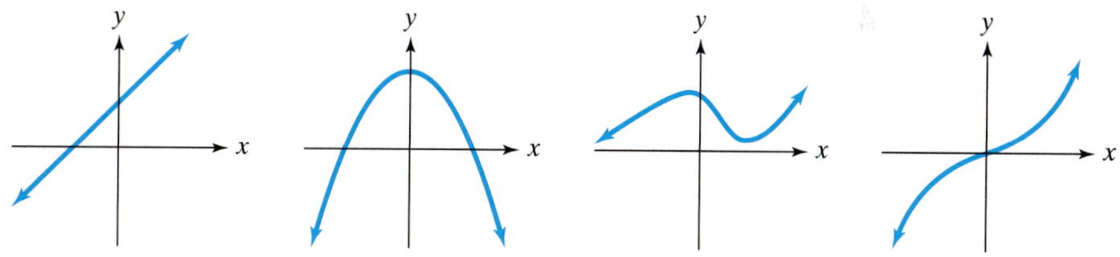

(a) (b) (c) (d)

Solution Notice that horizontal lines can be drawn in parts (b) and (c) that intersect the graphs more than once. These graphs do not pass the horizontal line test. These are not the graphs of functions with inverse functions. By contrast, no horizontal line can be drawn in parts (a) and (d) that intersects the graphs more than once. These graphs pass the horizontal line test. Thus, the graphs in parts (a) and (d) represent functions that have inverse functions.

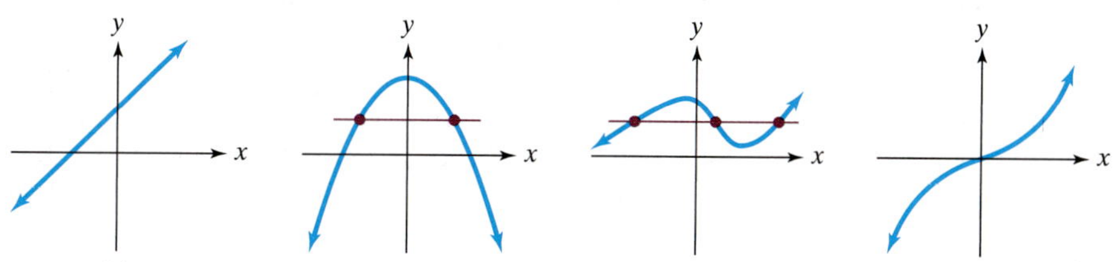

(a) Has an inverse function

(b) No inverse function

(c) No inverse function

(d) Has an inverse function ■

✔ CHECK POINT 6 Which of the following graphs represent functions that have inverse functions?

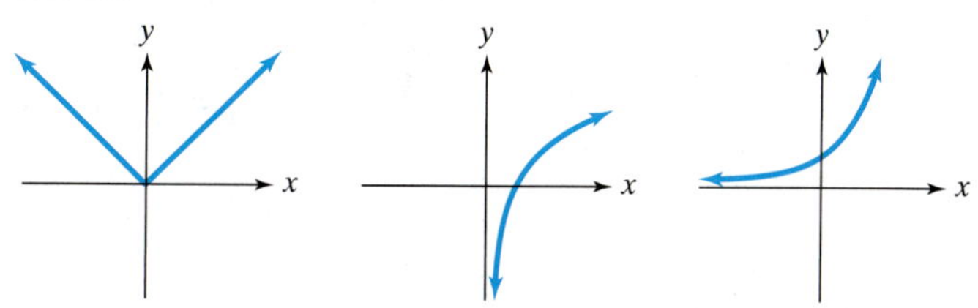

(a) (b) (c)

A function passes the horizontal line test when no two different ordered pairs have the same second component. This means that if $x_1 \neq x_2$, then $f(x_1) \neq f(x_2)$. Such a function is called a **one-to-one function**. Thus, a one-to-one function is a function in which no two different ordered pairs have the same second component. Only one-to-one functions have inverse functions. Any function that passes the horizontal line test is a one-to-one function. Any one-to-one function has a graph that passes the horizontal line test.

5 Use the graph of a one-to-one function to graph its inverse function.

Graphs of f and f^{-1}

There is a relationship between the graph of a one-to-one function, f, and its inverse, f^{-1}. Because inverse functions have ordered pairs with the coordinates reversed, if the point (a, b) is on the graph of f, then the point (b, a) is on the graph of f^{-1}. The points (a, b) and (b, a) are symmetric with respect to the line $y = x$. Thus, **the graph of f^{-1} is a reflection of the graph of f about the line $y = x$.** This is illustrated in Figure 9.10.

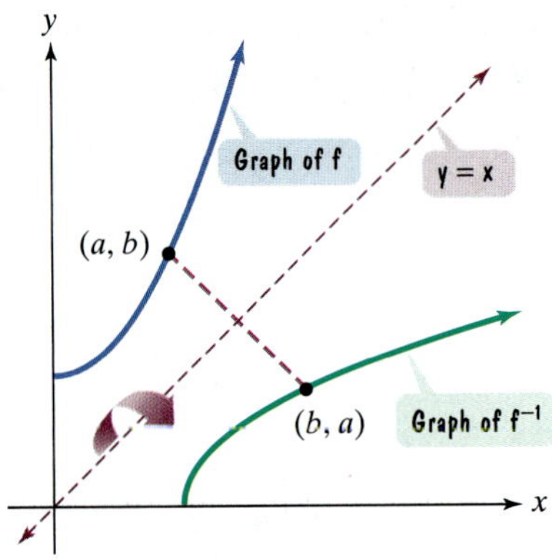

Figure 9.10 The graph of f^{-1} is a reflection of f about $y = x$.

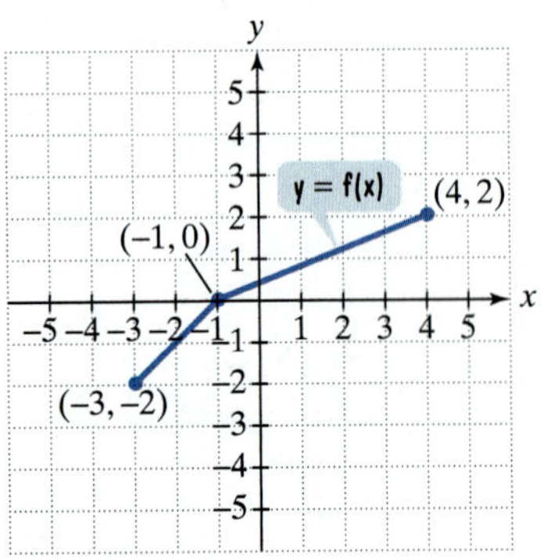

Figure 9.11

EXAMPLE 7 Graphing the Inverse Function

Use the graph of f in Figure 9.11 to draw the graph of its inverse function.

Solution We begin by noting that no horizontal line intersects the graph of f at more than one point, so f does have an inverse function. Because the points $(-3, -2)$, $(-1, 0)$, and $(4, 2)$ are on the graph of f, the graph of the inverse function, f^{-1}, has points with these ordered pairs reversed. Thus, $(-2, -3)$, $(0, -1)$, and $(2, 4)$ are on the graph of f^{-1}. We can use these points to graph f^{-1}. The graph of f^{-1} is shown in green in Figure 9.12. Note that the green graph of f^{-1} is the reflection of the blue graph of f about the line $y = x$.

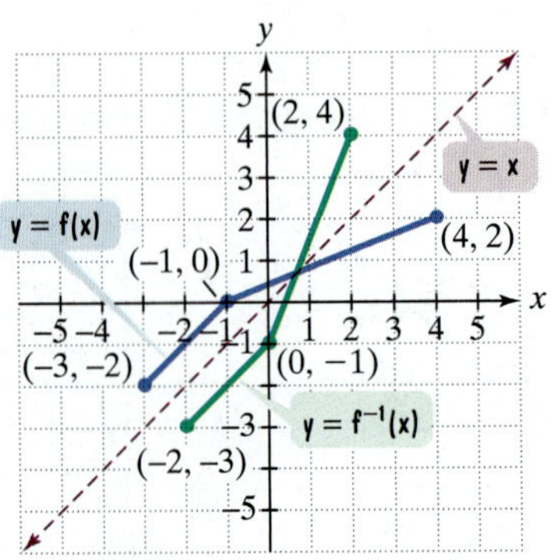

Figure 9.12 The graphs of f and f^{-1}

CHECK POINT 7 Use the graph of f in the figure below to draw the graph of its inverse function.

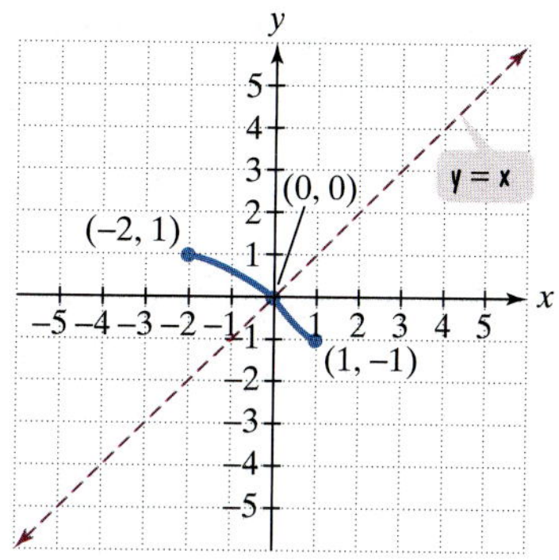

EXERCISE SET 9.2

Practice Exercises

In Exercises 1–14, find

a. $(f \circ g)(x)$;
b. $(g \circ f)(x)$;
c. $(f \circ g)(2)$.

1. $f(x) = 2x, \quad g(x) = x + 7$

2. $f(x) = 3x, \quad g(x) = x - 5$

3. $f(x) = x + 4, \quad g(x) = 2x + 1$

4. $f(x) = 5x + 2, \quad g(x) = 3x - 4$

5. $f(x) = 4x - 3, \quad g(x) = 5x^2 - 2$

6. $f(x) = 7x + 1, \quad g(x) = 2x^2 - 9$

7. $f(x) = x^2 + 2, \quad g(x) = x^2 - 2$

8. $f(x) = x^2 + 1, \quad g(x) = x^2 - 3$

9. $f(x) = \sqrt{x}, \quad g(x) = x - 1$

10. $f(x) = \sqrt{x}, \quad g(x) = x + 2$

11. $f(x) = 2x - 3, \quad g(x) = \dfrac{x + 3}{2}$

12. $f(x) = 6x - 3, \quad g(x) = \dfrac{x + 3}{6}$

13. $f(x) = \dfrac{1}{x}, \quad g(x) = \dfrac{1}{x}$

14. $f(x) = \dfrac{1}{x}, \quad g(x) = \dfrac{2}{x}$

In Exercises 15–24, find $f(g(x))$ and $g(f(x))$ and determine whether each pair of functions f and g are inverses of each other.

15. $f(x) = 4x$ and $g(x) = \dfrac{x}{4}$

16. $f(x) = 6x$ and $g(x) = \dfrac{x}{6}$

17. $f(x) = 3x + 8$ and $g(x) = \dfrac{x - 8}{3}$

18. $f(x) = 4x + 9$ and $g(x) = \dfrac{x - 9}{4}$

19. $f(x) = 5x - 9$ and $g(x) = \dfrac{x + 5}{9}$

20. $f(x) = 3x - 7$ and $g(x) = \dfrac{x + 3}{7}$

21. $f(x) = \dfrac{3}{x - 4}$ and $g(x) = \dfrac{3}{x} + 4$

22. $f(x) = \dfrac{2}{x - 5}$ and $g(x) = \dfrac{2}{x} + 5$

23. $f(x) = -x$ and $g(x) = -x$

24. $f(x) = \sqrt[3]{x - 4}$ and $g(x) = x^3 + 4$

The functions in Exercises 25–44 are all one-to-one. For each function:

a. *Find an equation for $f^{-1}(x)$, the inverse function.*
b. *Verify that your equation is correct by showing that $f(f^{-1}(x)) = x$ and $f^{-1}(f(x)) = x$.*

25. $f(x) = x + 3$

26. $f(x) = x + 5$

27. $f(x) = 2x$

28. $f(x) = 4x$

29. $f(x) = 2x + 3$

30. $f(x) = 3x - 1$

31. $f(x) = x^3 + 2$

32. $f(x) = x^3 - 1$

33. $f(x) = (x + 2)^3$

34. $f(x) = (x - 1)^3$

35. $f(x) = \dfrac{1}{x}$

36. $f(x) = \dfrac{2}{x}$

37. $f(x) = \sqrt{x}$

38. $f(x) = \sqrt[3]{x}$

39. $f(x) = x^2 + 1, \quad \text{for } x \geq 0$

40. $f(x) = x^2 - 1, \quad \text{for } x \geq 0$

41. $f(x) = \dfrac{2x + 1}{x - 3}$

42. $f(x) = \dfrac{2x - 3}{x + 1}$

43. $f(x) = \sqrt[3]{x - 4} + 3$

44. $f(x) = x^{\frac{3}{5}}$

Which graphs in Exercises 45–50 represent functions that have inverse functions?

45.

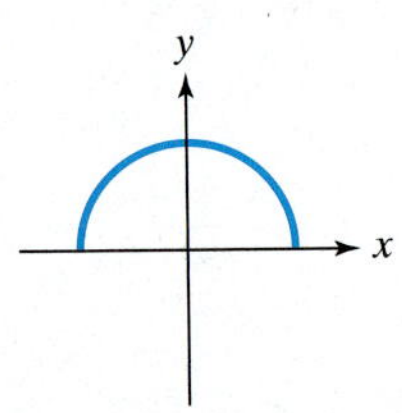

46.

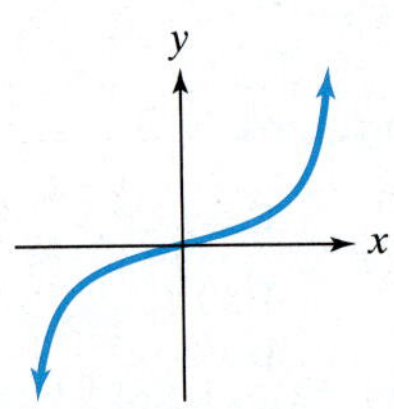

47.

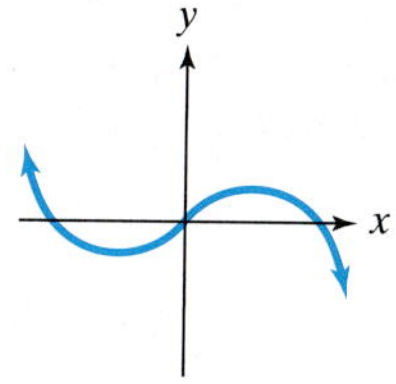

48.

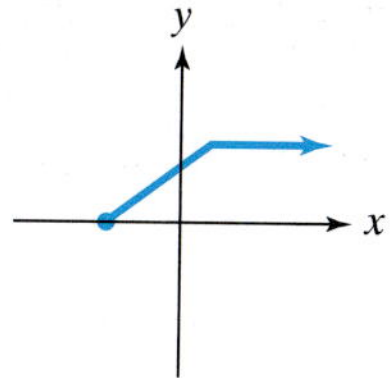

49.

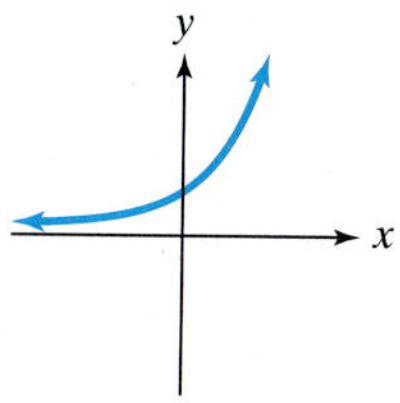

50.

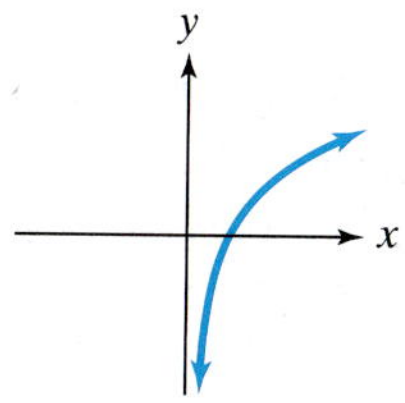

In Exercises 51–54, use the graph of f to draw the graph of its inverse function.

51.

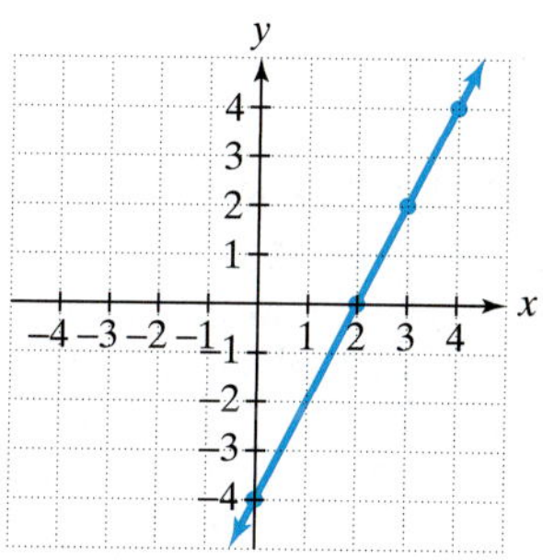

52.

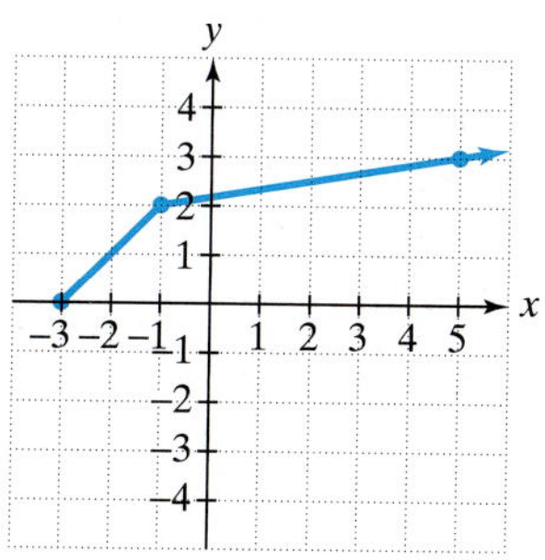

53.

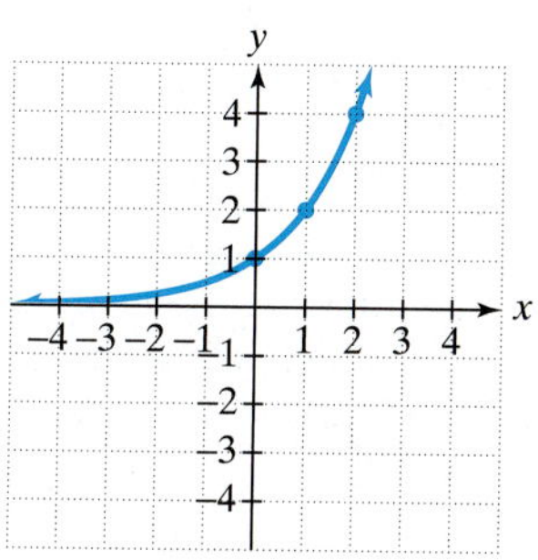

54.

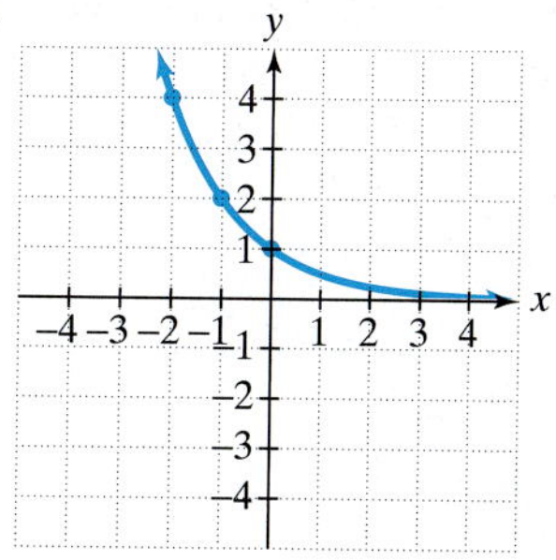

Application Exercises

55. The regular price of a computer is x dollars. Let $f(x) = x - 400$ and $g(x) = 0.75x$.
   a. Describe what the functions f and g model in terms of the price of the computer.
   b. Find $(f \circ g)(x)$ and describe what this models in terms of the price of the computer.
   c. Repeat part (b) for $(g \circ f)(x)$.
   d. Which composite function models the greater discount on the computer, $f \circ g$ or $g \circ f$? Explain.
   e. Find f^{-1} and describe what this models in terms of the price of the computer.

56. The regular price of a pair of jeans is x dollars. Let $f(x) = x - 5$ and $g(x) = 0.6x$.
   a. Describe what functions f and g model in terms of the price of the jeans.
   b. Find $(f \circ g)(x)$ and describe what this models in terms of the price of the jeans.
   c. Repeat part (b) for $(g \circ f)(x)$.
   d. Which composite function models the greater discount on the jeans, $f \circ g$ or $g \circ f$? Explain.
   e. Find f^{-1} and describe what this models in terms of the price of the jeans.

57. The graph represents the probability of two people in the same room sharing a birthday as a function of the number of people in the room. Call the function f.

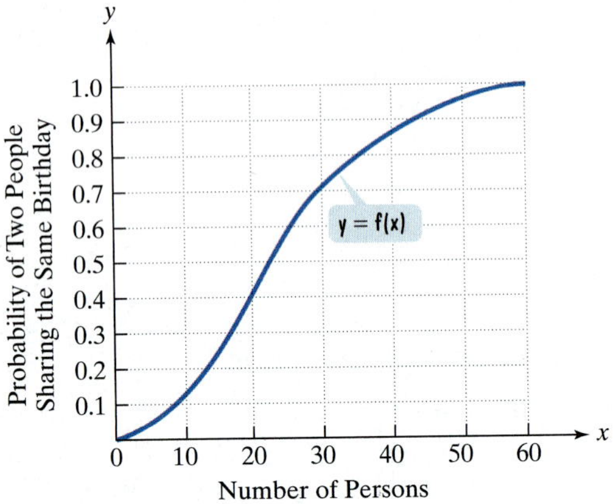

a. Explain why f has an inverse that is a function.

b. Describe in practical terms the meanings of $f^{-1}(0.25)$, $f^{-1}(0.5)$, and $f^{-1}(0.7)$.

58. The line graph shown below is based on data from the World Health Organization.

a. Explain why f has an inverse that is a function.

b. Describe in practical terms the meaning of $f^{-1}(20)$.

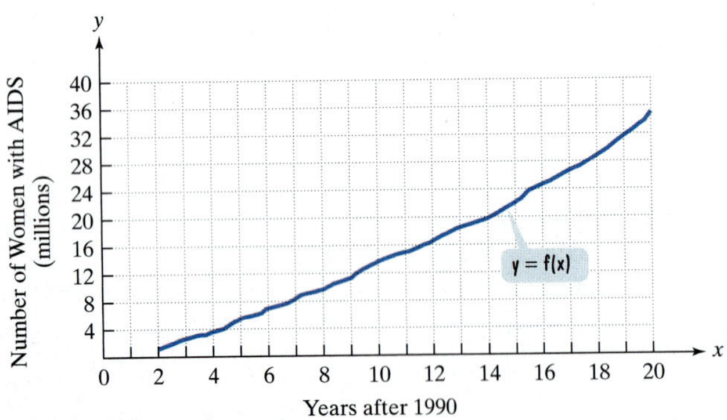

Source: Boston Globe

The graph shows the average age at which women in the United States married for the first time over a 110-year period. Use the graph to solve Exercises 59–60.

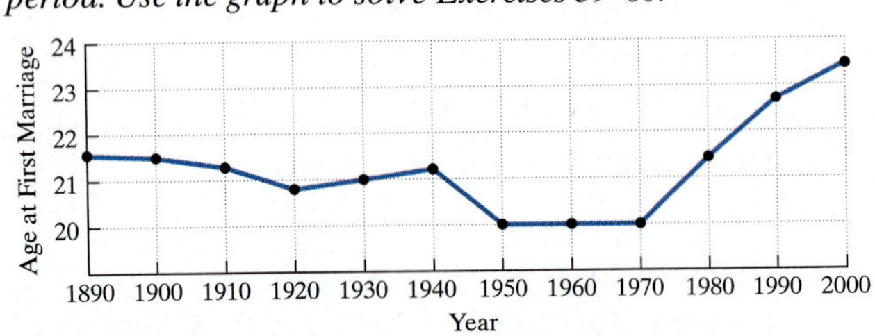

Source: U.S. Census Bureau

59. Does this graph have an inverse that is a function? What does this mean about the average age at which U.S. women married for the first time during the period shown?

60. Identify two or more years in which U.S. women married for the first time at the same average age. What is this average age?

Writing in Mathematics

61. Describe a procedure for finding $(f \circ g)(x)$.

62. Explain how to determine if two functions are inverses of each other.

63. Describe how to find the inverse of a one-to-one function.

64. What is the horizontal line test and what does it indicate?

65. Describe how to use the graph of a one-to-one function to draw the graph of its inverse function.

66. How can a graphing utility be used to visually determine if two functions are inverses of each other?

67. Consider the following function:

{(The Beatles, 20), (Elvis Presley, 18),
(Michael Jackson, 13)
(Mariah Carey, 13), (The Supremes, 12)}.

The domain is the set of the five recording artists with the most number 1 singles in the United States. The range is the set of the number of #1 singles for each artist. Reverse each of the five ordered pairs. Is the resulting relation a function? Describe what this means in terms of whether or not the given function is one-to-one. (*Source*: *The Popular Music Database*)

Technology Exercises

In Exercises 68–75, use a graphing utility to graph the function. Use the graph to determine whether the function has an inverse that is a function (that is, whether the function is one-to-one).

68. $f(x) = x^2 - 1$

69. $f(x) = \sqrt[3]{2 - x}$

70. $f(x) = \dfrac{x^3}{2}$

71. $f(x) = \dfrac{x^4}{4}$

72. $f(x) = |x - 2|$

73. $f(x) = (x - 1)^3$

74. $f(x) = -\sqrt{16 - x^2}$

75. $f(x) = x^3 + x + 1$

In Exercises 76–78, use a graphing utility to graph f and g in the same viewing rectangle. In addition, graph the line $y = x$ and visually determine if f and g are inverses.

76. $f(x) = 4x + 4, \quad g(x) = 0.25x - 1$

77. $f(x) = \frac{1}{x} + 2, \quad g(x) = \frac{1}{x - 2}$

78. $f(x) = \sqrt[3]{x} - 2, \quad g(x) = (x + 2)^3$

Critical Thinking Exercises

79. Which one of the following is true?

a. The inverse of $\{(1, 4), (2, 7)\}$ is $\{(2, 7), (1, 4)\}$.

b. The function $f(x) = 5$ is one-to-one.

c. If $f(x) = 3x$, then $f^{-1}(x) = \frac{1}{3x}$.

d. The domain of f is the same as the range of f^{-1}.

80. If $h(x) = \sqrt{3x^2 + 5}$, find functions f and g so that $h(x) = (f \circ g)(x)$.

81. If $f(x) = 3x$ and $g(x) = x + 5$, find $(f \circ g)^{-1}(x)$ and $(g^{-1} \circ f^{-1})(x)$.

82. Show that

$$f(x) = \frac{3x - 2}{5x - 3}$$

is its own inverse.

83. Consider the two functions defined by $f(x) = m_1x + b_1$ and $g(x) = m_2x + b_2$. Prove that the slope of the composite function of f with g is equal to the product of the slopes of the two functions.

Review Exercises

84. Divide and write the quotient in scientific notation:

$$\frac{4.3 \times 10^5}{8.6 \times 10^{-4}}.$$

(Section 1.7, Example 4)

85. Graph: $\quad f(x) = x^2 - 4x + 3$.
(Section 8.3, Example 3)

86. Solve: $\quad \sqrt{x + 4} - \sqrt{x - 1} = 1$.
(Section 7.6. Example 4)

▸ SECTION 9.3 Logarithmic Functions

Objectives

1. Change from logarithmic to exponential form.
2. Change from exponential to logarithmic form.
3. Evaluate logarithms.
4. Use basic logarithmic properties.
5. Graph logarithmic functions.
6. Find the domain of a logarithmic function.
7. Use common logarithms.
8. Use natural logarithms.

SSM PH Tutor Center CD-ROM Video

The earthquake that ripped through northern California on October 17, 1989, measured 7.1 on the Richter scale, killed more than 60 people, and injured more than 2400. Shown here is San Francisco's Marina district, where shock waves tossed houses off their foundations and into the street.

A higher measure on the Richter scale is more devastating than it seems because for each increase in one unit on the scale, there is a tenfold increase in the intensity of an earthquake. In this section our focus is on the inverse of the exponential function, called the logarithmic function. The logarithmic function will help you to understand diverse phenomena, including earthquake intensity, human memory, and the pace of life in large cities.

The Definition of Logarithmic Functions

No horizontal line can be drawn that intersects the graph of an exponential function at more than one point. This means that the exponential function is one-to-one and has an inverse. The inverse function of the exponential function with base b is called the **logarithmic function with base b**.

Definition of the Logarithmic Function

For $x > 0$ and $b > 0, b \neq 1$,

$$y = \log_b x \text{ is equivalent to } b^y = x.$$

The function $f(x) = \log_b x$ is the **logarithmic function with base b.**

The equations

$$y = \log_b x \quad \text{and} \quad b^y = x$$

are different ways of expressing the same thing. The first equation is in **logarithmic form** and the second equivalent equation is in **exponential form.**

Notice that a **logarithm**, y, **is an exponent**. You should learn the location of the base and exponent in each form.

Location of Base and Exponent in Exponential and Logarithmic Forms

Exponent — Exponent

Logarithmic Form: $y = \log_b x$ — Exponential Form: $b^y = x$

Base — Base

EXAMPLE 1 Changing from Logarithmic to Exponential Form

1 Change from logarithmic to exponential form.

Write each equation in its equivalent exponential form:

a. $2 = \log_5 x$ **b.** $3 = \log_b 64$ **c.** $\log_3 7 = y$.

Solution We use the fact that $y = \log_b x$ means $b^y = x$.

a. $2 = \log_5 x$ means $5^2 = x$. **b.** $3 = \log_b 64$ means $b^3 = 64$.

Logarithms are exponents. — Logarithms are exponents.

c. $\log_3 7 = y$ or $y = \log_3 7$ means $3^y = 7$. ■

✔ **CHECK POINT 1** Write each equation in its equivalent exponential form:

a. $3 = \log_7 x$ **b.** $2 = \log_b 25$ **c.** $\log_4 26 = y$.

EXAMPLE 2 Changing From Exponential to Logarithmic Form

2 Change from exponential to logarithmic form.

Write each equation in its equivalent logarithmic form:

a. $12^2 = x$ **b.** $b^3 = 8$ **c.** $e^y = 9$.

Solution We use the fact that $b^y = x$ means $y = \log_b x$.

a. $12^2 = x$ means $2 = \log_{12} x$. **b.** $b^3 = 8$ means $3 = \log_b 8$.

Logarithms are exponents. — Logarithms are exponents.

c. $e^y = 9$ means $y = \log_e 9$. ■

CHECK POINT 2 Write each equation in its equivalent logarithmic form:

a. $2^5 = x$ **b.** $b^3 = 27$ **c.** $e^y = 33.$

3 Evaluate logarithms.

Remembering that logarithms are exponents makes it possible to evaluate some logarithms by inspection. The logarithm of x with base b, $\log_b x$, is the exponent to which b must be raised to get x. For example, suppose we want to evaluate $\log_2 32$. We ask, 2 to what power gives 32? Because $2^5 = 32$, $\log_2 32 = 5$.

EXAMPLE 3 Evaluating Logarithms

Evaluate:

a. $\log_2 16$ **b.** $\log_3 9$ **c.** $\log_{25} 5$.

Solution

Logarithmic Expression	Question Needed for Evaluation	Logarithmic Expression Evaluated
a. $\log_2 16$	2 to what power gives 16?	$\log_2 16 = 4$ because $2^4 = 16$.
b. $\log_3 9$	3 to what power gives 9?	$\log_3 9 = 2$ because $3^2 = 9$.
c. $\log_{25} 5$	25 to what power gives 5?	$\log_{25} 5 = \frac{1}{2}$ because $25^{\frac{1}{2}} = \sqrt{25} = 5$.

■

CHECK POINT 3 Evaluate:

a. $\log_{10} 100$ **b.** $\log_3 3$ **c.** $\log_{36} 6$.

4 Use basic logarithmic properties.

Basic Logarithmic Properties

Because logarithms are exponents, they have properties that can be verified using properties of exponents.

Basic Logarithmic Properties Involving One

1. $\log_b b = 1$ because 1 is the exponent to which b must be raised to obtain b. $(b^1 = b)$

2. $\log_b 1 = 0$ because 0 is the exponent to which b must be raised to obtain 1. $(b^0 = 1)$

EXAMPLE 4 Using Properties of Logarithms

Evaluate:

a. $\log_7 7$ **b.** $\log_5 1$.

Solution

a. Because $\log_b b = 1$, we conclude $\log_7 7 = 1$.

b. Because $\log_b 1 = 0$, we conclude $\log_5 1 = 0$. ■

CHECK POINT 4 Evaluate:

a. $\log_9 9$ **b.** $\log_8 1$.

The inverse of the exponential function is the logarithmic function. Thus, if $f(x) = b^x$, then $f^{-1}(x) = \log_b x$. In Section 9.2, we saw how inverse functions "undo" one another. In particular,

$$f(f^{-1}(x)) = x \quad \text{and} \quad f^{-1}(f(x)) = x.$$

Applying these relationships to exponential and logarithmic functions, we obtain the following **inverse properties of logarithms**:

Inverse Properties of Logarithms

For $b > 0$ and $b \neq 1$,

$\log_b b^x = x$ The logarithm with base b of b raised to a power equals that power.

$b^{\log_b x} = x$ b raised to the logarithm with base b of a number equals that number.

EXAMPLE 5 Using Inverse Properties of Logarithms

Evaluate:

a. $\log_4 4^5$ **b.** $6^{\log_6 9}$.

Solution

a. Because $\log_b b^x = x$, we conclude $\log_4 4^5 = 5$.

b. Because $b^{\log_b x} = x$, we conclude $6^{\log_6 9} = 9$. ■

CHECK POINT 5 Evaluate:

a. $\log_7 7^8$ **b.** $3^{\log_3 17}$.

5 Graph logarithmic functions.

Graphs of Logarithmic Functions

How do we graph logarithmic functions? We use the fact that the logarithmic function is the inverse of the exponential function. This means that the logarithmic function reverses the coordinates of the exponential function. It also means that the graph of the logarithmic function is a reflection of the graph of the exponential function about the line $y = x$.

EXAMPLE 6 Graphs of Exponential and Logarithmic Functions

Graph $f(x) = 2^x$ and $g(x) = \log_2 x$ in the same rectangular coordinate system.

Solution We first set up a table of coordinates for $f(x) = 2^x$. Reversing these coordinates gives the coordinates for the inverse function $g(x) = \log_2 x$.

x	-2	-1	0	1	2	3
$f(x) = 2^x$	$\frac{1}{4}$	$\frac{1}{2}$	1	2	4	8

Reverse coordinates.

x	$\frac{1}{4}$	$\frac{1}{2}$	1	2	4	8
$g(x) = \log_2 x$	-2	-1	0	1	2	3

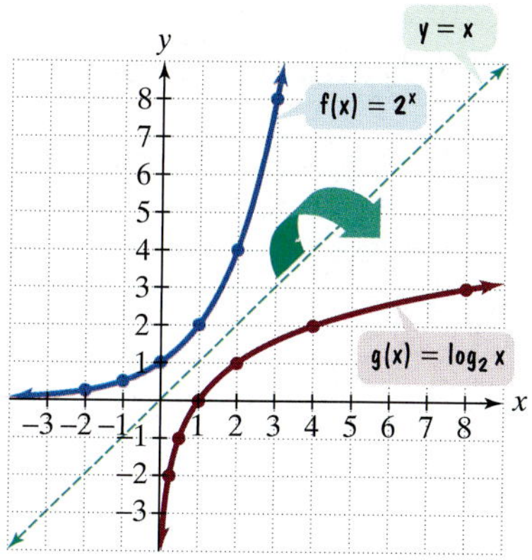

Figure 9.13 The graphs of $f(x) = 2^x$ and its inverse function

We now plot the ordered pairs in both tables, connecting them with smooth curves. Figure 9.13 shows the graphs of $f(x) = 2^x$ and its inverse function $g(x) = \log_2 x$. The graph of the inverse can also be drawn by reflecting the graph of $f(x) = 2^x$ about the line $y = x$. ■

CHECK POINT 6 Graph $f(x) = 3^x$ and $g(x) = \log_3 x$ in the same rectangular coordinate system.

Figure 9.14 illustrates the relationship between the graph of the exponential function, shown in blue and its inverse, the logarithmic function, shown in red, for bases greater than 1 and for bases between 0 and 1.

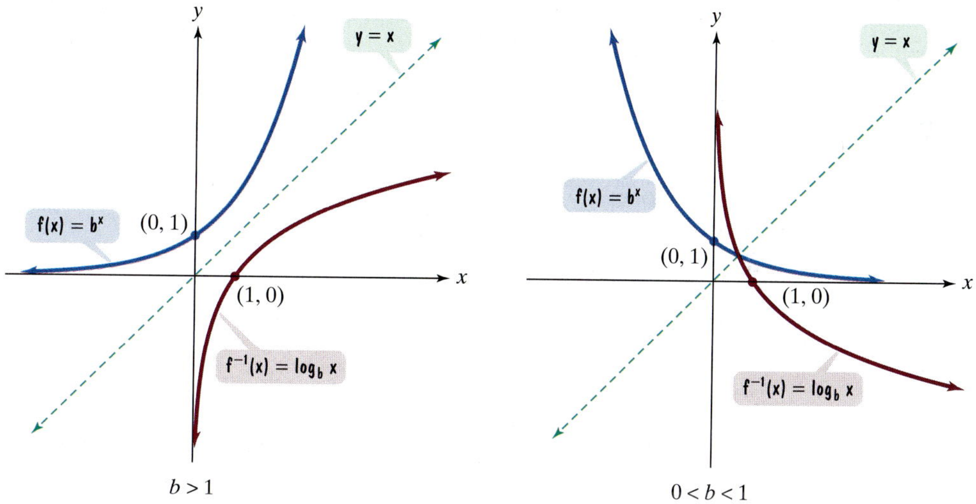

Figure 9.14 Graphs of exponential and logarithmic functions

Discover for Yourself

Verify each of the four characteristics in the box for the red graphs in Figure 9.14

Characteristics of the Graphs of Logarithmic Functions of the Form $f(x) = \log_b x$

- The x-intercept is 1. There is no y-intercept.
- The y-axis is a vertical asymptote.
- If $b > 1$, the function is increasing. If $0 < b < 1$, the function is decreasing.
- The graph is smooth and continuous. It has no sharp corners or gaps.

6 Find the domain of a logarithmic function.

The Domain of a Logarithmic Function

In Section 9.1, we learned that the domain of an exponential function of the form $f(x) = b^x$ includes all real numbers and its range is the set of positive real numbers. Because the logarithmic function reverses the domain and the range of the exponential function, the **domain of a logarithmic function of the form $f(x) = \log_b x$ is the set of all positive real numbers.** Thus, $\log_2 8$ is defined because the value of x in the logarithmic expression, 8, is greater than zero and therefore is included in the domain of the logarithmic function $f(x) = \log_2 x$. However, $\log_2 0$ and $\log_2(-8)$ are not defined because 0 and -8 are not positive real numbers and therefore are excluded from the domain of the logarithmic function $f(x) = \log_2 x$. In general, the domain of $f(x) = \log_b(x + c)$ consists of all x for which $x + c > 0$.

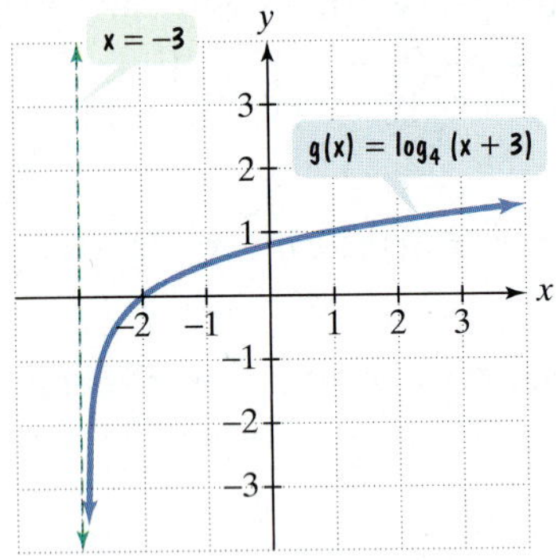

Figure 9.15 The domain of $g(x) = \log_4(x + 3)$ is $(-3, \infty)$.

EXAMPLE 7 Finding the Domain of a Logarithmic Function

Find the domain of $g(x) = \log_4(x + 3)$.

Solution The domain of g consists of all x for which $x + 3 > 0$. Solving this inequality for x, we obtain $x > -3$. Thus, the domain of g is $\{x | x > -3\}$ or $(-3, \infty)$. This is illustrated in Figure 9.15. The vertical asymptote is $x = -3$, and all points on the graph of g have x-coordinates that are greater than -3. ■

✔ **CHECK POINT 7** Find the domain of $h(x) = \log_4(x - 5)$.

7 Use common logarithms.

Common Logarithms

The logarithmic function with base 10 is called the **common logarithmic function.** The function $f(x) = \log_{10} x$ is usually expressed as $f(x) = \log x$. A calculator with a [LOG] key can be used to evaluate common logarithms. Here are some examples:

Logarithm	Most Scientific Calculator Keystrokes	Most Graphing Calculator Keystrokes	Display (or Approximate Display)
$\log 1000$	1000 [LOG]	[LOG] 1000 [ENTER]	3
$\log \frac{5}{2}$	[(] 5 [÷] 2 [)] [LOG]	[LOG] [(] 5 [÷] 2 [)] [ENTER]	0.39794
$\frac{\log 5}{\log 2}$	5 [LOG] [÷] 2 [LOG] [=]	[LOG] 5 [÷] [LOG] 2 [ENTER]	2.32192
$\log(-3)$	3 [+/−] [LOG]	[LOG] [(−)] 3 [ENTER]	[ERROR]

The error message given by many calculators for $\log(-3)$ is a reminder that the domain of every logarithmic function, including the common logarithmic function, is the set of positive real numbers.

Many real-life phenomena start with rapid growth, and then the growth begins to level off. This type of behavior can be modeled by logarithmic functions.

EXAMPLE 8 Modeling Height of Children

The percentage of adult height attained by a boy who is x years old can be modeled by

$$f(x) = 29 + 48.8 \log(x + 1)$$

where x represents the boy's age and $f(x)$ represents the percentage of his adult height. Approximately what percentage of his adult height is a boy at age eight?

Solution We substitute the boy's age, 8, for x and evaluate the function at 8.

$$f(x) = 29 + 48.8 \log(x + 1) \quad \text{This is the given function.}$$

$$f(8) = 29 + 48.8 \log(8 + 1) \quad \text{Substitute 8 for x.}$$

$$= 29 + 48.8 \log 9$$

Graphing calculator keystrokes: 29 [+] 48.8 [×] [LOG] 9 [ENTER]

$$\approx 76$$

Thus, an 8-year-old boy is approximately 76% of his adult height. ■

✔ CHECK POINT 8 Use the function in Example 8 to answer this question: Approximately what percentage of his adult height is a boy at age ten?

The basic properties of logarithms that were listed earlier in this section can be applied to common logarithms.

Properties of Common Logarithms

General Properties	Common Logarithms
1. $\log_b 1 = 0$	**1.** $\log 1 = 0$
2. $\log_b b = 1$	**2.** $\log 10 = 1$
3. $\log_b b^x = x$	**3.** $\log 10^x = x$
4. $b^{\log_b x} = x$	**4.** $10^{\log x} = x$

The property $\log 10^x = x$ can be used to evaluate common logarithms involving powers of 10. For example,

$$\log 100 = \log 10^2 = 2, \quad \log 1000 = \log 10^3 = 3, \text{ and } \log 10^{7.1} = 7.1.$$

EXAMPLE 9 Earthquake Intensity

The magnitude, R, on the Richter scale of an earthquake of intensity I is given by

$$R = \log \frac{I}{I_0}$$

where I_0 is the intensity of a barely felt zero-level earthquake. The earthquake that destroyed San Francisco in 1906 was $10^{8.3}$ times as intense as a zero-level earthquake. What was its magnitude on the Richter scale?

Solution Because the earthquake was $10^{8.3}$ times as intense as a zero-level earthquake, the intensity, I, is $10^{8.3}I_0$.

$$R = \log \frac{I}{I_0}$$ This is the formula for magnitude on the Richter scale.

$$R = \log \frac{10^{8.3}I_0}{I_0}$$ Substitute $10^{8.3}I_0$ for I.

$$= \log 10^{8.3}$$ Simplify.

$$= 8.3$$ Use the property $\log 10^x = x$.

San Francisco's 1906 earthquake registered 8.3 on the Richter scale. ■

✔ CHECK POINT 9 Use the formula in Example 9 to solve this problem. If an earthquake is 10,000 times as intense as a zero-level quake $(I = 10{,}000I_0)$, what is its magnitude on the Richter scale?

8 Use natural logarithms.

Natural Logarithms

The logarithmic function with base e is called the **natural logarithmic function.** The function $f(x) = \log_e x$ is usually expressed as $f(x) = \ln x$, read "el en of x." A calculator with an LN key can be used to evaluate natural logarithms.

Like the domain of all logarithmic functions, the domain of the natural logarithmic function is the set of all positive real numbers. Thus, the domain of $f(x) = \ln(x + c)$ consists of all x for which $x + c > 0$.

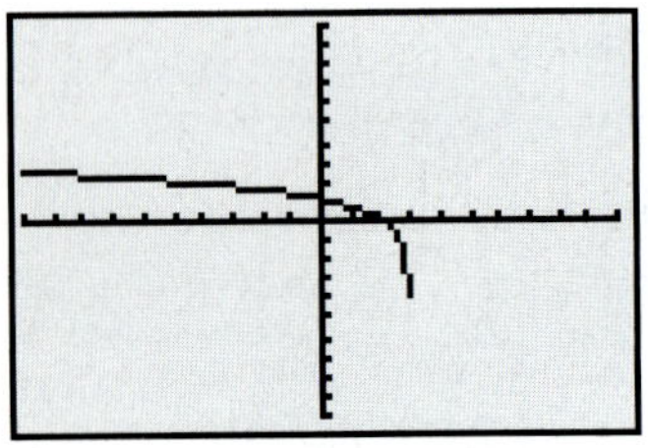

Figure 9.16 The domain of $f(x) = \ln(3 - x)$ is $(-\infty, 3)$.

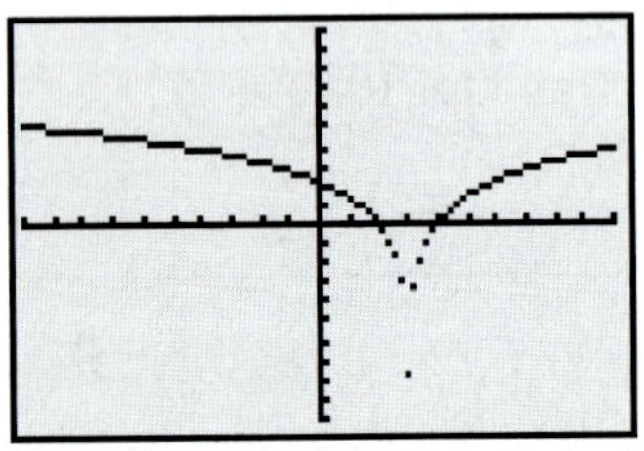

Figure 9.17 3 is excluded from the domain of $g(x) = \ln(x - 3)^2$.

EXAMPLE 10 Finding Domains of Natural Logarithmic Functions

Find the domain of each function:

a. $f(x) = \ln(3 - x)$ **b.** $g(x) = \ln(x - 3)^2$.

Solution

a. The domain of f consists of all x for which $3 - x > 0$. Solving this inequality for x, we obtain $x < 3$. Thus, the domain of f is $\{x|x < 3\}$ or $(-\infty, 3)$. This is verified by the graph in Figure 9.16.

b. The domain of g consists of all x for which $(x - 3)^2 > 0$. It follows that the domain of g is all real numbers except 3. Thus, the domain of g is $\{x|x \neq 3\}$ or $(-\infty, 3) \cup (3, \infty)$. This is shown by the graph in Figure 9.17. To make it more obvious that 3 is excluded from the domain, we used a DOT format. ■

✔ **CHECK POINT 10** Find the domain of each function:

a. $f(x) = \ln(4 - x)$

b. $g(x) = \ln x^2$.

The basic properties of logarithms that were listed earlier in this section can be applied to natural logarithms.

Properties of Natural Logarithms

General Properties	Natural Logarithms
1. $\log_b 1 = 0$	**1.** $\ln 1 = 0$
2. $\log_b b = 1$	**2.** $\ln e = 1$
3. $\log_b b^x = x$	**3.** $\ln e^x = x$
4. $b^{\log_b x} = x$	**4.** $e^{\ln x} = x$

The property $\ln e^x = x$ can be used to evaluate natural logarithms involving powers of e. For example,

$$\ln e^2 = 2, \quad \ln e^3 = 3, \quad \ln e^{7.1} = 7.1, \quad \text{and} \quad \ln \frac{1}{e} = \ln e^{-1} = -1.$$

EXAMPLE 11 Using Inverse Properties

Use inverse properties to simplify:

a. $\ln e^{7x}$ **b.** $e^{\ln 4x^2}$.

Solution

a. Because $\ln e^x = x$, we conclude that $\ln e^{7x} = 7x$.

b. Because $e^{\ln x} = x$, we conclude that $e^{\ln 4x^2} = 4x^2$. ■

✔ **CHECK POINT 11** Use inverse properties to simplify:

a. $\ln e^{25x}$ **b.** $e^{\ln \sqrt{x}}$.

EXAMPLE 12 Walking Speed and City Population

As the population of a city increases, the pace of life also increases. The formula

$$W = 0.35 \ln P + 2.74$$

models average walking speed, W, in feet per second, for a resident of a city whose population is P thousand. Find the average walking speed for people living in New York City with a population of 7323 thousand.

Solution We use the formula and substitute 7323 for P, the population in thousands.

$W = 0.35 \ln P + 2.74$ — This is the given formula.

$W = 0.35 \ln 7323 + 2.74$ — Substitute 7323 for P.

≈ 5.9 — Graphing calculator keystrokes: 0.35 [×] [LN] 7323 [+] 2.74 [ENTER]. (On some calculators, a parenthesis is needed after 7323.)

The average walking speed in New York City is approximately 5.9 feet per second. ■

✔ **CHECK POINT 12** Use the formula $W = 0.35 \ln P + 2.74$ to find the average walking speed in Jackson, Mississippi with a population of 197 thousand.

EXERCISE SET 9.3

Practice Exercises

In Exercises 1–8, write each equation in its equivalent exponential form.

1. $4 = \log_2 16$
2. $6 = \log_2 64$
3. $2 = \log_3 x$
4. $2 = \log_9 x$
5. $5 = \log_b 32$
6. $3 = \log_b 27$
7. $\log_6 216 = y$
8. $\log_5 125 = y$

In Exercises 9–20, write each equation in its equivalent logarithmic form.

9. $2^3 = 8$
10. $5^4 = 625$
11. $2^{-4} = \frac{1}{16}$
12. $5^{-3} = \frac{1}{125}$
13. $\sqrt[3]{8} = 2$
14. $\sqrt[3]{64} = 4$
15. $13^2 = x$
16. $15^2 = x$
17. $b^3 = 1000$
18. $b^3 = 343$
19. $7^y = 200$
20. $8^y = 300$

In Exercises 21–38, evaluate each expression without using a calculator.

21. $\log_4 16$
22. $\log_7 49$
23. $\log_2 64$
24. $\log_3 27$
25. $\log_7 \sqrt{7}$
26. $\log_6 \sqrt{6}$
27. $\log_2 \frac{1}{8}$
28. $\log_3 \frac{1}{9}$
29. $\log_{64} 8$
30. $\log_{81} 9$
31. $\log_5 5$
32. $\log_{11} 11$
33. $\log_4 1$
34. $\log_6 1$
35. $\log_5 5^7$
36. $\log_4 4^6$
37. $8^{\log_8 19}$
38. $7^{\log_7 23}$

39. Graph $f(x) = 4^x$ and $g(x) = \log_4 x$ in the same rectangular coordinate system.

40. Graph $f(x) = 5^x$ and $g(x) = \log_5 x$ in the same rectangular coordinate system.

41. Graph $f(x) = \left(\frac{1}{2}\right)^x$ and $g(x) = \log_{\frac{1}{2}} x$ in the same rectangular coordinate system.

42. Graph $f(x) = \left(\frac{1}{4}\right)^x$ and $g(x) = \log_{\frac{1}{4}} x$ in the same rectangular coordinate system.

In Exercises 43–48, find the domain of each logarithmic function.

43. $f(x) = \log_5(x + 4)$

44. $f(x) = \log_5(x + 6)$

45. $f(x) = \log(2 - x)$

46. $f(x) = \log(7 - x)$

47. $f(x) = \ln(x - 2)^2$

48. $f(x) = \ln(x - 7)^2$

In Exercises 49–62, evaluate each expression without using a calculator.

49. $\log 100$

50. $\log 1000$

51. $\log 10^7$

52. $\log 10^8$

53. $10^{\log 33}$

54. $10^{\log 53}$

55. $\ln 1$

56. $\ln e$

57. $\ln e^6$

58. $\ln e^7$

59. $\ln \frac{1}{e^6}$

60. $\ln \frac{1}{e^7}$

61. $e^{\ln 125}$

62. $e^{\ln 300}$

In Exercises 75–80, use inverse properties of logarithms to simplify each expression.

63. $\ln e^{9x}$

64. $\ln e^{13x}$

65. $e^{\ln 5x^2}$

66. $e^{\ln 7x^2}$

67. $10^{\log \sqrt{x}}$

68. $10^{\log \sqrt[3]{x}}$

Application Exercises

The percentage of adult height attained by a girl who is x years old can be modeled by

$$f(x) = 62 + 35\log(x - 4)$$

where x represents the girl's age (from 5 to 15) and f(x) represents the percentage of her adult height. Use the formula to solve Exercises 69–70.

69. Approximately what percentage of her adult height is a girl at age 13?

70. Approximately what percentage of her adult height is a girl at age ten?

71. The annual amount that we spend to attend sporting events can be modeled by

$$f(x) = 2.05 + 1.3\ln x$$

where x represents the number of years after 1984 and $f(x)$ represents the total annual expenditures for admission to spectator sports, in billions of dollars. In 2000, approximately how much was spent on admission to spectator sports?

72. The percentage of U.S. households with cable television can be modeled by

$$f(x) = 18.32 + 15.94\ln x$$

where x represents the number of years after 1979 and $f(x)$ represents the percentage of U.S. households with cable television. What percentage of U.S. households had cable television in 1990?

The loudness level of a sound, D, in decibels, is given by the formula

$$D = 10\log(10^{12}I)$$

where I is the intensity of the sound, in watts per meter². Decibel levels range from 0, a barely audible sound, to 160, a sound resulting in a ruptured eardrum. Use the formula to solve Exercises 73–74.

73. The sound of a blue whale can be heard 500 miles away, reaching an intensity of 6.3×10^6 watts per meter^2. Determine the decibel level of this sound. At close range, can the sound of a blue whale rupture the human eardrum?

74. What is the decibel level of a normal conversation, 3.2×10^{-6} watt per meter^2?

75. Students in a psychology class took a final examination. As part of an experiment to see how much of the course content they remembered over time, they took equivalent forms of the exam in monthly intervals thereafter. The average score for the group, $f(t)$, after t months was modeled by the function

$$f(t) = 88 - 15\ln(t + 1), \quad 0 \le t \le 12.$$

a. What was the average score on the original exam?

b. What was the average score after 2 months? 4 months? 6 months? 8 months? 10 months? one year?

c. Sketch the graph of f (either by hand or with a graphing utility). Describe what the graph indicates in terms of the material retained by the students.

Writing in Mathematics

76. Describe the relationship between an equation in logarithmic form and an equivalent equation in exponential form.

77. What question can be asked to help evaluate $\log_3 81$?

78. Explain why the logarithm of 1 with base b is 0.

79. Describe the following property using words: $\log_b b^x = x$.

80. Explain how to use the graph of $f(x) = 2^x$ to obtain the graph of $g(x) = \log_2 x$.

81. Explain how to find the domain of a logarithmic function.

82. New York City is one of the world's great walking cities. Use the formula in Example 12 on page 649 to describe what frequently happens to tourists exploring the city by foot.

83. Logarithmic models are well suited to phenomena in which growth is initially rapid but then begins to level off. Describe something that is changing over time that can be modeled using a logarithmic function.

84. Suppose that a girl is 4'6" at age 10. Explain how to use the function in Exercises 69–70 to determine how all she can expect to be as an adult.

Technology Exercises

In Exercises 85–88, graph f and g in the same viewing rectangle. Then describe the relationship of the graph of g to the graph of f.

85. $f(x) = \ln x$, $g(x) = \ln(x + 3)$

86. $f(x) = \ln x$, $g(x) = \ln x + 3$

87. $f(x) = \log x$, $g(x) = -\log x$

88. $f(x) = \log x$, $g(x) = \log(x - 2) + 1$

89. Students in a mathematics class took a final examination. They took equivalent forms of the exam in monthly intervals thereafter. The average score, $f(t)$, for the group after t months was modeled by the human memory function $f(t) = 75 - 10\log(t + 1)$, where $0 \le t \le 12$. Use a graphing utility to graph the function. Then determine how many months will elapse before the average score falls below 65.

90. Graph f and g in the same viewing rectangle.

a. $f(x) = \ln(3x)$, $g(x) = \ln 3 + \ln x$

b. $f(x) = \log(5x^2)$, $g(x) = \log 5 + \log x^2$

c. $f(x) = \ln(2x^3)$, $g(x) = \ln 2 + \ln x^3$

d. Describe what you observe in parts (a)–(c). Generalize this observation by writing an equivalent expression. In each case, the graphs of f and g are the same for $\log_b(MN)$, where $M > 0$ and $N > 0$.

e. Complete this statement: The logarithm of a product is equal to ____________.

91. Graph each of the following functions in the same viewing rectangle and then place the functions in order from the one that increases most slowly to the one that increases most rapidly.

$$y = x,\ y = \sqrt{x},\ y = e^x,\ y = \ln x,\ y = x^x,\ y = x^2$$

Critical Thinking Exercises

92. Which one of the following is true?

a. $\dfrac{\log_2 8}{\log_2 4} = \dfrac{8}{4}$

b. $\log(-100) = -2$.

c. The domain of $f(x) = \log_2 x$ is $(-\infty, \infty)$.

d. $\log_b x$ is the exponent to which b must be raised to obtain x.

93. Without using a calculator, find the exact value of

$$\frac{\log_3 81 - \log_\pi 1}{\log_{2\sqrt{2}} 8 - \log 0.001}.$$

94. Solve for x: $\log_4[\log_3(\log_2 x)] = 0$.

95. Without using a calculator, determine which is the greater number: $\log_4 60$ or $\log_3 40$.

Review Exercises

96. Solve the system:

$$2x = 11 - 5y$$
$$3x - 2y = -12.$$

(Section 3.1, Example 5)

97. Factor completely:

$$6x^2 - 8xy + 2y^2.$$

(Section 5.4, Example 9)

98. Solve: $x + 3 \le -4$ or $2 - 7x \le 16$.

(Section 4.2, Example 6)

▸ SECTION 9.4 Properties of Logarithms

Objectives

1. Use the product rule.
2. Use the quotient rule.
3. Use the power rule.
4. Expand logarithmic expressions.
5. Condense logarithmic expressions.
6. Use the change-of-base property.

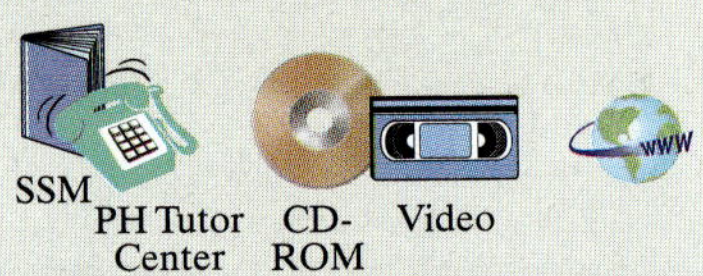

We all learn new things in different ways. In this section, we consider important properties of logarithms. What would be the most effective way for you to learn about these properties? Would it be helpful to use your graphing utility and discover one of these properties for yourself? To do so, work Exercise 90 in Exercise Set 9.3 before continuing. Would the properties become more meaningful if you could see exactly where they come from? If so, you will find details of the proofs of many of these properties in the appendix. The remainder of our work in this chapter will be based on the properties of logarithms that you learn in this section.

1 Use the product rule.

The Product Rule

Properties of exponents correspond to properties of logarithms. For example, when we multiply with the same base, we add exponents:

$$b^M \cdot b^N = b^{M+N}.$$

This property of exponents, coupled with an awareness that a logarithm is an exponent, suggests the following property, called the **product rule**:

The Product Rule

Let b, M, and N be positive real numbers with $b \neq 1$.

$$\log_b(MN) = \log_b M + \log_b N$$

The logarithm of a product is the sum of the logarithms.

When we use the product rule to write a single logarithm as the sum of two logarithms, we say that we are **expanding a logarithmic expression**. For example, we can use the product rule to expand $\ln(4x)$:

$$\ln(7x) = \ln 7 + \ln x.$$

The logarithm of a product | is | the sum of the logarithms.

EXAMPLE 1 Using the Product Rule

Use the product rule to expand each logarithmic expression:

a. $\log_4(7 \cdot 5)$ **b.** $\log(10x)$.

Solution

a. $\log_4(7 \cdot 5) = \log_4 7 + \log_4 5$ The logarithm of a product is the sum of the logarithms.

b. $\log(10x) = \log 10 + \log x$ The logarithm of a product is the sum of the logarithms. These are common logarithms with base 10 understood.

$= 1 + \log x$ Because $\log_b b = 1$, then $\log_{10} 10 = 1$. ■

✔ **CHECK POINT 1** Use the product rule to expand each logarithmic expression:

a. $\log_6(7 \cdot 11)$ **b.** $\log(100x)$.

2 Use the quotient rule.

The Quotient Rule

When we divide with the same base, we subtract exponents:

$$\frac{b^M}{b^N} = b^{M-N}.$$

This property suggests the following property of logarithms, called the **quotient rule**:

> **The Quotient Rule**
>
> Let b, M, and N be positive real numbers with $b \neq 1$.
>
> $$\log_b\left(\frac{M}{N}\right) = \log_b M - \log_b N$$
>
> The logarithm of a quotient is the difference of the logarithms.

When we use the quotient rule to write a single logarithm as the difference of two logarithms, we say that we are **expanding a logarithmic expression**. For example, we can use the quotient rule to expand $\log \frac{x}{2}$:

$$\log \frac{x}{2} = \log x - \log 2.$$

The logarithm of a quotient is the difference of the logarithms.

EXAMPLE 2 Using the Quotient Rule

Use the quotient rule to expand each logarithmic expression:

a. $\log_7\left(\frac{19}{x}\right)$ **b.** $\ln\left(\frac{e^3}{7}\right)$.

Solution

a. $\log_7\left(\frac{19}{x}\right) = \log_7 19 - \log_7 x$ The logarithm of a quotient is the difference of the logarithms.

b. $\ln\left(\frac{e^3}{7}\right) = \ln e^3 - \ln 7$ The logarithm of a quotient is the difference of the logarithms. These are natural logarithms with base e understood.

$= 3 - \ln 7$ Because $\ln e^x = x$, then $\ln e^3 = 3$. ■

CHECK POINT 2 Use the quotient rule to expand each logarithmic expression:

a. $\log_8\left(\frac{23}{x}\right)$ **b.** $\ln\left(\frac{e^5}{11}\right)$.

3 Use the power rule.

The Power Rule

When an exponential expression is raised to a power, we multiply exponents:

$$(b^M)^p = b^{Mp}.$$

This property suggests the following property of logarithms, called the **power rule**:

The Power Rule

Let b, M, and N be positive real numbers with $b \neq 1$, and let p be any real number.

$$\log_b M^p = p \log_b M$$

The logarithm of a number with an exponent is the product of the exponent and the logarithm of that number.

When we use the power rule to "pull the exponent to the front," we say that we are **expanding a logarithmic expression**. For example, we can use the power rule to expand $\ln x^2$:

$$\ln x^2 = 2 \ln x.$$

The logarithm of a number with an exponent | is | the product of the exponent and the logarithm of that number.

Figure 9.18 shows the graphs of $y = \ln x^2$ and $y = 2 \ln x$. Are $\ln x^2$ and $2 \ln x$ the same? The graphs illustrate that $y = \ln x^2$ and $y = 2 \ln x$ have different domains. The graphs are only the same if $x > 0$. Thus, we should write

$$\ln x^2 = 2 \ln x \quad \text{for} \quad x > 0.$$

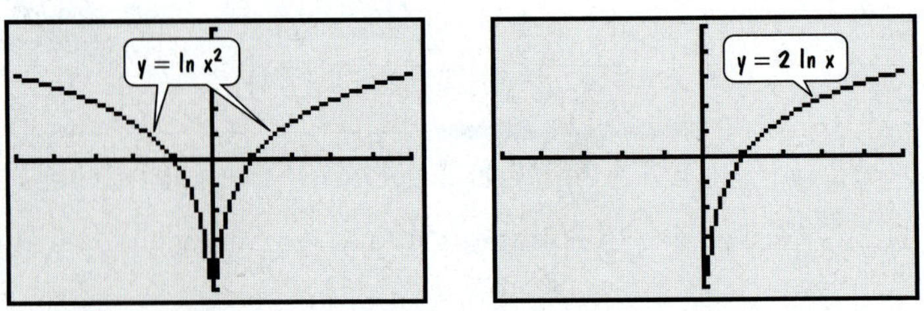

Domain: $(-\infty, 0) \cup (0, \infty)$ Domain: $(0, \infty)$

Figure 9.18 $\ln x^2$ and $2 \ln x$ have different domains.

When expanding a logarithmic expression, you might want to determine whether the rewriting has changed the domain of the expression. For the rest of this section, assume that all variables and variable expressions represent positive numbers.

4 Expand logarithmic expressions.

EXAMPLE 3 Using the Power Rule

Use the power rule to expand each logarithmic expression:

a. $\log_5 7^4$ **b.** $\ln \sqrt{x}$.

Solution

a. $\log_5 7^4 = 4\log_5 7$ The logarithm of a number with an exponent is the exponent times the logarithm of that number.

b. $\ln \sqrt{x} = \ln x^{\frac{1}{2}}$ Rewrite the radical using a rational exponent.

$= \frac{1}{2}\ln x$ Use the power rule to bring the exponent to the front. ■

CHECK POINT 3 Use the power rule to expand each logarithmic expression:

a. $\log_6 8^9$ **b.** $\ln \sqrt[3]{x}$.

Study Tip

The graphs show

$$y_1 = \ln(x+3)$$

and

$$y_2 = \ln x + \ln 3.$$

The graphs are not the same. Thus, we see that

$$\ln(x+3) \neq \ln x + \ln 3.$$

In general,

$$\log_b(M+N) \neq \log_b M + \log_b N.$$

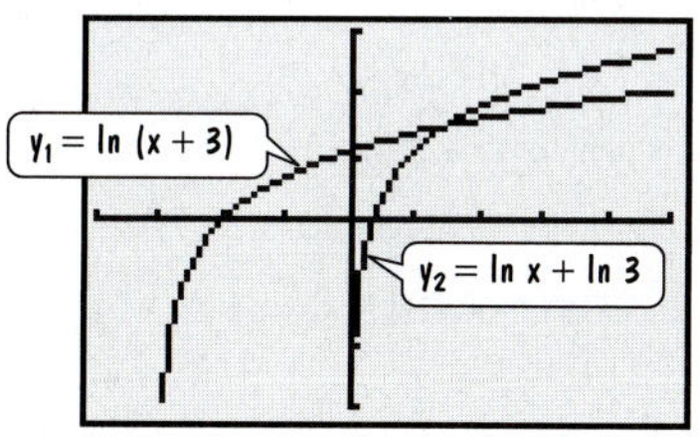

Try to avoid the following errors.

INCORRECT

$\log_b(M+N) = \log_b M + \log_b N$

$\log_b(M-N) = \log_b M - \log_b N$

$\log_b(M \cdot N) = \log_b M \cdot \log_b N$

$\log_b\left(\frac{M}{N}\right) = \frac{\log_b M}{\log_b N}$

$\frac{\log_b M}{\log_b N} = \log_b M - \log_b N$

Expanding Logarithmic Expressions

It is sometimes necessary to use more than one property of logarithms when you expand a logarithmic expression. Properties for expanding logarithmic expressions are as follows:

Properties for Expanding Logarithmic Expressions

1. $\log_b(MN) = \log_b M + \log_b N$ Product rule
2. $\log_b\left(\frac{M}{N}\right) = \log_b M - \log_b N$ Quotient rule
3. $\log_b M^p = p\log_b M$ Power rule

In all cases, $M > 0$ and $N > 0$.

EXAMPLE 4 Expanding Logarithmic Expressions

Use logarithmic properties to expand each expression as much as possible:

a. $\log_b x^2\sqrt{y}$ **b.** $\log_6\left(\frac{\sqrt[3]{x}}{36y^4}\right)$.

Solution We will have to use two or more of the properties for expanding logarithms in each part of this example.

a. $\log_b x^2\sqrt{y} = \log_b x^2 y^{\frac{1}{2}}$ Use exponential notation.

$= \log_b x^2 + \log_b y^{\frac{1}{2}}$ Use the product rule.

$= 2\log_b x + \frac{1}{2}\log_b y$ Use the power rule.

b. $\log_6\left(\frac{\sqrt[3]{x}}{36y^4}\right) = \log_6 \frac{x^{\frac{1}{3}}}{36y^4}$ — Use exponential notation.

$= \log_6 x^{\frac{1}{3}} - \log_6 36y^4$ — Use the quotient rule.

$= \log_6 x^{\frac{1}{3}} - (\log_6 36 + \log_6 y^4)$ — Use the product rule on $\log_6 36y^4$.

$= \frac{1}{3}\log_6 x - (\log_6 36 + 4\log_6 y)$ — Use the power rule.

$= \frac{1}{3}\log_6 x - \log_6 36 - 4\log_6 y$ — Apply the distributive property.

$= \frac{1}{3}\log_6 x - 2 - 4\log_6 y$ — $\log_6 36 = 2$ because 2 is the power to which we must raise 6 to get 36. $(6^2 = 36)$ ■

✔ **CHECK POINT 4** Use logarithmic properties to expand each expression as much as possible:

a. $\log_b x^4\sqrt[3]{y}$ **b.** $\log_5\left(\frac{\sqrt{x}}{25y^3}\right)$.

5 Condense logarithmic expressions.

Condensing Logarithmic Expressions

To **condense a logarithmic expression**, we write the sum or difference of two or more logarithmic expressions as a single logarithmic expression. We use the properties of logarithms to do so.

Properties for Condensing Logarithmic Expressions

1. $\log_b M + \log_b N = \log_b(MN)$ — Product rule
2. $\log_b M - \log_b N = \log_b\left(\frac{M}{N}\right)$ — Quotient rule
3. $p\log_b M = \log_b M^p$ — Power rule

In all cases, $M > 0$ and $N > 0$.

EXAMPLE 5 Condensing Logarithmic Expressions

Write as a single logarithm:

a. $\log_4 2 + \log_4 32$ **b.** $\log(4x - 3) - \log x$.

Solution

a. $\log_4 2 + \log_4 32 = \log_4(2 \cdot 32)$ — Use the product rule.

$= \log_4 64$ — We now have a single logarithm. However, we can simplify.

$= 3$ — $\log_4 64 = 3$ because $4^3 = 64$.

b. $\log(4x - 3) - \log x = \log\left(\frac{4x - 3}{x}\right)$ — Use the quotient rule. ■

✔ **CHECK POINT 5** Write as a single logarithm:

a. $\log 25 + \log 4$ **b.** $\log(7x + 6) - \log x$.

Coefficients of logarithms must be 1 before you can condense them using the product and quotient rules. For example, to condense

$$2 \ln x + \ln(x + 1),$$

the coefficient of the first term must be 1. We use the power rule to rewrite the coefficient as an exponent:

1. Make the number in front an exponent.

$$2 \ln x + \ln(x + 1) = \ln x^2 + \ln(x + 1) = \ln x^2(x + 1).$$

2. Use the product rule. The sum of logarithms with coefficients 1 is the logarithm of the product.

EXAMPLE 6 Condensing Logarithmic Expressions

Write as a single logarithm:

a. $\frac{1}{2}\log x + 4\log(x - 1)$ **b.** $3\ln(x + 7) - \ln x$.

Solution

a. $\frac{1}{2}\log x + 4\log(x - 1)$

$= \log x^{\frac{1}{2}} + \log(x - 1)^4$ Use the power rule so that all coefficients are 1.

$= \log x^{\frac{1}{2}}(x - 1)^4$ Use the product rule.

$= \log \sqrt{x}(x - 1)^4$

b. $3\ln(x + 7) - \ln x$

$= \ln(x + 7)^3 - \ln x$ Use the power rule so that all coefficients are 1.

$= \ln \frac{(x + 7)^3}{x}$ Use the quotient rule. ■

✔ **CHECK POINT 6** Write as a single logarithm:

a. $2\ln x + \frac{1}{3}\ln(x + 5)$ **b.** $2\log(x - 3) - \log x$.

6 Use the change-of-base property.

The Change-of-Base Property

We have seen that calculators give the values of both common logarithms (base 10) and natural logarithms (base e). To find a logarithm with any other base, we can use the following change-of-base property:

The Change-of-Base Property

For any logarithmic bases a and b, and any positive number M,

$$\log_b M = \frac{\log_a M}{\log_a b}.$$

The logarithm of M with base b is equal to the logarithm of M with any new base divided by the logarithm of b with that new base.

In the change-of-base property, base b is the base of the original logarithm. Base a is a new base that we introduce. Thus, the change-of-base property allows us to change from base b to *any* new base a, as long as the newly introduced base is a positive number not equal to 1.

The change-of-base property is used to write a logarithm in terms of quantities that can be evaluated with a calculator. Because calculators contain keys for common (base 10) and natural (base e) logarithms, we will frequently introduce base 10 or base e.

Change-of-Base Property

$$\log_b M = \frac{\log_a M}{\log_a b}$$

a is the new introduced base.

Introducing Common Logarithms

$$\log_b M = \frac{\log_{10} M}{\log_{10} b}$$

10 is the new introduced base.

Introducing Natural Logarithms

$$\log_b M = \frac{\log_e M}{\log_e b}$$

e is the new introduced base.

Using the notations for common logarithms and natural logarithms, we have the following results:

The Change-of-Base Property: Introducing Common and Natural Logarithms

Introducing Common Logarithms

$$\log_b M = \frac{\log M}{\log b}$$

Introducing Natural Logarithms

$$\log_b M = \frac{\ln M}{\ln b}$$

EXAMPLE 7 Changing Base to Common Logarithms

Use common logarithms to evaluate $\log_5 140$.

Solution Because $\log_b M = \dfrac{\log M}{\log b}$,

$$\log_5 140 = \frac{\log 140}{\log 5} \approx 3.07.$$

Use a calculator: 140 [LOG] [÷] 5 [LOG] [=] or [LOG] 140 [÷] [LOG] 5 [ENTER].

This means that $\log_5 140 \approx 3.07$. ■

Discover for Yourself

Find a reasonable estimate of $\log_5 140$ to the nearest whole number: 5 to what power is 140? Compare your estimate to the value obtained in Example 7.

✔ **CHECK POINT 7** Use common logarithms to evaluate $\log_7 2506$.

EXAMPLE 8 Changing Base to Natural Logarithms

Use natural logarithms to evaluate $\log_5 140$.

Solution Because $\log_b M = \dfrac{\ln M}{\ln b}$,

$$\log_5 140 = \frac{\ln 140}{\ln 5} \approx 3.07.$$

Use a calculator: 140 [LN] [÷] 5 [LN] [=] or [LN] 140 [÷] [LN] 5 [ENTER].

We have again shown that $\log_5 140 \approx 3.07$. ■

CHECK POINT 8 Use natural logarithms to evaluate $\log_7 2506$.

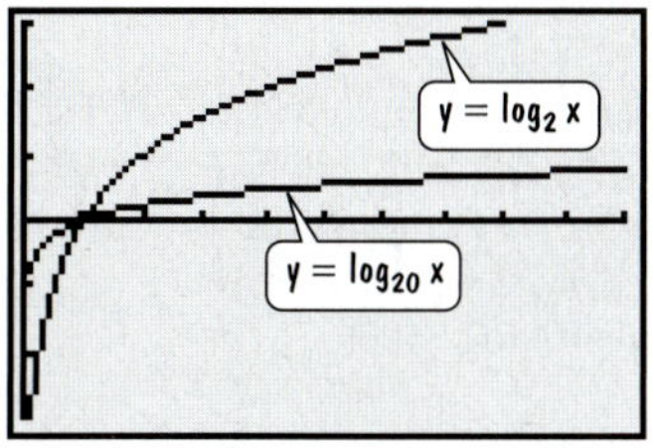

Figure 9.19 Using the change-of-base property to graph logarithmic functions

We can use the change-of-base property to graph logarithmic functions with bases other than 10 or e on a graphing utility. For example, Figure 9.19 shows the graphs of

$$y = \log_2 x \quad \text{and} \quad y = \log_{20} x$$

in a $[0, 10, 1]$ by $[-3, 3, 1]$ viewing rectangle. Because $\log_2 x = \frac{\ln x}{\ln 2}$ and $\log_{20} x = \frac{\ln x}{\ln 20}$ the functions are entered as

$$y_1 = \boxed{\text{LN}}\, x \,\boxed{\div}\,\boxed{\text{LN}}\, 2$$

$$\text{and} \quad y_2 = \boxed{\text{LN}}\, x \,\boxed{\div}\,\boxed{\text{LN}}\, 20.$$

EXERCISE SET 9.4

Practice Exercises

In all exercises, assume that all variables and variable expressions represent positive numbers.

In Exercises 1–32, use properties of logarithms to expand each logarithmic expression as much as possible. Where possible, evaluate logarithmic expressions without using a calculator.

1. $\log_5(7 \cdot 3)$

2. $\log_8(13 \cdot 7)$

3. $\log_7(7x)$

4. $\log_9(9x)$

5. $\log(1000x)$

6. $\log(10{,}000x)$

7. $\log_7\left(\frac{7}{x}\right)$

8. $\log_9\left(\frac{9}{x}\right)$

9. $\log\left(\frac{x}{100}\right)$

10. $\log\left(\frac{x}{1000}\right)$

11. $\log_4\left(\frac{64}{y}\right)$

12. $\log_5\left(\frac{125}{y}\right)$

13. $\ln\left(\frac{e^2}{5}\right)$

14. $\ln\left(\frac{e^4}{8}\right)$

15. $\log_b x^3$

16. $\log_b x^7$

17. $\log N^{-6}$

18. $\log M^{-8}$

19. $\ln \sqrt[5]{x}$

20. $\ln \sqrt[7]{x}$

21. $\log_b x^2 y$

22. $\log_b xy^3$

23. $\log_4\left(\frac{\sqrt{x}}{64}\right)$

24. $\log_5\left(\frac{\sqrt{x}}{25}\right)$

25. $\log_6\left(\frac{36}{\sqrt{x+1}}\right)$

26. $\log_8\left(\frac{64}{\sqrt{x+1}}\right)$

27. $\log_b\left(\frac{x^2 y}{z^2}\right)$

28. $\log_b\left(\frac{x^3 y}{z^2}\right)$

29. $\log \sqrt{100x}$

30. $\ln \sqrt{ex}$

31. $\log \sqrt[3]{\frac{x}{y}}$

32. $\log \sqrt[5]{\frac{x}{y}}$

In Exercises 33–52, use properties of logarithms to condense each logarithmic expression. Write the expression as a single logarithm whose coefficient is 1. Where possible, evaluate logarithmic expressions.

33. $\log 5 + \log 2$

34. $\log 250 + \log 4$

35. $\ln x + \ln 7$

36. $\ln x + \ln 3$

37. $\log_2 96 - \log_2 3$

38. $\log_3 405 - \log_3 5$

39. $\log(2x + 5) - \log x$

40. $\log(3x + 7) - \log x$

41. $\log x + 3 \log y$

42. $\log x + 7 \log y$

43. $\frac{1}{2} \ln x + \ln y$

44. $\frac{1}{3} \ln x + \ln y$

45. $2 \log_b x + 3 \log_b y$

46. $5 \log_b x + 6 \log_b y$

47. $5 \ln x - 2 \ln y$

48. $7 \ln x - 3 \ln y$

49. $3 \ln x - \frac{1}{3} \ln y$

50. $2 \ln x - \frac{1}{2} \ln y$

51. $4 \ln(x + 6) - 3 \ln x$

52. $8 \ln(x + 9) - 4 \ln x$

In Exercises 53–60, use common logarithms or natural logarithms and a calculator to evaluate to four decimal places

53. $\log_5 13$ **54.** $\log_6 17$

55. $\log_{14} 87.5$ **56.** $\log_{16} 57.2$

57. $\log_{0.1} 17$ **58.** $\log_{0.3} 19$

59. $\log_\pi 63$ **60.** $\log_\pi 400$

Application Exercises

61. The loudness level of a sound can be expressed by comparing the sound's intensity to the intensity of a sound barely audible to the human ear. The formula

$$D = 10(\log I - \log I_0)$$

describes the loudness level of a sound, D, in decibels, where I is the intensity of the sound, in watts per meter^2, and I_0 is the intensity of a sound barely audible to the human ear.

a. Express the formula so that the expression in parentheses is written as a single logarithm.

b. Use the form of the formula from part (a) to answer this question. If a sound has an intensity 100 times the intensity of a softer sound, how much larger on the decibel scale is the loudness level of the more intense sound?

62. The formula

$$t = \frac{1}{c}[\ln A - \ln(A - N)]$$

describes the time, t, in weeks, that it takes to achieve mastery of a portion of a task, where A is the maximum learning possible, N is the portion of the learning that is to be achieved, and c is a constant used to measure an individual's learning style.

a. Express the formula so that the expression in brackets is written as a single logarithm.

b. The formula is also used to determine how long it will take chimpanzees and apes to master a task. For example, a typical chimpanzee learning sign language can master a maximum of 65 signs. Use the form of the formula from part (a) to answer this question. How many weeks will it take a chimpanzee to master 30 signs if c for that chimp is 0.03?

Writing in Mathematics

63. Describe the product rule for logarithms and give an example.

64. Describe the quotient rule for logarithms and give an example.

65. Describe the power rule for logarithms and give an example.

66. Without showing the details, explain how to condense $\ln x - 2\ln(x + 1)$.

67. Describe the change-of-base property and give an example.

68. Explain how to use your calculator to find $\log_{14} 283$.

69. You overhear a student talking about a property of logarithms in which division becomes subtraction. Explain what the student means by this.

70. Find $\ln 2$ using a calculator. Then calculate each of the following: $1 - \frac{1}{2}$; $1 - \frac{1}{2} + \frac{1}{3}$; $1 - \frac{1}{2} + \frac{1}{3} - \frac{1}{4}$; $1 - \frac{1}{2} + \frac{1}{3} - \frac{1}{4} + \frac{1}{5}$; .... Describe what you observe.

Technology Exercises

71. a. Use a graphing utility (and the change-of-base property) to graph $y = \log_3 x$.

b. Graph $y = 2 + \log_3 x$, $y = \log_3(x + 2)$, and $y = -\log_3 x$ in the same viewing rectangle as $y = \log_3 x$. Then describe the change or changes that need to be made to the graph of $y = \log_3 x$ to obtain each of these three graphs.

72. Graph $y = \log x$, $y = \log(10x)$, and $y = \log(0.1x)$ in the same viewing rectangle. Describe the relationship among the three graphs. What logarithmic property accounts for this relationship?

73. Use a graphing utility and the change-of-base property to graph $y = \log_3 x$, $y = \log_{25} x$, and $y = \log_{100} x$ in the same viewing rectangle.

a. Which graph is on the top in the interval $(0, 1)$? Which is on the bottom?

b. Which graph is on the top in the interval $(1, \infty)$? Which is on the bottom?

c. Generalize by writing a statement about which graph is on top, which is on the bottom, and in which intervals, using $y = \log_b x$ where $b > 1$.

Disprove each statement in Exercises 74–78 by

a. *letting y equal a positive constant of your choice.*

b. *using a graphing utility to graph the function on each side of the equal sign. The two functions should have different graphs, showing that the equation is not true in general.*

74. $\log(x + y) = \log x + \log y$

75. $\log \frac{x}{y} = \frac{\log x}{\log y}$

76. $\ln(x - y) = \ln x - \ln y$

77. $\ln(xy) = (\ln x)(\ln y)$

78. $\frac{\ln x}{\ln y} = \ln x - \ln y$

Critical Thinking Exercises

79. Which one of the following is true?

a. $\frac{\log_7 49}{\log_7 7} = \log_7 49 - \log_7 7$

b. $\log_b(x^3 + y^3) = 3\log_b x + 3\log_b y$

c. $\log_b(xy)^5 = (\log_b x + \log_b y)^5$

d. $\ln\sqrt{2} = \frac{\ln 2}{2}$

80. Use the change-of-base property to prove that

$$\log e = \frac{1}{\ln 10}.$$

81. If $\log 3 = A$ and $\log 7 = B$, find $\log_7 9$ in terms of A and B.

82. Write as a single term that does not contain a logarithm:

$$e^{\ln 8x^5 - \ln 2x^2}.$$

Review Exercises

83. Graph: $5x - 2y > 10$. (Section 4.4, Example 1)

84. Solve: $x - 2(3x - 2) > 2x - 3$. (Section 4.1, Example 4)

85. Divide and simplify: $\frac{\sqrt[3]{40x^2y^6}}{\sqrt[3]{5xy}}$.

(Section 7.4, Example 5)

SECTION 9.5 Exponential and Logarithmic Equations

Objectives

1. Solve exponential equations.
2. Solve logarithmic equations.
3. Solve applied problems involving exponential and logarithmic equations.

SSM PH Tutor Center CD-ROM Video

Is an early retirement awaiting you?

You inherited \$30,000. You'd like to put aside \$25,000 and eventually have over half a million dollars for early retirement. Is this possible? In this section you will see how techniques for solving equations with variable exponents provide an answer to this question.

1 Solve exponential equations.

Exponential Equations

An **exponential equation** is an equation containing a variable in an exponent. Examples of exponential equations include

$$2^{3x-8} = 16, \quad 4^x = 15, \quad \text{and} \quad 40e^{0.6x} = 240.$$

Some exponential equations can be solved by expressing each side of the equation as a power of the same base. All exponential functions are one-to-one—that is, no two different ordered pairs have the same second component. Thus, if b is a positive number other than 1 and $b^M = b^N$, then $M = N$.

Using Technology

The graphs of

$$y_1 = 2^{3x-8}$$

and $y_2 = 16$

have an intersection point whose x-coordinate is 4. This verifies that $\{4\}$ is the solution set for $2^{3x-8} = 16$.

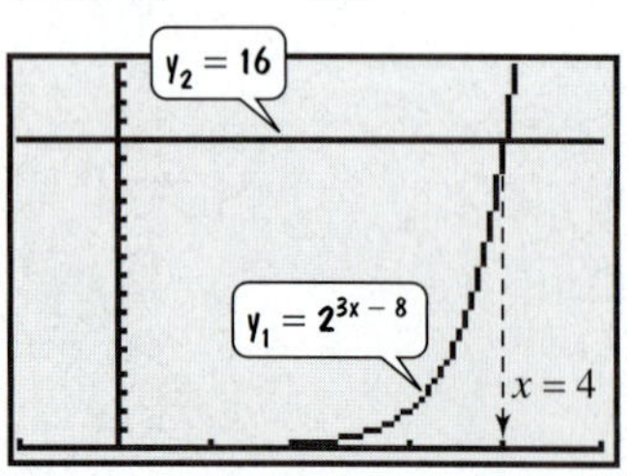

$[-1, 5, 1]$ by $[0, 20, 1]$

EXAMPLE 1 Solving Exponential Equations

Solve: **a.** $2^{3x-8} = 16$ **b.** $16^x = 64$.

Solution In each equation, express both sides as a power of the same base. Then set the exponents equal to each other.

a. Because 16 is 2^4, we express each side in terms of base 2.

$2^{3x-8} = 16$ This is the given equation.

$2^{3x-8} = 2^4$ Write each side as a power of the same base.

$3x - 8 = 4$ If $b^M = b^N$, $b > 0$ and $b \neq 1$, then $M = N$.

$3x = 12$ Add 8 to both sides.

$x = 4$ Divide both sides by 3.

CHECK 4:

$$2^{3x-8} = 16$$
$$2^{3\cdot 4-8} \stackrel{?}{=} 16$$
$$2^4 \stackrel{?}{=} 16$$
$$16 = 16, \quad \text{true}$$

The solution is 4 and the solution set is $\{4\}$.

b. Because $16 = 4^2$ and $64 = 4^3$, we express each side in terms of base 4.

$16^x = 64$ This is the given equation.

$(4^2)^x = 4^3$ Write each side as a power of the same base.

$4^{2x} = 4^3$ When an exponential expression is raised to a power, multiply exponents.

$2x = 3$ If two powers of the same base are equal, then the exponents are equal.

$x = \frac{3}{2}$ Divide both sides by 2.

CHECK $\frac{3}{2}$:

$$16^x = 64$$
$$16^{\frac{3}{2}} \stackrel{?}{=} 64$$
$$(\sqrt{16})^3 \stackrel{?}{=} 64 \qquad b^{\frac{m}{n}} = (\sqrt[n]{b})^m$$
$$4^3 \stackrel{?}{=} 64$$
$$64 = 64, \quad \text{true}$$

The solution is $\frac{3}{2}$ and the solution set is $\{\frac{3}{2}\}$. ■

Discover for Yourself

The equation $16^x = 64$ can also be solved by writing each side in terms of base 2. Do this. Which solution method do you prefer?

CHECK POINT 1 Solve:

a. $5^{3x-6} = 125$ **b.** $4^x = 32$.

When both sides of an exponential equation cannot be written as a power of the same base, logarithms are used in solving the equation. The solution begins with isolating the exponential expression and taking the natural logarithm on both sides. Why can we do this? All logarithmic functions are one-to-one—that is, no two different ordered pairs have the same second component. Thus, if M and N are positive real numbers and $M = N$, then $\log_b M = \log_b N$.

Using Natural Logarithms to Solve Exponential Equations

1. Isolate the exponential expression.
2. Take the natural logarithm on both sides of the equation.
3. Simplify using one of the following properties:

$$\ln b^x = x \ln b \quad \text{or} \quad \ln e^x = x.$$

4. Solve for the variable.

EXAMPLE 2 Solving an Exponential Equation

Solve: $4^x = 15$.

Solution Because the exponential expression, 4^x, is already isolated on the left, we begin by taking the natural logarithm on both sides of the equation.

$$4^x = 15 \quad \text{This is the given equation.}$$

$$\ln 4^x = \ln 15 \quad \text{Take the natural logarithm on both sides.}$$

$$x \ln 4 = \ln 15 \quad \text{Use the power rule and bring the variable exponent to the front: } \ln b^x = x \ln b.$$

$$x = \frac{\ln 15}{\ln 4} \quad \text{Solve for } x \text{ by dividing both sides by } \ln 4.$$

We now have an exact value for x. We use the exact value for x in the equation's solution set. Thus, the equation's solution is $\frac{\ln 15}{\ln 4}$ and the solution set is $\left\{\frac{\ln 15}{\ln 4}\right\}$. We can obtain a decimal approximation by using a calculator: $x \approx 1.95$. Because $4^2 = 16$, it seems reasonable that the solution to $4^x = 15$ is approximately 1.95. ■

✔ **CHECK POINT 2** Solve: $5^x = 134$. Find the solution, and then use a calculator to obtain a decimal approximation to two decimal places for the solution.

EXAMPLE 3 Solving an Exponential Equation

Solve: $40e^{0.6x} = 240$.

Solution

Solution We begin by dividing both sides by 40 to isolate the exponential expression, $e^{0.6x}$. Then we take the natural logarithm on both sides of the equation.

$$40e^{0.6x} = 240 \quad \text{This is the given equation.}$$

$$e^{0.6x} = 6 \quad \text{Isolate the exponential factor by dividing both sides by 40.}$$

$$\ln e^{0.6x} = \ln 6 \quad \text{Take the natural logarithm on both sides.}$$

$$0.6x = \ln 6 \quad \text{Use the inverse property } \ln e^x = x \text{ on the left.}$$

$$x = \frac{\ln 6}{0.6} \approx 2.99 \quad \text{Divide both sides by 0.6.}$$

Thus, the solution of the equation is $\frac{\ln 6}{0.6} \approx 2.99$. Try checking this approximate solution in the original equation, verifying that $\left\{\frac{\ln 6}{0.6} \approx 2.99\right\}$ is the solution set. ■

✔ **CHECK POINT 3** Solve: $7e^{2x} = 63$. Find the solution set, and then use a calculator to obtain a decimal approximation to two decimal places for the solution.

2 Solve logarithmic equations.

Logarithmic Equations

A **logarithmic equation** is an equation containing a variable in a logarithmic expression. Examples of logarithmic equations include

$$\log_4(x+3) = 2 \quad \text{and} \quad \ln 2x = 3.$$

If a logarithmic equation is in the form $\log_b x = c$, we can solve the equation by rewriting it in its equivalent exponential form $b^c = x$. Example 4 illustrates how this is done.

Using Technology

The graphs of

$$y_1 = \log_4(x+3) \quad \text{and} \quad y_2 = 2$$

have an intersection point whose x-coordinate is 13. This verifies that $\{13\}$ is the solution set for $\log_4(x+3) = 2$.

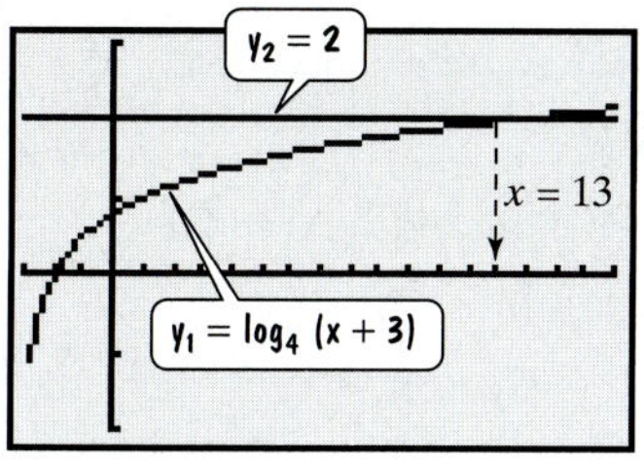

$[-3, 17, 1]$ by $[-2, 3, 1]$

Note:
Because

$$\log_b x = \frac{\ln x}{\ln b}$$

(change-of-base property),

we entered y_1 using

$$y_1 = \log_4(x+3) = \frac{\ln(x+3)}{\ln 4}.$$

EXAMPLE 4 Solving a Logarithmic Equation

Solve: $\log_4(x+3) = 2$.

Solution We first rewrite the equation as an equivalent equation in exponential form using the fact that $\log_b x = c$ means $b^c = x$.

$$\log_4(x+3) = 2 \quad \text{means} \quad 4^2 = x + 3.$$

Logarithms are exponents.

Now we solve the equivalent equation for x.

$4^2 = x + 3$ This is the equivalent equation.

$16 = x + 3$ Square 4.

$13 = x$ Subtract 3 from both sides.

CHECK 13:

$\log_4(x+3) = 2$ This is the given logarithmic equation.

$\log_4(13+3) \stackrel{?}{=} 2$ Substitute 13 for x.

$\log_4 16 \stackrel{?}{=} 2$

$2 = 2$, true $\quad \log_4 16 = 2$ because $4^2 = 16$.

This true statement indicates that the solution is 13 and the solution set is $\{13\}$. ■

✔ CHECK POINT 4 Solve: $\log_2(x-4) = 3$.

Logarithmic expressions are defined only for logarithms of positive real numbers. Always check proposed solutions of a logarithmic equation in the original equation. Exclude from the solution set any proposed solution that produces the logarithm of a negative number or the logarithm of 0.

To rewrite the logarithmic equation $\log_b x = c$ in the equivalent exponential form $b^c = x$, we need a single logarithm whose coefficient is one. It is sometimes necessary to use properties of logarithms to condense logarithms into a single logarithm. In the next example, we use the product rule for logarithms to obtain a single logarithmic expression on the left side.

EXAMPLE 5 Using the Product Rule to Solve a Logarithmic Equation

Solve: $\log_2 x + \log_2(x - 7) = 3$.

Solution

$\log_2 x + \log_2(x-7) = 3$	This is the given equation.
$\log_2 x(x-7) = 3$	Use the product rule to obtain a single logarithm: $\log_b M + \log_b N = \log_b(MN)$.
$2^3 = x(x-7)$	$\log_b x = c$ means $b^c = x$.
$8 = x^2 - 7x$	Apply the distributive property on the right.
$0 = x^2 - 7x - 8$	Set the equation equal to 0.
$0 = (x-8)(x+1)$	Factor.
$x - 8 = 0$ or $x + 1 = 0$	Set each factor equal to 0.
$x = 8$ $\quad x = -1$	Solve for x.

Check 8:

$$\log_2 x + \log_2(x-7) = 3$$
$$\log_2 8 + \log_2(8-7) \stackrel{?}{=} 3$$
$$\log_2 8 + \log_2 1 \stackrel{?}{=} 3$$
$$3 + 0 \stackrel{?}{=} 3$$
$$3 = 3, \quad \text{true}$$

Check −1:

$$\log_2 x + \log_2(x-7) = 3$$
$$\log_2(-1) + \log_2(-1-7) \stackrel{?}{=} 3$$

The number −1 does not check. Negative numbers do not have logarithms.

The solution is 8 and the solution set is $\{8\}$. ■

✔ CHECK POINT 5 Solve: $\log x + \log(x - 3) = 1$.

EXAMPLE 6 Using the Quotient Rule to Solve a Logarithmic Equation

Solve: $\log_4(x + 3) - \log_4 x = 2$.

Solution

$\log_4(x+3) - \log_4 x = 2$	This is the given equation.
$\log_4 \frac{x+3}{x} = 2$	Use the quotient rule to obtain a single logarithm: $\log_b M - \log_b N = \log_b \frac{M}{N}$.
$4^2 = \frac{x+3}{x}$	$\log_b x = c$ means $b^c = x$.
$16 = \frac{x+3}{x}$	Square 4: $4^2 = 16$.
$16x = x + 3$	Multiply both sides by x.
$15x = 3$	Subtract x from both sides.
$x = \frac{3}{15} = \frac{1}{5}$	Divide both sides by 15.

Check by substituting $\frac{1}{5}$ in the original equation. In this equation, both x and $x + 3$ must be positive. The proposed solution, $\frac{1}{5}$, satisfies these conditions, and the solution set is $\{\frac{1}{5}\}$. ■

CHECK POINT 6 Solve: $\log_5(x + 1) - \log_5 x = 2$

Equations involving natural logarithms can be solved using the inverse property $e^{\ln x} = x$. For example, to solve

$$\ln x = 5$$

we write both sides of the equation as exponents on base e:

$$e^{\ln x} = e^5.$$

This is called **exponentiating both sides** of the equation. Using the inverse property $e^{\ln x} = x$, we simplify the left side of the equation and obtain the solution:

$$x = e^5.$$

EXAMPLE 7 Solving an Equation with a Natural Logarithm

Solve: $3 \ln 2x = 12$.

Solution

$$3 \ln 2x = 12$$ This is the given equation.

$$\ln 2x = 4$$ Divide both sides by 3.

$$e^{\ln 2x} = e^4$$ Exponentiate both sides.

$$2x = e^4$$ Use the inverse property to simplify the left side: $e^{\ln x} = x$.

$$x = \frac{e^4}{2} \approx 27.30$$ Divide both sides by 2.

CHECK $\frac{e^4}{2}$:

$$3 \ln 2x = 12$$ This is the given logarithmic equation.

$$3 \ln 2\left(\frac{e^4}{2}\right) \stackrel{?}{=} 12$$ Substitute $\frac{e^4}{2}$ for x.

$$3 \ln e^4 \stackrel{?}{=} 12$$ Simplify: $\frac{\cancel{2}}{1} \cdot \frac{e^4}{\cancel{2}} = e^4$.

$$3 \cdot 4 \stackrel{?}{=} 12$$ Because $\ln e^x = x$, we conclude $\ln e^4 = 4$.

$$12 = 12, \quad \text{true}$$

This true statement indicates that the solution is $\frac{e^4}{2}$ and the solution set is $\left\{\frac{e^4}{2}\right\}$. ■

CHECK POINT 7 Solve: $4 \ln 3x = 8$.

3 Solve applied problems involving exponential and logarithmic equations.

Applications

Our first applied example provides a mathematical perspective on the old slogan "Alcohol and driving don't mix." In California, where 38% of fatal traffic crashes involve drinking drivers, it is illegal to drive with a blood alcohol concentration of 0.08 or higher. At these levels, drivers may be arrested and charged with driving under the influence.

EXAMPLE 8 Alcohol and Risk of a Car Accident

Medical research indicates that the risk of having a car accident increases exponentially as the concentration of alcohol in the blood increases. The risk is modeled by

$$R = 6e^{12.77x}$$

where x is the blood alcohol concentration and R, given as a percent, is the risk of having a car accident. What blood alcohol concentration corresponds to a 17% risk of a car accident?

Solution For a risk of 17%, we let $R = 17$ in the equation and solve for x, the blood alcohol concentration.

$$R = 6e^{12.77x}$$ This is the given equation.

$$6e^{12.77x} = 17$$ Substitute 17 for R and (optional) reverse the two sides of the equation.

$$e^{12.77x} = \frac{17}{6}$$ Isolate the exponential factor by dividing both sides by 6.

$$\ln e^{12.77x} = \ln\left(\frac{17}{6}\right)$$ Take the natural logarithm on both sides.

$$12.77x = \ln\left(\frac{17}{6}\right)$$ Use the inverse property $\ln e^x = x$ on the left side.

$$x = \frac{\ln\left(\frac{17}{6}\right)}{12.77} \approx 0.08$$ Divide both sides by 12.77.

For a blood alcohol concentration of 0.08, the risk of a car accident is 17%. In many states, it is illegal to drive at this blood alcohol concentration. ■

✔ **CHECK POINT 8** Use the formula in Example 8 to solve this problem. What blood alcohol concentration corresponds to a 7% risk of a car accident? (In many states, drivers under the age of 21 can lose their license for driving at this level.)

Suppose that you inherit $30,000. Is it possible to invest $25,000 and have over half a million dollars for early retirement? Our next example illustrates the power of compound interest.

EXAMPLE 9 Revisiting the Formula for Compound Interest

The formula

$$A = P\left(1 + \frac{r}{n}\right)^{nt}$$

describes the accumulated value, A, of a sum of money, P, the principal, after t years at annual percentage rate r (in decimal form) compounded n times a year. How long will it take \$25,000 to grow to \$500,000 at 9% annual interest compounded monthly?

Solution

$$A = P\left(1 + \frac{r}{n}\right)^{nt}$$ This is the given formula.

$$500{,}000 = 25{,}000\left(1 + \frac{0.09}{12}\right)^{12t}$$ A(the desired accumulated value) = 500,000, P(the principal) = 25,000, r(the interest rate) = 9% = 0.09, and n = 12 (monthly compounding).

Our goal is to solve the equation for t. Let's reverse the two sides of the equation and then simplify within parentheses.

$$25{,}000\left(1 + \frac{0.09}{12}\right)^{12t} = 500{,}000$$

$$25{,}000(1 + 0.0075)^{12t} = 500{,}000$$ Divide within parentheses: $\frac{0.09}{12} = 0.0075$.

$$25{,}000(1.0075)^{12t} = 500{,}000$$ Add within parentheses.

$$(1.0075)^{12t} = 20$$ Divide both sides by 25,000.

$$\ln(1.0075)^{12t} = \ln 20$$ Take the natural logarithm on both sides.

$$12t\ln(1.0075) = \ln 20$$ Use the power rule to bring the exponent to the front: $\ln b^x = x \ln b$.

$$t = \frac{\ln 20}{12 \ln 1.0075}$$ Solve for t, dividing both sides by $12 \ln 1.0075$.

$$\approx 33.4$$ Use a calculator.

After approximately 33.4 years, the \$25,000 will grow to an accumulated value of \$500,000. If you set aside the money at age 20, you can begin enjoying a life of leisure at about age 53. ■

CHECK POINT 9 How long, to the nearest tenth of a year, will it take \$1000 to grow to \$3600 at 8% annual interest compounded quarterly?

ENRICHMENT ESSAY

Playing Doubles: Interest Rates and Doubling Time

One way to calculate what your savings will be worth at some point in the future is to consider doubling time. The following table shows how long it takes for your money to double at different annual interest rates subject to continuous compounding.

Annual Interest Rate	Years to Double
5%	13.9 years
7%	9.9 years
9%	7.7 years
11%	6.3 years

Of course, the first problem is collecting some money to invest. The second problem is finding a reasonably safe investment with a return of 9% or more.

EXAMPLE 10 The Growth of the Environmental Industry

The formula

$$N = 461.87 + 299.4 \ln x$$

models the thousands of workers, N, in the environmental industry in the United States x years after 1979. By which year will there be 1,500,000, or 1500 thousand, U.S. workers in the environmental industry?

Solution We substitute 1500 for N and solve for x, the number of years after 1979.

$$N = 461.87 + 299.4 \ln x \quad \text{This is the given formula.}$$

$$461.87 + 299.4 \ln x = 1500 \quad \text{Substitute 1500 for N and reverse the two sides of the equation.}$$

Our goal is to isolate $\ln x$. We can then find x by exponentiating both sides of the equation, using the inverse property $e^{\ln x} = x$.

$$299.4 \ln x = 1038.13 \quad \text{Subtract 461.87 from both sides.}$$

$$\ln x = \frac{1038.13}{299.4} \quad \text{Divide both sides by 299.4.}$$

$$e^{\ln x} = e^{1038.13/299.4} \quad \text{Exponentiate both sides.}$$

$$x = e^{1038.13/299.4} \quad e^{\ln x} = x$$

$$\approx 32 \quad \text{Use a calculator.}$$

Approximately 32 years after 1979, in the year 2011, there will be 1.5 million U.S. workers in the environmental industry. ■

CHECK POINT 10 Use the formula in Example 10 to find by which year there will be two million, or 2000 thousand, U.S. workers in the environmental industry.

EXERCISE SET 9.5

Practice Exercises

Solve each exponential equation in Exercises 1–12 by expressing each side as a power of the same base and then equating exponents.

1. $2^x = 64$

2. $3^x = 81$

3. $5^x = 125$

4. $5^x = 625$

5. $2^{2x-1} = 32$

6. $3^{2x+1} = 27$

7. $4^{2x-1} = 64$

8. $5^{3x-1} = 125$

9. $32^x = 8$

10. $4^x = 32$

11. $9^x = 27$

12. $125^x = 625$

Solve each exponential equation in Exercises 13–34 by taking the natural logarithm on both sides. Express the solution set in terms of natural logarithms. Then use a calculator to obtain a decimal approximation, correct to two decimal places, for the solution.

13. $10^x = 3.91$

14. $10^x = 8.07$

15. $e^x = 5.7$

16. $e^x = 0.83$

17. $5^x = 17$

18. $19^x = 143$

19. $5e^x = 25$

20. $9e^x = 99$

21. $3e^{5x} = 1977$

22. $4e^{7x} = 10{,}273$

23. $e^{0.7x} = 13$

24. $e^{0.08x} = 4$

25. $1250e^{0.055x} = 3750$

26. $1250e^{0.065x} = 6250$

27. $30 - (1.4)^x = 0$

28. $135 - (4.7)^x = 0$

29. $e^{1-5x} = 793$

30. $e^{1-8x} = 7957$

31. $7^{x+2} = 410$

32. $5^{x-3} = 137$

33. $2^{x+1} = 5^x$

34. $4^{x+1} = 9^x$

In Exercises 35–58, solve each logarithmic equation. Be sure to reject any value of x that produces the logarithm of a negative number or the logarithm of zero in the original equation.

35. $\log_3 x = 4$

36. $\log_5 x = 3$

37. $\log_2 x = -4$

38. $\log_2 x = -5$

39. $\log_9 x = \frac{1}{2}$

40. $\log_{25} x = \frac{1}{2}$

41. $\log x = 2$

42. $\log x = 3$

43. $\ln x = -3$

44. $\ln x = -4$

45. $\log_4(x + 5) = 3$

46. $\log_5(x - 7) = 2$

47. $\log_3(x - 4) = -3$

48. $\log_7(x + 2) = -2$

49. $\log_4(3x + 2) = 3$

50. $\log_2(4x + 1) = 5$

51. $\log_5 x + \log_5(4x - 1) = 1$

52. $\log_6(x + 5) + \log_6 x = 2$

53. $\log_3(x - 5) + \log_3(x + 3) = 2$

54. $\log_2(x - 1) + \log_2(x + 1) = 3$

55. $\log_2(x + 2) - \log_2(x - 5) = 3$

56. $\log_4(x + 2) - \log_4(x - 1) = 1$

57. $\log(3x - 5) - \log 5x = 2$

58. $\log(2x - 1) - \log x = 2$

Exercises 59–66 involve equations with natural logarithms. Solve each equation by isolating the natural logarithm and exponentiating both sides. Express the answer in terms of e. Then use a calculator to obtain a decimal approximation, correct to two decimal places, for the solution.

59. $\ln x = 2$

60. $\ln x = 3$

61. $5 \ln 2x = 20$

62. $6 \ln 2x = 30$

63. $6 + 2 \ln x = 5$

64. $7 + 3 \ln x = 6$

65. $\ln \sqrt{x + 3} = 1$

66. $\ln \sqrt{x + 4} = 1$

Application Exercises

Use the formula $R = 6e^{12.77x}$, where x is the blood alcohol concentration and R, given as a percent, is the risk of having a car accident, to solve Exercises 67–68.

67. What blood alcohol concentration corresponds to certainty, or a 100% risk, of a car accident?

68. What blood alcohol concentration corresponds to a 50% risk of a car accident?

69. The function $f(t) = 18.2e^{0.001t}$ models the population of New York State, $f(t)$, in millions, t years after 1994.

a. What was the population of New York in 1994?

b. When will the population of New York reach 18.5 million?

70. The function $f(t) = 14e^{0.168t}$ models the population of Florida, $f(t)$, in millions, t years after 1994.

a. What was the population of Florida in 1994?

b. When will the population of Florida reach 18.5 million?

In Exercises 71–74, complete the table for a savings account subject to n compoundings yearly $\left(A = P\left(1 + \frac{r}{n}\right)^{nt}\right)$. *Round answers to one decimal place.*

	Amount Invested	Number of Compounding Periods	Annual Interest Rate	Accumulated Amount	Time t in Years
71.	\$12,500	4	5.75%	\$20,000	
72.	\$7250	12	6.5%	\$15,000	
73.	\$1000	360		\$1400	2
74.	\$5000	360		\$9000	4

In Exercises 75–78, complete the table for a savings account subject to continuous compounding $(A = Pe^{rt})$. *Round answers to one decimal place.*

	Amount Invested	Annual Interest Rate	Accumulated Amount	Time t in Years
75.	\$8000	8%	Double the amount invested	
76.	\$8000		\$12,000	2
77.	\$2350		Triple the amount invested	7
78.	\$17,425	4.25%	\$25,000	

79. The function $f(x) = 15{,}557 + 5259 \ln x$ models the average cost of a new car, $f(x)$, in dollars, x years after 1989. When was the average cost of a new car \$25,000?

80. The function $f(x) = 68.41 + 1.75 \ln x$ models the life expectancy, $f(x)$, in years, for African-American females born x years after 1969. In which birth year was life expectancy 73.7 years? Round to the nearest year.

The function $P(x) = 95 - 30 \log_2 x$ *models the percentage,* $P(x)$*, of students who could recall the important features of a classroom lecture as a function of time, where x represents the number of days that have elapsed since the lecture was given. The figure shows the graph of the function. Use this information to solve Exercises 81–82. Round answers to one decimal place.*

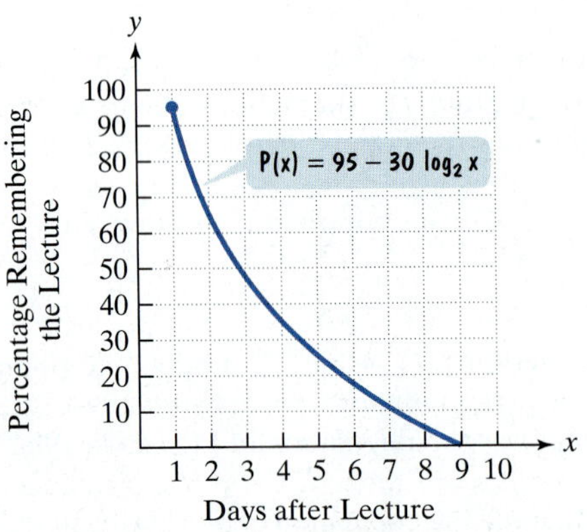

81. After how many days do only half the students recall the important features of the classroom lecture? (Let $P(x) = 50$ and solve for x.) Locate the point on the graph that conveys this information.

82. After how many days have all students forgotten the important features of the classroom lecture? (Let $P(x) = 0$ and solve for x.) Locate the point on the graph that conveys this information.

The pH of a solution ranges from 0 to 14. An acid solution has a pH less than 7. Pure water is neutral and has a pH of 7. Normal, unpolluted rain has a pH of about 5.6. The pH of a solution is given by

$$\text{pH} = -\log x$$

where x represents the concentration of the hydrogen ions in the solution, in moles per liter. Use the formula to solve Exercises 83–84.

83. An environmental concern involves the destructive effects of acid rain. The most acidic rainfall ever had a pH of 2.4. What was the hydrogen ion concentration? Express the answer as a power of 10, and then round to the nearest thousandth.

84. The figure at the top of the next page shows very acidic rain in the northeast United States. What is the hydrogen ion concentration of rainfall with a pH of 4.2? Express the answer as a power of 10, and then round to the nearest hundred-thousandth.

Acid Rain Over Canada and the United States

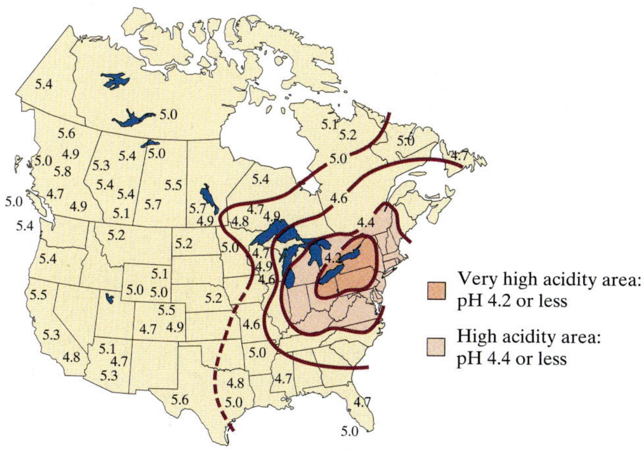

Source: National Atmospheric Program

Writing in Mathematics

85. What is an exponential equation?

86. Explain how to solve an exponential equation when both sides can be written as a power of the same base.

87. Explain how to solve an exponential equation when both sides cannot be written as a power of the same base. Use $3^x = 140$ in your explanation.

88. What is a logarithmic equation?

89. Explain how to solve a logarithmic equation. Use $\log_3(x - 1) = 4$ in your explanation.

90. In many states, a 17% risk of a car accident with a blood alcohol concentration of 0.08 is the lowest level for charging a motorist with driving under the influence. Do you agree with the 17% risk as a cutoff percentage, or do you feel that the percentage should be lower or higher? Explain your answer. What blood alcohol concentration corresponds to what you believe is an appropriate percentage?

91. Have you purchased a new or used car recently? If so, describe if the function in Exercise 79 accurately models what you paid for your car. If there is a big difference between the figure given by the function and the amount that you paid, how can you explain this difference?

Technology Exercises

In Exercises 92–99, use your graphing utility to graph each side of the equation in the same viewing rectangle. Then use the x-coordinate of the intersection point to find the equation's solution set. Verify this value by direct substitution into the equation.

92. $2^{x+1} = 8$

93. $3^{x+1} = 9$

94. $\log_3(4x - 7) = 2$

95. $\log_3(3x - 2) = 2$

96. $\log(x + 3) + \log x = 1$

97. $\log(x - 15) + \log x = 2$

98. $3^x = 2x + 3$

99. $5^x = 3x + 4$

Hurricanes are one of nature's most destructive forces. These low-pressure areas often have diameters of over 500 miles. The function $f(x) = 0.48\ln(x + 1) + 27$ models the barometric air pressure, $f(x)$, in inches of mercury, at a distance of x miles from the eye of a hurricane. Use this function to solve Exercises 100–101.

100. Graph the function in a $[0, 500, 50]$ by $[27, 30, 1]$ viewing rectangle. What does the shape of the graph indicate about barometric air pressure as the distance from the eye increases?

101. Use an equation to answer this question: How far from the eye of a hurricane is the barometric air pressure 29 inches of mercury? Use the TRACE and ZOOM features or the intersect command of your graphing utility to verify your answer.

102. The function $P(t) = 145e^{-0.092t}$ models a runner's pulse, $P(t)$, in beats per minute, t minutes after a race, where $0 \le t \le 15$. Graph the function using a graphing utility. TRACE along the graph and determine after how many minutes the runner's pulse will be 70 beats per minute. Round to the nearest tenth of a minute. Verify your observation algebraically.

103. The function $W(t) = 2600(1 - 0.51e^{-0.075t})^3$ models the weight, $W(t)$, in kilograms, of a female African elephant at age t years. (1 kilogram $\approx$ 2.2 pounds) Use a graphing utility to graph the function. Then TRACE along the curve to estimate the age of an adult female elephant weighing 1800 kilograms.

Critical Thinking Exercises

104. Which one of the following is true?

a. If $\log(x + 3) = 2$, then $e^2 = x + 3$.

b. If $\log(7x + 3) - \log(2x + 5) = 4$, then in exponential form $10^4 = (7x + 3) - (2x + 5)$.

c. If $x = \frac{1}{k}\ln y$, then $y = e^{kx}$.

d. Examples of exponential equations include $10^x = 5.71$, $e^x = 0.72$, and $x^{10} = 5.71$.

105. If \$4000 is deposited into an account paying 3% interest compounded annually and at the same time \$2000 is deposited into an account paying 5% interest compounded annually, after how long will the two accounts have the same balance? Round to the nearest year.

Solve each equation in Exercises 106–108. Check each proposed solution by direct substitution or with a graphing utility.

106. $(\ln x)^2 = \ln x^2$

107. $(\log x)(2\log x + 1) = 6$

108. $\ln(\ln x) = 0$

Review Exercises

109. Solve: $\sqrt{x+4} - \sqrt{x-1} = 1$

(Section 7.6, Example 4)

110. Solve:

$$\frac{3}{x+1} - \frac{5}{x} = \frac{19}{x^2 + x}.$$

(Section 6.6, Example 5)

111. Simplify: $(-2x^3y^{-2})^{-4}$.

(Section 1.6, Example

▸ SECTION 9.6 Exponential Growth and Decay; Modeling Data

Objectives

1. Model exponential growth and decay.
2. Model data with exponential and logarithmic functions.

SSM PH Tutor Center CD-ROM Video

The most casual cruise on the Internet shows how people disagree when it comes to making predictions about the effects of the world's growing population. Some argue that there is a recent slowdown in the growth rate, economies remain robust, and famines in Biafra and Ethiopia are aberrations rather than signs of the future. Others say that the 6 billion people on Earth is twice as many as can be supported in middle-class comfort, and the world is running out of arable land and fresh water. Debates about entities that are growing exponentially can be approached mathematically: We can create functions that model data and use these functions to make predictions. In this section we will show you how this is done.

1 Model exponential growth and decay.

Exponential Growth and Decay

One of algebra's many applications is to predict the behavior of variables. This can be done with **exponential growth** and **decay models**. With exponential growth and decay, quantities grow or decay at a rate directly proportional to their size. Populations that are growing exponentially grow extremely rapidly as they get larger because there are more adults to have offspring. For example, the **growth rate** for world population is 1.3%, or 0.013. This means that each year world population is 1.3% more than what it was in the previous year. In 1999, world population was 6 billion. Thus, we compute the world population in 2000 as follows:

$$6 \text{ billion} + 1.3\% \text{ of } 6 \text{ billion} = 6 + (0.013)(6) = 6.078.$$

This computation suggests that 6.078 billion people populated the world in 2000. The 0.078 billion represents an increase of 78 million people from 1999 to 2000, the equivalent of the population of Germany. Using 1.3% as the annual growth rate, world population for 2001 is found similarly:

$$6.078 \text{ billion} + 1.3\% \text{ of } 6.078 \text{ billion} = 6.078 + (0.013)(6.078) \approx 6.157.$$

This computation suggests that approximately 6.157 billion people populated the world in 2001.

The explosive growth of world population may remind you of the growth of money in an account subject to compound interest. Just as the growth rate for world population is multiplied by the population plus any increase in the population, a compound interest rate is multiplied by your original investment plus any accumulated interest. The balance in an account subject to continuous compounding and world population are special cases of an *exponential growth model*.

Exponential Growth and Decay Models

The mathematical model for **exponential growth** or **decay** is given by

$$f(t) = A_0 e^{kt} \quad \text{or} \quad A = A_0 e^{kt}.$$

- **If $k > 0$, the function models the amount or size of a *growing* entity.** A_0 is the original amount or size of the growing entity at time $t = 0$, A is the amount at time t, and k is a constant representing the growth rate.
- **If $k < 0$ the function models the amount or size of a *decaying* entity.** A_0 is the original amount or size of the decaying entity at time $t = 0$, A is the amount at time, t, and k is a constant representing the decay rate.

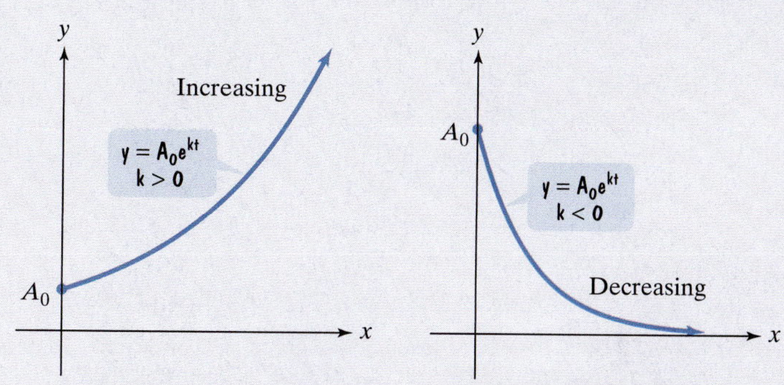

(a) Exponential growth (b) Exponential decay

Sometimes we need to use given data to determine k, the rate of growth or decay. After we compute the value of k, we can use the formula $A = A_0 e^{kt}$ to make predictions. This idea is illustrated in our first two examples.

EXAMPLE 1 Modeling the Growth of the Minimum Wage

The graph in Figure 9.20 shows the growth of the minimum wage from 1970 through 2000. In 1970, the minimum wage was \$1.60 per hour. By 2000, it had grown to \$5.15 per hour.

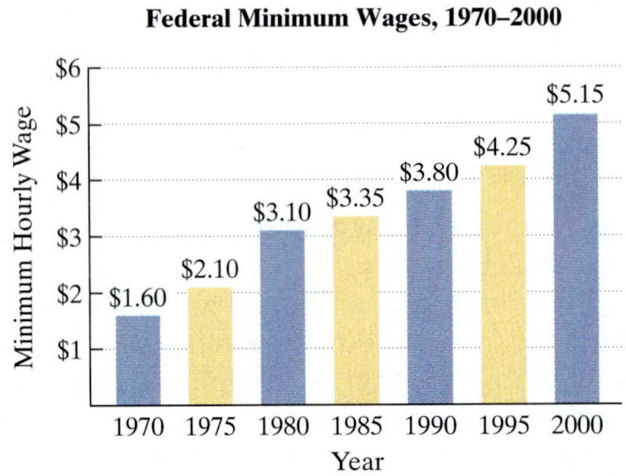

Figure 9.20
Source: U.S. Employment Standards Administration

a. Find the exponential growth function that models the data for 1970 through 2000.

b. By which year will the minimum wage reach \$7.50 per hour?

Solution

a. We use the exponential growth model

$$A = A_0e^{kt}$$

in which t is the number of years after 1970. This means that 1970 corresponds to $t = 0$. At that time the minimum wage was \$1.60, so we substitute 1.6 for A_0 in the growth model:

$$A = 1.6e^{kt}.$$

We are given that \$5.15 is the minimum wage in 2000. Because 2000 is 30 years after 1970, when $t = 30$ the value of A is 5.15. Substituting these numbers into the growth model will enable us to find k, the growth rate. We know that $k > 0$ because the problem involves growth.

$$A = 1.6e^{kt}$$ Use the growth model with $A_0 = 1.6$.

$$5.15 = 1.6e^{k\cdot 30}$$ When $t = 30$, $A = 5.15$. Substitute these numbers into the model.

$$e^{30k} = \frac{5.15}{1.6}$$ Isolate the exponential factor by dividing both sides by 1.6. We also reversed the sides.

$$\ln e^{30k} = \ln\frac{5.15}{1.6}$$ Take the natural logarithm on both sides.

$$30k = \ln\frac{5.15}{1.6}$$ Simplify the left side using $\ln e^x = x$.

$$k = \frac{\ln\frac{5.16}{1.6}}{30} \approx 0.039$$

We substitute 0.039 for k in the growth model to obtain the exponential growth function for the minimum wage. It is

$$A = 1.6e^{0.039t}$$

where t is measured in years after 1970.

b. To find the year in which the minimum wage will reach \$7.50 per hour, we substitute 7.5 for A in the model from part (a) and solve for t.

$$A = 1.6e^{0.039t} \quad \text{This is the model from part (a).}$$

$$7.5 = 1.6e^{0.039t} \quad \text{Substitute 7.5 for A.}$$

$$e^{0.039t} = \frac{7.5}{1.6} \quad \text{Divide both sides by 1.6.}$$

$$\ln e^{0.039t} = \ln \frac{7.5}{1.6} \quad \text{Take the natural logarithm on both sides.}$$

$$0.039t = \ln \frac{7.5}{1.6} \quad \text{Simplify on the left using } \ln e^x = x.$$

$$t = \frac{\ln \frac{7.5}{1.6}}{0.039} \approx 40 \quad \text{Solve for t by dividing both sides by 0.039.}$$

Because 40 is the number of years after 1970, the model indicates that the minimum wage will reach \$7.50 by 1970 + 40, or in the year 2010. ■

ENRICHMENT ESSAY

Lying with Statistics

Benjamin Disraeli, Queen Victoria's prime minister, stated that there are "lies, damned lies, and statistics." The problem is not that data lie, but rather that liars use data. For example, the data in Example 1 create the impression that wages are on the rise and workers are better off each year. The graph in Figure 9.21 is more effective in creating an accurate picture. Why? It is adjusted for inflation and measured in constant 1996 dollars. Something else to think about: In predicting a minimum wage of \$7.50 by 2010, are we using the best possible model for the data? We return to this issue in the exercise set.

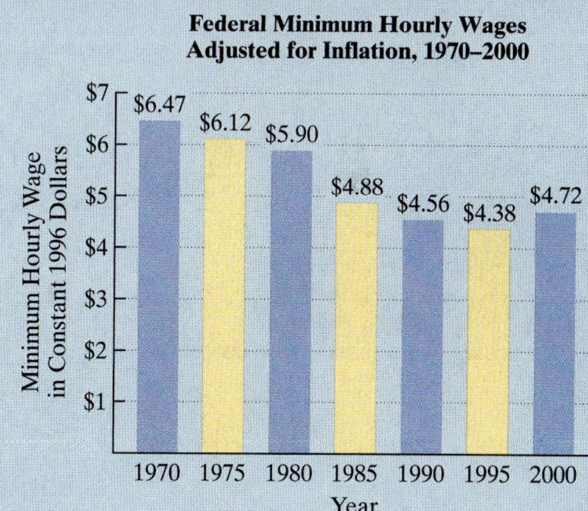

Figure 9.21

Source: U.S. Employment Standards Administration

✔ **CHECK POINT 1** In 1980, the population of Africa was 491 million and by 1990 it had grown to 643 million.

a. Use the exponential growth model $A = A_0e^{kt}$, in which t is the number of years after 1980, to find the exponential growth function that models the data.

b. By which year will Africa's population reach 1000 million, or one billion?

ENRICHMENT ESSAY

Carbon Dating and Artistic Development

The artistic community was electrified by the discovery in 1995 of spectacular cave paintings in a limestone cavern in France. Carbon dating of the charcoal from the site showed that the images, created by artists of remarkable talent, were 30,000 years old, making them the oldest cave paintings ever found. The artists seemed to have used the cavern's natural contours to heighten a sense of perspective. The quality of the painting suggests that the art of early humans did not mature steadily from primitive to sophisticated in any simple linear fashion.

Our next example involves exponential decay and its use in determining the age of fossils and artifacts. The method is based on considering the percentage of carbon-14 remaining in the fossil or artifact. Carbon-14 decays exponentially with a *half-life* of approximately 5715 years. The **half-life** of a substance is the time required for half of a given sample to disintegrate. Thus, after 5715 years a given amount of carbon-14 will have decayed to half the original amount. Carbon dating is useful for artifacts or fossils up to 80,000 years old. Older objects do not have enough carbon-14 left to date age accurately.

EXAMPLE 2 Carbon-14 Dating: The Dead Sea Scrolls

a. Use the fact that after 5715 years a given amount of carbon-14 will have decayed to half the original amount to find the exponential decay model for carbon-14.

b. In 1947, earthenware jars containing what are known as the Dead Sea Scrolls were found by an Arab Bedouin herdsman. Analysis indicated that the scroll wrappings contained 76% of their original carbon-14. Estimate the age of the Dead Sea Scrolls.

Solution We begin with the exponential decay model $A = A_0e^{kt}$. We know that $k < 0$ because the problem involves the decay of carbon-14. After 5715 years $(t = 5715)$, the amount of carbon-14 present, A, is half the original amount, A_0. Thus, we can substitute $\frac{A_0}{2}$ for A in the exponential decay model. This will enable us to find k, the decay rate.

a. $A = A_0e^{kt}$ — Begin with the exponential decay model.

$\frac{A_0}{2} = A_0e^{k5715}$ — After 5715 years $(t = 5715)$, $A = \frac{A_0}{2}$ (because the amount present, A, is half the original amount, A_0).

$\frac{1}{2} = e^{5715k}$ — Divide both sides of the equation by A_0.

$\ln\frac{1}{2} = \ln e^{5715k}$ — Take the natural logarithm on both sides.

$\ln\frac{1}{2} = 5715k$ — $\ln e^x = x$

$k = \frac{\ln\frac{1}{2}}{5715} \approx -0.000121$ — Solve for k.

Substituting for k in the decay model, the model for carbon-14 is $A = A_0e^{-0.000121t}$.

b. $A = A_0e^{-0.000121t}$ — This is the decay model for carbon-14.

$0.76A_0 = A_0e^{-0.000121t}$ — A, the amount present, is 76% of the original amount, so $A = 0.76A_0$.

$0.76 = e^{-0.000121t}$ — Divide both sides of the equation by A_0.

$\ln 0.76 = \ln e^{-0.000121t}$ — Take the natural logarithm on both sides.

$\ln 0.76 = -0.000121t$ — $\ln e^x = x$

$t = \frac{\ln 0.76}{-0.000121} \approx 2268$ — Solve for t.

The Dead Sea Scrolls are approximately 2268 years old plus the number of years between 1947 and the current year. ■

✔ **CHECK POINT 2** Strontium-90 is a waste product from nuclear reactors. As a consequence of fallout from atmospheric nuclear tests, we all have a measurable amount of strontium-90 in our bones.

a. Use the fact that after 28 years a given amount of strontium-90 will have decayed to half the original amount to find the exponential decay model for strontium-90.

b. Suppose that a nuclear accident occurs and releases 60 grams of strontium-90 into the atmosphere. How long will it take for strontium-90 to decay to a level of 10 grams?

The Art of Modeling

2 Model data with exponential and logarithmic functions.

Throughout this chapter, we have been working with models that were given. However, we can create functions that model data by observing patterns in scatter plots. Figure 9.22 shows scatter plots for data that are exponential or logarithmic.

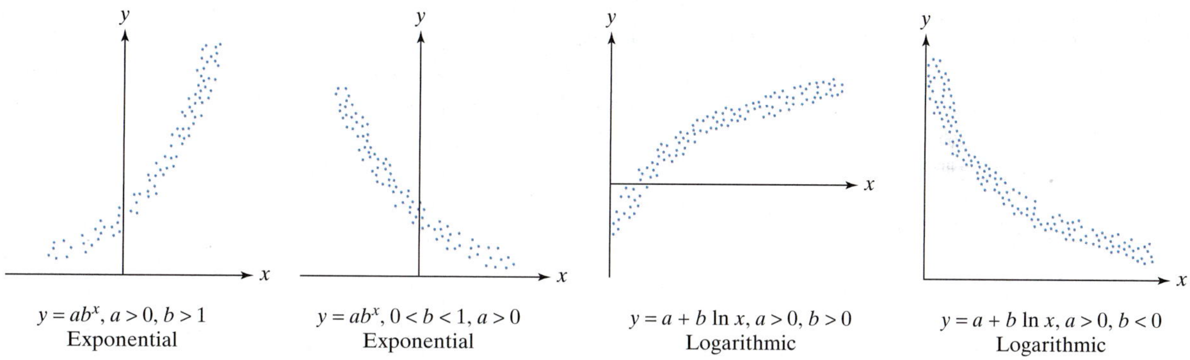

Figure 9.22 Scatter plots for exponential or logarithmic models

Graphing utilities can be used to find the equation of a function that is derived from data. For example, earlier in the chapter we encountered a function that modeled the size of a city and the average walking speed, in feet per second, of pedestrians. The function was derived from the data in Table 9.1. The scatter plot is shown in Figure 9.23.

Table 9.1

x, Population (thousands)	y, Walking Speed (feet per second)
5.5	3.3
14	3.7
71	4.3
138	4.4
342	4.8

Source: Mark and Helen Bornstein, "The Pace of Life"

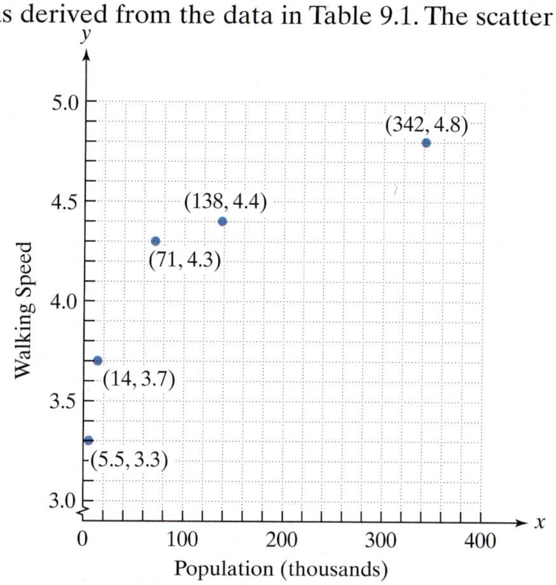

Figure 9.23 Scatter plot for data in Table 9.1

Because the data in this scatter plot increase rapidly at first and then begin to level off a bit, the shape suggests that a logarithmic model might be a good choice.

Figure 9.24 A logarithmic model for the data in Table 9.1 on page 679

A graphing utility fits the data in Table 9.1 on the previous page to a logarithmic model of the form $y = a + b \ln x$ by using the logarithmic REGression option (see Figure 9.24). From the figure, we see that the logarithmic model of the data, with numbers rounded to three decimal places, is

$$y = 2.735 + 0.352 \ln x.$$

The number r that appears in Figure 9.24 is called the **correlation coefficient** and is a measure of how well the model fits the data. The value of r is such that $-1 \leq r \leq 1$. A positive r means that as the x-values increase, so do the y-values. A negative r means that as the x-values increase, the y-values decrease. **The closer that r is to -1 or 1, the better the model fits the data.** Because r is approximately 0.996, the model

$$y = 2.735 + 0.352 \ln x$$

fits the data very well.

Now let's look at data whose scatter plot suggests an exponential model. The data in Table 9.2 indicate world population for six years. The scatter plot is shown in Figure 9.25.

Table 9.2

x, Year	y, World Population (billions)
1950	2.6
1960	3.1
1970	3.7
1980	4.5
1989	5.3
1999	6.0

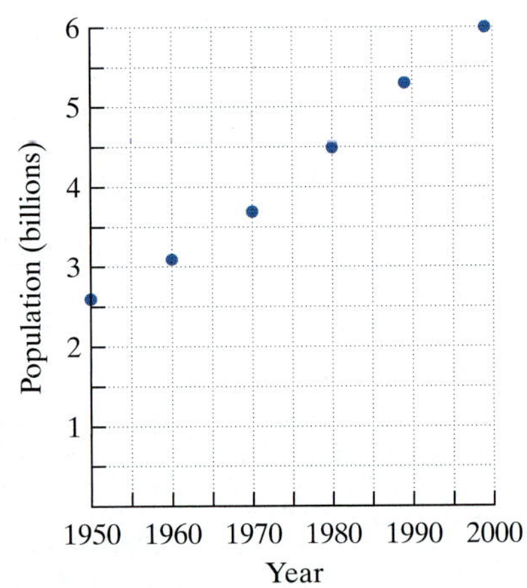

Figure 9.25 A scatter plot for the data in Table 9.2

Because the data in this scatter plot have a rapidly increasing pattern, the shape suggests that an exponential model might be a good choice. (You might also want to try a linear model.) If you select the exponential option, you will use a graphing utility's Exponential REGression option. With this feature, a graphing utility fits the data to an exponential model of the form $y = ab^x$.

When computing an exponential model of the form $y = ab^x$, a graphing utility rewrites the equation using logarithms. Because the domain of the logarithmic function is the set of positive numbers, **zero must not be a value for x.** What does this mean for our data for world population that starts in the year 1950? We must start values of x after 0. Thus, we'll assign x to represent the number of years after 1949. This gives us the data shown in Table 9.3. Using the Exponential REGression option, we obtain the equation in Figure 9.26.

Table 9.3

x, Numbers of Years after 1949	*y*, World Population (billions)
1 (1950)	2.6
11 (1960)	3.1
21 (1970)	3.7
31 (1980)	4.5
40 (1989)	5.3
50 (1999)	6.0

Figure 9.26 An exponential model for the data in Table 9.3

From Figure 9.26, we see that the exponential model of the data for world population x years after 1949, with numbers rounded to three decimal places, is

$$y = 2.570(1.018)^x.$$

The correlation coefficient, r, is close to 1, indicating that the model fits the data very well.

When using a graphing utility to model data, begin with a scatter plot, drawn either by hand or with the graphing utility, to obtain a general picture for the shape of the data. It might be difficult to determine what model best fits the data—linear, logarithmic, exponential, quadratic, or something else. If necessary, use your graphing utility to fit several models to the data. The best model is the one that yields the value r, the correlation coefficient, closest to 1 or -1. Finding a proper fit for data can be almost as much art as it is mathematics. In this era of technology, the process of creating models that best fit data is one that involves more decision making than computation.

ENRICHMENT ESSAY

The Future of World Population

In the future, a new world order looms. The percentage of world population in less developed countries will increase, North America's will remain relatively stable, and Europe's will decrease.

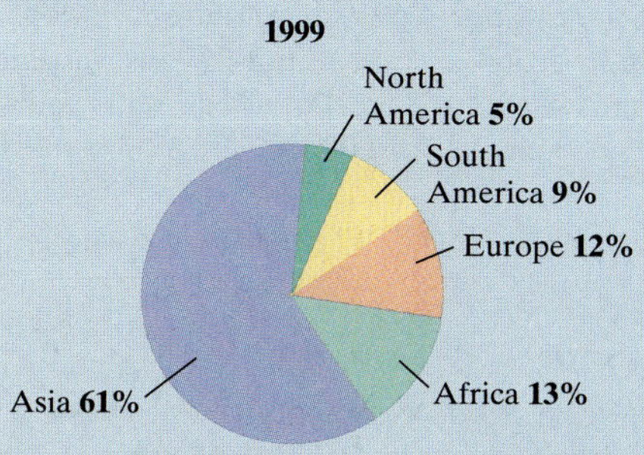

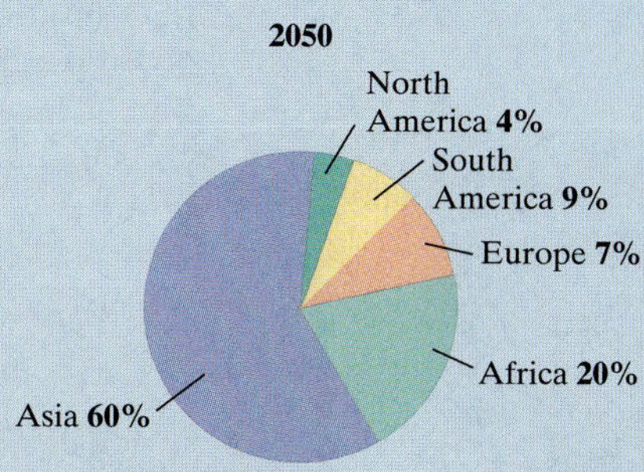

Most Populous Countries, 2050

1. India	1529 million
2. China	1478 million
3. U.S.	349 million
4. Pakistan	345 million
5. Indonesia	312 million
6. Nigeria	244 million
7. Brazil	244 million
8. Bangladesh	212 million
9. Ethiopia	169 million
10. Congo	160 million

Source: United Nations Population Fund

EXERCISE SET 9.6

Practice and Application Exercises

The exponential growth model $A = 208e^{0.008t}$ describes the population of the United States, in millions, t years after 1970. Use this model to solve Exercises 1–4.

1. What was the population of the United States in 1970?

2. By what percentage is the population of the United States increasing each year?

3. When will the U.S. population be 300 million?

4. When will the U.S. population be 350 million?

India is currently one of the world's fastest-growing countries. By 2040, the population of India will be larger than the population of China; by 2050, nearly one-third of the world's population will live in these two countries alone. The exponential growth model $A = 574e^{0.026t}$ describes the population of India, in millions, t years after 1974. Use this model to solve Exercises 5–8.

5. By what percentage is the population of India increasing each year?

6. What was the population of India in 1974?

7. When will India's population be 1624 million?

8. When will India's population be 2732 million?

The value of houses in a neighborhood follows a pattern of exponential growth. In the year 2000, you purchased a house in this neighborhood. The value of your house, in thousands of dollars, t years after 2000 is given by the exponential growth model $V = 140e^{0.068t}$. Use this model to solve Exercises 9–12.

9. What did you pay for your house?

10. By what percentage is the price of houses in your neighborhood increasing each year?

11. When will your house be worth $200,000?

12. When will your house be worth $300,000

13. Through the end of 1991, 200,000 cases of AIDS had been reported to the Centers for Disease Control in the United States. By the end of 1998, the number had grown to 680,000. The exponential growth function $A = 200e^{kt}$ describes the thousands of AIDS cases in the United States t years after 1991. Use the fact that 7 years after 1991 there were 680 thousand cases to find k to three decimal places. Then write the exponential growth function. According to your model, by what percentage is the number of AIDS cases in the United States increasing each year?

14. In 1980, China's population was 983 million; in 1990, it was 1154 million. The exponential growth function $A = 983e^{kt}$ describes the population of China, in millions, t years after 1980. Use the fact that 10 years after 1980 the population was 1154 million to find k to three decimal places. Then write the exponential growth function. According to your model, by what percentage is the population of China increasing each year?

An artifact originally had 16 grams of carbon-14 present. The decay model $A = 16e^{-0.000121t}$ describes the amount of carbon-14 present after t years. Use this model to solve Exercises 15–16.

15. How many grams of carbon-14 will be present in 5715 years?

16. How many grams of carbon-14 will be present in 11,430 years?

17. The half-life of the radioactive element krypton-91 is 10 seconds. If 16 grams of krypton-91 are initially present, how many grams are present after 10 seconds? 20 seconds? 30 seconds? 40 seconds? 50 seconds?

18. The half-life of the radioactive element plutonium-239 is 25,000 years. If 16 grams of plutonium-239 are initially present, how many grams are present after 25,000 years? 50,000 years? 75,000 years? 100,000 years? 125,000 years?

Use the exponential decay model for carbon-14, $A = A_0e^{-0.000121t}$, to solve Exercises 19–20.

19. Prehistoric cave paintings were discovered in a cave in France. The paint contained 15% of the original carbon-14. Estimate the age of the paintings.

20. Skeletons were found at a construction site in San Francisco in 1989. The skeletons contained 88% of the expected amount of carbon-14 found in a living person. In 1989, how old were the skeletons?

21. The August 1978 issue of *National Geographic* described the 1964 find of dinosaur bones of a newly discovered dinosaur weighing 170 pounds, measuring 9 feet, with a 6-inch claw on one toe of each hind foot. The age of the dinosaur was estimated using potassium-40 dating of rocks surrounding the bones.

a. Potassium-40 decays exponentially with a half-life of approximately 1.31 billion years. Use the fact that after 1.31 billion years a given amount of potassium-40 will have decayed to half the original amount to show that the decay model for potassium-40 is given by $A = A_0e^{-0.52912t}$, where t is in billions of years.

b. Analysis of the rocks surrounding the dinosaur bones indicated that 94.5% of the original amount of potassium-40 was still present. Let $A = 0.945A_0$ in the model in part (a) and estimate the age of the bones of the dinosaur.

22. A bird species in danger of extinction has a population that is decreasing exponentially $(A = A_0e^{kt})$. Five years ago the population was at 1400 and today only 1000 of the birds are alive. Once the population drops below 100, the situation will be irreversible. When will this happen?

23. Use the exponential growth model, $A = A_0e^{kt}$, to show that the time it takes a population to double (to grow from A_0 to $2A_0$) is given by $t = \frac{\ln 2}{k}$.

24. Use the exponential growth model, $A = A_0e^{kt}$, to show that the time it takes a population to triple (to grow from A_0 to $3A_0$) is given by $t = \frac{\ln 3}{k}$.

Use the formula $t = \frac{\ln 2}{k}$ that gives the time for a population with a growth rate k to double to solve Exercises 25–26. Express each answer to the nearest whole year.

25. China is growing at a rate of 1.1% per year. How long will it take China to double its population?

26. Japan is growing at a rate of 0.3% per year. How long will it take Japan to double its population?

Writing in Mathematics

27. Nigeria has a growth rate of 0.031 or 3.1%. Describe what this means.

28. How can you tell if an exponential model describes exponential growth or exponential decay?

29. Suppose that a population that is growing exponentially increases from 800,000 people in 1997 to 1,000,000 people in 2000. Without showing the details, describe how to obtain the exponential growth function that models this data.

30. What is the half-life of a substance?

31. Describe the shape of a scatter plot that suggests modeling the data with an exponential function.

32. You take up weightlifting and record the maximum number of pounds you can lift at the end of each week. You start off with rapid growth in terms of the weight you can lift from week to week, but then the growth begins to level off. Describe how to obtain a function that models the number of pounds you can lift at the end of each week. How can you use this function to predict what might happen if you continue the sport?

33. Would you prefer that your salary be modeled exponentially or logarithmically? Explain your answer.

34. One problem with all exponential growth models is that nothing can grow exponentially forever. Describe factors that might limit the size of a population.

Technology Exercises

In Example 1 on page 676, we used two data points and an exponential function to model federal minimum wages that were not adjusted for inflation from 1970 through 2000. The data are shown again in the table. Use all seven data points to solve Exercises 35–39.

x, Number of Years after 1969	y, Federal Minimum Wage
1	1.60
6	2.10
11	3.10
16	3.35
21	3.80
26	4.25
31	5.15

35. Use your graphing utility's Exponential REGression option to obtain a model of the form $y = ab^x$ that fits the data. How well does the correlation coefficient, r, indicate that the model fits the data?

36. Use your graphing utility's logarithmic REGression option to obtain a model of the form $y = a + b \ln x$ that fits the data. How well does the correlation coefficient, r, indicate that the model fits the data?

37. Use your graphing utility's Linear REGression option to obtain a model of the form $y = ax + b$ that fits the data. How well does the correlation coefficient, r, indicate that the model fits the data?

38. Use your graphing utility's Power REGression option to obtain a model of the form $y = ax^b$ that fits the data. How well does the correlation coefficient, r, indicate that the model fits the data?

39. Use the value of r in Exercises 35–38 to select the model of best fit. Use this model to predict by which year the minimum wage will reach \$7.50. How does this answer compare to the year we found in Example 1, namely 2010? If you obtained a different year, how do you account for this difference?

40. The figure shows the number of people in the United States age 65 and over, with projected figures for the year 2000 and beyond.

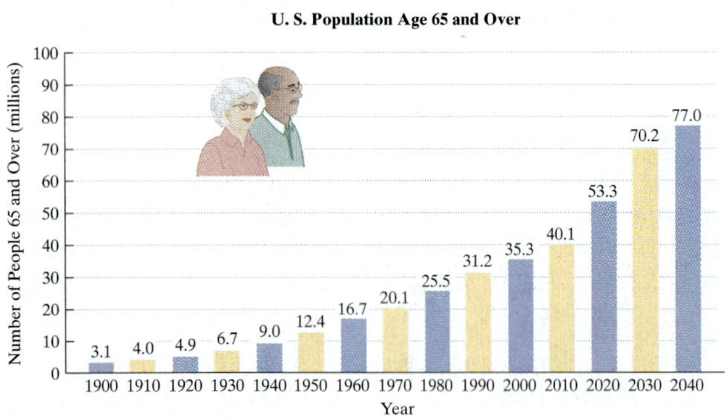

Source: U. S. Bureau of the Census

Let x represent the number of years after 1899 and let y represent the U.S. population, in millions. Use your graphing utility to find the model that best fits the data in the bar graph. Then use the model to find the U.S. population age 65 and over in 2050.

Critical Thinking Exercises

41. The World Health Organization makes predictions about the number of AIDS cases based on a compromise between a linear model and an exponential growth model. Explain why the World Health Organization does this.

42. a. Show that $y = ab^x$ is equivalent to $y = ae^{(\ln b)\cdot x}$. *Hint*: Use the fact that $b = e^{\ln b}$.

b. In this section, we used the Exponential REGression feature of a graphing utility and obtained an exponential model for world population. World population, y, in billions, x years after 1949 is modeled by

$$y = 2.57(1.018)^x.$$

Use the equivalence from part (a) and express the model as

$$y = 2.57e^{0.018x}.$$

c. Express the model for world population in part (b) so that it has the same letters as those in the exponential growth model $A = A_0e^{kt}$. According to the model, what is the growth rate for world population?

d. The actual growth rate for world population is now 1.3%. Describe what might happen if you use the model in part (c) to make projections about world population in the future.

Review Exercises

43. Divide:

$$\frac{x^2 - 9}{2x^2 + 7x + 3} \div \frac{x^2 - 3x}{2x^2 + 11x + 5}.$$

(Section 6.1, Example 7)

44. Solve: $x^{\frac{2}{3}} + 2x^{\frac{1}{3}} - 3 = 0.$
(Section 8.4, Example 5)

45. Simplify: $6\sqrt{2} - 2\sqrt{50} + 3\sqrt{98}$.
(Section 7.4, Example 2)

CHAPTER 9 GROUP PROJECTS

1. This group exercise involves exploring the way we grow. Group members should create a graph for the function that models the percentage of adult height attained by a boy who is x years old, $f(x) = 29 + 48.8\log(x + 1)$. Let $x = 1, 2, 3, \ldots, 12$, find function values, and connect the resulting points with a smooth curve. Then create a graph for the function that models the percentage of adult height attained by a girl who is x years old, $g(x) = 62 + 35\log(x - 4)$. Let $x = 5, 6, 7, \ldots, 15$, find function values, and connect the resulting points by a smooth curve. Group members should then discuss similarities and differences in the growth patterns for boys and girls based on the graphs.
2. Research applications of logarithmic functions as mathematical models and plan a seminar based on your group's research. Each group member should research one of the following areas or any other area of interest: pH (acidity of solutions), intensity of sound (decibels), brightness of stars, consumption of natural resources, human memory, progress over time in a sport, profit over time. For the area that you select, explain how logarithmic functions are used and provide examples.
3. This activity is intended for three or four people who would like to take up weightlifting. Each person in the group should record the maximum number of pounds that he or she can lift at the end of each week for the first 10 consecutive weeks. Use the logarithmic REGression option of a graphing utility to obtain a model showing the amount of weight that group members can lift from week 1 through week 10. Graph each of the models in the same viewing rectangle to observe similarities and differences among weight-growth patterns of each member. Use the functions to predict the amount of weight that group members will be able to lift in the future. If the group continues to work out together, check the accuracy of these predictions.
4. Each group member should consult an almanac, newspaper, magazine, or the Internet to find data that can be modeled by exponential or logarithmic functions. Group members should select the two sets of data that are most interesting and relevant. For each data set selected, find a model that best fits the data. Each group member should make one prediction based on the model and then discuss a consequence of this prediction. What factors might change the accuracy of each prediction?
5. In Tom Stoppard's play *Arcadia*, the characters dream and talk about mathematics, including ideas involving graphing, composite functions, symmetry, and lack of symmetry in things that are tangled, mysterious, and unpredictable. Group members should rent and view the movie. Present a report on the ideas discussed by the characters that are related to concepts involving functions, their graphs, and composite functions. Bring in a copy of the video and show appropriate excerpts.

CHAPTER SUMMARY, REVIEW, AND TEST

SUMMARY

DEFINITIONS AND CONCEPTS | **EXAMPLES**

Section 9.1 Exponential Functions

The exponential function with base b is defined by $f(x) = b^x$, where $b > 0$ and $b \neq 1$. The graph contains the point $(0, 1)$. When $b > 1$, the graph rises from left to right. When $0 < b < 1$, the graph falls from left to right. The x-axis is a horizontal asymptote. The domain is $(-\infty, \infty)$; the range is $(0, \infty)$. The natural exponential function is $f(x) = e^x$, $e \approx 2.71828$.

Graph $f(x) = 2^x$ and $g(x) = 2^{x-1}$.

x	$f(x) = 2^x$	$g(x) = 2^{x-1}$
-2	$2^{-2} = \frac{1}{4}$	$2^{-3} = \frac{1}{8}$
-1	$2^{-1} = \frac{1}{2}$	$2^{-2} = \frac{1}{4}$
0	$2^0 = 1$	$2^{-1} = \frac{1}{2}$
1	$2^1 = 2$	$2^0 = 1$
2	$2^2 = 4$	$2^1 = 2$

The graph of g is the graph of f shifted one unit to the right.

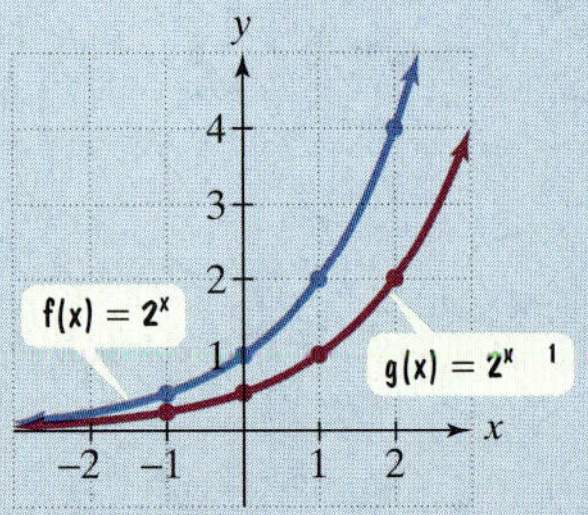

Formulas for Compound Interest
After t years, the balance, A, in an account with principal P and annual interest rate r is given by the following formulas:

1. For n compoundings per year: $A = P\left(1 + \frac{r}{n}\right)^{nt}$
2. For continuous compounding: $A = Pe^{rt}$

Select the better investment for \$4000 over 6 years:

- 6% compounded semiannually

$$A = P\left(1 + \frac{r}{n}\right)^{nt}$$
$$= 4000\left(1 + \frac{0.06}{2}\right)^{2 \cdot 6} \approx \$5703$$

- 5.9% compounded continuously

$$A = Pe^{rt} = 4000e^{0.059(6)} \approx \$5699$$

The first investment is better.

Section 9.2 Composite and Inverse Functions

Composite Functions
The composite function $f \circ g$ is defined by

$$(f \circ g)(x) = f(g(x)).$$

The composite function $g \circ f$ is defined by

$$(g \circ f)(x) = g(f(x)).$$

Let $f(x) = x^2 + x$ and $g(x) = 2x + 1$.

- $(f \circ g)(x) = f(g(x)) = (g(x))^2 + g(x)$

Replace x by $g(x)$.

$$= (2x + 1)^2 + (2x + 1) = 4x^2 + 4x + 1 + 2x + 1$$
$$= 4x^2 + 6x + 2$$

- $(g \circ f)(x) = g(f(x)) = 2f(x) + 1$

Replace x by $f(x)$.

$$= 2(x^2 + x) + 1 = 2x^2 + 2x + 1$$

DEFINITIONS AND CONCEPTS	EXAMPLES

Section 9.2 Composite and Inverse Functions (continued)

Inverse Functions

If $f(g(x)) = x$ and $g(f(x)) = x$, function g is the inverse of function f, denoted f^{-1} and read "f inverse." The procedure for finding a function's inverse uses a switch-and-solve strategy. Switch x and y, then solve for y.

If $f(x) = 2x - 5$, find $f^{-1}(x)$.

$y = 2x - 5$ Replace f(x) with y.

$x = 2y - 5$ Exchange x and y.

$x + 5 = 2y$ Solve for y.

$\frac{x+5}{2} = y$

$f^{-1}(x) = \frac{x+5}{2}$ Replace y with $f^{-1}(x)$.

The Horizontal Line Test for Inverse Functions

A function, f, has an inverse that is a function, f^{-1}, if there is no horizontal line that intersects the graph of f at more than one point. A one-to-one function is one in which no two different ordered pairs have the same second component. Only one-to-one functions have inverse functions. If the point (a, b) is on the graph of f, then the point (b, a) is on the graph of f^{-1}. The graph of f^{-1} is a reflection of the graph of f about the line $y = x$.

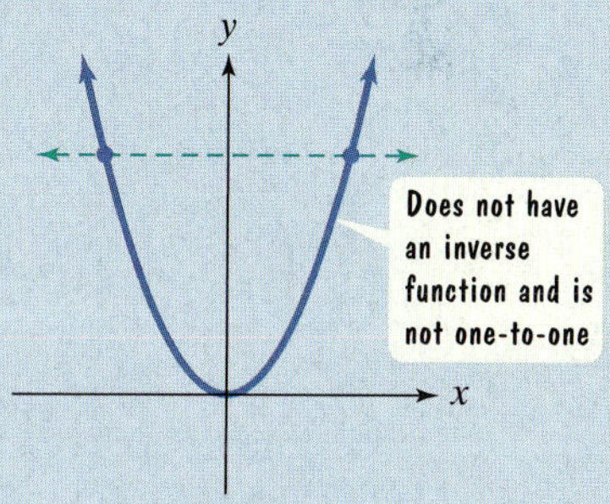

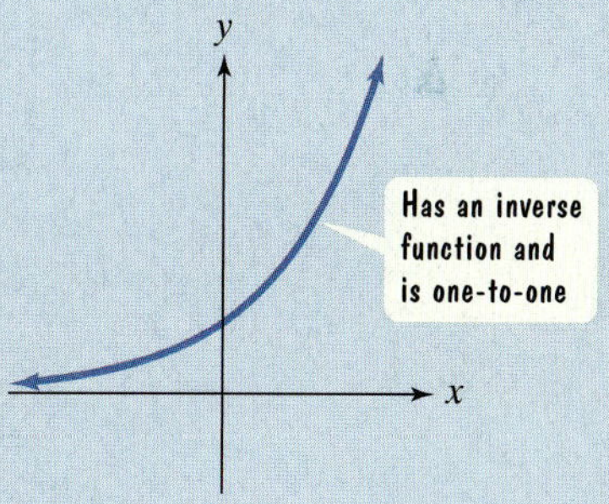

Section 9.3 Logarithmic Functions

Definition of the logarithmic function: For $x > 0$ and $b > 0, b \neq 1, y = \log_b x$ is equivalent to $b^y = x$. The function $f(x) = \log_b x$ is the logarithmic function with base b. This function is the inverse function of the exponential function with base b.

- Write $\log_2 32 = 5$ in exponential form.

  $2^5 = 32$ $y = \log_b x$ means $b^y = x$.

- Write $\sqrt{49} = 7$, or $49^{\frac{1}{2}} = 7$, in logarithmic form.

  $\frac{1}{2} = \log_{49} 7$ $b^y = x$ means $y = \log_b x$.

DEFINITIONS AND CONCEPTS	EXAMPLES

Section 9.3 Logarithmic Functions (continued)

The graph of $f(x) = \log_b x$ can be obtained from $f(x) = b^x$ by reversing coordinates. The graph of $f(x) = \log_b x$ contains the point $(1, 0)$. If $b > 1$, the graph rises from left to right. If $0 < b < 1$, the graph falls from left to right. The y-axis is a vertical asymptote. The domain is $(0, \infty)$; the range is $(-\infty, \infty)$. $f(x) = \log x$ means $f(x) = \log_{10} x$ and is the common logarithmic function. $f(x) = \ln x$ means $f(x) = \log_e x$ and is the natural logarithmic function. The domain of $f(x) = \log_b(x + c)$ consists of all x for which $x + c > 0$.

- Graph $f(x) = \log_3 x$.

x	-1	0	1	2
$y = 3^x$	$\frac{1}{3}$	1	3	9

x	$\frac{1}{3}$	1	3	9
$f(x) = \log_3 x$	-1	0	1	2

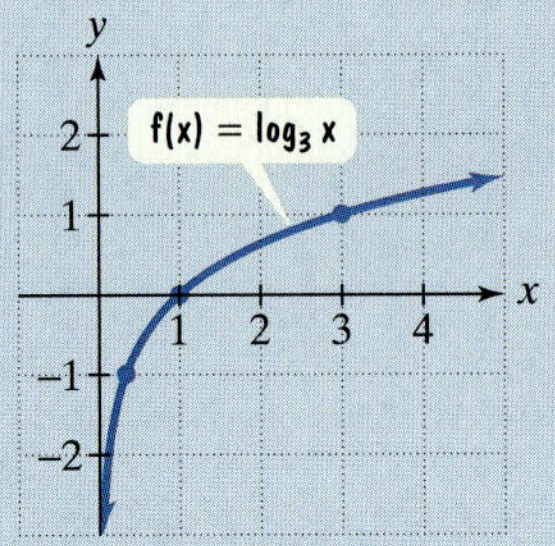

- Find the domain: $f(x) = \log_6(4 - x)$.

$$4 - x > 0$$
$$4 > x \quad (\text{or } x < 4)$$

The domain is $\{x | x < 4\}$ or $(-\infty, 4)$.

Basic Logarithmic Properties

Base b ($b > 0, b \neq 1$)	**Base 10 (Common Logarithms)**	**Base e (Natural Logarithms)**
$\log_b 1 = 0$	$\log 1 = 0$	$\ln 1 = 0$
$\log_b b = 1$	$\log 10 = 1$	$\ln e = 1$
$\log_b b^x = x$	$\log 10^x = x$	$\ln e^x = x$
$b^{\log_b x} = x$	$10^{\log x} = x$	$e^{\ln x} = x$

- $\log_8 1 = 0$ because $\log_b 1 = 0$.
- $\log_4 4 = 1$ because $\log_b b = 1$.
- $\ln e^{8x} = 8x$ because $\ln e^x = x$.
- $e^{\ln \sqrt[3]{x}} = \sqrt[3]{x}$ because $e^{\ln x} = x$.
- $\log_t t^{25} = 25$ because $\log_b b^x = x$.

Section 9.4 Properties of Logarithms

Properties of Logarithms

1. *The Product Rule*: $\log_b(MN) = \log_b M + \log_b N$
2. *The Quotient Rule*: $\log_b\left(\frac{M}{N}\right) = \log_b M - \log_b N$
3. *The Power Rule*: $\log_b M^p = p \log_b M$
4. *The Change-of Base Property*:

The General Property	**Introducing Common Logarithms**	**Introducing Natural Logarithms**
$\log_b M = \frac{\log_a M}{\log_a b}$	$\log_b M = \frac{\log M}{\log b}$	$\log_b M = \frac{\ln M}{\ln b}$

- Expand: $\log_3(81x^7)$.

$$= \log_3 81 + \log_3 x^7$$
$$= 4 + 7\log_3 x$$

- Write as a single logarithm:

$$7 \ln x - 4 \ln y.$$
$$= \ln x^7 - \ln y^4 = \ln \frac{x^7}{y^4}$$

- Evaluate: $\log_6 92$.

$$\log_6 92 = \frac{\ln 92}{\ln 6} \approx 2.5237$$

DEFINITIONS AND CONCEPTS	EXAMPLES
Section 9.5 Exponential and Logarithmic Equations	
An exponential equation is an equation containing a variable in an exponent. Some exponential equations can be solved by expressing both sides as a power of the same base. Then set the exponents equal to each other.	Solve: $4^{2x-1} = 64$. $4^{2x-1} = 4^3$; $2x - 1 = 3$; $2x = 4$; $x = 2$. The solution is 2 and the solution set is $\{2\}$.
If both sides of an exponential equation cannot be expressed as a power of the same base, isolate the exponential expression and take the natural logarithm on both sides. Simplify using $\ln b^x = x \ln b$ or $\ln e^x = x$.	Solve: $7^x = 103$. $\ln 7^x = \ln 103$; $x \ln 7 = \ln 103$; $x = \frac{\ln 103}{\ln 7}$. The solution is $\frac{\ln 103}{\ln 7}$ and the solution set is $\left\{\frac{\ln 103}{\ln 7}\right\}$.
A logarithmic equation is an equation containing a variable in a logarithmic expression. Logarithmic equations in the form $\log_b x = c$ can be solved by rewriting as $b^c = x$. When checking logarithmic equations, reject proposed solutions that produce the logarithm of a negative number or the logarithm of zero in the original equation.	Solve: $\log_2(3x - 1) = 5$. $2^5 = 3x - 1$; $32 = 3x - 1$; $33 = 3x$; $11 = x$. The solution is 11 and the solution set is $\{11\}$.
Equations involving natural logarithms are solved by isolating the natural logarithm with coefficient 1 on one side and exponentiating both sides. Simplify using $e^{\ln x} = x$.	Solve: $3 \ln 2x = 15$. $\ln 2x = 5$; $e^{\ln 2x} = e^5$; $2x = e^5$; $x = \frac{e^5}{2}$. The solution is $\frac{e^5}{2}$ and the solution set is $\left\{\frac{e^5}{2}\right\}$.
Section 9.6 Exponential Growth and Decay; Modeling Data	
Exponential growth and decay models are given by $A = A_0 e^{kt}$ in which t represents time, A_0 is the amount present at $t = 0$, and A is the amount present at time t. If $k > 0$, the model describes growth and k is the growth rate. If $k < 0$, the model describes decay and k is the decay rate. Scatter plots for exponential and logarithmic models are shown in Figure 9.22 on page 679 When using a graphing utility to model data, the closer that the correlation coefficient r is to -1 or 1, the better the model fits the data.	The 1970 population of Mexico City was 9.4 million; in 1990, it was 20.2 million. Write the exponential growth function that describes the population, in millions, t years after 1970. Begin with $A = A_0 e^{kt}$. $A = 9.4e^{kt}$ (In 1970 ($t = 0$), population was 9.4 million.) $20.2 = 9.4e^{k \cdot 20}$ (When $t = 20$ (in 1990), $A = 20.2$.) $e^{20k} = \frac{20.2}{9.4}$; $\ln e^{20k} = \ln \frac{20.2}{9.4}$; $20k = \ln \frac{20.2}{9.4}$ and $k = \frac{\ln \frac{20.2}{9.4}}{20} \approx 0.038$. Growth function is $A = 9.4e^{0.038t}$. (Growth rate is 0.038 or 3.8%.)

Review Exercises

9.1 *In Exercises 1–4, set up a table of coordinates for each function. Select integers from −2 to 2 for x. Then use the table of coordinates to match the function with its graph. [The graphs are labeled (a) through (d).]*

1. $f(x) = 4^x$

2. $f(x) = 4^{-x}$

3. $f(x) = -4^{-x}$

4. $f(x) = -4^{-x} + 3$

a.

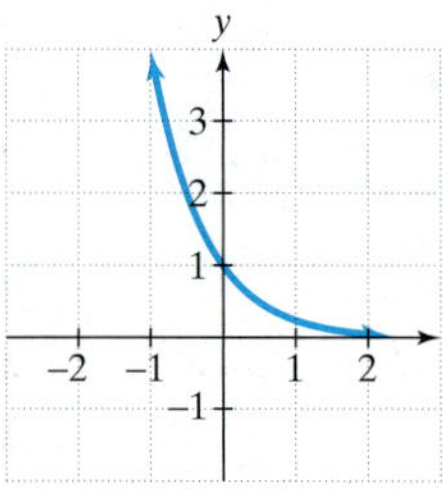

b.

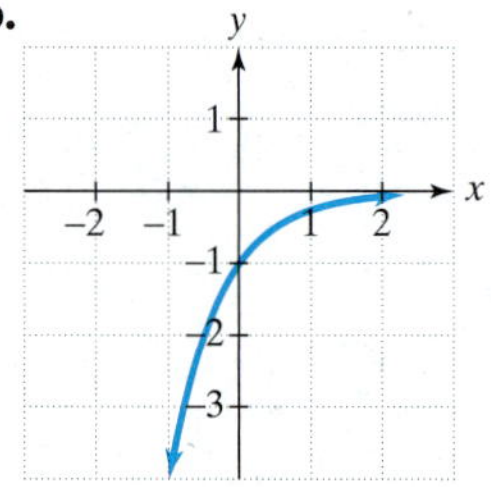

c.

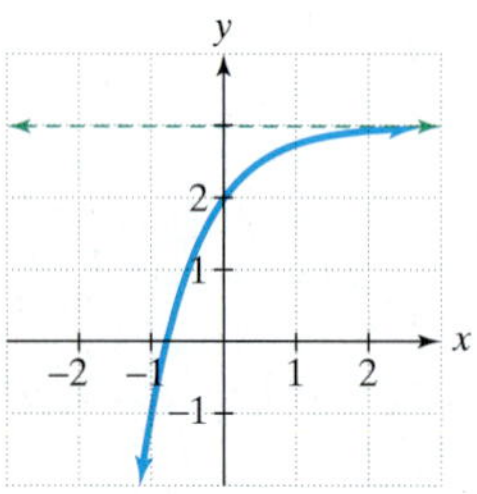

d.

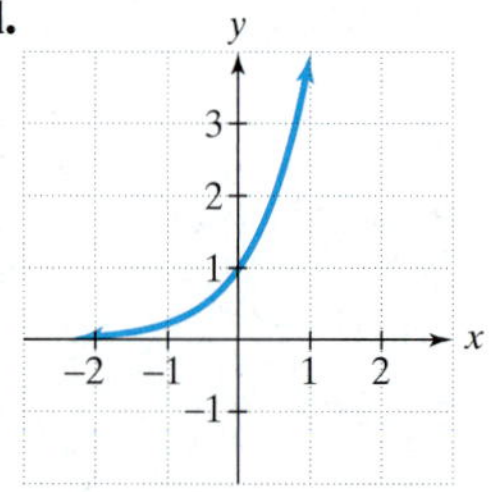

In Exercises 5–8, graph functions f and g in the same rectangular coordinate system. Select integers from −2 to 2 for x. Then describe how the graph of g is related to the graph of f. If applicable, use a graphing utility to confirm your hand-drawn graphs.

5. $f(x) = 2^x$ and $g(x) = 2^{x-1}$

6. $f(x) = 2^x$ and $g(x) = \left(\frac{1}{2}\right)^x$

7. $f(x) = 3^x$ and $g(x) = 3^x - 1$

8. $f(x) = 3^x$ and $g(x) = -3^x$

Use the compound interest formulas

$$A = P\left(1 + \frac{r}{n}\right)^{nt} \quad \text{and} \quad A = Pe^{rt}$$

to solve Exercises 9–10.

9. Suppose that you have \$5000 to invest. Which investment yields the greater return over 5 years: 5.5% compounded semiannually or 5.25% compounded monthly?

10. Suppose that you have \$14,000 to invest. Which investment yields the greater return over 10 years: 7% compounded monthly or 6.85% compounded continuously?

11. A cup of coffee is taken out of a microwave oven and placed in a room. The temperature, T, in degrees Fahrenheit, of the coffee after t minutes is modeled by the function $T = 70 + 130e^{-0.04855t}$. The graph of the function is shown in the figure. Use the graph to answer each of the following questions.

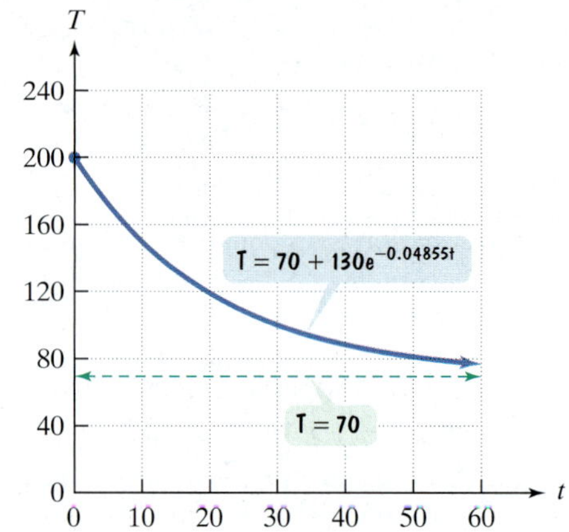

a. What was the temperature of the coffee when it was first taken out of the microwave?

b. What is a reasonable estimate of the temperature of the coffee after 20 minutes? Use your calculator to verify this estimate.

c. What is the limit of the temperature to which the coffee will cool? What does this tell you about the temperature of the room?

9.2 *In Exercises 12–13, find* **a.** $(f \circ g)(x)$; **b.** $(g \circ f)(x)$; **c.** $(f \circ g)(3)$.

12. $f(x) = x^2 + 3, \quad g(x) = 4x - 1$

13. $f(x) = \sqrt{x}, \quad g(x) = x + 1$

In Exercises 14–15, find $f(g(x))$ and $g(f(x))$ and determine whether each pair of functions f and g are inverses of each other.

14. $f(x) = \frac{3}{5}x + \frac{1}{2}$ and $g(x) = \frac{5}{3}x - 2$

15. $f(x) = 2 - 5x$ and $g(x) = \dfrac{2 - x}{5}$

The functions in Exercises 16–18 are all one-to-one. For each function:

- **a.** *Find an equation of $f^{-1}(x)$, the inverse function.*
- **b.** *Verify that your equation is correct by showing that $f(f^{-1}(x)) = x$ and $f^{-1}(f(x)) = x$.*

16. $f(x) = 4x - 3$

17. $f(x) = \sqrt{x + 2}$

18. $f(x) = 8x^3 + 1$

Which graphs in Exercises 19–22 represent functions that have inverse functions?

19.

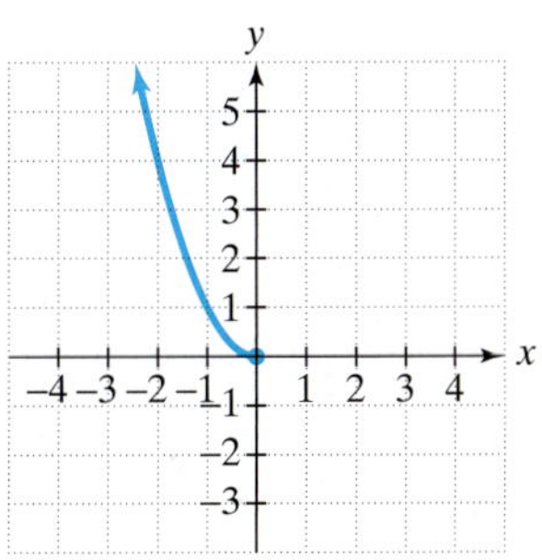

20.

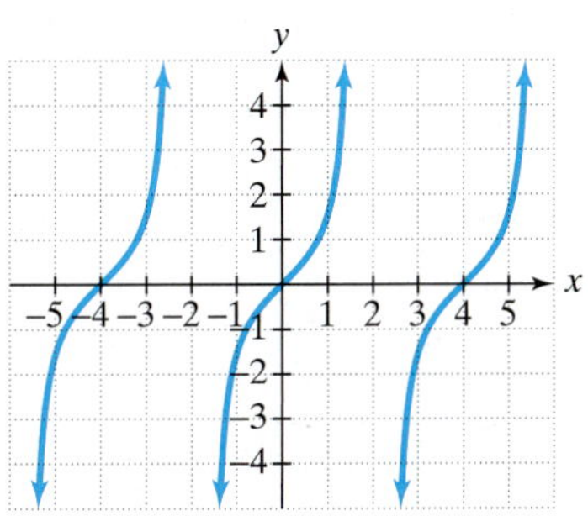

21.

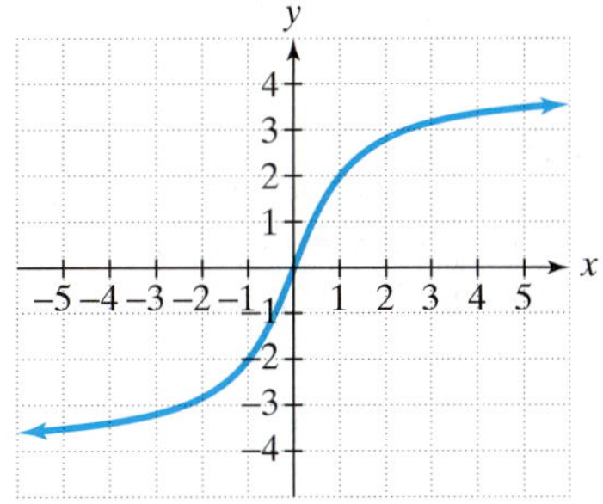

22.

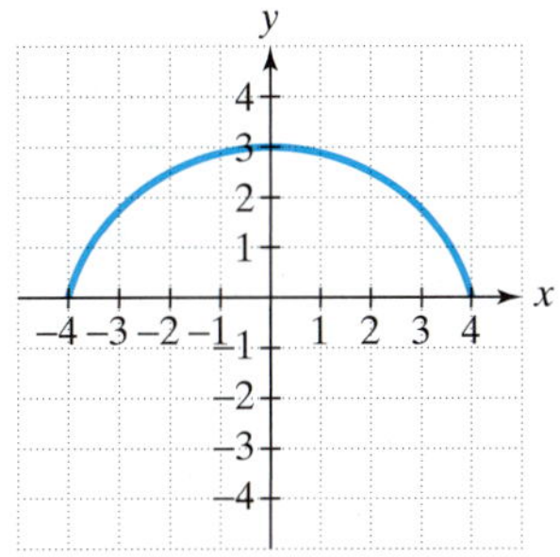

23. Use the graph of f in the figure shown to draw the graph of its inverse function.

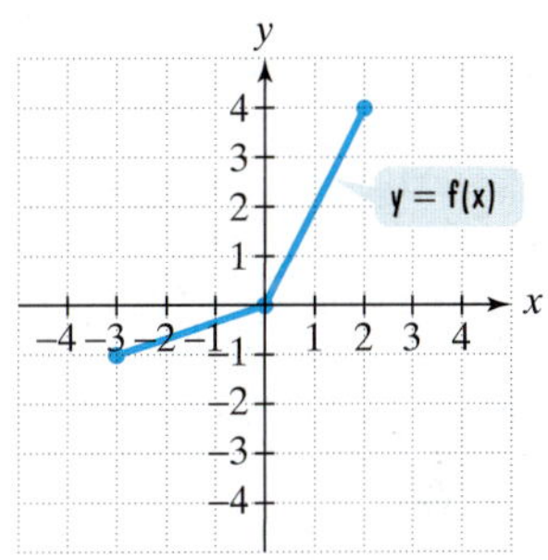

9.3 *In Exercises 24–26, write each equation in its equivalent exponential form.*

24. $\dfrac{1}{2} = \log_{49} 7$

25. $3 = \log_4 x$

26. $\log_3 81 = y$

In Exercises 27–29, write each equation in its equivalent logarithmic form.

27. $6^3 = 216$

28. $b^4 = 625$

29. $13^y = 874$

In Exercises 30–37, evaluate each expression without using a calculator. If evaluation is not possible, state the reason.

30. $\log_4 64$

31. $\log_5 \frac{1}{25}$

32. $\log_3(-9)$

33. $\log_{16} 4$

34. $\log_{17} 17$

35. $\log_3 3^8$

36. $\ln e^5$

37. $\log_3(\log_8 8)$

38. Graph $f(x) = 2^x$ and $g(x) = \log_2 x$ in the same rectangular coordinate system.

39. Graph $f(x) = (\frac{1}{3})^x$ and $g(x) = \log_{\frac{1}{3}} x$ in the same rectangular coordinate system.

In Exercises 40–42, find the domain of each logarithmic function.

40. $f(x) = \log_8(x + 5)$

41. $f(x) = \log(3 - x)$

42. $f(x) = \ln(x - 1)^2$

In Exercises 43–45, use inverse properties of logarithms to simplify each expression.

43. $\ln e^{6x}$

44. $e^{\ln \sqrt{x}}$

45. $10^{\log 4x^2}$

46. On the Richter scale, the magnitude, R, of an earthquake of intensity I is given by $R = \log \frac{I}{I_0}$, where I_0 is the intensity of a barely felt zero-level earthquake. If the intensity of an earthquake is $1000I_0$, what is its magnitude on the Richter scale?

47. Students in a psychology class took a final examination. As part of an experiment to see how much of the course content they remembered over time, they took equivalent forms of the exam in monthly intervals thereafter. The average score, $f(t)$, for the group after t months is modeled by the function $f(t) = 76 - 18\log(t + 1)$, where $0 \le t \le 12$.

a. What was the average score when the exam was first given?

b. What was the average score after 2 months? 4 months? 6 months? 8 months? one year?

c. Use the results from parts (a) and (b) to graph f. Describe what the shape of the graph indicates in terms of the material retained by the students.

48. The formula

$$t = \frac{1}{c}\ln\left(\frac{A}{A - N}\right)$$

describes the time, t, in weeks, that it takes to achieve mastery of a portion of a task. In the formula, A represents maximum learning possible, N is the portion of the learning that is to be achieved, and c is a constant used to measure an individual's learning style. A 50-year-old man decides to start running as a way to maintain good health. He feels that the maximum rate he could ever hope to achieve is 12 miles per hour. How many weeks will it take before the man can run 5 miles per hour if $c = 0.06$ for this person?

9.4 *In Exercises 49–52, use properties of logarithms to expand each logarithmic expression as much as possible. Where possible, evaluate logarithmic expressions without using a calculator. Assume that all variables represent positive numbers.*

49. $\log_6(36x^3)$

50. $\log_4 \frac{\sqrt{x}}{64}$

51. $\log_2\left(\frac{xy^2}{64}\right)$

52. $\ln \sqrt[3]{\frac{x}{e}}$

In Exercises 53–56, use properties of logarithms to condense each logarithmic expression. Write the expression as a single logarithm whose coefficient is 1.

53. $\log_b 7 + \log_b 3$

54. $\log 3 - 3\log x$

55. $3\ln x + 4\ln y$

56. $\frac{1}{2}\ln x - \ln y$

In Exercises 57–58, use common logarithms or natural logarithms and a calculator to evaluate to four decimal places.

57. $\log_6 72{,}348$

58. $\log_4 0.863$

9.5 *Solve each exponential equation in Exercises 59–61 by expressing each side as a power of the same base and then equating exponents.*

59. $2^{4x-2} = 64$

60. $125^x = 25$

61. $9^x = \frac{1}{27}$

Solve each exponential equation in Exercises 62–64 by taking the natural logarithm on both sides. Express the solution in terms of natural logarithms. Then use a calculator to obtain a decimal approximation, correct to two decimal places, for the solution.

62. $8^x = 12{,}143$

63. $9e^{5x} = 1269$

64. $30e^{0.045x} = 90$

In Exercises 65–71, solve each logarithmic equation.

65. $\log_5 x = -3$

66. $\log x = 2$

67. $\log_4(3x - 5) = 3$

68. $\log_2(x + 3) + \log_2(x - 3) = 4$

69. $\log_3(x - 1) - \log_3(x + 2) = 2$

70. $\ln x = -1$

71. $3 + 4\ln 2x = 15$

72. The function $f(t) = 10.1e^{0.005t}$ models the population, $f(t)$, of Los Angeles, California, in millions, t years after 1992. If the growth rate continues into the future, when will the population reach 13 million?

73. The amount of carbon dioxide in the atmosphere, measured in parts per million, has been increasing as a result of the burning of oil and coal. The buildup of gases and particles traps heat and raises the planet's temperature, a phenomenon called the greenhouse effect. Carbon dioxide accounts for about half of the warming. The function $f(t) = 364(1.005)^t$ projects carbon dioxide concentration, $f(t)$, in parts per million, t years after 2000. Using the projections given by the function, when will the carbon dioxide concentration be double the preindustrial level of 280 parts per million?

74. The function $\overline{C}(x) = 15{,}557 + 5259\ln x$ models the average cost of a new car, $\overline{C}(x)$, x years after 1989. When will the average cost of a new car be \$30,000?

75. Use the compound interest formula

$$A = P\left(1 + \frac{r}{n}\right)^{nt}$$

to solve this problem. How long, to the nearest tenth of a year, will it take \$12,500 to grow to \$20,000 at 6.5% annual interest compounded quarterly?

Use the compound interest formula

$$A = Pe^{rt}$$

to solve Exercises 76–77.

76. How long, to the nearest tenth of a year, will it take \$50,000 to triple in value at 7.5% annual interest compounded continuously?

77. What interest rate is required for an investment subject to continuous compounding to triple in 5 years?

9.6

78. According to the U.S. Bureau of the Census, in 1980 there were 14.6 million residents of Hispanic origin living in the United States. By 1997, the number had increased to 29.3 million. The exponential growth function $A = 14.6e^{kt}$ describes the U.S. Hispanic population, in millions, t years after 1980.

a. Find k, correct to three decimal places.

b. Use the resulting model to project the Hispanic resident population in 2005.

c. In which year will the Hispanic resident population reach 50 million?

79. Use the exponential decay model for carbon-14, $A = A_0e^{-0.000121t}$, to solve this exercise. Prehistoric cave paintings were discovered in the Lascaux cave in France. The paint contained 15% of the original carbon-14. Estimate the age of the paintings at the time of the discovery.

80. The figure shows world population projections through the year 2150. The data are from the United Nations Family Planning Program and are based on optimistic or pessimistic expectations for successful control of human population growth. Suppose that you are interested in modeling these data using exponential, logarithmic, linear, and quadratic functions. Which function would you use to model each of the projections? Explain your choices. For the choice corresponding to a quadratic model, would your formula involve one with a positive or negative leading coefficient? Explain.

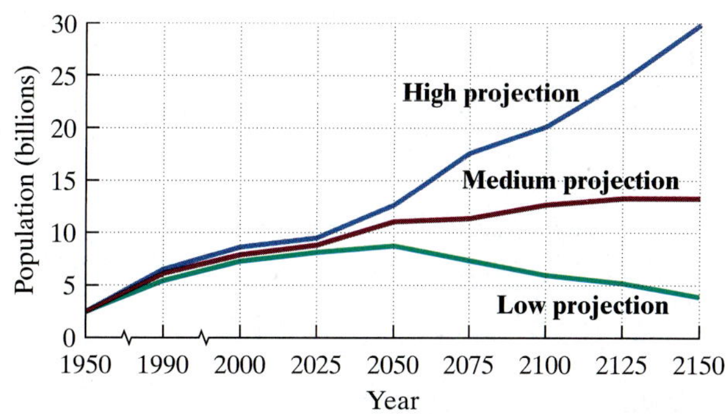

Chapter 9 Test

1. Graph $f(x) = 2^x$ and $g(x) = 2^{x+1}$ in the same rectangular coordinate system.

2. Use $A = P\left(1 + \frac{r}{n}\right)^{nt}$ and $A = Pe^{rt}$ to solve this problem. Suppose you have \$3000 to invest. Which investment yields the greater return over 10 years: 6.5% compounded semiannually or 6% compounded continuously? How much more (to the nearest dollar) is yielded by the better investment?

3. If $f(x) = x^2 + x$ and $g(x) = 3x - 1$, find $(f \circ g)(x)$ and $(g \circ f)(x)$.

4. If $f(x) = 5x - 7$, find $f^{-1}(x)$.

5. A function f models the amount given to charity as a function of income. The graph of f is shown in the figure.

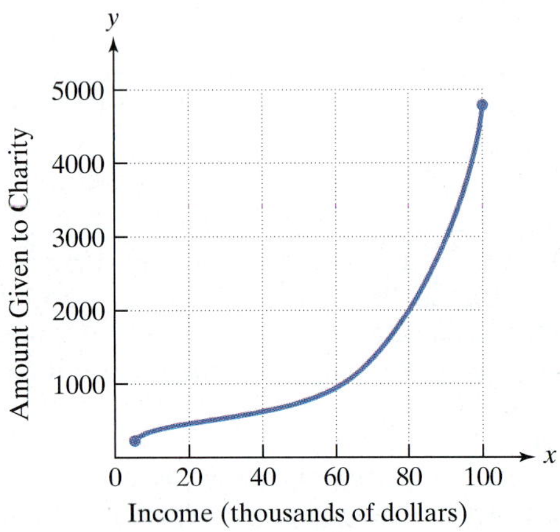

a. Explain why f has an inverse that is a function.

b. Find $f(80)$.

c. Describe in practical terms the meaning of $f^{-1}(2000)$.

6. Write in exponential form: $\log_5 125 = 3$.
7. Write in logarithmic form: $\sqrt{36} = 6$.
8. Graph $f(x) = 3^x$ and $g(x) = \log_3 x$ in the same rectangular coordinate system.

In Exercises 9–11, simplify each expression.

9. $\ln e^{5x}$

10. $\log_b b$

11. $\log_6 1$

12. Find the domain: $f(x) = \log_5(x - 7)$.

13. On the decibel scale, the loudness of a sound, in decibels, is given by $D = 10 \log \frac{I}{I_0}$, where I is the intensity of the sound, in watts per meter^2, and I_0 is the intensity of a sound barely audible to the human ear. If the intensity of a sound is $10^{12}I_0$, what is its loudness in decibels? (Such a sound is potentially damaging to the ear.)

In Exercises 14–15, use properties of logarithms to expand each logarithmic expression as much as possible. Where possible, evaluate logarithmic expressions without using a calculator.

14. $\log_4(64x^5)$

15. $\log_3 \frac{\sqrt[3]{x}}{81}$

In Exercises 16–17, write each expression as a single logarithm.

16. $6 \log x + 2 \log y$

17. $\ln 7 - 3 \ln x$

18. Use a calculator to evaluate $\log_{15} 71$ to four decimal places.

In Exercises 19–25, solve each equation.

19. $3^{x-2} = 81$

20. $5^x = 1.4$

21. $400e^{0.005x} = 1600$

22. $\log_{25} x = \frac{1}{2}$

23. $\log_6(4x - 1) = 3$

24. $\log x + \log(x + 15) = 2$

25. $2 \ln 3x = 8$

26. The function

$$P(t) = 89.18e^{-0.004t}$$

models the percentage, $P(t)$, of married men in the United States who were employed t years after 1959.

a. What percentage of married men were employed in 1959?

b. Is the percentage of married men who are employed increasing or decreasing? Explain.

c. In which year were 77% of U.S. married men employed?

Use the formulas

$$A = P\left(1 + \frac{r}{n}\right)^{nt} \quad \text{and} \quad A = Pe^{rt}$$

to solve Exercises 27–28.

27. How long, to the nearest tenth of a year, will it take \$4000 to grow to \$8000 at 5% annual interest compounded quarterly?

28. What interest rate is required for an investment subject to continuous compounding to double in 10 years?

29. The 1980 population of Europe was 484 million; in 1990, it was 509 million. Write the exponential growth function that describes the population of Europe, in millions, t years after 1980.

30. Use the exponential decay model for carbon-14, $A = A_0 e^{-0.000121t}$, to solve this exercise. Bones of a prehistoric man were discovered and contained 5% of the original amount of carbon-14. How long ago did the man die?

Cumulative Review Exercises (Chapters 1–9)

In Exercises 1–8, solve each equation, inequality, or system.

1. $\sqrt{2x + 5} - \sqrt{x + 3} = 2$

2. $(x - 5)^2 = -49$

3. $x^2 + x > 6$

4. $6x - 3(5x + 2) = 4(1 - x)$

5. $\dfrac{2}{x - 3} - \dfrac{3}{x + 3} = \dfrac{12}{x^2 - 9}$

6. $3x + 2 < 4$ and $4 - x > 1$

7. $3x - 2y + z = 7$
$2x + 3y - z = 13$
$x - y + 2z = -6$

8. $\log_9 x + \log_9 (x - 8) = 1$

In Exercises 9–11, graph each function, equation, or inequality in a rectangular coordinate system.

9. $f(x) = (x + 2)^2 - 4$

10. $y < -3x + 5$

11. $f(x) = 3^{x-2}$

In Exercises 12–15, perform the indicated operations, and simplify, if possible.

12. $\dfrac{2x + 1}{x - 5} - \dfrac{4}{x^2 - 3x - 10}$

13. $\dfrac{\dfrac{1}{x - 1} + 1}{\dfrac{1}{x + 1} - 1}$

14. $\dfrac{6}{\sqrt{5} - \sqrt{2}}$

15. $8\sqrt{45} + 2\sqrt{5} - 7\sqrt{20}$

16. Rationalize the denominator: $\dfrac{5}{\sqrt[3]{2x^2 y}}$.

17. Factor completely: $5ax + 5ay - 4bx - 4by$.

18. Write as a single logarithm: $5 \log x - \dfrac{1}{2} \log y$.

19. Solve for p: $\dfrac{1}{p} + \dfrac{1}{q} = \dfrac{1}{f}$.

In Exercises 20–22, find the domain of each function.

20. $f(x) = \dfrac{2}{x^2 + 2x - 15}$

21. $f(x) = \sqrt{2x - 6}$

22. $f(x) = \ln(1 - x)$

23. The length of a rectangular garden is 2 feet more than twice its width. If 22 feet of fencing is needed to enclose the garden, what are its dimensions?

24. With a 6% raise, you will earn \$19,610 annually. What is your salary before this raise?

25. The function $F(t) = 1 - k \ln(t + 1)$ models the fraction of people, $F(t)$, who remember all the words in a list of nonsense words t hours after memorizing the list. After 3 hours, only half the people could remember all the words. Determine the value of k and then predict the fraction of people in the group who will remember all the words after 6 hours. Round to three decimal places and then express the fraction with a denominator of 1000.

One minute you're in class, enjoying the lecture. Then a sharp pain radiates down your side. The next minute you're being diagnosed with, of all things, a kidney stone. It took your cousin six weeks to recover from kidney stone surgery, but your doctor assures you there is nothing to worry about. A new procedure, based on a curve that looks like the cross section of a football, will dissolve the stone painlessly and let you return to class in a day or two. How can this be?

From ripples in water to the path on which humanity journeys through space, certain curves occur naturally throughout the universe. Over 2000 years ago the ancient Greeks studied these curves, called *conic sections*, without regard to their immediate usefulness simply because the study elicited ideas that were exciting, challenging, and interesting. The ancient Greeks could not have imagined the applications of these curves in the twenty-first century. Overwhelmed by the choices on satellite television? Blame it on a conic section! In this chapter, we use the rectangular coordinate system to study the conic sections and the mathematics behind their surprising applications.

Chapter 10

Conic Sections and Systems of Nonlinear Equations

▸ SECTION 10.1 Distance and Midpoint Formulas; Circles

Objectives

1. Find the distance between two points.
2. Find the midpoint of a line segment.
3. Write the standard form of a circle's equation.
4. Give the center and radius of a circle whose equation is in standard form.
5. Convert the general form of a circle's equation to standard form.

SSM PH Tutor Center CD-ROM Video

It's a good idea to know your way around a circle. Clocks, angles, maps, and compasses are based on circles. Circles occur everywhere in nature: in ripples on water, patterns on a butterfly's wings, and cross sections of trees. Some consider the circle to be the most pleasing of all shapes.

The rectangular coordinate system gives us a unique way of knowing a circle. It enables us to translate a circle's geometric definition into an algebraic equation. To do this, we must first develop a formula for the distance between any two points in rectangular coordinates.

1 Find the distance between two points.

The Distance Formula

Using the Pythagorean Theorem, we can find the distance between the two points $P_1(x_1, y_1)$ and $P_2(x_2, y_2)$ in the rectangular coordinate system. The two points are illustrated in Figure 10.1.

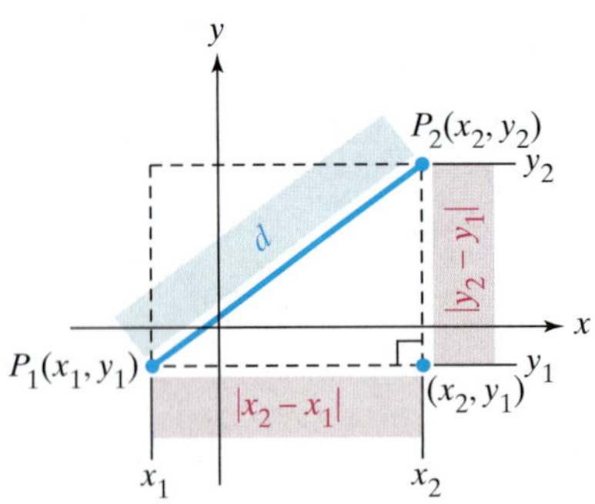

Figure 10.1

The distance that we need to find is represented by d and shown in blue. Notice that the distance between two points on the dashed horizontal line is the absolute value of the difference between the x-coordinates of the two points. This distance, $|x_2 - x_1|$, is shown in pink. Similarly, the distance between two points on the dashed vertical line is the absolute value of the difference between the y-coordinates of the two points. This distance, $|y_2 - y_1|$, is also shown in pink.

Because the dashed lines are horizontal and vertical, a right triangle is formed. Thus, we can use the Pythagorean Theorem to find distance d. By the Pythagorean Theorem,

$$d^2 = |x_2 - x_1|^2 + |y_2 - y_1|^2$$
$$d = \sqrt{|x_2 - x_1|^2 + |y_2 - y_1|^2}$$
$$d = \sqrt{(x_2 - x_1)^2 + (y_2 - y_1)^2}.$$

This result is called the **distance formula**.

The Distance Formula

The distance, d, between the points (x_1, y_1) and (x_2, y_2) in the rectangular coordinate system is

$$d = \sqrt{(x_2 - x_1)^2 + (y_2 - y_1)^2}.$$

When using the distance formula, it does not matter which point you call (x_1, y_1) and which you call (x_2, y_2).

EXAMPLE 1 Using the Distance Formula

Find the distance between $(-1, -3)$ and $(2, 3)$.

Solution Letting $(x_1, y_1) = (-1, -3)$ and $(x_2, y_2) = (2, 3)$, we obtain

$$\begin{aligned}
d &= \sqrt{(x_2 - x_1)^2 + (y_2 - y_1)^2} && \text{Use the distance formula.}\\
&= \sqrt{[2 - (-1)]^2 + [3 - (-3)]^2} && \text{Substitute the given values.}\\
&= \sqrt{(2 + 1)^2 + (3 + 3)^2} && \text{Perform subtractions within the grouping symbols.}\\
&= \sqrt{3^2 + 6^2} &&\\
&= \sqrt{9 + 36} && \text{Square 3 and 6.}\\
&= \sqrt{45} && \text{Add.}\\
&= 3\sqrt{5} \approx 6.71. && \sqrt{45} = \sqrt{9 \cdot 5} = \sqrt{9}\sqrt{5} = 3\sqrt{5}
\end{aligned}$$

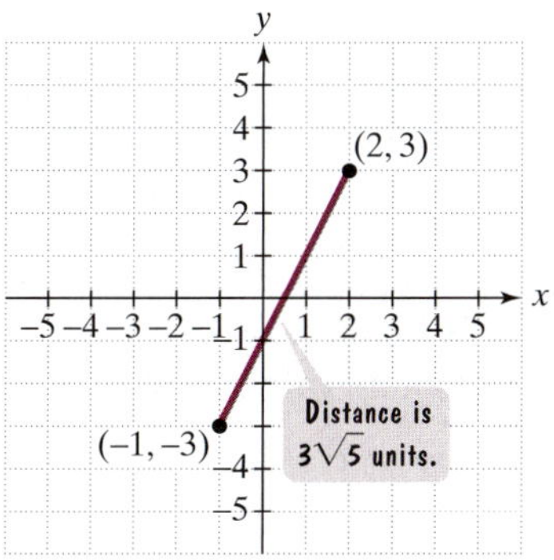

Figure 10.2 Finding the distance between two points

The distance between the given points is $3\sqrt{5}$ units, or approximately 6.71 units. The situation is illustrated in Figure 10.2. ■

CHECK POINT 1 Find the distance between $(2, -2)$ and $(5, 2)$.

The Midpoint Formula

The distance formula can be used to derive a formula for finding the midpoint of a line segment between two given points. The formula is given in the box on page 700.

2 Find the midpoint of a line segment.

The Midpoint Formula

Consider a line segment whose endpoints are (x_1, y_1) and (x_2, y_2). The coordinates of the segment's midpoint are

$$\left(\frac{x_1 + x_2}{2}, \frac{y_1 + y_2}{2}\right).$$

To find the midpoint, take the average of the two x-coordinates and of the two y-coordinates.

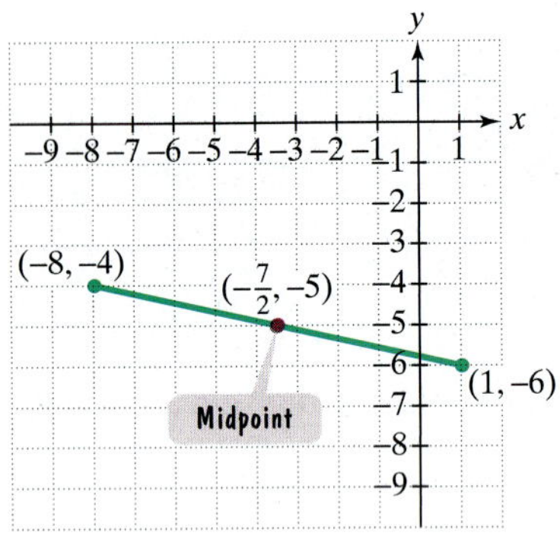

Figure 10.3 Finding a line segment's midpoint

EXAMPLE 2 Using the Midpoint Formula

Find the midpoint of the line segment with endpoints $(1, -6)$ and $(-8, -4)$.

Solution To find the coordinates of the midpoint, we average the coordinates of the endpoints.

$$\text{Midpoint} = \left(\frac{1 + (-8)}{2}, \frac{-6 + (-4)}{2}\right) = \left(\frac{-7}{2}, \frac{-10}{2}\right) = \left(-\frac{7}{2}, -5\right)$$

Average the x-coordinates. Average the y-coordinates.

Figure 10.3 illustrates that the point $(-\frac{7}{2}, -5)$ is midway between the points $(1, -6)$ and $(-8, -4)$. ■

✔ CHECK POINT 2 Find the midpoint of the line segment with endpoints $(1, 2)$ and $(7, -3)$.

Circles

Our goal is to translate a circle's geometric definition into an equation. We begin with this geometric definition.

Definition of a Circle

A **circle** is the set of all points in a plane that are equidistant from a fixed point, called the **center**. The fixed distance from the circle's center to any point on the circle is called the **radius**.

Figure 10.4 is our starting point for obtaining a circle's equation. We've placed the circle into a rectangular coordinate system. The circle's center is (h, k) and its radius is r. We let (x, y) represent the coordinates of any point on the circle.

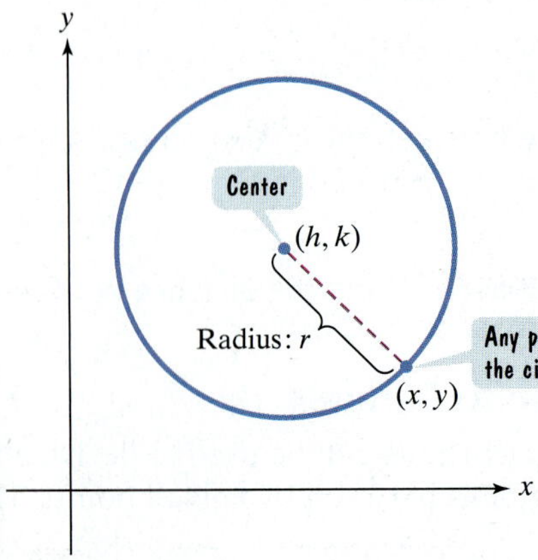

Figure 10.4 A circle centered at (h, k) with radius r

What does the geometric definition of a circle tell us about the point (x, y) in Figure 10.4? The point is on the circle if and only if its distance from the center is r. We can use the distance formula to express this idea algebraically:

The distance between (x, y) and (h, k) is always r.

$$\sqrt{(x-h)^2 + (y-k)^2} = r.$$

Squaring both sides of this equation yields the **standard form of the equation of a circle**.

The Standard Form of the Equation of a Circle

The **standard form of the equation of a circle** with center (h, k) and radius r is

$$(x-h)^2 + (y-k)^2 = r^2.$$

3 Write the standard form of a circle's equation.

EXAMPLE 3 Finding the Standard Form of a Circle's Equation

Write the standard form of the equation of the circle with center $(0, 0)$ and radius 2. Graph the circle.

Solution The center is $(0, 0)$. Because the center is represented as (h, k) in the standard form of the equation, $h = 0$ and $k = 0$. The radius is 2, so we will let $r = 2$ in the equation.

$$(x-h)^2 + (y-k)^2 = r^2$$ This is the standard form of a circle's equation.

$$(x-0)^2 + (y-0)^2 = 2^2$$ Substitute 0 for h, 0 for k, and 2 for r.

$$x^2 + y^2 = 4$$ Simplify.

The standard form of the equation of the circle is $x^2 + y^2 = 4$. Figure 10.5 shows the graph. ■

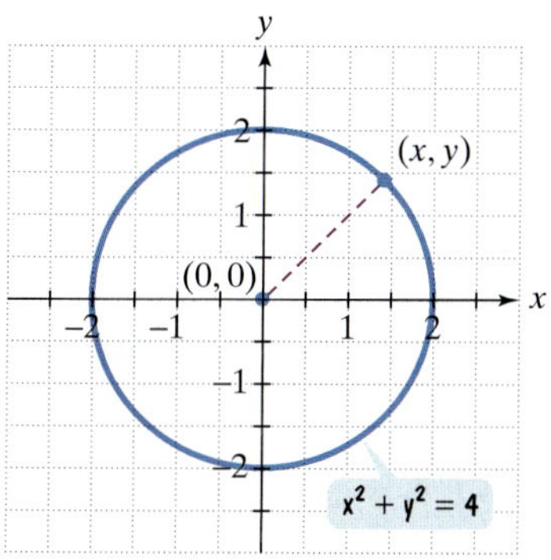

Figure 10.5 The graph of $x^2 + y^2 = 4$

✔ **CHECK POINT 3** Write the standard form of the equation of the circle with center $(0, 0)$ and radius 4.

Using Technology

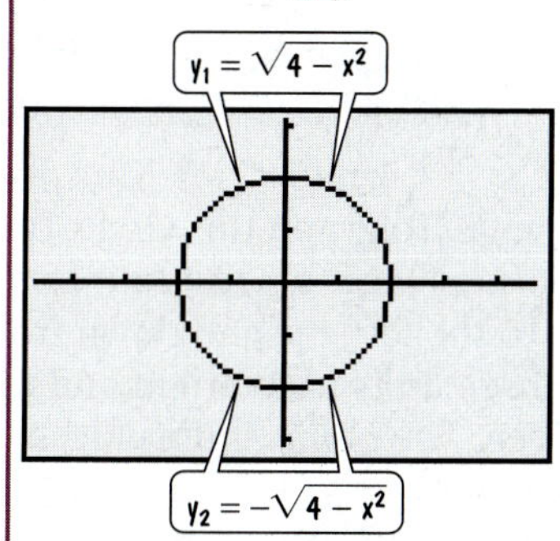

To graph a circle with a graphing utility, first solve the equation for y.

$$x^2 + y^2 = 4$$
$$y^2 = 4 - x^2$$
$$y = \pm\sqrt{4 - x^2}$$

Graph the two equations

$$y_1 = \sqrt{4 - x^2} \quad \text{and} \quad y_2 = -\sqrt{4 - x^2}$$

in the same viewing rectangle. The graph of $y_1 = \sqrt{4 - x^2}$ is the top semicircle because y is always positive. The graph of $y_2 = -\sqrt{4 - x^2}$ is the bottom semicircle because y is always negative. Use a ZOOM SQUARE setting so that the circle looks like a circle. (Many graphing utilities have problems connecting the two semicircles because the segments directly across from the center become nearly vertical.)

Example 3 and Check Point 3 involved circles centered at the origin. The standard form of the equation of all such circles is $x^2 + y^2 = r^2$, where r is the circle's radius. Now, let's consider a circle whose center is not at the origin.

EXAMPLE 4 Finding the Standard Form of a Circle's Equation

Write the standard form of the equation of the circle with center $(-2, 3)$ and radius 4.

Solution The center is $(-2, 3)$. Because the center is represented as (h, k) in the standard form of the equation, $h = -2$ and $k = 3$. The radius is 4, so we will let $r = 4$ in the equation.

$$(x - h)^2 + (y - k)^2 = r^2$$ This is the standard form of a circle's equation.

$$[x - (-2)]^2 + (y - 3)^2 = 4^2$$ Substitute −2 for h, 3 for k, and 4 for r.

$$(x + 2)^2 + (y - 3)^2 = 16$$ Simplify.

The standard form of the equation of the circle is $(x + 2)^2 + (y - 3)^2 = 16$ ■

✔ **CHECK POINT 4** Write the standard form of the equation of the circle with center $(5, -6)$ and radius 10.

4 Give the center and radius of a circle whose equation is in standard form.

EXAMPLE 5 Using the Standard Form of a Circle's Equation to Graph the Circle

Find the center and radius of the circle whose equation is

$$(x - 2)^2 + (y + 4)^2 = 9$$

and graph the equation.

Solution In order to graph the circle, we need to know its center, (h, k), and its radius, r. We can find the values for h, k, and r by comparing the given equation to the standard form of the equation of a circle.

$$(x - 2)^2 + (y + 4)^2 = 9$$

$$(x - 2)^2 + (y - (-4))^2 = 3^2$$

This is $(x - h)^2$, with $h = 2$. This is $(y - k)^2$, with $k = -4$. This is r^2, with $r = 3$.

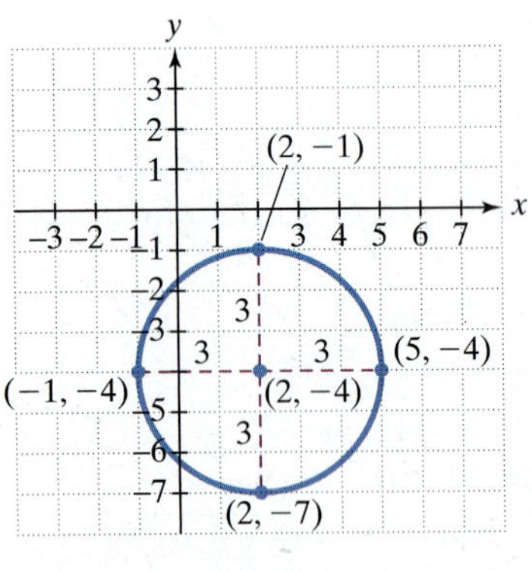

Figure 10.6 The graph of $(x - 2)^2 + (y + 4)^2 = 9$

We see that $h = 2$, $k = -4$, and $r = 3$. Thus, the circle has center $(h, k) = (2, -4)$ and a radius of 3 units. To graph this circle, first plot the center $(2, -4)$. Because the radius is 3, you can locate at least four points on the circle by going out three units to the right, to the left, up, and down from the center.

The points three units to the right and to the left of $(2, -4)$ are $(5, -4)$ and $(-1, -4)$, respectively. The points three units up and down from $(2, -4)$ are $(2, -1)$ and $(2, -7)$, respectively.

Using these points, we obtain the graph in Figure 10.6. ■

✔ **CHECK POINT 5** Find the center and radius of the circle whose equation is

$$(x + 3)^2 + (y - 1)^2 = 4$$

and graph the equation.

If we square $x - 2$ and $y + 4$ in the standard form of the equation of Example 5, we obtain another form for the circle's equation.

$$(x - 2)^2 + (y + 4)^2 = 9$$ This is the standard form of the equation from Example 5.

$$x^2 - 4x + 4 + y^2 + 8y + 16 = 9$$ Square $x - 2$ and $y + 4$.

$$x^2 + y^2 - 4x + 8y + 20 = 9$$ Combine numerical terms and rearrange terms.

$$x^2 + y^2 - 4x + 8y + 11 = 0$$ Subtract 9 from both sides.

This result suggests that an equation in the form $x^2 + y^2 + Dx + Ey + F = 0$ can represent a circle. This is called the **general form of the equation of a circle**.

The General Form of the Equation of a Circle

The **general form of the equation of a circle** is

$$x^2 + y^2 + Dx + Ey + F = 0.$$

5 Convert the general form of a circle's equation to standard form.

We can convert the general form of the equation of a circle to the standard form $(x - h)^2 + (y - k)^2 = r^2$. We do so by completing the square on x and y. Let's see how this is done.

EXAMPLE 6 Converting the General Form of a Circle's Equation to Standard Form and Graphing the Circle

Write in standard form and graph: $x^2 + y^2 + 4x - 6y - 23 = 0$.

Solution Because we plan to complete the square on both x and y, let's rearrange terms so that x-terms are arranged in descending order, y-terms are arranged in descending order, and the constant term appears on the right.

$$x^2 + y^2 + 4x - 6y - 23 = 0$$ This is the given equation.

$$(x^2 + 4x \quad\;) + (y^2 - 6y \quad\;) = 23$$ Rewrite in anticipation of completing the square.

$$(x^2 + 4x + 4) + (y^2 - 6y + 9) = 23 + 4 + 9$$ Complete the square on x: $\frac{1}{2} \cdot 4 = 2$ and $2^2 = 4$, so add 4 to both sides. Complete the square on y: $\frac{1}{2}(-6) = -3$ and $(-3)^2 = 9$, so add 9 to both sides.

Remember that numbers added on the left side must also be added on the right side.

$$(x + 2)^2 + (y - 3)^2 = 36$$ Factor on the left and add on the right.

This last equation is in standard form. We can identify the circle's center and radius by comparing this equation to the standard form of the equation of a circle, $(x - h)^2 + (y - k)^2 = r^2$.

$$(x + 2)^2 + (y - 3)^2 = 36$$

$$(x - (-2))^2 + (y - 3)^2 = 6^2$$

This is $(x - h)^2$, with $h = -2$. This is $(y - k)^2$, with $k = 3$. This is r^2, with $r = 6$.

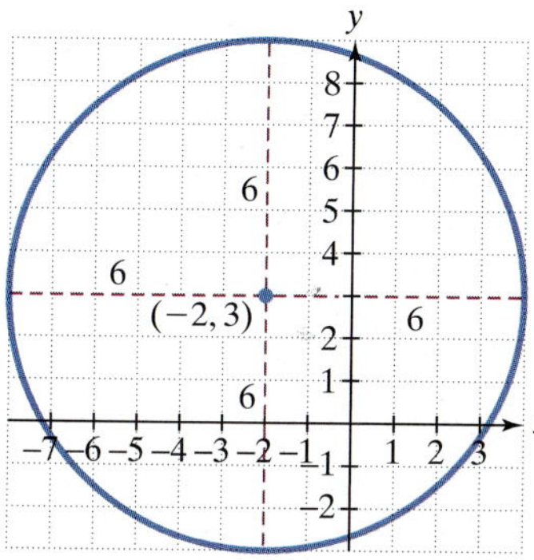

Figure 10.7 The graph of $(x + 2)^2 + (y - 3)^2 = 36$

We use the center, $(h, k) = (-2, 3)$, and the radius, $r = 6$, to graph the circle. The graph is shown in Figure 10.7. ■

Using Technology

To graph $x^2 + y^2 + 4x - 6y - 23 = 0$, rewrite the equation as a quadratic equation in y.

$$y^2 - 6y + (x^2 + 4x - 23) = 0$$

Now solve for y using the quadratic formula, with $a = 1, b = -6$, and $c = x^2 + 4x - 23$.

$$y = \frac{-b \pm \sqrt{b^2 - 4ac}}{2a} = \frac{-(-6) \pm \sqrt{(-6)^2 - 4 \cdot 1(x^2 + 4x - 23)}}{2 \cdot 1} = \frac{6 \pm \sqrt{36 - 4(x^2 + 4x - 23)}}{2}$$

Because we will enter these equations, there is no need to simplify further. Enter

$$y_1 = \frac{6 + \sqrt{36 - 4(x^2 + 4x - 23)}}{2}$$

and

$$y_2 = \frac{6 - \sqrt{36 - 4(x^2 + 4x - 23)}}{2}.$$

Use a ZOOM SQUARE setting. The graph is shown on the right.

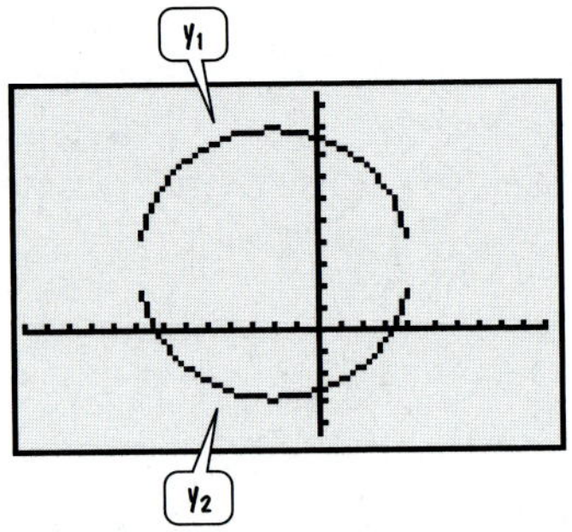

CHECK POINT 6 Write in standard form and graph:

$$x^2 + y^2 + 4x - 4y - 1 = 0.$$

EXERCISE SET 10.1

Practice Exercises

In Exercises 1–18, find the distance between each pair of points. If necessary, round answers to two decimal places.

1. $(2, 3)$ and $(14, 8)$
2. $(5, 1)$ and $(8, 5)$
3. $(4, 1)$ and $(6, 3)$
4. $(2, 3)$ and $(3, 5)$
5. $(0, 0)$ and $(-3, 4)$
6. $(0, 0)$ and $(3, -4)$
7. $(-2, -6)$ and $(3, -4)$
8. $(-4, -1)$ and $(2, -3)$
9. $(0, -3)$ and $(4, 1)$
10. $(0, -2)$ and $(4, 3)$
11. $(3.5, 8.2)$ and $(-0.5, 6.2)$
12. $(2.6, 1.3)$ and $(1.6, -5.7)$
13. $(0, -\sqrt{3})$ and $(\sqrt{5}, 0)$
14. $(0, -\sqrt{2})$ and $(\sqrt{7}, 0)$
15. $(3\sqrt{3}, \sqrt{5})$ and $(-\sqrt{3}, 4\sqrt{5})$
16. $(2\sqrt{3}, \sqrt{6})$ and $(-\sqrt{3}, 5\sqrt{6})$
17. $\left(\frac{7}{3}, \frac{1}{5}\right)$ and $\left(\frac{1}{3}, \frac{6}{5}\right)$
18. $\left(-\frac{1}{4}, -\frac{1}{7}\right)$ and $\left(\frac{3}{4}, \frac{6}{7}\right)$

In Exercises 19–30, find the midpoint of each line segment with the given endpoints.

19. $(6, 8)$ and $(2, 4)$
20. $(10, 4)$ and $(2, 6)$
21. $(-2, -8)$ and $(-6, -2)$
22. $(-4, -7)$ and $(-1, -3)$
23. $(-3, -4)$ and $(6, -8)$
24. $(-2, -1)$ and $(-8, 6)$
25. $\left(-\frac{7}{2}, \frac{3}{2}\right)$ and $\left(-\frac{5}{2}, -\frac{11}{2}\right)$
26. $\left(-\frac{2}{5}, \frac{7}{15}\right)$ and $\left(-\frac{2}{5}, -\frac{4}{15}\right)$
27. $(8, 3\sqrt{5})$ and $(-6, 7\sqrt{5})$
28. $(7\sqrt{3}, -6)$ and $(3\sqrt{3}, -2)$
29. $(\sqrt{18}, -4)$ and $(\sqrt{2}, 4)$
30. $(\sqrt{50}, -6)$ and $(\sqrt{2}, 6)$

In Exercises 31–40, write the standard form of the equation of the circle with the given center and radius.

31. Center $(0, 0), r = 7$
32. Center $(0, 0), r = 8$
33. Center $(3, 2), r = 5$
34. Center $(2, -1), r = 4$
35. Center $(-1, 4), r = 2$
36. Center $(-3, 5), r = 3$
37. Center $(-3, -1), r = \sqrt{3}$
38. Center $(-5, -3), r = \sqrt{5}$
39. Center $(-4, 0), r = 10$
40. Center $(-2, 0), r = 6$

In Exercises 41–48, give the center and radius of the circle described by the equation and graph each equation.

41. $x^2 + y^2 = 16$

42. $x^2 + y^2 = 49$

43. $(x - 3)^2 + (y - 1)^2 = 36$

44. $(x - 2)^2 + (y - 3)^2 = 16$

45. $(x + 3)^2 + (y - 2)^2 = 4$

46. $(x + 1)^2 + (y - 4)^2 = 25$

47. $(x + 2)^2 + (y + 2)^2 = 4$

48. $(x + 4)^2 + (y + 5)^2 = 36$

In Exercises 49–56, complete the square and write the equation in standard form. Then give the center and radius of each circle and graph the equation.

49. $x^2 + y^2 + 6x + 2y + 6 = 0$

50. $x^2 + y^2 + 8x + 4y + 16 = 0$

51. $x^2 + y^2 - 10x - 6y - 30 = 0$

52. $x^2 + y^2 - 4x - 12y - 9 = 0$

53. $x^2 + y^2 + 8x - 2y - 8 = 0$

54. $x^2 + y^2 + 12x - 6y - 4 = 0$

55. $x^2 - 2x + y^2 - 15 = 0$

56. $x^2 + y^2 - 6y - 7 = 0$

Application Exercises

57. A rectangular coordinate system with coordinates in miles is placed on the map in the figure shown. Bangkok has coordinates $(-115, 170)$ and Phnom Penh has coordinates $(65, 70)$. How long will it take a plane averaging 400 miles per hour to fly directly from one city to the other? Round to the nearest tenth of an hour. Approximately how many minutes is the flight?

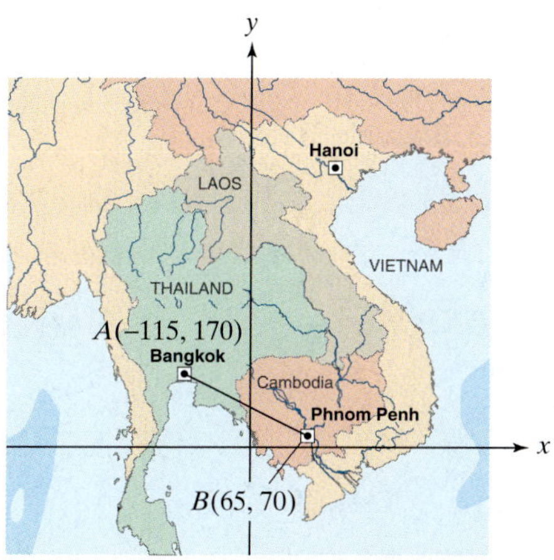

58. We refer to the driveway in the figure shown as being *circular*, meaning that it is bounded by two circles. The figure indicates that the radius of the larger circle is 52 feet and the radius of the smaller circle is 38 feet.

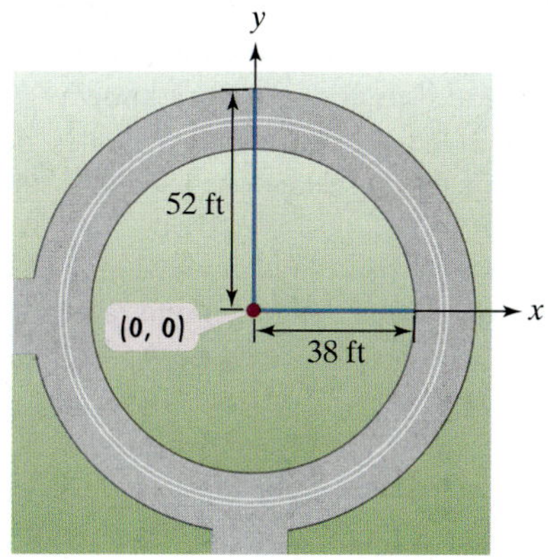

a. Use the coordinate system shown to write the equation of the smaller circle.

b. Use the coordinate system shown to write the equation of the larger circle.

59. The ferris wheel in the figure has a radius of 68 feet. The clearance between the wheel and the ground is 14 feet. The rectangular coordinate system shown has its origin on the ground directly below the center of the wheel. Use the coordinate system to write the equation of the circular wheel.

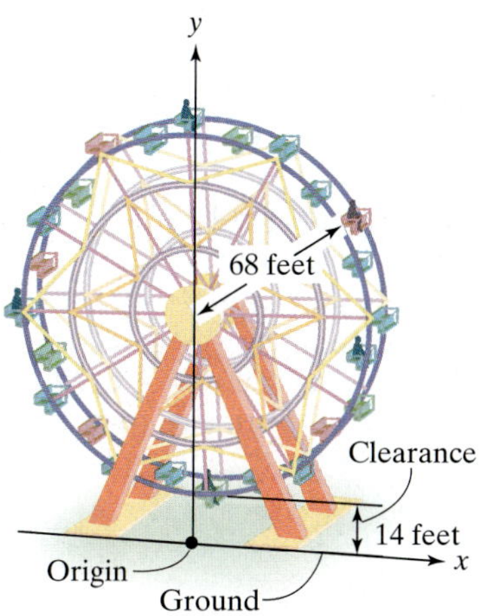

60. The circle formed by the middle lane of a circular running track can be described algebraically by $x^2 + y^2 = 4$, where all measurements are in miles. If you run around the track's middle lane twice, approximately how many miles have you covered?

Writing in Mathematics

61. In your own words, describe how to find the distance between two points in the rectangular coordinate system.

62. In your own words, describe how to find the midpoint of a line segment if its endpoints are known.

63. What is a circle? Without using variables, describe how the definition of a circle can be used to obtain a form of its equation.

64. Give an example of a circle's equation in standard form. Describe how to find the center and radius for this circle.

65. How is the standard form of a circle's equation obtained from its general form?

66. Does $(x - 3)^2 + (y - 5)^2 = 0$ represent the equation of a circle? If not, describe the graph of this equation.

67. Does $(x - 3)^2 + (y - 5)^2 = -25$ represent the equation of a circle? What sort of set is the graph of this equation?

Technology Exercises

In Exercises 68–70, use a graphing utility to graph each circle whose equation is given.

68. $x^2 + y^2 = 25$

69. $(y + 1)^2 = 36 - (x - 3)^2$

70. $x^2 + 10x + y^2 - 4y - 20 = 0$

Critical Thinking Exercises

71. Which one of the following is true?

a. The equation of the circle whose center is at the origin with radius 16 is $x^2 + y^2 = 16$.

b. The graph of $(x - 3)^2 + (y + 5)^2 = 36$ is a circle with a radius 6 centered at $(-3, 5)$.

c. The graph of $(x - 4) + (y + 6) = 25$ is a circle with a radius 5 centered at $(4, -6)$.

d. None of the above is true.

72. Show that the points $A(1, 1 + d)$, $B(3, 3 + d)$, and $C(6, 6 + d)$ are collinear (lie along a straight line) by showing that the distance from A to B plus the distance from B to C equals the distance from A to C.

73. Prove the midpoint formula by using the following procedure.

a. Show that the distance between (x_1, y_1) and $\left(\frac{x_1 + x_2}{2}, \frac{y_1 + y_2}{2}\right)$ is equal to the distance between (x_2, y_2) and $\left(\frac{x_1 + x_2}{2}, \frac{y_1 + y_2}{2}\right)$.

b. Use the procedure from Exercise 72 and the distances from part (a) to show that the points (x_1, y_1), $\left(\frac{x_1 + x_2}{2}, \frac{y_1 + y_2}{2}\right)$, and (x_2, y_2) are collinear.

In Exercises 74–75, write the standard form and the general form of the equation of each circle.

74. Center at $(3, -5)$ and passing through the point $(-2, 1)$

75. Passing through $(-7, 2)$ and $(1, 2)$; these points lie on the circle and also on the line that passes through the circle's center.

76. Find the area of the donut-shaped region bounded by the graphs of $(x - 2)^2 + (y + 3)^2 = 25$ and $(x - 2)^2 + (y + 3)^2 = 36$.

77. A **tangent line** to a circle is a line that intersects the circle at exactly one point. The tangent line is perpendicular to the radius of the circle at this point of contact. Write the point-slope equation of a line tangent to the circle whose equation is $x^2 + y^2 = 25$ at the point $(3, -4)$.

Review Exercises

78. If $f(x) = x^2 - 2$ and $g(x) = 3x + 4$, find $f(g(x))$ and $g(f(x))$. (Section 9.2. Example 1)

79. Solve: $2x = \sqrt{7x - 3} + 3$. (Section 7.6. Example 3)

80. Solve: $|2x - 5| < 10$. (Section 4.3. Example 3)

▸ SECTION 10.2 The Ellipse

Objectives

1. Graph ellipses centered at the origin.
2. Graph ellipses not centered at the origin.
3. Solve applied problems involving ellipses.

Tunnel cut in rock wall. © Kevin Fleming/CORBIS.

You took on a summer job driving a truck, delivering books that were ordered online. You're an avid reader, so just being around books sounded appealing. However, now you're feeling a bit shaky driving the truck for the first time. It's 10 feet wide and 9 feet high; compared to your compact car, it feels like you're behind the wheel of a tank. Up ahead you see a sign at the semielliptical entrance to a tunnel: Caution! Tunnel is 10 Feet High at Center Peak. Then you see another sign: Caution! Tunnel is 40 Feet Wide. Will your truck clear the opening of the tunnel's archway?

The mathematics of your world is present in the movements of planets, bridge and tunnel construction, navigational systems used to keep track of a ship's location, manufacture of lenses for telescopes, and even a procedure for disintegrating kidney stones. The mathematics behind these applications involves conic sections. **Conic sections** are curves that result from the intersection of a right circular cone and a plane. Figure 10.8 illustrates the four conic sections: the circle, the ellipse, the parabola, and the hyperbola.

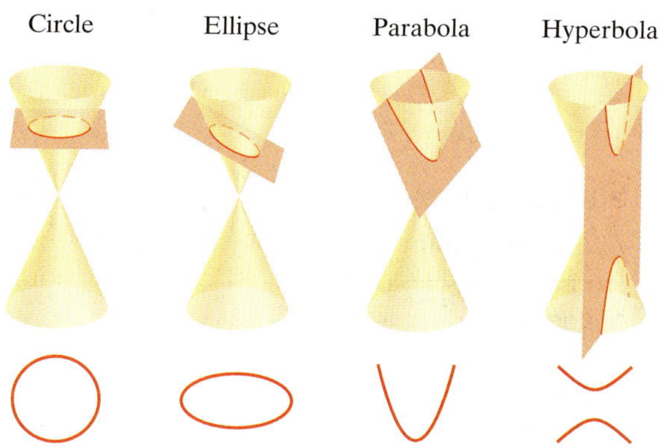

Figure 10.8 Obtaining the conic sections by intersecting a plane and a cone

In this section, we study the symmetric oval-shaped curve known as the ellipse. We will use a geometric definition for an ellipse to derive its equation. With this equation, we will determine if your delivery truck will clear the tunnel's entrance.

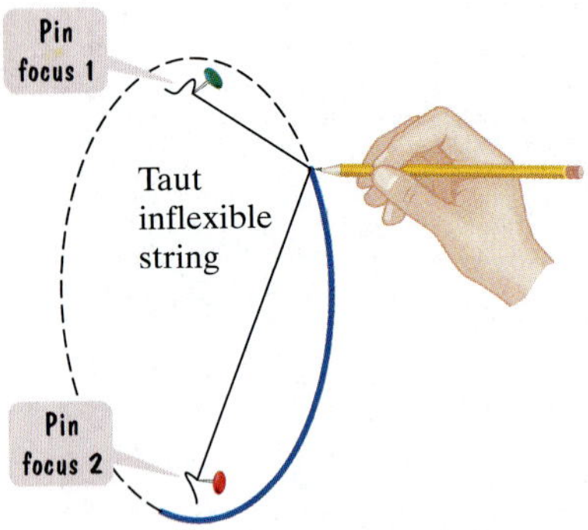

Figure 10.9 Drawing an ellipse

Definition of an Ellipse

Figure 10.9 illustrates how to draw an ellipse. Place pins at two fixed points, each of which is called a focus (plural: foci). If the ends of a fixed length of string are fastened to the pins and we draw the string taut with a pencil, the path traced by the pencil will be an ellipse. Notice that the sum of the distances of the pencil point from the foci remains constant because the length of the string is fixed. This procedure for drawing an ellipse illustrates its geometric definition.

Definition of an Ellipse

An **ellipse** is the set of all points, P, in a plane the sum of whose distances from two fixed points, F_1 and F_2, is constant (see Figure 10.10). These two fixed points are called the **foci** (plural of **focus**). The midpoint of the segment connecting the foci is the **center** of the ellipse.

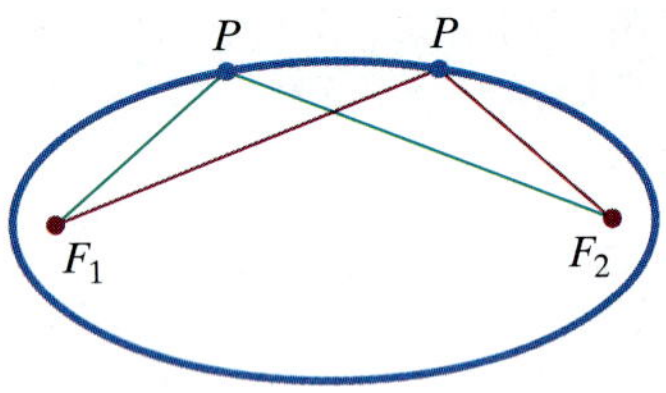

Figure 10.10

Figure 10.11 illustrates that an ellipse can be elongated horizontally or vertically. The line through the foci intersects the ellipse at two points, called the **vertices** (singular: **vertex**). The line segment that joins the vertices is the **major axis**. Notice that the midpoint of the major axis is the center of the ellipse. The line segment whose endpoints are on the ellipse and that is perpendicular to the major axis at the center is called the **minor axis** of the ellipse.

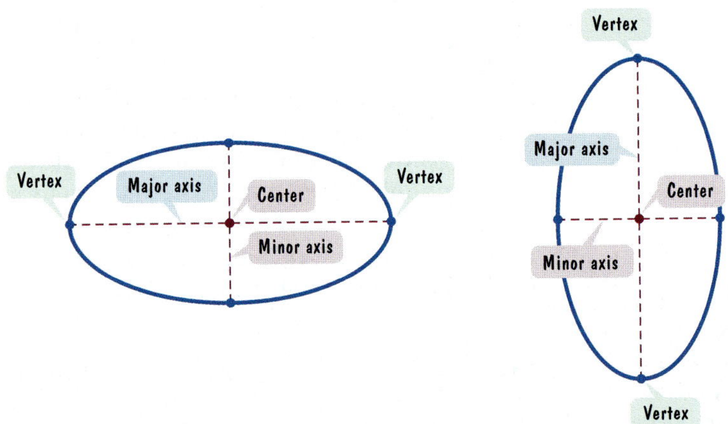

Figure 10.11 Horizontal and vertical elongations of an ellipse

Standard Form of the Equation of an Ellipse

The rectangular coordinate system gives us a unique way of describing an ellipse. It enables us to translate an ellipse's geometric definition into an algebraic equation.

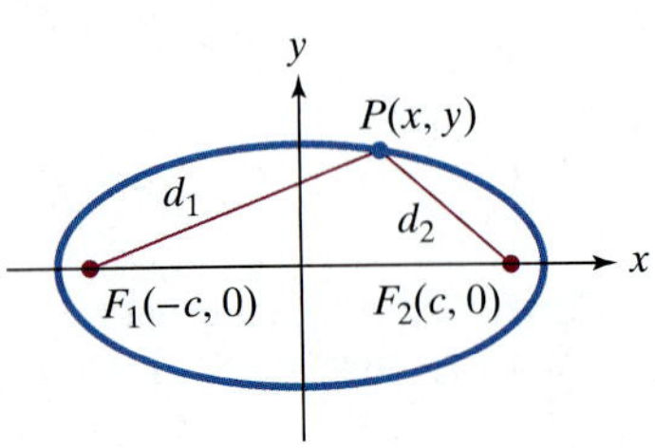

Figure 10.12

We start with Figure 10.12 to obtain an ellipse's equation. We've placed an ellipse that is elongated horizontally into a rectangular coordinate system. The foci are on the x-axis at $(-c, 0)$ and $(c, 0)$, as in Figure 10.12. In this way, the center of the ellipse is at the origin. We let (x, y) represent the coordinates of any point, P, on the ellipse.

What does the definition of an ellipse tell us about the point (x, y) in Figure 10.12? For any point (x, y) on the ellipse, the sum of the distances to the two foci, $d_1 + d_2$, must be constant. We denote this constant by $2a$. Thus, the point (x, y) is on the ellipse if and only if

$$d_1 + d_2 = 2a.$$

$$\sqrt{(x+c)^2 + y^2} + \sqrt{(x-c)^2 + y^2} = 2a \quad \text{Use the distance formula.}$$

Discover for Yourself

Perform the algebra mentioned on the right by eliminating radicals and obtaining the equation shown.

After eliminating radicals and simplifying, we obtain

$$(a^2 - c^2)x^2 + a^2y^2 = a^2(a^2 - c^2).$$

In order to simplify this equation, let $b^2 = a^2 - c^2$. Substituting b^2 for $a^2 - c^2$, we obtain

$$b^2x^2 + a^2y^2 = a^2b^2$$

$$\frac{b^2x^2}{a^2b^2} + \frac{a^2y^2}{a^2b^2} = \frac{a^2b^2}{a^2b^2} \quad \text{Divide both sides by } a^2b^2.$$

$$\frac{x^2}{a^2} + \frac{y^2}{b^2} = 1 \quad \text{Simplify.}$$

This last equation is the **standard form of the equation of an ellipse**. There are two such equations, one for a horizontal major axis and one for a vertical major axis.

Standard Forms of the Equations of an Ellipse

The **standard form of the equation of an ellipse** with center at the origin, and major and minor axes of lengths $2a$ and $2b$ (where a and b are positive, and $a^2 > b^2$) is

$$\frac{x^2}{a^2} + \frac{y^2}{b^2} = 1 \quad \text{or} \quad \frac{x^2}{b^2} + \frac{y^2}{a^2} = 1.$$

Figure 10.13 illustrates that the vertices are on the major axis, a units from the center. The foci are are on the major axis, c units from the center.

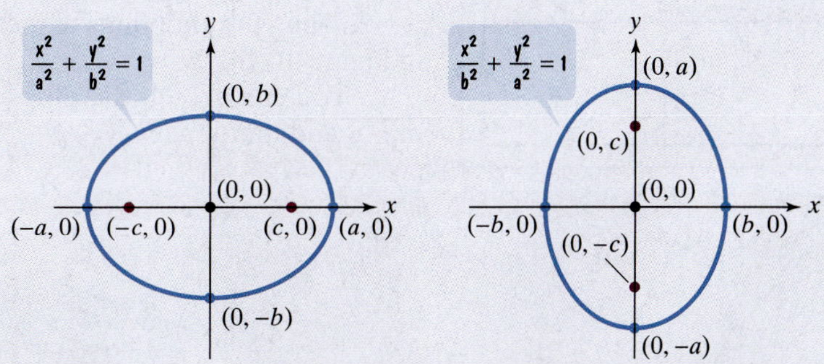

Figure 10.13 **(a)** Major axis is horizontal with length $2a$. **(b)** Major axis is vertical with length $2a$.

Using the Standard Form of the Equation of an Ellipse

We can use the standard form of an ellipse's equation to graph the ellipse.

1 Graph ellipses centered at the origin.

EXAMPLE 1 Graphing an Ellipse Centered at the Origin

Graph the ellipse: $\frac{x^2}{9} + \frac{y^2}{4} = 1.$

Solution The given equation is the standard form of an ellipse's equation with $a^2 = 9$ and $b^2 = 4$.

$$\frac{x^2}{9} + \frac{y^2}{4} = 1$$

$a^2 = 9$. This is the larger of the two denominators.

$b^2 = 4$. This is the smaller of the two denominators.

Because the denominator of the x^2-term is greater than the denominator of the y^2-term, the major axis is horizontal. Based on the standard form of the equation, we know the vertices are $(-a, 0)$ and $(a, 0)$. Because $a^2 = 9$, $a = 3$. Thus, the vertices are $(-3, 0)$ and $(3, 0)$, shown in Figure 10.14.

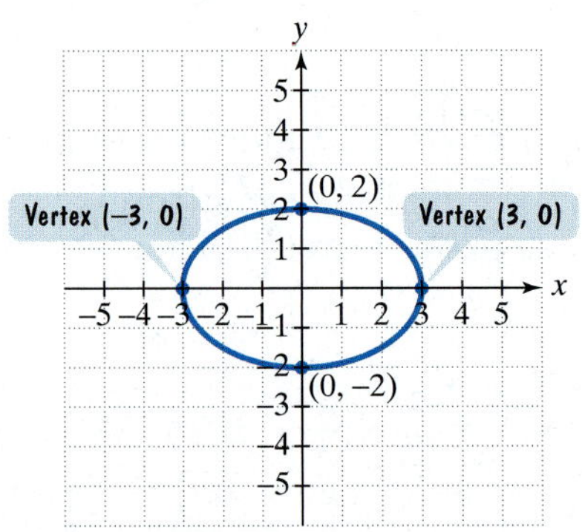

Figure 10.14 The graph of $\frac{x^2}{9} + \frac{y^2}{4} = 1$

Now let us find the endpoints of the vertical minor axis. According to the standard form of the equation, these endpoints are $(0, -b)$ and $(0, b)$. Because $b^2 = 4$, $b = 2$. Thus, the endpoints of the minor axis are $(0, -2)$ and $(0, 2)$. They are shown in Figure 10.14.

You can sketch the ellipse in Figure 10.14 by locating endpoints on the major and minor axes.

$$\frac{x^2}{3^2} + \frac{y^2}{2^2} = 1$$

Endpoints of the major axis are 3 units to the left and right of the center.

Endpoints of the minor axis are 2 units up and down from the center.

■

✔ CHECK POINT 1 Graph the ellipse: $\frac{x^2}{36} + \frac{y^2}{9} = 1.$

Using Technology

We graph $\frac{x^2}{9} + \frac{y^2}{4} = 1$ with a graphing utility by solving for y and defining two functions.

$$\frac{y^2}{4} = 1 - \frac{x^2}{9}$$

$$y^2 = 4\left(1 - \frac{x^2}{9}\right)$$

$$y = \pm 2\sqrt{1 - \frac{x^2}{9}}$$

Enter

$y_1 = 2$ [√] (1 [−] x [∧] 2 [÷] 9)

and

$y_2 = -y_1.$

To see the true shape of the ellipse, use the [ZOOM SQUARE] feature so that one unit on the x-axis is the same length as one unit on the y-axis.

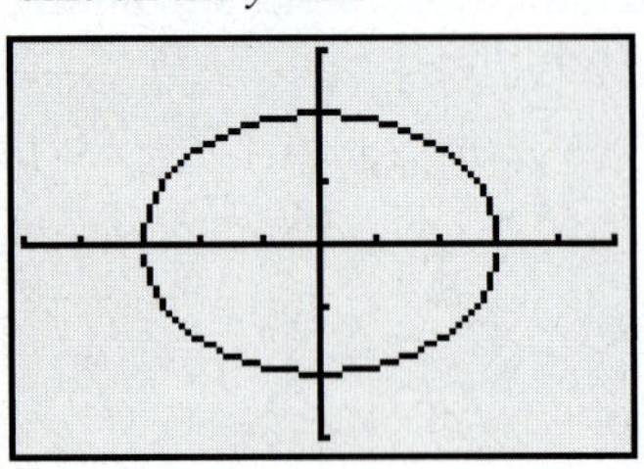

EXAMPLE 2 Graphing an Ellipse Centered at the Origin

Graph the ellipse: $25x^2 + 16y^2 = 400$.

Solution We begin by expressing the equation in standard form. Because we want 1 on the right side, we divide both sides by 400.

$$\frac{25x^2}{400} + \frac{16y^2}{400} = \frac{400}{400}$$

$$\frac{x^2}{16} + \frac{y^2}{25} = 1$$

$b^2 = 16$. This is the smaller of the two denominators.

$a^2 = 25$. This is the larger of the two denominators.

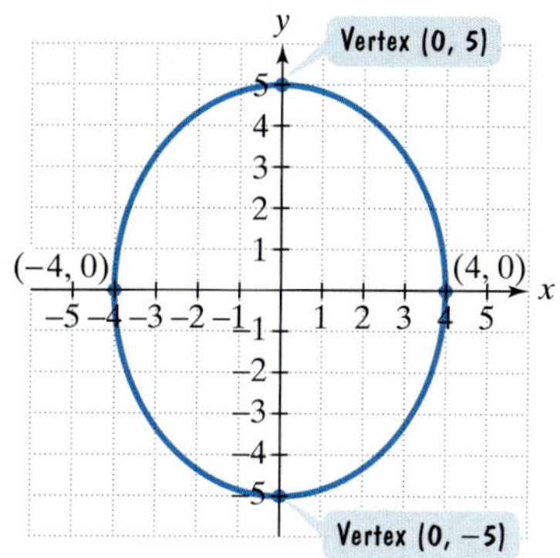

Figure 10.15 The graph of $\frac{x^2}{16} + \frac{y^2}{25} = 1$

The equation is the standard form of an ellipse's equation with $a^2 = 25$ and $b^2 = 16$. Because the denominator of the y^2-term is greater than the denominator of the x^2-term, the major axis is vertical. Based on the standard form of the equation, we know the vertices are $(0, -a)$ and $(0, a)$. Because $a^2 = 25$, $a = 5$. Thus, the vertices are $(0, -5)$ and $(0, 5)$, shown in Figure 10.15.

Now let us find the endpoints of the horizontal minor axis. According to the standard form of the equation, these endpoints are $(-b, 0)$ and $(b, 0)$. Because $b^2 = 16$, $b = 4$. Thus, the endpoints of the minor axis are $(-4, 0)$ and $(4, 0)$. They are shown in Figure 10.15.

You can sketch the ellipse in Figure 10.15 by locating endpoints on the major and minor axes:

$$\frac{x^2}{4^2} + \frac{y^2}{5^2} = 1.$$

Endpoints of the minor axis are 4 units to the left and right of the center.

Endpoints of the major axis are 5 units up and down from the center. ■

CHECK POINT 2 Graph the ellipse: $16x^2 + 9y^2 = 144$.

2 Graph ellipses not centered at the origin.

Ellipses Centered at (h, k)

Horizontal and vertical translations can be used to graph ellipses that are not centered at the origin. Figure 10.16 illustrates that the graphs of

$$\frac{(x-h)^2}{a^2} + \frac{(y-k)^2}{b^2} = 1 \quad \text{and} \quad \frac{x^2}{a^2} + \frac{y^2}{b^2} = 1$$

have the same size and shape. However, the graph of the first equation is centered at (h, k) rather than at the origin.

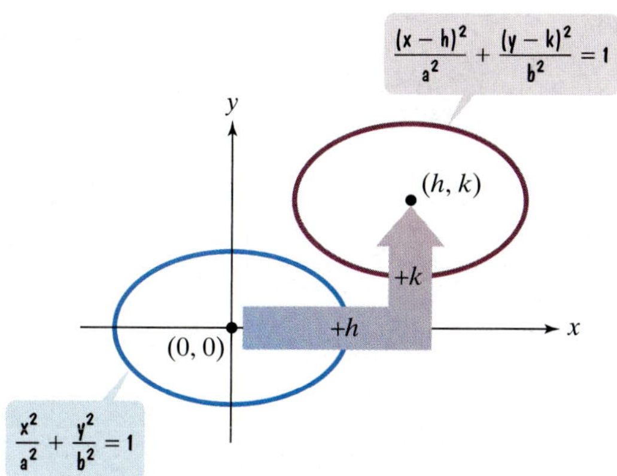

Figure 10.16 Translating an ellipse's graph

Table 10.1 gives the standard forms of equations of ellipses centered at (h, k). Figure 10.17 shows their graphs.

Table 10.1 Standard Forms of Equations of Ellipses Centered at (*h*, *k*)

Equation	Center	Major Axis	Vertices
$\frac{(x-h)^2}{a^2} + \frac{(y-k)^2}{b^2} = 1$, $a^2 > b^2$	(h, k)	Parallel to the x-axis, horizontal	$(h - a, k)$ $(h + a, k)$
$\frac{(x-h)^2}{b^2} + \frac{(y-k)^2}{a^2} = 1$, $a^2 > b^2$	(h, k)	Parallel to the y-axis, vertical	$(h, k - a)$ $(h, k + a)$

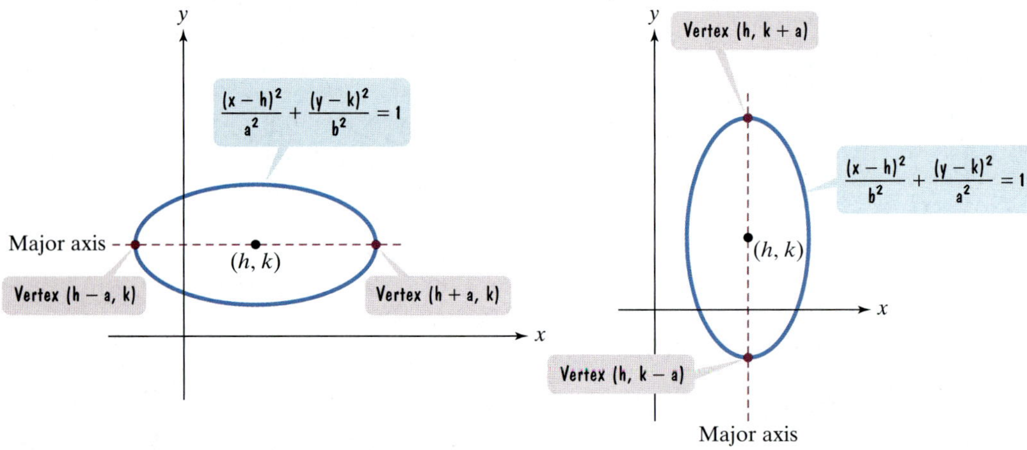

Figure 10.17 Graphs of ellipses centered at (h, k)

EXAMPLE 3 Graphing an Ellipse Centered at (*h*, *k*)

Graph the ellipse: $\frac{(x-1)^2}{4} + \frac{(y+2)^2}{9} = 1.$

Solution In order to graph the ellipse, we need to know its center, (h, k). In the standard forms of equations centered at (h, k), h is the number subtracted from x and k is the number subtracted from y.

$$\frac{(x-1)^2}{4} + \frac{(y-(-2))^2}{9} = 1$$

This is $(x - h)^2$ with $h = 1$

This is $(y - k)^2$ with $k = -2$.

We see that $h = 1$ and $k = -2$. Thus, the center of the ellipse, (h, k), is $(1, -2)$. We can graph the ellipse by locating endpoints on the major and minor axes. To do this, we must identify a^2 and b^2.

$$\frac{(x-1)^2}{4} + \frac{(y+2)^2}{9} = 1$$

$b^2 = 4$. This is the smaller of the two denominators.

$a^2 = 9$. This is the larger of the two denominators.

The larger number is under the expression involving y. This means that the major axis is vertical and parallel to the y-axis. Because $a^2 = 9$, $a = 3$ and the vertices lie three units above and below the center. Also, because $b^2 = 4$, $b = 2$ and the endpoints of the minor axis lie two units to the right and left of the center. We categorize these observations as follows:

Center	Vertices: three units above and below center	Endpoints of Minor Axis: two units right and left of center
$(1, -2)$	$(1, -2 + 3) = (1, 1)$	$(1 + 2, -2) = (3, -2)$
	$(1, -2 - 3) = (1, -5)$	$(1 - 2, -2) = (-1, -2)$

Using the center and these four points, we can sketch the ellipse shown in Figure 10.18.

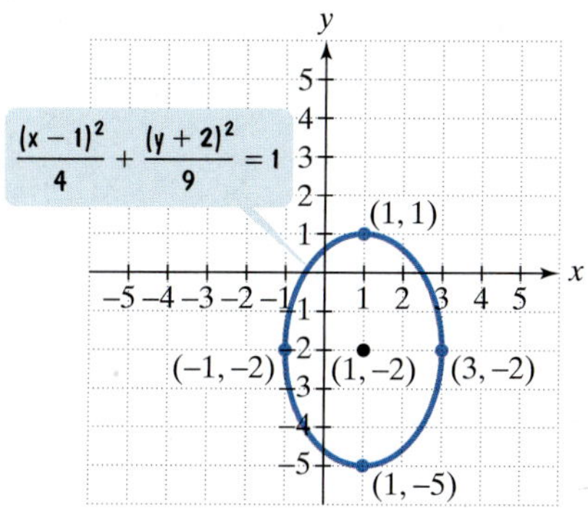

Figure 10.18 The graph of an ellipse centered at $(1, -2)$ ■

✔ CHECK POINT 3 Graph the ellipse: $\frac{(x+1)^2}{9} + \frac{(y-2)^2}{4} = 1$.

3 Solve applied problems involving ellipses.

Applications

Ellipses have many applications. German scientist Johannes Kepler (1571–1630) showed that the planets in our solar system move in elliptical orbits, with the sun at a focus. Earth satellites also travel in elliptical orbits, with Earth at a focus.

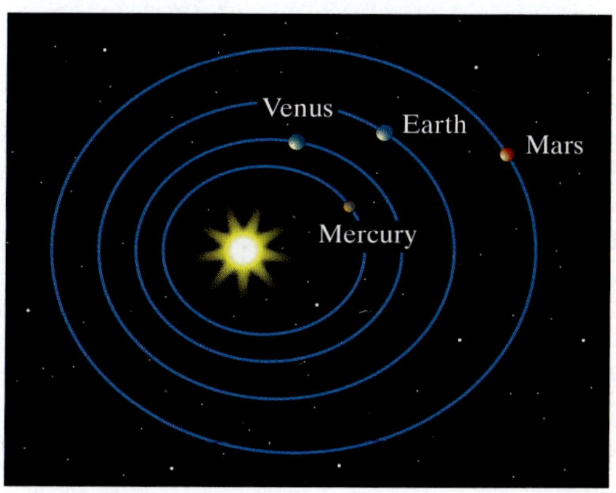

Planets move in elliptical orbits.

One intriguing aspect of the ellipse is that a ray of light or a sound wave emanating from one focus will be reflected from the ellipse to exactly the other focus. A whispering gallery is an elliptical room with an elliptical, dome-shaped ceiling. People standing at the foci can whisper and hear each other quite clearly, while persons in other locations in the room cannot hear them. Statuary Hall in the U.S. Capitol Building is elliptical. President John Quincy Adams, while a member of the House of Representatives, was aware of this acoustical phenomenon. He situated his desk at a focal point of the elliptical ceiling, easily eavesdropping on the private conversations of other House members located near the other focus.

The elliptical reflection principle is used in a procedure for disintegrating kidney stones. The patient is placed within a device that is elliptical in shape. The patient is at one focus, while ultrasound waves from the other focus hit the walls and are reflected to the kidney stone. The convergence of the ultrasound waves at the kidney stone causes vibrations that shatter it into fragments. The small pieces can then be passed painlessly through the patient's system. The patient recovers in days, as opposed to up to six weeks if surgery is used instead.

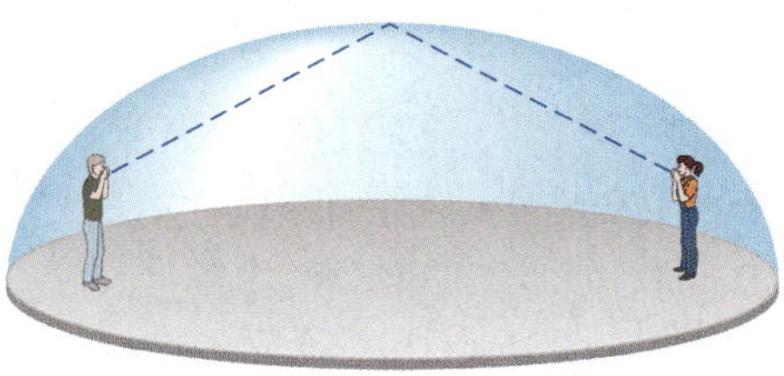

Whispering in an elliptical dome

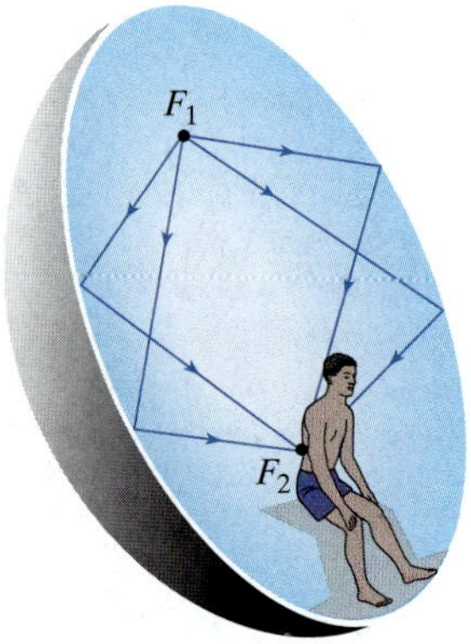

Disintegrating kidney stones

Ellipses are often used for supporting arches of bridges and in tunnel construction. This application forms the basis of our next example.

EXAMPLE 4 An Application Involving an Ellipse

A semielliptical archway over a one-way road has a height of 10 feet and a width of 40 feet (see Figure 10.19). Your truck has a width of 10 feet and a height of 9 feet. Will your truck clear the opening of the archway?

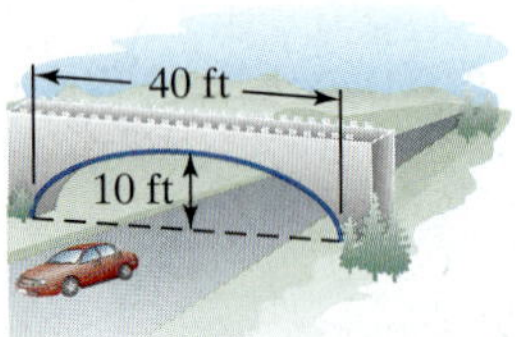

Figure 10.19 A semielliptical archway

Solution To determine the clearance, we must find the height of the archway 5 feet from the center. If that height is 9 feet or less, the truck will not clear the opening.

ENRICHMENT ESSAY

Halley's Comet

Halley's Comet has an elliptical orbit with the sun at one focus. The comet returns every 76.3 years. The first recorded sighting was in 239 B.C. It was last seen in 1986. At that time, spacecraft went close to the comet, measuring its nucleus to be 7 miles long and 4 miles wide. By 2024, Halley's Comet will have reached the farthest point in its elliptical orbit before returning to be next visible from Earth in 2062.

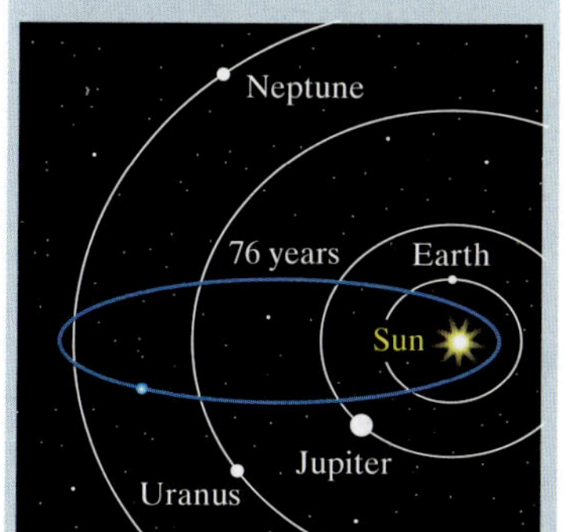

The elliptical orbit of Halley's Comet

In Figure 10.20, we've constructed a coordinate system with the x-axis on the ground and the origin at the center of the archway. Also shown is the truck, whose height is 9 feet.

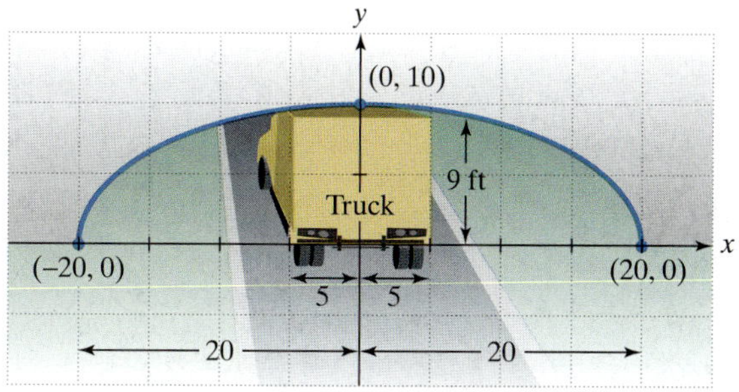

Figure 10.20

Using the equation $\frac{x^2}{a^2} + \frac{y^2}{b^2} = 1$, we can express the equation of the blue archway in Figure 10.20 as $\frac{x^2}{20^2} + \frac{y^2}{10^2} = 1$ or $\frac{x^2}{400} + \frac{y^2}{100} = 1$.

As shown in Figure 10.20, the edge of the 10-foot-wide truck corresponds to $x = 5$. We find the height of the archway 5 feet from the center by substituting 5 for x and solving for y.

$$\frac{5^2}{400} + \frac{y^2}{100} = 1 \quad \text{Substitute 5 for } x.$$

$$\frac{25}{400} + \frac{y^2}{100} = 1 \quad \text{Square 5.}$$

$$400\left(\frac{25}{400} + \frac{y^2}{100}\right) = 400(1) \quad \text{Clear fractions by multiplying both sides by 400.}$$

$$25 + 4y^2 = 400 \quad \text{Use the distributive property and simplify.}$$

$$4y^2 = 375 \quad \text{Subtract 25 from both sides.}$$

$$y^2 = \frac{375}{4} \quad \text{Divide both sides by 4.}$$

$$y = \sqrt{\frac{375}{4}} \quad \text{Take only the positive square root. The archway is above the } x\text{-axis, so } y \text{ is nonnegative.}$$

$$\approx 9.68$$

Thus, the height of the archway 5 feet from the center is approximately 9.68 feet. Because your truck's height is 9 feet, there is enough room for the truck to clear the archway. ■

✔ CHECK POINT 4 Will a truck that is 12 feet wide and has a height of 9 feet clear the opening of the archway described in Example 4?

EXERCISE SET 10.2

Practice Exercises

In Exercises 1–16, graph each ellipse.

1. $\frac{x^2}{16} + \frac{y^2}{4} = 1$
2. $\frac{x^2}{25} + \frac{y^2}{16} = 1$
3. $\frac{x^2}{9} + \frac{y^2}{36} = 1$
4. $\frac{x^2}{16} + \frac{y^2}{49} = 1$
5. $\frac{x^2}{25} + \frac{y^2}{64} = 1$
6. $\frac{x^2}{49} + \frac{y^2}{36} = 1$
7. $\frac{x^2}{49} + \frac{y^2}{81} = 1$
8. $\frac{x^2}{64} + \frac{y^2}{100} = 1$
9. $25x^2 + 4y^2 = 100$
10. $9x^2 + 4y^2 = 36$
11. $4x^2 + 16y^2 = 64$
12. $16x^2 + 9y^2 = 144$
13. $25x^2 + 9y^2 = 225$
14. $4x^2 + 25y^2 = 100$
15. $x^2 + 2y^2 = 8$
16. $12x^2 + 4y^2 = 36$

In Exercises 17–20, find the standard form of the equation of each ellipse.

17.

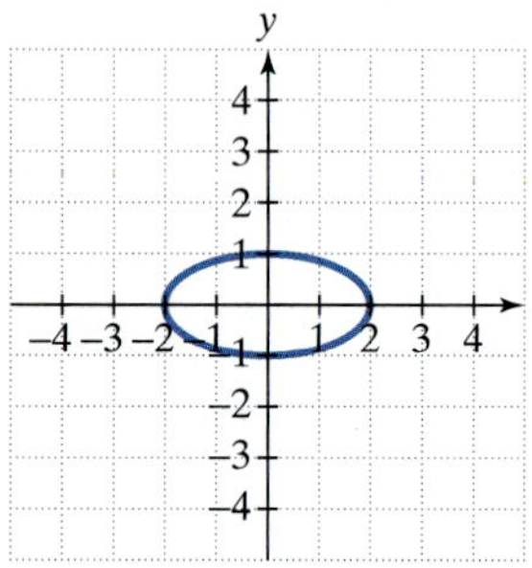

18.

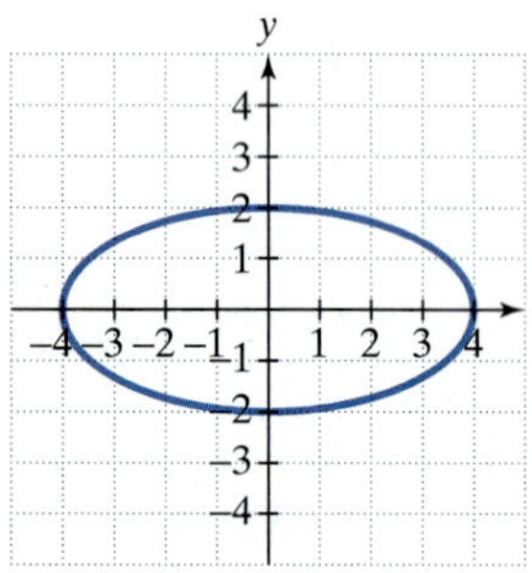

19.

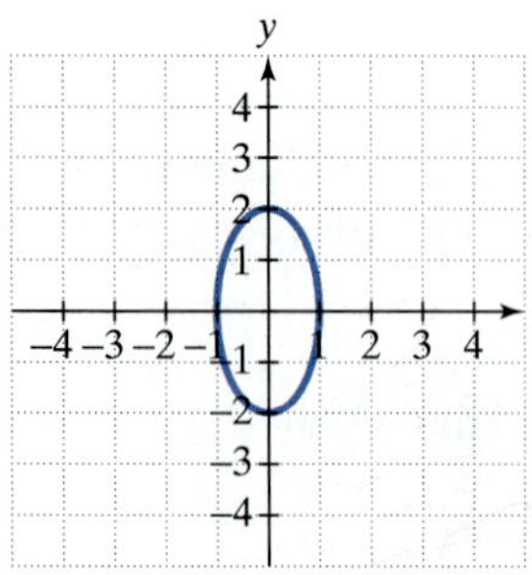

20.

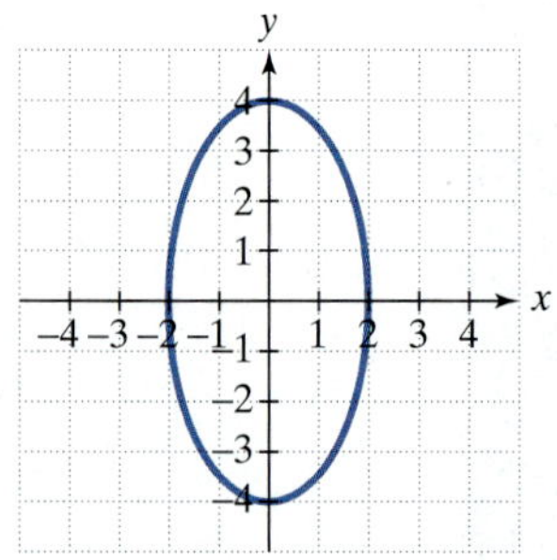

In Exercises 21–32, graph each ellipse.

21. $\frac{(x-2)^2}{9} + \frac{(y-1)^2}{4} = 1$
22. $\frac{(x-1)^2}{16} + \frac{(y+2)^2}{9} = 1$
23. $(x+3)^2 + 4(y-2)^2 = 16$
24. $(x-3)^2 + 9(y+2)^2 = 18$
25. $\frac{(x-4)^2}{9} + \frac{(y+2)^2}{25} = 1$
26. $\frac{(x-3)^2}{9} + \frac{(y+1)^2}{16} = 1$
27. $\frac{x^2}{25} + \frac{(y \quad 2)^2}{36} = 1$
28. $\frac{(x-4)^2}{4} + \frac{y^2}{25} = 1$
29. $\frac{(x+3)^2}{9} + (y-2)^2 = 1$
30. $\frac{(x+2)^2}{16} + (y-3)^2 = 1$
31. $9(x-1)^2 + 4(y+3)^2 = 36$
32. $36(x+4)^2 + (y+3)^2 = 36$

Application Exercises

33. Will a truck that is 8 feet wide carrying a load that reaches 7 feet above the ground clear the semielliptical arch on the one-way road that passes under the bridge shown in the figure?

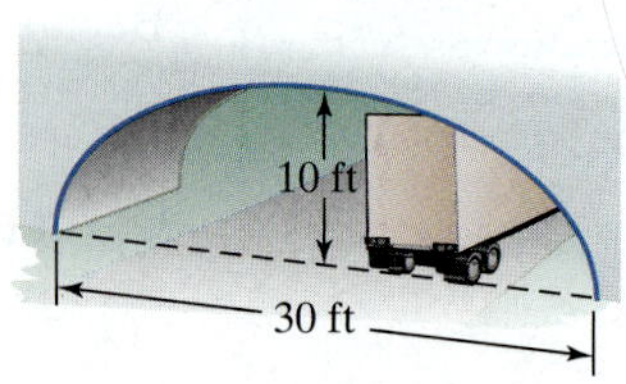

34. A semielliptic archway has a height of 20 feet and a width of 50 feet, as shown in the figure. Can a truck 14 feet high and 10 feet wide drive under the archway without going into the other lane?

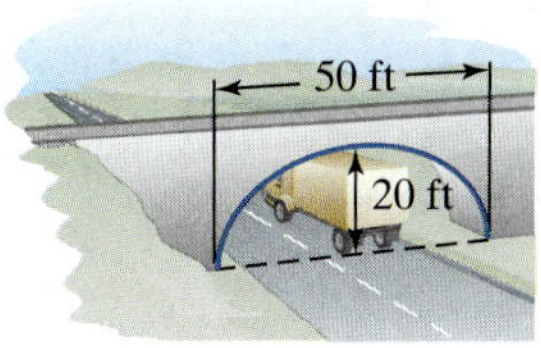

35. The elliptical ceiling in Statuary Hall in the U.S. Capitol Building is 96 feet long and 23 feet tall.

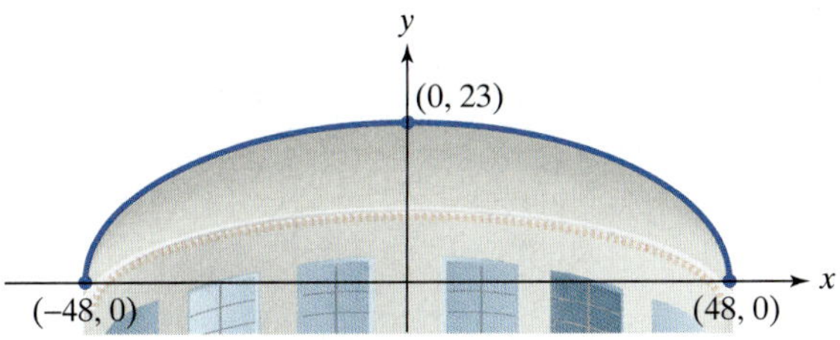

a. Using the rectangular coordinate system in the figure shown, write the standard form of the equation of the elliptical ceiling.

b. John Quincy Adams discovered that he could overhear the conversations of opposing party leaders near the left side of the chamber if he situated his desk at the focus at the right side of the chamber. How far from the center of the ellipse along the major axis did Adams situate his desk? (Round to the nearest foot.)

36. If an elliptical whispering room has a height of 30 feet and a width of 100 feet, where should two people stand if they would like to whisper back and forth and be heard?

Writing in Mathematics

37. What is an ellipse?

38. Describe how to graph $\frac{x^2}{25} + \frac{y^2}{16} = 1$.

39. Describe one similarity and one difference between the graphs of $\frac{x^2}{25} + \frac{y^2}{16} = 1$ and $\frac{x^2}{16} + \frac{y^2}{25} = 1$.

40. Describe one similarity and one difference between the graphs of $\frac{x^2}{25} + \frac{y^2}{16} = 1$ and $\frac{(x-1)^2}{25} + \frac{(y-1)^2}{16} = 1$.

41. An elliptipool is an elliptical pool table with only one pocket. A pool shark places a ball on the table, hits it in what appears to be a random direction, and yet it bounces off the edge, falling directly into the pocket. Explain why this happens.

Technology Exercises

42. Use a graphing utility to graph any five of the ellipses that you graphed by hand in Exercises 1–16.

43. Use a graphing utility to graph any three of the ellipses that you graphed by hand in Exercises 21–32. First solve the given equation for y by using the square root method. Enter each of the two resulting equations to produce each half of the ellipse.

Critical Thinking Exercises

44. Which one of the following is true?

a. The graphs of $x^2 + y^2 = 16$ and $\frac{x^2}{4} + \frac{y^2}{9} = 1$ do not intersect.

b. Some ellipses have equations that define y as a function of x.

c. The graph of $\frac{x^2}{9} + \frac{y^2}{4} = 0$ is an ellipse.

d. None of the above is true.

45. Find the standard form of the equation of an ellipse with vertices at $(0, -6)$ and $(0, 6)$, passing through $(2, -4)$.

In Exercises 46–47, convert each equation to standard form by completing the square on x and y. Then graph the ellipse.

46. $9x^2 + 25y^2 - 36x + 50y - 164 = 0$

47. $4x^2 + 9y^2 - 32x + 36y + 64 = 0$

48. An Earth satellite has an elliptical orbit described by

$$\frac{x^2}{(5000)^2} + \frac{y^2}{(4750)^2} = 1.$$

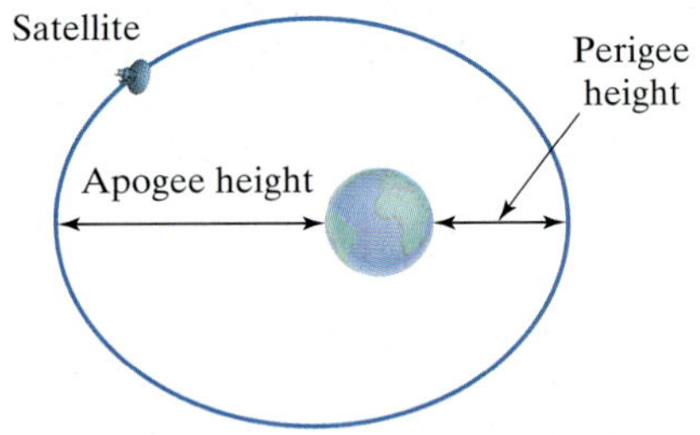

(All units are in miles.) The coordinates of the center of Earth are (16, 0).

a. The perigee of the satellite's orbit is the point that is nearest Earth's center. If the radius of Earth is approximately 4000 miles, find the distance of the perigee above Earth's surface.

b. The apogee of the satellite's orbit is the point that is the greatest distance from Earth's center. Find the distance of the apogee above Earth's surface.

49. The equation of the red ellipse in the figure at the top of the next column is

$$\frac{x^2}{25} + \frac{y^2}{9} = 1.$$

Write the equation for each circle shown in the figure.

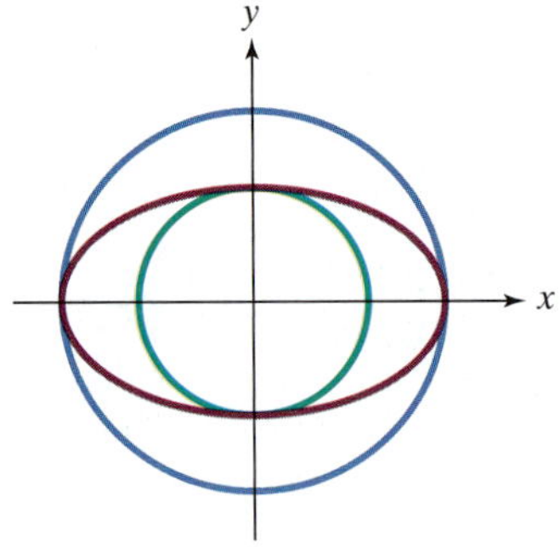

50. What happens to the shape of the graph of $\frac{x^2}{a^2} + \frac{y^2}{b^2} = 1$ as $\frac{c}{a}$ is close to zero?

Review Exercises

51. Factor completely: $x^3 + 2x^2 - 4x - 8$.
(Section 5.5, Example 4)

52. Simplify: $\sqrt[3]{40x^4y^7}$.
(Section 7.3, Example 5)

53. Solve: $\frac{2}{x+2} + \frac{4}{x-2} = \frac{x-1}{x^2-4}$.
(Section 6.6, Example 5)

▸ SECTION 10.3 The Hyperbola

Objectives

1. Locate a hyperbola's vertices.
2. Graph hyperbolas centered at the origin.

SSM PH Tutor Center CD-ROM Video

St. Mary's Cathedral

Conic sections are often used to create unusual architectural designs. The top of St. Mary's Cathedral in San Francisco is a 2135-cubic-foot dome with walls rising 200 feet above the floor and supported by four massive concrete pylons that extend 94 feet into the ground. Cross sections of the roof are parabolas and hyperbolas. In this section, we study the curve with two parts known as the hyperbola.

Figure 10.21 Casting hyperbolic shadows

Definition of a Hyperbola

Figure 10.21 shows a cylindrical lampshade casting two shadows on a wall. These shadows indicate the distinguishing feature of hyperbolas: Their graphs contain two disjoint parts, called **branches**. Although each branch might look like a parabola, its shape is actually quite different.

The definition of a hyperbola is similar to that of the ellipse. For the ellipse, the *sum* of the distances to the foci is a constant. By contrast, for a hyperbola the *difference* of the distances to the foci is a constant.

Definition of a Hyperbola

A **hyperbola** is the set of points in a plane the difference of whose distances from two fixed points (called foci) is a constant.

Figure 10.22 illustrates the two branches of a hyperbola. The line through the foci intersects the hyperbola at two points, called the **vertices**. The line segment that joins the vertices is the **transverse axis**. The midpoint of the transverse axis is the **center** of the hyperbola. Notice that the center lies midway between the vertices, as well as midway between the foci.

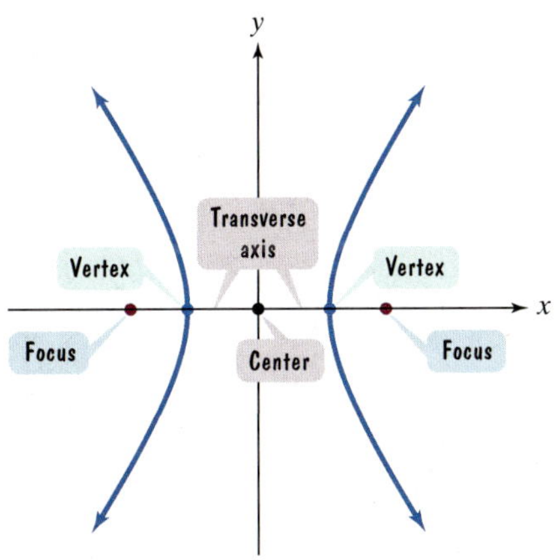

Figure 10.22 The two branches of a hyperbola

Standard Form of the Equation of a Hyperbola

The rectangular coordinate system enables us to translate a hyperbola's geometric definition into an algebraic equation. Figure 10.23 is our starting point for obtaining an equation. We place the foci, F_1 and F_2, on the x-axis at the points $(-c, 0)$ and $(c, 0)$. Note that the center of this hyperbola is at the origin. We let (x, y) represent the coordinates of any point, P, on the hyperbola.

What does the definition of a hyperbola tell us about the point (x, y) in Figure 10.23? For any point (x, y) on the hyperbola, the absolute value of the difference of the distances from the two foci, $|d_2 - d_1|$, must be constant. We denote this constant by $2a$, just as we did for the ellipse. Thus, the point (x, y) is on the hyperbola if and only if

$$|d_2 - d_1| = 2a$$

$$\left|\sqrt{(x + c)^2 + y^2} - \sqrt{(x - c)^2 + y^2}\right| = 2a \quad \text{Use the distance formula.}$$

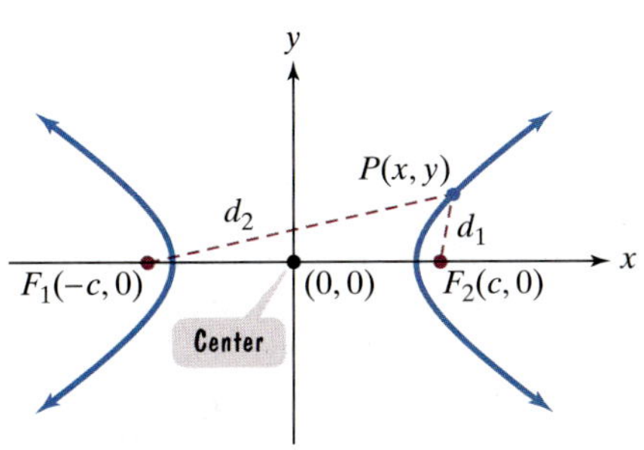

Figure 10.23

After eliminating radicals and simplifying, we obtain

$$(c^2 - a^2)x^2 - a^2y^2 = a^2(c^2 - a^2).$$

For convenience, let $b^2 = c^2 - a^2$. Substituting b^2 for $c^2 - a^2$ in the preceding equation, we obtain

$$b^2x^2 - a^2y^2 = a^2b^2.$$

$$\frac{b^2x^2}{a^2b^2} - \frac{a^2y^2}{a^2b^2} = \frac{a^2b^2}{a^2b^2} \quad \text{Divide both sides by } a^2b^2.$$

$$\frac{x^2}{a^2} - \frac{y^2}{b^2} = 1. \quad \text{Simplify.}$$

This last equation is called the **standard form of the equation of a hyperbola**. There are two such equations. The first is for a hyperbola in which the transverse axis lies on the x-axis. The second is for a hyperbola in which the transverse axis lies on the y-axis.

Standard Forms of the Equations of a Hyperbola

The **standard form of the equation of a hyperbola** with center at the origin is

$$\frac{x^2}{a^2} - \frac{y^2}{b^2} = 1 \quad \text{or} \quad \frac{y^2}{a^2} - \frac{x^2}{b^2} = 1.$$

Figure 10.24 illustrates that for the equation on the left, the transverse axis lies on the x-axis. For the equation on the right, the transverse axis lies on the y-axis. The vertices are a units from the center and the foci are c units from the center.

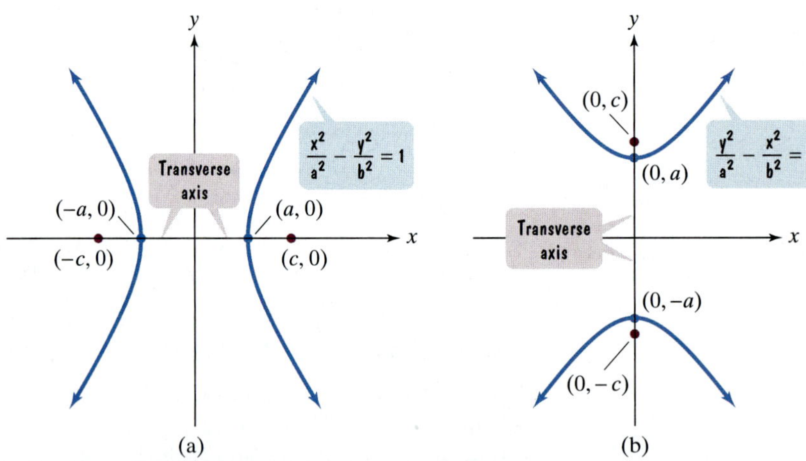

Figure 10.24 **(a)** Transverse axis lies on the x-axis. **(b)** Transverse axis lies on the y-axis.

1 Locate a hyperbola's vertices.

Using the Standard Form of the Equation of a Hyperbola

We can use the standard form of the equation of a hyperbola to find its vertices. Because the vertices are a units from the center, begin by identifying a^2 in the equation. In the standard form of a hyperbola's equation, $\boldsymbol{a^2}$ **is the number under the variable whose term is preceded by a plus sign** $(+)$. If the x^2-term is preceded by a plus sign, the transverse axis lies along the x-axis. Thus, the vertices are a units to the left and right of the origin. If the y^2-term is preceded by a plus sign, the transverse axis lies along the y-axis. Thus, the vertices are a units above and below the origin.

EXAMPLE 1 Finding Vertices from a Hyperbola's Equation

Find the vertices for each of the following hyperbolas with the given equation:

a. $\frac{x^2}{16} - \frac{y^2}{9} = 1$ **b.** $\frac{y^2}{9} - \frac{x^2}{16} = 1.$

Solution Both equations are in standard form. We begin by identifying a^2 and b^2 in each equation.

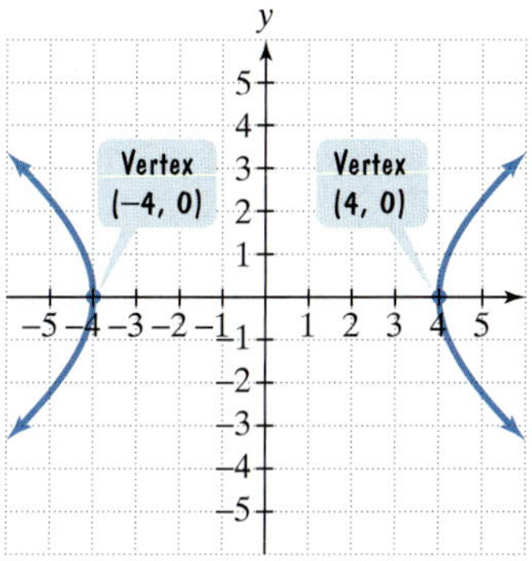

Figure 10.25 The graph of $\frac{x^2}{16} - \frac{y^2}{9} = 1$

a. The first equation is in the form $\frac{x^2}{a^2} - \frac{y^2}{b^2} = 1.$

$$\frac{x^2}{16} - \frac{y^2}{9} = 1$$

$a^2 = 16$. This is the number in the denominator of the term preceded by a plus sign.

$b^2 = 9$. This is the number in the denominator of the term preceded by a minus sign.

Because the x^2-term is preceded by a plus sign, the transverse axis lies along the x-axis. Thus, the vertices are a units to the *left* and *right* of the origin. Based on the standard form of the equation, we know the vertices are $(-a, 0)$ and $(a, 0)$. Because $a^2 = 16$, $a = 4$. Thus, the vertices are $(-4, 0)$ and $(4, 0)$, shown in Figure 10.25.

b. The second given equation is in the form $\frac{y^2}{a^2} - \frac{x^2}{b^2} = 1.$

$$\frac{y^2}{9} - \frac{x^2}{16} = 1$$

$a^2 = 9$. This is the number in the denominator of the term preceded by a plus sign.

$b^2 = 16$. This is the number in the denominator of the term preceded by a minus sign.

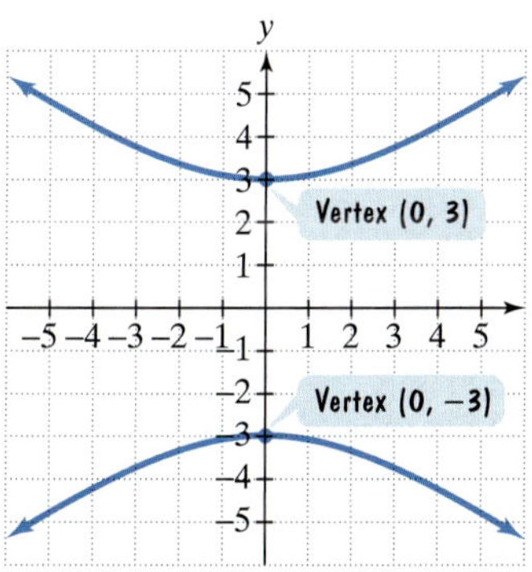

Figure 10.26 The graph of $\frac{y^2}{9} - \frac{x^2}{16} = 1$

Because the y^2-term is preceded by a plus sign, the transverse axis lies along the y-axis. Thus, the vertices are a units *above* and *below* the origin. Based on the standard form of the equation, we know the vertices are $(0, -a)$ and $(0, a)$. Because $a^2 = 9$, $a = 3$. Thus, the vertices are $(0, -3)$ and $(0, 3)$, shown in Figure 10.26. ■

✔ **CHECK POINT 1** Find the vertices for each of the following hyperbolas with the given equation:

a. $\frac{x^2}{25} - \frac{y^2}{16} = 1$

b. $\frac{y^2}{25} - \frac{x^2}{16} = 1.$

The Asymptotes of a Hyperbola

As x and y get larger, the two branches of the graph of a hyperbola approach a pair of intersecting lines, called **asymptotes**. The asymptotes pass through the center of the hyperbola and are helpful in graphing hyperbolas.

Figure 10.27 shows the asymptotes for the graphs of hyperbolas centered at the origin. The asymptotes pass through the corners of a rectangle. Note that the dimensions of this rectangle are $2a$ by $2b$.

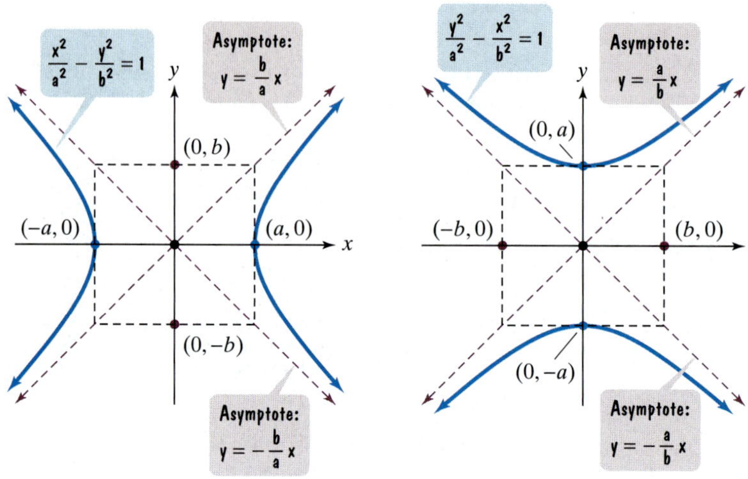

Figure 10.27 Asymptotes of a hyperbola

Why are $y = \pm\frac{b}{a}x$ the asymptotes for a hyperbola whose transverse axis is horizontal? The proof can be found in the appendix.

2 Graph hyperbolas centered at the origin.

Graphing Hyperbolas Centered at the Origin

Hyperbolas are graphed using vertices and asymptotes.

Graphing Hyperbolas

1. Locate the vertices.
2. Draw the rectangle centered at the origin with sides parallel to the axes, crossing one axis at $\pm a$ and the other at $\pm b$.
3. Draw the diagonals of this rectangle and extend them to obtain the asymptotes.
4. Draw the two branches of the hyperbola by starting at each vertex and approaching the asymptotes.

The rectangle in step 2 and the asymptotes in step 3 are drawn using dashed lines to show that they are not part of the hyperbola.

EXAMPLE 2 Graphing a Hyperbola

Graph the hyperbola: $\frac{x^2}{25} - \frac{y^2}{16} = 1.$

Solution

Step 1 Locate the vertices. The given equation is in the form $\frac{x^2}{a^2} - \frac{y^2}{b^2} = 1$, with $a^2 = 25$ and $b^2 = 16$.

$$\frac{x^2}{25} - \frac{y^2}{16} = 1$$

$a^2 = 25$ $b^2 = 16$

Based on the standard form of the equation with the transverse axis on the x-axis, we know that the vertices are $(-a, 0)$ and $(a, 0)$. Because $a^2 = 25$, $a = 5$. Thus, the vertices are $(-5, 0)$ and $(5, 0)$, shown in Figure 10.28.

Step 2 Draw a rectangle. Because $a^2 = 25$ and $b^2 = 16$, $a = 5$ and $b = 4$. We construct a rectangle to find the asymptotes, using -5 and 5 on the x-axis (the vertices are located here) and -4 and 4 on the y-axis. The rectangle passes through these four points, shown using dashed lines in Figure 10.28.

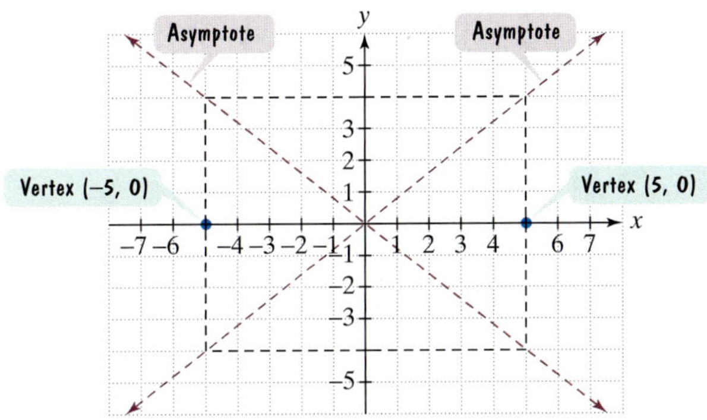

Figure 10.28 Preparing to graph $\frac{x^2}{25} - \frac{y^2}{16} = 1$

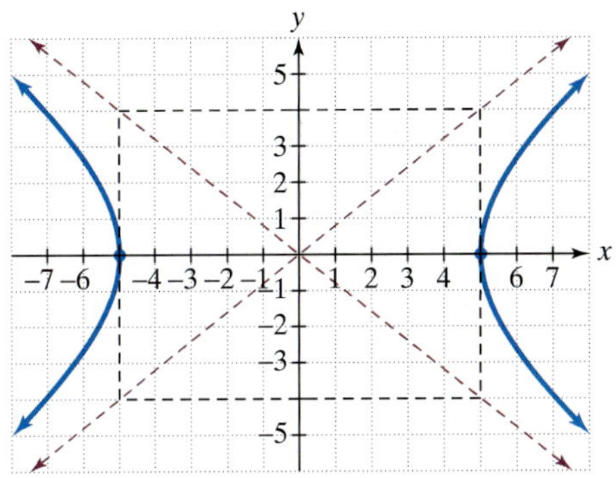

Figure 10.29 The graph of $\frac{x^2}{25} - \frac{y^2}{16} = 1$

Using Technology

Graph $\frac{x^2}{25} - \frac{y^2}{16} = 1$ by solving for y:

$$y_1 = \frac{\sqrt{16x^2 - 400}}{5}$$

$$y_2 = -\frac{\sqrt{16x^2 - 400}}{5} = -y_1.$$

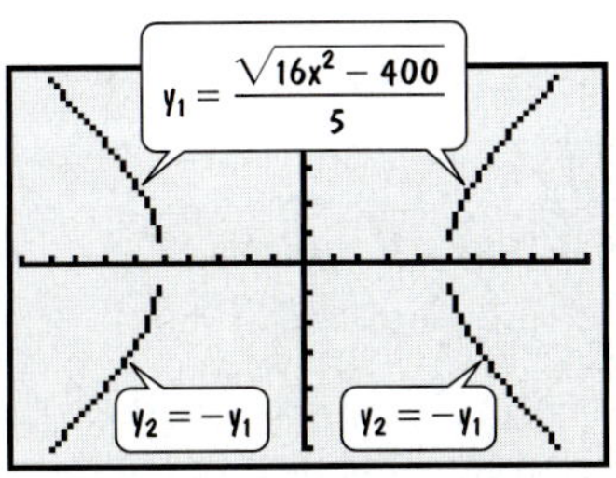

$[-10, 10, 1]$ by $[-6, 6, 1]$

Step 3 Draw extended diagonals for the rectangle to obtain the asymptotes. We draw dashed lines through the opposite corners of the rectangle, shown in Figure 10.28, to obtain the graph of the asymptotes.

Step 4 Draw the two branches of the hyperbola by starting at each vertex and approaching the asymptotes. The hyperbola is shown in Figure 10.29. ■

✔ **CHECK POINT 2** Graph the hyperbola: $\frac{x^2}{36} - \frac{y^2}{9} = 1$.

EXAMPLE 3 Graphing a Hyperbola

Graph the hyperbola: $9y^2 - 4x^2 = 36$.

Solution We begin by writing the equation in standard form. The right side should be 1, so we divide both sides by 36.

$$\frac{9y^2}{36} - \frac{4x^2}{36} = \frac{36}{36}$$

$$\frac{y^2}{4} - \frac{x^2}{9} = 1 \quad \text{Simplify. The right side is now 1.}$$

Now we are ready to use our four-step procedure for graphing hyperbolas.

Step 1 Locate the vertices. The equation that we obtained is in the form $\frac{y^2}{a^2} - \frac{x^2}{b^2} = 1$, with $a^2 = 4$ and $b^2 = 9$.

$$\frac{y^2}{4} - \frac{x^2}{9} = 1$$

$a^2 = 4$ $b^2 = 9$

Based on the standard form of the equation with the transverse axis on the y-axis, we know that the vertices are $(0, -a)$ and $(0, a)$. Because $a^2 = 4$, $a = 2$. Thus, the vertices are $(0, -2)$ and $(0, 2)$, shown in Figure 10.30.

Step 2 Draw a rectangle. Because $a^2 = 4$ and $b^2 = 9$, $a = 2$ and $b = 3$. We construct a rectangle to find the asymptotes, using -2 and 2 on the y-axis (the vertices are located here) and -3 and 3 on the x-axis. The rectangle passes through these four points, shown using dashed lines in Figure 10.30.

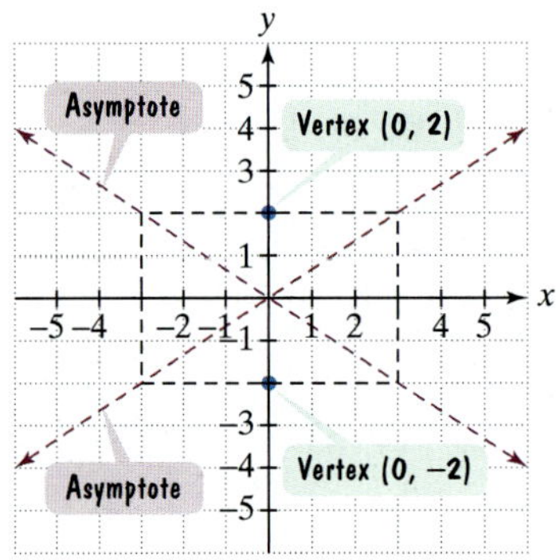

Figure 10.30 Preparing to graph $\frac{y^2}{4} - \frac{x^2}{9} = 1$

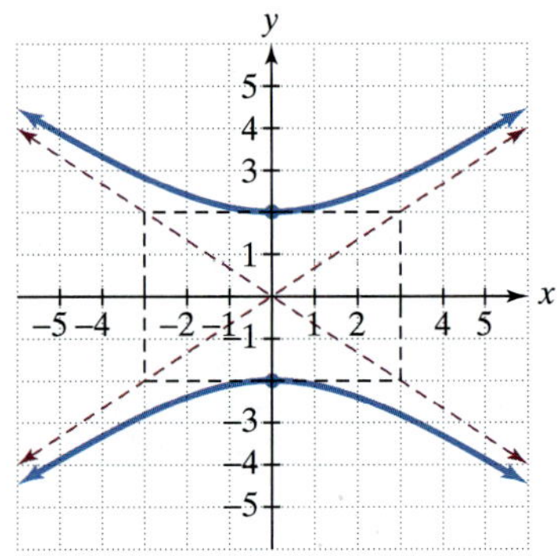

Figure 10.31 The graph of $\frac{y^2}{4} - \frac{x^2}{9} = 1$

Step 3 Draw extended diagonals of the rectangle to obtain the asymptotes. We draw dashed lines through the opposite corners of the rectangle, shown in Figure 10.30, to obtain the graph of the asymptotes.

Step 4 Draw the two branches of the hyperbola by starting at each vertex and approaching the asymptotes. The hyperbola is shown in Figure 10.31. ■

✔ CHECK POINT 3 Graph the hyperbola: $y^2 - 4x^2 = 4$.

Applications

Hyperbolas have many applications. When a jet flies at a speed greater than the speed of sound, the shock wave that is created is heard as a sonic boom. The wave has the shape of a cone. The shape formed as the cone hits the ground is one branch of a hyperbola.

ENRICHMENT ESSAY

Where Exactly Am I?

The hyperbola is the basis for the navigational system LORAN (for long-range navigation), used by a ship or aircraft to determine its location. The measured time-of-arrival difference between signals transmitted from two ground stations determines the hyperbola on which the ship or aircraft is located. The process is then repeated by taking a similar time-difference reading from a second pair of stations, determining a second hyperbola. The point of intersection of the two hyperbolas is the location of the ship or aircraft.

LORAN will eventually be replaced by the Global Positioning System. Using 24 satellites that orbit at 11,000 miles above Earth, the system is able to show you your exact position on Earth anytime, in any weather, anywhere.

Halley's Comet, a permanent part of our solar system, travels around the sun in an elliptical orbit. Other comets pass through the solar system only once, following a hyperbolic path with the sun as a focus.

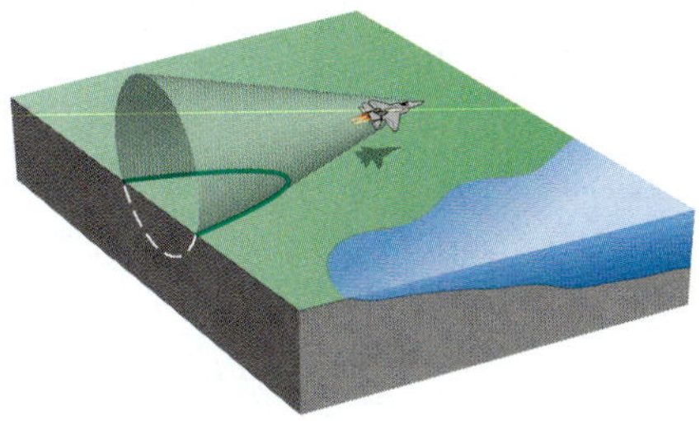

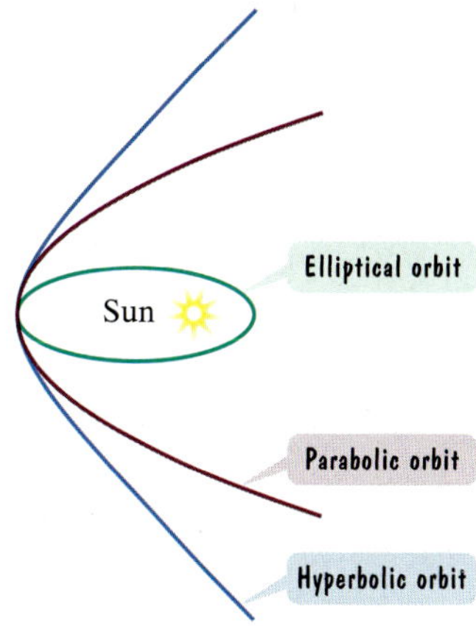

Hyperbolas are of practical importance in fields ranging from architecture to navigation. Cooling towers used in the design for nuclear power plants have cross sections that are both ellipses and hyperbolas. Three-dimensional solids whose cross sections are hyperbolas are used in some rather unique architectural creations, including the TWA building at Kennedy Airport and the St. Louis Science Center Planetarium.

EXERCISE SET 10.3

Practice Exercises

In Exercises 1–4, find the vertices of each hyperbola with the given equation. Then match each equation to one of the graphs that are shown and labeled (a)–(d).

1. $\dfrac{x^2}{4} - \dfrac{y^2}{1} = 1$

2. $\dfrac{x^2}{1} - \dfrac{y^2}{4} = 1$

3. $\dfrac{y^2}{4} - \dfrac{x^2}{1} = 1$

4. $\dfrac{y^2}{1} - \dfrac{x^2}{4} = 1$

(a)

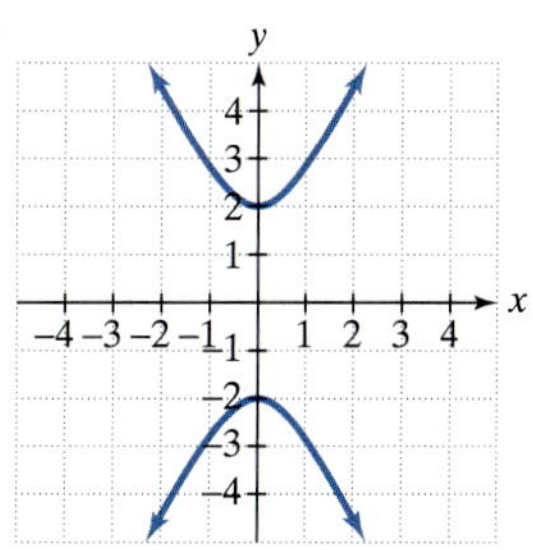

(b)

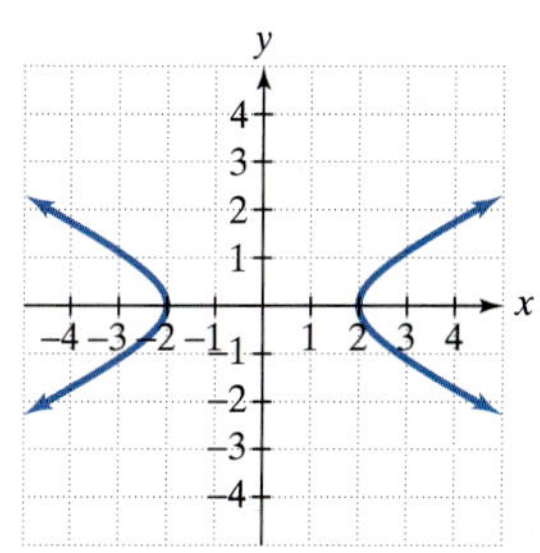

(c)

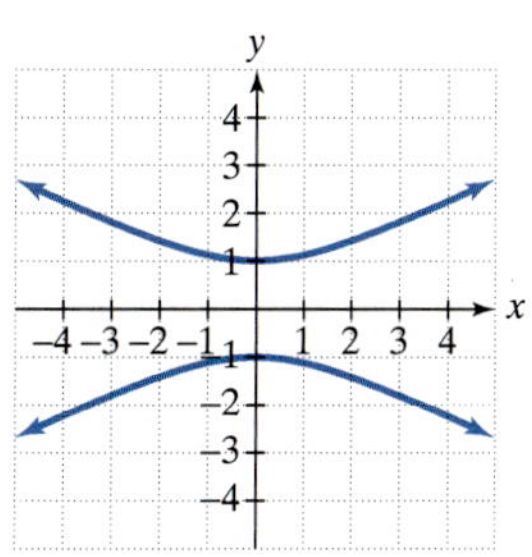

d.

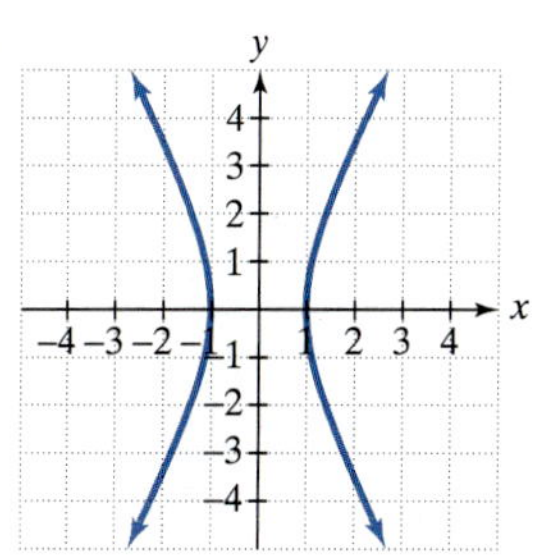

In Exercises 5–18, use vertices and asymptotes to graph each hyperbola.

5. $\frac{x^2}{9} - \frac{y^2}{25} = 1$

6. $\frac{x^2}{16} - \frac{y^2}{25} = 1$

7. $\frac{x^2}{100} - \frac{y^2}{64} = 1$

8. $\frac{x^2}{144} - \frac{y^2}{81} = 1$

9. $\frac{y^2}{16} - \frac{x^2}{36} = 1$

10. $\frac{y^2}{25} - \frac{x^2}{64} = 1$

11. $\frac{y^2}{36} - \frac{x^2}{25} = 1$

12. $\frac{y^2}{100} - \frac{x^2}{49} = 1$

13. $9x^2 - 4y^2 = 36$

14. $4x^2 - 25y^2 = 100$

15. $9y^2 - 25x^2 = 225$

16. $16y^2 - 9x^2 = 144$

17. $4x^2 = 4 + y^2$

18. $25y^2 = 225 + 9x^2$

In Exercises 19–22, find the standard form of the equation of each hyperbola.

19.

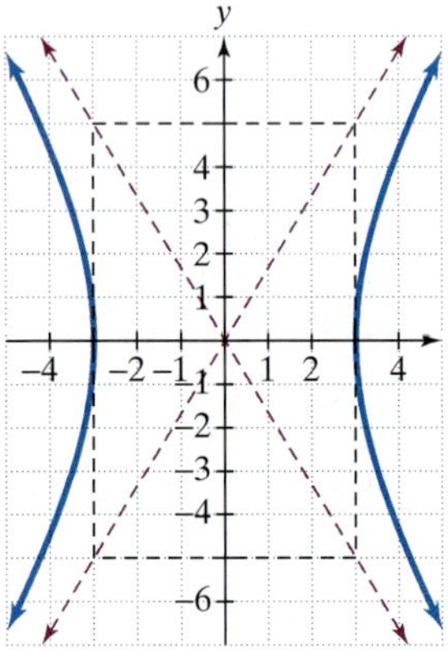

20.

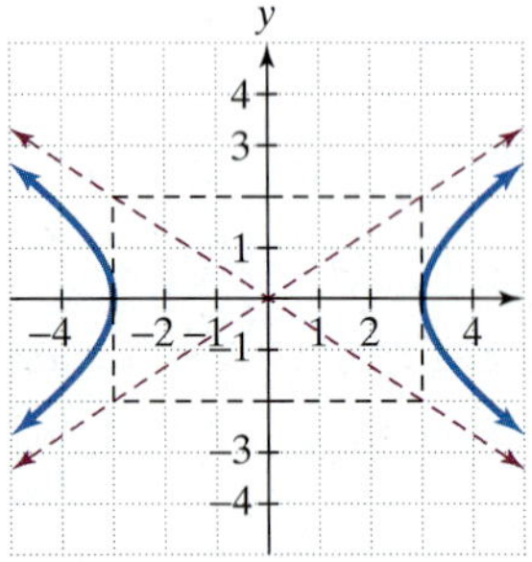

21.

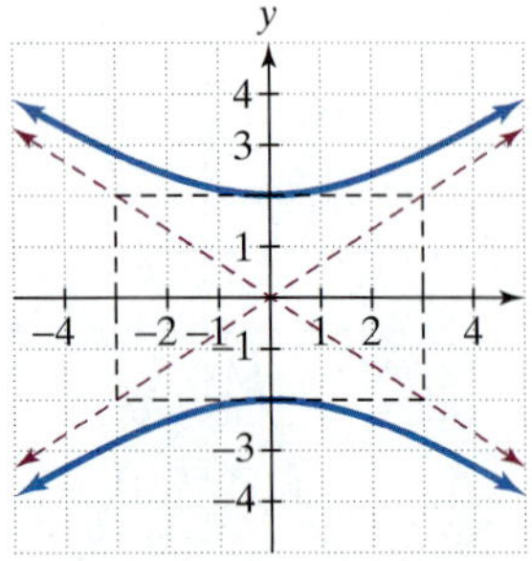

22.

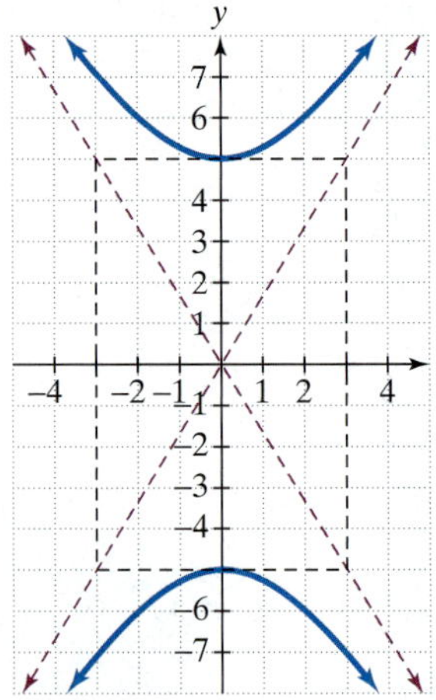

Application Exercises

23. An architect designs two houses that are shaped and positioned like a part of the branches of the hyperbola whose equation is $625y^2 - 400x^2 = 250{,}000$, where x and y are in yards. How far apart are the houses at their closest point?

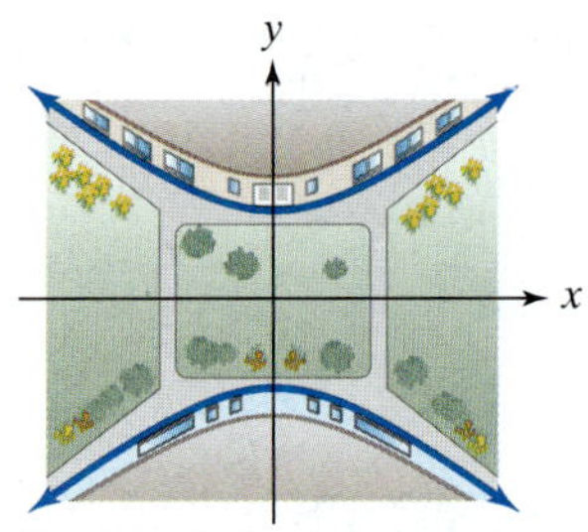

24. Scattering experiments, in which moving particles are deflected by various forces, led to the concept of the nucleus of an atom. In 1911, the physicist Ernest Rutherford (1871–1937) discovered that when alpha particles are directed toward the nuclei of gold atoms, they are eventually deflected along hyperbolic paths, illustrated in the figure. If a particle gets as close as 3 units to the nucleus along a hyperbolic path with an asymptote given by $y = \frac{1}{2}x$, what is the equation of its path?

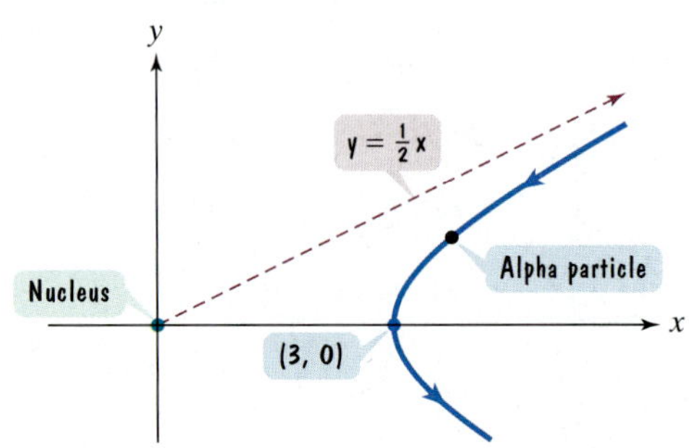

Writing in Mathematics

25. What is a hyperbola?

26. Describe how to graph $\frac{x^2}{9} - \frac{y^2}{1} = 1$.

27. Describe one similarity and one difference between the graphs of $\frac{x^2}{9} - \frac{y^2}{1} = 1$ and $\frac{y^2}{9} - \frac{x^2}{1} = 1$.

28. How can you distinguish an ellipse from a hyperbola by looking at their equations?

29. In 1992, a NASA team began a project called Spaceguard Survey, calling for an international watch for comets that might collide with Earth. Why is it more difficult to detect a possible "doomsday comet" with a hyperbolic orbit than one with an elliptical orbit?

Technology Exercises

30. Use a graphing utility to graph any five of the hyperbolas that you graphed by hand in Exercises 5–18.

31. Use a graphing utility to graph $\frac{x^2}{4} - \frac{y^2}{9} = 0$. Is the graph a hyperbola? In general, what is the graph of $\frac{x^2}{a^2} - \frac{y^2}{b^2} = 0$?

32. Graph $\frac{x^2}{a^2} - \frac{y^2}{b^2} = 1$ and $\frac{x^2}{a^2} - \frac{y^2}{b^2} = -1$ in the same viewing rectangle for values of a^2 and b^2 of your choice. Describe the relationship between the two graphs.

Critical Thinking Exercises

33. Which one of the following is true?

a. If one branch of a hyperbola is removed from a graph, then the branch that remains must define y as a function of x.

b. All points on the asymptotes of a hyperbola also satisfy the hyperbola's equation.

c. The graph of $\frac{x^2}{9} - \frac{y^2}{4} = 1$ does not intersect the line $y = -\frac{2}{3}x$.

d. Two different hyperbolas can never share the same asymptotes.

The graph of

$$\frac{(x-h)^2}{a^2} - \frac{(y-k)^2}{b^2} = 1$$

is the same as the graph of $\frac{x^2}{a^2} - \frac{y^2}{b^2} = 1$, *except the center is at* (h, k) *rather than at the origin. Use this information to graph the hyperbolas in Exercises 34–37.*

34. $\frac{(x-2)^2}{16} - \frac{(y-3)^2}{9} = 1$

35. $\frac{(x+2)^2}{9} - \frac{(y-1)^2}{25} = 1$

36. $(x-3)^2 - 4(y+3)^2 = 4$

37. $x^2 - y^2 - 2x - 4y - 4 = 0$

In Exercises 38–39, find the standard form of the equation of each hyperbola satisfying the given conditions.

38. vertices: $(6, 0), (-6, 0)$; asymptotes: $y = 4x, y = -4x$

39. vertices: $(0, 7), (0, -7)$; asymptotes: $y = 5x, y = -5x$

Review Exercises

40. Use intercepts and the vertex to graph the quadratic function: $y = -x^2 - 4x + 5$. (Section 8.3, Example 3)

41. Solve: $3x^2 - 11x - 4 \geq 0$. (Section 8.5, Example 2)

42. Solve: $\log_4(3x + 1) = 3$. (Section 9.5, Example 4)

▸ SECTION 10.4 The Parabola; Identifying Conic Sections

Objectives

1 Graph horizontal parabolas.

2 Identify conic sections by their equations.

SSM PH Tutor Center CD-ROM Video

At first glance, this image looks like columns of smoke rising from a fire into a starry sky. Those are, indeed, stars in the background, but you are not looking at ordinary smoke columns. These stand almost 6 trillion miles high and are 7000 light-years from Earth—more than 400 million times as far away as the sun.

This NASA photograph is one of a series of stunning images captured from the ends of the universe by the Hubble Space Telescope. The image shows infant star systems the size of our solar system emerging from the gas and dust that shrouded their creation. Using a parabolic mirror that is 94.5 inches in diameter, the Hubble is providing answers to many of the profound mysteries of the cosmos: How big and how old is the universe? How did the galaxies come to exist? Do other Earth-like planets orbit other sun-like stars? In this section, we study parabolas and their applications, including parabolic shapes that gather distant rays of light and focus them into spectacular images.

Definition of a Parabola

In Chapter 8, we studied parabolas, viewing them as graphs of quadratic functions in the form

$$y = a(x - h)^2 + k \quad \text{or} \quad y = ax^2 + bx + c.$$

Study Tip

Here is a summary of what you should already know about graphing parabolas.

Graphing $y = a(x - h)^2 + k$ and $y = ax^2 + bx + c$

1. If $a > 0$, the graph opens upward. If $a < 0$, the graph opens downward.
2. The vertex of $y = a(x - h)^2 + k$ is (h, k).
3. The x-coordinate of the vertex of $y = ax^2 + bx + c$ is $x = -\frac{b}{2a}$.

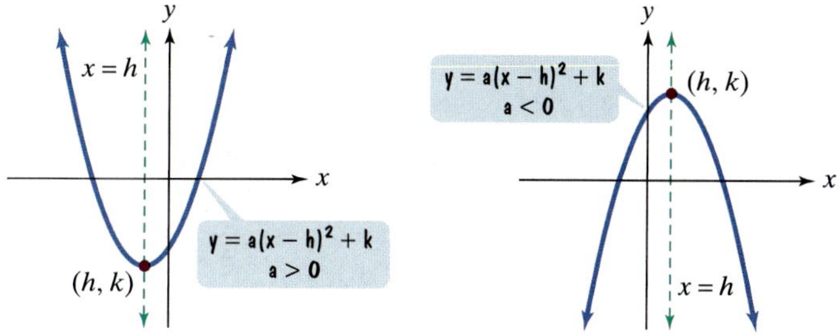

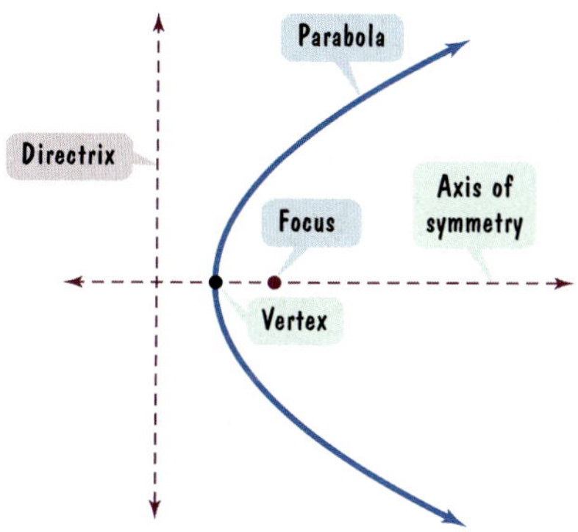

Figure 10.32

Parabolas can be given a geometric definition that enables us to include graphs that open to the left or to the right. The definitions of ellipses and hyperbolas involved two fixed points, the foci. By contrast, the definition of a parabola is based on one point and a line.

Definition of a Parabola

A **parabola** is the set of all points in a plane that are equidistant from a fixed line (the **directrix**) and a fixed point (the **focus**) that is not on the line (see Figure 10.32).

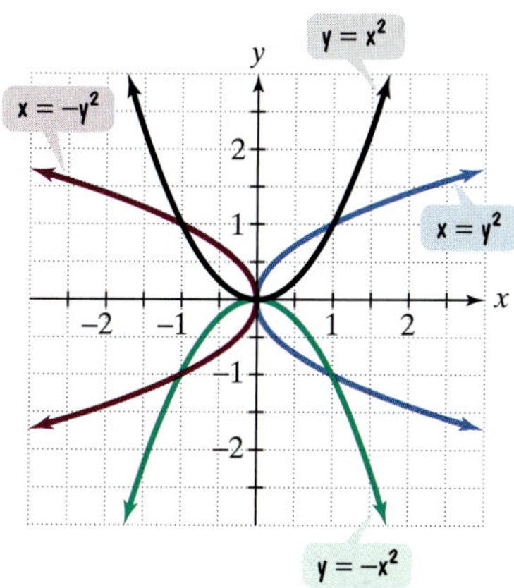

Figure 10.33

In Figure 10.32, the parabola is shown opening to the right. Find the line passing through the focus and perpendicular to the directrix. This is the **axis of symmetry** of the parabola. The point of intersection of the parabola with its axis of symmetry is the parabola's **vertex**. Notice that the vertex is midway between the focus and the directrix.

Parabolas can open to the left, right, upward, or downward. Figure 10.33 shows a basic "family" of four parabolas and their equations. Notice that the parabolas that open left or right are not functions of x because they fail the vertical line test—that is, it is possible to draw vertical lines that intersect these graphs at more than one point.

The equation $x = y^2$ interchanges the variables in the equation $y = x^2$. By interchanging x and y in the two forms of a parabola's equation,

$$y = a(x - h)^2 + k \quad \text{and} \quad y = ax^2 + bx + c,$$

we can obtain the forms of the equations of parabolas that open to the right or to the left.

Parabolas Opening to the Left or to the Right

The graphs of

$$x = a(y - k)^2 + h \quad \text{and} \quad x = ay^2 + by + c$$

are parabolas opening to the left or to the right.

1. If $a > 0$, the graph opens to the right. If $a < 0$, the graph opens to the left.
2. The vertex of $x = a(y - k)^2 + h$ is (h, k).
3. The y-coordinate of the vertex of $x = ay^2 + by + c$ is $y = -\dfrac{b}{2a}$.

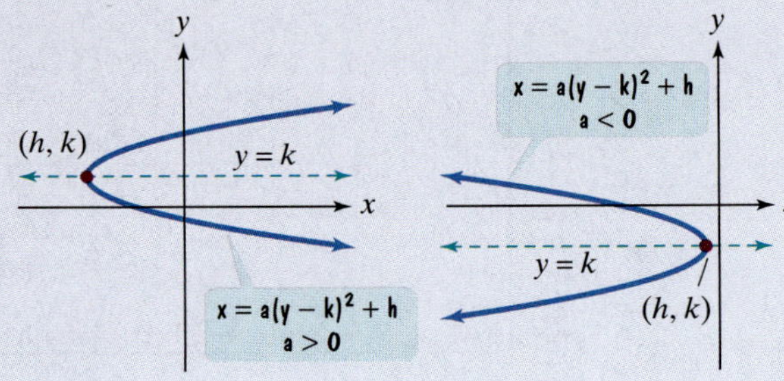

1 Graph horizontal parabolas.

Graphing Parabolas Opening to the Right or the Left

Here is a procedure for graphing horizontal parabolas that are not functions. Notice how this procedure is similar to the one that we used in Chapter 8 for graphing vertical parabolas.

Graphing Horizontal Parabolas

To graph $x = a(y - k)^2 + h$ or $x = ay^2 + by + c$:

1. Determine whether the parabola opens to the left or to the right. If $a > 0$, it opens to the right. If $a < 0$, it opens to the left.
2. Determine the vertex of the parabola. The vertex of $x = a(y - k)^2 + h$ is (h, k). The y-coordinate of the vertex of $x = ay^2 + by + c$ is $y = -\dfrac{b}{2a}$. Substitute to find the x-coordinate.
3. Find the x-intercept by replacing y with 0.
4. Find any y-intercepts by replacing x with 0. Solve the resulting quadratic equation for y.
5. Plot the intercepts and vertex. Connect these points with a smooth curve. If additional points are needed to obtain a more accurate graph, select values for y on each side of the axis of symmetry and compute values for x.

EXAMPLE 1 Graphing a Horizontal Parabola in the Form $x = a(y - k)^2 + h$

Graph: $x = -3(y - 1)^2 + 2$.

Solution We can graph this equation by following the steps in the preceding box. We begin by identifying values for a, k, and h.

$$x = a(y - k)^2 + h$$

$a = -3$ $k = 1$ $h = 2$

$$x = -3(y - 1)^2 + 2$$

Step 1 Determine how the parabola opens. Note that a, the coefficient of y^2, is -3. Thus, $a < 0$; this negative value tells us that the parabola opens to the left.

Step 2 Find the vertex. The vertex of the parabola is at (h, k). Because $k = 1$ and $h = 2$, the parabola has its vertex at $(2, 1)$.

Step 3 Find the x-intercept. Replace y with 0 in $x = -3(y - 1)^2 + 2$.

$$x = -3(0 - 1)^2 + 2 = -3(-1)^2 + 2 = -3(1) + 2 = -1$$

The x-intercept is -1. The parabola passes through $(-1, 0)$.

Step 4 Find the y-intercepts. Replace x with 0 in the given equation.

$$x = -3(y - 1)^2 + 2$$ This is the given equation.

$$0 = -3(y - 1)^2 + 2$$ Replace x with 0.

$$3(y - 1)^2 = 2$$ Solve for y. Add $3(y - 1)^2$ to both sides of the equation

$$(y - 1)^2 = \frac{2}{3}$$ Divide both sides by 3.

$$y - 1 = \sqrt{\frac{2}{3}} \quad \text{or} \quad y - 1 = -\sqrt{\frac{2}{3}}$$ Apply the square root property.

$$y = 1 + \sqrt{\frac{2}{3}} \qquad y = 1 - \sqrt{\frac{2}{3}}$$ Add 1 to both sides in each equation.

$$y \approx 1.8 \qquad y \approx 0.2$$ Use a calculator.

The y-intercepts are approximately 1.8 and 0.2. The parabola passes through approximately $(0, 1.8)$ and $(0, 0.2)$.

Step 5 Graph the parabola. With a vertex at $(2, 1)$, an x-intercept at -1, and y-intercepts at approximately 1.8 and 0.2, the graph of the parabola is shown in Figure 10.34. The axis of symmetry is the horizontal line whose equation is $y = 1$.

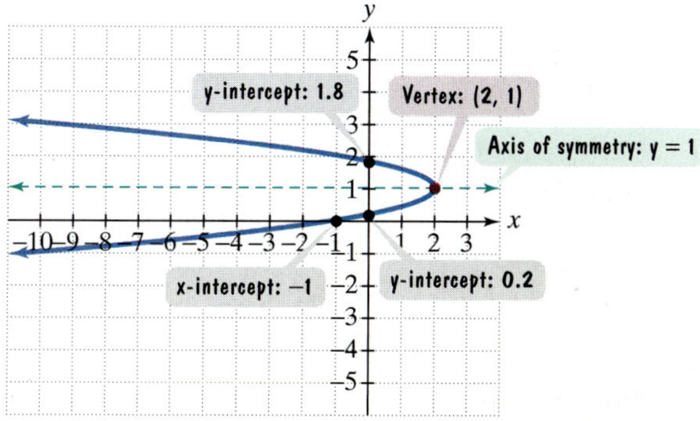

Figure 10.34 The graph of $x = -3(y - 1)^2 + 2$

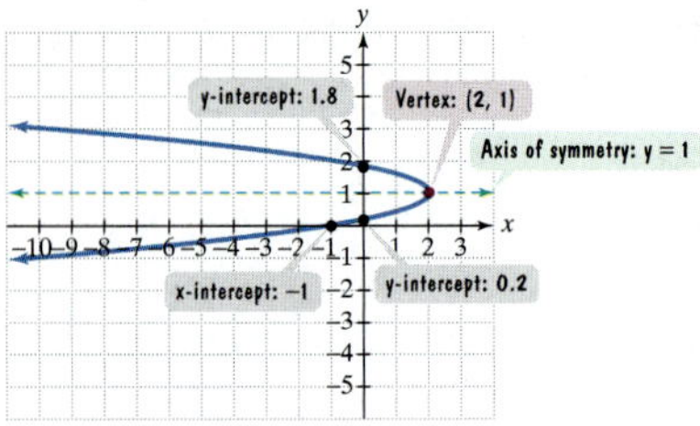

Figure 10.34, repeated

We can possibly improve our graph in Figure 10.34 by finding additional points on the parabola. Choose values of y on each side of the axis of symmetry, $y = 1$. We use $y = 2$ and $y = -1$. Then we compute values for x. The values in the table of coordinates show that $(-1, 2)$ and $(-10, -1)$ are points on the parabola. Locate each point in Figure 10.34.

x	y
−1	2
−10	−1

Choose values for y.

Compute values for x using

$x = -3(y-1)^2 + 2$:

$x = -3(2-1)^2 + 2 = -3 \cdot 1^2 + 2 = -1$

$x = -3(-1-1)^2 + 2 = -3(-2)^2 + 2 = -10$. ■

✔ CHECK POINT 1 Graph: $x = -(y-2)^2 + 1$.

EXAMPLE 2 Graphing a Horizontal Parabola in the Form $x = ay^2 + by + c$

Graph: $x = y^2 + 4y - 5$.

Solution

Step 1 Determine how the parabola opens. Note that a, the coefficient of y^2, is 1. Thus $a > 0$; this positive value tells us that the parabola opens to the right.

Step 2 Find the vertex. We know that the y-coordinate of the vertex is $y = -\frac{b}{2a}$. We identify a, b, and c in $x = ay^2 + by + c$.

$$x = y^2 + 4y - 5$$

$a = 1$ $b = 4$ $c = -5$

Substitute the values of a and b into the equation for the y-coordinate:

$$y = -\frac{b}{2a} = -\frac{4}{2 \cdot 1} = -2.$$

The y-coordinate of the vertex is −2. We substitute −2 for y in the parabola's equation, $x = y^2 + 4y - 5$, to find the x-coordinate:

$$x = (-2)^2 + 4(-2) - 5 = 4 - 8 - 5 = -9.$$

The vertex is at $(-9, -2)$.

Step 3 Find the x-intercept. Replace y with 0 in $x = y^2 + 4y - 5$.

$$x = 0^2 + 4 \cdot 0 - 5 = -5$$

The x-intercept is −5. The parabola passes through $(-5, 0)$.

Step 4 Find the y-intercepts. Replace x with 0 in the given equation.

$x = y^2 + 4y - 5$ — This is the given equation.

$0 = y^2 + 4y - 5$ — Replace x with 0.

$0 = (y - 1)(y + 5)$ — Use factoring to solve the quadratic equation.

$y - 1 = 0$ or $y + 5 = 0$ — Set each factor equal to 0.

$y = 1$ $y = -5$ — Solve.

The y-intercepts are 1 and −5. The parabola passes through $(0, 1)$ and $(0, -5)$.

Using Technology

To graph $x = y^2 + 4y - 5$ using a graphing utility, rewrite the equation as a quadratic equation in y.

$$y^2 + 4y + (-x - 5) = 0$$

$a = 1$ $b = 4$ $c = -x - 5$

Use the quadratic formula to solve for y and enter the resulting equations.

$$y_1 = \frac{-4 + \sqrt{16 - 4(-x - 5)}}{2}$$

$$y_2 = \frac{-4 - \sqrt{16 - 4(-x - 5)}}{2}$$

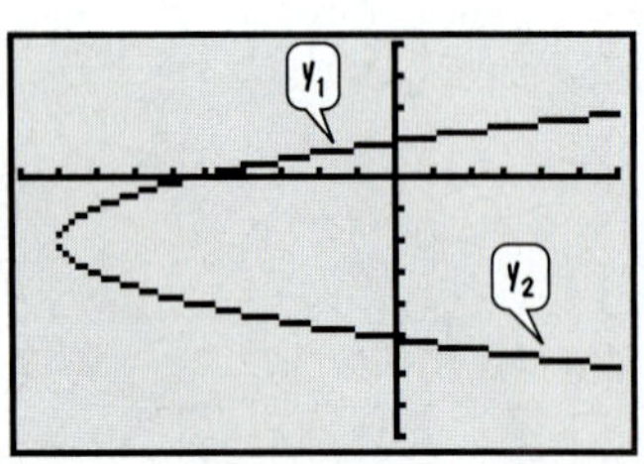

[−10, 6, 1] by [−8, 4, 1]

Step 5 Graph the parabola. With a vertex at $(-9, -2)$, an x-intercept at -5, and y-intercepts at 1 and -5, the graph of the parabola is shown in Figure 10.35. The axis of symmetry is the horizontal line whose equation is $y = -2$.

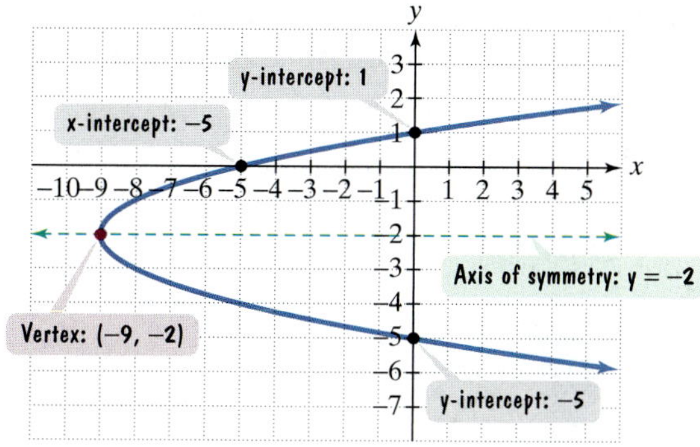

Figure 10.35 The graph of $x = y^2 + 4y - 5$

CHECK POINT 2 Graph: $x = y^2 + 8y + 7$.

Applications

Parabolas have many applications. Cables hung between structures to form suspension bridges form parabolas. Arches constructed of steel and concrete, whose main purpose is strength, are usually parabolic in shape.

Suspension bridge

Arch bridge

We have seen that comets in our solar system travel in orbits that are ellipses and hyperbolas. Some comets also follow parabolic paths. Only comets with elliptical orbits, such as Halley's Comet, return to our part of the galaxy.

A projectile, such as a baseball thrown directly upward, moves along a parabolic path, illustrated in Figure 10.36.

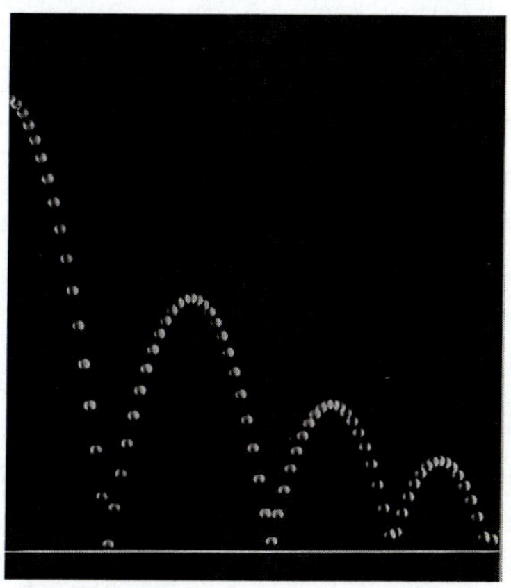

Figure 10.36 Multiflash photo showing the parabolic path of a ball thrown into the air ("Bouncing Ball" Cambridge, Massachusetts. © Berenice Abbot/Commerce Graphics, Ltd., Inc.)

If a parabola is rotated about its axis of symmetry, a parabolic surface is formed. Figure 10.37(a) shows how a parabolic surface can be used to reflect light. Light originates at the focus. Note how the light is reflected by the parabolic surface, so that the outgoing light is parallel to the axis of symmetry. The reflective properties of parabolic surfaces are used in the design of searchlights (see Figure 10.37(b)), automobile headlights, and parabolic microphones.

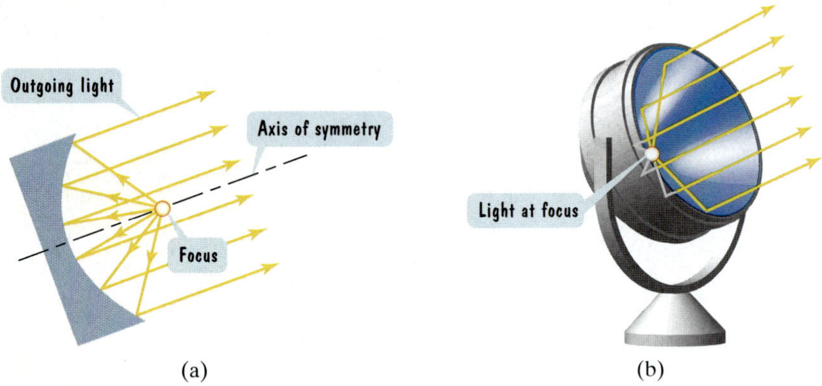

Figure 10.37 **(a)** Parabolic surface reflecting light **(b)** Light from the focus is reflected parallel to the axis of symmetry.

Figure 10.38(a) shows how a parabolic surface can be used to reflect *incoming* light. Note that light rays strike the surface and are reflected *to the focus*. This principle is used in the design of reflecting telescopes, radar, and television satellite dishes. Reflecting telescopes magnify the light from distant stars by reflecting the light from these bodies to the focus of a parabolic mirror (see Figure 10.38(b)).

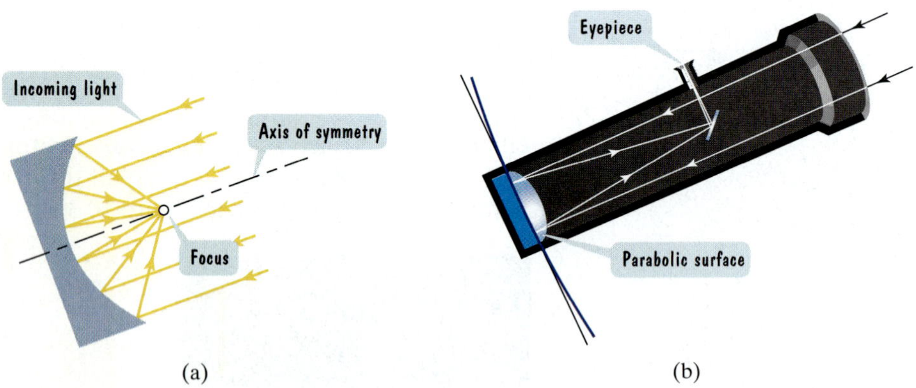

Figure 10.38 **(a)** Parabolic surface reflecting incoming light **(b)** Incoming light rays are reflected to the focus.

ENRICHMENT ESSAY

The Hubble Space Telescope

The Hubble Space Telescope

For decades, astronomers hoped to create an observatory above the atmosphere that would provide an unobscured view of the universe. This dream came true with the 1990 launching of the Hubble Space Telescope. The telescope initially had blurred vision due to problems with its parabolic mirror. The mirror had been ground two millionths of a meter smaller than design specifications. In 1993, astronauts from the Space Shuttle *Endeavor* equipped the telescope with optics to correct the blurred vision. "A small change for a mirror, a giant leap for astronomy," Christopher J. Burrows of the Space Telescope Science Institute said when clear images from the ends of the universe were presented to the public after the repair mission.

2 Identify conic sections by their equations.

Identifying Conic Sections by their Equations

What do the equations of the conic sections in this chapter have in common? They contain x^2-terms, y^2-terms, or both. Furthermore, they do not contain terms with exponents greater than 2. Here is how to identify conic sections by their equations:

Recognizing Conic Sections from Equations

Conic Section	How to Identity the Equation	Example
Circle	When x^2- and y^2-terms are on the same side, they have the same coefficient.	$x^2 + y^2 = 16$
Ellipse	When x^2- and y^2-terms are on the same side, they have different coefficients of the same sign.	$4x^2 + 16y^2 = 64$ or (dividing by 64) $\frac{x^2}{16} + \frac{y^2}{4} = 1$
Hyperbola	When x^2- and y^2-terms are on the same side, they have coefficients with opposite signs.	$9y^2 - 4x^2 = 36$ or (dividing by 36) $\frac{y^2}{4} - \frac{x^2}{9} = 1$
Parabola	Only one of the variables is squared.	$x = y^2 + 4y - 5$

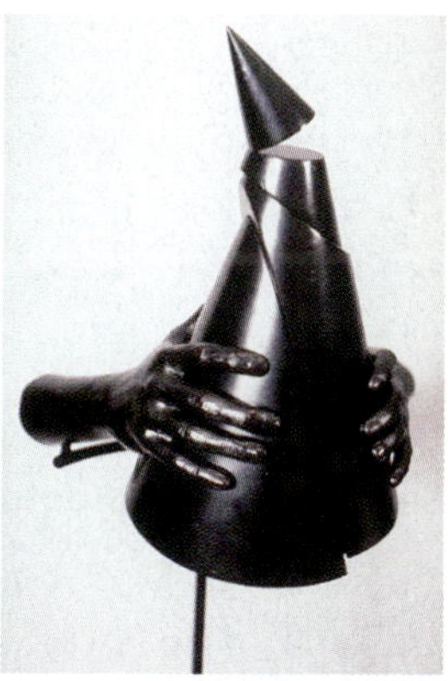

Richard E. Prince "The Cone of Apollonius" (detail), fiberglass, steel, paint, graphite. Equinox Gallery.

EXAMPLE 3 Recognizing Equations of Conic Sections

Indicate whether the graph of each equation is a circle, an ellipse, a hyperbola, or a parabola:

a. $4y^2 = 16 - 4x^2$ **b.** $x^2 = y^2 + 9$

c. $x + 7 - 6y = y^2$ **d.** $x^2 = 16 - 16y^2$.

Solution If both variables are squared, the graph of the equation is not a parabola. In this case, we get the x^2- and y^2-terms on the same side of the equation.

a. $4y^2 = 16 - 4x^2$

Both variables, x and y, are squared.

The graph is a circle, an ellipse, or a hyperbola. To see which one it is, we get the x^2- and y^2-terms on the same side. Add $4x^2$ to both sides. We obtain

$$4x^2 + 4y^2 = 16.$$

Because the coefficients of x^2 and y^2 are the same, namely 4, the equation's graph is a circle. This becomes more obvious if we divide both sides by 4.

$$x^2 + y^2 = 4$$

$$(x - 0)^2 + (y - 0)^2 = 2^2$$

The equation of the circle with center (h, k) and radius r is $(x - h)^2 + (y - k)^2 = r^2$.

The graph is a circle with center at the origin and radius 2.

b. $x^2 = y^2 + 9$

Both variables, x and y, are squared.

The graph cannot be a parabola. To see if it is a circle, an ellipse, or a hyperbola, we get the x^2- and y^2-terms on the same side. Subtract y^2 from both sides. We obtain

$$x^2 - y^2 = 9.$$

Because x^2- and y^2-terms have coefficients with opposite signs, the equation's graph is a hyperbola. This becomes more obvious if we divide both sides by 9 to obtain 1 on the right.

$$\frac{x^2}{9} - \frac{y^2}{9} = 1$$

$a^2 = 9$ $\quad$ $b^2 = 9$

The hyperbola's vertices are $(a, 0)$ and $(-a, 0)$, namely $(3, 0)$ and $(-3, 0)$.

c. $x + 7 - 6y = y^2$

Only one variable, y, is squared.

Because only one variable is squared, the graph of the equation is a parabola. We can express the equation of the horizontal parabola in the form $x = ay^2 + by + c$ by isolating x on the left. We obtain

$$x = y^2 + 6y - 7.$$

Because the coefficient of the y^2-term, 1, is positive, the graph of the horizontal parabola opens to the right.

d. $x^2 = 16 - 16y^2$

Both variables, x and y, are squared.

The graph cannot be a parabola. To see if it is a circle, an ellipse, or a hyperbola, we get the x^2- and y^2-terms on the same side. Add $16y^2$ to both sides. We obtain

$$x^2 + 16y^2 = 16.$$

Because the x^2- and y^2-terms have different coefficients of the same sign, namely 1 and 16, the equation's graph is an ellipse. This becomes more obvious if we divide both sides by 16 to obtain 1 on the right.

$$\frac{x^2}{16} + \frac{y^2}{1} = 1$$

$a^2 = 16$ $\quad$ $b^2 = 1$

An equation in the form $\frac{x^2}{a^2} + \frac{y^2}{b^2} = 1$ is an ellipse.

The vertices are $(4, 0)$ and $(-4, 0)$. ■

✔ CHECK POINT 3 Identify whether the graph of each equation is a circle, an ellipse, a hyperbola, or a parabola:

a. $x^2 = 4y^2 + 16$

b. $x^2 = 16 - 4y^2$

c. $4x^2 = 16 - 4y^2$

d. $x = -4y^2 + 16y.$

EXERCISE SET 10.4

Practice Exercises

In Exercises 1–6, the equation of a horizontal parabola is given. For each equation:
Determine how the parabola opens.
Find the parabola's vertex.
Use your results to identify the equation's graph. [The graphs are labeled (a) through (f).]

1. $x = (y - 2)^2 - 1$

2. $x = (y + 2)^2 - 1$

3. $x = (y + 2)^2 + 1$

4. $x = (y - 2)^2 + 1$

5. $x = -(y - 2)^2 + 1$

6. $x = -(y + 2)^2 + 1$

a.

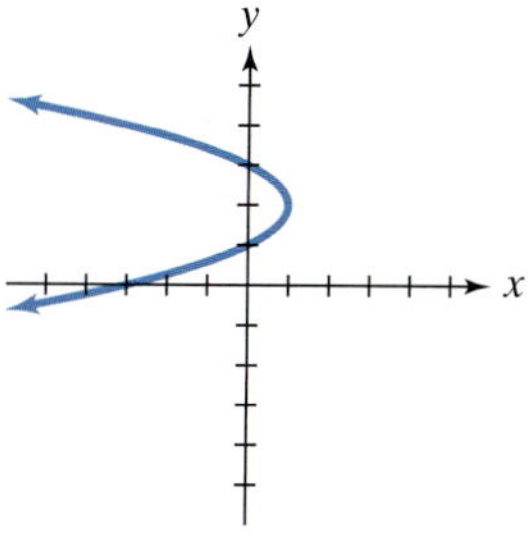

b.

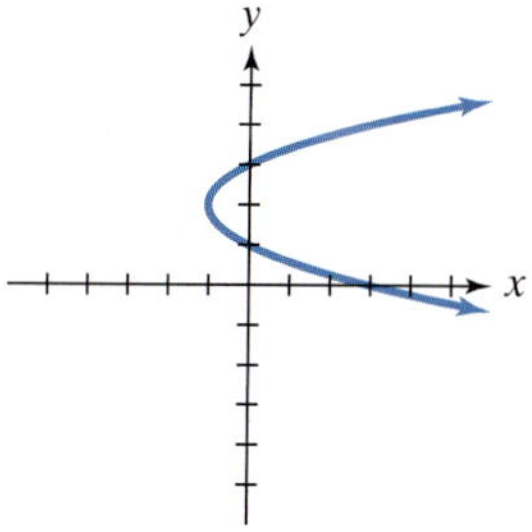

c.

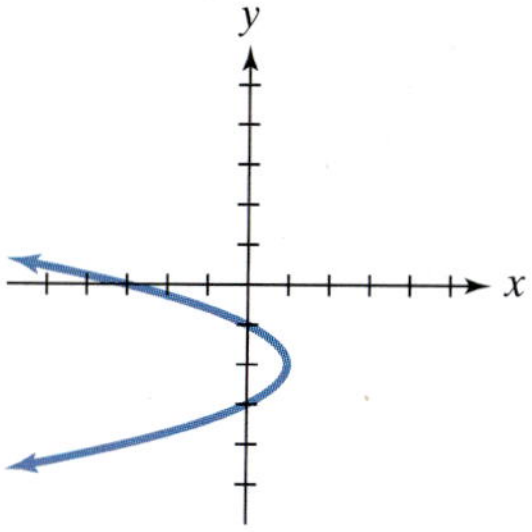

d.

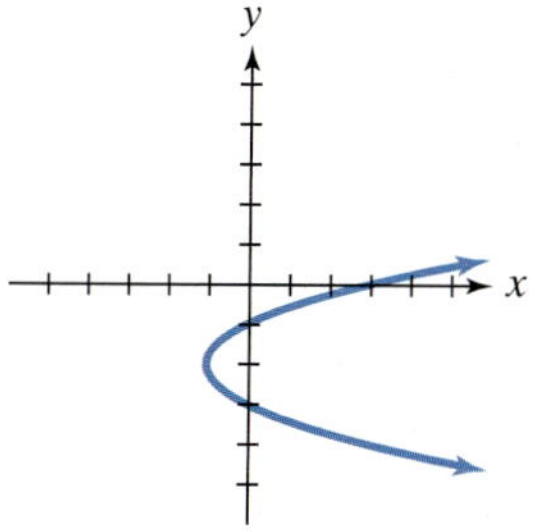

e.

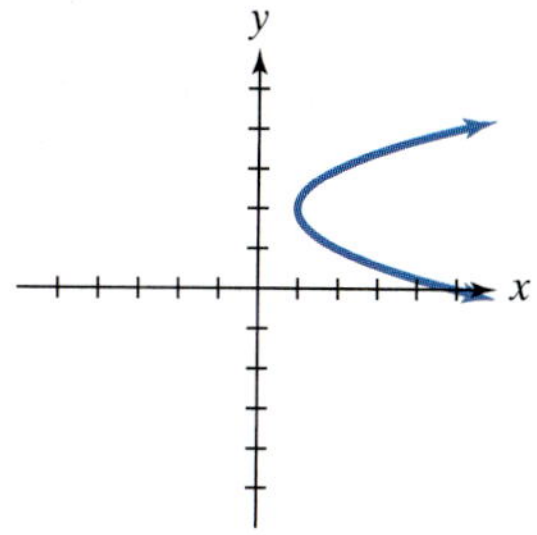

f.

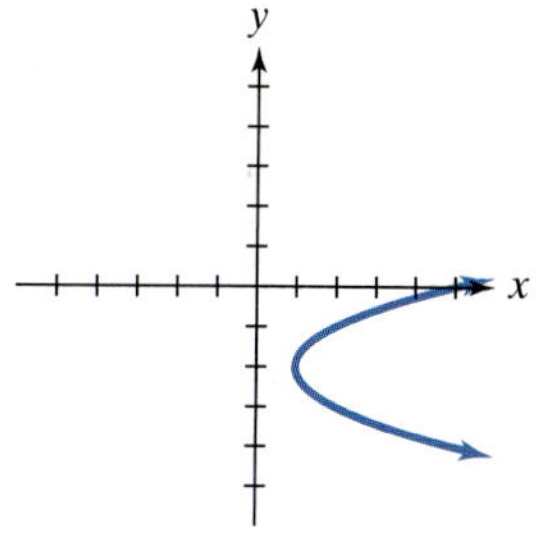

In Exercises 7–18, find the coordinates of the vertex for the horizontal parabola defined by the given equation.

7. $x = 2y^2$

8. $x = 4y^2$

9. $x = (y - 2)^2 + 3$

10. $x = (y - 3)^2 + 4$

11. $x = -4(y + 2)^2 - 1$

12. $x = -2(y + 5)^2 - 1$

13. $x = 2(y - 6)^2$

14. $x = 3(y - 7)^2$

15. $x = y^2 - 6y + 6$

16. $x = y^2 + 6y + 8$

17. $x = 3y^2 + 6y + 7$

18. $x = -2y^2 + 4y + 6$

In Exercises 19–42, use the vertex and intercepts to sketch the graph of each equation. Give the equation for the parabola's axis of symmetry. If needed, find additional points on the parabola by choosing values of y on each side of the axis of symmetry.

19. $x = (y - 2)^2 - 4$

20. $x = (y - 3)^2 - 4$

21. $x = (y - 3)^2 - 5$

22. $x = (y + 2)^2 - 3$

23. $x = -(y - 5)^2 + 4$

24. $x = -(y - 3)^2 + 4$

25. $x = (y - 4)^2 + 1$

26. $x = (y - 2)^2 + 3$

27. $x = -3(y - 5)^2 + 3$

28. $x = -2(y + 6)^2 + 2$

29. $x = -2(y + 3)^2 - 1$

30. $x = -3(y + 1)^2 - 2$

31. $x = \frac{1}{2}(y + 2)^2 + 1$

32. $x = \frac{1}{2}(y + 1)^2 + 2$

33. $x = y^2 + 2y - 3$

34. $x = y^2 - 6y + 8$

35. $x = y^2 - 4y + 5$

36. $x = -y^2 - 6y + 7$

37. $x = y^2 + 6y$

38. $x = y^2 + 4y$

39. $x = -2y^2 - 4y$

40. $x = -3y^2 - 6y$

41. $x = -2y^2 - 4y + 1$

42. $x = -2y^2 + 4y - 3$

In Exercises 43–54, the equation of a parabola is given. Determine:

a. *if the parabola is horizontal or vertical.*
b. *the way the parabola opens.*
c. *the vertex.*

43. $x = 2(y - 1)^2 + 2$

44. $x = 2(y - 3)^2 + 1$

45. $y = 2(x - 1)^2 + 2$

46. $y = 2(x - 3)^2 + 1$

47. $y = -(x + 3)^2 + 4$

48. $y = -(x + 1)^2 + 4$

49. $x = -(y + 3)^2 + 4$

50. $x = -(y + 1)^2 + 4$

51. $y = x^2 - 4x - 1$

52. $y = x^2 + 6x + 10$

53. $x = -y^2 + 4y + 1$

54. $x = -y^2 - 6y - 10$

In Exercises 55–64, indicate whether the graph of each equation is a circle, an ellipse, a hyperbola, or a parabola.

55. $x - 7 - 8y = y^2$

56. $x - 3 - 4y = 6y^2$

57. $4x^2 = 36 - y^2$

58. $4x^2 = 36 + y^2$

59. $x^2 = 36 + 4y^2$

60. $x^2 = 36 - 4y^2$

61. $3x^2 = 12 - 3y^2$

62. $3x^2 = 27 - 3y^2$

63. $3x^2 = 12 + 3y^2$

64. $3x^2 = 27 + 3y^2$

In Exercises 65–74, indicate whether the graph of each equation is a circle, an ellipse, a hyperbola, or a parabola. Then graph the conic section.

65. $x^2 - 4y^2 = 16$

66. $7x^2 - 7y^2 = 28$

67. $4x^2 + 4y^2 = 16$

68. $7x^2 + 7y^2 = 28$

69. $x^2 + 4y^2 = 16$

70. $4x^2 + y^2 = 16$

71. $x = (y - 1)^2 - 4$

72. $x = (y - 4)^2 - 1$

73. $(x - 2)^2 + (y + 1)^2 = 16$

74. $(x - 1)^2 + (y + 2)^2 = 16$

Application Exercises

75. The George Washington Bridge spans the Hudson River from New York to New Jersey. Its two towers are 3500 feet apart and rise 316 feet above the road. The cable between the towers has the shape of a parabola, and the cable just touches the sides of the road midway between the towers. The parabola is positioned in a rectangular coordinate system with its vertex at the origin. The point (1750, 316) lies on the parabola, as shown.

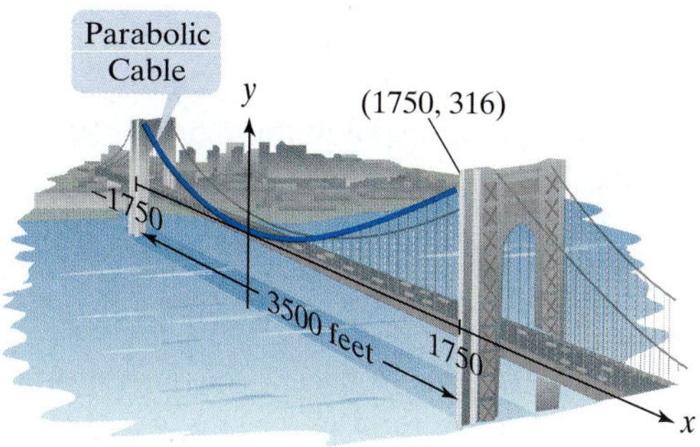

a. Write an equation in the form $y = ax^2$ for the parabolic cable. Do this by substituting 1750 for x and 316 for y and determining the value of a.

b. Use the equation in part (a) to find the height of the cable 1000 feet from a tower. Round to the nearest foot.

76. The towers of the Golden Gate Bridge connecting San Francisco to Marin County are 1280 meters apart and rise 160 meters above the road. The cable between the towers has the shape of a parabola, and the cable just touches the sides of the road midway between the towers. The parabola is positioned in a rectangular coordinate system with its vertex at the origin. The point (640, 140) lies on the parabola, as shown.

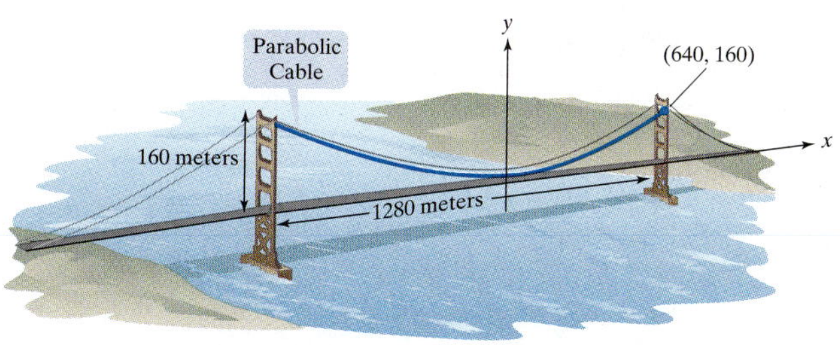

a. Write an equation in the form $y = ax^2$ for the parabolic cable. Do this by substituting 640 for x and 140 for y and determining the value of a.

b. Use the equation in part (a) to find the height of the cable 200 meters from a tower. Round to the nearest meter.

77. A satellite dish is in the shape of a parabolic surface. Signals coming from a satellite strike the surface of the dish and are reflected to the focus, where the receiver is located. The satellite dish shown has a diameter of 12 feet and a depth of 2 feet. The parabola is positioned in a rectangular coordinate system with its vertex at the origin.

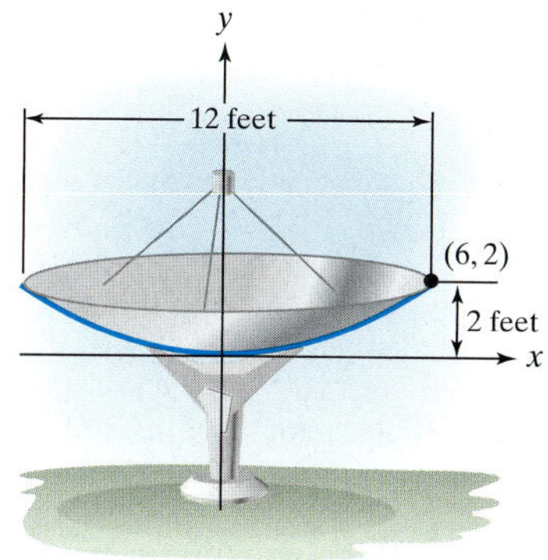

a. Write an equation in the form $y = ax^2$ for the parabola used to shape the dish.

b. The received should be placed at the focus $(0, p)$. The value of p is given by the equation $a = \frac{1}{4p}$. How far from the base of the dish should the received be placed?

78. An engineer is designing a flashlight using a parabolic reflecting mirror and a light source, as shown. The casting has a diameter of 4 inches and a depth of 2 inches. The parabola is positioned in a rectangular coordinate system with its vertex at the origin.

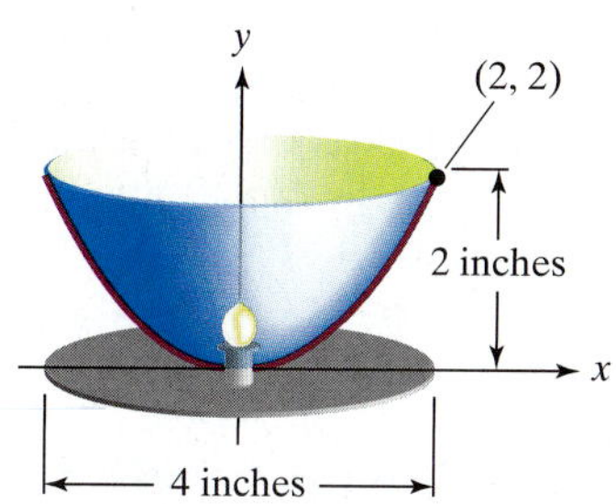

a. Write an equation in the form $y = ax^2$ for the parabola used to shape the mirror.

b. The light source should be placed at the focus $(0, p)$. The value of p is given by the equation $a = \frac{1}{4p}$. Where should the light source be placed relative to the mirror's vertex?

Writing in Mathematics

79. What is a parabola?

80. If you are given an equation of a parabola, explain how to determine if the parabola opens to the right, left, upward, or downward.

81. Explain how to use $x = 2(y + 3)^2 - 5$ to find the parabolas's vertex.

82. Explain how to use $x = y^2 + 8y + 9$ to find the parabolas's vertex.

83. Describe one similarity and one difference between the graphs of $x = 4y^2$ and $x = 4(y - 1)^2 + 2$.

84. How can you distinguish parabolas from other conic sections by looking at their equations?

85. How can you distinguish ellipses from hyperbolas by looking at their equations?

86. How can you distinguish ellipses from circles by looking at their equations?

Technology Exercises

Use a graphing utility to graph the parabolas in Exercises 87–88. Write the given equation as a quadratic equation in y and use the quadratic formula to solve for y. Enter each of the equations to produce the complete graph.

87. $y^2 + 2y - 6x + 13 = 0$

88. $y^2 + 10y - x + 25 = 0$

89. Use a graphing utility to graph any three of the parabolas that you graphed by hand in Exercises 19–42. First solve the given equation for y, possibly using the square root method. Enter each of the two resulting equations to produce the complete graph.

Critical Thinking Exercises

90. Which one of the following is true?

a. The parabola whose equation is $x = 2y - y^2 + 5$ opens to the right.

b. If the parabola whose equation is $x = ay^2 + by + c$ has its vertex at $(3, 2)$ and $a > 0$, then it has no y-intercepts.

c. Some parabolas that open to the right have equations that define y as a function of x.

d. The graph of $x = a(y - k) + h$ is a parabola with vertex at (h, k).

91. Look at the satellite dish shown in Exercise 77. Why must the receiver for a shallow dish be farther from the base of the dish than for a deeper dish of the same diameter?

92. The parabolic arch shown in the figure is 50 feet above the water at the center and 200 feet wide at the base. Will a boat that is 30 feet tall clear the arch 30 feet from the center?

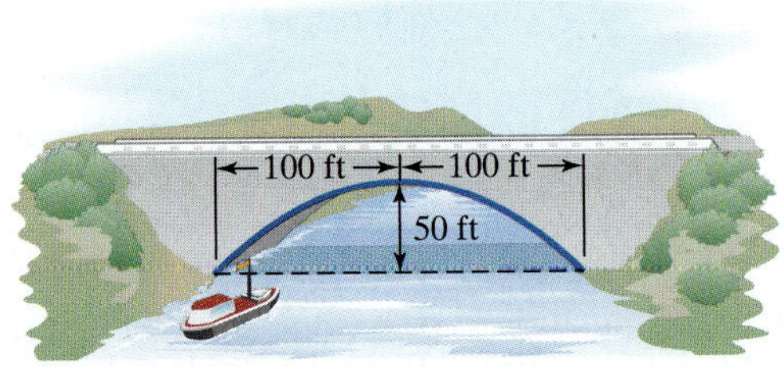

Review Exercises

93. Graph: $f(x) = 2^{1-x}$.
(Section 9.1, Example 4)

94. If $f(x) = \frac{1}{3}x - 5$, find $f^{-1}(x)$.
(Section 9.2, Example 4)

95. Solve the system:

$$4x - 3y = 12$$
$$3x - 4y = 2.$$

(Section 3.1, Example 6)

▸ SECTION 10.5 Systems of Nonlinear Equations in Two Variables

Objectives

1. Recognize systems of nonlinear equations in two variables.
2. Solve nonlinear systems by substitution.
3. Solve nonlinear systems by addition.
4. Solve problems using systems of nonlinear equations.

SSM | PH Tutor Center | CD-ROM | Video

Scientists debate the probability that a "doomsday rock" will collide with Earth. It has been estimated that an asteroid, a tiny planet that revolves around the sun, crashes into Earth about once every 250,000 years, and that such a collision would have disastrous results. In 1908, a small fragment struck Siberia, leveling thousands of acres of trees. One theory about the extinction of dinosaurs 65 million years ago involves Earth's collision with a large asteroid and the resulting drastic changes in Earth's climate.

Understanding the path of Earth and the path of a comet is essential to detecting threatening space debris. Orbits about the sun are not described by linear equations in the form $Ax + By = C$. The ability to solve systems that do not contain linear equations provides NASA scientists watching for troublesome asteroids with a possible collision point with Earth's orbit.

1 Recognize systems of nonlinear equations in two variables.

Systems of Nonlinear Equations and their Solutions

A **system of** two **nonlinear equations** in two variables contains at least one equation that cannot be expressed in the form $Ax + By = C$. Here are two examples:

$$x^2 = 2y + 10$$
$$3x - y = 9$$

Not in the form $Ax + By = C$. The term x^2 is not linear.

$$y = x^2 + 3$$
$$x^2 + y^2 = 9.$$

Neither equation is in the form $Ax + By = C$. The terms x^2 and y^2 are not linear.

A **solution** to a nonlinear system in two variables is an ordered pair of real numbers that satisfies all equations in the system. The **solution set** to the system is the set of all such ordered pairs. As with linear systems in two variables, the solution to a nonlinear system (if there is one) corresponds to the intersection point(s) of the graphs of the equations in the system. Unlike linear systems, the graphs can be circles, ellipses, hyperbolas, parabolas, or anything other than two lines. We will solve nonlinear systems using the substitution method and the addition method.

2 Solve nonlinear systems by substitution.

Eliminating a Variable Using the Substitution Method

The substitution method involves converting a nonlinear system to one equation in one variable by an appropriate substitution. The steps in the solution process are exactly the same as those used to solve a linear system by substitution. However, when you obtain an equation in one variable, this equation will not be linear. In our first example, this equation is quadratic.

EXAMPLE 1 Solving a Nonlinear System by the Substitution Method

Solve by the substitution method:

$$x^2 = 2y + 10 \quad \text{The graph is a parabola.}$$
$$3x - y = 9. \quad \text{The graph is a line.}$$

Solution

Step 1 Solve one of the equations for one variable in terms of the other. We begin by isolating one of the variables raised to the first power in either of the equations. By solving for y in the second equation, which has a coefficient of -1, we can avoid fractions.

$$3x - y = 9 \quad \text{This is the second equation in the given system.}$$
$$3x = y + 9 \quad \text{Add y to both sides.}$$
$$3x - 9 = y \quad \text{Subtract 9 from both sides.}$$

Step 2 Substitute the expression from step 1 into the other equation. We substitute $3x - 9$ for y in the first equation.

$$y = \boxed{3x - 9} \qquad x^2 = 2\boxed{y} + 10$$

This gives us an equation in one variable, namely

$$x^2 = 2(3x - 9) + 10.$$

The variable y has been eliminated.

Step 3 Solve the resulting equation containing one variable.

$$x^2 = 2(3x - 9) + 10 \quad \text{This is the equation containing one variable.}$$
$$x^2 = 6x - 18 + 10 \quad \text{Use the distributive property.}$$
$$x^2 = 6x - 8 \quad \text{Combine numerical terms on the right.}$$
$$x^2 - 6x + 8 = 0 \quad \text{Move all terms to one side and set the quadratic equation equal to 0.}$$
$$(x - 4)(x - 2) = 0 \quad \text{Factor.}$$
$$x - 4 = 0 \quad \text{or} \quad x - 2 = 0 \quad \text{Set each factor equal to 0.}$$
$$x = 4 \quad \text{or} \quad x = 2 \quad \text{Solve for x.}$$

Step 4 Back-substitute the obtained values into the equation from step 1. Now that we have the x-coordinates of the solutions, we back-substitute 4 for x and 2 for x in the equation $y = 3x - 9$.

If x is 4, $\quad y = 3(4) - 9 = 3,\quad$ so $(4, 3)$ is a solution.

If x is 2, $\quad y = 3(2) - 9 = -3,\quad$ so $(2, -3)$ is a solution.

Step 5 Check the proposed solutions in both of the system's given equations. We begin by checking (4, 3). Replace x with 4 and y with 3.

$x^2 = 2y + 10$	$3x - y = 9$	These are the given equations.
$4^2 \stackrel{?}{=} 2(3) + 10$	$3(4) - 3 \stackrel{?}{=} 9$	Let $x = 4$ and $y = 3$.
$16 \stackrel{?}{=} 6 + 10$	$12 - 3 \stackrel{?}{=} 9$	Simplify.
$16 = 16$, true	$9 = 9$, true	True statements result.

The ordered pair (4, 3) satisfies both equations. Thus, (4, 3) is a solution to the system.

Now let's check (2, −3). Replace x with 2 and y with −3 in both given equations.

$x^2 = 2y + 10$	$3x - y = 9$	These are the given equations.
$2^2 \stackrel{?}{=} 2(-3) + 10$	$3(2) - (-3) \stackrel{?}{=} 9$	Let $x = 2$ and $y = -3$.
$4 \stackrel{?}{=} -6 + 10$	$6 + 3 \stackrel{?}{=} 9$	Simplify.
$4 = 4$, true	$9 = 9$, true	True statements result.

The ordered pair (2, −3) also satisfies both equations and is a solution to the system. The solutions are (4, 3) and (2, −3), and the solution set is $\{(4, 3), (2, -3)\}$.

Figure 10.39 shows the graphs of the equations in the system and the solutions as intersection points. ■

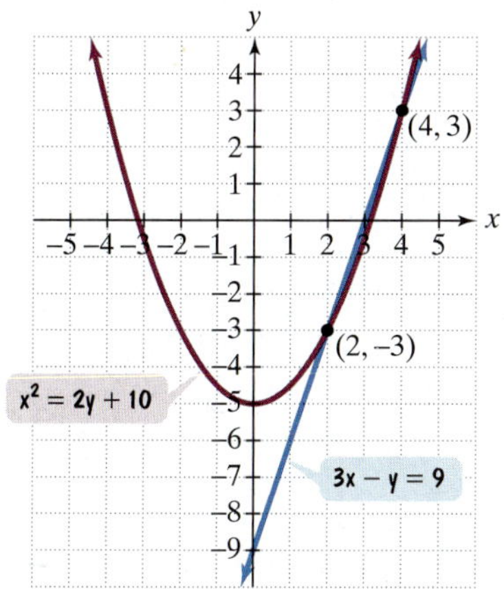

Figure 10.39 Points of intersection illustrate the nonlinear system's solutions.

✔ **CHECK POINT 1** Solve by the substitution method:

$$x^2 = y - 1$$
$$4x - y = -1.$$

EXAMPLE 2 Solving a Nonlinear System by the Substitution Method

Solve by the substitution method:

$$x - y = 3 \quad \text{The graph is a line.}$$
$$(x - 2)^2 + (y + 3)^2 = 4. \quad \text{The graph is a circle.}$$

Solution Graphically, we are finding the intersection of a line and a circle with center (2, −3) and radius 2.

Step 1 Solve one of the equations for one variable in terms of the other. We will solve for x in the linear equation – that is, the first equation. (We could also solve for y).

$$x - y = 3 \quad \text{This is the first equation in the given system.}$$
$$x = y + 3 \quad \text{Add } y \text{ to both sides.}$$

Step 2 Substitute the expression from step 1 into the other equation. We substitute $y + 3$ for x in the second equation.

$$x = \boxed{y + 3} \qquad (\boxed{x} - 2)^2 + (y + 3)^2 = 4$$

This gives an equation in one variable, namely

$$(y + 3 - 2)^2 + (y + 3)^2 = 4.$$

The variable x has been eliminated.

Step 3 Solve the resulting equation containing one variable.

$(y + 3 - 2)^2 + (y + 3)^2 = 4$	This is the equation containing one variable.
$(y + 1)^2 + (y + 3)^2 = 4$	Combine numerical terms in the first parentheses.
$y^2 + 2y + 1 + y^2 + 6y + 9 = 4$	Use the formula $(A + B)^2 = A^2 + 2AB + B^2$ to square $y + 1$ and $y + 3$.
$2y^2 + 8y + 10 = 4$	Combine like terms on the left.
$2y^2 + 8y + 6 = 0$	Subtract 4 from both sides and set the quadratic equation equal to 0.
$y^2 + 4y + 3 = 0$	Simplify by dividing both sides by 2.
$(y + 3)(y + 1) = 0$	Factor.
$y + 3 = 0 \quad \text{or} \quad y + 1 = 0$	Set each factor equal to 0.
$y = -3 \quad \text{or} \quad y = -1$	Solve for y.

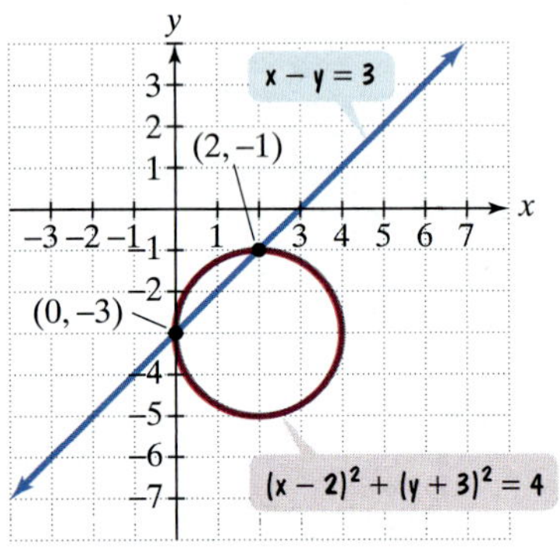

Figure 10.40 Points of intersection illustrate the nonlinear system's solutions.

Step 4 Back-substitute the obtained values into the equation from step 1. Now that we have the y-coordinates of the solutions, we back-substitute -3 for y and -1 for y in the equation $x = y + 3$.

If $y = -3$: $\quad x = -3 + 3 = 0,\quad$ so $(0, -3)$ is a solution.

If $y = -1$: $\quad x = -1 + 3 = 2,\quad$ so $(2, -1)$ is a solution.

Step 5 Check the proposed solution in both of the system's given equations. Take a moment to show that each ordered pair satisfies both equations. The solution are $(0, -3)$ and $(2, -1)$, and the solution set of the given system is $\{(0, -3), (2, -1)\}$.

Figure 10.40 shows the graphs of the equations in the system and the solutions as intersection points. ■

✔ **CHECK POINT 2** Solve by the substitution method:

$$x + 2y = 0$$
$$(x - 1)^2 + (y - 1)^2 = 5.$$

3 Solve nonlinear systems by addition.

Eliminating a Variable Using the Addition Method

In solving linear systems with two variables, we learned that the addition method works well when each equation is in the form $Ax + By = C$. For nonlinear systems, the addition method can be used when each equation is in the form $Ax^2 + By^2 = C$. If necessary, we will multiply either equation or both equations by appropriate numbers so that the coefficients of x^2 or y^2 will have a sum of 0. We then add equations. The sum will be an equation in one variable.

EXAMPLE 3 Solving a Nonlinear System by the Addition Method

Solve the system:

$4x^2 + y^2 = 13$	Equation 1 (The graph is an ellipse.)
$x^2 + y^2 = 10.$	Equation 2 (The graph is a circle.)

The given system:

$$4x^2 + y^2 = 13$$
$$x^2 + y^2 = 10,$$

repeated from page 744

Solution We can use the same steps that we did when we solved linear systems by the addition method.

Step 1 Write both equations in the form $Ax^2 + By^2 = C$. Both equations are already in this form, so we can skip this step.

Step 2 If necessary, multiply either equation or both equations by appropriate numbers so that the sum of the x^2-coefficients or the sum of the y^2-coefficients is 0. We can eliminate y^2 by multiplying Equation 2 by -1.

$$4x^2 + y^2 = 13 \xrightarrow{\text{No change}} 4x^2 + y^2 = 13$$
$$x^2 + y^2 = 10 \xrightarrow{\text{Multiply by } -1.} -x^2 - y^2 = -10$$

Steps 3 and 4 Add equations and solve for the remaining variable.

$$\begin{aligned} 4x^2 + y^2 &= 13 \\ -x^2 - y^2 &= -10 \\ \hline 3x^2 &= 3 && \text{Add equations.} \\ x^2 &= 1 && \text{Divide both sides by 3.} \\ x &= \pm 1 && \text{Use the square root method: If } x^2 = c \text{, then } x = \pm\sqrt{c}. \end{aligned}$$

Step 5 Back-substitute and find the values for the other variables. We must back-substitute each value of x into either one of the original equations. Let's use $x^2 + y^2 = 10$, Equation 2. If $x = 1$,

$$\begin{aligned} 1^2 + y^2 &= 10 && \text{Replace } x \text{ with 1 in Equation 2.} \\ y^2 &= 9 && \text{Subtract 1 from both sides.} \\ y &= \pm 3 && \text{Apply the square root method.} \end{aligned}$$

$(1, 3)$ and $(1, -3)$ are solutions. If $x = -1$,

$$\begin{aligned} (-1)^2 + y^2 &= 10 && \text{Replace } x \text{ with } -1 \text{ in Equation 2.} \\ y^2 &= 9 && \text{The steps are the same as before.} \\ y &= \pm 3 \end{aligned}$$

$(-1, 3)$ and $(-1, -3)$ are solutions.

Step 6 Check. Take a moment to show that each of the four ordered pairs satisfies Equation 1 and Equation 2. The solutions are $(1, 3)$, $(1, -3)$, $(-1, 3)$, and $(-1, -3)$, and the solution set of the given system is $\{(1, 3), (1, -3), (-1, 3), (-1, -3)\}$.

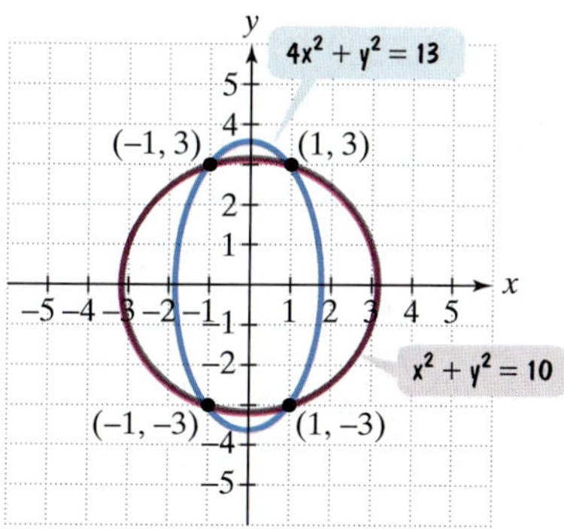

Figure 10.41 A system with four solutions

Figure 10.41 shows the graphs of the equations in the system, an ellipse and a circle, and the solutions as intersection points. ■

CHECK POINT 3 Solve the system:

$$3x^2 + 2y^2 = 35$$
$$4x^2 + 3y^2 = 48.$$

In solving nonlinear systems, we will include only ordered pairs with real numbers in the solution set. We have seen that each of these ordered pairs corresponds to a point of intersection of the system's graphs.

Study Tip

When solving nonlinear systems, extra solutions may be introduced that do not satisfy both equations in the system. Therefore, you should get into the habit of checking all proposed pairs in each of the system's two equations.

EXAMPLE 4 Solving a Nonlinear System by the Addition Method

Solve the system:

$$y = x^2 + 3 \quad \text{Equation 1 (The graph is a parabola.)}$$
$$x^2 + y^2 = 9. \quad \text{Equation 2 (The graph is a circle.)}$$

Solution We could use substitution because Equation 1 has y expressed in terms of x, but this would result in a fourth-degree equation. However, we can rewrite Equation 1 by subtracting x^2 from both sides and adding the equations to eliminate the x^2-terms.

$$-x^2 + y = 3 \quad \text{Subtract } x^2 \text{ from both sides of Equation 1.}$$
$$x^2 + y^2 = 9 \quad \text{This is Equation 2.}$$
$$y + y^2 = 12 \quad \text{Add the equations.}$$

We now solve this quadratic equation.

$$y + y^2 = 12 \quad \text{This is the equation containing one variable.}$$
$$y^2 + y - 12 = 0 \quad \text{Subtract 12 from both sides and set the quadratic equation equal to 0.}$$
$$(y + 4)(y - 3) = 0 \quad \text{Factor.}$$
$$y + 4 = 0 \quad \text{or} \quad y - 3 = 0 \quad \text{Set each factor equal to 0.}$$
$$y = -4 \quad \text{or} \quad y = 3 \quad \text{Solve for } y.$$

To complete the solution, we must back-substitute each value of y into either one of the original equations. We will use $y = x^2 + 3$, Equation 1. First, we substitute -4 for y.

$$-4 = x^2 + 3$$
$$-7 = x^2 \quad \text{Subtract 3 from both sides.}$$

Because the square of a real number cannot be negative, the equation $x^2 = -7$ does not have real-number solutions. We will not include the imaginary solutions, $x = \pm\sqrt{-7}$, or $\sqrt{7}i$ and $-\sqrt{7}i$, in the ordered pairs that make up the solution set. Thus, we move on to our other value for y, 3, and substitute this value into Equation 1.

$$y = x^2 + 3 \quad \text{This is Equation 1.}$$
$$3 = x^2 + 3 \quad \text{Back-substitute 3 for } y.$$
$$0 = x^2 \quad \text{Subtract 3 from both sides.}$$
$$0 = x \quad \text{Solve for } x.$$

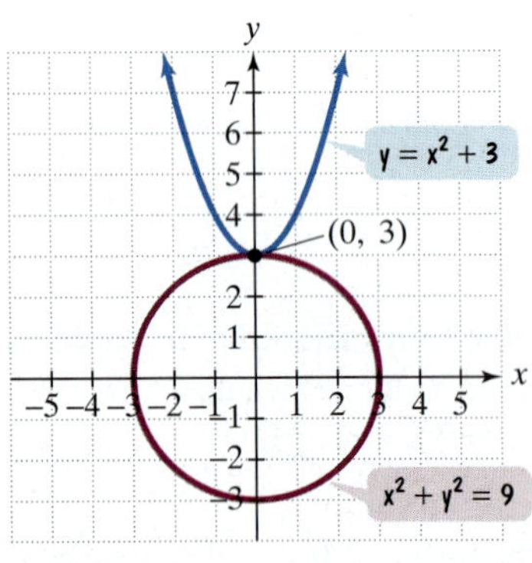

Figure 10.42 A system with one real solution

We showed that if $y = 3$, then $x = 0$. Thus, $(0, 3)$ is the solution with a real ordered pair. Take a moment to show that $(0, 3)$ satisfies Equation 1 and Equation 2. The solution set of the given system is $\{(0, 3)\}$. Figure 10.42 shows the system's graphs and the solution as an intersection point. ■

CHECK POINT 4 Solve the system:

$$y = x^2 + 5$$
$$x^2 + y^2 = 25.$$

4 Solve problems using systems of nonlinear equations.

Applications

Many geometric problems can be modeled and solved by the use of nonlinear systems of equations. We will use our step-by-step strategy for solving problems using mathematical models that are created from verbal models.

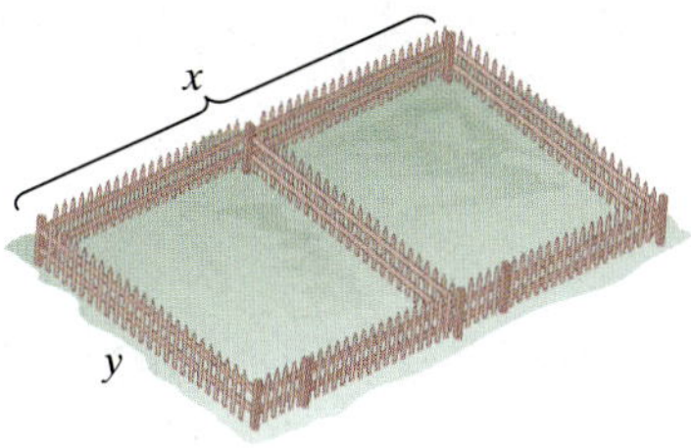

Figure 10.43 Building an enclosure

EXAMPLE 5 An Application of a Nonlinear System

You have 36 yards of fencing to build the enclosure in Figure 10.43. Some of this fencing is to be used to build an internal divider. If you'd like to enclose 54 square yards, what are the dimensions of the enclosure?

Solution

Step 1 Use variables to represent unknown quantities. Let x = the enclosure's length and y = the enclosure's width. These variables are shown in Figure 10.43.

Step 2 Write a system of equations describing the problem's conditions. The first condition is that you have 36 yards of fencing.

Fencing along both lengths	plus	fencing along both widths	plus	fencing for the internal divider	equals	36 yards.
$2x$	$+$	$2y$	$+$	y	$=$	36

Adding like terms, we can express the equation that models the verbal conditions for the fencing as $2x + 3y = 36$.

The second condition is that you'd like to enclose 54 square yards. The rectangle's area, the product of its length and its width, must be 54 square yards.

Length	times	width	is	54 square yards.
x	$\cdot$	y	$=$	54

Step 3 Solve the system and answer the problem's question. We must solve the system

$$2x + 3y = 36 \quad \text{Equation 1}$$
$$xy = 54. \quad \text{Equation 2}$$

We will use substitution. Because Equation 1 has no coefficients of 1 or -1, we will solve Equation 2 for y. Dividing both sides of $xy = 54$ by x, we obtain

$$y = \frac{54}{x}.$$

Now we substitute $\frac{54}{x}$ for y in Equation 1 and solve for x.

$$2x + 3y = 36 \quad \text{This is Equation 1.}$$

$$2x + 3 \cdot \frac{54}{x} = 36 \quad \text{Substitute } \frac{54}{x} \text{ for } y.$$

$$2x + \frac{162}{x} = 36 \quad \text{Multiply.}$$

$$x\left(2x + \frac{162}{x}\right) = 36 \cdot x \quad \text{Clear fractions by multiplying both sides by } x.$$

$$2x^2 + 162 = 36x \quad \text{Use the distributive property on the left side.}$$

$$2x^2 - 36x + 162 = 0 \quad \text{Subtract } 36x \text{ from both sides and set the quadratic equation equal to 0.}$$

$$x^2 - 18x + 81 = 0 \quad \text{Simplify by dividing both sides by 2.}$$

$$(x - 9)^2 = 0 \quad \text{Factor using } A^2 - 2AB + B^2 = (A - B)^2.$$

$$x - 9 = 0 \quad \text{Set the repeated factor equal to zero.}$$

$$x = 9 \quad \text{Solve for } x.$$

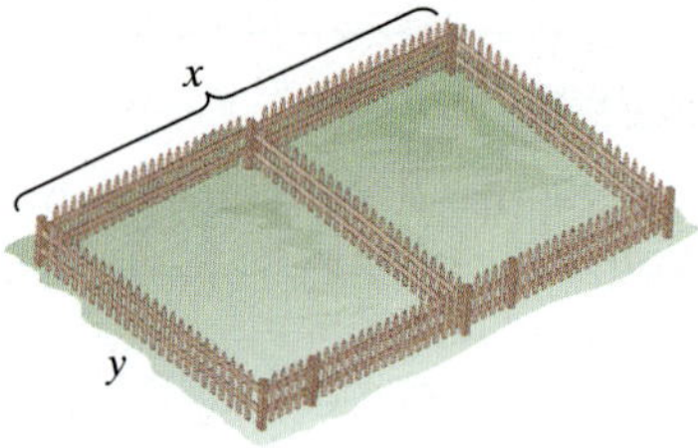

Figure 10.43, repeated

We back-substitute this value of x into $y = \frac{54}{x}$.

$$\text{If } x = 9, \quad y = \frac{54}{9} = 6.$$

This means that the dimensions of the enclosure in Figure 10.43 are 9 yards by 6 yards.

Step 4 Check the proposed solution in the original wording of the problem. With a length of 9 yards and a width of 6 yards, take a moment to check that this results in 36 yards of fencing and an area of 54 square yards. ■

CHECK POINT 5 Find the length and width of a rectangle whose perimeter is 20 feet and whose area is 21 square feet.

EXERCISE SET 10.5

Practice Exercises

In Exercises 1–18, solve each system by the substitution method.

1. $x + y = 2$
$y = x^2 - 4$

2. $x - y = -1$
$y = x^2 + 1$

3. $x + y = 2$
$y = x^2 - 4x + 4$

4. $2x + y = -5$
$y = x^2 + 6x + 7$

5. $y = x^2 - 4x - 10$
$y = -x^2 - 2x + 14$

6. $y = x^2 + 4x + 5$
$y = x^2 + 2x - 1$

7. $x^2 + y^2 = 25$
$x - y = 1$

8. $x^2 + y^2 = 5$
$3x - y = 5$

9. $xy = 6$
$2x - y = 1$

10. $xy = -12$
$x - 2y + 14 = 0$

11. $y^2 = x^2 - 9$
$2y = x - 3$

12. $x^2 + y = 4$
$2x + y = 1$

13. $xy = 3$
$x^2 + y^2 = 10$

14. $xy = 4$
$x^2 + y^2 = 8$

15. $x + y = 1$
$x^2 + xy - y^2 = -5$

16. $x + y = -3$
$x^2 + 2y^2 = 12y + 18$

17. $x + y = 1$
$(x - 1)^2 + (y + 2)^2 = 10$

18. $2x + y = 4$
$(x + 1)^2 + (y - 2)^2 = 4$

In Exercises 19–28, solve each system by the addition method.

19. $x^2 + y^2 = 13$
$x^2 - y^2 = 5$

20. $4x^2 - y^2 = 4$
$4x^2 + y^2 = 4$

21. $x^2 - 4y^2 = -7$
$3x^2 + y^2 = 31$

22. $3x^2 - 2y^2 = -5$
$2x^2 - y^2 = -2$

23. $3x^2 + 4y^2 - 16 = 0$
$2x^2 - 3y^2 - 5 = 0$

24. $16x^2 - 4y^2 - 72 = 0$
$x^2 - y^2 - 3 = 0$

25. $x^2 + y^2 = 25$
$(x - 8)^2 + y^2 = 41$

26. $x^2 + y^2 = 4$
$x^2 + (y - 3)^2 = 9$

27. $y^2 - x = 4$
$x^2 + y^2 = 4$

28. $x^2 - 2y = 8$
$x^2 + y^2 = 16$

In Exercises 29–42, solve each system by the method of your choice.

29. $3x^2 + 4y^2 = 16$
$2x^2 - 3y^2 = 5$

30. $x + y^2 = 4$
$x^2 + y^2 = 16$

31. $2x^2 + y^2 = 18$
$xy = 4$

32. $x^2 + 4y^2 = 20$
$xy = 4$

33. $x^2 + 4y^2 = 20$
$x + 2y = 6$

34. $3x^2 - 2y^2 = 1$
$4x - y = 3$

35. $x^3 + y = 0$
$x^2 - y = 0$

36. $x^3 + y = 0$
$2x^2 - y = 0$

37. $x^2 + (y - 2)^2 = 4$
$x^2 - 2y = 0$

38. $x^2 - y^2 - 4x + 6y - 4 = 0$
$x^2 + y^2 - 4x - 6y + 12 = 0$

39. $y = (x + 3)^2$
$x + 2y = -2$

40. $(x - 1)^2 + (y + 1)^2 = 5$
$2x - y = 3$

41. $x^2 + y^2 + 3y = 22$
$2x + y = -1$

42. $x - 3y = -5$
$x^2 + y^2 - 25 = 0$

In Exercises 43–46, let x represent one number and let y represent the other number. Use the given conditions to write a system of nonlinear equations. Solve the system and find the numbers.

43. The sum of two numbers is 10 and their product is 24. Find the numbers.

44. The sum of two numbers is 20 and their product is 96. Find the numbers.

45. The difference between the squares of two numbers is 3. Twice the square of the first number increased by the square of the second number is 9. Find the numbers.

46. The difference between the squares of two numbers is 5. Twice the square of the second number subtracted from three times the square of the first number is 19. Find the numbers.

Application Exercises

47. A planet's orbit follows an elliptical path described by $16x^2 + 4y^2 = 64$. A comet follows the parabolic path $y = x^2 - 4$. Where might the comet intersect the orbiting planet?

48. A system for tracking ships indicates that a ship lies on a hyperbolic path described by $2y^2 - x^2 = 1$. The process is repeated and the ship is found to lie on a hyperbolic path described by $2x^2 - y^2 = 1$. If it is known that the ship is located in the first quadrant of the coordinate system, determine its exact location.

49. Find the length and width of a rectangle whose perimeter is 36 feet and whose area is 77 square feet.

50. Find the length and width of a rectangle whose perimeter is 40 feet and whose area is 96 square feet.

Use the formula for the area of a rectangle and the Pythagorean Theorem to solve Exercises 51–52.

51. A small television has a picture with a diagonal measure of 10 inches and a viewing area of 48 square inches. Find the length and width of the screen.

52. The area of a rug is 108 square feet, and the length of its diagonal is 15 feet. Find the length and width of the rug.

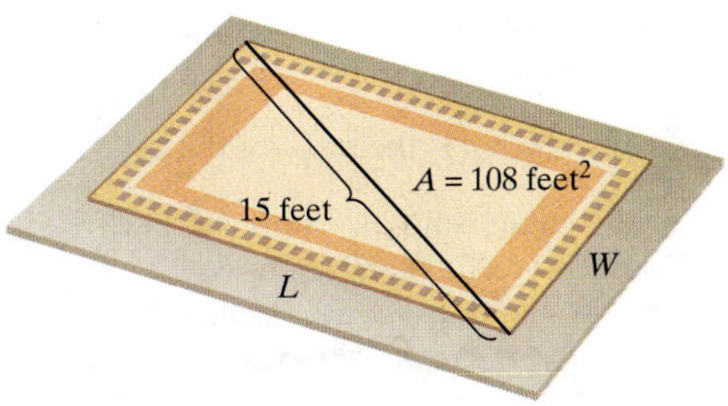

53. The figure shows a square floor plan with a smaller square area that will accommodate a combination fountain and pool. The floor with the fountain-pool area removed has an area of 21 square meters and a perimeter of 24 meters. Find the dimensions of the floor and the dimensions of the square that will accommodate the fountain and pool.

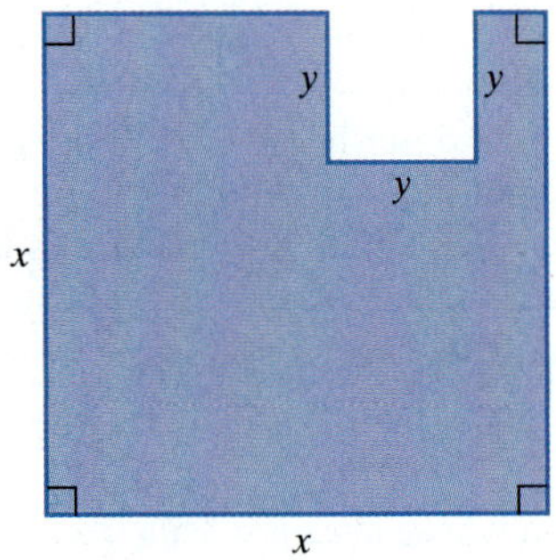

54. The area of the rectangular piece of cardboard shown on the left is 216 square inches. The cardboard is used to make an open box by cutting a 2-inch square from each corner and turning up the sides. If the box is to have a volume of 224 cubic inches, find the length and width of the cardboard that must be used.

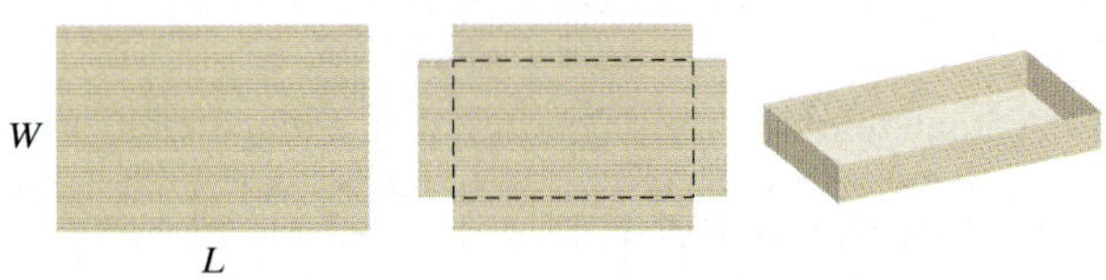

Writing in Mathematics

55. What is a system of nonlinear equations? Provide an example with your description.

56. Explain how to solve a nonlinear system using the substitution method. Use $x^2 + y^2 = 9$ and $2x - y = 3$ to illustrate your explanation.

57. Explain how to solve a nonlinear system using the addition method. Use $x^2 - y^2 = 5$ and $3x^2 - 2y^2 = 19$ to illustrate your explanation.

58. The daily demand and supply models for a carrot cake supplied by a bakery to a convenience store are given by the demand model $N = 40 - 3p$ and the supply model $N = \frac{p^2}{10}$, in which p is the price of the cake and N is the number of cakes sold or supplied each day to the convenience store. Explain how to determine the price at which supply and demand are equal. Then describe how to find how many carrot cakes can be supplied and sold each day at this price.

Technology Exercises

59. Verify your solutions to any five exercises from 1 through 42 by using a graphing utility to graph the two equations in the system in the same viewing rectangle. Then use the trace or intersection feature to verify the solutions.

60. Write a system of equations, one equation whose graph is a line and the other whose graph is a parabola, that has no ordered pairs that are real numbers in its solution set. Graph the equations using a graphing utility and verify that you are correct.

Critical Thinking Exercises

61. Which one of the following is true?

a. A system of two equations in two variables whose graphs are a circle and a line can have four real ordered-pair solutions.

b. A system of two equations in two variables whose graphs are a parabola and a circle can have four real ordered-pair solutions.

c. A system of two equations in two variables whose graphs are two circles must have at least two real ordered-pair solutions.

d. A system of two equations in two variables whose graphs are a parabola and a circle cannot have only one real ordered-pair solution.

62. Find a and b in this figure.

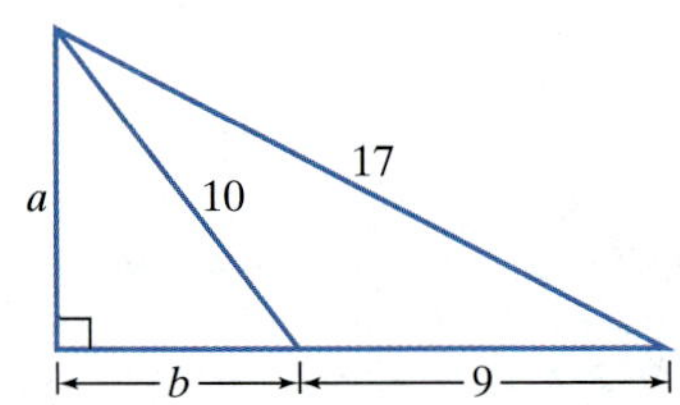

Solve the systems in Exercises 63–64.

63. $\log_y x = 3$
$\log_y(4x) = 5$

64. $\log x^2 = y + 3$
$\log x = y - 1$

Review Exercises

65. Graph: $3x - 2y \leq 6$.
(Section 4.4, Example 1)

66. Find the slope of the line passing through $(-2, -3)$ and $(1, 5)$.
(Section 2.3, Example 2)

67. Multiply: $(3x - 2)(2x^2 - 4x + 3)$.
(Section 5.2, Example 3)

CHAPTER 10 GROUP PROJECTS

MODELING PLANETARY MOTION

1. Polish astronomer Nicolaus Copernicus (1473–1543) was correct in stating that planets in our solar system revolve around the sun and not the Earth. However, he incorrectly believed that celestial orbits move in perfect circles, calling his system "the ballet of the planets."

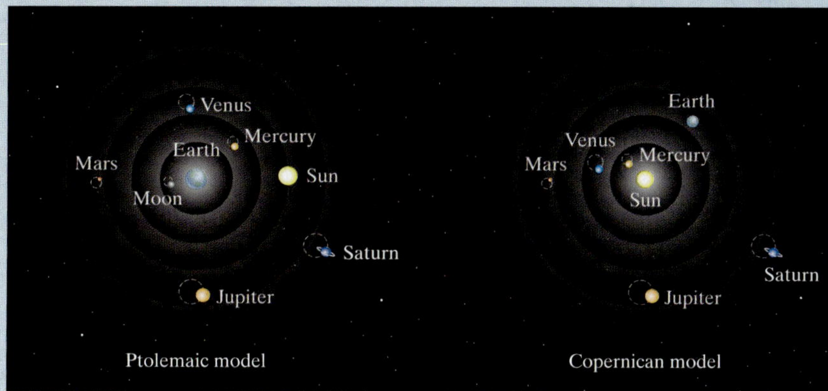

German scientist and mathematician Johannes Kepler (1571–1630) discovered that planets move in elliptical orbits with the sun at one focus. In this exercise, group members will write equations for two of these orbits and use a graphing utility to see what they look like. Use the following information:

Earth's orbit:	Length of major axis: 186 million miles
	Length of minor axis: 185.8 million miles
Mars's orbit:	Length of major axis: 283.5 million miles
	Length of minor axis: 278.5 million miles.

a. Group members should write equations in the form $\frac{x^2}{a^2} + \frac{y^2}{b^2} = 1$ for the elliptical orbits of Earth and Mars.

b. Use a graphing utility to graph the two ellipses in the same $[-300, 300, 50]$ by $[-200, 200, 50]$ viewing rectangle. Based on these graphs, explain why early astronomers incorrectly used the Copernican model to describe planetary motion.

c. Describing planetary orbits, Kepler wrote, "The heavenly motions are nothing but a continuous song for several voices, to be perceived by the intellect, not by the ear." Group members should discuss what Kepler meant by this statement.

2. Consult the research department of your library or the Internet to find an example of architecture that incorporates one or more conic sections in its design. Share this example with other group members. Explain precisely how conic sections are used. Do conic sections enhance the appeal of the architecture? In what ways?

3. Many public and private organizations and schools provide educational materials and information for the blind and visually impaired. Using your library, resources on the Worldwide Web, or local organizations, investigate how your group or college could make a contribution to enhance the study of conic sections for the blind and visually impaired. Group members should discuss how to create graphs in tactile, or touchable, form that show blind students the visual structure of each of the conic sections.

CHAPTER SUMMARY, REVIEW, AND TEST

SUMMARY

DEFINITIONS AND CONCEPTS	EXAMPLES
Section 10.1 Distance and Midpoint Formulas; Circles	
The Distance Formula The distance, d, between the points (x_1, y_1) and (x_2, y_2) is given by $d = \sqrt{(x_2 - x_1)^2 + (y_2 - y_1)^2}$.	Find the distance between $(-3, -5)$ and $(6, -2)$. $d = \sqrt{[6 - (-3)]^2 + [-2 - (-5)]^2}$ $= \sqrt{9^2 + 3^2} = \sqrt{81 + 9} = \sqrt{90} = 3\sqrt{10} \approx 9.49$
The Midpoint Formula The midpoint of the line segment whose endpoints are (x_1, y_1) and (x_2, y_2) is the point with coordinates $\left(\frac{x_1 + x_2}{2}, \frac{y_1 + y_2}{2}\right)$.	Find the midpoint of the line segment whose endpoints are $(-3, 6)$ and $(4, 1)$. midpoint $= \left(\frac{-3 + 4}{2}, \frac{1 + 6}{2}\right) = \left(\frac{1}{2}, \frac{7}{2}\right)$
A circle is the set of all points in a plane that are equidistant from a fixed point, the center. The distance from the center to any point on the circle is the radius. ***Standard Form of the Equation of a Circle*** The graph of $(x - h)^2 + (y - k)^2 = r^2$ is a circle with center (h, k) and radius r. ***General Form of the Equation of a Circle*** $x^2 + y^2 + Dx + Ey + F = 0$ Convert from the general form to the standard form by completing the square on x and y.	Find the center and radius and graph: $(x - 3)^2 + (y + 4)^2 = 16$. $(x - 3)^2 + (y - (-4))^2 = 4^2$ ($h = 3$, $k = -4$, $r = 4$) The center, (h, k), is $(3, -4)$ and the radius, r, is 4. 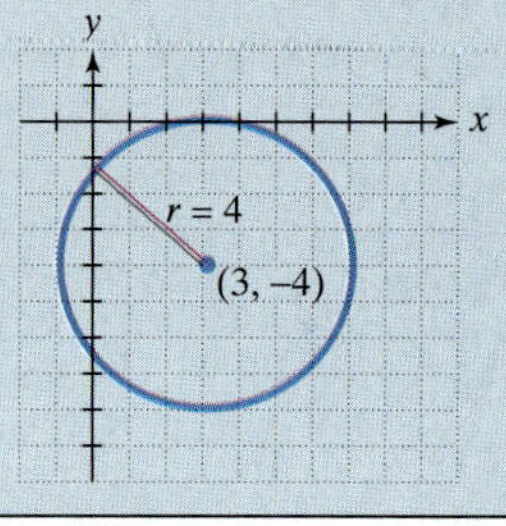
Section 10.2 The Ellipse	
An ellipse is the set of all points in a plane the sum of whose distances from two fixed points, the foci, is constant. The midpoint of the segment connecting the foci is the center of the ellipse. ***Standard Forms of the Equations of an Ellipse Centered at the Origin*** $(a^2 > b^2)$ Horizontal with vertices $(a, 0)$ and $(-a, 0)$: $\frac{x^2}{a^2} + \frac{y^2}{b^2} = 1$ (Endpoints of major axis a units left and right of center; minor axis b units up and down from center.) Vertical with vertices $(0, a)$ and $(0, -a)$: $\frac{x^2}{b^2} + \frac{y^2}{a^2} = 1$ (Endpoints of major axis a units up and down from center, minor axis b units left and right of center.) The equations $\frac{(x-h)^2}{a^2} + \frac{(y-k)^2}{b^2}$ and $\frac{(x-h)^2}{b^2} + \frac{(y-k)^2}{a^2} = 1$ $(a^2 > b^2)$ represent ellipses centered at (h, k).	Graph: $\frac{(x + 2)^2}{9} + \frac{(y + 4)^2}{25} = 1$. $\frac{(x - (-2))^2}{9} + \frac{(y - (-4))^2}{25} = 1$ ($b^2 = 9$, $a^2 = 25$) Center, (h, k), is $(-2, -4)$. With $a^2 = 25$, vertices are 5 units above and below the center. With $b^2 = 9$, endpoints of minor axis are 3 units to the left and right of the center. 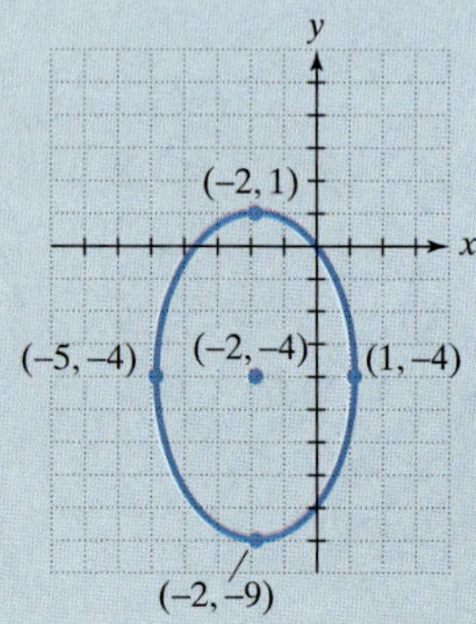

DEFINITIONS AND CONCEPTS	EXAMPLES

Section 10.3 The Hyperbola

A hyperbola is the set of all points in a plane the difference of whose distances from two fixed points, the foci, is constant.

Standard Forms of the Equations of a Hyperbola Centered at the Origin

$$\frac{x^2}{a^2} - \frac{y^2}{b^2} = 1 \qquad \frac{y^2}{a^2} - \frac{x^2}{b^2} = 1$$

Vertices are $(a, 0)$ and $(-a, 0)$.

Vertices are $(0, a)$ and $(0, -a)$.

As x and y get larger, the two branches of a hyperbola approach a pair of intersecting lines, called asymptotes. Draw the rectangle centered at the origin with sides parallel to the axes, crossing one axis at $\pm a$ and the other at $\pm b$. Draw the diagonals of this rectangle and extend them to obtain the asymptotes. Draw the two branches of the hyperbola by starting at each vertex and approaching the asymptotes.

Graph: $4x^2 - 9y^2 = 36$.

$$\frac{4x^2}{36} - \frac{9y^2}{36} = \frac{36}{36}$$

$$\frac{x^2}{9} - \frac{y^2}{4} = 1$$

$a^2 = 9$ $b^2 = 4$

Vertices are $(3, 0)$ and $(-3, 0)$. Draw a rectangle using -3 and 3 on the x-axis and -2 and 2 on the y-axis. Its extended diagonals are the asymptotes.

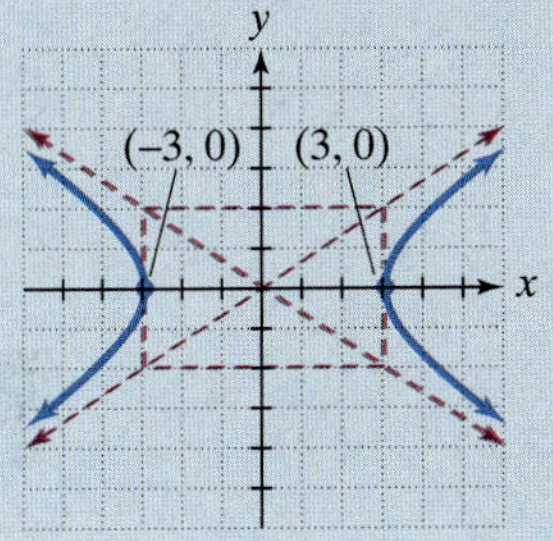

Section 10.4 The Parabola; Identifying Conic Sections

A parabola is the set of all points in a plane that are equidistant from a fixed line, the directrix, and a fixed point, the focus, that is not on the line. The line passing through the focus and perpendicular to the directrix is the axis of symmetry. The point of intersection of the parabola with its axis of symmetry is the vertex.

Equations of Horizontal Parabolas

$x = a(y - k)^2 + h$ — Vertex is (h, k).

$x = ay^2 + by + c$ — y-coordinate of vertex is $y = -\frac{b}{2a}$.

If $a > 0$, parabola opens to the right. If $a < 0$, parabola opens to the left.

Equations of Vertical Parabolas

$y = a(x - h)^2 + k$ — Vertex is (h, k).

$y = ax^2 + bx + c$ — x-coordinate of vertex is $x = -\frac{b}{2a}$.

If $a > 0$, the parabola opens upward. If $a < 0$, the parabola opens downward.

Find the vertex and graph:

$$x = -(y + 3)^2 + 4.$$

$$x = -1(y - (-3))^2 + 4$$

$a = -1$ $k = -3$ $h = 4$

Parabola opens to the left. Vertex, (h, k), is $(4, -3)$.

x-intercept: Let $y = 0$.

$$x = -(0 + 3)^2 + 4 = -9 + 4 = -5$$

y-intercepts: Let $x = 0$.

$$0 = -(y + 3)^2 + 4$$
$$(y + 3)^2 = 4$$
$$y + 3 = \pm\sqrt{4}$$
$$y = -1 \quad \text{or} \quad y = -5$$

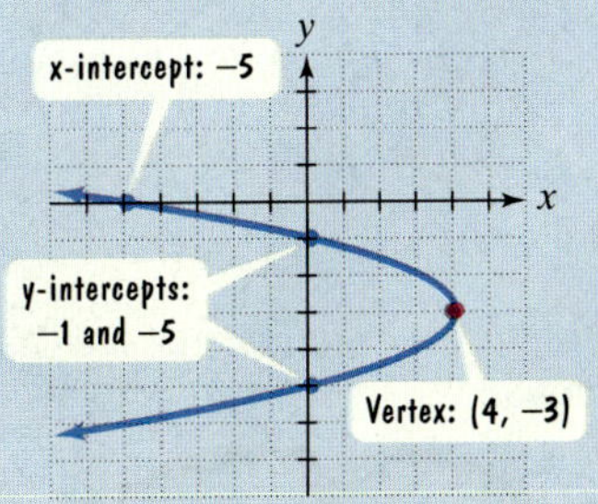

DEFINITIONS AND CONCEPTS	EXAMPLES
Section 10.4 The Parabola; Identifying Conic Sections (continued)	
Recognizing Conic Sections from Equations Conic sections result from the intersection of a cone and a plane. Their graphs contain x^2-terms, y^2-terms, or both. For parabolas, only one variable is squared. For circles, ellipses, and hyperbolas, both variables are squared. Put x^2- and y^2-terms on the same side of the equation to identify the graph: Circle: x^2 and y^2 have the same coefficient of the same sign. Ellipse: x^2 and y^2 have different coefficients of the same sign. Hyperbola: x^2 and y^2 have coefficients with opposite signs.	• $9x^2 + 4y^2 = 36$ Different coefficients of the same sign;ellipse • $9y^2 = 25x^2 + 225$ $9y^2 - 25x^2 = 225$ Coefficients with opposite signs; hyperbola • $x = -y^2 + 4y$ Only one variable, y, is squared; parabola • $\frac{x^2}{9} + \frac{y^2}{9} = 1$ $x^2 + y^2 = 9$ Same coefficient; circle
Section 10.5 Systems of Nonlinear Equations in Two Variables	
A system of two nonlinear equations in two variables, x and y, contains at least one equation that cannot be expressed in the form $Ax + By = C$. Systems can be solved by the substitution method or the addition method. Each solution corresponds to a point of intersection of the system's graphs.	Solve: $x^2 + y^2 = 25$ $x - 3y = -5$. Using substitution: $x = 3y - 5$ $(3y - 5)^2 + y^2 = 25$ $9y^2 - 30y + 25 + y^2 = 25$ $10y^2 - 30y = 0$ $10y(y - 3) = 0$ $y = 0$ or $y = 3$ If $y = 0$: $x = 3y - 5 = 3 \cdot 0 - 5 = -5$ If $y = 3$: $x = 3y - 5 = 3 \cdot 3 - 5 = 4$ The solution are $(-5, 0)$ and $(4, 3)$, and the solution set is $\{(-5, 0), (4, 3)\}$.

Review Exercises

10.1 *In Exercises 1–2, find the distance between each pair of points. If necessary, round answers to two decimal places.*

1. $(-2, -3)$ and $(3, 9)$
2. $(-4, 3)$ and $(-2, 5)$

In Exercises 3–4, find the midpoint of each line segment with the given endpoints.

3. $(2, 6)$ and $(-12, 4)$
4. $(4, -6)$ and $(-15, 2)$

In Exercises 5–6, write the standard form of the equation of the circle with the given center and radius.

5. Center $(0, 0)$, $r = 3$
6. Center $(-2, 4)$, $r = 6$

In Exercises 7–10, give the center and radius of each circle and graph its equation.

7. $x^2 + y^2 = 1$
8. $(x + 2)^2 + (y - 3)^2 = 9$

9. $x^2 + y^2 - 4x + 2y - 4 = 0$

10. $x^2 + y^2 - 4y = 0$

10.2 *In Exercises 11–16, graph each ellipse.*

11. $\frac{x^2}{36} + \frac{y^2}{25} = 1$

12. $\frac{x^2}{25} + \frac{y^2}{16} = 1$

13. $4x^2 + y^2 = 16$

14. $4x^2 + 9y^2 = 36$

15. $\frac{(x - 1)^2}{16} + \frac{(y + 2)^2}{9} = 1$

16. $\frac{(x + 1)^2}{9} + \frac{(y - 2)^2}{16} = 1$

17. A semielliptic archway has a height of 15 feet at the center and a width of 50 feet, as shown in the figure. The 50-foot width consists of a two-lane road. Can a truck that is 12 feet high and 14 feet wide drive under the archway without going into the other lane?

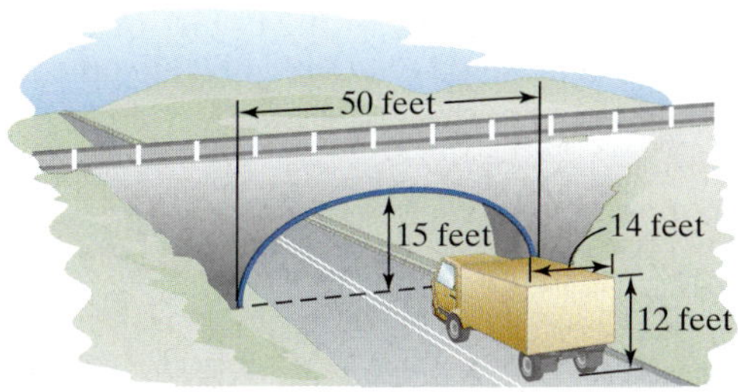

10.3 *In Exercises 18–21, use vertices and asymptotes to graph each hyperbola.*

18. $\frac{x^2}{16} - y^2 = 1$

19. $\frac{y^2}{16} - x^2 = 1$

20. $9x^2 - 16y^2 = 144$

21. $4y^2 - x^2 = 16$

10.4 *In Exercises 22–25, use the vertex and intercepts to sketch the graph of each equation. Give the equation for the horizontal parabola's axis of symmetry. If needed, find additional points on the parabola by choosing values of y on each side of the axis of symmetry.*

22. $x = (y - 3)^2 - 4$

23. $x = -2(y + 3)^2 + 2$

24. $x = y^2 - 8y + 12$

25. $x = -y^2 - 4y + 6$

In Exercises 26–32, indicate whether the graph of each equation is a circle, an ellipse, a hyperbola, or a parabola.

26. $x + 8y = y^2 + 10$

27. $16x^2 = 32 - y^2$

28. $x^2 = 25 + 25y^2$

29. $x^2 = 4 - y^2$

30. $36y^2 = 576 + 16x^2$

31. $\frac{(x + 3)^2}{9} + \frac{(y - 4)^2}{25} = 1$

32. $y = x^2 + 6x + 9$

In Exercises 33–41, indicate whether the graph of each equation is a circle, an ellipse, a hyperbola, or a parabola. Then graph the conic section.

33. $5x^2 + 5y^2 = 180$

34. $4x^2 + 9y^2 = 36$

35. $4x^2 - 9y^2 = 36$

36. $\frac{x^2}{25} + \frac{y^2}{1} = 1$

37. $x + 3 = -y^2 + 2y$

38. $y - 3 = x^2 - 2x$

39. $\frac{(x + 2)^2}{16} + \frac{(y - 5)^2}{4} = 1$

40. $(x - 3)^2 + (y + 2)^2 = 4$

41. $x^2 + y^2 + 6x - 2y + 6 = 0$

42. An engineer is designing headlight units for automobiles. The unit has a parabolic surface with a diameter of 12 inches and a depth of 3 inches. The situation is illustrated in the figure, where a coordinate system has been superimposed.

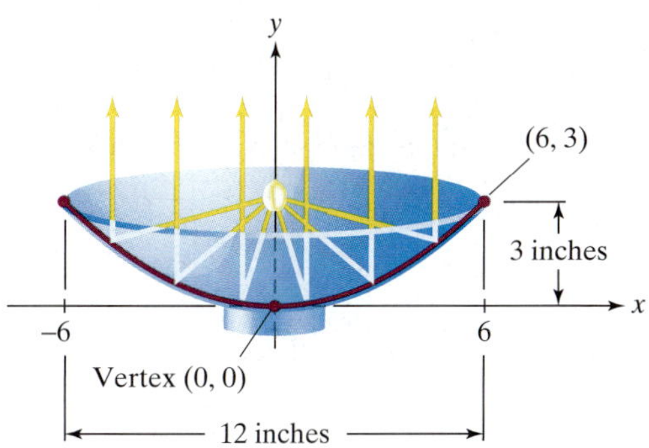

a. Use the point (6, 3) to write an equation in the form $y = ax^2$ for the parabola used to design the headlight.

b. The light source should be placed at the focus $(0, p)$. The value of p is given by the equation $a = \frac{1}{4p}$. Where should the light source be placed? Describe this placement relative to the vertex.

10.5 *In Exercises 43–53, solve each system by the method of your choice.*

43. $5y = x^2 - 1$
$x - y = 1$

44. $y = x^2 + 2x + 1$
$x + y = 1$

45. $x^2 + y^2 = 2$
$x + y = 0$

46. $2x^2 + y^2 = 24$
$x^2 + y^2 = 15$

47. $xy - 4 = 0$
$y - x = 0$

48. $y^2 = 4x$
$x - 2y + 3 = 0$

49. $x^2 + y^2 = 10$
$y = x + 2$

50. $xy = 1$
$y = 2x + 1$

51. $x + y + 1 = 0$
$x^2 + y^2 + 6y - x = -5$

52. $x^2 + y^2 = 13$
$x^2 - y = 7$

53. $2x^2 + 3y^2 = 21$
$3x^2 - 4y^2 = 23$

54. The perimeter of a rectangle is 26 meters, and its area is 40 square meters. Find its dimensions.

55. Find the coordinates of all points (x, y) that lie on the line whose equation is $2x + y = 8$, so that the area of the rectangle shown in the figure is 6 square units.

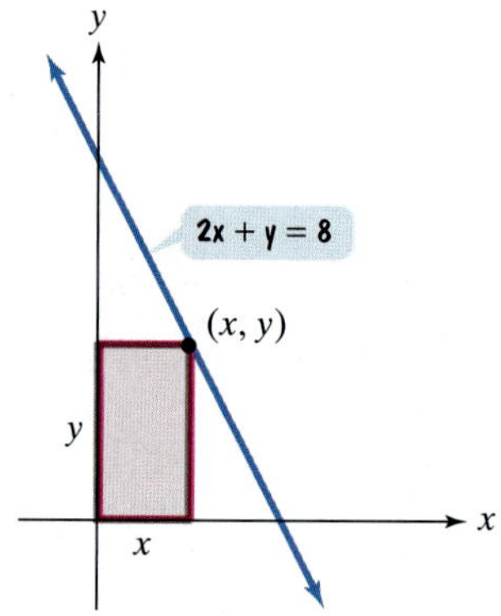

56. Two adjoining square fields with an area of 2900 square feet are to be enclosed with 240 feet of fencing. The situation is represented in the figure. Find the length of each side where a variable appears.

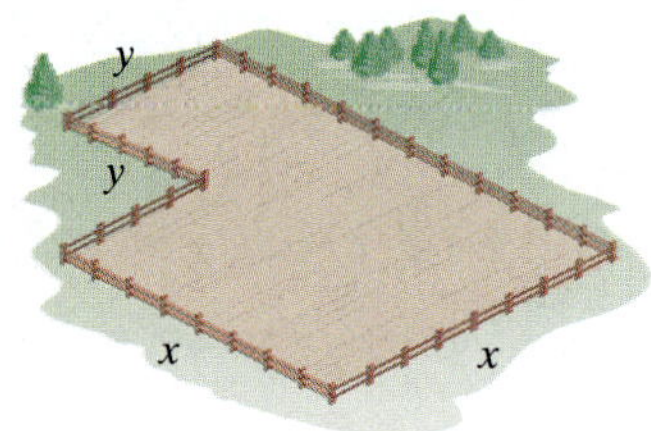

Chapter 10 Test

1. Find the distance between $(-1, 5)$ and $(2, -3)$. If necessary, round the answer to two decimal places.

2. Find the midpoint of the line segment whose endpoints are $(-5, -2)$ and $(12, -6)$.

3. Write the standard form of the equation of the circle with center $(3, -2)$ and radius 5.

In Exercises 4–5, give the center and radius of each circle.

4. $(x - 5)^2 + (y + 3)^2 = 49$

5. $x^2 + y^2 + 4x - 6y - 3 = 0$

In Exercises 6–7, give the coordinates of the vertex for each parabola.

6. $x = -2(y + 3)^2 + 7$

7. $x = y^2 + 10y + 23$

In Exercises 8–16, indicate whether the graph of each equation is a circle, an ellipse, a hyperbola, or a parabola. Then graph the conic section.

8. $\frac{x^2}{4} - \frac{y^2}{9} = 1$

9. $4x^2 + 9y^2 = 36$

10. $x = (y + 1)^2 - 4$

11. $16x^2 + y^2 = 16$

12. $25y^2 = 9x^2 + 225$

13. $x = -y^2 + 6y$

14. $\frac{(x - 2)^2}{16} + \frac{(y + 3)^2}{9} = 1$

15. $(x + 1)^2 + (y + 2)^2 = 9$

16. $\frac{x^2}{4} + \frac{y^2}{4} = 1$

In Exercises 17–18, solve each system.

17. $x^2 + y^2 = 25$
$x + y = 1$

18. $2x^2 - 5y^2 = -2$
$3x^2 + 2y^2 = 35$

19. The rectangular plot of land shown in the Figure is to be fenced along three sides using 39 feet of fencing. No fencing is to be placed along the river's edge. The area of the plot is 180 square feet. What are its dimensions?

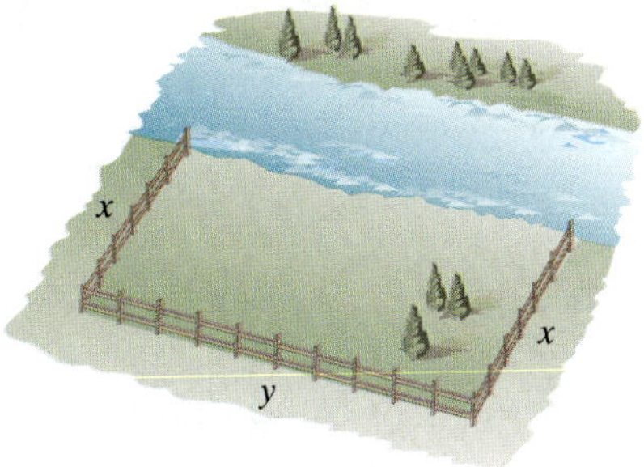

20. A rectangle has a diagonal of 5 feet and a perimeter of 14 feet. Find the rectangle's dimensions.

Cumulative Review Exercises (Chapters 1–10)

In Exercises 1–7, solve each equation, inequality, or system.

1. $3x + 7 > 4 \quad \text{or} \quad 6 - x < 1$

2. $x(2x - 7) = 4$

3. $\dfrac{5}{x - 3} = 1 + \dfrac{30}{x^2 - 9}$

4. $3x^2 + 8x + 5 < 0$

5. $3^{2x-1} = 81$

6. $30e^{0.7x} = 240$

7. $3x^2 + 4y^2 = 39$
$5x^2 - 2y^2 = -13$

In Exercises 8–11, graph each function, equation, or inequality in a rectangular coordinate system.

8. $f(x) = -\dfrac{2}{3}x + 4$

9. $3x - y > 6$

10. $x^2 + y^2 + 4x - 6y + 9 = 0$

11. $9x^2 - 4y^2 = 36$

In Exercises 12–15, perform the indicated operations, and simplify, if possible.

12. $-2(3^2 - 12)^3 - 45 \div 9 - 3$

13. $(3x^3 - 19x^2 + 17x + 4) \div (3x - 4)$

14. $\sqrt[3]{4x^2y^5} \cdot \sqrt[3]{4xy^2}$

15. $(2 + 3i)(4 - i)$

In Exercises 16–17, factor completely.

16. $12x^3 - 36x^2 + 27x$

17. $x^3 - 2x^2 - 9x + 18$

18. Find the domain: $f(x) = \sqrt{6 - 3x}$.

19. Rationalize the denominator: $\dfrac{1 - \sqrt{x}}{1 + \sqrt{x}}$.

20. Write as a single logarithm: $\dfrac{1}{3}\ln x + 7\ln y$.

21. Divide using synthetic division:

$(3x^3 - 5x^2 + 2x - 1) \div (x - 2)$.

22. Write a quadratic equation whose solution set is $\{-2\sqrt{3}, 2\sqrt{3}\}$.

23. Two cars leave from the same place at the same time, traveling in opposite directions. The rate of the faster car exceeds that of the slower car by 10 miles per hour. After 2 hours, the cars are 180 miles apart. Find the rate of each car.

24. Rent-a-Truck charges a daily rental rate of \$39 plus \$0.16 per mile. A competing agency, Ace Truck Rentals, charges \$25 a day plus \$0.24 per mile for the same truck. How many miles must be driven in a day to make the daily cost of both agencies the same? What will be the cost?

25. Three apples and two bananas provide 354 calories, and two apples and three bananas provide 381 calories. Find the number of calories in one apple and one banana.

Something incredible has happened. Your college roommate, a gifted athlete, has been given a six-year contract with a professional baseball team. He will be playing against the likes of Mark McGwire and Sammy Sosa. Management offers him three options. One is a beginning salary of $1,700,000 with annual increases of $70,000 per year starting in the second year. A second option is $1,700,000 the first year with an annual increase of 2% per year beginning in the second year. The third offer involves less money the first year—$1,500,000—but there is an annual increase of 9% yearly after that. Which option offers the most money over the six-year contract?

We often save for the future by investing small amounts at periodic intervals. To understand how our savings accumulate, we need to understand properties of lists of numbers that are related to each other by a rule. Such lists are called *sequences*. Learning about properties of sequences will show you how to make your financial goals a reality. Your knowledge of sequences will enable you to inform your college roommate of the best of the three appealing offers.

Chapter 11

Sequences, Series, and the Binomial Theorem

▸ SECTION 11.1 Sequences and Summation Notation

Objectives

1. Find particular terms of a sequence from the general term.
2. Use factorial notation.
3. Use summation notation.

Sequences

Many creations in nature involve intricate mathematical designs, including a variety of spirals. For example, the arrangement of the individual florets in the head of a sunflower forms spirals. In some species, there are 21 spirals in the clockwise direction and 34 in the counterclockwise direction. The precise numbers depend on the species of sunflower: 21 and 34, or 34 and 55, or 55 and 89, or even 89 and 144.

This observation becomes even more interesting when we consider a sequence of numbers investigated by Leonardo of Pisa, also known as Fibonacci, an Italian mathematician of the thirteenth century. The **Fibonacci sequence** of numbers is an infinite sequence that begins as follows:

$$1, 1, 2, 3, 5, 8, 13, 21, 34, 55, 89, 144, 233 \ldots.$$

The first two terms are 1. Every term thereafter is the sum of the two preceding terms. For example, the third term, 2, is the sum of the first and second terms: $1 + 1 = 2$. The fourth term, 3, is the sum of the second and third terms: $1 + 2 = 3$, and so on. Did you know that the number of spirals in a daisy or a sunflower, 21 and 34, are two Fibonacci numbers? The number of spirals in a pinecone, 8 and 13, and a pineapple, 8 and 13, are also Fibonacci numbers.

We can think of the Fibonacci sequence as a function. The terms of the sequence

$$1, 1, 2, 3, 5, 8, 13, 21, 34, 55, 89, 144, 233, \ldots$$

are the range values for a function whose domain is the set of positive integers.

Domain:	1,	2,	3,	4,	5,	6,	7,	...
	↓	↓	↓	↓	↓	↓	↓	
Range:	1,	1,	2,	3,	5,	8,	13,	...

Thus, $f(1) = 1$, $f(2) = 1$, $f(3) = 2$, $f(4) = 3$, $f(5) = 5$, $f(6) = 8$, $f(7) = 13$, and so on.

The letter a with a subscript is used to represent function values of a sequence, rather than the usual function notation. The subscripts make up the domain of the sequence, and they identify the location of a term. Thus, a_1 represents the first term of the sequence, a_2 represents the second term, a_3 the third term, and so on. This notation is shown for the first six terms of the Fibonacci sequence:

1,	1,	2,	3,	5,	8.
$a_1 = 1$	$a_2 = 1$	$a_3 = 2$	$a_4 = 3$	$a_5 = 5$	$a_6 = 8$

ENRICHMENT ESSAY

Fibonacci Numbers on the Piano Keyboard

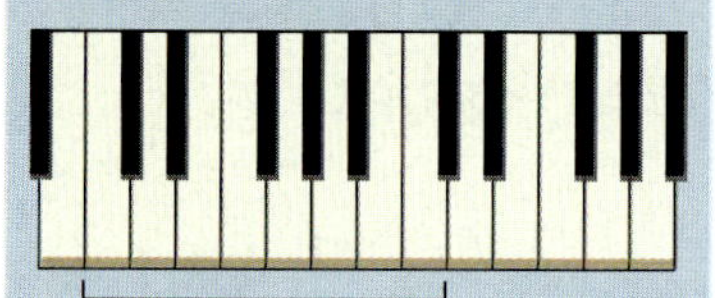

Numbers in the Fibonacci sequence can be found in an octave on the piano keyboard. The octave contains 2 black keys in one cluster, 3 black keys in another cluster, 5 black keys, 8 white keys, and a total of 13 keys altogether. The numbers 2, 3, 5, 8, and 13 are the third through seventh terms of the Fibonacci sequence.

The notation a_n represents the nth term, or **general term**, of a sequence. The entire sequence is represented by $\{a_n\}$.

Definition of a Sequence

An **infinite sequence** $\{a_n\}$ is a function whose domain is the set of positive integers. The function values, or **terms**, of the sequence are represented by

$$a_1, a_2, a_3, a_4, \ldots, a_n, \ldots.$$

Sequences whose domains consist only of the first n positive integers are called **finite sequences**.

1 Find particular terms of a sequence from the general term.

EXAMPLE 1 Writing Terms of a Sequence from the General Term

Write the first four terms of the sequence whose nth term, or general term, is given:

a. $a_n = 3n + 4$ **b.** $a_n = \dfrac{(-1)^n}{3^n - 1}$.

Solution

a. We need to find the first four terms of the sequence whose general term is $a_n = 3n + 4$. To do so, we replace n in the formula by 1, 2, 3, and 4.

a_1, 1st term: $3 \cdot 1 + 4 = 3 + 4 = 7$ a_2, 2nd term: $3 \cdot 2 + 4 = 6 + 4 = 10$

a_3, 3rd term: $3 \cdot 3 + 4 = 9 + 4 = 13$ a_4, 4th term: $3 \cdot 4 + 4 = 12 + 4 = 16$

The first four terms are 7, 10, 13, and 16. The sequence defined by $a_n = 3n + 4$ can be written as

$$7, 10, 13, \ldots, 3n + 4, \ldots.$$

b. We need to find the first four terms of the sequence whose general term is $a_n = \dfrac{(-1)^n}{3^n - 1}$. To do so, we replace each occurrence of n in the formula by 1, 2, 3, and 4.

a_1, 1st term: $\dfrac{(-1)^1}{3^1 - 1} = \dfrac{-1}{3 - 1} = -\dfrac{1}{2}$ a_2, 2nd term: $\dfrac{(-1)^2}{3^2 - 1} = \dfrac{1}{9 - 1} = \dfrac{1}{8}$

a_3, 3rd term: $\dfrac{(-1)^3}{3^3 - 1} = \dfrac{-1}{27 - 1} = -\dfrac{1}{26}$ a_4, 4th term: $\dfrac{(-1)^4}{3^4 - 1} = \dfrac{1}{81 - 1} = \dfrac{1}{80}$

The first four terms are $-\frac{1}{2}, \frac{1}{8}, -\frac{1}{26}, \frac{1}{80}$. The sequence defined by $\dfrac{(-1)^n}{3^n - 1}$ can be written as

$$-\frac{1}{2}, \frac{1}{8}, -\frac{1}{26}, \ldots, \frac{(-1)^n}{3^n - 1}, \ldots.$$

■

Study Tip

The factor $(-1)^n$ in the general term of a sequence causes the signs of the terms to alternate between positive and negative, depending on whether n is even or odd.

CHECK POINT 1 Write the first four terms of the sequence whose nth term, or general term, is given:

a. $a_n = 2n + 5$ **b.** $a_n = \dfrac{(-1)^n}{2^n + 1}$.

Using Technology

Graphing utilities can write the terms of a sequence and graph them. For example, to find the first six terms of $\{a_n\} = \left\{\frac{1}{n}\right\}$, enter

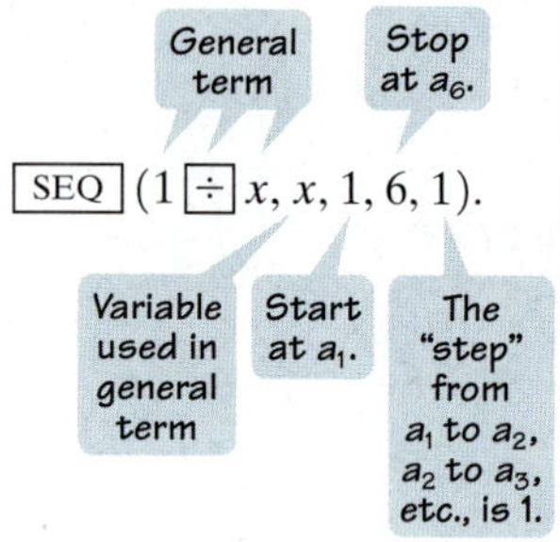

The first few terms of the sequence are shown in the viewing rectangle. By pressing the right arrow key to scroll right, you can see the remaining terms.

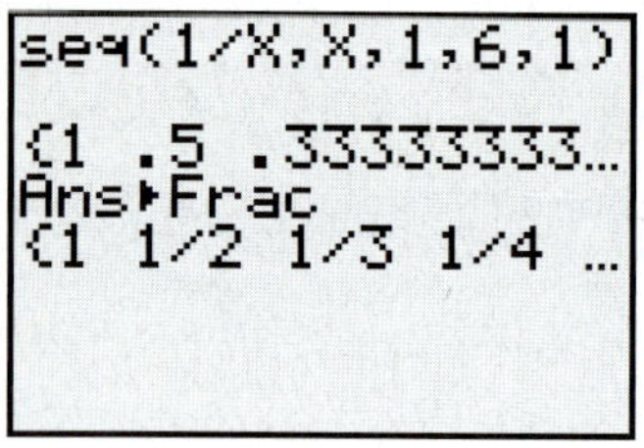

Although sequences are usually named with the letter a, any lowercase letter can be used. For example, the first four terms of the sequence $\{b_n\} = \{(\frac{1}{2})^n\}$ are $b_1 = \frac{1}{2}$, $b_2 = \frac{1}{4}$, $b_3 = \frac{1}{8}$, and $b_4 = \frac{1}{16}$.

Because a sequence is a function whose domain is the set of positive integers, the **graph of a sequence** is a set of discrete points. For example, consider the sequence whose general term is $a_n = \frac{1}{n}$. How does the graph of this sequence differ from the graph of the rational function $f(x) = \frac{1}{x}$? The graph of $f(x) = \frac{1}{x}$ is shown in Figure 11.1(a) for positive values of x. To obtain the graph of the sequence $\{a_n\} = \{\frac{1}{n}\}$, remove all the points from the graph of f except those whose x-coordinates are positive integers. Thus, we remove all points except $(1, 1)$, $(2, \frac{1}{2})$, $(3, \frac{1}{3})$, $(4, \frac{1}{4})$, and so on. The remaining points are the graph of the sequence $\{a_n\} = \{\frac{1}{n}\}$, shown in Figure 11.1(b). Notice that the horizontal axis is labeled n and the vertical axis a_n.

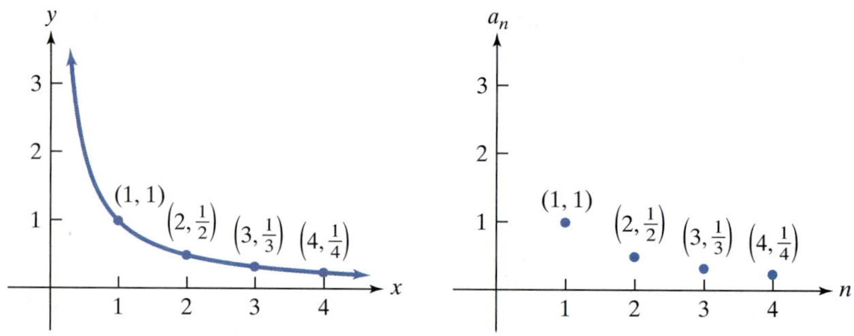

(a) The graph of $f(x) = \frac{1}{x}$, $x > 0$ (b) The graph of $\{a_n\} = \{\frac{1}{n}\}$

Figure 11.1 Comparing a continuous graph to the graph of a sequence

Factorial Notation

2 Use factorial notation.

Products of consecutive positive integers occur quite often in sequences. These products can be expressed in a special notation, called **factorial notation**.

Factorial Notation

If n is a positive integer, the notation $n!$ (read "n factorial") is the product of all positive integers from n down through 1.

$$n! = n(n-1)(n-2)\ldots(3)(2)(1)$$

0! (zero factorial), by definition, is 1.

$$0! = 1$$

The values of $n!$ for the first six positive integers are

$$1! = 1$$
$$2! = 2 \cdot 1 = 2$$
$$3! = 3 \cdot 2 \cdot 1 = 6$$
$$4! = 4 \cdot 3 \cdot 2 \cdot 1 = 24$$
$$5! = 5 \cdot 4 \cdot 3 \cdot 2 \cdot 1 = 120$$
$$6! = 6 \cdot 5 \cdot 4 \cdot 3 \cdot 2 \cdot 1 = 720.$$

Factorials affect only the number or variable that they follow unless grouping symbols appear. For example,

$$2 \cdot 3! = 2(3 \cdot 2 \cdot 1) = 2 \cdot 6 = 12$$

whereas

$$(2 \cdot 3)! = 6! = 6 \cdot 5 \cdot 4 \cdot 3 \cdot 2 \cdot 1 = 720.$$

In this sense, factorials are similar to exponents.

Using Technology

Most calculators have factorial keys. To find 5 factorial, most calculators use one of the following:

Scientific Calculators

5 [x!]

Graphing Calculators

5 [!] [ENTER].

Because $n!$ becomes quite large as n increases, your utility will display these larger values in scientific notation.

EXAMPLE 2 Finding Terms of a Sequence Involving Factorials

Write the first four terms of the sequence whose nth term is

$$a_n = \frac{2^n}{(n-1)!}.$$

Solution We need to find the first four terms of the sequence. To do so, we replace each n in the formula by 1, 2, 3, and 4.

a_1, 1st term: $$\frac{2^1}{(1-1)!} = \frac{2}{0!} = \frac{2}{1} = 2$$

a_2, 2nd term: $$\frac{2^2}{(2-1)!} = \frac{4}{1!} = \frac{4}{1} = 4$$

a_3, 3rd term: $$\frac{2^3}{(3-1)!} = \frac{8}{2!} = \frac{8}{2 \cdot 1} = 4$$

a_4, 4th term: $$\frac{2^4}{(4-1)!} = \frac{16}{3!} = \frac{16}{3 \cdot 2 \cdot 1} = \frac{16}{6} = \frac{8}{3}$$

The first four terms are 2, 4, 4, $\frac{8}{3}$. ■

CHECK POINT 2 Write the first four terms of the sequence whose nth term is

$$a_n = \frac{20}{(n+1)!}.$$

3 Use summation notation.

Summation Notation

It is sometimes useful to find the sum of the first n terms of a sequence. For example, consider the number of AIDS cases diagnosed in the United States for each year from 1991 through 1999, shown in Table 11.1.

Table 11.1 AIDS Cases Diagnosed in the United States, 1991–1999

Year	1991	1992	1993	1994	1995	1996	1997	1998	1999
Cases Diagnosed	60,472	79,477	79,752	72,684	69,172	59,832	47,439	38,587	25,434

Source: U.S. Department of Health and Human Services

We can let a_n represent the number of AIDS cases diagnosed in year n, where $n = 1$ corresponds to 1991, $n = 2$ to 1992, $n = 3$ to 1993, and so on. The terms of the finite sequence in Table 11.1 are given as follows:

$$\underset{a_1}{60{,}472},\ \underset{a_2}{79{,}477},\ \underset{a_3}{79{,}752},\ \underset{a_4}{72{,}684},\ \underset{a_5}{69{,}172},\ \underset{a_6}{59{,}832},\ \underset{a_7}{47{,}439},\ \underset{a_8}{38{,}587},\ \underset{a_9}{25{,}434}.$$

Why might we want to add the terms of this sequence? We do this to find the number of AIDS cases diagnosed from 1991 through 1999. Thus,

$$\begin{aligned} &a_1 + a_2 + a_3 + a_4 + a_5 + a_6 + a_7 + a_8 + a_9 \\ &= 60{,}472 + 79{,}477 + 79{,}752 + 72{,}684 + 69{,}172 + 59{,}832 + 47{,}439 + 38{,}587 + 25{,}434 \\ &= 532{,}849 \end{aligned}$$

We see that there were 532,849 AIDS cases diagnosed in the United States from 1991 through 1999.

There is a compact notation for expressing the sum of the first n terms of a sequence. For example, rather than write

$$a_1 + a_2 + a_3 + a_4 + a_5 + a_6 + a_7 + a_8 + a_9,$$

we can use **summation notation** to express the sum as

$$a_1 + a_2 + a_3 + a_4 + a_5 + a_6 + a_7 + a_8 + a_9 = \sum_{i=1}^{9} a_i.$$

We read the expression on the right as "the sum as i goes from 1 to 9 of a_i." The letter i is called the **index of summation** and is not related to the use of i to represent $\sqrt{-1}$.

You can think of the symbol Σ (the uppercase Greek letter sigma) as an instruction to add up terms of a sequence.

Summation Notation

The sum of the first n terms of a sequence is represented by the **summation notation**

$$\sum_{i=1}^{n} a_i = a_1 + a_2 + a_3 + a_4 + \cdots + a_n$$

where i is the **index of summation,** n is the **upper limit of summation,** and 1 is the **lower limit of summation.**

Any letter can be used for the index of summation. The letters i, j, and k are used commonly. Furthermore, the lower limit of summation can be an integer other than 1.

When we write out a sum that is given in summation notation, we are **expanding the summation notation**. Example 3 shows how to do this.

EXAMPLE 3 Using Summation Notation

Expand and evaluate the sum:

a. $\sum_{i=1}^{6} (i^2 + 1)$ **b.** $\sum_{k=4}^{7} [(-2)^k - 5]$ **c.** $\sum_{i=1}^{5} 3.$

Using Technology

Graphing utilities can calculate the sum of a sequence. For example, to find the sum of the sequence in Example 3a, enter

[SUM] [SEQ] $(x^2 + 1, x, 1, 6, 1)$.

Then press [ENTER]; 97 should be displayed. Use this capability to verify Example 3b.

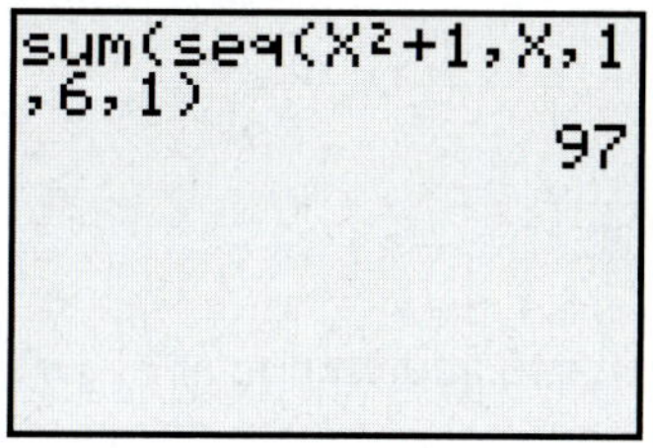

Solution

a. We must replace i in the expression $i^2 + 1$ with all consecutive integers from 1 to 6 inclusively. Then we add.

$$\begin{aligned}\sum_{i=1}^{6}(i^2+1) &= (1^2+1)+(2^2+1)+(3^2+1)+(4^2+1)\\ &\quad +(5^2+1)+(6^2+1)\\ &= 2+5+10+17+26+37\\ &= 97\end{aligned}$$

b. This time the index of summation is k. First we evaluate $(-2)^k - 5$ for all consecutive integers from 4 through 7 inclusively. Then we add.

$$\begin{aligned}\sum_{k=4}^{7}[(-2)^k-5] &= [(-2)^4-5]+[(-2)^5-5]\\ &\quad +[(-2)^6-5]+[(-2)^7-5]\\ &= (16-5)+(-32-5)+(64-5)+(-128-5)\\ &= 11+(-37)+59+(-133)\\ &= -100\end{aligned}$$

c. To find $\sum_{i=1}^{5} 3$, we observe that every term of the sum is 3. The notation $i = 1$ through 5 indicates that we must add the first five 3s from a sequence in which every term is 3.

$$\sum_{i=1}^{5} 3 = 3+3+3+3+3 = 15$$

■

✔ CHECK POINT 3 Expand and evaluate the sum:

a. $\sum_{i=1}^{6} 2i^2$

b. $\sum_{k=3}^{5} (2^k - 3)$

c. $\sum_{i=1}^{5} 4$.

For a given sum, we can vary the upper and lower limits of summation as well as the letter used for the index of summation. By doing so, we can produce different-looking summation notations for the same sum. For example, the sum of the squares of the first four positive integers, $1^2 + 2^2 + 3^2 + 4^2$, can be expressed in a number of equivalent ways:

$$\sum_{i=1}^{4} i^2 = 1^2+2^2+3^2+4^2 = 30$$

$$\begin{aligned}\sum_{i=0}^{3}(i+1)^2 &= (0+1)^2+(1+1)^2+(2+1)^2+(3+1)^2\\ &= 1^2+2^2+3^2+4^2 = 30\end{aligned}$$

$$\begin{aligned}\sum_{k=2}^{5}(k-1)^2 &= (2-1)^2+(3-1)^2+(4-1)^2+(5-1)^2\\ &= 1^2+2^2+3^2+4^2 = 30.\end{aligned}$$

EXAMPLE 4 Writing Sums in Summation Notation

Express each sum using summation notation:

a. $1^3 + 2^3 + 3^3 + \cdots + 7^3$ **b.** $1 + \frac{1}{3} + \frac{1}{9} + \frac{1}{27} + \cdots + \frac{1}{3^{n-1}}$.

Solution In each case, we will use 1 as the lower limit of summation and i for the index of summation.

a. The sum $1^3 + 2^3 + 3^3 + \cdots + 7^3$ has seven terms, each of the form i^3, starting at $i = 1$ and ending at $i = 7$. Thus,

$$1^3 + 2^3 + 3^3 + \cdots + 7^3 = \sum_{i=1}^{7} i^3.$$

b. The sum

$$1 + \frac{1}{3} + \frac{1}{9} + \frac{1}{27} + \cdots + \frac{1}{3^{n-1}}$$

has n terms, each of the form $\frac{1}{3^{i-1}}$, starting at $i = 1$ and ending at $i = n$.

Thus,

$$1 + \frac{1}{3} + \frac{1}{9} + \frac{1}{27} + \cdots + \frac{1}{3^{n-1}} = \sum_{i=1}^{n} \frac{1}{3^{i-1}}.$$

■

CHECK POINT 4 Express each sum using summation notation:

a. $1^2 + 2^2 + 3^2 + \cdots + 9^2$

b. $1 + \frac{1}{2} + \frac{1}{4} + \frac{1}{8} + \cdots + \frac{1}{2^{n-1}}$.

EXERCISE SET 11.1

Practice Exercises

In Exercises 1–16, write the first four terms of each sequence whose general term is given.

1. $a_n = 3n + 2$

2. $a_n = 4n - 1$

3. $a_n = 3^n$

4. $a_n = \left(\frac{1}{3}\right)^n$

5. $a_n = (-3)^n$

6. $a_n = \left(-\frac{1}{3}\right)^n$

7. $a_n = (-1)^n(n + 3)$

8. $a_n = (-1)^{n+1}(n + 4)$

9. $a_n = \frac{2n}{n + 4}$

10. $a_n = \frac{3n}{n + 5}$

11. $a_n = \frac{(-1)^{n+1}}{2^n - 1}$

12. $a_n = \frac{(-1)^{n+1}}{2^n + 1}$

13. $a_n = \frac{n^2}{n!}$

14. $a_n = \frac{(n + 1)!}{n^2}$

15. $a_n = 2(n + 1)!$

16. $a_n = -2(n - 1)!$

In Exercises 17–30, find each indicated sum.

17. $\sum_{i=1}^{6} 5i$

18. $\sum_{i=1}^{6} 7i$

19. $\sum_{i=1}^{4} 2i^2$

20. $\sum_{i=1}^{5} i^3$

21. $\sum_{k=1}^{5} k(k + 4)$

22. $\sum_{k=1}^{4} (k - 3)(k + 2)$

23. $\sum_{i=1}^{4} \left(-\frac{1}{2}\right)^i$

24. $\sum_{i=2}^{4} \left(-\frac{1}{3}\right)^i$

25. $\sum_{i=5}^{9} 11$

26. $\sum_{i=3}^{7} 12$

27. $\sum_{i=0}^{4} \frac{(-1)^i}{i!}$

28. $\sum_{i=0}^{4} \frac{(-1)^{i+1}}{(i+1)!}$

29. $\sum_{i=1}^{5} \frac{i!}{(i-1)!}$

30. $\sum_{i=1}^{5} \frac{(i+2)!}{i!}$

In Exercises 31–42, express eac§h sum using summation notation. Use 1 as the lower limit of summation and i for the index of summation.

31. $1^2 + 2^2 + 3^2 + \cdots + 15^2$

32. $1^4 + 2^4 + 3^4 + \cdots + 12^4$

33. $2 + 2^2 + 2^3 + \cdots + 2^{11}$

34. $5 + 5^2 + 5^3 + \cdots + 5^{12}$

35. $1 + 2 + 3 + \cdots + 30$

36. $1 + 2 + 3 + \cdots + 40$

37. $\frac{1}{2} + \frac{2}{3} + \frac{3}{4} + \cdots + \frac{14}{14+1}$

38. $\frac{1}{3} + \frac{2}{4} + \frac{3}{5} + \cdots + \frac{16}{16+2}$

39. $4 + \frac{4^2}{2} + \frac{4^3}{3} + \cdots + \frac{4^n}{n}$

40. $\frac{1}{9} + \frac{2}{9^2} + \frac{3}{9^3} + \cdots + \frac{n}{9^n}$

41. $1 + 3 + 5 + \cdots + (2n - 1)$

42. $a + ar + ar^2 + \cdots + ar^{n-1}$

In Exercises 43–48, express each sum using summation notation. Use a lower limit of summation of your choice and k for the index of summation.

43. $5 + 7 + 9 + 11 + \cdots + 31$

44. $6 + 8 + 10 + 12 + \cdots + 32$

45. $a + ar + ar^2 + \cdots + ar^{12}$

46. $a + ar + ar^2 + \cdots + ar^{14}$

47. $a + (a + d) + (a + 2d) + \cdots + (a + nd)$

48. $(a + d) + (a + d^2) + \cdots + (a + d^n)$

Application Exercises

49. The bar graph shows the number of compact discs (CDs) sold in the United States. Let a_n represent the number of CDs sold, in millions, in year n, where $n = 1$ corresponds to 1991, $n = 2$ to 1992, and so on.

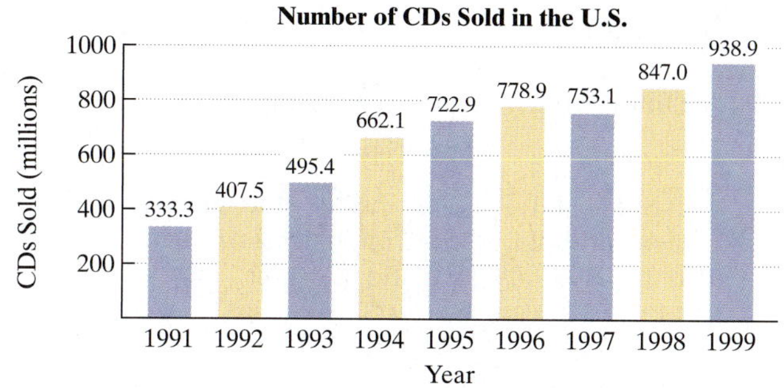

Source: Recording Industry Association of America

a. Find $\sum_{i=1}^{9} a_i$. What does this represent?

b. Find $\frac{1}{9}\sum_{i=1}^{9} a_i$. What does this represent?

50. The bar graph shows the number of vinyl long-playing records (LPs) sold in the United States. Let a_n represent the number of LPs sold, in millions, in year n, where $n = 1$ corresponds to 1991, $n = 2$ to 1992, and so on.

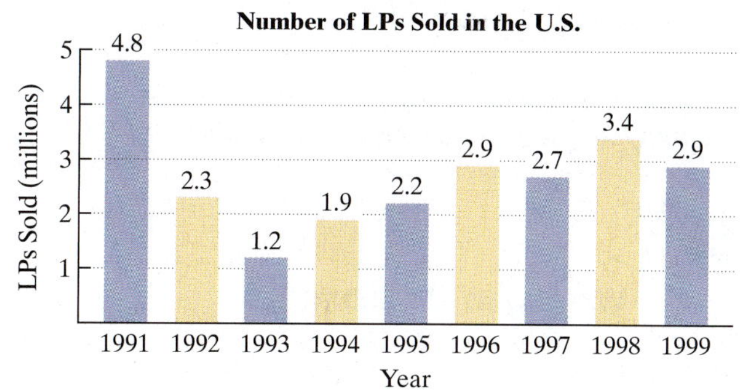

Source: Recording Industry Association of America

a. Find $\sum_{i=1}^{9} a_i$. What does this represent?

b. Find $\frac{1}{9}\sum_{i=1}^{9} a_i$. What does this represent?

51. The finite sequence whose general term is

$$a_n = 0.16n^2 - 1.04n + 7.39$$

where $n = 1, 2, 3, \ldots, 9$ models the total number of dollars, in billions, that Americans spent on recreational boating from 1991 through 1999. Find and interpret

$$\sum_{i=1}^{5} a_i.$$

52. The finite sequence whose general term is

$$a_n = 2.54e^{-0.09n}$$

where $n = 0, 1, 2, \ldots, 9$ models the number of new foreign cars sold in the United States, in millions, from 1990 through 1999. Find and interpret

$$\sum_{i=0}^{4} a_i.$$

Round to the nearest tenth of a million.

53. A deposit of \$6000 is placed in an account that earns 6% interest compounded quarterly. The balance in the account after n quarters is given by the sequence

$$a_n = 6000\left(1 + \frac{0.06}{4}\right)^n, \quad n = 1, 2, 3, \ldots.$$

Find the balance in the account after five years by computing a_{20}. Round to the nearest cent.

54. A deposit of \$10,000 is placed in an account that earns 8% interest compounded quarterly. The balance in the account after n quarters is given by the sequence

$$a_n = 10{,}000\left(1 + \frac{0.08}{4}\right)^n, \quad n = 1, 2, 3, \ldots.$$

Find the balance in the account after six years by computing a_{24}. Round to the nearest cent.

Writing in Mathematics

55. What is a sequence? Give an example with your description.

56. Explain how to write terms of a sequence if the formula for the general term is given.

57. What does the graph of a sequence look like? How is it obtained?

58. Explain how to find $n!$ if n is a positive integer.

59. What is the meaning of the symbol Σ? Give an example with your description.

60. You buy a new car for \$24,000. At the end of n years, the value of your car is given by the sequence

$$a_n = 24{,}000\left(\frac{3}{4}\right)^n, \quad n = 1, 2, 3, \ldots.$$

Find a_5 and write a sentence explaining what this value represents. Describe the nth term of the sequence in terms of the value of your car at the end of each year.

Technology Exercises

61. Use the SEQ (sequence) capability of a graphing utility to verify the terms of the sequences you obtained for any five sequences from Exercises 1–16.

62. Use the SUM SEQ (sum of the sequence) capability of a graphing utility to verify any five of the sums you obtained in Exercises 17–30.

63. As n increases, the terms of the sequence

$$a_n = \left(1 + \frac{1}{n}\right)^n$$

get closer and closer to the number e (where $e \approx 2.7183$). Use a calculator to find a_{10}, a_{100}, a_{1000}, $a_{10,000}$, and $a_{100,000}$, comparing these terms to the decimal approximation for e.

Many graphing utilities have a sequence-graphing mode that plots the terms of a sequence as points on a rectangular coordinate system. Consult your manual; if your graphing utility has this capability, use it to graph each of the sequences in Exercises 64–67. What appears to be happening to the terms of each sequence as n gets larger?

64. $a_n = \dfrac{n}{n+1}$; $n:[0, 10, 1]$ by $a_n:[0, 1, 0.1]$

65. $a_n = \dfrac{100}{n}$; $n:[0, 1000, 100]$ by $a_n:[0, 1, 0.1]$

66. $a_n = \dfrac{2n^2 + 5n - 7}{n^3}$; $n:[0, 10, 1]$ by $a_n:[0, 2, 0.2]$

67. $a_n = \dfrac{3n^4 + n - 1}{5n^4 + 2n^2 + 1}$; $n:[0, 10, 1]$ by $a_n:[0, 1, 0.1]$

Critical Thinking Exercises

68. Which one of the following is true?

a. $\sum_{i=1}^{2} (-1)^i 2^i = 0$

b. $\sum_{i=1}^{2} a_i b_i = \sum_{i=1}^{2} a_i \sum_{i=1}^{2} b_i$

c. $\sum_{i=1}^{4} 3i + \sum_{i=1}^{4} 4i = \sum_{i=1}^{4} 7i$

d. $\sum_{i=0}^{6} (-1)^i (i+1)^2 = \sum_{j=1}^{7} (-1)^j j^2$

In Exercises 69–70, expand and write the answer as a single logarithm with a coefficient of 1.

69. $\sum_{i=1}^{4} \log 2i$

70. $\sum_{i=2}^{4} 2i \log x$

71. If $a_1 = 7$ and $a_n = a_{n-1} + 5$ for $n \geq 2$, write the first four terms of the sequence.

72. Evaluate without using a calculator: $\frac{600!}{599!}$.

Review Exercises

73. Simplify: $\sqrt[3]{40x^4y^7}$. (Section 7.3, Example 5)

74. Factor: $27x^3 - 8$. (Section 5.5, Example 9)

75. Solve: $\frac{6}{x} + \frac{6}{x+2} = \frac{5}{2}$. (Section 6.6, Example 4)

▸ SECTION 11.2 Arithmetic Sequences

Objectives

1. Find the common difference for an arithmetic sequence.
2. Write terms of an arithmetic sequence.
3. Use the formula for the general term of an arithmetic sequence.
4. Use the formula for the sum of the first n terms of an arithmetic sequence.

Your grandmother and her financial counselor are looking at options in case nursing home care is needed in the future. The good news is that your grandmother's total assets are \$350,000. The bad news is that yearly nursing home costs average \$49,730, increasing by \$1800 each year. In this section, we will see how sequences can be used to describe your grandmother's situation and help her to identify realistic options.

Arithmetic Sequences

A mathematical model for the average annual salaries of major league baseball players generates the following data.

Year	1991	1992	1993	1994	1995	1996	1997	1998	1999
Salary	801,000	892,000	983,000	1,074,000	1,165,000	1,256,000	1,347,000	1,438,000	1,529,000

From 1991 to 1992, salaries increased by \$892,000 − \$801,000 = \$91,000. From 1992 to 1993, salaries increased by \$983,000 − \$892,000 = \$91,000. If we make these computations for each year, we find that the yearly salary increase is \$91,000. The sequence of annual salaries shows that each term after the first, 801,000, differs from the preceding term by a constant amount, namely 91,000. The sequence of annual salaries

$$801{,}000,\ 892{,}000,\ 983{,}000,\ 1{,}074{,}000,\ 1{,}165{,}000,\ 1{,}256{,}000, \ldots$$

is an example of an *arithmetic sequence*.

Definition of an Arithmetic Sequence

An **arithmetic sequence** is a sequence in which each term after the first differs from the preceding term by a constant amount. The difference between consecutive terms is called the **common difference** of the sequence.

1 Find the common difference for an arithmetic sequence.

The common difference, d, is found by subtracting any term from the term that directly follows it. In the following examples, the common difference is found by subtracting the first term from the second term, $a_2 - a_1$.

ARITHMETIC SEQUENCE	COMMON DIFFERENCE
801,000, 892,000, 983,000, 1,074,000, ...	$d = 892{,}000 - 801{,}000 = 91{,}000$
2, 6, 10, 14, 18, ...	$d = 6 - 2 = 4$
−2, −7, −12, −17, ...	$d = -7 - (-2) = -5$

If the first term of an arithmetic sequence is a_1, each term after the first is obtained by adding d, the common difference, to the previous term.

2 Write terms of an arithmetic sequence.

EXAMPLE 1 Writing the Terms of an Arithmetic Sequence Using the First Term and the Common Difference

Write the first six terms of the arithmetic sequence with first term 6 and common difference −2.

Solution To find the second term, we add −2 to the first term, 6, giving 4. For the next term, we add −2 to 4, and so on.

$$a_1 \text{ (first term)} = 6$$

$$a_2 \text{ (second term)} = 6 + (-2) = 4$$

$$a_3 \text{ (third term)} = 4 + (-2) = 2$$

$$a_4 \text{ (fourth term)} = 2 + (-2) = 0$$

$$a_5 \text{ (fifth term)} = 0 + (-2) = -2$$

$$a_6 \text{ (sixth term)} = -2 + (-2) = -4$$

The first six terms are

$$6, 4, 2, 0, -2, \text{ and } -4.$$

■

✔ **CHECK POINT 1** Write the first six terms of the arithmetic sequence with first term 100 and common difference −30.

3 Use the formula for the general term of an arithmetic sequence.

The General Term of an Arithmetic Sequence

Consider an arithmetic sequence whose first term is a_1 and whose common difference is d. We are looking for a formula for the general term, a_n. Let's begin by writing the first six terms. The first term is a_1. The second term is $a_1 + d$. The third term is $a_1 + d + d$, or $a_1 + 2d$. Thus, we start with a_1 and add d to each successive term. The first six terms are

$$a_1, \quad a_1 + d, \quad a_1 + 2d, \quad a_1 + 3d, \quad a_1 + 4d, \quad a_1 + 5d.$$

a_1, first term	a_2, second term	a_3, third term	a_4, fourth term	a_5, fifth term	a_6, sixth term

Compare the coefficient of d and the subscript of a denoting the term number. Can you see that the coefficient of d is 1 less than the subscript of a denoting the term number?

a_3: third term $= a_1 + 2d$ — 2 is one less than 3.

a_4: fourth term $= a_1 + 3d$ — 3 is one less than 4.

Thus, the formula for the nth term is

a_n: nth term $= a_1 + (n - 1)d.$ — $n - 1$ is one less than n.

General Term of an Arithmetic Sequence

The nth term (the general term) of an arithmetic sequence with first term a_1 and common difference d is

$$a_n = a_1 + (n - 1)d.$$

EXAMPLE 2 Using the Formula for the General Term of an Arithmetic Sequence

Find the eighth term of the arithmetic sequence whose first term is 4 and whose common difference is −7.

Solution To find the eighth term, a_8, we replace n in the formula with 8, a_1 with 4, and d with −7.

$$a_n = a_1 + (n - 1)d$$

$$a_8 = 4 + (8 - 1)(-7) = 4 + 7(-7) = 4 + (-49) = -45$$

The eighth term is −45. We can check this result by writing the first eight terms of the sequence:

$$4, -3, -10, -17, -24, -31, -38, -45.$$ ■

✔ **CHECK POINT 2** Find the ninth term of the arithmetic sequence whose first term is 6 and whose common difference is −5.

EXAMPLE 3 Using an Arithmetic Sequence to Model Teachers' Earnings

According to the National Education Association, teachers in the United States earned an average of \$21,700 in 1984. This amount has increased by approximately \$1472 per year.

a. Write a formula for the nth term of the arithmetic sequence that describes teachers' average earnings n years after 1983.

b. How much will U.S. teachers earn by the year 2005?

Solution

a. We can express teachers' earnings by the following arithmetic sequence:

21,700, 23,172, 24,644, 26,116, ...

a_1: earnings in 1984, 1 year after 1983 | a_2: earnings in 1985, 2 years after 1983 | a_3: earnings in 1986, 3 years after 1983 | a_4: earnings in 1987, 4 years after 1983

In this sequence a_1, the first term, represents the amount teachers earned in 1984. Each subsequent year this amount increases by \$1472, so $d = 1472$. We use the formula for the general term of an arithmetic sequence to write the nth term of the sequence that describes teachers' earnings n years after 1983.

$$a_n = a_1 + (n-1)d$$ This is the formula for the general term of an arithmetic sequence.

$$a_n = 21{,}700 + (n-1)1472$$ $a_1 = 21{,}700$ and $d = 1472$.

$$a_n = 21{,}700 + 1472n - 1472$$ Distribute 1472 to each term in parentheses.

$$a_n = 1472n + 20{,}228$$ Simplify.

Thus, teachers' earnings n years after 1983 can be described by $a_n = 1472n + 20{,}228$.

b. Now we need to find teachers' earnings in 2005. The year 2005 is 22 years after 1983: That is, $2005 - 1983 = 22$. Thus, $n = 22$. We substitute 22 for n in $a_n = 1472n + 20{,}228$.

$$a_{22} = 1472 \cdot 22 + 20{,}228 = 52{,}612$$

The 22nd term of the sequence is 52,612. Therefore, U.S. teachers are predicted to earn an average of \$52,612 by the year 2005. ■

✔ **CHECK POINT 3** According to the U.S. Bureau of Economic Analysis, U.S. travelers spent \$12,808 million in other countries in 1984. This amount has increased by approximately \$2350 million yearly.

a. Write a formula for the nth term of the arithmetic sequence that describes what U.S. travelers spend in other countries n years after 1983.

b. How much will U.S. travelers spend in other countries by the year 2010?

4 Use the formula for the sum of the first n terms of an arithmetic sequence.

The Sum of the First *n* Terms of an Arithmetic Sequence

The sum of the first n terms of an arithmetic sequence, denoted by S_n, and called the ***n*th partial sum**, can be found without having to add up all the terms. Let

$$S_n = a_1 + a_2 + a_3 + \cdots + a_n$$

be the sum of the first n terms of an arithmetic sequence. Because d is the common difference between terms, S_n can be written forward and backward as follows.

Forward: Start with the first term. Keep adding d.

Backward: Start with the last term. Keep subtracting d.

$$\begin{aligned} S_n &= a_1 + (a_1 + d) + (a_1 + 2d) + \cdots + a_n \\ S_n &= a_n + (a_n - d) + (a_n - 2d) + \cdots + a_1 \\ \hline 2S_n &= (a_1 + a_n) + (a_1 + a_n) + (a_1 + a_n) + \cdots + (a_1 + a_n) \end{aligned}$$

Add the two equations.

Because there are n sums of $(a_1 + a_n)$ on the right side, we can express this side as $n(a_1 + a_n)$. Thus, the last equation can be simplified:

$$2S_n = n(a_1 + a_n).$$

$$S_n = \frac{n}{2}(a_1 + a_n)$$ Solve for S_n, dividing both sides by 2.

We have proved the following result:

The Sum of the First *n* Terms of an Arithmetic Sequence

The sum, S_n, of the first n terms of an arithmetic sequence is given by

$$S_n = \frac{n}{2}(a_1 + a_n)$$

in which a_1 is the first term and a_n is the nth term.

To find the sum of the terms of an arithmetic sequence, we need to know the first term, a_1, the last term, a_n, and the number of terms, n. The following examples illustrate how to use this formula.

EXAMPLE 4 Finding the Sum of *n* Terms of an Arithmetic Sequence

Find the sum of the first 100 terms of the arithmetic sequence: $1, 3, 5, 7, \ldots$.

Solution We are finding the sum of the first 100 odd numbers. To find the sum of the first 100 terms, S_{100}, we replace n in the formula with 100.

$$S_n = \frac{n}{2}(a_1 + a_n)$$

$$S_{100} = \frac{100}{2}(a_1 + a_{100})$$

The first term, a_1, is 1.

We must find a_{100}, the 100th term.

We use the formula for the general term of a sequence to find a_{100}. The common difference, d, of $1, 3, 5, 7, \ldots$, is 2.

$$a_n = a_1 + (n - 1)d$$

This is the formula for the nth term of an arithmetic sequence. Use it to find the 100th term.

$$a_{100} = 1 + (100 - 1) \cdot 2$$

Substitute 100 for n, 2 for d, and 1 (the first term) for a_1.

$$= 1 + 99 \cdot 2$$

$$= 1 + 198 = 199$$

Now we are ready to find the sum of the 100 terms $1, 3, 5, 7, \ldots, 199$.

$$S_n = \frac{n}{2}(a_1 + a_n)$$

$$S_{100} = \frac{100}{2}(1 + 199) = 50(200) = 10{,}000$$

Use the formula for the sum of the first n terms of an arithmetic sequence. Let $n = 100$, $a_1 = 1$, and $a_{100} = 199$.

The sum of the first 100 odd numbers is 10,000. Equivalently, the 100th partial sum of the sequence $1, 3, 5, 7, \ldots$ is 10,000. ■

✔ **CHECK POINT 4** Find the sum of the first 15 terms of the arithmetic sequence: 3, 6, 9, 12, ...

Using Technology

To find

$$\sum_{i=1}^{25}(5i-9)$$

on a graphing utility, enter:

[SUM] [SEQ] $(5x - 9, x, 1, 25, 1)$.

Then press [ENTER].

EXAMPLE 5 Using S_n to Evaluate a Summation

Find the following sum: $\sum_{i=1}^{25}(5i-9)$.

Solution

$$\sum_{i=1}^{25}(5i-9) = (5\cdot 1-9)+(5\cdot 2-9)+(5\cdot 3-9)+\cdots+(5\cdot 25-9)$$

$$= -4 + 1 + 6 + \cdots + 116$$

By evaluating the first three terms and the last term, we see that $a_1 = -4$; d, the common difference, is $1 - (-4)$ or 5; and a_{25}, the last term, is 116.

$$S_n = \frac{n}{2}(a_1 + a_n)$$ Use the formula for the sum of the first n terms of an arithmetic sequence. Let $n = 25$, $a_1 = -4$, and $a_{25} = 116$.

$$S_{25} = \frac{25}{2}(-4 + 116) = \frac{25}{2}(112) = 1400$$

Thus,

$$\sum_{i=1}^{25}(5i-9) = 1400.$$

■

✔ **CHECK POINT 5** Find the following sum: $\sum_{i=1}^{30}(6i-11)$.

EXAMPLE 6 Modeling Total Nursing Home Costs over a Six-Year Period

Your grandmother has assets of \$350,000. One option that she is considering involves nursing home care for a six-year period beginning in 2001. The model

$$a_n = 1800n + 49{,}730$$

describes yearly nursing home costs n years after 2000. Does your grandmother have enough to pay for the facility?

Solution We must find the sum of an arithmetic sequence. The first term of the sequence corresponds to nursing home costs in the year 2001. The last term corresponds to nursing home costs in the year 2006. Because the model describes costs n years after 2000, $n = 1$ describes the year 2001 and $n = 6$ describes the year 2006.

$$a_n = 1800n + 49{,}730$$ This is the given formula for the general term of the sequence.

$$a_1 = 1800\cdot 1 + 49{,}730 = 51{,}530$$ Find a_1 by replacing n with 1.

$$a_6 = 1800\cdot 6 + 49{,}730 = 60{,}530$$ Find a_6 by replacing n with 6.

The first year the facility will cost \$51,530. By year six, the facility will cost \$60,530. Now we must find the sum of the costs for all six years. We focus on the sum of the first six terms of the arithmetic sequence

$$51{,}530, \quad 53{,}330, \ldots, \quad 60{,}530.$$

We find this sum using the formula for the sum of the first n terms of an arithmetic sequence. We are adding 6 terms: $n = 6$. The first term is 51,530: $a_1 = 51{,}530$. The last term—that is, the sixth term—is 60,530: $a_6 = 60{,}530$.

$$S_n = \frac{n}{2}(a_1 + a_n)$$

$$S_6 = \frac{6}{2}(51{,}530 + 60{,}530) = 3(112{,}060) = 336{,}180$$

Total nursing home costs for your grandmother are predicted to be \$336,180. Because your grandmother's assets are \$350,000, she has enough to pay for the facility. ■

CHECK POINT 6 In Example 6, how much would it cost for nursing home care for a ten-year period beginning in 2001?

EXERCISE SET 11.2

Practice Exercises

In Exercises 1–6, find the common difference for each arithmetic sequence.

1. 2, 6, 10, 14, …

2. 3, 8, 13, 18, …

3. −7, −2, 3, 8, …

4. −10, −4, 2, 8, …

5. 714, 711, 708, 705, …

6. 611, 606, 601, 596, …

In Exercises 7–16, write the first six terms of each arithmetic sequence with the given first term, a_1, and common difference, d.

7. $a_1 = 200, d = 20$

8. $a_1 = 300, d = 50$

9. $a_1 = -7, d = 4$

10. $a_1 = -8, d = 5$

11. $a_1 = 300, d = -90$

12. $a_1 = 200, d = -60$

13. $a_1 = \frac{5}{2}, d = -\frac{1}{2}$

14. $a_1 = \frac{3}{4}, d = -\frac{1}{4}$

15. $a_1 = -0.4, d = -1.6$

16. $a_1 = -0.3, d = -1.7$

In Exercises 17–24, use the formula for the general term (the nth term) of an arithmetic sequence to find the indicated term of each sequence with the given first term, a_1, and common difference, d.

17. Find a_6 when $a_1 = 13, d = 4$.

18. Find a_{16} when $a_1 = 9, d = 2$.

19. Find a_{50} when $a_1 = 7, d = 5$.

20. Find a_{60} when $a_1 = 8, d = 6$.

21. Find a_{200} when $a_1 = -40, d = 5$.

22. Find a_{150} when $a_1 = -60, d = 5$.

23. Find a_{60} when $a_1 = 35, d = -3$.

24. Find a_{70} when $a_1 = -32, d = 4$.

In Exercises 25–34, write a formula for the general term (the nth term) of each arithmetic sequence. Then use the formula for a_n to find a_{20}, the 20th term of the sequence.

25. 1, 5, 9, 13, …

26. 2, 7, 12, 17, …

27. $7, 3, -1, -5, \ldots$

28. $6, 1, -4, -9, \ldots$

29. $-20, -24, -28, -32, \ldots$

30. $-70, -75, -80, -85, \ldots$

31. $a_1 = -\frac{1}{3}, d = \frac{1}{3}$

32. $a_1 = -\frac{1}{4}, d = \frac{1}{4}$

33. $a_1 = 4, d = -0.3$

34. $a_1 = 5, d = -0.2$

35. Find the sum of the first 20 terms of the arithmetic sequence: $4, 10, 16, 22, \ldots$.

36. Find the sum of the first 25 terms of the arithmetic sequence: $7, 19, 31, 43, \ldots$.

37. Find the sum of the first 50 terms of the arithmetic sequence: $-10, -6, -2, 2, \ldots$.

38. Find the sum of the first 50 terms of the arithmetic sequence: $-15, -9, -3, 3, \ldots$.

39. Find $1 + 2 + 3 + 4 + \ldots + 100$, the sum of the first 100 natural numbers.

40. Find $2 + 4 + 6 + 8 + \ldots + 200$, the sum of the first 100 positive even integers.

41. Find the sum of the first 60 positive even integers.

42. Find the sum of the first 80 positive even integers.

43. Find the sum of the even integers between 21 and 45.

44. Find the sum of the odd integers between 30 and 54.

For Exercises 45–50, write out the first three terms and the last term. Then use the formula for the sum of the first n terms of an arithmetic sequence to find the indicated sum.

45. $\sum_{i=1}^{17} (5i + 3)$

46. $\sum_{i=1}^{20} (6i - 4)$

47. $\sum_{i=1}^{30} (-3i + 5)$

48. $\sum_{i=1}^{40} (-2i + 6)$

49. $\sum_{i=1}^{100} 4i$

50. $\sum_{i=1}^{50} (-4i)$

Application Exercises

51. According to the U.S. Bureau of Labor Statistics, in 1990 there were 126,424 thousand employees in the United States. This number has increased by approximately 1265 thousand employees per year.

a. Write the general term for the arithmetic sequence modeling the thousands of employees in the United States n years after 1989.

b. How many thousands of employees will there be by the year 2005?

52. According to the National Center for Education Statistics, the total enrollment in U.S. public elementary and secondary schools in 1985 was 39.05 million. Enrollment has increased by approximately 0.45 million per year.

a. Write the general term for the arithmetic sequence modeling the millions of students enrolled in U.S. public elementary and secondary schools n years after 1984.

b. How many millions of students will be enrolled by the year 2005?

53. Company A pays \$24,000 yearly with raises of \$1600 per year. Company B pays \$28,000 yearly with raises of \$1000 per year. Which company will pay more in year 10? How much more?

54. Company A pays \$23,000 yearly with raises of \$1200 per year. Company B pays \$26,000 yearly with raises of \$800 per year. Which company will pay more in year 10? How much more?

55. According to the Environmental Protection Agency, in 1960 the United States recovered 3.78 million tons of solid waste. Due primarily to recycling programs, this amount has increased by approximately 0.576 million ton per year.

a. Write the general term for the arithmetic sequence modeling the amount of solid waste recovered in the United States n years after 1959.

b. What is the total amount of solid waste recovered from 1960 through 2000?

56. According to the Environmental Protection Agency, in 1960 the United States generated 87.1 million tons of solid waste. This amount has increased by approximately 3.14 million tons per year.

a. Write the general term for the arithmetic sequence modeling the amount of solid waste generated in the United States n years after 1959.

b. What is the total amount of solid waste generated from 1960 through 2000?

57. A company offers a starting yearly salary of \$33,000 with raises of \$2500 per year. Find the total salary over a ten-year period.

58. You are considering two job offers. Company A will start you at \$19,000 a year and guarantee you a raise of \$2600 per year. Company B will start you at a higher salary, \$27,000 a year, but will only guarantee a raise of \$1200 per year. Find the total salary that each company will pay you over a ten-year period. Which company pays the greater total amount?

59. A theater has 30 seats in the first row, 32 seats in the second row, increasing by 2 seats each row for a total of 26 rows. How many seats are there in the theater?

60. A section in a stadium has 20 seats in the first row, 23 seats in the second row, increasing by 3 seats each row for a total of 38 rows. How many seats are in this section of the stadium?

Writing in Mathematics

61. What is an arithmetic sequence? Give an example with your explanation.

62. What is the common difference in an arithmetic sequence?

63. Explain how to find the general term of an arithmetic sequence.

64. Explain how to find the sum of the first n terms of an arithmetic sequence without having to add up all the terms.

65. Teachers' earnings n years after 1983 can be described by $a_n = 1472n + 20{,}228$. According to this model, what will teachers earn in 2083? Describe two possible circumstances that would render this predicted salary incorrect.

Technology Exercises

66. Use the SEQ (sequence) capability of a graphing utility and the formula you obtained for a_n to verify the value you found for a_{20} in any five exercises from Exercises 25–34.

67. Use the capability of a graphing utility to calculate the sum of a sequence to verify any five of your answers to Exercises 45–50.

Critical Thinking Exercises

68. Give examples of two different arithmetic sequences whose fourth term, a_4, is 10.

69. In the sequence 21,700, 23,172, 24,644, 26,116, ..., which term is 314,628?

70. A *degree-day* is a unit used to measure the fuel requirements of buildings. By definition, each degree that the average daily temperature is below 65°F is 1 degree-day. For example, a temperature of 42°F constitutes 23 degree-days. If the average temperature on January 1 was 42°F and fell 2°F for each subsequent day up to and including January 10, how many degree-days are included from January 1 to January 10?

71. Show that the sum of the first n positive odd integers,

$$1 + 3 + 5 + \cdots + (2n - 1),$$

is n^2.

Review Exercises

72. Solve: $\log(x^2 - 5) - \log(x + 5) = 3$. (Section 9.5, Example 6)

73. Solve: $x^2 + 3x \le 10$. (Section 8.5, Example 1)

74. Solve for P: $A = \dfrac{Pt}{P + t}$. (Section 6.7, Example 1)

SECTION 11.3 Geometric Sequences and Series

Objectives

1. Find the common ratio of a geometric sequence.
2. Write terms of a geometric sequence.
3. Use the formula for the general term of a geometric sequence.
4. Use the formula for the sum of the first n terms of a geometric sequence.
5. Find the value of an annuity.
6. Use the formula for the sum of an infinite geometric series.

SSM PH Tutor Center CD-ROM Video

Here we are at the closing moments of a job interview. You're shaking hands with the manager. You managed to answer all the tough questions without losing your poise, and now you've been offered a job. As a matter of fact, your qualifications are so terrific that you've been offered two jobs—one just the day before, with a rival company in the same field! One company offers \$30,000 the first year, with increases of 6% per year for four years after that. The other offers \$32,000 the first year, with annual increases of 3% per year after that. Over a five-year period, which is the better offer?

If salary raises amount to a certain percent each year, the yearly salaries over time form a geometric sequence. In this section, we investigate geometric sequences and their properties. After studying the section, you will be in a position to decide which job offer to accept: you will know which company will pay you more over five years.

Geometric Sequences

Figure 11.2 shows a sequence in which the number of squares is increasing. From left to right, the number of squares is 1, 5, 25, 125, and 625. In this sequence, each term after the first, 1, is obtained by multiplying the preceding term by a constant amount, namely 5. This sequence of increasing number of squares is an example of a *geometric sequence*.

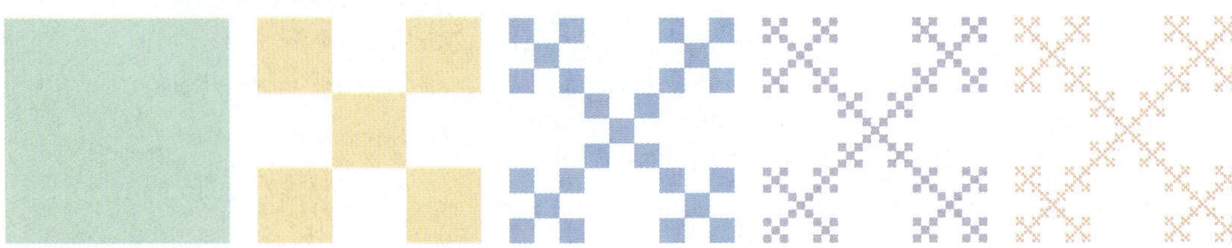

Figure 11.2 A geometric sequence of squares

Definition of a Geometric Sequence

A **geometric sequence** is a sequence in which each term after the first is obtained by multiplying the preceding term by a fixed nonzero constant. The amount by which we multiply each time is called the **common ratio** of the sequence.

1 Find the common ratio of a geometric sequence.

The common ratio, r, is found by dividing any term after the first term by the term that directly precedes it. In the following examples, the common ratio is found by dividing the second term by the first term, $\frac{a_2}{a_1}$.

GEOMETRIC SEQUENCE	COMMON RATIO
$1, 5, 25, 125, 625, \ldots$	$r = \frac{5}{1} = 5$
$4, 8, 16, 32, 64, \ldots$	$r = \frac{8}{4} = 2$
$6, -12, 24, -48, 96, \ldots$	$r = \frac{-12}{6} = -2$
$9, -3, 1, -\frac{1}{3}, \frac{1}{9}, \ldots$	$r = \frac{-3}{9} = -\frac{1}{3}$

Study Tip

When the common ratio of a geometric sequence is negative, the signs of the terms alternate.

2 Write terms of a geometric sequence.

How do we write out the terms of a geometric sequence when the first term and the common ratio are known? We multiply the first term by the common ratio to get the second term, multiply the second term by the common ratio to get the third term, and so on.

EXAMPLE 1 Writing the Terms of a Geometric Sequence

Write the first six terms of the geometric sequence with first term 6 and common ratio $\frac{1}{3}$.

Solution The first term is 6. The second term is $6 \cdot \frac{1}{3}$, or 2. The third term is $2 \cdot \frac{1}{3}$, or $\frac{2}{3}$. The fourth term is $\frac{2}{3} \cdot \frac{1}{3}$, or $\frac{2}{9}$, and so on. The first six terms are

$$6, 2, \frac{2}{3}, \frac{2}{9}, \frac{2}{27}, \frac{2}{81}.$$

■

CHECK POINT 1 Write the first six terms of the geometric sequence with first term 12 and common ratio $\frac{1}{2}$.

3 Use the formula for the general term of a geometric sequence.

The General Term of a Geometric Sequence

Consider a geometric sequence whose first term is a_1, and whose common ratio is r. We are looking for a formula for the general term, a_n. Let's begin by writing the first six terms. The first term is a_1. The second term is a_1r. The third term is $a_1r \cdot r$, or a_1r^2. The fourth term is $a_1r^2 \cdot r$, or a_1r^3, and so on. Starting with a_1 and multiplying each successive term by r, the first six terms are

$a_1,$	$a_1r,$	$a_1r^2,$	$a_1r^3,$	$a_1r^4,$	$a_1r^5.$
a_1, first term	a_2, second term	a_3, third term	a_4, fourth term	a_5, fifth term	a_6, sixth term

Compare the exponent on r and the subscript of a denoting the term number. Can you see that the exponent on r is 1 less than the subscript of a denoting the term number?

a_3: third term $= a_1r^2$ (2 is one less than 3.)

a_4: fourth term $= a_1r^3$ (3 is one less than 4.)

Thus, the formula for the nth term is

$$a_n = a_1r^{n-1}.$$

$n - 1$ is one less than n.

General Term of a Geometric Sequence

The nth term (the general term) of a geometric sequence with first term a_1 and common ratio r is

$$a_n = a_1r^{n-1}.$$

Study Tip

Be careful with the order of operations when evaluating

$$a_1r^{n-1}.$$

First find r^{n-1}. Then multiply the result by a_1.

EXAMPLE 2 Using the Formula for the General Term of a Geometric Sequence

Find the eighth term of the geometric sequence whose first term is −4 and whose common ratio is −2.

Solution To find the eighth term, a_8, we replace n in the formula with 8, a_1 with −4, and r with −2.

$$a_n = a_1r^{n-1}$$
$$a_8 = -4(-2)^{8-1} = -4(-2)^7 = -4(-128) = 512$$

The eighth term is 512. We can check this result by writing the first eight terms of the sequence:

$$-4, 8, -16, 32, -64, 128, -256, 512.$$

✔ CHECK POINT 2 Find the seventh term of the geometric sequence whose first term is 5 and whose common ratio is −3.

In Chapter 9, we studied exponential functions of the form $f(x) = b^x$ and the explosive exponential growth of world population. In our next example, we consider Florida's geometric population growth. Because **a geometric sequence is an exponential function whose domain is the set of positive integers**, geometric and exponential growth mean the same thing. (By contrast, an arithmetic sequence is a *linear function* whose domain is the set of positive integers.)

EXAMPLE 3 Geometric Population Growth

The population of Florida from 1980 through 1987 is shown in the following table.

Year	1980	1981	1982	1983	1984	1985	1986	1987
Population (in millions)	9.75	10.03	10.32	10.62	10.93	11.25	11.58	11.92

ENRICHMENT ESSAY

Geometric Population Growth

Economist Thomas Malthus (1766–1834) predicted that population growth would increase as a geometric sequence and food production would increase as an arithmetic sequence. He concluded that eventually population would exceed food production. If two sequences, one geometric and one arithmetic, are increasing, the geometric sequence will eventually overtake the arithmetic sequence, regardless of any head start that the arithmetic sequence might initially have.

a. Show that the population is increasing geometrically.

b. Write the general term for the geometric sequence describing population growth for Florida n years after 1979.

c. Estimate Florida's population, in millions, for the year 2000.

Solution

a. First, we divide the population for each year by the population in the preceding year.

$$\frac{10.03}{9.75} \approx 1.029, \quad \frac{10.32}{10.03} \approx 1.029, \quad \frac{10.62}{10.32} \approx 1.029$$

Continuing in this manner, we will keep getting approximately 1.029. This means that the population is increasing geometrically with $r \approx 1.029$. In this situation, the common ratio is the growth rate, indicating that the population of Florida in any year shown in the table is approximately 1.029 times the population the year before.

b. The sequence of Florida's population growth is

$$9.75, 10.03, 10.32, 10.62, 10.93, 11.25, 11.58, 11.92, \ldots.$$

Because the population is increasing geometrically, we can find the general term of this sequence using

$$a_n = a_1 r^{n-1}.$$

In this sequence, $a_1 = 9.75$ and r [from part (a)] ≈ 1.029. We substitute these values into the formula for the general term. This gives the general term for the geometric sequence describing Florida's population n years after 1979.

$$a_n = 9.75(1.029)^{n-1}$$

c. We can use the formula for the general term, a_n, in part (b) to estimate Florida's population for the year 2000. The year 2000 is 21 years after 1979—that is, $2000 - 1979 = 21$. Thus, $n = 21$. We substitute 21 for n in $a_n = 9.75(1.029)^{n-1}$.

$$a_{21} = 9.75(1.029)^{21-1} = 9.75(1.029)^{20} \approx 17.27$$

The formula indicates that Florida had a population of approximately 17.27 million in the year 2000. ■

✔ **CHECK POINT 3** Write the general term for the geometric sequence

$$3, 6, 12, 24, 48, \ldots.$$

Then use the formula for the general term to find the eighth term.

4 Use the formula for the sum of the first n terms of a geometric sequence.

The Sum of the First n Terms of a Geometric Sequence

The sum of the first n terms of a geometric sequence, denoted by S_n, and called the ***n*th partial sum**, can be found without having to add up all the terms. Recall that the first n terms of a geometric sequence are

$$a_1, a_1 r, a_1 r^2, \ldots, a_1 r^{n-2}, a_1 r^{n-1}.$$

We proceed as follows:

$$S_n = a_1 + a_1r + a_1r^2 + \cdots + a_1r^{n-2} + a_1r^{n-1}$$

S_n is the sum of the first n terms of the sequence.

$$rS_n = a_1r + a_1r^2 + a_1r^3 + \cdots + a_1r^{n-1} + a_1r^n$$

Multiply both sides of the equation by r.

$$S_n - rS_n = a_1 - a_1r^n$$

Subtract the second equation from the first equation.

$$S_n(1 - r) = a_1(1 - r^n)$$

Factor out S_n on the left and a_1 on the right.

$$S_n = \frac{a_1(1 - r^n)}{1 - r}.$$

Solve for S_n by dividing both sides by $1 - r$ (assuming that $r \neq 1$).

We have proved the following result:

Study Tip

If the common ratio is 1, the geometric sequence is

$$a_1, a_1, a_1, a_1, \ldots .$$

The sum of the first n terms of this sequence is na_1:

$$S_n = \underbrace{a_1 + a_1 + a_1 + \cdots + a_1}_{\text{There are } n \text{ terms.}}$$
$$= na_1.$$

The Sum of the First *n* Terms of a Geometric Sequence

The sum, S_n, of the first n terms of a geometric sequence is given by

$$S_n = \frac{a_1(1 - r^n)}{1 - r}$$

in which a_1 is the first term and r is the common ratio $(r \neq 1)$.

To find the sum of the terms of a geometric sequence, we need to know the first term, a_1, the common ratio, r, and the number of terms, n. The following examples illustrate how to use this formula.

EXAMPLE 4 Finding the Sum of the First *n* Terms of a Geometric Sequence

Find the sum of the first 18 terms of the geometric sequence: $2, -8, 32, -128, \ldots$.

Solution To find the sum of the first 18 terms, S_{18}, we replace n in the formula with 18.

$$S_n = \frac{a_1(1 - r^n)}{1 - r}$$

$$S_{18} = \frac{a_1(1 - r^{18})}{1 - r}$$

The first term, a_1, is 2.

We must find r, the common ratio.

We can find the common ratio by dividing the second term by the first term.

$$r = \frac{a_2}{a_1} = \frac{-8}{2} = -4$$

Now we are ready to find the sum of the first 18 terms of 2, −8, 32, −128, ...

$$S_n = \frac{a_1(1 - r^n)}{1 - r}$$ Use the formula for the sum of the first n terms of a geometric sequence.

$$S_{18} = \frac{2(1 - (-4)^{18})}{1 - (-4)}$$ a_1 (the first term) $= 2$, $r = -4$, and $n = 18$ because we want the sum of the first 18 terms.

$$= -27{,}487{,}790{,}694$$ Use a calculator.

The sum of the first 18 terms is −27,487,790,694. Equivalently, this number is the 18th partial sum of the sequence 2, −8, 32, −128, ... ■

✔ **CHECK POINT 4** Find the sum of the first nine terms of the geometric sequence: 2, −6, 18, −54, ...

EXAMPLE 5 Using S_n to Evaluate a Summation

Find the following sum: $\sum_{i=1}^{10} 6 \cdot 2^i$.

Solution Let's write out a few terms in the sum.

$$\sum_{i=1}^{10} 6 \cdot 2^i = 6 \cdot 2 + 6 \cdot 2^2 + 6 \cdot 2^3 + \cdots + 6 \cdot 2^{10}$$

Can you see that each term after the first is obtained by multiplying the preceding term by 2? To find the sum of the 10 terms ($n = 10$), we need to know the first term, a_1, and the common ratio, r. The first term is $6 \cdot 2$ or 12: $a_1 = 12$. The common ratio is 2.

$$S_n = \frac{a_1(1 - r^n)}{1 - r}$$ Use the formula for the sum of the first n terms of a geometric sequence.

$$S_{10} = \frac{12(1 - 2^{10})}{1 - 2}$$ a_1 (the first term) $= 12$, $r = 2$, and $n = 10$ because we are adding ten terms.

$$= 12{,}276$$ Use a calculator.

Thus,

$$\sum_{i=1}^{10} 6 \cdot 2^i = 12{,}276.$$ ■

Using Technology

To find

$$\sum_{i=1}^{10} 6 \cdot 2^i$$

on a graphing utility, enter

[SUM] [SEQ] $(6 \times 2^x, x, 1, 10, 1)$.

Then press [ENTER].

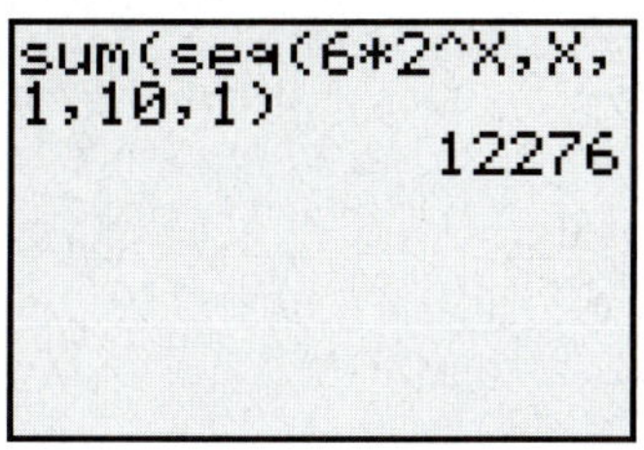

✔ **CHECK POINT 5** Find the following sum: $\sum_{i=1}^{8} 2 \cdot 3^i$.

Some of the exercises in the previous exercise set involved situations in which salaries increase by a fixed amount each year. A more realistic situation is one in which salary raises increase by a certain percent each year. Example 6 shows how such a situation can be described using a geometric series.

EXAMPLE 6 Computing a Lifetime Salary

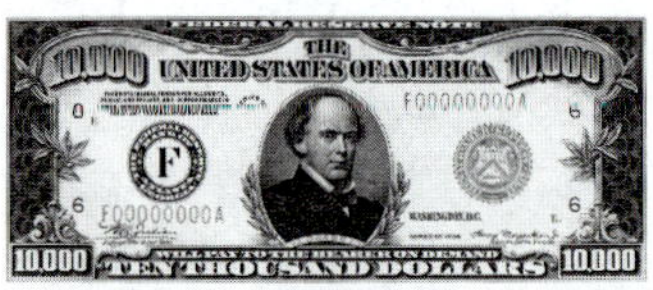

A union contract specifies that each worker will receive a 5% pay increase each year for the next 30 years. One worker is paid \$20,000 the first year. What is this person's total lifetime salary over a 30-year period?

Solution The salary for the first year is \$20,000. With a 5% raise, the second-year salary is computed as follows:

$$\text{Salary for year 2} = 20{,}000 + 20{,}000(0.05) = 20{,}000(1.05).$$

Each year, the salary is 1.05 times what it was in the previous year. Thus, the salary for year 3 is 1.05 times 20,000(1.05), or $20{,}000(1.05)^2$. The salaries for the first five years are given in the table.

Yearly Salaries					
Year 1	Year 2	Year 3	Year 4	Year 5	...
20,000	20,000(1.05)	$20{,}000(1.05)^2$	$20{,}000(1.05)^3$	$20{,}000(1.05)^4$	...

The numbers in the bottom row form a geometric sequence with $a_1 = 20{,}000$ and $r = 1.05$. To find the total salary over 30 years, we use the formula for the sum of the first n terms of a geometric sequence, with $n = 30$.

$$S_n = \frac{a_1(1 - r^n)}{1 - r}$$

$$S_{30} = \frac{20{,}000(1 - (1.05)^{30})}{1 - 1.05}$$

Total salary over 30 years

$$= \frac{20{,}000(1 - (1.05)^{30})}{-0.05}$$

$$\approx 1{,}328{,}777 \quad \text{Use a calculator.}$$

The total salary over the 30-year period is approximately \$1,328,777. ■

CHECK POINT 6 A job pays a salary of \$30,000 the first year. During the next 29 years, the salary increases by 6% each year. What is the total lifetime salary over the 30-year period?

5 Find the value of an annuity.

Annuities

The compound interest formula

$$A = P(1 + r)^t$$

gives the future value, A, after t years, when a fixed amount of money, P, the principal, is deposited in an account that pays an annual interest rate r (in decimal form) compounded once a year. However, money is often invested in small amounts at periodic intervals. For example, to save for retirement, you might decide to place \$1000 into an Individual Retirement Account (IRA) at the end of each year until you retire. An **annuity** is a sequence of equal payments made at equal time periods. An IRA is an example of an annuity.

Suppose P dollars is deposited into an account at the end of each year. The account pays an annual interest rate, r, compounded annually. At the end of the first year, the account contains P dollars. At the end of the second year, P dollars is deposited again. At the time of this deposit, the first deposit has received interest earned during the second year. The **value of the annuity** is the sum of all deposits made plus all interest paid. Thus, the value of the annuity after two years is

$$P + P(1 + r).$$

Deposit of P dollars at end of second year

First-year deposit of P dollars with interest earned for a year

The value of the annuity after three years is

$$P \quad + \quad P(1 + r) \quad + \quad P(1 + r)^2.$$

Deposit of P dollars at end of third year

Second-year deposit of P dollars with interest earned for a year

First-year deposit of P dollars with interest earned over two years

The value of the annuity after t years is

$$P + P(1 + r) + P(1 + r)^2 + P(1 + r)^3 + \cdots + P(1 + r)^{t-1}.$$

Deposit of P dollars at end of year t

First-year deposit of P dollars with interest earned over $t - 1$ years

This is the sum of the terms of a geometric sequence with first term P and common ratio $1 + r$. We use the formula

$$S_n = \frac{a_1(1 - r^n)}{1 - r}$$

to find the sum of the terms:

$$S_n = \frac{P(1 - (1 + r)^t)}{1 - (1 + r)} = \frac{P(1 - (1 + r)^t)}{-r} = P\frac{(1 + r)^t - 1}{r}.$$

This formula gives the value of an annuity after t years if interest is compounded once a year. We can adjust the formula to find the value of an annuity if equal payments are made at the end of each of n yearly compounding periods.

Value of an Annuity: Interest Compounded n Times per Year

If P is the deposit made at the end of each compounding period for an annuity at r percent annual interest compounded n times per year, the value, A, of the annuity after t years is

$$A = P\frac{\left(1 + \frac{r}{n}\right)^{nt} - 1}{\frac{r}{n}}.$$

EXAMPLE 7 Determining the Value of an Annuity

To save for retirement, you decide to deposit \$1000 into an IRA at the end of each year for the next 30 years. If the interest rate is 10% per year compounded annually, find the value of the IRA after 30 years.

Solution The annuity involves 30 year-end deposits of $P = \$1000$. The interest rate is 10%: $r = 0.10$. Because the deposits are made once a year and the interest is compounded once a year, $n = 1$. The number of years is 30: $t = 30$. We replace the variables in the formula for the value of an annuity with these numbers.

$$A = P\frac{\left(1+\frac{r}{n}\right)^{nt}-1}{\frac{r}{n}}$$

$$A = 1000\frac{\left(1+\frac{0.10}{1}\right)^{1\cdot 30}-1}{\frac{0.10}{1}} \approx 164{,}494$$

The value of the IRA at the end of 30 years is approximately \$164,494. ■

✔ CHECK POINT 7 If \$3000 is deposited into an IRA at the end of each year for 40 years and the interest rate is 10% per year compounded annually, find the value of the IRA after 40 years.

6 Use the formula for the sum of an infinite geometric series.

Geometric Series

An infinite sum of the form

$$a_1 + a_1 r + a_1 r^2 + a_1 r^3 + \cdots + a_1 r^{n-1} + \cdots$$

with first term a_1 and common ratio r is called an **infinite geometric series.** How can we determine which infinite geometric series have sums and which do not? We look at what happens to r^n as n gets larger in the formula for the sum of the first n terms of this series, namely

$$S_n = \frac{a_1(1-r^n)}{1-r}.$$

If r is any number between -1 and 1, that is, $-1 < r < 1$, the term r^n approaches 0 as n gets larger. For example, consider what happens to r^n for $r = \frac{1}{2}$:

$$\left(\frac{1}{2}\right)^1 = \frac{1}{2} \quad \left(\frac{1}{2}\right)^2 = \frac{1}{4} \quad \left(\frac{1}{2}\right)^3 = \frac{1}{8} \quad \left(\frac{1}{2}\right)^4 = \frac{1}{16} \quad \left(\frac{1}{2}\right)^5 = \frac{1}{32} \quad \left(\frac{1}{2}\right)^6 = \frac{1}{64}.$$

These numbers are approaching 0 as n gets larger.

Take another look at the formula for the sum of the first n terms of a geometric sequence.

$$S_n = \frac{a_1(1 - r^n)}{1 - r}$$

If $-1 < r < 1$, r^n approaches 0 as n gets larger.

Let us replace r^n with 0 in the formula for S_n. This change gives us a formula for the sum of infinite geometric series with common ratios between -1 and 1.

The Sum of an Infinite Geometric Series

If $-1 < r < 1$ (equivalently, $|r| < 1$), then the sum of the infinite geometric series

$$a_1 + a_1r + a_1r^2 + a_1r^3 + \cdots$$

in which a_1 is the first term and r is the common ratio is given by

$$S = \frac{a_1}{1 - r}.$$

If $|r| \geq 1$, the infinite series does not have a sum.

To use the formula for the sum of an infinite geometric series, we need to know the first term and the common ratio. For example, consider

$$\frac{1}{2} + \frac{1}{4} + \frac{1}{8} + \frac{1}{16} + \frac{1}{32} + \cdots.$$

First term, a_1, is $\frac{1}{2}$.

Common ratio, r, is $\frac{a_2}{a_1}$.
$r = \frac{1}{4} \div \frac{1}{2} = \frac{1}{4} \cdot 2 = \frac{1}{2}$

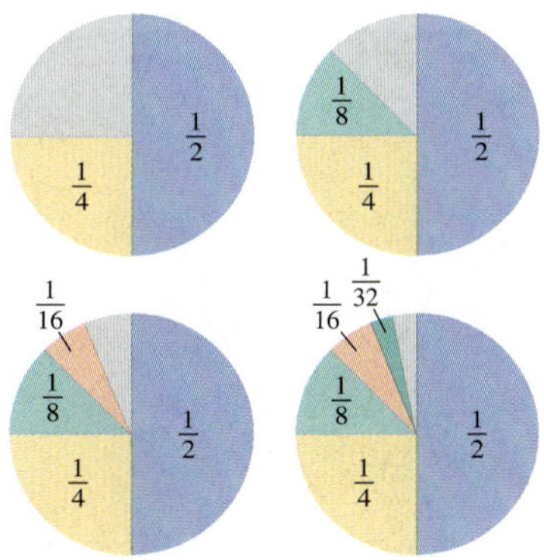

Figure 11.3 The sum $\frac{1}{2} + \frac{1}{4} + \frac{1}{8} + \frac{1}{16} + \frac{1}{32} + \cdots$ is approaching 1.

With $r = \frac{1}{2}$, the condition that $|r| < 1$ is met, so the infinite geometric series has a sum given by $S = \frac{a_1}{1 - r}$. The sum of the series is found as follows:

$$\frac{1}{2} + \frac{1}{4} + \frac{1}{8} + \frac{1}{16} + \frac{1}{32} + \cdots = \frac{a_1}{1 - r} = \frac{\frac{1}{2}}{1 - \frac{1}{2}} = \frac{\frac{1}{2}}{\frac{1}{2}} = 1.$$

Thus, the sum of the infinite geometric series is 1. Notice how this is illustrated in Figure 11.3. As more terms are included, the sum is approaching the area of one complete circle.

EXAMPLE 8 Finding the Sum of an Infinite Geometric Series

Find the sum of the infinite geometric series: $\frac{3}{8} - \frac{3}{16} + \frac{3}{32} - \frac{3}{64} + \cdots$.

Solution Before finding the sum, we must find the common ratio.

$$r = \frac{a_2}{a_1} = \frac{-\frac{3}{16}}{\frac{3}{8}} = -\frac{3}{16} \cdot \frac{8}{3} = -\frac{1}{2}$$

The infinite geometric series

$\frac{3}{8} - \frac{3}{16} + \frac{3}{32} - \frac{3}{64} + \cdots,$

repeated

Because $r = -\frac{1}{2}$, the condition that $|r| < 1$ is met. Thus, the infinite geometric series has a sum.

$$S = \frac{a_1}{1 - r}$$

This is the formula for the sum of an infinite geometric series. Let $a_1 = \frac{3}{8}$ and $r = -\frac{1}{2}$.

$$= \frac{\frac{3}{8}}{1 - \left(-\frac{1}{2}\right)} = \frac{\frac{3}{8}}{\frac{3}{2}} = \frac{3}{8} \cdot \frac{2}{3} = \frac{1}{4}$$

Thus, the sum of this infinite geometric series is $\frac{1}{4}$. Put in an informal way, as we continue to add more and more terms, the sum is approximately $\frac{1}{4}$. ■

✔ **CHECK POINT 8** Find the sum of the infinite geometric series: $3 + 2 + \frac{4}{3} + \frac{8}{9} + \cdots$.

We can use the formula for the sum of an infinite series to express a repeating decimal as a fraction in lowest terms.

EXAMPLE 9 Writing a Repeating Decimal as a Fraction

Express $0.\overline{7}$ as a fraction in lowest terms.

Solution

$$0.\overline{7} = 0.7777\ldots = \frac{7}{10} + \frac{7}{100} + \frac{7}{1000} + \frac{7}{10{,}000} + \cdots$$

Observe that $0.\overline{7}$ is an infinite geometric series with first term $\frac{7}{10}$ and common ratio $\frac{1}{10}$. Because $r = \frac{1}{10}$, the condition that $|r| < 1$ is met. Thus, we can use our formula to find the sum. Therefore,

$$0.\overline{7} = \frac{a_1}{1 - r} = \frac{\frac{7}{10}}{1 - \frac{1}{10}} = \frac{\frac{7}{10}}{\frac{9}{10}} = \frac{7}{10} \cdot \frac{10}{9} = \frac{7}{9}.$$

An equivalent fraction for $0.\overline{7}$ is $\frac{7}{9}$. ■

✔ **CHECK POINT 9** Express $0.\overline{9}$ as a fraction in lowest terms.

Infinite geometric series have many applications, as illustrated in Example 10.

EXAMPLE 10 Tax Rebates and the Multiplier Effect

A tax rebate that returns a certain amount of money to taxpayers can have a total effect on the economy that is many times this amount. In economics, this phenomenon is called the **multiplier effect**. Suppose, for example, that the government reduces taxes so that each consumer has \$2000 more income. The government assumes that each person will spend 70% of this (= \$1400). The individuals and businesses receiving this \$1400 in turn spend 70% of it (= \$980), creating extra income for other people to spend, and so on. Determine the total amount spent on consumer goods from the initial \$2000 tax rebate.

Solution The total amount spent is given by the infinite geometric series

$$1400 + 980 + 686 + \cdots.$$

70% of 1400 | 70% of 980

The first term is 1400: $a_1 = 1400$. The common ratio is 70%, or 0.7: $r = 0.7$. Because $r = 0.7$, the condition that $|r| < 1$ is met. Thus, we can use our formula to find the sum. Therefore,

$$1400 + 980 + 686 + \cdots = \frac{a_1}{1 - r} = \frac{1400}{1 - 0.7} \approx 4667.$$

This means that the total amount spent on consumer goods from the initial $2000 rebate is approximately $4667. ■

✔ CHECK POINT 10 Rework Example 10 and determine the total amount spent on consumer goods with a $1000 tax rebate and 80% spending down the line.

EXERCISE SET 11.3

Practice Exercises

In Exercises 1–8, find the common ratio for each geometric sequence.

1. $5, 15, 45, 135, \ldots$

2. $5, 10, 20, 40, \ldots$

3. $-15, 30, -60, 120, \ldots$

4. $-2, 6, -18, 54, \ldots$

5. $3, \frac{9}{2}, \frac{27}{4}, \frac{81}{8}, \ldots$

6. $4, \frac{8}{3}, \frac{16}{9}, \frac{32}{27}, \ldots$

7. $4, -0.4, 0.04, -0.004, \ldots$

8. $7, -0.7, 0.07, -0.007, \ldots$

In Exercises 9–16, write the first five terms of each geometric sequence with the given first term, a_1, and common ratio, r.

9. $a_1 = 2, r = 3$

10. $a_1 = 2, r = 4$

11. $a_1 = 20, r = \frac{1}{2}$

12. $a_1 = 24, r = \frac{1}{3}$

13. $a_1 = -4, r = -10$

14. $a_1 = -3, r = -10$

15. $a_1 = -\frac{1}{4}, r = -2$

16. $a_1 = -\frac{1}{16}, r = -4$

In Exercises 17–24, use the formula for the general term (the nth term) of a geometric sequence to find the indicated term of each sequence with the given first term, a_1, and common ratio, r.

17. Find a_8 when $a_1 = 6, r = 2$.

18. Find a_8 when $a_1 = 5, r = 3$.

19. Find a_{12} when $a_1 = 5, r = -2$.

20. Find a_{12} when $a_1 = 4, r = -2$.

21. Find a_6 when $a_1 = 6400, r = -\frac{1}{2}$.

22. Find a_6 when $a_1 = 8000, r = -\frac{1}{2}$.

23. Find a_8 when $a_1 = 1{,}000{,}000, r = 0.1$.

24. Find a_8 when $a_1 = 40{,}000, r = 0.1$.

In Exercises 25–32, write a formula for the general term (the nth term) of each geometric sequence. Then use the formula for a_n to find a_7, the seventh term of the sequence.

25. $3, 12, 48, 192, \ldots$

26. $3, 15, 75, 375, \ldots$

27. $18, 6, 2, \frac{2}{3}, \ldots$

28. $12, 6, 3, \frac{3}{2}, \ldots$

29. $1.5, -3, 6, -12, \ldots$

30. $5, -1, \frac{1}{5}, -\frac{1}{25}, \ldots$

31. $0.0004, -0.004, 0.04, -0.4. \ldots$

32. $0.0007, -0.007, 0.07, -0.7, \ldots$

Use the formula for the sum of the first n terms of a geometric sequence to solve Exercises 33–38.

33. Find the sum of the first 12 terms of the geometric sequence: $2, 6, 18, 54 \ldots$.

34. Find the sum of the first 12 terms of the geometric sequence: $3, 6, 12, 24, \ldots$.

35. Find the sum of the first 11 terms of the geometric sequence: $3, -6, 12, -24, \ldots$.

36. Find the sum of the first 11 terms of the geometric sequence: $4, -12, 36, -108, \ldots$.

37. Find the sum of the first 14 terms of the geometric sequence: $-\frac{3}{2}, 3, -6, 12, \ldots$.

38. Find the sum of the first 14 terms of the geometric sequence: $-\frac{1}{24}, \frac{1}{12}, -\frac{1}{6}, \frac{1}{3}, \ldots$.

In Exercises 39–44, find the indicated sum. Use the formula for the sum of the first n terms of a geometric sequence.

39. $\sum_{i=1}^{8} 3^i$

40. $\sum_{i=1}^{6} 4^i$

41. $\sum_{i=1}^{10} 5 \cdot 2^i$

42. $\sum_{i=1}^{7} 4(-3)^i$

43. $\sum_{i=1}^{6} \left(\frac{1}{2}\right)^{i+1}$

44. $\sum_{i=1}^{6} \left(\frac{1}{3}\right)^{i+1}$

In Exercises 45–52, find the sum of each infinite geometric series.

45. $1 + \frac{1}{3} + \frac{1}{9} + \frac{1}{27} + \cdots$

46. $1 + \frac{1}{4} + \frac{1}{16} + \frac{1}{64} + \cdots$

47. $3 + \frac{3}{4} + \frac{3}{4^2} + \frac{3}{4^3} + \cdots$

48. $5 + \frac{5}{6} + \frac{5}{6^2} + \frac{5}{6^3} + \cdots$

49. $1 - \frac{1}{2} + \frac{1}{4} - \frac{1}{8} + \cdots$

50. $3 - 1 + \frac{1}{3} - \frac{1}{9} + \cdots$

51. $\sum_{i=1}^{\infty} 26(-0.3)^{i-1}$

52. $\sum_{i=1}^{\infty} 51(-0.7)^{i-1}$

In Exercises 53–58, express each repeating decimal as a fraction in lowest terms.

53. $0.\overline{5} = \frac{5}{10} + \frac{5}{100} + \frac{5}{1000} + \frac{5}{10{,}000} + \cdots$

54. $0.\overline{1} = \frac{1}{10} + \frac{1}{100} + \frac{1}{1000} + \frac{1}{10{,}000} + \cdots$

55. $0.\overline{47} = \frac{47}{100} + \frac{47}{10{,}000} + \frac{47}{1{,}000{,}000} + \cdots$

56. $0.\overline{83} = \frac{83}{100} + \frac{83}{10{,}000} + \frac{83}{1{,}000{,}000} + \cdots$

57. $0.\overline{257}$

58. $0.\overline{529}$

In Exercises 59–64, the general term of a sequence is given. Determine whether the sequence is arithmetic, geometric, or neither. If the sequence is arithmetic, find the common difference; if it is geometric, find the common ratio.

59. $a_n = n + 5$

60. $a_n = n - 3$

61. $a_n = 2^n$

62. $a_n = \left(\frac{1}{2}\right)^n$

63. $a_n = n^2 + 5$

64. $a_n = n^2 - 3$

Application Exercises

Use the formula for the general term (the nth term) of a geometric sequence to solve Exercises 65–68.

In Exercises 65–66, suppose you save \$1 the first day of a month, \$2 the second day, \$4 the third day, and so on. That is, each day you save twice as much as you did the day before.

65. What will you put aside for savings on the fifteenth day of the month?

66. What will you put aside for savings on the thirtieth day of the month?

67. A professional baseball player signs a contract with a beginning salary of \$3,000,000 for the first year with an annual increase of 4% per year beginning in the second year. That is, beginning in year 2, the athlete's salary will be 1.04 times what it was in the previous year. What is the athlete's salary for year 7 of the contract? Round to the nearest dollar.

68. You are offered a job that pays \$30,000 for the first year with an annual increase of 5% per year beginning in the second year. That is, beginning in year 2, your salary will be 1.05 times what it was in the previous year. What can you expect to earn in your sixth year on the job? Round to the nearest dollar.

69. The population of Iraq from 1995 through 1998 is shown in the following table.

Year	1995	1996	1997	1998
Population in millions	20.60	21.36	22.19	23.02

Source: U.N. Population Division

a. Divide the population for each year by the population in the preceding year. Round to two decimal places and show that Iraq's population is increasing geometrically.

b. Write the general term of the geometric sequence describing population growth for Iraq n years after 1994.

c. Estimate Iraq's population, in millions, for the year 2005. Round to two decimal places.

70. The population of China from 1995 through 1998 is shown in the following table.

Year	1995	1996	1997	1998
Population in millions	1218.80	1232.21	1245.76	1259.46

Source: U.N. Population Division

a. Divide the population for each year by the population in the preceding year. Round to two decimal places and show that China's population is increasing geometrically.

b. Write the general term of the geometric sequence describing population growth for China n years after 1994.

c. Estimate China's population, in millions, for the year 2005. Round to two decimal places.

Use the formula for the sum of the first n terms of a geometric sequence to solve Exercises 71–76.

In Exercises 71–72, you save \$1 the first day of a month, \$2 the second day, \$4 the third day, continuing to double your savings each day.

71. What will your total savings be for the first 15 days?

72. What will your total savings be for the first 30 days?

73. A job pays a salary of \$24,000 the first year. During the next 19 years, the salary increases by 5% each year. What is the total lifetime salary over the 20-year period? Round to the nearest dollar.

74. You are investigating two employment opportunities. Company A offers \$30,000 the first year. During the next four years, the salary is guaranteed to increase by 6% per year. Company B offers \$32,000 the first year, with guaranteed annual increases of 3% per year after that. Which company offers the better total salary for a five-year contract? By how much? Round to the nearest dollar.

75. A pendulum swings through an arc of 20 inches. On each successive swing, the length of the arc is 90% of the previous length.

$$20, \quad 0.9(20), \quad 0.9^2(20), \quad 0.9^3(20), \quad \ldots$$

1st swing — 2nd swing — 3rd swing — 4th swing

After 10 swings, what is the total length of the distance the pendulum has swung? Round to the nearest hundredth of an inch.

76. A pendulum swings through an arc of 16 inches. On each successive swing, the length of the arc is 96% of the previous length.

$$16, \quad 0.96(16), \quad (0.96)^2(16), \quad (0.96)^3(16), \quad \ldots$$

1st swing — 2nd swing — 3rd swing — 4th swing

After 10 swings, what is the total length of the distance the pendulum has swung? Round to the nearest hundredth of an inch.

Use the formula for the value of an annuity to solve Exercises 77–80. Round answers to the nearest dollar.

77. To save for retirement, you decide to deposit \$2500 into an IRA at the end of each year for the next 40 years. If the interest rate is 9% per year compounded annually, find the value of the IRA after 40 years.

78. You decide to deposit \$100 at the end of each month into an account paying 8% interest compounded monthly to save for your child's education. How much will you save over 16 years?

79. You contribute \$600 at the end of each quarter to a Tax Sheltered Annuity (TSA) paying 8% annual interest compounded quarterly. Find the value of the TSA after 18 years.

80. To save for a new home, you invest \$500 per month in a mutual fund with an annual rate of return of 10% compounded monthly. How much will you have saved after four years?

Use the formula for the sum of an infinite geometric series to solve Exercises 81–83.

81. A new factory in a small town has an annual payroll of \$6 million. It is expected that 60% of this money will be spent in the town by factory personnel. The people in the town who receive this money are expected to spend 60% of what they receive in the town, and so on. What is the total of all this spending, called the *total economic impact* of the factory, on the town each year?

82. How much additional spending will be generated by a \$10 billion tax rebate if 60% of all income is spent?

83. If the shading process shown in the figure is continued indefinitely, what fractional part of the largest square is eventually shaded?

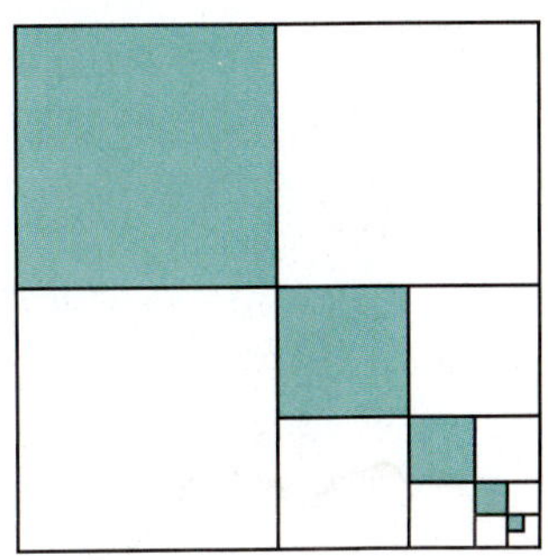

Writing in Mathematics

84. What is a geometric sequence? Give an example with your explanation.

85. What is the common ratio in a geometric sequence?

86. Explain how to find the general term of a geometric sequence.

87. Explain how to find the sum of the first n terms of a geometric sequence without having to add up all the terms.

88. What is an annuity?

89. What is the difference between a geometric sequence and an infinite geometric series?

90. How do you determine if an infinite geometric series has a sum? Explain how to find the sum of an infinite geometric series.

91. Would you rather have \$10,000,000 and a brand new BMW or 1¢ today, 2¢ tomorrow, 4¢ on day 3, 8¢ on day 4, 16¢ on day 5, and so on, for 30 days? Explain.

92. For the first 30 days of a flu outbreak, the number of students on your campus who become ill is increasing. Which is worse: the number of students with the flu is increasing arithmetically or is increasing geometrically? Explain your answer.

Technology Exercises

93. Use the SEQ (sequence) capability of a graphing utility and the formula you obtained for a_n to verify the value you found for a_7 in any three exercises from Exercises 25–32.

94. Use the capability of a graphing utility to calculate the sum of a sequence to verify any three of your answers to Exercises 39–44.

Critical Thinking Exercises

97. Which one of the following is true?

a. The sequence 2, 6, 24, 120, ... is an example of a geometric sequence.

b. The sum of the geometric series $\frac{1}{2} + \frac{1}{4} + \frac{1}{8} + \cdots + \frac{1}{512}$ can only be estimated without knowing precisely what terms occur between $\frac{1}{8}$ and $\frac{1}{512}$.

c. $10 - 5 + \frac{5}{2} - \frac{5}{4} + \cdots = \dfrac{10}{1 - \frac{1}{2}}$

d. If the nth term of a geometric sequence is $a_n = 3(0.5)^{n-1}$, the common ratio is $\frac{1}{2}$.

98. In a pest-eradication program, sterilized male flies are released into the general population each day. Ninety percent of those flies will survive a given day. How many flies should be released each day if the long-range goal of the program is to keep 20,000 sterilized flies in the population?

99. You are now 25 years old and would like to retire at age 55 with a retirement fund of \$1,000,000. How much should you deposit at the end of each month for the next 30 years in an IRA paying 10% annual interest compounded monthly to achieve your goal? Round to the nearest dollar.

Review Exercises

100. Simplify: $\sqrt{28} - 3\sqrt{7} + \sqrt{63}$. (Section 7.4, Example 2)

101. Solve: $2x^2 = 4 - x$. (Section 8.2, Example 2)

102. Rationalize the denominator: $\dfrac{6}{\sqrt{3} - \sqrt{5}}$. (Section 7.5, Example 5)

In Exercises 95–96, use a graphing utility to graph the function. Determine the horizontal asymptote for the graph of f and discuss its relationship to the sum of the given series.

	Function	**Series**
95.	$f(x) = \dfrac{2\left[1 - \left(\frac{1}{3}\right)^x\right]}{1 - \frac{1}{3}}$	$2 + 2\left(\frac{1}{3}\right) + 2\left(\frac{1}{3}\right)^2 + 2\left(\frac{1}{3}\right)^3 + \cdots$
96.	$f(x) = \dfrac{4[1 - (0.6)^x]}{1 - 0.6}$	$4 + 4(0.6) + 4(0.6)^2 + 4(0.6)^3 + \cdots$

▸ SECTION 11.4 The Binomial Theorem

Objectives

1. Recognize patterns in binomial expansions.
2. Evaluate a binomial coefficient.
3. Expand a binomial raised to a power.
4. Find a particular term in a binomial expansion.

SSM PH Tutor Center CD-ROM Video

Galaxies are groupings of billions of stars bound together gravitationally. Some galaxies, such as the Centaurus galaxy shown here, are elliptical in shape.

Is mathematics discovered or invented? For example, planets revolve in elliptical orbits. Does that mean that the ellipse is out there, waiting for the mind to discover it? Or do people create the definition of an ellipse just as they compose a song? And is it possible for the same mathematics to be discovered/invented by independent researchers separated by time, place, and culture? This is precisely what occurred when mathematicians attempted to find efficient methods for raising binomials to higher and higher powers, such as

$$(x + 2)^3, (x + 2)^4, (x + 2)^5, (x + 2)^6,$$

and so on. In this section, we study higher powers of binomials and a method first discovered/invented by great minds in Eastern and Western culture working independently.

Patterns in Binomial Expansions

1 Recognize patterns in binomial expansions.

When we write out the *binomial expression* $(a + b)^n$, where n is a positive integer, several patterns begin to appear.

$$(a + b)^1 = a + b$$
$$(a + b)^2 = a^2 + 2ab + b^2$$
$$(a + b)^3 = a^3 + 3a^2b + 3ab^2 + b^3$$
$$(a + b)^4 = a^4 + 4a^3b + 6a^2b^2 + 4ab^3 + b^4$$
$$(a + b)^5 = a^5 + 5a^4b + 10a^3b^2 + 10a^2b^3 + 5ab^4 + b^5$$

Discover for Yourself

Each expanded form of the binomial expression is a polynomial. Study the five polynomials and answer the following questions:

1. For each polynomial, describe the pattern for the exponents on a. What is the largest exponent on a? What happens to the exponent on a from term to term?
2. Describe the pattern for the exponents on b. What is the exponent on b in the first term? What is the exponent on b in the second term? What happens to the exponent on b from term to term?
3. Find the sum of the exponents on the variables in each term for the polynomials in the five rows. Describe the pattern.
4. How many terms are there in the polynomials on the right in relation to the power of the binomial?

Expansions of binomials

$(a + b)^1 = a + b$

$(a + b)^2 = a^2 + 2ab + b^2$

$(a + b)^3 = a^3 + 3a^2b + 3ab^2 + b^3$

$(a + b)^4 = a^4 + 4a^3b + 6a^2b^2 + 4ab^3 + b^4$

$(a + b)^5 = a^5 + 5a^4b + 10a^3b^2 + 10a^2b^3 + 5ab^4 + b^5$,

repeated from page 793.

How many of the following patterns were you able to discover?

1. The first term is a^n. The exponent on a decreases by 1 in each successive term.
2. The exponents on b increase by 1 in each successive term. In the first term, the exponent on b is 0. (Because $b^0 = 1$, b is not shown in the first term.) The last term is b^n.
3. The sum of the exponents on the variables in any term is equal to n, the exponent on $(a + b)^n$.
4. There is one more term in the polynomial expansion than there is in the power of the binomial, n. There are $n + 1$ terms in the expanded form of $(a + b)^n$.

Using these observations, the variable parts of the expansion of $(a + b)^6$ are

$$a^6,\quad a^5b,\quad a^4b^2,\quad a^3b^3,\quad a^2b^4,\quad ab^5,\quad b^6.$$

The first term is a^6, with the exponent on a decreasing by 1 in each successive term. The exponents on b increase from 0 to 6, with the last term being b^6. The sum of the exponents in each term is equal to 6.

We can generalize from these observations to obtain the variable parts of the expansion of $(a + b)^n$. They are

$$a^n,\quad a^{n-1}b,\quad a^{n-2}b^2,\quad a^{n-3}b^3, \ldots,\quad ab^{n-1},\quad b^n.$$

Exponents on a are decreasing by 1. Exponents on b are increasing by 1.

Sum of exponents: $n - 1 + 1 = n$

Sum of exponents: $n - 3 + 3 = n$

Sum of exponents: $1 + n - 1 = n$

Let's now establish a pattern for the coefficients of the terms in the binomial expansion. Notice that each row in the figure shown below begins and ends with 1. Any other number in the row can be obtained by adding the two numbers immediately above it.

Coefficients for $(a + b)^1$.	1 1
Coefficients for $(a + b)^2$.	1 2 1
Coefficients for $(a + b)^3$.	1 3 3 1
Coefficients for $(a + b)^4$.	1 4 6 4 1
Coefficients for $(a + b)^5$.	1 5 10 10 5 1

This triangular array of coefficients is called **Pascal's triangle**. If we continue with the sixth row, the first and last numbers are 1. Each of the other numbers is obtained by finding the sum of the two closest numbers above it in the fifth row.

1 1

1 2 1

1 3 3 1

1 4 6 4 1

1 5 10 10 5 1

1 + 5 | 5 + 10 | 10 + 10 | 10 + 5 | 5 + 1

1 6 15 20 15 6 1

We can use the numbers in the sixth row and the variable parts we found to write the expansion for $(a + b)^6$. It is

$$(a + b)^6 = a^6 + 6a^5b + 15a^4b^2 + 20a^3b^3 + 15a^2b^4 + 6ab^5 + b^6.$$

Study Tip

We have not shown the number in the top row of Pascal's triangle on the right. The top row is *row zero* because it corresponds to $(a + b)^0 = 1$. With row zero, the triangle appears as

1

1 1

1 2 1

1 3 3 1

1 4 6 4 1

etc.

2 Evaluate a binomial coefficient.

Binomial Coefficients

Pascal's triangle becomes cumbersome when a binomial is raised to a relatively large power. Therefore, the coefficients in a binomial expansion are instead given in terms of factorials. The coefficients are written in a special notation, which we define next.

Definition of a Binomial Coefficient $\binom{n}{r}$

For nonnegative integers n and r, with $n \geq r$, the expression $\binom{n}{r}$ (read "n above r") is called a **binomial coefficient** and is defined by

$$\binom{n}{r} = \frac{n!}{r!(n-r)!}.$$

The symbol ${}_nC_r$ is often used in place of $\binom{n}{r}$ to denote binomial coefficients.

Can you see that the definition of a binomial coefficient involves a fraction with factorials in the numerator and the denominator? When evaluating such an expression, try to reduce the fraction before performing the multiplications. For example, consider $\frac{26!}{21!}$. Rather than write out 26! as the product of all integers from 26 down to 1, we can express 26! as

$$26! = 26 \cdot 25 \cdot 24 \cdot 23 \cdot 22 \cdot 21!.$$

In this way, we can divide both the numerator and the denominator by the common factor, 21!.

$$\frac{26!}{21!} = \frac{26 \cdot 25 \cdot 24 \cdot 23 \cdot 22 \cdot \cancel{21!}}{\cancel{21!}} = 26 \cdot 25 \cdot 24 \cdot 23 \cdot 22 = 7{,}893{,}600$$

EXAMPLE 1 Evaluating Binomial Coefficients

Evaluate: **a.** $\binom{6}{2}$ **b.** $\binom{3}{0}$ **c.** $\binom{9}{3}$ **d.** $\binom{4}{4}$.

Solution In each case, we apply the definition of the binomial coefficient.

a. $\binom{6}{2} = \frac{6!}{2!(6-2)!} = \frac{6!}{2!4!} = \frac{6 \cdot 5 \cdot \cancel{4!}}{2 \cdot 1 \cdot \cancel{4!}} = 15$

b. $\binom{3}{0} = \frac{3!}{0!(3-0)!} = \frac{\cancel{3!}}{0!\cancel{3!}} = \frac{1}{1} = 1$

Remember that 0! = 1.

c. $\binom{9}{3} = \frac{9!}{3!(9-3)!} = \frac{9!}{3!6!} = \frac{9 \cdot 8 \cdot 7 \cdot \cancel{6!}}{3 \cdot 2 \cdot 1 \cdot \cancel{6!}} = 84$

d. $\binom{4}{4} = \frac{4!}{4!(4-4)!} = \frac{\cancel{4!}}{\cancel{4!}0!} = \frac{1}{1} = 1$ ■

Using Technology

Graphing utilities can compute binomial coefficients. For example, to find $\binom{6}{2}$, many utilities require the sequence

6 [nCr] 2 [ENTER].

The graphing utility will display 15. Consult your manual and verify the other evaluations in Example 1.

✔ CHECK POINT 1 Evaluate:

a. $\binom{6}{3}$ **b.** $\binom{6}{0}$

c. $\binom{8}{2}$ **d.** $\binom{3}{3}$.

3 Expand a binomial raised to a power.

The Binomial Theorem

If we use binomial coefficients and the pattern for the variable part of each term, a formula called the **Binomial Theorem** can be used to expand any positive integral power of a binomial.

A Formula for Expanding Binomials: The Binomial Theorem

For any positive integer n,

$$(a+b)^n = \binom{n}{0}a^n + \binom{n}{1}a^{n-1}b + \binom{n}{2}a^{n-2}b^2 + \binom{n}{3}a^{n-3}b^3 + \cdots + \binom{n}{n}b^n.$$

ENRICHMENT ESSAY

The Universality of Mathematics

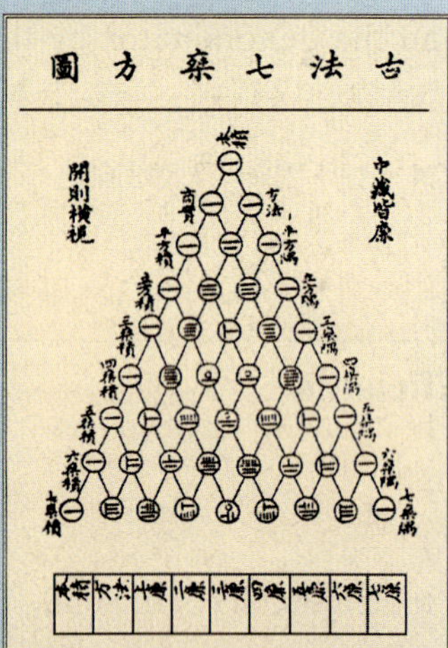

"Pascal's" triangle, credited to French mathematician Blaise Pascal (1623–1662), appeared in a Chinese document written in 1303. The Binomial Theorem was known in Eastern cultures prior to its discovery in Europe. The same mathematics is often discovered/invented by independent researchers separated by time, place, and culture.

EXAMPLE 2 Using the Binomial Theorem

Expand: $(x+2)^4$.

Solution We use the Binomial Theorem

$$(a+b)^n = \binom{n}{0}a^n + \binom{n}{1}a^{n-1}b + \binom{n}{2}a^{n-2}b^2 + \binom{n}{3}a^{n-3}b^3 + \cdots + \binom{n}{n}b^n$$

to expand $(x+2)^4$. In $(x+2)^4$, $a = x$, $b = 2$, and $n = 4$.

$$(x+2)^4 = \binom{4}{0}x^4 + \binom{4}{1}x^3 \cdot 2 + \binom{4}{2}x^2 \cdot 2^2 + \binom{4}{3}x \cdot 2^3 + \binom{4}{4}2^4$$

These binomial coefficients are evaluated using $\binom{n}{r} = \frac{n!}{r!(n-r)!}$.

$$= \frac{4!}{0!4!}x^4 + \frac{4!}{1!3!}x^3 \cdot 2 + \frac{4!}{2!2!}x^2 \cdot 4 + \frac{4!}{3!1!}x \cdot 8 + \frac{4!}{4!0!} \cdot 16$$

$\frac{4!}{2!2!} = \frac{4 \cdot 3 \cdot \cancel{2!}}{\cancel{2!} \cdot 2 \cdot 1} = \frac{12}{2} = 6$

Take a few minutes to verify the other factorial evaluations.

$$= 1 \cdot x^4 + 4x^3 \cdot 2 + 6x^2 \cdot 4 + 4x \cdot 8 + 1 \cdot 16$$

$$= x^4 + 8x^3 + 24x^2 + 32x + 16$$ ■

✔ CHECK POINT 2 Expand: $(x+1)^4$.

EXAMPLE 3 Using the Binomial Theorem

Expand: $(2x - y)^5$.

Solution Because the Binomial Theorem involves the addition of two terms raised to a power, we rewrite $(2x - y)^5$ as $[2x + (-y)]^5$. We use the Binomial Theorem

$$(a+b)^n = \binom{n}{0}a^n + \binom{n}{1}a^{n-1}b + \binom{n}{2}a^{n-2}b^2 + \binom{n}{3}a^{n-3}b^3 + \cdots + \binom{n}{n}b^n$$

to expand $[2x + (-y)]^5$. In $[2x + (-y)]^5$, $a = 2x$, $b = -y$, and $n = 5$.

$$(2x - y)^5 = [2x + (-y)]^5$$

$$= \binom{5}{0}(2x)^5 + \binom{5}{1}(2x)^4(-y) + \binom{5}{2}(2x)^3(-y)^2 + \binom{5}{3}(2x)^2(-y)^3 + \binom{5}{4}(2x)(-y)^4 + \binom{5}{5}(-y)^5$$

Evaluate binomial coefficients using $\binom{n}{r} = \frac{n!}{r!(n-r)!}$.

$$= \frac{5!}{0!5!}(2x)^5 + \frac{5!}{1!4!}(2x)^4(-y) + \frac{5!}{2!3!}(2x)^3(-y)^2 + \frac{5!}{3!2!}(2x)^2(-y)^3 + \frac{5!}{4!1!}(2x)(-y)^4 + \frac{5!}{5!0!}(-y)^5$$

$\frac{5!}{2!3!} = \frac{5 \cdot 4 \cdot \cancel{3!}}{2 \cdot 1 \cdot \cancel{3!}} = 10$
Take a few minutes to verify the other factorial evaluations.

$$= 1(2x)^5 + 5(2x)^4(-y) + 10(2x)^3(-y)^2 + 10(2x)^2(-y)^3 + 5(2x)(-y)^4 + 1(-y)^5$$

Raise both factors in these parentheses to the indicated powers.

$$= 1(32x^5) + 5(16x^4)(-y) + 10(8x^3)(-y)^2 + 10(4x^2)(-y)^3 + 5(2x)(-y)^4 + 1(-y)^5$$

Now raise $-y$ to the indicated powers.

$$= 1(32x^5) + 5(16x^4)(-y) + 10(8x^3)y^2 + 10(4x^2)(-y^3) + 5(2x)y^4 + 1(-y^5)$$

Multiplying factors in each of the six terms gives us the desired expansion:

$$(2x - y)^5 = 32x^5 - 80x^4y + 80x^3y^2 - 40x^2y^3 + 10xy^4 - y^5.$$ ■

✔ **CHECK POINT 3** Expand: $(x - 2y)^5$.

4 Find a particular term in a binomial expansion.

Finding a Particular Term in a Binomial Expansion

In the formula for expanding binomials, $\binom{n}{0}a^nb^0$ is the first term, $\binom{n}{1}a^{n-1}b$ is the second term, $\binom{n}{2}a^{n-2}b^2$ is the third term, and so on. Generalizing these observations provides a formula for finding a particular term without writing the entire expansion.

Finding a Particular Term in a Binomial Expansion

The rth term of the expansion of $(a + b)$ is

$$\binom{n}{r-1}a^{n-r+1}b^{r-1}.$$

EXAMPLE 4 Finding a Single Term of a Binomial Expansion

Find the fourth term in the expansion of $(3x + 2y)^7$.

Solution We will use the formula for the rth term of the expansion of $(a + b)^n$,

$$\binom{n}{r-1}a^{n-r+1}b^{r-1}$$

to find the fourth term of $(3x + 2y)^7$. For the fourth term, $n = 7$, $r = 4$, $a = 3x$, and $b = 2y$. Thus, the fourth term is

$$\binom{7}{4-1}(3x)^{7-4+1}(2y)^{4-1} = \binom{7}{3}(3x)^4(2y)^3 = \frac{7!}{3!(7-3)!}(3x)^4(2y)^3.$$

We use $\binom{n}{r} = \frac{n!}{r!(n-r)!}$ to evaluate $\binom{7}{3}$.

Now we need to evaluate the factorial expression and raise $3x$ and $2y$ to the indicated powers. We obtain

$$\frac{7!}{3!4!}(81x^4)(8y^3) = \frac{7 \cdot 6 \cdot 5 \cdot \cancel{4!}}{3 \cdot 2 \cdot 1 \cdot \cancel{4!}}(81x^4)(8y^3) = 35(81x^4)(8y^3) = 22{,}680x^4y^3.$$

The fourth term of $(3x + 2y)^7$ is $22{,}680x^4y^3$. ■

CHECK POINT 4 Find the fifth term in the expansion of $(2x + y)^9$.

EXERCISE SET 11.4

Practice Exercises

In Exercises 1–8, evaluate the given binomial coefficient.

1. $\binom{8}{3}$
2. $\binom{7}{2}$
3. $\binom{12}{1}$
4. $\binom{11}{1}$
5. $\binom{6}{6}$
6. $\binom{15}{2}$
7. $\binom{100}{2}$
8. $\binom{100}{98}$

In Exercises 9–30, use the Binomial Theorem to expand each binomial and express the result in simplified form.

9. $(x + 2)^3$
10. $(x + 4)^3$
11. $(3x + y)^3$
12. $(x + 3y)^3$
13. $(5x - 1)^3$
14. $(4x - 1)^3$
15. $(2x + 1)^4$
16. $(3x + 1)^4$
17. $(x^2 + 2y)^4$

18. $(x^2 + y)^4$
19. $(y - 3)^4$
20. $(y - 4)^4$
21. $(2x^3 - 1)^4$
22. $(2x^5 - 1)^4$
23. $(c + 2)^5$
24. $(c + 3)^5$
25. $(x - 1)^5$
26. $(x - 2)^5$
27. $(3x - y)^5$
28. $(x - 3y)^5$
29. $(2a + b)^6$
30. $(a + 2b)^6$

In Exercises 31–38, write the first three terms in each binomial expansion, expressing the result in simplified form.

31. $(x + 2)^8$
32. $(x + 3)^8$
33. $(x - 2y)^{10}$
34. $(x - 2y)^9$
35. $(x^2 + 1)^{16}$
36. $(x^2 + 1)^{17}$
37. $(y^3 - 1)^{20}$
38. $(y^3 - 1)^{21}$

In Exercises 39–46, find the term indicated in each expansion.

39. $(2x + y)^6$; third term
40. $(x + 2y)^6$; third term
41. $(x - 1)^9$; fifth term
42. $(x - 1)^{10}$; fifth term
43. $(x^2 + y^3)^8$; sixth term
44. $(x^3 + y^2)^8$; sixth term
45. $(x - \frac{1}{2})^9$; fourth term
46. $(x + \frac{1}{2})^8$; fourth term

Application Exercises

47. The percentage of people taking the SAT whose intended college major is engineering, $f(t)$, can be modeled by

$$f(t) = 0.002t^3 - 0.9t^2 + 1.27t + 6.76, \qquad 0 \le t \le 20,$$

where $t = 0$ represents 1975. How can we adjust this model so that $t = 0$ corresponds to 1985 rather than 1975? We shift the graph of f ten units to the left. We obtain $g(t) = f(t + 10)$. Use the Binomial Theorem to express $g(t)$ in descending powers of t.

48. The personal income per capita in the United States, $f(t)$, in constant 1992 dollars, can be modeled by

$$f(t) = 3.75t^3 - 115.23t^2 + 1229.81t + 16{,}025.65, \qquad 0 \le t \le 15,$$

where $t = 0$ represents 1979. How can we adjust this model so that $t = 0$ corresponds to 1989 rather than 1979? We shift the graph of f ten units to the left. We obtain $g(t) = f(t + 10)$. Use the Binomial Theorem to express $g(t)$ in descending powers of t.

Writing in Mathematics

49. Describe the pattern on the exponents on a in the expansion of $(a + b)^n$.
50. Describe the pattern on the exponents on b in the expansion of $(a + b)^n$.
51. What is true about the sum of the exponents on a and b in any term in the expansion of $(a + b)^n$?
52. How do you determine how many terms there are in a binomial expansion?
53. What is Pascal's triangle? How do you find the numbers in any row of the triangle?
54. Explain how to evaluate $\binom{n}{r}$. Provide an example with your explanation.
55. Explain how to use the Binomial Theorem to expand a binomial. Provide an example with your explanation.
56. Explain how to find a particular term in a binomial expansion without having to write out the entire expansion.
57. Are there situations in which it is easier to use Pascal's triangle than binomial coefficients? Describe these situations.

Technology Exercises

58. Use the [nCr] key on a graphing utility to verify your answers in Exercises 1–8.

In Exercises 59–60, graph each of the functions in the same viewing rectangle. Describe how the graphs illustrate the Binomial Theorem.

59. $f_1(x) = (x + 2)^3$
$f_2(x) = x^3$
$f_3(x) = x^3 + 6x^2$
$f_4(x) = x^3 + 6x^2 + 12x$
$f_5(x) = x^3 + 6x^2 + 12x + 8$
Use a $[-10, 10, 1]$ by $[-30, 30, 10]$ viewing rectangle.

60. $f_1(x) = (x + 1)^4$
$f_2(x) = x^4$
$f_3(x) = x^4 + 4x^3$
$f_4(x) = x^4 + 4x^3 + 6x^2$
$f_5(x) = x^4 + 4x^3 + 6x^2 + 4x$
$f_6(x) = x^4 + 4x^3 + 6x^2 + 4x + 1$
Use a $[-5, 5, 1]$ by $[-30, 30, 10]$ viewing rectangle.

In Exercises 61–63, use the Binomial Theorem to find a polynomial expansion for each function. Then use a graphing utility and an approach similar to the one in Exercises 59 and 60 to verify the expansion.

61. $f_1(x) = (x - 1)^3$

62. $f_1(x) = (x - 2)^4$

63. $f_1(x) = (x + 2)^6$

64. Graphing utilities capable of symbolic manipulation, such as the TI-92, will expand binomials. On the TI-92, to expand $(3a - 5b)^{12}$, input the following:

[EXPAND] ((3a [−] 5b) [∧] 12) [ENTER].

Use a graphing utility with this capability to verify any five of the expansions you performed by hand in Exercises 9–30.

Critical Thinking Exercises

65. Which one of the following is true?

a. The binomial expansion for $(a + b)^n$ contains n terms.

b. The Binomial Theorem can be written in condensed form as $(a + b)^n = \sum_{r=0}^{n} \binom{n}{r} a^{n-r} b^r$.

c. The sum of the binomial coefficients in $(a + b)^n$ cannot be 2^n.

d. There are no values of a and b such that $(a + b)^4 = a^4 + b^4$.

66. Use the Binomial Theorem to expand and then simplify the result: $(x^2 + x + 1)^3$. [*Hint*: Write $x^2 + x + 1$ as $x^2 + (x + 1)$].

67. Find the term in the expansion of $(x^2 + y^2)^5$ containing x^4 as a factor.

Review Exercises

68. If $f(x) = x^2 + 2x + 3$, find $f(a + 1)$. (Section 5.2, Example 11)

69. If $f(x) = x^2 + 5x$ and $g(x) = 2x - 3$, find $f(g(x))$ and $g(f(x))$. (Section 9.2, Example 1)

70. Subtract: $\dfrac{x}{x + 3} - \dfrac{x + 1}{2x^2 - 2x - 24}$. (Section 6.2, Example 7)

CHAPTER 11 GROUP PROJECTS

1. Group members serve as a financial team analyzing the three options given to the professional baseball player described in the chapter opener on page 758. As a group, determine which option provides the most amount of money over the six-year contract and which provides the least. Describe one advantage and one disadvantage to each option.
2. Enough curiosities involving the Fibonacci sequence exist to warrant a flourishing Fibonacci Association, which publishes a quarterly journal. Do some research on the Fibonacci sequence by consulting the Internet or the research department of your library, and find one property that interests you. After doing this research, get together with your group to share these intriguing properties.
3. Members of your group have been hired by the Environmental Protection Agency to write a report on whether we are making significant progress in recovering solid waste. Use the models from Exercises 55 and 56 in Exercise Set 11.2 on page 776 as the basis for your report. A graph of each model from 1960 through the present would be helpful. What percentage of solid waste generated is actually recovered on a year-to-year basis? Be as creative as you want in your report and then draw conclusions. The group should write up the report and perhaps even include suggestions as to how we might improve recycling progress.

CHAPTER SUMMARY, REVIEW, AND TEST

SUMMARY

DEFINITIONS AND CONCEPTS	EXAMPLES

Section 11.1 Sequences and Summation Notation

An infinite sequence $\{a_n\}$ is a function whose domain is the set of positive integers. The function values, or terms, are represented by

$$a_1, a_2, a_3, a_4, \ldots, a_n, \ldots.$$

General term: $a_n = \dfrac{(-1)^n}{n^3}$.

$$a_1 = \frac{(-1)^1}{1^3} = -1, \quad a_2 = \frac{(-1)^2}{2^3} = \frac{1}{8},$$

$$a_3 = \frac{(-1)^3}{3^3} = -\frac{1}{27}, \quad a_4 = \frac{(-1)^4}{4^3} = \frac{1}{64}$$

First four terms are $-1, \frac{1}{8}, -\frac{1}{27}, \frac{1}{64}$.

Factorial Notation

$$n! = n(n-1)(n-2)\cdots(3)(2)(1) \quad \text{and} \quad 0! = 1$$

$$6! = 6 \cdot 5 \cdot 4 \cdot 3 \cdot 2 \cdot 1 = 720$$

$$3! = 3 \cdot 2 \cdot 1 = 6$$

Summation Notation

$$\sum_{i=1}^{n} a_i = a_1 + a_2 + a_3 + a_4 + \cdots + a_n$$

In the summation shown here, i is the index of summation, n is the upper limit of summation, and 1 is the lower limit of summation.

$$\sum_{i=3}^{7} (i^2 - 4)$$
$$= (3^2 - 4) + (4^2 - 4) + (5^2 - 4) + (6^2 - 4) + (7^2 - 4)$$
$$= (9 - 4) + (16 - 4) + (25 - 4) + (36 - 4) + (49 - 4)$$
$$= 5 + 12 + 21 + 32 + 45$$
$$= 115$$

Section 11.2 Arithmetic Sequences

In an arithmetic sequence, each term after the first differs from the preceding term by a constant, the common difference. Subtract any term from the term that directly follows to find the common difference.

General Term of an Arithmetic Sequence
The nth term (the general term) of an arithmetic sequence with first term a_1 and common difference d is

$$a_n = a_1 + (n-1)d.$$

Find the general term and the tenth term:

$$3, 7, 11, 15, \ldots.$$

$a_1 = 3$ $\quad d = 7 - 3 = 4$

Using $a_n = a_1 + (n-1)d$:

$$a_n = 3 + (n-1)4 = 3 + 4n - 4 = 4n - 1.$$

The general term is $a_n = 4n - 1$.
The tenth term is $a_{10} = 4 \cdot 10 - 1 = 39$.

The Sum of the First n Terms of an Arithmetic Sequence
The sum, S_n, of the first n terms of an arithmetic sequence is given by

$$S_n = \frac{n}{2}(a_1 + a_n)$$

in which a_1 is the first term and a_n is the nth term.

Find the sum of the first ten terms:

$$2, 5, 8, 11, \ldots.$$

$a_1 = 2$ $\quad d = 5 - 2 = 3$

First find a_{10}, the 10th term. Using $a_n = a_1 + (n-1)d$,

$$a_{10} = 2 + (10 - 1) \cdot 3 = 2 + 9 \cdot 3 = 29.$$

Find the sum of the first ten terms using

$$S_n = \frac{n}{2}(a_1 + a_n).$$

$$S_{10} = \frac{10}{2}(a_1 + a_{10}) = 5(2 + 29) = 5(31) = 155$$

DEFINITIONS AND CONCEPTS	EXAMPLES
Section 11.3 Geometric Sequences and Series	
In a geometric sequence, each term after the first is obtained by multiplying the preceding term by a nonzero constant, the common ratio. Divide any term after the first by the term that directly precedes it to find the common ratio. ***General Term of a Geometric Sequence*** The nth term (the general term) of a geometric sequence with first term a_1 and common ratio r is $$a_n = a_1r^{n-1}.$$	Find the general term and the ninth term: $$12, -6, 3, -\frac{3}{2}, \ldots.$$ $a_1 = 12$; $r = \frac{-6}{12} = -\frac{1}{2}$ Using $a_n = a_1r^{n-1}$ $$a_n = 12\left(-\frac{1}{2}\right)^{n-1} \text{ is the general term.}$$ The ninth term is $$a_9 = 12\left(-\frac{1}{2}\right)^{9-1} = 12\left(-\frac{1}{2}\right)^8 = \frac{12}{256} = \frac{3}{64}.$$
The Sum of the First n Terms of a Geometric Sequence The sum, S_n, of the first n terms of a geometric sequence is given by $$S_n = \frac{a_1(1 - r^n)}{1 - r}$$ in which a_1 is the first term and r is the common ratio $(r \neq 1)$.	Find $\sum_{i=1}^{8} 4 \cdot 3^i$ $$= 4 \cdot 3 + 4 \cdot 3^2 + 4 \cdot 3^3 + \cdots + 4 \cdot 3^8.$$ $a_1 = 12$; $r = \frac{4 \cdot 3^2}{4 \cdot 3} = 3$ Using $S_n = \frac{a_1(1 - r^n)}{1 - r}$: $$S_8 = \frac{12(1 - 3^8)}{1 - 3} = 39{,}360.$$
The Sum of an Infinite Geometric Series If $-1 < r < 1$ (equivalently, $\|r\| < 1$), then the sum of the infinite geometric series $$a_1 + a_1r + a_1r^2 + a_1r^3 + \cdots$$ in which a_1 is the first term and r is the common ratio is given by $$S = \frac{a_1}{1 - r}.$$ If $\|r\| \geq 1$, the infinite series does not have a sum.	Find the sum: $$6 + \frac{6}{3} + \frac{6}{3^2} + \frac{6}{3^3} + \cdots.$$ $a_1 = 6$; $r = \frac{1}{3}$ Using $S = \frac{a_1}{1 - r}$, the sum is $$S = \frac{6}{1 - \frac{1}{3}} = \frac{6}{\frac{2}{3}} = 6 \cdot \frac{3}{2} = 9.$$
Section 11.4 The Binomial Theorem	
Definition of a Binomial Coefficient $$\binom{n}{r} = \frac{n!}{r!(n - r)!}$$	$$\binom{8}{3} = \frac{8!}{3!(8 - 3)!} = \frac{8!}{3!5!}$$ $$= \frac{8 \cdot 7 \cdot 6 \cdot \cancel{5!}}{3 \cdot 2 \cdot 1 \cdot \cancel{5!}} = 56$$

DEFINITIONS AND CONCEPTS	EXAMPLES
Section 11.4 The Binomial Theorem (continued)	
A Formula for Expanding Binomials: The Binomial Theorem For any positive integer n, $(a+b)^n = \binom{n}{0}a^n + \binom{n}{1}a^{n-1}b + \binom{n}{2}a^{n-2}b^2 + \binom{n}{3}a^{n-3}b^3 + \cdots + \binom{n}{n}b^n.$	Expand: $(3x-y)^4 = [3x+(-y)]^4.$ $= \binom{4}{0}(3x)^4 + \binom{4}{1}(3x)^3(-y) + \binom{4}{2}(3x)^2(-y)^2 + \binom{4}{3}(3x)^1(-y)^3 + \binom{4}{4}(-y)^4$ $= 1\cdot 81x^4 + 4\cdot 27x^3(-y) + 6\cdot 9x^2y^2 + 4\cdot 3x(-y^3) + 1\cdot y^4$ $= 81x^4 - 108x^3y + 54x^2y^2 - 12xy^3 + y^4$
Finding a Particular Term in a Binomial Expansion The rth term of the expansion of $(a+b)^n$ is $\binom{n}{r-1}a^{n-r+1}b^{r-1}.$	The 8th term of $(x+2y)^{10}$ is $\binom{10}{8-1}x^{10-8+1}(2y)^{8-1}$ $= \binom{10}{7}x^3(2y)^7 = 120x^3\cdot 128y^7$ $= 15{,}360x^3y.$

Review Exercises

11.1 *In Exercises 1–4, write the first four terms of each sequence whose general term is given.*

1. $a_n = 7n - 4$

2. $a_n = (-1)^n\dfrac{n+2}{n+1}$

3. $a_n = \dfrac{1}{(n-1)!}$

4. $a_n = \dfrac{(-1)^{n+1}}{2^n}$

In Exercises 5–6, find each indicated sum.

5. $\sum_{i=1}^{5}(2i^2 - 3)$ **6.** $\sum_{i=0}^{4}(-1)^{i+1}i!$

In Exercises 7–8, express each sum using summation notation. Use i for the index of summation.

7. $\frac{1}{3} + \frac{2}{4} + \frac{3}{5} + \cdots + \frac{15}{17}$

8. $4^3 + 5^3 + 6^3 + \cdots + 13^3$

11.2 *In Exercises 9–11, write the first six terms of each arithmetic sequence.*

9. $a_1 = 7, d = 4$

10. $a_1 = -4, d = -5$

11. $a_1 = \frac{3}{2}, d = -\frac{1}{2}$

In Exercises 12–14, use the formula for the general term (the nth term) of an arithmetic sequence to find the indicated term of each sequence.

12. Find a_6 when $a_1 = 5, d = 3$.

13. Find a_{12} when $a_1 = -8, d = -2$.

14. Find a_{14} when $a_1 = 14, d = -4$.

In Exercises 15–18, write a formula for the general term (the nth term) of each arithmetic sequence. Then use the formula for a_n to find a_{20}, the 20th term of the sequence.

15. $-7, -3, 1, 5, \ldots$

16. $a_1 = 200, d = -20$

17. $a_1 = -12, d = -\frac{1}{2}$

18. $15, 8, 1, -6, \ldots$

19. Find the sum of the first 22 terms of the arithmetic sequence: $5, 12, 19, 26, \ldots$.

20. Find the sum of the first 15 terms of the arithmetic sequence: $-6, -3, 0, 3, \ldots$.

21. Find $3 + 6 + 9 + \cdots + 300$, the sum of the first 100 positive multiples of 3.

In Exercises 22–24, use the formula for the sum of the first n terms of an arithmetic sequence to find the indicated sum.

22. $\sum_{i=1}^{16}(3i + 2)$ **23.** $\sum_{i=1}^{25}(-2i + 6)$

24. $\sum_{i=1}^{30}(-5i)$

25. In 1911, the world record for the men's mile run was 1043.04 seconds. The world record has decreased by approximately 0.4118 second each year since then.

a. Write the general term for the arithmetic sequence modeling record times for the men's mile run n years after 1910.

b. Use the model to predict the record time for the men's mile run for the year 2010.

26. A company offers a starting salary of \$31,500 with raises of \$2300 per year. Find the total salary over a ten-year period.

27. A theater has 25 seats in the first row and 35 rows in all. Each successive row contains one additional seat. How many seats are in the theater?

11.3 *In Exercises 28–31, write the first five terms of each geometric sequence.*

28. $a_1 = 3, r = 2$

29. $a_1 = \frac{1}{2}, r = \frac{1}{2}$

30. $a_1 = 16, r = -\frac{1}{4}$

31. $a_1 = -5, r = -1$

In Exercises 32–34, use the formula for the general term (the nth term) of a geometric sequence to find the indicated term of each sequence.

32. Find a_7 when $a_1 = 2, r = 3$.

33. Find a_6 when $a_1 = 16, r = \frac{1}{2}$.

34. Find a_5 when $a_1 = -3, r = 2$.

In Exercises 35–37, write a formula for the general term (the nth term) of each geometric sequence. Then use the formula for a_n to find a_8, the eighth term of the sequence.

35. $1, 2, 4, 8, \ldots$

36. $100, 10, 1, \frac{1}{10}, \ldots$

37. $12, -4, \frac{4}{3}, -\frac{4}{9}, \ldots$

38. Find the sum of the first 15 terms of the geometric sequence: $5, -15, 45, -135, \ldots$.

39. Find the sum of the first 7 terms of the geometric sequence: $8, 4, 2, 1, \ldots$.

In Exercises 40–42, use the formula for the sum of the first n terms of a geometric sequence to find the indicated sum.

40. $\sum_{i=1}^{6} 5^i$

41. $\sum_{i=1}^{7} 3(-2)^i$

42. $\sum_{i=1}^{5} 2\left(\frac{1}{4}\right)^{i-1}$

In Exercises 43–46, find the sum of each infinite geometric series.

43. $9 + 3 + 1 + \frac{1}{3} + \cdots$

44. $2 - 1 + \frac{1}{2} - \frac{1}{4} + \cdots$

45. $-6 + 4 - \frac{8}{3} + \frac{16}{9} - \cdots$

46. $\sum_{i=1}^{\infty} 5(0.8)^i$

In Exercises 47–48, express each repeating decimal as a fraction in lowest terms.

47. $0.\overline{6}$

48. $0.\overline{47}$

49. A job pays \$32,000 for the first year with an annual increase of 6% per year beginning in the second year. What is the salary in the sixth year? What is the total salary paid over this six-year period? Round answers to the nearest dollar.

50. You decide to deposit \$200 at the end of each month into an account paying 10% interest compounded monthly to save for your child's education. How much will you save over 18 years? Round to the nearest dollar.

51. A factory in an isolated town has an annual payroll of \$4 million. It is estimated that 70% of this money is spent within the town, that people in the town receiving this money will again spend 70% of what they receive in the town, and so on. What is the total of all this spending in the town each year?

11.4 *In Exercises 52–53, evaluate the given binomial coefficient.*

52. $\binom{11}{8}$

53. $\binom{90}{2}$

In Exercises 54–57, use the Binomial Theorem to expand each binomial and express the result in simplified form.

54. $(2x + 1)^3$

55. $(x^2 - 1)^4$

56. $(x + 2y)^5$

57. $(x - 2)^6$

In Exercises 58–59, write the first three terms in each binomial expansion, expressing the result in simplified form.

58. $(x^2 + 3)^8$

59. $(x - 3)^9$

In Exercises 60–61, find the term indicated in each expansion.

60. $(x + 2)^5$; fourth term

61. $(2x - 3)^6$; fifth term

Chapter 11 Test

1. Write the first five terms of the sequence whose general term is $a_n = \dfrac{(-1)^{n+1}}{n^2}$.

2. Find the indicated sum: $\sum_{i=1}^{5} (i^2 + 10)$.

3. Express the sum using summation notation. Use i for the index of summation.

$$\frac{2}{3} + \frac{3}{4} + \frac{4}{5} + \cdots + \frac{21}{22}$$

In Exercises 4–5, write a formula for the general term (the nth term) of each sequence. Then use the formula to find the twelfth term of the sequence.

4. 4, 9, 14, 19, ...

5. 16, 4, 1, $\frac{1}{4}$, ...

In Exercises 6–7, use the formula for the sum of the first n terms of an arithmetic sequence.

6. Find the sum of the first ten terms of the arithmetic sequence: −7, −14, −21, −28, ...

7. Find $\sum_{i=1}^{20} (3i - 4)$.

In Exercises 8–9, use the formula for the sum of the first n terms of a geometric sequence.

8. Find the sum of the first ten terms of the geometric sequence: 7, −14, 28, −56, ...

9. Find $\sum_{i=1}^{15} (-2)^i$.

10. Find the sum of the infinite geometric series:

$$4 + \frac{4}{2} + \frac{4}{2^2} + \frac{4}{2^3} + \cdots.$$

11. Express $0.\overline{73}$ in fractional notation.

12. A job pays \$30,000 for the first year with an annual increase of 4% per year beginning in the second year. What is the total salary paid over an eight-year period? Round to the nearest dollar.

13. Evaluate: $\dbinom{9}{2}$.

14. Use the Binomial Theorem to expand and simplify: $(x^2 - 1)^5$.

15. Use the Binomial Theorem to write the first three terms in the expansion and simplify: $(x + y^2)^8$.

Cumulative Review Exercises (Covering the Entire Book)

In Exercises 1–17, solve each equation, inequality, or system.

1. $9(x - 1) = 1 + 3(x - 2)$

2. $3x + 4y = -7$
$x - 2y = -9$

3. $x - y + 3z = -9$
$2x + 3y - z = 16$
$5x + 2y - z = 15$

4. $7x + 18 \le 9x - 2$

5. $4x - 3 < 13$ and $-3x - 4 \ge 8$

6. $2x + 4 > 8$ or $x - 7 \ge 3$

7. $|2x - 1| < 5$

8. $\left|\frac{2}{3}x - 4\right| = 2$

9. $\dfrac{4}{x - 3} - \dfrac{6}{x + 3} = \dfrac{24}{x^2 - 9}$

10. $\sqrt{x + 4} - \sqrt{x - 3} = 1$

11. $2x^2 = 5 - 4x$

12. $x^{\frac{2}{3}} - 5x^{\frac{1}{3}} + 6 = 0$

13. $2x^2 + x - 6 \le 0$

14. $\log_8 x + \log_8 (x + 2) = 1$

15. $5^{2x+3} = 125$

16. $2x^2 - 3y^2 = 5$
$3x^2 + 4y^2 = 16$

17. $2x^2 - y^2 = -8$
$x - y = 6$

In Exercises 18–24, graph each function, equation, or inequality in a rectangular coordinate system.

18. $x - 3y = 6$

19. $f(x) = \frac{1}{2}x - 1$

20. $3x - 2y > -6$

21. $f(x) = -2(x - 3)^2 + 2$

22. $\frac{x^2}{16} + \frac{y^2}{4} = 1$

23. $y = \log_2 x$

24. $x^2 - y^2 = 9$

In Exercises 25–35, perform the indicated operations, and simplify, if possible.

25. $4[2x - 6(x - y)]$

26. $(-5x^3y^2)(4x^4y^{-6})$

27. $(8x^2 - 9xy - 11y^2) - (7x^2 - 4xy + 5y^2)$

28. $(3x - 1)(2x + 5)$

29. $(3x^2 - 4y)^2$

30. $\frac{3x}{x + 5} - \frac{2}{x^2 + 7x + 10}$

31. $\frac{1 - \frac{9}{x^2}}{1 + \frac{3}{x}}$

32. $\frac{x^2 - 6x + 8}{3x + 9} \div \frac{x^2 - 4}{x + 3}$

33. $\sqrt{5xy} \cdot \sqrt{10x^2y}$

34. $4\sqrt{72} - 3\sqrt{50}$

35. $(5 + 3i)(7 - 3i)$

In Exercises 36–38, factor completely.

36. $81x^4 - 1$

37. $24x^3 - 22x^2 + 4x$

38. $x^3 + 27y^3$

In Exercises 39–42, let $f(x) = x^2 + 3x - 15$ and $g(x) = x - 2$. Find each indicated expression.

39. $(f - g)(x)$ and $(f - g)(5)$

40. $\left(\frac{f}{g}\right)(x)$ and the domain of $\frac{f}{g}$

41. $f(g(x))$

42. $g(f(x))$

43. If $f(x) = 7x - 3$, find $f^{-1}(x)$.

44. Find the distance between $(6, -1)$ and $(-3, -4)$. Round to two decimal places.

45. Divide using synthetic division:

$(3x^3 - x^2 + 4x + 8) \div (x + 2)$.

46. Solve for R: $I = \frac{R}{R + r}$.

47. Write the slope-intercept form of the equation of the line through $(-2, 5)$ and parallel to the line whose equation is $3x + y = 9$.

48. Evaluate the determinant: $\begin{vmatrix} -2 & -4 \\ 5 & 7 \end{vmatrix}$.

49. Write as a single logarithm whose coefficient is 1: $2 \ln x - \frac{1}{2} \ln y$.

50. Find the indicated sum: $\sum_{i=2}^{5} (i^3 - 4)$.

51. Find the sum of the first 30 terms of the arithmetic sequence: 2, 6, 10, 14, ...

52. Express $0.\overline{3}$ as a fraction in lowest terms.

53. Use the Binomial Theorem to expand and simplify: $(2x - y^3)^4$.

In Exercises 54–55, find the domain of each function.

54. $f(x) = \frac{x - 2}{x^2 - 3x + 2}$

55. $f(x) = \ln(2x - 8)$

56. The price of a computer is reduced by 30% to \$434. What was the original price?

57. The area of a rectangle is 52 square yards. The length of the rectangle is 1 yard longer than 3 times its width. Find the rectangle's dimensions.

58. You invested \$4000 in two stocks paying 12% and 14% annual interest, respectively. At the end of the year, the total interest from these investments was \$508. How much was invested at each rate?

59. Use the formula for continuous compounding, $A = Pe^{rt}$, to solve this problem. What interest rate is required for an investment of \$6000 subject to continuous compounding to grow to \$18,000 in 10 years?

60. The current, I, in amperes, flowing in an electrical circuit varies inversely as the resistance, R, in ohms, in the circuit. When the resistance of an electric percolator is 22 ohms, it draws 5 amperes of current. How much current is needed when the resistance is 10 ohms?

CHAPTER

2

TRIGONOMETRIC FUNCTIONS

Tidal Coastline and Pots of Water

In Florida, they post times of tides coming in and going out very precisely, like 11:23 A.M. How can they be so precise?

There is more to tides, the rise and fall of ocean waters, than the gravitational pull of the moon and sun.

These are the chief factors, of course. And because the movement of the Earth, sun and moon in relationship to each other are known precisely, the rhythm of tides rising and falling along coastlines is easy to predict.

Yet the time and heights of high and low tide may vary along different stretches of the same coast though they're reacting to similar driving forces and pressures.

Historic observation makes possible the exact timing of high tides and low tides along a particular section of coast for a month, a year, or far into the future.

The reason for the difference is oscillation. Think of pots and pans filled with varying levels of water on a table, says Charles O'Reilly, Chief of Tidal Analysis for the Geological Survey of Canada's hydrographic service in Dartmouth, N.S. Then kick the table.

"You'll notice the water in the pots and pans will slosh differently. That's their natural oscillation," he said. "If you kick the table rhythmically, you'll find each pot continues to slosh differently because it has its own rhythm.

"Now, if you join those pots and pans together, that's sort of like a coastal ocean. They're all feeling the same 'kick,' but they are all responding differently. In order to predict a tide, you have to measure them for some period of time."

SOURCE: *Toronto Star*, June 13, 2001, p. GT02

SEE CHAPTER PROJECT 1.

OUTLINE

For additional study help, go to

www.prenhall.com/sullivanegu3e

Materials include:

- Graphing Calculator Help
- Chapter Quiz
- Chapter Test
- PowerPoint Downloads
- Chapter Projects
- Student Tips

A Look Back, A Look Forward

In Chapter 1, we began our discussion of functions. We defined domain and range and independent and dependent variables; we found the value of a function and graphed functions. We continued our study of functions by listing properties that a function might have, like being even or odd, and we created a library of functions, naming key functions and listing their properties, including the graph.

In this chapter we define the trigonometric functions, six functions that have a wide application. We shall talk about their domain and range, how to find values, graph them, and develop a list of their properties.

There are two widely accepted approaches to the development of the trigonometric functions: one uses right triangles; the other uses circles, especially the unit circle. In this book, we develop the trigonometric functions using right triangles. In Section 2.5, we introduce trigonometric functions using the unit circle and show that this approach leads to the definition using right triangles.

PREPARING FOR THIS SECTION

Before getting started, review the following:

- ✓ Circumference and Area of a Circle (Section A.2, p. 568)
- ✓ Rectangular Coordinates (Section 1.1, pp. 2–3)

2.1 ANGLES AND THEIR MEASURE

OBJECTIVES

1. Convert between Degrees, Minutes, Seconds, and Decimal Forms for Angles
2. Find the Arc Length of a Circle
3. Convert from Degrees to Radians
4. Convert from Radians to Degrees
5. Find the Area of a Sector of a Circle
6. Find the Linear Speed of an Object Traveling in Circular Motion

A **ray**, or **half-line**, is that portion of a line that starts at a point V on the line and extends indefinitely in one direction. The starting point V of a ray is called its **vertex**. See Figure 1.

Figure 1

If two rays are drawn with a common vertex, they form an **angle**. We call one of the rays of an angle the **initial side** and the other the **terminal side**. The angle that is formed is identified by showing the direction and amount of rotation from the initial side to the terminal side. If the rotation is in the counterclockwise direction, the angle is **positive**; if the rotation is clockwise, the angle is **negative**. See Figure 2. Lowercase Greek letters, such as α (alpha), β (beta), γ (gamma), and θ (theta), will be used to denote angles. Notice in Figure 2(a) that the angle α is positive because the direction of the rotation from the initial side to the terminal side is counterclockwise. The angle β in Figure 2(b) is negative because the rotation is clockwise. The angle γ in Figure 2(c) is positive. Notice that the angle α in Figure 2(a) and the angle γ in Figure 2(c) have the same initial side and the same terminal side. However, α and γ are unequal, because the amount of rotation required to go from the initial side to the terminal side is greater for angle γ than for angle α.

Figure 2

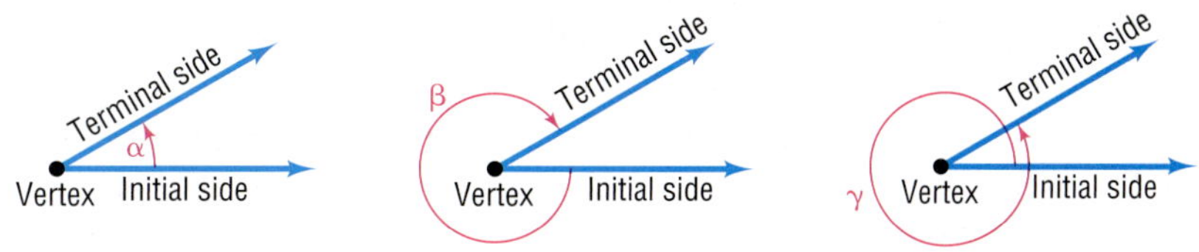

Counterclockwise rotation
Positive angle
(a)

Clockwise rotation
Negative angle
(b)

Counterclockwise rotation
Positive angle
(c)

An angle θ is said to be in **standard position** if its vertex is at the origin of a rectangular coordinate system and its initial side coincides with the positive x-axis. See Figure 3.

Figure 3

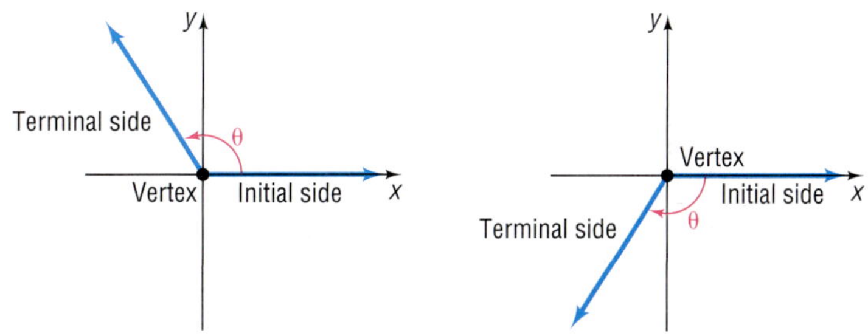

(a) θ in standard position; θ positive

(b) θ in standard position; θ negative

When an angle θ is in standard position, the terminal side will lie either in a quadrant, in which case we say that θ **lies in that quadrant**, or on the x-axis or the y-axis, in which case we say that θ is a **quadrantal angle**. For example, the angle θ in Figure 4(a) lies in quadrant II, the angle θ in Figure 4(b) lies in quadrant IV, and the angle θ in Figure 4(c) is a quadrantal angle.

Figure 4

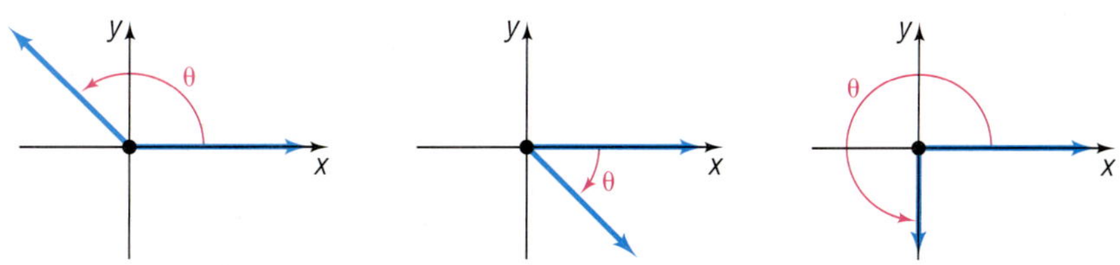

(a) θ lies in quadrant II

(b) θ lies in quadrant IV

(c) θ is a quadrantal angle

We measure angles by determining the amount of rotation needed for the initial side to become coincident with the terminal side. The two commonly used measures for angles are *degrees* and *radians*.

Degrees

The angle formed by rotating the initial side exactly once in the counterclockwise direction until it coincides with itself (1 revolution) is said to measure 360 degrees, abbreviated 360°. **One degree, 1°,** is $\frac{1}{360}$ revolution. A **right angle** is an angle that measures 90°, or $\frac{1}{4}$ revolution; a **straight angle** is an angle that measures 180°, or $\frac{1}{2}$ revolution. See Figure 5. As Figure 5(b) shows, it is customary to indicate a right angle by using the symbol ⊾.

Figure 5

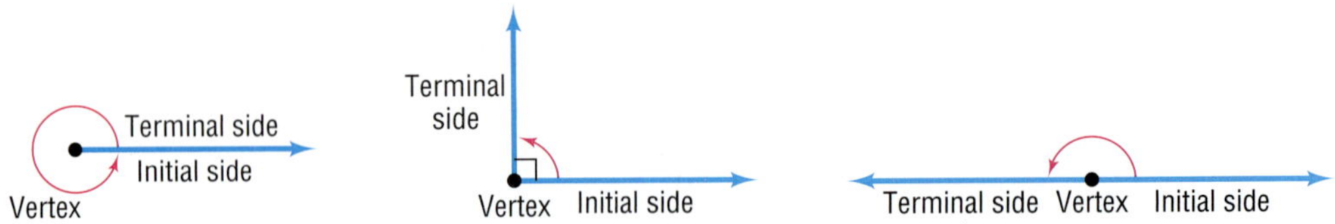

(a) 1 revolution counterclockwise, 360° **(b)** $\frac{1}{4}$ revolution counter-clockwise; right angle; 90° **(c)** $\frac{1}{2}$ revolution counter-clockwise; straight angle; 180°

It is also customary to refer to an angle that measures θ degrees as an angle of θ degrees.

EXAMPLE 1 Drawing an Angle

Draw each angle.

(a) 45° (b) −90° (c) 225° (d) 405°

Solution (a) An angle of 45° is $\frac{1}{2}$ of a right angle. See Figure 6.

(b) An angle of −90° is $\frac{1}{4}$ revolution in the clockwise direction. See Figure 7.

Figure 6

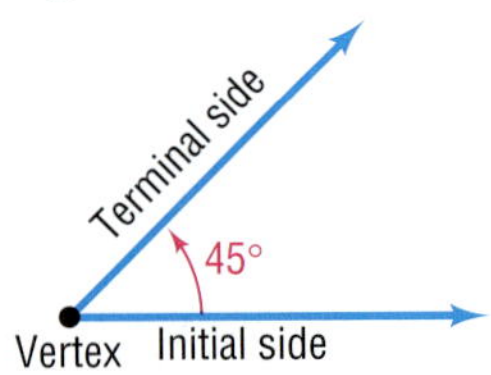

Figure 7

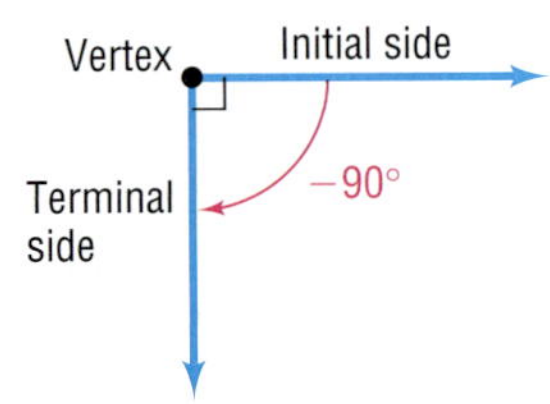

(c) An angle of 225° consists of a rotation through 180° followed by a rotation through 45°. See Figure 8.

(d) An angle of 405° consists of 1 revolution (360°) followed by a rotation through 45°. See Figure 9.

Figure 8

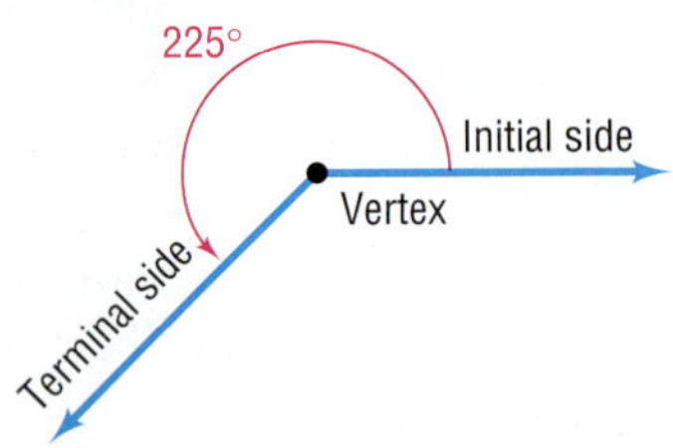

Figure 9

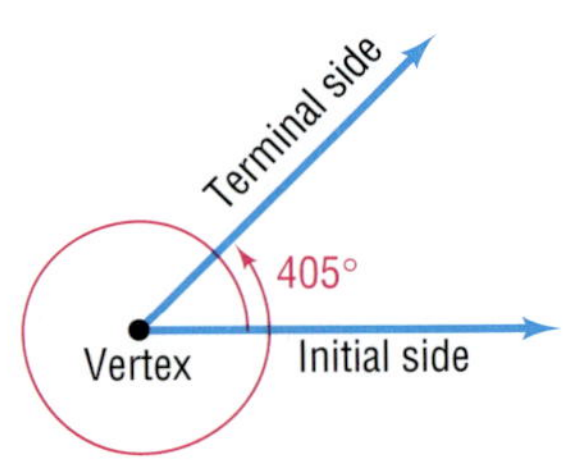

NOW WORK PROBLEM 1.

1 Although subdivisions of a degree may be obtained by using decimals, we also may use the notion of *minutes* and *seconds.* **One minute,** denoted by $\mathbf{1'}$, is defined as $\frac{1}{60}$ degree. **One second,** denoted by $\mathbf{1''}$, is defined as $\frac{1}{60}$ minute or, equivalently, $\frac{1}{3600}$ degree. An angle of, say, 30 degrees, 40 minutes, 10 seconds is written compactly as $30°40'10''$. To summarize:

$$\begin{gathered}1 \text{ counterclockwise revolution} = 360° \\ 1° = 60' \qquad 1' = 60''\end{gathered} \tag{1}$$

It is sometimes necessary to convert from the degree, minute, second notation ($D°M'S''$) to a decimal form, and vice versa. Check your calculator; it should be capable of doing the conversion for you.

Before getting started you must set the mode to degrees, because there are two common ways to measure angles: degree mode and radian mode. (We will define radians shortly.) Usually, a menu is used to change from one mode to another. Check your owner's manual to find out how your particular calculator works.

Now let's see how to convert by hand from the degree, minute, second notation ($D°M'S''$) to a decimal form, and vice versa, by looking at some examples: $15°30' = 15.5°$ because $30' = \frac{1}{2}^{\circ} = 0.5°$

$32.25° = 32°15'$ because $0.25° = \frac{1}{4}^{\circ} = \frac{1}{4}(60') = 15'$.

EXAMPLE 2 Converting between Degrees, Minutes, Seconds, and Decimal Forms

(a) Convert $50°6'21''$ to a decimal in degrees.

(b) Convert $21.256°$ to the $D°M'S''$ form.

Algebraic Solution

(a) Because $1' = \left(\frac{1}{60}\right)^{\circ}$ and $1'' = \left(\frac{1}{60}\right)' = \left(\frac{1}{60}\cdot\frac{1}{60}\right)^{\circ}$, we convert as follows:

$$\begin{aligned}50°6'21'' &= 50° + 6' + 21'' = 50° + 6\cdot\left(\frac{1}{60}\right)^{\circ} + 21\cdot\left(\frac{1}{60}\cdot\frac{1}{60}\right)^{\circ} \\ &\approx 50° + 0.1° + 0.005833° \\ &= 50.105833°\end{aligned}$$

(b) We start with the decimal part of $21.256°$, that is, $0.256°$.

$$0.256° = (0.256)(1°) \underset{1° = 60'}{=} (0.256)(60') = 15.36'$$

Now we work with the decimal part of $15.36'$, that is, $0.36'$.

$$0.36' = (0.36)(1') \underset{1' = 60''}{=} (0.36)(60'') = 21.6'' \approx 22''$$

Thus,

$$\begin{aligned}21.256° &= 21° + 0.256° = 21° + 15.36' = 21° + 15' + 0.36' \\ &= 21° + 15' + 21.6'' \approx 21°15'22''\end{aligned}$$

Graphing Solution (a) Figure 10 shows the solution using a TI-83 graphing calculator.

Figure 10

(b) Figure 11 shows the solution using a TI-83 graphing calculator.

Figure 11

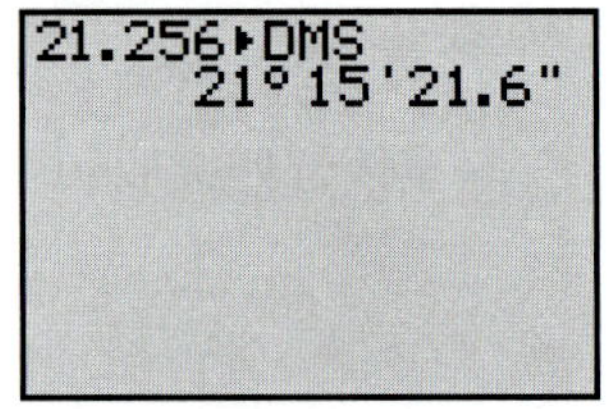

NOW WORK PROBLEMS 69 AND 75.

In many applications, such as describing the exact location of a star or the precise position of a boat at sea, angles measured in degrees, minutes, and even seconds are used. For calculation purposes, these are transformed to decimal form. In other applications, especially those in calculus, angles are measured using *radians.*

Radians

A **central angle** is an angle whose vertex is at the center of a circle. The rays of a central angle subtend (intersect) an arc on the circle. If the radius of the circle is r and the length of the arc subtended by the central angle is also r, then the measure of the angle is **1 radian.** See Figure 12(a).

For a circle of radius 1, the rays of a central angle with measure 1 radian would subtend an arc of length 1. For a circle of radius 3, the rays of a central angle with measure 1 radian would subtend an arc of length 3. See Figure 12(b).

Figure 12

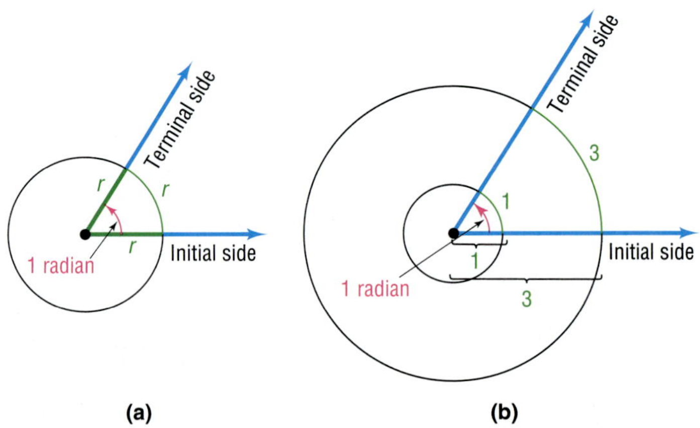

Figure 13
$\frac{\theta}{\theta_1} = \frac{s}{s_1}$

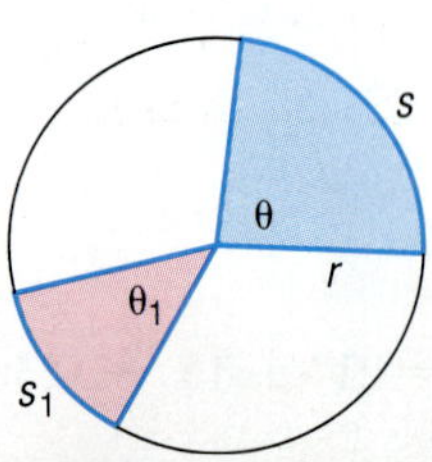

2 Now consider a circle of radius r and two central angles, θ and θ_1, measured in radians. Suppose that these central angles subtend arcs of lengths s and s_1, respectively, as shown in Figure 13. From geometry, we know that the ratio of the measures of the angles equals the ratio of the corresponding lengths of the arcs subtended by these angles; that is,

$$\frac{\theta}{\theta_1} = \frac{s}{s_1} \tag{2}$$

Suppose that $\theta_1 = 1$ radian. Refer again to Figure 12(a). The amount of arc s_1 subtended by the central angle $\theta_1 = 1$ radian equals the radius r of the circle. Then $s_1 = r$, so formula (2) reduces to

$$\frac{\theta}{1} = \frac{s}{r} \quad \text{or} \quad s = r\theta \tag{3}$$

Theorem

Arc Length

For a circle of radius r, a central angle of θ radians subtends an arc whose length s is

$$s = r\theta \tag{4}$$

Note: Formulas must be consistent with regard to the units used. In equation (4), we write

$$s = r\theta$$

To see the units, however, we must go back to equation (3) and write

$$\frac{\theta \text{ radians}}{1 \text{ radian}} = \frac{s \text{ length units}}{r \text{ length units}}$$

$$s \text{ length units} = r \text{ length units} \frac{\theta \text{ radians}}{1 \text{ radian}}$$

Since the radians cancel, we are left with

$$s \text{ length units} = (r \text{ length units})\theta \qquad s = r\theta$$

where θ appears to be "dimensionless" but, in fact, is measured in radians. So, in using the formula $s = r\theta$, the dimension for θ is radians, and any convenient unit of length (such as inches or meters) may be used for s and r.

EXAMPLE 3 Finding the Length of Arc of a Circle

Find the length of the arc of a circle of radius 2 meters subtended by a central angle of 0.25 radian.

Solution We use equation (4) with $r = 2$ meters and $\theta = 0.25$. The length s of the arc is

$$s = r\theta = 2(0.25) = 0.5 \text{ meter}$$

NOW WORK PROBLEM 37.

Relationship Between Degrees And Radians

Figure 14
1 revolution = 2π radians

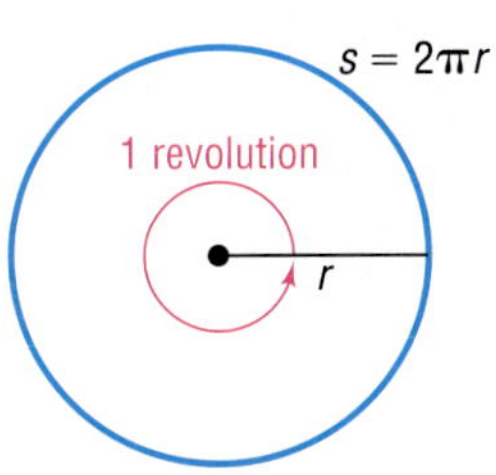

Consider a circle of radius r. A central angle of 1 revolution will subtend an arc equal to the circumference of the circle (Figure 14). Because the circumference of a circle equals $2\pi r$, we use $s = 2\pi r$ in equation (4) to find that, for an angle θ of 1 revolution,

$$s = r\theta$$

$$2\pi r = r\theta \qquad \theta = 1 \text{ revolution}; s = 2\pi r$$

$$\theta = 2\pi \text{ radians} \qquad \text{Solve for } \theta$$

From this we have,

$$1 \text{ revolution} = 2\pi \text{ radians} \tag{5}$$

so

$$360° = 2\pi \text{ radians}$$

or

$$180° = \pi \text{ radians} \tag{6}$$

Divide both sides of equation (6) by 180. Then

$$1 \text{ degree} = \frac{\pi}{180} \text{radian}$$

Divide both sides of (6) by π. Then

$$\frac{180}{\pi} \text{degrees} = 1 \text{ radian}$$

We have the following two conversion formulas:

$$1 \text{ degree} = \frac{\pi}{180} \text{radian} \qquad 1 \text{ radian} = \frac{180}{\pi} \text{degrees} \tag{7}$$

3 **EXAMPLE 4** **Converting from Degrees to Radians**

Convert each angle in degrees to radians.

(a) 60° (b) 150° (c) −45° (d) 90°

Solution (a) $60° = 60 \cdot 1 \text{ degree} = 60 \cdot \frac{\pi}{180} \text{radian} = \frac{\pi}{3} \text{radians}$

(b) $150° = 150 \cdot \frac{\pi}{180} \text{radian} = \frac{5\pi}{6} \text{radians}$

(c) $-45° = -45 \cdot \frac{\pi}{180} \text{radian} = -\frac{\pi}{4} \text{radian}$

(d) $90° = 90 \cdot \frac{\pi}{180} \text{radian} = \frac{\pi}{2} \text{radians}$

Example 4 illustrates that angles that are fractions of a revolution are expressed in radian measure as fractional multiples of π, rather than as decimals. For example, a right angle, as in Example 4(d), is left in the form $\frac{\pi}{2}$ radians, which is exact, rather than using the approximation $\frac{\pi}{2} \approx \frac{3.1416}{2} = 1.5708$ radians.

NOW WORK PROBLEM 13.

4

EXAMPLE 5 Converting Radians to Degrees

Convert each angle in radians to degrees.

(a) $\frac{\pi}{6}$ radian (b) $\frac{3\pi}{2}$ radians (c) $-\frac{3\pi}{4}$ radians (d) $\frac{7\pi}{3}$ radians

Solution (a) $\frac{\pi}{6} \text{ radian} = \frac{\pi}{6} \cdot 1 \text{ radian} = \frac{\pi}{6} \cdot \frac{180}{\pi} \text{ degrees} = 30°$

(b) $\frac{3\pi}{2} \text{ radians} = \frac{3\pi}{2} \cdot \frac{180}{\pi} \text{ degrees} = 270°$

(c) $-\frac{3\pi}{4} \text{ radians} = -\frac{3\pi}{4} \cdot \frac{180}{\pi} \text{ degrees} = -135°$

(d) $\frac{7\pi}{3} \text{ radians} = \frac{7\pi}{3} \cdot \frac{180}{\pi} \text{ degrees} = 420°$

NOW WORK PROBLEM 25.

Table 1 lists the degree and radian measures of some commonly encountered angles. You should learn to feel equally comfortable using degree or radian measure for these angles.

TABLE 1

Degrees	0°	30°	45°	60°	90°	120°	135°	150°	180°
Radians	0	$\frac{\pi}{6}$	$\frac{\pi}{4}$	$\frac{\pi}{3}$	$\frac{\pi}{2}$	$\frac{2\pi}{3}$	$\frac{3\pi}{4}$	$\frac{5\pi}{6}$	π
Degrees		210°	225°	240°	270°	300°	315°	330°	360°
Radians		$\frac{7\pi}{6}$	$\frac{5\pi}{4}$	$\frac{4\pi}{3}$	$\frac{3\pi}{2}$	$\frac{5\pi}{3}$	$\frac{7\pi}{4}$	$\frac{11\pi}{6}$	2π

EXAMPLE 6 Finding the Distance between Two Cities

See Figure 15(a). The latitude of a location L is the angle formed by a ray drawn from the center of Earth to the Equator and a ray drawn from the center of Earth to L. See Figure 15(b). Glasgow, Montana, is due north of Albuquerque, New Mexico. Find the distance between Glasgow (48°9′ north latitude) and Albuquerque (35°5′ north latitude). Assume that the radius of Earth is 3960 miles.

Figure 15

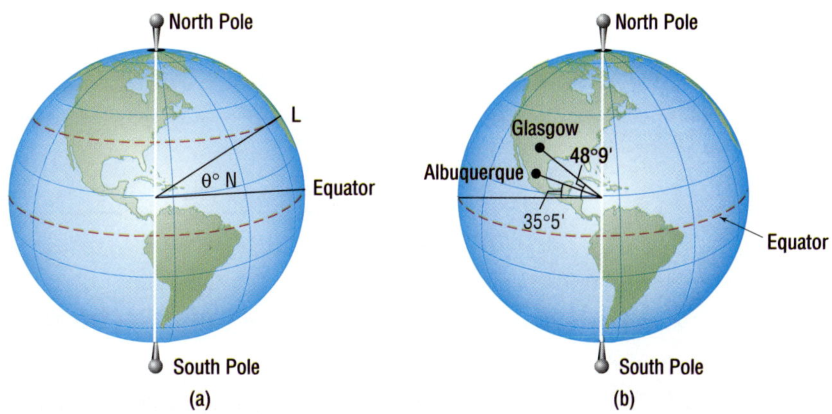

Solution The measure of the central angle between the two cities is $48°9' - 35°5' = 13°4'$. We use equation (4), $s = r\theta$, but first we must convert the angle of $13°4'$ to radians.

$$\theta = 13°4' \approx 13.0667° = 13.0667 \cdot \frac{\pi}{180} \text{ radian} \approx 0.228 \text{ radian}$$

We use $\theta = 0.228$ radian and $r = 3960$ miles in equation (4). The distance between the two cities is

$$s = r\theta = 3960 \cdot 0.228 \approx 903 \text{ miles}$$

When an angle is measured in degrees, the degree symbol will always be shown. However, when an angle is measured in radians, we will follow the usual practice and omit the word *radians*. So, if the measure of an angle is given as $\frac{\pi}{6}$ it is understood to mean $\frac{\pi}{6}$ radian.

NOW WORK PROBLEM 91.

Area of a Sector

Figure 16

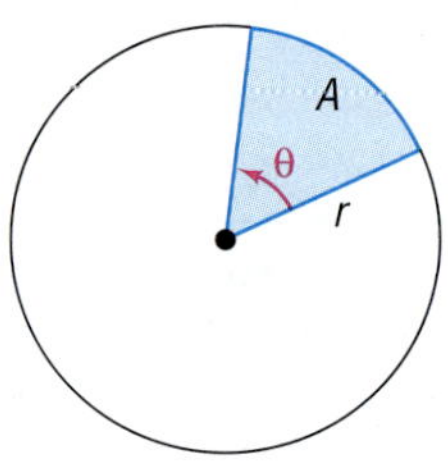

5 Consider a circle of radius r. Suppose that θ, measured in radians, is a central angle of this circle. See Figure 16. We seek a formula for the area A of the sector formed by the angle θ (shown in blue).

Now consider a circle of radius r and two central angles θ and θ_1, both measured in radians. See Figure 17. From geometry, we know the ratio of the measures of the angles equals the ratio of the corresponding areas of the sectors formed by these angles. That is,

Figure 17

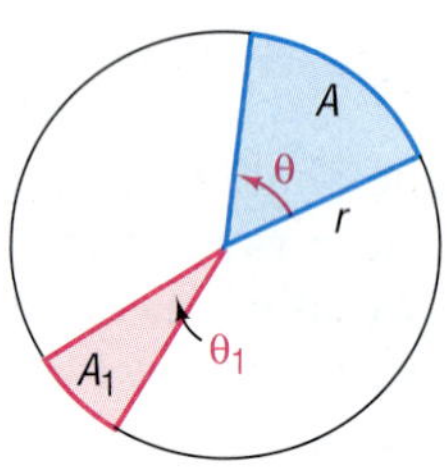

$$\frac{\theta}{\theta_1} = \frac{A}{A_1}$$

Suppose that $\theta_1 = 2\pi$ radians. Then A_1 = area of the circle $= \pi r^2$. Solving for A, we find

$$A = A_1 \frac{\theta}{\theta_1} = \pi r^2 \frac{\theta}{2\pi} = \frac{1}{2} r^2 \theta$$

Theorem

Area of a Sector

The area A of the sector of a circle of radius r formed by a central angle of θ radians is

$$A = \frac{1}{2} r^2 \theta \qquad (8)$$

EXAMPLE 7 **Finding the Area of a Sector of a Circle**

Find the area of the sector of a circle of radius 2 feet formed by an angle of 30°. Round the answer to two decimal places.

Solution We use equation (8) with $r = 2$ feet and $\theta = 30° = \frac{\pi}{6}$ radian. [Remember, in equation (8), θ must be in radians.] The area A of the sector is

$$A = \frac{1}{2}r^2\theta = \frac{1}{2}(2)^2\frac{\pi}{6} = \frac{\pi}{3}\text{square feet} \approx 1.05\text{ square feet}$$

rounded to two decimal places.

NOW WORK PROBLEM 45.

Circular Motion

6 We have already defined the average speed of an object as the distance traveled divided by the elapsed time. Suppose that an object moves around a circle of radius r at a constant speed. If s is the distance traveled in time t around this circle, then the **linear speed** v of the object is defined as

$$v = \frac{s}{t} \tag{9}$$

Figure 18
$v = \frac{s}{t}$; $\omega = \frac{\theta}{t}$

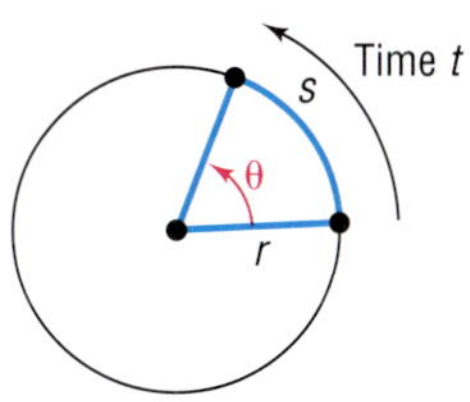

As this object travels around the circle, suppose that θ (measured in radians) is the central angle swept out in time t. See Figure 18. Then the **angular speed** ω (the Greek letter omega) of this object is the angle (measured in radians) swept out divided by the elapsed time; that is,

$$\omega = \frac{\theta}{t} \tag{10}$$

Angular speed is the way the turning rate of an engine is described. For example, an engine idling at 900 rpm (revolutions per minute) is one that rotates at an angular speed of

$$900\frac{\text{revolutions}}{\text{minute}} = 900\frac{\cancel{\text{revolutions}}}{\text{minute}} \cdot 2\pi\frac{\text{radians}}{\cancel{\text{revolution}}} = 1800\pi\frac{\text{radians}}{\text{minute}}$$

There is an important relationship between linear speed and angular speed:

$$\text{linear speed} = v \underset{(9)}{\underset{\uparrow}{=}} \frac{s}{t} \underset{s = r\theta}{\underset{\uparrow}{=}} \frac{r\theta}{t} = r\left(\frac{\theta}{t}\right)$$

Then, using equation (10), we obtain

$$v = r\omega \tag{11}$$

where ω is measured in radians per unit time.

When using equation (11), remember that $v = \frac{s}{t}$ (the linear speed) has the dimensions of length per unit of time (such as feet per second or miles per hour), r (the radius of the circular motion) has the same length dimension as s, and ω (the angular speed) has the dimensions of radians per unit of time. If the angular speed is given in terms of *revolutions* per unit of time (as is often the case), be sure to convert it to *radians* per unit of time before attempting to use equation (11).

EXAMPLE 8 Finding Linear Speed

A child is spinning a rock at the end of a 2-foot rope at the rate of 180 revolutions per minute (rpm). Find the linear speed of the rock when it is released.

Figure 19

Solution Look at Figure 19. The rock is moving around a circle of radius $r = 2$ feet. The angular speed ω of the rock is

$$\omega = 180\frac{\text{revolutions}}{\text{minute}} = 180\frac{\cancel{\text{revolutions}}}{\text{minute}}\cdot 2\pi\frac{\text{radians}}{\cancel{\text{revolution}}} = 360\pi\frac{\text{radians}}{\text{minute}}$$

From equation (11), the linear speed v of the rock is

$$v = r\omega = 2\text{ feet}\cdot 360\pi\frac{\text{radians}}{\text{minute}} = 720\pi\frac{\text{feet}}{\text{minute}} \approx 2262\frac{\text{feet}}{\text{minute}}$$

The linear speed of the rock when it is released is 2262 ft/min $\approx$ 25.7 mi/hr. ■

NOW WORK PROBLEM 87.

HISTORICAL FEATURE

Trigonometry was developed by Greek astronomers, who regarded the sky as the inside of a sphere, so it was natural that triangles on a sphere were investigated early (by Menelaus of Alexandria about AD 100) and that triangles in the plane were studied much later. The first book containing a systematic treatment of plane and spherical trigonometry was written by the Persian astronomer Nasîr Eddîn (about AD 1250).

Regiomontanus (1436–1476) is the person most responsible for moving trigonometry from astronomy into mathematics. His work was improved by Copernicus (1473–1543) and Copernicus's student Rhaeticus (1514–1576). Rhaeticus's book was the first to define the six trigonometric functions as ratios of sides of triangles, although he did not give the functions their present names. Credit for this is due to Thomas Finck (1583), but Finck's notation was by no means universally accepted at the time. The notation was finally stabilized by the textbooks of Leonhard Euler (1707–1783).

Trigonometry has since evolved from its use by surveyors, navigators, and engineers to present applications involving ocean tides, the rise and fall of food supplies in certain ecologies, brain wave patterns, and many other phenomena.

2.1 Concepts and Vocabulary

In Problems 1–3, fill in the blanks.

1. An angle θ is in __________ __________ if its vertex is at the origin of a rectangular coordinate system and its initial side coincides with the positive x-axis.

2. On a circle of radius r, a central angle of θ radians subtends an arc of length $s =$ __________; the area of the sector formed by this angle θ is $A =$ __________.

3. An object travels around a circle of radius r with constant speed. If s is the distance traveled in time t around the circle and θ is the central angle (in radians) swept out in time t, then the linear speed of the object is $v =$ __________ and the angular speed of the object is $\omega =$ __________.

In Problems 4–6, answer True or False to each statement.

4. $\pi = 180$.

5. $180° = \pi$ radians

6. On the unit circle, if s is the length of the arc subtended by a central angle θ, measured in radians, then $s = \theta$.

7. What is 1 radian?

8. Which angle has the larger measure: 1 degree or 1 radian? Or are they equal?

9. Explain the difference between linear speed and angular speed.

2.1 Exercises

In Problems 1–12, draw each angle.

1. $30°$ **2.** $60°$ **3.** $135°$ **4.** $-120°$ **5.** $450°$ **6.** $540°$

7. $\frac{3\pi}{4}$ **8.** $\frac{4\pi}{3}$ **9.** $-\frac{\pi}{6}$ **10.** $-\frac{2\pi}{3}$ **11.** $\frac{16\pi}{3}$ **12.** $\frac{21\pi}{4}$

In Problems 13–24, convert each angle in degrees to radians. Express your answer as a multiple of π.

13. $30°$ **14.** $120°$ **15.** $240°$ **16.** $330°$ **17.** $-60°$ **18.** $-30°$

19. $180°$ **20.** $270°$ **21.** $-135°$ **22.** $-225°$ **23.** $-90°$ **24.** $-180°$

In Problems 25–36, convert each angle in radians to degrees.

25. $\frac{\pi}{3}$ **26.** $\frac{5\pi}{6}$ **27.** $-\frac{5\pi}{4}$ **28.** $-\frac{2\pi}{3}$ **29.** $\frac{\pi}{2}$ **30.** 4π

31. $\frac{\pi}{12}$ **32.** $\frac{5\pi}{12}$ **33.** $-\frac{\pi}{2}$ **34.** $-\pi$ **35.** $-\frac{\pi}{6}$ **36.** $-\frac{3\pi}{4}$

In Problems 37–44, s denotes the length of the arc of a circle of radius r subtended by the central angle θ. Find the missing quantity. Round answers to three decimal places, if necessary.

37. $r = 10$ meters, $\theta = \frac{1}{2}$ radian, $s = ?$

38. $r = 6$ feet, $\theta = 2$ radians, $s = ?$

39. $\theta = \frac{1}{3}$ radian, $s = 2$ feet, $r = ?$

40. $\theta = \frac{1}{4}$ radian, $s = 6$ centimeters, $r = ?$

41. $r = 5$ miles, $s = 3$ miles, $\theta = ?$

42. $r = 6$ meters, $s = 8$ meters, $\theta = ?$

43. $r = 2$ inches, $\theta = 30°$, $s = ?$

44. $r = 3$ meters, $\theta = 120°$, $s = ?$

In Problems 45–52, A denotes the area of the sector of a circle of radius r formed by the central angle θ. Find the missing quantity. Round answers to three decimal places, if necessary.

45. $r = 10$ meters, $\theta = \frac{1}{2}$ radian, $A = ?$

46. $r = 6$ feet, $\theta = 2$ radians, $A = ?$

47. $\theta = \frac{1}{3}$ radian, $A = 2$ square feet, $r = ?$

48. $\theta = \frac{1}{4}$ radian, $A = 6$ square centimeters, $r = ?$

49. $r = 5$ miles, $A = 3$ square miles, $\theta = ?$

50. $r = 6$ meters, $A = 8$ square meters, $\theta = ?$

51. $r = 2$ inches, $\theta = 30°$, $A = ?$

52. $r = 3$ meters, $\theta = 120°$, $A = ?$

In Problems 53–56, find the length s and area A. Round answers to three decimal places.

53.

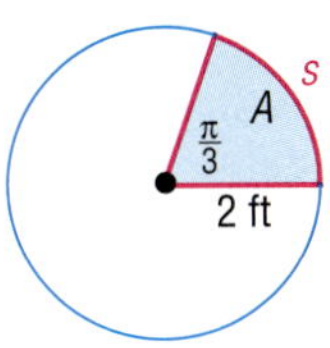

54.

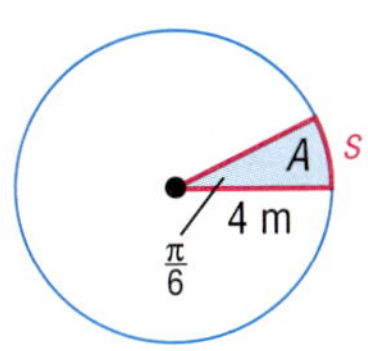

55.

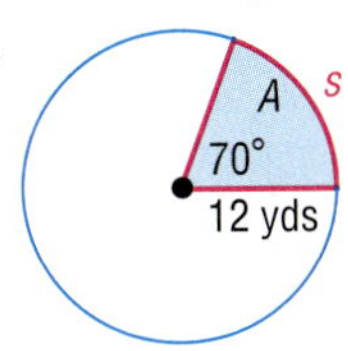

56.

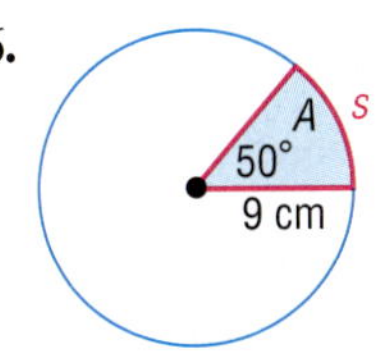

In Problems 57–62, convert each angle in degrees to radians. Express your answer in decimal form, rounded to two decimal places.

57. $17°$ **58.** $73°$ **59.** $-40°$ **60.** $-51°$ **61.** $125°$ **62.** $350°$

In Problems 63–68, convert each angle in radians to degrees. Express your answer in decimal form, rounded to two decimal places.

63. 3.14 **64.** 0.75 **65.** 2 **66.** 3 **67.** 6.32 **68.** $\sqrt{2}$

In Problems 69–74, convert each angle to a decimal in degrees. Round your answer to two decimal places.

69. $40°10'25''$ **70.** $61°42'21''$ **71.** $1°2'3''$ **72.** $73°40'40''$ **73.** $9°9'9''$ **74.** $98°22'45''$

In Problems 75–80, convert each angle to D°M'S" form. Round your answer to the nearest second.

75. $40.32°$ **76.** $61.24°$ **77.** $18.255°$ **78.** $29.411°$ **79.** $19.99°$ **80.** $44.01°$

81. Minute Hand of a Clock The minute hand of a clock is 6 inches long. How far does the tip of the minute hand move in 15 minutes? How far does it move in 25 minutes?

82. Movement of a Pendulum A pendulum swings through an angle of 20° each second. If the pendulum is 40 inches long, how far does its tip move each second?

83. Area of a Sector Find the area of the sector of a circle of radius 4 meters formed by an angle of 45°. Round the answer to two decimal places.

84. Area of a Sector Find the area of the sector of a circle of radius 3 centimeters formed by an angle of 60°. Round the answer to two decimal places.

85. Watering a Lawn A water sprinkler sprays water over a distance of 30 feet while rotating through an angle of 135°. What area of lawn receives water?

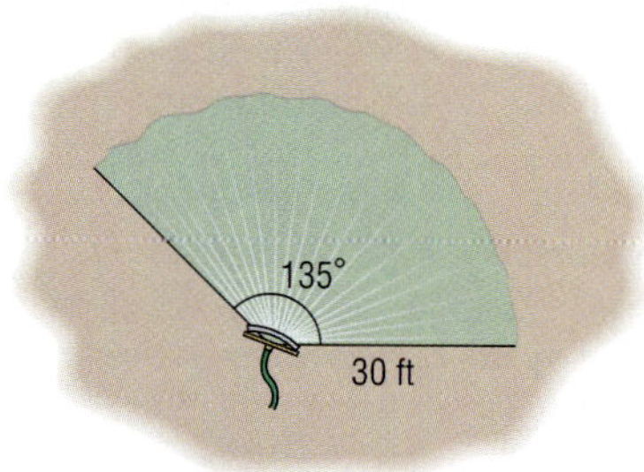

86. Designing a Water Sprinkler An engineer is asked to design a water sprinkler that will cover a field of 100 square yards that is in the shape of a sector of a circle of radius 50 yards. Through what angle should the sprinkler rotate?

87. Motion on a Circle An object is traveling around a circle with a radius of 5 centimeters. If in 20 seconds a central angle of $\frac{1}{3}$ radian is swept out, what is the angular speed of the object? What is its linear speed?

88. Motion on a Circle An object is traveling around a circle with a radius of 2 meters. If in 20 seconds the object travels 5 meters, what is its angular speed? What is its linear speed?

89. Bicycle Wheels The diameter of each wheel of a bicycle is 26 inches. If you are traveling at a speed of 35 miles per hour on this bicycle, through how many revolutions per minute are the wheels turning?

90. Car Wheels The radius of each wheel of a car is 15 inches. If the wheels are turning at the rate of 3 revolutions per second, how fast is the car moving? Express your answer in inches per second and in miles per hour.

In Problems 91–94, the latitude of a location L is the angle formed by a ray drawn from the center of Earth to the Equator and a ray drawn from the center of Earth to L. See the figure.

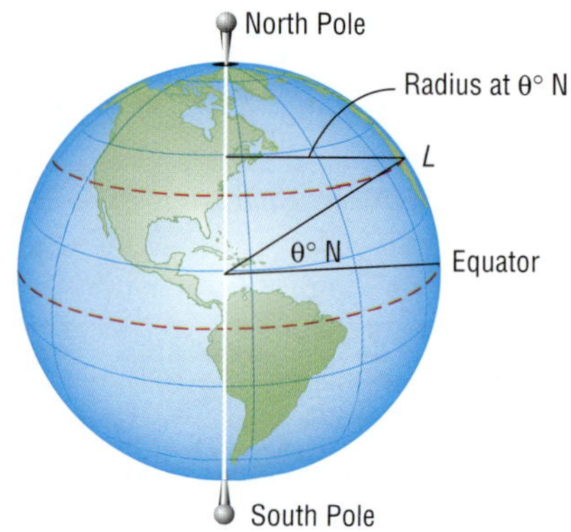

91. Distance between Cities Memphis, Tennessee, is due north of New Orleans, Louisiana. Find the distance between Memphis (35°9′ north latitude) and New Orleans (29°57′ north latitude). Assume that the radius of Earth is 3960 miles.

92. Distance between Cities Charleston, West Virginia, is due north of Jacksonville, Florida. Find the distance between Charleston (38°21′ north latitude) and Jacksonville (30°20′ north latitude). Assume that the radius of Earth is 3960 miles.

93. Linear Speed on Earth Earth rotates on an axis through its poles. The distance from the axis to a location on Earth 30° north latitude is about 3429.5 miles. Therefore, a location on Earth at 30° north latitude is spinning on a circle of radius 3429.5 miles. Compute the linear speed on the surface of Earth at 30° north latitude.

94. Linear Speed on Earth Earth rotates on an axis through its poles. The distance from the axis to a location on Earth 40° north latitude is about 3033.5 miles. Therefore, a location on Earth at 40° north latitude is spinning on a circle of radius 3033.5 miles. Compute the linear speed on the surface of Earth at 40° north latitude.

95. Speed of the Moon The mean distance of the Moon from Earth is 2.39×10^5 miles. Assuming that the orbit of the Moon around Earth is circular and that 1 revolution takes 27.3 days, find the linear speed of the Moon. Express your answer in miles per hour.

96. Speed of Earth The mean distance of Earth from the Sun is 9.29×10^7 miles. Assuming that the orbit of Earth around the Sun is circular and that 1 revolution takes 365 days, find the linear speed of Earth. Express your answer in miles per hour.

97. Pulleys Two pulleys, one with radius 2 inches and the other with radius 8 inches, are connected by a belt. (See the figure on page 111.) If the 2-inch pulley is caused to rotate at 3 revolutions per minute, determine the revolutions per minute of the 8-inch pulley.

[**Hint:** The linear speeds of the pulleys, that is, the speed of the belt, are the same.]

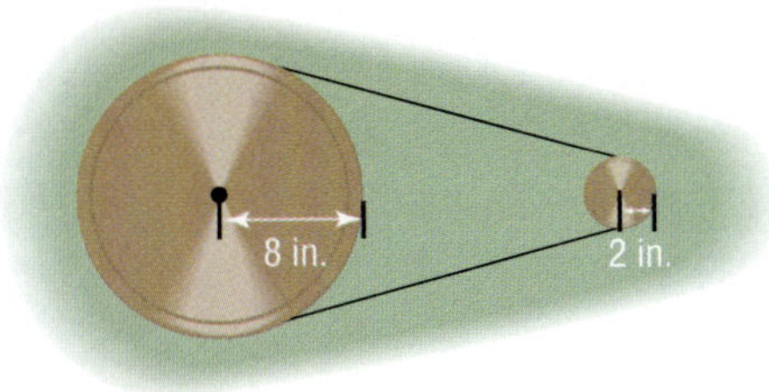

98. Ferris Wheels A neighborhood carnival has a Ferris wheel whose radius is 30 feet. You measure the time it takes for one revolution to be 70 seconds. What is the linear speed (in feet per second) of this Ferris wheel? What is the angular speed in radians per second?

99. Computing the Speed of a River Current To approximate the speed of the current of a river, a circular paddle wheel with radius 4 feet is lowered into the water. If the current causes the wheel to rotate at a speed of 10 revolutions per minute, what is the speed of the current? Express your answer in miles per hour.

100. Spin Balancing Tires A spin balancer rotates the wheel of a car at 480 revolutions per minute. If the diameter of the wheel is 26 inches, what road speed is being tested? Express your answer in miles per hour. At how many revolutions per minute should the balancer be set to test a road speed of 80 miles per hour?

101. The Cable Cars of San Francisco At the Cable Car Museum you can see the four cable lines that are used to pull cable cars up and down the hills of San Francisco. Each cable travels at a speed of 9.55 miles per hour, caused by a rotating wheel whose diameter is 8.5 feet. How fast is the wheel rotating? Express your answer in revolutions per minute.

102. Difference in Time of Sunrise Naples, Florida, is approximately 90 miles due west of Ft. Lauderdale. How much sooner would a person in Ft. Lauderdale first see the rising Sun than a person in Naples?

[**Hint:** Consult the figure. When a person at Q sees the first rays of the Sun, a person at P is still in the dark. The person at P sees the first rays after Earth has rotated so that P is at the location Q. Now use the fact that at the latitude of Ft. Lauderdale in 24 hours a length of arc of $2\pi(3559)$ miles is subtended.]

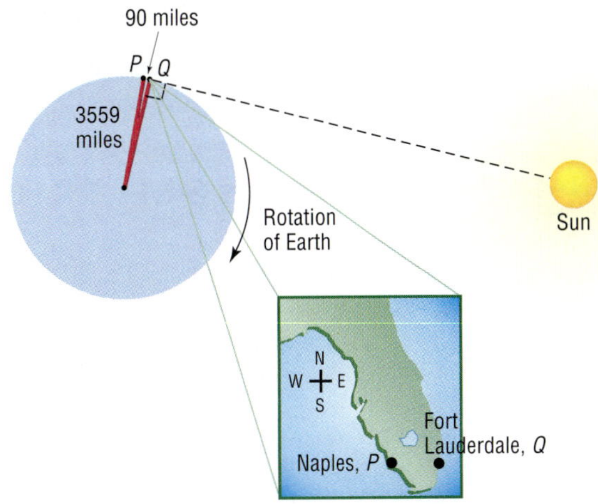

103. Keeping Up with the Sun How fast would you have to travel on the surface of Earth at the equator to keep up with the Sun (that is, so that the Sun would appear to remain in the same position in the sky)?

104. Nautical Miles A **nautical mile** equals the length of arc subtended by a central angle of 1 minute on a great circle* on the surface of Earth. (See the figure.) If the radius of Earth is taken as 3960 miles, express 1 nautical mile in terms of ordinary, or **statute,** miles.

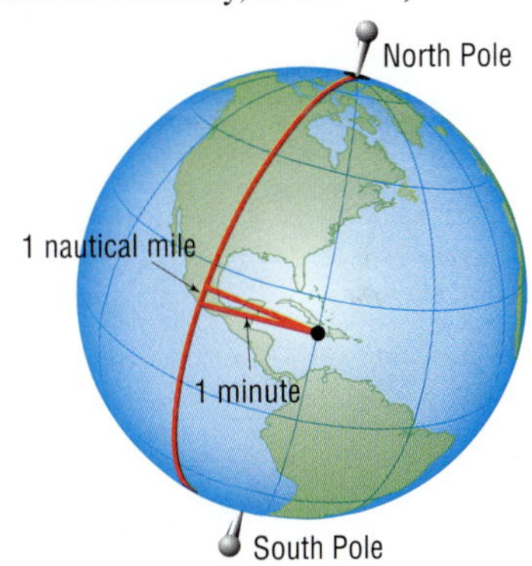

105. Pulleys Two pulleys, one with radius r_1 and the other with radius r_2, are connected by a belt. The pulley with radius r_1 rotates at ω_1 revolutions per minute, whereas the pulley with radius r_2 rotates at ω_2 revolutions per minute. Show that $\dfrac{r_1}{r_2} = \dfrac{\omega_2}{\omega_1}$.

106. Do you prefer to measure angles using degrees or radians? Provide justification and a rationale for your choice.

107. Discuss why ships and airplanes use nautical miles to measure distance. Explain the difference between a nautical mile and a statute mile.

108. Investigate the way that speed bicycles work. In particular, explain the differences and similarities between 5-speed and 9-speed derailleurs. Be sure to include a discussion of linear speed and angular speed.

* Any circle drawn on the surface of Earth that divides Earth into two equal hemispheres.

PREPARING FOR THIS SECTION

Before getting started, review the following:

- ✓ Pythagorean Theorem (Section A.2, pp. 565–567)
- ✓ Functions (Section 1.4, pp. 33–44)

2.2 RIGHT TRIANGLE TRIGONOMETRY

OBJECTIVES

1. Find the Value of Trigonometric Functions of Acute Angles
2. Use the Fundamental Identities
3. Find the Remaining Trigonometric Functions Given the Value of One of Them
4. Use the Complementary Angle Theorem

A triangle in which one angle is a right angle $(90°)$ is called a **right triangle.** Recall that the side opposite the right angle is called the **hypotenuse,** and the remaining two sides are called the **legs** of the triangle. In Figure 20 we have labeled the hypotenuse as c to indicate that its length is c units, and, in a like manner, we have labeled the legs as a and b. Because the triangle is a right triangle, the Pythagorean Theorem tells us that

Figure 20

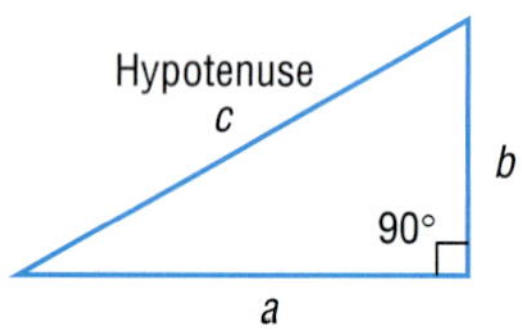

$$c^2 = a^2 + b^2$$

Now, suppose that θ is an **acute angle;** that is, $0° < \theta < 90°$ (if θ is measured in degrees) and $0 < \theta < \frac{\pi}{2}$ (if θ is measured in radians). See Figure 21(a). Using this acute angle θ, we can form a right triangle, like the one illustrated in Figure 21(b), with hypotenuse of length c and legs of lengths a and b. Using the three sides of this triangle, we can form exactly six ratios:

$$\frac{b}{c},\quad \frac{a}{c},\quad \frac{b}{a},\quad \frac{c}{b},\quad \frac{c}{a},\quad \frac{a}{b}$$

Figure 21

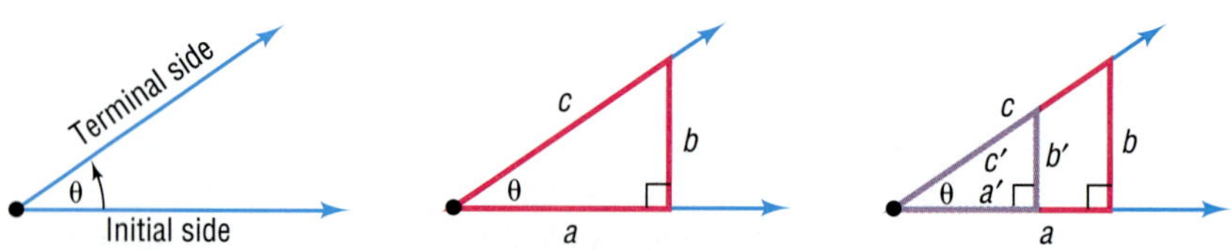

(a) Acute angle **(b)** Right triangle **(c)** Similar triangles

In fact, these ratios depend only on the size of the angle θ and not on the triangle formed. To see why, look at Figure 21(c). Any two right triangles formed using the angle θ will be similar and, hence, corresponding ratios will be equal. As a result,

$$\frac{b}{c} = \frac{b'}{c'} \quad \frac{a}{c} = \frac{a'}{c'} \quad \frac{b}{a} = \frac{b'}{a'} \quad \frac{c}{b} = \frac{c'}{b'} \quad \frac{c}{a} = \frac{c'}{a'} \quad \frac{a}{b} = \frac{a'}{b'}$$

Because the ratios depend only on the angle θ and not on the triangle itself, we give each ratio a name that involves θ: sine of θ, cosine of θ, tangent of θ, cosecant of θ, secant of θ, and cotangent of θ.

The six ratios of a right triangle are called **trigonometric functions of acute angles** and are defined as follows:

Function Name	Abbreviation	Value
sine of θ	$\sin\theta$	$\frac{b}{c}$
cosine of θ	$\cos\theta$	$\frac{a}{c}$
tangent of θ	$\tan\theta$	$\frac{b}{a}$
cosecant of θ	$\csc\theta$	$\frac{c}{b}$
secant of θ	$\sec\theta$	$\frac{c}{a}$
cotangent of θ	$\cot\theta$	$\frac{a}{b}$

Figure 22

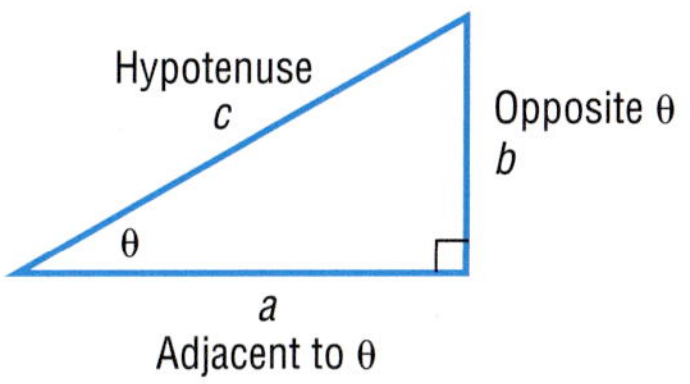

As an aid to remembering these definitions, it may be helpful to refer to the lengths of the sides of the triangle by the names *hypotenuse (c), opposite (b),* and *adjacent (a).* See Figure 22. In terms of these names, we have the following ratios:

$$\sin\theta = \frac{\text{opposite}}{\text{hypotenuse}} = \frac{b}{c} \quad \cos\theta = \frac{\text{adjacent}}{\text{hypotenuse}} = \frac{a}{c} \quad \tan\theta = \frac{\text{opposite}}{\text{adjacent}} = \frac{b}{a}$$
$$\csc\theta = \frac{\text{hypotenuse}}{\text{opposite}} = \frac{c}{b} \quad \sec\theta = \frac{\text{hypotenuse}}{\text{adjacent}} = \frac{c}{a} \quad \cot\theta = \frac{\text{adjacent}}{\text{opposite}} = \frac{a}{b} \qquad (1)$$

Notice that each of the trigonometric functions of the acute angle θ is positive.

1 **EXAMPLE 1** **Finding the Value of Trigonometric Functions**

Figure 23

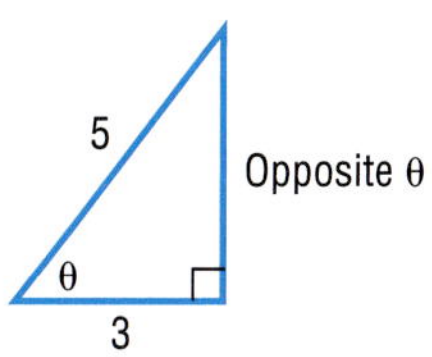

Find the value of each of the six trigonometric functions of the angle θ in Figure 23.

Solution We see in Figure 23 that the two given sides of the triangle are

$$c = \text{hypotenuse} = 5 \qquad a = \text{adjacent} = 3$$

To find the length of the opposite side, we use the Pythagorean Theorem.

$$\begin{aligned}(\text{adjacent})^2 + (\text{opposite})^2 &= (\text{hypotenuse})^2\\ 3^2 + (\text{opposite})^2 &= 5^2\\ (\text{opposite})^2 &= 25 - 9 = 16\\ \text{opposite} &= 4\end{aligned}$$

Now that we know the lengths of the three sides, we use the ratios in (1) to find the value of each of the six trigonometric functions:

$$\sin\theta = \frac{\text{opposite}}{\text{hypotenuse}} = \frac{4}{5} \qquad \cos\theta = \frac{\text{adjacent}}{\text{hypotenuse}} = \frac{3}{5} \qquad \tan\theta = \frac{\text{opposite}}{\text{adjacent}} = \frac{4}{3}$$

$$\csc\theta = \frac{\text{hypotenuse}}{\text{opposite}} = \frac{5}{4} \qquad \sec\theta = \frac{\text{hypotenuse}}{\text{adjacent}} = \frac{5}{3} \qquad \cot\theta = \frac{\text{adjacent}}{\text{opposite}} = \frac{3}{4}$$

NOW WORK PROBLEM 1.

Fundamental Identities

2 You may have observed some relationships that exist among the six trigonometric functions of acute angles. For example, the **reciprocal identities** are

Reciprocal Identities

$$\csc\theta = \frac{1}{\sin\theta} \qquad \sec\theta = \frac{1}{\cos\theta} \qquad \cot\theta = \frac{1}{\tan\theta} \tag{2}$$

Two other fundamental identities that are easy to see are the **quotient identities.**

Quotient Identities

$$\tan\theta = \frac{\sin\theta}{\cos\theta} \qquad \cot\theta = \frac{\cos\theta}{\sin\theta} \tag{3}$$

If $\sin\theta$ and $\cos\theta$ are known, formulas (2) and (3) make it easy to find the values of the remaining trigonometric functions.

EXAMPLE 2 Finding the Value of the Remaining Trigonometric Functions, Given $\sin\theta$ and $\cos\theta$

Given $\sin\theta = \frac{\sqrt{5}}{5}$ and $\cos\theta = \frac{2\sqrt{5}}{5}$, find the value of each of the four remaining trigonometric functions of θ.

Solution Based on formula (3), we have

$$\tan\theta = \frac{\sin\theta}{\cos\theta} = \frac{\frac{\sqrt{5}}{5}}{\frac{2\sqrt{5}}{5}} = \frac{1}{2}$$

Then we use the reciprocal identities from formula (2) to get

$$\csc\theta = \frac{1}{\sin\theta} = \frac{1}{\frac{\sqrt{5}}{5}} = \frac{5}{\sqrt{5}} = \sqrt{5} \qquad \sec\theta = \frac{1}{\cos\theta} = \frac{1}{\frac{2\sqrt{5}}{5}} = \frac{5}{2\sqrt{5}} = \frac{\sqrt{5}}{2} \qquad \cot\theta = \frac{1}{\tan\theta} = \frac{1}{\frac{1}{2}} = 2$$

NOW WORK PROBLEM 11.

Figure 24
$a^2 + b^2 = c^2$

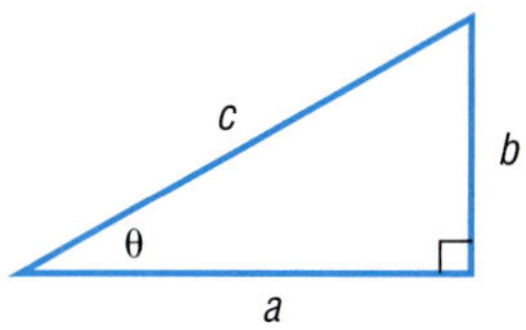

Refer now to the triangle in Figure 24. The Pythagorean Theorem states that $a^2 + b^2 = c^2$, which we write as

$$b^2 + a^2 = c^2$$

Dividing each side by c^2, we get

$$\frac{b^2}{c^2} + \frac{a^2}{c^2} = 1 \quad \text{or} \quad \left(\frac{b}{c}\right)^2 + \left(\frac{a}{c}\right)^2 = 1$$

In terms of trigonometric functions of the angle θ, this equation states that

$$(\sin\theta)^2 + (\cos\theta)^2 = 1 \qquad (4)$$

Equation (4) is, in fact, an identity, since the equation is true for any acute angle θ.

It is customary to write $\sin^2\theta$ instead of $(\sin\theta)^2$, $\cos^2\theta$ instead of $(\cos\theta)^2$, and so on. With this notation, we can rewrite equation (4) as

$$\sin^2\theta + \cos^2\theta = 1 \qquad (5)$$

Another identity can be obtained from equation (5) by dividing each side by $\cos^2\theta$.

$$\frac{\sin^2\theta}{\cos^2\theta} + 1 = \frac{1}{\cos^2\theta}$$

Now use formulas (2) and (3) to get

$$\tan^2\theta + 1 = \sec^2\theta \qquad (6)$$

Similarly, by dividing each side of equation (5) by $\sin^2\theta$, we get

$$1 + \cot^2\theta = \csc^2\theta \qquad (7)$$

Collectively, the identities in equations (5), (6), and (7) are referred to as the **Pythagorean identities.**

Let's pause here to summarize the fundamental identities.

Fundamental Identities

$$\tan\theta = \frac{\sin\theta}{\cos\theta} \qquad \cot\theta = \frac{\cos\theta}{\sin\theta}$$

$$\csc\theta = \frac{1}{\sin\theta} \qquad \sec\theta = \frac{1}{\cos\theta} \qquad \cot\theta = \frac{1}{\tan\theta}$$

$$\sin^2\theta + \cos^2\theta = 1 \qquad \tan^2\theta + 1 = \sec^2\theta \qquad 1 + \cot^2\theta = \csc^2\theta$$

EXAMPLE 3 Finding the Exact Value of a Trigonometric Expression Using Identities

Find the exact value of each expression. Do not use a calculator.

(a) $\tan 20° - \dfrac{\sin 20°}{\cos 20°}$ (b) $\sin^2\dfrac{\pi}{12} + \dfrac{1}{\sec^2\dfrac{\pi}{12}}$

Solution (a) $\tan 20° - \dfrac{\sin 20°}{\cos 20°} = \tan 20° - \tan 20° = 0$

$\uparrow$ $\dfrac{\sin\theta}{\cos\theta} = \tan\theta$

$$\text{(b) } \sin^2\frac{\pi}{12} + \frac{1}{\sec^2\frac{\pi}{12}} \underset{\substack{\uparrow \\ \cos\theta = \frac{1}{\sec\theta}}}{=} \sin^2\frac{\pi}{12} + \cos^2\frac{\pi}{12} \underset{\substack{\uparrow \\ \sin^2\theta + \cos^2\theta = 1}}{=} 1$$

NOW WORK PROBLEM 29.

3 Once the value of one trigonometric function is known, it is possible to find the value of each of the remaining five trigonometric functions.

EXAMPLE 4 Finding the Value of the Remaining Trigonometric Functions, Given $\sin\theta$, θ Acute

Given that $\sin\theta = \frac{1}{3}$ and θ is an acute angle, find the exact value of each of the remaining five trigonometric functions of θ.

Solution We solve this problem in two ways: The first way uses the definition of the trigonometric functions; the second method uses the fundamental identities.

Solution 1 Using the Definition We draw a right triangle with acute angle θ, opposite side of length $b = 1$, and hypotenuse of length $c = 3$ $\left(\text{because } \sin\theta = \frac{1}{3} = \frac{b}{c}\right)$. See Figure 25. The adjacent side a can be found by using the Pythagorean Theorem.

Figure 25

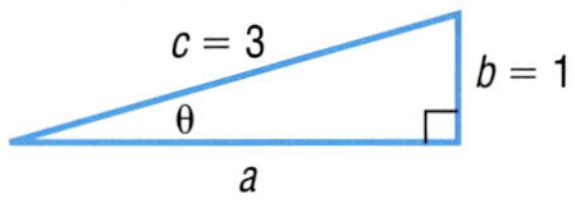

$$a^2 + 1^2 = 3^2$$
$$a^2 + 1 = 9$$
$$a^2 = 8$$
$$a = 2\sqrt{2}$$

Now the definitions given in equation (1) can be used to find the value of each of the remaining five trigonometric functions. (Refer back to the method used in Example 1.) Using $a = 2\sqrt{2}$, $b = 1$, and $c = 3$, we have

$$\cos\theta = \frac{a}{c} = \frac{2\sqrt{2}}{3} \qquad \tan\theta = \frac{b}{a} = \frac{1}{2\sqrt{2}} = \frac{\sqrt{2}}{4}$$

$$\csc\theta = \frac{c}{b} = \frac{3}{1} = 3 \qquad \sec\theta = \frac{c}{a} = \frac{3}{2\sqrt{2}} = \frac{3\sqrt{2}}{4} \qquad \cot\theta = \frac{a}{b} = \frac{2\sqrt{2}}{1} = 2\sqrt{2}$$

Solution 2 Using Identities We begin by seeking $\cos\theta$, which can be found by using the Pythagorean identity from equation (5).

$$\sin^2\theta + \cos^2\theta = 1$$
$$\frac{1}{9} + \cos^2\theta = 1 \qquad \sin\theta = \frac{1}{3}$$
$$\cos^2\theta = 1 - \frac{1}{9} = \frac{8}{9}$$

Because $\cos\theta > 0$ for an acute angle θ, we have

$$\cos\theta = \sqrt{\frac{8}{9}} = \frac{2\sqrt{2}}{3}$$

Now we know that $\sin\theta = \frac{1}{3}$ and $\cos\theta = \frac{2\sqrt{2}}{3}$, so we can proceed as we did in Example 2.

$$\tan\theta = \frac{\sin\theta}{\cos\theta} = \frac{\frac{1}{3}}{\frac{2\sqrt{2}}{3}} = \frac{1}{2\sqrt{2}} = \frac{\sqrt{2}}{4} \qquad \cot\theta = \frac{1}{\tan\theta} = \frac{1}{\frac{\sqrt{2}}{4}} = \frac{4}{\sqrt{2}} = 2\sqrt{2}$$

$$\sec\theta = \frac{1}{\cos\theta} = \frac{1}{\frac{2\sqrt{2}}{3}} = \frac{3}{2\sqrt{2}} = \frac{3\sqrt{2}}{4} \qquad \csc\theta = \frac{1}{\sin\theta} = \frac{1}{\frac{1}{3}} = 3$$

■

Finding the Values of the Trigonometric Functions When One Is Known

Given the value of one trigonometric function of an acute angle θ, the exact value of each of the remaining five trigonometric functions of θ can be found in either of two ways.

Method 1 Using the Definition

STEP 1: Draw a right triangle showing the acute angle θ.

STEP 2: Two of the sides can then be assigned values based on the given trigonometric function.

STEP 3: Find the length of the third side by using the Pythagorean Theorem.

STEP 4: Use the definitions in equation (1) to find the value of each of the remaining trigonometric functions.

Method 2 Using Identities

Use appropriately selected identities to find the value of each of the remaining trigonometric functions.

EXAMPLE 5 Given One Value of a Trigonometric Function, Find the Remaining Ones

Given $\tan\theta = \frac{1}{2}$, θ an acute angle, find the exact value of each of the remaining five trigonometric functions of θ.

Solution 1 Using the Definition

Figure 26
$\tan\theta = \frac{1}{2}$

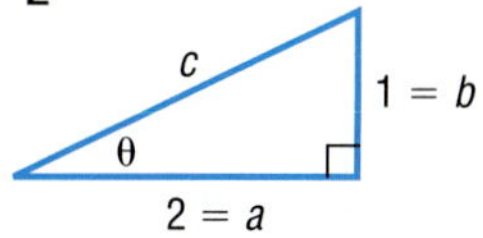

Figure 26 shows a right triangle with acute angle θ, where

$$\tan\theta = \frac{1}{2} = \frac{\text{opposite}}{\text{adjacent}} = \frac{b}{a}$$

Choose $b = 1$ and $a = 2$. The hypotenuse c can be found by using the Pythagorean Theorem.

$$c^2 = a^2 + b^2 = 2^2 + 1^2 = 5$$
$$c = \sqrt{5}$$

Now apply the definitions using $a = 2$, $b = 1$, and $c = \sqrt{5}$.

$$\sin\theta = \frac{b}{c} = \frac{1}{\sqrt{5}} = \frac{\sqrt{5}}{5} \qquad \cos\theta = \frac{a}{c} = \frac{2}{\sqrt{5}} = \frac{2\sqrt{5}}{5}$$

$$\csc\theta = \frac{c}{b} = \frac{\sqrt{5}}{1} = \sqrt{5} \qquad \sec\theta = \frac{c}{a} = \frac{\sqrt{5}}{2} \qquad \cot\theta = \frac{a}{b} = \frac{2}{1} = 2$$

Solution 2 Using Identities We use the Pythagorean identity that involves $\tan\theta$:

$$\tan^2\theta + 1 = \sec^2\theta$$

$$\left(\frac{1}{2}\right)^2 + 1 = \sec^2\theta \qquad \tan\theta = \frac{1}{2}$$

$$\sec^2\theta = \frac{1}{4} + 1 = \frac{5}{4} \qquad \text{Proceed to solve for } \sec\theta.$$

$$\sec\theta = \frac{\sqrt{5}}{2} \qquad \sec\theta > 0,\ \theta \text{ acute}$$

Now

$$\cos\theta = \frac{1}{\sec\theta} = \frac{1}{\frac{\sqrt{5}}{2}} = \frac{2}{\sqrt{5}} = \frac{2\sqrt{5}}{5}$$

$$\tan\theta = \frac{\sin\theta}{\cos\theta}, \quad \text{so} \quad \sin\theta = (\tan\theta)(\cos\theta) = \frac{1}{2}\cdot\frac{2\sqrt{5}}{5} = \frac{\sqrt{5}}{5}$$

$$\csc\theta = \frac{1}{\sin\theta} = \frac{1}{\frac{\sqrt{5}}{5}} = \sqrt{5}$$

$$\cot\theta = \frac{1}{\tan\theta} = \frac{1}{\frac{1}{2}} = 2$$

NOW WORK PROBLEM 15.

Complementary Angles; Cofunctions

4 Two acute angles are called **complementary** if their sum is a right angle. Because the sum of the angles of any triangle is 180°, it follows that, for a right triangle, the two acute angles are complementary.

Figure 27

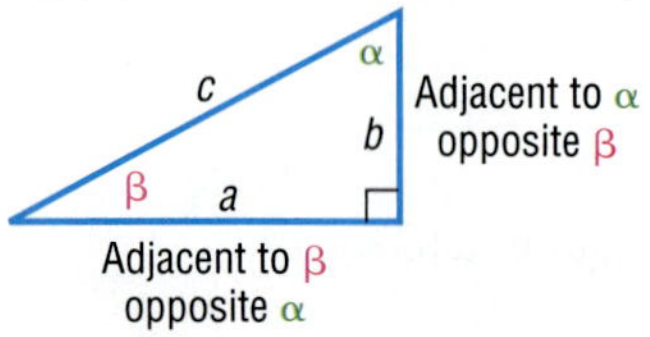

Refer now to Figure 27; we have labeled the angle opposite side b as β and the angle opposite side a as α. Notice that side a is adjacent to angle β and is opposite angle α. Similarly, side b is opposite angle β and is adjacent to angle α. As a result,

$$\begin{aligned} \sin\beta &= \frac{b}{c} = \cos\alpha & \cos\beta &= \frac{a}{c} = \sin\alpha & \tan\beta &= \frac{b}{a} = \cot\alpha \\ \csc\beta &= \frac{c}{b} = \sec\alpha & \sec\beta &= \frac{c}{a} = \csc\alpha & \cot\beta &= \frac{a}{b} = \tan\alpha \end{aligned} \tag{8}$$

Because of these relationships, the functions sine and cosine, tangent and cotangent, and secant and cosecant are called **cofunctions** of each other. The identities (8) may be expressed in words as follows:

Theorem

Complementary Angle Theorem

Cofunctions of complementary angles are equal.

Here are examples of this theorem.

Complementary angles	Complementary angles	Complementary angles
$\sin 30° = \cos 60°$	$\tan 40° = \cot 50°$	$\sec 80° = \csc 10°$
Cofunctions	Cofunctions	Cofunctions

If an angle θ is measured in degrees, we will use the degree symbol when writing a trigonometric function of θ, as, for example, in $\sin 30°$ and $\tan 45°$. If an angle θ is measured in radians, then no symbol is used when writing a trigonometric function of θ, as, for example, in $\cos \pi$ and $\sec \frac{\pi}{3}$.

If θ is an acute angle measured in degrees, the angle $90° - \theta$ (or $\frac{\pi}{2} - \theta$, if θ is in radians) is the angle complementary to θ. Table 2 restates the preceding theorem on cofunctions.

TABLE 2

θ (Degrees)	θ (Radians)
$\sin\theta = \cos(90° - \theta)$	$\sin\theta = \cos\left(\frac{\pi}{2} - \theta\right)$
$\cos\theta = \sin(90° - \theta)$	$\cos\theta = \sin\left(\frac{\pi}{2} - \theta\right)$
$\tan\theta = \cot(90° - \theta)$	$\tan\theta = \cot\left(\frac{\pi}{2} - \theta\right)$
$\csc\theta = \sec(90° - \theta)$	$\csc\theta = \sec\left(\frac{\pi}{2} - \theta\right)$
$\sec\theta = \csc(90° - \theta)$	$\sec\theta = \csc\left(\frac{\pi}{2} - \theta\right)$
$\cot\theta = \tan(90° - \theta)$	$\cot\theta = \tan\left(\frac{\pi}{2} - \theta\right)$

Although the angle θ in Table 2 is acute, we will see later (Section 3.4) that these results are valid for any angle θ.

EXAMPLE 6 Using the Complementary Angle Theorem

(a) $\sin 62° = \cos(90° - 62°) = \cos 28°$

(b) $\tan\frac{\pi}{12} = \cot\left(\frac{\pi}{2} - \frac{\pi}{12}\right) = \cot\frac{5\pi}{12}$

(c) $\cos\frac{\pi}{4} = \sin\left(\frac{\pi}{2} - \frac{\pi}{4}\right) = \sin\frac{\pi}{4}$

(d) $\csc\frac{\pi}{6} = \sec\left(\frac{\pi}{2} - \frac{\pi}{6}\right) = \sec\frac{\pi}{3}$

EXAMPLE 7 **Using the Complementary Angle Theorem**

Find the exact value of each expression. Do not use a calculator.

(a) $\sec 28^\circ - \csc 62^\circ$ (b) $\dfrac{\sin 35^\circ}{\cos 55^\circ}$

Solution (a) $\sec 28^\circ - \csc 62^\circ = \csc(90^\circ - 28^\circ) - \csc 62^\circ$
$= \csc 62^\circ - \csc 62^\circ = 0$

(b) $\dfrac{\sin 35^\circ}{\cos 55^\circ} = \dfrac{\cos(90^\circ - 35^\circ)}{\cos 55^\circ} = \dfrac{\cos 55^\circ}{\cos 55^\circ} = 1$

NOW WORK PROBLEMS 33 AND 47.

HISTORICAL FEATURE

The name *sine* for the sine function is due to a medieval confusion. The name comes from the Sanskrit word *jīva* (meaning chord), first used in India by Araybhata the Elder (AD 510). He really meant half-chord, but abbreviated it. This was brought into Arabic as *jība,* which was meaningless. Because the proper Arabic word *jaib* would be written the same way (short vowels are not written out in Arabic), *jība* was pronounced as *jaib,* which meant bosom or hollow, and *jaib* remains as the Arabic word for sine to this day. Scholars translating the Arabic works into Latin found that the word *sinus* also meant bosom or hollow, and from *sinus* we get the word *sine.*

The name *tangent,* due to Thomas Finck (1583), can be understood by looking at Figure 28. The line segment $\overline{DC}$ is tangent to the circle at C. If $d(O, B) = d(O, C) = 1$, then the length of the line segment $\overline{DC}$ is

$$d(D, C) = \frac{d(D, C)}{1} = \frac{d(D, C)}{d(O, C)} = \tan\alpha$$

The old name for the tangent is *umbra versa* (meaning turned shadow), referring to the use of the tangent in solving height problems with shadows.

Figure 28

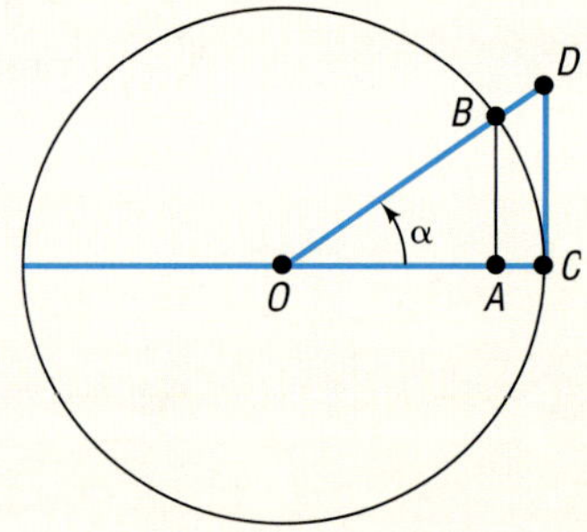

The names of the cofunctions came about as follows. If α and β are complementary angles, then $\cos\alpha = \sin\beta$. Because β is the complement of α, it was natural to write the cosine of α as *sin co* α. Probably for reasons involving ease of pronunciation, the *co* migrated to the front, and then cosine received a three-letter abbreviation to match sin, sec, and tan. The two other cofunctions were similarly treated, except that the long forms *cotan* and *cosec* survive to this day in some countries.

2.2 Concepts and Vocabulary

In Problems 1–3, fill in the blanks.

1. Two acute angles whose sum is a right angle are called ________.

2. The sine and ________ functions are cofunctions.

3. $\tan 28^\circ = \cot$ ________.

In Problems 4–6, answer True or False to each statement.

4. $1 + \tan^2\theta = \csc^2\theta$.

5. If θ is an acute angle and $\sec\theta = 3$, then $\cos\theta = \dfrac{1}{3}$.

6. $\tan\dfrac{\pi}{5} = \cot\dfrac{4\pi}{5}$.

7. Write down the Reciprocal Identities.

8. Explain the Complementary Angle Theorem. Give an example that uses it.

9. If you know the value of $\tan\theta$, θ an acute angle, how would you find $\sec\theta$?

10. Name the two functions that are reciprocals and cofunctions of each other.

2.2 Exercises

In Problems 1–10, find the value of the six trigonometric functions of the angle θ in each figure.

1.
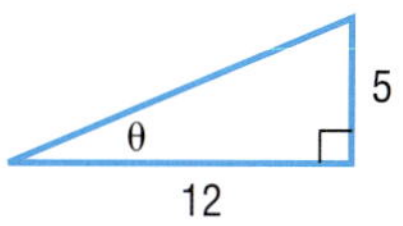

2.
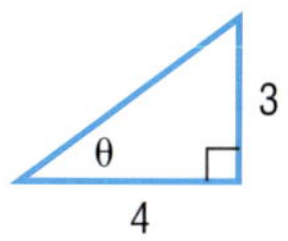

3.
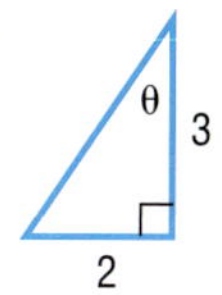

4.
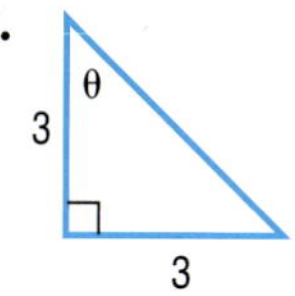

5.
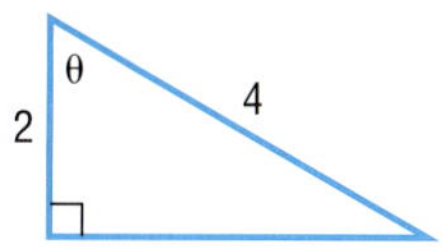

6.
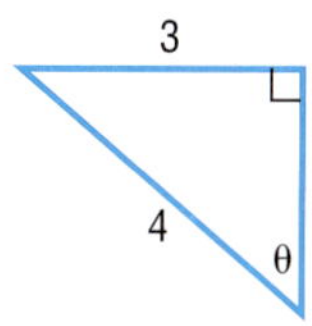

7.
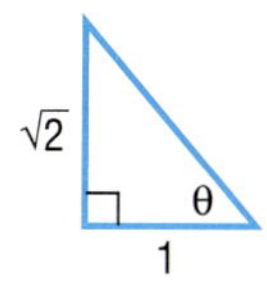

8.
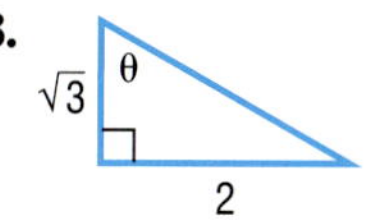

9.
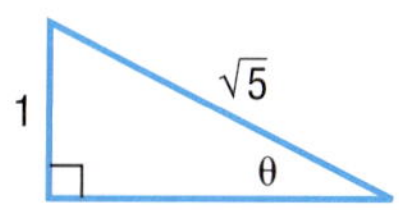

10.
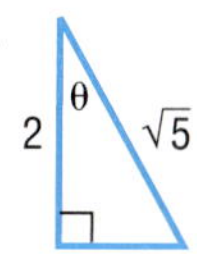

In Problems 11–14, use identities to find the exact value of each of the four remaining trigonometric functions of the acute angle θ.

11. $\sin\theta = \frac{1}{2}, \quad \cos\theta = \frac{\sqrt{3}}{2}$

12. $\sin\theta = \frac{\sqrt{3}}{2}, \quad \cos\theta = \frac{1}{2}$

13. $\sin\theta = \frac{2}{3}, \quad \cos\theta = \frac{\sqrt{5}}{3}$

14. $\sin\theta = \frac{1}{3}, \quad \cos\theta = \frac{2\sqrt{2}}{3}$

In Problems 15–26, use the definition or identities to find the exact value of each of the remaining five trigonometric functions of the acute angle θ.

15. $\sin\theta = \frac{\sqrt{2}}{2}$

16. $\cos\theta = \frac{\sqrt{2}}{2}$

17. $\cos\theta = \frac{1}{3}$

18. $\sin\theta = \frac{\sqrt{3}}{4}$

19. $\tan\theta = \frac{1}{2}$

20. $\cot\theta = \frac{1}{2}$

21. $\sec\theta = 3$

22. $\csc\theta = 5$

23. $\tan\theta = \sqrt{2}$

24. $\sec\theta = \frac{5}{3}$

25. $\csc\theta = 2$

26. $\cot\theta = 2$

In Problems 27–44, use Fundamental Identities and/or the Complementary Angle Theorem to find the exact value of each expression. Do not use a calculator.

27. $\sin^2 20° + \cos^2 20°$

28. $\sec^2 28° - \tan^2 28°$

29. $\sin 80° \csc 80°$

30. $\tan 10° \cot 10°$

31. $\tan 50° - \frac{\sin 50°}{\cos 50°}$

32. $\cot 25° - \frac{\cos 25°}{\sin 25°}$

33. $\sin 38° - \cos 52°$

34. $\tan 12° - \cot 78°$

35. $\frac{\cos 10°}{\sin 80°}$

36. $\frac{\cos 40°}{\sin 50°}$

37. $1 - \cos^2 20° - \cos^2 70°$

38. $1 + \tan^2 5° - \csc^2 85°$

39. $\tan 20° - \frac{\cos 70°}{\cos 20°}$

40. $\cot 40° - \frac{\sin 50°}{\sin 40°}$

41. $\tan 35° \cdot \sec 55° \cdot \cos 35°$

42. $\cot 25° \cdot \csc 65° \cdot \sin 25°$

43. $\cos 35° \sin 55° + \cos 55° \sin 35°$

44. $\sec 35° \csc 55° - \tan 35° \cot 55°$

45. Given $\sin 30° = \frac{1}{2}$, use trigonometric identities to find the exact value of

(a) $\cos 60°$ (b) $\cos^2 30°$

(c) $\csc \frac{\pi}{6}$ (d) $\sec \frac{\pi}{3}$

46. Given $\sin 60° = \frac{\sqrt{3}}{2}$, use trigonometric identities to find the exact value of

(a) $\cos 30°$ (b) $\cos^2 60°$

(c) $\sec \frac{\pi}{6}$ (d) $\csc \frac{\pi}{3}$

47. Given $\tan \theta = 4$, use trigonometric identities to find the exact value of

(a) $\sec^2 \theta$ (b) $\cot \theta$

(c) $\cot\left(\frac{\pi}{2} - \theta\right)$ (d) $\csc^2 \theta$

48. Given $\sec \theta = 3$, use trigonometric identities to find the exact value of

(a) $\cos \theta$ (b) $\tan^2 \theta$

(c) $\csc(90° - \theta)$ (d) $\sin^2 \theta$

49. Given $\csc \theta = 4$, use trigonometric identities to find the exact value of

(a) $\sin \theta$ (b) $\cot^2 \theta$

(c) $\sec(90° - \theta)$ (d) $\sec^2 \theta$

50. Given $\cot \theta = 2$, use trigonometric identities to find the exact value of

(a) $\tan \theta$ (b) $\csc^2 \theta$

(c) $\tan\left(\frac{\pi}{2} - \theta\right)$ (d) $\sec^2 \theta$

51. Given the approximation $\sin 38° \approx 0.62$, use trigonometric identities to find the approximate value of

(a) $\cos 38°$ (b) $\tan 38°$

(c) $\cot 38°$ (d) $\sec 38°$

(e) $\csc 38°$ (f) $\sin 52°$

(g) $\cos 52°$ (h) $\tan 52°$

52. Given the approximation $\cos 21° \approx 0.93$, use trigonometric identities to find the approximate value of

(a) $\sin 21°$ (b) $\tan 21°$

(c) $\cot 21°$ (d) $\sec 21°$

(e) $\csc 21°$ (f) $\sin 69°$

(g) $\cos 69°$ (h) $\tan 69°$

53. If $\sin \theta = 0.3$, find the exact value of $\sin \theta + \cos\left(\frac{\pi}{2} - \theta\right)$.

54. If $\tan \theta = 4$, find the exact value of $\tan \theta + \tan\left(\frac{\pi}{2} - \theta\right)$.

55. Find an acute angle θ that satisfies the equation $\sin \theta = \cos(2\theta + 30°)$.

56. Find an acute angle θ that satisfies the equation $\tan \theta = \cot(\theta + 45°)$.

57. Calculating the Time of a Trip From a parking lot you want to walk to a house on the ocean. The house is located 1500 feet down a paved path that parallels the beach, which is 500 feet wide. Along the path you can walk 300 feet per minute, but in the sand on the beach you can only walk 100 feet per minute. See the illustration.

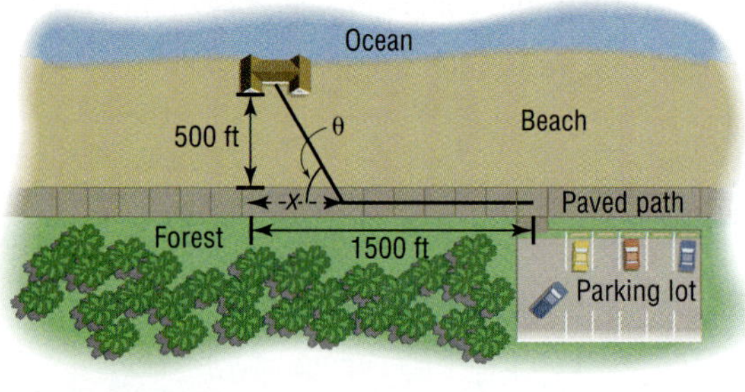

(a) Calculate the time T if you walk 1500 feet along the paved path and then 500 feet in the sand to the house.

(b) Calculate the time T if you walk in the sand directly toward the ocean for 500 feet and then turn left and walk along the beach for 1500 feet to the house.

(c) Express the time T to get from the parking lot to the beachhouse as a function of the angle θ shown in the illustration.

(d) Calculate the time T if you walk directly from the parking lot to the house.
[**Hint:** $\tan \theta = 500/1500$]

(e) Calculate the time T if you walk 1000 feet along the paved path and then walk directly to the house.

(f) Use a graphing utility to graph $T = T(\theta)$. For what angle θ is T least? What is x for this angle? What is the minimum time?

(g) Explain why $\tan \theta = \frac{1}{3}$ gives the smallest angle θ that is possible.

58. Carrying a Ladder around a Corner A ladder of length L is carried horizontally around a corner from a hall 3 feet wide into a hall 4 feet wide. See the illustration. Find the length L of the ladder as a function of the angle θ shown in the illustration.

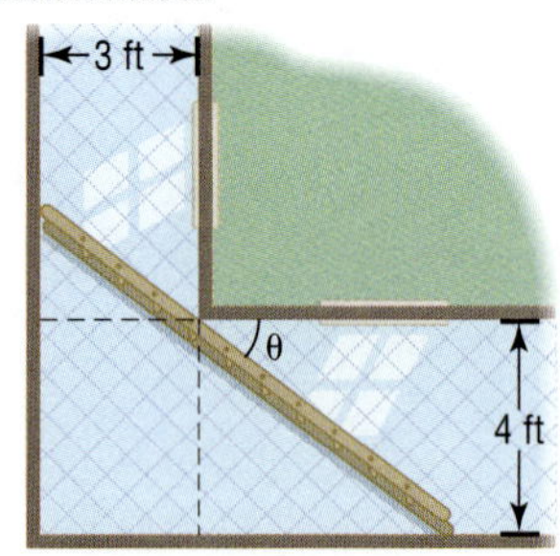

59. Suppose that the angle θ is a central angle of a circle of radius 1 (see the figure). Show that

(a) Angle $OAC = \frac{\theta}{2}$

(b) $|CD| = \sin \theta$ and $|OD| = \cos \theta$

(c) $\tan \frac{\theta}{2} = \frac{\sin \theta}{1 + \cos \theta}$

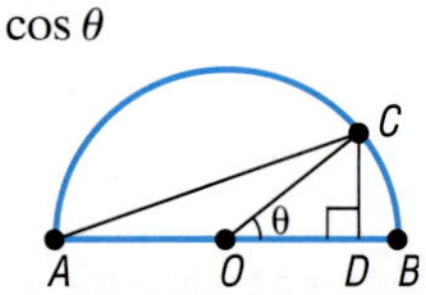

60. Show that the area A of an isosceles triangle is $A = a^2 \sin\theta\cos\theta$, where a is the length of one of the two equal sides and θ is the measure of one of the two equal angles (see the figure).

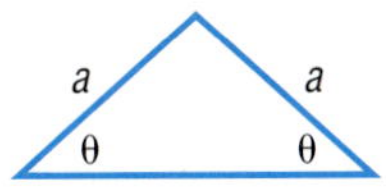

61. Let $n \geq 1$ be any real number and let θ be any angle for which $0 < n\theta < \frac{\pi}{2}$. Then we can draw a triangle with the angles θ and $n\theta$ and included side of length 1 (do you see why?) and place it on the unit circle as illustrated. Now, drop the perpendicular from C to $D = (x, 0)$ and show that

$$x = \frac{\tan(n\theta)}{\tan\theta + \tan(n\theta)}$$

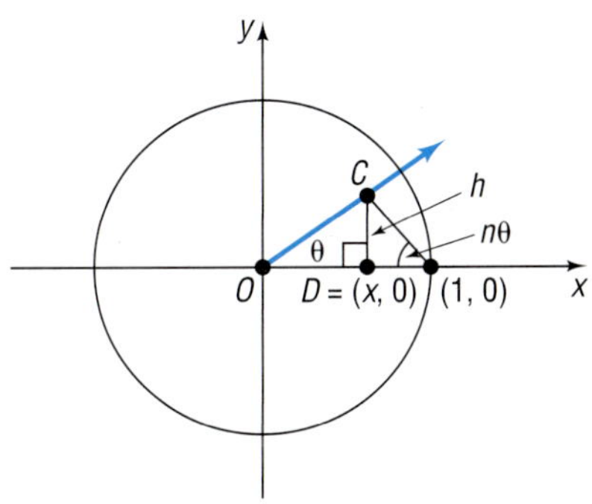

62. Refer to the figure. The smaller circle, whose radius is a, is tangent to the larger circle, whose radius is b. The ray $\overrightarrow{OA}$ contains a diameter of each circle, and the ray $\overrightarrow{OB}$ is tangent to each circle. Show that

$$\cos\theta = \frac{\sqrt{ab}}{\frac{a+b}{2}}$$

(This shows that $\cos\theta$ equals the ratio of the geometric mean of a and b to the arithmetic mean of a and b.)

[**Hint:** First show that $\sin\theta = \frac{b-a}{b+a}$.]

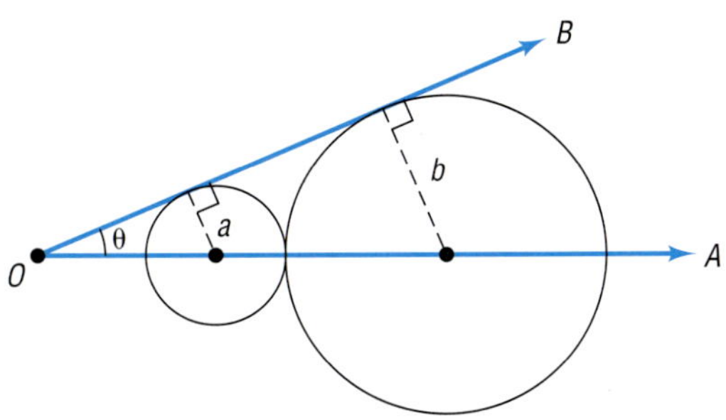

63. Refer to the figure. If $|OA| = 1$, show that

(a) Area $\Delta OAC = \frac{1}{2}\sin\alpha\cos\alpha$

(b) Area $\Delta OCB = \frac{1}{2}|OB|^2 \sin\beta\cos\beta$

(c) Area $\Delta OAB = \frac{1}{2}|OB|\sin(\alpha + \beta)$

(d) $|OB| = \frac{\cos\alpha}{\cos\beta}$

(e) $\sin(\alpha + \beta) = \sin\alpha\cos\beta + \cos\alpha\sin\beta$

[**Hint:** Area ΔOAB = Area ΔOAC + Area ΔOCB]

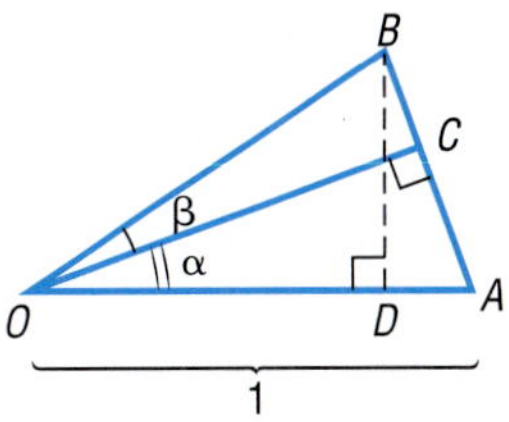

64. Refer to the figure, where a unit circle is drawn. The line DB is tangent to the circle.

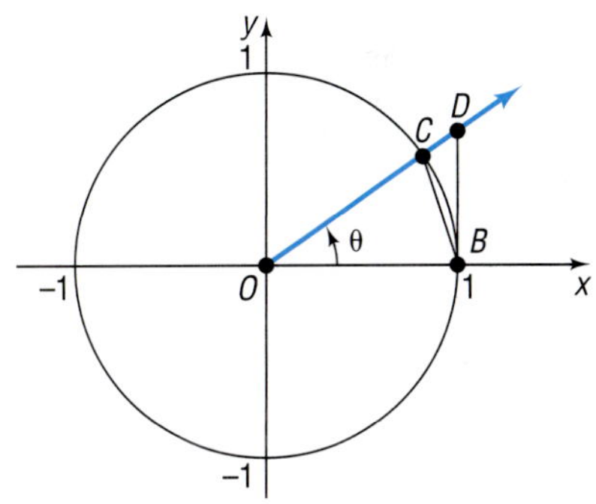

(a) Express the area of ΔOBC in terms of $\sin\theta$ and $\cos\theta$. [**Hint:** Use the altitude from C to the base $\overline{OB} = 1$.]

(b) Express the area of ΔOBD in terms of $\sin\theta$ and $\cos\theta$.

(c) The area of the sector of the circle OBC is $\frac{1}{2}\theta$, where θ is measured in radians. Use the results of parts (a) and (b) and the fact that

$$\text{Area } \Delta OBC < \text{Area } O\overset{\frown}{BC} < \text{Area } \Delta OBD$$

to show that

$$1 < \frac{\theta}{\sin\theta} < \frac{1}{\cos\theta}$$

65. If $\cos\alpha = \tan\beta$ and $\cos\beta = \tan\alpha$, where α and β are acute angles, show that

$$\sin\alpha = \sin\beta = \sqrt{\frac{3 - \sqrt{5}}{2}}$$

66. If θ is an acute angle, explain why $\sec\theta > 1$.

67. If θ is an acute angle, explain why $0 < \sin\theta < 1$.

68. How would you explain the meaning of the sine function to a fellow student who has just completed college algebra?

2.3 COMPUTING THE VALUES OF TRIGONOMETRIC FUNCTIONS OF GIVEN ANGLES

OBJECTIVES

1. Find the Exact Value of the Trigonometric Functions of $\frac{\pi}{4} = 45°$
2. Find the Exact Value of the Trigonometric Functions of $\frac{\pi}{6} = 30°$ and $\frac{\pi}{3} = 60°$
3. Use a Calculator to Approximate the Value of the Trigonometric Functions of Acute Angles

In the previous section, we developed ways to find the value of each trigonometric function of an acute angle when one of the functions is known. In this section, we discuss the problem of finding the value of each trigonometric function of an acute angle when the angle is given.

For three special acute angles, we can use some results from plane geometry to find the exact value of each of the six trigonometric functions.

1 Trigonometric Functions of $\frac{\pi}{4} = 45°$

EXAMPLE 1 Finding the Exact Value of the Trigonometric Functions of $\frac{\pi}{4} = 45°$

Find the exact value of the six trigonometric functions of $\frac{\pi}{4} = 45°$.

Figure 29

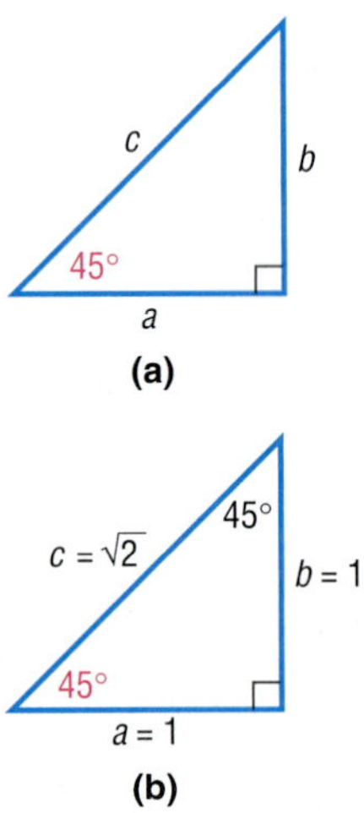

Solution Form the right triangle in Figure 29(a), in which one of the angles is $\frac{\pi}{4} = 45°$. It follows that the other acute angle is also $\frac{\pi}{4} = 45°$, and hence the triangle is isosceles. As a result, side a and side b are equal in length. Because the values of the trigonometric functions of an angle depend only on the angle and not on the size of the triangle, we may choose to use the triangle for which

$$a = b = 1$$

Then, by the Pythagorean Theorem,

$$c^2 = a^2 + b^2 = 1 + 1 = 2$$
$$c = \sqrt{2}$$

As a result, we have the triangle in Figure 29(b), from which we find

$$\sin\frac{\pi}{4} = \sin 45° = \frac{b}{c} = \frac{1}{\sqrt{2}} = \frac{\sqrt{2}}{2} \qquad \cos\frac{\pi}{4} = \cos 45° = \frac{a}{c} = \frac{1}{\sqrt{2}} = \frac{\sqrt{2}}{2}$$

Using Quotient and Reciprocal Identities, we find

$$\tan\frac{\pi}{4} = \tan 45° = \frac{\sin 45°}{\cos 45°} = \frac{\frac{\sqrt{2}}{2}}{\frac{\sqrt{2}}{2}} = 1 \qquad \cot\frac{\pi}{4} = \cot 45° = \frac{1}{\tan 45°} = \frac{1}{1} = 1$$

$$\sec\frac{\pi}{4} = \sec 45° = \frac{1}{\cos 45°} = \frac{1}{\frac{1}{\sqrt{2}}} = \sqrt{2} \qquad \csc\frac{\pi}{4} = \csc 45° = \frac{1}{\sin 45°} = \frac{1}{\frac{1}{\sqrt{2}}} = \sqrt{2}$$

EXAMPLE 2 Finding the Exact Value of a Trigonometric Expression

Find the exact value of each expression.

(a) $(\sin 45°)(\tan 45°)$ (b) $\left(\sec\frac{\pi}{4}\right)\left(\cot\frac{\pi}{4}\right)$

Solution We use the results obtained in Example 1.

(a) $(\sin 45°)(\tan 45°) = \frac{\sqrt{2}}{2}\cdot 1 = \frac{\sqrt{2}}{2}$

(b) $\left(\sec\frac{\pi}{4}\right)\left(\cot\frac{\pi}{4}\right) = \sqrt{2}\cdot 1 = \sqrt{2}$

NOW WORK PROBLEMS 1 AND 13.

2 Trigonometric Functions of $\frac{\pi}{6} = 30°$ and $\frac{\pi}{3} = 60°$

EXAMPLE 3 Finding the Exact Value of the Trigonometric Functions of $\frac{\pi}{6} = 30°$ and $\frac{\pi}{3} = 60°$

Figure 30

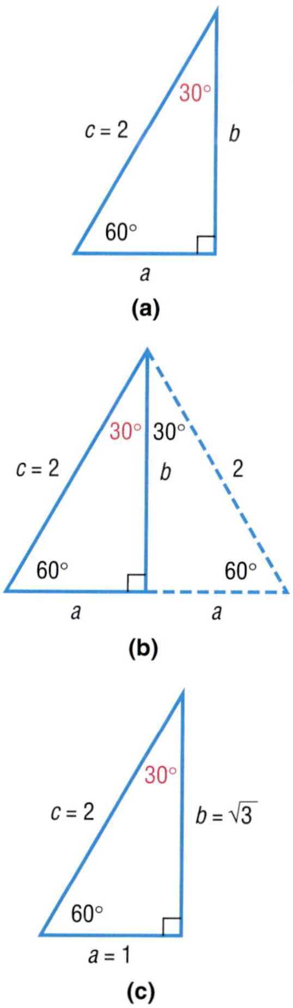

Find the exact value of the six trigonometric functions of $\frac{\pi}{6} = 30°$ and $\frac{\pi}{3} = 60°$.

Solution Form a right triangle in which one of the angles is $\frac{\pi}{6} = 30°$. It then follows that the other angle is $\frac{\pi}{3} = 60°$. Figure 30(a) illustrates such a triangle with hypotenuse of length 2. Our problem is to determine a and b.

We begin by placing next to the triangle in Figure 30(a) another triangle congruent to the first, as shown in Figure 30(b). Notice that we now have a triangle whose angles are each 60°. This triangle is therefore equilateral, so each side is of length 2. In particular, the base is $2a = 2$, so $a = 1$. By the Pythagorean Theorem, b satisfies the equation $a^2 + b^2 = c^2$, so we have

$$a^2 + b^2 = c^2$$
$$1^2 + b^2 = 2^2 \qquad a = 1, c = 2$$
$$b^2 = 4 - 1 = 3$$
$$b = \sqrt{3}$$

Using the triangle in Figure 30(c) and the fact that $\frac{\pi}{6} = 30°$ and $\frac{\pi}{3} = 60°$ are complementary angles, we find

$$\sin\frac{\pi}{6} = \sin 30° = \frac{\text{opposite}}{\text{hypotenuse}} = \frac{1}{2} \qquad \cos\frac{\pi}{3} = \cos 60° = \frac{1}{2}$$

$$\cos\frac{\pi}{6} = \cos 30° = \frac{\text{adjacent}}{\text{hypotenuse}} = \frac{\sqrt{3}}{2} \qquad \sin\frac{\pi}{3} = \sin 60° = \frac{\sqrt{3}}{2}$$

$$\tan\frac{\pi}{6} = \tan 30° = \frac{\sin 30°}{\cos 30°} = \frac{\frac{1}{2}}{\frac{\sqrt{3}}{2}} = \frac{1}{\sqrt{3}} = \frac{\sqrt{3}}{3} \qquad \cot\frac{\pi}{3} = \cot 60° = \frac{\sqrt{3}}{3}$$

$$\csc\frac{\pi}{6} = \csc 30^\circ = \frac{1}{\sin 30^\circ} = \frac{1}{\frac{1}{2}} = 2 \qquad \sec\frac{\pi}{3} = \sec 60^\circ = 2$$

$$\sec\frac{\pi}{6} = \sec 30^\circ = \frac{1}{\cos 30^\circ} = \frac{1}{\frac{\sqrt{3}}{2}} = \frac{2}{\sqrt{3}} = \frac{2\sqrt{3}}{3} \qquad \csc\frac{\pi}{3} = \csc 60^\circ = \frac{2\sqrt{3}}{3}$$

$$\cot\frac{\pi}{6} = \cot 30^\circ = \frac{1}{\tan 30^\circ} = \frac{1}{\frac{\sqrt{3}}{3}} = \frac{3}{\sqrt{3}} = \sqrt{3} \qquad \tan\frac{\pi}{3} = \tan 60^\circ = \sqrt{3}$$

Table 3 summarizes the information just derived for the angles $\frac{\pi}{6} = 30^\circ$, $\frac{\pi}{4} = 45^\circ$, and $\frac{\pi}{3} = 60^\circ$. Until you memorize the entries in Table 3, you should draw the appropriate triangle to determine the values given in the table.

TABLE 3

θ (Radians)	θ (Degrees)	$\sin\theta$	$\cos\theta$	$\tan\theta$	$\csc\theta$	$\sec\theta$	$\cot\theta$
$\frac{\pi}{6}$	30°	$\frac{1}{2}$	$\frac{\sqrt{3}}{2}$	$\frac{\sqrt{3}}{3}$	2	$\frac{2\sqrt{3}}{3}$	$\sqrt{3}$
$\frac{\pi}{4}$	45°	$\frac{\sqrt{2}}{2}$	$\frac{\sqrt{2}}{2}$	1	$\sqrt{2}$	$\sqrt{2}$	1
$\frac{\pi}{3}$	60°	$\frac{\sqrt{3}}{2}$	$\frac{1}{2}$	$\sqrt{3}$	$\frac{2\sqrt{3}}{3}$	2	$\frac{\sqrt{3}}{3}$

EXAMPLE 4 Finding the Exact Value of a Trigonometric Expression

Find the exact value of each expression.

(a) $\sin 45^\circ \cos 30^\circ$ (b) $\tan\frac{\pi}{4} - \sin\frac{\pi}{3}$ (c) $\tan^2\frac{\pi}{6} + \sin^2\frac{\pi}{4}$

Solution

(a) $\sin 45^\circ \cos 30^\circ = \frac{\sqrt{2}}{2}\cdot\frac{\sqrt{3}}{2} = \frac{\sqrt{6}}{4}$

(b) $\tan\frac{\pi}{4} - \sin\frac{\pi}{3} = 1 - \frac{\sqrt{3}}{2} = \frac{2-\sqrt{3}}{2}$

(c) $\tan^2\frac{\pi}{6} + \sin^2\frac{\pi}{4} = \left(\frac{\sqrt{3}}{3}\right)^2 + \left(\frac{\sqrt{2}}{2}\right)^2 = \frac{1}{3} + \frac{1}{2} = \frac{5}{6}$

NOW WORK PROBLEMS 5 AND 15.

The exact values of the trigonometric functions for the angles $\frac{\pi}{6} = 30^\circ$, $\frac{\pi}{4} = 45^\circ$, and $\frac{\pi}{3} = 60^\circ$ are relatively easy to calculate, because the triangles that contain such angles have "nice" geometric features. For most other angles, we can only approximate the value of each trigonometric function. To do this, we will need a calculator.

Using a Calculator to Find Values of Trigonometric Functions

3 Before getting started, you must first decide whether to enter the angle in the calculator using radians or degrees and then set the calculator to the correct MODE. (Check your instruction manual to find out how your calculator handles degrees and radians.) Your calculator has the keys marked [sin], [cos], and [tan]. To find the values of the remaining three trigonometric functions (secant, cosecant, and cotangent), we use the reciprocal identities.

$$\sec\theta = \frac{1}{\cos\theta} \qquad \csc\theta = \frac{1}{\sin\theta} \qquad \cot\theta = \frac{1}{\tan\theta}$$

EXAMPLE 5 Using a Calculator to Approximate the Value of Trigonometric Functions

Use a calculator to find the approximate value of:

(a) $\cos 48°$ (b) $\csc 21°$ (c) $\tan\dfrac{\pi}{12}$

Express your answer rounded to two decimal places.

Solution (a) First, we set the MODE to receive degrees. See Figure 31(a). Figure 31(b) shows the solution using a TI-83 graphing calculator.

Figure 31

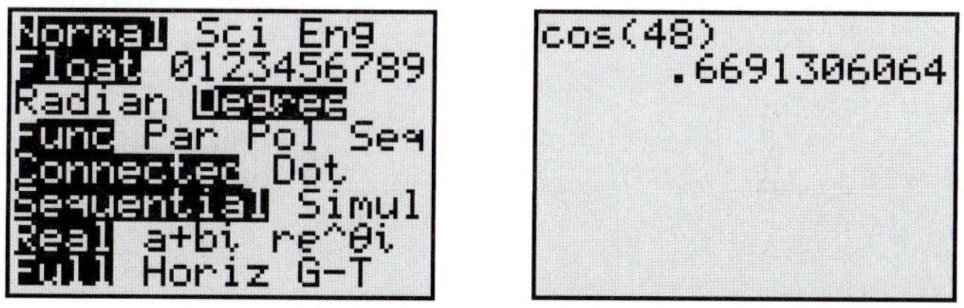

(a) (b)

Rounded to two decimal places,

$$\cos 48° = 0.67$$

(b) Most calculators do not have a csc key. The manufacturers assume the user knows some trigonometry. To find the value of $\csc 21°$, we use the fact that $\csc 21° = \dfrac{1}{\sin 21°}$. Figure 32 shows the solution using a TI-83 graphing calculator. Rounded to two decimal places,

Figure 32

$$\csc 21° = 2.79$$

(c) Set the MODE to receive radians. Figure 33 shows the solution using a TI-83 graphing calculator. Rounded to two decimal places,

Figure 33

$$\tan\frac{\pi}{12} = 0.27$$

NOW WORK PROBLEM 25.

EXAMPLE 6 Constructing a Rain Gutter

A rain gutter is to be constructed of aluminum sheets 12 inches wide. After marking off a length of 4 inches from each edge, this length is bent up at an angle θ. See Figure 34.

Figure 34

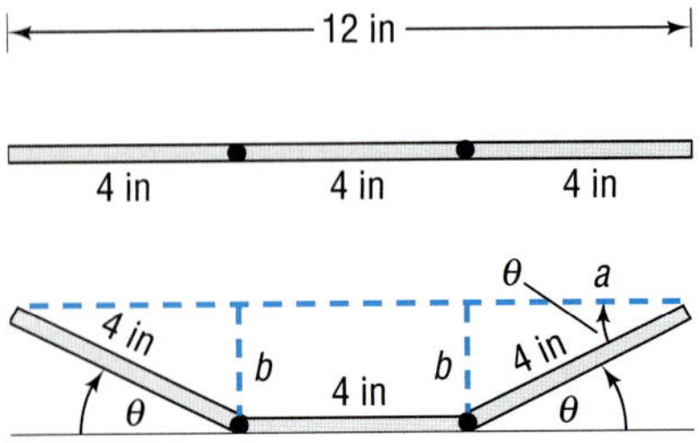

(a) Express the area A of the opening as a function of θ.

[**Hint:** Let b denote the vertical height of the bend.]

(b) Find the area A of the opening for $\theta = 30°, \theta = 45°, \theta = 60°$, and $\theta = 75°$.

(c) Graph $A = A(\theta)$. Find the angle θ that makes A largest. (This bend will allow the most water to flow through the gutter.)

Solution (a) Look again at Figure 34. The area A of the opening is the sum of the areas of two congruent right triangles and one rectangle. Look at Figure 35, showing the triangle in Figure 34 redrawn. We see that

Figure 35

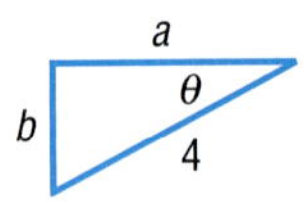

$$\cos\theta = \frac{a}{4} \quad \text{so} \quad a = 4\cos\theta \qquad \sin\theta = \frac{b}{4} \quad \text{so} \quad b = 4\sin\theta$$

The area of the triangle is

$$\text{area} = \frac{1}{2}(\text{base})(\text{height}) = \frac{1}{2}ab = \frac{1}{2}(4\cos\theta)(4\sin\theta) = 8\sin\theta\cos\theta$$

So the area of the two triangles is $16\sin\theta\cos\theta$.

The rectangle has length 4 and height b, so its area is

$$4b = 4(4\sin\theta) = 16\sin\theta$$

The area A of the opening is

$$A = \text{area of the two triangles} + \text{area of the rectangle}$$
$$A(\theta) = 16\sin\theta\cos\theta + 16\sin\theta = 16\sin\theta(\cos\theta + 1)$$

(b) For $\theta = 30°$: $\quad A(30°) = 16\sin 30°(\cos 30° + 1)$

$$= 16\left(\frac{1}{2}\right)\left(\frac{\sqrt{3}}{2} + 1\right) = 4\sqrt{3} + 8$$

The area of the opening for $\theta = 30°$ is about 14.9 square inches.

For $\theta = 45°$: $\quad A(45°) = 16\sin 45°(\cos 45° + 1)$

$$= 16\left(\frac{\sqrt{2}}{2}\right)\left(\frac{\sqrt{2}}{2} + 1\right) = 8 + 8\sqrt{2}$$

The area of the opening for $\theta = 45°$ is about 19.3 square inches.

For $\theta = 60°$: $\quad A(60°) = 16\sin 60°(\cos 60° + 1)$

$$= 16\left(\frac{\sqrt{3}}{2}\right)\left(\frac{1}{2} + 1\right) = 12\sqrt{3}$$

The area of the opening for $\theta = 60°$ is about 20.8 square inches.

For $\theta = 75°$: $\quad A(75°) = 16\sin 75°(\cos 75° + 1) \approx 19.5$

The area of the opening for $\theta = 75°$ is about 19.5 square inches.

Figure 36

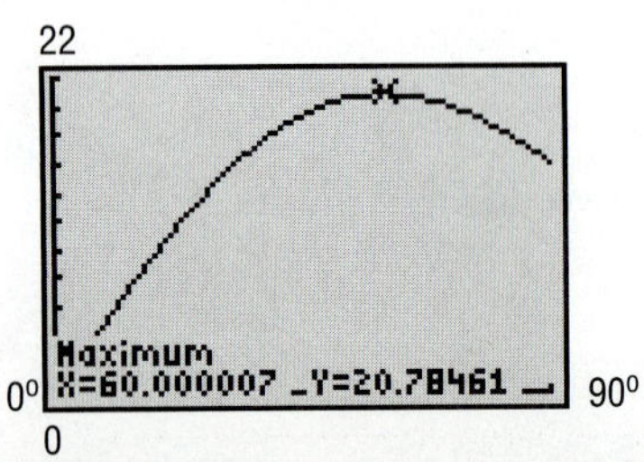

(c) Figure 36 shows the graph of $A = A(\theta)$. Using MAXIMUM, the angle θ that makes A largest is 60°. ■

2.3 Concepts and Vocabulary

In Problems 1 and 2, fill in the blanks.

1. $\tan\frac{\pi}{4} + \sin 30° =$ ________.

2. Using a calculator, $\sin 2 =$ ________, rounded to two decimal places.

In Problems 3 and 4, answer True or False to each statement.

3. Exact values can be found for the trigonometric functions of 60°.

4. Exact values can be found for the sine of any angle.

5. Explain how you remember that $\sin 45° = \frac{\sqrt{2}}{2}$.

6. Explain how you remember that $\sin\frac{\pi}{6} = \frac{1}{2}$.

2.3 Exercises

1. Write down the exact value of each of the six trigonometric functions of 45°.

2. Write down the exact value of each of the six trigonometric functions of 30° and of 60°.

In Problems 3–12, find the exact value of each expression if $\theta = 60°$. Do not use a calculator.

3. $\sin\theta$ **4.** $\cos\theta$ **5.** $\sin\frac{\theta}{2}$ **6.** $\cos\frac{\theta}{2}$ **7.** $(\sin\theta)^2$

8. $(\cos\theta)^2$ **9.** $2\sin\theta$ **10.** $2\cos\theta$ **11.** $\frac{\sin\theta}{2}$ **12.** $\frac{\cos\theta}{2}$

In Problems 13–24, find the exact value of each expression. Do not use a calculator.

13. $4\cos 45° - 2\sin 45°$ **14.** $2\sin 45° + 4\cos 30°$ **15.** $6\tan 45° - 8\cos 60°$

16. $\sin 30° \cdot \tan 60°$ **17.** $\sec\frac{\pi}{4} + 2\csc\frac{\pi}{3}$ **18.** $\tan\frac{\pi}{4} + \cot\frac{\pi}{4}$

19. $\sec^2\frac{\pi}{6} - 4$ **20.** $4 + \tan^2\frac{\pi}{3}$ **21.** $\sin^2 30° + \cos^2 60°$

22. $\sec^2 60° - \tan^2 45°$ **23.** $1 - \cos^2 30° - \cos^2 60°$ **24.** $1 + \tan^2 30° - \csc^2 45°$

In Problems 25–42, use a calculator to find the approximate value of each expression. Round the answer to two decimal places.

25. $\sin 28°$ **26.** $\cos 14°$ **27.** $\tan 21°$ **28.** $\cot 70°$ **29.** $\sec 41°$ **30.** $\csc 55°$

31. $\sin\frac{\pi}{10}$ **32.** $\cos\frac{\pi}{8}$ **33.** $\tan\frac{5\pi}{12}$ **34.** $\cot\frac{\pi}{18}$ **35.** $\sec\frac{\pi}{12}$ **36.** $\csc\frac{5\pi}{13}$

37. $\sin 1$ **38.** $\tan 1$ **39.** $\sin 1°$ **40.** $\tan 1°$ **41.** $\tan 0.3$ **42.** $\tan 0.1$

Projectile Motion The path of a projectile fired at an inclination θ to the horizontal with initial speed v_0 is a parabola (see the figure). The range R of the projectile, that is, the horizontal distance that the projectile travels, is found by using the formula

$$R = \frac{2v_0^2 \sin\theta\cos\theta}{g}$$

where $g \approx 32.2$ feet per second per second ≈ 9.8 meters per second per second is the acceleration due to gravity.

The maximum height H of the projectile is

$$H = \frac{v_0^2 \sin^2\theta}{2g}$$

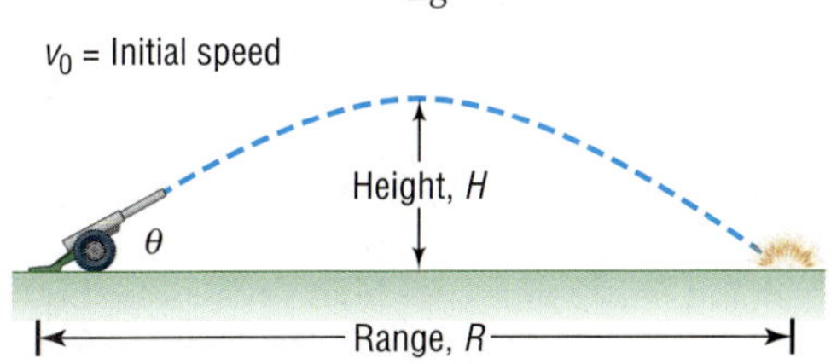

In Problems 43–46, find the range R and maximum height H. Round answers to two decimal places.

43. The projectile is fired at an angle of 45° to the horizontal with an initial speed of 100 feet per second.

44. The projectile is fired at an angle of 30° to the horizontal with an initial speed of 150 meters per second.

45. The projectile is fired at an angle of 25° to the horizontal with an initial speed of 500 meters per second.

46. The projectile is fired at an angle of 50° to the horizontal with an initial speed of 200 feet per second.

47. Inclined Plane If friction is ignored, the time t (in seconds) required for a block to slide down an inclined plane (see the figure) is given by the formula

$$t = \sqrt{\frac{2a \sin\theta}{g \cos\theta}}$$

where a is the length (in feet) of the base and $g \approx 32$ feet per second per second is the acceleration of gravity. How long does it take a block to slide down an inclined plane with base $a = 10$ feet when

(a) $\theta = 30°$? (b) $\theta = 45°$? (c) $\theta = 60°$?

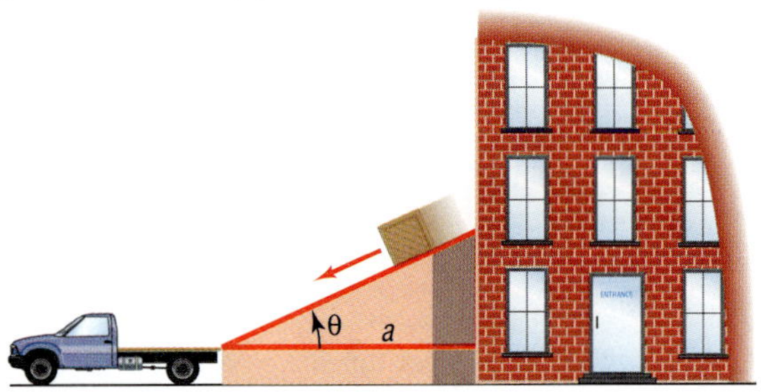

48. Piston Engines In a certain piston engine, the distance x (in meters) from the center of the drive shaft to the head of the piston is given by

$$x = \cos\theta + \sqrt{16 + 0.5(2\cos^2\theta - 1)}$$

where θ is the angle between the crank and the path of the piston head (see the figure). Find x when $\theta = 30°$ and when $\theta = 45°$.

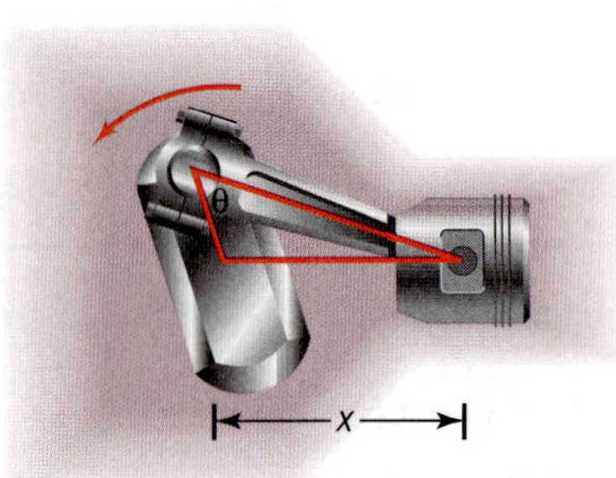

49. Calculating the Time of a Trip Two oceanfront homes are located 8 miles apart on a straight stretch of beach, each a distance of 1 mile from a paved road that parallels the ocean. Sally can jog 8 miles per hour along the paved road, but only 3 miles per hour in the sand on the beach. Because of a river between the two houses, it is necessary to jog on the sand to the road, continue on the road, and then jog on the sand to get from one house to the other. See the illustration.

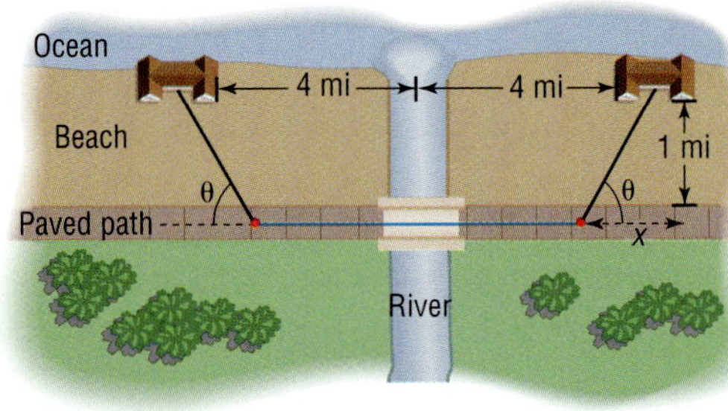

(a) Express the time T to get from one house to the other as a function of the angle θ shown in the illustration.

(b) Calculate the time T for $\theta = 30°$. How long is Sally on the paved road?

(c) Calculate the time T for $\theta = 45°$. How long is Sally on the paved road?

(d) Calculate the time T for $\theta = 60°$. How long is Sally on the paved road?

(e) Calculate the time T for $\theta = 90°$. Describe the path taken.

(f) Calculate the time T for $\tan\theta = \frac{1}{4}$. Describe the path taken. Explain why θ must be larger than 14°.

(g) Use a graphing utility to graph $T = T(\theta)$. What angle θ results in the least time? What is the least time? How long is Sally on the paved road?

50. Designing Fine Decorative Pieces A designer of decorative art plans to market solid gold spheres encased in clear crystal cones. Each sphere is of fixed radius R and will be enclosed in a cone of height h and radius r. See the illustration. Many cones can be used to enclose the sphere, each having a different slant angle θ.

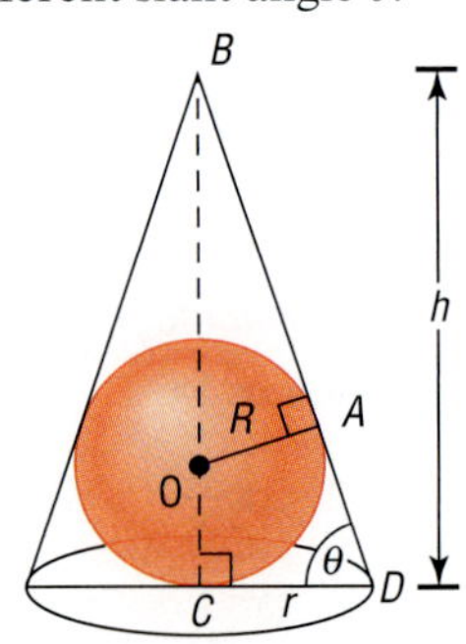

(a) Express the volume V of the cone as a function of the slant angle θ of the cone.
[**Hint:** The volume V of a cone of height h and radius r is $V = \frac{1}{3}\pi r^2 h$.]

(b) What volume V is required to enclose a sphere of radius 2 centimeters in a cone whose slant angle θ is 30°? 45°? 60°?

(c) What slant angle θ should be used for the volume V of the cone to be a minimum? (This choice minimizes the amount of crystal required and gives maximum emphasis to the gold sphere.)

51. Use a calculator set in radian mode to complete the following table. What can you conclude about the ratio $\frac{\sin\theta}{\theta}$ as θ approaches 0?

θ	0.5	0.4	0.2	0.1	0.01	0.001	0.0001	0.00001
$\sin\theta$								
$\frac{\sin\theta}{\theta}$								

52. Use a calculator set in radian mode to complete the following table. What can you conclude about the ratio $\frac{\cos\theta - 1}{\theta}$ as θ approaches 0?

θ	0.5	0.4	0.2	0.1	0.01	0.001	0.0001	0.00001
$\cos\theta - 1$								
$\frac{\cos\theta - 1}{\theta}$								

53. Find the exact value of $\tan 1° \cdot \tan 2° \cdot \tan 3° \cdot \ldots \cdot \tan 89°$.

54. Find the exact value of $\cot 1° \cdot \cot 2° \cdot \cot 3° \cdot \ldots \cdot \cot 89°$.

55. Find the exact value of $\cos 1° \cdot \cos 2° \cdot \ldots \cdot \cos 45° \cdot \csc 46° \cdot \ldots \cdot \csc 89°$.

56. Find the exact value of $\sin 1° \cdot \sin 2° \cdot \ldots \cdot \sin 45° \cdot \sec 46° \cdot \ldots \cdot \sec 89°$.

57. Write a brief paragraph that explains how to quickly compute the trigonometric functions of 30°, 45°, and 60°.

2.4 TRIGONOMETRIC FUNCTIONS OF GENERAL ANGLES

OBJECTIVES

1. Find the Exact Value of the Trigonometric Functions for General Angles
2. Use Coterminal Angles to Find the Exact Value of a Trigonometric Function
3. Determine the Sign of the Trigonometric Functions of an Angle in a Given Quadrant
4. Find the Reference Angle of a General Angle
5. Use the Theorem on Reference Angles
6. Find the Exact Value of Trigonometric Functions of an Angle Given One of them and the Quadrant of the Angle

Figure 37

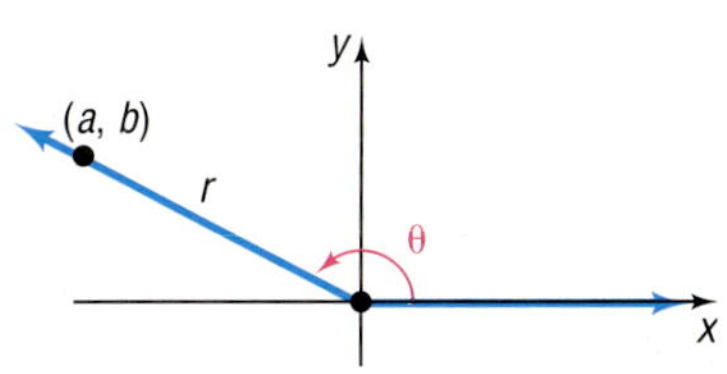

To extend the definitions of the trigonometric functions to include angles that are not acute, we employ a rectangular coordinate system and place the angle in the standard position so that its vertex is at the origin and its initial side is along the positive x-axis. See Figure 37.

Let θ be any angle in standard position, and let (a, b) denote the coordinates of any point, except the origin $(0, 0)$, on the terminal side of θ. If $r = \sqrt{a^2 + b^2}$ denotes the distance from $(0, 0)$ to (a, b), then the **six trigonometric functions of θ** are defined as the ratios

$$\sin\theta = \frac{b}{r} \qquad \cos\theta = \frac{a}{r} \qquad \tan\theta = \frac{b}{a}$$

$$\csc\theta = \frac{r}{b} \qquad \sec\theta = \frac{r}{a} \qquad \cot\theta = \frac{a}{b}$$

provided no denominator equals 0. If a denominator equals 0, that trigonometric function of the angle θ is not defined.

Figure 38

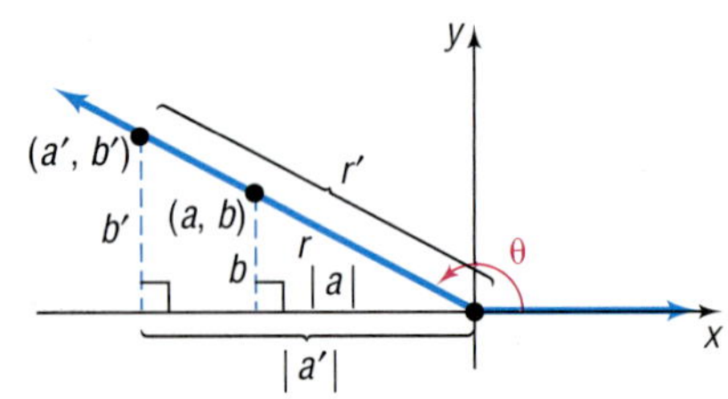

Notice in the preceding definitions that if $a = 0$, that is, the point $(0, b)$ is on the y-axis, then the tangent function and the secant function are undefined. Also, if $b = 0$, that is, the point $(a, 0)$ is on the x-axis, then the cosecant function and the cotangent function are undefined.

By constructing similar triangles, you should be convinced that the values of the six trigonometric functions of an angle θ do not depend on the selection of the point (a, b) on the terminal side of θ, but rather depend only on the angle θ itself. See Figure 38 for an illustration when θ lies in quadrant II. Since the triangles are similar, the ratio $\frac{b}{r}$ equals the ratio $\frac{b'}{r'}$, the common value being $\sin\theta$. Also, the ratio $\frac{|a|}{r}$ equals the ratio $\frac{|a'|}{r'}$, so $\frac{a}{r} = \frac{a'}{r'}$, the common value being $\cos\theta$. And so on.

Also, observe that if θ is acute these definitions reduce to the right triangle definitions given in Section 2.2, as illustrated in Figure 39.

Finally, from the definition of the six trigonometric functions of a general angle, we see that the Quotient and Reciprocal Identities hold. Also, using $r^2 = a^2 + b^2$ and dividing each side by r^2, we can derive the Pythagorean Identities for general angles.

Figure 39

$$\sin\theta = \frac{b}{r} = \frac{\text{opposite}}{\text{hypotenuse}}$$

$$\cos\theta = \frac{a}{r} = \frac{\text{adjacent}}{\text{hypotenuse}} \text{ and so on.}$$

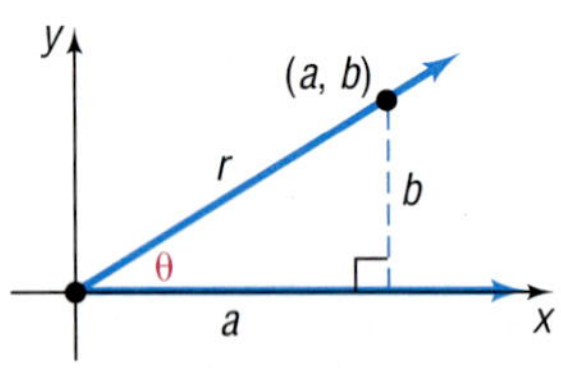

1 **EXAMPLE 1** **Finding the Exact Value of the Six Trigonometric Functions of θ Given a Point on the Terminal Side**

Find the exact value of each of the six trigonometric functions of a positive angle θ if $(4, -3)$ is a point on its terminal side.

Solution Figure 40 illustrates the situation. For the point $(a, b) = (4, -3)$, we have $a = 4$ and $b = -3$. Then $r = \sqrt{a^2 + b^2} = \sqrt{16 + 9} = 5$, so that

Figure 40

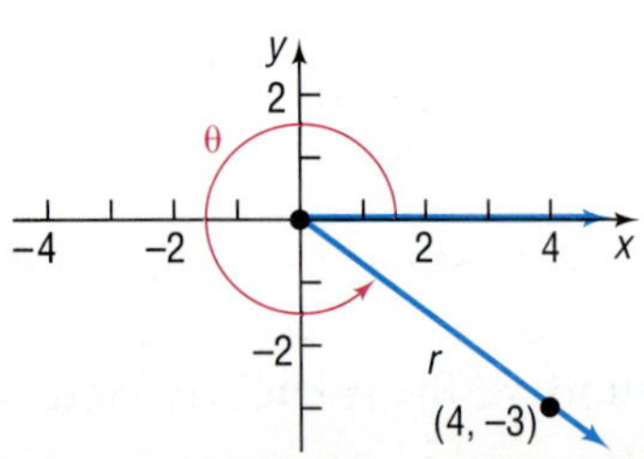

$$\sin\theta = \frac{b}{r} = -\frac{3}{5} \qquad \cos\theta = \frac{a}{r} = \frac{4}{5} \qquad \tan\theta = \frac{b}{a} = -\frac{3}{4}$$

$$\csc\theta = \frac{r}{b} = -\frac{5}{3} \qquad \sec\theta = \frac{r}{a} = \frac{5}{4} \qquad \cot\theta = \frac{a}{b} = -\frac{4}{3}$$

NOW WORK PROBLEM 1.

In the next example, we find the exact value of each of the six trigonometric functions at the quadrantal angles $0, \frac{\pi}{2}, \pi,$ and $\frac{3\pi}{2}$.

EXAMPLE 2 Finding the Exact Value of the Six Trigonometric Functions of Quadrantal Angles

Find the exact value of each of the six trigonometric functions of

(a) $\theta = 0 = 0°$ (b) $\theta = \frac{\pi}{2} = 90°$

(c) $\theta = \pi = 180°$ (d) $\theta = \frac{3\pi}{2} = 270°$

Solution

Figure 41
$\theta = 0 = 0°$

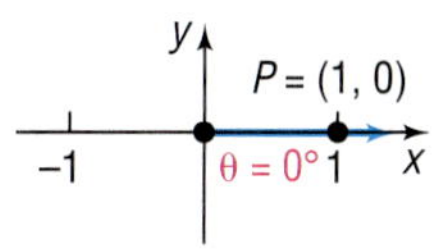

(a) The point $P = (1, 0)$ is on the terminal side of $\theta = 0 = 0°$ and is a distance of $r = 1$ unit from the origin. See Figure 41. Then,

$$\sin 0 = \sin 0° = \frac{0}{1} = 0 \qquad \cos 0 = \cos 0° = \frac{1}{1} = 1$$

$$\tan 0 = \tan 0° = \frac{0}{1} = 0 \qquad \sec 0 = \sec 0° = \frac{1}{1} = 1$$

Since the y-coordinate of P is 0, $\csc 0$ and $\cot 0$ are not defined.

Figure 42
$\theta = \frac{\pi}{2} = 90°$

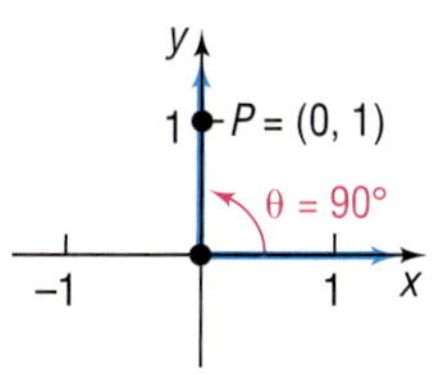

(b) The point $P = (0, 1)$ is on the terminal side of $\theta = \frac{\pi}{2} = 90°$ and is a distance of $r = 1$ unit from the origin. See Figure 42. Then

$$\sin \frac{\pi}{2} = \sin 90° = \frac{1}{1} = 1 \qquad \cos \frac{\pi}{2} = \cos 90° = \frac{0}{1} = 0$$

$$\csc \frac{\pi}{2} = \csc 90° = \frac{1}{1} = 1 \qquad \cot \frac{\pi}{2} = \cot 90° = \frac{0}{1} = 0$$

Since the x-coordinate of P is 0, $\tan \frac{\pi}{2}$ and $\sec \frac{\pi}{2}$ are not defined.

Figure 43
$\theta = \pi = 180°$

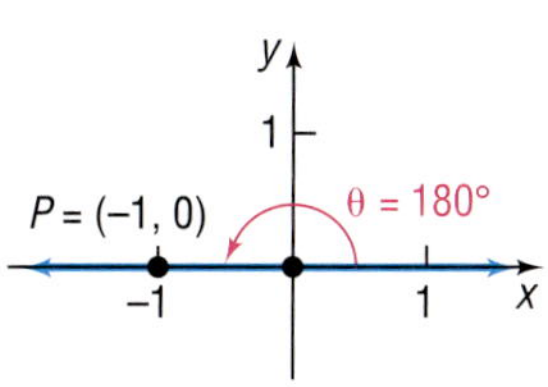

(c) The point $P = (-1, 0)$ is on the terminal side of $\theta = \pi = 180°$ and is a distance of $r = 1$ unit from the origin. See Figure 43. Then,

$$\sin \pi = \sin 180° = \frac{0}{1} = 0 \qquad \cos \pi = \cos 180° = \frac{-1}{1} = -1$$

$$\tan \pi = \tan 180° = \frac{0}{-1} = 0 \qquad \sec \pi = \sec 180° = \frac{1}{-1} = -1$$

Since the y-coordinate of P is 0, $\csc \pi$ and $\cot \pi$ are not defined.

Figure 44
$\theta = \frac{3\pi}{2} = 270°$

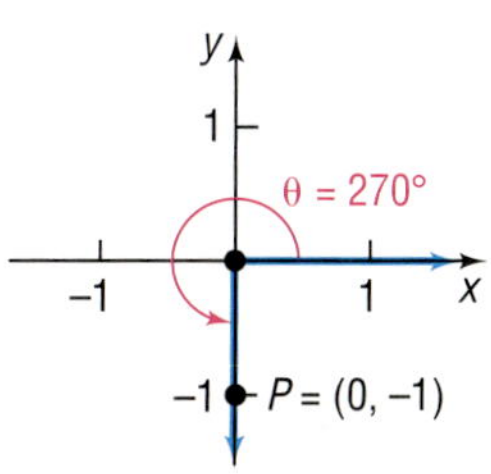

(d) The point $P = (0, -1)$ is on the terminal side of $\theta = \frac{3\pi}{2} = 270°$ and is a distance of $r = 1$ unit from the origin. See Figure 44. Then,

$$\sin \frac{3\pi}{2} = \sin 270° = \frac{-1}{1} = -1 \qquad \cos \frac{3\pi}{2} = \cos 270° = \frac{0}{1} = 0$$

$$\csc \frac{3\pi}{2} = \csc 270° = \frac{1}{-1} = -1 \qquad \cot \frac{3\pi}{2} = \cot 270° = \frac{0}{-1} = 0$$

Since the x-coordinate of P is 0, $\tan \frac{3\pi}{2}$ and $\sec \frac{3\pi}{2}$ are not defined. ■

Table 4 summarizes the values of the trigonometric functions found in Example 2.

TABLE 4

θ (Radians)	θ (Degrees)	$\sin\theta$	$\cos\theta$	$\tan\theta$	$\csc\theta$	$\sec\theta$	$\cot\theta$
0	0°	0	1	0	Not defined	1	Not defined
$\frac{\pi}{2}$	90°	1	0	Not defined	1	Not defined	0
π	180°	0	−1	0	Not defined	−1	Not defined
$\frac{3\pi}{2}$	270°	−1	0	Not defined	−1	Not defined	0

There is no need to memorize Table 4. To find the value of a trigonometric function of a quadrantal angle, draw the angle and apply the definition as we did in Example 2.

Coterminal Angles

Two angles in standard position are said to be **coterminal** if they have the same terminal side.

See Figure 45.

Figure 45

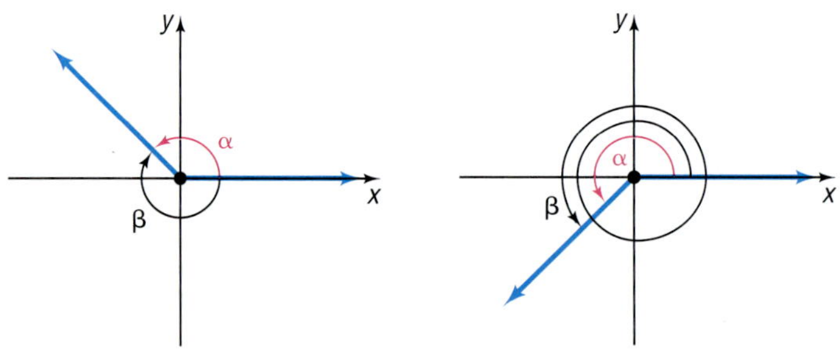

(a) α and β are coterminal **(b)** α and β are coterminal

In general, if θ is an angle measured in degrees, then $\theta \pm 360°k$, where k is any integer, is an angle coterminal with θ. If θ is measured in radians, then $\theta \pm 2\pi k$, where k is any integer, is an angle coterminal with θ.

2 Because coterminal angles have the same terminal side, it follows that the values of the trigonometric functions of coterminal angles are equal. We use this fact in the next example.

EXAMPLE 3 **Using the Coterminal Angle to Find the Exact Value of a Trigonometric Function**

Find the exact value of each of the following:

(a) $\sin 390°$ (b) $\cos 420°$ (c) $\tan\frac{9\pi}{4}$ (d) $\sec\left(-\frac{7\pi}{4}\right)$ (e) $\csc(-270°)$

Solution (a) It is best to sketch the angle first. See Figure 46. The angle 390° is coterminal with 30°.

$$\sin 390° = \sin(360° + 30°) = \sin 30° = \frac{1}{2}$$

(b) See Figure 47. The angle 420° is coterminal with 60°.

$$\cos 420° = \cos(360° + 60°) = \cos 60° = \frac{1}{2}$$

Figure 46

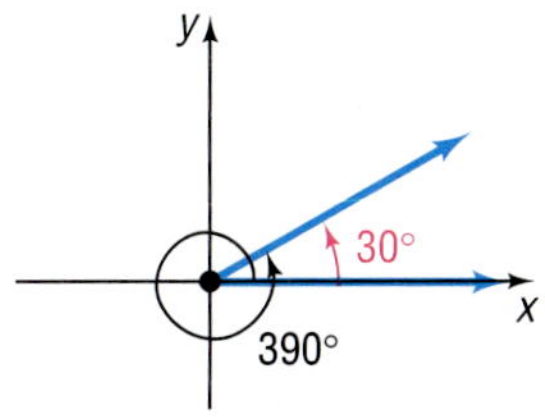

Figure 47

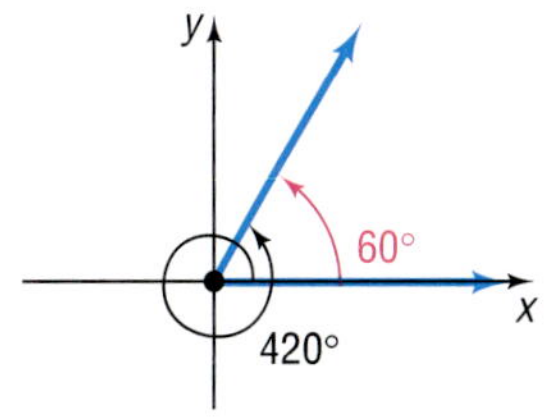

Figure 48

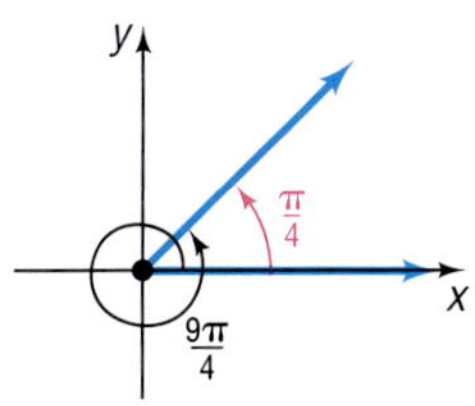

(c) The angle $\frac{9\pi}{4}$ is coterminal with $\frac{\pi}{4}\left(\frac{9\pi}{4} = \frac{8\pi}{4} + \frac{\pi}{4} = 2\pi + \frac{\pi}{4}\right)$. See Figure 48.

$$\tan\frac{9\pi}{4} = \tan\left(2\pi + \frac{\pi}{4}\right) = \tan\frac{\pi}{4} = 1$$

Figure 49

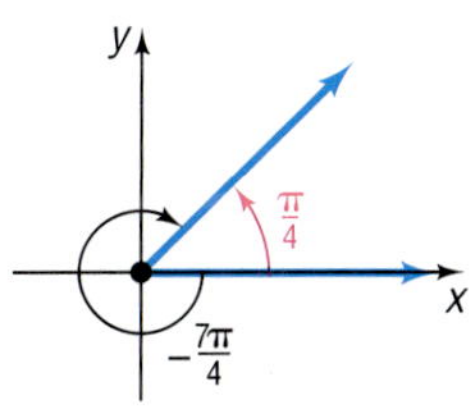

(d) See Figure 49. The angle $-\frac{7\pi}{4}$ is coterminal with $\frac{\pi}{4}$.

$$\sec\left(-\frac{7\pi}{4}\right) = \sec\left(-2\pi + \frac{\pi}{4}\right) = \sec\frac{\pi}{4} = \sqrt{2}$$

(e) See Figure 50. The angle −270° is coterminal with 90°.

$$\csc(-270°) = \csc(-360° + 90°) = \csc 90° = 1$$

Figure 50

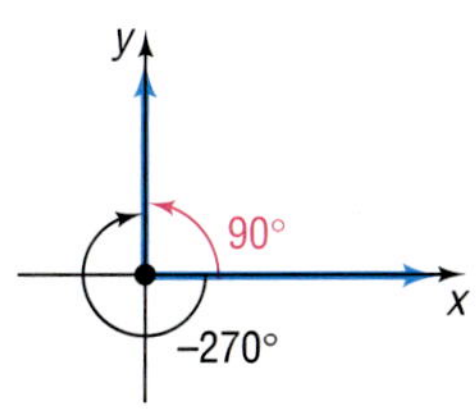

As Example 3 illustrates, the value of a trigonometric function of any angle is equal to the value of the same trigonometric function of an angle θ coterminal to it where $0° \le \theta < 360°$ (or $0 \le \theta < 2\pi$). Because the angles θ and $\theta \pm 360°k$ (or $\theta \pm 2\pi k$), where k is any integer, are coterminal, and because the values of the trigonometric functions are equal for coterminal angles, it follows that

$$\begin{aligned}
\sin(\theta \pm 360°k) &= \sin\theta & \sin(\theta \pm 2\pi k) &= \sin\theta \\
\cos(\theta \pm 360°k) &= \cos\theta & \cos(\theta \pm 2\pi k) &= \cos\theta \\
\tan(\theta \pm 360°k) &= \tan\theta & \tan(\theta \pm 2\pi k) &= \tan\theta \\
\csc(\theta \pm 360°k) &= \csc\theta & \csc(\theta \pm 2\pi k) &= \csc\theta \\
\sec(\theta \pm 360°k) &= \sec\theta & \sec(\theta \pm 2\pi k) &= \sec\theta \\
\cot(\theta \pm 360°k) &= \cot\theta & \cot(\theta \pm 2\pi k) &= \cot\theta
\end{aligned} \tag{1}$$

These formulas show that the values of the trigonometric functions repeat themselves every 360° (or 2π radians).

NOW WORK PROBLEM 37.

The Signs of the Trigonometric Functions

3 If θ is not a quadrantal angle, then it will lie in a particular quadrant. In such a case, the signs of the x-coordinate and the y-coordinate of a point (a, b) on the terminal side of θ are known. Because $r = \sqrt{a^2 + b^2} > 0$, it follows that the signs of the trigonometric functions of an angle θ can be found if we know in which quadrant θ lies.

For example, if θ lies in quadrant II, as shown in Figure 51, then a point (a, b) on the terminal side of θ has a negative x-coordinate and a positive y-coordinate. Thus,

Figure 51
θ in quadrant II; $a < 0$, $b > 0$, $r > 0$

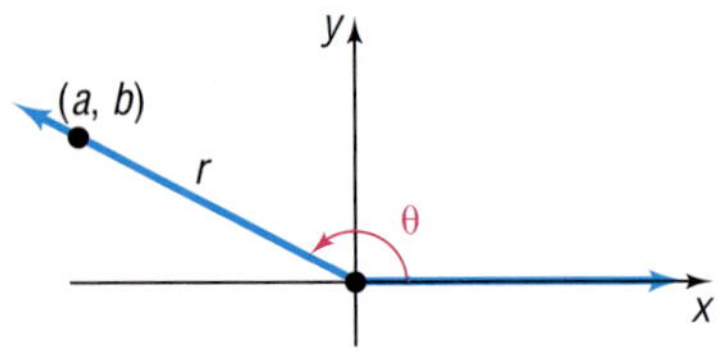

$$\sin\theta = \frac{b}{r} > 0 \qquad \cos\theta = \frac{a}{r} < 0 \qquad \tan\theta = \frac{b}{a} < 0$$
$$\csc\theta = \frac{r}{b} > 0 \qquad \sec\theta = \frac{r}{a} < 0 \qquad \cot\theta = \frac{a}{b} < 0$$

Table 5 lists the signs of the six trigonometric functions for each quadrant. See also Figure 52.

TABLE 5

Quadrant of θ	$\sin\theta$, $\csc\theta$	$\cos\theta$, $\sec\theta$	$\tan\theta$, $\cot\theta$
I	Positive	Positive	Positive
II	Positive	Negative	Negative
III	Negative	Negative	Positive
IV	Negative	Positive	Negative

Figure 52

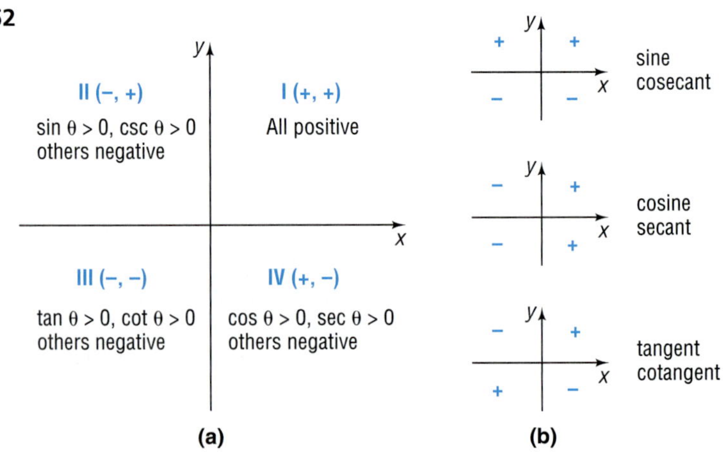

EXAMPLE 4 **Finding the Quadrant in Which an Angle Lies**

If $\sin\theta < 0$ and $\cos\theta < 0$, name the quadrant in which the angle θ lies.

Solution If $\sin\theta < 0$, then θ lies in quadrant III or IV. If $\cos\theta < 0$, then θ lies in quadrant II or III. Therefore, θ lies in quadrant III. ■

NOW WORK PROBLEM 11.

Reference Angle

4 Once we know in which quadrant an angle lies, we know the sign of each trigonometric function of this angle. The use of a certain reference angle may help us to evaluate the trigonometric functions of such an angle.

> Let θ denote a nonacute angle that lies in a quadrant. The acute angle formed by the terminal side of θ and either the positive x-axis or the negative x-axis is called the **reference angle** for θ.

Figure 53 illustrates the reference angle for some general angles θ. Note that a reference angle is always an acute angle, that is, an angle whose measure is between 0° and 90°.

Figure 53

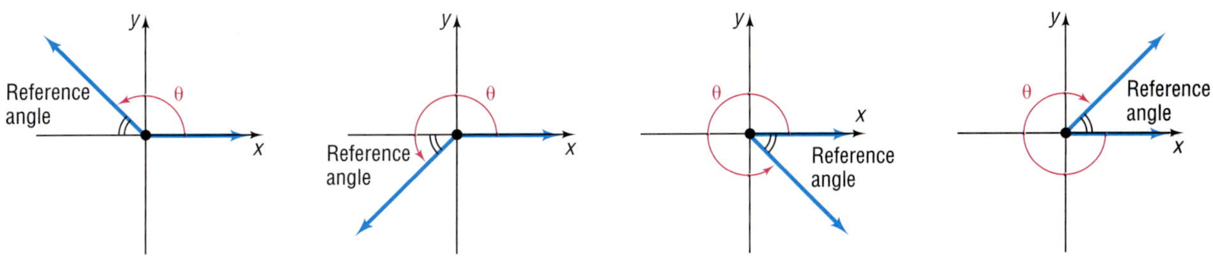

(a) (b) (c) (d)

Although formulas can be given for calculating reference angles, usually it is easier to find the reference angle for a given angle by making a quick sketch of the angle.

EXAMPLE 5 **Finding Reference Angles**

Find the reference angle for each of the following angles:

(a) 150° (b) −45° (c) $\frac{9\pi}{4}$ (d) $-\frac{5\pi}{6}$

Solution (a) Refer to Figure 54. The reference angle for 150° is 30°.

Figure 54

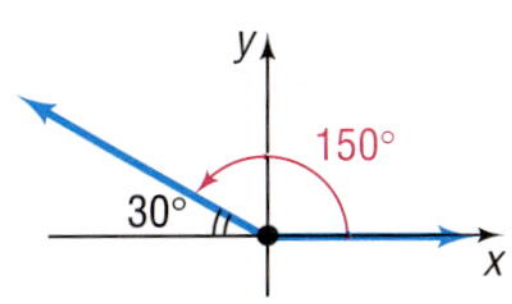

(b) Refer to Figure 55. The reference angle for −45° is 45°.

Figure 55

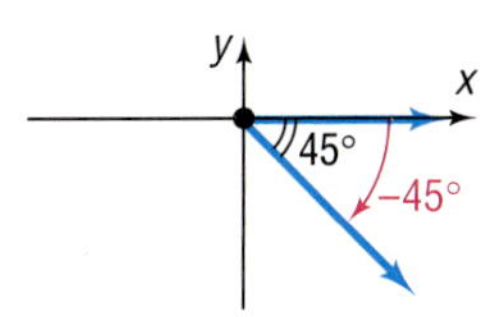

(c) Refer to Figure 56. The reference angle for $\frac{9\pi}{4}$ is $\frac{\pi}{4}$.

Figure 56

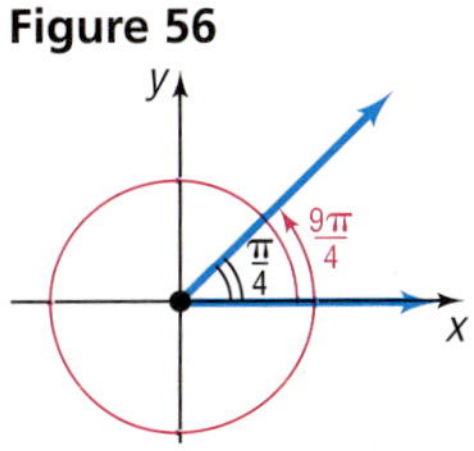

(d) Refer to Figure 57. The reference angle for $-\frac{5\pi}{6}$ is $\frac{\pi}{6}$.

Figure 57

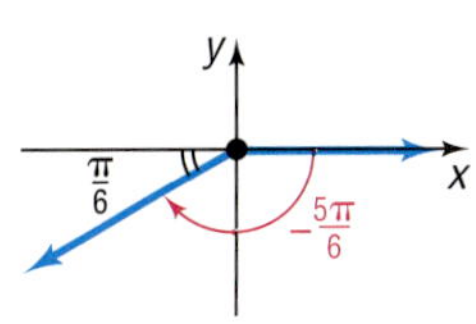

NOW WORK PROBLEM 19.

5 The advantage of using reference angles is that, except for the correct sign, the values of the trigonometric functions of a general angle θ equal the values of the trigonometric functions of its reference angle.

Theorem

Reference Angles

If θ is an angle that lies in a quadrant and if α is its reference angle, then

$$\begin{array}{lll} \sin\theta = \pm\sin\alpha & \cos\theta = \pm\cos\alpha & \tan\theta = \pm\tan\alpha \\ \csc\theta = \pm\csc\alpha & \sec\theta = \pm\sec\alpha & \cot\theta = \pm\cot\alpha \end{array} \qquad (2)$$

where the + or − sign depends on the quadrant in which θ lies.

Figure 58
$\sin\theta = b/r$, $\sin\alpha = b/r$,
$\cos\theta = a/r$, $\cos\alpha = |a|/r$

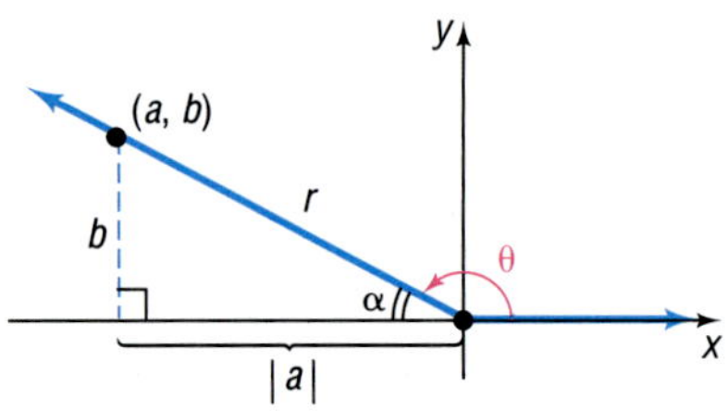

For example, suppose that θ lies in quadrant II and α is its reference angle. See Figure 58. If (a, b) is a point on the terminal side of θ and if $r = \sqrt{a^2 + b^2}$, we have

$$\sin\theta = \frac{b}{r} = \sin\alpha \qquad \cos\theta = \frac{a}{r} = \frac{-|a|}{r} = -\cos\alpha$$

and so on.

The next example illustrates how the theorem on reference angles is used.

EXAMPLE 6 Using the Reference Angle to Find the Exact Value of a Trigonometric Function

Find the exact value of each of the following trigonometric functions using reference angles.

(a) $\sin 135°$ (b) $\cos 240°$ (c) $\cos\frac{5\pi}{6}$ (d) $\tan\left(-\frac{\pi}{3}\right)$

Solution (a) Refer to Figure 59. The angle 135° is in quadrant II, where the sine function is positive. The reference angle for 135° is 45°.

$$\sin 135° = \sin 45° = \frac{\sqrt{2}}{2}$$

Figure 59

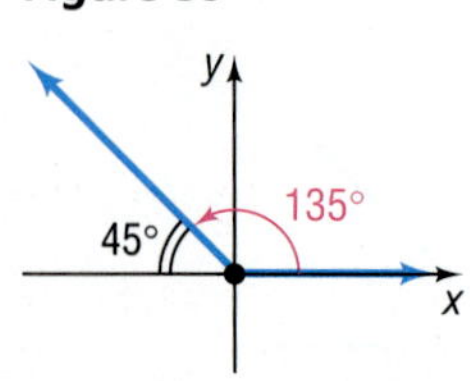

(b) Refer to Figure 60. The angle 240° is in quadrant III, where the cosine function is negative. The reference angle for 240° is 60°.

$$\cos 240° = -\cos 60° = -\frac{1}{2}$$

Figure 60

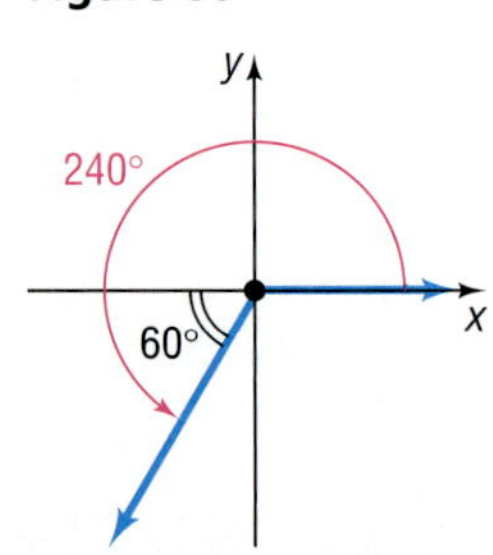

(c) Refer to Figure 61. The angle $\frac{5\pi}{6}$ is in quadrant II, where the cosine function is negative. The reference angle for $\frac{5\pi}{6}$ is $\frac{\pi}{6}$.

$$\cos\frac{5\pi}{6} = -\cos\frac{\pi}{6} = -\frac{\sqrt{3}}{2}$$

Figure 61

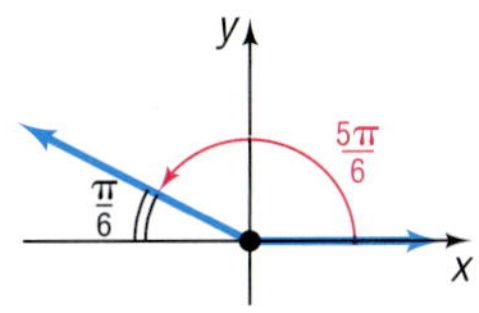

(d) Refer to Figure 62. The angle $-\frac{\pi}{3}$ is in quadrant IV, where the tangent function is negative. The reference angle for $-\frac{\pi}{3}$ is $\frac{\pi}{3}$.

$$\tan\left(-\frac{\pi}{3}\right) = -\tan\frac{\pi}{3} = -\sqrt{3}$$

Figure 62

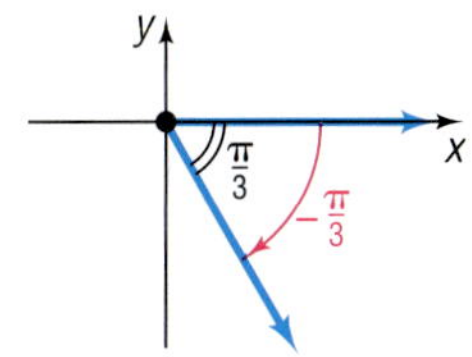

NOW WORK PROBLEM 53.

6 **EXAMPLE 7** **Finding the Exact Value of Trigonometric Functions**

Given that $\cos\theta = -\frac{2}{3}$, $\frac{\pi}{2} < \theta < \pi$, find the exact value of each of the remaining trigonometric functions.

Solution The angle θ lies in quadrant II, so we know that $\sin\theta$ and $\csc\theta$ are positive, whereas the other trigonometric functions are negative. If α is the reference angle for θ, then $\cos\alpha = \frac{2}{3}$. The values of the remaining trigonometric functions of the angle α can be found by drawing the appropriate triangle and using the Pythagorean Theorem. We use Figure 63 to obtain

Figure 63

$\cos\alpha = \frac{2}{3}$

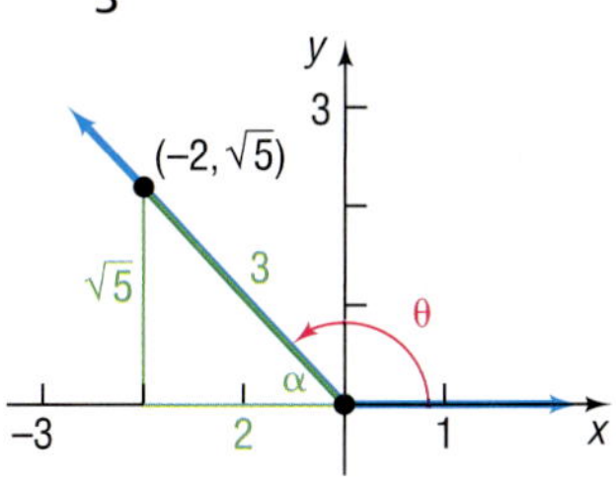

$$\sin\alpha = \frac{\sqrt{5}}{3} \qquad \cos\alpha = \frac{2}{3} \qquad \tan\alpha = \frac{\sqrt{5}}{2}$$

$$\csc\alpha = \frac{3}{\sqrt{5}} = \frac{3\sqrt{5}}{5} \qquad \sec\alpha = \frac{3}{2} \qquad \cot\alpha = \frac{2}{\sqrt{5}} = \frac{2\sqrt{5}}{5}$$

Now we assign the appropriate signs to each of these values to find the values of the trigonometric functions of θ.

$$\sin\theta = \frac{\sqrt{5}}{3} \qquad \cos\theta = -\frac{2}{3} \qquad \tan\theta = -\frac{\sqrt{5}}{2}$$

$$\csc\theta = \frac{3\sqrt{5}}{5} \qquad \sec\theta = -\frac{3}{2} \qquad \cot\theta = -\frac{2\sqrt{5}}{5}$$

NOW WORK PROBLEM 79.

EXAMPLE 8 **Finding the Exact Value of Trigonometric Functions**

If $\tan\theta = -4$ and $\sin\theta < 0$, find the exact value of each of the remaining trigonometric functions of θ.

Solution Since $\tan\theta = -4 < 0$ and $\sin\theta < 0$, it follows that θ lies in quadrant IV. If α is the reference angle for θ, then $\tan\alpha = 4$. See Figure 64. Then

Figure 64
$\tan\alpha = 4$

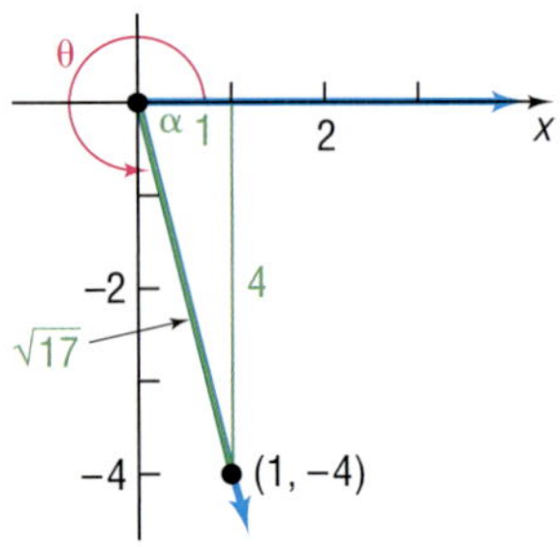

$$\sin\alpha = \frac{4}{\sqrt{17}} = \frac{4\sqrt{17}}{17} \qquad \cos\alpha = \frac{1}{\sqrt{17}} = \frac{\sqrt{17}}{17} \qquad \tan\alpha = \frac{4}{1} = 4$$

$$\csc\alpha = \frac{\sqrt{17}}{4} \qquad \sec\alpha = \frac{\sqrt{17}}{1} = \sqrt{17} \qquad \cot\alpha = \frac{1}{4}$$

We assign the appropriate sign to each of these to obtain the values of the trigonometric functions of θ.

$$\sin\theta = -\frac{4\sqrt{17}}{17} \qquad \cos\theta = \frac{\sqrt{17}}{17} \qquad \tan\theta = -4$$

$$\csc\theta = -\frac{\sqrt{17}}{4} \qquad \sec\theta = \sqrt{17} \qquad \cot\theta = -\frac{1}{4}$$

NOW WORK PROBLEM 89.

SUMMARY

To find the values of the trigonometric functions of a general angle:

STEP 1: If the angle θ is a quadrantal angle, draw the angle, pick a point on its terminal side, and apply the definitions of the trigonometric functions.

STEP 2: If the angle θ lies in a quadrant, determine the correct signs of the trigonometric functions in that quadrant and determine the reference angle α for θ. Now express each trigonometric function of θ in terms of the same value (except for the sign) of the trigonometric function of α, an acute angle. Finally, apply the correct sign to each function.

2.4 Concepts and Vocabulary

In Problems 1–3, fill in the blanks.

1. For an angle θ that lies in quadrant III, the trigonometric functions __________ and __________ are positive.

2. Two angles in standard position that have the same terminal side are __________.

3. The reference angle of 240° is __________.

In Problems 4–6, answer True or False to each statement.

4. $\sin 182° = \cos 2°$.

5. $\tan\frac{\pi}{2}$ is not defined.

6. The reference angle is always an acute angle.

7. Explain how you would find the reference angle of 240°.

8. In which quadrants is the cosine function positive?

9. If $0 \le \theta < 2\pi$, for what angles θ, if any, is $\tan\theta$ undefined?

10. Find an acute angle that is coterminal with −130°.

2.4 Exercises

In Problems 1–10, a point on the terminal side of an angle θ is given. Find the exact value of each of the six trigonometric functions of θ.

1. $(-3, 4)$
2. $(5, -12)$
3. $(2, -3)$
4. $(-1, -2)$
5. $(-3, -3)$
6. $(2, -2)$
7. $\left(\frac{\sqrt{3}}{2}, \frac{1}{2}\right)$
8. $\left(-\frac{1}{2}, \frac{\sqrt{3}}{2}\right)$
9. $\left(\frac{\sqrt{2}}{2}, -\frac{\sqrt{2}}{2}\right)$
10. $\left(-\frac{\sqrt{2}}{2}, -\frac{\sqrt{2}}{2}\right)$

In Problems 11–18, name the quadrant in which the angle θ lies.

11. $\sin\theta > 0, \quad \cos\theta < 0$
12. $\sin\theta < 0, \quad \cos\theta > 0$
13. $\sin\theta < 0, \quad \tan\theta < 0$
14. $\cos\theta > 0, \quad \tan\theta > 0$
15. $\cos\theta > 0, \quad \cot\theta < 0$
16. $\sin\theta < 0, \quad \cot\theta > 0$
17. $\sec\theta < 0, \quad \tan\theta > 0$
18. $\csc\theta > 0, \quad \cot\theta < 0$

In Problems 19–36, find the reference angle of each angle.

19. $-30°$
20. $60°$
21. $120°$
22. $300°$
23. $210°$
24. $330°$
25. $\frac{5\pi}{4}$
26. $\frac{5\pi}{6}$
27. $\frac{8\pi}{3}$
28. $\frac{7\pi}{4}$
29. $-135°$
30. $-240°$
31. $-\frac{2\pi}{3}$
32. $-\frac{7\pi}{6}$
33. $440°$
34. $490°$
35. $-\frac{3\pi}{4}$
36. $-\frac{11\pi}{6}$

In Problems 37–78, find the exact value of each expression. Do not use a calculator.

37. $\sin 405°$
38. $\cos 420°$
39. $\tan 405°$
40. $\sin 390°$
41. $\csc 450°$
42. $\sec 540°$
43. $\cot 390°$
44. $\sec 420°$
45. $\cos\frac{33\pi}{4}$
46. $\sin\frac{9\pi}{4}$
47. $\tan(21\pi)$
48. $\csc\frac{9\pi}{2}$
49. $\sec\frac{17\pi}{4}$
50. $\cot\frac{17\pi}{4}$
51. $\tan\frac{19\pi}{6}$
52. $\sec\frac{25\pi}{6}$
53. $\sin 150°$
54. $\cos 210°$
55. $\cos 315°$
56. $\sin 120°$
57. $\sec 240°$
58. $\csc 300°$
59. $\cot 330°$
60. $\tan 225°$
61. $\sin\frac{3\pi}{4}$
62. $\cos\frac{2\pi}{3}$
63. $\cot\frac{7\pi}{6}$
64. $\csc\frac{7\pi}{4}$
65. $\cos(-60°)$
66. $\tan(-120°)$
67. $\sin\left(-\frac{2\pi}{3}\right)$
68. $\cot\left(-\frac{\pi}{6}\right)$
69. $\tan\frac{14\pi}{3}$
70. $\sec\frac{11\pi}{4}$
71. $\csc(-315°)$
72. $\sec(-225°)$
73. $\sin(8\pi)$
74. $\cos(-2\pi)$
75. $\tan(7\pi)$
76. $\cot(5\pi)$
77. $\sec(-3\pi)$
78. $\csc\left(-\frac{5\pi}{2}\right)$

In Problems 79–96, find the exact value of each of the remaining trigonometric functions of θ.

79. $\sin\theta = \frac{12}{13}$, θ in quadrant II
80. $\cos\theta = \frac{3}{5}$, θ in quadrant IV
81. $\cos\theta = -\frac{4}{5}$, θ in quadrant III
82. $\sin\theta = -\frac{5}{13}$, θ in quadrant III
83. $\sin\theta = \frac{5}{13}$, $90° < \theta < 180°$
84. $\cos\theta = \frac{4}{5}$, $270° < \theta < 360°$
85. $\cos\theta = -\frac{1}{3}$, $180° < \theta < 270°$
86. $\sin\theta = -\frac{2}{3}$, $180° < \theta < 270°$
87. $\sin\theta = \frac{2}{3}$, $\tan\theta < 0$
88. $\cos\theta = -\frac{1}{4}$, $\tan\theta > 0$
89. $\sec\theta = 2$, $\sin\theta < 0$
90. $\csc\theta = 3$, $\cot\theta < 0$
91. $\tan\theta = \frac{3}{4}$, $\sin\theta < 0$
92. $\cot\theta = \frac{4}{3}$, $\cos\theta < 0$
93. $\tan\theta = -\frac{1}{3}$, $\sin\theta > 0$
94. $\sec\theta = -2$, $\tan\theta > 0$
95. $\csc\theta = -2$, $\tan\theta > 0$
96. $\cot\theta = -2$, $\sec\theta > 0$

97. Find the exact value of $\sin 45° + \sin 135° + \sin 225° + \sin 315°$.

98. Find the exact value of $\tan 60° + \tan 150°$.

99. If $\sin\theta = 0.2$, find $\sin(\theta + \pi)$.

100. If $\cos\theta = 0.4$, find $\cos(\theta + \pi)$.

101. If $\tan\theta = 3$, find $\tan(\theta + \pi)$.

102. If $\cot\theta = -2$, find $\cot(\theta + \pi)$.

103. If $\sin\theta = \frac{1}{5}$, find $\csc\theta$.

104. If $\cos\theta = \frac{2}{3}$, find $\sec\theta$.

105. Find the exact value of

$$\sin 1° + \sin 2° + \sin 3° + \cdots + \sin 358° + \sin 359°$$

106. Find the exact value of

$$\cos 1° + \cos 2° + \cos 3° + \cdots + \cos 358° + \cos 359°$$

---

107. Projectile Motion An object is propelled upward at an angle θ, $45° < \theta < 90°$, to the horizontal with an initial velocity of v_0 feet per second from the base of a plane that makes an angle of 45° with the horizontal. See the illustration. If air resistance is ignored, the distance R that it travels up the inclined plane is given by

$$R = \frac{v_0^2\sqrt{2}}{32}[\sin(2\theta) - \cos(2\theta) - 1]$$

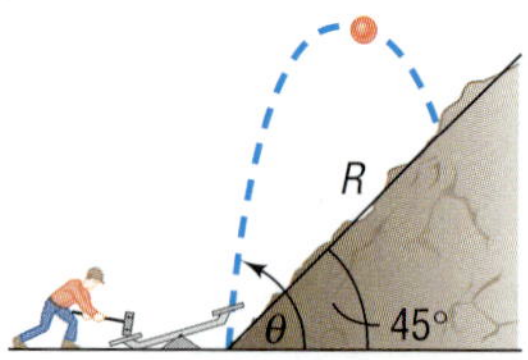

(a) Find the distance R that the object travels along the inclined plane if the initial velocity is 32 feet per second and $\theta = 60°$.
(b) Use a graphing utility to graph $R = R(\theta)$ if the initial velocity is 32 feet per second.
(c) What value of θ makes R largest?

108. Give three examples that demonstrate how to use the theorem on reference angles.

109. Write a brief paragraph that explains how to quickly compute the value of the trigonometric functions of 0°, 90°, 180°, and 270°.

PREPARING FOR THIS SECTION

Before getting started, review the following:

✓ Unit Circle (Section 1.3, p. 27)
✓ Symmetry (Section 1.3, pp. 20–22)
✓ Functions (Section 1.4, pp. 33–41)
✓ Even and Odd Functions (Section 1.5, pp. 55–57)

2.5 PROPERTIES OF THE TRIGONOMETRIC FUNCTIONS; UNIT CIRCLE APPROACH

OBJECTIVES

1. Find the Exact Value of the Trigonometric Functions Using the Unit Circle
2. Know the Domain and Range of the Trigonometric Functions
3. Use the Periodic Properties to Find the Exact Value of the Trigonometric Functions
4. Use Even-Odd Properties to Find the Exact Value of the Trigonometric Functions

In this section, we develop important properties of the trigonometric functions. We begin by introducing the trigonometric functions using the unit circle. This approach will lead to the definition given earlier of the trigonometric functions of a general angle.

The Unit Circle

1 Recall that the unit circle is a circle whose radius is 1 and whose center is at the origin of a rectangular coordinate system. Also recall that any circle of radius r has circumference of length $2\pi r$. Therefore, the unit circle (radius = 1) has a circumference of length 2π. In other words, for 1 revolution around the unit circle the length of arc is 2π units.

The following discussion sets the stage for defining the trigonometric functions.

Let $t \geq 0$ be any real number and let s be the distance from the origin to t on the real number line. See the red portion of Figure 65(a). Now look at the unit circle in Figure 65(a). Beginning at the the point $(1, 0)$ on the unit circle, travel $s = t$ units in the counterclockwise direction along the circle to arrive at the point $P = (a, b)$. In this sense, the length $s = t$ units is being **wrapped** around the unit circle.

Figure 65

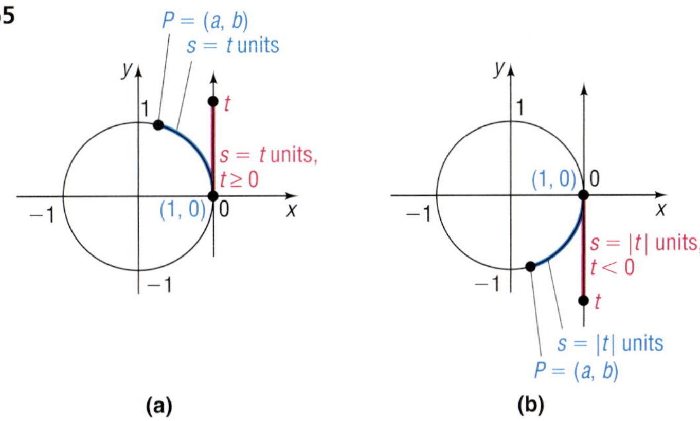

If $t < 0$, we begin at the point $(1, 0)$ on the unit circle and travel $s = |t|$ units in the clockwise direction to arrive at the point $P = (a, b)$. See Figure 65(b).

If $t > 2\pi$ or if $t < -2\pi$, it will be necessary to travel around the unit circle more than once before arriving at point P. Do you see why?

Let's describe this process another way. Picture a string of length $s = |t|$ units being wrapped around a circle of radius 1 unit. We start wrapping the string around the circle at the point $(1, 0)$. If $t \geq 0$, we wrap the string in the counterclockwise direction; if $t < 0$, we wrap the string in the clockwise direction. The point $P = (a, b)$ is the point where the string ends.

This discussion tells us that, for any real number t, we can locate a unique point $P = (a, b)$ on the unit circle. This point is called **the point *P* on the unit circle that corresponds to *t*.** This is the important idea here. No matter what real number t is chosen, there is a unique point P on the unit circle corresponding to it. We use the coordinates of the point $P = (a, b)$ on the unit circle corresponding to the real number t to define the **six trigonometric functions of *t*.**

Let t be a real number and let $P = (a, b)$ be the point on the unit circle that corresponds to t.

The **sine function** associates with t the y-coordinate of P and is denoted by

$$\sin t = b$$

The **cosine function** associates with t the x-coordinate of P and is denoted by

$$\cos t = a$$

If $a \neq 0$, the **tangent function** is defined as

$$\tan t = \frac{b}{a}$$

If $b \neq 0$, the **cosecant function** is defined as

$$\csc t = \frac{1}{b}$$

If $a \neq 0$, the **secant function** is defined as

$$\sec t = \frac{1}{a}$$

If $b \neq 0$, the **cotangent function** is defined as

$$\cot t = \frac{a}{b}$$

Once again, notice in these definitions that if $a = 0$ [that is, the point $P = (0, b)$ is on the y-axis] the tangent function and the secant function are undefined. Also, if $b = 0$ [that is, the point $P = (a, 0)$ is on the x-axis] the cosecant function and the cotangent function are undefined.

Because we use the unit circle in these definitions of the trigonometric functions, they are also sometimes referred to as **circular functions**.

EXAMPLE 1 Finding the Value of the Trigonometric Functions Using a Point on the Unit Circle

Figure 66

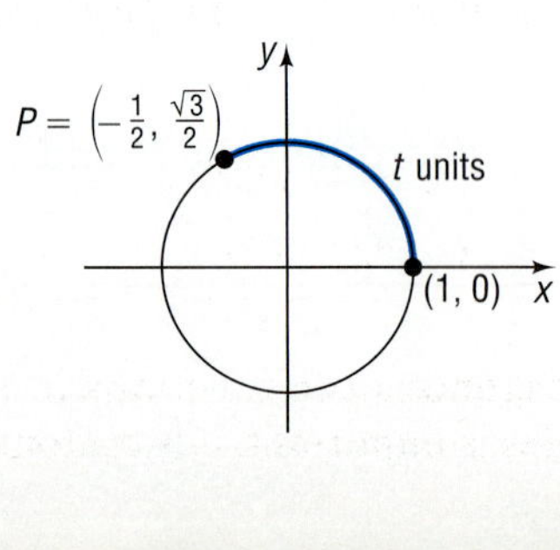

Find the value of $\sin t$, $\cos t$, $\tan t$, $\csc t$, $\sec t$, and $\cot t$ if $P = \left(-\frac{1}{2}, \frac{\sqrt{3}}{2}\right)$ is the point on the unit circle that corresponds to the real number t.

Solution See Figure 66. We follow the definition of the six trigonometric functions using $P = \left(-\frac{1}{2}, \frac{\sqrt{3}}{2}\right) = (a, b)$. Then, with $a = -\frac{1}{2}$, $b = \frac{\sqrt{3}}{2}$, we have

$$\sin t = b = \frac{\sqrt{3}}{2} \qquad \cos t = a = -\frac{1}{2} \qquad \tan t = \frac{b}{a} = \frac{\frac{\sqrt{3}}{2}}{-\frac{1}{2}} = -\sqrt{3}$$

$$\csc t = \frac{1}{b} = \frac{1}{\frac{\sqrt{3}}{2}} = \frac{2}{\sqrt{3}} = \frac{2\sqrt{3}}{3} \qquad \sec t = \frac{1}{a} = \frac{1}{-\frac{1}{2}} = -2 \qquad \cot t = \frac{a}{b} = \frac{-\frac{1}{2}}{\frac{\sqrt{3}}{2}} = -\frac{1}{\sqrt{3}} = -\frac{\sqrt{3}}{3}$$

NOW WORK PROBLEM 1.

Trigonometric Functions of Angles

Let $P = (a, b)$ be the point on the unit circle corresponding to the real number t. See Figure 67(a). Let θ be the angle in standard position, measured in radians, whose terminal side is the ray from the origin through P. See Figure 67(b). Since the unit circle has radius 1 unit, from the formula for arc length, $s = r\theta$, we find that

$$s = r\theta \underset{\substack{\uparrow \\ r = 1}}{=} \theta$$

So, if $s = |t|$ units, then $\theta = t$ radians. See Figures 67(c) and (d).

Figure 67

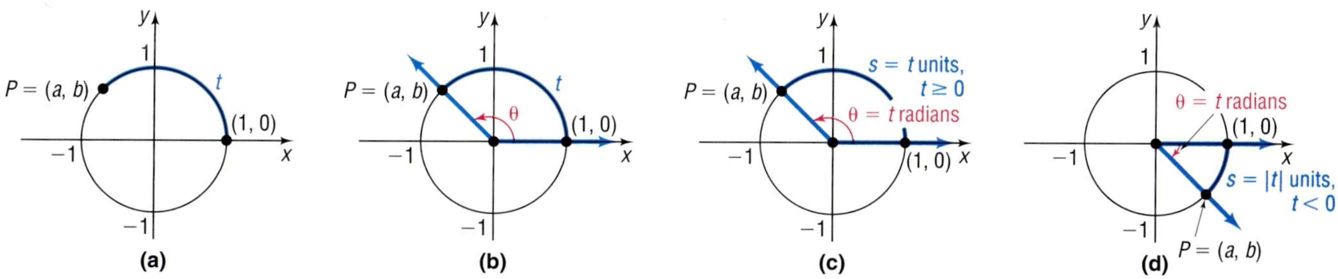

The point $P = (a, b)$ on the unit circle that corresponds to the real number t is the point P on the terminal side of the angle $\theta = t$ radians. As a result, we can say that

$$\underset{\substack{\uparrow \\ \text{Real number}}}{\sin t} = \underset{\substack{\uparrow \\ \theta = t \text{ radians}}}{\sin \theta}$$

and so on. We can now define the trigonometric functions of the angle θ.

If $\theta = t$ radians, the six **trigonometric functions of the angle θ** are defined as

$$\sin\theta = \sin t \qquad \cos\theta = \cos t \qquad \tan\theta = \tan t$$
$$\csc\theta = \csc t \qquad \sec\theta = \sec t \qquad \cot\theta = \cot t$$

Even though the distinction between trigonometric functions of real numbers and trigonometric functions of angles is important, it is customary to refer to trigonometric functions of real numbers and trigonometric functions of angles collectively as *the trigonometric functions*. We will follow this practice from now on.

Since the values of the trigonometric functions of an angle θ are determined by the coordinates of the point $P = (a, b)$ on the unit circle corresponding to θ, the units used to measure the angle θ are irrelevant. For example, it does not matter whether we write $\theta = \frac{\pi}{2}$ radians or $\theta = 90°$. The point on the unit circle corresponding to this angle is $P = (0, 1)$. Hence,

$$\sin\frac{\pi}{2} = \sin 90° = 1 \quad \text{and} \quad \cos\frac{\pi}{2} = \cos 90° = 0$$

To find the exact value of a trigonometric function of an angle θ requires that we locate the corresponding point $P^* = (a^*, b^*)$ on the unit circle. In fact, though, any circle whose center is at the origin can be used.

Figure 68

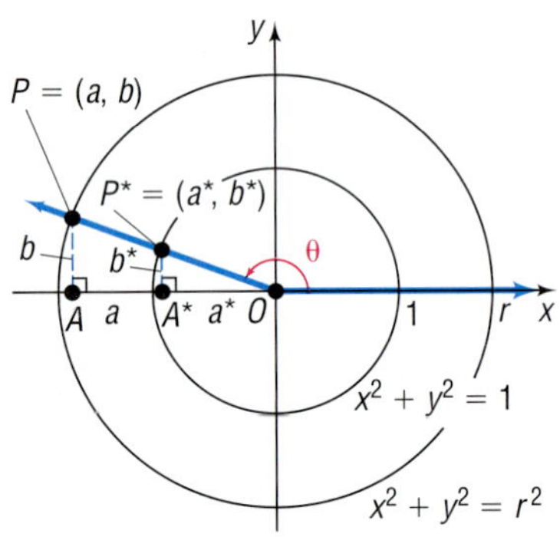

Let θ be any nonquadrantal angle placed in standard position. Let $P = (a, b)$ be the point on the circle $x^2 + y^2 = r^2$ that corresponds to θ. See Figure 68.

Notice that the triangles OA^*P^* and OAP are similar so that the ratios of the corresponding sides are equal.

$$\frac{b^*}{1} = \frac{b}{r} \qquad \frac{a^*}{1} = \frac{a}{r} \qquad \frac{b^*}{a^*} = \frac{b}{a}$$
$$\frac{1}{b^*} = \frac{r}{b} \qquad \frac{1}{a^*} = \frac{r}{a} \qquad \frac{a^*}{b^*} = \frac{a}{b}$$

These results lead us to formulate the following theorem:

Theorem

For an angle θ in standard position, let $P = (a, b)$ be any point on the terminal side of θ that is also on the circle $x^2 + y^2 = r^2$. Then

$$\sin\theta = \frac{b}{r} \qquad \cos\theta = \frac{a}{r} \qquad \tan\theta = \frac{b}{a}, \quad a \neq 0$$
$$\csc\theta = \frac{r}{b}, \quad b \neq 0 \qquad \sec\theta = \frac{r}{a}, \quad a \neq 0 \qquad \cot\theta = \frac{a}{b}, \quad b \neq 0$$

This result coincides with the definition given in Section 2.4 for the six trigonometric functions of a general angle θ.

NOW WORK PROBLEM 7.

Figure 69

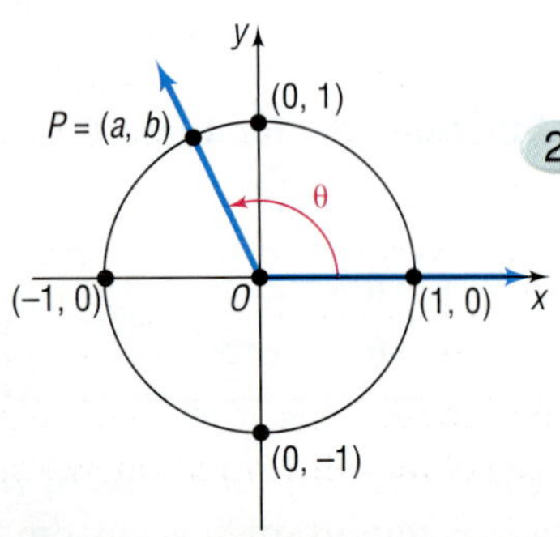

Domain and Range of the Trigonometric Functions

2 Let θ be an angle in standard position, and let $P = (a, b)$ be the point on the unit circle that corresponds to θ. See Figure 69. Then, by definition,

$$\sin\theta = b \qquad \cos\theta = a \qquad \tan\theta = \frac{b}{a}, \quad a \neq 0$$
$$\csc\theta = \frac{1}{b}, \quad b \neq 0 \qquad \sec\theta = \frac{1}{a}, \quad a \neq 0 \qquad \cot\theta = \frac{a}{b}, \quad b \neq 0$$

For $\sin\theta$ and $\cos\theta$, θ can be any angle, so it follows that the domain of the sine function and cosine function is the set of all real numbers.

> The domain of the sine function is the set of all real numbers.
>
> The domain of the cosine function is the set of all real numbers.

If $a = 0$, then the tangent function and the secant function are not defined. That is, for the tangent function and secant function, the x-coordinate of $P = (a, b)$ cannot be 0. On the unit circle, there are two such points $(0, 1)$ and $(0, -1)$. These two points correspond to the angles $\frac{\pi}{2}\,(90°)$ and $\frac{3\pi}{2}\,(270°)$ or, more generally, to any angle that is an odd multiple of $\frac{\pi}{2}\,(90°)$, such as $\frac{\pi}{2}\,(90°)$, $\frac{3\pi}{2}\,(270°)$, $\frac{5\pi}{2}\,(450°)$, $-\frac{\pi}{2}\,(-90°)$, and $-\frac{3\pi}{2}\,(-270°)$. Such angles must therefore be excluded from the domain of the tangent function and secant function.

> The domain of the tangent function is the set of all real numbers, except odd multiples of $\frac{\pi}{2}\,(90°)$.
>
> The domain of the secant function is the set of all real numbers, except odd multiples of $\frac{\pi}{2}\,(90°)$.

If $b = 0$, then the cotangent function and the cosecant function are not defined. For the cotangent function and cosecant function, the y-coordinate of $P = (a, b)$ cannot be 0. On the unit circle, there are two such points, $(1, 0)$ and $(-1, 0)$. These two points correspond to the angles $0\,(0°)$ and $\pi\,(180°)$ or, more generally, to any angle that is an integral multiple of $\pi\,(180°)$, such as $0\,(0°)$, $\pi\,(180°)$, $2\pi\,(360°)$, $3\pi\,(540°)$, and $-\pi\,(-180°)$. Such angles must therefore be excluded from the domain of the cotangent function and cosecant function.

> The domain of the cotangent function is the set of all real numbers, except integral multiples of $\pi\,(180°)$.
>
> The domain of the cosecant function is the set of all real numbers, except integral multiples of $\pi\,(180°)$.

Next, we determine the range of each of the six trigonometric functions. Refer again to Figure 69. Let $P = (a, b)$ be the point on the unit circle that corresponds to the angle θ. It follows that $-1 \le a \le 1$ and $-1 \le b \le 1$. Since $\sin\theta = b$ and $\cos\theta = a$, we have

$$-1 \le \sin\theta \le 1 \quad \text{and} \quad -1 \le \cos\theta \le 1$$

The range of both the sine function and the cosine function consists of all real numbers between -1 and 1, inclusive. Using absolute value notation, we have $|\sin\theta| \le 1$ and $|\cos\theta| \le 1$.

If θ is not a multiple of π (180°), then $\csc\theta = \frac{1}{b}$. Since $b = \sin\theta$ and $|b| = |\sin\theta| \le 1$, it follows that $|\csc\theta| = \frac{1}{|\sin\theta|} = \frac{1}{|b|} \ge 1$. The range of the cosecant function consists of all real numbers less than or equal to -1 or greater than or equal to 1. That is,

$$\csc\theta \le -1 \quad \text{or} \quad \csc\theta \ge 1$$

Using absolute value notation, we have $|\csc\theta| \ge 1$.

If θ is not an odd multiple of $\frac{\pi}{2}$ (90°), then, by definition, $\sec\theta = \frac{1}{a}$. Since $a = \cos\theta$ and $|a| = |\cos\theta| \le 1$, it follows that $|\sec\theta| = \frac{1}{|\cos\theta|} = \frac{1}{|a|} \ge 1$. The range of the secant function consists of all real numbers less than or equal to -1 or greater than or equal to 1. That is,

$$\sec\theta \le -1 \quad \text{or} \quad \sec\theta \ge 1$$

Using absolute value notation, we have $|\sec\theta| \ge 1$.

The range of both the tangent function and the cotangent function consists of all real numbers. That is,

$$-\infty < \tan\theta < \infty \quad \text{and} \quad -\infty < \cot\theta < \infty$$

You are asked to prove this in Problems 81 and 82.

Table 6 summarizes these results.

TABLE 6

Function	Symbol	Domain	Range
sine	$f(\theta) = \sin\theta$	All real numbers	All real numbers from -1 to 1, inclusive
cosine	$f(\theta) = \cos\theta$	All real numbers	All real numbers from -1 to 1, inclusive
tangent	$f(\theta) = \tan\theta$	All real numbers, except odd multiples of $\frac{\pi}{2}$ (90°)	All real numbers
cosecant	$f(\theta) = \csc\theta$	All real numbers, except integral multiples of π (180°)	All real numbers greater than or equal to 1 or less than or equal to -1
secant	$f(\theta) = \sec\theta$	All real numbers, except odd multiples of $\frac{\pi}{2}$ (90°)	All real numbers greater than or equal to 1 or less than or equal to -1
cotangent	$f(\theta) = \cot\theta$	All real numbers, except integral multiples of π(180°)	All real numbers

NOW WORK PROBLEMS 53 AND 57.

Period of the Trigonometric Functions

Figure 70

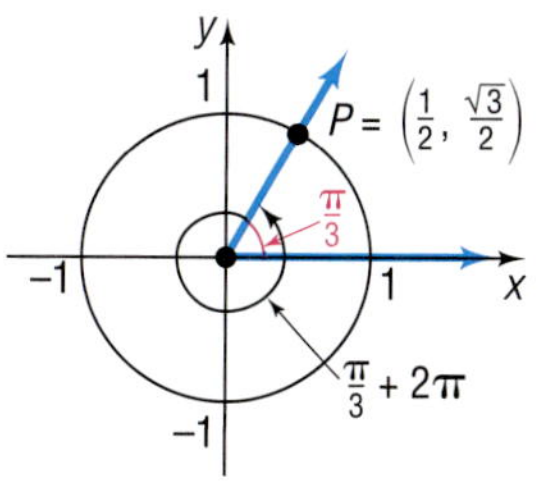

3 Look at Figure 70. This figure shows that for an angle of $\frac{\pi}{3}$ radians the corresponding point P on the unit circle is $\left(\frac{1}{2}, \frac{\sqrt{3}}{2}\right)$. Notice that for an angle of $\frac{\pi}{3} + 2\pi$ radians the corresponding point P on the unit circle is also $\left(\frac{1}{2}, \frac{\sqrt{3}}{2}\right)$. Thus,

$$\sin\frac{\pi}{3} = \frac{\sqrt{3}}{2} \quad \text{and} \quad \sin\left(\frac{\pi}{3} + 2\pi\right) = \frac{\sqrt{3}}{2}$$
$$\cos\frac{\pi}{3} = \frac{1}{2} \quad \text{and} \quad \cos\left(\frac{\pi}{3} + 2\pi\right) = \frac{1}{2}$$

This example illustrates a more general situation. For a given angle θ, measured in radians, suppose that we know the corresponding point $P = (a, b)$ on the unit circle. Now add 2π to θ. The point on the unit circle corresponding to $\theta + 2\pi$ is identical to the point P on the unit circle corresponding to θ. See Figure 71. The values of the trigonometric functions of $\theta + 2\pi$ are equal to the values of the corresponding trigonometric functions of θ.

Figure 71

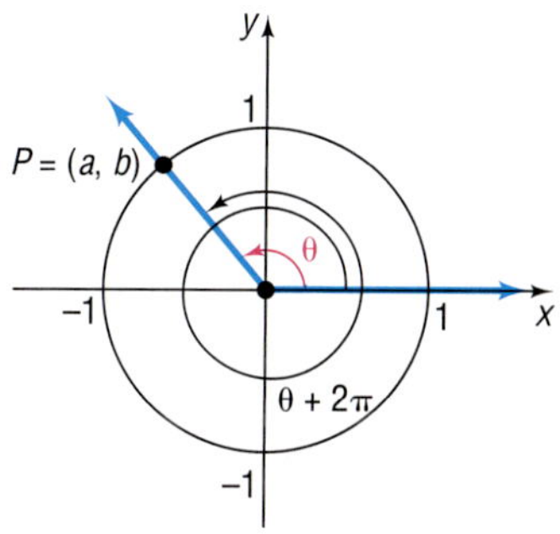

If we add (or subtract) integral multiples of 2π to θ, the trigonometric values remain unchanged. That is, for all θ,

$$\sin(\theta + 2\pi k) = \sin\theta \qquad \cos(\theta + 2\pi k) = \cos\theta \qquad (1)$$
where k is any integer

Functions that exhibit this kind of behavior are called *periodic functions*.

A function f is called **periodic** if there is a positive number p such that, whenever θ is in the domain of f, so is $\theta + p$, and

$$f(\theta + p) = f(\theta)$$

If there is a smallest such number p, this smallest value is called the **(fundamental) period** of f.

Based on equation (1), the sine and cosine functions are periodic. In fact, the sine and cosine functions have period 2π. You are asked to prove this fact in Problems 83 and 84. The secant and cosecant functions are also periodic with period 2π; and the tangent and cotangent functions are periodic with period π. You are asked to prove these statements in Problems 85 through 88.

These facts are summarized as follows:

Periodic Properties

$$\sin(\theta + 2\pi) = \sin\theta \quad \cos(\theta + 2\pi) = \cos\theta \quad \tan(\theta + \pi) = \tan\theta$$
$$\csc(\theta + 2\pi) = \csc\theta \quad \sec(\theta + 2\pi) = \sec\theta \quad \cot(\theta + \pi) = \cot\theta$$

Because the sine, cosine, secant, and cosecant functions have period 2π, once we know their values for $0 \le \theta < 2\pi$, we know all their values; similarly, since the tangent and cotangent functions have period π, once we know their values for $0 \le \theta < \pi$, we know all their values.

EXAMPLE 2 Using Periodic Properties to Find Exact Values

Find the exact value of: (a) $\sin 420°$ (b) $\tan \dfrac{5\pi}{4}$ (c) $\cos \dfrac{11\pi}{4}$

Solution (a) $\sin 420° = \sin(360° + 60°) = \sin 60° = \dfrac{\sqrt{3}}{2}$

(b) $\tan \dfrac{5\pi}{4} = \tan\left(\pi + \dfrac{\pi}{4}\right) = \tan \dfrac{\pi}{4} = 1$

(c) $\cos \dfrac{11\pi}{4} = \cos\left(\dfrac{3\pi}{4} + \dfrac{8\pi}{4}\right) = \cos\left(\dfrac{3\pi}{4} + 2\pi\right) = \cos \dfrac{3\pi}{4} = -\dfrac{\sqrt{2}}{2}$ ■

NOW WORK PROBLEMS 13 AND 71.

The periodic properties of the trigonometric functions will be very helpful to us when we study their graphs in the next section.

Even–Odd Properties

4 Recall that a function f is even if $f(-\theta) = f(\theta)$ for all θ in the domain of f; a function f is odd if $f(-\theta) = -f(\theta)$ for all θ in the domain of f. We will now show that the trigonometric functions sine, tangent, cotangent, and cosecant are odd functions, whereas the functions cosine and secant are even functions.

Theorem **Even–Odd Properties**

$$\sin(-\theta) = -\sin\theta \qquad \cos(-\theta) = \cos\theta \qquad \tan(-\theta) = -\tan\theta$$
$$\csc(-\theta) = -\csc\theta \qquad \sec(-\theta) = \sec\theta \qquad \cot(-\theta) = -\cot\theta$$

■

Figure 72

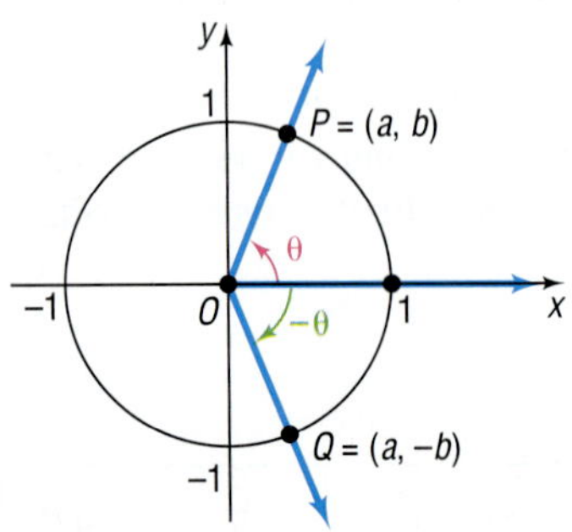

Proof Let $P = (a, b)$ be the point on the unit circle that corresponds to the angle θ. See Figure 72. The point Q on the unit circle that corresponds to the angle $-\theta$ will have coordinates $(a, -b)$. Using the definition for the trigonometric functions, we have

$$\sin\theta = b \qquad \cos\theta = a \qquad \sin(-\theta) = -b \qquad \cos(-\theta) = a$$

so

$$\sin(-\theta) = -b = -\sin\theta \qquad \cos(-\theta) = a = \cos\theta$$

Now, using these results and some of the Fundamental Identities, we have

$$\tan(-\theta) = \frac{\sin(-\theta)}{\cos(-\theta)} = \frac{-\sin\theta}{\cos\theta} = -\tan\theta$$

$$\cot(-\theta) = \frac{1}{\tan(-\theta)} = \frac{1}{-\tan\theta} = -\cot\theta$$

$$\sec(-\theta) = \frac{1}{\cos(-\theta)} = \frac{1}{\cos\theta} = \sec\theta$$

$$\csc(-\theta) = \frac{1}{\sin(-\theta)} = \frac{1}{-\sin\theta} = -\csc\theta$$

EXAMPLE 3 Finding Exact Values Using Even–Odd Properties

Find the exact value of:

(a) $\sin(-45°)$ (b) $\cos(-\pi)$ (c) $\cot\left(-\frac{3\pi}{2}\right)$ (d) $\tan\left(-\frac{37\pi}{4}\right)$

Solution (a) $\sin(-45°) = -\sin 45° = -\frac{\sqrt{2}}{2}$ (Odd function) (b) $\cos(-\pi) = \cos\pi = -1$ (Even function)

(c) $\cot\left(-\frac{3\pi}{2}\right) = -\cot\frac{3\pi}{2} = 0$ (Odd function)

(d) $\tan\left(-\frac{37\pi}{4}\right) = -\tan\frac{37\pi}{4} = -\tan\left(\frac{\pi}{4} + 9\pi\right) = -\tan\frac{\pi}{4} = -1$ (Odd function; Period is π)

SEEING THE CONCEPT To see that the cosine function is even, use a graphing utility to graph $Y_1 = \cos x$. Do you see the symmetry? Why would you conjecture that the cosine function is even? Clear the screen. Now graph $Y_1 = \sin x$. Do you see the symmetry? Do you conjecture that the sine function is odd?

NOW WORK PROBLEMS 29 AND 65.

2.5 Concepts and Vocabulary

In Problems 1–3, fill in the blanks.

1. The sine, cosine, cosecant, and secant functions have period ________; the tangent and cotangent functions have period ________.
2. The domain of the tangent function is ________.
3. The range of the sine function is ________.

In Problems 4–6, answer True or False to each statement.

4. The only even trigonometric functions are the cosine and secant functions.
5. All the trigonometric functions are periodic, with period 2π.
6. The range of the secant function is the set of positive real numbers.

7. Explain how you would find the value of sin 390° using periodic properties.
8. Explain how you would find the value of $\cos(-45°)$ using even–odd properties.

2.5 Exercises

In Problems 1–6, the point P on the unit circle that corresponds to a real number t is given. Find $\sin t$, $\cos t$, $\tan t$, $\csc t$, $\sec t$, *and* $\cot t$.

1. $\left(\frac{\sqrt{3}}{2}, -\frac{1}{2}\right)$ **2.** $\left(-\frac{\sqrt{3}}{2}, -\frac{1}{2}\right)$ **3.** $\left(-\frac{\sqrt{2}}{2}, -\frac{\sqrt{2}}{2}\right)$

4. $\left(\frac{\sqrt{2}}{2}, -\frac{\sqrt{2}}{2}\right)$ **5.** $\left(\frac{\sqrt{5}}{3}, \frac{2}{3}\right)$ **6.** $\left(-\frac{\sqrt{5}}{5}, \frac{2\sqrt{5}}{5}\right)$

In Problems 7–12, the point P on the circle $x^2 + y^2 = r^2$ *that is also on the terminal side of an angle* θ *in standard position is given. Find* $\sin\theta$, $\cos\theta$, $\tan\theta$, $\csc\theta$, $\sec\theta$, *and* $\cot\theta$.

7. $(3, -4)$ **8.** $(4, -3)$ **9.** $(-2, 3)$ **10.** $(2, -4)$ **11.** $(-1, -1)$ **12.** $(-3, 1)$

In Problems 13–28, use the fact that the trigonometric functions are periodic to find the exact value of each expression. Do not use a calculator.

13. $\sin 405°$ **14.** $\cos 420°$ **15.** $\tan 405°$ **16.** $\sin 390°$

17. $\csc 450°$ **18.** $\sec 540°$ **19.** $\cot 390°$ **20.** $\sec 420°$

21. $\cos\frac{33\pi}{4}$ **22.** $\sin\frac{9\pi}{4}$ **23.** $\tan(21\pi)$ **24.** $\csc\frac{9\pi}{2}$

25. $\sec\frac{17\pi}{4}$ **26.** $\cot\frac{17\pi}{4}$ **27.** $\tan\frac{19\pi}{6}$ **28.** $\sec\frac{25\pi}{6}$

In Problems 29–46, use the even-odd properties to find the exact value of each expression. Do not use a calculator.

29. $\sin(-60°)$ **30.** $\cos(-30°)$ **31.** $\tan(-30°)$ **32.** $\sin(-135°)$

33. $\sec(-60°)$ **34.** $\csc(-30°)$ **35.** $\sin(-90°)$ **36.** $\cos(-270°)$

37. $\tan\left(-\frac{\pi}{4}\right)$ **38.** $\sin(-\pi)$ **39.** $\cos\left(-\frac{\pi}{4}\right)$ **40.** $\sin\left(-\frac{\pi}{3}\right)$

41. $\tan(-\pi)$ **42.** $\sin\left(-\frac{3\pi}{2}\right)$ **43.** $\csc\left(-\frac{\pi}{4}\right)$ **44.** $\sec(-\pi)$

45. $\sec\left(-\frac{\pi}{6}\right)$ **46.** $\csc\left(-\frac{\pi}{3}\right)$

In Problems 47–52, find the exact value of each expression. Do not use a calculator.

47. $\sin(-\pi) + \cos(5\pi)$ **48.** $\tan\left(-\frac{5\pi}{6}\right) - \cot\frac{7\pi}{2}$ **49.** $\sec(-\pi) + \csc\left(-\frac{\pi}{2}\right)$

50. $\tan(-6\pi) + \cos\frac{9\pi}{4}$ **51.** $\sin\left(-\frac{9\pi}{4}\right) - \tan\left(-\frac{9\pi}{4}\right)$ **52.** $\cos\left(-\frac{17\pi}{4}\right) - \sin\left(-\frac{3\pi}{2}\right)$

53. What is the domain of the sine function?

54. What is the domain of the cosine function?

55. For what numbers θ is $f(\theta) = \tan\theta$ not defined?

56. For what numbers θ is $f(\theta) = \cot\theta$ not defined?

57. For what numbers θ is $f(\theta) = \sec\theta$ not defined?

58. For what numbers θ is $f(\theta) = \csc\theta$ not defined?

59. What is the range of the sine function?

60. What is the range of the cosine function?

61. What is the range of the tangent function?

62. What is the range of the cotangent function?

63. What is the range of the secant function?

64. What is the range of the cosecant function?

65. Is the sine function even, odd, or neither? Is its graph symmetric? With respect to what?

66. Is the cosine function even, odd, or neither? Is its graph symmetric? With respect to what?

67. Is the tangent function even, odd, or neither? Is its graph symmetric? With respect to what?

68. Is the cotangent function even, odd, or neither? Is its graph symmetric? With respect to what?

69. Is the secant function even, odd, or neither? Is its graph symmetric? With respect to what?

70. Is the cosecant function even, odd, or neither? Is its graph symmetric? With respect to what?

71. If $\sin\theta = 0.3$, find the value of

$$\sin\theta + \sin(\theta + 2\pi) + \sin(\theta + 4\pi)$$

72. If $\cos\theta = 0.2$, find the value of

$$\cos\theta + \cos(\theta + 2\pi) + \cos(\theta + 4\pi)$$

73. If $\tan\theta = 3$, find the value of

$$\tan\theta + \tan(\theta + \pi) + \tan(\theta + 2\pi).$$

74. If $\cot\theta = -2$, find the value of

$$\cot\theta + \cot(\theta - \pi) + \cot(\theta - 2\pi).$$

In Problems 75–80, use the periodic and even-odd properties.

75. If $f(x) = \sin x$ and $f(a) = \dfrac{1}{3}$, find the exact value of:
(a) $f(-a)$ (b) $f(a) + f(a + 2\pi) + f(a + 4\pi)$

76. If $f(x) = \cos x$ and $f(a) = \dfrac{1}{4}$, find the exact value of:
(a) $f(-a)$ (b) $f(a) + f(a + 2\pi) + f(a - 2\pi)$

77. If $f(x) = \tan x$ and $f(a) = 2$, find the exact value of:
(a) $f(-a)$ (b) $f(a) + f(a + \pi) + f(a + 2\pi)$

78. If $f(x) = \cot x$ and $f(a) = -3$, find the exact value of:
(a) $f(-a)$ (b) $f(a) + f(a + \pi) + f(a + 4\pi)$

79. If $f(x) = \sec x$ and $f(a) = -4$, find the exact value of:
(a) $f(-a)$ (b) $f(a) + f(a + 2\pi) + f(a + 4\pi)$

80. If $f(x) = \csc x$ and $f(a) = 2$, find the exact value of:
(a) $f(-a)$ (b) $f(a) + f(a + 2\pi) + f(a + 4\pi)$

81. Show that the range of the tangent function is the set of all real numbers.

82. Show that the range of the cotangent function is the set of all real numbers.

83. Show that the period of $f(\theta) = \sin\theta$ is 2π.

[**Hint:** Assume that $0 < p < 2\pi$ exists so that $\sin(\theta + p) = \sin\theta$ for all θ. Let $\theta = 0$ to find p. Then let $\theta = \dfrac{\pi}{2}$ to obtain a contradiction.]

84. Show that the period of $f(\theta) = \cos\theta$ is 2π.

85. Show that the period of $f(\theta) = \sec\theta$ is 2π.

86. Show that the period of $f(\theta) = \csc\theta$ is 2π.

87. Show that the period of $f(\theta) = \tan\theta$ is π.

88. Show that the period of $f(\theta) = \cot\theta$ is π.

89. If $\theta\,(0 < \theta < \pi)$ is the angle between a horizontal ray directed to the right (say, the positive x-axis) and a nonhorizontal, nonvertical line L, show that the slope m of L equals $\tan\theta$. The angle θ is called the **inclination** of L.

[**Hint:** See the illustration, where we have drawn the line L^* parallel to L and passing through the origin. Use the fact that L^* intersects the unit circle at the point $(\cos\theta, \sin\theta)$.]

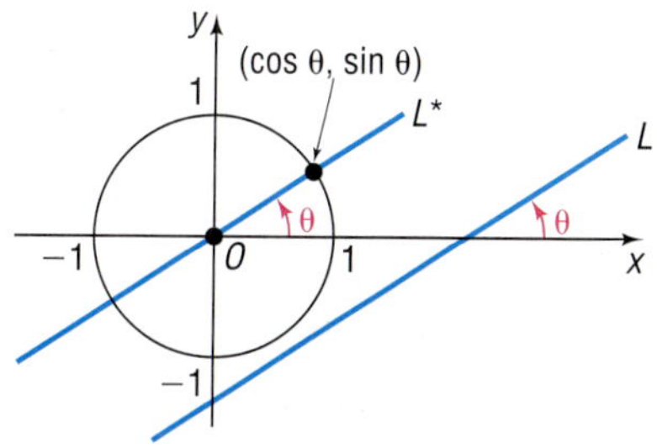

90. Write down five properties of the tangent function. Explain the meaning of each.

91. Describe your understanding of the meaning of a periodic function.

PREPARING FOR THIS SECTION

Before getting started, review the following:

- ✓ Functions (Section 1.4, pp. 41–44)
- ✓ Graphing Techniques: Transformations (Section 1.6, pp. 64–73)

2.6 GRAPHS OF THE SINE AND COSINE FUNCTIONS

OBJECTIVES

1. Graph Transformations of the Sine Function
2. Graph Transformations of the Cosine Function
3. Determine the Amplitude and Period of Sinusoidal Functions
4. Graph Sinusoidal Functions: $y = A\sin(\omega x)$
5. Find an Equation for a Sinusoidal Graph

Since we want to graph the trigonometric functions in the xy-plane, we shall use the traditional symbols x for the independent variable (or argument) and y for the dependent variable (or value at x) for each function. So we write the six trigonometric functions as

$$y = f(x) = \sin x \qquad y = f(x) = \cos x \qquad y = f(x) = \tan x$$
$$y = f(x) = \csc x \qquad y = f(x) = \sec x \qquad y = f(x) = \cot x$$

Here the independent variable x represents an angle, measured in radians. In calculus, x will usually be treated as a real number. As we said earlier, these are equivalent ways of viewing x.

The Graph of $y = \sin x$

Since the sine function has period 2π, we need to graph $y = \sin x$ only on the interval $[0, 2\pi]$. The remainder of the graph will consist of repetitions of this portion of the graph.

We begin by constructing Table 7, which lists some points on the graph of $y = \sin x$, $0 \le x \le 2\pi$. As the table shows, the graph of $y = \sin x$, $0 \le x \le 2\pi$, begins at the origin. As x increases from 0 to $\frac{\pi}{2}$, the value of $y = \sin x$ increases from 0 to 1; as x increases from $\frac{\pi}{2}$ to π to $\frac{3\pi}{2}$, the value of y decreases from 1 to 0 to -1; as x increases from $\frac{3\pi}{2}$ to 2π, the value of y increases from -1 to 0. If we plot the points listed in Table 7 and connect them with a smooth curve, we obtain the graph shown in Figure 73.

TABLE 7

x	$y = \sin x$	(x, y)
0	0	$(0, 0)$
$\frac{\pi}{6}$	$\frac{1}{2}$	$\left(\frac{\pi}{6}, \frac{1}{2}\right)$
$\frac{\pi}{2}$	1	$\left(\frac{\pi}{2}, 1\right)$
$\frac{5\pi}{6}$	$\frac{1}{2}$	$\left(\frac{5\pi}{6}, \frac{1}{2}\right)$
π	0	$(\pi, 0)$
$\frac{7\pi}{6}$	$-\frac{1}{2}$	$\left(\frac{7\pi}{6}, -\frac{1}{2}\right)$
$\frac{3\pi}{2}$	-1	$\left(\frac{3\pi}{2}, -1\right)$
$\frac{11\pi}{6}$	$-\frac{1}{2}$	$\left(\frac{11\pi}{6}, -\frac{1}{2}\right)$
2π	0	$(2\pi, 0)$

Figure 73

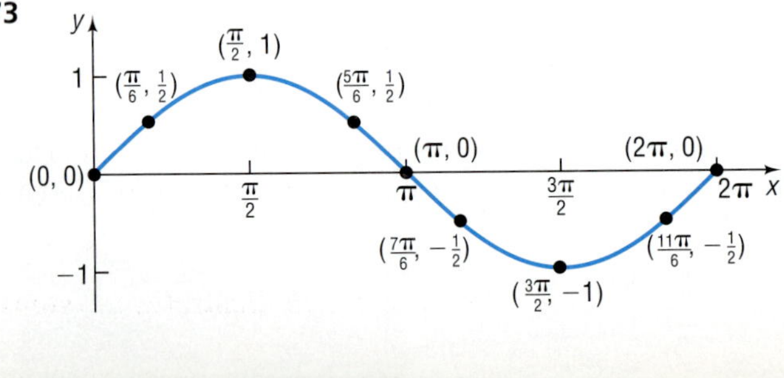

The graph in Figure 73 is one period, or **cycle**, of the graph of $y = \sin x$. To obtain a more complete graph of $y = \sin x$, we repeat this period in each direction, as shown in Figure 74(a). Figure 74(b) shows the graph on a TI-83 graphing calculator.

Figure 74
$y = \sin x, -\infty < x < \infty$

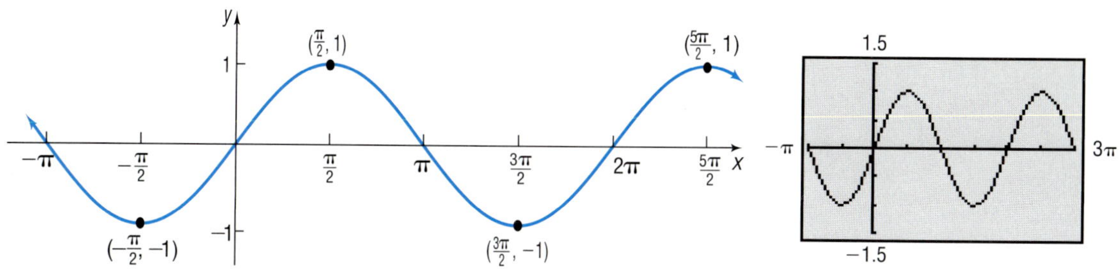

(a) (b)

The graph of $y = \sin x$ illustrates some of the facts that we already know about the sine function.

Properties of the Sine Function

1. The domain is the set of all real numbers.
2. The range consists of all real numbers from -1 to 1, inclusive.
3. The sine function is an odd function, as the symmetry of the graph with respect to the origin indicates.
4. The sine function is periodic, with period 2π.
5. The x-intercepts are $\ldots, -2\pi, -\pi, 0, \pi, 2\pi, 3\pi, \ldots$; the y-intercept is 0.
6. The maximum value is 1 and occurs at $x = \ldots, -\frac{3\pi}{2}, \frac{\pi}{2}, \frac{5\pi}{2}, \frac{9\pi}{2}, \ldots$; the minimum value is -1 and occurs at $x = \ldots, -\frac{\pi}{2}, \frac{3\pi}{2}, \frac{7\pi}{2}, \frac{11\pi}{2}, \ldots$.

NOW WORK PROBLEMS **1, 3,** AND **5.**

1 The graphing techniques introduced in Chapter 1 may be used to graph functions that are transformations of the sine function (refer to Section 1.6).

EXAMPLE 1 Graphing Variations of $y = \sin x$ Using Transformations

Use the graph of $y = \sin x$ to graph $y = \sin\left(x - \frac{\pi}{4}\right)$.

Solution Figure 75 illustrates the steps.

Figure 75

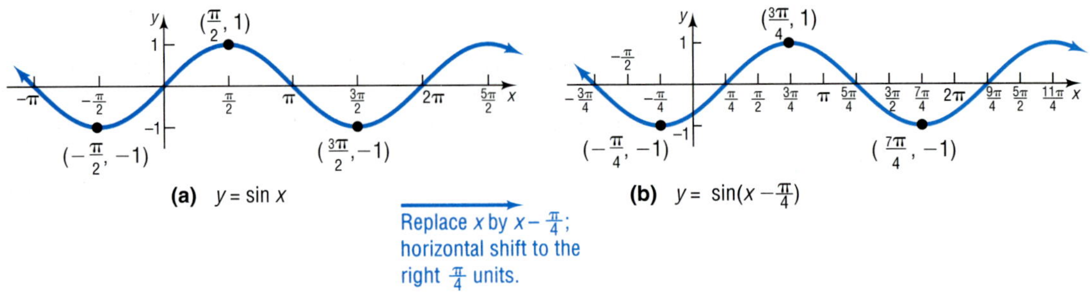

(a) $y = \sin x$ (b) $y = \sin(x - \frac{\pi}{4})$

✔ CHECK: Figure 76 shows the graph of $Y_1 = \sin\left(x - \dfrac{\pi}{4}\right)$.

Figure 76

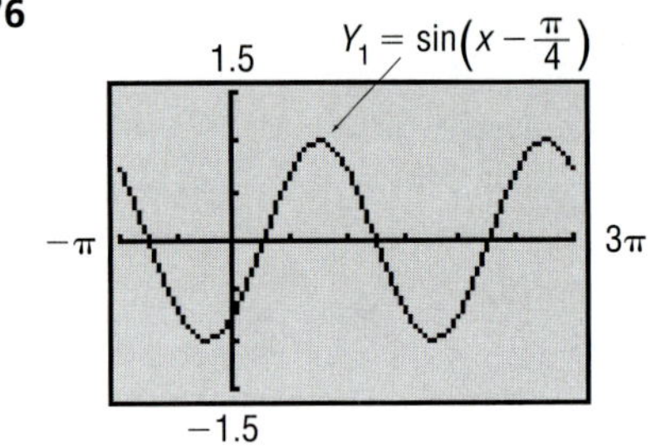

EXAMPLE 2 Graphing Variations of $y = \sin x$ Using Transformations

Use the graph of $y = \sin x$ to graph $y = -\sin x + 2$.

Solution Figure 77 illustrates the steps.

Figure 77

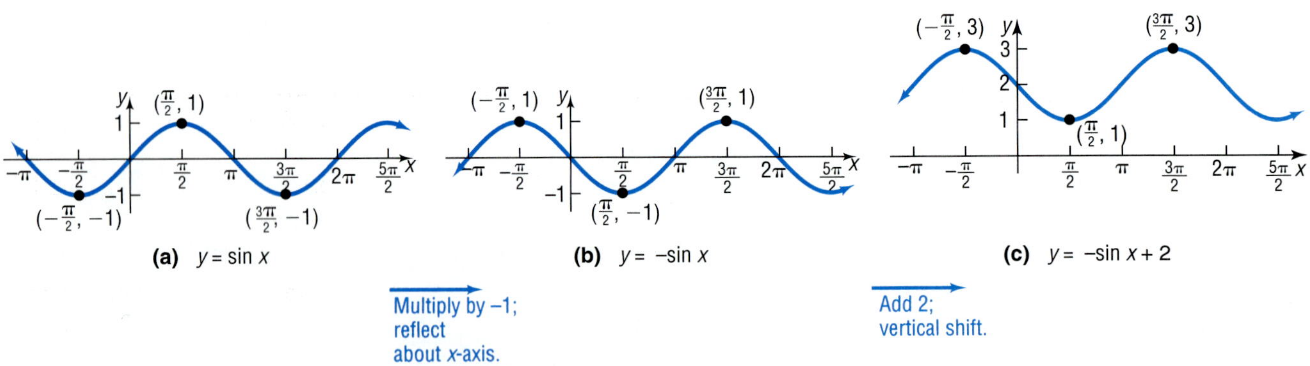

(a) $y = \sin x$ (b) $y = -\sin x$ (c) $y = -\sin x + 2$

✔ CHECK: Figure 78 shows the graph of $Y_1 = -\sin x + 2$.

Figure 78

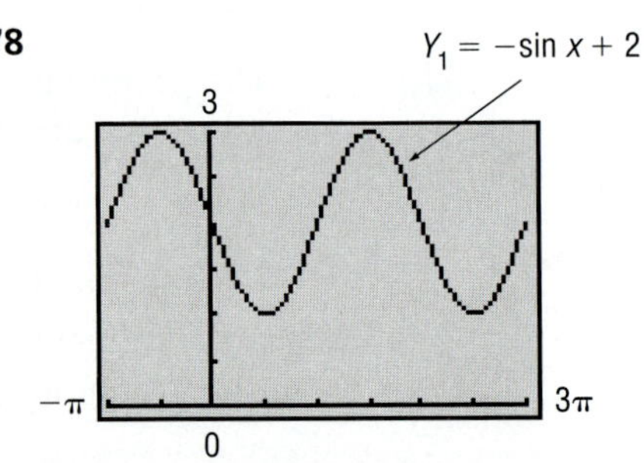

NOW WORK PROBLEM 17.

TABLE 8

x	$y = \cos x$	(x, y)
0	1	$(0, 1)$
$\frac{\pi}{3}$	$\frac{1}{2}$	$\left(\frac{\pi}{3}, \frac{1}{2}\right)$
$\frac{\pi}{2}$	0	$\left(\frac{\pi}{2}, 0\right)$
$\frac{2\pi}{3}$	$-\frac{1}{2}$	$\left(\frac{2\pi}{3}, -\frac{1}{2}\right)$
π	-1	$(\pi, -1)$
$\frac{4\pi}{3}$	$-\frac{1}{2}$	$\left(\frac{4\pi}{3}, -\frac{1}{2}\right)$
$\frac{3\pi}{2}$	0	$\left(\frac{3\pi}{2}, 0\right)$
$\frac{5\pi}{3}$	$\frac{1}{2}$	$\left(\frac{5\pi}{3}, \frac{1}{2}\right)$
2π	1	$(2\pi, 1)$

The Graph of $y = \cos x$

The cosine function also has period 2π. We proceed as we did with the sine function by constructing Table 8, which lists some points on the graph of $y = \cos x$, $0 \le x \le 2\pi$. As the table shows, the graph of $y = \cos x$, $0 \le x \le 2\pi$, begins at the point $(0, 1)$. As x increases from 0 to $\frac{\pi}{2}$ to π, the value of y decreases from 1 to 0 to -1; as x increases from π to $\frac{3\pi}{2}$ to 2π, the value of y increases from -1 to 0 to 1. As before, we plot the points in Table 8 to get one period or cycle of the graph. See Figure 79.

Figure 79
$y = \cos x$, $0 \le x \le 2\pi$

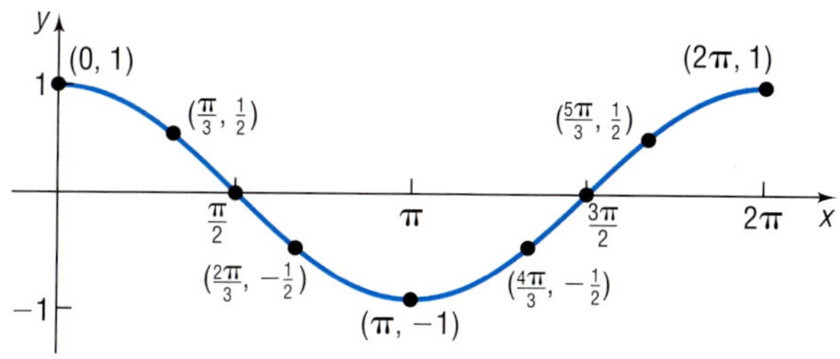

A more complete graph of $y = \cos x$ is obtained by repeating this period in each direction, as shown in Figure 80(a). Figure 80(b) shows the graph on a TI-83 graphing calculator.

Figure 80
$y = \cos x$, $-\infty < x < \infty$

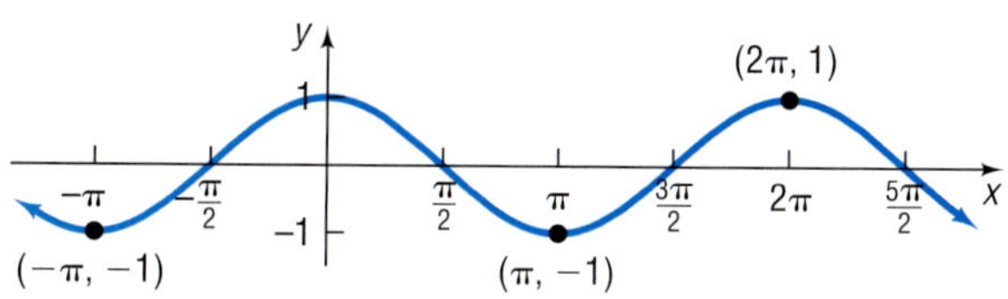

(a)

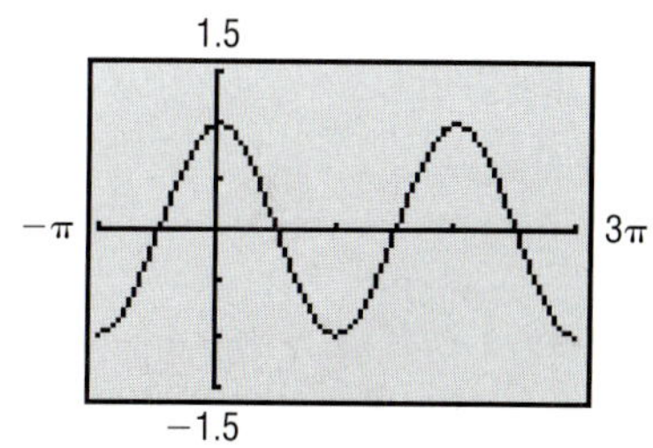

(b)

The graph of $y = \cos x$ illustrates some of the facts that we already know about the cosine function.

> ***Properties of the Cosine Function***
>
> 1. The domain is the set of all real numbers.
> 2. The range consists of all real numbers from -1 to 1, inclusive.
> 3. The cosine function is an even function, as the symmetry of the graph with respect to the y-axis indicates.
> 4. The cosine function is periodic, with period 2π.
> 5. The x-intercepts are $\dots, -\frac{3\pi}{2}, -\frac{\pi}{2}, \frac{\pi}{2}, \frac{3\pi}{2}, \frac{5\pi}{2}, \dots$; the y-intercept is 1.
> 6. The maximum value is 1 and occurs at $x = \dots, -2\pi, 0, 2\pi, 4\pi, 6\pi, \dots$; the minimum value is -1 and occurs at $x = \dots, -\pi, \pi, 3\pi, 5\pi, \dots$.

2 Again, the graphing techniques from Chapter 1 may be used to graph transformations of the cosine function.

EXAMPLE 3 Graphing Variations of $y = \cos x$ Using Transformations

Use the graph of $y = \cos x$ to graph $y = 2\cos x$.

Solution Figure 81 illustrates the steps.

Figure 81

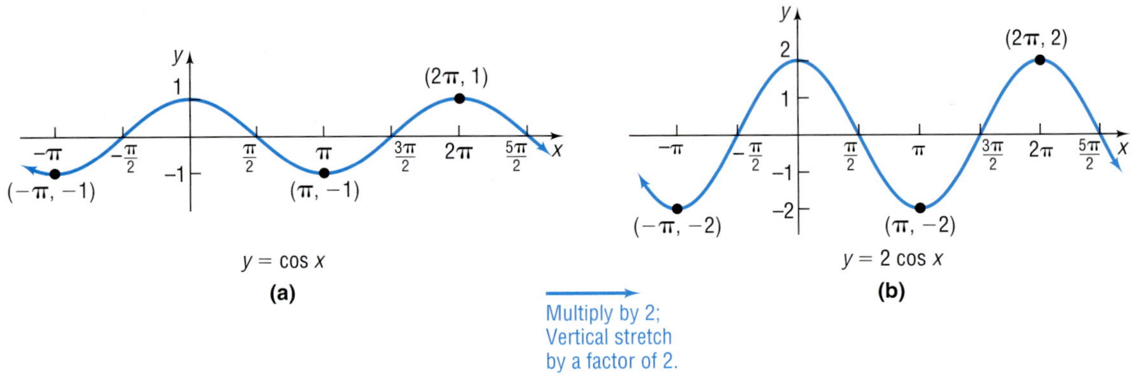

✔ CHECK: Figure 82 shows the graph of $Y_1 = 2\cos x$.

Figure 82

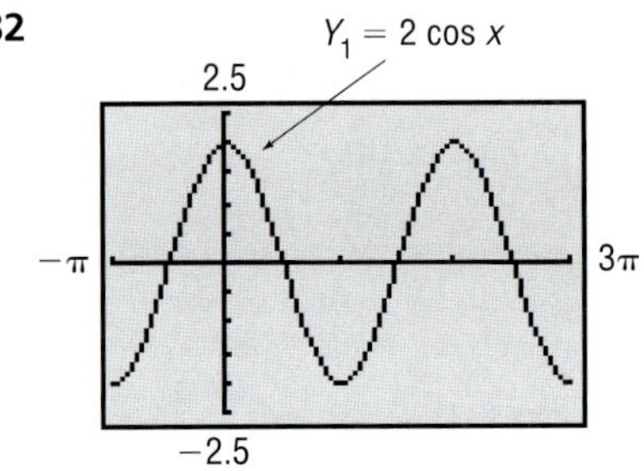

EXAMPLE 4 Graphing Variations of $y = \cos x$ Using Transformations

Use the graph of $y = \cos x$ to graph $y = \cos(3x)$.

Solution Figure 83 illustrates the graph, which is a horizontal compression of the graph of $y = \cos x$. (Multiply each x-coordinate by $\frac{1}{3}$.) Notice that, due to this compression, the period of $y = \cos(3x)$ is $\frac{2\pi}{3}$, whereas the period of $y = \cos x$ is 2π.

Figure 83

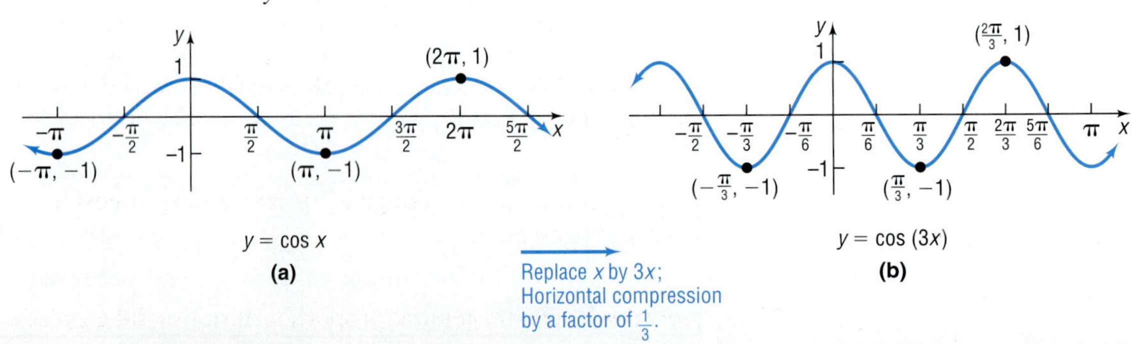

✔ CHECK: Figure 84 shows the graph of $Y_1 = \cos(3x)$.

Figure 84

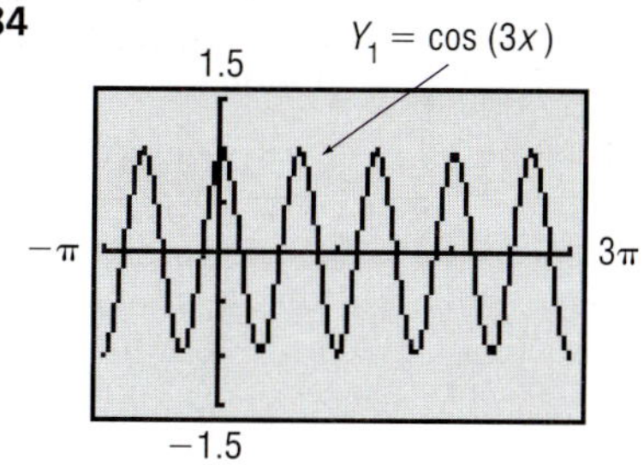

SEEING THE CONCEPT Graph $Y_1 = \cos(3x)$ with $X\text{min} = 0$, $X\text{max} = \frac{2\pi}{3}$ and $X\text{scl} = \frac{\pi}{6}$ to verify that the period is $\frac{2\pi}{3}$.

NOW WORK PROBLEM 25.

Sinusoidal Graphs

The graph of $y = \cos x$, when compared to the graph of $y = \sin x$, suggests that the graph of $y = \sin x$ is the same as the graph of $y = \cos\left(x - \frac{\pi}{2}\right)$. See Figure 85.

Figure 85

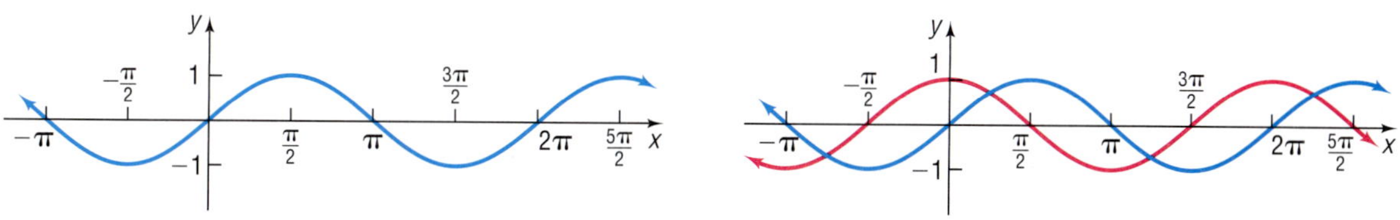

(a) $y = \sin x$ (b) $y = \cos x$ $y = \cos\left(x - \frac{\pi}{2}\right)$

Based on Figure 85, we conjecture that

$$\sin x = \cos\left(x - \frac{\pi}{2}\right)$$

(We shall prove this fact in Chapter 3.) Because of this similarity, the graphs of sine functions and cosine functions are referred to as **sinusoidal graphs**.

SEEING THE CONCEPT Graph $Y_1 = \sin x$ and $Y_2 = \cos\left(x - \frac{\pi}{2}\right)$. How many graphs do you see?

Let's look at some general properties of sinusoidal graphs.

3 In Example 3 we obtained the graph of $y = 2\cos x$, which we reproduce in Figure 86. Notice that the values of $y = 2\cos x$ lie between -2 and 2, inclusive.

Figure 86
$y = 2\cos x$

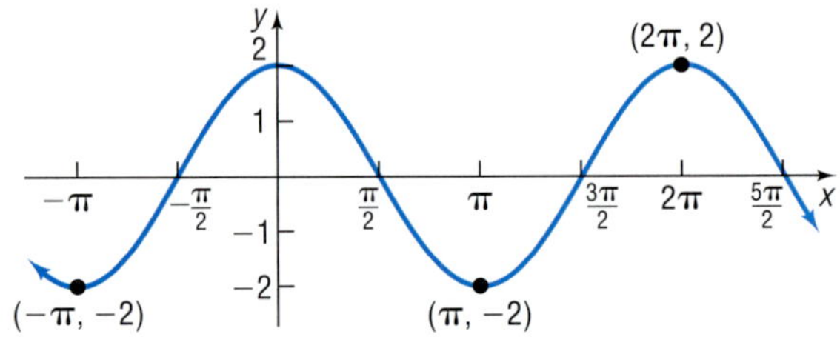

In general, the values of the functions $y = A\sin x$ and $y = A\cos x$, where $A \neq 0$, will always satisfy the inequalities

$$-|A| \leq A\sin x \leq |A| \quad \text{and} \quad -|A| \leq A\cos x \leq |A|$$

respectively. The number $|A|$ is called the **amplitude** of $y = A\sin x$ or $y = A\cos x$. See Figure 87.

Figure 87

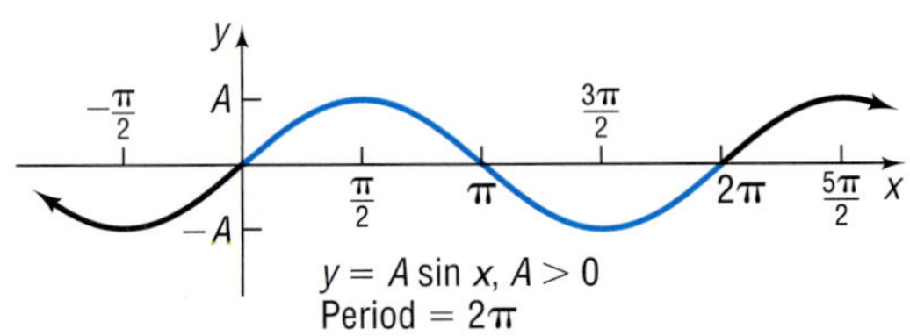

In Example 4, we obtained the graph of $y = \cos(3x)$, which we reproduce in Figure 88. Notice that the period of this function is $\dfrac{2\pi}{3}$.

Figure 88
$y = \cos(3x)$

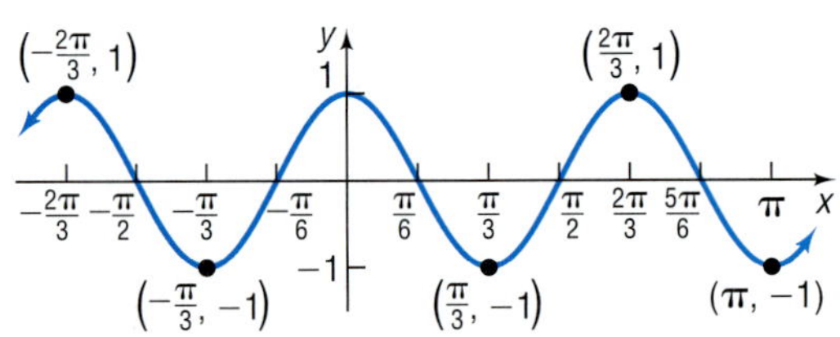

In general, if $\omega > 0$, the functions $y = \sin(\omega x)$ and $y = \cos(\omega x)$ will have period $T = \dfrac{2\pi}{\omega}$. To see why, recall that the graph of $y = \sin(\omega x)$ is obtained from the graph of $y = \sin x$ by performing a horizontal compression or stretch by a factor $\dfrac{1}{\omega}$. This horizontal compression replaces the interval $[0, 2\pi]$, which contains one period of the graph of $y = \sin x$, by the interval $\left[0, \dfrac{2\pi}{\omega}\right]$, which contains one period of the graph of $y = \sin(\omega x)$. The period of the functions $y = \sin(\omega x)$ and $y = \cos(\omega x)$, $\omega > 0$, is $\dfrac{2\pi}{\omega}$.

For example, for the function $y = \cos(3x)$, graphed in Figure 88, $\omega = 3$, so the period is $\dfrac{2\pi}{\omega} = \dfrac{2\pi}{3}$.

One period of the graph of $y = \sin(\omega x)$ or $y = \cos(\omega x)$ is called a **cycle**. Figure 89 illustrates the general situation. The blue portion of the graph is one cycle.

Figure 89

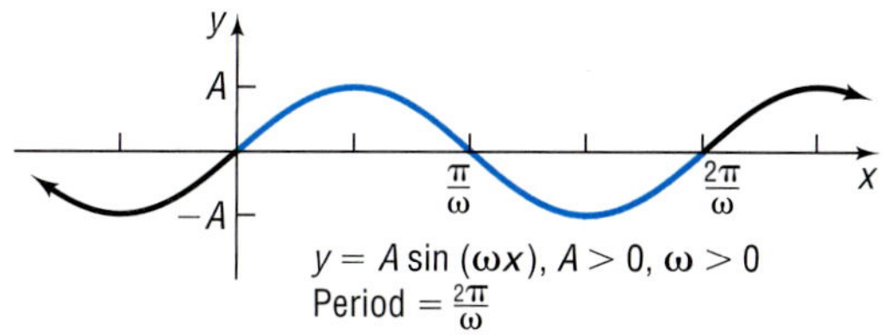

If $\omega < 0$ in $y = \sin(\omega x)$ or $y = \cos(\omega x)$, we use the even-odd properties of the sine and cosine functions as follows:

$$\sin(-\omega x) = -\sin(\omega x) \quad \text{and} \quad \cos(-\omega x) = \cos(\omega x)$$

This gives us an equivalent form in which the coefficient of x is positive. For example,

$$\sin(-2x) = -\sin(2x) \quad \text{and} \quad \cos(-\pi x) = \cos(\pi x)$$

Theorem

If $\omega > 0$, the amplitude and period of $y = A\sin(\omega x)$ and $y = A\cos(\omega x)$ are given by

$$\text{Amplitude} = |A| \qquad \text{Period} = T = \frac{2\pi}{\omega} \tag{1}$$

EXAMPLE 5 Finding the Amplitude and Period of a Sinusoidal Function

Determine the amplitude and period of $y = 3\sin(4x)$.

Solution Comparing $y = 3\sin(4x)$ to $y = A\sin(\omega x)$, we find that $A = 3$ and $\omega = 4$. From equation (1),

$$\text{Amplitude} = |A| = 3 \qquad \text{Period} = T = \frac{2\pi}{\omega} = \frac{2\pi}{4} = \frac{\pi}{2}$$

NOW WORK PROBLEM 33.

4 Earlier, we graphed sine and cosine functions using transformations. We now introduce another method that can be used to graph these functions.

Figure 90 shows one cycle of the graphs of $y = \sin x$ and $y = \cos x$ on the interval $[0, 2\pi]$. Notice that each graph consists of four parts corresponding to the four subintervals:

$$\left[0, \frac{\pi}{2}\right], \quad \left[\frac{\pi}{2}, \pi\right], \quad \left[\pi, \frac{3\pi}{2}\right], \quad \left[\frac{3\pi}{2}, 2\pi\right]$$

Each of these subintervals is of length $\frac{\pi}{2}$ (the period 2π divided by 4) and the endpoints of these intervals give rise to five key points, as shown in Figure 90.

Figure 90

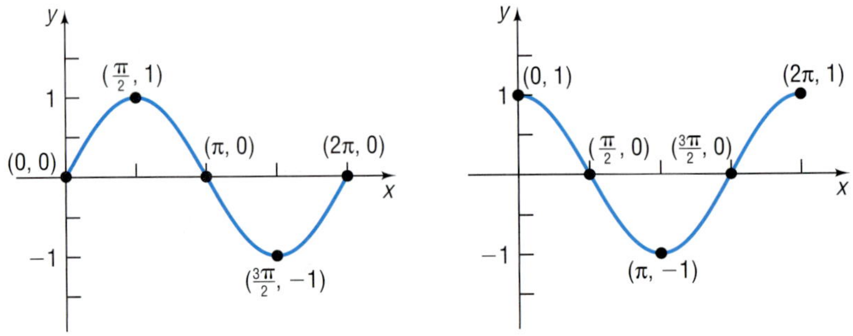

When graphing a sinusoidal function of the form $y = A\sin(\omega x)$ or $y = A\cos(\omega x)$ by hand, we use the amplitude to determine the maximum and minimum values of the function. The period is used to divide the x-axis into four subintervals. The endpoints of the subintervals give rise to five key points on the graph, which are used to sketch one cycle. Finally, extend the graph in either direction to make it complete.

To graph a sinusoidal function using a graphing utility, we use the amplitude to set Ymin and Ymax and use the period to set Xmin and Xmax.

Let's look at an example.

EXAMPLE 6 Graphing a Sinusoidal Function

Graph: $y = 3\sin(4x)$

Solution From Example 5, the amplitude is 3 and the period is $\frac{\pi}{2}$. The graph of $y = 3\sin(4x)$ will lie between -3 and 3 on the y-axis. One cycle will begin at $x = 0$ and end at $x = \frac{\pi}{2}$.

We divide the interval $\left[0, \frac{\pi}{2}\right]$ into four subintervals, each of length $\frac{\pi}{2} \div 4 = \frac{\pi}{8}$:

$$\left[0, \frac{\pi}{8}\right], \quad \left[\frac{\pi}{8}, \frac{\pi}{4}\right], \quad \left[\frac{\pi}{4}, \frac{3\pi}{8}\right], \quad \left[\frac{3\pi}{8}, \frac{\pi}{2}\right]$$

The endpoints of these intervals give rise to five key points on the graph:

$$(0, 0), \quad \left(\frac{\pi}{8}, 3\right), \quad \left(\frac{\pi}{4}, 0\right), \quad \left(\frac{3\pi}{8}, -3\right), \quad \left(\frac{\pi}{2}, 0\right)$$

We plot these five points and fill in the graph of the sine curve as shown in Figure 91(a). If we extend the graph in either direction, we obtain the complete graph shown in Figure 91(b).

Figure 91

Figure 91(c) shows the graph using a graphing utility.

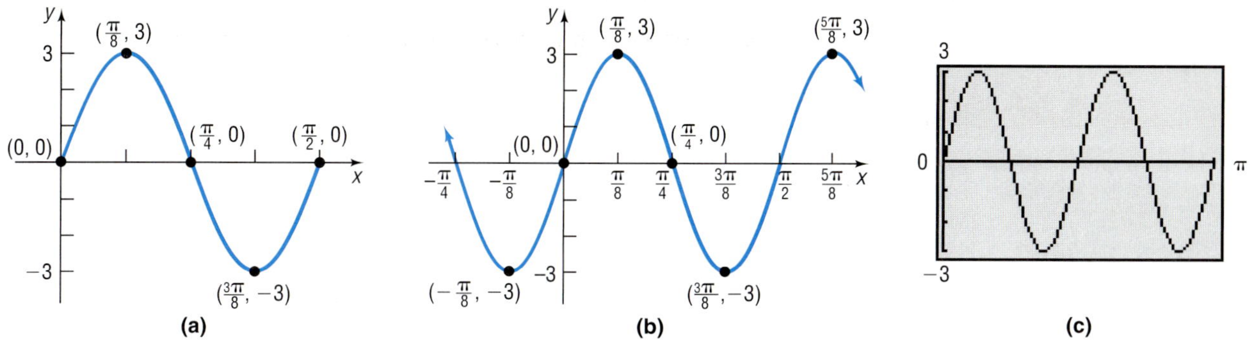

✔ CHECK: Graph $y = 3\sin(4x)$ by hand using transformations. Which graphing method do you prefer?

NOW WORK PROBLEM 39.

EXAMPLE 7 Finding the Amplitude and Period of a Sinusoidal Function and Graphing It

Determine the amplitude and period of $y = -4\cos(\pi x)$, and graph the function.

Solution Comparing $y = -4\cos(\pi x)$ with $y = A\cos(\omega x)$, we find that $A = -4$ and $\omega = \pi$. The amplitude is $|A| = |-4| = 4$, and the period is $T = \dfrac{2\pi}{\omega} = \dfrac{2\pi}{\pi} = 2$.

The graph of $y = -4\cos(\pi x)$ will lie between -4 and 4 on the y-axis. One cycle will begin at $x = 0$ and end at $x = 2$. We divide the interval $[0, 2]$ into four subintervals, each of length $2 \div 4 = \dfrac{1}{2}$:

$$\left[0, \frac{1}{2}\right], \quad \left[\frac{1}{2}, 1\right], \quad \left[1, \frac{3}{2}\right], \quad \left[\frac{3}{2}, 2\right].$$

The five key points on the graph are

$$(0, -4), \quad \left(\frac{1}{2}, 0\right), \quad (1, 4), \quad \left(\frac{3}{2}, 0\right), \quad (2, -4).$$

We plot these five points and fill in the graph of the cosine function as shown in Figure 92(a). Extending the graph in either direction, we obtain Figure 92(b).

Figure 92(c) shows the graph using a graphing utility.

Figure 92
$y = -4\cos(\pi x)$

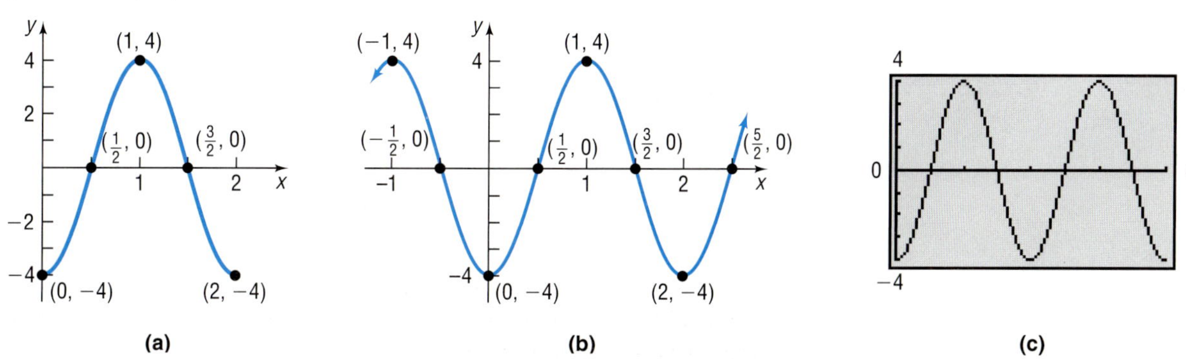

✔ CHECK: Graph $y = -4\cos(\pi x)$ by hand using transformations. Which graphing method do you prefer?

EXAMPLE 8 Finding the Amplitude and Period of a Sinusoidal Function and Graphing It

Determine the amplitude and period of $y = 2\sin\left(-\frac{\pi}{2}x\right)$, and graph the function.

Solution Since the sine function is odd, we use the equivalent form:

$$y = -2\sin\left(\frac{\pi}{2}x\right)$$

Comparing $y = -2\sin\left(\frac{\pi}{2}x\right)$ to $y = A\sin(\omega x)$, we find that $A = -2$ and $\omega = \frac{\pi}{2}$. The amplitude is $|A| = 2$, and the period is $T = \frac{2\pi}{\omega} = \frac{2\pi}{\frac{\pi}{2}} = 4$.

The graph of $y = -2\sin\left(\frac{\pi}{2}x\right)$ will lie between -2 and 2 on the y-axis. One cycle will begin at $x = 0$ and end at $x = 4$. We divide the interval $[0, 4]$ into four subintervals, each of length $4 \div 4 = 1$:

$$[0, 1], \quad [1, 2], \quad [2, 3], \quad [3, 4]$$

The five key points on the graph are

$$(0, 0), \quad (1, -2), \quad (2, 0), \quad (3, 2), \quad (4, 0)$$

We plot these five points and fill in the graph of the sine function as shown in Figure 93(a). Extending the graph in either direction, we obtain Figure 93(b).

Figure 93(c) shows the graph using a graphing utility.

Figure 93

$y = 2\sin\left(-\frac{\pi}{2}x\right)$

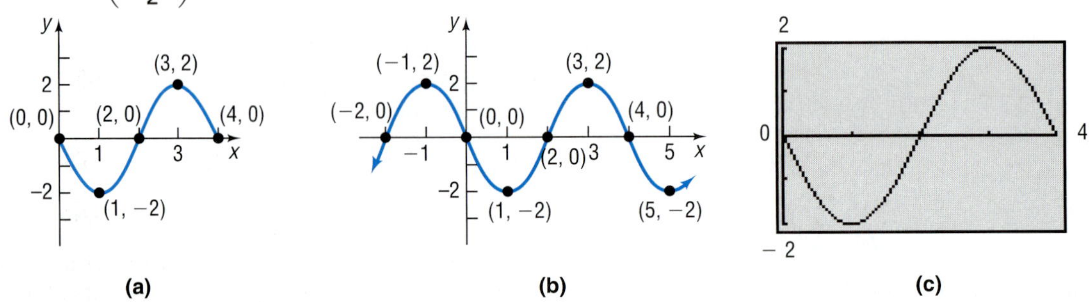

✔ CHECK: Graph $y = 2\sin\left(-\frac{\pi}{2}x\right)$ by hand using transformations. Which graphing method do you prefer?

NOW WORK PROBLEM 53.

5 We can also use the ideas of amplitude and period to identify a sinusoidal function when its graph is given.

EXAMPLE 9 Finding an Equation for a Sinusoidal Graph

Find an equation for the graph shown in Figure 94.

Figure 94

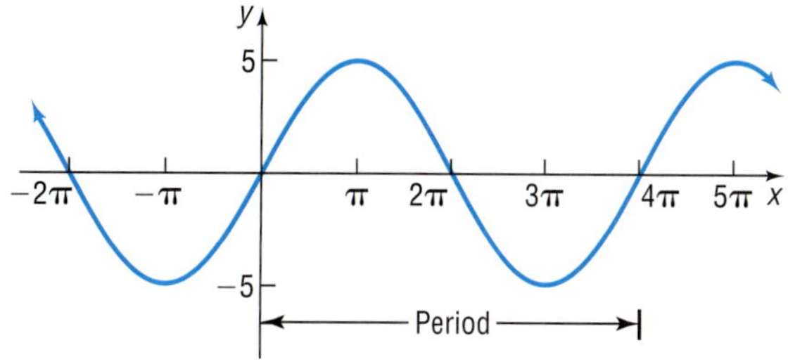

Solution This graph can be viewed as the graph of a sine function* with amplitude $A = 5$. The period T is observed to be 4π. By equation (1),

$$T = \frac{2\pi}{\omega}$$

$$4\pi = \frac{2\pi}{\omega}$$

$$\omega = \frac{2\pi}{4\pi} = \frac{1}{2}$$

The sine function whose graph is given in Figure 94 is

$$y = A\sin(\omega x) = 5\sin\left(\frac{1}{2}x\right)$$

✔ CHECK: Graph $Y_1 = 5\sin\left(\frac{1}{2}x\right)$ and compare the result with Figure 94.

EXAMPLE 10 Finding an Equation for a Sinusoidal Graph

Find an equation for the graph shown in Figure 95.

Figure 95

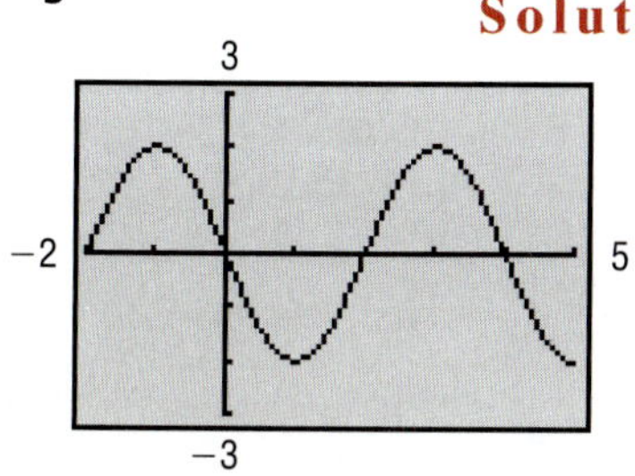

Solution The graph is sinusoidal, with amplitude $|A| = 2$. The period is 4, so $\frac{2\pi}{\omega} = 4$ or $\omega = \frac{\pi}{2}$. Since the graph passes through the origin, it is easiest to view the equation as a sine function, but notice that the graph is actually the reflection of a sine function about the x-axis (since the graph is decreasing near the origin). As a result, $A = -2$ and we have

$$y = A\sin(\omega x) = -2\sin\left(\frac{\pi}{2}x\right)$$

*The equation could also be viewed as a cosine function with a horizontal shift, but viewing it as a sine function is easier.

✔ CHECK: Graph $Y_1 = -2\sin\left(\frac{\pi}{2}x\right)$ and compare the result with Figure 95.

NOW WORK PROBLEMS 63 AND 67.

2.6 Concepts and Vocabulary

In Problems 1–3, fill in the blanks.

1. The maximum value of $y = \sin x$ is ________ and occurs at $x =$ ________.
2. The function $y = A\sin(\omega x)$, $A > 0$, has amplitude 3 and period 2; then $A =$ ________ and $\omega =$ ________.
3. The function $y = 3\cos(6x)$ has amplitude ________ and period ________.

In Problems 4–6, answer True or False to each statement.

4. The graphs of $y = \sin x$ and $y = \cos x$ are identical except for a horizontal shift.
5. For $y = 2\sin(\pi x)$, the amplitude is 2 and the period is $\frac{\pi}{2}$.
6. The graph of the sine function has infinitely many x-intercepts.

7. Draw a quick sketch of $y = \sin x$. Be sure to label at least five points.
8. Explain how you would scale the x-axis and the y-axis before graphing $y = 3\cos(\pi x)$.
9. Explain the term *amplitude* as it relates to the graph of a sinusoidal function.

2.6 Exercises

In Problems 1–10, if necessary, refer to the graphs to answer each question.

1. What is the y-intercept of $y = \sin x$?
2. What is the y-intercept of $y = \cos x$?
3. For what numbers x, $-\pi \le x \le \pi$, is the graph of $y = \sin x$ increasing?
4. For what numbers x, $-\pi \le x \le \pi$, is the graph of $y = \cos x$ decreasing?
5. What is the largest value of $y = \sin x$?
6. What is the smallest value of $y = \cos x$?
7. For what numbers x, $0 \le x \le 2\pi$, does $\sin x = 0$?
8. For what numbers x, $0 \le x \le 2\pi$, does $\cos x = 0$?
9. For what numbers x, $-2\pi \le x \le 2\pi$, does $\sin x = 1$? What about $\sin x = -1$?
10. For what numbers x, $-2\pi \le x \le 2\pi$, does $\cos x = 1$? What about $\cos x = -1$?

In Problems 11 and 12, match the graph to a function. Three answers are possible.

A. $y = -\sin x$ B. $y = -\cos x$ C. $y = \sin\left(x - \frac{\pi}{2}\right)$

D. $y = -\cos\left(x - \frac{\pi}{2}\right)$ E. $y = \sin(x + \pi)$ F. $y = \cos(x + \pi)$

11.

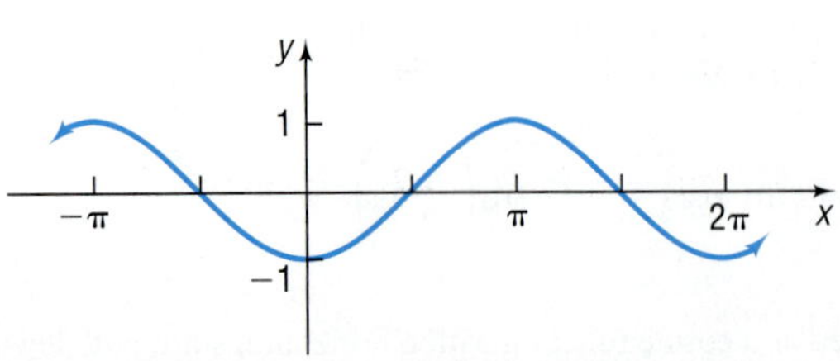

12.

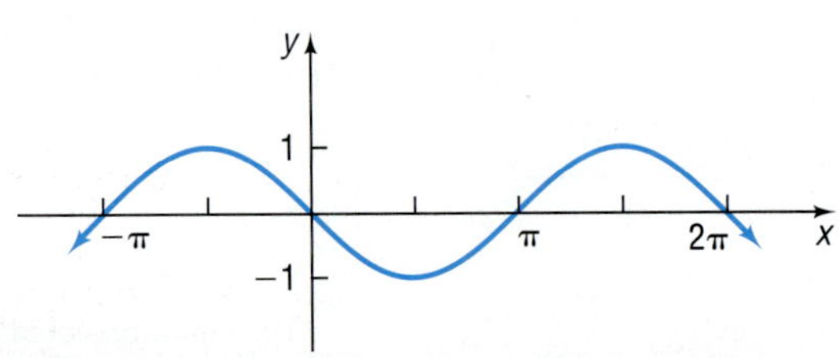

In Problems 13–28, use transformations to graph each function.

13. $y = 3\sin x$ **14.** $y = 4\cos x$ **15.** $y = \cos\left(x + \frac{\pi}{4}\right)$ **16.** $y = \sin(x - \pi)$

17. $y = \sin x - 1$ **18.** $y = \cos x + 1$ **19.** $y = -2\sin x$ **20.** $y = -3\cos x$

21. $y = \sin(\pi x)$ **22.** $y = \cos\left(\frac{\pi}{2}x\right)$ **23.** $y = 2\sin x + 2$ **24.** $y = 3\cos x + 3$

25. $y = -2\cos\left(x - \frac{\pi}{2}\right)$ **26.** $y = -3\sin\left(x + \frac{\pi}{2}\right)$ **27.** $y = 3\sin(\pi - x)$ **28.** $y = 2\cos(\pi - x)$

In Problems 29–38, determine the amplitude and period of each function without graphing.

29. $y = 2\sin x$ **30.** $y = 3\cos x$ **31.** $y = -4\cos(2x)$ **32.** $y = -\sin\left(\frac{1}{2}x\right)$

33. $y = 6\sin(\pi x)$ **34.** $y = -3\cos(3x)$ **35.** $y = -\frac{1}{2}\cos\left(\frac{3}{2}x\right)$ **36.** $y = \frac{4}{3}\sin\left(\frac{2}{3}x\right)$

37. $y = \frac{5}{3}\sin\left(-\frac{2\pi}{3}x\right)$ **38.** $y = \frac{9}{5}\cos\left(-\frac{3\pi}{2}x\right)$

In Problems 39–48, match the given function to one of the graphs (A)–(J).

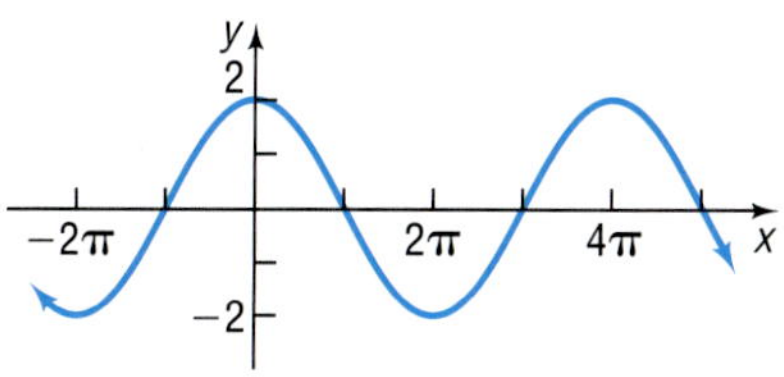

(A)

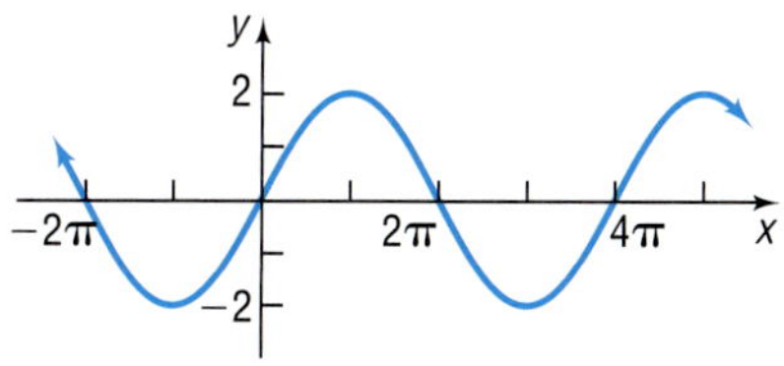

(B)

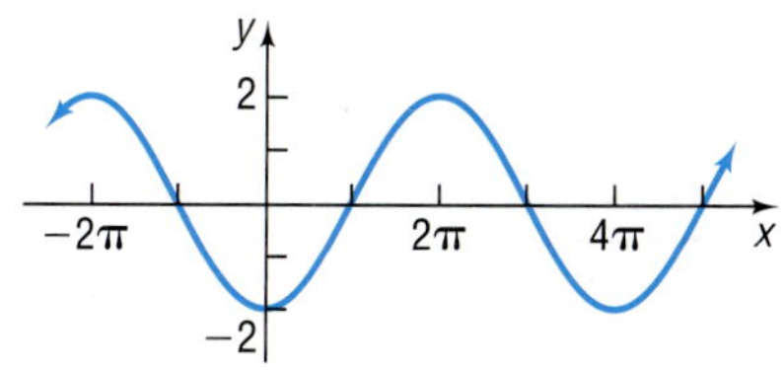

(C)

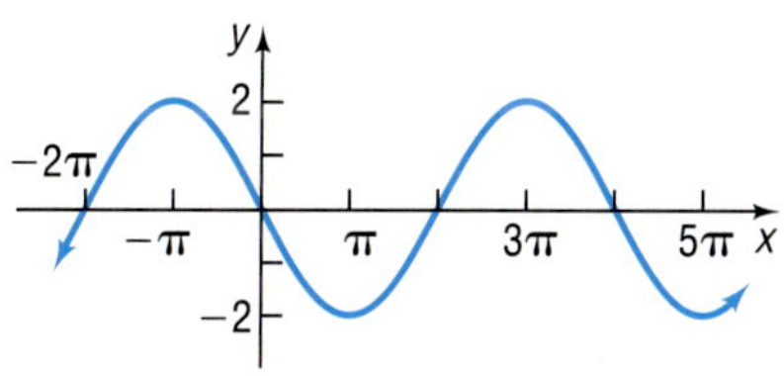

(D)

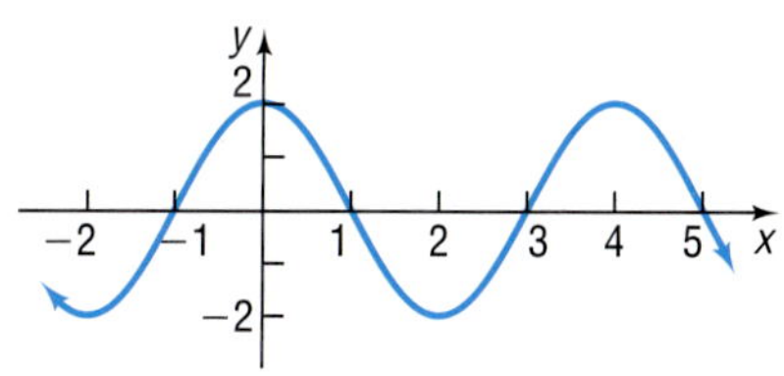

(E)

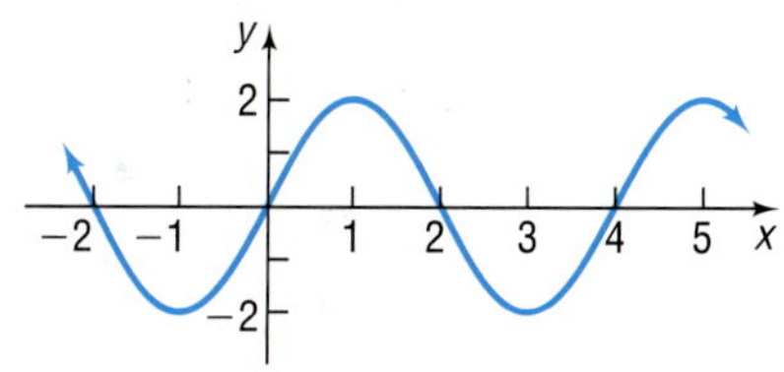

(F)

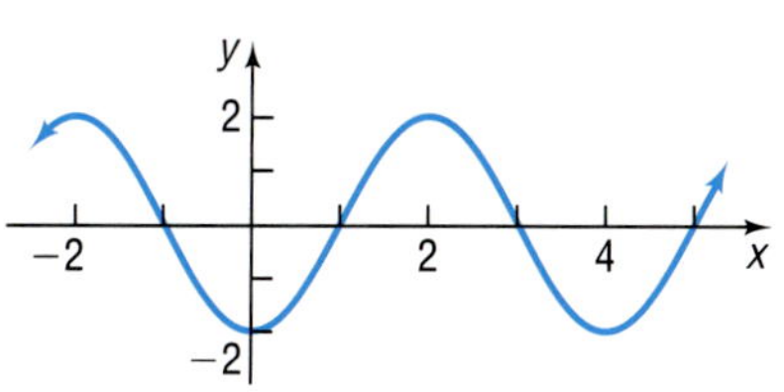

(G)

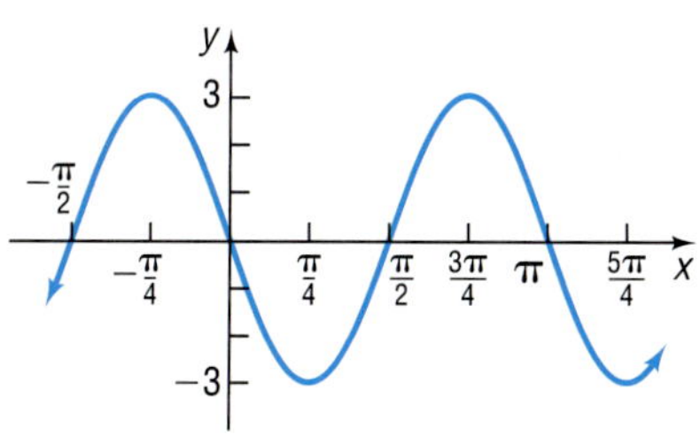

(H)

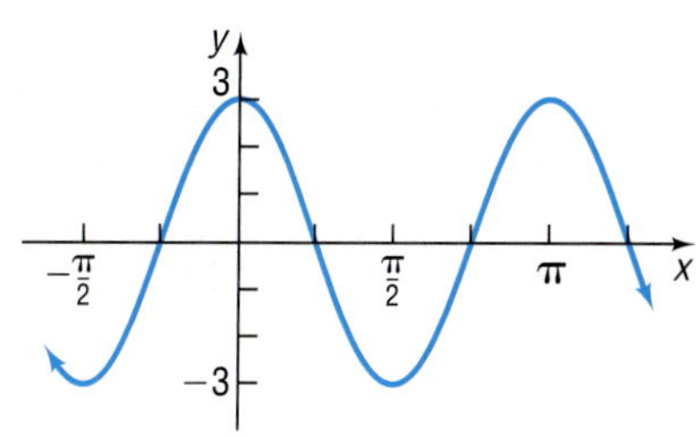

(I)

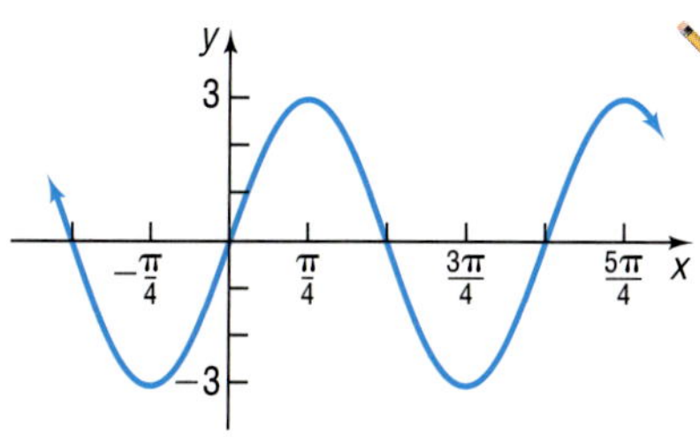

(J)

39. $y = 2\sin\left(\frac{\pi}{2}x\right)$ **40.** $y = 2\cos\left(\frac{\pi}{2}x\right)$ **41.** $y = 2\cos\left(\frac{1}{2}x\right)$

42. $y = 3\cos(2x)$ **43.** $y = -3\sin(2x)$ **44.** $y = 2\sin\left(\frac{1}{2}x\right)$

45. $y = -2\cos\left(\frac{1}{2}x\right)$ **46.** $y = -2\cos\left(\frac{\pi}{2}x\right)$ **47.** $y = 3\sin(2x)$

48. $y = -2\sin\left(\frac{1}{2}x\right)$

In Problems 49–52, match the given function to one of the graphs (A)–(D).

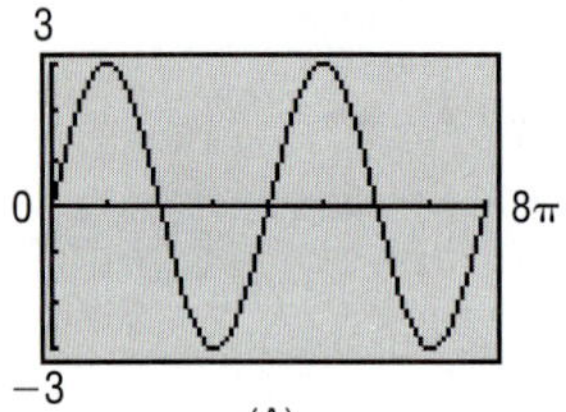

(A)

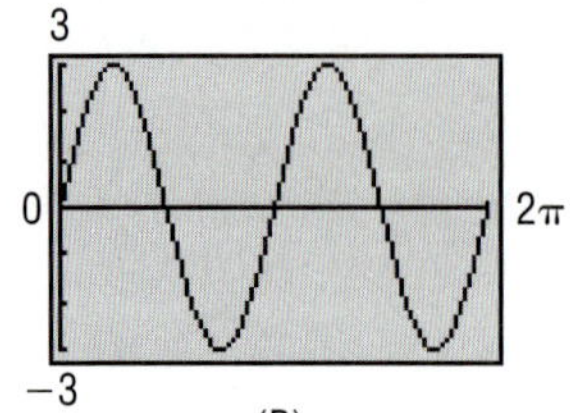

(B)

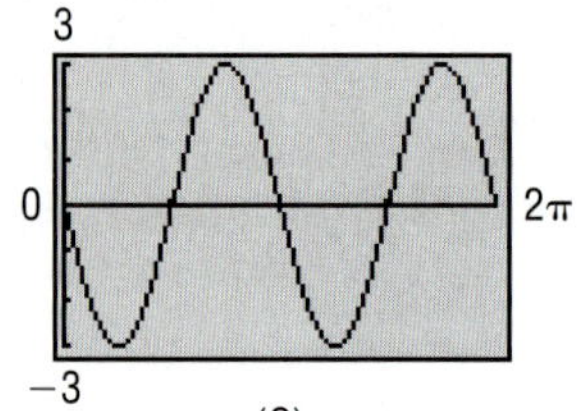

(C)

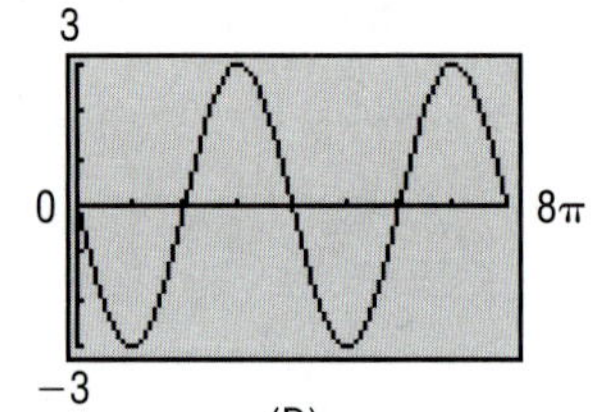

(D)

49. $y = 3\sin\left(\frac{1}{2}x\right)$

50. $y = -3\sin(2x)$

51. $y = 3\sin(2x)$

52. $y = -3\sin\left(\frac{1}{2}x\right)$

In Problems 53–62, graph each sinusoidal function.

53. $y = 5\sin(4x)$

54. $y = 4\cos(6x)$

55. $y = 5\cos(\pi x)$

56. $y = 2\sin(\pi x)$

57. $y = -2\cos(2\pi x)$

58. $y = -5\cos(2\pi x)$

59. $y = -4\sin\left(\frac{1}{2}x\right)$

60. $y = -2\cos\left(\frac{1}{2}x\right)$

61. $y = \frac{3}{2}\sin\left(-\frac{2}{3}x\right)$

62. $y = \frac{4}{3}\cos\left(-\frac{1}{3}x\right)$

In Problems 63–66, write the equation of a sine function $y = A\sin(\omega x)$, $A > 0$, that has the given characteristics.

63. Amplitude: 3
Period: π

64. Amplitude: 2
Period: 4π

65. Amplitude: 3
Period: 2

66. Amplitude: 4
Period: 1

In Problems 67–78, find an equation for each graph.

67.

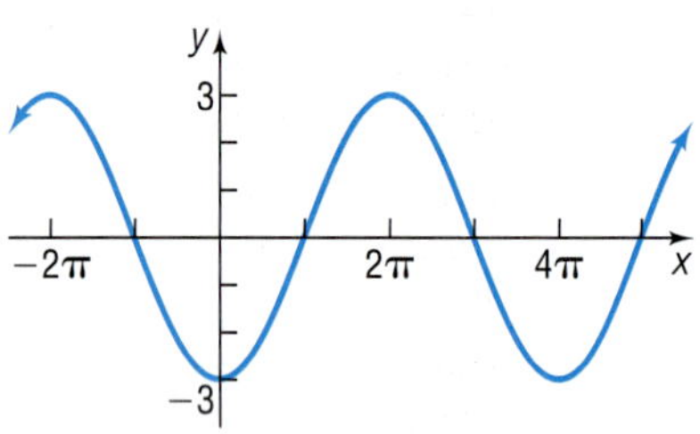

68.

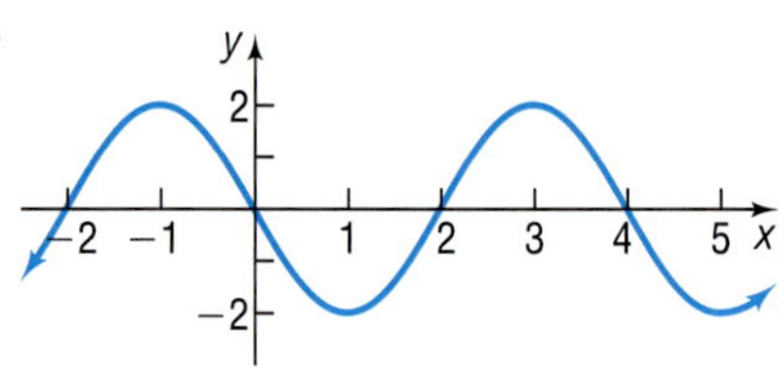

69.

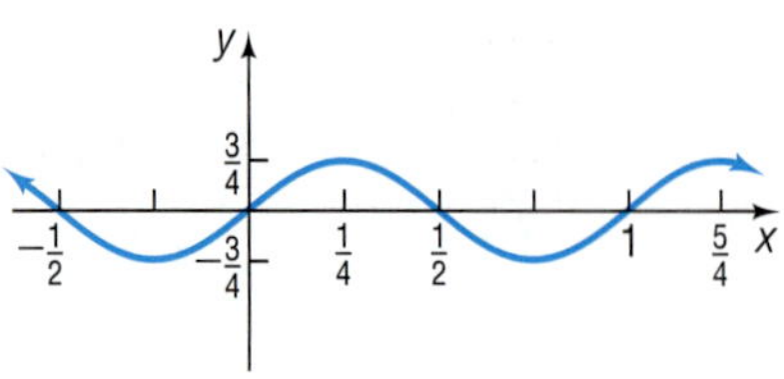

70.

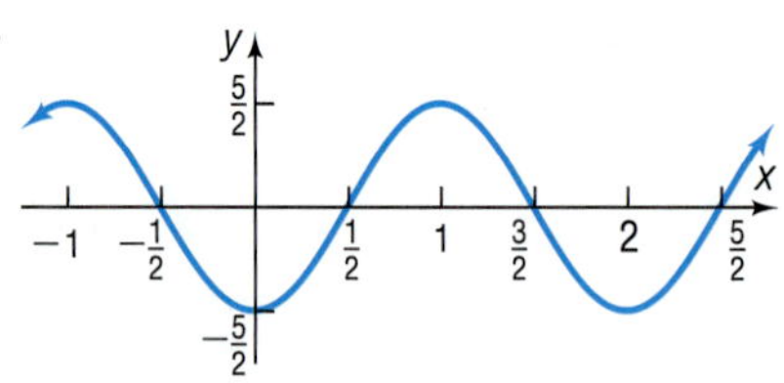

71.

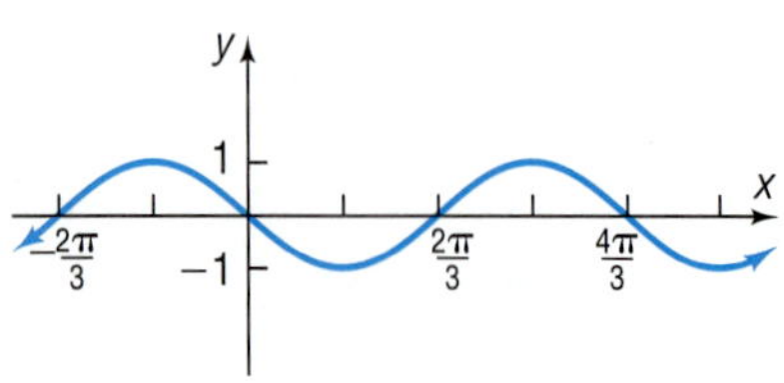

72.

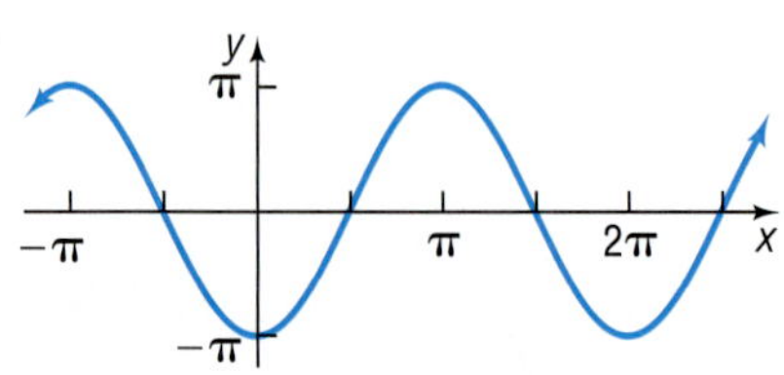

73.

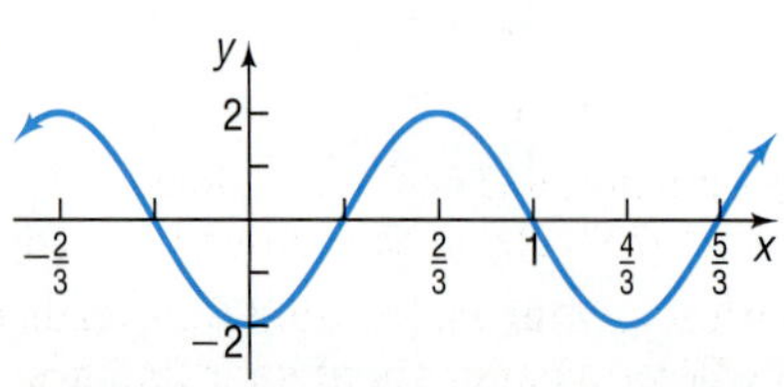

74.

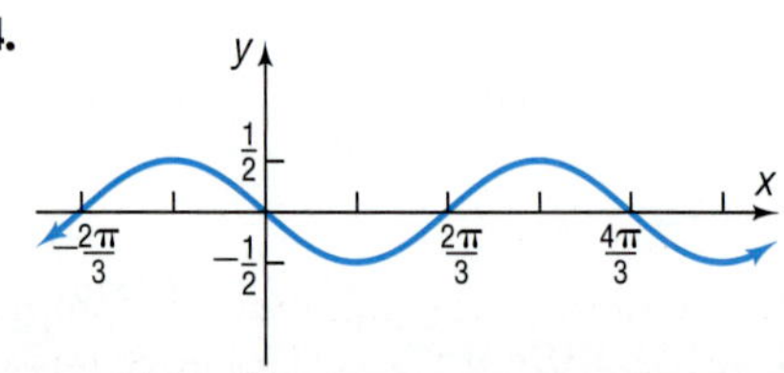

75.

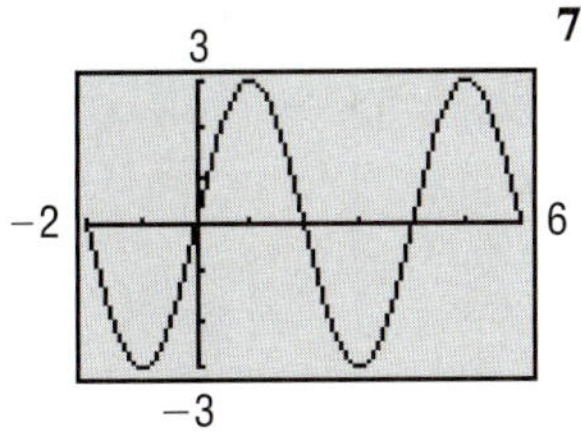

76.

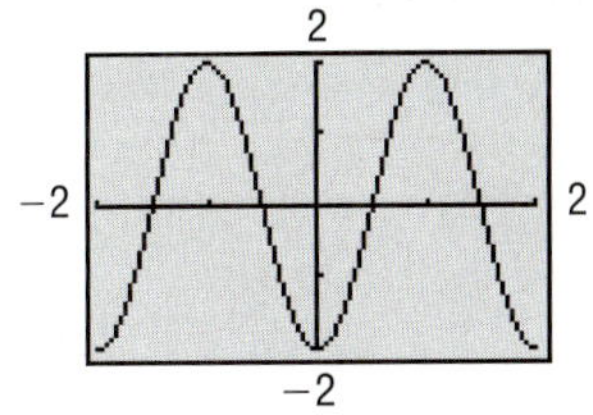

77.

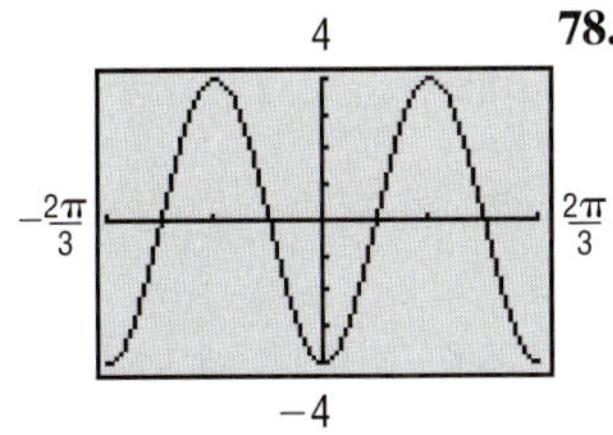

78.

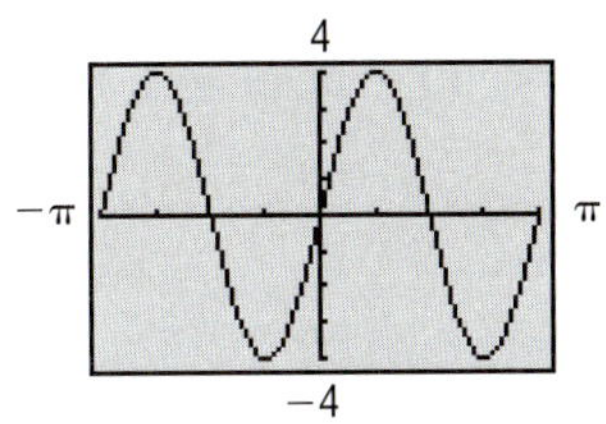

79. Alternating Current (ac) Circuits The current I, in amperes, flowing through an ac (alternating current) circuit at time t is

$$I = 220\sin(60\pi t), \qquad t \geq 0$$

What is the period? What is the amplitude? Graph this function over two periods.

80. Alternating Current (ac) Circuits The current I, in amperes, flowing through an ac (alternating current) circuit at time t is

$$I = 120\sin(30\pi t), \qquad t \geq 0$$

What is the period? What is the amplitude? Graph this function over two periods.

81. Alternating Current (ac) Generators The voltage V produced by an ac generator is

$$V = 220\sin(120\pi t)$$

(a) What is the amplitude? What is the period?
(b) Graph V over two periods, beginning at $t = 0$.
(c) If a resistance of $R = 10$ ohms is present, what is the current I?
[**Hint:** Use Ohm's Law, $V = IR$.]
(d) What is the amplitude and period of the current I?
(e) Graph I over two periods, beginning at $t = 0$.

82. Alternating Current (ac) Generators The voltage V produced by an ac generator is

$$V = 120\sin(120\pi t)$$

(a) What is the amplitude? What is the period?
(b) Graph V over two periods, beginning at $t = 0$.
(c) If a resistance of $R = 20$ ohms is present, what is the current I?
[**Hint:** Use Ohm's Law, $V = IR$.]
(d) What is the amplitude and period of the current I?
(e) Graph I over two periods, beginning at $t = 0$.

83. Alternating Current (ac) Generators The voltage V produced by an ac generator is sinusoidal. As a function of time, the voltage V is

$$V = V_0\sin(2\pi f t)$$

where f is the **frequency**, the number of complete oscillations (cycles) per second. [In the United States and Canada, f is 60 hertz (Hz).] The **power** P delivered to a resistance R at any time t is defined as

$$P = \frac{V^2}{R}$$

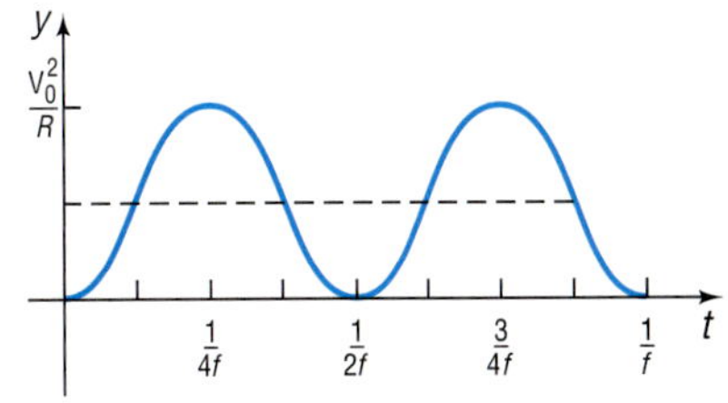

Power in an ac generator

(a) Show that $P = \frac{V_0^2}{R}\sin^2(2\pi f t)$.
(b) The graph of P is shown in the figure. Express P as a sinusoidal function.
(c) Deduce that

$$\sin^2(2\pi f t) = \frac{1}{2}[1 - \cos(4\pi f t)]$$

84. Biorhythms In the theory of biorhythms, a sine function of the form

$$P = 50\sin(\omega t) + 50$$

is used to measure the percent P of a person's potential at time t, where t is measured in days and $t = 0$ is the person's birthday. Three characteristics are commonly measured:

Physical potential: period of 23 days
Emotional potential: period of 28 days
Intellectual potential: period of 33 days

(a) Find ω for each characteristic.
(b) Use a graphing utility to graph all three functions.
(c) Is there a time t when all three characteristics have 100% potential? When is it?
(d) Suppose that you are 20 years old today ($t = 7305$ days). Describe your physical, emotional, and intellectual potential for the next 30 days.

85. Graph $y = |\cos x|, -2\pi \leq x \leq 2\pi$.

86. Graph $y = |\sin x|, -2\pi \leq x \leq 2\pi$.

87. Explain how the amplitude and period of a sinusoidal graph are used to establish the scale on each coordinate axis.

88. Find an application in your major field that leads to a sinusoidal graph. Write a paper about your findings.

PREPARING FOR THIS SECTION

Before getting started, review the following:

- ✓ Functions (Section 1.4, pp. 41–44)
- ✓ Graphing Techniques: Transformations (Section 1.6, pp. 64–73)

2.7 GRAPHS OF THE TANGENT, COTANGENT, COSECANT, AND SECANT FUNCTIONS

OBJECTIVES

1. Graph Transformations of the Tangent Function and Cotangent Function
2. Graph Transformations of the Cosecant Function and Secant Function

The Graphs of $y = \tan x$ and $y = \cot x$

1 Because the tangent function has period π, we only need to determine the graph over some interval of length π. The rest of the graph will consist of repetitions of that graph. Because the tangent function is not defined at $\dots, -\frac{3\pi}{2}, -\frac{\pi}{2}, \frac{\pi}{2}, \frac{3\pi}{2}, \dots$, we will concentrate on the interval $\left(-\frac{\pi}{2}, \frac{\pi}{2}\right)$, of length π, and construct Table 9, which lists some points on the graph of $y = \tan x$, $-\frac{\pi}{2} < x < \frac{\pi}{2}$. We plot the points in the table and connect them with a smooth curve. See Figure 96 for a partial graph of $y = \tan x$, where $-\frac{\pi}{3} \le x \le \frac{\pi}{3}$.

TABLE 9

x	$y = \tan x$	(x, y)
$-\frac{\pi}{3}$	$-\sqrt{3}$	$\left(-\frac{\pi}{3}, -\sqrt{3}\right)$
$-\frac{\pi}{4}$	-1	$\left(-\frac{\pi}{4}, -1\right)$
$-\frac{\pi}{6}$	$-\frac{\sqrt{3}}{3}$	$\left(-\frac{\pi}{6}, -\frac{\sqrt{3}}{3}\right)$
0	0	$(0, 0)$
$\frac{\pi}{6}$	$\frac{\sqrt{3}}{3}$	$\left(\frac{\pi}{6}, \frac{\sqrt{3}}{3}\right)$
$\frac{\pi}{4}$	1	$\left(\frac{\pi}{4}, 1\right)$
$\frac{\pi}{3}$	$\sqrt{3}$	$\left(\frac{\pi}{3}, \sqrt{3}\right)$

Figure 96

$y = \tan x$, $-\frac{\pi}{3} \le x \le \frac{\pi}{3}$

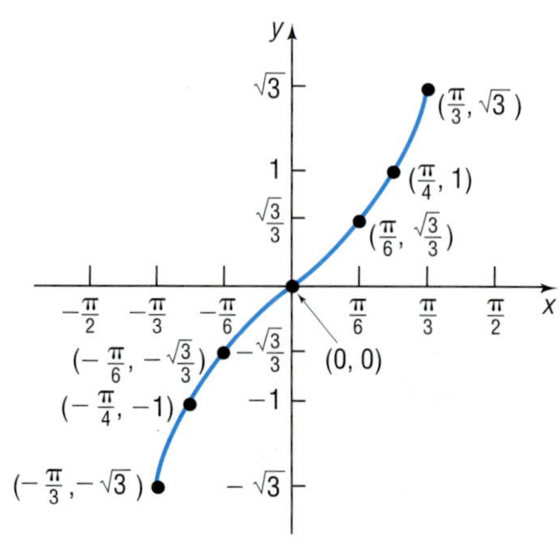

To complete one period of the graph of $y = \tan x$, we need to investigate the behavior of the function as x approaches $-\frac{\pi}{2}$ and $\frac{\pi}{2}$. We must be careful, though, because $y = \tan x$ is not defined at these numbers. To determine this behavior, we use the identity

$$\tan x = \frac{\sin x}{\cos x}$$

If x is close to $\frac{\pi}{2} \approx 1.5708$, but remains less than $\frac{\pi}{2}$, then $\sin x$ will be close to 1 and $\cos x$ will be positive and close to 0. (Refer back to the graphs of the sine function and the cosine function.) Hence, the ratio $\frac{\sin x}{\cos x}$ will be positive and large. In fact, the closer x gets to $\frac{\pi}{2}$, the closer $\sin x$ gets to 1 and $\cos x$ gets to 0, so $\tan x$ approaches ∞ $\left(\lim_{x \to \frac{\pi}{2}^-} \tan x = \infty\right)$. In other words, the vertical line $x = \frac{\pi}{2}$ is a vertical asymptote to the graph of $y = \tan x$. See Table 10.

TABLE 10

	X	Y1
	1.5	14.101
	1.57	1255.8
	1.5707	10381
$\frac{\pi}{2}$ →	1.5708	ERROR
	-1.5	-14.1
	-1.57	-1256
$-\frac{\pi}{2}$ →	-1.571	ERROR

Y1 = tan(X)

If x is close to $-\frac{\pi}{2}$, but remains greater than $-\frac{\pi}{2}$, then $\sin x$ will be close to -1 and $\cos x$ will be positive and close to 0. Hence, the ratio $\frac{\sin x}{\cos x}$ approaches $-\infty$ $\left(\lim_{x \to -\frac{\pi}{2}^+} \tan x = -\infty\right)$. In other words, the vertical line $x = -\frac{\pi}{2}$ is also a vertical asymptote to the graph.

With these observations, we can complete one period of the graph. We obtain the complete graph of $y = \tan x$ by repeating this period, as shown in Figure 97(a).

Figure 97(b) shows the graph of $y = \tan x$, $-\infty < x < \infty$ using a graphing utility. Notice we used dot mode when graphing $y = \tan x$. Do you know why?

Figure 97
$y = \tan x$, $-\infty < x < \infty$, x not equal to odd multiples of $\frac{\pi}{2}$

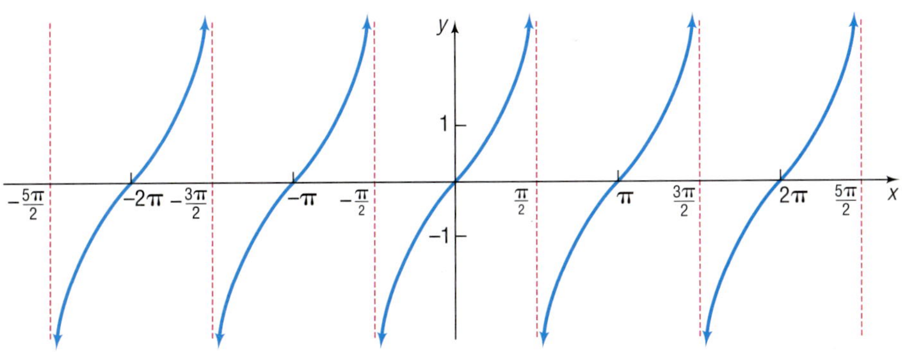

(a)

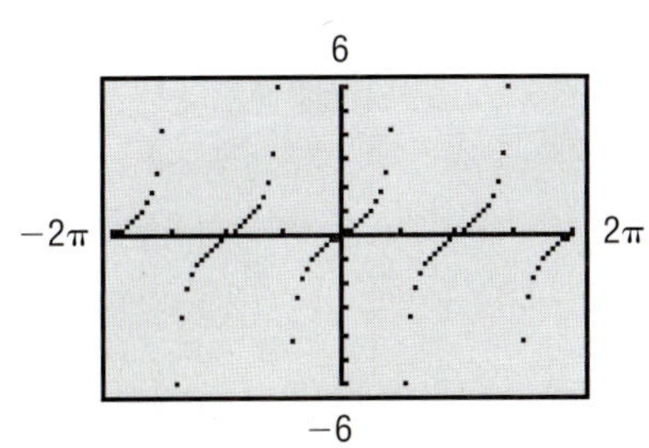

(b)

The graph of $y = \tan x$ illustrates some facts that we already know about the tangent function.

Properties of the Tangent Function

1. The domain is the set of all real numbers, except odd multiples of $\frac{\pi}{2}$.
2. The range consists of all real numbers.
3. The tangent function is an odd function, as the symmetry of the graph with respect to the origin indicates.
4. The tangent function is periodic, with period π.
5. The x-intercepts are $\dots, -2\pi, -\pi, 0, \pi, 2\pi, 3\pi, \dots$; the y-intercept is 0.
6. Vertical asymptotes occur at $x = \dots, -\frac{3\pi}{2}, -\frac{\pi}{2}, \frac{\pi}{2}, \frac{3\pi}{2}, \dots$.

NOW WORK PROBLEMS 1 AND 9.

EXAMPLE 1 Graphing Variations of $y = \tan x$ Using Transformations

Graph: $y = 2\tan x$

Solution We start with the graph of $y = \tan x$ and vertically stretch it by a factor of 2. See Figure 98.

Figure 98

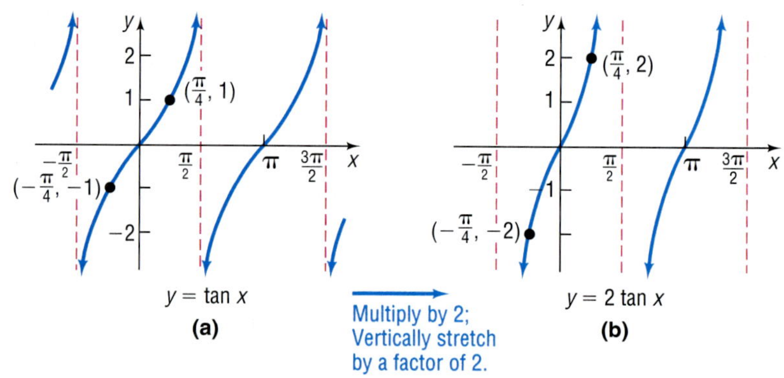

Figure 99 shows the graph using a graphing utility.

Figure 99

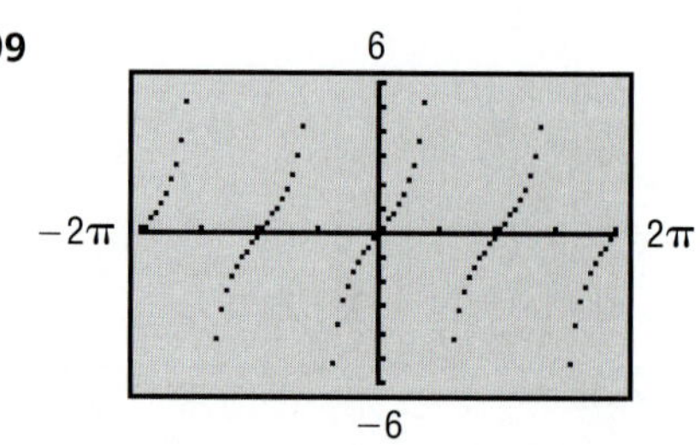

EXAMPLE 2 Graphing Variations of $y = \tan x$ Using Transformations

Graph: $y = -\tan\left(x + \frac{\pi}{4}\right)$

Solution We start with the graph of $y = \tan x$. See Figure 100.

Figure 100

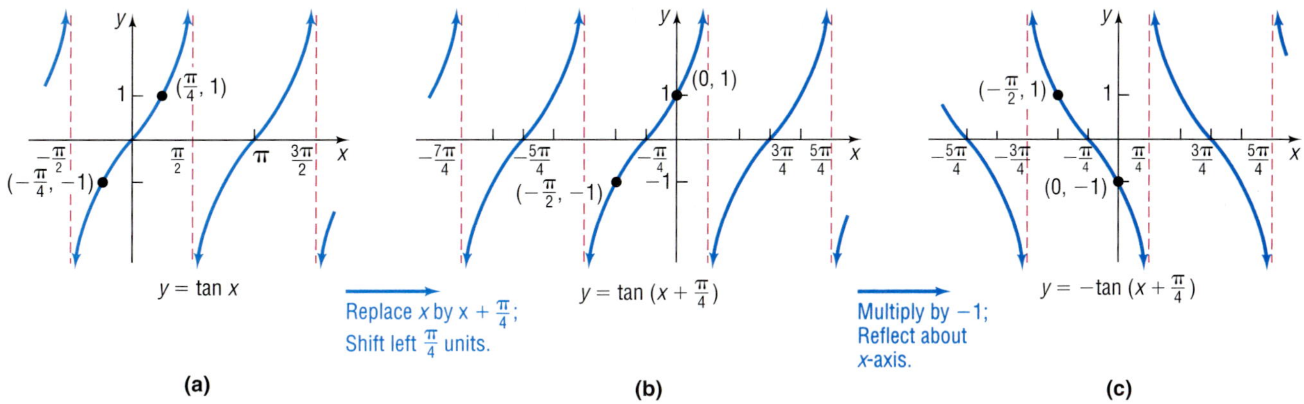

✔ CHECK: Graph $Y_1 = -\tan\left(x + \dfrac{\pi}{4}\right)$ and compare the result to Figure 100(c).

NOW WORK PROBLEM 19.

TABLE 11

x	$y = \cot x$	(x, y)
$\frac{\pi}{6}$	$\sqrt{3}$	$\left(\frac{\pi}{6}, \sqrt{3}\right)$
$\frac{\pi}{4}$	1	$\left(\frac{\pi}{4}, 1\right)$
$\frac{\pi}{3}$	$\frac{\sqrt{3}}{3}$	$\left(\frac{\pi}{3}, \frac{\sqrt{3}}{3}\right)$
$\frac{\pi}{2}$	0	$\left(\frac{\pi}{2}, 0\right)$
$\frac{2\pi}{3}$	$-\frac{\sqrt{3}}{3}$	$\left(\frac{2\pi}{3}, -\frac{\sqrt{3}}{3}\right)$
$\frac{3\pi}{4}$	-1	$\left(\frac{3\pi}{4}, -1\right)$
$\frac{5\pi}{6}$	$-\sqrt{3}$	$\left(\frac{5\pi}{6}, -\sqrt{3}\right)$

We obtain the graph of $y = \cot x$ as we did the graph of $y = \tan x$. The period of $y = \cot x$ is π. Because the cotangent function is not defined for integral multiples of π, we will concentrate on the interval $(0, \pi)$. Table 11 lists some points on the graph of $y = \cot x, 0 < x < \pi$. As x approaches 0, but remains greater than 0, the value of $\cos x$ will be close to 1 and the value of $\sin x$ will be positive and close to 0. Hence, the ratio $\dfrac{\cos x}{\sin x} = \cot x$ will be positive and large; so as x approaches 0, with $x > 0$, $\cot x$ approaches ∞ $\left(\lim_{x \to 0^+} \cot x = \infty\right)$. Similarly, as x approaches π, but remains less than π, the value of $\cos x$ will be close to -1, and the value of $\sin x$ will be positive and close to 0. Hence, the ratio $\dfrac{\cos x}{\sin x} = \cot x$ will be negative and will approach $-\infty$ as x approaches π $\left(\lim_{x \to \pi^-} \cot x = -\infty\right)$. Figure 101 shows the graph.

Figure 101
$y = \cot x, -\infty < x < \infty$, x not equal to integral multiples of π, $-\infty < y < \infty$

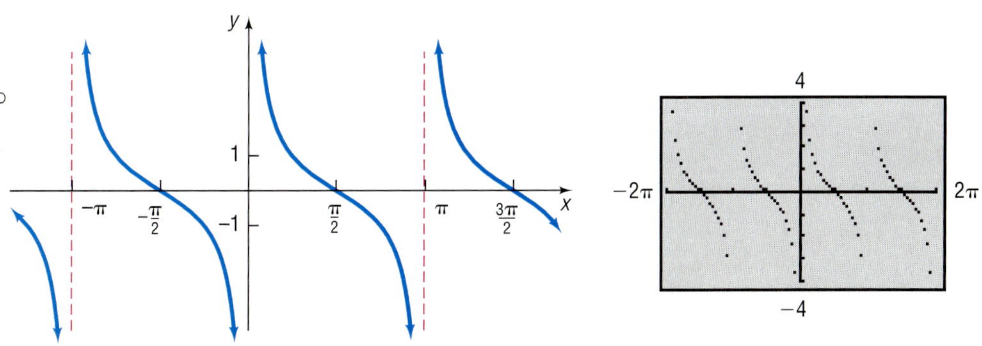

NOW WORK PROBLEM 25.

The Graphs of $y = \csc x$ and $y = \sec x$

2 The cosecant and secant functions, sometimes referred to as **reciprocal functions**, are graphed by making use of the reciprocal identities

$$\csc x = \frac{1}{\sin x} \quad \text{and} \quad \sec x = \frac{1}{\cos x}$$

For example, the value of the cosecant function $y = \csc x$ at a given number x equals the reciprocal of the corresponding value of the sine function, provided that the value of the sine function is not 0. If the value of $\sin x$ is 0, then, at such numbers x, (integral multiples of π) the cosecant function is not defined. In fact, the graph of the cosecant function has vertical asymptotes at integral multiples of π. Figure 102 shows the graph.

Figure 102
$y = \csc x$, $-\infty < x < \infty$, x not equal to integral multiples of π, $|y| \geq 1$

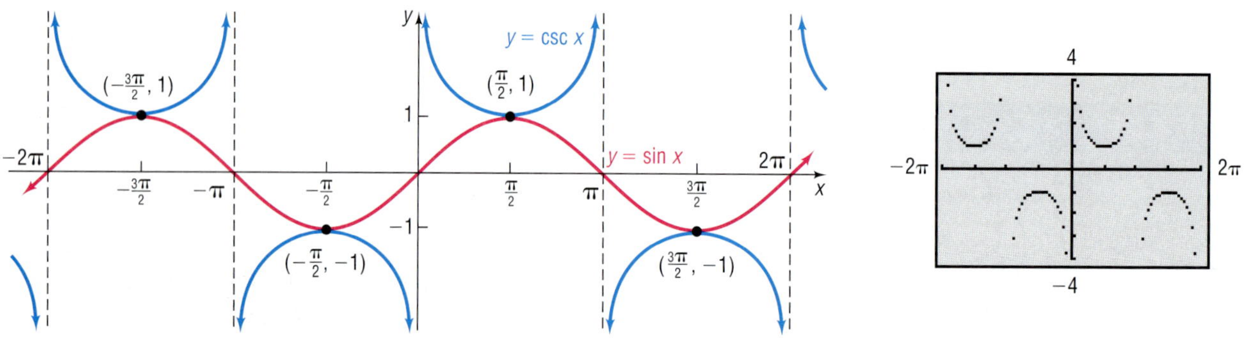

EXAMPLE 3 Graphing Variations of $y = \csc x$ Using Transformations

Graph: $y = 2\csc\left(x - \dfrac{\pi}{2}\right)$

Solution Figure 103 shows the required steps.

Figure 103

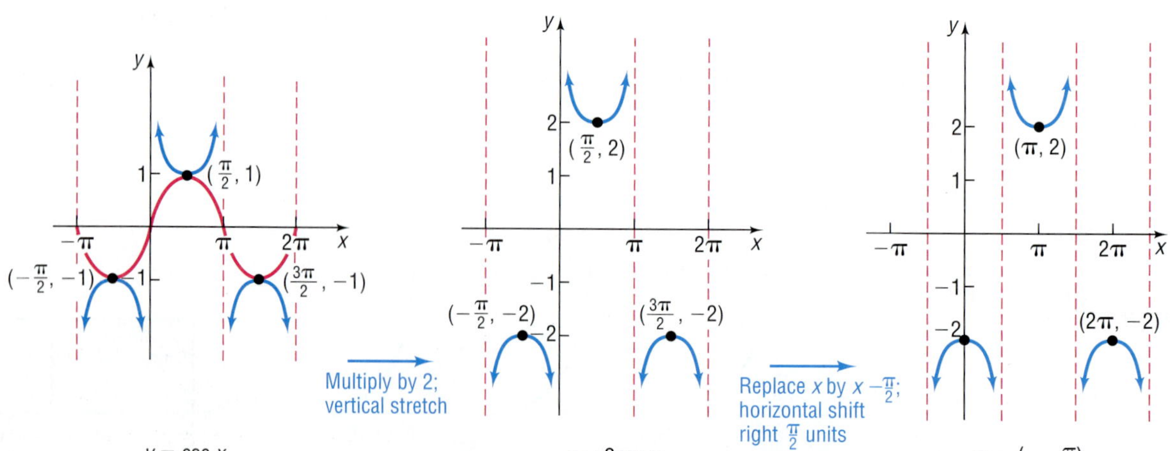

$y = \csc x$
(a)

$y = 2\csc x$
(b)

$y = 2\csc\left(x - \frac{\pi}{2}\right)$
(c)

✔ CHECK: Graph $Y_1 = 2\csc\left(x - \dfrac{\pi}{2}\right)$ and compare the result with Figure 103(c).

NOW WORK PROBLEM 31.

Using the idea of reciprocals, we can similarly obtain the graph of $y = \sec x$. See Figure 104.

Figure 104

$y = \sec x$, $-\infty < x < \infty$, x not equal to odd multiples of $\frac{\pi}{2}$, $|y| \geq 1$

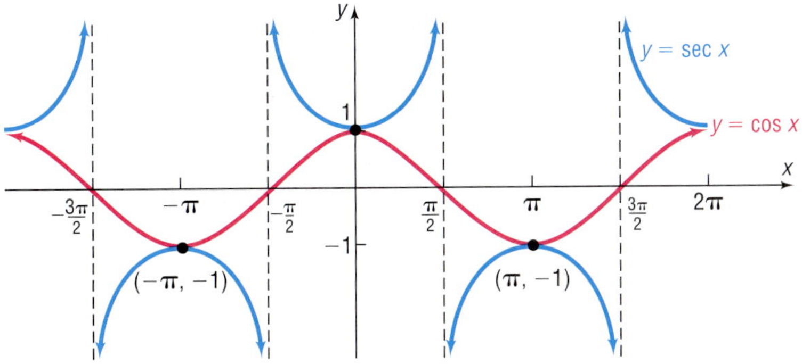

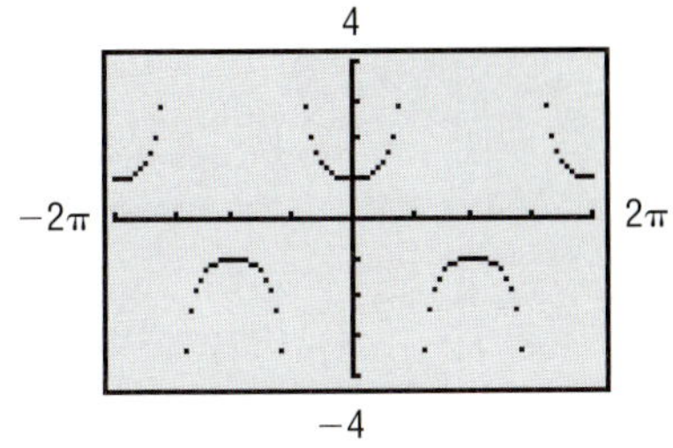

2.7 Concepts and Vocabulary

In Problems 1–3, fill in the blanks.

1. The graph of $y = \tan x$ is symmetric with respect to the ________ and has vertical asymptotes at ________.
2. The graph of $y = \sec x$ is symmetric with respect to the ________ and has vertical asymptotes at ________.
3. It is easiest to graph $y = \sec x$ by first sketching the graph of ________.

In Problem 4, answer True of False.

4. The graphs of $y = \tan x$, $y = \cot x$, $y = \sec x$, and $y = \csc x$ each have infinitely many vertical asymptotes.

5. Explain how you would graph $y = \csc x$ using the idea of a reciprocal function.
6. Sketch the graph of $y = \tan x$.

2.7 Exercises

In Problems 1–10, if necessary, refer to the graphs to answer each question.

1. What is the y-intercept of $y = \tan x$?
2. What is the y-intercept of $y = \cot x$?
3. What is the y-intercept of $y = \sec x$?
4. What is the y-intercept of $y = \csc x$?
5. For what numbers x, $-2\pi \leq x \leq 2\pi$, does $\sec x = 1$? What about $\sec x = -1$?
6. For what numbers x, $-2\pi \leq x \leq 2\pi$, does $\csc x = 1$? What about $\csc x = -1$?
7. For what numbers x, $-2\pi \leq x \leq 2\pi$, does the graph of $y = \sec x$ have vertical asymptotes?
8. For what numbers x, $-2\pi \leq x \leq 2\pi$, does the graph of $y = \csc x$ have vertical asymptotes?
9. For what numbers x, $-2\pi \leq x \leq 2\pi$, does the graph of $y = \tan x$ have vertical asymptotes?
10. For what numbers x, $-2\pi \leq x \leq 2\pi$, does the graph of $y = \cot x$ have vertical asymptotes?

In Problems 11–14, match each function to its graph.

A. $y = -\tan x$ B. $y = \tan\left(x + \frac{\pi}{2}\right)$ C. $y = \tan(x + \pi)$ D. $y = -\tan\left(x - \frac{\pi}{2}\right)$

11.

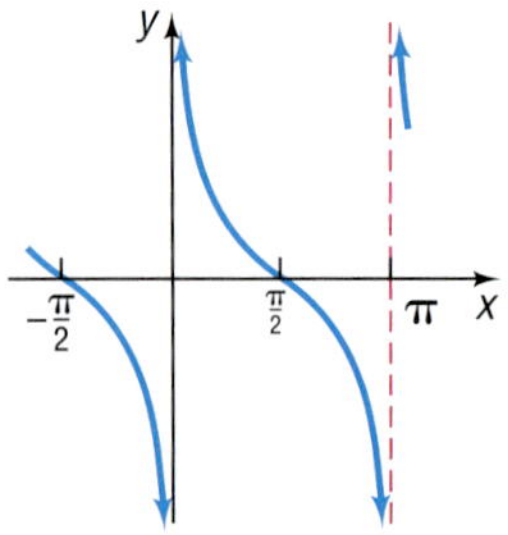

12.

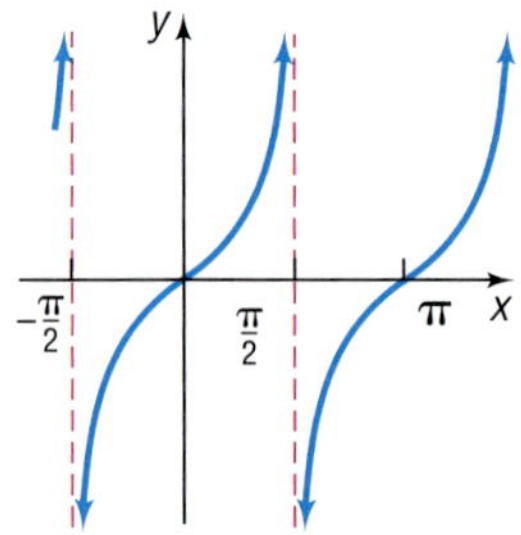

13.

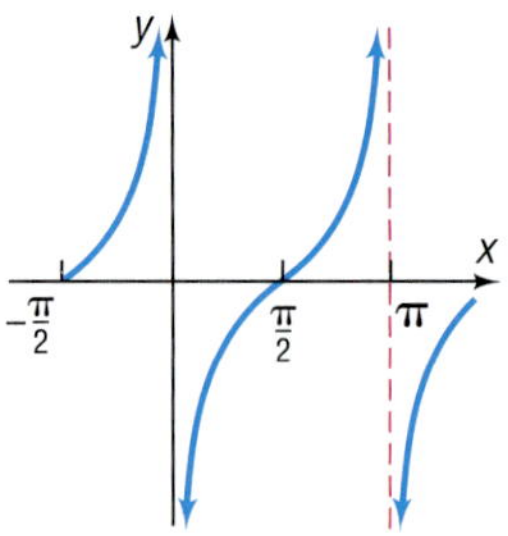

14.

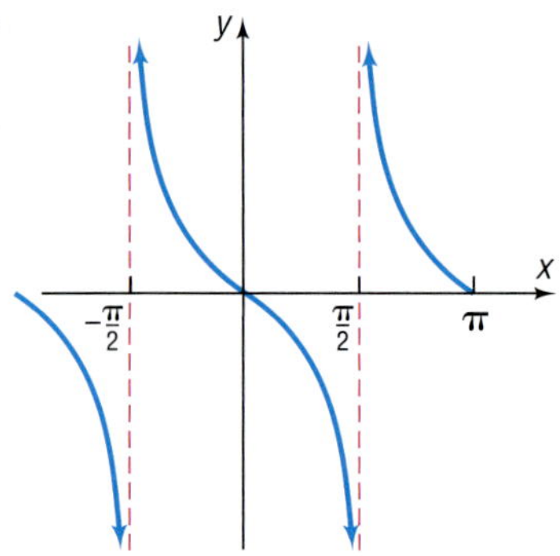

In Problems 15–34, use tranformations to graph each function. Verify your result using a graphing utility.

15. $y = -\sec x$ **16.** $y = -\cot x$ **17.** $y = \sec\left(x - \frac{\pi}{2}\right)$ **18.** $y = \csc(x - \pi)$

19. $y = \tan(x - \pi)$ **20.** $y = \cot(x - \pi)$ **21.** $y = 3\tan(2x)$ **22.** $y = 4\tan\left(\frac{1}{2}x\right)$

23. $y = \sec(2x)$ **24.** $y = \csc\left(\frac{1}{2}x\right)$ **25.** $y = \cot(\pi x)$ **26.** $y = \cot(2x)$

27. $y = -3\tan(4x)$ **28.** $y = -3\tan(2x)$ **29.** $y = 2\sec\left(\frac{1}{2}x\right)$ **30.** $y = 2\sec(3x)$

31. $y = -3\csc\left(x + \frac{\pi}{4}\right)$ **32.** $y = -2\tan\left(x + \frac{\pi}{4}\right)$ **33.** $y = \frac{1}{2}\cot\left(x - \frac{\pi}{4}\right)$ **34.** $y = 3\sec\left(x + \frac{\pi}{2}\right)$

---

35. Carrying a Ladder around a Corner A ladder of length L is carried horizontally around a corner from a hall 3 feet wide into a hall 4 feet wide. See the illustration below.

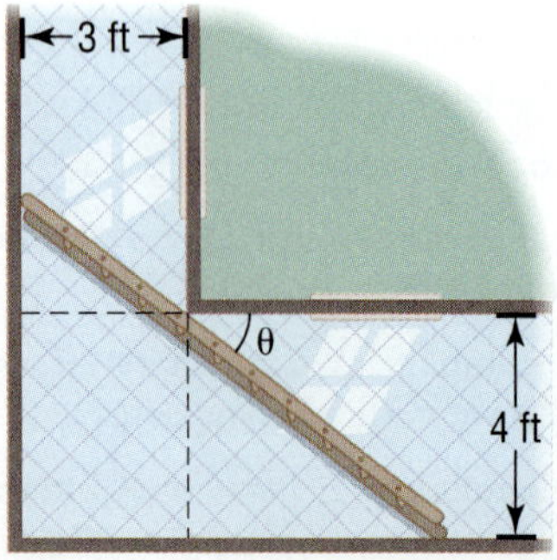

(a) Show that the length L of the ladder as a function of the angle θ is

$$L(\theta) = 3\sec\theta + 4\csc\theta$$

(b) Use a graphing utility to graph $L, 0 < \theta < \frac{\pi}{2}$.

(c) For what value of θ is L the least?

(d) What is the length of the longest ladder that can be carried around the corner? Why is this also the least value of L?

36. Exploration Graph

$$y = \tan x \quad \text{and} \quad y = -\cot\left(x + \frac{\pi}{2}\right)$$

Do you think that $\tan x = -\cot\left(x + \frac{\pi}{2}\right)$?

2.8 PHASE SHIFT; SINUSOIDAL CURVE FITTING

OBJECTIVES

1 Determine the Phase Shift of a Sinusoidal Function

2 Graph Sinusoidal Functions: $y = A\sin(\omega x - \phi)$

3 Find a Sinusoidal Function from Data

Phase Shift

1 We have seen that the graph of $y = A\sin(\omega x)$, $\omega > 0$, has amplitude $|A|$ and period $T = \frac{2\pi}{\omega}$. One cycle can be drawn as x varies from 0 to $\frac{2\pi}{\omega}$ or, equivalently, as ωx varies from 0 to 2π. See Figure 105.

Figure 105
One cycle
$y = A\sin(\omega x)$, $A > 0$, $\omega > 0$

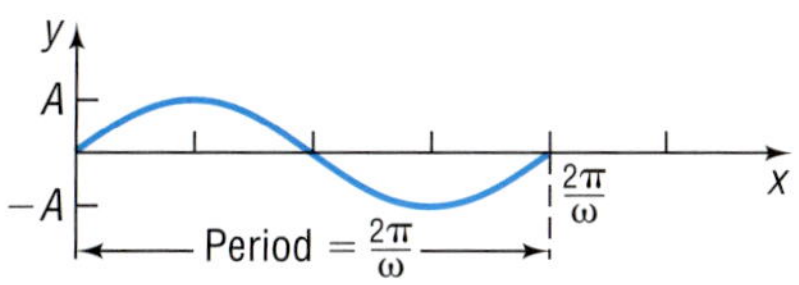

We now want to discuss the graph of

$$y = A\sin(\omega x - \phi) = A\sin\left[\omega\left(x - \frac{\phi}{\omega}\right)\right]$$

where $\omega > 0$ and ϕ (the Greek letter phi) are real numbers. The graph will be a sine curve of amplitude $|A|$. As $\omega x - \phi$ varies from 0 to 2π, one period will be traced out. This period will begin when

$$\omega x - \phi = 0 \quad \text{or} \quad x = \frac{\phi}{\omega}$$

and will end when

$$\omega x - \phi = 2\pi \quad \text{or} \quad x = \frac{2\pi}{\omega} + \frac{\phi}{\omega}$$

Figure 106
One cycle
$y = A\sin(\omega x - \phi)$, $A > 0$, $\omega > 0$, $\phi > 0$

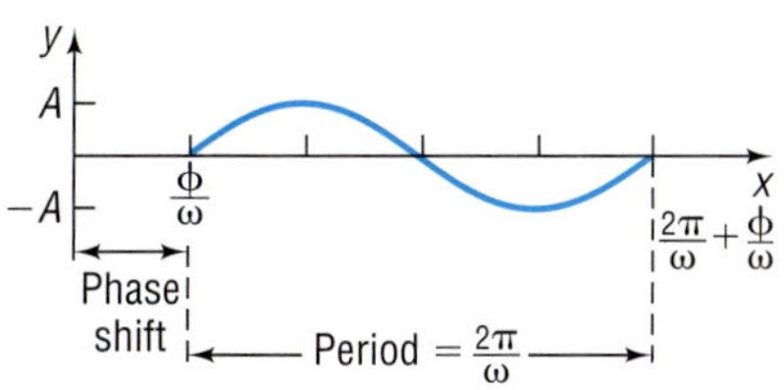

See Figure 106.

We see that the graph of $y = A\sin(\omega x - \phi) = A\sin\left[\omega\left(x - \frac{\phi}{\omega}\right)\right]$ is the same as the graph of $y = A\sin(\omega x)$, except that it has been shifted $\frac{\phi}{\omega}$ units (to the right if $\phi > 0$ and to the left if $\phi < 0$). This number $\frac{\phi}{\omega}$ is called the **phase shift** of the graph of $y = A\sin(\omega x - \phi)$.

For the graphs of $y = A\sin(\omega x - \phi)$ or $y = A\cos(\omega x - \phi)$, $\omega > 0$,

$$\text{Amplitude} = |A| \qquad \text{Period} = T = \frac{2\pi}{\omega} \qquad \text{Phase shift} = \frac{\phi}{\omega}$$

The phase shift is to the left if $\phi < 0$ and to the right if $\phi > 0$.

2 **EXAMPLE 1** **Finding the Amplitude, Period, and Phase Shift of a Sinusoidal Function and Graphing It**

Find the amplitude, period, and phase shift of $y = 3\sin(2x - \pi)$, and graph the function.

Solution Comparing

$$y = 3\sin(2x - \pi) = 3\sin\left[2\left(x - \frac{\pi}{2}\right)\right]$$

to

$$y = A\sin(\omega x - \phi) = A\sin\left[\omega\left(x - \frac{\phi}{\omega}\right)\right]$$

we find that $A = 3, \omega = 2$, and $\phi = \pi$. The graph is a sine curve with amplitude $|A| = 3$, period $T = \frac{2\pi}{\omega} = \frac{2\pi}{2} = \pi$, and phase shift $= \frac{\phi}{\omega} = \frac{\pi}{2}$.

The graph of $y = 3\sin(2x - \pi)$ will lie between -3 and 3 on the y-axis. One cycle will begin at $x = \frac{\phi}{\omega} = \frac{\pi}{2}$ and end at $x = \frac{2\pi}{\omega} + \frac{\phi}{\omega} = \pi + \frac{\pi}{2} = \frac{3\pi}{2}$. We divide the interval $\left[\frac{\pi}{2}, \frac{3\pi}{2}\right]$ into four subintervals, each of length $\pi \div 4 = \frac{\pi}{4}$:

$$\left[\frac{\pi}{2}, \frac{3\pi}{4}\right], \quad \left[\frac{3\pi}{4}, \pi\right], \quad \left[\pi, \frac{5\pi}{4}\right], \quad \left[\frac{5\pi}{4}, \frac{3\pi}{2}\right]$$

The five key points on the graph are

$$\left(\frac{\pi}{2}, 0\right), \quad \left(\frac{3\pi}{4}, 3\right), \quad (\pi, 0), \quad \left(\frac{5\pi}{4}, -3\right), \quad \left(\frac{3\pi}{2}, 0\right)$$

We plot these five points and fill in the graph of the sine function as shown in Figure 107(a). Extending the graph in either direction, we obtain Figure 107(b).

Figure 107

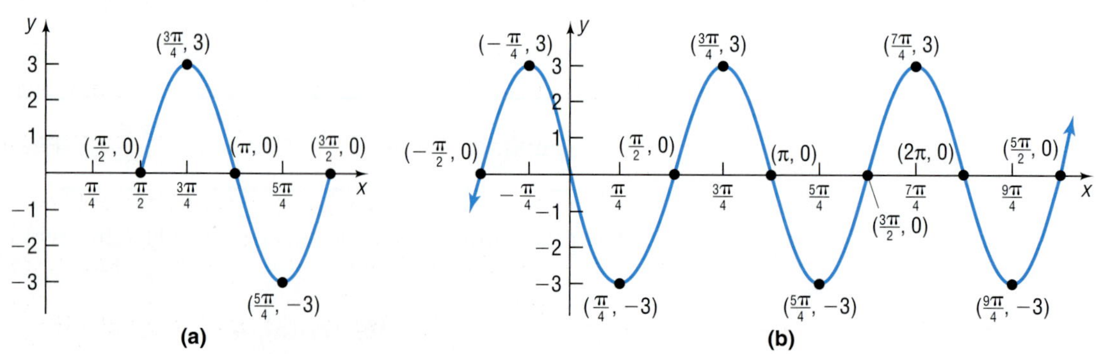

The graph of $y = 3\sin(2x - \pi) = 3\sin\left[2\left(x - \frac{\pi}{2}\right)\right]$ may also be obtained using transformations. See Figure 108.

Figure 108

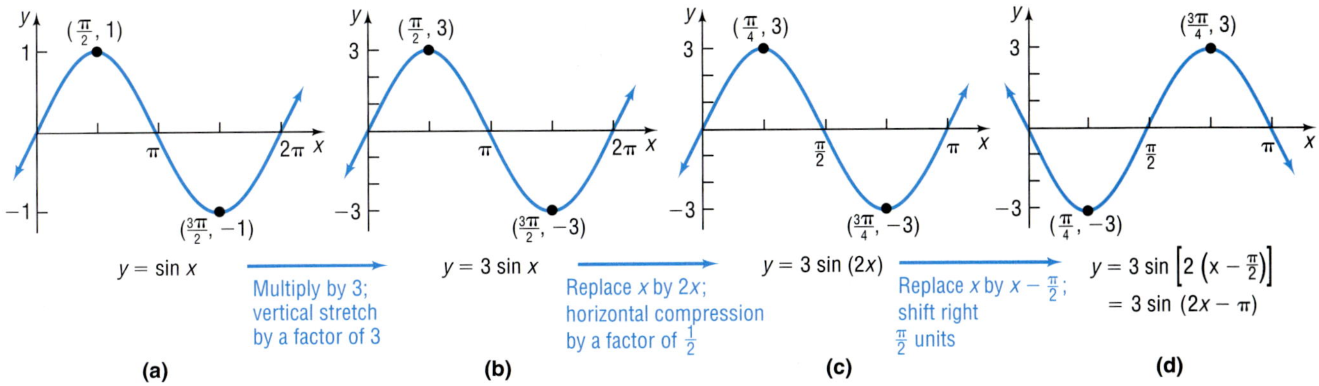

✔ CHECK: Figure 109 shows the graph of $Y_1 = 3\sin(2x - \pi)$ using a graphing utility.

Figure 109

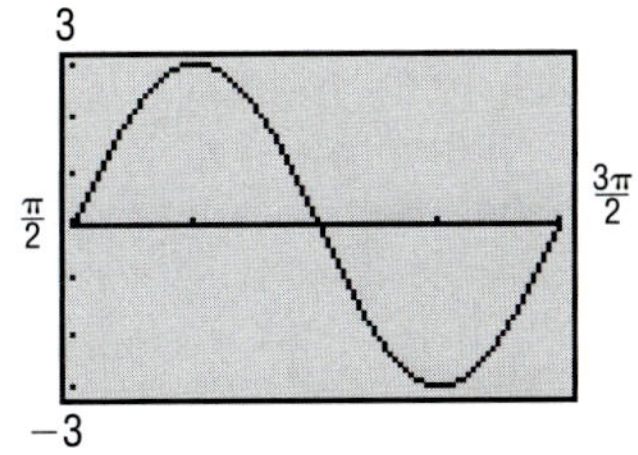

EXAMPLE 2 Finding the Amplitude, Period, and Phase Shift of a Sinusoidal Function and Graphing It

Find the amplitude, period, and phase shift of $y = 2\cos(4x + 3\pi)$, and graph the function.

Solution Comparing

$$y = 2\cos(4x + 3\pi) = 2\cos\left[4\left(x + \frac{3\pi}{4}\right)\right]$$

to

$$y = A\cos(\omega x - \phi) = A\cos\left[\omega\left(x - \frac{\phi}{\omega}\right)\right]$$

we see that $A = 2, \omega = 4$, and $\phi = -3\pi$. The graph is a cosine curve with amplitude $|A| = 2$, period $T = \frac{2\pi}{\omega} = \frac{2\pi}{4} = \frac{\pi}{2}$, and phase shift $= \frac{\phi}{\omega} = -\frac{3\pi}{4}$.

The graph of $y = 2\cos(4x + 3\pi)$ will lie between -2 and 2 on the y-axis. One cycle will begin at $x = \frac{\phi}{\omega} = -\frac{3\pi}{4}$ and end at $x = \frac{2\pi}{\omega} + \frac{\phi}{\omega} = \frac{\pi}{2} + \left(-\frac{3\pi}{4}\right) = -\frac{\pi}{4}$. We divide the interval $\left[-\frac{3\pi}{4}, -\frac{\pi}{4}\right]$ into four subintervals, each of the length $\frac{\pi}{2} \div 4 = \frac{\pi}{8}$:

$$\left[-\frac{3\pi}{4}, -\frac{5\pi}{8}\right], \quad \left[-\frac{5\pi}{8}, -\frac{\pi}{2}\right], \quad \left[-\frac{\pi}{2}, -\frac{3\pi}{8}\right], \quad \left[-\frac{3\pi}{8}, -\frac{\pi}{4}\right]$$

The five key points on the graph are

$$\left(-\frac{3\pi}{4}, 2\right), \left(-\frac{5\pi}{8}, 0\right), \left(-\frac{\pi}{2}, -2\right), \left(-\frac{3\pi}{8}, 0\right), \left(-\frac{\pi}{4}, 2\right)$$

We plot these five points and fill in the graph of the cosine function as shown in Figure 110(a). Extending the graph in either direction, we obtain Figure 110(b).

Figure 110

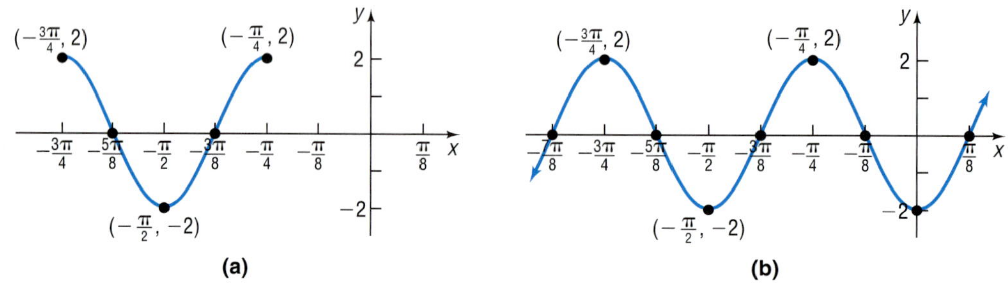

The graph of $y = 2\cos(4x + 3\pi) = 2\cos\left[4\left(x + \frac{3\pi}{4}\right)\right]$ may also be obtained using transformations. See Figure 111.

Figure 111

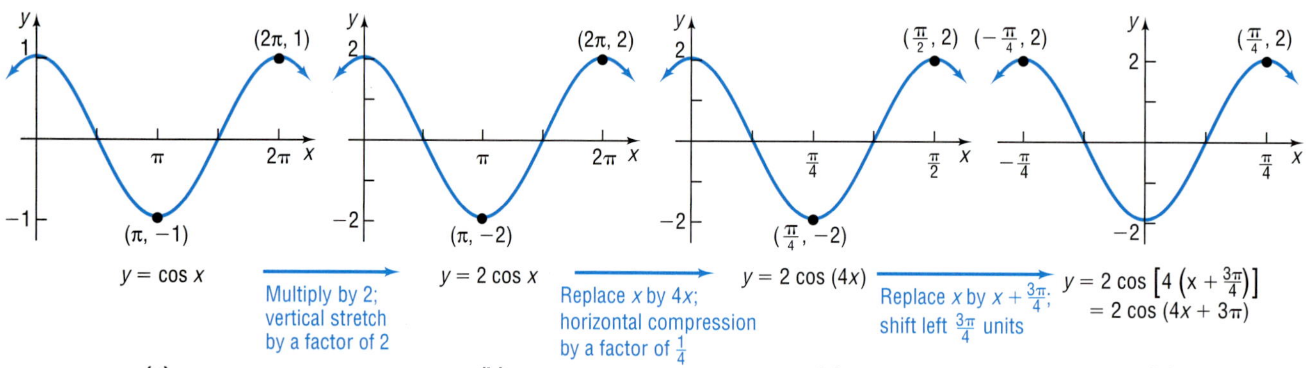

✔ CHECK: Graph $Y_1 = 2\cos(4x + 3\pi)$ using a graphing utility.

NOW WORK PROBLEM 1.

SUMMARY Steps for Graphing Sinusoidal Functions

To graph sinusoidal functions of the form $y = A\sin(\omega x - \phi)$ or $y = A\cos(\omega x - \phi)$:

STEP 1: Determine the amplitude $|A|$ and period $T = \frac{2\pi}{\omega}$.

STEP 2: Determine the starting point of one cycle of the graph, $\frac{\phi}{\omega}$.

STEP 3: Determine the ending point of one cycle of the graph, $\frac{2\pi}{\omega} + \frac{\phi}{\omega}$.

STEP 4: Divide the interval $\left[\frac{\phi}{\omega}, \frac{2\pi}{\omega} + \frac{\phi}{\omega}\right]$ into four subintervals, each of length $\frac{2\pi}{\omega} \div 4$.

STEP 5: Use the endpoints of the subintervals to find the five key points on the graph.

STEP 6: Fill in one cycle of the graph.

STEP 7: Extend the graph in each direction to make it complete.

Finding Sinusoidal Functions from Data

3 Scatter diagrams of data sometimes take the form of a sinusoidal function. Let's look at an example.

The data given in Table 12 represent the average monthly temperatures in Denver, Colorado. Since the data represent average monthly temperatures collected over many years, the data will not vary much from year to year and so will essentially repeat each year. In other words, the data are periodic. Figure 112 shows the scatter diagram of these data repeated over two years, where $x = 1$ represents January, $x = 2$ represents February, and so on.

TABLE 12

Month, *x*	Average Monthly Temperature, °F
January, 1	29.7
February, 2	33.4
March, 3	39.0
April, 4	48.2
May, 5	57.2
June, 6	66.9
July, 7	73.5
August, 8	71.4
September, 9	62.3
October, 10	51.4
November, 11	39.0
December, 12	31.0

Source: U.S. National Oceanic and Atmospheric Administration

Figure 112

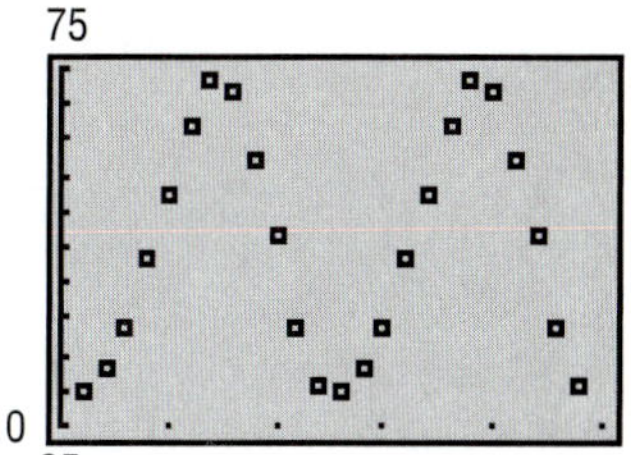

Notice that the scatter diagram looks like the graph of a sinusoidal function. We choose to fit the data to a sine function of the form

$$y = A\sin(\omega x - \phi) + B$$

where A, B, ω, and ϕ are constants.

EXAMPLE 3 Finding a Sinusoidal Function from Temperature Data

Fit a sine function to the data in Table 12.

Figure 113

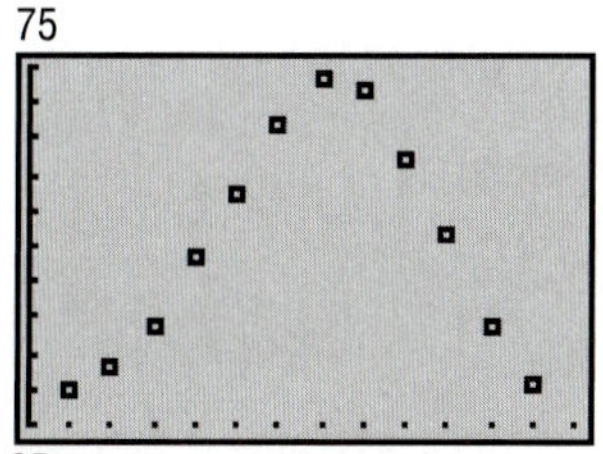

Solution We begin with a scatter diagram of the data for one year. See Figure 113. The data will be fitted to a sine function of the form

$$y = A\sin(\omega x - \phi) + B$$

STEP 1: To find the amplitude A, we compute

$$\text{Amplitude} = \frac{\text{largest data value} - \text{smallest data value}}{2}$$

$$= \frac{73.5 - 29.7}{2} = 21.9$$

To see the remaining steps in this process, we superimpose the graph of the function $y = 21.9 \sin x$, where x represents months, on the scatter diagram. Figure 114 shows the two graphs.

Figure 114

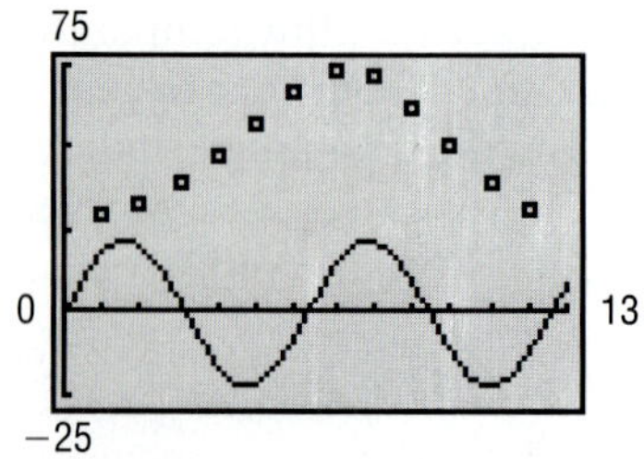

To fit the data, the graph needs to be shifted vertically, shifted horizontally, and stretched horizontally.

STEP 2: We determine the vertical shift by finding the average of the highest and lowest data value.

$$\text{Vertical shift} = \frac{73.5 + 29.7}{2} = 51.6$$

Now we superimpose the graph of $y = 21.9 \sin x + 51.6$ on the scatter diagram. See Figure 115.

Figure 115

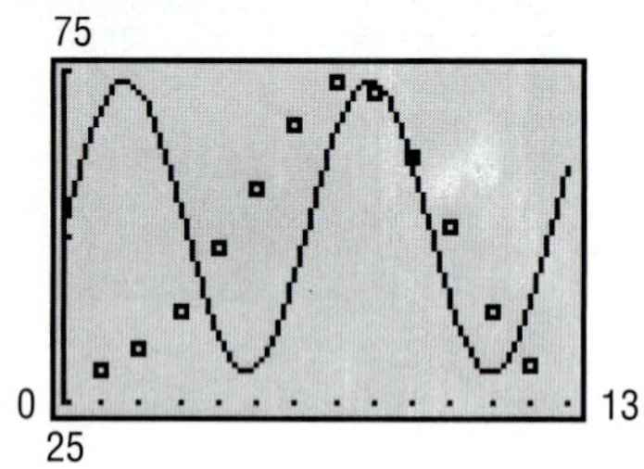

We see that the graph needs to be shifted horizontally and stretched horizontally.

STEP 3: It is easier to find the horizontal stretch factor first. Since the temperatures repeat every 12 months, the period of the function is $T = 12$. Since $T = \frac{2\pi}{\omega} = 12$,

$$\omega = \frac{2\pi}{12} = \frac{\pi}{6}$$

Now we superimpose the graph of $y = 21.9 \sin\left(\frac{\pi}{6}x\right) + 51.6$ on the scatter diagram. See Figure 116.

Figure 116

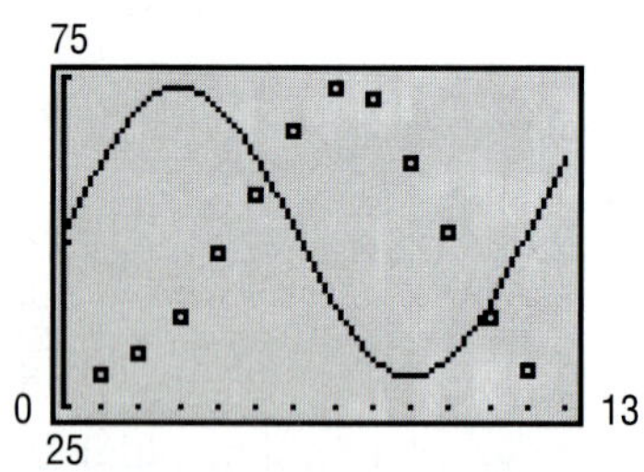

We see that the graph still needs to be shifted horizontally.

STEP 4: To determine the horizontal shift, we solve the equation

$$y = 21.9 \sin\left(\frac{\pi}{6}x - \phi\right) + 51.6$$

for ϕ by letting $y = 29.7$ and $x = 1$ (the average temperature in Denver in January).*

$$29.7 = 21.9 \sin\left(\frac{\pi}{6} \cdot 1 - \phi\right) + 51.6$$

$$-21.9 = 21.9 \sin\left(\frac{\pi}{6} - \phi\right) \qquad \text{Subtract 51.6 from both sides of the equation.}$$

$$-1 = \sin\left(\frac{\pi}{6} - \phi\right) \qquad \text{Divide both sides of the equation by 21.9.}$$

$$\frac{\pi}{6} - \phi = -\frac{\pi}{2} \qquad \sin\theta = -1 \text{ when } \theta = -\frac{\pi}{2}.$$

$$\phi = \frac{2\pi}{3} \qquad \text{Solve for } \phi.$$

The sine function that fits the data is

$$y = 21.9 \sin\left(\frac{\pi}{6}x - \frac{2\pi}{3}\right) + 51.6$$

*The data point selected here to find ϕ is arbitrary. Selecting a different data point will usually result in a different value for ϕ. To maintain consistency, we will always choose the data point for which y is smallest (in this case, January gives the lowest temperature).

Figure 117

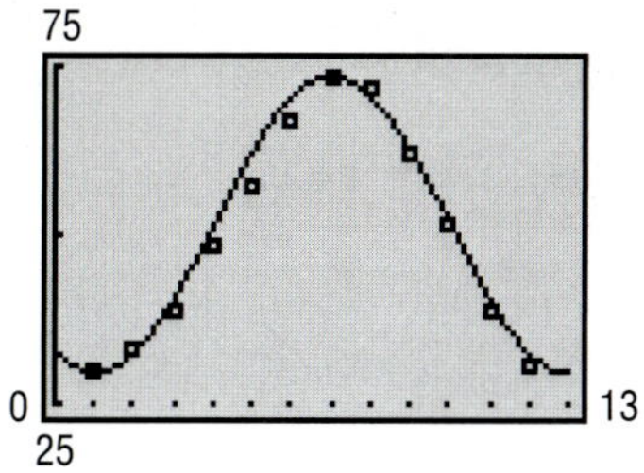

The graph of $y = 21.9\sin\left(\frac{\pi}{6}x - \frac{2\pi}{3}\right) + 51.6$ and the scatter diagram of the data are shown in Figure 117. ■

The steps to fit a sine function

$$y = A\sin(\omega x - \phi) + B$$

to sinusoidal data follow:

Steps for Fitting Data to a Sine Function $y = A\sin(\omega x - \phi) + B$

STEP 1: Determine A, the amplitude of the function.

$$\text{Amplitude} = \frac{\text{largest data value} - \text{smallest data value}}{2}$$

STEP 2: Determine B, the vertical shift of the function.

$$\text{Vertical shift} = \frac{\text{largest data value} + \text{smallest data value}}{2}$$

STEP 3: Determine ω. Since the period T, the time it takes for the data to repeat, is $T = \frac{2\pi}{\omega}$, we have

$$\omega = \frac{2\pi}{T}$$

STEP 4: Determine the horizontal shift of the function by solving the equation

$$y = A\sin(\omega x - \phi) + B$$

for ϕ by choosing an ordered pair (x, y) from the data. Since answers will vary depending on the ordered pair selected, we will always choose the ordered pair for which y is smallest in order to maintain consistency.

NOW WORK PROBLEM 19(a)–(c).

Certain graphing utilities (such as a TI-83 and TI-86) have the capability of finding the sine function of best fit for sinusoidal data. At least four data points are required for this process.

EXAMPLE 4 Finding the Sine Function of Best Fit

Use a graphing utility to find the sine function of best fit for the data in Table 12. Graph this function with the scatter diagram of the data.

Figure 118

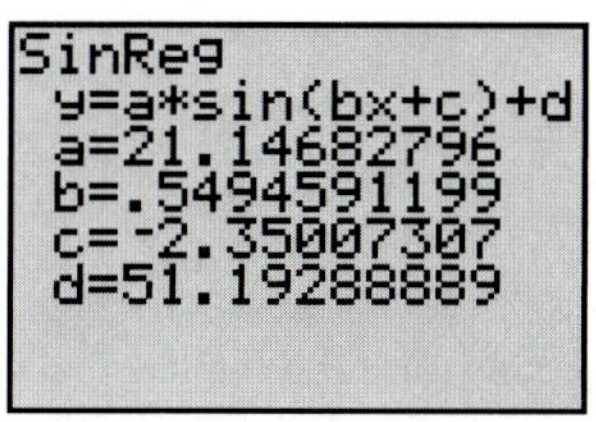

Solution Enter the data from Table 12 and execute the SINe REGression program. The result is shown in Figure 118.

The output that the utility provides shows us the equation

$$y = a\sin(bx + c) + d$$

The sinusoidal function of best fit is

$$y = 21.15\sin(0.55x - 2.35) + 51.19$$

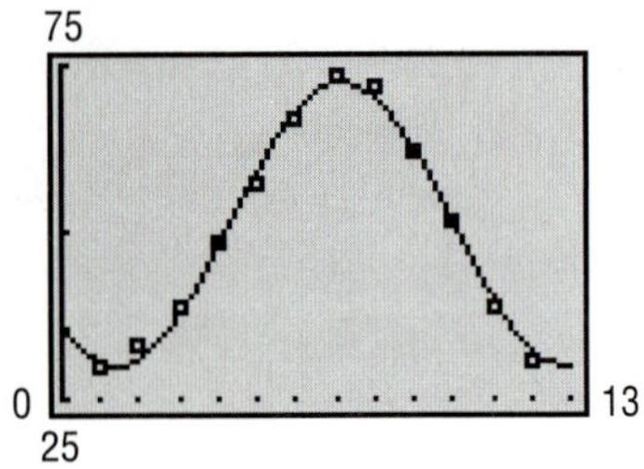

Figure 119

where x represents the month and y represents the average temperature.

Figure 119 shows the graph of the sinusoidal function of best fit on the scatter diagram. ■

NOW WORK PROBLEM 19(d) AND (e).

Since the number of hours of sunlight in a day cycles annually, the number of hours of sunlight in a day for a given location can be modeled by a sinusoidal function.

The longest day of the year (in terms of hours of sunlight) occurs on the day of the summer solstice. The summer solstice is the time when the sun is farthest north (for locations in the northern hemisphere). In 2001, the summer solstice occurred on June 21 (the 172nd day of the year) at 3.38 AM EDT. The shortest day of the year occurs on the day of the winter solstice. The winter solstice is the time when the sun is farthest south (again, for locations in the northern hemisphere). In 2001, the winter solstice occurred on December 21 (the 355th day of the year) at 2:22 PM (EST).

EXAMPLE 5 Finding a Sinusoidal Function for Hours of Daylight

According to the *Old Farmer's Almanac*, the number of hours of sunlight in Boston on the summer solstice is 15.283 and the number of hours of sunlight on the winter solstice is 9.067.

(a) Find a sinusoidal function of the form $y = A\sin(\omega x - \phi) + B$ that fits the data.*

(b) Use the function found in part (a) to predict the number of hours of sunlight on April 1, the 91st day of the year.

(c) Draw a graph of the function found in part (a).

(d) Look up the number of hours of sunlight for April 1 in the *Old Farmer's Almanac* and compare the actual hours of daylight to the results found in part (b).

Solution (a) **STEP 1:** Amplitude $= \dfrac{\text{largest data value} - \text{smallest data value}}{2}$

$$= \frac{15.283 - 9.067}{2} = 3.108$$

STEP 2: Vertical shift $= \dfrac{\text{largest data value} + \text{smallest data value}}{2}$

$$= \frac{15.283 + 9.067}{2} = 12.175$$

STEP 3: The data repeat every 365 days. Since $T = \dfrac{2\pi}{\omega} = 365$, we find

$$\omega = \frac{2\pi}{365}$$

So far, we have $y = 3.108\sin\left(\dfrac{2\pi}{365}x - \phi\right) + 12.175$.

*Notice that only two data points are given, so a graphing utility cannot be used to find the sine function of best fit.

STEP 4: To determine the horizontal shift, we solve the equation

$$y = 3.108 \sin\left(\frac{2\pi}{365}x - \phi\right) + 12.175$$

for ϕ by letting $y = 9.067$ and $x = 355$ (the number of hours of daylight in Boston on December 21).

$$9.067 = 3.108 \sin\left(\frac{2\pi}{365} \cdot 355 - \phi\right) + 12.175$$

$$-3.108 = 3.108 \sin\left(\frac{2\pi}{365} \cdot 355 - \phi\right)$$ Subtract 12.175 from both sides of the equation.

$$-1 = \sin\left(\frac{2\pi}{365} \cdot 355 - \phi\right)$$ Divide both sides of the equation by 3.108.

$$\frac{2\pi}{365} \cdot 355 - \phi = -\frac{\pi}{2}$$ $\sin\theta = -1$ when $\theta = -\frac{\pi}{2}$.

$$\phi = \frac{357}{146}\pi$$ Solve for ϕ.

The function that provides the number of hours of daylight in Boston for any day, x, is given by

$$y = 3.108 \sin\left(\frac{2\pi}{365}x - \frac{357}{146}\pi\right) + 12.175$$

(b) To predict the number of hours of daylight on April 1, we let $x = 91$ in the function found in part (a) and obtain

$$y = 3.108 \sin\left(\frac{2\pi}{365} \cdot 91 - \frac{357}{146}\pi\right) + 12.175$$

$$\approx 12.69$$

So we predict that there will be about 12.69 hours of sunlight on April 1 in Boston.

Figure 120

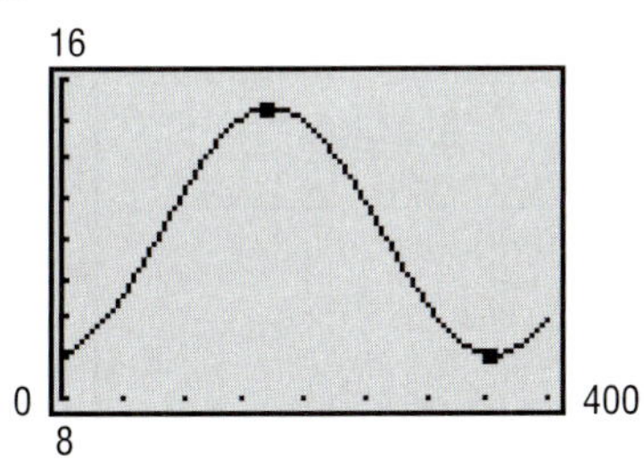

(c) The graph of the function found in part (a) is given in Figure 120.

(d) According to the *Old Farmer's Almanac*, there will be 12 hours 43 minutes of sunlight on April 1 in Boston. Our prediction of 12.69 hours converts to 12 hours 41 minutes.

NOW WORK PROBLEM 25.

2.8 Concepts and Vocabulary

In Problem 1, fill in the blank.

1. For the graph of $y = A\sin(\omega x - \phi)$, the number $\frac{\phi}{\omega}$ is called the __________.

In Problem 2, answer True or False.

2. Only two data points are required by a graphing utility to find the sine function of best fit.

2.8 Exercises

In Problems 1–12, find the amplitude, period, and phase shift of each function. Graph each function by hand. Show at least one period. Verify the result using a graphing utility.

1. $y = 4\sin(2x - \pi)$
2. $y = 3\sin(3x - \pi)$
3. $y = 2\cos\left(3x + \frac{\pi}{2}\right)$
4. $y = 3\cos(2x + \pi)$
5. $y = -3\sin\left(2x + \frac{\pi}{2}\right)$
6. $y = -2\cos\left(2x - \frac{\pi}{2}\right)$
7. $y = 4\sin(\pi x + 2)$
8. $y = 2\cos(2\pi x + 4)$
9. $y = 3\cos(\pi x - 2)$
10. $y = 2\cos(2\pi x - 4)$
11. $y = 3\sin\left(-2x + \frac{\pi}{2}\right)$
12. $y = 3\cos\left(-2x + \frac{\pi}{2}\right)$

In Problems 13–16, write the equation of a sine function, $y = A\sin(\omega x - \phi)$, $A > 0$, that has the given characteristics.

13. Amplitude: 2
    Period: π
    Phase shift: $\frac{1}{2}$
14. Amplitude: 3
    Period: $\frac{\pi}{2}$
    Phase shift: 2
15. Amplitude: 3
    Period: 3π
    Phase shift: $-\frac{1}{3}$
16. Amplitude: 2
    Period: π
    Phase shift: -2

17. **Alternating Current (ac) Circuits** The current I, in amperes, flowing through an ac (alternating current) circuit at time t is

$$I = 120\sin\left(30\pi t - \frac{\pi}{3}\right), \qquad t \geq 0$$

What is the period? What is the amplitude? What is the phase shift? Graph this function over two periods.

18. **Alternating Current (ac) Circuits** The current I, in amperes, flowing through an ac (alternating current) circuit at time t is

$$I = 220\sin\left(60\pi t - \frac{\pi}{6}\right), \qquad t \geq 0$$

What is the period? What is the amplitude? What is the phase shift? Graph this function over two periods.

19. **Monthly Temperature** The following data represent the average monthly temperatures for Juneau, Alaska.

Month, x	Average Monthly Temperature, °F
January, 1	24.2
February, 2	28.4
March, 3	32.7
April, 4	39.7
May, 5	47.0
June, 6	53.0
July, 7	56.0
August, 8	55.0
September, 9	49.4
October, 10	42.2
November, 11	32.0
December, 12	27.1

Source: U.S. National Oceanic and Atmospheric Administration.

(a) Use a graphing utility to draw a scatter diagram of the data for one period.
(b) By hand, find a sinusoidal function of the form $y = A\sin(\omega x - \phi) + B$ that fits the data.
(c) Draw the sinusoidal function found in part (b) on the scatter diagram.
(d) Use a graphing utility to find the sinusoidal function of best fit.
(e) Draw the sinusoidal function of best fit on the scatter diagram.

20. **Monthly Temperature** The following data represent the average monthly temperatures for Washington, D.C.

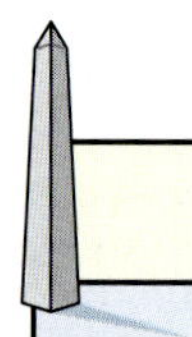

Month, x	Average Monthly Temperature, °F
January, 1	34.6
February, 2	37.5
March, 3	47.2
April, 4	56.5
May, 5	66.4
June, 6	75.6
July, 7	80.0
August, 8	78.5
September, 9	71.3
October, 10	59.7
November, 11	49.8
December, 12	39.4

Source: U.S. National Oceanic and Atmospheric Administration.

(a) Use a graphing utility to draw a scatter diagram of the data for one period.

(b) By hand, find a sinusoidal function of the form $y = A\sin(\omega x - \phi) + B$ that fits the data.
(c) Draw the sinusoidal function found in part (b) on the scatter diagram.
(d) Use a graphing utility to find the sinusoidal function of best fit.
(e) Graph the sinusoidal function of best fit on the scatter diagram.

21. Monthly Temperature The following data represent the average monthly temperatures for Indianapolis, Indiana.

Month, x	Average Monthly Temperature, °F
January, 1	25.5
February, 2	29.6
March, 3	41.4
April, 4	52.4
May, 5	62.8
June, 6	71.9
July, 7	75.4
August, 8	73.2
September, 9	66.6
October, 10	54.7
November, 11	43.0
December, 12	30.9

Source: U.S. National Oceanic and Atmospheric Administration.

(a) Use a graphing utility to draw a scatter diagram of the data for one period.
(b) By hand, find a sinusoidal function of the form $y = A\sin(\omega x - \phi) + B$ that fits the data.
(c) Draw the sinusoidal function found in part (b) on the scatter diagram.
(d) Use a graphing utility to find the sinusoidal function of best fit.
(e) Graph the sinusoidal function of best fit on the scatter diagram.

22. Monthly Temperature The data at the top of the right column represent the average monthly temperatures for Baltimore, Maryland.

(a) Use a graphing utility to draw a scatter diagram of the data for one period.
(b) By hand, find a sinusoidal function of the form $y = A\sin(\omega x - \phi) + B$ that fits the data.
(c) Draw the sinusoidal function found in part (b) on the scatter diagram.
(d) Use a graphing utility to find the sinusoidal function of best fit.
(e) Graph the sinusoidal function of best fit on the scatter diagram.

Month, x	Average Monthly Temperature, °F
January, 1	31.8
February, 2	34.8
March, 3	44.1
April, 4	53.4
May, 5	63.4
June, 6	72.5
July, 7	77.0
August, 8	75.6
September, 9	68.5
October, 10	56.6
November, 11	46.8
December, 12	36.7

Source: U.S. National Oceanic and Atmospheric Administration.

23. Tides Suppose that the length of time between consecutive high tides is approximately 12.5 hours. According to the National Oceanic and Atmospheric Administration, on Saturday, June 28, 1997, in Savannah, Georgia, high tide occurred at 3:38 AM (3.6333 hours) and low tide occurred at 10:08 AM (10.1333 hours). Water heights are measured as the amounts above or below the mean lower low water. The height of the water at high tide was 8.2 feet and the height of the water at low tide was −0.6 foot.
(a) Approximately when will the next high tide occur?
(b) Find a sinusoidal function of the form $y = A\sin(\omega x - \phi) + B$ that fits the data.
(c) Draw a graph of the function found in part (b).
(d) Use the function found in part (b) to predict the height of the water at the next high tide.

24. Tides Suppose that the length of time between consecutive high tides is approximately 12.5 hours. According to the National Oceanic and Atmospheric Administration, on Saturday, June 28, 1997, in Juneau, Alaska, high tide occurred at 8:11 AM (8.1833 hours) and low tide occurred at 2:14 PM (14.2333 hours). Water heights are measured as the amounts above or below the mean lower low water. The height of the water at high tide was 13.2 feet and the height of the water at low tide was 2.2 feet.
(a) Approximately when will the next high tide occur?
(b) Find a sinusoidal function of the form $y = A\sin(\omega x - \phi) + B$ that fits the data.
(c) Draw a graph of the function found in part (b).
(d) Use the function found in part (b) to predict the height of the water at the next high tide.

25. Hours of Daylight According to the *Old Farmer's Almanac*, in Miami, Florida, the number of hours of sunlight on the summer solstice is 12.75 and the number of hours of sunlight on the winter solstice is 10.583.
(a) Find a sinusoidal function of the form $y = A\sin(\omega x - \phi) + B$ that fits the data.

(b) Draw a graph of the function found in part (a).
(c) Use the function found in part (a) to predict the number of hours of sunlight on April 1, the 91st day of the year.
(d) Look up the number of hours of sunlight for April 1 in the *Old Farmer's Almanac* and compare the actual hours of daylight to the results found in part (c).

26. **Hours of Daylight** According to the *Old Farmer's Almanac*, in Detroit, Michigan, the number of hours of sunlight on the summer solstice is 13.65 and the number of hours of sunlight on the winter solstice is 9.067.
(a) Find a sinusoidal function of the form $y = A\sin(\omega x - \phi) + B$ that fits the data.
(b) Draw a graph of the function found in part (a).
(c) Use the function found in part (a) to predict the number of hours of sunlight on April 1, the 91st day of the year.
(d) Look up the number of hours of sunlight for April 1 in the *Old Farmer's Almanac* and compare the actual hours of daylight to the results found in part (c).

27. **Hours of Daylight** According to the *Old Farmer's Almanac*, in Anchorage, Alaska, the number of hours of sunlight on the summer solstice is 16.233 and the number of hours of sunlight on the winter solstice is 5.45.
(a) Find a sinusoidal function of the form $y = A\sin(\omega x - \phi) + B$ that fits the data.
(b) Draw a graph of the function found in part (a).
(c) Use the function found in part (a) to predict the number of hours of sunlight on April 1, the 91st day of the year.
(d) Look up the number of hours of sunlight for April 1 in the *Old Farmer's Almanac* and compare the actual hours of daylight to the results found in part (c).

28. **Hours of Daylight** According to the *Old Farmer's Almanac*, in Honolulu, Hawaii, the number of hours of sunlight on the summer solstice is 12.767 and the number of hours of sunlight on the winter solstice is 10.783.
(a) Find a sinusoidal function of the form $y = A\sin(\omega x - \phi) + B$ that fits the data.
(b) Draw a graph of the function found in part (a).
(c) Use the function found in part (a) to predict the number of hours of sunlight on April 1, the 91st day of the year.
(d) Look up the number of hours of sunlight for April 1 in the *Old Farmer's Almanac* and compare the actual hours of daylight to the results found in part (c).

Chapter Review

Things To Know

Definitions

Angle in standard position (p. 99)	Vertex is at the origin; initial side is along the positive x-axis
Degree $(1°)$ (p. 99)	$1° = \frac{1}{360}$ revolution
Radian (1) (p. 102)	The measure of a central angle of a circle whose rays subtend an arc whose length is the radius of the circle
Acute angle (p. 112)	An angle θ whose measure is $0° < \theta < 90°\left(\text{or } 0 < \theta < \frac{\pi}{2}\right)$
Trigonometric functions (p. 113)	$P = (a, b)$ is the point on the terminal side of θ a distance r from the origin: $\sin\theta = \frac{b}{r}$, $\cos\theta = \frac{a}{r}$, $\tan\theta = \frac{b}{a},\ a \neq 0$; $\csc\theta = \frac{r}{b},\ b \neq 0$, $\sec\theta = \frac{r}{a},\ a \neq 0$, $\cot\theta = \frac{a}{b},\ b \neq 0$
Complementary angles (p. 118)	Two acute angles whose sum is $90°\left(\frac{\pi}{2}\right)$
Cofunction (p. 118)	The following pairs of functions are cofunctions of each other: sine and cosine; tangent and cotangent; secant and cosecant
Complementary Angle Theorem (p. 119)	Cofunctions of complementary angles are equal
Reference angle of θ (p. 137)	The acute angle formed by the terminal side of θ and either the positive or negative x-axis
Periodic function (p. 149)	$f(\theta + p) = f(\theta)$, for all θ, $p > 0$, where the smallest such p is the fundamental period

Formulas

1 revolution counterclockwise $= 360° = 2\pi$ (p. 101)
counterclockwise $= 2\pi$ radians (p. 104)

$s = r\theta$ (p. 103) — θ is measured in radians; s is the length of arc subtended by the central angle θ of the circle of radius r; A is the area of the sector

$A = \frac{1}{2}r^2\theta$ (p. 106)

$v = r\omega$ (p. 107) — v is the linear speed along the circle of radius r; ω is the angular speed (measured in radians per unit time)

TABLE OF VALUES

θ (Radians)	θ (Degrees)	$\sin\theta$	$\cos\theta$	$\tan\theta$	$\csc\theta$	$\sec\theta$	$\cot\theta$
0	0°	0	1	0	Not defined	1	Not defined
$\frac{\pi}{6}$	30°	$\frac{1}{2}$	$\frac{\sqrt{3}}{2}$	$\frac{\sqrt{3}}{3}$	2	$\frac{2\sqrt{3}}{3}$	$\sqrt{3}$
$\frac{\pi}{4}$	45°	$\frac{\sqrt{2}}{2}$	$\frac{\sqrt{2}}{2}$	1	$\sqrt{2}$	$\sqrt{2}$	1
$\frac{\pi}{3}$	60°	$\frac{\sqrt{3}}{2}$	$\frac{1}{2}$	$\sqrt{3}$	$\frac{2\sqrt{3}}{3}$	2	$\frac{\sqrt{3}}{3}$
$\frac{\pi}{2}$	90°	1	0	Not defined	1	Not defined	0
π	180°	0	-1	0	Not defined	-1	Not defined
$\frac{3\pi}{2}$	270°	-1	0	Not defined	-1	Not defined	0

Fundamental Identities (p. 115)

$$\tan\theta = \frac{\sin\theta}{\cos\theta}, \quad \cot\theta = \frac{\cos\theta}{\sin\theta}$$

$$\cot\theta = \frac{1}{\tan\theta}, \quad \sec\theta = \frac{1}{\cos\theta}, \quad \csc\theta = \frac{1}{\sin\theta}$$

$$\sin^2\theta + \cos^2\theta = 1, \quad \tan^2\theta + 1 = \sec^2\theta, \quad 1 + \cot^2\theta = \csc^2\theta$$

Properties of the Trigonometric Functions

$y = \sin x$ (p. 155)
Domain: $-\infty < x < \infty$
Range: $-1 \le y \le 1$
Periodic: period $= 2\pi(360°)$
Odd function

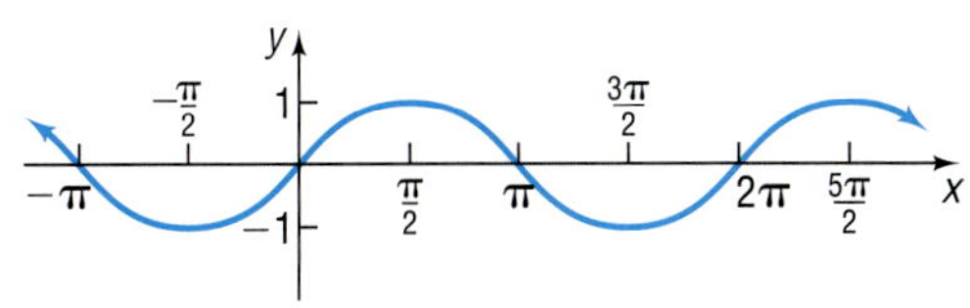

$y = \cos x$ (p. 157)
Domain: $-\infty < x < \infty$
Range: $-1 \le y \le 1$
Periodic: period $= 2\pi(360°)$
Even function

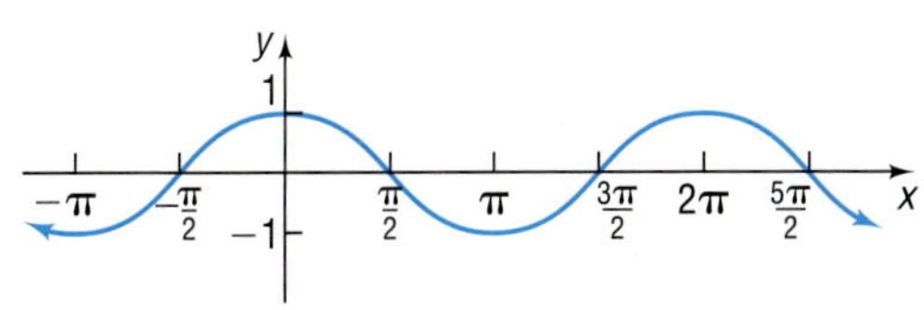

$y = \tan x$ (p. 172)

Domain: $-\infty < x < \infty$, except odd multiples of $\frac{\pi}{2}(90°)$
Range: $-\infty < y < \infty$
Periodic: period $= \pi(180°)$
Odd function

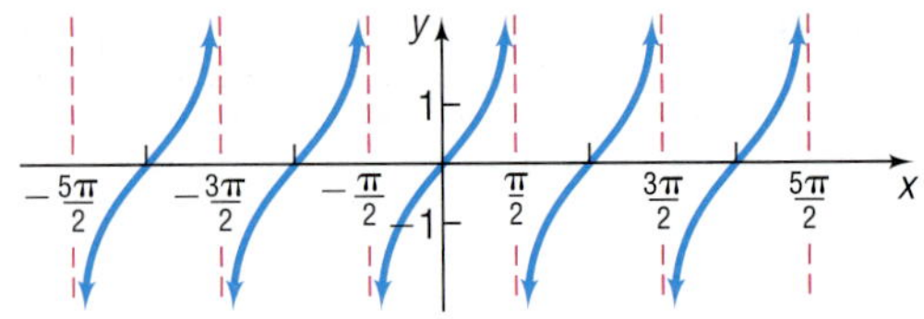

$y = \cot x$ (p. 173)

Domain: $-\infty < x < \infty$, except integral multiples of $\pi(180°)$
Range: $-\infty < y < \infty$
Periodic: period $= \pi(180°)$
Odd function

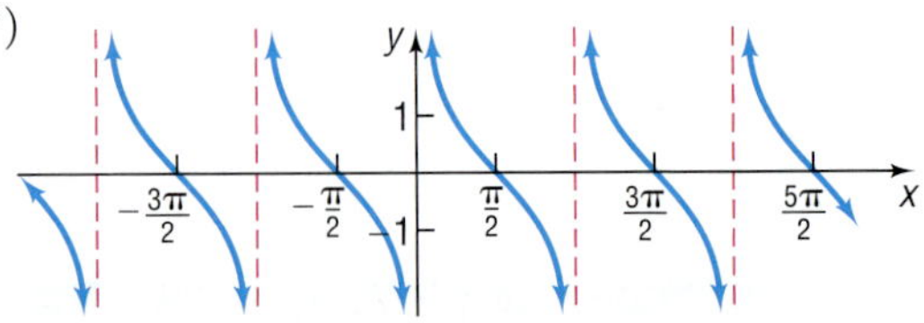

$y = \csc x$ (p. 174)

Domain: $-\infty < x < \infty$, except integral multiples of $\pi(180°)$
Range: $|y| \geq 1$
Periodic: period $= 2\pi(360°)$
Odd function

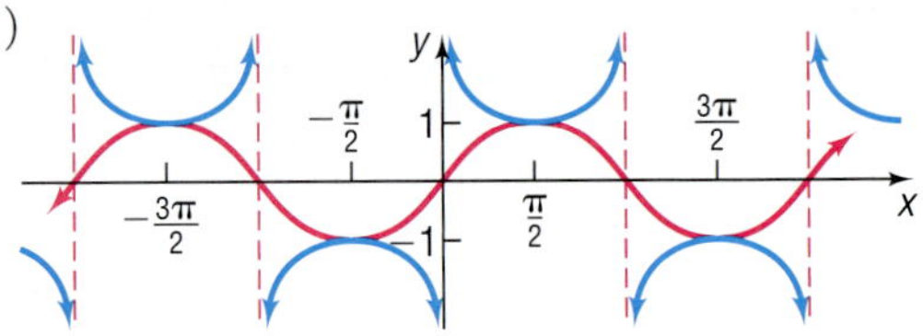

$y = \sec x$ (p. 175)

Domain: $-\infty < x < \infty$, except odd multiples of $\frac{\pi}{2}(90°)$
Range: $|y| \geq 1$
Periodic: period $= 2\pi(360°)$
Even function

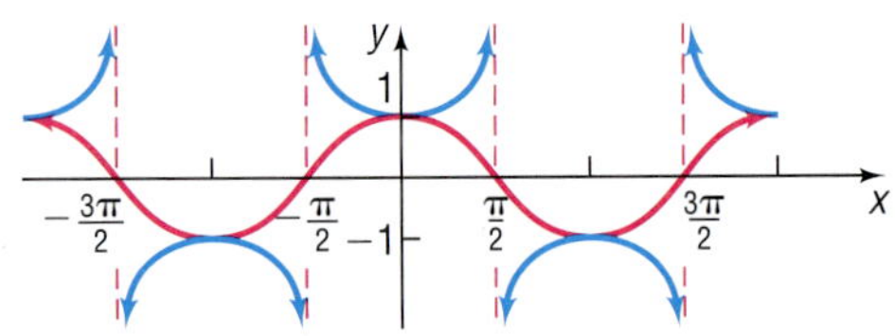

Sinusoidal graphs (pp. 161 and 177)

$y = A\sin(\omega x), \quad \omega > 0$

$y = A\cos(\omega x), \quad \omega > 0$

$y = A\sin(\omega x - \phi) = A\sin\left[\omega\left(x - \frac{\phi}{\omega}\right)\right]$

$y = A\cos(\omega x - \phi) = A\cos\left[\omega\left(x - \frac{\phi}{\omega}\right)\right]$

Period $= \frac{2\pi}{\omega}$

Amplitude $= |A|$

Phase shift $= \frac{\phi}{\omega}$

Objectives

Section		You should be able to:	Review Exercises
2.1	1	Convert between degrees, minutes, seconds, and decimal forms for angles (p. 101)	82
	2	Find the arc length of a circle (p. 102)	83–84
	3	Convert from degrees to radians (p. 104)	1–4
	4	Convert from radians to degrees (p. 105)	5–8
	5	Find the area of a sector of a circle (p. 106)	83
	6	Find the linear speed of an object traveling in circular motion (p. 107)	85–88
2.2	1	Find the value of trigonometric functions of acute angles (p. 113)	75
	2	Use the fundamental identities (p. 114)	21–24
	3	Find the remaining trigonometric functions given the value of one of them (p. 116)	31–32
	4	Use the complementary angle theorem (p. 118)	25–26

Section	Objective	Review Exercises
2.3	1 Find the exact value of the trigonometric functions of $\frac{\pi}{4} = 45°$ (p. 124)	9–12
	2 Find the exact value of the trigonometric functions of $\frac{\pi}{6} = 30°$ and $\frac{\pi}{3} = 60°$ (p. 125)	9–12
	3 Use a calculator to approximate the value of the trigonometric functions of acute angles (p. 127)	76
2.4	1 Find the exact value of the trigonometric functions for general angles (p. 132)	77
	2 Use coterminal angles to find the exact value of a trigonometric function (p. 134)	19–20
	3 Determine the sign of the trigonometric functions of an angle in a given quadrant (p. 136)	78
	4 Find the reference angle of a general angle (p. 137)	79
	5 Use the theorem on reference angles (p. 138)	13–16, 19–20
	6 Find the exact value of the trigonometric functions of an angle given one of them and the quadrant of the angle (p. 139)	31–46
2.5	1 Find the exact value of the trigonometric functions using the unit circle (p. 142)	80
	2 Know the domain and range of the trigonometric functions (p. 146)	81
	3 Use the periodic properties to find the exact values of the trigonometric functions (p. 149)	19–20
	4 Use even-odd properties to find the exact value of the trigonometric functions (p. 150)	13, 15–16, 18–20, 27–30
2.6	1 Graph transformations of the sine function (p. 155)	47, 50
	2 Graph transformations of the cosine function (p. 158)	48, 49
	3 Determine the amplitude and period of sinusoidal functions (p. 160)	59–64, 89
	4 Graph sinusoidal functions: $y = A\sin(\omega x)$ (p. 161)	47, 48, 63, 64, 89
	5 Find an equation for a sinusoidal graph (p. 165)	71–74
2.7	1 Graph transformations of the tangent function and cotangent function (p. 170)	51–56
	2 Graph transformations of the cosecant function and secant function (p. 174)	57–58
2.8	1 Determine the phase shift of a sinusoidal function (p. 177)	65–70, 90
	2 Graph sinusoidal functions: $y = A\sin(\omega x - \phi)$ (p. 177)	65–70, 90
	3 Find a sinusoidal function from data (p. 181)	91–94

Review Exercises

Blue problem numbers indicate the authors' suggestions for use in a Practice Test.

In Problems 1–4, convert each angle in degrees to radians. Express your answer as a multiple of π.

1. 135° **2.** 210° **3.** 18° **4.** 15°

In Problems 5–8, convert each angle in radians to degrees.

5. $\frac{3\pi}{4}$ **6.** $\frac{2\pi}{3}$ **7.** $-\frac{5\pi}{2}$ **8.** $-\frac{3\pi}{2}$

In Problems 9–30, find the exact value of each expression. Do not use a calculator.

9. $\tan\frac{\pi}{4} - \sin\frac{\pi}{6}$

10. $\cos\frac{\pi}{3} + \sin\frac{\pi}{4}$

11. $3\sin 45° - 4\tan\frac{\pi}{6}$

12. $4\cos 60° + 3\tan\frac{\pi}{3}$

13. $6\cos\frac{3\pi}{4} + 2\tan\left(-\frac{\pi}{3}\right)$

14. $3\sin\frac{2\pi}{3} - 4\cos\frac{5\pi}{2}$

15. $\sec\left(-\frac{\pi}{3}\right) - \cot\left(-\frac{5\pi}{4}\right)$

16. $4\csc\frac{3\pi}{4} - \cot\left(-\frac{\pi}{4}\right)$

17. $\tan\pi + \sin\pi$

18. $\cos\frac{\pi}{2} - \csc\left(-\frac{\pi}{2}\right)$

19. $\cos 540° - \tan(-45°)$

20. $\sin 630° + \cos(-180°)$

21. $\sin^2 20° + \frac{1}{\sec^2 20°}$

22. $\frac{1}{\cos^2 40°} - \frac{1}{\cot^2 40°}$

23. $\sec 50° \cos 50°$

24. $\tan 10° \cot 10°$

25. $\frac{\sin 50°}{\cos 40°}$

26. $\frac{\tan 20°}{\cot 70°}$

27. $\frac{\sin(-40°)}{\cos 50°}$

28. $\tan(-20°)\cot 20°$

29. $\sin 400° \sec(-50°)$

30. $\cot 200° \cot(-70°)$

In Problems 31–46, find the exact value of each of the remaining trigonometric functions.

31. $\sin\theta = \frac{4}{5}$, θ acute

32. $\cos\theta = \frac{3}{5}$, θ acute

33. $\tan\theta = \frac{12}{5}$, $\sin\theta < 0$

34. $\cot\theta = \frac{12}{5}$, $\cos\theta < 0$

35. $\sec\theta = -\frac{5}{4}$, $\tan\theta < 0$

36. $\csc\theta = -\frac{5}{3}$, $\cot\theta < 0$

37. $\sin\theta = \frac{12}{13}$, θ in quadrant II

38. $\cos\theta = -\frac{3}{5}$, θ in quadrant III

39. $\sin\theta = -\frac{5}{13}$, $\frac{3\pi}{2} < \theta < 2\pi$

40. $\cos\theta = \frac{12}{13}$, $\frac{3\pi}{2} < \theta < 2\pi$

41. $\tan\theta = \frac{1}{3}$, $180° < \theta < 270°$

42. $\tan\theta = -\frac{2}{3}$, $90° < \theta < 180°$

43. $\sec\theta = 3$, $\frac{3\pi}{2} < \theta < 2\pi$

44. $\csc\theta = -4$, $\pi < \theta < \frac{3\pi}{2}$

45. $\cot\theta = -2$, $\frac{\pi}{2} < \theta < \pi$

46. $\tan\theta = -2$, $\frac{3\pi}{2} < \theta < 2\pi$

In Problems 47–58, graph each function. Each graph should contain at least one period.

47. $y = 2\sin(4x)$

48. $y = -3\cos(2x)$

49. $y = -2\cos\left(x + \frac{\pi}{2}\right)$:

50. $y = 3\sin(x - \pi)$

51. $y = \tan(x + \pi)$

52. $y = -\tan\left(x - \frac{\pi}{2}\right)$

53. $y = -2\tan(3x)$

54. $y = 4\tan(2x)$

55. $y = \cot\left(x + \frac{\pi}{4}\right)$

56. $y = -4\cot(2x)$

57. $y = \sec\left(x - \frac{\pi}{4}\right)$

58. $y = \csc\left(x + \frac{\pi}{4}\right)$

In Problems 59–62, determine the amplitude and period of each function without graphing.

59. $y = 4\cos x$

60. $y = \sin(2x)$

61. $y = -8\sin\left(\frac{\pi}{2}x\right)$

62. $y = -2\cos(3\pi x)$

In Problems 63–70, find the amplitude, period, and phase shift of each function. Graph each function. Show at least one period.

63. $y = 4\sin(3x)$

64. $y = 2\cos\left(\frac{1}{3}x\right)$

65. $y = 2\sin(2x - \pi)$

66. $y = -\cos\left(\frac{1}{2}x + \frac{\pi}{2}\right)$

67. $y = \frac{1}{2}\sin\left(\frac{3}{2}x - \pi\right)$

68. $y = \frac{3}{2}\cos(6x + 3\pi)$

69. $y = -\frac{2}{3}\cos(\pi x - 6)$

70. $y = -7\sin\left(\frac{\pi}{3}x + \frac{4}{3}\right)$

In Problems 71–74, find a function whose graph is given.

71.

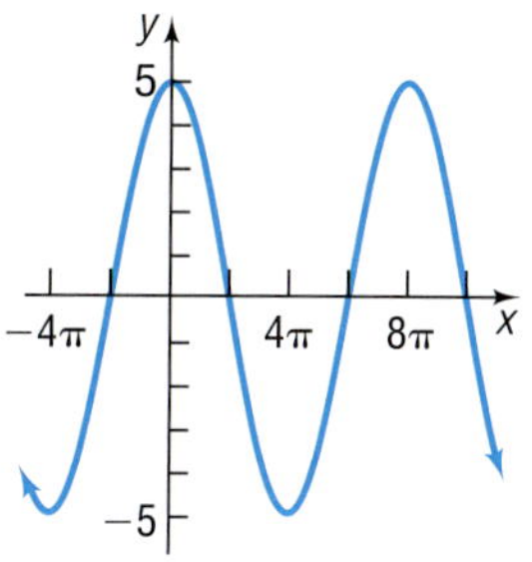

72.

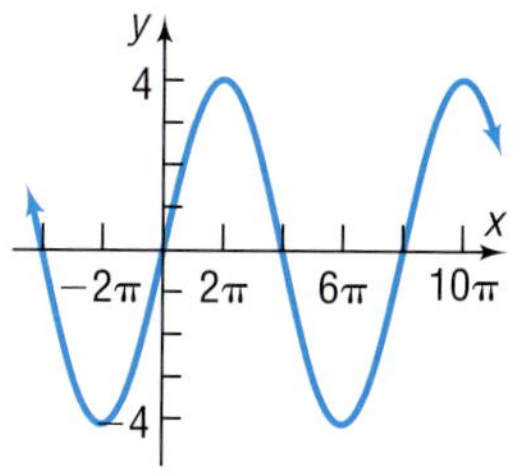

73.

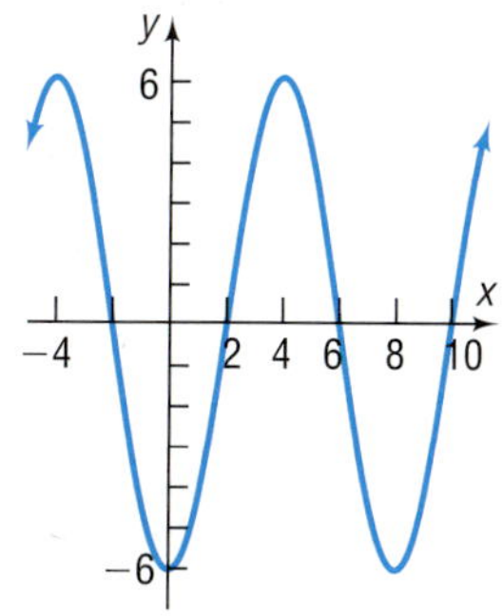

74.

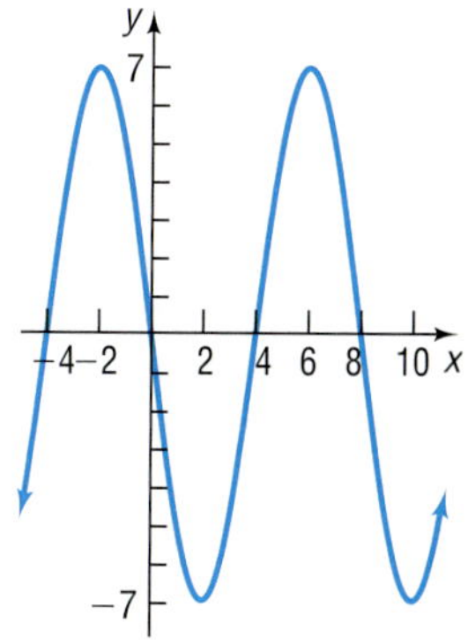

75. Find the value of each of the six trigonometric functions of the angle θ in the illustration.

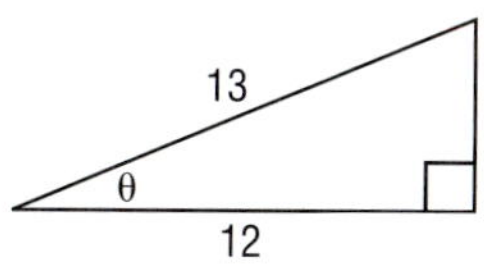

76. Use a calculator to approximate $\sec 10°$. Round the answer to two decimal places.

77. Find the exact value of each of the six trigonometric functions of an angle θ if $(3, -4)$ is a point on the terminal side of θ.

78. Name the quadrant θ lies in if $\cos\theta > 0$ and $\tan\theta < 0$.

79. Find the reference angle of $-\dfrac{4\pi}{5}$.

80. Find the exact value of $\sin t$, $\cos t$, and $\tan t$ if $P = \left(-\dfrac{3}{5}, \dfrac{4}{5}\right)$ is the point on the unit circle that corresponds to t.

81. Find the domain and range of the secant function.

82. (a) Convert the angle $32°20'35''$ to a decimal in degrees. Round the answer to two decimal places.
(b) Convert the angle $63.18°$ to $D°M'S''$ form. Express the answer to the nearest second.

83. Find the length of arc subtended by a central angle of $30°$ on a circle of radius 2 feet. What is the area of the sector?

84. The minute hand of a clock is 8 inches long. How far does the tip of the minute hand move in 30 minutes? How far does it move in 20 minutes?

85. Angular Speed of a Race Car A race car is driven around a circular track at a constant speed of 180 miles per hour. If the diameter of the track is $\dfrac{1}{2}$ mile, what is the angular speed of the car? Express your answer in revolutions per hour (which is equivalent to laps per hour).

86. Merry-Go-Rounds A neighborhood carnival has a merry-go-round whose radius is 25 feet. If the time for one revolution is 30 seconds, how fast is the merry-go-round going?

87. Lighthouse Beacons The Montauk Point Lighthouse on Long Island has dual beams (two light sources opposite each other). Ships at sea observe a blinking light every 5 seconds. What rotation speed is required to do this?

88. Spin Balancing Tires The radius of each wheel of a car is 16 inches. At how many revolutions per minute should a spin balancer be set to balance the tires at a speed of 90 miles per hour? Is the setting different for a wheel of radius 14 inches? If so, what is this setting?

89. Alternating Voltage The electromotive force E, in volts, in a certain ac circuit obeys the equation

$$E = 120\sin(120\pi t), \quad t \geq 0$$

where t is measured in seconds.
(a) What is the maximum value of E?
(b) What is the period?
(c) Graph this function over two periods.

90. Alternating Current The current I, in amperes, flowing through an ac (alternating current) circuit at time t is

$$I = 220\sin\left(30\pi t + \frac{\pi}{6}\right), \quad t \geq 0$$

(a) What is the period?
(b) What is the amplitude?
(c) What is the phase shift?
(d) Graph this function over two periods.

91. Monthly Temperature The following data represent the average monthly temperatures for Phoenix, Arizona.

Month, *m*	Average Monthly Temperature, T
January, 1	51
February, 2	55
March, 3	63
April, 4	67
May, 5	77
June, 6	86
July, 7	90
August, 8	90
September, 9	84
October, 10	71
November, 11	59
December, 12	52

Source: U.S. National Oceanic and Atmospheric Administration.

(a) Use a graphing utility to draw a scatter diagram of the data for one period.
(b) By hand, find a sinusoidal function of the form $y = A\sin(\omega x - \phi) + B$ that fits the data.
(c) Draw the sinusoidal function found in part (b) on the scatter diagram.
(d) Use a graphing utility to find the sinusoidal function of best fit.
(e) Graph the sinusoidal function of best fit on the scatter diagram.

92. Monthly Temperature The following data represent the average monthly temperatures for Chicago, Illinois.

Month, *m*	Average Monthly Temperature, T
January, 1	25
February, 2	28
March, 3	36
April, 4	48
May, 5	61
June, 6	72
July, 7	74
August, 8	75
September, 9	66
October, 10	55
November, 11	39
December, 12	28

Source: U.S. National Oceanic and Atmospheric Administration.

(a) Use a graphing utility to draw a scatter diagram of the data for one period.
(b) By hand, find a sinusoidal function of the form $y = A\sin(\omega x - \phi) + B$ that fits the data.
(c) Draw the sinusoidal function found in part (b) on the scatter diagram.
(d) Use a graphing utility to find the sinusoidal function of best fit.
(e) Graph the sinusoidal function of best fit on the scatter diagram.

93. Hours of Daylight According to the *Old Farmer's Almanac*, in Las Vegas, Nevada, the number of hours of sunlight on the summer solstice is 13.367 and the number of hours of sunlight on the winter solstice is 9.667.
(a) Find a sinusoidal function of the form

$$y = A\sin(\omega x - \phi) + B$$

that fits the data.
(b) Draw a graph of the function found in part (a).
(c) Use the function found in part (a) to predict the number of hours of sunlight on April 1, the 91st day of the year.
(d) Look up the number of hours of sunlight for April 1 in the *Old Farmer's Almanac* and compare the actual hours of daylight to the results found in part (c).

94. Hours of Daylight According to the *Old Farmer's Almanac*, in Seattle, Washington, the number of hours of sunlight on the summer solstice is 13.967 and the number of hours of sunlight on the winter solstice is 8.417.
(a) Find a sinusoidal function of the form

$$y = A\sin(\omega x - \phi) + B$$

that fits the data.
(b) Draw a graph of the function found in part (a).
(c) Use the function found in part (a) to predict the number of hours of sunlight on April 1, the 91st day of the year.
(d) Look up the number of hours of sunlight for April 1 in the *Old Farmer's Almanac* and compare the actual hours of daylight to the results found in part (c).

Chapter Projects

1. **Tides** A partial tide table for September, 2001, for Sabine Pass along the Texas Gulf Coast is given in the table on page 195:

(a) On September 15, when was the tide high? This is called "high tide". On September 19, when was the tide low? This is called "low tide". Most days will have two low tides and two high tides.
(b) Why do you think there is a negative height for the low tide on September 14? What is the tide height measured against?
(c) On your graphing utility, draw a scatter diagram for the data in the table. Let T (time) be the independent variable, with $T = 0$ being 12:00 A.M. on September 1, $T = 24$ being 12:00 A.M. on September 2, etc. Remember that there are 60 minutes in an hour. Let H be the height in feet when converting the times. Also, make sure your graphing utility is in radian mode.
(d) What shape does the data take? What is the period of the data? What is the amplitude? Is the amplitude constant? Explain.

Sept	High Tide Time	Ht(ft)	High Tide Time	Ht(ft)	Low Tide Time	Ht(ft)	Low Tide Time	Ht(ft)	Sun/Moon phase Rise/Set
F 14	03:08a	2.4	11:12a	2.2	08:14a	2.0	07:19p	−0.1	7:00a/7:23p
S 15	03:33a	2.4	12:56p	2.2	08:15a	1.9	08:13p	0.0	7:00a/7:22p
S 16	03:57a	2.3	02:17p	2.3	08:45a	1.6	09:05p	0.3	7:01a/7:20p
M17	04:20a	2.2	03:33p	2.3	09:24a	1.4	09:54p	0.5	7:01a/7:19p
T18	04:41a	2.2	04:47p	2.3	10:08a	1.0	10:43p	1.0	7:02a/7:08p
W19	05:01a	2.0	06:04p	2.3	10:54a	0.7	11:32p	1.4	7:02a/7:17p
T20	05:20a	2.0	07:27p	2.3	11:44a	0.4			7:03a/7:15p

Date from: *www.harbortides.com*

(e) Using Steps 1–4, given on page 183, fit a sine curve to the data. Let the amplitude be the average of the amplitudes that you found in part (c), unless the amplitude was constant. Is there a vertical shift? Is there a phase shift?

(f) Using your graphing utility, find the sinusoidal function of best fit. How does it compare to your equation?

(g) Using the equation found in part (e) and using the sinusoidal equation of best fit found in part (f), predict the high tides and the low tides on September 21.

(h) Looking at the times of day that the low tides occur, what do you think causes the low tides to vary so much each day? Explain. Does this seem to have the same type of effect on the high tides? Explain.

2. **Identifying Mountain Peaks in Hawaii** Suppose that you are standing on the southeastern shore of Oahu and you see three mountain peaks on the horizon. You want to determine which mountains are visible from Oahu. The possible mountain peaks that can be seen from Oahu and the height (above sea level) of their peaks are:

Island	Distance (miles)	Mountain	Height (feet)
Lanai	65	Lanaihale	3,370
Maui	110	Haleakala	10,023
Hawaii	190	Mauna Kea	13,796
Molokai	40	Kamakou	4,961

(a) To determine which of these mountain peaks would be visible from Oahu, consider that you are standing on the shore and looking "straight out" so that your line of sight is tangent to the surface of Earth at the point you're standing. Make a sketch of the right triangle formed by your sight line, the radius from the center of Earth to the point you are standing, and the line from the center of Earth through Lanai.

(b) Assuming that the radius of Earth is 3960 miles, determine the angle formed at the center of Earth.

(c) Determine the length of the hypotenuse of the triangle. Is Lanaihale visible from Oahu?

(d) Repeat parts (a)–(d) for the other three islands.

(e) Which three mountains are visible from Oahu?

3. **CBL Experiment** Using a CBL, the microphone probe and a tuning fork, record the amplitude, frequency, and period of the sound from the graph of the sound created by the tuning fork over time. Repeat the experiment for different tuning forks.

Cumulative Review

1. Find the real solutions, if any, of the equation $2x^2 + x - 1 = 0$.
2. Find an equation for the line with slope -3, containing the point $(-2, 5)$.
3. Find an equation for the circle of radius 4 and center at the point $(0, -2)$.
4. Graph the equation $2x - 3y = 12$.
5. Graph the equation $x^2 + y^2 - 2x + 4y - 4 = 0$.
6. Use transformations to graph the function $y = (x - 3)^2 + 2$.
7. Sketch a graph of each of the following functions. Label at least three points on each graph.
   (a) $y = x^2$
   (b) $y = x^3$
   (c) $y = \sin x$
   (d) $y = \tan x$
8. Find the inverse function of $f(x) = 3x - 2$.
9. Find the exact value of $(\sin 14°)^2 + (\sin 76°)^2 - 3$.
10. Graph $y = 3\sin(2x)$.
11. Find the exact value of $\tan\frac{\pi}{4} - 3\cos\frac{\pi}{6} + \csc\frac{\pi}{6}$.
12. Find a sinusoidal function for the following graph.

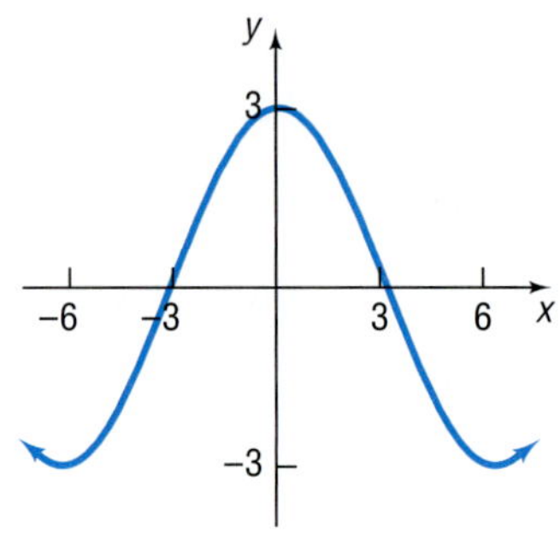

APPLICATIONS OF TRIGONOMETRIC FUNCTIONS

CHAPTER 4

New maps pinpoint Lewis and Clark's journey through Missouri

KANSAS CITY, Mo.—Nearly two centuries ago, Congress commissioned Meriwether Lewis and William Clark to explore trade routes to the West.

Their long-documented journey took them through Missouri, but their exact route has been up for debate.

Now, modern computer technology combined with 19th century land surveys may provide a precise picture. The latest computer-generated maps of Lewis and Clark's expedition were unveiled here this week by Missouri Secretary of State Matt Blunt.

The maps combine the terrain of the early 19th century with contemporary geographical markers. They are the most precise to date of Lewis and Clark's journeys through Missouri in 1804 and 1806, said the project's lead researcher, Jim Harlan.

"We've known what they've done, but not with this much certainty," said Harlan, geographic resource project director at the University of Missouri-Columbia. "I think we need more information and less speculation. That's what this is all about."

SOURCE: Sophia Maines, "New Maps Pinpoint Lewis and Clark's Journey Through Missouri," *The Kansas City Star*, August 1, 2001 Distributed by Knight Ridder/Tribune Information Services..

SEE CHAPTER PROJECT 1.

OUTLINE

For additional study help, go to

www.prenhall.com/sullivanegu3e

Materials include:

- Graphing Calculator Help
- Chapter Quiz
- Chapter Test
- PowerPoint Downloads
- Chapter Projects
- Student Tips

A Look Back, A Look Forward

In Chapter 2, we defined the six trigonometric functions using right triangles and then extended the definition to include general angles. In particular, we learned to evaluate the trigonometric functions. We also learned how to graph sinusoidal functions.

In this chapter, we use the trigonometric functions to solve applied problems. The first four sections deal with applications involving right triangles and *oblique triangles*, triangles that do not have a right angle. To solve problems involving oblique triangles, we will develop the Law of Sines and the Law of Cosines. We will also develop formulas for finding the area of a triangle.

The final section deals with applications of sinusoidal functions involving simple harmonic motion and damped motion.

PREPARING FOR THIS SECTION

Before getting started, review the following:

- ✓ Pythagorean Theorem (Section A.2, pp. 565–567)
- ✓ Complementary Angle Theorem (Section 2.2, pp. 118–120)
- ✓ Trigonometric Equations (I) (Section 3.7, pp. 247–250)

4.1 APPLICATIONS INVOLVING RIGHT TRIANGLES

OBJECTIVES

1. Solve Right Triangles
2. Solve Applied Problems

Solving Right Triangles

1 In the discussion that follows, we will always label a right triangle so that side a is opposite angle α, side b is opposite angle β, and side c is the hypotenuse, as shown in Figure 1. **To solve a right triangle** means to find the missing lengths of its sides and the measurements of its angles. We shall follow the practice of expressing the lengths of the sides rounded to two decimal places and of expressing angles in degrees rounded to one decimal place. (Be sure that your calculator is in degree mode.)

Figure 1

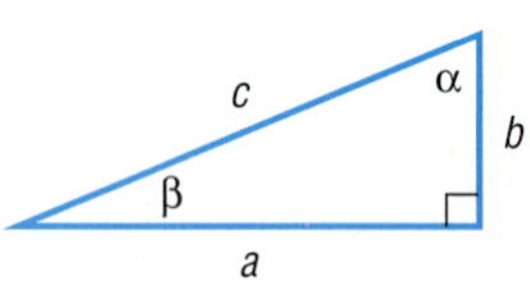

To solve a right triangle, we need to know one of the acute angles α or β and a side, or else two sides. Then we make use of the Pythagorean Theorem and the fact that the sum of the angles of a triangle is 180°. The sum of the angles α and β in a right triangle is therefore 90°.

For the right triangle shown in Figure 1, we have

$$c^2 = a^2 + b^2, \qquad \alpha + \beta = 90^\circ$$

Figure 2

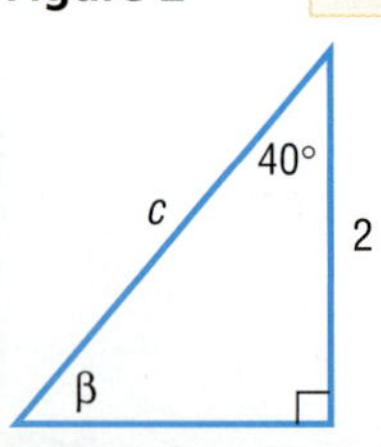

EXAMPLE 1 **Solving a Right Triangle**

Use Figure 2. If $b = 2$ and $\alpha = 40^\circ$, find a, c, and β.

Solution Since $\alpha = 40^\circ$ and $\alpha + \beta = 90^\circ$, we find that $\beta = 50^\circ$. To find the sides a and c, we use the facts that

$$\tan 40^\circ = \frac{a}{2} \quad \text{and} \quad \cos 40^\circ = \frac{2}{c}$$

Now solve for a and c.

$$a = 2\tan 40° \approx 1.68 \quad \text{and} \quad c = \frac{2}{\cos 40°} \approx 2.61$$

NOW WORK PROBLEM 1.

EXAMPLE 2 Solving a Right Triangle

Use Figure 3. If $a = 3$ and $b = 2$, find c, α, and β.

Figure 3

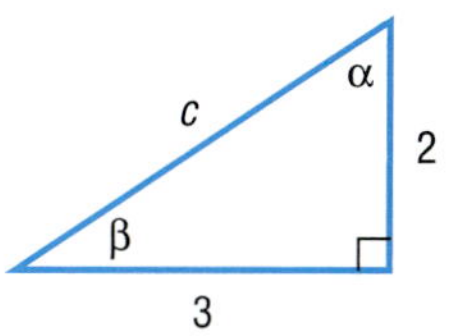

Solution Since $a = 3$ and $b = 2$, then, by the Pythagorean Theorem, we have

$$c^2 = a^2 + b^2 = 3^2 + 2^2 = 9 + 4 = 13$$
$$c = \sqrt{13} \approx 3.61$$

To find angle α, we use the fact that

$$\tan\alpha = \frac{3}{2} \quad \text{so} \quad \alpha = \tan^{-1}\frac{3}{2}$$

Set the mode on your calculator to degrees. Then, rounded to one decimal place, we find that $\alpha = 56.3°$. Since $\alpha + \beta = 90°$, we find that $\beta = 33.7°$.

Note: To avoid round-off errors when using a calculator, we will store unrounded values in memory for use in subsequent calculations.

NOW WORK PROBLEM 11.

Applications

2 One common use for trigonometry is to measure heights and distances that are either awkward or impossible to measure by ordinary means.

EXAMPLE 3 Finding the Width of a River

A surveyor can measure the width of a river by setting up a transit* at a point C on one side of the river and taking a sighting of a point A on the other side. Refer to Figure 4. After turning through an angle of 90° at C, the surveyor walks a distance of 200 meters to point B. Using the transit at B, the angle β is measured and found to be 20°. What is the width of the river rounded to the nearest meter?

Figure 4

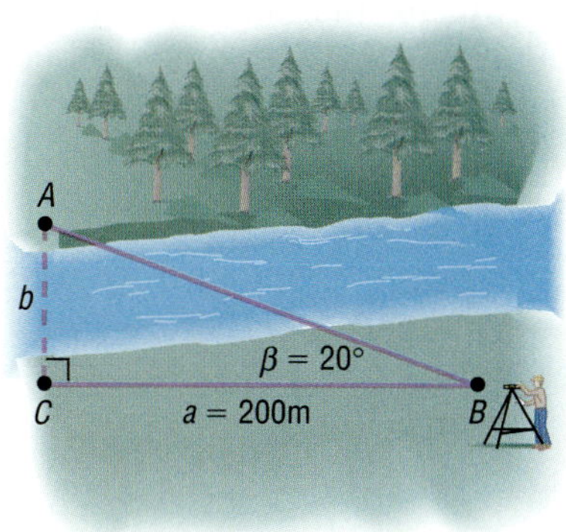

Solution We seek the length of side b. We know a and β, so we use the fact that

$$\tan\beta = \frac{b}{a}$$

to get

$$\tan 20° = \frac{b}{200}$$
$$b = 200\tan 20° \approx 72.79 \text{ meters}$$

The width of the river is 73 meters, rounded to the nearest meter.

NOW WORK PROBLEM 21.

* An instrument used in surveying to measure angles.

EXAMPLE 4 Finding the Inclination of a Mountain Trail

A straight trail leads from the Alpine Hotel, elevation 8000 feet, to a scenic overlook, elevation 11,100 feet. The length of the trail is 14,100 feet. What is the inclination of the trail? That is, what is the angle β in Figure 5?

Figure 5

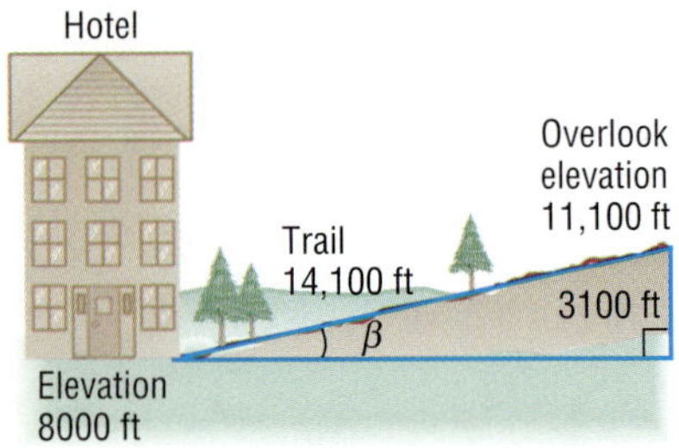

Solution As Figure 5 illustrates, the angle β obeys the equation

$$\sin\beta = \frac{3100}{14{,}100}$$

Using a calculator,

$$\beta = \sin^{-1}\frac{3100}{14{,}100} \approx 12.7°$$

The inclination of the trail is approximately 12.7°. ■

Vertical heights can sometimes be measured using either the angle of elevation or the angle of depression. If a person is looking up at an object, the acute angle measured from the horizontal to a line-of-sight observation of the object is called the **angle of elevation**. See Figure 6(a).

If a person is standing on a cliff looking down at an object, the acute angle made by the line-of-sight observation of the object and the horizontal is called the **angle of depression**. See Figure 6(b).

Figure 6

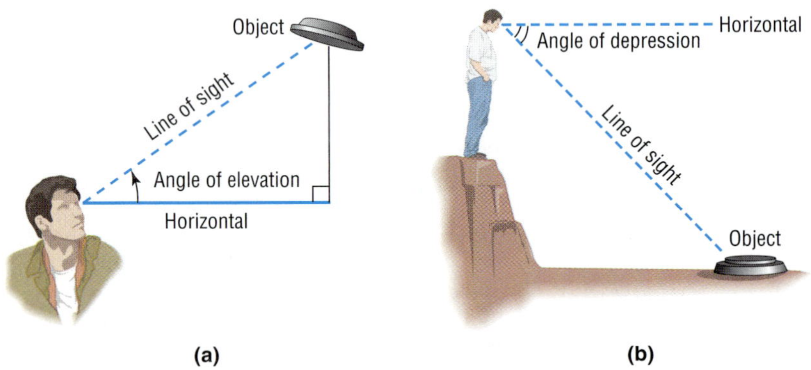

EXAMPLE 5 Finding the Height of a Cloud

Meteorologists find the height of a cloud using an instrument called a **ceilometer**. A ceilometer consists of a **light projector** that directs a vertical light beam up to the cloud base and a **light detector** that scans the cloud to detect the light beam. See Figure 7(a). On December 1, 2001, at Midway Airport in Chicago, a ceilometer with a base of 300 feet was employed to find the height of the cloud cover. If the angle of elevation of the light detector is 75°, what is the height of the cloud cover?

Figure 7

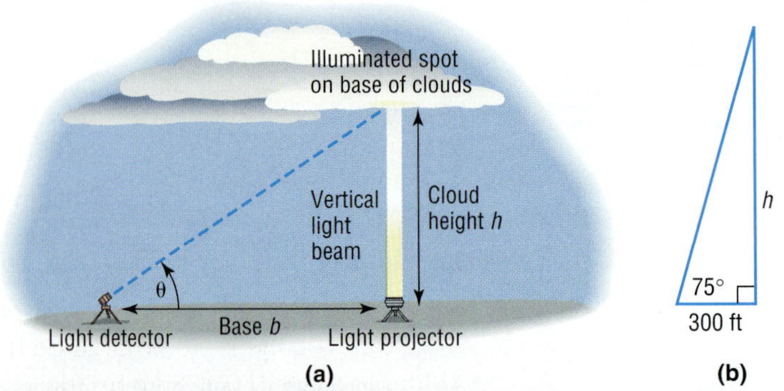

Solution Figure 7(b) illustrates the situation. To find the height h, we use the fact that $\tan 75^\circ = \frac{h}{300}$, so

$$h = 300 \tan 75^\circ \approx 1120 \text{ feet}$$

The ceiling (height to the base of the cloud cover) is approximately 1120 feet.

NOW WORK PROBLEM 23.

The idea behind Example 5 can also be used to find the height of an object with a base that is not accessible to the horizontal.

EXAMPLE 6 Finding the Height of a Statue on a Building

Adorning the top of the Board of Trade building in Chicago is a statue of Ceres, the Greek goddess of wheat. From street level, two observations are taken 400 feet from the center of the building. The angle of elevation to the base of the statue is found to be 45°; the angle of elevation to the top of the statue is 47.2°. See Figure 8(a). What is the height of the statue?

Figure 8

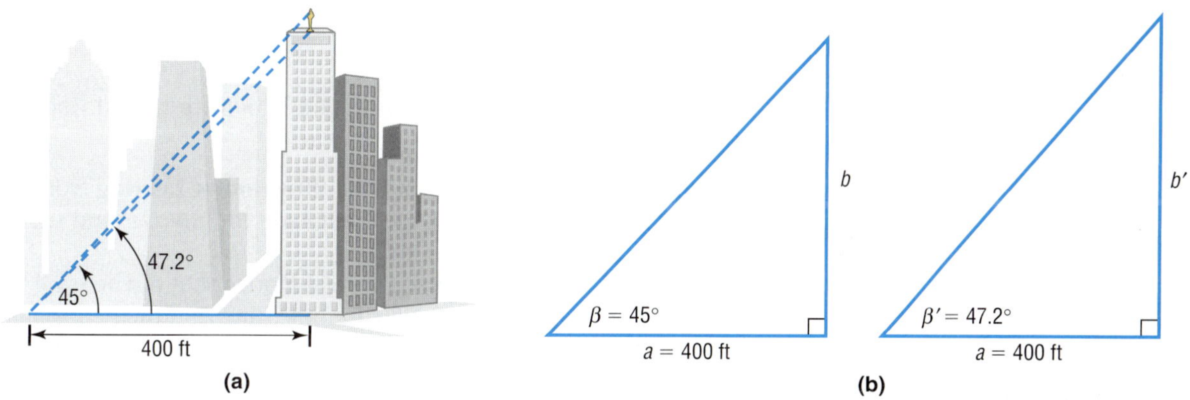

Solution Figure 8(b) shows two triangles that replicate Figure 8(a). The height of the statue of Ceres will be $b' - b$. To find b and b', we refer to Figure 8(b).

$$\tan 45^\circ = \frac{b}{400} \qquad \tan 47.2^\circ = \frac{b'}{400}$$

$$b = 400 \tan 45^\circ = 400 \qquad b' = 400 \tan 47.2^\circ \approx 432$$

The height of the statue is approximately $432 - 400 = 32$ feet.

NOW WORK PROBLEM 31.

EXAMPLE 7 The Gibb's Hill Lighthouse, Southampton, Bermuda

In operation since 1846, the Gibb's Hill Lighthouse stands 117 feet high on a hill 245 feet high, so its beam of light is 362 feet above sea level. A brochure states that the light can be seen on the horizon about 26 miles distant. Verify the accuracy of this statement.

Solution Figure 9 illustrates the situation. The central angle θ, positioned at the center of Earth, radius 3960 miles, obeys the equation

Figure 9

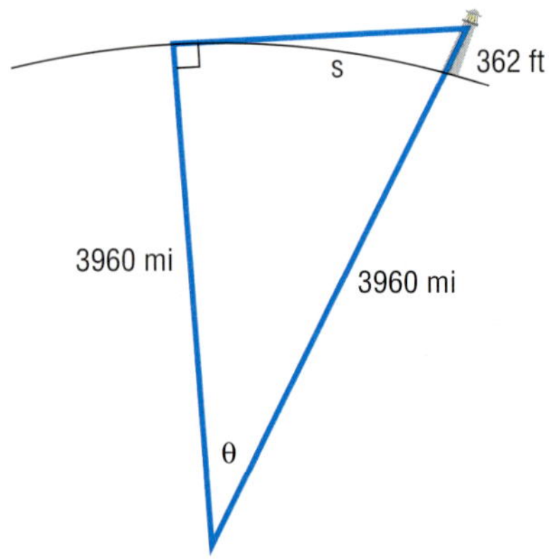

$$\cos\theta = \frac{3960}{3960 + \dfrac{362}{5280}} \approx 0.999982687 \qquad \text{1 mile} = \text{5280 feet}$$

Solving for θ, we find

$$\theta \approx 0.33715^\circ \approx 20.23'$$

The brochure does not indicate whether the distance is measured in nautical miles or statute miles. Let's calculate both distances.

The distance s in nautical miles (refer to Problem 104, p. 111) is the measure of angle θ in minutes, so $s = 20.23$ nautical miles.

The distance s in statute miles is given by the formula $s = r\theta$, where θ is measured in radians. Then, since

$$\theta = 20.23' \underset{\substack{\uparrow \\ 1' = \frac{1}{60}^\circ}}{=} 0.33715^\circ \underset{\substack{\uparrow \\ 1^\circ = \frac{\pi}{180}\text{ radian}}}{=} 0.00588 \text{ radian}$$

we find that

$$s = r\theta = (3960)(0.00588) = 23.3 \text{ miles}$$

In either case, it would seem that the brochure overstated the distance somewhat. ■

Figure 10

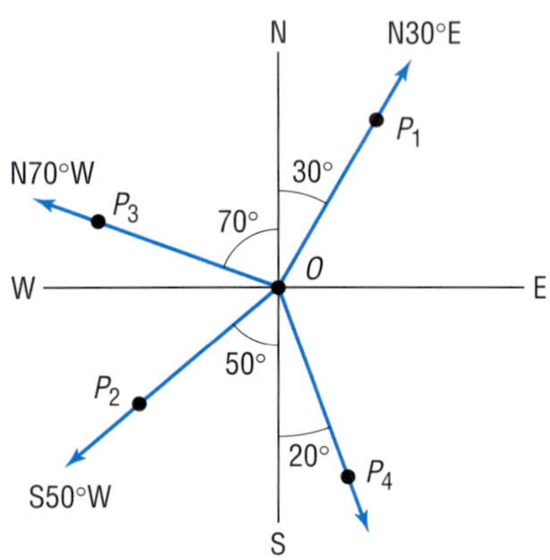

In navigation and surveying, the **direction** or **bearing** from a point O to a point P equals the acute angle θ between the ray OP and the vertical line through O, the north–south line.

Figure 10 illustrates some bearings. Notice that the bearing from O to P_1 is denoted by the symbolism N30°E, indicating that the bearing is 30° east of north. In writing the bearing from O to P, the direction north or south always appears first, followed by an acute angle, followed by east or west. In Figure 10, the bearing from O to P_2 is S50°W, and from O to P_3 it is N70°W.

EXAMPLE 8 Finding the Bearing of an Object

In Figure 10, what is the bearing from O to an object at P_4?

Solution The acute angle between the ray OP_4 and the north–south line through O is given as 20°. The bearing from O to P_4 is S20°E. ■

EXAMPLE 9 Finding the Bearing of an Airplane

A Boeing 777 aircraft takes off from O'Hare Airport on runway 2 LEFT which has a bearing of N20°E.* After flying for 1 mile, the pilot of the air craft requests permission to turn 90° and head toward the northwest. The request is granted. After the plane goes 2 miles in this direction, what bearing should the control tower use to locate the aircraft?

*In air navigation, the term **azimuth** is employed to denote the positive angle measured clockwise from the north (N) to a ray OP. In Figure 10, the azimuth from O to P_1 is 30°; the azimuth from O to P_2 is 230°; the azimuth from O to P_3 is 290°. In naming runways, the units digit is left off the azimuth. Runway 2 LEFT means the left runway with a direction of azimuth 20° (bearing N20°E). Runway 23 is the runway with azimuth 230° and bearing S50°W.

Figure 11

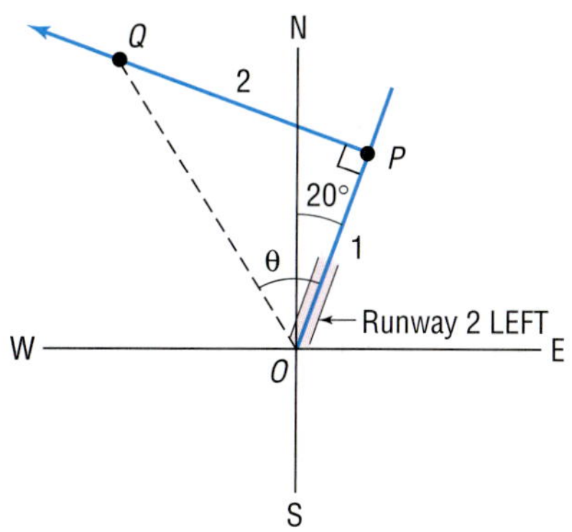

Solution Figure 11 illustrates the situation. After flying 1 mile from the airport O (the control tower), the aircraft is at P. After turning 90° toward the northwest and flying 2 miles, the aircraft is at the point Q. In triangle OPQ, the angle θ obeys the equation

$$\tan\theta = \frac{2}{1} = 2 \quad \text{so} \quad \theta = \tan^{-1} 2 \approx 63.4°$$

The acute angle between the north-south line and the ray OQ is $63.4° - 20° = 43.4°$. The bearing of the aircraft from O to Q is N43.4°W. ■

NOW WORK PROBLEM 39.

4.1 Concepts and Vocabulary

In Problems 1 and 2, fill in the blanks.

1. When you look up at an object, the acute angle measured from the horizontal to a line-of-sight observation of the object is called the ________ ________ ________.
2. When you look down at an object, the acute angle described in Problem 1 is called the ________ ________ ________.

In Problems 3 and 4, answer True or False to each statement.

3. In a right triangle, if two sides are known, we can solve the triangle.
4. In a right triangle, if we know the two acute angles, we can solve the triangle.

5. Explain how you would measure the width of the Grand Canyon from a point on its ridge.
6. Explain how you would measure the height of a TV tower that is on the roof of a tall building.

4.1 Exercises

In Problems 1–14, use the right triangle shown in the margin. Then, using the given information, solve the triangle.

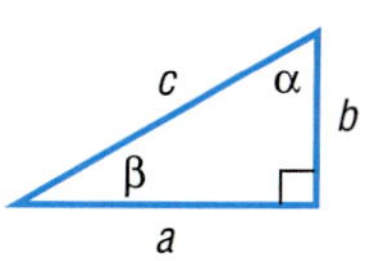

1. $b = 5$, $\beta = 20°$; find a, c, and α
2. $b = 4$, $\beta = 10°$; find a, c, and α
3. $a = 6$, $\beta = 40°$; find b, c, and α
4. $a = 7$, $\beta = 50°$; find b, c, and α
5. $b = 4$, $\alpha = 10°$; find a, c, and β
6. $b = 6$, $\alpha = 20°$; find a, c, and β
7. $a = 5$, $\alpha = 25°$; find b, c, and β
8. $a = 6$, $\alpha = 40°$; find b, c, and β
9. $c = 9$, $\beta = 20°$; find b, a, and α
10. $c = 10$, $\alpha = 40°$; find b, a, and β
11. $a = 5$, $b = 3$; find c, α, and β
12. $a = 2$, $b = 8$; find c, α, and β
13. $a = 2$, $c = 5$; find b, α, and β
14. $b = 4$, $c = 6$; find a, α, and β

15. **Geometry** A right triangle has a hypotenuse of length 8 inches. If one angle is 35°, find the length of each leg.
16. **Geometry** A right triangle has a hypotenuse of length 10 centimeters. If one angle is 40°, find the length of each leg.
17. **Geometry** A right triangle contains a 25° angle. If one leg is of length 5 inches, what is the length of the hypotenuse?
[**Hint:** Two answers are possible.]

18. Geometry A right triangle contains an angle of $\pi/8$ radian. If one leg is of length 3 meters, what is the length of the hypotenuse?

[**Hint:** Two answers are possible.]

19. Geometry The hypotenuse of a right triangle is 5 inches. If one leg is 2 inches, find the degree measure of each angle.

20. Geometry The hypotenuse of a right triangle is 3 feet. If one leg is 1 foot, find the degree measure of each angle.

21. Finding the Width of a Gorge Find the distance from A to C across the gorge illustrated in the figure.

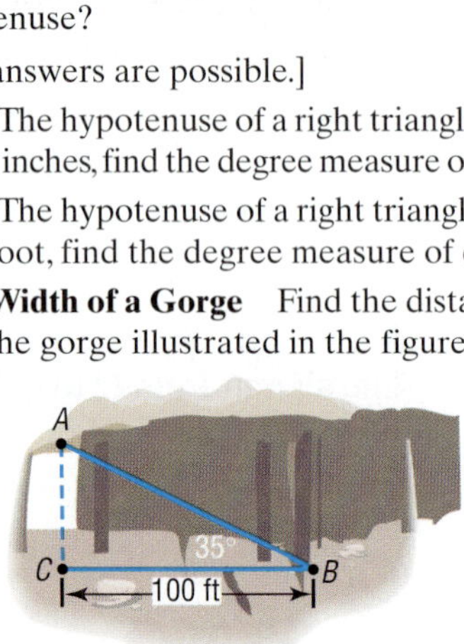

22. Finding the Distance across a Pond Find the distance from A to C across the pond illustrated in the figure.

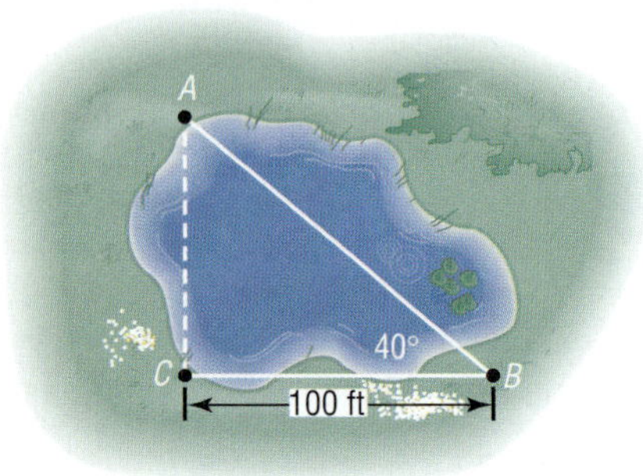

23. The Eiffel Tower The tallest tower built before the era of television masts, the Eiffel Tower was completed on March 31, 1889. Find the height of the Eiffel Tower (before a television mast was added to the top) using the information given in the illustration.

24. Finding the Distance of a Ship from Shore A ship, offshore from a vertical cliff known to be 100 feet in height, takes a sighting of the top of the cliff. If the angle of elevation is found to be 25°, how far offshore is the ship?

25. Finding the Distance to a Plateau Suppose that you are headed toward a plateau 50 meters high. If the angle of elevation to the top of the plateau is 20°, how far are you from the base of the plateau?

26. Statue of Liberty A ship is just offshore of New York City. A sighting is taken of the Statue of Liberty, which is about 305 feet tall. If the angle of elevation to the top of the statue is 20°, how far is the ship from the base of the statue?

27. Finding the Reach of a Ladder A 22-foot extension ladder leaning against a building makes a 70° angle with the ground. How far up the building does the ladder touch?

28. Finding the Height of a Building To measure the height of a building, two sightings are taken a distance of 50 feet apart. If the first angle of elevation is 40° and the second is 32°, what is the height of the building?

29. Directing a Laser Beam A laser beam is to be directed through a small hole in the center of a circle of radius 10 feet. The origin of the beam is 35 feet from the circle (see the figure). At what angle of elevation should the beam be aimed to ensure that it goes through the hole?

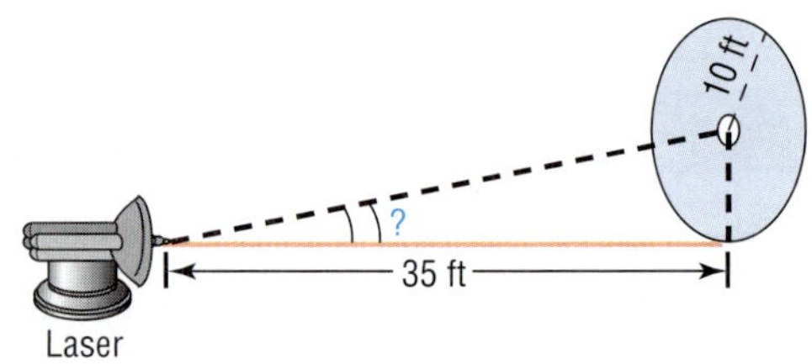

30. Finding the Angle of Elevation of the Sun At 10 AM on April 26, 2000, a building 300 feet high casts a shadow 50 feet long. What is the angle of elevation of the Sun?

31. Mt. Rushmore To measure the height of Lincoln's caricature on Mt. Rushmore, two sightings 800 feet from the base of the mountain are taken. If the angle of elevation to the bottom of Lincoln's face is 32° and the angle of elevation to the top is 35°, what is the height of Lincoln's face?

32. Finding the Distance between Two Objects A blimp, suspended in the air at a height of 500 feet, lies directly over a line from Soldier Field to the Adler Planetarium on Lake Michigan (see the figure). If the angle of depression from the blimp to the stadium is 32° and from the blimp to the planetarium is 23°, find the distance between Soldier Field and the Adler Planetarium.

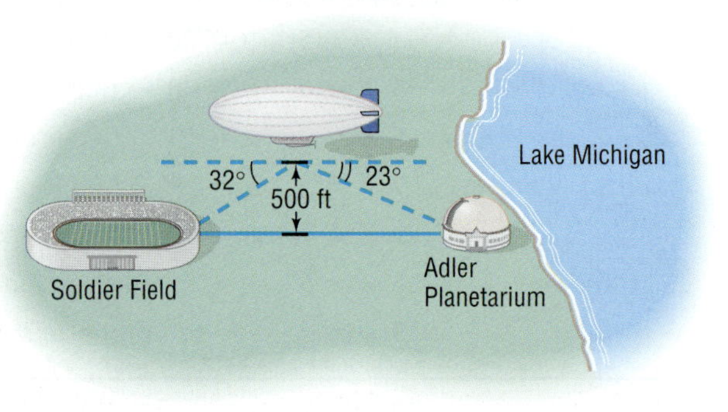

33. Finding the Length of a Guy Wire A radio transmission tower is 200 feet high. How long should a guy wire be if it is to be attached to the tower 10 feet from the top and is to make an angle of 21° with the ground?

34. Finding the Height of a Tower A guy wire 80 feet long is attached to the top of a radio transmission tower, making an angle of 25° with the ground. How high is the tower?

35. Washington Monument The angle of elevation to the top of the Washington Monument is 35.1° at the instant it casts a shadow 789 feet long. Use this information to calculate the height of the monument.

36. Finding the Length of a Mountain Trail A straight trail with an inclination of 17° leads from a hotel at an elevation of 9000 feet to a mountain lake at an elevation of 11,200 feet. What is the length of the trail?

37. Finding the Speed of a Truck A state trooper is hidden 30 feet from a highway. One second after a truck passes, the angle θ between the highway and the line of observation from the patrol car to the truck is measured. See the illustration.

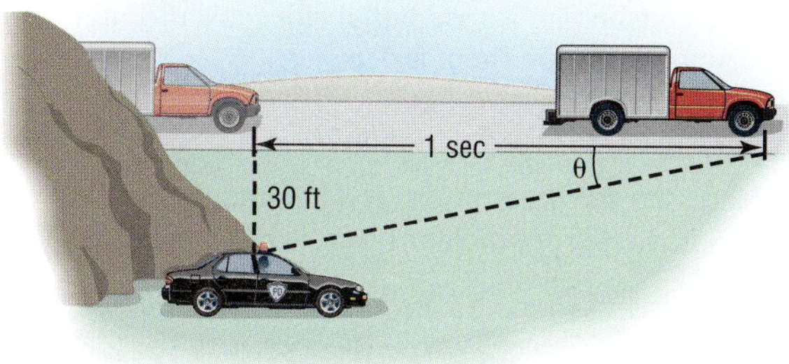

(a) If the angle measures 15°, how fast is the truck traveling? Express the answer in feet per second and in miles per hour.

(b) If the angle measures 20°, how fast is the truck traveling? Express the answer in feet per second and in miles per hour.

(c) If the speed limit is 55 miles per hour and a speeding ticket is issued for speeds of 5 miles per hour or more over the limit, for what angles should the trooper issue a ticket?

38. Security A security camera in a neighborhood bank is mounted on a wall 9 feet above the floor. What angle of depression should be used if the camera is to be directed to a spot 6 feet above the floor and 12 feet from the wall?

39. Finding the Bearing of an Aircraft A DC-9 aircraft leaves Midway Airport from runway 4 RIGHT, whose bearing is N40°E. After flying for 1/2 mile, the pilot requests permission to turn 90° and head toward the southeast. The permission is granted. After the airplane goes 1 mile in this direction, what bearing should the control tower use to locate the aircraft?

40. Finding the Bearing of a Ship A ship leaves the port of Miami with a bearing of S80°E and a speed of 15 knots. After 1 hour, the ship turns 90° toward the south. After 2 hours, maintaining the same speed, what is the bearing to the ship from port?

41. Shooting Free Throws in Basketball The eyes of a basketball player are 6 feet above the floor. The player is at the free-throw line, which is 15 feet from the center of the basket rim (see the figure). What is the angle of elevation from the player's eyes to the center of the rim?

[**Hint:** The rim is 10 feet above the floor.]

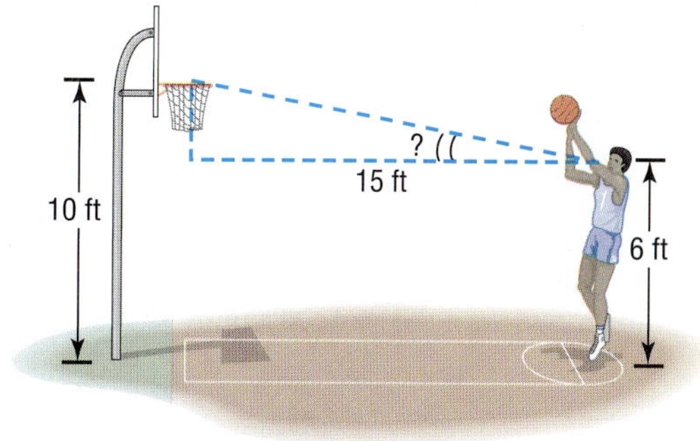

42. Finding the Pitch of a Roof A carpenter is preparing to put a roof on a garage that is 20 feet by 40 feet by 20 feet. A steel support beam 46 feet in length is positioned in the center of the garage. To support the roof, another beam will be attached to the top of the center beam (see the figure). At what angle of elevation is the new beam? In other words, what is the pitch of the roof?

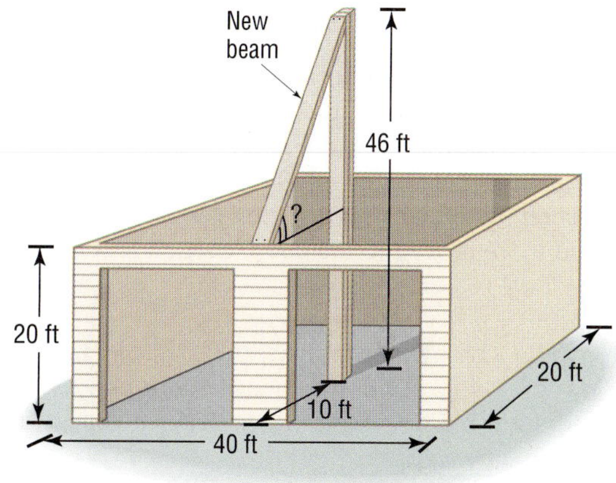

43. Constructing a Highway A highway whose primary directions are north–south is being constructed along the west coast of Florida. Near Naples, a bay obstructs the straight path of the road. Since the cost of a bridge is prohibitive, engineers decide to go around the bay. The illustration shows the path that they decide on and the measurements taken. What is the length of highway needed to go around the bay?

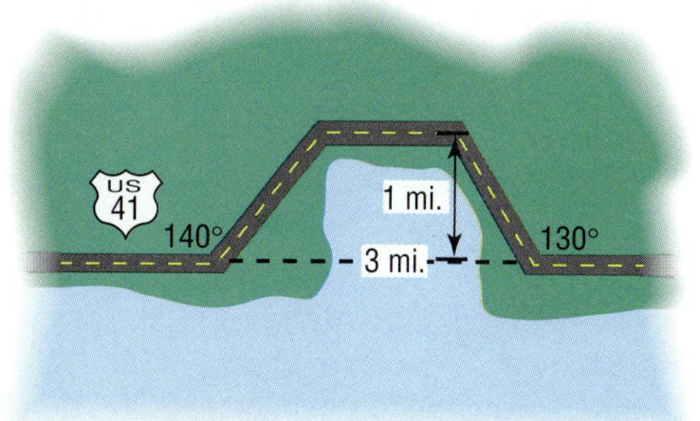

44. Surveillance Satellites A surveillance satellite circles Earth at a height of h miles above the surface. Suppose that d is the distance, in miles, on the surface of Earth that can be observed from the satellite. See the illustration.

(a) Find an equation that relates the central angle θ to the height h.

(b) Find an equation that relates the observable distance d and θ.

(c) Find an equation that relates d and h.

(d) If d is to be 2500 miles, how high must the satellite orbit above Earth?

(e) If the satellite orbits at a height of 300 miles, what distance d on the surface can be observed?

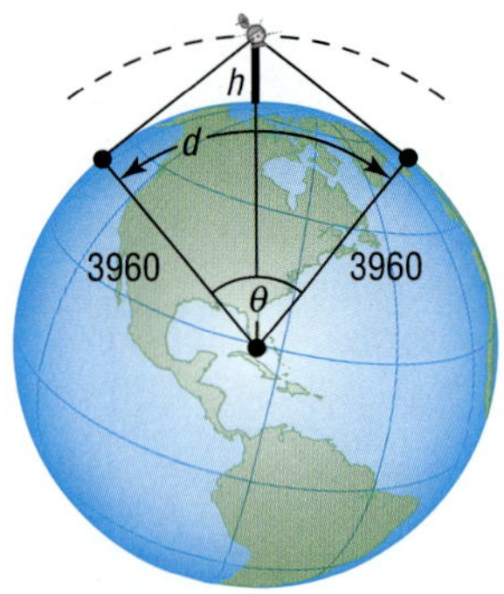

45. Photography A camera is mounted on a tripod 4 feet high at a distance of 10 feet from George, who is 6 feet tall. See the illustration. If the camera lens has angles of depression and elevation of 20°, will George's feet and head be seen by the lens? If not, how far back will the camera need to be moved to include George's feet and head?

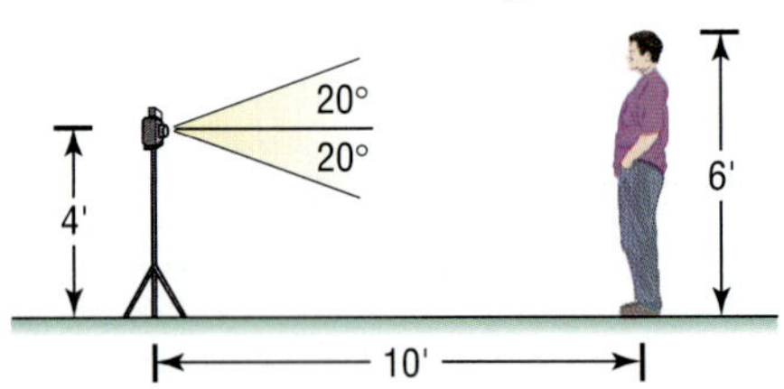

46. Construction A ramp for wheelchair accessibility is to be constructed with an angle of elevation of 15° and a final height of 5 feet. How long is the ramp?

47. Geometry A rectangle is inscribed in a semicircle of radius 1. See the illustration.

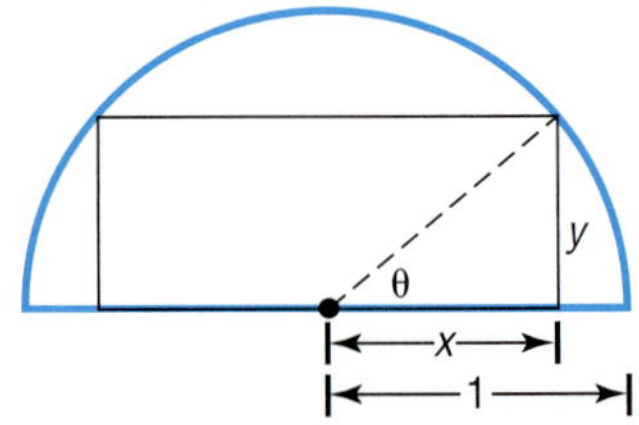

(a) Express the area A of the rectangle as a function of the angle θ shown in the illustration.

(b) Show that $A = \sin(2\theta)$.

(c) Find the angle θ that results in the largest area A.

(d) Find the dimensions of this largest rectangle.

48. Area of an Isosceles Triangle Show that the area A of an isosceles triangle, whose equal sides are of length s and the angle between them is θ, is

$$A = \frac{1}{2}s^2 \sin\theta$$

[**Hint:** See the illustration. The height h bisects the angle θ and is the perpendicular bisector of the base.]

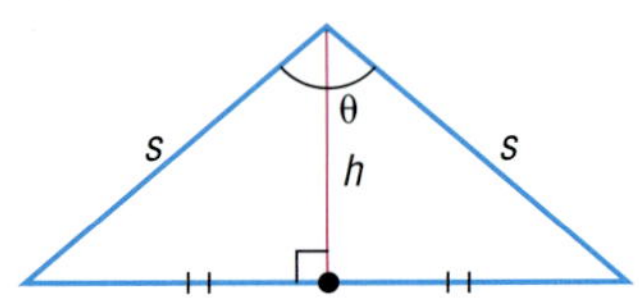

49. The Gibb's Hill Lighthouse, Southampton, Bermuda In operation since 1846, the Gibb's Hill Lighthouse stands 117 feet high on a hill 245 feet high, so its beam of light is 362 feet above sea level. A brochure states that ships 40 miles away can see the light and planes flying at 10,000 feet can see it 120 miles away. Verify the accuracy of these statements. What assumption did the brochure make about the height of the ship?

PREPARING FOR THIS SECTION

Before getting started, review the following:

✓ Trigonometric Equations (I) (Section 3.7, pp. 247–250) ✓ Difference Formulas for Sine (Section 3.4, p. 225)

4.2 THE LAW OF SINES

OBJECTIVES

1. Solve SAA or ASA Triangles
2. Solve SSA Triangles
3. Solve Applied Problems Using the Law of Sines

If none of the angles of a triangle is a right angle, the triangle is called **oblique**. An oblique triangle will have either three acute angles or two acute angles and one obtuse angle (an angle between 90° and 180°). See Figure 12.

Figure 12

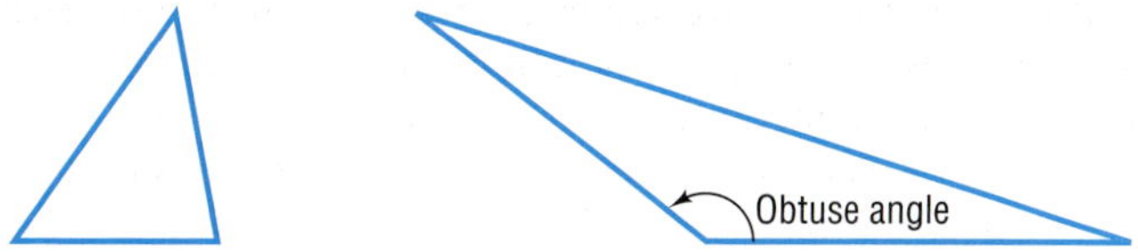

(a) All angles are acute **(b)** Two acute angles and one obtuse angle

Figure 13

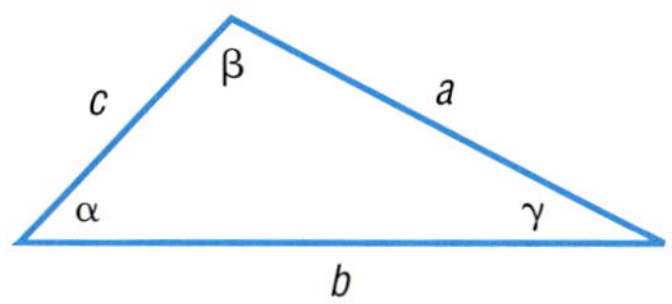

In the discussion that follows, we will always label an oblique triangle so that side a is opposite angle α, side b is opposite angle β, and side c is opposite angle γ, as shown in Figure 13.

To **solve an oblique triangle** means to find the lengths of its sides and the measurements of its angles. To do this, we need to know the length of one side along with: (1) two angles or (2) the other two sides or (3) one angle and one other side. Knowing three angles of a triangle determines a family of *similar triangles,* that is, triangles that have the same shape but different sizes. There are four possibilities to consider:

CASE 1: One side and two angles are known (ASA or SAA).
CASE 2: Two sides and the angle opposite one of them are known (SSA).
CASE 3: Two sides and the included angle are known (SAS).
CASE 4: Three sides are known (SSS).

Figure 14 illustrates the four cases.

Figure 14

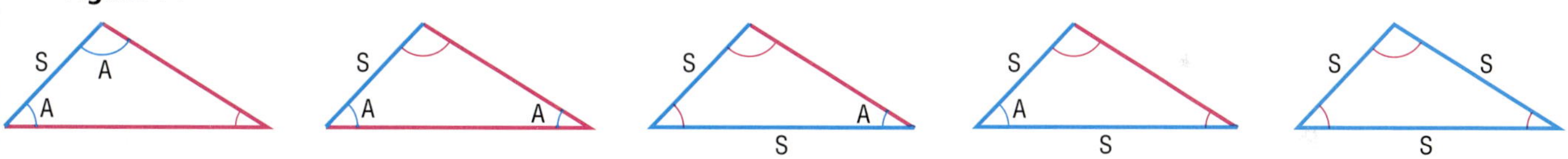

Case 1: ASA Case 1: SAA Case 2: SSA Case 3: SAS Case 4: SSS

The **Law of Sines** is used to solve triangles for which Case 1 or 2 holds. Cases 3 and 4 are considered when we study the Law of Cosines in Section 4.3.

Theorem

Law of Sines

For a triangle with sides a, b, c and opposite angles α, β, γ, respectively,

$$\frac{\sin \alpha}{a} = \frac{\sin \beta}{b} = \frac{\sin \gamma}{c} \qquad (1)$$

Proof To prove the Law of Sines, we construct an altitude of length h from one of the vertices of a triangle. Figure 15(a) shows h for a triangle with three acute angles, and Figure 15(b) shows h for a triangle with an obtuse angle. In each case, the altitude is drawn from the vertex at β. Using either illustration, we have

$$\sin \gamma = \frac{h}{a}$$

Figure 15

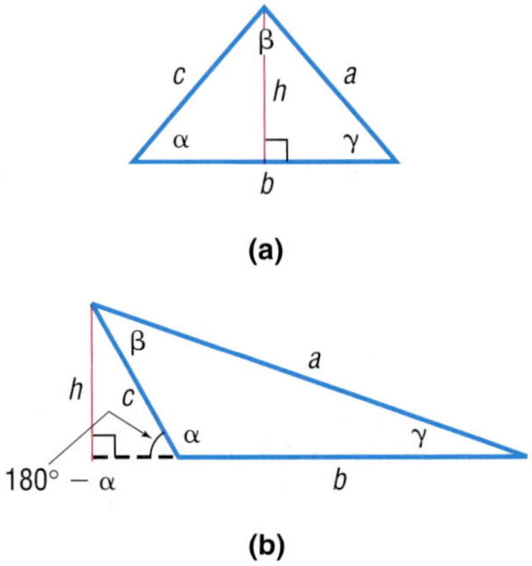

from which

$$h = a \sin \gamma \tag{2}$$

From Figure 15(a), it also follows that

$$\sin \alpha = \frac{h}{c}$$

from which

$$h = c \sin \alpha \tag{3}$$

From Figure 15(b), it follows that

$$\sin(180^\circ - \alpha) \underset{\uparrow}{=} \sin \alpha = \frac{h}{c}$$

Difference formula

which again gives

$$h = c \sin \alpha$$

So, whether the triangle has three acute angles or has two acute angles and one obtuse angle, equations (2) and (3) hold. As a result, we may equate the expressions for h in equations (2) and (3) to get

$$a \sin \gamma = c \sin \alpha$$

from which

$$\frac{\sin \alpha}{a} = \frac{\sin \gamma}{c} \tag{4}$$

In a similar manner, by constructing the altitude h' from the vertex of angle α as shown in Figure 16, we can show that

Figure 16

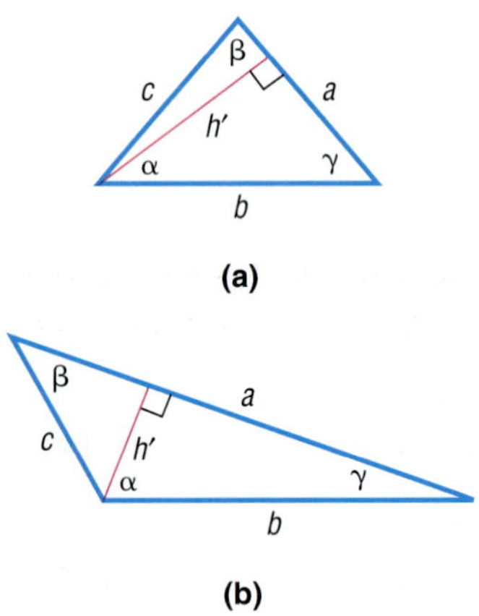

$$\sin \beta = \frac{h'}{c} \quad \text{and} \quad \sin \gamma = \frac{h'}{b}$$

Equating the expressions for h', we find that

$$h' = c \sin \beta = b \sin \gamma$$

from which

$$\frac{\sin \beta}{b} = \frac{\sin \gamma}{c} \tag{5}$$

When equations (4) and (5) are combined, we have equation (1), the Law of Sines. ■

In applying the Law of Sines to solve triangles, we use the fact that the sum of the angles of any triangle equals 180°; that is,

$$\alpha + \beta + \gamma = 180^\circ \tag{6}$$

1 Our first two examples show how to solve a triangle when one side and two angles are known (Case 1: SAA or ASA).

EXAMPLE 1 Using the Law of Sines to Solve a SAA Triangle

Solve the triangle: $\alpha = 40°, \beta = 60°, a = 4$

Solution Figure 17 shows the triangle that we want to solve. The third angle γ is found using equation (6).

Figure 17

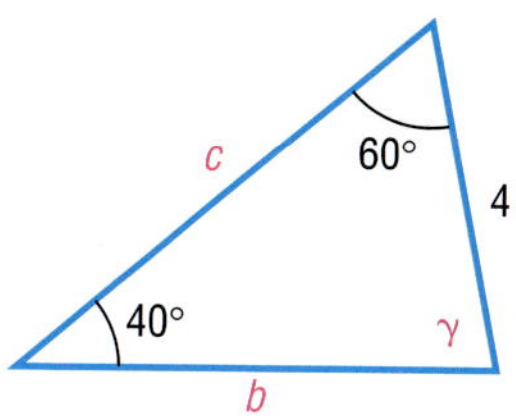

$$\alpha + \beta + \gamma = 180°$$
$$40° + 60° + \gamma = 180°$$
$$\gamma = 80°$$

Now we use the Law of Sines (twice) to find the unknown sides b and c.

$$\frac{\sin \alpha}{a} = \frac{\sin \beta}{b} \qquad \frac{\sin \alpha}{a} = \frac{\sin \gamma}{c}$$

Because $a = 4, \alpha = 40°, \beta = 60°$, and $\gamma = 80°$, we have

$$\frac{\sin 40°}{4} = \frac{\sin 60°}{b} \qquad \frac{\sin 40°}{4} = \frac{\sin 80°}{c}$$

Solving for b and c, we find that

$$b = \frac{4 \sin 60°}{\sin 40°} \approx 5.39 \qquad c = \frac{4 \sin 80°}{\sin 40°} \approx 6.13$$

Notice in Example 1 that we found b and c by working with the given side a. This is better than finding b first and working with a rounded value of b to find c.

NOW WORK PROBLEM 1.

EXAMPLE 2 Using the Law of Sines to Solve an ASA Triangle

Solve the triangle: $\alpha = 35°, \beta = 15°, c = 5$

Solution Figure 18 illustrates the triangle that we want to solve. Because we know two angles ($\alpha = 35°$ and $\beta = 15°$), we find the third angle using equation (6).

Figure 18

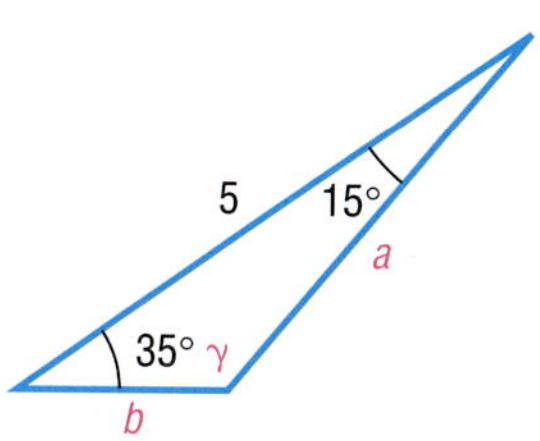

$$\alpha + \beta + \gamma = 180°$$
$$35° + 15° + \gamma = 180°$$
$$\gamma = 130°$$

Now we know the three angles and one side ($c = 5$) of the triangle. To find the remaining two sides a and b, we use the Law of Sines (twice).

$$\frac{\sin \alpha}{a} = \frac{\sin \gamma}{c} \qquad \frac{\sin \beta}{b} = \frac{\sin \gamma}{c}$$

$$\frac{\sin 35°}{a} = \frac{\sin 130°}{5} \qquad \frac{\sin 15°}{b} = \frac{\sin 130°}{5}$$

$$a = \frac{5 \sin 35°}{\sin 130°} \approx 3.74 \qquad b = \frac{5 \sin 15°}{\sin 130°} \approx 1.69$$

NOW WORK PROBLEM 15.

The Ambiguous Case

2 Case 2 (SSA), which applies to triangles for which two sides and the angle opposite one of them are known, is referred to as the **ambiguous case**, because the known information may result in one triangle, two triangles, or no triangle at all. Suppose that we are given sides a and b and angle α, as illustrated in Figure 19. The key to determining the possible triangles, if any, that may be formed from the given information lies primarily with the height h and the fact that $h = b \sin \alpha$.

Figure 19

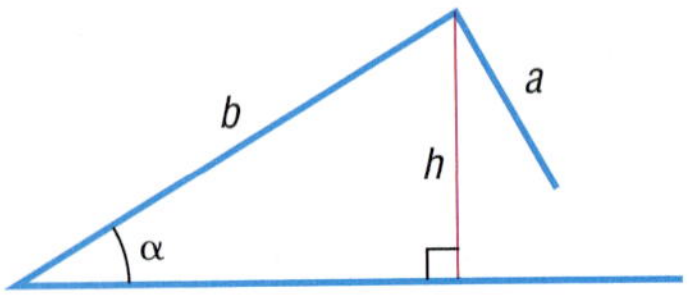

No Triangle If $a < h = b \sin \alpha$, then side a is not sufficiently long to form a triangle. See Figure 20.

One Right Triangle If $a = h = b \sin \alpha$, then side a is just long enough to form a right triangle. See Figure 21.

Figure 20
$a < h = b \sin \alpha$

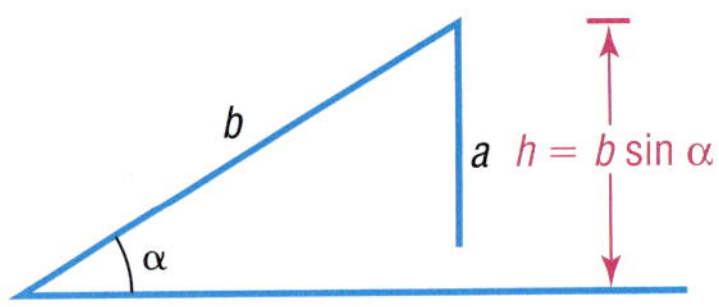

Figure 21
$a = b \sin \alpha$

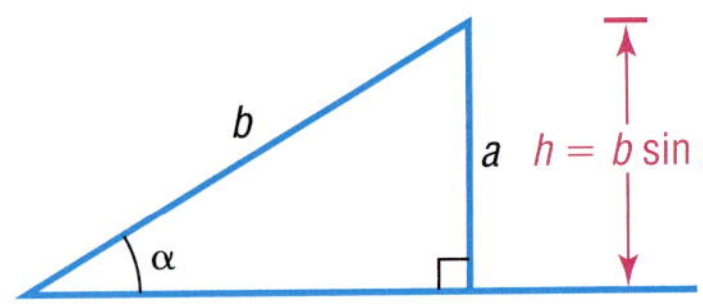

Two Triangles If $a < b$ and $h = b \sin \alpha < a$, then two distinct triangles can be formed from the given information. See Figure 22.

One Triangle If $a \geq b$, then only one triangle can be formed. See Figure 23.

Figure 22
$b \sin \alpha < a$ and $a < b$

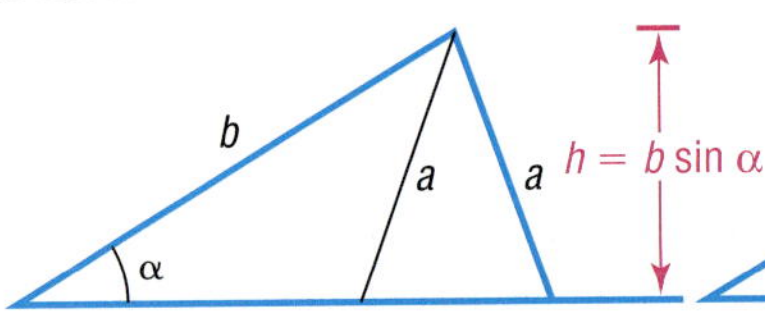

Figure 23
$a \geq b$

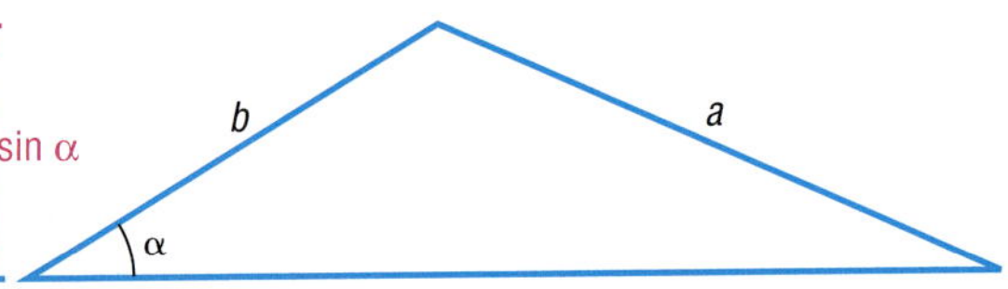

Fortunately, we do not have to rely on an illustration to draw the correct conclusion in the ambiguous case. The Law of Sines will lead us to the correct determination. Let's see how.

EXAMPLE 3 Using the Law of Sines to Solve a SSA Triangle (One Solution)

Solve the triangle: $a = 3, b = 2, \alpha = 40°$

Solution See Figure 24(a). Because $a = 3, b = 2$, and $\alpha = 40°$ are known, we use the Law of Sines to find the angle β.

Figure 24(a)

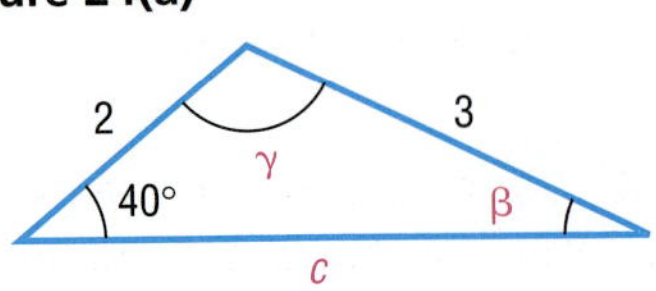

$$\frac{\sin \alpha}{a} = \frac{\sin \beta}{b}$$

Then

$$\frac{\sin 40°}{3} = \frac{\sin \beta}{2}$$

$$\sin \beta = \frac{2 \sin 40°}{3} \approx 0.43$$

There are two angles $\beta, 0° < \beta < 180°$, for which $\sin \beta \approx 0.43$.

$$\beta_1 \approx 25.4° \quad \text{and} \quad \beta_2 \approx 154.6°$$

Note: Here we computed β using the stored value of $\sin \beta$. If you use the rounded value, $\sin \beta \approx 0.43$, you will obtain slightly different results.

The second possibility, $\beta_2 \approx 154.6°$, is ruled out, because $\alpha = 40°$, making $\alpha + \beta_2 \approx 194.6° > 180°$. Now, using $\beta_1 \approx 25.4°$, we find that

$$\gamma = 180° - \alpha - \beta_1 \approx 180° - 40° - 25.4° = 114.6°$$

The third side c may now be determined using the Law of Sines.

$$\frac{\sin \alpha}{a} = \frac{\sin \gamma}{c}$$

$$\frac{\sin 40°}{3} = \frac{\sin 114.6°}{c}$$

$$c = \frac{3 \sin 114.6°}{\sin 40°} \approx 4.24$$

Figure 24(b)

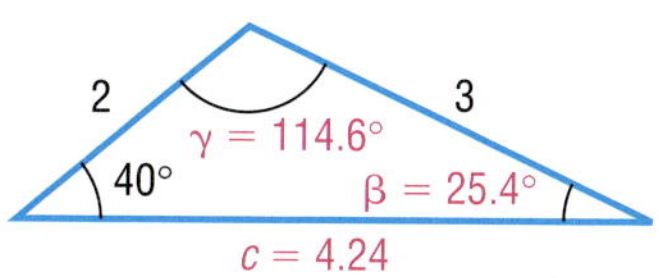

Figure 24(b) illustrates the solved triangle. ■

EXAMPLE 4 **Using the Law of Sines to Solve a SSA Triangle (Two Solutions)**

Solve the triangle: $a = 6, b = 8, \alpha = 35°$

Solution See Figure 25(a). Because $a = 6, b = 8$, and $\alpha = 35°$ are known, we use the Law of Sines to find the angle β.

Figure 25(a)

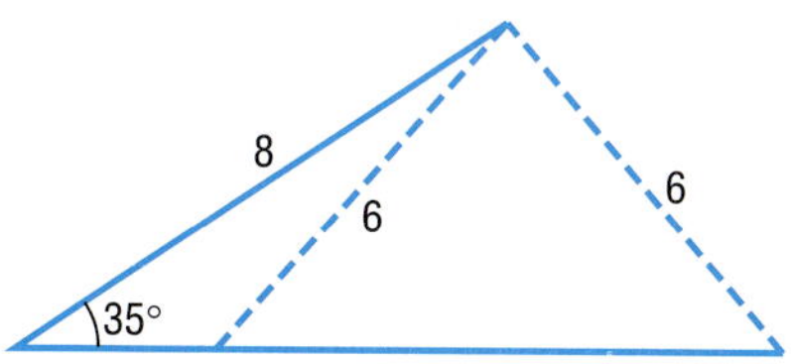

$$\frac{\sin \alpha}{a} = \frac{\sin \beta}{b}$$

Then

$$\frac{\sin 35°}{6} = \frac{\sin \beta}{8}$$

$$\sin \beta = \frac{8 \sin 35°}{6} \approx 0.76$$

$$\beta_1 \approx 49.9° \quad \text{or} \quad \beta_2 \approx 130.1°$$

For both choices of β, we have $\alpha + \beta < 180°$. There are two triangles one containing the angle $\beta_1 \approx 49.9°$ and the other containing the angle $\beta_2 \approx 130.1°$. The third angle γ is either

$$\gamma_1 = 180° - \alpha - \beta_1 \approx 95.1° \quad \text{or} \quad \gamma_2 = 180° - \alpha - \beta_2 \approx 14.9°$$

$\uparrow$ $\alpha = 35°$, $\beta_1 = 49.9°$ $\qquad$ $\uparrow$ $\alpha = 35°$, $\beta_2 = 130.1°$

Figure 25(b)

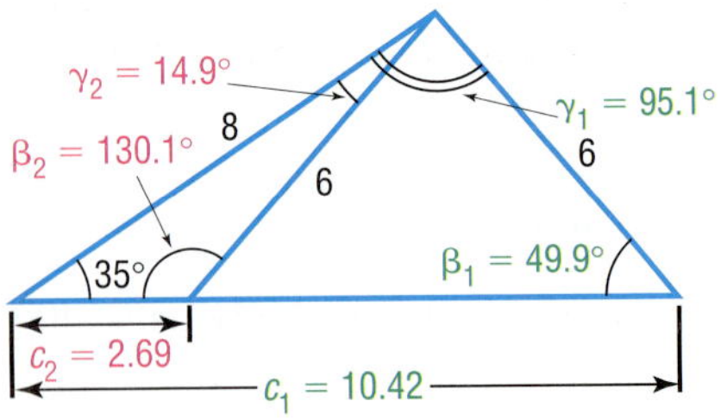

The third side c obeys the Law of Sines, so we have

$$\frac{\sin \alpha}{a} = \frac{\sin \gamma_1}{c_1} \qquad \frac{\sin \alpha}{a} = \frac{\sin \gamma_2}{c_2}$$

$$\frac{\sin 35°}{6} = \frac{\sin 95.1°}{c_1} \qquad \frac{\sin 35°}{6} = \frac{\sin 14.9°}{c_2}$$

$$c_1 = \frac{6 \sin 95.1°}{\sin 35°} \approx 10.42 \qquad c_2 = \frac{6 \sin 14.9°}{\sin 35°} \approx 2.69$$

The two solved triangles are illustrated in Figure 25(b).

EXAMPLE 5 Using the Law of Sines to Solve a SSA Triangle (No Solution)

Solve the triangle: $a = 2, c = 1, \gamma = 50°$

Solution Because $a = 2, c = 1$, and $\gamma = 50°$ are known, we use the Law of Sines to find the angle α.

$$\frac{\sin \alpha}{a} = \frac{\sin \gamma}{c}$$

$$\frac{\sin \alpha}{2} = \frac{\sin 50°}{1}$$

$$\sin \alpha = 2 \sin 50° \approx 1.53$$

Figure 26

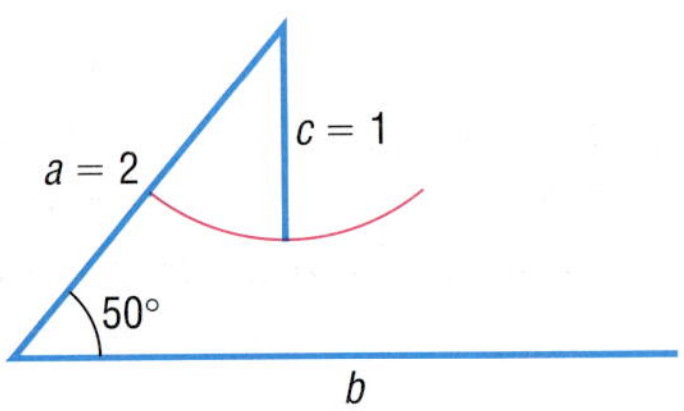

Since there is no angle α for which $\sin \alpha > 1$, there can be no triangle with the given measurements. Figure 26 illustrates the measurements given. Notice that, no matter how we attempt to position side c, it will never touch side b to form a triangle.

NOW WORK PROBLEMS **17** AND **23**.

Applications

3 The Law of Sines is particularly useful for solving certain applied problems.

EXAMPLE 6 Finding the Height of a Mountain

To measure the height of a mountain, a surveyor takes two sightings of the peak at a distance 900 meters apart on a direct line to the mountain.* See Figure 27(a). The first observation results in an angle of elevation of 47°, whereas the second results in an angle of elevation of 35°. If the transit is 2 meters high, what is the height h of the mountain?

Figure 27

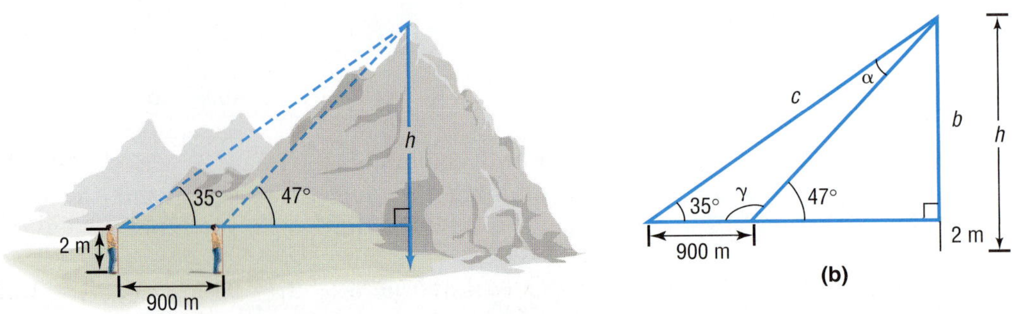

*For simplicity, we assume that these sightings are at the same level.

Solution Figure 27(b) shows the triangles that replicate the illustration in Figure 27(a). Since $\gamma + 47° = 180°$, we find that $\gamma = 133°$. Also, since $\alpha + \beta + \gamma = 180°$, we find that $\alpha = 180° - 35° - 133° = 12°$. We use the Law of Sines to find c.

$$\frac{\sin \alpha}{a} = \frac{\sin \gamma}{c} \qquad \alpha = 12°, \gamma = 133°, a = 900$$

$$c = \frac{900 \sin 133°}{\sin 12°} = 3165.86$$

Using the larger right triangle, we have

$$\sin 35° = \frac{b}{c} \qquad c = 3165.86$$

$$b = 3165.86 \sin 35° = 1815.86 \approx 1816 \text{ meters}$$

The height of the peak from ground level is approximately $1816 + 2 = 1818$ meters. ■

NOW WORK PROBLEM 31.

EXAMPLE 7 Rescue at Sea

Coast Guard Station Zulu is located 120 miles due west of Station X-ray. A ship at sea sends an SOS call that is received by each station. The call to Station Zulu indicates that the bearing of the ship from Zulu is N40°E (40° east of north). The call to Station X-ray indicates that the bearing of the ship from X-ray is N30°W (30° west of north).

(a) How far is each station from the ship?

(b) If a helicopter capable of flying 200 miles per hour is dispatched from the nearest station to the ship, how long will it take to reach the ship?

Solution (a) Figure 28 illustrates the situation. The angle γ is found to be

$$\gamma = 180° - 50° - 60° = 70°$$

Figure 28

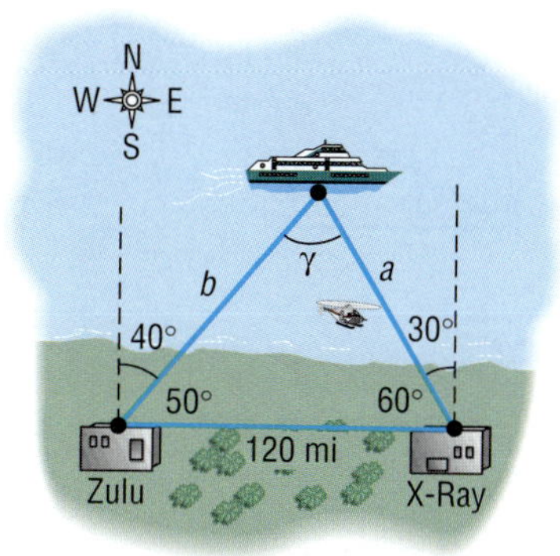

The Law of Sines can now be used to find the two distances a and b that we seek.

$$\frac{\sin 50°}{a} = \frac{\sin 70°}{120}$$

$$a = \frac{120 \sin 50°}{\sin 70°} \approx 97.82 \text{ miles}$$

$$\frac{\sin 60°}{b} = \frac{\sin 70°}{120}$$

$$b = \frac{120 \sin 60°}{\sin 70°} \approx 110.59 \text{ miles}$$

Station Zulu is about 111 miles from the ship, and Station X-ray is about 98 miles from the ship.

(b) The time t needed for the helicopter to reach the ship from Station X-ray is found by using the formula

$$(\text{Velocity}, v)(\text{Time}, t) = \text{Distance}, a$$

Then

$$t = \frac{a}{v} = \frac{97.82}{200} \approx 0.49 \text{ hour} \approx 29 \text{ minutes}$$

It will take about 29 minutes for the helicopter to reach the ship. ■

NOW WORK PROBLEM 29.

4.2 Concepts and Vocabulary

In Problems 1 and 2, fill in the blanks.

1. If none of the angles of a triangle is a right angle, the triangle is called __________.
2. For a triangle with sides a, b, c and opposite angles α, β, γ, the Law of Sines states that __________.

In Problems 3–5, answer True or False to each statement.

3. An oblique triangle in which two sides and an angle are given always results in at least one triangle.
4. The sum of the angles of any triangle equals 180°.
5. The ambiguous case refers to the fact that, when two sides and the angle opposite one of them is given, sometimes the Law of Sines cannot be used.

6. What do you do first if you are asked to solve a triangle and are given one side and two angles?
7. What do you do first if you are asked to solve a triangle and are given two sides and the angle opposite one of them?

4.2 Exercises

In Problems 1–8 solve each triangle.

1.
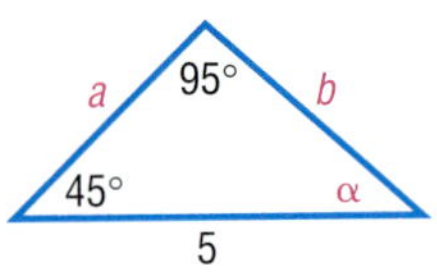

2.
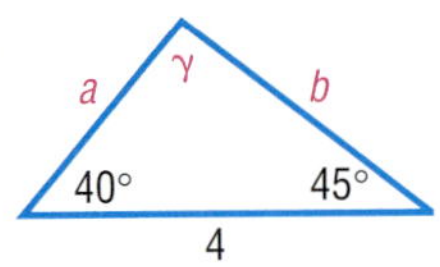

3.
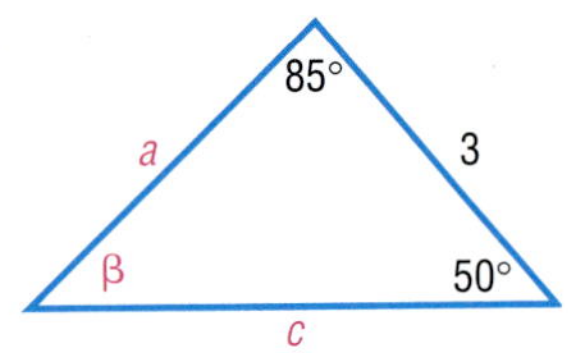

4.
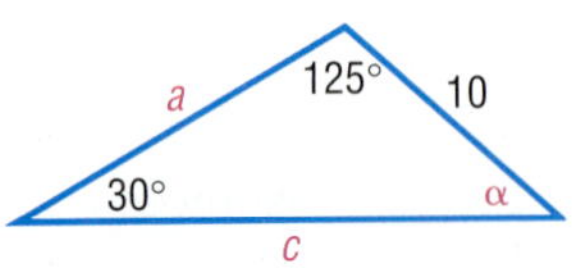

5.
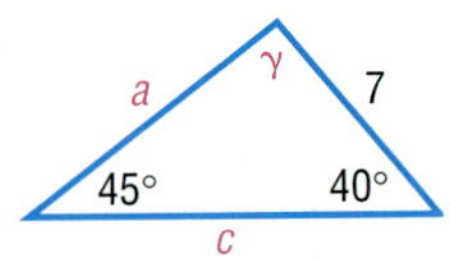

6.
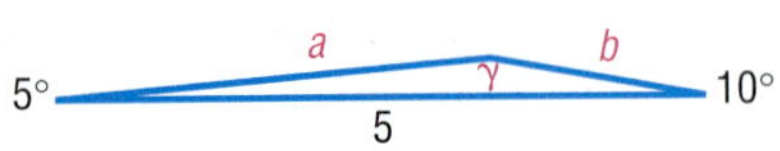

7.
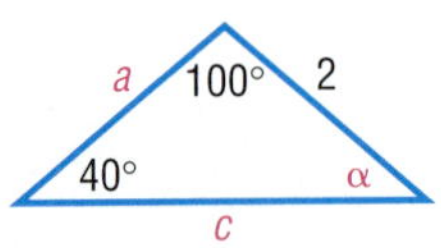

8.
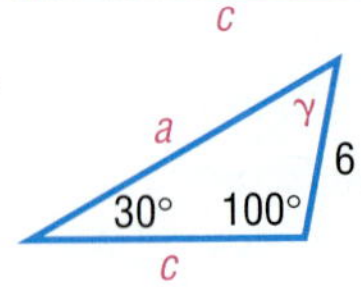

In Problems 9–14 solve each triangle.

9. $\alpha = 40°, \quad \beta = 20°, \quad a = 2$
10. $\alpha = 50°, \quad \gamma = 20°, \quad a = 3$
11. $\beta = 70°, \quad \gamma = 10°, \quad b = 5$
12. $\alpha = 70°, \quad \beta = 60°, \quad c = 4$
13. $\alpha = 110°, \quad \gamma = 30°, \quad c = 3$
14. $\beta = 10°, \quad \gamma = 100°, \quad b = 2$
15. $\alpha = 40°, \quad \beta = 40°, \quad c = 2$
16. $\beta = 20°, \quad \gamma = 70°, \quad a = 1$

In Problems 17–28, two sides and an angle are given. Determine whether the given information results in one triangle, two triangles, or no triangle at all. Solve any resulting triangle(s).

17. $a = 3, \quad b = 2, \quad \alpha = 50°$
18. $b = 4, \quad c = 3, \quad \beta = 40°$
19. $b = 5, \quad c = 3, \quad \beta = 100°$
20. $a = 2, \quad c = 1, \quad \alpha = 120°$
21. $a = 4, \quad b = 5, \quad \alpha = 60°$
22. $b = 2, \quad c = 3, \quad \beta = 40°$
23. $b = 4, \quad c = 6, \quad \beta = 20°$
24. $a = 3, \quad b = 7, \quad \alpha = 70°$
25. $a = 2, \quad c = 1, \quad \gamma = 100°$
26. $b = 4, \quad c = 5, \quad \beta = 95°$
27. $a = 2, \quad c = 1, \quad \gamma = 25°$
28. $b = 4, \quad c = 5, \quad \beta = 40°$

29. Rescue at Sea Coast Guard Station Able is located 150 miles due south of Station Baker. A ship at sea sends an SOS call that is received by each station. The call to Station Able indicates that the ship is located N55°E; the call to Station Baker indicates that the ship is located S60°E.

(a) How far is each station from the ship?

(b) If a helicopter capable of flying 200 miles per hour is dispatched from the nearest station to the ship, how long will it take to reach the ship?

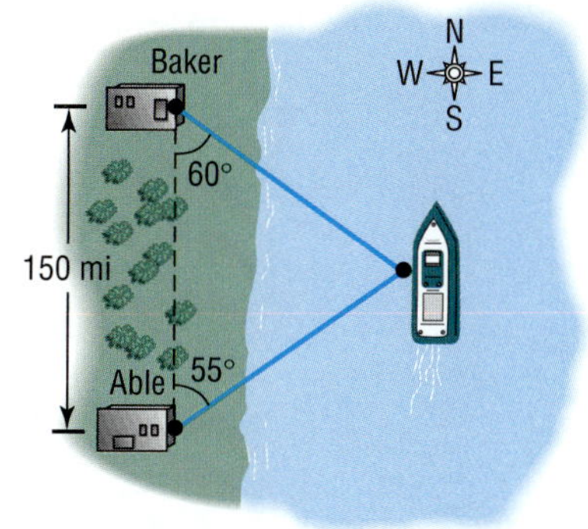

30. Surveying Consult the figure. To find the distance from the house at A to the house at B, a surveyor measures the angle BAC to be 40° and then walks off a distance of 100 feet to C and measures the angle ACB to be 50°. What is the distance from A to B?

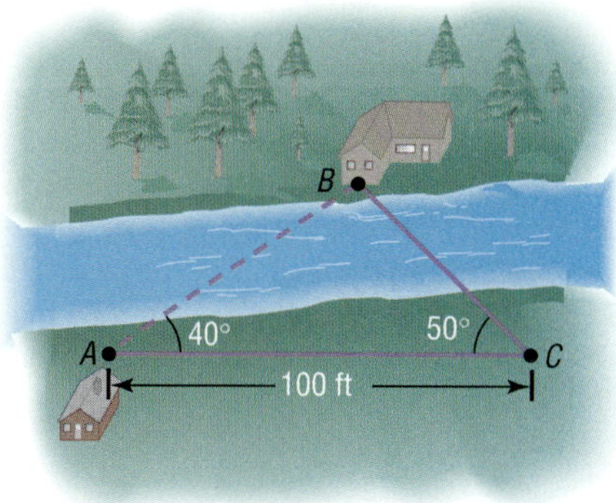

31. Finding the Length of a Ski Lift Consult the figure. To find the length of the span of a proposed ski lift from A to B, a surveyor measures the angle DAB to be 25° and then walks off a distance of 1000 feet to C and measures the angle ACB to be 15°. What is the distance from A to B?

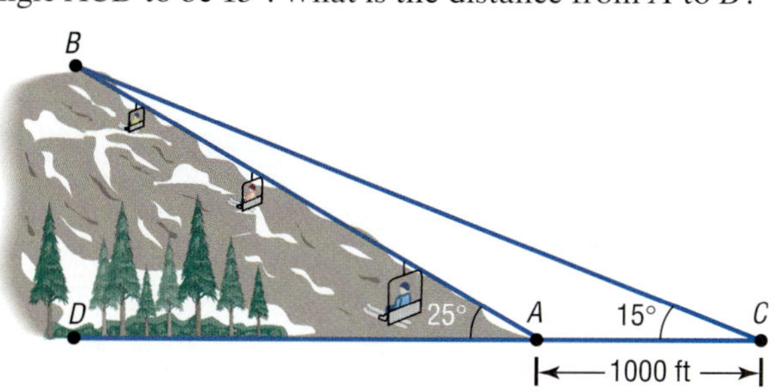

32. Finding the Height of a Mountain Use the illustration in Problem 31 to find the height BD of the mountain at B.

33. Finding the Height of an Airplane An aircraft is spotted by two observers who are 1000 feet apart. As the airplane passes over the line joining them, each observer takes a sighting of the angle of elevation to the plane, as indicated in the figure. How high is the airplane?

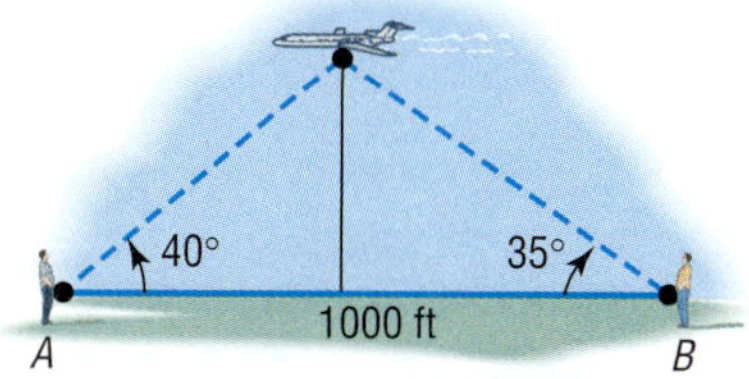

34. Finding the Height of the Bridge over the Royal Gorge The highest bridge in the world is the bridge over the Royal Gorge of the Arkansas River in Colorado.* Sightings to the same point at water level directly under the bridge are taken from each side of the 880-foot-long bridge, as indicated in the figure. How high is the bridge?

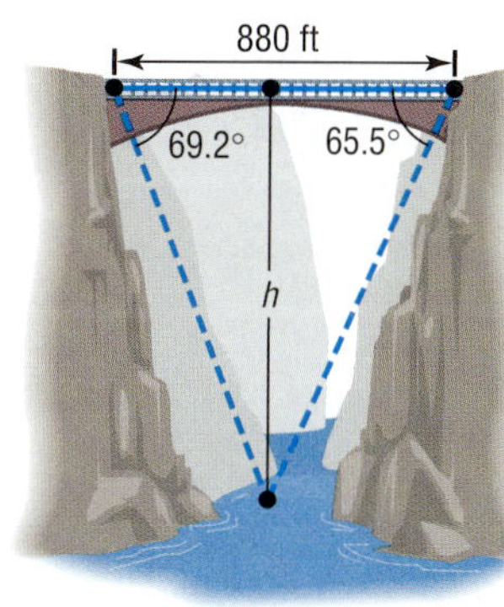

35. Navigation An airplane flies from city A to city B, a distance of 150 miles, and then turns through an angle of 40° and heads toward city C, as shown in the figure.

(a) If the distance between cities A and C is 300 miles, how far is it from city B to city C?

(b) Through what angle should the pilot turn at city C to return to city A?

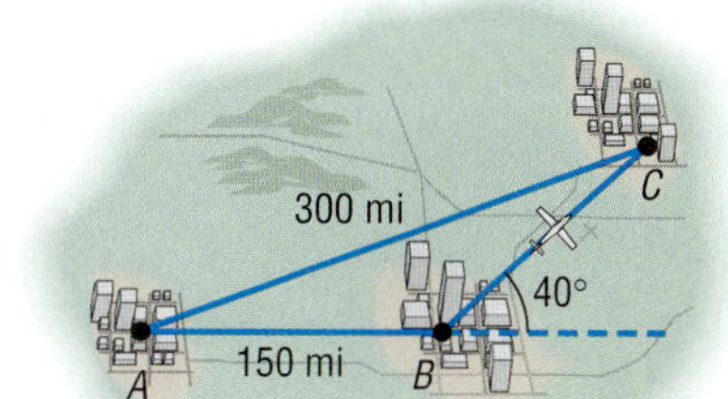

*Source: *Guinness Book of World Records.*

36. Time Lost due to a Navigation Error In attempting to fly from city A to city B, an aircraft followed a course that was 10° in error, as indicated in the figure. After flying a distance of 50 miles, the pilot corrected the course by turning at point C and flying 70 miles farther. If the constant speed of the aircraft was 250 miles per hour, how much time was lost due to the error?

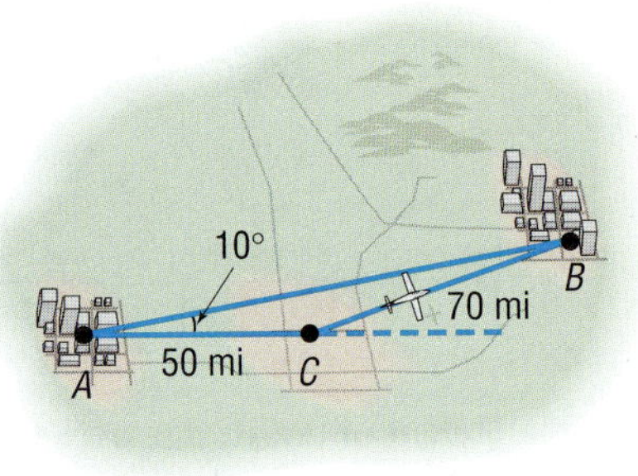

37. Finding the Lean of the Leaning Tower of Pisa The famous Leaning Tower of Pisa was originally 184.5 feet high. At a distance of 123 feet from the base of the tower, the angle of elevation to the top of the tower is found to be 60°. Find the angle CAB indicated in the figure. Also, find the perpendicular distance from C to AB.

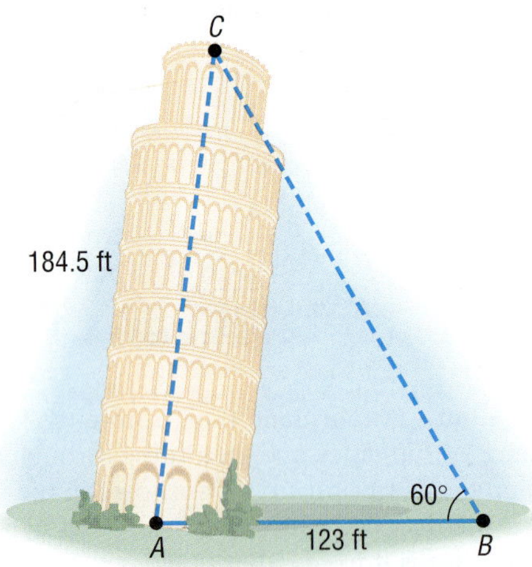

38. Crankshafts on Cars On a certain automobile, the crankshaft is 3 inches long and the connecting rod is 9 inches long (see the figure). At the time when the angle OPA is 15°, how far is the piston (P) from the center (O) of the crankshaft?

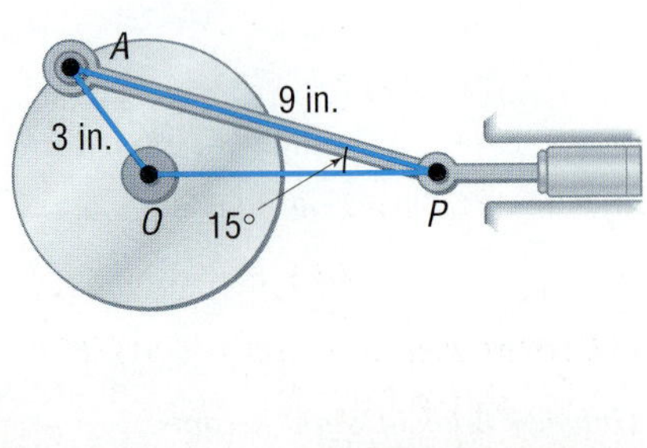

39. Constructing a Highway U.S. 41, a highway whose primary directions are north–south, is being constructed along the west coast of Florida. Near Naples, a bay obstructs the straight path of the road. Since the cost of a bridge is prohibitive, engineers decide to go around the bay. The illustration shows the path that they decide on and the measurements taken. What is the length of highway needed to go around the bay?

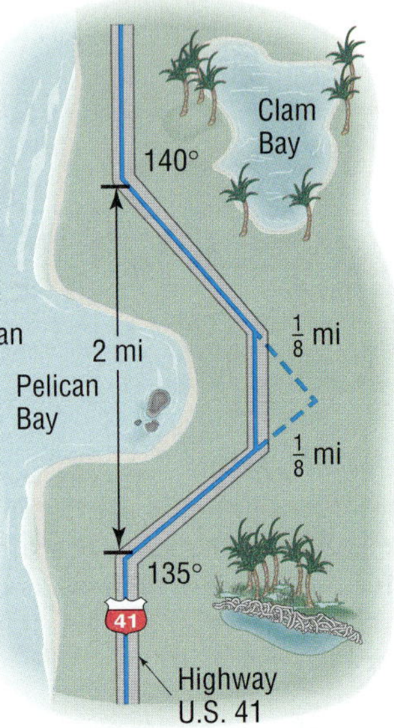

40. Determining Distances at Sea The navigator of a ship at sea spots two lighthouses that she knows to be 3 miles apart along a straight seashore. She determines that the angles formed between two line-of-sight observations of the lighthouses and the line from the ship directly to shore are 15° and 35°. See the illustration.

(a) How far is the ship from lighthouse A?
(b) How far is the ship from lighthouse B?
(c) How far is the ship from shore?

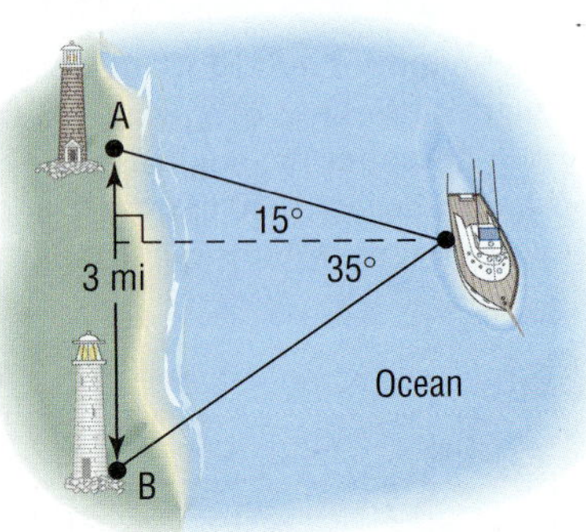

41. Calculating Distance at Sea The navigator of a ship at sea has the harbor in sight at which the ship is to dock. She spots a lighthouse that she knows is 1 mile up the coast from the mouth of the harbor, and she measures the angle between the line-of-sight observations of the harbor and lighthouse to be 20°. With the ship heading directly toward the harbor, she repeats this measurement after 5 minutes of traveling at 12 miles per hour. If the new angle is 30°, how far is the ship from the harbor?

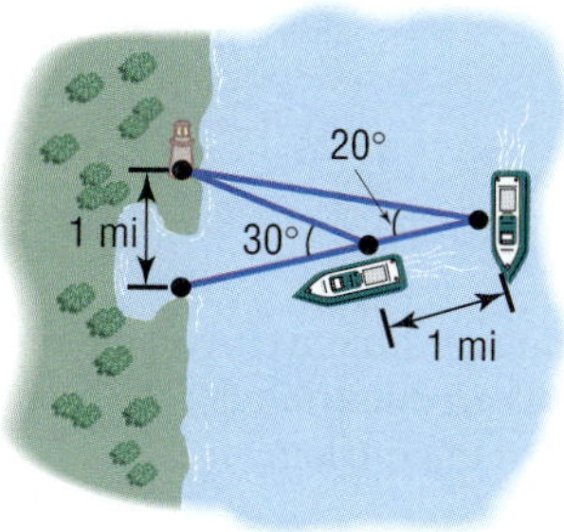

42. Finding Distances A forest ranger is walking on a path inclined at 5° to the horizontal directly toward a 100-foot-tall fire observation tower. The angle of elevation from the path to the top of the tower is 40°. How far is the ranger from the tower at this time?

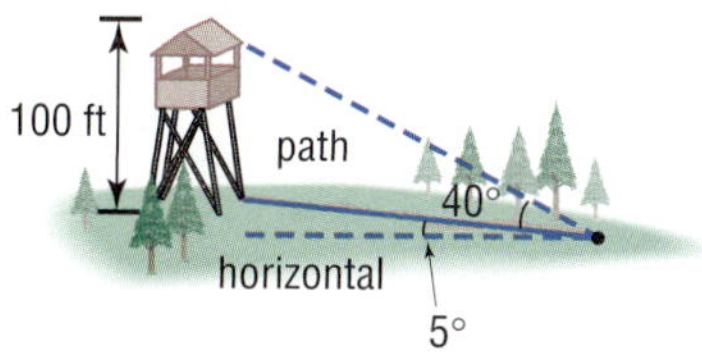

43. Great Pyramid of Cheops One of the original Seven Wonders of the World, the Great Pyramid of Cheops was built about 2580 BC. Its original height was 480 feet 11 inches, but due to the loss of its topmost stones, it is now shorter.* Find the current height of the Great Pyramid, using the information given in the illustration.

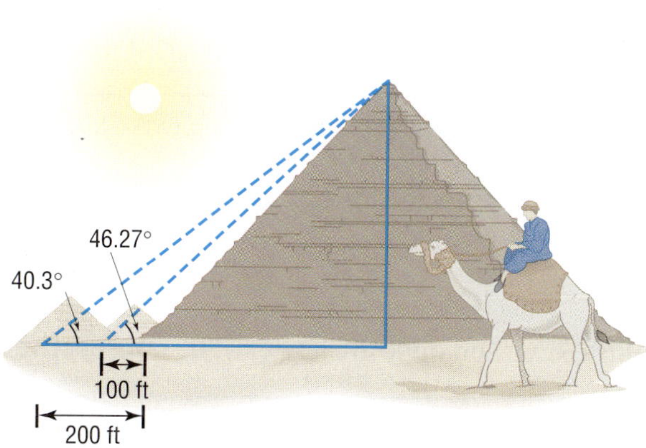

44. Determining the Height of an Aircraft Two sensors are spaced 700 feet apart along the approach to a small airport. When an aircraft is nearing the airport, the angle of elevation from the first sensor to the aircraft is 20°, and from the second sensor to the aircraft it is 15°. Determine how high the aircraft is at this time.

45. Landscaping Pat needs to determine the height of a tree before cutting it down to be sure that it will not fall on a nearby fence. The angle of elevation of the tree from one position on a flat path from the tree is 30°, and from a second position 40 feet farther along this path it is 20°. What is the height of the tree?

46. Construction A loading ramp 10 feet long that makes an angle of 18° with the horizontal is to be replaced by one that makes an angle of 12° with the horizontal. How long is the new ramp?

47. Finding the Height of a Helicopter Two observers simultaneously measure the angle of elevation of a helicopter. One angle is measured as 25°, the other as 40° (see the figure). If the observers are 100 feet apart and the helicopter lies over the line joining them, how high is the helicopter?

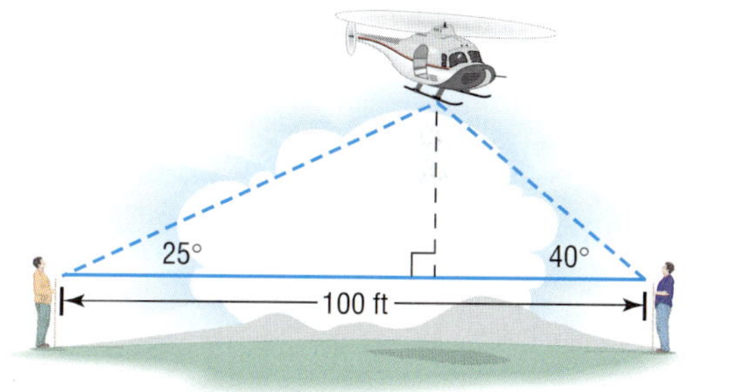

48. Mollweide's Formula For any triangle, Mollweide's Formula (named after Karl Mollweide, 1774–1825) states that

$$\frac{a+b}{c} = \frac{\cos\left[\frac{1}{2}(\alpha - \beta)\right]}{\sin\left(\frac{1}{2}\gamma\right)}$$

Derive it.

[**Hint:** Use the Law of Sines and then a sum-to-product formula. Notice that this formula involves all six parts of a triangle. As a result, it is sometimes used to check the solution of a triangle.]

49. Mollweide's Formula Another form of Mollweide's Formula is

$$\frac{a-b}{c} = \frac{\sin\left[\frac{1}{2}(\alpha - \beta)\right]}{\cos\left(\frac{1}{2}\gamma\right)}$$

Derive it.

50. For any triangle, derive the formula

$$a = b\cos\gamma + c\cos\beta$$

[**Hint:** Use the fact that $\sin\alpha = \sin(180° - \beta - \gamma)$.]

*Source: *Guinness Book of World Records*.

51. Law of Tangents For any triangle, derive the Law of Tangents.

$$\frac{a-b}{a+b} = \frac{\tan\left[\frac{1}{2}(\alpha - \beta)\right]}{\tan\left[\frac{1}{2}(\alpha + \beta)\right]}$$

[**Hint:** Use Mollweide's Formula.]

52. Circumscribing a Triangle Show that

$$\frac{\sin\alpha}{a} = \frac{\sin\beta}{b} = \frac{\sin\gamma}{c} = \frac{1}{2r}$$

where r is the radius of the circle circumscribing the triangle ABC whose sides are a, b, and c, as shown in the figure.

[**Hint:** Draw the diameter AB'. Then β = angle ABC = angle $AB'C$, and angle $ACB' = 90°$.]

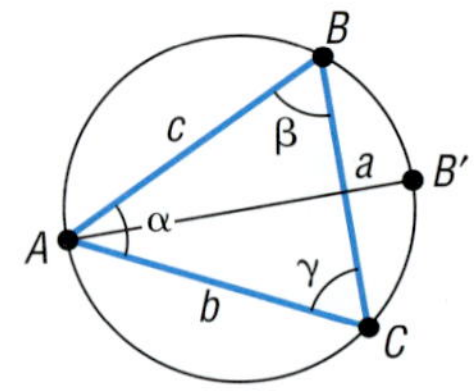

53. Make up three problems involving oblique triangles. One should result in one triangle, the second in two triangles, and the third in no triangle.

PREPARING FOR THIS SECTION

Before getting started, review the following:

✓ Trigonometric Equations (I) (Section 3.7, pp. 247–250) ✓ The Distance Formula (Section 1.1, pp. 5–7)

4.3 THE LAW OF COSINES

OBJECTIVES

1. Solve SAS Triangles
2. Solve SSS Triangles
3. Solve Applied Problems Using the Law of Cosines

In Section 4.2, we used the Law of Sines to solve Case 1 (SAA or ASA) and Case 2 (SSA) of an oblique triangle. In this section, we derive the Law of Cosines and use it to solve the remaining cases, 3 and 4.

CASE 3: Two sides and the included angle are known (SAS).
CASE 4: Three sides are known (SSS).

Theorem

Law of Cosines

For a triangle with sides a, b, c and opposite angles α, β, γ, respectively,

$$c^2 = a^2 + b^2 - 2ab\cos\gamma \quad (1)$$
$$b^2 = a^2 + c^2 - 2ac\cos\beta \quad (2)$$
$$a^2 = b^2 + c^2 - 2bc\cos\alpha \quad (3)$$

Proof We will prove only formula (1) here. Formulas (2) and (3) may be proved using the same argument.

Figure 29

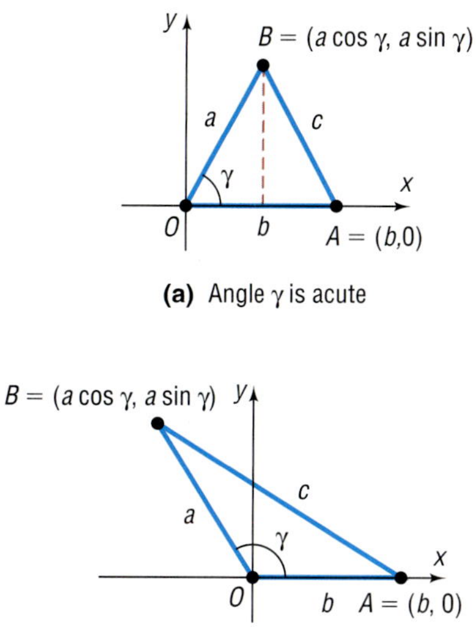

(a) Angle γ is acute

(b) Angle γ is obtuse

We begin by strategically placing a triangle on a rectangular coordinate system so that the vertex of angle γ is at the origin and side b lies along the positive x-axis. Regardless of whether γ is acute, as in Figure 29(a), or obtuse, as in Figure 29(b), the vertex B has coordinates $(a\cos\gamma, a\sin\gamma)$. Vertex A has coordinates $(b, 0)$.

We can now use the distance formula to compute c^2.

$$\begin{aligned} c^2 &= (b - a\cos\gamma)^2 + (0 - a\sin\gamma)^2 \\ &= b^2 - 2ab\cos\gamma + a^2\cos^2\gamma + a^2\sin^2\gamma \\ &= b^2 - 2ab\cos\gamma + a^2(\cos^2\gamma + \sin^2\gamma) \\ &= a^2 + b^2 - 2ab\cos\gamma \end{aligned}$$

■

Each of formulas (1), (2), and (3) may be stated in words as follows:

Theorem

Law of Cosines

The square of one side of a triangle equals the sum of the squares of the other two sides minus twice their product times the cosine of their included angle.

■

Observe that if the triangle is a right triangle (so that, say, $\gamma = 90°$) then formula (1) becomes the familiar Pythagorean Theorem: $c^2 = a^2 + b^2$. Thus, the Pythagorean Theorem is a special case of the Law of Cosines.

1 Let's see how to use the Law of Cosines to solve Case 3 (SAS), which applies to triangles for which two sides and the included angle are known.

EXAMPLE 1 Using the Law of Cosines to Solve a SAS Triangle

Solve the triangle: $a = 2, b = 3, \gamma = 60°$.

Figure 30

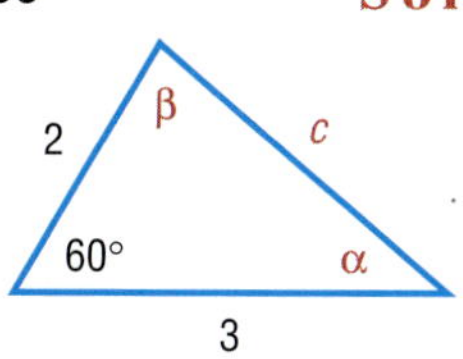

Solution See Figure 30. The Law of Cosines makes it easy to find the third side, c.

$$\begin{aligned} c^2 &= a^2 + b^2 - 2ab\cos\gamma \\ &= 4 + 9 - 2\cdot 2\cdot 3\cdot\cos 60° \\ &= 13 - \left(12\cdot\frac{1}{2}\right) = 7 \\ c &= \sqrt{7} \end{aligned}$$

Side c is of length $\sqrt{7}$. To find the angles α and β, we may use either the Law of Sines or the Law of Cosines. It is preferable to use the Law of Cosines, since it will lead to an equation with one solution. Using the Law of Sines would lead to an equation with two solutions that would need to be checked to determine which solution fits the given data. We choose to use formulas (2) and (3) of the Law of Cosines to find α and β.

For α:

$$a^2 = b^2 + c^2 - 2bc\cos\alpha$$
$$2bc\cos\alpha = b^2 + c^2 - a^2$$
$$\cos\alpha = \frac{b^2 + c^2 - a^2}{2bc} = \frac{9 + 7 - 4}{2 \cdot 3\sqrt{7}} = \frac{12}{6\sqrt{7}} = \frac{2\sqrt{7}}{7}$$
$$\alpha \approx 40.9°$$

For β:

$$b^2 = a^2 + c^2 - 2ac\cos\beta$$
$$\cos\beta = \frac{a^2 + c^2 - b^2}{2ac} = \frac{4 + 7 - 9}{4\sqrt{7}} = \frac{1}{2\sqrt{7}} = \frac{\sqrt{7}}{14}$$
$$\beta \approx 79.1°$$

Notice that $\alpha + \beta + \gamma = 40.9° + 79.1° + 60° = 180°$, as required. ■

NOW WORK PROBLEM **1.**

2 The next example illustrates how the Law of Cosines is used when three sides of a triangle are known, Case 4 (SSS).

EXAMPLE 2 Using the Law of Cosines to Solve a SSS Triangle

Solve the triangle: $a = 4, b = 3, c = 6$.

Solution See Figure 31. To find the angles α, β, and γ, we proceed as we did in the latter part of the solution to Example 1.

Figure 31

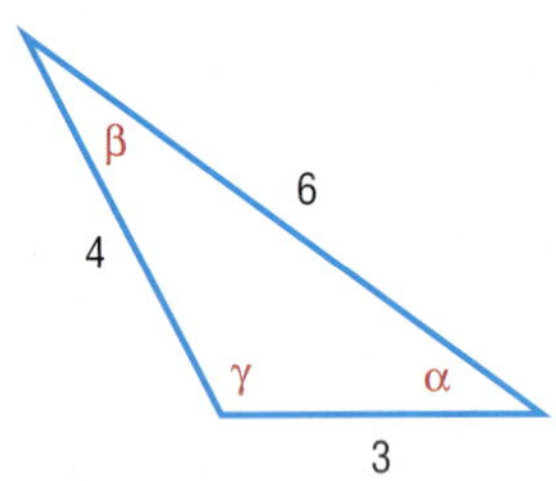

For α:

$$\cos\alpha = \frac{b^2 + c^2 - a^2}{2bc} = \frac{9 + 36 - 16}{2 \cdot 3 \cdot 6} = \frac{29}{36}$$
$$\alpha \approx 36.3°$$

For β:

$$\cos\beta = \frac{a^2 + c^2 - b^2}{2ac} = \frac{16 + 36 - 9}{2 \cdot 4 \cdot 6} = \frac{43}{48}$$
$$\beta \approx 26.4°$$

Since we know α and β,

$$\gamma = 180° - \alpha - \beta \approx 180° - 36.3° - 26.4° = 117.3°$$ ■

NOW WORK PROBLEM **7.**

3 EXAMPLE 3 Correcting a Navigational Error

A motorized sailboat leaves Naples, Florida, bound for Key West, 150 miles away. Maintaining a constant speed of 15 miles per hour, but encountering heavy crosswinds and strong currents, the crew finds, after 4 hours, that the sailboat is off course by 20°.

(a) How far is the sailboat from Key West at this time?

(b) Through what angle should the sailboat turn to correct its course?

(c) How much time has been added to the trip because of this? (Assume that the speed remains at 15 miles per hour.)

Figure 32

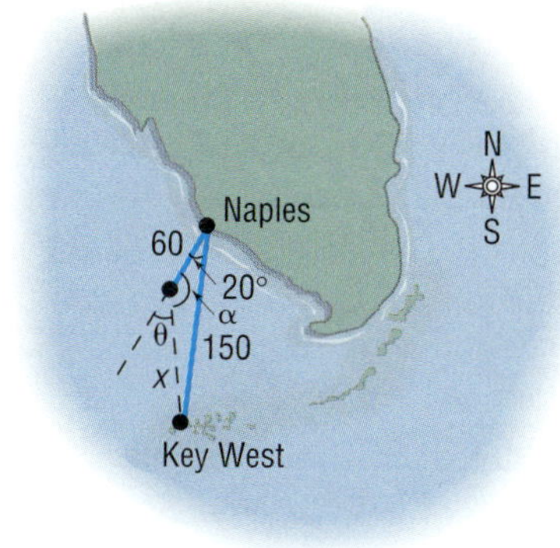

Solution See Figure 32. With a speed of 15 miles per hour, the sailboat has gone 60 miles after 4 hours. We seek the distance x of the sailboat from Key West. We also seek the angle θ that the sailboat should turn through to correct its course.

(a) To find x, we use the Law of Cosines, since we know two sides and the included angle.

$$x^2 = 150^2 + 60^2 - 2(150)(60)\cos 20° \approx 9186$$
$$x \approx 95.8$$

The sailboat is about 96 miles from Key West.

(b) We now know three sides of the triangle, so we can use the Law of Cosines again to find the angle α opposite the side of length 150 miles.

$$150^2 = 96^2 + 60^2 - 2(96)(60)\cos\alpha$$
$$9684 = -11{,}520\cos\alpha$$
$$\cos\alpha \approx -0.8406$$
$$\alpha \approx 147.2°$$

The sailboat should turn through an angle of

$$\theta = 180° - \alpha \approx 180° - 147.2° = 32.8°$$

The sailboat should turn through an angle of about 33° to correct its course.

(c) The total length of the trip is now $60 + 96 = 156$ miles. The extra 6 miles will only require about 0.4 hour or 24 minutes more if the speed of 15 miles per hour is maintained. ■

NOW WORK PROBLEM **27.**

HISTORICAL FEATURE

The Law of Sines was known vaguely long before it was explicitly stated by Nasîr Eddîn (about AD 1250). Ptolemy (about AD 150) was aware of it in a form using a chord function instead of the sine function. But it was first clearly stated in Europe by Regiomontanus, writing in 1464.

The Law of Cosines appears first in Euclid's *Elements* (Book II), but in a well-disguised form in which squares built on the sides of triangles are added and a rectangle representing the cosine term is subtracted. It was thus known to all mathematicians because of their familiarity with Euclid's work. An early modern form of the Law of Cosines, that for finding the angle when the sides are known, was stated by François Viète (in 1593).

The Law of Tangents (see Problem 51 of Exercise 4.2) has become obsolete. In the past it was used in place of the Law of Cosines, because the Law of Cosines was very inconvenient for calculation with logarithms or slide rules. Mixing of addition and multiplication is now very easy on a calculator, however, and the Law of Tangents has been shelved along with the slide rule.

4.3 Concepts and Vocabulary

In Problems 1–3, fill in the blanks.

1. If three sides of a triangle are given, the Law of __________ is used to solve the triangle.
2. If one side and two angles of a triangle are given, the Law of __________ is used to solve the triangle.
3. If two sides and the included angle of a triangle are given, the Law of __________ is used to solve the triangle.

In Problems 4–6, answer True or False to each statement.

4. Given only the three sides of a triangle, there is insufficient information to solve the triangle.
5. Given two sides and the included angle, the first thing to do to solve the triangle is to use the Law of Sines.
6. A special case of the Law of Cosines is the Pythagorean Theorem.

7. What do you do first if you are asked to solve a triangle and are given two sides and the included angle?
8. What do you do first if you are asked to solve a triangle and are given three sides?
9. Make up an applied problem that requires using the Law of Cosines.

4.3 Exercises

In Problems 1–8, solve each triangle.

1.

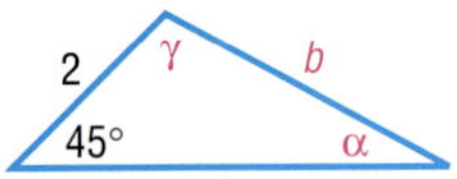

2.

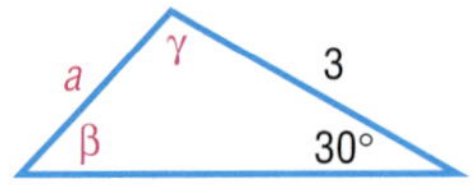

3.

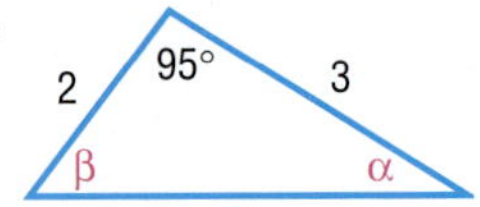

4.

5.

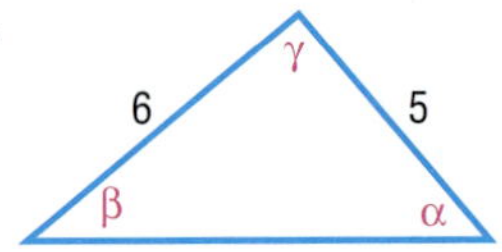

6.

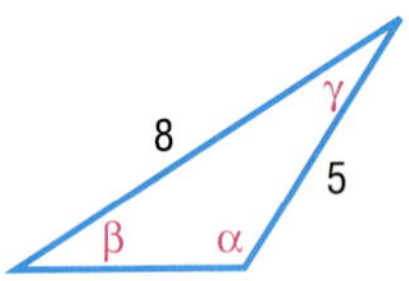

7.

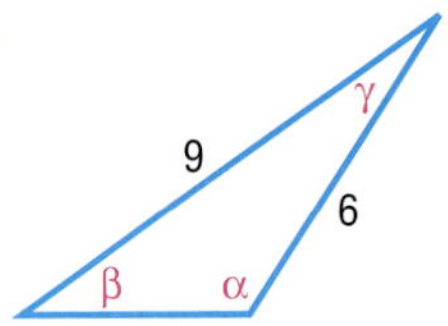

8.

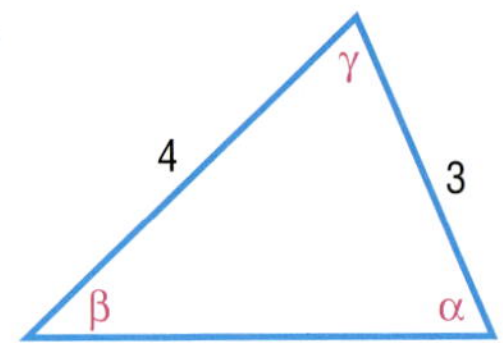

In Problems 9–24, solve each triangle.

9. $a = 3, \quad b = 4, \quad \gamma = 40°$

10. $a = 2, \quad c = 1, \quad \beta = 10°$

11. $b = 1, \quad c = 3, \quad \alpha = 80°$

12. $a = 6, \quad b = 4, \quad \gamma = 60°$

13. $a = 3, \quad c = 2, \quad \beta = 110°$

14. $b = 4, \quad c = 1, \quad \alpha = 120°$

15. $a = 2, \quad b = 2, \quad \gamma = 50°$

16. $a = 3, \quad c = 2, \quad \beta = 90°$

17. $a = 12, \quad b = 13, \quad c = 5$

18. $a = 4, \quad b = 5, \quad c = 3$

19. $a = 2, \quad b = 2, \quad c = 2$

20. $a = 3, \quad b = 3, \quad c = 2$

21. $a = 5, \quad b = 8, \quad c = 9$

22. $a = 4, \quad b = 3, \quad c = 6$

23. $a = 10, \quad b = 8, \quad c = 5$

24. $a = 9, \quad b = 7, \quad c = 10$

---

25. Surveying Consult the figure. To find the distance from the house at A to the house at B, a surveyor measures the angle ACB, which is found to be 70°, and then walks off the distance to each house, 50 feet and 70 feet, respectively. How far apart are the houses?

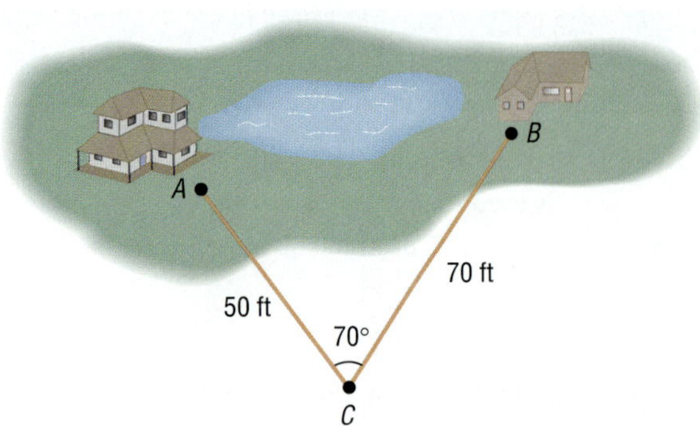

26. Navigation An airplane flies from Ft. Myers to Sarasota, a distance of 150 miles, and then turns through an angle of 50° and flies to Orlando, a distance of 100 miles (see the figure).

(a) How far is it from Ft. Myers to Orlando?

(b) Through what angle should the pilot turn at Orlando to return to Ft. Myers?

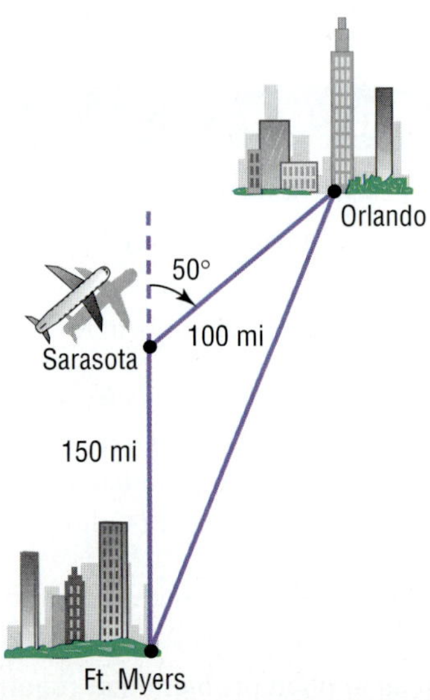

27. Revising a Flight Plan In attempting to fly from Chicago to Louisville, a distance of 330 miles, a pilot inadvertently took a course that was 10° in error, as indicated in the figure.

(a) If the aircraft maintains an average speed of 220 miles per hour and if the error in direction is discovered after 15 minutes, through what angle should the pilot turn to head toward Louisville?

(b) What new average speed should the pilot maintain so that the total time of the trip is 90 minutes?

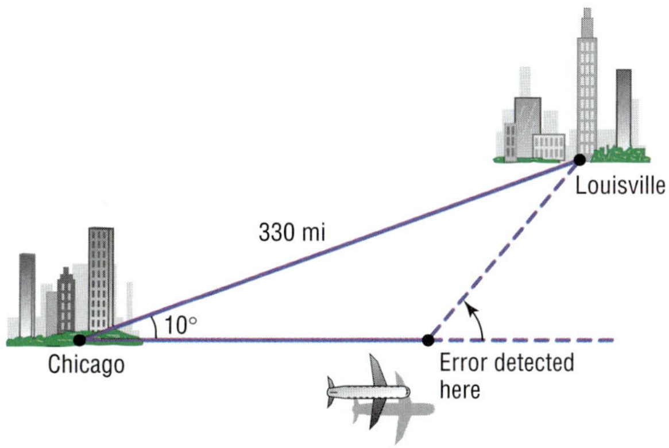

28. Avoiding a Tropical Storm A cruise ship maintains an average speed of 15 knots in going from San Juan, Puerto Rico, to Barbados, West Indies, a distance of 600 nautical miles. To avoid a tropical storm, the captain heads out of San Juan in a direction of 20° off a direct heading to Barbados. The captain maintains the 15-knot speed for 10 hours, after which time the path to Barbados becomes clear of storms.

(a) Through what angle should the captain turn to head directly to Barbados?

(b) Once the turn is made, how long will it be before the ship reaches Barbados if the same 15-knot speed is maintained?

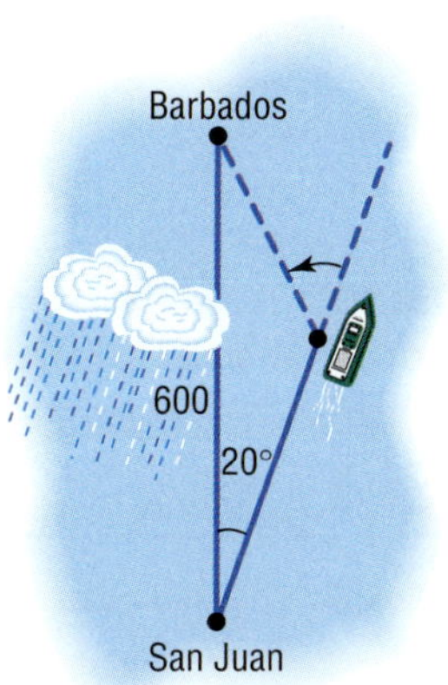

29. Major League Baseball Field A Major League baseball diamond is actually a square 90 feet on a side. The pitching rubber is located 60.5 feet from home plate on a line joining home plate and second base.

(a) How far is it from the pitching rubber to first base?

(b) How far is it from the pitching rubber to second base?

(c) If a pitcher faces home plate, through what angle does he need to turn to face first base?

30. Little League Baseball Field According to Little League baseball official regulations, the diamond is a square 60 feet on a side. The pitching rubber is located 46 feet from home plate on a line joining home plate and second base.

(a) How far is it from the pitching rubber to first base?

(b) How far is it from the pitching rubber to second base?

(c) If a pitcher faces home plate, through what angle does he need to turn to face first base?

31. Finding the Length of a Guy Wire The height of a radio tower is 500 feet, and the ground on one side of the tower slopes upward at an angle of 10° (see the figure).

(a) How long should a guy wire be if it is to connect to the top of the tower and be secured at a point on the sloped side 100 feet from the base of the tower?

(b) How long should a second guy wire be if it is to connect to the middle of the tower and be secured at a point 100 feet from the base on the flat side?

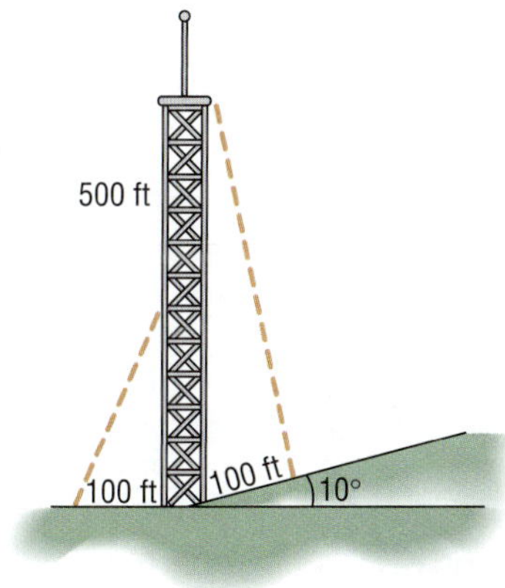

32. Finding the Length of a Guy Wire A radio tower 500 feet high is located on the side of a hill with an inclination to the horizontal of 5° (see the figure). How long should two guy wires be if they are to connect to the top of the tower and be secured at two points 100 feet directly above and directly below the base of the tower?

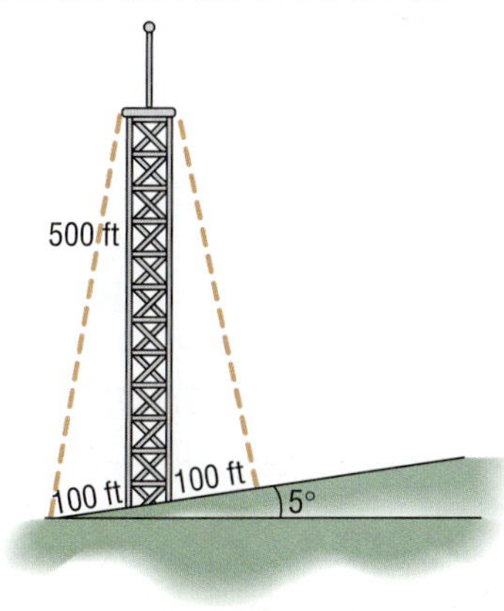

33. Wrigley Field, Home of the Chicago Cubs The distance from home plate to the fence in dead center in Wrigley Field is 400 feet (see the figure). How far is it from the fence in dead center to third base?

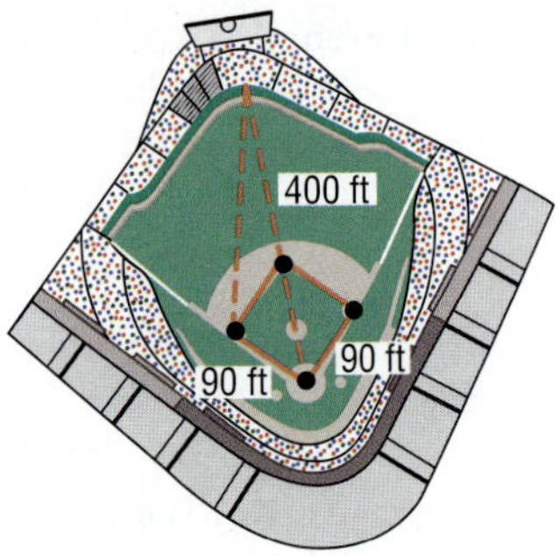

34. Little League Baseball The distance from home plate to the fence in dead center at the Oak Lawn Little League field is 280 feet. How far is it from the fence in dead center to third base?

[**Hint:** The distance between the bases in Little League in 60 feet.]

35. Rods and Pistons Rod OA (see the figure) rotates about the fixed point O so that point A travels on a circle of radius r. Connected to point A is another rod AB of length $L > 2r$, and point B is connected to a piston. Show that the distance x between point O and point B is given by

$$x = r\cos\theta + \sqrt{r^2\cos^2\theta + L^2 - r^2}$$

where θ is the angle of rotation of rod OA.

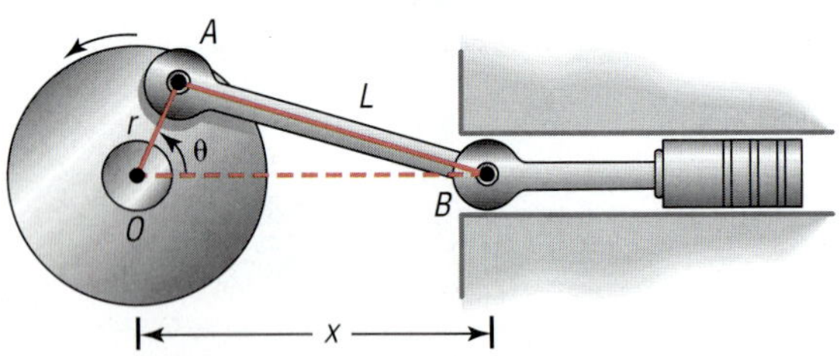

36. Geometry Show that the length d of a chord of a circle of radius r is given by the formula

$$d = 2r\sin\frac{\theta}{2}$$

where θ is the central angle formed by the radii to the ends of the chord (see the figure). Use this result to derive the fact that $\sin\theta < \theta$, where $\theta > 0$ is measure in radians.

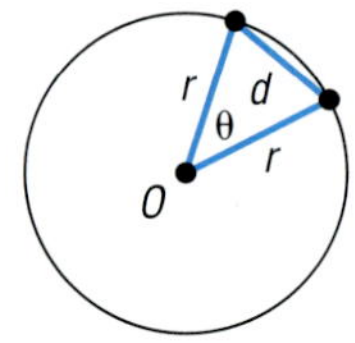

37. For any triangle, show that

$$\cos\frac{\gamma}{2} = \sqrt{\frac{s(s-c)}{ab}}$$

where $s = \frac{1}{2}(a + b + c)$.

[**Hint:** Use a Half-angle Formula and the Law of Cosines.]

38. For any triangle show that

$$\sin\frac{\gamma}{2} = \sqrt{\frac{(s-a)(s-b)}{ab}}$$

where $s = \frac{1}{2}(a + b + c)$.

39. Use the Law of Cosines to prove the identity

$$\frac{\cos\alpha}{a} + \frac{\cos\beta}{b} + \frac{\cos\gamma}{c} = \frac{a^2 + b^2 + c^2}{2abc}$$

40. Write down your strategy for solving an oblique triangle.

PREPARING FOR THIS SECTION

Before getting started, review the following:

✓ Geometry Review (Section A.2, pp. 565–568)

4.4 AREA OF A TRIANGLE

OBJECTIVES

1. Find the Area of SAS Triangles
2. Find the Area of SSS Triangles

In this section, we will derive several formulas for calculating the area A of a triangle. The most familiar of these is the following:

APPENDIX

Where Did That Come From? Selected Proofs

▸ SECTION 9.4 Properties of Logarithms

The Product Rule

Let b, and N be positive real numbers with $b \neq 1$.

$$\log_b(MN) = \log_b M + \log_b N$$

Proof

We begin by letting $\log_b M = R$ and $\log_b N = S$.

Now we write each logarithm in exponential form.

$$\log_b M = R \quad \text{means} \quad b^R = M.$$

$$\log_b N = S \quad \text{means} \quad b^S = N.$$

By substituting and using a property of exponents, we see that

$$MN = b^R b^S = b^{R+S}$$

Now we change $MN = b^{R+S}$ to logarithmic form.

$$MN = b^{R+S} \quad \text{means} \quad \log_b(MN) = R + S.$$

Finally, substituting $\log_b M$ for R and $\log_b N$ for S gives us

$$\log_b(MN) = \log_b M + \log_b N,$$

The property that we wanted to prove.

The quotient and power rules for logarithms are proved using similar procedures.

The Change-of-Base Property

For any logarithmic bases a and b, and any positive number M,

$$\log_b M = \frac{\log_a M}{\log_a b}.$$

Proof

To prove tha change-of-base property, we let x equal the logarithm on the left side:

$$\log_b M = x.$$

Now we rewrite this logarithm in exponential form.

$$\log_b M = x \quad \text{means} \quad b^x = M.$$

Because b^x and M are equal, the logarithms with base a for each of these expressions must be equal. This means that

$$\log_a b^x = \log_a M$$

$$x \log_a b = \log_a M$$ Apply the power rule for logarithms on the left side.

$$x = \frac{\log_a M}{\log_a b}$$ Solve for x by dividing both sides by $\log_a b$.

In our first step we let x equal $\log_b M$. Replacing x on the left side by $\log_b M$ gives us

$$\log_b M = \frac{\log_a M}{\log_a b},$$

which is the change-of-base property.

▸ SECTION 10.3 The Hyperbola

The Asymptotes of a Hyperbola Centered at the Origin

The hyperbola

$$\frac{x^2}{a^2} - \frac{y^2}{b^2} = 1$$

with a horizontal transverse axis has the two asymptotes

$$y = \frac{b}{a}x \quad \text{and} \quad y = -\frac{b}{a}x.$$

Proof

Begin by solving the hyperbola's equation for y.

$$\frac{x^2}{a^2} - \frac{y^2}{b^2} = 1$$ This is the standard form of the equation of a hyperbola.

$$\frac{y^2}{b^2} = \frac{x^2}{a^2} - 1$$ We isolate the term involving y^2 to solve for y.

$$y^2 = \frac{b^2x^2}{a^2 - b^2}$$ Multiply both sides by b^2.

$$y^2 = \frac{b^2x^2}{a^2}\left(1 - \frac{a^2}{x^2}\right)$$ Factor out $\frac{b^2x^2}{a^2}$ on the right. Verify that this result is correct by multiplying using the distributive property and obtaining the previous step.

$$y = \pm\sqrt{\frac{b^2x^2}{a^2}\left(1 - \frac{a^2}{x^2}\right)}$$ Solve for y using the square root method: If $u^2 = d$, $u = \pm\sqrt{d}$.

$$y = \pm\frac{b}{a}x\sqrt{1 - \frac{a^2}{x^2}}$$ Simplify.

As x increases or decreases without bound, the value of $\frac{a^2}{x^2}$ approaches 0. Consequently, the value of y can be approximated by

$$y = \pm \frac{b}{a} x.$$

This means that the lines whose equations are $y = \frac{b}{a} x$ and $y = -\frac{b}{a} x$ are asymptotes for the graph of the hyperbola.

Answers to Selected Exercises

CHAPTER 1

Section 1.1

Check Point Exercises

1. a. $8x + 5$ **b.** $\frac{x}{7} - 2x$ **2.** 51; In 1950, approximately 51% of American adults smoked cigarettes. **3.** 608
4. 2311; The formula models the data quite well. **5. a.** true **b.** true **6. a.** -8 is less than -2.; true **b.** 7 is greater than -3.; true
c. -1 is less than or equal to -4.; false **d.** 5 is greater than or equal to 5.; true **e.** 2 is greater than or equal to -14.; true

Exercise Set 1.1

1. $x + 5$ **3.** $x - 4$ **5.** $4x$ **7.** $2x + 10$ **9.** $6 - \frac{1}{2}x$ **11.** $\frac{4}{x} - 2$ **13.** $\frac{3}{5 - x}$ **15.** 57 **17.** 10 **19.** 88 **21.** 10
23. 44 **25.** 46 **27.** $\{1, 2, 3, 4\}$ **29.** $\{-7, -6, -5, -4\}$ **31.** $\{8, 9, 10, \ldots\}$ **33.** $\{1, 3, 5, 7, 9\}$ **35.** true **37.** true
39. false **41.** true **43.** false **45.** true **47.** false **49.** -6 is less than -2.; true **51.** 5 is greater than -7.; true
53. 0 is less than -4.; false **55.** -4 is less than or equal to 1.; true **57.** -2 is less than or equal to -6.; false
59. -2 is less than or equal to -2.; true **61.** -2 is greater than or equal to -2.; true **63.** 2 is less than or equal to $-\frac{1}{2}$.; false
65. 82 years; Formula models data quite well. **67.** 10; Formula models data quite well. **69.** 10°C **71.** 60 ft
73–89. Answers will vary. **91.** c **93.** $(2 \cdot 3 + 3) \cdot 5 = 45$ **95.** -6 and -5

Section 1.2

Check Point Exercises

1. a. 6 **b.** 4.5 **c.** 0 **2. a.** -28 **b.** 0.7 **c.** $-\frac{1}{10}$ **3. a.** 8 **b.** $-\frac{1}{3}$ **4. a.** -3 **b.** 10.5 **c.** -6
5. a. 25 **b.** -25 **c.** -64 **d.** 81 **6. a.** -8 **b.** $\frac{8}{15}$ **7.** 74 **8.** -4 **9.** addition: $9 + 4x$; multiplication: $x \cdot 4 + 9$
10. a. $(6 + 12) + x = 18 + x$ **b.** $(-7 \cdot 4)x = -28x$ **11.** $-28x - 8$ **12.** $14x + 15x^2$ or $15x^2 + 14x$ **13.** $12x - 40$
14. $42 - 4x$

Exercise Set 1.2

1. 7 **3.** 4 **5.** 7.6 **7.** $\frac{\pi}{2}$ **9.** $\sqrt{2}$ **11.** -11 **13.** -4 **15.** -4.5 **17.** $\frac{2}{15}$ **19.** -8.2 **21.** -12.4 **23.** 0
25. -11 **27.** 5 **29.** 0 **31.** -12 **33.** 18 **35.** -15 **37.** $-\frac{1}{4}$ **39.** 5.5 **41.** $\sqrt{2}$ **43.** -90 **45.** 33 **47.** $-\frac{15}{13}$
49. 0 **51.** -8 **53.** 48 **55.** 100 **57.** -100 **59.** -8 **61.** 1 **63.** -1 **65.** -3 **67.** 45 **69.** 0 **71.** undefined
73. $\frac{9}{14}$ **75.** -15 **77.** -2 **79.** -24 **81.** 45 **83.** 1 **85.** 14 **87.** $-\frac{8}{3}$ **89.** 31
91. addition: $10 + 4x$; multiplication: $x \cdot 4 + 10$ **93.** addition: $-5 + 7x$; multiplication: $x \cdot 7 - 5$ **95.** $(4 + 6) + x = 10 + x$
97. $(-7 \cdot 3)x = -21x$ **99.** $\left(-\frac{1}{3} \cdot -3\right)y = y$ **101.** $6x + 15$ **103.** $-14x - 21$ **105.** $-3x + 6$ **107.** $12x$ **109.** $5x^2$
111. $10x + 12x^2$ **113.** $18x - 40$ **115.** $8y - 12$ **117.** $16y - 25$ **119.** $12x^2 + 11$ **121.** \$0.63 **123.** \$1.74
125. \$477 billion **127. a.** $132 - 0.6a$ **b.** 120 beats per min **129–145.** Answers will vary. **147.** $(8 - 2) \cdot 3 - 4 = 14$
149. -7 **150.** $\frac{10}{x} - 4x$ **151.** 42 **152.** true

Section 1.3

Check Point Exercises

1.

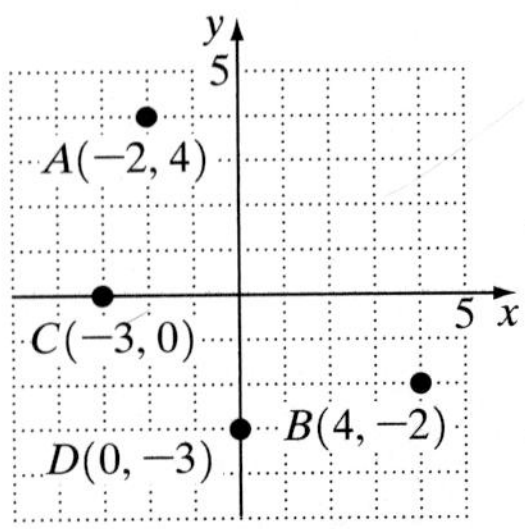

2.

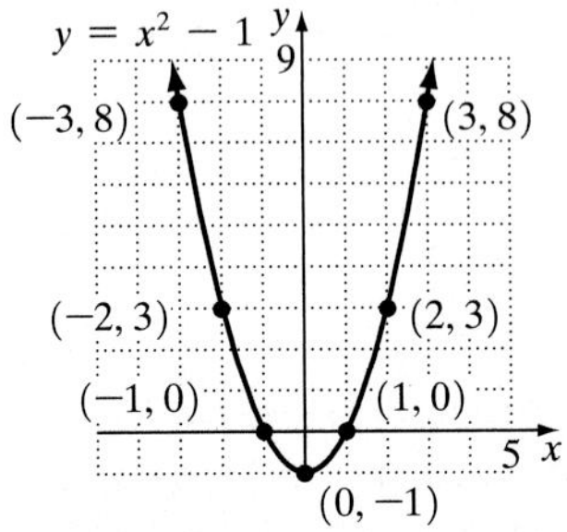

3.

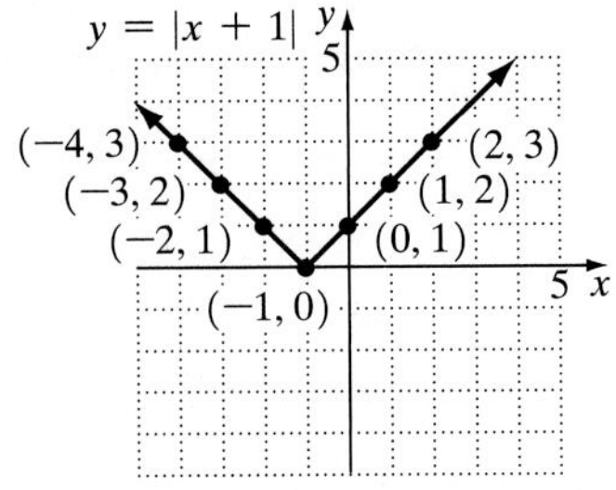

4. a. 0 to 3 hr **b.** 3 to 13 hr **c.** 0.05 mg per 100 ml; after 3 hr **d.** The drug concentration is 0 mg per 100 ml.

5. minimum x-value: −100; maximum x-value: 100; distance between tick marks on x-axis: 50; minimum y-value: −100; maximum y-value: 100; distance between tick marks on y-axis: 10

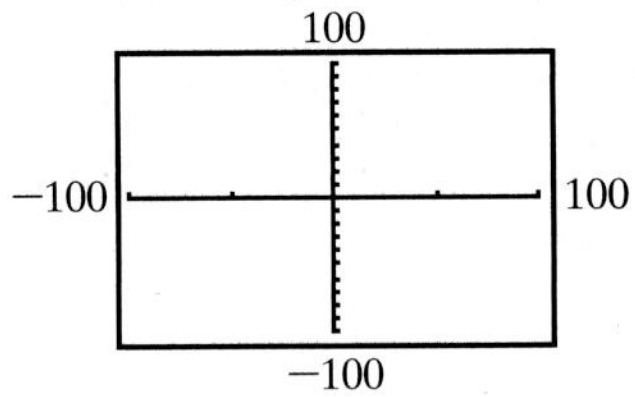

Exercise Set 1.3

1–9.

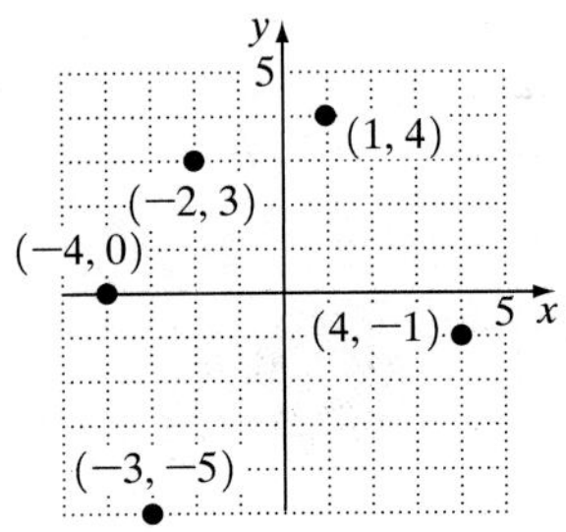

11.

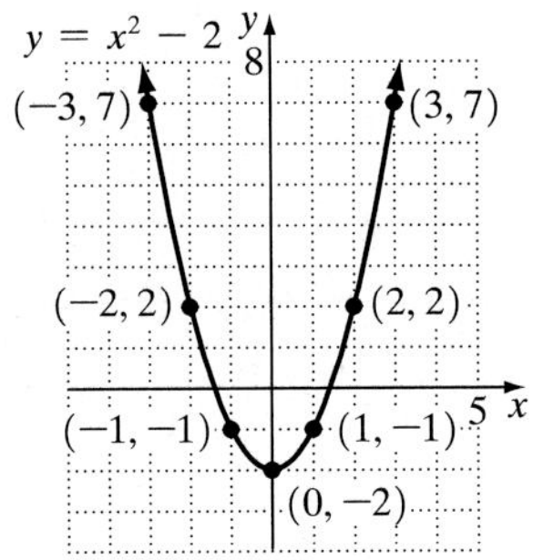

13.

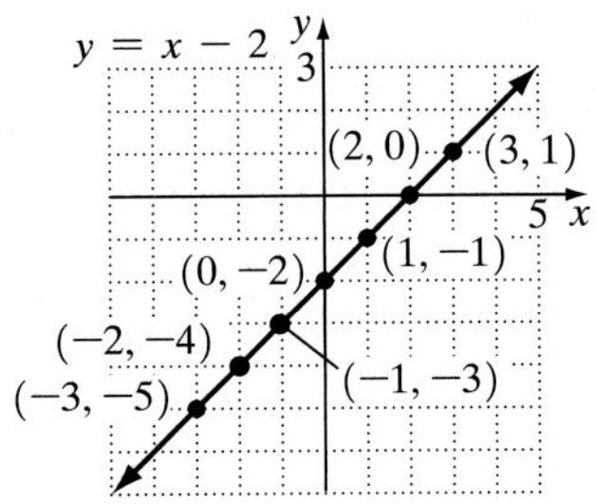

15.

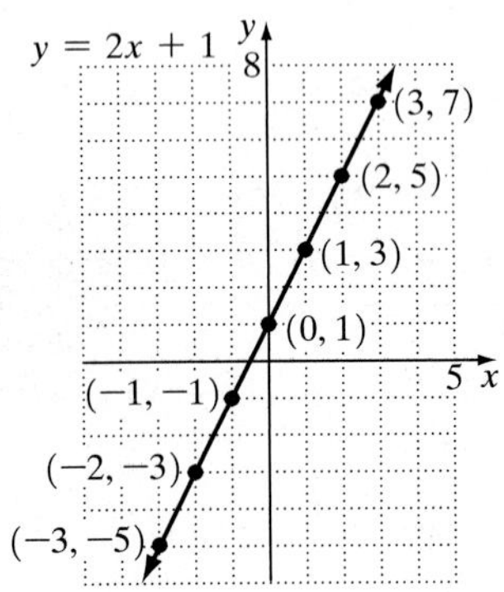

17.

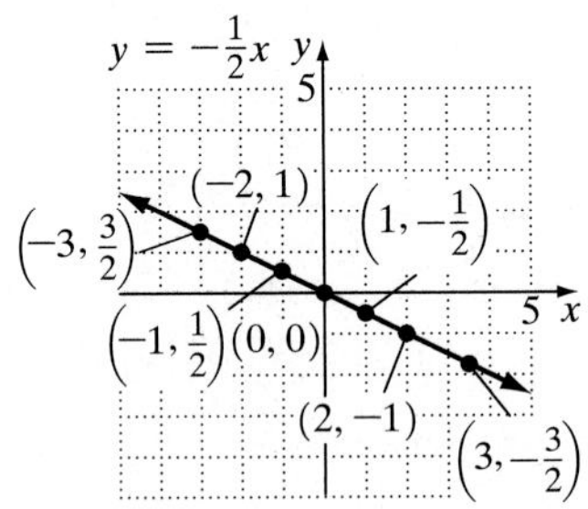

19.

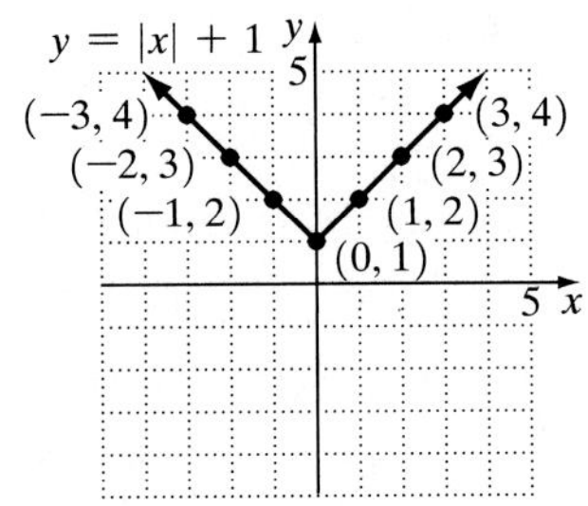

21.

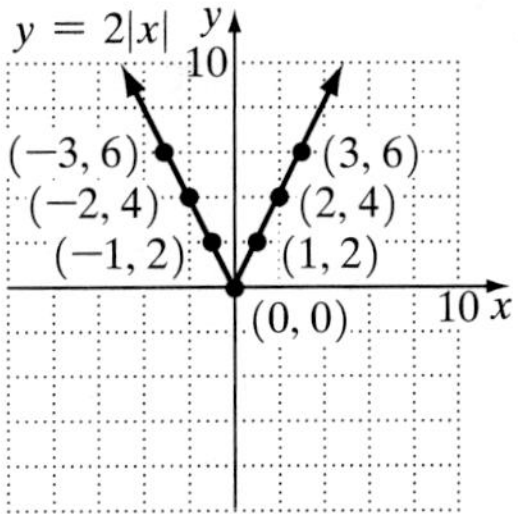

23.

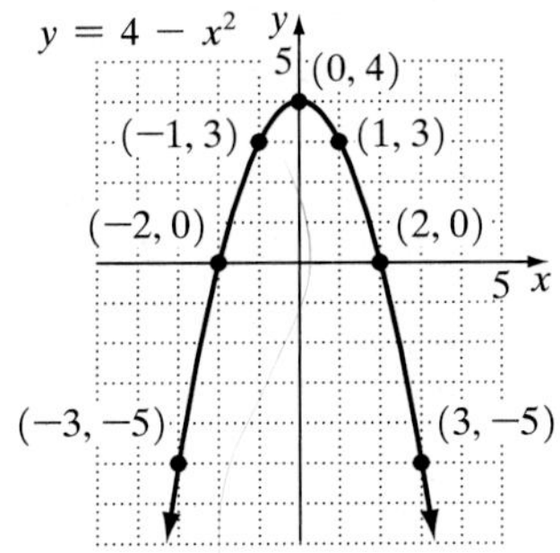

25.

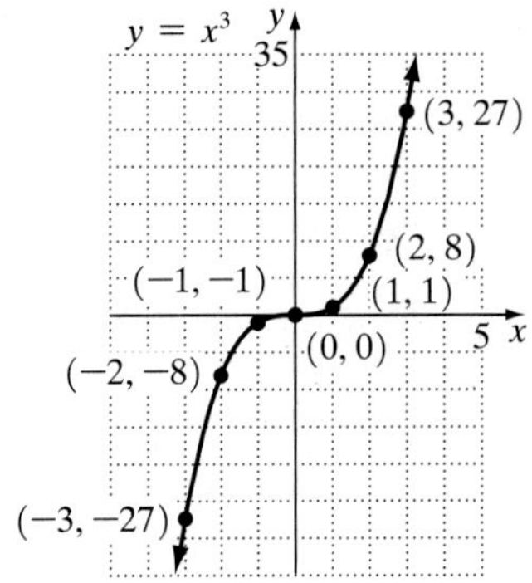

27. c **29.** b **31.** 0 to 4 years **33.** 4th year; about 8.3% **35.** 1832; 20% **37.** 52%
39. a **41.** b **43.** b **45.** 181 lb **47.** 135 lb **49–55.** Answers will vary.

59. c;

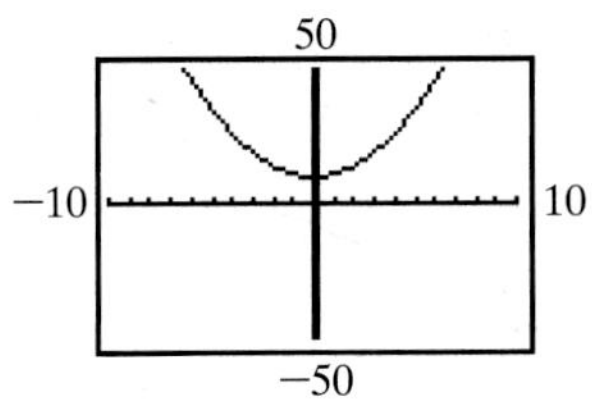

61. c;

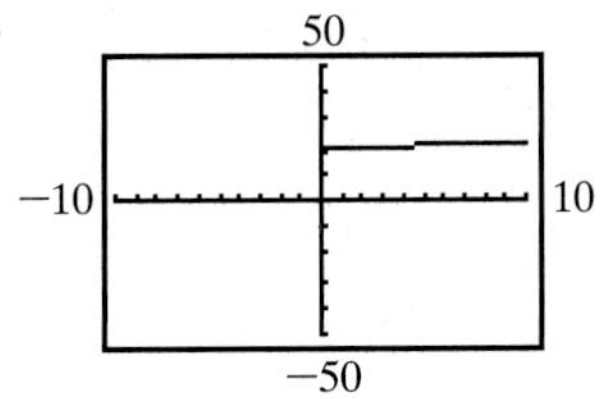

63. d
65. 8.8 sec
66. 14.3
67. 3
68. $-14x - 25$

Section 1.4

Check Point Exercises

1. 6 or $\{6\}$ **2.** -1 or $\{-1\}$ **3.** 6 or $\{6\}$ **4.** 1 or $\{1\}$ **5.** no solution; inconsistent **6.** all real numbers; identity **7.** 2011

Exercise Set 1.4

1. 3 **3.** 11 **5.** 11 **7.** 7 **9.** 13 **11.** 2 **13.** -4 **15.** 9 **17.** -5 **19.** 6 **21.** 19 **23.** 12 **25.** 24
27. -15 **29.** 5 **31.** 13 **33.** -12 **35.** $\frac{46}{5}$ **37.** all real numbers; identity **39.** no solution; inconsistent
41. 0; conditional **43.** 0; conditional **45.** no solution; inconsistent **47.** 0; conditional **49.** 40 years old; Answers will vary.
51. 2000 **53.** 2001 **55–67.** Answers will vary.

69. inconsistent

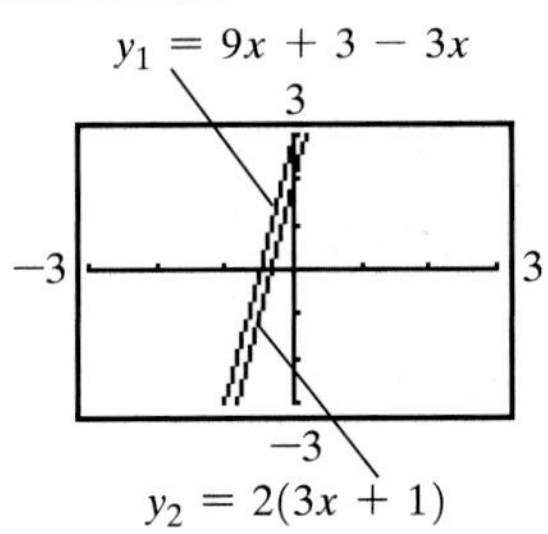

71. conditional; -5

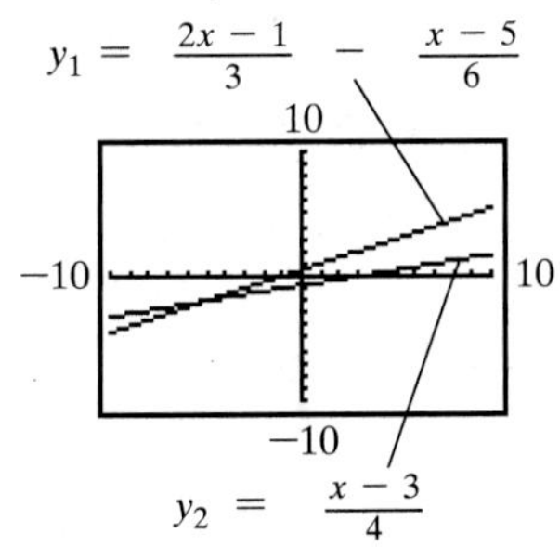

73. $x = \frac{c - b}{a}$ **75.** Answers will vary.

77. $\frac{3}{10}$ **78.** -60

79.

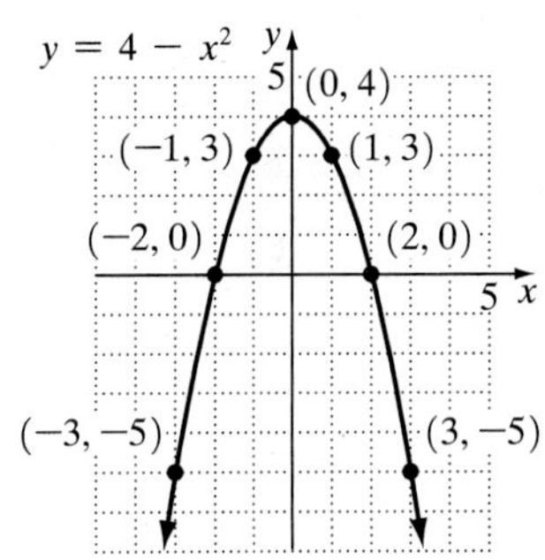

Section 1.5

Check Point Exercises

1. *Purple Rain*: 13 million; *Saturday Night Fever*: 15 million; *The Bodyguard*: 16 million **2.** 2010 **3.** 300 min **4.** \$1400
5. 50 ft by 94 ft **6.** $l = \frac{P - 2w}{2}$ **7.** $h = \frac{V}{lw}$ **8.** $W = 106 + 6H$ or $W = 6H + 106$ **9.** $C = \frac{P}{1 + M}$

Exercise Set 1.5

1. 6 **3.** 25 **5.** 120 **7.** 320 **9.** 19 and 45 **11.** Hurricane Hugo: \$7 billion; Northridge earthquake: \$12.5 billion; Hurricane Andrew: \$20 billion **13.** 100°, 50°, and 30° **15.** 2050 **17.** 2005 **19.** 2004 **21.** \$600 of merchandise; \$580 **23.** 28 times **25.** \$420 **27.** \$150 **29.** \$31,000 **31.** \$467.20 **33.** 50 yd by 100 yd **35.** 36 ft by 78 ft **37.** length: 2 ft; height: 5 ft **39.** 5 ft 7 in. **41.** $l = \frac{A}{w}$ **43.** $b = \frac{2A}{h}$ **45.** $P = \frac{I}{rt}$ **47.** $p = \frac{T - D}{m}$ **49.** $a = \frac{2A}{h} - b$ **51.** $h = \frac{3V}{\pi r^2}$ **53.** $m = \frac{y - y_1}{x - x_1}$ **55.** $d_1 = Vt + d_2$ **57.** $x = \frac{C - By}{A}$ **59.** $v = \frac{2s - at^2}{2t}$ **61.** $n = \frac{L - a}{d} + 1$ **63.** $l = \frac{A - 2wh}{2w + 2h}$ **65.** $I = \frac{E}{R + r}$ **67–71.** Answers will vary.

73.

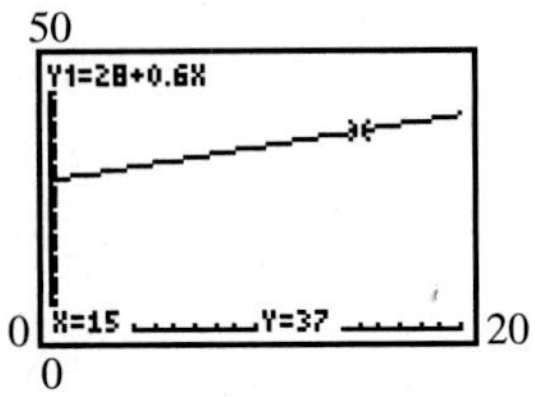

75. \$200 **77.** Mrs. Ricardo: \$4000; boy: \$8000; girl: \$2000 **79.** $C = \frac{LV - SN}{L - N}$ **80.** −6 is less than or equal to −6.; true **81.** $\frac{7}{3}$ **82.** −8

Section 1.6

Check Point Exercises

1. a. b^{11} **b.** $40x^5y^{10}$ **2. a.** b^{12} **b.** $9x^{11}y^3$ **3. a.** 1 **b.** 1 **c.** −1 **d.** 10 **e.** 1 **4. a.** $\frac{1}{25}$ **b.** $\frac{1}{81}$ **c.** 16 **d.** $\frac{3y^4}{x^6}$ **5. a.** $\frac{4^3}{7^2} = \frac{64}{49}$ **b.** $\frac{x^2}{5}$ **6. a.** x^{15} **b.** $\frac{1}{y^{14}}$ **c.** b^{12} **7. a.** $16x^4$ **b.** $-27y^6$ **c.** $\frac{y^2}{16x^{10}}$ **8. a.** $\frac{x^{15}}{64}$ **b.** $\frac{16}{x^{12}y^8}$ **9. a.** $-12y^9$ **b.** $\frac{4y^{14}}{x^6}$

Exercise Set 1.6

1. b^{11} **3.** x^4 **5.** 32 **7.** $6x^6$ **9.** $20y^{12}$ **11.** $100x^{10}y^{12}$ **13.** b^9 **15.** $5x^5$ **17.** x^5y^5 **19.** $10xy^3$ **21.** $-8a^{11}b^8c^4$ **23.** 1 **25.** 1 **27.** −1 **29.** 13 **31.** 1 **33.** $\frac{1}{3^2} = \frac{1}{9}$ **35.** $\frac{1}{(-5)^2} = \frac{1}{25}$ **37.** $-\frac{1}{5^2} = -\frac{1}{25}$ **39.** $\frac{x^2}{y^3}$ **41.** $\frac{8y^3}{x^7}$ **43.** $5^3 = 125$ **45.** $(-3)^4 = 81$ **47.** $\frac{y^5}{x^2}$ **49.** $\frac{b^7c^3}{a^4}$ **51.** x^{60} **53.** $\frac{1}{b^{12}}$ **55.** 7^{20} **57.** $64x^3$ **59.** $9x^{14}$ **61.** $8x^3y^6$ **63.** $9x^4y^{10}$ **65.** $-\frac{x^6}{27}$ **67.** $\frac{y^8}{25x^6}$ **69.** $\frac{x^{20}}{16y^{16}z^8}$ **71.** $\frac{16}{x^4}$ **73.** $\frac{x^6}{25}$ **75.** $\frac{81x^4}{y^4}$ **77.** $\frac{x^{24}}{y^{12}}$ **79.** x^9y^{12} **81.** a^8b^{12} **83.** $\frac{1}{x^6}$ **85.** $-\frac{4}{x}$ **87.** $\frac{2}{x^7}$ **89.** $\frac{10a^2}{b^5}$ **91.** $\frac{1}{x^9}$ **93.** $-\frac{6}{a^2}$ **95.** $3x^{10}$ **97.** $\frac{1}{x^{10}}$ **99.** $-\frac{5y^8}{x^6}$ **101.** x^{16} **103.** $-\frac{27b^{15}}{a^{18}}$ **105.** 1 **107.** 125 kg; not safe **109.** 0.55 astronomical unit **111.** 1; Earth is 1 astronomical unit from the sun. **113–119.** Answers will vary.

121.

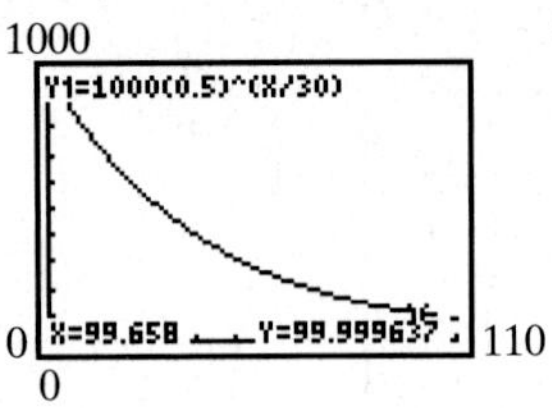

; about 100 years or 2086 **123.** d **125.** x^{4n+3} **127.** x^6

129.

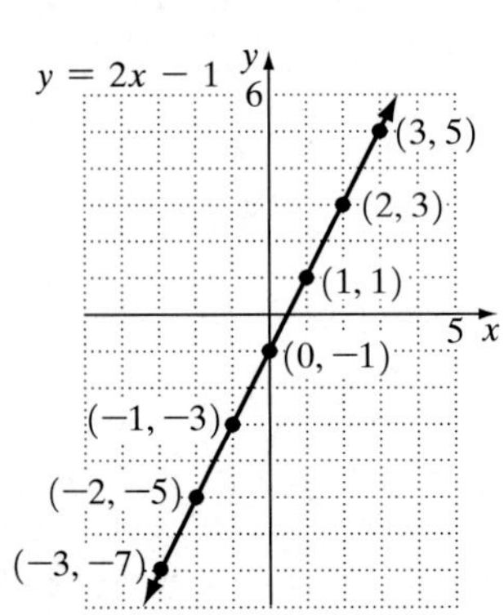

130. $y = \frac{C - Ax}{B}$ **131.** 40 m by 75 m

Section 1.7

Check Point Exercises

1. a. 2,600,000,000 **b.** 0.000003017 **2. a.** 7.41×10^9 **b.** 9.2×10^{-8} **3.** 4.1×10^9 **4. a.** 3.55×10^{-1} **b.** 4×10^8
5. \$20,000 **6.** 3.1×10^7 mi

Exercise Set 1.7

1. 380 **3.** 0.0006 **5.** 7,160,000 **7.** 1.4 **9.** 0.79 **11.** 0.00415 **13.** 60,000,100,000 **15.** 3.2×10^4 **17.** 6.38×10^{17}
19. 3.17×10^2 **21.** 5.716×10^3 **23.** 2.7×10^{-3} **25.** 5.04×10^{-9} **27.** 7×10^{-3} **29.** 3.14159×10^0 **31.** 6.3×10^7
33. 6.4×10^4 **35.** 1.22×10^{-11} **37.** 2.67×10^{13} **39.** 2.1×10^3 **41.** 4×10^5 **43.** 2×10^{-8} **45.** 5×10^3
47. 4×10^{15} **49.** 9×10^{-3} **51.** 5.33×10^7 people **53.** 4.17×10^7 people **55.** \$6800 **57.** $\$1.12 \times 10^{12}$
59. 1.06×10^{-18} g **61–67.** Answers will vary. **69.** d **71.** Answers will vary. **72.** $85x - 26$ **73.** 4 **74.** $\frac{y^6}{64x^8}$

Review Exercises

1. $2x - 10$ **2.** $6x + 4$ **3.** $\frac{9}{x} + \frac{1}{2}x$ **4.** 34 **5.** 60 **6.** 15 **7.** $\{1, 2\}$ **8.** $\{-3, -2, -1, 0, 1\}$ **9.** false **10.** true
11. true **12.** −5 is less than 2.; true **13.** −7 is greater than or equal to −3.; false **14.** −7 is less than or equal to −7.; true
15. 124 ft **16.** 9.7 **17.** 5.003 **18.** 0 **19.** −7.6 **20.** −4.4 **21.** 13 **22.** 60 **23.** −240 **24.** 16 **25.** −32
26. $-\frac{5}{12}$ **27.** 7 **28.** −9.1 **29.** 7 **30.** 9 **31.** −2 **32.** −18 **33.** 55 **34.** 1 **35.** −4 **36.** −13 **37.** $17x - 15$
38. $9x^2 + x$ **39.** $5y - 17$ **40.** $10x$ **41.** $-3x - 8$

42–44.

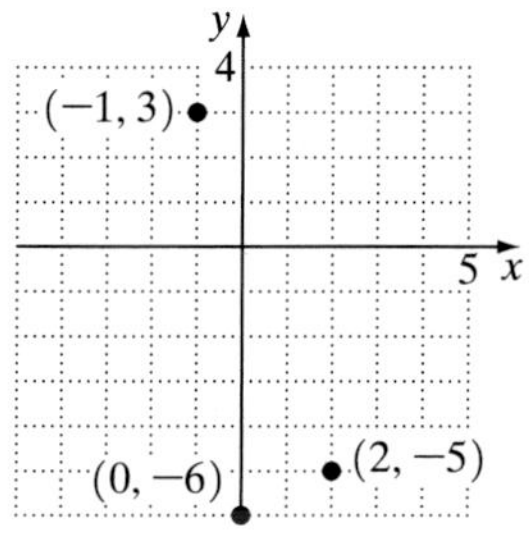

45.

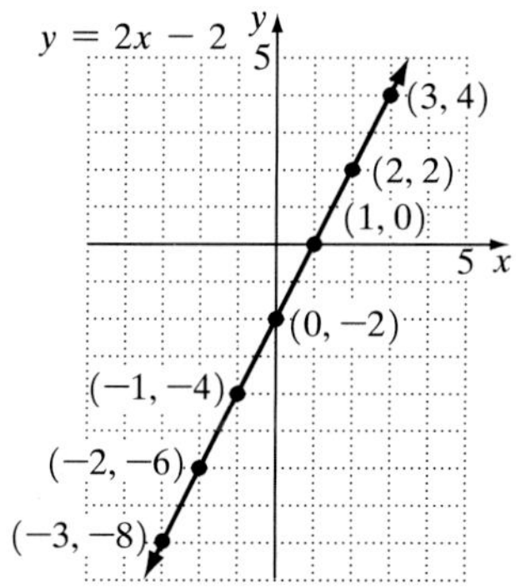

46.

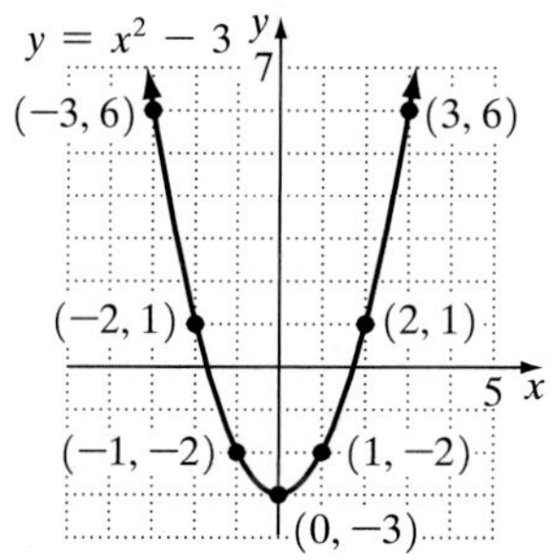

47.

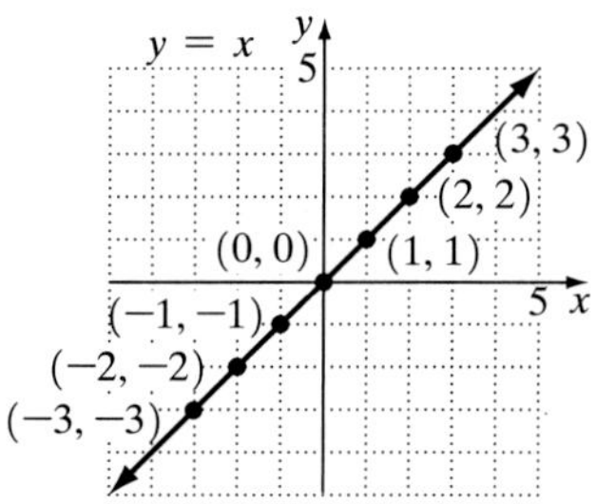

48.

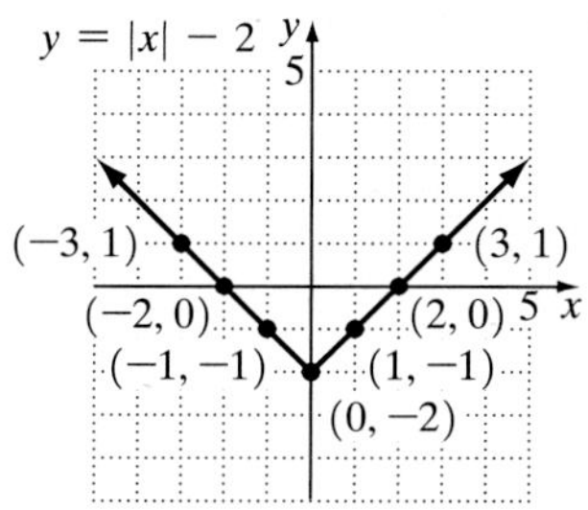

49. minimum x-value: −20;
maximum x-value: 40;
distance between tick marks on x-axis: 10;
minimum y-value: −5;
maximum y-value: 5;
distance between tick marks on y-axis: 1

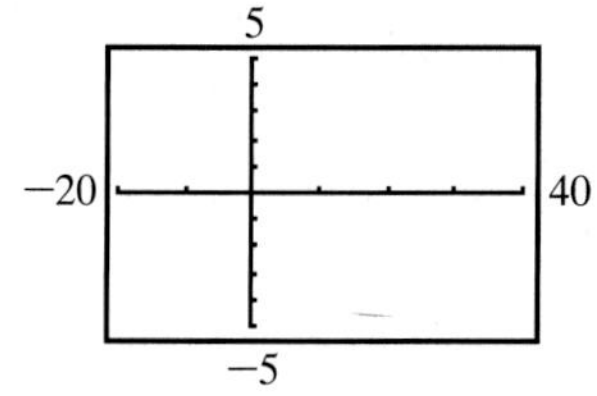

50. 20% **51.** 85 years old **52.** Answers will vary. **53.** c **54.** 6 **55.** −10 **56.** 5 **57.** −13 **58.** −3 **59.** −1
60. 2 **61.** 2 **62.** $\frac{72}{11}$ **63.** −12 **64.** $\frac{77}{15}$ **65.** no solution; inconsistent **66.** all real numbers; identity **67.** 0; conditional
68. 1997 **69.** low income: \$174 thousand; middle income: \$237 thousand; high income: \$345 thousand **70.** 2005 **71.** 500 min
72. \$60 **73.** \$10,000 in sales **74.** 44 yd by 126 yd **75.** $h = \frac{3V}{B}$ **76.** $x = \frac{y - y_1}{m} + x_1$ **77.** $R = \frac{E}{I} - r$
78. $F = \frac{9C + 160}{5}$ **79.** $g = \frac{s - vt}{t^2}$ **80.** $g = \frac{T}{r + vt}$ **81.** $15x^{13}$ **82.** $\frac{x^2}{y^5}$ **83.** $\frac{x^4y^7}{9}$ **84.** $\frac{1}{x^{18}}$ **85.** $49x^6y^2$ **86.** $-\frac{8}{y^7}$

87. $-\frac{12}{x^7}$ **88.** $3x^{10}$ **89.** $-\frac{a^8}{2b^5}$ **90.** $-24x^7y^4$ **91.** $\frac{3}{4}$ **92.** $\frac{y^{12}}{125x^6}$ **93.** $-\frac{6x^9}{y^5}$ **94.** 7,160,000 **95.** 0.000107 **96.** 4.1×10^{13} **97.** 8.09×10^{-3} **98.** 1.26×10^{8} **99.** 2.5×10^{1} **100.** $\$4.2 \times 10^{10}$

Chapter 1 Test

1. $4x - 5$ **2.** 170 **3.** $\{-4, -3, -2, -1\}$ **4.** true **5.** -3 is greater than -1.; false **6.** 616 convictions **7.** 17.9 **8.** -7.6 **9.** $\frac{1}{4}$ **10.** -60 **11.** $\frac{1}{8}$ **12.** 3.1 **13.** -3 **14.** 6 **15.** -4 **16.** $-5x - 18$ **17.** $6y - 27$

18.

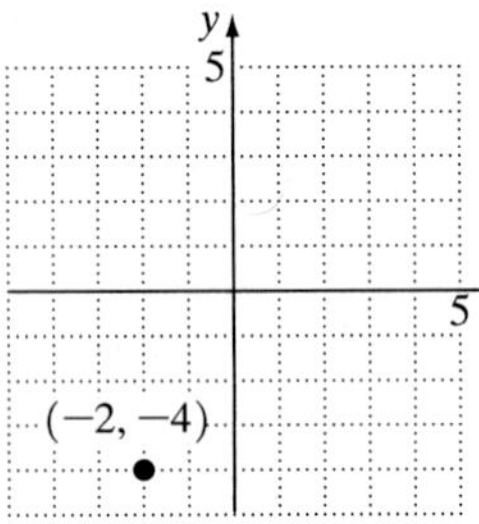

19.

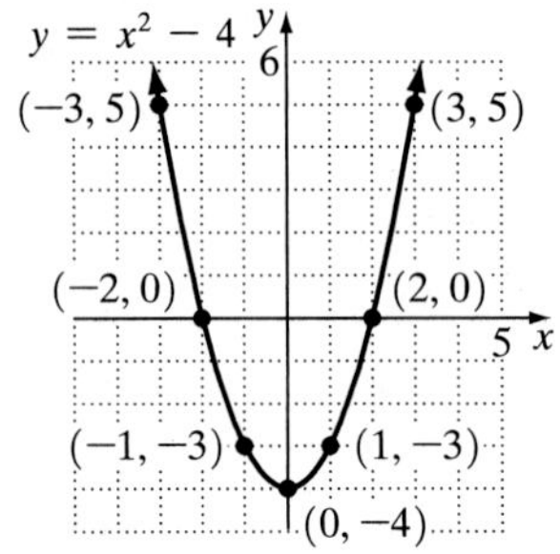

20. 2 **21.** -6 **22.** no solution; inconsistent **23.** 23 and 49 **24.** 5 yr **25.** \$50 **26.** 120 yd by 380 yd **27.** $h = \frac{3V}{lw}$ **28.** $y = \frac{C - Ax}{B}$ **29.** $-\frac{14}{x^5}$ **30.** $\frac{x^6}{4y^3}$ **31.** $\frac{x^{15}}{64y^6}$ **32.** 0.0000038 **33.** 4.07×10^{11} **34.** 5×10^{3} **35.** 1.216×10^{10}

CHAPTER 2

Section 2.1

Check Point Exercises

1. domain: {5, 10, 15, 20, 25}; range: {12.8, 16.2, 18.9, 20.7, 21.8} **2. a.** not a function **b.** function **3. a.** 29 **b.** 65 **c.** 46 **d.** $6a + 6h + 9$ **4.**

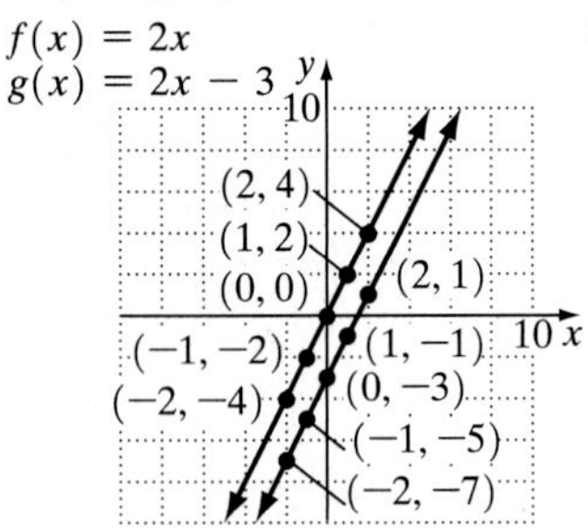

; The graph of g is the graph of f shifted down 3 units. **5. a.** function **b.** function **c.** not a function **6. a.** 16 **b.** 8

Exercise Set 2.1

1. function; domain: {1, 3, 5}; range: {2, 4, 5} **3.** not a function; domain: {3, 4}; range: {4, 5} **5.** function; domain: {−3, −2, −1, 0}; range: {−3, −2, −1, 0} **7.** not a function; domain: {1}; range: {4, 5, 6} **9. a.** 1 **b.** 6 **c.** -7 **d.** $2a + 1$ **e.** $a + 3$ **11. a.** -2 **b.** -17 **c.** 0 **d.** $12b - 2$ **e.** $3b + 10$ **13. a.** 5 **b.** 8 **c.** 53 **d.** 32 **e.** $48b^2 + 5$ **15. a.** -1 **b.** 26 **c.** 19 **d.** $2b^2 + 3b - 1$ **e.** $50a^2 + 15a - 1$ **17. a.** $\frac{3}{4}$ **b.** -3 **c.** $\frac{11}{8}$ **d.** $\frac{13}{9}$ **e.** $\frac{2a + 2h - 3}{a + h - 4}$ **f.** Denominator would be zero.

19.

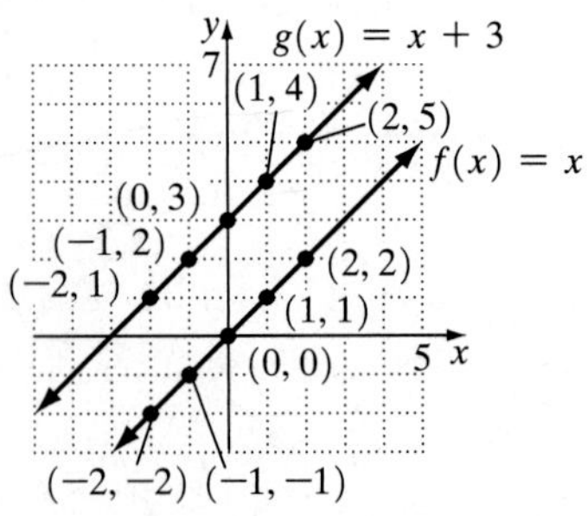

The graph of g is the graph of f shifted up 3 units.

21.

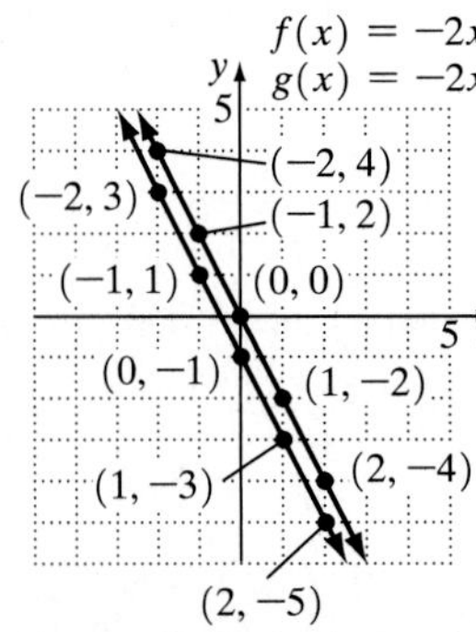

The graph of g is the graph of f shifted down 1 unit.

23.

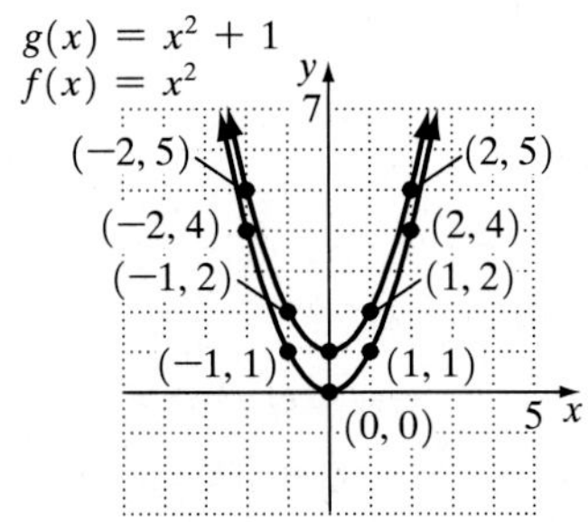

The graph of g is the graph of f shifted up 1 unit.

25. $f(x) = |x|$
$g(x) = |x| - 2$

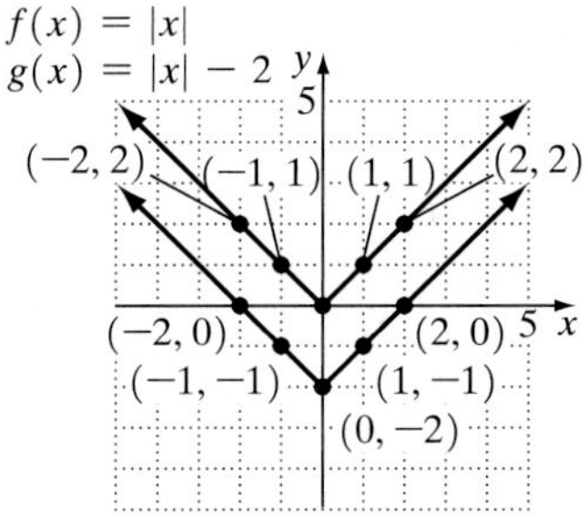

The graph of g is the graph of f shifted down 2 units.

27. $f(x) = x^3$
$g(x) = x^3 + 2$

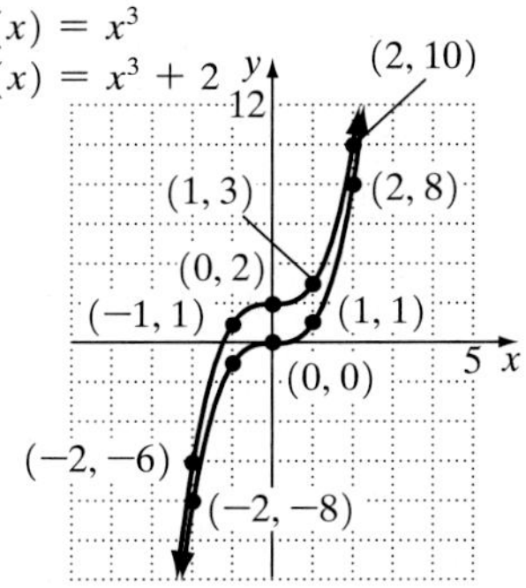

The graph of g is the graph of f shifted up 2 units.

29. function **31.** function **33.** not a function **35.** function **37.** −4 **39.** 4 **41.** 0 **43.** 2 **45.** 2 **47.** −2 **49.** $\{(1, 31), (2, 53), (3, 70), (4, 86), (5, 86)\}$; domain: $\{1, 2, 3, 4, 5\}$; range: $\{31, 53, 70, 86\}$; function; Answers will vary. **51.** 5.22; In 2000 there were 5.22 million women enrolled in U.S. colleges.; (2000, 5.22) **53.** 1.4; In 2004 there will be 1.4 million more women than men enrolled in U.S. colleges. **55** 440; For 20-year-old people there are 440 accidents per 50 million miles driven.; (20, 440)

57. $x = 45$; $y = 190$; The minimum number of accidents is 190 per 50 million miles driven and is attributed to 45-year-old drivers. **59.** 3.1; In 1960 about 3.1% of the U.S. population was Jewish American. **61.** 19 and 64; In 1919 and in 1962 3% of the U.S. population was Jewish American. **63.** 1940; 3.7% **65.** Each year corresponds to only one percentage. **67.** 0.76; It costs \$0.76 to mail a 3-ounce first-class letter. **69.** \$0.55 **71–77.** Answers will vary.

79.

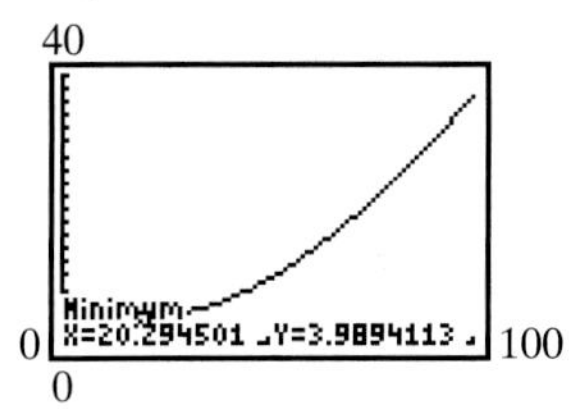

; The number of annual visits decreases until about age 20 and then increases.; (20.3, 4.0); The minimum is at about age 20 when there are about 4 annual physician visits.

81. 3 **83.** $f(2) = 6$; $f(3) = 9$; $f(4) = 12$; no **84.** 0 **85.** $\dfrac{y^{10}}{9x^4}$ **86.** −15

Section 2.2

Check Point Exercises

1. a. $\{x|x$ is a real number$\}$ **b.** $\{x|x$ is a real number and $x \neq -5\}$ **2. a.** $3x^2 + 6x + 6$ **b.** 78 **3. a.** $\dfrac{5}{x} + \dfrac{7}{x-8}$ **b.** $\{x|x$ is a real number and $x \neq 0$ and $x \neq 8\}$ **4. a.** 23 **b.** $x^2 - 3x - 3$; 1 **c.** $\dfrac{x^2 - 2x}{x+3}$; $\dfrac{7}{2}$ **d.** −24

Exercise Set 2.2

1. $\{x|x$ is a real number$\}$ **3.** $\{x|x$ is a real number and $x \neq -4\}$ **5.** $\{x|x$ is a real number and $x \neq 3\}$ **7.** $\{x|x$ is a real number and $x \neq 5\}$ **9.** $\{x|x$ is a real number and $x \neq -7$ and $x \neq 9\}$ **11. a.** $5x - 5$ **b.** 20 **13. a.** $3x^2 + x - 5$ **b.** 75 **15. a.** $2x^2 - 2$ **b.** 48 **17.** $\{x|x$ is a real number$\}$ **19.** $\{x|x$ is a real number and $x \neq 5\}$ **21.** $\{x|x$ is a real number and $x \neq 0$ and $x \neq 5\}$ **23.** $\{x|x$ is a real number and $x \neq 2$ and $x \neq -3\}$ **25.** $\{x|x$ is a real number and $x \neq 2\}$ **27.** $\{x|x$ is a real number$\}$ **29.** $x^2 + 3x + 2$; 20 **31.** 0 **33.** $x^2 + 5x - 2$; 48 **35.** −8 **37.** −16 **39.** −135 **41.** $\dfrac{x^2 + 4x}{2 - x}$; 5 **43.** −1 **45.** $\{x|x$ is a real number$\}$ **47.** $\{x|x$ is a real number and $x \neq 2\}$ **49.** 20; In 2000 veterinary costs in the U.S. for dogs and cats were about \$20 billion. **51.** $\{1983, 1987, 1991, 1996, 2000\}$ **53.** the total world population h **55.** 5.9 billion **57.** −200,000; 0; 200,000; The company has profits of −200,000, 0, and 200,000 dollars when 20,000, 30,000 and 40,000 radios are produced respectively. **59–63.** Answers will vary.

65.

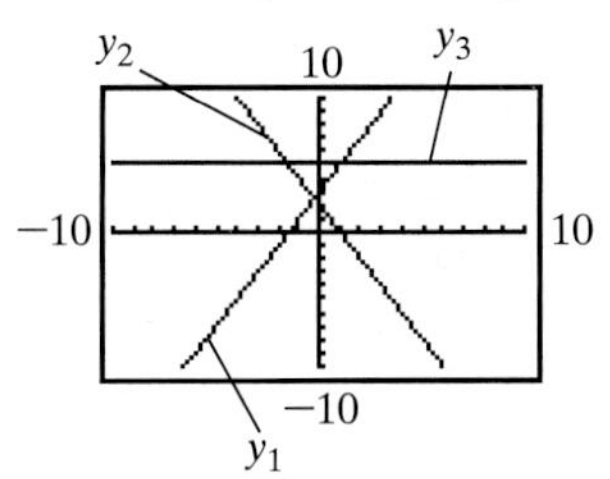

67.

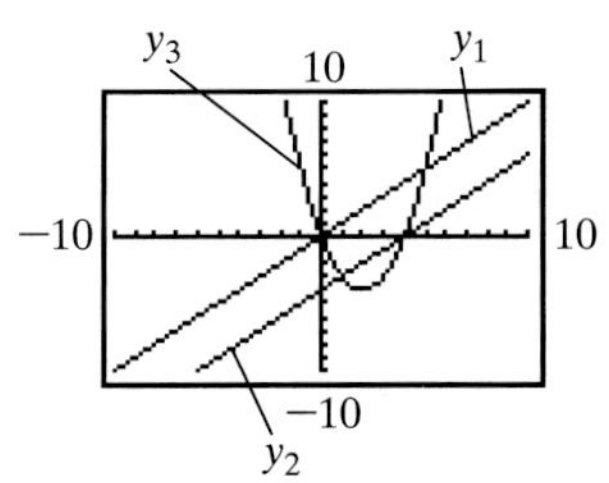

69. No y-value is displayed.; y_3 is undefined at $x = 0$.

71.

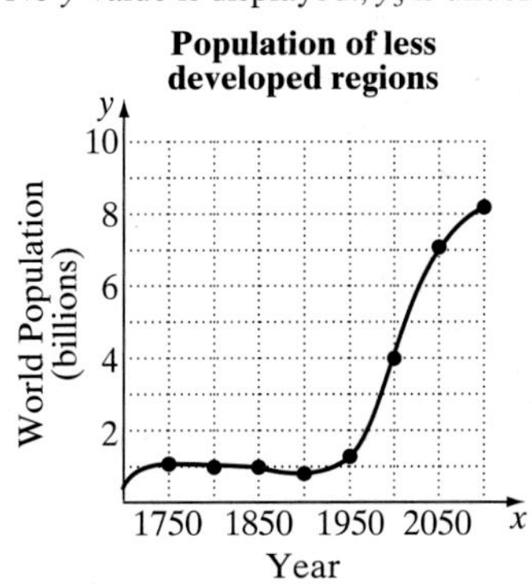

72. $b = \dfrac{R - 3a}{3}$ or $b = \dfrac{R}{3} - a$ **73.** 7 **74.** $6b + 8$

Section 2.3

Check Point Exercises

1.

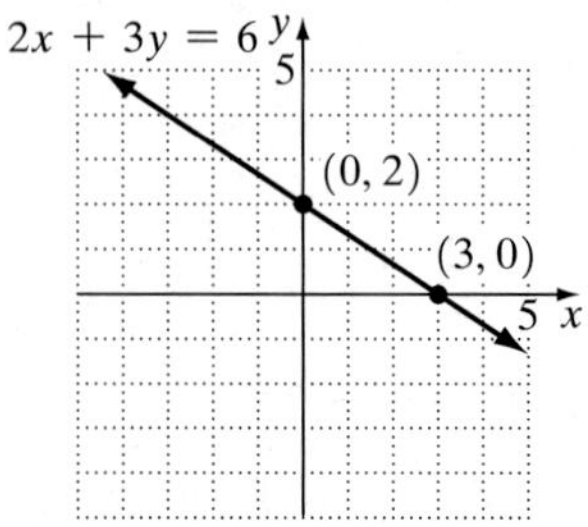

2. a. 6 **b.** $-\frac{7}{5}$

3.

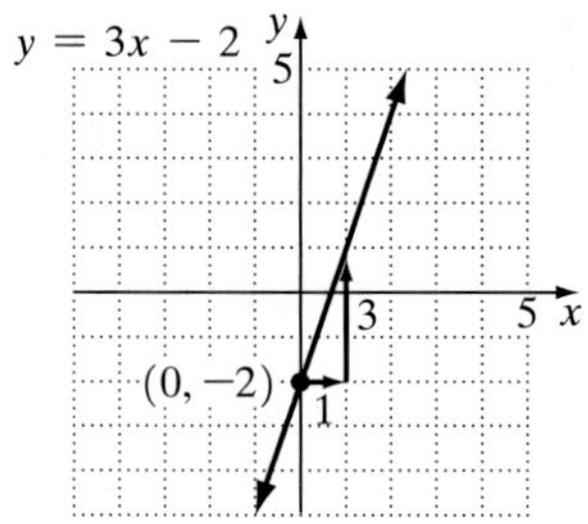

4.

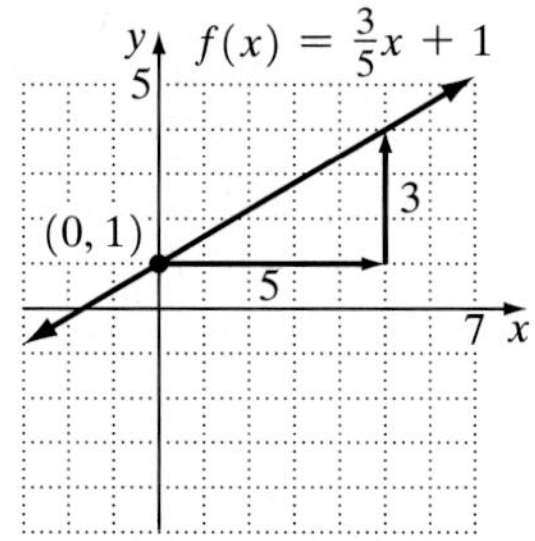

5.

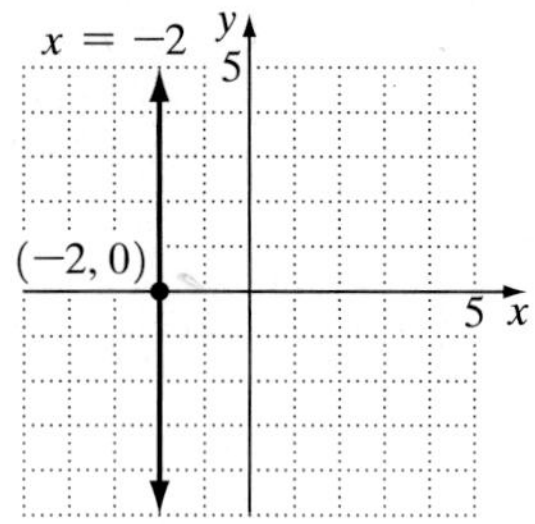

6. 0.01 mg per 100 ml per hr

Exercise Set 2.3

1.

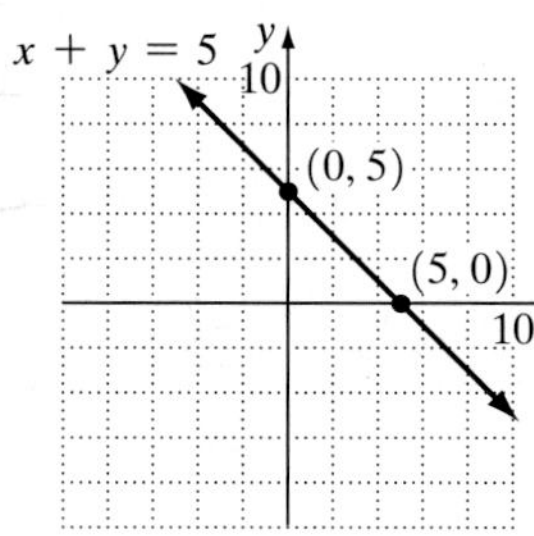

3.

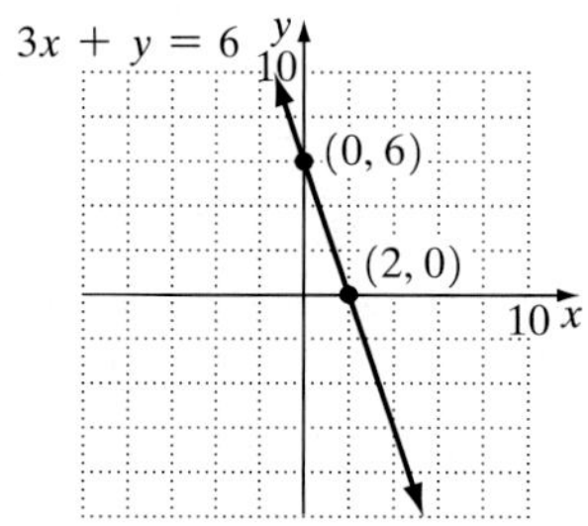

5.

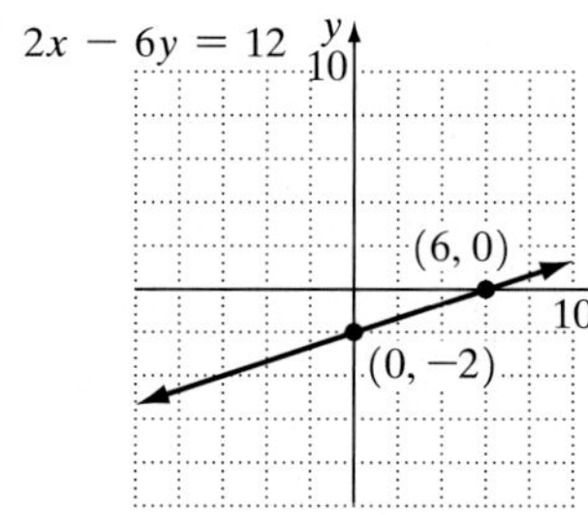

7.

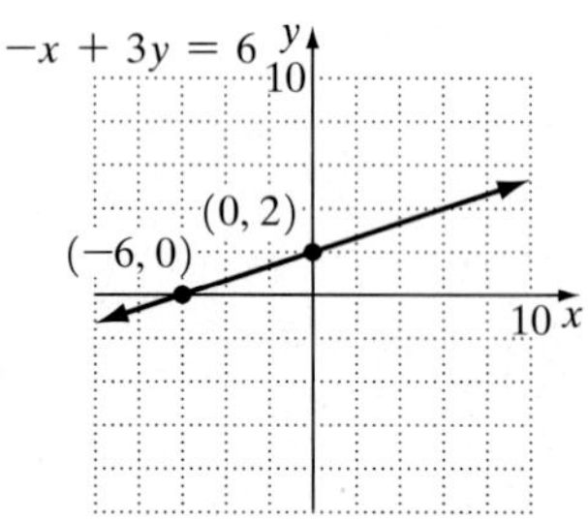

9.

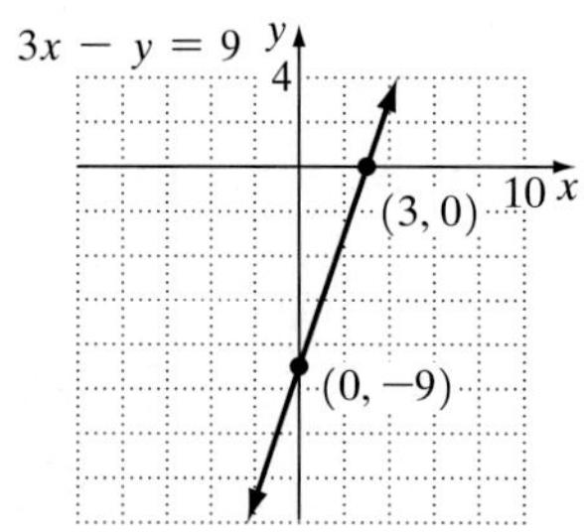

11.

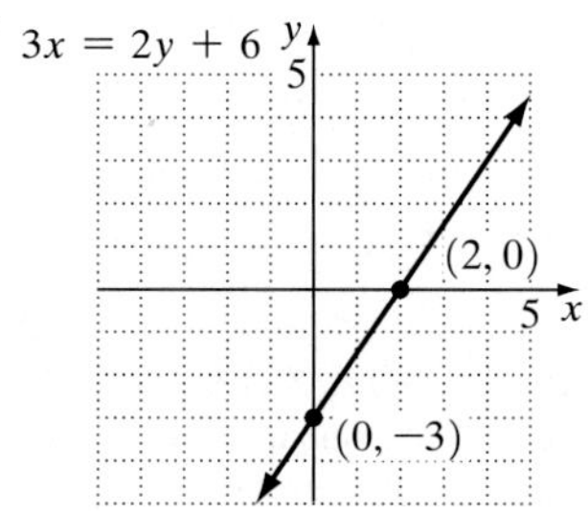

13.

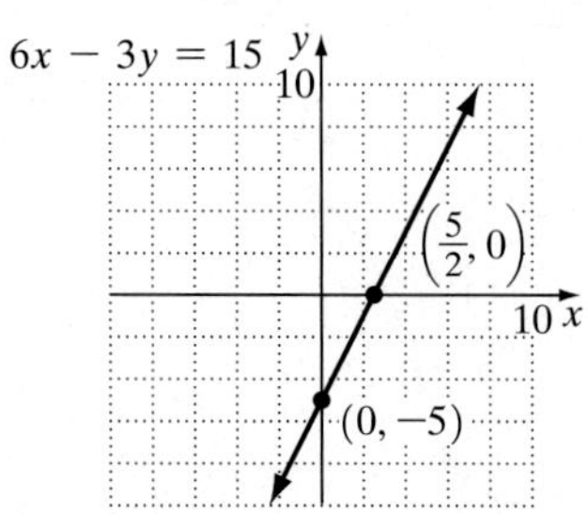

15. $\frac{3}{4}$; rises **17.** $\frac{1}{4}$; rises **19.** 0; horizontal **21.** −5; falls

23. undefined; vertical

25. $m = 2; b = 1;$

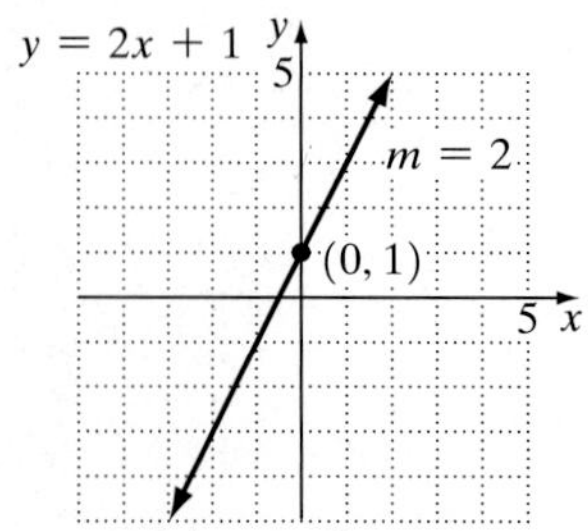

27. $m = -2; b = 1;$

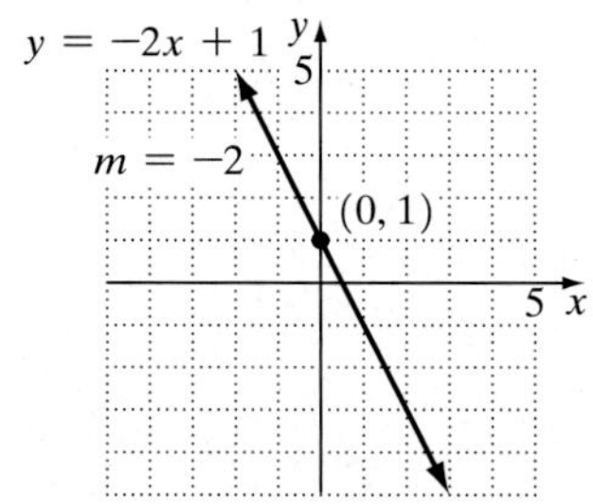

29. $m = \frac{3}{4}; b = -2;$

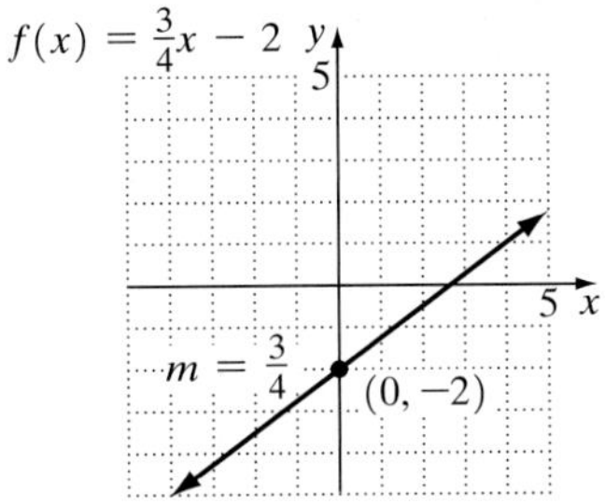

31. $m = -\frac{3}{5}; b = 7;$

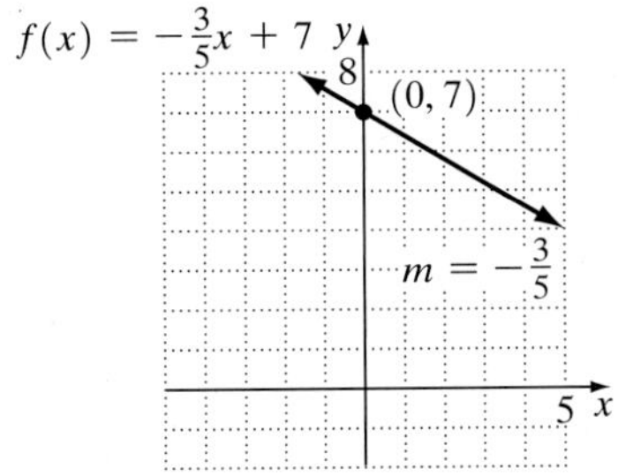

33. $m = -\frac{1}{2}; b = 0;$

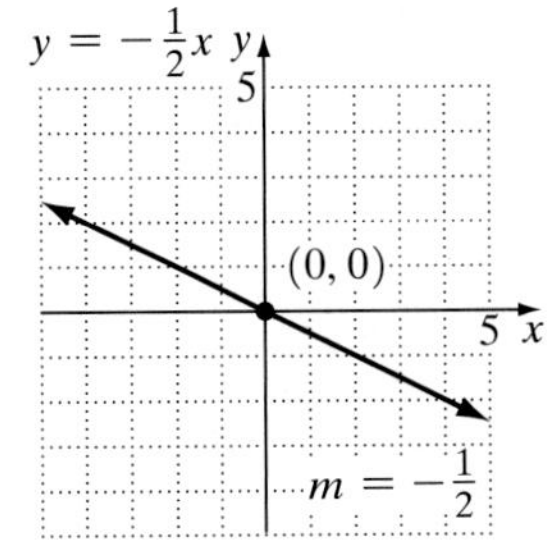

35. a. $y = -2x$

b. $m = -2; b = 0$

c.

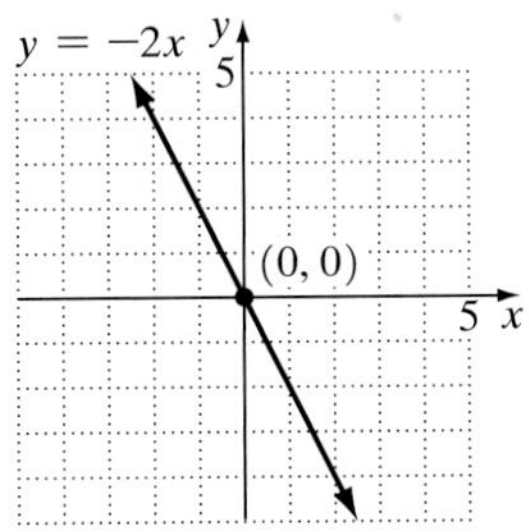

37. a. $y = \frac{5}{4}x$

b. $m = \frac{5}{4}; b = 0$

c.

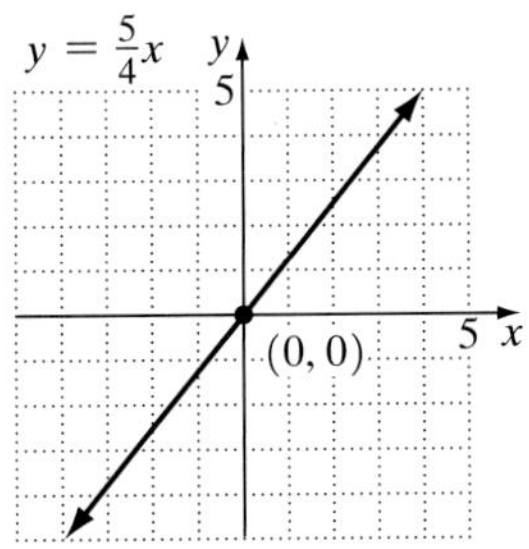

39. a. $y = -3x + 4$

b. $m = -3; b = 4$

c.

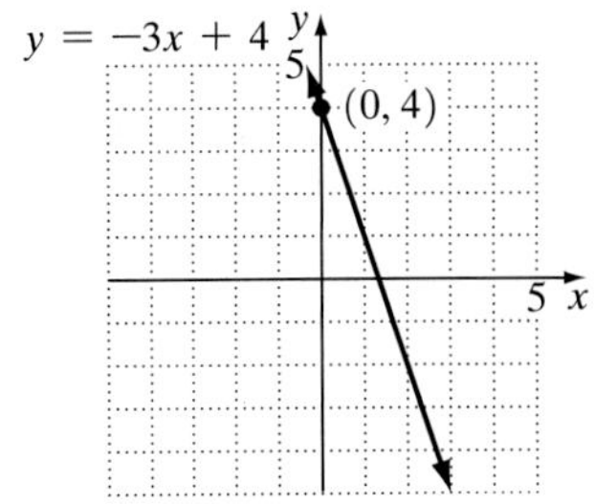

41. a. $y = -\frac{5}{3}x + 5$

b. $m = -\frac{5}{3}; b = 5$

c.

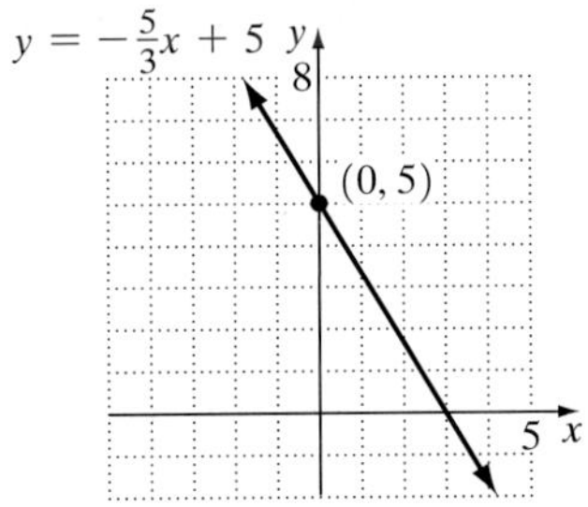

43.

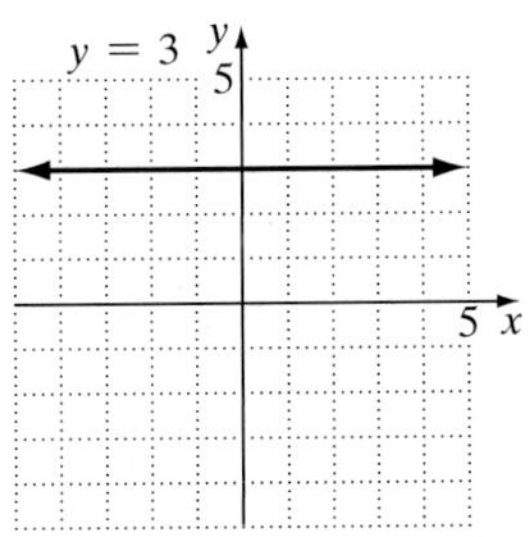

45.

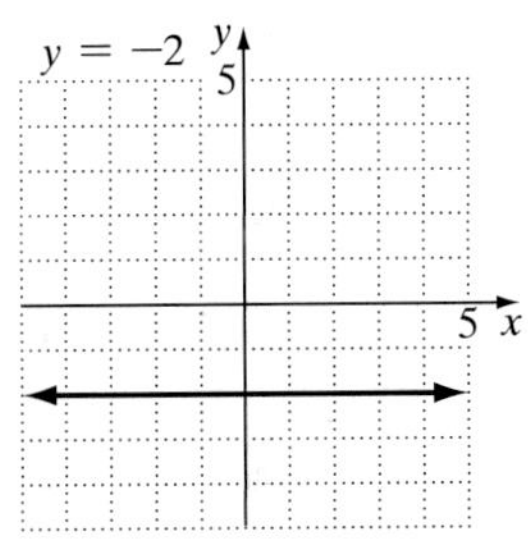

47.

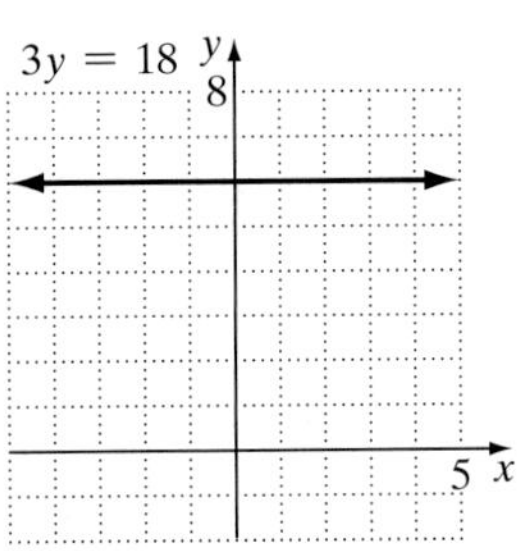

49.

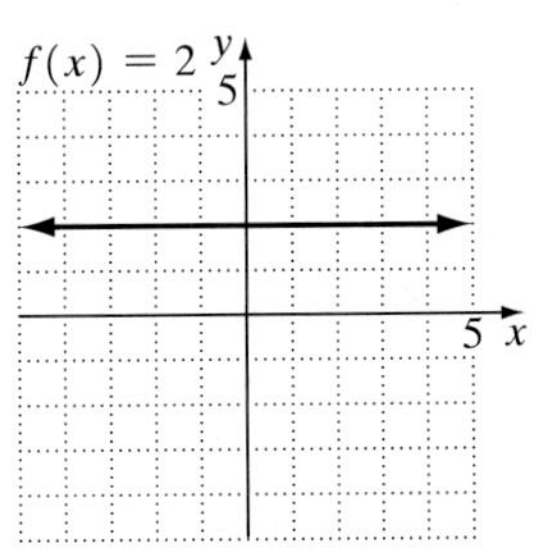

51.

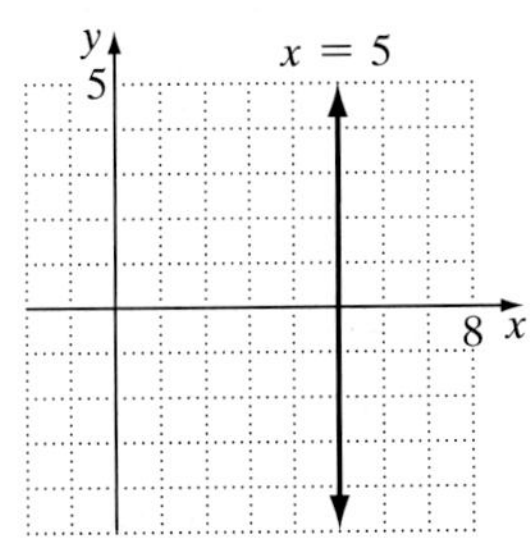

53.

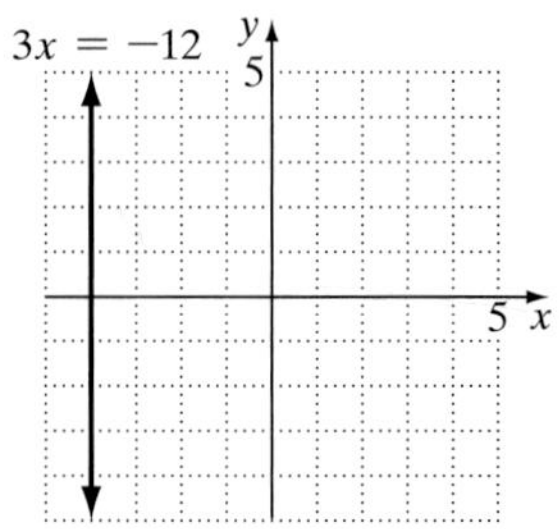

55.

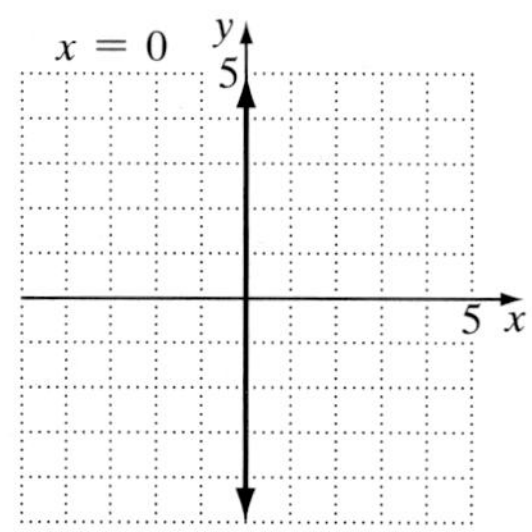

57. $\frac{47}{500} = 0.094$ billion people per year
59. -108; Consumption was decreasing at an average rate of 108 cigarettes per year per year.
61. 0 cigarettes per year per year
63. about 300 cigarettes per year per year; Answers will vary depending upon point estimation.
65. $y = 15$ **67.** $1 billion per year

69. 2; The average cost of a retail drug prescription has increased at a rate of $2 per year. **71.** -0.7; The percentage of U.S. adults who read a newspaper has decreased at a rate of 0.7% per year since 1995. **73–89.** Answers will vary.

91.

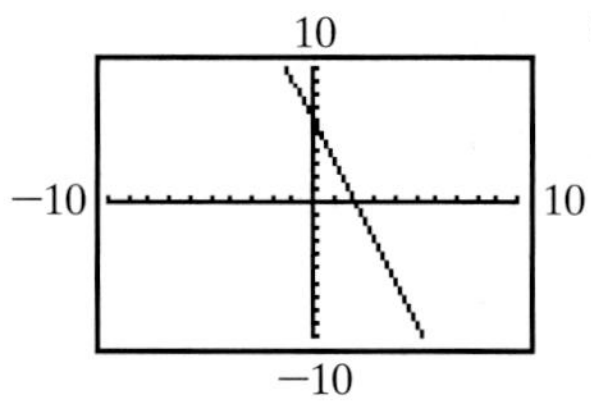

; $m = -3$

93.

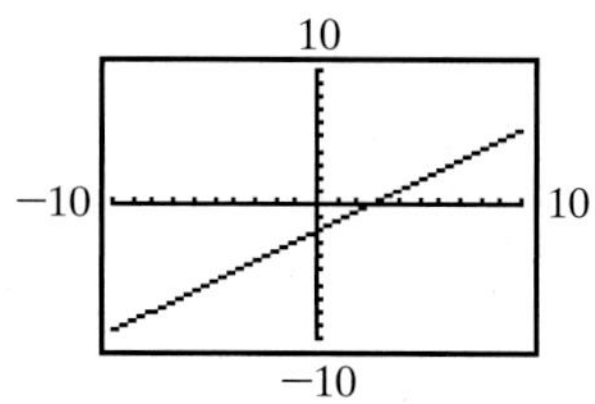

; $m = \frac{3}{4}$

95. coefficient of x: -6; coefficient of y: 3
97. Answers will vary; estimating sales as $16 billion in 1993 and $23 billion in 2000, the average price increased $0.50 per year.
98. $16x^4y^6$ **99.** 3.2×10^{-3}
100. $3x + 17$

Section 2.4

Check Point Exercises

1. $y + 5 = 6(x - 2)$; $y = 6x - 17$ or $f(x) = 6x - 17$ **2. a.** $y + 1 = -5(x + 2)$ or $y + 6 = -5(x + 1)$ **b.** $y = -5x - 11$ or $f(x) = -5x - 11$ **3.** $y = 2.35x + 179.5$ or $f(x) = 2.35x + 179.5$; 297 million people **4.** $y - 5 = 3(x + 2)$; $y = 3x + 11$ or $f(x) = 3x + 11$ **5.** 3

Exercise Set 2.4

1. $y - 5 = 2(x - 3)$; $f(x) = 2x - 1$ **3.** $y - 5 = 6(x + 2)$; $f(x) = 6x + 17$ **5.** $y + 3 = -3(x + 2)$; $f(x) = -3x - 9$ **7.** $y - 0 = -4(x + 4)$; $f(x) = -4x - 16$ **9.** $y + 2 = -1\left(x + \frac{1}{2}\right)$; $f(x) = -x - \frac{5}{2}$ **11.** $y - 0 = \frac{1}{2}(x - 0)$; $f(x) = \frac{1}{2}x$ **13.** $y + 2 = -\frac{2}{3}(x - 6)$; $f(x) = -\frac{2}{3}x + 2$ **15.** $y - 2 = 2(x - 1)$ or $y - 10 = 2(x - 5)$; $f(x) = 2x$ **17.** $y - 0 = 1(x + 3)$ or $y - 3 = 1(x - 0)$; $f(x) = x + 3$ **19.** $y + 1 = 1(x + 3)$ or $y - 4 = 1(x - 2)$; $f(x) = x + 2$ **21.** $y + 2 = \frac{4}{3}(x + 3)$ or $y - 6 = \frac{4}{3}(x - 3)$; $f(x) = \frac{4}{3}x + 2$ **23.** $y + 1 = 0(x + 3)$ or $y + 1 = 0(x - 4)$; $f(x) = -1$ **25.** $y - 4 = 1(x - 2)$ or $y - 0 = 1(x + 2)$; $f(x) = x + 2$ **27.** $y - 0 = 8\left(x + \frac{1}{2}\right)$ or $y - 4 = 8(x - 0)$; $f(x) = 8x + 4$ **29. a.** 5 **b.** $-\frac{1}{5}$ **31. a.** -7 **b.** $\frac{1}{7}$ **33. a.** $\frac{1}{2}$ **b.** -2 **35. a.** $-\frac{2}{5}$ **b.** $\frac{5}{2}$ **37. a.** -4 **b.** $\frac{1}{4}$ **39. a.** $-\frac{1}{2}$ **b.** 2 **41. a.** $\frac{2}{3}$ **b.** $-\frac{3}{2}$ **43. a.** undefined **b.** 0 **45.** $y - 2 = 2(x - 4)$; $y = 2x - 6$ or $f(x) = 2x - 6$ **47.** $y - 4 = -\frac{1}{2}(x - 2)$; $y = -\frac{1}{2}x + 5$ or $f(x) = -\frac{1}{2}x + 5$ **49.** $y + 10 = -4(x + 8)$; $y = -4x - 42$ or $f(x) = -4x - 42$ **51.** $y + 3 = -5(x - 2)$; $y = -5x + 7$ or $f(x) = -5x + 7$ **53.** $y - 2 = \frac{2}{3}(x + 2)$; $y = \frac{2}{3}x + \frac{10}{3}$ or $f(x) = \frac{2}{3}x + \frac{10}{3}$ **55.** $y + 7 = -2(x - 4)$; $y = -2x + 1$ or $f(x) = -2x + 1$ **57. a.** $y - 34 = \frac{13}{2}(x - 2)$ or $y - 60 = \frac{13}{2}(x - 6)$ **b.** $f(x) = \frac{13}{2}x + 21$ **c.** 125 accidents **59.** Answers will vary; an example is: (0, 0.5) and (15, 1.5); $y - 0.5 = \frac{1}{15}(x - 0)$; 2.5 million

61. a. & b.

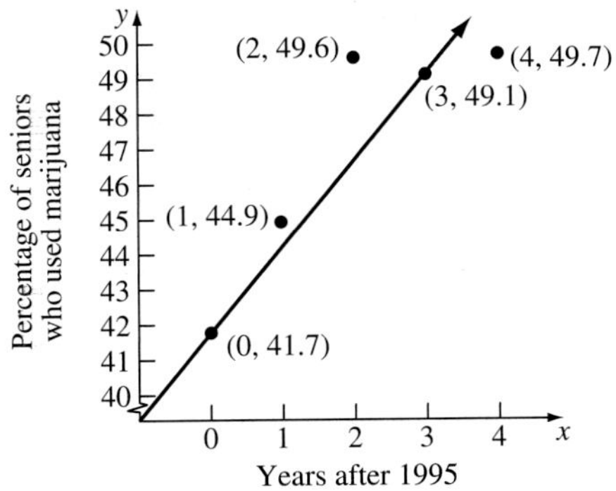

c. Answers will vary; an example is: using points (0, 41.7) and (3, 49.1), $y - 41.7 = \frac{37}{15}(x - 0)$ and $y = \frac{37}{15}x + 41.7$.

d. 78.7%

63. a. $\frac{1}{5}$ **b.** $\frac{2}{15}$ **c.** no; The yearly increase for women is greater. **65–71.** Answers will vary.

73. a.

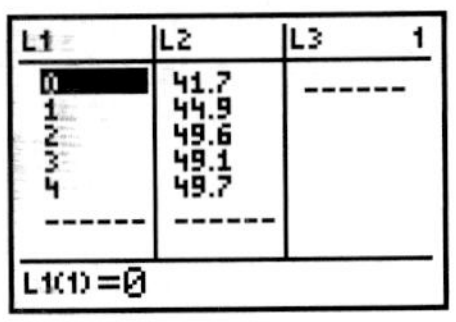

b. & d.

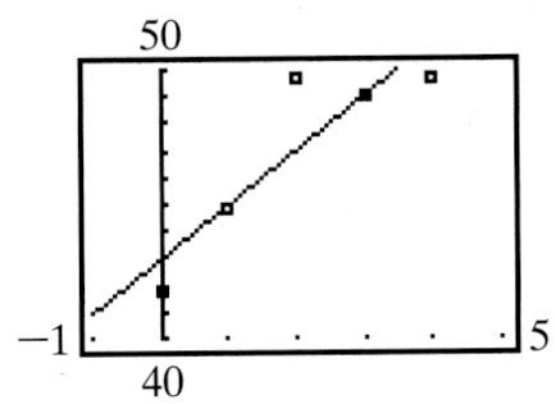

c.

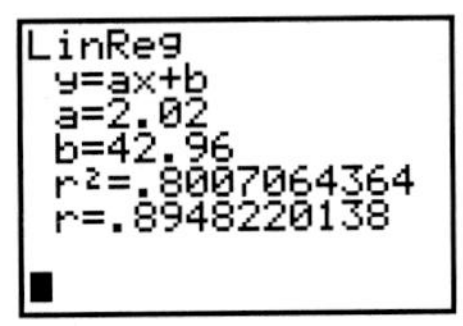

75. $y - 0 = 3(x + 3)$; $y = 3x + 9$ or $f(x) = 3x + 9$

77. $y - 0 = -\frac{b}{a}(x - a)$ ← Point-slope form

$y = -\frac{b}{a}x + b$ ← Slope-intercept form

$\frac{y}{b} = -\frac{x}{a} + 1$ ← Divide by b.

$\frac{x}{a} + \frac{y}{b} = 1$ ← Intercept form; Answers will vary.

78. 33 **79.** −56 **80.** 40°, 60°, and 80°

Review Exercises

1. function; domain: {3, 4, 5}; range: {10} **2.** function; domain: {1, 2, 3, 4}; range: {12, 100, π, −6}
3. not a function; domain: {13, 15}; range: {14, 16, 17} **4. a.** −5 **b.** 16 **c.** −75 **d.** $14a - 5$ **e.** $7a + 9$
5. a. 2 **b.** 52 **c.** 70 **d.** $3b^2 - 5b + 2$ **e.** $48a^2 - 20a + 2$

6. $f(x) = x^2$
$g(x) = x^2 - 1$

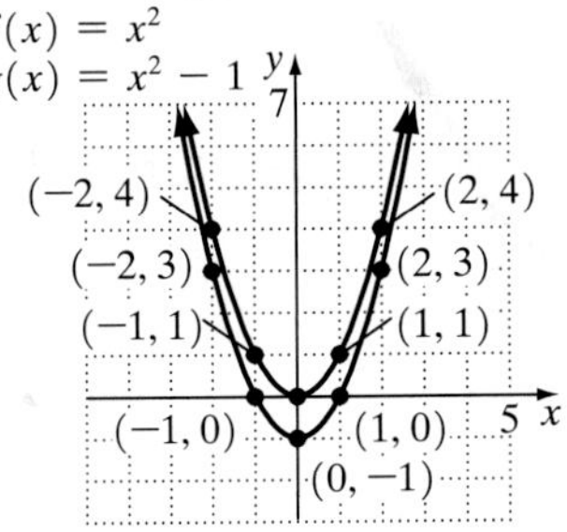

7.

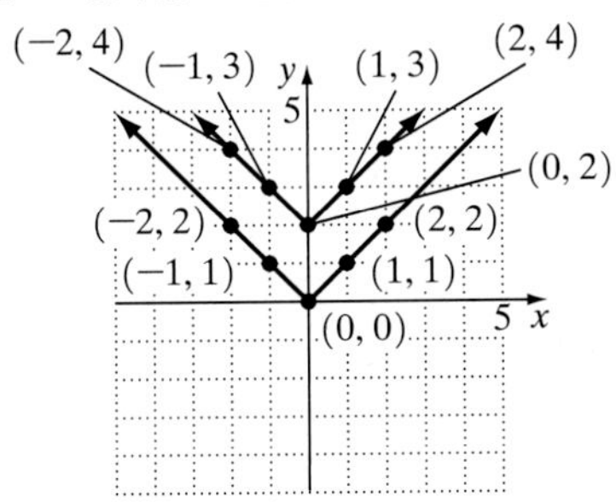

$f(x) = |x|$
$g(x) = |x| + 2$

8. not a function **9.** function
10. function **11.** not a function
12. −3 **13.** −2 **14.** 3
15. a. For each time, there is only one height.
b. 0; The vulture was on the ground after 15 seconds.
c. 45 m
d. 7 and 22; After 7 and 22 seconds the vulture's height is 20 meters.
e. Answers will vary.
16. $\{x | x$ is a real number$\}$

17. $\{x | x$ is a real number and $x \neq -8\}$ **18.** $\{x | x$ is a real number and $x \neq 5\}$ **19. a.** $6x - 4$ **b.** 14 **20. a.** $5x^2 + 1$ **b.** 46
21. $\{x | x$ is a real number and $x \neq 4\}$ **22.** $\{x | x$ is a real number and $x \neq -6$ and $x \neq -1\}$ **23.** $x^2 - x - 5$; 1 **24.** 1
25. $x^2 - 3x + 5$; 3 **26.** 9 **27.** −120 **28.** $\frac{x^2 - 2x}{x - 5}$; −8 **29.** $\{x | x$ is a real number$\}$ **30.** $\{x | x$ is a real number and $x \neq 5\}$

31.

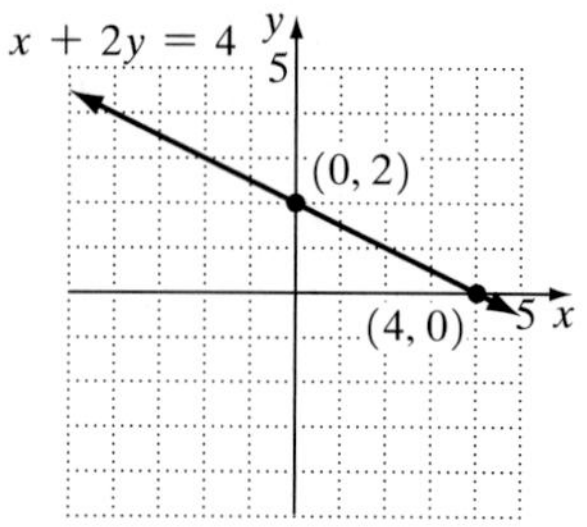

32.

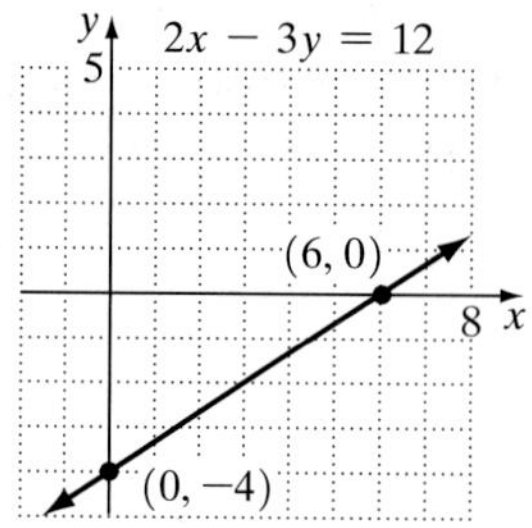

33.

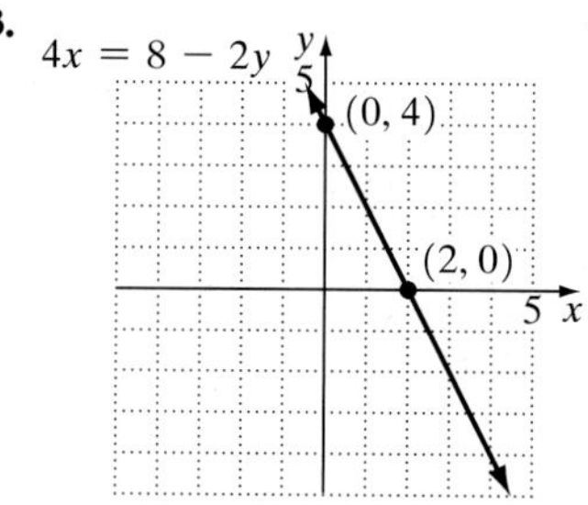

34. 2; rises **35.** $-\frac{2}{3}$; falls **36.** undefined; vertical **37.** 0; horizontal

38. $m = 2; b = -1;$

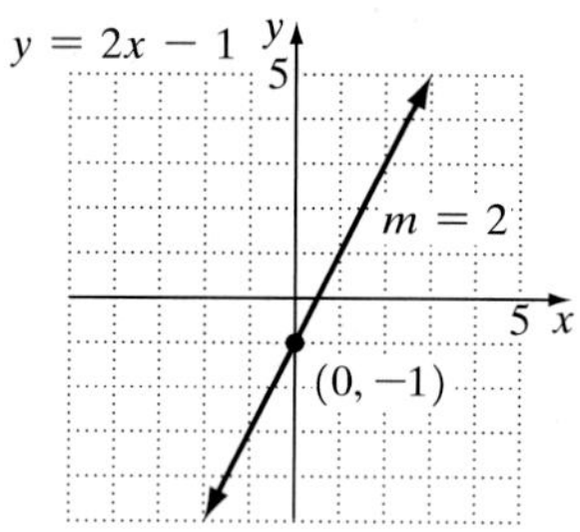

39. $m = -\frac{1}{2}; b = 4;$

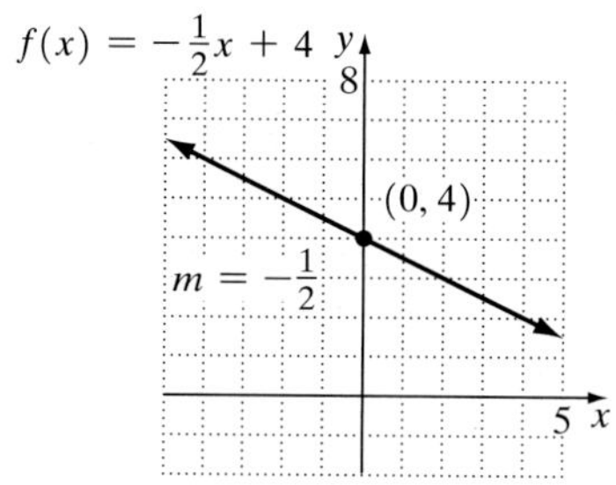

40. $m = \frac{2}{3}; b = 0;$

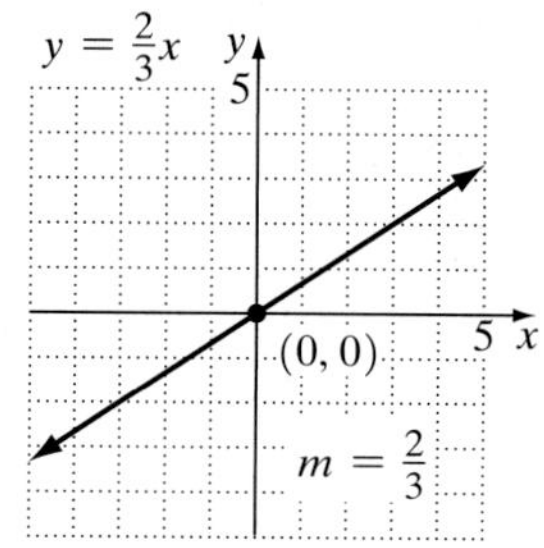

41. $y = -2x + 4; m = -2; b = 4$ **42.** $y = -\frac{5}{3}x; m = -\frac{5}{3}; b = 0$ **43.** $y = -\frac{5}{3}x + 2; m = -\frac{5}{3}; b = 2$

44.

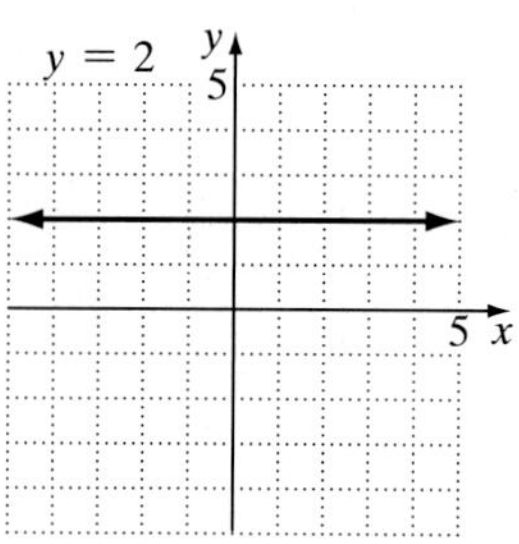

45.

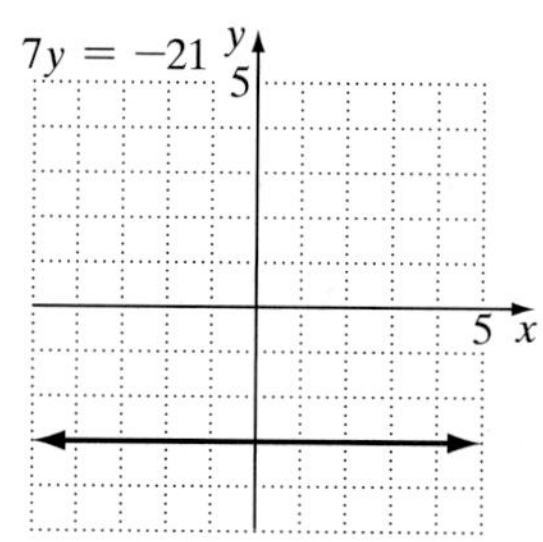

46.

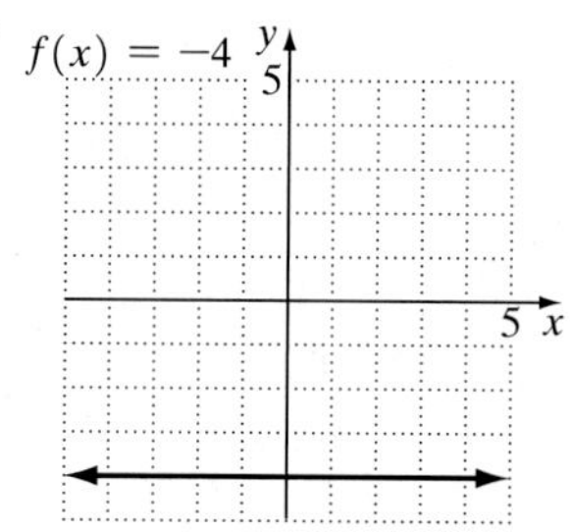

47.

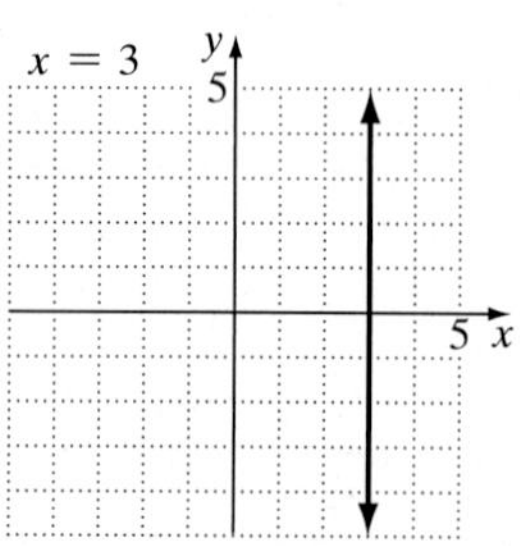

48.

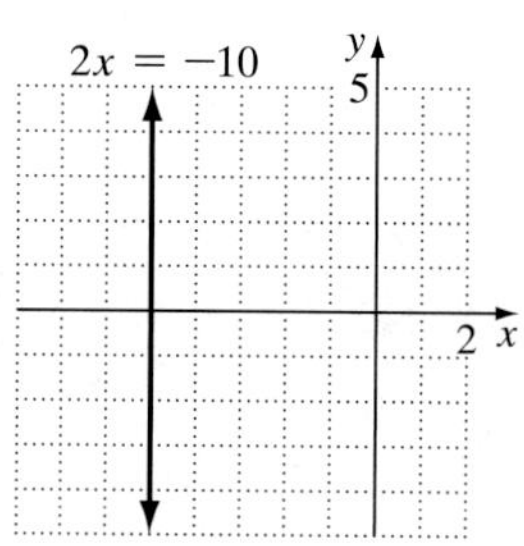

49. $\frac{3500}{3} \approx 1167$ dollars per year

50. −0.27; Record time has been decreasing at a rate of 0.27 seconds per year since 1900.

51. $y - 2 = -6(x + 3)$; $y = -6x - 16$ or $f(x) = -6x - 16$

52. $y - 6 = 2(x - 1)$; $y = 2x + 4$ or $f(x) = 2x + 4$

53. $y + 7 = -3(x - 4)$; $y = -3x + 5$ or $f(x) = -3x + 5$

54. $y - 6 = -3(x + 2)$; $y = -3x$ or $f(x) = -3x$ **55. a.** $y - 162 = 1(x - 2)$ or $y - 168 = 1(x - 8)$ **b.** $f(x) = x + 160$ **c.** 175 lb

Chapter 2 Test

1. function; domain: {1, 3, 5, 6}; range: {2, 4, 6} **2.** not a function; domain: {2, 4, 6}; range: {1, 3, 5, 6} **3.** $3a + 10$ **4.** 28

5.

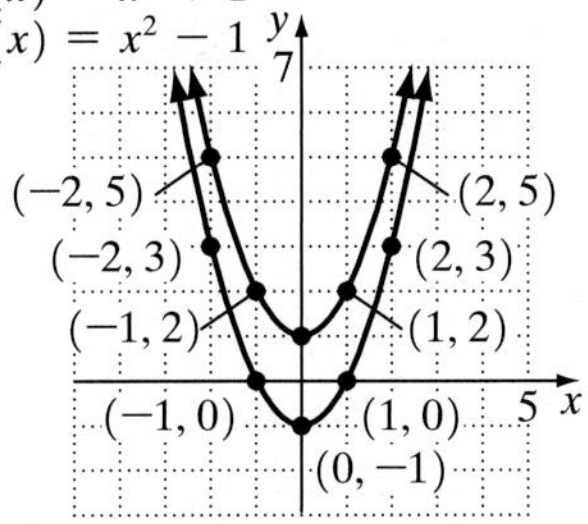

The graph of g is the graph of f shifted up 2 units.

6. function **7.** not a function **8.** −3 **9.** −5
10. $\{x|x \text{ is a real number and } x \neq 10\}$ **11.** $x^2 + 5x + 2; 26$
12. $x^2 + 3x - 2; -4$ **13.** −15 **14.** $\frac{x^2 + 4x}{x + 2}; 3$
15. $\{x|x \text{ is a real number and } x \neq -2\}$

16.

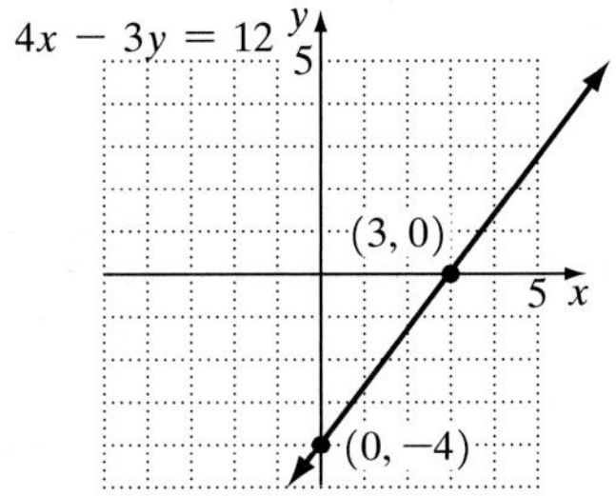

17.

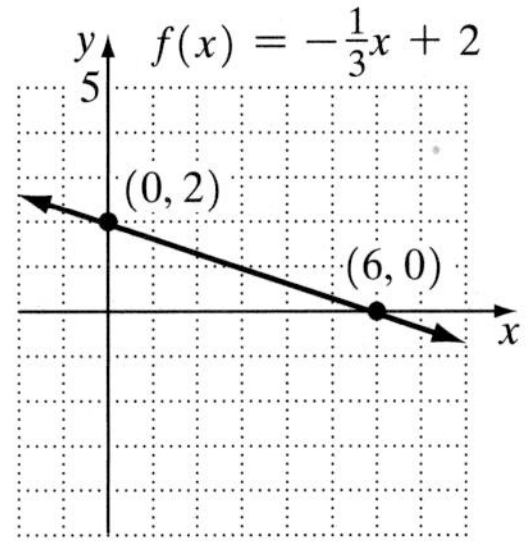

18.

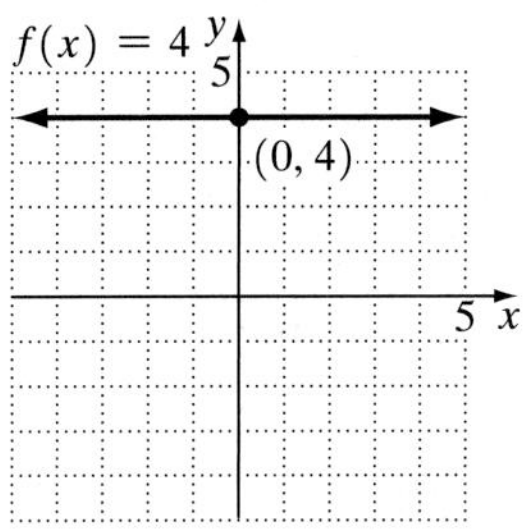

19. $-\frac{1}{2}$; falls **20.** undefined; vertical
21. 176; In 2005 the number of Super Bowl viewers will be 176 million.
22. 3.6; The number of Super Bowl viewers is increasing at a rate of 3.6 million per year.
23. $y + 3 = 1(x + 1)$ or $y - 2 = 1(x - 4)$; $y = x - 2$ or $f(x) = x - 2$
24. $y - 3 = 2(x + 2)$; $y = 2x + 7$ or $f(x) = 2x + 7$
25. a. $y - 320 = 42(x - 0)$ or $y - 530 = 42(x - 5)$
b. $f(x) = 42x + 320$
c. \$866

Cumulative Review Exercises (Chapters 1–2)

1. {0, 1, 2, 3} **2.** false **3.** 7 **4.** 15 **5.** $3x + 7$ **6.** −4 **7.** all real numbers; identity **8.** −6 **9.** \$2250
10. $t = \frac{A - p}{pr}$ **11.** $\frac{y^{10}}{9x^8}$ **12.** $\frac{9x^{10}}{y^{12}}$ **13.** 2.1×10^{-5} **14.** function; domain: {1, 2, 3, 4, 6}; range: {5}

15.

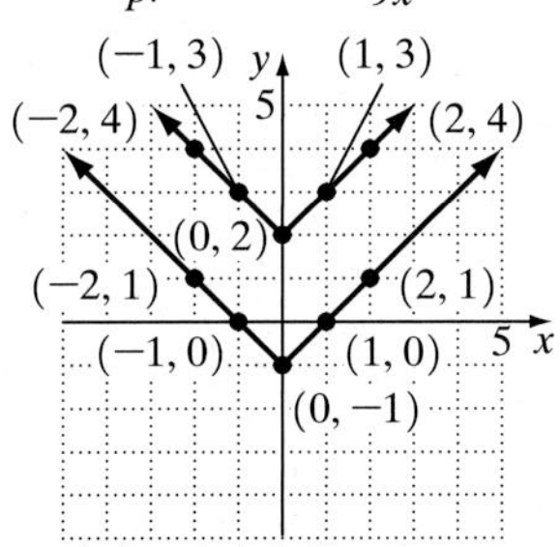

$g(x) = |x| + 2$
$f(x) = |x| - 1$
The graph of g is the graph of f shifted up 3 units.

16. $\{x|x \text{ is a real number and } x \neq 15\}$ **17.** $2x^2 + x + 5; 6$

18.

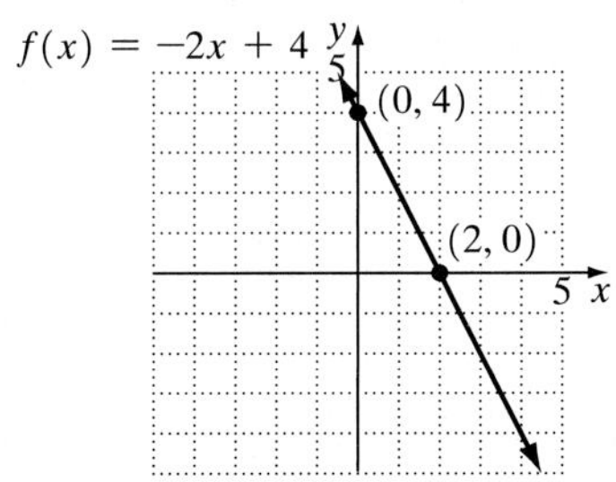

19.

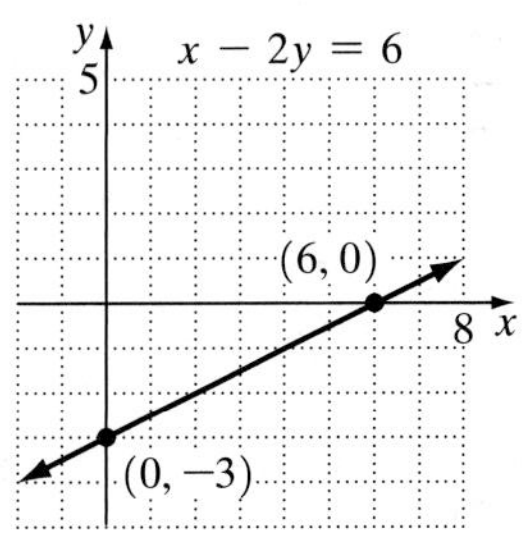

20. $y + 5 = 4(x - 3)$; $y = 4x - 17$ or $f(x) = 4x - 17$

CHAPTER 3

Section 3.1

Check Point Exercises

1. solution **2.** (1, 4) or {(1, 4)} **3.** (6, 11) or {(6, 11)} **4.** (5, 4) or {(5, 4)} **5.** (2, −1) or {(2, −1)}
6. (2, −1) or {(2, −1)} **7.** $\left(\frac{37}{7}, \frac{19}{7}\right)$ or $\left\{\left(\frac{37}{7}, \frac{19}{7}\right)\right\}$ **8.** no solution or ∅ **9.** $\{(x, y)|y = 4x - 4\}$ or $\{(x, y)|4x + 2y = 6\}$

Exercise Set 3.1

1. solution **3.** not a solution **5.** solution **7.** $(3, 1)$ **9.** $\left(\frac{1}{2}, 3\right)$ **11.** $(4, 3)$ **13.** $\{(x, y)|2x + 3y = 6\}$ or $\{(x, y)|4x = -6y + 12\}$ **15.** $(1, 0)$ **17.** $\varnothing$ **19.** $(1, 2)$ **21.** $(3, 1)$ **23.** $\varnothing$ **25.** $(2, 4)$ **27.** $(3, 1)$ **29.** $(2, 1)$ **31.** $(2, -3)$ **33.** $\varnothing$ **35.** $(3, -2)$ **37.** $\varnothing$ **39.** $(-5, -1)$ **41.** $\left\{(x, y) \middle| y = \frac{2}{5}x - 2\right\}$ or $\{(x, y)|2x - 5y = 10\}$ **43.** $(5, 2)$ **45.** $(2, -3)$ **47.** $(-1, 1)$ **49.** $(-2, -7)$ **51.** $(7, 2)$ **53.** $(4, -1)$ **55.** $\{(x, y)|2x + 6y = 8\}$ or $\{(x, y)|3x + 9y = 12\}$ **57.** $\left(\frac{29}{22}, -\frac{5}{11}\right)$ **59.** $(-2, -1)$ **61.** $(1, -3)$ **63.** $(1, -3)$ **65.** $(4, 3)$ **67.** $\{(x, y)|x = 3y - 1\}$ or $\{(x, y)|2x - 6y = -2\}$ **69.** $\varnothing$ **71.** $(5, 1)$ **73.** $\left(\frac{32}{7}, -\frac{20}{7}\right)$ **75.** $(-5, 7)$ **77.** $\varnothing$ **79.** $\{(x, y)|x + 2y - 3 = 0\}$ or $\{(x, y)|12 = 8y + 4x\}$
81. $(0, 0)$ **83.** 2004; about 12.24 deaths per 100,000 persons; The lines intersect at approximately the point (2004, 12.24).
85. 2032; about 0.61 deaths per 1000 live births **87. a.** 526 videos sold; 466 videos supplied **b.** \$22; 474 copies
89–97. Answers will vary. **99.** chicken: 0.94; red meat: 0.74; Chicken consumption is increasing; red meat consumption is decreasing.; The lines are not parallel.; Chicken consumption is greater than red meat consumption. **101–103.** Answers will vary. **105.** $\frac{10}{9}$
106. $-128x^{19}y^{8}$ **107.** 11

Section 3.2

Check Point Exercises

1. pan pizza: 1120 calories; burrito: 430 calories **2.** \$5000 at 6%; \$11,000 at 8% **3.** 4 of 18% solution; 8 of 45% solution
4. boat: 35 mph; current: 7 mph **5.** $P(x) = 125x - 160{,}000$ **6. a.** $C(x) = 300{,}000 + 30x$ **b.** $R(x) = 80x$
c. (6000, 480,000); The company will break even when it produces and sells 6000 shoes. At this level, both revenue and cost are \$480,000.

Exercise Set 3.2

1. 3 and 4 **3.** 5 and 2 **5. a.** $P(x) = 17x - 25{,}500$ **b.** 1500 units; \$48,000 **7. a.** $P(x) = 140x - 70{,}000$ **b.** 500 units; \$122,500
9. 3.1 million in the 40–44 age group; 5.4 million in the 45–54 age group **11.** sponge cake: 162 mg; pound cake: 68 mg
13. \$12,500 at 7%; \$7500 at 9% **15.** \$220,000 at 8%; \$30,000 at 18% **17.** \$6000 at 12%; \$2000 at a 5% loss
19. 12 g of 45% fat content; 18 g of 20% fat content **21.** north campus: 300 students; south campus: 900 students
23. 36 lb of the \$6-per-pound tea; 108 lb of the \$8-per-pound tea **25.** plane: 130 mph; wind: 30 mph
27. crew: 6 km/hr; current: 2 km/hr **29.** speed in still water: 4.5 mph; current: 1.5 mph **31.** 100 ft by 80 ft
33. 50 rooms with kitchen facilities; 150 rooms without kitchen facilities **35.** 3 apples and 2 avocados **37.** 500 radios
39. −6000; When the company produces and sells 200 radios, the profit loss is \$6000. **41.** \$190,000 **43. a.** $C(x) = 18{,}000 + 20x$
b. $R(x) = 80x$ **c.** (300, 24,000); When 300 canoes are produced and sold, both revenue and cost are \$24,000.
45. a. $C(x) = 30{,}000 + 2500x$ **b.** $R(x) = 3125x$ **c.** (48, 150,000); For 48 sold-out performances, both cost and revenue are \$150,000.
47–53. Answers will vary. **55.**

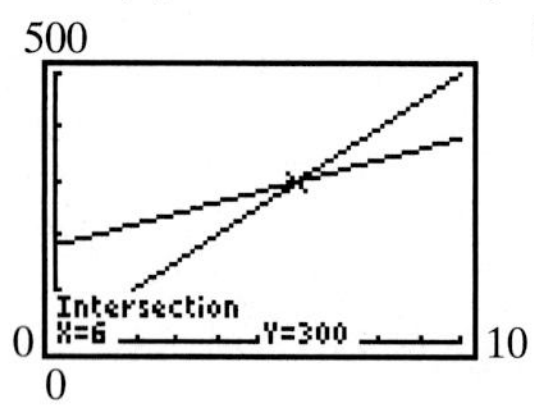

; (6, 300) **57.** Answers will vary.
59. $x°$ is 55° and $y°$ is 35°.
61. \$45 for mangos and \$22 for avocados
62. $y - 5 = -2(x + 2)$ or $y - 13 = -2(x + 6)$; $y = -2x + 1$ or $f(x) = -2x + 1$
63. $y - 0 = 1(x + 3)$; $y = x + 3$ or $f(x) = x + 3$
64. $\{x|x \text{ is a real number and } x \neq 3\}$

Section 3.3

Check Point Exercises

1. $(-1) - 2(-4) + 3(5) = 22$; $2(-1) - 3(-4) - 5 = 5$; $3(-1) + (-4) - 5(5) = -32$ **2.** $(1, 4, -3)$ or $\{(1, 4, -3)\}$
3. $(4, 5, 3)$ or $\{(4, 5, 3)\}$ **4.** $y = 3x^2 - 12x + 13$ or $f(x) = 3x^2 - 12x + 13$

Exercise Set 3.3

1. not a solution **3.** solution **5.** $(2, 3, 3)$ **7.** $(2, -1, 1)$ **9.** $(1, 2, 3)$ **11.** $(3, 1, 5)$ **13.** $(1, 0, -3)$ **15.** $(1, -5, -6)$ **17.** no solution or $\varnothing$ **19.** infinitely many solutions; dependent equations **21.** $\left(\frac{1}{2}, \frac{1}{3}, -1\right)$ **23.** $y = 2x^2 - x + 3$ or $f(x) = 2x^2 - x + 3$ **25.** $y = 2x^2 + x - 5$ or $f(x) = 2x^2 + x - 5$ **27.** 7, 4, and 5 **29. a.** (0, 5.4), (10, 4.7), (20, 6.2)
b. $0a + 0b + c = 5.4$; $100a + 10b + c = 4.7$; $400a + 20b + c = 6.2$ **31. a.** $y = -16x^2 + 40x + 200$
b. 0; After 5 seconds the ball hits the ground. **33.** chemical: \$42,758; mechanical: \$39,852; electrical: \$38,811
35. \$1200 at 8%; \$2000 at 10%; \$3500 at 12% **37.** 200 \$8 tickets; 150 \$10 tickets; 50 \$12 tickets

39. 4 oz of Food A; 0.5 oz of Food B; 1 oz of Food C **41–45.** Answers will vary. **49.** c **51.** 13 triangles, 21 rectangles, and 6 pentagons

53.
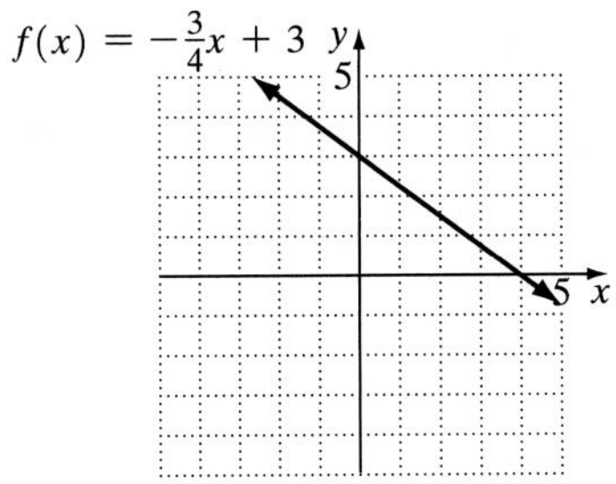

54.
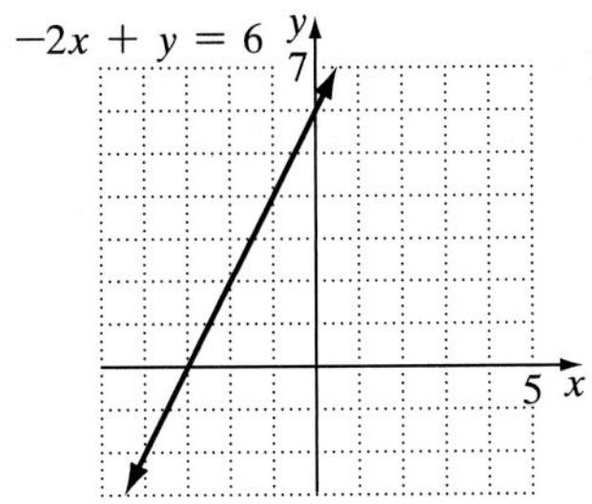

55.
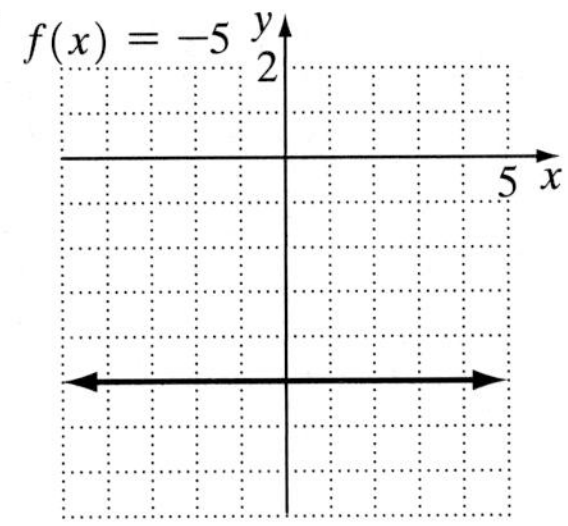

Section 3.4

Check Point Exercises

1. $(4, -3, 1)$ or $\{(4, -3, 1)\}$ **2. a.** $\left[\begin{array}{ccc|c} 1 & 6 & -3 & 7 \\ 4 & 12 & -20 & 8 \\ -3 & -2 & 1 & -9 \end{array}\right]$ **b.** $\left[\begin{array}{ccc|c} 1 & 3 & -5 & 2 \\ 1 & 6 & -3 & 7 \\ -3 & -2 & 1 & -9 \end{array}\right]$ **c.** $\left[\begin{array}{ccc|c} 4 & 12 & -20 & 8 \\ 1 & 6 & -3 & 7 \\ 0 & 16 & -8 & 12 \end{array}\right]$

3. $(-1, 2)$ or $\{(-1, 2)\}$ **4.** $(5, 2, 3)$ or $\{(5, 2, 3)\}$

Exercise Set 3.4

1. $x - 3y = 11; y = -3; (2, -3)$ **3.** $x - 3y = 1; y = -1; (-2, -1)$ **5.** $x - 4z = 5; y - 12z = 13; z = -\frac{1}{2}; \left(3, 7, -\frac{1}{2}\right)$

7. $x + \frac{1}{2}y + z = \frac{11}{2}; y + \frac{3}{2}z = 7; z = 4; (1, 1, 4)$ **9.** $\left[\begin{array}{cc|c} 1 & -\frac{3}{2} & 5 \\ 2 & 2 & 5 \end{array}\right]$ **11.** $\left[\begin{array}{cc|c} 1 & -\frac{4}{3} & 2 \\ 3 & 5 & -2 \end{array}\right]$ **13.** $\left[\begin{array}{cc|c} 1 & -3 & 5 \\ 0 & 12 & -6 \end{array}\right]$

15. $\left[\begin{array}{cc|c} 1 & -\frac{3}{2} & \frac{7}{2} \\ 0 & \frac{17}{2} & -\frac{17}{2} \end{array}\right]$ **17.** $\left[\begin{array}{ccc|c} 1 & -3 & 2 & 5 \\ 1 & 5 & -5 & 0 \\ 3 & 0 & 4 & 7 \end{array}\right]$ **19.** $\left[\begin{array}{ccc|c} 1 & -3 & 2 & 0 \\ 0 & 10 & -7 & 7 \\ 2 & -2 & 1 & 3 \end{array}\right]$ **21.** $\left[\begin{array}{ccc|c} 1 & 1 & -1 & 6 \\ 0 & -3 & 3 & -15 \\ 0 & -4 & 2 & -14 \end{array}\right]$ **23.** $(4, 2)$

25. $(3, -3)$ **27.** $(2, -5)$ **29.** no solution or $\varnothing$ **31.** infinitely many solutions; dependent equations **33.** $(1, -1, 2)$
35. $(3, -1, -1)$ **37.** $(1, 2, -1)$ **39.** $(1, 2, 3)$ **41.** no solution or $\varnothing$ **43.** infinitely many solutions; dependent equations
45. $(-1, 2, -2)$ **47. a.** $y = 2x^2 + 22x + 320$ **b.** 2780 thousand, or 2,780,000, inmates **c.** Answers will vary.
49. 35% are under 30, 49% are 30–49 and 16% are 50 and over. **51–57.** Answers will vary. **59.** d
61. a. Matrices show that the system of equations for this problem is inconsistent. **b.** Matrices show that the system has infinitely many solutions. There are many combinations of foods that satisfy the new requirements. **62.** $-6a + 13$ **63.** $6x^2 - 9x; 15$ **64.** $-\frac{x^{11}}{3y^{36}}$

Section 3.5

Check Point Exercises

1. a. -4 **b.** -17 **2.** $(4, -2)$ or $\{(4, -2)\}$ **3.** 80 **4.** $(2, -3, 4)$ or $\{(2, -3, 4)\}$

Exercise Set 3.5

1. 1 **3.** -29 **5.** 0 **7.** 33 **9.** $-\frac{7}{16}$ **11.** $(5, 2)$ **13.** $(2, -3)$ **15.** $(3, -1)$ **17.** $(4, 0)$ **19.** $(4, 2)$ **21.** $(7, 4)$
23. inconsistent **25.** infinitely many solutions; dependent equations **27.** 72 **29.** -75 **31.** 0 **33.** $(-5, -2, 7)$

35. $(2, -3, 4)$ **37.** $(3, -1, 2)$ **39.** $(2, 3, 1)$ **41.** 28 sq units **43.** yes **45.** $\begin{vmatrix} x & y & 1 \\ 3 & -5 & 1 \\ -2 & 6 & 1 \end{vmatrix} = 0; y = -\frac{11}{5}x + \frac{8}{5}$

47–55. Answers will vary. **57.** d **59.** The value is multiplied by -1. **61.** Answers will vary.
62. all real numbers or $\{x|x \text{ is a real number}\}$ or $\mathbb{R}$. **63.** $y = \frac{2x + 7}{3}$ **64.** 0

Review Exercises

1. not a solution **2.** solution **3.** $(2, 3)$ **4.** $\{(x, y)|3x - 2y = 6\}$ or $\{(x, y)|6x - 4y = 12\}$ **5.** $(-5, -6)$ **6.** $\varnothing$ **7.** $(3, 4)$ **8.** $(23, -43)$ **9.** $(-4, 2)$ **10.** $\left(3, \frac{1}{2}\right)$ **11.** $\{(x, y)|y = 4 - x\}$ or $\{(x, y)|3x + 3y = 12\}$ **12.** $\left(3, \frac{8}{3}\right)$ **13.** $\varnothing$ **14.** TV: \$350; stereo: \$370 **15.** \$3000 at 8%; \$7000 at 10% **16.** 10 ml of 34%; 90 ml of 4% **17.** plane: 630 mph; wind: 90 mph **18.** 9 ft by 5 ft **19.** profit loss of \$4500 **20.** $P(x) = 45x - 22{,}500$ **21.** $(500, 42{,}500)$; When 500 calculators are produced and sold, both cost and revenue are \$42,500. **22. a.** $C(x) = 60{,}000 + 200x$ **b.** $R(x) = 450x$ **c.** $(240, 108{,}000)$; When 240 desks are produced and sold, both cost and revenue are \$108,000. **23.** no **24.** $(0, 1, 2)$ **25.** $(2, 1, -1)$ **26.** infinitely many solutions; dependent equations **27.** $y = 3x^2 - 4x + 5$ **28.** Japan: 16%; Germany: 15%; France: 14% **29.** $2x + 3y = -10$; $y = -6$; $(4, -6)$

30. $x + y + 3z = 12$; $y - 2z = -4$; $z = 3$; $(1, 2, 3)$ **31.** $\left[\begin{array}{cc|c} 1 & -8 & 3 \\ 0 & 1 & -2 \end{array}\right]$ **32.** $\left[\begin{array}{cc|c} 1 & -3 & 1 \\ 0 & 7 & -7 \end{array}\right]$ **33.** $\left[\begin{array}{ccc|c} 1 & -1 & \frac{1}{2} & -\frac{1}{2} \\ 1 & 2 & -1 & 2 \\ 6 & 4 & 3 & 5 \end{array}\right]$

34. $\left[\begin{array}{ccc|c} 1 & 2 & 2 & 2 \\ 0 & 1 & -1 & 2 \\ 0 & 0 & 9 & -9 \end{array}\right]$ **35.** $(-5, 3)$ **36.** no solution or $\varnothing$ **37.** $(1, 3, -4)$ **38.** $(-2, -1, 0)$ **39.** 17 **40.** 4 **41.** -86

42. -236 **43.** $\left(\frac{7}{4}, -\frac{25}{8}\right)$ **44.** $(2, -7)$ **45.** $(23, -12, 3)$ **46.** $(-3, 2, 1)$ **47.** $y = \frac{5}{8}x^2 - 50x + 1150$; 30-year-olds are involved in 212.5 accidents daily, and 50-year-olds are involved in 212.5 accidents daily.

Chapter 3 Test

1. $(2, 4)$ **2.** $(6, -5)$ **3.** $(1, -3)$ **4.** $\{(x, y)|4x = 2y + 6\}$ or $\{(x, y)|y = 2x - 3\}$ **5.** shrimp: 42 mg; scallops: 15 mg **6.** \$4000 at 9%; \$2000 at 6% **7.** 40 oz of 20%; 20 oz of 50% **8.** plane: 725 km/hr; wind: 75 km/hr **9.** $P(x) = 85x - 350{,}000$ **10.** $C(x) = 360{,}000 + 850x$ **11.** $R(x) = 1150x$ **12.** $(1200, 1{,}380{,}000)$; When 1200 computers are produced and sold, both cost and revenue are \$1,380,000. **13.** $(1, 3, 2)$ **14.** $\left[\begin{array}{ccc|c} 1 & 0 & -4 & 5 \\ 0 & -1 & 26 & -20 \\ 2 & -1 & 4 & -3 \end{array}\right]$ **15.** $(4, -2)$ **16.** $(-1, 2, 2)$ **17.** 17 **18.** -10 **19.** $(-1, -6)$ **20.** $(4, -3, 3)$

Cumulative Review Exercises (Chapters 1–3)

1. 1 **2.** $15x - 7$ **3.** -6 **4.** -15 **5.** no solution or $\varnothing$ **6.** \$2000 **7.** $-\frac{x^8}{4y^{30}}$ **8.** $-4a - 3$ **9.** $\{x|x$ is a real number and $x \neq -3\}$ **10.** $x^2 - 3x - 1$; -1

11.

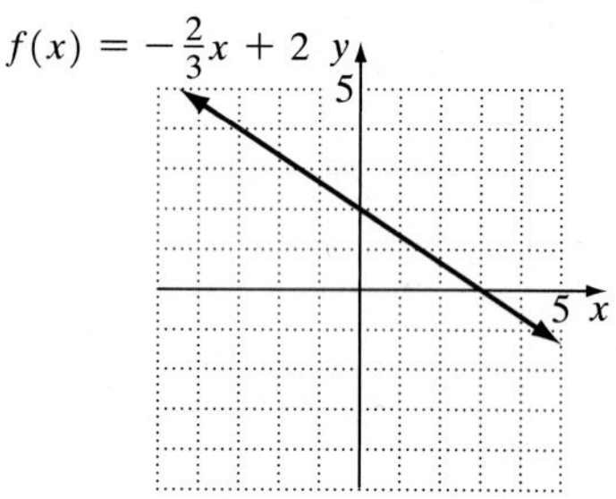

12.

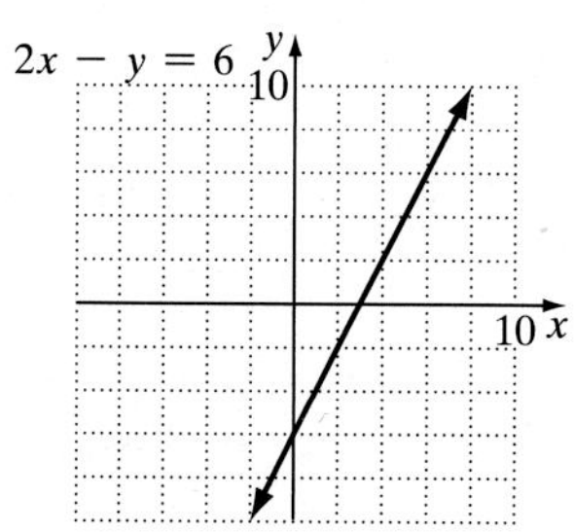

13. $y - 4 = -3(x - 2)$ or $y + 2 = -3(x + 4)$; $y = -3x + 10$ or $f(x) = -3x + 10$
14. $y - 0 = -3(x + 1)$; $y = -3x - 3$ or $f(x) = -3x - 3$
15. $\left(7, \frac{1}{3}\right)$
16. $(3, 2, 4)$
17. pad: \$0.80; pen: \$0.20
18. 23 **19.** $(3, -2, 1)$
20. $(-3, 2)$

CHAPTER 4

Section 4.1

Check Point Exercises

1. a.

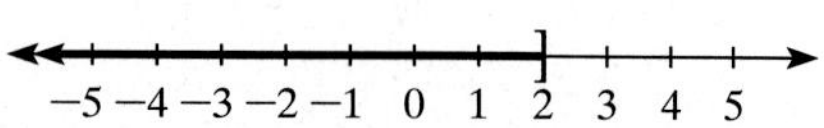

b.

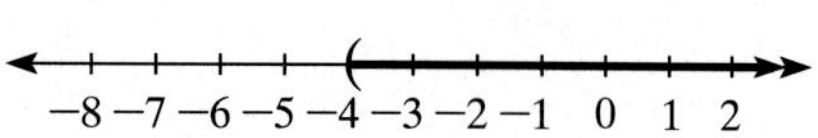

c.

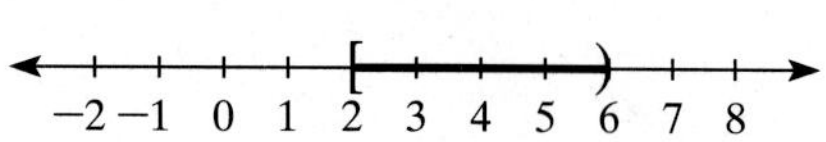

2. a. $\{x|-2 \leq x < 5\}$

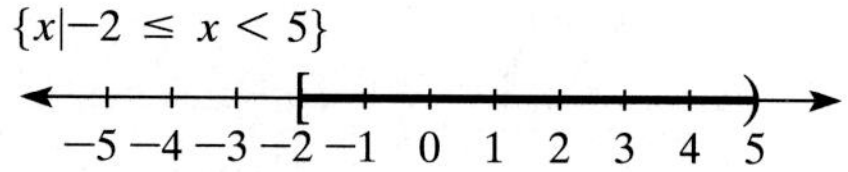

b. $\{x|1 \leq x \leq 3.5\}$

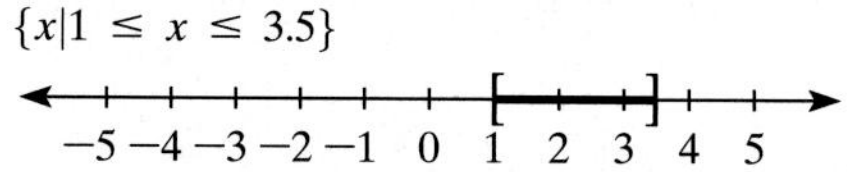

c. $\{x|x < -1\}$

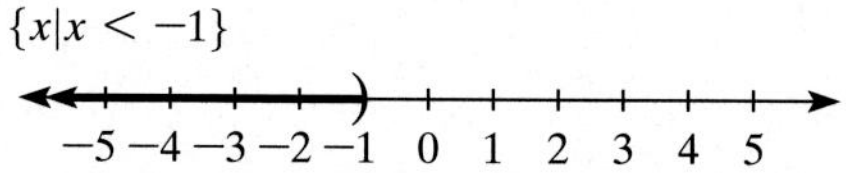

3. $\{x|x > -5\}$ or $(-5, \infty)$

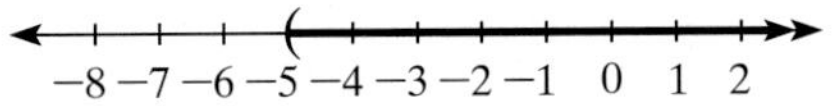

4. $\{x|x \le 1\}$ or $(-\infty, 1]$

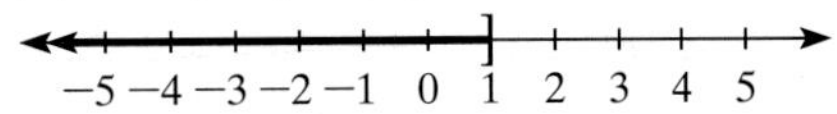

5. $\{x|x > 13\}$ or $(13, \infty)$

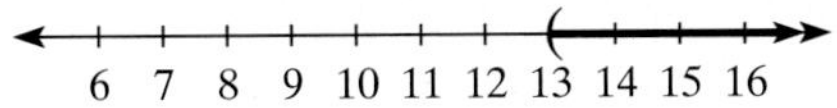

6. **a.** $\{x|x \text{ is a real number}\}$ or $\mathbb{R}$ or $(-\infty, \infty)$
b. $\varnothing$

7. more than 720 miles

Exercise Set 4.1

1.
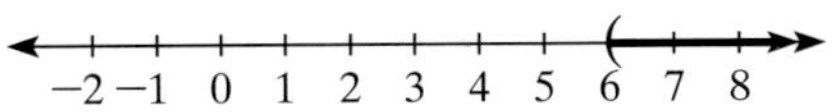

3.
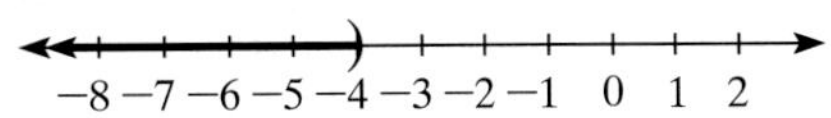

5.
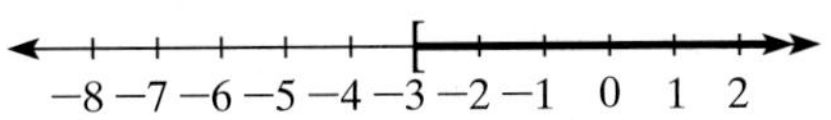

7.
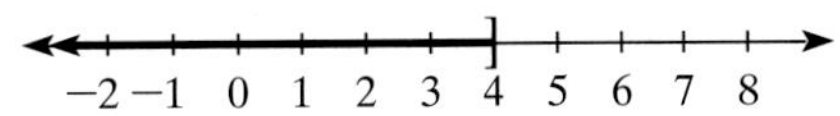

9.
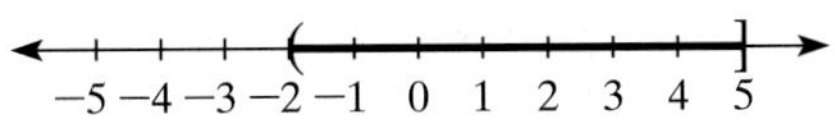

11.
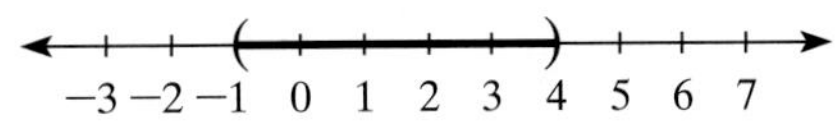

13. $\{x|1 < x \le 6\}$

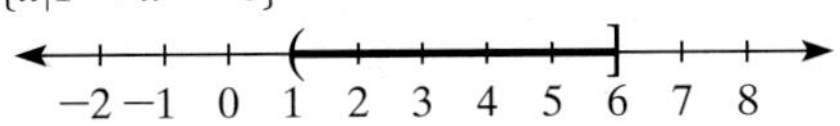

15. $\{x|-5 \le x < 2\}$

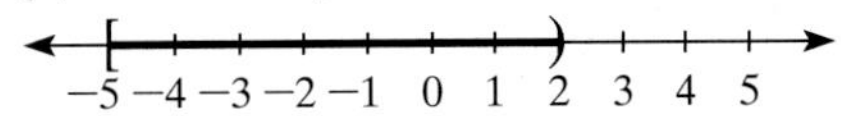

17. $\{x|-3 \le x \le 1\}$

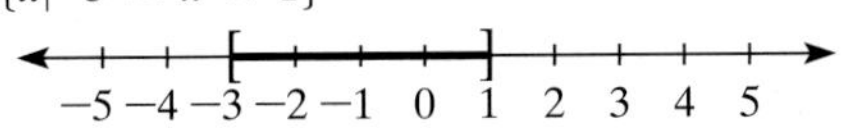

19. $\{x|x > 2\}$

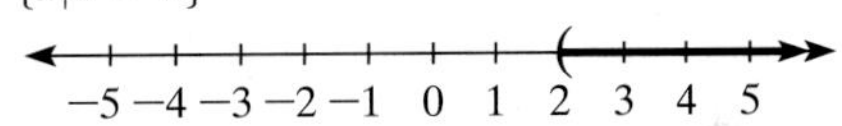

21. $\{x|x \ge -3\}$

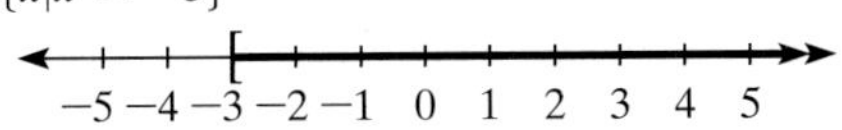

23. $\{x|x < 3\}$

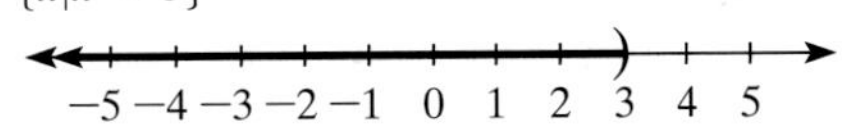

25. $\{x|x < 5.5\}$

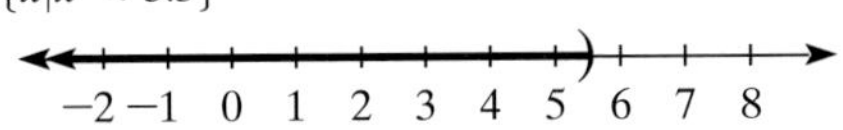

27. $\{x|x < 3\}$ or $(-\infty, 3)$

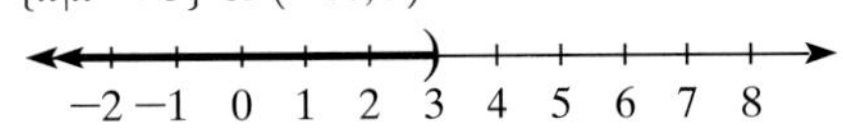

29. $\{x|x \ge 7\}$ or $[7, \infty)$

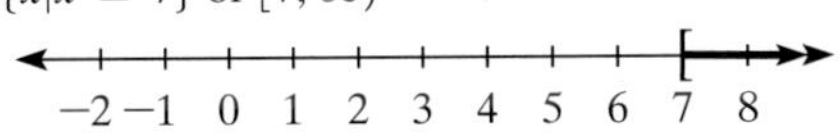

31. $\{x|x \le -4\}$ or $(-\infty, -4]$

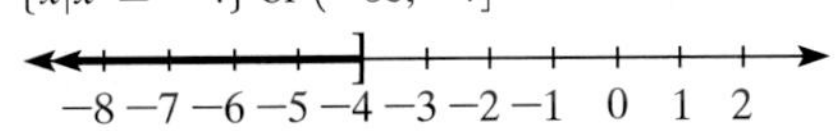

33. $\left\{x \middle| x \le -\frac{2}{5}\right\}$ or $\left(-\infty, -\frac{2}{5}\right]$

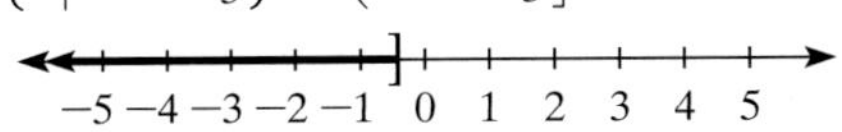

35. $\{x|x \ge 0\}$ or $[0, \infty)$

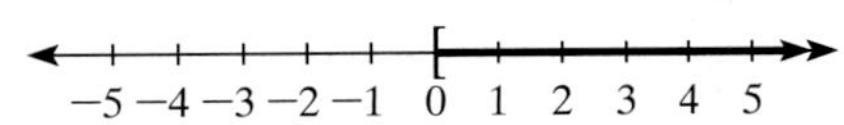

37. $\{x|x < 1\}$ or $(-\infty, 1)$

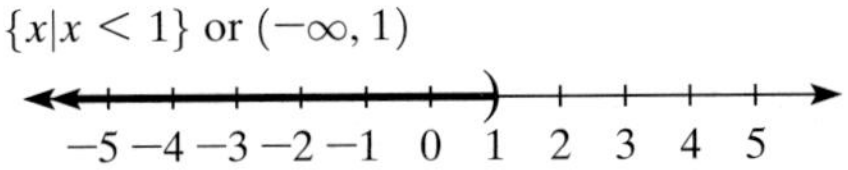

39. $\{x|x \ge 6\}$ or $[6, \infty)$

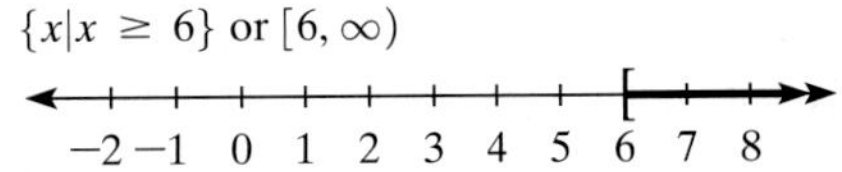

41. $\{x|x \ge -6\}$ or $[-6, \infty)$

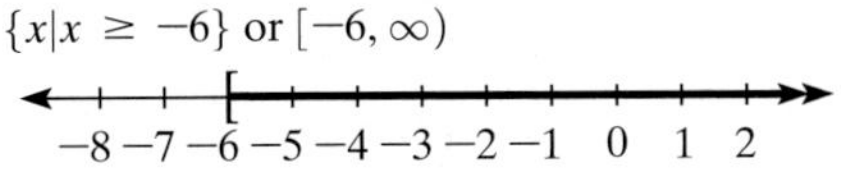

43. $\{x|x < -6\}$ or $(-\infty, -6)$

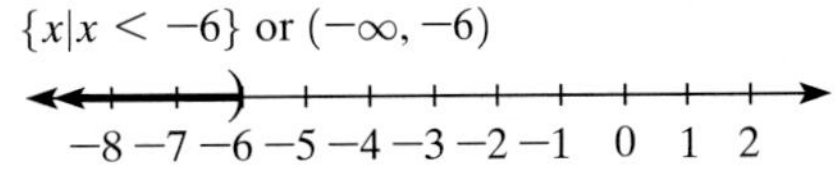

45. $\{x|x \ge 13\}$ or $[13, \infty)$

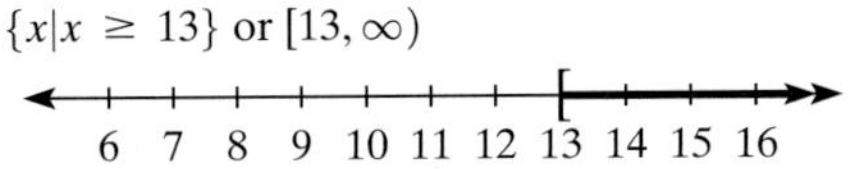

47. $\{x|x \text{ is a real number}\}$ or $\mathbb{R}$ or $(-\infty, \infty)$

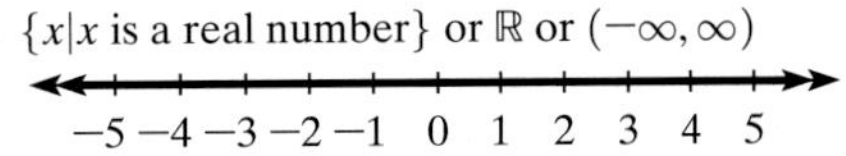

49. $\{x|x \text{ is a real number}\}$ or $\mathbb{R}$ or $(-\infty, \infty)$

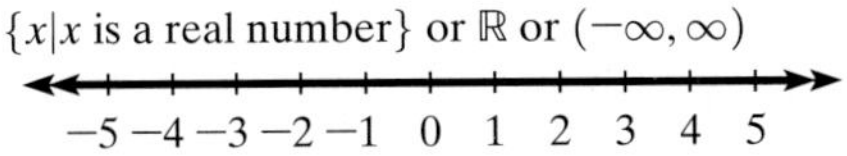

51. $\varnothing$
53. playing sports and sports events
55. exercise, movies, gardening, and amusement parks

57. movies and gardening **59.** gardening, amusement parks, and home improvement **61.** after the year 2005 **63.** $\{t > 175\}$ or $(175, \infty)$; The women's speed skating times will be less than the men's after the year 2075. **65.** more than 100 miles per day **67.** greater than \$32,000 **69.** more than 6250 tapes **71.** 39 bags or fewer **73–79.** Answers will vary.

81. 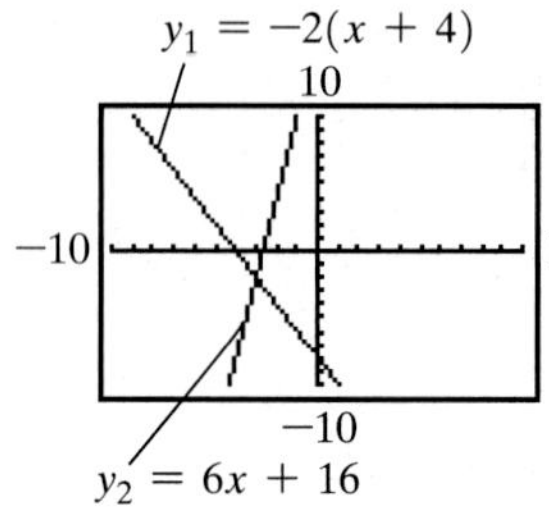

; $\{x|x > -3\}$ or $(-\infty, -3)$

83. $\{x|x$ is a real number$\}$ or $\mathbb{R}$ or $(-\infty, \infty)$
85. d
87. $\{x|-10 < x < -8\}$ or $(-10, -8)$
88. 29
89. $(-1, -1, 2)$
90. $\frac{x^9}{8y^{15}}$

Section 4.2

Check Point Exercises

1. $\{3, 7\}$ **2.** $\{x|x < 1\}$ or $(-\infty, 1)$ **3.** $\varnothing$ **4.** $\{x|-1 \le x < 4\}$ or $[-1, 4)$; 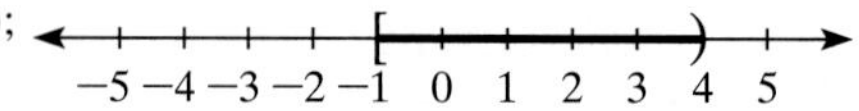

5. $\{3, 4, 5, 6, 7, 8, 9\}$ **6.** $\{x|x \le 1$ or $x > 3\}$ or $(-\infty, 1] \cup (3, \infty)$ **7.** $\{x|x$ is a real number$\}$ or $\mathbb{R}$ or $(-\infty, \infty)$

Exercise Set 4.2

1. $\{2, 4\}$ **3.** $\varnothing$

5. $\{x|x > 6\}; (6, \infty)$

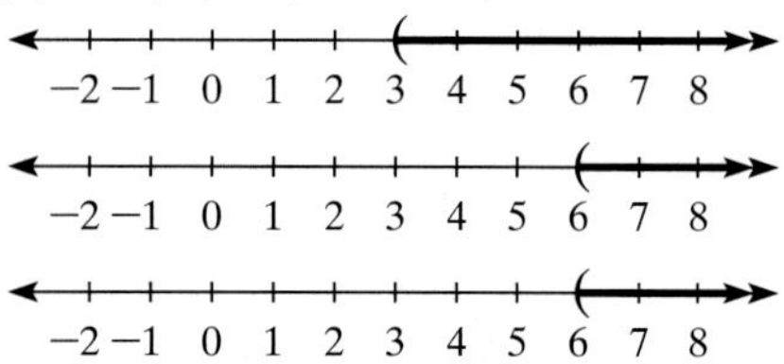

7. $\{x|x \le 1\}; (-\infty, 1]$

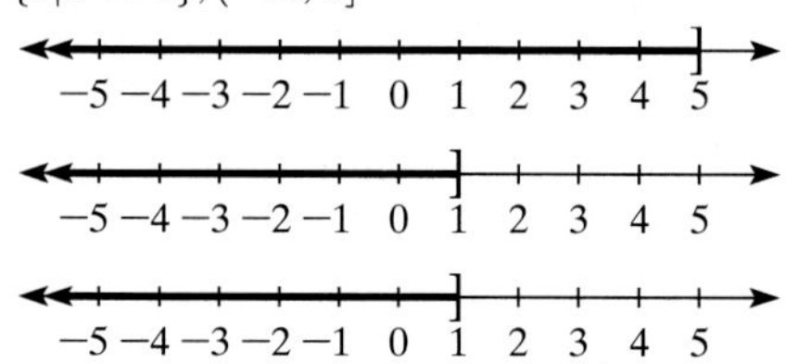

9. $\{x|-1 \le x < 2\}; [-1, 2)$

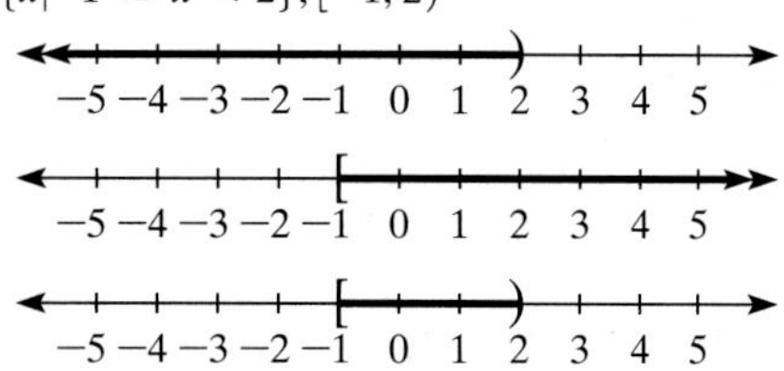

11. $\varnothing$

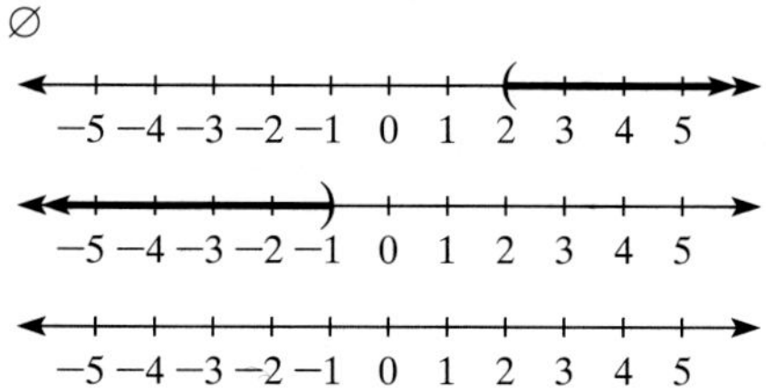

13. $\{x|-6 < x < -4\}; (-6, -4)$

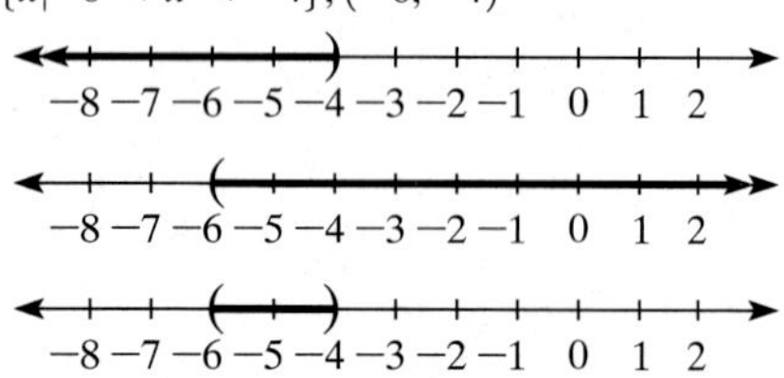

15. $\{x|-3 < x \le 6\}; (-3, 6]$

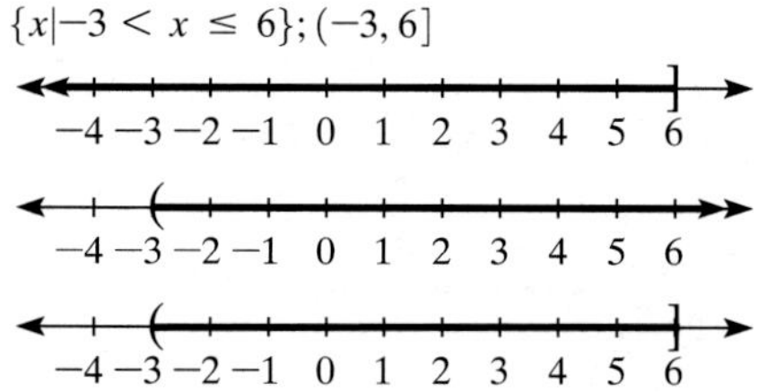

17. $\{x|2 < x < 5\}; (2, 5)$

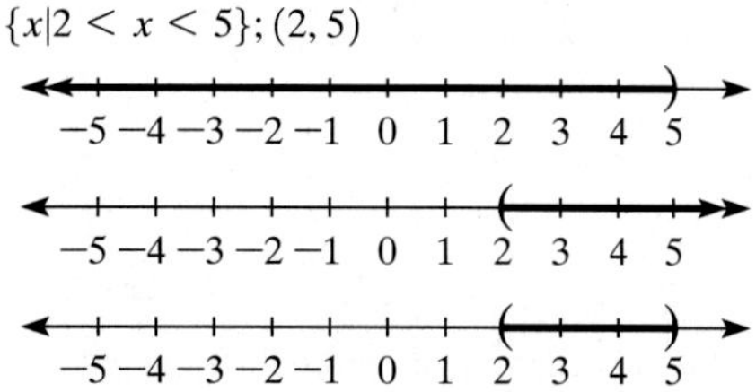

19. $\varnothing$

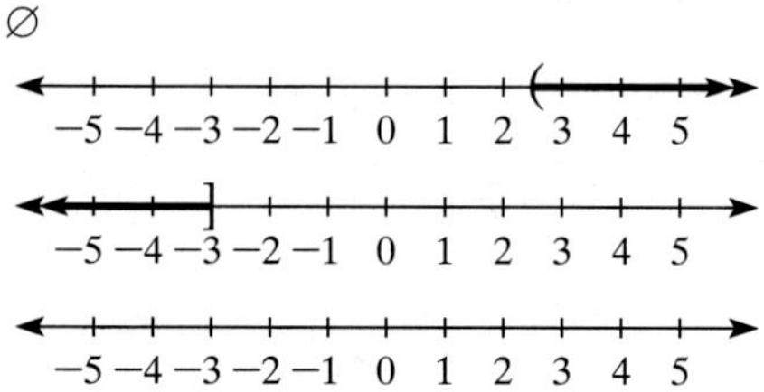

21. $\{x|0 \le x < 2\}; [0, 2)$

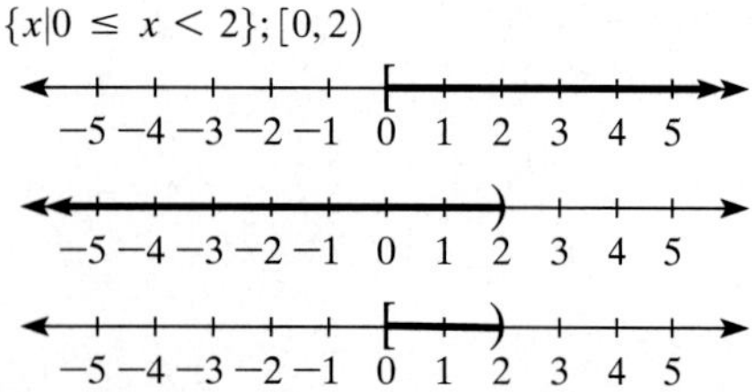

23. $\{x|3 < x < 5\}; (3, 5)$

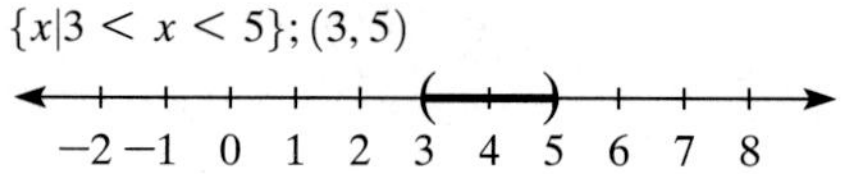

25. $\{x|-1 \le x < 3\}; [-1, 3)$

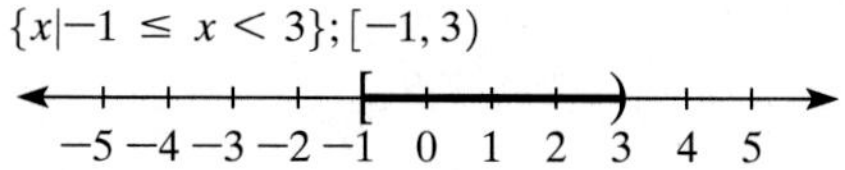

27. $\{x|-5 < x \le -2\}; (-5, -2]$

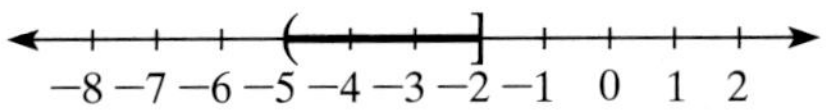

29. $\{x|3 \le x < 6\}; [3, 6)$

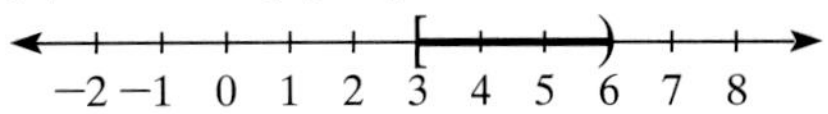

31. $\{1, 2, 3, 4, 5\}$ **33.** $\{1, 2, 3, 4, 5, 6, 7, 8, 10\}$

35. $\{x|x > 3\}; (3, \infty)$

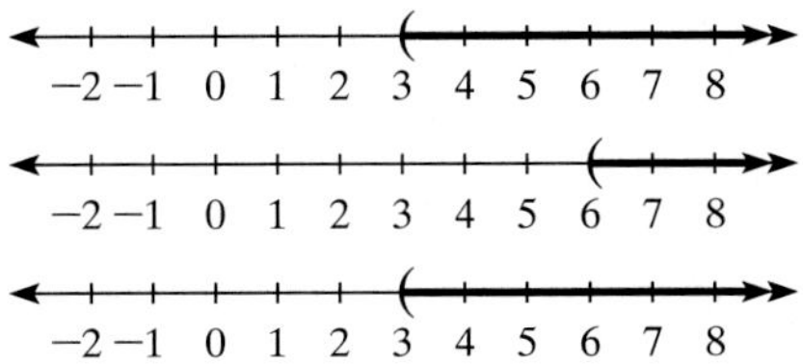

37. $\{x|x \le 5\}; (-\infty, 5]$

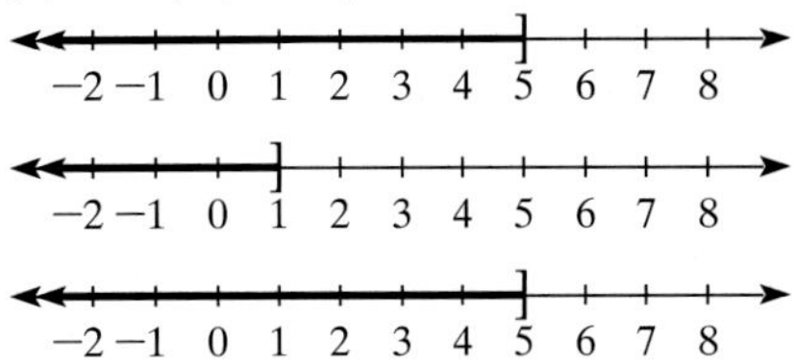

39. $\{x|x$ is a real number$\}; (-\infty, \infty)$

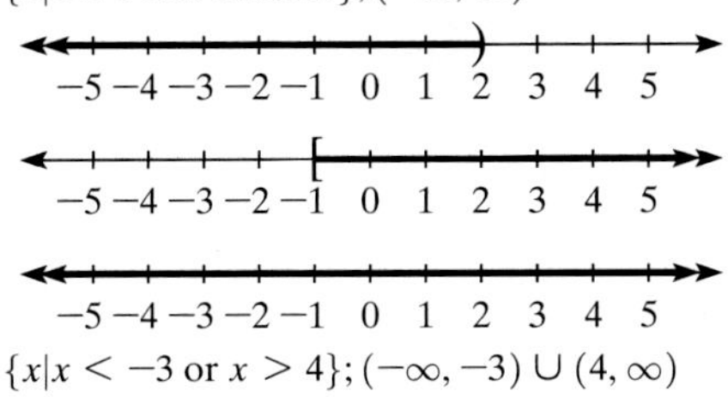

41. $\{x|x < -1 \text{ or } x \ge 2\}; (-\infty, -1) \cup [2, \infty)$

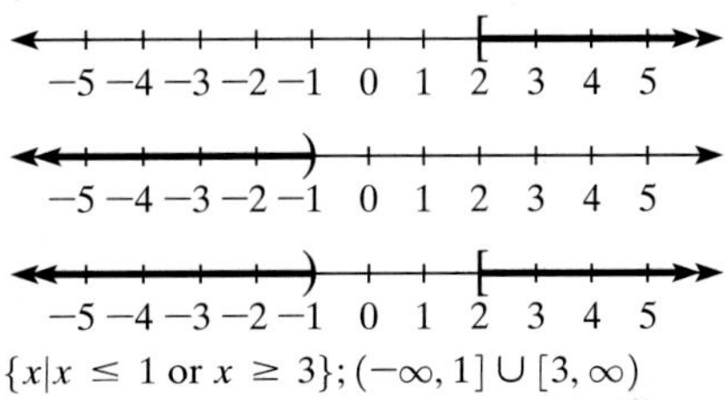

43. $\{x|x < -3 \text{ or } x > 4\}; (-\infty, -3) \cup (4, \infty)$

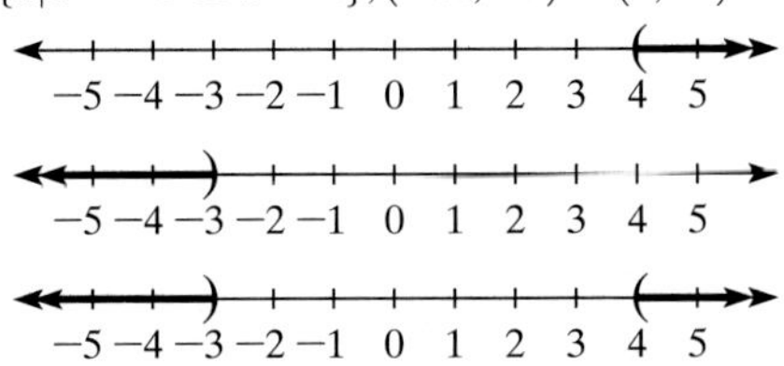

45. $\{x|x \le 1 \text{ or } x \ge 3\}; (-\infty, 1] \cup [3, \infty)$

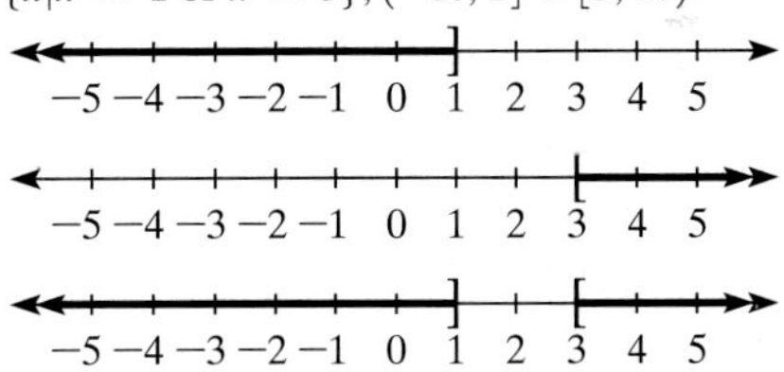

47. $\{x|x$ is a real number$\}; (-\infty, \infty)$

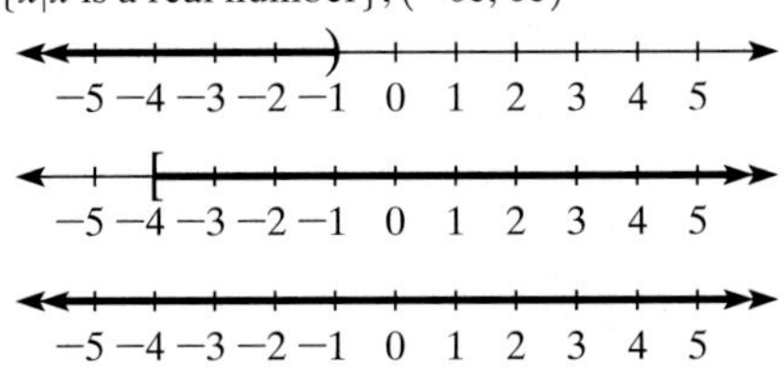

49. $\{x|x < 2\}; (-\infty, 2)$

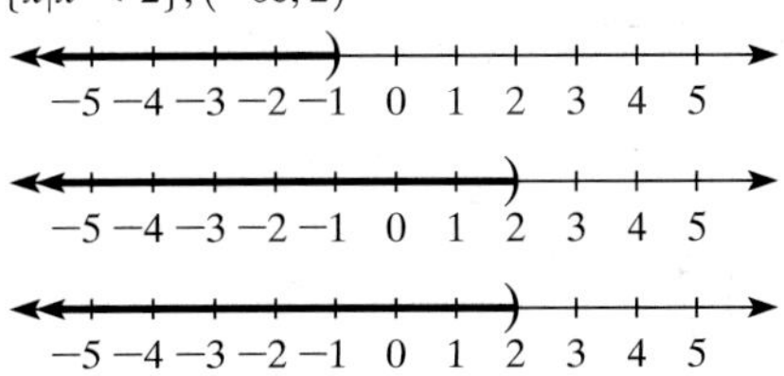

51. spatial-temporal toys, sports equipment, and toy cars and trucks **53.** doll houses, spatial-temporal toys, sports equipment, and toy cars and trucks **55.** none of the toys **57.** from 1905 to 1955 **59.** [76, 126); If the highest grade is 100, then [76, 100]. **61.** more than 3 and less than 15 crossings per 3-month period **63–69.** Answers will vary.

71. $\{x|-6 < x < 2\}$ or $(-6, 2)$

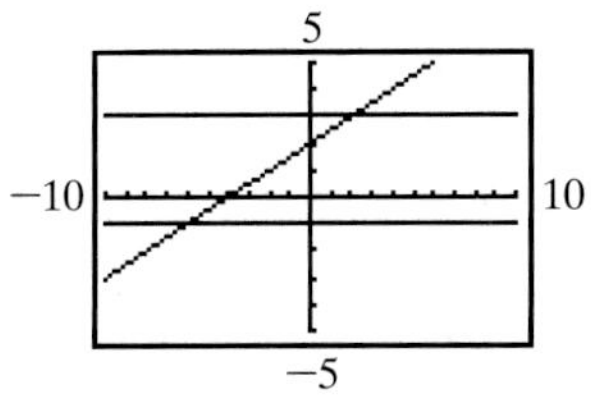

73. $\{x|-3 \le x \le 2\}$ or $[-3, 2]$

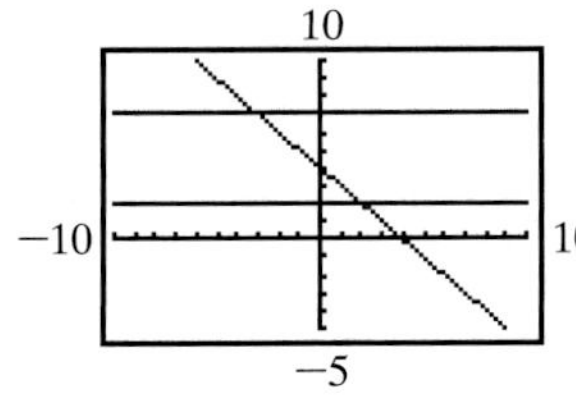

75. $(-1, 5]$

77.

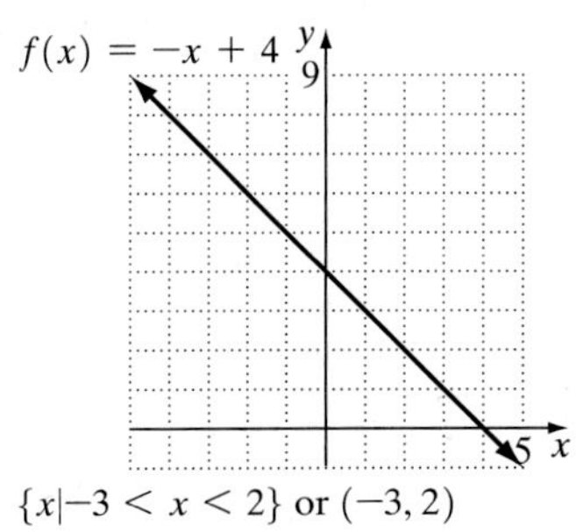

$\{x|-3 < x < 2\}$ or $(-3, 2)$

79. $(-\infty, 4]$ **81.** $[-1, 4]$ **83.** least: 4; greatest: 7 **84.** $-x^2 + 5x - 9; -15$ **85.** $f(x) = -\frac{1}{2}x + 4$ **86.** $17 - 2x$

Section 4.3

Check Point Exercises

1. -2 and 3, or $\{-2, 3\}$ **2.** $\frac{4}{3}$ and 10, or $\left\{\frac{4}{3}, 10\right\}$

3. $\{x|-3 < x < 7\}$ or $(-3, 7)$

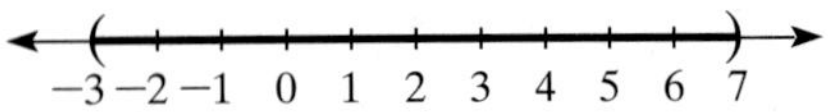

4. $\{x|x \le 1 \text{ or } x \ge 4\}$ or $(-\infty, 1] \cup [4, \infty)$

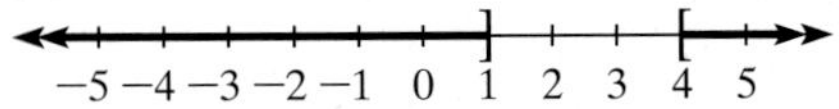

5. $\{x|8.1 \le x \le 13.9\}$ or $[8.1, 13.9]$; The percentage of children in the population who think that not being able to do anything they want is a bad thing is between a low of 8.1% and a high of 13.9%.

Exercise Set 4.3

1. $\{-8, 8\}$ **3.** $\{-5, 9\}$ **5.** $\{-3, 4\}$ **7.** $\{-1, 2\}$ **9.** $\varnothing$ **11.** $\{-3\}$ **13.** $\{-11, -1\}$ **15.** $\{-3, 4\}$ **17.** $\left\{-\frac{13}{3}, 5\right\}$ **19.** $\left\{-\frac{2}{5}, \frac{2}{5}\right\}$ **21.** $\varnothing$ **23.** $\varnothing$ **25.** $\left\{\frac{1}{2}\right\}$ **27.** $\left\{\frac{3}{4}, 5\right\}$ **29.** $\left\{\frac{5}{3}, 3\right\}$ **31.** $\{0\}$ **33.** $\{4\}$ **35.** $\{4\}$ **37.** $\{-1, 15\}$

39. $\{x|-3 < x < 3\}$ or $(-3, 3)$

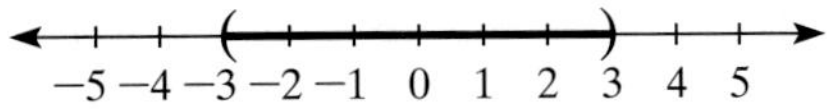

41. $\{x|1 < x < 3\}$ or $(1, 3)$

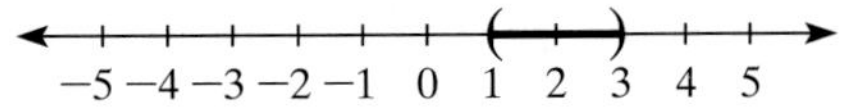

43. $\{x|-3 \le x \le -1\}$ or $[-3, -1]$

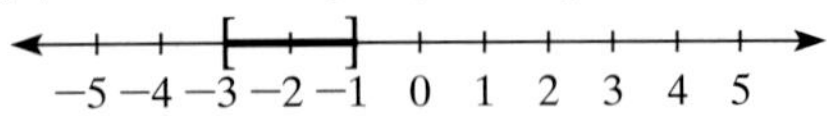

45. $\{x|-1 < x < 7\}$ or $(-1, 7)$

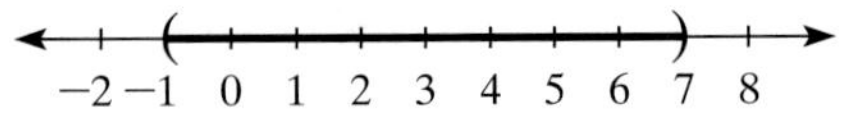

47. $\{x|x < -3 \text{ or } x > 3\}$ or $(-\infty, -3) \cup (3, \infty)$

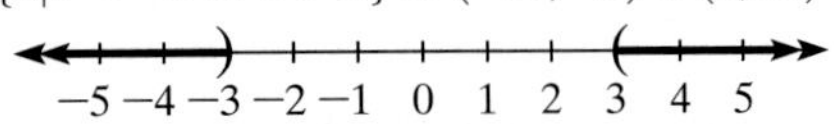

49. $\{x|x < -4 \text{ or } x > -2\}$ or $(-\infty, -4) \cup (-2, \infty)$

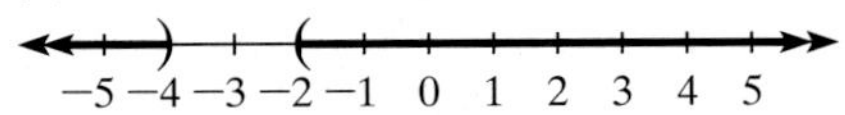

51. $\{x|x \le 2 \text{ or } x \ge 6\}$ or $(-\infty, 2] \cup [6, \infty)$

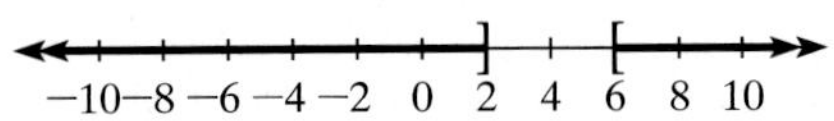

53. $\left\{x \,\middle|\, x < \frac{1}{3} \text{ or } x > 5\right\}$ or $\left(-\infty, \frac{1}{3}\right) \cup (5, \infty)$

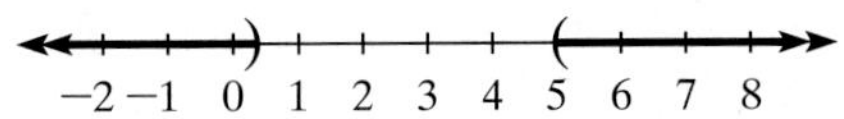

55. $\{x|-5 \le x \le 3\}$ or $[-5, 3]$

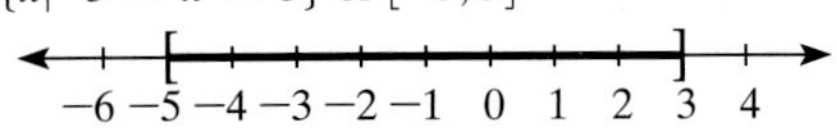

57. $\{x|-6 < x < 0\}$ or $(-6, 0)$

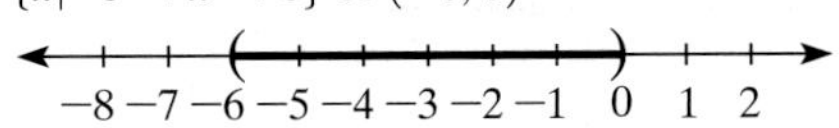

59. $\{x|x \le -5 \text{ or } x \ge 3\}$ or $(-\infty, -5] \cup [3, \infty)$

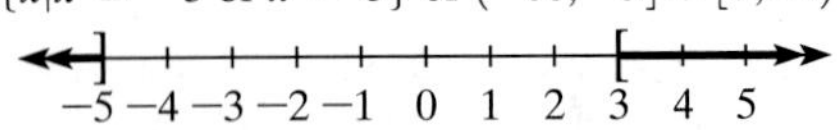

61. $\{x|x < -3 \text{ or } x > 12\}$ or $(-\infty, -3) \cup (12, \infty)$

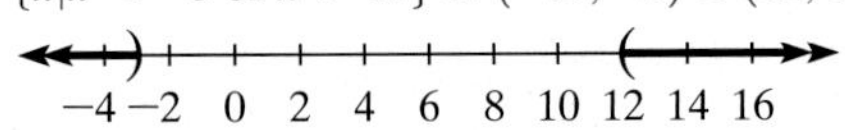

63. $\varnothing$

65. $\{x|x \text{ is a real number}\}$ or $\mathbb{R}$ or $(-\infty, \infty)$

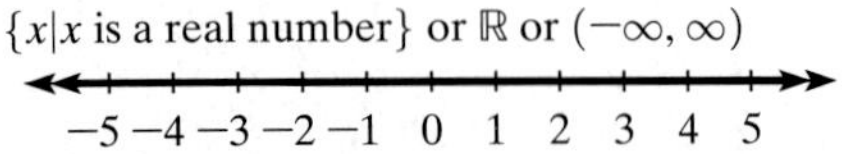

67. $\{x|-9 \le x \le 5\}$ or $[-9, 5]$

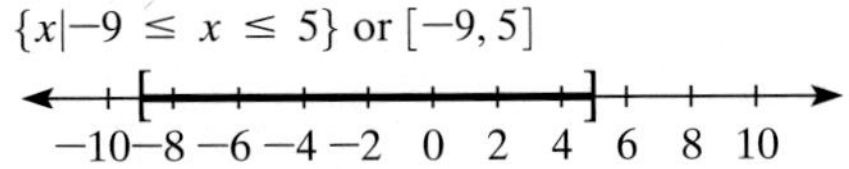

69. $\{x|x < 1 \text{ or } x > 2\}$ or $(-\infty, 1) \cup (2, \infty)$

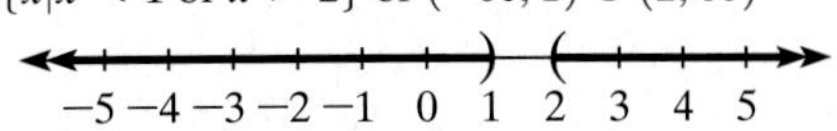

71. $\{x|58.6 \le x \le 61.8\}$ or $[58.6, 61.8]$; The percentage of the U.S. population that watched M*A*S*H is between a low of 58.6% and a high of 61.8%.; 1.6%

73. $\{T|50 \le T \le 64\}$ or $[50, 64]$; The monthly average temperature for San Francisco, CA is between a low of 50°F and a high of 64°F. **75.** $\{x|8.59 \le x \le 8.61\}$ or $[8.59, 8.61]$; A machine part that is supposed to be 8.6 centimeters is acceptable between a low of 8.59 and a high of 8.61 centimeters. **77.** If the number of outcomes that result in heads is 41 or less or 59 or more, then the coin is unfair.

79–85. Answers will vary.

87.

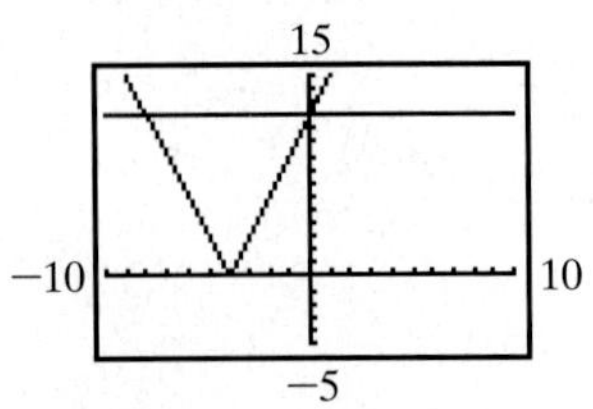

89.

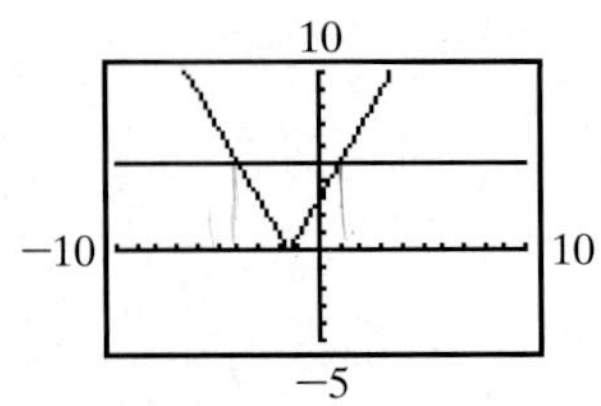

91.

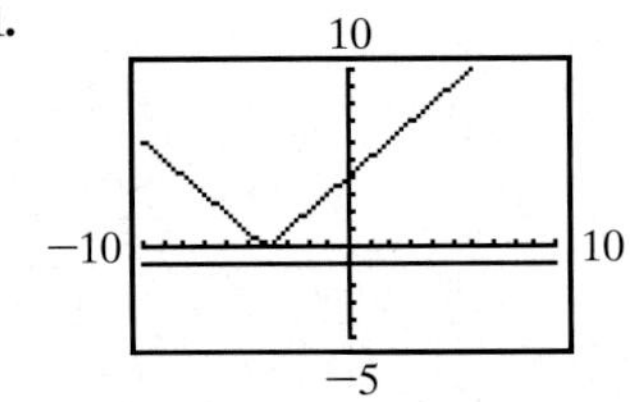

93.

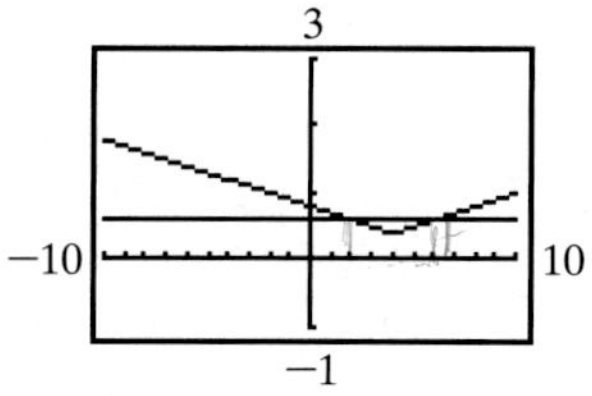

95. Answers will vary. **97.** 3 and 5

99. The cost for refunds will be between a low of $500 and a high of $2500.

101.

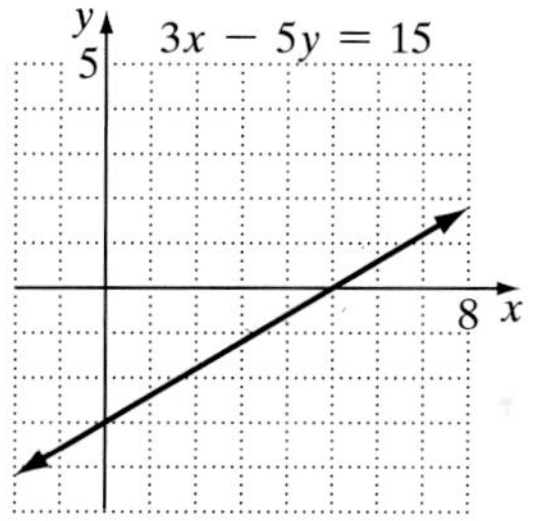

102.

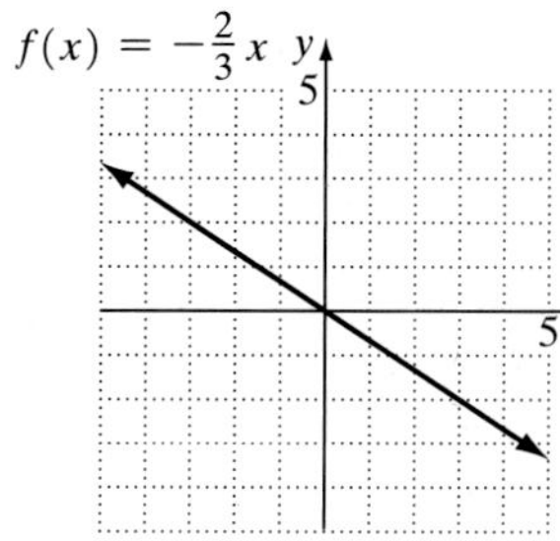

103.

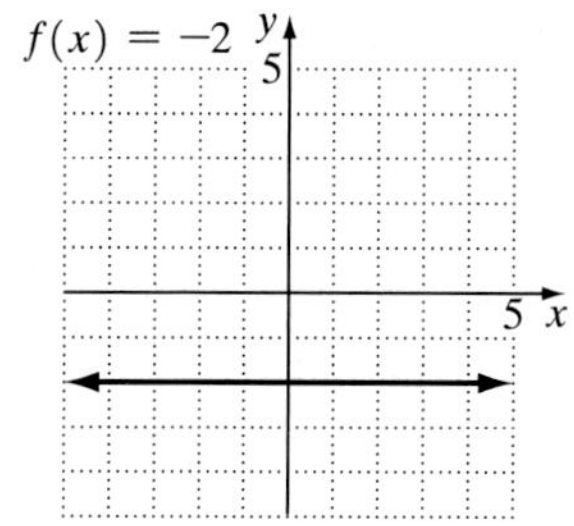

Section 4.4

Check Point Exercises

1.

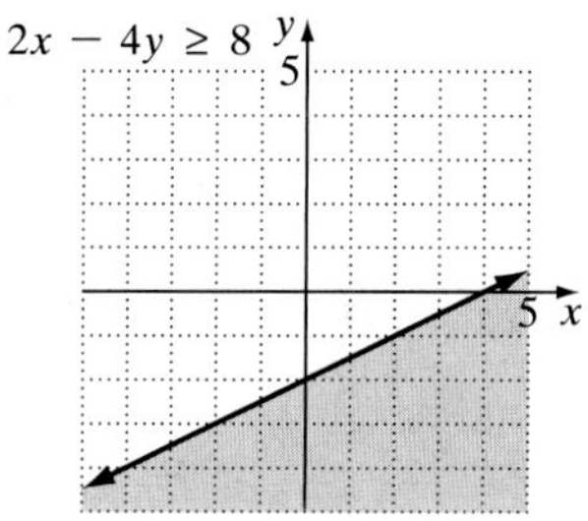

2.

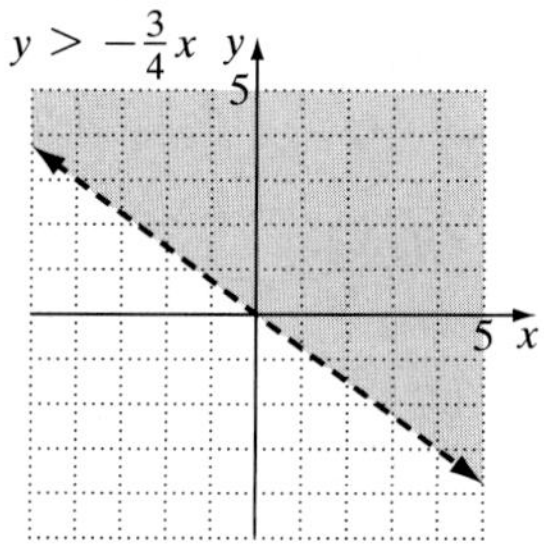

3.

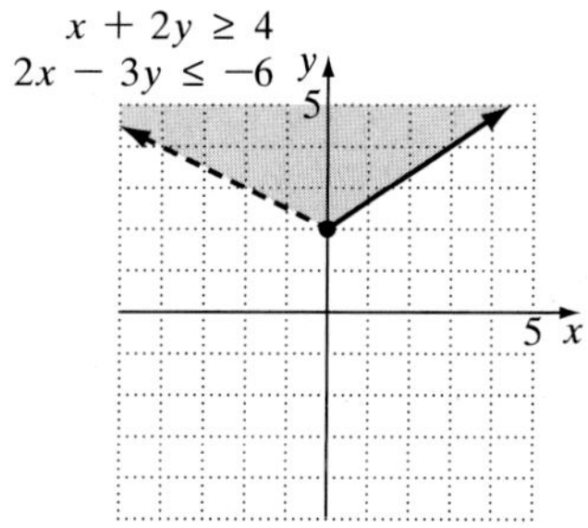

4.

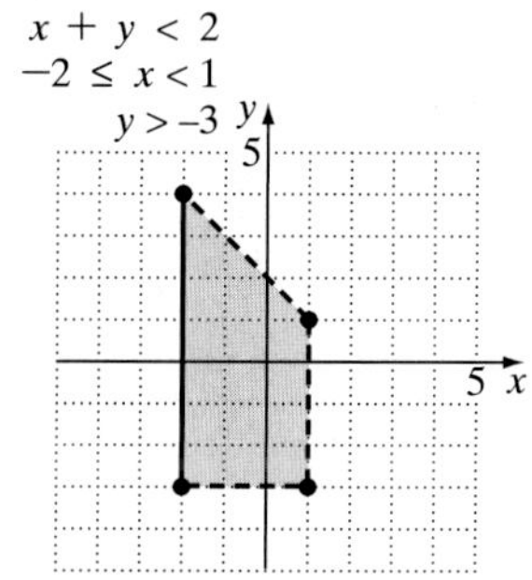

5. $B = (60, 20)$; Using $T = 60$ and $P = 20$, each of the three inequalities for grasslands is true: $60 \geq 5$, true; $5(60) - 7(20) \geq 70$, true; $3(60) - 35(20) \leq -140$, true.

Exercise Set 4.4

1.

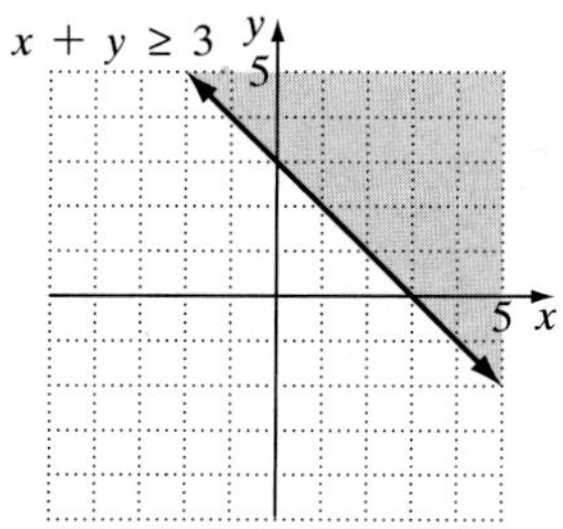

3.

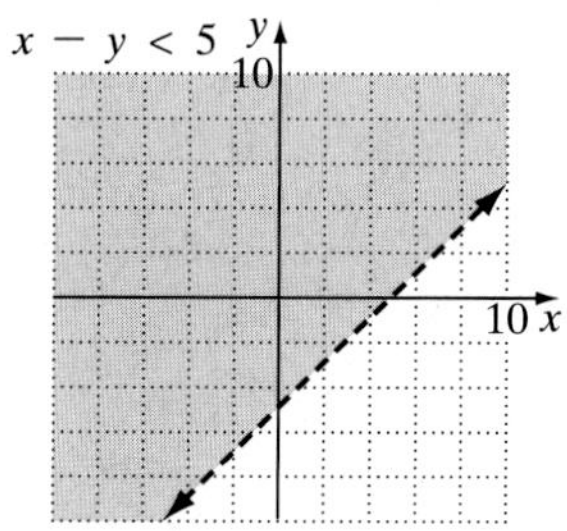

5.

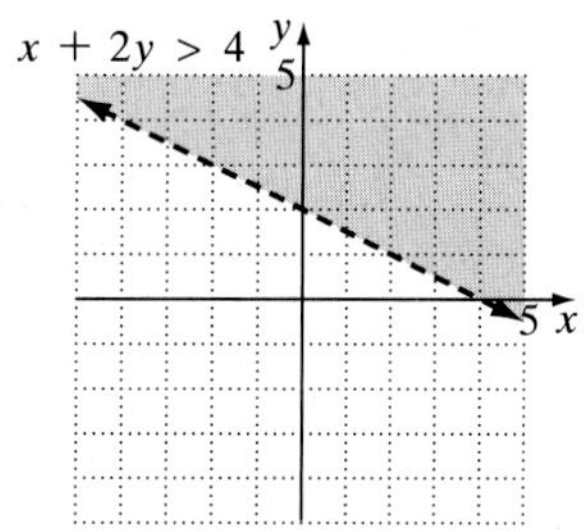

7.

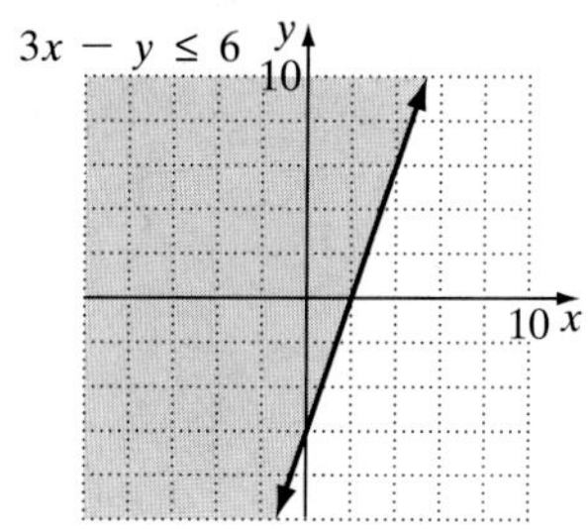

9.

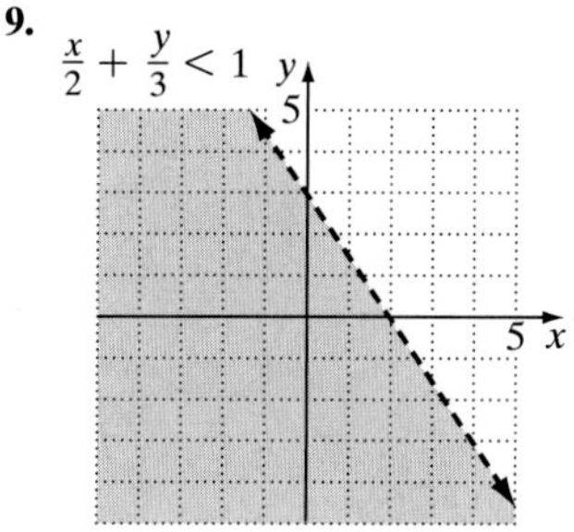

11.

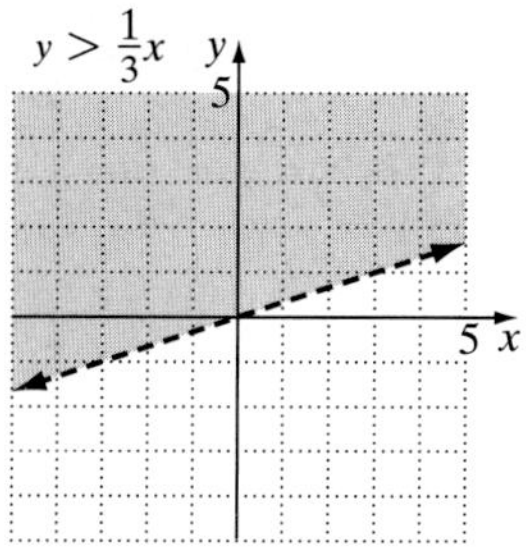

13.

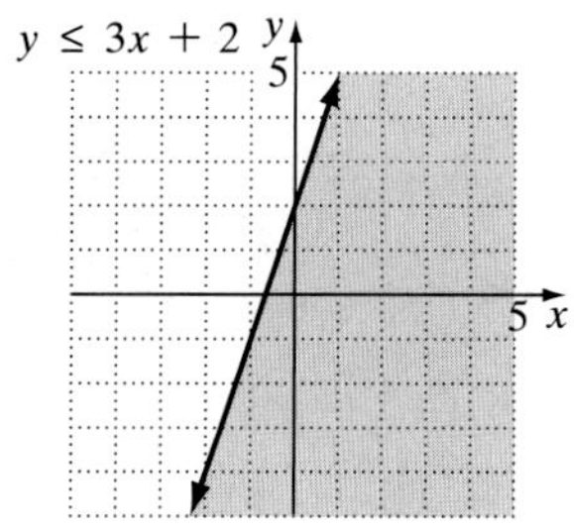

15.

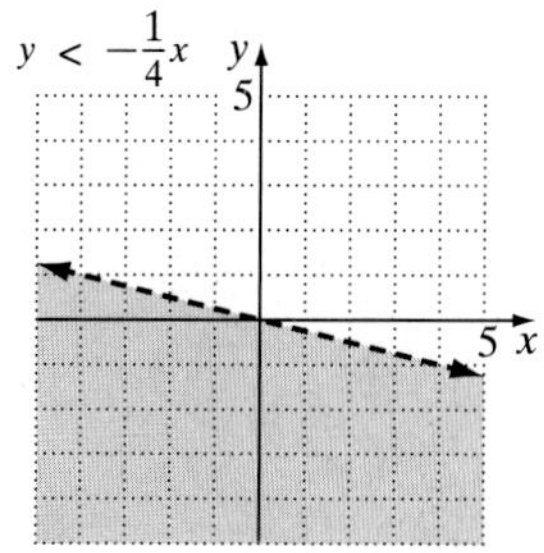

17. $x \le 2$

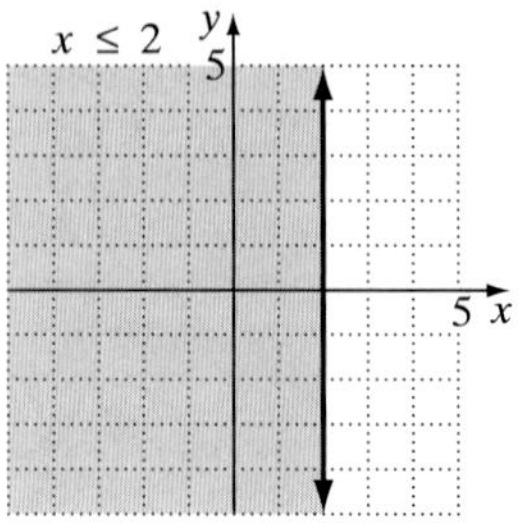

19. $y > -4$

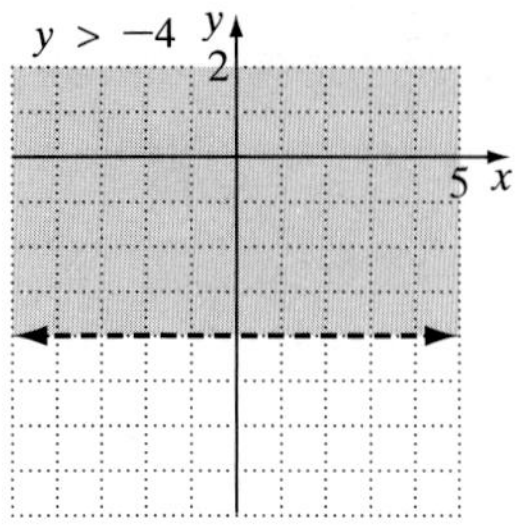

21. $y \ge 0$

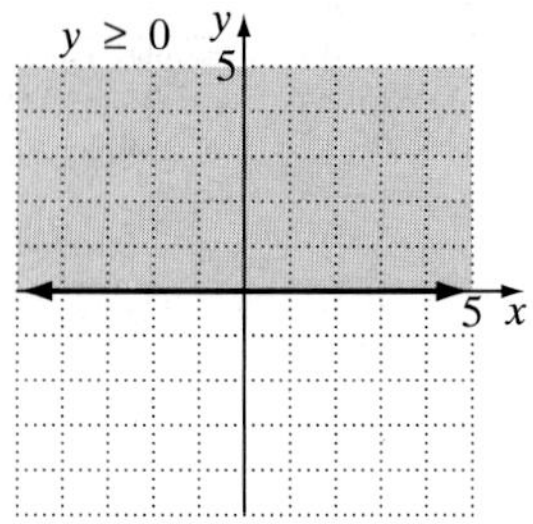

23. $3x + 6y \le 6$
$2x + y \le 8$

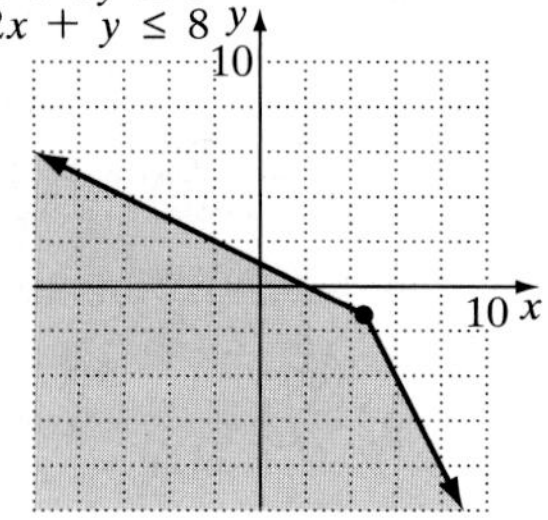

25. $2x - 5y \le 10$
$3x - 2y > 6$

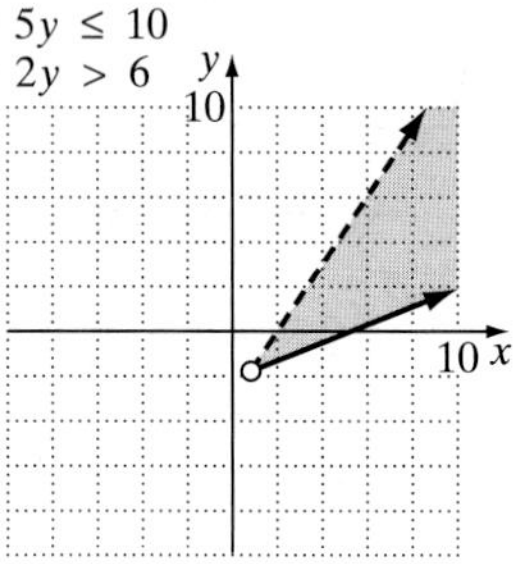

27. $y > 2x - 3$
$y < -x + 6$

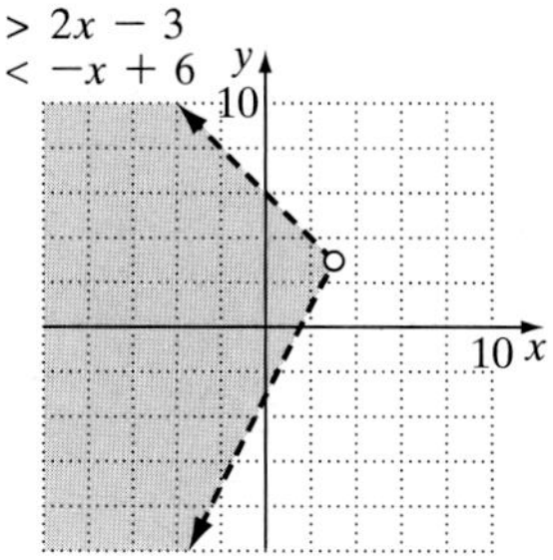

29. $x + 2y \le 4$
$y \ge x - 3$

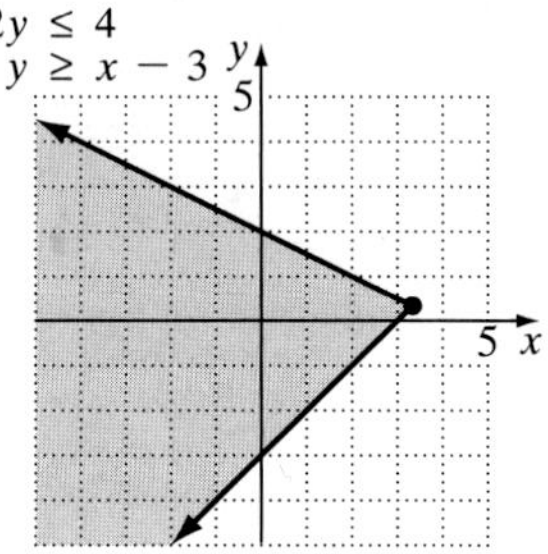

31. $x \le 2$
$y \ge -1$

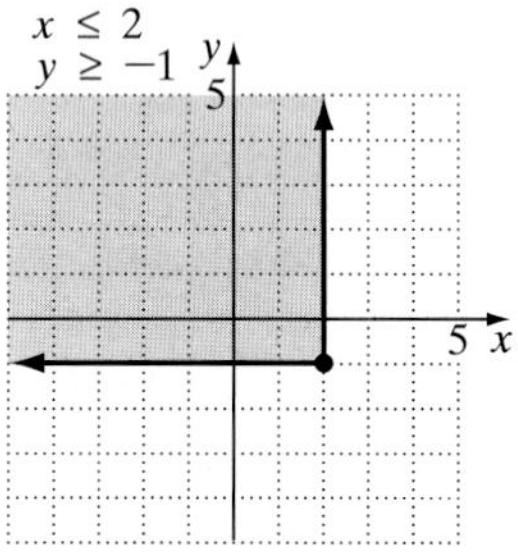

33. $-2 \le x < 5$

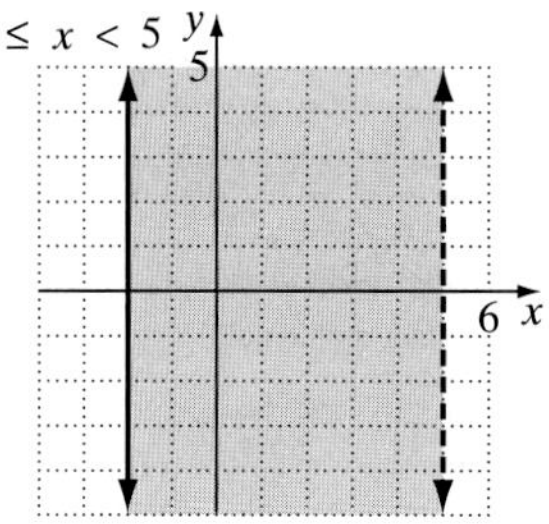

35. $x - y \le 1$
$x \ge 2$

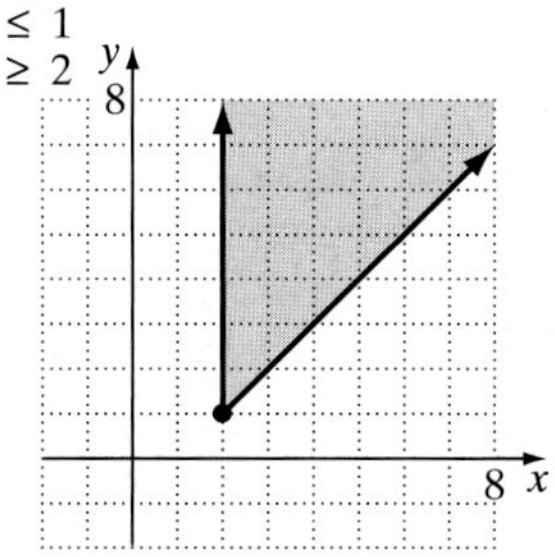

37. $\varnothing$

39. $x + y > 4$
$x + y > -1$

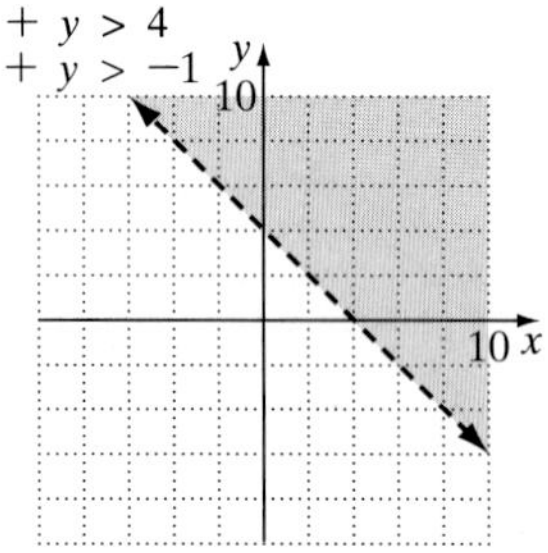

41. $x - y \le 2$
$x \ge -2$
$y \le 3$

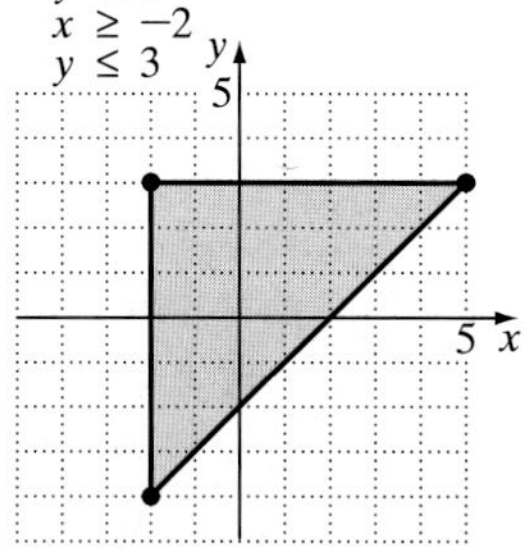

43. $x \ge 0$
$y \ge 0$
$2x + 5y \le 10$
$3x + 4y \le 12$

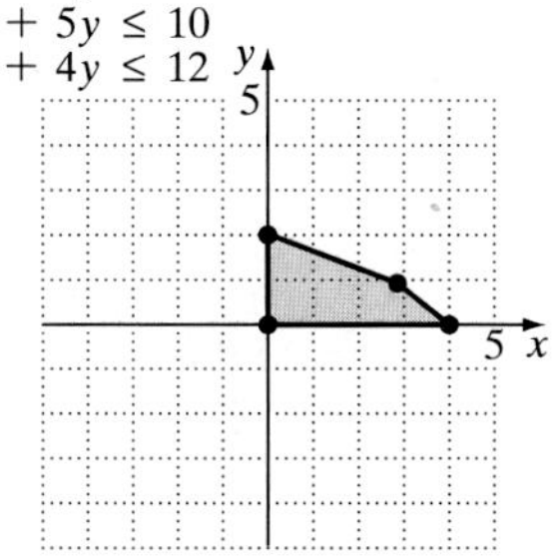

45. $3x + y \le 6$
$2x - y \le -1$
$x \ge -2$
$y \le 4$

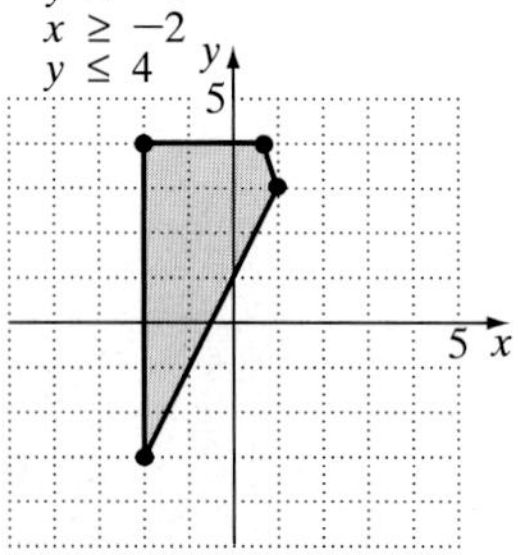

47. a. (20, 160); 20 year-old with a pulse rate of 160 beats per minute is within the target zone.
b. $10 \le 20 \le 70$, true; $2(20) + 3(160) \ge 450$, true; $20 + 160 \le 190$, true.

49. Answers will vary.

51. 190 lb

53. $7w - 25h < -800$; Using (100, 70) then $7(100) - 25(70) < -800$ is true.

55. $7w - 25h \ge -800$, $w - 5h \le -170$; Using (150, 70) then $7(150) - 25(70) \ge -800$ is true and $150 - 5(70) \le -170$ is true.

57–65. Answers will vary.

67.

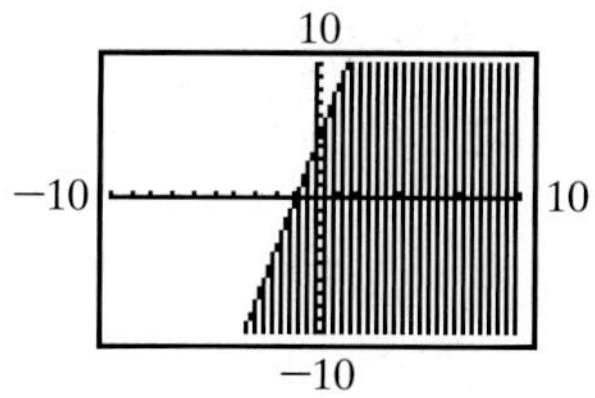

69.

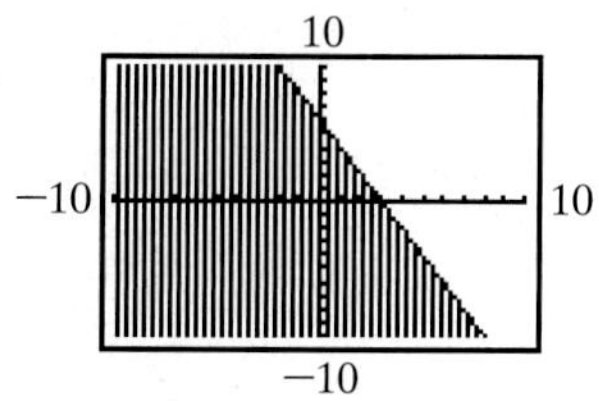

71–75. Answers will vary. **77.**

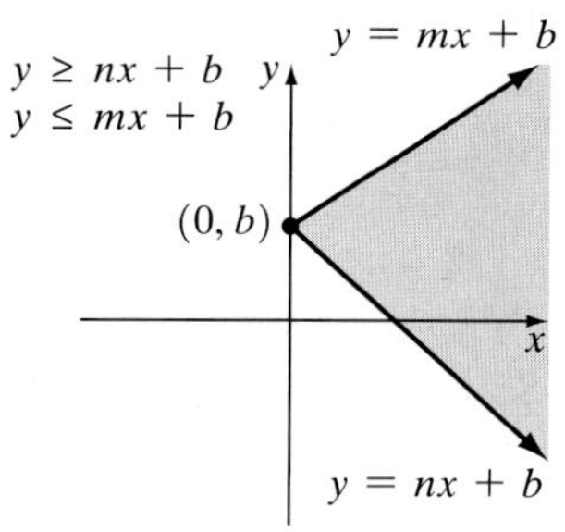

78. $(3, 1)$ or $\{(3, 1)\}$ **79.** $(2, 4)$ or $\{(2, 4)\}$ **80.** 165

Section 4.5

Check Point Exercises

1. $z = 25x + 155y$ **2.** $x + y \leq 80$ **3.** $30 \leq x \leq 80, 10 \leq y \leq 30; z = 25x + 55y, x + y \leq 80, 30 \leq x \leq 80, 10 \leq y \leq 30$
4. 50 bookshelves and 30 desks; \$2900 **5.** 30

Exercise Set 4.5

1. $(1, 2)$: 17; $(2, 10)$: 70; $(7, 5)$: 65; $(8, 3)$: 58; maximum: 70; minimum: 17
3. $(0, 0)$: 0; $(0, 8)$: 400; $(4, 9)$: 610; $(8, 0)$: 320; maximum: 610; minimum: 0

5. a.

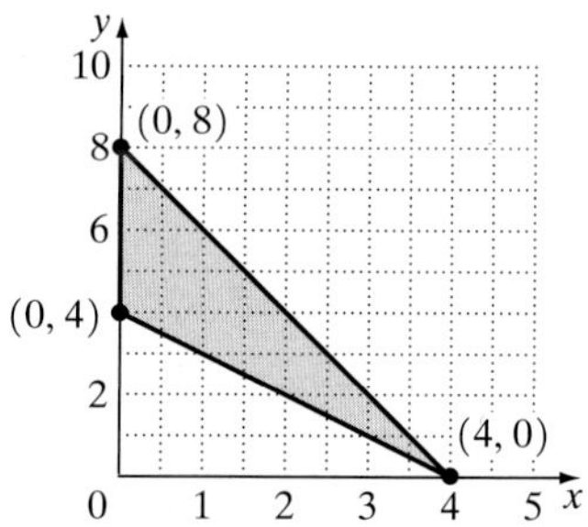

b. $(0, 4)$: 8; $(0, 8)$: 16; $(4, 0)$: 12
c. maximum: 16; at $(0, 8)$

7. a.

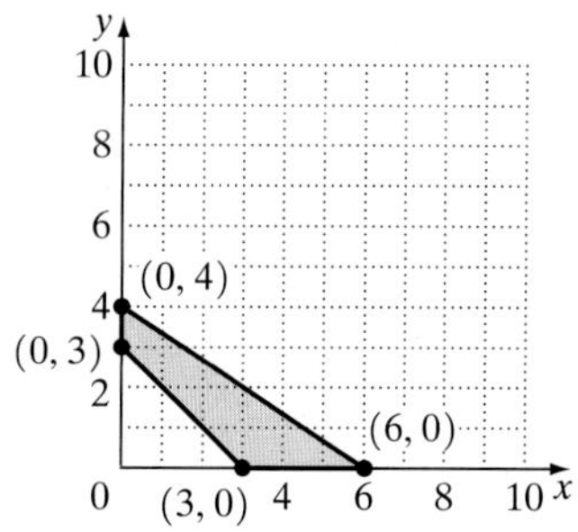

b. $(0, 3)$: 3; $(0, 4)$: 4; $(6, 0)$: 24; $(3, 0)$: 12
c. maximum: 24; at $(6, 0)$

9. a.

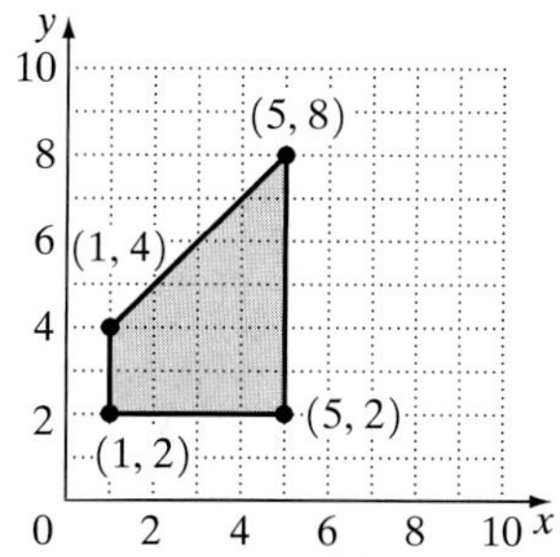

b. $(1, 2)$: −1; $(1, 4)$: −5; $(5, 8)$: −1; $(5, 2)$: 11

c. maximum: 11; at $(5, 2)$

11. a.

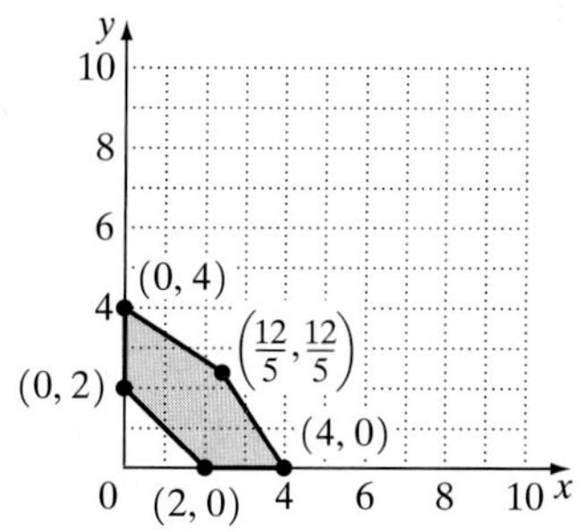

b. $(0, 2)$: 4; $(0, 4)$: 8; $\left(\frac{12}{5}, \frac{12}{5}\right)$: $\frac{72}{5} = 14.4$; $(4, 0)$: 16; $(2, 0)$: 8
c. maximum: 16; at $(4, 0)$

13. a.

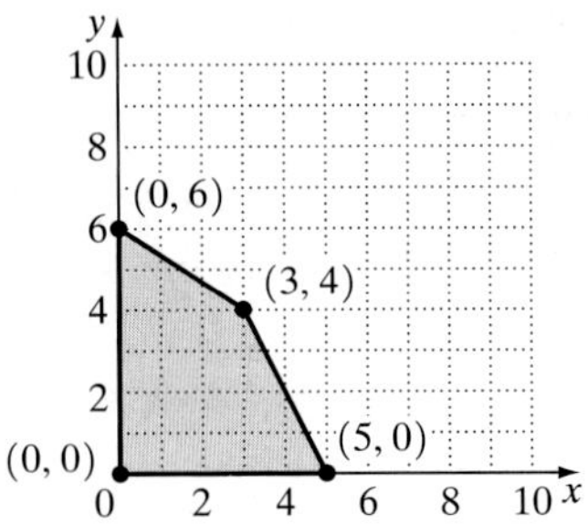

b. (0, 0): 0; (0, 6): 72; (3, 4): 78; (5, 0): 50
c. maximum: 78; at (3, 4)

15. a. $z = 125x + 200y$
b. $x \leq 450$; $y \leq 200$; $600x + 900y \leq 360{,}000$
c.

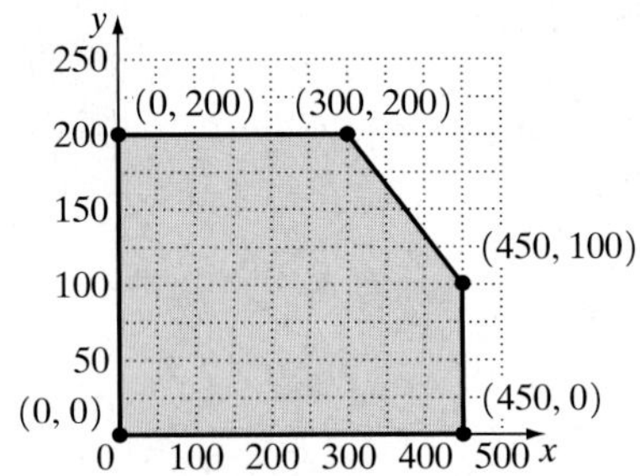

d. 0; 40,000; 77,500; 76,250; 56,250
e. 300; 200; 77,500

17. 40 of model A and 0 of model B **19.** 300 boxes of food and 200 boxes of clothing **21.** 50 parents and 100 students **23.** 10 Boeing 727s and 42 Falcon 20s **25–27.** Answers will vary. **29.** \$5000 in stocks and \$5000 in bonds **31.** $54x^7y^{15}$ **32.** $L = \dfrac{12P + W}{2}$ **33.** 10

Review Exercises

1.

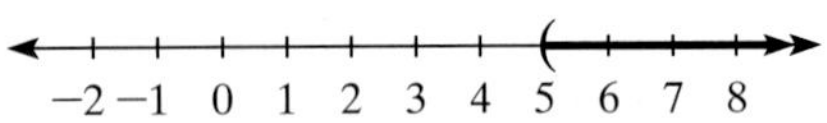

2.

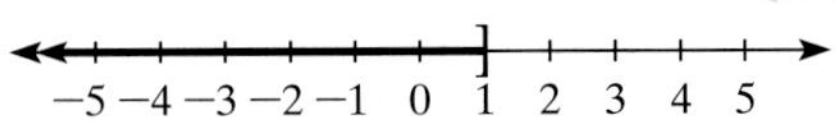

3.

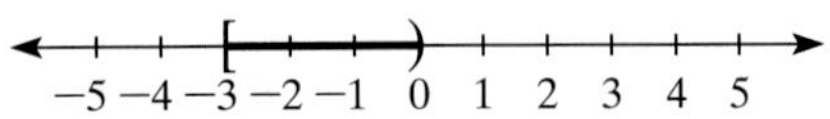

4. $\{x|-2 < x \leq 3\}$

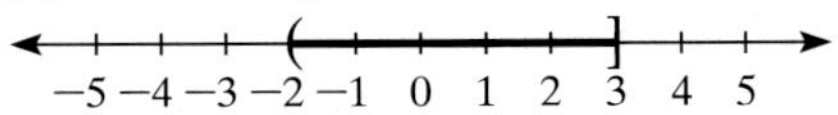

5. $\{x|-1.5 \leq x \leq 2\}$

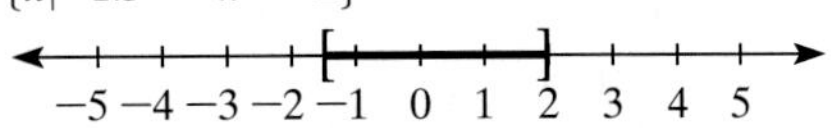

6. $\{x|x > -1\}$

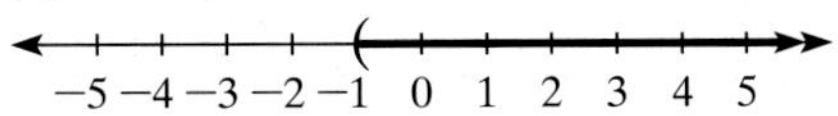

7. $\{x|x \geq -2\}$; $[-2, \infty)$

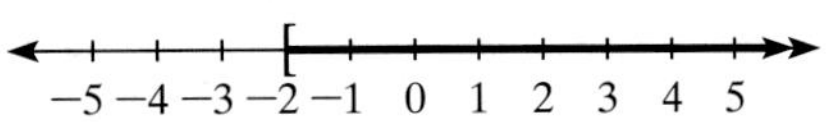

8. $\left\{x \middle| x \geq \frac{3}{5}\right\}$; $\left[\frac{3}{5}, \infty\right)$

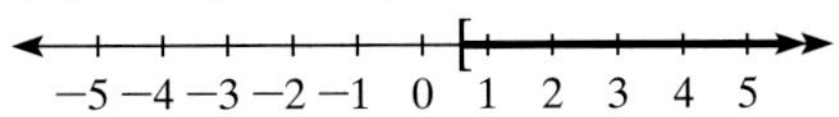

9. $\left\{x \middle| x < -\frac{21}{2}\right\}$; $\left(-\infty, -\frac{21}{2}\right)$

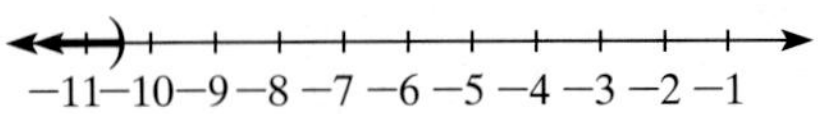

10. $\{x|x > -3\}$; $(-3, \infty)$

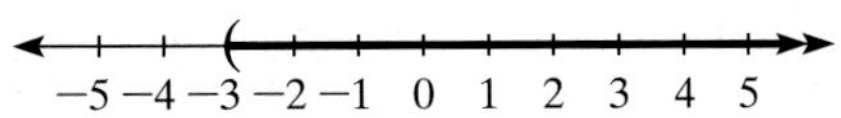

11. $\{x|x \leq -2\}$; $(-\infty, -2]$

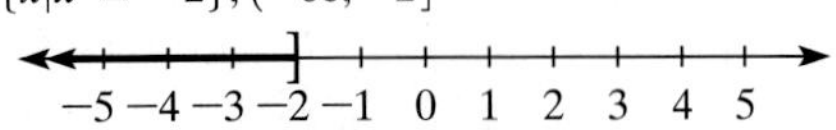

12. $\varnothing$
13. more than 50 checks
14. more than \$13,500 in sales

15. $\{a, c\}$ **16.** $\{a\}$ **17.** $\{a, b, c, d, e\}$ **18.** $\{a, b, c, d, f, g\}$

19. $\{x|x \leq 3\}$; $(-\infty, 3]$

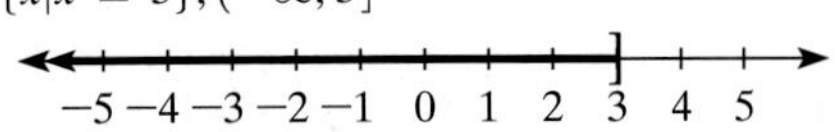

20. $\{x|x < 6\}$; $(-\infty, 6)$

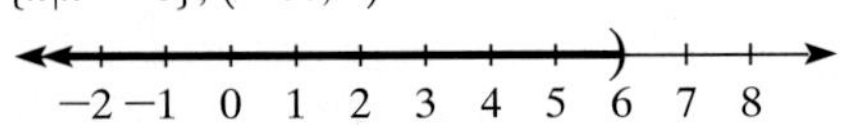

21. $\{x|6 < x < 8\}$; $(6, 8)$

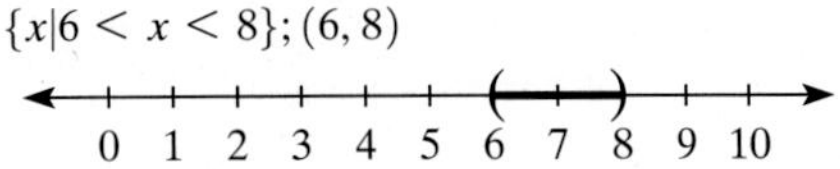

22. $\{x|x \leq 1\}$; $(-\infty, 1]$

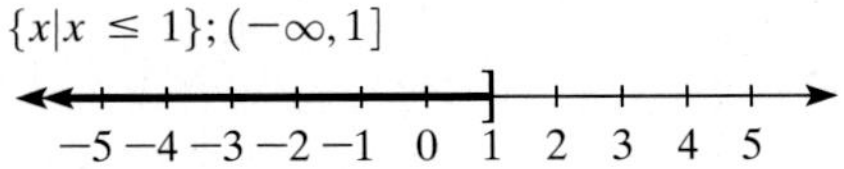

23. $\varnothing$ **24.** $\{x|x < 1 \text{ or } x > 2\}$; $(-\infty, 1) \cup (2, \infty)$

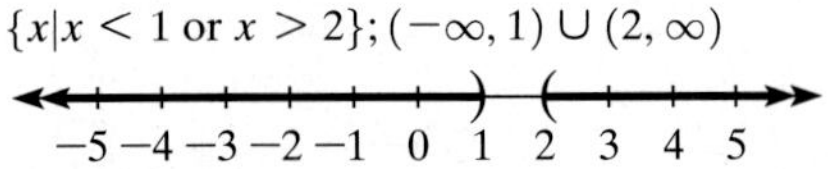

25. $\{x|x \leq -4 \text{ or } x > 2\}$; $(-\infty, -4] \cup (2, \infty)$

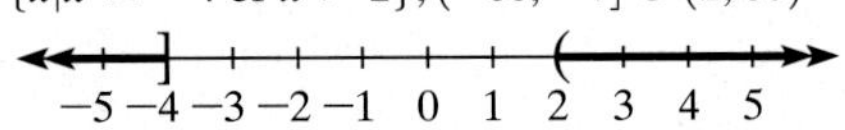

26. $\{x|x < -2\}$; $(-\infty, -2)$

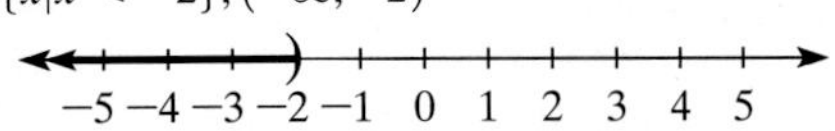

27. $\{x|x \text{ is a real number}\}$; $(-\infty, \infty)$

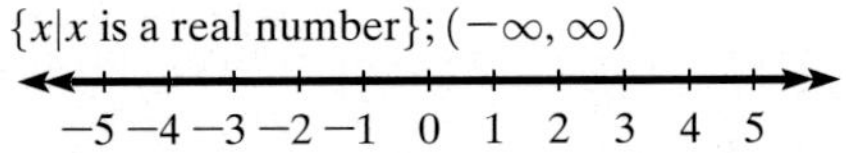

28. $\{x|-5 < x \leq 2\}; (-5, 2]$

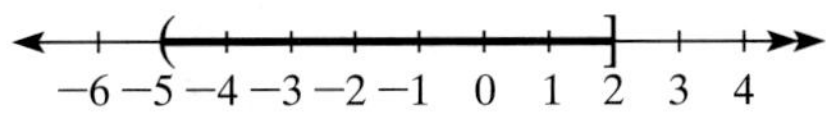

29. $\left\{x \middle| -\frac{3}{4} \leq x \leq 1\right\}; \left[-\frac{3}{4}, 1\right]$

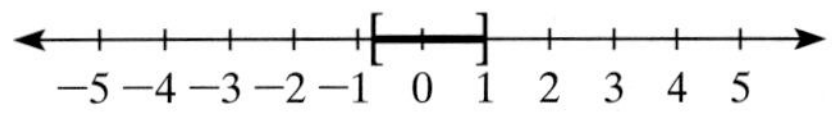

30. $[81, 131)$; If the highest grade is 100, then $[81, 100]$. **31.** $\{-4, 3\}$ **32.** $\varnothing$ **33.** $\left\{-\frac{11}{2}, \frac{23}{2}\right\}$ **34.** $\left\{-4, -\frac{6}{11}\right\}$

35. $\{x|-9 \leq x \leq 6\}; [-9, 6]$

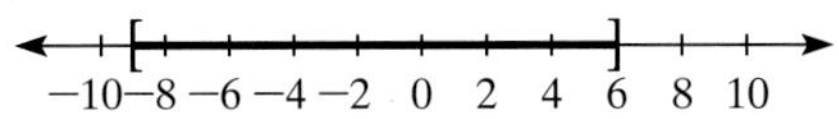

36. $\{x|x < -6 \text{ or } x > 0\}; (-\infty, -6) \cup (0, \infty)$

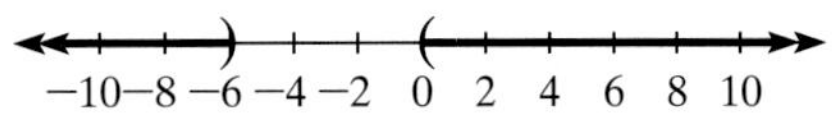

37. $\{x|-3 < x < -2\}; (-3, -2)$

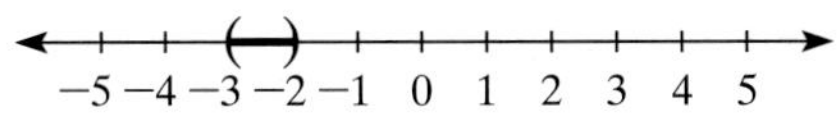

38. $\varnothing$

39. Approximately 90% of the population sleeps between 5.5 hours and 7.5 hours daily.

40.

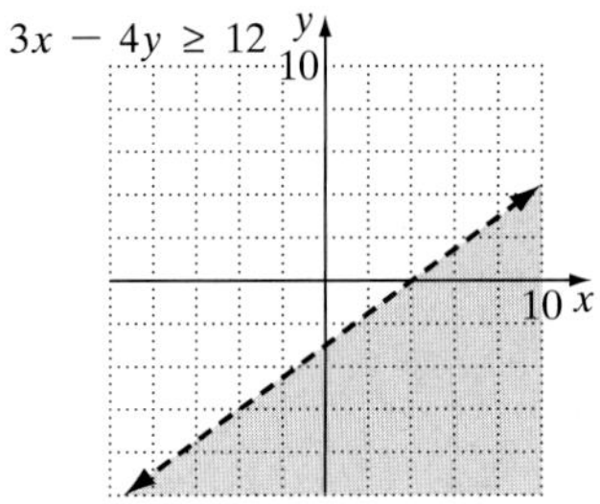

41.

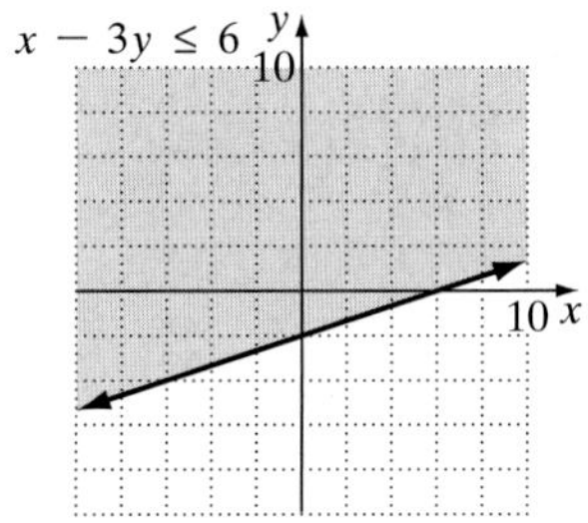

42.

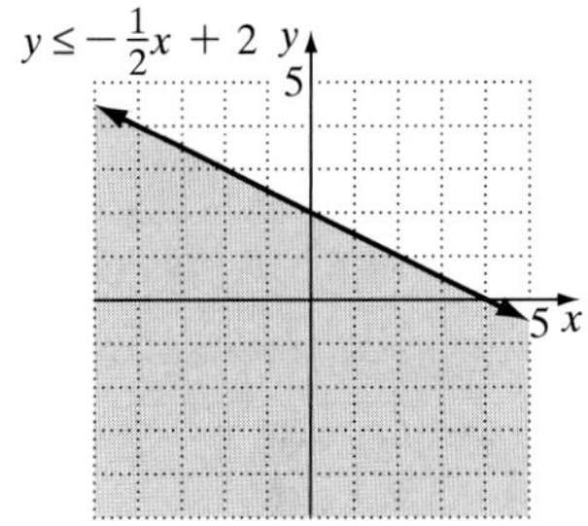

43.

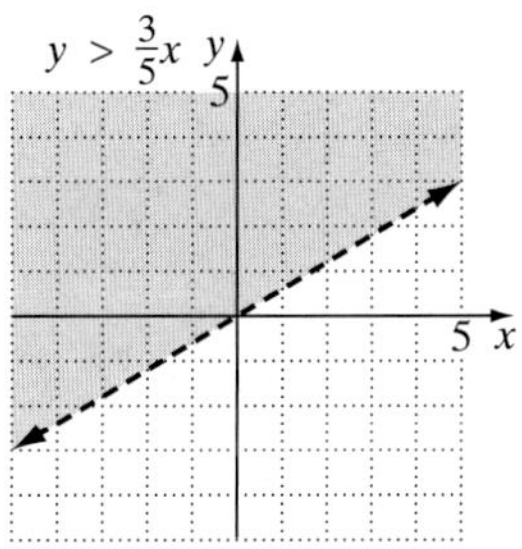

44.

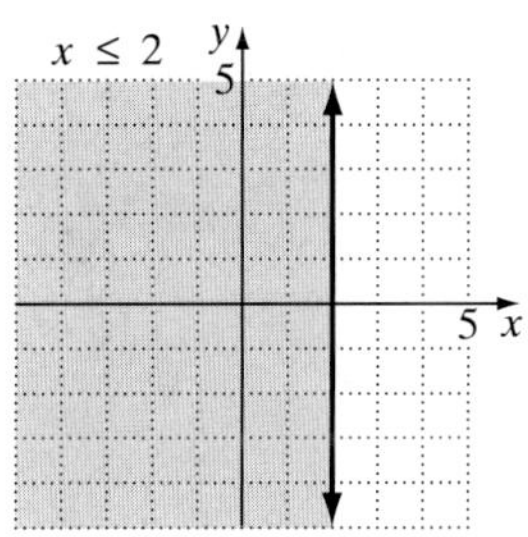

45.

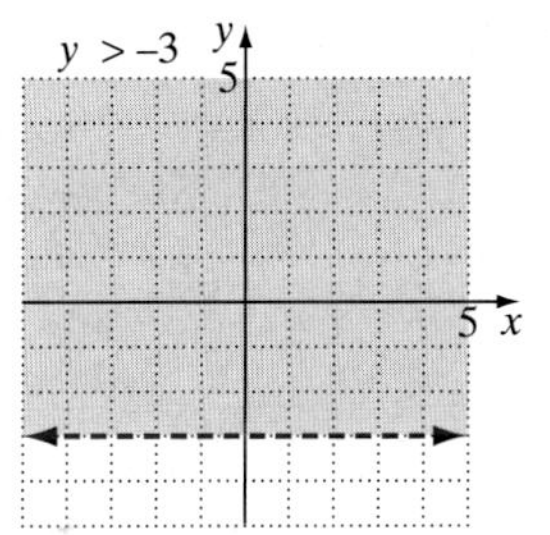

46.

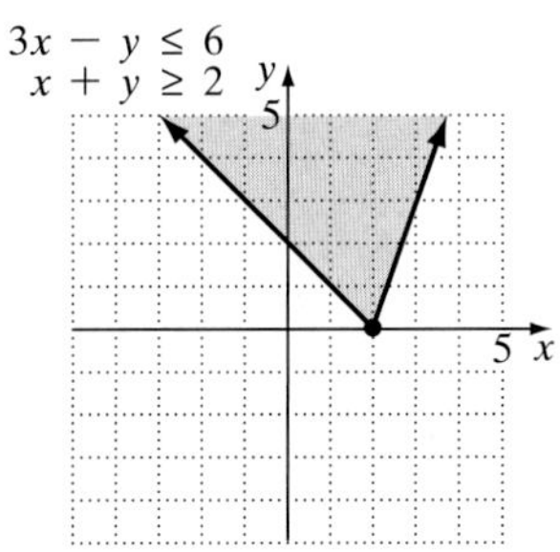

47.

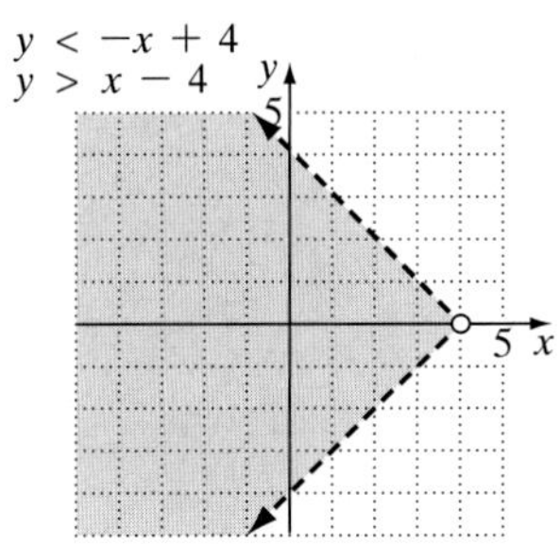

48.

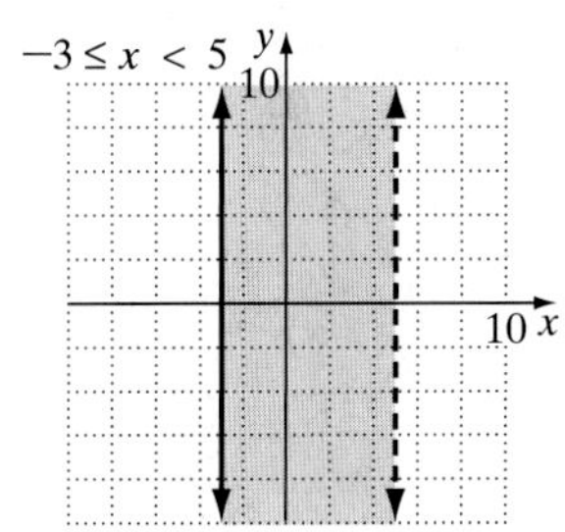

49.

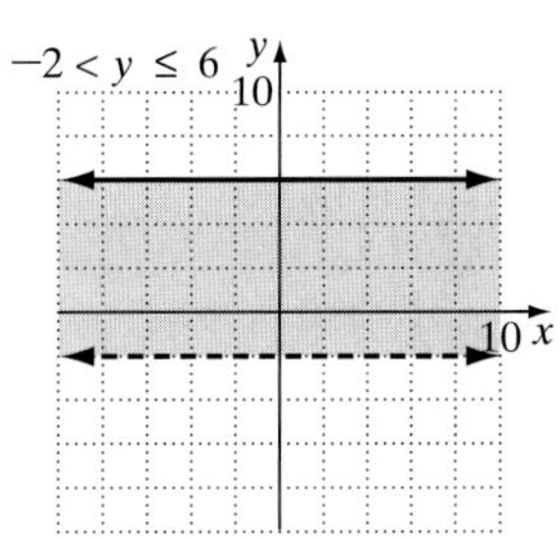

50.

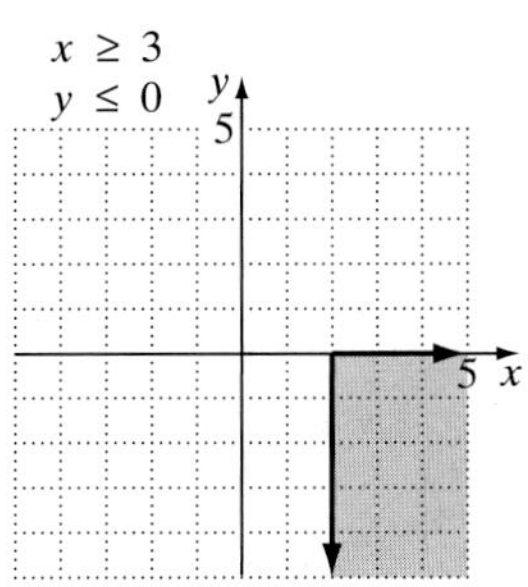

51.

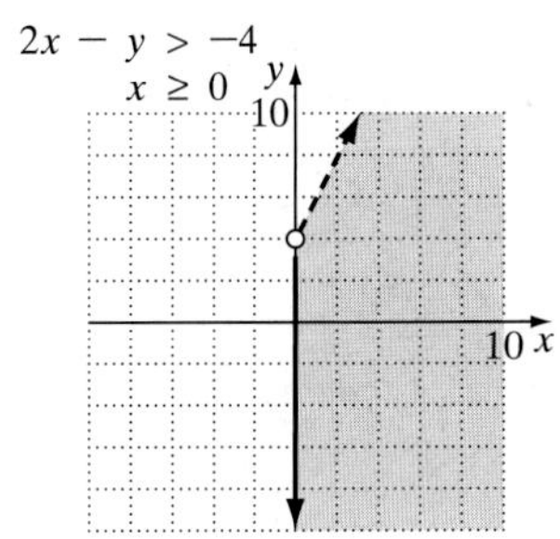

52. $x + y \le 6$
$y \ge 2x - 3$

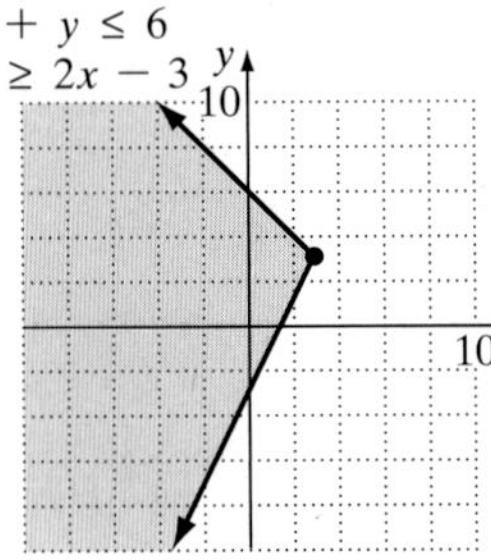

53. $3x + 2y \ge 4$
$x - y \le 3$
$x \ge 0,\ y \ge 0$

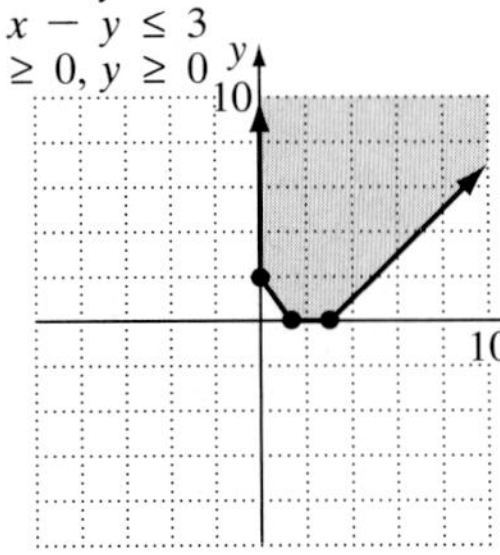

54. $\varnothing$

55. $\left(\frac{1}{2}, \frac{1}{2}\right): \frac{5}{2}$; $(2, 2)$: 10; $(4, 0)$: 8; $(1, 0)$: 2; maximum: 10; minimum: 2

56.

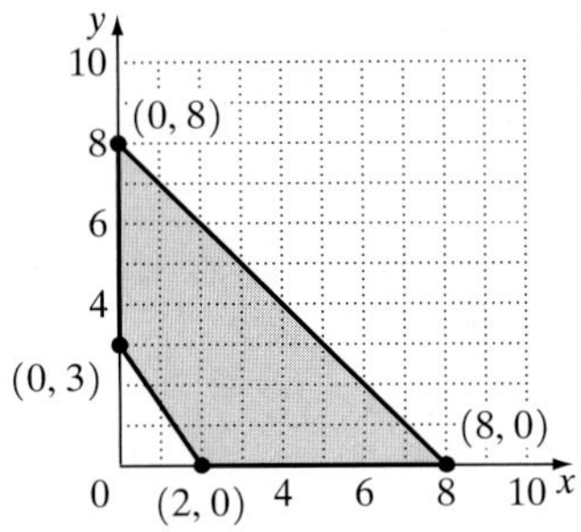

maximum: 24

57.

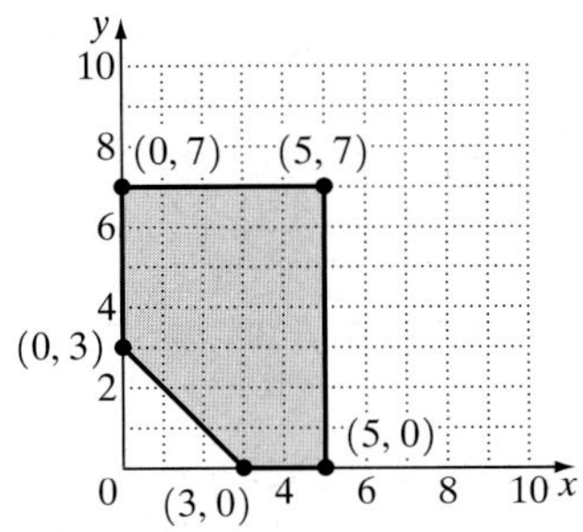

maximum: 33

58.

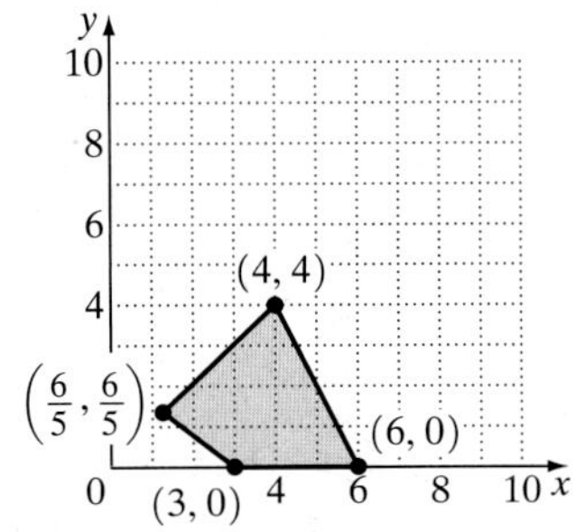

maximum: 44

59. a. $z = 500x + 350y$ **b.** $x + y \le 200$; $x \ge 10$; $y \ge 80$

c.

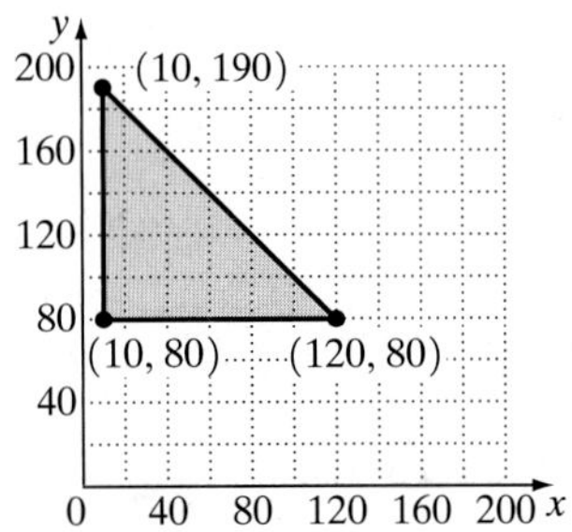

d. (10, 80): 33,000; (10, 190): 71,500; (120, 80): 88,000

e. 120; 80; 88,000

60. 480 of model A and 240 of model B

Chapter 4 Test

1. $\{x \mid -3 \le x < 2\}$

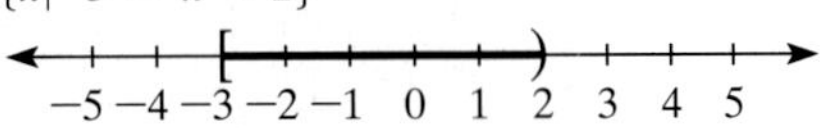

2. $\{x \mid x \le -1\}$

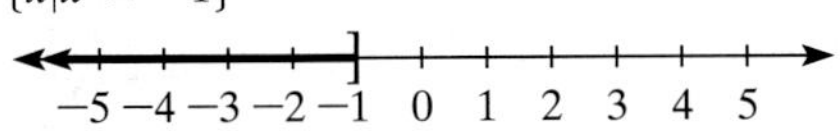

3. $\{x \mid x \le 12\}$; $(-\infty, 12]$

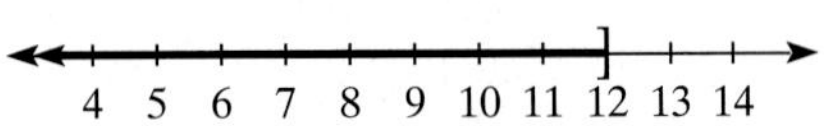

4. $\left\{x \,\middle|\, x \ge \frac{21}{8}\right\}$; $\left[\frac{21}{8}, \infty\right)$

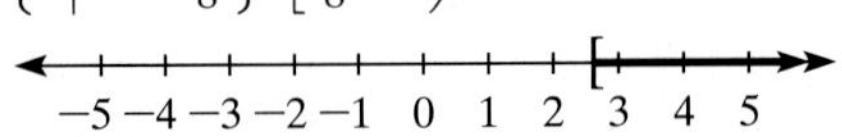

5. more than 200 calls **6.** $\{4, 6\}$ **7.** $\{2, 4, 6, 8, 10, 12, 14\}$

8. $\{x \mid -2 < x < -1\}$; $(-2, -1)$

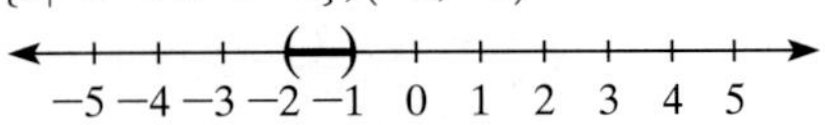

9. $\{x \mid x \ge -2\}$; $[-2, \infty)$

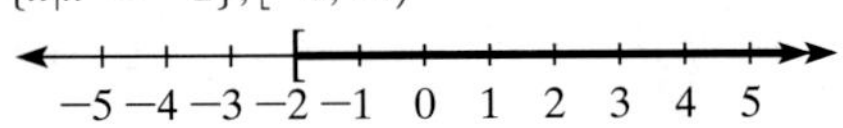

10. $\{x \mid x < 4\}$; $(-\infty, 4)$

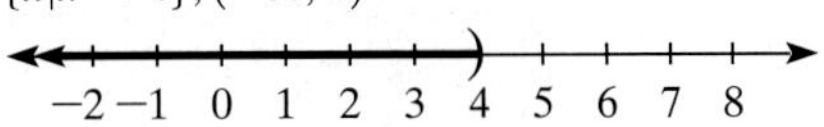

11. $\{x \mid x \le -4 \text{ or } x > 2\}$; $(-\infty, -4] \cup (2, \infty)$

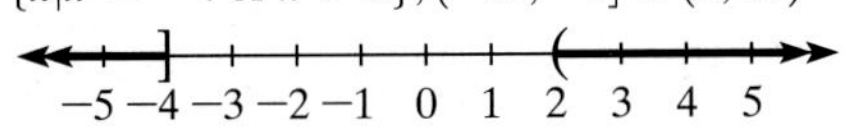

12. $\left\{x \,\middle|\, -7 \le x < \frac{13}{2}\right\}$; $\left[-7, \frac{13}{2}\right)$

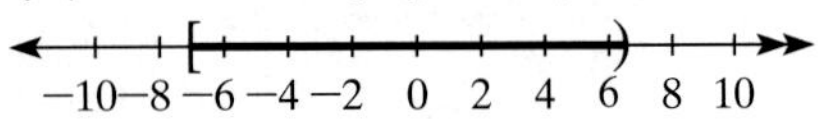

13. $\left\{-2, \frac{4}{5}\right\}$

14. $\left\{-\frac{8}{5}, 7\right\}$

15. $\{x|-3 < x < 4\}$; $(-3, 4)$

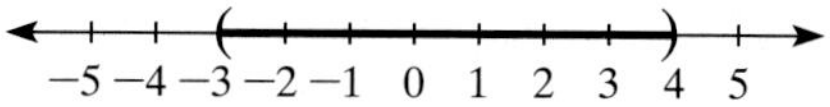

16. $\{x|x \le -1 \text{ or } x \ge 4\}$; $(-\infty, -1] \cup [4, \infty)$

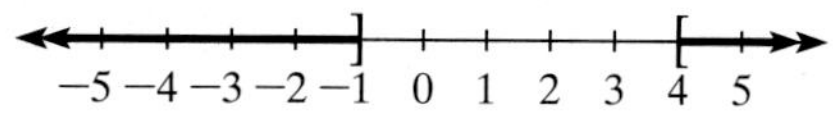

17. $\{T|66 \le T \le 82\}$ or $[66, 82]$; The monthly average temperature for Miami, FL is between a low of 66°F and a high of 82°F.

18. $3x - 2y < 6$

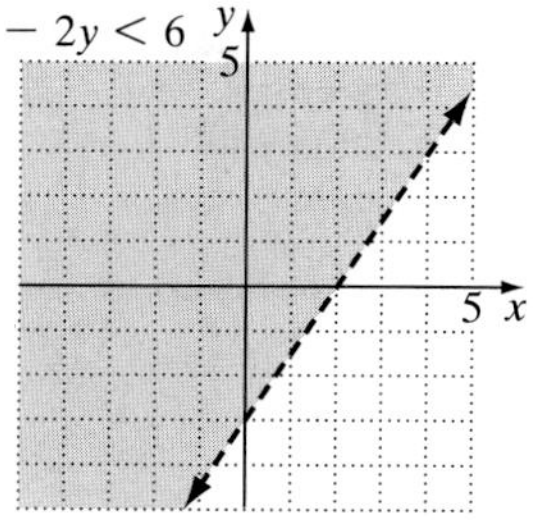

19. $y \ge \frac{1}{2}x - 1$

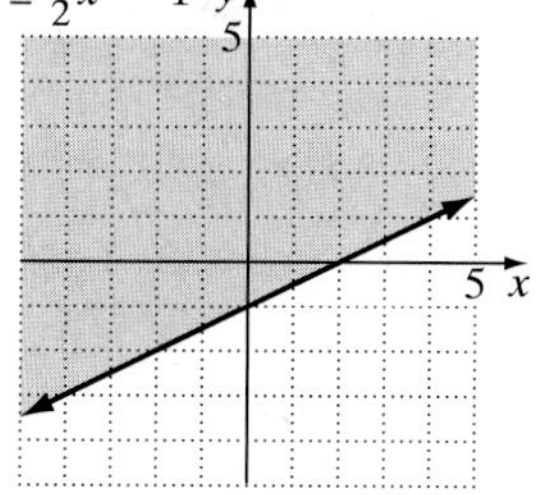

20. $y \le -1$

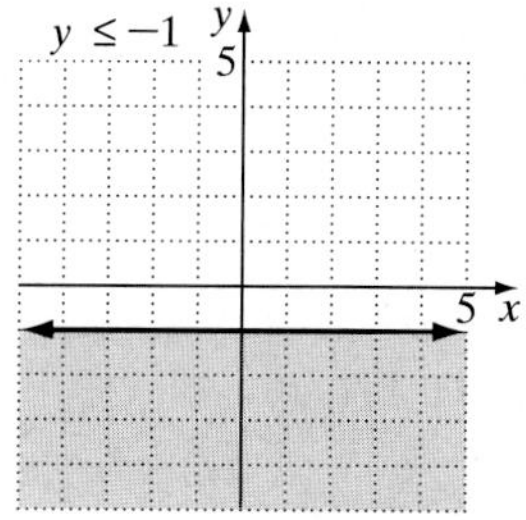

21. $x + y \ge 2$, $x - y \ge 4$

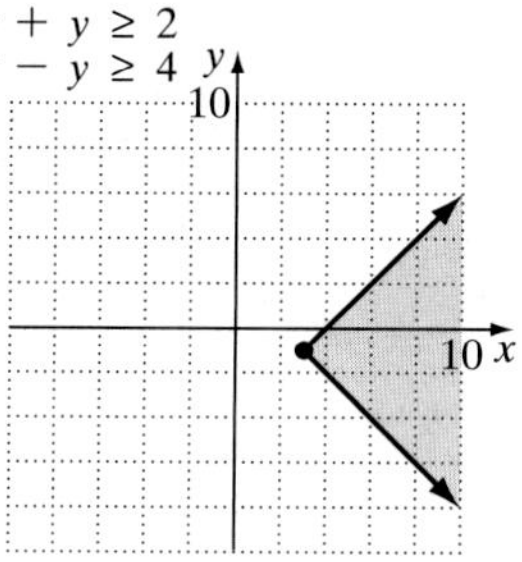

22. $3x + y \le 9$, $2x + 3y \ge 6$, $x \ge 0$, $y \ge 0$

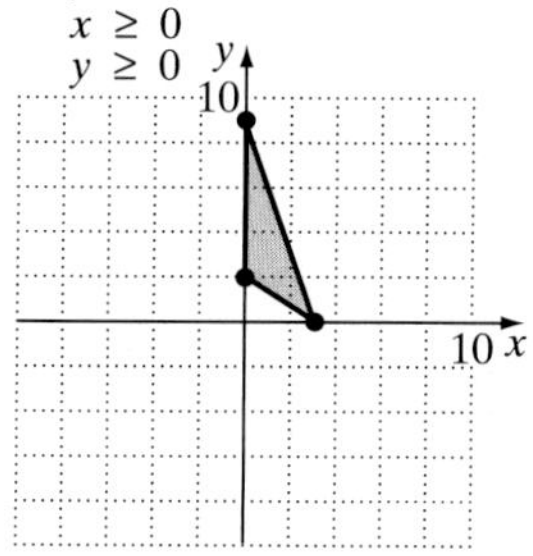

23. $-2 < x \le 4$

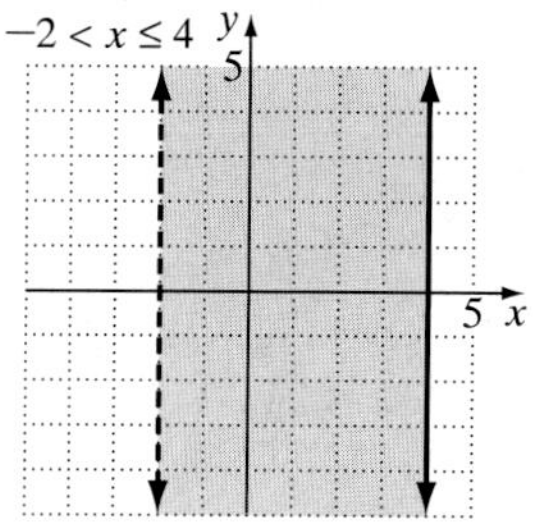

24. maximum: 26 **25.** 50 regular and 100 deluxe; $35,000

Cumulative Review Exercises (Chapters 1–4)

1. -1 **2.** 8 **3.** $-\frac{2y^7}{3x^5}$ **4.** 22; $4a^2 - 6a + 4$ **5.** $2x^2 + x + 2$; 12 **6.** $f(x) = -\frac{1}{2}x + 4$

7. $f(x) = 2x + 1$

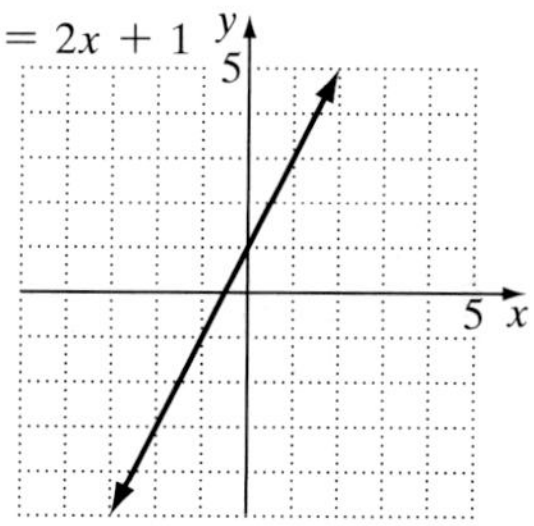

8. $y > 2x$

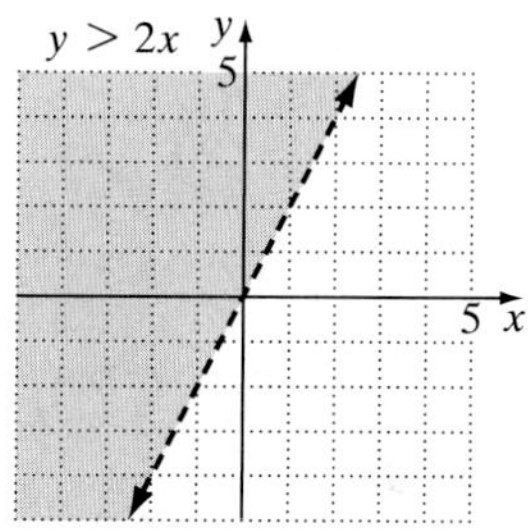

9. $2x - y \ge 6$

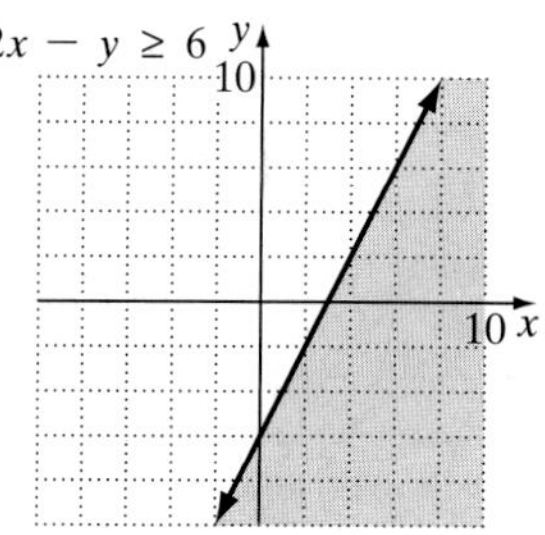

10. $f(x) = -1$

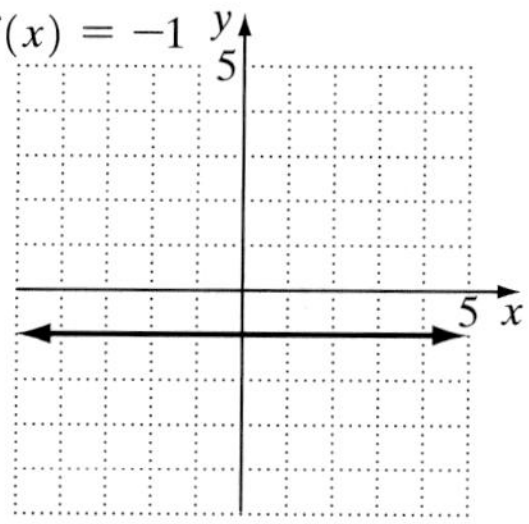

11. $(-4, 2, -1)$ **12.** $(-1, 2)$ **13.** -17

14. 46 rooms with kitchen facilities and 14 without kitchen facilities

15. a. and b. are functions.

16. $\{x|x \ge -7\}$; $[-7, \infty)$;

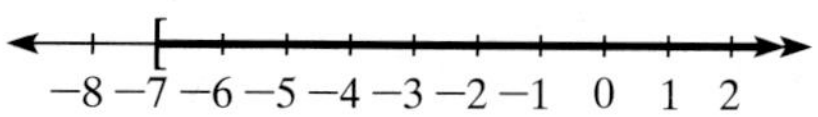

17. $\{x|x < -6\}$; $(-\infty, -6)$;

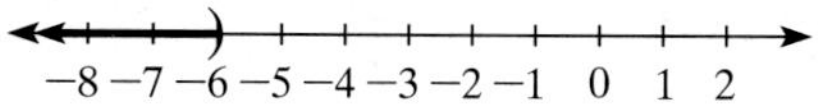

18. $\{x|x \le 3 \text{ or } x \ge 5\}$; $(-\infty, 3] \cup [5, \infty)$;

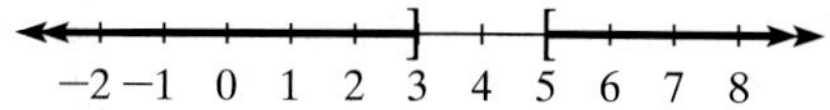

19. $\{x|-10 \le x \le 7\}$; $[-10, 7]$

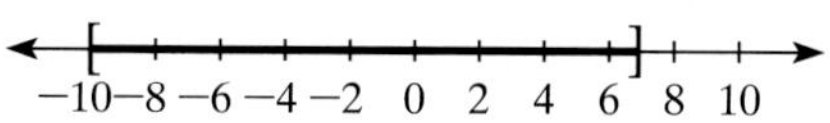

20. $\left\{x \middle| x < \frac{1}{3} \text{ or } x > 5\right\}$; $\left(-\infty, \frac{1}{3}\right) \cup (5, \infty)$

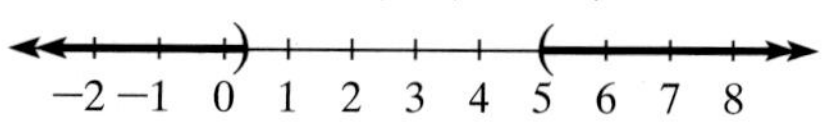

CHAPTER 5

Section 5.1

Check Point Exercises

1.

Term	Coefficient	Degree
$8x^4y^5$	8	9
$-7x^3y^2$	-7	5
$-x^2y$	-1	3
$-5x$	-5	1
11	11	0

The degree of the polynomial is 9, the leading term is $8x^4y^5$, and the leading coefficient is 8.

2. 16 **3.** The graph rises to the left and to the right.
4. This would not be appropriate over long time periods. Since the graph falls to the right, at some point the ratio would be negative which is not possible.
5. The graph does not show the end behavior of the function. The graph should fall to the left.
6. $-3x^3 + 10x^2 - 10$ **7.** $9xy^3 + 3xy^2 - 15y - 9$
8. $10x^3 - 2x^2 + 8x - 10$ **9.** $13x^2y^5 + 2xy^3 - 10$

Exercise Set 5.1

1.

Term	Coefficient	Degree
$-x^4$	-1	4
x^2	1	2

The degree of the polynomial is 4, the leading term is $-x^4$, and the leading coefficient is -1.

3.

Term	Coefficient	Degree
$5x^3$	5	3
$7x^2$	7	2
$-x$	-1	1
9	9	0

The degree of the polynomial is 3, the leading term is $5x^3$, and the leading coefficient is 5.

5.

Term	Coefficient	Degree
$3x^2$	3	2
$-7x^4$	-7	4
$-x$	-1	1
6	6	0

The degree of the polynomial is 4, the leading term is $-7x^4$, and the leading coefficient is -7.

7.

Term	Coefficient	Degree
x^3y^2	1	5
$-5x^2y^7$	-5	9
$6y^2$	6	2
-3	-3	0

The degree of the polynomial is 9, the leading term is $-5x^2y^7$, and the leading coefficient is -5.

9.

Term	Coefficient	Degree
x^5	1	5
$3x^2y^4$	3	6
$7xy$	7	2
$9x$	9	1
-2	-2	0

The degree of the polynomial is 6, the leading term is $3x^2y^4$, and the leading coefficient is 3.

11. 0 **13.** 12 **15.** 56
17. -29 **19.** -1
21–23. Graph #23 is not that of a polynomial function.
25. falls to the left and falls to the right; graph (b)
27. rises to the left and rises to the right; graph (a)
29. $11x^3 + 7x^2 - 12x - 4$

31. $-\frac{2}{5}x^4 + x^3 + \frac{3}{8}x^2$ **33.** $9x^2y - 6xy$ **35.** $2x^2y + 15xy + 15$ **37.** $-9x^4y^2 - 6x^2y^2 - 5x^2y + 2xy$
39. $12x^3 + 4x^2 + 12x - 14$ **41.** $22y^5 + 9y^4 + 7y^3 - 13y^2 + 3y - 5$ **43.** $-5x^3 + 8xy - 9y^2$ **45.** $x^4y^2 + 8x^3y + y - 6x$
47. $8a^2b^4 + 3ab^2 + 8ab$ **49.** $5x^3 + 3x^2y - xy^2 - 4y^3$ **51.** 8.14; There are 8.14 deaths per year per thousand people among 50 year olds. **53.** (50, 8.14) **55.** 57 years old **57.** 55 human years **59.** 2 dog years **61.** falls to the right; The viral particles eventually decrease as the days increase. **63.** falls to the right; Eventually the elk population will be extinct. **65–75.** Answers will vary.
77. Answers will vary; an example is $-x^2 + x + 1$. **79.** Answers will vary; an example is $x^3 + x + 1$.

81.

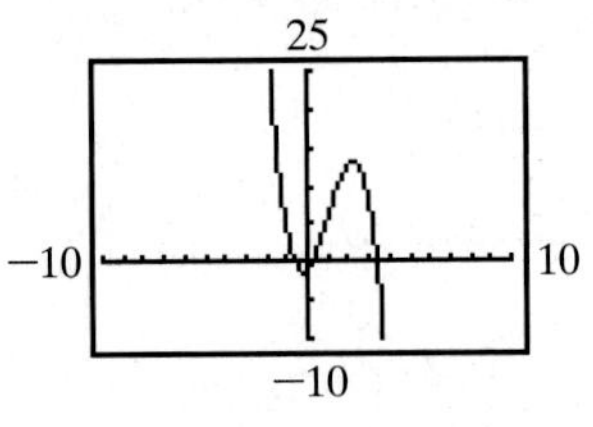

83.

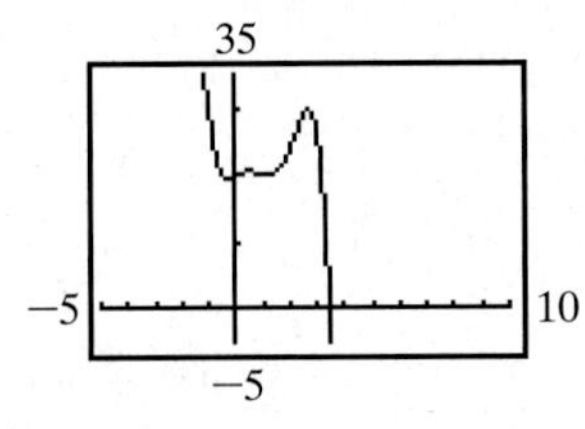

85.

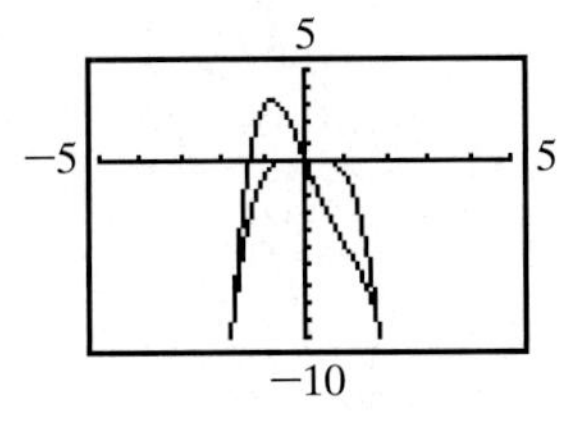

87. $3x^{2n} - x^n + 4$ **89.** $9x^2 - 3x + 5$ **90.** $\frac{2}{3}$ **91.**

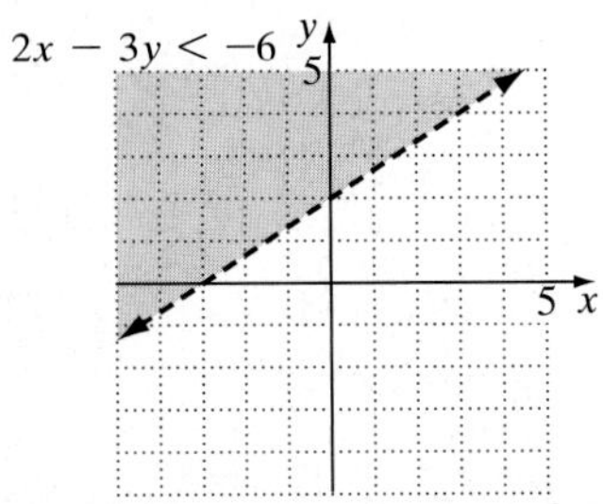

92. $y - 5 = 3(x + 2)$; $y = 3x + 11$ or $f(x) = 3x + 11$

Section 5.2

Check Point Exercises

1. a. $-18x^7y^{11}$ **b.** $30x^{10}y^6z^8$ **2. a.** $12x^9 - 18x^6 + 24x^4$ **b.** $10x^5y^9 - 8x^7y^7 - 10x^4y^3$ **3.** $5x^3 - 18x^2 + 7x + 6$ **4.** $12x^2y^6 - 8x^2y^4 + 6xy^5 + 2y^2$ **5. a.** $x^2 + 8x + 15$ **b.** $14x^2 + xy - 4y^2$ **c.** $4x^6 - 12x^4 - 5x^3 + 15x$ **6. a.** $x^2 + 20x + 100$ **b.** $16x^2 + 40xy + 25y^2$ **7. a.** $x^2 - 10x + 25$ **b.** $4x^2 - 24xy^4 + 36y^8$ **8. a.** $x^2 - 9$ **b.** $25x^2 - 49y^2$ **c.** $25a^2b^4 - 16a^2$ **9. a.** $9x^2 + 12x + 4 - 25y^2$ **b.** $4x^2 + 4xy + y^2 + 12x + 6y + 9$ **10. a.** $x^2 - 10x + 21$ **b.** 5 **11. a.** $a^2 + a - 2$ **b.** $2ah + h^2 - 5h$

Exercise Set 5.2

1. $15x^6$ **3.** $15x^3y^{11}$ **5.** $-6x^2y^9z^9$ **7.** $12x^3 + 8x^2$ **9.** $2y^3 - 10y^2$ **11.** $10x^8 - 20x^5 + 45x^3$ **13.** $28x^2y + 12xy^2$ **15.** $18a^3b^5 + 15a^2b^3$ **17.** $-12x^6y^3 + 28x^3y^4 - 24x^2y$ **19.** $x^3 - x^2 - x - 15$ **21.** $x^3 - 1$ **23.** $a^3 - b^3$ **25.** $x^4 + 5x^3 + x^2 - 11x + 4$ **27.** $x^3 - 4x^2y + 4xy^2 - y^3$ **29.** $x^3y^3 + 8$ **31.** $x^2 + 11x + 28$ **33.** $y^2 - y - 30$ **35.** $10x^2 + 11x + 3$ **37.** $6y^2 - 11y + 4$ **39.** $15x^2 - 22x + 8$ **41.** $2x^2 + xy - 21y^2$ **43.** $14x^2y^2 - 19xy - 3$ **45.** $x^3 - 5x - 4x^2 + 20$ or $x^3 - 4x^2 - 5x + 20$ **47.** $8x^5 - 40x^3 + 3x^2 - 15$ **49.** $x^2 + 6x + 9$ **51.** $y^2 - 10y + 25$ **53.** $4x^2 + 4xy + y^2$ **55.** $25x^2 - 30xy + 9y^2$ **57.** $4x^4 + 12x^2y + 9y^2$ **59.** $16x^2y^4 - 8x^2y^3 + x^2y^2$ **61.** $x^2 - 16$ **63.** $25x^2 - 9$ **65.** $16x^2 - 49y^2$ **67.** $y^6 - 4$ **69.** $1 - y^{10}$ **71.** $49x^2y^4 - 100y^2$ **73.** $4x^2 + 12x + 9 - 16y^2$ **75.** $x^2 + 2xy + y^2 - 9$ **77.** $25x^2 + 70xy + 49y^2 - 4$ **79.** $25y^2 - 4x^2 - 12x - 9$ **81.** $x^2 + 2xy + y^2 + 2x + 2y + 1$ **83.** $x^4 - 1$ **85. a.** $x^2 + 4x - 12$ **b.** -15 **c.** -12 **87. a.** $x^3 - 27$ **b.** -35 **c.** -27 **89. a.** $a^2 + a + 5$ **b.** $2ah + h^2 - 3h$ **91. a.** $3a^2 + 14a + 15$ **b.** $6ah + 3h^2 + 2h$ **93. a.** $x^2 + 6x + 4x + 24$ **b.** $(x + 6)(x + 4) = x^2 + 10x + 24$ **95. a.** $x^2 + 7x + 12$ **b.** $x^2 + 3x + 2$ **c.** $4x + 10$ **97. a.** $4x^2 - 36x + 80$ **b.** $4x^3 - 36x^2 + 80x$ **99–107.** Answers will vary.

109.

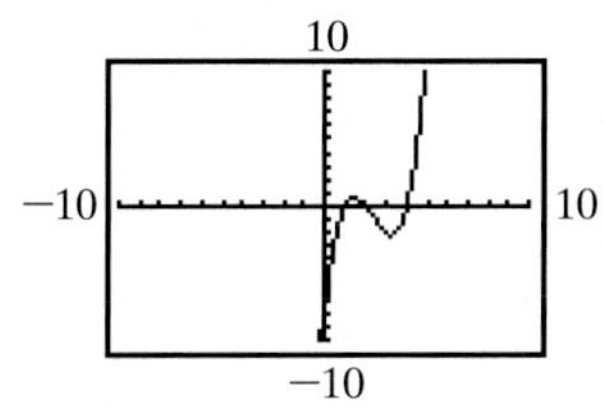

conclusion: $y_1 = y_2$

111.

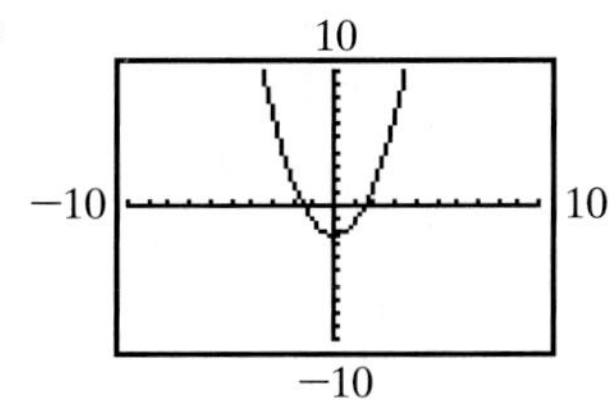

conclusion: $y_1 = y_2$

113. d **115.** $x^3 + 7x^2 - 3x$ **117.** $6y^n - 13$ **118.** $\left\{x \middle| x \le -\frac{14}{3} \text{ or } x \ge 2\right\}$ or $\left(-\infty, -\frac{14}{3}\right] \cup [2, \infty)$ **119.** $\{x|x \ge -3\}$ or $[-3, \infty)$ **120.** 8.034×10^9

Section 5.3

Check Point Exercises

1. $10x(2x + 3)$ **2. a.** $3x^2(3x^2 + 7)$ **b.** $5x^3y^2(3 - 5xy)$ **c.** $4x^2y^3(4x^2y^2 - 2xy + 1)$ **3.** $-2x(x^2 - 5x + 3)$ **4. a.** $(x - 4)(3 + 7a)$ **b.** $(a + b)(7x - 1)$ **5.** $(x - 4)(x^2 + 5)$ **6.** $(x + 5)(4x - 3y)$

Exercise Set 5.3

1. $2x(5x + 2)$ **3.** $y(y - 4)$ **5.** $x^2(x + 5)$ **7.** $4x^2(3x^2 - 2)$ **9.** $2x^2(16x^2 + x + 4)$ **11.** $2xy(2xy^2 + 3)$ **13.** $10xy^2(3xy - 1)$ **15.** $2x(6y - 3z + 2w)$ **17.** $3x^2y^4(5xy^2 - 3x^2 + 4y)$ **19.** $-4(x - 3)$ **21.** $-8(x + 6)$ **23.** $-2(x^2 - 3x + 7)$ **25.** $-5(y^2 - 8x)$ **27.** $-4x(x^2 - 8x + 5)$ **29.** $-1(x^2 + 7x - 5)$ **31.** $(x + 3)(4 + a)$ **33.** $(y - 6)(x - 7)$ **35.** $(x + y)(3x - 1)$ **37.** $(3x - 1)(4x^2 + 1)$ **39.** $(x + 3)(2x + 1)$ **41.** $(x + 3)(x + 5)$ **43.** $(x + 7)(x - 4)$ **45.** $(x - 3)(x^2 + 4)$ **47.** $(y - 6)(x + 2)$ **49.** $(y + 1)(x - 7)$ **51.** $(5x - 6y)(2x + 7y)$ **53.** $(4x - 1)(x^2 - 3)$ **55.** $(x - a)(x - b)$ **57.** $(x - 3)(x^2 + 4)$ **59.** $(a - b)(y - x)$ **61. a.** 16; The ball is 16 feet above the ground after 2 seconds. **b.** 0; The ball is on the ground after 2.5 seconds. **c.** $-8t(2t - 5)$; $f(t) = -8t(2t - 5)$ **d.** 16; 0; yes; Answers will vary. **63. a.** $(x - 0.4x)(1 - 0.4) = (0.6x)(0.6) = 0.36x$ **b.** no; 36% **65.** $A = P + Pr + (P + Pr)r = P(1 + r) + Pr(1 + r) = (1 + r)(P + Pr) = P(1 + r)(1 + r) = P(1 + r)^2$ **67.** $A = r(\pi r + 2l)$ **69–73.** Answers will vary.

75.

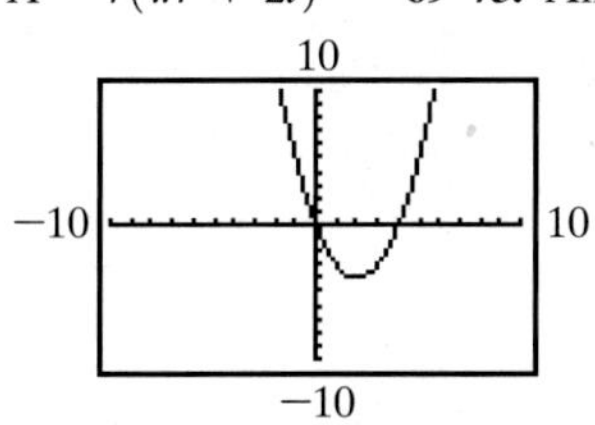

Graphs coincide.; factored correctly

77.

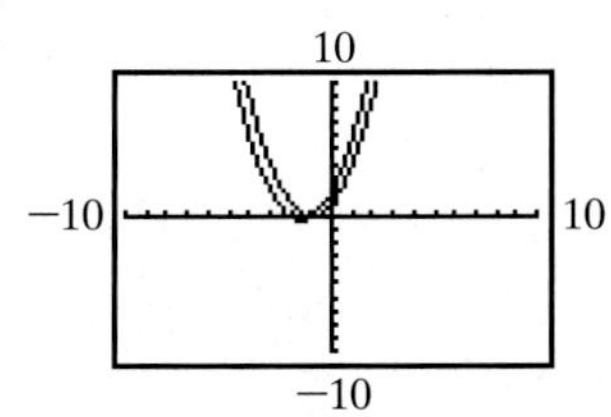

Graphs do not coincide.; $x^2 + 2x + x + 2 = (x + 2)(x + 1)$

79.

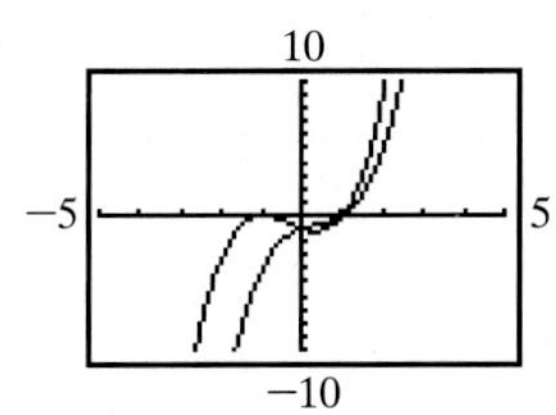

Graphs do not coincide.; $x^3 + x^2 - x - 1 = (x + 1)^2(x - 1)$

81. $x^{2n}(x^{2n} + 1 + x^n)$ **83.** $4y^{2n}(2y^4 + 4y^3 - 3)$ **85.** Answers will vary; an example is: $6x^2 - 4x + 9x - 6$.
86. $\left(\frac{20}{11}, -\frac{14}{11}\right)$ or $\left\{\left(\frac{20}{11}, -\frac{14}{11}\right)\right\}$ **87. a.** function **b.** not a function **88.** length: 8 ft; width: 3 ft

Section 5.4

Check Point Exercises

1. $(x + 4)(x + 2)$ or $(x + 2)(x + 4)$ **2.** $(x - 5)(x - 4)$ **3.** $(y + 22)(y - 3)$ **4.** $(x - 3y)(x - 2y)$
5. $3x(x - 7)(x + 2)$ **6.** $(x^3 - 5)(x^3 - 2)$ **7.** $(3x - 14)(x - 2)$ **8.** $x^4(3x - 1)(2x + 7)$ **9.** $(2x - y)(x - 3y)$
10. $(3y^2 - 2)(y^2 + 4)$ **11.** $(2x - 5)(4x - 1)$

Exercise Set 5.4

1. $(x + 3)(x + 2)$ **3.** $(x + 6)(x + 2)$ **5.** $(x + 5)(x + 4)$ **7.** $(y + 8)(y + 2)$ **9.** $(x - 3)(x - 5)$ **11.** $(y - 2)(y - 10)$
13. $(a + 7)(a - 2)$ **15.** $(x + 6)(x - 5)$ **17.** $(x + 4)(x - 7)$ **19.** $(y + 4)(y - 9)$ **21.** prime **23.** $(x - 2y)(x - 7y)$
25. $(x + 5y)(x - 6y)$ **27.** prime **29.** $(a - 10b)(a - 8b)$ **31.** $3(x + 3)(x - 2)$ **33.** $2x(x - 3)(x - 4)$
35. $3y(y - 2)(y - 3)$ **37.** $2x^2(x + 3)(x - 16)$ **39.** $(x^3 + 2)(x^3 - 3)$ **41.** $(x^2 - 6)(x^2 + 1)$ **43.** $(x + 6)(x + 2)$
45. $(3x + 5)(x + 1)$ **47.** $(x + 11)(5x + 1)$ **49.** $(y + 8)(3y - 2)$ **51.** $(y + 2)(4y + 1)$ **53.** $(2x + 3)(5x + 2)$
55. $(4x - 3)(2x - 3)$ **57.** $(y - 3)(6y - 5)$ **59.** prime **61.** $(x + y)(3x + y)$ **63.** $(2x + y)(3x - 5y)$
65. $(5x - 2y)(3x - 5y)$ **67.** $(3a - 7b)(a + 2b)$ **69.** $5x(3x - 2)(x - 1)$ **71.** $2x^2(3x + 2)(4x - 1)$ **73.** $y^3(5y + 1)(3y - 1)$
75. $3(8x + 9y)(x - y)$ **77.** $2b(a + 3)(3a - 10)$ **79.** $2y(2x - y)(3x - 7y)$ **81.** $13x^3y(y + 4)(y - 1)$
83. $(2x^2 - 3)(x^2 + 1)$ **85.** $(2x^3 + 5)(x^3 + 3)$ **87.** $(2y^5 + 3)(y^5 + 1)$ **89.** $(5x + 12)(x + 2)$ **91.** $(2x - 13)(x - 2)$
93. a. 32; The diver is 32 feet above the water after 1 second. **b.** 0; The diver hits the water after 2 seconds.
c. $-16(t - 2)(t + 1)$; $f(t) = -16(t - 2)(t + 1)$ **d.** 32; 0 **95. a.** $x^2 + x + x + x + 1 + 1 = x^2 + 3x + 2$
b. $(x + 2)(x + 1)$ **c.** Answers will vary. **97–103.** Answers will vary.

105.

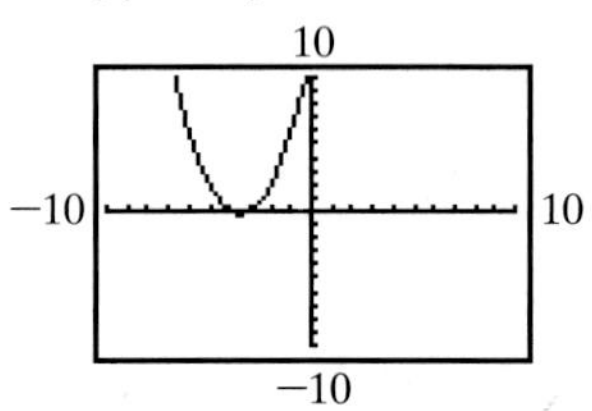

Graphs coincide.; factored correctly

107.

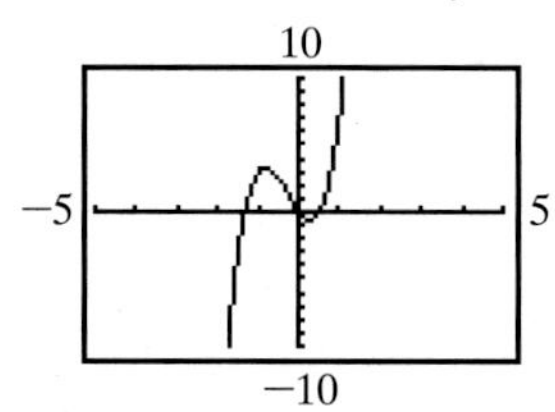

Graphs coincide.; factored correctly

109. b
111. −16, −8, 8, 16
113. $(4x^n - 5)(x^n - 1)$
114. $\{x|x > 5\}$ or $(5, \infty)$
115. $(2, -1, 3)$ or $\{(2, -1, 3)\}$

116. $x - y \geq -4$
$x + 2y \geq 2$

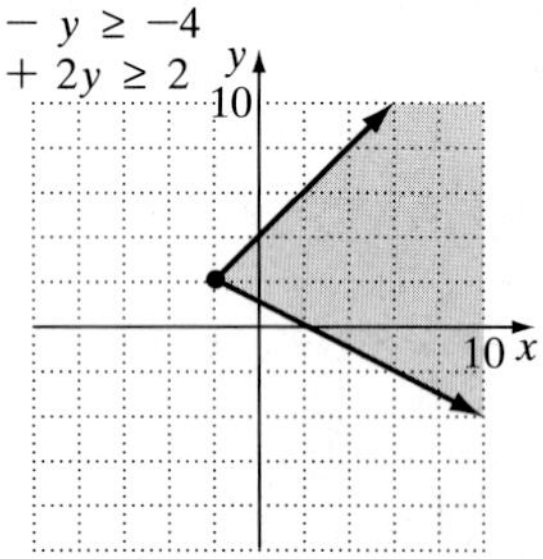

Section 5.5

Check Point Exercises

1. a. $(4x + 5)(4x - 5)$ **b.** $(10y^3 + 3x^2)(10y^3 - 3x^2)$ **2.** $6y(1 + xy^3)(1 - xy^3)$ **3.** $(4x^2 + 9)(2x + 3)(2x - 3)$
4. $(x + 7)(x + 2)(x - 2)$ **5. a.** $(x + 3)^2$ **b.** $(4x + 5y)^2$ **c.** $(2y^2 - 5)^2$ **6.** $(x + 5 + y)(x + 5 - y)$
7. $(a + b - 2)(a - b + 2)$ **8. a.** $(x + 3)(x^2 - 3x + 9)$ **b.** $(x^2 + 10y)(x^4 - 10x^2y + 100y^2)$
9. a. $(x - 2)(x^2 + 2x + 4)$ **b.** $(1 - 3xy)(1 + 3xy + 9x^2y^2)$

Exercise Set 5.5

1. $(x + 2)(x - 2)$ **3.** $(3x + 5)(3x - 5)$ **5.** $(3 + 5y)(3 - 5y)$ **7.** $(6x + 7y)(6x - 7y)$ **9.** $(xy + 1)(xy - 1)$
11. $(3x^2 + 5y^3)(3x^2 - 5y^3)$ **13.** $(x^7 + y^2)(x^7 - y^2)$ **15.** $(x - 3 + y)(x - 3 - y)$ **17.** $(a + b - 2)(a - b + 2)$
19. $2x(x + 2)(x - 2)$ **21.** $2(5 + y)(5 - y)$ **23.** $8(x + y)(x - y)$ **25.** $2xy(x + 3)(x - 3)$ **27.** $a(ab + 7c)(ab - 7c)$
29. $5y(1 + xy^3)(1 - xy^3)$ **31.** $8(x^2 + y^2)$ **33.** prime **35.** $(x^2 + 4)(x + 2)(x - 2)$ **37.** $(9x^2 + 1)(3x + 1)(3x - 1)$
39. $2x(x^2 + y^2)(x + y)(x - y)$ **41.** $(x + 3)(x + 2)(x - 2)$ **43.** $(x - 7)(x + 1)(x - 1)$ **45.** $(x + 2)^2$ **47.** $(x - 5)^2$
49. $(x^2 - 2)^2$ **51.** $(3y + 1)^2$ **53.** $(8y - 1)^2$ **55.** $(x - 6y)^2$ **57.** prime **59.** $(3x + 8y)^2$ **61.** $(x - 3 + y)(x - 3 - y)$

63. $(x + 10 + x^2)(x + 10 - x^2)$ **65.** $(3x - 5 + 6y)(3x - 5 - 6y)$ **67.** $(x^2 + x + 1)(x^2 - x - 1)$ **69.** $(z + x - 2y)(z - x + 2y)$ **71.** $(x + 4)(x^2 - 4x + 16)$ **73.** $(x - 3)(x^2 + 3x + 9)$ **75.** $(2y + 1)(4y^2 - 2y + 1)$ **77.** $(5x - 2)(25x^2 + 10x + 4)$ **79.** $(xy + 3)(x^2y^2 - 3xy + 9)$ **81.** $x(4 - x)(16 + 4x + x^2)$ **83.** $(x^2 + 3y)(x^4 - 3x^2y + 9y^2)$ **85.** $(5x^2 - 4y^2)(25x^4 + 20x^2y^2 + 16y^4)$ **87.** $(x + 1)(x^2 - x + 1)(x^6 - x^3 + 1)$ **89.** $x^2 - 16 = (x + 4)(x - 4)$ **91.** $9x^2 - 16 = (3x + 4)(3x - 4)$ **93.** $4a^3 - 4ab^2 = 4a(a + b)(a - b)$ **95–97.** Answers will vary.

99.

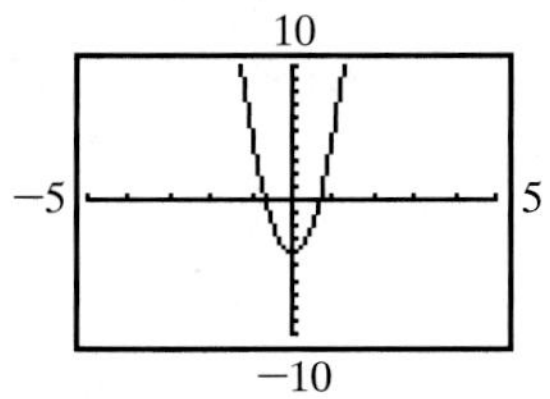

Graphs coincide.; factored correctly

101.

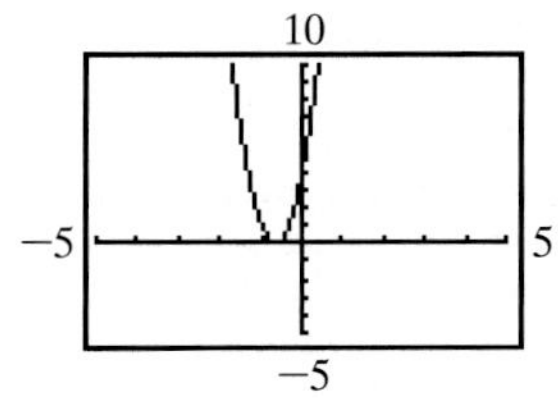

Graphs coincide.; factored correctly

103.

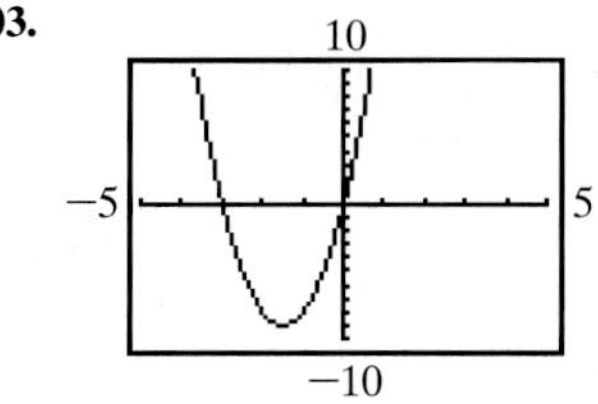

Graphs coincide.; factored correctly

105.

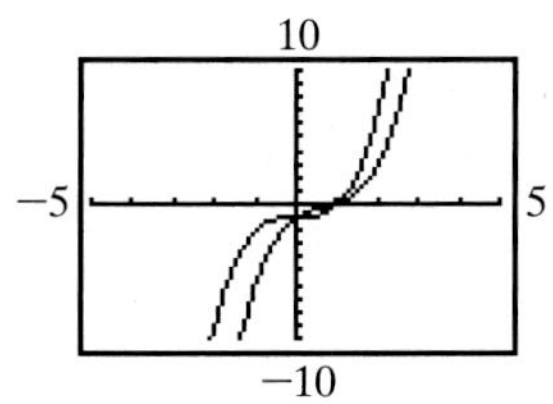

Graphs do not coincide.; $x^3 - 1 = (x - 1)(x^2 + x + 1)$

107. d **109.** $(6x^n + y^n)(6x^n - y^n)$ **111.** $(2x^n - 5y^m)^2$ **113.** 16 **115.** $\{x|5 \le x \le 11\}$ or $[5, 11]$ **116.** $(-2, 1)$ or $\{(-2, 1)\}$ **117.** $\frac{y^6}{16x^5}$

Section 5.6

Check Point Exercises

1. $3x(x - 5)^2$ **2.** $3y(x + 2)(x - 6)$ **3.** $(x + y)(4a + 5)(4a - 5)$ **4.** $(x + 10 + 6a)(x + 10 - 6a)$ **5.** $x(x + 2)(x^2 - 2x + 4)(x^6 - 8x^3 + 64)$

Exercise Set 5.6

1. $x(x + 4)(x - 4)$ **3.** $3(x + 3)^2$ **5.** $3(3x - 1)(9x^2 + 3x + 1)$ **7.** $(x + 4)(x - 4)(y - 2)$ **9.** $2b(2a + 5)(a - 3)$ **11.** $(y + 2)(y - 2)(a - 4)$ **13.** $11x(x^2 + y)(x^2 - y)$ **15.** $4x(x^2 + 4)(x + 2)(x - 2)$ **17.** $(x - 4)(x + 3)(x - 3)$ **19.** $2x^2(x + 3)(x^2 - 3x + 9)$ **21.** $3y(x^2 + 4y^2)(x + 2y)(x - 2y)$ **23.** $3x(2x + 3y)^2$ **25.** $(x - 6 + 7y)(x - 6 - 7y)$ **27.** prime **29.** $12xy(x + y)(x - y)$ **31.** $6b(x^2 + y^2)$ **33.** $(x + y)(x - y)(x^2 + xy + y^2)$ **35.** $(x + 6 + 2a)(x + 6 - 2a)$ **37.** $(x^3 - 2)(x^3 + 7)$ **39.** $(2x - y)(2 + x^2)$ **41.** $2(3x - 2y)(9x^2 + 6xy + 4y^2)$ **43.** $(x + 5 + y)(x + 5 - y)$ **45.** $(x^4 + y^4)(x^2 + y^2)(x + y)(x - y)$ **47.** $xy(x + 4y)(x - 4y)$ **49.** $x(1 + 2x)(1 - 2x + 4x^2)$ **51.** $2(4y + 1)(2y - 1)$ **53.** $y(14y^2 + 7y - 10)$ **55.** $3(3x + 2y)^2$ **57.** $3x(4x^2 + y^2)$ **59.** $x^3y^3(xy - 1)(x^2y^2 + xy + 1)$ **61. a.** $x(x + y) - y(x + y)$ **b.** $(x + y)(x - y)$ **63. a.** $xy + xy + xy + 3x(x) = 3xy + 3x^2$ **b.** $3x(y + x)$ **65.** Answers will vary.

67.

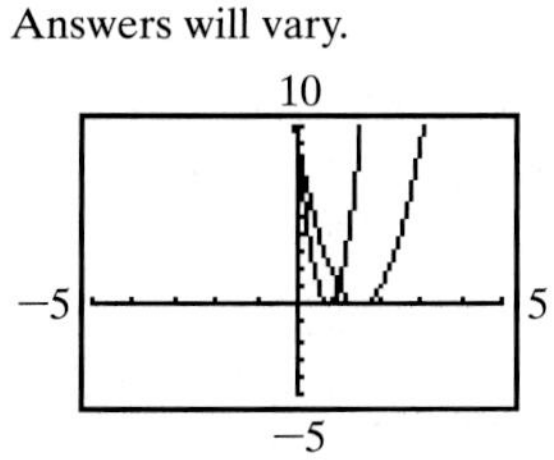

Graphs do not coincide.; $4x^2 - 12x + 9 = (2x - 3)^2$

69.

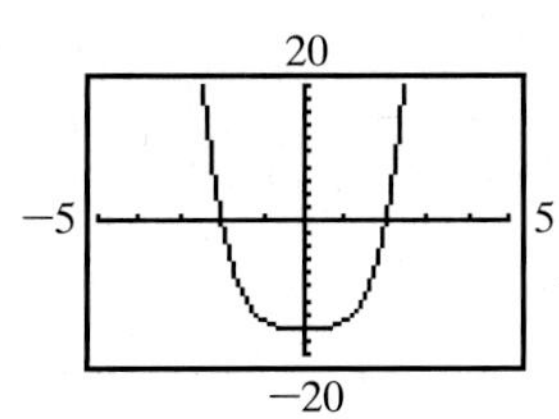

Graphs coincide.; factored correctly

71. b **73.** $(x + 4)(x - 4)(x + 2)(x - 2)$ **75.** $x(x^{2n} + y^{2n})(x^n + y^n)(x^n - y^n)$ **77.** $(x^2 + 2x + 2)(x^2 - 2x + 2)$ **79.** -1 or $\{-1\}$ **80.** $\frac{8x^9}{y^3}$ **81.** 52

Section 5.7

Check Point Exercises

1. $-\frac{1}{2}$ and 5, or $\left\{-\frac{1}{2}, 5\right\}$ **2. a.** 0 and $\frac{2}{3}$, or $\left\{0, \frac{2}{3}\right\}$ **b.** 5 or $\{5\}$ **c.** -4 and 3, or $\{-4, 3\}$ **3.** $-\frac{3}{2}$, -2, and 2, or $\left\{-\frac{3}{2}, -2, 2\right\}$ **4.** after 1 second and 2 seconds; $(1, 192)$ and $(2, 192)$ **5.** 2 ft **6.** 5 ft

Exercise Set 5.7

1. −4 and 3, or $\{-4, 3\}$ **3.** −7 and 1, or $\{-7, 1\}$ **5.** −4 and $\frac{2}{3}$, or $\left\{-4, \frac{2}{3}\right\}$ **7.** $\frac{3}{5}$ and 1, or $\left\{\frac{3}{5}, 1\right\}$ **9.** −2 and $\frac{1}{3}$, or $\left\{-2, \frac{1}{3}\right\}$ **11.** 0 and 8, or $\{0, 8\}$ **13.** 0 and $\frac{5}{3}$, or $\left\{0, \frac{5}{3}\right\}$ **15.** −2 or $\{-2\}$ **17.** 7 or $\{7\}$ **19.** $\frac{5}{3}$ or $\left\{\frac{5}{3}\right\}$ **21.** −5 and 5, or $\{-5, 5\}$ **23.** $-\frac{10}{3}$ and $\frac{10}{3}$, or $\left\{-\frac{10}{3}, \frac{10}{3}\right\}$ **25.** −3 and 6, or $\{-3, 6\}$ **27.** −3 and −2, or $\{-3, -2\}$ **29.** 4 or $\{4\}$ **31.** −2 and 8, or $\{-2, 8\}$ **33.** −1 and $\frac{2}{5}$, or $\left\{-1, \frac{2}{5}\right\}$ **35.** −6 and −3, or $\{-6, -3\}$ **37.** −5, −4, and 5, or $\{-5, -4, 5\}$ **39.** −5, 1, and 5, or $\{-5, 1, 5\}$ **41.** −4, 0, and 4, or $\{-4, 0, 4\}$ **43.** 0 and 2, or $\{0, 2\}$ **45.** −5, −3, and 0, or $\{-5, -3, 0\}$ **47.** 2 and 4; d **49.** −4 and −2; c

51.

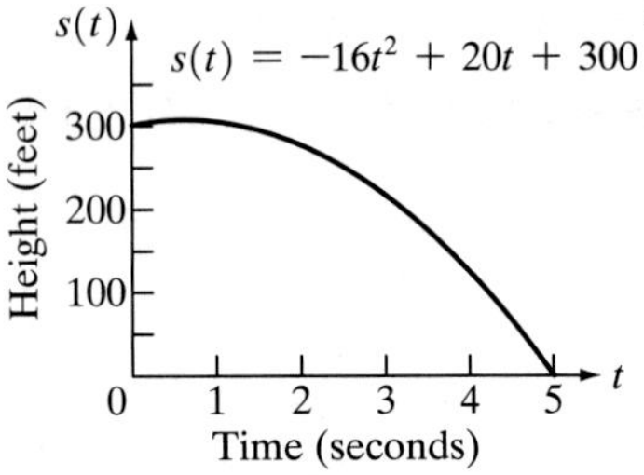

; 5 seconds

53. after 2 seconds; (2, 276) **55.** 1994 **57.** (4, 27); This is the highest point on the graph; it signifies that during this time period the greatest number of recipients was 27 million in 1994. **59.** length: 9 feet; width: 6 ft **61.** 5 in. **63.** 5 m **65. a.** $4x^2 + 44x$ **b.** 3 ft **67.** length: 10 in.; width: 10 in. **69.** length: 12 ft; width: 5 ft **71.** $30\frac{1}{8}$ ft **73–83.** Answers will vary.

85.

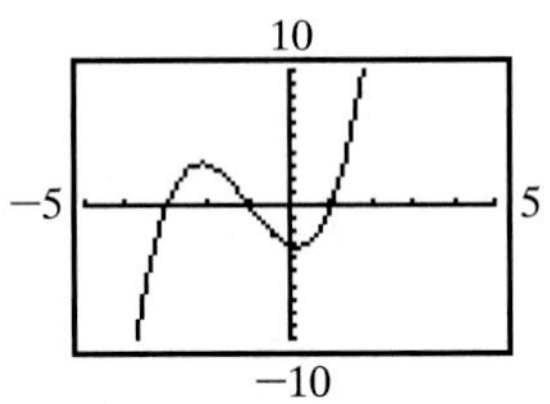

−3, −1, and 1, or $\{-3, -1, 1\}$

87.

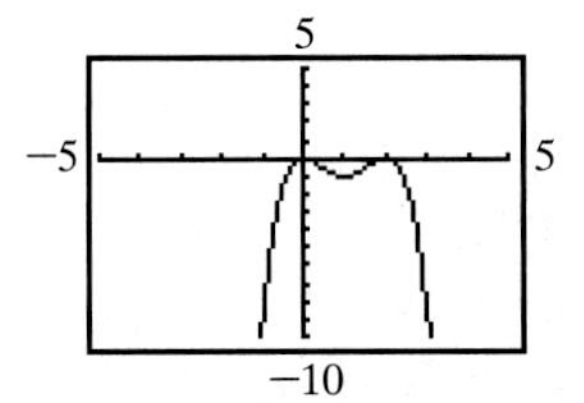

0 and 2, or $\{0, 2\}$

89. $\{x|x$ is a real number and $x \neq -9$ and $x \neq 5\}$ or $(-\infty, -9) \cup (-9, 5) \cup (5, \infty)$ **91.** Answers will vary; an example is: $x^2 - 4x - 21$. **93.** −2 and $\frac{10}{3}$, or $\left\{-2, \frac{10}{3}\right\}$ **94.** 12 **95.** \$1700 at 5% and \$1300 at 8%

Review Exercises

1.

Term	Coefficient	Degree
$-5x^3$	−5	3
$7x^2$	7	2
$-x$	−1	1
2	2	0

The degree of the polynomial is 3, the leading term is $-5x^3$, and the leading coefficient is −5.

2.

Term	Coefficient	Degree
$8x^4y^2$	8	6
$-7xy^6$	−7	7
$-x^3y$	−1	4

The degree of the polynomial is 7, the leading term is $-7xy^6$, and the leading coefficient is −7.

3. −31 **4.** 63.3; There were 63.3 thousand, or 63,000, people waiting for transplants in the United States in 2000. **5.** rises to the left and falls to the right; c **6.** rises to the left and to the right; b **7.** falls to the left and rises to the right; a **8.** falls to the left and to the right; d **9.** The graph of this function falls to the right which indicates that the percentage would eventually be negative, which is not possible. **10.** $x^3 - 6x^2 - x - 9$ **11.** $12x^3y - 2x^2y - 14y - 17$ **12.** $15x^3 - 10x^2 + 11x - 8$ **13.** $-2x^3y^2 - 4x^3y - 8$ **14.** $3x^3 + 5x^2y - xy^2 - 8y^3$ **15.** $-12x^6y^2z^7$ **16.** $2x^8 - 24x^5 - 12x^3$ **17.** $21x^5y^4 - 35x^2y^3 - 7xy^2$ **18.** $6x^3 + 29x^2 + 27x - 20$ **19.** $x^4 + 4x^3 + 4x^2 - x - 2$ **20.** $12x^2 - 23x + 5$ **21.** $15x^2y^2 + 2xy - 8$ **22.** $9x^2 + 42xy + 49y^2$ **23.** $x^4 - 10x^2y + 25y^2$ **24.** $4x^2 - 49y^2$ **25.** $9x^2y^4 - 16x^2$ **26.** $x^2 + 6x + 9 - 25y^2$ **27.** $x^2 + 2xy + y^2 + 8x + 8y + 16$ **28.** $2x^2 - x - 15$; 21 **29. a.** $a^2 - 9a + 10$ **b.** $2ah + h^2 - 7h$ **30.** $8x^2(2x + 3)$ **31.** $2x(1 - 18x)$ **32.** $7xy(3xy - 2y + 1)$ **33.** $9x^2y(2xy - 3)$ **34.** $-4(3x^2 - 2x + 12)$ **35.** $-1(x^2 + 11x - 14)$ **36.** $(x - 1)(x^2 - 2)$ **37.** $(y - 3)(x - 5)$ **38.** $(x - 3y)(5a + 2b)$ **39.** $(x + 5)(x + 3)$ **40.** $(x + 20)(x - 4)$ **41.** $(x + 17y)(x - y)$ **42.** $3x(x - 1)(x - 11)$ **43.** $(x + 7)(3x + 1)$ **44.** $(3x - 2)(2x - 3)$ **45.** $(5x + 4y)(x - 2y)$ **46.** $x(3x + 4)(2x - 1)$ **47.** $(x + 3)(2x + 5)$ **48.** $(x^3 + 6)(x^3 - 5)$ **49.** $(x^2 + 3)(x^2 - 13)$ **50.** $(x + 11)(x + 9)$ **51.** $(5x^3 + 2)(x^3 + 3)$ **52.** $(2x + 5)(2x - 5)$ **53.** $(1 + 9xy)(1 - 9xy)$ **54.** $(x^4 + y^3)(x^4 - y^3)$ **55.** $(x - 1 + y)(x - 1 - y)$ **56.** $(x + 8)^2$ **57.** $(3x - 1)^2$ **58.** $(5x + 2y)^2$ **59.** prime **60.** $(5x - 4y)^2$ **61.** $(x + 9 + y)(x + 9 - y)$ **62.** $(z + 5x - 1)(z - 5x + 1)$ **63.** $(4x + 3)(16x^2 - 12x + 9)$ **64.** $(5x - 2)(25x^2 + 10x + 4)$ **65.** $(xy + 1)(x^2y^2 - xy + 1)$ **66.** $3x(5x + 1)$ **67.** $3x^2(2x + 1)(2x - 1)$ **68.** $4x(5x^3 - 6x^2 + 7x - 3)$ **69.** $x(x - 2)(x - 13)$ **70.** $-2y^2(y - 3)(y - 9)$ **71.** $(3x - 5)^2$ **72.** $5(x + 3)(x - 3)$ **73.** $(2x - 1)(x + 3)(x - 3)$ **74.** $(3x - y)(2x - 7y)$ **75.** $2y(y + 3)^2$ **76.** $(x + 3 + 2a)(x + 3 - 2a)$ **77.** $(2x - 3)(4x^2 + 6x + 9)$ **78.** $x(x^2 + 1)(x + 1)(x - 1)$ **79.** $(x^2 - 3)^2$ **80.** prime **81.** $4(a + 2)(a^2 - 2a + 4)$ **82.** $(x^2 + 9)(x + 3)(x - 3)$ **83.** $(a + 3b)(x - y)$ **84.** $(3x - 5y)(9x^2 + 15xy + 25y^2)$ **85.** $2xy(x + 3)(5x - 4)$ **86.** $(2x^3 + 5)(3x^3 - 1)$ **87. a.** $2xy + 2y^2$ **b.** $2y(x + y)$ **88. a.** $x^2 - 4y^2$ **b.** $(x + 2y)(x - 2y)$

89. −5 and −1, or {−5, −1} **90.** $\frac{1}{3}$ and 7, or $\left\{\frac{1}{3}, 7\right\}$ **91.** −8 and 7, or {−8, 7} **92.** 0 and 4, or {0, 4}
93. −5, −3, and 3, or {−5, −3, 3} **94.** 9 seconds **95.** 1990 **96.** length: 9 ft; width: 6 ft **97.** 2 in. **98.** 50 yd, 120 yd, and 130 yd

Chapter 5 Test

1. degree: 3; leading coefficient: −6 **2.** degree: 9; leading coefficient: 7 **3.** 6; 4 **4.** falls to the left and to the right
5. falls to the left and rises to the right **6.** $7x^3y - 18x^2y - y - 9$ **7.** $11x^2 - 13x - 6$ **8.** $35x^7y^3$ **9.** $x^3 - 4x^2y + 2xy^2 + y^3$
10. $21x^2 - 20xy - 9y^2$ **11.** $4x^2 - 25y^2$ **12.** $16y^2 - 56y + 49$ **13.** $x^2 + 4x + 4 - 9y^2$ **14.** $3x^2 + x - 10$; 60
15. $2ah + h^2 - 5h$ **16.** $x^2(14x - 15)$ **17.** $(9y + 5)(9y - 5)$ **18.** $(x + 3)(x + 5)(x - 5)$ **19.** $(5x - 3)^2$
20. $(x + 5 + 3y)(x + 5 - 3y)$ **21.** prime **22.** $(y + 2)(y - 18)$ **23.** $(2x + 5)(7x + 3)$ **24.** $5(x - 1)(x^2 + x + 1)$
25. $3(2x + y)(2x - y)$ **26.** $2(3x - 1)(2x - 5)$ **27.** $3(x^2 + 1)(x + 1)(x - 1)$ **28.** $(x^4 + y^4)(x^2 + y^2)(x + y)(x - y)$
29. $4x^2y(3y^3 + 2xy - 9)$ **30.** $(x^3 + 2)(x^3 - 14)$ **31.** $(x^2 - 6)(x^2 + 4)$ **32.** $3y(4x - 1)(x - 2)$ **33.** $y(y - 3)(y^2 + 2)$
34. $-\frac{1}{3}$ and 2, or $\left\{-\frac{1}{3}, 2\right\}$ **35.** −1 and $\frac{6}{5}$, or $\left\{-1, \frac{6}{5}\right\}$ **36.** 0 and $\frac{1}{3}$, or $\left\{0, \frac{1}{3}\right\}$ **37.** −1, 1, and 4, or {−1, 1, 4} **38.** 7 seconds
39. length: 5 yd; width: 3 yd **40.** 12 units, 9 units, and 15 units

Cumulative Review Exercises (Chapters 1–5)

1. 4 or {4} **2.** $\left(-4, \frac{1}{2}\right)$ or $\left\{\left(-4, \frac{1}{2}\right)\right\}$ **3.** (2, 1, −1) or {(2, 1, −1)} **4.** $\{x \mid 2 < x < 3\}$ or (2, 3)
5. $\{x \mid x \le -2 \text{ or } x \ge 7\}$ or $(-\infty, -2] \cup [7, \infty)$ **6.** 1 and $\frac{5}{2}$, or $\left\{1, \frac{5}{2}\right\}$ **7.** −5, 0, and 2, or {−5, 0, 2} **8.** $x = \frac{b}{c - a}$
9. $f(x) = 2x + 1$ **10.** winner: 1480 votes; loser: 1320 votes

11.
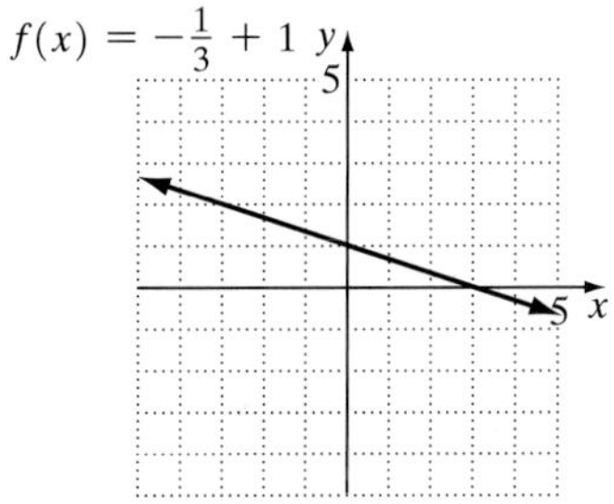

12.
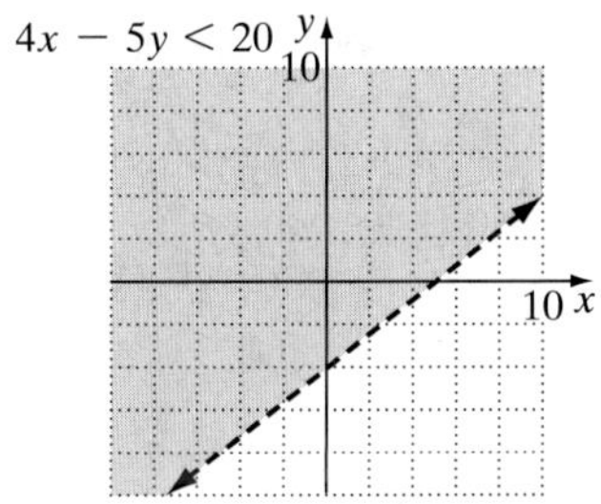

13.
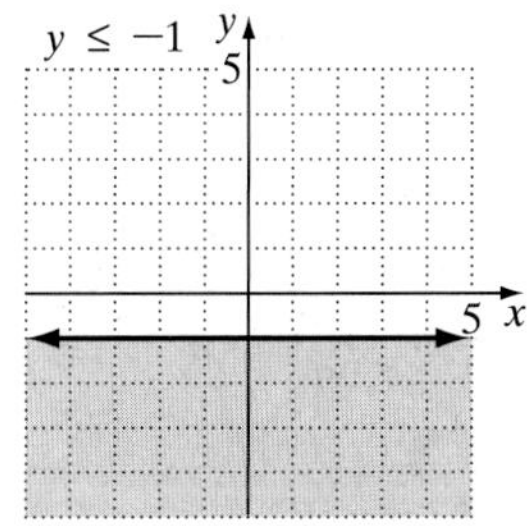

14. $-\frac{y^{10}}{2x^6}$ **15.** 7.06×10^{-5} **16.** $9x^4 - 6x^2y + y^2$ **17.** $9x^4 - y^2$ **18.** $(x + 3)(x - 3)^2$ **19.** $x^2(x^2 + 1)(x + 1)(x - 1)$
20. $14x^3y^2(1 - 2x)$

CHAPTER 6

Section 6.1

Check Point Exercises

1. a. 80; The cost to remove 40% of the lake's pollutants is \$80 thousand.; (40, 80) **b.** 180; The cost to remove 60% of the lake's pollutants is \$180 thousand.; (60, 180) **2.** $\left\{x \,\middle|\, x \text{ is a real number and } x \neq -3 \text{ and } x \neq \frac{1}{2}\right\}$ or $(-\infty, -3) \cup \left(-3, \frac{1}{2}\right) \cup \left(\frac{1}{2}, \infty\right)$
3. $x + 5$ **4. a.** $\frac{x - 5}{3x - 1}$ **b.** $\frac{x + 4y}{3x(x + y)}$ **5.** $\frac{x + 3}{x - 4}$ **6.** $\frac{-4(x + 2)}{3x(x - 2)}$ or $-\frac{4(x + 2)}{3x(3x - 2)}$ **7. a.** $9(3x + 7)$ **b.** $\frac{x + 3}{5}$

Exercise Set 6.1

1. −5; −3; 2 **3.** 0; $-\frac{3}{2}$; $-\frac{21}{2}$ **5.** $-\frac{7}{2}$; −5; $\frac{11}{5}$ **7.** $\{x \mid x \text{ is a real number and } x \neq 5\}$ or $(-\infty, 5) \cup (5, \infty)$ **9.** $\{x \mid x$ is a real number and $x \neq -3$ and $x \neq 1\}$ or $(-\infty, -3) \cup (-3, 1) \cup (1, \infty)$ **11.** $\{x \mid x \text{ is a real number and } x \neq -5\}$ or $(-\infty, -5) \cup (-5, \infty)$
13. $\{x \mid x \text{ is a real number and } x \neq 3 \text{ and } x \neq 5\}$ or $(-\infty, 3) \cup (3, 5) \cup (5, \infty)$ **15.** $\left\{x \mid x \text{ is a real number and } x \neq -\frac{4}{3} \text{ and } x \neq 2\right\}$ or $\left(-\infty, -\frac{4}{3}\right) \cup \left(-\frac{4}{3}, 2\right) \cup (2, \infty)$ **17.** 4 **19.** $\{x \mid x \text{ is a real number and } x \neq -2 \text{ and } x \neq 2\}$ or $(-\infty, -2) \cup (-2, 2) \cup (2, \infty)$
21. As x decreases, the function values are approaching 3.; $y = 3$ **23.** There is no point on the graph with x-coordinate −2.

25. The graph is not continuous. Furthermore, it neither rises nor falls to either the left or the right. **27.** $x + 2$ **29.** $\frac{1}{x - 3}$ **31.** $\frac{4}{x}$
33. $\frac{4}{y + 5}$ **35.** $-\frac{1}{3x + 5}$ **37.** $\frac{y + 7}{y - 7}$ **39.** $\frac{x + 9}{x - 1}$ **41.** cannot be simplified **43.** $-\frac{x + 3}{x + 4}$ **45.** $\frac{x + 5y}{3x - y}$
47. $\frac{x^2 + 2x + 4}{x + 2}$ **49.** $x^2 - 3$ **51.** $\frac{3}{2}$ **53.** $\frac{x + 7}{x}$ **55.** $\frac{x + 3}{x - 2}$ **57.** $\frac{x + 4}{x + 2}$ **59.** $-\frac{2}{y(y + 3)}$ **61.** $\frac{y^2 + 2y + 4}{2y}$
63. $\frac{x^2 + x + 1}{x - 2}$ **65.** 4 **67.** $\frac{4(x + y)}{3(x - y)}$ **69.** $\frac{9}{28}$ **71.** $\frac{7}{10}$ **73.** $\frac{x}{3}$ **75.** $\frac{y - 5}{2}$ **77.** $\frac{(x^2 + 4)(x - 4)}{x - 1}$ **79.** $\frac{y - 7}{y - 5}$
81. $(x - 1)(2x - 1)$ **83.** $\frac{1}{x - 2y}$ **85.** $3x^2$ **87.** 195; The cost to inoculate 60% of the population is \$195 million.; (60, 195)
89. 100; This indicates that we cannot inoculate 100% of the population. **91.** 7; About 7% of people with 10 years of education are unemployed. ; (10, 7) **93.** The function values are approaching 0.; no; The function values never actually reach 0.
95. 90; Lung cancer has a incidence ratio of 10 between the ages of 55 and 64. 90% of deaths from lung cancer in this group are smoking related.; (10, 90) **97.** $y = 100$; As the incidence ratio increases the percentage of smoking-related deaths is approaching 100%.
99–109. Answers will vary.

111.

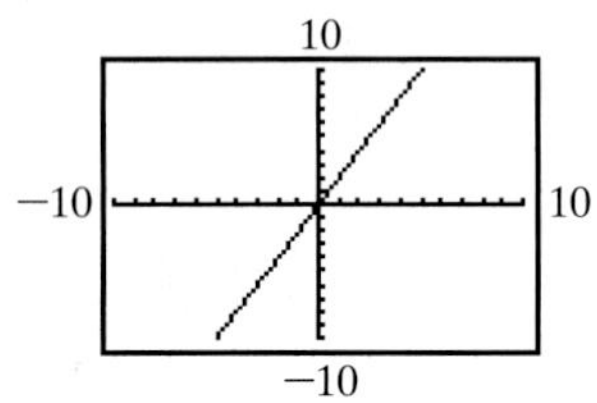

Graphs coincide.; multiplied correctly

113.

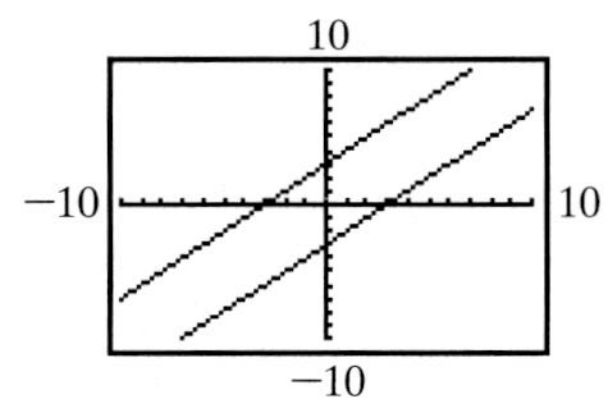

Graphs do not coincide.; Right side should be $x + 3$.

115. a.

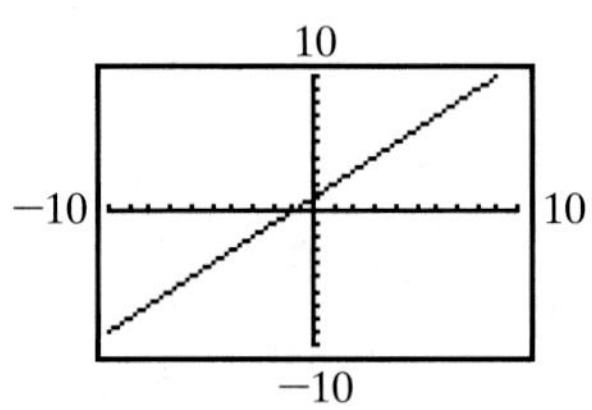

; The graphs coincide. **b.** $\frac{x^2 - x - 2}{x - 2} = x + 1$; no; $g(2) = 3$ but $f(2)$ is undefined.
c. When tracing the graph of f, for $x = 2$, no y-value is given; however, on the graph of g, for $x = 2$, the y-value 3 is displayed.; f and g are not equal when $x = 2$.

117. d **119.** $2a + h - 5$ **121.** $\frac{y^n - 1}{y^n + 4}$ **122.**

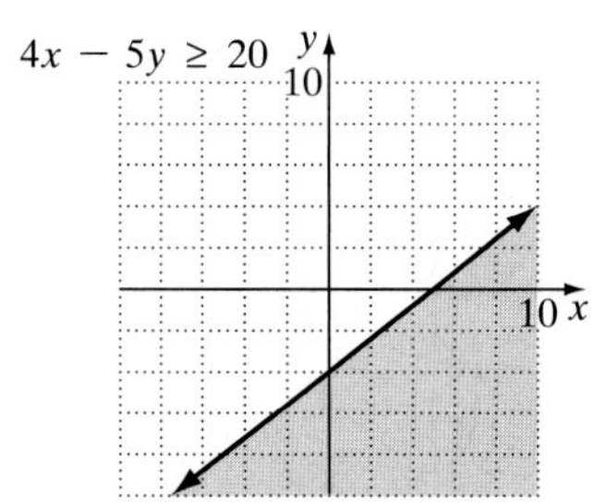

123. $2x^3 - 11x^2 + 3x + 30$
124. $x(x - 3)(x^2 + 2)$

Section 6.2

Check Point Exercises

1. $\frac{x - 5}{x + 3}$ **2.** $\frac{1}{x - y}$ **3.** $18x^2$ **4.** $5x(x + 3)(x + 3)$ or $5x(x + 3)^2$ **5.** $\frac{21 + 4x}{18x^2}$ or $\frac{4x + 21}{18x^2}$
6. $\frac{2x^2 - 2x + 8}{(x - 4)(x + 4)}$ or $\frac{2(x^2 - x + 4)}{(x - 4)(x + 4)}$ **7.** $\frac{x^2 - 3x + 5}{(x - 3)(x + 1)(x - 2)}$ **8.** $\frac{3}{y + 2}$ **9.** $\frac{3x - 5y}{x - 3y}$

Exercise Set 6.2

1. $\frac{4}{3x}$ **3.** $\frac{10x + 7}{x - 4}$ **5.** $\frac{2x - 1}{x + 3}$ **7.** $\frac{y - 3}{y + 3}$ **9.** $\frac{x + 1}{4x - 3}$ **11.** $\frac{x - 3}{x - 1}$ **13.** $\frac{4y + 3}{2y - 1}$ **15.** $\frac{x^2 + xy + y^2}{x + y}$ **17.** $175x^2$
19. $(x - 5)(x + 5)$ **21.** $y(y + 10)(y - 10)$ **23.** $(x + 4)(x - 4)(x - 4)$ **25.** $(y - 5)(y - 6)(y + 1)$ **27.** $\frac{3 + 50x}{5x^2}$
29. $\frac{7x + 1}{(x - 2)(x + 3)}$ **31.** $\frac{x^2 + x + 12}{(x - 3)(x + 2)(x + 3)}$ **33.** $\frac{2y^2 - 6y + 65}{(y + 4)(y - 7)}$ **35.** $\frac{3x + 20}{(x + 4)(x - 4)}$ **37.** $\frac{13}{(y - 3)(y - 2)}$

39. $\frac{1}{x+3}$ **41.** $\frac{6x^2+14x+10}{(x+1)(x+4)(x+3)}$ **43.** $-\frac{x^2-3x-13}{(x+4)(x-2)(x+1)}$ **45.** $\frac{4x-11}{x-3}$ **47.** $\frac{2y-14}{(y-4)(y+4)}$ **49.** $\frac{x^2-2x-6}{3(x+2)(x-2)}$ **51.** $\frac{16}{x-4}$ **53.** $\frac{10x+3y}{(x+y)(x-y)}$ **55.** $\frac{x^2-18x-30}{(5x+6)(x-2)}$ **57.** $\frac{3x^2+16xy-13y^2}{(x+5y)(x-5y)(x-4y)}$ **59.** $\frac{8x^3-32x^2+23x+6}{(x+2)(x-2)(x-2)(x-1)}$ **61.** $\frac{16a^2-12ab-18b^2}{(3a+4b)(3a-4b)(2a-b)}$ **63.** 11; If you average 0 miles per hour over the speed limits, the total driving time is about 11 hours.; (0, 11) **65.** $\frac{720x+48{,}050}{(x+70)(x+65)}$; 11 **67.** 12 mph; Answers will vary. **69.** $\frac{2x(2x+9)}{(x+4)(x+5)}$ **71.** $\frac{-5x^3+27{,}680x-388{,}150}{x^2+9}$ **73–75.** Answers will vary. **77.** Answers will vary.; $\frac{b+a}{ab}$

79.

As age increases the number of arrests per 100,000 drivers increases until about age 25 and then decreases.

81. $\frac{x^2+5x+8}{(x+2)(x+1)}$ **83.** $\frac{x-y+1}{(x-y)(x-y)}$ **85.** $-\frac{2x(3x^2+8x-6)}{(x+3)(x-3)(x+6)}$ **87.** $\frac{3x(x+3)}{x+6}$ **88.** $\frac{y^{10}}{9x^4}$ **89.** $\left\{x \middle| -\frac{13}{3} < x \le 5\right\}$ or $\left[-\frac{13}{3}, 5\right]$ **90.** $2x(5x+3)(5x-3)$

Section 6.3

Check Point Exercises

1. $\frac{y}{x+y}$ **2.** $-\frac{1}{x(x+7)}$ **3.** $\frac{2x}{x^2+1}$ **4.** $\frac{x+2}{x-5}$

Exercise Set 6.3

1. $\frac{4x+2}{x-3}$ **3.** $\frac{x^2+9}{(x-3)(x+3)}$ **5.** $\frac{y+x}{y-x}$ **7.** $\frac{4-x}{5x-3}$ **9.** $\frac{1}{x-3}$ **11.** $-\frac{1}{x(x+5)}$ **13.** $-x$ **15.** $-\frac{x+1}{x-1}$ **17.** $\frac{(x+y)(x+y)}{xy}$ **19.** $\frac{4x}{x^2+4}$ **21.** $\frac{2y^3+5x^2}{y^3(5-3x^2)}$ **23.** $-\frac{12}{5}$ **25.** $\frac{3ab(b+a)}{(2b+3a)(2b-3a)}$ **27.** $-\frac{x+10}{5x-2}$ **29.** $\frac{2y}{3y+7}$ **31.** $\frac{2b+a}{b-2a}$ **33.** $\frac{4(7x+5)}{69(x+5)}$ **35.** $\frac{3x-2y}{13y+5x}$ **37.** $x(x+1)$ **39. a.** $\frac{Pi(1+i)^n}{(1+i)^n-1}$ **b.** \$527 per month **41.** $R=\frac{R_1R_2R_3}{R_2R_3+R_1R_3+R_1R_2}$; about 2.18 **43–45.** Answers will vary. **47.** Graphs coincide.; simplified correctly **49.** Graphs do not coincide.; $\frac{\frac{1}{x}+\frac{1}{3}}{\frac{1}{3x}}=x+3$ **51.** $\frac{1}{(x+h+1)(x+1)}$ **53.** $\frac{x^4+3x^2+1}{x(x^2+2)}$ **55.** $-\frac{2a+h}{a^2(a+h)^2}$ **56.** 3 and 9, or $\{3, 9\}$ **57.** $16x^4-8x^2y+y^2$ **58.** $\{x \mid 1 < x < 5\}$ or $(1, 5)$

Section 6.4

Check Point Exercises

1. $4x^2-8x+\frac{1}{2}+\frac{1}{x}$ **2.** $3x^2y^2-xy+\frac{2}{y}$ **3.** $x+5$ **4.** $2x^2-x+3-\frac{6}{2x-1}$ **5.** $2x^2+7x+14+\frac{21x-10}{x^2-2x}$

Exercise Set 6.4

1. $5x^4-3x^2+2$ **3.** $6x^2+2x-3-\frac{2}{x}$ **5.** $7x-\frac{7}{4}-\frac{4}{x}$ **7.** $-5x^3+10x^2-\frac{3}{5}x+8$ **9.** $2a^2b-a-3b$ **11.** $6x-\frac{3}{y}-\frac{2}{xy^2}$ **13.** $x+3$ **15.** x^2+x-2 **17.** $x-2+\frac{2}{x-5}$ **19.** $x+5-\frac{10}{2x+3}$ **21.** $x^2+2x+3+\frac{1}{x+1}$ **23.** $2y^2+3y-1$ **25.** $3x^2-3x+1+\frac{2}{3x+2}$ **27.** $2x^2+4x+5+\frac{9}{2x-4}$ **29.** $2-\frac{2}{2y-1}$ **31.** $2y^3+3y^2-4y+1$ **33.** $4x^2+3x-8+\frac{18}{x^2+3}$ **35.** $5x^2+x+3+\frac{x+7}{3x^2-1}$ **37.** 70; At the tax rate of 30% the government tax revenue is \$70 tens of billions or \$700 billion.; (30, 70). **39.** $f(x) = 80 + \frac{800}{x - 110}$; 70; yes; Answers will vary. **41–45.** Answers will vary.

47.

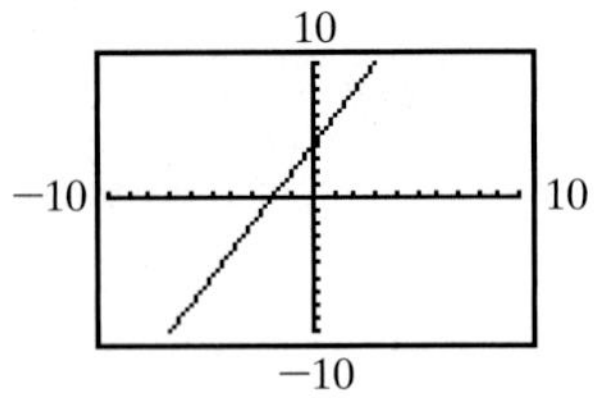

Graphs coincide.; division correct

49.

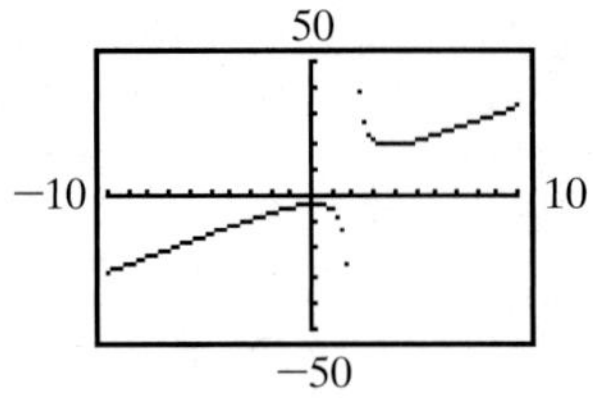

Graphs coincide.; division correct

51. d **53.** $x^{2n} - x^n + 1$ **55.** −12

56. $\left\{x \middle| x < -\frac{1}{2} \text{ or } x > \frac{7}{2}\right\}$ or $\left(-\infty, -\frac{1}{2}\right) \cup \left(\frac{7}{2}, \infty\right)$ **57.** 4.061×10^7 **58.** $22x + 12$

Section 6.5

Check Point Exercises

1. $x^2 - 2x - 3$ **2.** −105 **3.** The remainder is zero.; $-1, -\frac{1}{3}$, and $\frac{2}{5}$, or $\left\{-1, -\frac{1}{3}, \frac{2}{5}\right\}$

Exercise Set 6.5

1. $2x + 5$ **3.** $3x - 8 + \frac{20}{x + 5}$ **5.** $4x^2 + x + 4 + \frac{3}{x - 1}$ **7.** $6x^4 + 12x^3 + 22x^2 + 48x + 93 + \frac{187}{x - 2}$
9. $x^3 - 10x^2 + 51x - 260 + \frac{1300}{5 + x}$ **11.** $3x^2 + 3x - 3$ **13.** $x^4 + x^3 + 2x^2 + 2x + 2$ **15.** $x^3 + 4x^2 + 16x + 64$
17. $2x^4 - 7x^3 + 15x^2 - 31x + 64 - \frac{129}{x + 2}$ **19.** −25 **21.** −133 **23.** 240 **25.** 4729 **27.** The remainder is 0.; −1, 2, and 3, or {−1, 2, 3} **29.** The remainder is 0.; $-\frac{1}{2}$, 1, and 2, or $\left\{-\frac{1}{2}, 1, 2\right\}$ **31.** The remainder is 0.; $-5, \frac{1}{3}$, and $\frac{1}{2}$, or $\left\{-5, \frac{1}{3}, \frac{1}{2}\right\}$
33. a. The remainder is 0. **b.** 3 mm **35–39.** Answers will vary. **41.** The remainder is 0.; −2, −1, 2, and 5, or {−2, −1, 2, 5}
42. $\{x|x > -10\}$ or $(-10, \infty)$ **43.** −5 and −1,or {−5, −1} **44.** (5, 3) or {(5, 3)}

Section 6.6

Check Point Exercises

1. 6 or {6} **2.** 4 or {4} **3.** no solution or ∅ **4.** 4 and 6, or {4, 6} **5.** 1 and 7, or {1, 7} **6.** 50%

Exercise Set 6.6

1. 1 or {1} **3.** 3 or {3} **5.** 4 or {4} **7.** −4 or {−4} **9.** $\frac{9}{7}$ or $\left\{\frac{9}{7}\right\}$ **11.** no solution or ∅ **13.** 2 or {2} **15.** 2 or {2}
17. 5 or {5} **19.** −7 and −1, or {−7, −1} **21.** −6 and 3, or {−6, 3} **23.** −1 or {−1} **25.** no solution or ∅ **27.** 2 or {2}
29. no solution or ∅ **31.** $-\frac{4}{3}$ or $\left\{-\frac{4}{3}\right\}$ **33.** 60% **35.** 10 days; (10, 8) **37.** $y = 5$; On average the students remembered 5 words over an extended period of time. **39.** 11 learning trials; (11, 0.95) **41.** As the number of learning trials increases, the proportion of correct responses increases.; Initially, the proportion of correct responses increases rapidly but slows down as time increases.
43. either 50 or approximately 67 cases; (50, 350) or $\left(66\frac{2}{3}, 350\right)$ **45–51.** Answers will vary.

53.

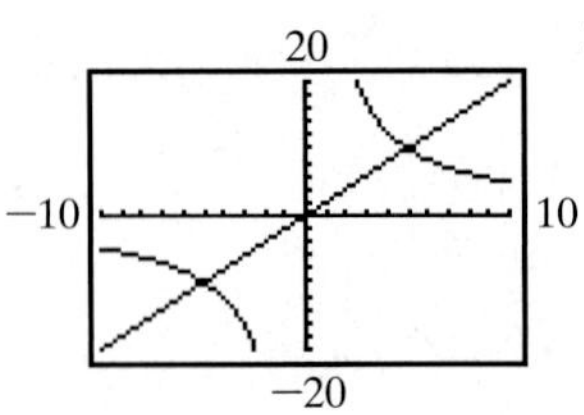

; −5 and 5, or {−5, 5}

55.

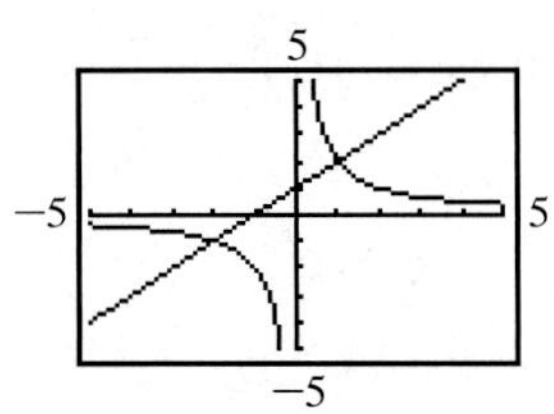

; −2 and 1, or {−2, 1} **57.** d

59. Answers will vary. **61.** $x + 2y \geq 2$
$x - y \geq -4$

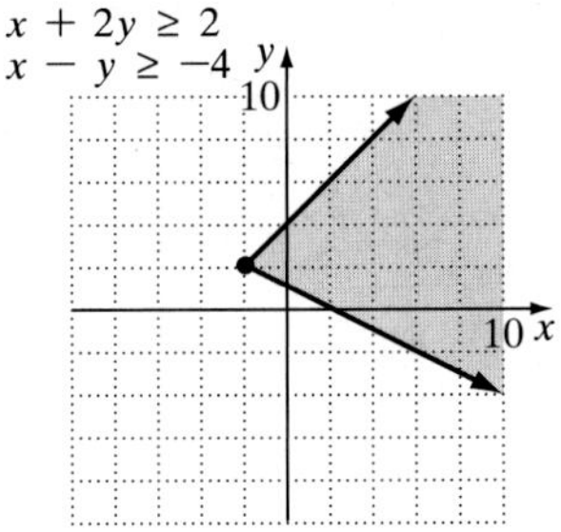

62. 15 or {15} **63.** $F = \frac{9C + 160}{5}$

Section 6.7

Check Point Exercises

1. $x = \frac{yz}{y - z}$ **2. a.** $C(x) = 500{,}000 + 400x$ **b.** $\overline{C}(x) = \frac{500{,}000 + 400x}{x}$ **c.** 10,000 wheelchairs **3.** 12 mph
4. $2\frac{2}{3}$ hr or 2 hr and 40 min **5.** experienced carpenter: 8 hr; apprentice: 24 hr

Exercise Set 6.7

1. $P_1 = \frac{P_2V_2}{V_1}$ **3.** $f = \frac{pq}{q + p}$ **5.** $r = \frac{A - P}{P}$ **7.** $m_1 = \frac{Fd^2}{Gm_2}$ **9.** $x = \overline{x} + zs$ **11.** $R = \frac{E - Ir}{I}$ **13.** $f_1 = \frac{ff_2}{f_2 - f}$
15. 50,000 wheelchairs **17.** $y = 400$; Average cost is nearing \$400 as production level increases. **19. a.** $C(x) = 100{,}000 + 100x$
b. $\overline{C}(x) = \frac{100{,}000 + 100x}{x}$ **c.** 500 bikes **21.** 5 mph **23.** The time increases as the walking rate is close to zero miles per hour.
25. bicycle: 12 mph; motorcycle: 32 mph **27.** 6 mph **29.** 2 mph **31.** 4 mph **33.** 4.2 ft per sec **35.** 12 miles
37. 17 min; no **39.** 2 hr **41.** 12 hr **43.** 10 hr **45.** 10 hr **47.** 10 min **49–55.** Answers will vary.
57. For Exercise 45:

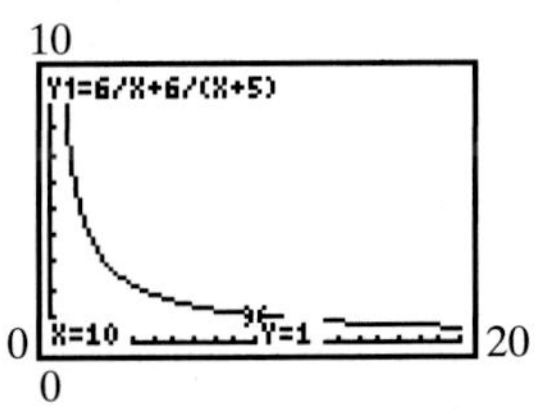

For Exercise 46:

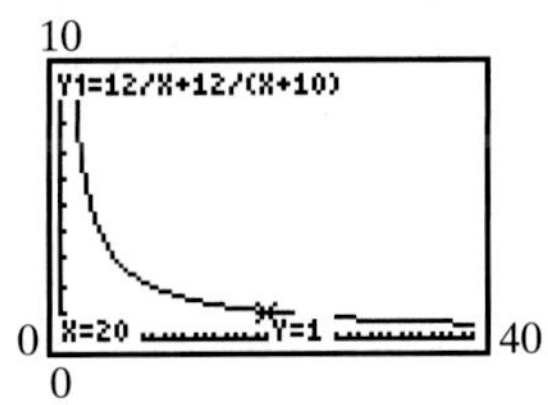

59. c
61. 54 mph
63. $(x + 2 + 3y)(x + 2 - 3y)$
64. $(5, -3)$ or $\{(5, -3)\}$
65. $(3, -1, 2)$ or $\{(3, -1, 2)\}$

Section 6.8

Check Point Exercises

1. 137.5 pounds per square inch **2.** about 556 ft **3.** \$26.00 per barrel **4.** 24 min **5.** 96π cubic feet

Exercise Set 6.8

1. 84 **3.** 25 **5.** $\frac{5}{6}$ **7.** 240 **9. a.** $G = kW$ **b.** $G = 0.02W$ **c.** 1.04 in. **11.** \$60.00 **13.** 2442 mph **15.** about 607 lb
17. 0.5 hr **19.** 6.4 lb **21.** 32; not in the desirable range **23.** about 11.1 foot-candles **25.** 72 ergs **27.** 3983 calls
29. Yes, the wind will exert a force of 360 pounds on the window. **31–35.** Answers will vary.
37. z varies jointly as the square of x and the square root of y. **39.** Answers will vary. **41.** It is decreased to $\frac{1}{4}$ of the former intensity.
43. Distance is increased by $\sqrt{50}$ or about 7.07 for the space telescope. **44.** -2 **45.** $(x + 3)(x - 3)(y - 3)$ **46.** 4

Review Exercises

1. a. $\frac{7}{4}$ **b.** $\frac{3}{4}$ **c.** does not exist **d.** 0 **2.** $\{x|x$ is a real number and $x \neq -4$ and $x \neq 3\}$ or $(-\infty, -4) \cup (-4, 3) \cup (3, \infty)$
3. $\{x|x$ is a real number and $x \neq -2$ and $x \neq 1\}$ or $(-\infty, -2) \cup (-2, 1) \cup (1, \infty)$ **4.** $\frac{x^2 - 7}{3x}$ **5.** $\frac{x - 1}{x - 7}$ **6.** $\frac{3x + 2}{x - 5}$
7. cannot be simplified **8.** $\frac{x^2 + 2x + 4}{x + 2}$ **9.** $\frac{x + 1}{3}$ **10.** $\frac{2x - 1}{x + 1}$ **11.** $\frac{x - 7}{x^2}$ **12.** $\frac{1}{3(x + 3)}$ **13.** $\frac{1}{2}$

14. $\dfrac{(y+4)(y+2)}{(y-6)(y^2+4y+16)}$ **15. a.** 50 deer **b.** 150 deer **c.** $y = 225$; The population will approach 225 deer over time.
16. 4 **17.** $\dfrac{1}{x+3}$ **18.** $3x+2$ **19.** $36x^3$ **20.** $(x+7)(x-5)(x+2)$ **21.** $\dfrac{3x-5}{x(x-5)}$ **22.** $\dfrac{5x-2}{(x-3)(x+2)(x-2)}$
23. $\dfrac{2x-8}{(x-3)(x-5)}$ or $\dfrac{2(x-4)}{(x-3)(x-5)}$ **24.** $\dfrac{4x}{(3x+4)(x-4)}$ **25.** $\dfrac{y-3}{(y+1)(y+3)}$ **26.** $\dfrac{2x^2-9}{(x+3)(x-3)}$ **27.** $3(x+y)$
28. $\dfrac{3}{8}$ **29.** $\dfrac{x}{x-5}$ **30.** $\dfrac{3x+8}{3x+10}$ **31.** $\dfrac{4(x-2)}{2x+5}$ **32.** $\dfrac{1+x}{1-x}$ **33.** $3x-6+\dfrac{2}{x}-\dfrac{2}{5x^2}$ **34.** $6x^3y+2xy-10x$
35. $3x-7+\dfrac{26}{2x+3}$ **36.** $2x^2-4x+1-\dfrac{10}{5x-3}$ **37.** $x^5+5x^4+8x^3+16x^2+33x+63+\dfrac{128}{x-2}$ **38.** $2x^2+3x-1$
39. $4x^2-7x+5-\dfrac{4}{x+1}$ **40.** $3x^3+6x^2+10x+10$ **41.** $x^3-4x^2+16x-64+\dfrac{272}{x+4}$ **42.** 3 **43.** 396 **44.** solution
45. not a solution **46.** The remainder is 0.; $-1, \frac{1}{3}$, and $\frac{1}{2}$, or $\left\{-1, \frac{1}{3}, \frac{1}{2}\right\}$. **47.** 6 or {6} **48.** 52 or {52} **49.** 7 or {7}
50. $-\frac{9}{2}$ or $\left\{-\frac{9}{2}\right\}$ **51.** $-\frac{1}{2}$ and 3, or $\left\{-\frac{1}{2}, 3\right\}$ **52.** −23 and 2, or {−23, 2} **53.** −3 and 2, or {−3, 2} **54.** 80%
55. $C = R - nP$ **56.** $T_1 = \dfrac{P_1V_1T_2}{P_2V_2}$ **57.** $P = \dfrac{A}{rT+1}$ **58.** $R = \dfrac{R_1R_2}{R_2+R_1}$ **59.** $n = \dfrac{IR}{E-Ir}$
60. a. $C(x) = 50{,}000 + 25x$ **b.** $\overline{C}(x) = \dfrac{50{,}000+25x}{x}$ **c.** 5000 graphing calculators **61.** 12 mph **62.** 9 mph **63.** 2 hr; no
64. faster crew: 36 hr; slower crew: 45 hr **65.** 240 min or 4 hr **66.** \$154 **67.** 1600 ft **68.** 5 hr **69.** 112 decibels
70. 16 hr **71.** 800 cubic feet

Chapter 6 Test

1. $\{x \mid x$ is a real number and $x \neq 2$ and $x \neq 5\}$ or $(-\infty, 2) \cup (2, 5) \cup (5, \infty)$; $\dfrac{x}{x-5}$ **2.** $\dfrac{x}{x-4}$ **3.** $\dfrac{(x+3)(x-1)}{x+1}$ **4.** $\dfrac{1}{3x+5}$
5. $\dfrac{x^2+2x+15}{(x+3)(x-3)}$ **6.** $\dfrac{3x-4}{(x+2)(x-3)}$ **7.** $\dfrac{5x^2-7x+4}{(x+2)(x-1)(x-2)}$ **8.** $\dfrac{x+3}{x+5}$ **9.** $\dfrac{x-2}{x-10}$ **10.** $\dfrac{x-2}{8}$ **11.** $1-x$
12. $3x^2y^2+4y^2-\frac{5}{2}y$ **13.** $3x^2-3x+1+\dfrac{2}{3x+2}$ **14.** $3x^2+2x+3+\dfrac{9-6x}{x^2-1}$ **15.** $3x^3-4x^2+7$ **16.** 12
17. solution **18.** $\frac{1}{3}$ or $\left\{\frac{1}{3}\right\}$ **19.** −1 or {−1} **20.** 3 years **21.** $a = \dfrac{Rs}{s-R}$ or $a = -\dfrac{Rs}{s-R}$
22. a. $C(x) = 300{,}000 + 10x$ **b.** $\overline{C}(x) = \dfrac{300{,}000+10x}{x}$ **c.** 20,000 televisions **23.** 12 hr **24.** 4 mph **25.** 45 foot-candles

Cumulative Review Exercises (Chapters 1–6)

1. $\{x \mid x < -6\}$ or $(-\infty, -6)$ **2.** $-\frac{1}{2}$ and 4, or $\left\{-\frac{1}{2}, 4\right\}$ **3.** $(-2, 0, 4)$ or $\{(-2, 0, 4)\}$ **4.** $\left\{x \middle| -2 \le x \le \frac{14}{3}\right\}$ or $\left[-2, \frac{14}{3}\right]$
5. 12 or {12} **6.** $s = \dfrac{2R - Iw}{2I}$ **7.** (3, 2) or {(3, 2)} **8.** $f(x) = -3x - 2$

9.

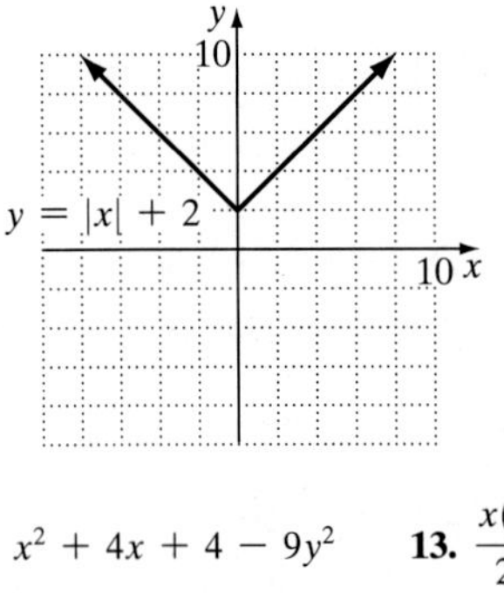

10.

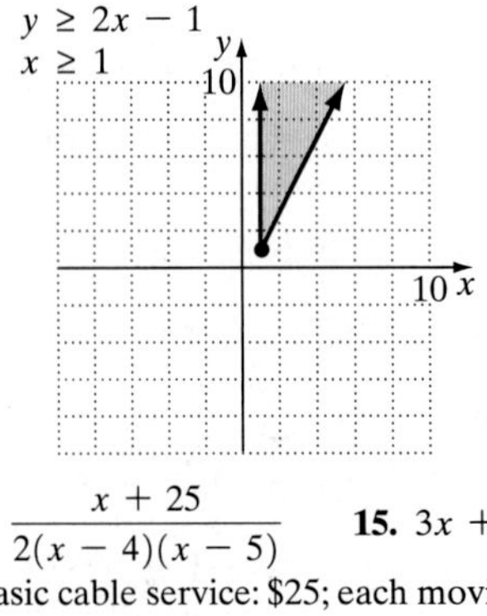

11.

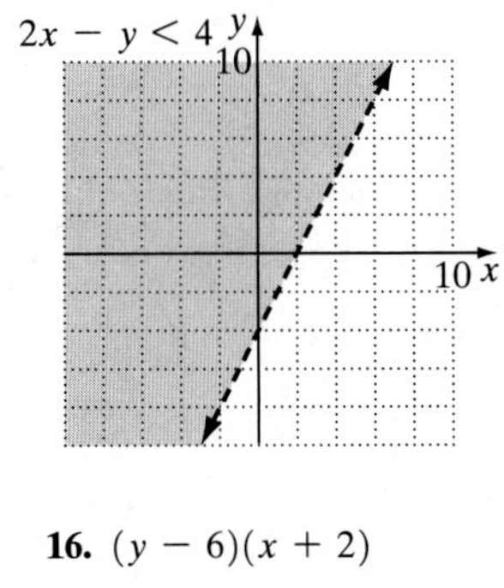

12. $x^2+4x+4-9y^2$ **13.** $\dfrac{x(x+1)}{2x+5}$ **14.** $\dfrac{x+25}{2(x-4)(x-5)}$ **15.** $3x+4+\dfrac{2}{x+2}$ **16.** $(y-6)(x+2)$
17. $2xy(2x+3)(6x-5)$ **18.** 4 sec **19.** basic cable service: \$25; each movie channel: \$10 **20.** 1 ft

CHAPTER 7

Section 7.1

Check Point Exercises

1. a. 8 **b.** -7 **c.** $\frac{4}{5}$ **d.** 0.09 **e.** 5 **f.** 7 **2. a.** 4 **b.** $-\sqrt{24} \approx -4.90$ **3.** $\{x|x \geq 3\}$ or $[3, \infty)$
4. \$25.5 billion; The model describes the data well. **5. a.** 7 **b.** $|x + 8|$ **c.** $|7x^5|$ **d.** $|x - 3|$ **6. a.** 3 **b.** -2 **7.** $-3x$
8. a. 2 **b.** -2 **c.** not a real number **d.** -1 **9. a.** $|x + 6|$ **b.** $3x - 2$ **c.** 8

Exercise Set 7.1

1. 6 **3.** -6 **5.** not a real number **7.** $\frac{1}{5}$ **9.** -0.2 **11.** 3 **13.** 1 **15.** not a real number **17.** 4; 1; 0; not a real number
19. -5; -2.24; -1; not a real number **21.** 4; 2; 1; 6 **23.** $\{x|x \geq 3\}$ or $[3, \infty)$; c **25.** $\{x|x \geq -5\}$ or $[-5, \infty)$; d
27. $\{x|x \leq 3\}$ or $(-\infty, 3]$; e **29.** 5 **31.** 4 **33.** $|x - 1|$ **35.** $|6x^2|$ or $6x^2$ **37.** $-|10x^3|$ or $-10|x^3|$ **39.** $|x + 6|$
41. $-|x - 4|$ **43.** 3 **45.** -3 **47.** $\frac{1}{5}$ **49.** 3; 2; -1; -4 **51.** -2; 0; 2 **53.** 1 **55.** 2 **57.** -2 **59.** not a real number
61. -1 **63.** not a real number **65.** -4 **67.** 2 **69.** -2 **71.** x **73.** $|y|$ **75.** $-2x$ **77.** -5 **79.** 5 **81.** $|x + 3|$
83. $-2(x - 1)$ **85.** 40.2 in.; The model describes actual data well. **87.** 70 mph; The officer should not believe the motorist.; Answers will vary. **89–99.** Answers will vary.
101.

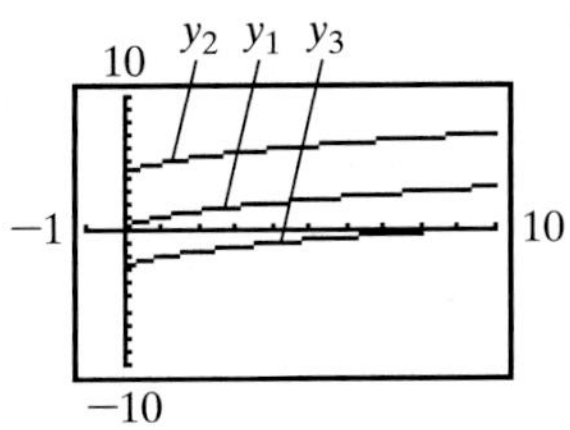

; Answers will vary.

103.

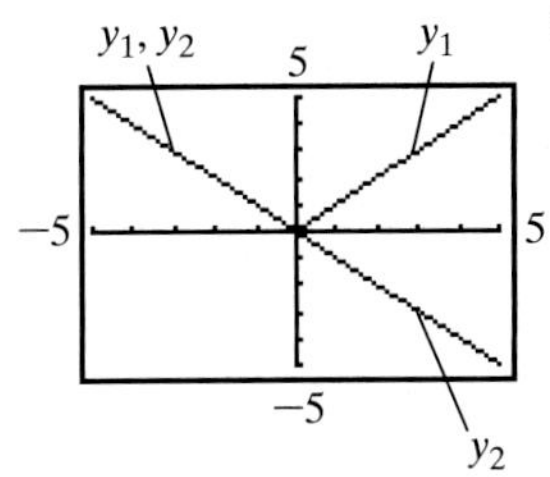

; **a.** $\{x|x \leq 0\}$ **b.** $\{x|x > 0\}$

105. Answers will vary; an example is $f(x) = \sqrt{15 - 3x}$. **107.** $|(2x + 3)^5|$ **109.** The graph of h is the graph of f shifted left 3 units.
110. $7x + 30$ **111.** $\frac{x^8}{9y^6}$ **112.** $\left\{x \middle| x < -\frac{7}{3} \text{ or } x > 5\right\}$ or $\left(-\infty, -\frac{7}{3}\right) \cup (5, \infty)$

Section 7.2

Check Point Exercises

1. a. 5 **b.** -2 **c.** $\sqrt[4]{5xy^2}$ **2. a.** $(5xy)^{1/4}$ **b.** $\left(\frac{a^3b}{2}\right)^{1/5}$ **3. a.** 81 **b.** 8 **c.** -8 **4. a.** $6^{4/3}$ **b.** $(2xy)^{7/5}$
5. a. $\frac{1}{25^{1/2}} = \frac{1}{5}$ **b.** $\frac{1}{64^{1/3}} = \frac{1}{4}$ **c.** $\frac{1}{32^{4/5}} = \frac{1}{16}$ **d.** $\frac{1}{(3xy)^{5/9}}$ **6.** \$126.5 thousand or \$126,500 **7. a.** $7^{5/6}$ **b.** $\frac{5}{x}$ **c.** $9.1^{3/10}$
d. $\frac{y^{1/12}}{x^{1/5}}$ **8. a.** $\sqrt{x}$ **b.** $2a^4$ **c.** $\sqrt[4]{x^2y}$ **d.** $\sqrt[6]{x}$ **e.** $\sqrt[6]{x}$

Exercise Set 7.2

1. $\sqrt{49} = 7$ **3.** $\sqrt[3]{-27} = -3$ **5.** $-\sqrt[4]{16} = -2$ **7.** $\sqrt[3]{xy}$ **9.** $\sqrt[5]{2xy^3}$ **11.** $(\sqrt{81})^3 = 729$ **13.** $(\sqrt[3]{125})^2 = 25$
15. $(\sqrt[5]{-32})^3 = -8$ **17.** $(\sqrt[3]{27})^2 + (\sqrt[4]{16})^3 = 17$ **19.** $\sqrt[7]{(xy)^4}$ **21.** $7^{1/2}$ **23.** $5^{1/3}$ **25.** $(11x)^{1/5}$ **27.** $x^{3/2}$ **29.** $x^{3/5}$
31. $(x^2y)^{1/5}$ **33.** $(19xy)^{3/2}$ **35.** $(7xy^2)^{5/6}$ **37.** $2xy^{2/3}$ **39.** $\frac{1}{49^{1/2}} = \frac{1}{7}$ **41.** $\frac{1}{27^{1/3}} = \frac{1}{3}$ **43.** $\frac{1}{16^{3/4}} = \frac{1}{8}$ **45.** $\frac{1}{8^{2/3}} = \frac{1}{4}$
47. $\left(\frac{27}{8}\right)^{1/3} = \frac{3}{2}$ **49.** $\frac{1}{(-64)^{2/3}} = \frac{1}{16}$ **51.** $\frac{1}{(2xy)^{7/10}}$ **53.** $\frac{5x}{z^{1/3}}$ **55.** 3 **57.** 4 **59.** $x^{5/6}$ **61.** $x^{3/5}$ **63.** $\frac{1}{x^{5/12}}$ **65.** 25
67. $\frac{1}{y^{1/6}}$ **69.** $32x$ **71.** $5x^2y^3$ **73.** $\frac{x^{1/4}}{y^{3/10}}$ **75.** 3 **77.** $27y^{2/3}$ **79.** $\sqrt[5]{x}$ **81.** $2a^2$ **83.** x^2y^3 **85.** x^6y^6 **87.** $\sqrt[5]{3y}$
89. $\sqrt[3]{4a^2}$ **91.** $\sqrt[3]{x^2y}$ **93.** $\sqrt[6]{2^5}$ or $\sqrt[6]{32}$ **95.** $\sqrt[20]{x}$ **97.** $\sqrt{y}$ **99.** $\sqrt[8]{x}$ **101.** $\sqrt[4]{x^2y}$ **103.** $\sqrt[12]{2x}$ **105.** x^9y^{15}
107. 58 species of plants **109.** 1.9 hours **111.** 4.25% **113.** about 1872 calories per day
115. a. $L + 1.25S^{1/2} - 9.8D^{1/3} \leq 16.296$ **b.** eligible **117–127.** Answers will vary.

129. a.

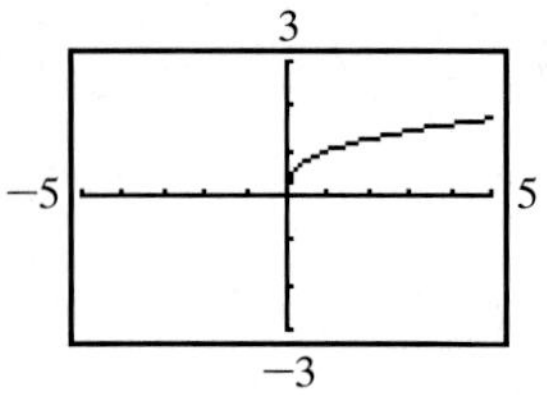

b.

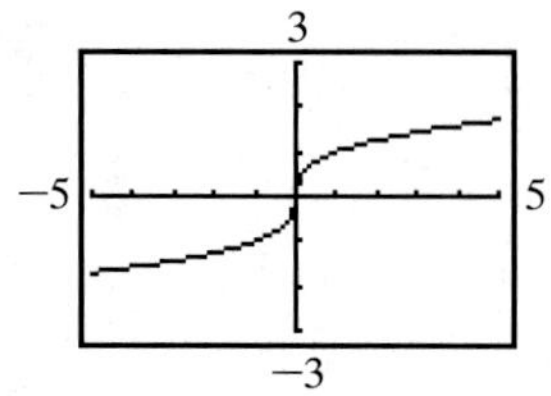

c. no; The 6th root of a negative number is undefined; thus, $(x^{1/6})^2$ is undefined for negative numbers.

131. 8 years old **133.** 61 **135.** $y = -2x + 11$ or $f(x) = -2x + 11$
136.

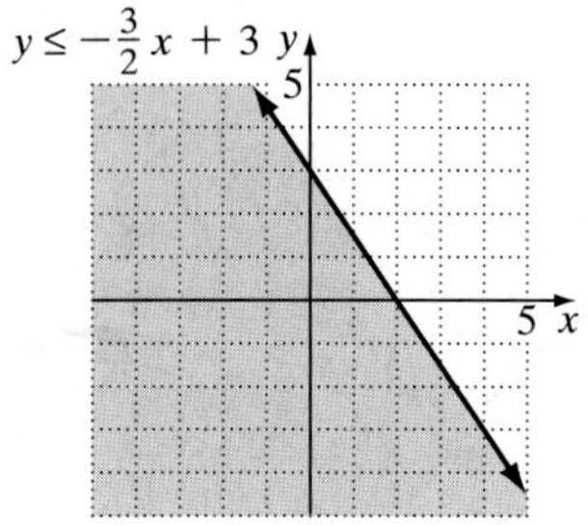

137. $(3, 4)$ or $\{(3, 4)\}$

Section 7.3

Check Point Exercises

1. a. $\sqrt{55}$ **b.** $\sqrt{x^2 - 16}$ **c.** $\sqrt[3]{60}$ **d.** $\sqrt[7]{12x^4}$ **2. a.** $4\sqrt{5}$ **b.** $2\sqrt[3]{5}$ **c.** $2\sqrt[4]{2}$ **d.** $10|x|\sqrt{2y}$ **3.** $f(x) = \sqrt{3}|x - 2|$
4. $x^4y^5z\sqrt{xyz}$ **5.** $2x^3y^4\sqrt[3]{5xy^2}$ **6.** $2x^2z\sqrt[5]{x^2y^2z^3}$ **7. a.** $2\sqrt{3}$ **b.** $100\sqrt[3]{4}$ **c.** $2x^2y\sqrt[4]{2}$

Exercise Set 7.3

1. $\sqrt{15}$ **3.** $\sqrt[3]{18}$ **5.** $\sqrt[4]{33}$ **7.** $\sqrt{33xy}$ **9.** $\sqrt[5]{24x^4}$ **11.** $\sqrt{x^2 - 9}$ **13.** $\sqrt[6]{(x - 4)^5}$ **15.** $\sqrt{x}$ **17.** $\sqrt[4]{\frac{3x}{7y}}$
19. $\sqrt[7]{77x^5y^3}$ **21.** $5\sqrt{2}$ **23.** $3\sqrt{5}$ **25.** $5\sqrt{3x}$ **27.** $2\sqrt[3]{2}$ **29.** $3x$ **31.** $-2y\sqrt[3]{2x^2}$ **33.** $6|x + 2|$ **35.** $2(x + 2)\sqrt[3]{4}$
37. $|x - 1|\sqrt{3}$ **39.** $x^3\sqrt{x}$ **41.** $x^4y^4\sqrt{y}$ **43.** $4x\sqrt{3x}$ **45.** $y^2\sqrt[3]{y^2}$ **47.** $x^4y\sqrt[3]{x^2z}$ **49.** $3x^2y^2\sqrt[3]{3x^2}$ **51.** $(x + y)\sqrt[3]{(x + y)^2}$
53. $y^3\sqrt[5]{y^2}$ **55.** $2xy^3\sqrt[5]{2xy^2}$ **57.** $2x^2\sqrt[4]{5x^2}$ **59.** $(x - 3)^2\sqrt[4]{(x - 3)^2}$ **61.** $2\sqrt{6}$ **63.** $5\sqrt{2xy}$ **65.** $6x$ **67.** $10xy\sqrt{2y}$
69. $60\sqrt{2}$ **71.** $2\sqrt[3]{6}$ **73.** $2x^2\sqrt{10x}$ **75.** $5xy^4\sqrt[4]{x^2y^2}$ **77.** $2xyz\sqrt[4]{x^2z^3}$ **79.** $2xy^2\sqrt[5]{2y^3}$ **81.** $(x - y)^2\sqrt[3]{(x - y)^2}$
83. $8\sqrt{2}$ mph; 11.3 mph **85.** $2\sqrt{5}$ sec; 4.5 sec **87. a.** $\frac{7.644}{2\sqrt[4]{2}} = \frac{3.822}{\sqrt[4]{2}}$ **b.** 3.21 liters of blood per minute per square meter; (32, 3.21)
89–93. Answers will vary. **95.** Simplification is correct. **97.** Graphs are the same; simplification is correct. **99.** d
101. The number must be multiplied by 27. **103.**

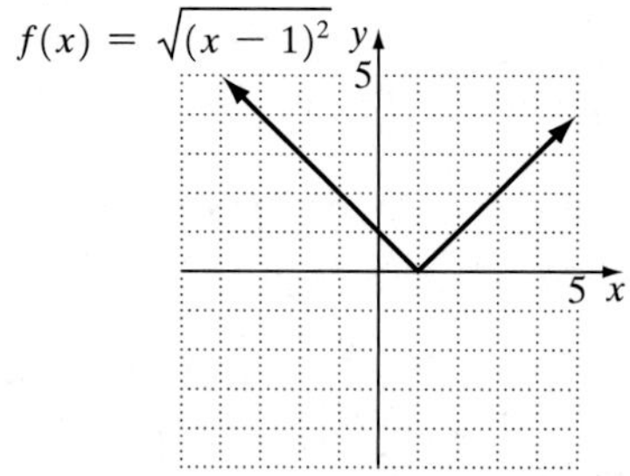

104. $\{x|5 \leq x \leq 11\}$ or $[5, 11]$
105. $(2, -4)$ or $\{(2, -4)\}$
106. $(4x - 3)(16x^2 + 12x + 9)$

Section 7.4

Check Point Exercises

1. a. $10\sqrt{13}$ **b.** $(21 - 6x)\sqrt[3]{7}$ **c.** $5\sqrt[4]{3x} + 2\sqrt[3]{3x}$ **2. a.** $21\sqrt{5}$ **b.** $-12\sqrt{3x}$ **c.** cannot be simplified
3. a. $-9\sqrt[3]{3}$ **b.** $(5 + 3xy)\sqrt[3]{x^2y}$ **4. a.** $\frac{2\sqrt[3]{3}}{5}$ **b.** $\frac{3x\sqrt{x}}{y^5}$ **c.** $\frac{2y^2\sqrt[3]{y}}{x^4}$ **5. a.** $2x^2\sqrt{5}$ **b.** $\frac{5\sqrt{xy}}{2}$ **c.** $2x^2y$

Exercise Set 7.4

1. $11\sqrt{5}$ **3.** $7\sqrt[3]{6}$ **5.** $2\sqrt[5]{2}$ **7.** $\sqrt{13}+2\sqrt{5}$ **9.** $7\sqrt{5}+2\sqrt[3]{x}$ **11.** $4\sqrt{3}$ **13.** $19\sqrt{3}$ **15.** $6\sqrt{2x}$ **17.** $13\sqrt[3]{2}$ **19.** $(9x+1)\sqrt{5x}$ **21.** $7y\sqrt[3]{2x}$ **23.** $(3x-2)\sqrt[3]{2x}$ **25.** $4\sqrt{x-2}$ **27.** $5x\sqrt[3]{xy^2}$ **29.** $\frac{\sqrt{11}}{2}$ **31.** $\frac{\sqrt[3]{19}}{3}$ **33.** $\frac{x}{6y^4}$ **35.** $\frac{2x\sqrt{2x}}{5y^3}$ **37.** $\frac{x\sqrt[3]{x}}{2y}$ **39.** $\frac{x^2\sqrt[3]{50x^2}}{3y^4}$ **41.** $\frac{y\sqrt[4]{9y^2}}{x^2}$ **43.** $\frac{2x^2\sqrt[5]{2x^3}}{y^4}$ **45.** $2\sqrt{2}$ **47.** 2 **49.** $3x$ **51.** x^2y **53.** $2x^2\sqrt{5}$ **55.** $3\sqrt{xy}$ **57.** $2xy$ **59.** $2x^2y^2\sqrt[4]{y^2}$ **61.** $\sqrt[3]{x+3}$

63. a. $R_f\frac{\sqrt{c^2-v^2}}{\sqrt{c^2}}=R_f\sqrt{\frac{c^2-v^2}{c^2}}=R_f\sqrt{\frac{c^2}{c^2}-\frac{v^2}{c^2}}=R_f\sqrt{1-\frac{v^2}{c^2}}=R_f\sqrt{1-\left(\frac{v}{c}\right)^2}$

b. $R_f\sqrt{1-\left(\frac{c}{c}\right)^2}=0$; Aging rate is 0.; A person moving at the speed of light does not age.

65. $P=18\sqrt{5}$ ft; $A=100$ sq ft **67–73.** Answers will vary. **75.** Graphs are not the same.; $\sqrt{16x}-\sqrt{9x}=\sqrt{x}$ **77.** d **79.** $-\frac{31\sqrt{5}}{12}$ **81.** 0 or {0} **82.** $(x-2y)(x-6y)$ **83.** $\frac{3x^2+8x+6}{(x+3)^2(x+2)}$

Section 7.5

Check Point Exercises

1. a. $x\sqrt{6}+2\sqrt{15}$ **b.** $y-\sqrt[3]{7y}$ **c.** $36-18\sqrt{10}$ **2. a.** $11+2\sqrt{30}$ **b.** 1 **c.** $a-7$ **3. a.** $\frac{\sqrt{21}}{7}$ **b.** $\frac{\sqrt[3]{6}}{3}$ **4. a.** $\frac{\sqrt{14xy}}{7y}$ **b.** $\frac{\sqrt[3]{3xy^2}}{3y}$ **c.** $\frac{3\sqrt[5]{4x^3y}}{y}$ **5.** $12\sqrt{3}-18$ **6.** $\frac{3\sqrt{5}+3\sqrt{2}+\sqrt{35}+\sqrt{14}}{3}$ **7.** $\frac{1}{\sqrt{x+3}+\sqrt{x}}$

Exercise Set 7.5

1. $x\sqrt{2}+\sqrt{14}$ **3.** $7\sqrt{6}-6$ **5.** $12\sqrt{2}-6$ **7.** $\sqrt[3]{12}+4\sqrt[3]{10}$ **9.** $2x\sqrt[3]{2}-\sqrt[3]{x^2}$ **11.** $32+11\sqrt{2}$ **13.** $34-15\sqrt{5}$ **15.** $117-36\sqrt{7}$ **17.** $\sqrt{6}+\sqrt{10}+\sqrt{21}+\sqrt{35}$ **19.** $\sqrt{6}-\sqrt{10}-\sqrt{21}+\sqrt{35}$ **21.** $-48+7\sqrt{6}$ **23.** $8+2\sqrt{15}$ **25.** $3x-2\sqrt{3xy}+y$ **27.** -44 **29.** -71 **31.** 6 **33.** $6-5\sqrt{x}+x$ **35.** $\sqrt[3]{x^2}+\sqrt[3]{x}-20$ **37.** $2x^2+x\sqrt[3]{y^2}-y\sqrt[3]{y}$ **39.** $\frac{\sqrt{10}}{5}$ **41.** $\frac{\sqrt{11x}}{x}$ **43.** $\frac{3\sqrt{3y}}{y}$ **45.** $\frac{\sqrt[3]{4}}{2}$ **47.** $3\sqrt[3]{2}$ **49.** $\frac{\sqrt[3]{18}}{3}$ **51.** $\frac{4\sqrt[3]{x^2}}{x}$ **53.** $\frac{\sqrt[3]{2y}}{y}$ **55.** $\frac{7\sqrt[3]{4x}}{2x}$ **57.** $\frac{\sqrt[3]{2x^2y}}{xy}$ **59.** $\frac{3\sqrt[4]{x^3}}{x}$ **61.** $\frac{3\sqrt[5]{4x^2}}{x}$ **63.** $x\sqrt[5]{8x^3y}$ **65.** $8\sqrt{5}-16$ **67.** $\frac{13\sqrt{11}+39}{2}$ **69.** $3\sqrt{5}-3\sqrt{3}$ **71.** $\frac{a+\sqrt{ab}}{a-b}$ **73.** $25\sqrt{2}+15\sqrt{5}$ **75.** $4+\sqrt{15}$ **77.** $\frac{x-2\sqrt{x}-3}{x-9}$ **79.** $\frac{3\sqrt{6}+4}{2}$ **81.** $\frac{4\sqrt{xy}+4x+y}{y-4x}$ **83.** $\frac{3}{\sqrt{6}}$ **85.** $\frac{2x}{\sqrt[3]{2x^2y}}$ **87.** $\frac{x-9}{x-3\sqrt{x}}$ **89.** $\frac{a-b}{a-2\sqrt{ab}+b}$ **91.** $\frac{1}{\sqrt{x+5}+\sqrt{x}}$ **93.** $\frac{1}{(x+y)(\sqrt{x}-\sqrt{y})}$ **95.** 33; In 2001, 33% of U.S. households were online.; Answer models data well. **97.** 3.5% **99.** 3.43%; This answer models the actual yearly increase well. **101. a.** $\frac{6.85}{\sqrt{t+h}+\sqrt{t}}$ **b.** $\frac{6.85}{2\sqrt{t}}$ or $\frac{3.425}{\sqrt{t}}$ **c.** 1.7% **103. a.** 18% **b.** $p(x)=\frac{13\sqrt{x}+x}{5}$ **105.** $P=8\sqrt{2}$ in.; $A=7$ sq in. **107.** $\frac{7\sqrt{2\cdot 2\cdot 3}}{6}=\frac{7\cdot 2\sqrt{3}}{6}=\frac{14\sqrt{3}}{6}=\frac{7\sqrt{3}}{3}=\frac{7}{3}\sqrt{3}$ **109–117.** Answers will vary. **119.** Graphs are not the same.; $(\sqrt{x}-1)(\sqrt{x}-1)=x-2\sqrt{x}+1$ **121.** Graphs are not the same.; $(\sqrt{x}+1)^2=x+2\sqrt{x}+1$ **123.** c **125.** 6

127. $3\left(\frac{3+\sqrt{3}}{3}\right)^2+2=6\left(\frac{3+\sqrt{3}}{3}\right)$ Substitute $\frac{3+\sqrt{3}}{3}$ for x.

$$\frac{(3+\sqrt{3})^2}{3}+2=2(3+\sqrt{3})$$
$$(3+\sqrt{3})^2+6=6(3+\sqrt{3})$$
$$(9+6\sqrt{3}+3)+6=18+6\sqrt{3}$$
$$18+6\sqrt{3}=18+6\sqrt{3}\quad \text{True}$$

128. $\frac{2x+7}{x^2-4}$ **129.** $-\frac{7}{2}$ or $\left\{-\frac{7}{2}\right\}$ **130.** 13

Section 7.6

Check Point Exercises

1. 20 or {20} **2.** no solution or ∅ **3.** −1 and 3, or {−1, 3} **4.** 4 or {4} **5.** −12 or {−12} **6.** 2013

Exercise Set 7.6

1. 6 or $\{6\}$ **3.** 17 or $\{17\}$ **5.** no solution or $\varnothing$ **7.** 8 or $\{8\}$ **9.** 0 and 3, or $\{0, 3\}$ **11.** 1 and 3, or $\{1, 3\}$ **13.** 3 and 7, or $\{3, 7\}$ **15.** 9 or $\{9\}$ **17.** 8 or $\{8\}$ **19.** 35 or $\{35\}$ **21.** 16 or $\{16\}$ **23.** 2 or $\{2\}$ **25.** $\frac{5}{2}$ or $\left\{\frac{5}{2}\right\}$ **27.** 5 or $\{5\}$ **29.** no solution or $\varnothing$ **31.** 2 and 6, or $\{2, 6\}$ **33.** 0 and 10, or $\{0, 10\}$ **35.** 8 or $\{8\}$ **37.** 2018 **39.** 2002 **41.** 36 years old; (36, 40,000) **43.** 27 sq mi **45.** 6078 million km **47–53.** Answers will vary.

55. 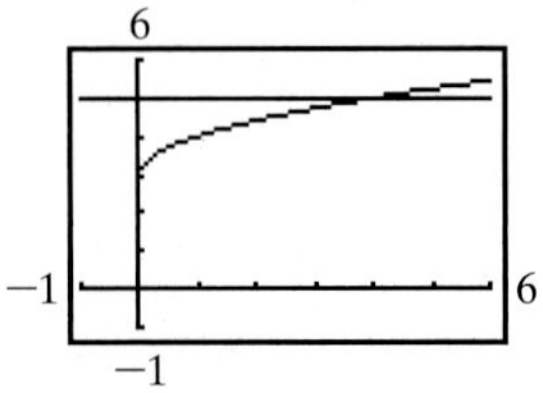

; 4 or $\{4\}$

57. 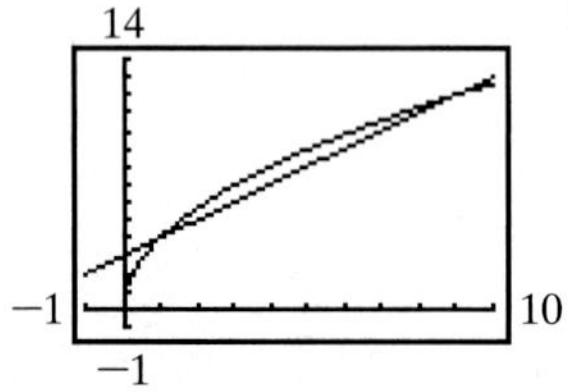

; 1 and 9, or $\{1, 9\}$

59. c **61.** 81 or $\{81\}$ **63.** 129 or $\{129\}$ **64.** $4x^3 - 15x^2 + 47x - 142 + \frac{425}{x + 3}$ **65.** $\frac{x + 2}{2(x + 3)}$ **66.** $(y - 3 + 5x)(y - 3 - 5x)$

Section 7.7

Check Point Exercises

1. a. $4i$ **b.** $\sqrt{5}i$ **c.** $5\sqrt{2}i$ **2. a.** $8 + i$ **b.** $-10 + 10i$ **3. a.** $63 + 14i$ **b.** $58 - 11i$ **4.** $-\sqrt{35}$ **5.** $\frac{18}{25} + \frac{26}{25}i$ **6.** $-\frac{1}{2} - \frac{3}{4}i$ **7. a.** 1 **b.** i **c.** $-i$

Exercise Set 7.7

1. $7i$ **3.** $\sqrt{17}i$ **5.** $5\sqrt{3}i$ **7.** $2\sqrt{7}i$ **9.** $-5\sqrt{6}i$ **11.** $7 + 4i$ **13.** $5 + \sqrt{5}i$ **15.** $6 - 3\sqrt{2}i$ **17.** $8 + 3i$ **19.** $8 - 2i$ **21.** $5 + 3i$ **23.** $-1 - 7i$ **25.** $-2 + 9i$ **27.** $8 + 15i$ **29.** $-14 + 17i$ **31.** $9 + 5\sqrt{3}i$ **33.** $-6 + 10i$ **35.** $-21 - 15i$ **37.** $-35 - 14i$ **39.** $7 + 19i$ **41.** $-1 - 31i$ **43.** $3 + 36i$ **45.** 34 **47.** 34 **49.** 11 **51.** $-5 + 12i$ **53.** $21 - 20i$ **55.** $-\sqrt{14}$ **57.** -6 **59.** $-5\sqrt{7}$ **61.** $-2\sqrt{6}$ **63.** $\frac{3}{5} - \frac{1}{5}i$ **65.** $1 + i$ **67.** $\frac{28}{25} + \frac{21}{25}i$ **69.** $-\frac{12}{13} + \frac{18}{13}i$ **71.** $0 + i$ or i **73.** $\frac{3}{10} - \frac{11}{10}i$ **75.** $\frac{11}{13} - \frac{16}{13}i$ **77.** $-\frac{23}{58} + \frac{43}{58}i$ **79.** $0 - \frac{7}{3}i$ or $-\frac{7}{3}i$ **81.** $-\frac{5}{2} - 4i$ **83.** $-\frac{7}{3} + \frac{4}{3}i$ **85.** -1 **87.** $-i$ **89.** -1 **91.** 1 **93.** i **95.** 1 **97.** $-i$ **99.** 0 **101.** $(47 + 13i)$ volts **103.** $(5 + \sqrt{15}i) + (5 - \sqrt{15}i) = 10$; $(5 + \sqrt{15}i)(5 - \sqrt{15}i) = 25 - 15i^2 = 25 + 15 = 40$ **105–115.** Answers will vary. **117.** $\sqrt{-9} + \sqrt{-16} = 3i + 4i = 7i$ **119.** d **121.** $\frac{14}{25} - \frac{2}{25}i$ **123.** 0 **124.** $\frac{x^2}{y^2}$ **125.** $x = \frac{yz}{y - z}$ **126.** $\frac{21}{11}$ or $\left\{\frac{21}{11}\right\}$

Review Exercises

1. 9 **2.** $-\frac{1}{10}$ **3.** -3 **4.** not a real number **5.** -2 **6.** 5; 1.73; 0; not a real number **7.** 2; −2; −4 **8.** $\{x \mid x \geq 2\}$ or $[2, \infty)$ **9.** $\{x \mid x \leq 25\}$ or $(-\infty, 25]$ **10.** $5|x|$ **11.** $|x + 14|$ **12.** $|x - 4|$ **13.** $4x$ **14.** $2|x|$ **15.** $-2(x + 7)$ **16.** $\sqrt[3]{5xy}$ **17.** $(\sqrt{16})^3 = 64$ **18.** $(\sqrt[5]{32})^4 = 16$ **19.** $(7x)^{1/2}$ **20.** $(19xy)^{5/3}$ **21.** $\frac{1}{8^{2/3}} = \frac{1}{4}$ **22.** $\frac{3x}{(ab)^{4/5}} = \frac{3x}{\sqrt[5]{(ab)^4}}$ **23.** $x^{7/12}$ **24.** $5^{1/6}$ **25.** $2x^2y$ **26.** $\frac{y^{1/8}}{x^{1/3}}$ **27.** x^3y^4 **28.** $y\sqrt[3]{x}$ **29.** $\sqrt[6]{x^5}$ **30.** $\sqrt[6]{x}$ **31.** $\sqrt[15]{x}$ **32.** \$3150 million **33.** $\sqrt{21xy}$ **34.** $\sqrt[5]{77x^3}$ **35.** $\sqrt[6]{(x - 5)^5}$ **36.** $f(x) = \sqrt{7}|x - 1|$ **37.** $2x\sqrt{5x}$ **38.** $3x^2y^2\sqrt[3]{2x^2}$ **39.** $2y^2\sqrt[4]{2x^3y^3}$ **40.** $2x^2\sqrt{6x}$ **41.** $2xy\sqrt[3]{2y^2}$ **42.** $xy\sqrt[5]{16y^4}$ **43.** $\sqrt{x^2 - 1}$ **44.** $8\sqrt[3]{3}$ **45.** $9\sqrt{2}$ **46.** $(3x + y^2)\sqrt[3]{x}$ **47.** $-8\sqrt[3]{6}$ **48.** $\frac{2}{5}\sqrt[3]{2}$ **49.** $\frac{x\sqrt{x}}{10y^2}$ **50.** $\frac{y\sqrt[4]{3y}}{2x^5}$ **51.** $2\sqrt{6}$ **52.** $2\sqrt[3]{2}$ **53.** $2x\sqrt[4]{2x}$ **54.** $10x^2\sqrt{xy}$ **55.** $6\sqrt{2} + 12\sqrt{5}$ **56.** $5\sqrt[3]{2} - \sqrt[3]{10}$ **57.** $-83 + 3\sqrt{35}$ **58.** $\sqrt{xy} - \sqrt{11x} - \sqrt{11y} + 11$ **59.** $13 + 4\sqrt{10}$ **60.** $22 - 4\sqrt{30}$ **61.** -6 **62.** 4 **63.** $\frac{2\sqrt{6}}{3}$ **64.** $\frac{\sqrt{14}}{7}$ **65.** $4\sqrt[3]{3}$ **66.** $\frac{\sqrt{10xy}}{5y}$ **67.** $\frac{7\sqrt[3]{4x}}{x}$ **68.** $\frac{\sqrt[4]{189x^3}}{3x}$ **69.** $\frac{5\sqrt[5]{xy^4}}{2xy}$ **70.** $3\sqrt{3} + 3$ **71.** $\frac{\sqrt{35} - \sqrt{21}}{2}$ **72.** $10\sqrt{5} + 15\sqrt{2}$ **73.** $\frac{x + 8\sqrt{x} + 15}{x - 9}$ **74.** $\frac{5 + \sqrt{21}}{2}$ **75.** $\frac{3\sqrt{2} + 2}{7}$ **76.** $\frac{2}{\sqrt{14}}$ **77.** $\frac{3x}{\sqrt[3]{9x^2y}}$ **78.** $\frac{7}{\sqrt{35} + \sqrt{21}}$

79. $\frac{4}{10 - 2\sqrt{21}}$ **80.** 16 or $\{16\}$ **81.** no solution or $\varnothing$ **82.** 2 or $\{2\}$ **83.** 8 or $\{8\}$ **84.** -4 and -2, or $\{-4, -2\}$
85. 256 feet **86.** 84 years old **87.** $9i$ **88.** $3\sqrt{7}i$ **89.** $-2\sqrt{2}i$ **90.** $12 + 2i$ **91.** $-9 + 4i$ **92.** $-12 - 8i$
93. $29 + 11i$ **94.** $-7 - 24i$ **95.** 113 **96.** $-2\sqrt{6}$ **97.** $\frac{15}{13} - \frac{3}{13}i$ **98.** $\frac{1}{5} + \frac{11}{10}i$ **99.** $\frac{1}{3} - \frac{5}{3}i$ **100.** 1 **101.** $-i$

Chapter 7 Test

1. a. 6 **b.** $\{x \mid x \le 4\}$ or $(-\infty, 4]$ **2.** $\frac{1}{81}$ **3.** $\frac{5\sqrt[8]{y}}{\sqrt[4]{x}}$ **4.** $\sqrt{x}$ **5.** $\sqrt[20]{x^9}$ **6.** $5|x|\sqrt{3}$ **7.** $|x - 5|$ **8.** $2xy^2\sqrt[3]{xy^2}$
9. $-\frac{2}{x^2}$ **10.** $\sqrt[3]{50x^2y}$ **11.** $2x\sqrt[4]{2y^3}$ **12.** $-7\sqrt{2}$ **13.** $(2x + y^2)\sqrt[3]{x}$ **14.** $2x\sqrt[3]{x}$ **15.** $12\sqrt{2} - \sqrt{15}$ **16.** $26 + 6\sqrt{3}$
17. $52 - 14\sqrt{3}$ **18.** $\frac{\sqrt{5x}}{x}$ **19.** $\frac{\sqrt[3]{25x}}{x}$ **20.** $-5 + 2\sqrt{6}$ **21.** 6 or $\{6\}$ **22.** 16 or $\{16\}$ **23.** -3 or $\{-3\}$ **24.** 49 months
25. $5\sqrt{3}i$ **26.** $-1 + 6i$ **27.** $26 + 7i$ **28.** -6 **29.** $\frac{1}{5} + \frac{7}{5}i$ **30.** $-i$

Cumulative Review Exercises (Chapters 1–7)

1. $(-2, -1, -2)$ or $\{(-2, -1, -2)\}$ **2.** $-\frac{1}{3}$ and 4, or $\left\{-\frac{1}{3}, 4\right\}$ **3.** $\left\{x \,\middle|\, x > \frac{1}{3}\right\}$ or $\left(\frac{1}{3}, \infty\right)$ **4.** $\frac{3}{4}$ or $\left\{\frac{3}{4}\right\}$ **5.** -1 or $\{-1\}$

6. $x + 2y < 2$
$2y - x > 4$

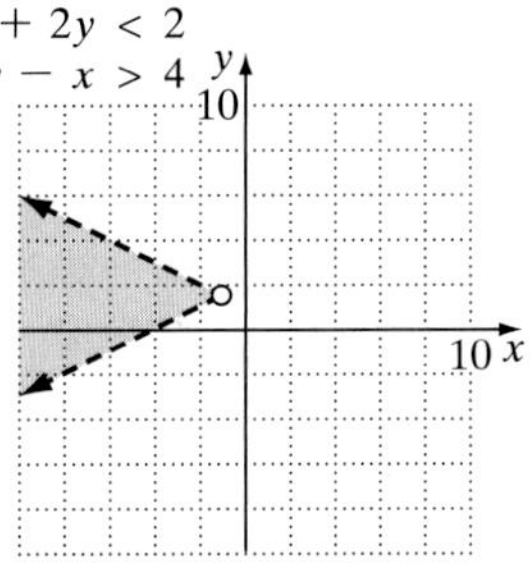

7. $\frac{x^2}{15(x + 2)}$ **8.** $\frac{x}{y}$ **9.** $8x^3 - 22x^2 + 11x + 6$
10. $\frac{5x - 6}{(x - 5)(x + 3)}$ **11.** -64 **12.** 0
13. $-\frac{16 - 9\sqrt{3}}{13}$ **14.** $2x^2 + x + 5 + \frac{6}{x - 2}$ **15.** $-34 - 3\sqrt{6}$
16. $2(3x + 2)(4x - 1)$ **17.** $(4x^2 + 1)(2x + 1)(2x - 1)$
18. about 53 lumens **19.** \$1500 at 7% and \$4500 at 9% **20.** 2650 students

CHAPTER 8

Section 8.1

Check Point Exercises

1. $\pm\sqrt{3}$ or $\{\pm\sqrt{3}\}$ **2.** $\pm\frac{\sqrt{10}}{2}$ or $\left\{\pm\frac{\sqrt{10}}{2}\right\}$ **3.** $\pm\frac{3}{2}i$ or $\left\{\pm\frac{3}{2}i\right\}$ **4.** $2 \pm \sqrt{7}$ or $\{2 \pm \sqrt{7}\}$ **5. a.** $x^2 + 10x + 25 = (x + 5)^2$
b. $x^2 - 3x + \frac{9}{4} = \left(x - \frac{3}{2}\right)^2$ **c.** $x^2 + \frac{3}{4}x + \frac{9}{64} = \left(x + \frac{3}{8}\right)^2$ **6.** $5 \pm \sqrt{7}$ or $\{5 \pm \sqrt{7}\}$ **7.** $\frac{5 \pm 3\sqrt{3}}{2}$ or $\left\{\frac{5 \pm 3\sqrt{3}}{2}\right\}$
8. 20% **9.** $10\sqrt{21}$ ft; 45.8 ft

Exercise Set 8.1

1. ±5 or $\{\pm5\}$ **3.** $\pm\sqrt{6}$ or $\{\pm\sqrt{6}\}$ **5.** $\pm\frac{5}{4}$ or $\left\{\pm\frac{5}{4}\right\}$ **7.** $\pm\frac{\sqrt{6}}{3}$ or $\left\{\pm\frac{\sqrt{6}}{3}\right\}$ **9.** $\pm\frac{4}{5}i$ or $\left\{\pm\frac{4}{5}i\right\}$
11. -10 and -4, or $\{-10, -4\}$ **13.** $3 \pm \sqrt{5}$ or $\{3 \pm \sqrt{5}\}$ **15.** $-2 \pm 2\sqrt{2}$ or $\{-2 \pm 2\sqrt{2}\}$ **17.** $5 \pm 3i$ or $\{5 \pm 3i\}$
19. $\frac{-3 \pm \sqrt{11}}{4}$ or $\left\{\frac{-3 \pm \sqrt{11}}{4}\right\}$ **21.** -3 and 9, or $\{-3, 9\}$ **23.** $x^2 + 2x + 1 = (x + 1)^2$ **25.** $x^2 - 14x + 49 = (x - 7)^2$
27. $x^2 + 7x + \frac{49}{4} = \left(x + \frac{7}{2}\right)^2$ **29.** $x^2 - \frac{1}{2}x + \frac{1}{16} = \left(x - \frac{1}{4}\right)^2$ **31.** $x^2 + \frac{4}{3}x + \frac{4}{9} = \left(x + \frac{2}{3}\right)^2$
33. $x^2 - \frac{9}{4}x + \frac{81}{64} = \left(x - \frac{9}{8}\right)^2$ **35.** -4 and -2, or $\{-4, -2\}$ **37.** $-3 \pm \sqrt{7}$ or $\{-3 \pm \sqrt{7}\}$ **39.** $-2 \pm \sqrt{3}$ or $\{-2 \pm \sqrt{3}\}$
41. $-1 \pm i$ or $\{-1 \pm i\}$ **43.** $\frac{-3 \pm \sqrt{13}}{2}$ or $\left\{\frac{-3 \pm \sqrt{13}}{2}\right\}$ **45.** $\frac{7 \pm \sqrt{37}}{2}$ or $\left\{\frac{7 \pm \sqrt{37}}{2}\right\}$ **47.** $\frac{-1 \pm \sqrt{5}}{2}$ or $\left\{\frac{-1 \pm \sqrt{5}}{2}\right\}$
49. $\frac{1}{2}$ and 1, or $\left\{\frac{1}{2}, 1\right\}$ **51.** $\frac{-5 \pm \sqrt{2}}{2}$ or $\left\{\frac{-5 \pm \sqrt{2}}{2}\right\}$ **53.** $\frac{1 \pm \sqrt{13}}{3}$ or $\left\{\frac{1 \pm \sqrt{13}}{3}\right\}$ **55.** $\frac{1}{4} \pm \frac{1}{4}i$ or $\left\{\frac{1}{4} \pm \frac{1}{4}i\right\}$ **57.** 20%
59. 3.69% **61.** 2003 **63.** $10\sqrt{3}$ sec; 17.3 sec **65.** $3\sqrt{5}$ mi; 6.7 mi **67.** $5\sqrt{7}$ ft; 13.2 ft

69. $50\sqrt{2}$ ft; 70.7 ft **70.** $70\sqrt{2}$ ft; 99 ft **71.** $c = \sqrt{\frac{E}{m}}$ or $c = \frac{\sqrt{mE}}{m}$ **73.** 8 m

75–83. Answers will vary. **85.**

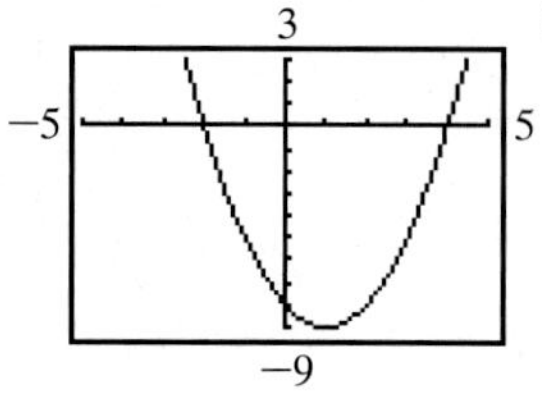

; −2 and 4, or {−2, 4} **87.** d **89.** $\frac{-1 \pm \sqrt{1-4c}}{2}$ or $\left\{\frac{-1 \pm \sqrt{1-4c}}{2}\right\}$

91. $\pm\sqrt{5}$ and $\pm\sqrt{3}$, or $\{\pm\sqrt{5}, \pm\sqrt{3}\}$

92. $4 - 2x$ **93.** $(1 - 2x)(1 + 2x + 4x^2)$

94. $x^3 - 2x^2 - 4x - 12 - \frac{42}{x-3}$

Section 8.2

Check Point Exercises

1. −5 and $\frac{1}{2}$, or $\left\{-5, \frac{1}{2}\right\}$ **2.** $\frac{3 \pm \sqrt{7}}{2}$ or $\left\{\frac{3 \pm \sqrt{7}}{2}\right\}$ **3.** $-1 \pm \frac{\sqrt{6}}{3}i$ or $\left\{-1 \pm \frac{\sqrt{6}}{3}i\right\}$ **4. a.** 0; one real solution

b. 81; two rational solutions **c.** −44; two imaginary solutions **5. a.** $20x^2 + 7x - 3 = 0$ **b.** $x^2 + 49 = 0$

6. 1995; actual number: 340 thousand; The function describes the situation well for that year.

Exercise Set 8.2

1. −6 and −2, or {−6, −2} **3.** 1 and $\frac{5}{2}$, or $\left\{1, \frac{5}{2}\right\}$ **5.** $\frac{-3 \pm \sqrt{89}}{2}$ or $\left\{\frac{-3 \pm \sqrt{89}}{2}\right\}$ **7.** $\frac{7 \pm \sqrt{85}}{6}$ or $\left\{\frac{7 \pm \sqrt{85}}{6}\right\}$

9. $\frac{1 \pm \sqrt{7}}{6}$ or $\left\{\frac{1 \pm \sqrt{7}}{6}\right\}$ **11.** $\frac{3}{8} \pm \frac{\sqrt{87}}{8}i$ or $\left\{\frac{3}{8} \pm \frac{\sqrt{87}}{8}i\right\}$ **13.** $2 \pm 2i$ or $\{2 \pm 2i\}$ **15.** $\frac{4}{3} \pm \frac{\sqrt{5}}{3}i$ or $\left\{\frac{4}{3} \pm \frac{\sqrt{5}}{3}i\right\}$

17. 52; two irrational solutions **19.** 4; two rational solutions **21.** −23; two imaginary solutions **23.** 36; two rational solutions

25. −60; two imaginary solutions **27.** 0; one real solution **29.** $-\frac{2}{3}$ and 2, or $\left\{-\frac{2}{3}, 2\right\}$ **31.** $1 \pm \sqrt{2}$ or $\{1 \pm \sqrt{2}\}$

33. $\frac{3 \pm \sqrt{65}}{4}$ or $\left\{\frac{3 \pm \sqrt{65}}{4}\right\}$ **35.** 0 and $\frac{8}{3}$; or $\left\{0, \frac{8}{3}\right\}$ **37.** $\frac{-6 \pm 2\sqrt{6}}{3}$ or $\left\{\frac{-6 \pm 2\sqrt{6}}{3}\right\}$ **39.** $\frac{2 \pm \sqrt{10}}{3}$ or $\left\{\frac{2 \pm \sqrt{10}}{3}\right\}$

41. $2 \pm \sqrt{10}$ or $\{2 \pm \sqrt{10}\}$ **43.** 1 and $\frac{5}{2}$, or $\left\{1, \frac{5}{2}\right\}$ **45.** $x^2 - 2x - 15 = 0$ **47.** $12x^2 + 5x - 2 = 0$ **49.** $x^2 + 36 = 0$

51. $x^2 - 2 = 0$ **53.** $x^2 - 20 = 0$ **55.** $x^2 - 2x + 2 = 0$ **57.** $x^2 - 2x - 1 = 0$ **59.** about 2002

61. 19 year olds and 72 year olds; Answers will vary. **63.** 3.8 sec **65.** 4.7 m by 7.7 m **67.** 2.3 ft by 3.3 ft

69. 9.3 in. and 0.7 in. **71–79.** Answers will vary.

81.

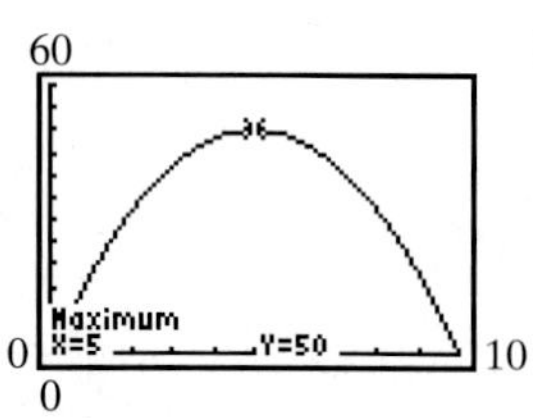

; depth: 5 in.; maximum area: 50 sq in.; Answers will vary. **83.** $\frac{3\sqrt{2} \pm \sqrt{10}}{2}$ or $\left\{\frac{3\sqrt{2} \pm \sqrt{10}}{2}\right\}$

85. 2.4 m; yes

87. −3 and $\frac{1}{4}$, or $\left\{-3, \frac{1}{4}\right\}$

88. 3 and 7, or {3, 7} **89.** $\frac{5\sqrt{3} - 5x}{3 - x^2}$

Section 8.3

Check Point Exercises

1.

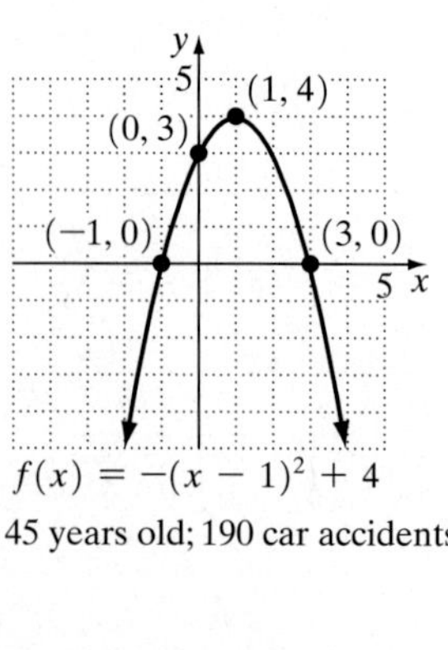

2.

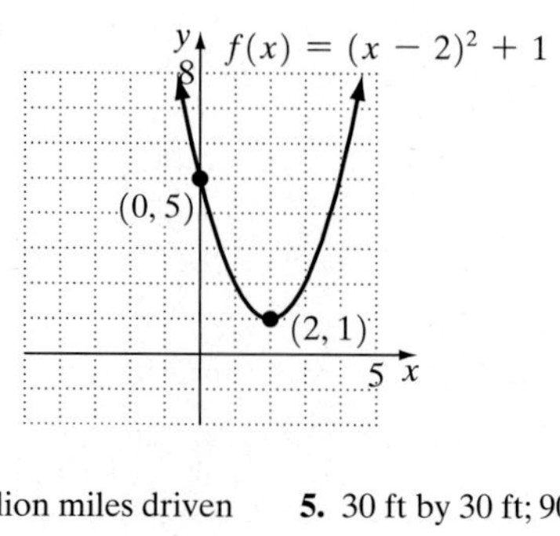

3.

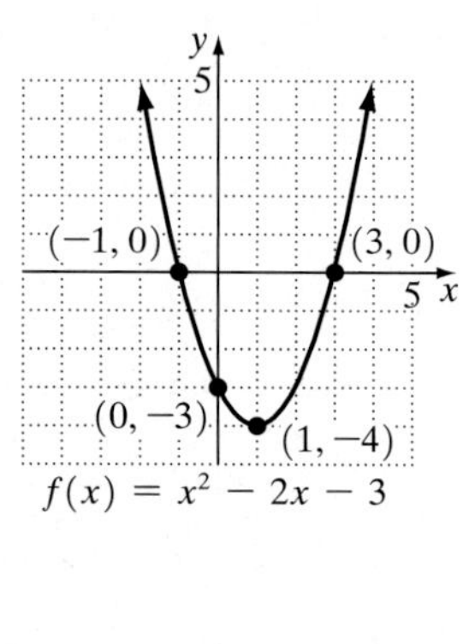

4. 45 years old; 190 car accidents per 50 million miles driven **5.** 30 ft by 30 ft; 90 sq ft

Exercise Set 8.3

1. $h(x) = (x - 1)^2 + 1$ **3.** $j(x) = (x - 1)^2 - 1$ **5.** $h(x) = x^2 - 1$ **7.** $g(x) = x^2 - 2x + 1$ **9.** $(3, 1)$ **11.** $(-1, 5)$
13. $(2, -5)$ **15.** $(-1, 9)$

17.

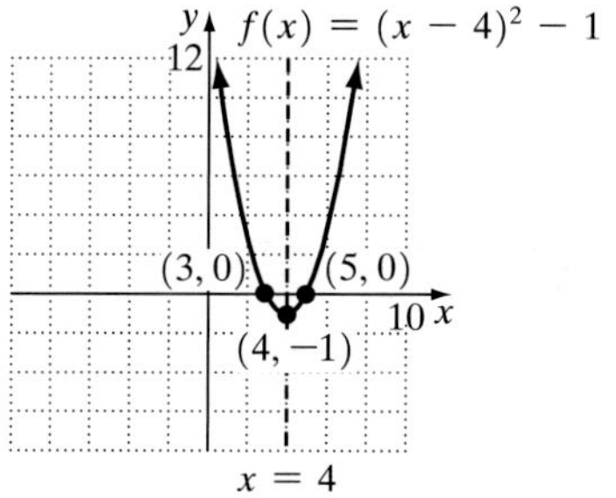

19.

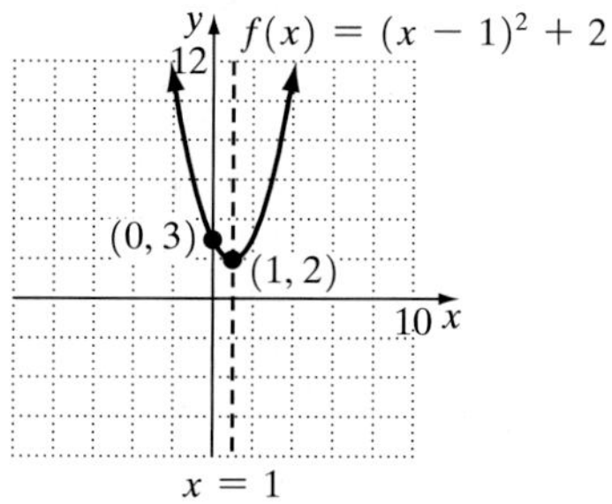

21.

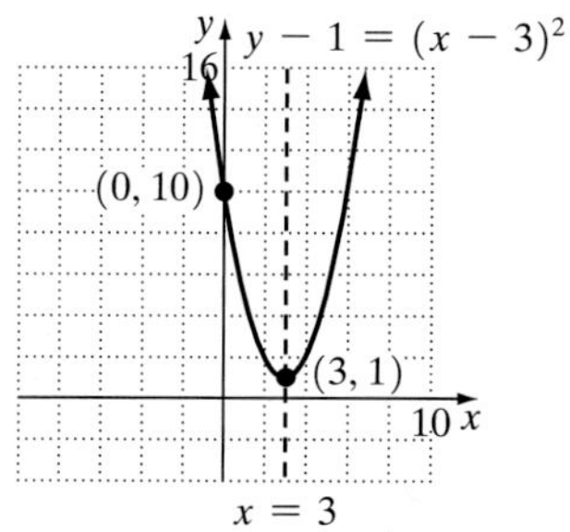

23.

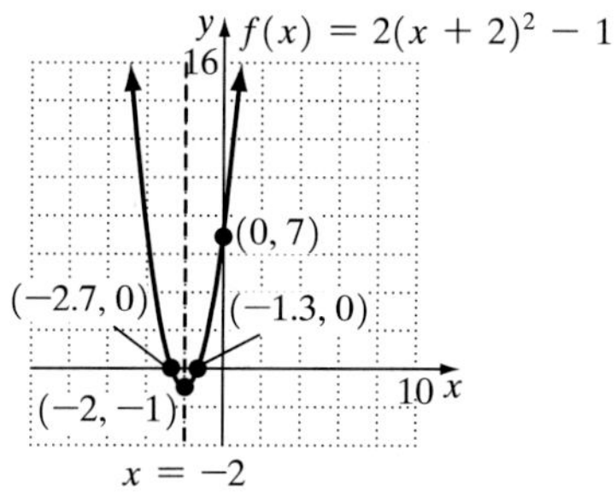

25.

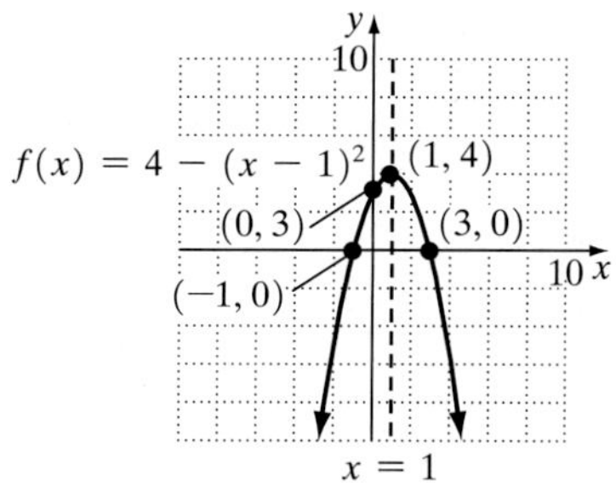

27.

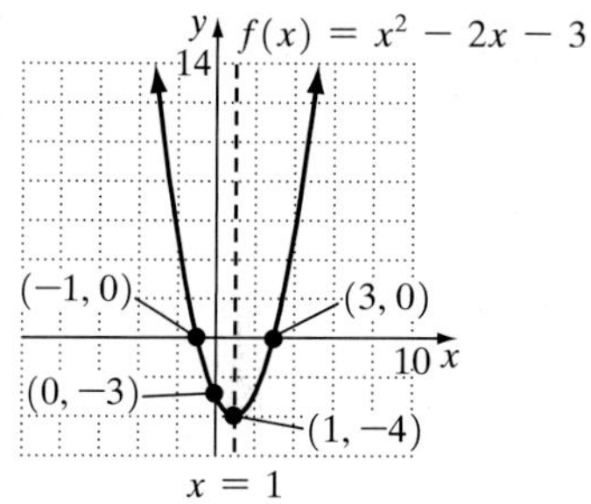

29.

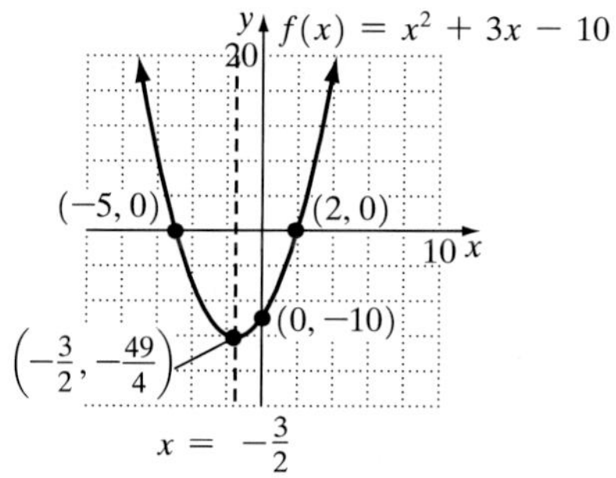

31.

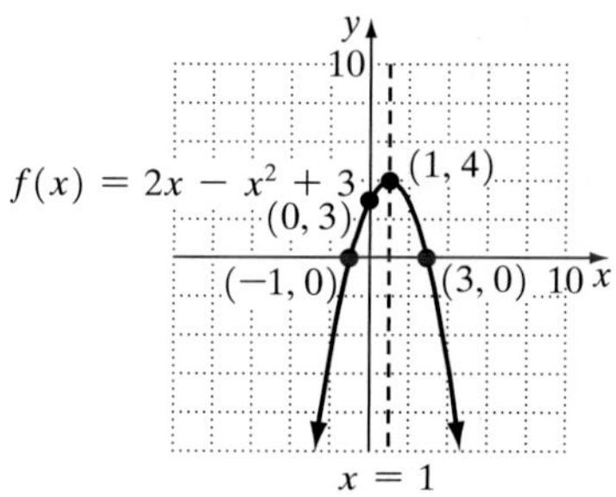

33.

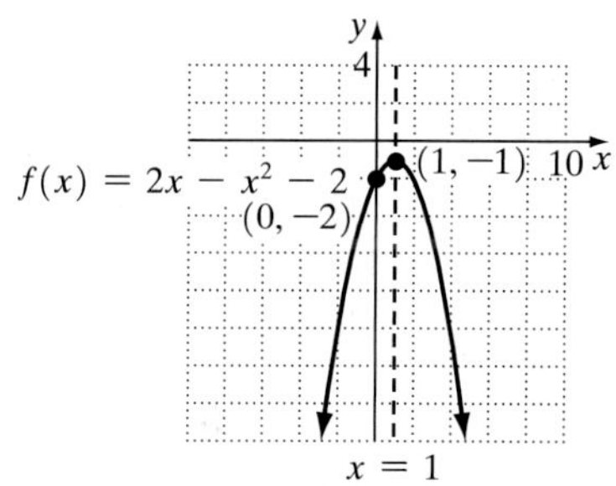

35. minimum; $(2, -13)$ **37.** maximum;t $(1, 1)$ **39.** minimum; $\left(\frac{1}{2}, -\frac{5}{4}\right)$ **41.** 1968; 4237 cigarettes per person

43. 7.2 hr; 622 per 100,000 males

45. **a.** 2 sec; 224 ft
b. 5.7 sec
c. 160; 160 feet is the height of the building.
d.

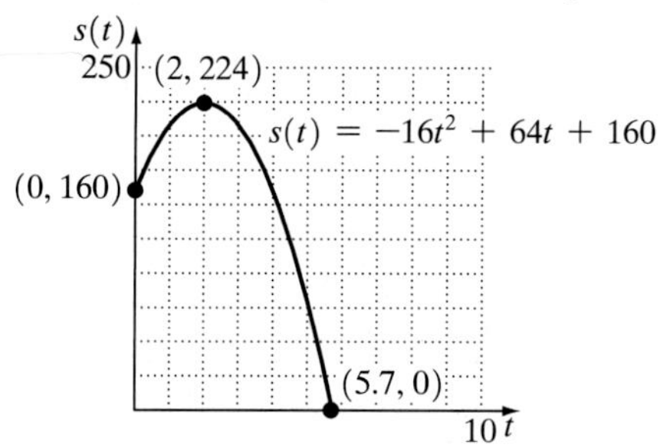

47. $(15, 41.1)$; What cost \$100 in 1967 cost \$41.10 in 1935.
49. length: 60 ft; width: 30 ft; 1800 sq ft
51. 8 and 8; 64 **53.** 5 and −5; −25
55. 12.5 yd by 12.5 yd; 156.25 sq yd **57.** 5 in.; 50 sq in.
59. **a.** $C(x) = 525 + 0.55x$
b. $P(x) = -0.001x^2 + 2.45x - 525$
c. 1225 sandwiches; \$975.63
61–67. Answers will vary.
69. **a.**

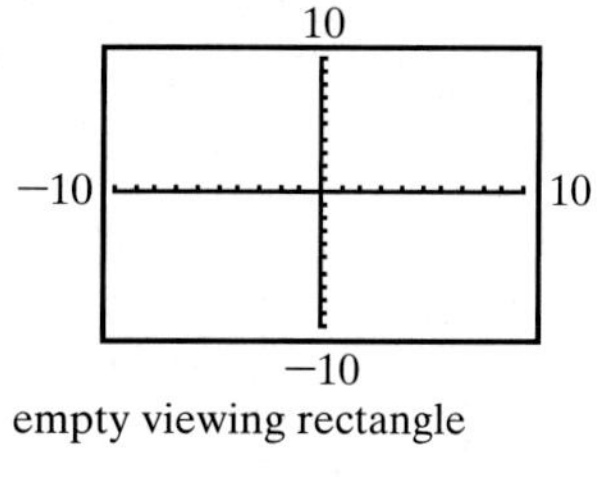

empty viewing rectangle

b. $(20.5, -120.5)$
c.

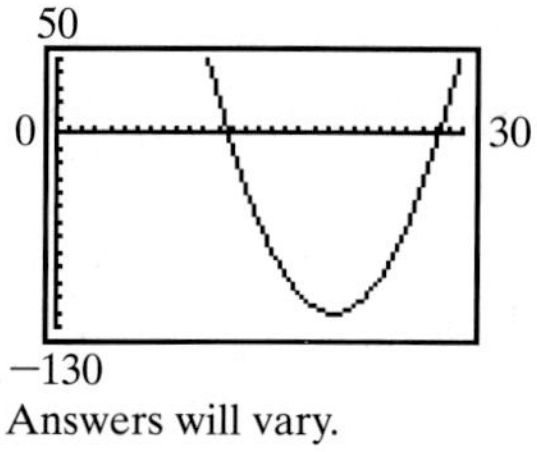

d. Answers will vary.

71. (2.5, 185)

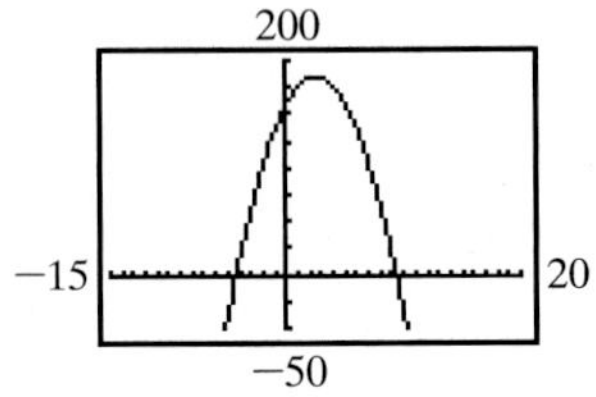

73. (−30, 91)

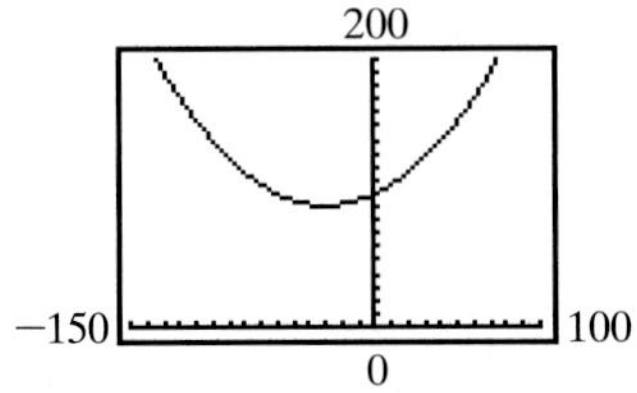

75. (4.4, 3.9); The minimum number of people holding more than one job was 3.9 million in 1975.

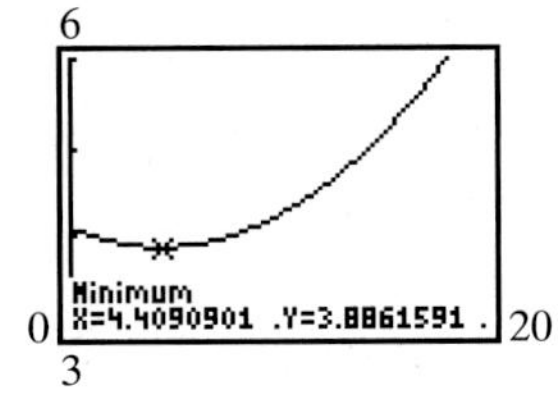

77. a **79.** $x = 3$; (0, 11) **81.** 65 trees; 16,900 lb **82.** 7 or {7} **83.** $\frac{x}{x-2}$ **84.** (6, −2) or {(6, −2)}

Section 8.4

Check Point Exercises

1. −4, 4, −1, and 1, or {−4, 4, −1, 1} **2.** 16 or {16} **3.** $-\sqrt{6}, -1, 1$, and $\sqrt{6}$, or $\{-\sqrt{6}, -1, 1, \sqrt{6}\}$ **4.** −1 and 2, or {−1, 2}
5. $-\frac{1}{27}$ and 64, or $\left\{-\frac{1}{27}, 64\right\}$

Exercise Set 8.4

1. −2, 2, −1, and 1, or {−2, 2, −1, 1} **3.** $-3, 3, -\sqrt{2}$, and $\sqrt{2}$, or $\{-3, 3, -\sqrt{2}, \sqrt{2}\}$ **5.** $-2i, 2i, -\sqrt{2}$, and $\sqrt{2}$, or $\{-2i, 2i, -\sqrt{2}, \sqrt{2}\}$
7. 1 or {1} **9.** 49 or {49} **11.** 25 and 64, or {25, 64} **13.** 2 and 12, or {2, 12} **15.** $-\sqrt{3}, \sqrt{3}$, and 0, or $\{-\sqrt{3}, \sqrt{3}, 0\}$
17. −5, −2, −1, and 2, or {−5, −2, −1, 2} **19.** $-\frac{1}{4}$ and $\frac{1}{5}$, or $\left\{-\frac{1}{4}, \frac{1}{5}\right\}$ **21.** $\frac{1}{3}$ and 2, or $\left\{\frac{1}{3}, 2\right\}$ **23.** $\frac{-2 \pm \sqrt{7}}{3}$ or $\left\{\frac{-2 \pm \sqrt{7}}{3}\right\}$
25. −8 and 27, or {−8, 27} **27.** −243 and 32, or {−243, 32} **29.** 1 or {1} **31.** −8, −2, 1, and 4, or {−8, −2, 1, 4}
33. −2, 2, −1, and 1; c **35.** 1; e **37.** 2 and 3; f **39.** ages 20 and 55; The function models the data well. **41–43.** Answers will vary.
45. −1 and 2, or {−1, 2} **47.** −1, 1, −3, and 3, or {−1, 1, −3, 3} **49.** $-\frac{3}{2}$ and 2, or $\left\{-\frac{3}{2}, 2\right\}$ **51.** 1 or {1} **53.** b
55. $\sqrt[3]{-2}$ and $\frac{\sqrt[3]{225}}{5}$, or $\left\{\sqrt[3]{-2}, \frac{\sqrt[3]{225}}{5}\right\}$ **57.** $\frac{1}{5x-1}$ **58.** $\frac{1}{2} + \frac{3}{2}i$ **59.** (4, −2) or {(4, −2)}

Section 8.5

Check Point Exercises

1. $\{x|-3 < x < 1\}$ or $(-3, 1)$

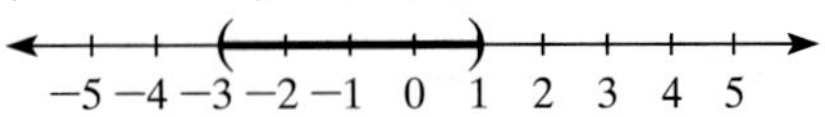

2. $\{x|x \leq -4$ or $x \geq 5\}$ or $(-\infty, -4] \cup [5, \infty)$

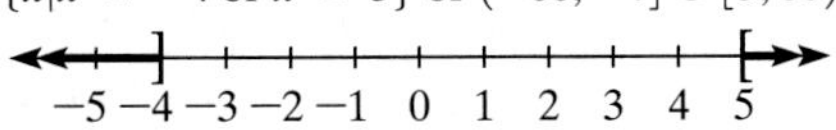

3. $\{x|x < -2$ or $x > 5\}$ or $(-\infty, -2) \cup (5, \infty)$

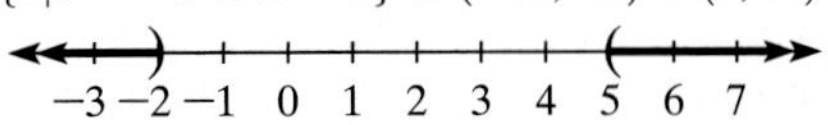

4. $\{x|-1 < x \leq 1\}$ or $(-1, 1]$

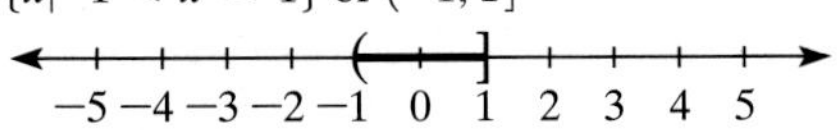

5. between 1 and 4 seconds, excluding $t = 1$ and $t = 4$

Exercise Set 8.5

1. $\{x|x < -2$ or $x > 4\}$ or $(-\infty, -2) \cup (4, \infty)$

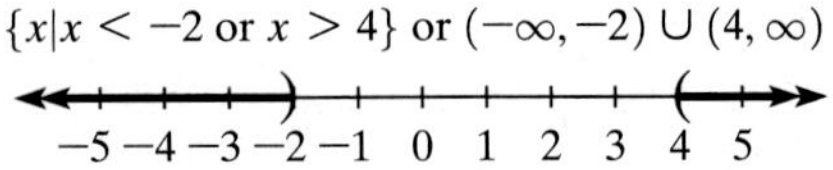

3. $\{x|-3 \leq x \leq 7\}$ or $[-3, 7]$

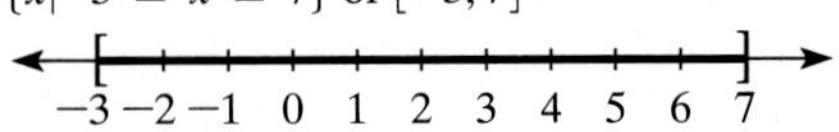

5. $\{x|x < 1$ or $x > 4\}$ or $(-\infty, 1) \cup (4, \infty)$

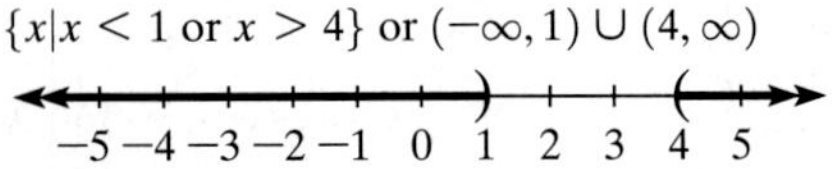

7. $\{x|x < -4$ or $x > -1\}$ or $(-\infty, -4) \cup (-1, \infty)$

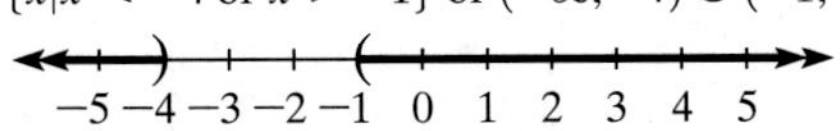

9. $\{x|2 \leq x \leq 4\}$ or $[2, 4]$

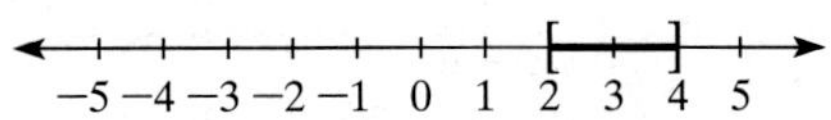

11. $\left\{x \middle| -4 \leq x \leq \frac{2}{3}\right\}$ or $\left[-4, \frac{2}{3}\right]$

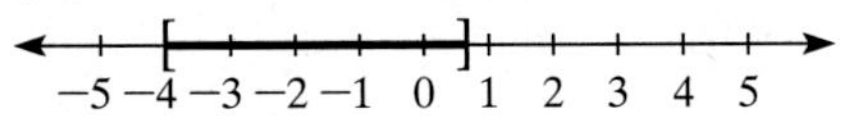

13. $\left\{x \middle| -3 < x < \frac{5}{2}\right\}$ or $\left(-3, \frac{5}{2}\right)$

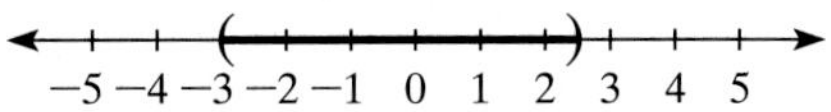

15. $\left\{x \middle| -1 < x < -\frac{3}{4}\right\}$ or $\left(-1, -\frac{3}{4}\right)$

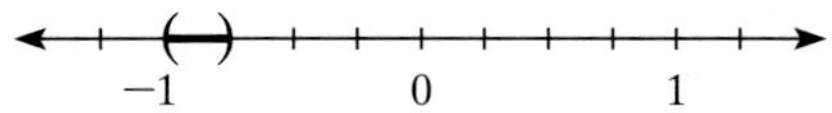

17. $\{x|x \le 0 \text{ or } x \ge 4\}$ or $(-\infty, 0] \cup [4, \infty)$

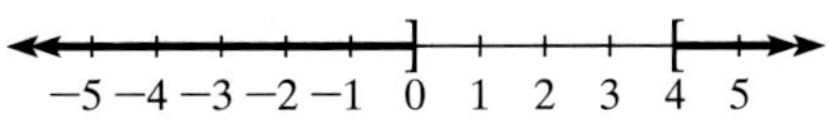

19. $\left\{x \middle| x < -\frac{3}{2} \text{ or } x > 0\right\}$ or $\left(-\infty, -\frac{3}{2}\right) \cup (0, \infty)$

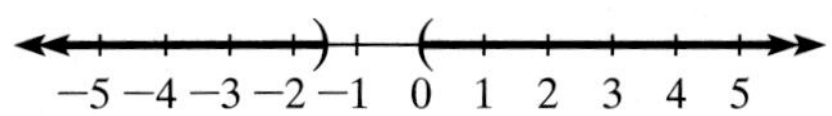

21. $\{x|0 \le x \le 1\}$ or $[0, 1]$

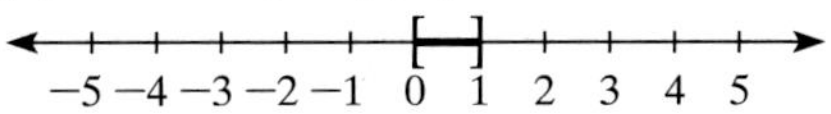

23. $\{x|2 - \sqrt{2} \le x \le 2 + \sqrt{2}\}$ or $[2 - \sqrt{2}, 2 + \sqrt{2}]$

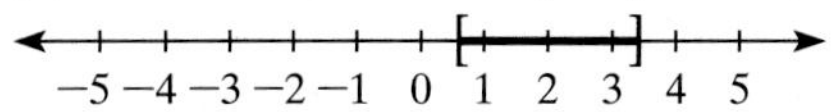

25. no solution or $\varnothing$

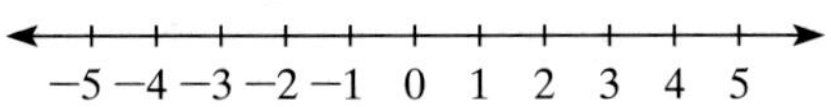

27. $\{x|x < -3 \text{ or } x > 4\}$ or $(-\infty, -3) \cup (4, \infty)$

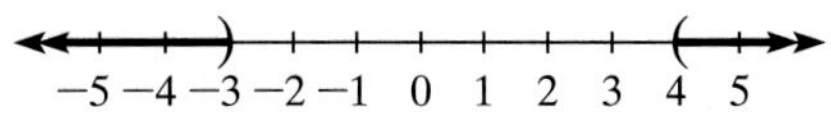

29. $\{x|-4 < x < -3\}$ or $(-4, -3)$

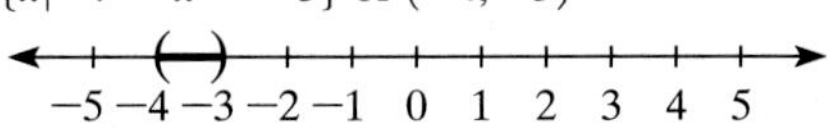

31. $\{x|2 \le x < 4\}$ or $[2, 4)$

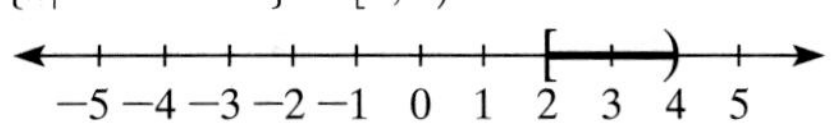

33. $\left\{x \middle| x < -\frac{4}{3} \text{ or } x \ge 2\right\}$ or $\left(-\infty, -\frac{4}{3}\right) \cup [2, \infty)$

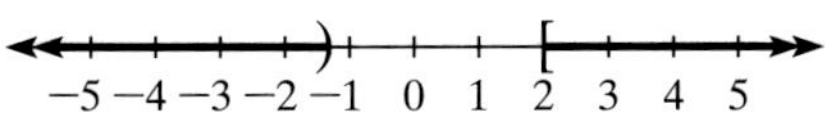

35. $\{x|x < 0 \text{ or } x > 3\}$ or $(-\infty, 0) \cup (3, \infty)$

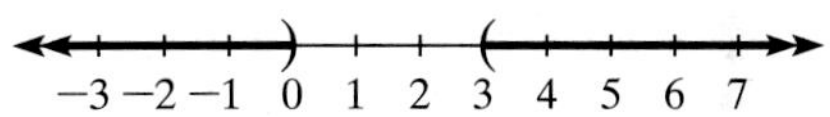

37. $\{x|x < -5 \text{ or } x > -3\}$ or $(-\infty, -5) \cup (-3, \infty)$

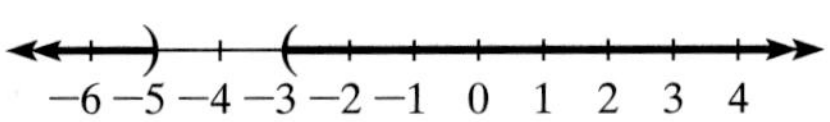

39. $\left\{x \middle| x < \frac{1}{2} \text{ or } x \ge \frac{7}{5}\right\}$ or $\left(-\infty, \frac{1}{2}\right) \cup \left(\frac{7}{5}, \infty\right)$

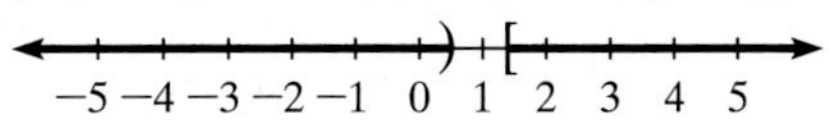

41. $\{x|x \le -6 \text{ or } x > -2\}$ or $(-\infty, -6] \cup (-2, \infty)$

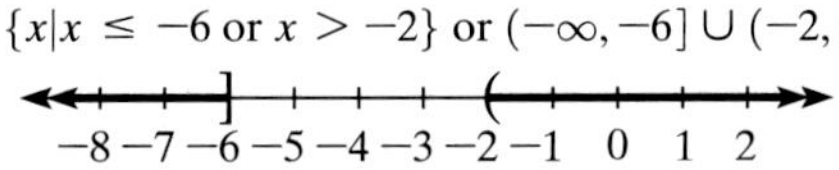

43. between 0 and 3 seconds, excluding $t = 0$ and $t = 3$
45. f: \$379 billion; g: \$379.8 billion; The functions model the value in the graph well.

47. after 2007 **49.** 1300.2 thousand or 1,300,200; The function models the actual number for 1998 well. **51.** years after 2010
53. The company's production level must be at least 20,000 wheelchairs per month. For values of x greater than or equal to 20,000, the graph lies on or below the line $y = 425$. **55–57.** Answers will vary. **59.** $\{x|x < -5 \text{ or } x > 2\}$ or $(-\infty, -5) \cup (2, \infty)$
61. $\{x|1 < x \le 4\}$ or $(1, 4]$ **63.** $\{x|-4 < x < -1 \text{ or } x \ge 2\}$ or $(-4, -1) \cup [2, \infty)$ **65.** d
67. Answers will vary; an example is: $\frac{x - 3}{x + 4} \ge 0$. **69.** $\{2\}$ **71.** $\{x|x < 2 \text{ or } x > 2\}$ or $(-\infty, 2) \cup (2, \infty)$
73. $27 - 3x^2 \ge 0$; $\{x|-3 \le x \le 3\}$ or $[-3, 3]$ **74.** $\{x|-19 < x < 29\}$ or $(-19, 29)$ **75.** $\frac{2(x - 1)}{(x + 4)(x - 3)}$
76. $(x^2 + 4y^2)(x + 2y)(x - 2y)$

Review Exercises

1. ± 8 or $\{\pm 8\}$ **2.** $\pm 5\sqrt{2}$ or $\{\pm 5\sqrt{2}\}$ **3.** $\pm\frac{\sqrt{6}}{3}$ or $\left\{\pm\frac{\sqrt{6}}{3}\right\}$ **4.** $4 \pm 3\sqrt{2}$ or $\{4 \pm 3\sqrt{2}\}$ **5.** $-7 \pm 6i$ or $\{-7 \pm 6i\}$
6. $x^2 + 20x + 100 = (x + 10)^2$ **7.** $x^2 - 3x + \frac{9}{4} = \left(x - \frac{3}{2}\right)^2$
8. 3 and 9, or $\{3, 9\}$ **9.** $\frac{7 \pm \sqrt{53}}{2}$ or $\left\{\frac{7 \pm \sqrt{53}}{2}\right\}$ **10.** $\frac{-3 \pm \sqrt{41}}{4}$ or $\left\{\frac{-3 \pm \sqrt{41}}{4}\right\}$ **11.** 8% **12.** 20 weeks
13. $60\sqrt{5}$ m; 134.2 m **14.** $1 \pm \sqrt{5}$ or $\{1 \pm \sqrt{5}\}$ **15.** $1 \pm 3\sqrt{2}i$ or $\{1 \pm 3\sqrt{2}i\}$ **16.** $\frac{-2 \pm \sqrt{10}}{2}$ or $\left\{\frac{-2 \pm \sqrt{10}}{2}\right\}$
17. -36; two imaginary solutions **18.** 81; two rational solutions **19.** 40; two irrational solutions **20.** $\frac{1}{2}$ and 5, or $\left\{\frac{1}{2}, 5\right\}$
21. -2 and $\frac{10}{3}$, or $\left\{-2, \frac{10}{3}\right\}$ **22.** $\frac{7 \pm \sqrt{37}}{6}$ or $\left\{\frac{7 \pm \sqrt{37}}{6}\right\}$ **23.** -3 and 3, or $\{-3, 3\}$ **24.** $3 \pm 2\sqrt{2}$ or $\{3 \pm 2\sqrt{2}\}$
25. $\frac{1}{6} \pm \frac{\sqrt{23}}{6}i$ or $\left\{\frac{1}{6} \pm \frac{\sqrt{23}}{6}i\right\}$ **26.** $4 \pm \sqrt{5}$ or $\{4 \pm \sqrt{5}\}$ **27.** $15x^2 - 4x - 3 = 0$ **28.** $x^2 + 81 = 0$
29. $x^2 - 48 = 0$ **30.** 2001 **31.** 8.8 sec

32.

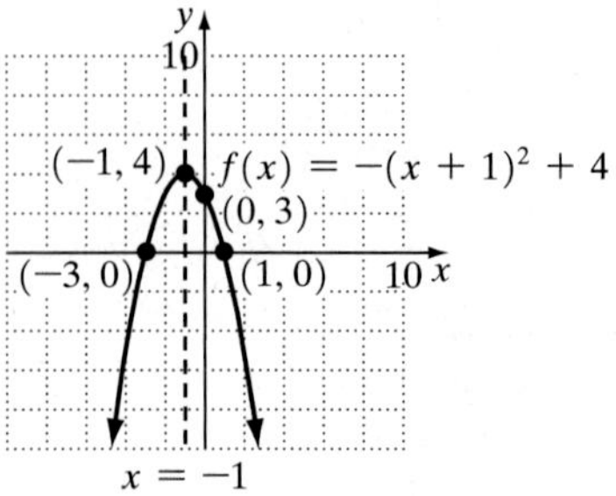

33.

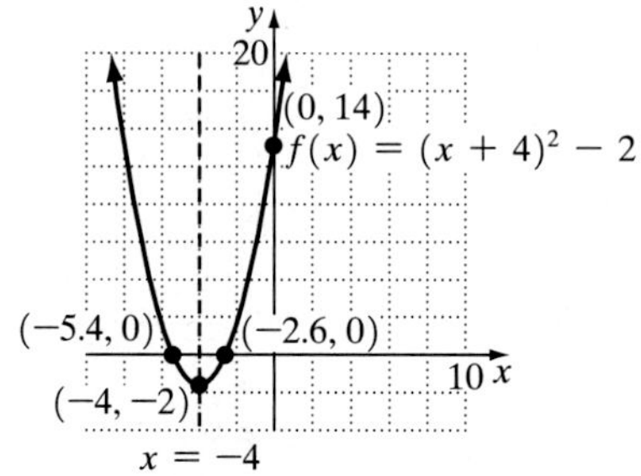

34.

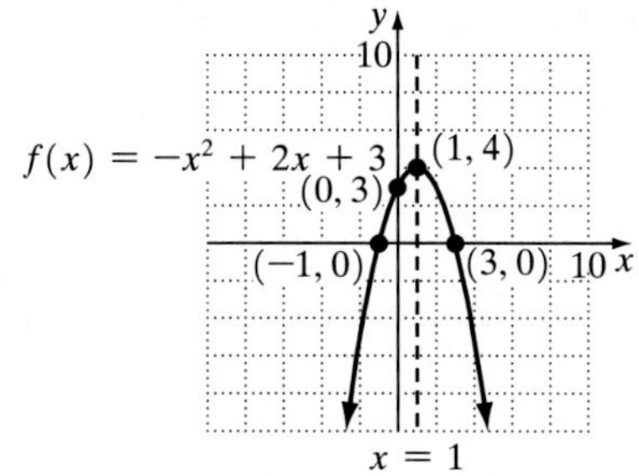

35.

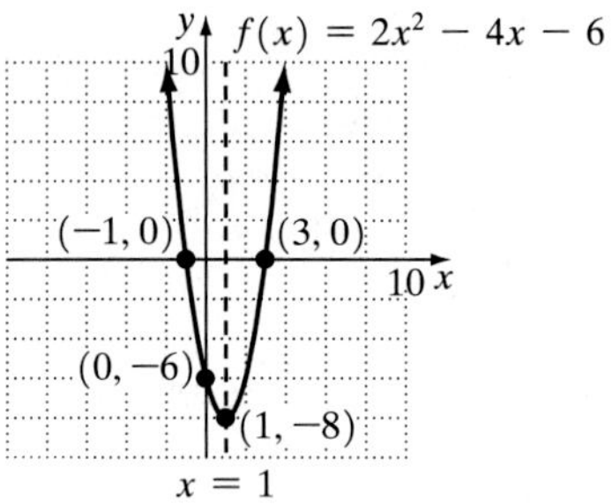

36. 25 in. of rainfall per year; 13.5 in. of growth
37. 12.5 sec; 2540 feet
38. (20, 5.4); The maximum divorce rate of about 5.4 per 1000 population was in 1980.
39. 250 yd by 500 yd; 125,000 sq yard
40. −7 and 7; −49 **41.** $-\sqrt{2}, \sqrt{2}, -2$, and 2, or $\{-\sqrt{2}, \sqrt{2}, -2, 2\}$ **42.** 1 or {1}
43. −5, −1, and 3, or {−5, −1, 3} **44.** $-\frac{1}{8}$ and $\frac{1}{7}$, or $\left\{-\frac{1}{8}, \frac{1}{7}\right\}$
45. −27 and 64, or {−27, 64} **46.** 16 or {16}
47. $\left\{x \middle| -3 < x < \frac{1}{2}\right\}$ or $\left(-3, \frac{1}{2}\right)$;

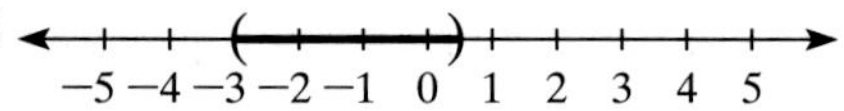

48. $\left\{x \middle| x \le -4 \text{ or } x \ge -\frac{1}{2}\right\}$ or $(-\infty, -4] \cup \left[-\frac{1}{2}, \infty\right)$;

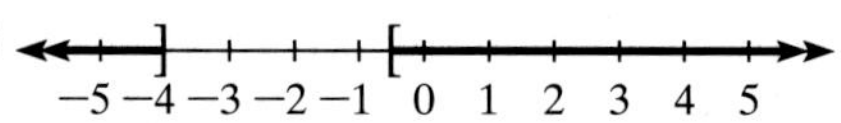

49. $\{x | x < -2 \text{ or } x > 6\}$ or $(-\infty, -2) \cup (6, \infty)$;

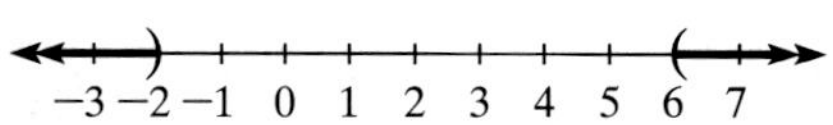

50. $\left\{x \middle| x < 4 \text{ or } x \ge \frac{23}{4}\right\}$ or $(-\infty, 4) \cup \left[\frac{23}{4}, \infty\right)$;

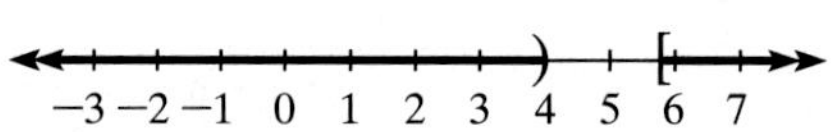

51. between 1 and 2 seconds, excluding $t = 1$ and $t = 2$ **52. a.** 200 beats per minute
b. between 0 and 4 minutes and more than 12 minutes after the workout; between 0 and 4 minutes; Answers will vary.

Chapter 8 Test

1. $\pm\frac{\sqrt{10}}{2}$ or $\left\{\pm\frac{\sqrt{10}}{2}\right\}$ **2.** $3 \pm 2\sqrt{5}$ or $\{3 \pm 2\sqrt{5}\}$ **3.** $x^2 - 16x + 64 = (x - 8)^2$ **4.** $x^2 + \frac{2}{5}x + \frac{1}{25} = \left(x + \frac{1}{5}\right)^2$
5. $3 \pm \sqrt{2}$ or $\{3 \pm \sqrt{2}\}$
6. $50\sqrt{2}$ ft **7.** 40; two irrational solutions. **8.** −16; two imaginary solutions.
9. −5 and $\frac{1}{2}$, or $\left\{-5, \frac{1}{2}\right\}$ **10.** $-4 \pm \sqrt{11}$ or $\{-4 \pm \sqrt{11}\}$ **11.** $-2 \pm 5i$ or $\{-2 \pm 5i\}$ **12.** $\frac{3}{2} \pm \frac{1}{2}i$ or $\left\{\frac{3}{2} \pm \frac{1}{2}i\right\}$
13. $x^2 - 4x - 21 = 0$ **14.** $x^2 + 100 = 0$
15. 1990 and 1998
16.

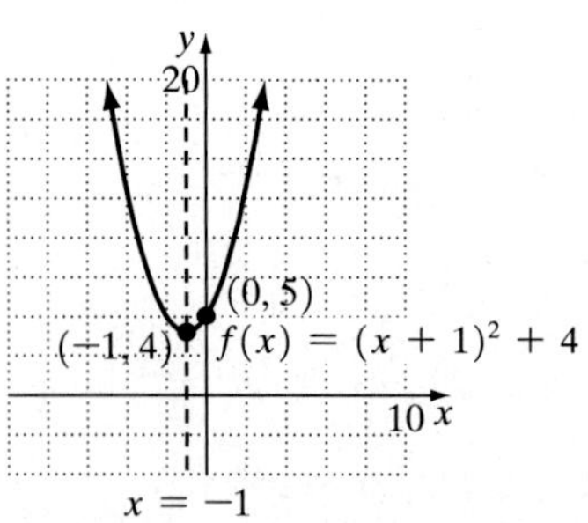

17.

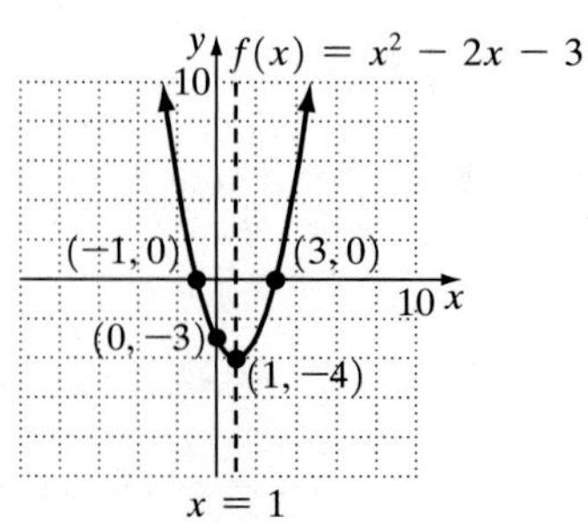

18. after 2 sec; 69 ft
19. 4.1 sec
20. 23 computers; \$169 hundreds or \$16,900
21. 1 and 2, or {1, 2}
22. −3, 3, −2, and 2, or {−3, 3, −2, 2}
23. 1 and 512, or {1, 512}

24. $\{x | -3 < x < 4\}$ or $(-3, 4)$

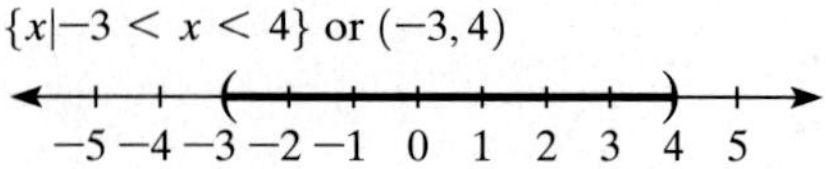

25. $\{x | x < 3 \text{ or } x \ge 10\}$ or $(-\infty, 3) \cup [10, \infty)$

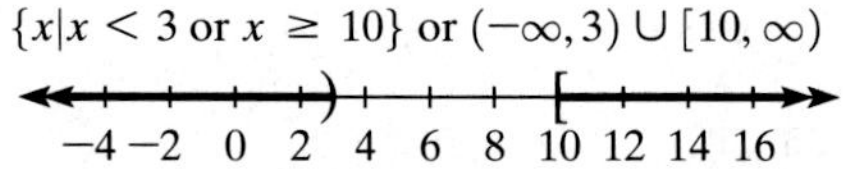

Cumulative Review Exercises (Chapters 1–8)

1. 4 or $\{4\}$ **2.** $(2, 3)$ or $\{(2, 3)\}$ **3.** $(2, 0, 3)$or $\{(2, 0, 3)\}$ **4.** $\{x|x < -2 \text{ or } x > 4\}$ or $(-\infty, -2) \cup (4, \infty)$ **5.** 5 or $\{5\}$
6. $\{x|x \geq 4\}$ or $[4, \infty)$ **7.** $\frac{3}{4} \pm \frac{\sqrt{7}}{4}i$ or $\left\{\frac{3}{4} \pm \frac{\sqrt{7}}{4}i\right\}$

8.

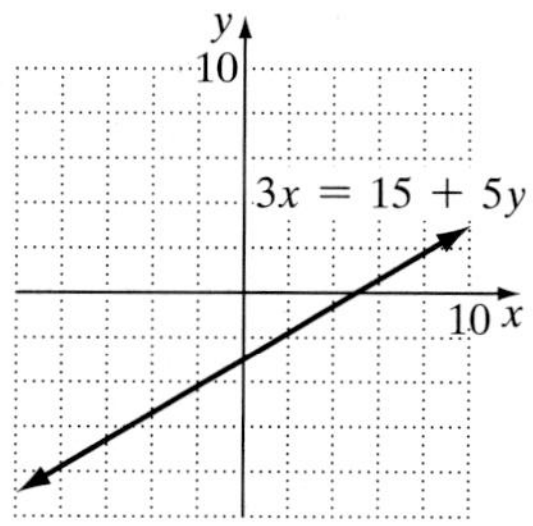

9.

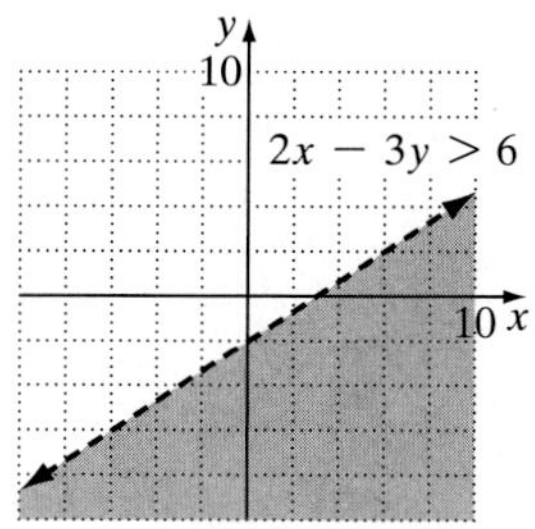

10.

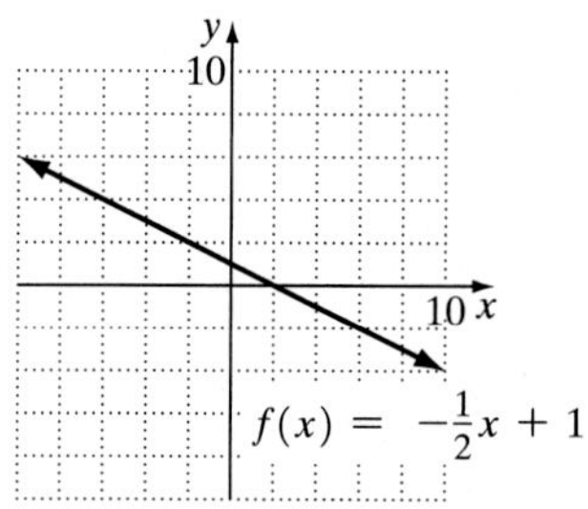

11.

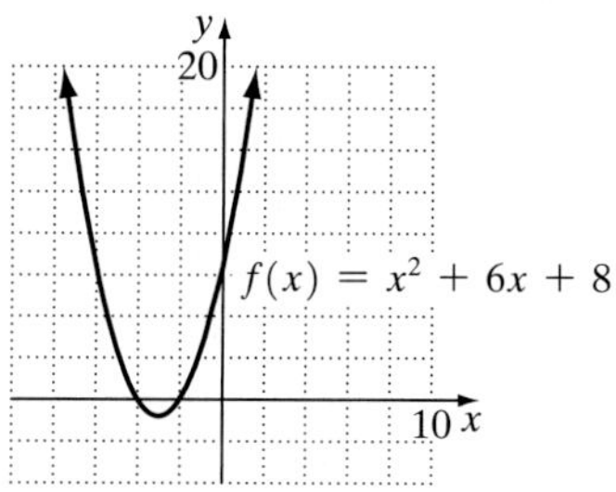

12.

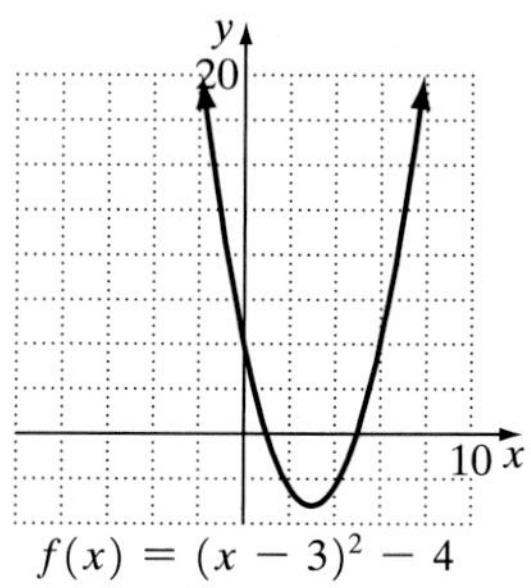

13. -6 **14.** $c = \frac{Ad}{d - A}$
15. $y = \frac{1}{2}x + 5$ or $f(x) = \frac{1}{2}x + 5$
16. $-\frac{y^9}{3x}$ **17.** $16x^4 - 40x^2y + 25y^2$
18. $x^2 - 5x + 1 + \frac{8}{5x + 1}$
19. $2y^2\sqrt[3]{2y^2}$ **20.** $\frac{4x - 13}{(x - 3)(x - 4)}$
21. $(x - 4)(x + 2)(x^2 - 2x + 4)$
22. $2(x + 3y)^2$ **23.** length: 12 ft; width: 4 ft **24.** $\frac{6}{5}$ hr or 1 hr and 12 min **25.** 5 mph

CHAPTER 9

Section 9.1

Check Point Exercises

1. one
2.

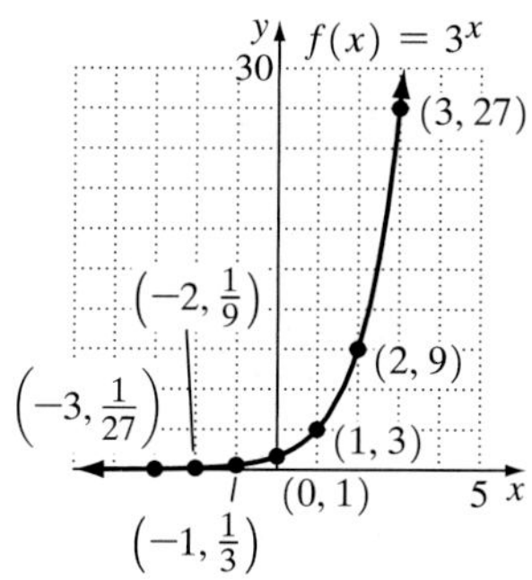

3.

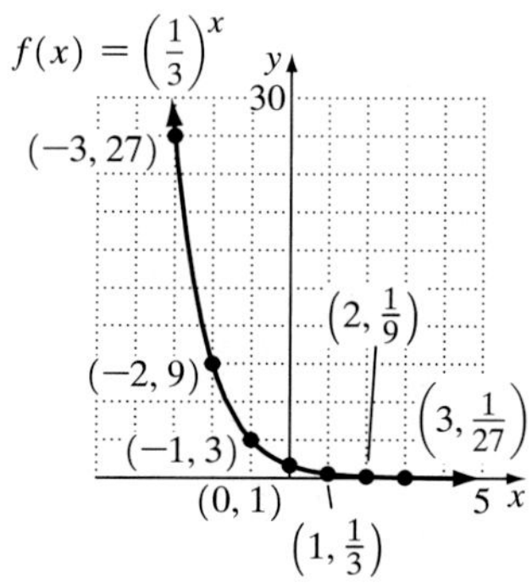

4.

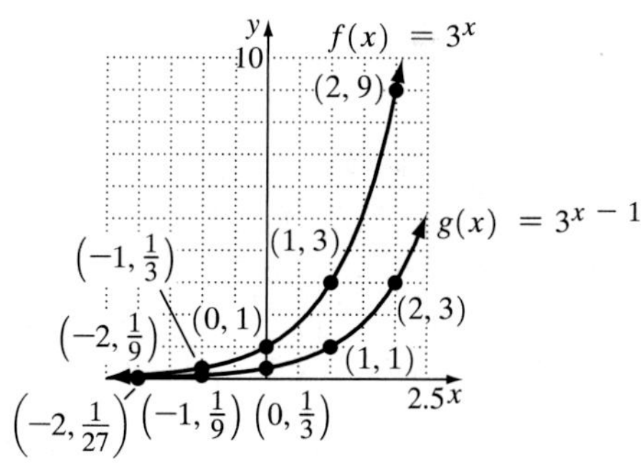

The graph of g is the graph of f shifted 1 unit to the right.

5.

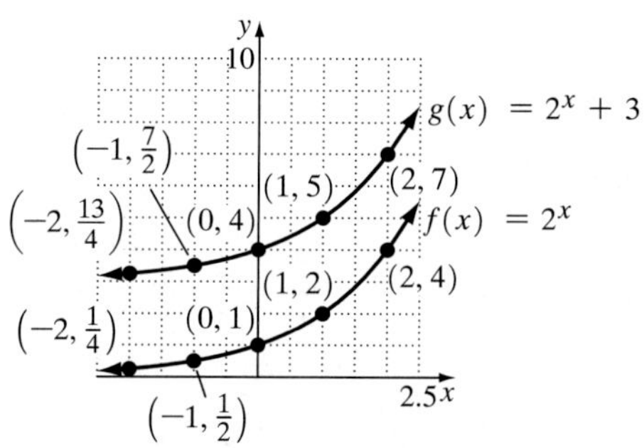

The graph of g is the graph of f shifted up 3 units.

6. approximately 11.49 billion
7. a. $14,859.47
b. $14,918.25

Exercise Set 9.1

1. 10.556 **3.** 11.665 **5.** 0.125 **7.** 9.974 **9.** 0.387

11. ; d

x	$f(x)$
-2	$\frac{1}{9}$
-1	$\frac{1}{3}$
0	1
1	3
2	9

13. ; e

x	$f(x)$
-2	$-\frac{8}{9}$
-1	$-\frac{2}{3}$
0	0
1	2
2	8

15. ; f

x	$h(x)$
-2	9
-1	3
0	1
1	$\frac{1}{3}$
2	$\frac{1}{9}$

17.

x	$f(x)$
-2	$\frac{1}{16}$
-1	$\frac{1}{4}$
0	1
1	4
2	16

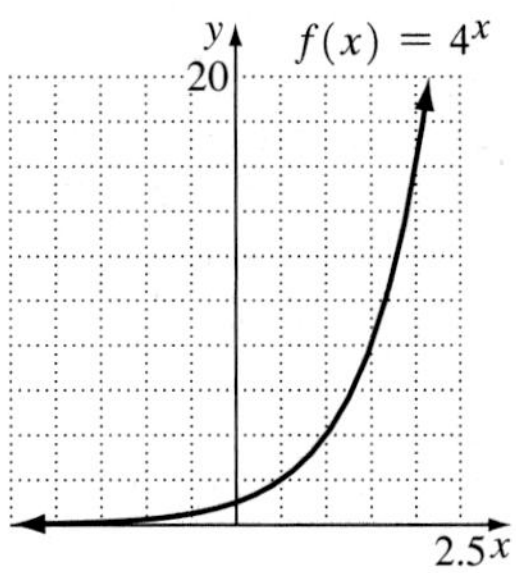

19.

x	$g(x)$
-2	$\frac{4}{9}$
-1	$\frac{2}{3}$
0	1
1	$\frac{3}{2}$
2	$\frac{9}{4}$

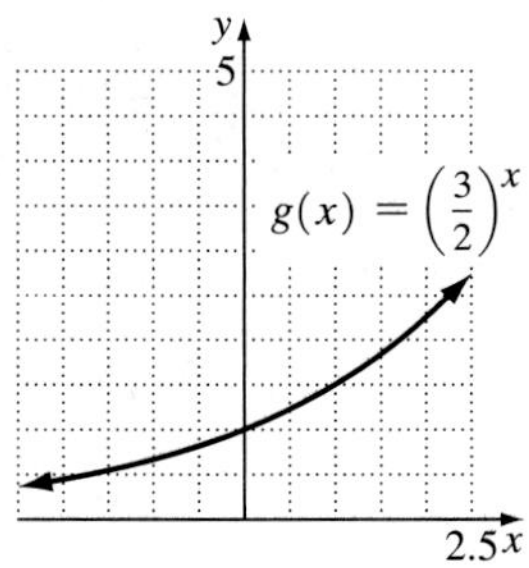

21.

x	$h(x)$
-2	4
-1	2
0	1
1	$\frac{1}{2}$
2	$\frac{1}{4}$

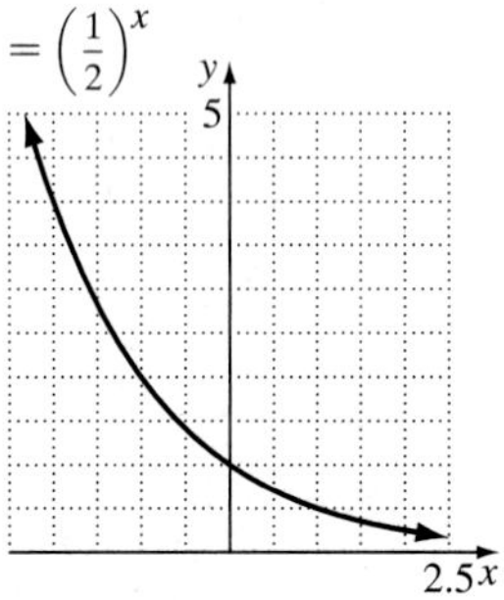

23.

x	$f(x)$
-2	2.78
-1	1.67
0	1
1	0.6
2	0.36

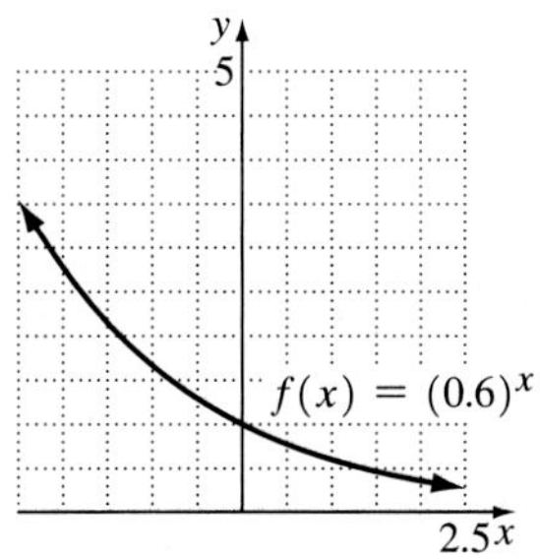

25.

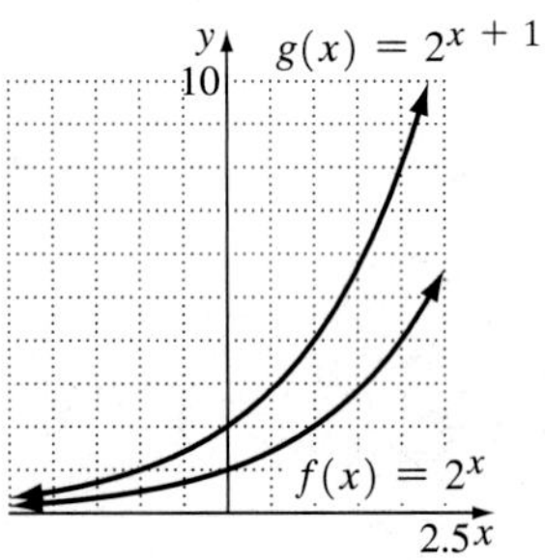

The graph of g is the graph of f shifted 1 unit to the left.

27.

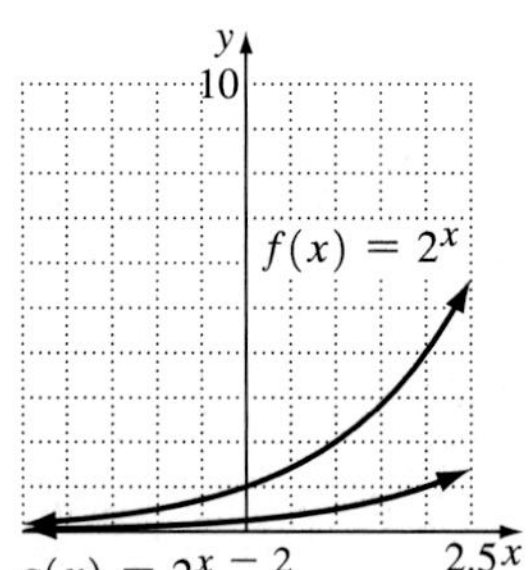

The graph of g is the graph of f shifted 2 units to the right.

29.

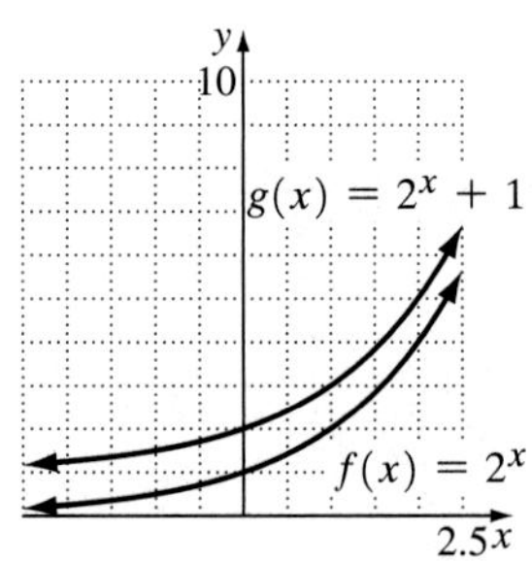

The graph of g is the graph of f shifted up 1 unit.

31.

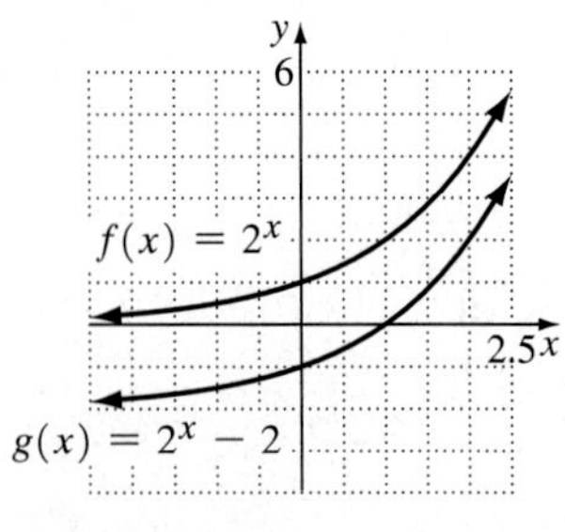

The graph of g is the graph of f shifted down 2 units.

33.

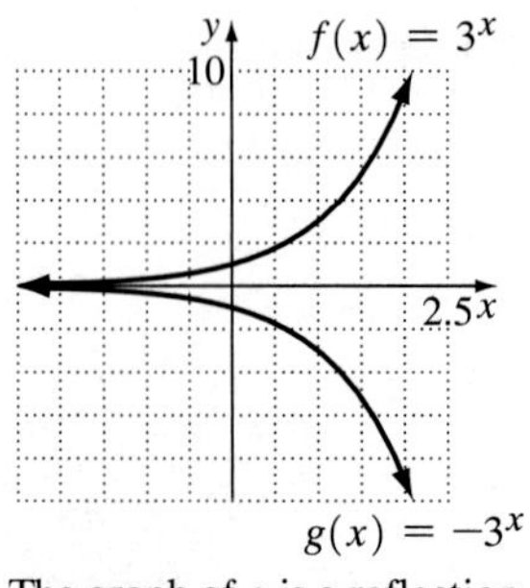

The graph of g is a reflection of the graph of f across the x-axis.

35.

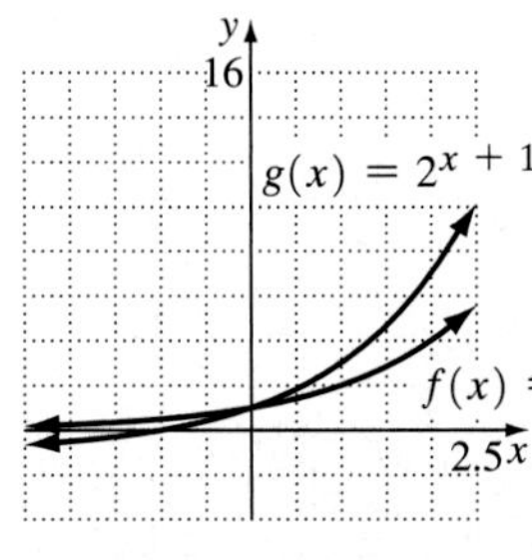

The graph of g is the graph of f shifted 1 unit to the left and 1 unit down.

37.

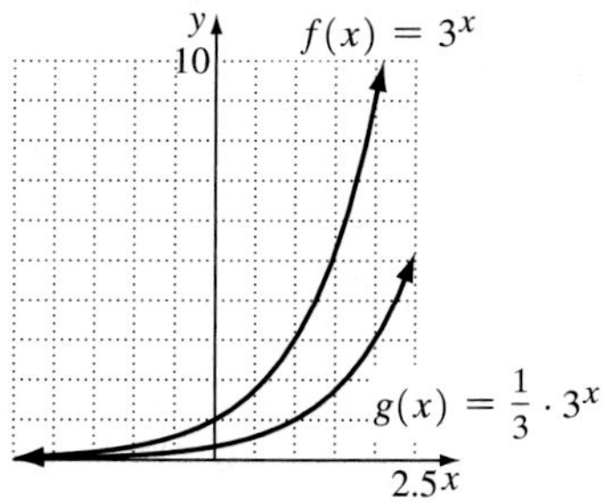

The graph of g is the graph of f shifted 1 unit to the right.

39. a. \$13,116.51 **b.** \$13,157.04 **c.** \$13,165.31
41. \$7% compounded monthly
43. a. 67.38 million **b.** 134.74 million **c.** 269.46 million **d.** 538.85 million **e.** It appears to double.
45. \$116,405.10
47. $2^{1.7} \approx 3.249010$; $2^{1.73} \approx 3.317278$; $2^{1.732} \approx 3.321880$; $2^{1.73205} \approx 3.321995$; $2^{1.7320508} \approx 3.321997$; $2^{\sqrt{3}} \approx 3.321997$; Answers will vary.
49. 241,786; There were about 241,786 AIDS cases in the United States among intravenous drug users in 2000.
51. a. 100% **b.** about 68.5% **c.** about 30.8% **d.** about 20%
53. about 0.07; About 7 out of 100 people, 25 years old have some coronary heart disease.
55. a. about 1429 people **b.** about 24,546 people
c. The number of people cannot exceed the population.; The asymptote indicates that the number of ill people will not exceed 30,000, the population of the town.

57–61. Answers will vary. **63.**

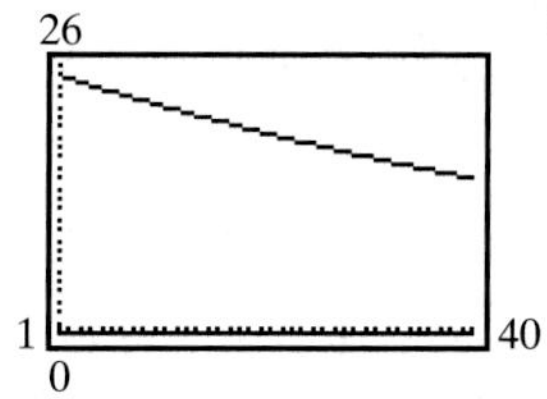

; 1973

65. a.

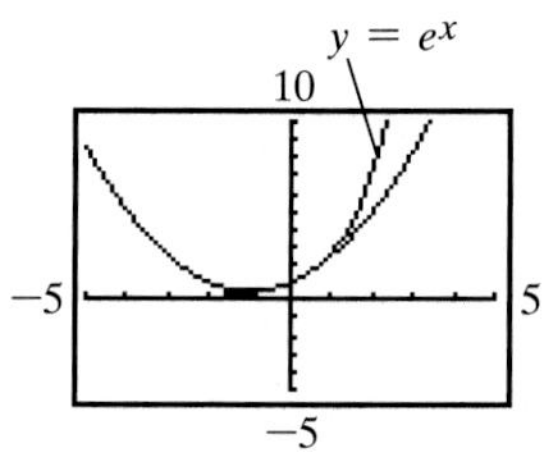

b.

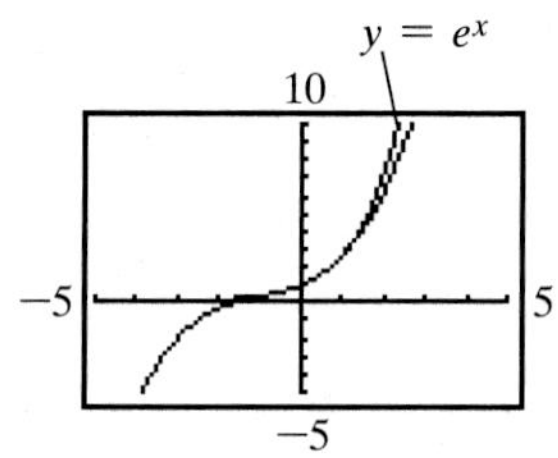

c.

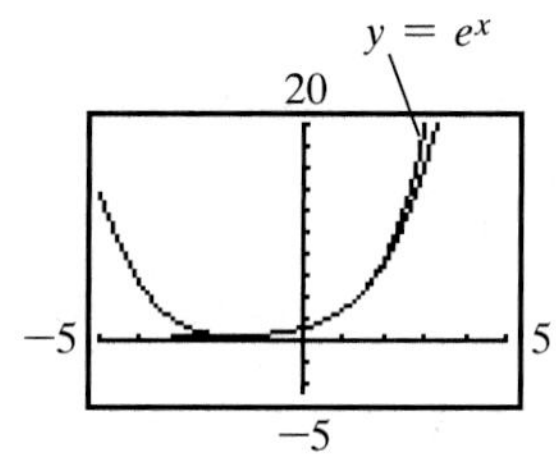

d. Answers will vary.

67. a. $y = \left(\frac{1}{3}\right)^x$ **b.** $y = \left(\frac{1}{5}\right)^x$ **c.** $y = 5^x$ **d.** $y = 3^x$; Answers will vary. **69.** $b = \frac{Da}{a - D}$ **70.** -1
71. -2 and 5, or $\{-2, 5\}$

Section 9.2

Check Point Exercises

1. a. $5x^2 + 1$ **b.** $25x^2 + 60x + 35$ **2.** $f(g(x)) = 7\left(\frac{x}{7}\right) = x$; $g(f(x)) = \frac{7x}{7} = x$
3. $f(g(x)) = 4\left(\frac{x + 7}{4}\right) - 7 = x + 7 - 7 = x$; $g(f(x)) = \frac{(4x - 7) + 7}{4} = \frac{4x}{4} = x$ **4.** $f^{-1}(x) = \frac{x - 7}{2}$ **5.** $f^{-1}(x) = \sqrt[3]{\frac{x + 1}{4}}$
6. (b) and (c) **7.**

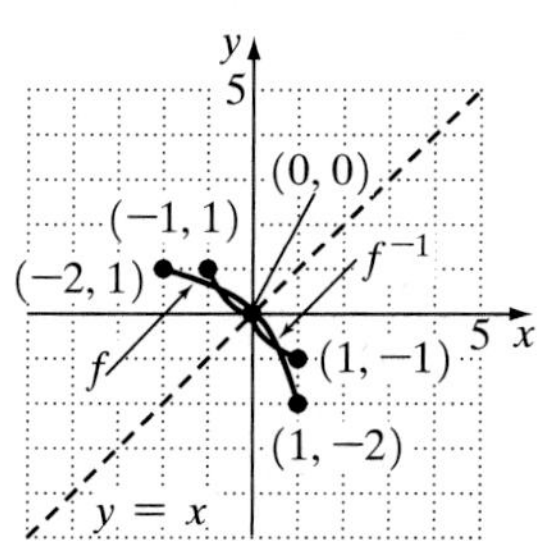

Exercise Set 9.2

1. a. $2x + 14$ **b.** $2x + 7$ **c.** 18 **3. a.** $2x + 5$ **b.** $2x + 9$ **c.** 9 **5. a.** $20x^2 - 11$ **b.** $80x^2 - 120x + 43$ **c.** 69
7. a. $x^4 - 4x^2 + 6$ **b.** $x^4 + 4x^2 + 2$ **c.** 6 **9. a.** $\sqrt{x - 1}$ **b.** $\sqrt{x} - 1$ **c.** 1 **11. a.** x **b.** x **c.** 2
13. a. x **b.** x **c.** 2 **15.** $f(g(x)) = x$; $g(f(x)) = x$; inverses **17.** $f(g(x)) = x$; $g(f(x)) = x$; inverses
19. $f(g(x)) = \frac{5x - 4}{9}$; $g(f(x)) = \frac{5x - 56}{9}$; not inverses **21.** $f(g(x)) = x$; $g(f(x)) = x$; inverses **23.** $f(g(x)) = x$; $g(f(x)) = x$;

inverses **25. a.** $f^{-1}(x) = x - 3$ **b.** $f(f^{-1}(x)) = (x - 3) + 3 = x$ and $f^{-1}(f(x)) = (x + 3) - 3 = x$

27. a. $f^{-1}(x) = \frac{x}{2}$ **b.** $f(f^{-1}(x)) = 2\left(\frac{x}{2}\right) = x$ and $f^{-1}(f(x)) = \frac{2x}{2} = x$ **29. a.** $f^{-1}(x) = \frac{x - 3}{2}$

b. $f(f^{-1}(x)) = 2\left(\frac{x - 3}{2}\right) + 3 = x$ and $f^{-1}(f(x)) = \frac{(2x + 3) - 3}{2} = x$ **31. a.** $f^{-1}(x) = \sqrt[3]{x - 2}$

b. $f(f^{-1}(x)) = (\sqrt[3]{x - 2})^3 + 2 = x$ and $f^{-1}(f(x)) = \sqrt[3]{(x^3 + 2) - 2} = x$ **33. a.** $f^{-1}(x) = \sqrt[3]{x} - 2$

b. $f(f^{-1}(x) = ((\sqrt[3]{x} - 2) + 2)^3 = x$ and $f^{-1}(f(x)) = \sqrt[3]{(x + 2)^3} - 2 = x$ **35. a.** $f^{-1}(x) = \frac{1}{x}$

b. $f(f^{-1}(x)) = \frac{1}{\frac{1}{x}} = x$ and $f^{-1}(f(x)) = \frac{1}{\frac{1}{x}} = x$ **37. a.** $f^{-1}(x) = x^2, x \geq 0$ **b.** $f(f^{-1}(x)) = \sqrt{x^2} = x$ and $f^{-1}(f(x) = (\sqrt{x})^2 = x$

39. a. $f^{-1}(x) = \sqrt{x - 1}$ **b.** $f(f^{-1}(x)) = (\sqrt{x - 1})^2 + 1 = x$ and $f^{-1}(f(x)) = \sqrt{(x^2 + 1) - 1} = x$ **41. a.** $f^{-1}(x) = \frac{3x + 1}{x - 2}$

b. $f(f^{-1}(x)) = \dfrac{2\left(\frac{3x + 1}{x - 2}\right) + 1}{\left(\frac{3x + 1}{x - 2}\right) - 3} = x$ and $f^{-1}(f(x)) = \dfrac{3\left(\frac{2x + 1}{x - 3}\right) + 1}{\left(\frac{2x + 1}{x - 3}\right) - 2} = x$ **43. a.** $f^{-1}(x) = (x - 3)^3 + 4$

b. $f(f^{-1}(x)) = \sqrt[3]{((x - 3)^3 + 4) - 4} + 3 = x$ and $f^{-1}(f(x)) = ((\sqrt[3]{x - 4} + 3) - 3)^3 + 4 = x$ **45.** no inverse **47.** no inverse

49. inverse function

51.

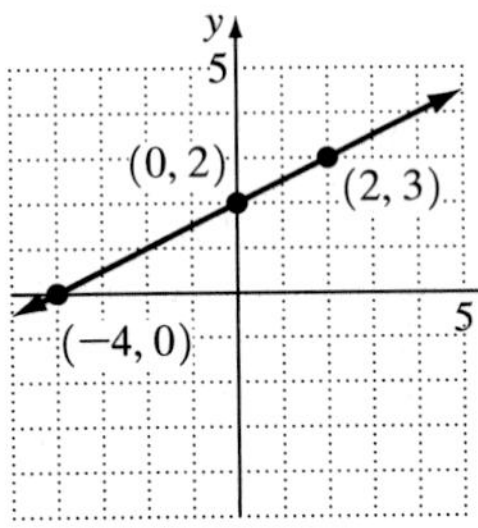

53.

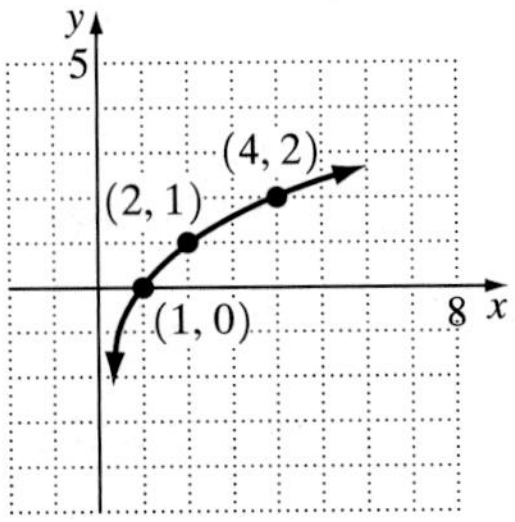

55. a. f represents the price after a \$400 discount, and g represents the price after a 25% discount (75% of the regular price).
b. $0.75x - 400$; $f \circ g$ represents an additional \$400 discount on a price that has already been reduced by 25%.
c. $0.75(x - 400) = 0.75x - 300$; $g \circ f$ represents an additional 25% discount on a price that has already been reduced \$400.
d. $f \circ g$; $0.75x - 400 < 0.75x - 300$; so $f \circ g$ represents the lower price after the two discounts.
e. $f^{-1}(x) = x + 400$; f^{-1} represents the regular price, since the value of x here is the price after a \$400 discount.

57. a. No horizontal line intersects the graph of f in more than one point. **b.** $f^{-1}(0.25)$, or 15, represents the number of people who would have to be in the room so that the probability of two sharing a birthday would be 0.25; $f^{-1}(0.5)$, or 23, represents the number of people so that the probability would be 0.5; $f^{-1}(0.7)$, or 30, represents the number of people so that the probability would be 0.7.

59. no; There are at least two years for which the average age at which women married is the same. **61–67.** Answers will vary.

69.

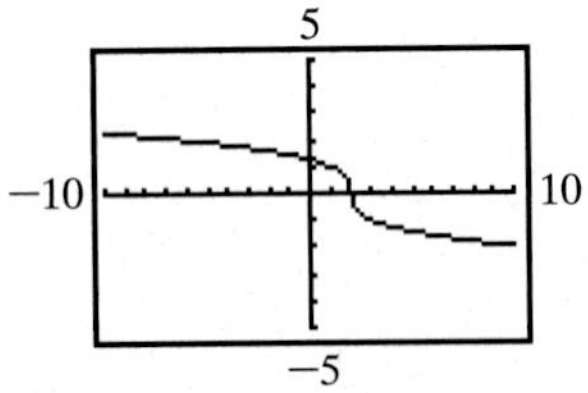

inverse function

71.

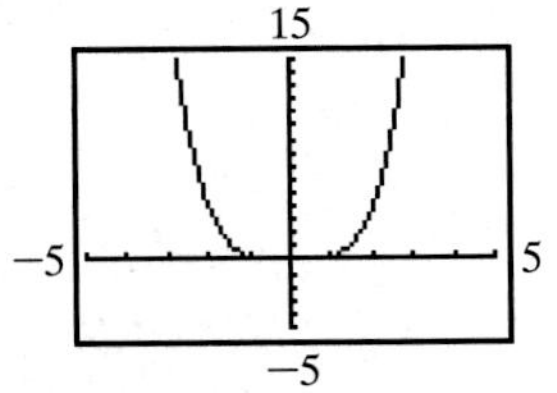

no inverse function

73.

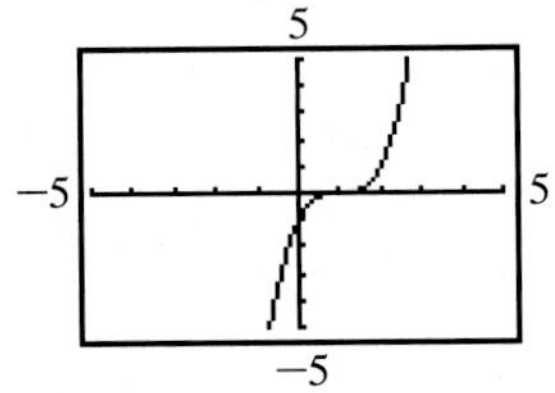

inverse function

75.

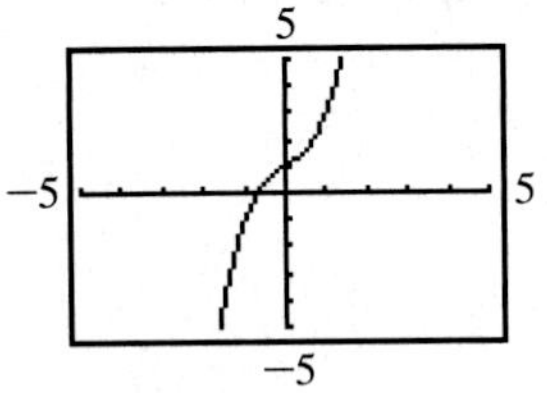

inverse function

77.

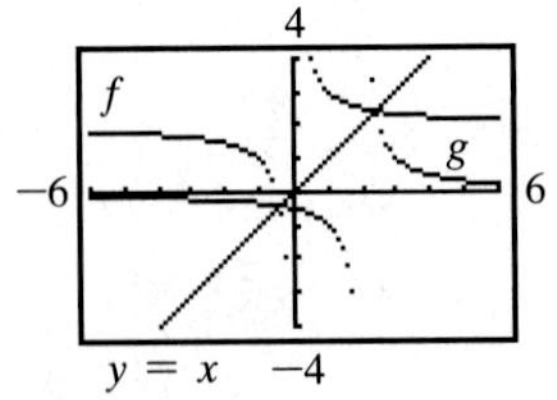

inverses

79. d **81.** $(f \circ g)^{-1}(x) = \frac{x - 15}{3}$; $(g^{-1} \circ f^{-1})(x) = \frac{x}{3} - 5 = \frac{x - 15}{3}$

83. $(f \circ g)(x) = m_1(m_2x + b_2) + b_1 = m_1m_2x + m_1b_2 + b_1$; The slope of $f \circ g$ is m_1m_2, which is the product of the slopes of f and g.

84. 5×10^8 **85.**

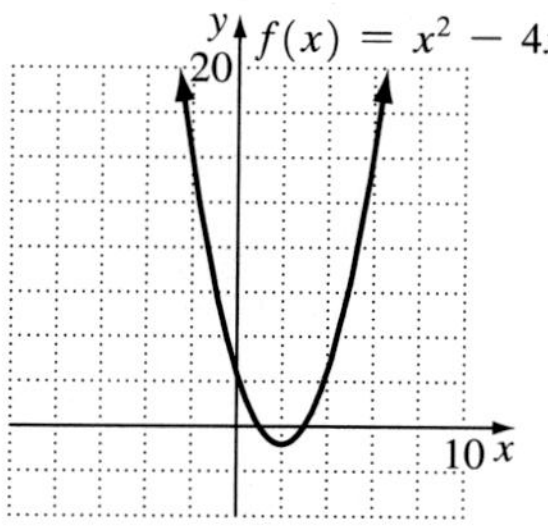

86. 5 or $\{5\}$

Section 9.3

Check Point Exercises

1. a. $7^3 = x$ **b.** $b^2 = 25$ **c.** $4^y = 26$ **2. a.** $5 = \log_2 x$ **b.** $3 = \log_b 27$ **c.** $y = \log_e 33$ **3.** 2 **b.** 1 **b.** $\frac{1}{2}$
4. a. 1 **b.** 0 **5. a.** 8 **b.** 17

6.

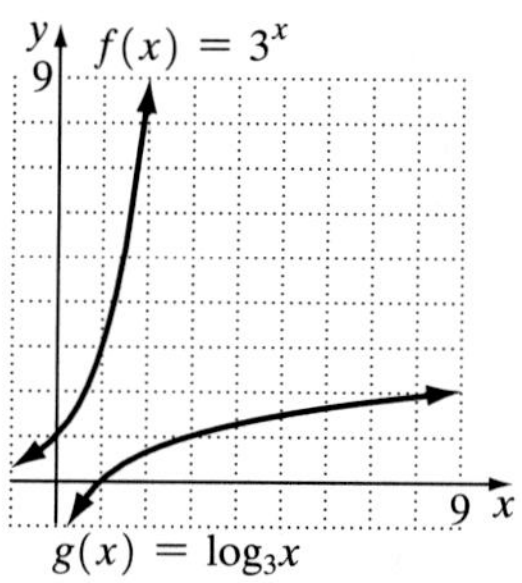

7. $\{x|x > 5\}$ or $(5, \infty)$
8. approximately 80%
9. 4
10. a. $\{x|x < 4\}$ or $(-\infty, 4)$
b. $\{x|x$ is a real number and $x \neq 0\}$ or $(-\infty, 0) \cup (0, \infty)$
11. a. $25x$
b. $\sqrt{x}$
12. approximately 4.6 feet per second

Exercise Set 9.3

1. $2^4 = 16$ **3.** $3^2 = x$ **5.** $b^5 = 32$ **7.** $6^y = 216$ **9.** $\log_2 8 = 3$ **11.** $\log_2 \frac{1}{16} = -4$ **13.** $\log_8 2 = \frac{1}{3}$ **15.** $\log_{13} x = 2$
17. $\log_b 1000 = 3$ **19.** $\log_7 200 = y$ **21.** 2 **23.** 6 **25.** $\frac{1}{2}$ **27.** -3 **29.** $\frac{1}{2}$ **31.** 1 **33.** 0 **35.** 7 **37.** 19

39.

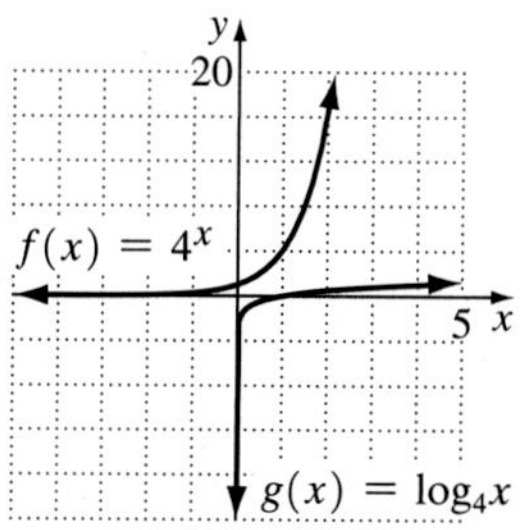

41.

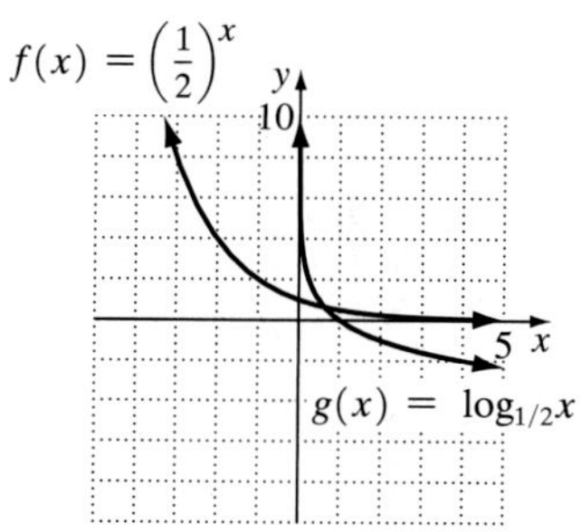

43. $\{x|x > -4\}$ or $(-4, \infty)$
45. $\{x|x < 2\}$ or $(-\infty, 2)$
47. $\{x|x$ is a real number and $x \neq 2\}$ or $(-\infty, 2) \cup (2, \infty)$
49. 2 **51.** 7 **53.** 33 **55.** 0
57. 6 **59.** -6 **61.** 125
63. $9x$ **65.** $5x^2$ **67.** $\sqrt{x}$
69. approximately 95.4%
71. approximately \$5.7 billion
73. approximately 188 decibels; yes

75. a. 88
b. 71.5; 63.9; 58.8; 55.0; 52.0; 49.5
c.

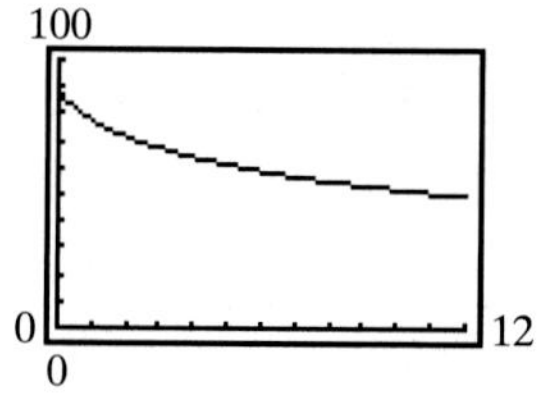

The students remembered less of the material over time.

77–83. Answers will vary.

85.

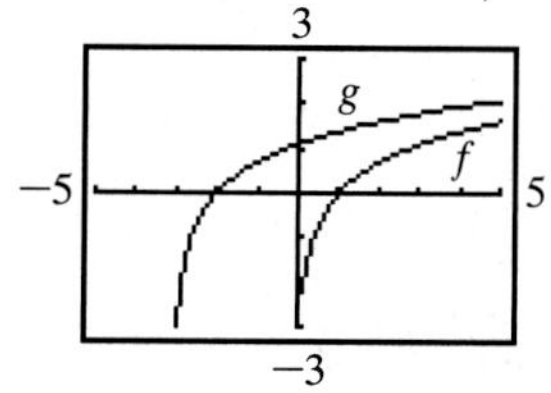

The graph of g is the graph of f shifted 3 units to the left.

87.

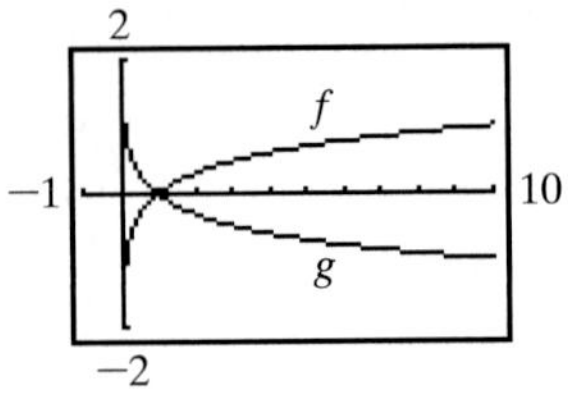

The graph of g is the reflection of the graph of f across the x-axis.

89.

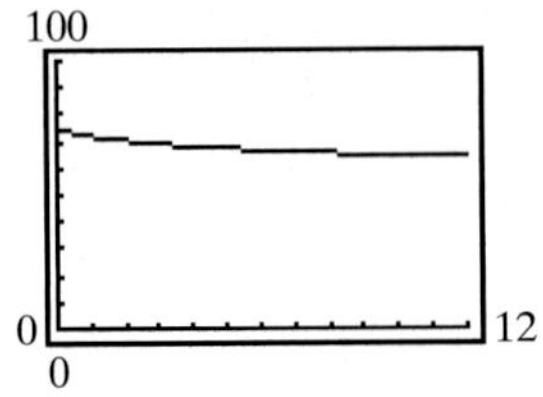

9 months

91.

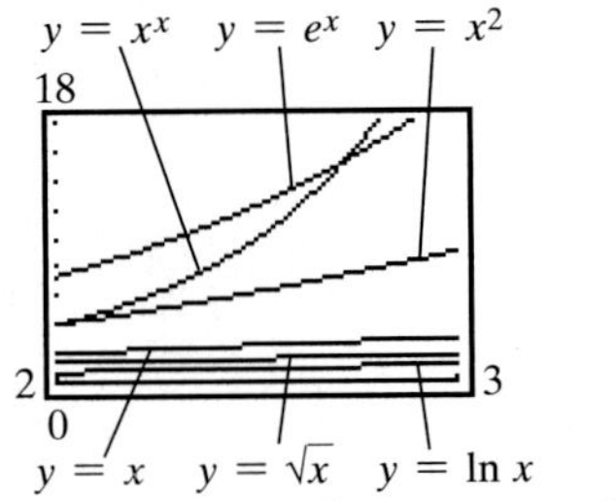

$y = \ln x$; $y = \sqrt{x}$; $y = x$; $y = x^2$; $y = e^x$; $y = x^x$

93. $\frac{4}{5}$ **95.** $\log_3 40$ **96.** $(-2, 3)$ or $\{(-2, 3)\}$ **97.** $2(3x - y)(x - y)$ **98.** $\{x | x \le -7 \text{ or } x \ge -2\}$ or $(-\infty, -7] \cup [-2, \infty)$

Section 9.4

Check Point Exercises

1. a. $\log_6 7 + \log_6 11$ **b.** $2 + \log x$ **2. a.** $\log_8 23 - \log_8 x$ **b.** $5 - \ln 11$ **3. a.** $9 \log_6 8$ **b.** $\frac{1}{3} \ln x$

4. a. $4 \log_b x + \frac{1}{3} \log_b y$ **b.** $\frac{1}{2} \log_5 x - 2 - 3 \log_5 y$ **5. a.** $\log 100 = 2$ **b.** $\log \frac{7x + 6}{x}$

6. a. $\ln x^2 \sqrt[3]{x + 5}$ **b.** $\log \frac{(x - 3)^2}{x}$ **7.** $\frac{\log 2506}{\log 7} \approx 4.02$ **8.** $\frac{\ln 2506}{\ln 7} \approx 4.02$

Exercise Set 9.4

1. $\log_5 7 + \log_5 3$ **3.** $1 + \log_7 x$ **5.** $3 + \log x$ **7.** $1 - \log_7 x$ **9.** $\log x - 2$ **11.** $3 - \log_4 y$ **13.** $2 - \ln 5$

15. $3 \log_b x$ **17.** $-6 \log N$ **19.** $\frac{1}{5} \ln x$ **21.** $2 \log_b x + \log_b y$ **23.** $\frac{1}{2} \log_4 x - 3$ **25.** $2 - \frac{1}{2} \log_6(x + 1)$

27. $2 \log_b x - \log_b y - 2 \log_b z$ **29.** $1 + \frac{1}{2} \log x$ **31.** $\frac{1}{3} \log x - \frac{1}{3} \log y$ **33.** $\log 10 = 1$ **35.** $\ln 7x$ **37.** $\log_2 32 = 5$

39. $\log \frac{2x + 5}{x}$ **41.** $\log xy^3$ **43.** $\ln y\sqrt{x}$ **45.** $\log_b x^2 y^3$ **47.** $\ln \frac{x^5}{y^2}$ **49.** $\ln \frac{x^3}{\sqrt[3]{y}}$ **51.** $\ln \frac{(x + 6)^4}{x^3}$ **53.** 1.5937

55. 1.6944 **57.** −1.2304 **59.** 3.6193

61. a. $D = 10 \log \frac{I}{I_0}$ **b.** 20 decibels **63–69.** Answers will vary.

71. a. & b.

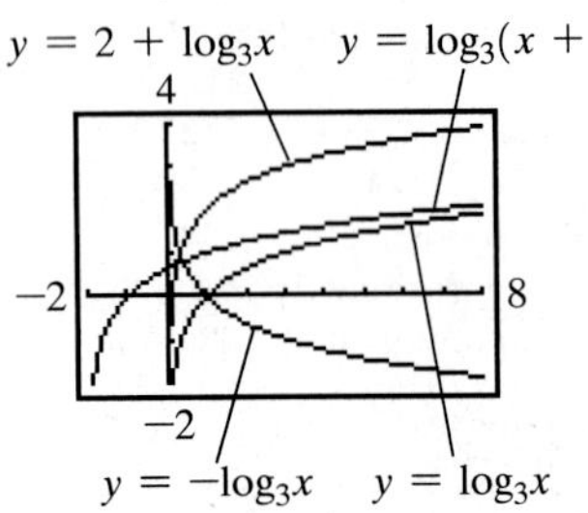

; $y = \log_3 x$ is shifted up 2 units to obtain $y = 2 + \log_3 x$, $y = \log_3 x$ is shifted to the left 2 units to obtain $y = \log_3(x + 2)$, and $y = \log_3 x$ is reflected across the x-axis to obtain $y = -\log_3 x$.

73.

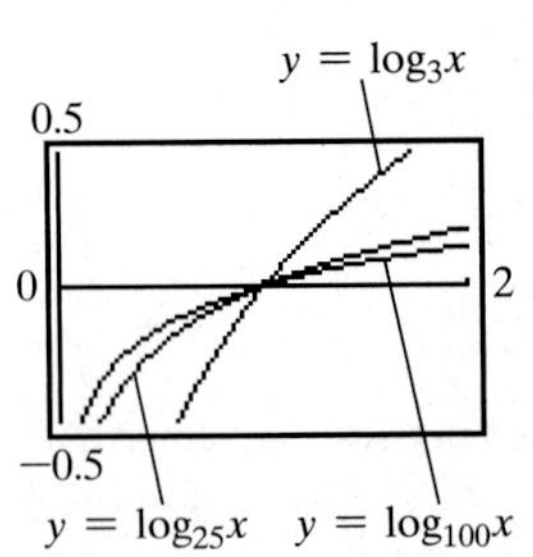

a. $y = \log_{100} x$ is on the top, and $y = \log_3 x$ is on the bottom.
b. $y = \log_3 x$ is on the top, and $y = \log_{100} x$ is on the bottom.
c. If $y = \log_b x$ is graphed for two different values of b, the graph of the one with the larger base will be on top in the interval $(0, 1)$ and the one with the smaller base will be on top in the interval $(0, \infty)$.

79. d **81.** $\frac{2A}{B}$ **83.**

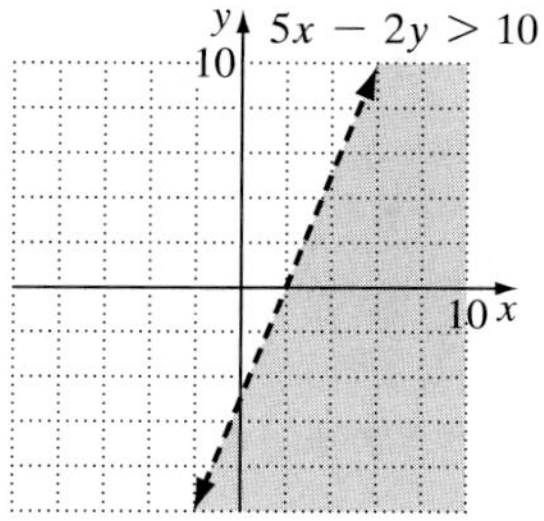

84. $\{x|x < 1\}$ or $(-\infty, 1)$ **85.** $2y\sqrt[3]{xy^2}$

Section 9.5

Check Point Exercises

1. a. 3 or {3} **b.** $\frac{5}{2}$ or $\left\{\frac{5}{2}\right\}$ **2.** $\frac{\ln 134}{5} \approx 3.04$ or $\left\{\frac{\ln 134}{\ln 5} \approx 3.04\right\}$ **3.** $\frac{\ln 9}{2} = \ln 3 \approx 1.10$ or $\{\ln 3 \approx 1.10\}$ **4.** 12 or {12} **5.** 5 or {5} **6.** $\frac{1}{24}$ or $\left\{\frac{1}{24}\right\}$ **7.** $\frac{e^2}{3}$ or $\left\{\frac{e^2}{3}\right\}$ **8.** blood alcohol concentration of 0.01 **9.** 16.2 years **10.** 2149

Exercise Set 9.5

1. 6 or {6} **3.** 3 or {3} **5.** 3 or {3} **7.** 2 or {2} **9.** $\frac{3}{5}$ or $\left\{\frac{3}{5}\right\}$ **11.** $\frac{3}{2}$ or $\left\{\frac{3}{2}\right\}$ **13.** $\frac{\ln 3.91}{\ln 10} \approx 0.59$ or $\left\{\frac{\ln 3.91}{\ln 10} \approx 0.59\right\}$ **15.** $\ln 5.7 \approx 1.74$ or $\{\ln 5.7 \approx 1.74\}$ **17.** $\frac{\ln 17}{\ln 5} \approx 1.76$ or $\left\{\frac{\ln 17}{\ln 5} \approx 1.76\right\}$ **19.** $\ln 5 \approx 1.61$ or $\{\ln 5 \approx 1.61\}$ **21.** $\frac{\ln 659}{5} \approx 1.30$ or $\left\{\frac{\ln 659}{5} \approx 1.30\right\}$ **23.** $\frac{\ln 13}{0.7} \approx 3.66$ or $\left\{\frac{\ln 13}{0.7} \approx 3.66\right\}$ **25.** $\frac{\ln 3}{0.055} \approx 19.97$ or $\left\{\frac{\ln 3}{0.055} \approx 19.97\right\}$ **27.** $\frac{\ln 30}{\ln 1.4} \approx 10.11$ or $\left\{\frac{\ln 30}{\ln 1.4} \approx 10.11\right\}$ **29.** $\frac{1 - \ln 793}{5} \approx -1.14$ or $\left\{\frac{1 - \ln 793}{5} \approx -1.14\right\}$ **31.** $\frac{\ln 410}{\ln 7} - 2 \approx 1.09$ or $\left\{\frac{\ln 410}{\ln 7} - 2 \approx 1.09\right\}$ **33.** $\frac{\ln 2}{\ln 5 - \ln 2} \approx 0.76$; $\left\{\frac{\ln 2}{\ln 5 - \ln 2} \approx 0.76\right\}$ **35.** 81 or {81} **37.** $\frac{1}{16}$ or $\left\{\frac{1}{16}\right\}$ **39.** 3 or {3} **41.** 100 or {100} **43.** $\frac{1}{e^3}$ or $\left\{\frac{1}{e^3}\right\}$ **45.** 59 or {59} **47.** $\frac{109}{27}$ or $\left\{\frac{109}{27}\right\}$ **49.** $\frac{62}{3}$ or $\left\{\frac{62}{3}\right\}$ **51.** $\frac{5}{4}$ or $\left\{\frac{5}{4}\right\}$ **53.** 6 or {6} **55.** 6 or {6} **57.** no solution or ∅ **59.** $e^2 \approx 7.39$ or $\{e^2 \approx 7.39\}$ **61.** $\frac{e^4}{2} \approx 27.30$ or $\left\{\frac{e^4}{2} \approx 27.30\right\}$ **63.** $e^{-1/2} \approx 0.61$ or $\{e^{-1/2} \approx 0.61\}$ **65.** $e^2 - 3 \approx 4.39$ or $\{e^2 - 3 \approx 4.39\}$ **67.** a blood alcohol concentration of 0.22 **69. a.** 18.2 million **b.** 2010 **71.** 8.2 **73.** 16.8% **75.** 8.7 **77.** 15.7% **79.** 1995 **81.** about 2.8 days; (2.8, 50) **83.** $10^{-2.4}$ or 0.004 moles per liter **85–91.** Answers will vary.

93.

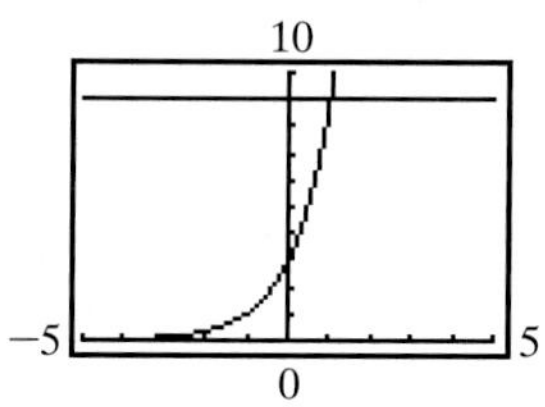

; 1 or {1}

95.

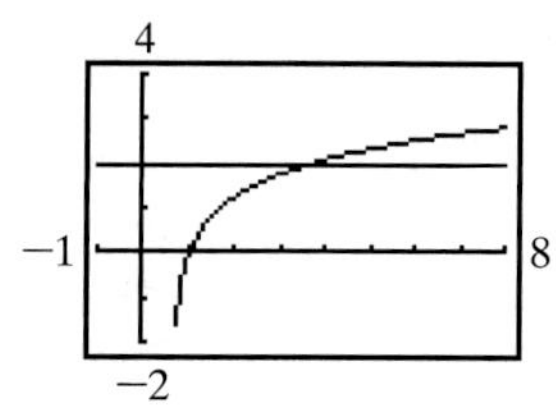

; $\frac{11}{3}$ or $\left\{\frac{11}{3}\right\}$

97.

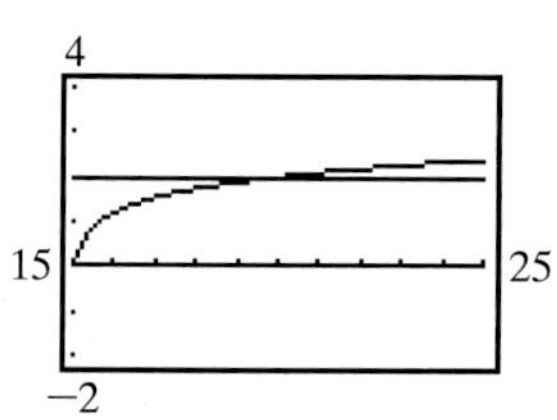

; 20 or {20}

99.

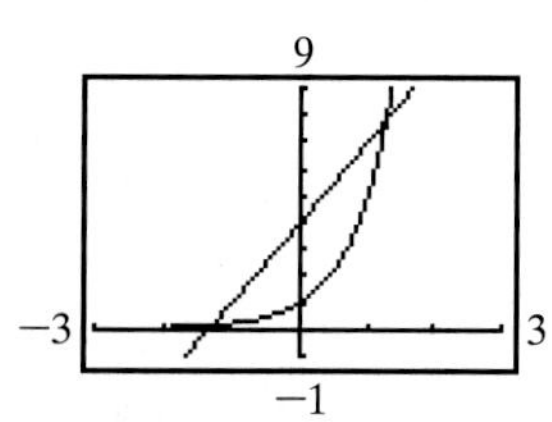

; −1.29 and 1.28, or {−1.29, 1.28}

101. 63.5 miles **103.** 20 years old **105.** about 36 years **107.** 10^{-2} and $10^{3/2}$, or $\{10^{-2}, 10^{3/2}\}$ **109.** 5 or {5} **110.** −12 or {−12} **111.** $\frac{y^8}{16x^{12}}$

Section 9.6

Check Point Exercises

1. a. $A = 491e^{0.027t}$ **b.** in 2006 **2. a.** $A = A_0 e^{-0.0248t}$ **b.** about 72.2 years

Exercise Set 9.6

1. 208 million **3.** 2016 **5.** 2.6% per year **7.** 2014 **9.** \$140 thousand or \$140,000 **11.** about 2005
13. $k = 0.175$; $A = 200e^{0.175t}$; 17.5% per year **15.** approximately 8 grams **17.** 8 grams after 10 seconds; 4 grams after 20 seconds; 2 grams after 30 seconds; 1 gram after 40 seconds; 0.5 grams after 50 seconds **19.** approximately 15,679 years old

21. a. $\frac{1}{2} = e^{1.31k}$ yields $k = \frac{\ln \frac{1}{2}}{1.31} \approx -0.52912$. **b.** about 0.1069 billion or 106,900,000 years old

23. $$\begin{aligned} 2A_0 &= A_0e^{kt} \\ 2 &= e^{kt} \\ \ln 2 &= \ln e^{kt} \\ \ln 2 &= kt \\ \frac{\ln 2}{k} &= t \end{aligned}$$

25. about 63 years
27–33. Answers will vary.
35. $y = 1.74(1.037)^x$; $r \approx 0.97$; Since r is close to 1, the model fits the data well.
37. $y = 1.547 + 0.112x$; $r \approx 0.99$; Since r is close to 1, the model fits the data well.
39. $y = 1.547 + 0.112x$; 2022; Answers will vary.

41. Answers will vary. **43.** $\frac{x+5}{x}$ **44.** -27 and 1, or $\{-27, 1\}$ **45.** $17\sqrt{2}$

Review Exercises

1.

x	$f(x)$
-2	$\frac{1}{16}$
-1	$\frac{1}{4}$
0	1
1	4
2	16

; d

2.

x	$f(x)$
-2	16
-1	4
0	1
1	$\frac{1}{4}$
2	$\frac{1}{16}$

; a

3.

x	$f(x)$
-2	-16
-1	-4
0	-1
1	$-\frac{1}{4}$
2	$-\frac{1}{16}$

; b

4.

x	$f(x)$
-2	-13
-1	-1
0	2
1	$\frac{11}{4}$
2	$\frac{47}{16}$

; c

5.

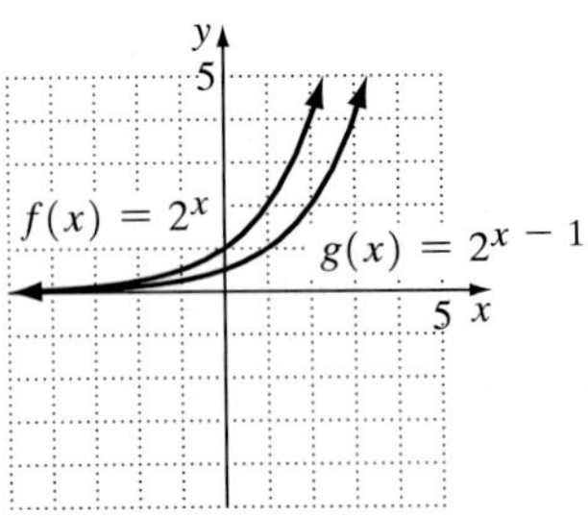

The graph of g is the graph of f shifted to the right 1 unit.

6.

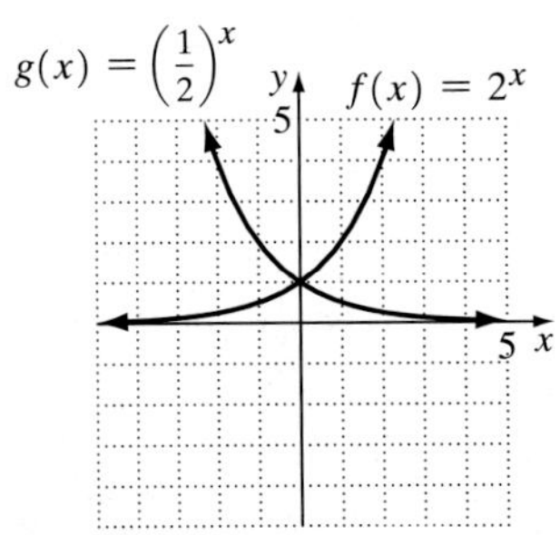

The graph of g is the reflection of f across the y-axis.

7.

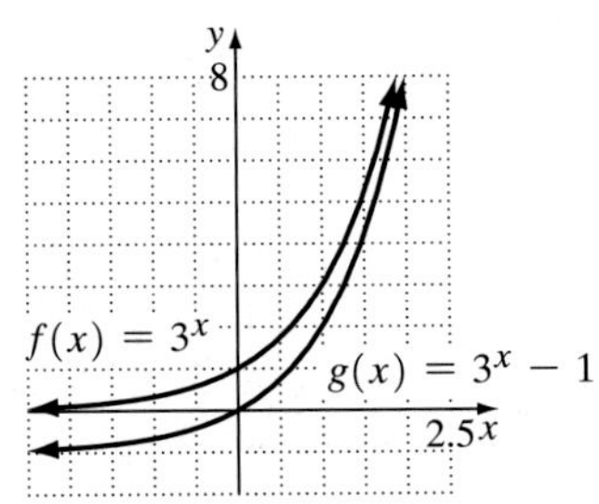

The graph of g is the graph of f shifted down 1 unit.

8.

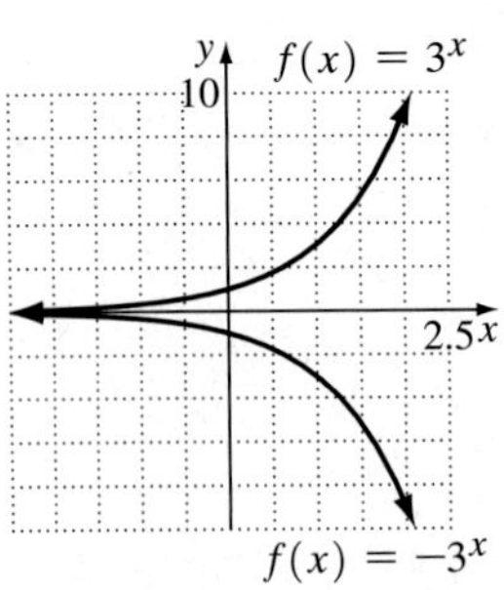

The graph of g is the reflection of f across the x-axis.

9. 5.5% compounded semiannually
10. 7% compounded monthly
11. a. 200°F
b. 119°F
c. The temperature of the room is 70°F.
12. a. $16x^2 - 8x + 4$ **b.** $4x^2 + 11$ **c.** 124
13. a. $\sqrt{x+1}$ **b.** $\sqrt{x} + 1$ **c.** 2
14. $f(g(x)) = x - \frac{7}{10}$; $g(f(x)) = x - \frac{7}{6}$; not inverses
15. $f(g(x)) = x$; $g(f(x)) = x$; inverses

16. a. $f^{-1}(x) = \frac{x+3}{4}$ **b.** $f(f^{-1}(x)) = 4\left(\frac{x+3}{4}\right) - 3 = x$ and $f^{-1}(f(x)) = \frac{(4x-3)+3}{4} = x$ **17. a.** $f^{-1}(x) = x^2 - 2$
b. $f(f^{-1}(x)) = \sqrt{(x^2-2)+2} = x$ and $f^{-1}(f(x)) = (\sqrt{x+2})^2 - 2 = x$ **18. a.** $f^{-1}(x) = \frac{\sqrt[3]{x-1}}{2}$
b. $f(f^{-1}(x)) = 8\left(\frac{\sqrt[3]{x-1}}{2}\right)^3 + 1 = x$ and $f^{-1}(f(x)) = \frac{\sqrt[3]{(8x^3+1)-1}}{2} = x$ **19.** inverse function
20. no inverse function **21.** inverse function **22.** no inverse function
23.

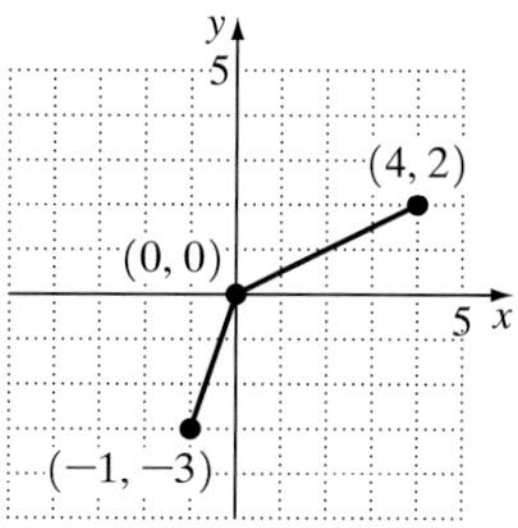

24. $49^{1/2} = 7$ **25.** $4^3 = x$ **26.** $3^y = 81$ **27.** $\log_6 216 = 3$ **28.** $\log_b 625 = 4$
29. $\log_{13} 874 = y$ **30.** 3 **31.** −2 **32.** −9 is not in the domain of $y = \log_3 x$.
33. $\frac{1}{2}$ **34.** 1 **35.** 8 **36.** 5 **37.** 0
38.

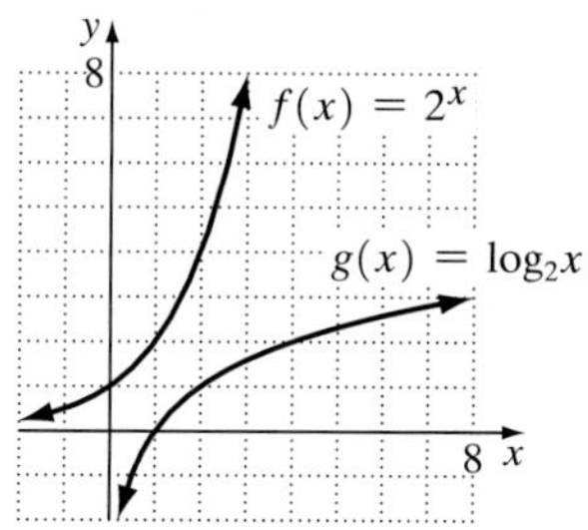

39.

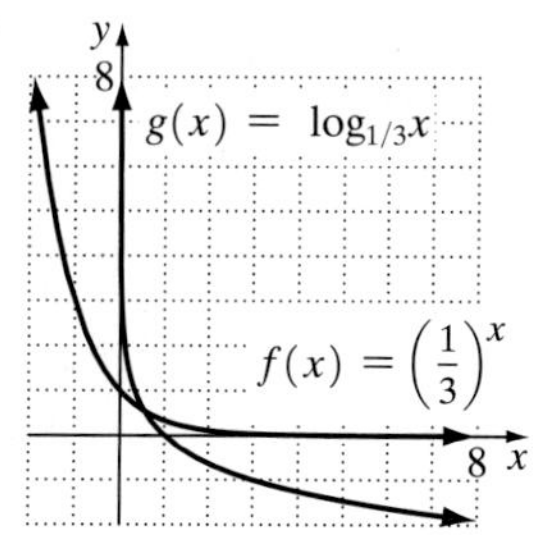

40. $\{x|x > -5\}$ or $(-5, \infty)$
41. $\{x|x < 3\}$ or $(-\infty, 3)$
42. $\{x|x < 1$ or $x > 1\}$ or $(-\infty, 1) \cup (1, \infty)$
43. $6x$
44. $\sqrt{x}$ **45.** $4x^2$ **46.** 3

47. a. 76
b. 67.4 after 2 months; 63.4 after 4 months; 60.8 after 6 months; 58.8 after 8 months; 55.9 after one year
c.

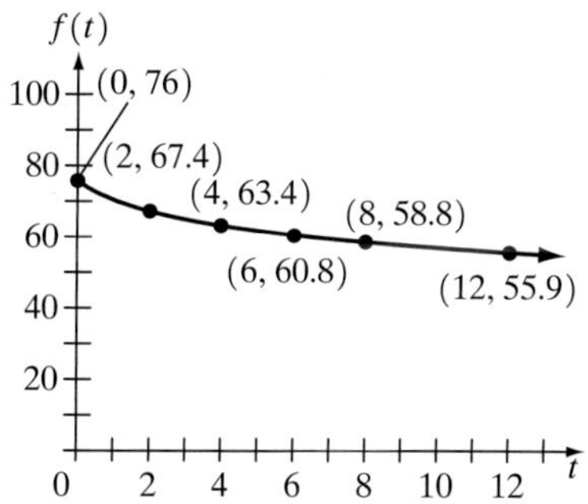

As time increases the amount of material retained by the students decreases.

48. about 9 weeks **49.** $2 + 3\log_6 x$ **50.** $\frac{1}{2}\log_4 x - 3$
51. $\log_2 x + 2\log_2 y - 6$ **52.** $\frac{1}{3}\ln x - \frac{1}{3}$ **53.** $\log_b 21$
54. $\log \frac{3}{x^3}$ **55.** $\ln(x^3y^4)$ **56.** $\ln \frac{\sqrt{x}}{y}$ **57.** 6.2448
58. −0.1063 **59.** 2 or {2} **60.** $\frac{2}{3}$ or $\left\{\frac{2}{3}\right\}$ **61.** $-\frac{3}{2}$ or $\left\{-\frac{3}{2}\right\}$
62. $\frac{\ln 12{,}143}{\ln 8} \approx 4.52$ or $\left\{\frac{\ln 12{,}143}{\ln 8} \approx 4.52\right\}$
63. $\frac{\ln 141}{5} \approx 0.99$ or $\left\{\frac{\ln 141}{5} \approx 0.99\right\}$
64. $\frac{\ln 3}{0.045} \approx 24.41$ or $\left\{\frac{\ln 3}{0.045} \approx 24.41\right\}$ **65.** $\frac{1}{125}$ or $\left\{\frac{1}{125}\right\}$
66. 100 or {100} **67.** 23 or {23} **68.** 5 or {5}
69. no solution or ∅

70. $\frac{1}{e}$ or $\left\{\frac{1}{e}\right\}$ **71.** $\frac{e^3}{2}$ or $\left\{\frac{e^3}{2}\right\}$ **72.** 2042 **73.** approximately 2086 **74.** 2005 **75.** 7.3 years **76.** 14.6 years **77.** 22%
78. a. $k = 0.041$ **b.** about 40.7 million **c.** 2010 **79.** 15,679 years old **80.** Answers will vary.

Chapter 9 Test

1.

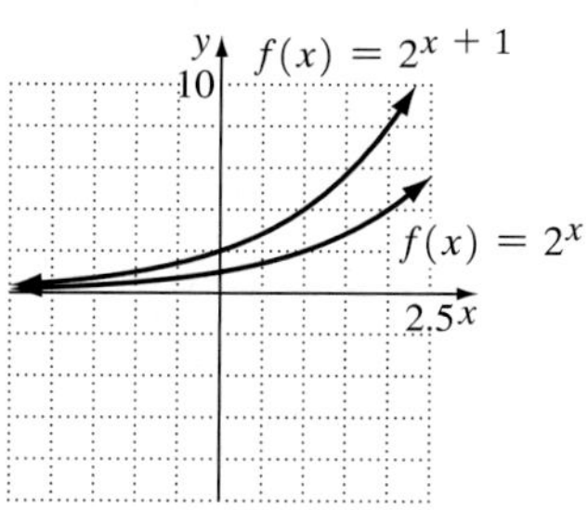

2. 6.5% compounded semiannually
3. $(f \circ g)(x) = 9x^2 - 3x$; $(g \circ f)(x) = 3x^2 + 3x - 1$
4. $f^{-1}(x) = \frac{x+7}{5}$
5. a. No horizontal line intersects the graph of f in more than one point. **b.** 2000
c. $f^{-1}(2000)$ represents the income, $80 thousand, of a family that gives $2000 to charity.
6. $5^3 = 125$ **7.** $\log_{36} 6 = \frac{1}{2}$

8.

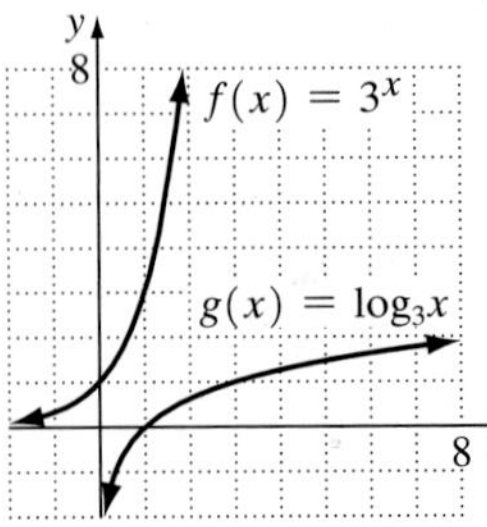

9. $5x$ **10.** 1 **11.** 0 **12.** $\{x|x > 7\}$ or $(7, \infty)$ **13.** 120 decibels **14.** $3 + 5\log_4 x$ **15.** $\frac{1}{3}\log_3 x - 4$ **16.** $\log x^6y^2$ **17.** $\ln\frac{7}{x^3}$ **18.** 1.5741 **19.** 6 or $\{6\}$ **20.** $\frac{\ln 1.4}{\ln 5}$ or $\left\{\frac{\ln 1.4}{\ln 5}\right\}$ **21.** $\frac{\ln 4}{0.005}$ or $\left\{\frac{\ln 4}{0.005}\right\}$ **22.** 5 or $\{5\}$ **23.** $\frac{217}{4}$ or $\left\{\frac{217}{4}\right\}$ **24.** 5 or $\{5\}$ **25.** $\frac{e^4}{3}$ or $\left\{\frac{e^4}{3}\right\}$ **26. a.** 89.18% **b.** decreasing; The growth rate, -0.004, is negative. **c.** 1996 **27.** 13.9 years **28.** about 6.9% **29.** $A = 484e^{0.005t}$ **30.** 24,758 years ago

Cumulative Review Exercises (Chapters 1–9)

1. 22 or $\{22\}$ **2.** $5 \pm 7i$ or $\{5 \pm 7i\}$ **3.** $\{x|x < -3 \text{ or } x > 2\}$ or $(-\infty, -3) \cup (2, \infty)$ **4.** -2 or $\{-2\}$ **5.** no solution or $\varnothing$ **6.** $\left\{x \middle| x < \frac{2}{3}\right\}$ or $\left(-\infty, \frac{2}{3}\right)$ **7.** $(4, 0, -5)$ or $\{(4, 0, -5)\}$ **8.** 9 or $\{9\}$

9.

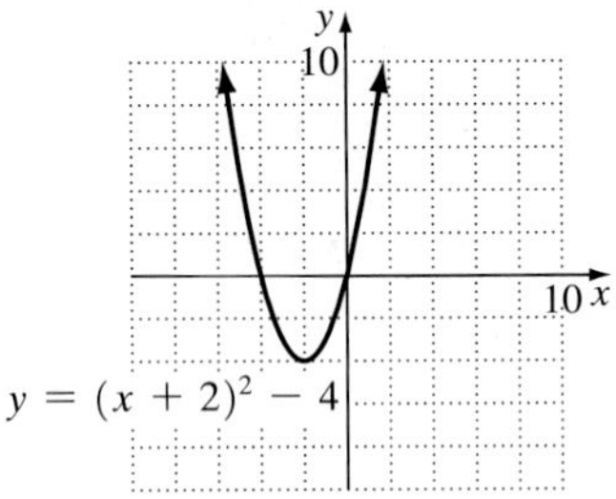

10.

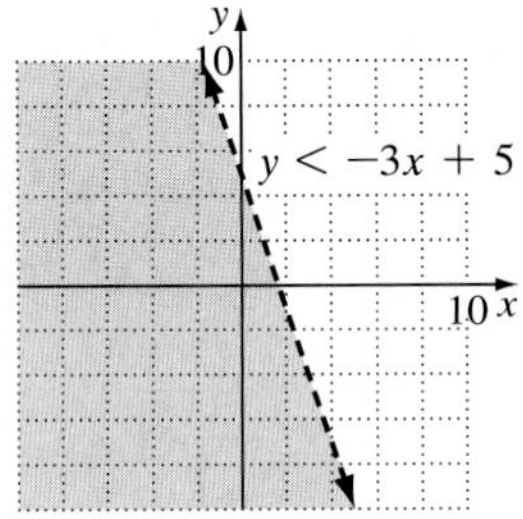

11.

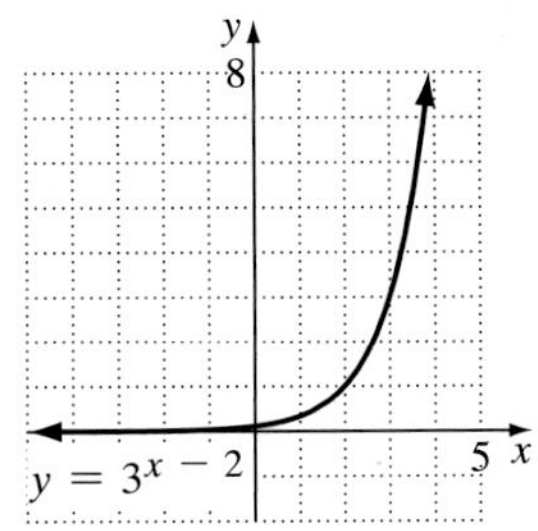

12. $\frac{2x^2 + 5x - 2}{(x - 5)(x + 2)}$ **13.** $-\frac{x + 1}{x - 1}$ **14.** $2\sqrt{5} + 2\sqrt{2}$ **15.** $12\sqrt{5}$ **16.** $\frac{5\sqrt[3]{4xy^2}}{2xy}$ **17.** $(x + y)(5a - 4b)$ **18.** $\frac{\log x^5}{\log\sqrt{y}}$ **19.** $p = \frac{qf}{q - f}$ **20.** $\{x|x < -5 \text{ or } x > 3\}$ or $(-\infty, -5) \cup (3, \infty)$ **21.** $\{x|x > 3\}$ or $(3, \infty)$ **22.** $\{x|x < 1\}$ or $(-\infty, 1)$ **23.** 8 ft by 3 ft **24.** \$18,500 **25.** $k = \frac{1}{2\ln 4} \approx 0.3607$; about 0.298 or $\frac{298}{1000}$

CHAPTER 10

Section 10.1

Check Point Exercises

1. 5 units **2.** $\left(4, -\frac{1}{2}\right)$ **3.** $x^2 + y^2 = 16$ **4.** $(x - 5)^2 + (y + 6)^2 = 100$ **5.** center: $(-3, 1)$; radius: 2 units **6.** $(x + 2)^2 + (y - 2)^2 = 9$; center: $(-2, 2)$; $r = 3$

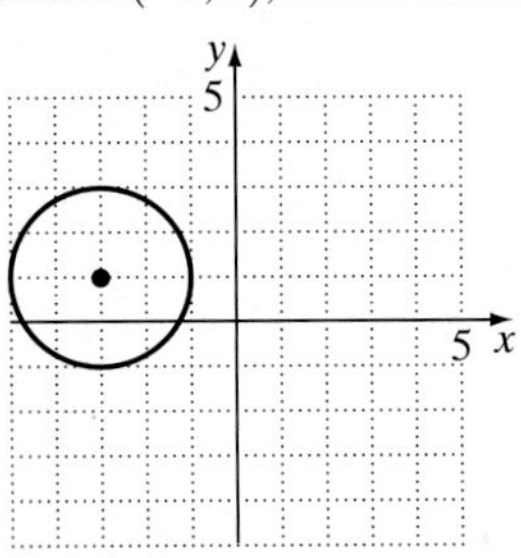

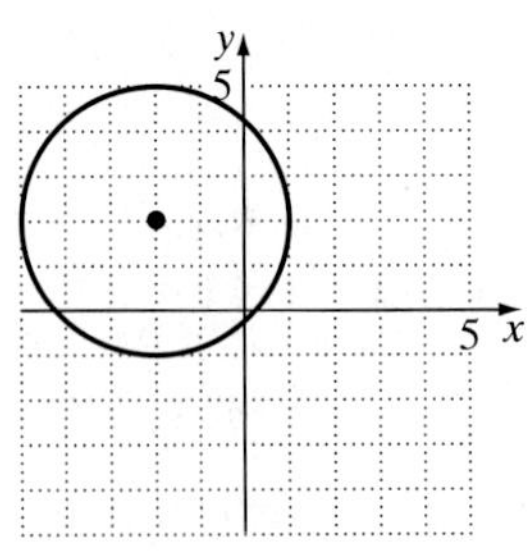

Exercise Set 10.1

1. 13 units **3.** $2\sqrt{2}$ or 2.83 units **5.** 5 units **7.** $\sqrt{29}$ or 5.39 units **9.** $4\sqrt{2}$ or 5.66 units **11.** $2\sqrt{5}$ or 4.47 units **13.** $2\sqrt{2}$ or 2.83 units **15.** $\sqrt{93}$ or 9.64 units **17.** $\sqrt{5}$ or 2.24 units **19.** $(4, 6)$ **21.** $(-4, -5)$ **23.** $\left(\frac{3}{2}, -6\right)$ **25.** $(-3, -2)$ **27.** $(1, 5\sqrt{5})$ **29.** $(2\sqrt{2}, 0)$ **31.** $x^2 + y^2 = 49$ **33.** $(x - 3)^2 + (y - 2)^2 = 25$ **35.** $(x + 1)^2 + (y - 4)^2 = 4$ **37.** $(x + 3)^2 + (y + 1)^2 = 3$ **39.** $(x + 4)^2 + y^2 = 100$

41. center: $(0, 0)$; $r = 4$

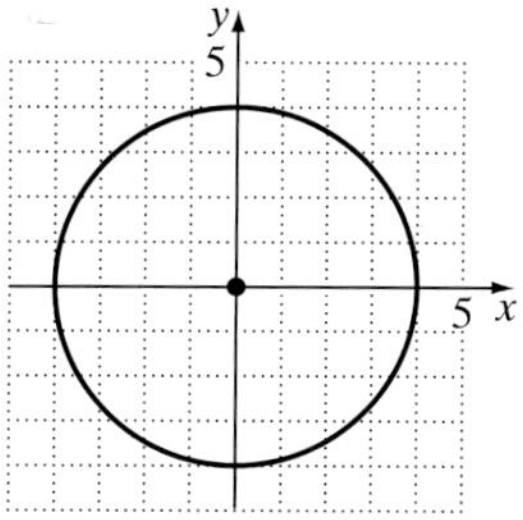

43. center: $(3, 1)$; $r = 6$

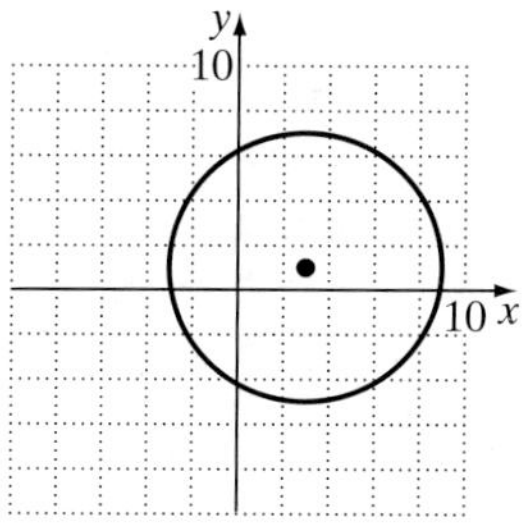

45. center: $(-3, 2)$; $r = 2$

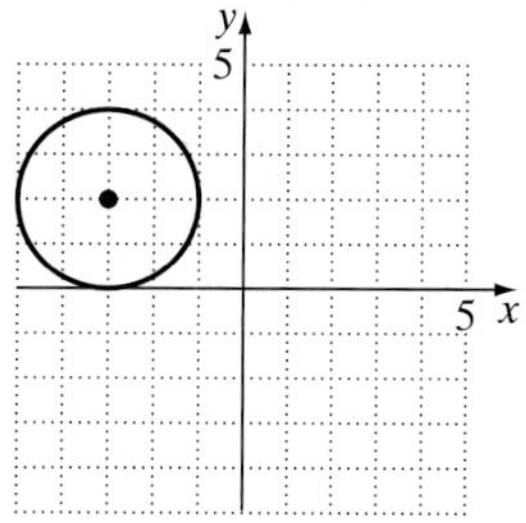

47. center: $(-2, -2)$; $r = 2$

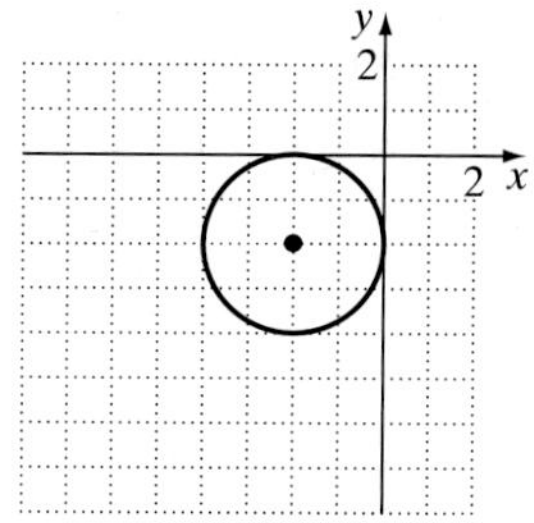

49. $(x + 3)^2 + (y + 1)^2 = 4$;
center: $(-3, -1)$; $r = 2$

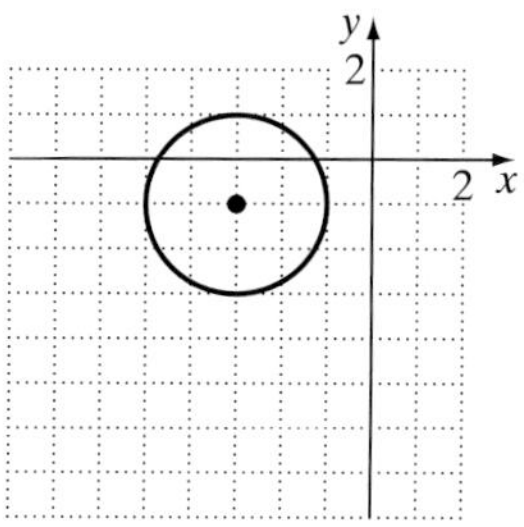

51. $(x - 5)^2 + (y - 3)^2 = 64$;
center: $(5, 3)$; $r = 8$

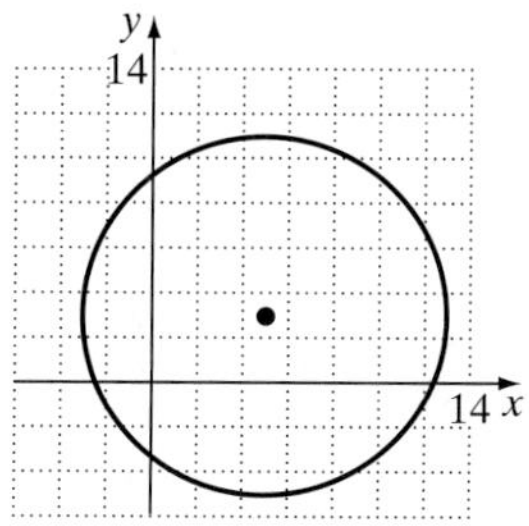

53. $(x + 4)^2 + (y - 1)^2 = 25$
center: $(-4, 1)$; $r = 5$

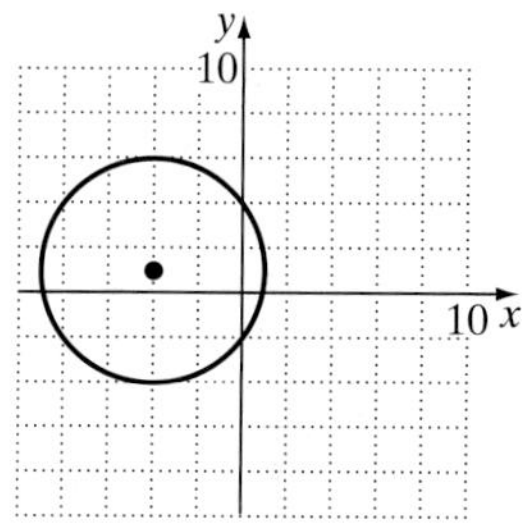

55. $(x - 1)^2 + y^2 = 16$;
center: $(1, 0)$; $r = 4$

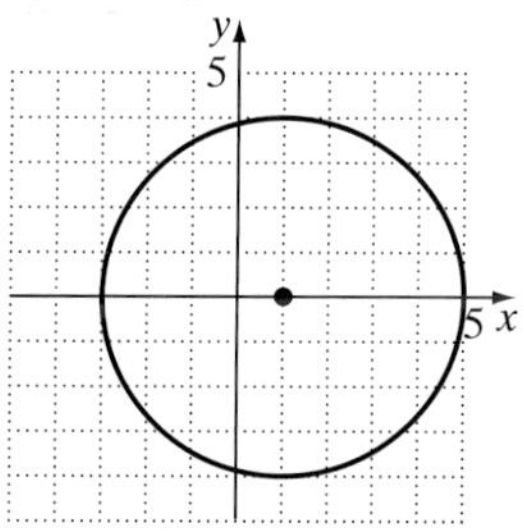

57. 0.5 hour or 30 minutes

59. $x^2 + (y - 82)^2 = 4624$ **61–67.** Answers will vary.

69.

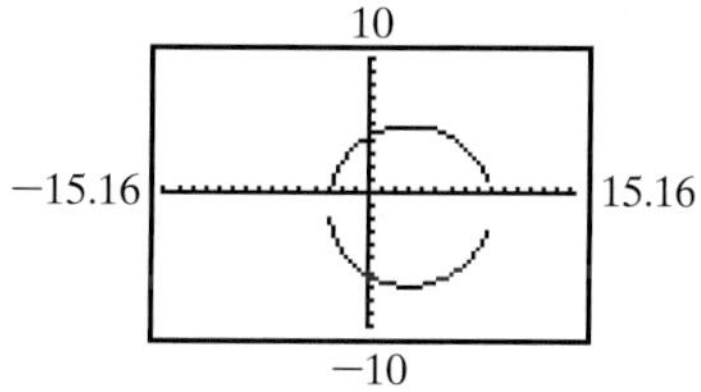

71. d

73. a. $\sqrt{\left(x_1 - \frac{x_1 + x_2}{2}\right)^2 + \left(y_1 - \frac{y_1 + y_2}{2}\right)^2} = \frac{1}{2}\sqrt{(x_1 - x_2)^2 + (y_1 - y_2)^2}$

$\sqrt{\left(x_2 - \frac{x_1 + x_2}{2}\right)^2 + \left(y_2 - \frac{y_1 + y_2}{2}\right)^2} = \frac{1}{2}\sqrt{(x_1 - x_2)^2 + (y_1 - y_2)^2}$

b. The distance from (x_1, y_1) to (x_2, y_2) is $\sqrt{(x_1 - x_2)^2 + (y_1 - y_2)^2}$, which equals the sum of the distances from part (a).

75. $(x + 3)^2 + (y - 2)^2 = 16; x^2 + y^2 + 6x - 4y - 3 = 0$ **77.** $y + 4 = \frac{3}{4}(x - 3)$

78. $f(g(x)) = 9x^2 + 24x + 14; g(f(x)) = 3x^2 - 2$ **79.** $\{4\}$ **80.** $\left\{x \middle| -\frac{5}{2} < x < \frac{15}{2}\right\}$ or $\left(-\frac{5}{2}, \frac{15}{2}\right)$

Section 10.2

Check Point Exercises

1.

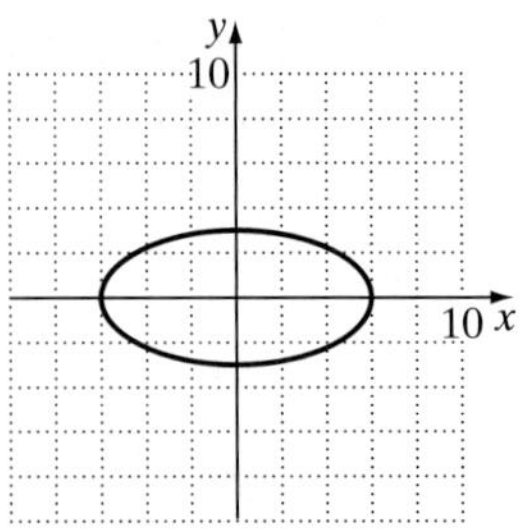

2.

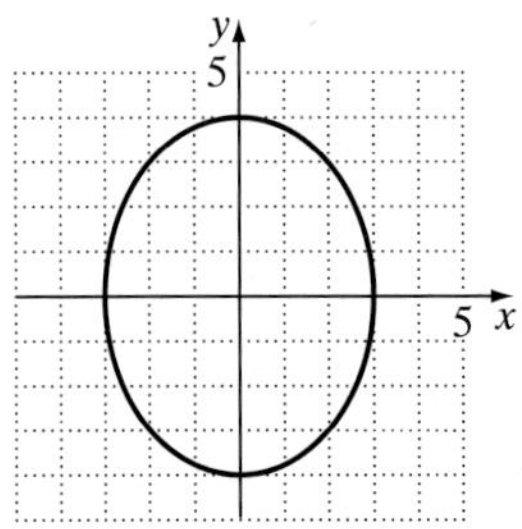

3.

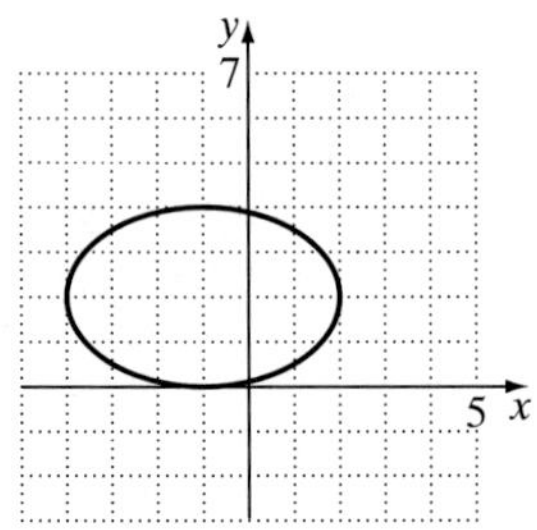

4. Yes, the height of the archway 6 feet from the center is approximately 9.54 feet.

Exercise Set 10.2

1.

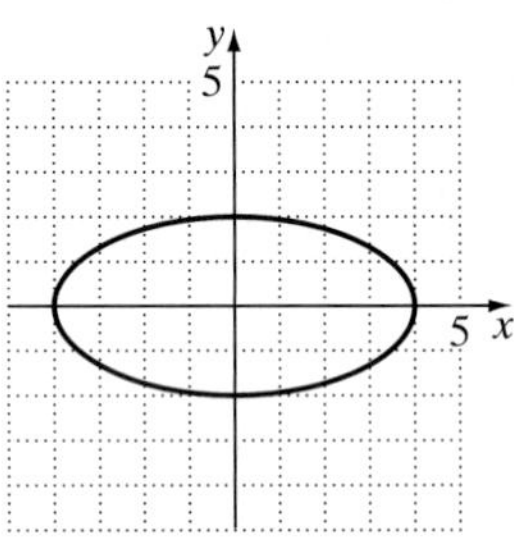

3.

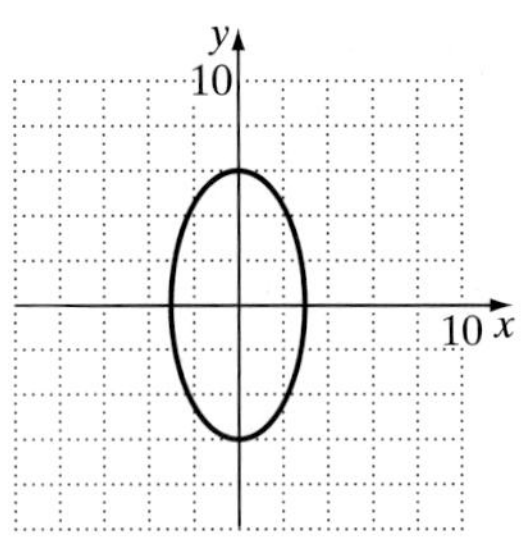

5.

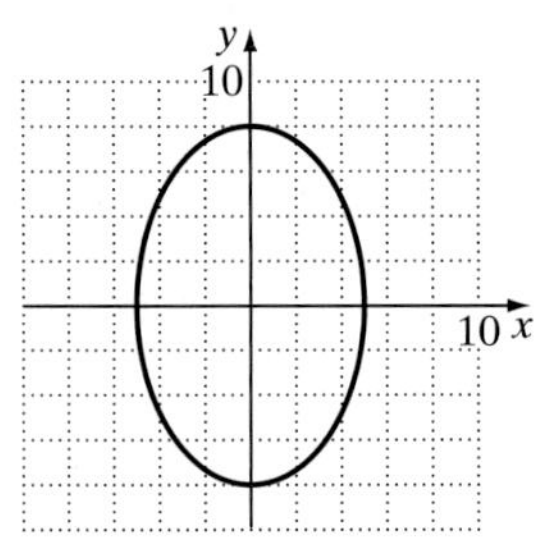

7.

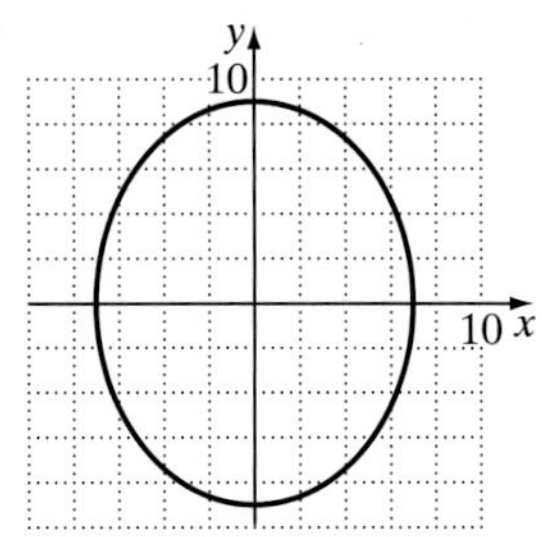

9.

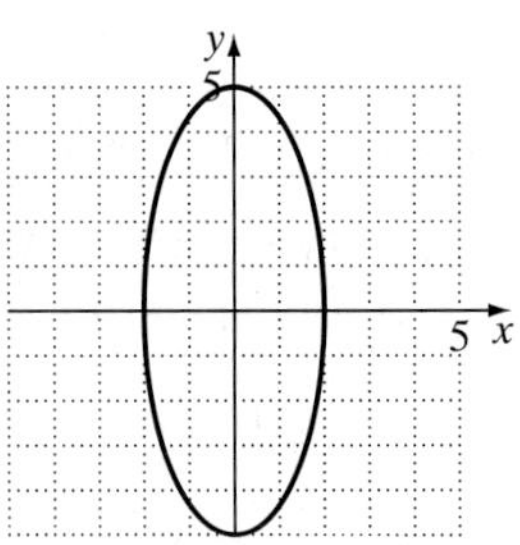

11.

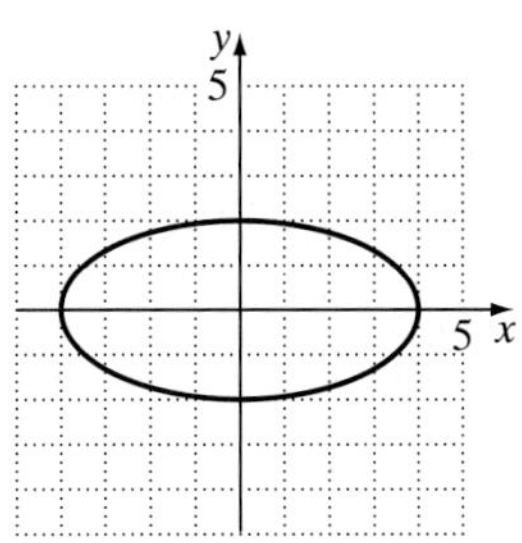

13.

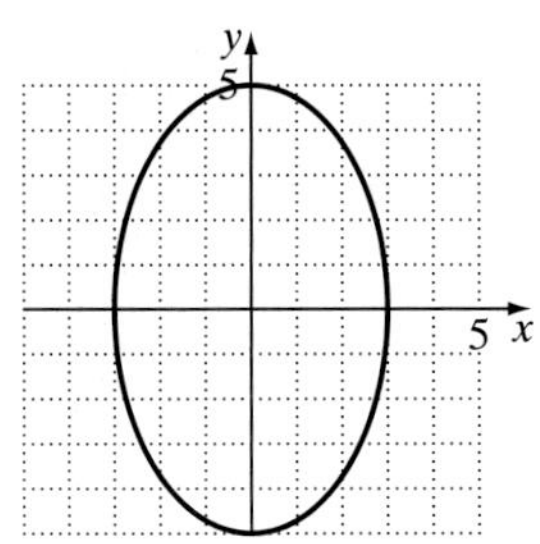

15.

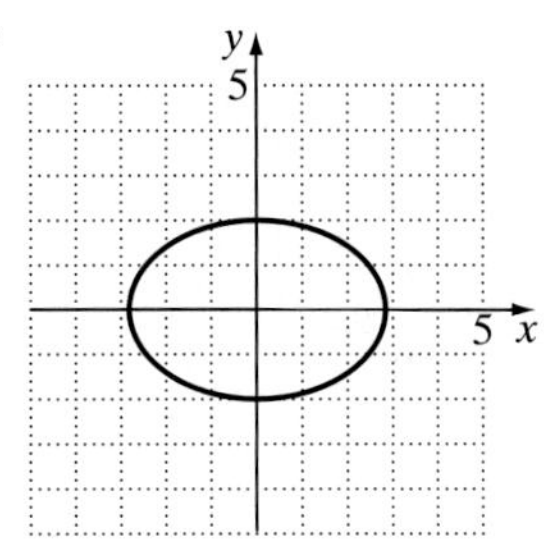

17. $\frac{x^2}{4} + \frac{y^2}{1} = 1$ or $\frac{x^2}{4} + y^2 = 1$

19. $\frac{x^2}{1} + \frac{y^2}{4} = 1$ or $x^2 + \frac{y^2}{4} = 1$

21.

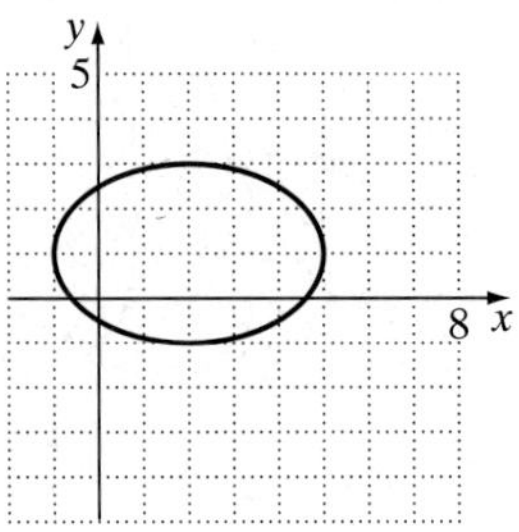

23.

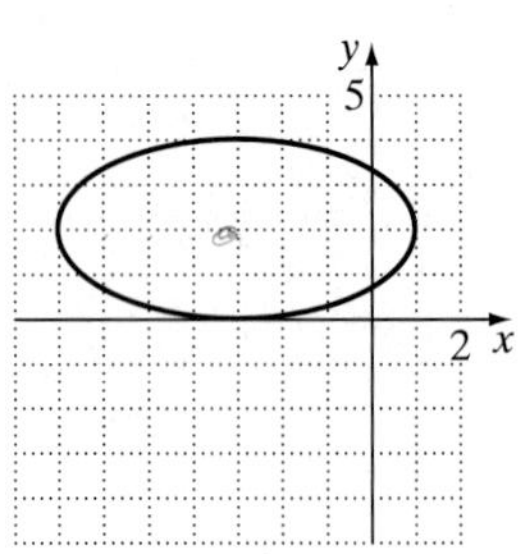

25.

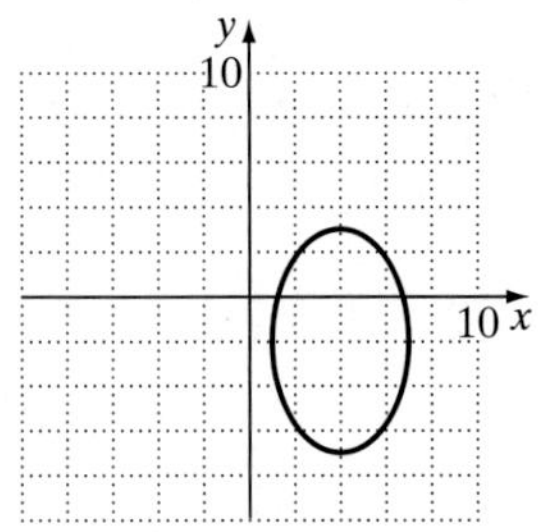

27.

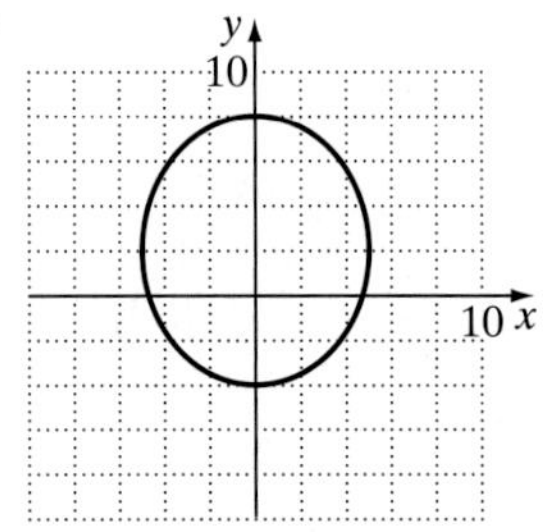

29.

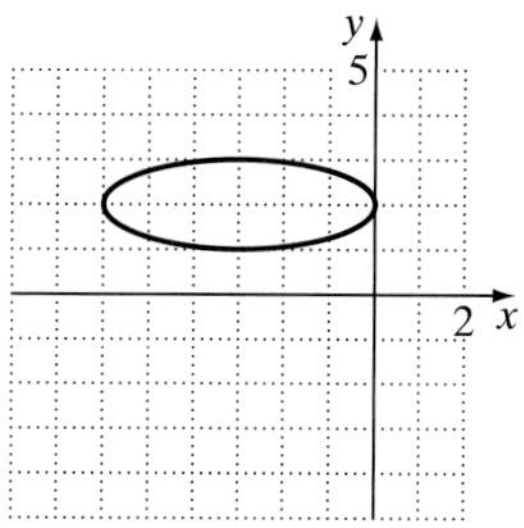

31.

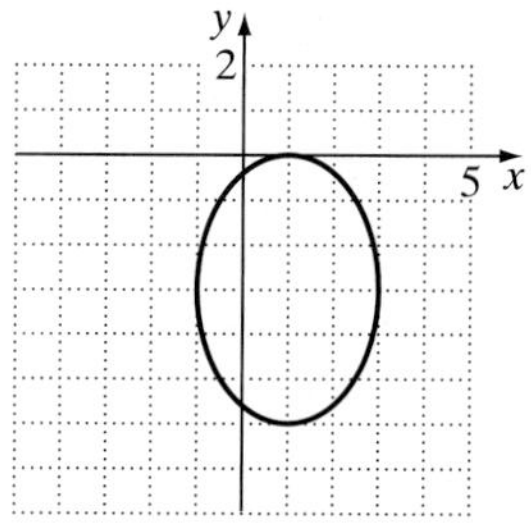

33. Yes, the height of the archway 4 feet from the center is approximately 9.96 feet.

35. a. $\frac{x^2}{48^2} + \frac{y^2}{23^2} = 1$ or $\frac{x^2}{2304} + \frac{y^2}{529} = 1$

b. 42 feet

37–43. Answers will vary.

45. $\frac{x^2}{\frac{36}{5}} + \frac{y^2}{36} = 1$

47.

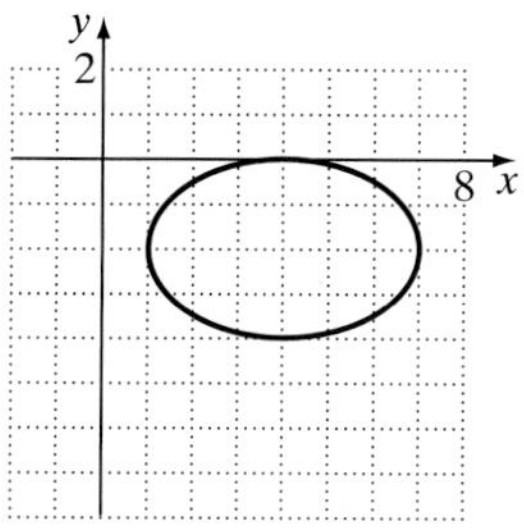

49. small circle: $x^2 + y^2 = 9$; large circle: $x^2 + y^2 = 25$

51. $(x + 2)(x + 2)(x - 2)$

52. $2xy^2\sqrt[3]{5xy}$

53. -1 or $\{-1\}$

Section 10.3

Check Point Exercises

1. a. $(-5, 0)$ and $(5, 0)$ **b.** $(0, -5)$ and $(0, 5)$

2.

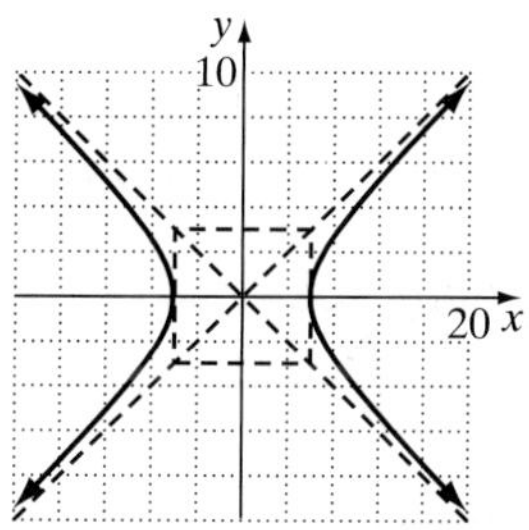

3.

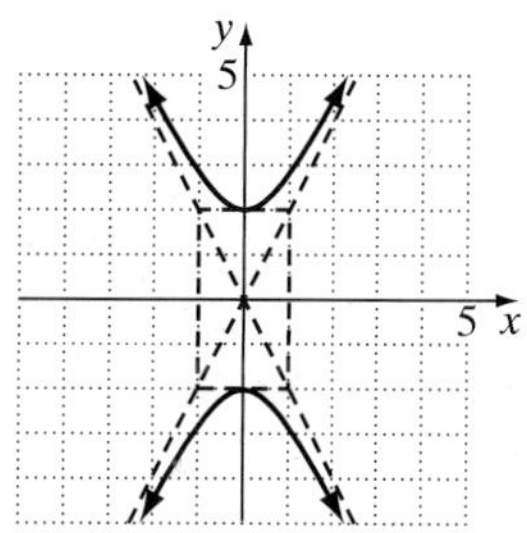

Exercise Set 10.3

1. $(-2, 0)$ and $(2, 0)$; b **3.** $(0, -2)$ and $(0, 2)$; a

5.

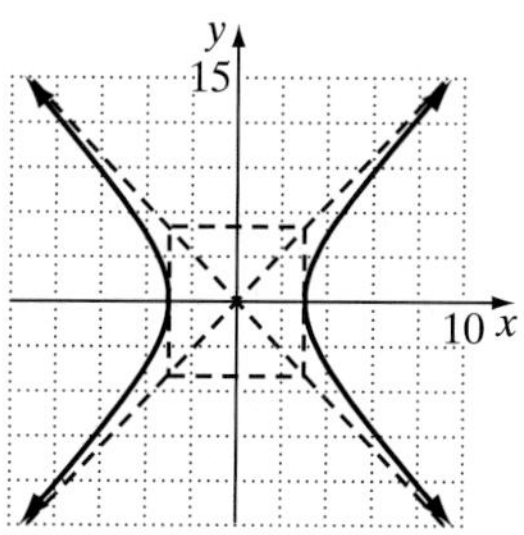

7.

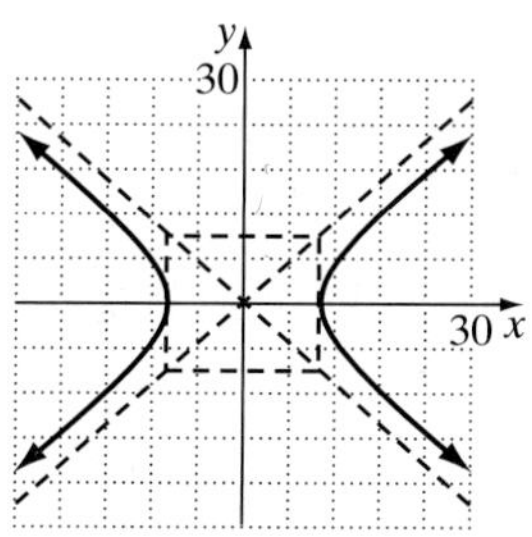

9.

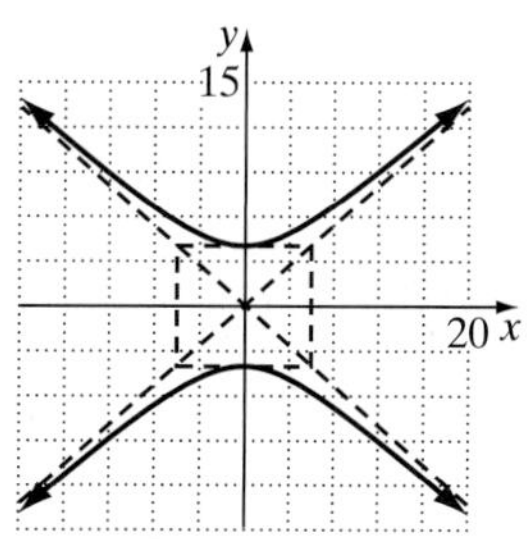

11.

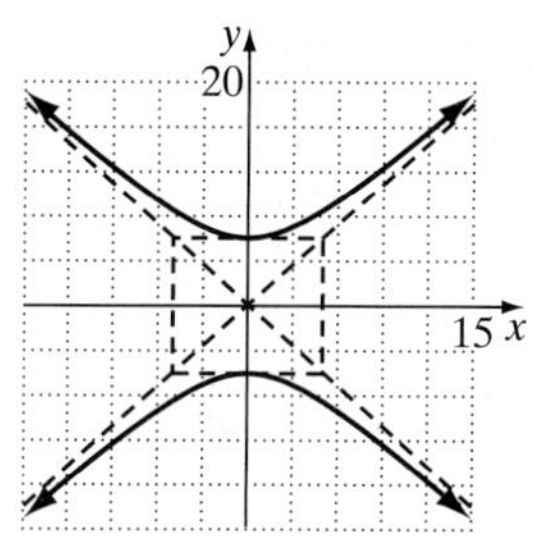

13.

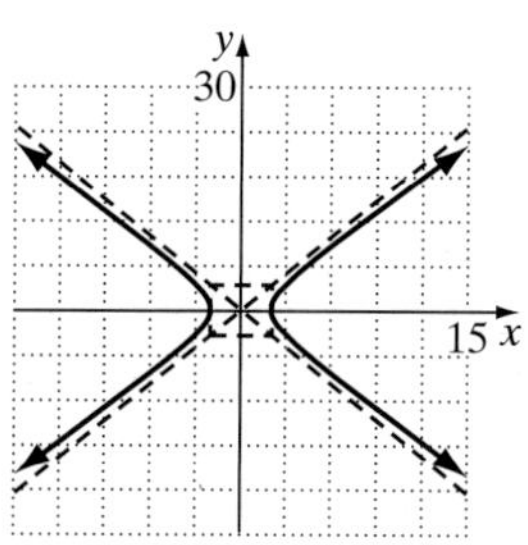

15.

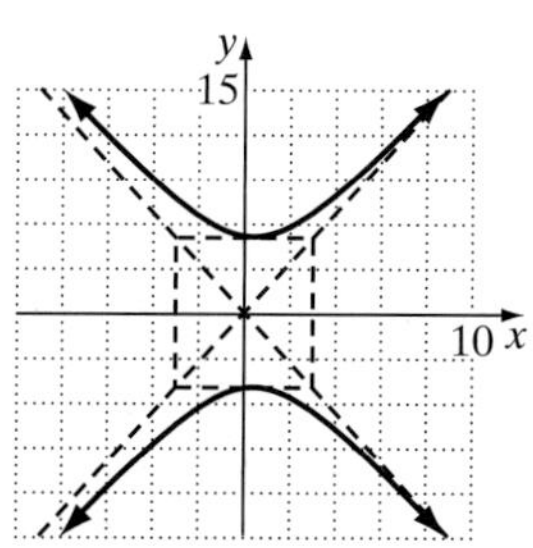

17.

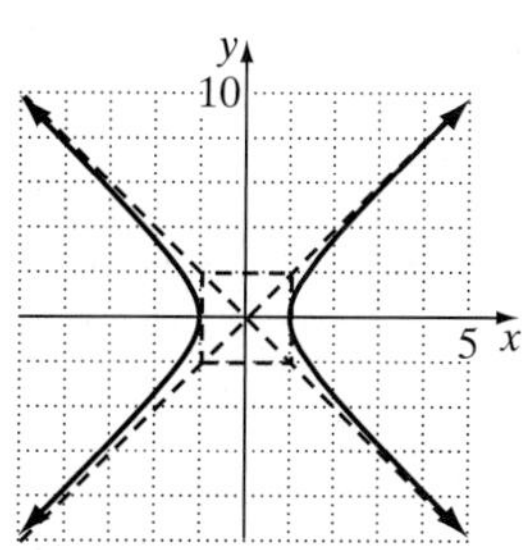

19. $\frac{x^2}{9} - \frac{y^2}{25} = 1$

21. $\frac{y^2}{4} - \frac{x^2}{9} = 1$

23. 40 yards

25–29. Answers will vary.

31.

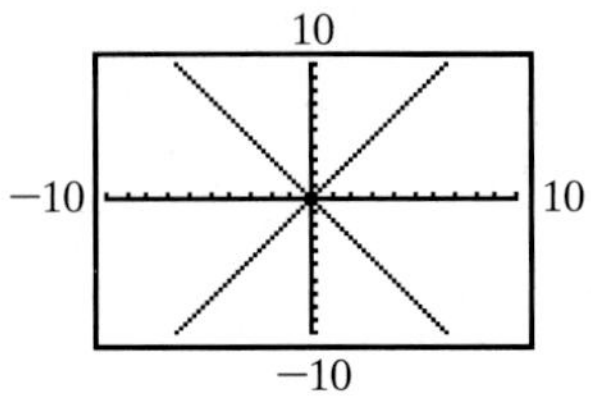

no; two lines; $y = \frac{b}{a}x$ and $y = -\frac{b}{a}x$

33. c

35.

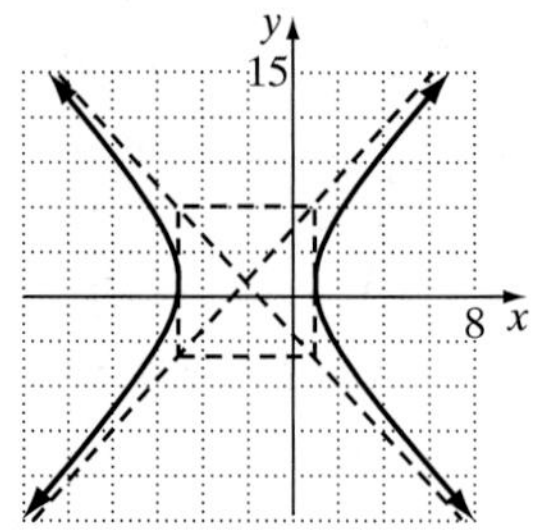

37.

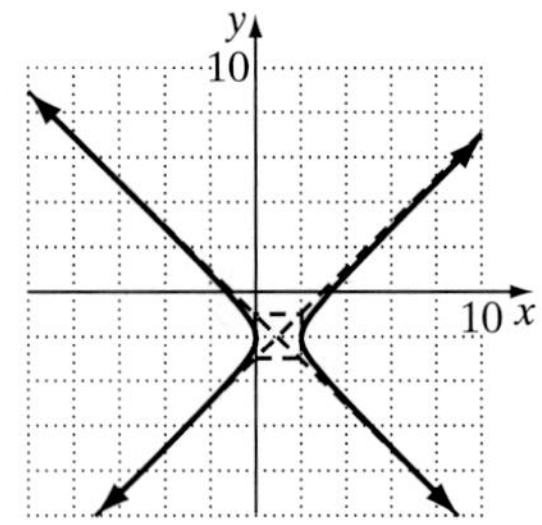

39. $\frac{y^2}{49} - \frac{x^2}{\frac{49}{25}} = 1$

40.

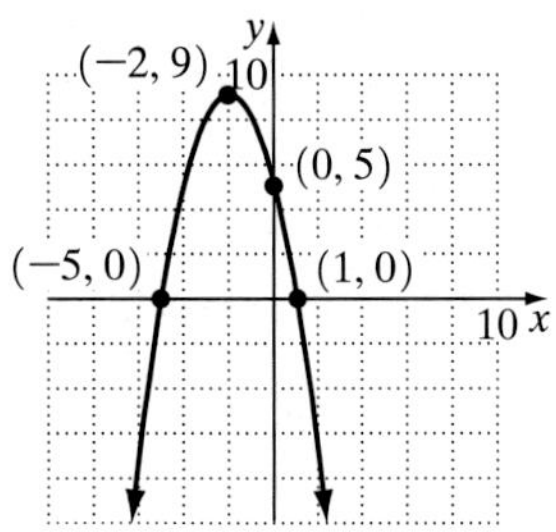

41. $\left\{x \middle| x \leq -\frac{1}{3} \text{ or } x \geq 4\right\}$ or $\left(-\infty, -\frac{1}{3}\right] \cup [4, \infty)$

42. 21 or {21}

Section 10.4

Check Point Exercises

1.

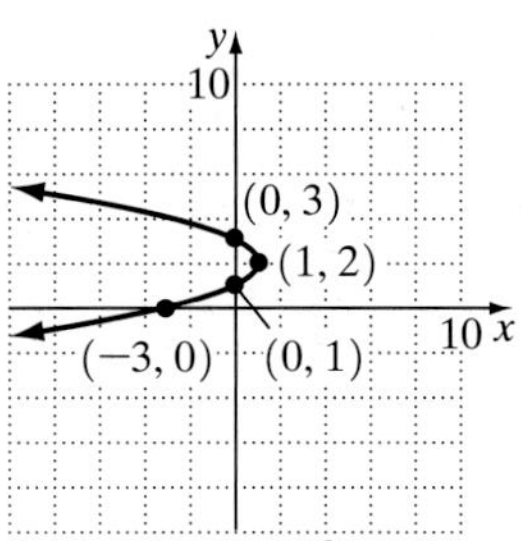

$x = -(y - 2)^2 + 1$

2.

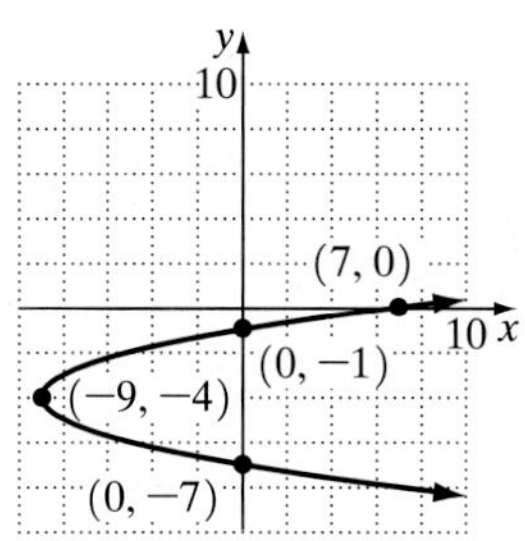

$x = y^2 + 8y + 7$

3. **a.** hyperbola
b. ellipse
c. circle
d. parabola

Exercise Set 10.4

1. opens to right; (−1, 2); b **3.** opens to right; (1, −2); f **5.** opens to left; (1, 2); a
7. (0, 0) **9.** (3, 2) **11.** (−1, −2) **13.** (0, 6) **15.** (−3, 3) **17.** (4, −1)

19.

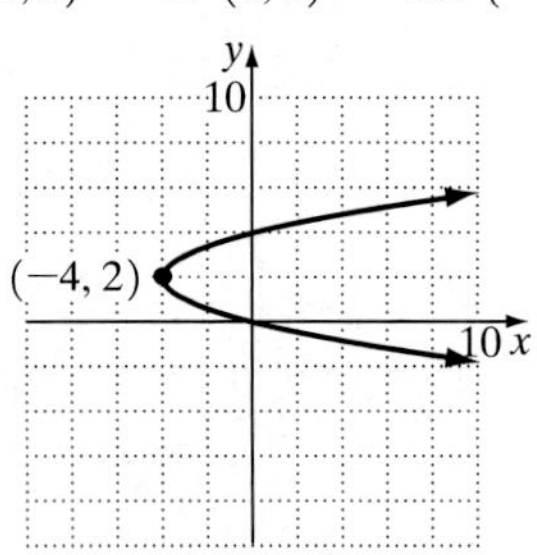

$x = (y - 2)^2 - 4$
axis of symmetry: $y = 2$

21.

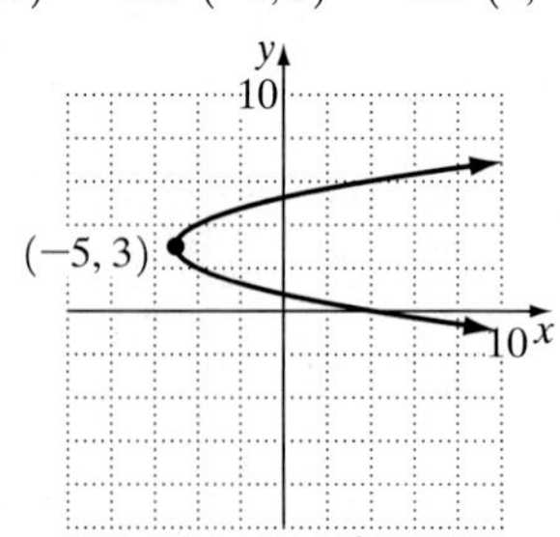

$x = (y - 3)^2 - 5$
axis of symmetry: $y = 3$

23.

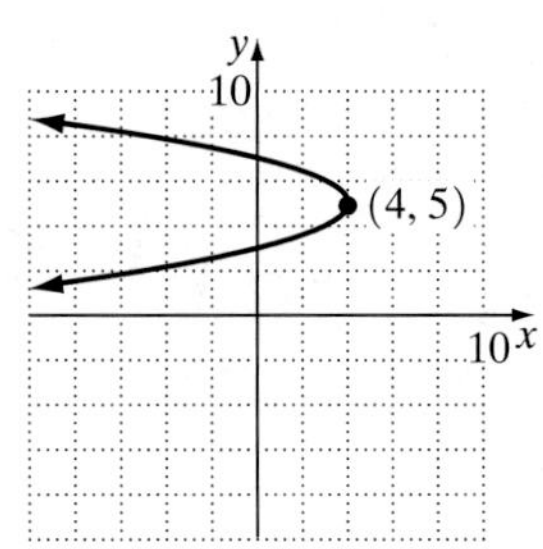

$x = -(y - 5)^2 + 4$
axis of symmetry: $y = 5$

25.

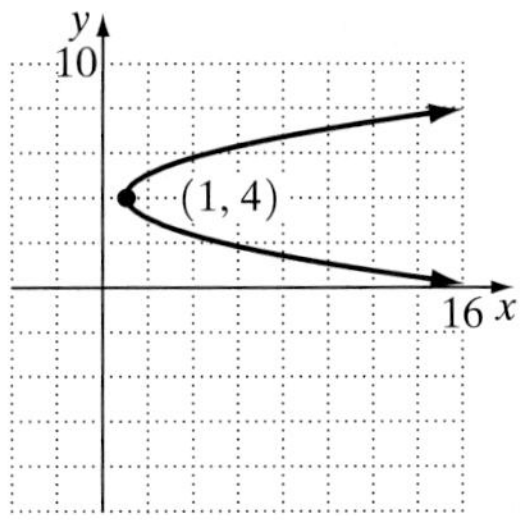

$x = (y - 4)^2 + 1$

axis of symmetry: $y = 4$

27.

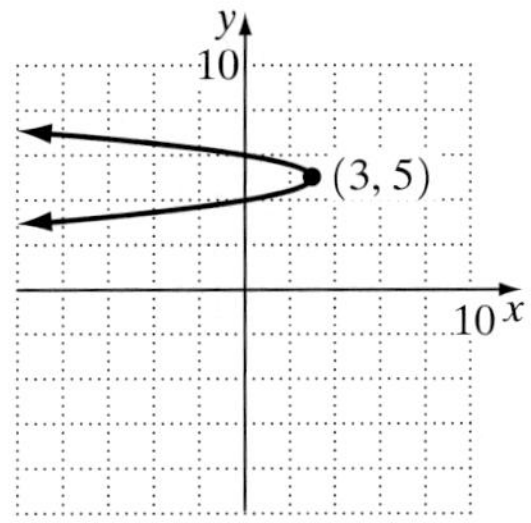

$x = -3(y - 5)^2 + 3$

axis of symmetry: $y = 5$

29.

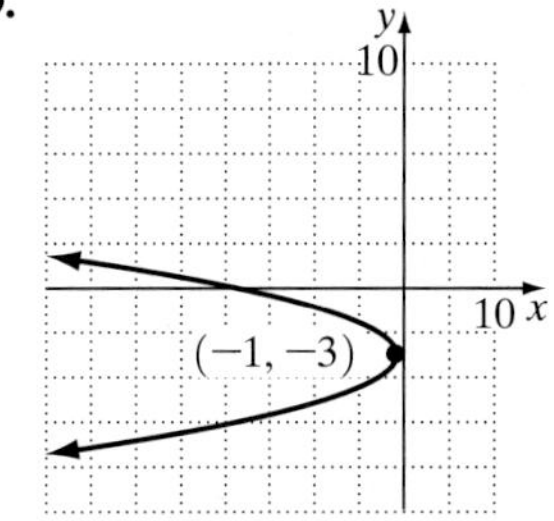

$x = -2(y + 3)^2 - 1$

axis of symmetry: $y = -3$

31.

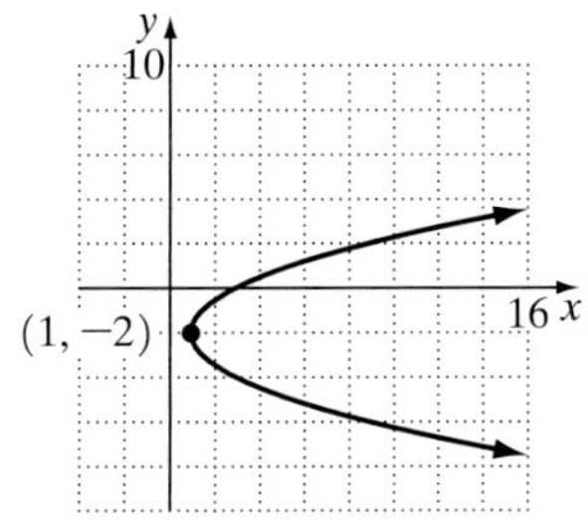

$x = \frac{1}{2}(y + 2)^2 + 1$

axis of symmetry: $y = -2$

33.

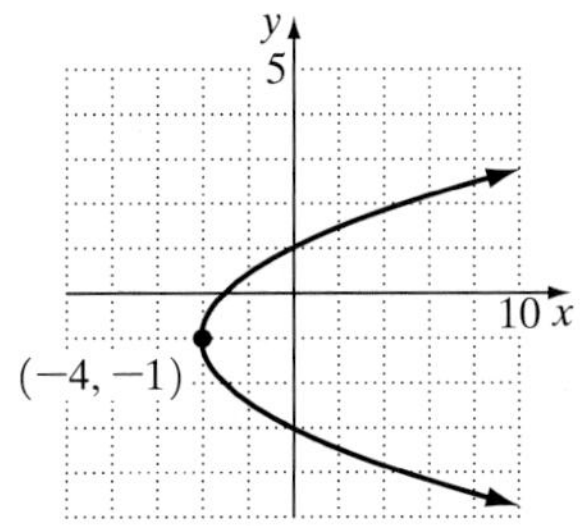

$x = y^2 + 2y - 3$

axis of symmetry: $y = -1$

35.

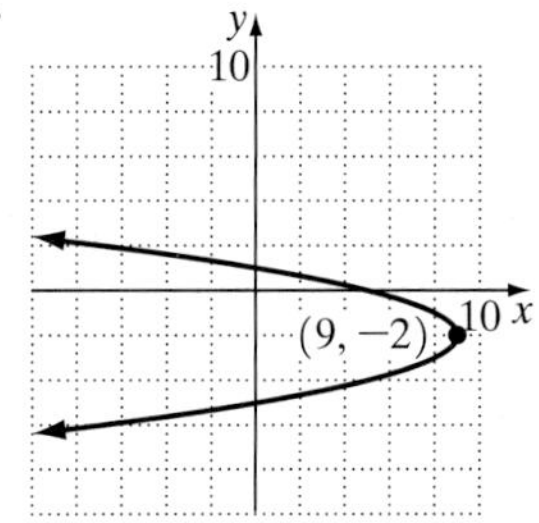

$x = -y^2 - 4y + 5$

axis of symmetry: $y = -2$

37.

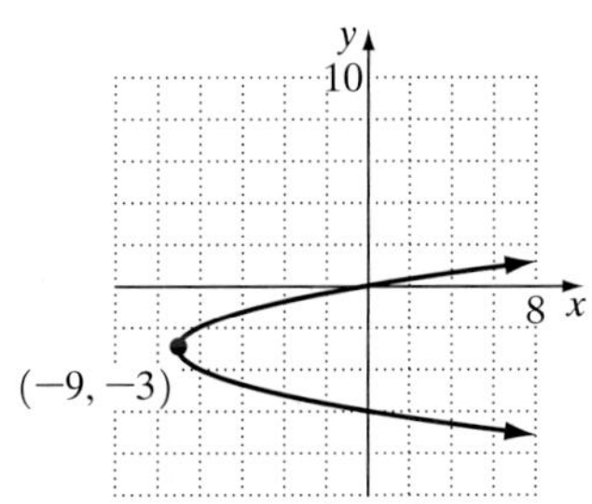

$x = y^2 + 6y$

axis of symmetry: $y = -3$

39.

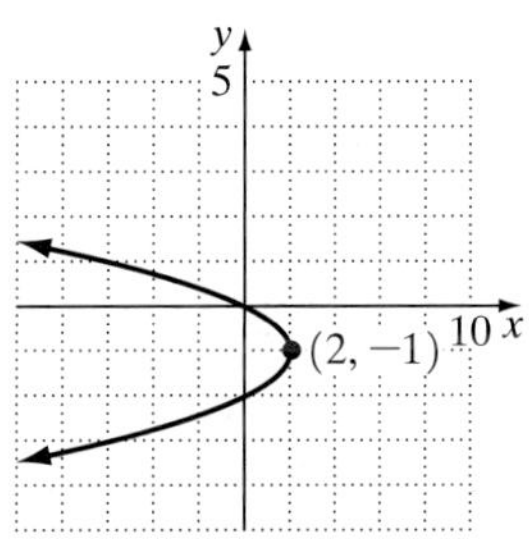

$x = -2y^2 - 4y$

axis of symmetry: $y = -1$

41.

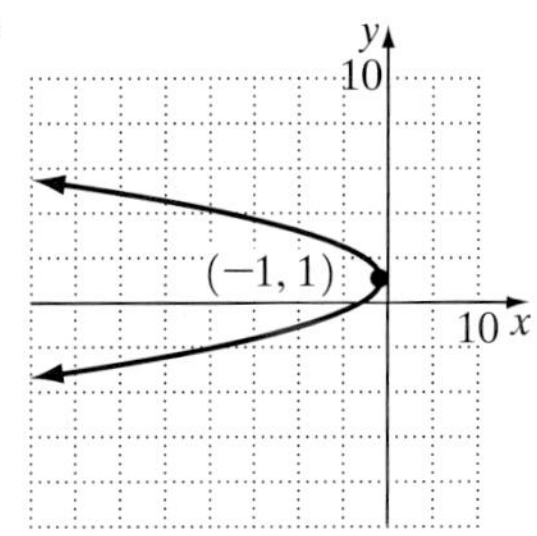

$x = -2y^2 - 4y + 1$

axis of symmetry: $y = -1$

43. a. horizontal **b.** to the right **c.** (2, 1) **45. a.** vertical **b.** upward **c.** (1, 2) **47. a.** vertical **b.** downward **c.** (−3, 4) **49. a.** horizontal **b.** to the left **c.** (4, −3) **51. a.** vertical **b.** upward **c.** (2, −5) **53.** horizontal **b.** to the left **c.** (5, 2) **55.** parabola **57.** ellipse **59.** hyperbola **61.** circle **63.** hyperbola

65. hyperbola

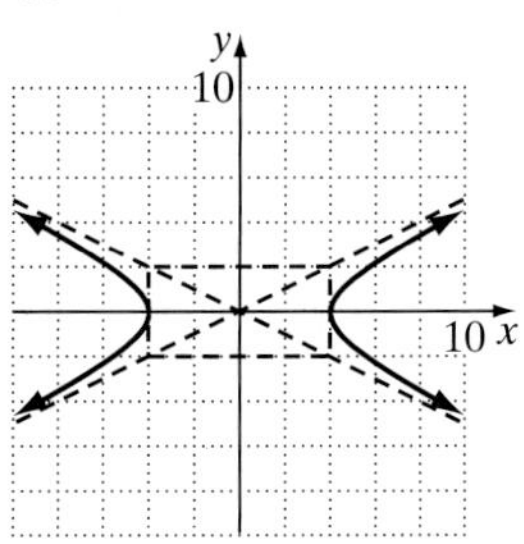

67. circle

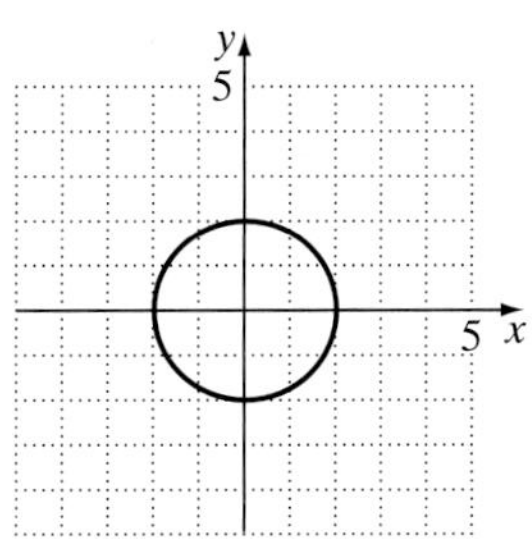

69. ellipse

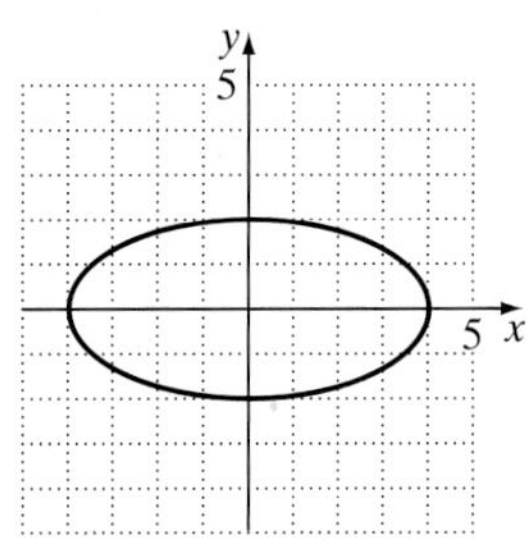

71. parabola

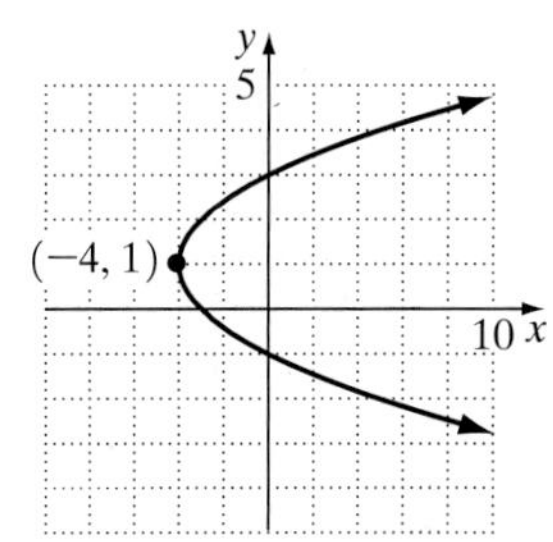

73. circle

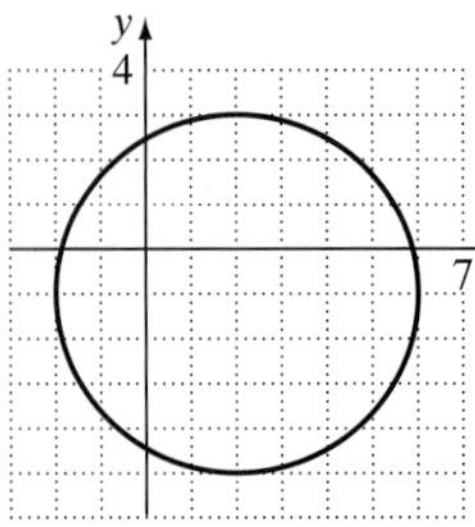

75. a. $y = 0.000342x^2$ **b.** 58 ft **77. a.** $y = \frac{1}{18}x^2$ **b.** 4.5 ft **79–85.** Answers will vary.

87. $y^2 + 2y + (-6x + 13) = 0$; $y = -1 \pm \sqrt{6x - 2}$

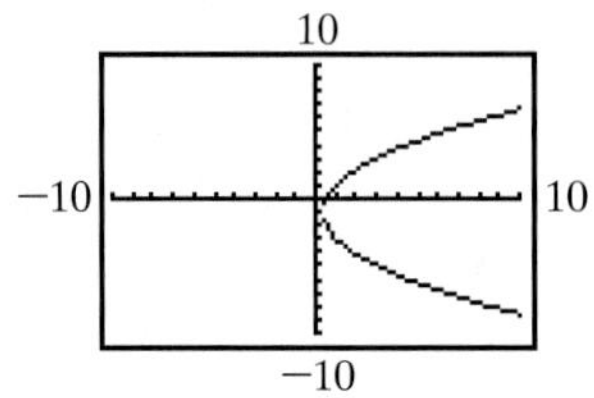

89–91. Answers will vary.

93.

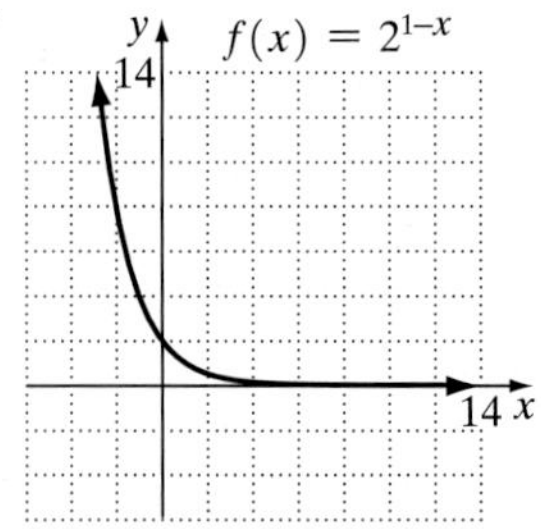

94. $f^{-1}(x) = 3x + 15$
95. (6, 4) or {(6, 4)}

Section 10.5

Check Point Exercises

1. (0, 1) and (4, 17), or {(0, 1), (4, 17)} **2.** (2, −1) and $\left(-\frac{6}{5}, \frac{3}{5}\right)$, or $\left\{(2, -1), \left(-\frac{6}{5}, \frac{3}{5}\right)\right\}$
3. (3, 2), (−3, 2), (3, −2), and (−3, −2), or {(3, 2), (−3, 2), (3, −2), (−3, −2)} **4.** (0, 5) or {(0, 5)} **5.** length: 7 feet; width: 3 feet

Exercise Set 10.5

1. (2, 0) and (−3, 5), or {(2, 0), (−3, 5)} **3.** (2, 0) and (1, 1), or {(2, 0), (1, 1)} **5.** (−3, 11) and (4, −10), or {(−3, 11), (4, −10)}
7. (−3, −4) and (4, 3), or {(−3, −4), (4, 3)} **9.** (2, 3) and $\left(-\frac{3}{2}, -4\right)$, or $\left\{(2, 3), \left(-\frac{3}{2}, -4\right)\right\}$ **11.** (3, 0) and (−5, −4), or {(3, 0), (−5, −4)} **13.** (3, 1), (−1, −3), (1, 3), and (−3, −1), or {(3, 1), (−1, −3), (1, 3), (−3, −1)} **15.** (4, −3) and (−1, 2), or {(4, −3), (−1, 2)} **17.** (4, −3) and (0, 1), or {(4, −3), (0, 1)} **19.** (3, 2), (−3, 2), (3, −2), and (−3, −2), or {(3, 2), (−3, 2), (3, −2), (−3, −2)} **21.** (3, 2), (−3, 2), (3, −2), and (−3, −2), or {(3, 2), (−3, 2), (3, −2), (−3, −2)}
23. (2, 1), (−2, 1), (2, −1), and (−2, −1) or {(2, 1), (−2, 1), (2, −1), (−2, −1)} **25.** (3, 4) and (3, −4), or {(3, 4), (3, −4)}
27. (0, 2), (0, −2), $(-1, \sqrt{3})$, and $(-1, -\sqrt{3})$, or $\{(0, 2), (0, -2), (-1, \sqrt{3}), (-1, -\sqrt{3})\}$ **29.** (2, 1), (−2, 1), (2, −1), and (−2, −1), or {(2, 1), (−2, 1), (2, −1), (−2, −1)} **31.** (−1, −4), (1, 4), $(2\sqrt{2}, \sqrt{2})$, and $(-2\sqrt{2}, -\sqrt{2})$, or $\{(-1, -4), (1, 4), (2\sqrt{2}, \sqrt{2}), (-2\sqrt{2}, -\sqrt{2})\}$
33. (4, 1) and (2, 2), or {(4, 1), (2, 2)} **35.** (0, 0) and (−1, 1), or {(0, 0), (−1, 1)} **37.** (0, 0), (2, 2), and (−2, 2), or {(0, 0), (2, 2), (−2, 2)} **39.** (−4, 1) and $\left(-\frac{5}{2}, \frac{1}{4}\right)$, or $\left\{(-4, 1), \left(-\frac{5}{2}, \frac{1}{4}\right)\right\}$ **41.** (−2, 3) and $\left(\frac{12}{5}, -\frac{29}{5}\right)$, or $\left\{(-2, 3), \left(\frac{12}{5}, -\frac{29}{5}\right)\right\}$ **43.** $x + y = 10$; $xy = 24$; 6 and 4 **45.** $x^2 - y^2 = 3$; $2x^2 + y^2 = 9$; 1 and 2, −2 and 1, −2 and −1 or 2 and −1
47. (0, −4), (2, 0), (−2, 0) **49.** length: 11 feet; width: 7 feet **51.** length: 8 inches; width: 6 inches
53. large square: 5 meters by 5 meters; small square: 2 meters by 2 meters **55–59.** Answers will vary. **61.** b **63.** (8, 2) or {(8, 2)}

65.

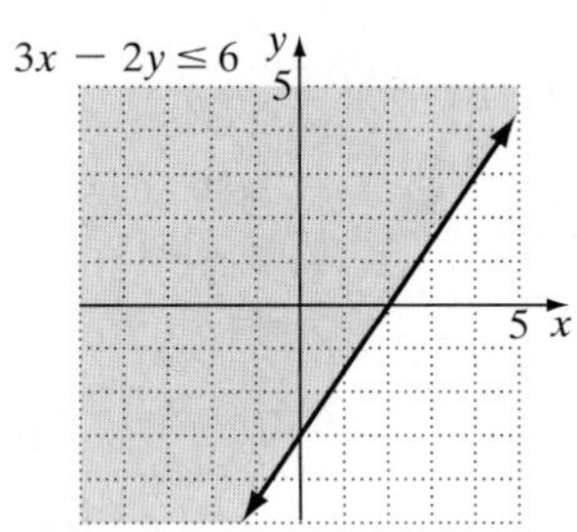

66. $m = \frac{8}{3}$

Review Exercises

1. 13 units **2.** $2\sqrt{2} \approx 2.83$ units **3.** $(-5, 5)$ **4.** $\left(-\frac{11}{2}, -2\right)$ **5.** $x^2 + y^2 = 9$ **6.** $(x + 2)^2 + (y - 4)^2 = 36$

7. center: $(0, 0)$; $r = 1$

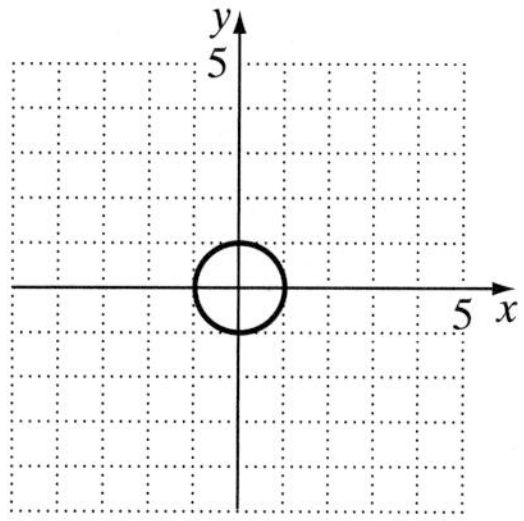

8. center: $(-2, 3)$; $r = 3$

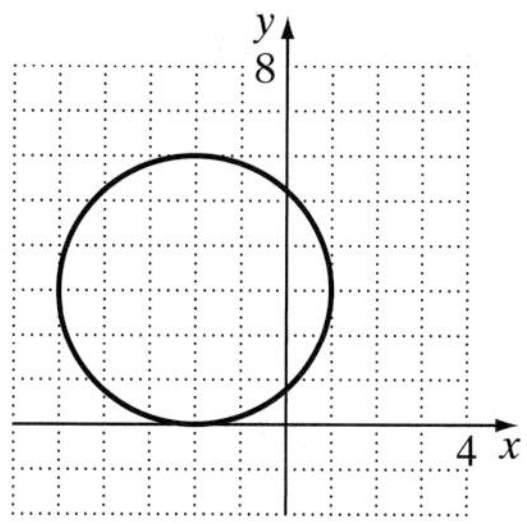

9. center: $(2, -1)$; $r = 3$

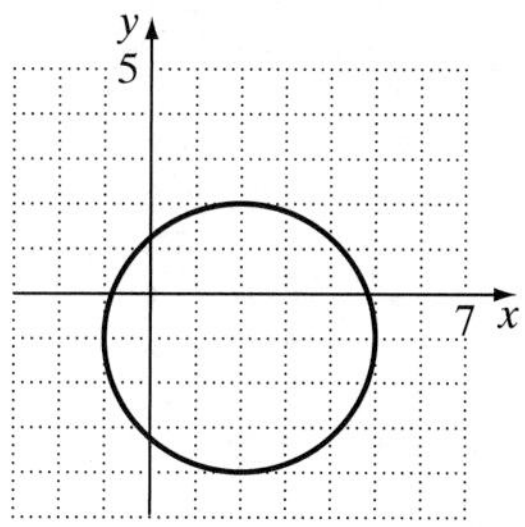

10. center: $(0, 2)$; $r = 2$

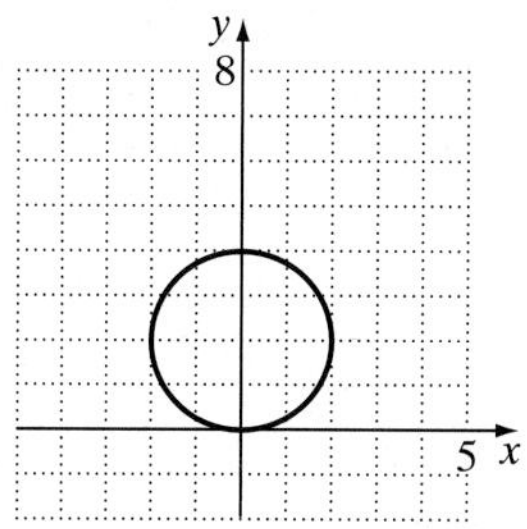

11.

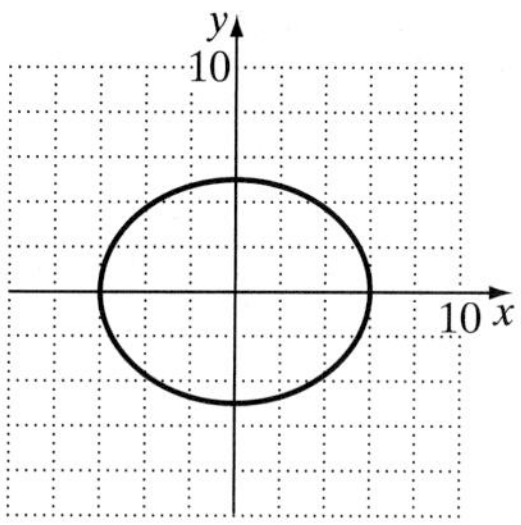

12.

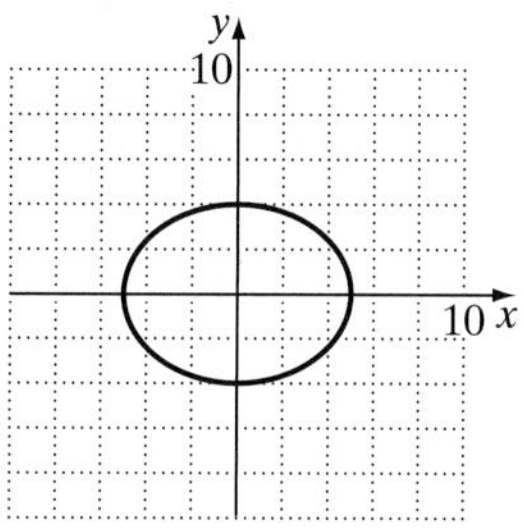

13.

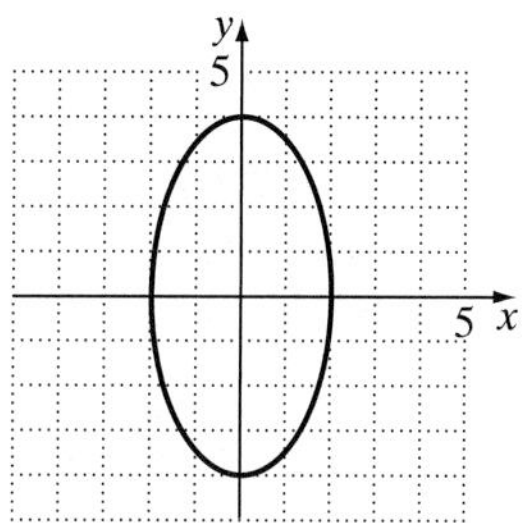

14.

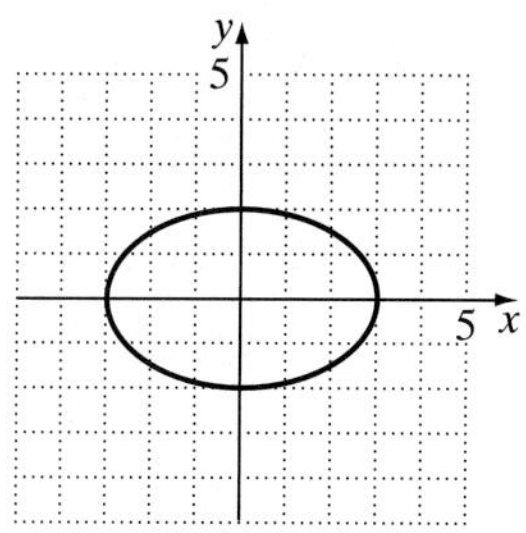

15.

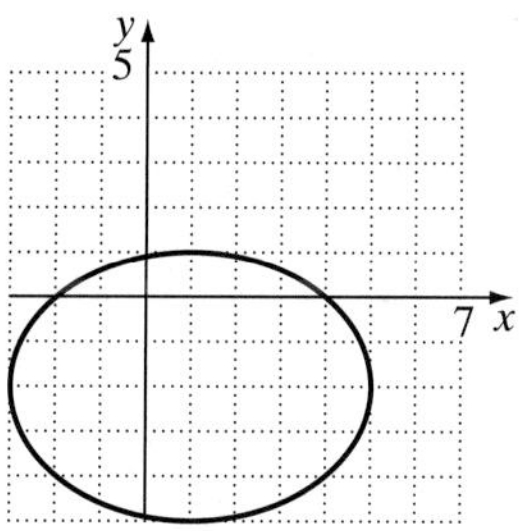

16.

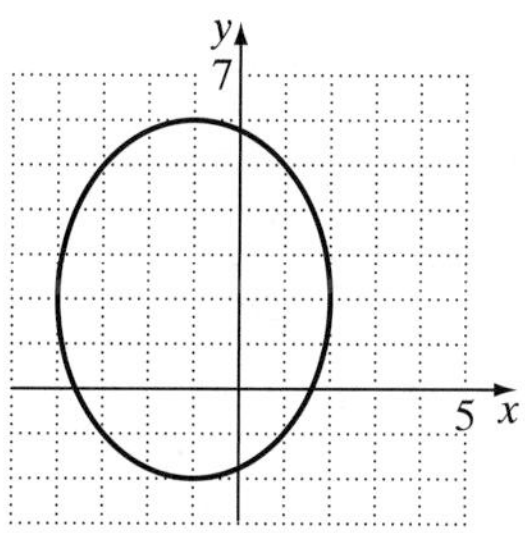

17. Yes, the height of the archway 14 feet from the center is approximately 12.43 feet.

18.

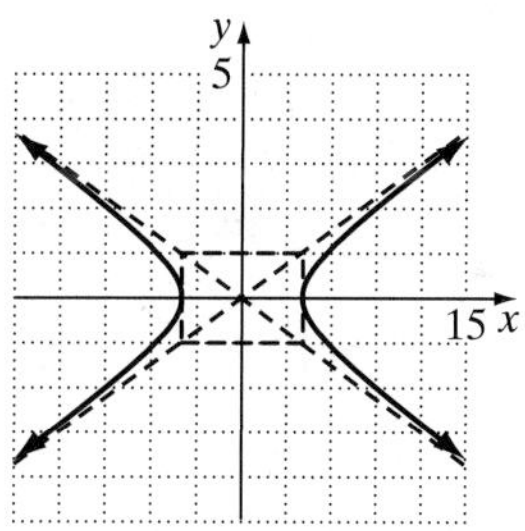

19.

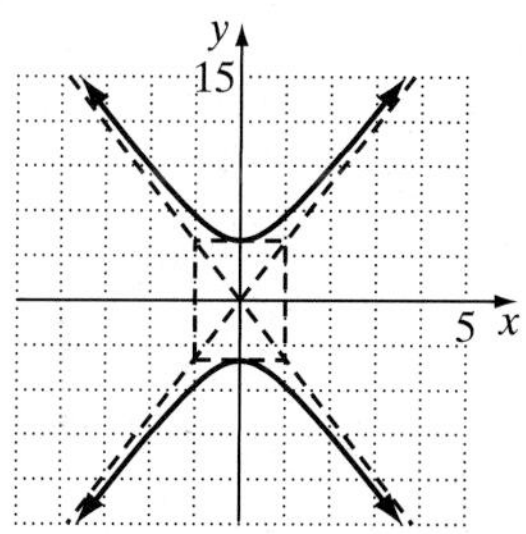

20.

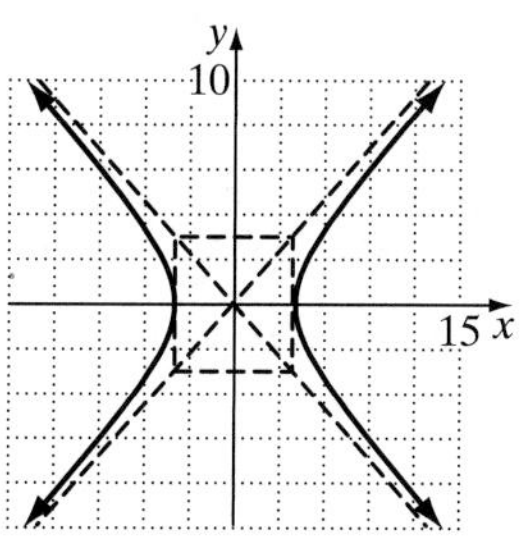

21.

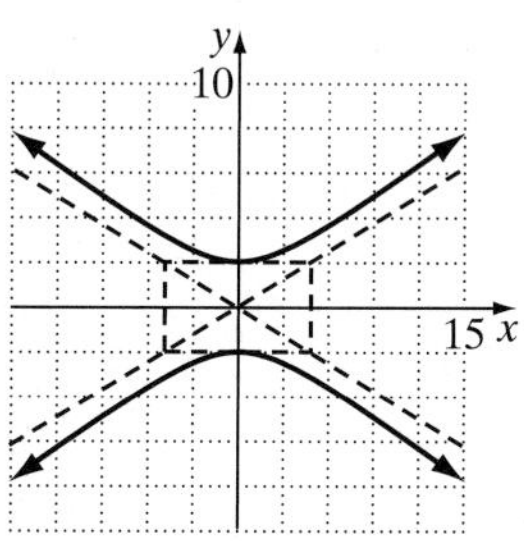

22.

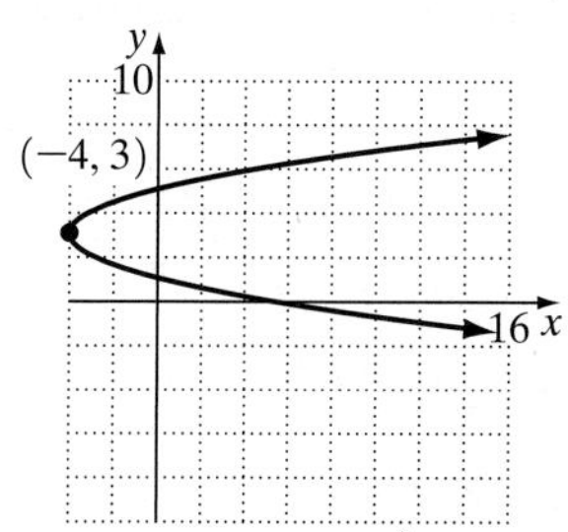

$x = (y - 3)^2 - 4$

axis of symmetry: $y = 3$

23.

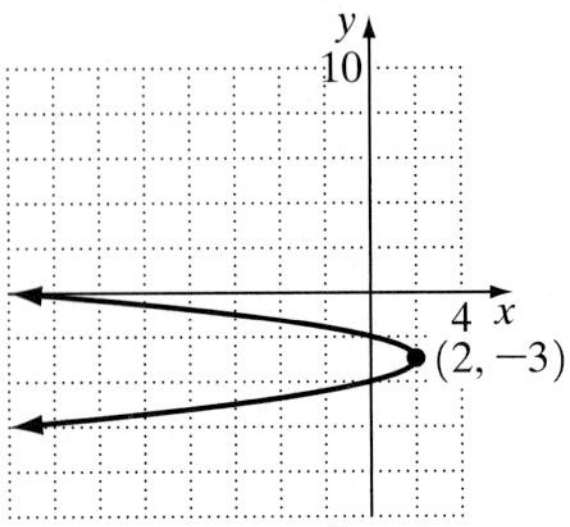

$x = -2(y + 3)^2 + 2$
axis of symmetry: $y = -3$

24.

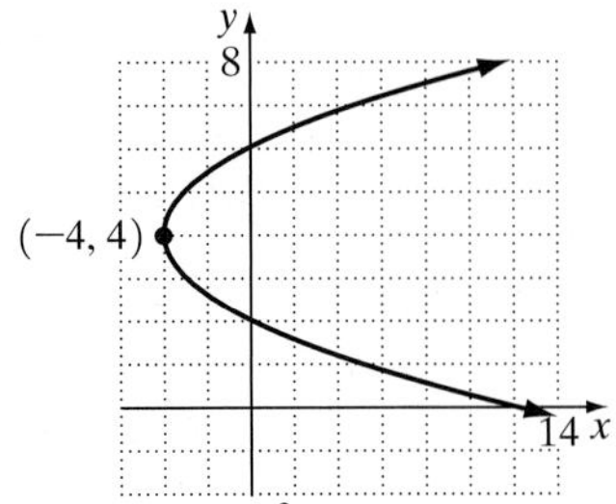

$x = y^2 - 8y + 12$
axis of symmetry: $y = 4$

25.

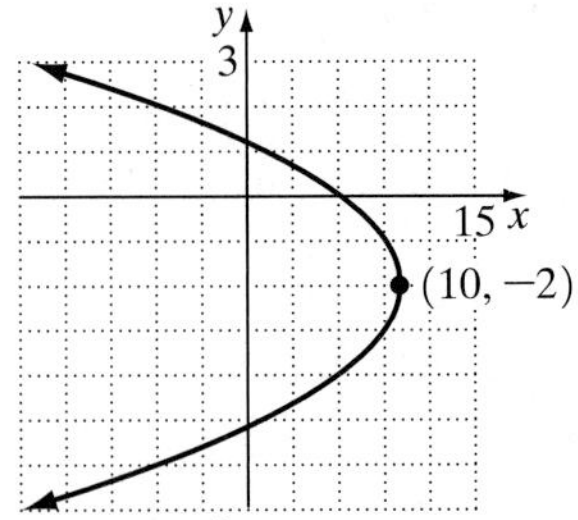

$x = -y^2 - 4y + 6$
axis of symmetry: $y = -2$

26. parabola **27.** ellipse **28.** hyperbola **29.** circle **30.** hyperbola **31.** ellipse **32.** parabola

33. circle

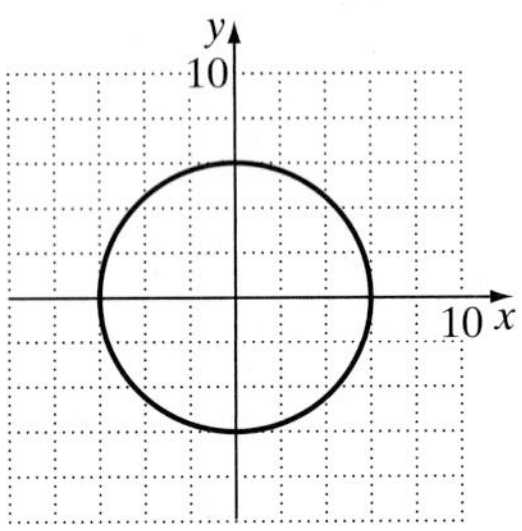

34. ellipse

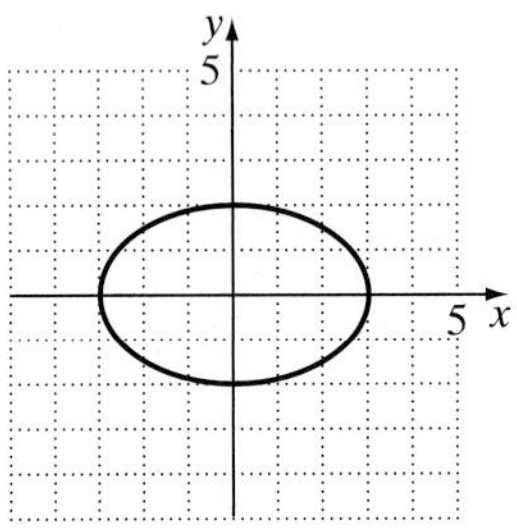

35. hyperbola

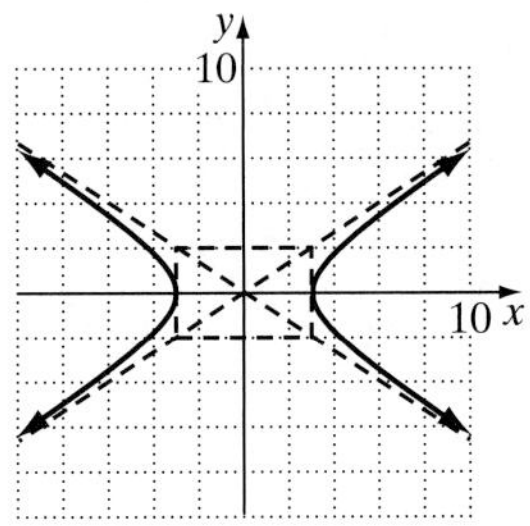

36. ellipse

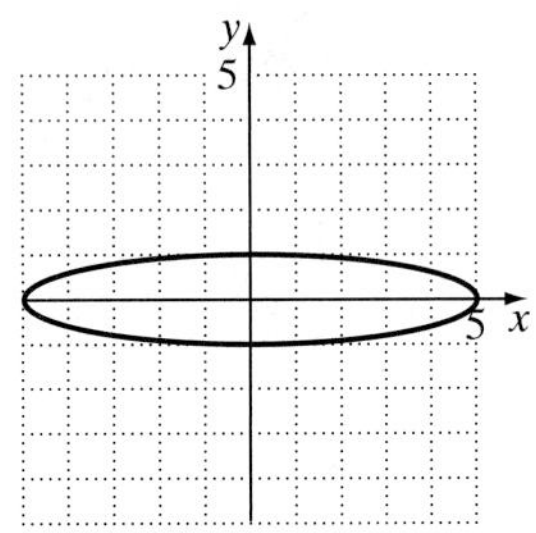

37. parabola

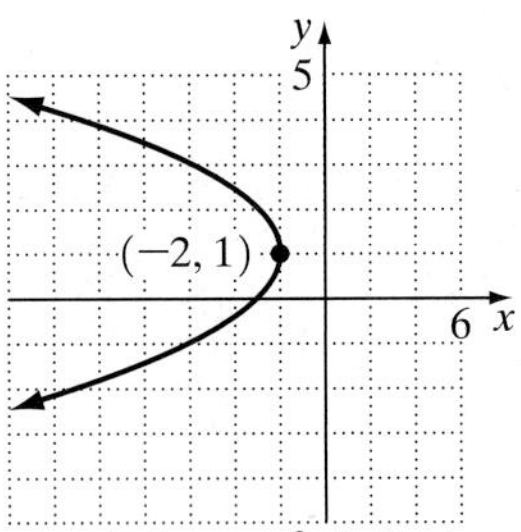

$x + 3 = -y^2 + 2y$

38. parabola

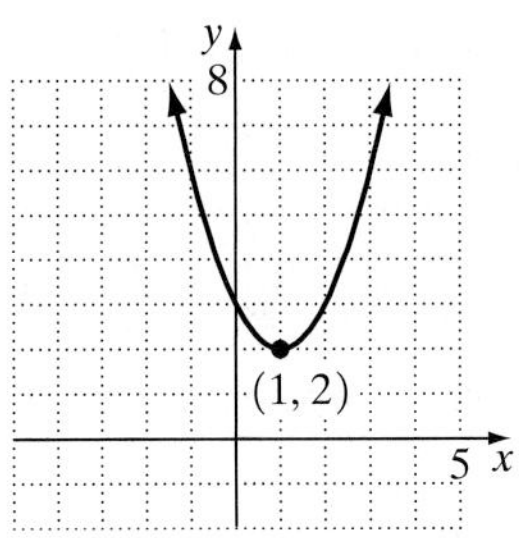

$y - 3 = x^2 - 2x$

39. ellipse

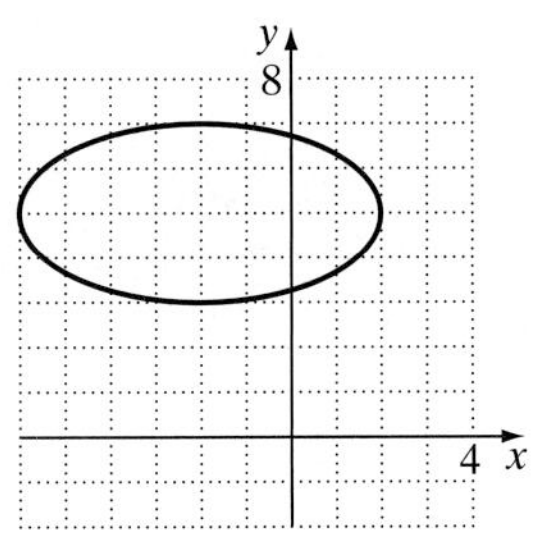

40. circle

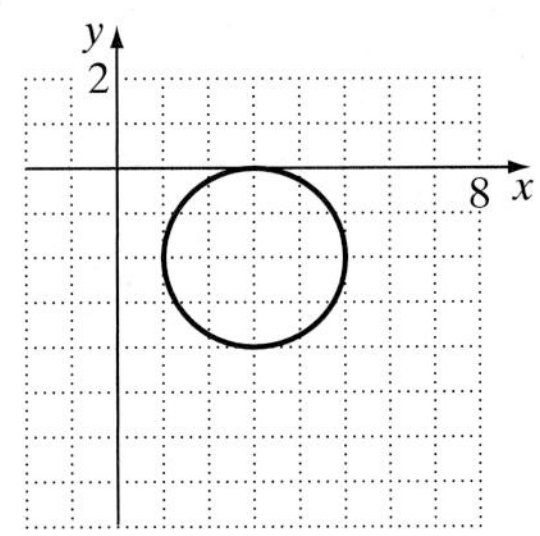

41. circle

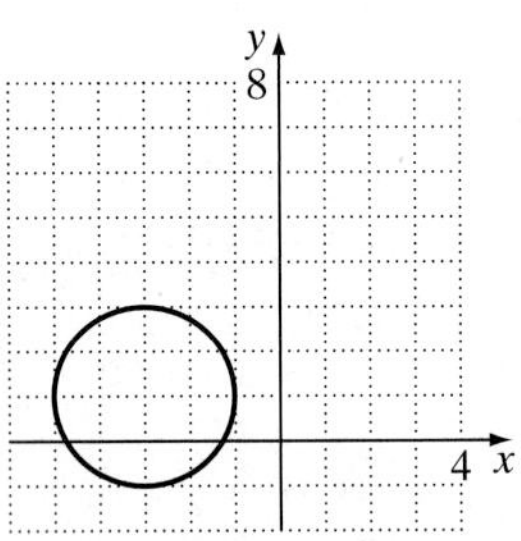

42. **a.** $y = \frac{1}{12}x^2$ **b.** $(0, 3)$; 3 inches above the vertex

43. $(1, 0)$ and $(4, 3)$, or $\{(1, 0), (4, 3)\}$ **44.** $(0, 1)$ and $(-3, 4)$, or $\{(0, 1), (-3, 4)\}$

45. $(-1, 1)$ and $(1, -1)$, or $\{(-1, 1), (1, -1)\}$

46. $(3, \sqrt{6}), (-3, \sqrt{6}), (3, -\sqrt{6})$, and $(-3, -\sqrt{6})$, or $\{(3, \sqrt{6}), (-3, \sqrt{6}), (3, -\sqrt{6}), (-3, -\sqrt{6})\}$

47. $(2, 2)$ and $(-2, -2)$, or $\{(2, 2), (-2, -2)\}$ **48.** $(1, 2)$ and $(9, 6)$, or $\{(1, 2), (9, 6)\}$

49. $(-3, -1)$ and $(1, 3)$, or $\{(-3, -1), (1, 3)\}$ **50.** $(-1, -1)$ and $\left(\frac{1}{2}, 2\right)$, or $\left\{(-1, -1), \left(\frac{1}{2}, 2\right)\right\}$

51. $(0, -1)$ and $\left(\frac{5}{2}, -\frac{7}{2}\right)$, or $\left\{(0, -1), \left(\frac{5}{2}, -\frac{7}{2}\right)\right\}$

52. $(3, 2), (-3, 2), (2, -3)$, and $(-2, -3)$, or $\{(3, 2), (-3, 2), (2, -3), (-2, -3)\}$

53. $(3, 1), (-3, 1), (3, -1)$, and $(-3, -1)$ or $\{(3, 1), (-3, 1), (3, -1)$, and $(-3, -1)\}$ **54.** 8 meters by 5 meters **55.** $(1, 6)$ and $(3, 2)$

56. x: 50 ft; y: 20 ft

Chapter 10 Test

1. $\sqrt{73} \approx 8.54$ units **2.** $\left(\frac{7}{2}, -4\right)$ **3.** $(x-3)^2 + (y+2)^2 = 25$ **4.** center: $(5, -3)$; $r = 7$ **5.** center: $(-2, 3)$; $r = 4$
6. center: $(7, -3)$ **7.** $(-2, -5)$

8. hyperbola

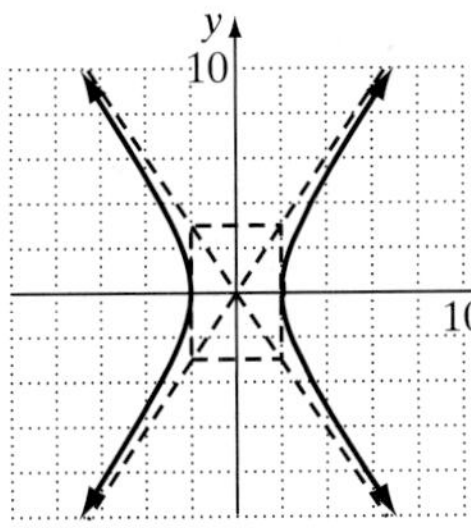

9. ellipse

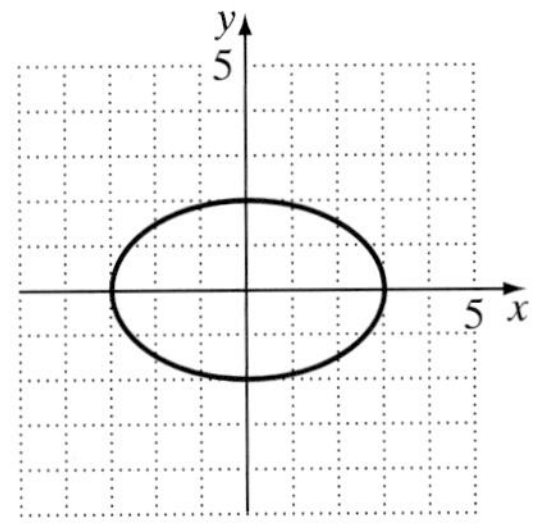

10. parabola

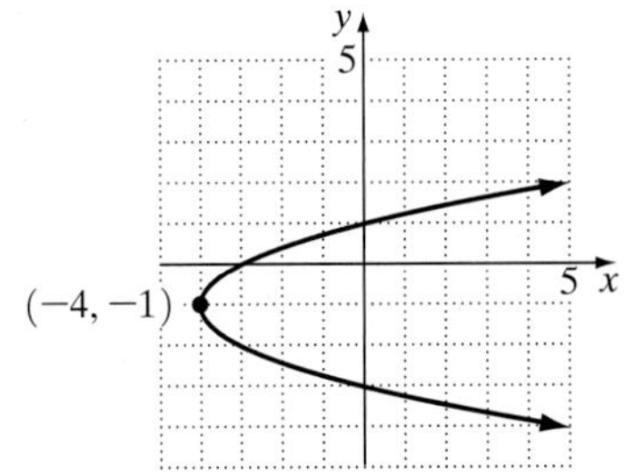

$x = (y+1)^2 - 4$

11. ellipse

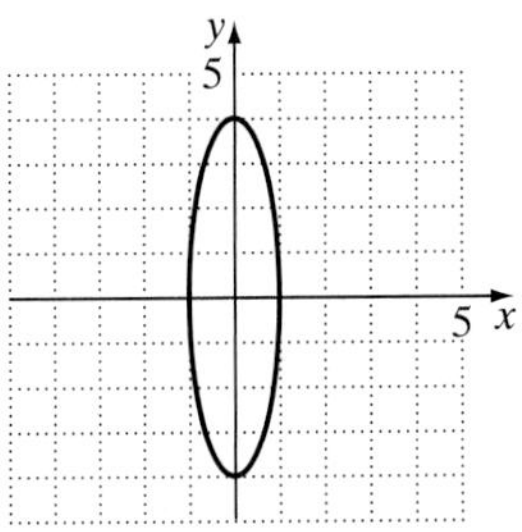

12. hyperbola

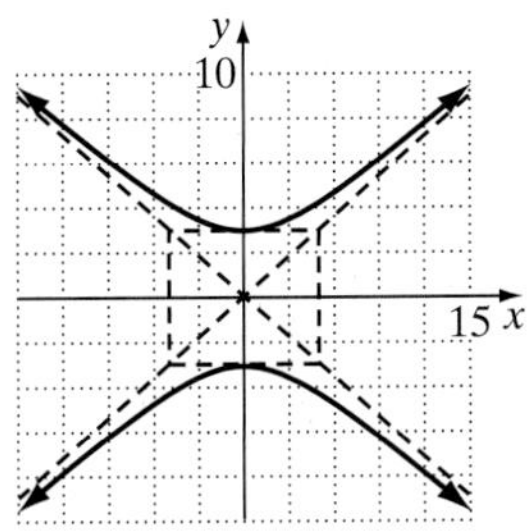

13. parabola

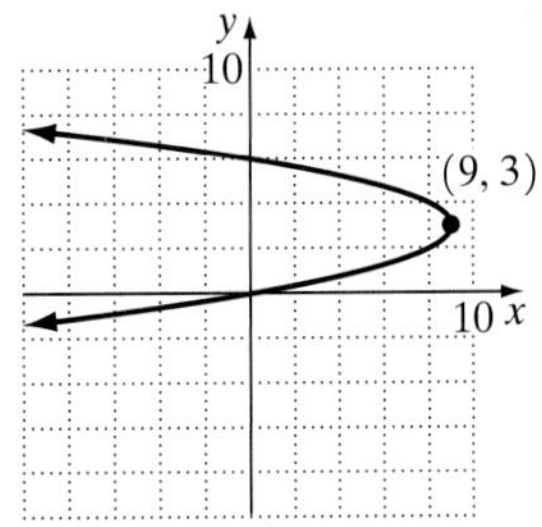

$x = -y^2 + 6y$

14. ellipse

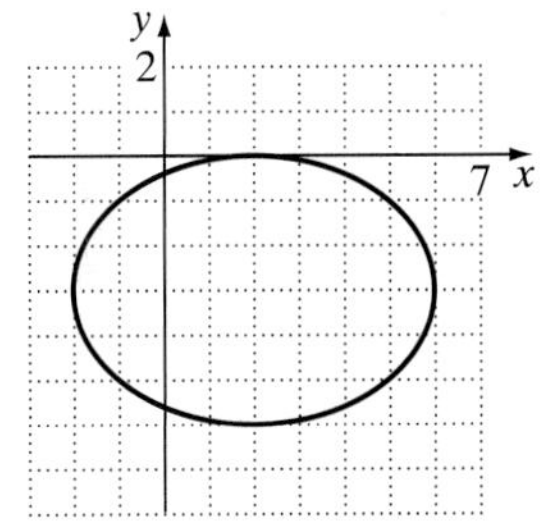

15. circle

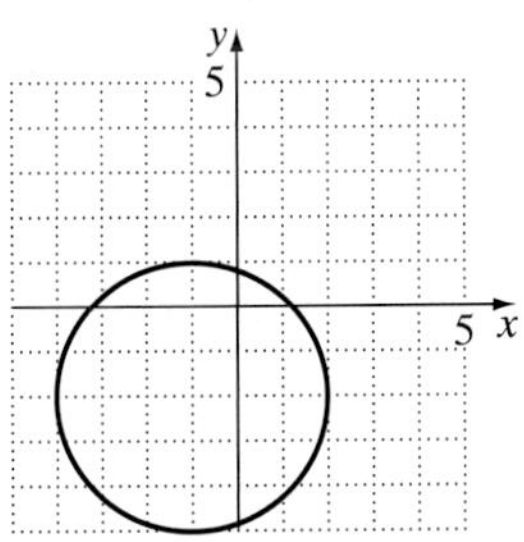

16. circle

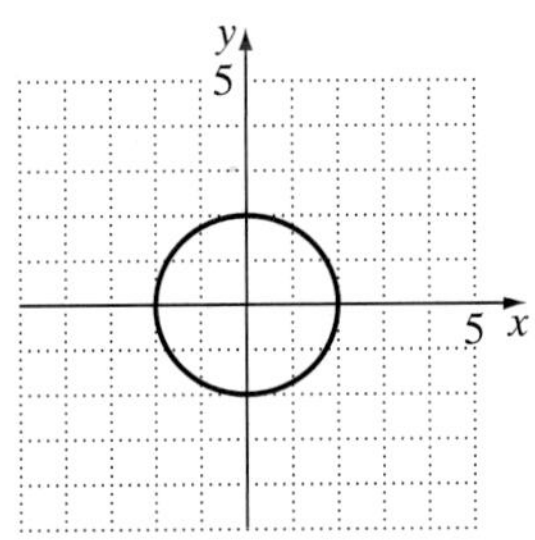

17. $(4, -3)$ and $(-3, 4)$, or $\{(4, -3), (-3, 4)\}$
18. $(3, 2), (-3, 2), (3, -2)$, and $(-3, -2)$, or $\{(3, 2), (-3, 2), (3, -2), (-3, -2)\}$
19. 15 feet by 12 feet or 24 feet by 7.5 feet
20. 4 feet by 3 feet

Cumulative Review Exercises (Chapters 1–10)

1. $\{x|x > -1\}$ or $(-1, \infty)$ **2.** $-\frac{1}{2}$ and 4, or $\left\{-\frac{1}{2}, 4\right\}$ **3.** 2 or $\{2\}$ **4.** $\left\{x \middle| -\frac{5}{3} < x < -1\right\}$ or $\left(-\frac{5}{3}, -1\right)$ **5.** $\frac{5}{2}$ or $\left\{\frac{5}{2}\right\}$
6. $\frac{\ln 8}{0.7} \approx 2.97$ or $\left\{\frac{\ln 8}{0.7} \approx 2.97\right\}$ **7.** $(1, 3), (-1, 3), (1, -3)$, and $(-1, -3)$, or $\{(1, 3), (-1, 3), (1, -3), (-1, -3)\}$

8.

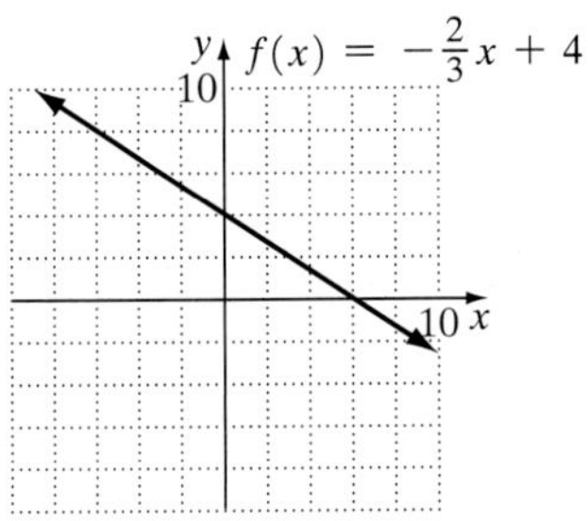

9.

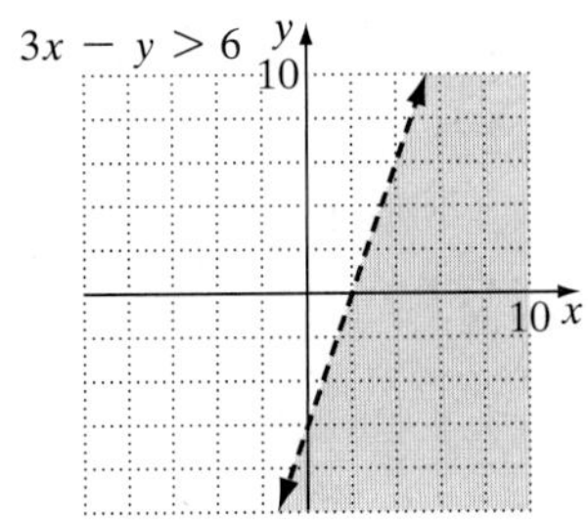

10.

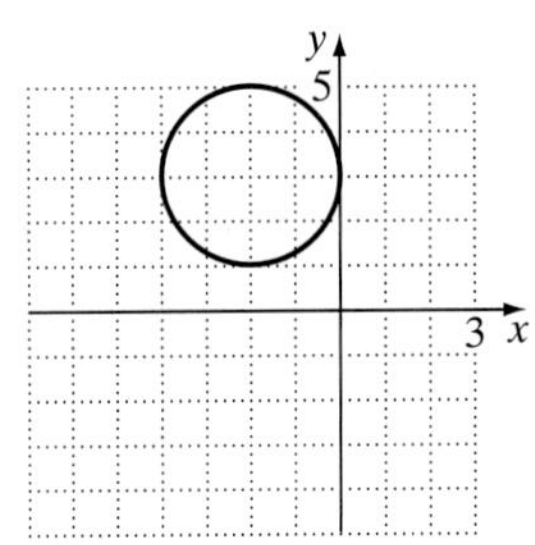

11.
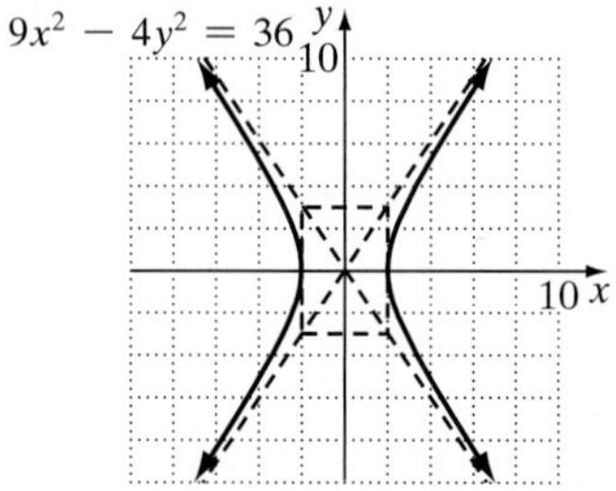

12. 46 **13.** $x^2 - 5x - 1$ **14.** $2xy^2\sqrt[3]{2y}$ **15.** $11 + 10i$ **16.** $3x(2x - 3)^2$ **17.** $(x - 2)(x + 3)(x - 3)$ **18.** $\{x|x \le 2\}$ or $(-\infty, 2]$ **19.** $\dfrac{(1 - \sqrt{x})^2}{1 - x}$ or $\dfrac{1 - 2\sqrt{x} + x}{1 - x}$ **20.** $\ln x^{1/3}y^7$ **21.** $3x^2 + x + 4 + \dfrac{7}{x - 2}$ **22.** $x^2 - 12$ **23.** faster car: 50 mph; slower car: 40 mph **24.** 175 miles; $67 **25.** apple: 60 calories; banana: 87 calories

CHAPTER 11

Section 11.1

Check Point Exercises

1. a. $7, 9, 11, 13$ **b.** $-\frac{1}{3}, \frac{1}{5}, -\frac{1}{9}, \frac{1}{17}$ **2.** $10, \frac{10}{3}, \frac{5}{6}, \frac{1}{6}$ **3. a.** $2(1)^2 + 2(2)^2 + 2(3)^2 + 2(4)^2 + 2(5)^2 + 2(6)^2 = 182$ **b.** $(2^3 - 3) + (2^4 - 3) + (2^5 - 3) = 47$ **c.** $4 + 4 + 4 + 4 + 4 = 20$ **4. a.** $\sum_{i=1}^{9} i^2$ **b.** $\sum_{i=1}^{n} \frac{1}{2^{i-1}}$

Exercise Set 11.1

1. $5, 8, 11, 14$ **3.** $3, 9, 27, 81$ **5.** $-3, 9, -27, 81$ **7.** $-4, 5, -6, 7$ **9.** $\frac{2}{5}, \frac{2}{3}, \frac{6}{7}, 1$ **11.** $1, -\frac{1}{3}, \frac{1}{7}, -\frac{1}{15}$ **13.** $1, 2, \frac{3}{2}, \frac{2}{3}$ **15.** $4, 12, 48, 240$ **17.** 105 **19.** 60 **21.** 115 **23.** $-\frac{5}{16}$ **25.** 55 **27.** $\frac{3}{8}$ **29.** 15 **31.** $\sum_{i=1}^{15} i^2$ **33.** $\sum_{i=1}^{11} 2^i$ **35.** $\sum_{i=1}^{30} i$ **37.** $\sum_{i=1}^{14} \frac{i}{i + 1}$ **39.** $\sum_{i=1}^{n} \frac{4^i}{i}$ **41.** $\sum_{i=1}^{n} (2i - 1)$ **43.** Answers will vary; examples are: $\sum_{k=1}^{14} (2k + 3)$ or $\sum_{k=2}^{15} (2k + 1)$. **45.** Answers will vary; an example is: $\sum_{k=0}^{12} ar^k$. **47.** Answers will vary; an example is: $\sum_{k=0}^{n} (a + kd)$. **49. a.** 5939.1; From 1991 to 1999 a total of 5939.1 million CDs were sold. **b.** 659.9; From 1991 to 1999 an average of 659.9 million CDs were sold each year. **51.** 30.15; From 1991 to 1995 American's spent a total of $30.15 billion on recreational boating. **53.** $8081.13 **55–61.** Answers will vary. **63.** $a_{10} \approx 2.5934$; $a_{100} \approx 2.70481$; $a_{1000} \approx 2.71692$; $a_{10,000} \approx 2.71815$; $a_{100,000} \approx 2.71827$ **65.** As n gets larger, the terms get closer to 0. **67.** As n gets larger, the terms get closer to $\frac{3}{5}$ or 0.6. **69.** $\log 2 + \log 4 + \log 6 + \log 8 = \log 384$ **71.** $7, 12, 17, 22$ **73.** $2xy^2\sqrt[3]{5xy}$ **74.** $(3x - 2)(9x^2 + 6x + 4)$ **75.** $-\frac{6}{5}$ and 4, or $\left\{-\frac{6}{5}, 4\right\}$

Section 11.2

Check Point Exercises

1. $100, 70, 40, 10, -20, -50$ **2.** -34 **3. a.** $a_n = 2350n + 10{,}458$ **b.** $73,908 million or $73,908,000,000 **4.** 360 **5.** 2460

Exercise Set 11.2

1. 4 **3.** 5 **5.** -3 **7.** $200, 220, 240, 260, 280, 300$ **9.** $-7, -3, 1, 5, 9, 13$ **11.** $300, 210, 120, 30, -60, -150$ **13.** $\frac{5}{2}, 2, \frac{3}{2}, 1, \frac{1}{2}, 0$ **15.** $-0.4, -2, -3.6, -5.2, -6.8, -8.4$ **17.** 33 **19.** 252 **21.** 955 **23.** -142 **25.** $a_n = 4n - 3$; $a_{20} = 77$ **27.** $a_n = 11 - 4n$; $a_{20} = -69$ **29.** $a_n = -4n - 16$; $a_{20} = -96$ **31.** $a_n = \frac{1}{3}n - \frac{2}{3}$; $a_{20} = 6$ **33.** $a_n = 4.3 - 0.3n$; $a_{20} = -1.7$ **35.** 1220 **37.** 4400 **39.** 5050 **41.** 3660 **43.** 396 **45.** $8 + 13 + 18 + \cdots + 88 = 816$ **47.** $2 + (-1) + (-4) + \cdots + (-85) = -1245$ **49.** $4 + 8 + 12 + \cdots + 400 = 20{,}200$ **51. a.** $a_n = 1265n + 125{,}159$ **b.** 145,399 thousand or 145,399,000 employees **53.** company A; $1400 **55. a.** $a_n = 0.576n + 3.204$ **b.** 627.3 million or 627,300,000 tons **57.** $442,500 **59.** 1430 seats **61–67.** Answers will vary. **69.** 200th term **71.** $S_n = \frac{n}{2}[1 + (2n - 1)] = \frac{n}{2}(2n) = n^2$ **72.** 1005 or {1005} **73.** $\{x|-5 \le x \le 2\}$ or $[-5, 2]$ **74.** $P = \dfrac{At}{t - A}$

Section 11.3

Check Point Exercises

1. $12, 6, 3, \frac{3}{2}, \frac{3}{4}, \frac{3}{8}$ **2.** 3645 **3.** $a_n = 3(2)^{n-1}; a_8 = 384$ **4.** 9842 **5.** 19,680 **6.** \$2,371,746 **7.** \$1,327,778 **8.** 9
9. 1 **10.** \$4000

Exercise Set 11.3

1. $r = 3$ **3.** $r = -2$ **5.** $r = \frac{3}{2}$ **7.** $r = -0.1$ **9.** 2, 6, 18, 54, 162 **11.** $20, 10, 5, \frac{5}{2}, \frac{5}{4}$ **13.** −4, 40, −400, 4000, −40,000
15. $-\frac{1}{4}, \frac{1}{2}, -1, 2, -4$ **17.** $a_8 = 768$ **19.** $a_{12} = -10{,}240$ **21.** $a_6 = -200$ **23.** $a_8 = 0.1$ **25.** $a_n = 3(4)^{n-1}; a_7 = 12{,}288$
27. $a_n = 18\left(\frac{1}{3}\right)^{n-1}; a_7 = \frac{2}{81}$ **29.** $a_n = 1.5(-2)^{n-1}; a_7 = 96$ **31.** $a_n = 0.0004(-10)^{n-1}; a_7 = 400$ **33.** 531,440 **35.** 2049
37. $\frac{16{,}383}{2}$ **39.** 9840 **41.** 10,230 **43.** $\frac{63}{128}$ **45.** $\frac{3}{2}$ **47.** 4 **49.** $\frac{2}{3}$ **51.** 20 **53.** $\frac{5}{9}$ **55.** $\frac{47}{99}$ **57.** $\frac{257}{999}$
59. arithmetic; $d = 1$ **61.** geometric; $r = 2$ **63.** neither **65.** \$16,384 **67.** \$3,795,957 **69. a.** 1.04; 1.04; 1.04; $r = 1.04$
b. $a_n = 20.60(1.04)^{n-1}$ **c.** 30.49 million or 30,490,000 **71.** \$32,767 **73.** \$793,583 **75.** 130.26 in. **77.** \$844,706
79. \$94,834 **81.** \$9 million **83.** $\frac{1}{3}$ **85–93.** Answers will vary.
95.

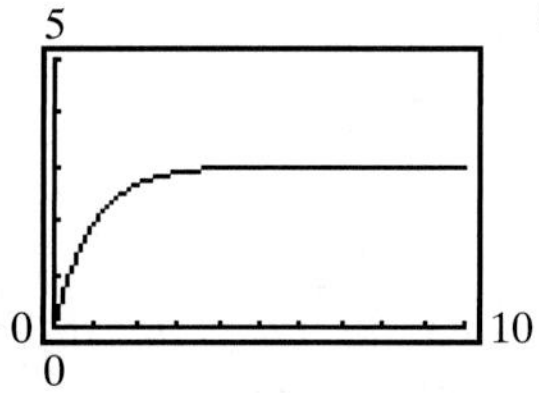

; horizontal asymptote: $y = 3$; sum of series: 3

97. d
99. \$442
100. $2\sqrt{7}$
101. $\frac{-1 \pm \sqrt{33}}{4}$ or $\left\{\frac{-1 \pm \sqrt{337}}{4}\right\}$
102. $-3(\sqrt{3} + \sqrt{5})$

Section 11.4

Check Point Exercises

1. a. 20 **b.** 1 **c.** 28 **d.** 1 **2.** $x^4 + 4x^3 + 6x^2 + 4x + 1$ **3.** $x^5 - 10x^4y + 40x^3y^2 - 80x^2y^3 + 80xy^4 - 32y^5$
4. $4032x^5y^4$

Exercise Set 11.4

1. 56 **3.** 12 **5.** 1 **7.** 4950 **9.** $x^3 + 6x^2 + 12x + 8$ **11.** $27x^3 + 27x^2y + 9xy^2 + y^3$ **13.** $125x^3 - 75x^2 + 15x - 1$
15. $16x^4 + 32x^3 + 24x^2 + 8x + 1$ **17.** $x^8 + 8x^6y + 24x^4y^2 + 32x^2y^3 + 16y^4$ **19.** $y^4 - 12y^3 + 54y^2 - 108y + 81$
21. $16x^{12} - 32x^9 + 24x^6 - 8x^3 + 1$ **23.** $c^5 + 10c^4 + 40c^3 + 80c^2 + 80c + 32$ **25.** $x^5 - 5x^4 + 10x^3 - 10x^2 + 5x - 1$
27. $243x^5 - 405x^4y + 270x^3y^2 - 90x^2y^3 + 15xy^4 - y^5$ **29.** $64a^6 + 192a^5b + 240a^4b^2 + 160a^3b^3 + 60a^2b^4 + 12ab^5 + b^6$
31. $x^8 + 16x^7 + 112x^6$ **33.** $x^{10} - 20x^9y + 180x^8y^2$ **35.** $x^{32} + 16x^{30} + 120x^{28}$ **37.** $y^{60} - 20y^{57} + 190y^{54}$ **39.** $240x^4y^2$
41. $126x^5$ **43.** $56x^6y^{15}$ **45.** $-\frac{21}{2}x^6$ **47.** $g(t) = 0.002t^3 - 0.84t^2 - 16.13t - 68.54$ **49–57.** Answers will vary.
59.

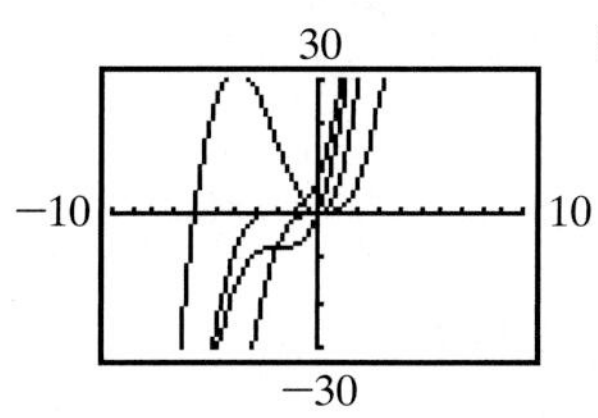

; f_2, f_3, and f_4 are approaching $f_1 = f_5$.

61. $x^3 - 3x^2 + 3x - 1$
63. $x^6 + 12x^5 + 60x^4 + 160x^3 + 240x^2 + 192x + 64$
65. b **67.** $10x^4y^6$ **68.** $a^2 + 4a + 6$
69. $f(g(x)) = 4x^2 - 2x - 6; g(f(x)) = 2x^2 + 10x - 3$
70. $\frac{2x^2 - 9x - 1}{(x - 4)(x + 3)}$

Review Exercises

1. 3, 10, 17, 24 **2.** $-\frac{3}{2}, \frac{4}{3}, -\frac{5}{4}, \frac{6}{5}$ **3.** $1, 1, \frac{1}{2}, \frac{1}{6}$ **4.** $\frac{1}{2}, -\frac{1}{4}, \frac{1}{8}, -\frac{1}{16}$ **5.** 95 **6.** −20 **7.** Answers will vary; an example is:
$\sum_{i=1}^{15} \frac{i}{i + 2}$. **8.** Answers will vary; examples are: $\sum_{i=4}^{13} i^3$ or $\sum_{i=1}^{10} (i + 3)^3$. **9.** 7, 11, 15, 19, 23, 27 **10.** −4, −9, −14, −19, −24, −29
11. $\frac{3}{2}, 1, \frac{1}{2}, 0, -\frac{1}{2}, -1$ **12.** 20 **13.** −30 **14.** −38 **15.** $a_n = 4n - 11; a_{20} = 69$ **16.** $a_n = 220 - 20n; a_{20} = -180$

17. $a_n = -\frac{23}{2} - \frac{1}{2}n; a_{20} = -\frac{43}{2}$ **18.** $a_n = 22 - 7n; a_{20} = -118$ **19.** 1727 **20.** 225 **21.** 15,150 **22.** 440 **23.** −500
24. −2325 **25. a.** $1043.458 - 0.418n$ **b.** 1001.658 sec **26.** $418,500 **27.** 1470 seats **28.** 3, 6, 12, 24, 48
29. $\frac{1}{2}, \frac{1}{4}, \frac{1}{8}, \frac{1}{16}, \frac{1}{32}$ **30.** $16, -4, 1, -\frac{1}{4}, \frac{1}{16}$ **31.** −5, 5, −5, 5, −5 **32.** $a_7 = 1458$ **33.** $a_6 = \frac{1}{2}$ **34.** $a_5 = -48$
35. $a_n = 1(2)^{n-1}$ or $a_n = 2^{n-1}; a_8 = 128$ **36.** $a_n = 100\left(\frac{1}{10}\right)^{n-1}; a_8 = \frac{1}{100{,}000} = 0.00001$ **37.** $a_n = 12\left(-\frac{1}{3}\right)^{n-1}; a_8 = -\frac{4}{729}$
38. 17,936,135 **39.** $\frac{127}{8}$ or 15.875 **40.** 19,530 **41.** −258 **42.** $\frac{341}{128}$ **43.** $\frac{27}{2}$ **44.** $\frac{4}{3}$ **45.** $-\frac{18}{15}$ **46.** 20 **47.** $\frac{2}{3}$
48. $\frac{47}{99}$ **49.** $42,823; $223,210 **50.** $120,113 **51.** $9.3 million **52.** 165 **53.** 4005 **54.** $8x^3 + 12x^2 + 6x + 1$
55. $x^8 - 4x^6 + 6x^4 - 4x^2 + 1$ **56.** $x^5 + 10x^4y + 40x^3y^2 + 80x^2y^3 + 80xy^4 + 32y^5$
57. $x^6 - 12x^5 + 60x^4 - 160x^3 + 240x^2 - 192x + 64$ **58.** $x^{16} + 24x^{14} + 252x^{12}$ **59.** $x^9 - 27x^8 + 324x^7$ **60.** $80x^2$ **61.** $4860x^2$

Chapter 11 Test

1. $1, -\frac{1}{4}, \frac{1}{9}, -\frac{1}{16}, \frac{1}{25}$ **2.** 105 **3.** Answers will vary; examples are: $\sum_{i=2}^{21} \frac{i}{i+1}$ or $\sum_{i=1}^{20} \frac{i+1}{i+2}$. **4.** $a_n = 5n - 1; a_{12} = 59$
5. $a_n = 16\left(\frac{1}{4}\right)^{n-1}; a_{12} = \frac{1}{262{,}144}$ **6.** −385 **7.** 550 **8.** −2387 **9.** −21,846 **10.** 8 **11.** $\frac{73}{99}$ **12.** $276,427 **13.** 36
14. $x^{10} - 5x^8 + 10x^6 - 10x^4 + 5x^2 - 1$ **15.** $x^8 + 8x^7y^2 + 28x^6y^4$

Cumulative Review Exercises (Covering the Entire Book)

1. $\frac{2}{3}$ or $\left\{\frac{2}{3}\right\}$ **2.** $(-5, 2)$ or $\{(-5, 2)\}$ **3.** $(1, 4, -2)$ or $\{(1, 4, -2)\}$ **4.** $\{x|x \geq 10\}$ or $[10, \infty)$ **5.** $\{x|x \leq -4\}$ or $(-\infty. -4]$
6. $\{x|x > 2\}$ or $(2, \infty)$ **7.** $\{x|-2 < x < 3\}$ or $(-2, 3)$ **8.** 3 and 9, or $\{3, 9\}$ **9.** no solution or $\varnothing$ **10.** 12 or $\{12\}$
11. $\frac{-2 \pm \sqrt{14}}{2}$ or $\left\{\frac{-2 \pm \sqrt{14}}{2}\right\}$ **12.** 8 and 27, or $\{8, 27\}$ **13.** $\left\{x \middle| -2 \leq x \leq \frac{3}{2}\right\}$ or $\left[-2, \frac{3}{2}\right]$ **14.** 2 or $\{2\}$ **15.** 0 or $\{0\}$
16. $(2, 1), (-2, 1), (2, -1)$, and $(-2, -1)$, or $\{(2, 1), (-2, 1), (2, -1), (-2, -1)\}$ **17.** $(-14, -20)$ and $(2, -4)$, or $\{(-14, -20), (2, -4)\}$

18.
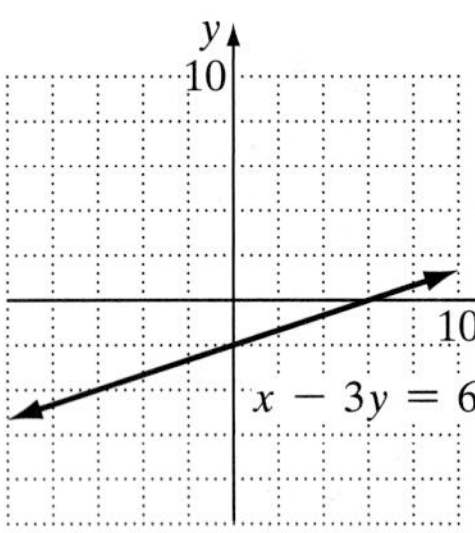

19.
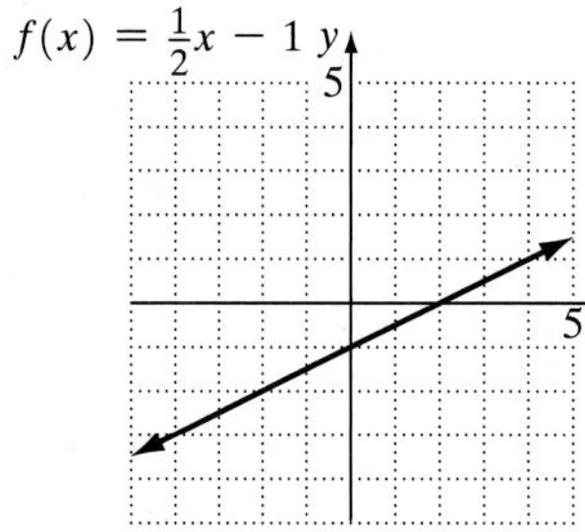

20.
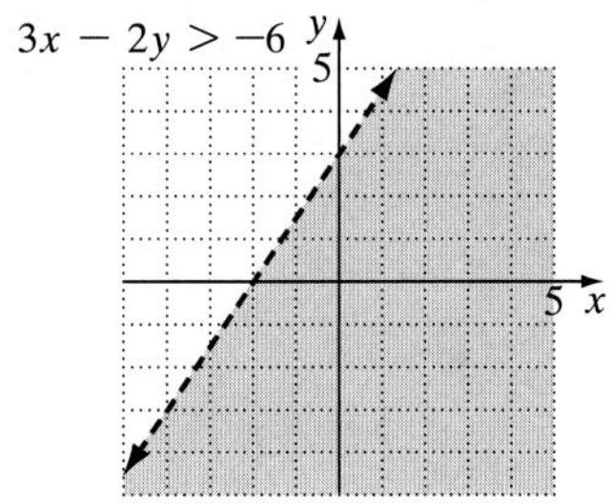

21.
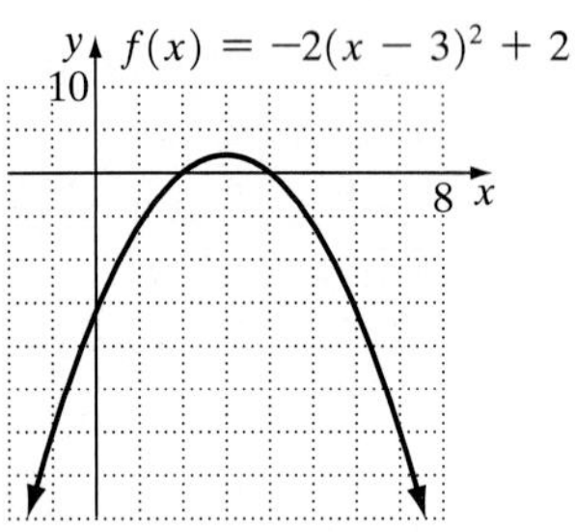

22.
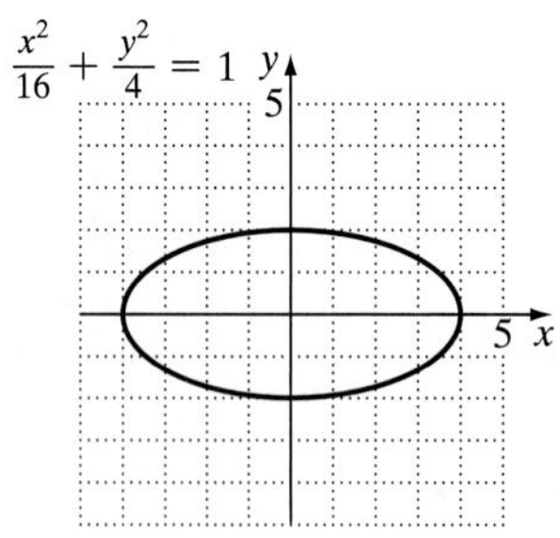

23.
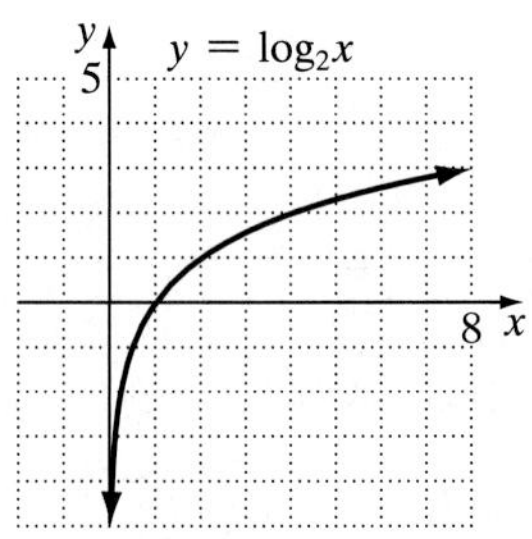

24.

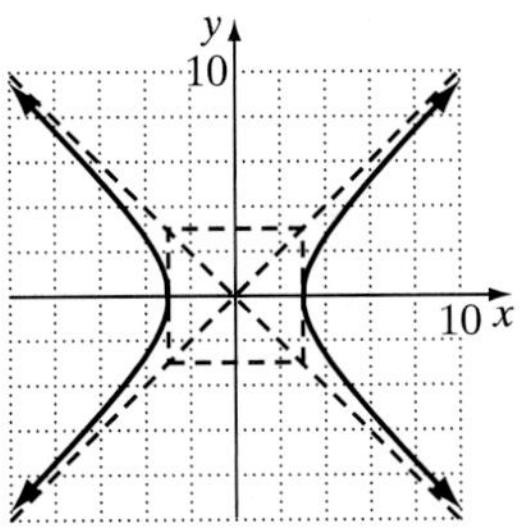

25. $-16x + 24y$ **26.** $-\frac{20x^7}{y^4}$ **27.** $x^2 - 5xy - 16y^2$ **28.** $6x^2 + 13x - 5$

29. $9x^4 - 24x^2y + 16y^2$ **30.** $\frac{3x^2 + 6x - 2}{(x + 5)(x + 2)}$ **31.** $\frac{x - 3}{x}$ **32.** $\frac{x - 4}{3x + 6}$

33. $5xy\sqrt{2x}$ **34.** $9\sqrt{2}$ **35.** $44 + 6i$ **36.** $(9x^2 + 1)(3x + 1)(3x - 1)$

37. $2x(4x - 1)(3x - 2)$ **38.** $(x + 3y)(x^2 - 3xy + 9y^2)$ **39.** $x^2 + 2x - 13; 22$

40. $x + 5$; $\{x|x$ is a real number and $x \neq 2\}$ or $(-\infty, 2) \cup (2, \infty)$

41. $x^2 - x - 17$ **42.** $x^2 + 3x - 17$ **43.** $f^{-1}(x) = \frac{x + 3}{7}$

44. $3\sqrt{10} \approx 9.49$ units **45.** $3x^2 - 7x + 18 - \frac{28}{x + 2}$ **46.** $R = -\frac{Ir}{I - 1}$ or $R = \frac{Ir}{1 - I}$ **47.** $y = -3x - 1$ or $f(x) = -3x - 1$

48. 6 **49.** $\ln \frac{x^2}{\sqrt{y}}$ **50.** 208 **51.** 1800 **52.** $\frac{1}{3}$ **53.** $16x^4 - 32x^3y^3 + 24x^2y^6 - 8xy^9 + y^{12}$

54. $\{x|x$ is a real number and $x \neq 1$ and $x \neq 2\}$ or $(-\infty, 1) \cup (1, 2) \cup (2, \infty)$ **55.** $\{x|x > 4\}$ or $(4, \infty)$ **56.** \$620

57. 13 yd by 4 yd **58.** \$2600 at 12% and \$1400 at 14% **59.** approximately 11% **60.** 11 amps

Subject Index

D

E

H

I

J

K

L

87.

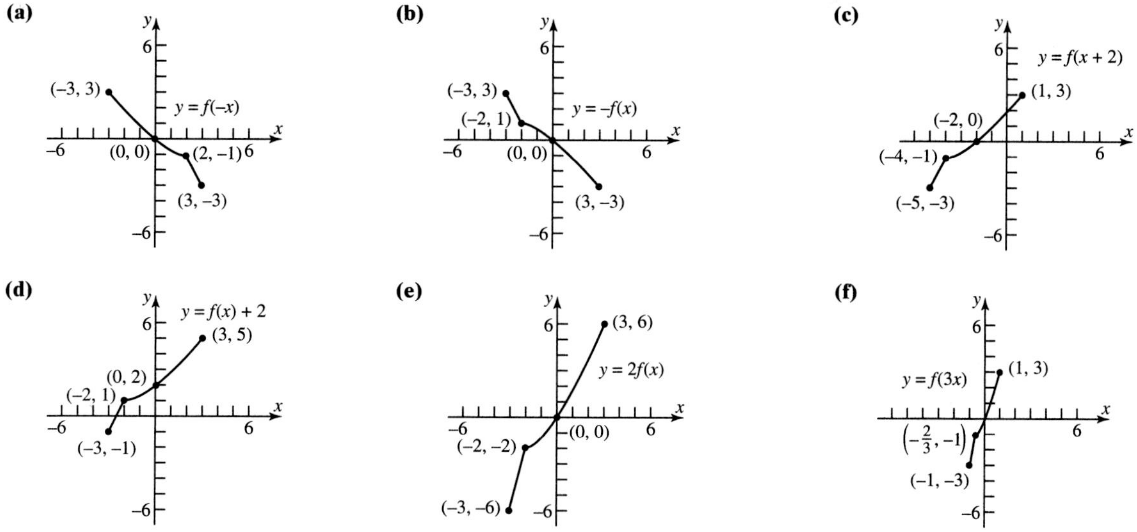

89. $d(A, B) = \sqrt{13}; d(B, C) = \sqrt{13}$ **91.** Center: $(1, -2)$; radius $= 4\sqrt{2}$; $x^2 + y^2 - 2x + 4y - 27 = 0$

CHAPTER 2 Trigonometric Functions

2.1 Concepts and Vocabulary *(page 108)*

1. Standard position **2.** $r\theta; \frac{1}{2}r^2\theta$ **3.** $\frac{s}{t}; \frac{\theta}{t}$ **4.** F **5.** T **6.** T **7.** On a circle of radius r, if the length of arc subtended by a central angle is also r, then the measure of the angle is 1 radian. **8.** 1 radian is larger. **9.** When an object travels around a circle, linear speed measures the length of the arc traveled per unit time; angular speed measures the angle swept out per unit time.

2.1 Exercises *(page 109)*

1. 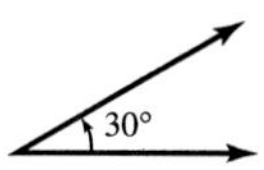

3. 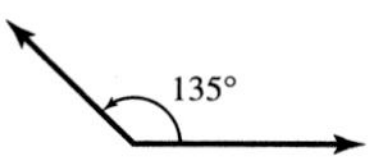

5. 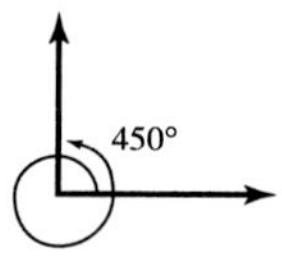

7. 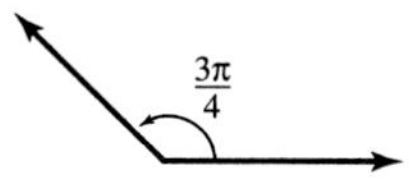

9. 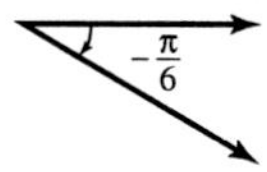

11. 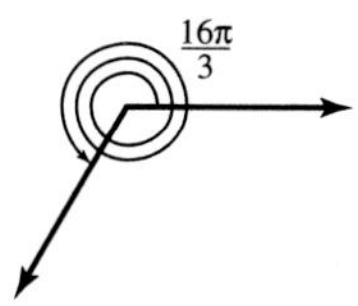

13. $\frac{\pi}{6}$ **15.** $\frac{4\pi}{3}$ **17.** $-\frac{\pi}{3}$ **19.** π **21.** $-\frac{3\pi}{4}$ **23.** $-\frac{\pi}{2}$ **25.** 60° **27.** −225° **29.** 90° **31.** 15° **33.** −90° **35.** −30° **37.** 5 m **39.** 6 ft **41.** 0.6 radian **43.** $\frac{\pi}{3} \approx 1.047$ in. **45.** 25 m^2 **47.** $2\sqrt{3} \approx 3.464$ ft **49.** 0.24 radian **51.** $\frac{\pi}{3} \approx 1.047$ in^2 **53.** $s = 2.094$ ft; $A = 2.094$ ft^2 **55.** $s = 14.661$ yd; $A = 87.965$ yd^2 **57.** 0.30 **59.** −0.70 **61.** 2.18 **63.** 179.91° **65.** 114.59° **67.** 362.11° **69.** 40.17° **71.** 1.03° **73.** 9.15° **75.** 40°19′12″ **77.** 18°15′18″ **79.** 19°59′24″ **81.** $3\pi \approx 9.42$ in.; $5\pi \approx 15.71$ in. **83.** $2\pi \approx 6.28$ m^2 **85.** $\frac{675\pi}{2} \approx 1060.29$ ft^2 **87.** $\omega = \frac{1}{60}$ radian/sec; $v = \frac{1}{12}$ cm/sec **89.** Approximately 452.49 rpm **91.** Approximately 359.40 mi **93.** Approximately 897.84 mi/hr **95.** Approximately 2291.94 mi/hr **97.** $\frac{3}{4}$ rpm **99.** Approximately 2.86 mi/hr **101.** Approximately 31.47 rpm **103.** Approximately 1036.73 mi/hr **105.** $v_1 = r_1\omega_1, v_2 = r_2\omega_2$, and $v_1 = v_2$ so $r_1\omega_1 = r_2\omega_2 \Rightarrow \frac{r_1}{r_2} = \frac{\omega_2}{\omega_1}$.

2.2 Concepts and Vocabulary *(page 120)*

1. Complementary **2.** Cosine **3.** 62° **4.** F **5.** T **6.** F **7.** $\csc\theta = \frac{1}{\sin\theta}; \sec\theta = \frac{1}{\cos\theta}; \cot\theta = \frac{1}{\tan\theta}$ **10.** Tangent, cotangent

2.2 Exercises *(page 121)*

1. $\sin\theta = \frac{5}{13}$; $\cos\theta = \frac{12}{13}$; $\tan\theta = \frac{5}{12}$; $\csc\theta = \frac{13}{5}$; $\sec\theta = \frac{13}{12}$; $\cot\theta = \frac{12}{5}$ **3.** $\sin\theta = \frac{2\sqrt{13}}{13}$; $\cos\theta = \frac{3\sqrt{13}}{13}$; $\tan\theta = \frac{2}{3}$; $\csc\theta = \frac{\sqrt{13}}{2}$; $\sec\theta = \frac{\sqrt{13}}{3}$; $\cot\theta = \frac{3}{2}$ **5.** $\sin\theta = \frac{\sqrt{3}}{2}$; $\cos\theta = \frac{1}{2}$; $\tan\theta = \sqrt{3}$; $\csc\theta = \frac{2\sqrt{3}}{3}$; $\sec\theta = 2$; $\cot\theta = \frac{\sqrt{3}}{3}$ **7.** $\sin\theta = \frac{\sqrt{6}}{3}$; $\cos\theta = \frac{\sqrt{3}}{3}$; $\tan\theta = \sqrt{2}$; $\csc\theta = \frac{\sqrt{6}}{2}$; $\sec\theta = \sqrt{3}$; $\cot\theta = \frac{\sqrt{2}}{2}$ **9.** $\sin\theta = \frac{\sqrt{5}}{5}$; $\cos\theta = \frac{2\sqrt{5}}{5}$; $\tan\theta = \frac{1}{2}$; $\csc\theta = \sqrt{5}$; $\sec\theta = \frac{\sqrt{5}}{2}$; $\cot\theta = 2$

11. $\tan\theta = \frac{\sqrt{3}}{3}$; $\csc\theta = 2$; $\sec\theta = \frac{2\sqrt{3}}{3}$; $\cot\theta = \sqrt{3}$ **13.** $\tan\theta = \frac{2\sqrt{5}}{5}$; $\csc\theta = \frac{3}{2}$; $\sec\theta = \frac{3\sqrt{5}}{5}$; $\cot\theta = \frac{\sqrt{5}}{2}$

15. $\cos\theta = \frac{\sqrt{2}}{2}$; $\tan\theta = 1$; $\csc\theta = \sqrt{2}$; $\sec\theta = \sqrt{2}$; $\cot\theta = 1$ **17.** $\sin\theta = \frac{2\sqrt{2}}{3}$; $\tan\theta = 2\sqrt{2}$; $\csc\theta = \frac{3\sqrt{2}}{4}$; $\sec\theta = 3$; $\cot\theta = \frac{\sqrt{2}}{4}$

19. $\sin\theta = \frac{\sqrt{5}}{5}$; $\cos\theta = \frac{2\sqrt{5}}{5}$; $\csc\theta = \sqrt{5}$; $\sec\theta = \frac{\sqrt{5}}{2}$; $\cot\theta = 2$ **21.** $\sin\theta = \frac{2\sqrt{2}}{3}$; $\cos\theta = \frac{1}{3}$; $\tan\theta = 2\sqrt{2}$; $\csc\theta = \frac{3\sqrt{2}}{4}$; $\cot\theta = \frac{\sqrt{2}}{4}$

23. $\sin\theta = \frac{\sqrt{6}}{3}$; $\cos\theta = \frac{\sqrt{3}}{3}$; $\csc\theta = \frac{\sqrt{6}}{2}$; $\sec\theta = \sqrt{3}$; $\cot\theta = \frac{\sqrt{2}}{2}$ **25.** $\sin\theta = \frac{1}{2}$; $\cos\theta = \frac{\sqrt{3}}{2}$; $\tan\theta = \frac{\sqrt{3}}{3}$; $\sec\theta = \frac{2\sqrt{3}}{3}$; $\cot\theta = \sqrt{3}$

27. 1 **29.** 1 **31.** 0 **33.** 0 **35.** 1 **37.** 0 **39.** 0 **41.** 1 **43.** 1 **45.** **(a)** $\frac{1}{2}$ **(b)** $\frac{3}{4}$ **(c)** 2 **(d)** 2 **47.** **(a)** 17 **(b)** $\frac{1}{4}$ **(c)** 4 **(d)** $\frac{17}{16}$

49. **(a)** $\frac{1}{4}$ **(b)** 15 **(c)** 4 **(d)** $\frac{16}{15}$ **51.** **(a)** 0.78 **(b)** 0.79 **(c)** 1.27 **(d)** 1.28 **(e)** 1.61 **(f)** 0.78 **(g)** 0.62 **(h)** 1.27 **53.** 0.6 **55.** 20°

57. **(a)** 10 min **(b)** 20 min **(c)** $T(\theta) = 5\left(1 - \frac{1}{3\tan\theta} + \frac{1}{\sin\theta}\right)$

(d) Approximately 15.81 min **(e)** Approximately 10.40 min

(f)

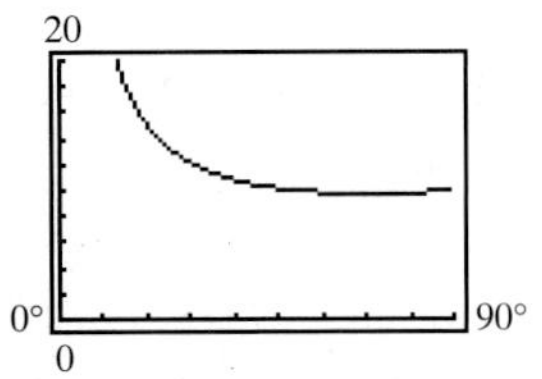

Approximately 70.53°
Approximately 176.78 ft
Approximately 9.71 min

59. **(a)** $|OA| = |OC| = 1$; angle OAC = angle OCA
angle OAC + angle OCA + $(180° - \theta) = 180°$
$2(\text{angle } OAC) - \theta = 0$; angle $OAC = \frac{\theta}{2}$

(b) $\sin\theta = \frac{|CD|}{|OC|} = |CD|$; $\cos\theta = \frac{|OD|}{|OC|} = |OD|$

(c) $\tan\frac{\theta}{2} = \frac{|CD|}{|AD|} = \frac{\sin\theta}{1 + |OD|} = \frac{\sin\theta}{1 + \cos\theta}$

61. $h = x\tan\theta$ and $h = (1 - x)\tan(n\theta)$; thus, $x\tan\theta = (1 - x)\tan(n\theta)$ so $x = \frac{\tan(n\theta)}{\tan\theta + \tan(n\theta)}$.

63. **(a)** Area $\triangle OAC = \frac{1}{2}|AC||OC| = \frac{1}{2}\cdot\frac{|AC|}{1}\cdot\frac{|OC|}{1} = \frac{1}{2}\sin\alpha\cos\alpha$

(b) Area $\triangle OCB = \frac{1}{2}|BC|\,|OC| = \frac{1}{2}|OB|^2\frac{|BC|}{|OB|}\cdot\frac{|OC|}{|OB|} = \frac{1}{2}|OB|^2\sin\beta\cos\beta$

(c) Area $\triangle OAB = \frac{1}{2}|BD|\,|OA| = \frac{1}{2}|OB|\frac{|BD|}{|OB|} = \frac{1}{2}|OB|\sin(\alpha + \beta)$ **(d)** $\frac{\cos\alpha}{\cos\beta} = \frac{\frac{|OC|}{1}}{\frac{|OC|}{|OB|}} = |OB|$

(e) Area $\triangle OAB$ = area $\triangle OAC$ + area $\triangle OCB$

$$\frac{1}{2}|OB|\sin(\alpha + \beta) = \frac{1}{2}\sin\alpha\cos\alpha + \frac{1}{2}|OB|^2\sin\beta\cos\beta$$

$$\sin(\alpha + \beta) = \frac{\sin\alpha\cos\alpha + |OB|^2\sin\beta\cos\beta}{|OB|}$$

$$\sin(\alpha + \beta) = \frac{\sin\alpha(|OB|\cos\beta) + |OB|^2\sin\beta\left(\frac{\cos\alpha}{|OB|}\right)}{|OB|}$$

$$\sin(\alpha + \beta) = \sin\alpha\cos\beta + \cos\alpha\sin\beta$$

65. $\sin\alpha = \tan\alpha\cos\alpha = \cos\beta\cos\alpha = \cos\beta\tan\beta = \sin\beta$;

$$\sin^2\alpha + \cos^2\alpha = 1$$
$$\sin^2\alpha + \tan^2\beta = 1$$
$$\sin^2\alpha + \frac{\sin^2\beta}{\cos^2\beta} = 1$$
$$\sin^2\alpha + \frac{\sin^2\alpha}{1-\sin^2\alpha} = 1$$
$$\sin^2\alpha - \sin^4\alpha + \sin^2\alpha = 1 - \sin^2\alpha$$
$$\sin^4\alpha - 3\sin^2\alpha + 1 = 0$$
$$\sin^2\alpha = \frac{3 \pm \sqrt{5}}{2}$$
$$\sin^2\alpha = \frac{3 - \sqrt{5}}{2}$$
$$\sin\alpha = \sqrt{\frac{3-\sqrt{5}}{2}}$$

Note that $\frac{3+\sqrt{5}}{2} > 1$.

67. Assume (a, b) is a point on the terminal side of θ. Since θ is acute, $a > 0$ and $b > 0$. Since $a^2 + b^2 = c^2$, then $0 < b^2 < c^2$, so $0 < b < c$.

Thus, $0 < \frac{b}{c} < 1$ and $0 < \sin\theta < 1$.

2.3 Concepts and Vocabulary *(page 129)*

1. $\frac{3}{2}$ **2.** 0.91 **3.** T **4.** F

2.3 Exercises *(page 129)*

1. $\sin 45^\circ = \frac{\sqrt{2}}{2}$; $\cos 45^\circ = \frac{\sqrt{2}}{2}$; $\tan 45^\circ = 1$; $\csc 45^\circ = \sqrt{2}$; $\sec 45^\circ = \sqrt{2}$; $\cot 45^\circ = 1$ **3.** $\frac{\sqrt{3}}{2}$ **5.** $\frac{1}{2}$ **7.** $\frac{3}{4}$ **9.** $\sqrt{3}$ **11.** $\frac{\sqrt{3}}{4}$ **13.** $\sqrt{2}$ **15.** 2 **17.** $\sqrt{2} + \frac{4\sqrt{3}}{3}$ **19.** $-\frac{8}{3}$ **21.** $\frac{1}{2}$ **23.** 0 **25.** 0.47 **27.** 0.38 **29.** 1.33 **31.** 0.31 **33.** 3.73 **35.** 1.04 **37.** 0.84 **39.** 0.02 **41.** 0.31 **43.** $R \approx 310.56$ ft; $H \approx 77.64$ ft **45.** $R \approx 19{,}541.95$ m; $H \approx 2278.14$ m **47.** **(a)** 0.60 sec **(b)** 0.79 sec **(c)** 1.04 sec

49. **(a)** $T(\theta) = 1 + \frac{2}{3\sin\theta} - \frac{1}{4\tan\theta}$
(b) 1.90 hr; 0.57 hr
(c) 1.69 hr; 0.75 hr
(d) 1.63 hr; 0.86 hr
(e) 1.67 hr **(f)** 2.75 hr
(g)

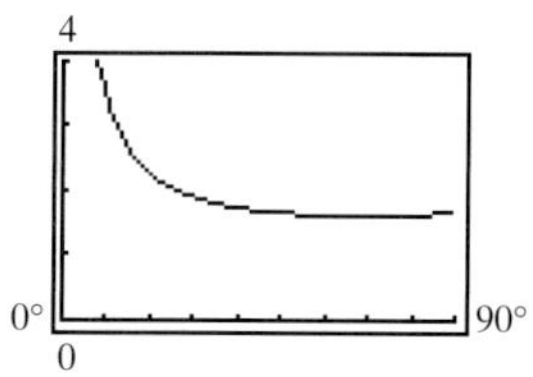

67.98°; 1.62 hr; 0.90 hr

51.

θ	0.5	0.4	0.2	0.1	0.01	0.001	0.0001	0.00001
$\sin\theta$	0.4794	0.3894	0.1987	0.0998	0.0100	0.0010	0.0001	0.00001
$\frac{\sin\theta}{\theta}$	0.9589	0.9735	0.9933	0.9983	1.0000	1.0000	1.0000	1.0000

$\frac{(\sin\theta)}{\theta}$ approaches 1 as $\theta \to 0$.

53. 1 **55.** $\frac{\sqrt{2}}{2}$

2.4 Concepts and Vocabulary *(page 140)*

1. Tangent; cotangent **2.** Coterminal **3.** 60° **4.** F **5.** T **6.** T **8.** I, IV **9.** $\frac{\pi}{2}, \frac{3\pi}{2}$ **10.** 50°

2.4 Exercises *(page 141)*

1. $\sin\theta = \frac{4}{5}$; $\cos\theta = -\frac{3}{5}$; $\tan\theta = -\frac{4}{3}$; $\csc\theta = \frac{5}{4}$; $\sec\theta = -\frac{5}{3}$; $\cot\theta = -\frac{3}{4}$ **3.** $\sin\theta = -\frac{3\sqrt{13}}{13}$; $\cos\theta = \frac{2\sqrt{13}}{13}$; $\tan\theta = -\frac{3}{2}$; $\csc\theta = -\frac{\sqrt{13}}{3}$; $\sec\theta = \frac{\sqrt{13}}{2}$; $\cot\theta = -\frac{2}{3}$ **5.** $\sin\theta = -\frac{\sqrt{2}}{2}$; $\cos\theta = -\frac{\sqrt{2}}{2}$; $\tan\theta = 1$; $\csc\theta = -\sqrt{2}$; $\sec\theta = -\sqrt{2}$; $\cot\theta = 1$ **7.** $\sin\theta = \frac{1}{2}$; $\cos\theta = \frac{\sqrt{3}}{2}$; $\tan\theta = \frac{\sqrt{3}}{3}$; $\csc\theta = 2$; $\sec\theta = \frac{2\sqrt{3}}{3}$; $\cot\theta = \sqrt{3}$ **9.** $\sin\theta = -\frac{\sqrt{2}}{2}$; $\cos\theta = \frac{\sqrt{2}}{2}$; $\tan\theta = -1$; $\csc\theta = -\sqrt{2}$; $\sec\theta = \sqrt{2}$; $\cot\theta = -1$ **11.** II **13.** IV **15.** IV **17.** III **19.** 30° **21.** 60° **23.** 30° **25.** $\frac{\pi}{4}$ **27.** $\frac{\pi}{3}$ **29.** 45° **31.** $\frac{\pi}{3}$ **33.** 80°

35. $\frac{\pi}{4}$ **37.** $\frac{\sqrt{2}}{2}$ **39.** 1 **41.** 1 **43.** $\sqrt{3}$ **45.** $\frac{\sqrt{2}}{2}$ **47.** 0 **49.** $\sqrt{2}$ **51.** $\frac{\sqrt{3}}{3}$ **53.** $\frac{1}{2}$ **55.** $\frac{\sqrt{2}}{2}$ **57.** -2 **59.** $-\sqrt{3}$ **61.** $\frac{\sqrt{2}}{2}$ **63.** $\sqrt{3}$ **65.** $\frac{1}{2}$ **67.** $-\frac{\sqrt{3}}{2}$ **69.** $-\sqrt{3}$ **71.** $\sqrt{2}$ **73.** 0 **75.** 0 **77.** -1 **79.** $\cos\theta = -\frac{5}{13}$; $\tan\theta = -\frac{12}{5}$; $\csc\theta = \frac{13}{12}$; $\sec\theta = -\frac{13}{5}$; $\cot\theta = -\frac{5}{12}$ **81.** $\sin\theta = -\frac{3}{5}$; $\tan\theta = \frac{3}{4}$; $\csc\theta = -\frac{5}{3}$; $\sec\theta = -\frac{5}{4}$; $\cot\theta = \frac{4}{3}$ **83.** $\cos\theta = -\frac{12}{13}$; $\tan\theta = -\frac{5}{12}$; $\csc\theta = \frac{13}{5}$; $\sec\theta = -\frac{13}{12}$; $\cot\theta = -\frac{12}{5}$ **85.** $\sin\theta = -\frac{2\sqrt{2}}{3}$; $\tan\theta = 2\sqrt{2}$; $\csc\theta = -\frac{3\sqrt{2}}{4}$; $\sec\theta = -3$; $\cot\theta = \frac{\sqrt{2}}{4}$ **87.** $\cos\theta = -\frac{\sqrt{5}}{3}$; $\tan\theta = -\frac{2\sqrt{5}}{5}$; $\csc\theta = \frac{3}{2}$; $\sec\theta = -\frac{3\sqrt{5}}{5}$; $\cot\theta = -\frac{\sqrt{5}}{2}$ **89.** $\sin\theta = -\frac{\sqrt{3}}{2}$; $\cos\theta = \frac{1}{2}$; $\tan\theta = -\sqrt{3}$; $\csc\theta = -\frac{2\sqrt{3}}{3}$; $\cot\theta = -\frac{\sqrt{3}}{3}$ **91.** $\sin\theta = -\frac{3}{5}$; $\cos\theta = -\frac{4}{5}$; $\csc\theta = -\frac{5}{3}$; $\sec\theta = -\frac{5}{4}$; $\cot\theta = \frac{4}{3}$ **93.** $\sin\theta = \frac{\sqrt{10}}{10}$; $\cos\theta = -\frac{3\sqrt{10}}{10}$; $\csc\theta = \sqrt{10}$; $\sec\theta = -\frac{\sqrt{10}}{3}$; $\cot\theta = -3$ **95.** $\sin\theta = -\frac{1}{2}$; $\cos\theta = -\frac{\sqrt{3}}{2}$; $\tan\theta = \frac{\sqrt{3}}{3}$; $\sec\theta = -\frac{2\sqrt{3}}{3}$; $\cot\theta = \sqrt{3}$ **97.** 0 **99.** -0.2 **101.** 3 **103.** 5 **105.** 0 **107. (a)** Approximately 16.56 ft
(b)

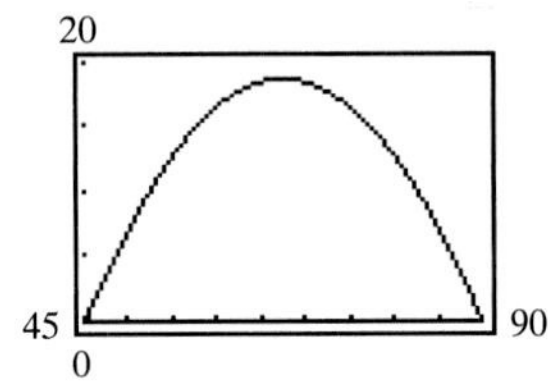

(c) 67.50°

2.5 Concepts and Vocabulary *(page 151)*

1. 2π; π **2.** All real numbers except odd multiples of $\frac{\pi}{2}$ **3.** All real numbers from -1 to 1 inclusive **4.** T **5.** F **6.** F

2.5 Exercises *(page 152)*

1. $\sin t = -\frac{1}{2}$; $\cos t = \frac{\sqrt{3}}{2}$; $\tan t = -\frac{\sqrt{3}}{3}$; $\csc t = -2$; $\sec t = \frac{2\sqrt{3}}{3}$; $\cot t = -\sqrt{3}$ **3.** $\sin t = -\frac{\sqrt{2}}{2}$; $\cos t = -\frac{\sqrt{2}}{2}$; $\tan t = 1$; $\csc t = -\sqrt{2}$; $\sec t = -\sqrt{2}$; $\cot t = 1$ **5.** $\sin t = \frac{2}{3}$; $\cos t = \frac{\sqrt{5}}{3}$; $\tan t = \frac{2\sqrt{5}}{5}$; $\csc t = \frac{3}{2}$; $\sec t = \frac{3\sqrt{5}}{5}$; $\cot t = \frac{\sqrt{5}}{2}$ **7.** $\sin\theta = -\frac{4}{5}$; $\cos\theta = \frac{3}{5}$; $\tan\theta = -\frac{4}{3}$; $\csc\theta = -\frac{5}{4}$; $\sec\theta = \frac{5}{3}$; $\cot\theta = -\frac{3}{4}$ **9.** $\sin\theta = \frac{3\sqrt{13}}{13}$; $\cos\theta = -\frac{2\sqrt{13}}{13}$; $\tan\theta = -\frac{3}{2}$; $\csc\theta = \frac{\sqrt{13}}{3}$; $\sec\theta = -\frac{\sqrt{13}}{2}$; $\cot\theta = -\frac{2}{3}$ **11.** $\sin\theta = -\frac{\sqrt{2}}{2}$; $\cos\theta = -\frac{\sqrt{2}}{2}$; $\tan\theta = 1$; $\csc\theta = -\sqrt{2}$; $\sec\theta = -\sqrt{2}$; $\cot\theta = 1$ **13.** $\frac{\sqrt{2}}{2}$ **15.** 1 **17.** 1 **19.** $\sqrt{3}$ **21.** $\frac{\sqrt{2}}{2}$ **23.** 0 **25.** $\sqrt{2}$ **27.** $\frac{\sqrt{3}}{3}$ **29.** $-\frac{\sqrt{3}}{2}$ **31.** $-\frac{\sqrt{3}}{3}$ **33.** 2 **35.** -1 **37.** -1 **39.** $\frac{\sqrt{2}}{2}$ **41.** 0 **43.** $-\sqrt{2}$ **45.** $\frac{2\sqrt{3}}{3}$ **47.** -1 **49.** -2 **51.** $\frac{2-\sqrt{2}}{2}$ **53.** All real numbers **55.** Odd multiples of $\frac{\pi}{2}$ **57.** Odd multiples of $\frac{\pi}{2}$ **59.** $[-1, 1]$ **61.** $(-\infty, \infty)$ **63.** $(-\infty, -1]$ or $[1, \infty)$ **65.** Odd; yes; origin **67.** Odd; yes; origin **69.** Even; yes; y-axis **71.** 0.9 **73.** 9 **75. (a)** $-\frac{1}{3}$ **(b)** 1 **77. (a)** -2 **(b)** 6 **79. (a)** -4 **(b)** -12

81. Let $P = (x, y)$ be the point on the unit circle that corresponds to t. Consider the equation $\tan t = \frac{y}{x} = a$. Then $y = ax$. But $x^2 + y^2 = 1$ so that $x^2 + a^2x^2 = 1$. Thus, $x = \pm\frac{1}{\sqrt{1+a^2}}$ and $y = \pm\frac{a}{\sqrt{1+a^2}}$; that is, for any real number a, there is a point $P = (x, y)$ on the unit circle for which $\tan t = a$. In other words, $-\infty < \tan t < \infty$, and the range of the tangent function is the set of all real numbers.

83. Suppose there is a number p, $0 < p < 2\pi$, for which $\sin(\theta + p) = \sin\theta$ for all θ. If $\theta = 0$, then $\sin(0 + p) = \sin p = \sin 0 = 0$; so that $p = \pi$. If $\theta = \frac{\pi}{2}$, then $\sin\left(\frac{\pi}{2} + p\right) = \sin\left(\frac{\pi}{2}\right)$. But $p = \pi$. Thus, $\sin\left(\frac{3\pi}{2}\right) = -1 = \sin\left(\frac{\pi}{2}\right) = 1$. This is impossible. Therefore, the smallest positive number p for which $\sin(\theta + p) = \sin\theta$ for all θ is $p = 2\pi$.

85. $\sec\theta = \frac{1}{\cos\theta}$; since $\cos\theta$ has period 2π, so does $\sec\theta$.

31.

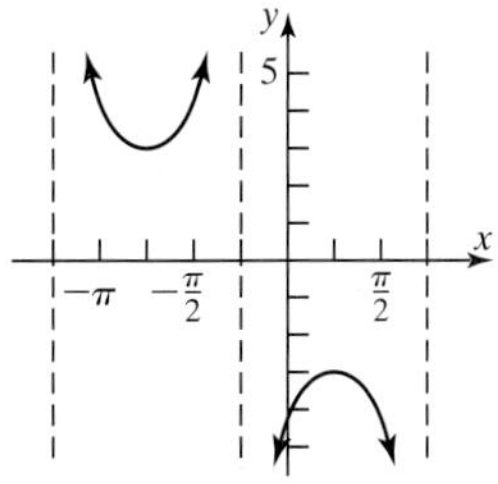

33.

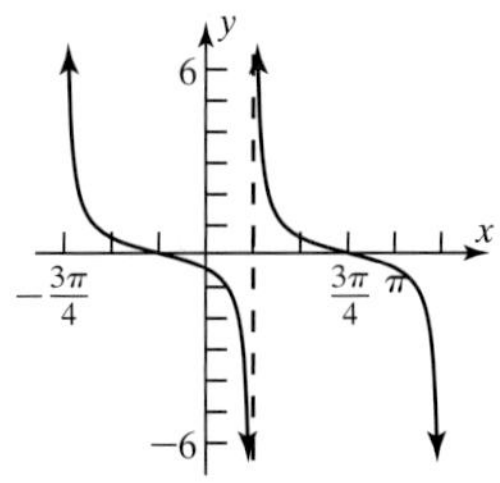

35. **(a)** $L(\theta) = \dfrac{3}{\cos\theta} + \dfrac{4}{\sin\theta} = 3\sec\theta + 4\csc\theta$

(b)

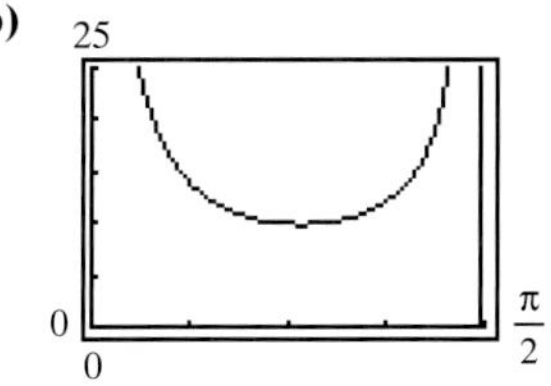

(c) 0.83

(d) 9.87 ft

2.8 Concepts and Vocabulary *(page 185)*

1. Phase shift **2.** F

2.8 Exercises *(page 186)*

1. Amplitude $= 4$

Period $= \pi$

Phase shift $= \dfrac{\pi}{2}$

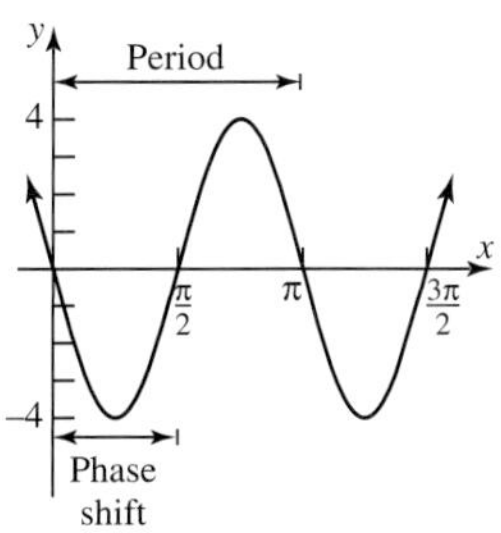

3. Amplitude $= 2$

Period $= \dfrac{2\pi}{3}$

Phase shift $= -\dfrac{\pi}{6}$

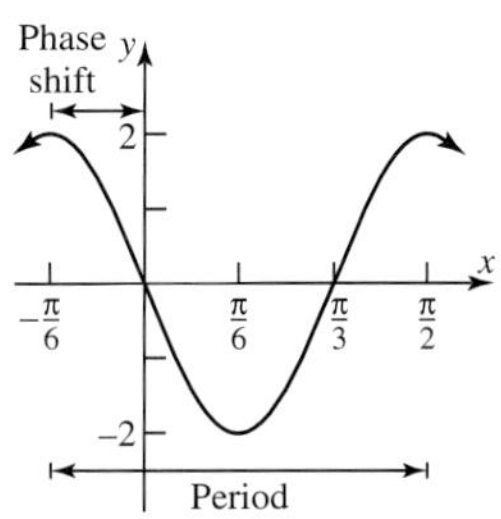

5. Amplitude $= 3$

Period $= \pi$

Phase shift $= -\dfrac{\pi}{4}$

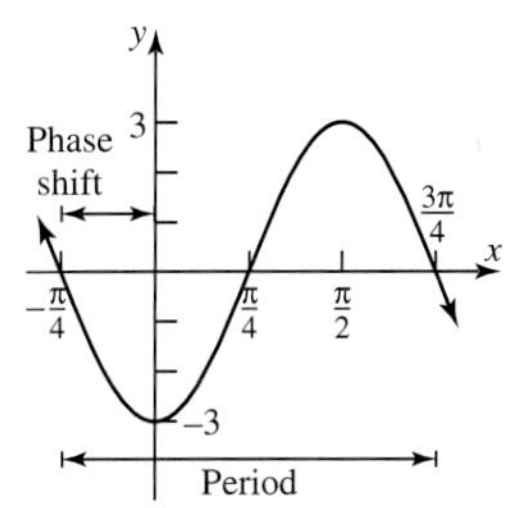

7. Amplitude $= 4$

Period $= 2$

Phase shift $= -\dfrac{2}{\pi}$

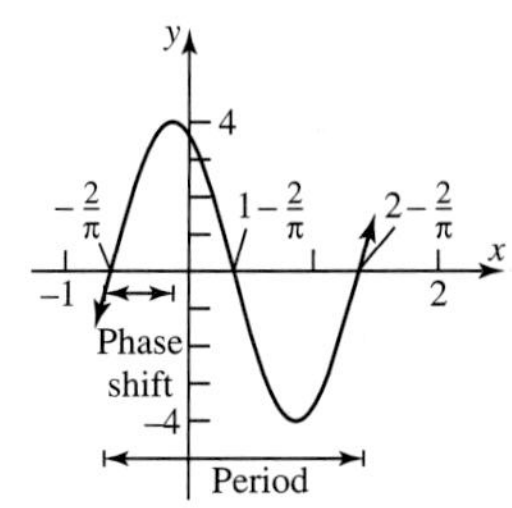

9. Amplitude $= 3$

Period $= 2$

Phase shift $= \dfrac{2}{\pi}$

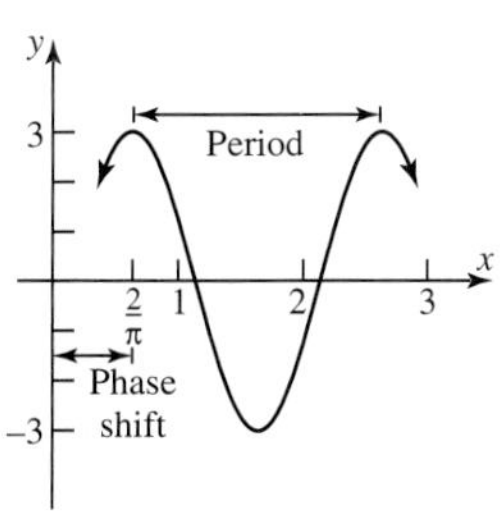

11. Amplitude $= 3$

Period $= \pi$

Phase shift $= \dfrac{\pi}{4}$

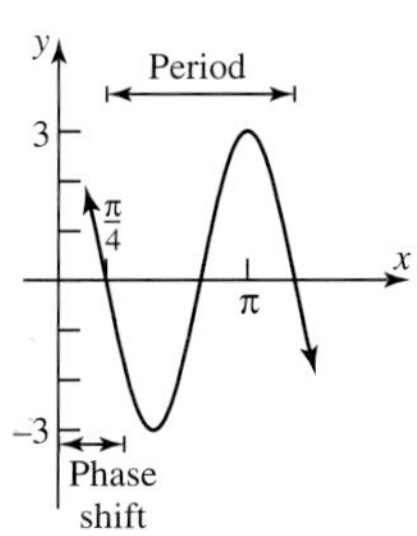

13. $y = \pm 2\sin\left[2\left(x - \dfrac{1}{2}\right)\right]$ or $y = \pm 2\sin(2x - 1)$

15. $y = \pm 3\sin\left[\dfrac{2}{3}\left(x + \dfrac{1}{3}\right)\right]$ or $y = \pm 3\sin\left(\dfrac{2}{3}x + \dfrac{2}{9}\right)$

17. Period $= \dfrac{1}{15}$; amplitude $= 120$; phase shift $= \dfrac{1}{90}$

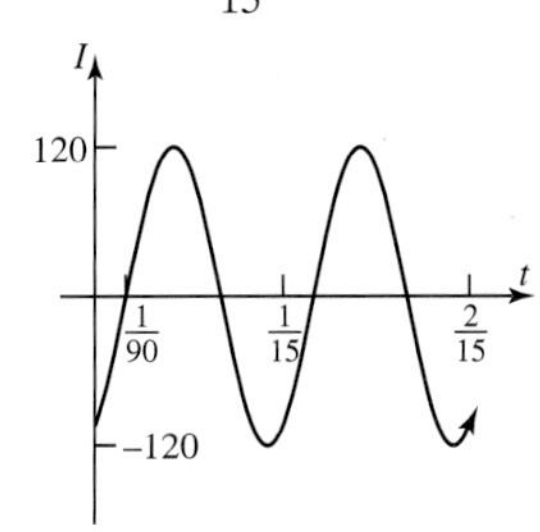

19. **(a)**

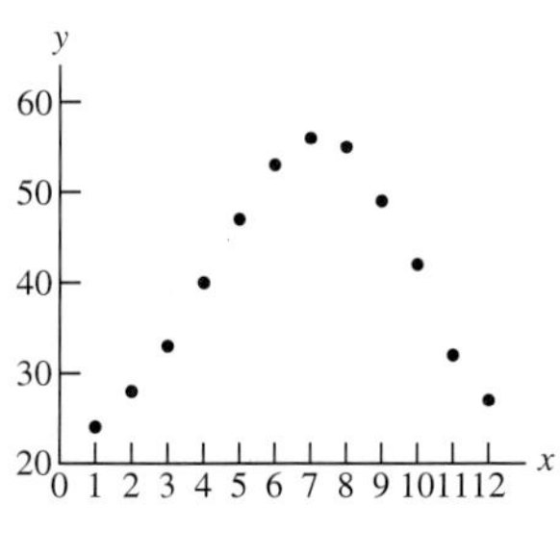

(b) $y = 15.9\sin\left(\dfrac{\pi}{6}x - \dfrac{2\pi}{3}\right) + 40.1$

(c)

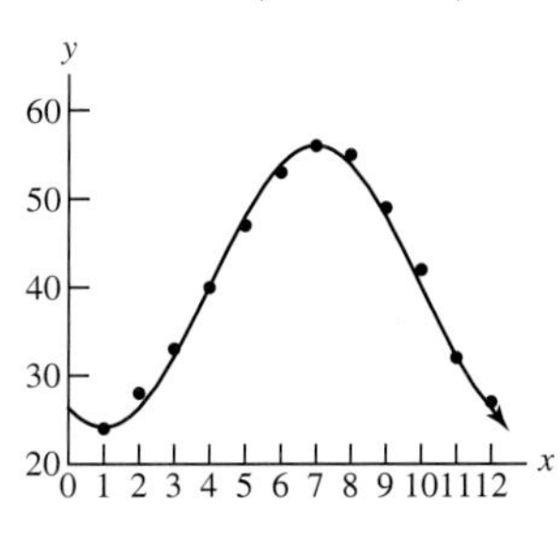

(d) $y = 15.62\sin(0.517x - 2.096) + 40.377$

(e)

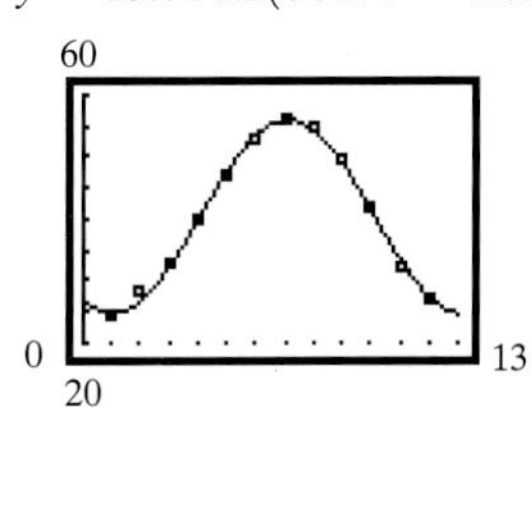

21. (a)

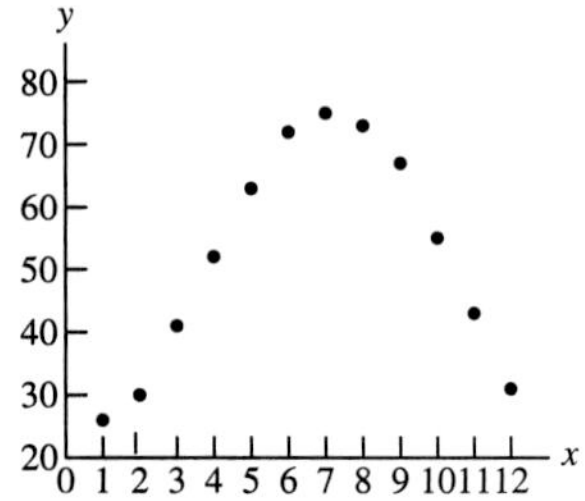

(b) $y = 24.95\sin\left(\dfrac{\pi}{6}x - \dfrac{2\pi}{3}\right) + 50.45$

(c)

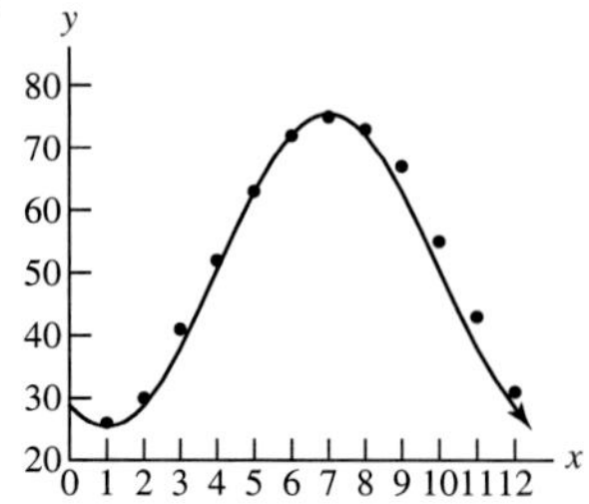

(d) $y = 25.693\sin(0.476x - 1.814) + 49.854$

(e)

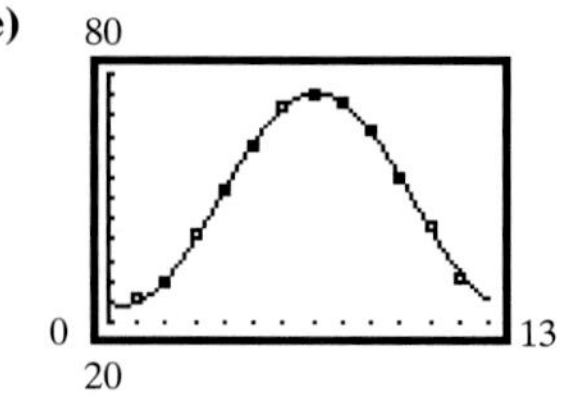

23. (a) 4:08 P.M.

(b) $y = 4.4\sin\left(\dfrac{4\pi}{25}x - 6.66\right) + 3.8$

(c)

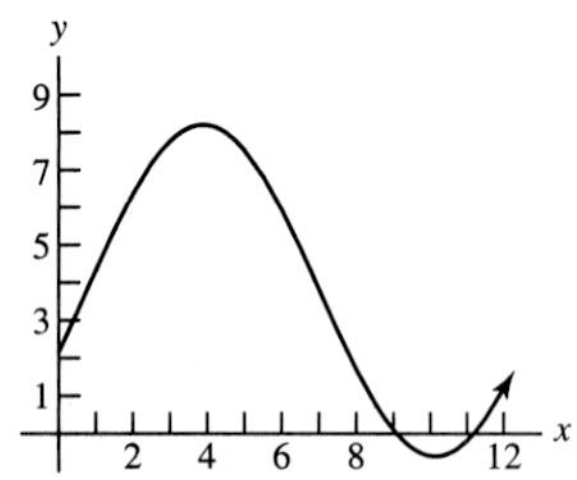

(d) 8.2 ft

25. (a) $y = 1.0835\sin\left(\dfrac{2\pi}{365}x - 2.45\pi\right) + 11.6665$

(b)

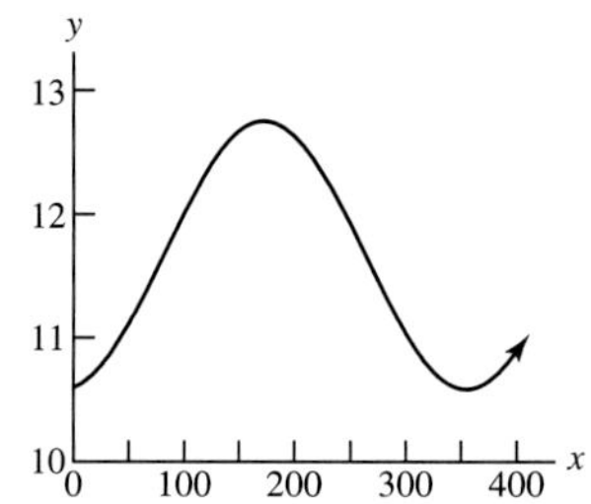

(c) 11.83 hr

27. (a) $y = 5.3915\sin\left(\dfrac{2\pi}{365}x - 2.45\pi\right) + 10.8415$

(b)

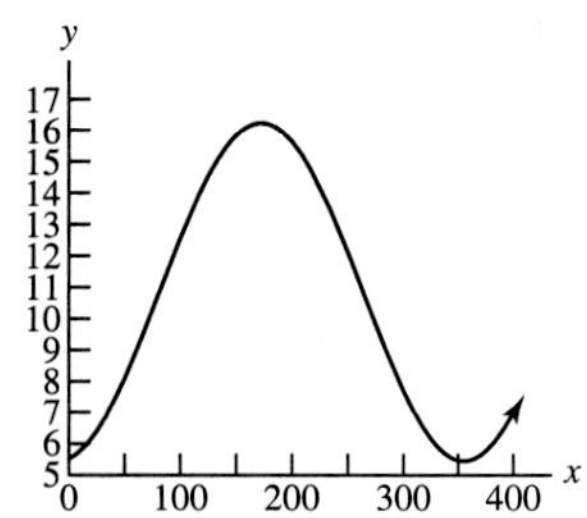

(c) 11.66 hr

Review Exercises *(page 191)*

1. $\dfrac{3\pi}{4}$ **3.** $\dfrac{\pi}{10}$ **5.** 135° **7.** −450° **9.** $\dfrac{1}{2}$ **11.** $\dfrac{3\sqrt{2}}{2} - \dfrac{4\sqrt{3}}{3}$ **13.** $-3\sqrt{2} - 2\sqrt{3}$ **15.** 3 **17.** 0 **19.** 0 **21.** 1 **23.** 1 **25.** 1 **27.** −1

29. 1 **31.** $\cos\theta = \dfrac{3}{5}$; $\tan\theta = \dfrac{4}{3}$; $\csc\theta = \dfrac{5}{4}$; $\sec\theta = \dfrac{5}{3}$; $\cot\theta = \dfrac{3}{4}$ **33.** $\sin\theta = -\dfrac{12}{13}$; $\cos\theta = -\dfrac{5}{13}$; $\csc\theta = -\dfrac{13}{12}$; $\sec\theta = -\dfrac{13}{5}$; $\cot\theta = \dfrac{5}{12}$

35. $\sin\theta = \dfrac{3}{5}$; $\cos\theta = -\dfrac{4}{5}$; $\tan\theta = -\dfrac{3}{4}$; $\csc\theta = \dfrac{5}{3}$; $\cot\theta = -\dfrac{4}{3}$ **37.** $\cos\theta = -\dfrac{5}{13}$; $\tan\theta = -\dfrac{12}{5}$; $\csc\theta = \dfrac{13}{12}$; $\sec\theta = -\dfrac{13}{5}$; $\cot\theta = -\dfrac{5}{12}$

39. $\cos\theta = \dfrac{12}{13}$; $\tan\theta = -\dfrac{5}{12}$; $\csc\theta = -\dfrac{13}{5}$; $\sec\theta = \dfrac{13}{12}$; $\cot\theta = -\dfrac{12}{5}$ **41.** $\sin\theta = -\dfrac{\sqrt{10}}{10}$; $\cos\theta = -\dfrac{3\sqrt{10}}{10}$; $\csc\theta = -\sqrt{10}$; $\sec\theta = -\dfrac{\sqrt{10}}{3}$; $\cot\theta = 3$ **43.** $\sin\theta = -\dfrac{2\sqrt{2}}{3}$; $\cos\theta = \dfrac{1}{3}$; $\tan\theta = -2\sqrt{2}$; $\csc\theta = -\dfrac{3\sqrt{2}}{4}$; $\cot\theta = -\dfrac{\sqrt{2}}{4}$

45. $\sin\theta = \dfrac{\sqrt{5}}{5}$; $\cos\theta = -\dfrac{2\sqrt{5}}{5}$; $\tan\theta = -\dfrac{1}{2}$; $\csc\theta = \sqrt{5}$; $\sec\theta = -\dfrac{\sqrt{5}}{2}$

47.

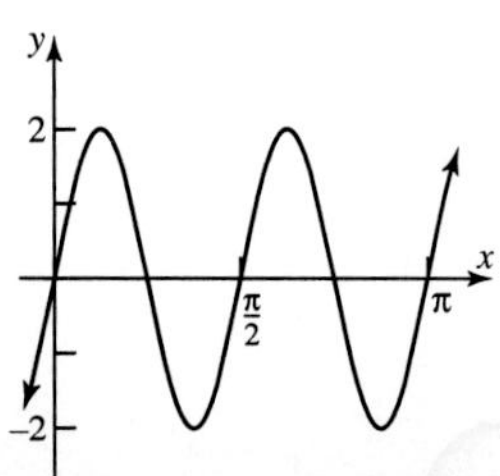

49.

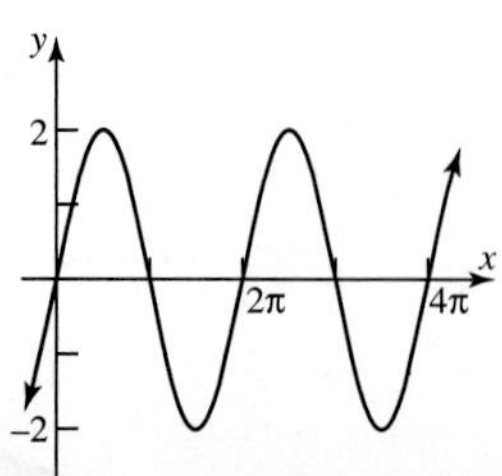

51.

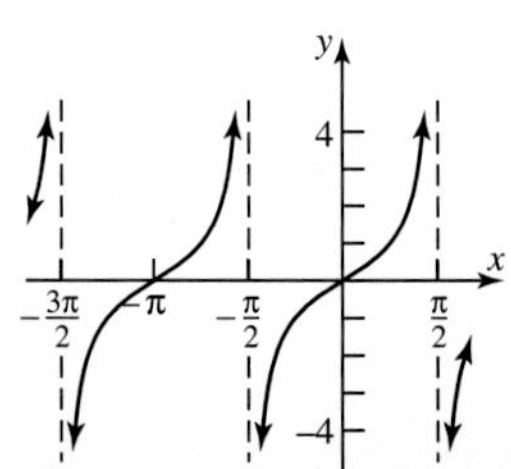

53.

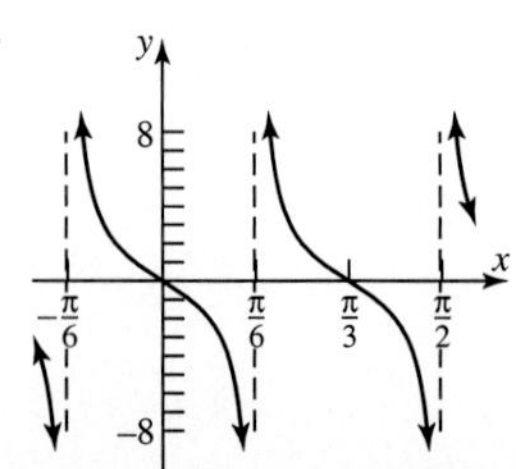

55.

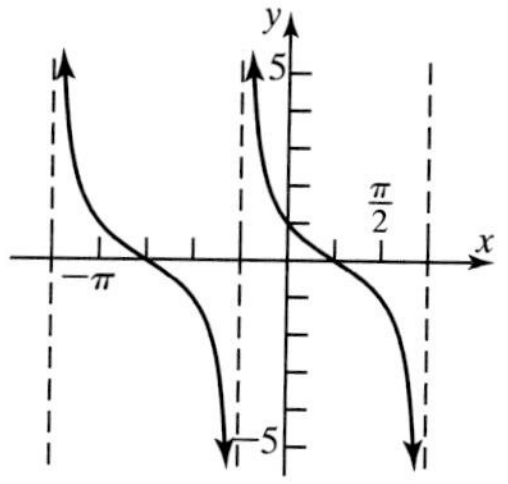

57.

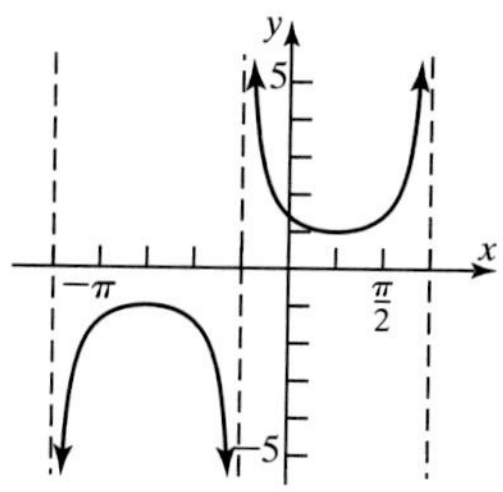

59. Amplitude $= 4$; period $= 2\pi$

61. Amplitude $= 8$; period $= 4$

63. Amplitude $= 4$

Period $= \frac{2\pi}{3}$

Phase shift $= 0$

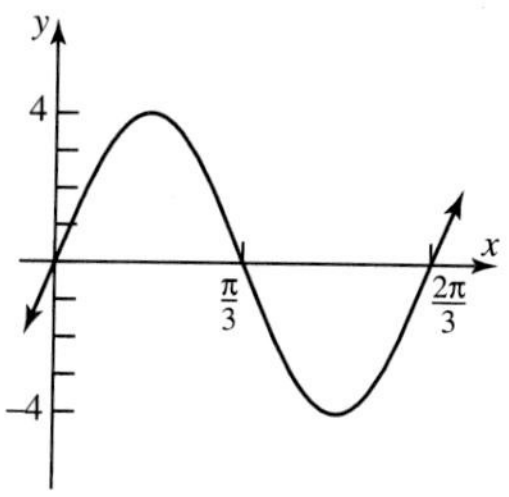

65. Amplitude $= 2$

Period $= \pi$

Phase Shift $= \frac{\pi}{2}$

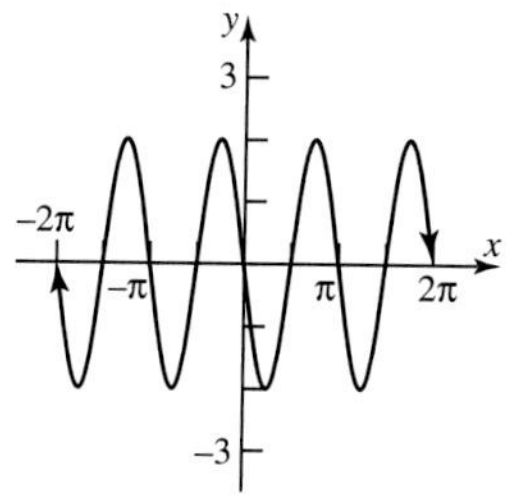

67. Amplitude $= \frac{1}{2}$

Period $= \frac{4\pi}{3}$

Phase shift $= \frac{2\pi}{3}$

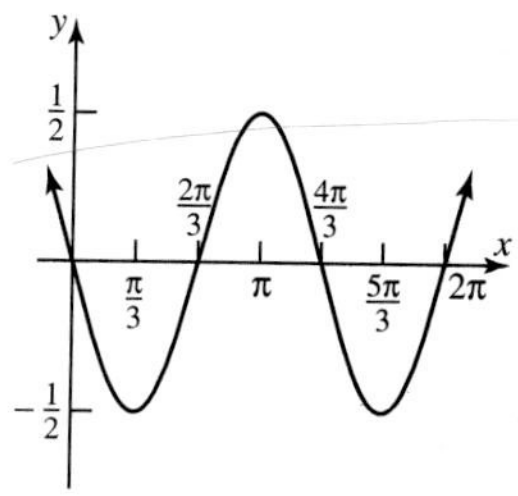

69. Amplitude $= \frac{2}{3}$

Period $= 2$

Phase shift $= \frac{6}{\pi}$

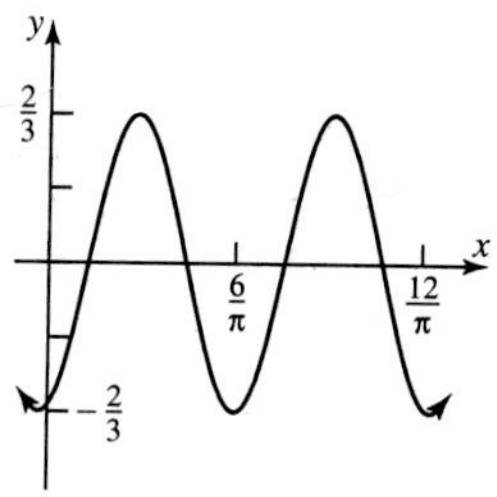

71. $y = 5\cos\frac{x}{4}$ **73.** $y = -6\cos\left(\frac{\pi}{4}x\right)$ **75.** $\sin\theta = \frac{5}{13}, \cos\theta = \frac{12}{13}, \tan\theta = \frac{5}{12}, \csc\theta = \frac{13}{5}, \sec\theta = \frac{13}{12}, \cot\theta = \frac{12}{5}$

77. $\sin\theta = -\frac{4}{5}, \cos\theta = \frac{3}{5}, \tan\theta = -\frac{4}{3}, \csc\theta = -\frac{5}{4}, \sec\theta = \frac{5}{3}, \cot\theta = -\frac{3}{4}$ **79.** $\frac{\pi}{5}$

81. Domain: $\left\{x \middle| x \neq \text{odd multiple of } \frac{\pi}{2}\right\}$; range: $\{y \,|\, |y| \geq 1\}$ **83.** $\frac{\pi}{3} \approx 1.05$ ft; $\frac{\pi}{3} \approx 1.05$ ft^2 **85.** Approximately 114.59 revolutions/hr

87. 0.1 revolution/sec $= \frac{\pi}{5}$ radian/sec **89.** **(a)** 120 **(b)** $\frac{1}{60}$ **(c)**

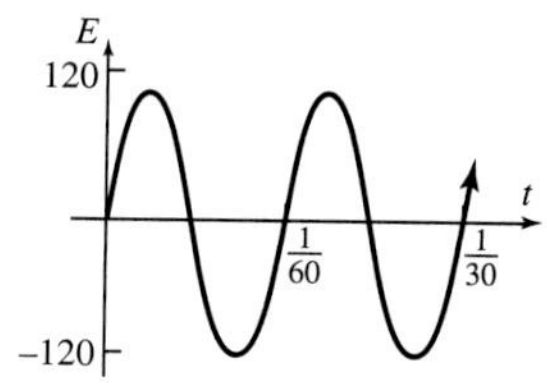

91. **(a)**

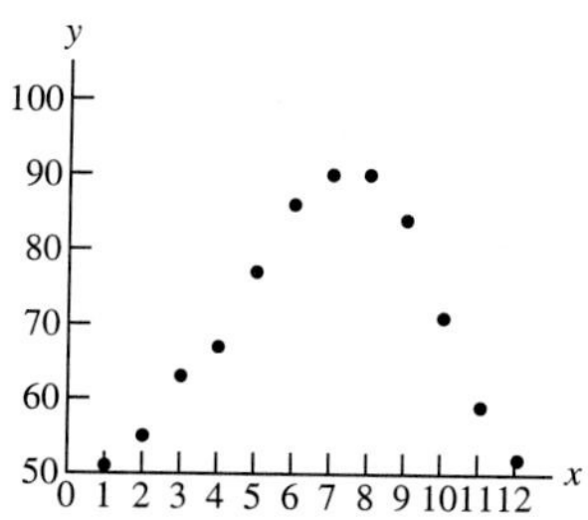

(b) $y = 19.5\sin\left(\frac{\pi}{6}x - \frac{2\pi}{3}\right) + 70.5$

(c)

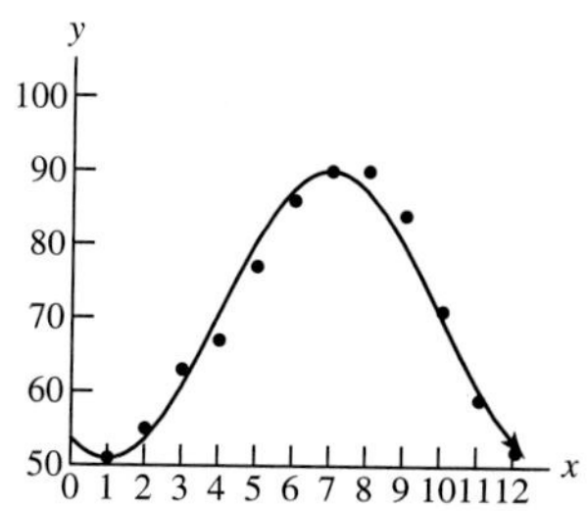

(d) $y = 19.52\sin(0.54x - 2.28) + 71.01$

(e)

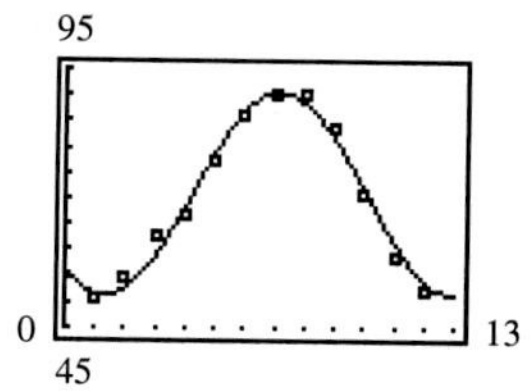

93. (a) $y = 1.85\sin\left(\frac{2\pi}{365}x - 2.45\pi\right) + 11.517$ **(b)**

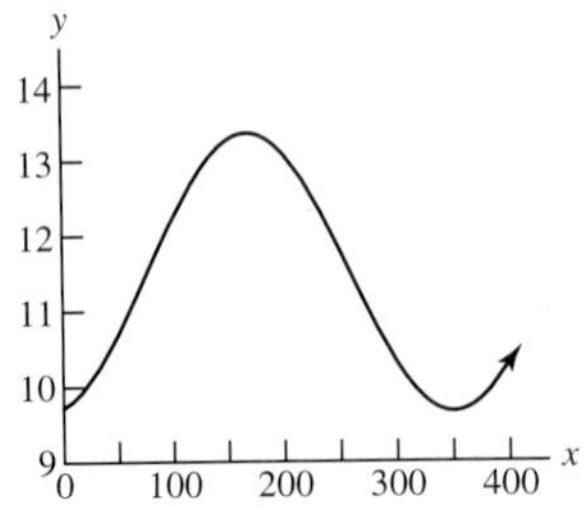

(c) 11.80 hr

Cumulative Review Exercises *(page 196)*

1. $\left\{-1, \frac{1}{2}\right\}$ **2.** $y - 5 = -3(x + 2)$ or $y = -3x - 1$ **3.** $x^2 + (y + 2)^2 = 16$

4.

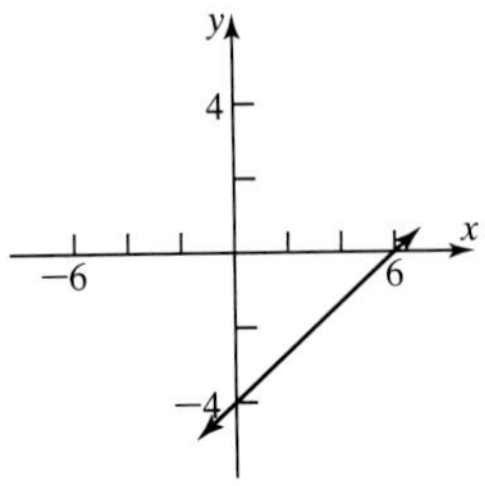

5.

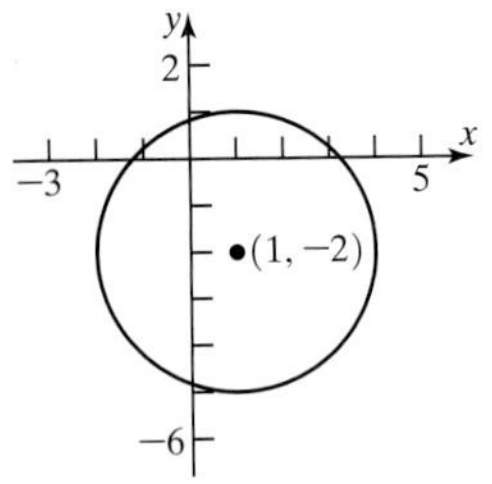

6.

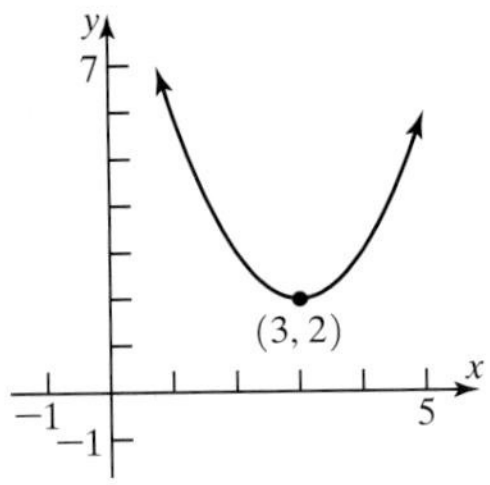

7. (a)

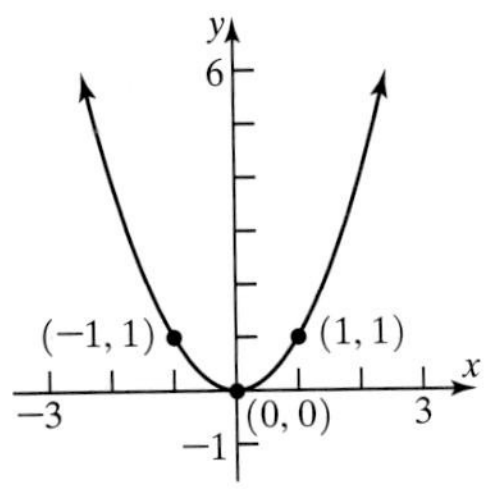

(b)

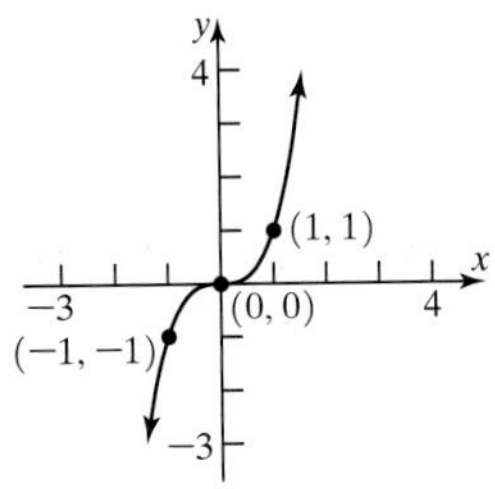

(c)

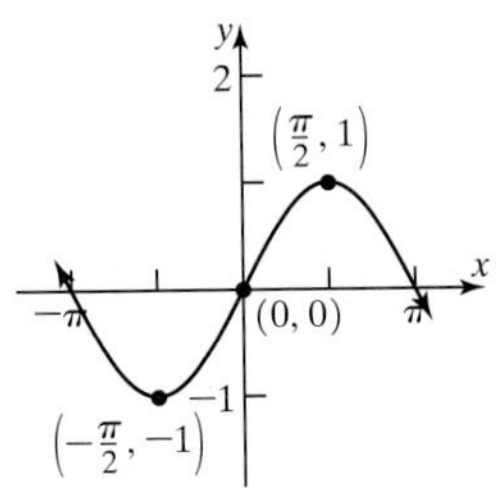

(d)

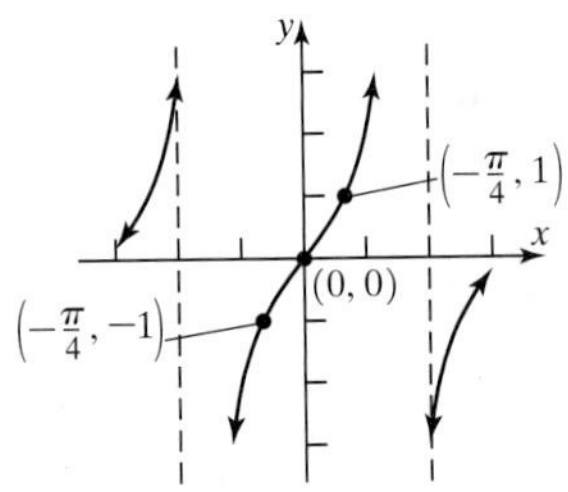

8. $f^{-1}(x) = \frac{1}{3}(x + 2)$ **9.** -2 **10.**

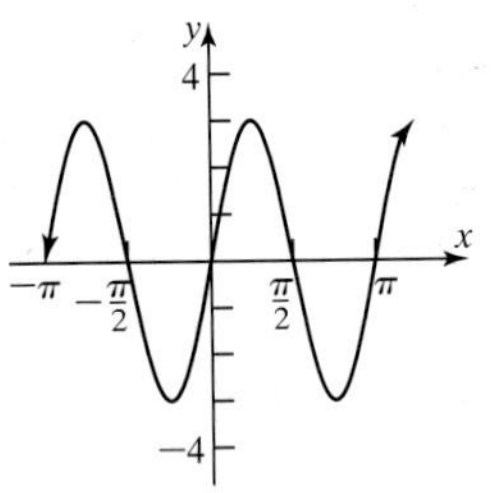

11. $3 - \frac{3\sqrt{3}}{2}$ **12.** $y = 3\cos\left(\frac{\pi}{6}x\right)$

CHAPTER 3 Analytic Trigonometry

3.1 Concepts and Vocabulary *(page 208)*

1. $x = \sin y$ **2.** 0 **3.** $\frac{\pi}{5}$ **4.** F **5.** T **6.** T **8.** $-\frac{\pi}{2} \le x \le \frac{\pi}{2}$ **9.** $-1 \le x \le 1$

3.1 Exercises *(page 208)*

1. 0 **3.** $-\frac{\pi}{2}$ **5.** 0 **7.** $\frac{\pi}{4}$ **9.** $\frac{\pi}{3}$ **11.** $\frac{5\pi}{6}$ **13.** 0.10 **15.** 1.37 **17.** 0.51 **19.** -0.38 **21.** -0.12 **23.** 1.08 **25.** 0.54 **27.** $\frac{4\pi}{5}$ **29.** -3.5 **31.** $-\frac{3\pi}{7}$ **33.** Yes; $-\frac{\pi}{6}$ lies in the interval $\left[-\frac{\pi}{2}, \frac{\pi}{2}\right]$. **35.** No; 2 is not in the domain of $\sin^{-1} x$. **37.** No; $-\frac{\pi}{6}$ does not lie in the interval $[0, \pi]$. **39.** Yes, $-\frac{1}{2}$ is in the domain of $\cos^{-1} x$. **41.** Yes; $-\frac{\pi}{3}$ lies in the interval $\left(-\frac{\pi}{2}, \frac{\pi}{2}\right)$. **43.** Yes; 2 is in the domain of $\tan^{-1} x$. **45. (a)** 13.92 hr or 13 hr, 55 min **(b)** 12 hr **(c)** 13.85 hr or 13 hr, 51 min **47. (a)** 13.3 hr or 13 hr, 18 min **(b)** 12 hr **(c)** 13.26 hr or 13 hr, 15 min **49. (a)** 12 hr **(b)** 12 hr **(c)** 12 hr **(d)** It's 12 hr. **51.** 3.35 min

Cumulative Review Exercises (page 265)

1. $\left\{\frac{-1-\sqrt{13}}{6}, \frac{-1+\sqrt{13}}{6}\right\}$ **2.** $y+1=-1(x-4)$ or $x+y=3$; $6\sqrt{2}$; $(1,2)$ **3.** x-axis symmetry; $(0,-3)$, $(0,3)$, $(3,0)$

4.

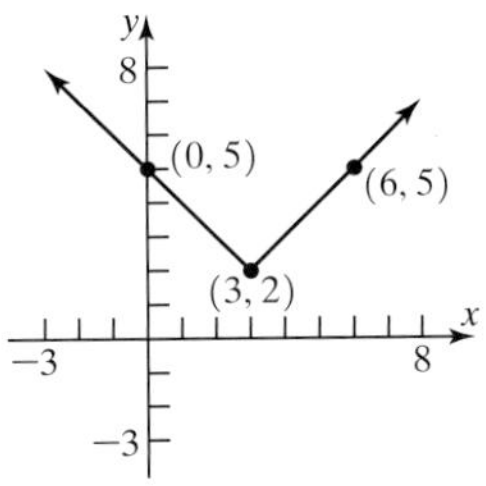

5.

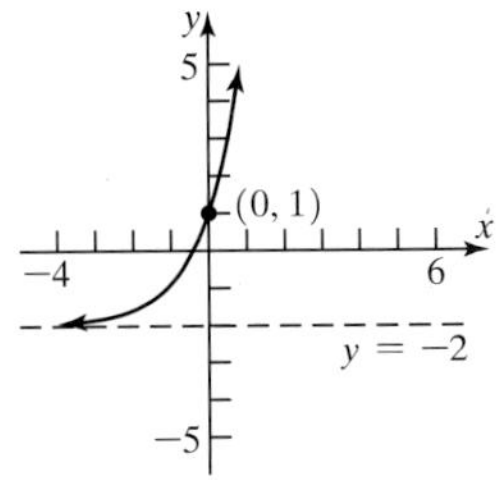

6.

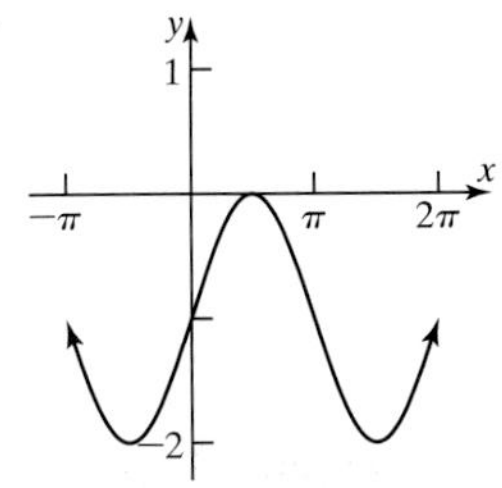

7. (a)

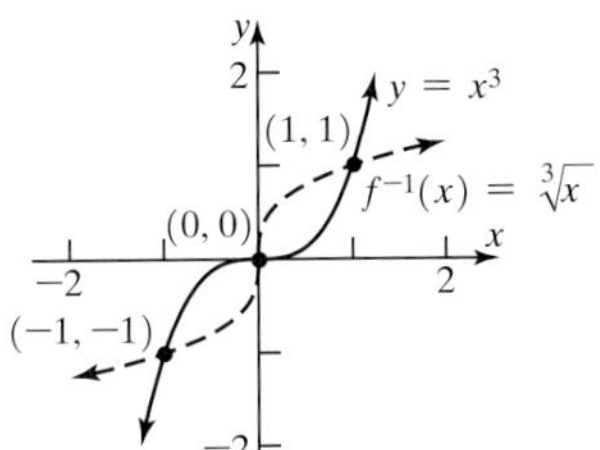

(b)

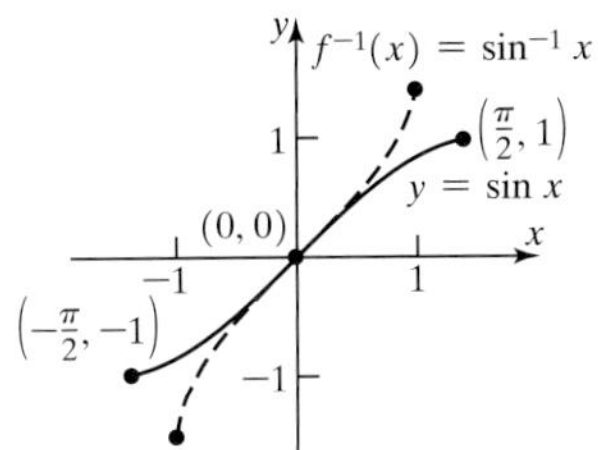

(c)

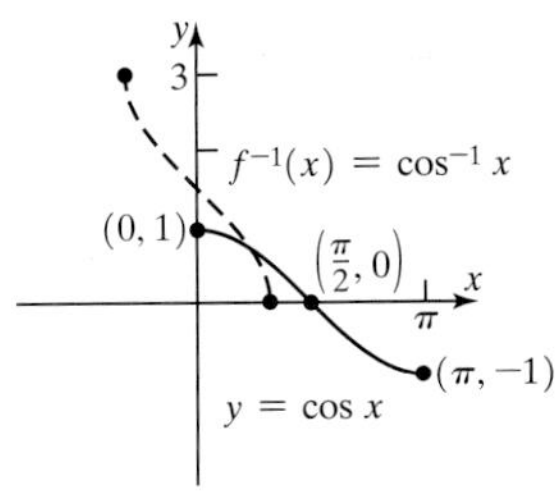

8. (a) $-\frac{2\sqrt{2}}{3}$ **(b)** $\frac{\sqrt{2}}{4}$ **(c)** $\frac{4\sqrt{2}}{9}$ **(d)** $\frac{7}{9}$ **(e)** $\sqrt{\frac{3+2\sqrt{2}}{6}}$ **(f)** $\sqrt{\frac{3-2\sqrt{2}}{6}}$ **9.** $\frac{\sqrt{5}}{5}$

10. (a) $-\frac{2\sqrt{2}}{3}$ **(b)** $-\frac{2\sqrt{2}}{3}$ **(c)** $\frac{7}{9}$ **(d)** $\frac{4\sqrt{2}}{9}$ **(e)** $\frac{\sqrt{6}}{3}$ **11.** 7.28

CHAPTER 4 Applications of Trigonometric Functions

4.1 Concepts and Vocabulary (page 273)

1. Angle of elevation **2.** Angle of depression **3.** T **4.** F

4.1 Exercises (page 273)

1. $a \approx 13.74, c \approx 14.62, \alpha = 70°$ **3.** $b \approx 5.03, c \approx 7.83, \alpha = 50°$ **5.** $a \approx 0.71, c \approx 4.06, \beta = 80°$ **7.** $b \approx 10.72, c \approx 11.83, \beta = 65°$ **9.** $b \approx 3.08, a \approx 8.46, \alpha = 70°$ **11.** $c \approx 5.83, \alpha \approx 59.0°, \beta = 31.0°$ **13.** $b \approx 4.58, \alpha \approx 23.6°, \beta = 66.4°$ **15.** 4.59 in., 6.55 in. **17.** 5.52 in. or 11.83 in. **19.** 23.6° and 66.4° **21.** 70.02 ft **23.** 985.91 ft **25.** 137.37 m **27.** 20.67 ft **29.** 15.9° **31.** 60.27 ft **33.** 530.18 ft **35.** 554.52 ft **37. (a)** 111.96 ft/sec or 76.3 mi/hr **(b)** 82.42 ft/sec or 56.2 mi/hr **(c)** 18.8° or less **39.** S76.6°E **41.** 14.9° **43.** 3.83 mi **45.** No; Move the tripod back about 1 ft.

47. (a) $A(\theta) = 2\sin\theta\cos\theta$ **(b)** From double-angle formula, since $2\sin\theta\cos\theta = \sin(2\theta)$ **(c)** $\theta = 45°$ **(d)** $\frac{\sqrt{2}}{2}$ by $\sqrt{2}$

4.2 Concepts and Vocabulary (page 284)

1. Oblique **2.** $\frac{\sin\alpha}{a} = \frac{\sin\beta}{b} = \frac{\sin\gamma}{c}$ **3.** F **4.** T **5.** F

4.2 Exercises (page 284)

1. $a \approx 3.23, b \approx 3.55, \alpha = 40°$ **3.** $a \approx 3.25, c \approx 4.23, \beta = 45°$ **5.** $\gamma = 95°, c \approx 9.86, a \approx 6.36$ **7.** $\alpha = 40°, a = 2, c \approx 3.06$ **9.** $\gamma = 120°, b \approx 1.06, c \approx 2.69$ **11.** $\alpha = 100°, a \approx 5.24, c \approx 0.92$ **13.** $\beta = 40°, a \approx 5.64, b \approx 3.86$ **15.** $\gamma = 100°, a \approx 1.31, b \approx 1.31$ **17.** One triangle; $\beta \approx 30.7°, \gamma \approx 99.3°, c \approx 3.86$ **19.** One triangle; $\gamma \approx 36.2°, \alpha \approx 43.8°, a \approx 3.51$ **21.** No triangle **23.** Two triangles; $\gamma_1 \approx 30.9°, \alpha_1 \approx 129.1°, a_1 \approx 9.07$ or $\gamma_2 \approx 149.1°, \alpha_2 \approx 10.9°, a_2 \approx 2.20$ **25.** No triangle **27.** Two triangles; $a_1 \approx 57.7°$, $\beta_1 \approx 97.3°, b_1 \approx 2.35$ or $\alpha_2 \approx 122.3°, \beta_2 \approx 32.7°, b_2 \approx 1.28$ **29. (a)** Station Able is about 143.33 mi from the ship; Station Baker is about 135.58 mi from the ship. **(b)** Approximately 41 min **31.** 1490.48 ft **33.** 381.69 ft **35. (a)** 169.18 mi **(b)** 161.3° **37.** 84.7°; 183.72 ft **39.** 2.64 mi **41.** 1.88 mi or 1.53 mi **43.** 449.36 ft **45.** 39.39 ft **47.** 29.97 ft

49. $\frac{a-b}{c} = \frac{a}{c} - \frac{b}{c} = \frac{\sin\alpha}{\sin\gamma} - \frac{\sin\beta}{\sin\gamma} = \frac{\sin\alpha - \sin\beta}{\sin\gamma} = \frac{2\sin\left(\frac{\alpha-\beta}{2}\right)\cos\left(\frac{\alpha+\beta}{2}\right)}{2\sin\frac{\gamma}{2}\cos\frac{\gamma}{2}} = \frac{\sin\left(\frac{\alpha-\beta}{2}\right)\cos\left(\frac{\pi}{2}-\frac{\gamma}{2}\right)}{\sin\frac{\gamma}{2}\cos\frac{\gamma}{2}} = \frac{\sin\left(\frac{\alpha-\beta}{2}\right)}{\cos\frac{\gamma}{2}}$

51. $\dfrac{a-b}{a+b} = \dfrac{\dfrac{a-b}{c}}{\dfrac{a+b}{c}} = \dfrac{\dfrac{\sin\left[\frac{1}{2}(\alpha-\beta)\right]}{\cos\frac{\gamma}{2}}}{\dfrac{\cos\left[\frac{1}{2}(\alpha-\beta)\right]}{\sin\frac{\gamma}{2}}} = \dfrac{\tan\left[\frac{1}{2}(\alpha-\beta)\right]}{\cot\frac{\gamma}{2}} = \dfrac{\tan\left[\frac{1}{2}(\alpha-\beta)\right]}{\tan\left(\frac{\pi}{2}-\frac{\gamma}{2}\right)} = \dfrac{\tan\left[\frac{1}{2}(\alpha-\beta)\right]}{\tan\left[\frac{1}{2}(\alpha+\beta)\right]}$

4.3 Concepts and Vocabulary *(page 291)*

1. Cosines **2.** Sines **3.** Cosines **4.** F **5.** F **6.** T

4.3 Exercises *(page 292)*

1. $b \approx 2.95, \alpha \approx 28.7°, \gamma \approx 106.3°$ **3.** $c \approx 3.75, \alpha \approx 32.1°, \beta \approx 52.9°$ **5.** $\alpha \approx 48.5°, \beta \approx 38.6°, \gamma \approx 92.9°$ **7.** $\alpha \approx 127.2°, \beta \approx 32.1°, \gamma \approx 20.7°$ **9.** $c \approx 2.57, \alpha \approx 48.6°, \beta \approx 91.4°$ **11.** $a \approx 2.99, \beta \approx 19.2°, \gamma \approx 80.8°$ **13.** $b \approx 4.14, \alpha \approx 43.0°, \gamma \approx 27.0°$ **15.** $c \approx 1.69, \alpha \approx 65.0°, \beta \approx 65.0°$ **17.** $\alpha \approx 67.4°, \beta = 90°, \gamma \approx 22.6°$ **19.** $\alpha = 60°, \beta = 60°, \gamma = 60°$ **21.** $\alpha \approx 33.6°, \beta \approx 62.2°, \gamma \approx 84.3°$ **23.** $\alpha \approx 97.9°, \beta \approx 52.4°, \gamma \approx 29.7°$ **25.** 70.75 ft **27. (a)** 12.0° **(b)** 220.8 mph **29. (a)** 63.7 ft **(b)** 66.8 ft **(c)** 92.8° **31. (a)** 492.6 ft **(b)** 269.3 ft **33.** 342.33 ft

35. Using the Law of Cosines:
$L^2 = x^2 + r^2 - 2rx\cos\theta$
$x^2 - 2rx\cos\theta + r^2 - L^2 = 0$
Then, using the quadratic formula:
$x = r\cos\theta + \sqrt{r^2\cos^2\theta + L^2 - r^2}$

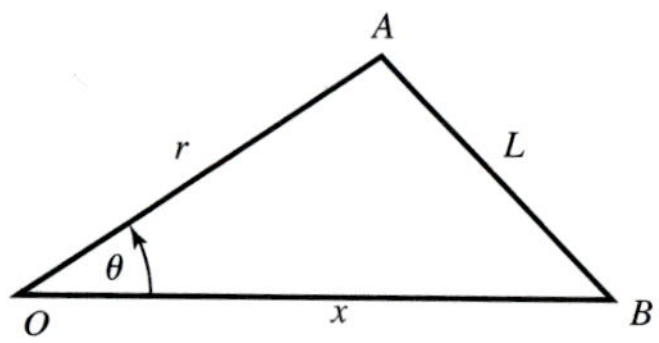

37. $\cos\dfrac{\gamma}{2} = \sqrt{\dfrac{1+\cos\gamma}{2}} = \sqrt{\dfrac{1+\dfrac{a^2+b^2-c^2}{2ab}}{2}} = \sqrt{\dfrac{2ab+a^2+b^2-c^2}{4ab}} = \sqrt{\dfrac{(a+b)^2-c^2}{4ab}} = \sqrt{\dfrac{(a+b+c)(a+b-c)}{4ab}}$
$= \sqrt{\dfrac{2s(2s-2c)}{4ab}} = \sqrt{\dfrac{s(s-c)}{ab}}$

39. $\dfrac{\cos\alpha}{a} + \dfrac{\cos\beta}{b} + \dfrac{\cos\gamma}{c} = \dfrac{b^2+c^2-a^2}{2abc} + \dfrac{a^2+c^2-b^2}{2abc} + \dfrac{a^2+b^2-c^2}{2abc} = \dfrac{b^2+c^2-a^2+a^2+c^2-b^2+a^2+b^2-c^2}{2abc}$
$= \dfrac{a^2+b^2+c^2}{2abc}$

4.4 Concepts and Vocabulary *(page 297)*

1. Heron's **2.** F **3.** T

4.4 Exercises *(page 298)*

1. 2.83 **3.** 2.99 **5.** 14.98 **7.** 9.56 **9.** 3.86 **11.** 1.48 **13.** 2.82 **15.** 1.53 **17.** 30 **19.** 1.73 **21.** 19.90 **23.** 19.81 **25.** 9.03 sq ft **27.** \$5446.38 **29.** 9.26 sq cm **31.** $A = \dfrac{1}{2}ab\sin\gamma = \dfrac{1}{2}a\sin\gamma\left(\dfrac{a\sin\beta}{\sin\alpha}\right) = \dfrac{a^2\sin\beta\sin\gamma}{2\sin\alpha}$ **33.** 0.92 **35.** 2.27 **37.** 5.44

39. $A = \dfrac{1}{2}r^2(\theta + \sin\theta)$ **41.** 31,145.15 sq ft

43. $h_1 = 2\dfrac{K}{a}, h_2 = 2\dfrac{K}{b}, h_3 = 2\dfrac{K}{c}$. Then $\dfrac{1}{h_1} + \dfrac{1}{h_2} + \dfrac{1}{h_3} = \dfrac{a}{2K} + \dfrac{b}{2K} + \dfrac{c}{2K} = \dfrac{a+b+c}{2K} = \dfrac{2s}{2K} = \dfrac{s}{K}$.

45. Angle AOB measures $180° - \left(\dfrac{\alpha}{2} + \dfrac{\beta}{2}\right) = 180° - \dfrac{1}{2}(180° - \gamma) = 90° + \dfrac{\gamma}{2}$, and $\sin\left(90° + \dfrac{\gamma}{2}\right) = \cos\left(-\dfrac{\gamma}{2}\right) = \cos\dfrac{\gamma}{2}$ since cosine is an even function. Therefore, $r = \dfrac{c\sin\dfrac{\alpha}{2}\sin\dfrac{\beta}{2}}{\sin\left(90° + \dfrac{\gamma}{2}\right)} = \dfrac{c\sin\dfrac{\alpha}{2}\sin\dfrac{\beta}{2}}{\cos\dfrac{\gamma}{2}}$.

47. $\cot\dfrac{\alpha}{2} + \cot\dfrac{\beta}{2} + \cot\dfrac{\gamma}{2} = \dfrac{s-a}{r} + \dfrac{s-b}{r} + \dfrac{s-c}{r} = \dfrac{3s-(a+b+c)}{r} = \dfrac{3s-2s}{r} = \dfrac{s}{r}$

Exponential and Logarithmic Functions

1. Exponential Function: $f(x) = b^x, b > 0, b \neq 1$
Graphs:

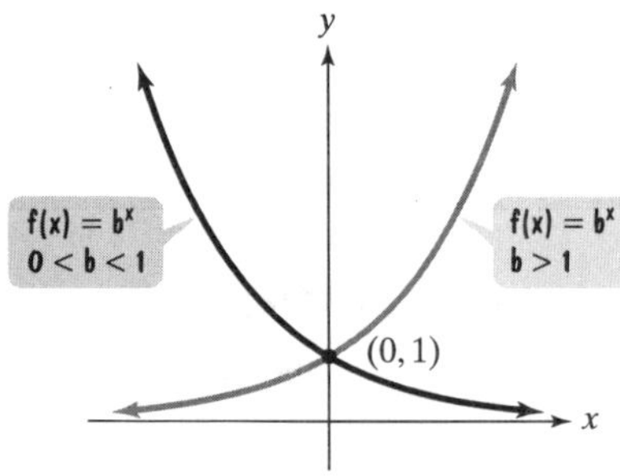

2. Logarithmic Function: $f(x) = \log_b x, b > 0, b \neq 1$
$y = \log_b x$ is equivalent to $x = b^y$.

Graphs:

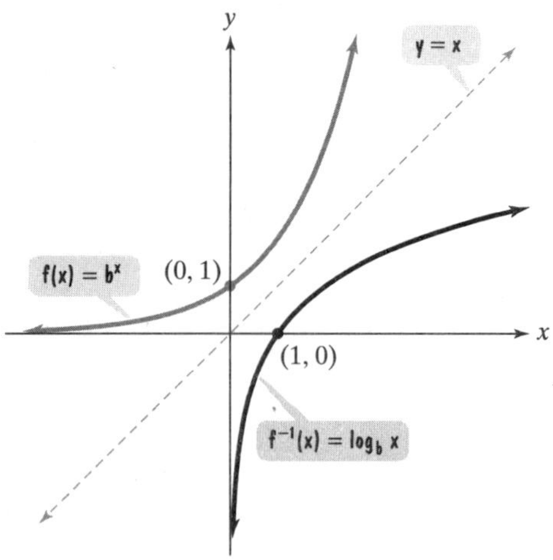

$b > 1$

3. Properties of Logarithms

a. $\log_b(MN) = \log_b M + \log_b N$

b. $\log_b\left(\frac{M}{N}\right) = \log_b M - \log_b N$

c. $\log_b M^p = p \log_b M$

d. $\log_b M = \frac{\log_a M}{\log_a b} = \frac{\ln M}{\ln b} = \frac{\log M}{\log b}$

e. $\log_b b^x = x; \log 10^x = x; \ln e^x = x$

f. $b^{\log_b x} = x; 10^{\log x} = x; e^{\ln x} = x$

Distance and Midpoint Formulas

1. The distance from (x_1, y_1) to (x_2, y_2) is

$$\sqrt{(x_2 - x_1)^2 + (y_2 - y_1)^2}.$$

2. The midpoint of the line segment with endpoints (x_1, y_1) and (x_2, y_2) is

$$\left(\frac{x_1 + x_2}{2}, \frac{y_1 + y_2}{2}\right).$$

Conic Sections

Circle

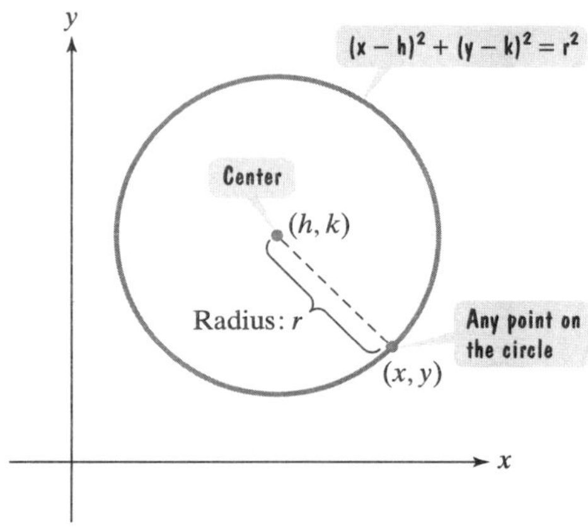

Ellipse

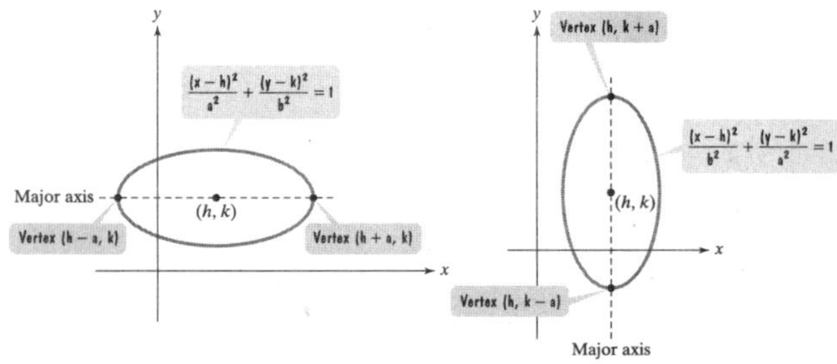

Hyperbola

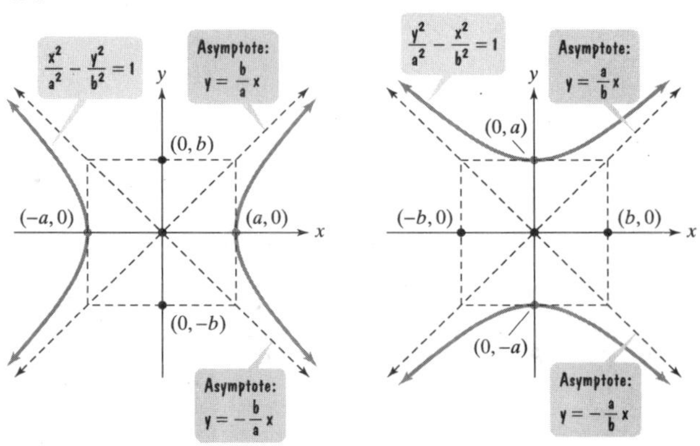